Weak and Electromagnetic
Interactions in Nuclei

Weak and Electromagnetic Interactions in Nuclei

Proceedings of the International Symposium
Heidelberg, July 1–5, 1986

Editor: H. V. Klapdor

With 555 Figures

Springer-Verlag Berlin Heidelberg New York
London Paris Tokyo

Professor Dr. Hans Volker Klapdor
Max-Planck-Institut für Kernphysik, D-6900 Heidelberg, Fed. Rep. of Germany

International Advisory Committee:
A. Arima (Tokyo), D. A. Bromley (Yale), G. E. Brown (Stony Brook), E. Fiorini (Milano), P. W. M. Glaudemans (Utrecht), H. Horie (Tokyo), R. Mößbauer (München), A. Pomanski (Moscow), W. Priester (Bonn), A. Richter (Darmstadt), V. G. Soloviev (Dubna), B. Stech (Heidelberg), D. Wilkinson (Sussex)

Organizing Committee:
H. V. Klapdor *(Chairman)*, J. Metzinger *(Scientific Secretary)*, A. Piepke, B. Povh, U. Schmidt-Rohr (Max-Planck-Institut für Kernphysik, Heidelberg), P. Bock, W. Bühring, D. Dubbers (Physikalisches Institut of the University of Heidelberg)

Sponsored by:
International Union of Pure and Applied Physics,
European Physical Society

Photos on pages XII–XVI: Schloßphoto Heidelberg and MPI
Cover illustration: S. L. Glashow, Annette Whelan

ISBN 3-540-17255-6 Springer-Verlag Berlin Heidelberg New York
ISBN 3-387-17255-6 Springer-Verlag New York Berlin Heidelberg

Library of Congress Cataloging-in-Publication Data. Weak and electromagnetic interactions in nuclei. Includes index. 1. Weak interactions (Nuclear physics) – Congresses. 2. Electromagnetic interactions – Congresses. 3. Nuclear structure – Congresses. 4. Astrophysics – Congresses. I. Klapdor, H.V. (Hans Volker), 1942-QC794.8.W4W36 1986 539.7'54 86-29713

This work is subject to copyright. All rights are reserved, whether the whole or part of the material is concerned, specifically those of translation, reprinting reuse of illustrations, broadcasting, reproduction by photocopying machine of similar means, and storage in data banks. Under § 54 of the German Copyright Law where copies are made for other than private use, a fee is payable to "Verwertungsgesellschaft Wort", Munich.

© Springer-Verlag Berlin Heidelberg 1986
Printed in Germany

The use of registered names, trademarks, etc. in this publication does not imply, even in the absence of a specific statement, that such names are exempt from the relevant protective laws and regulations and therefore free for general use.

Offset printing: Weihert-Druck GmbH, D-6100 Darmstadt

Preface

Nuclear physics is presently experiencing a thrust towards fundamental physics questions. Low-energy experiments help in testing beyond today's standard models of particle physics. The search for finite neutrino masses and neutrino oscillations, for proton decay, rare and forbidden muon and pion decays, for an electric dipole moment of the neutron denote some of the efforts to test today's theories of grand unification (GUTs, SUSYs, Superstrings, ...) complementary to the search for new particles and symmetries in high-energy experiments.

The close connections between the laws of microphysics, astrophysics and cosmology open further perspectives. This concerns, to mention some of them, properties of exotic nuclei and nuclear matter, and star evolution; the neutrino and the dark matter in the universe; relations between grand unification and evolution of the early universe.

The International Symposium on Weak and Electromagnetic Interactions in Nuclei (W.E.I.N. 1986), held in Heidelberg 1–5 July 1986, in conjunction with the 600th anniversary of the University of Heidelberg, brought together experts in the fields of nuclear and particle physics, astrophysics and cosmology.

The organization of these Proceedings reflects the concept and structure of the Symposium. About one-third of the volume deals with current nuclear physics problems such as weak and electromagnetic nuclear properties, high spin spectroscopy, nuclei far from stability, spin-isospin excitations, symmetries in nuclei; one chapter concerns electroweak interactions and subnucleonic structure. The major part is devoted to low-energy experiments contributing to our knowledge of electroweak theories and theories of grand unification and to connections between microphysics, astrophysics and cosmology.

The meeting – and this book – show that nuclear physics, in a broader sense, will play an important role in future investigation of fundamental problems in nature.

Heidelberg, October 1986 *H.V. Klapdor*

Acknowledgements

I would like to extend my sincere appreciation to the Rector of the University, Prof. G. zu Putlitz, under whose patronage the Symposium was held, for his generous support.

I am deeply indebted to the President of the International Union of Pure and Applied Physics, Prof. A. Bromley, for his encouragement and strong support. The sponsorship by IUPAP and the European Physical Society is gratefully acknowledged. I am also grateful to Prof. Th. Mayer-Kuckuk, Vice-President of IUPAP, and to Prof. J. Trümper, President of the German Physical Society, for valuable advice and support. I thank Mr. S. Alber, Vice-President of the European Parliament, for his interest and active participation in the Symposium.

I would like to thank the members of the International Advisory Committee and the Organizing Committee for their suggestions and assistance. I owe special thanks to Prof. P. Brix, Prof. A. Faessler, Prof. R. Mößbauer, Prof. A. Pomanski, Prof. B. Povh, Prof. A. Richter, and Prof. B. Stech.

At various stages of my work I enjoyed the collaboration of many colleagues and friends whose contributions have been important to the Symposium and are highly appreciated. I would like to thank especially Prof. T. Oda, Prof. R. Madey, Dr. I. Kondurow, Dr. K. Grotz, and my colleagues from the University of Heidelberg.

I personally am especially grateful to Dr. J. Metzinger for the endless hours spent doing so many things that led to the success of the meeting.

Without the generous support of the many Institutional and Corporate Sponsors the Symposium could not have taken place.

I would like to thank Dr. H.-U. Daniel and Mr. R. Michels of Springer-Verlag for their concern and their efficient cooperation. Finally, the generous permission of Prof. Sheldon Glashow to use his artistic representation, illustrating the philosophy of this Conference, on the front cover of these Proceedings is gratefully acknowledged.

H.V. Klapdor

Institutional Sponsors:
Deutsche Forschungsgemeinschaft (DFG)
Bundesministerium für Forschung und Technologie (BMFT)
Land Baden-Württemberg
Max-Planck-Gesellschaft zur Förderung der Wissenschaften e.V.
Universität Heidelberg

Corporate Sponsors:
KWU, Erlangen; Badische Landesbausparkasse Karlsruhe; Daimler-Benz AG, Stuttgart; IBM-Deutschland, Stuttgart; Control Data GmbH, Frankfurt; Vakuumschmelze GmbH, Hanau; Walter De Gruyter u. Co, Berlin; Canberra Elektronik GmbH, Frankfurt; EG u. G ORTEC, München; Nuclear Data Inc., Frankfurt; PGT Europa GmbH, Wiesbaden; Digital Equipment GmbH; Eltec Electronik, Nürnberg; Harshaw Chemie GmbH, Wermelskirchen; Heidelberger Volksbank, Heidelberg; Springer-Verlag GmbH, Heidelberg; Physics Trust Publications, Bristol; Dialog GmbH, Mannheim; Edwards Kniese u. Co., Marburg; Baden-Württembergische Bank AG, Heidelberg; North Holland Physics Publishing Co., Amsterdam; B.G. Teubner GmbH, Stuttgart; Tennelec GmbH, Unterhaching; Ariadne GmbH, Wiesbaden; CJT Vakuum Technik Produktions u. Vertriebs GmbH, Ramelsbach; Frank Elektronik, Nürnberg; VCH Verlagsgesellschaft, Weinheim; Vieweg u. Sohn Verlagsgesellschaft mbH, Wiesbaden; Fa. Janssen, Heidelberg

Opening Addresses

Opening by the Chairman of the Symposium
Prof. H.V. Klapdor

Dear colleagues, ladies and gentlemen.

It is a great pleasure for me to open this Symposium on Weak and Electromagnetic Interactions in Nuclei, and to welcome on this occasion so many participants from many countries. In particular, I have the honor to welcome Magnifizenz Professor zu Putlitz, Rector of the University of Heidelberg, Professor Mayer-Kuckuk, Vice President of the International Union of Pure and Applied Physics, Professor Trümper, President of the German Physical Society, and for the Ministry of Science and Arts of the Land Baden-Württemberg, Doctor Dederer.

The present symposium was conceived as a successor to the Symposium on Electromagnetic Properties of Atomic Nuclei held in Tokyo 1983. We have broadened, however, the scope and it is one of the ideas of this symposium to elucidate connections between nuclear physics and the neighboring fields particle physics, astrophysics, and cosmology. The organizers consider this meeting further as a modest contribution to the celebrations of the 600th anniversary of the University of Heidelberg which is held in this year. I hope that we will have a successful meeting, lively discussions, and I wish you all a pleasant time in Heidelberg.

Welcome by the Rector of the University
Magnifizenz Prof. G. zu Putlitz

Professor Klapdor, Doctor Dederer, ladies and gentlemen, dear colleagues.

It is a big pleasure for me to welcome you this morning on the occasion of this international meeting on weak and electromagnetic interactions in nuclei. As you know, Heidelberg has been chosen as a conference site also because the University celebrates its 600th anniversary. Actually, physics in Heidelberg has a long tradition. In 1386 when the University was founded, Heilmann von Wunnenberg taught already physics here. Of course, this was the physics of Aristotle, essentially the interpretation of the books of Aristotle in the limits given by the Church. The start of physics really can be attributed to Christian Mayer at the end of the 18th century. He started experimental physics and he also gave first indications for the existence of double stars. In 1803 this University and this countryside became part of Baden. For the University this was a big advantage because from then on the University was put on solid financial grounds. In 1817 physics was separated from chemistry and chemistry continued to grow and to flourish with such famous names like Leopold Gmelin, Robert Bunsen, and later Kuhn, Freudenberg and Wittig. In physics, Jolly installed the first laboratory for students. In the middle of the century, Kirchhoff together with Bunsen discovered spectral analysis and used this technique to explain the spectra of stars. Von Helmholtz taught here as well as Quincke, and at the turn of the century Philipp Lenard, who was professor of physics, received the Nobel Prize for cathode rays but was unfortunately better known later on for his anti-Semitism and his National Socialistic attitude.

Physics in the Third Reich as a consequence of Lenard's influence was not existent anymore. It is a particular pleasure to note that after the war men were here who reinstalled physics--men like Walther Bothe together with Heinz Maier-Leibnitz and also you were here, Professor Mößbauer at that time, like Otto Haxel, like Johannes Jensen, Wolfgang Gentner, and Hans Kopfermann.

Today you are here to have a scientific meeting on the topic of electromagnetic and weak interactions in nuclei. Indeed this field has a good anchor place in Heidelberg because of all the activities in the University but also in institutions around Heidelberg like the Max Planck Institute for Nuclear Physics or the Heavy Ion Laboratory in Darmstadt.

All that we do as scientists has ultimately the purpose to understand Nature, to improve the quality of life. When our colleague Quincke in 1885 as Pro-Rector of the University gave his annual speech, he said the following. I will read it first in German and then translate it!

'Wie aber auch die Zukunft sich gestalten mag, so lange die Quelle der Wissenschaft in Heidelberg fließt, wird die Physik als ebenbürtige Schwester der Wissenschaften aller Fakultäten nicht zurückstehen im Streben nach der Wahrheit bei dem Studium alter und neuer Prozesse in den ewigen Akten der Natur, oder der Arbeit für das Wohl das leidenden Menschengeschlechts.'

And this means in a free translation

'Wherever the future goes, as long as the force of sciences flows in Heidelberg, physics will be an equal sister of the sciences of all faculties in the quest for truth, in the study of old and new processes in the eternal books of Nature and with respect to the work for the benefit of human mankind.'

Thank you very much.

Address by Prof. T. Oda,
Member of the Organizing Committee of the Tokyo 1983 Symposium,

Ladies and Gentleman,

As a member of the organizing committee of the previous symposium in 1983 in Tokyo I am very pleased to offer best wishes for a successful symposium. We are pleased that this symposium is being held here in Heidelberg in conjunction with the 600th anniversary of the University of Heidelberg, which is one of the oldest and best known universities in the world. We are very grateful to Professor Klapdor and his collaborators for making this symposium a reality under such good circumstances and in such a beautiful place. Thank you very much.

Address by Prof. T. Mayer-Kuckuk, Vice-President of the International Union of Pure and Applied Physics (IUPAP)

Ladies and gentlemen.

Communication among scientists is something very essential for the progress in physics. The general acceptance by the international community of scientists is the criterion for a theory or a result to be established and a constant dialogue is the basis for going further. This dialogue may take place at several levels: At the blackboard of a physicist's office, at an institute seminar, or at an international conference like this, which is the true and ultimate forum of international contest. You all know what you take home from such a meeting. Perspectives have become much clearer, one knows better what is worth to do and what should better be avoided, and one learns about all the technical developments by other groups. Indeed, the money for such a conference is extremely well spent if one takes into account that research funds are better and more efficiently used after such an exchange of information. For these reasons, the International Union of Pure and Applied Physics is promoting the free and unbiased exchange of ideas on an international basis and it is prepared to defend, wherever necessary, the right of free circulation of scientists. This is in fact a very important, but by no means a trivial and unconditionally accepted principle in our world.

Ladies and gentlemen, it is a great pleasure for me to welcome you here on behalf of IUPAP which sponsors this conference. Topic and program show impressively that once again nuclear physics in a broader sense plays an important role in the study of fundamental laws of physics as it has done under various aspects since the beginning of the century. Thus it is in the best tradition of IUPAP to sponsor this conference, and in particular the President, Professor Allan Bromley, has asked me to convey his greetings to you. For my person, I wish you pleasant days here in Heidelberg and many fruitful discussions.

Address by Prof. J. Trümper, President of the German Physical Society (DPG)

Magnifizenz, ladies and gentlemen, dear colleagues, lieber Herr Klapdor!

On behalf of the German Physical Society I'd like to express a warm welcome to the participants of the International Symposium on Weak and Electromagnetic Interactions in Nuclei. It is a great pleasure for me to be with you here this morning, especially since the subject of this conference in general and the astrophysical aspects in particular are very close to my own scientific interests. It is also a great pleasure for me to be here in Heidelberg again during the year of the 600th anniversary of the University. Only three months ago the German Physical Society held its big annual meeting, the 50th one actually, here in Heidelberg and enjoyed the atmosphere of this most beautiful city and the hospitality of its university. I hope very much, Magnifizenz, that you enjoyed the company of the more than 2,000 physicists as well. At any rate, best wishes again for you and for your great university, which is the oldest one on German territory.

Let me make a few remarks on the German Physical Society, which, of course, is rather young in comparison. Actually it was founded in 1845 by a group of six young physicists in Berlin, and because of the rather complex German history it underwent several transmutations before it found its present shape after World War II. At the present time the Society is in a phase of very rapid growth. In February of this year the 10,000th member joined the Society, another young physicist from Berlin. In the meantime our membership has grown to more than 11,000 and the growth rate is still increasing. The German Physical Society is organized in 13 divisions, and the nuclear and elementary particle physics divisions are two of the larger ones. In addition we have 16 joint working groups dealing with interdisciplinary subjects in collaboration with other societies such as, for instance, the Astronomical Society.

One of the most important scientific developments in our century is the formation of strong links between micro- and cosmophysics. Several very important aspects of these relations will be discussed at this meeting. As a high-energy astrophysicist and X-ray astronomer I have mainly dealt with white dwarfs and neutron stars, which are almost ideal manifestations of this situation. The most general aspect, of course, is the evolution of our universe at large. Atomic physics has enabled us to observe the chemical evolution of matter. Nuclear physics allows us to understand the transformation processes which are at work. Elementary particle physics may provide the key for an understanding of the early evolution of the universe. The discovery of the underlying relationship between the large-scale properties of the universe and the laws of microphysics could represent a fundamental breakthrough in physics. I wish you a successful meeting. Thank you.

Grußwort von Dr. S. Dederer, Ministerialdirigent im Ministerium für Wissenschaft und Kunst des Landes Baden-Württemberg

Herr Professor Klapdor, Magnifizenz, meine sehr verehrten Damen und Herren.

Ich habe heute die Ehre und die Freude, im Namen und Auftrag von Herrn Wissenschaftsminister Professor Dr. Engler die Teilnehmer dieses Symposiums der Kernphysik in Heidelberg herzlich zu grüßen und willkommen zu heißen. Wir freuen uns sehr, daß eine so große Zahl von Wissenschaftlern von hohem Rang aus dem In- und Ausland der Einladung zu dieser Tagung in der wunderschönen Stadt Heidelberg gefolgt ist. Das Max-Planck-Institut für Kernphysik hat mit der Durchführung dieses Symposiums einen guten Beitrag zum Jubiläumsjahr der Universität geleistet. Dafür sei dem Institut und insbesondere Herr Professor Dr. Klapdor herzlich gedankt. Das Ministerium für Wissenschaft und Kunst sieht in der Tatsache, daß so viele namhafte Wissenschaftler sich heute hier zusammengefunden haben, den Beweis, daß das Max-Planck-Institut für Kernphysik und die Universität Heidelberg auf dem Gebiet der Wissenschaft und der Forschung Hervorragendes leisten und weltweit angesehen sind. Forschung und Entwicklung ist für uns alle, besonders aber auch für das rohstoffarme Land Baden-Württemberg lebensnotwendig. So schrieb Professor Dr. Theodor Heuss den Satz: "Der Aufstieg des Landes kann sich nicht mit dem billigen Massengut vollziehen, auch nicht mit gewerblichem Halbzeug, sondern nur mit Fertigfabrikaten und hochwertigen Sonderleistungen." Die Einschätzung der Entwicklungsmöglichkeiten unseres Landes ist auch die Grundlage für die Forschungs- und Technologiepolitik der Landesregierung von Baden-Württemberg. Wachstum, Beschäftigung, Wohlstand und sozialer Friede hängen davon ab, wie sich unsere Wirtschaft entwickelt und im internationalen Wettbewerb behauptet. Diese Wettbewerbsfähigkeit unserer Wirtschaft beruht zu einem wesentlichen Teil auf der Leistungkraft der Forschung in unseren Universitäten und in den Instituten der Max-Planck-Gesellschaft und der Fraunhofer-Gesellschaft. Die Landesregierung ist daher bemüht, den Boden und das Umfeld für eine gute und erfolgreiche Forschung zu bereiten. So erhalten die neun Universitäten in den vergangenen Jahren und in den kommenden Jahren jährlich 30 Millionen D Mark im Rahmen eines Forschungsschwerpunktprogrammes zusätzlich. Daneben ist ein Forschungspool eingerichtet, mit dem zur Vorbereitung neuer Sonderforschungsbereiche Personal- und Sachmittel bereitgestellt werden. Die Max-Planck-Gesellschaft hat 13 Max-Planck-Institute in Baden-Württemberg eingerichtet. Dreien dieser Max-Planck-Institute hat die Landesregierung zur Ausweitung der Forschungstätigkeit Sonderzuwendungen in erheblichem Umfange in den letzten Jahren gewährt. Mit diesen und weiteren Maßnahmen hat das Land eindeutig der Forschung eine besondere Priorität eingeräumt. Sie sind, meine Damen und Herren, in ein forschungsfreundliches Land gekommen. Mit dem heute beginnenden fünftägigen Kongress werden Sie der Forschung neue Impulse geben. Ich wünsche allen Teilnehmern der Tagung einen guten wissenschaftlichen Gewinn aus den Vorträgen und den sich anschließenden Diskussionen. Ich wünsche Ihnen und Ihren sehr verehrten Damen auch viele schöne Stunden außerhalb des Tagungsprogramms in der romantischen Stadt Heidelberg und in der reizenden Umgebung. Dem Symposium wünsche ich einen erfolgreichen Verlauf.

Translation

Mr. Chairman, ladies and gentlemen.

I have the honor and pleasure today of greeting and welcoming you, the participants of this symposium of nuclear physicists in Heidelberg most sincerely and on behalf of the Minister of Science, Professor Engler. We are delighted to see that such a great number of distinguished scientists from home and abroad have accepted the invitation to this conference in the beautiful city of Heidelberg. By organizing this symposium, the Max Planck Institute for Nuclear Physics has made an excellent contribution to the University's anniversary year. For this we owe thanks to the Institute and especially to Professor Klapdor. The Ministry for Science and Arts considers the fact that so many first-class scientists have gathered here today to be proof of the Max Planck Institute's and the University of Heidelberg's outstanding achievements in the field of research and science and of importance to all of us, and especially to the Land Baden-Württemberg, which is not rich in raw materials. It was Doctor Theodor Heuss, who once wrote, "The upswing of the country cannot be achieved with cheap mass goods. Not with commercial semifinished goods either. Only with finished goods and high-quality special performance." This view on Baden-Württemberg's potential for development is also the basis for the development and technology policy of Baden-Württemberg's state government. The competitiveness of our industry depends to a great extent on the quality of research at our universities and the institutes of the Max Planck Society and the Fraunhofer Society. That is why the state government of Baden-Württemberg tries to lay the foundations for good and successful research. For example, in the past few years the nine universities have received an additional 30 million German Marks per year within the framework of a special program focusing on research and will continue to receive the same amount in the coming years. The Max Planck Society has established 13 Max Planck institutes in Baden-Württemberg. Three of them were given considerable special funds for further research projects in the past few years. With these and other measures, Baden-Württemberg has clearly given priority to research. Ladies and gentlemen, you have come to a research-minded state of Germany. With the five-day congress beginning today, you will give fresh impetus to research. I would like to wish all participants in the symposium good scientific gains from the papers presented here and the ensuing discussions. I would also like to wish you and the ladies here present many pleasant hours in the beautiful city of Heidelberg and its splendid surroundings that you can spend away from the conference program. And then finally I would like to wish the symposium successful days of work. Thank you.

WEAK AND
ELECTROMAGNETIC
INTERACTIONS
IN NUCLEI
1–5 July 1986

63
27

Office
F
ELECTROMAGNE
INTERACTIONS

International Symposium on
WEAK AND
ELECTROMAGNETIC
INTERACTIONS

Contents

Part 1 Weak and Electromagnetic Nuclear Properties and Excitations

1.1 Nuclei at Low Excitation

1.2 Nuclei in Highly Excited States

1.2.1 High Spin States

1.2.2 Giant Resonances and Sum Rules

1.3 Exotic Nuclei and Beta Decay far from Stability

1.4 Spin-Isospin Excitations in Nuclei

Part 3 Status and Test of Electroweak Theories and GUT's

3.1 Status of Electroweak Theory

3.2 Electroweak Interactions and Symmetries in Baryons, Nuclei and Atoms

3.2.1 Nuclear Beta Decay and Weak Coupling Constants

3.2.2 Hyperons and Hypernuclei

3.3 Lepton Number Violation and Neutrino Mass

3.3.1 Double Beta Decay

3.3.2 Solar Neutrinos

3.3.3 Reactor Neutrino Oscillation Experiments

3.5 GUT's, SUSY's, Superstrings

3.5.1 Further Basic Experiments for GUT's

3.5.2 Theory of GUT's, SUSY's, Superstrings

Part 4 **Weak Interaction in Astrophysics and Cosmology**

4.1 Weak Interaction in Astrophysics

4.2 Evolution and Structure of the Universe

Part 5 **Summary**

Part 6 **Appendix**

Part 1

Weak and Electromagnetic Nuclear Properties and Excitations

Nuclear Moments: A Test of the Shell Model?

P.W.M. Glaudemans

Fysisch Laboratorium, Rijksuniversiteit Utrecht, P.O. Box 80.000,
NL-3508 TA Utrecht, The Netherlands

It is well-known that even the extreme single-particle model (Schmidt values) can quite well reproduce the observed magnetic dipole moments of several nuclei. Recent extensive investigations on the structure of p-shell nuclei in terms of large-scale shell-model calculations show, however, that certain calculated magnetic-dipole and electric-quadrupole moments are very sensitive to the choice made for the effective nucleon-nucleon interaction.

The results obtained sofar in small and large model spaces for p-shell nuclei with various interactions based on the Reid soft-core and Paris potential, on the Sussex matrix elements and on empirical Talmi integrals will be discussed.

1. INTRODUCTION

Many nuclear models are based on the shell model. They are often developed to provide a reasonable way of truncation for the unmanageable large set of basis states that should be required for a somewhat realistic model space. When describing nuclear structure properties in the conventional shell model one usually tries to overcome this computational problem of a too large model space by assuming that part of the nucleons forms closed shells. The excitations of this core of closed shells are ignored or to some extend included in some approximation.

Because of developments in present day (super)computers and related programs, ZWARTS [1], the assumption of closed shells does no longer have to be made for light nuclei. Hence, all particles are considered as active and thus core polarization effects are automatically included in the model space. As a result there are no single-particle energies (representing the interaction of a particle with the core) that must be determined separately. The hamiltonian can then be expressed in terms of two-body matrix elements only. There is still a limitation in the number of high-lying shells that can be considered as active, however.

In this paper we discuss some of the results that have been obtained recently, WOLTERS et al. [2], from a large basis shell-model treatment of the A = 4-16 nuclei. The observables e.g. spectra, rms charge radii and electromagnetic moments obtained with various interactions will be compared with the data. For earlier reports on this project see [3-6].

2. THE MODEL SPACE

All $2\hbar\omega$ excitations with respect to the $(p^n_{3/2}\ p^m_{1/2})$ configurations are included in the present model space. This means that one particle can be excited from the 1s shell into the 2s1d-shell or from the 1p-shell into the 1f2p-shell. Moreover, two particles can both be excited from the 1s-shell into the 1p-shell or from the 1p-shell into the 2s1d-shell.

This complete $2\hbar\omega$ model space allows us to remove exactly the spurious states due to excitation of the center-of-mass if, moreover, a translational invariant interaction is used, VAN HEES and GLAUDEMANS [7]. It should be remarked that the (unphysical) spurious-state contributions must be removed, in particular when one deals with light nuclei, since their effects are proportional with 1/A.

The hamiltonian matrices that must be constructed and diagonalized in a $2\hbar\omega$ space do not exceed order 2000. This does not create severe computational problems for modern computers and programs, however. The difficulty lies in the determination of an adequate effective two-body interaction. This subject will be briefly discussed in the next section.

3. THE EFFECTIVE INTERACTION

The hamiltonian required for a treatment of p-shell nuclei in a $2\hbar\omega$ model space is completely determined by 671 two-body matrix elements. This number is far too large to use a pure phenomenological approach, i.e. to treat all two-body matrix elements as parameters and determine their values from a fit of calculated spectra to the observed spectra. However, the requirement of translational invariance (see sect. 2) reduces the number of parameters from 671 to 51. This number is still too large for an unambiguous set of parameters. A further parameter reduction can be achieved when one makes use of some potential or some set of matrix elements that has been derived from other experimental data. When investigating this problem several approaches have been followed.

In this paper some of the results obtained with four different interactions will be presented. The parameters of each set have been obtained from a least-squares fit of calculated binding energies of ground states and excited states to the experimental counterparts. One and the same set of 76 selected experimental levels of predominant $0\hbar\omega$ nature in A = 4-16 nuclei has been used in all cases.

The Reid soft-core interaction

The bare Reid soft-core interaction (RSC), see [8], can not be used without renormalizations, because e.g. the actual calculation must take place in a finite model space. The RSC requires two-nucleon wave functions which are approximately zero for small internucleon distance, due to the short-range repulsion. To solve this problem we have used the following approach.

Let us assume that the radial part of the relative two-body matrix elements has a central singlet (S = 0) and triplet (S = 1) term, an LS term and a tensor term which are isospin dependent. One can then express the relative matrix elements needed in the $2\hbar\omega$ model space in terms of 29 Talmi integrals [2]. The repulsive short-range part of the potential affects the lower order Talmi integrals most. Hence, values of the 9 Talmi integrals with order 0 or 1 have been determined empirically from a fit to the selected set of experimental levels. The remaining 20 Talmi integrals have been derived from the RSC potential.

The Paris potential

The use of the Paris potential [9] gives rise to the same problems as those of the Reid soft-core potential. Hence, a similar procedure as described above, has been followed, i.e. 9 Talmi integrals have been obtained empirically and the remaining 20 values have been derived from the Paris potential. Moreover, a term proportional to A^3 has been added to all diagonal matrix elements of the hamiltonian. This term was found to further improve the agreement with experimental binding energies.

The Sussex interaction
In this approach the Sussex matrix elements [10] have been used. We present here the results obtained from the set of relative matrix elements for the harmonic-oscillator size parameter b = 1.80 fm. The required renormalization of this interaction is performed by an overall scaling factor and an additional term proportional to A^3 which are both empirically determined. This set of Sussex matrix elements is used also to compare results obtained in a $0\hbar\omega$ space with those in a $2\hbar\omega$ model space.

Empirical Talmi integrals
Finally we present some results of an interaction that is determined completely by empirically determined Talmi integrals. In this approach the 13 most important Talmi integrals have been determined from a fit to the selected set of experimental levels, whereas the remaining 16 Talmi integrals are set equal to zero.

4. WAVE FUNCTIONS

An impression of the structure of the p-shell ground-state wave functions for a $2\hbar\omega$ model space is given in fig. 1. Here the result for the Reid soft-core potential is shown, but a similar behaviour is found also for the other interactions discussed in this paper. Note that the total intensity of the $2\hbar\omega$ admixtures is nearly mass independent for A = 4 to 16.

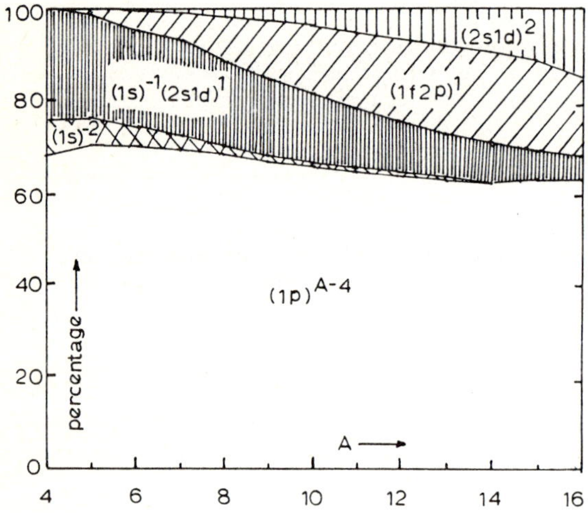

Fig. 1. The structure of the ground-state wave functions of p-shell nuclei in a $2\hbar\omega$ model space derived from the Reid soft-core potential with b = 2.0 fm

5. SPECTRA

The spectrum of ^{7}Li calculated with three different interactions (Sussex, Paris and Talmi integrals) is shown in fig. 2. This spectrum illustrates the typical agreement between theory and experiment for the $2\hbar\omega$ model space. For comparison, the results are also given for the $0\hbar\omega$ space with the Sussex interaction. It is seen from fig. 2 that all interactions produce the correct sequence of levels for ^{7}Li. The lower part of fig. 2 shows the rms deviation for each interaction separately. The rms deviation, defined as $(\Sigma[E_b(th) - E_b(exp)]^2/(N-n))^{\frac{1}{2}}$ with N = 76 and n = number of parameters, is calculated for the set of 76 selected levels used in the fitting procedure.

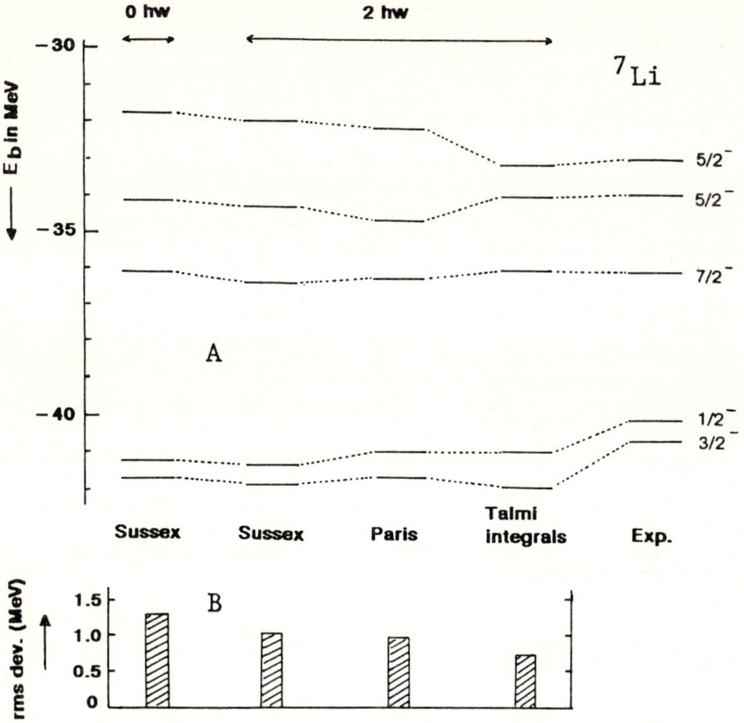

Fig. 2. The spectra of ^{7}Li obtained from various calculations in a $0\hbar\omega$ and $2\hbar\omega$ model space are compared with experiment (A). The rms deviation for all 76 levels used in the least-squares fit for each of the calculations (B)

An important result is that the rms deviation for the Sussex interaction in the $2\hbar\omega$ space is significantly lower than that for the same interaction in the $0\hbar\omega$ space. Moreover, it is seen that the rms deviation improves for the calculations from left to right from 1.3 MeV to 0.7 MeV, see fig. 2.

6. RMS CHARGE RADII

The effect of $2\hbar\omega$ admixtures on the nuclear charge radii as well as their dependence on the choice of the effective interaction have been investigated. It is important to reproduce the nuclear charge radii in particular for the interpretation of (e,e') form factors.

The results for the calculated rms charge radii, taking the proton finite size into account, are compared to the experimental data in figs. 3A and 3B. It follows from fig. 3A that $2\hbar\omega$ admixtures reduce the rms radii by about 5% with respect to the $0\hbar\omega$ values, when the same interaction and size parameter (Sussex, b = 1.80 fm) are used. This reduction is due to interference effects between $0\hbar\omega$ and $2\hbar\omega$ components in the wave functions. Note that for ^{4}He the effect is almost 15%. Hence, the ratio (radius ^{6}Li)/(radius ^{4}He) assumes the values of 1.17 and 1.29 for the $0\hbar\omega$ and $2\hbar\omega$ model space, respectively. This should be compared to the experimental ratio of 1.48 ± 0.06.

In fig. 3B the results for three interactions (Sussex, Paris, Talmi integrals) in a $2\hbar\omega$ model space for b = 1.80 fm are presented. It is seen that all interactions produce the same rms charge radii to within 3% except for He where the differences are somewhat larger. It would be interesting to investigate whether the differences between theory and experiment for He, Li, Be and O can be explained by effects of $N\hbar\omega$ admixtures with $N \geq 4$.

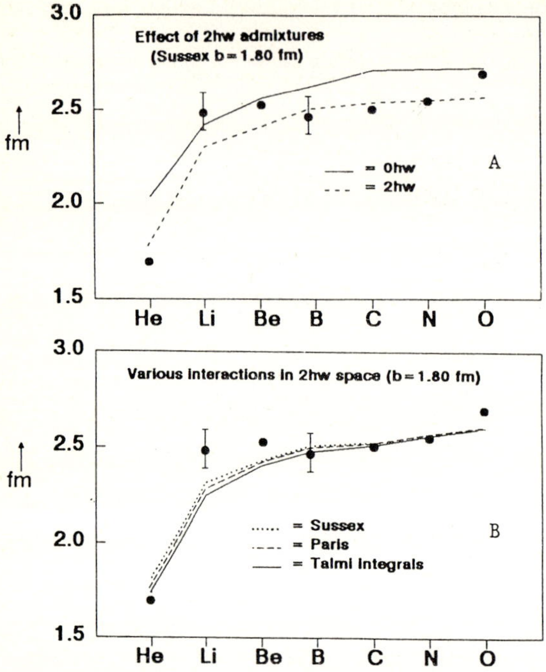

Fig. 3. The rms charge radii for a $0\hbar\omega$ and $2\hbar\omega$ model space, respectively, as obtained from the Sussex interaction with b = 1.80 fm (A). The radii in a $2\hbar\omega$ space from different interactions with b = 1.80 fm (B). The experimental values for the same Z-value have been averaged

7. MAGNETIC DIPOLE MOMENTS

A good test of the shell-model description of nuclear states is provided by the isovector part of g-factors. It is well-known that the isoscalar part of g-factors is not very sensitive to the details of a wave function [11]. The measured isoscalar g-factor, to be denoted by g(0), can be determined simply from the sum of the g-factors of corresponding states in mirror nuclei, whereas the isovector part g(1) follows from their difference.

It is illustrated in fig. 4A, which shows all cases that can be compared with experiment, that the g(0)-factors derived from the Sussex interaction in a $0\hbar\omega$ space as well as from several interactions in a $2\hbar\omega$ space differ very little and, moreover, are in good agreement with the experimental data. The largest deviation occurs for ^{14}N with the Sussex-$0\hbar\omega$ wave function.

The results for the isovector part g(1) are shown in fig. 4B. Only five cases can be compared with experiment because for $T_z = 0$ nuclei one has g(1) = 0. It follows from figs. 4A and 4B that the variations in calculated g-factors for the various interactions are considerable more pronounced for g(1) than for g(0). In particular the measured positive g(1)-factor for the ground state of ^{12}B cannot be reproduced at all by any of the interactions discussed in this paper. The approach with the 13 empirical Talmi integrals (see sect. 3) still gives the best overall agreement.

8. ELECTRIC QUADRUPOLE MOMENTS

The Q-pole moments are also found to provide a severe test of the present p-shell wave functions. In order to make a clear comparison possible between the various theoretical approaches, all calculations are performed with the same effective

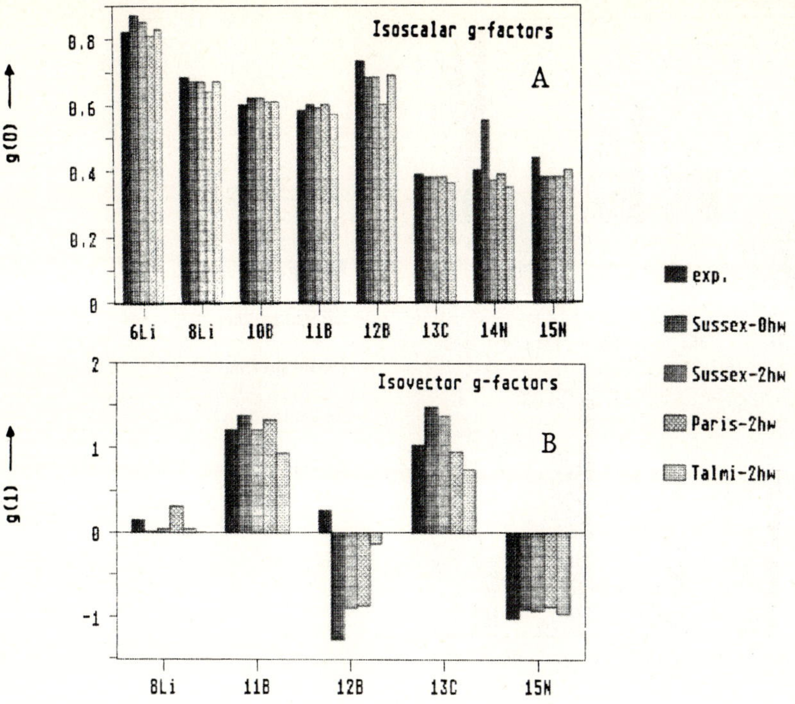

Fig. 4. The isoscalar (A) and isovector (B) g-factors for several interactions and model spaces are compared with the experimental data

charge $\Delta e = 0.1$ e, i.e. $e_p = 1.1$ e and $e_n = 0.1$ e. A proper choice for the harmonic-oscillator size parameter b is important since quadrupole moments are proportional to b^2. In this paper b is assumed to be mass-independent and fixed to b = 1.80 fm for all $2\hbar\omega$ calculations. This value of b is used also to reproduce the rms charge radii, see fig. 3B. The Q-pole moments in the $0\hbar\omega$ model space with the Sussex interaction are obtained for b = 1.67 fm in order to reproduce here also the rms charge radii as well as possible, see fig. 3A.

It follows from figs. 5A and 5B that the Q-values obtained from the Sussex interaction in the $0\hbar\omega$ space which quite well reproduces radii (see above) underestimate the Q-pole moments quite severely in most cases, see e.g. ^{7}Li, ^{9}Be, ^{10}B, ^{11}B, ^{11}C and ^{12}B. The same interaction in a $2\hbar\omega$ model space reproduces the radii and yields much better agreement for Q-pole moments. Large differences between the various interactions in a $2\hbar\omega$ space are found for ^{11}C and ^{12}B. The approach with the 13 empirical Talmi integrals yields the best agreement with the experimental Q-values.

9. SUMMARY AND CONCLUSIONS

A systematic study is discussed of different effective interactions based on the Sussex matrix elements, Reid soft-core potential, Paris potential and Talmi integrals, respectively, in a large shell-model space. The calculations are performed in a complete $2\hbar\omega$ no-core model space for A = 4-16 nuclei. Hence, the hamiltonian can be expressed in terms of two-body matrix elements without the need to introduce single-particle energies. The calculated spectra, rms charge radii, magnetic dipole and electric quadrupole moments are compared with the

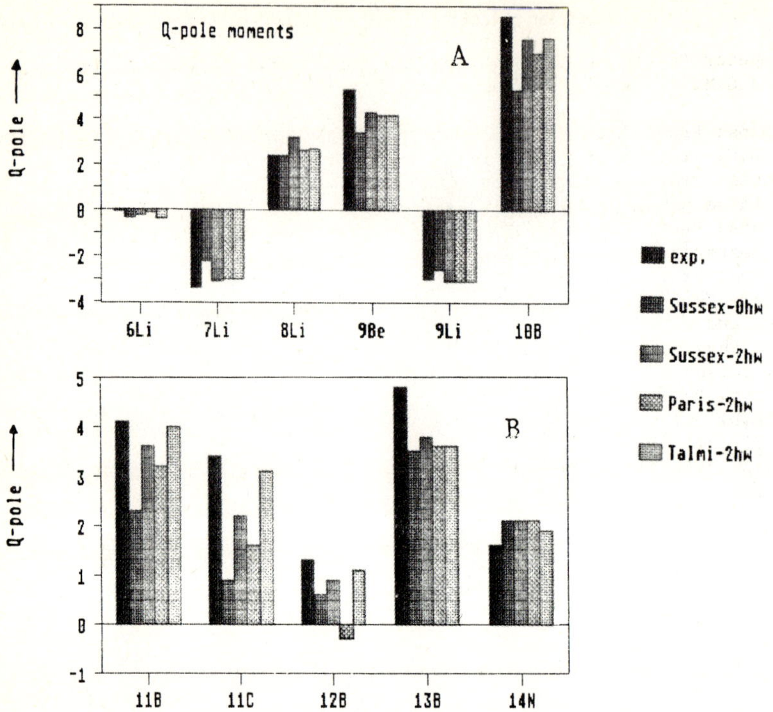

Fig. 5. The quadrupole moments (efm^2) for various interactions and model spaces calculated with an effective charge of $\Delta e = 0.1$ e and a b-value that reproduces the radii are compared with the experimental data

available experimental data. It is shown that the intensity of the $2\hbar\omega$ admixtures is of the order of 25%. These admixtures often make a more consistent description of charge radii and quadrupole moments possible, if for the latter only the small effective charge of $\Delta e = 0.1$ e is used. An effective charge which is as small as possible will be desirable in order to interpret (e,e') form factors, since effective charges might be q dependent.

From the present results it follows that the approach in which the effective interaction is completely expressed in terms of empirically determined Talmi integrals seems to be the most promising one, although up till now some quite serious discrepancies with the experimental data still exist.

It is important to realize that a static multipole moment depends on the wave function of a single state only. This aspect can be very helpful in the search for an improved effective interaction which later on can be used to study less well-known nuclear structure phenomena. However, it turns out that only a very limited number of these moments has been measured experimentally in particular for odd-odd nuclei, where the present sets of theoretical results differ most strongly from each other.

Finally I would like to thank my collaborators dr. A.G.M. van Hees, A.A. Wolters, N.A.F.M. Poppelier and P.H. Bruinsma who have provided the material presented in this paper.

10. REFERENCES

1. D. Zwarts: Computer Physics Communications 38, 365 (1985)
2. A.A. Wolters, A.G.M. van Hees, N.A.F.M. Poppelier and P.W.M. Glaudemans: to be publ.
3. P.W.M. Glaudemans: Proc. Int. Symp. on Nuclear Shell Models, eds. M. Vallieres and B.H. Wildenthal, World Scientific Publ. Comp., 2 (1985)
4. P.W.M. Glaudemans: Proc. Int. Symp. on Nuclear Structure, Bloomington, October 1985, to be publ.
5. P.W.M. Glaudemans: Proc. of the fourth Miniconference on Nuclear Structure in the 1p shell, Amsterdam, November 1985, 1.
6. P.W.M. Glaudemans: Proc. Int. Conf. Sorrento on Microscopic Approaches to Nuclear Structure Calculations, May 1986, to be publ.
7. A.G.M. van Hees and P.W.M. Glaudemans: Z. Phys. A315, 223 (1984)
8. R.V. Reid: Ann. Phys. 50, 411 (1968)
9. M. Lacombe, B. Loiseau, J.M. Richard, R. Vinh Mau, J. Côté, P. Pirès and R. de Tourreil: Phys. Rev. C21, 861 (1980)
10. J.P. Elliott, A.D. Jackson, H.A. Mavromatis, E.A. Sanderson and B. Singh: Nucl. Phys. A121, 241 (1968)
11. P.C. Zalm, J.F.A. van Hienen and P.W.M. Glaudemans: Z. Phys. A287, 255 (1978)

Nuclear Isoscalar Magnetic Moments and Gauge Invariance

W. Bentz

Department of Physics, University of Tokyo, Hongo, Bunkyo-ku, Tokyo 113, Japan

1. Introduction

The study of nuclear magnetic moments has a long and fascinating history, and is still an exciting field attracting considerable interest [1]-[3]. One of the reasons for this is that nuclear magnetic moments continue to provide us with touchstones for new theoretical developments. Historically, they gave evidence and support for the ideas of configuration mixing [4],[5] and exchange currents [6],[7]. In particular, isoscalar magnetic moments of nuclei with simple structures have shown very clearly the importance of the second order configuration mixing [5]. In these cases, meson exchange currents, although not yet fully explored, seemed to give only minor contributions [2].

Recently an approach to nuclear structure based on relativistic quantum field theory [8] (in its simplest form called the $\sigma\omega$-model) is discussed very much in the physical community. Using rather simple physical terms, it seemed to explain nuclear matter properties, properties of closed shell nuclei, and various observables in nucleon-nucleus scattering fairly well. However, it was noted in several works [9]-[11] that the description of magnetic moments seems to pose a serious problem for this model. In particular, relativistic 'corrections' to isoscalar magnetic moments for A=15 of roughly 100% were reported, which is, of course, in sharp disagreement with experiment. It also apparently disagrees with the above mentioned fact that meson exchange currents are expected to play a minor role for isoscalar quantities, since it is intuitively clear that the corrections to the Schmidt values obtained in the relativistic approach must have to do with mesonic degrees of freedom.

In this contribution we investigate the reason why the calculations employing the Dirac equation failed in describing the magnetic moments. As a basis we use the Ward identity study of Bentz et al.[12]. Furthermore we investigate the precise connection between traditionally considered meson exchange currents and the corrections arising from the small components of the nucleon spinor. We then present the recent results of Ichii et al.[13],[14] for meson exchange current contributions to isoscalar magnetic moments and discuss their importance compared with the second order configuration mixing. We will also make contact to recent suggestions [15] that isoscalar meson exchange currents could constitute the major part of the corrections to the Schmidt values.

2. The electromagnetic current in a relativistic description

Let us consider the electromagnetic current of a quasinucleon in interaction with particles filling an infinitely extended Fermi sea. On the basis of the Ward-Takahashi identity it can be shown [12] that for zero momentum transfer the current takes the form

$$\vec{j}(q=0)=\lim_{q_0\to 0}\lim_{\vec{q}\to 0}\bar{f}(p')\vec{\Gamma}(p',p)f(p) = \frac{\vec{p}}{\varepsilon_p}\frac{1+\tau^z}{2} - \tau^z\,\hat{p}\,\frac{p^2}{3\pi^2}\,(f_1-f_1') \,. \tag{1}$$

Here f(p) is the quasiparticle spinor, Γ is the full electromagnetic vertex, ε_p is the quasiparticle energy and f_1 and f_1' are two Landau-Migdal parameters [12]. It must be noted that eq.(1) is exact. The isoscalar part of the current (1) differs from the free nucleon current only by the replacement $E_p=\sqrt{p^2+M^2}\to\varepsilon_p$ (M is the nucleon mass). The difference $E_p-\varepsilon_p>0$ must be small compared with the nucleon mass in any model which gives a reasonable nucleon separation energy. We thus conclude that the isoscalar angular momentum g-factor obtained from (1),

$$g_\ell^{(0)} = \frac{1}{2}\frac{E_p}{\varepsilon_p} \tag{2}$$

is slightly enhanced relative to its free value 0.5. The enhancement can be interpreted as an effect of the nucleon binding and should not exceed a few percent.

Let us analyze this result in more detail for the special case of the relativistic $\sigma\omega$-model in the mean field approximation [8]. Here the spinor f of eq.(1) is a solution of the Dirac equation with the self energy $\Sigma=\Sigma_s+\gamma^0\Sigma_v$, where the attractive scalar part Σ_s arises from σ-meson exchange and the repulsive vector part Σ_v from ω-meson exchange (s.fig.1).
The spectrum in this case is given by

$$\varepsilon_p = E_p^* + \Sigma_v \tag{3}$$

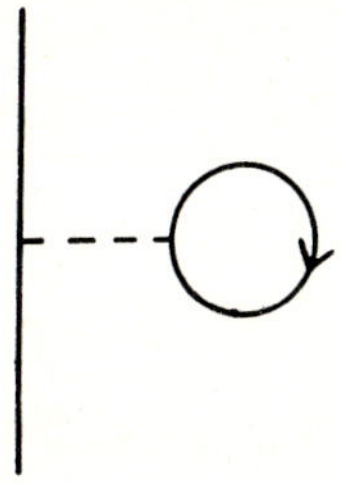

Fig.1 Nucleon self energy in the mean field approximation. The dashed line denotes the exchanged meson (σ or ω)

with $E_p^* = \sqrt{M^{*2}+p^2}$, $M^* = M+\Sigma_s$. The vertex Γ in (1) consists of its free part plus the vertex correction. Due to gauge invariance the latter is obtained by inserting a photon line into the self energy in all possible ways. Applying this to the self consistent self energy of fig.1 we obtain the RPA-type vertex correction shown in fig.2.

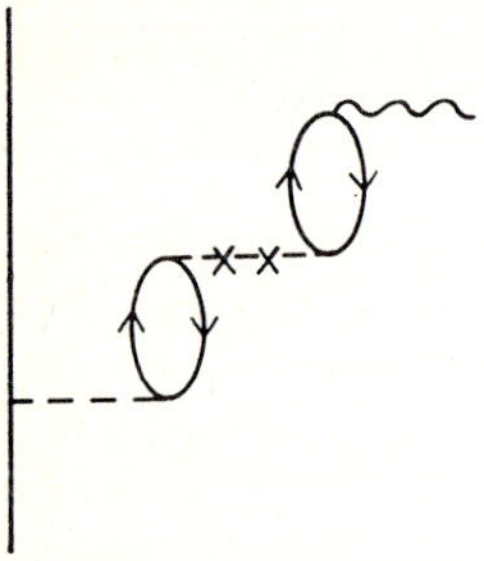

Fig.2 RPA-type vertex correction. The two crosses indicate that the bubbles, which represent particle-hole or nucleon-antinucleon excitations, are summed to all orders.

Note that in the limit $q \to 0$, as defined in eq.(1), there survive only the $N\bar{N}$ excitations due to ω-meson exchange. It is a purely isoscalar correction. We thus can split up the current (1) into two pieces,

$$\vec{j}(0) = \vec{j}_i(0) + \vec{j}_v(0) \ , \qquad (4)$$

where

$$\vec{j}_i(0) = \bar{f}(p)\ \vec{\gamma}\ \frac{1+\tau^z}{2}\ f(p) = \frac{\vec{p}}{E_p^*}\ \frac{1+\tau^z}{2} \qquad (5)$$

is the current in the impuls approximation, and j_v is due to the vertex correction of fig.2. In the σω-model, M^* is very small (M^*=0.56 M), and therefore the current (5) is strongly enhanced. Note that this enhancement comes only from the σ-meson. The enhancement factor for $g_\ell^{(0)}$ as obtained from eq.(5) amounts to E_p/E_p^*= 1.65 for normal nuclear matter density ρ_0 (or 1.23 for $\rho_0/2$). However, if we include the piece j_v due to ω-meson exchange, this large enhancement factor is cut down to E_p/ε_p=1.08 for ρ_0 (or 1.02 for $\rho_0/2$). We thus see clearly the reason for the difficulties encountered in recent calculations [9]-[11] of magnetic moments in the relativistic mean field approximation: The vertex corrections were neglected, and consequently an unreasonably large enhancement of isoscalar magnetic moments due to the small effective mass was obtained. We conclude that a consistent calculation must include these vertex corrections, since both in the energy (3) and in the isoscalar part of the current (1) there occur basically the same cancellations between the attractive σ-meson contributions and the repulsive ω-meson contributions.

3. Connection with exchange currents

How does the more conventional approach based on meson exchange currents [2],[3],[7] fit with the general constraints discussed in sect.2 ? To understand this, let us first note that the spinor f is related to the free spinor u in lowest order perturbation theory by [12],[23]

$$f(p) = u(p) + \Sigma(p)S_{-}(p)u(p) + \ldots \quad , \tag{6}$$

where S_{-} is the negative frequency part of the free nucleon propagator. Using this we can represent the corrections to the free current implied by j_i and j_v of eq.(4) by figs. 3a and 3b, respectively. Our arguments of sect.2 imply strong cancellations between these two pieces.

In the exchange current approach to isoscalar magnetic moments one considers the two-body pair current graphs shown in fig.4 besides the gauge invariant $\rho\pi\gamma$ dissociation current [13],[14]. We see that if we average the direct terms of these pair currents with respect to one of the two nucleons over the states in the Fermi sea, we automatically include all four graphs of fig.3, which simply arise from the four different time orderings of the two-body pair current operators. Thus, the cancellations due to the vertex corrections discussed in sect.2 are automatically respected in the more traditional approach starting from two-body

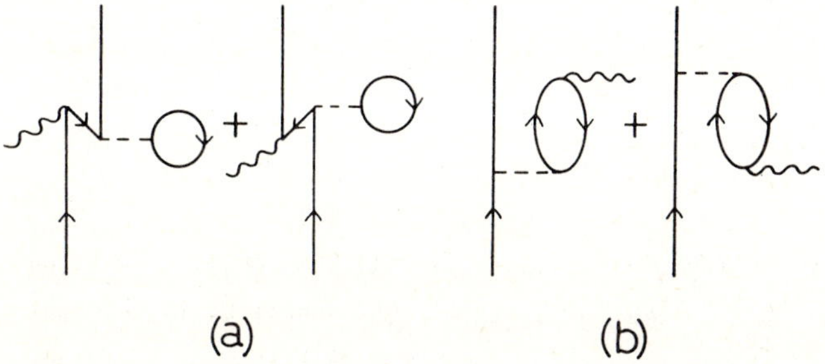

Fig.3 Corrections to the nucleon current in lowest order perturbation theory due to the admixture of free negative-energy solutions in the quasiparticle spinor (a) and due to the RPA-type vertex correction (b).

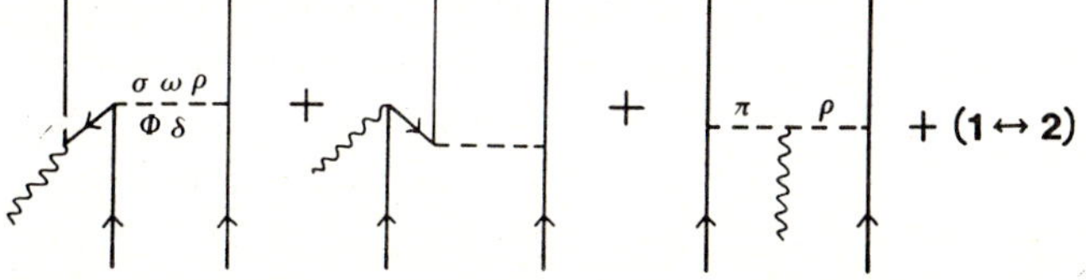

Fig.4 Exchange current processes contributing to isoscalar magnetic moments. The first two figures are the pair currents and the third one is the $\rho\pi\gamma$ dissociation current.

exchange currents. This explains why small corrections have been obtained in exchange current calculations, why very large ones in the σω-model neglecting the vertex corrections. The above arguments were discussed first by Arima et al.[14].

4. Model calculations

We have seen that a consistent description of currents in a relativistic framework requires the inclusion of vertex corrections. A fully relativistic calculation along these lines is now in progress [16]. Here we report on the recent exchange current calculation of Ichii et al.[13] based on the diagrams of fig.4. The calculations have been done for both nuclear matter and finite nuclei with one particle or one hole outside an LS closed core. It has been shown that the nuclear matter estimates employing a Fermi momentum corresponding to about half of the normal nuclear matter density can account for the results in finite nuclei fairly well. Here we show the results for finite nuclei in tables 1 and 2. Table 1 shows the deviations of the isoscalar g-factors from their free values $g_s^{(0)}=0.88$, $g_\ell^{(0)}=0.5$ and $g_p^{(0)}=0$, i.e; the deviation of the isoscalar magnetic moment from the Schmidt value is written in the form

$$\delta\mu^{(0)} = \delta g_s^{(0)} s + \delta g_\ell^{(0)} \ell + \delta g_p^{(0)} [Y^{(2)} x s]^{(1)} . \qquad (7)$$

For the meson masses and coupling constants we use the parameters of the Bonn potential [17]. A short range correlation function extracted from three-body calculations [18] is included. From table 1 we see that the exchange current contributions, listed in the columns ex, enhance both $g_s^{(0)}$ and $g_\ell^{(0)}$ by about 3% to 4%. These values are added to the contributions from second order configuration mixing [1],[5] (columns cm) to give the total corrections (tot). We see that for $g_s^{(0)}$ configuration mixing is clearly the dominant effect, while we obtain comparable contributions for $g_\ell^{(0)}$. The reasonably small, but non-negligible corrections due to exchange currents reflect the cancellation between the processes represented by figs.3a and 3b, the individual contributions of which can be written in nuclear matter as [1],[13]

$$\vec{j}_\sigma(p',p) = -\frac{\Sigma_s}{M}\frac{1+\tau^z}{2}\left(\frac{\vec{p}+\vec{p}'}{2M} + i\,\frac{\vec{\sigma}x\vec{q}}{2M}\right),$$

$$\vec{j}_\omega(p',p) = -\frac{\Sigma_v}{M}\frac{1}{2}\frac{m_\omega^2}{\vec{q}^2+m_\omega^2}\left(\frac{\vec{p}+\vec{p}'}{2M} + i\,\frac{\vec{\sigma}x\vec{q}}{2M}\right).$$

(Remember that $\Sigma_s<0$ and $\Sigma_v>0$). It is interesting to observe that our overall corrections to the g-factors for A=17 agree well with the empirical values derived by Brown and Wildenthal [19] for sd shell nuclei. Table 2 shows the calculatedand experimentally observed corrections to isoscalar magnetic moments and the

Table 1. Contributions of exchange currents (ex) and second order configuration mixing (cm) to the effective isoscalar g-factors

	$\delta g_{\ell}^{(0)}$			$\delta g_{s}^{(0)}$			$\delta g_{p}^{(0)}$		
A	cm	ex	tot	cm	ex	tot	cm	ex	tot
15	0.016	0.017	0.033	-0.180	0.035	-0.145	-0.015	0.016	0.001
17	0.011	0.013	0.024	-0.129	0.022	-0.107	0.000	0.016	0.016
39	0.010	0.018	0.028	-0.200	0.037	-0.163	-0.012	0.018	0.006
41	0.007	0.014	0.021	-0.148	0.026	-0.122	0.008	0.017	0.025

Table 2. Contributions of exchange currents and second order configuration mixing to isoscalar magnetic moments and the expectation values of the spin expressed as a pertentage of the Schmidt values

	$\delta\mu^{(0)}/\mu_S^{(0)}$ [%]				$\delta\tilde{s}_{obs}/s_S$[%]	$\delta s_{obs}/s_S$[%]	$\delta s_{calc}/s_S$[%]
A	cm	ex	tot	exp			
15	22.85	1.81	24.66	16.7	-49.26	-43.94	-67.36
17	-2.95	2.63	-0.23	-1.8	-13.63	-33.58	-22.37
39	12.49	3.01	15.50	11.1	-61.93	-45.13	-69.68
41	-2.70	2.89	0.19	-1.0	-10.21	-39.75	-27.62

corrections to the expectation values of the spin. The value $\tilde{s}_{obs}$ is extracted directly from the experimental isoscalar magnetic moment $\mu_{exp}^{(0)}$ assuming no exchange currents to be present. s_{obs} is obtained by subtracting our exchange current contributions from $\mu_{exp}^{(0)}$ and is compared with the value s_{calc} calculated from the second order configuration mixing [1],[5]. We see that for hole states (A=15,39) exchange current contributions are small compared with the second order configuration mixing effect, but that they give comparable contributions for particle states. However, these exchange current contributions are quite model dependent due to the delicate cancellation between σ-meson and ω-meson exchange contributions, which we discussed earlier. Thus, rather than presenting a final quantitative conclusion, we wish to point out that isoscalar exchange current effects, which somehow escaped a closer investigation so far, are present and should be taken into account besides the second order configuration mixing.

Finally we would like to add a comment on recent calculations [15],[20] in which an isoscalar exchange current is obtained by performing a minimal substitution in the two-body potential rather than by considering the relevant Feynman diagrams as done in this work. The connection between these two methods is discussed in [1],[13], and the result is as follows: The method of minimal substitution is consistent with current conservation only if a one-body current obtained via a FW transformation is employed. This current differs from the usual form $\bar{u}\vec{\gamma}u$ by a piece $\delta\vec{j}$, which affects the one-body off shell matrix elements [21],[22]. In the presence of a second nucleon, this piece $\delta\vec{j}$ may therefore be conveniently included in the definition of the exchange current, as noted first by Stichel and Werner [23]. The resulting exchange current, i.e; the current obtained by minimal substitution plus the piece $\delta\vec{j}$, satisfies the same proper continuity equation as does the exchange current obtained from Feynman diagrams [24]. In contrast to the work of Stichel and Werner, the piece $\delta\vec{j}$ has not been considered in refs.[15],[20].

5. Concluding remarks

We pointed out that a relativistic description of currents necessarily must include vertex corrections. These contributions are taken into account automatically in more traditional approaches starting from two-body exchange currents. From the traditional approach we learned, however, that these exchange current contributions to magnetic moments are in many cases subordinate to nuclear structure effects (configuration mixing). The inclusion of such effects in a relativistic framework is a formidable task. These problems will have to be solved in order to fully appreciate the role of nuclear magnetic moments as a touchstone for the recently developed relativistic many-body theories [8],[25].

References

1) A. Arima, K. Shimizu, W. Bentz and H. Hyuga, 'Nuclear Magnetic Properties and Gamow-Teller Transitions', to be published in Adv. Nucl. Phys.

2) A. Arima and H. Hyuga, in Mesons in Nuclei, ed. D.H. Wilkinson and M. Rho (North-Holland, Amsterdam, 1979) 685.

3) I.S. Towner and F.C. Khanna, Nucl. Phys. A399 (1983) 334.

4) A. Arima and H. Horie, Progr. Theor. Phys. 11 (1954) 509, 12 (1954) 623.

5) K. Shimizu, M. Ichimura and A. Arima, Nucl. Phys. A226 (1974) 282.

6) T. Yamazaki, in Mesons in Nuclei, ed. D.H. Wilkinson and M. Rho (North-Holland, Amsterdam 1979) 651.

7) H. Hyuga, A. Arima and K. Shimizu, Nucl. Phys. A336 (1980) 363.

8) B. Serot and J.D. Walecka, Adv. Nucl. Phys. 16 (1986) 1.

9) L.D. Miller, Ann. Phys. 91 (1975) 40.

10) A. Bouyssy, S. Marcos and J.F. Mathiot, Nucl. Phys. A415 (1984) 497.

11) M. Bawin, C. Hughes and G. Strobel, Phys. Rev. C28 (1983) 456.

12) W. Bentz, A. Arima, H. Hyuga, K. Shimizu and K. Yazaki, Nucl. Phys. A436 (1985) 593.

13) S. Ichii, W. Bentz and A. Arima, 'Isoscalar Currents and Nuclear Magnetic Moments', submitted to Nucl. Phys.

14) A. Arima, W. Bentz and S. Ichii, in Rationale of Beings, ed. K. Ishiwawa, Y. Kawazoe, M. Matsuzaki and K. Takahashi (World Scientific, Singapure, 1986)205.

15) D.O. Riska, Phys. Script. 31 (1985) 107.

16) S. Ichii, A. Arima, W. Bentz and T. Suzuki, in preparation.

17) K. Holinde, Phys. Rep. 68 (1981) 122.

18) E. Hadjimichael, S.N. Yang and G.E. Brown, Phys. Lett. 39B (1972) 594.

19) B.A. Brown and B.H. Wildenthal, Phys. Rev. C28 (1983) 2397.

20) A. Buchmann, W. Leidemann and H. Arenhoevel, Nucl. Phys. A443 (1985) 726.

21) K. Ohta, Phys. Rev. C19 (1979) 965.

22) K. Ohta and M. Ichimura, 'Gauge Invariance and the Nuclear Electromagnetic Interaction', Univ. of Tokyo preprint, 1980.

23) P. Stichel and E. Werner, Nucl. Phys. A145 (1970) 257.

24) W. Bentz, Nucl. Phys. A448 (1986) 669.

25) M.R. Anastasio, L.S. Celenza, W.S. Pong and C.M. Shakin, Phys. Rep. 100 (1983) 327.

Shell-Model Analyses of Weak and Electromagnetic Data: The Interplay of Many-Body and Single-Nucleonic Features

B.H. Wildenthal

Department of Physics and Atmospheric Science, Drexel University, Philadelphia, PA 19104, USA

1. Introduction

Our understanding of nuclear structure suggests that the values of electromagnetic and weak observables measured in nuclear physics experiments reflect an interplay between "universal" properties of neutrons and protons as they exist in finite nuclei and highly state-specific coherent features of the many-body wave functions which describe how these neutrons and protons are combined into individual nuclear eigenstates. Only for the "single-particle" and "single-hole" nuclei adjacent to doubly-magic nuclei can the many-body aspects of such observables be approximately factored out in a trivial fashion. For the great majority of nuclei, the implications of experimental data about the general features of nuclear structure can emerge only after the idiosyncrasies of the different many-body aspects of the measured matrix elements are understood and extracted. Since we do not have, a priori, quantitatively accurate theories for many-body nuclear structure, confidence in the factorization of the many-body features of nuclear matrix elements from the single-nucleonic elements is obtained only by internally consistent analyses of extensive sets of data.

Such analyses have been carried out for a variety (M1, Gamow-Teller, E2, E4) of nuclear observables in the A=17-39 region. A single family of shell-model wave functions, each member of which spans the full d5/2-s1/2-d3/2 model space, was used in these analyses. The resultant factorizations indicate that the properties of the neutrons and protons in this model context are quite invariant to the particular mass and state in which they are measured. Moreover, these properties are quantitatively consistent with current theoretical predictions and with the measured properties of the "single-particle" and "single-hole" examples. With the confidence obtained from these surveys, data from nuclei far from stability and from extended ranges of excitation energy and momentum transfer can now be examined to determine the extent to which such "extreme" conditions might modify nucleonic properties or/and the consistency between these additional many-body solutions for nuclear systems and those for previously studied "conventional" nuclei. Experiments on very neutron-rich nuclei and the results of (p,n) and (e,e') reactions all bear on this issue.

The shell-model wave functions [1] used in our analyses represent the latest stage in a long evolution during which increasing technical ability to handle large dimensions in shell-model calculations [2,3] has driven a search to discover the optimum modifications [4,5] to shell-model Hamiltonians based [6] on the features of the nucleon-nucleon systems. The "USD" wave functions of the present study have many features in common with their immediate ancestors, but are the first to describe the entire mass region in the context of a single Hamiltonian formulation. The essential feature of the new calculations which made this comprehensive treatment of the sd-shell possible is a scaling of the two-body matrix elements by a factor of A to the power -0.3. These calculations, with their internally consistent treatment of all sd-shell nuclei, create a more secure foundation for the study of shell-wide systematics than was previously available.

2. Analysis of Weak and Electromagnetic Matrix Elements for Nuclear States Occuring at Low Excitation Energies

The study of electric quadrupole (E2) matrix elements with the USD wave functions is at an intermediate stage. Static moments [7] and selected ground-state transitions [8,9] have been analysed, the latter in the context of inelastic electron scattering form factors. Detailed analyses of all measured matrix elements are under way. [10] No major modifications of the conclusions obtained in earlier studies [11,12] with Chung-Wildenthal wave functions [5] have emerged clearly. Key issues are the quantitative degree to which an effective-charge model can account for the relationships between the data and the shell-model wave functions and the detailed specifications for such renormalization models.

The conclusion of previous studies [11], that mass-independent, constant effective charges of 1.35e and 0.35e for the proton and neutron, respectively, provide a very economical renormalization of theoretical E2 strengths to experimental values in a harmonic oscillator context, continues to serve quite well with the new USD wave functions. The values of 1.15e and 0.45e for the proton and neutron in the Saxon-Woods prescription [12] also appear to be consistent with the newer results. Remaining issues which need to be elucidated include the degree to which mass and orbit dependence of the E2 effective charge are consistent with experiment and the current wave functions and, with the aid of more complete and accurate electron scattering data, the optimum prescription for the radial dependence of single-nucleon wave functions and of the effective charge or its equivalent. These empirical results will then need to be compared with new theoretical estimates of these effects and the foundations of the E2 renormalization established securely.

Electric hexadecupole data, completely in the province of electron scattering experiments, also seem to be encompassed within the context of shell-wide constant effective charges. For this operator, the optimum values are close to 1.5e and 0.5e for the proton and the neutron, respectively [8,13].

Gamow-Teller beta decay data have been compared with the predictions of the USD wavefunctions in the contexts of the subset of transitions which connect mirror states [14] and the entire body of transitions in which the daughter states can be matched one-to-one with model eigenstates [15]. The clear, mutually consistent, conclusions of the comparisons are that the predicted rates are too large if the values of the weak-interactions coupling constants are taken to be consistent with the decay of the free neutron and pure Fermi decay. A very concise renormalization of the predicted strengths, reducing them by a factor of 0.56, greatly improves the match between theory and experiment. In finer detail, independent renormalizations for the various individual single-nucleon matrix elements improve the agreement of the theory with experiment still further.

The empirical values of the renormalized single-nucleon matrix elements of the Gamow-Teller matrix elements which are determined in these shell-model analyses can be compared with predictions of various higher-order corrections to the "one major shell" shell-model picture. The empirical values are consistent with recent predictions [16,17] which are based on consideration of several competing correction mechanisms, such as two particle - two hole configuration mixing over many major shells via the mediation of the tensor force, mesonic exchange currents, and the excitation of the internal degrees of freedom of the nucleons.

The same family of model wave functions and the same nuclear states in many cases can also be used in an parallel analysis of magnetic dipole data. The M1 matrix elements are more complicated than their Gamow-Teller counterparts because they include the orbital operator in addition to the spin operator. The M1 data also are complicated in general by the mixture of isoscalar with isovector strength. Finally, the experimental values tend to be less complete and precise than the typical Gamow-Teller data.

The M1 data consisting of pairs of moments of mirror levels are, however, very precise and their individual isoscalar and isovector components can be separated neatly. Analysis of this subset of data in combination with the matching Gamow-Teller data [14] shows that, while the individual single-nucleon renormalizations of the isovector M1 operator are significant, their net effect on total multiparticle matrix elements is relatively small in comparison to the corresponding Gamow-Teller examples. This can be understood theoretically in the context of different phases between different components of the renormalizations, which arise because of the different extra-model excitations available with the weak and electromagnetic operators.

This same analysis also yielded empirical values for the renormalizations of the single-nucleon matrix elements of the isoscalar M1 operator. The conventional view is that isoscalar magnetic moments are easily reproduced to high accuracy in shell-model calculations. This conclusion ignores the fact that the experimental and calculated numbers are dominated by the expectation value of J, which, of course, any calculation not technically incorrect must yield exactly. The nontrivial residues of isoscalar magnetic moment matrix elements are reproduced at about the same level of accuracy with our wave functions as are the isovector matrix elements. While not perfect, this level of accuracy is sufficient to reveal that the single-nucleon matrix elements of the spin component of the isoscalar M1 operator need empirical renormalizations which are comparable to the isovector renormalizations.

This point is significant in the context of identifying the dominant source of the corrections to the conventional shell-model representation. Configuration mixing over many oscillator shells affects isovector and isoscalar M1 matrix elements in the same way, but the excitation of the nucleon into the delta resonance, an isovector excitation, would affect isovector matrix elements only. The evidence adduced for isoscalar quenching is, accordingly, suggestive of a important role for higher-order configuration mixing in all aspects of quenching of spin-related observables.

Analysis of M1 transitions [18] can provide a better fix on the renormalization of the spin-flip and s-state components of the isovector M1 operator than that available from the moments alone, but the fragmentary data from T=0 nuclei do not appear on close analysis to yield additional information on the isoscalar components. Even with more and better data, a thorough treatment of isospin mixing might be required before any conclusions about isoscalar renormalizations could safely be reached.

In summary, analysis of weak and electromagnetic moments and transition rates for sd-shell nuclei with the USD wave functions suggests that these model predictions are an efficient vehicle with which to account for observed matrix elements of complex nuclear states in terms of a many-body component, which varies over a couple of orders of magnitude from state to state, and a single-nucleonic component in which the properties of the model nucleons are virtually state independent. The properties of the model nucleons are expressed in terms of single-particle matrix elements which, while different from the values to be expected from the measured properties of the free neutron and proton in theoretically understood ways, are independent of nuclear state and mass to a good approximation.

3. Extensions to New Regions of Neutron-Proton Ratio, Excitation Energy, and Momentum Transfer

The USD Hamiltonian and its wave functions are deeply linked with the energy levels of stable and near-stable nuclei, from which the dominant portion of our experimental knowledge of nuclear structure is drawn. The energies themselves are used as the criterion by which the Hamiltonian is "renormalized". Measured values of the moments of these levels and of the strengths of transitions between them are compared with model predictions to validate the wavefunction amplitudes which emerge from diagonalizing the Hamiltonian. The results of these comparisons seem to indicate that the sd-shell model space as deployed with the USD Hamiltonian gives a systematically sound accounting of the "major" components of the wave functions of these energy levels. By this we mean that it seems likely that the relative magnitudes and phases of the largest wave function components are systematically correct to within "reasonable" uncertainty. If this were not the case, the model's systematically good theoretical reproduction of observed single-nucleon spectroscopic factors, as well as the comprehensively good accounting for weak and electromagnetic matrix elements discussed above, would be impossible to understand.

Despite these successes, fundamental questions about the range of validity of the model space and the efficacy of the present Hamiltonian remain to be answered. What are the "natural" limits of agreement between model predictions and experiment? How accurately estimated are the "small" components of the model wave functions? How well do the model predictions agree with experimental data from regimes of neutron-proton number, excitation energy and momentum transfer which are different from those characteristic of the levels used to renormalize the Hamiltonian?

The correctness of the small-amplitude components in the model wave functions is difficult to assess. Their influence is not apparent in the great majority of matrix elements, and comes to the fore only in instances in which selection rules suppress the contributions of the large-amplitude components. At the same time, the influence of systematic trends in small-amplitude components can compete easily with the effects of the renormalization of the single-nucleonic components. Consequently, it is essential to extend checks of the model wave functions over as wide and varied a range of data as possible.

One route to expanding the data set for model checks is to bring new nuclei into consideration. In practice, this means creating systems with large asymmetries in neutron-proton ratios and measuring their properties. Recent work with complex heavy-ion transfer reactions [19-21] and with isotope-separator techniques [22-24] has provided beautiful new tests of the USD predictions for the energy-level sequences of neutron-rich isotopes of Ne, Mg, and Si. A primary virtue of the shell-model formulation of nuclear structure is the intrinsic association of wave functions with predicted energies. Hence the energy predictions of the USD Hamiltonian are accompanied by predictions of nucleon-transfer widths, beta-decay rates and gamma-branching probabilities. Such predictions [25] make possible much more thorough tests of the model predictions and facilitate the planning of new experiments.

In addition to new combinations of neutrons and protons, the realm of experimental data can be expanded in terms of increasing the range of excitation energies considered. In the case of yrast and near-yrast levels, higher excitation energies mostly compensate for the trivial effects of higher angular momentum. For lower spin states, higher excitation energies are associated with the onset of "intruder" levels, extra states of the same spin-parity which naturally occur at higher energies by excitation of configurations which break the model-space limits. Intruder states, coupled with experimental difficulties, make it impossible to track the correspondence of model with experimental levels on a one-to-one basis up through the dozens of levels which are predicted to exist for low-spin states in the region of 5-15 MeV excitation energy.

At the same time, vital phenomena are predicted to occur at these energies, particularly in the area of spin excitations. The peak of the Gamow-Teller and M1 strength is typically predicted in the 8-12 MeV range. Comparisons of model predictions with experiment is most meaningful when the dominant fraction of the model strength is in play. Then, the basic parameters of the theory are at test, not just the fine details. M1 and Gamow-Teller matrix elements taken from states at low excitation energy tend to be small. It is critical to confirm conclusions based on comparisons of the model predictions with experiment in the low-energy "tails" of the distributions of spin-excitation strength by examining the "peaks". The experimental techniques which make such examinations possible are back-angle inelastic electron scattering [26-27], photon scattering [28-30], large-Q-value Gamow-Teller decays [31-33] and (p,n)-like reactions in appropriate configurations [34-36]. Data from such experiments are available for only a few nuclei in the sd-shell. It is very important that the fragmentary information from the existing samples be significantly enlarged by exploiting the superlative facilaties now in operation for such studies.

The existing data on M1 excitations, predominantly for even-even nuclei [26-28] is consistent with the analysis of states at low excitation energy. The detailed distribution of strength over the final J=1 states is not always predicted accurately, but the centroid of the strength and its integrated magnitude are. As with the low-lying data there is no clear-cut evidence yet for strong quenching relative to the model predictions. The complementary picture on Gamow-Teller excitations is comparably sparse. Studies of the A=26 and A=32 systems are complete [34,35] and some [n,p] data have been obtained [36]. Here, the present analysis is again consistent with that for strength at lower energies, namely, the model strengths must be quenched by a factor of approximately 0.55 to 0.60. Experiments with large-Q-value beta decays offer a different route to sample the dominant portion of Gamow-Teller strength. They offer the advantage of a better understood mechanism but they typically still have difficulty covering the entire strenth distribution. Again, shell-model studies of these data, including analyses [37] with the Chung-Wildenthal as well as the USD wave functions, to date yield quenching factors of 0.50 to 0.60.

In summary, the existing picture of spin excitation strengths in the vicinities of the peaks of the strenth distributions, while fragmentary, is consistent with the model-experiment relationships in the tails of the distributions. This is evidence that the model predictions are valid in a significantly larger domain than might have been expected and it gives increased confidence in the universality of the renormalizations of the single-nucleonic components of the theory.

The final dimension of an expanded realm for model-experimental comparisons is that of variable momentum transfer. We are still at a beginning stage in analysing modern data with modern wave functions. The analysis of quadrupole and hexadecupole data from even-even nuclei indicates [8,9,13] that internally consistent predictions agree with data up to 3 inverse fermis. High-resolution data on 27Al [38] are more detailed and complex, and the model predictions [39] are equivalently successful in reproducing their features. The 19F nucleus is so close to the shell boundary that intruder states dominate much of its spectrum, but for states clearly within the model's compass, predictions agree with experiment [9] in a fashion consistent with 27Al and the even-even systems.

4. Conclusion

The existence of comprehensive and internally consistent model predictions for a very wide range of phenomena in sd-shell nuclei creates a positive feedback loop for experimental studies on these systems. Through the model wave functions, every matrix element is linked to every other. Every measurement is thus important and meaningful in the advance towards increasing our understanding of nuclear structure on the whole and in improving the accuracy and reliability of model wave functions. The final limits of accuracy in the type of approach discussed here are not easily estimated at present, but the current achievements can certainly be surpassed if the appropriate experimental edifice is built on the existing foundations.

References

1. B. H. Wildenthal: Progress in Particle and Nuclear Physics (Pergamon, New York, 1983), Vol. 11, p. 5.
2. J. B. French, E. C. Halbert, J. B. McGrory and S. S. M. Wong: Advances in Nuclear Physics (Plenum Press, New York, 1969), Vol. 3, ch. 3
3. R. R. Whitehead, A. Watt, B. J. Cole and I. Morrison: Advances in Nuclear Physics (Plenum Press, New York, 1977), Vol. 8, ch. 3
4. B. M. Preedom and B. H. Wildenthal: Phys. Rev. C6, 1633 (1972)
5. W. Chung: Ph.D. Thesis, Michigan State University, unpublished (1976)
6. T. T. S. Kuo and G. E. Brown: Nucl. Phys. 85, 40 (1966)
7. M. C. Carchidi, B. H. Wildenthal and B. A. Brown: unpublished

8. B. A. Brown, R. Radhi and B. H. Wildenthal: Phys. Reports 101, 313 (1983)
9. B. A. Brown et al.: Phys. Rev. C32, 1127 (1985)
10. B. H. Wildenthal, B. A. Brown and J. Keinonen, unpublished
11. D. Schwalm, E. K. Warburton and J. W. Olness: Nucl. Phys. A293, 425 (1977)
12. B. A. Brown et al.: Phys. Rev. C26, 2247 (1982)
13. B. H. Wildenthal, B. A. Brown and I. Sick: Phys. Rev. C32, 2185 (1985)
14. B. A. Brown and B. H. Wildenthal: Phys. Rev. C28, 2397 (1983)
15. B. A. Brown and B. H. Wildenthal: Atomic Data and Nuclear Data Tables 33,347 (1985)
16. H. Hyuga, A. Arima and K. Shimizu: Nucl. Phys. A336, 363 (1980)
17. I. S. Towner and F. C. Khanna: Nucl. Phys. A399, 334 (1983)
18. M. C. Etchegoyen, A. Etchegoyen, B. H. Wildenthal and B. A. Brown: unpublished
19. L. K. Fifield et al.: Nucl. Phys. A437, 141 (1985)
20. L. K. Fifield et al.: Nucl. Phys. A441, 531 (1985)
21. D. Guillemaud-Mueller et al.: Nucl. Phys. A426, 37 (1984)
22. P. Baumann et al.: Contribution to this Symposium
23. A. C. Mueller: Contribution to this Symposium
24. J. P. Dufour: Contribution to this Symposium
25. B. H. Wildenthal, M. S. Curtin and B. A. Brown: Phys. Rev. C28, 1343 (1983)
26. L. W. Fagg: Rev. Mod. Phys. 47, 683 (1975)
27. R. W. Schneider et al: Nucl. Phys. A323, 13 (1979)
28. U. E. P. Berg et al.: Phys. Lett. 140B, 191 (1984)
29. R. Vodhanel et al.: Phys. Rev. C29, 409 (1984)
30. R. Moreh: Private Communication
31. J. Aysto et al.: Phys. Rev. C32, 1700 (1985)
32. T. Bjornstad et al.: Nucl. Phys. A443, 283 (1985)
33. G. Walter: Private Communication
34. R. Madey et al.: Phys. Rev. C, to be published
35. B. D. Anderson et al: Phys. Rev. C, to be published
36. O. Hausser: Contribution to this Symposium
37. W. Muller et al.: Nucl. Phys. A430, 61 (1884)
38. P. J. Hicks et al.: Phys. Rev. C27, 2515 (1983)
39. R. Radhi, Ph.D. Thesis, Michigan State University, (1983), unpublished

Global Set of Quadrupole Deformation Parameters for Even-Even Nuclei

S. Raman and C.W. Nestor, Jr.

Oak Ridge National Laboratory, Oak Ridge, TN 37831, USA

We have completed a compilation of experimental results for the reduced electric quadrupole transition probability [$B(E2)\uparrow$] between the 0^+ ground state and the first 2^+ state in even-even nuclei. This compilation together with certain simple relationships noted by other authors can be used to make reasonable predictions of unmeasured $B(E2)\uparrow$ values. The quadrupole deformation parameter β_2 immediately follows, because β_2 is proportional to $[B(E2)\uparrow]^{1/2}$.

We have collected experimental results for the reduced electric-quadrupole transition probability, $B(E2)\uparrow$, between the 0^+ ground state and the first 2^+ state in even-even nuclides [1]. These $B(E2)\uparrow$ values represent rather basic nuclear information complementary to our knowledge of the energies of low-lying levels in these nuclides. Generally larger than expected from the single-particle model, they have emphasized the widespread occurrence of quadrupole distortions in nuclides.

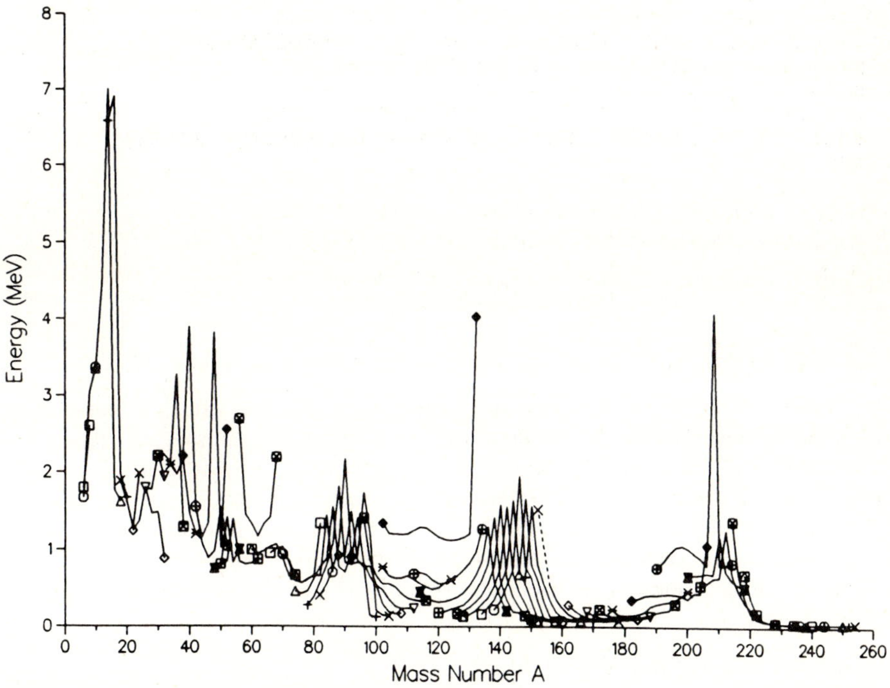

Fig. 1. Energy of the first-excited 2^+ state.

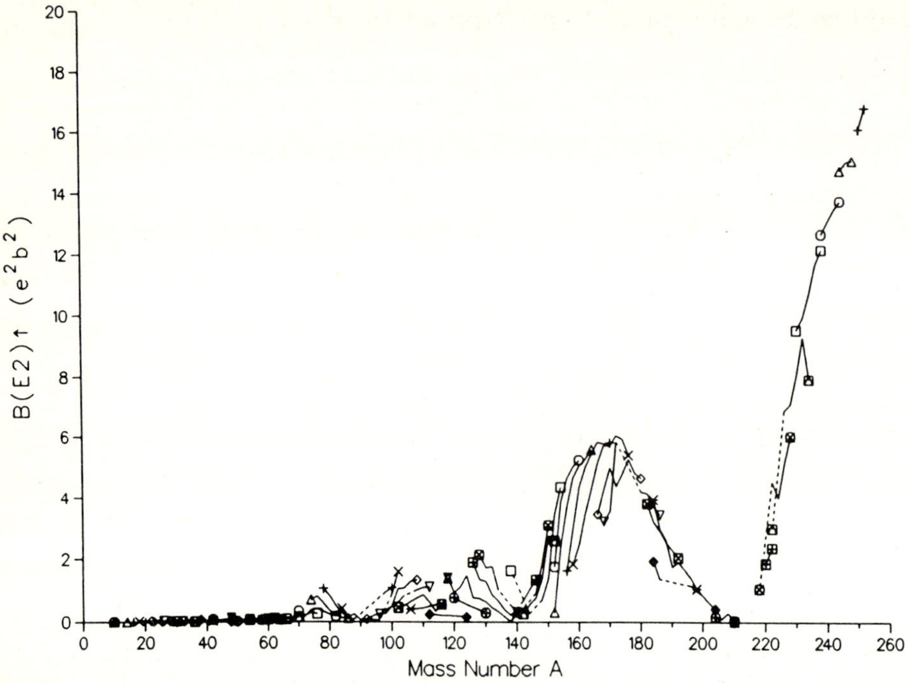

Fig. 2. Reduced transition probability, $B(E2)\uparrow$.

Our starting point was a previous $B(E2)\uparrow$ compilation [2], which contained 476 measured $B(E2)\uparrow$ values from 133 references leading to adopted $B(E2)\uparrow$ values for 155 nuclides. The current compilation contains 1765 entries from 793 references, leading to adopted $B(E2)\uparrow$ values for 281 nuclides. The energies of the first 2^+ states are now known for 457 nuclides. An overall view of the data in this compilation is shown graphically in Figs. 1 and 2.

Assuming a uniform charge distribution out to the distance $R(\theta, \phi)$ and zero charge beyond, the quadrupole deformation parameter β_2 is related to $B(E2)$ by the formula

$$\beta_2 = (4\pi/3ZR_0^2)\ [B(E2)\uparrow/e^2]^{1/2}\,. \qquad (1)$$

We have taken R_0 to be 1.2 $A^{1/3}$ fermis and the single-particle $\beta_{2\,(\mathrm{sp})}$ to be $1.59/Z$.

For nuclides without an experimentally determined $B(E2)\uparrow$ value, a reasonable prediction based on systematics can be obtained by adopting the approaches suggested by GRODZINS [3]; HAMAMOTO [4]; ROSS and BHADURI [5]; PATNAIK, PATRA, and SATPATHY [6]; and CASTEN [7]. We have first tested these approaches with the data in our compilation. In general, predictions of $B(E2)\uparrow$ values for nuclei off the stability line can then be made to an accuracy of $\pm 25\%$. Such predictions can be used to deduce systematic trends in deformation as was done recently [8] in the case of the Ce, Nd, Sm, and Gd isotopes.

GRODZINS [3] noted first that the γ-ray mean life τ_γ was approximately proportional to E^{-4}, where E is the energy of the $2^+ \rightarrow 0^+$ transition. He then suggested the following empirical relationship (τ_γ in psec and E in keV):

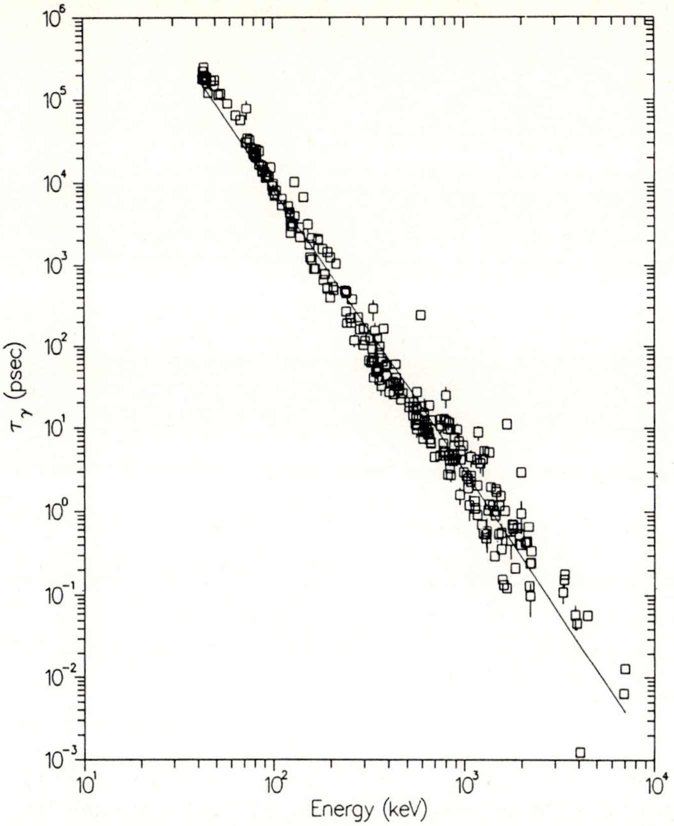

Fig. 3. Gamma-ray mean life as a function of E. This plot is similar to Fig. 2 of Ref. [2].

$$\tau_\gamma \approx 3.33 \times 10^{13}\ E^{-4}\ Z^{-2}\ A\ . \tag{2}$$

Our equivalent plot is shown in Fig. 3. A least-squares fit to the current data yields

$$\tau_\gamma \approx 5.96 \times 10^{10}\ E^{-3.6}\ Z^{-8.4}\ A^{7.1}\ . \tag{2}$$

HAMAMOTO [4] plotted the quantity $[\beta_2/\beta_{2(sp)}]/N_pN_n$ as a function of Z, where the valence number of protons (neutrons) $N_p(N_n)$ is defined as the number of particles below midshell and the number of holes past midshell. She noted that the above quantity was roughly constant over the entire mass region (both spherical and deformed) except for the regions near closed shells. Our equivalent plot is shown in Fig. 4.

After defining $F(N,Z) \equiv [E \times B(E2)]^{-1}$, ROSS and BHADURI [5] found that the value of $F(N,Z)$ for the anchor N,Z nucleus is related to the values for the three neighboring nuclides by the difference equation

$$F(N,Z) + F(N+2,Z+2) - F(N+2,Z) - F(N,Z+2) \sim 0\ . \tag{4}$$

PATNAIK, PATRA, and SATPATHY [6] further noted that a similar difference equation is also satisfied by E, the energy of the first 2^+ state, and by the reduced transition probability $B(E2)\uparrow$.

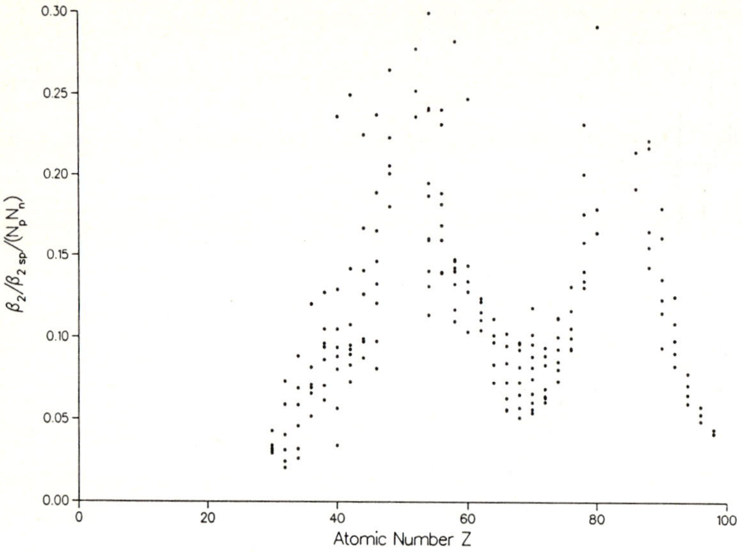

Fig. 4. Quadrupole deformation parameter, β_2 divided by $\beta_{2(sp)}$ ($\equiv 1.59/Z$) as a function of Z. This plot is similar to Fig. 2 of Ref. [4].

$$E(N,Z) + E(N+2,Z+2) - E(N+2,Z) - E(N,Z+2) \sim 0 \ . \tag{5}$$

$$B(E2)[N,Z] + B(E2)[N+2,Z+2] - B(E2)[N+2,Z] - B(E2)[N,Z+2] \sim 0 \ . \tag{6}$$

We have tested the difference equations (4) and (6) with the data in our compilation. The results are shown in Fig. 5. Except near closed shells, both approaches yield deviations of the difference equations from zero that are typically $\pm 25\%$. Even though they were proposed more than a decade ago, these difference equations have not received the attention that they probably deserve.

Finally, CASTEN [7] has recently attempted a systemization of a considerable amount of spectroscopic data, especially in the five nuclear transition regions [$A \approx 100$, $A \approx 130$, A ≈ 150 (Z $<$ 64), $A \approx 150$ ($Z \geqslant 64$), and $A \approx 190$] in terms of the valence nucleon product N_pN_n. He has shown that the $B(E2)$ values in the $A \approx 150$ (Z $\leqslant$ 64), $A \approx 150$ ($Z \geqslant 66$), and $A \approx 100$ regions fall on relatively smooth curves. Some of CASTEN's conclusions are based on *a posteriori* counting of valence nucleons. Separating the $A \approx 150$ region into $Z \leqslant 64$ and $Z \geqslant 66$ parts is just one example, for which he has provided ample justification in terms of a subshell closure. Our equivalent plot for the $A \approx 100$ region is shown in Fig. 6. We use his prescription that if $N < 60$, the proton shell is 38 to 50, and if $N \geqslant 60$, the proton shell is 28 to 50.

Returning to HAMAMOTO's [4] original idea, we have plotted in Fig. 7 the quantity $\beta_2/\beta_{2(sp)}$ vs N_pN_n for the isotopes of eight even-Z elements with $64 < Z < 82$. The curve drawn there accounts rather well for 61 data points.

We have shown that there are several ways of systematizing the $B(E2)$ values and the related quantities τ_γ and $\beta_2/\beta_{2(sp)}$. It is relatively straightforward then to arrive at a global set of recommended $B(E2)$ and β_2 values by combining measured values with predictions. We are currently in the process of arriving at such a set.

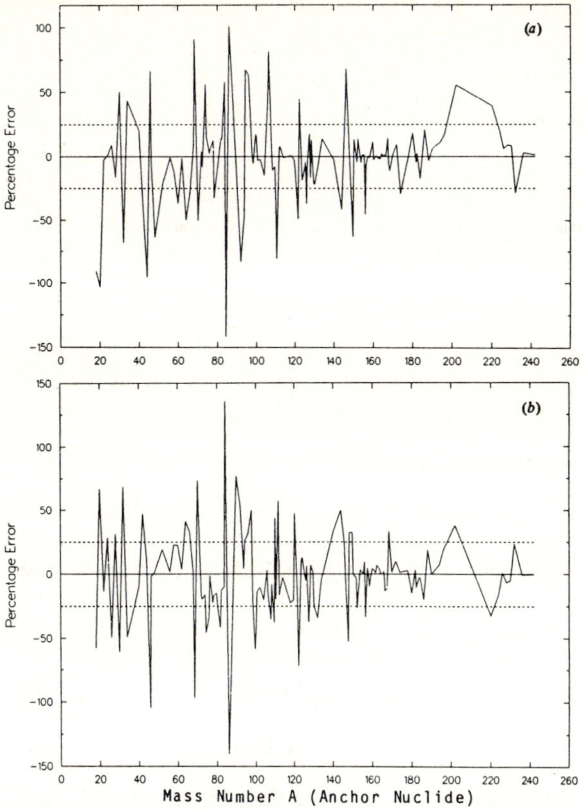

Fig. 5. Test of the difference equations of (a) ROSS and BHADURI [5]—see Eq. (3), and (b) PATNAIK, PATRA, and SATPATHY [6]—see Eq. (5). The deviation from zero of either Eq. (3) or Eq. (5) is divided by the average of the four values, expressed as a percentage, and plotted along the ordinate.

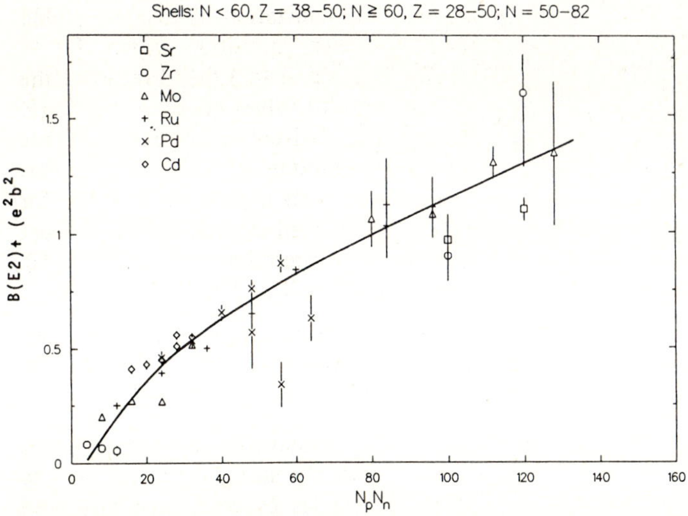

Fig. 6. $B(E2)$ systematics for the $A \approx 100$ region. This plot is similar to Fig. 25 of Ref. [7]. The curve is drawn to guide the eye.

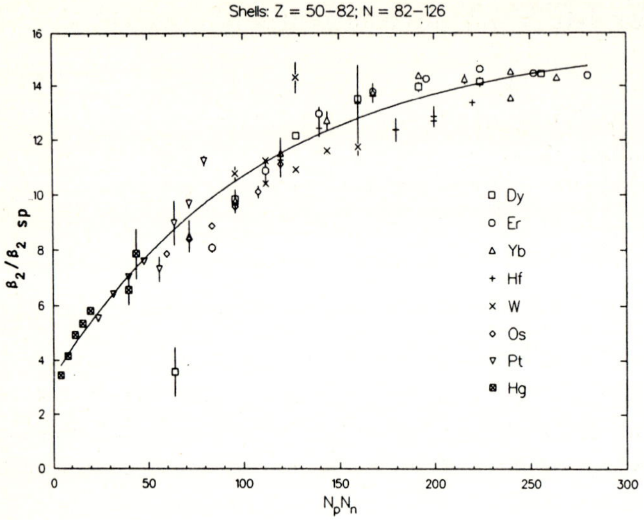

Fig. 7. Quadrupole deformation parameter, β_2, divided by $\beta_{2(sp)}$ ($\equiv 1.59/Z$) as a function of the quantity N_pN_n for even-Z isotopes in the $64 < Z < 82$ region. The curve is drawn to guide the eye.

This research was sponsored by the Division of Nuclear Physics, Office of High Energy and Nuclear Physics, U.S. Department of Energy, under Contract No. DE-AC05-84OR21400 with the Martin Marietta Energy Systems, Inc.

References

[1] S. Raman, C. H. Malarkey, W. T. Milner, C. W. Nestor, Jr., and P. H. Stelson, At. Data Nucl. Data Tables, to be published.
[2] P. H. Stelson and L. Grodzins, *Nucl. Data* **A1**, 21 (1965).
[3] L. Grodzins, *Phys. Lett.* **2**, 88 (1962).
[4] I. Hamamoto, *Nucl. Phys.* **73**, 225 (1965).
[5] C. K. Ross and R. K. Bhaduri, *Nucl. Phys.* **A196**, 369 (1972).
[6] R. Patnaik, R. Patra, and L. Satpathy, *Phys. Rev.* C **12**, 2038 (1975).
[7] R. F. Casten, *Nucl. Phys.* **A443,** 1 (1985).
[8] C. J. Lister et al., *Phys. Rev. Lett.* **55**, 810 (1985).

Core Polarization for Magnetic Transitions (λ=1,2,4)

W. Andrejtscheff[1], *Ch. Stoyanov*[1], *and A.I. Vdovin*[2]

[1]Bulgarian Academy of Sciences, Institute for Nuclear Research and Nuclear Energy, 1784 Sofia, Bulgaria
[2]Joint Institute of Nuclear Research, Laboratory of Theoretical Physics, Dubna 141980, USSR

The systematics of ℓ-forbidden $M1$ as well as of allowed $M2$ and $M4$ transitions in non-deformed nuclei is considered under emphasis of the core-polarization effect. Reported is an A-dependence of the ℓ-forbidden $M1$ strength. $M2$ and $M4$ transition probabilities are calculated within the quasiparticle-phonon model.

1. Mass-Number Dependence of ℓ-Forbidden $M1$ Strengths

This class of $M1$ transitions is of special interest because they might serve as a very sensitive measure of subnuclear effects [1]. However, in order to extract information on such effects a clear separation of all possible nuclear interactions is required. Here, we concentrate only on an A-dependence of the ℓ-forbidden $M1$ strength not reported so far in the literature.

Magnetic dipole transitions with $\Delta\ell = 2$ are forbidden in the pure shell model (ℓ-forbiddenness). Generally, they occur due to: (i) shell-model configuration admixtures arising from core polarization of first order [2] and/or of higher orders [3]. The influence of such admixtures depends on position and strength of the 1^+ core states to which the levels under discussion in the odd-A nucleus are coupled; ii) mesonic exchange currents [4] and (iii) isobaric effects, e.g., possible admixtures from the Δ-resonance [1,4]. It is possible to extend the bar $M1$ operator by a tensor term $g_p[Y_2 s]$ which allows $\Delta\ell = 2$ transitions. In the case of ℓ-forbidden $M1$ transitions the value of g_p becomes crucial. Attempts were undertaken to extract it from a fit of the data [5] or to calculate it taking into account the effects (i–iii) [6].

In nuclei with an LS shell closure (e.g. $Z = 20$ or $N = 20$), first-order core polarization (which we would naively expect to be one of the strongest effects) is exactly forbidden. Thus, in nuclei approaching $Z = 20$ or $N = 20$ an appreciable decrease of the transition strength should be observed. There are some indications in this sense which are stronger in odd-neutron nuclei than in odd-proton ones [7]. At the same time several experimental $B(M1)$ values indicate that first-order core-polarization does not necessarily provide the largest contributions. An example is ^{39}K with one proton-hole in the ^{40}Ca core. First-order core-polarization is there exactly forbidden. Nevertheless the $B(M1)$ value is in the same range as in several other nuclei. Extensive theoretical efforts have been undertaken in the $A = 39$ region [5,6]. Of particular interest here is the calculation [6] of the ℓ-forbidden $M1$ strength $d_{3/2} \to s_{1/2}$ in ^{39}K with the extended $M1$ operator including all the effects (i–iii) above mentioned. The theoretical value $B(M1)_{\mathrm{th}}$ is about seven times lower than the experimental one. Even of more interest is the formation of the theoretical value of g_p $[B(M1, d_{3/2} \to s_{1/2}) = (0.053 g_p)^2]$. There, by far the largest contribution comes from the isobaric currents, i.e., from subnuclear effects. Magnetic dipole ℓ-forbidden transition probabilities in nuclei with $90 < A < 150$ were recently calculated within the theory of finite Fermi systems. A reasonable agreement with the experimental data is reported [8].

In general, the ℓ-forbidden $B(M1)$ values in the vicinity of closed shells are of the size $10^{-3} \ldots 10^{2} \mu_N^2$, except for those around N, $Z = 20$ (LS closure) where they might become lower [7]. Away from closed shells the mixing in the wave functions considerably increases. It appears hard to predict a priori what would be the behaviour of the corresponding $B(M1)$ data. Within the particle-vibration coupling model it has been shown [9] that even in case of strong configuration mixing away from closed shells ℓ-forbiddeness continues to persist. The main contribution comes from the single-particle matrix element of the $[Y_2 \boldsymbol{s}]$ term while contributions including phonon admixtures cancel each other. Correspondingly, similar $B(M1)$ values should be expected in the vicinity and away from closed shells. These conclusions resemble the validity of the Ward identity in quantum electrodynamics. This expectation has been experimentally observed [10] in some odd-A nuclei away from closed shells considered within different models: in ^{101,103,105}Pd treated within a slightly-deformed-rotor-plus-quasiparticle model and in ^{71,73}As where the $M1$ transition rate is satisfactorily reproduced by the cluster-vibrational model. Magnetic dipole transitions with $\Delta\ell = 2$ might occur also between multiquasiparticle configurations in odd or even nuclei (^{91}Nb, ^{134}Te, ^{144}Sm, ^{144}Eu). There, the ℓ-forbidden $M1$ strength is in the same order of magnitude as that between corresponding one-quasiparticle configurations [10–12].

The reduced transition probabilities $B(M1)$ of the (most) known ℓ-forbidden $M1$ transitions in odd-A nuclei are presented in Fig. 1 versus the mass number A. This systematics has some pronounced features. (a) In the vicinity of N, $Z \approx 20$ ($A \approx 50$) some particularly low values occur. This is the region of LS closure discussed above. (b) For a given mass number, fluctuations within one or sometimes even two orders of magnitude are typical. They are due to the influence of the nuclear and subnuclear effects mentioned above (i–iii). (c) A general slowly decreasing tendency with increasing mass number can be nevertheless recognized (note the logarithmic scale). The physical background of this decrease is hard to understand.

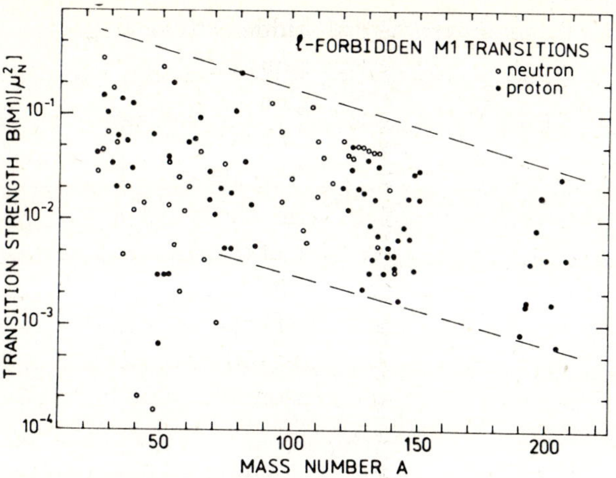

Fig. 1. Experimental $B(M1)$ values of ℓ-forbidden transitions versus the mass number A. The error of the values is about 20%

First we look at the position and strength of 1^+ (even-even) core states with increasing mass number. The trend in Fig. 1 could be understood if with increasing mass number the average position of such states increase and/or their $M1$ strength decrease. However, such a tendency for the 1^+ states can be hardly expected from the available (insufficient) data. From our present knowledge, a general slope of the (average) 1^+ energies with $\sim A^{-1/3}$ appears rather probable [1,13] although further data are needed to firmly establish the A-dependence of $E_{\rm av}(1^+)$. A mass-number dependence of the Δ-isobar influence on the magnetic properties of low-lying nuclear states is also not likely [14].

A possible idea to roughly understand the physical background (c) of Fig. 1 is offered by the model of a particle coupled to quadrupole vibrations. As already mentioned, within this model only the matrix element of the $[Y_s\boldsymbol{s}]$ term between the (main) $\Delta\ell = 2$ shell-model components does not vanish while the phonon contributions cancel each other [9]. With increasing mass the quadrupole vibrational excitations become more collective and lower in energy, i.e., their contributions to the wave functions of the odd-A nuclei (phonon contributions) should increase. Correspondingly, the (main) $\Delta\ell = 2$ shell-model components slowly decrease. This might finally produce the right trend of Fig. 1. Of course, this qualitative picture is quite rough, several points should be carefully considered: the role of shell effects, pairing correlations as well as the above mentioned phenomena (i–iii).

We believe that a deeper understanding of the presented A-dependence will provide access to important details in the formation of the $M1$ matrix elements.

2. Core Polarization for $M2$ Transitions in Odd Spherical Nuclei

Reduced transition probabilities for magnetic quadrupole transitions between states with $J^{\pi} = 11/2^{-}$ (predominantly $h_{11/2}$) and $7/2^{+}$ (predominantly $g_{7/2}$) have been determined experimentally in more than thirty odd-A non-deformed nuclei. In a single isotopic chain (e.g. Sn) they might vary smoothly offering a chance for the systematic study of the different effects involved. It seemed worthwhile to perform systematic calculations of $M2$ transition rates in the frame of the quasiparticle-phonon model [15]. This model has been successful in the description of many characteristics of spherical nuclei. Calculations of $M2$ transition probabilities were carried out in a series of Sn, Sb, Pr, Pm and Eu isotopes.

The wave functions of the states include quasiparticle and quasiparticle-plus-phonon components and have the form:

$$\Psi_{\nu}(JM) = C_{J\nu}\Big\{\alpha^{+}_{JM} + \sum_{\lambda ij} D^{\lambda i}_{j}(J\nu)[\alpha^{+}_{jm}Q^{+}_{\lambda\mu i}]_{JM}\Big\}\Psi_{0}$$

where α^{+}_{jm} and $Q^{+}_{\lambda\mu i}$ are the creation operators of a quasiparticle and a phonon, respectively. The subscripts are the corresponding quantum numbers; Ψ_0 is the ground state of the even-even core (quasiparticle and phonon vacuum). The amplitudes $C_{J\nu}$ and $D^{\lambda i}_{j}$ are found by solving the model equations [15]. The following parameters were used in the model Hamiltonian: the parameters of the Woods-Saxon single-particle potential, the BCS pairing constants G_N and G_Z and the constants of the isoscalar and isovector effective separable multipole and spinmultipole forces ($\kappa_0^{(\lambda)}$, $\kappa_1^{(\lambda)}$ and $\kappa_0^{(\lambda L)}$, $\kappa_1^{(\lambda L)}$, respectively). The matrix element $\langle J_2\|M\lambda\|J_1\rangle$ consists of two parts. The first part $\langle J_2\|M\lambda_{\mathrm{q.p.}}\|J_1\rangle$ is proportional to the pairing factor $V^{(+)}_{j_1j_2}$ and represents the contribution of one-quasiparticle components. The second one $\langle J_2\|M\lambda_{\mathrm{ph}}\|J_1\rangle$ takes into account the influence of the quasiparticle-plus-phonon components, i.e. this is the core polarization term.

In Fig. 2 we have shown the distribution of the $M2$ strength in the even core with two main regions of concentration. The first one is around 10 MeV. The orbital term of the $M2$ operator covers an essential part of the $M2$ strength [16] and this group of states depends only weakly on the constant of the spin-dipole forces. The states around 20 MeV are "spin" states and their position depends stronger on the spin-dipole strength constant $\kappa_1^{(12)}$. These states are very collective and interact very strongly with the odd quasiparticle. Their contribution to the structure of low-lying $11/2^{-}$ and $7/2^{+}$ states is around 0.1+0.4% but due to that strong interaction they are responsible for the main part of the $M2$ reduction. Around the closed shell $Z = 50$, the structure of the collective $M2$ states at $\approx$20 MeV consists of proton configurations. Thus, the odd proton interacts stronger with the $Z = 50$ even core than the odd neutron.

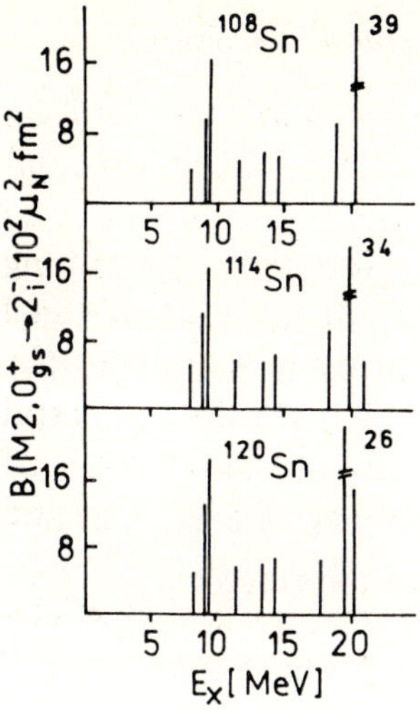

Fig. 2. $M2$ distribution in even tin nuclei (theory)

In detail, the numerical results on $M2$ transitions are given in [17]. Here, we point at two main features: (i) The core polarization clearly reduces the $B(M2)$ values. The degree of reduction depends on the number of 2^- states included in the calculation and on the ratio of the isoscalar and isovector spin-dipole constants. This means that the core polarization effect is different for odd-Z and odd-N nuclei. (ii) Some specific features in the variation of the $B(M2)$ values in an isotopic chain are associated with the pairing effect. A part of the results is presented in Table 1 ($\kappa_0^{(12)}/\kappa_1^{(12)} = 0.1$). The table illustrates in a clear way both features mentioned above. In spite of the strong reduction due to core polarization, the calculated transition rates $B(M2)$ are still by 4–5 times larger than the experimental values. In order to describe the data, the use of an effective spin gyromagnetic factor $g_{\mathrm{s\,eff}} = 0.5\, g_{\mathrm{s\,free}}$ is needed. This value

Table 1. Experimental and theoretical transition rates $B(M2, 11/2^- \to 7/2^+)$ $[\mu_N^2 fm^2]$. Abbreviation: c.p. = core polarization

Nuclei	Experiment	Theory	
		no c.p.	with c.p.
^{113}Sb	—	194	51.1
^{115}Sb	8.97±0.8	147	43.1
^{117}Sb	11.8±0.8	148	42.9
^{119}Sb	8.78±1.2	150	42.4
^{121}Sb	—	171	48.1
^{109}Sn	8.2±1.2	143	33.3
^{111}Sn	6.1±0.8	127	30.4
^{113}Sn	4.8±0.1	117	28.2
^{115}Sn	4.68±0.04	98	24.2
^{117}Sn	4.7±1.3	96	23.6
^{119}Sn	5.4±1.9	99	25.2
^{121}Sn	8.6±2.1	111	30.9

is about two times higher than that used in [18] where only the interaction of the odd particle with vibrations is taken into account.

3. Magnetic Hexadecapole Transitions

$M4$ transition probabilities are studied in the framework of the quasiparticle-phonon model. The distribution of the $M4$ strength in some non-deformed nuclei of different mass regions is illustrated in Fig. 3. The $M4$ strength distribution is similar to that obtained in [19]. The calculations were performed with following model parameters. The isovector spin-octupole strength parameter $\kappa^{(34)}$ has been chosen to be equal to the octupole parameter $\kappa_1^{(3)}$. The isoscalar parameter is ten times smaller than the isovector one. The calculated $B(M4)$ values do not essentially change by variations of the isoscalar parameter $\kappa_0^{(34)}$ as it is in the case of $M2$ transitions. Further, the theoretical values $B(M4)$ are only weakly influenced by pairing $(v^2 \sim 1)$ and the single-particle amplitudes are close to unity $(C^2 \sim 1)$. Numerical results are shown in Table 2.

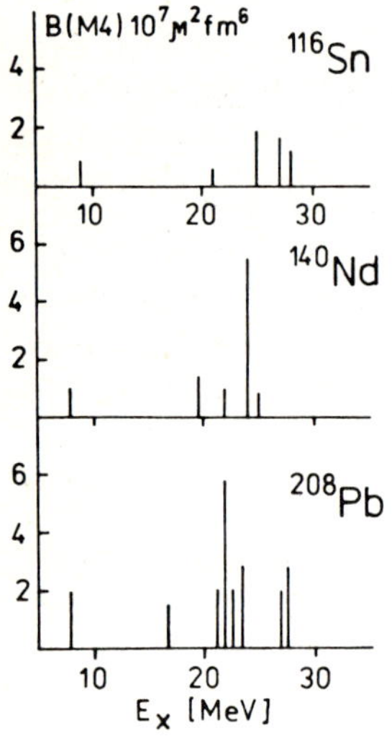

Fig. 3. $M4$ strength distribution in some even-even nuclei

The experimental quantities $B(M4)$ are reproduced with lower values of $g_{s\,\text{eff}}$ than the quantities $B(M2)$ in the same nuclei. This is due to our truncation of the single-particle space which does not include the single-particle continuum. This stronger influences transitions with higher λ. Thus the reduction of $M4$ transitions is weaker than that for $M2$.

Following remark appears of importance. The $M2$ transition probability in ^{209}Pb is by 3–8 times larger than that in other nuclei. Similarly, the $M4$ rate in ^{207}Tl and ^{207}Pb is larger by 5–10 times than $M4$ rates in other nuclei. Theoretical calculations (e.g. [19]) satisfactorily reproduce the $B(M\lambda)$ rates in

Table 2. Experimental and theoretical transition rates $B(M4)$ $10^5[\mu_N^2 fm^6]$. In ^{207}Pb, the transition included is of the type $13/2 \rightarrow 5/2^-$, in all the other nuclei the transition type is $11/2^- \rightarrow 3/2^+$. Presented are values with and without core polarization (c.p.) included. The shell model $B(M4)$ values are 13.4 for ^{207}Tl, 14.3 for ^{207}Pb and 6.4 for the other nuclei (in $[10^5 \mu_N^2 fm^6]$). The last two columns for ^{207}Tl and ^{207}Pb present results calculated with effective g factors used in the literature: (a) $g_s^p = 0.9\, g^p_{\text{s free}}$, $g_l^p = 1.04$; $g^n_{\text{s eff}} = 0.85\, g^n_{\text{s free}}$, $g_l^n = -0.04$; (b) $g^p_{\text{s eff}} = 0.81\, g_{\text{s free}}$, $g^p_{\text{l eff}} = 1.12$; $g^n_{\text{s eff}} = 0.72\, g^n_{\text{s free}}$, $g^n_{\text{l eff}} = -0.03$

Nuclei	Experiment	Theory				
		no c.p.	with c.p.			
			$g_{\text{s free}}$	$g_{\text{s eff}} = 0.4\, g_{\text{s free}}$		
^{117}Sn	0.414±0.008	5.2	2.4	0.37		
^{119}Sn	0.429±0.008	5.6	2.4	0.36		
^{121}Sn	0.375±0.043	6.0	2.3	0.32		
^{139}Ce	0.259±0.011	5.1	3.4	0.55		
^{141}Nd	0.227±0.012	6.0	4.3	0.69		
^{143}Sm	0.218±0.007	4.8	3.3	0.53		
					(a)	(b)
^{207}Tl	3.1±0.3	12.0	5.6		3.39	2.75
^{207}Pb	2.6±0.8	12.9	6.0		3.44	2.41

these nuclei with the usual values of $g_{\text{s eff}} \approx 0.7 - 0.9\, g_{\text{s free}}$. In this sense, our calculations are not an exception. However, in the other nuclei the effects we take into account appear not sufficient for the description of the $B(M\lambda)$ data. It seems unlikely that the renormalization of the g factors varies strongly between neighbouring nuclei. Therefore, some structure effects might be responsible for the decrease of the $B(M\lambda)$ rates. At this point it is not clear which are these effects. They might be associated with anharmonicity weak in the neighbourhood of ^{208}Pb, stronger in the Sn isotopes and more important in the region $A \sim 140$ (except for semimagic nuclei).

References

1. A. Richter: Darmstadt preprint IKDA 83/16 (1983)
2. A. Arima, H. Horie: Prog. Theor. Phys. **11**, 509 (1954)
3. M. Ichimura, K. Yazaki: Nucl. Phys. **63**, 401 (1965)
   H.A. Mavromatis, L. Zamick, G.E. Brown; Nucl. Phys. **80**, 545 (1966); Nucl. Phys. A**104**, 17 (1967); Phys. Lett. **20**, 171 (1966)
4. A. Arima, H. Hyuga: In *Mesons in Nuclei*, ed. M. Rho and D. Wilkinson (North Holland, Amsterdam 1979), p. 685
5. B.A. Brown, B.H. Wildenthal: Phys. Rev. C**28**, 2397 (1983)
6. I.S. Towner, F.C. Khanna: J. Phys. (Paris) **45**, C4–519 (1984); Nucl. Phys. A**399**, 334 (1983)
7. W. Andrejtscheff, L. Zamick, N.Z. Marupov, K.M. Muminov, T.M. Muminov: Nucl. Phys. A**351**, 54 (1981); A**368**, 45 (1981)
8. N.A. Bontch-Osmolovskaya, M.A. Dolgopolov, I.V. Kopytin, V.A. Morozov: Dubna reports JINR-**P4-85-759**; **P6-85-868**
9. V. Paar, S. Brant: Phys. Lett. **74B**, 297 (1978); Nucl. Phys. A**303**, 96 (1978)

10. W. Andrejtscheff, L.K. Kostov, L.G. Kostova, P. Petkov, M. Senba, N. Tsoupas, Z.Z. Ding, C. Tuniz: Nucl. Phys. A**445**, 515 (1985)
11. K. Heyde, J. San, E.A. Henry, R.A. Meyer: Phys. Rev. C**25**, 3193 (1982)
12. L.K. Kostov, W. Andrejtscheff, L.G. Kostova, P. Petkov, W. Enghard, H. Prade, L. Käubler, H. Rotter, F. Stary: To be published
13. C. Djalali, M. Morlet: Orsay preprint IPNO-DRE **85.13** (1985)
14. A. Bohr, B. Mottelson: Phys. Lett. **100**B, 10 (1981)
15. V.G. Soloviev et al.: Sov. J. Part. Nuclei **9**, 580 (1978); **11**, 301 (1980); **14**, 327 (1983); **16**, 245 (1985)
16. V. Ponomarev: J. Phys. G**10**, L177 (1984)
17. A.I. Vdovin, Ch. Stoyanov, W. Andrejtscheff: Nucl. Phys. A**440**, 437 (1985); Bull. Acad. Sci. USSR, Phys. Ser. **49**, 2173 (1985)
18. H. Eijri, J.I. Fujita: Phys. Rep. C**38**, 87 (1978)
19. J. Speth: Phys. Rep. **33**, 127 (1977)

Evidence for Three Microscopically Different Kinds of E1 Transitions in Lead-Region Nuclei

T. Lönnroth

Department of Physics, University of Jyväskylä, Seminaarinkatu 15, SF-40100 Jyväskylä, Finland

One outstanding feature of the shell model in heavy nuclei is that within a main shell the single-particle transitions of electric dipole (E1) character are forbidden. This leads to the experimentally observed fact that the E1 transitions are usually hindered by a factor of about 10^6 with respect to the Weisskopf estimate. But since there is a dispersion of about two or three orders of magnitude among these hindrance factors, the isomeric E1 transitions have not been used to elucidate the mocroscopic properties of the involved states, as has been done for the E3 transitions of this nuclear region [1]. The E3 transitions often proceed via admixtures of the low-lying collective octupole state, whereas the E1 transitions usually are explained to proceed via minute admixtures of the giant dipole, or RPA 1^- state [2].

The present report concerns the survey of electric-dipole transition rates, i.e. B(E1) values, in lead-region nuclei with Z = 82-88 and N $\sim$ 112-134. It is, indeed, found that the major part of the E1 transitions are hindered by factors ranging 10^6 - 10^{10} with respect to the Weisskopf estimate. But there are clearly a large number of E1 transitions, the hindrances of which are substantially smaller. In table 1 we have collected a number of selected E1 transitions to illustrate the situation.

This table clearly shows that a number of the E1 transitions are much faster than the 'typical' value of 10^{-6} W.u., and it will be shown below that these fast transitions are of two kinds. It is emphasized that in the mass region A $\sim$ 195-212 some transitions are as much as three orders of magnitude faster than the average. A closer look at table 1 reveals that all of them involve the proton $\pi i_{13/2}$ orbital in the initial state, and the $\pi h_{9/2}$ orbital in the final state. If one allows an admixture of the proton $\pi i_{11/2}$ orbital coupled to a 2^+ component in the initial state the transition would proceed as

$$|\pi i_{11/2} \otimes 2^+\rangle \rightarrow |\pi h_{9/2} \otimes 2^+\rangle,$$

and this is an <u>allowed</u> E1 transition. In contrast, the slow E1 transitions proceeding via the giant-dipole admixtures can be pictured as

$$|\Psi_{final}\rangle \otimes |1^-\rangle \rightarrow |\Psi_{final}\rangle,$$

irrespective of the valence-nucleon configuration Ψ. The excitation energy of the

Table 1. Sample hindrances F_W over the Weisskopf estimate for lead-region E1 transitions

Nucleus	$J_i \rightarrow J_f$	Main configuration a)	$F_W(\cdot 10^{-3})$	Reference
^{195}Bi	$29/2^- \rightarrow 27/2^+$	$\pi h_{9/2} \otimes \nu 12^+ \rightarrow \pi h_{9/2} \otimes \nu 9^-$	1,300	3
^{196}Pb	$5^- \rightarrow 4^+$	$\nu i_{13/2} f_{5/2} \rightarrow \nu f_{5/2} p_{3/2}$	200	4
^{197}Bi	$13/2^+ \rightarrow 11/2^-$	$\pi i_{13/2} \rightarrow h_{9/2} \otimes \nu 2^+$	12	5
^{198}Pb	$5^- \rightarrow 4^+$	$\nu i_{13/2} f_{5/2} \rightarrow \nu f_{5/2} p_{3/2}$	2,000	4
^{199}Bi	$13/2^+ \rightarrow 11/2^-$	$\pi i_{13/2} \rightarrow h_{9/2} \otimes \nu 2^+$	18	6
^{200}Pb	$5^- \rightarrow 4^+$	$\nu i_{13/2} f_{5/2} \rightarrow \nu f_{5/2} p_{3/2}$	500	4
^{200}Po	$12^+ \rightarrow 11^-$	$\nu i^2_{13/2} \rightarrow \pi h_{9/2} i_{13/2}$	10,000	7
^{202}Po	$12^+ \rightarrow 11^-$	$\nu i^2_{13/2} \rightarrow \pi h_{9/2} i_{13/2}$	6,600	8
^{203}At	$13/2^+ \rightarrow 11/2^-$	$\pi i_{13/2} \rightarrow \pi h_{9/2} \otimes \nu 2^+$	100	9
^{204}Po	$11^- \rightarrow 10^+$	$\pi h_{9/2} i_{13/2} \rightarrow \pi h^2_{9/2} \nu 2^+$	1.5	10
^{206}Po	$9^- \rightarrow 8^+$	$\nu i_{13/2} f_{5/2} \rightarrow \pi h^2_{9/2}$	$1.3 \cdot 10^7$	11
^{206}Po	$11^- \rightarrow 10^+$	$\pi h_{9/2} i_{13/2} \rightarrow \pi h^2_{9/2} \otimes \nu 2^+$	< 70	11
^{208}Po	$11^- \rightarrow 10^+$	$\pi h_{9/2} i_{13/2} \rightarrow \pi h^2_{9/2} \otimes \nu 2^+$	190	12
^{212}Rn	$11^- \rightarrow 10^+$	$\pi h^3_{9/2} i_{13/2} \rightarrow \pi h^4_{9/2}$ b)	≦ 2	13
^{216}Ra	$19^- \rightarrow 18^+$	$\pi h_{9/2} i_{13/2} \rightarrow \pi h^2_{9/2}$ c)	2.9	14
^{218}Ra	many	collective d)	∿ 100	15
A≧216	many	collective d)	100-200	e)

a) Only active configurations given, spectators omitted. b) A seniority-2 → seniority-4 transition. c) Spectator neutron configurations omitted. The E1 transition is placed differently in another work [16]. d) Many dipole transitions connecting positive- and negative-parity quadrupole bands are observed. e) See e.g. the systematics in refs. [17,18].

giant dipole 1^- state is about 11 MeV in ^{208}Pb [2] and should stay at a fairly constant energy, whereas that of the $\pi i_{11/2}$ orbital is lower [19].
The collective E1 transitions observed in all the heaviest lead-region nuclides are supposed to differ completely from the above-mentioned ones. It has been argued [15,20] that the light actinides develop a permanent dipole degree of freedom, in the form of alpha-particle clusters, and that this intrinsic structure is responsible for the collective E1 transitions.

In conclusion, it is proposed that there exist three microscopically different kinds of electric dipole transitions in lead-region nuclei, namely:

1) giant-dipole coupled transitions, which are the most frequently occurring ones ('normal' E1 transitions),
2) single-particle transitions ($\pi i_{13/2}$) probably involving admixtures of the $\pi i_{11/2}$ orbital, and
3) collective transitions involving an intrinsic dipole degree of freedom, the structure of which is not definitely known.

A detailed account of the present work is in preparation and will be published elsewhere [21].

References

1. I. Bregström and B. Fant: Phys. Scripta 31,26(1985)
2. V. Gillet et al.: Nucl. Phys 88,321(1966)
3. T. Lönnroth et al.: Phys. Rev. C 33,1641(1986)
4. M. Pautrat et al.:Nucl. Phys. A201,449(1973);A303,521(1978)
5. T. Chapuran et al.: Phys. Rev. C33,130(1986)
6. W.F. Piel, Jr. et al.: Phys. Rev. C 31,2087(1985)
7. T. Weckström et al.: Z. Physik A321,231(1985)
8. H. Beuscher et al.: Phys. Rev. Lett. 36,1128(1976)
9. K. Dybdal et al.: Phys. Rev. C 28,1171(1983)
10. V. Rahkonen and T. Lönnroth:Nucl. Phys. (submitted)
11. V. Rahkonen et al.: Phys. Scripta (in press)
12. V. Rahkonen and T. Lönnroth: Z. Physik A322,333(1985)
13. T. Lönnroth et al.: Phys. Scripta (to be published)
14. Y. Itoh et al.: Nucl. Phys. A410,156(1983)
15. M. Gai et al.: Phys. Rev. Lett. 51,646(1983)
16. T. Lönnroth et al.: Phys. Rev. C 27,180(1983)
17. W. Bonin et al.: Z. Physik A322,59(1985)
18. J.F. Shriner et al.: Phys. Rev. C 32,1888(1985)
19. J. Speth et al.: Phys. Rep. 33C,127(1977)
20. H.J. Daley and F. Iachello: Ann. Phys.(N.Y.) 167,73(1986)
21. T. Lönnroth: Z. Physik (to be published)

Electromagnetic Transitions in Octupolly Deformed Nuclei*

K. Böhning[1], *Z. Patyk*[1], *A. Sobiczewski*[1], *and P. Rozmej*[2,**]

[1]Institute for Nuclear Studies, Hoza 69, PL-00-681 Warszawa, Poland
[2]GSI, D-6100 Darmstadt, F. R. Germany

In the present contribution, we concentrate on even-even nuclei in the radium region. Collective spectra of these nuclei indicate that some of them are probably octupolly deformed [1]. Also theoretical calculations of the collective potential energy of the nuclei suggest such deformation (e.g. [2-4]). It is, however, hard to say how large the deformation is as no quantity has been measured which would be directly related to it. The reduced octupole-transition probability B(E3) seems to be a good candidate for such a quantity.

The scope of this contribution is to look at systematics of B(E3), calculated for few isotopes of radium. The calculation is an extension of the studies [5,6]. The lowest collective states of a nucleus are treated as large-amplitude quadrupole and octupole vibrations, coupled to each other by the collective potential energy. The potential is calculated in a microscopic way. The mass parameters are considered as independent of deformation, for reason of simplicity. Their values are taken the same as in refs. [5,6]. The collective wave functions, with the help of which the probability B(E3) is calculated, are obtained by numerical solution of the Schrödinger equation. The Nilsson quadrupole-, ϵ_2, and octupole-deformation, ϵ_3, parameters [7] are considered as dynamical variables.

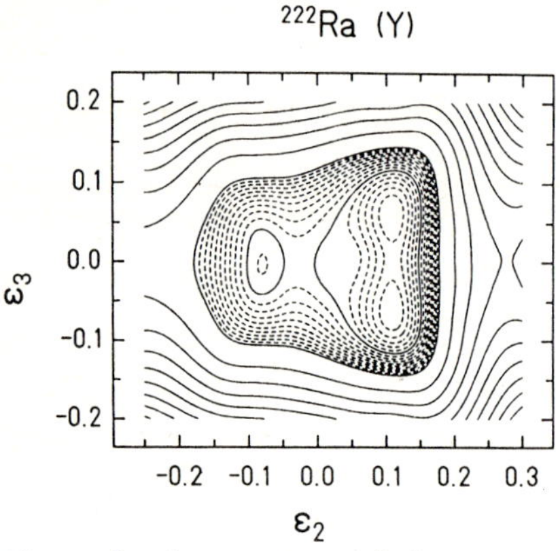

Figure 1. Contour map of the potential energy $V(\epsilon_2,\epsilon_3)$ for ^{222}Ra. The difference in energy between neighbouring solid lines is 1 MeV and between dashed lines 0.1 MeV.

---

* Supported in part by the Polish-US Maria Sklodowska-Curie Fund, Grant No. P-F7F037P

** Permanent address: Institute of Physics, MCS University, Lublin, Poland.

An example of the potential energy $V(\epsilon_2,\epsilon_3)$ is given in Fig. 1 for ^{222}Ra. It is obtained by the macroscopic-microscopic method. The macroscopic part of V is calculated from the Yukawa-plus-exponential model with the parameters of ref. [3]. The microscopic part is the Strutinski shell correction, based on the Nilsson potential. The parameters of the potential are taken from ref. [8]. They allow to better reproduce the ground-state properties of odd radium isotopes and lead to better quadrupole moments of even-even isotopes than the standard "A=225" parameters of ref. [7]. The values of the parameters are $\kappa_p = 0.0580$, $\mu_p = 0.630$ for protons and $\kappa_n = 0.0526$, $\mu_n = 0.457$ for neutrons.

One can see in Fig. 1 that the potential V, which is symmetric in ϵ_3, $V(-\epsilon_3) = V(\epsilon_3)$, has the minimum at the point $\epsilon_2^0 \approx 0.11$, $\epsilon_3^0 \approx 0.06$. The gain in the energy, due to the octupole deformation ϵ_3, is, however, rather small, $\Delta V_3 \approx 0.15$ MeV. The calculated quadrupole moment, $Q_2 = 6.10$ b, is rather close to the experimental value $Q_2^{exp} = (6.74 \pm 0.26)$ b [9]. It is obtained dynamically. i.e. with the use of the collective wave function of the ground state.

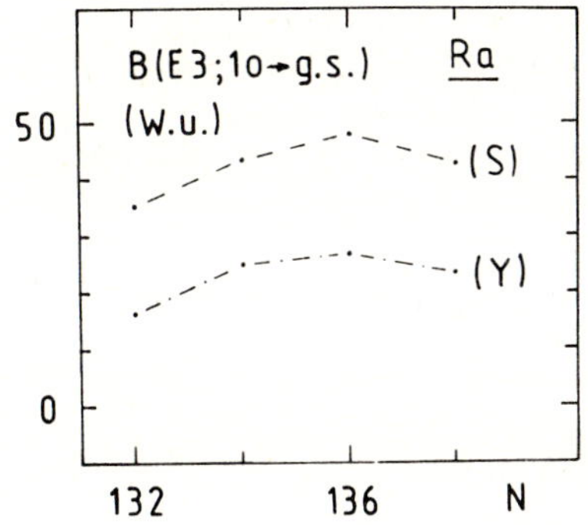

Figure 2. Reduced octupole-transition probability B(E3) calculated for two variants of the potential, (Y) and (S), described in text.

Figure 2 shows the values of B(E3; 10 → g.s.) calculated for the octupole transition from the lowest negative-parity state (i.e. the first octupole excitation 10) to the ground state (g.s.) for four even-even isotopes of radium: $^{220-226}$Ra. The values are given in the Weisskopf units and are plotted as a function of the neutron number N of an isotope. They are calculated with two variants of the potential energy V. One variant, (Y), is the same in which the energy of Fig. 1 is obtained. The other, (S), is the potential energy calculated as a sum of the single-particle energies smoothed by the pairing interaction. This energy is expected to have proper dependence on deformation only for the deformations not too far from the equilibrium value. The energy (S) usually leads to a larger gain ΔV_3 than the energy (Y). For example, for ^{222}Ra: $\Delta V_3 \approx$ 0.8 MeV for (S) while $\Delta V_3 \approx 0.15$ MeV for (Y), as already mentioned above. One can see in Fig. 2 that B(E3) depends much on the potential energy V. For the variant (S), it is about two times larger than for the variant (Y). For both variants, the largest B(E3) is obtained for ^{224}Ra, i.e. for the isotope for which the experimental energy of the first octupole state is smallest.

References

1. W. Kurcewicz et al.: Nucl. Phys. A356, 15(1981)
2. A. Gyurkovich et al.: Phys. Lett. 105B, 95(1981)
3. P. Möller and J.R. Nix: Nucl. Phys. A361, 117(1981)
4. G.A. Leander et al.: Nucl. Phys. A388, 452(1982)
5. K. Böning et al.: Phys. Lett. 161B, 231(1985)
6. R. Baranowski et al.: Proc. Int. Conf. on nuclear structure, reactions and symmetries, Dubrovnik 1986, in press
7. S.G. Nilsson et al.: Nucl. Phys. A131, 1(1969)
8. P. Rozmej et al.: Proc. 24th Int. Winter Meeting on nuclear physics, Bormio 1986, in press
9. A.S. Goldhaber and G. Scharff-Goldhaber: Phys. Rev. 17, 1171(1978)

Observation of Full Sets of Coexisting Intruder Excitations in the N=58 Isotones and Isotopes of Zr: Evidence for Alpha Correlated Excitations?

R.A. Meyer[1,2], *A. Aprahamian*[1], *E.A. Henry*[1], *G. Lhersonneau*[2], *K. Heyde*[3], *L.G. Mann*[1], *G. Menzen*[2], *N. Roy*[1], *and K. Sistemich*[2]

[1] Nuclear Chemistry Division, L.L.N.L., Livermore, CA 94550, USA
[2] Institut für Kernphysik, KFA-Jülich GmbH, D-5170 Jülich, F. R. Germany
[3] Institute of Nuclear Physics, Proeftuinstraat 86, B-9000 Gent, Belgium

We have performed in-beam gamma-ray singles and coincidence spectroscopy studies with the $^{96}Zr(t,p)^{98}Zr$ reaction using multiparameter (proton-gamma-gamma-time) coincidence techniques at the Lawrence Livermore National Laboratory's in-beam spectroscopy system at the Los Alamos Ion Beam Facility. Analyses of these data have revealed a number of previously unobserved levels. Our data shows that the excited states of ^{98}Zr below 3.3 MeV can be placed into two sets of levels. The first set has its main deexciting transitions leading to the ^{98}Zr ground state while the second set has its main deexciting transitons leading to the 854-keV 0^+ level.

Using unified shell-model calculations, Federman and Pittel [1] have shown that for the excited 0^+ state, a larger than expected occupancy of the proton $1g_{9/2}$ orbital will occur due to the promotion of extra protons from below the Z=38 subshell closure. Because of the larger occupation of the $1g_{9/2}$ orbital, the neutron $1g_{7/2}$ orbital has been shown to be lowered with respect to the $2d_{5/2}$ orbital through the residual p-n interaction [1]. This lowering will facilitate the promotion of neutron pairs across the $2d_{5/2}$ subshell closure, leading to a larger $1g_{7/2}$ neutron occupation thus giving rise to alpha-like correlations in the excited 0^+ state. Thus, from the view of the number of nucleon pairs available (particle plus hole pairs), we might expect that $^{98}Zr^*$ should look like ^{102}Ru. Our in-beam spectroscopy results show that there is a one-to-one correspondance between the first 8 levels of both sets.

The picture of the 0^+ excited state, in which the occupations of the $1g_{9/2}$-proton and the $1g_{7/2}$-neutron orbitals are enhanced at the expense of the orbitals below the respective gaps, is typical of the ground-state structure in the heavier Mo isotopes [1]. Thus, when the Mo isotopes are used as targets for the $(d,^6Li)$ reaction, the alpha-like correlations are evidenced by large cross sections into the excited 0^+ states of the even-even Zr nuclei, where the depleted population of the orbitals below the gaps is preserved, and conversely by small $(d,^6Li)$ cross sections to the ground states. These arguments are further supported by the very low energy needed to excite alpha pairing vibrations in this mass region. This energy can be estimated by using the quantity: $S_x = 2BE(A,Z) - BE(A-4,Z-2) - BE(A+4,Z+2)$ and gives values on the order of 1 to 0.5 MeV for the $^{96,98}Zr$ nuclei. As seen in Fig.1, even lower energies seem to occur in the nucleus ^{100}Mo which is just two protons outside a Z=40, N=56 core. We show that the alpha seperation energy can be taken as an indication of the ease of formation of specific alpha clustering. The alpha separation energy does have a rather small value of ~4 MeV in this A~100 mass region, although the smallest values of the alpha separation energy of ~2 MeV result in the Ru nuclei with N=52.

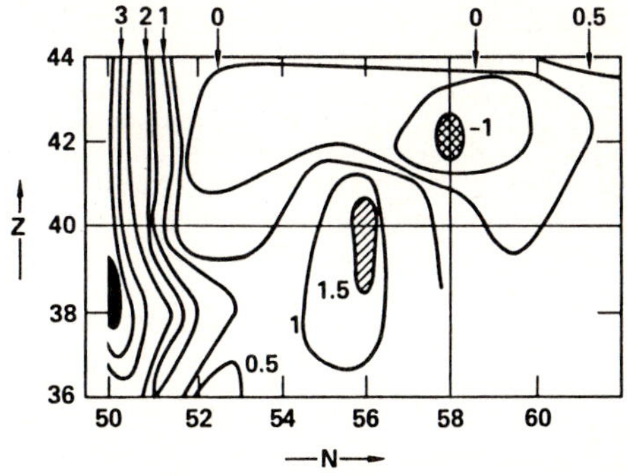

Fig.1 Alpha-correlation energies (S_x) for nuclei with $49<N<63$ and $35<Z<45$.

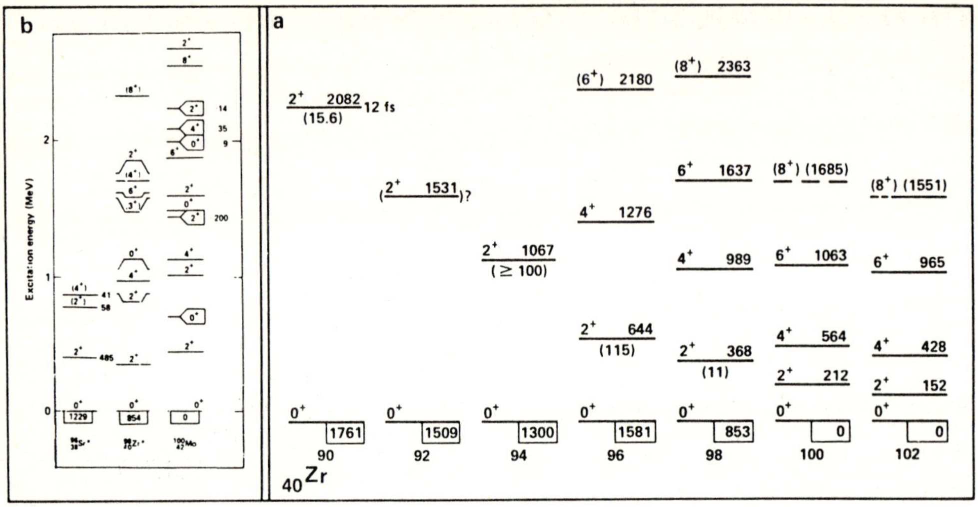

Fig. 2. Systematics of alpha-correlated structures in nuclei with a) Z=40 and b) N=58.

In performing IBM-2 calculations to describe ^{98}Zr*, a good estimate of the proton and neutron boson numbers is difficult to make due to the presence of subshell occurrence for both protons and neutrons. For ^{98}Zr, neutron pair excitation across the N=56 subshell gap actually does not change the neutron boson number N_ν, counting from N=50, since this merely results in a redistribution of the neutron 0^+ pair distribution (N_ν=4). The effect of increased neutron $1g_{7/2}$-orbital occupation through the proton-neutron interaction, on the other hand, can change the proton pair distribution by lifting extra proton pairs from the Z=38 shell into the $1g_{9/2}$ proton orbital. This can be taken to lead to a larger effective N_p value (ie. we use N_p(eff.)=3). Thus, the increased collectivity associated with the ^{98}Zr* excitation is simulated in a phenomenological way. We show that the calculations agree well with both the known ^{102}Ru levels and the set of levels built on the first excited 0^+ state in ^{98}Zr.

Comparison of the isotopes and isotones of Zr suggests that the alpha correlated structures extend over a wide range of nuclei. When we juxtapose the 0^+ excited state and associated 2^+ states for the A≤98 Zr nuclei to the ground-state bands for the A>98 Zr nuclei, we find that the levels of this series display a steady progression from i) the closed neutron shell nucleus ^{90}Zr that has its first alpha-correlated 0^+ level at 1761 keV which has an associated 2^+ level at 2081 keV higher in excitation energy; to ii) ^{102}Zr which has a fully developed rotational ground state band that possess a moment-of-inertia parameter of $h^2/2I$=25.3 keV. The characterization of ^{98}Zr* and its association with ^{102}Ru fits well into the progression from vibrational via gamma-soft deformed to fully rotational as the neutron $1g_{7/2}$ orbital occupancy grows. This comparison is a further manifestation of the increasing collectivity as the $1g_{7/2}$ neutron orbital is populated first by polarizaton and eventually at N>58 where it becomes the ground state configuration (see discussion by Federman and Pittel [1]). Thus, the "sudden" onset of deformation previously thought to occur in this mass region may be an artifact of correlating different structural types. Similar alpha-correlated structures occur in the isotones of $^{98}Zr_{58}$. The alpha-correlation energy contour suggests that while these alpha-correlated structures should be expected at higher excitation energies in $^{96}Sr_{58}$; they should become the ground-state set in $^{100}Mo_{58}$. We can identify a set of levels in ^{96}Sr built on the 1229-keV 0^+ level which behaves in a way similar to those we observe in ^{98}Zr. Similarily, using the results of recent in-beam gamma-ray specctroscopy studies in combination with on-line recoil mass-separator (JOSEF) results we show that ^{100}Mo has two sets of levels. However, in accordance with our alpha-correlation energy predictions, the ground state set in ^{100}Mo corresponds to the excited state set in ^{96}Sr and ^{98}Zr.

1] P. Federman and S. Pittel, Phys. Rev. C 20 (1979) 820.

On the β-Transition of ^{205}Tl to ^{205}Pb

E. Braun and I. Talmi
The Weizmann Institute of Science, Rehovot, Israel

Some first forbidden β-decays in nuclei near ^{208}Pb have low ft-values. The large matrix elements corresponding to these were interpreted long ago as due to purity of shell model configurations at the closed proton and neutron shells [1]. On the basis of this information it was suggested to build a low energy neutrino detector [2]. A neutrino with energy higher than 62 keV could be absorbed by ^{205}Tl nucleus in its $\frac{1}{2}^+$ ground state resulting in a $\frac{1}{2}^-$ state, 2.3 keV above the ^{205}Pb ground state.

BAHCALL [3] pointed out, however, that the transition will not proceed at the rate of simple $3p_{\frac{1}{2}} \rightarrow 3s_{\frac{1}{2}}$ transitions. In the lowest simple shell model configuration of 124 neutrons none is in the $3p_{\frac{1}{2}}$ orbit. The transition can take place only via admixtures of other configurations. Hence, the expected rate should be multiplied by the probability of the configuration with $(3p_{\frac{1}{2}})^2$ neutrons in the ground state of ^{205}Tl The success of the proposed detector depends crucially on this probability.

This probability can be estimated by considering direct reactions on the nearby ^{206}Pb nucleus. Picking up a neutron from ^{206}Pb leads also to the $\frac{1}{2}^-$ state (2.3 keV) of ^{205}Pb indicating the presence of $p^2_{\frac{1}{2}}$ neutrons in the ^{206}Pb ground state. The probability of a $p^2_{\frac{1}{2}}$ pair extracted from this reaction is 20-30% with a large margin of error. The expected log ft-values for the $^{205}Tl \rightarrow {}^{205}Pb$ transition would thus expected to be 5.5-6.

Another approach to obtain the expected rate is to consider experimental data about β-decays in nearby nuclei. To do that it is important to find out whether single nucleon wave functions can yield the measured ft-values. Only if this is so can we use the dominant components of shell model wave functions for such calculations. Had transitions taken place only via small components the experimental information would have been irrelevant for our purpose. It was found that certain radial shapes of $3p$ and $3s$ wave functions reproduce rather well the observed β-decay rates between the single hole nuclei ^{207}Tl and ^{207}Pb [4]. Repeating the calculation we could also reproduce the observed ft-values 5.11 to the $\frac{1}{2}^-$ ground state of ^{207}Pb and 6.22 to the $\frac{3}{2}^-$ excited state.

Our calculations show that the rates are extremely sensitive to the radial shapes adopted. This is not surprising since both proton and neutron radial functions have several nodes. It follows that much care must be taken in comparing ft-values in different nuclei. This may account for the rather ambiguous conclusions that can be drawn from ft-values observed in A=206 nuclei.

The most direct information may come from the β-decay of the $\frac{1}{2}^-$ ground state of ^{205}Hg to $\frac{1}{2}^+$ states of ^{205}Tl in which the $3p_{\frac{1}{2}}$ neutron becomes a $3s_{\frac{1}{2}}$ proton. The transition can take place only to the state with no $p_{\frac{1}{2}}$ neutrons. This state would have been the ^{205}Tl ground state in the simplest shell model picture. Still the decay proceeds to three $\frac{1}{2}^+$ states, the ground state and levels at 1.219 MeV and 1.435 MeV with log ft-values of 5.27, 7.1 and 5.69 respectively. Thus, the 1.435 MeV state has a fairly large component without $p^2_{\frac{1}{2}}$ neutrons and the ground state contains a fairly large admixture of states with $p^2_{\frac{1}{2}}$ neutrons. Under the assumption of the same proton and neutron radial functions in these two states we obtain from the measured ft-values 27% for the probability of $p^2_{\frac{1}{2}}$ neutrons in the ^{205}Tl ground state. This figure agrees well with the results of pick-up reactions.

The probability of 27% for $p^2_{\frac{1}{2}}$ neutrons in the ^{205}Tl ground state leads to log ft=5.7 for the transition to ^{205}Pb. There is, however, another uncertainty in that conclusion. The question is how much of that total strength goes into the 2.3 keV level. Some of it may belong to higher $\frac{1}{2}^-$ state in ^{205}Pb. Results of pick up experiments [5] indicate that most of the strength is indeed concentrated in the $\frac{1}{2}^-$ state 2.3 keV above the ^{205}Pb ground state.

1. A. de-Shalit and M. Goldhaber, Phys. Rev. 92 (1953 1211.
2. M.S. Freedman, in Proceedings of Conference on Status and Future of Solar Neutrino Research, 1978, Brookhaven National Laboratory Report BNL 50879, Vol. I p. 313
3. J.N. Bahcall quoted in ref. [2]
4. J. Damgaard and A. Winther, Nucl. Phys. 54 (1964) 615.
5. W.A. Lanford, Phys. Rev. C11 (1975) 815.

Study of the 1S Component of the Internal Bremsstrahlung Accompanying the (1u)-Forbidden Electron-Capture Decay of ^{41}Ca

P. Hornshøj[1] *and M. Pfützner*[2]

[1]Institute of Physics, Aarhus University, Denmark
[2]Institute of Experimental Physics, Warsaw University, Poland

The modern theory of IB in allowed transitions was developed by GLAUBER and MARTIN [1,2]. This theory, approximately valid also for the 1-forbidden nonunique transitions, was verified in several experiments [3]. ZON and RAPOPORT [4] extended the theory to transitions of higher-order of forbidenness. The validity of this formalism is still a question because of the lack of experimental data.

A very good case to test this latter theory is the first-forbidden unique (1u) decay of 41-Ca which proceeds 100% between ground states with a Q value of 421.3 keV and a half-life of $1x10^5$ years. Our aim is to study the shape and absolute intensity of the 1S component of the 41-Ca IB decay by KX-IBγ coincidences.

The experiment was performed at the Institute of Physics of Aarhus University (Denmark). The IB photon spectrum was measured by a Ge(Li) spectrometer, in coincidence with 3.3 keV potassium KX-rays detected by a Si(Li) spectrometer. The 7-mm-diameter source was prepared by sedimentation of $CaCo_3$ containing 0.47 mg 41-Ca on a plexiglass backing. This source was sandwiched between the two detectors in a very close geometry. During a 106-hour run, data were collected by standard slow-fast coincidence setup and stored on magnetic tape in the list mode. The efficiency and the response function of the Ge(Li) detector were determined by sources (226-Ra, 54-Mn, 137-Cs, 198-Au, 51-Cr, 139-Ce, 57-Co) measured in the 41-Ca source geometry. The experiment was repeated in a 182-hour run with another Ge(Li) detector having slightly better peak-to-Compton ratio.

The IBγ-K-X time spectrum collected during the first coincidence run is shown in Fig.1A. Photon spectra colleted in coincidence with KX-rays and the appropriate time window are shown in Fig.1B. After subtracting the random events, the resulting spectrum was corrected for the detector-response function. The spectra of monoenergetic sources were fitted with a simple nine-parameter function. The variation of these parameters with energy was established, enabling us to calculate the response of the detector to any monoenergetic radiation in the region of interest.

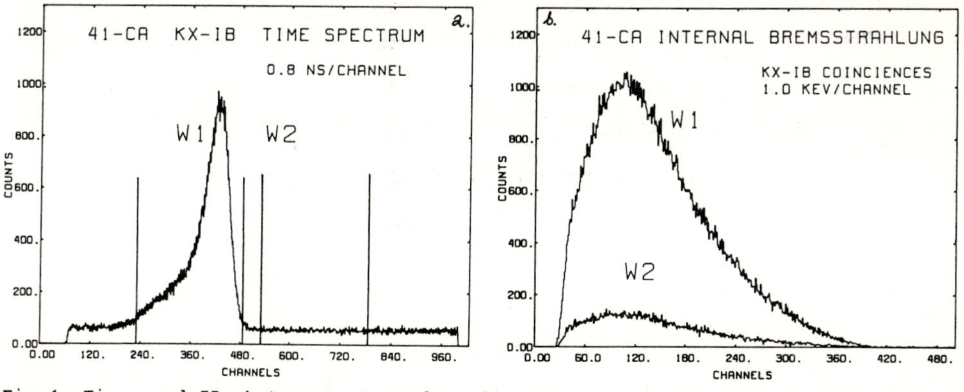

Fig.1. Time- and IB-photon spectrum from 41-Ca KX-IB coincidences. 'Real' and random windows, W1 and W2, are shown in a).

The IB spectrum was integrated in 10 keV bins and unfolded, starting from the high -energy end. Finally, the correction for the detector efficiency was applied.

In the simplest theoretical model, the probability of emitting the IB photon in the energy region (k,k+dk), normalized to the ordinary K-capture rate, is given by the MORRISON and SCHIFF formula [5],

$$\left.\frac{dw_{1S}}{w_K}\right|_{MS} = \frac{\alpha}{\pi(mc^2)^2}\frac{k(k_0-k)^2}{k_0^2}\,dk\ , \tag{1}$$

where α is the fine-structure constant, mc^2 is the electron rest energy, and k_0 is a maximum photon energy $k_0=Q-B_{1S}$. Q is the decay energy, B_{1S} is the binding energy of the 1S electron in the final atom. For 41-Ca, $k_0=417.7$ keV.

In all versions of realistic theories, the deviations of the IB spectrum from this simple model are described by the shape factor $R_{1S}(k)$:

$$\frac{dw_{1S}}{w_K} = \left.\frac{dw_{1S}}{w_K}\right|_{MS} R_{1S}(k)\ . \tag{2}$$

To compare with theory, the corrected experimental IB spectra were normalized to the number of KX-rays registered by the Si(Li) detector. Further division by the Morrison and Schiff formula yields the shape. The results of the experiments are shown in Fig.2. Together with the experimental shape factor, the theoretical predictions are shown. The detailed calculations were carried out for this case by ZON [6].

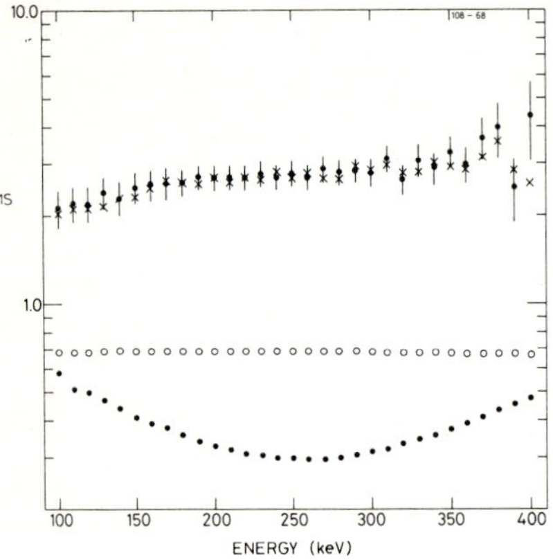

Fig. 2. The experimental S-shape factors from run 1 (black points with error bars) and run 2 (crosses, no error bars) on 41-Ca compared to the theoretical predictions of Glauber and Martin (open circles), and Zon and Rapoport (closed circles).

The total probability of emitting an IB photon in the energy region 95-421 keV per K capture is found to be $(2.6\pm0.3)\times10^{-4}$ in the first $(2.5\pm0.3)\times10^{-4}$ in the second run. The theoretical predictions are 0.69×10^{-4} in the version of the theory of GLAUBER and MARTIN and 0.38×10^{-4} in the theory of ZON and RAPOPORT.

We have found that the intensity of the IB associated with the K-capture decay of 41-Ca is considerably higher than the theoretical predictions. Furthermore, the shape of the spectrum is different from the shape derived from the Zon and Rapoport theory. It is remarkable that the calculations of ZON exhibit larger deviations from the experiment than the theory of allowed transitions. The reason for this is not understood at present.

We would like to thank Jan Zylicz for making us interested in this subject and for numerous valuable discussions.

References

1. R.J. Glauber and P.C. Martin, Phys.Rev. 104(1956)158
2. P.C. Martin and R.J. Glauber, Phys.Rev. 109(1956)1307
3. W. Bambynek et al., Rev.Mod.Phys. 49(1977)77
4. B.A. Zon and L.P. Rapoport, J.Nucl.Phys. 7(1968)528
5. P. Morrison and L. Schiff, Phys.Rev. 58(1940)24
6. B.A. Zon, Izv.Akad.Nauk SSSR Ser.Fiz. 37(1973)1978

Continuity-Equation Constraint and the Non-Uniqueness of the Vector Potential Decompositions

M. Gmitro[1,2], *J. Kvasil*[3], *and J. Řizek*[2]

[1]Laboratory of Theoretical Physics, JINR Dubna 141980, USSR
[2]Institute of Nuclear Physics, ČSAV, 25068 Řež, Czechoslovakia
[3]Dept. of Nucl. Physics, Charles University, Prague, Czechoslovakia

The use of the continuity-equation constraint (CEC) has proven highly useful in calculations of the deuteron photodisintegration [1,2], electric-type form factors in (e, e') scattering [3,4], and even more complicated radiative muon capture [5]. It is motivated as an extension of the Siegert hypothesis: the nuclear impulse-approximation transition operator is via the continuity equation and by-parts integration transcribed in a form which mainly depends on the nuclear charge density for which the omitted meson exchange corrections are of order $(v/c)^2$. The terms which depend on the nuclear current MEC of order v/c go as small corrections at low momentum transfers.

To perform the above program technically, one needs a decomposition of the electromagnetic vector potential of the form

$$\boldsymbol{A} = \boldsymbol{\varepsilon}_\lambda \exp(i\boldsymbol{k}\boldsymbol{\varrho}) = \boldsymbol{B} + \nabla s. \tag{1}$$

Several forms of such a decomposition are known in literature [6–9]. If used under the impulse approximation, different forms of (1) provide indeed different results for the calculated physical quantities. To find an optimal form of (1) might be especially important in the deuteron photodisintegration work, where sensitive and complicated relativistic corrections are added to the basic impulse-approximation result [2].

To provide a basis for more quantitative studies of the CEC effects, Řizek [10] suggested a class of decompositions of the vector potential (1) which depend on one real parameter. Here we put forward an example of the CEC calculation of the E2 form factor, with such decompositions.

The multipole fields are used in the Coulomb gauge. The L-th multipole of the electric type electromagnetic potential is [8]

$$\boldsymbol{A}_{LM} = \frac{1}{x}\left[\frac{d}{dx}(x j_L(x)) \cdot \boldsymbol{Y}^{(1)}_{LM} + \sqrt{L(L+1)}\, j_L(x) \boldsymbol{Y}^{(-1)}_{LM}\right] \tag{2}$$

where $\boldsymbol{Y}^{(a)}_{LM}$ are vector spherical harmonics [2], $x = k\varrho$, k being the photon transferred momentum. Taking the L-multipole of $\boldsymbol{B}$ in (1) in the form $\boldsymbol{B}_{LM} = f_L(x)(p\boldsymbol{Y}^{(1)}_{LM} + q\boldsymbol{Y}^{(-1)}_{LM})$ Řizek [10] has obtained

$$S_{LM}(x) = \frac{1}{k\sqrt{L(L+1)}}\Big[\frac{d}{dx}(xj_L(x)) + xg_L(x)\Big]\cdot Y_{LM}$$
$$B_{LM}(x) = g_L(x)\Big[\frac{r}{\sqrt{L(L+1)}}\mathbf{Y}^{(-1)}_{LM} - \mathbf{Y}^{(1)}_{LM}\Big] \tag{3}$$

where $r \geq 0$ is the mentioned parameter (connected with the above choice of B_{LM}) and

$$g_L(x) = x^{-r-1}\int_0^x y^{r+1} j_L(y)dy + C^{-r-1} \quad .$$

Here C is an integration constant. The "old" decompositions are obtained as follows

r	... Decomposition	Ref.
0	Foldy	[6]
$L+1$	Eisenberg and Greiner	[7]
∞	Rose	[8]

Řizek [10] has also shown that the result of [9] repeats that by *Foldy* [6] though in a very different form.

We have used the decompositions (3) to calculate with CEC the transverse form factor of the 2^+, $T = 0$, 4.44 MeV level in ^{12}C. The results for the three decompositions are shown in Fig. 1. We have checked numerically that taking $r \geq 0$ one obtains the transverse form factor which continuously deforms between the dot-dashed line $r = 0$ and the dotted line in Fig. 1. For $r = 500$ the results

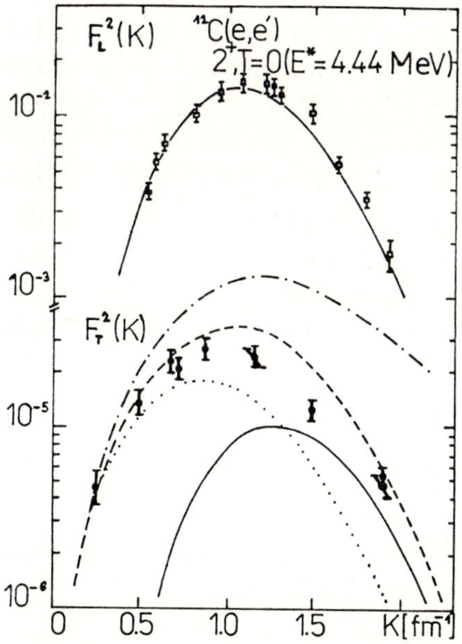

Fig. 1. The inelastic form factors of the 4.44 MeV level in ^{12}C. Wave functions of the Cohen-Kurath [11] type are used, results are scaled by a factor of 2. Full line – no CEC; dot-dashed, dashed, and dotted lines include the continuity-equation constraint via the decompositions of [6, 7, 8], respectively. They were calculated with the value of parameter r equal to 0, $L+1$, and 500, respectively. Data are from [12]

differ only by 1–2% from those which correspond to the decomposition by Rose ($r \to \infty$).

Obviously further work is needed to formulate a criterion of an optimal choice of r in (3).

References

1. F. Partovi: Ann. of Phys. **27**, 79 (1964)
2. Workshop on radiative processed in few-nucleon systems, Vancover, Can. J. Phys. **62**, 1019–1128 (1984)
3. D. Cha: Phys. Rev. C**21**, 1672 (1980)
4. R.A. Eramzhyan, M. Gmitro, S.S. Kamalov, T.D. Kaipov, R. Mach: Nucl. Phys. A**429**, 403 (1984)
5. M. Gmitro, A.A. Ovchinnikova, T.V. Tetereva: Nucl. Phys. A**453**, 685 (1986)
6. L.L. Foldy: Phys. Rev. **92**, 178 (1953)
7. J.M. Eisenberg, W. Greiner: Nuclear theory, Vol. 2 (North-Holland, Amsterdam 1970)
8. M.E. Rose: Multipole fields (Wiley, New York 1955)
9. J.L. Friar, S. Fallieros: Phys. Rev. C**29**, 1645 (1984)
10. J. Řizek: Report INP Řež, 1986
11. S. Cohen, D. Kurath: Nucl Phys. **73**, 1 (1965); **101**, 1 (1967)
12. J.B. Flanz et al.: Phys. Rev. Lett. **41**, 1642 (1978)

Nuclear Clustering Effects in Colliding N=Z and N≠Z Nuclei

R.K. Gupta, S.S. Malik, and R. Sultana

Physics Department, Panjab University, Chandigarh-160014, India

1. Introduction

Reactions between various N=Z and N≠Z s-d shell nuclei ($^{16}O+^{40}Ca$, $^{16}O+^{44}Ca$, $^{24}Mg+^{24}Mg$, $^{28}Si+^{28}Si$, $^{28}Si+^{30}Si$, $^{30}Si+^{30}Si$ and $^{32}S+^{32}S$) have been carried out recently [1-3]. These experiments are made at bombarding energies of 1.5 to 2 times the Coulomb barrier where the experiments do not distinguish between inelastic and transfer reactions. At these energies, though the transfer process is presumably very weak, some data are still recorded [1,2]. These data clearly indicate an explicit preference for alpha-cluster transfer in reactions using N=Z, alpha-particle nuclei and that as neutrons are added to the N=Z, alpha particle target, projectile or to both the reaction partners, then the occurrence of resonance-like structure for alpha-particle transfer gets suppressed. In particular, whereas at 75 and 80.6 MeV, the $^{16}O+^{40}Ca$ reactions show an explicit preference for alpha-particle transfer [1,2], the 80.6 MeV $^{16}O+^{44}Ca$ reaction does not indicate any such preferred alpha resonance structure [2]. In a recent paper, one of the authors and collaborators [4] have shown that the preferred alpha-cluster transfer in the reaction of 75 MeV ^{16}O on ^{40}Ca is a collective mass fragmentation process and in this contribution we show that the same dynamical fragmentation theory also gives the other result of the experiments.

2. Theory

Within the dynamical fragmentation theory, we define the mass fragmentation potential $V(\eta)$ at the touching configuration ($R=R_1+R_2$) of two nuclei as

$$V(\eta) = -B_1(A_1,Z_1) - B_2(A_2,Z_2) + \frac{Z_1Z_2e^2}{R} + V_P \qquad (1)$$

where $\eta = (A_1-A_2)/(A_1+A_2)$ is the collective mass asymmetry coordinate, $B_i(A_i,Z_i)$ are the experimental binding energies and V_P is the nuclear proximity potential [5]. The charges Z_i are fixed by minimizing the sum of the binding energies in the charge asymmetry coordinate $\eta_Z = (Z_1-Z_2)/(Z_1+Z_2)$.

Considering that the particle transfer occurs in steps of one nucleon, $\Delta\eta = 2/A$, the stationary Schrödinger equation in η:

$$\left[-\frac{\hbar^2}{2(B_{\eta\eta})^{1/2}}\frac{\partial}{\partial\eta}\frac{1}{2(B_{\eta\eta})^{1/2}}\frac{\partial}{\partial\eta} + V(\eta)\right]\Psi_R^{(\nu)}(\eta) = E_R^{(\nu)}\Psi_R^{(\nu)}(\eta) \qquad (2)$$

at fixed R, is solved numerically. On proper scaling, the solution of (2) gives the fractional mass distribution yields

$$Y(A_1) = |\Psi_R(\eta(A_1))|^2 \, (B_{\eta\eta}(A_1))^{1/2} \, 4/A. \tag{3}$$

The nuclear temperature effects are also included through

$$|\Psi_R|^2 = \sum_{\nu=0}^{\infty} |\Psi_R^{(\nu)}|^2 \quad \exp(-E_R^{(\nu)}/\theta) \tag{4}$$

where θ (in MeV) is related to the excitation energy

$$E^* = (1/9)A\theta^2 - \theta \tag{5}$$

3. Calculations

The fragmentation potentials, calculated for the composite system ^{48}Cr, ^{56}Ni and ^{64}Ge, formed due to N=Z, alpha-particle nuclei $^{24}Mg+^{24}Mg$, $^{16}O+^{40}Ca$ (or equivalently $^{28}Si+^{28}Si$) and $^{32}S+^{32}S$, show deep minima only at the alpha-particle transfer of the nucleons. As two neutrons are added to form ^{50}Cr and ^{58}Ni, the depths of the minima at two-nucleon or non-alpha-particle transfer are of the same order as for the alpha-particle transfer. On further increasing the N/Z ratios to obtain ^{52}Cr and ^{60}Ni, the alpha-clustering effect is almost lost. Similarly, in agreement with the experimental data on $^{16}O+^{40}Ca \rightarrow {}^{56}Ni$ and $^{16}O+^{44}Ca \rightarrow {}^{60}Ni$, the calculated yields for the fission of $^{56,58}Ni$ (and $^{48,50}Cr$) show the resonance-like structure for the alpha-cluster transfer, with greatly reduced amplitude in the cases of ^{58}Ni (and ^{50}Cr) and this resonance structure gets completely lost for ^{60}Ni (and ^{52}Cr). This effect is more vivid when $B_{\eta\eta}(\eta)$, rather than a constant value, is used. Some further details of the calculations can be seen in Ref.6.

4. Conclusion

We have shown that the alpha-cluster transfer during the collision between two N=Z alpha-particle s-d shell nuclei, which reduces gradually for N>Z nuclei, is a collective mass transfer process.

References

1. R.R. Betts: Lecture Notes in Physics, Vol. 156 p185 (Berlin: Springer) 1981.
2. R.R. Betts: Proc. 5th Adriatic Int. Conf. on Nuclear Physics, Hvar, Croatia, Yugoslavia, ed. N. Cindro et. al. (Singapore: World Sc.) 1984 p33.
3. P.H. Kutt, S.F. Pate, A.H. Wuosmaa, R.W. Zurmühle, O. Hansen, R.R. Betts, and S. Saini: Phys. Lett. 155B, 27 (1985).
4. D.R. Saroha, N. Malhotra and R.K. Gupta: J. Phys. G: Nucl. Phys. 11, L27 (1985).
5. J. Blocki, J. Randrup, W.J. Swiatecki and C.F. Tsang: Ann. Phys. N.Y. 105, 427 (1977).
6. S.S. Malik and R.K. Gupta: J. Phys. G: Nucl. Phys. (1986)-in press.

Cluster Distortion Effects in Electron Scattering from ^{6}Li

A.T. Kruppa

Institute of Nuclear Research of the Hungarian Academy of Sciences,
P.O.B. 51., H-4001 Debrecen, Hungary and
Kernforschungszentrum Karlsruhe, Institut für Kernphysik III,
P.O.B. 3640, D-7500 Karlsruhe, F. R. Germany

It is firmly established that the dominant clustering component of ^{6}Li is α+d. The effect of the distortion of the weakly bound d cluster on αd scattering [1] and on the energy of ^{6}Li [2] has already been investigated. We have now made a thorough analysis of the cluster distortion effects on the electromagnetic properties, which enter into the description of electron scattering from this nucleus.

We use a generator coordinate (GC) model, in which the trial wave function of ^{6}Li is

$$\Psi = \sum_{i=1}^{N_\alpha} \sum_{j=1}^{N_d} \sum_{k=1}^{N} F_{ijk} P_{JM} P_{K=0} \mathcal{A}_{\alpha d}(\Phi_\alpha^i(S_\alpha^k)\ \Phi_d^j(S_d^k)) \tag{1}$$

where $\mathcal{A}_{\alpha d}$ is the intercluster antisymmetrizer, P_K and P_{JM} are the linear and angular momentum projection operators, respectively. The wave function $\Phi_c^i(S_c^k)$, centered at the point S_c^k, is a Slater determinant representing cluster c(c=α or d) with the lowest harmonic oscillater shell model configuration of width parameter β_c^i. The main distinctive feature of the model is that not only the separation but also the widths of the shell model potentials are treated as GC's. In this way, in addition to the molecule-like collective motion, the correlation that corresponds to the breathing vibration of the clusters is also treated explicitly.

The discretized values of β_c^i were determined by the cluster stability condition, i.e. by minimizing the cluster energies with the trial function

$$\Psi_c^{(p)} = \sum_{i=1}^{N_c} F_c^{p,i} P_{JM} P_{K=0}\ \Phi_c^i \tag{2}$$

where p=1 corresponds to the ground state. The amplitudes F_{ijk} and $F_c^{p,i}$ were calculated by a Hill-Wheeler diagonalization of the respective microscopic hamiltonians. Using (2), we can rewrite (1) in the more plausible form

$$\Psi = \sum_{p=1}^{N_\alpha} \sum_{q=1}^{N_d} \sum_{k=1}^{N} P_{JM} \mathcal{A}_{\alpha d}(\Psi_\alpha^{(p)} \Psi_d^{(q)} \chi_{pqk}) \tag{3}$$

where χ_{pqk} are appropriate relative-motion functions. Thus, in the breathing cluster model (BCM) specified by (1) or (3) the specific distortion is taken into account through the inclusion of breathing excited states of α and d in the wave function (3). A simple cluster model (SCM), in which the clusters have only one common width parameters, was also considered. From among the NN interactions used the results calculated with the Brink-Boeker force B1 [3] are presented. In the hamiltonian we took into account the Coulomb force as well.

The elastic charge form factor F_{C0} and the inelastic form factor F_{C2} of the excitation of the 2.18 MeV state, which appear in the PWBA of electron scattering [4], are shown in Fig. 1. In the SCM the charge form factor deviates badly from the experimental data even in the small momentum transfer region $q^2 \leq 7$ fm^{-2}, dominated by the nucleonic degrees of freedom. The BCM, on the contrary, reproduces the data in this region excellently. The lack of the diffraction dip in the BCM must be due to our neglect of the short range NN correlation and mesonic currents [5]. The freezing of α and/or d in their ground states (no summation over p and/or

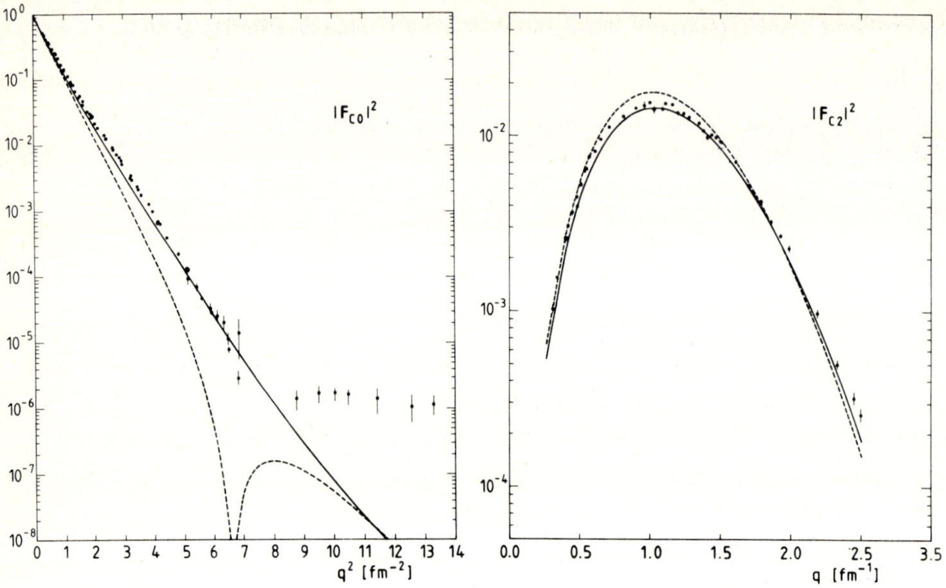

Fig.1. The elastic F_{C0} and inelastic F_{C2} form factors calculated with the BCM (solid line) and SCM (dashed line)

q in (3)) slightly influences the behaviour of F_{C0} below the diffraction dip but the distortion of α and d become more and more important at increasing q values. The effect of the d distortion is even more significant for the shape of F_{C2}. The diffraction minimum in the elastic magnetic form factor (not shown here) is reproduced in each model and the agreement with experiment is satisfactory. The calculated (and experimental) rms charge radius and reduced transition strength $B(E2,3^+\rightarrow1^+)$ are 2.65 fm (2.56±0.1 fm) and 9.02 e^2fm^4 (10.9±2.1 e^2fm^4) in the BCM.

To sum up, in the small momentum transfer region the main effect comes from the improved description of the cluster ground states, while in the high momentum transfer region the distortion of the clusters is also important.

References

1. H. Kanada, T. Kaneko, S. Saito and Y.C. Tang: Nucl. Phys. A444, 209 (1985)
2. R. Beck, F. Dickmann and A.T. Kruppa: Phys. Rev. C30, 1044 (1984)
3. D.M. Brink and E. Boeker, Nucl. Phys. A91, 1 (1967)
4. T. deForest, Jr. and J.D. Walecka: Adv. in Phys. 15, 1 (1966)
5. M.A.K. Lodhi and R.B. Hamilton: Phys. Rev. Lett. 54, 646 (1985)

The Effect of Hidden Colour on the Helium Isotropes

A. Abbas

Institut für Kernphysik, Technische Hochschule,
D-6100 Darmstadt, F. R. Germany

From the elastic scattering of electrons on the nuclei ^3He and ^4He it has been deduced [1] that these nuclei appear to have a pronounced depression in the central charge density. This hole could not be accounted for by any of the leading nuclear models. Here we seek an explanation in terms of the quark model (for further details see REF. 2).

A group-theoretical analysis of the 6, 9, and 12 quark systems has been done by MATVEEV and SORBA [3]. These multiquark states while being colour singlet were assigned to the completely antisymmetric representation of the SU(12) group which contains the direct product of the SU(4) spin-isospin group and the SU(3) colour group. In each case quarks were supposed to stand on the same energy shell with spin j=1/2. For 6q system the fully antisymmetric state (1^6) which is also colour singlet is represented as (50,1). This is built from the product $(1^3) \times (1^3)$ only for $(20,1) \times (20,1) = (50+..,1)$ and $(20',8) \times (20',8) = (50+..,1+..)$ (4). The latter corresponds to the hidden colour channel. It was determined that the 6q dibaryonic state consists of 80% hidden colour components[3]. This involved so much labour that the hidden colour components in the 9q and the 12q systems could not be obtained[3].

The realization that there exists a simple relation between the CFP of the unitary group and the isoscalar factors (ISF) of the permutation group ie. the $SU(mn) \supset SU(m) \times SU(n)$ CFP are at the same time the ISF of the group chain $S(f_1+f_2) \supset S(f_1) \times S(f_2)$ greatly simplifies the algebra[4]. We use this method to determine the $SU(12) \supset SU(4) \times SU(3)$ CFP for the tribaryonic 9q and the tetrabaryonic 12q systems. These are given in Table 1 for 9q states. From here one calculates that the 9q state consists of 97.6% hidden colour components. The 12q case is listed in Table 2. Note that this system is all coloured (99.8%).

It has been pointed out by BRODSKY [5] that the deuteron which is basically N-N at large distances must go over to the dibaryonic 6q system when the transverse separation of quarks goes to zero. This consists of 80% hidden colour. So when the two nucleons approach each other the system must do work to change the 6q state to a predominantly hidden colour

TABLE 1: The tribaryonic 9q system CFP.

1-baryonic state	x	2-baryonic states	CFP
(20,1)	x	(50,1)	$(5/42)^{1/2}$
(20',8)	x	(64,8)	$(32/42)^{1/2}$
$(\bar{4},10)$	x	$(\overline{10},\overline{10})$	$(5/42)^{1/2}$

TABLE 2: The tetrabaryonic 12q system CFP.

1-bar.state	x	3-bar.state	CFP	dibar.state	x	dibar.state	CFP
(20,1)	x	$(\overline{20},1)$	$(42/462)^{1/2}$	(50,1)	x	(50,1)	$(25/462)^{1/2}$
(20',8)	x	(20',8)	$(336/462)^{1/2}$	(64,8)	x	(64,8)	$(256/462)^{1/2}$
$(\bar{4},10)$	x	$(\bar{4},10)$	$(84/462)^{1/2}$	(6,27)	x	(6,27)	$(81/462)^{1/2}$
				$(\overline{10},\overline{10})$	x	(10,10)	$(50/462)^{1/2}$
				(10,10)	x	$(\overline{10},\overline{10})$	$(50/462)^{1/2}$

configuration. In other words QCD requires that the N-N potential must be repulsive at short distances.

^{3}He looks like 3N at large separations. It would evolve into the tribaryonic 9q state at small relative distances. In doing this it has to go over to a state which has predominant (97.6%) hidden colour components. The system has to do work against colour forces at short relative distances; or so to say there will be a "colour repulsion" creating a hole at the centre of ^{3}He. Similarly for ^{4}He.

In short, on the basis of QCD, it appears that the multibaryonic 9q and 12q systemsshould provide a viable description of the central region of ^{3}He and ^{4}He respectively. Since these multiquark systems have dominant hidden colour components the multinucleonic system will have to do work against colour forces. This colour repulsion will lead to creation of hole in the centre of ^{3}He and ^{4}He - exactly as seen in the experiments.

This work was supported by BMFT

REFERENCES

1. I. Sick, Lect. Notes in Phys. 87, 236 (Springer-Verlag, Berlin, 1978)
2. A. Abbas, Phys. Lett. 167B, 150(1986)
3. V.A. Matveev and P. Sorba, Nuovo Cimento 45, 257(1978)
4. J.Q. Chen, J. Math. Phys. 22, 1(1981)
5. S.J. Brodsky, Comm. Nucl. Part. Phys. 12, 213(1984); Nucl. Phys. A416, 3C(1984)

^{4}He D-State Effects in the ^{2}H(d,γ)^{4}He Reaction at Low Energies

A. Arriaga, A.M. Eiró, and F.D. Santos

Centro de Fisica Nuclear da Universidade de Lisboa, Av. Gama Pinto 2, 1699 Lisboa Codex, Portugal

Large tensor force effects have recently been reported in the ^{2}He(d, γ)^{4}He reaction induced by polarized deuterons [1,2,3] . For deuteron energies E_d < 20 MeV the reaction is dominated by the E2 transition. Among the allowed E2 amplitudes $A = \langle {}^4He|E2|{}^1_4D_2 \rangle$ is essentially determined by the ^{4}He S-state, and $B = \langle {}^4He|E2|{}^5S_2 \rangle$, $C = \langle {}^4He|E2|{}^5D_2 \rangle$, $D = \langle {}^4He|E2|{}^5G_2 \rangle$ by the ^{4}He D-state. It is well known [2] that the angular distributions of the tensor analyzing powers (TAP) are very sensitive to the value of the asymptotic D/S state ratio ρ in the d+d cluster configuration of ^{4}He. This effect can be observed for deuteron energies near 10 MeV where the dominant amplitudes are A and C. As the deuteron energy decreases, the transition amplitude from the $^5S_2$ scattering state becomes dominant due to the absence of the centrifugal barrier. The amplitudes B and A become of the same order of magnitude while C is negligible. Therefore large D state effects are also expected at low energies and the sensitivity to ρ can be observed either in the TAP or in the cross section [4].

To describe the initial scattering state we use separable potentials that reproduce the energy dependence of RGM phase shifts [2]. Coulomb distortion effects were also included. Calculations were performed using a point deuteron approximation in the E2 operator. In this approximation the structure of the α particle wave function comes into the transition amplitudes trough its projection into two deuteron clusters $\langle \phi_\alpha | \phi_d \phi_d \rangle$. To describe the S and D components of this overlap, we used Wood-Saxon wells with the same geometry for the S and D states and constrained to give equal binding energy. That calculation was improved using the momentum distribution of the d+d configuration in ^{4}He obtained by SCHIAVILLA et al [5], with the Urbana N-N interaction. This wave function corresponds to ρ=-0.13. Figure 1 shows the results of calculations for the angular distributions of the differential cross section $\sigma(\theta)$ at E_d=2 MeV and for A_{yy} at E_d=10 MeV. The curves correspond to different wave functions with the same value of ρ (full line and dash-dotted line), and to different values of ρ with Wood-Saxon wave functions (full line and dotted line). The deviation of the measured angular distribution $\sigma(\theta)$ from the $\sin^2 2\theta$ shape is particularly noticeable at $\theta = \pi/2$, and is essentially due to the presence of the B amplitude. The agreement obtained with the A_{yy} data is reasonable in the angular regions of $\theta = \pi/4$ and $3\pi/4$. The calculations indicate that the transition matrix elements have some sensitivity to the interior region of the overlap.

Calculations were also performed for the full E2 operator using S-state gaussian wave functions which reproduce the deuteron and α-particle rms radius. The α-particle D-state wave function was generated in a perturbative way under the condition that it gives the same D_2 parameter as the d+d amplitude of SCHIAVILLA et al. The results of the calculations show that the main effect of the extended E2 operator is in the B amplitude which has approximately the same modulus but changes sign. The calculated $\sigma(\theta)$ is almost unchanged relative to the cluster description. However there is a noticeable effect in $A_{yy}(\theta)$ which changes sign at $\theta = \pi/2$ (broken line in Fig.1).

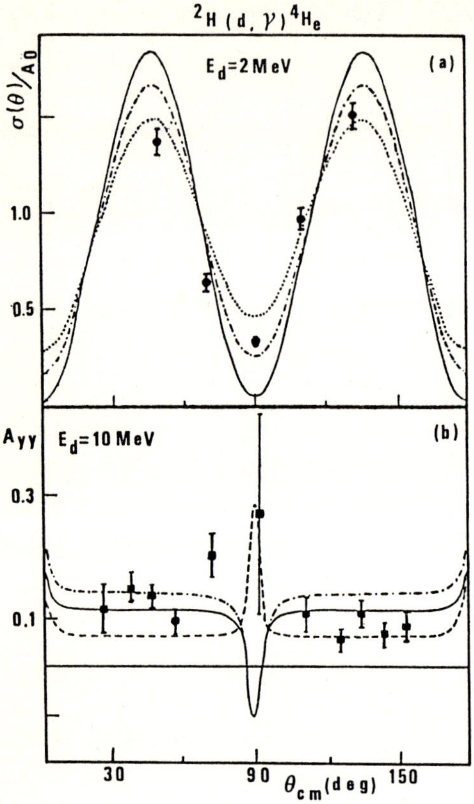

Fig.1
Angular distributions of the differential cross section and A_{yy}. The full and dotted curves correspond to Wood-Saxon wave functions with ρ=-0.13 and -0.30 respectively and the dash-dotted curve to the wave function of ref. [5]. The broken curve is the result of a microscopic calculation with gaussian wave functions. In (a) the full and broken curves coincide. Experimental data are from ref. 4 (a) and from ref. 3 (b). ($Ao = \sigma_{total}/4\pi$)

It can be argued that to analyse the effects of the tensor force in the α particle we should also consider these effects in the scattering states. We find that in the cluster model the mixing of J=2 states through the tensor force changes only the reduced matrix elements and not the angular dependences of the amplitudes. Although the cross section and A_{yy} can be reasonably described with a pure E2 transition, the angular distribution of A_y obtained by MELLEMA et al [3] where $A_y(\theta) \cong A_y(\pi - \theta)$, can only be explained by the presence of the M2 multipole. Microscopic calculations using realistic wave functions and taking into account both E2 and M2 transitions are needed.

1 - H.R.Weller et al., Phys.Rev.Lett. 53, 1325 (1984)
2 - F.D.Santos et al., Phys.Rev. C31, 707 (1985)
3 - S. Mellema et al., Phys.Lett. 166B, 282 (1986)
4 - H.R. Weller et al., Phys.Rev. C in press
5 - R.Schiavilla, V.R. Pandharipande, R.B.Wiringa,Nucl.Phys. A449 (1986) 219

Electromagnetic Structure of the Deuteron in the Skyrme Model

E.M. Nyman and D.O. Riska

Department of Physics, University of Helsinki, SF-00170 Helsinki, Finland

The Skyrme model for the nucleon [1] has, despite its simplicity, been shown to provide a qualitatively correct explanation for the properties of the nucleon and the delta-resonances. It alsoaccounts correctly for the key features, i.e., the long- and short-range behaviours, of the nucleon-nucleon interaction. We have developed methods for dealing with the static electromagnetic properties of the deuteron, using the Skyrme model with a product ansatz for the two-nucleon system. As the deuteron has no isospin, its electromagnetic response is generated solely by the anomalous current, related to the baryon current and the Wess-Zumino interaction. The anomalous current is of a particular interest because it is related to the topological aspects of the soliton. Thus, the corresponding exchange current is obtained entirely within the model, while in a conventional (meson-exchange) picture this current would have to be introduced as a new, independent interaction, the most important term involving a pi-rho-gamma-vertex.

The isoscalar part of the electromagnetic current operator is in the Skyrme model one half of the anomalous baryon current. We use the usual hedgehog ansatz, and for the the two-nucleon system we use a product approximation. The baryon current splits into a sum of terms, a one-body current for each nucleon and an irreducible two-soliton exchange-current operator. The charge and quadrupole form factors of the deuteron are obtained from the Fourier transform of the time component of the baryon current. The magnetic dipole form factor is obtained from the spatial part of the isoscalar baryon current.

When computing the exchange contributions, the single-nucleon solitons are obtained using the parameters of ref. [2]. Certain technical improvements compared to the published version of our work [3] have been introduced, giving minor modifications in the static results. Using the deuteron wave function given by the Paris potential the contribution to the square of the charge radius is 0.0015 fm^2. The reason for this small value is that the wave function is quite depleted near the origin and not that the operator itself is small. The exchange contribution to the quadrupole moment is -0.0015 fm^2. Again, we are dealing with a very small correction. This time the value is uncertain, however, beacuse the neglected D-state components are more important for the quadrupole moment than for the radius. (Work is in progress to incorporate the D-state.) The magnetic moment of the deuteron is in the impulse approximation determined by the isoscalar magnetic moment of the nucleons and the D-state probability in the deuteron. The exchange contribution is 0.036 n.m., which is slightly larger than what would be needed to cover the gap between the impulse approximation and the empirical value.

The main conclusion regarding the static properties of the deuteron is that the exchange contributions are small enough not to disturb the relatively good agreement with experiments obtainable without them. The calculations have given the correct signs and a reasonable general size.

We are in the process of computing the form factors at higher momentum transfers, where the exchange contributions dominate. As Skyrme's original version of

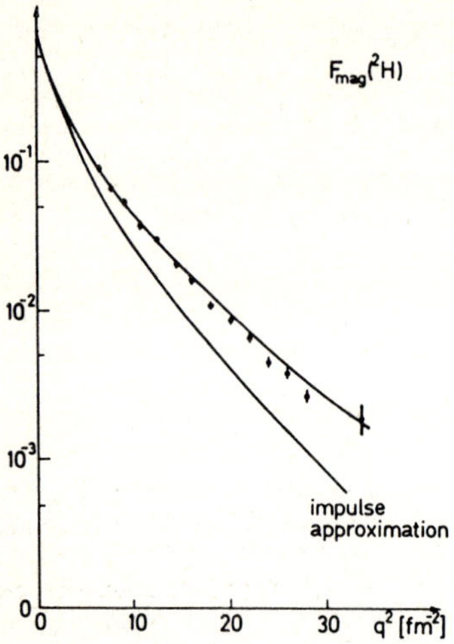

Fig 1. Magnetic form factor of the deuteron as function of the momentum transfer, calculated from the isoscalar electric form factor of the single nucleon

the model is not applicable at high q, however, we have adopted an approach which is independent of the Lagrangian: We determine the chiral angle theta from the single-nucleon isoscalar electric form factor. As seen in the figure, this gives a very good fit to the deuteron magnetic form factor.

References:

1. T.H.R. Skyrme: Proc. Roy. Soc. A260, 127 (1961)
2. G.S. Adkins and C.R. Nappi: Nucl. Phys. B233, 251 (1984)
3. E.M. Nyman and D.O. Riska: "Static Electromagnetic properties of the Deuteron in the Skyrme Model", University of Helsinki preprint HU-TFT-84-47 (1984)

The Electromagnetic Radii of the Deuteron

L. Črepinšek[1] *and H.F.K. Zingl*[2]

[1]Technical Faculty, University Maribor, Maribor, Yugoslavia
[2]Institut für Theoretische Physik, University Graz, A-8010 Graz, Austria

A few years ago, HAFTEL et al. [1] discussed the importance of measurements of the tensor polarisation in electron-deuteron scattering - even at low momentum transfer to the deuteron - to learn about important features of the off-shell nucleon-nucleon force. This paper initiated many investigations, both experimentally and theoretically.

We have calculated the complete analytical expressions for the radii of the deuteron, consisting of contributions from nonrelativistic impulse approximation, relativistic corrections and mesonic currents. In the case of the charge form factor of the deuteron our expressions could be used to decide on the controversial interpretations of the experimental results [2,3]. The slopes of the other form factors are not known experimentally at the moment, the value given by GOLDEMBERG and SCHAERF [4] for the radius of the magnetic moment has to be questioned nowadays [5]. Also unknown is the quadrupole form factor, which is the important quantity to decide experimentally on the D-state probability of the deuteron [6,7].

In our representation we use the notation of HAFTEL et al. [1], which yields to the following static limits for the deuteron form factors F_C, F_Q and F_M in the impulse approximation:

$$F_C^{IA}(0) = 0,\ F_Q^{IA}(0) = M_D{}^2 Q^{IA},\quad F_M^{IA}(0) = 2\mu_D^{IA}\ , \text{ with}$$

$$Q^{IA} \equiv Q_1 + Q_2 = (\sqrt{2}/10) \int_0^\infty \psi_o(r)\psi_2(r)r^4 dr - (1/20)\int_o^\infty \psi_2^2(r)r^4 dr\ ,$$

$\mu_D^{IA} = (\mu_P + \mu_N)(1 - (3/2)p_D) + (3/4)p_D$, μ_P and μ_N are the magnetic moments in n.m. of the proton and neutron respectively, p_D is the D-state probability and M_D the mass of the deuteron.

The corresponding radii are:

$$(r_C^{IA})^2 = r_{CP}^2 - 6\, G'_{CN}(0) + r_D^2\ ,\ (r_Q^{IA})^2 = r_{CP}^2 - 6\, G'_{CN}(0) + \frac{3\sqrt{2}\, f}{560\, Q^{IA}}\ ,$$

$$(r_M^{IA})^2 = (\mu_D^{IA})^{-1} [\mu_P r_{MP}^2 + \mu_N r_{MN}^2)(1 - (3/2)p_D) +$$

$$+ (\mu_P + \mu_N)(r_D^2 - Q^{IA}/2 + 9\, Q_2) + (3/4)p_D(r_{CP}^2 - 6\, G'_{CN}(0)) - (9/4)Q_2]\ .$$

r_{CP}, r_{MP}, $G'_{CN}(0)$, r_{MN} give the slopes of the nucleon form factors at $q = 0$, r_D is the internal deuteron radius defined by

$$r_D^2 = (1/4)\int_0^\infty (\psi_o^2(r) + \psi_2^2(r)r^4 dr \text{ and } f = 2\int_o^\infty \psi_2(r)(\psi_o(r) - 8^{-1/2}\psi_2(r))r^6\, dr$$

is the socalled Levinger factor investigated by MATHELITSCH and ZINGL [6] and DONNELLY and SICK [7]. It was already observed by LEVINGER [8] that f is inversely proportional to p_D.

The numerical values calculated with the Paris potential [9] give r_D^2=3,88754 fm^2, f=144,873 fm^4, p_D=0,0577621, Q_1=0,297626 fm^2, Q_2=-0,018197 fm^2 and therefore r_C^{IA}=2,124 fm, r_Q^{IA}=2,133 fm, r_M^{IA}=2,108 fm.

The well-known Darwin-term adds 0,033 fm^2, both to $(r_C^{IA})^2$ and $(r_Q^{IA})^2$.

The other most important contributions - we shall discuss these in more detail elsewhere - are the spin-orbit relativistic contribution and the pair-term, the $\rho\pi\gamma$-term and the retardation as mesonic corrections. Some of these quantities are calculated also by KOHNO [10] and our agreement is very good. For the Paris potential we have the final result: r_C=2,135 fm, r_Q=2,143 fm and r_M=2,113 fm.

This result is in contradiction to GOLDEMBERG and SCHAERF [4] , who found $r_M > r_C$.

This work was supported by the Fonds zur Förderung der wissenschaftlichen Forschung in Austria; project 5212.

References:

1 M.I. Haftel, L. Mathelitsch and H.F.K. Zingl, Phys. Rev. C22, 1285, 1980.
2 T.E.O. Ericson and M. Rosa-Clot, J. Phys. G: Nucl. Phys. 10, L201, 1984.
3 K. Klarsfeld, J. Martorell and D.W.L. Sprung, J. Phys. G: Nucl. Phys. 10, L205, 1984.
4 J. Goldemberg and C. Schaerf, Phys. Rev. Lett. 12, 298, 1964.
5 L. Mathelitsch, K. Schwarz, H.F.K. Zingl and M. Kohno, J. Phys. G: Nucl. Phys. 11, L151, 1985.
6 L. Mathelitsch and H.F.K. Zingl, Phys. Lett. 69B, 134, 1977, Il Nuovo Cim. 44A, 81, 1977.
7 T.W. Donnelly and I. Sick, Rev. of Mod. Phys. 56, 461, 1984.
8 J.S. Levinger, Acta Phys. Acad. Sci. Hung. 33 (2), 135, 1973.
9 M. Lacombe, B. Loiseau, R. Vinh Mau, J. Coté, P. Pirès and R. de Tourreil, Phys. Lett. 101B, 139, 1981.
10 M. Kohno, J. Phys. G: Nucl. Phys. 9L 85, 1983.

Asymmetry and Angular Distribution of Deuteron Photodisintegration in the 20–60 MeV Range

M.L. Rustgi, L.N. Pandey, and A. Kassaee

Physics Department, State University of New York at Buffalo, Buffalo, NY 14260, USA

Although deuteron photodisintegration has been studied extensively both experimentally and theoretically in the last fifty years, it continues to provide vital testing ground between new measurements and recent calculations incorporating corrections due to meson exchanges, isobar effects, and relativistic corrections, etc. The differences between theory and experimental measurements point to a need for new measurements employing polarized gamma rays. Recently such measurements on the cross section $\sigma(\theta,\phi)$ and polarization asymmetry $\Sigma(\theta)$ have been reported by De PASCALE et al. [1] in the 20-60 MeV gamma-ray energy range. We have attempted to fit their data by using the amplitude method of RUSTGI et al. [2]. The interaction Hamiltonian H' for E1, M1, and E2 multipoles is written as

$$H' = -eE_x A, \tag{1}$$

$$\begin{aligned} A = {} & \frac{1}{2}(\vec{I}_E\cdot\vec{r}) + \frac{i}{8}(\vec{k}\cdot\vec{r})(\vec{I}_E\cdot\vec{r}) + \frac{f^2}{4m}\Big[\frac{1}{3}(\phi_0\vec{\sigma}_1\cdot\vec{\sigma}_2+\phi_2 S_{12})\vec{r} \\ & + (2\mu_v\phi-\phi_1)[\vec{\sigma}_1(\vec{\sigma}_2\cdot\vec{r})+\vec{\sigma}_2(\vec{\sigma}_1\cdot\vec{r})]\Big]\cdot\vec{I}_E \\ & + \frac{3}{2}\frac{f^2}{M}\phi\alpha_s[\vec{\sigma}_1(\vec{r}\cdot\vec{\sigma}_2)-\vec{\sigma}_2(\vec{r}\cdot\vec{\sigma}_1)]\cdot\vec{I}_E + \frac{f^2}{2M}\frac{1}{\mu} \\ & \times\Big[\Phi_0[(\vec{\sigma}_2\cdot\vec{\nabla})(\vec{\sigma}_1\cdot\vec{r})+(\vec{\sigma}_1\cdot\vec{\nabla})(\vec{\sigma}_2\cdot\vec{r})]\vec{r} + \frac{1}{3}[(3\Phi_1+\Phi_2)(\vec{\sigma}_1\cdot\vec{\sigma}_2)+\Phi_2 S_{12}]\cdot\nabla\Big]\cdot\vec{I}_E \\ & - \frac{1}{8M^2}[(2\mu_v-1)(\vec{\sigma}_1+\vec{\sigma}_2)+2(\mu_s-1)(\vec{\sigma}_1-\vec{\sigma}_2)]\times\vec{P}_r\cdot\vec{I}_E \\ & + \frac{h}{2Mc}\Big[(\frac{1}{2}\mu_v+g_1+h_1)(\vec{\sigma}_1-\vec{\sigma}_2)+\frac{1}{2}(\mu_s-\frac{1}{2})(\vec{\sigma}_1+\vec{\sigma}_2)+(g_{II}+h_{II})T_{12}^{(-)}\Big]\cdot\vec{I}_H. \end{aligned} \tag{2}$$

The notation and symbols used are the same as in PANDEY et al. [3]. The differential cross section may be written as

$$\begin{aligned} \sigma(\theta,\phi) = {} & a + b\sin^2\theta + c\cos\theta + d\cos\theta\sin^2\theta + e\sin^4\theta \\ & + \cos 2\phi\,(f\sin^2\theta + g\cos\theta\sin^2\theta + h\sin^4\theta). \end{aligned} \tag{3}$$

The numerical calculations have been performed for supersoft core [SSC-B solid line] and Paris (dashed line) potentials. The results are compared in Figs. 1 and 2. The quality of agreement obtained by us is the same as that obtained by others as reported in Ref. [1]. Our work suggests that improved fit to the coefficients d and g can be obtained [see dot-dashed curve in the Figs.] if the E2 radial matrix elements are phenomenologically reduced by about 15%. The physical process responsible for this reduction is not clear.

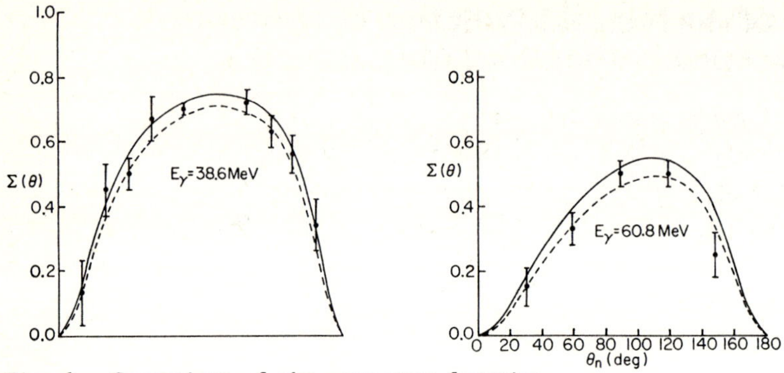

Fig. 1 Comparison of the asymmetry function

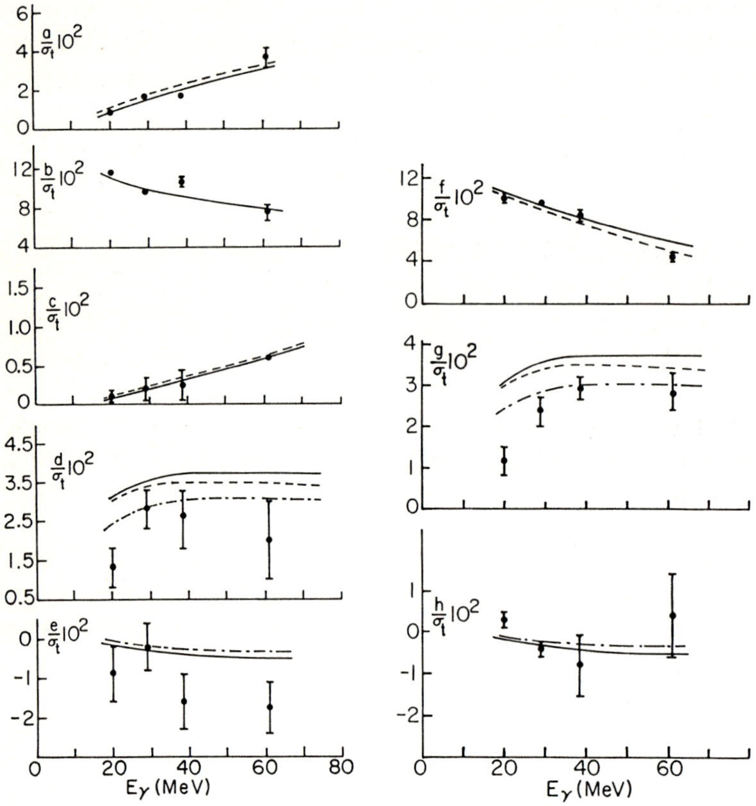

Fig. 2 Comparison of the angular distribution parameters for the supersoft core (solid) and Paris (dotted) potentials

References

1. M.P. De Pascale et al., Phys. Rev. C 32, 1830 (1985)
2. M.L. Rustgi, W. Zernik, G. Breit, and D.J. Andrews, Phys. Rev. 120, 1181 (1960)
3. L.N. Pandey and M.L. Rustgi, Phys. Rev. C 32, 1842 (1985)

Measurement of the Natural Widths of Thulium Atomic Levels in the Decay of ^{169}Yb

V.N. Pokrovskii[1], *A.Kh. Inoyatov*[1], *I.A. Prostakov*[1], *Ch. Briançon*[2], *Tz. Vylov*[1], *B. Legrand*[2], *A. Minkova*[3], *and A.A. Pas'ko*[1]

[1]Joint Institute of Nuclear Research, Dubna 141980, USSR
[2]Centre de Spectrométrie Nucléaire et de Spectrométrie de Masse, F-Orsay, France
[3]Faculty of Physics, University of Sofia, Sofia, Bulgaria

To extract the neutrino mass from a beta-spectrum the accurate knowledge of the resolution function (RF) is of crucial importance. Usually some known internal conversion lines are used to determine the instrumental RF. For this purpose one has to know the natural width of the internal conversion lines.

In recent neutrino mass experiments M-20.7 conversion lines in ^{169}Yb-decay are often used for RF determination. Unfortunately, there are no data available for the natural widths of M-levels of ^{169}Tm obtained by direct measurements so far.

The low energy conversion lines in the ^{169}Yb-decay have been measured with a resolution of about 5 and 7 eV in an electrostatic electron spectrometer[1]. The applicability of the empirical relation $W^n = \Gamma^n + R^n$ where W is the measured line width, Γ the natural line width and R the resolution, has been investigated by the convolution technique. The natural line widths of some atomic levels in ^{169}Tm are determined by a method based on a comparison between the convolution of a Lorentzian with the RF (Gaussian) and the high energy slope of the experimental line. The line K-14.4 in ^{57}Fe has been used as a reference line with known natural line width to obtain the spectrometer resolution. The procedure has been tested on the L_1-lines of two gamma transitions in ^{201}Hg (^{201}Tl-decay). Figure 1 shows the measured M_1-20.7 conversion line of ^{169}Tm. A χ^2-minimum of $\chi^2 = 0.87$ is obtained for $\Gamma = 14.2(3)$ eV at spectrometer resolution $R = 6.6$ eV.

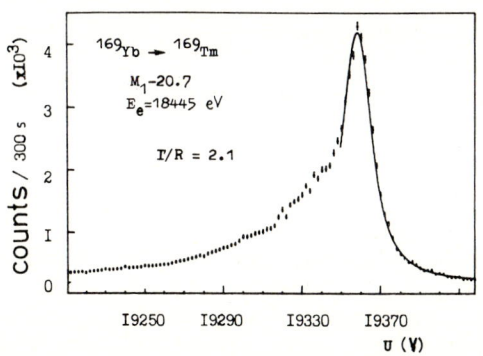

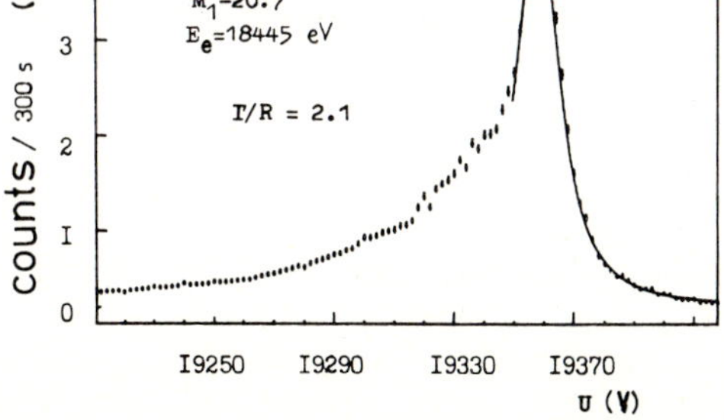

Fig. 1. The measured M_1-20.7 conversion line of ^{169}Tm

Table 1. Natural line widths for $Z = 69$ (^{169}Tm) in eV. The overall uncertainty is given in brackets

Atomic level	This work	Bennett et al.[2] (experiment)	Krause et al.[3] (semiempirical)	ITEP-83[4]
K	35.5(3)	-	30.1(10)	-
L_1	4.4(3)	-	5.47(80)	5.0
L_2	4.0(4)	-	4.49(50)	-
L_3	4.0(5)	-	4.48(40)	3.0
M_1	13.9(3)	14(3)	-	14.7
M_2	9.2(8)	7.1(20)	-	6.3
M_3	9.5(12)	7.7(14)	-	6.6

The results are summarized in Table 1. They are in good agreement with the results from X-ray transition measurements and consistent with the semiempirical data.

References

1. Briançon Ch. et al.: NIM in Phys. Res. **221**, 547 (1984)
2. Bennett C.L. et al.: Phys. Rev. C**31**, 197 (1985)
3. Krause M.O. et al.: J. Phys. Chem. Ref. Data **8**, 329 (1979)
4. Unpublished, quoted in Ref. 2

Measurements of Proton Strength Functions and Comparisons with Theory

E. Arai and Y. Ozawa

Nuclear Reactor Research Laboratory, Tokyo Institute of Technology, Oh-okayama 2-12, 152 Tokyo Meguro-ku, Japan

Using a high-resolution proton beam of the Tokyo Institute of Technology Van de Graaff, precise measurements of elastic and inelastic scattering cross sections have been performed in the past 15 years. By directly observing individual proton resonances, their spins, parities and proton decay widths were deduced. From these experiments we have evaluated (1) proton strength functions in terms of target mass number and of incident proton energy and (2) Coulomb matrix elements for split analogue resonances.

1. Proton strength functions

Theoretical values of strength functions were calculated according to the following procedure. A compound-nucleus formation cross section σ_c can be expressed with an average proton width $<\Gamma_p>$ and an average level distance $<D>$:

$$\sigma_c = 2\pi^2 \lambda\!\!\!^{-2} g_J <\Gamma_p>/<D>. \qquad (1)$$

On the other hand, σ_c is also described by employing a proton transmission coefficient $T_{\ell J}$ in the manner of compound nuclear statistical theory:

$$\sigma_c = \pi \lambda\!\!\!^{-2} g_J T_{\ell J} \qquad (2)$$

Reduced proton widths γ^2_p and proton width Γ_p are related through Eq.(3):

$$\gamma^2_p = \Gamma_p/(2P_\ell) \qquad (3)$$

Coulomb penetrabilities for incident protons P_ℓ were obtained from the R-matrix programme for scattering-experiment analyses. By combining the above three equation, proton strength functions are given as follows:

$$S = <\gamma_p^2>/<D> = T_{\ell J}/(4\pi P_\ell) \qquad (4)$$

The proton transmission coefficients $T_{\ell J}$ were calculated by using the computer code "DWUCK 4". We adopted Perey's parameter set for low proton energy and for the mass region of 30-100:

In Fig. 1, we display as an example the $p_{1/2}$ result. The region of target mass is restricted to A=50-64 mainly because of the availability of experimental data for E_p higher than 3 MeV. The experimental data are reproduced within error bars with some exceptions. The form of the curve is, however, generally well explained by the optical-model prediction. Detailed descriptions will be published elsewhere [1,2].

2. Coulomb matrix elements of analogue states

Proton strength functions were evaluated in terms of proton energy around analogue states with a rather small energy interval $\varepsilon(<<w_\lambda)$.

$$S(E) = \frac{1}{\varepsilon}\Sigma\gamma^2_\nu(\nu \text{ in } \varepsilon). \qquad (5)$$

The broadening of fine structure of an analogue state w_λ was deduced by fitting the resonance parameters to eq.(6):

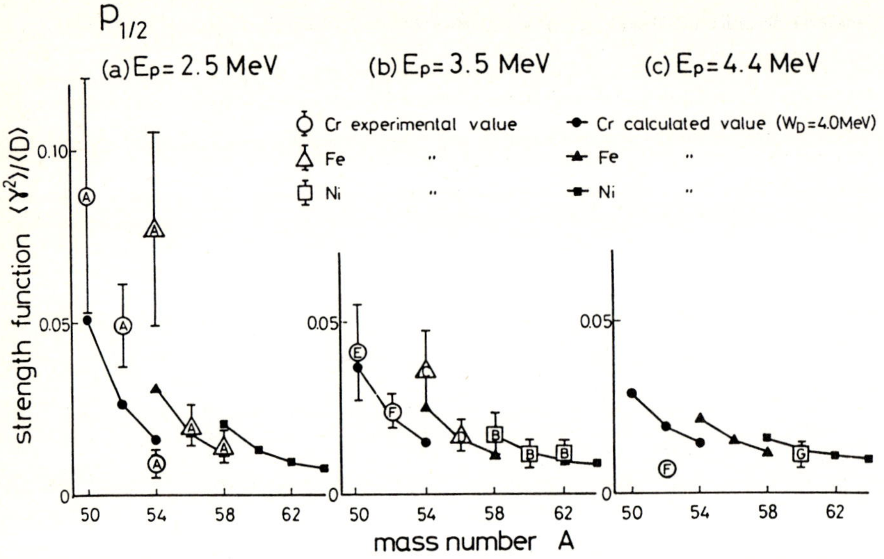

Fig.1 The dependence of $p_{1/2}$ strength functions on the incident proton energy and on the target mass number. The experimental values are deduced from the response parameters of $p_{1/2}$-wave resonances in the energy range of (a) 2-3 MeV, (b) 3-4 MeV and (c) 4-4.75 MeV. A: Bilpuch et al., Phys. Rev. 28C, 145 (1976); B-F: Data taken by the Tokyo Institute of Technology group

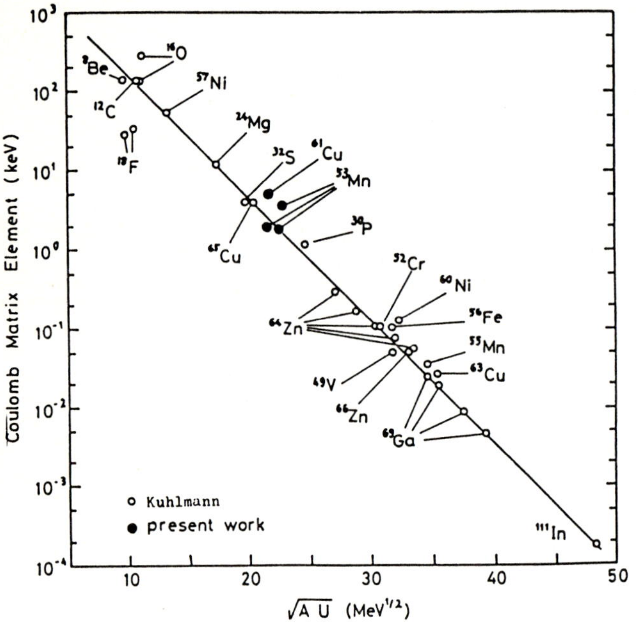

Fig.2 The dependence of the Coulomb matrix element on $\sqrt{A \cdot U}$. The straight line is the result of the semi-empirical formula given by Kuhlmann

Table 1 Coulomb matrix elements

Analogue		$\bar{E}_\lambda$ (MeV)	ρ_η (MeV^{-1})	present work W_λ (keV)	H_c (keV)	
^{31}P	$3/2^+$	3.411	23	$2^{+1.2}_{-0.2}$	$3.7^{+1.0}_{-0.2}$	(?)
^{53}Mn	$3/2^-$	3.688	272	7^{+7}_{-4}	$2.0^{+0.9}_{-0.7}$	
^{53}Mn	$1/2^-$	4.088	202	10^{+15}_{-10}	$2.8^{+1.6}_{-2.8}$	(?)
^{53}Mn	$5/2^+$	4.481	496	10^{+8}_{-5}	1.8 ± 0.6	
^{53}Mn	$5/2^+$	4.592	530	50 ± 10	3.9 ± 0.4	
^{61}Cu	$5/2^+$	4.286	307	55 ± 15	5.3 ± 0.7	
^{61}Cu	$1/2^+$	4.673	197	0.4 ± 0.1	0.6 ± 0.1	(?)

$$S_c(E)=S_c^{(pot)} - \frac{(2S_c^{(pot)}\Delta_{\lambda c})(\bar{E}_\lambda - E)}{(\bar{E}_\lambda - E)^2+\frac{1}{4}w_\lambda^2} + \frac{\tilde{\gamma}^2_{\lambda c}w_\lambda/2\pi}{(\bar{E}_\lambda - E)^2+\frac{1}{4}w_\lambda^2}, \quad (6)$$

where E_λ=energy of analogue state and $\Delta_{\lambda c}$=parameter for the asymmetry of analogue fine structure. The first term represents the background from $T_<$ states, the second and the third terms denote the symmetry and the asymmetry terms of fine structures, respectively. Coulomb matrix elements H_c ($=\langle\eta|V_c|\lambda)$) were calculated through eq. (7).

$$w_\lambda = 2\pi H_c^2 \rho_n, \quad (7)$$

where ρ_n was calculated by means of the Fermi gas model. Results are listed in Table 1 for seven analogue states in three compound nuclei. In Fig. 2, these values (·) are plotted with other data (°) to be compared with the semi-empirical formula derived by Kuhlmann /3/

$$H_c = 6.26 \exp(-0.36\sqrt{AU}) \quad /MeV/. \quad (8)$$

References

1. Y. Ozawa and E. Arai: Z. Physik A, to be published
2. Y. Ozawa: Doctor's thesis, Tokyo Institute of Technology (1986)
3. E. Kuhlmann: Phys. Rev. C20, 415(1979)

Low Energy Photofission of Actinides*

W. Wilke[1,+], *H. Ströher*[1], *U. Kneissl*[1], *R.D. Heil*[1], *K. Huber*[1], *K. Schaschek*[1], *U. Seemann*[1], *T. Weber*[1], *L.S. Cardman*[2], *P.T. Debevec*[2], *R.I. Jones*[2], *and A.M. Nathan*[2]

[1]Institut für Kernphysik, Justus Liebig Universität, Leihgesterner Weg 217, D-6300 Giessen, F. R. Germany
[2]Nucl. Phys. Lab., Univ. of Illinois, 23 Stadium Drive, Champaign, IL 61820, USA

The measurement of fragment angular and mass distributions in photofission represents an important tool to investigate the properties of the double humped fission barrier, in particular the spectrum of transition states, which provides direct information on nuclear shapes and possible collective modes. Photoexcitation has the unique feature to be specific to low spin states. The ratio of quadrupole to dipole photofission is directly connected to the relative heights of the bumps of the double humped fission barrier.

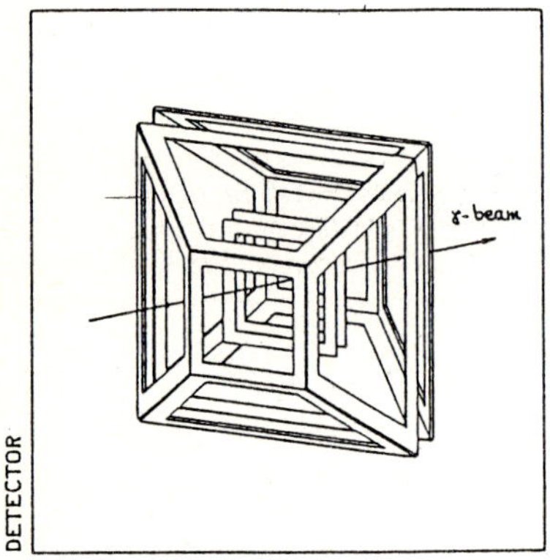

Fig. 1: 4π arrangement of position sensitive PPAC

Up to now all fragment angular and mass distributions in photofission experiments have been measured using continuous bremsstrahlung. For first experiments with monochromatic tagged photons at the MUSL 2 accelerator of the University of Illinois we constructed a special 4π arrangement of position sensitive parallel plate avalanche counters (PPAC), which is shown schematically in Fig. 1. The whole device is built of 12 PPAC's, each consisting of 3 electrodes: the two outer ones are wire planes (distance between wires 1 mm) with a delay-line read out which provides the position information. The inner electrode consists of an aluminized hostaphan foil which gives a fast timing signal. The performance of these PPAC's will be discussed in more detail in a forthcoming paper. The colinear emission of the two fragments allows to calculated the position of the fission event on the target from the detected points of incidence on the outer detectors. Angular resolutions < 1° can be achieved even at extended beam spot diameters of 7 cm. The two inner detectors are used as start detectors for fragment time of flight measurements and as an on line position and profile monitor.

---

*Supported by the Deutsche Forschungsgemeinschaft and NATO
+Excerpt from D26

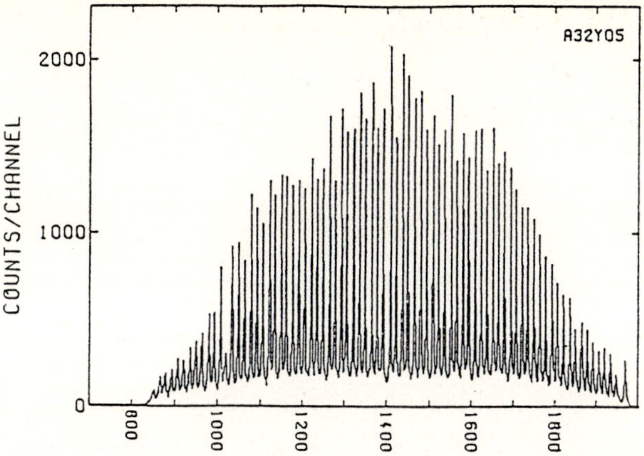

Fig. 2: Typical position spectrum

First experiments on ^{238}U have been performed with the detector set up at the tagged photon facility at Illinois in the energy range near the barrier and of the second chance fission threshold. At the bremsstrahlung facility of the Giessen Linac data have been taken within the energy range near the barrier for ^{232}Th, $^{236,238}U$. The data have been analysed with respect to the fragment angular and mass distribution.

In the following figures first results are presented. Fig. 3 shows typical fragment TOF-difference spectra for ^{232}Th at bremsstrahlung end point energies E_{BS} of 7.75 MeV and 16 MeV respectively and for ^{236}U (E_{BS}= 16 MeV). An increase of the symmetric fission yield at the higher bremsstrahlung energy can nicely be stated. For ^{236}U (lower figure) a less pronounced mass splitting has been observed.

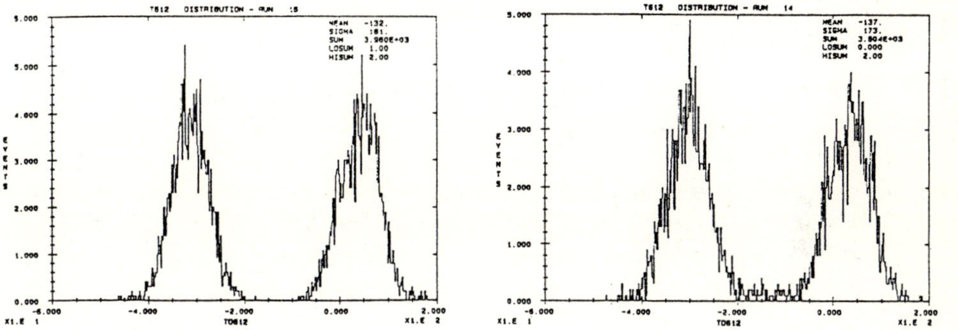

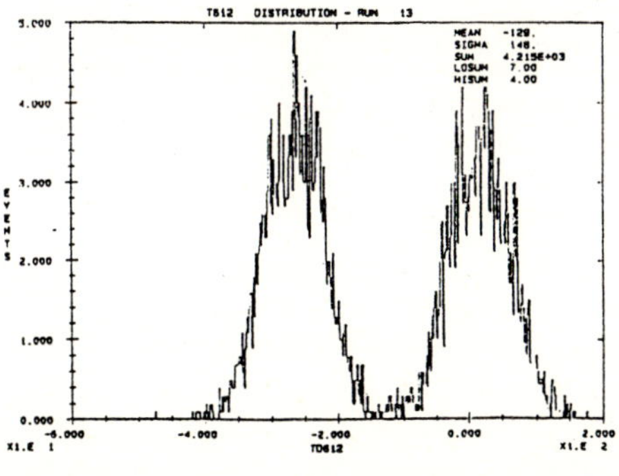

Fig. 3: Fragment TOF difference-spectra (flight path ≈ 12 cm) for ^{232}Th (E_{BS}= 7.75 MeV and 16 MeV) and ^{236}U (E_{BS}= 16 MeV)

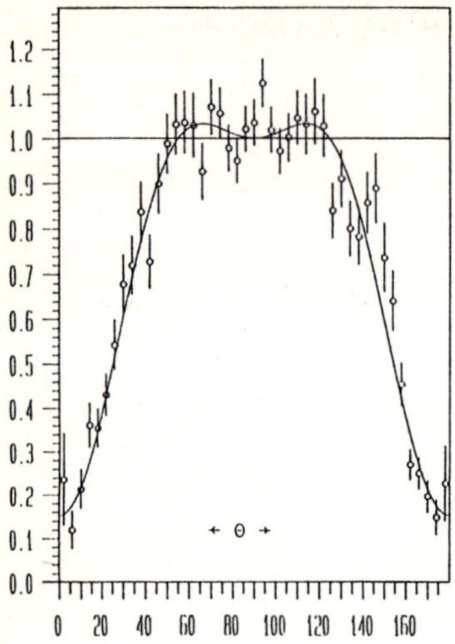

Fig. 4: Fragment angular distribution (E_{BS}= 6.5 MeV, ^{238}U)

As an example in Fig. 4 a fragment angular distribution (^{238}U, E_{BS}= 6.5 MeV) is plotted.

Effects of the Vacuum-Polarization on Sub-Coulomb $^{12}C-^{12}C$-Scattering

D. Vetterli, G. Baur, W. Boeglin, P. Egelhof, R. Henneck, A. Klein, H. Muehry, G.R. Plattner, F. Roesel, I. Sick, D. Trautmann, A. Weller**, and M. Yaskola*

Institut für Physik, Universität Basel, CH-4056 Basel, Switzerland

For many years QED-effects have been investigated in atomic and particle physics mainly in leptonic systems. Little information only is available on QED-effects in Coulomb-interaction between hadrons.[1,2] Experiments performed on the proton-proton system[1] yield results which disagree with theoretical predictions by an amount of about 50% of the vacuumpolarization effect.

The aim of the present experiment is to detect the influence of the vacuum polarization on sub-Coulomb-scattering of Heavy Ions. The angular distribution for scattering of identical spin zero particles shows a pronounced interference structure which is symmetric around $\Theta_{cm} = 90°$ (see Fig.1). Calculations using the program 'Vacpol' predict small angular shifts of the interference minima due to vacuum polarization. These shifts amount to deviations from the Mott-scattering cross section of a few percent in the region around the minima. The expected effects are shown in Fig.1 for the case of ^{12}C-^{12}C-scattering at $E_{cm} = 2$ MeV, which is far below the Coulomb barrier. Contributions due to additional processes such as the nuclear interaction, nuclear polarizability, relativistic effects, screening and higher order QED-effects have been calculated or estimated. They are found to be at least one to two orders of magnitude smaller than the effect of the vacuum polarization. An accurate measurement of the angular distribution therefore allows a sensitive test of the vacuum polarization contribution to the electromagnetic interaction of heavy ions.

The experiment is performed at the MPI Heidelberg using the EN-Tandem accelerator. The experimental set-up consists of a scattering table housed in a scattering chamber of 1.2 m diameter. Besides moveable detectors (range $38° < \Theta_{lab} < 52°$) a set of fixed monitor detectors for calibration and monitoring of beam position, target position etc. is used. The positions of entrance and exit-slits of the beam and the detector slits are known to an accuracy of a few microns. The energy calibration of the beam is done using a resonance in $^{16}O\ (\alpha,\alpha)\ ^{16}O$ scattering with α's having the same magnetic rigidity as the ^{12}C- beam.

In a first experiment data have been taken at 4 angles (marked with arrows in Fig.1). In order to reduce systematic errors due to the change of target thickness, target position, beam alignment etc. many short runs have been taken. After each cycle of crossection measurements a couple of test runs are done to measure the amount of smallangle scattering and the target position via a cinematical coincidence technique as well as the target thickness.

In the lower part of Fig.1 the measured data (result of a preliminary data analysis) are compared to a calculation for pure Mott-scattering (dashed line)

* present address: Institut für Physik, Universität Mainz, D-6500 Mainz, FRG
** Max-Planck-Institut für Kernphysik, D-6900 Heidelberg, FRG

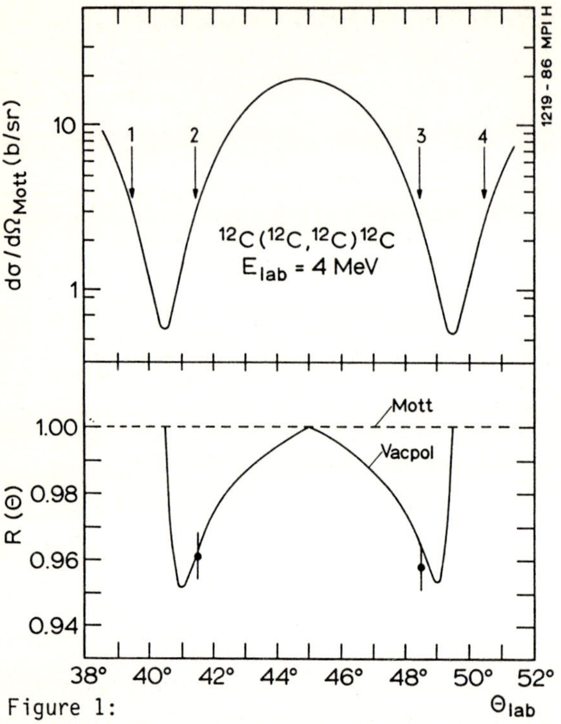

Figure 1:

Angular distribution of the differential cross section for Mott-scattering of ^{12}C from ^{12}C at E_{lab} = 4 MeV (upper part). The lower part shows the cross section ratio $R(\Theta)$ for angles symmetric to the minima [for example $R(\Theta_2) = \sigma(\Theta_2)/\sigma(\Theta_1)$]. The angles are chosen such that the ratio is one for Mott-scattering (dashed line). The solid line shows the prediction for $R(\Theta)$ where the contribution of vacuum polarization is taken into account.

and a prediction using the program 'Vacpol' where the effect of vacuum polarization is included. The error bars of the data points are purely statistical. The systematic error is estimated to be at present $\pm$ 0.008. Within the accuracy reached the data are in good agreement with the QED-prediction. For the first time the effect of vacuumpolarization could be established for the Heavyion-Heavyion-Coulomb-interaction. After a further reduction of the error bars the effects of screening and higher order vacuum polarization, which are being included in the theoretical calculation should be detectable.

References

1 H. Wassmer, H. Muehry: Helv.Phys. Acta 46(1973)627;
Ch. Thomann et al.: Nucl.Phys. A303(1978)457.

2 W.G. Lynch et al.: Phys.Rev.Lett. 48(1982)979.

1.2 Nuclei in Highly Excited States

1.2.1 High Spin States

Electromagnetic Properties of Nuclei at High Spins

G.A. Leander

UNISOR, Oak Ridge Associated Universities, Oak Ridge, TN 37830, USA

1. Introduction

A photon emitted by an excited state is likely to carry away, at most, 1 or $2\hbar$ of angular momentum. Therefore, a profusion of photons is needed to deexcite the rapidly rotating states of nuclei formed by heavy-ion reactions. The study of electromagnetic properties has become the primary source of information on nuclear structure at high spins and, also, at the warm temperatures present in the initial stage of the electromagnetic cascade process. The purpose of this paper is a review of the E1, M1, and E2 properties of such highly excited states, in order to set the stage with theoretical props for the section on this topic in the present volume. This review does not aspire to completeness or a fair distribution of credit for past achievements, but does attempt to highlight the major current research topics in the field. A chart of these topics in the plane of energy versus angular momentum is given in Fig. 1. They involve both the properties of the rotating but "cold" states on or near the yrast line, which can be experimentally resolved as discrete states, and of the increasingly closely spaced levels at higher temperature. The former reveal the response of nuclear structure to rotation, and the latter provide unique insight into the transition of a quantum system from order to chaos.

2. Dipole Resonances.

2.1. E1 Giant Resonance

The vibration of protons against neutrons superposed on any nuclear state has a collective resonance, according to a famous hypothesis discovered by Axel in the thesis of BRINK [1]. Empirically the resonance on low-lying states occurs at $\hbar\omega \approx 78\ A^{-1/3}$ MeV, about 15 MeV in a medium heavy nucleus. Therefore, the resonances even on the ground-state and low-energy excited states occur in a region of high level density. The resonances become damped, or fragmented onto their many neighbors, and the E1 transition rates from all states at energies $U \gtrsim \hbar\omega$ are more or less enhanced by a share of the collectivity of the resonances.

The emission of gamma-rays at the energy of the E1 giant resonance in heavy-ion fusion-evaporation reactions is most likely to occur from the compound nucleus prior to any evaporation of particles. This follows from elementary statistical physics. Transition rates depend on the level density ρ at the initial and final states: $T \propto \rho(U_f)/\rho(U_i)$. According to the Fermi gas formula, ρ is the product of a relatively slowly varying factor with $\exp(2\sqrt{aU})$, or equivalently $\exp(2U/t)$ where the temperature $t = \sqrt{U/a}$. Inserting $U_i - U_f = 15$ MeV for the gamma ray and $U_i - U_f = 8 + 2t$ MeV for a neutron with binding energy 8 MeV, we get $T_\gamma/T_n \propto \exp(-14/t)$ which is a rapidly increasing function of temperature. Thus, when NEWTON et al. [2] observed a bump on the high-energy tail of the γ spectrum from ^{40}Ar induced fusion reactions, and were able to reproduce this bump in a statistical model calculation using the E1 strength function of the resonance on the ground state, they confirmed the Axel-Brink hypothesis for states of very high energy and angular momentum (Fig. 1).

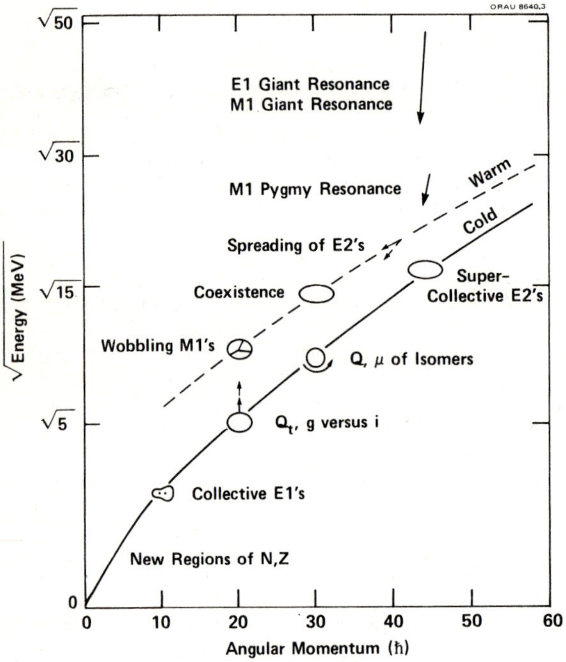

Fig. 1. Hot topics in warm and cold rotating nuclei

Continued interest in the E1 giant resonance rests on the fact that its shape is sensitive specifically to the shape of the nucleus. In a deformed nucleus there are several resonance frequencies, proportional to the reciprocal lengths along the principal axes. The angular momentum dependence has been investigated theoretically in the cranking model. It is found that at fixed shape there is little effect on either the centroid or the splitting of the resonance [3]. The inclusion of temperature in the calculations also has little effect. Even the damping width is unchanged at any rate on the RPA level of the theory [4]. Thus, there is the prospect of studying the shape of the giant E1 resonance to find out what happens to spherical and deformed shell structure at high spin and temperature [5]. A splitting into two peaks could be obtained for the "superdeformed" shapes that might occur at very high spins. Since the relative strength of the signal increases with temperature as discussed above, the E1 resonance might be used to probe the very existence of nuclei at extremely high temperatures. WONG [6] has suggested that rotating toroidal shapes might be favored under these conditions, and that the E1 resonance going around the torus would appear at an unmistakably low frequency.

2.2. M1 Resonances

It is not known whether M1 giant resonances play any role in the deexcitation of highly excited nuclei. It has been suggested, however, that a concentration of M1 transition strength in some energy interval can sometimes occur [7]. The resulting bump in the gamma spectrum should be observable when this energy lies above the energy of the yrast-like transitions. Such M1 pygmy resonances, to use the terminology of the corresponding phenomenon in β decay, are expected to arise when several of the strong M1 single-particle transitions near the Fermi level in the Nilsson scheme cluster around the same energy. In deformed nuclei this can happen for transitions of the type $[N\ n_z\ \Lambda\ \Omega] \to [N\ n_z{+}\ 1\ \Lambda{-}1\ \Omega{-}1]$ or $[N\ n_z\ \Lambda\ \Omega = \Lambda{-}1/2] \to [N\ n_z\ \Lambda\ \Omega = \Lambda{+}1/2]$. Experimental evidence for dipole bumps at the appropriate energies have been obtained from a light-ion reaction leading to ^{161}Dy [8] and from a study of angular distributions following a (HI, xn) reaction leading to ^{158}Yb [9].

3. E2 Spreading Widths

When a nucleus formed by heavy-ion fusion has cooled down to a temperature of about 1/2 MeV, E2 transitions along collective rotational bands running roughly parallel to the yrast line are expected to become competitive with the dipole transitions [10,11]. Since there are many closely spaced interacting bands above the yrast line, the collective E2 strength from each initial state is probably spread over several final states, in analogy with the damping of the E1 giant resonance discussed above. The behavior of the spreading width will depend strongly on the nature of the mechanisms that contribute to it [12]. One estimate, based on the fluctuations of rotation-aligned particle angular momenta in the cranked harmonic oscillator model, gives a spreading width proportional to I^2/U [13], in other words, decreasing with increasing intrinsic excitation energy U contrary to what might be expected intuitively. An increase of the spreading width with increasing temperature, on the other hand, might result because the potential-energy-of-deformation surface generally becomes flatter with increasing temperature so that the interacting bands have a wider spread in moment of inertia [14].

Experimentally, it is clear that the upper limit on the E2 spreading width is about 1/2 MeV, since the spectra from rotational nuclei exhibit a quadrupole bump with a clear-cut edge that moves up in energy with increasing multiplicity and total energy of the cascades. The component of the average E2 strength function with a width less than some tens of keV must be rather small, of the order of 10%, because at best only weak ridge-valley structures are observed in E_γ-E_γ correlation plots [15]. Similar conclusions are obtained by looking at the change in the spectrum that comes from requiring a coincidence with some specified gamma ray energy E_γ [14,16]. The width of this narrow component is constant or slowly increasing with increasing E_γ, and its fraction of the total strength decreases, but both the width and the relative strength of the narrow component

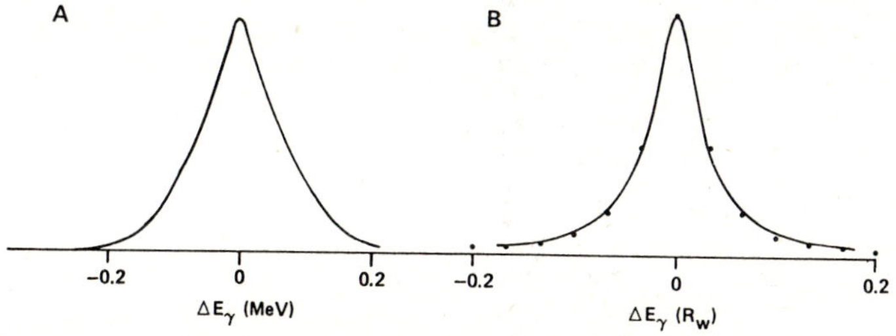

Fig. 2(A). A possible phenomenological E2 strength function obtained by superimposing a broad and a narrow Gaussian (see text). (B). An E2 strength function from the model of Ref. [12]. The dots show a Breit-Wigner line shape

are insensitive to the total energy and multiplicity of the cascades. The evaluation of these experimental findings is still somewhat speculative. Figure 2A shows a phenomenological E2 strength function that might be suggested by the data. It consists of two Gaussians of width σ = 40 keV and 150 keV, respectively, with 10% of the strength in the narrow component. Figure 2B shows the strength function obtained from consecutive transitions using the model of interacting bands described in Ref. [12]. The distribution of B(E2; i → j) B(E2; j → k) is plotted versus the spread in energy, $(E_i - E_j) - (E_j - E_k)$. The result appears to follow a Breit-Wigner distribution, indicated by dots in Fig. 2A, which is the shape usually associated with damping. This result is obtained from the model even when the initial states i are constrained to be from the region where the level density increases rapidly with energy. The peaks of the distributions in Figs. 2A and B are quite similar, but the tails are radically different. Taking into account the factor $E_\gamma{}^5$ in transition rates, the bulk of all transitions would go deep into the low-energy tail of the Breit-Wigner like distribution, contrary to the empirical evidence. Before the aptness of any damping mechanism for the E2 spreading can be established, it will be necessary to resolve this difficulty.

4. Shape Coexistence

The coexistence of different nuclear shapes at high spins has long been predicted by theory; see, for example, the summary of the cranked Nilsson-Strutinsky model calculations by the Lund group during the years 1975-1980 in Ref. [17]. At the highest spins the increase of the bulk 'liquid drop' energy with increasing deformation is counteracted by a decrease of the bulk rotational energy; the potential energy surface becomes flatter, and shape-dependent shell closures over a wide region of deformation space come into play. It is becoming increasingly evident from experiment that a richness of structure stemming from coexisting shapes and symmetries of the nucleus does indeed occur and can be studied in the gamma-ray spectrum (e.g., the contributions of Sharpey-Schafer, Khoo and de Voigt to this volume).

With the existence of coexistence well established, new questions arise. Most importantly, what is the degree of order, as opposed to chaos, in the

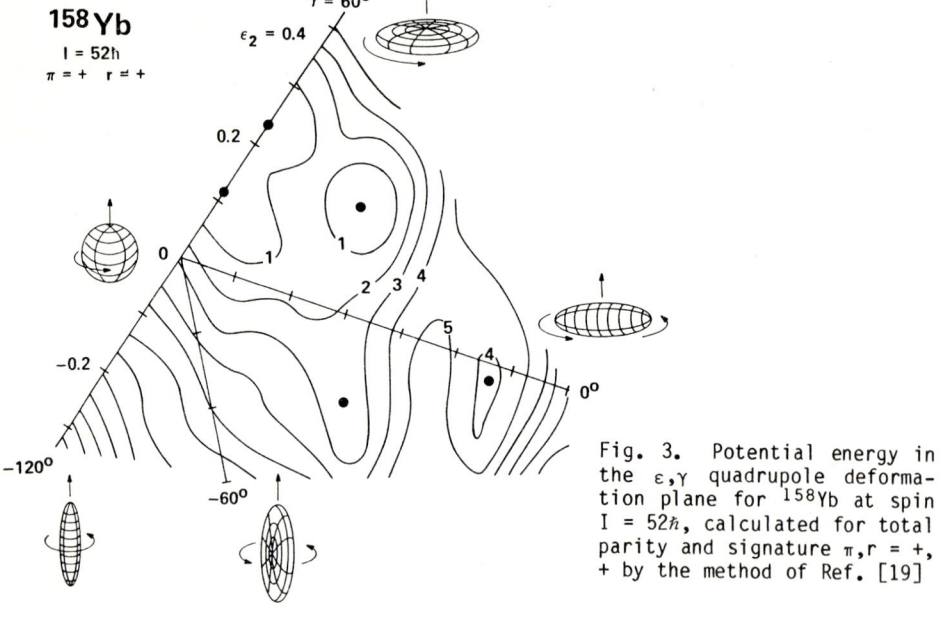

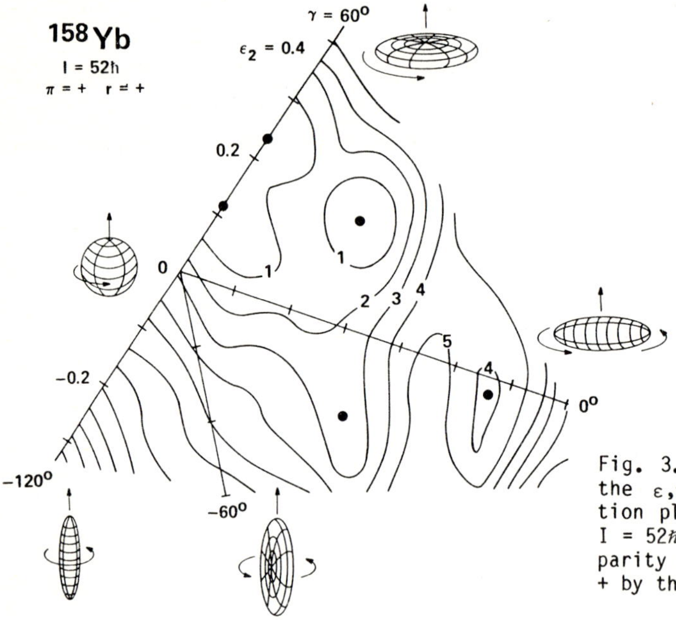

Fig. 3. Potential energy in the ε,γ quadrupole deformation plane for ^{158}Yb at spin I = 52ħ, calculated for total parity and signature π,r = +, + by the method of Ref. [19]

structure of the coexisting configurations? There is a hierarchy of quantum numbers which are present in the theoretical models, or at any rate imposed on them, which can be used to classify configurations and to calculate potential energy surfaces for the classes. At the extreme of internal chaos, there are the traditional potential-energy surfaces [17] and their extensions to finite temperature [18]. Various degrees of order were introduced in the calculations by T. BENGTSSON and RAGNARSSON [19]. For example, in Fig. 3 the total parity π and total signature r are conserved. The minima then tend to become more distinct, with higher potential barriers between them, than without these requirements. Slightly different surfaces are obtained for other combinations of π and r. The highest degree of order used by Bengtsson and Ragnarsson is conservation of the number of particles with each combination of π, r, N_r, τ, where π and r now refer to the single-particle parity and signature, N_r is the major oscillator quantum number in the rotating basis, and τ is isospin. Then, for example, superdeformed and weakly deformed minima do not appear in the same potential energy surfaces. At low spins the onset of pairing certainly scatters pairs of particles between the π, r, N_r groups, however, and allows different classes of bands to interact. Data like those recently obtained for the superdeformed band in ^{152}Dy and the onset of its electromagnetic decay into the weakly oblate configurations [20] are highly interesting in this context.

Some types of coexisting bands which should exist according to Fig. 3 have not yet been positively identified at high spins, namely, the well-deformed triaxial and oblate bands. Wobbling bands associated with triaxial configurations might be characterized by the M1 and E2 transitions "leaking" between such bands [21,22]. High-K bands associated with well-deformed oblate configurations could be characterized by $\Delta I = 1$ transitions.

5. Electromagnetic Moments of Discrete States

Structure theories of the more strongly populated yrast states can be tested by measuring their electromagnetic moments.

5.1. E2 Moments

One of the early predictions of the cranking model was the occurrence of high-spin isomers, with a shape made oblate by the ring-shaped orbits of the aligned high-j particles, in the regions of nuclei around A ~ 150 and ^{208}Pb [23]. Measured electric quadrupole moments of the high-spin isomers in ^{147}Gd support this picture [24,25]. In the lead region, however, the issue has been clouded by a recent measurement for the $63/2^-$ isomer of ^{211}Rn which gave no evidence for deformation [26]. In collectively rotating nuclei, the ring shaped orbits of aligned high-j particles in the S-band are expected to polarize the nuclear shape toward positive γ (c.f. Fig. 3), thereby reducing the collectivity of the rotation [27,28]. Evidence for this has been found in the in-band E2 transition moments, which are reduced above the backbend in nuclei around ^{158}Er [29]. The rotation-aligned high-j quasiparticles obtained when the Fermi level is nearer the middle of a high-j shell are expected to have other orbital shapes, however, and may even drive the γ deformation in the opposite direction so as to enhance the collectivity of the rotation [27,30]. Evidence for this has not yet been found. The nucleus ^{172}W would seem to be a relevant test case: the Fermi level lies higher in the neutron $i_{13/2}$ shell than for ^{160}Yb, and a cranking calculation predicts roughly constant E2 transition moments along the yrast line (the dashed curve in Fig. 4). A recent experiment on ^{172}W did give information about the high spin structure, but of an unexpected kind which has not yet been fully understood [31]. Namely, the measured E2 transition moments in Fig. 4 reveal a drop which is surprisingly sharp considering the smoothness of the upbend in ^{172}W, and closer analysis of the data establishes that the yrast upbend is due to a three-band crossing (cf. the inset in Fig. 4). Energy level systematics appear to link the mysterious third band to the proton $h_{9/2}$ shell, which would indeed have ring shaped aligned orbitals.

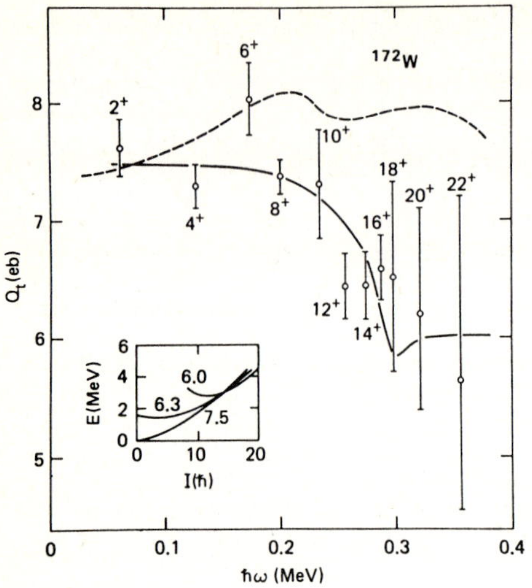

Fig. 4. Measured Q_t values [31] for yrast levels of ^{172}W. Mixing of the three unperturbed bands shown in the inset can reproduce the yrast energies. The unperturbed intrinsic quadrupole moments indicated (in eb) in the inset were obtained from the calculated change in γ deformation assuming first $(\pi h_{9/2})^2$ and then $(\nu i_{13/2})^2$ rotational alignment. The Q_t values that result from the mixing are shown by the solid curve

Overall, whereas earlier studies of deformed nuclei by Coulomb excitation had shown that rotor-like E2 moments could persist up to high spins (e.g. Ref. [32]), the recent lifetime measurements on neutron-deficient nuclei from Ce to W indicate a more consistent trend towards reduction of the E2 transition moments in the S-band than is expected from theory.

5.2. M1 Moments

Magnetic moments probe the microscopy of nuclear rotations. The magnetic g-factors in collective bands can indicate whether backbending phenomena are due to neutrons or protons [33]. Direct measurements on collective states have so far been carried out only for a couple of actinides [34], the quasicontinuum in a few rare-earth nuclei [35], and the nucleus ^{168}W which has a retarded yrast transition due to weak interaction at the backbend [36]. In $\Delta I = 1$ bands, indirect information on the M1 transition moments is more readily available from branching and mixing ratios. The signature splitting of the M1 transitions rates in such bands is highly sensitive to the triaxiality of the nuclear shape ([37] and references therein). The magnetic moments of high-spin isomers can put some constraints on their configurations. One interesting experiment that has not yet been made is to measure an M1 moment of a very high spin isomer together with the M1 transition moment of a rotational band built on that isomer.

5.3. E1 Transition Moments

In recent years, it has been established that some nuclei are characterized by reflection asymmetric intrinsic shapes [38]. With the reflection symmetry broken, the nucleus can sustain a collective intrinsic E1 moment. This E1 moment cannot be observed as a static moment in the laboratory frame due to parity conservation, but the E1 transition moment leads to enhanced E1 transitions within the parity-doubled rotational bands of such nuclei. Recently, several bands have been found where the E1 cascade transitions are competitive with the E2 cross-over transitions.

Early theoretical work had anticipated such bands. The assumption that the E1 moments could be estimated from the leading order term of the liquid drop model was not borne out by the data, however. This term is proportional to $\beta_2\beta_3$ so the B(E1)/B(E2) branching ratios would depend only on the effective value of $\beta_3{}^2$. The experimental data, however, exhibit a large variation of the B(E1)/B(E2) branching ratios over a sequence of Ra and Th isotopes (Fig. 5) whose equilibrium β_3 deformations are calculated to be quite similar. This behavior is

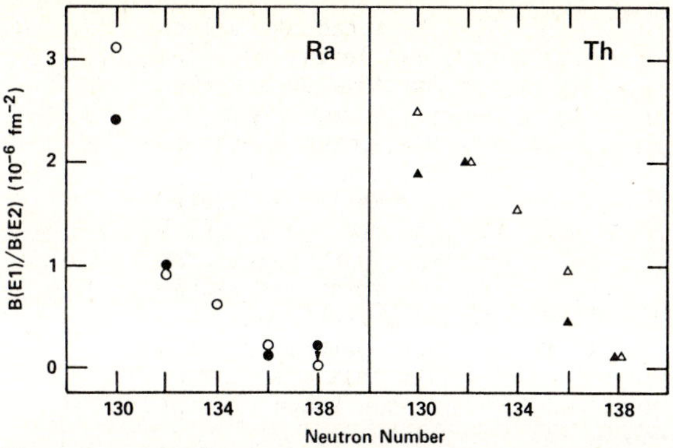

Fig. 5. B(E1)/B(E2) ratios at moderately high spins in the radium and light thorium isotopes, from experiment (solid symbols) and from a cranked shell model calculation [39] (open symbols). The theoretical value for ^{218}Ra is an average of the values at the prolate and oblate minima

explained by a strong shell effect on the E1 moment [39]. The center of gravity for both protons and neutrons in an octupole-deformed mean field is displaced toward the thick end of the "pear shape" near the spherical magic numbers, and toward the pointed end at mid-shell. Varying the neutron number while keeping the proton number fixed at an intermediate value relative to the shell closure, the shell correction to the E1 moment then changes sign. The B(E1)/B(E2) ratios are large for the lower neutron numbers because the shell correction and the liquid drop contribution have the same sign, and small for the higher neutron numbers because the two terms have opposite signs and cancel. Fig. 5 shows that there is quantitative agreement between the data and results from a cranked Woods-Saxon Bogolyubov Strutinsky model calculation of equilibrium deformations, including the isovector dipole deformation.

There are both experimental and theoretical indications that octupole deformation may occur at high spins in regions of nuclei where it does not occur at low spins [38-40]. The observation of two or more consecutive members of the $\Delta I = 1$ E1 cascade, as in ^{148,149,150}Sm and ^{148}Nd [41], provides strong evidence for such breaking of the reflection symmetry at high spins.

6. New Regions of Z and N. The electromagnetic properties of rotational states are helpful in studying the structure of nuclei far from stability. A single example of this will be given from the region of Z and N to be considered in another contribution to this volume. In the $Z \sim N \sim 36$ region, the search is on for a unique, strongly oblate band structure. These oblate bands were predicted a long time ago [42] to coexist with the strongly prolate bands that have since been observed in this region. Three $3/2^-$ bands are expected in the low-energy spectrum of ^{73}Kr: the prolate 3/2[312] and 3/2[301] bands, and the oblate 3/2[321] band. They could be distinguished by the cascade to cross over branching ratio from their $7/2^-$ band members, which would be 1:10, 2:1, and 8:1, respectively.

In conclusion, the study of electromagnetic properties in rotating nuclei has shed light on several new phenomena during the last few years, and rapid experimental progress is encouraging the theorists to ask new or more refined questions.

UNISOR is a consortium of ten institutions. It is supported by these institutions and by the U.S. Department of Energy under contract number DE-AC05-76OR00033 with Oak Ridge Associated Universities.

References

1. D.M. Brink: Ph.D. Thesis, University of Oxford, 1955
2. J.O. Newton, B. Herskind, R.M. Diamond, E.L. Dines, J.E. Draper, K.H. Lindenberger, C. Schuck, S. Shih and F.S. Stephens: Phys. Rev. Lett. 46, 1383 (1981)
3. P. Ring, L.M. Robledo, J.L. Egido and M. Faber, Nucl. Phys. A419 , 261 (1984)
4. I. Gallardo, M. Diebel, T. Døssing and R. Broglia: Nucl. Phys. A443, 415 (1985)
5. J.J. Gaardhøje: in Nuclear Structure 1985, edited by R. Broglia, G.B. Hagemann and B. Herskind (North Holland, Amsterdam, 1985) p. 519
6. C.Y. Wong: Phys. Rev. C, in press
7. Y.S. Chen and G.A. Leander: in High Angular Momentum Properties of Nuclei, edited by N.R. Johnson (Harwood, New York, 1983) p. 327
8. M. Guttormsen, J. Rekstad, A. Henriquez, F. Ingebretsen and T.F. Thorsteinsen: Phys. Rev. Lett. 52, 102 (1984)
9. Y. Schutz, C. Baktash, I.Y. Lee, F.K. McGowan, N.R. Johnson, M.L. Halbert, D.C. Hensley, L. Courtney, A.J. Larabee, L.L. Riedinger, D.G. Sarantites and Y.S. Chen: in Oak Ridge National Laboratory Physics Division Annual Report 1985, ORNL-6233, p. 80
10. R.J. Liotta and R.A. Sorensen: Nucl. Phys. A297, 136 (1978)
11. M. Wakai and A. Faessler: Nucl. Phys. A307, 349 (1978)
12. G. Leander, Phys. Rev. C25, 2780 (1982)
13. T. Døssing, in Nuclear Structure 1985, edited by R.A. Broglia, G.B. Hagemann and B. Herskind (North Holland, Amsterdam, 1985) p. 379
14. F.S. Stephens, J.E. Draper, J.L. Egido, J.C. Bacelar, E.M. Beck, M.A. Deleplanque and R.M. Diamond: preprint LBL-21288, submitted to Physical Review Letters
15. D.J.G. Love, A.H. Nelson, P.J. Nolan and P.J. Twin: Phys. Rev. Lett. 54, 1361 (1985)
16. I.Y. Lee: Proc. Int. Symp. on Physics at Tandem, Beijing, 1986 (World Scientific, Singapore) in press
17. S. Åberg: Phys. Scr. 25, 113 (1982)
18. A.V. Ignatyuk, I.N. Mikhailov, L.H. Molina, R.G. Nazmitdinov and K. Pomorski: Nucl. Phys. A346, 191 (1980)
19. T. Bengtsson and I. Ragnarsson: Phys. Lett. 115B ,431 (1982); Nucl. Phys. A436, 14 (1985)
20. P.J. Twin, B.M. Nyakó, A.H. Nelson, J. Simpson, M.A. Bentley, H.W. Cranmer-Gordon, P.D. Forsyth, D. Howe, A.R. Makhtar, J.D. Morrison, J.F. Sharpey-Schafer and G. Sletten: submitted to Physical Review Letters.
21. A. Bohr and B.R. Mottelson, in Nuclear Structure, vol. 2 (Benjamin, New York, 1975) pp. 190ff
22. R.J. Liotta, Phys. Scr. 21, 135 (1980)
23. G. Andersson, S.E. Larsson, G. Leander, P. Möller, S.G. Nilsson, I. Ragnarsson, S. Åberg, R. Bengtsson, J. Dudek, B. Nerlo-Pomorska, K. Pomorski and Z. Szymanski: Nucl. Phys. A268, 205 (1976)
24. O. Häusser, H.-E. Mahnke, J. F. Sharpey-Schafer, M.L. Swanson, P. Taras, D. Ward, H.R. Andrews and T.K. Alexander: Phys. Rev. Lett. 44, 132 (1980)
25. T. Døssing, K. Neergard and H. Sagawa: Phys. Scr. 24, 258 (1981)
26. E. Dafni, M. Hass, E. Naim, M.H. Rafailovich. A. Berger, H. Grawe and H.-E. Mahnke: Phys. Rev. Lett. 55, 1269 (1985)
27. S. Frauendorf and F.R. May: Phys. Lett. 125B, 245 (1983)
28. R. Bengtsson, Y.S. Chen, J.-y. Zhang and S. Åberg: Nucl. Phys. A405, 211 (1983)
29. M. Oshima, N.R. Johnson, F.K. McGowan, I.Y. Lee, C. Baktash, R.V. Ribas, Y. Schutz and J.C. Wells: Phys. Rev. C33, 1988 (1986)
30. G.A. Leander, S. Frauendorf and F.R. May: in High Angular Momentum Properties of Nuclei, edited by N.R. Johnson (Harwood, New York, 1983) p. 281

31. M.N. Rao, N.R. Johnson, F.K. McGowan, I.Y. Lee, C. Baktash, M. Oshima, J.W. McConnell, J.C. Wells, A. Larabee, L. L. Riedinger, R. Bengtsson, Z. Xing, Y.S. Chen, P.B. Semmes and G.A. Leander: submitted to Physical Review Letters.
32. E. Grosse, A. Balanda, H. Emling, F. Folkmann, P. Fuchs, R.B. Piercey, D. Schwalm, R.S. Simon, H.J. Wollersheim, D. Evers and H. Ower: Phys. Scr. 24, 337 (1981)
33. S. Frauendorf: Phys. Lett. 100B, 219 (1981)
34. O. Häusser, H. Gräf, L. Grodzins, E. Jaeschke, V. Metag, D. Habs, D. Pelte, H. Emling, E. Grosse, R. Kulessa, D. Schwalm, R.S. Simon and J. Keinonen, Phys. Rev. Lett. 48, 383 (1982)
35. O. Häusser, D. Ward, H.R. Andrews, P. Taras, B. Haas, M.A. Deleplanque, R.M. Diamond, E.L. Dines, A.O. Macchiavelli, R. McDonald, F.S. Stephens and C.V. Stager: Phys. Lett. 144B, 341 (1984)
36. Current work at Stony Brook.
37. I. Hamamoto and B.R. Mottelson: Phys. Lett. 167B, 370 (1986)
38. G.A. Leander: in Nuclear Structure 1985, edited by R.A. Broglia, G.B. Hagemann and B. Herskind (North Holland, Amsterdam) p. 249; W. Nazarewicz, ibid., p. 263.
39. G.A. Leander, W. Nazarewicz, G.F. Bertsch and J. Dudek: Nucl. Phys. A453 (1986) 58.
40. J. Dudek, W. Nazarewicz and G.A. Leander: in Precommunications to Niels Bohr Cent. Symp. on Nuclear Structure, Copenhagen, 1985, p. 40
41. W. Urban: in Proc. Workshop on Nuclear Structure at High Spin, Risø, 1986.
42. I. Ragnarsson and S.G. Nilsson: in Proc. Colloquium on Transitional Nuclei, Orsay, 1971, p.112

Shape Coexistence and Discrete Superdeformed States up to 60$\hbar$ in ^{152}Dy

J.F. Sharpey-Schafer

Oliver Lodge Laboratory, The University, P.O. Box 147, Liverpool, L69 3BX, U.K.

Abstract: Evidence is presented for the existence of a series of 20 discrete superdeformed states ($\varepsilon \approx 0.6$) extending up to a spin of at least 60$\hbar$ in ^{152}Dy. The existence of a discrete band of this kind raises the prospects of being able to make a detailed spectroscopic study of the properties of prolate deformed nuclei having a major to minor axis ratio of 2 to 1. The implications of the present results, techniques for searching for other superdeformed bands and the physics we may address by these studies are discussed.

1. Introduction

I had been asked to give a review talk on "γ-ray spectroscopy as a probe of nuclear structure and shape at high spins". However, I think that, in the short time available, it would be more interesting to concentrate on our recent discovery of a discrete superdeformed band in ^{152}Dy extending up to a spin of about 60$\hbar$ and to discuss a few of the implications of this discovery.

The first firm evidence for the existence of superdeformed ($\varepsilon \approx 0.6$) structures in ^{152}Dy was the observation [1] of ridges 47 keV from the diagonal of the $E_{\gamma 1} - E_{\gamma 2}$ correlation plot [2] for data taken with our original Total Energy and Suppression Shield Array [3] TESSA2 using the ^{108}Pd (^{48}Ca, 4n) ^{152}Dy reaction. These ridges could be seen up to a γ-ray energy of about 1350 keV and down to 800 keV where they disappeared into complex formations caused by the coincidences between many discrete lines. The decay scheme of the previously known discrete γ-rays in ^{152}Dy is shown in figure 1. The yrast γ-rays come from states that are formed by aligned single particle states [4] and the complex decay scheme [5] on the right hand side of figure 1 comes from near yrast levels formed by minor rearrangements of the yrast configurations. A reexamination of our original data [1] showed that the 60ns 17^+ isomer, through which all the oblate aligned states decay, is bypassed [6] by the decay of a sequence of positive parity, even spin (positive signature) states shown on the lefthand side of figure 1. This band is very similar [7] to the yrast levels in ^{154}Dy and therefore presumably has a normal deformation of $\varepsilon \approx 0.2$. It can be seen from figure 1 that the band coexists but does not interact with the known oblate yrast states although it is up to 1.5 MeV higher in excitation energy. This band could

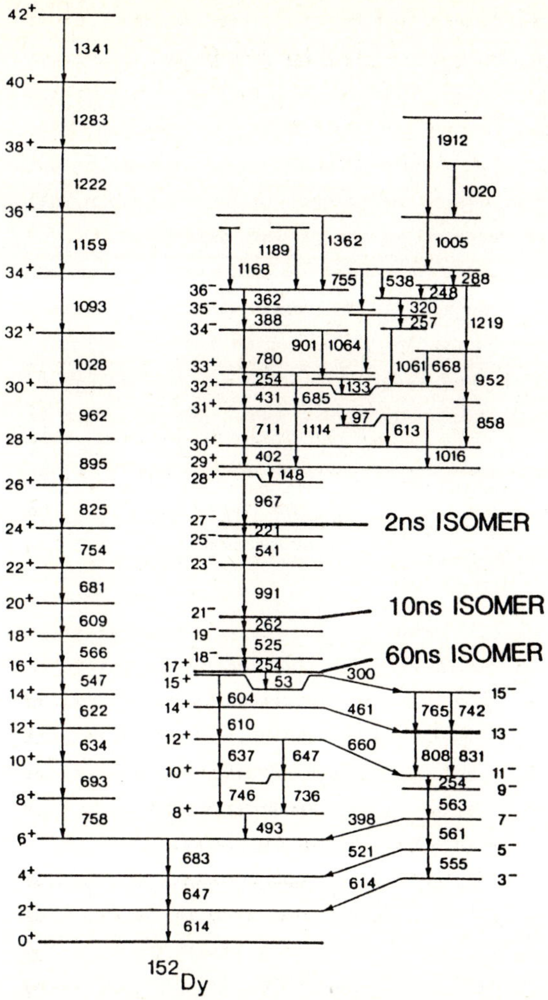

Figure 1. Decay scheme for ^{152}Dy showing the oblate aligned single particle states on the righthand side and a prolate rotational band on the left.

form a pathway by which superdeformed states could decay and bypass the 60ns isomer.

2. Discrete Superdeformed Band

At the end of March 1986 we carried out a new experiment to try to extend the coexistence band to higher spins and to investigate whether the superdeformed structures decayed through the 60ns isomer or whether they bypassed it via the low deformation band. The same ^{108}Pd (^{48}Ca, 4n) ^{152}Dy reaction was used as previously at an energy of 205 MeV. Our new γ-ray spectrometer TESSA3 was used to take the data and obtain greatly improved statistics. TESSA3 consists of the original 50 element BGO crystal ball but with 12 BGO suppressed [8] n-type hyper-pure Ge detectors instead of the 6 NaI suppressed detectors of TESSA2. The target consisted of 4 thin foils of enriched ^{108}Pd so that all γ-rays had the full Doppler shift. The geometry of the spectrometer is such that only γ-rays emitted within about 25mm of the target (equivalent to 3ns after the reaction in this experiment) are seen by the Ge detectors. A 15mg cm^{-2} gold foil 50mm down beam from the target caught recoiling reaction products within the volume of the BGO ball. The electronics was arranged so that a high multiplicity delayed pulse from the BGO, subsequent to the usual Ge-Ge-BGO coincidence, could be used to tag the decay of the 60ns isomer. This tag had an efficiency of approximately 25%. About 120 x 10^6 coincidence events were recorded in the present experiment.

Careful examination of data, gated by the sum and fold signals from the BGO ball to select ^{152}Dy, revealed [9] a sequence of 19 γ-rays all in coincidence with each other and separated in γ-ray energy by 47 keV which was the original separation observed [1] for the superdeformed ridge. In figure 2 a spectrum of these γ-rays is shown that is gated by the 60ns isomer tag. The spectrum is gated on the transitions in the band which are underlined and shows that the superdeformed ridge originally observed contains a discrete band extending from a γ-ray energy of 602 keV up to 1449 keV. This band decays with roughly equal intensity through the 60ns isomer and bypassing the isomer. In figure 2, the data triggered by the 17^+ isomer, γ-rays from the oblate yrast states can be seen at 254 (18^-), 525 (19^-), 262 (21^-), 991 (23^-) and 541 keV (25^-). The mean yrast spin that the superdeformed band decays to is 21.5 $\hbar$. Most of the intensity in the superdeformed band is lost after the 693 and 647 in-band decays. Candidates for γ-rays between the lower end of the superdeformed band and the yrast levels are seen at 571, 627, 770 and 1005 keV. Our data does not show any one strong path connecting the superdeformed to the yrast states so neither the spins nor the excitation energies of the superdeformed levels are known at present. If it is assumed that any path to the yrast line contains between 2 and 4 γ-rays and that theory is correct in predicting that the lowest superdeformed states have

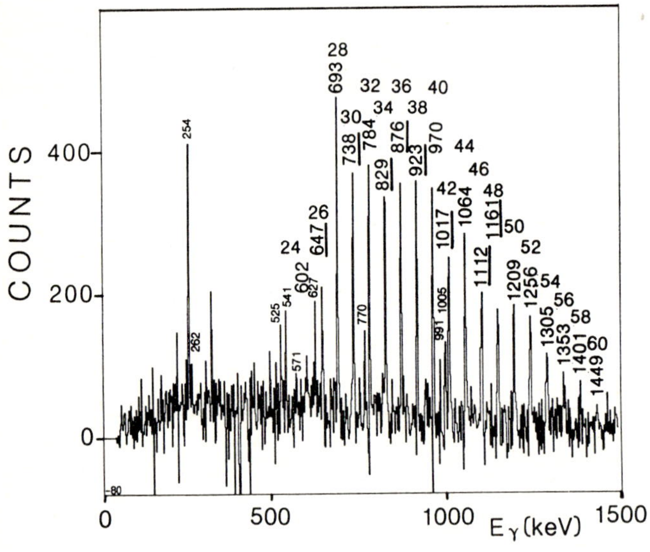

Figure 2. Spectrum of γ-rays from discrete superdeformed states in ^{152}Dy.

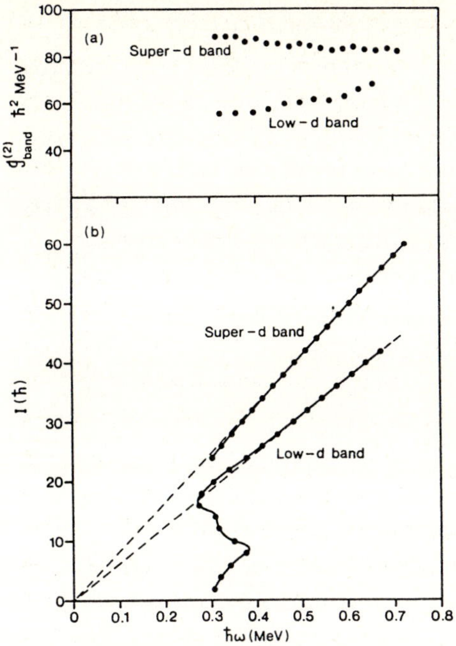

Figure 3. a) Plot of moment of inertia $\Theta^{(2)}$ versus frequency for the superdeformed band and the low deformation band in ^{152}Dy. b) Plot of spin against frequency assuming the spins shown in fig. 2 for the superdeformed band. The linear part of each graph is extrapolated back towards the origin.

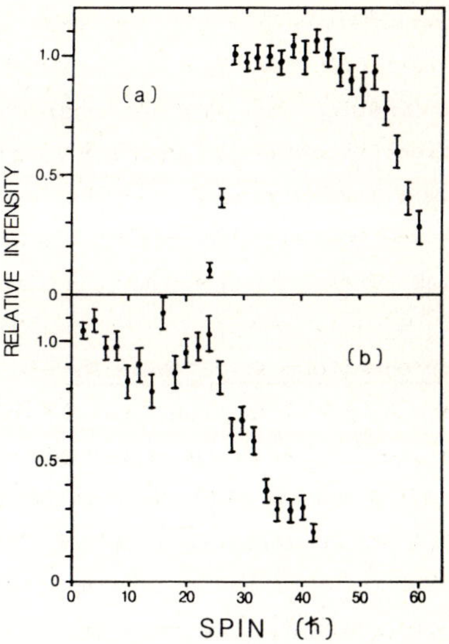

Figure 4. Relative intensity of population of a) the superdeformed band, b) the low deformation band.

even spin and positive parity, then 26^+ is the lowest reasonable spin for the state from which the 647 keV γ-ray decays. This assumption gives an average spin loss between superdeformation and the yrast line of about 3 to 4 $\hbar$.

A plot of the moment of inertia $\Theta^{(2)}$ for the superdeformed band, and also the lower deformation band of figure 1, is shown in figure 3(a). In figure 3(b) the spin is plotted against rotational frequency for both bands. In figure 4 the intensity in each band is plotted as a function of the spin. It can be seen from figure 4(a) that the feeding of the superdeformed band occurs mostly above spin 50. This may be compared with a mean input spin for the ^{152}Dy channel of about 55 calculated using normal statistical model assumptions. Further experiments are needed to ascertain how the intensity/spin distribution depends on the input angular momentum and excitation energy, the structure of ^{152}Dy at these high spins and the fission limit for superdeformed states. We might conjecture that the superdeformed band is fed when it is yrast which would indicate that this occurs near spin 50.

The feeding out of the superdeformed band occurs very rapidly after the transitions labelled 28^+ and 26^+ in figure 2. Calculations suggest [10, 11] that the superdeformed minimum becomes shallower and shallower as the spin decreases so that the sudden decay out of the band is probably due to the vanishing of the potential barrier between the superdeformed and other shapes.

3. Open Questions at 2:1 Axis Ratio

The intensity of about 2% of the total channel intensity to ^{152}Dy that we observe for the discrete superdeformed band makes spectroscopy possible in this second minimum that could not be contemplated for fission isomers. The physics for the 2 to 1 axis ratio deformations is not identical for the actinides and rare earths due to the strong spin dependence of the superdeformed potential in the lighter nuclei. However, our observation of a strong unambiguous signal for superdeformation opens up many exciting questions which demand experimental answers.

(i) The very large quadrupole moment should give rise to very fast E2 transitions [12] within the superdeformed band that will show enhancements of up to 2000 times the single particle rates. We have just completed an experiment to measure the lifetimes of the superdeformed states in ^{152}Dy using the same reaction as before but stopping the recoils in a gold target backing. We hope to have a measure of the superdeformed quadrupole moment in the near future.

(ii) The systematics of superdeformed structures in neighbouring nuclei should be explored to find out how the intensity, $\Theta^{(2)}$, I_{max}, I_{min}, E_x etc. varies with proton and neutron numbers Z and N. In searches for other superdeformed discrete bands strategies will be needed to look for these structures. One such strategy might be based on the following procedure. In figure 5(a) the original [1] $E_{\gamma 1}$ - $E_{\gamma 2}$ correlation plot is shown. If a cut is made along the ridge at 47±2 keV from the diagonal then any discrete γ-rays from superdeformed states will show up as peaks at 47 keV intervals. Such a cut is shown for our recent data [9] in figure 5(b) and peaks are labelled with their energies. A template consisting of lines separated by 47 keV may then be slid up and down the spectrum looking for a match to peaks along the cut. Clearly the 829 and 876 peaks are separated by the correct energy and they have many other candidates for companion superdeformed γ-rays. These are of course two of the γ-rays in the superdeformed band all of which are indicated by arrows in figure 5(b). If the spaces between the known superdeformed γ-rays at 784, 829, 876 and 923 are examined it can be seen that any other excited superdeformed band must consist of γ-rays less than about 10%

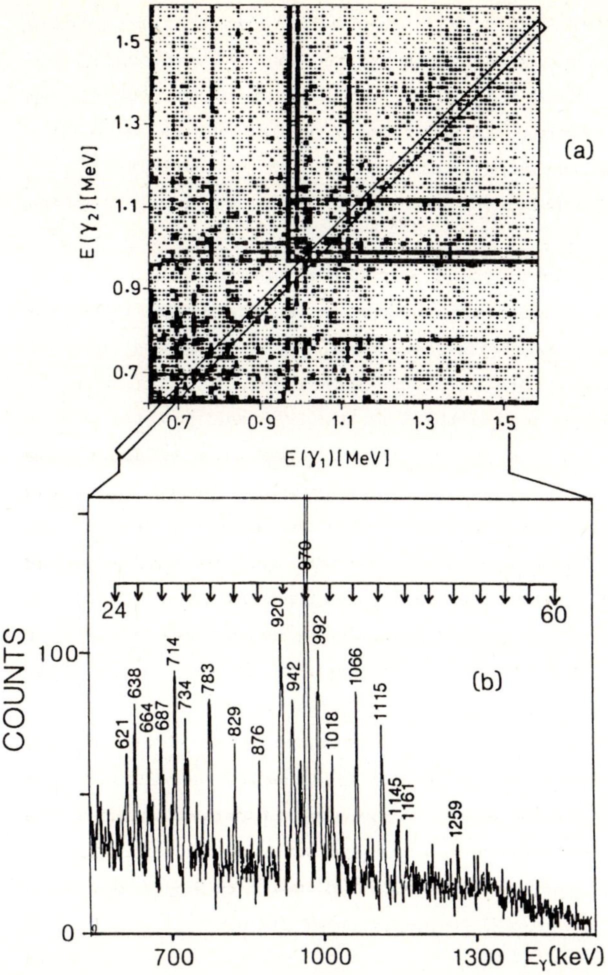

Figure 5. a) $E_{\gamma 1}$-$E_{\gamma 2}$ correlation plot for the ^{152}Dy data of ref. [1]. b) spectrum of γ- rays in a cut along the ridge at (47 ± 2) keV from the diagonal for the new data taken with TESSA3. A template with arrows spaced at 47 keV intervals is shown positioned on peaks corresponding to γ-rays in the superdeformed band.

of the intensity of the band observed. It should be noted that as the superdeformed minimum gets shallower the excited deformed bands will experience the loss of the containing barrier before the lowest observed band. Hence it might be expected that excited superdeformed bands will end at a higher spin than the observed band.

(iii) It is especially important to discover superdeformed states in the odd proton and odd neutron neighbours of ^{152}Dy. Such data can be used to check that calculations have the correct single particle orbitals at the Fermi surface of nuclear matter with a 2 to 1 axis ratio. It is important to know if any of the standard model constants are seriously affected by the extreme nuclear shape. Other intriguing questions are: How does the odd nucleon affect the core? How does it change the decay back to normal deformations? How is the fission limit affected? We have just completed an experiment to look for superdeformation in ^{153}Ho using the ^{109}Ag (^{48}Ca, 4n) reaction and hope to answer some of these questions in the near future.

(iv) The decay out of the superdeformed band is probably very complex. However at least one path between the superdeformed and the yrast states is needed in order to establish the spins and excitation energies of superdeformation. A brute force attempt to find such a path might be made using the largest γ-ray spectrometers. A 30 detector array, called the European Suppression Spectrometer Array (ESSA30), will be assembled at Daresbury early in 1987 as a collaborative venture between several European laboratories. ESSA30 will have no inner BGO ball and the 30 escape suppressed spectrometers will be held in a 32 faced polyhedral frame. However there is no guarantee that even such a powerful array will have sufficient sensitivity to uniquely determine the decay paths out of the superdeformed band. It is therefore sensible to look for subtler techniques which may be used to enhance the signal of superdeformation.

In figure 6 a schematic diagram is shown of the different paths by which the nucleus ^{152}Dy can loose angular momentum. Transitions down the yrast oblate states are very slow [17] taking many hundreds of picoseconds. Transitions down the low deformation band will be by enhanced (~ 200 spu) E2 decays which bypass the 60ns isomer. The decay of the superdeformed states will be by very enhanced (~ 2000 spu) E2 transitions and then, at the bottom of the band, by transitions

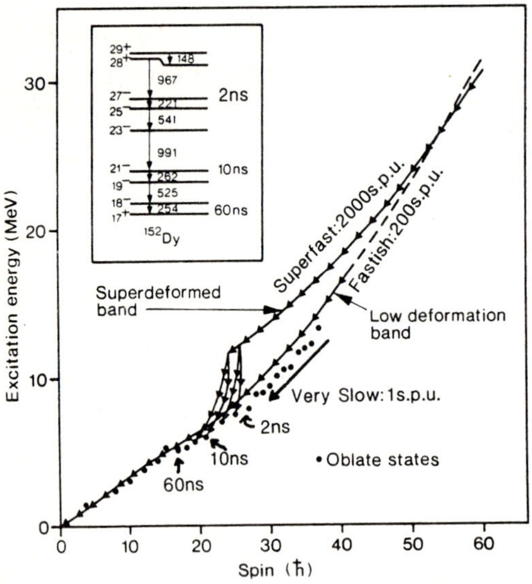

Figure 6. Schematic diagram of the decay paths of the prolate collective bands in ^{152}Dy. The insert shows the oblate states in ^{152}Dy just above the 60ns 17^+ isomer.

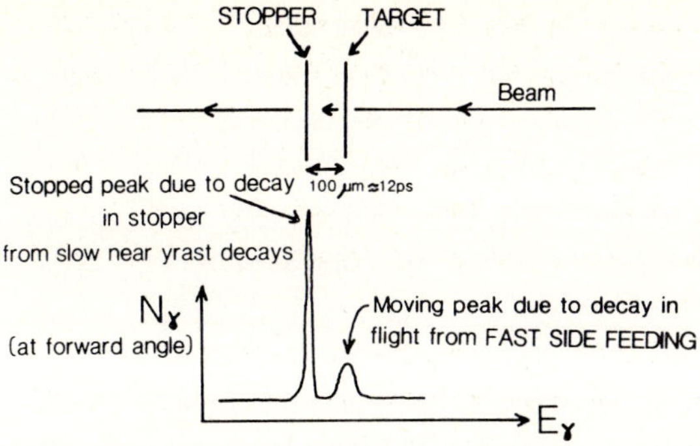

Figure 7. Proposed target/stopper geometry to select γ-rays side feeding levels below long lived isomeric yrast states.

which must compete with inband transition rates. Thus decay from superdeformed states will be extremely rapid and will only slow up somewhat in the last one or two transitions before the yrast line is fed between the 27^- 2ns isomer and the 17^+ 60ns isomer (see inset to figure 6). Thus a signature of superdeformation has to be very fast sidefeeding, especially at the highest input angular momentum, of the yrast levels between 27^- and 17^+. Normal (98%) decays down the yrast line will be delayed by the 2ns and 10ns isomers compared to the fast sidefeeding which should take a few picoseconds at most.

The fast side feeding may be selected preferentially using the target, recoil catcher arrangement shown in figure 7. If a thin target is used as usual but a gold foil recoil catcher is placed about 100μm behind it, then the lifetimes of the 25^-, 23^-, 19^- and 18^- levels, which decay by 541, 991, 525 and 254 keV γ-rays respectively, are such that they will decay in flight with the full Doppler shift when fed by the fast superdeformed levels. However they will be caught by the recoil catcher before they are fed via the yrast 2ns isomer and all subsequent decays will be unshifted. Thus, in this fixed distance recoil distance or "plunger" method, the yrast γ-rays in coincidence with the superdeformation will be fully Doppler shifted while the normal yrast γ-rays will be stopped and show no shift. This separation will improve the statistical sensitivity by removing the large numbers of yrast sequence decays from the coincidence spectra. It will also distinguish between pathways which come into the yrast line on either side of the 10ns isomer. It will be interesting to look for fast side feeding in the higher yrast states above the 2ns isomer to see if they are fed by excited superdeformed bands. This technique might be useful generally in the search for weakly populated superdeformed structures.

(v) Further work needs to be carried out to determine how the superdeformed population depends on the input angular momentum and on target/projectile asymmetry. Now that a definite signature for the population of superdeformed states has been found it should be able to resolve open [13,14] questions regarding the effect of superdeformation on the final channel cross-sections in fusion reactions.

(vi) There are difficulties in understanding how the superdeformed band gets fed in such strength and why other superdeformed bands are relatively weak. There is currently a conjecture [15] that statistical E1 transitions feeding the superdeformed states will be greatly enhanced by the severe lowering of one component of the Giant Dipole Resonance which is split by the prolate deformation. Experiments are underway [16] to look for the anomalously large splitting of the GDR built on superdeformed states.

In summary, the observation of discrete γ-rays from superdeformed states up to spin 60 in ^{152}Dy has not only set a new record for the highest spin for a discrete state but it also opens up the possibility of detailed studies of the properties of deformed nuclear matter with a major to minor axis ratio of 2 to 1. Many physicists should have a great deal of fun attempting to answer the many open questions generated by this discovery.

4. Acknowledgements

The discrete superdeformed band was discovered in the data by Peter Twin and Barna Nyakó. Others involved in the hard work on the experiment and the analysis were Mike Bentley, Debbie Howe, Andy Nelson, John Simpson, Howard Cranmer-Gordon, Peter Forsyth, Rahman Mokhtar, Dave Morrison and Geirr Sletten. This experiment could only be carried out because of the superlative performance of the TESSA3 spectrometer designed by Paul Nolan and Don Gifford and constructed by a large and very able team under their direction. "Number crunching" the data was possible because of the efforts of John Cresswell, John Lisle and the Daresbury software group. The research reported here was supported by grants from the United Kingdom Science and Engineering Research Council.

References

1. B M Nyakó, J R Cresswell, P D Forsyth, D Howe, P J Nolan, M A Riley, J F Sharpey-Schafer, J Simpson, N J Ward and P J Twin; Phys. Rev. Lett. 52 (1984) 507.
2. B Herskind; J. Phys. (Paris), Colloq. 41 (1980) C10-106.
3. P J Twin, P J Nolan, R Aryaeinajad, D J G Love, A H Nelson and A Kirwan; Nucl. Phys. A409 (1983) 343c.
4. T L Khoo, R K Smither, B Haas, O Häusser, H R Andrews, D Horn and D Ward; Phys. Rev. Lett. 41 (1978) 1027.
5. D Howe; Ph.D. thesis, Univ. of Liverpool (1986).
6. B M Nyakó, J Simpson, P J Twin, D Howe, P D Forsyth and J F Sharpey-Schafer; Phys. Rev. Lett. 56 (1986) 2680.
7. J Styczen, Y Nagai, M Piiparinen, A Ercan and P Kleinheinz; Phys. Rev. Lett. 50 (1986) 2680.
8. P J Nolan, D W Gifford and P J Twin, Nucl. Instr. A236 (1985) 95.

9. P J Twin, B M Nyako A H Nelson, J Simpson, M A Bentley, H W Cranmer-Gordon, P D Forsyth, D Howe, A R Mokhtar, J D Morrison, J F Sharpey-Schafer and G Sletten; Phys. Rev. Lett. (in press).

10. S Åberg; Proc. Workshop on Nucl Str., NBI, Copenhagen, May 1986, p.9.

11. J Dudek; ibid, p.145.

12. P J Twin, A H Nelson, B M Nyakó, D Howe, H W Cranmer-Gordon, D Elenkov, P D Forsyth, J K Jabber, J F Sharpey-Schafer, J Simpson and G Sletten; Phys. Rev. Lett. 55 (1985) 1380.

13. W Kühn, P Chowdhury, R V F Janssens, T L Khoo, F Haas, J Kasagi and R M Ronningen; Phys. Rev. lett. 51 (1983) 1858.

14. A Ruckelhausen et al; Phys. Rev. Lett. 56 (1986) 2356.

15. B Herskind; Private communciation.

16. B Herskind; J J Gaardhøe et al.

17. B Haas et al; Phys. Lett. 84B (1979) 178.

Evolution of Nuclear Structure with Spin and Temperature

T.L. Khoo

Argonne National Laboratory, Argonne, IL 60439, USA

1. INTRODUCTION

In this talk the decay of the compound nucleus (CN) -- from formation to ground state -- will be described. The shape and nuclear structure evolution in this process will be traced, in which spin I and excitation energy E^* decrease. Three topics will be discussed: (a) shape relaxation following formation; (b) γ-decay towards the yrast line; and (c) γ-decay along the cold yrast line.

A structural evolution with increasing temperature T (corresponding to increasing $E^* - E_{yrast}$) is to be expected. This is because the occupation of orbits changes with T, leading to an alteration of the shell effects which so profoundly influence nuclear structure. In fact, thermal fluctuations will give rise to admixtures of different types of structures (and shapes); the mixing of these structures and its relationship to rotational damping constitute interesting topics to study. Another important question relates to the damping of shell effects with increasing T. In other words, at what temperatures are the broken symmetries (induced by shell effects) restored? A question which still remains unanswered concerns the nature of the collective strutures built on high-j oblate aligned states.

The experimental methods for investigating nuclear structure depend on T. For E^*-E_{yrast} ~0-2 MeV, where T is low, we use γ- and, in certain cases, α-spectroscopy. For E^*-E_{yrast} ~2-15 MeV, where T ~ 1 MeV, the information comes from continuum γ rays; to date only the non-statistical γ's have been used, but the statistical γ rays, which reflect the level densities, may also contain information on nuclear structure. For E^*-E_{yrast} ~15-100 MeV, where T ~1-2.5 MeV, one depends on the properties of neutrons and charged particles emitted in the evaporative stage, viz. the multiplicity distributions, energy spectra and angular correlations. Here one aims at a complete characterization of the CN decay; this requires, among other things, measurements of the ℓ-distribution in the reaction and observation of the decay from specified regions of E^* and ℓ. In this regime, very high energy γ's in the giant resonance region is another tool [1].

2. INITIAL STAGE OF COMPOUND NUCLEUS DECAY

The first phase following contact of the interacting ions consists of shape relaxation towards the equilibrium shape at high temperature. During this process the kinetic energy of the cold incoming ions is converted into heat, deformation (surface) energy and rotational energy. The temperature should increase and reach a maximum when the equilibrium shape is attained. Normally one expects that particle evaporation occurs only after this point is reached. However, if the shape relaxation time is sufficiently long, then evaporation may occur during relaxation. One possible such case will be discussed. The pertinent experimental data consists of neutron multiplicity distributions [2-4] (including distributions for individual ℓ's), ℓ-distributions [3] and evaporation residue cross-sections σ_{ER} [4].

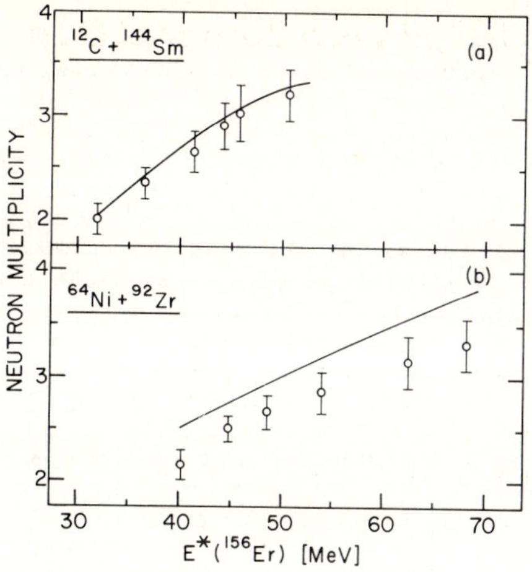

Fig. 1. Average neutron multiplicity $\bar{n}$ as a function of E^* in the CN ^{156}Er in the reaction (a) $^{12}C + ^{144}Sm$ and (b) $^{64}Ni + ^{92}Zr$. The solid line shows $\bar{n}$ from CASCADE calculations with ℓ-distributions derived from measured values of σ_{ER} (see Ref. 4 and caption in Fig. 2).

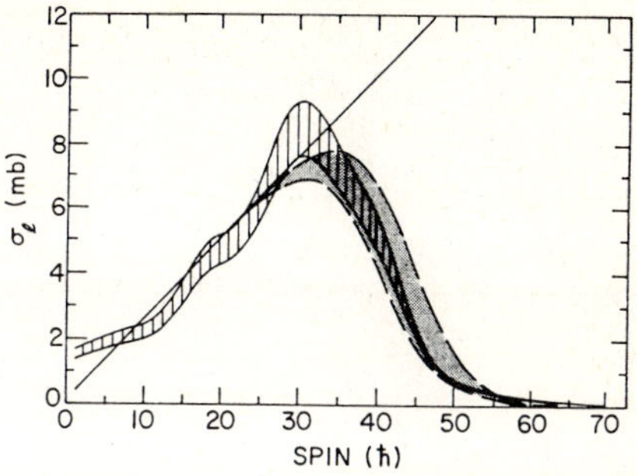

Fig. 2. Partial wave cross-section. The solid lines are the ℓ-distributions measured with the Crystal Ball [3], normalized to the ER cross-section s_{ER} and to 0.85 σ_{ER}, i.e. $\Sigma\sigma_\ell = \sigma_{ER}$ and $\Sigma\sigma_\ell = 0.85\ \sigma_{ER}$. The spread between the two solid lines reflects half of the ±15% uncertainty in σ_{ER}. The dashed lines are obtained using the form $\sigma_\ell = \pi\bar{\lambda}^2(2\ell + 1)T_\ell$; $T_\ell = \frac{1}{1 + \exp(\ell-\ell_o)/d}$, with d=4 chosen to reproduce the tail from the Ball measurement, and ℓ_o adjusted to satisfy the above normalization conditions.

The decay of $^{156}E^*$ is characterized by an inhibition of neutron emission when it is formed in a nearly mass symmetric channel, $^{64}Ni + ^{92}Zr$, but no inhibition when formed in a mass-asymmetric channel, $^{12}C + ^{144}Sm$ [2,4]. In other words, statistical model calculations (and straight forward considerations of available energy) can account for the average neutron multiplicity in the mass-asymmetric reaction, but not in the other -- see Fig. 1. The suppression of neutron emission in one case cannot be attributed to a lack of knowledge of the yrast line, γ-strength function, level density [2,4] or ℓ-distribution leading to evaporation residues (ER) [3,4]. The partial cross-sections for individual partial waves σ_ℓ (Fig. 2) have been determined by combining measurements of ER cross-sections at Argonne [4] and of ℓ-distributions with the Darmstadt-Heidelberg crystal ball [3]. The deduced σ_ℓ's, which have been used in the obtaining the calculated neutron multiplicities $\bar{n}$ of Fig. 1, distinctly show no anomalously high-ℓ components which could account for the neutron suppression.

From the ℓ-distributions of individual channels, one can compare the ^{156}Er decay modes, for identical E^* and ℓ, when formed in the two reactions. Fig. 3 shows the ratio σ_{2n}/σ_{3n} as a function of ℓ and suggests that the decay of the CN depends on the entrance channel for $\ell \gtrsim 26$, contrary to expectations based on the Bohr hypothesis [5]. Furthermore, the ratios for the $^{12}C + ^{144}Sm$ reaction can be properly described by statistical calculations [3], but not that for $^{64}Ni + ^{92}Zr$ when $\ell \gtrsim 26$.

A possible clue to the explanation of these observations is the initial shape of the CN. As shown in Fig. 4, the initial shape is more elongated in the more mass-svmmetric channel. One possible suggestion is that the energy is tied up in deformation during the neutron emission period, for times perhaps as long as 10^{-16}-10^{-17} s, when γ emission then sets in. This time scale is clearly several orders of magnitude longer than that shown in Fig. 4, which has been computed using Feldmeier's particle exchange model [7]. The exact mechanism for the

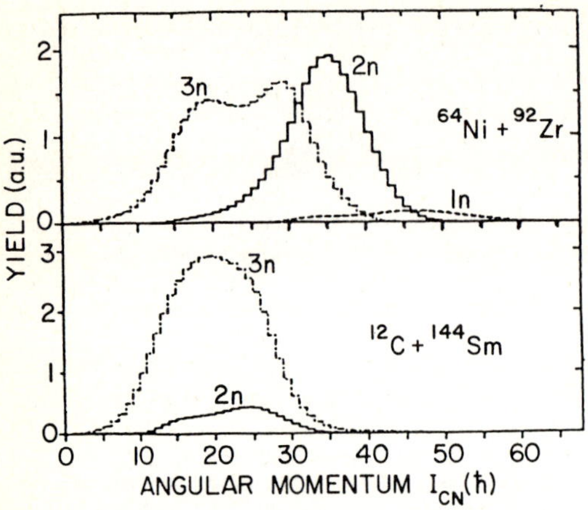

Fig. 3. (a) CN spin distributions for the channels corresponding to 2 and 3 neutron emission following the ^{64}Ni + ^{92}Zr and ^{12}C + ^{144}Sm reactions at E^*=47 MeV. The uncertainty in spin is ± 10%. (b) Ratio of σ_{2n}/σ_{3n} for the indicated reactions as function of CN spin. The errors include effects of increasing the measured spins of only the 2n channel by 10%, to simulate the worst case situation. For $\ell \gtrsim 26$ it appears that neutron emission is dependent on the entrance channel. The dotted line, from a statistical calculation using the code HIVAP (6), agrees with ratio from the C + Sm channel but with not that from the Ni + Zr channel.

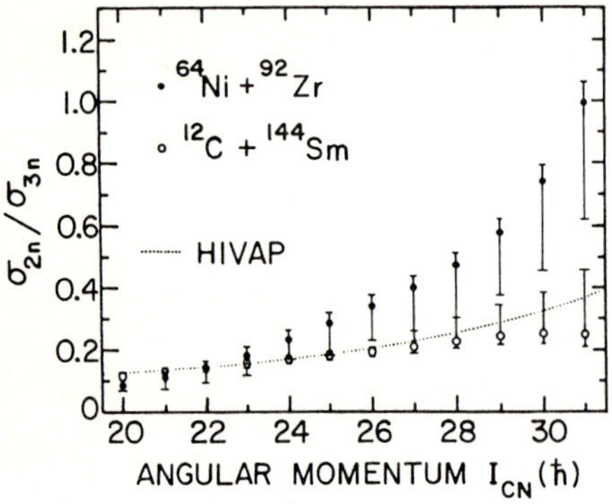

Fig. 4. Trajectories and shapes following fusion of (a) C + Sm and (b) Ni + Zr, calculated as a function of time with Feldmeier's particle exchange model [7]. Note that the initial quadrupole-like deformation is larger in Ni + Zr fusion. ►

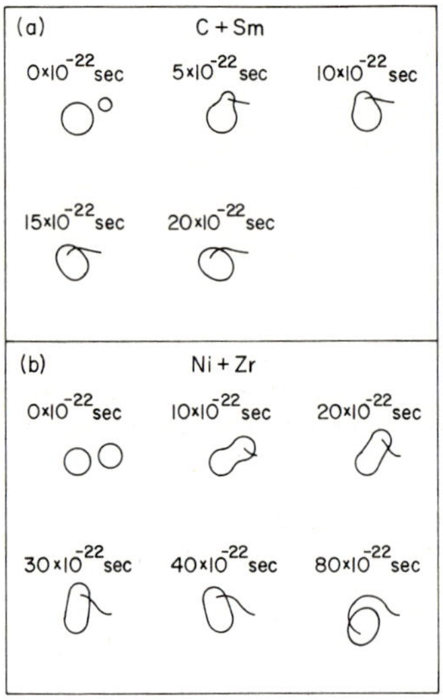

retardation in shape relaxation is not known at present but one speculation, which merits further investigation, is that the CN is trapped for a short time (10^{-16}-10^{17}s) in a superdeformed secondary minimum. (While one would normally not expect this trapping to persist for times of the order of 10^{-14} s, characteristic of the decay of the Daresbury superdeformed band [8], the possibility should not be ignored.) If this suggestion turns out to be correct, it would imply a persistence of shell structure effects for T ~1-2 MeV.

3. γ CASCADE FROM THE CONTINUUM INTO THE YRAST LINE

Following neutron decay the average nuclear excitation is typically 7-12 MeV above the yrast line. The subsequent γ cascade cools the nucleus towards the yrast line and removes angular momentum. If the deexcitation pathway is delineated then the structure of states through which each stage of the cascade proceeds can be probed and any accompanying structural evolution can be

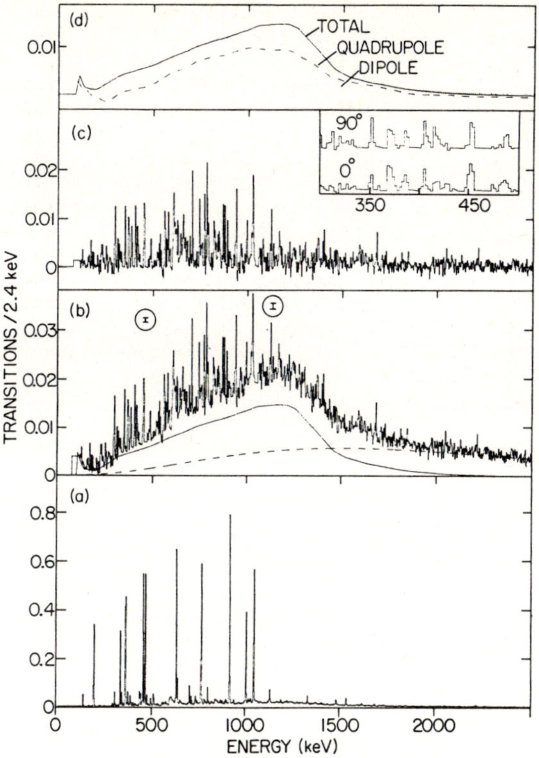

Fig. 5. (a) Total γ spectrum from ^{153}Ho, following the ^{120}Sn(^{37}Cl, 4n) reaction. The strong yrast transitions (trees) dominate. (b) Spectrum of γ rays from the pre-yrast cascade. (c) Grass transitions; insert shows that these transitions are not Doppler shifted. (d) Non-statistical smooth component (soil) decomposed into quadrupole and dipole parts.

studied. One method for tracing the γ cascade consists of identifying and determining the properties of the different components of the total γ spectrum, utilizing the power of Compton-suppressed Ge detectors [9]. For this type of study it cannot be overemphasized that it is imperative to isolate the spectrum of one and only one nucleus.

I shall discuss the γ spectra for $^{153}_{67}Ho_{86}$ and $^{152}_{66}Dy_{86}$, isotones with oblate aligned particle yrast configurations. The spectra in Fig. 5 show at least four identifiable components [9]. The sharp discrete lines in Fig. 5(a) are the yrast transitions-referred to as 'trees'. After removal of the tree transitions the remaining spectrum represents the pre-yrast cascade (Fig. 5b) which is clearly not continuous but consists of weak sharp peaks called 'grass' (isolated in Fig. 5c) which is superimposed on a smooth background. The smooth structure can be further decomposed into a statistical part -dashed line in Fig. 5b -- and a non-statistical bump -- referred to as 'soil'. A further decomposition into quadrupole and dipole parts can be done.

From the Doppler shifts, the time order of the different components can be ascertained. That the quadrupole soil is fast is immediately clear from the opposite Doppler shifts in forward and backward detectors (Fig. 6), indicating that the decay occurs mainly before the recoiling residue slows down in the stopping foil. In fact, from such spectra it is possible to deduce that this quadrupole component exhibits significant collectivity with $4 < Q_0 < 10$ eb. The dipole soil component occurs later (~ 1 ps) and the grass transitions represent the last stage ($\gtrsim$ 8ps) of feeding into the yrast line.

The energy and spin removed by each component can be deduced and, together with the time sequence information, used to construct the average decay pathway to the yrast line (Fig. 7). The relative ordering of the statistical and

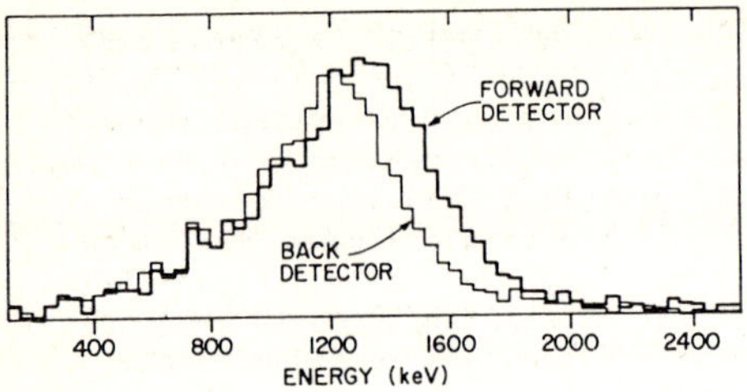

Fig. 6. Quadrupole soil component in forward (0°) and backward (146°) detectors from the $^{76}Ge(^{80}Se,4n)$ ^{152}Dy reaction, clearly illustrating the Doppler shifts of high energy edge.

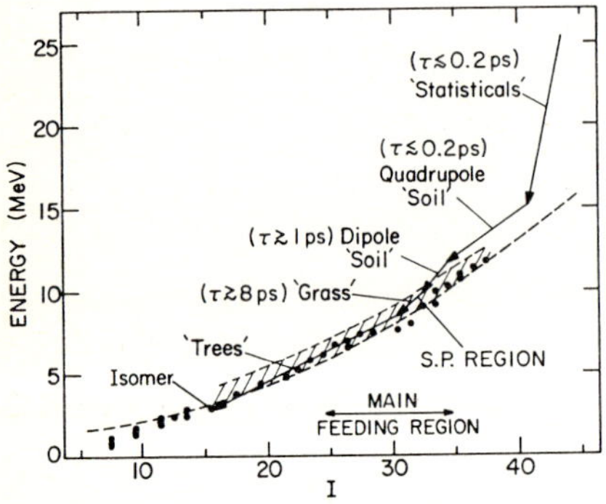

Fig. 7. Gamma deexcitation pathway. The relative order of the quadrupole soil and statistical components is not unambiguous and some intermixing is possible.

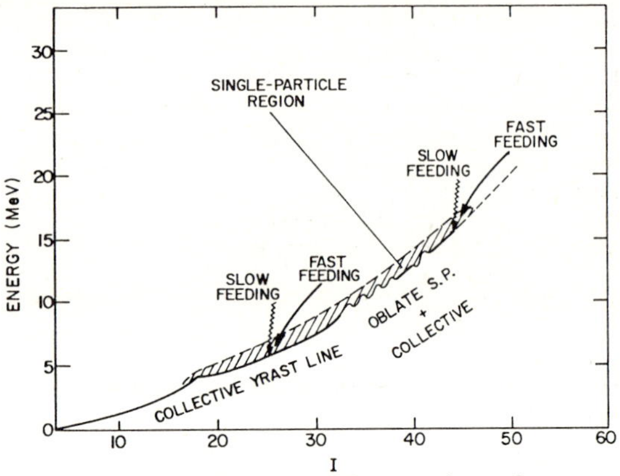

Fig. 8. Schematic illustration of changing nature of yrast line and of fast (collective) and slow (single-particle) feeding of yrast states in ^{154}Dy. Smooth and wriggly portions of yrast line denote collective and aligned particle character.

quadrupole parts is not yet completely clear and some intermixing of these 2 components may occur, particulary after the first 2 or 3 statistical γ's are emitted.

The decay times and irregular transition energies of the grass lines demonstrate that they, like the yrast transitions, are of single-particle character. The dipole soil component is probably an extension of the grass part to higher excitation energy above the yrast line, where the higher level densities result in a transition to a smooth spectrum.

Thus within 1-2 MeV of the yrast line the oblate coupling scheme is preserved before a transition to a region dominated by collective states. The change in structure along the decay pathway reflects an evolution with not only spin but also temperature. The low- and super-deformation discrete bands in ^{152}Dy [8, 10], which lie above the aligned-particle yrast line, are another manifestation of this evolution.

4. DECAY ALONG THE YRAST REGION

The last stage of γ decay occurs along the cold yrast line. Whether further structural changes occur in this stage depends on the nucleus. In transitional nuclei changes are expected. An example is provided by ^{154}Dy which, with 88 neutrons, lies between the oblate (or spherical) nuclei with $N \leqslant 86$ and the deformed region beginning at N = 90.

An extensive body of data exists for ^{154}Dy: the level scheme [11-13], lifetimes and feeding times [11,12,14] have been measured up to spin 44. (Much of the information for the highest spin states is preliminary as analysis of results [12] from the Argonne-Notre Dame Compton-suppressed Ge detector array is still in progress.)

The information on ^{154}Dy is schematically summarized in Fig. 8. The yrast line is collective at low spins, manifests backbending due to rotational alignment, and for $I \gtrsim 32$ exhibits an onset of the oblate coupling scheme. At the highest spins the preliminary lifetime results [12] suggest an alternation of both single-particle and collective structures, as might be expected in the region where terminating bands [15] emerge in the yrast region. The feeding of yrast states exhibit both fast (collective) and slow (~ 10 ps) components [12,14], implying the coexistence of aligned particle and collective configurations in the vicinity of the yrast line. For $I > 44$ the decay pathway seems to suddenly fragment into many pieces, and the decay times may indicate a predominance of collectivity. (Whether there is a connection between these two phenomena is not clear.) Some of the features described here, in particular the transition from prolate to oblate shapes at high spin, have also been observed in other transitional nuclei [15,16].

5. SUMMARY

The different stages of formation and decay of a compound nucleus are depicted in Fig. 9. The approximate temperature, time scale, mode of spin generation and relative orientation of the spin and symmetry axes are also indicated.

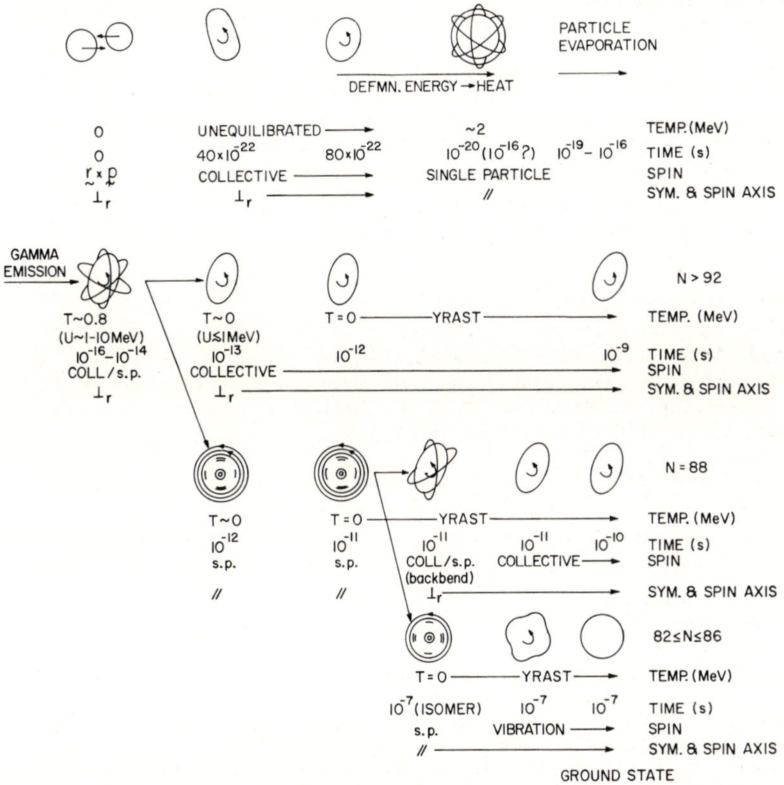

Fig. 9. Schematic representation of different stages of CN decay showing shape evolution, approximate temperature, time scale, spin modes and relative orientation of spin and symmetry axes.

Loosely speaking the kinetic energy of the fusing ions is converted to thermal, deformation and rotational energy. The thermal excitations impart energy to individual nucleons, eventually leading to particle evaporation and concomitant cooling. Subsequent cooling to the cold yrast line occurs primarily through emission of statistical γ rays.

The shape of the composite system undergoes many dramatic changes. The somewhat prolate-like object following fusion relaxes to the equilibrium shape, which is expected to be oblate at high spin when shell effects damp out at high temperature. In ^{64}Ni + ^{92}Zr fusion there may be a hint of trapping in a secondary superdeformed minimum before the oblate shape is attained. (This discussion should be qualified by the recognition that large shape fluctuations are expected at high T.) As the nucleus cools towards the yrast line, highly deformed (prolate or triaxial) shapes are encountered, which rapidly dissipate angular momentum through E2 γ emission. This collective mode is encountered in, it seems, all nuclei, regardless of whether the cold yrast states are of single-particle or collective nature. The energy domain (with respect to the yrast line) of this stage is not clearly established in all cases; where the yrast line is of aligned-particle nature it starts ~1.5 MeV above the yrast line. The subsequent shape evolution along the yrast line depends sensitively on shell structure and, thus, on the neutron number in the mass 150 region. For N > 92, the prolate shape persists to the ground state. For lower N, a transition to the oblate coupling scheme occurs in the vicinity of the yrast line ($E^* - E_{yrast} \lesssim 1.5$ MeV). For $82 \leqslant N \leqslant 86$, a slight oblate shape persists until vibrational states are encountered before a final spherical ground state is attained. For N = 88 there is a transition from oblate to prolate shape, perhaps through a triaxial regime.

This transition is accompanied by a change in the mode of angular momentum generation, from one where spin is obtained mainly from particle alignment to one where it is a parallel coupling of particle and collective angular momentum, i.e. rotational alignment. In addition, the orientation of the symmetry axis with respect to the spin vector flips from parallel to perpendicular. Such an orientation flip may also occur in the early stage of shape relaxation.

The decay of a compound nucleus from its formation to the ground state is indeed a process accompanied by dramatic structural changes.

6. ACKNOWLEDGEMENT

It is a pleasure to acknowledge the contributions of my collaborators in the work discussed here. The names of all participants are found in Refs. 2-4, 9 and 12, but I would like to particularly mention here the contributions of H. Emling, D. Habs, W. Henning, R. Holzmann, R. V. F. Janssens, W. Kühn, W. C. Ma, V. Metag, M. Quader, D. C. Radford and A. Ruckelshausen.

This research was supported by the U. S. Department of Energy, Nuclear Physics Division, under Contract W-31-109-Eng-38.

7. REFERENCES

1. K. Snover: these proceedings
2. W. Kühn, P. Chowdhury, R.V.F. Janssens, T.L. Khoo, F. Haas, J. Kasagi and R.M. Ronningen: Phys. Rev. Lett. 51, 1858 (1983)
3. A. Ruckelshäusen, R.D. Fisher, W. Kühn, V. Metag, R. Mülhans, R. Novotny, T.L. Khoo, R.V.F. Janssens, H. Gröger, D. Habs, H.W. Heyng, R. Repnow, D. Schwalm, G. Duchene, R.M. Freeman, B. Haas, F. Haas, S. Hlavac and R.S. Simon: Phys. Rev. Lett. 56, 2356 (1986)
4. R. V. F. Janssens, R. Holzmann, W. Henning. T.L. Khoo, K.T. Lesko, G.S.F. Stephans, D.C. Radford, A.M. van den Berg, W. Kühn and R.M. Ronningen: Phys. Lett. in press
5. N. Bohr: Nature (London) 137, 334 (1936)

6. W. Reisdorf: Z. Phys. A300, 227 (1981)
7. H. Feldmeier: Nuclear Structure and Heavy Ion Dynamics; ed. L. Moretto and R.A. Ricci: North-Holland (1984), p. 274
8. J.F. Sharpey-Schafer: these proceedings; P. Twin et al.: Daresbury preprint 1986
9. D.C. Radford, I. Ahmad, R. Holzmann, R.V.F. Janssens, T.L. Khoo, M.L. Drigert, U. Garg, H. Helppi: Phys. Rev. Lett. 55, 1727 (1985)
10. B.M. Nyakó, J. Simpson, P.J. Twin, D. Howe, P.D. Forsyth and J.F. Sharpey-Schafer: Phys. Rev. Lett. 56, 2680 (1986)
11. A. Pakkanen, Y.H. Chung, P.J. Daly, S.R. Faber, H. Helppi, J. Wilson, P. Chowdhury, T. L. Khoo, I. Ahmad, J. Borggreen, Z.W. Grabowski and D.C. Radford: Phys. Rev. Lett. 48, 1530 (1982)
12. M. Quader, W. Ma, H.E. Emling, T.L. Khoo, I. Ahmad, B. Dichter, R. Holzmann, R. V. F. Janssens, P.J. Daly, Z. Grabowski, M. Piiparinen, V. Trzaska, M. Drigert and U. Garg: to be published
13. H.W. Cranmer-Gordon et al.: to be published
14. F. Azgui, H. Emling, F. Grosse, C. Michel, R.S. Simon, W. Spreng, H.J. Wollersheim, T.L. Khoo, P. Chowdhury, D. Frekers, R.V.F. Janssens, A. Pakkanen, P.J. Daly, M. Kortelahti, D. Schwalm and G. Seiler-Clark: Nucl. Phys. A439, 573 (1985)
15. I. Ragnarsson, Z. Xing, T. Bengtsson and M.A. Riley: Physica Scripta in press
16. F.S. Stephens et al.: Phys. Rev. Lett. 54, 2584 (1985); P. Tjøm et al., ibid, 2405 (1985); and references therein

Nuclear Rotation in the $N \cong Z \cong 36$ Region*

K.P. Lieb

II. Physikalisches Institut, Universität Göttingen,
D-3400 Göttingen, F. R. Germany

1. Introduction

Heavy ion fusion reactions have provided evidence for pronounced variations of the nuclear moment of inertia as function of the rotational frequency $\hbar\omega$ and nucleon number A. These variations have been linked to the alignment of particles in high j orbits, reduction of pairing correlations, shape coexistence or other changes of the collective and/or single-particle structure. For several reasons, the $N \cong Z$ = 34 - 40 region offers an excellent testing ground for such effects: these nuclei are among the most deformed of the periodic table with deformation parameters β_2 = 0.3 - 0.4; they often are γ-unstable, triaxial or shape coexistent; particle alignment involving the $g_{9/2}$ proton and/or neutron orbit occur at $\hbar\omega \cong 0.5$ MeV, e.g. in a spin region easily accessible to standard γ-ray spectroscopy on discrete transitions.

In this article, I shall present results on level energies and lifetimes in evaporation residues excited in ^{36}Ar and ^{40}Ca induced fusion reactions on ^{40}Ca targets. The experiments were performed with the 105 - 130 MeV ^{36}Ar beam of VICKSI/Berlin and the 120 - 170 MeV ^{40}Ca beams provided by the Brookhaven and Daresbury tandem accelerators. In choosing N = Z projectiles on ^{40}Ca targets, the neutron deficient isotopes with $N \cong Z$ are being populated by proton and α-particle evaporation (e.g. $^{72,73}Br$, $^{70,72}Se$, $^{74,76}Kr$, ^{77}Rb), while neutron emission is strongly suppressed and the spectroscopy of the N = Z isotopes ^{72}Kr and ^{76}Sr is still very scarse, in spite of the use of sophisticated recoil spectrometers and filtering techniques.

2. Rotational alignment in ^{76}Kr

This nucleus was excited in the $^{40}Ca(^{40}Ca,4p)$ reaction at 155 MeV beam energy. Five BGO Compton suppressed Ge detectors positioned at 135° and 0° to the beam axis, a ΔE-E telescope for charged particle identification and a neutron multiplicity filter were operated in $\gamma\gamma$ and particle-γ multiparameter coincidence. Based on the previously known high spin states |1,2| reaching up to 12^+ resp. 11^-, we extended the positive-parity yrast band up to spin 22^+ and the signature $\alpha = \pm 1$ negative parity yrast line up to spin 17^- and 18^- |3|. The band structure is presented in Fig. 1, while the projection $I_x(\omega)$ of the spin on the rotational axis is plotted in Fig. 2 versus the rotational frequency $\hbar\omega$. A strong up-bending effect is clearly visible at $\hbar\omega$ = 0.65 MeV related to an aligned single-particle angular momentum $i_x = 6.6\ \hbar$.

Although $g_{9/2}$ (2qp) alignment can make up for $i_x \leq 8$, (2qp) $g_{9/2}$ proton alignment in the neighboring nuclei ^{75}Br |4| and ^{78}Kr |5| has been observed to give only i_x = 3 - 4 $\hbar$, so that the high i_x value in ^{76}Kr is surprising. Recent Hartree Fock Bogoliubov calculations by NAZAREWICZ, et al. |6| and LÜHMANN, et al. |4|, however, indicate that in ^{76}Kr $g_{9/2}$ proton and neutron alignment are very close in energy and may be difficult to separate experimentally. Figure 3 illustrates the calculated total energy surface of the ^{76}Kr ground state which exhibits a γ-unstable minimum near $\beta_2 = 0.36$, $\gamma = 0°$, as well as the quasi-particle Routhians of protons and neutrons for N = 40 and Z = 36 calculated for a somewhat smaller deformation taken from $B(E2,2^+ \rightarrow 0^+)$ |2|.

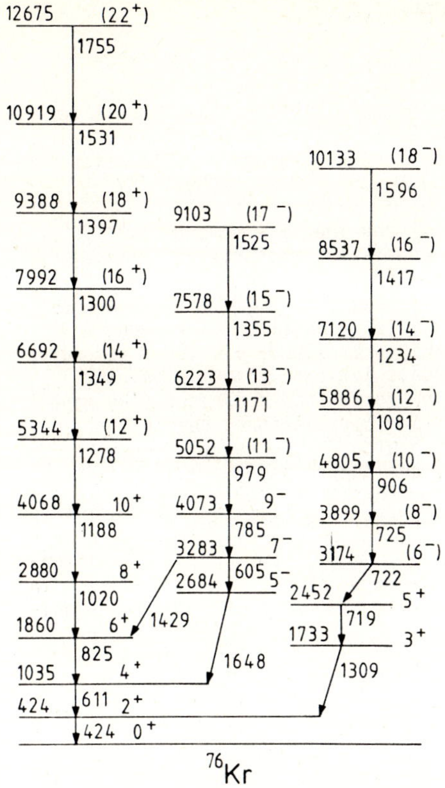

Fig. 1: High spin states in ^{76}Kr |2|

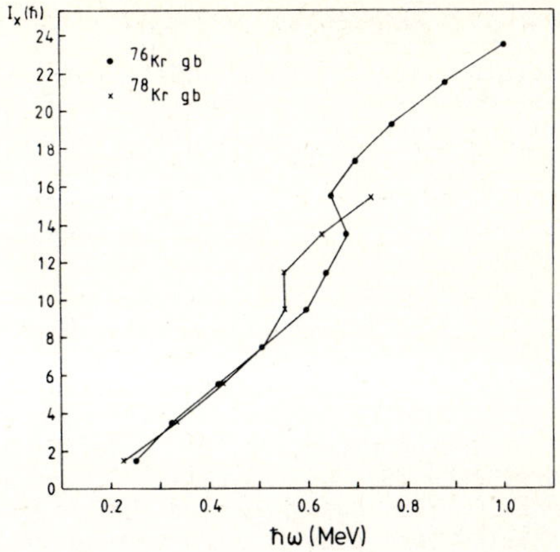

Fig. 2: $I_x(\omega)$ of the positive-parity yrast bands in ^{76}Kr and ^{78}Kr |2,5|

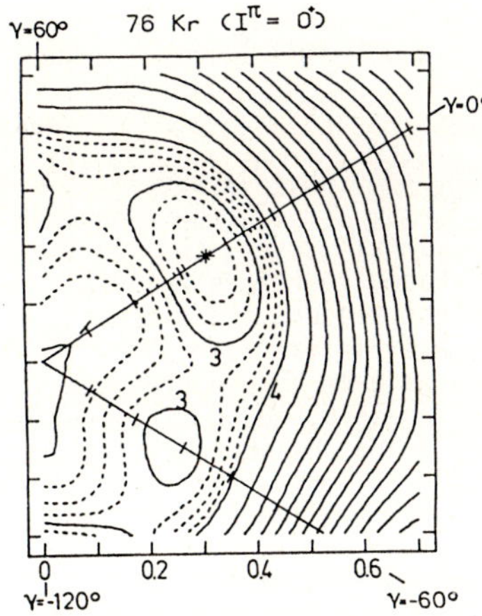

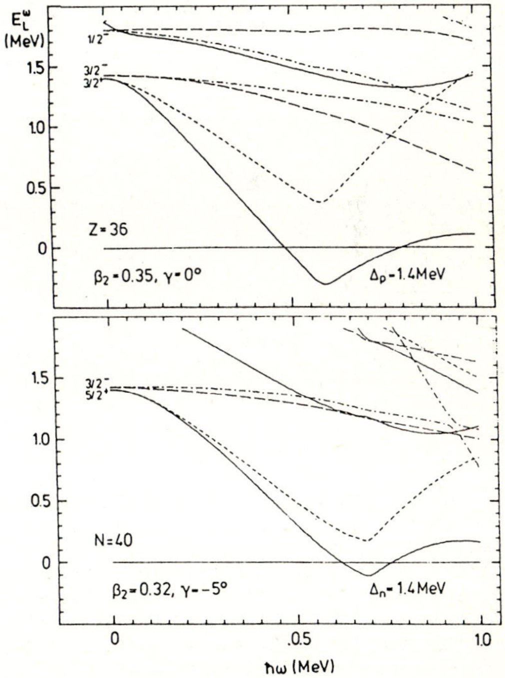

Fig. 3: Total energy surface of the ground state and quasi-particle Routhians in ^{76}Kr |4|

3. Rigid rotation in ^{77}Rb

High spin states in this nucleus were established via the reaction ^{40}Ca(^{40}Ca,3p) at 122 MeV beam energy and the α = 1/2 yrast bands were extended up to spin I^{π} = 37/2$^+$ and 37/2$^-$ |7|. Extensive recoil distance and DSA lifetime measurements were performed for the states up to spin 25/2$^+$ resp. 21/2$^-$; a systematic study of side feeding times in this reaction revealed a nearly linear dependence of the side feeding time τ_t with the excitation energy: τ_f(ps) = 1.56 - 0.32 E_x(MeV) |8|.

The deduced transitional quadrupole moments $|Q_t|$ and inertial moments $\mathfrak{I}^{(1)}/\hbar^2$ and $\mathfrak{I}^{(2)}/\hbar^2$ are displayed in Fig. 4; they are consistent with rigid, triaxial rotation in the $g_{9/2}$ decoupled proton band in the frequency range 0.4 $\leq \hbar\omega \leq$ 0.7 MeV, the deformation parameters being β_2 = 0.37(2) and γ = -25(7)^o. The average $|Q_t|$ value of the negative-parity band (α = ±1/2) is in excellent agreement with that of the ground state, Q = 3.48(16) b, determined at ISOLDE via laser resonance fluorescence |9|. On the basis of Hartree Fock calculations including pairing, NAZAREWICZ |10| has suggested that rigid rotation in the odd-Z isotope ^{77}Rb may be related to a drastic reduction of proton pairing correlations, due to the single-particle structure near β_2 = 0.4. As the proton Fermi energy lies right below the Z = 38 shell gap, pairing correlations are suppressed already at low spin value I ≅ 21/2. The calculated proton pairing gap Δ_p in ^{77}Rb is indeed reduced from Δ_p = 1.4 MeV at β_2 = 0 to Δ_p = 0.6 MeV at β_2 = 0.4 (see Fig. 5).

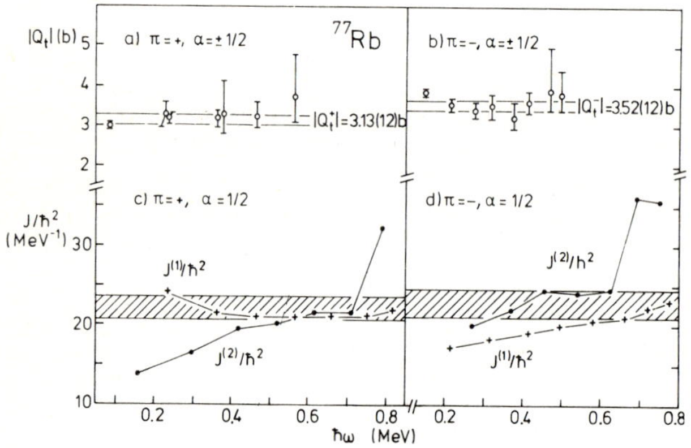

Fig. 4: Transitional quadrupole moments $|Q_t|$ and moments of inertia of the α = ±1/2 yrast bands in ^{77}Rb

Although the other odd-Z nuclei in this mass region, e.g. ^{79,81}Rb, ^{73,75}Br |4,11-13|, do not strictly follow rigid rotational pattern ($\mathfrak{I}^{(1)} \neq \mathfrak{I}^{(2)}$), their quadrupole and inertial moments vary slowly as function of $\hbar\omega$. In all cases, the additional $g_{9/2}$ proton thus stabilizes the collective prolate shape, therefore removes shape coexistence and blocks (2qp) $g_{9/2}$ alignment. For that reason, the simple triaxial rotor plus quasiparticle model works surprisingly well in many odd-A isotopes |11-14|, but has only limited success in the description of even-even nuclei, in particular concerning the E2 strengths. The same conclusion is borne out from Hartree Fock calculations with a deformed basis done by Bhatt and collaborators |15|.

4. High spin states in ^{73}Br

We have also recently investigated the decoupled $g_{9/2}$ proton band in ^{73}Br up to spin 29/2$^+$, via the reactions ^{40}Ca(^{36}Ar,3p) and ^{40}Ca(^{40}Ca,α3p) |11|. Fig. 6 illustrates the transitional quadrupole moments deduced from the measured lifetimes (recoil distance and DSA technique) in comparison with that of the ^{70}Se and ^{72}Se ground band |16|. While $|Q_t|$ is large and nearly constant in ^{73}Br, it varies rapidly as function of spin, in the even-even Se isotopes. HEESE, et al. |16| have interpreted the pronounced decrease of $|Q_t|$ in ^{70}Se as due to a transition from an oblate ground state deformation of β_2 = -0.30(3) to a prolate deformation above

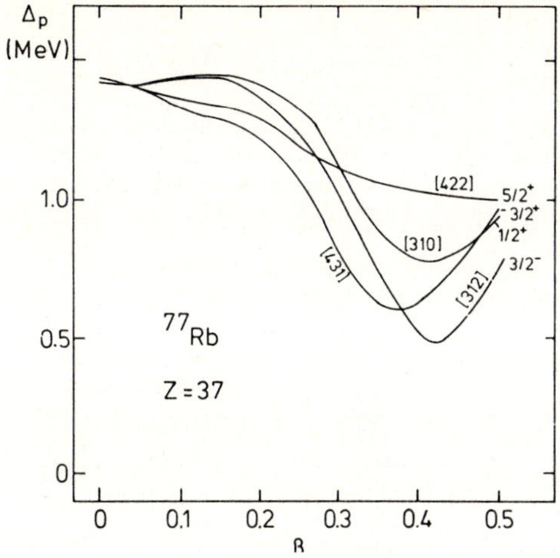

Fig. 5: Hartree Fock Bogoliubov calculations of the proton pairing gap Δ_p in ^{77}Rb |10|. Note the minima of the |431 3/2⁺| and |312 3/2⁻| (1qp) Nilsson states near β_2 = 0.4; $\gamma = 0^0$ was used

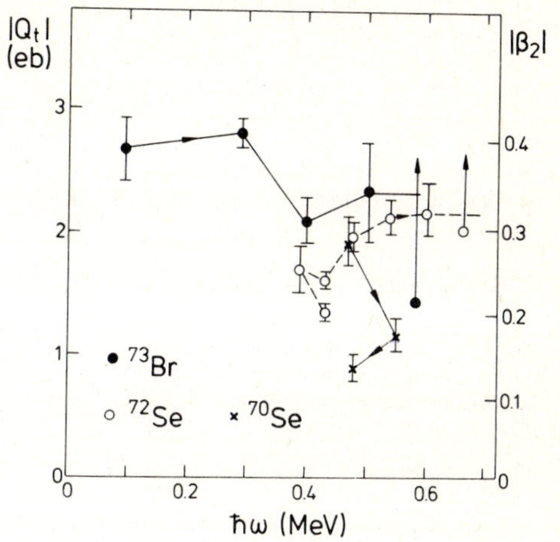

Fig. 6: Transitional quadrupole moment $|Q_t|$ and deformation parameter β_2 of the α = 1/2 $g_{9/2}$ band in ^{73}Br, and the ground bands in ^{70}Se and ^{72}Se |11,16,17|. The arrows mark the evolution for increasing spin value. All $|Q_t|$ and $|\beta_2|$ values have been derived from the corresponding B(E2) values assuming axial symmetry.

spin 8⁺, similarly as in ^{72}Se |17,11|. There seems to be a significant difference of B(E2) values at low spin: Whereas the ground state in ^{72}Se feature true coexistence of prolate and oblate deformation leading to a reduction in average deformation from β_2 = 0.31 above spin 8⁺ to $|\beta_2| \cong 0.20$ near the ground state, the decreasing B(E2) values along the ^{70}Se ground band may indicate a real change in the sign of the deformation. Clearly, lifetime measurements of the higher yrast states are required to test this point.- The slight decrease of $|Q_t|$ above $\hbar\omega$ = 0.35 MeV visible in the $g_{9/2}$ band in ^{73}Br has also been observed at the same rotational frequency in the positive-parity and negative-parity bands in ^{75}Br |4|. Hartree Fock calculations in this nucleus suggest that this decrease is connected to changes of γ from triaxial ($\gamma \cong -20^0$) to symmetric prolate rotation ($\gamma \cong 0^0$). These calculations suggest γ driving forces both by the collective core and the aligned particles.

5. Future experiments

Nucleon correlations producing deformation as well as pair correlations or pairing fluctuations often vary strongly as function of the nucleon number and angular momentum |18|. The few examples alluded to in this article indeed highlight the fact that although the average quadrupole deformation β_2 smoothly increases toward N ≅ Z ≅ 38, the individual nuclei differ widely in their structure.

There is no doubt that triaxiality plays an essential role although reliable data on the asymmetry parameter γ are still scarce. It has been emphasized |19| that the E2 strengths of the signature changing ΔI = -1 transitions in odd-A nuclei sensitively depend on γ and should be measured as carefully as the stretched E2 transition strengths. Furthermore, the ΔI = -1 M1 components reveal the nature(s) and alignment pattern of particles involved as spectators of the collective motion. In the mass A = 80 region, this has been nicely illustrated by the Rossendorf group for the (1qp) and (3qp) states in ^{81}Kr |20|.

BCS cranking calculations |21| predict that in the spin range $I = 10 - 30\ \hbar$ variations of γ at constant β_2 value may be the dominant structure effect, besides the possibility that non-collective quasi-particle states become yrast. The few isotopes studied so far by means of the new generation of Compton suppressed multidetector systems (TESSA, OSIRIS) reveal different shapes above the (2qp) $g_{9/2}$ proton or neutron alignment: rigid triaxial rotation in ^{77}Rb |7|, non-ridig triaxial rotation in ^{80}Sr and probably rigid symmetric rotation in ^{84}Zr |22,23|. These experiments indicate the excitement which remains in the spectroscopy of this region.

I am indebted to my collaborators L. Lühmann, J. Heese, F. Raether, B. Wörmann (Göttingen), W. Gelletly, C. J. Lister, B. J. Varley (Manchester) and D. Alber, H. Grawe and B. Spellmeyer (HMI Berlin) for the permission to quote some unpublished results and to W. Nazarewicz, J. Garrett and A. Faessler for fruitful discussions. This work has been funded by Deutsches Bundesministerium für Forschung und Technologie.

6. References

1 R.B. Piercey, et al.: Phys. Rev. Lett. 47 , 1524 (1981)
2 B. Wörmann, et al.: Nucl. Phys. A431 , 170 (1984)
3 F.J. Bergmeister, W. Gelletly, K.P. Lieb, C.J. Lister, L. Lühmann, R. Moskrop, H.G. Price, B.J. Varley and B. Wörmann, to be publ.
4 L. Lühmann, et al.: Phys. Rev. C31 , 828 (1985)
5 H.P. Hellmeister, et al.: Nucl. Phys. A332 , 241 (1979)
6 W. Nazarewicz, et al.: Nucl. Phys. A435 , 397 (1985)
7 L. Lühmann, K.P. Lieb, C.J. Lister, J.W. Olness, H.G. Price, and B.J. Varley: Europhys. Lett. 1 , 623 (1986)
8 L. Lühmann, doctoral thesis Göttingen (1985) unpubl.
9 C. Thibault, et al.: Phys. Rev. C27 , 2720 (1981)
10 W. Nazarewicz: private communication
11 B. Wörmann, et al.: Z. Phys. A332 , 171 (1985)
12 J. Panqueva, et al.: Nucl. Phys. A376 , 367 (1982)
13 J. Panqueva, et al.: Nucl. Phys. A389 , 424 (1982)
14 H. Toki and A. Faessler: Phys. Lett. 63B , 121 (1976); A. Petrovici and A. Faessler, Nucl. Phys. A395 , 44 (1983)
15 D.P. Ahalpara, K.H. Bhatt and R. Sahu: J. Phys. G10 , 735 (1985)
16 J. Heese, K.P. Lieb, L. Lühmann, F. Raether, B. Wörmann, D. Alber, H. Grawe, J. Eberth and T. Mylaeus: Z. Phys. in press
17 K.P. Lieb and J.J. Kolata, Phys. Rev. C15 , 939 (1977); H.P. Hellmeister, et al: Phys. Rev. C17 , 2113 (1978); J.H. Hamilton, et al.: Phys. Rev. Lett. 36,340 (76)
18 J.D. Garrett, G.B. Hagemann and B. Herskind: Ann. Rev. Nucl. Part. Sci. in press
19 I. Hamamoto and B.R. Mottelson: Phys. Lett. 132B , 7 (1982); 167B , 370 (1986)
20 L. Funke, et al.: Phys. Lett. 120B , 301 (1983); F. Dönau: NBI-ZFK preprint 85-36
21 R. Bengtsson and W. Nazarewicz: XIX Winter School on Physics, Zakopane (1984)
22 H.G. Price, et al.: Phys. Rev. Lett. 51 , 1842 (1983)
23 R.F. Davie, et al.: to be publ.

Cooling of Hot Rotating Nuclei by Electric and Magnetic Dipole Radiation

H.P. Morsch, B. Bochev, T. Kutsarova, R.M. Lieder, W. Gast, G. Hebbinghaus, A. Krämer-Flecken, W. Urban, and J.P. Didelez

Institut für Kernphysik, Kernforschungsanlage Jülich, D-5170 Jülich, F. R. Germany

In this talk results from the study of particle-γ coincidences are presented which are related to giant resonance decay of highly excited compound nuclei. The α + ^{159}Tb reaction was investigated in which compound nuclei of excitation energy up to about 90 MeV are produced. In the spectra a splitting of the electric giant dipole resonance built on highly excited states of Dy nuclei is clearly observed which is different from the splitting in cold nuclei. Also, indication for an enhanced E1 strength at low energies is found. For the first time evidence for M1 decay is obtained in these compound systems.

Giant resonances built on nuclear ground states have been investigated intensely over the last decade. In addition to the well known isovector giant dipole resonance (GDR) discovered already in 1947 isoscalar giant resonances have been found in the 70's. These modes have been studied in scattering experiments using many different projectiles. Among these resonances, the isoscalar compressional modes are of particular importance because they give information on compressional features of the nucleus. A summary of recent results is given in ref. [1].

An interesting problem already raised by Brink in 1955 [2] is the question of the existence of giant resonances on excited states. This should allow to obtain additional structure information on the properties of these states. Already for quite some time the GDR on low lying excited states in light nuclei has been investigated in the (p,γ) capture reaction [3]. A new development in this field started in 1981 when in the study of continuous γ-ray spectra a high energy contribution was found which was interpreted as decay of the GDR in highly excited compound nuclei [4]. This result has strong impact on the study of nuclear properties: it opens up the possibility to study nuclear properties at high temperature and fast rotation. From a splitting of the GDR information on the nuclear shape can be obtained whereas the resonance width determines nuclear damping mechanisms.

There are many questions related to the GDR in hot nuclei; e.g. is the width of the resonance strongly increased and is there a shift in excitation energy with respect to cold system. Another interesting problem is the Coriolis splitting in rotating systems. For all these questions we have theoretical predictions from a variety of different investigations mainly based on temperature dependent RPA [5]. From these, the temperature dependence of the resonance energy and width should be small up to temperatures of $\sim$ 3 MeV. Also the Coriolis splitting should be rather small (part of an MeV) up to spins of 30 - 40 $\hbar$. The experimental situation, however, is rather confusing. In most of the experiments strongly increased resonance widths have been found, further shifts in the centroid energy. A summary of results from heavy ion induced reactions is given in the talk by K.A. Snover.

An important question is related to the origin of the high energy γ-rays. An interesting investigation of the reaction mechanism was performed recently [6] in which neutron-γ coincidences were measured. The results indicate that the high energy γ-rays arise mainly from the early stage of the reaction. This is supported by our coincidence experiments discussed below.

We have studied particle-γ coincidences in α induced reactions on Tb at an α energy of 110 MeV. In comparison with recent heavy ion induced reactions in which compound nuclei of moderate excitation energy but rather high rotation are produced, in our system we populated compound nuclei of rather high excitation energy but lower angular momentum ($I \leq 26\,\hbar$). By the particle trigger used different excitation regions of the compound nucleus can be studied. Gating on protons at different energies E_p coincident γ ray spectra are obtained which are displayed in fig. 1. In this figure exponential slopes describing the statistical transitions emitted after neutron evaporation are given by the solid lines. Quite clearly, we see the enhancement of high energy γ rays. Its yield decreases rapidly for lower energies of the compound nucleus supporting the picture that GDR radiation is strongest in the early state of the reaction.

In order to analyze the giant resonance in the γ decay spectra statistical decay calculations were performed using the Monte Carlo program JULIAN-PACE. This code calculates the statistical decay cascade using level densities and ma-

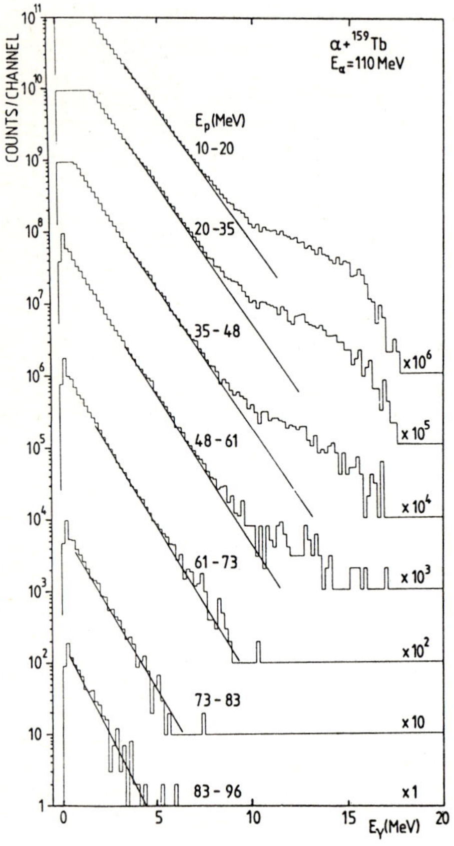

Figure 1:
γ-ray spectra coincident with outgoing protons of different energy from the α bombardement of ^{159}Tb. The slopes of the statistical decay contributions are given by the solid lines.

trix elements for different type of transitions. Using a level density parameter a = A/8 and a constant E1 matrix element of 1 W.U. yields a good description of the low energy part of the γ spectra in fig. 1. The initial ℓ distribution of the decaying compound system was proportional to $(2\ell+1)$ in the limits 10 - 26 $\hbar$ which gave best results on the description of mass distributions discussed below. Compared to the simple exponential behaviour plotted in fig. 1 the calculated curves become flatter at high γ energies indicating cooling of the highly excited compound nuclei by γ radiation.

To show details of the γ spectra it is useful to eliminate the exponential slope in the data. This can be done by dividing the experimental spectra by $\exp(-E_\gamma/T_c)$, T_c is the temperature of the compound system. A better way is to divide the experimental data by the calculated decay spectra. In such a representation we have the advantage to produce rather flat spectra at low energies, further in the comparison of calculations including the giant resonance response uncertainties in the calculations, e.g. level densities and ℓ-window effects cancel out to some extent. Divided spectra for the two bins of highest excitation energy are shown in fig. 2. These spectra show, for the first time, clear signature for a splitting of the GDR in hot nuclei. Further structure is observed in the 5 to 10 MeV region with a rather pronounced structure above 9 MeV.

Assuming an E1 strength distribution determined from cold nuclei (two-Lorentzian distribution with parameters from ref. [7]) we obtain decay spectra given by the dashed lines in fig. 2. The description of the data is not significantly improved. We obtain a GDR bump, however, the details of the data are not reproduced. Furthermore, below 10 MeV the calculated cross sections are too low (at 8 MeV only half of the spectral height is reproduced). To obtain a better description of the experimental data resonance parameters have to be assumed which are significantly different from those for cold nuclei with a large width of 5 MeV for the low energy component and a small width of 2.9 MeV for the high energy component. This clearly indicates that the shape of highly excited nuc-

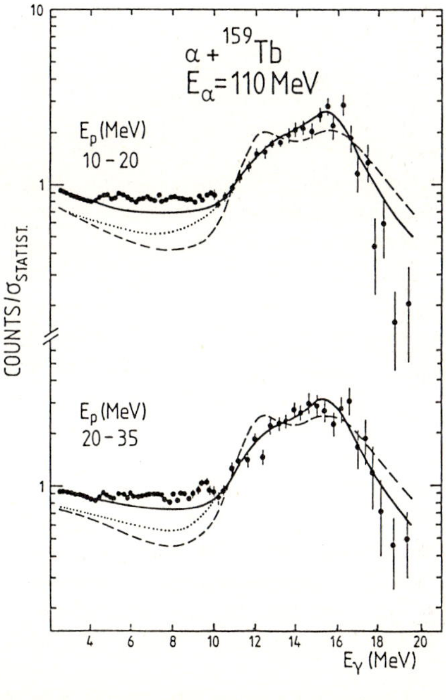

Figure 2:
Coincident γ-ray spectra for the proton energy bins 10-20 MeV and 20-35 MeV divided by calculated γ decay spectra assuming a constant E1 matrix element. The different lines are described in the text.

lei (E_x ~ 70 - 90 MeV) is significantly different from that of cold nuclei which may indicate a transition to oblate shapes. Still, the low energy spectrum is not well reproduced (dotted line in fig. 2). In order to reproduce the low energy spectrum a constant matrix element has to be added up to energies of 8 MeV. The final fit is given by the solid lines in fig. 2.

In the region of 4.5 to 10 MeV there is structure which is not described by these calculations. This is the region where M1 strength has been found in cold nuclei in (p,p') scattering experiments [8]. Thus, is is conceivable to assign a M1 character to this structure. The possibility to study M1 strength in hot nuclei is important because it allows to get unique information on magnetic properties of heated nuclei. A quantitative estimate from the present data is difficult because of ambiguities due to the addition of a constant matrix element. Assuming a total E1 strength of 130 % of the energy weighted sum rule (EWSR) which is the same as the E1 strength in cold nuclei [8] yields a M1 strength ≥ 1.5 times the M1 strength estimated for cold Dy nuclei.

From the statistical cascade the residual mass distribution can be obtained also. Under similar experimental conditions the distribution of final nuclei has been measured by particle-γ coincidences [9]. These data show in addition to bell shape equilibrium mass distributions preequilibrium tails towards lower particle energies. Extracted equilibrium mass distributions are given in fig. 3.

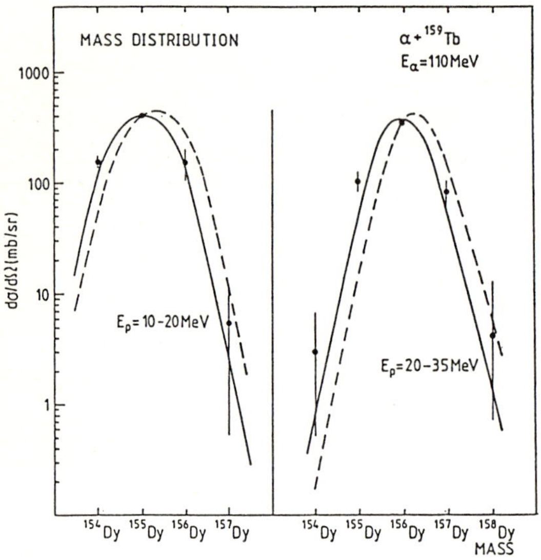

Figure 3: Equilibrium decay mass distributions from the data of ref. 9 in comparison with statistical cascade calculations. Solid lines: inclusion of GDR response, dashed lines: Constant E1 matrix element.

The solid lines correspond to our calculations which gave a good fit to the data in fig. 2. The mass distributions arequite well reproduced. It is interesting to note that these distributions are different from the ones obtained by using a constant matrix element without giant resonance effect (dashed lines). This indicates that the giant resonance decay affects significantly the compound nucleus decay cascade.

In summary giant resonance decay in hot nuclei is discussed. E1 decay yields information on deformation and damping mechanisms of highly excited compound nuclei whereas M1 decay gives insight into magnetic properties of hot nuclei. This appears as a very interesting new spectroscopic tool which can enable us to study nuclei under extreme condidtions as e.g. at very large angular momenta or critically high nuclear temperatures. However, many detailed and more systematic studies are absolutely necessary to obtain a reliable description of nuclear systems far away from the ground state.

References:

1. H.P. Morsch: proceedings of the Niels Bohr Centennial Conference on Nuclear Structure (1985) 433
2. D.M. Brink: Ph.D. thesis, University of Oxford (1955)
3. see e.g. K.A. Snover: Journal de Phys. C4 (1984) 337
4. J.O. Newton et al.: Phys. Rev. Lett. 46 (1981) 1383
5. M.E. Faber, J.L. Egido and P. Ring: Phys. Lett. 127B (1983) 3; M. Barranco et al.: Phys. Lett. 154B (1985) 96; J. Bar-Touv: Phys. Rev. C32 (1985) 1369
6. T. Arctaedius et al.: Phys. Lett. 158B (1985) 205
7. B.L. Berman and S.C. Fultz: Rev. Mod. Phys. 47 (1975) 713
8. C. Djalali et al.: Nucl. Phys. A388 (1982) 1
9. B. Bochev et al.: accepted in Nucl. Phys.

Energy Correlations of γ-Rays from Superdeformed States in Z=66 and 68 Isotopes; Multipolarities in ^{152}Dy

M.J.A. de Voigt[1], *H. Riezebos*[1], *J. van Klinken*[1], *A. Balanda*[1,3], *Z. Sujkowski*[1,4], *J.C. Bacelar*[2], *E.M. Beck*[2], *M.A. Deleplanque*[2], *R.M. Diamond*[2], *J. Draper*[2,5], and *F.S. Stephens*[2]

[1]Kernfysisch Versneller Instituut, NL-9747 AA Groningen, The Netherlands
[2]Lawrence Berkeley Laboratory, Berkeley, CA 94720, USA
[3]Uniwersytet Jagielloński, Instytut Fizyki, Ul. Reymonta 4, PL-30-059 Kraków, Poland
[4]Instytut Problemów Jadrowych, PL-05-400 Otwock, Swierk, Poland
[5]University of California, Davis, CA 95616, USA

1. Search for Superdeformed States in Z=66 and 68 Isotopes

The search has been carried out at LBL by the first two authors and the co-authors at Berkeley using the High Energy Resolution Array (HERA), which consists of 21 BGO Compton-suppressed Ge detectors for the measurement of $E_\gamma(1) - E_\gamma(2)$ correlations. Thin targets of enriched ^{114,116}Cd and ^{118,120}Sn were bombarded with 180 MeV ^{40}Ar beams. The recoils were stopped in a lead catcher, positioned at a distance of 20 cm behind the target which corresponds to a flight time of $\sim$28 ns in the case of ^{151,152}Dy, and at a distance of 13 cm in the other cases. The cylindrical BGO detector at 0°, surrounding the beam pipe with the catcher foil, provided a signal on delayed γ-ray emission from the recoils. All recorded energy signals of the Ge detectors were corrected for the Doppler shifts as to be gain matched on-line. In addition, the time spectrum obtained from the 0° BGO crystal with respect to the OR-signal (first time signal of all detectors) of the Ge detectors has been recorded for each event. Only triple coincidences and higher folds were written on magnetic tape. Typical event rates amounted to 2000 s^{-1}, with a beam current of 15 nA. Two days of running time did yield $\sim$200 million triple and higher-fold events resulting in more than 600 million double coincidences in the case of ^{152}Dy and about 1/3 of that in the other searches.

The correlation matrices obtained in the four reactions leading to the main evaporation products ^{152}Dy (a), ^{150}Dy (b), ^{154}Er (c) and ^{156}Er (d) can be compared in Fig. 1. The valley-ridge structure is most clearly seen in the matrix for ^{152}Dy. The average distance between the central ridges over the whole region is 94$\pm$4 keV, yielding an average moment of inertia of $F^{(2)} = 85\pm4\,\mathrm{MeV}^{-1}\hbar^2$, which is about the same value as reported in [1] and as the value for a recently observed [2] discrete SD band up to spin 60.

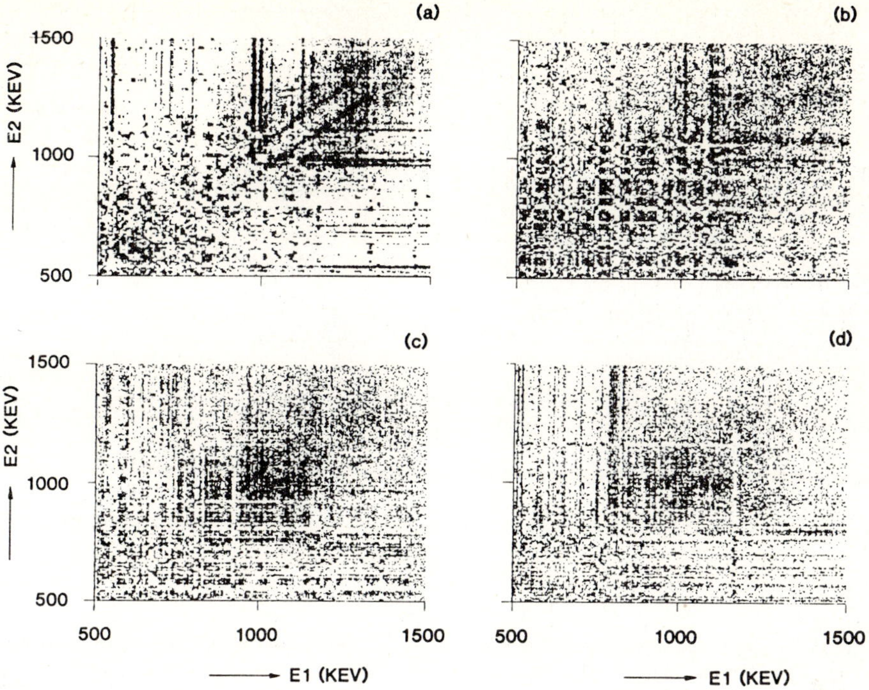

Fig. 1. Symmetrized $E_\gamma(1) - E_\gamma(2)$ correlation matrices, corrected for uncorrelated events, for the energy region of E_γ = 500–1500 keV (2 keV/ch). Matrix (a) contains 192 million events in the displayed energy region corresponding mainly to transitions in ^{152}Dy. The two strong horizontal and vertical lines in the center correspond to the known 968 and 991 discrete transitions in ^{152}Dy. The matrix for ^{150}Dy (b) contains 73 mill. events, for ^{154}Er (c) 64 mill. and for ^{156}Er (d) 76 mill. events

Projections in the energy region of 800–1340 keV perpendicular to the $E_1 = E_2$ diagonal in the matrices, corrected for uncorrelated background, for the full matrix (raw data) as well as those gated on discrete lines above and below the 17^+ isomer in ^{152}Dy are displayed in Fig. 2a,b and c, respectively. In the spectrum of Fig. 2c ridges are seen with the correct distance of 94 keV which indicates that a significant fraction of the γ-ray flow bypasses the isomeric state.

This was also found from the intensity ratio of the ridge with that of known discrete lines above the isomer, which was much larger in the case that no time restriction was applied than with a long time delay of γ-rays detected in the 0° BGO around the catcher.

A projection perpendicular to the $E_1 = E_2$ diagonal and gated on discrete lines in ^{151}Dy is displayed in Fig. 2d. The valley seems to be present as well as the ridges, 94 keV apart. Although this is consistent with the existence of SD bands in ^{151}Dy, the poor statistics in the matrix gated on discrete lines does not allow definite statements in this respect. The valley-ridge structure in ^{150}Dy and in ^{154}Er is weak, and absent for ^{156}Er (see Fig. 1b,c,d). It seems

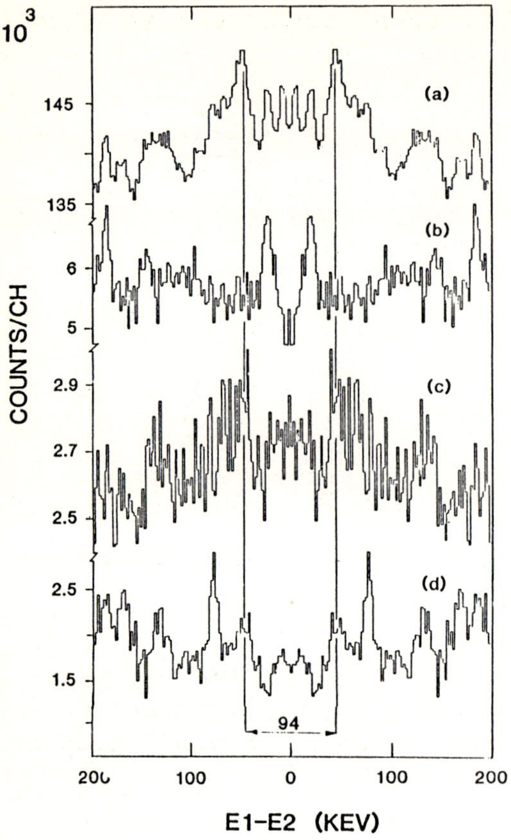

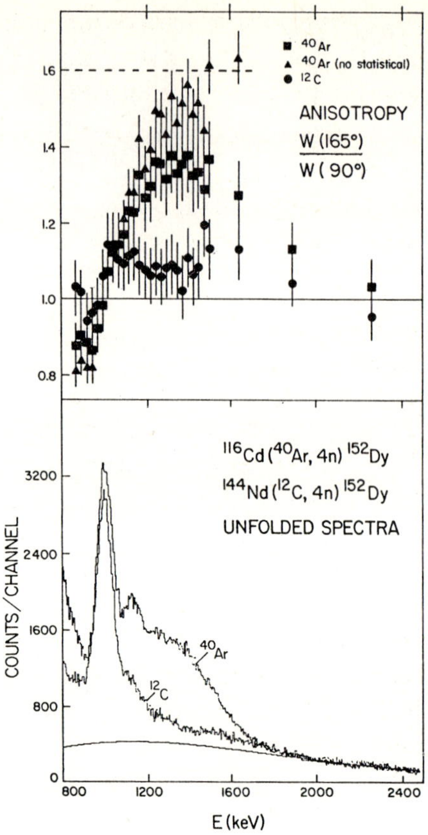

Fig. 2. Projections (in 2 keV/ch) perpendicular to the diagonal in the raw correlation matrices of ^{151,152}Dy imposing the following conditions: (a) E_γ=800–1340 keV, all data no selections. (b) E_γ=800–1360 keV gates on discrete transitions in ^{152}Dy above and (c) below the 17^+, 60 ns isomeric state. The strong peaks in the center (ΔE_γ=23 keV) of spectrum (b) are due to the 968–991 keV coincidences (see Fig. 1). (d) E_γ=900–1360 keV, gates on discrete lines in ^{151}Dy

Fig. 3. Unfolded γ-ray spectra measured with a NaI(Tl) detector (below), with the corresponding γ-ray anisotropies defined as the γ-ray yield at 165° divided by the corresponding yield at 90° (above). Triangle points show the anisotropy of the quasicontinuum for the ^{40}Ar reaction in case the statisticals are subtracted (assumed isotropically distributed) and the circles are for the ^{12}C data

that in ^{150}Dy the valley is present, but there are almost no ridges. In ^{154}Er, however, the valley is shallower and the ridges are more smeared than in ^{152}Dy. It is tempting to interprete a sharp ridge as being due to collective rotation corresponding to a distinct minimum in the potential energy at large deformation. The smearing out effect may be due to a fluctuation of the moment of inertia associated with a flatter superdeformed potential pocket, for instance a SD region at higher temperature. A shallower valley may be caused by a larger damping of the rotational motion [3,4]. This would mean that the four nuclei in Fig. 1 represent sort of limiting cases, because various combinations of valley-ridge structures occur, pointing to large structural effects in this transitional region even in the quasicontinuum spectrum corresponding to high spins and possibly high temperatures.

The measured moment of inertia $F^{(2)}_{\rm band}=85\pm4\,{\rm MeV}^{-1}\hbar^2$ for the SD bands in ^{152}Dy corresponds to a deformation with a ratio of the long to the short nuclear semiaxes of $\sim$1.7, indeed not too far from the ratio 2 for the earlier observed superdeformed (SD) fission isomers.

2. Multipolarity of γ Rays from Collective Quasicontinuum States in ^{152}Dy

This experiment has been carried out by the authors of the KVI, using 180 MeV ^{40}Ar and 82 MeV ^{12}C beams of the Groningen AVF cyclotron. The ^{40}Ar ions transfer enough angular momentum (max. 75 $\hbar$) to excite the SD states in the same way as in the investigation described above, while the ^{12}C ions do not bring in enough angular momentum to reach the high-spin region beyond 40 $\hbar$. A sum spectrometer has been used to enhance the high-spin region of ^{152}Dy in the spectra along with the catcher foil technique. Four small NaI(Tl) detectors served to trigger on delayed γ rays emitted from the stopper. Three large NaI(Tl) detectors were used to measure γ-ray spectra and angular anisotropies and two Ge detectors with BGO anti-Compton shields served to check the selection of evaporation channels with the sum spectrometer.

The difference between the γ-ray spectra and the anisotropy for ^{40}Ar and ^{12}C induced reactions, including the energy region where the γ rays from SD states have been found, is shown in Fig. 3. The anisotropy in the case of ^{40}Ar induced reactions reaches the value 1.6 of stretched E2 transitions when the statistical γ rays are subtracted. The ^{12}C data show a considerably smaller anisotropy, thus confirming that the collectivity (stretched E2) is only present in ^{152}Dy at spins >30. The multiplicities obtained from the large NaI detectors show a maximum around 1200 keV, also indicating enhanced collectivity in that γ-energy region.

The conversion electron yield has been determined from the spectra taken with different transmission settings (5A and 6A) of the Mini-Orange (MO) spectrometer. The background was determined with a transmission outside the energy region of interest after normalization with a ^{56}Co source. The resulting internal conversion coefficients (ICC) are shown in Fig. 4 (bottom) along with the theoretical values for $E1$, $E2$, and $M1$. In the case of ^{12}C (small Doppler broadening) it was possible to deduce the ICC for the 968 keV $E1$ and the 991 keV $E2$ transitions, as indicated. Their agreement with the theoretical values prove that the method is reliable, both for the ^{12}C and the ^{40}Ar data which have been obtained in the same experiment. The high-energy points have been measured with a different MO transmission setting (6A). The ^{40}Ar data have

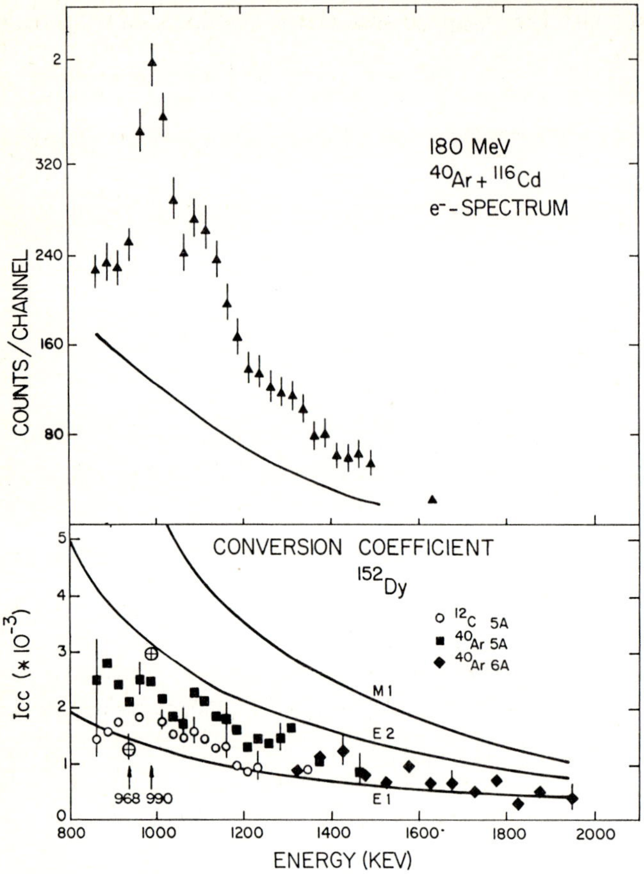

Fig. 4. Measured internal conversion coefficients (below) for the ^{116}Cd(^{40}Ar,4n) reaction, squares for the 5A MO setting and triangles for the 6A setting and the ^{144}Nd(^{12}C,4n) reaction (circles). Positions and measured ICC's for the 968($E1$) and 990($E2$) keV transitions are indicated (in the ^{12}C induced reaction). An electron spectrum is given (above) along with a "smoothed" background, measured for a transmission outside the region of interest

been processed with a gate on the high-sum energy as to enhance the highest spin region above 40 and to select predominantly ^{152}Dy. This sum energy selection was not necessary for the ^{12}C data.

The results show strong $E2$ admixtures in the γ-ray energy region 0.8–1.4 MeV where the decay from SD states has been observed. Although $M1$ transitions will also produce high ICC, the observed anisotropy (Fig. 3) allows only nonstretched $M1$ transitions, which seems unlikely for high-energy γ rays (>1 MeV) in the high-spin region. The ^{12}C data show lower ICC, which is probably due to more $E1$ admixture, as expected for non-collective excitation. Therefore we attribute the high ICC, exclusively obtained with the ^{40}Ar beam, to the strongly deformed quasi continuum in ^{152}Dy.

Calculations, in terms of the cranking approximation using the Woods-Saxon potential [5], indicate the existence of various collective bands in the high-spin region (>30). Up to spin 60 triaxial shapes appear close to the yrast line and above spin 60 the SD shape (β=0.65) becomes yrast. The present data of predominant $E2$ radiation in the γ-ray energy region corresponding to spin 30–50 confirm these results.

References

1. B.M. Nyako et al.: Phys. Rev. Lett. **52**, 507 (1984)
2. P.J. Twin et al.: Submitted to Phys. Rev. Lett.
3. J. Bacelar et al.: Phys. Rev. Lett. **55**, 1858 (1985)
4. J.E. Draper et al.: Phys. Rev. Lett. **56**, 309 (1985)
5. J. Dudek, W. Nazarewicz: Phys. Rev. C**31**, 298 (1985)

High Spin States Around A∼150

A. Piepke, N. Mansour, H. Sanchez, A. Moussavi, J. Metzinger, H. Strecker, and H.V. Klapdor

Max-Planck-Institut für Kernphysik, D-6900 Heidelberg, F. R. Germany

1. Introduction

The investigation of highly rotating nuclei in the mass region of $A \approx 150$ through gamma spectroscopic methods offers a good possibility to study spin-induced nuclear shape changes at highest spins. In this transitional mass region [$N \approx 82$-86, $Z \approx 62$-68] between the doubly closed $^{146}_{62}Gd_{84}$ core at one end and the well-deformed nuclei on the other, Strutinsky-cranked shell model calculations predict for the even-even isotopes Sm, Dy, Er, etc., a shape change from spherical or slightly oblate deformation ($\gamma = 60^{\circ}$, $\beta \approx 0.1$) at low spins to superdeformed prolate ($\gamma = 0^{\circ}$, $\beta \approx 0.6$) or triaxial rotors in the spin region of $J = 60\ \hbar$, as a function of proton and neutron number [1]. The search for superdeformed nuclei is urged with big experimental effort by several groups, so Sharpey-Schafer and co-workers reported [2,3] evidence for a superdeformed structure of $^{152}_{66}Dy_{86}$, the neighbouring even-even isotone of $^{153}_{67}Ho_{86}$, up to a spin $J = 60\ \hbar$. Janssens et al. [4] investigated the behaviour of ^{153}Ho at highest spins via spectroscopy of unresolved γ-radiation and reported for states in the spin region of $J \approx 40\ \hbar$ the onset of a collective E2-component in the continuum γ-spectra. We extended in this work the level scheme of ^{153}Ho [5] up to a state with $E_x = 14.7$ MeV. $J = (89/2\ \hbar)$ in order to resolve the so-called "M1-quasicontinuum" and to look for collective states at highest spins. For the odd-odd nuclide ^{152}Ho there existed no level scheme. Only some γ-ray energies were reported in ref. [6,7] to belong to this nucleus. We propose in this work a level scheme for ^{152}Ho, the only nucleus in the $A \approx 146$-154 mass region for which no detailed spectroscopic information was available.

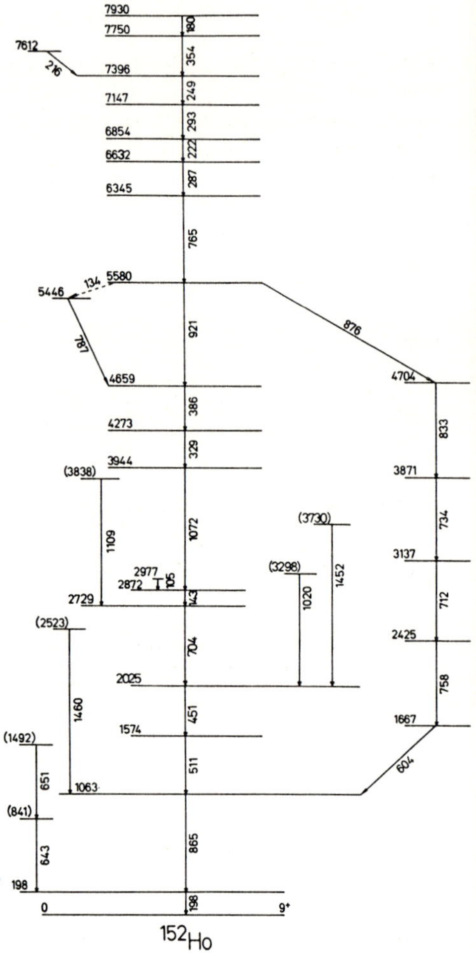

Fig. 1: The proposed level scheme of ^{152}Ho.

2. Reaction and Experimental Procedure

The high spin states in ^{152,153}Ho were populated by the ^{122}Sn(^{35}Cl,xn), x = 5,4 reaction at E_{lab} = 167 MeV. The target consisted of 98% enriched ^{122}Sn evaporated on a gold backing. The ^{35}Cl beam was extracted from the MP-Tandem and post-accelerator of the Max-Planck-Institut für Kernphysik in Heidelberg. The γ-radiation was detected by the Heidelberg double anti-Compton spectrometer consisting of two HP Ge detectors with active NaJ shielding. The present study involved four separate measurements: relative γ-ray excitation functions, γ-ray angular distributions, γ-γ-coincidence measurements and γ-ray timing measurements.

3. Results and Discussion

Figs. 1,3 show the suggested level schemes of ^{152,153}Ho. The main criterion for the level ordering was the intensity of the γ transitions so that for the highly excited states and also for these ones in ^{152}Ho which are contaminated with ^{153}Ho and ^{152}Dy lines (E_γ = 604 keV, 712 keV, 734 keV, 758 keV) as well as the E_γ = 511 keV line, the level ordering can only be tentative.

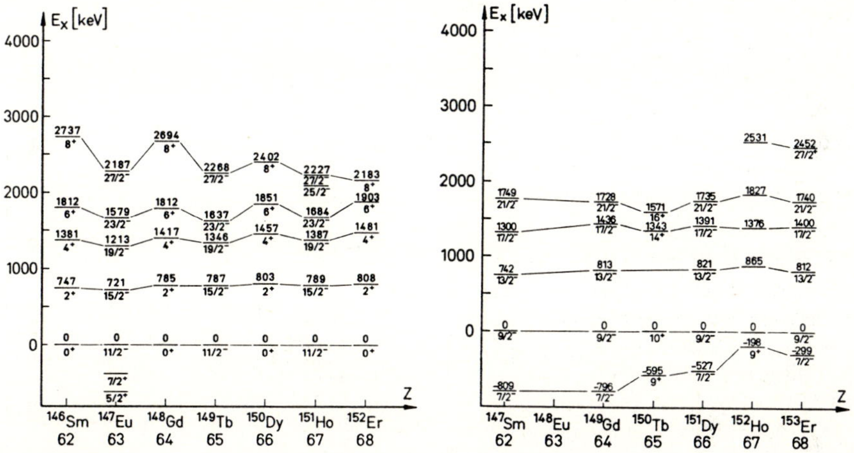

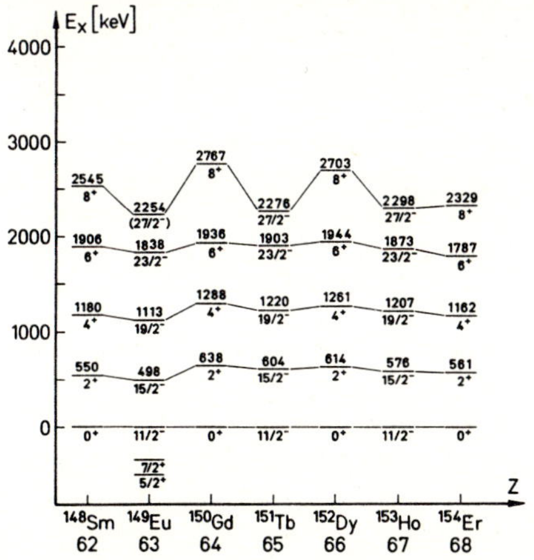

Fig. 2a,b,c: Comparison of the level schemes of odd and even Z isotones in the mass range A ≈ 150.

3.1 The nuclide ^{152}Ho

Because of the lack of calculations the proposed level scheme of ^{152}Ho will be discussed in the light of a comparison to the experimental systematics of the N = 85 isotones. For $^{152}_{67}Ho_{85}$ with three protons and three neutrons outside the doubly closed $^{146}_{64}Gd_{82}$ core even the ground state spin was unknown. Ref. [8] reports two longlived α-unstable isomers, one with $J^{\pi} = 9^{+}$ and $T_{1/2} = 52$ s and the other with $J^{\pi} = 3^{+}$ and $T_{1/2} = 2.4$ m. The ground state spin of ^{152}Ho was chosen for this work to $J^{\pi} = 9^{+}$ in accordance with the neighbouring odd-odd isotone $^{150}_{65}Tb_{85}$ [19] and odd-odd isotope $^{150}_{67}Ho_{83}$ [9]. This would correspond to a fully aligned $\pi h^{3}_{11/2} \otimes \nu f^{3}_{7/2}$ shell model con-figuration which can also be found in $^{148}_{63}Eu_{85}$ [14]. The striking similarities for energy spacing and spin of the states of nuclei belonging to isotone chains with Z = 62-68 in the mass region of A ≈ 150 is demonstra-ted in Fig. 2a,b,c where three chains are displayed with N = 84, 85, 86. The experi-mental information on $^{146-148}Sm$, $^{147-149}Eu$, $^{148-150}Gd$, $^{149-151}Tb$, $^{150-153}Dy$ and $^{152-154}Er$ was taken from ref.[3,5,10-25]. For all even Z members of this chain $\nu f^{3}_{7/2}$ and $\nu h_{9/2}f^{2}_{7/2}$ shell model multiplets are reported to feed directly into the ground state. For $^{150}_{65}Tb_{85}$ which is the only odd-odd N = 85 nuclide with known level pattern ref. [19] reports a similar structure, namely a $J^{\pi} = 9^{+}$ ground state fed by $\pi h_{11/2} \otimes \nu h_{9/2}f^{2}_{7/2}$ and $\pi h_{11/2} \otimes \nu f^{3}_{7/2}$ multiplets. In this situation it seems reasonable to compare the ground state band of ^{152}Ho with these well-established level schemes. In Fig. 2b the ^{152}Ho states with E_x = 198 keV, 1063 keV, 1574 keV, 2025 keV and 2729 keV are compared with the $\nu h_{9/2}f^{2}_{7/2}$ structures of the even Z isotones and the $\pi h_{11/2} \otimes \nu h_{9/2}f^{2}_{7/2}$ multiplet in $^{150}_{65}Tb_{85}$. This comparison shows that the energy spacing from the first excited state onward shows good agreement with the neighbouring isotones.

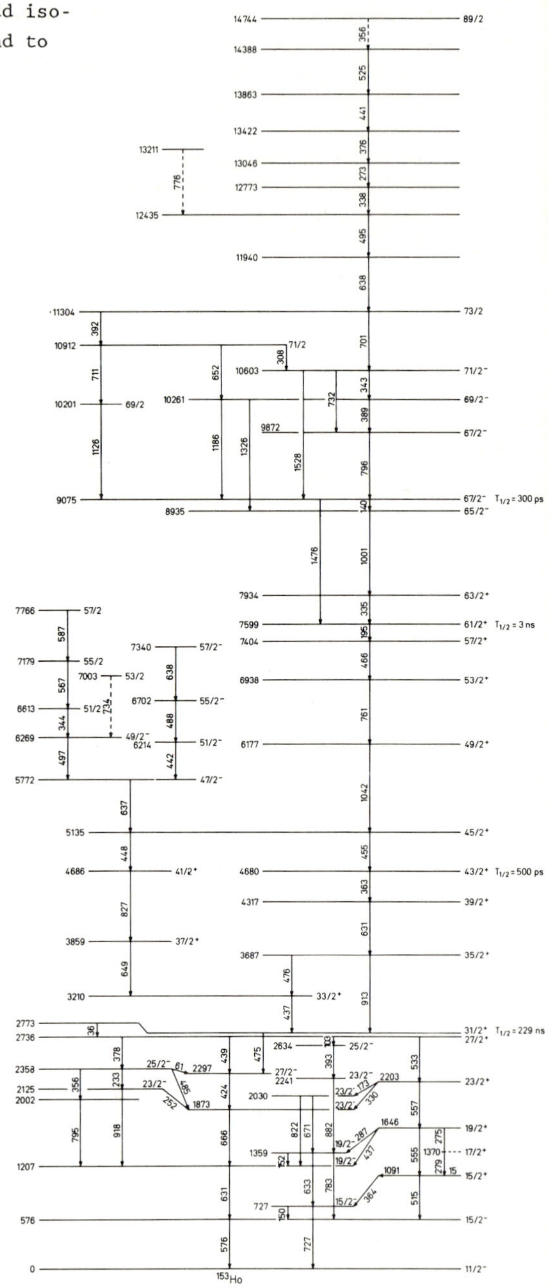

Fig. 3: The proposed extension of the level scheme of $^{153}_{67}Ho_{86}$.

Also the decline of the energy of the first excited state with growing Z fits into this systematics if one takes into account that nuclei in this mass region are reported to exhibit increasing oblate ($\epsilon \approx -0.1 \triangleq \beta \approx 0.11$) deformation when protons are added to the core [26]. The Nilsson diagram shows that the energy gap between the $\nu f_{7/2}$ and $\nu h_{9/2}$ states is decreasing with growing oblate deformation that would ease the promotion of a neutron into the $\nu h_{9/2}$ shell and lower the energy of the first excited state if it belongs to a $\pi h^3_{11/2} \otimes \nu h_{9/2} f^2_{7/2}$ multiplet. The assumption of an oblate deformation of the ^{152}Ho nucleus is supported by the irregular energy spacing and the occurrence of ns isomers which are indications for a single particle excitation mode. The spin assignments to ^{152}Ho deduced from the preliminary evaluation of the γ-ray angular distributions are consistent with the systematics of the nuclei in the N = 85 isotone chain.

The level lifetimes of ref. [7] were confirmed in this work. Together with the information from the coincidence spectra the measured $T_{1/2}$ indicates an isomeric state at E_x = 2729 keV with $T_{1/2} \approx$ 50 ns ± 7. Since γ transitions placed above the E_x = 2729 keV state were also found to have lifetimes a second isomer at higher excitation energy can be suggested but not located in the level scheme due to the big statistical error.

3.2 The nuclide ^{153}Ho:

The spherical or slightly oblate shape along the yrast line in the mass region of A ≈ 150 manifests itself in an excitation by alignment of individual nucleons along the symmetry axis. Theoretical calculations predict the coexistence of these oblate yrast states with prolate deformed states ~ 3 MeV above the yrast line. This prolate "phase" is reported to get yrast in the spin region of J ≈ 60 $\hbar$ [1]. The spectroscopy of these collective bands is aggravated by the fact that no interaction [3] of the oblate yrast states and the prolate rotational bands is expected in the low spin region. Since the yrast states are preferentially populated in evaporation-fusion reactions the intensity of the γ-transitions between the collective states is very low. So the only chance to prove the existence of these strongly deformed bands with discrete γ-γ-coincidence measurements is to enlarge the spectroscopic information along the yrast line up to highest spins and to find interactions of the oblate and the prolate phase in the region where the rotational band gets yrast. Ref. [4] reported for the spin range of $\langle J \rangle \approx 44\ \hbar$ - $49\ \hbar$ an onset of a collective E2 component in the continuum spectra with comparable intensity to the M1 component originating from the noncollective yrast states.

In this work the level scheme of ^{153}Ho Fig. (3) was extended up to a state with E_x = 14.744 MeV and J = (89/2 $\hbar$). More than 20 new transitions were added to the level scheme of Radford et. al. [3]. Up to those highest states an irregular energy spacing of the excited states can be observed which manifests an excitation by single particle alignment. A strong interaction of the oblate and prolate states up to this excitation energy seems unlikely.

References

[1] J. Dudek and W. Nazarewicz, Phys. Rev. C31, 298, (1985).

[2] B.M. Nyakó, J. Simpson, P.J. Twin, D. Howe, P.D. Forsyth and J.F. Sharpey-Schafer, Phys. Rev. Lett. 56, 2680 (1986) and 57, 811 (1986).

[3] J.F. Sharpey-Schafer, contribution to this volume.
[4] R.V.F. Janssens, D.C. Radford, I. Ahmad and A.M. Van den Berg, Phys. Lett. 152B, 167, (1985).
[5] D.C. Radford, M.S. Rosenthal, P.D. Parker, J.A. Cizewski, J.H. Thomas, B.Haas, F.A. Beck, T. Byrski, J.C. Merdinger, A. Nourredine, J. Schutz, J.P. Vivinen, J.S. Dionsio and Ch. Vieu, Phys. Lett. 126B, 24 (1983).
[6] D.C.J.M. Hageman, M.J.A. de Voigt and J.F.W. Jansen, Phys. Lett. 84B, 301, (1979).
[7] J. Jastrzębski, R. Kossakowski, J. Łsukasiak, M. Moszyński, Z. Preibisz, S. André, J. Genevey, A. Gizon and J. Gizon, Phys. Lett. 97B, 50 (1980).
[8] W.D. Schmidt-Ott, K.S. Toth, E. Newman and C.R. Bingham, Phys. Rev. C10, 296, (1974).
[9] J. McNeill, R. Broda, Y.H. Chung, P.J. Daly, Z.W. Grabowski, H. Helppi, M. Kortelahti, R.V.F. Janssens, T.L. Khoo, R.D. Lawson, D.C. Radford and J. Blomqvist, Z. Phys. A325, 27, (1986).
[10] C.H. King, B.A. Brown and T.L. Khoo, Phys. Rev. C5, 2127 (1978).
[11] M. Piiparinen, Y. Nagai, J. Styczen and P. Kleinheinz, Proc. of "Int. Conf. on Nucl. Behaviour at High Ang. Momentum", 53, Strasbourg (1980).
[12] E. Hammarén, E. Liukkonen, M. Piiparinen, J. Kownacki, Z. Sujkowski, Th. Lindblad and H. Ryde, Nuc. Phys. A321, 71, (1979).
[13] J.G. Fleissner, E.G. Funk, F.P. Venezia and J.W. Mihelich, Phys. Rev. C16, 227, (1977).
[14] M. Piiparinen, R. Broda, Y. Nagai, P. Kleinheinz and A. Pakkanen, Z. Phys. A301, 231, (1981).
[15] P. Kleinheinz, Proc. "Symp. on High-Spin Phenomena in Nuclei", 125, Argonne, (1979).
[16] M. Piiparinen, R. Pengo, Y. Nagai, E. Hammarén, P. Kleinheinz, N. Roy, L. Carlén, H. Ryde, Th. Lindblad, A. Johnson, S.A. Hjorth and J. Blomqvist, Z. Phys. A 300, 133, (1981).
[17] D.R. Haenni and T.T. Sugihara, Phys. Rev. C16, 120, (1977).
[18] N.C. Singhal, M.W. Johns and J.V. Thompson, Can. J. Phys. 57, 1959 (1979).
[19] R. Broda, M. Ogawa, P. Kleinheinz, R.K. Sheline, L. Richter, Proc. "Symp. on High-Spin Phenomena in Nuclei", 401, Argonne, (1979).
[20] P. Kemnitz, L. Funke, F. Stary, E. Will, G. Winter, S. Elfström, S.A. Hjorth, A. Johnson and Th. Lindblad, Nuc. Phys. A311, 11, (1978).
[21] M. Piiparinen, P. Kleinheinz, S. Lunardi, H. Backe, J. Blomqvist, Proc. "Symp. on High-Spin Phenomena in Nuclei", 407, Argonne, (1979).
[22] J. Gizon, A. Gizon, S. André, J. Genevey, J. Jastrzebski, R. Kossakowski, M. Moszynski and Z. Preibisz, Z. Phys. A301, 67, (1981).
[23] G. Bastin, A. Peghaire, J.P. Thibaud, S. Andre, D. Barneoud, C. Foin and P. Aguer, Nuc. Phys. A345, 302, (1980).
[24] L. Carlén, S. Jónsson, J. Krumlinde, J. Lyttkens, N. Roy, H. Ryde, S. Strömberg, W. Waluś, G.B. Hagemann and B. Herskind, Nucl. Phys. A381, 155, (1982).
[25] P. Aguer, G. Bastin, A. Charmant, Y. El Masri, P. Hubert, R. Janssens, C. Michel, J.P. Thibaud and J. Vervier, Phys. Lett. 82B, 55, (1979).
[26] G. Andersson, S.E. Larsson, G. Leander, P. Möller, S.G. Nilsson, I. Ragnarsson, S. Åberg, R. Bengtsson, J. Dudek, B. Nerlo-Pomorska, K. Posmorski and Z. Szymański, Nucl. Phys. A268, 205, (1976).

The Nucleus ^{136}Sm and the Transition to a Strong Deformed Region at N=74–76

F. Soramel, S. Lunardi, W. Meczynski, and M. Morando*

Dipartimento di Fisica, Università di Padova and INFN-Sezione di Padova, I-35131 Padova, Italy

Since some years interest has been concentrated on the nuclei with $50 \leq N, Z \leq 82$ where a strong permanent ground state deformation is expected [1]. The study of these nuclei is, however, not easy since they lie very far from the stability line and close to the proton dripline; therefore the experimental results are still scarce and incomplete. Some neutron-deficient rare earth isotopes can, nevertheless, be populated via heavy ion induced fusion reactions.

We have initiate an experimental program to investigate the neutron deficient nuclei close to Z=64 using the beams delivered by the Tandem XTU at Legnaro. The first results on N=78 and N=76 nuclei [2,3] show, together with excitations of single particle character, an increasing collectivity when going far away from the N=82 shell closure.

Through the reaction ^{107}Ag+^{32}S at beam energies ranging from 125 up to 150 MeV we have now studied the N=74 nucleus ^{136}Sm and established its level scheme which is shown in Fig.1. Lifetimes with plunger technique have been measured too and halflife values up to spin 10^+ have been obtained for ^{136}Sm as given in Table 1. Unlike the two heavier isotopes ^{138,140}Sm, no sharp discontinuity either in level

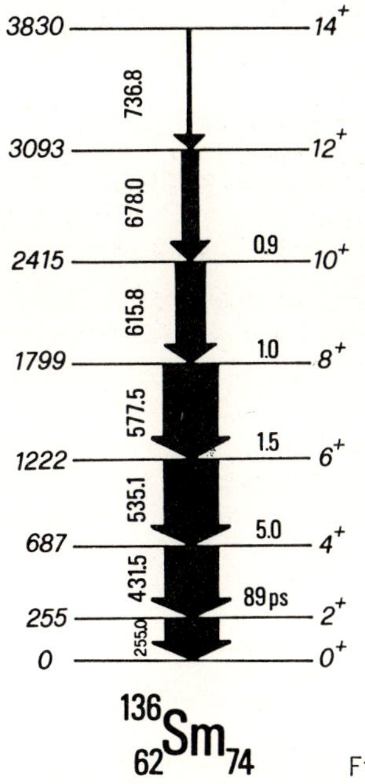

Figure 1. Level scheme of ^{136}Sm

* Permanent address: Institute of Nuclear Physics 31-342 Cracow (Poland)

Table 1. Halflives, experimental B(E2) values, and calculated rigid rotor B(E2)'s for the ^{136}Sm levels

E_γ [keV]	I^π	$T_{\frac{1}{2}}$ [ps]	B(E2) [fm^4e^2]	$B(E2)^{rotor}$ [fm^4e^2]
255	2^+	89.0(2.0)	5158(116)	5158
432	4^+	5.0(0.3)	7473(428)	7369
535	6^+	1.5(0.8)	8721(4546)	8116
578	8^+	1.0(0.3)	9191(2952)	8496
616	10^+	0.9(0.3)	7258(2362)	8726

spacings or B(E2) values is observed at 10^+ in ^{136}Sm. This differs also from the other N=74 isotones up to Nd which exhibit strong backbending: the ^{136}Sm nucleus shows only a small upbending. Moreover all the experimental B(E2) values are consistent with those calculated in the frame of an axial rigid rotor. These facts are an indication of a more stable rotational character for ^{136}Sm.

From the lifetime of the 2^+ state we have deduced a B(E2) of 124(3) W.u., which is one of the largest values measured sofar in this nuclear region. The corresponding deformation parameter is β=0.30(1) and is much larger than predicted [1] (β=0.23). We can note that this large β value is similar to that of ^{152}Sm nucleus which lies in the symmetric position of ^{136}Sm with respect to the N=82 closed shell, whereas the deformation is stronger in ^{138}Sm than in the corresponding ^{150}Sm.

Hence the transition to the strong deformed region which we observe at N=74 is not so abrupt as for the heavy rare earth nuclei at N=90; this is probably due to the Z=64 shell closure whose effects are evident for N>82 but disappear rapidly when subtracting nucleons from N=82.

1 G.A.Leander, P.Möller: Phys. Lett. 110B, 17 (1982).
2 S.Lunardi et al.: Proceedings of "Conference on Nuclear Structure with Heavy Ions" Legnaro, 27-31 May 1985, edited by R.A.Ricci and C.Villi; Editrice Compositori-Bologna; p. 145.
3 S.Lunardi et al.: Z.Physik A321, 177 (1985).

Excitation of Stretched Particle-Hole States in Closed Shell Nuclei

A. Yokoyama[1] *and H. Horie*[2]

[1]Laboratory of Physics, School of Medicine, Teikyo University, Hachioji, Tokyo, 192 Japan

[2]Faculty of Engineering, Kanto Gakuin University, Yokohama, Kanagawa, 236 Japan

High-spin states of unnatural parity have recently been observed in nuclei over a wide range of the nuclear charts by using various hadronic and leptonic reactions [1]. The high-spin states are expected to be generated by a simple particle-hole (p-h) configuration $j_p j_h^{-1}$, where $j_p=\ell_p+1/2$ and $j_h=\ell_h+1/2$ are the largest angular momenta found in the first open shell and the last filled shell, respectively. When the j_p and j_h couple to the angular momentum of the maximum value $J_{max}=j_p+j_h=\ell_p+\ell_h+1$, the high-spin states are often referred to as stretched p-h states. A number of experimental data indicate that the excitation strengths of the stretched and nearly stretched p-h states are suppressed by a factor of 1/5-1/2 compared to those predicted by the single-particle shell model. Subnucleonic effects such as the isobar-hole excitations and the mesonic exchange currents are expected to be rather weak in the stretched transitions due to higher multipole transitions[2]. The observed quenching of the strengths might be therefore attributed mainly to nuclear structure effects, such as ground state correlations and higher-shell configuration mixing effects. Since the stretched state in a closed shell nucleus is a unique p-h state with the $1\hbar\omega$ excitation, and since the number of configurations is severely limited due to high angular momentum of the state, a perturbative method would be applicable in order to estimate the nuclear structure effects. We attempt to make a systematic configuration mixing calculation of the transition strengths for the high-spin p-h states in the closed shell nuclei from ^{12}C to ^{208}Pb. Several calculations[3,4] have so far been made to interpret inelastic electron scattering form factors of the 14^- and 12^- states in ^{208}Pb.

A perturbation expansion of Rayleigh-Schrödinger type up to first order is assumed, and then an effective transition operator $f^{(\lambda)}_{eff}$ is defined by

$$f^{(\lambda)}_{eff} = f^{(\lambda)} + V \frac{Q}{E_0 - H} f^{(\lambda)} + f^{(\lambda)} \frac{Q}{E_0 - H} V \qquad (1)$$

Here, $f^{(\lambda)}$ is the free-nucleon one-particle transition operator of tensorial rank λ, V is the residual interaction, and Q is the projection operator taking account of the Pauli principle. As for the mixing interaction V, we assume, as a first step, phenomenological central forces with the delta-function, the Yukawa and the Gaussian radial shape, the mixture parameters of which were fitted so as to reproduce some observed levels[5]. By using these mixing interactions, we calculate the corrected matrix element $M^{corr} = \langle j_p j_h^{-1}; \lambda \| f^{(\lambda)}_{eff} \| 0 \rangle$ both for the $M\lambda$ transitions and for the transverse magnetic form factors of inelstic electron scattering. Each process appeared in M^{corr} is shown in Fig. 1 diagrammatically. In calculating the matrix elements of the transition operator and the two-body interaction, we assume

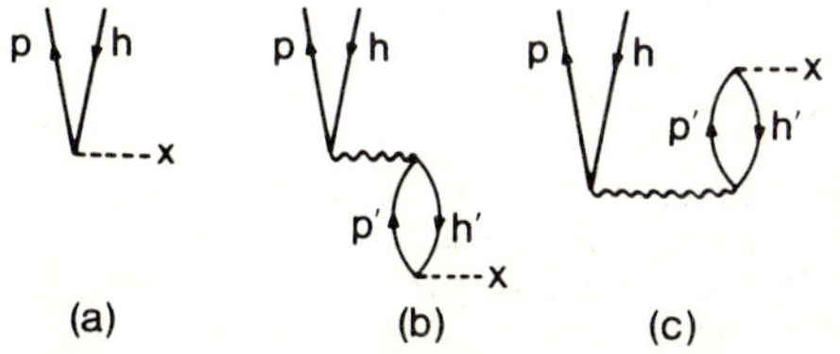

Fig. 1 Diagrammatical representation of the unperturbed and perturbed matrix elements considered in the present calculations: (a) unperturbed matrix element, (b) correction from the final state, and (c) correction from the initial state.

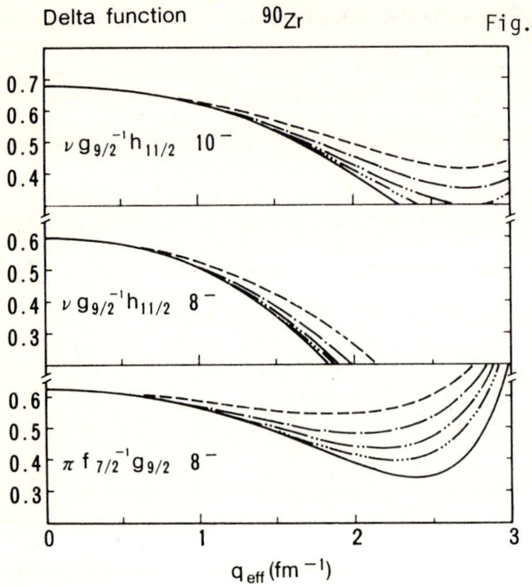

Fig. 2 The squared ratio of the corrected to the unperturbed matrix element as a function of momentum transfer q. The dashed line shows the calculation with $\Delta n=0$, the dot-dashed one with $\Delta n=2$, the double-dot-dashed one with $\Delta n=4$, the triple-dot-dashed one with $\Delta n=6$, and the solid line with $\Delta n=10$.

the harmonic oscillator radial wave functions with the oscillator constant $\nu = 0.96 \times A^{-1/3}$ [fm^{-2}]. Only the single-particle energy is assumed to calculate the energy denominators appeared in (1), and it is calculated by using the empirical formula given in the previous configuration mixing calculation[6]. We then define a ratio $R(q) = M^{corr}/M^{unp}$ as a measure of suppression, where $M^{unp} = \langle j_p j_h^{-1}; \lambda \| f^{(\lambda)} \| 0 \rangle$ and q is the momentum transfer.

The radial functions in the operator for transverse magnetic form factors of electron scattering are $j_L(qr)$ with $L=\lambda+1$, $\lambda-1$ and λ rather than $r^{\lambda-1}$ in the $M\lambda$ transitions, and therefore the configuration mixing which contributes to the form factors is not restricted up to the $(\lambda-1)\hbar\omega$ p-h excitations. We perform calculations with configurations up to $(\lambda-1+\Delta n)\hbar\omega$ excitations, where we take $\Delta n=0,2,\ldots,10$. In order to obtain sufficient convergence for most cases, we adopt $\Delta n=6$, and this value is used throughout the present calculations. The squared ratios $R(q)^2$ for the $\nu 10^-$, $\nu 8^-$ and $\pi 8^-$ states in ^{90}Zr are exemplified in Fig. 2.

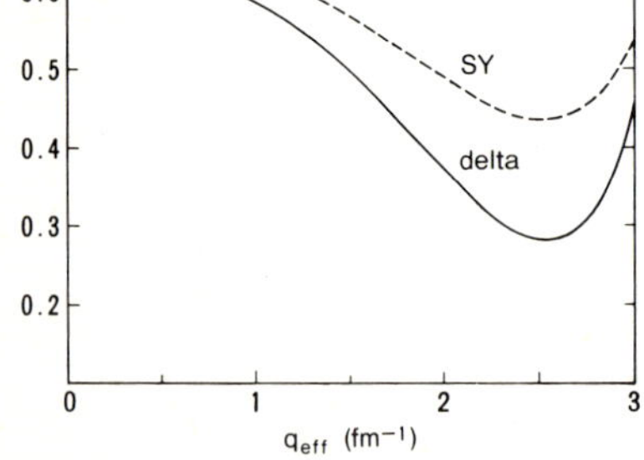

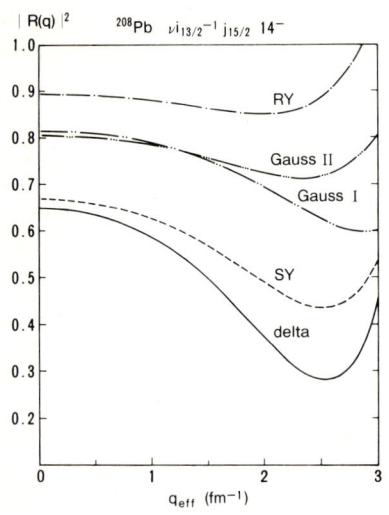

Fig. 3 The squared ratio of the corrected to the unperturbed form factor as a function of momentum transfer q. The effective interactions used here are the delta-function(delta), the Serber-Yukawa(SY), the Rosenfeld-Yukawa(RY), the Gauss I (attractive singlet-odd force), and the Gauss II (repulsive singlet-odd force).

Figure 3 shows the squared ratios of the corrected to the unperturbed matrix element as a function of q for the $\nu 14^-$ state in ^{208}Pb. It turns out that the q-dependence of this ratio is appreciably large, though it depends on the mixing interactions employed. The ratios calculated both for the proton and neutron unperturbed p-h states are always a declining function of q, at least up to around the first maximum of the form factors. By comparing the ratio obtained with the delta-function interaction to that obtained with the SY interaction, it is clearly seen that the delta function affects more strongly the form factors in the high-q region than the SY interaction. This may be readily understood from the radial shape in momentum space: $1/(m^2 + q^2)$ for the Yukawa and 1 (constant) for the delta function. The even-state components of the interaction, which are well known to be attractive, always play a role reducing the unperturbed matrix element. On the other hand, the odd-state components are not yet well understood and give an enhancement for the unperturbed matrix element, if they are repulsive. The difference in the squared ratios obtained with the finite range interactions comes mainly from the odd-state components, as demonstrated in Fig. 3.

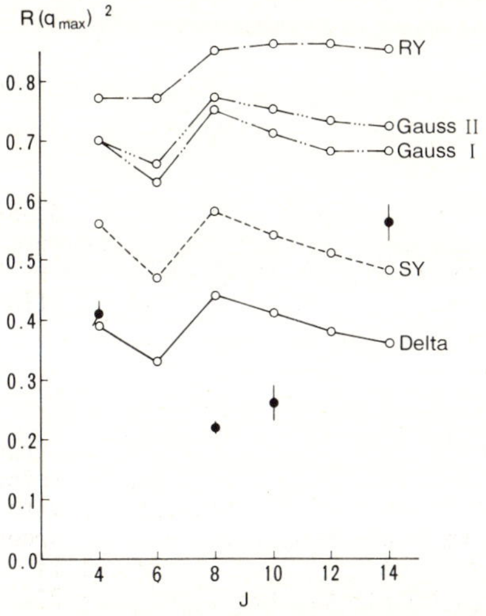

Fig. 4 The squared ratios of the corrected to the unperturbed matrix element at $q=q_{max}$, where q_{max} is the momentum transfer which gives the first maximum of the unperturbed form factors. The entries are the isovector $d_{5/2}p_{3/2}^{-1}$ 4^- state in ^{16}O, the isovector $f_{7/2}d_{5/2}^{-1}$ 6^- state in ^{40}Ca, the neutron $g_{9/2}f_{7/2}^{-1}$ 8^- state in ^{48}Ca, the neutron $h_{11/2}g_{9/2}^{-1}$ 10^- state in ^{90}Zr, the neutron $i_{13/2}h_{11/2}^{-1}$ 12^- state in ^{146}Gd, and the neutron $j_{15/2}i_{13/2}^{-1}$ 14^- state in ^{208}Pb. Experimental data, indicated by the filled circle, are taken from ref. 1.

The squared ratios at $q=q_{max}$ are presented in Fig. 4 for the stretched transitions of the isovector and the neutron p-h states in the good closed shell nuclei. The abscissa is taken to be the spin value of the p-h states, but it also corresponds approximately to the mass of the nucleus. The present calculations predict that the ratios are almost constant as they go from the light to the heavy nuclei. The observed data, on the other hand, seem to indicate a mass dependence; relatively larger suppression in the medium-light nuclei. The corrected transition strengths amount to 40-80% of the unperturbed strengths, though they depend considerably on the mixing interactions employed.

References:

1 A. Lindgren: Journal de Physique, Supplement C4, 433 (1984)
2 T. Suzuki, S. Krewald and J. Speth: Phys. Lett. 107B, 9 (1981)
3 I. Hamamoto, J. Lichtenstadt and G. F. Bertsch: Phys. Lett. 93B, 213 (1980)
4 T. Suzuki, M. Oka, H. Hyuga and A. Arima: Phys. Rev. C26, 750 (1982)
5 A. de Shalit and H. Feshbach: Theoretical Nuclear Physics (John Wiley & Sons, 1974)
6 H. Noya, A. Arima and H. Horie: Prog. Theor. Phys., Supplement 8, 33 (1958)

1.2.2 Giant Resonances and Sum Rules

Photon Decay of Giant Resonances

F.E. Bertrand, J.R. Beene, and M.L. Halbert †

Oak Ridge National Laboratory*, Oak Ridge, TN 37831, USA

We have determined the total gamma-decay probability, the ground-state gamma branching ratio, and the branching ratios to a number of low-lying states as a function of excitation energy in ^{208}Pb to ~ 15 MeV. The total yield of ground-state E2 gamma radiation in ^{208}Pb can only be understood if decay of compound states is considered. Other observations in ^{208}Pb include the absence of a significant branch from the giant quadrupole resonance (GQR) to the low-lying collective states at 2.6 MeV and 4.08 MeV, and a strong branch to a 3^- state at 4.97 MeV.

Over the past decade several new giant resonances (GR) have been observed and classified [1]. Most of the observation of these new resonances has been accomplished through the use of light mass hadronic probes (protons, alphas, etc.), utilizing either inelastic scattering or charge exchange reactions. Recently, some advantages to excitation of isoscalar giant resonances using inelastic scattering of medium energy heavy ions have been investigated [2].

While some selectivity in GR excitation is obtained in inelastic scattering by selection of the incident particle in general inelastic scattering is not selective among the various multipoles. It is of considerable importance to find a method to study the complicated GR structure that would provide multipole selectivity and would at the same time provide a model independent measure of the transition strength in the resonance. A measurement of the photon decay of the giant resonances can provide such information and we report here the results from such measurements.

The giant electric multipole resonances in heavy nuclei are simple nuclear states embedded in a dense spectrum of more complex states, with which they mix. The consequent damping of the giant resonances offers an excellent test of our

† Collaborators (ORNL) on work reported here are: R.L. Auble, E.E. Gross, D.C. Hensley, D.J. Horen, R.L. Robinson, R.O. Sayer, D. Shapira, and T.P. Sjoreen.

*Operated by Martin Marietta Energy Systems, Inc. under contract DE-AC05-84OR21400 with the U.S. Department of Energy.

understanding of many-body physics in atomic nuclei. The questions now being asked [3,4] concerning the microscopic structure and the damping of these resonances require more detailed experiments than those which have served to build up the systematic catalog of gross properties of the resonances over the last decade [1]. The data required are coincidence data on the particle and gamma decay of the resonances, which can probe aspects of the resonance structure not addressed by the existing systematics.

The GR are described microscopically as a coherent superposition of one-particle, one-hole excitations relative to the ground state [1,5,6]. This coherent state is connected — by definition — to the ground state by a strong electromagnetic matrix element. Observation of the corresponding electromagetic decay deexciting the GR is of great importance, because of its direct relationship to the concept of a GR, since it offers the possibility of a determination of the resonance strength independent of that provided by analysis of inelastic scattering data with reaction models. Unfortunately, the electric GR lies above particle emission thresholds, with the consequence that the γ decay, in heavy nuclei, occurs for only about one in 10^4 decays.

The particle-hole states that make up the resonance can decay directly into the continuum, producing a free particle and the A-1 nucleus in the corresponding hole state. This is considered a direct decay process, and the corresponding width $\Gamma^{\uparrow}$ is called the escape width. Observation of the distribution of hole states left behind after such decays would provide detailed information about the microscopic structure of the resonances. Unfortunately, in a heavy nucleus such direct particle decays are also rare and difficult to isolate from more common processes [7,8]. The resonances in a heavy nucleus typically lie in a region of very high level density. The simple 1p-1h states of the resonance are consequently mixed or damped into the more complex np-nh states which exist at the same excitaiton energy.

This mixing or damping can be thought of as an alternative decay process for the coherent state [3,7,8]. From this point of view, the GR is excited as a primary doorway state in the inelastic scattering process. This state decays directly via $\Gamma^{\uparrow}$ or by gamma emission, or it "decays" into the continuum of more complex (compound) states. These states then decay statistically, usually by particle emission. A width $\Gamma^{\downarrow}$, the spreading width, is associated with this decay into the continuum. The observed width of the GR state is thus $\Gamma_T = \Gamma^{\uparrow} + \Gamma^{\downarrow}$ (we can safely neglect Γ_γ). For heavy nuclei, $\Gamma_T \sim \Gamma^{\downarrow}$ [7,8]. A microscopic understanding of this damping process is the focus of current theoretical work on giant resonances. Decay studies can provide insight into this process too, if, as has been suggested, the most important states involved in the mixing process are the 2p-2h states formed by coupling the 1p-1h states of the resonance to low-lying surface vibrations [4]. Evidence for the importance of such couplings should appear in the particle or gamma decay to the low-lying collective states.

The measurements were carried out by exciting the giant resonances using 380 MeV ^{17}O inelastic scattering and detecting the γ-decay (in coincidence with the inelastically scattered ^{17}O) in a 4π, γ-ray spectrometer. The use of ~ 22 MeV/nucleon ^{17}O inelastic scattering provides very large cross sections and excellent peak-to-continuum ratios for the GQR. This is pointed out in Fig. 1 where we show a comparison between the giant resonance structure observed in ^{208}Pb as excited by 376-MeV ^{17}O ions [2b] and 334-MeV protons [9]. The large peak located at 10.6 MeV in the ^{17}O spectrum is from excitation of the isoscalar giant quadrupole resonance. The ^{17}O spectrum which was obtained with ~ 200-keV energy resolution shows the existence of fine structure at excitation energies between ~ 7 MeV and the GQR. These peaks are observed also in the (p,p') spectrum [Fig. 1(b)] which was obtained with about 70-keV resolution. The most pronounced difference between the ^{17}O and proton spectra is near 14 MeV, in the region of the giant dipole and giant monopole resonances. This is expected because at the incident energies utilized, proton scattering provides stronger excitation of these resonances. The considerable similarity between the spectra from proton and ^{17}O

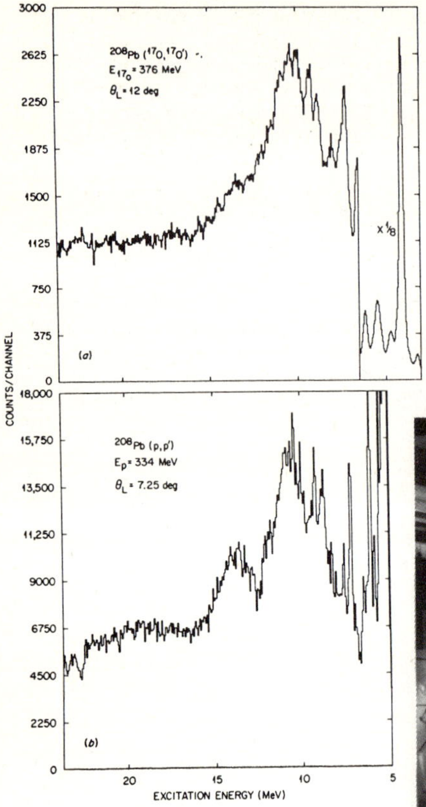

Fig. 1. Inelastic scattering spectra for excitation energies between ~ 3 MeV and ~ 24 MeV. (a) (^{17}O, ^{17}O'), 12 degrees [Ref. 2b] and (b) (p,p'), 7.25 degrees [Ref. 9].

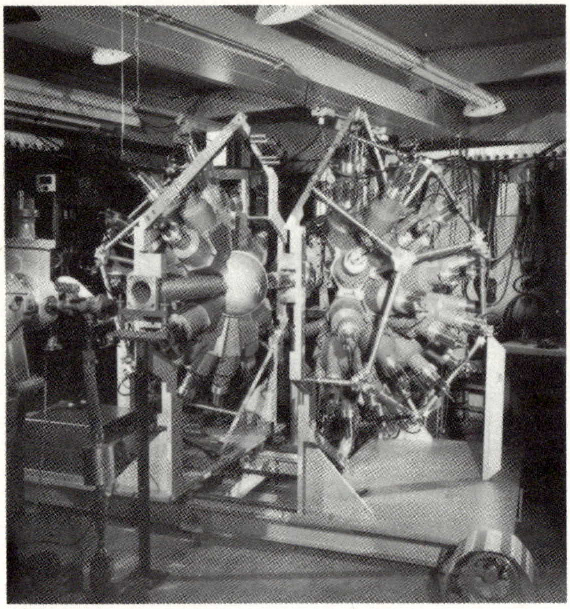

Fig. 2. ORNL Spin Spectrometer. The spectrometer is shown with one half pulled back to expose the spherical scattering chamber.

inelastic scattering is surprising since different types of states could be excited by the two different probes. The ^{17}O probe excites predominantly isoscalar, non spin-flip, states whereas in medium-energy proton scattering contributions from spin-flip excitations should be present. The similarity of the fine structure peaks in the ^{17}O and proton spectra strongly suggests that the peaks arise mainly from excitation of isoscalar states.

There are primarily two experimental capabilities available at ORNL that contributed to our successful γ-decay measurements. The first, discussed above, is the use of ~ 25 MeV/nucleon heavy ions that excite the giant resonances with large cross sections and yield large resonance peak-to-continuum ratios. We chose ^{17}O because the particle thresholds are very low and thus the projectile excitation cross section near the GR region in ^{208}Pb in coincidence with outgoing ^{17}O is negligible. The second feature is the existence at ORNL of the Spin Spectrometer [10], a crystal ball device, which is a 4π, segmented NaI gamma ray spectrometer consisting of 72 NaI detectors (see Fig. 2). Each detector is 17.8 cm thick and ~ 7.6 cm in diameter at the front and 15.2 cm diameter at the back. In the present experiment, the NaI elements at 0° and 180° (relative to the beam direction) were removed for the beam entrance and exit pipes. The Spin Spectrometer with its nearly 4π geometry provides high efficiency detection [10] for both gamma radiation and neutrons. Neutrons and gamma rays were distinguished by time of flight. The flight path is too short to permit resolution of neutron decay to individual levels in ^{207}Pb. However, the residual excitation energy in ^{207}Pb following neutron emission is accurately determined from the total gamma-ray energy in the Spin Spectrometer.

Charged reaction products were detected in six Si surface barrier detector telescopes each consisting of a 500 μm thick ΔE and a 1500 μm thick E detector. These detector telescopes provided excellent mass separation. Each telescope was covered with a trapezoidal collimator having an opening angle of $\Delta\theta = 3°$ and $\Delta\phi = 9°$, yielding a total solid angle for the array of 22.6 msr.

Events which involved pure γ decays were isolated by specifying two criteria. (a) No neutron pulse was seen by the spectrometer, and (b) the total energy carried away by gamma radiation accounted, within the resolution of the detectors involved, for the total excitation energy of ^{208}Pb in the event, as determined by the energy of the inelastically scattered ^{17}O.

This isolation of gamma decay events is illustrated in Fig. 3 which shows a two-parameter histogram of events in which NaI pulses were detected in coincidence

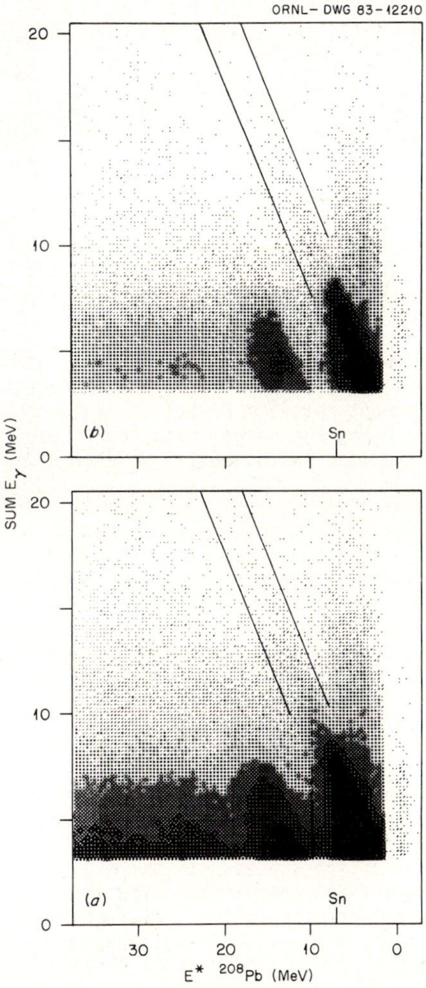

Fig. 3. (a) Is a plot of all events in which no delayed pulse (neutron) was observed. The solid lines indicate the boundaries of the region for which sum $E_\gamma \sim E^*\ ^{208}$Pb (they should extend to sum $E_\gamma = 0$). In (b) the additional constraint $V > 0.95$ has been applied.

with a charged particle identified as ^{17}O in one of the telescopes. The abscissa is the excitation energy in the initial ^{208}Pb nucleus derived from the energy of the ^{17}O. The ordinate is the sum of the gamma ray energies detected in the spectrometer. These should be events in which no neutron pulse was detected, but since virtually all the GR decay is via neutrons [above $E^*(^{208}Pb) \sim 8$ MeV], and since the neutron detection efficiency is less than 100%, the requirement of the absence of a neutron pulse still leaves a substantial background of n-decay events. However, these background events are well separated from pure γ-decay events because of the neutron separation energy, S_n. The pure gamma-decay events should be found in the region outlined on Fig. 3, for which the sum E_γ is approximately equal to $E^*(^{208}Pb)$. In order to avoid confusion from the detection of high energy particles from the sequential decay of ^{18}O and ^{18}F back to ^{17}O following transfer reactions, an event was considered for further analysis only if the largest pulse height occurred in a NaI element at $\theta_{lab} > 66°$. Figure 3a shows all γ rays that fulfill the above requirements. The yield of these events is found to fall off approximately exponentially above S_n. The total gamma branching ratio at 11 MeV is $\sim 2 \times 10^{-3}$.

It is important to select those gamma events which decay directly to the ground state. Unfortunately the number (k) of gamma detectors which are triggered in an event is not useful for this selection. Calibration experiments with 25-MeV protons on ^{12}C show that a single 15.1 MeV gamma ray triggers, on the average, about three detectors and has a significant probability to trigger as many as five. We have used the parameter

$$V = \left| \sum_{i=1}^{k} \vec{h}_i \right| / \sum_{i=1}^{k} \left| h_i \right|$$

to identify ground state gamma decays. The h_i are the individual gamma ray pulse heights recorded in an event. These pulse heights can be assigned a direction as well as a magnitude by noting the position in the Spin Spectrometer array of the detector which produced them; hence, a "vector pulse height," $\vec{h}$, (or apparent photon momentum vector) is obtained for each triggered detector. V is the ratio of the magnitude of the vector sum of pulse heights to the scaler sum. For an event resulting from a single gamma ray this quantity should be near one since only adjacent detectors are triggered. For a cascade decay involving multiple gamma rays V should approach zero as the number of gamma-rays increases. Figure 3b is the same plot as 3a, subject to the additional requirement that $V > 0.95$. It is clear that the rarity of the ground state, GR γ-branch among the large "background" of high-multiplicity cascade γ-ray events requires a device having many γ detectors and 4π geometry like the Spin Spectrometer.

Figure 4 shows the sum gamma-ray spectra obtained from the two-dimensional plots such as Fig. 3. The results shown in Fig. 4 are from those events located between the masks (diagonal lines) on Fig. 3. The solid curve on Fig. 4 is the γ-ray spectrum for all values of V, i.e. all gammas, and corresponds to the data on Fig. 3a. The dashed curve corresponds to γ-events for which $V > 0.98$ (Fig. 3b) and consists only of gamma rays from ground state transitions. The peak at 2.61 MeV from the 3^- state decay has the same number of counts in both spectra. This is of course expected since the state decays 100% to the ground state. On the other hand, in the region above ~ 10 MeV the total γ-branch exceeds the ground state γ-branch by factors of 5-10.

Figure 5 shows the ratio of the solid and dashed gamma-ray spectra in Fig. 4, which is equal to the ground state gamma ray branching ratio, $\Gamma\gamma_0/\Gamma\gamma_{Total}$. Figure 5 shows the regions of excitation in ^{208}Pb which have strong electromagnetic matrix elements to the ground state, i.e. very collective states. In the high excitation energy region such states are defined as giant resonances. The spectrum shows the 2.61 MeV, 3^-, state which has a branching to the ground state of 100%. The peak at ~ 4 MeV arises from excitation of the 2^+ and 4^+ states in ^{208}Pb. It is not completely clear what provides the strong ground state enhance-

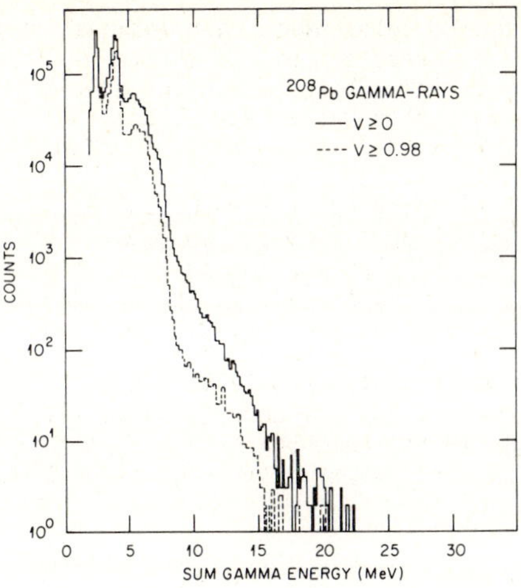

Fig. 4. Gamma-ray spectra from ^{208}Pb for V ≥ 0 (all gamma rays) and V ≥ 0.98 (only ground state gamma rays.)

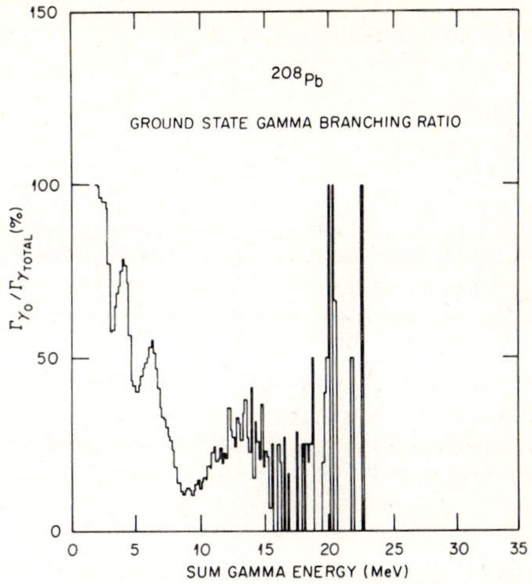

Fig. 5. Ground state gamma-ray branching ratio (%) as a function of sum gamma-ray energy (or excitation energy in ^{208}Pb).

ments in the 6 MeV region other than a group of 1^- states in that energy region. The ground state branching ratio then falls rapidly at the neutron separation energy but begins to rise again near 10 MeV. An obvious broad structure is observed in the 10-17 MeV energy region. Two peaks are found in this region, one at ~ 11 MeV, the other at ~ 13.5 MeV. These energies correspond with the known energies of the giant quadrupole and giant dipole resonances, respectively. It is to be noted that any L = 4 or 6 strength in the GQR region would not have an observable ground state decay. Furthermore, the giant monopole resonance would not have a ground state gamma branch. Thus, the peaks at 11 and 13.5 MeV are from "clean" excitations of the GQR and GDR.

Using angular distributions of the ground state gamma rays we can establish that the region of 9.5 to 11.5 MeV consists of (70 ± 10)% quadrupole radiation. The spectrum of ground-state gamma rays and the singles spectrum were fit using resonance parameters from Ref. 9. By dividing the ground-state gamma-ray yield by the singles yield of scattered particles populating the GQR, we obtain an experimental ground-state branch

$$R_{\gamma 0} = (3.2 \pm 0.4) \times 10^{-4} \qquad (^{208}\text{Pb,GQR}) .$$

This value has been corrected for instrumental efficiency and for the fraction of quadrupole radiation in the region obtained from fits to the photon angular distribution.

The results for $\Gamma_{\gamma 0}$ can be compared to expectation based on the energy-weighted sum rule (EWSR). If we consider the ground-state gamma decay to occur directly from the GR doorway state, then it should be considered as occurring in competition with the damping process, characterized by $\Gamma^{\downarrow}$, which we identify with the experimentally observed resonance width Γ_{exp}. The ground-state gamma width for a state exhausting 100% of the isoscalar L = 2 EWSR is [1,11,12]

$$\Gamma_{\gamma 0}(\text{EWSR}) = 8.07 \times 10^{-7}\, E_{\gamma 0}^{5}\, B(E2\downarrow)_{EWSR} \ \text{MeV} ;$$

$$B(E2\uparrow)_{EWSR} = 5B(E2\downarrow)_{EWSR} = \frac{49.88\ Z^2R^2}{A\ E_{GQR}}\ e^2\text{fm}^4 .$$

Using these expressions we find that for 100% of the EWSR the expected branch is

$$\left(\frac{\Gamma_{\gamma 0}}{\Gamma^{\downarrow}}\right)_{\text{DIRECT}} = 8.62 \times 10^{-5} .$$

This accounts for only about 25% of the observed branch. In ^{208}Pb the compound 2^+ states into which the GQR is damped decay almost exclusively by neutron emission. It has been pointed out [13] that the neutron widths of the compound states in ^{208}Pb are unusually small, leading to a large contribution to the ground-state gamma decay from the compound states. In order to relate the observed ground-state branch to the sum rule strength in the GQR region, a quantitative estimate of this effect is required. The desired quantity is $\langle\sigma_{\gamma 0}^{CN}(E)/\sigma(E)\rangle$, where the denominator is the total cross section for excitation of the GQR and $\sigma_{\gamma 0}^{CN}(E)$ is the cross section for ground-state gamma production in the reaction. This ratio can be expressed [14-16] as

$$R_{\gamma 0}^{CN} = \frac{\sigma_{\gamma 0}^{CN}(E)}{\sigma(E)} \approx S\,\frac{\langle\Gamma_{\gamma 0}(E)\rangle_{CN}}{\langle\Gamma_n(E)\rangle_{CN}} .$$

The quantity $\langle\Gamma_N(E)\rangle_{CN}$ is the average neutron ($\simeq$ total) width of the compound states; $\langle\Gamma_{\gamma 0}(E)\rangle$ is calculated as in Ref. [13] and is proportional to $\Gamma_{\gamma 0}$ (EWSR). The ratio of compound widths can be calculated with reasonable confidence [13], employing experimental neutron strength functions [13,17,18] for $\ell = 0$, 1, and 2 neutron emission up to ~ 3 MeV above threshold, and the Hauser-Feshbach formula with optical model transmission coefficients for larger ℓ and excitation energies. We find the ratio of average widths in the above expression averaged over the GQR between 9.5 and 11.5 MeV to be between 8.0×10^{-5} and 1.1×10^{-4}, depending on the optical potential employed. The enhancement factor [14-16] (S) in the expression for the compound branch arises because of the properties of the Porter-Thomas distributions which the individual $\Gamma_{\gamma 0}$ are assumed to follow [15,16], and because of the strong correlation between the excitation and ground-state gamma-decay process [16]. In the usual model of the inelastic excitation process, the excitation matrix elements are proportional to the ground-state decay matrix elements. Hence, the treatment of compound decay following inelastic excitation is very similar to compound-elastic gamma-ray scattering. Assuming that the total neutron width $\langle\Gamma_n\rangle_{CN}$ does not fluctuate rapidly and using the fact that $\langle\Gamma_n\rangle \gg \langle\Gamma_\gamma\rangle$, we obtain $S \approx 3.0$ [Refs. 15,16]. Thus, the theoretical ground-state branch, assuming 100% of the EWSR strength is in the range

$$R_{\gamma 0\ \text{Theory}} = 3.1 - 4.2 \times 10^{-4} ,$$

so that the experimental branch corresponds to between 79% and 105% of the EWSR. This value is in excellent agreement with the value from inelastic scattering [1,9].

It is also of great interest to see if gamma-decay branches other than the ground-state decay can be identified. In particular, direct decays to the low-lying collective states, the 3^- state at 2.61 MeV and the 2^+ state at 4.085 are of interest. Figure 6 shows the relative strength of gamma-ray branches to a number of low-lying states. Figures 6b and 6c are for direct decays to the 3^-, 2.61 and 2^+, 4.08 states, respectively. Figure 6d is the relative strengths for decays populating the 4.97-MeV, 3^- state. The yield distributions in Fig. 6, other than the ground-state yield, must be considered semiquantitative, especially where they indicate very small strengths, since adequate background subtraction has not been done. Nevertheless, they are valuable to indicate general features. A few of the more striking aspects include the marked absence of strength to the 2.61 and 4.08 MeV states across the resonance region. A strong yield of decays to the 3^- state at 4.97 MeV (thought to be a noncollective state dominated by a single 1p-1h configuration) is seen to appear at ~ 9 MeV and remains significant across the GQR region. A very similar, though weaker, strength distribution to that shown in

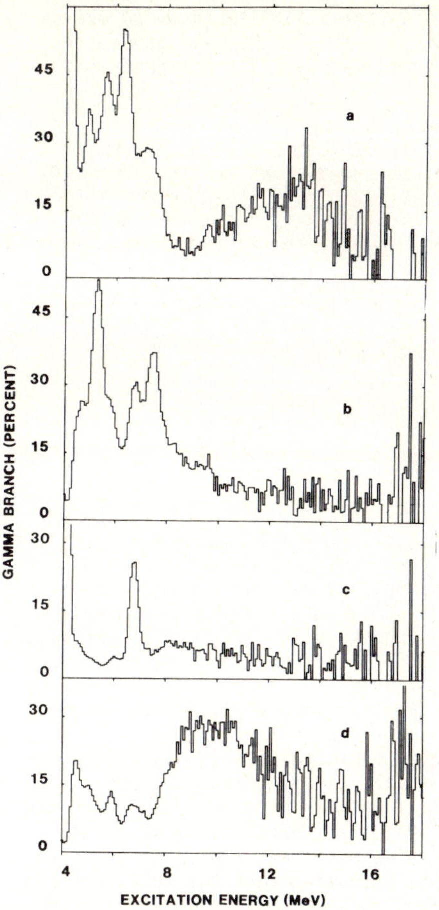

Fig. 6. Relative gamma-decay strengths for transitions to a number of low-lying levels in ^{207}Pb: (a) for ground-state decays; for transitions to the 2.61-MeV, 3^- state; (c) the 4.08-MeV, 2^+ state; (d) the 4.97-MeV, 3^- state.

Fig. 6d is seen for decays to a 5^- state at 3.9 MeV. This indicates the existence of high-spin strength underlying the GQR. A more quantitative treatment of decay branches from the GQR region (i.e., a bin from 9.5 to 11.5 MeV) is shown in Table I. It should be noted that the absence of decay to the 2.6-MeV, 3^- state, which appears remarkable at first sight, agrees with recent calculations [19,20].

The observed ground-state E2 gamma decay strength from the 9.5- to 11.5-MeV region in ^{208}Pb can be accounted for quantitatively using the properties of the GQR obtained from hadron scattering experiments, provided both compound and direct decays are considered. Compound decays dominate in ^{208}Pb, but the relative importance of compound and direct decays can vary widely, even for neighboring nuclei. While the direct decay branch varies slowly from nucleus to nucleus, the average compound neutron widths, which determine the compound gamma contribution, vary greatly. For example, calculations predict that the compound contribution of E2 gamma emission from the GQR in ^{209}Bi should be almost an order of magnitude smaller than in ^{208}Pb.

The very small branch observed to the 2.61-MeV, 3^- state from the GQR region is also worthy of note. Statistical estimates give a 30-40% E1 branch to this state from a 2^+ state at 10.7 MeV. As mentioned earlier, this suppression appears to be well understood [19,20]. However, neither of these calculations considers damping beyond 2p-2h states. The data show that the mechanisms responsible for the suppression of this decay survive to the compound states.

TABLE I. Relative gamma branching to low-lying states in ^{208}Pb from an excitation energy region 9.5-11.5 MeV [E(GQR) ± Γ(GQR)/2]. The 5-7 MeV, 1^- states refers to a group of 1^- states in that region known from (γ,γ') experiments.

Final state energy	Final state spin	Decay branch relative to g.s.		
		Experiment	Ref. 19	Ref. 20
0.0	0^+	1.0	1.0	1.0
2.61	3^-	0.04 ± 0.04	0.027	0.035
4.085	2^+	$0.02^{+0.05}_{-0.02}$		9×10^{-3}
5-7	1^-	1.80 ± 0.04		0.53
4.97	3^-	1.80 ± 0.04		0.53 (~ 4.7) 1.6 (~ 5.5) 0.2 (~ 6.3)

References

[1] Fred E. Bertrand, Annual Review of Nuclear Science 26, 457 (1976). "Giant Multipole Resonances," Proceedings of the Giant Multipole Resonance Topical Conference, Oak Ridge, Tennessee, October 1979, ed. Fred E. Bertrand (Harwood Academic Publishers, New York, 1980). Fred E. Bertrand, Nucl. Phys. A354, 129c (1981).

[2] (a) T. P. Sjoreen, F. E. Bertrand, R. L. Auble, E. E. Gross, D. J. Horen, D. Shapira and B. Wright, Phys. Rev. C 29, 1370 (1984). (b) "Excitation of the High Energy Nuclear Continuum in ^{208}Pb by 22 MeV/Nucleon ^{17}O and ^{32}S," F. E. Bertrand et al., submitted for publication in Phys. Rev. C.

[3] G. F. Bertsch, P. F. Bortignon, and R. A. Broglia, Rev. Mod. Phys. 55, 287 (1983).

[4] P. F. Bortignon and R. A. Broglia, Nucl. Phys. A371, 405 (1981).

[5] G. R. Satchler, Phys. Rep. 14, 99 (1974).

[6] K. Goeke and J. Speth, Annu. Rev. Nucl. Sci. 32, 65 (1982).

[7] G. J. Wagner in Giant Multipole Resonances, ed. F. E. Bertrand (Harwood Academic, New York, 1980), pp. 251-74.

[8] L. S. Cardman, Nucl. Phys. A354, 173c (1981).

[9] F. E. Bertrand et al., Physical Review C, to be published.

[10] M. Jääskeläinen et al., Nucl. Instrum. Methods 204, 385 (1983).

[11] A. Bohr and B. R. Mottelson, Nuclear Structure, Vol. I (Benjamin, Reading, Mass., 1969).

[12] A. Bohr and B. R. Mottelson, Nuclear Structure, Vol. II (Benjamin, Reading, Mass., 1975).

[13] J. R. Beene et al., Phys. Lett. 164B, 19 (1985).

[14] P. A. Moldauer, Phys. Rev. C 11, 426 (1974).

[15] J. E. Lynn, Theory of Neutron Resonance Cross Sections (Oxford University Press, Oxford, 1968).

[16] P. Axel et al., Phys. Rev. C 2, 689 (1970).

[17] S. G. Mughabghab, M. Divadeenam, and N. E. Holden, Neutron Cross Sections (Academic Press, New York, 1981).

[18] D. J. Horen, J. A. Harvey, and N. W. Hill, Phys. Rev. C 18, 722 (1978).

[19] P. F. Bortignon, R. A. Broglia, and G. F. Bertsch, Phys. Lett. 148B, 20 (1984).

[20] J. Speth et al., Phys. Rev. C 31, 2310 (1985).

Coincidence Electron Scattering (e,e′c) in the Giant Resonance Region of ^{28}Si

Th. Kihm[1], *K.T. Knöpfle*[1], *H. Riedesel*[1], *M. Spahn*[1], *P. Voruganti*[1], *H.J. Emrich*[2], *G. Fricke*[2], *R. Neuhausen*[2], *R.K.M. Schneider*[2], and *J.R. Calarco*[3]

[1]Max-Planck-Institut für Kernphysik, D-6900 Heidelberg, F. R. Germany
[2]Institut für Kernphysik, Universität Mainz, D-6500 Mainz, F. R. Germany
[3]University of New Hamsphire, Durham, NH 03824, USA

In the nuclear continuum region, the inherent power of inelastic electron scattering to map out the Fourier transforms of the transition charge and current densities is completely exploited only if the inelastically scattered electron is detected in coincidence with a nuclear decay product c. The coincidence requirement effectively eliminates the strong elastic radiative tail which hitherto plagued the analysis of single arm (e,e′) experiments.

We report on (e,e′c) coincidence studies of the giant resonance (GR) region of ^{28}Si (E_x=14-22 MeV) which have been performed at an incident energy of 183.5 MeV at the Mainz microtron MAMI A. Inelastically scattered electrons were detected at 25°, 30° and 45° with a resolution of 80 keV in the Mainz 180° magnetic spectrometer; the corresponding momentum transfers of q = 0.39, 0.47 and 0.68 fm^{-1} cover the maximum of the E1 form factor and the increasing slope of the E2/E0 form factors. Secondary charged decay products c=p,α were detected in coincidence by 8 ΔE-E surface barrier detector telescopes mounted in a plane rotated 45° around the q axis from the scattering plane; measured angular correlation functions (ACFs) cover the range from -10° to +200° with respect to the direction of q in 10° steps.

Taking advantage of the clean preparation of the nuclear response in the (e,e′c) reaction, a model-independent multipole analysis [1,2] of the 4π angle integrated coincidence spectra is developed which yields not only the decomposition of the E1 and E2/E0 strength distributions (Fig.1) in each measured decay channel c

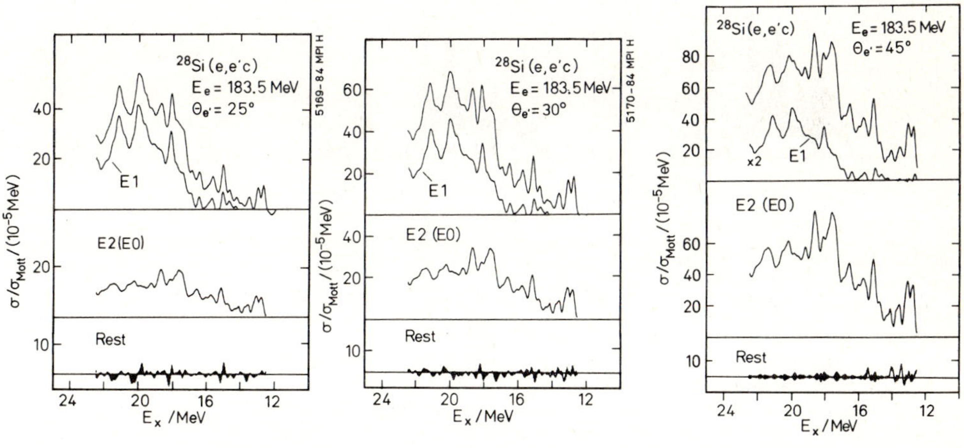

Fig. 1: The 4π integrated ^{28}Si(e,e′c) coincidence cross sections at the three measured momentum transfers in the total c=p,α decay channel (top) and their decomposition into the respective E1 and E2/E0 components. "Rest" denotes the difference spectra between measured yields and the sum of deduced E1 and E2/E0 cross sections.

but also the extraction of the respective form factors. These form factors are largely independent of excitation energy and well reproduced by the Goldhaber-Teller and Tassie model, respectively. The deduced E1 strengths agree well with the results from photodisintegration experiments. The total extracted E2/E0 strength, however, amounts to (35.5±2)% of the electromagnetic E2 energy-weighted sum rule which is about twice the value deduced from studies with isoscalar projectiles. The excess of E2/E0 strength is found predominantly in the proton channels [Γ_p/Γ_α~2.5 and 1; in (e,e'c) v. (α,α'c)] indicating the presence of large isovector E2/E0 contributions in the region of the isoscalar E2/E0 GRs.

The analysis of the (e,e'α_0) angular correlation functions (ACFs) separates E0, E1 and E2 excitations unambiguously only if one assumes that the E0 form factor increases more strongly with q than does the E1 form factor. For overlapping resonances of <u>pure</u> Coulomb character the α_0 ACF is given in the centre-of-mass system of the decaying nucleus by [3]

$$W_\lambda(\theta,q,\omega) = | \Sigma_\lambda \hat{\lambda}\cdot C\lambda(q,\omega)\cdot P_\lambda(\cos\theta) |^2 \quad \text{with} \quad \hat{\lambda} = (2\lambda + 1)^{1/2}.$$

Here, $P_\lambda(\cos\theta)$ is the Legendre polynomial of order λ where the polar angle θ is measured with respect to the direction of momentum transfer. The quantity $|C\lambda(q,\omega)|$ represents--apart from known kinematical factors--the absolute value of the reduced Coulomb matrix element of multipole order λ. In the present analysis the sum is extended over λ values of 0, 1, and 2. The fit to the experimental ACFs thus determines in each 80 keV wide energy bin ω five parameters: the absolute values of the three complex amplitudes $C\lambda(q,\omega)$ and two relative phases, e.g., $\delta_{01}(\omega)$ and $\delta_{02}(\omega)$ between C0 and C1 and C2, respectively. Due to the non-linear nature of the theoretical ACF, only $|C2(q,\omega)|$ is uniquely determined, while several different solutions do exist [4] for $C0(q,\omega)$ and $C1(q,\omega)$ which can be found analytically from a first solution obtained in the fit procedure. In general, we obtain two solutions for the E0 and E1 components, respectively (Fig. 2). Combining the results obtained at $\theta_{e'}$=25° and 30°, one E0 and E1 solution, respectively, can be discarded by

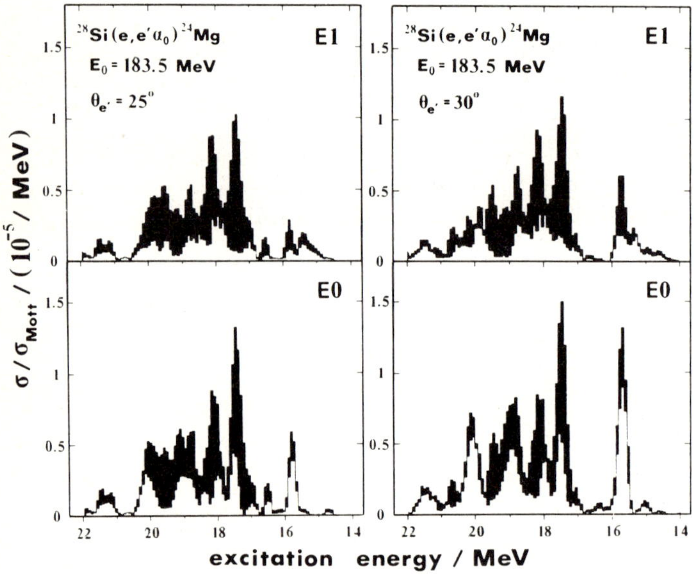

Fig. 2: The two solutions (separated by the black areas) for the E1 and E0 strength distributions deduced from the analysis of the (e,e'α_0) angular correlation functions measured at electron scattering angles of 25° (left) and 30° (right) [4].

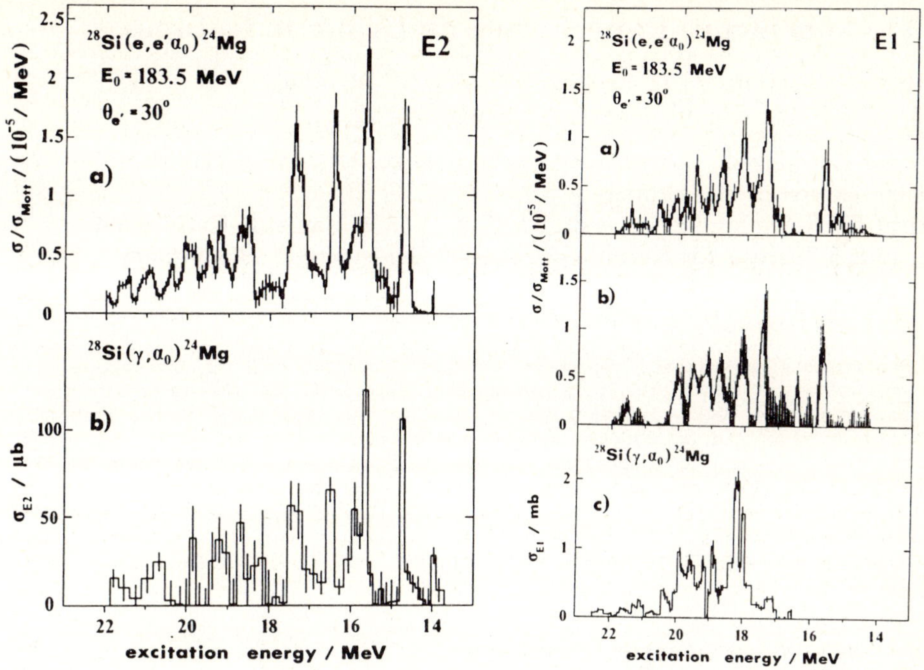

Fig. 3: a) The E2 and E1 strength distributions in ^{28}Si from the analysis of the (e,e'α_o) ACFs. The E2 result is compared with the E2 strength distribution deduced from the ^{24}Mg(α,γ_o)^{28}Si capture reaction [5]. The E1 result is compared with b) the corresponding result from a multipole analysis of the 4π integrated (e,e'α_o) coincidence cross sections and c) the E1 distribution from a ^{28}Si(γ,α_o)^{24}Mg study [6].

demanding that the E1 form factor increase more slowly with increasing q than does the E0 form factor. The resulting E1 distribution (Fig. 3) is in excellent agreement with the E1 strength distribution deduced from the multipole analysis mentioned above and a ^{28}Si(γ,α_o) study with tagged photons [6]. The corresponding E0 strength distributions are represented by the lower yield curves in Fig. 2. Between $14 < E_x < 22$ MeV, about 20% and 50% of the total observed (e,e'α_o) coincidence cross section arise from E0 and E2 excitations (Fig. 3), respectively. In terms of the respective energy-weighted sum rules, the E0 and E2 giant resonances exhibit thus in ^{28}Si a coupling to the α_o channel of comparable strength. Both being $2\hbar\omega$ excitations, this result points to a similar microscopic structure of both resonances.

1 Th. Kihm, Ph.D. Thesis, Heidelberg (1985).
2 Th. Kihm, K.T. Knöpfle, H. Riedesel, P. Voruganti, H.J. Emrich, G. Fricke, R. Neuhausen, and R.K.M. Schneider, Phys. Rev. Lett. 56, 2789 (1986).
3 W.E. Kleppinger and J.D. Walecka, Ann. Phys. 146, 349 (1983).
4 M. Spahn, Diplomarbeit, Heidelberg (1986), and to be published.
5 E. Kuhlmann, K.A. Snover, G. Feldmann, and M. Hindi, Phys. Rev. C27, 948 (1983).
6 R. L. Gulbranson L.S. Cardman, A. Doron, A. Erell, K.R. Lindgren, and A.I. Yavin, Phys. Rev. C27, 470 (1983).

(e,e'f)-Coincidence Experiments on Uranium Isotopes

Th. Weber[1], *R.D. Heil*[1], *U. Kneissl*[1], *W. Pecho*[1], *W. Wilke*[1], *H.J. Emrich*[2], *Th. Kihm*[3], *and K.T. Knöpfle*[3]

[1]Inst. für Kernphysik, Universität Giessen, Leihgesterner Weg 217, D-6300 Giessen, F. R. Germany
[2]Institut für Kernphysik, Universität Mainz, D-6500 Mainz, F. R. Germany
[3]Max-Planck-Institut für Kernphysik, D-6900 Heidelberg, F. R. Germany

(e,e'f)-coincidence experiments represent the most powerful tool to investigate the decay properties of giant multipole resonances, especially of the isoscalar giant quadrupole resonance (GQR), in heavy nuclei. Besides the advantages of the inelastic electron scattering, the coincidence between the fission fragments and the scattered electron causes a complete suppression of the huge radiation tail. The study of the fission decay of giant resonances in heavy nuclei provides interesting information about the coupling of the collective phenomena of fission and giant resonances. In particular the fission decay of the GQR has been subject of controversial experimental studies, using hadrons [1] and electrons (inclusive (e,f) [2,3] and exclusive (e,e'f) [4,5]) as probes.

Our exclusive electro-fission experiments have been performed at the 183 MeV stage of the CW-Mainz-Microtron (MAMI) using currents of typical 20 μA. A fission fragment detector device (PPAC-Ball [6]), consisting of 32 parallel plate avalanche counters covering nearly 2π with sufficient angular resolution, enabled a systematic, high resolution study of fission fragment angular distributions in a wide range of excitation energy ω and momentum transfer q. Our experiments have been performed for 4 q-values (0.62, 0.47, 0.235, 0.138 fm^{-1}) corresponding to the maxima of the E3, E2/E0, E1 form factors and a q-value near to the photon point. Furthermore, detailed information

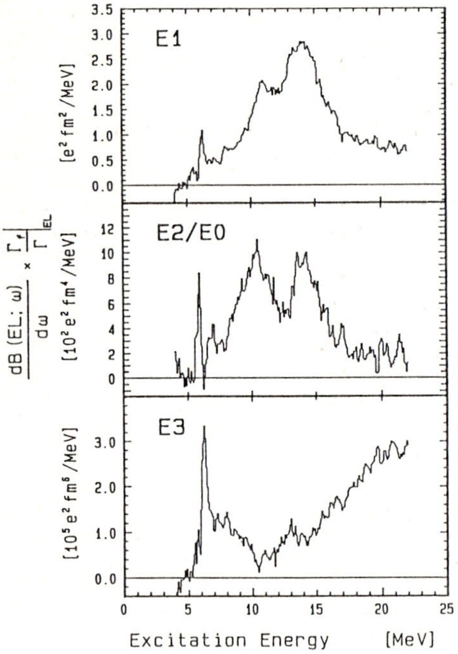

Fig. 1: Decomposed E1, E2/E0, and E3 strength distribution times the corresponding branching ratio for the fission decay of ^{238}U

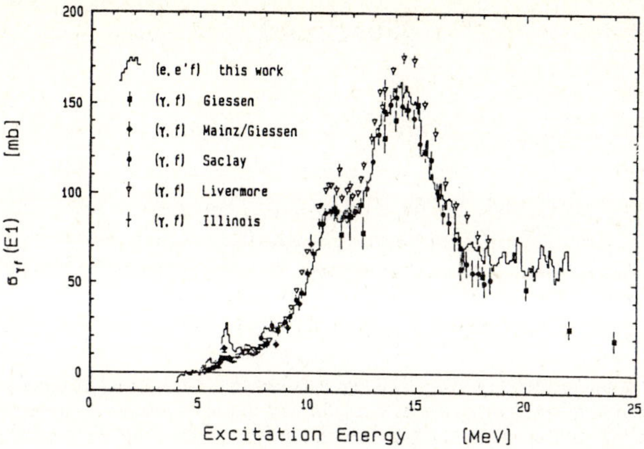

Fig. 2: Comparison of the photo-fission cross section, determined directly from the decomposed E1 strength distribution with photo-fission data from literature [8-11]

about the mass asymmetry of the fission fragments has been obtained by time-of-flight measurements.

The coincidence cross sections for ^{238}U measured for excitation energies from 4 to 22 MeV show a pronounced anisotropic fission fragment angular distribution near the barrier at about 6 MeV and distinct increases above the onsets of the higher chance fission thresholds. The prominent fission barrier peak shows an angular distribution very similar to the pattern one would expect for the quantum numbers L=2, M=0, K=0, which could well be understood in the channel formalism of fission.

With a model independent decomposition procedure [7] we extracted multipole strength distributions for E1, E2/E0, and E3, shown in fig. 1. From the determined E1 strength function we could directly calculate the corresponding E1 photo-fission cross section, which is in very good agreement with several photo-fission data from the literature [8-11], as indicated in fig. 2. The pronounced bump arround 10 MeV in the E2/E0 strength function shows a concentration of E2/E0 strength, which exhausts about 28 % of the E2 energy weighted sum rule (EWSR) between 8 MeV and 12.2 MeV, assuming that all observed strength corresponds to E2, since E2 and E0 excitations cannot be disentangled in our analysis due to their similar form factor shapes. This exhaustion of the E2 sum rule is higher than the results of previous (e,e'f)-experiments [4,5] using the traditional, model dependent multipole decomposition analysis. However, we have strong confidence in our analysis, since the extrapolation to the photon point results for the E1 strength distribution in an excellent agreement with photo-fission data [8-11].

First experiments on ^{235}U have been performed, which show a pronounced odd-even effect near the barrier and second chance fission threshold as compared to ^{238}U. A detailed analysis of the fragment angular distributions and mass distributions is not yet completed.

1 J.v.d. Plicht et al., Nucl. Phys. A346 (1980) 349
2 J.D.T. Arruda Neto et al., Nucl. Phys. A349 (1980) 483
3 H. Ströher et al., Phys. Rev. Lett. 47 (1981) 318
4 D.H. Dowell et al., Phys. Rev. Lett. 49 (1082) 113
5 K.A. Griffioen et al., Phys. Rev. Lett. 53 (1984) 2382
6 J. Drexler et al., Nucl. Instr. and Meth. 220 (1984) 409
7 Th. Kihm, K.T. Knöpfle, priv. communication
8 H. Ries et al., Phys. Rev. C 29 (1984) 2346
9 A. Veyssière et al., Nucl. Phys. A199 (1973) 45
10 J.T. Caldwell et al., Phys. Rev. C 21 (1980) 1215
11 P.A. Dickey, P. Axel, Phys. Rev. Lett. 35 (1975) 501

The Isovector E2 Resonance in ^{90}Y Observed in Neutron Radiative Capture

L. Nilsson[1], *S. Crona*[1], *A. Håkansson*[1], *A. Likar*[1,*], *A. Lindholm*[1], *N. Olsson*[1,+], *I. Bergqvist*[2], *and R. Zorro*[2]

[1]Tandem Accelerator Laboratory, Box 533, S-751 21 Uppsala, Sweden
[2]Department of Physics, University of Lund, S-Lund, Sweden

Giant resonances in nuclei are most commonly studied by inelastic hadron scattering. In this kind of experiments isoscalar resonances are strongly populated in contrast to isovector resonances, for which other means of excitation are normally required. The charge exchange reaction offers a good possibility to selectively excite isovector excitations. For the investigation of the isovector E2 resonance nucleon radiative capture has turned out to be a valuable tool. This has been shown in previous work, see e g [1-3] and will be illustrated in this report by the ^{89}Y(n,γ)^{90}Y reaction.

Monoenergetic neutron beams in the energy range 12 - 27 MeV are produced by the ^{2}H(d,n)^{3}He and ^{3}H(d,n)^{4}He reactions in gas targets at the 6 MV tandem accelerator at Uppsala. The neutron beam at forward angles irradiates a cylindrical yttrium sample, 6 cm in diameter and 8 cm long, placed about 15 cm from the gas target. Time-of-flight techniques with pulsed beam (frequency 2 MHz, pulse width 2 ns) are used to distinguish capture gamma rays from scattered neutrons in the gamma-ray detector, which is positioned in a heavy shielding of lead and borated paraffin about 1 m from the sample. The gamma-ray detector is a large (35 cm long by 24 cm in diameter) anti-coincidence shielded NaI crystal with excellent energy resolution (1.5 % at 22 MeV gamma-ray energy).

A signature for radiative capture via the isovector E2 resonance is the fore-aft asymmetry in the angular distribution of the gamma radiation to low-lying single-particle states in the residual nucleus. The asymmetry is caused by the interference between the resonant E2 amplitude and the E1 amplitude from the high-energy tail of the giant dipole resonance (and direct capture). By measuring three-point angular distributions (θ = 55°, 90° and 125°) the intensity ratios

$$A_1 = \frac{I(55^\circ) - I(125^\circ)}{I(55^\circ) + I(125^\circ)} \quad \text{and} \qquad (1)$$

$$A_2 = 2 - \frac{4I(90^\circ)}{I(55^\circ) + I(125^\circ)} \qquad (2)$$

are determined, where $I(\theta)$ is the measured intensity at the reaction angle θ. These ratios obtained from experiments at seven neutron energies in the range 12 - 27 MeV are compared with calculations based on the direct-semidirect capture model [4,5] with complex particle-vibration coupling [6]. The intercomparison of the data and the calculations is shown in Fig. 1 for different final states. The experimental data are compatible with a resonance energy of 26±1 MeV, a width of 11±2 MeV and with complete exhaustion of the lower isospin ($T_<$) part (88 %) of the isovector E2 sum rule.

---

* Permanent address: Institute Jozef Stefan, University E Kardelj Ljubljana, Yugoslavia
\+ Permanent address: Studsvik Science Research Laboratory, Nyköping Sweden

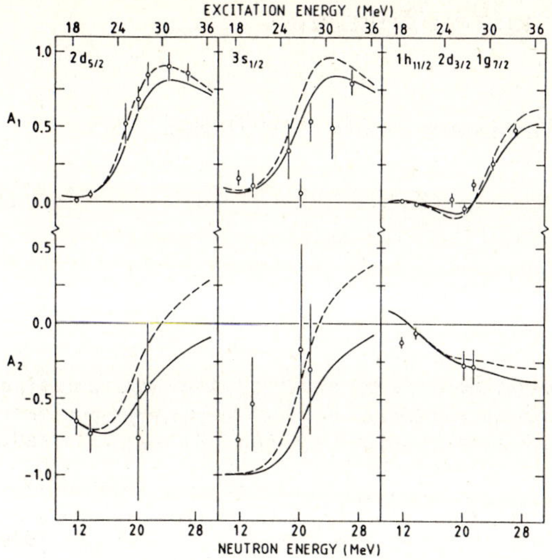

Fig. 1. Angular distribution ratios A_1 and A_2 (defined by eq. (1) and (2)) for the $^{89}Y(n,\gamma)^{90}Y$ reaction to various single-particle states in the residual nucleus. Open circles represent the experimental data. The curves are calculated on the basis of the direct-semidirect capture model with resonance parameters given in the text and with a resonance strength of 100 % (full curve) and 150 % (dashed curve) of the $T_<$ part of the isovector E2 sum rule.

References:

1. F S Dietrich, D W Heikkinen, K A Snover and K Ebisawa, Phys Rev Letters 47, 156 (1977)
2. D M Drake, S Joly, L Nilsson, S A Wender, K Aniol, I Halpern and D Storm, Phys Rev Letters 47, 1581 (1981)
3. I Bergqvist, R Zorro, A Håkansson, A Lindholm, L Nilsson, N Olsson and A Likar, Nucl Phys A419, 509 (1984)
4. G E Brown, Nucl Phys 57, 339 (1964)
5. C F Clement, A M Lane and J R Rook, Nucl Phys 66, 273, 293 (1965)
6. M Potokar, Phys Letters 46B, 346 (1973)

Giant Dipole Resonances in Excited Nuclei

K.A. Snover

Department of Physics, University of Washington, Seattle, WA 98195, USA

Since 1981 [1], many experiments have demonstrated the universal nature of the giant dipole resonance (GDR) in the high energy γ-decay of highly excited, equilibrated nuclei formed in heavy ion reactions. Typically the GDR is excited along with other degrees of freedom, so that its decay leaves the nucleus in an excited state. The signature of GDR decay is high energy (E_γ ~ 10–30 MeV) γ-emission. As in photoabsorption in the giant resonance region, other multipoles contribute only weakly, while nonstatistical emission processes such as pre-equilibrium giant resonance excitation and/or nuclear bremstrahlung are appreciable only at higher bombarding energies. The effect of GDR decay shows up as a "bump" for $E_\gamma \sim E_D$ (~ $80A^{-1/3}$ MeV in medium-heavy nuclei) on an otherwise exponentially decaying spectrum, and the shape of the bump reflects the average shape of the GDR built on excited states, which may be extracted by a statistical model analysis of the measured spectrum shape. Statistical evaporation of high energy γ rays with $E_\gamma \sim E_D > E_n \cong B_n + 2T$ (B_n=neutron binding energy, T=temperature) is favored to occur at the earliest stage of the evaporation process; as a result, the spectrum shape is characteristic of the properties of the highly excited nucleus. Nuclear deformation plays a key role in determining the GDR shape, so that GDR shape analyses of decays of highly excited nuclei provide unique information about deformation of hot nuclei.

In this paper I concentrate on recent experiments, mostly from Seattle, which demonstrate the systematic behavior of the GDR in excited nuclei at moderate temperature and spin. I also briefly discuss the issue of nonstatistical γ-ray emission in light complex-particle induced reactions. For a general review of giant resonances in excited nuclei, including (p,γ) studies, the reader is referred to ref. 2.

1. Systematics of GDR Decay in Excited Nuclei

Although a number of experimental results of GDR decay in excited, equilibrated nuclei have been published [see ref. 2], only recently has an understanding of the systematics emerged from a series of experiments at moderate temperature and spin performed at Seattle [3]. Decays of compound nuclei from $^{46}Ti^*$ to $^{166}Er^*$ were produced in various heavy ion reactions, using primarily ^{12}C and ^{18}O

projectiles. Inclusive γ-ray singles spectra were measured in a large NaI spectrometer, with pulsed-beam time-of-flight used to separate γ-rays produced in the target from background events. The NaI gain was stabilized to better than 1% using a stabilized LED, and was calibrated using high energy γ-ray lines from $^{11}B(p,\gamma)$. The E_γ dependence of the NaI lineshape and absolute efficiency was measured to better than ±10% using the $^{12,13}C(^2He,p\gamma)$ and $^{12}C(p,\gamma)^{13}N$ reactions. These considerations together with a consistent statistical model analysis of all the data, including least-square fitting of GDR parameters, resulted in a reliable extraction of GDR properties for the wide variety of reactions studied.

Many of the important features of these reactions are illustrated in Fig. 1, which shows decays of $^{76}Se^*$ (45.2 MeV) and $^{90}Zr^*$ (49.9 MeV) compound nuclei formed in $^{12}C+^{64}Ni$ and $^{18}O+^{72}Ge$ reactions, respectively [4]. The top panels show the measured spectral shapes along with the result of Cascade statistical model calculations including a GDR strength function of the form

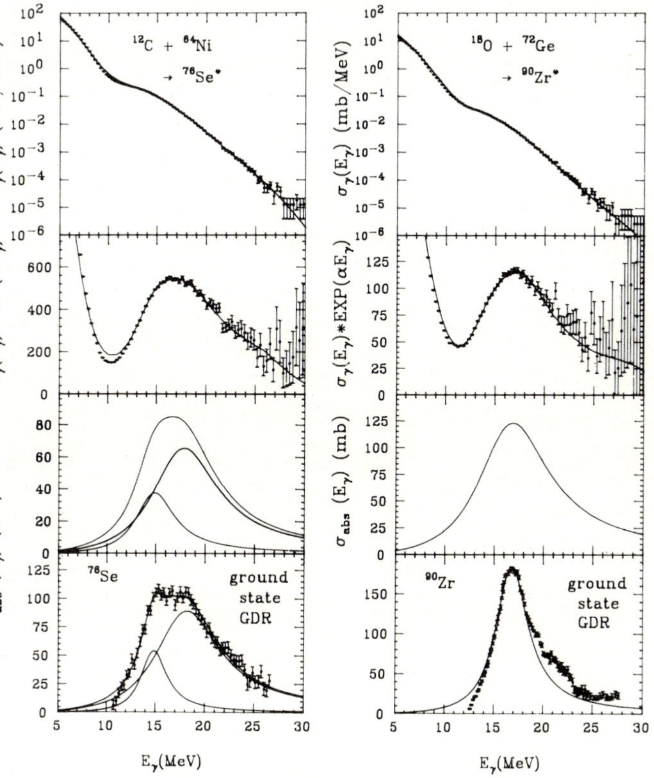

Fig. 1 Top panel: measured spectra and fitted Cascade calculations for the indicated reactions. Second panel: spectra and fitted calculations multiplied by $\exp(\alpha E_\gamma)$, α^{-1}=1.60 MeV for $^{76}Se^*$ and 1.71 MeV for $^{90}Zr^*$. Third panel: fitted $\sigma_{abs}(E_\gamma)$. Bottom panel: g.s.-GDR results for $\sigma_{abs}(E_\gamma)$

$$d\Gamma_\gamma/dE_\gamma = (\pi\hbar c)^{-2}\sigma_{abs}(E_\gamma)\ (E_\gamma^2/3)\rho_f/\rho_i$$

where ρ_f/ρ_i is the ratio of final to initial level density, and $\sigma_{abs}(E_\gamma)$, the average cross section for the inverse process of γ absorption by the ensemble of excited states in the parent and nearby daughter nuclei populated by high energy γ decay, is assumed to have a 1- or 2-component Lorentzian form:

$$\sigma_{abs}\left[E_\gamma\right] = 60\ \text{MeV-mb}(2/\pi)(NZ/A)\ \sum_{j=1}^{2} S_j\Gamma_j E_\gamma^2\left[\left[E_\gamma^2-E_D^2\right]^2+E_\gamma^2\Gamma^2\right]^{-1}$$

The calculated curve folded with the detector response is least-squares fitted to the data in the high energy region by varying the GDR energy, width and strength. Fig. 1 (second panel) shows both data and calculated spectra multiplied by $\exp(\alpha E_\gamma)$ for display purposes - the exponential factor α has been adjusted so that the shape of the multiplied spectra is similar to the shape of $\sigma_{abs}(E_\gamma)$ determined from the fit to the data. The actual fitted $\sigma_{abs}(E_\gamma)$ is shown in the third panel, while the bottom panel shows $\sigma_{abs}(E_\gamma)$ results for the g.s.-GDR.

Now ^{90}Zr and neighboring nuclides are spherical at low energy. The g.s.-GDR in ^{90}Zr shows a narrow ($\Gamma \sim 4$ MeV) single-Lorentzian shape (excluding the small $T_>$ peak), while the decays of $^{90}Zr^*$ (49.9 MeV) show a strongly broadened (Γ=8.75 MeV) GDR with about the same resonance energy and strength. On the other hand, ^{76}Se is a nucleus for which strong quadrupole deformation is important in the low-energy level spectra, and for which the g.s.-GDR is split into 2 components due to deformation. The average GDR shape deduced from decays of $^{76}Se^*$ also requires 2 components and, remarkably, agrees within experimental error with the g.s.-GDR, both in overall shape and in the shape and strength of the individual components (see Fig. 1).

The effect of strong deformation on the GDR shape in excited nuclei is clearly evident in the decays of $^{166}Er^*$ (49.2 MeV) and $^{160}Er^*$ (43.2 MeV) nuclei formed in $^{12}C+^{154,148}Sm$ reactions [5]. These rare earth nuclei behave at low energy like good rotators, with prolate deformations β=0.30 and 0.25, respectively. Again, within experimental error the GDR shapes deduced from decays at high excitation energy are the same as the g.s.-GDR shapes measured in photoabsorption on rare-earth nuclei with similar low-energy deformation. These results have several important implications: 1) the deformation of these nuclei remain prolate and essentially unchanged in magnitude up to temperatures $T \cong \sqrt{E_f/a} \sim 1$ MeV and spins $I \sim 0\text{-}25\ \hbar$, 2) all GDR properties, such as centroid energy, strength and width remain unchanged as well. In particular, the spreading width due to mixing of the GDR with more complicated configurations has not changed appreciably compared to the g.s.-GDR, even though the excited nucleus decays involve much higher excitation energy. Recent GDR spreading width calculations in ^{10}Zr at comparable temperature are consistent with this conclusion [6].

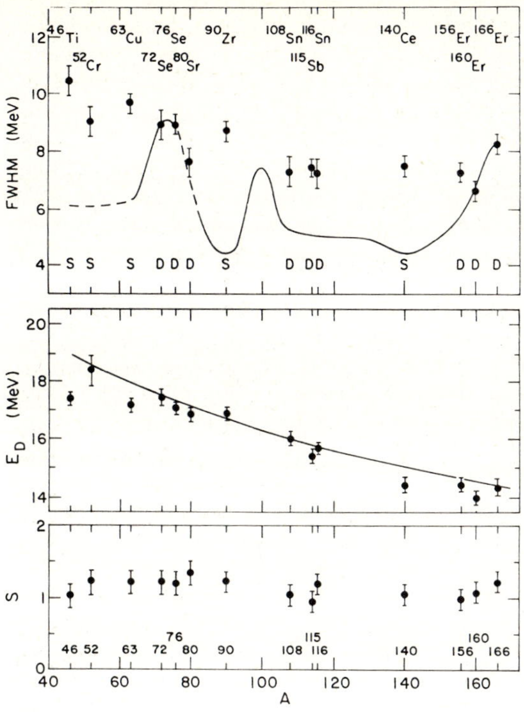

Fig. 2 Systematics of the GDR in excited nuclei at T=1-2 MeV and I=0-25 ℏ as a function of mass. Bottom: resonance strengths in units of the classical E1 sum rule. Middle: resonance energies together with a curve representing g.s.-GDR energies. Top: resonances widths (FWHM) together with a curve approximating the g.s.-GDR width for nuclei of similar mass and isotope. S and D indicate whether a single or double Lorentzian shape was required to fit the excited-nuclei data. The measured compound nuclei are shown.

A summary [3] of some of the basic features of the GDR in excited nuclei from A=46 to 166 is shown in Fig. 2. Most of the compound nuclei were formed with spins ~ 0-25 ℏ and initial excitation energies E_i ~ 40-52 MeV, corresponding to mean final-state temperatures (following γ-decay) ranging from ~ 1 MeV in the heavy cases to ~ 2 MeV in the lightest case. Spectra which could be fitted with a single Lorentzian are indicted by an S; other cases required a double Lorentzian as indicated by a D. The plotted quantities are the total GDR strength, the mean resonance energy and the FWHM of the fitted $\sigma_{abs}(E_\gamma)$ (for a single Lorentzian fit these are S, E_D and Γ).

In the high-energy region of the γ-ray spectra, the yield scales as the product $\sigma_c \cdot S$ of the fusion cross section σ_c and the sum rule strength fraction S. The uncertainty in the extracted strengths is estimated to be ±15-20%, due to ±10-15% uncertainty in σ_c and ±10% for detector efficiency. The strengths for A=46 to 166 have been multiplied by factors of 1.6 to 1.07 to account for the neglect of isospin in the statistical model fits, as estimated from calculations with an isospin-dependent version of CASCADE [7]. The measured strengths are essentially all consistent with 1 classical dipole sum rule, in good agreement with ground state systematics, indicating that within experimental error, these resonances fully equilibrate before decay, and essentially all of the E1 strength is accounted for by the assumed Lorentzian shapes.

The GDR resonance energies are shown along with a curve which fits ground-state GDR energies fairly well in this mass region. The measured energies agree quite well with the curve except for the light masses, where isospin splitting becomes important. The heavy-ion entrance channels always have isospin $T=T_3$, while the GDR built on a $T=T_3$ level will have components with $T=T_3$ and T_3+1. Ground-state systematics lead to estimated isospin shifts of ~ 1 MeV for A=46-63, which explains most of deviations from the curve for the light masses.

In the width plot (top of Fig. 2) the smooth curve is an approximation to the ground-state GDR FWHM, which is well-known to be narrow in the region of spherical nuclei and broadened primarily by quadrupole deformation in other mass regions. The observed GDR FWHM in excited nuclei varies more or less smoothly with mass, even though the single- and the double-Lorentzian shapes are significantly different in the various nuclei. The widths shown in Fig. 2 are generally broader than the ground-state GDR widths observed in the same or in neighboring nuclei, except for the three strongly deformed cases - $^{166}Er^*$, $^{160}Er^*$, and $^{76}Se^*$. The $^{166}Er^*$ and $^{160}Er^*$ cases, as discussed above, are clear examples of the persistence of ground-state-like deformation at high energies. Thus it is tempting to infer that nuclear quadrupole deformation plays a fundamental role in determining the GDR shapes in other excited nuclei as well.

Strong support for this conjecture is contained in the calculations of GALLARDO _et al_. [8]. The basic idea is that the ensemble of excited nuclear states characteristic of a nucleus in thermal equilibrium at finite temperature contains a distribution of deformations, and thermal averaging over such a deformation distribution leads to substantial broadening of the GDR. An essential element in this argument is that the nuclear potential energy as a function of the quadrupole shape degrees of freedom becomes rather flat at moderate temperature and spin for nuclei which are spherical or near-spherical at low energies. This appears to be the case for $^{108}Sn^*$, where the calculated [8] and measured [9] broadening appear comparable.

For all of the double-Lorentzian examples in Fig. 2, the fitted strength ratios favor prolate-like shapes ($S_{upper}/S_{lower}>1$). On the other hand, higher angular momenta should tend to drive the nuclear shape toward oblate deformation and increased temperature is expected to hasten this process. Indeed, evidence has been presented for oblate-like shapes in Er isotopes at somewhat higher spin [9]. For the shapes represented in Fig. 2, where angular momentum probably plays a minor role, the gradual increase in the FWHM as A decreases appears to be related to the higher temperatures for the light cases.

The results shown in Fig. 2 for the GDR strength and energy are in good agreement with recent theoretical calculations. The GDR strength is given essentially by the classical dipole sum rule, which is determined by the number of neutrons and protons in the nucleus, and hence is not expected to change in an excited nucleus. The exchange current enhancement factor in the sum rule is calculated to depend very weakly on temperature [11]. From finite temperature RPA calculations, the GDR energy is expected to drop with increasing T, due primarily to the softening of the nucleus surface. However, the calculated drop is only 2-3% at T=2 MeV [11,12].

All results for GDR parameters in excited nuclei depend on assumed level densities. Our results have been obtained using a prescription similar to that of PUHLHOFER [13] and are sensitive primarily to the assumed level density in the "high energy" region $E_x > 80A^{-1/3}$ MeV, for which the Δ's are taken from the Myers droplet model mass formula without the Wigner term and with shell and pairing corrections removed, and the a parameter is given by a=A/8. In spite of years of study, the nuclear level density in the energy range of interest is not firmly established. However, the above choice of parameters is a common one (see e.g., Ref. 14), and in the present work it leads to consistent results for GDR properties in excited nuclei over a wide mass range. Given the growing theoretical consensus that the GDR energy and strength should be stable over a wide range of temperatures and spin, one can turn the problem around and use the agreement between excited-state and g.s.-GDR strengths and/or energies to place limits on acceptable level density parameters. For example, in the heavier nuclei a change of a to A/7 or A/9 leads to ~ ±0.5 MeV change in E_D and ±30% change in S, and both of these changes are about at the limit of acceptability in comparison with g.s.-values. One should note that any level density with the same slope $d\rho/dE$ as that used in the present analyses would yield similar results.

2. The GDR in $^{63}Cu^*$ over a Wide Range of Temperature and Spin

Here I discuss decays of a particular compound nucleus, $^{63}Cu^*$, over a wide range of temperature and spins [15]. The purpose of this experiment was to search for a possible dependence of the GDR energy, width and strength on nuclear temperature and/or spin. Use of different entrance channels permitted the formation of $^{63}Cu^*$ with a variety of different spins and initial excitation energies, ranging from $^{4}He+^{59}Co$ forming $^{63}Cu^*$ at E_i=22.5 MeV and $\ell_0 \cong 4.5$, to $^{18}O+^{45}Sc$ at E_i=77 MeV and ℓ_0=34, where ℓ_0 is the grazing ℓ-value ($\ell_{mean} \cong 2\ell_0/3$). $^{6}Li+^{57}Fe$ and $^{12}C+^{51}V$ reactions were also employed, with several reaction pairs that formed the compound nucleus at the same energy but different spin.

Results of least-squares fitting of Cascade statistical model calculations to the measured spectra are shown in Fig. 3. All spectra were consistent with statistical decay except for $^{4}He+^{59}Co$ and $^{6}Li+^{57}Fe$ reactions at the highest bombarding energies, which were omitted from the figure. The spectra were fitted adequately by a single Lorentzian, for which the deduced strengths, energies

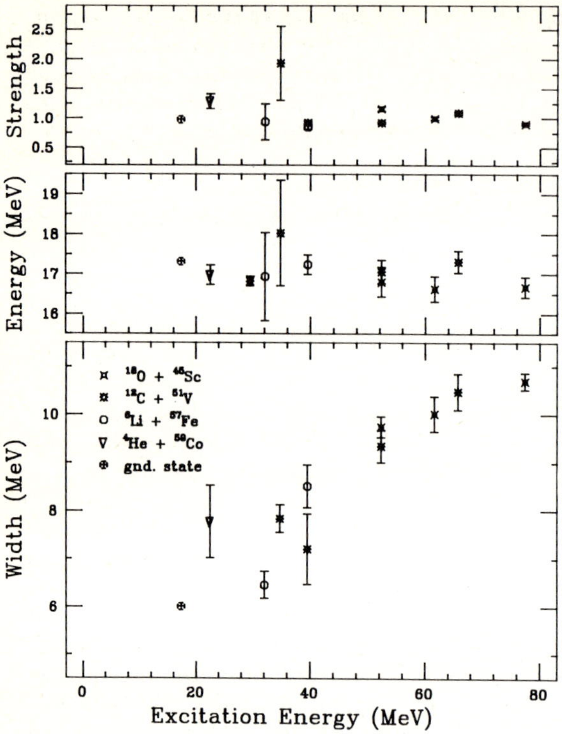

Fig. 3 Fitted GDR energies, strengths and widths for decays of the $^{63}Cu^*$ compound nucleus, as a function of initial excitation energy.

and widths are shown in Fig. 3 as a function of initial excitation energy. Also shown are the parameters for the g.s.-GDR. To a good approximation the GDR strengths and energies are the same for all reactions and energies, and are in agreement with ground-state values. The width, however, increases smoothly by nearly a factor of two in going to the highest excitation energy, which corresponds to T≅2 MeV.

In order to get a better idea of the relative importance of temperature and spin in broadening the GDR, the width is shown in Fig. 4 as a function of (final-state) temperature and initial grazing angular momentum ℓ_0. This plot shows that at large T ~ 1.6-1.9 MeV, the contours of constant Γ appear to be roughly independent of ℓ_0 suggesting that the dominant contribution to the broadening may be due to temperature.

The extensive range of initial excitation energies examined in this study permitted a closer examination of the possible level densities compatible with the measured spectra. Several different prescriptions [14] were judged to be adequate over the energy range studied, and the error bars in Figs. 3 and 4 include the variations due to these different choices.

The absence of any appreciable variations in S or E_D as a function of spin in decays of $^{63}Cu^*$ may be contrasted with the results of a study of decays of $^{162}Er^*$ performed with the

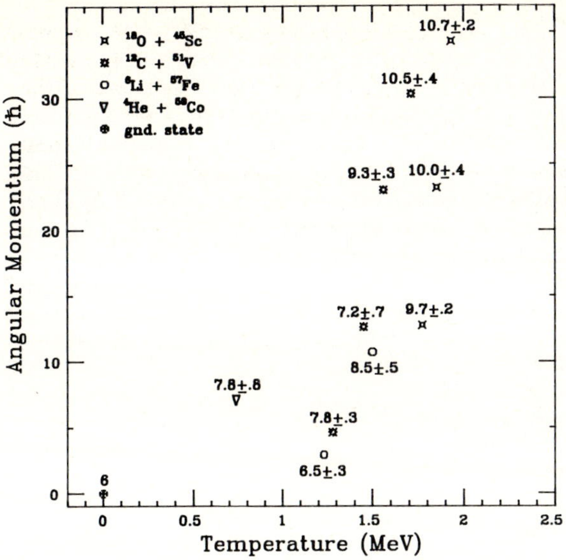

Fig. 4 Fitted GDR width as a function of (final state) temperature and initial grazing angular momentum for decays of $^{63}Cu^*$.

Heidelberg NaI crystal ball [16]. In that experiment, a strong downward shift of E_D of ≈2 MeV was observed as a function of increasing angular momentum from 10 to 50 ħ. The Heidelberg experiment is the only one of its kind, and the result is incompatible with our current theoretical understanding of the physics of rapidly rotating nuclei. Although the $^{63}Cu^*$ decay experiment does not extend to as high a spin as the $^{162}Er^*$ study (<40 ħ compared with 50 ħ), the rotational frequencies in the $^{63}Cu^*$ study are over twice as large as in the $^{162}Er^*$ study, so that the effects of angular momentum on the properties of $^{63}Cu^*$ might be expected to be at least as great as in $^{162}Er^*$.

3. Nonstatistical Emission of High Energy Gamma Rays in ^{3}He- and ^{4}He-Induced Reactions

Heavy ion reactions near the Coulomb barrier produce high energy γ-rays predominantly by statistical decay of the GDR. On the other hand, proton capture is dominated by strong nonstatistical direct and semidirect GDR emission processes. At much higher bombarding energies, heavy ion collisions produce hard photons from what is perhaps a nuclear bremstrahlung process (for recent references, see [2]). The transition between statistical and nonstatistical emission of high energy (E_γ ~ 5-30 MeV) γ-rays has been studied in ^{3}He- and ^{4}He-induced reactions at bombarding energies of 4-9 MeV/nucleon [2,17]. Evidence of nonstatistical behavior in these reactions consists of two types: an enhancement of the high energy γ-ray cross section compared to statistical model predictions, and angular distribution asymmetries about θ_γ=90°.

For decays of an equilibrated system, the angular distribution of the decay products is required to be symmetric about θ_γ=90° in the center-of-mass. In general, an asymmetry about 90° can arise only from interfering channels of opposite parity such as E1 - E2 interference in γ decay. The symmetry requirement for equilibrated decays follows from the fact that opposite parity radiations populating a given final state must come from opposite parity and hence different initial compound states. The assumed absence of phase correlations among different states of an equilibrated system implies that such interferences disappear due to averaging over initial states. The observation of an angular distribution asymmetry in the center of mass thus requires both the presence of interfering radiations of opposite parity plus a pre-equilibrium reaction mechanism with phase coherence between different reaction amplitudes.

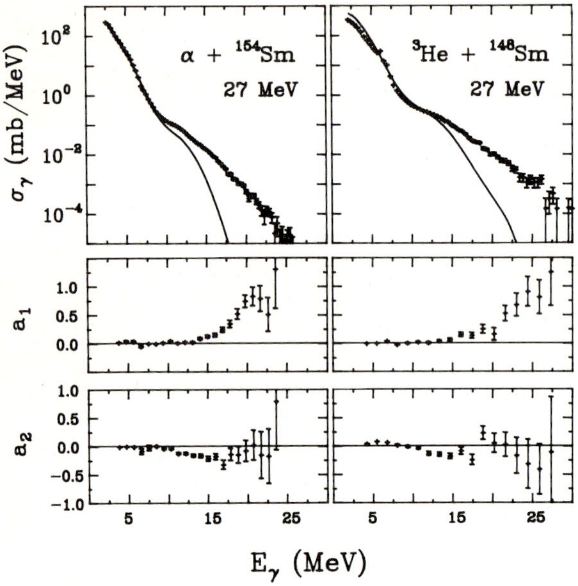

Fig. 5 Measured γ-ray spectra, a_1 and a_2 coefficients for the $^4He+^{154}Sm$ and $^3He+^{148}Sm$ reactions at a bombarding energy E=27 MeV. The solid curves in the top panel are the result of a statistical model calculation.

In light nuclei at low bombarding energies E≈12 MeV, ^{3}He-induced reactions appear statistical whereas at E=27 MeV in both light and heavy nuclei strong nonstatistical contributions are apparent at high E_γ. Alpha-induced reactions appear statistical in light nuclei up to E_α≈27 Mev, while nonstatistical behavior is again evident in heavy nuclei. The situation for ^{3,4}He + Sm at E_α=27 MeV is illustrated in Fig. 5. At high E_γ, these results show both a strong enhancement of the γ yield relative to a statistical model calculation, and a strong forward peaking of the angular distributions in the center-of-mass as indicated by the large positive a_1 coefficient.

The very strong degree of phase coherence required to produce the large observed a_1 coefficients in ^{3}He and α-induced reactions implies that the nonstatistical γ-ray production mechanism most likely involves radiative emission at an early stage of the reaction process. Because the spectral enhancement occurs in the giant resonance region, it seems likely that the GDR and, perhaps, the isovector GQR play a role. Direct and semidirect (one-step doorway) emission, as discussed in Section 2 for nucleon-induced reactions, are possible processes for He-induced reactions. However, recent calculations [17] for 27 MeV $^{3,4}He$ + ^{124}Sn indicate cross sections for $E_\gamma \approx E_D$ which are 1-2 orders of magnitude too low at high E_γ to explain the data, which are similar to the Sm data shown in Fig. 5. Also the calculated cross sections, which are dominated by the direct amplitude, have a relatively flat energy dependence in contrast to the exponential-like fall of the data with increasing E_γ. The amplitude for γ decay of the projectile excited into its GDR by the collision process was also estimated and found to be small. Photon production following projectile breakup, which might be important, has not been calculated.

In conclusion, at sufficiently high bombarding energy, all nucleus-nucleus reactions may be expected to exhibit nonstatistical γ-emission processes. High-energy γ emission may occur by decay of giant resonances produced both in equilibrium and pre-equilibrium processes, as well as by nuclear bremsstrahlung.

I am indebted to my colleagues, including J.A. Behr, G.A. Feldman, E.F. Garman, C.A. Gossett, J.H. Gundlach, H. Glatzel, and M. Kicinska-Habior for their valuable collaborations and for their contributions of results prior to publication.

References:

1. J.O. Newton, B. Herskind, R.M. Diamond, E.L. Dines, J.E. Draper, K.H. Lindenberger, C. Schück, S. Shih, and F.S. Stephens, Phys. Rev. Lett. 46, 1383 (1981).
2. K.A. Snover, Ann. Rev. of Nucl. Part. Phys., Vol. 36, 1986, in press.
3. C.A. Gossett, K.A. Snover, J.H. Gundlach, G. Feldman, J. Behr, and M. Kicinska-Habior, to be published.
4. J.H. Gundlach et al., to be published.
5. C.A. Gossett, K.A. Snover, J.A. Behr, G. Feldman, and J.L. Osborne, Phys. Rev. Lett. 54, 1486 (1985).
6. P.F. Bortignon and R.A. Broglia, Proc. of the Topical Meeting on Phase Space Approach to Nuclear Dynamics, Trieste, Italy, Singapore: World Sci. Pub. Co. (1986); P.F. Bortignon, R.A. Broglia, G.F. Bertsch, and J. Pacheco, to be published (1986).
7. M.N. Harakeh, D.H. Dowell, G. Feldman, E.F. Garman, R. Loveman, J.L. Osborne, and K.A. Snover, Phys. Lett., in press.
8. M. Gallardo, M. Diebel, T. Dossing, and R.A. Broglia, Nucl. Phys. A 443, 415 (1985).
9. J.J. Gaardhøje, C. Ellegaard, B. Herskind, R.M. Diamond, M.A. Delaplangue, G. Dines, A.O. Macchiavelli, and F.S. Stephens, Phys. Rev. Lett. 56, 1783 (1986).

10. J.J. Gaardhøje, C. Ellegaard, B. Herskind, and S.G. Steadman, Phys. Rev. Lett. 53, 148 (1984); J.J. Gaardhøje, Nuclear Structure, 1985 R. Broglia, G. Hagemann, B. Herskind, eds. Amsterdam: North-Holland pp. 519-534 (1985).
11. H. Sagawa and G.F. Bertsch, Phys. Lett. B 146, 138 (1984).
12. M. Barranco, A. Polls, S. Marcos, J. Navarro, and J. Treiner, Phys. Lett. B 154, 96 (1985).
13. F. Puhlhofer, Nucl. Phys. A 280:267 (1977).
14. G. Nebbia et al., to be published.
15. M. Kicinska-Habior, E.F. Garman, C.A. Gossett, K.A. Snover, J. Behr, G. Feldman, H. Glatzel, and J.H. Gundlach, to be published.
16. D. Schwalm, 1984 Erice Summer School Proceedings (1985).
17. J.A. Behr, G. Feldman, C.A. Gossett, J. Gundlach, and K.A. Snover, to be published (1986).

Giant Resonances on Excited States in Deformed Nuclei

V. Kitipova

Institute of Nuclear Research and Nuclear Energy,
Bulgarian Academy of Science, 1784 Sofia, Bulgaria

The interest in giant multipole resonances (GMR) built on excited states increased during the last years. In some recent papers [1–3] experimental results about such giant resonances of deformed nuclei are considered. The object of this work is the theoretical investigation of the giant dipole, quadrupole and octupole resonances built on the lowest excited states of deformed nuclei.

The calculations are performed in the framework of the quasi-particle-phonon model (QPM) [4]. To avoid the description of every level at high energy (whose density grows sharply with the increase of the energy) the method of the strength function [5] is used. An investigation of the GMR requires to take into account complex configurations. Here the wave functions are considered up to two-phonon components

$$|\psi_n(K^\pi)\rangle = \Big(\sum_g C_g^n Q_g^+ + (1/\sqrt{2}) \sum_{g_1 g_2} D^n_{g_1 g_2} Q^+_{g_1} Q^+_{g_2}\Big) |\psi_0\rangle, \tag{1}$$

where C_g^n and $D^n_{g_1 g_2}$ are the coefficients of the one-phonon and two-phonon components. Q_g^+ is the creation operator of the phonon $g = \lambda\mu j$ with energy ω_g, where j is the number of a phonon with multipolarity $\lambda\mu$. ψ_0 is the phonon vacuum. The Hamiltonian of the model is given by [5]

$$H = \sum_g \omega_g Q_g^+ Q_g - (1/\sqrt{2}) \sum_g \sum_{qq'} [\Gamma^g_{qq'} B(qq')(Q_g^+ + Q_g) + \text{h.c.}]. \tag{2}$$

The explicite expressions of the operators $B(qq')$ are given in eq. (5), and the values $\Gamma^g_{qq'}$ which characterize the phonon with energy ω_g can be found in [5]. The reduced probability of an $E\lambda$-transition between the initial state $(I^\pi K)_i$ and the final state $(I^\pi K)_f$ can be written as

$$B_{i\to f}(E\lambda) = (I_i K_i \lambda\mu | I_f K_f)^2 (\langle \psi_f(K^\pi)|\mathcal{M}|\psi_i(K^\pi)\rangle)^2, \tag{3}$$

where $\mathcal{M}$ is the operator of the $E\lambda$-transition and

$$\begin{aligned}\mathcal{M} = \sum_{qq'} \{ &\langle q+|\Gamma(E\lambda)|q'+\rangle [v_{qq'} B(qq') + (1/\sqrt{2}) u_{qq'} (A^+(qq') + A(qq'))] \\ &+ \langle q+|\Gamma(E\lambda)|q'-\rangle [v_{qq'} \overline{B}(qq') \\ &+ (1/\sqrt{2}) u_{qq'} (\overline{A}^+(qq') + \overline{A}(qq'))]\}.\end{aligned} \tag{4}$$

Here the following expressions are used:

$$v_{qq'} = u_q u_{q'} - v_q v_{q'} \qquad u_{qq'} = u_q v_{q'} + u_{q'} v_q$$
$$A(qq') = (1/\sqrt{2}) \sum_{\sigma} \sigma \alpha_{q'\sigma} \alpha_{q-\sigma} \quad B(qq') = \sum_{\sigma} \alpha^{+}_{q\sigma} \alpha_{q'\sigma}$$
$$\overline{A}(qq') = (1/\sqrt{2}) \sum_{\sigma} \alpha_{q\sigma} \alpha_{q'\sigma} \quad \overline{B}(qq') = \sum_{\sigma} \sigma \alpha^{+}_{q-\sigma} \alpha_{q'\sigma} \tag{5}$$

where u_q and v_q are the Bogolubov transformation coefficients, α^{+}_{q} is the creation operator of the quasiparticles and $\Gamma(E\lambda)$ determines the single-particle matrix elements (m.e.). One can write $\mathcal{M}$ as

$$\mathcal{M} = \mathcal{M}_1 + \Delta \mathcal{M} \tag{6}$$

where $\Delta\mathcal{M}$ includes the terms containing $B(qq')$ and $\overline{B}(qq')$. Usually $\Delta\mathcal{M}$ is neglected in the QPM [4,5]. Its contribution to the transition probability will be considered later. For $E\lambda$-transitions between states of the type (1) one obtains [6]

$$\begin{aligned} \langle \psi_f(K^\pi)|\mathcal{M}|\psi_i(K^\pi)\rangle &= (1/\sqrt{2}) \sum_{g} \Big(\sum_{g_f} C^{f}_{g_f} D^{i}_{gg_f} + \sum_{g_i} C^{i}_{g_i} D^{f}_{gg_i} \Big) \\ &\times [(1/Y_g)(2-\delta_{\mu 0})]^{1/2} [e^{(\lambda)}_{eff}(n) X_g(n) + e^{(\lambda)}_{eff}(p) y_g(p) X_g(p)] \\ &= \sqrt{2} \sum_{g} \Big(\sum_{g_f} C^{f}_{g_f} D^{i}_{gg_f} + \sum_{g_i} C^{i}_{g_i} D^{f}_{gg_i} \Big) L_g, \end{aligned} \tag{7}$$

where $e^{(\lambda)}_{eff}(n)$ and $e^{(\lambda)}_{eff}(p)$ are the effective charges and the values Y_i, y^{p}_{i}, $X^{\lambda\mu i}_{p}$, $X^{\lambda\mu i}_{n}$, characterizing the phonons, are given in [4]. Further the strength function $b_{i\to f}(E\lambda, \eta)$ is introduced

$$\begin{aligned} b_{i\to f}(E\lambda, \eta) &= \sum_{i} B_{i\to f}(E\lambda) \varrho(\eta - \eta_i) \\ &= (I_i K_i \lambda\mu | I_f K_f)^2 P_{i\to f}(E\lambda, \eta), \end{aligned} \tag{8}$$

where

$$\varrho(\eta - \eta_i) = \Delta\{2\pi[(\eta-\eta_i)^2 + (\Delta/2)^2]\}^{-1} \tag{9}$$

and Δ is a parameter of averaging. Using a variational principle [4] and the analytical properties of the function (9) one obtains that [6,7]

$$b_{i\to f}(E\lambda, \eta) = (I_i K_i \lambda\mu | I_f K_f)(1/\pi) \mathrm{Im} \left. \frac{\left| \begin{array}{c|c} -K_{00} & -K_{0g_i} \\ \hline -K_{g_i 0} & \|(\omega_g - z)\delta_{g_i g'_i} - K_{g_i g'_i}\| \end{array} \right|}{|(\omega_{g_i} \delta_{g_i g'_i} - K_{g_i g'_i}|} \right|_{z=\eta+i(\Delta/2)} \tag{10}$$

In a similar way one obtains for transitions to the ground state

$$b_{i\to 0^+}(E\lambda,\eta) = (1/\pi)\mathrm{Im}\frac{\left|\begin{array}{c|c} 0 & L_g \\ \hline L_g & \|(\omega_g - z)\delta_{gg'} - K_{gg'}\| \end{array}\right|}{|(\omega_g - z)\delta_{gg'} - K_{gg'}|}\Bigg|_{z=\eta+i(\Delta/2)} \tag{11}$$

where

$$K_{gg'} = \sum_{g_1 g_2} U^g_{g_1 g_2} U^{g'}_{g_1 g_2}(\omega_{g_1} + \omega_{g_2} - \eta)^{-1} \quad ,$$
$$U^g_{g_1 g_2} = (1/2)\langle Q_g H Q^+_{g_1} Q^+_{g_2}\rangle. \tag{12}$$

As is known the Axel-Brink hypothesis states that a GMR can be built on every excited state with the same form as that of the GMR built on the ground state, but shifted upward in the energy of the excited state [8]. Therefore comparing the strength function (10) with the strength function describing the GMR built on the ground state (11) one can verify the Axel-Brink hypothesis.

Numerical calculations of the strength functions for $E\lambda$-transitions ($\lambda<4$) from highly excited states to the lowest vibrational states and also to the ground states of the nuclei ^{176}Hf and ^{156}Gd are performed. The parameters of the average field and the constants of the pairing and multipole-multipole interactions are taken from [9, 10]. The 30 most collective phonons of every multipolarity are taken into account. The strength function (10) for $E1$-transitions to the ground states and also from initial states with $(I^\pi K)_i$ = 3^-3, 3^-2, 3^-1, 1^-1 to the

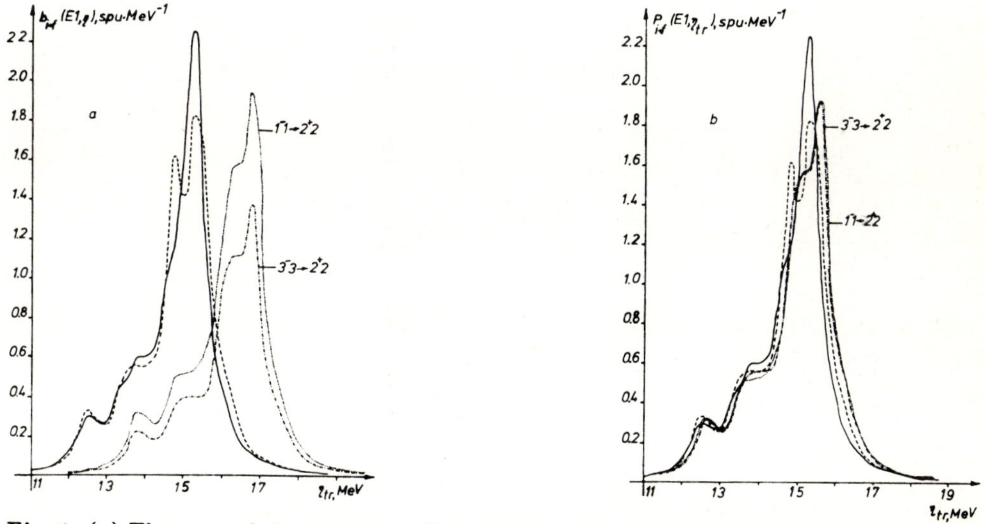

Fig. 1. (a) The strength functions $b_{i\to f}(E1,\eta)$ of the reduced probabilities of $E1$-transitions to the ground state 0^+ (the solid curve shows the calculation taking into account the anharmonicity, the dashed – the calculation in the RPA) and $E1$-decays $3^+3\to 2^+2$, $1^+1\to 2^+2$; (b) the strength functions $P_{i\to f}(E1,\eta_{\mathrm{tr}})$, depending on the transition energy $\eta_{\mathrm{tr}} = \eta - \eta_f$

lowest state with $(I^\pi K)_f = 2^+2$ are calculated. Some of the results for ^{176}Hf are shown in Fig. 1a. One can see that the giant dipole resonance calculated in the RPA also taking into account the anharmonicity, described by $b_{i\to 2^+2}(E1,\eta)$, is moved approximately to 1.2 MeV with respect to $b_{i\to 0}+(E1,\eta)$ (the energy of the state $(I^\pi K)_f = 2^+2$ is about 1.2 MeV). It is difficult to compare the forms of the both curves because of that energy shift and of the difference in the Clebsch-Gordon coefficients. For to make easier the comparison, the strengh functions $P_{i\to f}(E1,\eta_{\rm tr})$ and $P_{i\to 0^+}(E1,\eta_{\rm tr})$ are shown in Fig. 1b. Here $\eta_{\rm tr} = \eta - \eta_f$ is the transition energy. One can see that the curve $P_{i\to 2^+2}$ is near to the $P_{i\to 0^+}(E1,\eta_{\rm tr})$. The results for the strength function of $E2$-transitions to the ground state and to the low-lying excited states with $I^\pi K = 2^+2$, 3^-0 are shown in Fig. 2a. The curves have similar forms. The same result is obtained for the strength functions of $E3$-transitions for one of the projections of the octupole resonance ($K = 0$) built on the ground state and on the low-lying excited state 2^+0 of ^{176}Hf. They are shown in Fig. 2b.

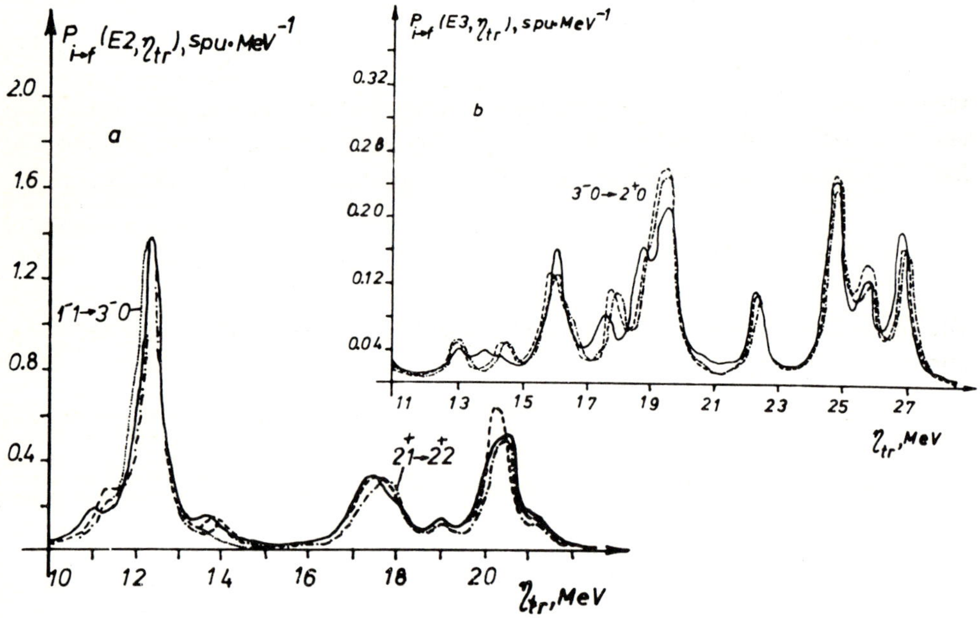

Fig. 2a,b. The strength functions $P_{i\to f}(E\lambda,\eta_{\rm tr})$ of $E\lambda$-transitions to the ground state 0^+ (the indications are the same as in Fig. 1) and to the low-lying excited states $(I^\pi K)_f$: (a) $\lambda = 2$ and $(I^\pi K)_f = 2^+2, 3^-3$; (b) $\lambda = 3$ and $(I^\pi K) = 2^+0$

The numerical estimates of the values of $\langle\psi_f|\mathcal{M}_1|\psi_i\rangle$ and $\langle\psi_f|\Delta\mathcal{M}|\psi_i\rangle$ have shown that the contribution of the second expression to the probabilities for $E\lambda$-transitions is about 4% [11].

From these results one can conclude that for the even-even deformed nuclei for $E\lambda$-transitions going to low-lying excited states with approximately one-phonon structure the hypothesis of Axel-Brink is fulfiled with a good accuracy. Applying that hypothesis to the isobaric analog states one can obtain estimations for the first-forbidden (f.f.) β-decay [1]. So an investigation of such β-decay will give information about the properties of $E1$ charge-exchange modes. The precise inclusion of the pairing correlations is important for the correct description of the transition probabilities [12]. Here the influence of the pairing correlations on the m.e. of the f.f. β_--decay of transitional even-even nuclei with a small deformation is investigated [13]. These nuclei have many quasiparticles in the ground state, therefore the condition for applying of the RPA is violated. Because of that the calculations are performed in the framework of the quasiparticle model [4]. The transition operators of the f.f. β_--decay [14] can be reduced with the help of some approximations to the following two:

$$M(j_v, \lambda = 1) = (1/\sqrt{3})(E_f - E_i - \Delta E_c)M(\varrho_v, \lambda = 1)$$

$$M(\varrho_A, \lambda = 0) = (E_f - E_i - \Delta E_c)M(j_A, \lambda = 0) \ , \tag{13}$$

where ΔE_c is the averaged Coulomb energy and $E_i - E_f$ is the transition energy. Averaging (13) one obtains the following expressions for the relation between the quasiparticle and the single-particle m.e. of $M(\varrho_v, \lambda = 1)$:

$$\langle\psi_f|M(\varrho_v, \lambda = 1)|\psi_i\rangle^2_{\text{q.p.}} = u^2_{q_1}u^2_{r_2}\langle\psi_f|M(\varrho_v, \lambda = 1)|\psi_i\rangle^2_{\text{s.p.}} \ . \tag{14}$$

For to determine the contribution of the pairing correlations to the $M(j_A, \lambda)$ one has to compare the expressions

$$\langle\psi_f|M(j_A, \lambda)|\psi_i\rangle^2_{\text{s.p.}} = 5D^2(m_{l_1}, m_{l_1}) + 18D^2(m_{l_1}, m_{l_1-1}) + 1.17D^2(m_{l_1}, m_{l_1+1})$$

$$\langle\psi_f|M(j_A, \lambda)|\psi_i\rangle^2_{\text{q.p.}} = u^2_{s_1}u^2_{r_2}4.5D^2(m_{l_1}, m_{l_1}) + 16D^2(m_{l_1}, m_{l_1-1}) + 1.17D^2(m_{l_1}, m_{l_1+1}) \ , \tag{15}$$

where the coefficients $D(m_{l_1}, m_{l_2})$ are algebraic expressions of the quantum numbers of the states $|\psi_i\rangle$ and $|\psi_f\rangle$. Here s_1 are the quantum numbers characterizing the state of the neutron taking part in the β_--decay. r_2 are the quantum numbers of the odd proton in the daughter nucleus. A numerical estimate of the influence of the pairing correlations on the m.e. of the f.f. β_--decay of ^{144}Ba is done. It is shown that the transition to the quasiparticle basis leads to a reduction of the β_--decay probabilities by approximately 75%.

References

1. H.V. Klapdor, C.O. Wene: Izv. Akad. Nauk SSSR, Ser. Fiz. **44,** 2 (1980)
2. J.O. Newton et al.: Phys. Rev. Lett. **46,** 1383 (1981)
3. J.L. Egido, P. Ring: Phys. Rev. C**25,** 3239 (1982)
4. V.G. Soloviev: *Theory of Complex Nuclei* (Pergamon Press, Oxford 1976)
5. V.G. Soloviev: Fizika Element. Chastits i Atomn. Yadra **9,** 580 (1978)
6. V. Kitipova et al.: Izv. Akad. Nauk SSSR, Ser. Fiz. **44,** 1915 (1980)
7. V. Kitipova et al.: Izv. Akad. Nauk SSSR, Ser. Fiz. **45,** 1923 (1981)
8. G.A. Bartholomew et al.: Adv. Nucl. Phys. **7,** 229 (1973)
9. G. Kyrchev, L.A. Malov: Izv. Akad. Nauk SSSR, Ser. Fiz **43,** 107 (1979)
10. F.A. Gareev et al.: Fizika Element. Chastits i Atom. Yadra **4,** 357 (1973)
11. V. Kitipova: Bulg. J. Phys. (in press)
12. T. Kondoh, T. Tachibana, M. Yamada: Progr. Theor. Phys. **74,** 708 (1985)
13. V. Kitipova: Bulg. J. Phys. **11,** 582 (1984)
14. A. Bohr, B.R. Mottelson: *Nuclear Structure*, Vol. 1 (Benjamin, New York 1969)

Direct-Semidirect Proton Capture in the GDR Region

T. Rzaca-Urban[1], *G. Szeflinska*[1], *Z. Szeflinski*[1], *Z. Wilhelmi*[1], *H.V. Klapdor*[2], *K. Grotz*[2], and *J. Metzinger*[2]

[1]Institute of Experimental Physics, Warsaw University, Hoza 69, 00-681 Warsaw, Poland

[2]Max-Planck-Institut für Kernphysik, D-6900 Heidelberg, F. R. Germany

Proton radiative capture excites the giant dipole resonance (GDR) predominantly through direct or semidirect processes. At excitation energies below the GDR region contributions from compound nucleus capture are expected to complicate this picture. The recent work of DOWELL et al. [1] has introduced a simple model, which finds the inverse photoproton strength integrated over the GDR to a good approximation proportional to the proton transfer spectroscopic factors.

The aim of the present work is to study the mechanism of radiative capture in medium mass nuclei.

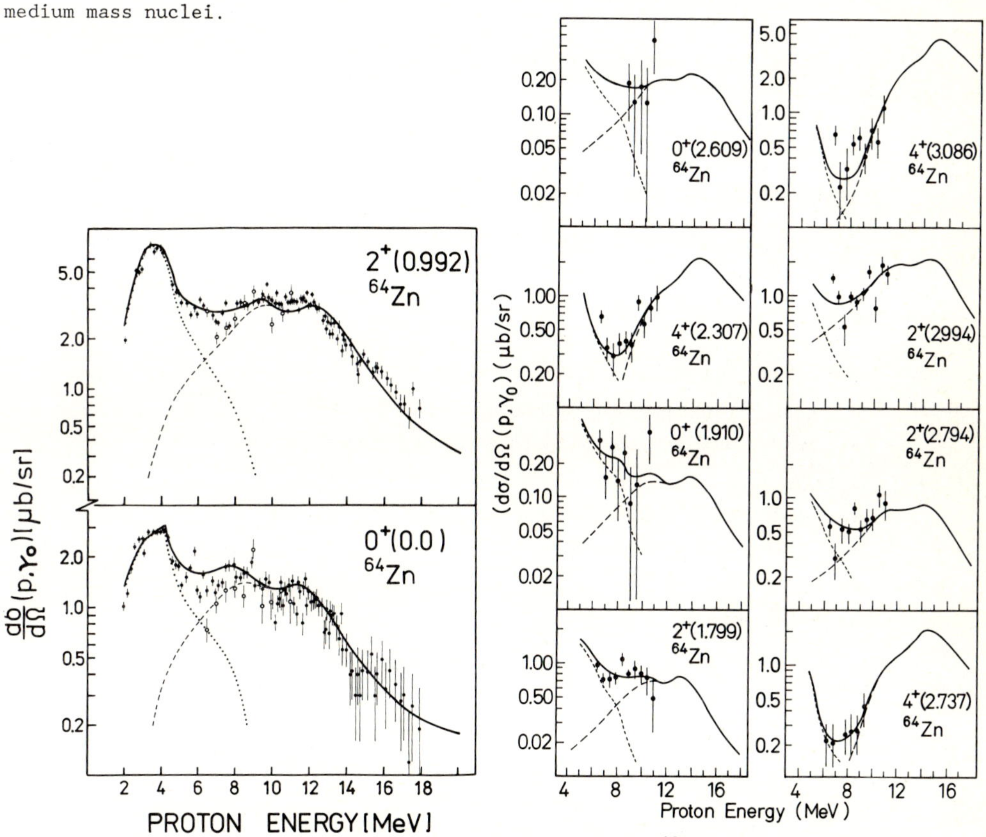

Fig. 1. Differential cross sections at $\Theta = 90°$ for the $^{63}Cu(p,\gamma_i)$ reaction. Calculations of the statistical and DSD contributions are given by the dotted and dashed lines, respectively. The full curve is the sum of both.

The gamma-ray spectra were measured at the Heidelberg EN-Tandem using a Ge(Li) detector surrounded by a big NaI(Tl), optically divided into two parts, operating in the pair spectrometer mode. The differential cross sections for the $^{63}Cu(p,\gamma)^{64}Zn$ reaction measured at $\Theta = 90^{\circ}$ are displayed in Fig. 1. The experimental cross sections are compared with the predictions of statistical and direct-semidirect (DSD) model calculations. The partial cross sections (p,γ_i) were calculated as a sum of statistical and DSD cross sections. The parameters of GDR's built on excited states in ^{64}Zn were assumed to be the same as those determined by LEPRETRE et al. [2] for the ground state of ^{64}Zn.

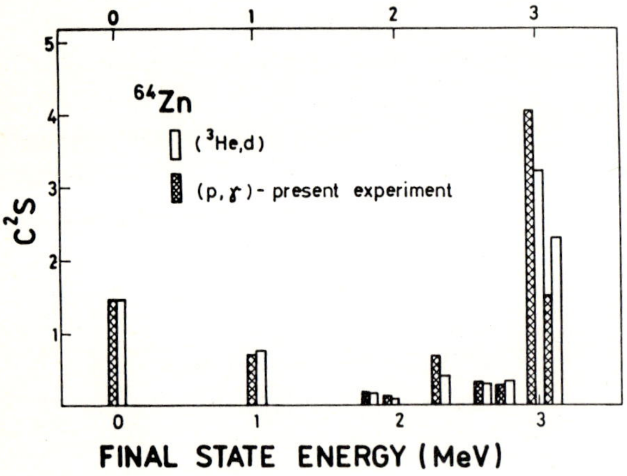

Fig. 2. Comparison of spectroscopic factors obtained from the present experiment with those from the (^{3}He,d) reaction [3].

For proton energies below 4 MeV the statistical process is found to dominate, while above E_p = 8 MeV the DSD capture prevails. Proton transfer spectroscopic factors were determined and compared to those derived from (^{3}He,d) reaction by FORD et al. [3] (Fig. 2).

In order to describe radiative proton capture in the GDR region of medium mass nuclei it is thus necessary to apply model calculations which account for both the DSD and the statistical mechanism of the reaction. The calculations account properly for the contributions of direct interactions to the radiative capture, which allows us to extract the proton transfer spectroscopic factors.

References

1. D.H. Dowell et al.: Phys. Rev. Lett. 50 1191 (1983)
2. A. Lepretre et al.: Nucl. Phys. A175, 609 (1971)
3. J.L.C. Ford et al.: Nucl. Phys. A103, 525 (1967)

Dispersion Relation Analysis of Photonuclear Data*

J. Ahrens[1], *L.S. Ferreira*[2], *and W. Weise*[3]

[1]Nuclear Physics Div., Max Planck Institute for Chemistry, D-6500 Mainz, F. R. Germany
[2]Department of Physics, University of Coimbra, Portugal
[3]Institute of Theoretical Physics, University of Regensburg, D-8400 Regensburg, F. R. Germany

1. Introduction

An important nuclear physics issue is the question whether the intrinsic properties of free nucleons change inside nuclei. One such property is the strong isovector M1 transition connecting the nucleon with the $\Delta(1232)$. It dominates the nuclear electromagnetic response at excitation energies in the range ω = 200 - 400 MeV.

In this context, a detailed comparison of the nuclear photon scattering amplitude with the one for A free nucleons is of some interest. This is done here in a dispersion relation framework. In fact, such a comparison is the basic idea behind the dispersion relation sum rule of GELL-MANN, GOLDBERGER and THIRRING (GGT) [1], and its refined version [2] which accounts for characteristic shadowing effects at high energies.

2. Basic Dispersion Relation

This paper reports on a systematic dispersion relation analyis of the forward Compton amplitude for a series of nuclei from A = 9 to A = 238. Such a global analysis has now become possible by the existence of total γA cross section data covering the low energy giant dipole resonance region, the region of Δ-exciations in nuclei, and the high energy shadowing region up to about 200 GeV [3]. Typical cross sections $\sigma_{\gamma A}$ obtained by interpolation of the data are shown in Fig. 1. It is particularly interesting to note that the Δ-resonance region shows a universal behaviour of $\sigma_{\gamma A}/A$ for all nuclei.

The basic dispersion relation for the nuclear forward Compton scattering amplitude $F_{\gamma A}(\omega)$ is

$$\mathrm{Re}\, F_{\gamma A}(\omega) = -\frac{Z^2}{A}\frac{e^2}{M} + \frac{\omega^2}{2\pi^2}\, P\int_0^\infty d\omega' \frac{\sigma_{\gamma A}(\omega')}{\omega'^2-\omega^2} \tag{1}$$

with

$$\sigma_{\gamma A}(\omega) = \frac{4\pi}{\omega}\,\mathrm{Im}\, F_{\gamma A}(\omega). \tag{2}$$

Using the available data of $\sigma_{\gamma A}$ and suitable interpolations between them, the real part of $F_{\gamma A}$ can be evaluated at all energies ω. A typical result is shown in Fig. 2

+) Work supported in part by DFG and by DAAD.

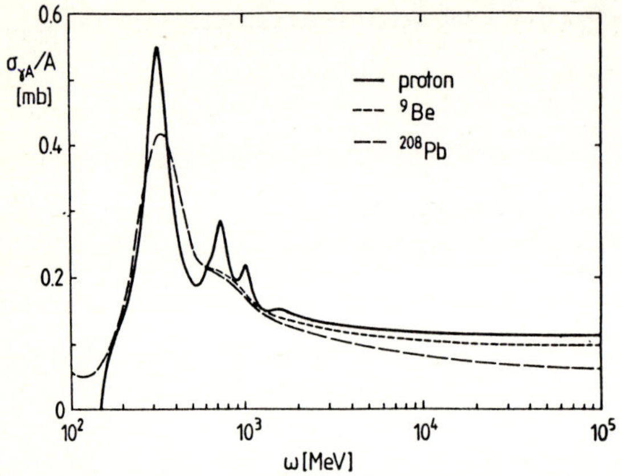

Figure 1: Total photon cross sections per nucleon for the proton, ^{9}Be and ^{208}Pb.

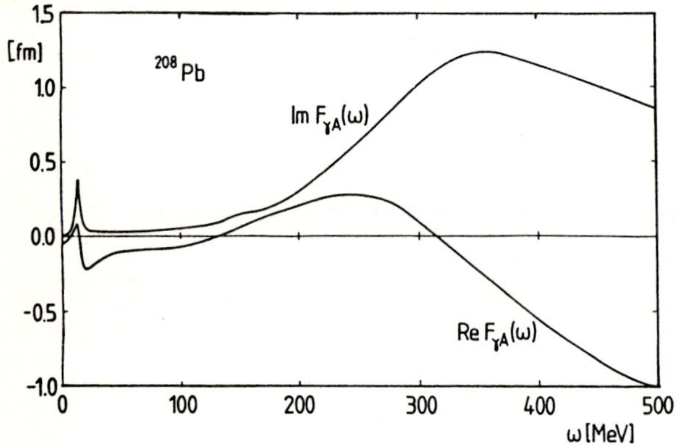

Figure 2: Real and imaginary parts of the forward Compton amplitude for ^{208}Pb.

for the case of ^{208}Pb. One notes that the zero of Re $F_{\gamma A}(\omega)$ close to the Δ resonace is at

$$\omega_o \simeq 314 \text{ MeV}, \tag{3}$$

very close to the corresponding zero in the real part of the free proton amplitude $F_{\gamma p}(\omega)$. In fact, the value (3) for ω_o turns out to be again universal for all nuclei within narrow limits. Hence the position of the Δ(1232) resonance does not change much in nuclei: the main effect of the nuclear medium is a moderate increase of its decay width.

3. The GGT Sum Rule, Revisited

Combining the dispersion relations (1) for a given nucleus and for A times the average nucleon amplitude $F_{\gamma N} = (Z/A)F_{\gamma p} + (N/A)F_{\gamma n}$, one obtains

$$\int_0^{m_\pi} d\omega\, \sigma_{\gamma A}(\omega) = \frac{NZ}{A} S + \omega_o^2\, P \int_{m_\pi}^{\infty} d\omega \frac{A\sigma_{\gamma N}(\omega) - \sigma_{\gamma A}(\omega)}{\omega_o^2 - \omega^2} + \delta(\omega_o) \tag{4a}$$

$$\equiv \frac{NZ}{A} S\,(1 + K)\ , \tag{4b}$$

where $S = 2\pi^2 e^2/M$, and

$$\delta(\omega_o) = 2\pi^2 [A\, \mathrm{Re}\, F_{\gamma N}(\omega_o) - \mathrm{Re}\, F_{\gamma A}(\omega_o)] - P\int_0^{m_\pi} d\omega \frac{\omega^2 \sigma_{\gamma A}(\omega)}{\omega_o^2 - \omega^2}\ . \tag{5}$$

The sum rule (4a) converges for any finite ω_o. The optimal choice is the ω_o of eq. (3) for which the first term on the r.h.s. of eq. (5) vanishes, while the second term is relatively small and of order $(\bar{\omega}/\omega_o)^2$, where $\bar{\omega}$ is an average energy around $\sim$ 100 MeV. The evaluation of the enhancement K over the classical dipole sum with S = 60 MeV mb (see eq. (4b)) from the intermediate and high energy cross sections gives K = 0.75 for ^{208}Pb, to be compared with the empirical K = 0.78 ± 0.15 obtained by integrating the photoabsorption cross section up to $\omega = m_\pi$ [4]. We also mention that a recent accurate measurement [5] of the enhancement K_{dip} of the sum rule in the giant resonance region (i.e. up to about 20 MeV) for ^{209}Bi gives $K_{dip} = 0.46 \pm 0.05$. The remaining $\Delta K \simeq 0.3$ is then located in the quasi-deuteron absorption region between the giant dipole resonance and the pion threshold.

4. Concluding Remarks

A universal feature of photonuclear cross sections in the Δ resonance region is that their maximum almost coincides with the position of the free Δ(1232) as seen in the γp total cross section. The same universality for all nuclei holds for the zero of Re $F_{\gamma A}$ at $\omega_o \simeq$ 314 MeV which coincides with the corresponding zero of Re $F_{\gamma p}$. This observation confirms that the Δ(1232) survives in nuclei as a distinct baryonic species: its interactions with surrounding nucleons are at most of a similar strength as the binding potential experienced by the nucleons themselves. These and other features, such as the moderate broadening of the Δ width in a nuclear medium, are quite well described in Δ-hole models [6,7].

References

[1] M. Gell-Mann, M.L. Goldberger and W. Thirring: Phys. Rev. 95, 1612 (1954)
[2] W. Weise: Phys. Rev. Lett. 31, 773 (1973); Phys. Reports 13, 53 (1974)
[3] J. Ahrens: Nucl. Phys. A 446, 229c (1985), and refs. therein
[4] A. Lepetre et al.: Nucl. Phys. A 367, 237 (1981); P. Carlos et al: Nucl. Phys. A 431, 431 (1984)
[5] R. Nolte, A. Baumann, K.W. Rose and M. Schumacher, Univ. of Göttingen preprint (1985)
[6] E. Oset, H. Toki and W. Weise, Phys. Reports 83, 281 (1982)
[7] J.H. Koch et al.: Ann. of Phys. 154, 99 (1984)

1.3 Exotic Nuclei and Beta Decay far from Stability

Search for Superheavy Elements – A Status Report

G. Herrmann

Institut für Kernchemie, Universität Mainz, D-6500 Mainz, and
GSI Darmstadt, D-6100 Darmstadt, F. R. Germany

Abstract: A survey is given of nuclear and chemical properties in the predicted island of spherical superheavy nuclei around element 114 and of recent attempts to produce such nuclei by transfer and fusion reactions.

1. Introduction

The search for new chemical elements was a major impetus in the development of chemistry over centuries. The University of Heidelberg where this symposium is held at the occasion of the 600th anniversary, played an important role in this field. With a new physical detection technique, spectral analysis, developed here in a collaboration of the physicist Gustav Kirchhoff and the chemist Robert Bunsen [1], two new elements, the alkali metals cesium [2] and rubidium [3], were discovered by Bunsen in natural spring water. This discovery marks a new area in the history of the chemical elements, the application of sensitive physical detection techniques, such as optical and X-ray spectroscopy or radioactivity measurements, in the discovery of new elements in natural samples and during their chemical enrichment.

In the following eight decades the existing gaps in the periodic table of the elements shown in Figure 1 were filled, the last one with promethium (element 61) discovered 1945 in fission products. Already earlier, in 1940, the first step beyond the heaviest natural element, uranium, was made by the synthesis of neptunium (element 93). The periodic table was further extended element-by-element by synthetic elements but it became obvious that the decreasing stability of such heavy nuclei as manifested in decreasing half-lives and production cross sections would make further progress more and more difficult unless strong shell effects would dramatically increase the stability of still heavier nuclei. Such shell closures were predicted in 1966/67 to occur at proton number 114 and neutron number 184 [4,5], leading to an island of superheavy elements not too far from the then heaviest element, 106. Since that time, numerous attempts were made to reach this island by nuclear reactions, in addition to the extension of the periodic table element-by-element.

This paper gives an outline of nuclear and chemical properties around element 114 and of recent attempts to jump into the island. Comprehensive reviews on superheavy elements can be found elsewhere [6-8]. Recent progress in the

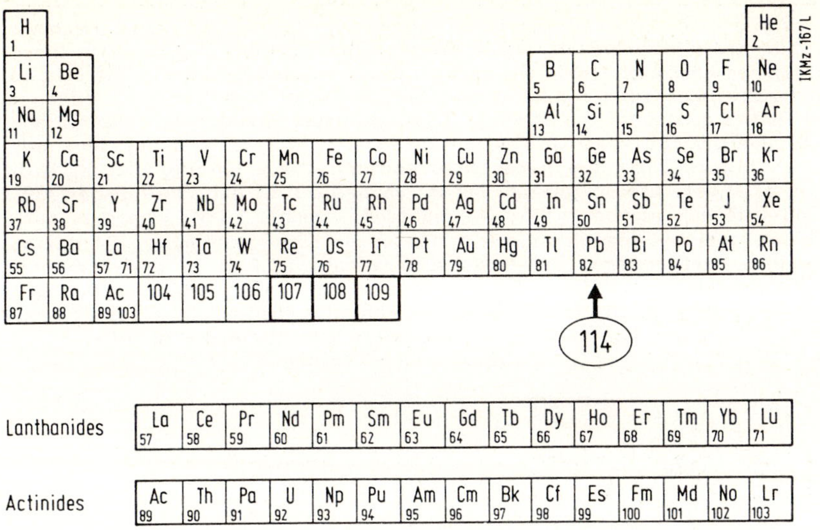

Fig.1: Present periodic table of the elements

step-by-step extension of the periodic table is surveyed in another contribution to this conference [9].

2. Nuclear and Chemical Properties around Element 114

Figure 2 shows the present upper end of the chart of nuclides in the region of the heavy actinide and the transactinide elements (lower left corner) and the location of the predicted island of superheavy nuclides (upper right corner). Two versions of the island are indicated by contour lines connecting nuclides with 1 μs half-life. In one version [10], diagonally hatched from right to left, the island extends much farther toward neutron deficient nuclei than in the other version [11] hatched from left to right. Both calculations predict the longest half-life for element 110 with 184 neutrons (cross-hatched) with values of 10^9 y [10] and 10^5 y [11].

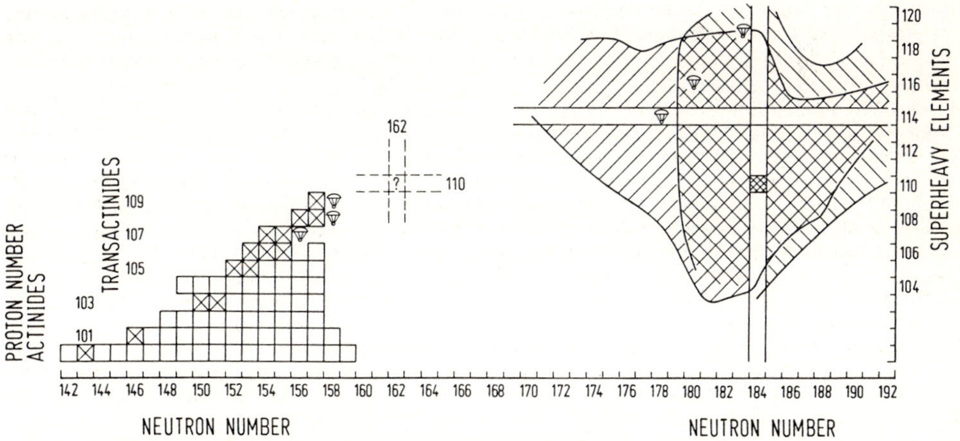

Fig.2: Chart of the nuclides with the heaviest known nuclides (lower left corner) and two versions [10,11] of the predicted island of spherical superheavy nuclides (upper right corner)

The major problem for the synthesis of such superheavy elements becomes immediately evident from Figure 2: Whereas their proton numbers overlap with those of the heaviest known elements, their neutron numbers are much larger. There is no combination of easily accessible nuclei which can be fused together to produce a nucleus located in the center of the island. Even the most favourable combinations using calcium-48 as a projectile lead to compound nuclei on the neutron-deficient side as is indicated in Figure 2 by the parachutes showing these landing places. With plutonium-244 as a target, atomic number 114 is reached but at neutron number 178, with curium-248 atomic number 116 at neutron number 180 whereas with einsteinium-254 one gets by one unit close to the magic neutron number 184 but at atomic number 119. Furthermore, as shown schematically in Fig. 3, the compound nuclei formed will predominantly decay by fission and only a very small fraction will undergo one or several neutron evaporation steps finally leading to a superheavy nucleus in its ground state. The branching ratio $\Gamma n/\Gamma f$ is in the order of 0.01.

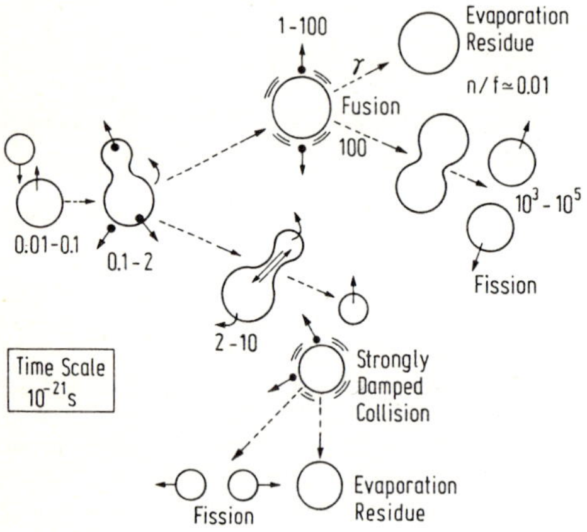

Fig.3: Two approaches to the synthesis of superheavy nuclei by heavy-ion reactions: fusion (upper row) and transfer reaction (lower row), together with characteristic reaction times in 10^{-21}s.

Alternatively, strongly damped collisions between two very heavy nuclei may be used for the synthesis of superheavy nuclei, as is also shown in Figure 3. In spite of the short life-time of the dinuclear system formed in the entrance channel, one of the reaction partners may take up many nucleons from the other partner to grow into a superheavy nucleus. After separation into two fragments, the superheavy fragment will again fission predominantly and only a very small fraction will survive.

For detection, spontaneous fission should be the most characteristic decay mode of superheavy nuclei, either directly or after one or several alpha-transitions. Theoretical estimates predict an average total kinetic energy of 200-230 MeV for the fission fragments and the evaporation of about ten neutrons per fission. These numbers should be compared with 170 MeV kinetic energy and 2.5 neutrons in fission of uranium-238. Alpha particle energies are much less characteristic since they fall in line with many known alpha-emitters. Hence, the observation of fission events with unusually high kinetic energy and neutron multiplicity is a very characteristic fingerprint for superheavy nuclei. Extremely sensitive measurements can be performed by fission-fragment fission-neutron coincidence counting [12]. Chemically isolated samples are placed between pairs of surface barrier detectors located in the central hole of a moderator cylinder which is equipped with helium-3 detectors for neutron counting. The system registers in the surface barrier detectors fission events as single and coincident fragments and records the fragment energies. It registers in addition the number of neutrons appearing in coincidence with such events and their time- and spatial-distribution. The system is operated below ground and is further shielded against cosmic-ray induced events by plastic scintillator sheets. The background rate for coincident events is less than one event per year at a detection efficiency of about 60%.

Since chemical separation procedures play an important role in searches for superheavy elements predictions of their chemical properties are required. The position of element 114 in the periodic table is shown in Figure 1. Detailed quantum mechanical calculations have indicated that the electronic structures of the atomic ground states of superheavy elements are quite similar to those of their homologous elements. Thus at element 112, eka-mercury, the 6d shell should be completed in analogy to the completion of the 5d shell at mercury. Element 114, eka-lead, should possess two 7p(1/2) electrons analogous to the two 6p(1/2) electrons of lead. At element 118, eka-radon, the 7p(3/2) shell should be completed with four electrons as it is the case for the 6p(3/2) shell at radon. Hence, many chemical properties of superheavy elements [13] can be predicted by extrapolation, e.g., within the vertical columns of the periodic table. Deviations from such systematic trends may be caused, however, by relativistic effects in the electron energies which should lead to a considerable splitting between the 7p(1/2) and the 7p(3/2) electron orbitals so that the 7p(1/2) subshell filled at element 114 may obtain the character of a closed shell as well as the 7s configuration filled at element 112. Elements 112 and 114 may, thus, behave like chemically inert, noble gases or liquids [14] in addition to the superheavy noble gas, element 118, eka-radon.

On the basis of such predicted chemical properties several separation procedures were worked out, designed as group separations in order to collect together as many superheavy elements as possible. If evidence for such elements would show up more specific procedures would then be applied in order to identify the element. Three approaches were followed: (i) Element 112 through 116 are predicted to be volatile at higher temperatures [15]: these are the 'lead-like' elements separated by gas-phase chemistry. (ii) Some elements such as 112, 114, and 118 may even be volatile at room temperature and may be chemically inert [14]: these 'radon-like' elements are isolated by another version of gas-phase chemistry. (iii) Elements 108 through 116 are thought to form in aqueous solutions negatively charged complexes with hydrobromic acid [16]: these 'platinum-like' elements are separated by solution chemistry. Off -line as well as on-line versions of such procedures were developed [17,18]. In the off-line approach the chemical separation is performed after irradiation, followed by counting of the isolated samples. This provides the highest sensitivity for long-lived nuclides with a cross-section limit of 10^{-33} cm^2 for a half-life region of hours to years. With on-line procedures, chemical separation and counting are performed during irradiation in order to cover short half-lives, down to a few seconds at a somewhat lower sensitivity of about 10^{-34} cm^2.

In addition to chemical separation techniques, various fast transport techniques have been applied, as is shown below. Furthermore, two recoil-fragment separators were used in recent fusion experiments to extend the half-life region covered down to microseconds: the gas-filled magnetic separator SASSY [19] at Berkeley and the velocity filter SHIP [9] at Darmstadt.

3. Results: Multinucleon Transfer Reactions

When the UNILAC heavy-ion accelerator at Darmstadt went into operation, a large number of fusion reactions had already been tried for the synthesis of superheavy elements without any success [6]. Hence it appeared to be most appealing to follow the alternative approach, the transfer of many nucleons between very heavy nuclei. With the availability of the uranium beam a reaction such as

$$^{238}_{92}U^{146} + {}^{238}_{92}U^{146} \rightarrow {}^{298}_{114}X^{184} + {}^{178}_{70}Yb^{108}$$

was within reach. Already the first studies of the element distribution for fragments originating from interactions between two uranium nuclei [20,21] indicated massive multiple nucleon transfer to occur. In particular, evidence for the production of complementary fragments such as ytterbium isotopes at a cross section level as high as 10^{-28} cm^2 was found in radiochemical studies [20]. Thus, the transfer proceeds into the region of superheavy nuclei. Systematic work on the neutron-to-proton ratios of heavy fragments in multinucleon transfer reactions [22] indicated that even very neutron-rich superheavy nuclei close to the 184 neutron shell may be formed in good yield. Furthermore, counter experiments [21,23] showed that superheavy nuclei may be formed with relatively little excitation energy and, hence, the chance for their escape from prompt fission seemed to be large enough for the formation of detectable quantities.

These results encouraged several groups to perform a direct search for superheavy elements in the $^{238}U + ^{238}U$ reaction. The most sensitive experiment was carried out by a GSI-Mainz collaboration [24] using off-line chemistry and low-level counting techniques. No evidence for superheavy elements was obtained. The upper limits of the production cross sections are shown in Figure 4 by the curve labeled CHEM. In the evaluation of the data, the half-life of the superheavy nuclide is a free parameter leading to a parabolic or trapezoidal shape of such curves. For a given detection technique there is an optimum half-life range. Below this range the sensitivity decreases because shorter-lived nuclides decay already before counting,

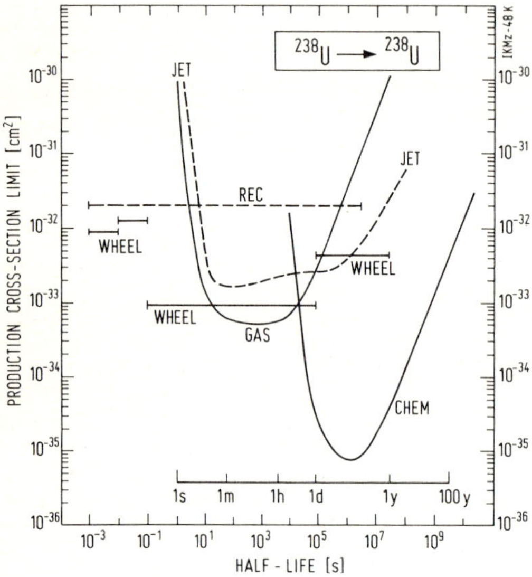

Fig. 4: Upper limits for the production cross sections of superheavy elements in collisions of two uranium-238 nuclei as a function of the half-life. Data labeled CHEM refer to radiochemical off-line experiments [23], GAS to on-line searches for very volatile elements [24], WHEEL to measurements with a rotating wheel system [23], JET to gas-jet experiments [25,26] and REC to measurements of recoil atoms implanted into surface barrier detectors [20]

whereas beyond this range longer-lived nuclides escape detection. In the curve denoted CHEM the highest sensitivity, 7×10^{-36} cm^2 as a cross section limit, is reached for half-lives of a few weeks. In addition to radiochemical techniques the same group applied fission track detection with a rotating wheel system in order to cover the region of short half-lives. The resulting limits denoted by WHEEL are set by a background of spontaneously fissioning actinide nuclei present in the chemically unseparated mixture of reaction products. Since some fission fragment pulses were observed for "radon-like" elements in this and other transfer reactions, a Mainz-GSI collaboration [25] recently reinvestigated this fraction with an improved technique but again without positive results, as shown by the curve GAS. Also given in Figure 4 are the results of measurements of unseparated reaction product mixtures after transport by a gas jet from the target to the detector system, JET [26,27], and after direct implantation into a surface barrier detector, REC [21].

After the failure of this reaction the next attempt was the use of a heavier target, curium-248, in transfer reactions with uranium-238 projectiles. Substantially increased production cross sections for heavy actinides were found [28] giving hope for a production of superheavy nuclei in detectable quantities. A series of radiochemical search experiments were performed by a GSI-Mainz-Livermore-Berkeley-Oak Ridge collaboration [29]. Special precautions had to be made including studies of target failure mechanisms [30] to prevent contamination by the strong spontaneous fission activity of the target, 10^8 to 10^9 fissions per day, during bombardment and chemistry. Again these experiments produced negative results as shown in Figure 5.

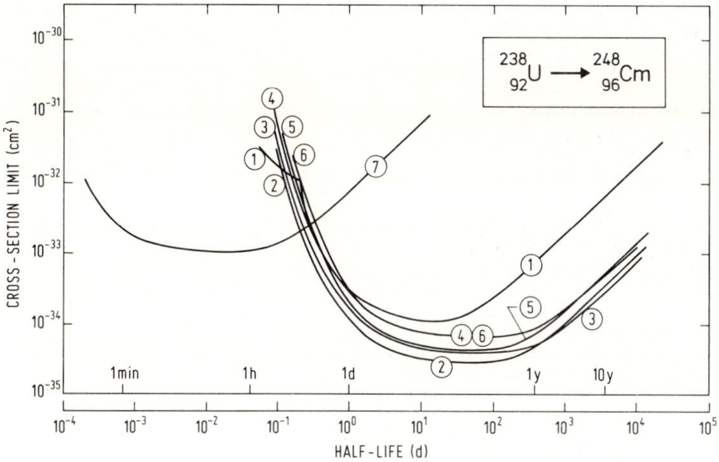

Fig. 5: Upper limits for the production cross sections of superheavy nuclei in the reaction of uranium-238 with curium-248. Curves 1 and 7 refer to radon-like elements separated off-line and on-line, respectively, curves 2 to 6 to different fractions containing lead-like and platinum-like elements [29]

Also negative remained experiments of a Marburg-Giessen-GSI collaboration [31] to synthesize superheavy elements by instantaneous fission in the reaction of lead-208 with uranium-238. During close contact uranium-238 may fission and one of the fission fragments may fuse with lead-208 forming a superheavy nucleus. Figure 6 gives the results. No evidence for delayed fission fragments from superheavy nuclei was found by a München group [32] in the reaction of lead-208 with xenon-136 applying fragment detection with counters for half-lives down to 10^{-12} s at 10^{-30} cm^2 cross section.

4. Results: Fusion Reactions

Fusion of calcium-48 with curium-248 according to

$$^{48}_{20}Ca^{28} + {}^{248}_{96}Cm^{152} \rightarrow {}^{296}_{116}X^{180}$$

had already been attempted for the production of superheavy elements in elaborate experiments by a Livermore-Berkeley collaboration [33-36] and by the Dubna group [37]. In these earlier experiments an extra push of projectile energy was thought to be necessary to overcome entrance channel limitations of the fusion process. The interaction energy chosen exceeded the calculated fusion barrier by about 20 MeV introducing to the compound nucleus an excitation energy of at least 45 MeV.

Since more recent systematic work on fusion reactions gave evidence, however, that the dynamic hindrance of fusion may be small in this reaction, a new search for

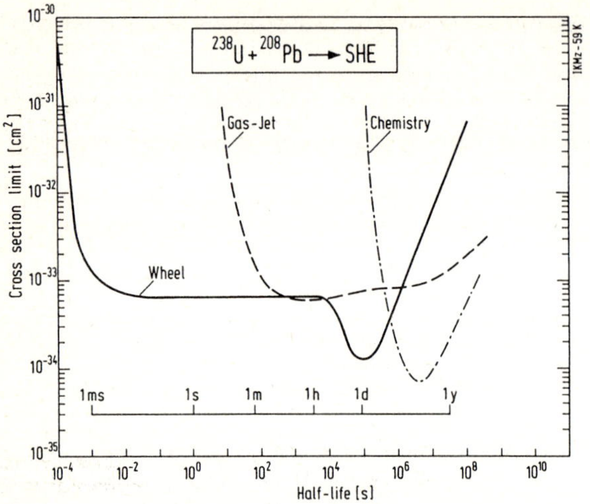

Fig. 6: Upper limits for the production cross sections of superheavy nuclei in the reaction of uranium-238 with lead-208 as obtained with a rotating wheel, by a gas-jet on-line transport technique and by off-line chemistry [31]

superheavy elements in the calcium-48 on curium-248 reaction was undertaken by a GSI-Berkeley-Mainz-Los Alamos-Bern-Göttingen collaboration [38] at projectile energies much closer to the barrier. In this series of experiments the excitation energy of the compound nucleus was kept between 16 and 40 MeV. A variety of chemical and physical techniques was used for the isolation and detection of superheavy nuclei in order to cover a half-life range range of 14 orders of magnitude, from 1 µs to 10 y, at production rates of a few atoms per day. This reaction represents the system most extensively studied so far for the production of superheavy nuclei. No positive

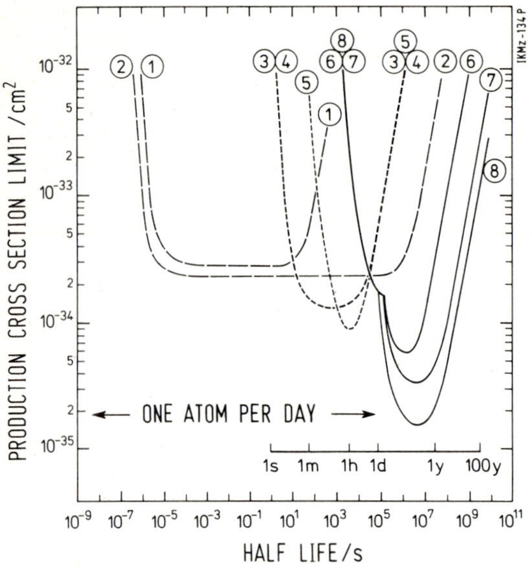

Fig. 7: Upper limits for the production cross sections of superheavy nuclei in the reaction of calcium-48 with curium-248 as a function of the half-life [38]. The data result from experiments with recoil-fragment separators (curves 1 and 2), with fast on-line chemistry (3,4, and 5), and with off-line chemistry (6,7, and 8)

evidence was found, although a production rate of a few atoms per day would be sufficient for detection. Figure 7 summarizes the results.

Irradiations were performed at the Berkeley SuperHILAC and the UNILAC. Radiochemical methods were applied in on-line (curves 3,4,5) and off-line versions (6,7,8) for radon-like (3,6), lead-like (4,8) and platinum-like elements (5,7). The region of very short half-lives was covered by experiments with recoil-fragment separators (1,2).

Further work with fusion reactions may be conducted into two directions. A closer approach to the 184 neutron shell may help to reach a region of increased stability against fission. This can be achieved by fusion of calcium-48 with einsteinium-254, the heaviest nuclide available in microgram quantities. An exploratory radiochemical experiment using this reaction was performed at Berkeley by a Livermore-Berkeley-GSI collaboration with upper cross section limits of 10^{-31} cm^2 [39]. It should be noted that very recent theoretical calculations [40] indicate a shift of the island toward lower neutron numbers, 178 and 180.

The other line of direction is the production of nuclei in the region of element 110 with neutron numbers around 162 where a local stability region is postulated to exist in these recent calculations [40], see Figure 2 and Ref. [9]. These predictions are supported by the decay properties of the heaviest known nuclei [9]. The region may be reached by fusion of argon-40 with uranium-233, -235 and -238. In exploratory experiments by a GSI-Mainz-Würenlingen group [41] a search was made for the decay products descendant from such nuclei. Two techniques were applied, (i) on-line transport of the reaction products by a gas-jet followed by detection of time-correlated alpha- and spontaneous fission events with a rotating wheel system, and (ii) chemical isolation of long-lived decay products such as 20-h fermium-255. The experiments remained negative at a cross section level of about 10^{-33} cm^2. Other experiments in this direction are described in another contribution to this conference [9].

5. Conclusions

The reasons for the failure of so many attempts to produce superheavy elements may be manyfold. Either the stability of superheavy nuclei is grossly overestimated, or there are unexpected difficulties with the synthesis reactions due to the too high excitation energy imposed on the nuclei formed or due to hindrances in the fusion process. In view of the surprising stability of the heaviest known nuclei one is tempted to conclude that the limitations lie in the production processes. Hence we may still be convinced that spherical superheavy elements around atomic number 114 should exist but it may be difficult to make them.

References

1. G. Kirchhoff and R. Bunsen, J. prakt. Chemie 80, 449 (1860)
2. R. Bunsen, J. prakt. Chemie 80, 477 (1860)
3. R. Bunsen, J. prakt. Chemie 83 198 (1861)
4. W.D. Myers and W.J. Swiatecki, Nucl. Phys. 81, 1 (1966)
5. H. Meldner, Arkiv Fysik 36, 593 (1967)
6. G. Herrmann, Nature 280, 543 (1979)
7. J.V. Kratz, Radiochim. Acta 32, 25 (1983)
8. G.N. Flerov and G.M. Ter-Akopyan, Rep. Progr. Phys. 46, 817 (1983)
9. S. Hoffmann, Contrib. to this Conference
10. E.O. Fiset and J.R. Nix, Nucl. Phys. A193, 647 (1972)
11. J. Randrup et al., Phys. Scr. 10A, 60 (1974), and S. Aberg, priv. comm. (1979)
12. P. Peuser et al., Nucl. Instr. Meth. A239, 529 (1985)
13. O.L. Keller, Radiochim. Acta 37, 169 (1984)
14. K.S. Pitzer, J. Chem. Phys. 63, 1032 (1975)
15. B. Eichler, Kernenergie 19, 307 (1976)
16. J.V. Kratz et al., Inorg. Nucl. Chem. Lett. 10, 951 (1974)

17. G. Herrmann, Pure & Applied Chemistry 53, 949 (1981)
18. K. Sümmerer et al., Contrib. XXII. Int. Meeting on Nuclear Physics, Bormio 1984, GSI Darmstadt Preprint GSI-84-17
19. M. Leino, S. Yashita and A. Ghiorso, Phys. Rev. C24, 2370 (1981)
20. M. Schädel et al., Phys. Rev. Lett. 41, 469 (1978)
21. K.D. Hildenbrand et al., Phys. Rev. Lett. 39, 1065 (1977)
22. H. Freiesleben and J.V. Kratz, Phys. Rep. 106, 1 (1984)
23. H. Freiesleben et al., Z. Physik A292, 171 (1979)
24. H. Gäggeler et al. Phys. Rev. Lett. 45, 1824 (1980)
25. N. Hildebrand et al., GSI Darmstadt Rep. GSI-85-1, p. 89, and to be published
26. D.C. Aumann et al., Phys. Lett. 82B, 361 (1979)
27. H. Jungclas et al., Phys. Lett. 79B, 58 (1978), and D. Hirdes et al., GSI Darmstadt Rep. GSI-79-11, p. 71
28. M. Schädel et al., Phys. Rev. Lett. 48, 852 (1982)
29. J.V. Kratz et al., Phys. Rev. C33, 504 (1986)
30. R.W. Lougheed et al., Nucl. Instr. Meth. 200, 71 (1982)
31. T. Lund et al., Z. Physik A303, 115 (1981)
32. P. Sperr et al., Z. Physik 287, 57 (1978)
33. E.K. Hulet et al., Phys. Rev. Lett. 39, 385 (1977)
34. J.D. Illige et al., Phys. Lett. 78B, 209 (1978)
35. R.J. Otto et al., J. Inorg. Nucl. Chem. 40, 589 (1978)
36. A. Ghiorso et al., LBL Berkeley Rep. LBL 6575 (1977), p. 242
37. Yu.Ts. Oganessian et al., Nucl. Phys. A294, 213 (1978)
38. P. Armbruster et al., Phys. Rev. Lett. 54, 406 (1985)
39. R.W. Lougheed et al., Phys. Rev. C32, 1760 (1985)
40. P. Möller, G.A. Leander and J.R. Nix, Z. Physik A323, 41 (1986)
41. K. Sümmerer et al., GSI Darmstadt Rep. GSI-86-1, p. 30

Heavy Elements – Experiments on Synthesis and Decay

S. Hofmann, P. Armbruster, G. Berthes, H. Folger, U. Gollerthan, E. Hanelt, F.P. Heßberger, M.E. Leino, G. Münzenberg, K. Poppensieker, B. Quint, W. Reisdorf, K.-H. Schmidt, H.-J. Schött, K. Sümmerer, and I. Zychor***

Gesellschaft für Schwerionenforschung mbH, P.O. Box 110541,
D-6100 Darmstadt, F. R. Germany

Attempts were made to identify new elements in heavy ion fusion reactions. After an effective separation by the velocity filter SHIP the ions passed a pair of thin secondary electron detectors and were finally implanted into position sensitive Si-detectors. With these techniques the odd elements 105, 107, and 109 have been investigated in the years 1981 and 1982. The even elements 106 and 108 have been studied in the year 1984, and in 1985 attempts were made to produce element 110. So far in most experiments ^{208}Pb and ^{209}Bi targets have been irradiated with beams of ^{50}Ti, ^{54}Cr, ^{58}Fe, and ^{64}Ni. In a recent experiment (April/May, 1986) the decay of the ee nucleus ^{264}108 was measured. This isotope was produced in the reaction ^{58}Fe+^{207}Pb with a cross section of $(3.2^{+6.1}_{-2.6})$ pb. The nucleus ^{264}108 decays by α emission with a half-life of (76^{+365}_{-35}) μs. A deduced α energy is $(10.8^{+0.1}_{-0.3})$ MeV. Also ^{260}106 is decaying by α emission with a branching of 50 %. The decay sequence $^{264}108(\alpha) \rightarrow {}^{260}106(\alpha) \rightarrow {}^{256}104(sf)$ demonstrates that the heavier element isotopes ^{264}108 and ^{260}106 are more stable against fission than the lighter element isotope ^{256}104. The mass excesses of ^{264}108 and ^{260}106 are (120.0 ± 0.3) MeV and (106.58+0.10) MeV, respectively. The values are obtained by α chain summation. The investigated isotopes are situated near the center of a shell stabilized region of deformed nuclei. The fission barriers for the discovered isotopes are mainly due to shell correction energies. Values as large as 5 MeV have been deduced from the data. The liquid drop fission barriers are negligible, the stabilization of the nuclei observed is due to shell corrections. In this sense the nuclei are deformed superheavy nuclei.

1. Introduction

The heaviest known atomic nucleus identified unambiguously by parent-daughter correlation is the isotope with mass number 266 of element 109. In the decay chain of the only one observed atom the three decay modes alpha decay, beta decay, and fission, occur.

$$^{266}109 \xrightarrow[T_{1/2}\,=\,3\,\text{ms}]{\alpha} {}^{262}107 \xrightarrow[T_{1/2}\,=\,5\,\text{ms}]{\alpha} {}^{258}105 \xrightarrow[T_{1/2}\,=\,4\,\text{s}]{\text{EC}} {}^{258}104 \xrightarrow[T_{1/2}\,=\,13\,\text{ms}]{\text{sf}} A_1 + A_2$$

The prediction of the liquid drop model gives a vanishing fission barrier near element 110. Already for elements beyond Md half-lives of less than 10^{-6} sec were expected. However, nuclear structure effects change the height and shape of the fission barrier and thus prevent these heavy nuclei from an immediate fission [1,2]. The unexpected deviation of the measured spontaneous fission half-lives beyond element 102 from early systematics proves that these microscopic effects cannot be extrapolated with sufficient accuracy [3]. Experimental data are needed to establish the ground-state properties of the heaviest isotopes.

---

*) On leave from University of Helsinki, Finland

**) On leave from Warsaw University, Poland

2. Experimental Method Used to Synthesize and to Identify Elements 106 to 109

2.1 General Guidlines

Since the first operation of SHIP [4] the sensitivity of experiments to search for rare isotopes could be steadily improved. Developments in ion source techniques and accelerator acceptance increased the beam currents of titanium to nickel ions up to 5×10^{12}/s. Thin lead or bismuth targets (400 μg/cm^2) mounted on a wheel of 30 cm diameter are moved with a velocity of 2 cm/ms through the beam spot of 5 mm diameter. Radiative cooling prevents the targets from melting.

Using targets of lead and bismuth, compound nuclei of the highly fissionable heavy elements can be produced in fusion reactions with a minimum of excitation energy of 15 to 20 MeV (cold fusion) [5]. After evaporation of 1 or 2 neutrons the residues, if α-emitting, belong to one and the same decay chain and unknown heavier elements can be identified by genetic correlations to their known daughters.

The biggest α branching ratios could be expected for odd-odd isotopes, whereas the even-even isotopes were expected to decay by fission with half-lives decreasing below 1 μs (the separation time of SHIP) with increasing element number [6]. Therefore, negative results in search experiments for even elements could have two reasons, either too low cross-sections or too short half-lives. As there is no enhanced cross section for even-even isotopes to be expected, the detection of odd elements not suffering from the above ambiguities would help to design later experiments aiming at even elements. To start our program of heavy element synthesis experiments on odd elements were chosen.

2.2 Detection System

The velocity filter SHIP is a two-stage filter consisting of spatially separated electrostatic and magnetic dipole fields, and two magnetic quadrupole lenses [4]. The separated ions pass a pair of time-of-flight detectors and are finally implanted into an array of position sensitive silicon detectors. The detection of single decay chains has become possible by correlation analysis. Time and position of subsequent α- or fission-decays are correlated to the time and position signals obtained from the implantation process [7]. Half-lives can be determined from the time difference between two subsequent signals even in case of only one decaying atom. The energy and position resolution of Si-surface barrier detectors cooled to 260° K is $\Delta E(FWHM) = 18$ keV and $\Delta Y(FWHM) = 210$ μm, respectively. These values increase to about 35 keV and 400 μm after implantation of 10^9 particles. Since the implantation depth of the ions is smaller than the range of the decay alphas, there is a chance of about 45 % that an alpha particle is observed with only a part of its full energy.

3. Results on Elements 106-109

3.1 The Odd Elements 105, 107, and 109 with N-Z = 48

The isotopes $^{258}105$, $^{262}107$, and $^{266}109$ have been observed in the reactions of ^{50}Ti, ^{54}Cr, and ^{58}Fe projectiles with ^{209}Bi targets [8-10]. The measured total half-lives are 4.4 s, 5 ms, and 3 ms, respectively, and the main decay mode is α emission. The number of observed atoms was 129, 6, and 1, respectively, in measuring times of 2, 6, and 13 days resulting in cross-sections of 2.9 nb, 200 pb, and 16 pb at projectile energies between 4.7 and 5.1 MeV/u. A compilation of decay and production data is given in [11].

The cross-sections are shown in Fig. 1. An exponential function fitted to the data gives a cross-section decrease of a factor of 0.37 per element. An extrapolation to element 111 results in $\sigma \sim 1$ pb. Cross-section limits for element 111 are published by the Dubna group [12] (s. Fig. 1). They do not contradict our

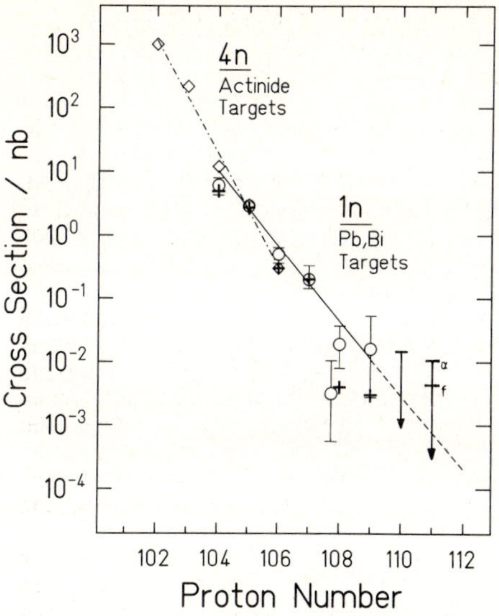

Fig. 1: Comparison of cross-sections for production of heavy elements (◇ Berkeley data, + Dubna data, o SHIP data). The 3 pb cross section was measured for $^{264}108$. Only cross-section limits are known from experiments to produce element 110 (SHIP) and 111 (Dubna) in cold fusion reactions.

extrapolations. Results of our last experiment to produce $^{261}107$ in a 2n evaporation channel gave similar decay properties as for the neighbouring isotope $^{262}107$ [13]. Both are α-emitters. An 80 % spontaneous fission branch in the ms-region discussed by the Dubna group [14] has not been seen in our experiment.

3.2 The Even Elements 106 to 108

Replacing ^{209}Bi by ^{208}Pb should not significantly change the reaction process and an interpolation of the odd element cross sections is, therefore, a good base to plan experiments to produce isotopes of even elements. These experiments have been carried out in March, 1984 (Z = 106 and 108), and in March, 1985 (Z = 110). In May, 1986, we performed an experiment to identify $^{264}108$ in the reaction $^{58}Fe+^{207}Pb$ [29] and to search for isotopes of element 110 in the $^{40}Ar+^{235}U$ reaction.

The even-even isotope $^{260}106$ was produced at 4.92 MeV/u in the reaction $^{208}Pb(^{54}Cr,2n)^{260}106$. It decays by emission of α particles to the spontaneous fissioning nucleus $^{256}104$ ($T_{1/2}$ = 7.4 ms). Additionally, a spontaneous fission branch of 50 % has been measured. The half-life of $^{260}106$ is $(3.6^{+0.9}_{-0.6})$ ms, giving a partial sf half-life of 7 ms. When going from $^{256}104$ to $^{260}106$ the sf half-life does not decrease, a finding which contradicts the theoretical predictions which reproduce the Z = 104 half-lives, but is in agreement with the analysis of Demin et al. [15]. The production cross-section of the 2n-reaction channel is within a factor 2 equal to the 1n-reaction channel leading to $^{261}106$ (0.5 nb).

Oganessian et al. [17] found in experiments to produce element 108 a 6 ms fission activity. Assigning the fission activity to $^{256}104$ they concluded that the even-even nucleus $^{264}108$ was produced with a cross section of 5 pb in the reaction $^{58}Fe+^{207}Pb$ and decayed predominantly by α-emission. This observation, the discovery of the α-emitter $^{265}108$ [16], and the high α branching of $^{260}106$ was not expected after measurements of the short fission half-lives of the even-even isotopes of element 104. Now it seemed to be possible that the fission probabilities of the heavier elements could follow a new systematic. By extrapolation, relatively stable nuclei with respect to fission can be expected even for elements beyond 109. Irradiating ^{207}Pb targets (410 μg/cm^2) with ^{58}Fe projectiles of 5.04 MeV/u we tried to produce the even-even isotope ^{264}108 and to measure the decay properties. In an 13 days experiment we achieved an integral dose of 1.2 x 10^{18} projectiles on target. Figure 2 shows the only one decay chain that could be assigned to a compound nucleus formation. After implantation of an evaporation residue into the detector - this event is characterized by a time-of-flight signal and an implantation energy corresponding to a compound nucleus formation - we measured at the same detector position after 108 μs a 1.6 MeV signal. After a time interval of 10 272 μs a signal of 4.8 MeV and after 18 839 μs a high energy signal of about 235 MeV. We interprete the three signals following the implantation as being due to a decay chain of two α decays which both escape into backward direction depositing only part of their energy and a fission process. The deduced half-lives of the implanted nucleus and the daughter products are (76^{+365}_{-35}) μs, $(7.1^{+34.1}_{-3.3})$ ms, and (13^{+62}_{-6}) ms, respectively (a time of 1.7 μs to pass SHIP was considered in the half-life calculation of the implanted nucleus). Decay modes and half-lives of the daughter products are in agreement with the known decay data of ^{260}106(50% α) → ^{256}104(98% sf). Consequently, the implanted nucleus is ^{264}108 decaying by α emission with 76 μs half-life. The production cross section is $(3.2^{+6.1}_{-2.6})$ pb.

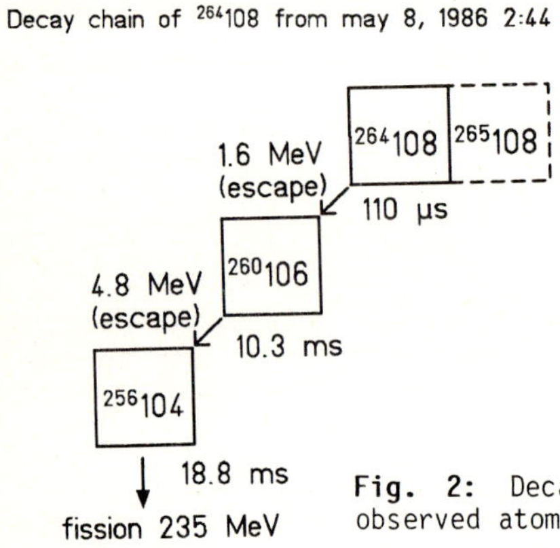

Fig. 2: Decay chain of the only one observed atom of ^{264}108.

All measured 1n cross sections of reactions with lead and bismuth targets decrease logarithmically with increasing element number. An extrapolation of the cross sections to element 110 results in values between 1 and 5 pb. In a 15 days' irradiation of ^{208}Pb a dose of 1 x 10^{18} ^{64}Ni projectiles (E = 5.06 MeV/u) was collected. No decay was observed that could be assigned to an evaporation residue of the compound nucleus ^{272}110. A cross-section limit is 12 pb, calculated with 95 % confidence. The experimental cross-section limit does not contradict the expectations and no further conclusions can be drawn except that the cross sections do not increase.

4. Discussion of Ground State Properties of the Heaviest Nuclei

4.1 Experimental Masses and Shell Corrections

The mass excesses of the even-even nucleus ^{260}106, and its daughter nucleus, ^{256}104, as determined in our experiments are given in Table 1. Best agreement is obtained with the mass predictions of Liran and Zeldes [22]. A mass excess of ^{264}108 can be deduced with an α energy calculated from the measured half-life and a reduced α width relative to the polonium α decay of 1. The w_α value of the daughter nucleus ^{260}106 amounts to about 1. The deduced α energy is $(10.8^{+0.1}_{-0.3})$ MeV. The errors are due to the half-life uncertainties.

Table 1: Experimental mass excess (MeV) of $^{264}108$, $^{260}106$, and $^{256}104$ compared to predictions of different mass tables.

Nucleus	Ref. 19	20	21	22	23	this work
$^{264}108$	121.28	120.40	120.4	120.27	--	120.0±0.3
$^{260}106$	108.27	107.29	107.7	106.94	108.13	106.58±0.10
$^{256}104$	95.90	94.84	95.6	94.37	95.77	94.24±0.07

From the experimental mass excesses and the liquid drop (spherical) part of the Møller-Nix mass formula [23] we determined shell corrections for the N-Z = 48 nuclei (Fig. 3). Shell effects are included for the odd-mass nuclei with the same N-Z. The shell effects show a continuous increase with proton number, up to Z = 107, from 0.5 MeV for uranium to more than 5 MeV for Z > 104 elements.

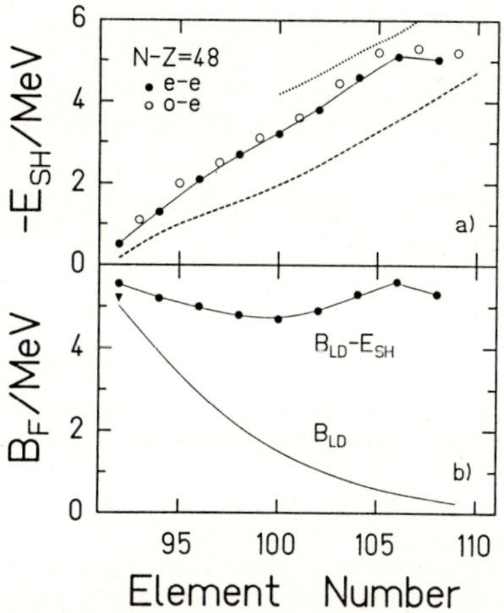

Fig. 3: a) Ground state shell effects of N-Z = 48 nuclei. Dashed line: Microscopic part of the Møller-Nix mass formula [23]. Dotted line: Calculations from Cwiok et al. [24].
b) Liquid drop fission barrier (B_{LD}) [25] and experimental fission barriers ($B_{LD}-E_{SH}$) [26].

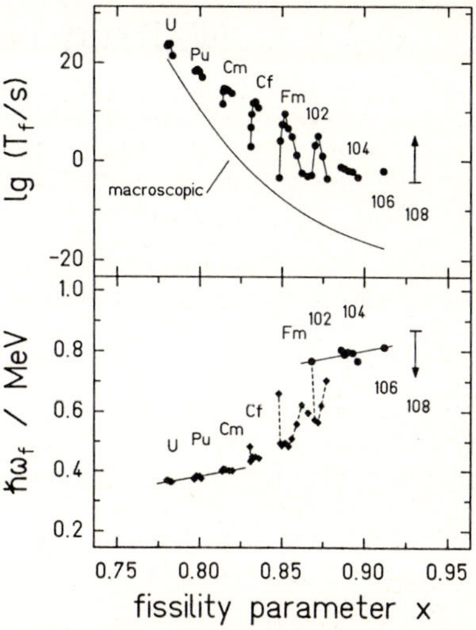

Fig. 4: The upper part shows the fission half-lives of all known even-even nuclei together with the half-lives expected for nuclei with macroscopic fission barriers [25]. The lower part of the figure shows the curvature parameter $\hbar\omega_f$. Diamonds belong to nuclei with a double-humped barrier and dots are nuclei with a predicted single-humped barrier.

Experimental data cannot yet answer the question whether the shell effects continue to increase slowly further to heavier elements. The N-Z = 48 nuclei rather suggest a flattening. This would confirm calculations [24,27] predicting a region of deformed nuclei of higher stability extended in front of the classical spherical island of superheavy nuclei. The latter gain their stability from spherical shell closures at Z = 114 and N = 184 [1]. Shell corrections increase up to a maximum value of nearly 9 MeV for the nuclei Z = 114, N = 178 [27]. The island around Z = 109, N = 162 consists of nuclei which are deformed (ε_2 = + 0.20, ε_4 = + 0.09). In Fig. 5 we have plotted the results of the recent shell model calculations. The figure suggests that the investigated N-Z = 48 and 49 nuclei are

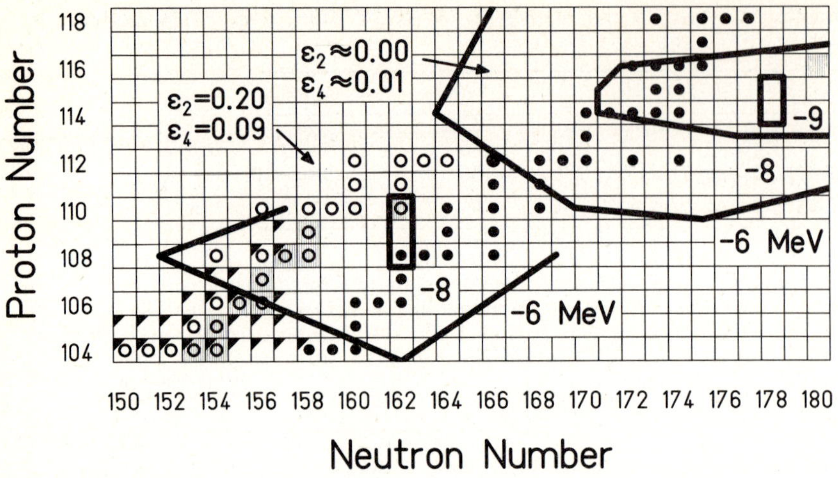

Fig. 5: Ground state shell corrections as calculated in [24] (lower left part) and [27] (upper right part). Open circles: Compound nuclei of reactions (Ti to Zn) + ^{208}Pb → Z = 104 to 112. Dots: Compound nuclei of reactions (Mg to Fe) + ^{238}U → Z = 104 to 118. Full triangles: Known nuclei. Shaded: Compound systems investigated with SHIP.

crossing the island of increased stability west from the maximum, and ^{266}109 is already on the side of decreasing shell corrections.

4.2 Fission Barriers and Half-lives

Fission barriers of N-Z = 48 nuclei have been determined as a sum of the liquid drop barrier [25] and the experimental ground state shell effects (Fig. 3b). Although the liquid drop barrier vanishes for the heavy elements the increasing shell effects keep the fission barriers constant near 5 MeV.

Figure 4a shows the systematic decrease of experimental fission half-lives up to Z = 102 by 2 or 3 orders of magnitude per element at equal N-Z. The overlying structure with maximum half-lives of N = 152 isotones is due to a subshell closure at this neutron number. As the barrier height is constant for the N-Z = 48 nuclei the decrease of the half-lives by more than 20 orders of magnitude is due to the decrease of the liquid drop barrier widths. This trend changes at element 104 from where on the constant half-lives suggest that beside the barrier heights also the shapes of the barriers remain constant. For N-Z = 50 nuclei the shapes have changed from a double humped barrier to a less broad single humped barrier between element 102 and element 104 [3].

In Fig. 4b the curvature $\hbar\omega_f$ of the fission barrier is plotted as a function of the fissility parameter x. $\hbar\omega_f$ was extracted from a Hill-Wheeler barrier transmission formula using the experimental values for the spontaneous fission half-lives and the fission barriers [28]. For elements 104 and 106 high values of $\hbar\omega_f$ point to a narrow barrier the width of which is about 50 % of the values obtained for liquid drop fission barriers near uranium. Moreover, constant half-lives and fission barriers point to a barrier which is nearly independent of the fissility x of the nucleus.

Recent calculations of the fission half-lives [27] of elements 106 to 110 give increasing values with increasing neutron number up to N = 162, the isotones with the highest shell effects. However, in case of ^{260}106 the experimental partial fission half-life is 7 ms and still 71 times longer than the calculated value (98 μs). This trend is continued by a sf half-life limit of ^{264}108 ($T_{1/2,sf}$ > 62 μs) deduced from our experiment. Also this value is 150 times longer than the calculated value of 0.4 μs.

5. Outlook Beyond Element 109

Future studies on fusion reactions may allow to find better strategies for the time consuming search experiments. Theory predicts an island of increased stability of deformed nuclei around Z = 109, N = 162. Most of the known isotopes of elements above 104 gain their stability from these nuclear structure effects. Future experiments will take into account the island character and will proof the border lines. Further neutron deficient isotopes of the known elements can be investigated in reactions with lead and bismuth targets. Experiments to produce nuclei of the neutron-rich part of the island with targets around uranium will be continued with a systematic study of cross sections. If the theoretical predictions are correct, the increased stability of deformed nuclei up to about 162 neutrons against fission may help to keep the compound nuclei stable during the evaporation process, and may help us again to identify new isotopes of the heaviest elements by detection of α decay chains offering a direct possibility to measure nuclear binding energies of the heaviest nuclei. New data on the synthesis and decay of lighter elements and improvements of the experimental set-up will possibly help to synthesize nuclei of elements beyond Z = 109.

References

1. W.D. Myers, W.J. Swiatecki: Nucl. Phys. 81, 1 (1966)
2. S.G. Nilsson et al.: Nucl. Phys. A131, 1 (1969)
3. Y.T. Oganessian: Lect. Notes Phys. 33, 221 (1974)
4. G. Münzenberg et al.: Nucl. Instr. Meth. 161, 65 (1979)
5. Y.T. Oganessian et al.: JETP Lett. 20, 265 (1974)
6. J. Randrup et al.: Phys. Rev. C13, 229 (1976)
7. S. Hofmann et al.: Nucl. Instr. Meth. 223, 312 (1984)
8. F.P. Heßberger et al.: GSI Annual Report 1981, p. 66 (1982); Z. Phys.
9. G. Münzenberg et al.: Z. Phys. A300, 107 (1981)
10. G. Münzenberg et al.: Z. Phys. A309, 89 (1982)
11. P. Armbruster: Ann. Rev. of Nucl. and Part. Science 35, 135 (1985)
12. Y.T. Oganessian et al.: Radiochemica Acta 37, 113 (1984)
13. G. Münzenberg et al.: GSI Annual Report 1985, p. 31 (1986)
14. Y.T. Oganessian: Int. School-Seminar on Heavy Ion Physics, Alushta, JINR D7-83-644, p. 55, Dubna (1983)
15. A.G. Demin et al.: Z. Phys. A315, 197 (1984)
16. G. Münzenberg et al.: Z. Phys. A317, 235 (1984)
17. Y.T. Oganessian et al.: Z. Phys. A319, 215 (1984)
18. Y.T. Oganessian: private communication
19. W.D. Myers: Droplet Model of Atomic Nuclei, IFI/Plenum, NY (1973)
20. H.v. Groote et al.: At. Data and Nucl. Data Tables 17, 418 (1976)
21. P.A. Seeger and W.M. Howard: Nucl. Phys. A238, 49 (1975)
22. S. Liran and N. Zeldes: At. Data and Nucl. Data Tables 17, 431 (1976)
23. P. Møller and J.R. Nix: At. Data and Nucl. Data Tables 26, 165 (1981)
24. S. Cwiok et al.: Nucl. Phys. A410, 254 (1983)
25. M. Dahlinger et al.: Nucl. Phys. A376, 94 (1982)
26. P. Armbruster: The Int. School of Physics "Enrico Fermi", Varenna (1984)
27. G.A. Leander et al.: Proceedings of the 7th International Conf. on Atomic Masses and Fundamental Constants, AMCO-7, Darmstadt-Seeheim, p. 466 (1984) and Z. Phys. A323, 41 (1986)
28. F.P. Heßberger: Thesis, TH Darmstadt (1984)
29. G. Münzenberg et al.: to be published

What is the Source of the Narrow Positron Peaks Observed in Superheavy Collision Systems?

J. Schweppe and J.S. Greenberg

Yale University, A.W. Wright Nuclear Structure Laboratory, P.O. Box 6666, New Haven, CT 06511, USA

We have observed that the positrons associated with a narrow peak in the positron spectrum from U+Th collisions are correlated with the simultaneous emission of electrons whose energy spectrum also contains a narrow peak. The mean energies and widths of the two peaks are equal within measurement errors. Neither the coincidence-peak intensity nor the energy distributions of the positrons and electrons can be accounted for by known nuclear internal conversion processes. Similar observations have also been made in the Th+Th and Th+Cm collision systems. New measurements confirm the emission of correlated positrons and electrons and establish the existence of more than one set of correlated energies.

1 Introduction

This report describes the latest developments in a series of experiments performed at GSI Darmstadt by the EPOS collaboration[1-3] to study the production of positrons in collisions of very heavy ions and atoms. The original motivation for this effort was an experimental search for spontaneous positron production[4-6], a new QED process predicted to occur in the large electromagnetic fields produced transiently in these collsions. It is not necessary to describe spontaneous positron production in any greater detail here because that was apparently not what was found.

What was found was the following. Shown in Figure 1 is the energy distribution of positrons emitted in six very-heavy-ion collision systems[2]. The six systems cover a range of combined nuclear charge, $Z_u \equiv Z_{proj} + Z_{targ}$, the important parameter for spontaneous positron production, from $Z_u = 188$ down to $Z_u = 163$. For the purposes of spontaneous positron production, the heavier systems in parts

*The experiments described here were carried out by the EPOS collaboration at the Gesellschaft für Schwerionenforschung, Darmstadt, and reflect the combined efforts of its members: Hartmut Backe, Klaus Bethge, Helmut Bokemeyer, Tom Cowan, Helmut Folger, Kiyo Sakaguchi, Dirk Schwalm, Kurt E. Stiebing, and Paul Vincent.

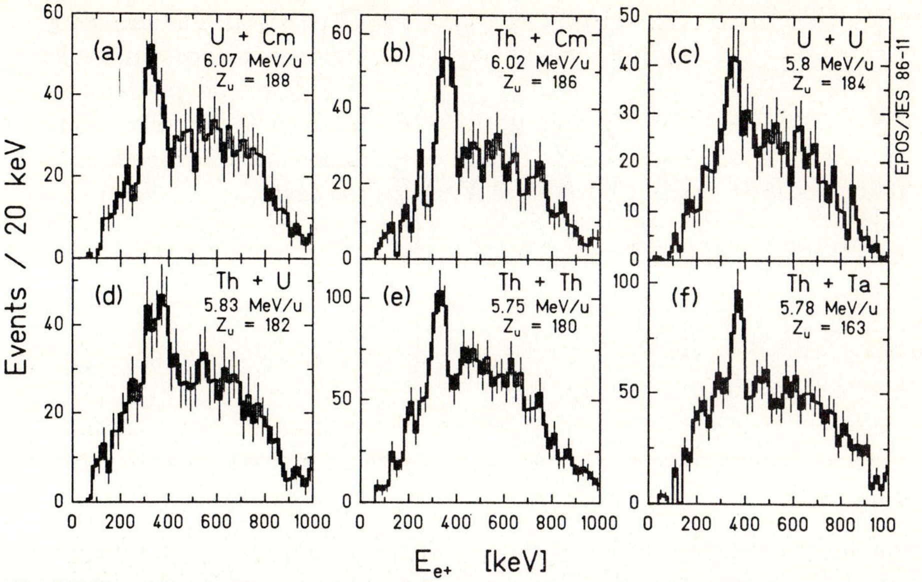

FIGURE 1. Positron energy distributions for the six collision systems and projectile energies indicated. Kinematic constraints have been chosen as discussed in the text.

(a) through (e) are supercritical, while the Th+Ta collision system shown in part (f) is well below the critical point. The bombarding energies of the collisions are indicated and can be summarized as being collisions near the Coulomb barrier, corresponding in each case to a partial overlap of the nuclear density distributions. In addition to the smooth background of dynamically produced positrons, a narrow peak is clearly visible in each spectra at a energy between 300 and 400 keV (see also References 7 and 8). The spectra correspond to kinematic conditions chosen empirically to enhance the peak with respect to the background by taking advantage of an apparent difference in the heavy-ion scattering-angle correlations for the events associated with the positron peak with respect to the dominant Rutherford-scattering events[1,2]. The width of the peak in each case is about 70 keV. This is consistent with the Doppler broadening expected for monoenergetic emission of positrons from a system moving with the velocity of the center of mass of the heavy-ion collision system. It also corresponds by the Uncertainty Relationship to emission from a system surviving more than 10^{-20} seconds, or at least an order of magnitude longer than the Rutherford scattering time.

Perhaps the most striking feature of the data is the near constancy of the energy of the peak, despite the range of collision systems studied. This is summarized in Figure 2, where the peak position is plotted as a function of the combined nuclear charge, Z_u. This constancy is in direct contradiction to the predictions[9] of the simplest model of spontaneous positron production in which it is assumed that the nuclear charge configuration of the two colliding nuclei does not change

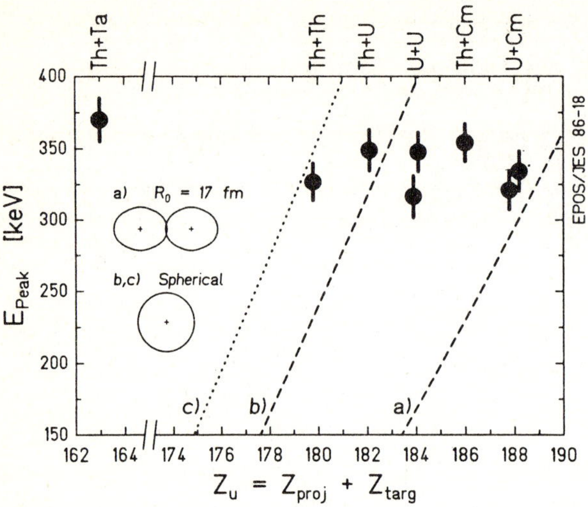

FIGURE 2. The mean energies of the positron peaks in Figure 1 are plotted as a function of Z_u. The calculated[9] lines are described in the text.

appreciablely during the collision. This prediction[9] is shown by the dashed line (a) in Figure 2. The curves (b) and (c) show the effect of assuming that the two nuclei coalesce into a spherical form and that additionally all electrons are removed during the collision, respectively. It is obviously difficult to explain the data within a model based on spontaneous positron production unless one assumes that the nuclear charge configuration and perhaps also the electronic charge state track with the combined charge of the collision system in just such a way as to maintain a nearly constant binding energy for the $1s\sigma$ electrons of the quasimolecule. Furthermore, the coalesced system must somehow reform a binary system with kinematics resembling that of Rutherford scattering in order to match the experimental detection trigger.

The difficulties with an explanation based on spontaneous positron production led us to consider other possible sources for the similar, approximately monoenergetic, positron emission found in all these collsion systems. An earlier analysis of the positron line shape had already shown us the difficulty of explaining the narrow positron structures on the basis of the relatively broad energy distribution of positrons from the internal pair conversion of excited nuclear states populated during the collsion. Indeed, three-body decay processes in general, in which the positron shares kinetic energy statstically with other particles, produce an energy distributions of positrons which is too wide to match the observed peaks. As a result, we began to consider the possibility that the experimentally-observed positron peaks are narrow because the positron is the product of the two-body decay of a neutral state into a positron-electron pair[2]. In this case , of course, the symmetric positron and electron would be emitted back to back in the rest frame of the decaying state and share equally the available energy. Obviously the signature of such a decay process would be the emission of monoenergetic electrons of approximately the same energy as the observed positron peaks in coincidence with these events.

2 Apparatus

In order to look for coincident positron-electron emission in heavy-ion collisions, we modified our positron spectrometer, EPOS[10], to include the detection of electrons. The spectrometer is shown in Figure 3 below in three views: from the side in part (a), in three cross-sectional views through the new electron counter, through the target chamber, and through the positron detector in part (b), and as a perspective view of the main elements in part (c). The original positron spectrometer, built for the detection of positrons emitted in heavy-ion collisions, is based on a solenoidal-magnetic-field transport system which combines a large positron detection efficiency with good separation of the positrons from the overwhelming backgrounds of gamma rays and delta electrons copiously produced in heavy-ion collisions. The positrons are transported at right angles to the beam direction by the $\sim$ 2 kilogauss magnetic field to an axially positioned, 10 mm $\times$ 100 mm, cylindrical Si(Li) detector some 83 cm away from the target. The intrinsic resolution of the liquid-nitrogen-cooled Si(Li) counter is $\sim$ 12 keV. The axial-focussing property of the solenoidal field for charged particles produces the separation of positrons from gamma rays, and a baffle system suppresses electrons. Backgrounds are further suppressed by the detection of the characteristic positron annihilation radiation in a cylindrical arrangement of NaI(Tl) crystals positioned around the positron detector.

The kinematic parameters for each positron-producing collision are determined by measuring the scattering angles of both heavy nuclei in a pair of position-sensitive, parallel-plate, avalanche counters located in the target chamber. They cover the laboratory polar-angular region from 25° to 65° with respect to the beam axis and a total azimuthal-angular region of 120°.

A major background of positrons, that due to the internal pair conversion of excited nuclear states, was determined by folding the gamma-ray yield monitored in two 3"×3" NaI(Tl) crystals, positioned just outside the target chamber, with theoretical internal-pair-conversion coefficients according to an empirical method based on measurements of low-Z collision systems where this source of positrons dominates.

The new electron detector is installed in the opposite arm of the solenoid from the positron detector, about 112 cm from the target. The electron detector consists of two $65 \times 32 \times 3$ mm^3 planar Si(Li) crystals. Their arrangement[11] is shown in Figure 3 (b). As opposed to the radially symmetric and axially positioned positron-detection system, the electron detectors are positioned symmetrically off the solenoid axis to allow the large flux of low-energy delta electrons to spiral past between the two detectors. They are oriented with their sensitive faces parallel to the solenoid axis and facing inward to suppress the detection of oppositely spiralling positrons. The intrinsic resolution of the alcohol-cooled Si(Li) counters was $\sim$ 35 keV.

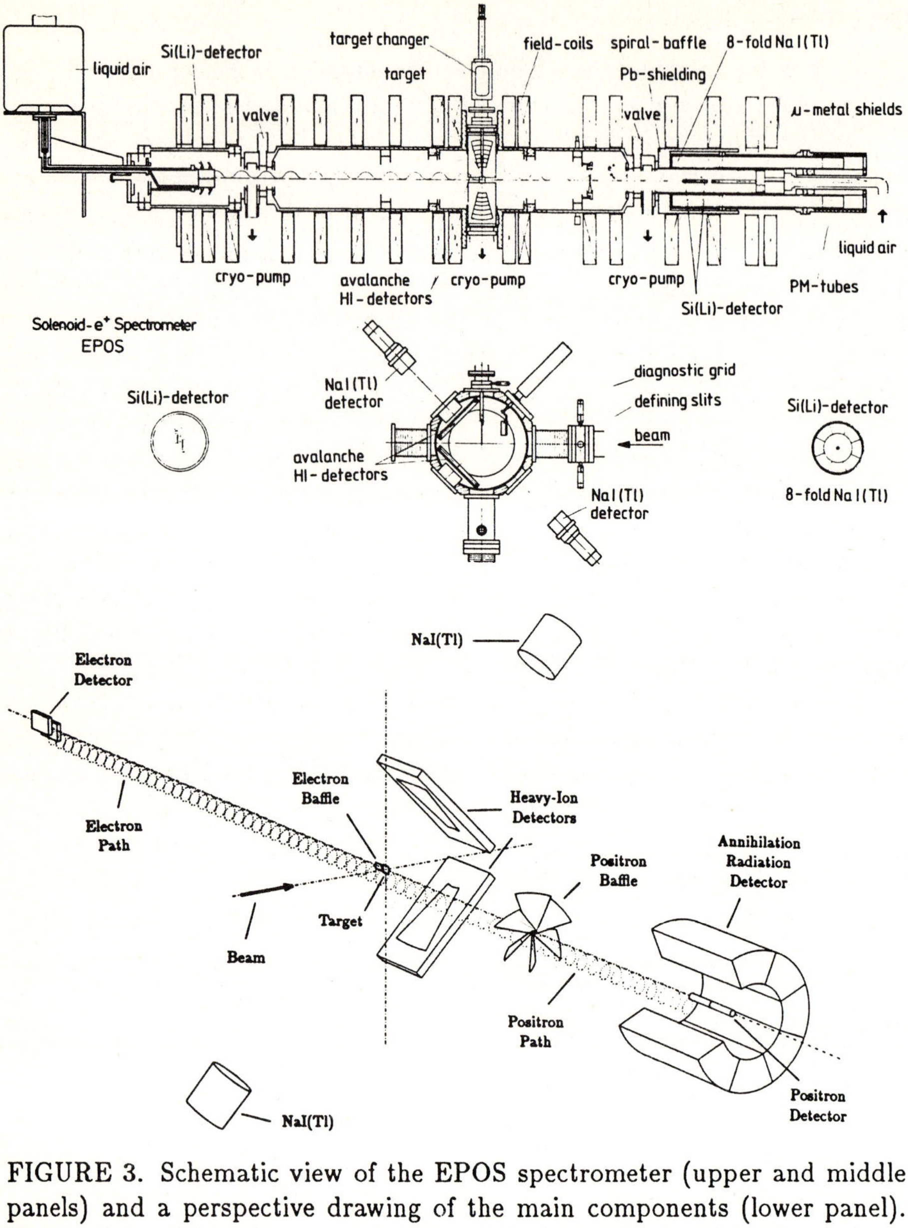

(a)

FIGURE 3. Schematic view of the EPOS spectrometer (upper and middle panels) and a perspective drawing of the main components (lower panel). (Two of the eight segments of the annihilation-radiation detector are not drawn in the lower panel in order to allow a better view of the positron detector.)

3 Monte Carlo Simulations

In order to gain an understanding of the response of our apparatus to different electron-positron production processes, Monte Carlo simulations were performed[12]. Of these, two important processes will be discussed here. The simplest process in many ways is the previously mentioned, two-body decay of a neutral state into an electron-positron pair. The results of a simulation[12] of this process are shown in Figure 4 for the case that a neutral state existing at rest in the center of mass of the colliding heavy nuclei decays into an electron and a positron, which are emitted back to back in this reference frame, each claiming half of the available total energy of 1.8 MeV (chosen to reproduce the energy of the peaks in the positron spectra of Figure 1). The input into the Monte Carlo calculation is not only the calculated kinematics of such a decay from the center of mass of a heavy-ion collision as observed in the laboratory reference frame, but also the measured transport efficiencies and detector response matrices for the detection of the positron and the electron. Figure 4 (a) shows the calculated distribution of positron energies measured in singles. Part (b) shows the two-dimensional distribution of coincidence events as a function of the kinetic energy of the emitted positron (horizontal scale) and of the electron (vertical scale). As expected, in the laboratory reference frame both the positron and electron show the full Doppler broadening of ~ 70 keV. This is seen even more clearly in Figure 4 (c) and (d) for positrons and electrons, respectively. In anticipation of the comparison with the measured data below, the peak has been isolated in these projections of Figure 4

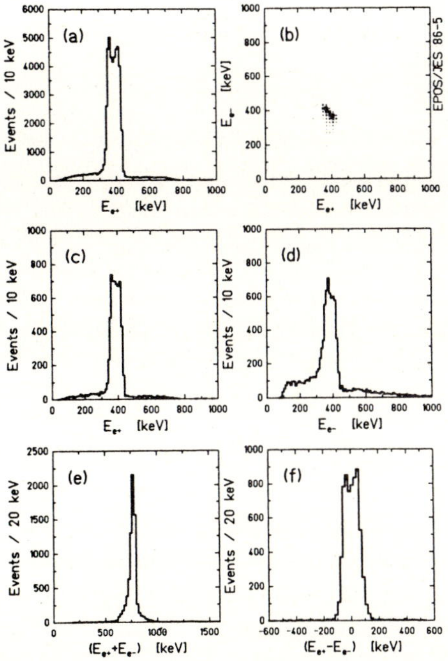

FIGURE 4. Monte Carlo simulation of the two-body decay of a neutral state of mass 1.8 MeV into an electron-positron pair in the EPOS spectrometer.

(b) by gating in each case on the peak in the other lepton channel in the energy interval 340 $< E <$ 420 keV. These are the two gates marked A and B, respectively, in Figure 4 (b).

As can be seen in Figure 4 (b), because of the particularly simple form of back-to-back emission in the center-of-mass rest frame, the Doppler broadening has a simple, diagonal form in the laboratory rest frame. As a result, a more natural coordinate system is the sum and the difference of the positron and electron energies. This gives new coordinate axes at 45° to the E_{e^+} and E_{e^-} axes. The projections onto these variables are shown in parts (e) and (f), respectively. As above, the peak area has been picked out by the gates marked C and D in Figure 4 (b). Gate C is wedge shaped to compensate approximately for the increase in Doppler broadening for larger positron and electron energies. The naturalness of these projections can be seen most clearly in Figure 4 (c) where, due to the event-by-event cancellation of the Doppler shifts of the positron and the electron, the width of the peak is determined essentially by the resolution of the detection systems and is significantly less than the Doppler-broadened widths measured in the individual positron and electron spectra of parts (c) and (d).

For comparison, Figure 5 shows the same projections of a Monte Carlo simulation[12] of the most obvious source of electron-positron pairs in heavy-ion collisions, the internal pair conversion of excited nuclear states. An E1 transition for $Z = 92$ is assumed and the energy of the transition is again 1.8 MeV. This three-body decay process is representative in general of normal positron-producing processes

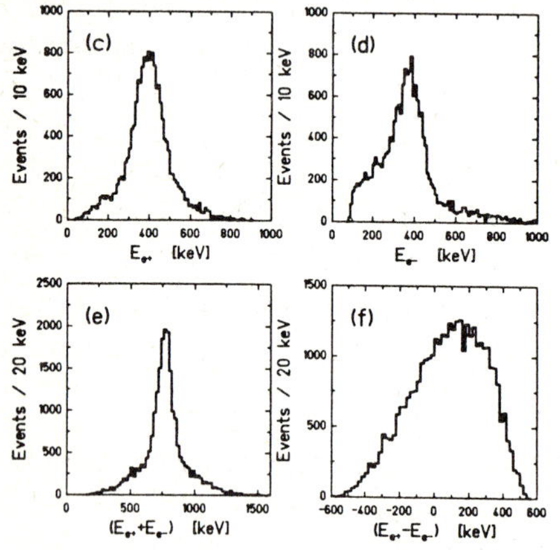

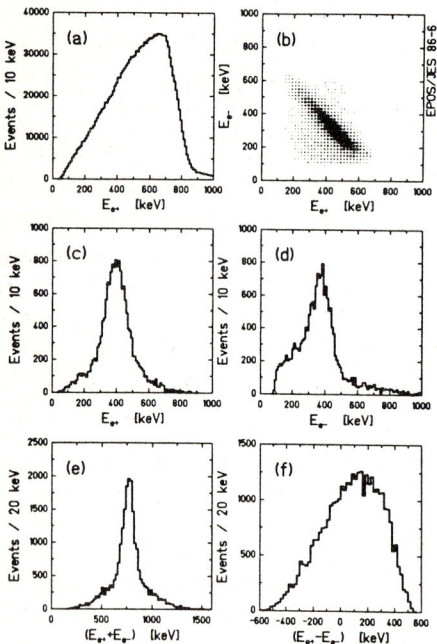

FIGURE 5. Monte Carlo simulation of the internal pair conversion of an E1 nuclear transition for $Z = 92$ in the EPOS spectrometer.

in heavy-ion collisions. The differences between it and the two-body decay of Figure 4 are clear. The singles positron spectrum in part (a) is much broader and the distribution of coincidence events in the plane of positron and electron energies of part (b) is also much larger. This last point is important not only for the form of the various projections, but also for the relative detection efficiency of such events within the narrow gates A through D. Although one can recover relatively narrow peaks by placing narrow gates on the electron or positron energies, as shown in parts (c) and (d), respectively, the projections onto the energy sum and difference axes, shown in parts (e) and (f), respectively, once again bring out the large difference to the two-body decay considered above. The energy-sum spectrum of Figure 5 (e) is about three times as wide as the corresponding projection for two-body decay, shown in Figure 4 (e) above and the difference distribution in part (f) is about four times as wide as that in Figure 4 (f). Clearly these two types of production processes can be distinguished.

Of course, the peak events are only a small fraction of the electrons and positrons produced in heavy-ion collsions, so these events have to be added to the smooth backgrounds of electrons and positrons from dynamic quasimolecular processes and from nuclear processes. This sum is shown below in Figure 7, parts (e) through (l), for the two processes and the four projections discussed above in Figures 4 and 5. The ratio of peak events to dynamic and nuclear background is chosen for comparison to the measured data.

4 Data

These two Monte Carlo simulations form a basis for examining the measured data. The first measurements of coincident electron-positron emission in heavy-ion collisions with the new apparatus were performed in July, 1985[3]. The U+Th collision system was measured at a bombarding energy of 5.83 MeV/u. The $\sim$ 1830 detected coincident events are shown in Figure 6 (a) as a function of the measured kinetic energy of the positron and the electron. The total projections onto the positron and electron energy axes are shown in parts (b) and (c), respectively. Also indicated on the two-dimensional plot of part (a) are the same four gates A through D discussed above in Figure 4. The projections of the measured coincident events onto the positron-energy, electron-energy, energy-sum, and energy-difference axes using gates A through D are shown in Figure 7, parts (a) through (d), respectively. In each measured spectrum, a peak is clearly visable above the smooth background. The peak in the energy-sum spectrum of part (c) contains 35.3 ± 9.4 counts above the continuous background. A statistical fluctuation of this size is excluded at the $\sim 6\sigma$ level.

The results of the two Monte Carlo simulations described above are shown below the data in parts (e) through (l) of Figure 6. The peak events have been added to a Monte Carlo calculation[12] of the dynamic and nuclear background of

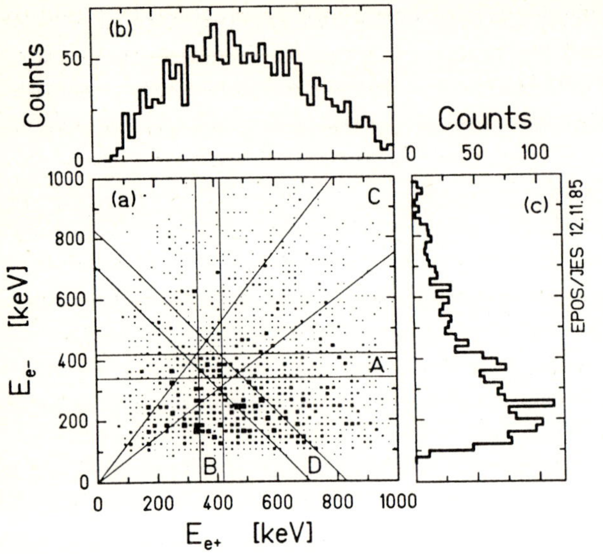

FIGURE 6. Measured coincident electron positron data for the U+Th collision system at a bombarding energy of 5.83 MeV/u. Part (a) is the distribution of coincident events as a function of the positron and the electron kinetic energies. Parts (b) and (c) are the total projections of this data onto the positron-energy and the electron-energy axes, respectively.

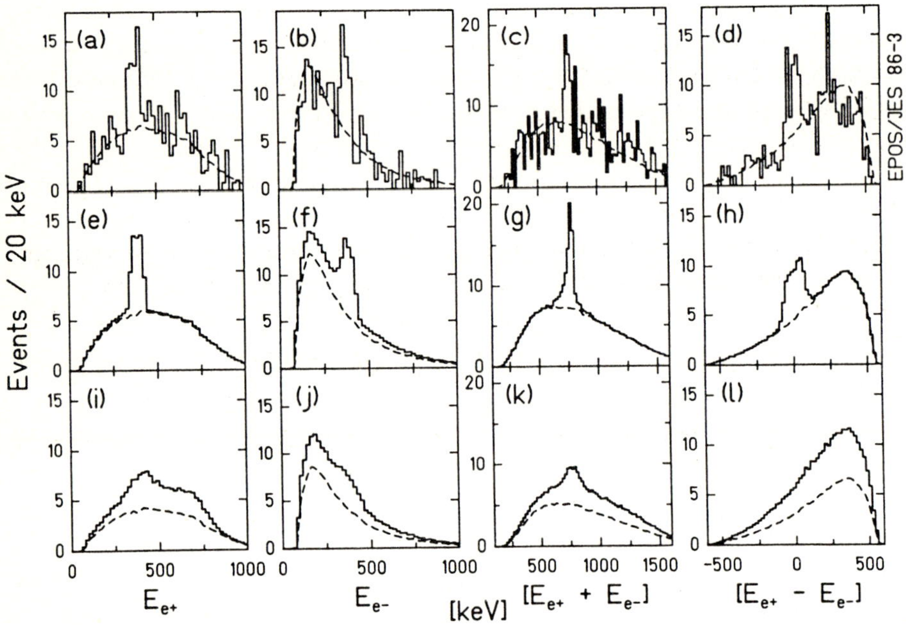

FIGURE 7. Comparison of projections (described in the text) of a measurement of the U+Th collision system at a bombarding energy of 5.83 MeV/u in parts (a) through (d) with the Monte Carlo simulation of the decay of a neutral state in parts (e) through (h) and of the internal pair conversion of a nuclear transition in parts (i) through (l).

positrons and electrons with a single, overall normalization to the total number of detected events. This part of the calculation is indicated by the dashed line in all of the panels of Figure 6. Parts (e) through (h) show the two-body decay of a neutral state, whereby it has been assumed that 3% of the positrons emitted in the heavy-ion collisions are due to this decay process in order to reproduce the observed peak intensity in the measured coincident spectra. Parts (i) through (l) show the internal pair conversion of an E1 nuclear transition. Here it is necessary to assume that fully 30% of the positron yield is due to the internal pair conversion of this single nuclear transition in order to produce the peak intensity shown. This is due to the reduced detection efficiency mentioned above for detecting these events within the narrow gates used to produce these plots. (To reproduce the 35 counts measured in the peak would actually require more like 60% of *all* measured positrons to come from *one* nuclear transition, in contradiction with the observation that *all* nuclear internal pair conversion accounts for only 20–30% of the positron yield.) Clearly the agreement between the measured spectra and the simulation of a two-body decay process is striking.

A study has been made to check that the observed peaks in the four projections shown in Figure 7 (a) through (d) correspond to events truly concentrated in one spot in the two-dimensional distribution of Figure 6 (a), and not manufactured by the narrow gates A through D. As an example, Figure 8 shows three projections onto the energy-sum axis. The three gates used are indicated in the two-dimensional plot of part (c). Part (b) repeats Figure 7 (c) above. Parts (a) and (d) are projections of the two neighboring regions. No significant peak is obvious in either distribution. As in Figure 6 above, the dashed lines are the same projections of a Monte Carlo simulation. Figure 9 shows projections onto neighboring regions for the other three coordinate axes: positron energy, electron energy, and energy-difference. In each case, the two neighboring projections, indicated in Figure 9 (a), are added together. Again, no prominent structure is evident in these neighboring regions.

Measurements were also made on two other collision systems, Th+Th and Th+Cm, at similar bombarding energies[3]. Similar correlated structures are seen in both systems, although at a lower level of statistical certainty. There is one additional feature worth noting which appears in these data: In the Th+Th system, peaks were observed in the positron spectra both at $\sim$ 310 and at $\sim$ 370 keV. There is evidence that, in both cases, an electron peak accompanies the positron peak with $E_{e^-} \simeq E_{e^+}$. This observation may be related to possible differences in the energies of the positron peaks shown in Figure 1 above, reflecting the possibilty of more than one structure.

These measurements were repeated in the Spring and the Summer of 1986. For these runs, the resolution of the electron counters was improved to $\sim$ 10 keV by cooling with liquid nitrogen. The collision system U+Th was measured again at bombarding energies around 5.87 MeV/u. The analysis of the data collected

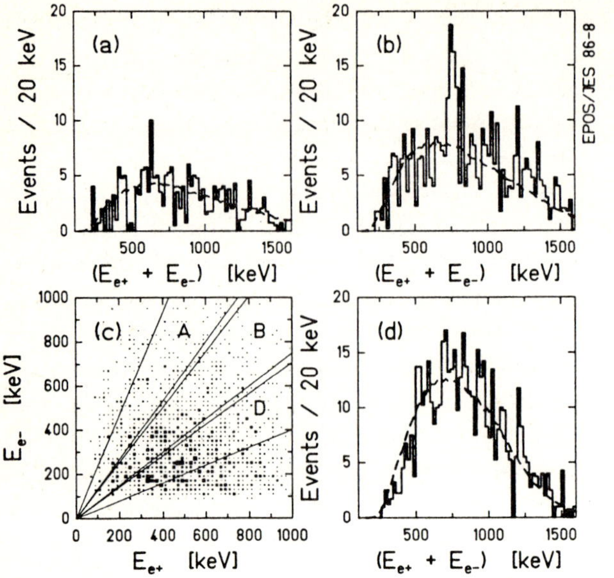

FIGURE 8. Three projections of a measurement of the U+Th collision system at a bombarding energy of 5.83 MeV/u onto the energy-sum axis. The projections shown in parts (a), (b), and (d) of the data in part (c) are gated on the regions marked A, B, and D in part (c), respectively.

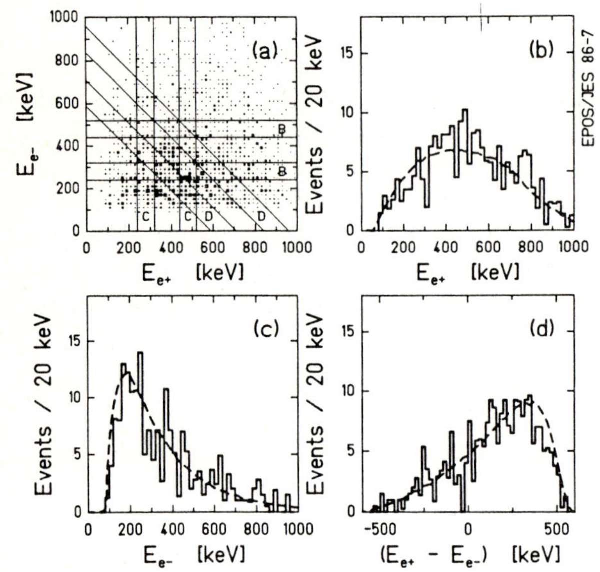

FIGURE 9. Projections of a measurement of the U+Th collision system at a bombarding energy of 5.83 MeV/u onto the positron-energy, electron-energy, and energy-difference axes. The projections shown in parts (b), (c), and (d) of the data in part (a) are gated on the pairs of regions marked B, C, and D in part (a), respectively.

is still proceeding, but a preliminary analysis has reaffirmed the existence of the coincident electron-positron events, indicates that the energy-sum lines are significantly narrower than than the corresponding positron and electron lines, and confirms the evidence of the earlier measurement that there are more than one set of discrete energies. Some preliminary results are shown in Figure 10. The upper panels (a) and (b) and the lower panels (c) and (d) show energy-sum and -difference spectra for two subsets of the data gated on beam energy, heavy-ion scattering angle, and positron/electron time of flight chosen to enhance these two sum lines, respectively.

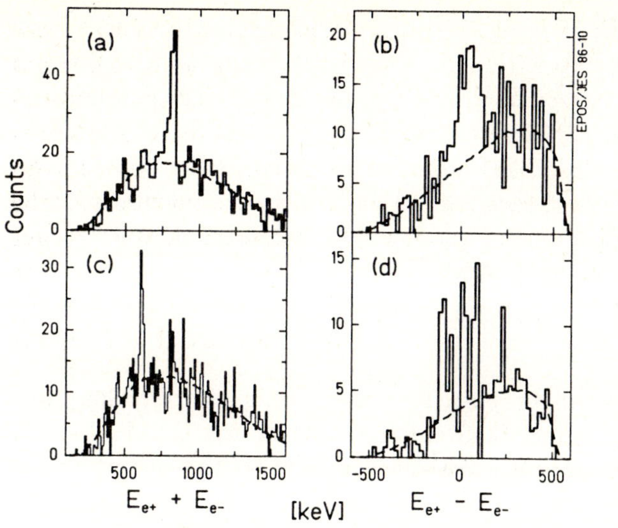

FIGURE 10. Results of a preliminary analysis of measurements in the U+Th collision system at bombarding energies around 5.87 MeV/u. The upper panels and lower panels show energy sum and energy difference distributions for two subsets of the data gated on beam energy, heavy-ion scattering angle, and positron/electron time of flight chosen to enhance these two sum lines, respectively.

5 Conclusions

The experimental situation can be briefly summarized as follows. A narrow structure is observed in the distribution of electron energies correlated to the narrow peaks already reported in the energy distribution of positrons emitted in very-heavy-ion collisions near the Coulomb barrier. The energy of the electron structure is approximately that of the correlated positron structure. The distribution of the event-by-event sum of the positron and electron energies also contains a structure which is significantly narrower than that in either the positron or electron distributions. This seems to indicate that the Doppler shifts of the positron and the electron partially cancel event by event, implying that the positron and the electron are emitted approximately back to back in the rest frame of their production. The measured widths of the peaks in the four projections of Figure 7 (a) through (d) are consistent with a velocity for the source of the correlated electrons and positrons comparable to the velocity of the center of mass of the colliding heavy ions. The coincidence intensity and the shape of the spectra cannot be reproduced by known nuclear internal-conversion processes. There are at least two sets of correlated electron-positron energies, and the lines have been seen in three different collision systems: U+Th, Th+Th, and Th+Cm. There is also some evidence for a dependence of the correlated electron-positron production on the projectile energy, the scattering angles of the colliding nuclei, and the measured time of flight of the positron and the electron in the spectrometer. The last point may indicate the existence in the production process of either a time delay or a dependence on the emission angle of the positron and electron.

Theoretical efforts to understand this experimental effect have have included considerations of conventional atomic physics effects[13] as well as speculations on the breakdown of the QED vacuum[14], the production of a new, previously undetected, neutral particle[2,3,15] with a mass of $\sim 3m_{e^-}$, in particular the axion[16], the formation of polyelectronic complexes[17], or new bound states of electron-positron pairs[18], or solitons in the QED vacuum[19]. So far, no completely consistent description has been found which can account for all of the observed properties — in particular the multiplicity of lines, the observed intensity, or the apparent low velocity of the decaying state. It appears that an extensive experimental program studying the various characteristics listed above will be necessary in order to shed light on this unexplained experimental effect.

Acknowledgements

As stated at the begining of this report, the experiments described here were carried out by the members of the EPOS collaboration at the Gesellschaft für Schwerionenforschung, Darmstadt. We are very grateful to the operating and engineering staff at the GSI UNILAC for their dedication in operating the accelerator and assistence in performing these measurements, and to the Transplutonium Program of the U.S Department of Energy for the loan of the curium material. Particular appreciation is due to Werner Kreuzer for his dedication and ingenuity in the design and construction of our spectrometer. This work was supported in part by the Bundesministerium für Forschung und Technologie of the Federal Republic of Germany, by the U.S. Department of Energy through Contracts No. DE–AC02–76ER03074 and No. DE–AC02–76CH00016, and by the U.S. National Science Foundation.

References

(1) J. Schweppe *et al.*, Phys. Rev. Lett. **51**, 2261 (1983).
(2) T. Cowan *et al.*, Phys. Rev. Lett. **54**, 1761 (1985).
(3) T. Cowan *et al.*, Phys. Rev. Lett. **56**, 444 (1986).
(4) V.V. Voronkov and N.N. Kolesnikov, ZETF **39**, 189 (1960) (Sov. Phys. JETP **12**, 136 (1961)).
(5) W. Pieper and W. Greiner, Z. Phys. **218**, 327 (1969)
(6) S.S. Gershtein and Ya.B. Zel'dovich, ZETF **57**, 654 (1969) (Sov. Phys. JETP **30**, 358 (1970)); Lett. Nuovo Cimento **1**, 835 (1969)).
(7) M. Clemente *et al.*, Phys. Lett. **B137**, 41 (1984).
(8) H. Tsertos *et al.*, Phys. Lett. **B162**, 372 (1985).
(9) J. Reinhardt *et al.*, Z. Phys. **A303**, 173 (1981).
(10) H. Bokemeyer *et al.*, in *Quantum Electrodynamics of Strong Fields*, edited by W. Greiner (Plenum, New York, 1983), pp. 273–292; J.S. Greenberg and P. Vincent, in *Treatise on Heavy-Ion Science*, edited by D.A. Bromeley (Plenum, New York, 1985), pp. 139–421.
(11) H. Backe *et al.*, Phys. Rev. Lett. **50**, 1838 (1983).

(12) T. Cowan, to be published in *Physics of Strong Fields*, edited by W. Greiner (Plenum, New York).
(13) W. Lichten and A. Robatino, Phys. Rev. Lett. **54**, 781 (1985); P. Schlüter *et al.*, Z. Phys. **A323**, 139 (1986); P. Schlüter *et al.*, Phys. Rev. **C33**, 1816 (1986).
(14) G. Soff *et al.*, Nuc. Ins. Meth. Phys. Res. **B9**, 747 (1985); **B10** 214 (1985)
(15) A.B. Balentekin *et al.*, Phys. Rev. Lett. **55**, 461 (1985); A. Schäfer *et al.*, J. Phys. **G11**, L69 (1985); J. Reinhardt *et al.*, Phys. Rev. **C33**, 194 (1986); A. Chodos and L.C.R. Wijewardhana, Phys. Rev. Lett. **56**, 302 (1986); K. Lane, Phys. Lett. **B169**, 97 (1986); B. Müller and J. Reinhardt, Phys. Rev. Lett. **56**, 2108 (1986); E. Ma, Phys. Rev. **D34**, 293 (1986); D. Carrier *et al.*, Phys. Rev. **D34**, 1332 (1986); G. Mageras *et al.*, Phys. Rev. Lett. **56**, 2672 (1986); T. Bowcock *et al.*, Phys. Rev. Lett. **56**, 2676 (1986); A. Zee, Phys. Lett. **B172**, 377 (1986); A. Schäfer *et al.*, Mod. Phys. Lett. **A1**, 1 (1986); A. Schäfer *et al.*, Z. Phys. **A324**, 243 (1986); U.E. Schröder, Mod. Phys. Lett. **A1**, 157 (1986).
(16) N.C. Mukhopadhyay and A. Zehnder, Phys. Rev. Lett. **56**, 206 (1986); R.D. Peccei *et al.*, Phys. Lett. **B172**, 435 (1986); L.M. Kraus and F. Wilczek, Phys. Lett. **B173**, 189 (1986); M. Suzuki, Phys. Lett. **B175**, 364 (1986); L.M. Kraus and M.B. Wise, Phys. Lett. **B176**, 483 (1986).
(17) C.-Y. Wong, Phys. Rev. Lett. **56**, 1047 (1986); M.-C. Chu and V. Pönisch, Phys. Rev. **C33**, 2222 (1986).
(18) B. Müller *et al.*, J. Phys. **G12**, L109 (1986).
(19) L.S. Celenza *et al.*, Phys. Rev. Lett. **57**, 55 (1986).

New Information on Nuclear Structure in the Cd-In-Sn Region from Laser Spectroscopy and the Question of Core Polarization Contribution to Nuclear Radii

E.W. Otten

Institut für Physik der Universität Mainz, D-6500 Mainz, F. R. Germany

Nuclear spin, moments and isotope shifts of charge radii have been measured by laser spectroscopy for about 70 nuclear states in the range $48 \leqq Z \leqq 50$, $54 \leqq N \leqq 78$. $1/2^-$-states in heavy In-isotopes cross the Schmidt line, indicating complex nuclear structure. Magnetic as well as spectroscopic quadrupole moments of most of the odd odd In-isotopes can be reproduced satisfactorily by coupling the respective experimental moments of odd even and even odd neighbouring nuclei. The isotope shift of all three elements exhibits a parabolic shape, which is superimposed to the almost linear droplet model expectation. The shape can be fitted quantitatively to Talmis core polarization model. The curvature of the parabola which peaks in the middle of the neutron shell is a direct measure of the collective contribution to the charge radius. It corresponds to deformation parameters $\langle\beta^2\rangle$ twice as large as observed from $B(E2, 0^+ \rightarrow 2_1^+)$ values. One suggests, therefore, that the latter one does not exhaust the full collectivity of the nuclear ground state. By comparing to charge radii in the Rb and Cs region one derives isotonic and isobaric shifts in accordance with the droplet model. The isobaric shifts may be interpreted as a build up of a neutron skin in a simple way.

I. Introduction

Recent laser spectroscopic experiments at the ISOLDE-facility/CERN, the GSI/Darmstadt and the KFZ/Karlsruhe concentrated on the investigation of nuclear spins, moments and charge radii in the region of medium-heavy elements Cd [1] - In [2] - Sn [3,4]. The chains investigated cover typically 20 mass numbers for each element. Different laser spectroscopic methods have been used: 1. Doppler-limited, laser-induced fluorescence from an atomic vapor in a resonance cell for the case of cadmium; 2. Doppler-free, laser-excited fluorescence from an atomic beam for the case of tin; 3. collinear laser spectroscopy for the cases of indium and tin.

Whereas in the two latter, more recent experiments the relevant spectroscopic parameters, namely the hyperfine splitting (hfs) and the isotope shift (IS) by far exceed the residual line width of roughly 50 MHz, the analysis of the Cd results, which date back a few years, was handicapped (particularly in the presence of isomers) by overlapping hfs components having Gaussian profiles of approximately 3 GHz width. In the evaluation of IS's this problem played a minor role only. The atomic beam technique applied off-line by the Karlsruhe group is described in detail in ref. [5], the on-line collinear technique in ref. [6]. Figure 1 shows a hyperfine pattern obtained in collinear laser spectroscopy for the case of ^{122}In.

The measurements in the three adjacent elements Cd, In, Sn cover an almost complete net in the range $48 \leqq Z \leqq 50$ and $54 \leqq N \leqq 78$. The proton co-ordinate remains close to the shell closure $Z = 50$, whereas the neutron co-ordinate spans over most of the shell $50 \leqq N \leqq 82$. The region is characterized, therefore, by spherical shell model states near the double shell closures at the far neutron-deficient ard neutron-rich ends of the nuclear chart and by moderately deformed nuclei near midshell in the stable valley. In this paper we will discuss, how nuclear magnetic moments (μ_I), spectroscopic quadrupole moments (Q_s) and the isotopic variation of charge radii ($\delta\langle r^2\rangle$) systematically respond to this

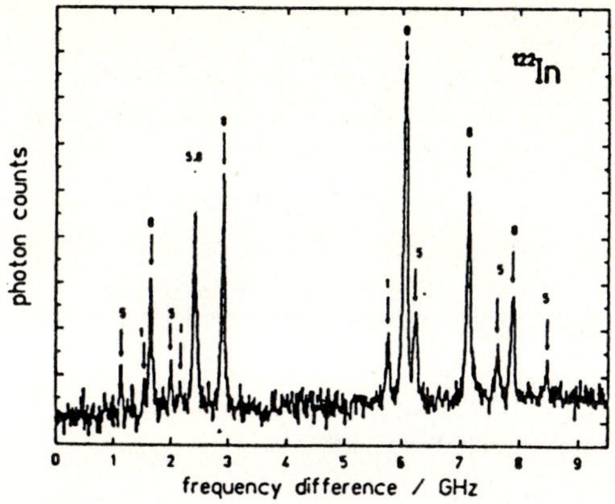

Fig. 1:
Hfs pattern of ^{122}In ground state (I=1) and two isomers (I=5), (I=8) observed in the optical resonance line 5p $^2P_{3/2}$ → 6s $^2S_{1/2}$, (λ=451 nm) by collinear laser spectroscopy.

situation. Moreover, we discuss isobaric and isotonic shifts by comparing to charge radii in the Rb and Cs region.

II. Spins and Magnetic Moments

The discussion of nuclear spins and moments is particularly fruitful for the case of In. In the mass range 104 ≦ A ≦ 127 37 nuclear states have been investigated with spin values I^P = $9/2^+$, $1/2^-$, 3, 4, 5, 8, including many isomers. Figure 2 displays the course of μ_I as a function of N for the odd-even isotopes, Fig. 3 for the odd-odd ones.

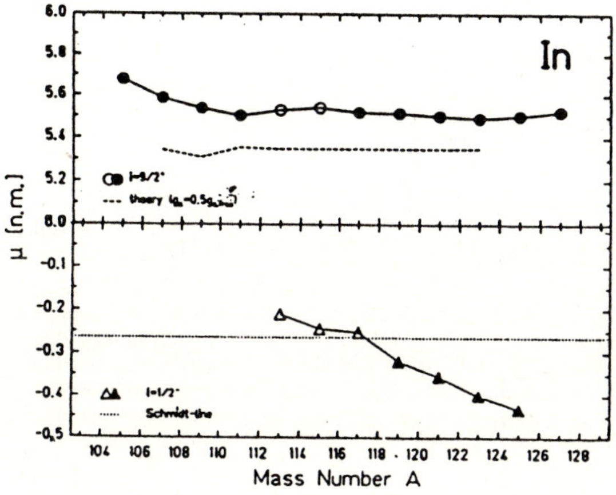

Fig. 2:
μ_I-values of odd-even In-isotopes [2]. Upper half: $9/2^+$-states, theory from ref. [7], [8]. Lower half: $1/2^-$-states, crossing the Schmidt line.

Fig. 3:

μ_I-values of odd-odd In-isotopes [2]. Equal spins are connected by lines. Full symbols represent experiment, open ones the empirical calculation according equ. (1).

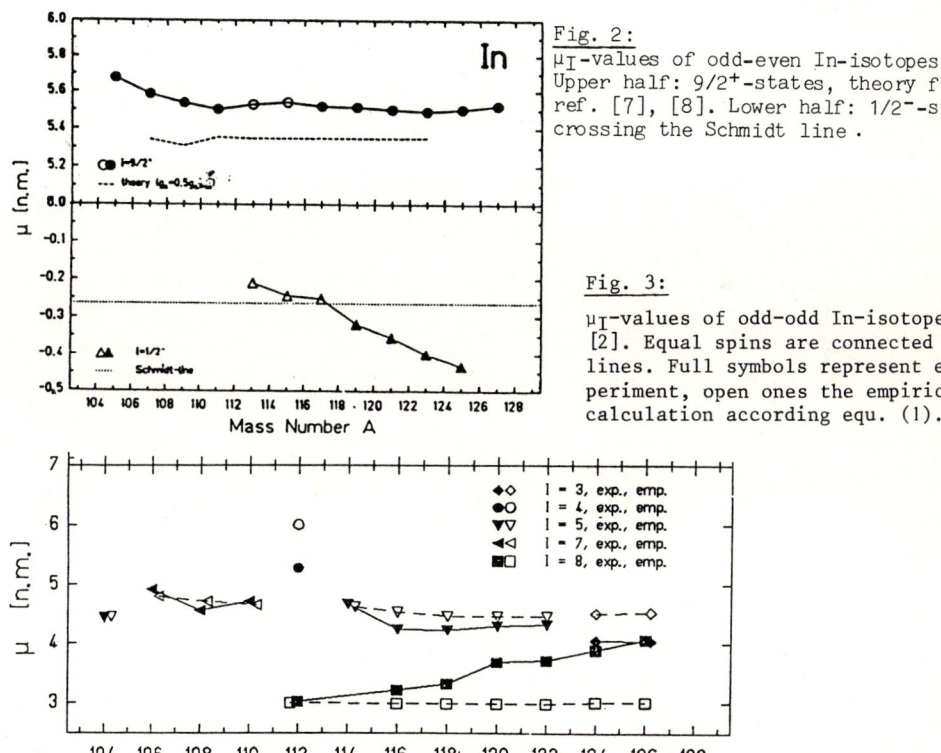

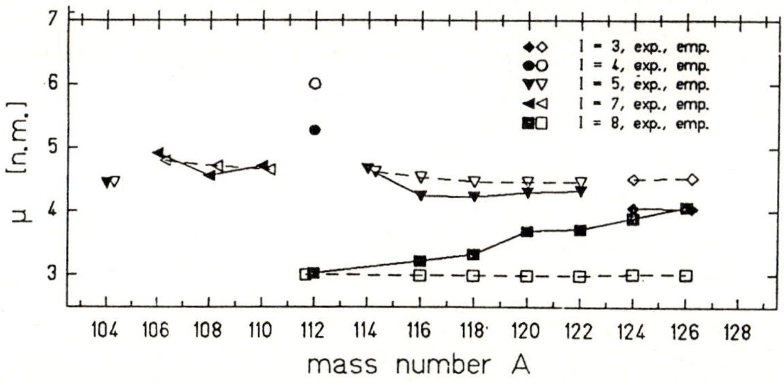

The almost constant $\mu_I \approx 5.5$ nm, observed in the $9/2^+$ ground state series, is reproduced [7,8] within a few percent in a hole plus vibrating core model [9] choosing $g_s = 0.5\ g_{s,free}$, $g_R = Z/A$. A striking feature of the $1/2^-$ moments is their overshooting over the Schmidt line by a factor 1.7 for the heaviest isotope. This result is all the more surprising as it is known that $p_{1/2}$ states are insensitive to core polarization in first order. A satisfactory explanation is difficult to find therefore; the moments indicate very complex nuclear structure.

The set of odd-odd In moments together with the data for their isotopic and isotonic neighbours offers an excellent chance for checking systematically the odd group coupling model which predicts $\mu_{I,odd,odd}$ to be the vector sum of the two adjacent odd proton group (p) and odd neutron group (n) moments with the same configuration. The relation reads

$$\mu_{I,odd,odd} = \frac{I}{2}\ [g_p + g_n + (g_p - g_n)\ \frac{I_p(I_p+1)-I_n(I_n+1)}{I(I+1)}\ ]\ . \qquad (1)$$

These "empirical" moments are plotted together with the experimental ones in Fig. 3. The configurations involved are $(\pi g_{9/2})^{-1} \otimes (\nu d_{5/2})$, $(\nu g_{7/2})$, $(\nu s_{1/2})$, $(\nu d_{3/2})$, $(\nu h_{11/2})$. For $I < 8$ the agreement varies from fair ($I = 3$) to very good ($I = 5$). For the discrepancy at $I = 8$ see below.

III. Spectroscopic Quadrupole Moments

The odd group coupling model may be applied to Q_s as well as to μ_I, although the former case is hardly discussed in the literature. The odd group moments $Q_{s,p}$ and $Q_{s,n}$ are added with tensor coupling coefficients q_1, q_2, defined in ref. [10], to the resulting moment

$$Q_{s,odd,odd} = q_1\ Q_{s,p} + q_2\ Q_{s,n}\ . \qquad (2)$$

The perfect agreement of these empirical moments with the measured ones is a great surprise, indeed (see Fig. 4) [2].

However, the $I = 8$ states form an exception again. The experimental moments rise much faster than the empirical ones. It seems that in this high spin state, build on a $(\pi g_{9/2})^{-1} \otimes (\nu h_{11/2})^n$ configuration, the neutrons decouple from the prolate core into the smallest, unfilled Ω projection in order to maximize their overlap with the prolate core. In this case Q_s is related to the intrinsic quadrupole moment Q_o by

$$Q_s = [3\ \Omega^2 - I\ (I+1)]\ /\ [(I+1)\ (2I+3)]\ \ Q_o\ . \qquad (3)$$

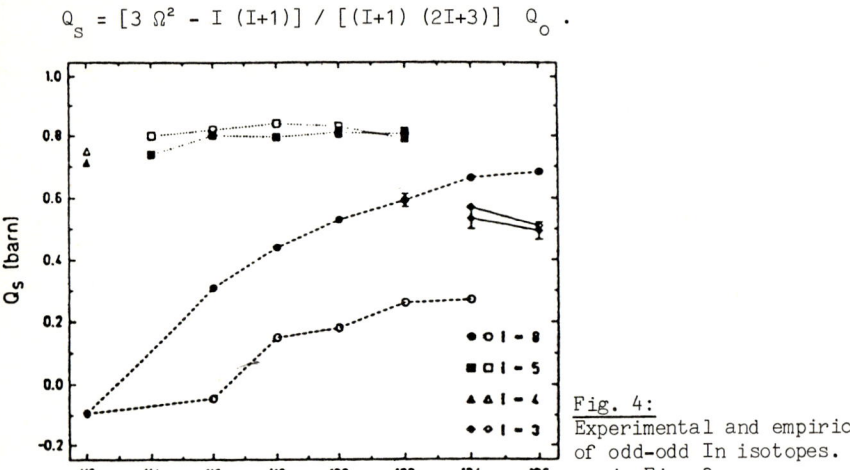

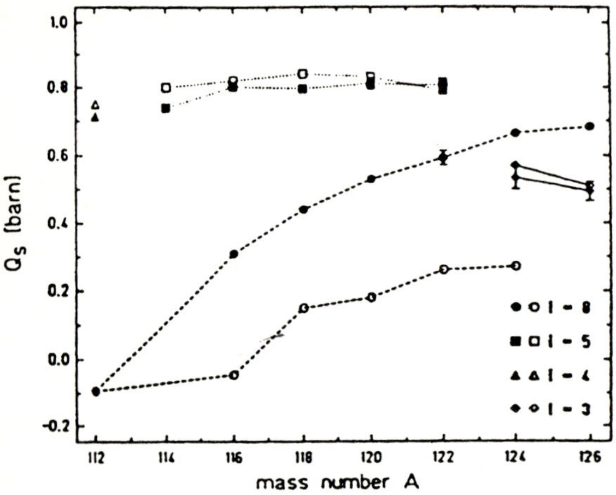

Fig. 4: Experimental and empirical Q_s values of odd-odd In isotopes. Symbols defined as in Fig. 3.

With increasing Ω the fore factor of (3) changes sign from minus to plus in the middle of a j-shell which, according to Fig. 4, would be located around A = 112. This model has been successfully applied to the ($\nu i_{13/2}$) isomers of Hg already [11,12]. On the other hand one should note that a core polarization induced by a $(j)^n$ shell model configuration produces exactly the same signature of Q_s as a function of n.

IV. Isotope Shift and Nuclear Deformation

The course of $<r^2>$ as a function of N is very regular and similar in all three elements (compare Fig. 5 and 6). Usually one decomposes $\delta<r^2>$ into a volume and a deformation part:

$$\delta<r^2> = \delta<r^2>_{vol} + \frac{5}{4\pi} <r^2> \quad \delta<\beta^2> . \qquad (4)$$

The volume part is varying with $A^{1/3}$, i.e. practically linear in the limited range considered. It seems to be well described by Myers droplet model [13]. The deformation part of $\delta<r^2>$ depends on the shell structure. It can exhibit sharp shape transistions, but also a smooth N-dependence like observed in the Cd-Sn region in the neighbourhood of the Z = 50 proton shell closure. The $\delta<r^2>$ curve has obviously a parabolic shape which maximizes in the middle of the neutron shell at N = 66. The curvature and hence $<\beta^2>$ increase with the distance from the proton shell closure.

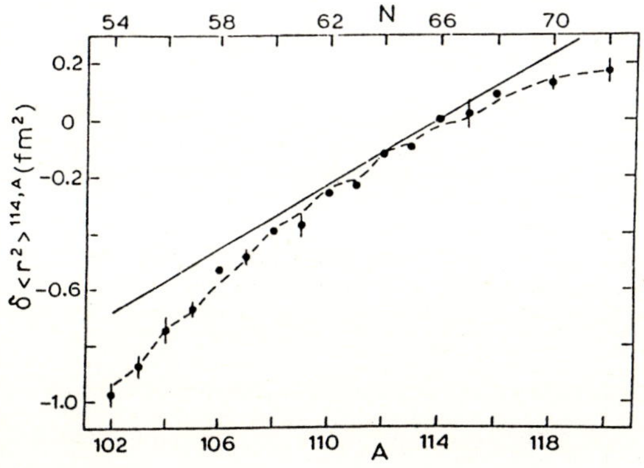

Fig. 5:
$\delta<r^2>$ for Cd isotopes [1]. The full line corresponds to the droplet model [13], the dotted line to a parabolic fit according to Talmis core polarization model [14] (equ. (5)). It includes odd-even staggering.

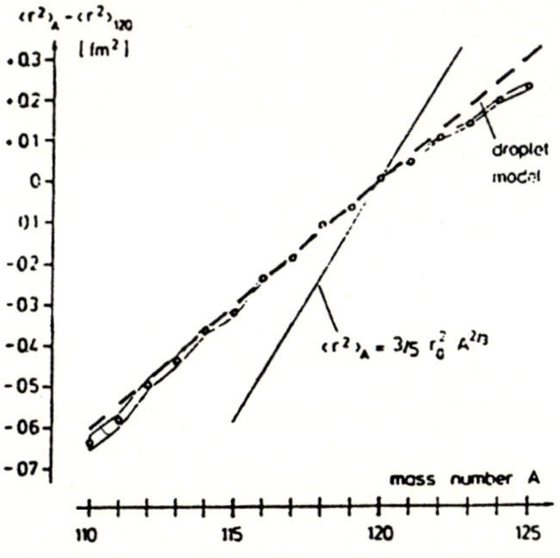

Fig. 6:
$\delta<r^2>$ for Sn isctopes [3] in comparison with the droplet model and the naive liquid drop model (full line). The envelope around the data points indicates a systematic scaling error predominantly.

Talmi has recently applied a core polarization model to the similar case of the IS of Ca [14]. He shows that in a single (ν_j^n) shell model configuration the polarization contribution is proportional to the number of particles (n) times the number of holes (2j+1-n). The odd-even staggering of $\langle r^2\rangle$ is contained in the model as a constant offset of the odd N isotopes. The model can be generalized for semi-magic nuclei also to a major N-shell; it then predicts $\delta\langle r^2\rangle$ to be a general quadratic form in the particle number plus the odd-even staggering term. Looking to the data and from theoretical arguments it obviously makes sense to retain the symmetry in particles and holes also with respect to a major shell. This leads to a parametrization of (4) in the form [1]:

$$\delta\langle r^2\rangle^{50,N} = a(N - 50) + b(N - 50)(82 - N) + (1/2)d\,(1 - (-1)^N). \qquad (5)$$

The first term in (5) is now identified with the volume, the second with the deformation contribution.

Fitting (5) to the data yields a(Cd) = 0.047 fm^2, a(In) = 0.060 fm^2, a(Sn) = 0.065 fm^2; the droplet model predicts a ≈ 0.055 fm^2 in fair agreement with the average of the data. The deformation part vanishes at the shell closures by definition of the model (5) and peaks in the center at N = 66. By fitting the parameter b one obtains the center deformations $\langle \beta^2\rangle^{N=66}$ = 0.078, 0.048, 0.032 for Cd [1], In [2], Sn [4].

Figure 7 compares for the case of Cd $\langle\beta^2\rangle_{IS}$ derived from $\delta\langle r^2\rangle$ through equations (4), (5) with those obtained from quadrupole strengths through the closure relation:

$$\langle\beta^2\rangle_{BE2} = \left(\frac{4\pi}{3ZR_o^2}\right)^2 \sum_i B(E2,\, 0^+\rightarrow 2_i^+). \qquad (6)$$

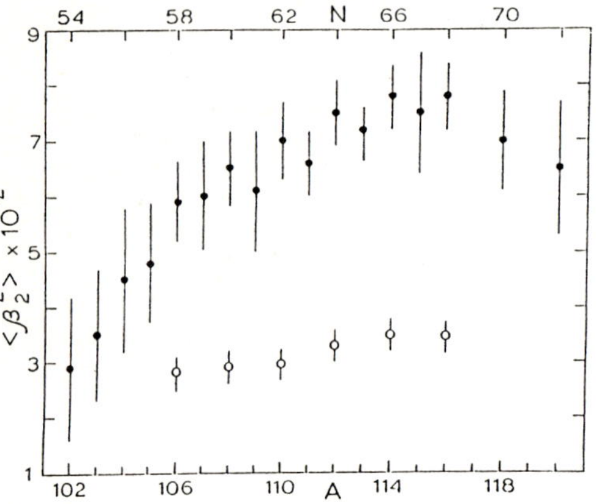

Fig. 7: $\langle\beta^2\rangle$ of Cd ground states from $\delta\langle r^2\rangle$ analyzed by equ. (5) (full dots) compared to those derived from B(E2, 0^+-2_1^+) values (open circles) [1].

For harmonic vibrators as well as rigid rotators the sum is exhausted by its first term. This truncation was also used in the Cd case. From Fig. 7 one reads that the deformation parameter as derived from $\delta\langle r^2\rangle$ reaches the astonishing large value of $\langle\beta^2\rangle_{IS}^{1/2}$ = 0.28 whereas $\langle\beta^2\rangle_{BE2}^{1/2}$ is in the familiar range of 0.18 to 0.20. The same proportion is observed in In between $\langle\beta^2\rangle_{IS}^{1/2}$ and $\langle\beta\rangle_{Q_s}$, where the latter is derived from Q_s values by equation (3) in the strong coupling limit Ω = I and the relation

$$\langle\beta\rangle = \sqrt{5\pi}\; Q_o \,/\, (4Zr_o^2\, A^{2/3}) \qquad (7)$$

with r_o = 1.2 fm (compare Fig. 8).

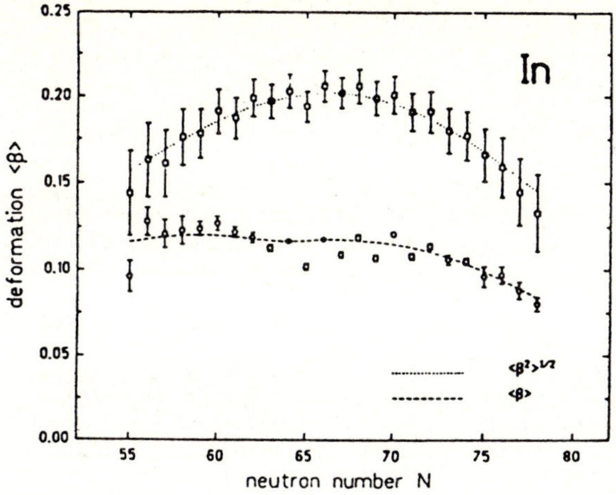

Fig. 8:
$<\beta^2>^{1/2}$ values from $\delta<r^2>$ analyzed by equ. (5) (dotted line) in comparison with <ß> values derived from Q_S values (dashed line) [2]. The error bars on $<\beta^2>^{1/2}$ increase towards the wings of the curve due to systematic scaling errors.

In both cases the IS reveals a much stronger collective contribution to $<r^2>$ than expected from the simple model of vibrational or static quadrupole deformation. In the latter case one could argue that the contribution of zero point motion to Q_S and hence to $<\beta>_{Q_S}$ cancels. This is not the case in $<\beta^2>_{IS}$ and also not in $<\beta^2>_{BE2}$. The truncation of (6) is certainly unjustified in case of anharmonic vibration. Strong anharmonicity of a soft mode is postulated in the Jahn-Teller model of nuclear deformation [15]. Moreover, the question is open to what extent multipoles other than quadrupole may contribute to the ground-state collectivity. But they hardly make up for the whole discrepancy.

V. Isobaric Shift and Neutron Skin

IS measruements extending far off stability enable also the evaluation of isotonic and isobaric shifts over large distances. In addition one uses absolute $<r^2>$ measurements of stable elements, obtained from muonic spectra, e.g., in order to link the data across the Z coordinate. Tables of isotonic and isobaric shifts have been prepared in ref. [1] for the Cd region linked to the Rb and to the Cs. Figure 9 gives an example of the isobaric shift in A = 118 between Cd, Sn and Cs. After correcting for the deformation effect on $\delta<r^2>$ the agreement with the prediction from the droplet model is perfect. Note that in a naive, homogeneously charged liquid drop model of radius $R_o = r_o A^{1/3}$ the isobaric shift would vanish identically. What does it tell us then?

We can answer this question easily under the schematic, but plausible assumption that the mean-squared nuclear matter radius

$$<r^2>_A = \frac{Z}{A} <r^2>_Z + \frac{N}{A} <r^2>_N \qquad (8)$$

is constant for isobars (incompressible nuclear matter). One then derives [1] for two isobars $N_1 + Z_1 = N_o + Z_o = A$ a difference of neutron radii

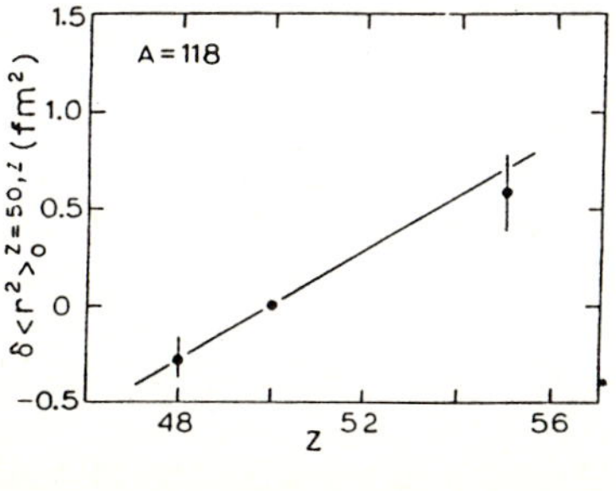

Fig. 9:
Isobaric shift in the A = 118 chain, corrected for deformation effect [1]. The line represents the prediction from the droplet model [13].

$$\langle r^2\rangle_{N_1} - \langle r^2\rangle_{N_o} = \frac{Z_1}{N_1} (\langle r^2\rangle_{Z_o} - \langle r^2\rangle_{Z_1})$$

$$- \frac{A}{N_o N_1} (N_1 - N_o) (\langle r^2\rangle_A - \langle r^2\rangle_{Z_o}) \qquad (9)$$

where the first term on the right dominates. Thus (9) relates the isobaric shift of the charge radius directly to the one of the neutrons. For the case considered follows from the data

$$\langle r^2\rangle_N ({}^{118}_{48}Cd^{70}) - \langle r^2\rangle_N ({}^{118}_{55}Cs^{63}) = 0.59\ fm^2. \qquad (10)$$

In terms of r.m.s. radii (10) corresponds to a swelling of the neutron radius by $\delta\langle r^2\rangle_N^{1/2} = 0.063$ fm on cost of a shrinking of the charge radius by $\delta\langle r^2\rangle_Z^{1/2} = -0.092$ fm. In the same region Angeli et al. [16] have calculated with a BCS-Hartree-Fock code $\delta\langle r^2\rangle_N^{1/2} - \delta\langle r^2\rangle_Z^{1/2} = 0.12$ fm for an exchange of 4 nucleons instead of 7. Scaling this number by (7/4) yields 0.196 fm in fair **agreement** with our result of 0.155 fm. The introduction of a neutron skin, on the other hand, was a major achievement in the development from the liquid drop to the droplet model.

References

[1] F. Buchinger, P. Dabkiewicz, H.-J. Kluge, A.C. Mueller and E.W. Otten, Hyperfine Interactions 9, 165 (1981) and submitted to Nucl. Phys. A

[2] U. Dinger, J. Eberz, G. Huber, H. Lochmann, R. Menges, R. Neugart, R. Kirchner, O. Klepper, T. Kühl, D. Marx, G. Ulm and K. Wendt, submitted to Nucl. Phys. A

[3] M. Anselment, K. Bekk, H. Hanser, H. Hoeffgen, G. Meisel, S. Göring, H. Rebel and G. Schatz, submitted to Phys. Rev. C

[4] H. Lochmann, U. Dinger, J. Eberz, G. Huber, G. Ulm, R. Kirchner, O. Klepper, T. Kühl and D. Marx, submitted to Z. Phys.

[5] G. Nowicki, K. Bekk, S. Göring, A. Hanser, H. Rebel and G. Schatz, Phys. Rev. C18, 2369 (1978)

[6] A.C. Mueller, F. Buchinger, W. Klempt, E.W. Otten, R. Neugart, C. Ekström and J. Heinemeier, Nucl. Phys. A403, 234 (1983)

[7] J. Van Maldeghem, K. Heyde and J. Sau, Phys. Rev. C32, 1067 (1985)

[8] K. Heyde, Gent, private communication

[9] K. Heyde et al., Phys. Rep. 102, 291 (1983) and references therein

[10] A. Bohr and B.R. Mottelson, Nuclear Structure , Vol. 1, Benjamin, New York (1969)

[11] E.I. Volmyanski and V.G. Dubro, Izv. Akad. Nauk SSSR, Ser. Fiz. 41, 1252 (1977)

[12] I. Ragnarsson, in Proc. Int. Symp. Future directions in studies of nuclei far from stability, Nashville 1979, eds. J.H. Hamilton et al., North Holland, Amsterdam (1980)

[13] W.D. Myers and K.H. Schmidt, Nucl. Phys. A410, 61 (1983)

[14] I. Talmi, Nucl. Phys. A423, 189 (1984)

[15] P.G. Reinhard and E.W. Otten, Nucl. Phys. A420, 173 (1984)

[16] I. Angeli, M. Beiner, R.J. Lombard and D. Mas, J. Phys. G6, 303 (1980)

Beta Decay of Neutron-Rich Transuranic Nuclei*

R.W. Hoff

Nuclear Chemistry Division, Lawrence Livermore National Laboratory, Livermore, CA 94550, USA

The intense neutron flux generated in a thermonuclear explosion is of interest to nuclear physicists because, among other reasons, the atoms exposed to this flux will undergo many successive capture reactions during the brief period the flux exists. If 238U target atoms are given such exposure, the phenomenon becomes a way to probe the nuclear properties of very neutron-rich uranium nuclei and to produce heavy isotopes of the transuranium elements that survive the chains of beta decay following the initial phase of the reaction. This unique method of transuranium element production was investigated by U.S. scientists in a series of thermonuclear explosions during the period 1952-1969, first as an unexpected result of the Mike explosion [1] that occurred in November 1952 and later in a series of underground explosions where the explosive devices were designed specifically for this purpose [2,3]. The heaviest species identified in this work was 257Fm, which implies the production of 257U during the capture phase. While nuclei this neutron-rich cannot be produced by any other man-made technique, it is known, of course, that such nuclei are generated by the astrophysical r process. The neutron-capture path for this process is thought to pass through a region around 262U[4]. Given the clear relationship between these processes, it is the purpose of this paper to see what insights into recent r-process calculations can be gained from a consideration of the experimental data for heavy element production in thermonuclear explosions.

Experimentally determined mass-yield curves for several underground thermonuclear explosions are shown in Fig. 1. The curves are all similar in form, each with the following characteristics: 1) A regular, exponential decrease with increasing A; 2) Data points often missing for either A=249 or A=251, which is due to the circumstances of the experimental measurements; 3) No data shown for A < 244 (or perhaps 242). Although such nuclides can be readily determined, it was often impossible to deduce the fraction of observed atoms arising from multiple neutron capture processes in a 238U target due to interference from large quantities of residual plutonium isotopes; 4) Termination of the yield curve data at A=257; 5) Some odd-even variation within the curve. For all of the data in the mass range A=244-248, the observed yields of the even-A products are systematically higher than those for the odd-A products. Beginning somewhere near A=250, this odd-even variation is seen to be reversed at higher masses. The only exception to this observation is the Hutch mass yield curve where no reversal is observed up to A=255.

As a starting point for calculating product yields, we will consider a simple model in which the assumed capture sequence begins with a single target isotope and consists of isotopes of the same element, all with identical capture cross sections(σ). Losses due to processes other than capture (e.g. neutron-induced fission) are neglected. For such a system the yield of the nth member in the sequence is given by the following expression:

$$N_n = N_0(0) \exp(-\sigma\Phi) (\sigma\Phi)^n / n! \qquad (1)$$

* Work performed under the auspices of the U.S. Department of Energy by the Lawrence Livermore National Laboratory under Contract No. W-7405-Eng-48.

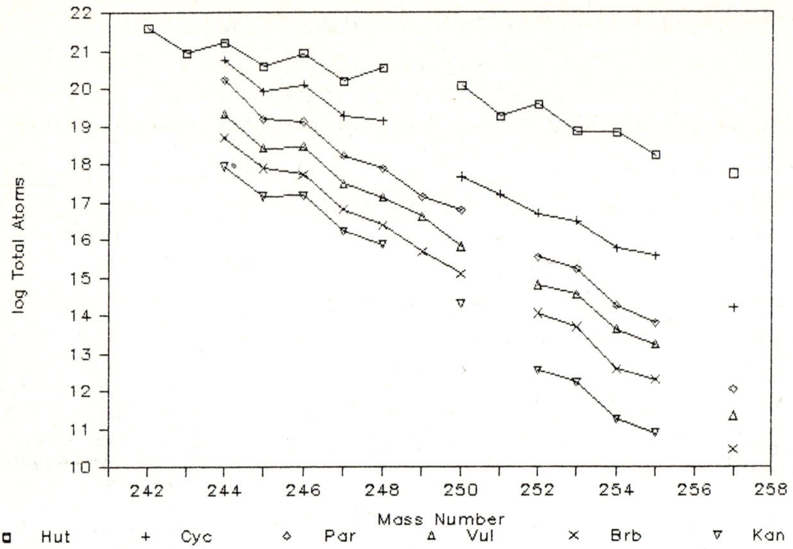

Fig.1 Mass-yield curves for several underground explosions. Except for Hutch, the relative positions of the curves on the ordinate have been adjusted to facilitate comparison; the true position of each curve is given by the absolute amounts of mass 244 atoms produced, in units of 10^{20} atoms: Hutch 17, Cyclamen 3.8, Par 8.7, Vulcan 2.7, Barbel 0.10, and Kankakee 11.

where $N_0(0)$ is the amount of target material at zero time, and Φ is the neutron exposure, i.e. the product of neutron flux (n/cm^2-s) and time (t).

This expression describes a family of smooth mass-yield curves, all concave downwards and whose slopes at any given mass number tend to decrease with increasing values of neutron exposure.

By departing from these simple assumptions, the odd-even variation within a given curve is readily explained by a corresponding variation in capture cross section for odd-mass and even-mass isotopes. This is exactly the expected behavior for the capture cross sections because of an odd-even alternation in neutron separation energies, a natural manifestation of the nucleon pairing phenomenon. On the other hand, understanding the linear nature of the observed yield curves presents a somewhat greater challenge. It was this feature of the mass-yield curve from the Mike event that led CAMERON [5] to suggest a revised form of the mass equation for neutron-rich nuclei which provided for a less rapid decrease in the neutron separation energy with increasing neutron number. Nevertheless, since it is known that the overall trend for separation energies is one of decrease and that eventually neutrons are no longer bound (at the neutron drip line), it is not possible, within the simple model considerations discussed here, to produce a calculated mass-yield curve that is linear over an extended range of masses. For this reason, the observed mass yield curves in Fig. 1 were thought to be the envelope of several separate curves.

This idea has been invoked, in fact, to explain the existence of the reversal in the odd-even alternation observed in the majority of the data (Fig.1). It was proposed that the nuclides observed beyond A=250 were produced by sequential neutron capture in an odd-Z element. The odd-even effect is reversed because now capture cross sections for even-mass isotopes (which are odd-odd species) will be larger than for odd-mass isotopes. BELL [6] has shown that an initial production of 238Pa and/or 239Np can occur by (n,p) and (d,n) reactions on the 238U target before moderation of the source neutrons. Although these odd-Z nuclides would be present in much lower abundance (approx. 0.01) than the 238U, the larger cross sections of the capture sequence eventually cause the yields from the odd-Z chain to surpass those from the uranium chain. INGLEY [3] has made quantitative fits to most of the data of Fig. 1, deriving best values for effective capture cross sections (at 20 keV), neutron exposure values for each event, and ratios of initial amounts of 238Pa and 238U required to reproduce the mass-yield curves.

Another variable studied in the underground explosion series was that of the composition of the initial target material. It was found that the greatest success was obtained with 238U, although various other heavy nuclides such as 243Am, 242Pu, 237Np, and 232Th were incorporated in the targets of various test devices [2,3]. Neutron-induced fission presumably was the cause of low production for some of the heavier alternate target materials.

A disappointing feature of the experimental study of these products from thermonuclear explosions was the inability to detect evidence for nuclides with A>257. Certainly the trends of yields for the higher exposures (e.g. in Hutch and Cyclamen) were such that one would predict production of detectable amounts of species up to A=260 or even higher, depending upon specific decay properties. This absence of heavier isotopes can be explained by at least three loss mechanisms: 1) Neutron-induced fission competing with capture, 2) Beta-delayed fission in the decay chain, and 3) Short half-lives of the isotopes at the end of the decay chain. This last mechanism apparently can account for the inability to detect species with A=256, 258, and 259. The beta-stable nuclides at the end of these decay chains are now known to have very short spontaneous fission (SF) half lives [7-9], viz. 12-m 256Cf, 0.4-ms 258Fm, and 1.5-s 259Fm.

Of greater interest to this question has been the recent discovery [10] of a 32-day SF activity assigned to 260Md. Although it appears that the experimental techniques used in the 1960's would have permitted detection of 260Md, had it been present in the samples then, there are two arguments suggesting it was not present: 1) The end of the mass 260 beta minus decay chain is calculated to be 260Fm; 2) Even if 260Fm proved to be unstable towards beta minus decay, it is unlikely that the recently discovered 260Md activity would have been produced. The argument here is that the relatively long half life of the observed 260Md, coupled with a decay energy of 0.5-1.0 MeV, suggests considerable hindrance of its beta decay (log ft >8.5), probably due to its high spin. If this is so, the 32-day activity will not be populated at the end of a beta-decay chain because the successive decays, which are populating even-even and odd-odd nuclides alternately, will immediately revert to populating only low-spin levels once the decay has passed through the first 0^+ even-even ground state in the chain.

In the past 10-15 years, advances in the formulation of beta-strength functions and in the extension of mass formulae and fission barrier calculations away from beta stability have resulted in several papers that treat the phenomenon of beta-delayed fission (BDF) and beta-delayed neutron emission in neutron-rich nuclides. With the availability of quantitative estimates of these phenomena, allowance has been made for their effect on the calculated abundances of nuclides in beta-decay chains in the astrophysical r-process, and, less frequently, on the observed abundances of products from thermonuclear explosions. We will consider here the effect of BDF in this latter application by assessing the results of four separate calculations, the elements of which are listed in Table 1. Beta-delayed neutron emission will not be considered, at least partially because its effect is to shift atoms from one beta-decay chain to the next and, thus, to some extent to nullify any loss by replacement of atoms from the next higher chain. We have calculated correction factors for each mass chain by taking the product of the survival fraction for each decay step in the chain. These correction factors, which are shown in Fig. 2, were then applied to the Hutch mass yield curve with the results shown in Fig. 3.

Table 1. Calculations of beta-delayed fission and r-process abundances

Beta-Strength Function	Mass Formula	Fission Barrier
WENE and JOHANSSON [11] Constant, i.e. prop. to daughter level density	JOHANSSON and WENE [12]	Statistical, employing fission, neutron,and gamma widths
KODAMA and TAKAHASHI [13] Gross theory of beta decay (no level density formula required)	MYERS and SWIATECKI [14]	Systematic formulation based on exp. data
THIELEMANN [4] Tamm-Dancoff approx. with long-range GT residual interaction	HILF [15], (Fig. 5a [4])	HOWARD and MOELLER [16]
	VON GROOTE [17]	
	HOWARD and MOELLER [16], (Fig. 5c [4])	
MEYER [18] Nilsson RPA treatment with infinite range GT interaction [19]	HOWARD and MOELLER [16]	HOWARD and MOELLER [16]

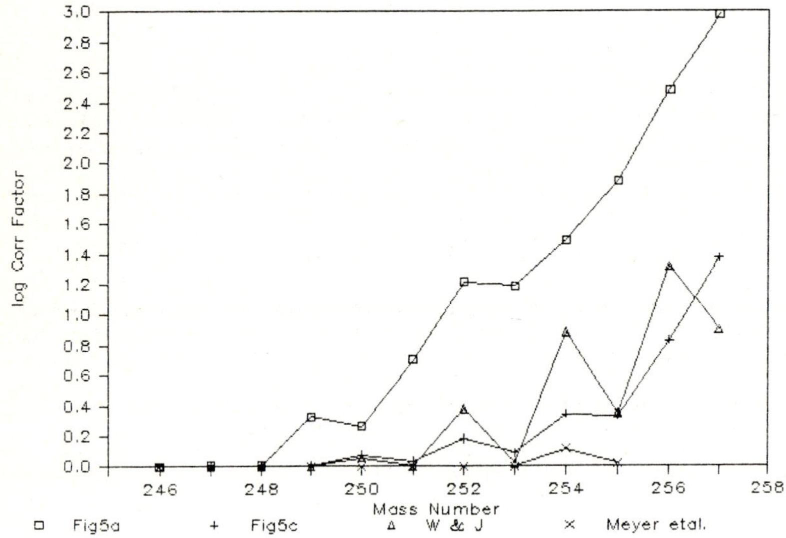

Fig.2 Beta-delayed fission correction factors for decay chains beginning with uranium isotopes.

The largest amount of BDF is calculated by THIELEMANN [4], hereafter referred to as TMK, using the mass formula of HILF [15]. When these BDF survival factors are applied as corrections to the Hutch mass-yield curve, the new data set (labelled as "Fig.5a" in the legend of Fig. 3, referring to a figure in TMK) indicates substantially greater yields than before for nuclides with A>250. The correction to the A=257 point is so great that its yield is second only to that for A=252 for all masses with A>246. The capture cross-section values required to produce a calculated fit to these data will be much larger for A>250 isotopes than deduced previously. In fact, we expect these revisions to require cross-section values so large as to be unphysical. There is nothing in the trend of calculated neutron binding energies that would suggest enough perturbation to permit large increases in the calculated cross sections.

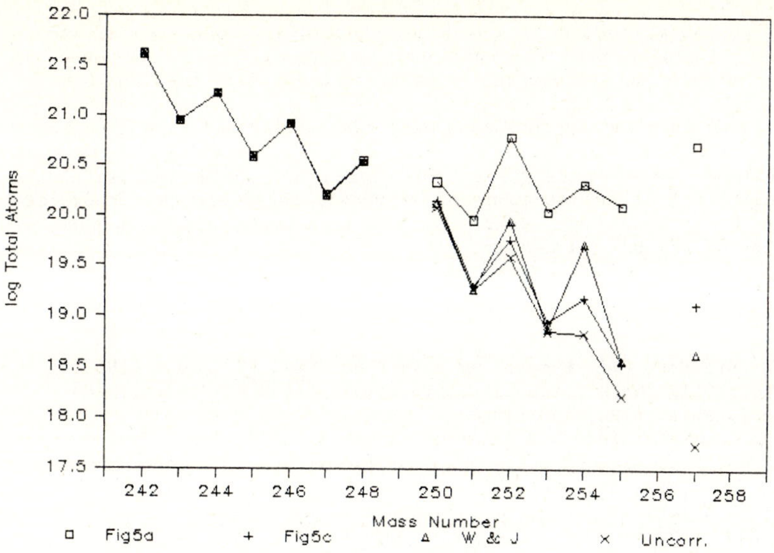

Fig.3 Hutch mass-yield curve corrected for beta-delayed fission.

A second set of results from TMK (labelled in the legend as "Fig.5c") is also shown in Fig. 2; these can be compared with the calculations of MEYER [18] where common elements are the mass formulae and fission barrier calculations. One can see the correction factors of MEYER [18] are negligible on the scale of Fig.2, except for A=254, while those of TMK for A=252-255 are as large as a factor of 2 and rise rapidly to even greater values for A=256, 257. The differences here, which are a function only of the assumed form of the beta-strength function, show that TMK's treatment tends to place a greater fraction of the beta strength in excited levels near or above the fission barrier than does that of MEYER [18].

Those BDF calculations that produce large even-odd variation, e.g. see the data labelled "W&J" in Fig. 2 (from Wene and Johansson [11]), can account for the reversal of this variation in the mass-yield curves of Fig. 1. However, large even-odd variations are not a feature of all BDF calculations. In all cases, the TMK calculations show that the correction factors become extremely large (>1000) for decay chains with A>or=258; in fact, their calculations indicate the beginning of a region at A=258 where most nuclei will exhibit 100% BDF, and atoms entering this region will not survive. On this basis, they predict no production of superheavy elements by the r process. KODAMA and TAKAHASHI [13] found that BDF was negligible for Z<94, but was appreciable in heavier nuclei. For the only examples given in the paper, they indicate little significant BDF for neutron-rich species with A=266-270, excepting 269Lr and 270No, for which BDF branching was 35-40%.

Thus, the four papers cited in Table 1 include calculated estimates of beta-delayed fission with widely varying magnitudes, an indication of the degree of uncertainty inherent in extrapolating the various basic parameters. Using the effect of BDF corrections on the mass-yield curves considered in this paper as a general indication of their accuracy, we reach the following conclusions: 1) The TMK treatment with the Hilf mass formula definitely overestimates BDF for uranium decay chains in the A=244-257 region. Paradoxically, TMK considered the Hilf mass formula most realistic for neutron-rich nuclides and used these BDF survival factors to modify their r-process calculations. Some of the other theoretical estimates summarized in Table 1 may also overestimate BDF; 2) The effect of BDF can be invoked to explain two features of the thermonuclear explosion mass-yield curves, the reversal of the odd-even yield variation around A=250 and the termination of the curves at A=257. On the other hand, since there are alternate possibilities that explain the observations adequately, the presence of substantial amounts of BDF is not required in order to understand the data.

Since the results of BDF calculations show much variability and, thus, there is no indication that they are accurate, it seems wise to assess their effect on r-process calculations rather carefully. Therefore, it would be useful to determine the sensitivity of calculations of r-process abundances and r-process production ratios for chronometric pairs to the magnitudes and structure of various calculations of the BDF phenomenon.

References

1. A. Ghiorso, M.H. Studier, C.I. Browne, et al.: Phys. Rev. 99 , 1048 (1955)
2. R.W. Hoff: Lawrence Livermore National Laboratory Report UCRL-81566 (1978)
3. J.S. Ingley: Nucl. Phys. A124, 130 (1969)
4. F.-K. Thielemann, J. Metzinger, H.V. Klapdor: Z. Phys. A309, 301 (1983)
5. A.G.W.Cameron, R. Elkin: Can. J. Phys. 43, 1288 (1965)
6. G.I. Bell: Rev. Mod. Phys. 39, 59 (1967)
7. D.C. Hoffman, J.B. Wilhelmy, J. Weber, W.R. Daniels, E.K. Hulet, R.W. Lougheed, J.H. Landrum, J.F. Wild, R.J. Dupzyk: Phys. Rev. C21 , 972 (1980)
8. E.K. Hulet,J.F. Wild, R.W. Lougheed, J.E.Evans, B.J. Qualheim, M. Nurmia, A. Ghiorso: Phys. Rev. Lett. 26, 523 (1971)
9. E.K. Hulet, R.W. Lougheed, J.H. Landrum, J.F. Wild, D.C. Hoffman, J. Weber, J.B. Wilhelmy: Phys. Rev. C21, 966 (1980)
10. R.W. Lougheed, E.K. Hulet, R.J. Dougan, J.F. Wild, R.J. Dupzyk, C.M. Henderson, K.J. Moody, R.L. Hahn, K. Suemmerer, G. Bethune: Proceedings of the Actinides 85 Conference, Aix-en-Provence, France, Sept. 1985.
11. C.-O. Wene, S.A.E. Johansson: Phys. Scr. 10A, 156 (1974)
12. S.A. Johansson, C.-O. Wene: Arkiv Fysik 36, 353 (1967)
13. T. Kodama, K. Takahashi: Nucl. Phys. A239, 489 (1975)
14. W.D. Myers, W.J. Swiatecki: Nucl. Phys. 81, 1 (1966)
15. E.R. Hilf, H. von Groote, K. Takahashi: CERN 76-13, 142 (1976)
16. W.M. Howard, P. Moeller: At. Data Nucl. Data Tables 25, 219 (1980)
17. H. von Groote, E.R. Hilf, K. Takahashi: At. Data Nucl. Data Tables 17, 418 (1976)
18. B.S. Meyer, W.M. Howard, G.J. Mathews, P. Moeller, K. Takahashi: Lawrence Livermore National Laboratory Report UCRL-93519 (1985)
19. J. Krumlinde, P. Moeller: Nucl. Phys. A417, 419 (1984)

Studies of Heavy-Ion Produced Proton-Rich and Neutron-Rich Nuclei

O. Klepper

GSI Darmstadt, D-6100 Darmstadt, F. R. Germany

Abstract: Recent studies of medium-mass nuclei far-off the β-stability line are surveyed that were performed at the GSI on-line mass separator. Measurements of half-lives of new neutron-rich nuclei and of the strength of Gamow-Teller β-decays of proton-rich nuclei are presented in some detail and compared to related experiments at other facilities.

1. Introduction

In this contribution recent spectroscopic investigations at the GSI mass separator on-line to the heavy-ion accelerator UNILAC will be described. Proton-rich or neutron-rich nuclei of the elements in the range from chromium (Z=24) to neptunium (Z=93) have been produced by fusion-evaporation or multi-nucleon transfer reactions. In both cases on-line mass-separation has been applied successfully in many areas of the nuclear chart for nuclei with half-lives longer than about 100 ms.

At first, half-lives of new neutron-rich nuclei produced outside the classical fission-product region by heavy-ion reactions will be presented in comparison with theoretical predictions, together with recent experimental results obtained from other production schemes. As a second topic the reduced strength of Gamow-Teller (GT) β-decays of proton-rich nuclei relative to shell-model predictions will be discussed. This concerns the nucleus ^{48}Mn and even-mass magic nuclei around ^{100}Sn and above ^{146}Gd which were produced by fusion evaporation reactions with ^{40}Ca and ^{58}Ni projectiles. For other recent experiments, not dealt with here, the reader is referred to the literature or other contributions to this symposium. For instance in the ^{146}Gd region, by investigating the N=81 isotones ^{149}Er and ^{151}Yb the four positive-parity proton single-particle states s1/2, d3/2, d5/2, g7/2 have been located in the odd-Z N=82 isotones ^{149}Ho and ^{151}Tm. The results indicate a decrease of the Z=64 gap when more protons are added to ^{147}Tb [1]. In the ^{100}Sn region information also on electromagnetic moments and mean square charge radii was obtained by collinear laser spectroscopy of indium and tin isotopes. The indium measurements were extended to neutron-rich isotopes at ISOLDE/CERN. For the whole range of $^{105-127}$In the ground-state magnetic dipole and essentially also the electric quadrupole moments for I=9/2 show a surprisingly constant value which is about 80% of the Schmidt-value for the magnetic moments. On the other hand, $^{115-125}$In the magnetic moments of the I=1/2 isomers drop more and more below the lower Schmidt-limit with increasing mass. The charge radii for tin and indium show a parabolic behaviour in the middle between the N=50 and 82 shell closures. The curvature is about two times bigger than expected from static or dynamic deformation derived from quadrupole moments or B(E2) values in this region [2,3].

Broader reviews of the spectroscopy of nuclei far from stability in general or the work at the GSI on-line mass separator in particular are given in [4] and [5], respectively. For the application of heavy-ion reactions to investigate heavy or superheavy elements or to produce lighter exotic nuclei with projectiles at energies above 20 MeV/u see other contributions to this conference.

2. Half-Lives of New Neutron-Rich Nuclei

In the past, spectroscopy of β^--decaying nuclei has been concentrated mainly in those regions where fission processes or spallation of heavy nuclei by high-energy protons could supply nuclei in sufficient quantities. A new way to approach these nuclei is provided by deep-inelastic reactions induced by energetic heavy-ions. With the availability of heavy-ion beams with energies ranging up to 20 MeV/u at the UNILAC, the Göttingen-GSI-Mainz-Warsaw collaboration studied the question whether neutron-rich isotopes in the areas below the light and above the heavy fragments of asymetric fission ($Z \leq 29$ and $Z \geq 66$, respectively) can be produced with sufficient yields to allow mass separation and β-, γ-, and X-ray spectroscopy. For these investigations, targets of ^{nat}W/Ta were bombarded by the neutron-rich isotopes ^{76}Ge, ^{82}Se, ^{136}Xe, ^{186}W and ^{238}U with energies between 9 and 15 MeV/u and beam intensities of 4 - 20 particle•nA. In order to obtain sufficiently high production rates, the target thickness was chosen as to decelerate the projectiles from their initial energy down to the Coulomb barrier within the target and therefore to include both deep-inelastic and quasi-elastic collisions. Target-like fragments in the heavy-lanthanide region and projectile-like fragments in the chromium to zinc, heavy-lanthanide and francium to neptunium region have been investigated so far (see [6,7] and references therein).

The assignment of a new β-coincident γ-activity was often faciliated by the knowledge of excited states in their β-decay daughters. In this aspect it should be pointed out that measurements of excited states (and ground-state masses) with charge-exchange reactions and magnetic spectrographs and/or ΔE-E techniques turned out to be very helpful and complementary to the β-decay studies described here (e.g. [8]). Production rates of 1 to 10 mass-separated atoms/s obtained for isotopes at the actual border-line of known nuclei can be considered as the present limit for decay studies. In general partial decay schemes have been established. A review of the obtained spectroscopic results and references can be found in [9], while application of this technique to an astrophysical problem concerning the solar abundance of ^{180m}Ta is presented in [10]. In the following, only the aspect of comparing the newly established β half-lives with predictions from the gross theory of β-decay [11] and a microscopic model [12] is discussed (Fig. 1). The new neptunium isotopes presented in Fig. 1 were produced in reactions with 16 MeV/u ^{136}Xe

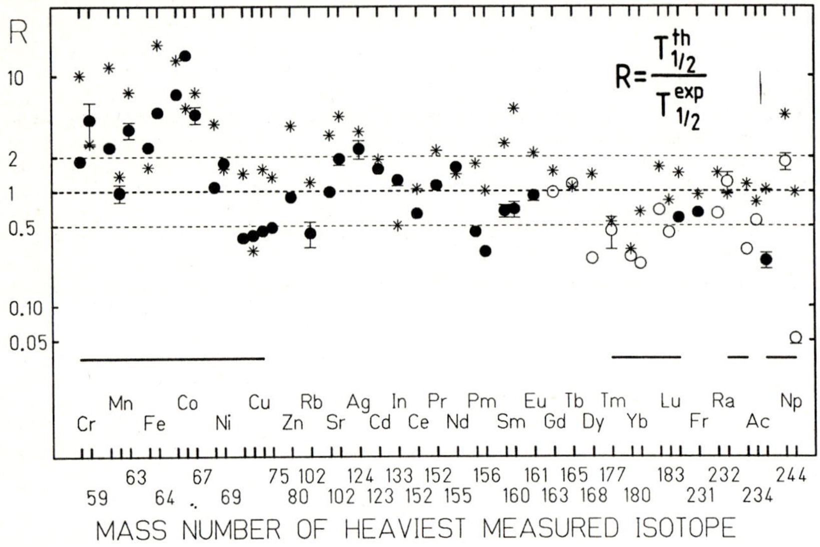

Fig. 1 Ratios R of predicted ($T_{1/2}^{th}$) to measured ($T_{1/2}^{exp}$) new β-decay half-lives. Circles indicate calculations from a microscopic model [12] and stars ones from the gross theory of β-decay [11] using decay-energies from the mass formula of [13] in the latter case. Full and open circles correspond to half-lives shorter or longer than 60 s, respectively. (Note: inbetween the row of displayed nickel and copper isotopes the still unknown β-decays of ^{68}Ni and ^{74}Cu are missing.) The horizontal bars indicate the regions where the new isotopes were produced in multi-nucleon transfer reactions [6,9,14]. Other results are from [15-26].

ions and ^{244}Pu targets and were found by on-line fast chemistry [14]. Besides the 26 new half-lives from heavy-ion reactions, also 22 values of nuclei produced by other techniques (thermal-neutron or spontaneous fission, spallation or fission by high energy protons) since the publication of Ref. [12] are shown. (Nuclei with $A \geq 55$ not cited in [27] or in the 1980 edition of [15] were considered. The sample may be incomplete.)

As generally found, the microscopic model gives lower values than the gross theory. The predictive power of the two theories is tested in Fig. 1, as the displayed experimental values were not yet available, when the models were developed. At that time 65 % (58 %) of the known β-decays with half-lives ≤ 60 s (≤ 1000 s) were reproduced within a factor of two by the more refined microscopic model [12]. This holds also in general for data shown in Fig.1. For the chromium to iron region , however, a systematic trend of both predictions towards too long half-lives seems to evolve. The peak in the cobalt region (Z=27) is consistent with the observation already mentioned in [12] that the predicted half-life for 0.30-s ^{64}Co is 33 times too long (not shown in Fig.1). This was the only nucleus with an experimental half-life smaller than 1s which was not reproduced within a factor of 10.

Half-life predictions in the iron-cobalt region are important, for instance, for the calculation of the astrophysical rapid neutron-capture process (r-process) that has its starting point here. The so far emerging trend to shorter half-lives, however, seems not to change the predicted r-process isotopic abundances drastically, and can therefore not account for discrepancies between calculated and observed solar abundances [6].

A higher sensitivity to detect even more neutron-rich nuclei in the chromium to nickel region against the isobaric background may be achieved by counting β-delayed neutrons. The search for the β-decay of the doubly magic ^{68}Ni nucleus [28] will be continued. A new chemically selective ion source with cold-trap technique [29] may help to separate this nucleus from the isobaric copper and gallium background which prevented the measurement up to now.

3. Strength of Gamow-Teller Beta Decays of Proton-Rich Nuclei

48-Manganese In the ^{56}Ni region new β-delayed and direct proton emitters were searched for and the Fermi and GT β-decay of the new isotope ^{48}Mn was investigated via spectroscopy of β-delayed protons and γ-rays [30]. A branching ratio of $2.7(12)\cdot 10^{-3}$ was deduced for the emission of β-delayed protons of ^{48}Mn. This isotope with a half-life of 150(10) ms represents the heaviest member of the A=4n, T_Z=-1 family of β-delayed proton precursors known to date. The decay proceeds mainly via the superallowed transition to the isobaric analogue state at 5792.4(6) keV in ^{48}Cr. The experimental GT strength distribution, deduced for the ^{48}Cr excitation energies in the interval 0-5.3 MeV from γ data is compared to shell-model calculations with the OXBASH code [31]. As present preliminary conclusion follows that the observed strength is about 60% of the predicted value; this quenching factor agrees with the general trend observed in β-decay of mirror nuclei in the fp shell (e.g. [32]), the global GT quenching of 0.58(5) observed for the middle of the sd-shell [33] and the reduction factor 0.49(5) recently obtained for the β-decay of ^{32}Ar [34].

100-Sn Region Nuclear-structure studies [2] around the expected double shell-closure at ^{100}Sn have recently been focused to investigate the quenching of GT β-transitions. From β- and γ-spectroscopy Q_{EC} values and half-lives of the N=50 isotone ^{96}Pd and of the Z=50 isotopes ^{104,106,108}Sn have been obtained (e.g. [35]). The observed fast β-decays with summed strengths corresponding to log ft values between 3.1 and 3.5 (Fig.2). (The errors do not include systematic uncertainties due to the non-observation of 1^+ states at higher excitation energies in the daughter nucleus because of unobserved weak transitions or of the limiting Q_{EC} window.) They are interpreted as allowed GT transitions where a proton in the single-particle orbit g9/2 transforms into a neutron in the unoccupied spin-orbit partner g7/2 . In this condition, where both the proton and the neutron Fermi energies lie between the same spin-orbit partners, the total available GT strength should be exhausted by this transition. As expected, the experimental GT strength tends to increase with the number of active protons n, the values, however, are 4 to 14 times smaller than expected from the extreme single-particle shell model (Fig. 2). This missing strength indicates that individual nuclear structure and configuration mixing going beyond this simple model have to be considered. This approach was undertaken by I.S. Towner [41] for the N=50 and N=82 isotones. The smearing out of the shell closures was accounted for by a standard pairing approach and the configuration mixing of the

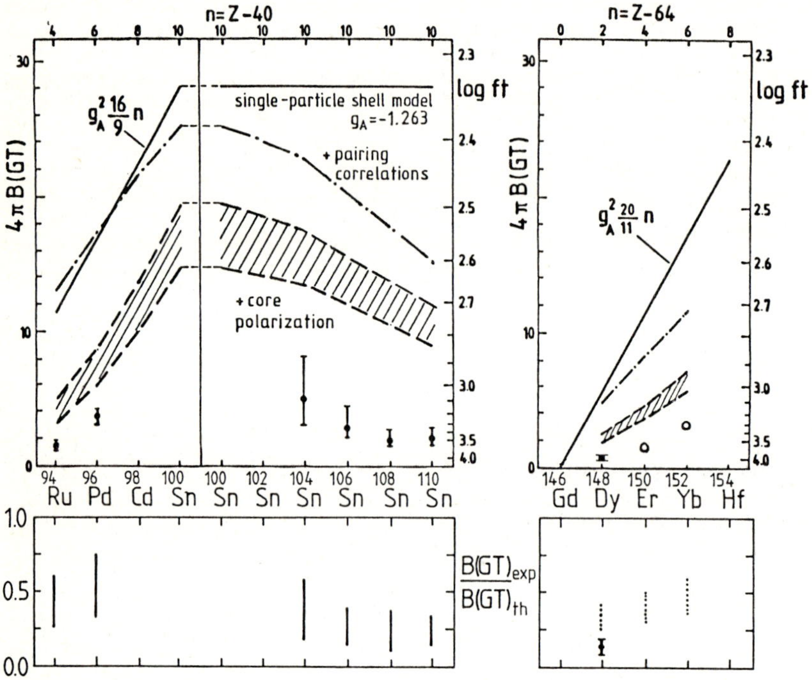

Fig. 2 Top Synopsis of the experimentally observed total GT transition strength B(GT) of even-mass N=50 isotones, tin isotopes and N=82 isotones and predictions. The corresponding log ft values are given at the right hand scale. For references concerning the left hand figure see [36], except for the improved value for ^{104}Sn [37]. The experimental points for the N=82 isotones (right hand side) are from [38] (full dot) and from [39] (open dots).
The full line represents predictions from the extreme single-particle shell model as function of the number n of protons in the g9/2 shell (left) or the h11/2 shell (right), assuming the free-neutron value of the coupling constant g_A and, in the case of the tin isotopes, an empty g7/2 neutron shell. Lower transition probabilities are obtained by taking into account, at first, pairing correlations of protons and neutrons (dotted-dashed lines; from model II of [40] and from [41] for the left and right hand figure, resp.) and then first order core polarization from particle-hole interactions. The shaded region is obtained for different residual forces [41].
Bottom The fraction of strength ("quenching factor") observed experimentally relative to the calculated $B(GT)_{th}$ in the shaded areas is indicated by vertical bars (full or dotted). Their lengths include also the experimental error. The dotting of the bars for the N=82 isotones should indicate that the reduction of the transition strength due to pairing is probably overestimated in [41] and that more precise experimental studies are in progress in the cases of ^{150}Er and ^{152}Yb. For ^{148}Dy the observed strength from [38] is given relative to the calculations from [41] (dotted bar) and to those from [38] (dot with error bar).The error bar of this latter value includes besides the experimental uncertainty also the estimated value of 30% for the total theoretical analysis.

spin orbit partners due to particle-hole interaction by first order perturbation theory ("core polarization" or "ground state correlations"). Recent more extended pairing calculations [40] for the ^{100}Sn region show much less retardation from pairing than given in [41] and even lead to a light enhancement in the cases of ^{94}Ru and ^{96}Pd. Fig. 2 displays the single-particle shell model values corrected for pairing correlations due to DOBACZEWSKI et al. [40] and for core polarization according to TOWNER [41] applying in this case the same correction for all tin isotopes. The fraction of observed strength relative to these so derived theoretical ones is presented in the left bottom part of Fig. 2.

N=82 Isotones above 146Gd Fig. 9 displays also the observed GT transition strength for the β-decays of even-mass of N=82 isotones above ^{146}Gd. Here the h11/2 protons transform to neutrons in the empty h9/2 neutron orbit. In contrast to the analogous case of the N=50 isotones, however, here the β-decays seem to feed mainly just one 1^+ level in the daughter nucleus, e.g. 97 % in the case of ^{148}Dy [38]. The very detailed investigation of this nucleus at ISOCELE/Orsay is presently being extended to the isotones ^{150}Er and ^{152}Yb at GSI. Earlier in-beam results from [39] for these three isotones were based solely on the identification of a single γ-ray; they agree, however, quite well with the new ^{148}Dy measurement [38]. As in the ^{100}Sn region, the retardation due to pairing correlations seems to be overestimated in [41]. Pairing calculations for ^{148}Dy, which were adjusted to experimental occupation numbers for h11/2 protons in ^{144}Sm, yield more then two h11/2 protons in ^{148}Dy and therefore an enhancement of the GT transition strength and, together with corrections for ground-state correlations in ^{148}Dy, a surprisingly low quenching factor of 0.15 in Fig. 2 (for details see [38] and the contribution by P. Kleinheinz to this symposium).

More clearly than in the cases of $^{106,108,110}Sn$, it seems that the GT quenching for ^{148}Dy is stronger than in the sd and fp shells, as discussed above. It is also stronger than expected from charge-exchange reactions such as (p,n), where the observed fraction of the GT sumrule strength is 50 to 60% for heavier nuclei with $A \geq 90$ [42]. In the magic nuclei around ^{100}Sn and ^{146}Gd the retardation of the GT strength due to pairing correlations and first order core polarization is much smaller than in other nuclei and of comparable size to the one due to "higher order terms", so that here the effect of GT quenching in medium mass nuclei can be studied more effectively. Investigations are in progress (to search for and) to study the β-decays of $^{98,100}Cd$, ^{102}Sn, ^{150}Er and ^{152}Yb, partially in collaboration with ISOLDE at CERN.

REFERENCES

1. P. Kleinheinz et al., Z. Phys. A322 705 (1985).
2. U. Dinger et al., "Spins, Moments and Charge Radii of Indium Isotopes in the Range of $^{104-127}In$ Determined by Laser Spectroscopy", to be submitted to Nucl. Phys. A; J. Eberz et al., Z. Phys. A323 119 (1986); H. Lochmann et al., GSI Scientific Report 1985, GSI 1986-1, p. 44
3. E.W. Otten, contribution to this symposium
4. J.H. Hamilton et al., Rep. Prog. Phys. 48 631 (1985)
5. O. Klepper in Proc. Symposium "10 Years of Uranium Beam at the UNILAC", GSI Darmstadt, Fed. Rep. of Germany, April 1986
6. U. Bosch et al, Phys. Lett. 164B 22 (1985)
7. K.-L. Gippert et al., Nucl. Phys. A453 1 (1986)
8. M. Bernas et al., Nucl. Phys. A413 363 (1984)
9. K. Rykaczewski et al., Proc. XIII Int. Winter Meeting on Nuclear Physics, Bormio, Italy, Jan. 1985, p. 764.
10. W. Eschner et al., Z. Phys. A317 281 (1984)
11. K. Takahashi et al., At. Nucl. Data Tables 12 101 (1973).
12. H. V. Klapdor et al., At. Nucl. Data Tables 31 81 (1984) and Z. Phys. A309 91 (1982).
13. P. Möller, J. R. Nix, Nucl. Phys. A361 117 (1981).
14. H. Tetzlaff et al., GSI Scientific Report 1985, GSI 86-1, p. 33.
15. Y. Yoshizawa et al., "Chart of the Nuclides" (Jap. At. Energy Res. Inst., Tokai, Japan), 1984 edition

16. P.L. Reeder et al., Phys. Rev. C31 1029 (1985)
17. R.L. Gill et al., Phys. Rev. Lett. 56 1874 (1986)
18. K.-L. Kratz, Z. Phys. A312 263 (1983)
19. F.K. Wohn et al., Phys.Rev. Lett. 51 873 (1983)
20. J.C. Hill et al., Phys. Rev. C33 1727 (1986)
21. P.L. Reeder et al. Phys. Rev. C27 3002 (1983)
22. T. Björnsted et al., "Double Closed Shell Nucleus ^{132}Sn and its Valence Nuclei", (ISOLDE Workshop, Zinal, Switzerland, 1984), Abstract C9 and to be published
23. K. Okano et al. "The Half-life of a New Isotope ^{155}Nd and its Comparison with Predicted Values" (Res. Reactor Inst., Kyoto Univ., Japan), Preprint 1986
24. H. Mach et al., Phys. Rev. Lett. 56 1547 (1986)
25. P. Hill et al. Z. Phys. A320 531 (1985)
26. T. Karlewski et al., GSI Scientific Report 1983, GSI 1984-1, p. 72
27. W. Seelmann-Eggebert et al. "Karlsruhe Chart of Nuclides", (KFZ Karlsruhe, Fed. Rep. of Germany) 5th edition 1981
28. M. Bernas et al., J. Physique Lett. 45 L-851 (1984)
29. R. Kirchner et al., Nucl. Instr. Meth. A247 265 (1986)
30. T. Sekine et al., "The Decay of ^{48}Mn", to be submitted to Nucl. Phys. A; see also: GSI Scientific Report 1985, GSI 1986-1, p. 41.
31. B.A. Brown et al., (Michigan State Univ., USA) Int. Rep. MSUCL-524, 1985
32. Y.Arai et al., Nucl. Phys. A420 193 (1984).
33. B.A. Brown et al., At. Data Nucl. Data Tables 33 347 (1985).
34. T. Björnstad et al., Nucl. Phys. A443 283 (1985)
35. K. Rykaczewski et al., Z. Phys. A322 263 (1985).
36. O. Klepper, in Proc. XXIII Intern. Winter Meeting on Nucl. Phys., Bormio (1985), p. 747.
37. G.-E. Rathke, private communication.
38. P. Kleinheinz et al., Phys. Rev. Lett. 55 2664 (1985) and contribution to this symposium
39. W. Habenicht et al., in Proc. 7th Int. Conf. on At. Masses and Fundamental Constants (AMCO-7), Darmstadt-Seeheim (1984), ed. O.Klepper, p. 244.
40. J. Dobaczewski et al., contribution to this symposium
41. I. S. Towner, Nucl. Phys. A444 402 (1985) and private communication.
42. C. Gaarde, Nucl. Phys. A396 127c (1983).

Study of Properties of Nuclei far from Stability at GANIL

A.C. Mueller

GANIL, BP. 5027, F-14021 Caen Cedex, France

1. Introductory Remarks

There is certainly no need, at this place, to stress the interest of systematic studies of nuclei as far away from stability as possible. One may mention, just for citing a prominent example, the renormalization of the axial-vector strength in nuclear β-decay by the spectroscopic study of the T_Z=-2 nucleus ^{32}Ar[1]. The experiments on isotopes situated in the borderland off nuclear stability are however difficult due to the sharp fall-off of the production cross sections. The remedy of bombarding very thick targets with intense beams of high-energy protons (as done, for example, at the ISOLDE facility at CERN) is often of only limited success : the diffusion time out of such targets [2] may become prohibitive for the short lifetimes of very exotic species. Then, the inverse method, i.e. the fragmentation of a heavy-ion beam, even of moderate intensity may prove competitive as has been demonstrated by the pioneering experiments at the BEVALAC by Symons et al. [3].

In fact, the fragmentation of a heavy projectile of relativistic energy exhibits two interesting features : since there is only very small momentum transfer between the projectile and the target, the produced fragments are moving close to the projectile velocity and their angular distribution is peaked around 0°. These properties allow for in-flight magnetic separation.

It has been shown [4] that the mechanism of projectile fragmentation remains predominant in the intermediate-energy (20-100 MeV/n) regime which is covered by the accelerator facility GANIL at Caen. Comparatively high heavy-ion beam intensities (see Table 1) are available at GANIL with excellent focal qualities which have allowed for some recent advance in producing isotopes at both drip-lines for light elements.

For these experiments is used the 0° magnetic analyser LISE (angular acceptance = 1 msr) which recently has been constructed at GANIL[5, 6]. It mainly

Table 1: Beams currently available at GANIL.

ION	CHARGE STATE	ENERGY (MeV/u)	CURRENT AT TARGET(μA)
^{14}N	7^+	60	0.8
^{16,18}O	8^+, 8^+	60, 94	2.0, 1.5
^{20}Ne	9^+, 10^+	40, 60	1.0, 0.45
^{40}Ar	14^+,15^+,16^+,16^+	27,35,44,60	0.08,0.4,1.0,0.9
^{40}Ca	16^+, 18^+	60, 77	0.1, 0.1
^{58}Ni	21^+, 23^+	30, 55	0.1, 0.05
^{84,86}Kr	26^+, 29^+	35, 44	0.15, 0.25
^{129}Xe	32^+, 35^+	23, 27	0.02, 0.1

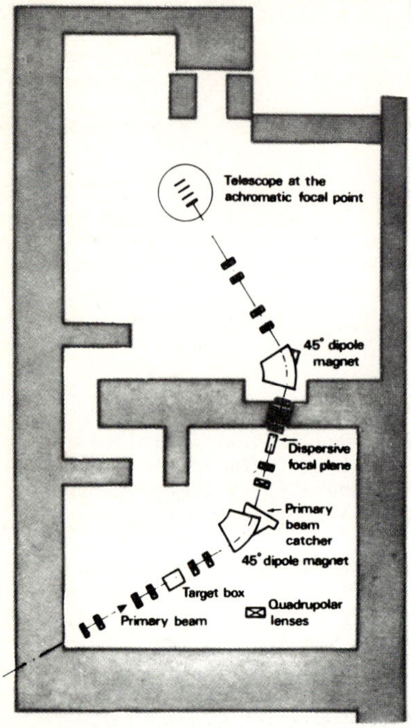

Figure 1 : Schematic layout of LISE

consists of two identical dipole magnets (see Fig. 1), the first, dispersive, plays the rôle of an A/Z separator of the projectile fragments, whereas the second compensates the dispersion. Such a system is doubly achromatic, in angle as well as in position. This, in turn, ensures a practically constant flight path length ($\Delta l/l \sim 10^{-4}$, l = 18 m) between the initial and the final focal points, i.e. the production target and the detector position. Thus, the time-of-flight becomes an important particle identification parameter. In addition, LISE has been designed for maintaining its optical properties after a change of the energy provoked by an energy degrader installed at the intermediate focal plane between the two magnets. This allows for isotopic separation since the Z-dependence of the energy-loss in the degrader combines with the A/Z separation of the first dipole.

This feature has been used by J.P. Dufour et al. for the study of β-γ decay properties of light neutron-rich nuclei for which, so far, only the existence has been known experimentally. The results of these experiments are subject to a special contribution to this conference [7].

2. Search for new isotopes at the drip-lines

These experiments have been carried out, so far, by fragmentation of Ca and Ni for the production of neutron-deficient isotopes and Ar for neutron-rich ones. (It may be noted that also several new neutron-rich isotopes with Z≥18 have been found in the fragmetation of Kr, albeit still far away from the drip-line [8]). For the identification of the fragments a four stage solid-state detector telescope (ΔE_1, ΔE_2, E, E) is mounted at the final focal point of LISE thus taking benefit of the achromatic refocusing properties of the spectrometer. The information of this telescope combined to the time-of-flight (TOF) and the magnetic rigidity setting of the spectrometer readily ensures a overdetermined identification in A and Z.

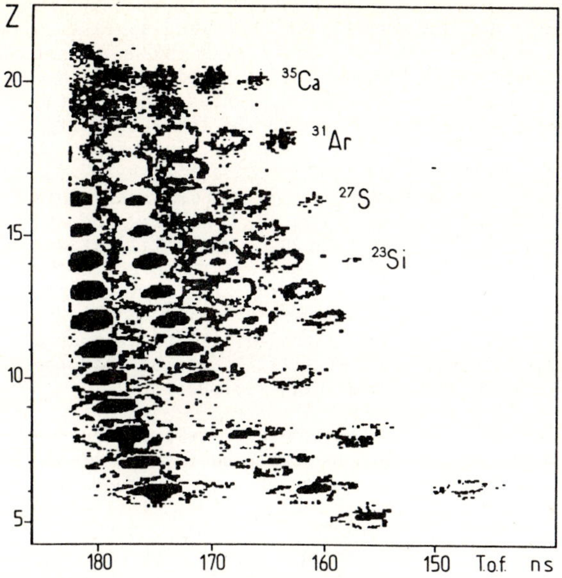

Figure 2 : see text

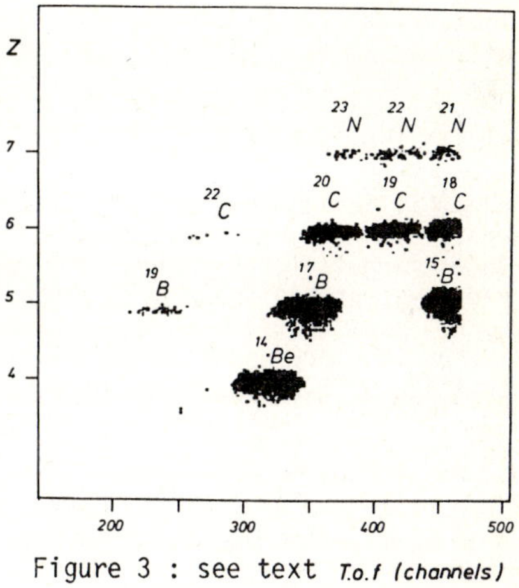

Figure 3 : see text

Fig. 2, as an example, shows the fragments obtained by the interaction of a 77 MeV/u ^{40}Ca beam with a 92 mg/cm^2 Ni target at a spectrometer setting of $B\rho$ = 2.10 Tm. Here the bidimensional representation of $\sqrt{\Delta E_1}$/T.O.F. (proportional to Z) versus T.O.F (proportional to A/Z) is shown. Characteristic curves, which are labelled by their isospin projection T_Z become visible. The whole series of the T_Z = -5/2 nuclei predicted to represent the proton drip-line up to Z = 20 [9], i.e. ^{23}Si, ^{27}S, ^{31}Ar and ^{35}Ca (with the possible exception of ^{22}Si predicted "unbound" by 16keV [9]).

A recent experiment with the 55 MeV/u Ni beam on a Ni target is still in the state of analysis. Preliminarily, the new isotopes ^{43}V, $^{46-47}$Mn, ^{48}Fe, $^{50-51-52}$Co, ^{53}Ni and, by using an Al target, $^{55-56}$Cu are identified [10]. This would indicate that the proton drip-line is also reached for the odd-Z elements Z = 23, 25, 27, 29. The observation of the Cu isotopes points to the interesting fact that transfer reactions, even at GANIL energies may prove effective for the production of nuclei far from stability [11].

On the neutron-rich side, the fragmentation of 44 MeV/u ^{40}Ar projectiles allowed for the first observation of ^{22}C [12], and ^{23}N, ^{29}Ne, ^{30}Ne [13]. Fig. 3 shows, for example, the results of the fragmentation of the 44 MeV/u Ar beam on a 168 mg/cm^2 Ta target at a magnetic field optimized for very neutron-rich B and C isotopes. The new isotope ^{22}C (T_Z = +5) is readily identified whereas the absence of ^{21}C confirms the odd-even effects which are predicted by current mass formulae. This is also nicely shown by the absence of ^{18}B and presence of ^{19}B in Fig. 3 which confirms an experiment by Musser and Stevenson [14] at a level of much improved statistics. According to the mass formulae [15] ^{19}B, ^{22}C and ^{23}N should be the last particle-bound nuclei. Surprises are however possible, and our observation of the isotope ^{29}Ne, which is predicted unbound by most formulae may point in this direction.

3. Future Projects

Obviously , the proof of its existence is only the first step in getting knowledge on the properties of a new isotope. The observation of ^{31}Ar, for example, which is predicted stable against 1p- but slightly unbound against 2p emission may point to a possible case of the new decay mode of a direct two-proton radioactivity [12]. Its detection is however not so straightforward : low-energy protons have to be detected from a deeply implanted source (the energy of the radioactive nucleus is close to the full energy of the accelerator beam) in the presence of an unavoidable background of positrons. Therefore, it is planned to try a detection in the following way : the radioactive heavy-ion will be stopped, at the exit of the spectrometer LISE, in a relatively thin semiconductor detector (thus minimizing the energy signal of the heavy ion and of positrons) preceeded by ΔE detectors for slowing down and for identification. Still, the arrival of nucleus will be marked by a huge signal as compared to the low energy protons. This will be cared for by switching the gain of the preamplifier by a special electronics which is developed actually [16]. The whole semiconductor telescope will be surrounded by an anticoincident β-counter to distinguish the protons from the positrons. The set-up will also be used for β-delayed proton spectroscopy for the other new proton-rich isotopes.

For the very neutron-rich light isotopes, the decay mode of β-delayed neutron emission will become prepondarate. Therefore, we are presently building a liquid scintillator type neutron counter destinated to surround the semiconductor telescope. This detector will be similar to the one we used for the detection of β-delayed neutrons from Na and Mg isotopes [17] produced by a mass separator connected to the CERN PS. Considering the current production rates at GANIL we plan the extension of these experiments to elements up to Ar in the near future.

Another project consist in the measurement of nuclear g-factors. Here, the idea is that there might be spin alignment of the projectile-like fragment in a heavy-ion collision at intermediate energies [18]. This alignment is detected by implanting the fragment in a Pt stopper which is kept in a static magnetic field and observing the anisotropy of the γ-rays following the β-decay. The g-factor may then be measured by a NMR-technique : a radiofrequency is generated around the stopper region which will destroy the alignement when at resonance with the Larmor frequency of the implanted nucleus. A very first test has just been performed on ^{14}B produced with LISE by projectile fragmentation of ^{18}O at 65 MeV/u. It is much too early for a final conclusion ; yet it seemed on-line, that alignment of a few percent was present. If this is confirmed, the method is applicable to several light isotopes under the present conditions.

4. Direct Mass Measurements

A very original method of the measurements of nuclear masses of light neutron-rich isotopes has been developed by W. Mittig et al. [19,20]. This time, the production target for the projectile fragments is located just at the exit of the accelerator. The achromatic design of the GANIL beam transport lines is used to provide a flight path of 116 m down-stream to the high-precision spectrograph SPEG [21]. Thus, knowing the speed and magnetic rigidity with precision ($\Delta v/v \leqslant 5\times10^{-4}$; $\Delta B/B \leqslant 10^{-4}$), the mass of light neutron-rich reaction products is directly deduced.

Presently, the errors are about 500 KeV. Some of the first results are shown in fig. 4 where the experimental two-neutron separation energies are compared to calculations by Uno and Yamada [22] and Möller and Nix [23]. The predictions agree well with the experimental data for N and F isotopes, but a disagreement of several MeV is found for the most neutron-rich O isotopes corresponding to a more bound behaviour than predicted.

5. Concluding Remarks

The favourable production conditions for light exotic nuclei at GANIL led to the start of several experimental programs which have briefly been sketched above. First promising results have been obtained already. This may justify optimism

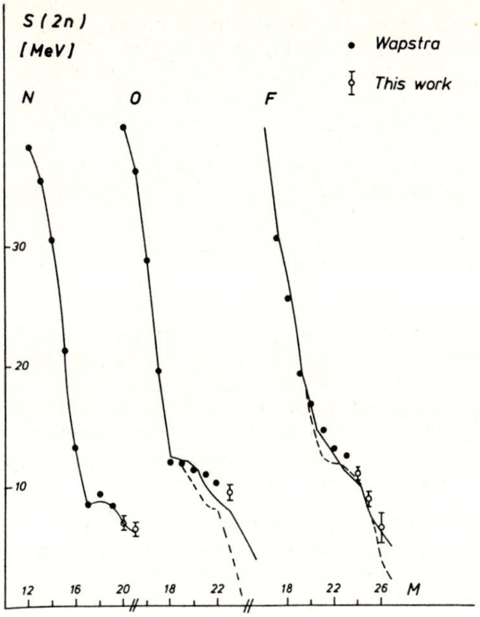

Figure 4 : Two neutron separation energies for N, O and F isotopes. The solid and the dashed lines correspond to the predictions by [22] and [23], respectively.

for the future, especially in view of the upgrading of GANIL which is planned as well for the energy as well as for the intensity of the beams.

REFERENCES

1. T. Björnstad, M.J.G. Borge, P. Dessagne, R.D. von Dincklage, G.T. Ewan, P.G. Hansen, A. Huck, B. Jonson, G. Klotz, A. Knipper, P.O. Larsson, G. Nyman, H.L. Ravn, C. Richard-Serre, K. Riisager, D. Schardt and G. Walter.
Nucl. Phys. A443 (1985) 283

2. H.L. Ravn, Accelerated Radioactive Beams Workshop, Vancouver Island, September 1985, Preprint CERN-EP/85-173

3. T.J.M. Symons, Y.P. Viyogi, G.D. Westfall, P. Doll, D.E. Greiner, H. Faraggi, P.J. Lindstrom and D.K. Scott, H.J. Crawford and C. McParland Phys. Rev. Lett 42 (1979) 40

4. D. Guerreau, V. Borrel, D. Jacquet, J. Galin, B. Gatty and X. Tarrago Phys. Lett. 131B (1983) 293

5. A.C. Mueller, Proceedings of the 7th International Conference on Atomic Masses and Fundamental Constants, ed. O. Klepper, GSI LIbrary, GSI Darmstadt, 1984, p. 696

6. M. Langevin and R. Anne in Instrumentation for Heavy Ion Nuclear Research, ed. D. Shapira, Vol. 7 (Harwood Academic Publishers, Chur, London, Paris, New York (1985) p. 191

7. J.P. Dufour see proceedings of this Conference

8. D. Guillemaud-Mueller, A.C. Mueller, D. Guerreau, F. Pougheon, R. Anne, M. Bernas, J. Galin, J.C. Jacmart, M. Langevin, F. Naulin, E. Quiniou and C. Détraz
Z. Phys. A322 (1985) 415

9. M. Langevin, A.C. Mueller, D. Guillemaud-Mueller, M.G. Saint-Laurent, R. Anne, M. Bernas, J. Galin, D. Guerreau, J.C. Jacmart, S.D. Hoath, F. Naulin, F. Pougheon, E. Quiniou and C. Détraz
Nucl. Phys. A455 (1986) 149

10. To be published by the same group of authors as ref. 12

11. D. Guerreau, Proceedings of the International Conference on Heavy Ion Nuclear Collisions in the Fermi Energy Domain, May 1986, Caen, France (Preprint GANIL P.86-07)

12. F. Pougheon, D. Guillemaud-Mueller, E. Quiniou, M.G. Saint-Laurent, R. Anne, D. Bazin, M. Bernas, D. Guerreau, J.C. Jacmart, S.D. Hoath, A.C. Mueller and C. Détraz
Europhysics Letters in press and Preprint IPN Orsay, IPNO-DRE 86-05

13. M. Langevin, E. Quiniou, M. Bernas, J. Galin, J.C. Jacmart, F. Naulin, F. Pougheon, R. Anne, C. Détraz, D. Guerreau, D. Guillemaud-Mueller and A.C. Mueller
Phys. Lett. 150B (1985) 71

14. J.A. Musser and J.D. Stevenson Phys. Rev. Lett. 53 (1984) 2544

15. See, for example, references in [12]

16. A. Richard, IPN Orsay, private communication

17. M. Langevin, C. Détraz, D. Guillemaud-Mueller, A.C. Mueller, C. Thibault, F. Touchard and M. Epherre.
Nucl. Phys. A414 (1984) 151

18. M. Ishihara, K. Asahi, T. Ichihara, K. Ishida, R. Anne, D. Bazin, D. Guillemaud-Mueller, A.C. Mueller and R. Bimbot
Proposal to the Program Committee GANIL, P-77

19. W. Mittig, L. Bianchi, A. Cunsolo, B. Fernandez, A. Foti, J. Gastebois, A. Gillibert, C. Grégoire, Y. Schutz, C. Stephan and A. Peghaire
XXIV International Winter Meeting on Nuclear Physics, Bormio, Italy, 1986 and GANIL Preprint P.86-03

20. A. Gillibert, A. Bianchi, A. Cunsolo, B. Fernandez, A. Foti, J. Gastebois, Ch. Grégoire, W. Mittig, A. Péghaire, Y. Schutz and C. Stephan
Submitted to Physics Letters

21. P. Birien and S. Valero, CEN-Saclay, DPhN/MF, Mai 1981, CEA-N-2215

22. M. Uno, M. Yamada INS Report NUMA 40 (1982)

23. P. Möller and J.R. Nix, At. Data and Nucl. Data Tables 26 (1981) 165

Beta Decay of Twelve Light Neutron-Rich Isotopes from ^{17}C to ^{40}S *

J.P. Dufour[1], *R. Del Moral*[1], *A. Fleury*[1], *F. Hubert*[1], *D. Jean*[1], *M.S. Pravikoff*[1], *H. Delagrange*[2], *H. Geissel*[3], *and K.-H. Schmidt*[3]

[1]Centre d'Etudes Nucléaires de Bordeaux-Gradignan, Le Haut-Vigneau, F-33170 Gradignan, France
[2]GANIL, B.P. 5027, F-14021 Caen Cedex, France
[3]Gesellschaft für Schwerionenforschung, D-6100 Darmstadt, F. R. Germany

The target fragmentation induced by particule beams is since long recognized as a powerful tool to produce exotic nuclei. Nearly all nuclei with neutron and proton numbers less than that of the target are produced. The drawback of such a wide production is that it implies the use of efficient Z and A selective devices in order to be able to carry detailed spectroscopic measurements. Although very productive, the well-known ISOL facilities do not cover the full range of elements and even in a favorable case like cesium they may fail to separate short-lived isotopes like ^{113}Cs which decays by proton emission. An interesting alternative to ISOL techniques is now available with the fragmentation of heavy ion beams which is essentially equivalent to target fragmentation except for the inverse kinematic. At energies higher than 40Mev/n projectile fragments are forward focused with velocities close to that of the beam. As was already shown by SYMONS in 1979 [1] these favorable conditions result in a high efficiency and very fast separation (according to A/Z) when the nuclei go through a simple magnet.

At GANIL the high intensity beams have also a low emittance and this opens new separation possibilities by allowing the use of a good quality spectrometer. The work presented here has been made at the LISE spectrometer. In one of the modes of operation this device allows to separate the nuclei in two steps: i) the nuclei are selected according to their magnetic rigidities at the exit of the target (A/Z criterium) ii) the nuclei selected by the first dipole are slowed-down in a solid

---

* work performed at the GANIL national facility in CAEN (France)

material and are again selected by a second magnet ($A^{2.5}/Z^{1.5}$ criterium). The second selection should not be understood as an energy-loss discrimination insensitive towards A since it rather involves relative momentum-loss which is both A and Z sensitive. The details of this Projectile Fragments Isotopic Separation (PFIS) are given in ref 2. In the intermediate energy domain one can obtain analytical formulae giving the correspondence between the two magnetic rigidities and the A and Z of the nucleus transmitted with the highest transmission. Neglecting relativistic effects, the following equations are obtained when the projectile velocity is supposed to be conserved in the nuclear reaction:

$$A = \frac{k_1}{B\rho_1^{\lambda-2}} \; \frac{B\rho_1^{\lambda} - B\rho_2^{\lambda}}{d}$$

$$Z = \frac{k_2}{B\rho_1^{\lambda-1}} \; \frac{B\rho_1^{\lambda} - B\rho_2^{\lambda}}{d}$$

where d is the thickness of the degrader , k_1, k_2, λ are constants charaterizing the slowing-down process in the material ($\lambda \simeq 3.5$ for Al) at intermediate energies.

The mass resolution at fixed Z is proportional to d/R, R being the range of the fragment at the exit of the target . In the case of LISE the highest values of d/R were set to 0.5 due to a limitation in the way of lowering the field in the second magnet. We measured a mass resolution $A/\Delta A \simeq 100$ due to the spectrometer. However, the use of a non-adjustable exit collimator reduced the experimental mass resolution to about 50.

The results reported here have been obtained with an ^{40}Ar beam at 60Mev/n reacting on a 190mg/cm² Be target. The separated nuclei were implanted in a catcher foil placed in between a thin scintillator detecting the betas and a high volume (174cm³) Ge detector; only gammas in coincidence with betas were recorded. The half-life measurements have been made by pulsing the beam. Due to the short transport time (≃200ns) and the low background at the exit of LISE the gamma spectroscopy was carried on during both growth and decay of the activity. We thus used two different ways of determining the half-life: i) by the ratio of the beam-off over beam-on activities. ii) by the exponential decay during the beam-off cycles. The first determination may be inaccurate if the beam intensity is not constant, in average, during the beam-on period. In practice, both types of measurement are consistent, and the first type of information has been used when the statistics in the beam-off cycles was low, as in the case of ^{26}Ne. The check of the procedure was made on the following isotopes: ^{36}P, ^{32}Al, ^{30}Mg, ^{25}Ne, ^{27}Na and ^{17}C for which the results of this work are given and compared to previous determinations in table 1.

Table 1. Half-life [s] values measured in this work for the isotopes which have been used for a check of the procedure.

ISOTOPE	^{25}Ne	^{27}Na	^{30}Mg	^{32}Al	^{36}P	^{17}C
This work	0,62(5)	0,295(20)	0,34(2)	0,031(6)	5,33(53)	0,22(8)
Prev.Meas. [ref]	0,602(8) [3]	0,304(5) [4]	0,325(20) [4]	0,035(5) [5]	5,9(4) [6]	0,202(20) [7]

These nuclei were identified through their characteristic gammas and some of them were used to calibrate the $(B\rho_1, B\rho_2)$ tuning to the optimum transmission of a given isotope. Using the formulae given in ref. 2 the calibration constants were set for ^{30}Mg and allowed subsequent "a priori" tunings to a precision better than 0.1 mass unit on other elements with Z ranging from 6 to 15. The mass precision checks were made on ^{15}C, ^{21}O, ^{25}Ne and ^{37}P. Due to the finite resolution, the tuning of the spectrometer for the ^{A}Z isotope allowed some contaminants to be also transmitted at levels ranging from a few percent to three times the abundance of the ^{A}Z nuclide depending on the "exoticity" of the observed nucleus. Extreme exemples are ^{36}P and ^{40}S, the first being highly produced and clearly separated, the second being produced with a low cross-section by double charge exchange and the neighbouring isotopes, although strongly suppressed, are observed as contaminants due to their much higher cross-sections.

The attribution of a given gamma ray to a given ^{A}Z is based on the variation of the counting-rate for different $B\rho_1$-$B\rho_2$ tunings. When observed as a contaminant a given

Table 2. List of the energies, relative intensities and half-lives [s] of the beta-delayed gammas observed in the decay of the studied nuclei.

^{40}S ; $T_{1/2}$=8,8(2,2)

211,6 (4)	100 (10)
431,9 (4)	51 (10)
677,5 (5)	38 (10)
888,6 (4)	50 (12)

^{38}P ; $T_{1/2}$=0,64(14)

1292,3 (4)	100 (6)
2224,3 (10)	23 (4)
3516,0 (10)	13 (4)
3698,0 (10)	11 (3)
4713,3 (10)	10 (3)

^{37}P ; $T_{1/2}$=2,31(13)

* 646,2 (3)	100 (4)
* 751,2 (3)	7,2(6)
*1582,9 (4)	74,4(30)
2100,8 (4)	6,1(8)
2254,1 (4)	8,2(10)

26Ne ; $T_{1/2}$=0,25(20)

233,4 (5)	100

^{36}Si ; $T_{1/2}$=0,54(15)

174,7 (3)	106, (6)
250,3 (4)	100, (5)
424,8 (4)	60, (10)
878,2 (5)	64, (12)
921,8 (5)	19, (7)
1856,2 (6)	45, (10)

^{35}Si ; $T_{1/2}$=0,87(17)

241,4 (3)	100 (4)
392,3 (3)	59 (5)
633,7 (5)	27 (4)
1473,4 (5)	19 (6)
1714,7 (6)	24 (6)
1994,8 (6)	36 (6)
2386,4 (6)	127 (12)
3173,5 (10)	41 (7)
3349,1 (10)	46 (6)
3590,0 (11)	60 (7)
3859,5 (10)	117 (9)
4100,8 (10)	146 (10)

^{34}Al ; $T_{1/2}$=0,050(25)

123,8 (4)	100

^{32}Al ; $T_{1/2}$=0,031(6)

1941,4 (5)	100 (7)
2289,4 (8)	11 (3)
3042,1 (10)	36 (5)
3844,0 (15)	10 (3)
4230,0 (15)	14 (3)

^{17}C ; $T_{1/2}$=0,22(8)

475,6 (4)	22 (7)
619,6 (5)	18 (8)
1374,7 (5)	100 (20)
1848,7 (6)	75 (13)
1906,0 (6)	49 (13)

^{24}F ; $T_{1/2}$=0,34(8)

1981,6 (4)	100

^{36}P ; $T_{1/2}$=5,33(53)

185,8 (4)	2,1(1)
579,8 (4)	0,4(1)
757,5 (4)	1,6(2)
812,0 (4)	4,9(3)
826,9 (4)	15,7(3)
902,7 (4)	71,0(20)
1012,3 (4)	0,7(2)
1058,8 (4)	5,2(3)
1256,9 (4)	4,5(3)
1284,3 (4)	4,1(3)
1439,6 (5)	0,6(2)
1638,3 (4)	35,5(10)
1729,0 (5)	0,9(2)
1960,9 (5)	14,7(5)
2066,1 (5)	0,7(2)
2019,3 (5)	5,0(3)
2251,4 (5)	1,5(2)
2320,1 (5)	1,8(3)
2540,3 (5)	20,0(8)
3076,6 (5)	3,0(10)
3290,3 (5)	100,0(30)
3681,5 (6)	0,8(2)

^{22}O ; $T_{1/2}$=2,9(15) or $T_1 \simeq T_2 \simeq$ 0,8(4)

637,5 (4)	100 (8)
917,9 (4)	43 (7)
1862,0 (5)	67 (10)

^{19}N ; $T_{1/2}$=0,32(10)

96,0 (10)	100(10)
709,2 (8)	63(21)
3137,8 (10)	76(21)

isotope is transmitted at a level always less than 30% of that observed for the nominal tuning. Confirmation in the identification is obtained by the intensity ratios which are clearly constant for different $B\rho_1$-$B\rho_2$ tunings when taken for two gammas attributed to the same isotope.

The results for the gamma energies and the half-lives of the observed isotopes are given in table 2. In the case of ^{22}O the beam-off activity in the 637.5 kev peak exceeds the beam-on one by 30%. If due to statistical fluctuations this observation has a less than 1% probability and the half-life yielded by a fit of the beam-off decay is $T_{1/2}$=2.9±1.5s, a value significantly different from the measurement by MURPHY [5], $T_{1/2}$=0.91±0.35s. A more likely explanation implies the existence of an isomer in ^{22}O having a half-life in the range 0.4 to 1.2s and close to that of the ground-state. Then both beam-on and beam-off data are well described. A further experiment is planned to study the radioactivity of ^{22}O in more details.

References:

1- T.J.M. Symons et al., Phys.Rev.Lett 42 (1979) 40.

2- J.P.Dufour et al.,to appear in Nucl.Inst.Meth. (1986)

3- D.R. Goosman et al., Phys.Rev. C7 (1973) 1133.

4- D. Guillemaud-Mueller et al., Nucl.Phys. A426 (1984) 36.

5- M.J. Murphy et al., Phys.Rev.Lett. 49 (1982) 455.

6- J.C. Hill et al., Phys.Rev. C25 (1982) 3104.

7- M.S. Curtin et al., Phys.Rev.Lett. 56 (1986) 34.

Beta Decay far from Stability and the Decay Heat of Nuclear Reactors

H.V. Klapdor, J. Metzinger, and K. Grotz

Max-Planck-Institut für Kernphysik, D-6900 Heidelberg, F. R. Germany

1. Beta-decay half-lives

The interest in the β-decay properties of nuclides far (and not so far) from stability has increased in recent years. Since the calculation of the half-lives of neutron-rich nuclei published in [1], about 70 new isotopes have been discovered, and this number will increase rapidly with new experimental possibilities (see, e.g.[2-5]). The theoretical predictions [1] thus can be tested now on a rather large number of nuclei unknown at the time of the calculations. Fig. 1 compares the half-lives of the latter nuclides with the predictions of [1]. The theoretical values are found reliable within the expected limits. This is of importance in particular for the astrophysical conclusions on element synthesis, age of galaxy and more recently the value of the cosmological constant and the corresponding energy density of the vacuum [6-8]. The new data allow on the other hand to improve the calculations.

First results of a second generation microscopic calculation which we denote by KGM, in contrast to the older KMO calculations [1] (which were actually done in 1981), are shown in Fig. 2. It is seen that the new calculations which use an RPA approach treating the odd particle by perturbation theory, can give an improved description of the half-lives, particularly also in the chromium to cobalt region, where some systematic deviation seemed to be present (see also Fig. 1 in [4]). We find that the special problem in this region is a particular sensitivity of the half-lives to small changes in deformation, which lead to large changes in the g.s. configurations of the parent nuclei. Fig. 2 shows as examples the effect of a change of the deformation within δ = 0.05 and 0.1 for the nuclides ^{66}Co and ^{67}Co by the corresponding "error" bars.

2. The Decay Heat of Nuclear Reactors.

The improved predictability of nuclear β-decay leads in addition to improved reactor antineutrino spectra from fissile material as ^{235}U, ^{239}Pu, ^{241}Pu [34-38] which are important for the interpretation of reactor neutrino oscillation experiments, to important applications in nuclear technology [38,39].

In this section the results are given of the first attempt at all of a summation calculation of the reactor decay heat basing on a microscopic nuclear structure calculation of the β-decay of the fission products. A detailed description of these new decay heat calculations and of their consequences is given in [38,39].

Basis is a new program, THOR-I (Theory of Heat of Reactors-I), which calculates the isotopic inventory of a reactor as function of time during reactor op-

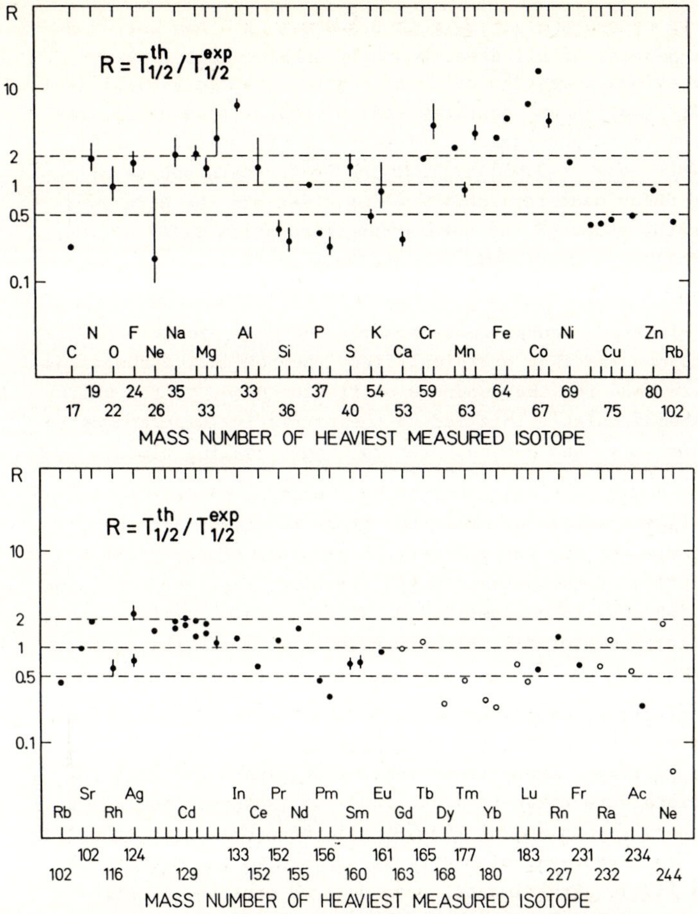

Fig. 1: Ratios R of predicted ($T^{th}_{1/2}$) to measured ($T^{exp}_{1/2}$) new β-decay half-lives [4,9-33] unknown at the time of calculation. Full and open circles correspond to measured half-lives shorter or longer than 60 s, respectively.

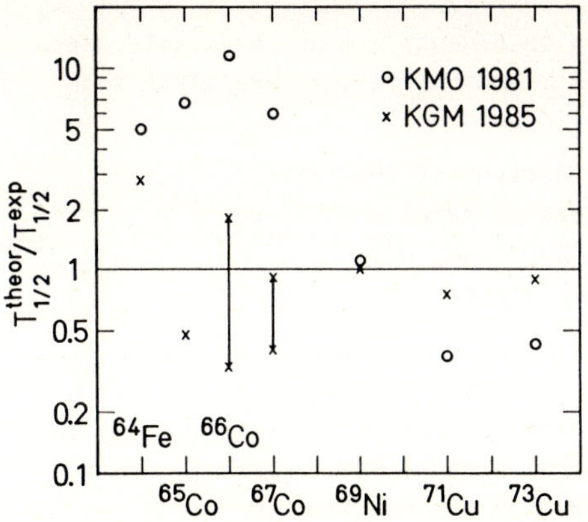

Fig. 2: First results of the second generation calculations (KGM) compared to the older calculations (KMO) of [1]. Ratios of predicted ($T^{th}_{1/2}$) new β-decay half-lives to experimental half-lives ($T^{exp}_{1/2}$).

eration and after shutdown by the analytical method and use of a new set of β-decay data, in which the β-decay of all experimentally unknown nuclei is microscopically calculated. Previous summation calculations suffered in general from the problem of lacking information on the beta decay of short-lived nuclei far from stability which resulted in uncertain predictions for the first several 100 sec after shutdown (see, e.g. [40,41]). That the by far main source of uncertainty of predicted decay heat for short cooling times are the β decay energies or in other words the shape of the beta strength function $S_\beta(E)$ of short-lived fission products follows already from the analysis of [42] and has been discussed in detail by [43].

Besides deficiencies in experimental decay schemes the main problem in summation calculations up to now was that oversimplified 'theoretical' assumptions on the shape of $S_\beta(E)$ were made for the hundreds of fission products for which no experimental decay schemes exist. This led to the persisting discrepancy between calculated and measured γ- and β-decay heat for short cooling times.

An improvement of the decay heat calculations by improving the assumption on $S_\beta(E)$ has been tried by [44] and [45] by using the gross theory of beta decay to calculate $\overline{E}_\beta$ and $\overline{E}_\gamma$. In view of the general serious problems of the gross theory to give a correct description of the beta strength function (see the discussion in [43,6] the result of [45] has to be considered, however, as an improvement obtained by a kind of multiparameter fit which in principle limits the predictive power of the approach.

In the present calculations, experimental beta decay half-lives were taken from the Karlsruhe chart of the nuclides [46] complemented by values from the literature up to the end of 1984. Experimental beta and gamma decay data including isomeric transitions were taken from the Table of Isotopes, Nuclear Data Sheets and recent references (up to end of 1984). All these data were standardized in the sense that they were recalculated to be adapted to the New Atomic Mass Table (Wapstra 1985 [47]). Experimental beta-delayed neutron emission rates (P_n values) were taken from the ORIGEN-2 decay data set [48]. For the neutron capture cross sections the spectrum-averaged cross sections of the ORIGEN-2 code [48] typical for light water reactors (LWRs) were taken. For the fission yields we used the values of Rider [49].

For all nuclei where experimental results on β-decay are not available, beta decay half lives, beta branching ratios and P_{1n} and P_{2n} values were taken from the microscopic calculations described above in section 1.

In Figs. 3a-d we show the result of calculations of the total decay heat (from β and γ decay) of short time irradiations of samples of ^{235}U, ^{239}Pu, ^{241}Pu, ^{233}U with thermal and fast neutrons and compare them with the in our eyes most precise short time irradiation measurements [40,50,51].

The experimental data shown in Figs. 3a-d consist of overlapping sets corresponding to different irradiation times in each of the measurements. For each set of the measurements there exists correspondingly one theoretical curve. It is seen that the calculations even reproduce the details of the irradiation and detection history. The calculations are performed - corresponding to the measurements - for thermal fission in the case of the Dickens experiment [40,50] and for fast fission in the case of the Akiyama experiment [51]. Both sets of experiments overlap within the error bars and the calculations reproduce the ex-

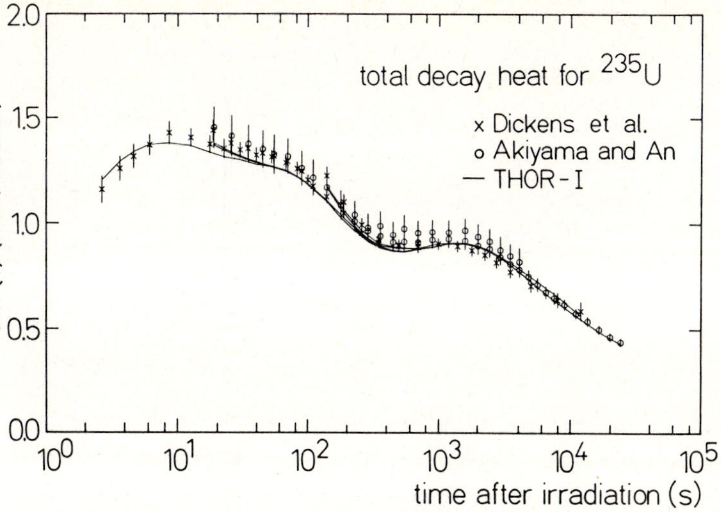

Fig. 3a: Measured total ($\beta + \gamma$) decay heat power f(t) as function of time t after irradiation for short-time irradiation (different irradiation times) of ^{235}U by thermal neutrons (crosses, Dickens et al. [40] and by fast neutrons (circles, Akiyama and An [51]) and the corresponding calculations by THOR-I.

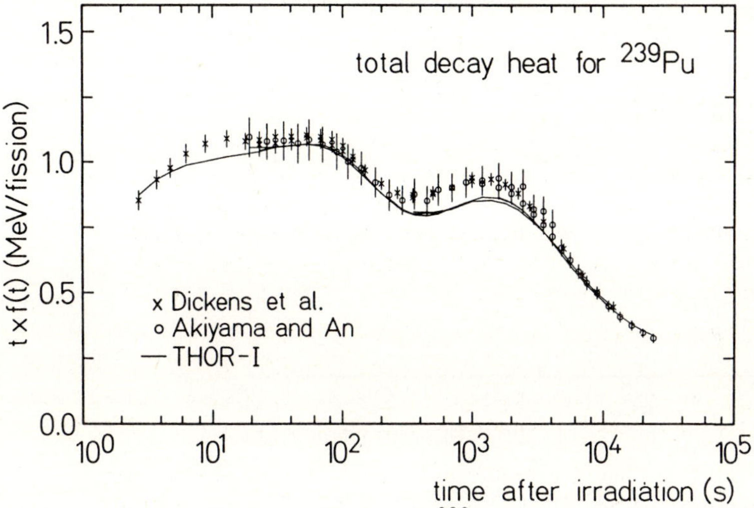

Fig. 3b: Same as in Fig. 3a, but for ^{239}Pu. The Dickens et al. data are from [50].

periments reasonably. Still better agreement is obtained between calculation and measurement when considering only the part of the decay heat produced by β-decay alone.

Figures 3e,f demonstrate the effect of the experimentally unknown nuclides far from stability on the decay heat. The calculation in these cases considers only experimentally known decay schemes. Comparison of Figs. 3e,f with Figs. 3a,c may give a feeling for the quality of the description of the β decay of un-

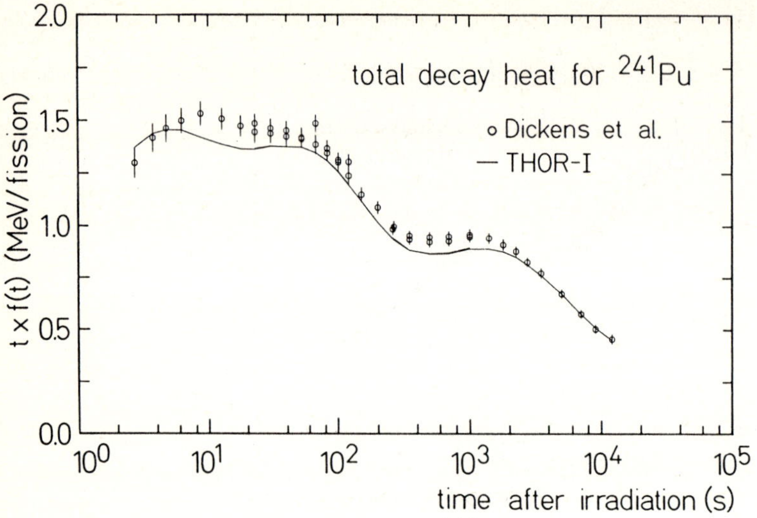

Fig. 3c: Same as in Fig. 3a, but for ^{241}Pu. The circles denote the experiment by Dickens et al. [50].

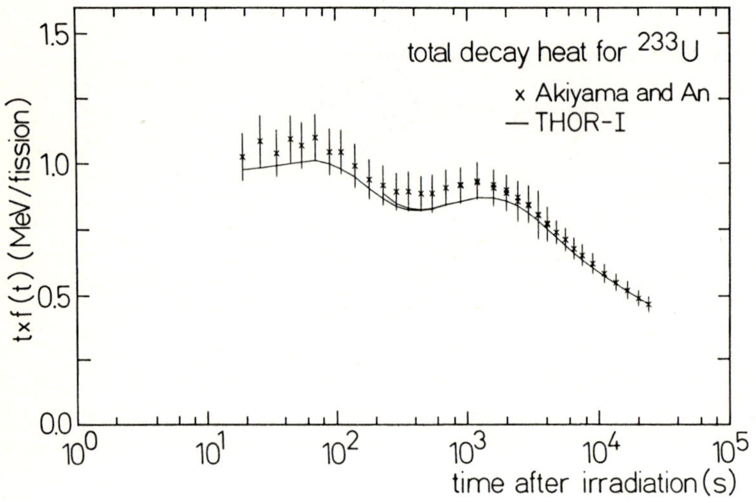

Fig. 3d: Same as in Fig. 3a, but for fast fission of ^{233}U. The crosses denote the experiment of Akiyama and An [41].

known nuclei. Comparison of Figs. 3f and 3c shows further that the disagreement between experiment and calculation around times after irradiation of 1000 s is an effect of the experimentally 'known' nuclei or of the decay heat measurements. The same is true for the case of ^{239}Pu (Fig. 3b).

We have further calculated the total decay heat power for the following types of reactors under realistic reactor operation conditions using our new program THOR-I: pressurized water reactors, boiling water reactors, advanced pressurized water reactors, CANDU. The result is that in all cases present de-

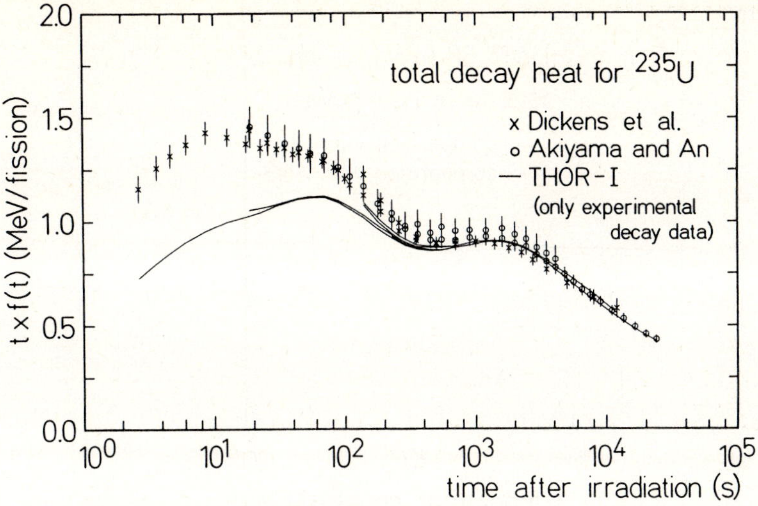

Fig. 3e: The same experimental results as in Fig. 3a, but the calculation performed taking into account only experimentally known decay schemes.

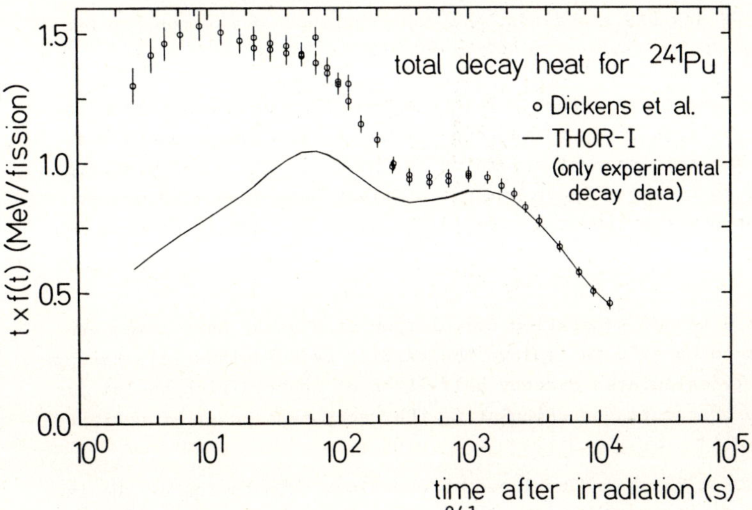

Fig. 3f: Same as in Fig. 3e, but for ^{241}Pu. The experimental data are those of [50] (same as in Fig. 3c).

cay heat standards [52,53] overestimate the decay heat, the amount being different for different reactors [38,39].

As example we shall discuss the situation for light water reactors in some detail. Fig. 4 shows the total decay heat power as function of time after shutdown calculated by THOR-I for a realistic light water reactor operation cycle with a total burnup of 38 MWd/kghm in comparison to the ANS and DIN standards. The assumed cycle was 335-30-335-30-335 days (reactor on and off, respectively). The initial enrichment of ^{235}U was taken to be 3.4%. A constant average neutron

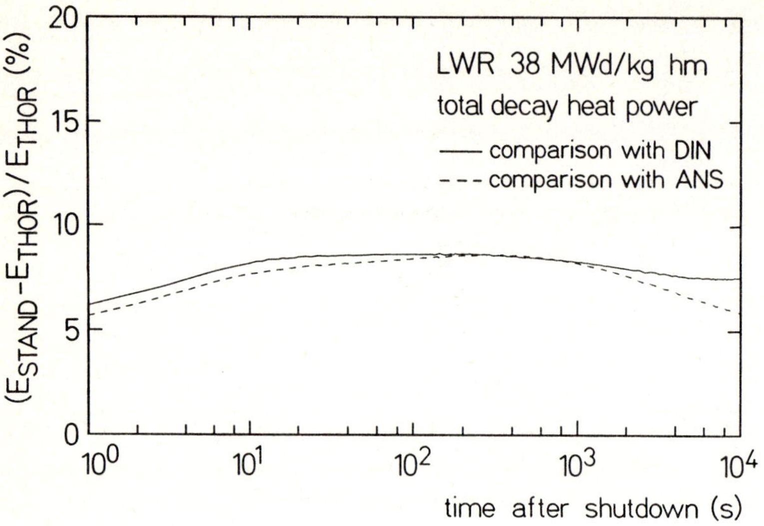

Fig. 4: Total decay heat power integrated up to time t after shutdown as function of t calculated by THOR-I for a realistic light water reactor operation cycle (see text) with a total burnup of 38 MWd/kghm in comparison to the predictions of the ANS and DIN standards. All contributions of neutron capture are included.

flux of $3.25\text{x}10^{14}$ $\text{cm}^{-2}\text{sec}^{-1}$ was assumed. In the calculation of the ANS (DIN) expectations the contributions from neutron capture by the fission products and the actinides are taken into account according to the different given prescriptions. Fig. 4 shows, that according to our calculations both standards overestimate the decay heat power by about 6 - 8% in the first 10^4 sec after shutdown.

3. Conclusion

First results of a second generation calculation of β-decay half-lives are presented which seem to be able to improve the earlier calculations [1]. Basing on the microscopically calculated β-decay half-lives of neutron-rich nuclei far from stability a new procedure for calculating the decay heat power of nuclear reactors is presented.

A new program THOR-I has been developed which allows calculation of the isotopic inventory of a reactor at any time during reactor operation and after shutdown by the analytical method. Part of THOR-I is a new set of β-decay data, in which β-decay of experimentally uninvestigated nuclei is described by microscopic calculations - the first attempt of this kind to our knowledge.

The internal consistency of this data set is reflected in the fact that it allows simultaneous description of decay heat power, β-decay half-lives and rates for β-delayed neutron emission and the shape (and absolute yields) of electron and antineutrino spectra produced in the reactor core. No other set presently in use for the description of experimentally uninvestigated nuclei is capable to do this.

We have calculated by THOR-I the total decay heat power as function of time after shutdown for realistic reactor operation cycles and have compared the result with the ANS and DIN standards. It is found that according to our calculations both standards overestimate the decay heat power by about 6 - 8% in the first 10^4 sec after shutdown.

In view of the progress in the accuracy of summation calculations demonstrated in this paper it might be worthwhile to take such calculations into serious consideration as a basis of future standards. To the extent that the present standards allow the users the option of computing the decay heat power by their own programs and justifying the calculations, the results presented here might be of direct economic benefit for operators of light water reactors. The progress in the predictability of the decay heat power for arbitrary fissile materials should be of importance further for reactor technologies at present under development such as high-converting LWRs and fast breeders and also for the handling of burnt fuels.

References

1. H.V. Klapdor, J. Metzinger, T. Oda, At. Data Nucl. Data Tables 31 (1984) 81.
2. A.C. Mueller, this volume.
3. J.P. Dufour et al., this volume.
4. O. Klepper, this volume.
5. R.W. Hoff, this volume.
6. H.V. Klapdor, Fortschr. d. Physik 33 (1985) 1.
7. H.V. Klapdor, Invited lecture presented at Internat. School of Nucl. Phys.: The Early Universe and its Evolution, Erice, Sicily, 2-14 April 1986, in press in Progr. Part. Nucl. Phys.
8. H.V. Klapdor, K. Grotz, Astrophys. J. 301 (1986) L39 and this volume.
9. M.J. Murphy et al., Phys. Rev. Lett. 49, 455 (1982).
10. J.D. Baker et al., J. Radioanal. Chem. 74, 117 (1982).
11. R. Kirchner et al., Nucl. Phys. A378, 549 (1982).
12. F.K. Wohn et a., Phys. Rev. Lett. 51, 873 (1983).
13. R.C. Greenwood et al., Phys. Rev. C27, 1266 (1983).
14. M. Langevin et al., Phys. Lett. 125B, 116 (1983).
15. E. Runte et al., Nucl. Phys. A399, 63 (1983).
16. J.C. Hill et al., Phys. Rev. C27, 2857 (1983).
17. M. Langevin et al., Phys. Lett. 130B, 251 (1983).
19. J.C. Hill et al., Phys. Rev. C29, 1078 (1984).
18. K. Rykaczewski et al., Z. Phys. A309, 273 (1983).
20. M. Langevin et al., Nucl. Phys. A414, 151 (1984).
21. E. Runte et al., Nucl. Phys. A441, 237 (1985).
22. P. Hill et al., Z. Phys. A320, 531 (1985).
23. U. Bosch et al., Phys. Lett. 164B, 22 (1985).
24. P.L. Reeder et al., Phys. Rev. C31, 1029 (1985).
25. H. Göktürk et al., Z. Phys. A324, 117 (1986).
26. J.P. Dufour et al., Z. Phys. A324, 487 (1986).
27. M.S. Curtin et al., Phys. Rev. Lett. 56, 34 (1986).
28. M. Mach et al., Phys. Rev. C34, 1117 (1986).
29. M. Mach et al., Phys. Rev. Lett. 56, 1547 (1986).
30. R.L. Gill et al., Phys. Rev. Lett. 56, 1874 (1986).
31. J.C. Hill et al., Phys. Rev. C33, 1727 (1986).
32. M.J.G. Borge et al., CERN-EP/86-89.
33. P. Taskinen et al., this volume.
34. H.V. Klapdor, J. Metzinger Phys. Rev. Lett. 48 (1982) 127 and Phys. Lett. 112B (1982) 22.
35. F. von Feilitzsch et al., Phys. Lett. 118B (1982) 162.
36. K. Schreckenbach et al., Phys. Lett. 160B (1985) 325.
37. K. Schreckenbach, this volume.
38. H.V. Klapdor, Proceed. KTG/ENS-Internat. "State of the Art"- Seminar on Nuclear Data, Cross Section Libraries and Their Application in Nuclear Technology, Bonn, October 1-2, 1985, p. 419-158.

39. J. Metzinger, H.V. Klapdor, Proceed. ASM Internat. Conf. on Nuclear Power Plant Aging, Availability Factor and Reliability Analysis, San Diego, California, 8-12 July, 1985.
40. J.K. Dickens, T.A. Love, J.W. McConnell, R.W. Peele, Nucl. Sci. Eng. 74 (1980) 106
41. A. Tobias, Progr. Nucl. Energy 5 (1980) 1
42. F. Schmittroth, R.W. Schenter, Nucl. Sci. Eng. 63 (1977) 276
43. H.V. Klapdor, Progr. Part. Nucl. Phys. 10 (1983) 131
44. T. Yoshida, Nucl. Sci. Eng. 63 (1976) 376
45. T. Yoshida, R. Nakasima, J. Nucl. Sci. Techn. 18 (1981) 393
46. Karlsruhe Chart of the Nuclides, 5th edition 1981, W. Seelmann-Eggebert, G. Pfennig, H. Münzel, H. Klewe-Nebenius
47. A.H. Wapstra, G. Audi, The 1983 Atomic Mass Table, Nucl. Phys. A 432 (1985) 1
48. A.G. Croff, Nucl. Technology 62 (1982) 335 and ORNL-RSIC CCC 371 A
49. B.F. Rider, Compil. of Fiss. Prod. ENDF/B-VI (1981)
50. J.K. Dickens, T.A. Love, J.W. McConnell, R.W. Peele, Nucl. Sci. Eng. 78 (1981) 126
51. M. Akiyama, S. An, Nucl. Data Sci. Technology 237-244, ed. K.H. Böckhoff, 1983, ECSC, EEC, EAEC, Brussels and Luxembourg
52. ANSI/ANS-5.1-1979, American National Standard for Decay Heat Power in Light Water Reactors (1979)
53. DIN 25463, German Standard for the Decay Heat Power in Light Water Reactors (1982)

Gamow-Teller Resonance in β^+-Decay of Heavy Nuclei and Delayed Proton Emission

G.D. Alkhazov, L.H. Batist, A.A. Bykov, V.D. Wittmann, and S.Yu. Orlov

Leningrad Nuclear Physics Institute of the USSR Academy of Sciences, Gatchina, Leningrad district, 188350, USSR

The states with large components of charge-exchange excitations are populated most intensively in nuclear β-decay. Studies of β-decay strength functions S_β using a total γ-absorption spectrometry are in progress at the IRIS isotope separator working on-line to the 1 GeV proton synchrocyclotron. By this technique proposed in [1], the β-feeds I_β may be determined without knowledge of the level scheme and the intensity balances. In our spectrometer a large NaI crystal of 200 mm · 200 mm dimension with a well of Ø 40 mm · 100 mm is used. Mass-separated radioactive sources are delivered into the well by a tape transport system. Details of this technique may be found elsewhere [2].

In nuclei far from stability besides the Gamow-Teller (pn^{-1})-resonance [4] there exists a charge-conjugated branch of spin-isospin excitations - the Gamow-Teller (np^{-1})-resonance. This resonance in contrast with the first one is energetically accessible in the β-decay. In order to investigate its properties in heavy spherical nuclei, S_β for nuclides close to the shell N = 82 were measured using the total γ-absorption spectrometer [3]. It was found that the well-defined resonant structure of S is a representative feature of nuclei in Z ~ 64, N ~ 82 region. With the neutron magic number crossing, the energy of the resonance E increases by ~ 3 MeV, but the transition energy (Q - E) changes slightly. The reduced probability of the β-transition smoothly increases as neutron excess decreases. Such behaviour typifies collective charge-exchange excitations.

The S_β of delayed proton precursor 147-Dy has been measured (Fig. 1) with a proton detector placed inside of the total γ-absorption spectrometer. As it was found in special coincidence experiment, the intensity of the proton transitions to excited states of the daughter 146-Gd nucleus is negligible. Therefore the proton intensities $I_p = I_\beta \cdot (\Gamma_p/\Gamma)$, so one can obtain the value of the com-

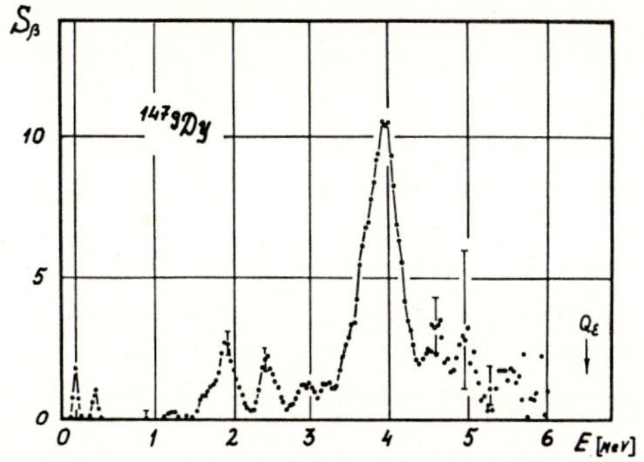

Fig. 1. The experimental β-strength function of 147-Dy decay in units of $[10^{-5}MeV^{-1}s^{-1}]$ obtained by using the total γ-absorption spectrometer.

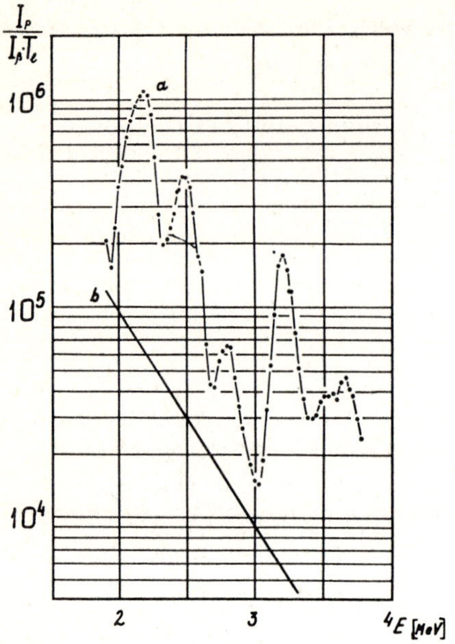

Fig. 2. The average values of the reduced competition factors versus the delayed proton energies: a)- obtained from the experiment with the energy resolution 0.2 MeV; b)- the predictions from the statistical model of compound-nuclei where the reduced proton widths are in inverse proportion to the level density.

petition factor $R = \Gamma_p/\Gamma$ using the experimental values of I_p and I_β. It is convenient to introduce the reduced variables $\gamma_p = \Gamma/T_\ell$ and $r = R/T_\ell$ where T_ℓ is the transmission coefficient calculated from the optical model potential. Figure 2 presents the average values of the reduced competition factors $r = I_p/(T_\ell \cdot I_\beta)$ obtained from the experimental data and calculated in the routine statistical model. As it is seen from Fig. 2, the statistical approach has very little in common with the experiment. The experimental curve exhibits pronounced resonant structure, the resonances being correlated to those of S_β. The appearance of such correlations may be elucidated in a simple model as follows. The ground state, $1/2^+$, of the precursor is identified to the $s_{1/2}$ neutron-hole configuration, while the ground state, $1/2^+$, of the emitter should have a large contribution of the proton orbital $s_{1/2}$. Therefore the β-strength in the region of the resonance is due to transitions to 3-quasiparticle states, the proton decay of these states to the ground state of the even daughter nucleus being forbidden. However, an admixture of the proton single-particle $s_{1/2}$ state appears in the high-excited states due to the spin-isospin residual interaction $H = G(\tau \cdot \tau)(\sigma \cdot \sigma)$. The mixing parameter δ in the wave functions is given by

$$\delta = G \frac{A \cdot B(E)}{E}$$

where B(E), A are the reduced matrix elements of the GT-β-transitions to the state at the energy E and to the ground state, respectively. Then Γ_p may be estimated as $\Gamma_p = \delta^2 \cdot T_\ell \cdot \gamma_0$, γ_0 being a single particle estimation of the proton width. As a result, Γ_p becomes proportional to the β-strength, whereas I_p becomes proportional to the β-strength squared. Figure 3 suggests that the direct proton emission in the decay of 147-Dy predominates over the compound process, since the S_β obtained in this approach is close to that measured through the use of the total γ-absorption technique.

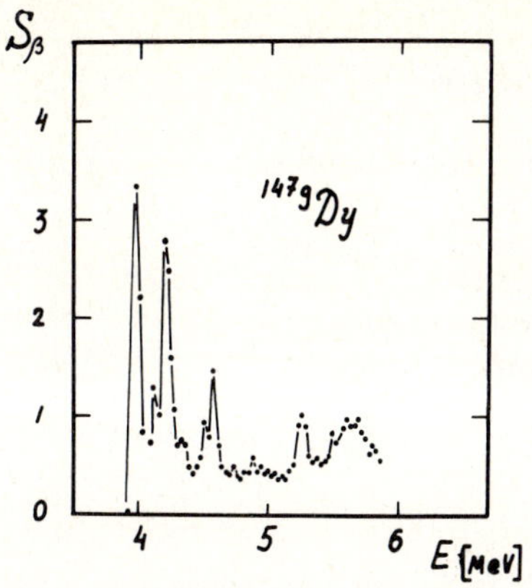

Fig. 3. The β-strength function of 147g-Dy decay extracted from the delayed proton spectra using the model of the direct proton emission where the reduced proton widths are proportional to the β-strength. The S_β and the excitation energy are in units of $[10^{-4}\ MeV^{-1}\ s^{-1}]$ and [MeV], respectively. S_β at $E < 3.9$ MeV is below the registration threshold.

The high energy resolution in the delayed proton spectrum makes it possible to investigate the fine structure of S_β. To our surprise we have found the energies of the fine structure terms as follows: 4.0, 4.18, 4.55, 5.26 MeV, that is the energy intervals between the terms are the constant multiple of 0.18 MeV. The existence of such fine structure is not an exceptional feature of 147-Dy. The analyses of the total γ-absorption data as well as the available β-decay schemes lead us to the conclusion that periodical oscillations of S_β occur in the β-decay of many nuclides, the oscillation period being dependent upon the atomic number only:

$$\Delta E = \frac{(55 \pm 5) \cdot 2^{-n}}{A}\ \text{MeV}, n = 0,1,2; \qquad 15 < A < 200$$

1. C.L. Duke et al.: Nucl. Phys. A151, 609 (1970).
2. G.D. Alkhazov et al.: Proc. of 4th Intern. Conf. on Nuclei Far from Stab., CERN 81-09, 238 (1981),
   G.D. Alkhazov et al.: Nucl. Phys. A438, 482 (1985),
   G.D. Alkhazov et al.: Phys. Lett. B159, 350, (1985).
3. G.D. Alkhazov et al.: Proc. of AMCO-7 Intern. Conf., Darmstadt 1984, p. 612.
4. C. Gaarde et al.: Nucl. Phys. A369, 258, (1981).

GT Beta Decay of ^{29}Na – Comparison with Shell Model Predictions

P. Baumann[1], *Ph. Dessagne*[1], *A. Huck*[1], *G. Klotz*[1], *Ch. Miehé*[1], *A. Knipper*[1], *M. Ramdane*[1], *G. Walter*[1], *G. Marguier*[2], *and C. Richard-Serre*[3]

[1]Centre de Recherches Nucléaires, F-67037 Strasbourg, France
[2]Institut de Physique Nucléaire, F-60622 Villeurbanne, France
[3]IN2P3/CERN, CH-1211 Geneva 23, Switzerland and
the ISOLDE Collaboration

For very neutron rich nuclei (31Na, 32Mg), anomalies in the low energy level structure have been interpreted with the contribution of the fp shell in the model space. As a part of a study of the different decay modes of 30Na, a new investigation of the level structure of 29Mg has been made. To know if the decay of 29Na and 30Na can be understood in terms of sd-shell systematics, we have compared detailed measurements with the precise theoretical predictions made by WILDENTHAL et al. |1| in the complete sd shell model space.

1 EXPERIMENT

The 29Na nuclei have been produced by fragmentation of a uranium carbide target with the 600 MeV proton beam of the CERN SC. The mass-separated sources delivered by the ISOLDE separator have been studied by gamma (Ge (Li) counters) and neutron (time of flight) spectroscopy using singles and coincidence measurements. Our data complement previous results |2,3| and allow a detailed comparison with the shell model calculations for neutron-rich nuclei.

2 RESULTS for BOUND STATES in 29Mg (E_x < 3.78 MeV)

The decay properties of 9 excited states (E_x = 55,1638,2192,2500,2615,3224,3227, 3674 and 3985 keV) can be compared with theoretical predictions and an excellent agreement is found for excitation energies, beta feeding and gamma branching ratios. In particular, the first excited state of 29Mg has been found at 54.6 keV and from the lifetime value (τ = 1.83 ns) deduced from γ-γ coincidences with BaF2 counters, it has been identified as the missing $1/2^+$ level, decaying to the $(3/2^+)$ G.S. with a B(M1) = 0.11 ± 0.01 W.u. strength. To illustrate the quality of the agreement between theory and experiment, we have reported in fig.1 the distribution of the transition strength with excitation energy up to 4 MeV, predicted by WILDENTHAL et al. |1| and deduced from our experiment. It should be noted that one level (E_x = 1095 keV) cannot be associated with shell model states in sd space configurations. Our results are consistent with $J^\pi = 3/2^-$ assignment proposed by FIFIELD et al. |3| for this intruder state.

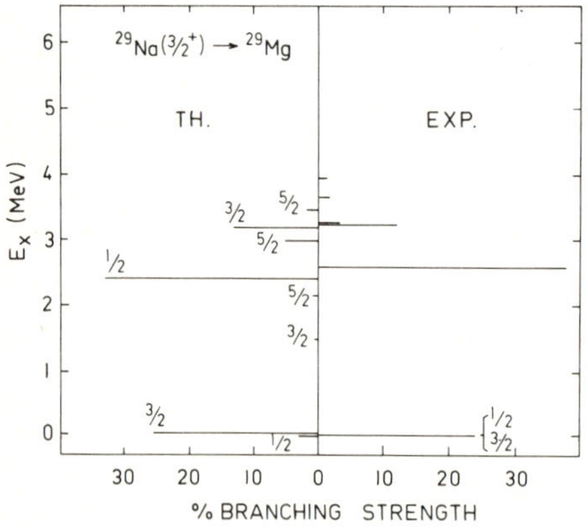

Fig.1
Branching strength (%) for E_x < 5 MeV

3 RESULTS for PARTICLE UNBOUND EXCITED STATES in 29Mg ($4 < E_x < 9$ MeV)

The time of flight of delayed neutrons (fig.2) reveals that the neutron emission in addition to the previously reported decays |4|, proceeds via high energy branches. From our experiment, 51% of the neutron emissions are found above 1.7 MeV and correspond to several strong branches depopulating high lying levels in the 1n channel. To assign initial and final levels in the delayed neutron emission, neutron-gamma coincidences have been studied. The experimental energy distribution of the GT strength is shown in fig.3 with shell-model predictions by WILDENTHAL |5|, carried out with the "free-nucleon" normalization for the GT single-nucleon matrix element and a number of final states taken into account considerably higher than in the previous case (ref.1). In fig.3, the GT strength measured and predicted to fall within each 200 keV energy bite is summed and these values are plotted as a histogram. It can be seen that the shell-model calculation in sd space predicts a significant strength up to 9 MeV when the number of final states involved in the calculation is high enough. However, above E_x = 8 MeV, the dominant strength deduced from the time of flight measurements is not reproduced . This anomaly can be related to the few discrepancies already discussed by WILDENTHAL et al. |1| for very neutron-rich sd-shell nuclei.

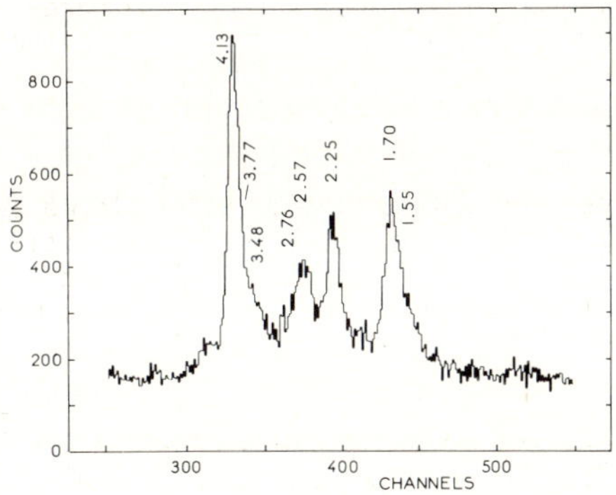

Fig.2
Neutron time of flight spectrum (E_n in MeV)

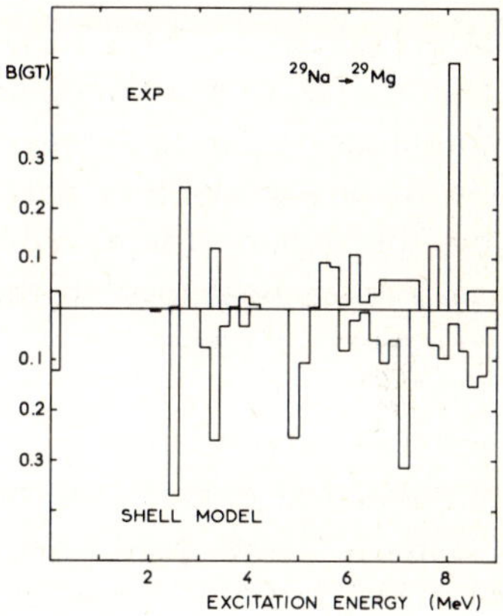

Fig.3
B (GT) strength for 29Na decay

1 B.H. Wildenthal et al.: Phys. Rev. C28, 1343 (1983)
2 D. Guillemaud-Mueller et al.: Nucl. Phys. A426 37 (1984)
3 L.K. Fifield et al.: Nucl. Phys. A437 141 (1985)
4 W. Ziegert et al.: Helsingör, 1981, CERN 81-09, p. 327
5 B.H. Wildenthal: Private Communication (1986)

The Renormalization of the Axial-Vector Strength in Nuclei: Experiments on Superallowed Beta-Decay

B. Jonson[1], *M.J. G. Borge*[2], *P.G. Hansen*[2,3], *S. Mattsson*[1], *G. Nyman*[1], *A. Richter*[4], *and K. Riisager*[3]

[1]Dept. of Physics, Chalmers University of Technology, S-41296 Göteborg, Sweden
[2]EP Division, CERN, CH-1211 Geneva 23, Switzerland
[3]Institute of Physics, University of Aarhus, DK-8000 Aarhus C, Denmark
[4]Institut für Kernphysik, TH, D-6100 Darmstadt, F. R. Germany

It is known that the empirical coupling constant g_A' characterizing the axial-vector (or, Gamow–Teller) strength in nuclear beta-decay is reduced relative to the free-nucleon value g_A determined from the neutron decay. This reduction, which was first estimated by WILKINSON [1], arises largely from pionic contributions and hence is one of the examples of sub-nucleonic effects in nuclear physics. Very approximately one may say that one part of the effect comes from virtual hole-delta pairs and another part from second-order core polarization brought about by the tensor force, which itself is predominantly due to one-pion exchange. Papers dealing with these problems, and references to earlier work, can be found in publication [2].

Experiments to determine the quenching factor $(g_A'/g_A)^2$ directly in beta-decay are, however, rendered difficult by the existence of a highly collective spin-isospin mode ('Gamow–Teller giant resonance' GTGR), which is not accessible in the decay of nuclei with an excess of neutrons. In this case the rates of low-energy transitions reflect essentially the screening of the interaction due to the GTGR. Luckily a series of beautiful (p,n) experiments [3] at high energy have allowed the GTGR to be observed from neutron-rich targets, and indicate that only about 50–65% of the single-nucleon strength is observed in the heavier nuclei, rather independently, it seems, of mass and isospin.

The present paper charts the progress in a series of experiments to determine the quenching from the beta-decay of extremely proton-rich nuclei, for which the major part of the collective transitions ('the superallowed decay') are energetically possible. We have already [4] reported results for ^{32}Ar, found [5] as the first case with $Z - N = 4$, and now available in copious quantities thanks to a new argon-producing target and ion-source system [6] at CERN's ISOLDE Facility. These improved experimental conditions have allowed also a study of the ^{33}Ar strength, in more detail than the previous work of HARDY et al. [8]. A summary of the present experimental situation for all the proton-rich argon isotopes is given in Figs. 1 and 2 and in Table 1.

The theoretical scale is set by a large shell-model calculation [9] based on complete diagonalization in the *sd* shell with the Chung–Wildenthal parameters.

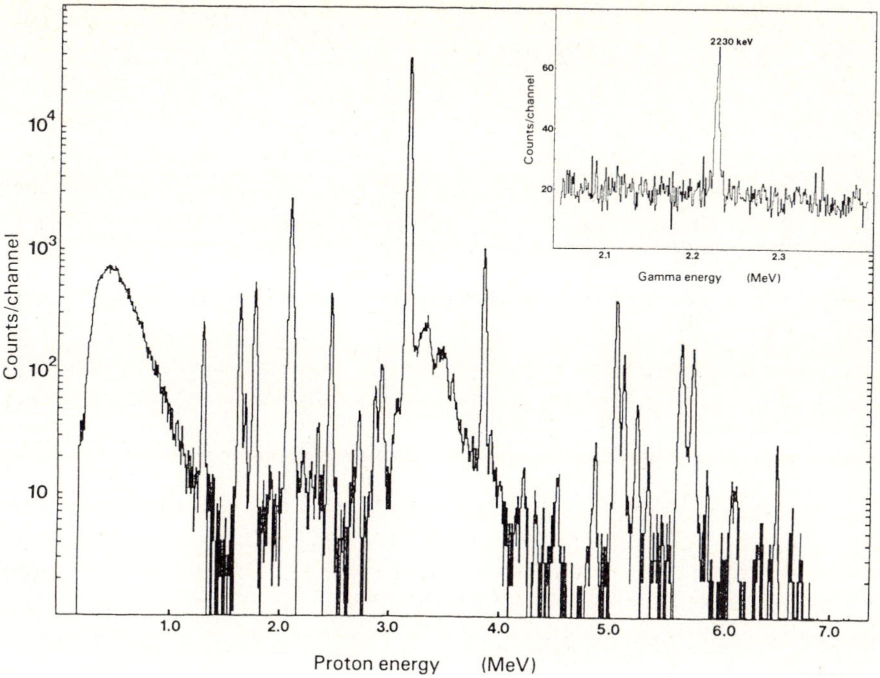

Fig. 1 Spectrum of beta-delayed protons [Ref. 7] from ^{33}Ar measured with a 500 μm, 300 mm^2 Si detector (FWHM = 25 keV at 5 MeV). The spectrum was measured with the condition that the β^+ particle preceding the proton emission was emitted in the opposite direction to the proton detector. This measuring technique was used to avoid β^+p summing in the proton detector. The proton spectrum has been corrected for the proton response function. The branch of delayed protons obtained from these data is 40.1 ± 1.0%. The inset shows the result of a separate experiment where coincidences between protons and gamma-rays were detected. The peak at 2230 keV corresponds to a (0.7 ± 0.2)% branch to the first excited state in ^{32}S.

Table 1

Nuclide	T_Z	$\Sigma B(GT)^{exp}$	[Ref.]	$(g'_A/g_A)^2$ [Ref. 9]
^{35}Ar	$-1/2$	0.3	[11]	0.56
^{34}Ar	-1	1.7	[12]	0.52
^{33}Ar	$-3/2$	2.7	[7]	0.54
^{32}Ar	-2	3.8	[4]	0.49

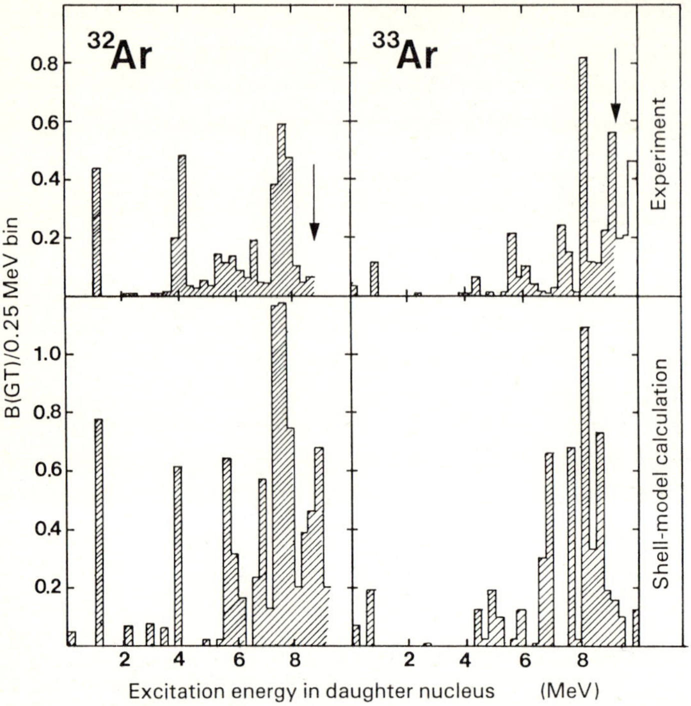

Fig. 2 Experimental and theoretical strength functions for ^{32}Ar and ^{33}Ar. The experimental strength functions are obtained from beta-delayed proton and gamma data (Refs. [6] and [7]). The experimental cut-off energies used for calculating the quenching factors are indicated by arrows. The lower part of the figure shows the theoretical strength function from a shell-model calculation by Müller et al. [9]. Note the excellent agreement between the theoretical and experimental energies. The over-all renormalization of the axial-vector strength in the energy regions below the arrows are $(g'_A/g_A)^2 = 0.49 \pm 0.05$ for ^{32}Ar and $(g'_A/g_A)^2 = 0.54 \pm 0.05$ for ^{33}Ar. The weak branch to the first excited state in ^{32}S, in the ^{33}Ar decay, gives rise to a relatively high uncertainty of the strength above 9.25 MeV excitation in ^{33}Cl.

This is the approach taken by BROWN and WILDENTHAL [10] and earlier work cited therein, but it seems as if the newer parameter set reproduces the argon energies less well. Part of this problem could be that argon is too close to the end of *sd* shell so that excitations into the *fp* shell have begun to play a role. Later we expect to be able to investigate also some cases near the beginning and middle of the *sd* shell. With the reservation that the error on the theoretical scale is still an open question, one sees from Table 1 that all the results for proton-rich argon isotopes are consistent with $(g'_A/g_A)^2 = 0.53 \pm 0.05$ (experimental error only).

REFERENCES

[1] D.H. Wilkinson: Phys. Rev. **C7**, 930 (1973); Nucl. Phys. **A209**, 470 (1973)
[2] Progr. Part. Nucl. Phys. **11**, 1-618 (1983)
[3] C. Gaarde *in: Nuclear Structure 1985*, eds. R. Broglia et al., (North-Holland, Amsterdam, 1985), p. 449
[4] T. Bjørnstad et al.: Nucl. Phys. **A433**, 283 (1985)
[5] E. Hagberg et al.: Phys. Rev. Lett. **49**, 792 (1977)
[6] T. Bjørnstad et al.: Methods for production of intense beams of unstable nuclei: new developments at ISOLDE, to be published in Physica Scripta
[7] M.J.G. Borge et al.: The decay of ^{33}Ar, in preparation
[8] J.C. Hardy et al.: Phys. Rev. **C3**, 700 (1971).
[9] W. Müller, B.C. Metsch, W. Knüpfer and A. Richter: Nucl. Phys. **A430**, 61 (1984)
[10] B.A. Brown and B.H. Wildenthal, Atomic and Nucl. Data Tables **33**, 347 (1985)
[11] H.S. Wilson et al.: Phys. Rev. **C22**, 1696 (1980)
[12] J.C. Hardy et al.: Nucl. Phys. **A223**, 157 (1974)

The $\pi g_{9/2} \rightarrow \nu g_{7/2}$ Gamow-Teller Beta Decay of Even Nuclei Near ^{100}Sn

J. Dobaczewski[1], *W. Nazarewicz*[2], *A. Plochocki*[1], *K. Rykaczewski*[1], *and J. Zylicz*[1,3]

[1]Department of Physics, Warsaw University, PL-00681 Warsaw, Poland
[2]Institute of Physics, Technical University, PL-00662 Warsaw, Poland
[3]GSI Darmstadt, D-6100 Darmstadt, F. R. Germany

1. Introduction

A typical even nucleus in the ^{100}Sn region decays by a few very fast $0^+ \rightarrow 1^+$ Gamow-Teller (GT) transitions [1]. These transitions result from the transformation of a $g_{9/2}$ proton into a $g_{7/2}$ neutron. The 1^+ states fed directly in the β-decay are expected to have a $(\pi g^{-1}_{9/2} \nu g_{7/2})_{1^+}$ component in their wave function which defines the β-transition strength. The latter can be deduced from the decay scheme if the decay energy Q(EC) and the half-life are determined.

The present work aims at the interpretation of existing experimental β-decay data and provides predictions for future experiments. For nuclei of interest, two nuclear-structure models are applied to calculate the Q(EC) value, the centre-of-gravity energy $E(1^+)$ of the $(\pi g^{-1}_{9/2} \nu g_{7/2})_{1^+}$ configuration and the reference β-strength S(GT). A comparison of S(GT) with the experimental total (summed over the individual $0^+ \rightarrow 1^+$ transitions) β-strength is a starting point for the analysis of the origin of the well known β-strength quenching phenomenon.

2. Models Applied

Model I is a self consistent Hartree-Fock-Bogolubov spherical theory which assumes a minimalization of the energy of an A-nucleon system of independent quasiparticles interacting with the Skyrme effective two-body forces [2]. The full HFB method is used instead of the BCS approach in the treatment of the pairing correlations.

Model II assumes a minimalization of the nuclear potential energy expressed as a sum of the macroscopic liquid-drop term of ref. [3] and the microscopic correction term based on the Woods -Saxon potential with parameters of ref. [4]. The microscopic correction term contains a shell correction evaluated by use of the Strutinsky method and a pairing correction obtained by use of the Lipkin-Nogami approach (the approximate particle-number projection before variation being employed to avoid sudden pairing collapse at the shell closure) [5]. The calculations have been performed first under an assumption of a spherical nuclear shape (II-sph) and, afterwards, by varying the nuclear deformation (II-def).

3. Results of Calculations and Experimental Data

Theoretical Q(EC) and $E(1^+)$ values,selected here for the mother nuclei with Z=50 and/or N=50, are compared in table 1 with the available experimental data (a more complete presentation of the results will be given in a forthcoming paper). The predictions of models I and II-sph are fairly close to the experiment, but a better, quite satisfactory, agreement is obtained with the model II-def.

For the ground states of ^{110}In through ^{104}In the model II-def predicts the deformation parameter β_2 ranging from 0.12 to 0.07 (with β_4 close to zero), which compares well with the results of the laser measurements [1]. However, for the 1^+ excited states fed in the β-decay the calculated β_2 values are close to 0. The energy splitting observed for these states can only partly be due to the deformation; to explain the observations one should take into account residual n-p interactions. The ground states of the mother nuclei are found in our calculations to be spherical, cf. also ref [3].

The pairing functions listed in table 2, v^2 for the $\pi g_{9/2}$ orbital and u^2 for the $\nu g_{7/2}$ orbital in the mother nucleus, can be used to obtain the reference β-strength:

$$S(GT) = (160/9)\ v_p^2\ u_n^2.$$

Here the factor 160/9 represents the strength calculated for the $^{100}Sn \rightarrow {}^{100}In$ decay within the extreme single-particle shell model. The experimental β-strength for nuclei near ^{100}Sn is found to be smaller than the reference value by a factor ≥ 5. According to Towner [6], the core polarization and higher order effects (including the non-nucleonic degrees of freedom) are responsible for this hindrance of the β-decay.

Compared to the results of our calculations (carried out with the models which reproduce quite well the Q(EC) and $E(1^+)$ data), the pairing factors v^2 for the N=50 isotones from ref. [6] and u^2 for the Z=50 isotopes from ref. [7] seem to overestimate the role of pairing in retardation of the GT-transitions.

This work was partially supported by the Polish Ministry of Science and High Education under the contract CPBP01.09.

Table 1. Predicted and experimental energy values

Beta decay	Q(EC) MeV				$E(1^+)$ MeV			
	I	II-sph	II-def	Exp [1]	I	II-sph	II-def	Exp [1]
$^{110}Sn \rightarrow {}^{110}In$	0.99	-0.16	1.12	0.58(3)	0.28	0	0.72	0.3
$^{108}Sn \rightarrow {}^{108}In$	2.34	1.04	1.89	2.059(25)	0.56	0	0.78	0.7
$^{106}Sn \rightarrow {}^{106}In$	3.49	2.48	3.35	3.19(6)	1.02	0	0.85	1.1
$^{104}Sn \rightarrow {}^{104}In$	5.15	4.12	4.51	4.15(30)	1.95	0.52	0.98	1.2
$^{102}Sn \rightarrow {}^{102}In$	6.85	5.75	6.09	-	2.80	0.95	1.32	-
$^{100}Sn \rightarrow {}^{100}In$	8.10	7.37	7.82	-	2.37	1.37	1.78	-
$^{98}Cd \rightarrow {}^{98}Ag$	6.22	5.29	5.53	-	2.46	1.46	1.53	-
$^{96}Pd \rightarrow {}^{96}Rh$	4.27	3.19	3.33	3.45(15)	2.54	1.54	1.34	1.3
$^{94}Ru \rightarrow {}^{94}Tc$	2.12	1.01	1.21	1.586(15)	2.56	1.63	1.32	0.7

Table 2. Results of pairing calculations for mother nuclei of table 1

Z	50	50	50	50	50	50	48	46	44
N	60	58	56	54	52	50	50	50	50
v^2(I)	1.00	1.00	1.00	1.00	1.00	1.00	0.80	0.61	0.43
u^2(I)	0.73	0.85	0.95	0.97	0.99	1.00	1.00	1.00	1.00
v^2(II-sph)	0.94	0.94	0.93	0.93	0.92	0.92	0.79	0.63	0.47
u^2(II-sph)	0.57	0.67	0.76	0.87	0.93	0.97	0.97	0.97	0.97

REFERENCES

[1] O. Klepper, contribution to this conference and references quoted there-in
[2] J. Dobaczewski et al.,Nucl. Phys.A422,103(1985)
[3] P. Möller and J.R. Nix, ADNDT 26,165(1981)
[4] J. Dudek et al., Phys Rev. C23,920(1981)
[5] W. Nazarewicz, Proceedings of the Nuclear Structure Conference in Dubrovnik, 5-14 June 1986, to be published
[6] I.S. Towner, Nucl. Phys. A444,402(1985)
[7] X. Campi et al.,Nucl. Phys. A223,541(1974)

Giant GT^+ Excitations of N=82 Nuclei Populated in β^+-Decay

P. Kleinheinz

Institut für Kernphysik, KFA Jülich, D-5170 Jülich, F. R. Germany

Whereas the GT^- giant excitation is energetically not accessible in β^--decay, the GT^+ giant state may be reached in β^+-decay of very neutron-deficient nuclei where the proton single-particle states are raised through the Coulomb repulsion. This becomes possible whenever the proton and neutron Fermi-energies are between the spin-orbit partners of a high-ℓ orbit, i.e. such that the $j_>$ state is partly filled by protons whereas the $j_<$ state for neutrons is (partly) empty. This condition is fulfilled for the 1g shell in the N≥50 nuclei below ^{100}Sn, and for the 1h orbitals in the N≥82 nuclei above ^{146}Gd, and in both regions such $\pi j_> \rightarrow \nu j_<$ GT β^+-decays have been identified. Since e.g. in nuclei with 82 neutrons all N=4 neutron states are filled, the Pauli principle only allows a $\sigma\tau_+$ transition of the N=5 $h_{11/2}$ protons to the unoccupied $h_{9/2}$ neutron orbit. This is the only possible $\sigma\tau_+$ transition which will exhaust the entire available strength and thus the $(\pi j_> \nu j_<)1^+$ state in the daughter nucleus is identical with the GT^+ giant state. The principal locations of the pertinent states for the N=82 region are shown in fig. 1, which refers to ^{148}Dy; this case has recently[1)] been investigated in a high sensitivity measurement. The figure also includes the isobaric-analog state at its estimated energy, and the GT^- resonance based on the $^{148}Gd_{84}$ ground state, which should lie about 2 MeV higher.

Already in earlier β^+-decay studies of ^{148}Dy and of the higher-lying isotones[2)] ^{150}Er and ^{152}Yb it was noted that the $0^+ \rightarrow 1^+$ GT transition strengths are strongly retarded compared to the shell model prediction for $(\pi h^n_{11/2})0^+ \rightarrow (\pi h^{n-1}_{11/2} \nu h_{9/2})1^+$ transitions. All these studies have however identified a single γ-ray only, viz., the $(1^+ \rightarrow 2^-, E1)$ ground state transition in the

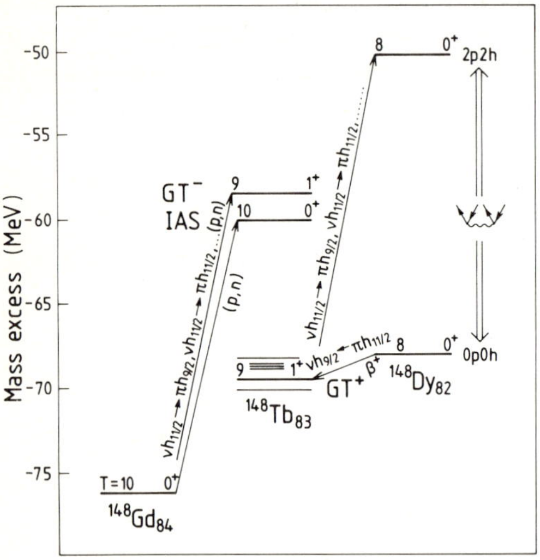

Fig. 1: The T=9 GT^+ and GT^- giant states in ^{148}Tb and the T=10 analog state of the ^{148}Gd ground state. The excited (2p-2h) 0^+ state in Dy created by the combined action of the $\sigma\tau_+$ and $\sigma\tau_-$ operators mixes into the ground state (ground state correlations), which causes a reduction of the transition matrix element to the GT^+ state in Tb.

odd-odd daughter, and therefore it remained unknown whether additional GT-strength lies at higher excitation within the Q_β window. The recent ^{148}Dy decay study[1] elucidates on this question by more precise measurements carried out at the ISOCELE II on-line mass separator at Orsay which allowed detection of γ-transitions with an intensity as low as 10^{-4} per decay. The experiments included decay-time measurements, γγ- and γ x-ray-coincidences, and conversion electron measurements providing α_K-values for transitions up to 1.3 MeV with intensities $\geq 2 \times 10^{-3}$ per decay. These data gave the ^{148}Dy decay scheme of fig. 2 where fourteen new ^{148}Tb levels with firm I^π assignments or with narrow limits are identified. It was found that 97 % of the β-decay intensity feeds the 620 keV 1^+ state in ^{148}Tb which is the GT^+ giant excitation on the ^{148}Dy ground state. The resulting log ft = 3.95(3) compared to the shell model estimate gives

$$B_{exp}(GT^+)/B_{SM}(GT^+) = 0.12(2)$$

assuming two $h_{11/2}$ protons to be present in the ^{148}Dy ground state. In this experiment the 1^+-strength-detection-sensitivity is $\leq 10^{-3}$ of the shell model prediction for $E_x \leq 2$ MeV (cf. fig. 2). The data also show that GT-strength mixing into higher-lying 1^+ states of other structural character is quite weak.

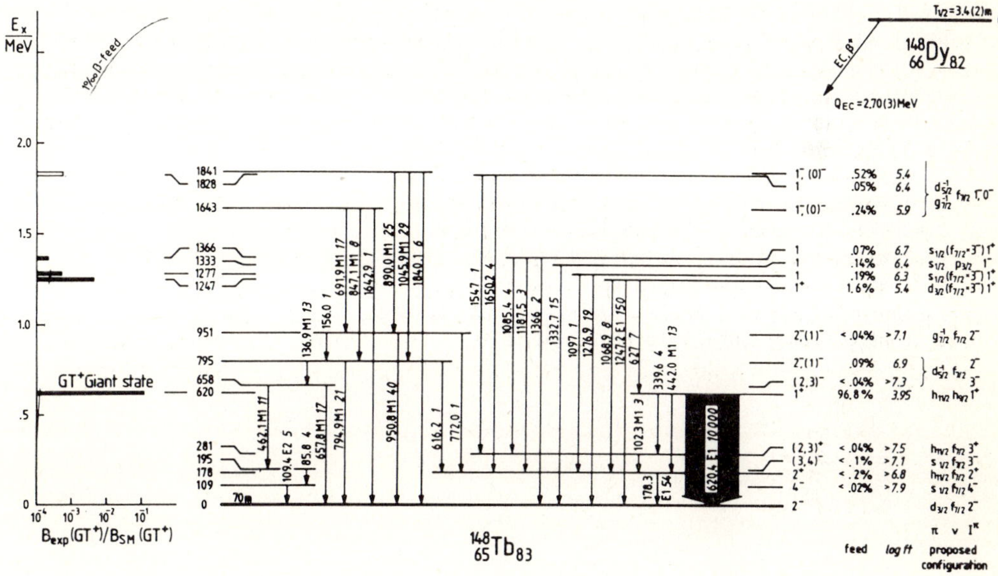

Fig. 2: Levels in ^{148}Tb observed in decay of ^{148}Dy. Transition multipolarities derived from α_K-values are given together with the γ-ray energies and intensities. The GT^+ transition strenths are plotted to the left where also the experimental detection sensitivity limit for 1^+ strength is included.

A more detailed analysis was made by comparison with a full QRPA calculation which takes into account proton pairing in ^{148}Dy as well as ground state correlations arising from mixing with 2p-2h states involving one $\nu^{+1}\pi^{-1}$ and one $\pi^{+1}\nu^{-1}$ excitation (cf. fig. 1). A zero-range force of the Migdal type plus a finite range contribution deduced from one-boson exchange potentials was used[3] to calculate the 1^+ states in ^{148}Tb, and as expected no significant 1^+ strength is predicted above 1 MeV excitation. Comparison with experiment gives

$$B_{exp}(GT^+)/B_{QRPA}(GT^+) = 0.15(2)$$

where the uncertainty of the theoretical analysis is believed to be of the order of 30 %.

The GT^+ strength retardation for ^{148}Tb is unexpectedly large, it agrees however with retardations quoted[2] for the isotones ^{150}Ho and ^{152}Tm. Since however in these nuclei only a single γ-ray was reported the uncertainty in the deduced strength is quite large. More precise studies of the ^{150}Er and ^{152}Yb decays[4] recently begun at the GSI on-line mass separator have confirmed the strongly-fed low-lying 1^+ level and have identified several additional states up to ~ 1.5 MeV excitation in the two daughter nuclei, involving deexcitation γ-rays with intensities up to 16 % of the intense $1^+ \rightarrow$ g.st. transition. Detailed conclusions must await α_K-measurements, but the new data might well support the earlier quoted retardations, and in any case provide more useful experimental errors for them.

The strength retardation factor of 0.15 in ^{148}Tb is significantly smaller than found for β^+-decays leading to the GT^+ giant state in the daughter nucleus in other regions. For the N=50 nuclei ^{96}Pd and ^{94}Ru the analogous retardation factors are[5,6] 0.55(12) and 0.49(5), and a study[7,8] of β^+-decay of the Z N nucleus ^{32}Ar gave 0.49(3).

References

1. P. Kleinheinz, K. Zuber, C. Conci, C. Protop, J. Zuber, C.F. Liang, P. Paris, J. Blomqvist, Phys. Rev. Lett. 55 2664 (1985)
2. E. Nolte, G. Korschinek, and Ch. Setzensack, Z. Phys. A309, 33 (1982)
3. C. Conci, Jül-Spez-241 (1984); D. Cha and J. Speth, Phys. Lett. 143B, 297 (1984)
4. D. Schardt, B. Rubio, A. Plochocki, R. Barden, O. Klepper, R. Kirchner, P. Kleinheinz et al., to be published
5. K. Rykaczewski et al., Z. Phys. A322, 263 (1985)
6. O. Klepper, contribution to this volume
7. T. Björnstad et al., Nucl. Phys. A443, 283 (1985)
8. B. Jonson, contribution to this volume

Experimental and Shell-Model Study of the Beta Decay of ^{43}Ti [1]

J. Honkanen, V. Koponen, H. Hyvönen[2]*, P. Taskinen, J. Äystö, K. Ogawa*[3]*, and K. Eskola*[4]

Department of Physics, University of Jyväskylä, SF-40100 Jyväskylä, Finland

The mirror nuclei in the $f_{7/2}$ shell offer an excellent possibility for systematic studies of various nuclear properties. These nuclei decay by a strong mixed Fermi and Gamow-Teller decay to the ground state of the daughter nucleus and by weak Gamow-Teller decays to the excited states. The accurate determination of the Gamow-Teller strength provides a sensitive test for the nuclear shell-model calculations.

The ^{43}Ti nuclides were produced by the $^{40}Ca(\alpha, n)^{43}Ti$ reaction with 18 MeV alpha beam from the MC-20 cyclotron of the University of Jyväskylä. Self-supporting 2-3 mg/cm^2 thick natural calcium targets were used. The He-jet technique and the recently developed ion-guide on-line isotope separation method together with beta-gamma coincidence techniques were utilized.

The half-life of ^{43}Ti was measured from the decay curve of the high energy portion of the β spectrum by using the electrostatic deflection of the separator beam. A least-squares fit gives a value of 509$\pm$5 ms for the ^{43}Ti half-life. This value agrees with the average of the previous measurements although they have poor mutual agreement and large error bars [1]. In this measurement the accuracy is improved and the interfering activities have been eliminated by mass separation. The beta branching ratios were determined by comparing the γ-ray yield to the yield of the 0.5 s component of the annihilation radiation. Seventeen γ lines were observed to deexcite ten levels of ^{43}Sc populated by the beta decay of ^{43}Ti. These earlier unobserved transitions correspond to a total beta branching of 9.8$\pm$0.8 %.

In this work the Gamow-Teller matrix elements were calculated by a shell model code [2], which applies ^{40}Ca as an inner core and in which one ore more of the valence nucleons can be lifted from the $f_{7/2}$ orbital to the higher $f_{5/2}, p_{3/2}$ and $p_{1/2}$ orbitals. Because ^{43}Ti has only three valence nucleons, all four fp levels could be treated as active orbitals. The effective interactions were calculated by using slightly modified Kuo-Brown matrix elements.

The experimental Gamow-Teller matrix elements were determined from the expression

---

[1] Financially supported by the Academy of Finland

[2] Present address: Finnish Centre for Radiation and Nuclear Safety, 00101 Helsinki, Finland

[3] Institute for Nuclear Study, University of Tokyo, Tokyo 188, Japan

[4] Department of Physics, University of Helsinki, 00170 Helsinki, Finland

$$ft = \frac{6163s}{\langle 1\rangle^2 + (g_A/g_V)^2\langle\sigma\tau\rangle^2},$$

where $\langle 1\rangle$ is the Fermi and $\langle\sigma\tau\rangle$ is the Gamow-Teller matrix element. Free nucleon values were used for the coupling constants $(g_A/g_V) = 1.2606\pm0.0075$ [3].

The experimental and calculated Gamow-Teller matrix elements have been plotted as a function of the excitation energy in Fig. 1. The figure shows that the levels of ^{43}Sc are not very well reproduced by the calculations. In addition the calculated Gamow-Teller strength is concentrated on the lowest level of each spin but experimentally the strength is more evenly distributed. However, the total observed Gamow-Teller strength for levels below 5 MeV is $\Sigma\langle\sigma\tau\rangle^2_{exp}$=1.35 and the theoretical prediction $\Sigma\langle\sigma\tau\rangle^2_{theor}$=1.97. This indicates a 30 % quenching of the strength, which is the same as observed in the systematic shell-model calculations in the sd shell [4] and also in the (p,n) reactions [5]. If a pure $(f_{7/2})^n$ model is used in the calculations the total strength is $\Sigma\langle\sigma\tau\rangle^2_{theor}$=1.56 and if only one particle is lifted to the higher shells a value of 2.21 is obtained.

It seems that the lower sd shell configurations are perturbing the level energies of ^{43}Sc calculated by the fp shell model but they are not affecting the beta decay strength more than the core polarisation is generally esimated to do.

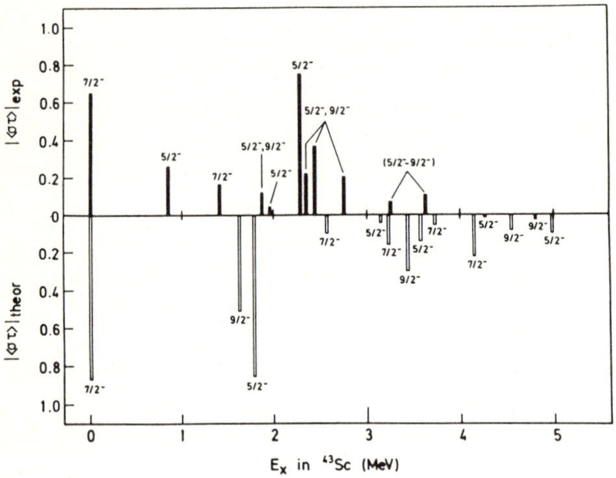

Fig.1 Experimental and calculated Gamow-Teller matrix elements $\langle\sigma\tau\rangle$ in the beta decay of ^{43}Ti. The spin and parity are indicated for each level.

References

[1] A. M. Aldridge, H. S. Plendl and J. P. Aldridge, Nucl. Phys. **A98** 329 (1967)
[2] K. Ogawa, INS Shell Model Code (unpublished)
[3] D. H. Wilkinson, Nucl. Phys. **A377** 474 (1982)
[4] B. A. Brown and B. H. Wildent hal, At. Data Nucl. Data tables **33** 347 (1985)
[5] C. Gaarde, Nucl. Phys. **A396** 127c (1983)

Energies for Superallowed ft-Values: $^{42}Sc(\beta^+)^{42}Ca$

P.H. Barker and V.T. Kirk

Auckland University, Auckland, New Zealand

Figure 1 shows the ft-values of those 0^+, T=1 superallowed Fermi beta decays which have been measured to high accuracy. The figure is essentially that given by Koslowsky et al. [1] with the inclusion of the calculations of Ormond and Brown [2] for the nuclear coulomb correction for ^{34}Cl and $^{26}Al^m$. Although the Conserved Vector Current theory predicts that these ft-values should be the same, it is obvious that they do not form a statistically self-consistent set, and so their use to extract a value for G_V, the weak interaction coupling constant, is open to some objection.

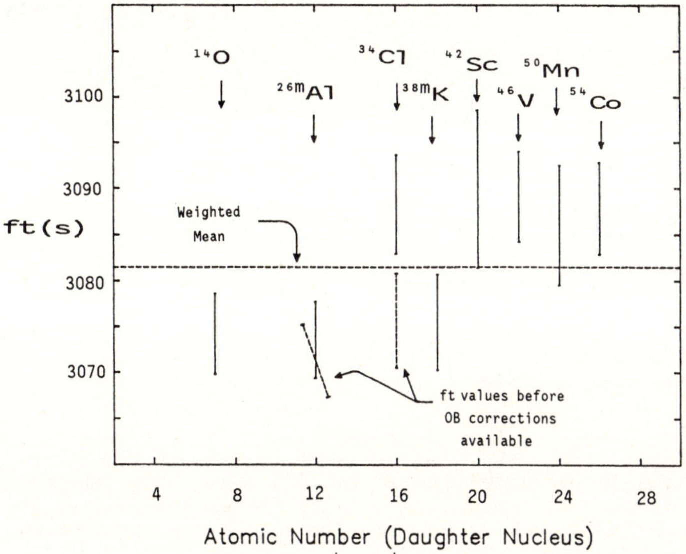

Fig. 1: ft-values of some $0^+ \rightarrow 0^+$, T=1, Fermi beta decays.

The experimental data on which the ^{42}Sc ft-value rests are the halflife and the energy release in the beta decay. Each of these is the mean of essentially two measurements and unfortunately in both cases there is marked disagreement.

Originally the beta decay energy release, Q_β had been determined from the threshold energy of the associated reaction $^{42}Ca(p,n)^{42}Sc$, [3], but when the value derived from subsequent ^{42}Ca $(^3He,t)^{42}Sc$ Q-value measurements, [4], [5], [6], was found to be 7-10 keV at variance with the (p,n) results the latter were abandoned by the compilers. However the two most precise results from $(^3He,t)$ measurements are not in good agreement, viz. Q_β = 5401.7 ± 0.4 keV [5] and Q_β = 5403.3 ± 0.2 keV [6].

In our laboratory, we have over the last ten years developed a system for measuring reaction Q-values to 10^{-5}, with the absolute energy calibration being based on a one-volt standard, [7]. For superallowed beta decay energies, the assigned error has normally depended solely on the statistics of the (p,n) reaction yield just above threshold.

In reference [3], no (p,n) yield curves are shown, the threshold energy being merely quoted. Figure 2 shows one of a series of yield curves of positrons from the ^{42}Ca(p,n)^{42}Sc(β^+)^{42}Ca reaction, around the threshold energy at 7.4 MeV, taken in 1977 in Auckland. The target was approximately 30 keV thick and the positrons were detected in a plastic scintillator while the proton beam was switched off. The position A shows the threshold energy which would agree with [3], whilst B is the position derived from the average of the (^{3}He,t) results. The difference is about 7 keV, and the data presented in fig.2 seem in accord with ref [3]. However there is obviously background below threshold.

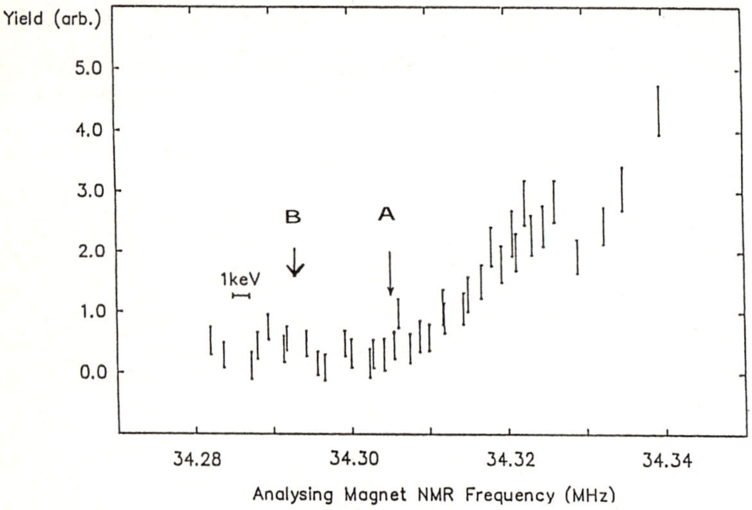

Fig. 2: Yield curve of ^{42}Ca(p,n).(see text)

The (p,n) threshold has recently been reinvestigated by us with a view to resolving the uncertainties in our knowledge of the ^{42}Sc beta decay energy. Figure 3 shows the yield of positrons from a thin ($\sim$ 2 keV) target of ^{42}Ca. Great care was taken to eradicate activities produced by the 7.4 MeV proton beam on trace amounts of lanthanum, fluorine and nitrogen in particular. The positrons were detected as event-mode data in a positron telescope and the reaction energy was altered by changing the electrical potential of the target whilst keeping the proton kinetic energy constant.

Several features may be noted. There is a prominent resonance in the thin target yield but below this there is still positron yield. An analysis of the yield from target voltage -2 kV to +12 kV gives a threshold at position B with an error ± 0.4 keV. The positions labelled A are approximately (i.e. to ± 0.5 keV) where the (^{3}He,t) Q-values would indicate the threshold should be. These are not precisely given because, since this experiment was aimed principally at elucidating the shape of the (p,n) yield curve, the absolute energy calibration was not rigorously performed.

We hope in the near future to complete our measurements of the ^{42}Ca(p,n)^{42}Sc Q-value, and hence to resolve the uncertainty in the ^{42}Sc superallowed β-decay energy.

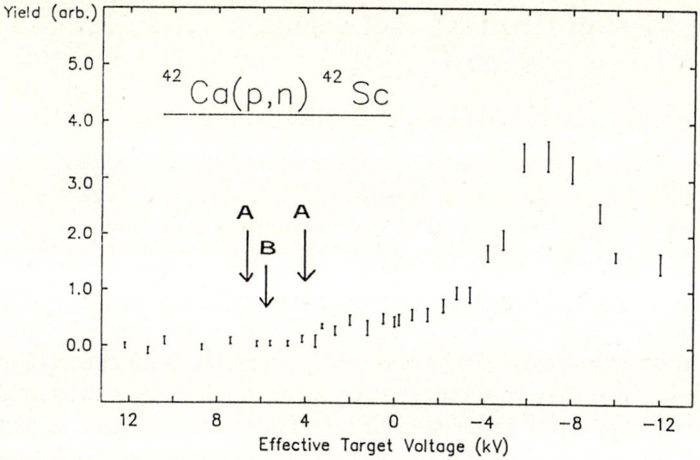

Fig. 3: Yield curve of ^{42}Ca(p,n). (see text)

References:

1. V.T. Koslowsky et al.: AMCO 7 Proceedings (T.H. Darmstadt), 572 (1984)
2. W.E. Ormond and B.A. Brown: This Conference
3. J.M. Freeman et al.: Phys. Lett. 17A, 317 (1965)
4. J.C. Hardy et al.: Phys. Rev. Lett. 33, 320 (1974)
5. H. Vonach et al.: Nucl. Phys. A278, 189 (1977)
6. V.T. Koslowsky et al.: AMCO 7 Proceedings (T.H. Darmstadt), 60 (1984)
7. See e.g. R.E. White et al.: Phys. Lett. 105B, 116 (1981)

Discovery of New Fission Product Activities in the A=110–118 Mass Region [1]

P. Taskinen, J. Honkanen, J. Äystö, P. Jauho, M. Yoshii[2]*, and J. Ärje*

University of Jyväskylä, Department of Physics, SF-40100 Jyväskylä, Finland

Neutron rich nuclei around the mass A=100 have recently attracted both theoretical and experimental interest since they are expected to form a new region of strong deformation. In contrary, very little experimental information is available in the adjacent transitional region A= 110-118. Due to expected isomerism in this region both low and high spin states can be studied via beta decay in many nuclei. The studies of these nuclides have many experimental difficulties. They are only weakly produced in thermal neutron induced fission and because all the elements in this mass region (Tc, Ru, Rh, Pd) have very high melting points they have not been available as beams of on-line isotope separators. Now the recently developed ion guide method [1] allows the very fast separation of both volatile and nonvolatile elements with equal efficiency.

In this work the fission fragments were produced by bombarding 12-25 mg/cm^2 thick ^{nat}U targets with 19 MeV proton beam from the University of Jyväskylä MC-20 cyclotron. The mass separated samples of very neutron rich nuclei were studied using $\beta-\gamma$, $\gamma-\gamma$ and $\beta-e^-$ coincidence techniques. The half-lifves were determined using the electrostatic deflection of the separator beam.

Several new activities were observed in the mass region A=110-118 and many new transitions were assigned to the β decay of known nuclides. One excample of the mass separated activities is presented in Fig. 1. It shows the $\beta-\gamma$ and $\beta-e^-$ coincidence spectra measured at the mass 114 position. Both the Z identification and the multipolarity of the transition can be deduced from the spectra. The ground state band of ^{114}Pd is already known from the prompt γ-ray studies of the spontaneous fission of ^{252}Cf [2]. A proposed level scheme is shown in the inset. The conversion electron spectrum indicates that the 317.2, 332.8 and 362.1 keV transitions have $M1$ or $E2$ multipolarity.

Summary of the first observations of the mass separated $^{112,114,116}Rh$ isotopes are shown in Table 1. The β-decay half-lives were determined from the decay curves of the lowest $2^+ \rightarrow 0^+$ transitions 348.9, 332.8 and 340.6 keV in ^{112}Pd, ^{114}Pd and ^{116}Pd, respectively. Although isomerism is expected in these nuclei, no strong other components were observed in these decay curves. A modification of the ^{112}Rh half-life may be necessary after the half-life of ^{112}Ru and its role as a source for ^{112}Rh activity are known. Several new activities were also observed at the mass 112 and

[1] Financially supported by the Academy of Finland

[2] Supported in part by the JSPS, Japan

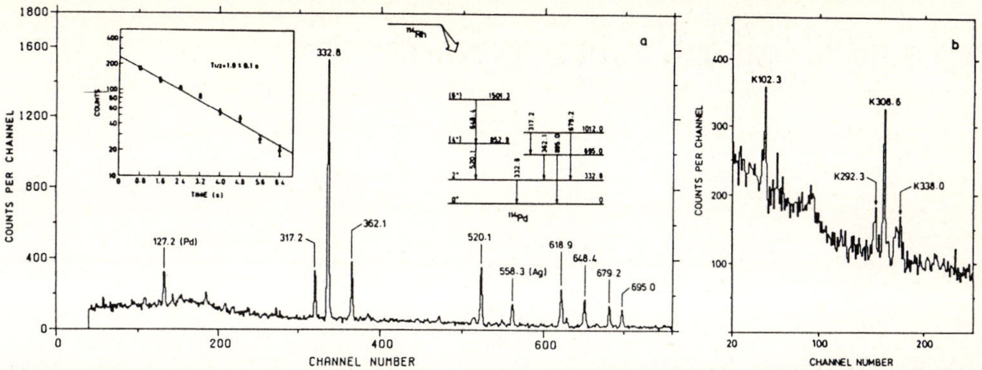

Fig.1a) Beta-gamma coincidence spectrum measured at mass 114 position. Decay curve of the strongest 332.8 keV γ -line and a suggested decay scheme of ^{114}Rh are shown in the inset.
b) Conversion electron spectrum.

Table 1. Summary of β-decay half-lives of $^{112,114,116}Rh$ isotopes.

Nuclide	$T_{1/2}$ (s) This work	$T_{1/2}$ (s) Prompt fission $^{a)}$	$T_{1/2}$ (s) Gross theory $^{b)}$
^{112}Rh	(5.3±0.3)	4.65±0.14	10
^{114}Rh	1.9±0.1	1.68±0.07	5
^{116}Rh	1.0±0.2	-	2

a) Ref. 3 b) Ref.4

114 position. The half-lives predicted by the Gross theory [4] are systematically by a factor of two longer than the measured values. However, a recent calculation of Klapdor et al. [5] gives a shorter value $T_{1/2}$=0.61 s for the ^{116}Rh half-life.

In summary, we have initiated a systematic study of the fission product activities using the ion guide method. It has proven an effective way of producing radioactive mass separated beams especially in the highly nonvolatile *Ru*-*Rh* region. In this work a new isotope ^{116}Rh was observed andthe half-lives were measured for $^{112,114,116}Rh$ and the lowest $2^+ \rightarrow 0^+$ transitions were confirmed in the corresponding even palladium isotopes.

[1] J. Ärje, J. Äystö, H. Hyvönen, P. Taskinen, V. Koponen, J. Honkanen, A. Hautojärvi and K. Vierinen, Phys. Rev. Lett. **54** 99 (1985)

[2] C. Cheifets, R. C. Jared, S. G. Thompson and J. B. Wilhelmy, Phys, Rev. Lett. **25** 38 (1970)

[3] J. B. Wilhelmy, S. G. Thompson, J. O. Rasmussen, J. T. Routti and J. E. Phillips, UCRL-19530 p.178 (1970)

[4] K. Takahashi, M. Yamada and T. Kondoh, At. Data Nucl. Data Tables **12** 101 (1973)

[5] H. V. Klapdor, J. Merzinger and T. Oda, At. Data Nucl. Data Tables **31** 81 (1984)

Δ-Excitations in Nuclei

C. Gaarde

The Niels Bohr Institute, University of Copenhagen, Blegdamsvej 17, DK-2100 Copenhagen Ø, Denmark

Introduction

The study of Δ-excitations in a nucleus is a natural extension of the study of nuclear structure. The Δ is however not identical with the nucleons in a nucleus, and the Pauli-principle will have a very different effect on the Δ. The properties of the nuclear medium in the presence of a Δ is therefore different from normal nuclear matter.

We shall in this paper discuss 2 aspects of Δ's in nuclei. The possible quenching of στ-strength in the low excitation energy region of the spectrum. The other aspect is the direct excitation of Δ's in nuclei as observed in charge-exchange reactions.

In fig.1 we show a spectrum from the (^{3}He,t) reaction on ^{12}C at 2 GeV [1]. The spectrum shows that the response to a probe like (^{3}He,t) is concentrated in 2 energy regions and that the Δ-excitation is indeed a simple mode in the nucleus.

I. Quenching of στ-Strength

The στ-matrix element between the nucleon and the Δ ⟨Δ|||στ|||N⟩ (reduced both in spin and isospin) is large.

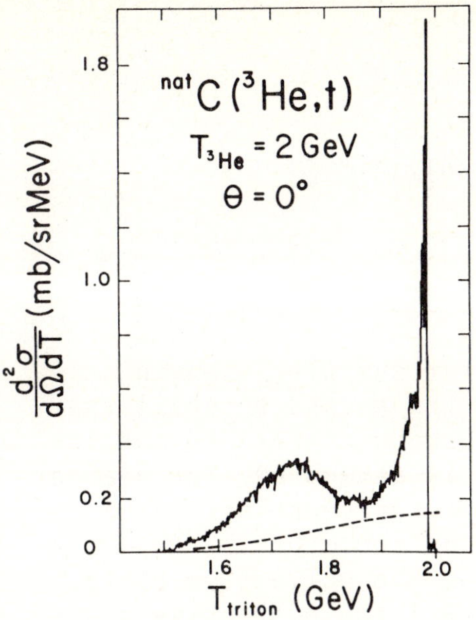

Figure 1. Triton spectrum at 2 GeV bombarding energy

We have no experimental number for this quantity, but even if different models give rather different values, it is consistently found that the coupling to the ΔN^- states has an important effect on the $\sigma\tau$-strength in the nucleon sector of the spectrum [2]. As an example of such a model the field approximation with the above matrix element taken from the constituent quark model finds that 35% of the Gamow-Teller is removed from the nucleon sector [3,4]. In this model there is a 7% amplitude of ΔN^{-1} states in normal ground states of nuclei.

The experimental situation is summarized in fig.2 and the corresponding caption [5]. This figure refers to the strength in the "shell model region", i.e. under and below the collective state. Above this region it is experimentally very difficult to determine how much of the cross section corresponds to Gamow-Teller strength. We can not from the (p,n) studies exclude that ℓ=0 strength could be there. The missing GT strength would then be explained as an effect of coupling to 2p-2h states. The more detailed studies where spin-flip probabilities are measured in $(\vec{p},\vec{n})$ [6] experiments do not show any GT strength in this energy region, but with the same uncertainty as above: that a small fraction of the cross section can not be excluded as ℓ=0 strength.

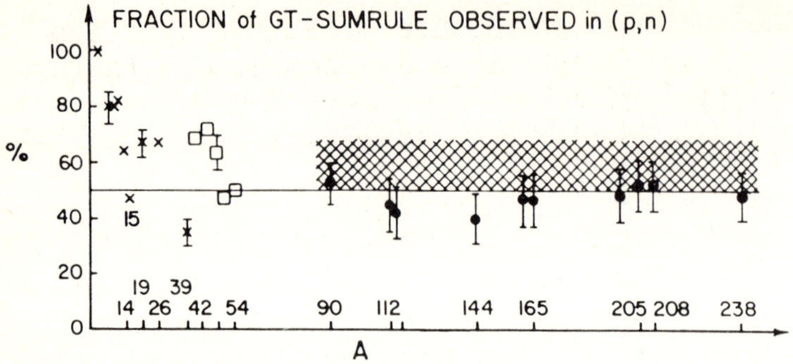

Figure 2. Fraction of Gamow-Teller sumrule strength observed in (p,n) reactions. In the p- and sd-shell the strength is most often in a few sharp states. In the fp-shell a multipole decomposition is attempted. For heavier nuclei the dots (with error bars) represent strengths in peaks (low lying + giant), whereas the cross hatched region also includes strength under the collective state. Possible strength above (larger E_x) the collective state is not included.

Still another set of data from the particle decay of the collective Gamow-Teller state in ^{208}Bi shows that there is very little 1 particle - 1 hole strength above the collective state [7].

In this connection we mention the very interesting data from the (n,p) experiments in Triumf and the (d,2p) studies at Saturne (see below). The GT strength in the β^+-channel is lying in a region of the spectrum where the density of 2p-2h states is lower and the coupling could therefore be quite different. We can hope that we soon can be more specific on the quenching of Gamow-Teller strength.

II. Direct Excitation of the Δ

In this section we discusss data from charge exchange reactions at intermediate energies as studied at Laboratoire National Saturne. The spectrum shown in fig.1 is typical of the findings in these experiments. A very selective excitation of spin-isospin modes in the nucleon sector and a strong excitation of the Δ.

II.1. The (^{3}He,t) reaction

The experiments are performed at Saturne with bombarding energies between 600 MeV and 2.3 GeV. The outgoing tritons are analyzed in a magnetic spectrometer SPES4 with a resolution of $\frac{\Delta E}{E} \sim 10^{-3}$. The spectrometer allows the study at small angles including zero degrees.

The spectrum shown in fig.1 is obtained from spectra at 6 different field settings of the spectrometer each taking 10-20 minutes with a beam current of 10-20 nA and target thicknesses of 50-200 mg/cm^2. An important test case for our study is the p(^{3}He,t)Δ^{++} reaction. The results for this case have been discussed in a recent publication [8]. It is found that the data can be explained in terms of one pion exchange between the target proton and a proton in the ^{3}He. The ^{3}He-t system is described by a formfactor. Such a model accounts well for the absolute cross section as well as details in the resonance shape.

We also have rather systematic data on the Δ-production in nuclei with the (^{3}He,t) reaction [1]. The results may be summarized as:

i) Energy dependence.
At 1500 MeV bombarding energy the Δ-cross section is quite small but increasing rapidly with energy. The energy dependence can be explained as an effect of the decrease in (four-)momentum transfer with increasing energy (for the same energy transfer). The formfactor for the ^{3}He-t system is therefore used at a different value of momentum transfer.

ii) A-dependence.
A large Δ-cross section is observed for all A with a resonance energy and shape that is independent of A for A$\geq$12. The cross section is going from 100 mb/sr for ^{12}C to 130 mb/sr for ^{40}Ca and then increasing rather slowly to 170 mb/sr for ^{208}Pb. The numbers refer to lab. cross sections at θ=0°.

iii) Angular distributions.
The angular distributions for the Δ-peak are very similar and also similar to that observed for the p$\rightarrow\Delta^{++}$ case.

iv) Energy shift relative to quasifree Δ-production.
The Δ-peak is found to be shifted around 35 MeV relative to the energy corresponding to quasifree Δ-formation.

The findings described above are consistent with quasi-free Δ-production except for the shift in energy. Several attempts have been made to account for the shift. A recent publication describes a surface-response for the (^{3}He,t) probe [9]. It is shown that ΔN^{-1} correlations of the RPA type only would give rise to energy shifts of 5-10 MeV. The observed shift is then described as an effect of the change of properties of the Δ in the nuclear medium, i.e. a smaller mass and a larger width.

II.2. The (d,2p) reaction.

The (d,2p) reaction with the 2 protons in a relative singlet S state [10] is at intermediate energies a very selective probe of spin-isospin modes in the β^+-direction. In the impulse approximation the (d,2p) is like the (n,p) reaction with spin transfer, so even more specific than the (n,p) reaction.

In the (d,2p) experiments performed at Saturne the 2 protons are recorded in the same magnetic spectrometer, SPES4. This means that the relative kinetic energy of the 2 protons has to be very small and that means that the very experimental set-up selects the singlet S state. We mention that the contribution from the ^{3}P is less than 1%. Data have been obtained at 650 MeV and 2 GeV deuteron bombarding energy and the results obtained do indeed confirm the expectations mentioned above. We shall not here go into a detailed discussion of the (d,2p) data, but show 2 figures that illustrate the spin transfer selectivity of the reaction.

In fig.3 the zero degree spectrum for ^{54}Fe(d,2p) is shown. Spectra at larger angles show that the peak dominating the spectrum corresponds to an ℓ=0 transition. The spectrum therefore shows that the β^+-strength is concentrated in a very narrow region of the spectrum. We have again no indication of GT strength in the tail region, and we note that in the β^+-channel we are at a much lower excitation energy, and the continuum is therefore rather different from that in the β^--channel. The other figure shows spectra at 2 GeV with CH_2 and C targets. The difference spectrum then gives us the p(d,2p) data. At 2 GeV the Δ-excitation is again a dominant feature of the spectra. We see also here a shift in Δ-energy between the

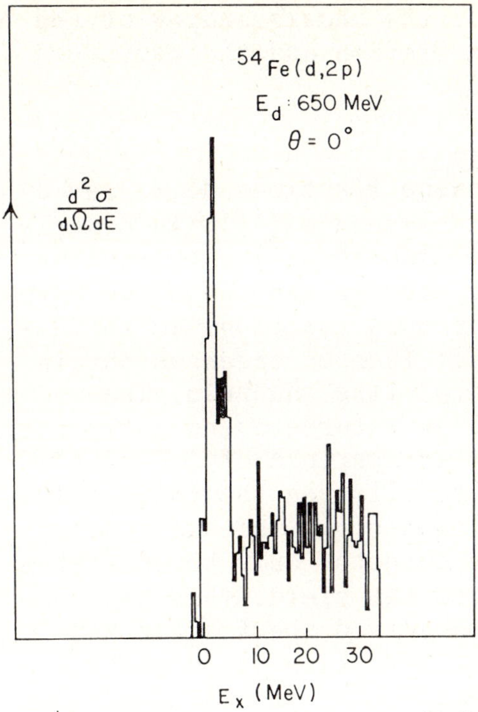

Figure 3. (d,2p) spectrum at zero degrees

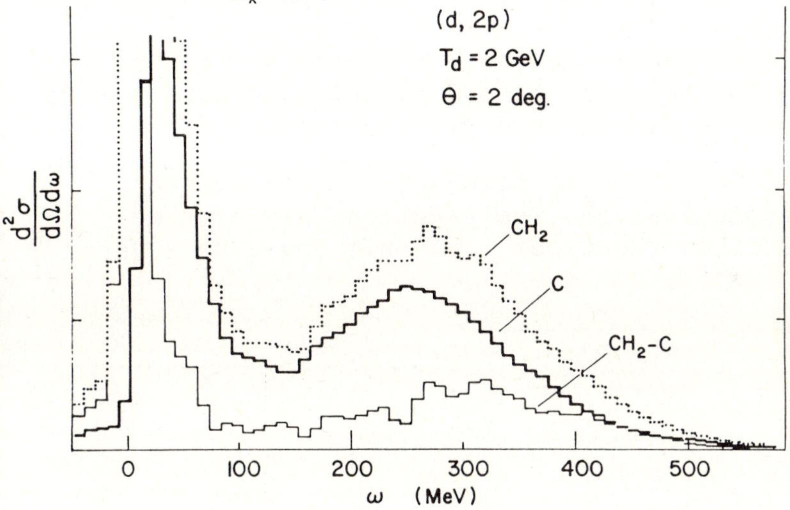

Figure 4. (d,2p) spectra with CH_2 and C targets. The difference spectrum is also shown

proton and ^{12}C nucleus, very similar to what is observed in the (^3He,t) reaction.

The experiments are performed with a polarized deuteron beam. Since the deuteron has S=1 we can in a simple cross section measurement determine the tensor analyzing power. We shall not go into any detail here, only mention that this quantity is very useful for the understanding of the spin structure of both the Δ-excitation and the excitations in the nucleon sector [10].

II.3. Heavy ion charge exchange

Also with heavy ion charge exchange reactions do we see a very strong excitation of the Δ-resonance [11]. An example is shown in fig.5 where spectra from the ($^{16}O,^{16}N$) reaction with ^{12}C and the proton as targets are shown. We have data for a number of targets for this reaction. We further have data for ^{12}C and ^{20}Ne beams. In both cases we obtain data for both the (p,n)- and (n,p)-like channels. The experiments have been performed at Saturne using again the SPES4 spectrometer at $\theta=0°$ with an aperture of 1.7°x3.4°. Within this angle range we practically see the total cross section in the charge exchange channels. Since we require definite mass and charge in our detector only bound states in the ejectile can contribute to the yield. This has some important consequences for the observed spectra. In say ($^{12}C,^{12}N$) only the g.s. and 0.96 MeV state would contribute, whereas in ($^{12}C,^{12}B$) several other states up to 3.4 MeV could be involved in the reactions observed in the detector. In the ^{20}Ne-beams the 2 channels are rather different. As for the (^3He,t) and (d,2p) reactions the data on the proton are an important test case.

The data are analyzed in terms of one-step processes in the impulse approximation. The reactions are extreme examples of surface reactions, and this could be a very

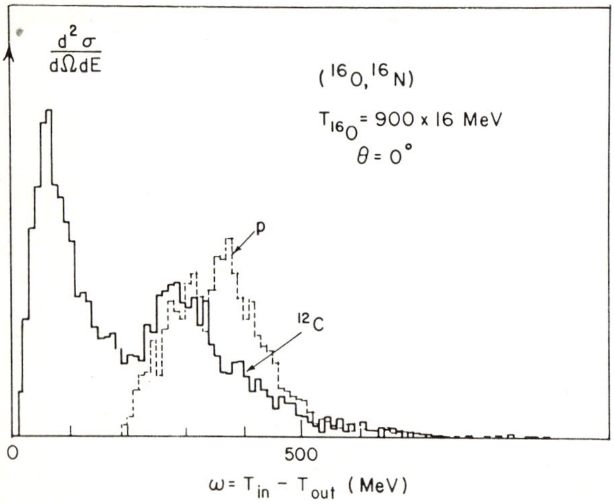

Figure 5. ($^{16}O,^{16}N$) spectra with ^{12}C and the proton as targets

interesting feature in probing the long range parts of the effective interaction.

We conclude this section by noting that also in heavy ion reactions at intermediate energies the Δ is a dominant feature of the spectra. The charge-exchange reactions with heavy ions seem to be extremely sensitive probes of isospin-spin modes in nuclei.

Acknowledgements
The material presented here is the result of contributions from many people at many laboratories. Most of the recent work is done at Saturne, and I am particularly indebted to the collaborators whose names are given in ref.[1]. This work has been supported in part by the Danish Natural Science Research Council.

References

1. D. Contardo, M. Bedjidian, J.Y. Grossiord, A. Guichard, R. Haroutunian, J.R. Pizzi, C. Ellegaard, C. Gaarde, J.S. Larsen, C. Goodman, I. Bergqvist, A. Brockstedt, L. Carlén, P. Ekström, D. Bachelier, J.L. Boyard, T. Hennino, J.C. Jourdain, M. Roy-Stephan, M. Boivin and P. Radvanyi, Phys. Lett. 168B, 331 (1986)
2. M. Ericsson, Ann. of Physics 63, 562 (1971)
3. A. Bohr and B. Mottelson, Phys. Lett. 100B, 10 (1981)
4. C. Gaarde et al., Nucl. Phys. A369, 258 (1981)
5. C. Gaarde et al., Proceedings Spin Excitations in Nuclei, Telluride 1982, eds. F. Petrovich, Plenum Press, New York, p.65
6. T. Taddeucci et al., Phys. Rev. C33, 746 (1986)
7. C. Gaarde et al., Phys. Rev. Lett. 46, 902 (1981)
8. C. Ellegaard et al., Phys. Lett. 154B, 110 (1985)
9. H. Esbensen and T.S.H. Lee, Phys.Rev. C32, 1966(1985)
10. D.V. Bugg and C. Wilkin, Phys. Lett. 152B, 37 (1985) and 154B, 243 (1985)
11. D. Bachelier et al., Phys. Lett. 172, 23 (1986)

Spin-Isospin Excitation by the (p, p′) Reaction

N. Marty

Institut de Physique Nucléaire, B.P. N° 1, F-91406 Orsay, France

1. Introduction

Many experimental and theoretical data have been obtained during the last years on the spin-isospin excitation mode in the nuclei. Charge exchange (p,n) reactions performed at the Indiana University Cyclotron facility [1] have settled the existence of Gamow-Teller transitions in all the nuclei studied. Inelastic scattering using different probes as (e,e') [2], (γ,γ') [3] and (p,p') [4] afford details on the structure of 1+ excitations. Intermediate energy protons from 200 to 400 MeV are especially suited for exciting spin-isospin modes as the ratio of the spin-isospin scattering amplitude $t_{\sigma\tau}$ on the t_{00} amplitude goes through a strong maximum at these energies. This paper will focus on some results obtained at Orsay by 201 MeV(p,p') inelastic scattering.

In inelastic scattering there is no model independent sum rule for the M1 strength. If the missing strength or quenching Q is defined as the ratio of the measured to the model predicted cross sections, quenchings of the order of 30 to 35 % are generally obtained in the (p,p') experiments (Fig. 1). Different explanations have been proposed for these quenchings as ground state correlations, core polarization [5], coupling with high lying 2p, 2h configurations [6] and possible Δ nucleon hole admixture [7].

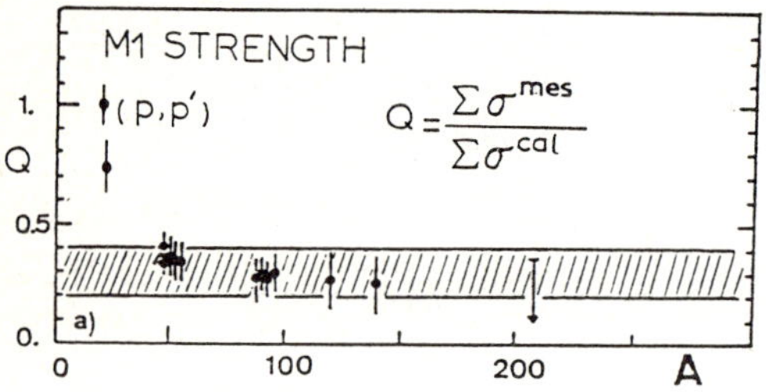

Fig. 1 Fraction of the M1 predicted strength observed in (pp') scattering at 201 MeV

As Q is strongly model dependent, it is important to study nuclei for which different theoretical predictions are available or nuclei for which a full space shell model calculation can be performed.

In order to test the models, one can compare experimental results obtained with different probes which interact differently with the nuclei ; (p,p') scattering at forward angles interacts only through the spin term $V_{\sigma\tau_3}$ of the central nucleon-nucleon interaction. The electromagnetic interaction includes both a spin and an orbital term :

$$B(M1) = |\langle j_f| e^{iqr}(g_\ell \underset{\sim}{\ell} + g_s \underset{\sim}{s})|j_i\rangle|^2 .$$

In the following section some examples will be discussed.

2. Some examples

2.1 A heavy nucleus ^{90}Zr

The existence of a giant M1 resonance was first established for ^{90}Zr [8]. For such heavy nuclei two problems appear. The first is experimental ; the resonance, 1.5 MeV wide is superposed on a continuum, partly physical (tail of the giant dipole resonance), partly due to the rescattering of the elastically scattered beam. This last background could be subtracted (Fig. 2) as its shape is the same inside and outside angular windows put on the trajectories of the scattered particles. The uncertainty on the physical continuum is then greatly reduced. Two possible "continua" are drawn on Fig. 2c. There is actually no theoretical prediction for the continuum.

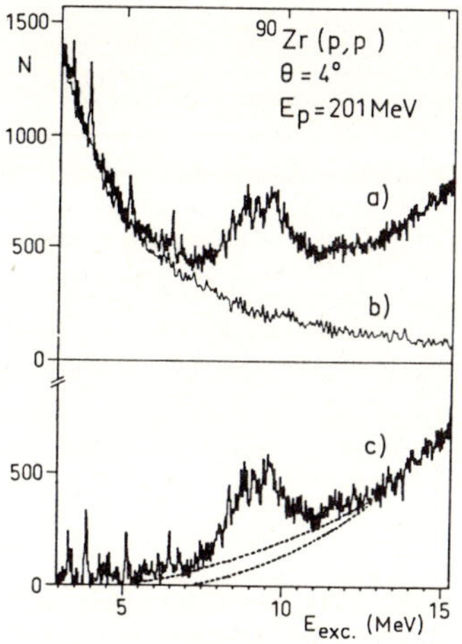

Fig. 2 Spectrum of protons inelastically scattered from ^{90}Zr at θ = 4° a) with the background b) shape of the background c) after background subtraction. The dotted curves are two limits of the continuum

The second problem is theoretical : only restricted shell model calculations can be performed for such heavy nuclei. Our results were first compared with the predictions of a particle-hole shell model calculation with wave functions $\nu(1\ g_{9/2},\ 2\ d_{5/2},\ 3\ s_{1/2},\ 2\ d_{3/2},\ 1\ g_{7/2})$ with 9 or 10 particles on the orbital $g_{9/2}$ [9], and RPA calculations [10]. The quenching obtained with both models is $Q = 0.35 \pm 0.06$.

In a recent calculation based on a quasi-particle phonon model (QPM) [11] which includes pairing interaction between protons and neutrons, the quenching factor which is 0.32 for a $\nu(1\ g_{7/2},\ 1\ g_{9/2}^{-1})$ transition raises to 0.48 in a TDA calculation and to 0.64 in a RPA calculation. When accounting in an approximate way of two phonon states, 20 % of the strength is shifted of the resonance region and Q = 0.79. This model predicts a small resonance at 11.5 MeV due to a $\pi(1\ g_{7/2},\ 1\ g_{9/2}^{-1})$ transition which is clearly seen on the (pp') experimental spectra.

The example of ^{90}Zr shows how the quenching factor Q is sensitive to configuration mixing and to coupling to two p two h excitations.

2.2 ^{48}Ca

^{48}Ca seems the most simple example for studying M1 transitions. A large $\nu(f_{5/2},\ f_{7/2}^{-1})$ transition is expected ; it has effectively been observed by (e,e') scattering at Darmstadt at an energy of 10.22 MeV [12] ; it is strongly excited in (p,p') [13] and its analogue in ^{48}Sc is also well known [14] In(pp') there is no problem about the continuum as the ratio(peak on continuum)is about 40 (Fig. 3).

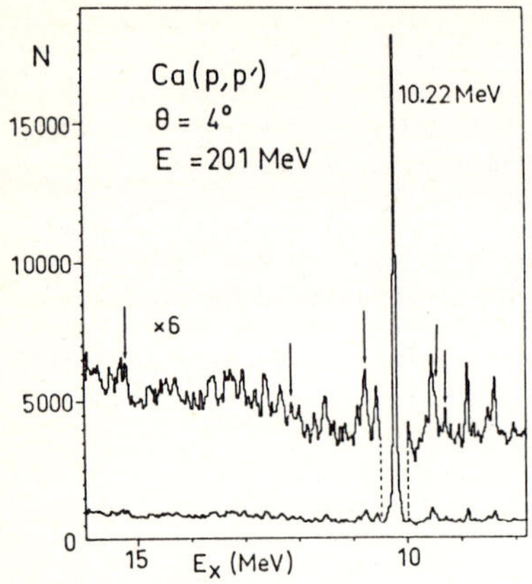

Fig. 3 Proton scattering spectrum on ^{48}Ca at $\theta = 4°$. Note the different scales for the whole spectrum and for the small 1+ peaks pointed by arrows

For a neutron spin-flip transition, as the isoscalar and isovector parts of the nucleon-nucleon interaction are well known, equivalent $\sigma(p,n)$ cross sections and B(M1) value can be extracted from the (p,p') data [15]. The results obtained are in perfect agreement with the (e,e') and (p,n) data.

In a full (1 f, 2 p) shell model calculation [16] the wave function is dominated by the configuration :

$$|1+\rangle = 0.89|\nu(f_{5/2}, f_{7/2}^{-1})\rangle + 0.11|\nu(f_{7/2}, f_{5/2}^{-1})\rangle$$

The quenching Q is 0.33 : it is 0.35 in the model of MUTO and HORIE [17]. The 2 p - 2 h component $\nu(f_{5/2}^{2}, f_{7/2}^{-2})$ plays a dominant role in reducing the strength : 1 % admixture reduces the total strength by 20 %. This last model predicts that 94 % of the total 1+ strength is concentrated on a single state in agreement with our results. Only five small peaks are 1+ candidates totalizing less than 12 % of the strength of the 10.22 MeV state. These peaks are indicated by arrows on Fig. 3.

In (e,e') data many more 1+ states have been detected. Recently in a (d,α) experiment [18] the admixture of a large $\pi(1\,f_{7/2})$ component has been found in the ground state of ^{48}Ca. An explanation of the discrepancy between the (e,e') and (p,p') results may be the role of orbital terms in the electromagnetic excitation of states which involve proton configuration admixture. Such effects have been clearly seen for a state in ^{50}Ti [19]. In a simple shell model ^{50}Ti differs from ^{48}Ca by the addition of two protons in the $(1\,f_{7/2})$ shell.

2.3 ^{20}Ne

^{20}Ne is a good example to test the effects of orbital and spin interactions in the electromagnetic excitation of 1+ states. A full (1 d, 2 s) shell model calculation [20] predicts a state at 11.25 MeV excited dominantly by orbital interaction, spin and orbital effects interfering coherently (Fig. 4 c). For the states predicted at higher energies, the interference is destructive and the expected B(M 1) strengths are less than 0.1 μ_N^2. B(σ) in Fig. 4 is the "spin" strength obtained by setting g_ℓ equal to 0.

^{20}Ne has been studied by electron scattering [21] and by nuclear fluorescence [22]. The strong 11.25 MeV state is seen at the energy and with the B(M 1) strength predicted. In the (e,e') experiment no other 1+ state is reported.

We have studied ^{20}Ne by (p,p') scattering to measure the quenching of the strength excited through the spin part of the interaction. A gaseous target at a pressure of 3.5 atm has been used ; three 1+ states are clearly excited, weak

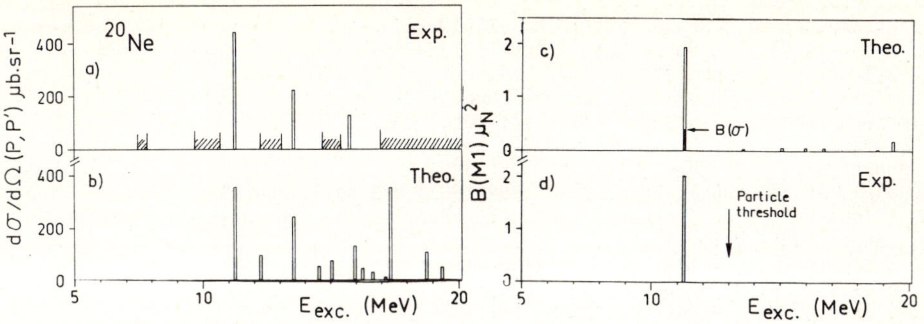

Fig. 4 Comparison between experimental and theoretical values for the (p,p') cross sections at 4° and the B(M 1) values for ^{20}Ne

states in the energy ranges marked by hatches on Fig. 4 a could have been obscured by strong peaks from the carbon and oxygen of the kapton windows of the target. The ratio of the experimental to the predicted cross sections summed over the three states is 1.0 ± 0.1. ^{20}Ne is then an example for which a full space shell model calculation predicts 1+ states excited differently by electromagnetic ans spin interaction. No quenching is found for both the B(M 1) strength and the (p,p') cross sections.

3. Summary and conclusion

Intermediate energy (p,p') scattering at forward angles is a specific tool for studying $\Delta L = 0$ spin-flip transitions. Except for light nuclei, only 30 to 35 % of the predicted strength is observed in the region of the resonance. A comparison between (p,p') and (e,e') or (γ,γ') data allows to measure the orbital contribution to the electromagnetic interaction and affords a severe test of the models. The strength of 1+ excitations appears to be very sensitive to correlation in the ground state, to pairing interactions and to coupling to high lying to 2 p, 2 h states. Large model space calculations are needed before any conclusion can be given about the role of non nuclear degrees of freedom to explain the M1 missing strength.

Aknowledgments

The (p,p') results discussed in this paper were all obtained at the Orsay synchrocyclotron in a M.S.U., I.P.N. collaboration involving N. Anantaraman, G.M. Crawley, A. Galonsky from M.S.U., C. Djalali, M. Morlet and A. Willis from Orsay.

References

1 C. Gaarde, Nucl. Phys. A396, 127C(1983) and references therein.
2 A. Richter, Proc. of the Int. Conf. Nucl. Phys. (Florence 1983) 189, Ed. P. Blasi and R.A. Ricci.
3 U.E.P. Berg, Proc. Int. Symp. HESANS 83 (J. Phys. Coll. C4, 359(1984).
4 C. Djalali, Proc. Int. Symp. HESANS 83 (J. Phys. Coll. C4, 375(1984).
5 I.S. Towner and F.C. Khanna, Nucl. Phys. A399, 334(1983).
6 G.F. Bertsch and I. Hamamoto, Phys. Rev. C26, 1323(1982).
7 A. Bohr and B. Mottelson, Phys. Let. 100B, 10(1981)
A. Harting et al.,Phys. Let. 104B, 261(1981).
8 N. Anantaraman et al., Phys. Rev. Let. 46, 1318(1981).
9 N. Anantaraman and B.H. Wildenthal, private communication.
10 H. Sagawa and Nguyen Van Giaï, Phys. Let. 113B, 119(1982) and private communication.
11 A.I. Vdovin et al., preprint submitted to Iadernaia Physica, 1986.
12 W. Steffen et al., Phys. Let. 95B, 23(1980).

[13] G.M. Crawley et al., Phys. Let. 127B, 322(1983).
[14] B.D. Anderson et al., Phys. Let. 114B, 15(1982)
C. Gaarde et al., unpublished.
[15] C. Djalali, thesis Orsay 1984, unpublished.
[16] J.B.Mc Grory and B.H. Wildenthal, Phys. Let. 103B, 173(1981).
[17] K. Muto and H. Horie, Nucl. Phys. A440, 254(1985).
[18] H. Nann et al., Phys. Rev. Let. 55, 578(1985).
[19] C. Djalali et al., Nucl. Phys. A410, 399(1983) and A417, 564(1984).
[20] B.H. Wildenthal, Proq. in Part. and Nucl. Phys. ed. by D.H. Wilkinson (Pergamon, London 1984) Vol. 11,5.
B.A. Brown and B.H. Wildenthal, private communication.
[21] W.L. Bendel et al., Phys. Rev. C3, 1821(1971)
[22] U.E.P. Berg et al., Phys. Let. 140B, 191(1984).

Universal Gamow-Teller Quenching in (n,p), $(\vec{p}, \vec{p}')$ and (p, n) Reactions

O. Häusser

Simon Fraser University, Burnaby, B.C., Canada, V5A 1S6 and
TRIUMF, 4004 Wesbrook Mall, Vancouver, B.C., Canada, V6T 2A3

1. INTRODUCTION

During the past two years significant improvements have occurred at the medium resolution spectrometer (MRS) at TRIUMF which have made possible many nuclear physics experiments over the full energy range (200-500 MeV) of the TRIUMF cyclotron. Using the dispersion matched spectrometer system nearly background-free (p,p') spectra with a typical resolution of 150 keV are observed at forward angles. Spin transfer experiments in (p,p') have been carried out with a new focal-plane polarimeter which covers a large region of nuclear excitations (~45 MeV). Finally, the addition of a compact sweeping magnet at the target location of the spectrometer has led to the development of a unique facility for studying both (n,p) and (p,n) charge exchange reactions with a typical resolution of 1 MeV.

In this talk recent results obtained with the new MRS facilities will be presented. A preliminary analysis of the small-angle data was carried out with two objectives in mind. First, we have investigated the suitability of the nucleon-nucleus interaction — as a substitute of the well-understood electroweak interaction — for deducing nuclear structure information. The dependence of cross sections on incident energy was observed for strong transitions which were selected to probe specific spin-isospin components of the effective N-nucleus interaction. In the TRIUMF energy regime the elementary NN cross sections are at a minimum and the microscopic distorted wave impulse approximation (DWIA) is expected to work well, especially above 350 MeV.

The second objective is concerned with spin-isospin excitations which provide the strongest and best understood features in the forward angle spectra. It is now well established [1] that in the (p,n) reaction on heavier targets (A>16) only 50-60% of the Gamow-Teller (GT) strength calculated in the shell model with free-nucleon operators is actually observed in the low-energy region. In the (e,e') and (p,p') reactions the 1^+ excitations do not appear to follow such a simple pattern; e.g. for ^{51}V both B(M1) values [2] and (p,p') cross sections [3] are much smaller than in other N=28 isotones in contrast to shell model calculations (see [4]) which predict little A-dependence. We show, at least for the (p,p') reaction, that small instrumental background and spin-flip information are essential for a reliable determination of GT strength at low excitation. Finally first (n,p) results from TRIUMF are discussed. In many nuclei the GT strength in the (n,p) channel is largely Pauli-blocked and extremely sensitive to details of the Fermi surface. In the examples chosen here the (n,p) channel contains a large fraction of the GT sum rule, and a quenching factor can be defined in a meaningful way.

2. The N-NUCLEUS INTERACTION at SMALL MOMENTUM TRANSFERS

2.1 Isoscalar and Isovector 1^+ Excitations in ^{28}Si(p,p')

The (p,p') experiments were carried out with the standard upgraded MRS, a 1.4 GeV/c QD spectrometer with a vertical bend angle of 60°. A noteworthy feature of the small-angle setup is a front-end wire chamber (FEC) 1.3 m from the target, and a copper beam stop between target and FEC which shadows scattering angles inside of

2.5°. Using the FEC information the solid angle (~1 msr) is accurately defined and instrumental background is reduced drastically by precise ray tracing to the target. A $^{24}Mg(p,p')$ spectrum obtained at 250 MeV and $\theta=2.9°$ is shown in Fig. 1. The 1^+, T=1 state at 10.71 MeV is the most prominent inelastic excitation. The 1^+ doublet at 9.83/9.97 MeV is probably of mixed isospin.

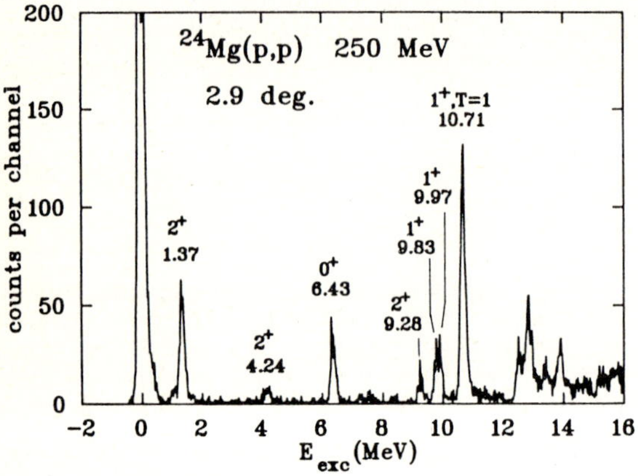

Fig. 1. Momentum spectrum of 250 MeV protons inelastically scattered from ^{24}Mg at $\theta=2.9°$.

In ^{28}Si a strong 1^+ state at 9.50 MeV is believed [5] to have a rather pure T=0 isospin. We have used it together with the dominant 1^+, T=1 state at 11.45 MeV to probe the energy dependence of isoscalar and isovector $\Delta S=1$ terms of the N-nucleus interaction. The quenching factors (QF), $\sigma_{experiment}/\sigma_{theory}$, for $\theta_{cm}=4°$ are shown in Fig. 2. The cross sections were calculated using a density-dependent interaction based on the Paris NN potential [6] and transition densities by Brown and Wildenthal [7]. The QF, a nuclear structure property, should be energy independent. For the isovector excitation this seems to be the case regardless of whether medium modifications are included (DD) or not (IA). The QF for the isoscalar transition increases with energy indicating a deficiency in the $\Delta S=1$, $\Delta T=0$ part of the interaction [6]. Similar conclusions are obtained for the Love-Franey interaction [8]. We notice that the QF for individual states may vary considerably, e.g. QF=0.3 and 0.7 for the strongest 1^+, T=1 states in ^{28}Si (at 11.45 MeV) and ^{24}Mg (at 10.71 MeV), respectively. Because of this we have, whenever possible, evaluated the GT strength over the entire low-energy region and then made the comparison with the integral GT strength from the shell model.

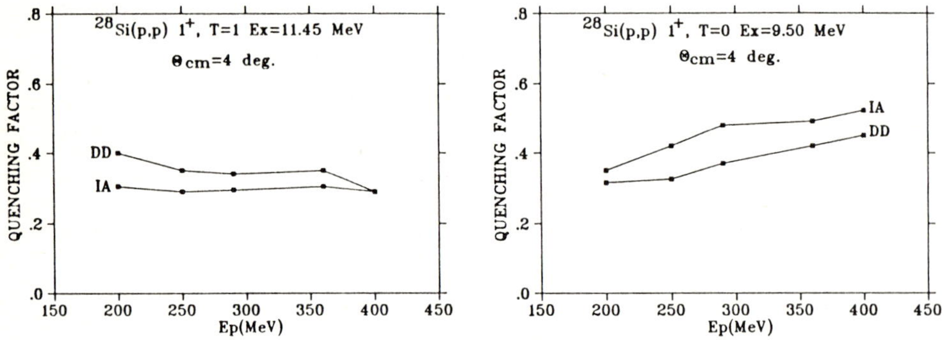

Fig. 2. Energy dependence of the ratio $\sigma_{experiment}/\sigma_{theory}$ at $\theta_{cm}=4°$ for pure isovector (left) and isoscalar (right) $\Delta S=1$ excitations in $^{28}Si(p,p')$. Calculations of the theoretical cross sections are explained in the text.

2.2 Isovector Effective Interaction Strengths from the $^{14}C(p,n)^{14}N$ Reaction

In the MRS facility for charge exchange reactions the incident proton beam passes through the primary target at T_{pn} and is then deflected through a 20° angle by a sweeping magnet. In the (n,p) mode, shown schematically in Fig. 3, a nearly monoenergetic neutron beam from the $^7Li(p,n)$ reaction interacts at 0° with the secondary (n,p) target located only 0.9 m downstream over the MRS pivot. Up to six target layers are separated by wire chambers which identify the origin of the proton from the (n,p) reaction. Up to 1 g/cm^2 thick targets can be used while maintaining a resolution of 1 MeV or less. Targets of CH_2 and C can be present simultaneously to determine shape and intensity of the incident neutron spectrum via H(n,p) elastic scattering.

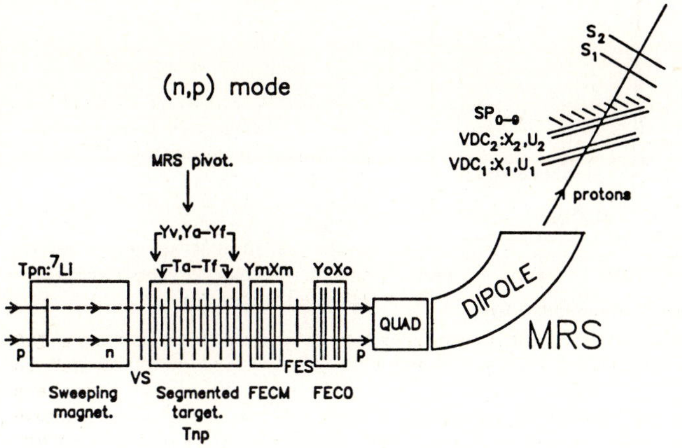

Fig. 3. Schematic side view of the MRS facility for (n,p) charge exchange reactions. The detection system consists of up to 19 wire planes and 5 veto and trigger scintillators.

In the (p,n) mode the primary target of interest is placed at T_{pn} above the MRS pivot and the segmented secondary targets are replaced by a scintillator which serves as an n → p converter and whose light output is used to correct the proton momentum determined by the MRS. A system resolution of 0.8-1 MeV was observed for the $^{14}C(p,n)^{14}N$ reaction between 200 and 450 MeV. This is sufficient to separate the Fermi transition to the 2.31 MeV 0^+, T=1 IAS from the strong GT transition to the 3.95 MeV 1^+, T=0 state in ^{14}N. From the cross section ratios and using the factorized DWIA the ratio of interaction strengths, $|J_{\sigma\tau}/J_\tau|^2$, has been deduced at q=0 [9] and is shown in Fig. 4. The TRIUMF results are in clear disagreement with the Love-Franey interaction [8]. Judging from the results of Fig. 2 the discrepancy can be mainly attributed to J_τ.

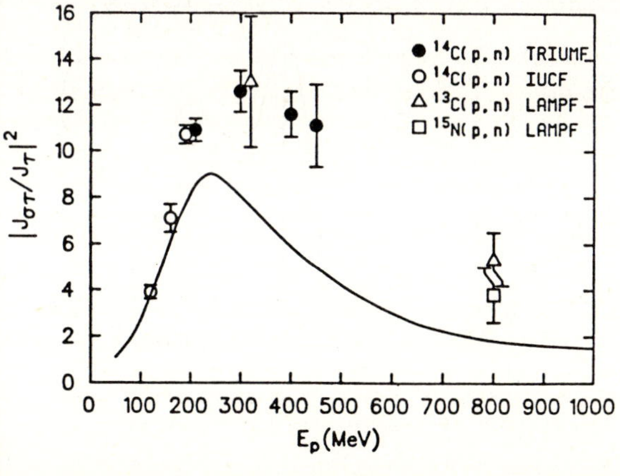

Fig. 4. Ratio of effective isovector interaction strengths derived from (p,n) reactions at incident energies from 100 to 800 MeV. From [9].

2.3 Normalization of B(GT+) and (n,p) Cross Sections at 200 MeV

For targets of ^{6}Li, ^{12}C and ^{13}C the B(GT+) strengths are known from the log(ft) values measured in β^- decay. The 0° cross sections at 200 MeV (Figure 5 shows the typical quality of the raw spectra) were extrapolated to q=0. The ratio $\sigma(0°,q=0)/B(GT+) = 9.8\pm0.7$ mb/sr is constant for the three transitions and anomalies reported for the (p,n) reaction [10] have not yet been observed in (n,p).

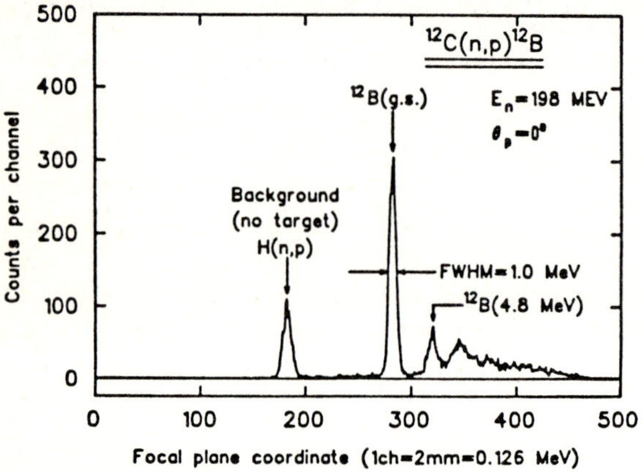

Fig. 5. Spectrum from the $^{12}C(n,p)^{12}B$ reaction at 200 MeV observed at 0°. Measurements with an 'empty' target stack show that the peak labelled H arises from mylar windows in the target chamber segments.

3. GAMOW-TELLER STRENGTH in the SHELL MODEL REGION

3.1 The $^{26}Mg(n,p)^{26}A\ell$ and $^{54}Fe(n,p)^{54}Mn$ Reactions

For the majority of targets the (n,p) reaction opens up uncharted spectroscopic territory. The TRIUMF facility has so far been used to measure angular distributions for targets of ^{19}F, ^{26}Mg, ^{90}Zr and ^{208}Pb at 200 MeV and for ^{54}Fe at 300 MeV. We discuss here ^{26}Mg and ^{54}Fe for which substantial GT+ strength has been observed. Such measurements, when combined with (p,n) data, provide a full test of the GT sum rule, $S = \Sigma\, B(GT-) - \Sigma\, B(GT+) = 3(N-Z)$.

The 0° spectrum for $^{26}Mg(n,p)$ is shown on the left of Fig. 6. Three GT groups at 0.16, 2.86 and 5.48 MeV are strongly forward peaked. On the right of Fig. 6 the

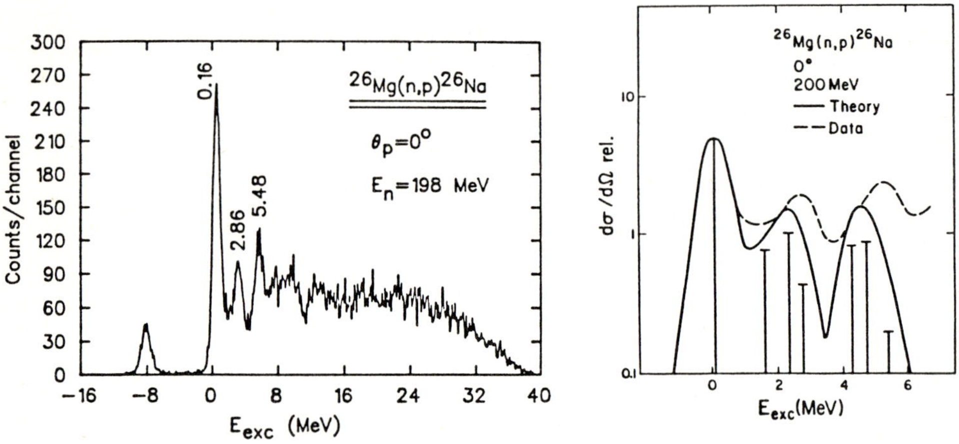

Fig. 6. Spectrum from the $^{26}Mg(n,p)$ reaction at 0° and E_p=200 MeV (left). The data are compared to a theoretical calculation of the GT component only (right).

data are compared to a full (sd) shell calculation of B(GT+) [11]. The theoretical GT operator has been renormalized by $(g_A^{eff})^2 = 0.6(g_A^{free})^2$, and the resulting strengths have been folded with the experimental lineshape for comparison with the data. Good qualitative agreement is observed although above E_{exc}=3 MeV careful analysis is required to assess the L>0 contributions to the data.

A test of the GT sum rule was also carried out for the $^{54}Fe(n,p)^{54}Mn$ reaction at 300 MeV. The 0° cross section shown in Fig. 7 amounts to 16 mb below E_{exc}=8 MeV. Angular distributions indicate that ~85% of this value arises from the L=0 multipole. The corresponding B(GT+)=5.1±0.6 is smaller than estimated in the shell model with a simple $(f_{7/2})^{14}$ ^{54}Fe ground state (10.3), or with a slightly expanded (fp) basis [12,13] (~9.2). The GT strength distribution calculated by Bloom and Fuller [12], shown as a cross-hatched area in Fig. 7, agrees in shape fairly well with experiment. Combining the (n,p) results with previous (p,n) data [14], B(GT−)=7.8±1.9, we obtain for the sum rule S_{exp}=2.7±2.0, smaller than the expected value of 6. This is a further indication that a sizeable fraction of the GT strength is shifted out of the low-energy shell model region. The (n,p) results are of considerable astrophysical relevance since electron capture on N~28, Z~28 nuclei plays an important role in the pre-supernova stage of stellar collapse [12,15].

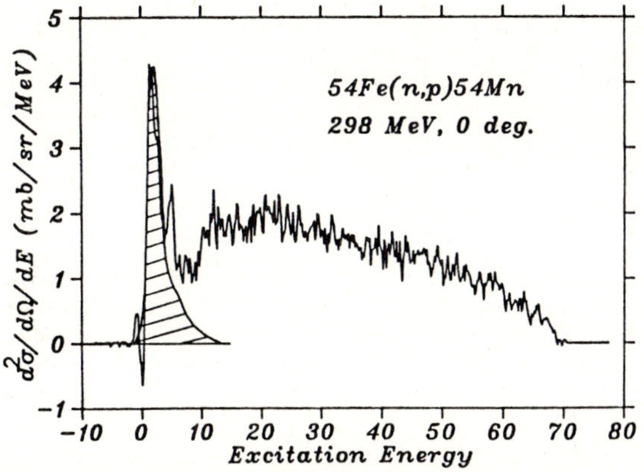

Fig. 7. Spectrum for the $^{54}Fe(n,p)^{54}Mn$ reaction at 0° and E_p=300 MeV. The cross hatched area is the GT distribution calculated by Bloom and Fuller [12] and multiplied with a quenching factor of ~0.6. The bin width is 0.18 MeV per channel.

3.2 Gamow-Teller strength deduced from (p,p') spin-flip cross sections

Small-angle measurements of inelastic proton scattering can be difficult to interpret because both ΔS=0 and ΔS=1 excitations may contribute to the cross section. We have performed double scattering experiments of the spin-flip cross section, σS_{nn}, which is a measure of the ΔS=1 excitations only [16,17] to deduce GT quenching factors for ^{24}Mg and ^{54}Fe(p,p'). The experiments were carried out with a newly constructed focal plane polarimeter [18] which utilizes inclusive scattering from carbon to determine the transverse polarization of momentum-analyzed protons. An important feature of the polarimeter is the large momentum acceptance which allows simultaneous measurements over a wide region of nuclear excitation.

In Fig. 8 the differential and spin-flip cross sections are shown for ^{24}Mg(p,p') at 2.9° and E_p=250 MeV. We observe a strong suppression of natural parity states with ΔS=0. The σS_{nn} strength in the shell model region between 9-16.2 MeV implies QF=0.55±0.10.

Small angle results for σ in ^{54}Fe and ^{51}V, and for σS_{nn} in ^{54}Fe at E_p=290 MeV are shown in Fig. 9. We identify GT strength at 7.4-12.6 MeV in ^{54}Fe and ^{51}V, and a non-spin-flip giant resonance (presumably E0/E1) at ~20 MeV. Since full (f,p) shell model calculations are not feasible we have estimated $\sigma(1^+)$ from the ratio, $B(M1,\text{full fp})/B(M1,f_{7/2}^{A-40}\text{g.s.}) = 0.75$, calculated in ^{48}Ca [19]. This ratio was used

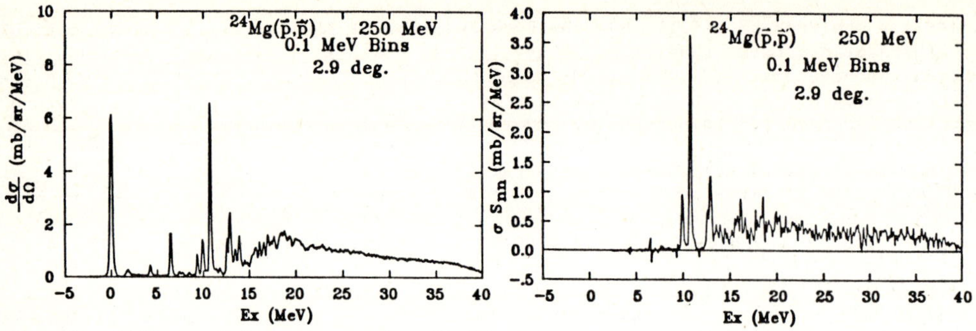

Fig. 8. Differential cross section (left) and spin-flip cross section (right) for ^{24}Mg(p,p') at 2.9° and E_p=250 MeV.

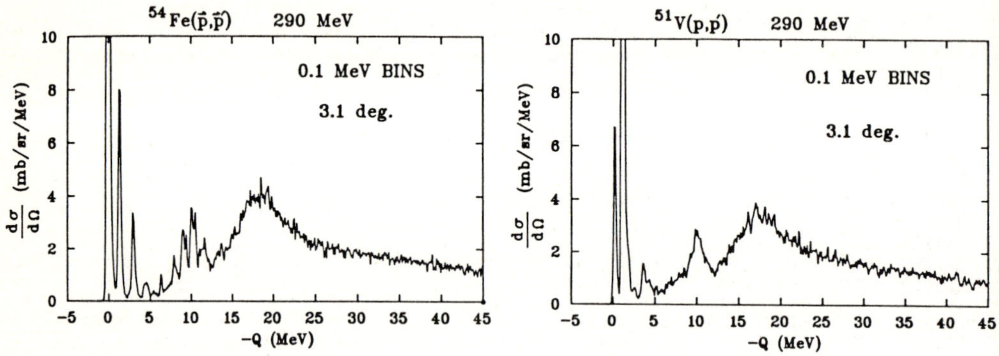

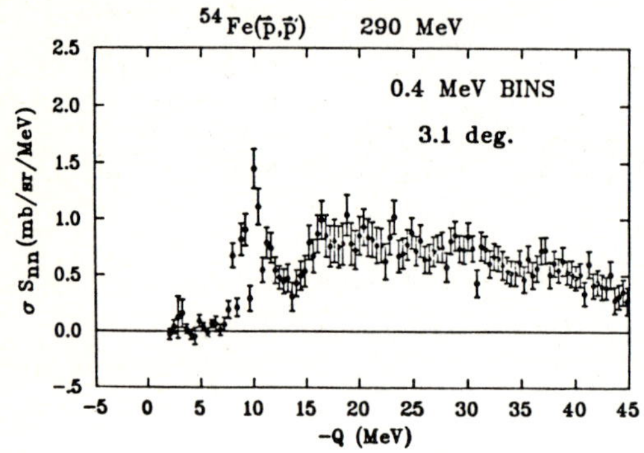

Fig. 9. Differential (p,p') cross section at E_p=290 MeV for ^{54}Fe (top,left) and ^{51}V (top right). The spin-flip cross section for ^{54}Fe is shown on the left.

to deduce the QF in the (f,p) shell shown in Table I which summarizes the present experiments. The results suggest universal values of QF~0.5-0.6 in both (s,d) and (f,p) shell nuclei. In contrast to [3] ^{51}V and ^{54}Fe appear to have similar GT strength.

4. CONCLUSIONS

We have shown that at small momentum transfer spin-isospin components of the N-nucleus interaction are well understood between 200-400 MeV. N-nucleus scattering at these energies provides a comprehensive, quantitative overview of spin-isospin excitations in nuclei. In (p,p') experiments spin-flip cross sections and small instrumental background are essential to locate all the GT strength. We obtain universal quenching factors of ~0.5-0.6 for both (p,p') and (n,p) reactions provided the theory is based on a realistic model of the Fermi surface (untruncated major shell). This supports the concept of an effective one-body operator which

Table I. Gamow-Teller quenching factors from (p,p') and (n,p) reactions at TRIUMF

reaction	measured quantity	energy region (MeV)	quenching factor
^{24}Mg(p,p')	σS_{nn}	9-16.2	0.55
^{26}Mg(n,p)	σ	0-3	~0.6
^{54}Fe(n,p)	σ	0-8	0.64
^{54}Fe(p,p')	σS_{nn}	7.4-12.6	0.62
	σ	8.6-11.8	0.62
^{51}V(p,p')	σ	8.6-11.8	0.51

applies to a wide range of nuclei [7]. Evaluation of GT strength in the high-energy tail is hampered by large L>0 "background" and requires meticulous analysis of the existing detailed angular distributions.

The (n,p) reaction is emerging as a powerful new spectroscopic tool for the study of spin-isospin distributions of $T_>$ states. It provides estimates of important weak interaction rates which are required in calculations of double beta decay and in models of stellar collapse.

The rapid development of the new TRIUMF facilities was made possible by the close collaboration of many individuals. These include R. Abegg, W.P. Alford, D. Frekers, R. Helmer, R. Henderson, K. Hicks, D.A. Hutcheon, K.P. Jackson, C.A. Miller, R. Sawafta, R. Schubank, M. Vetterli and S. Yen. We also thank B.A. Brown, R. Dymarz and F.C. Khanna for discussions of theoretical aspects of this work. The encouragement and financial support provided by the Director of TRIUMF and his associates is gratefully acknowledged. This work was supported by grants from the Natural Sciences and Engineering Research Council of Canada.

REFERENCES

1. C.D. Goodman: Spin Excitations in Nuclei (Plenum, New York, 1984), p.143
2. D. Bender et al.: Nucl. Phys. A398, 408 (1983)
3. C. Djalali et al.: Nucl. Phys. A410, 399 (1983)
4. J. Rapaport et al.: Nucl. Phys. A427, 332 (1984)
5. N. Anantaraman et al.: Phys. Rev. Lett. 52, 1409 (1984)
6. L. Rikus and H. von Geramb: Nucl. Phys. A426, 496 (1984); and refs. therein
7. B.A. Brown and B.H. Wildenthal: Phys. Rev. C28, 2397 (1983); and refs. therein
8. M.A. Franey and W.G. Love: Phys. Rev. C31, 488 (1985)
9. W.P. Alford et al.: Phys. Lett. B, in press
10. J.W. Watson et al.: Phys. Rev. Lett. 55, 1369 (1985)
11. B.A. Brown: private communication
12. S.D. Bloom and G.M. Fuller: Nucl. Phys. A440, 511 (1985)
13. K. Muto, Nucl. Phys: A451, 481 (1986)
14. J. Rapaport et al.: Nucl. Phys. A410, 371 (1983)
15. J. Wambach: contribution to this conference
16. J.M. Moss: Phys. Rev. C26, 727 (1982)
17. S.K. Nanda et al.: Phys. Rev. Lett. 51, 1526 (1983)
18. O. Häusser et al.: Nucl. Instrum. Methods, to be published
19. J.B. McGrory and B.H. Wildenthal: Phys. Lett. 103B, 173 (1981)

Spin-Isospin Excitations in Nuclei by the (p,n) Reaction

R. Madey, B.D. Anderson, B.S. Flanders, and J.W. Watson*

Kent State University, Kent, OH 44242, USA

The (p,n) reaction has been successful in delineating the Gamow-Teller strength function in the nuclear excitation spectrum; however, the total strength extracted from experiment is less than that expected from the sum rule if only nucleon spin and isospin degrees of freedom are important in nuclei. Apparently the missing strength is shifted to higher excitation energies primarily through mixing with both 2p-2h and Δ-h configurations; however, an experimental test is needed to decide the relative importance of these two principal mechanisms.

1. Gamow-Teller Strength Function

The (p,n) reaction is an invaluable complement to weak-interaction decays because of its ability to survey experimentally the dominant portion of Gamow-Teller (GT) transition strength. The results of (p,n) reaction studies on many 0^+ target nuclei reveal that the transition strength to 1^+ states is quenched in the sense that these peaks contain only a fraction of the sum-rule strength [1] expected if only nucleon spin and isospin degrees of freedom are important in nuclei. The strength of a GT transition relative to an absolute standard can be evaluated in terms of the sum rule for GT beta decay, which relates the sums of the strengths of β^- and β^+ transitions to the neutron excess N-Z of the target nucleus:

$$S(\beta^-) - S(\beta^+) = 3(N-Z) \tag{1}$$

This sum rule is independent of the structure of the ground state. The strength $S(\beta^{\pm})$ is given by the sum of the reduced transition probabilities $B(GT^{\pm})$ for all $\beta^{\pm}$ transitions from the ground state of the target nucleus to all possible final states in the residual nucleus. The strength $S(\beta^-)$ can be observed in (p,n) reactions; and $S(\beta^+)$, in (n,p) reactions. In cases where β^+ transitions are blocked or nearly blocked by the Pauli principle, $S(\beta^+)$ is zero or small. The reduced transition probabilities B(GT) are defined such that B(GT) = 3 for the beta decay of the free neutron.

The percentage of the 3(N-Z) sum rule extracted from measurements of cross sections for (p,n) reactions on most nuclei is typically 50 to 60% when $S(\beta^+)$ is assumed to be zero. Cross sections for the (p,n) reaction to discrete 1^+ states were normalized to known beta transition rates for nuclei with mass numbers from 13 to 90 [2]. In heavier nuclei, the GT strength was extracted [3] by normalizing to the Fermi transition strength which is concentrated in the isobaric analog state (IAS) transition and which exhausts the (N-Z) sum rule for Fermi transitions.

2. The $^{26}Mg(p,n)^{26}Al$ Reaction

Although the β^+ strength function $S(\beta^+)$ is weak, it is not always negligible--particularly in light nuclei. In a recent study of the $^{26}Mg(p,n)^{26}Al$ reaction [4], the magnitude of $S(\beta^+)$ was found to be a substantial fraction of 3(N-Z).

*Present address: University of Maryland, College Park, MD 20742

From a shell model calculation, this fraction was estimated to be 30%. For the $^{26}Mg(p,n)^{26}Al$ reaction, the effect of $S(\beta^+)$ is significant. When $S(\beta^+)$ is taken from the shell model, the 1^+ peaks in the ^{26}Al spectrum contain 57% of the sum rule given by (1); however, neglect of $S(\beta^+)$ leads to the erroneous conclusion that the 1^+ peaks contain 73% of the 3(N-Z) sum rule. The shell-model evaluation of $S(\beta^+)$ is supported by extraction of cross sections for T=2 peaks at high excitation energies. From the measured (p,n) cross section for the narrow peak at E_x=13.6 MeV, which corresponds to the lowest T=2, 1^+ state in the spectrum of ^{26}Al, a lower limit for $S(\beta^+)$ was found to be about 16% of 3(N-Z). If the peaks at 14.6 and 14.9 MeV are assumed to be pure T=2 peaks, then $S(\beta^+)$ in discrete states could amount to 32% of 3 (N-Z). Measurements with a recently constructed (n,p) facility at the TRIUMF laboratory in Vancouver is expected to yield $S(\beta^+)$ values for several nuclei.

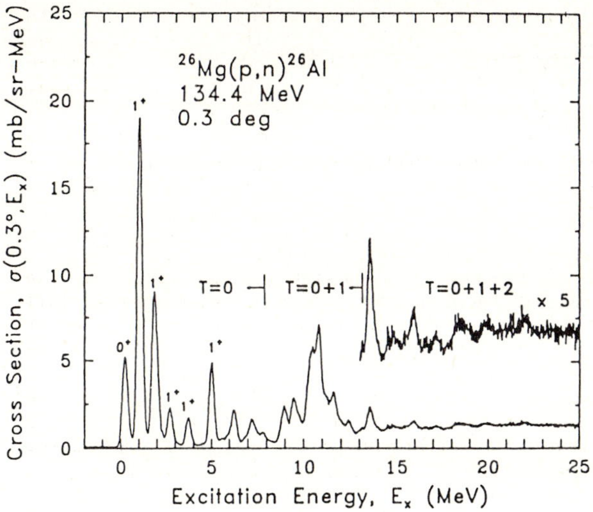

Fig. 1 Excitation energy spectrum at 0.3° for $^{26}Mg(p,n)$ at 134 MeV

The complex spectroscopy in the sd-shell is illustrated in Fig. 1. The ^{26}Mg nucleus is near the middle of the sd shell ($17 \leq A \leq 39$). The Hamiltonian of the multiconfiguration shell model describes the overall distribution of strength better than the detailed distribution among nearby levels. For comparison with the shell model predictions, the measured distribution was smoothened artificially as shown in Fig. 2. The concentration of strength at low excitation energies is associated with transitions that leave the particles in the same shells; whereas the concentration of strength at high excitation energies is dominated by transitions from $d_{5/2}$ to $d_{3/2}$ orbits. The relative strengths and energy distributions agree with theory when the calculated strength function is renormalized by the factor 0.57. This factor comes from a survey [5] of Gamow-Teller beta decay where it was found that the assumption of the free nucleon normalization of the G-T operator yielded theoretical strengths larger than experiment by a factor $B(GT)_{th,fn}/B(GT)_{exp}$ = 1.75 $\pm$ 0.14. The reciprocal of this factor gives the quenching factor of 0.57 which coincidentally is the same as that obtained from the $^{26}Mg(p,n)^{26}Al$ reaction study. The significance of this comparison is that the (p,n) reaction shows the same overall reduction in GT strength as that observed in beta decay. The problem of the missing strength in beta decays does not appear to be resolved by a redistribution of GT strength to the higher excitation energies seen in the (p,n) reaction. The predicted distribution of GT strength between the T=0,1 and 2 isospin channels is consistent with observations except that the relatively small fraction of strength in the T=2 channel is predicted to be too small on an absolute scale. Also, analysis of the shell-model results indicates that the sensitivity

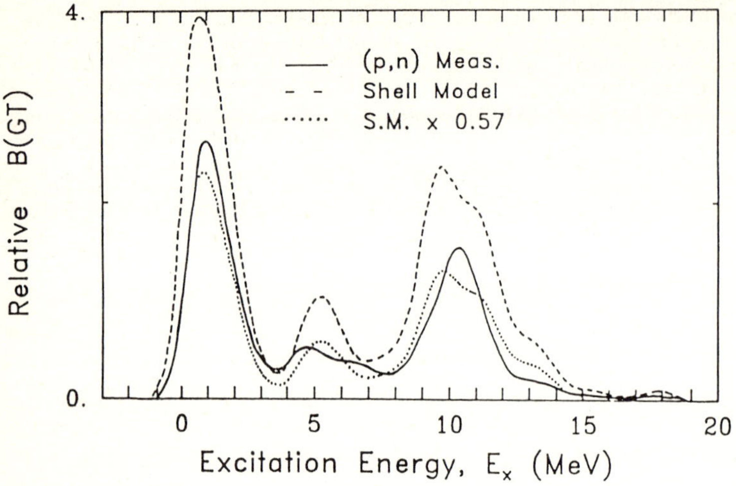

Fig. 2 Comparison of the theoretical and smeared (1 MeV width) experimental energy distributions of GT strength from ^{26}Mg to ^{26}Al

of the predicted strength to configuration mixing increases with increasing isospin; hence, the T=0 comparison should be most reliable in terms of extracting a quenching factor, while the T=2 results indicate that the shell-model wave functions may incorporate too much configuration mixing.

3. Polarization-Transfer Measurements

Recent advances in neutron polarimeters [6,7] have made possible the measurement of the excitation energy distribution of the spin-flip probability for $(\vec{p},\vec{n})$ reactions at 0°. The polarization P_n of a neutron emitted at 0° is related to the polarization P_p of the incident proton:

$$P_n = P_p \, K_y^{y'} \, (0°) \tag{2}$$

Here the symbol $K_y^{y'}$ denotes the transverse polarization-transfer coefficient (according to the Madison Convention [8]). The symbol D_{NN} is sometimes used for $K_y^{y'}$. The relation between $K_y^{y'}$ and the spin-flip probability S is:

$$2S = (1 - K_y^{y'}) \tag{3}$$

For a one-step, central, interaction with a single angular-momentum-transfer, CORNELIUS et al [9] showed that spin-flip probabilities are simple combinations of Clebsch-Gordan coefficients. Thus, the signature for an allowed ($\Delta L=0$) GT($\Delta S=1$) transition is an expected value of S = 2/3. For natural parity ($\Delta J = \Delta L$) transitions without spin-transfer, the spin-flip probability S = 0. The value S = 2/3 is the minimum value of S expected for an unnatural parity ($\Delta J \neq \Delta L$) transition; for example, the value of S = 7/10 for a $0^+ \rightarrow 2^-$ transition and S = 1 for a $0^+ \rightarrow 0^-$ transition. Shown in Fig. 3 is the spin-flip probability spectrum at 0° for the $^{48}Ca(\vec{p},\vec{n})^{48}Sc$ reaction at 135 MeV. The measured value [10] of S is approximately 2/3 for the entire excitation energy region of the spectrum where 1^+ strength was identified from previous cross-section measurements [11]. A multipole analysis of the cross-section measurements indicated that the apparent continuum up to 30 MeV of excitation energy is predominantly (~75%) $\Delta L = 0$ at 0°. This result taken together with the measured values of S = 2/3 up to $E_x \sim 20$ MeV shows that this entire region is predominantly 1^+ strength. Note that the spin-flip probability for the IAS state does not reach the theoretical value of zero because the IAS is superimposed on the low-energy side of the GTGR; hence, some of the cross section under the IAS

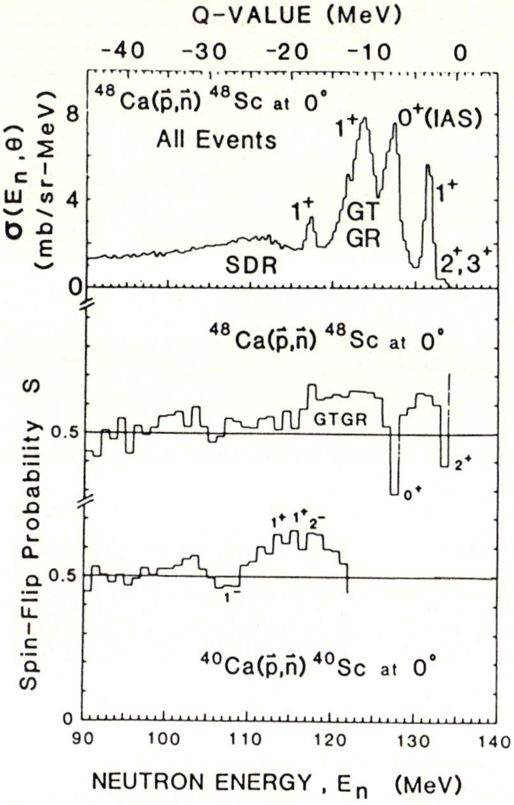

Fig. 3 Neutron energy (and Q-value) distribution of the 0° cross section and spin-flip probability for $^{48}Ca(\vec{p},\vec{n})$ at 135 MeV. The ground state Q-value is -0.49 MeV for $^{48}Ca(p,n)$ and -15.25 MeV for $^{40}Ca(p,n)$

has the spin-flip probability of 2/3 characteristic of a GT transition. The magnitude of the 1^+ strength in the continuum is estimated from the multipole decomposition to be about 26% of the 3(N-Z) sum rule. When this strength from the continuum is added to the 43% of the sum rule strength found [10] in peaks above the apparent continuum, the percentage of the sum rule found in the $^{48}Ca(p,n)^{48}Sc$ reaction is about 69%. If the apparent continuum in the GT region consisted of a background of transitions without spin-transfer, then such [S = 0] transitions would dilute the spin-flip and yield a value smaller than 2/3 for the overall spin-flip probability in the GT region. TADDEUCCI et al [12] obtained a similar result for the Zr(p,n) reaction, namely, that the cross sections in the apparent continuum correspond to $\Delta S = 1$ transitions.

4. Low-Lying GT Strength

Recently, FLANDERS [13] explored the structures in the GT strength distribution at excitation energies below the GTGR in medium- and heavy-mass targets. Theoretical understanding of the low-lying part of the GT strength distribution is relevant to problems in nuclear physics and in astrophysics such as double-beta decay, lepton-number conservation, the mass of the neutrino, neutrino oscillations and stellar nucleosynthesis. Neutron spectra were measured at the Indiana University Cyclotron Facility from 134 MeV protons on targets of Ge, Se, Te, Sr, Sn, Pb and U. The excitation energy spectrum at 0.3° is shown in Fig. 4 for the $^{128}Te(p,n)^{128}I$ reaction. Except for the U target, the spectra display the following features: (1) a dominant narrow peak corresponding to the excitation of the 0^+ IAS transition, (2) a large broad peak corresponding to the excitation of the Gamow-Teller Giant Resonance (GTGR), (3) a few narrow peaks at very low excitation energies, and (4)

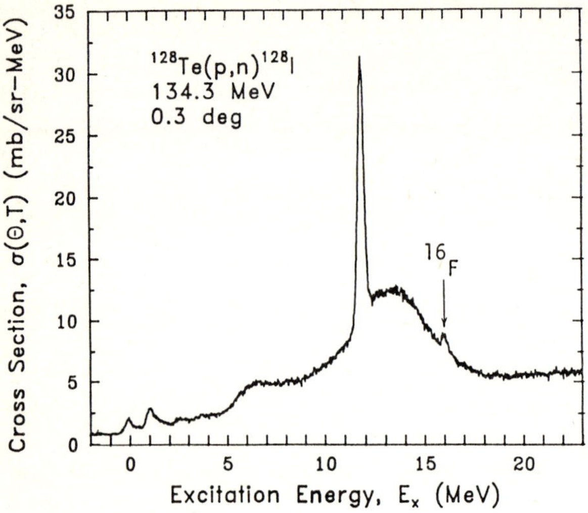

Fig. 4 Excitation-energy distribution of the cross section at 0.3 deg. in the c.m. system for $^{128}Te(p,n)$ at 134.3 MeV

a few broad-structured bumps of varying magnitude, which appear to be complexes of many unresolved states. The last two features were not seen in the U spectrum. The GT strength observed in peaks above a fitted polynomial background varied (as shown in Table 1) from 49 to 62% of the 3(N-Z) sum rule with $S(\beta^+)=0$. The portion of this GT strength in low-lying transitions below the GTGR varied from 15 to 32% as shown in Table 1 also.

Table 1. Percentage of the 3(N-Z) sum rule observed and the percentage of observed GT strength located in low-lying transitions below the GTGR

Target	^{76}Ge	^{82}Se	^{128}Te	^{130}Te	^{88}Sr	^{116}Sn	^{120}Sn	^{124}Sn	^{208}Pb
% of 3(N-Z) Sum Rule	55	52	60	62	56	49	53	52	55
% of Observed GT below GTGR	25	32	15	17	20	25	21	21	16

The low-lying GT strength functions for the four ßß-decay nuclei ^{76}Ge, ^{82}Se, ^{128}Te and ^{130}Te provide constraints on calculations of the matrix elements involved in the first-step of the ßß decay processes. The matrix elements are used to calculate lifetimes for ßß decay. Limits on the neutrino mass are set by comparing calculated lifetimes with measured lifetimes. Predictions for ßß decay lifetimes are consistent with β^- strength functions in the intermediate nuclei [14].

An estimate of the difference in the GT strengths from ^{128}Te and ^{130}Te is useful for testing the assumption that the matrix elements are the same for the double-beta decay of these two nuclei. The neutron spectra from $^{128,130}Te$ were obtained with an energy resolution of about 330 keV. A difference spectrum was obtained by normalizing the yield of the IAS to be proportional to N-Z. The resulting difference spectrum is plotted in Fig. 5 as the fractional difference $1-Y^*(^{128}Te)/Y(^{130}Te)$ in the normalized yields of the allowed ($\Delta\ell = 0$) GT strength versus the neutron time-of-flight. The yield $Y^*(^{128}Te) = (13/12)\ Y(^{128}Te)$. An excitation energy scale is shown. A peak with $\Delta\ell \neq 0$ at $E_x = 50$ keV in ^{130}I was removed from this difference spectrum because $\Delta\ell \neq 0$ strength does not contribute to the double-beta decay process. The fractional difference is small in the neighborhood of the GTGR; for the excitation energy region below the GTGR, this

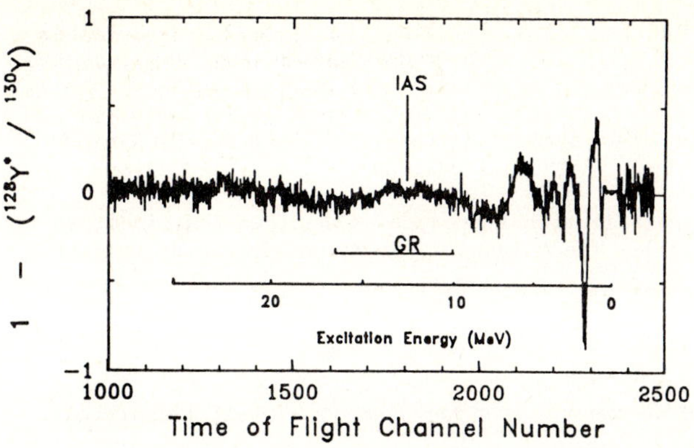

Fig. 5 The fractional difference 1-Y*(^{128}Te)/Y(^{130}Te) in the normalized yields of the allowed ($\Delta\ell$=0) GT strength versus the neutron time-of-flight.

difference is about 15%. The fractional difference over the entire excitation energy spectrum is about 3%; however, it increases to about 8% when the low-lying transitions are weighted by the reciprocal of the energy. The energy denominator in the transition amplitude emphasizes the low-lying transitions.

5. Conclusion

Explanations for the missing GT strength generally fall into two categories; one invokes subnuclear degrees of freedom [15], the other incorporates configuration mixing [16] which is estimated to transfer about 25% of the strength to continuum states at high excitation energies. In a recent review, BERTSCH and ESBENSEN [17] discuss the status of both of these categories and conclude that a convincing demonstration of this additional GT strength seems unlikely with experimental tools that are presently available.

This work was supported in part by the U.S. National Science Foundation under grant numbers PHY 85-01054 and PHY 83-40353.

References

1. K.I. Ikeda, S. Fujii and J.I. Fujita, Phys. Lett. 3, 271 (1963).

   C. Gaarde, J.S. Larsen, M.N. Harakeh, S.Y. van der Werf, M. Igarashi and A. Müller-Arnke, Nucl. Phys. A334, 248 (1980).

2. C.D. Goodman and S.D. Bloom in "Spin Excitations in Nuclei", ed. by F. Petrovich, G.E. Brown, G.T. Garvey, C.D. Goodman, R.A. Lindgren, W.G. Love, p. 143, Plenum, New York (1982).

3. C. Gaarde, Nuclear Structure, ed. by R. Broglia et al, p. 449, Elsevier Science Publishers (1985).

4. R. Madey, B.S. Flanders, B.D. Anderson, A.R. Baldwin, C. Lebo, J.W. Watson, S.M. Austin, A. Galonsky, B.A. Wildenthal and C.C. Foster, Phys. Rev. C (submitted).

5. B.A. Brown and B.H. Wildenthal, Atomic Data and Nuclear Data Tables 33, 347 (1985).

6. R. Madey, J.W. Watson, B.D. Anderson, A.R. Baldwin and P.J. Pella, Proceedings 1985 Summer Workshop, Continuous Electron Beam Accelerator Facility, ed. by Hall Crannel and Franz Gross, pp. 290-315 (1985).

7. T.N. Taddeucci, C.D. Goodman, R.C. Byrd, T.A. Carey, D.J. Horen, J. Rapaport and E. Sugarbaker, Nucl. Instrum. Methods A 241, 448 (1985).

8. The Madison Convention, in Polarization Phenomena in Nuclear Reactions, ed. by H.H. Barshall and W. Haeberli, The University of Wisconsin Press, Madison, Wisconsin (1970).

9. W.D. Cornelius, J.M. Moss and T. Yamaya, Phys. Rev. C23, 1364 (1980).

10. J.W. Watson, P.J. Pella, B.D. Anderson, A.R. Baldwin, T. Chittrakarn, B.S. Flanders, R. Madey, C.C. Foster and I.J. van Heerden, Phys. Lett (submitted).

11. B.D. Anderson, T. Chittrakarn, A.R. Baldwin, C. Lebo, R. Madey, P.C. Tandy, J.W. Watson, B.A. Brown and C.C. Foster, Phys. Rev. C 31, 1161 (1985).

12. T.N. Taddeucci, C.D. Goodman, R.C. Byrd, I.J. Van Heerden, T.A. Carey, D.J. Horen, J.S. Larsen, C. Gaarde, J. Rapaport, T.P. Welch, and E. Sugarbaker, Phys. Rev C33, 746 (1986).

13. B. Flanders, Ph.D. Dissertation, Kent State University (December 1985).

14. K. Grotz and H.V. Klapdor, Nuclear Physics (in press).

15. M. Ericson, A. Figereau and D. Thevenet, Phys. Lett. 45B, 19 (1973).
M. Rho, Nucl. Phys. A231, 493 (1974).
E. Oset and M. Rho, Phys. Rev. Lett. 42, 47 (1979).
A. Bohr and B.R. Mottleson, Phys. Lett. 100B, 10 (1981).
M. Rho, Ann. Rev. Nucl. & Part. Sci 34, 531 (1984).

16. K. Shimizu, M. Ichimura and A. Arima, Nucl. Phys. A226, 282 (1974).
I.S. Towner and F.C. Khanna, Phys. Rev. Lett. 42, 51 (1979).
G.F. Bertsch and I. Hamamoto, Phys. Rev C26, 1323 (1982).
K. Takayanagi, K. Shimizu and A. Arima, Nucl. Phys. A 444, 436 (1985).
A. Klein, W.G. Love and N. Auerbach, Phys. Rev. C31, 710 (1985).
F. Osterfeld, D. Cha and J. Speth, Phys. Rev C31, 372 (1985).
S. Drozdz, V. Klemt, J. Speth and J. Wambach, Phys. Lett. 166B, 18 (1986).

17. G.F. Bertsch and H. Esbensen, The (p,n) Reaction and the Nucleon-Nucleon Force, Argonne National Laboratory Report PHY-4803-TH-66 (May 1986).

Microscopic Description of (p, n) Reactions at Intermediate Energies

S.N. Ershov[1], *F.A. Gareev*[1], *N.I. Pyatov*[1], *and S.A. Fayans*[2]

[1]Joint Institute for Nuclear Research, Dubna 141980, USSR
[2]I.V. Kurchatov Institute of Atomic Energy, Moscow 123182, USSR

Recently,(p, n) reactions at intermediate proton energies ($E_p \gtrsim 100$ MeV) have turned out to be a fruitful tool for studying charge-exchange excitations of nuclei, in particular, in the spin-isospin channel. Thus, the Gamow - Teller resonance (GTR) was discovered and the quenching of the integral strength of spin-isospin transitions with a small momentum transfer in the low-energy range of the excitation spectrum ($0 \leq E_x \leq 30$ MeV) was established[1]. To obtain a quantitative estimate of the quenching factor in the continuous spectrum region, it was necessary to develop a microscopic approach for the description of observed inclusive neutron spectra and separation of contributions of various multipolarities. First estimates of a background below the GTR were made in papers[2] without taking into account the effective interactions of quasiparticles. Further development of this model and detailed calculations for ^{90}Zr are given in ref.[3]. Analogous calculations by the Hartree-Fock method with the Skyrme interactions were carried out in ref.[4]. We have developed a microscopic model[5] for the description of angular distributions for individual excitations and inclusive spectra of neutrons, which differs from the above-mentioned approaches, mainly, by the method of structure calculations.

At intermediate energies the direct one-step processes which are well-described in the distorted-wave impulse approximation (DWIA)[6], dominate in the forward-angle cross section up to excitation energies $E_x \lesssim E_p/2$. Usually, a free NN t-matrix parametrized in ref.[7] is used in the scattering calculations. The process of nucleon knock-on exchange associated with the central part of t-matrix interaction is taken into account in the pseudo-potential approximation[8]. Distorted waves necessary for the calculation were obtained using the optical potential with the parameters[9].

In our approach the structure calculations of charge-exchange excitations were carried out in the theory of finite Fermi systems (TFFS) using a complete particle-hole basis generated in the Woods-Saxon potential[10]. In the spin-isospin channel the effective interaction of quasiparticles includes a short-range repulsion with the Landau-Migdal constant $g_0' = 1.1$ ($G_0' = 330$ MeV·fm^3) and the renormalized one-pion exchange amplitude which contains the contribution from virtual Δ-isobar-hole excitations. In the isospin channel use is made of the effective density-dependent forces consistent with the isovector potential and isovector density. As shown in refs.[5,10], such an approach satisfactorily describes the energies of observed charge-exchange discrete and resonance states. For each state a transition density was calculated. In the continuous spectrum this density is normalized to the integral transition strength for a selected range of the excitation energy. Such a procedure allows us to exhaust sufficiently completely the total transition strength (sum rules) for each multipolarity L up to $E_x \approx 40$ MeV.

In the TFFS the effects of suppressing the low-energy $\sigma\tau$-transitions are taken into account phenomenologically by means of a local charge of quasiparticles $e_q[\sigma\tau]$, which at small momentum transfers is approximated by a constant. The charge $e_q[\sigma\tau]$ is included as an external factor into any matrix element of the $\sigma\tau$-transition and, consequently, into the multipole sum rules for the particle-hole branch of the $\sigma\tau$-excitations. Physically, the value of e_q^2 characterizes the contribution of the particle-hole excitation branch to the complete response of the nucleus to the $\sigma\tau$-field. With the help of this quantity the theory phenomenologically takes into account the effects associated with virtual excitations of baryon resonances (in particular, the excitations ΔN^{-1}), the meson-exchange currents and multipair excitations. In the DWIA calculations the $\sigma\tau$-components of the t-matrix are multiplied by $e_q[\sigma\tau]$, i.e. for $\sigma\tau$-transitions the reaction cross section is reduced by $e_q^2[\sigma\tau]$.

We have calculated differential cross sections and inclusive spectra of neutrons for the reactions $^{40}Ca(p, n)^{40}Sc$, $^{48}Ca(p, n)^{48}Sc$ at E_p = 134 and 160 MeV, as well as $^{90}Zr(p, n)^{90}Nb$ and $^{208}Pb(p, n)^{208}Bi$ at E_p = 200 MeV. The t-matrix normalization has been made[5] for the reaction $^{42}Ca(p, n)^{42}Sc$ (1^+, E_x =0.61 MeV), where the GT transition matrix element M_{GT}^2 = 2.57 is considered to be known from the β-decay of $^{42}Ti \to {}^{42}Sc$. Using the ph-transition density $(1\ f_{7/2})^2_{0^+} \to$ $\to (1\ f_{7/2})^2_{1^+}$, normalized to the above value of M^2_{GT}, we have obtained at E_p =160 MeV the cross section $\sigma(0^0)$ = 15.8 mb/sr which is in good agreement with the experimental value of (15.2±1.0) mb/sr.

The total neutron spectrum is a superposition of contributions of individual transitions with the multipolarities $0 \leq L \lesssim 4$, i.e. excitations with $J^\pi = 0^+$, 0^-, 1^+, 1^-, 2^+, 2^-,..., 5^+, since at small angles the contribution of higher multipolarities is small. The calculations have shown that at small θ the contribution of the spin-orbit and tensor components of the t-matrix to the cross section is rather small for the multipolarities $L \lesssim 3$. To obtain a continuous theoretical spectrum and to compare it with the experimental one, the cross sections calculated for each excitation were smoothed with a Breit-Wigner function which simulates a spreading width. The partial widths depend on the excitation energy and they have been chosen so as to reproduce the gross structures observed in the spectra (for example, for the GTR width $\Gamma \approx$ 4-5 MeV). The value of $e_q[\sigma\tau]$ was obtained by fitting the theoretical spectrum to the experimental one in the excitation energy range $E_x \lesssim$ 25 MeV, where the contribution of multistep processes is expected to be small.

In fig.1 some theoretical neutron spectra calculated at e_q = 0.8 are compared with the experimental data kindly presented by Dr. C.Gaarde. Partial distributions for the Fermi (0^+), Gamow-Teller (1^+), spin-dipole (L = 1) transitions are shown, as well as a summary background of the transitions with L > 1. It is clear that at small angles the contribution of the GT transitions dominates, and the background of the multipole transitions below the GTR is small at $\theta = 0^0$ but it quickly grows with θ. The integrated cross sections in the energy range from 0 to $-Q_{max}$ are given in Table 1. Comparing the results of calculations with the experimental data, one should bear in mind that the uncertainty of both is of the order of 10%. Within the limits of this uncertainty one may conclude that $e_q[\sigma\tau] \approx 0.85 \pm 0.05$, i.e. the strength of the low-energy ($E_x \lesssim$ 15-25 MeV) $\sigma\tau$-transitions is quenched by the factor $\approx 0.7 \pm 0.1$. Note that, according to our estimates with e_q = 0.8 in the low-energy part of the excitation spectrum only ≈ 50% of the sum rule 3(N - Z) for the GT transitions is exhausted in ^{48}Ca and ^{90}Zr; and ≈ 40%, in ^{208}Pb.

In conclusion, we have obtained the evidence that all $\sigma\tau$-vertices irreducible in the particle-hole channel are suppressed in a nucleus by a factor 0.8-0.9. In particular, the axial-vector weak coupling constant is renormalized in the nucleus $g_A \to G_A = e_q \cdot g_A \simeq 1.0$, which is confirmed by the data from the β-decay[11]. Analogously renormalized is the constant of the πNN coupling, which has already been taken into account in our calculations. A detailed analysis of the effects connected with the $\sigma\tau$-transition quenching is given in ref.[12].

The authors are grateful to Dr. C.Gaarde for providing us with a number of experimental data.

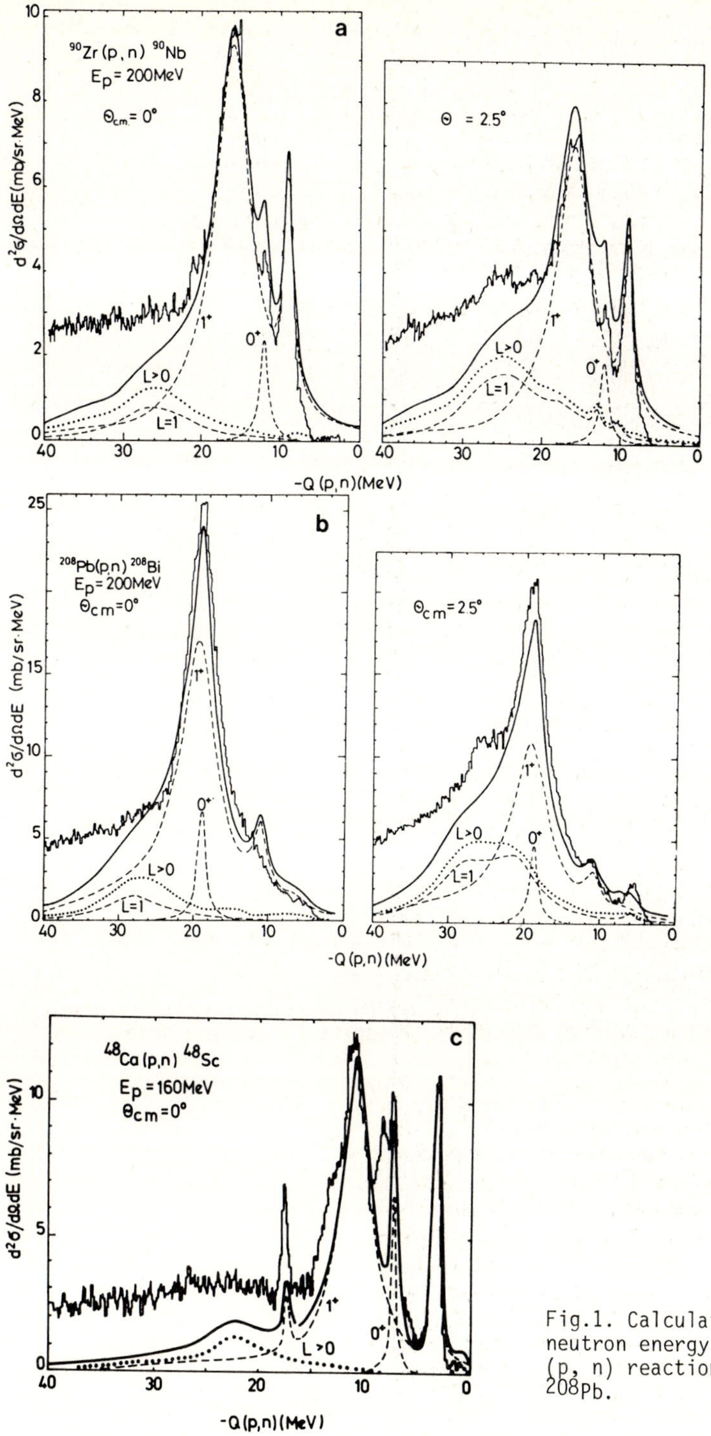

Fig.1. Calculated and experimental neutron energy spectra from the (p, n) reactions on ^{48}Ca, ^{90}Zr and ^{208}Pb.

Table 1. Energy-integrated up to $-Q_{max}$ partial and total cross sections calculated with $e_q[\sigma\tau]$ = 0.8

θ_{cm} (deg.)	$-Q_{max}$ (MeV)	$\sigma(0^+)$	$\sigma(1^+)$	$\sigma(L=1)$ (mb/sr)	$\sigma(L>1)$	σ_{tot}	σ_{exp}
			$^{48}Ca(p, n)^{48}Sc$,		E_p = 160 MeV		
0^0	15	5.1	60.0	0.9	1.7	67.7	81.6
	35	5.5	71.7	7.8	6.8	91.8	142.6
			$^{90}Zr(p, n)^{90}Nb$,		E_p = 200 MeV		
0^0	25	4.8	80.3	4.9	4.2	94.2	88
	40	4.9	86.3	10.3	9.2	110.7	130
2.5^0	25	3.9	60.6	15.9	3.8	84.2	71
	40	3.9	65.6	26.1	8.9	104.5	116
			$^{208}Pb(p, n)^{208}Bi$,		E_p = 200 MeV		
0^0	25	13.9	146.2	6.6	11.6	178.3	183
	40	14.2	177.1	18.4	20.4	230.1	270
2.5^0	25	9.7	91.6	37.5	8.9	147.7	165
	40	10.0	113.6	68.6	16.6	208.8	270

References

1. D.E.Bainum et al.: Phys.Rev.Lett. 44, 1751 (1980); C.Goodman: Nucl.Phys. A374, 241c (1982); C.Gaarde: Nucl.Phys. A396, 127c (1983) and references therein.
2. F.Osterfeld: Phys.Rev. C26, 762 (1982); F.Osterfeld, A.Schulte: Phys.Lett. 138B, 23 (1984).
3. F.Osterfeld et al.: Phys.Rev. C31, 372 (1985).
4. A.Klein et al.: Phys.Rev. C31, 710 (1985).
5. F.A.Gareev et al.: Yad.Fiz. 39, 1401 (1984); J.Bang et al.: Nucl.Phys. A440, 445 (1985).
6. H.C.Chiang, J.Hüfner: Nucl.Phys. A349, 466 (1980).
7. W.G.Love, M.A.Franey: Phys.Rev. C24, 1073 (1981); ibid. C31, 488 (1985).
8. W.G.Love: Nucl.Phys. A312, 160 (1978).
9. A.Nadasen et al.: Phys.Rev. C23, 1023 (1981).
10. S.A.Fayans, N.I.Pyatov: in Proc. 4th Int.Conf. on Nuclei Far from Stability, Helsingør, 1981 (CERN 81-09, Geneva, 1981), p.287; N.I.Pyatov, S.A.Fayans: Sov.J.Part.Nucl. 14, 401 (1983).
11. B.Buck, S.M.Perez: Phys.Rev.Lett. 50, 1975 (1983); S.V.Tolokonnikov, R.U.Khafizov: Phys.Lett. 153B, 353 (1985); I.S.Towner: Nucl.Phys. A444, 402 (1985).
12. F.A.Gareev et al.: in "Nuclear Structure", JINR D4-85-851, Dubna, 1985, p.124.

Charge-Exchange Resonances in Deformed Nuclei

L.A. Malov, V.G. Soloviev, and A.V. Sushkov

Joint Institute for Nuclear Research, Dubna 141980, USSR

In recent years the Gamow-Teller (GT) and spin-dipole charge-exchange resonances have widely been studied experimentally and theoretically in atomic nuclei including the deformed ones [1]. The study of the GT resonance is associated with the calculation of β-strength functions. The first calculations of the GT resonances in deformed nuclei have been performed in refs. [2,3] while studying β-decay of the rare-earch nuclei and rubidium isotopes and for some nuclei in the rare-earth and actinide region. The calculations were performed in the random phase approximation (RPA). The above approximation is well fulfilled for the deformed nuclei.

In this report within the quasiparticle-phonon nuclear model (QPNM) we give the results of calculations in RPA for the GT resonances and other 1^+ states, spin-dipole with $\lambda^\pi = 0^-, 1^-, 2^-$, E1 charge-exchange and $T_>$ resonances in deformed nuclei in the range of $156 \leq A \leq 168$ and $236 \leq A \leq 240$ [4-5]. Moreover, the developed formalism [4] is applied for describing allowed unhindered β-transitions in the nuclear region of $A \sim 165$. The role of tensor forces is analysed. The QPNM Hamiltonian consists of the Saxon-Woods potential for the average nuclear field and pairing interactions. It contains also the multipole-multipole and spin-multipole - spin-multipole isoscalar and isovector interactions. The characteristics of the model Hamiltonian for the deformed nuclei are given in ref. [6]. The model Hamiltonian may include also an isovector tensor interaction, which allows one to take into account the one-pion exchange contribution [7-8].

For the study of charge-exchange resonances the (p,n) and (n,p) transition strengths B(p,n) and B(n,p) are introduced [9]. These quantities are directly connected with the strength functions of the above-mentioned transitions $b(p, n; E)$ and $b(n, p; E)$, that are defined and widely used within the QPNM [4]. For example, $B(p,n) = S(p,n) = \int_0^\infty b(p, n; E)\,dE$, where E is the excitation energy. In particular, they are very useful for analyzing the properties of charge-exchange resonances with the use of the sum rules. For the GT resonances the sum rule is model independent

$$S(p, n) - S(n, p) = 3(N - Z)\,. \qquad (1)$$

In the calculations [4] the sum rule for GT resonance is satisfied by (97-99)%. For the E1 and spin-dipole resonances the model dependent sum rules are overestimated by (10-20)% [4].

The Gamow-Teller and charge-exchange spin-dipole and E1 resonances were calculated for a large number of nuclei in the ranges A=155, 165 and 240. The same parameters of the Saxon-Woods potential and the pairing constants were used as in ref. [6]. The isovector constant for the GT forces is $\kappa_1^{\ell=0} = 17/A$ MeV and for the spin-dipole forces is $\kappa_1^{\ell=1} = 0.75 \frac{4\pi}{\langle r^2 \rangle A} \kappa_1^{\ell=0}$. A typical example of the behaviour of the strength functions of (p,n) and (n,p) transitions with excitation of 1^+ states on ^{164}Er versus the excitation energy is shown in Fig. 1. The energy is reckoned from the ground state of a target-nucleus. For (p, n)-transitions in the strength distribution one can distinguish the low-energy region, region of maximum of the GT resonance and the high-energy region. The main strength of the resonance is distributed in the interval 5-40 MeV. The GT resonance has a maximum

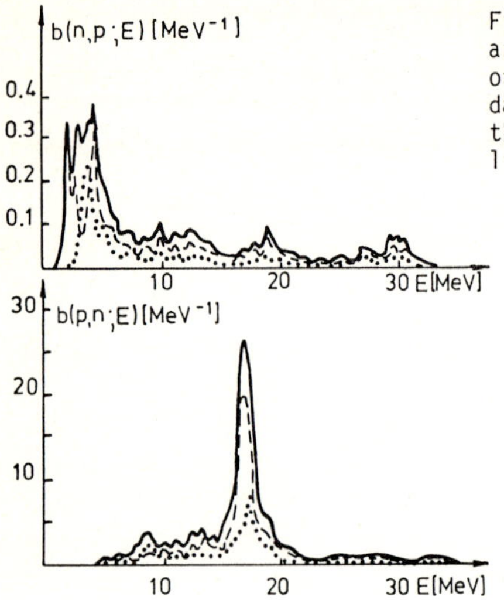

Fig. 1. Strength functions of (p, n) and (n, p) transitions with excitation of 1^+ states on ^{164}Er. The dotted and dashed lines denote transitions to the states with K=0 and 1; the solid line is their sum

at 18-20 MeV. In the region of maximum for the nuclei in A=165 and 240 ranges (60-70)% of the total (p,n) strength is exhausted (1/3 of this strength is for the state with $I^{\pi}K = 1^+0$ and the rest for 1^+1 states). For A=155 the maximum is splitted into two-three peaks, and it is localised in the energy interval of 8 MeV. The maximum for transitions to the 1^+0 states is by 0.1-0.3 MeV higher in comparison with transitions to the 1^+1 states. The strength of the (p, n) transitions to the low-energy states, partially observed in β-decay,is of about 20%.The centroid energy of this part is 10 MeV for the rare-earth nuclei and 12 MeV for the actinide region. The difference between the centroid of the low-energy region and the GT resonance maximum is 8 MeV. It is almost the same for all calculated nuclei. About 10% (for actinides ~15%) of the (p,n) transition strength is exhausted at an energy above 22 MeV.

The (n, p) strength is distributed rather uniformly within 2-35 though there are peaks at 5-8 MeV, 10-15 MeV and 25-30 MeV (the low-energy peak is most pronounced). With increasing A the strength function maximum shifts towards higher energies and the total strength S(n, p) decreases from 2-2.5% in the rare-earth nuclei up to 0.6-0.8% in actinides with respect to the total strength S(p, n) .

Thus, the calculations show that the GT resonance concentrated in a narrow energy region may be studied in detail experimentally. Its characteristics depend weakly on A and the resonance with components K=0 and 1 does not split.

The spin-dipole resonance consists of the states with $\lambda^{\pi} = 0^-, 1^-, 2^-$. The strength is distributed within 2-40 MeV, 87-90% of this strength being concentrated in the region of resonance maximum 14-33 MeV. The low-energy part (2-14 MeV) has only the states with $\lambda = 2$. For the states with $\lambda = 0$ there is one pronounced maximum at 30-32 MeV; for $\lambda = 1$, two maxima at 24-25 MeV and 29-30 MeV and for $\lambda = 2$ the strength maximum is distributed in a wide energy region of 14-30 MeV. The high-energy part of the resonance gives (3-4)% contribution to the total strength. The strength of (n, p) spin-dipole transitions is distributed within 2-25 MeV, has individual maxima and comprises (5-14)% of S(p, n).

The calculations have shown that though the distributions of the charge-exchange E1 and spin-dipole resonances overlap strongly, their mutual influence is small. The (p, n) transition strength is distributed within 15-40 MeV and has two pronounced maxima at 25-27 MeV and 29 MeV. These maxima correspond to the transitions to the states with K = 0 and 1. About 75% of the strength is concentrated in the region of maximum 24-31 MeV;and (17-20)%, in the low-energy region. Note that the charge-exchange E1 resonance in deformed nuclei is concentrated in a narrower interval in comparison with the E1 giant resonance.

Now we shall consider the effect of the inclusion of tensor forces (T) on the GT resonance characteristics. The calculations have been made for the attracting tensor interaction with different values of the constants κ_T. A simultaneous inclusion of the GT, spin-quadrupole and tensor forces shifts part of the GT strength to an energy region higher than 20 MeV and to the region of 6-10 MeV. The position of the centroid energy of (p, n) transitions with tensor forces included changes slightly whereas the maximum energy of the resonance falls. The centroid energy of (n, p) transition rises. The strength distribution for (p, n) and (n, p) transitions in some energy intervals for different constants of tensor forces κ_T and κ_{GT} = const is shown in Table 1 for ^{168}Er.

Table 1. Strength distribution of (p, n) - and (n, p) - GT transitions in excitation energy in ^{168}Er (in % of the total strength)

κ_T/κ_{GT} \ Energy interval (MeV)	0-16 (p,n)-transition	16-20	>20	0-7 (n,p)-transition	7-15	>15
0	25	60	15	41	28	31
-0.1	28	54	18	38	27	35
-0.2	32	46	22	32	27	41

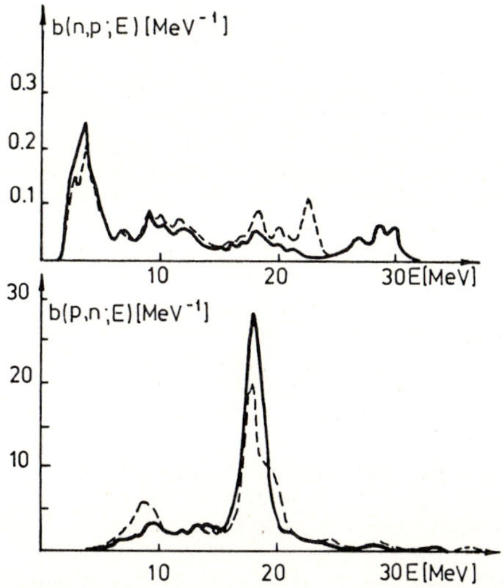

Fig. 2. The same as in Fig. 1 for ^{164}Er. The solid line is the inclusion of only GT forces; the dashed line is a simultaneous inclusion of GT, spin-quadrupole and tensor ($\kappa_T = -0.2\,\kappa_{GT}$) forces

Figure 2 shows the strength functions of (p, n) and (n, p) transitions for ^{168}Er, that have been obtained taking into account different residual forces; κ_{GT} is renormalized so that the position of the GT resonance maximum would not change.

The description of charge-exchange resonances in the QPNM has also been used for describing $T_>$ giant dipole resonances [5]. It was shown that energy centroids of $T_>$ resonances are placed at energies 24-28 MeV, their excitation cross sections in photonuclear reactions are considerably less than those of an electric giant resonance with $T_<$.

This formalism is applied for studying β-decay. In the rare-earth nuclei in A = 165 range there are allowed unhindered β-transitions p523↑ ↔ n523↓ with log ft ~ 4.8. Table 2 exemplifies the calculated results for the matrix elements of transitions between doubly even and doubly odd nuclei. Similar results have been obtained in ref. [10] for odd-A nuclei. It is seen that the inclusion of the residual GT interaction decreases two or three times the square of the GT transition matrix element in comparison with the one-quasiparticle estimates. However, the calculated values are yet on the average two-five times the experimental ones. The inclusion of more complex configurations (2p-2h), tensor forces and non-nuclear degrees of freedom ($\Delta\bar{N}$) may lead to a further descrease of theoretical estimates. It has been studying how the two-phonon states and a simultaneous in-

Table 2. Calculated and experimental matrix elements for transition p523↑ $7/2^-$ ↔ n523↓ $5/2^-$ between doubly even and doubly odd nuclei

Transition	$\|\langle 1^+ \| GT \| 0^+ \rangle\|^2$		
	Exp.	BCS	GT
$^{162}Ho \rightarrow ^{162}Dy$	0.06	0.43	0.10
$^{162}Yb \rightarrow ^{162}Tm$	0.04	0.73	0.25
$^{164}Ho \rightarrow ^{164}Dy$	0.16	0.61	0.14
$^{164}Yb \rightarrow ^{164}Tm$	0.03	0.62	0.20
$^{164}Tm \rightarrow ^{164}Er$	0.10	0.23	0.04
$^{164}Ho \rightarrow ^{164}Er$	0.02	0.39	0.13
$^{166}Dy \rightarrow ^{166}Ho$	0.01	0.65	0.18
$^{166}Yb \rightarrow ^{166}Tm$	0.03	0.47	0.16

clusion of the GT, spin-quadrupole and tensor forces influence the probability of β-transitions and GT resonances. The obtained results showed that the tensor forces increase the region of strength distribution of (p,n) GT transitions over excitation energy and decrease the probability of β-transitions.

References:

1. H.V.Klapdor: "The Shape of the Beta Strength Function and Consequences for Nuclear Physics and Astrophysics", In: Progr. in Part. and Nucl.Phys., v.10, pp. 131-225, Pergamon Press Ltd (1983)
2. S.I.Gabrakov et al.: Phys.Lett., 36B, 275 (1971)
   H.V.Klapdor et al.: Phys.Lett. 78B, 20 (1978)
   J.Krumlinde: Nucl.Phys. A413, 223 (1984)
3. V.G.Soloviev et al.: Pis'ma Zh.Eksp.Teor.Fiz. 38, 151 (1983)
4. V.G.Soloviev et al.: Z.Phys. A - At. and Nucl. 316, 65 (1984)
5. V.G.Soloviev, A.V.Sushkov: Izv. Akad. Nauk SSSR, Ser.Fiz. 48,1798 (1984)
6. L.A.Malov, V.G.Soloviev: Part. Nucl. 11, 301 (1980)
7. D.Cha, J.Speth: Phys.Lett. 143B, 297 (1984)
8. V.G.Soloviev: JINR preprint E4-85-706, Dubna (1985)
9. C.Gaarde et al.: Nucl.Phys. A369, 258 (1981)
10. J.Krumlinde, P.Möller: Nucl.Phys. A417, 419 (1984)

Quenching of Gamow-Teller Strength and a Microscopic Derivation of the Effective Δ_{33}-Nucleon Interaction

S. Krewald

Institut für Kernphysik, Kernforschungsanlage Jülich, D-5170 Jülich, F. R. Germany

Subnuclear degrees of freddom have been an exciting topic in nuclear physics during the last ten years. The present review concentrates on the possible role of the Δ_{33}-resonance in reducing magnetic strength in nuclei.

In the analysis of energetically low-lying nuclear excitations, the wave functions are usually constructed from purely nuclear degrees of freedom, such as particle-hole excitations. There is no general principle, however, that would prevent to mix a nucleon-hole excitation with a Δ_{33}-hole excitation. In fact, it is quite easy to obtain order-of-magnitude estimates, how large possible Δ_{33}-hole admixtures might be. Since the Δ_{33}-resonance has isospin 3/2, it has to be excited by a vector meson. This limits the possible meson-exchanges to pion- and rho-exchange. The bare nucleon-nucleon interaction mediated by one-pion and one-rho exchange is known fairly well (see e.g. ref. 1). In the nuclear medium, the bare interaction is modified by short-range correlations. For semi-quantitative purposes, short-range correlations may be summarized by a local correlation function which prevents two nucleons to have a vanishing distance. The Δ-nucleon interaction may be obtained from the nucleon-nucleon interaction by replacing the spin- and isospin-operators by the corresponding transition spin operators and by replacing the pion-nucleon coupling constant f_π by the Δ_{33}-nucleon constant f_π^*. An estimation of this kind has been performed in 1977 by Huber and his collaborators[2] for the electroexcitation of the 15.11 MeV state in ^{12}C (see fig. 1). The

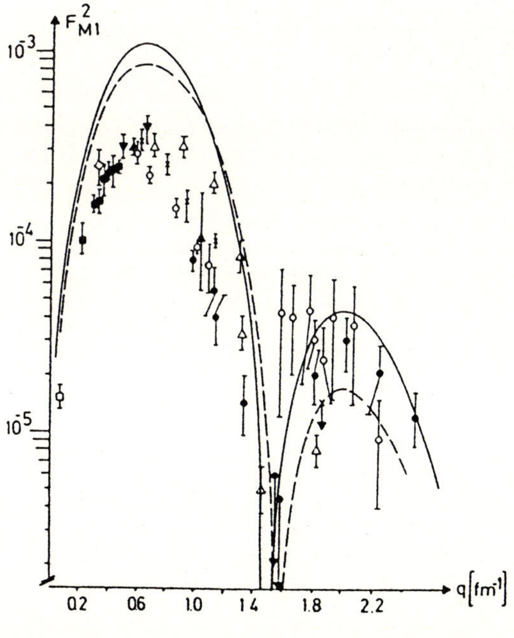

Fig. 1 Transition form factor of the electroexcitation of the 15.11 MeV, T=1, $J^\pi = 1^+$ state in ^{12}C, obtained within the random-phase approximation with (dashed line) and without (solid line) inclusion of the Δ_{33}-resonance.

Δ_{33}-resonance is found to reduce the cross section by about 25 %. Since the Δ_{33}-nucleon interaction is repulsive (after taking care of short-range correlations), one can easily understand that strength is shifted from the low-energy region (15.11 MeV) presumably into the Δ_{33}-region (approximately 300 MeV). The magnitude of this effect, however, is quite model-dependent.

With the experimental discovery of the Gamow-Teller resonance in 1980[3], the interest in the possible role of the Δ_{33}-resonance rose sharply. Ikeda's sum rule states that the difference between the 0^0 cross sections of the (p,n) and (n,p) reaction is proportional to the neutron excess. This provides a model-independent lower limit for the Gamow-Teller strength to be seen in the (p,n) reaction[4]. Much less strength than given by Ikeda's sum rule was found in the early analysis of the data (see ref. 5). Today we know that a careful background subtraction[6] restores most of the missing strength and the remaining question is to which extent the long tails of the Gamow-Teller resonance (see fig. 2) are actual Gamow-Teller strength. In this situation, a microscopic analysis of the Gamow-Teller mode may provide important independent information.

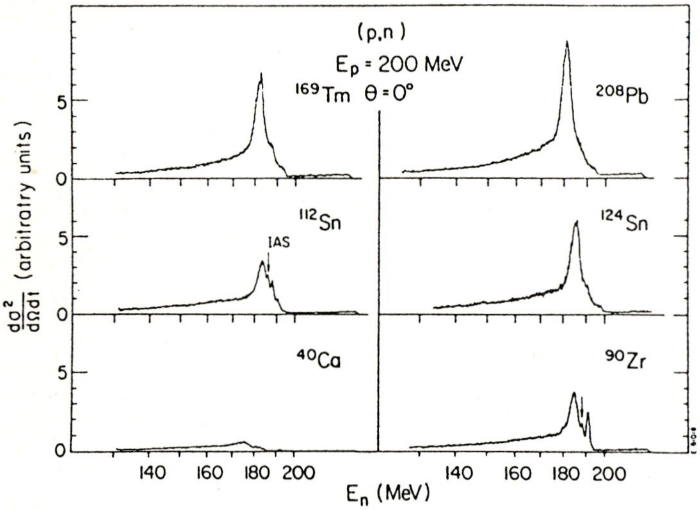

Fig. 2 Cross section for the (p,n) reaction at 0^0 exhibiting the Gamow-Teller resonance and its tail in several neutron excess nuclei (taken from ref. 5).

Any theoretical model for the Gamow-Teller strength needs the Δ_{33}-nucleon interaction as an input. It would be nice if direct experimental information on the Δ_{33}-nucleon interaction in the medium were available. The reaction (p,Δ^{++}) studied at Saturne[7] provides a possibility to study the Δ_{33}-nucleon interaction directly. Jain has analyzed the $^6Li(p,\Delta^{++})^6He$ reaction within the distorted-wave Born approximation[8] and has obtained a coupling parameter $g'_\Delta = 0.4$ C. In pionic units, the constant C is given by

$$C = 4\pi\ \hbar c\ f_\pi f_\pi^* \left(\frac{\hbar}{m_\pi c}\right)^2 \tag{1}$$

and has the numerical value C = 821 MeV fm^3. The coupling parameter obtained by Jain is of limited value in the present discussion, since it represents the Δ_{33}-nucleon coupling at a very large momentum transfer (0.11 (GeV/c)^2) and has been obtained for a proton energy of E_p = 1.04 GeV. In the present context, however, the Δ_{33}-nucleon interaction is needed for vanishing momentum transfer and for bound particles. Therefore we have to rely on many-body theory.

Starting from symmetry-arguments, M. Rho has suggested that the Δ_{33}-nucleon interaction should be obtained from the corresponding nucleon-nucleon interaction by a simple rescaling with the ratio of the bare coupling strength f^*/f (this is reviewed in ref. 9). This prescription does not hold when one treats short-range correlations in standard Brueckner-Hartree-Fock theory because of a strong cancellation between the Hartree and the Fock term[10,11]. The antisymmetrization of the G-matrix can be performed most easily in the usual partial-wave representation

$$G = \sum_{JST,L} G^{JST}_{LL} \; (1-(-)^{L+S+T}) \qquad (2)$$

Because of its spin and isospin 3/2, the Δ_{33}-resonance can only be excited via spin and isospin transfers S=1 and T=1. This suppresses all even partial wave contributions to the G-matrix, in particular the S-wave which is most important for short-range correlations. As a result, one obtains a Landau parameter of the order of $g'_\Delta = 0.4$ in nuclear matter which results in a reduction of the Gamow-Teller strength of less than ten percent only[10,12].

This observation does not yet close the issue of the quenching of the Gamow-Teller strength, however, since one has to check whether the underlying microscopic model gives a consistent description not only of spin-isospin modes, such as the Gamow-Teller resonance, but also of other nuclear excitations, such as density oscillations. If one describes density oscillations as small vibrations around the Brueckner-Hartree-Fock ground state, i.e. one performs a random-phase approximation (RPA) with the G-matrix as an effective interaction, one finds - at least in heavy nuclei - an instability which produces a negative excitation energy e.g. for the breathing mode or the energetically low-lying collective isoscalar 3^- mode[13]. The instability of density oscillations in a conventional RPA calculation based on G-matrices is very plausible when one recalls that nuclear matter saturates at about twice normal nuclear matter density, if one starts e.g. from a one-boson exchange potential[14]. A density oscillation of a finite nucleus will therefore be amplified in the conventional RPA and lead to a collapsed system of high density. Obviously, one has to use an improved many-body theory, before one can make firm conclusions about the reduction of the Gamow-Teller strength.

In order to obtain the correct saturation properties of nuclear matter, a relativistic treatment of the nucleon dynamics appears to be essential[15]. In the nuclear medium, the nucleon spinor $f(\vec{p},s)$ is determined by the Dirac equation

$$[\vec{\alpha}\cdot\vec{p} + \beta m + \Sigma(\vec{p})] \; f(\vec{p},s) = E \; f(\vec{p},s) \; . \qquad (3)$$

The self-energy $\Sigma(\vec{p})$ basically modifies the mass of the nucleon inside the nucleus and as a result, the repulsive ω-exchange is made even more repulsive while the attractive σ-exchange turns out to be less attractive. Thus a new saturation mechanism results, which pushes the saturation point of nuclear matter toward the empirical binding energy of -16 MeV and the empirical Fermi momentum $k_F = 1.36$ fm^{-1}.

Apart from the relativistic treatment of the nucleon, the long-range correlations of the nuclear medium have to be treated much better than in the conventional RPA framework. The importance of long-range correlations is well known in liquid helium[16]. Long-range correlations strongly modify the effective interaction in the medium by adding the so-called "induced interaction"[17] to the

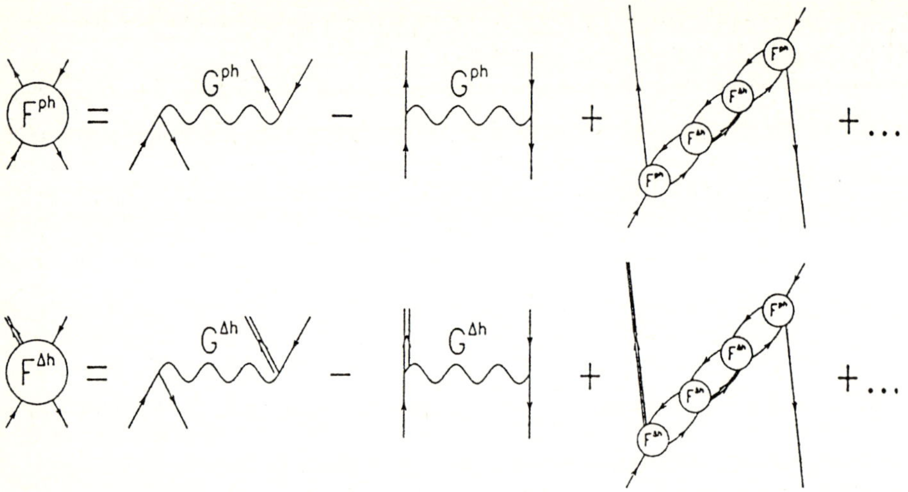

Fig. 3 Graphical representation of the Babu-Brown equation which embodies the effects due to long-range correlations into the effective particle-hole interaction in the medium.

Brueckner G-matrix which only includes the effects due to short-range correlations. A microscopic model for the induced interaction has been developed by Babu and Brown[18]. The induced interaction polarizes the meson being exchanged between two nucleons (see fig. 3). The effect of the induced interaction is to give a repulsive contribution to the effective particle-hole or Δ_{33}-hole interaction in the medium.

When one includes both the relativistic treatment of the nucleon and the influence of long-range correlations into the microscopic model for the description of nuclear excited states, one obtains e.g. excitation energies for the monopole mode which are thirty percent below the experimental values[19]. Given the fact that a conventional RPA produces no stable monopole mode at all, one has to conclude that microscopic many-body theory now incorporates the most essential physical effects, even though there is still plenty of room for quantitative improvements.

The effects of the long-range correlations on the Δ_{33}-nucleon coupling strength parameter g'_Δ is displayed in table 1. If one determines the value of g'_Δ by requiring that the entire reduction of the Gamow-Teller strength is due to the Δ_{33}-hole quenching mechanism, one obtains the value $g'_\Delta = 0.6$. Microscopic theories incorporating short-range correlations only give values of $g'_\Delta \approx 0.4$ which does not produce a substantial decrease of Gamow-Teller strength. The more sophisticated many-body theory outlined here results in a very small spin and isospin dependent Landau parameter f_0 which is in a reasonable agreement with the empirical value. The small value of f_0 is necessary in order to have a fair description of collective density oscillations. The same mechanism which pushes up the parameter f_0, leads to a more repulsive Δ_{33}-hole coupling parameter g'_0. Quantitatively, however, the enhancement of g'_Δ is not very large. The present calculation has been performed in nuclear matter. In a finite nucleus, the enhancement of g'_Δ due to long-range correlations may be even less than in nuclear matter since in finite nuclei the relevant spin-isospin modes are weaker than the density oscillation modes. This point has been investigated quantitatively by Czerski et al.[24] who analyzed the influence of the Δ_{33}-resonance on the reduction of the cross section for the electroexcitation of the 15.1 MeV state in ^{12}C.

Table 1 The Landau parameters f_0, g_0' and g_Δ' of the effective interaction in the nuclear medium are given in units of C = 821 MeV fm^3.

		f_0	g_0'	g_Δ'
model	ref.	1	$\sigma\cdot\sigma'$ $\tau\cdot\tau'$	$\sigma\cdot S$ $\tau\cdot T$
Δ_{33}-quenching	20	-	-	0.6
short-range correlations	10	-	-	0.35
	21	-	-	0.36
	22	-0.43	0.23	0.33
long-range correlations	22	-0.12	0.25	0.45
empirical	23	0.03	0.33	-
	8	-	-	(0.4)

The corresponding B(M1)-value was quenched by less than 10 % due to the Δ_{33}-admixture.

The conclusions to be drawn are quite clear: the Δ_{33}-resonance may very well admix to nuclear states in the low-energy region. The magnitude of the reduction due to Δ_{33}-admixtures, however, is of the order of ten percent according to the best microscopic many-body theories available. This makes the effect of the Δ_{33}-resonance in low energy nuclear physics comparable to the magnitude of meson-exchange current effects.

1. K. Holinde: Phys. Rep. 68, 121 (1981)
2. E. Grecksch, M. Dillig and M.G. Huber: Phys. Lett. 72B, 11 (1977)
3. D.E. Bainum, J. Rapaport, C.D. Goodman, D.J. Horen, D.D. Foster, M.B. Greenfield, C.A. Goulding: Phys. Rev. Lett. 44, 1751 (1980)
4. I. Ikeda, S. Fujii and J.J. Fujita: Phys. Lett. 3, 271 (1963)
5. C. Gaarde: Nucl. Phys. A396, 127 (1983)
6. F. Osterfeld: Phys. Rev. C26, 762 (1982);
   F. Osterfeld, D. Cha and J. Speth: Phys. Rev. C31, 372 (1985)
7. T. Hennino et al.: Phys. Rev. Lett. 48, 997 (1982)
8. B.V. Jain: Phys. Rev. C29, 1396 (1984)
9. M. Rho: Ann. Rev. Nucl. and Part. Science 34 (1984)
10. A. Arima, T. Cheon, K. Shimizu, H. Hyuga and T. Suzuki: Phys. Lett. 122B, 126 (1983)
11. Toru Suzuki, S. Krewald and J. Speth: Phys. Lett. 107B, 9 (1981)
12. W.H. Dickhoff, J. Meyer-ter-Vehn, H. Müther and A. Faessler: Phys. Rev. C23, 1154 (1981)
13. K. Nakayama, S. Krewald and J. Speth: Phys. Lett. 148B, 399 (1984)
14. B. Day: Phys. Rev. Lett. 47, 226 (1980)
15. J.D. Walecka: Ann. Phys. 83, 491 (1974);
    M.R. Anastasio, L.S. Celenza, W.S. Pong and C.M. Shakin: Phys. Rep. 100, 327 (1983)
16. R.D. Viollier and J.D. Walecka: Acta Phys. Polon. B8, 25 (1977)

17. J. Bardeen, G. Baym and D. Pines: Phys. Rev. 156, 207 (1967)
18. S. Babu and G.E. Brown, Ann. Phys. 78, 1 (1973)
19. K. Nakayama, S. Drozdz, S. Krewald and J. Speth: preprint
20. A. Harting, M. Kohno and W. Weise: Nucl. Phys. A420, 423 (1984)
21. H. Sagawa, T.S.H. Lee, K. Ohta: Phys. Rev. C33, 629 (1986)
22. S. Krewald: Habilitationsschrift, Bonn (1985), Jül-1983
23. J. Speth, E. Werner and W. Wild: Phys. Rep. 33, 127 (1977)
24. P. Czerski, W.H. Dickhoff, A. Faessler and H. Müther: Phys. Rev. C33, 1753 (1986)

Does the Delta Quench Gamow-Teller Strength in (p, n)- and ($\vec{p}$, $\vec{p}'$)-Reactions?

F. Osterfeld, A. Schulte, T. Udagawa, and M. Yabe

Institut für Kernphysik, Kernforschungsanlage Jülich,
D-5170 Jülich, F. R. Germany

Microscopic analyses of complete forward angle intermediate energy (p,n)-, (^{3}He,t)- and ($\vec{p}$,$\vec{p}'$)-spin-flip spectra are presented for the reactions ^{90}Zr(p,n), ^{90}Zr(^{3}He,t) and ^{90}Zr($\vec{p}$,$\vec{p}'$). It is shown that the whole spectra up to high excitation energies ($E_x \sim 50$ MeV) are the result of correlated one-particle-one-hole (1p1h) spin-isospin transitions only. The spectra reflect, therefore, the linear spin-isospin response of the target nucleus to the probing external hadronic fields. Our results suggest that the measured (p,n)-, (^{3}He,t)- and ($\vec{p}$,$\vec{p}'$)-cross sections are compatible with the transition strength predictions as obtained from random phase approximation (RPA) calculations. This means that the Δ isobar quenching mechanism is likely to be rather small.

1. Introduction

Recent (p,n) experiments on many nuclei throughout the periodic table[1,2] have led to the discovery that systematically a large fraction of the minimum Gamow-Teller (GT) strength of 3(N-Z) is missing in the excitation energy region where the shell model would predict it to be. Three physically different mechanisms have been discussed to explain this so-called quenching of the total GT strength. The first is that Δ(1232)-isobar-nucleon-hole (ΔN^{-1}) states couple into the proton-particle-neutron-hole (pn^{-1}) GT states and remove strength from the low-lying excitation spectrum[3]. Here the internal degrees of freedom of the nucleon, specifically the Δ, are made responsible for the quenching of the GT strength. The second mechanism is ordinary nuclear configuration mixing[4], where energetically high-lying two-particle-two-hole (2p2h) states mix with the low-lying one-particle-one-hole (1p1h) GT states and shift GT strength into the energy region far beyond the GTR. The third possibility[5-8], closely connected with the second mechanism, is that a large fraction of GT strength is actually located in the physical background below and beyond the giant GT state and is therefore escaping experimental detection.

The ambiguity in the extraction of the GT-strength from the data is due to the shape of the 0^0 (p,n) spectrum. The spectrum shows a couple of very prominent peaks identified as GT transitions, but also shows a broad continuum with a large and long high energy tail. This broad continuum, which in reality takes up a major portion of the spectrum, makes it difficult to distinguish between the GT and the background cross section due to other excitations.

Meanwhile, it has been first recognized by one of the present authors (F.O.) that the whole continuous spectrum is calculable within microscopic models[5]. In this contribution we present such microscopic model calculations for the ^{90}Zr(p,n)- and ^{90}Zr(^{3}He,t)-reactions at 200 MeV/nucleon and for the ^{90}Zr($\vec{p}$,$\vec{p}'$)-spin flip reaction at 319 MeV. We show that essentially all the measured cross section up to high excitation energies ($E_x \leqslant 50$ MeV) is produced by 1p1h spin-isospin excitations only and that this cross section therefore represents the linear spin-isospin response of the target nucleus to the probing hadronic fields.

2. The Model

To analyze the experimental spectra we have to make assumptions about the nuclear excitation spectrum and about the reaction mechanism. In our calculations we assumed that the nuclear excited states can be described reasonably well by microscopic RPA wave functions. The RPA calculations were performed in a model space[7] which included all $3\hbar\omega$ ph excitations in case of $^{90}Zr(p,n)$, and all $2\hbar\omega$ ph excitations in case of $^{90}Zr(^3He,t)$ and $^{90}Zr(\vec{p},\vec{p}\,')$ [10].

From the RPA wave functions we calculated the (p,n)- and $(\vec{p},\vec{p}\,')$-DWIA-cross sections with the code FROST-MARS and the $(^3He,t)$-cross sections with the newly developed code DCP (DWIA code for Composite Particle scattering)[9] which can handle the knock-on exchange process for composite particle scattering exactly. For the effective projectile-target nucleon interaction we used the free nucleon-nucleon (NN) t-matrix in the parametrization of Love and Franey[11]. The interaction was calibrated to the known β-decay transition $^{42}Ca(0^+)\rightarrow{}^{42}Sc(1^+$, E_x = 0.61 MeV) in order to guarantee a force independent analysis of the data[2].

3. Results and Discussion

3.1 Analysis of the 200 MeV $^{90}Zr(p,n)$-Spectrum

In the microscopic model already described we have calculated energy spectra at various scattering angles for the reaction $^{90}Zr(p,n)$ at 200 MeV incident energy. In fig. 1 we show the results for the 0^0 and 4.5^0 spectra. The 0^0 spectrum in fig. 1a is dominated by the GT 1^+ transitions. The theoretical spectrum is a incoherent sum of cross sections with multipolarities L=0 through L=4 (J^π = 0^+, $0^-,1^+,1^-,2^+,2^-,3^+,3^-,4^-,4^+,5^+$). From these states, the $0^-,1^-,2^-$ and $1^+,2^+,3^+$ states were calculated with RPA, while the $3^-,4^-,4^+,5^+$ states were treated within the unperturbed 1p1h doorway model of ref. 5 which includes the nuclear continuum exactly. The RPA model space included all $3\hbar\omega$ excitations so that the RPA states extend in excitation energies up to a Q value of Q = -40 MeV. The cross section

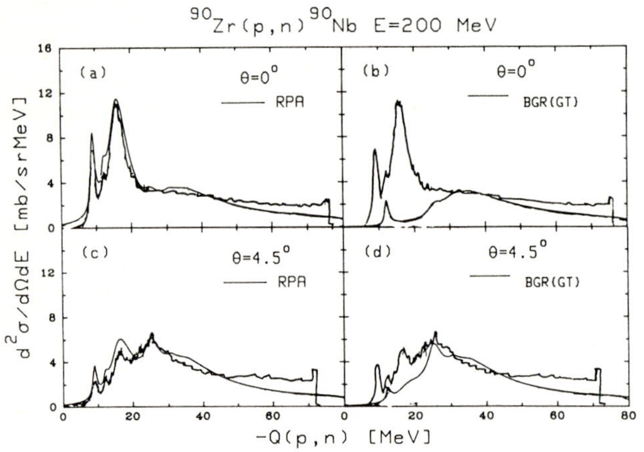

Fig. 1 Neutron spectra from the reaction $^{90}Zr(p,n)^{90}Nb$ at angles of θ = 0^0 (a) and (b), and θ = 4.5^0 (c) and (d). The data were taken from ref. 2. The complete theoretical spectra in (a) and (c) were calculated with RPA wave functions. Figs. (b) and (d) show the background (BGR) with respect to the GT-resonance.

beyond Q = -40 MeV is mainly due to states with $E_x > 3\hbar\omega$ which were again treated within the unperturbed 1p1h doorway model of ref. 5.

For all the RPA final states n we first calculated the differential cross sections, $d\sigma_n/d\Omega$. The continuous spectra, $d^2\sigma/dEd\Omega$, were then generated by folding each cross section $d\sigma_n/d\Omega$ into an asymmetric Breit-Wigner weight function with widths taken either from experiment or from microscopic 2p2h-calculations[12)].

The theoretical spectrum in fig. 1a slightly overestimates the experimental one in the Q-value range, 0 MeV > Q > -40 MeV. Considering this Q-value range we need a quenching of 15 % of the theoretical spectrum in order to bring experiment and theory into agreement. Unfortunately, we cannot decide whether this quenching should be due to the Δ isobar effect or due to additional spreading of both the GT strength and the strength of the multipoles with L > 0. A larger asymmetric spreading of these states would shift more strength to higher excitation energies. Such an additional shift would actually be welcome since the theory is underestimating the data at high negative Q values with the present widths.

Particularly important for our discussion of the quenching of the total GT strength are the results for the high scattering angles at $\theta = 9.5^0$, $\theta = 12.8^0$, and $\theta = 18.7^0$ which are shown in fig. 2. At these scattering angles the GT resonance gives a comparatively small contribution to the total (p,n)-spectrum. The shape and magnitude of these spectra are therefore mainly determined by states of other multipolarities. Note that the theoretical spectra calculated provide in all cases a good description of the experimental data. This leads us to the following important conclusion: At 200 MeV incident energy the whole (p,n) spectra up to E_x = 70 MeV are a result of one-step processes only. Two-step processes with explicit excitation of 2p2h states are suppressed.

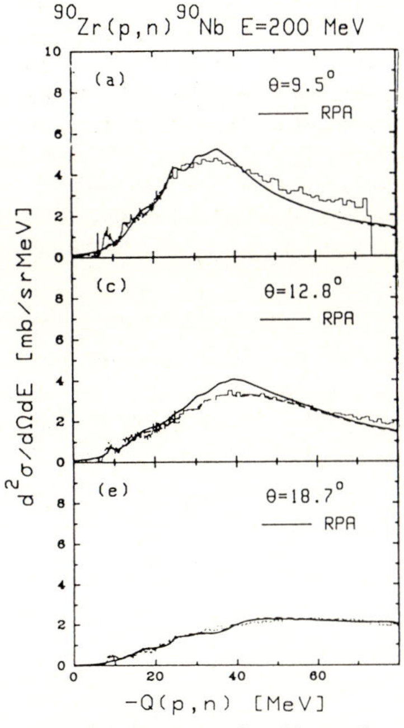

Fig. 2 Same as in figs. 1a and 1c, but now for the scattering angles $\theta = 9.5^0$, 12.8^0 and 18.7^0.

3.2 Analysis of the $^{90}Zr(^3He,t)$-Reaction at 600 MeV

The calculations[9] for the $^{90}Zr(^3He,t)$-reaction were performed in a similar manner as for $^{90}Zr(p,n)$ apart from the fact that we used here a smaller model space which only included all $2\hbar\omega$ excitations. The finite projectile size was taken into account in the analysis of the data by folding the effective projectile nucleon-target nucleon interaction[11] into the magnetic projectile density distribution which was taken from experiment[13]. Similarly the optical potentials in the incident and exit channels were generated from the 200 MeV proton optical potentials of ref. 14 by following the single folding procedure.

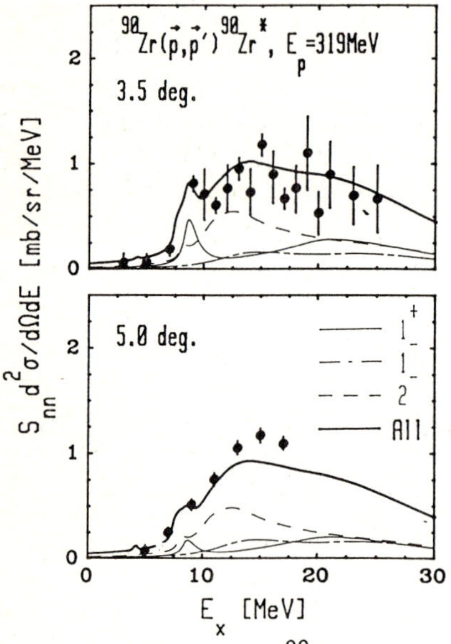

Fig. 3 Zero degree $^{90}Zr(p,n)$- and $^{90}Zr(^3He,t)$-spectra at the incident energy of 200 MeV/nucleon. The (p,n)-data (thick full line in (a)) were taken from ref. 2 and the $(^3He,t)$-data (thick full line in (b)) were taken from ref. 19. In each case the short dashed curve represents the complete theoretical spectrum while the dotted curve represents the cross section for the $2\hbar\omega$, $\Delta L=2$-excitation. The calculated $(^3He,t)$-spectrum was multiplied with a factor of N = 1.3 to reproduce the data.

In fig. 3b we show the results for the $^{90}Zr(^3He,t)$-spectrum at 0^0. The dotted curve represents the GT (L=0) and the long-dashed curve the $2\hbar\omega$, L=2 ($J^\pi = 1^+, 2^+, 3^+$) cross section distribution. Note that in case of the $(^3He,t)$-reaction the cross section in the high excitation energy region is dominated by the $\Delta L=2$-excitations. This is not the case for the (p,n)-reaction in which the $2\hbar\omega$ L=2-contribution is rather small (see long-dashed curve in fig. 3a). Actually the $(^3He,t)$ 0^0-cross section to these states is by roughly a factor of 10 larger than the corresponding (p,n) cross section. The enhancement of the $\Delta L=2$ $(^3He,t)$-cross section at 0^0 over the corresponding (p,n)-cross section is due to the strong surface character of the $(^3He,t)$-reaction. This reaction thus provides valuable nuclear structure informations which differ from those of the (p,n)-reaction.

3.3 Analysis of $^{90}Zr(\vec{p},\vec{p}')$-Spin-Flip Spectra

Recently, cross sections, analyzing powers and spin-flip probabilities have been measured at small angles in the polarized proton inelastic scattering on ^{90}Zr at E_p = 319 MeV [15]. These measurements reveal a large cross section for spin ($\Delta S=1$) excitations distributed roughly uniformly over the excitation energy region from about 8 to 25 MeV [15]. This is a very interesting result, in particular in connection with the question whether the quenching of the magnetic spin-flip strength observed in (p,p')-[15,16] and (e,e')-experiments[17] is due to the isobar-hole (ΔN^{-1})-effect[3] or due to ordinary nuclear configuration mixing[4]. In order to answer this question we analyzed the $^{90}Zr(\vec{p},\vec{p}')$-spectra in detail decomposing them into the various multipoles.

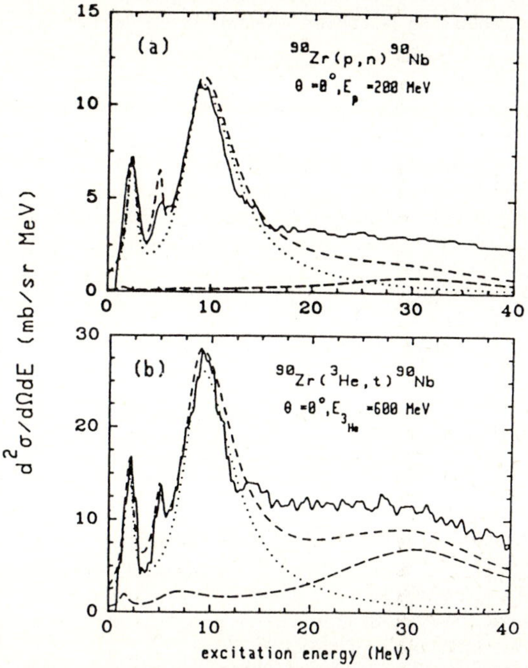

Fig. 4 Spin-flip spectra for the $^{90}Zr(\vec{p},\vec{p}')$-reaction at scattering angles of 3.5° and 5°, respectively. The solid circles are the experimental data[15] and the thick solid curve shows the result of our calculation. For comparison, also the cross section contributions to the 1^+-, 1^-- and 2^--states are given.

In fig. 4 we show the calculated spin-flip spectra for scattering angles of θ = 3.5° and 5°, respectively, along with the data of ref. 14. The theoretical spectra are incoherent sums of cross sections with multipolarities L=0 through L=3 (J^π = $0^-,1^+,1^-,2^+,2^-,3^+$). The theoretical spectra reproduce the experimental ones rather well. The largest contribution to the total spin-flip cross section comes from the 2^--states which roughly generate 40 % of the total theoretical cross section. Next to the 2^--states the 1^--, 1^+- and 0^--states produce the second largest contributions to the total cross section while the contributions of the 2^+- and 3^+- states are somewhat smaller but not negligible. We compared the 2^--strength distribution of fig. 2 with 2p2h-calculations performed by Drozdz et al.[12] and found that the shapes of both distributions are rather similar in the

5 MeV to 20 MeV excitation energy region although the 2p2h-calculations predict more strength in the high energy tail ($E_x > 20$ MeV). We mention that Esbensen and Bertsch[18] obtained a similar result to ours for the $^{90}Zr(\vec{p},\vec{p}')$-reaction using a semi-infinite nuclear slab model for the nuclear structure description.

4. Conclusions

In summary, we have shown that the measured $^{90}Zr(p,n)$-, the measured $^{90}Zr(^3He,t)$- and the measured $^{90}Zr(\vec{p},\vec{p}')$-spectra are in good agreement with the transition strength predictions as obtained from the RPA. This means that much of the strength which was supposed to be missing from the low-lying states like the GT- and M1-state is practically located in the continuous part of the low energy excitation spectrum. The result suggests that the Δ isobar quenching mechanism is likely to be rather small.

References

1. For reviews on the experimental and theoretical situation of Gamow-Teller resonances see: Proc. Int. Conf. on Spin Excitations in Nuclei, 1982, Telluride, CO, ed. by F. Petrovich, G.E. Brown, G.T. Garvey, C.D. Goodman, R.A. Lindgren, and W.G. Love (Plenum, New York 1984)
2. Gaarde, C. et al.: Nucl. Phys. A369, 258 (1981);
   Gaarde, C.: contribution to this conference
3. Ericson, M., Figureau, A., Thévenet, C.: Phys. Lett. 45B, 19 (1973);
   Rho, M.: Nucl. Phys. A231, 493 (1974);
   Ohta, K., Wakamatsu, M.: ibid A234, 445 (1974);
   Delorme, J., Ericson, M., Figureau, A., Thévenet, C.: Ann. Phys. (N.Y.) 102, 273 (1976);
   Oset, E., Rho, M.: Phys. Rev. Lett. 42, 42 (1979);
   Knüpfer, W., Dillig, M., Richter, A.: Phys. Lett. 95B, 349 (1980);
   Härting, A., Weise, W., Toki, H., Richter, A.: ibid 104B, 261 (1981);
   Toki, H., Weise, W.; ibid 97B, 12 (1980);
   Bohr, A., Mottelson, B.R.: ibid 100B, 10 (1981);
   Brown, G.E., Rho, M.: Nucl. Phys. A328, 397 (1981);
   Sagawa, H., Van Giai, N.: Phys. Lett. 118B, 167 (1982);
   Suzuki, T., Krewald, S., Speth, J.: Phys. Lett. 107B, 9 (1981);
   Osterfeld, F., Krewald, S., Speth, J., Suzuki, T.: Phys. Rev. Lett. 49, 11 (1982)
4. Towner, I.S., Khanna, F.C.: Phys. Rev. Lett. 42, 51 (1979);
   Arima, A., Hyuga, H.: Mesons in Nuclei, ed. by D. Wilkinson (North-Holland, Amsterdam 1979), p. 683;
   Shimizu, K., Ichimura, M., Arima, A.: Nucl. Phys. A226, 282 (1978);
   Bertsch, G.F., Hamamoto, I.: Phys. Rev. C26, 1323 (1982)
5. Osterfeld, F.: Phys. Rev. C26, 762 (1982)
6. Izumoto, T.: Nucl. Phys. A395, 189 (1983)
7. Osterfeld, F., Cha, D., Speth, J.: Phys. Rev. C31, 372 (1985)
8. Klein, A., Love, W.G., Auerbach, N.: Phys. Rev. C31, 710 (1983)
9. Schulte, A., Udagawa, T., Osterfeld, F., Cha, D.: to be published
10. Yabe, M., Osterfeld, F., Cha, D.: Phys. Lett. (in press)
11. Love, W.G., Franey, M.A.: Phys. Rev. C24, 1073 (1981), Phys. Rev. C32, 553 (1985)
12. Drozdz, S., Klemt, V., Speth, J., Wambach J.: Phys. Lett. 166B, 18 (1986)
13. McCarthy, J.S., Sick, I., Whitney, R.R.: Phys. Rev. C15, 1396 (1977)

14. Crawley, G.M. et al.: Phys. Rev. C26, 87 (1982)
15. Nanda, S.K. et al: Phys. Rev. Lett. 51, 1526 (1983);
Glashausser, C., Nanda, S.K.: AIP Conf Proc. (USA) No. 123, p. 1096
16. Anantaraman, N. et al.: Phys. Rev. Lett. 46, 1318 (1981);
Crawley, C.M. et al.: Phys. Lett. 127B, 322 (1983);
Crawley, C.M. et al.: Phys. Rev. C26, 87 (1982)
17. Steffen, W. et al.: Phys. Lett. 95B, 23 (1980);
see also: Richter, A.: Proc. Int. Conf. on Nucl. Phys., ed. P. Blasi and R.A. Ricci, Vol. 2 (1983)
18. Esbensen, H., Bertsch, G.F.: Phys. Rev. C32, 553 (1985)
19. Ellegaard, C. et al: Phys. Rev. Lett. 50, 1745 (1983)

Symmetry Violation and Interplay between Giant Resonances and Background in Finite Nuclei

I. Rotter

Zentralinstitut für Kernforschung, Rossendorf, DDR-8051 Dresden, GDR

In the standard nuclear structure theory, the nucleus is considered as a quantum mechanical many body problem within the space of bound single particle wavefunctions. But in reality, the nucleus is an open quantum mechanical system since most excited states are unstable against particle decay. The space of scattering states is the environment into which the discrete nuclear states are embedded.

The Rossendorf continuum shell model /1/ represents a method to describe the nucleus as an open quantum mechanical many body problem. The starting-point is a standard nuclear structure calculation with inclusion of the single particle resonances into the space of bound single particle wavefunctions up to a certain cut-off radius. The coupling to the continuum, including the tails of the single particle resonances beyond the cut-off radius, is taken into account by the same Hamiltonian $H = H_0 + V$ without additional approximations. From the point of view of selforganization, the nuclear states appear as structures in an open quantum mechanical system.

The corrections to the results of a standard nuclear structure calculation, which arise from the coupling to the continuum, are connected mainly with the correction term to the Hamiltonian,

$$H_{QQ}^{eff} = H_{QQ} + H_{QP}\, G_P^{(+)}\, H_{PQ} , \qquad (1)$$

which takes into account the coupling via the continuum. Here, the Q subspace is the function space of the standard nuclear structure calculation with the Hamiltonian $H_{QQ} \equiv QHQ$, the P subspace is the function space of scattering states with the Hamiltonian $H_{PP} \equiv PHP$ and one particle in the continuum, $G_P^{(+)}$ is the Green function for the motion of the particle in the continuum and $H_{QP} \equiv QHP$ etc. The approximation in the numerical calculations is P + Q = 1. The matrix elements of $H_{QQ}^{eff} - H_{QQ}$ are given by

$$< \phi_R | H_{QP}\, G_P^{(+)}\, H_{PQ} | \phi_{R'} > = \sum_c \int_{E_c}^{\infty} dE' \, \langle \phi_R | H | \xi_{E'}^{c(+)} \rangle \, (E^+ - E')^{-1} \langle \xi_{E'}^{c(+)} | H | \phi_{R'} \rangle \qquad (2)$$

where ϕ_R is eigenfunction of H_{QQ}, ξ_E^c is solution of the coupled channels equations

$$(E - H_{PP})\, \xi_E^{c(+)} = 0 \qquad (3)$$

while c denotes a channel with threshold energy E_c. The matrix elements (2) do not vanish also at energies where all channels are closed, i.e. for bound states. The eigenvalues of H_{QQ}^{eff} are real in this case, otherwise complex. The real part determines the energy, the imaginary part the width of the state. The matrix elements (2) determine the external mixing of the resonance states R and R' via the continuum which increases with increasing degree of overlapping of the states.

In the following, some corrections to the results of standard nuclear structure calculations will be outlined.

(i) The environment of the open quantum mechanical nuclear system (continuum) is different for neutrons and protons due to the Coulomb interaction. It causes a small charge dependence of the nuclear forces /2/ which exists only in finite nuclei and has nothing to do with the problem of charge symmetry of the nuclear forces in nuclear matter. The differences have, according to eq. (2), regularities with general many-body properties of the nuclei as, e.g., binding energy and shell closure. Their general behaviour in dependence on A is in accordance with the trends observed in the Coulomb energy anomaly.

(ii) A further consequence of the replacement of H_{QQ} by H_{QQ}^{eff} in an open quantum mechanical system is the appearance of another symmetry violation in the nuclear forces for finite nuclei which does also not exist in nuclear matter. The nucleus is, indeed, an open quantum mechanical system in relation to the nucleon degrees of freedom. But the threshold for the emission of Δ isobars is so high in energy that the nucleus may be considered as a closed system in relation to these particles. As a consequence, the NN, $\Delta\Delta$ and ΔN correlations in finite nuclei are different /3/ although they may be the same in nuclear matter. The differences correspond to the differences in the effective nuclear forces used, e.g., in the standard shell model and in the continuum shell model calculations.

In the usual ansatz for the spin-isospin dependent particle-hole interaction, all corrections other than those considered explicitely are summarized in the parameter g' /4/. This parameter is obtained from a fit to the experimental data and involves all the effects in finite nuclei which are connected with the coupling to the environment. This parameter should be assumed therefore to be smaller for isobars than for nucleons in finite nuclei although the universality of the spin-isospin correlations, suggested by the naive quark model, holds surely in nuclear matter /5/. The differences between the parameter g' for nucleons and Δ isobars are expected to be larger, but to show the same regularities with A as the deviations from charge symmetry, due to (2). They have similar properties for a wide range of nuclei with only smooth dependence on mass number and angular momentum.

(iii) The external mixing (2) cannot be neglected in the interplay between a giant resonance and its background of the same spin and parity. Numerical calculations /6/ with realistic wavefunctions for ^{16}O have shown that deviations from the statistical behaviour occur at the top of resonances due to the unitarity of the S-matrix the constraint by which correlates the underlying levels by means of the external mixing (2) in such a way that the maximal possible value of the cross section will not be passed over. The constraint by the unitarity of the S-matrix is strongest at that energy and in that channel where the contribution of the giant resonance to the cross section is maximal. Here, the contribution which the underlying levels as a whole can give to the cross section is reduced in comparison with their contribution at other energies and in other channels. This fact caused by the unitarity of the S-matrix should be taken into account in the analysis of the experimental data by a selective transparency.

By means of these results obtained numerically it is shown that some part of the missing spectroscopic strength of resonances, discussed in the experimental data, is in reality not "missing" but connected with disregarding the nonstatistical properties of the underlying levels near the top of resonances in the standard method of analysing the data.

It can be shown that the standard nuclear reaction theory resting on a statistical distribution of the spectroscopic factors gives, nevertheless, good results for the average cross section at high excitation energy. But it fails in the calculation of decay widths due to the cooperative interaction (2) /7/.

(iv) The Rossendorf continuum shell model /1/ allows to describe long-living states and short-living ones in a unified manner and can therefore be used in order to investigate the interplay between the short-living collective states and the long-living states of single-particle nature. The numerical results obtained show that the collective features of the nuclei may be identified with cooperative effects in selforganizing systems. The paradigm of synergetics /8/ seems to be valid also for nuclear structure studies although the mathematical formalism used is completely different. Moreover, the concept of a value parameter /9/ is included automatically into the nuclear structure studies /10/.

A proof of this statement consists in investigating processes which, although very improbable from a pure statistical point of view, may appear due to their large "selective value". A process of such a type is the emission of a high-energy particle in a nuclear reaction by leaving all the remaining nucleons in their energetical lowest state. The concentration of all the energy on one particle is the less probable the higher the available energy is and should be suppressed from a pure statistical point of view at high energies.

Another proof bases on the finite lifetime of the compound nucleus resonance states. All the memory of the entrance channel is lost in the limit of a very long lifetime of the resonance states as it has been assumed in the original definition of the compound nucleus. The opposite is true in the limit of a very small lifetime. The reaction proceeds in this case almost without formation of a compound nucleus. The existence of a finite lifetime of the compound nucleus, seen by means of entrance channel effects, points therefore to the fact that in the nuclear reaction considered (dissipative) structures appear by selforganization.

Other proofs consist in lifetime measurements of the compound nucleus states /10/ as well as in an identification of amplitude correlations in nuclear reaction cross sections /11/.

References

1. I. Rotter, Fiz. Elem. Chastits At. Yadra 15 (1984) 762 (translation: Sov. J. Part. Nucl. 15 (1984) 341 and ZfK-508 (Rossendorf 1983))

2. I. Rotter, J. Phys. G6 (1980) 185

3. I. Rotter, Annalen der Physik (Leipzig), Jubiläumsband G. Richter, 1986

4. A.B. Migdal, Rev. Mod. Phys. 50 (1978) 107

5. W. Weise, Nucl. Phys. A374 (1982) 505c

6. P. Kleinwächter and I. Rotter, J. Phys. G12 (1986) in press

7. P, Kleinwächter and I. Rotter, Phys. Rev. C32 (1985) 1742

8. H. Haken, Synergetics, Springer-Verlag 1978, and Advanced Synergetics, Springer-Verlag 1983

9. M. Eigen, Naturwissenschaften 58 (1971) 465

10. I. Rotter, J. Phys. G12 (1986) in press

11. I. Rotter, J. Phys. G11 (1985) L219

Properties of a New Magnetic Dipole Mode Discovered in Low Energy Electron Scattering*

D. Bohle, Th. Guhr, U. Hartmann, K.-D. Hummel, G. Kilgus, U. Milkau, and A. Richter

Institut für Kernphysik, Technische Hochschule Darmstadt,
D-6100 Darmstadt, F. R. Germany

In a large range of nuclei low lying $J^{\pi}=1^{+}$ states have been found that are excited predominantly by a new M1 mode. Four properties of the new mode will be discussed in detail. Firstly, from the excitation energy systematics observed the strength of the Majorana force of the interacting boson model (IBA) is deduced. Secondly, through the comparison of electron scattering and proton scattering experiments it is shown that the new mode is largely due to the orbital motion of protons with respect to neutrons. Thirdly, taking the nucleus ^{164}Dy as an example, g-factors and effective boson charges of the M1-, E2- and M3 IBA transition operators, respectively, are studied. The F-scalar magnetic octupol g-factor Ω_S is derived for the first time. Finally, the distribution of M1 strength in ^{156}Gd will be discussed in the light of recent theoretical calculations.

1. Introduction

Experimental information on low lying $J^{\pi}=1^{+}$ states from high resolution inelastic electron scattering is now available in three regions of the periodic table, i.e. the deformed rare earth nuclei [1,2], the $f_{7/2}$-shell nuclei [3] ^{46,48}Ti and the actinides [4]. These low lying states form the magnetic analogon of the E1 giant resonance which in nuclei with mass number A lies at an excitation energy of

$$E_x \approx 77A^{-1/3}\ \text{MeV}. \tag{1}$$

Both, macroscopic [5,6] and microscopic pictures (see ref. [7] and references therein) of the new M1 mode yield an equally simple formula

$$E_x \approx 66\delta A^{-1/3}\ \text{MeV} \tag{2}$$

where now the mass deformation δ enters into this scaling law. Therefore, in contrast to the E1 giant resonance, the states excited by the new M1 mode are bound states. With the exception of ^{46}Ti the states observed in inelastic electron scattering follow the excitation energy given by eq.(2) rather closely. In terms of the interacting boson model the excitation energy is determined by the strength of the Majorana force. We will show here that this strength is also related to the nuclear mass deformation in a simple way.

The transition strength connected with the new mode is typically $B(M1)\uparrow=1\text{-}3\mu_N^2$. Since the orbital single particle estimate [8] yields

$$B(M1)\uparrow \approx 0.1\ \mu_N^2 \tag{3}$$

the observed M1 transitions are highly collective. That indeed the convection current part of the M1 operator dominates over the spin part will be shown for light and heavy nuclei through the comparison of inelastic electron and proton scattering experiments. Furthermore, electron scattering form factors have been measured up to high momentum transfer. These form factors are in good agreement with all predic-

* Work supported by the Deutsche Forschungsgemeinschaft

tions that assume an orbital excitation mode and no deviations have been observed so far.

It has been pointed out [9] that the low lying $J^{\pi}=1^{+}$ states which are of mixed symmetry in the proton-neutron degree of freedom might be the first ones of an entire new class of mixed symmetric states occuring at that excitation energy and having collective properties. Taking ^{164}Dy as an example, the effective boson charges of the E2 transition operator are derived and a 2^{+} candidate for the member of the $K^{\pi}=1^{+}$ band has been found. Furthermore, the magnetic octupol g-factor Ω_s which is symmetric in the proton-neutron degree of freedom is determined for the first time. Based on theoretical arguments [10] predictions for the transition strength to mixed symmetric 3^{+} states are given.

Finally, the M1 strength distribution of ^{156}Gd will be discussed in the light of recent theoretical calculations [11,12].

2. Systematics of Excitation Energy and the Strength of the Majorana Force

Two examples for M1 transitions excited by the new mode are shown in figs.1 and 2 where inelastic electron scattering spectra on the light nuclei ^{46,48}Ti and the heavy deformed nucleus ^{164}Dy are displayed. While only one strong state is observed in ^{46}Ti two states are excited in ^{48}Ti, the second one roughly 1.8 MeV higher in excitation energy carrying about half as much strength as the lower one. This is just what has to be expected [13] if ^{48}Ti is an example of a triaxially deformed nucleus. In ^{164}Dy the M1 strength rests essentially in three equally strong levels [14]; the higher lying dublet is not resolved in the spectra shown in fig.2. The data taken on ^{164}Dy at high momentum transfer in Amsterdam [15] displayed in the lower half of fig.2 show - due to the high level density - many transitions of high multipolarity which lead to many states of higher spin in the excitation energy region of interest.

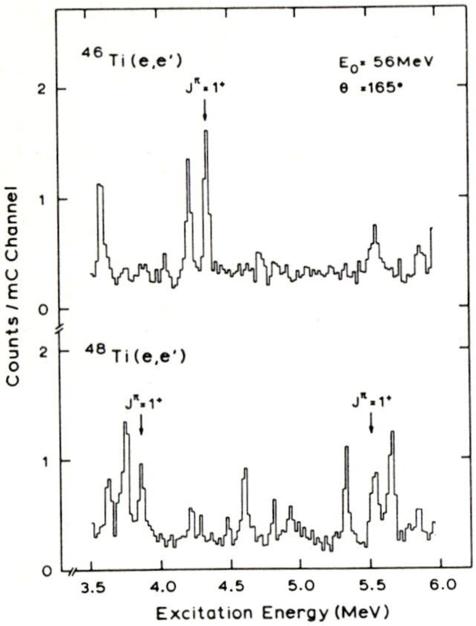

Fig.1 Inelastic electron scattering spectra on ^{46,48}Ti. In the triaxial nucleus ^{48}Ti two 1^{+} states are excited by the new M1 mode. They are separated by an energy of roughly 1.8 MeV.

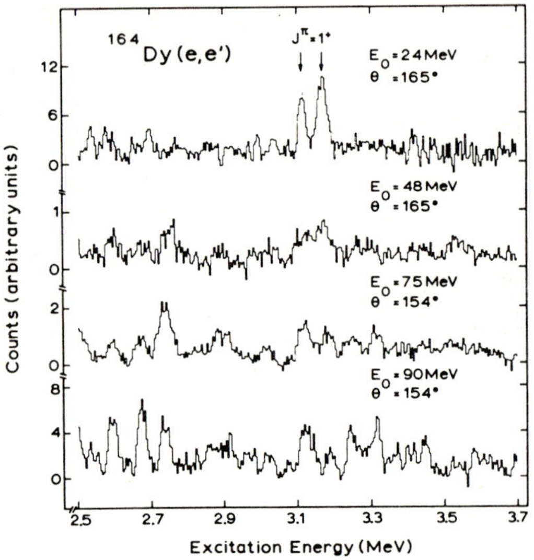

Fig.2 Inelastic electron scattering spectra on ^{164}Dy. In the 24 MeV spectrum two peaks corresponding to the excitation of three $J^{\pi}=1^{+}$ states are seen at $E_x\approx3.1$ MeV. These states are only weakly excited at 48 MeV. In the 75 and 90 MeV spectra the prominent peaks close to the excitation energy of the $J^{\pi}=1^{+}$ states correspond to transitions of higher multipolarity.

The experimental information obtained from our inelastic electron scattering work is used to describe the excitation energies of the $J^{\pi}=1^{+}$ states. The excitation energies and transition strengths observed are summerized in Table I. We use the interacting boson model to discuss these quantities without prejudice towards the other existing models, but the IBA-2 is well suited to study the general trend of excitation energy and transition strength.

Table I Excitation energies, transition strengths of the strongest $J^{\pi}=1^{+}$ states, parameters used in the IBA-2 description of the energy spectra, and mass deformations of the nuclei studied

Nucleus	E_x (MeV)	$B(M1)\uparrow$ (μ_N^2)	ε_d (MeV)	$\varkappa$ (keV)	χ	λ (MeV)	δ
^{46}Ti	4.3	1.0±0.2	1.16	-43	-0.63	0.64	0.16
^{110}Pd	(3.9)	$\lesssim$ 0.5	0.68	-26	-0.26	0.28	0.22
^{154}Sm	3.2	0.8±0.2	0.41	-20	-1.36	0.21	0.26
^{154}Gd	2.9	0.9±0.2	0.46	-17	-0.88	0.19	0.24
^{156}Gd	3.1	1.3±0.2	0.40	-20	-0.92	0.18	0.25
^{158}Gd	3.2	1.4±0.3	0.38	-20	-0.88	0.18	0.26
^{164}Dy	3.1 3.2	1.3±0.2 2.3±0.4	0.24	-22	-0.29	0.13	0.26
^{168}Er	3.4	0.9±0.2	0.34	-24	-0.32	0.15	0.27
^{232}Th	2.0	1.3±0.2	0.22	-14	-0.91	0.12	0.20
^{238}U	2.3	2.7±0.6	0.26	-12	-1.18	0.11	0.24

In order to determine the strength of the Majorana force we proceed as follows[16]. The IBM-Hamiltonian

$$H = \varepsilon_d n_d + \varkappa(Q_\pi+Q_\nu)\cdot(Q_\pi+Q_\nu) + \lambda M \tag{4}$$

is used to describe the spectra of the nuclei studied. The first term of the Hamiltonian accounts for the pairing-, the second one for the quadrupole-quadrupole- and the third one for the Majorana interaction. The operators Q and M are given in second quantized form using the boson creation and annihalation operators for bosons of angular momentum zero and two (s and d bosons) as

$$Q_\rho = (s_\rho^+\tilde{d}_\rho + d_\rho^+ s_\rho)^{(2)} + \chi_\rho(d_\rho^+\tilde{d}_\rho)^{(2)} , \tag{5}$$

$$M = (s_\pi^+ d_\nu^+ + d_\pi^+ s_\nu^+)^{(2)}\cdot(s_\nu\tilde{d}_\pi + \tilde{d}_\nu s_\pi)^{(2)} - 2\sum_{k=1,3}(\tilde{d}_\pi\tilde{d}_\nu)\cdot(d_\pi^+ d_\nu^+)^{(k)} \tag{6}$$

with $\rho=(\pi,\nu)$ denoting proton and neutron bosons, respectively.

If the structure constants of the quadrupol operator are equal, i.e. $\chi \equiv \chi_\pi = \chi_\nu$, the Hamiltonian is symmetric under the interchange of proton and neutron variables. This symmetry is related to the boson quantum number F. Bosons are assumed to have F-spin F = 1/2 with projection F_z = 1/2 for protons and F_z = -1/2 for neutron bosons. With the help of this new quantum number the boson states can be labeled according to their symmetry in the proton neutron degree of freedom. The low lying symmetric states have $F_{max} = (N_\pi+N_\nu)/2$ while the mixed symmetric $J^{\pi}=1^{+}$ states have $F = F_{max}-1$. The Majorana operator (6) used in the Hamiltonian (4) also reduces to a much simpler form, namely

$$M = F_{max}(F_{max} + 1) - F(F+1). \tag{7}$$

Clearly, the eigenvalues of the Majorana operator determine the energy splitting between symmetric and mixed symmetric states. It has to be pointed out, however, that this most simple choice of an IBA-2 Hamiltonian may have to be modified [17] when experimental information on mixed symmetric bands, other than $K^{\pi} = 1^{+}$ bands studied here, becomes available. In Table I the experimental excitation energies and transition strengths as well as the parameters of the Hamiltonian are given. The parameters of the pairing and quadrupole part have been fixed by the well known low energy spectra and the strength of the Majorana force has been determined from the excitation energy of the $J^{\pi} = 1^{+}$ state.

Concerning the parameters of the Hamiltonian listed in Table I several remarks are in order. Firstly, although the IBM is not so well suited for the description of nuclei where protons and neutrons occupy the same shell, nevertheless a calculation has been performed for $^{46}Ti(N_{\pi}=1,\ N_{\nu}=2)$. We have taken, however, in the calculation only those levels into account which were shown [18] to have shell model wave functions with a large overlap with IBM-2 wave functions. Secondly, with the further exception of ^{110}Pd (where at present only an indication for a 1^{+} state exists at $E_x \approx 3.9$ MeV) all nuclei are good rotors. This is reflected in the smoothly varying behaviour of the parameters with proton and neutron numbers. Furthermore, the ratio ε_d/N_{κ}, where $N = N_{\pi}+N_{\nu}$, is small and χ is close to the rotational value of -1.3. The deviations from this value observed in ^{164}Dy and ^{168}Er tell us that these nuclei - though good rotors - are transitional nuclei that belong to the transition region between good rotors and γ-unstable nuclei for which $\chi=0$ holds.

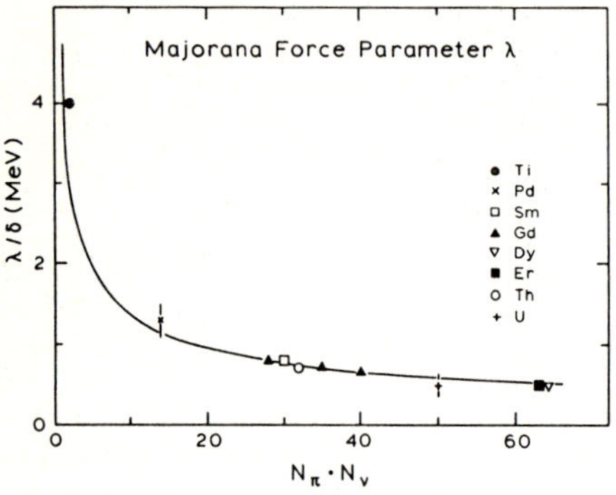

Fig.3 Majorana force paramter λ normalized to the mass deformation δ as a function of $N_{\pi} \cdot N_{\nu}$. Compared to the experimental data is the function eq.(8).

In fig.3 the result obtained for the strength of the Majorana force is displayed. The strength parameter λ has been normalized to the mass deformation δ since nuclei which have the same mass number but different deformations have the same value of λ/δ. The ratio λ/δ is plotted against the product of proton and neutron boson numbers as proposed by CASTEN [19]. This kind of plot yields a simple curve whenever the quantity under study is related to the proton neutron interaction which also generates the nuclear deformation. It had already been proposed by GREINER [20] a long time ago that proton and neutron deformations are responsible for the magnetic properties of nuclei.

The data shown in fig.3 are compared with the function

$$\lambda/\delta = 4.3 \cdot (N_{\pi} \cdot N_{\nu})^{-1/2} \quad . \tag{8}$$

The good description using formula eq.(8) hints at the possibility to use the new M1 mode to study nuclear deformations. Furthermore, using this formula, predictions of excitation energies of mixed symmetric states are now possible in the framework of the IBM.

3. The Orbital M1 Mode in Light and Heavy Nuclei

In order to prove that the new M1 mode indeed corresponds to the orbital motion of protons about neutrons as the classical picture of LO IUDICE and PALUMBO [5] suggests proton scattering experiments at E_0=200 MeV and small scattering angles have been performed in Orsay [3]. Under these kinematical conditions it is the spin part that dominates the effective nucleon-nucleon interaction. On the other hand, the electromagnetic transition operator contains an orbital and a spin term, so that

$$B(M1)\uparrow = (\pm\sqrt{B(\ell)} + \sqrt{B(\sigma)})^2 \quad . \tag{9}$$

Combining the results of electron and proton scattering experiments yields thus - within a certain model dependence on the side of the (p,p') experiments - a separation of the spin and orbital contributions B(σ), B(ℓ) to the transition strength B(M1)↑. One has to keep in mind, however, that in general these quantities may interfere both constructively and destructively. Since we are interested in a lower limit of the orbital strength we choose the positive sign in eq.(9). This method has been tested using the 10.23 MeV neutron spin flip transition in ^{48}Ca [3].

Figure 4 displays the low excitation energy part of the measured spectra where the $J^\pi=1^+$ state at E_x= 4.319 MeV is seen clearly in both experiments. The excitation energy around the known strong $J^\pi=1^+$ state at $E_x\approx$10.18 MeV is shown in fig.5 for comparison. This state, in analogy to the 10.23 MeV level in ^{48}Ca, should mainly be excited through the spin part of the transition operator though the (e,e') experiment reveals also a strong orbital contribution to this transition, nevertheless. The transition strength of the 4.319 MeV state was measured [3] to be B(M1)↑ = 1.0±0.2 μ_N^2 while the spin strength is B(σ) = 0.13±0.03 μ_N^2. From this values it follows that the ratio B(ℓ)/B(σ) is as large as 3. In a $f_{7/2}$-shell model ZAMICK who has been first to point out [21] that low lying $J^\pi=1^+$ states are a general property of nuclei calculated [22] that this ratio should be about 1. The measured ratio is somewhat larger but disagrees seriously with a recent calculation of CIVITARESE

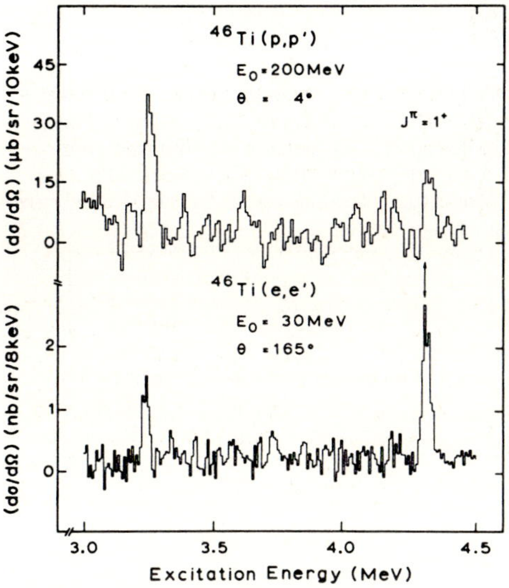

Fig.4 Comparison between (p,p') and (e,e') spectra on ^{46}Ti in the excitation energy region between E_x = 3.0-4.5 MeV. The $J^\pi=1^+$ state is largely excited by the orbital part of the M1 operator.

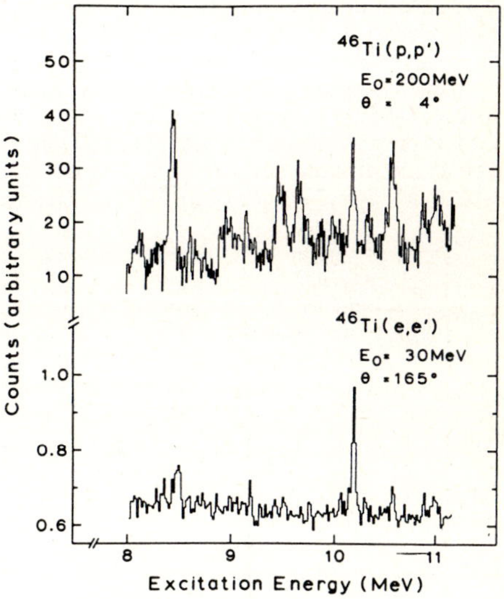

Fig.5 Comparison between (p,p')and (e,e') spectra in the excitation energy region between E_x=8.0-11.0 MeV. The $J^\pi=1^+$ state at 10.18 MeV should be excited mainly by the spin part of the M1 operator.

et al. [23] who predict a ratio of 0.18 which corresponds to the statement that spin magnetism would by far be dominating over orbital magnetism. This is clearly ruled out by experiment.

Since only upper limits [3] on the spin strength could be given for rare earth nuclei another proton scattering experiment on ^{156}Gd was undertaken [24]. This time the incident proton energy was chosen to be E_0= 25 MeV where an excellent energy resolution of ΔE=8 keV could be obtained. Again the new M1 mode was not excited. The upper limit is displayed together with two predictions for the angular distribution in fig.6. The fully drawn line has been obtained using the same wave function for the nuclear structure description that has been used to calculate the (e,e') form factor. The ratio $B(\ell)/B(\sigma)$ obtained with this wave function is 90. The dashed line results when a purely orbital excitation mode is assumed. The upper limit of the present experiment yields 45, i.e. a factor of 2 lower than the theoretical prediction. This result for $B(\ell)/B(\sigma)$ demonstrates, that the new M1 mode in deformed rare earth nuclei is of orbital character.

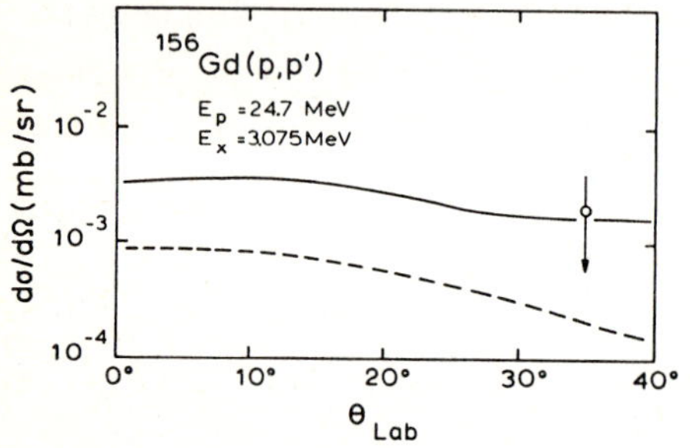

Fig. 6 DWBA calculation of the cross section for inelastic proton scattering at E_0=25 MeV. The dashed curve simulates a purely orbital mode, while the full one assumes that $B(\ell)/B(\sigma)$=90.

One further argument that the new M1 mode is a dominantly orbital mode is based on the electron scattering form factor, which is very sensitive to the nuclear structure description. The electron scattering experiments on ^{48}Ti and ^{164}Dy have therefore been extended in Amsterdam [15] to cover also the second maximum of the form factor. In fig.7 the transverse form factors of the $J^\pi=1^+$ states excited in ^{46}Ti and ^{48}Ti are displayed. Only the nucleus ^{48}Ti has been studied at higher momentum transfer. Compared to the experimental data is the prediction calculated using ZAMICK's wave functions [21]. No deviation that would indicate a large spin contribution to the cross section has been observed in agreement with what was found in proton scattering. Figure 8 shows the transverse form factor of the $J^\pi=1^+$ state at E_x= 3.11 MeV in ^{164}Dy. Compared to the experimental data is the IBM-2 pre-

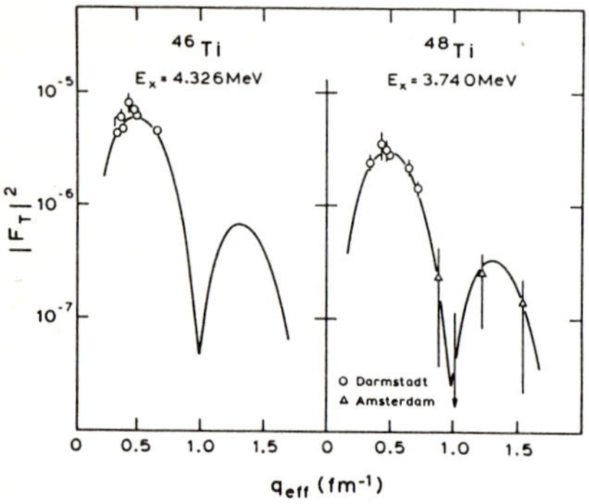

Fig.7 Transvers form factors of the transitions to the $J^\pi=1^+$ states in ^{46,48}Ti. The full drawn line is the form factor calculated using the wave functions of ZAMICK.

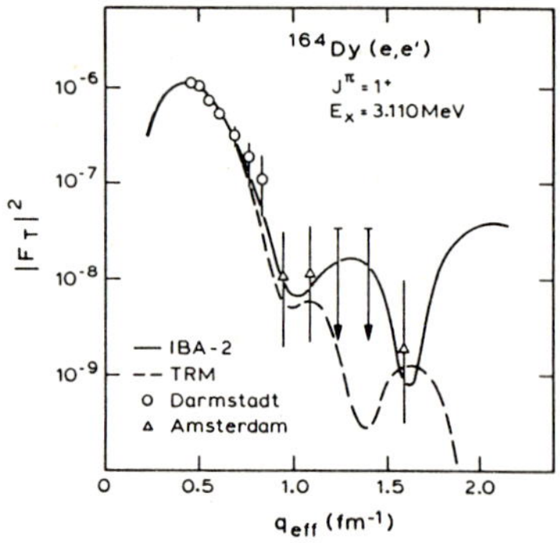

Fig.8 Transverse form factors of the transition to the $J^\pi=1^+$ state in ^{164}Dy at E_x=3.11 MeV. Compared to the data are the IBM and two rotor model predictions.

diction (full line) where $B(\ell)/B(\sigma) = 90$ is assumed. The dashed line gives the two rotor model [25] prediction that assumes that the new M1 mode is purely orbital. The experiment is in agreement with both predictions.

4. Boson g-Factors and Effective Charges

It has been pointed out by IACHELLO [9] that the mixed symmetric $J^\pi=1^+$ states may be just the first ones discovered of a new class of collective states. Taking ^{164}Dy as an example we will summarize what is known from inelastic electron scattering on mixed symmetric 1^+, 2^+ and 3^+ states, the magnetic dipole and octupole g-factors and the E2 effective boson charges of the IBM transition operators.

Because the excitation of the symmetric counterpart of the mixed symmetric $J^\pi=1^+$ state corresponds to the rotation of the nucleus as a whole it cannot be excited. This fact reflects itself in the microscopic [26] values of the magnetic dipole g-factors $g_\pi=1$ and $g_\nu=0$.

States of spin higher than 1 may be symmetric or mixed symmetric. It is a well known fact that the first excited symmetric $J^\pi=2^+_S$ state is very collective. And since close to the rotational limit [27]

$$B(E2,0^+\to 2^+_S) \simeq \frac{2N+3}{N} \quad (N_\pi e_\pi + N_\nu e_\nu)^2 \tag{10}$$

- where $e_\pi(e_\nu)$ is the proton (neutron) boson charge - it is immediately clear that the F-scalar boson charge [28] $e_S = 1/2(e_\pi+e_\nu)$ is large while the F-vector one has necessarily to be small. An adjustment of $(NB(E2;0^+\to 2^+_S)/2N+3)^{1/2}$ to the known strengths of the first excited $J^\pi=2^+$ states in the Dy isotopic chain yields

$$e_\pi \simeq 12.8\ efm^2 \text{ and } e_\nu \simeq 8.6\ efm^2 \tag{11}$$

in excellent agreement with our results [29] for ^{156}Gd. Using again the rotational limit estimate [27]

$$B(E2,0^+\to 2^+_M) \simeq (e_\pi - e_\nu)^2 \frac{3(N-1)}{N(2N-1)} N_\pi N_\nu \tag{12}$$

we get $B(E2)\uparrow = 102\ e^2fm^4$. The single particle estimate yields $270\ e^2fm^4$. Therefore due to the renormalisation using effective charges the E2 transition to the mixed symmetric $J^\pi=2^+$ state is no longer collective.

In fig.9 electron scattering form factors [15] of the transitions to the first excited 2^+ state, the 2^+ member of the γ-rotational band and a candidate for the mixed symmetric 2^+_M state are displayed. The tremendous differences in transition strength are clearly visible; for the transition to the mixed symmetric state we have $B(E2)\uparrow = 60\pm 15\ e^2fm^4$. There may be a small contribution to this form factor from a transition to a 3^- state which has not been resolved in the present experiment. Therefore, an extraction of transition densities has not jet been undertaken and the curves shown in fig.9 are to guide the eye only.

Looking thirdly at the excitation of 3^+ states the experimental as well as theoretical picture is much less clear than in the two previous cases. As in the 2^+ case F-scalar and F-vector excitations are possible.

The respective magnetic octupol g-factors (called Ω_S and Ω_V) are

$$\Omega_S = 1/2\ (\Omega_\pi+\Omega_\nu) \text{ and } \Omega_V = 1/2\ (\Omega_\pi-\Omega_\nu)\ . \tag{13}$$

Since very little is known on the relative sign of Ω_S vs. Ω_V it is difficult to predict whether the symmetric or mixed symmetric 3^+ states will be collectively excited.

While from generalized seniority scheme calculations [10,30] is argued that Ω_π and Ω_ν should have opposite signs thus making the mixed symmetric 3^+ state the

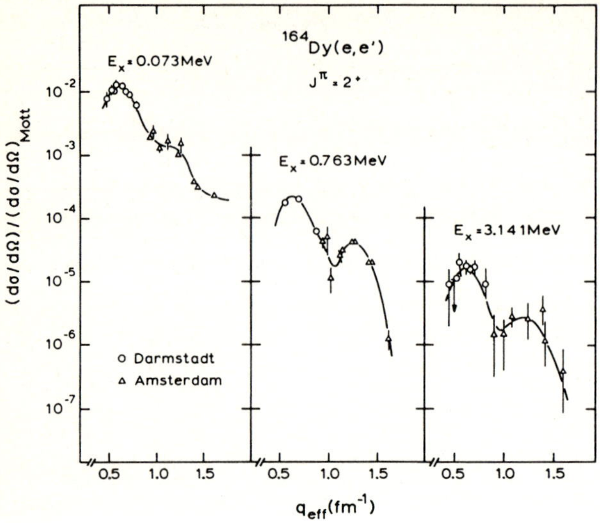

Fig.9 Electron scattering form factors to the $J^{\pi}=2^+_S$, 2^+_γ and 2^+_M levels observed in ^{164}Dy. A small contribution to the form factor of the 2^+_M state from an unresolved E3 transition cannot be ruled out as yet. The curves are to guide the eye only.

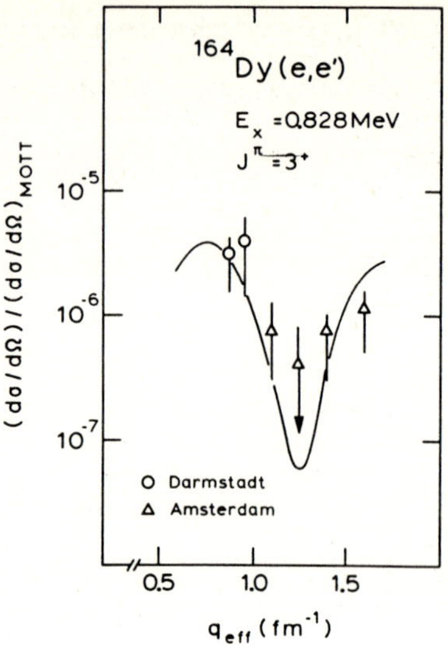

Fig.10 Inelastic electron scattering form factor of the transition to the first excited 3^+ state in ^{164}Dy. The fully drawn line is a IBA-2 prediction.

collective one also the other choice cannot be ruled out [31]. Furthermore, it has been pointed out that the usual choice [10]of the M3 transition operator may not be sufficient and higher order corrections may have to be taken into account [32]. Moreover microscopic calculations [10,30,33] suggest that the values of Ω_π and Ω_ν are highly nuclear structure dependent and that the spin contribution does not necessarily vanish. Therefore the M3 mode is not expected to be a dominantly orbital mode anymore.

Our search for M3 excitations in ^{164}Dy was spirited by the assumption that this nucleus having very collective M1 transitions might also exhibit strong M3 transitions. In fig.10 the form factor to the first excited 3^+ state belonging to the γ-band in ^{164}Dy is shown. It is compared to a DWBA calculation using a transition density which has been calculated by SCHOLTEN [10,34]. The M3 strength is found to be $B(M3)\uparrow = (0.3^{+0.1}_{-0.2})\ \mu_N^2 b^2$,

Using the rotational limit prediction [27]

$$B(M3,0^+\rightarrow 3^+_S) = \frac{35}{8\pi}\,(\Omega_\nu N_\nu+\Omega_\pi N_\pi)^2\ \frac{8(N-2)(N-1)}{3N(2N-3)(2N-1)} \qquad (14)$$

we get $\Omega_S = 0.08\pm0.05\ \mu_N b$ which is in excellent agreement with SCHOLTEN's prediction [10]$\Omega_S = 0.085\ \mu_N b$. This striking agreement tempted us to use his microscopic g-factors to calculate the M3 strength into the mixed symmetric states. With the formula given in ref.27 we find $B(M3;0^+\rightarrow 3^+_M,\ K=1) = 0.07\ \mu_N^2 b^2$ which would be too small to be detected in the present experiment.

The predicted strength into the mixed symmetric 3^+ state belonging to the $K^{\pi}=2^+$ band , however, is of the same order as the strength to the 3^+_S state,namely $B(M3)\uparrow = 0.24\ \mu_N^2 b^2$. So far we did not detect any M3 strength in this region.

5. M1 Strength Distribution in ^{156}Gd

The fragmentation of M1 strength in ^{156}Gd has been discussed in detail [2,7,35] elsewhere. Since, however, the detailed orbital strength distribution has been measured [35] making use of the comparison between inelastic electron scattering and nuclear resonance fluorescence it is of interest to discuss it again in the light of two recent quasiparticle RPA (QRPA) calculations [11,12]. In fig.11 the experimental M1 strength distribution is compared to the two QRPA calculations. The experimental M1 strength is displayed in the upper part of the figure.

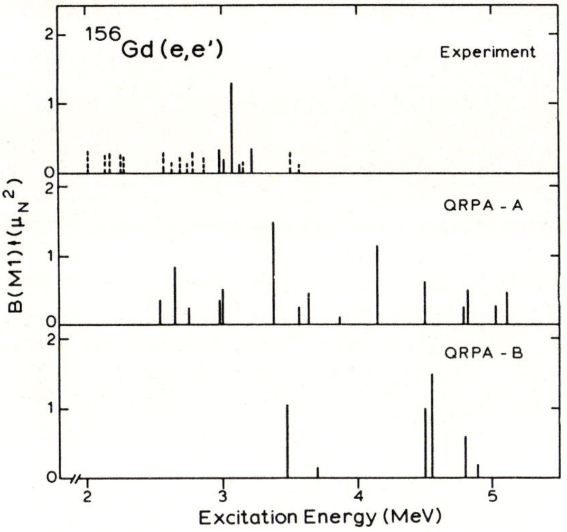

Fig.11 Experimental M1 strength distribution in ^{156}Gd (upper part) compared with two theoretical QRPA predictions [11,12].

The comparison of (e,e') and (γ,γ') experiments [35] has shown that only the full drawn lines correspond to $J^{\pi}=1^{+}$ states that are excited by the rotational M1 mode. The IBA-2 form factor has been used , however, to extrapolate to the photon point in every case.

The orbital M1 strength in ^{156}Gd rests in five levels closely centered around the strong collective state at E_x= 3.07 MeV. The estimate of the mixing matrix element derived by the assumption that the mixed symmetric $J^{\pi}=1^{+}$ state couples strongly to the background of symmetric g-boson $J^{\pi}=1^{+}$ levels [30] has to be lowered so that it becomes about 50 keV.

In the middle part of fig.11 the strength distribution calculated by HILTON et al.[11] is shown. All strong states displayed are mainly due to an orbital excitation. Spin-flip contributions to the M1 strength are always less than 0.4 μ_N^2 . We believe that the self consistent microscopic description of the dominance of orbital M1 strength at low excitation energies is a breakthrough. The theoretical distribution of orbital M1 strength has, however, a width that is much too large. The total orbital M1 strength predicted of B(M1)↑ = 5.8 μ_N^2 is much larger than the experimental value of B(M1)↑ = 2.3±0.5 μ_N^2 .

While in the approach described above a Skyrme interaction was used in a deformed oscillator basis the QRPA calculation of CIVITARESE et al. [12] shown in the lower part of fig.11 employs a quadrupol-quadrupol interaction and an axially symmetric Woods-Saxon potential.

The structure of the M1 mode in this approach is such that the wave functions of the isovector $J^{\pi}=1^{+}$ states are mainly built on one or two quasiproton pair configurations. Though the overall description of the M1 strength of the lowest $J^{\pi}=1^{+}$ state with a strength above 1 μ_N^2 in a number of nuclei besides ^{156}Gd is surprisingly good the strong states predicted at higher excitation energies have not yet been observed in the surveys we undertook so far in the nuclei ^{156}Gd, ^{174}Yb and ^{238}U.

Furthermore, in this approach it is predicted, that orbital and spin contributions to the transition matrix elements are comparable while experimentally the clear dominance of the orbital over the spin part has been established[3,24]

Acknowledgments

We would like to thank P. von Brentano, A.E.L. Dieperink, K. Heyde, F. Iachello, P. van Isacker, C.W. de Jager, F. Palumbo and O. Scholten for helpful discussions.

[1] D. Bohle, A. Richter, W. Steffen, A.E.L. Dieperink, N. Lo Iudice, F. Palumbo and O. Scholten: Phys. Lett. 137B, 27 (1984)
[2] D. Bohle, G. Küchler, A. Richter and W. Steffen: Phys. Lett. 148B, 260 (1984)
[3] C. Djalali, N. Marty, M. Morlet, A. Willis, J.C. Jourdain, D. Bohle, U. Hartmann, G. Küchler, A. Richter, G. Caskey, G.M. Crawley and A. Galonsky Phys. Lett. 164B, 269 (1985)
[4] R.D. Heil, U.E.P. Berg, A. Jung, U. Kneissl, H.H. Pitz, U. Seemann, R. Stock, F.J. Urban, B. Fischer, H. Hollick, D. Kollewe, K.-D. Hummel, G. Kilgus, D. Bohle, Th. Guhr, U. Hartmann, U. Milkau and A. Richter: to be published
[5] N. Lo Iudice and F. Palumbo: Phys. Rev. Lett. 41, 1532 (1978)
[6] E. Lipparini and S. Stringari: Phys. Lett. 130B, 139 (1983)
[7] A. Richter: Nuclear Structure 1985; R. Broglia, G.B. Hagemann and B.Herskind, (ed.). Elsevier, Amsterdam 1985
[8] T. Otsuka and J.N. Ginocchio: Phys. Rev. Lett. 54, 777 (1984)
[9] F. Iachello: Phys. Rev. Lett. 53, 1427 (1984)
[10] O. Scholten, A.E.L. Dieperink, K. Heyde and P. van Isacker: Phys. Lett. 149B, 279 (1984)
[11] R.R. Hilton, S. Iwasaki, H.J. Mang, P. Ring and M. Faber: Contribution to the Conference on Phase-Space Dynamics, Triest, Sept. 1985
[12] O. Civitarese, A. Faessler and R. Nojarov: preprint Tübingen
[13] N. Lo Iudice, E. Lipparini, S. Stringari, F. Palumbo and A. Richter: Phys.Lett. 161B, 18 (1985)
[14] C. Wesselborg: private communication
[15] D. Bohle, K. Alrutz-Ziemssen, K.-D. Hummel, G. Kilgus, A. Richter, C.W. de Jager and H. de Vries: to be published
[16] U. Hartmann, D. Bohle, Th. Guhr, K.-D. Hummel, G. Kilgus, U. Milkau and A. Richter: to be published
[17] A. Novoselski and I. Talmi: Phys. Lett. 160B, 13 (1985)
[18] J.A. Evans, J.P. Elliott and S. Szpikowski: Nucl. Phys. A435, 317 (1985)
[19] R.F. Casten: Phys. Lett. 152B, 145 (1985)
[20] W. Greiner: Nucl. Phys. 80 ,417 (1966)
[21] L. Zamick: Phys. Rev. C31, 1955 (1985); Phys. Rev. C33,691 (1986)
[22] L. Zamick: Phys. Lett. 167B, 1 (1986)
[23] O. Civitarese, A. Faessler and R. Nojarov: preprint Tübingen
[24] C. Wesselborg, K. Schiffer, K.O. Zell, P. von Brentano, D. Bohle, A. Richter, G.P.A. Berg, B. Brinkmöller, J.G.M. Römer, F. Osterfeld and M. Yabe: Zeitschr. f. Phys. A323, 485 (1986)
[25] G. de Franceschi, N. Lo Iudice and F. Palumbo: Phys. Rev. C29, 1436 (1984)
[26] M. Sambataro, D. Scholten, A.E.L. Dieperink and G. Piccitto: Nucl. Phys. A423, 333 (1984)
[27] P. van Isacker, K. Heyde, J. Jolie and A. Sevrin: Ann. Phys. (N.Y.), in press
[28] P. Sala, A. Gelberg and P. von Brentano: Zeitschr.f.Phys.A323, 281 (1986)
[29] D. Bohle, A. Richter, K. Heyde, P. van Isacker, J. Moreau and A. Sevrin: Phys. Rev.Lett. 55, 1661 (1985)
[30] O. Scholten, K. Heyde, P. van Isacker, J. Jolie, J. Moreau, M. Waroquier and J. Sau: Nucl. Phys. A438, 41 (1985)
[31] F. Iachello: private communication
[32] A. van Egmond, K. Allaart and A. Bonsignori: Nucl. Phys. A436, 458 (1985)
[33] K. Heyde and J. Sau: Phys. Rev. C30, 1355 (1984)
[34] O. Scholten: private communication
[35] D. Bohle, A. Richter, U.E.P. Berg, J. Drexler, R.D. Heil, U. Kneissl, H. Metzger, R. Stock, B. Fischer, H. Hollick and D. Kollewe: Nucl. Phys. A, in press

Mixed-Symmetry States in Proton-Neutron Systems

K. Heyde

Institute for Nuclear Physics, Proeftuinstraat 86, B-9000 Gent, Belgium

1 Introduction

Since the study of the electric giant dipole resonance which results from a collective motion of protons against neutrons, many more multipole resonances have been observed experimentally. As early as 1966 [1] , the possibility of isovector quadrupole vibrations was discussed for vibrational-like excitations with a resulting energy eigenvalue of $\hbar\omega \simeq 15$ MeV. Later on, rotational oscillations in a two-rotor model [2,3] or in a vibrating potential model [4] were studied, leading to $K^\pi=1^+$ excitations in deformed nuclei. Within these macroscopic geometric models, a large difficulty remains for a reliable calculation of the restoring force defining the energy of the particular proton-neutron modes [5].

Besides the studies pointed out above, macroscopic algebraic calculations, in the framework of the proton-neutron interacting boson model (IBM-2), have allowed for a systematic exploration of these new collective modes, called mixed-symmetry states since they correspond, group theoretically with the two-row representations of the combined proton-neutron $U_{\pi+\nu}(6)$ group, according to the group reduction

$$U_\pi(6) \otimes U_\nu(6) \supset U_{\pi+\nu}(6) \supset \ldots \qquad (1)$$

2 The IBM-2

We shall not discuss the IBM-2 here, but refer to the studies of Arima and Iachello [6,7], Scholten et al. [8] and P.Van Isacker et al. [9]. Here, we concentrate only onto the mixed-symmetry states in different dynamical symmetries U(5), SU(3), O(6) and their transitional regions concerning the energy spectra and M1 and E2 decay properties.

We start from the Hamiltonian

$$H = \varepsilon_d(n_{d_\pi} + n_{d_\nu}) + \kappa(Q_\pi + Q_\nu)\cdot(Q_\pi + Q_\nu) + aM \quad , \qquad (2)$$

with

$$Q_\rho \equiv (d^+s + s^+\tilde{d})^{(2)}_\rho + \chi_\rho(d^+\tilde{d})^{(2)}_\rho \quad , \qquad (3)$$

and M the Majorana operator [8,9].

2.1 Excitation energies

In fig.1, we show the calculated spectra, starting from the Hamiltonian (2), with parameters as discussed by Bijker et al.[10] for the Pt nuclei. Here, one clearly observes an almost linear rise of the 1^+, 3^+ mixed-symmetry states with increasing boson number since in the exact O(6) and SU(3) limits (between which the Pt nuclei are situated as a transitional region) the excitation energy of the 1^+_M becomes $(-2\kappa+a).N - 6\kappa$ and $(-3\kappa+a).N - 3/8\kappa.L(L+1)$, respectively. Similar results have been obtained for the U(5) → O(6) transitional region (Pd, Ru nuclei) [8] as well as for the U(5) → SU(3) transitional region (Sm, Gd nuclei) [8] . In the U(5) limit, $E_x(1^+_M) = 2\,\varepsilon_d + a.N$.

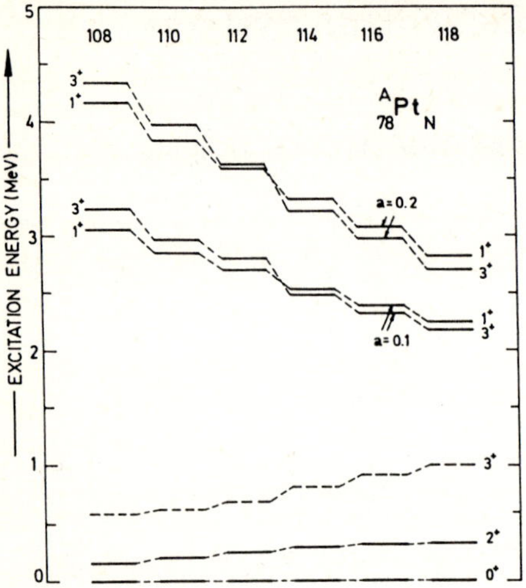

Fig.1 Energy spectra for the even-even Pt nuclei, using the parameters of ref. [10]. Besides the lowest, symmetric $J^\pi = 2^+, 3^+$ levels, we indicate the lowest mixed-symmetry $J^\pi = 1^+, 3^+$ levels for two different choices of the Majorana force strength a.

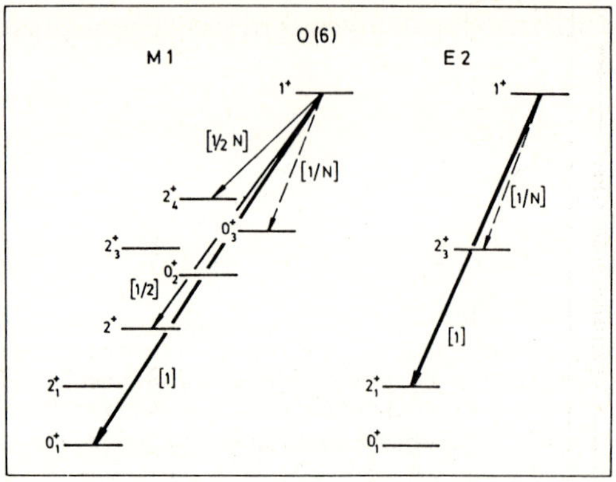

Fig 2 The M1 and E2 decay of the 1^+_M state in the O(6) limit. The B(M1) and B(E2) values are normalized to the strongest M1 or E2 transition, respectively. The results shown are valid in the limit $N \to \infty$.

2.2 M1-properties

Since the recent observation of 1^+ states in inelastic (e,e') scattering in different mass regions (light nuclei, $1f_{7/2}$ shell, deformed rare-earth nuclei, actinides) has clearly established the existence of mixed-symmetry 1^+ states [11-13], we now discuss some details on the decay of these states.

In fig 2, we show the γ-decay of the 1^+_M state in the O(6) limit for both M1 and E2 decay properties. The numbers given are correct in leading order in N (for $N \to \infty$) and give the branching rules in the γ-decay. These results should be a good guidance for experiments that are now being performed using photon inelastic scattering [14,15]. The results of fig 2 only apply to the extreme O(6) limit. Therefore, we show besides, in fig 3, a transitional nucleus ^{104}Pd(U(5) → O(6) transition). Although the 0^+_1 level is most strongly populated, the "two-phonon" 2^+_2 level is more strongly populated than the 2^+_1 level which could be a specific signature in order to identify the mixed-symmetry 1^+_M state near the O(6) region of nuclei. This particular feature is the most striking one and is typical for the heavier Ru, Pd nuclei in this U(5) → O(6) region since in the exact O(6) limit one has $B(M1;1^+_M \to 2^+_1) = 0$. Similar studies for the SU(3), U(5) limit and other transitional regions have been carried out [8,9].

2.3 E2-properties

Already a few years ago, it was suggested [16] that besides the mixed symmetry 1^+_M mode, also 2^+ mixed-symmetry states should occur. Within the U(5) limit, the 2^+_M level indeed results as the lowest-lying mixed-symmetry state. Since in a microscopic approach, most E2 strength resides in the $\Delta N=2$ (N : harmonic oscillator quantum number) mode, in the $\Delta N=0$ mode, which roughly corresponds to the IBM-2 model space, only a moderate total summed E2 strength remains. Some recent experiments however, indicate possible evidence for the observation of 2^+ levels with the properties of the 2^+_M states e.g. the 2^+_2, 2^+_3 levels in ^{56}Fe [17], the N=84 nuclei (Ba,Ce,Nd) [18] and recently, also ^{156}Gd [19]. Thus even though the ratio $R \equiv B(E2;0^+_1 \to 2^+_M)/B(E2;0^+_1 \to 2^+_1)$ is less favourable in the SU(3) limit ($\simeq 0.01$ for

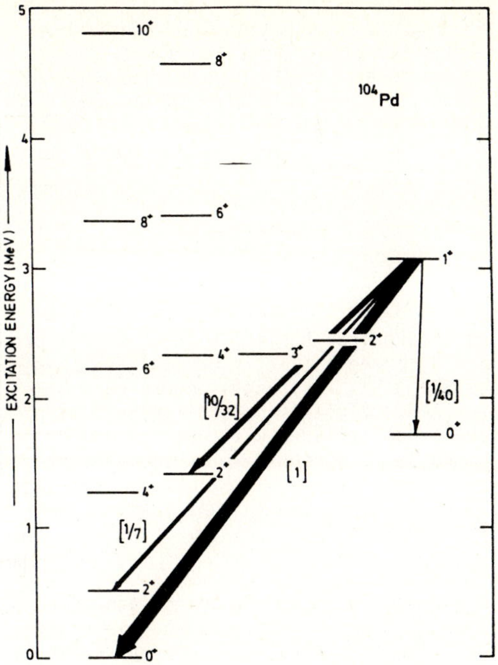

Fig 3 The total gamma-decay intensities in ^{104}Pd, normalized to the ground state M1 transition rate (the thickness of the arrow indicates the relative intensity T(M1)+T(E2) and compares with the experimental situation)

TABLE I. $B(T\lambda;J_i \rightarrow J_f)$ Values in the U(5) Limit

J_i^π	J_f^π	$T\lambda$	$B(T\lambda;J_i \rightarrow J_f)$
0_1^+	$2_1^+(d)$	E2	$(e_\nu N_\nu + e_\pi N_\pi)^2 \frac{5}{N}$
0_1^+	$2_M^+(d)$	E2	$(e_\nu - e_\pi)^2 \frac{5}{N} N_\nu N_\pi$
$1_M^+(d^2)$	$0_2^+(d^2)$	M1	$\frac{3}{4\pi}(g_\nu - g_\pi)^2 \frac{4}{(N-1)N} N_\nu N_\pi$
$1_M^+(d^2)$	$2_2^+(d^2)$	M1	$\frac{3}{4\pi}(g_\nu - g_\pi)^2 \frac{7}{(N-1)N} N_\nu N_\pi$
$1_M^+(d^2)$	$2_1^+(d)$	E2	$(e_\nu - e_\pi)^2 \frac{1}{N} N_\nu N_\pi$
$1_M^+(d^2)$	$2_2^+(d^2)$	E2	$(e_\nu \chi_\nu - e_\pi \chi_\pi)^2 \frac{1}{2(N-1)N} N_\nu N_\pi$
$1_M^+(d^2)$	$2_3^+(d^3)$	E2	0
$1_M^+(d^2)$	$2_M^+(d)$	E2	$(e_\nu N_\nu + e_\pi N_\pi)^2 \frac{1}{N}$
$1_M^+(d^2)$	$3_1^+(d^3)$	E2	0
$2_M^+(d)$	$2_1^+(d)$	M1	$\frac{3}{4\pi}(g_\nu - g_\pi)^2 \frac{6}{N^2} N_\nu N_\pi$
$2_M^+(d)$	$0_2^+(d^2)$	E2	$(e_\nu - e_\pi)^2 \frac{2}{5(N-1)N^2} N_\nu N_\pi$
$2_M^+(d)$	$2_1^+(d)$	E2	$(e_\nu \chi_\nu - e_\pi \chi_\pi)^2 \frac{1}{N^2} N_\nu N_\pi$
$2_M^+(d)$	$2_2^+(d^2)$	E2	$(e_\nu - e_\pi)^2 \frac{2}{(N-1)N^2} N_\nu N_\pi$
$2_M^+(d)$	$4_1^+(d)$	E2	$(e_\nu - e_\pi)^2 \frac{18}{5(N-1)N^2} N_\nu N_\pi$

Note. The $B(T\lambda)$ values which are zero by virtue of the $\Delta(n_1+n_2)$ selection rule, are not given.

$e_\nu = e_\pi/2$) compared to the U(5) limit ($\simeq 0.1$ for $e_\nu = e_\pi/2$), E2 strength can be observed in strongly deformed nuclei. In the light of the above discussion and because of the larger ratio R in the U(5) limit, we present in table I, the B(E2) and B(M1) values that are characteristic and should allow for a possible observation of the 2^+ mixed symmetry states in or near the U(5) limit.

3 Splitting and spreading of mixed-symmetry states

As becomes clear from the above discussion, the IBM-2 allows for a detailed and systematic study of the properties of mixed-symmetry states. Analytic expressions have been derived for $E_x(1_M^+, 2_M^+, \ldots)$ and for B(M1), B(E2) values in the dynamical symmetries of the IBM-2 [9] and moreover, numerical studies have been performed in transitional regions [8]. Since the IBM-2 is a collective model, most mixed-symmetry strength is highly concentrated. For the 1_M^+ state, in the SU(3) and O(6) limit, only one state contains all the 1^+ strength; realistic calculations give only a small mixing into other 1^+ levels. In the experimental data, more fragmentation has been observed varying from a splitting into two major peaks towards spreading into more levels above $E_x \gtrsim 3$ MeV (see fig 1 of [12]). From microscopic studies (shell-model in light nuclei, RPA calculations in deformed nuclei, ... for refs. see [8,20]), a rather important fragmentation results with as a general result, concentration of strength near $E_x \simeq 3$ MeV and near 4.5 MeV in deformed nuclei. The first group seems to be related to $j \rightarrow j$ transitions whereas for the second group, mainly spin flip $j^\uparrow \rightarrow j^\downarrow$ transitions contribute. The way to go from the microscopic into the macroscopic models is still not understood but some possibilities exist from schematic model studies.

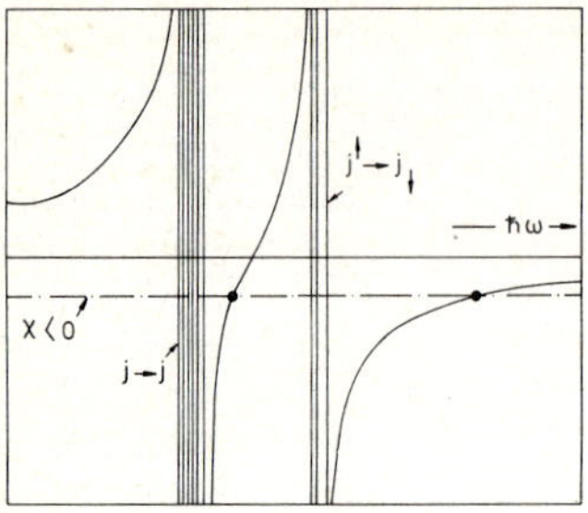

Fig 4 The RPA eigenvalue equation for a separable force in a schematic model where the $j \rightarrow j$ 1^+ 2qp states are taken degenerate in unperturbed energy ($\simeq 2\Delta$) and $j^{\uparrow} \rightarrow j^{\downarrow}$ 1^+ 2qp spin-flip states are placed at the energy of the shell-gap ($\simeq$ 4.5 MeV). The roots are given with a heavy dot on the dash-dot line.

If we depict the M1 strength in an almost spherical basis for the N=4 (for protons) and N=5 (for neutrons) shells, the $j \rightarrow j$ strength (partly orbital and spin) is concentrated around $E_x \simeq 2.5$ MeV whereas the $j\uparrow \rightarrow j\downarrow$ strength occurs near $E_x \simeq 4.5$ MeV. For a schematic two-level TDA (or RPA) calculation, and concentrating on the isovector M1 component, two prominent states will result : at the higher energy, the spin flip dominates and will concentrate strength in a Gamow-Teller resonance. In between the two groups of unperturbed states, a collective state will result where mainly the $j \rightarrow j$ strength becomes concentrated in what could be the analogon of the IBM-2 (or two-rotor model) collective 1^+_M mixed-symmetry state (fig 4).

A possible way now to relate the IBM-2 strength and its spreading (and eventual splitting) with the microscopic studies could be via the hexadecapole states that can occur, in a (sdg)boson model description in the relevant energy region [9] which could act as natural doorway states. For mixing 1^+ strengths in the underlying 2qp,4qp, ... background we have calculated a typical $\langle 1^+_M(\text{IBM-2})|H_{mix}|(sdg)1^+\rangle$ $1^+\rangle$ matrix element (in the vibrational limit) to be of the order of $\simeq$ 100 keV giving a spreading with (using the density of hexadecapole states in the region 3 MeV $< E_x <$ 4 MeV) of $\Gamma\downarrow \simeq$ 300 keV (fig 5). More work needs to be done to elucidate the mixing into the qp background and a similar calculation of $\langle H_{mix}\rangle$ for the SU(3) limit before understanding the relation with the microscopic RPA calculations. Up to now, most RPA type of calculations give $\Sigma B(M1;\uparrow) \simeq 10\text{-}15\ \mu_N^2$. This is a topic for both theoretical and experimental study i.e. how much M1 strength remains in the background grass in the region 3 MeV $< E_x <$ 10 MeV.

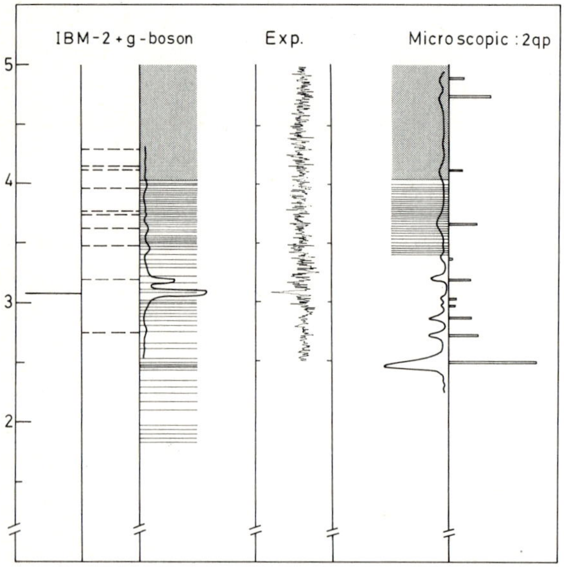

Fig 5 Schematic figure pointing out how from a microscopic 2qp calculation, taking into account the growing 4qp,.. density at $E_x \simeq 4$ MeV, M1 strength could get fragmented. On the left-hand side we show, starting from the IBM-2 and by introducing states increasing in complexity (g-bosons (dashed-line) and 2qp + 4qp +.. states (full lines and hatched region)), how M1 strength can be split and fragmented further in the background. We compare with the experimental data for ^{156}Gd [13].

4 Conclusion

Mixed-symmetry states have been shown to form a most interesting class of excitations in nuclei. After the first experimental evidence obtained in ^{156}Gd, it has been shown that mixed-symmetry states are a general feature common to all two-component systems eg IBM-2, two-rotor model, proton-neutron 2qp ; shell-model $(j_\pi)^2_{0^+,2^+} - (j_\nu)^2_{0^+,2^+}, \ldots$. Experimentally, detailed mapping of 1^+, 2^+ mixed-symmetry states has started (light nuclei, rare-earth nuclei, vibrational-like nuclei, actinides). Major questions remain as to the fragmentation and total M1 and E2 strength in the mixed-symmetry states and how to bridge the gap between the collective models on one side and the microscopic studies on the other side.

The work presented here grew out of a collaborative work over the last years in Gent with P.Van Isacker, J.Jolie and A.Sevrin on mixed-symmetry states in the IBM-2. He is most grateful to D.Bohle, A.E.L.Dieperink, A.Frank, A.Richter and O.Scholten for the generous exchange of information and intense discussion meetings in Gent.

1. A.Faessler, Nucl.Phys. 85, 653 (1966)
2. N.Lo Iudice and F.Palumbo, Phys.Rev.Lett. 41, 1532 (1978)
3. N.Lo Iudice and F.Palumbo, Nucl.Phys. A326, 193 (1979)
4. T.Suzuki and D.J.Rowe, Nucl.Phys. A289, 461 (1977)
5. A.Faessler, Z.Bochnacki and R.Nojarov, J.Phys. G12, L47 (1986)
6. A.Arima and F.Iachello, Adv. in Nucl.Phys. 13, 139 (1984)
7. A.Arima and F.Iachello, Ann.Rev.Nucl.Sci. 31, 75 (1981)
8. O.Scholten et al. Nucl.Phys. A438, 41 (1985)
9. P.Van Isacker, K.Heyde, J.Jolie and A.Sevrin, Ann.Phys. (N.Y.) to be publ.
10. R.Bijker, A.E.L.Dieperink, O.Scholten and R.Spanhoff, Nucl.Phys. A344, 207 (1980)
11. D.Bohle et al., Phys.Lett. 137B, 27 (1984)
12. D.Bohle, G.Küchler, A.Richter and W.Steffen, Phys.Lett. 148B, 260 (1984)
13. D.Bohle, Ph.D.Thesis, Technische Hochschule, Darmstadt, 1986 (Unpubl)
14. U.E.P.Berg et al., Phys.Lett. 149B, 59 (1984)
15. D.Bohle et al, preprint, to be publ.
16. F.Iachello, Phys.Rev.Lett. 53, 1427 (1984)
17. S.A.A.Eid, W.D.Hamilton and J.P.Elliott, Phys.Lett. 166B, 267 (1986)
18. W.D.Hamilton, A.Irbäck and J.P.Elliott, Phys.Rev.Lett. 53, 2469 (1984)
19. D.Bohle et al, Phys.Rev.Lett. 55, 1661 (1985)
20. K.Heyde, Proc. of the Int. Research Conf. on Nuclear Structure, Reactions and Symmetries, (Dubrovnik, 1986), (World Scientific Publ. Co).

Proton-Neutron Symmetry among Bosons

P. von Brentano

Institut für Kernphysik der Universität zu Köln,
D-5000 Köln 41, F.R. Germany

Properties of collective states which depend on the proton and neutron content are discussed in the frame of the F-spin concept. Experimental evidence for F-spin as data on F-spin multiplets are given. "Global" descriptions of groups of nuclei in the Interacting Boson Model are discussed.

Introduction:

The collective properties of low lying nuclear states have been successfully described both in geometric models based on the concept of mean field and in algebraic models based on the concept of bosons. It has been found, that many properties of the collective states can be described by one mean field in the geometric models or by using one kind of bosons in the algebraic models. The nucleus consists, however, of two kinds of particles namely protons and neutrons. Already in the sixties Faessler and Greiner (1-3) introduced separate mean proton and neutron fields in the geometric models and these ideas have been taken up recently as is discussed in the talk of Nojarov (4,5). In the algebraic model Arima et al (6-8) introduced the Interacting Boson Model with proton and neutron bosons(IBM-2). After the formulation of these models with a proton and neutron degree of freedom two important questions arise:

1) how can it be that the collective models with one kind of particles e.g.IBM-1 are so successful in describing many phenomena although the nucleus consists of protons and neutrons
2) which kind of nuclear phenomena will definitely need models with proton and neutron bosons.

The first problem, why IBM-1 is successful, has been answered by the introduction of the F-spin concept by Arima et al (6-8). F-spin is the isospin for systems of proton and neutron bosons. In particular it has been shown that the subset of the F=Fmax fully symmetric states of the IBM-2 correspond to the states of the IBM-1 (6-8,12). An even stronger result has been obtained by Harter et al (9), who have shown that to each IBM-2 Hamiltonian one can associate a projected IBM-1 Hamiltonian, which will reproduce exactly the energies of those states of the given IBM-2, which have a pure F-spin F=Fmax. The second question, concerning nuclear properties, which can be understood only in the IBM-2, has been answered in a spectacular way by the discovery of the 1^+(F=Fmax-1) levels in electron scattering by the Darmstadt group (10-12). A subject which is discussed in the talks by Bohle and Heyde, where extensive references are given. Other methods which probe the proton and neutron content of collective states are electromagnetic transitions and in particular M1-transitions, which are discussed in the talk by Gelberg. In this respect also transition densities from electron scattering must be mentioned. Another area where the proton-neutron model is needed are attempts to describe "global" aspects of collective states in many nuclei with one Hamiltonian with constant parameters. These "global" aspects will be the main topic of this talk. A third question is which phenomena are particularly related to the concept of F-spin. As F-spin is the isospin among a system of proton and neutron bosons, we will be guided by our knowledge on isospin. The following questions arise.

1) Are there F-spin multiplets with rather constant energies corresponding to isospin multiplets?

2) What are the energies of the lowest Fmax-1, Fmax-2 ... states? This is discussed by Bohle and Heyde.

3) What is the purity of F-spin and what are the important facts concerning its breaking? This is discussed in Gelbergs talk.

F-spin multiplets

The most direct way of observing spin is to look at the splitting of a level in a magnetic field. Similarly the most direct way to observe isospin and F-spin is to look for isospin and F-spin multiplets. An F-spin multiplet is a series of nuclei with a constant number of bosons N=Np+Nn, but varying numbers of proton bosons Np. As the low lying collective levels have the maximum F-spin Fmax = 0.5 (Np+Nn) it is much more easy to find F-spin multiplets from experiments than it is to find isospin multiplets. Fig. 1 shows an F-spin multiplet with N=13 in the rare earth region from reference (13). It is remarkable how constant the excitation energies of the groundband and the gammaband are in this F-spin multiplet, even though ^{158}Dy and ^{182}Pt differ by 24 mass units. One has to stress that without the

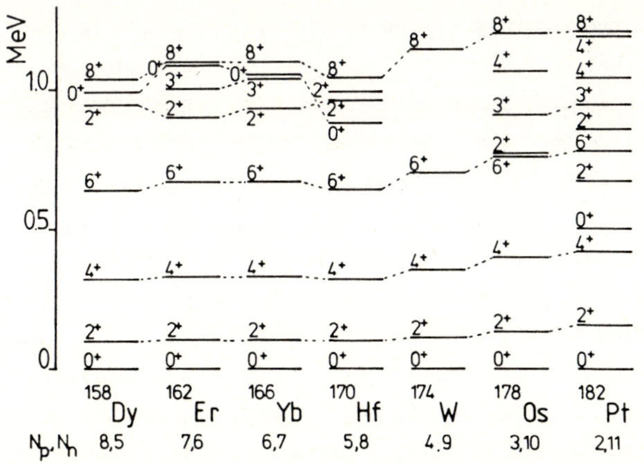

Figure 1. F-spin multiplet with F=13/2 Np and Nn are the proton and neutron boson numbers. We show the levels of the ground band, the quasi gamma band and the quasi beta band up to an excitation energy around 1 MeV.

P. von Brentano et al (13)

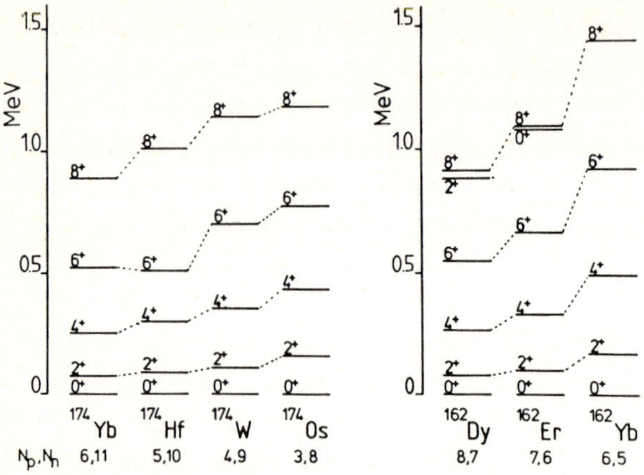

Figure 2. Levels of the ground band up to the 8^+ levels for several chains of isobaric nuclei with A=174 and A=162. Np and Nn are the proton and neutron boson numbers. We note that N=Np+Nn varies in these chains of isobars.
P. von Brentano et al (13)

Interacting Boson Model (IBM-2) and the F-spin concept there would have been no reason to examine this particular series of nuclei, which forms the F-spin multiplet. In order to test the significance of the observed constancy of the energies of the ground band within the multiplet, we show in fig. 2 several series of isobaric nuclei, for which the total boson number N=Np+Nn and the F-spin Fmax = 1/2 N vary strongly. One notes that the energies in these series of nuclei vary indeed strongly with N. The 6+ energy of ^{162}Yb is about 70 % higher than the 6+ energy in ^{162}Dy, although the two nuclei have the same mass number. The observed constancy of the energies in this F-spin multiplet is therefore a real effect.

"Global" fits with the Interacting Boson Model

What are the consequences of the observation of rather constant excitation energies in F-spin multiplets? The most simple explanation is to assume that the IBM-2 Hamiltonian is F-spin invariant and has constant coefficients for the multiplet (14). An elegant way in which this concept can be formulated is to postulate a dynamic U(12) symmetry as has been suggested by Frank and van Isacker (15). The problem with this suggestion is that it implies rigorously constant energies within the multiplet. As shown in fig. 3 the energies in the multiplets vary with Fo in general, however indicating the presence of F-vector and F-tensor components in the Hamiltonian. Thus the phenomenological boson interaction must contain not only F-scalar terms as e.g. terms like $(Qp+Qn)^2$ but also higher tensor operators as in Qp·Qp. A surprising result is that even with the Qp·Qn interaction one can obtain multiplets with rather constant energies in the rare earth region, however(13). This is been shown by Sala et al (16), who have fitted spectra of about 22 rare earth nuclei with 5 constant parameters, with an IBM-2 Hamiltonian which contains only the boson energy terms and proton neutron monopole and quadrupole terms. Examples of these fits are shown in fig. 4 and 5. In this work the energies of the groundband and the gammaband are well reproduced, whereas the beta band is not so well reproduced. We mention also a similar work by Casten et al

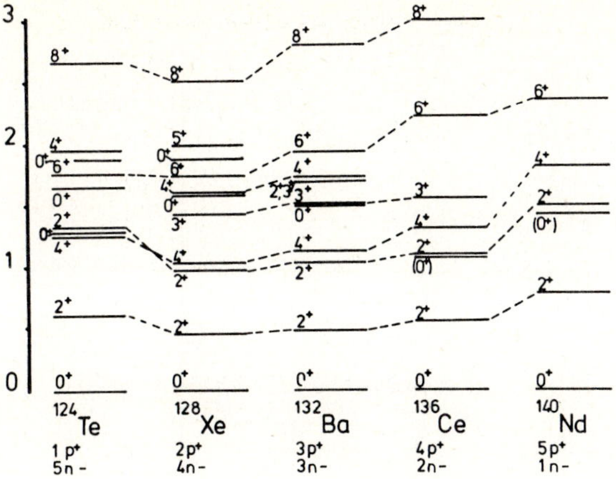

Figure 3. The figure shows a part of an F-spin multiplet for ^{124}Te-^{140}Nd. Nuclei with Z=50 and N=82, which are not described by the boson model, are deleted. All known positive parity levels up Ex-2 MeV and the ground and gamma bands up to the energy of the 8_1^+ level are shown.
From Harter et al (14)

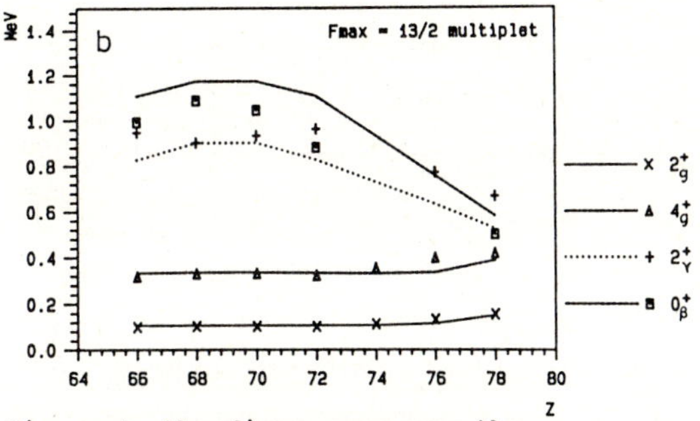

Figure 4. the figure compares the predictions of a "global" IBM-2 fit with 5 constant parameters to the experimental data.

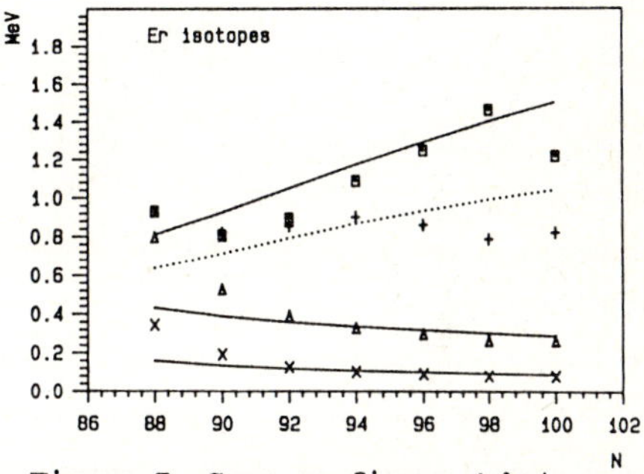

Figure 5. Same as figure 4 but for a sequence of Er isotoprs from ^{156}Er to ^{168}Er.
From P. Sala et al (16)

(17) in which a "global" fit to spectra of about 100 nuclei is achieved in the IBM-1 by using 6 constant parameters and a parametrized functional dependence of the boson-energy on the Casten scaling parameter Np·Nn. We want to mention in this respect also a recent "global" fit for the A=130 nuclei by Novoselsky. (19) The success of these "global" fits clearly shows the validity of the Interacting Boson Model.

Collectivity, F-spin, Np·Nn scaling and deformation

The F-spin concept allows also to discuss various models of nuclear collectivity in a unified frame. A comparison of three models is given in fig. 6 where the isoenergy contours for the energy of the first 2+ level in the N-Z-plane for the rare earth nuclei are shown. The lower left part of fig. 6 gives the prediction of an F-spin invariant IBM-2. The diamond lines are the lines of constant F-spin. The nuclei which are on these diamond lines form a four fold F-spin multiplet. This is

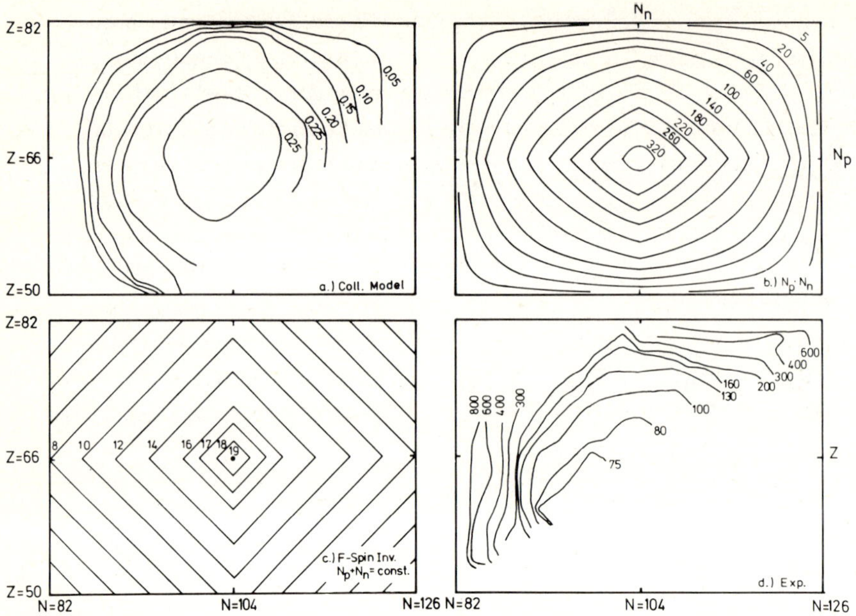

Figure 6. Comparison of constant parameter contours for three collective models with the empirical results. The contours are those of constant deformation, for the Strutinsky type calculations from the Lund group, of constant Np·Nn for the Np·Nn parametrization, of constant total valence nucleon number (boson number N and F-spin Fmax) for the F-spin approach and of the experimental excitation energies for the first 2^+ states.
From Casten et al (17)

surprising at the first moment, because one expects that a complete F-spin multiplet with F-spin F contains (2F+1) nuclei. The situation is more complex, in real nuclei, however, because the bosons are made of fermion pairs and these are counted as particles at the lower half of the shell and as holes in the upper part of the shell. Thus the boson number increases till the middle of the shell and decreases after it. If this fact is considered, one gets indeed the diamond lines for all nuclei which have constant F-spin Fmax and constant total boson number N=Np+Nn. The diamond lines are also lines of constant collective spectra in the most simple case of an F-spin invarant IBM-2 Hamiltonian. In fig. 6 the predictions of this schematic model are compared to the experimental 2+ energies. One notes that the diamond lines are not reproducing the data in the vicinity of the magic shells. It has been noted by Casten (e.g.17), that a rather general scaling of collective properties is obtained by a scaling factor Np.Nn. This scaling factor measures the overall strength of the neutron proton force. The contours of constant Np·Nn are also shown in fig. 6. It is interesting to note that the "global" IBM-2 fit discussed above obeys the Casten Np Nn scaling to a surprising good accuracy.

Finally we mention, that in fig. 6 also calculated isodeformation curves from Strutinsky type calculations by the Lund group are shown. Comparing the three models, we note that collectivity is strongly related to the proton neutron force as has been stressed since a long time by Talmi. We note also that the most simple way in which deformation phenomenon can be discussed is in terms of an F-spin invariant Interacting Boson Model.

Summary

It has been shown, that the F-spin concept is a very useful concept in the proton neutron boson models, which allows to discuss proton neutron symmetry and which unifies many different phenomena in collective models. In particular this concept explains the rather constant energies in F-spin multiplets and it suggests the possibility of "global" fits with the proton neutron Interacting Boson Model. It also gives a unified description of the new experiments on the 1+ states done by the Darmstadt group. Clearly much work is needed in future to investigate the purity of F-spin in collective nuclei.

Acknowledgements

The autor warmly thanks A. Gelberg H. Harter, B. Barrett, R. F. Casten, I. Morrison, P. Sala and J. Theuerkauf for stimulating discussions and cooperation. Supported by BMFT.

1) W. Greiner Phys. Rev. Lett. 599 14 (1965)

2) A. Faessler Nuclear Physics 653 85 (1966)

3) W. Greiner Nucl. Phys. 417 80 (1966)

4) S.G. Rohozinski and W. Greiner Preprint Frankfurt (1985)

5) A. Faessler, R. Nojarov and S. Zubik preprint Tübingen (1986)

6) A. Arima, T. Otsuka, F. Iachello and I. Talmi, Phys. Lett. 66B, 209, (1977)

7) T. Otsuka, A. Arima, F. Iachello and I. Talmi, Phys. Lett. 76B, 139, (1978)

8) T. Otsuka, A. Arima and F. Iachello Nucl. Phys. A309 1 (1978)

9) H. Harter, P. von Brentano, A. Gelberg, Phys. Lett. 157B 1 (1985)

10) A. Richter, in Nuclear Structure 1985, Niels Bohr Centennial Conference 1985 and references given therein

11) D. Bohle, G. Küchler, A. Richter, W. Steffen, Phys. Lett. 148B 260 (1984)

12) F. Iachello, Phys. Rev. Lett. 53 1427 (1984)

13) P. von Brentano, A. Gelberg, H. Harter, P. Sala, J. Phys. G. Nucl. Phys. 11 85 (1985)

14) H. Harter, P. von Brentano, A. Gelberg and R. F. Casten, Phys Rev. C32, 631 (1985)

15) A. Frank and P. van Isacker Phys. Rev. C32 1770 (1985)

16) P. Sala. P. von Brentano, H. Harter, B. Barrett and R. F. Casten, Nucl. Phys. (1986)

17) R. F. Casten, P. von Brentano, A. Gelberg, H. Harter, Journal of Phys. G (1986)

18) P. Sala, A. Gelberg, P. von Brentano, Z. Phys. A232 1 (1986)

19) A. Novoselsky, I. Talmi, Phys. Lett. B172 139 (1986)

F-Spin and Collective M1 Transitions*

A. Gelberg

Institut für Kernphysik der Universität zu Köln,
D-5000 Köln 41, F.R. Germany

The basic concepts of F-spin are discussed and applied to M1 transitions. An IBM-2 fit of energies and branching ratios in ^{128}Xe has been achieved. Information on the F-spin structure of eigenstates has been obtained.

Collective levels in many nuclei have been successfully described by the interacting boson model (IBM) [1]. In its proton-neutron version known as IBM-2 [1,2] separate proton and neutron bosons are introduced, viz. d_p, s_p and d_n,s_n bosons, as long as only bosons with L=0and 2 are considered. The proton-neutron symmetry of this model can be formalised by introducing the F-spin concept [3,4] which can be shortly characterized as boson isospin. The purpose of this work is to investigate and make use of the close relation between F-spin and the strength of collective M1 transitions.

Proton and neutron bosons are assigned an F-spin F=1/2 with projections F_0=1/2 and F_0= -1/2 respectively. The F-spin labels SU(2) representations in the proton-neutron boson space. The F-spin generators have been given in ref. [4] . A nucleus is characterized by $F_0=(N_p-N_n)/2$ which is a constant of motion; N_p and N_n are the numbers of proton and neutron bosons respectively, the maximum value of F is $F_{max}=(N_p+N_n)/2=N/2$. States with $F=F_{max}$ belong to the totally symmetric representation [N].

The boson creation and destruction operators can be easily written as F-spin tensors of rank 1/2 [5], which in turn can be coupled to higher rank tensors by Clebsch-Gordan coefficients. The IBM hamiltonian in its usual form consists of one- and two- body terms, i.e. of products of maximum four boson creation and destruction operators. Consequently, the hamiltonian can be decomposed into a scalar, a vector and a rank 2 tensor in F-space.

If the hamiltonian is scalar, it commutes with all F-spin generators. A lower degree of symmetry is possible in which the hamiltonian commutes only with $\vec{F}^2$; in this case F is still a good quantum number. The states with $F=F_{max}$ which have maximum symmetry are lowest in energy.

Let us analyse the F-spin structure of the M1 operator, which in IBM is defined as [2]

$$T(M1)=\sqrt{3/4\pi}\,(g_pL_p+g_nL_n) \tag{1}$$

where $L_i=\sqrt{10}\,[d^+\times\tilde{d}]^{(1)}$ (i=p,n) is the angular momentum operator and g_p and g_n are the proton and neutron boson gyromagnetic factors respectively. T(M1) can be rewritten

$$\sqrt{4\pi/3}\;T(M1)=g_s(L_p+L_n)+g_v(L_p-L_n) \tag{2}$$

* Supported by Bundesministerium für Forschung und Technologie

with $g_s=(g_p+g_n)/2$ and $g_v=(g_p-g_n)/2$. The first term in (2), which is proportional to the total angular momentum is an F-scalar and it contributes only to the static magnetic moment. The second term has non vanishing transition, i.e. non diagonal matrix elements [2] . This term is the zero component of an F-vector. Therefore the F-spin selection rule for M1 transitions is $\Delta F=0,\pm1$.

A typical example for this selection rule is the M1 transition from a collective, mixed symmetry 1+ state [2] to the ground state. This state has no $F=F_{max}$ component and has mainly an $F=F_{max}-1$ structure. Such states have been found in deformed nuclei by electron scattering [6] and subsequently by photon scattering [7].

Besides, it can be shown that there are no M1 transitions between states with $F=F_{max}$ [8] . Consequently M1 transitions between low energy states, e.g. those belonging to the ground state band (gsb), gamma band, etc. can appear only if components with $F=F_{max}-1$ are admixed to the main component $F=F_{max}$. The M1 transition strengths provide us with a tool for investigating the F-spin structure of states. As a matter of fact the F-spin selection rule constitutes only a necessary condition for the occurence of M1 transitions. The situation is complicated by the possible action of other selection rules, as will be shown below.

In order to investigate the connection of F-spin with M1 transitions, let us analyse the concrete case of ^{128}Xe. This nucleus has recently been investigated by means of in-beam gamma-ray spectroscopy [9] and Coulomb excitation [10] . Although only one E2/M1 mixing ratio has been accurately measured, the model parameters can be fitted to branching ratios of pairs of transitions with at least one mixed transition.

Xe and Ba isotopes with less than 82 neutrons can fairly well be described in the O(6) limit of IBM-1 [11] . In this version of IBM only one kind of bosons is considered and all states are totally symmetric, i.e. they are identical to states with $F=F_{max}$. We will use an IBM-2 hamiltonian which, in a first approximation generates eigenstates which are as close as possible to IBM-1 states. This has been done by using the projection method described in ref. [5] .

The following form of the hamiltonian has been chosen [12]

$$H=\varepsilon(n_p+n_n)+K(Q_p\cdot Q_n)+K'(Q_p+Q_n)^2-\lambda M \qquad (4)$$

where $Q_i=(d^+s+s^+\tilde{d}+[d^+x\tilde{d}]^{(2)})_i$, $n_d=d_i^+\cdot\tilde{d}_i (i=n,p)$ and M has the $M=1/4N(N+2)-F$ 5 . Two cases are considered which will be labelled as Q_pQ_n ($K'=0$) and $(Q_p+Q_n)^2$ (K=0). We will first put $\chi_p=\chi_n=0$. By fitting energies and B(E2) [5] in ^{128}Xe, following parameters were found: in the $(Q_p+Q_n)^2$ case $\varepsilon=0, K'=-0.07$ MeV, $\lambda=0.183$ MeV. In the Q_pQ_n case we found $\varepsilon=0.28$ MeV, K=-0.263 MeV, $\lambda=0.1$ MeV. The value of λ has been chosen so as to put the mixed symmetry 1+ level at 2.3 MeV. A comparison of experimental and and theoretical excitation energies is shown in fig. 1. It should be noticed that the $(Q_p+Q_n)^2$ hamiltonian is an F-scalar, therefore all the states have good F-spin and M1 transitions between low lying levels vanish.

In order to induce M1 transitions we have chosen to vary the parameter $\chi=(\chi_p-\chi_n)/2$ while keeping $\chi=(\chi_p+\chi_n)/2=0$. The parameter χ was fitted to the intensity of mixed M1+E2 transitions, or more exactly to branching ratios. Theoretical and experimental relative intensities for both hamiltonians are given in table 1 .

The choice for the g-factors has been $g_p=1$, $g_n=0$ which corresponds to a purely orbital origin of the magnetic moments [2]. Experimental systematics indicates $g_p=0.63(4)$ and $g_n=0.05(5)$ [16].

As far as mixing ratios are concerned δ(E2/M1) is relatively well known only for the $2_2 2_1$ transitions. The best experimental values are 6.1(5) [17] and 3.8+1.6-0.8 [10] . The theoretical value is 4.66 with the Q_pQ_n hamiltonian and $\chi=-0.22$.

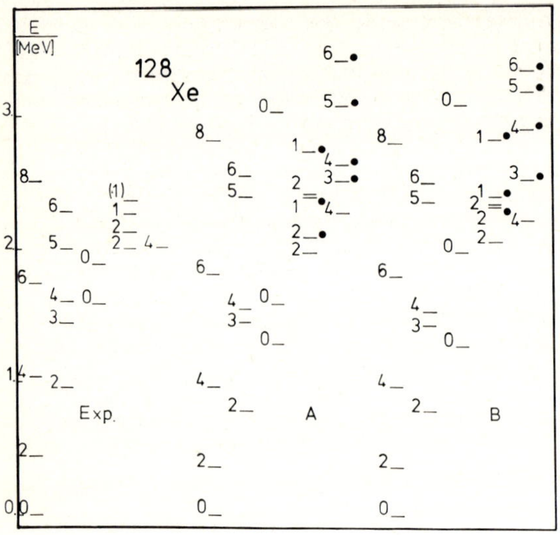

Fig. 1

Excitation energies in ^{128}Xe.

A: Q_pQ_n hamiltonian

B: $(Q_p+Q_n)^2$ hamiltonian

(the dots label those levels which are nearly pure $F_{max}-1$ states)

Table 1: Relative gamma ray intensities for ^{128}Xe. Effective charge obtained from $B(E2;2_g^+0_g)=1500e^2fm^4$ [13]. Experimental intensities were taken from ref. [13,14,15]. Parameters as above, but with $\chi'\neq 0$.

a. Q_pQ_n hamiltonian $\chi'=-0.22$

I_iI_f	Intensities Exp.	Theory
3_14_1	17.8(3)	21.9
3_12_2	100	100
4_24_1	78.6(15)	64.1
4_22_2	100	100
5_14_2	13.8(11)	15.4
5_16_1	4.1(4)	4.2
5_13_1	100	100

b. $(Q_p+Q_n)^2$ hamiltonian $\chi'=-0.155$

I_iI_f	Intensities Exp.	Theory
3_14_1	17.8(3)	22.9
3_12_2	100	100
4_24_1	78.6(15)	53.4
4_24_2	100	100
5_14_2	13.8(11)	15.3
5_16_1	4.1(4)	4.1
5_13_1	100	100

The next step is the analysis of the F-spin structure of the wave functions by means of the NPBOS-F code [18]. The amplitudes of the components with F_{max}, $F_{max}-1$ and $F_{max}-2$ are given in table 2, a, b . One can see that the amplitudes of $F=F_{max}$ components are pretty small, especially in the $(Q_p+Q_n)^2$ case.

Strangely enough, although the eigenstates of the Q_pQ_n hamiltonian are F-mixed even at $\chi'=0$ (fig.2), the M1 strength vanishes in this limit too. The decomposition of this hamiltonian into Casimir operators shows that it belongs to an intermediate case between the dynamic symmetries

$$O_p(6)\times O_n(6) \supset \begin{array}{c} O_p(5)\times O_n(5) \\ O_{p+n}(6) \end{array} \supset O_{p+n}(5)\supset O_{p+n}(3)$$

These chains have $O_{p+n}(5)$ in common. It has been shown [19] that an $O_{p+n}(5)$ selection rule leads to vanishing B(M1) between low lying states.

Table 2: Calculated F-spin amplitudes, (parameters as for table 1)

a. Q_pQ_n hamiltonian $\chi'=-0.22$				b. $(Q_p+Q_n)^2$ hamiltonian $\chi'=-0.155$			
	Amplitudes				Amplitudes		
I E (MeV)	F_{max}	$F_{max}-1$	$F_{max}-2$	I E (MeV)	F_{max}	$F_{max}-1$	$F_{max}-2$
2_1 0.443	-0.9827	0.1313	-0.1306	2_1 0.443	0.9995	0.0321	-0.0014
4_1 1.033	-0.9819	0.1461	-0.1209	4_1 1.033	-0.9994	0.0343	-0.0014
2_2 0.969	-0.9806	0.1547	-0.1207	2_2 0.969	-0.9982	0.0593	-0.0015
3_1 1.430	-0.9786	0.1776	-0.1049	3_1 1.430	-0.9944	0.1059	-0.0021
4_2 1.603	-0.9794	0.1727	-0.1043	4_2 1.603	-0.9986	0.0524	-0.0021
5_1 1.997	-0.9746	0.2090	-0.0808	5_1 1.997	-0.9915	0.1301	-0.0034

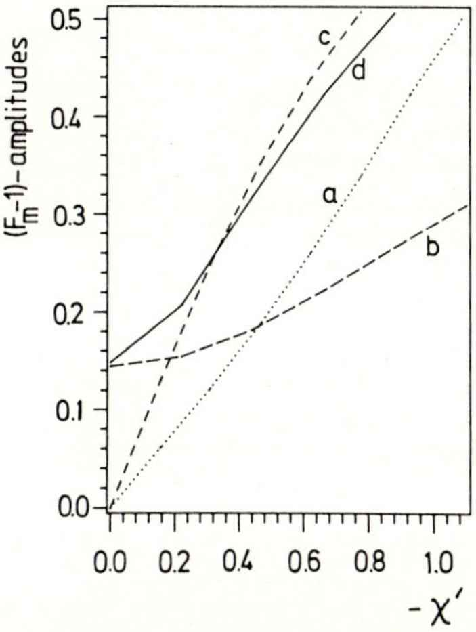

Fig. 2

$(F_{max}-1)$ amplitudes of $|I\rangle$ vs $\chi'=-1/2(\chi_p-\chi_n)$

a. $I_i=2$ $I_f=2$ $(Q_p+Q_n)^2$ hamiltonian

b. $I_i=2$ $I_f=2$ Q_pQ_n hamiltonian

c. $I_i=5$ $I_f=6$ $(Q_p+Q_n)^2$ hamiltonian

d. $I_i=5$ $I_f=6$ Q_pQ_n hamiltonian

It has been shown that one can obtain a satisfactory IBM-2 fit of energies and branching ratios for the low lying collective levels in ^{128}Xe. The fit to branching ratios which include M1 transitions is a very sensitive test of the F-spin structure of wave functions. The complete results of this calculation are being published elsewhere [20] .

The author warmly thanks P. von Brentano and H. Harter for stimulating discussions and cooperation.

References:

1. A. Arima and F. Iachello: Advances in Nuclear Physics, Vol.13 , Ed. I. W. Negele and E. Vogt (Plenum, 1984)
2. A. E. L. Dieperink and G. Wenes: Ann. Rev. Nucl. Part. Sci. 35, 77 (1985)
3. A. Arima, T. Otsuka, F. Iachello and I. Talmi: Phys. Lett. 66B, 205 (1977)
4. T. Otsuka, A. Arima and F. Iachello: Nucl. Phys. A309,1 (1978)
5. H. Harter, A. Gelberg and P. von Brentano: Phys. Lett. 157B (1985)
6. A. Richter, in Nuclear Structure 1985, Copenhagen 1985, ed. R. Broglia, G. Hagemann and B. Herskind, pag. 469 (North-Holland, Amsterdam 1986)

7. U. E. P. Berg, C. Bläsing, J. Drexler, R. D. Heil, U. Kneissel, W. Naatz, R. Ratzek, S. Schennach, R. Stock, T. Weber, H. Wickert, B. Fischer, H. Hollick, and D. Kollewe Phys. Lett. 149B, 59 (1984)
8. P. van Isackker, K. Heyde, J. Jolie and A. Severin (preprint, Gent 1986)
9. A. Dewald. R. Reinhardt, J. Panqueva, K. O. Zell and P. von Brentano: Z. Phy. A315, 77 (1984)
10. J. Srebrny, private communication
11. R. F. Casten and P. von Brentano: Phys. Lett.152 B, 22 (1985)
12. A. E. L. Dieperink and R. Bijker: Phys. Lett. 116B, 77 (1982)
13. L. Goettig, Ch. Droste and A. Dygo: Nucl. Phys. A357, 109 (1981)
14. R. Reinhardt, Ph. D. Thesis, Institut für Kernphysik Köln 1986 (unpublished)
15. C. M. Lederer and V. Shirley, Table of Isotopes (J. Wiley, New York, 1978)
16. A. Wolf, D. D. Warner and N. Benczer-Koller: Phys. Lett. 158B, 7 (1985)
17. I. Lange, K. Kumar, I. H. Hamilton: Rev. Mod. Phy. 59, 119 (1982)
18. H. Harter, Program NPBOS-F (unpublished)
19. P. van Isacker, private communication
20. H. Harter, P. von Brentano and A. Gelberg: due to appear in Phys. Rev. C

Microscopic Calculations for Low-Lying M1-Collective States in Deformed Nuclei

N.I. Pyatov [1] *and S.I. Gabrakov* [2]

[1] Joint Institute of Nuclear Research, Laboratory of Theoretical Physics, Dubna 141980, USSR
[2] Sofia University, Sofia, Bulgaria

Recently, low-lying (E_x = 3 MeV) M1-collective states in deformed nuclei have been discovered [1] experimentally. We would like to mention that probably for the first time collective M1-transitions in the energy region of 2 ÷ 4 MeV in deformed nuclei were predicted in our work [2]. We treated them in a microscopic model with pairing and residual spin-spin interactions in the framework of RPA. In particular, for $K^\pi = 1^+$ state in ^{168}Er the calculated energy was E_x = 3.18 MeV and $B(M1, 0^+ \to 1^+) \simeq 1.8\ \mu_N^2$. Later it was derived using a modified model in which the broken rotational invariance of the deformed self-consistent field was restored. These calculations have confirmed in general the predicted characteristics of the 1^+ - states in Ref. [2]. In Ref. [3] the calculations, for example, for ^{154}Sm and ^{166}Er clearly indicated the appearance of collective 1^+-excitations within the energy range E_x = 2.5 ÷ 4 MeV and with a typical value of $B(M1, 0^+ \to 1^+) \simeq 1 \div 2\ \mu_N^2$. But one should stress that the main strength of the M1-transitions is concentrated in the region of $E_x \approx 7 \div 10$ MeV. The low-lying 1^+ states exhausted no more than 10% of the energy weighted sum-rule.

The earliest experimental search for such states was undertaken in (γ,γ') resonance scattering [4]. In fact it was found in ^{168}Er that a 1^+ -state existed with the energy E_x = 3.39 MeV and the width Γ_0 = 113 ± 12 meV which corresponds to $B(M1, 0^+ \to 1^+) \simeq (0.75 \pm 0.08)\mu_N^2$.

Here we want to outline the model and to present some results of calculations for several deformed nuclei.

We start with the model Hamiltonian:

$$H = H_{qp} + H_\sigma + h \quad \ldots\ldots\ldots\ldots\ldots\ldots \quad (1)$$

Where H_{qp} is the quasiparticle Hamiltonian with pairing interaction, H_σ is the residual spin-spin interaction and h is the effective interaction restoring the rotational invariance of the quasi-particle Hamiltonian [3]. The strength parameters for the spin-interactions were chosen from calculations of magnetic moments [2] and β-decay rates [5].

The results obtained for the energies and probabilities of M1-transitions for three nuclei are shown in Fig. 1. The calculations indicate that the collective 1^+-phonon is a superposition of a large number of two-quasiparticle states,

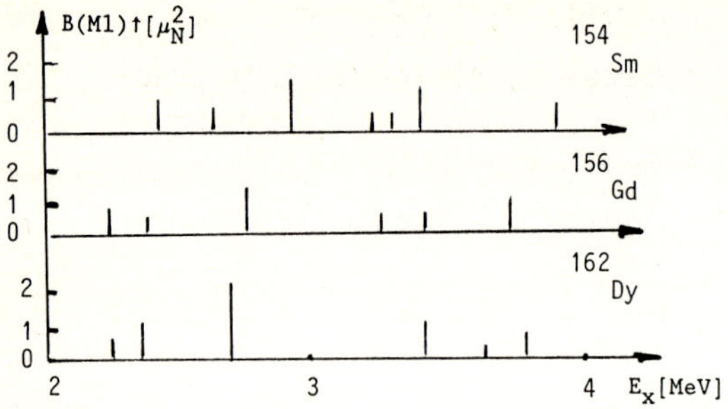

Fig. 1

but in wave-function is dominated by the contribution from transitions between states belonging to the same spherical j-shell which are split by the deformation. Such transitions arise only in deformed nuclei.

The large fragmentation of these states arises from our microscopic basis which is not the case in the macroscopic approach [6]. In the region of rare earths we find that the low-lying Ml-resonances are mainly due to the splitting of the spherical j-levels $2f_{7/2}$, $1h_{9/2}$, $1i_{13/2}$ (for neutrons) and $1g_{7/2}$, $1h_{11/2}$ (for protons). As a rule, 1^+-states with the largest B(M1) have strongly mixed neutron-proton wave functions. The calculated B(M1) could be reduced by the use of effective g-factors in nuclear matter instead of the free nucleon values.

1. D. Bohle et al.: Phys. Lett. 137B, 27(1984)
D. Bohle et al.: Phys. Lett. 148B, 260(1984)
U.E.P. Berg et al.: Phys. Lett. 149B, 59(1984)

2. S.I. Gabrakov, A.A. Kuliev and N.I. Pyatov: Yadernaja Fizika (in Russian) 12, 82(1970)
S.I. Gabrakov et al.: Nucl. Phys. A182, 625(1972)

3. A.A. Kuliev and N.I. Pyatov: Yadernaja Fizika (in Russian) 20, 297(1972)

4. F.R. Metzger: Phys. Rev. C13, 626(1976)

5. S.I. Gabrakov, A.A. Kuliev and N.I. Pyatov: Phys. Lett. 26B, 275(1971)

6. See for example, R.R. Hilton: Z. Phys. A316, 121(1984)

Low-Frequency Neutron-Proton Vibrations

A. Faessler, R. Nojarov, and Z. Bochnacki***

Institut für Theoretische Physik, Universität Tübingen,
D-7400 Tübingen 1, F. R. Germany

The low-frequency isovector neutron-proton vibrations, which have been identified recently in experiment [1] as the lowest "mixed-symmetry" 2^+ states in Ba-Ce nuclei near closed shells, were predicted a long time ago [2] in the framework of an extended vibrational model. Isospin-dependent collective coordinates allow the description of both isoscalar and isovector vibrations, coupled by the strength of the restoring force for isovector vibrations. The strength G is calculated microscopically [3,4] from the second (neutron-proton) derivative of a density-dependent symmetry energy using the wave functions of a spherical [3] (1) or and axially symmetric [4] (2) Woods-Saxon potential together with the BCS approximation:

$$G_{sph} = D \int \left[\rho_r^n \rho_r^p r^4/(\rho^n + \rho^p)^{1/3} \right] dr, \quad \rho_r^i = d\rho^i/dr, \qquad i = n,p$$

$$E_{iv}(2^+) = E_{is}(2^+) \sqrt{1 + 8G/C_{is}} \tag{1}$$

$$G_{def} = \pi D \iint \frac{\left[\rho_r^p \rho_r^n z^2 + \rho_z^p \rho_z^n r^2 - (\rho_z^p \rho_r^n + \rho_r^p \rho_z^n) rz \right]}{(\rho^n + \rho^p)^{1/3}} r\, drdz$$

$$\rho_r^i = \partial\rho^i/\partial r, \quad \rho_z^i = \partial\rho^i/\partial z, \quad E_{iv}(1^+) = 2\sqrt{G_{def}/J}, \tag{2}$$

$$B(M1;0^+ \rightarrow 1^+) = 3JE_{iv}(1^+)/16\pi,$$

$$k = D\rho^{2/3}, \qquad D = 91.6 \text{ MeV fm}^2.$$

The density-dependence of the symmetry energy coefficient k reduces about twice the energy of the isovector vibrations, bringing it close to the experimental values. The spherical limit is described by (1), which relate the energies of the isovector (E_{iv}) and isoscalar (E_{is}) 2^+ states. The isoscalar stiffness C_{is} is obtained from the experimental $B(E2;0^+ \rightarrow 2_1^+)$ value. The deformed limit is given by (2) in cylindrical coordinates r,z. J is the cranking moment of inertia, which is calculated microscopically.

The experimental energies of the isovector 2^+ states lying at about 2 MeV in ^{124}Te, ^{140}Ba, ^{142}Ce, and ^{144}Nd [1], are reproduced with an accuracy better than 10% and the 2^+ state at 2.193 MeV or 2.352 MeV in ^{128}Te is expected to be also an isovector vibrational one [3].

---

*) Fellow of the AvHumboldt Foundation. Institute of Nuclear Research and Nuclear Energy, Bulgarian Academy of Sciences, Sofia 1784, Bulgaria
**) Institute of Nuclear Physics, 31-342, Krakow, Poland

The magnetic dipole states with $K^{\pi} = 1^{+}$, which have been found at about 3 MeV in a number of deformed rare-earth nuclei [5], are described in our model [4] as scissor-like neutron-proton vibrations. Results for three of these nuclei are compared with the experimental data [5] in Table 1.

Table 1. Energies and reduced M1-transition probabilities for scissor isovector vibrations.

nucleus		^{154}Sm	^{158}Gd	^{164}Dy
$E(1^{+})$ [MeV]	exp.	3.2	3.2	3.11
	th.	3.197	3.207	3.126
$B(M1;0^{+} \rightarrow 1^{+})$ $[\mu_N^2]$	exp.	0.8(0.2)	1.4(0.3)	1.5(0.3)
	th.	5.1	4.9	5.1

These results were obtained under the assumption that only the outer particles, which give the main contribution (about 98%) to the nuclear moment of inertia, take part in the rotation-like isovector vibrations. These are about half of all the nucleons. The overestimation of B(M1) is a typical shortcoming of the collective model, since all the strength of the M1 mode is concentrated in one state only.

A possible splitting of the 1^{+} state due to triaxiality was studied [6] by obtaining the strength G of the restoring force with a triaxial Woods-Saxon potential. The results for ^{48}Ti (Table 2) were obtained with $\beta = 0.2$, $\gamma = 20°$.

Table 2. Splitting of the scissor vibrations in ^{48}Ti (exp. from [7])

	$E(1_1^{+})$ [MeV]	$E(1_2^{+})$	$B(M1;0^{+} \rightarrow 1_1^{+})$	$[\mu_N^2]$	$B(M1;0^{+} \rightarrow 1_2^{+})$
exp.	3.75	5.7	0.54(0.1)		-
th.	3.86	6.0	0.70		0.93

Thus, the low-energy isovector vibrations in spherical, axially symmetric and non-axial nuclei are described within a unified approach.

References:

1. W.D. Hamilton et al: Phys. Rev. Lett. 26, 2469 (1984)
2. A. Faessler: Nucl. Phys. 85, 653 (1966)
3. A. Faessler, R. Nojarov: Phys. Lett. 166B, 367 (1986)
   R. Nojarov, A. Faessler: to be published in J. Phys. G: Nucl. Phys.
4. A. Faessler et al: J. Phys. G: Nucl. Phys. 12, L47 (1986)
   R. Nojarov et al: Z. Phys. A324 (1986), in press
5. D. Bohle et al: Phys. Lett. 148B, 260 (1984)
6. A. Faessler et al: Z. Phys. A324 (1986), in press
7. D. Bohle et al: Bull. DPG 4, 592 (1986); priv. comm. Darmstadt

The Convection Current for the $0^+ \to 1^+$ Excitations in the Even-Even $f_{7/2}$ Shell Nuclei

T. Oda and K. Muto

Department of Physics, Tokyo Institute of Technology, Oh-okayama, Meguro-ku, Tokyo 152, Japan

The experimental puzzle —— a broad bump centered at 10.15 MeV in ^{51}V is observed in the proton inelastic scattering[1], while there are no strong M1 excitations in the (e,e') spectra[2] —— might suggest the destructive interference of spin and orbital contributions in this energy range, because intermediate energy proton scattering at small angles excites magnetic dipole states only through the spin part of the nucleon-nucleon interaction[3,4], whereas both spin and orbital parts of the electromagnetic interaction may contribute to inelastic electron scattering. In the large scale shell model calculation[5], however, appreciable orbital contribution for $T=T_Z$ states appears below 8.5 MeV, and no appreciable orbital contribution is seen in the excitation energy range where a strong M1 excitation is observed in the (p,p') reaction. MUTO and HORIE have shown that the strong fragmentation and continuous distribution of the calculated strength in the energy range between 8 and 12 MeV in ^{51}V could cause M1 amplitudes to be hardly detected by the (e,e') measurement, consistent with such an experimental condition that the peak-background ratio of (e,e') spectra is much smaller than that of (p,p')[6].

In Fig. 1, absolute values of matrix elements of convection (orbital) current and spin current for the single particle isovector M1 transitions are plotted. A large convection current contribution should occur through the single particle matrix elements between the same j orbits. We could expect, therefore, in the medium heavy and heavy nuclei that the M1 transition with large convection current contribution should excite the low-lying magnetic dipole state. M1 states at higher energies around 10 MeV are excited mainly by spin-flip transitions. As one can see from Fig. 1, orbital current could not contribute to excite such M1 states.

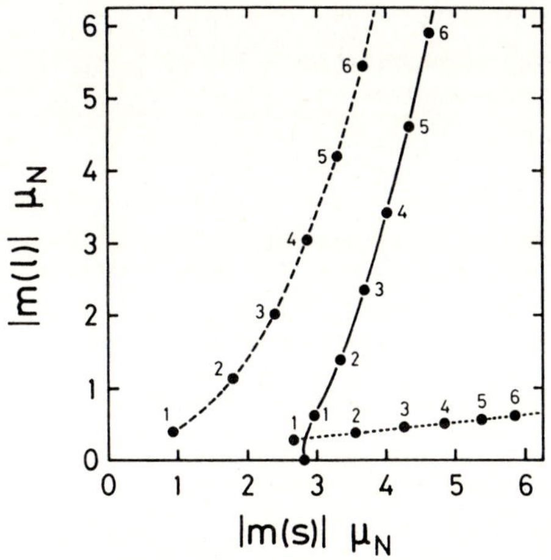

Fig. 1 $|m(\ell)|$ vs $|m(s)|$ for single-particle isovector M1 transitions $(j||M1||j') = m(s)+m(\ell)$; $j = j' = \ell+1/2$ (solid), $j = j' = \ell-1/2$ (dash) and $j = j'-1 = \ell-1/2$ (dot)

Isovector $K^{\pi}=1^{+}$ states, orthogonal to the spurious isoscalar $K^{\pi}=1^{+}$ states, have been predicted by the various collective models[7] also to be low-lying (2 ∿ 4 MeV). The very characteristic feature is that the M1 strength for electro-excitation of these 1^{+} states is assumed to come fully from convection current.

Recently, DJALALI et al.[8] performed high resolution inelastic proton scattering at 201 MeV on the heavy deformed nuclei (^{154}Sm, ^{156}Gd and ^{164}Dy) and the $f_{7/2}$ shell nucleus ^{46}Ti to clarify the nature of low-lying magnetic dipole states previously observed in inelastic electron scattering[9] and photoexcitation[10]. Combining the fact that intermediate energy proton scattering at small angles excites magnetic dipole states only through the spin part of the nucleon-nucleon interaction[3,4] and the backward angle cross sections for inelastic electron scattering, they separated the $0^{+}_{g.s.} \rightarrow 1^{+}_{1}$ transition strength $B(M1)\uparrow$ $(=M(\ell)\pm M(s))^{2}$) into its spin and convection current parts. They deduced the orbital-to-spin ratios $(M(\ell)/M(s))$ for ^{46}Ti and the lower limits of $M(\ell)/M(s)$ for the heavy deformed nuclei. For all four nuclei, they have shown that the spin contribution to the $B(M1)\uparrow$ is generally smaller than the orbital one[8].

ZAMICK[11] compared magnetic dipole excitations in the $f^{n}_{7/2}$ model with the new collective excitations in the heavy deformed nuclei. He showed by carrying out calculations for the titanium isotopes that there are systematic similarities in excitation energy and strength between those two types of magnetic dipole excitations. The $0^{+} \rightarrow 1^{+}$ M1 transition probability in the single j model is proportional to $(g_{\pi}-g_{\nu})^{2}$, and thus isovector type[11]. The orbiral-to-spin ratio is, therefore, fixed to be 0.64 if one uses the free g factors. The renormalized g factors generally accepted increase the ratio to 1.1[11]. Although IBA-2 formula[12] also has a factor $(\bar{g}_{\pi}-\bar{g}_{\nu})^{2}$, those g factors have no spin current contribution, ZAMICK[12] also showed in the $f_{7/2}$ shell that spin current contributes significantly even in a rotational model which uses the Nilsson intrinsic wave functions, and that the orbital-to-spin ratio does not exceed 1.8 even if one uses the large deformation parameter and the renormalized g factors.

In the pure $f_{7/2}$ model, most of the M1 strength is concentrated in the lowest 1^{+} state[12]. Since spin-flip transitions make main contribution both to the total B(M1) and to the M1 strength distribution up to higher energy (∿ 15 MeV), we have carried out the shell model calculations for the $0^{+}_{gr} \rightarrow 1^{+}$ M1 transitions of the $f_{7/2}$ shell nuclei in the full $f^{n}_{7/2} + f^{n-1}_{7/2}(f_{5/2}p_{3/2}p_{1/2})^{1}$ basis space in order to understand those M1 transitions in a unified way. The effective interactions obtained by YOKOYAMA and HORIE[13] are used. In Fig. 2, we show the calculated results of orbital current contribution $(M(\ell))$ versus spin current contribution $(M(s))$ to the $0^{+}_{gr} \rightarrow 1^{+}_{1}$ M1 transition probabilities. The full circles on the solid line, whose tangent is equal to 0.64, denote the values in the pure $f^{n}_{7/2}$ model. The full circle at the end of an arrow shows the value in the full $f^{n}_{7/2} + f^{n-1}_{7/2}(f_{5/2}p_{3/2}p_{1/2})^{1}$ configurations. The dashed line, the tangent of which is 1, is drawn for the guide to the eye. The orbital-to-spin ratios except the case of ^{44}Ti increase from 0.64 of the $f^{n}_{7/2}$ model to ∿ 1, when the model space is expanded to include the one nucleon excited configurations. This trend is caused by the effect of the first order configuration mixing. In the single j model $(j = \ell+1/2)$, the M1 transition amplitude is proportional to $g_{s}+2\ell g_{\ell}$. The first order configuration mixing gives rise to the spin-flip transition proportional to $g_{s}-g_{\ell}$. Therefore, the transition amplitude can be written as $\alpha(g_{s}+2\ell g_{\ell}) + \beta(g_{s}-g_{\ell}) = (\alpha+\beta)g_{s} + (2\ell\alpha-\beta)g_{\ell}$. Generally sign of α and β differ from each other, and magnitude of β is much smaller than that of α[14]. It can be expected that one nucleon configurations $(j^{n-1}j';j' = \ell-1/2)$ quench mainly spin current and enhance convection current only a little bit[15]. As a combined effect of these two, the orbital-to-spin ratios increase close to 1 from 0.64 of the pure $f^{n}_{7/2}$ model.

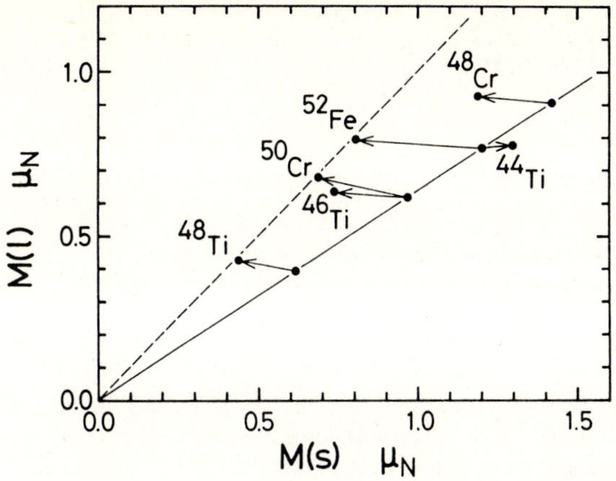

Fig. 2 M(ℓ) vs M(s) in the pure $f^n_{7/2}$ model and in the $f^n_{7/2}+f^{n-1}_{7/2}j'$ ($j' = p_{1/2}p_{3/2}$ or $f_{5/2}$) configurations

We have not shown in Fig. 2 the results for the N=28 isotones. Since 1^+ states do not exist in the identical nucleon j^n configurations, any 1^+ states of the N=28 isotones do not have large orbital configurations in the M1 transitions. EVANS et al.[16] pointed out using the IBM-3 that the 1^+ states in Fig. 2, the states of seniority 4 and reduced isospin 1 in the shell model classification, are the states of mixed symmetry as in the heavy nuclei.

If we switch off spin current and calculate orbital current B(M1)'s for all the 1^+ states, we have the largest orbital current B(M1) for the lowest 1^+ state — 60 ∿ 70% of the total B(M1) of orbital current (40% in the case of ^{48}Ti) —. Spin current contributes to the total B(M1) more than a factor 10 as orbital current does, and those 1^+ states, which have strong spin current contribution, are lying in the region higher than 7 ∿ 8 MeV. Although there is still comparable contribution from spin current for the lowest 1^+ B(M1), the lowest 1^+ states in Fig. 2 are considered to be precursors of the new M1 collective 1^+ states whose B(M1)'s are believed to come only from orbital current.

1. C. Djalali et al: Nucl. Phys. A388, 1 (1982)
2. D. Bender et al.: Nucl. Phys. A398, 408 (1983)
3. G. M. Crawley et al.: Phys. Rev. C26, 87 (1982)
4. C. Djalali: Proc. Int. Symp. on Highly Excited States, J. Phys. (Paris) C4, 375 (1984)
5. K. Muto and H. Horie: Nucl. Phys. A440, 254 (1985)
6. The same as [4].
7. N. Lo Iudice and F. Palumbo: Phys. Rev. Lett. 41, 1532 (1978)
   T. Suzuki and D. J. Rowe: Nucl. Phys. A289, 461 (1977)
   A. E. L. Dieperink: Prog. Part. Nucl. Phys. 9, 121 (1983)
   E. Lipparini and S. Stringari: Phys. Lett. 130B, 139 (1983)
   H. Kurasawa and T. Suzuki: Phys. Lett. 144B, 151 (1984)
   I. Hamamoto and S. Aberg: Phys. Lett. 145B, 163 (1984)
   O. Scholten et al.: Nucl. Phys. A438, 41 (1985)
   A. van Egmond et al.: Preprint Amsterdam University
8. C. Djalali et al.: Phys. Lett. 164B, 269 (1985)
9. D. Bohle et al.: Phys. Lett. 137B, 27 (1984)
   D. Bohle et al.: Phys. Lett. 148B, 260 (1984)
10. U. E. P. Berg et al.: Phys. Lett. 149B, 59 (1984)
11. L. Zamick: Phys. Rev. C31, 1955 (1985)
12. L. Zamick: Phys. Lett. 167B, 1 (1986)
13. A. Yokoyama and H. Horie: Phys. Rev. C31, 1012 (1985)
14. H. Noya, A. Arima and H. Horie: Prog. Theor. Phys. Suppl. No. 8, 33 (1958)
15. A. Arima: Proc. Int. Conf. on Nuclear Moments and Nuclear Structure (Osaka, 1972), edited by H. Horie and K. Sugimoto (Phys. Soc. Japan), p.205, 1973
16. J. A. Evans, J. P. Elliott and S. Szpikowski: Nucl. Phys. A435, 317 (1985)

Gamow-Teller Strength from Spin-Isospin Saturated Nuclei

B. Desplanques[1] *and S. Noguera*[2]

[1]Division de Physique Théorique*, Institut de Physique Nucléaire, F-91406 Orsay Cedex, France

[2]Departamento de Fisica Teorica, Facultad C. Fisica, c/Dr Moliner s/n, Burjasot (Valencia), Spain

The Gamow-Teller strength has mainly been discussed for nuclei with $N \neq Z$ where it should be different from 0 in the simplest model used for their description. With more refined descriptions, some strength is also expected for nuclei with $N = Z$ such as ^4He, ^{16}He, ^{40}Ca and ^{80}Zr. It then takes its origin in the ground-state correlations induced from those parts of the NN force which break the spin-isospin SU(4) symmetry. Its observation, which is not easy, may cast some light on these ground-state correlations as well as to some quenching mechanism of the Gamow-Teller strength at low energy /1/.

The present work improves upon a previous one by ADACHI, LIPPARINI and NGUYEN VAN GIAI /2/ in several respects. It includes the contribution of 2p-2h in the ground-state with an energy up to 150 MeV. This has partly been done to make some contact with nuclear matter calculations and to test the convergence of the results. Our study includes ^4He as a closed shell nucleus. This allows to check the reliability of the ingredients of our study as far as its D-state probability is concerned. This quantity is expected to be of the order of 10% from the most realistic calculations. We also consider various single-particle wave-functions. Beside those calculated with an harmonic oscillator or a Woods-Saxon potential, we use those derived from a Skyrme-type force. In this case, the main parameter relevant for our study is the effective mass which has been chosen to be 0.7 times the nucleon mass at the nuclear density of 0.15 fm^{-3}. Finally, we also consider the contribution of some force in odd parity states.

The Gamow-Teller strength which we are interested in here may be written as :

$$S_{GT}^{\beta^{\pm}} = \sum_{\substack{n \\ j=1,2,3}} |\langle n| \sum_{i=1} \sigma_i^j \tau_i^{\pm} |g.s.\rangle|^2$$

$$= \int dE\, G^{\sigma\tau^{\pm}}(E). \qquad (1)$$

The last equality defines the strength density which is introduced to provide some graphical representation of the results (after some smoothing). The calculation has been performed at the lowest order in the interaction, therefore neglecting the rearrangement of the strength in the final state, or the coupling between 2p-2h configurations in the final state.

Several forces have been employed in our study. Results are presented here for a force which is close to the one used by BERTSCH and HAMAMOTO /3/ for the even parity states. For the central part relevant to odd parity states,

---

* Laboratoire associé au C.N.R.S.

we added a ρ-exchange type contribution to the π-exchange one, while the tensor part was chosen to be identical to the one for even parity states /4/.

Results obtained with single-particle harmonic oscillator wave function (HO) for the strength density per nucleon are presented in Fig.1. For all nuclei under consideration, the high energy tail around 150 MeV is essentially determined by the contribution of the $^3S_1$-$^3D_1$ transition arising from the tensor force. The differences at lower energies are largely due to other contributions from the tensor and central forces in odd parity states. The integrated strength for ^{4}He would correspond to a D-state probability of about 25%, which is much larger than what is indicated by current calculations with realistic forces. The percentage of excitation of the $^3D_1$ component per nucleon through the $^3S_1$-$^3D_1$ tensor coupling in ^{40}Ca and ^{80}Zr is close to 15%, which is also somewhat higher than what is usually obtained in the most recent nuclear matter calculations (8%) /5/.

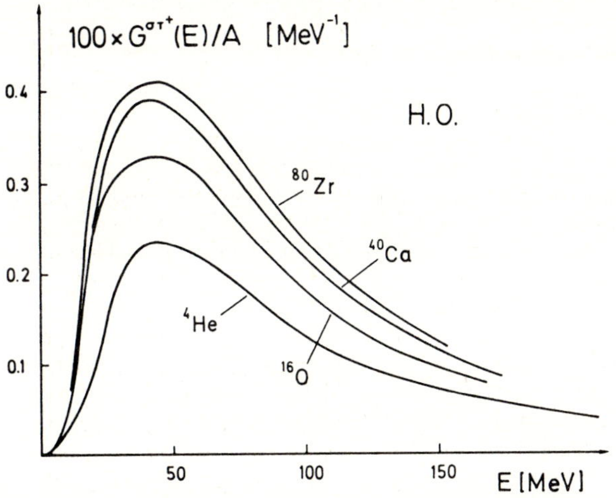

Figure 1 : Representation of the Gamow-Teller strength density calculated with harmonic oscillator wave functions for ^{4}He, ^{16}O, ^{40}Ca and ^{80}Zr

Results with single-particle wave-functions calculated from a Woods-Saxon potential (WS) or from a Skyrme force (SK) are presented in Fig.2 for ^{40}Ca. The structures which appear in the figure are due to the calculations which allow for a better energy resolution than in the previous case where the strength appearing in principle at 2, 4, 6 ... $\hbar\omega_0$ was spread somewhat uniformly over an energy range of $2\hbar\omega_0$. On the average, results with Woods-Saxon wave-functions are close to those obtained with harmonic oscillator wave-functions and the first structures observed in that case would roughly correspond to the ones expected for the harmonic oscillator model in absence of spreading of the strength.

A striking difference appears when single-particle wave-functions obtained from a Skyrme force are used. The strength is considerably reduced (a factor 2) and slightly shifted to higher energies. This is entirely due to the spatial non-locality of the mean-field, parametrized by an effective mass whose value is $M^* = 0.7$ M at the nuclear density of 0.15 fm^{-3}. The D-state probability in ^{4}He is now close to 10% while the percentage of excitation of the $^3D_1$ component per nucleon (through the $^3S_1$-$^3D_1$ tensor coupling) is about 8%, in

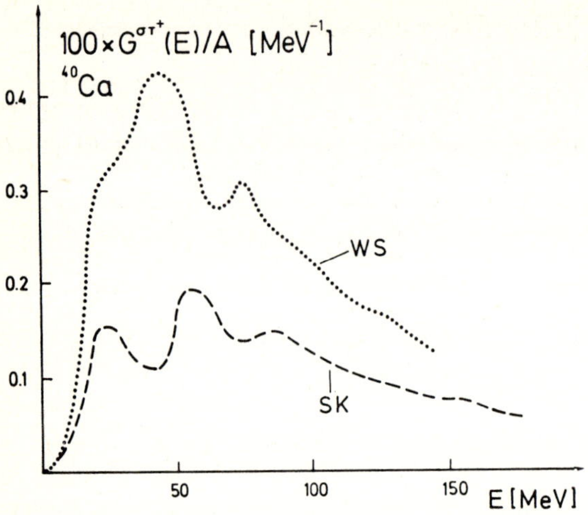

Figure 2 : Representation of the strength density calculated in ^{40}Ca for Woods-Saxon and velocity dependent potential wells

much better agreement with nuclear matter calculations. As this last set of results turns out to be quite reasonable in view of what is known from other fields, it is now possible to consider as representative ones the corresponding expectations for the total Gamow-Teller strength, $S_{GT}^{\beta\pm}$, of closed shell nuclei. They are respectively 0.45, 2.7, 8 and 15 (extrapolated) for ^{4}He, ^{16}O, ^{40}Ca and ^{80}Zr. For a local single-particle potential, they would have been 1.1, 5.4, 15.7 and 31. The comparison between both sets of numbers shows the importance of an effect which has been largely ignored in studies dealing with Gamow-Teller transitions and should reduce corrections to them due to 2nd order configuration mixing /3, 6, 7/.

The Gamow-Teller strength considered here should also affect the study of the Gamow-Teller strength in nuclei with N ≠ Z. Thus, to the strengths of 24 and 30 expected in the simplest models for ^{48}Ca and ^{90}Zr respectively ($\sigma\tau^+$ branch), one has to roughly add extra strengths whose values are likely to be close to those we calculated for ^{40}Ca and ^{80}Zr, 8 and 15. This point is important because part of the strength of the first kind (15-20% according to estimates with the same ingredients as in this paper) which is missing at low energy is expected to be found at higher energy. There, it should mix with the strength of the second kind, what has to be kept in mind in analyzing experiments.

Part of this work was supported by CAICYT.

/1/ B. Desplanques : Phys. Lett. **141B**, 285 (1984)
/2/ S. Adachi, E. Lipparini and Nguyen Van Giai : Nucl. Phys. **A438**, 1 (1985)
/3/ G.F. Bertsch and I. Hamamoto : Phys. Rev. **C26**, 1323 (1982)
/4/ B. Desplanques and S. Noguera : Phys. Lett. **173**, 23 (1986)
/5/ P. Grangé (private communication)
B.D. Day and R.B. Wiringa : Phys. Rev. **C32**, 1057 (1985)
/6/ K. Shimizu, M. Ichimura and A. Arima : Nucl. Phys. **A226**, 282 (1974)
/7/ I.S. Towner and F.C. Khanna : Nucl. Phys. **A399**, 334 (1983)

Calculation of the Gamow-Teller Resonance in Nuclear β-Decay: The Cases ^{34}Ar and ^{35}Ar

W. Knüpfer[1], *B. Metsch*[1*], *W. Müller*[1], *and A. Richter*[2**]

[1]Institut für Theoretische Physik, Universität Erlangen-Nürnberg, D-8520 Erlangen, F. R. Germany

[2]Institut für Kernphysik der Technischen Hochschule Darmstadt, D-6100 Darmstadt, F. R. Germany

Abstract: The strength functions for the ß+ decay of the proton rich nuclei ^{34}Ar and ^{35}Ar into a Gamow Teller resonance are studied using shell model wave functions. A resonance-like structure is calculated to be positioned about 3 MeV above the state that is the isobaric analogue state of the ground state of the decaying nucleus. Similarly to previous calculations for A = 32 and 36 these nuclei posses a spin-flip resonance based on the excited counterpart of the ground state.

The excitation of the $T_> = T_0 + 1$ states, where T_0 is the lowest isospin for a given nucleus, by strong magnetic dipole transitions has been the subject of continous interest during the past decade. In light A = 4 n nuclei these (J^π; T = 1+; 1) resonance states are in turn fed by strong M 1 transitions from the lowest (J^π; T = 0+;2) state which led Hanna [1] to the interpretation of giant M1 resonances built on giant M1 resonances as sketched in the left part of fig. 1. In view of the fact that the electromagnetic decay of the (J^π;T=0+;2) state mainly reveals the lowest part of the M1 distribution it seems to be an interesting question as to how the (J_i ; T_i = 0+; 2) -> (J_f; T_f = 1+; 1) spin-flip strength is distributed. Clearly one would expect this spin-flip strength to be localized at higher excitation energies above the (J^π; T = 0+, 2) state, which means that they are virtually inaccessible for electromagnetic investigation. However the analogue GT transitions from the (J^π; T = 0+; 2) ground state into the (J^π; T = 1+; 1) states of the daughter nucleus can be studied provided that the parent state lies energeti cally above the resonance states of the daughter nucleus (see right part of fig. 1). In practice this is realized in proton rich light nuclei, where the Coulomb interaction can shift the parent state (stable against heavier particle decay!) above the resonance states of daughter nucleus, thus allowing an investigation of the spin-isospin GT resonance by measuring ß+ decay rates.

Recently, detailed measurements of the ß+ strength functions have been reported for A = 32 from the ISOLDE collaboration at CERN [2]. Shell model predictions exist for the strength distributions $^{32}Ar \xrightarrow{\beta^+} {}^{32}Cl$, $^{33}Ar \xrightarrow{\beta^+} {}^{33}Cl$ and $^{36}Ca \xrightarrow{\beta^+} {}^{36}K$. [3] demonstrating that a large part of the strength of the spin-isospin flip transitions starting from the (J^π; T = 0+; 2) ground state of proton rich nuclei in the sd-shell is concentrated in a relatively narrow energy region, approximately 3 MeV above the state that is the isobaric analogue of the ß+ decaying state.

---

*) permanent address: Institut für theoretische Kernphysik der Universität Bonn

**) supported by Deutsche Forschungsgemeinschaft

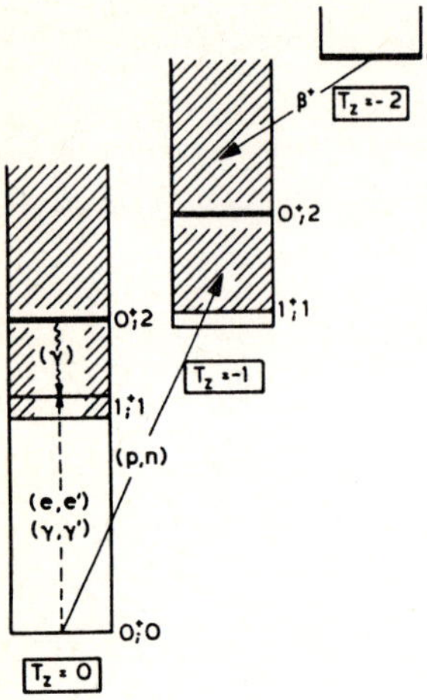

Fig. 1. Interrelations of spin-isospin excitations in light A=4n mass-systems. The $(J_i;T_i;\to J_f;T_f)=(0^+;0\to 1^+;1)$ excitations can be explored by (e,e) and (p,p) inelastic scattering or alternatively by the analogue GT excitation via (p,n). The $(J_i;T_i\to J_f;T_f)=(0^+;2\to 1^+;1)$ excitation may be studied through β^+ decay of the proton-rich isobar. The lowest part of this excitation can also be seen in γ-decay of the $(J^\pi;T=0^+;2)$ state.

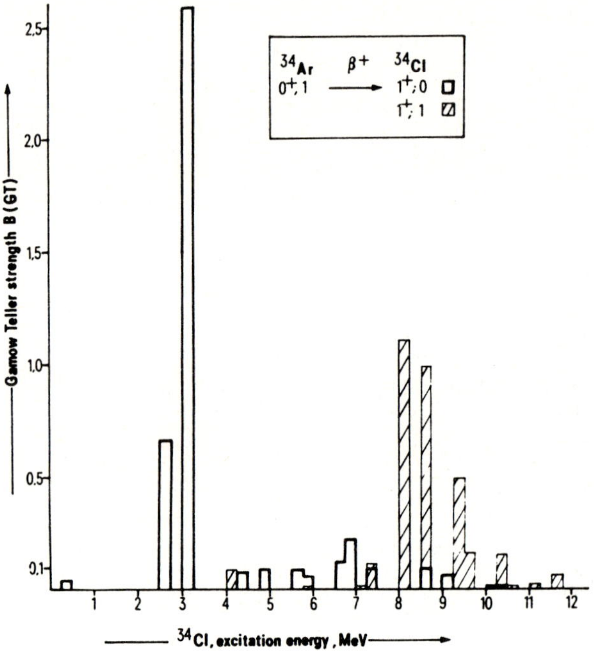

Figs. 2,3 : GT-strength distributions for the ^{34}Ar and ^{35}Ar β^+ decay. For ^{35}Ar all transitions to J^π, T states are included accessible by β^+ decay.

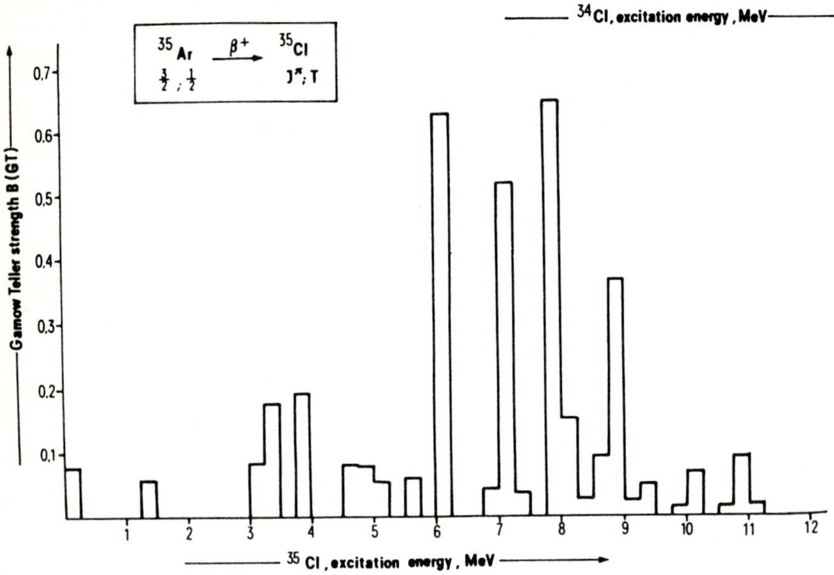

This GT resonance can be interpreted as the ($d5/2^{-1}$ d3/2) excitation of the (J^π; T = 0+; 2) state. For the experimentally observed GT strengths of the nuclei ^{32}Ar and ^{33}Ar a retardation is found of about half of what would be expected from the calculations. We therefore infer that this retardation is not much different in β^+ decay and in (p,n) reactions.

In the present calculations we extend the investigations to the nuclei 34 Ar and 35 Ar, respectively. The GT-strength distributions for these nuclei have been obtained by diagonalization of an empirically determined effective interaction in a model space that includes all distributions of A-16 particles over the sd-shell orbits [3].

For $^{34}Ar \xrightarrow{\beta^+} {}^{34}Cl$ (see fig. 2) the concentration of GT strength around 3 and 8 MeV reflects the isospin splitting between the T=0 and T=1 channel, respectively. Both resonances justify an interpretation as the $d5/2^{-1}$ d3/2 of the (J^π;T) = (0+, 1) state. The resulting strength distribution of $^{35}Ar \xrightarrow{\beta^+} {}^{35}Cl$ decay is shown in fig. 3. Again the concentration of strength around 8 MeV is of the same microscopic structure as for the other Ar-nuclei. Preliminary results of the ISOLDE collaboration on ^{34}Ar show approximately a similar retardation of the spin-flip strength than for ^{32}, ^{33}Ar.

In conclusion, the existence of resonance like spin-flip excitations on excited states |0>* with the spin of the ground state has been studied by calculating the Gamow Teller β^+ decay from the state that is the isobaric analogue state of this specific excited state |0>*. A comparison with the available data for A=32, 33 and 34 reveals a hindrance of strength in the resonance region of a factor 2-3, in accordance with informations from other spin-flip probes.

|1| S.S. Hanna, Nukleonika 19 (1974) 655
|2| T. Bjömstad et al., Nucl. Phys. A 443 (1985) 283
|3| W. Müller et al., Nucl. Phys. A 430 (1984) 61

Nuclear Spin-Isospin Excitations Studied by Photopion Productions

K. Shoda and A. Kagaya

Laboratory of Nuclear Science, Tohoku University, Mikamine, Sendai 982, Japan

Nuclear spin-isospin excitations have been studied by the electromagnetic interaction (e,e'), (π^-,γ) and $(\gamma,\pi^\pm)$ as well as high energy hadronic interaction (p,n) and (n,p) reactions. In these reactions, isospin transition is uniquely $\Delta T=1$ except (e,e'). We will discuss the general aspect and systematic comparisons for the angular distribution of $(\gamma,\pi^\pm)$ reactions. The momentum transfer to the residual nucleus can be changed by the pion emission angle and photon energy in this reaction.

The main term of the $(\gamma,\pi^\pm)$ interaction matrix is given by PWIA as $\langle f|\sum_i^A \vec{\sigma}\cdot\vec{\varepsilon}\tau^\pm e^{-i\vec{q}\cdot\vec{r}}|i\rangle$ where $|i\rangle$, $|f\rangle$ are the nuclear states. The multipole expansion of this operator is given as shown in table 1 (L is the transfer of the angular momentum). The M0 and electric transitions correspond to a single operator but magnetic ones double. The transitions within the p shell lead to L=0 and 2, then the $j_0\sigma Y_0$ type (M1) and $j_2\sigma Y_2$ type (M1, E2, M3) are possible. The transitions from p to d shell lead to L=1 and 3 then the $j_1\sigma Y_1$ (M0, E1, M2) and $j_3\sigma Y_3$ type (M2, E3, M4) are possible. The pion angular distrubutions in these transitions are approximately expressed by the spherical Bessel function as $|\langle f|j_L(q)Y_L(\hat{q})|i\rangle|^2$. There are two types of patterns in special magnetic transitions (M1 for p-p, M2 for p-d). Precise operator of the $(\gamma,\pi^\pm)$ transition is composed of several terms with spin-flip type and non--flip type in addition to the $\vec{\sigma}\cdot\vec{\varepsilon}\tau^\pm$ main term [1]. When the matrix element of the main term is very small, the other terms and/or other modes contribute apparently.

table 1. Multipole components of the main term $(\vec{\sigma}\cdot\vec{\varepsilon})$ in $(\gamma,\pi^\pm)$

L	operator	ΔJ^π	possible transition from s, p, d shell				
0	$j_0(q)\sigma Y_0$	1^+ (M1)	s-s	p-p		d-d	...
1	$j_1(q)\sigma Y_1$	0^- 1^- 2^- (M0, E1, M2)	s-p		p-d	d-f	...
2	$j_2(q)\sigma Y_2$	1^+ 2^+ 3^+ (M1, E2, M3)		p-p		d-d	...
3	$j_3(q)\sigma Y_3$	2^- 3^- 4^- (M2, E3, M4)			p-d	d-f	...
4	$j_4(q)\sigma Y_4$	3^+ 4^+ 5^+ (M3, E4, M5)				d-d	...
5	$j_5(q)\sigma Y_5$	4^- 5^- 6^- (M4, E5, M6)				d-f	...

There are many experimental examples of the π angular distributions of $(\gamma,\pi^\pm)$ reaction on p-shell nuclei ^{6}Li[2], ^{9}Be[3], ^{10}B[4], ^{12}C[5], ^{13}C[6,7], and ^{14}N[8,9] and s-d shell nuclei ^{24}Mg[10] and ^{28}Si[11]. Some of them are shown in fig.1. In the case of $^{13}C(\gamma,\pi^+)$, $^{13}C(\gamma,\pi^-)$ and $^{14}N(\gamma,\pi^+)^{14}C^*$ in fig.1 and also ^{6}Li, ^{12}C, ^{24}Mg and ^{28}Si, the angular distributions show forward (small q) peak which agrees with the theoretical estimates reflecting $\langle j_0(q)\rangle^2$ pattern for $j_0\sigma Y_0$ type. In the backward region (large q) the cross section of this pattern is usually small, therfore other mode of transition sometimes can be studied as shown in fig.1 a) b) c). In the case of M1 transition for $^{14}N(\gamma,\pi^+)^{14}C$g.s., the pattern of the angular distributions is quite different from others. In addition to the small $\langle j_0(q)\rangle^2$ component, $\langle j_2(q)\rangle^2$ contributes to this M1 transition. This pattern shows a clear contrast to the case of $^{14}N(\gamma,\pi^+)^{14}C^*$ in fig.1 c). In the electric transition, since

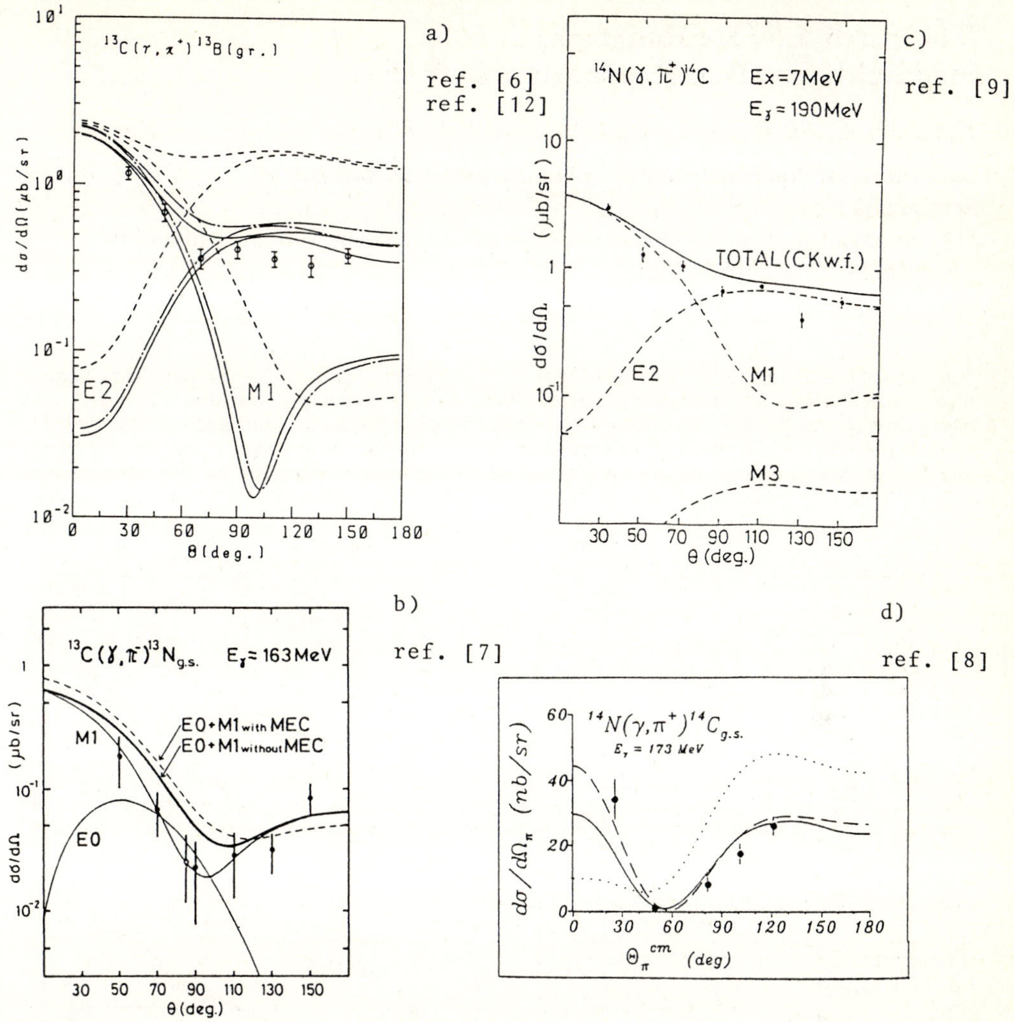

fig. 1. Examples of angular distributions.

the main term of operator is unique, the angular distribution shows sensitive dependence on |i> and |f>.

References
1 M.K. Singham and F. Tabakin, Annals of Physics 135 (1981) 71.
2 K. Shoda et al., Phys. Lett. 101B (1981) 124.
3 K. Shoda et al., Nucl. Phys. A403 (1983) 469.
4 M. Yamazaki et al., preprint (1985).
5 K. Shoda et al., Nucl. Phys. A305 (1980) 377.
6 K. Shoda et al., Phys. Rev. C27 (1983) 443.
7 K. Shoda et al., Phys. Lett. to be published (1986).
8 K. Rohrich et al., Phys. Lett. 153B (1985) 203; B.H. Cottman et al., Phys. Rev. Lett. 55 (1985) 684; K. Rohrich et al., private commiunication.
9 B.N. Sung et al. private communication (1986).
10 H. Tsubota et al. private communication (1986).
11 K. Shoda et al. Nucl. Phys. A439 (1985) 669.
12 T. Sato et al. Z.Phys. A320 (1985) 507.

$^{6,7}Li(\gamma,\pi^+)$ $^{6,7}He$ Reactions for Highly Excited Resonances in $^{6,7}He$

K. Shoda, O. Sasaki, S. Toyama, H. Tsubota, T. Kobayashi, and A. Kagaya*

Laboratory of Nuclear Science, Tohoku University, Mikamine, Sendai 982, Japan

*Department of Physics, College of General Education, Tohoku University, Kawauchi, Sendai 980, Japan

The excited states in ^{6}He and ^{7}He have not been well known. Two broad resonances have been studied in the energy region upto 25 MeV by (n,p) reaction [1] and radiative π^- capture [2]. The (n,p) study concludes these resonances correspond to the GDR with spin-non-flip isovector type. On the other hand, the (π^-,γ) results which show resonances similar to the (n,p) results are expected by the spin-isospin flip modes. Present experiments have been made for the study of these resonances.

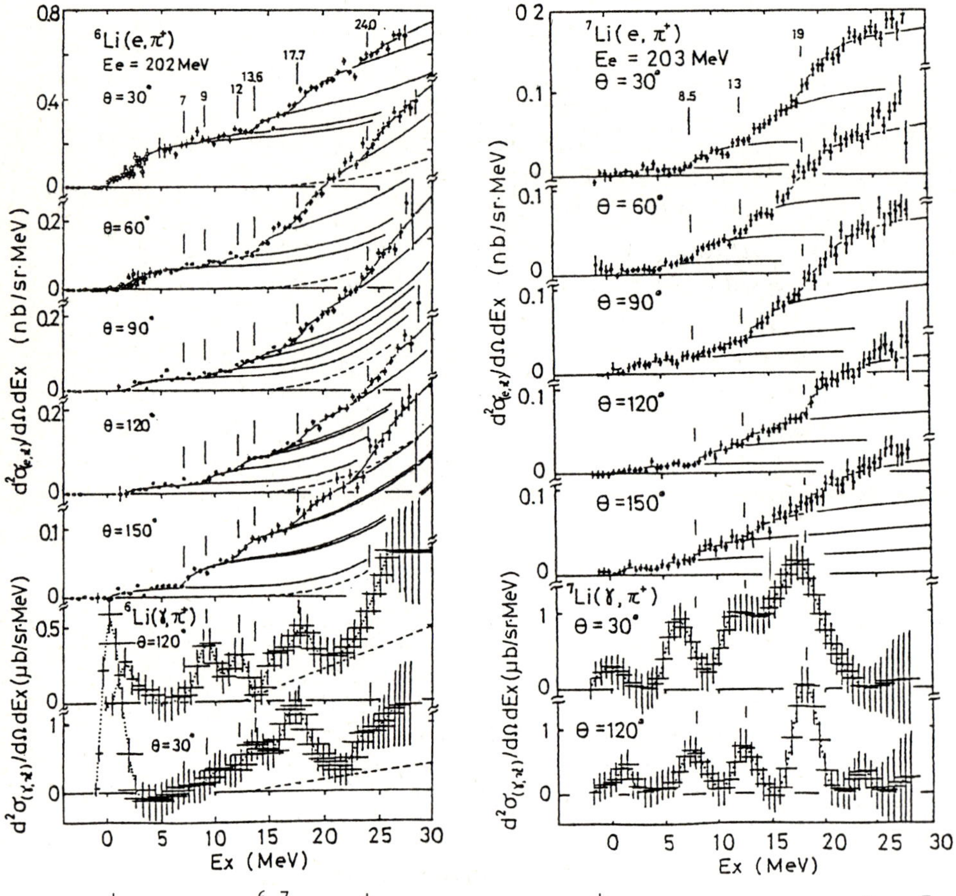

Fig.1. π^+ spectra of $^{6,7}Li(e,\pi^+)$ and unfolded (γ,π^+) cross sections. $Ex=T^\pi_{max}-T^\pi$ for spectra. Determined resonance energies are shown.

Isotopically enriched ^{6}Li and ^{7}Li targets 100-200 mg/cm^2 thick were bombarded by a 202 MeV and 203 MeV electron beam from the Tohoku University linear accelerator, respectively. The (e,π^+) spectra measured by a magnetic spectrometer are shown in fig.1. They are unfolded to give $\sigma(\gamma,\pi^+)$ as a function of the residual energy (Ex) by a method similar to the least structure method [3] which leads to correct absolute value but smeared out the structure. The results are shown in the lower parts of fig.1. The dashed curves show the continuum estimates from these results. Another photon difference method as well as the fitting method on the breaks photon difference method as well as the fitting method on the breaks in the spectra is applied to find the residual states for strong transitions. The results of the resonance energies in ^{6,7}He are indicated by virtical lines in the figure.

The least square fit to the spectra is made using the determined Ex mentioned above in order to deduce $d\sigma(\gamma,\pi^+)/d\Omega$. The results are shown by the angular distrubutions in fig.2. In the figure, comparison with the theoretical estimates is made. The estimates were made by DWIA calculation with a single particle shell model by harmonic oscillator potential. Though the absolute values are not discussed, the normalized patterns reproduce the data well as shown in the figure. In the case of ^{6}Li(γ,π^+), the theoretical results by Eramzhyan et al. [4] are also shown by the absolute cross section. Their energies Ex are modified here so as to give a best fit to the present experimental spectra.

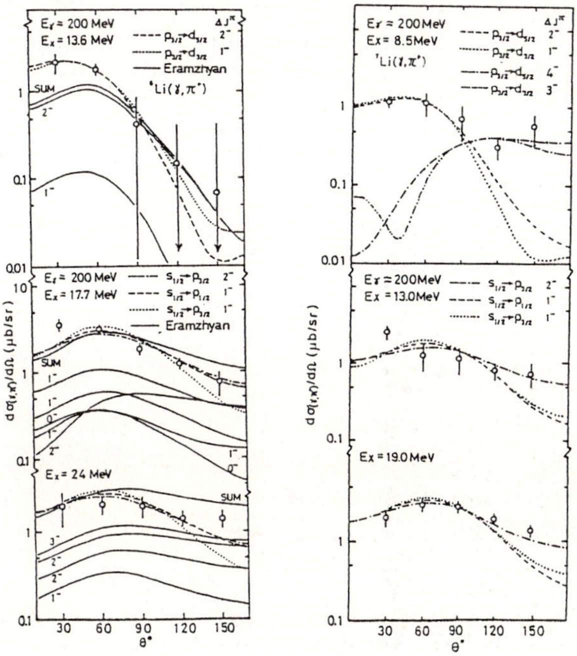

fig.2. Angular distributions of ^{6,7}Li(γ,π^+). Curves are theoretical estimates.

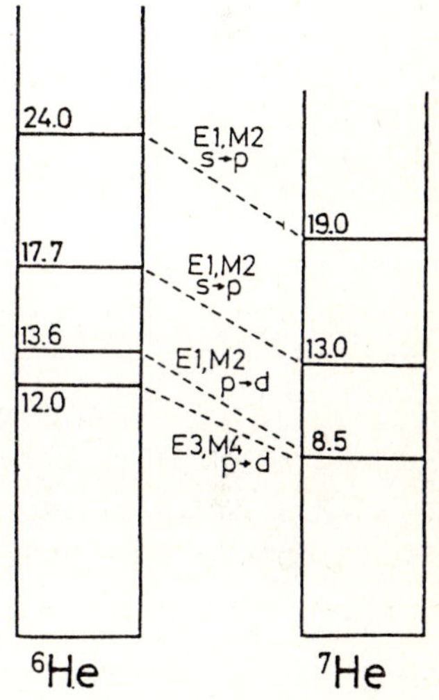

fig.3. Spin-isospin flip giant resonances in ^{6}He and ^{7}He.

The residual resonance energies and the deduced multipolarity and modes of the nuclear transitions are summarized in fig.3. Since the strong (γ,π^+) reaction relates to the spin-isospin flip type $(\vec{\sigma},\vec{\varepsilon}\tau^+)$, the resonance should be of spin--isospin flip mode. The estimated multipolarity shows mainly spin-dipole isovector giant resonances. The lower resonances correspond to the p→d transitions and higher ones to the s→p transitions. The excitation energies of these resonances are about 5 MeV low in ^{7}He than the corresponding ones in ^{6}He.

References

1 F.P. Brady et al., J. Phys. G:Nucl. Phys. 10 (1984) 363.
2 H.W. Baer et al., Phys. Rev. C8 (1973) 2029; D. Renker et al., Phys. Rev. Lett. 41 (1978) 1279.
3 B.C. Book, Nucl. Instrum. Methods 24 (1963) 256.
4 R.A. Eramzhyan et al. Z. Phys. A322 (1985) 321.

Part 2

Electroweak Interactions in Nuclei and Subnucleonic Structure

The Experimental Status of the EMC Effect

K. Rith

Max-Planck-Institut für Kernphysik, D-6900 Heidelberg, F. R. Germany

1. INTRODUCTION

The challenge for todays nuclear physics is to develop a fundamental theory of nuclei and of nuclear forces based on quarks, gluons and their interactions.

In this connection there are a lot of basic questions to be answered such as: are the properties of the nucleon, like its mass or size, affected by the presence of other nucleons in a nucleus? Do nucleons swell, but still keep their quark and gluon content? Do they (or at least some of them) overlap and form multiquark clusters or even one big bag, where there is free colour flow and the quarks can no longer be assigned to individual nucleons while due to the strong colour forces it is not possible to liberate them from free hadrons? Are the confinement conditions in a nucleus different from the free nucleon case and could therefore the nucleus be used as a laboratory to study and to try to understand confinement?

Furthermore, how are these colourless objects bound together? Is it by the exchange of other colourless composite objects like pions or generally speaking meson exchange currents, a mechanism which is currently not understood in the framework of QCD? Can at least the short range part of the nuclear forces be directly described by the fundamental strong force between the fundamental particles, that means by the exchange of quarks and gluons or some multigluon states? Can QCD be extended into a region where the strong coupling constant is large and is it possible to find a unique description of the strong colour forces and the nuclear forces?

2. THE TOOL

In my opinion the best way to attack these questions is to look at quark and gluons directly in deep inelastic lepton nucleus scattering experiments, which allow to probe the nucleus at distances which are two to three orders of magnitude smaller than the nucleon size, and to study the modifications of quark and gluon distributions $q(x,Q^2)$ and $g(x,Q^2)$ due to the nuclear environment.

These are obtained from the measurement of the structure function $F_2^N(x,Q^2)$ per nucleon which is defined as

$$F_2^N(x,Q^2) = x\sum_f z_f^2(q_f(x,Q^2) + \bar{q}_f(x,Q^2))$$

where the sum runs over the different quark flavours f, z_f is the charge of the quark (in units of $|e|$) and $q_f(x,Q^2)$ ($\bar{q}_f(x,Q^2)$) is the probability that a quark (antiquark) of flavour f carries the fraction x of the nucleon light cone momentum. x is given by $x=Q^2/2M\nu$, where Q^2 is the negative square of the four momentum carried by the virtual photon exchanged between lepton and nucleon, M is taken to be the proton mass and $\nu=E-E'$ is the energy transfered by the virtual photon from the lepton to the nucleon.

Nuclear effects in F_2^N were first observed by the European Muon Collaboration (EMC) at CERN [1], therefore the name EMC effect, and then by a series of other deep inelastic electron, muon and neutrino experiments. In this contribution I will summarise the main features of the existing data and some implications due to recent results, but I will not discuss the large variety of possible explanations since many reviews already exist [2] and the theoretical aspects will be covered by many other speakers at this conference.

3. THE DATA

3.1 EMC Results

3.1.1 Quark distributions

Figure 1 shows the original data of the EMC effect [1], the ratio of the nucleon structure functions for iron, $F_2^N(Fe)$, and deuterium, $F_2^N(D)$, plotted against x, together with the expectation of previous Fermi motion model calculations [3]. The data have been taken at muon energies between 200 and 280 GeV. The mean Q^2 of the data points at low x is about 16–20 GeV^2, it increases to about 80 GeV^2 for x=0.65. The shaded area indicates the range for the error on the slope of a straight line fit to the data, the systematic errors for each point are somewhat larger. In addition there is an overall normalisation uncertainty of ±7%. It is obvious that these errors are too large for detailed conclusions about the underlying physics and that dedicated simultaneous measurements on pairs of nuclei are needed to reduce normalisation errors and systematic errors to a minimum.

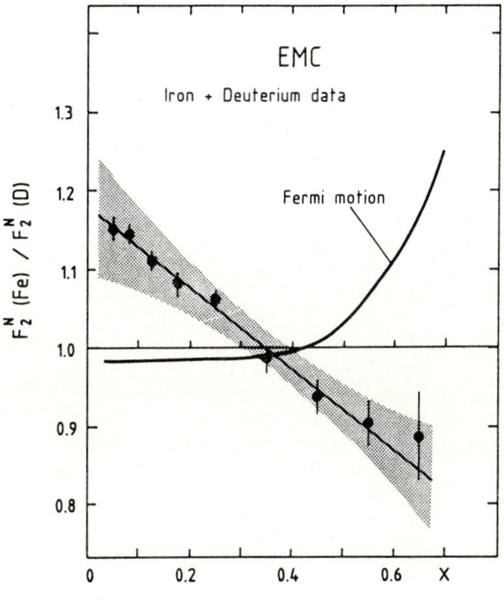

Fig. 1 Q^2 averaged ratio of the nucleon structure function F_2^N for iron and deuterium

Keeping in mind these restrictions the result tells us that at x>0.3 (valence quark region) less quarks with high momentum are found in a nucleus than in a free nucleon. At x=0.65 the difference to the expectation for iron is about 30–40%. At low x (sea quark region), where only little difference had been predicted, an enhancement is observed, suggesting an increase of the momentum fraction carried by seaquarks.

This behaviour of the data immediately provoked the question: does one observe an <u>additional sea component</u> in the nucleus? Are those the sea quarks responsible for the short range forces? Or does one observe the presence of extra pions (meson exchange currents) in the nucleus [4]?

To answer this important question obviously better experimental information is needed. Due to their large normalisation uncertainty the EMC data allow any value for an enhanced sea between 0 and 45%. For these studies, in principle neutrino/antineutrino experiments are superiour since they allow to extract the sea quark distributions directly, but unfortunately they will always suffer from too low statistics. The CDHS experiment [5a], for example, obtained for the ratio of the antiquark distributions in iron and in hydrogen a mean value of 1.10 ± 0.11 (stat.) ±0.07 (syst.), where the systematic error, in my opinion, is still underestimated. The results of the other neutrino experiments [5] are even more inconclusive. Hopefully Drell-Yan experiments on different nuclei [6], with incident proton energies around 800 GeV, which will be soon possible at FNAL, will prove to be better suited for these studies.

3.1.2 Gluon distribution

Quark and gluon distributions are related by the elementary QCD processes of gluon emission by quarks and the quark-antiquark creation out of gluon. Therefore, if the quark distributions are modified by nuclear effects, then the gluon distribution must also be affected.

One way to determine the gluon distribution is from the cross section for J/Ψ production studied via the $\mu^+\mu^-$ decay of the J/Ψ.

The EMC has found that the cross section per nucleon for J/Ψ production is much larger for iron than for hydrogen-deuterium [7]. In the framework of the photon-gluon fusion model [8] this implies that the gluon distribution is enhanced in iron compared to hydrogen and deuterium. The data are shown in fig. 2. They cover a range $0.026<x<0.086$. The average ratio of the cross sections, or the gluon distributions respectively, is 1.45 ± 0.12 (stat.) ±0.22 (syst.). However, these data have too low statistics to allow any conclusion about the magnitude and the x and A dependence of the the change of the gluon distribution. Obviously more and better data are needed.

3.2 SLAC Electron Data

Figure 3 shows the results of the SLAC experiment E139 [9], which has measured cross sections per nucleon for a series of nuclei with incident electron energies between 10 and 20 GeV. The data cover a Q^2 range of $2\leq Q^2\leq15$ GeV^2. Also plotted

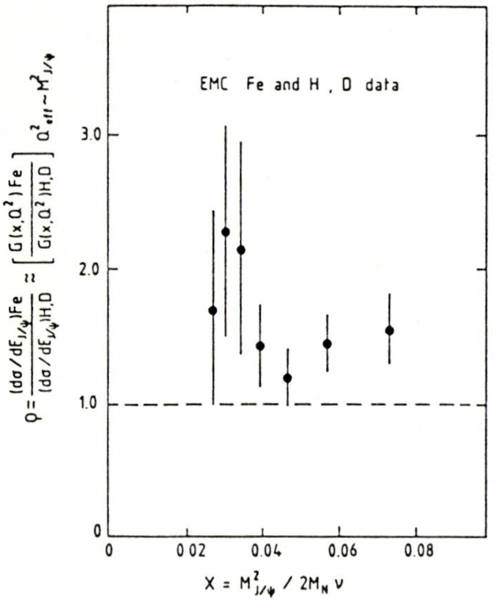

Fig. 2
The ratio of the cross sections per nucleon for J/Ψ production for iron and hydrogen-deuterium measured by EMC

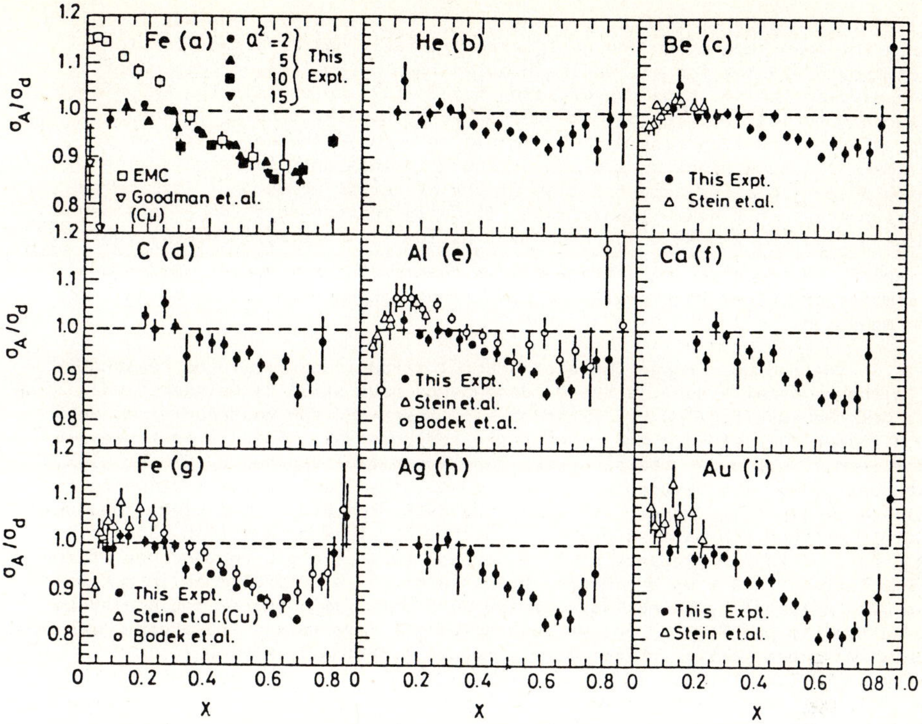

Fig. 3 A dependence of the EMC effect measured by SLAC experiment E139

are the recovered target wall data for aluminium and steel from the experiments E49B and E87 [10,11] the results of the shadowing experiment E61 [12] and the EMC data.

For all nuclei one observes that the nuclear cross section σ^A is reduced compared to the 'free nucleon' one, σ^D, in the region $0.3<x<0.8$. Obviously there is little Q^2 dependence of the effect, since there is good agreement between these data and the EMC result in shape as well as normalisation (see fig. 3a) although their Q^2 range differs by more than a magnitude. The rapid increase of the ratio at larger x is dominantly a kinematic effect since the free nucleon cross section vanishes for $x \to 1$. It is however, not clear whether this increase is entirely due to Fermi motion of nucleons or whether it is caused by collective phenomena such as nucleon–nucleon correlations or, for instance, the formation of multiquark clusters. The reduction in the region $0.3<x<0.8$ has a very characteristic shape which is similar for all nuclei, with a minimum around $x \sim 0.65$. The effect is already present for helium, its magnitude increases logarithmically with the atomic weight A. Since the average nuclear density $\rho(A)$ shows a similar A dependence, all models which relate the effect in some way to this quantity (increase of confinement size of nucleons, fraction of multiquark clusters, ...) are able to reproduce this A dependence. Also the number of extra pions per nucleon, as obtained from standard nuclear physics calculations [13] increases approximately like log A. Therefore, pion models can, in principle, describe as well an enhancement at low x as the A dependence of the cross section reduction at higher x, simply by assuming that the momentum carried by the pions has been taken from the valence quarks of the nucleons. Another quantity showing the same A dependence and which has therefore, been used by several model makers is the shift of the quasi-elastic peak ϵ_{meas} measured in A(e,e')A–1 reactions [14]. One should however, be aware of the fact that after coulomb corrections, which should be applied to the experimental data points [15], ε_{cc} is nearly constant for all A. Similarly the mean separation energy $\bar{E}_s$ experimentally has an almost constant value of around −25 MeV for all nuclei heavier than carbon [16]. This quantity has been used in the so called x-rescaling models [17] to explain the EMC effect as a standard nuclear physics phenomenon caused by nuclear binding, assuming that an effective nucleon mass $M'=M+\bar{E}_s$ should be used in the definition of x: $x \rightarrow x'=x/(M+\bar{E}_s)$. There might be a grain of truth in this approach and it might indeed be responsible for some fraction of the effect at medium x. Its ingredients are however, not unambiguous [18]. It fails to reproduce the A dependence at medium x [19] and a possible enhancement at low x, where F_2^D is nearly constant with x and F_2^A is therefore not affected by a change of the x-scale. Additional mechanisms are required to explain the remaining A dependence and to balance the missing momentum.

While in the medium x range the experimental situation seems to be reasonably clear (although details such as the Q^2 dependence have still to be clarified), this is not the case for low x. First of all, there are for most nuclei no data at x<0.2 from E139, secondly the results are controversal. This is most clearly seen for iron (fig.3 a,g): the structure function ratio measured by EMC at high Q^2 shows a continuous rise between x=0.3 and x=0.05, the low Q^2 cross section ratios from E139 are compatible with unity while the data from E61 show a clear enhancement with a maximum at x around 0.1–0.15 which seems to increase (fig. 3c,e,g,i) with A. Part of the discrepancy could be caused by a strong Q^2 dependence of the nuclear effects at low x or by $R=\sigma_L/\sigma_T$ being larger in a nucleus than in a free nucleon, as indicated by an analysis of the E139 data [20], but it could also be due to systematic problems in one or several of the data sets. Therefore, much more detailed experimental information is needed espectially at low x.

3.3 Neutrino Data

The data from a counter experiment [5a] and three bubble chamber experiments [5b–d] are summarised in fig. 4. The mean Q^2 values of these data are typically a factor of ten smaller than those of the EMC data. The results of all four experiments are, within their large error bars, compatible with the EMC data for x≳0.1

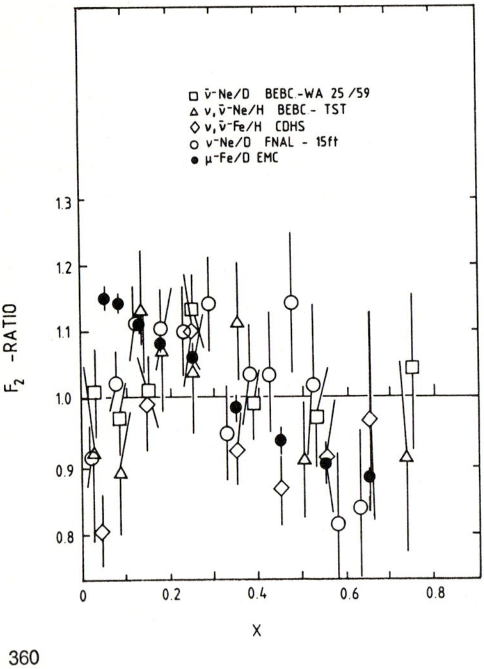

Fig. 4 Neutrino results for the EMC effect

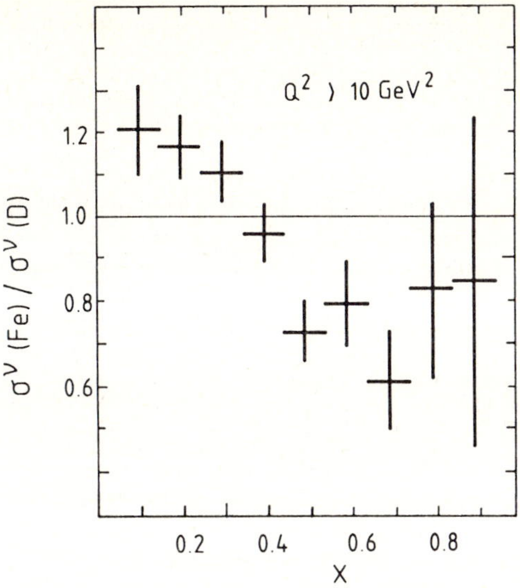

Fig. 5
Cross section ratios for Fe/D from FNAL neutrino experiment E545

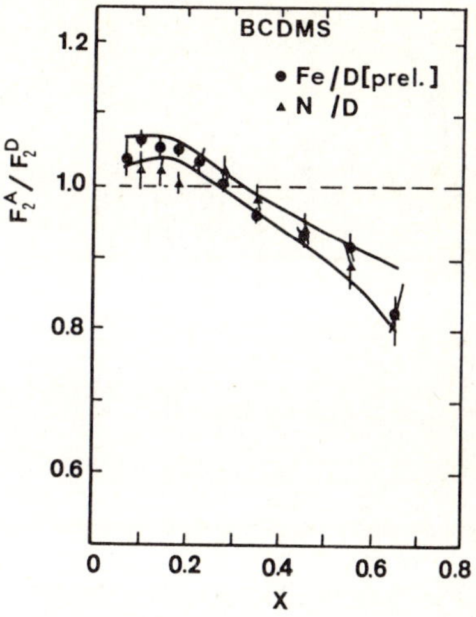

Fig. 6
Preliminary results for N/D and Fe/D from BCDMS. The curves indicate the range of systematic errors for the iron data.

showing some enhancement below $x \simeq 0.3$ and some depletion at higher x. At $x<0.1$ however, they decline below one. On the other hand two results from the 15ft bubble chamber at FNAL [21,22] show a clear rise at low x. In fig. 5 preliminary data [22] for iron/deuterium are plotted which have been obtained from a compari-ison of events originating from the steel walls of the bubble chamber and those from the deuterium filling. Since these data have been taken simultaneously many of the systematic errors should cancel.

3.4 New Muon Data From BCDMS and EMC

There are recent data from the BCDMS and EMC muon experiments at CERN. The BCDMS results for N/D and Fe/D are shown in fig. 6. Details of this experiment have been presented at this conference [23]. The Fe/D ratio shows a clear enhancement at $x<0.3$ with a maximum of ~ 1.06 at $x \simeq 0.1$. Taking into account the systematic errors, indicated by the two curves, and a possible normalisation difference of 3–4%, there is globally good agreement between this data set and the original EMC data. The point below $x=0.1$ indicates a possible drop towards low x. Only a small enhancement of 2–3% below $x=0.3$ is seen in the N/D ratio. The drop at larger x is very similar also in magnitude to that of the Fe/D data indicating that the A dependence at high values of Q^2 might be different from that extracted from the SLAC data. This in turn would point to a sizable Q^2 dependence of the effect.

During the measurements of the polarised proton structure functions in 1984/85 the EMC has performed a parasitic measurement from several nuclear targets (D, He, C, Cu, Sn), each approximately 8g/cm^2 thick. These were placed downstream of the polarised target, one at a time, and exchanged every few hours, thus minimising systematic errors due to variations in the acceptance and apparatus performance. The data are still being analysed. Preliminary results for C/D and Cu/D, based on approximately 10K events for each target are presented in fig. 7. These data also show an enhancement at $x<0.3$ which is more pronounced for Cu/D than for C/D. The Cu/D results are compatible with the original Fe/D data if one takes into

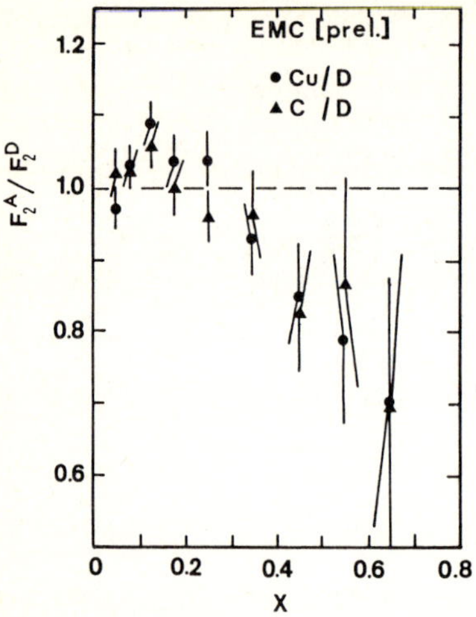

Fig. 7 Preliminary results for C/D and Cu/D from EMC

account the point to point systematic errors and allows for a normalisation shift of the Fe/D results downwards by about 3–4%. Only the point at x=0.05 differs by more than one standard deviation.

The Cu/D ratio shows a drop towards very small x which could be an indication that shadowing, undoubtably observed in photoproduction ($x=Q^2=0.0$), persists to relative high Q^2 values around 10 GeV^2. This would favour the quark parton model explanations [24] of shadowing.

At higher x the decrease of the cross section ratios Cu/D and C/D is steeper than that observed in the original Fe/D ratio and in the BCDMS results, but the error bars are too large to allow any further conclusion especially about a different A dependence at high values of Q^2.

3.5 Summary of the Experimental Results

Due to recent data the general trend of the nuclear effects has become much clearer. It might look very similar to that indicated in figs. 3c,e,g,i: there is shadowing at $x<0.05$, probably even at high Q^2 values, but this has to be explored in a dedicated experiment covering a large Q^2 range at very low x. In the range $0.05<x<0.3$ there is an enhancement of σ^A over σ^D with a maximum around x=0.1, its magnitude increasing with A. Details of the A dependence and of the Q^2 dependence, which appears to be small, are still unknown and require further high statistic investigations before one can relate it to effects like antishadowing, extra pions or an enhanced sea. Neither the large enhancement at $x\lesssim0.08$ in the original EMC data nor the nearly constant ratio around unity below $x\simeq0.3$ in the E139 data has been confirmed by recent experiments. Therefore there is little need for an A dependence of $R=\sigma_L/\sigma_T$, nevertheless this question has to be clarified in its own right to find out whether there are more pointlike spin 0 objects in a nucleus than in a free nucleon. At $0.3<x<0.8$ there is a reduction of the cross section ratio which has the same shape for all nulei. The effect has its maximum at $x\simeq0.65$ and increases approximately with log A, while nuclear binding corrections have nearly the same magnitude for all A. There are indications that this A dependence might look different at high Q^2. The Q^2 dependence of the effect in this region has to be measured with a precision in the level of a few per mille to find out whether the effect can be related to perturbative QCD [25] or whether quark–quark correlations between different nucleons in the nucleus (higher twist effects) play an important role [26]. At even higher x σ^A/σ^D rises again above one, essentially nothing is known about the behaviour of σ^A at $x>1$, the region where multiquark clusters and other nucleon nucleon correlation effects play the dominant role. The exploration of this kinematic regime might be essential to discriminate between models.

4. OUTLOOK

Many of the proposed theoretical approaches can describe some features of the existing data, which have just scratched the surface of this exciting field. However, much more detailed experimental information is needed until the final picture with all its facets emerges and one will find a unique explanation for the observed nuclear effects. Nearly all of the open experimental questions, indicated in the previous paragraphs, will be investigated during the following years by the New Muon Collaboration, NA37 at CERN [27], the aspect of the enhanced sea by the Drell-Yan experiment E772 at FNAL [6]. My hope is that in a few years we will understand better how and why the nuclear environment influences quark and gluon distributions and confinement and that the data will help to understand nuclear forces in a fundamental way in terms of quarks, gluons and their interactions.

References

1. EMC, J.J. Aubert et al., Phys. Lett. 123B (1983) 275.
2. Examples are:
O. Nachtmann, Proc. 11th Conf. on Neutrino Physics and Astrophysics, Nordkirchen near Dortmund, 1984, eds. K. Kleinknecht and E.A. Paschos, p.405.
I. Savin, Proc. of the XIIth Int. Conf. on High Energy Physics, Leipzig 1984, A. Meyer and E. Wieczorek ed., Vol.II, p.241.
C.H. Llewellyn-Smith, Proc. PANIC (Heidelberg) 1984, eds. B. Povh and G. zu Putlitz, Nucl. Phys. A434 (1985) 35C.
A. Krzywicki, Proc. 11th Europhysics Conf. 'Nuclear Physics with Electromagnetic Probes', Paris 1985, eds. A. Gerard and C. Samour, Nucl. Phys. A446 (1985) 135C.
3. A. Bodek and J.L. Ritchie, Phys. Rev. D23 (1981) 1070, D24 (1981) 140; L.L. Frankfurt and M.I. Strikman, Nucl. Phys. B181 (1981) 22.
4. C.H. Llewellyn-Smith, Phys. Lett. 128B (1983) 107;
M. Ericson and A.W. Thomas, Phys. Lett. 128B (1983) 112;
E.L. Berger et al., Phys. Rev. D29 (1984) 398.
5. a) H. Abramowicz et al., Z. Phys. C25 (1984) 29;
b) M.A. Parker et al., Nucl. Phys. B232 (1984) 1;
c) A.M. Cooper et al., Phys. Lett. 141B (1984) 133;
d) V.V. Ammosov et al., PISMA v.ZHETF 39 (1984) 327.
6. E.L. Berger, these proceedings.
7. EMC, J.J. Aubert et al., Phys. Lett. 152B (1985) 433.
8. J.P. Leveille and T. Weiler, Nucl. Phys. B147 (1979) 147;
M. Glück and E. Reya, Phys. Lett. 83B (1979) 98.
9. R.G. Arnold et al., Phys. Rev. Lett. 52 (1984) 727.
10. A. Bodek et al., Phys. Rev. Lett. 50 (1983) 1431.
11. A. Bodek et al., Phys. Rev. Lett. 51 (1983) 534.
12. S. Stein et al., Phys. Rev. D12 (1975) 1884.
13. B.L. Friman et al., Phys. Rev. Lett. 51 (1983) 763.
14. E.J. Moniz et al., Phys. Rev. Lett. 26 (1971) 445.
15. R. Rosenfelder, Ann. Phys. 128 (1980) 188.
16. S. Frullani and J. Mougey, Advances in Nuclear Physics 14 (1984) 1.
17. S.V. Akulinichev et al., Phys. Lett. 158B (1985) 485, Phys. Rev. Lett 55 (1985) 2239;
B.I. Birbrair et al., Phys. Lett. 166B (1986) 119;
G.V. Dunne and A.W. Thomas, Phys. Rev C (1986) (in print);
S. Shlomo, these proceedings.
18. M.I. Strikman, L.L. Frankfurt, Leningrad preprint, LINR 1197 (1986).
19. S. Gupta, Bombay preprint TIFR/TH/86-17.
20. R. Arnold et al., SLAC proposal E140 (Nov. 1984);
J. Gomez, SLAC-Pub-3552 (1985).
21. J. Hanlon et al., Phys. Rev. D32 (1985) 2441.
22. E545 data, presented by T. Kitagaki at the 11th Int. Conf. on Neutrino Physics and Astrophysics, Sendai, Japan, June, 1986.

23. BCDMS, G. Bari et al., Phys. Lett. 163B (1985) 282;
A. Milsztajn, these proceedings.
24. N.N. Nikolaev and V.J. Zakharov, Phys. Lett. B56 (1975) 397;
A.H. Müller, proceedings of the XVIIth Rencontre de Moriond, 1982 ed. J. Tran Thanh Van, p.13, and private communication.
25. The first publications proposing a QCD related explanation of the EMC effect were:
F.E. Close et al., Phys. Lett 129B (1983) 346;
O. Nachtmann and H. Pirner, Z. Phys. C21 (1984) 277.
26. E.V. Shuryak, Proc. 11th Europhysics Conf. 'Nuclear Physics with Electromagnetic Probes' Paris 1985, eds. A. Gerard and C. Samour, Nucl. Phys. A446 (1985) 259C.
27. NMC, 'Detailed Measurements of Structure Functions from Nucleons and Nuclei', CERN/SPSC/85-18, SPSC/P210.

QCD and Fermi Gas Model Interpretations of the E. M. C. Effect

F.E. Close

Rutherford Appleton Laboratory, Chilton, Didcot, Oxon, UK

There are "too many" explanations of the EMC effect, some of them emphasising models of nuclear binding (such as the Fermi-gas model) others emphasising the fundamental role of quarks and gluons ("rescaled QCD"). These apparently different descriptions of quark distributions in a nucleus may in fact be connected. A duality between the QCD approach and the conventional model of nucleon binding leads to nuclear properties being simply related to the anomalous dimensions of QCD. The possibility of underwriting nuclear models by QCD, leading to a QCD theory of nuclear binding will be discussed. Striking implications are obtained for the quark-gluon substructure of the nuclear pions.

RESCALED QCD AND THE FERMI GAS MODEL

In this symposium we will hear that nuclear theory, when applied carefully, can describe the data for inelastic scattering of leptons on nuclei (the "EMC effect"). I will leave the discussion to those later talks and merely state that I believe their gross features to be correct[1,,2].

I would only dispute one point and that is the claim that as a standard nuclear theory can accommodate the data, "there is no room left for rescaled QCD". There is no doubt that one can describe the data by emphasising the quark-gluon degrees of freedom, within rescaled QCD[3,4,]. The description in terms of nucleons and pions emphasises a rather different level of matter[1,2,5,6]. I find it exciting that both theories can successfully describe the nuclear data and so offer the promise of teaching us how to map quark-gluon physics onto the nucleon-pion models.

We are just beginning to study this mapping (4,7) and I will give a brief summary of what we are finding. Some of the results are rather interesting; and I suspect that they will cause a change in attitudes to the theme of "quarks in nuclei". We know that quarks are in nucleons and pions, and that these latter packages are in the nucleus; we can describe high energy data by appealing to either layer. I suspect that in the future there will be much interest in the role of gluons in nuclei, which are essential to rescaled QCD but do not appear explicitly in the nucleon-pion models.

In order to motivate the correspondence between quark-gluon and nucleon-pion physics, I will begin with the essentials of the convolution picture underlying nuclear models of the EMC effect. Details are in refs (7,8).

If $f_{N/A}(y)$ is the probability distribution to find a nucleon (N) in the nucleus (A) carrying fraction $y = \frac{p_+}{P_+}$ of the nuclear momentum and $f_{Q/A}(X_A)$, $f_{Q/N}(z)$ are defined analogously for quarks (Q) with $X_A \equiv AX = A\frac{k_+}{P_+}$; then the mathematical statement of the probability convolution is

$$f_{Q/A}(x) = \sum_{T=N,\pi} \iint dy\, dz\, f_{Q/T}(z)\, f_{T/A}(y)\, \delta(x-yz) \tag{1}$$

where T are the nucleon or pion packages in which the quarks are clustered.

Given a distribution of nucleons and pions in the nucleus, $f_{T/A}(y)$, nuclear physicists can deduce the quark distributions in the nucleus, $f_{Q/A}(x)$, and compare with those in the free nucleon, $f_{Q/N}(x)$. This ratio is what the EMC data display[9]. Nuclear effects are subsumed in $f_{T/A}(y)$ involving, in the Fermi-gas model for example, two parameters k_F (Fermi momentum) and ε (the fraction of nucleon momentum that goes into binding in the form of pion particles)[4,5,8]. For iron $k \equiv k_F/M \simeq 30\%$ and $\varepsilon \simeq 4\%-5\%$[1,2,5,6]. As shown in refs (2,5,6,10) and discussed elsewhere in this conference[1] the EMC effect is rather well described as a function of x.

Note that any Q^2 dependence of the distributions in $f_{Q/N}(x,Q^2)$ will also appear in $f_{Q/A}(x,Q^2)$. The ability to fit the x dependence is independent of the form of Q^2 dependence. It would be true even if the data were Q^2 independent ("scale invariant"). This remark, apparently trivial, will turn out to be more profound later.

Equation (1) takes on a simple factorised form if we integrate both sides over x. Define the moment

$$M_n \equiv \int d\beta\, \beta^{n-1}\, f(\beta) \tag{2}$$

then for arbitrary n we find

$$M_n^{Q/A}(Q^2) = \sum_T M_n^{Q/T}(Q^2)\, M_n^{T/A} \tag{3}$$

The QCD theory of quark-gluon interactions has a most useful property: it is well suited to describing the Q^2 dependence of moments. Indeed in ref (3) we showed how a "rescaled" QCD theory can describe the ratio of non-singlet moments $M_n^{Q/A}/M_n^{Q/N}(Q^2)$; (for Q ≡ quark non-singlet distributions, for which the binding quanta do not contribute). This implies that

$$M_n^{Q/A}(Q^2) \equiv M_n^{Q/N}(\xi Q^2) \tag{4}$$

at least for a range of $2 < n \lesssim 10$. The rescaling parameter is itself a function of Q^2. In ref (3) we defined ξ with respect to an explicit scale

$$\xi(Q^2) = \left(\frac{\mu_N^2}{\mu_A^2}\right)^{\alpha(\mu_N^2)/\alpha(Q^2)} \qquad (5)$$

where $\frac{\mu_A^3}{\mu_N^3} \equiv \frac{R_N^3}{R_A^3}$ is the ratio of the mean confinement volumes for quarks and gluons when in nuclei or nucleons.

A more general statement of rescaled QCD which makes no explicit reference to a scale μ_N follows from eq (5) and is

$$\alpha(Q^2) \log \xi(Q^2) = \text{const.} \simeq 0.15\text{-}0.2 \qquad (6)$$

The magnitude of the constant has been read off from the Q^2 dependence of the data. For iron it is is now well known that[3] $\xi \simeq 2$ to 3 when Q^2 is such that $\alpha(Q^2) \simeq 0.2$, hence 0.15-0.2 is the empirical magnitude of the combination in eq (6). This is a measure of the length scale change (hence statements like "the data show a 15% increase in the confinement scale for quarks in iron relative to free nucleons"). However, to identify it explicitly as such requires commitment to specific models, such as bag models, which involve additional assumptions. The minimal demand that there is an overall scale change, expressed in the Q^2 dependence of the moments, is stated mathematically by eq (6) and the magnitude read off from the data

The evolution equations of QCD, with eq (4) and (5) then give[3,4]

$$\log\left(\frac{M_n^{Q/A}(Q^2)}{M_n^{Q/N}(Q^2)}\right) = \frac{\gamma_o^n}{8\pi}\,\alpha(Q^2)\log\xi(Q^2) \qquad (7)$$

where γ_o^n are the (non-singlet) anomalous dimensions of QCD. ($\gamma_o^{(2)} = -64/9$ and $\gamma_o^{(3)}/\gamma_o^{(2)} = 25/16$), (see e.g. refs 4 and 7).

Note that rescaled QCD addresses the Q^2 dependences directly. Our ability to describe the EMC effect depends cruically upon the fact that the Q^2 dependence is in accord with QCD theory.

Now consider eq (3) which can link the nuclear physics (expressed in terms of moments, $M_n^{N/A}$) and rescaled QCD (expressed in eq (7)).

Taking the Fermi-gas model, to be explicit, one finds[4],

$$\log M_n^{N/A}(k,\varepsilon) = (n-1)\left[\varepsilon + \frac{n-2}{10}k^2\right] \qquad (8)$$

Thus in particular

$$\log M_2^{N/A} = \varepsilon \qquad (9)$$

$$\frac{\log M_3^{N/A}}{\log M_2^{N/A}} = 2 - \frac{k^2}{5\varepsilon} \qquad (10)$$

On comparing these with $\log(M_n^{Q/A}/M_n^{Q/N})$ some remarkable correspondences are seen. Thus for the second moment we have

$$\frac{\gamma_o^{(2)}}{8\pi}\,\alpha(Q^2)\,\log \xi(Q^2) = \varepsilon \qquad (11)$$

and inserting magnitudes in the left hand side

$$-\frac{64}{9} \times \frac{1}{8\pi} \times [0.15 - 0.2] = (4.2 - 5.6)\% \qquad (12)$$

This implies that some 4-5% of quark-energy-momentum in the free nucleon is shed when that nucleon enters a nucleus. Conversely, this amount has to be supplied when the nucleon is removed from a nucleus. This agrees quantitatively with the average removal energy of a nucleon[1,2] and with the pion theory of nuclear binding which implies that this amount of nucleon energy goes into pion particles[6].

Taking the ratio of third and second moments, the magnitude of rescaling cancels and leaves simply the ratio of QCD anomalous dimensions $\gamma_o^{(3)}/\gamma_o^{(2)}$ (eq 7). Comparing eq (7 and 10) requires that

$$\frac{\gamma_{(3)}}{\gamma_{(2)}} = 2 - \frac{k^2}{5\varepsilon} \qquad (13)$$

The left hand side has magnitude $\frac{25}{16}$ (ref 4) and hence

$$\varepsilon = \frac{16}{35}k^2 \qquad (14)$$

independent of nucleus A. This remarkable result is well satisfied[4] but is surely an accident: how can Fermi motion (a property of "pointlike" nucleons) depend upon the space-time properties of quark-gluon interactions (the QCD anomalous dimensions)?

The result at eq (12), however, is more likely to be "genuine". We know little about quark-gluon confinement in either free or bound nucleons. However, we know that these partons can leak from nucleons in their ability to bind several nucleons into nuclei - thus the average confinement volume of quarks or gluons is larger in nuclei than in free nucleons. The minimal assumption is that the "nasty" features of confinement are, in some sense, common to both free and bound nucleons and cancel out in ratios up to an overall change of scale. It is a gift of Nature that this appears to be a reasonable first approximation. Eq (11) is the mapping from rescaled QCD.

It is instructive to realise the qualitative implications of this result. The energy lost to binding is carried by quarks, antiquarks and gluons in rescaled QCD. These are the constituents of the nucleons and pions which are accepted as the quanta of nuclear binding. Thus we have obtained a description of binding with desired qualitative features and some quantitative agreement.

THE QUARK-GLUON CONTENT OF THE NUCLEAR BINDING QUANTA

The analysis so far has concentrated on the valence quarks, or "non-singlet" distributions. The contributions from sea quarks and antiquarks and from gluons cancel out; all we could say was that 4-5% of the nuclear energy is carried by them. If we perform a similar analysis for the gluons or the sum of quark and antiquark (the "singlet" distributions) we can make explicit statements about the quark-gluon content of the binding quanta.

One may anticipate the result immediately. Rescaling has implied that gluons and quark-antiquarks are shed when the nucleon is bound. Thus gluons are important in nuclear binding. Insofar as pions are conventionally believed to be responsible for binding at the "1 fermi" length scale, the correspondence between the two approaches requires that the pion contains a substantial gluon component (in the sense that gluons must carry over 50% of a pion's light cone momentum). The pion is "a glueball with isospin". This can be tested independently by good Drell-Yan annihilation data with pion beams, and if verified will surely have a profound effect on our understanding of the nature of the pion (and by implication the role of glue in its isoscalar partners, the η and η').

The details of this analysis will be found in ref (7). Here I will merely summarise some of the steps.

I will define $Q_{\pm} \equiv q \pm \bar{q}$, the sum and difference of quark (q) and antiquark ($\bar{q}$) distributions. Also for ease of notation I will define

$$Q^n/T(Q^2) \equiv M_n^{Q/T}(Q^2) \qquad (15)$$

for any package T. Thus eq (3) rewritten becomes

$$q_-^n/A(Q^2) = q_-^n/N(Q^2) \quad N^n/A \qquad (16)$$

$$q_+^n/A(Q^2) = q_+^n/N(Q^2) \quad N^n/A + q_+^n/M(Q^2) \; M^n/A \qquad (17)$$

where N,M refer to nucleon and meson, and where we have used

$$q^n/M(Q^2) \equiv \bar{q}^n/M(Q^2) \qquad (18)$$

in isoscalar nuclei. It is this feature that enabled the results of the previous section to be obtained since both eq (16) and the rescaled QCD equation

$$q_-^n/A(Q^2) \equiv q_-^n/N(\xi Q^2) = q_-/N(Q^2)\ \{1-\frac{\gamma^n}{8\pi}\ \alpha(Q^2)\ \log\ \xi(Q^2)\} \tag{19}$$

involved only q_- and not q_+ nor glue (g) distributions.
For ease of notation I will define

$$K \equiv \alpha(Q^2)\ \log\ \xi\ (Q^2)/8\pi\ . \tag{20}$$

Eqs (16) and (19) then led to the central equation of our duality

$$N/A = 1-\frac{\gamma^{(n)}}{8\pi}\ \alpha(Q^2)\ \log\ \xi(Q^2) \equiv 1 - \gamma^{(n)}K \tag{21}$$

(whose logarithm, for n=2, yields eq (11) for small K).

If rescaled QCD is a valid description of quarks in nuclei then it should also apply to eq (17) - the q_+ distributions. The evolution equation for the q_+ moments may be found in standard texts on QCD and gives

$$q_+^{(n)}/A(Q^2) \equiv q_+^{(n)}/N(\xi Q^2) = q_+^{(n)}/N(Q^2)\ \{1- \gamma^{(n)}K\} - g^{(n)}/N(Q^2)\gamma_{qg}K \tag{22}$$

This involves the gluon distributions in nucleons and the anomalous dimension γ_{qg} associated with the vertex for $g \to q\bar{q}$.

Compare eq (22) with the nuclear evolution eq (17) and notice the common factor of eq (21) whose left hand side appears in eq (17) and whose right hand side appears in eq (22). Eqs (17) and (22) thus imply

$$q_+^{(n)}/M(Q^2)\ M^{(n)}/A = -\ (\gamma_{qg}^{(n)}\ K)\ g^{(n)}/N(Q^2) \tag{23}$$

This relates the distribution of quarks in a meson (pion) to the glue in a nucleon; at first sight a most startling result.
Physically the L.H.S. is a measure of the "excess" $q\bar{q}$ sea in the nucleus ("excess" in the sense that it is generated by the binding - or rescaling - over and above that already in the nucleons). The source of this excess is the free nucleon's glue. Binding the nucleons effectively evolves the glue, generating additional $q\bar{q}$ by an amount controlled by the shift K and the anomalous dimension $(-\gamma_{qg})$ proportional to the $g \to q\bar{q}$ probability.

Notice that the equation implies that if there are extra $q\bar{q}$ in nuclear binding mesons viewed at some Q^2, then necessarily there has to be glue in the free nucleon at that Q^2. Conversely, if there is some Q^2 where the nucleon contains only valence quarks and no glue, then the nuclear binding cannot be due to extra $q\bar{q}$, (i.e. not simple "naive" pions). Here we have the first hint of our general conclusion: gluonic mesons must play an essential role in nuclear binding.

This mapping of rescaled QCD onto nulcear convolution models simplifies considerably when n=2, which describes the energy-momentum distributions within the packages. If there are F flavours of quark then the n=2 anomalous dimensions are related

$$\gamma_{qg}^{(2)} = \frac{3F}{16}\,\gamma^{(2)} \tag{24}$$

and the equation (23) becomes

$$g^{(2)}/N\,(Q^2) = \frac{16}{3F}\,(1 - g^{(2)}/M(Q^2))\;. \tag{25}$$

and so

$$\frac{d}{dQ^2}\; g^{(2)}/N(Q^2) = -\frac{16}{3F}\,\frac{d}{dQ^2}\; g^{(2)}/M(Q^2)\;. \tag{26}$$

Thus if, as is conventionally believed, the fraction of the nucleon's momentum carried by glue increases with Q^2, then that in the pion must <u>decrease</u> with Q^2 i.e. the pion contains glue that cannot be generated by evolution in Q^2. The QCD evolution equations only allow this if

$$g^{(2)}/M(Q^2) > 16/16 + 3F\;. \tag{27}$$

and so the pion is gluonic in the sense that gluons carry the majority of its momentum. (In general we may define a gluonic hadron to be one that satisfies eq 27).

Surprisingly, perhaps, data on the pion structure function are consistent with this within the considerable errors. There are independent theoretical reasons for suspecting that valence $q\bar{q}$ play the minor role in the pion(4,11,12). Independent of the QCD - nuclear duality, there is the crucial question of the pion's internal constituency and whether glue satisfies eq (27) or not (whether the pion is gluonic or quarkic).

If rescaled QCD is mappable onto nuclear convolution models to good approximation, then eq (23) must be satisfied for some range of n. We find that this places severe constraints on the quark and gluon distributions within pions and nucleons, and depends on the nuclear model for π/A.

If the π/A distributions of ref 5 are employed, we find consistency in eq (23) with "sensible" distributions of the form

$$q^{\pi}(x) \sim (1-x)$$

$$g^{N}(x) \sim (1-x)^{5\pm2} \text{ or } (1+9x)(1-x)^{5\pm1} \tag{28}$$

These are in remarkable agreement with data[13]; the solution has 65% of the pion's momentum carried by glue.

The π/A distributions of ref 6 are rather different from those in ref 5. They are harder[14] and so extend the $q^A(x)$ to higher x intrinsically. Eq (23) leads to a solution where $q^\pi(x) \simeq g^N(x)$, which is quite unlike theoretical prejudices and probably inconsistent with data.

If one uses "conventional" distributions in the sense of eq (24) with the Argonne π/A model, the mapping will fail. This is implicitly evident in the work of ref (14) who note that the $\bar{q}^A(x)$ in the Argonne model is very different from that in other approaches. However, if eq (23) is satisfied, as in the nuclear model of ref (5), the different theories map into one another and will lead to similar phenomenology. A "unified" approach to nuclear and quark-gluon physics may be possible.

This work was performed in collaboration with R.G. Roberts and G.G. Ross. I am also indebted to many colleagues for their comments and questions after many seminar presentations during the development of these ideas.

References

1. S. Shlomo, these proceedings.
F. Coester, these proceedings.
2. G.V. Dunne and A.W. Thomas, Phys. Rev. C (1986), in press.
3. F.E. Close, R.G. Roberts and G.G. Ross, Phys. Letters 129B, 346 (1983).
R.L. Jaffe, F.E. Close, R.G. Roberts and G.G. Ross, Phys. Letters 134B, 449 (1984).
F.E. Close, R.L. Jaffe, R.G. Roberts and G.G. Ross, Phys. Rev. D31, 1004 (1985).
F.E. Close, Nucl. Phys. A446, 273 (1985).
R.G. Roberts, "Quarks in Nuclei", p. 215 in Vol. 234, Lecture notes in physics, Springer-Verlag (1985).
C.H. Llewellyn Smith, Nucl. Phys. A434, 35c (1984).
4. F.E. Close, R.G. Roberts and G.G. Ross, Phys. Letters 168B, 400 (1986).
5. C.H. Llewellyn Smith, Phys. Lett. 128B, 107 (1983).
M. Ericson and A.W. Thomas, Phys. Lett. 128B, 112 (1983).
6. E. Berger, F. Coester and R.B. Wiringa, Phys. Rev. D29, 398 (1984).
7. F.E. Close, R.G. Roberts and G.G. Ross, RAL-report (1986) "Quark and Gluon distributions in nucleons, pions and nuclei" (in preparation).
8. R.L. Jaffe, MIT-CTP-1261, Lectures at 1985 Los Alamos School on Quark Nucleon Physics (1985).

9. K. Rith, these proceedings.
10. A. Kryzwicki, Nucl. Phys. A446, 135 (1985).
11. S.J. Brodsky, p. 99 "Quarks and Nucleon Forces", Vol 100, Springer Tracts in Modern Physics.
12. N. Isgur (private communication).
13. D.W. Duke and J.F. Owens, Phys. Rev. D30, 49 (1984).
14. R.P. Bickerstaffe, M.C. Birse and G.A. Miller, p. 144 in "Hadronic probes and nuclear interactions", American Institute of Physics, 1985; see also Phys. Rev. Letters 53, 2532 (1984).

Massive Lepton Pair Production – the Drell-Yan Process – with Nuclear Targets

E.L. Berger

High Energy Physics Division, Argonne National Laboratory,
Argonne, IL 60439, USA

The Drell-Yan process, $hA \to (\mu\bar{\mu})X$, in which a massive lepton-pair is produced inclusively from a nuclear target, offers the possibility for gaining unique insight into the A dependence of the quark and antiquark densities of nuclei, $q^A(x,Q^2)$ and $\bar{q}^A(x,Q^2)$. Measurements are crucial for testing whether the A dependence of deep-inelastic structure functions, known as the EMC effect, is a universal, process independent property of parton densities. Precise experiments with pion and nucleon beams on nuclear targets would extend knowledge of the A dependence of quark and antiquark densities of nuclei into important intervals of the fractional longitudinal momentum variable x which are not accessible easily in deep inelastic scattering. Models constructed to interpret the EMC effect are used to obtain predictions for the A dependence of massive lepton pair production. For tests of quantum chromodynamics, isolating nuclear dependence is necessary before definite conclusions can be drawn on the behavior of the pion structure function and on the magnitude and kinematic variation of predicted higher order terms (K factor). I indicate the sensitive regions of phase space and the increased level of experimental precision which is needed if a contribution is to be made to isolating the A dependence of quark and antiquark densities from the Drell-Yan process.

I. Introduction

In this paper I show that inclusive hadronic production of massive lepton pairs from nuclei, $hA \to (\ell\bar{\ell})X$, will provide unique information on the nuclear (A) dependence of the quark and antiquark density distributions in nuclei, $q^A(x)$ and $\bar{q}^A(x)$. This information is important for the development of models of the nucleus and of parton confinement, and for detailed tests of perturbative quantum chromodynamics. The variable x is the Bjorken scaling variable. It represents the fraction of the (light-front) momentum of the nucleus carried by the quark or antiquark.

Substantial interest in the nuclear dependence of the densities $q^A(x)$ and $\bar{q}^A(x)$ may be traced in part to the observation [1-3] that the deep inelastic structure function $F_2(x,Q^2)$ of an iron nucleus differs as a function of x from the structure function of deuterium. This phenomenon, known as the European Muon Collaboration [1] (EMC) effect, has led to a wide spectrum of theoretical interpretations [4-6]. In $\ell A \to \ell' X$, Q^2 is the usual square of the invariant four-momentum transfer from the initial to the final lepton.

Descriptions of the nuclear A dependence $F_2^A(x,Q^2)$ are sensitive only to the sum of the quark and antiquark densities measured in nuclei, $q^A(x,Q^2)$ and $\bar{q}^A(x,Q^2)$.

It is of interest to try to understand the potentially different A dependences of $q^A(x,Q^2)$ and $\bar{q}^A(x,Q^2)$. Moreover, different approaches [5,6] developed to explain the EMC effect in $F_2^A(x,Q^2)$ make significantly different statements [4] regarding the x dependence of the antiquark density $\bar{q}^A(x)$. Neutrino and antineutrino experiments should be a rich source of information on the ratio $\bar{q}^A(x,Q^2)/\bar{q}^N(x,Q^2)$. However, the present data sample [3] is limited, and it is unlikely that the required increase in statistical and systematic precision will be forthcoming soon at large Q^2 ($\tilde{>}$ 10 GeV^2).

The proton induced reaction $pA \to (\ell\bar{\ell})X$ is an especially valuable source of detailed information on the A dependence of the antiquark density $\bar{q}^A(x)$. It is noteworthy that the range in x spanned by lepton-pair data is greater in practice than in the neutrino case meaning that lepton pair data have more discriminating power. The nuclear dependence of the quark density, $q^A(x,Q^2)$, can be obtained from pion and antiproton data, e.g. $\pi^-A \to (\ell\bar{\ell})X$. This information is necessary

i) for establishing whether the A dependence of $q^A(x,Q^2)$ observed in deep inelastic lepton scattering is a process-independent, universal property of parton densities, as well as

ii) for extending the range in x over which $q^A(x,Q^2)$ is measured.

A more detailed discussion of these and other points may be found in [7].

In Sec. II, the Drell-Yan model is reviewed briefly, with particular emphasis on nuclear effects. Proton induced reactions, $pA \to \gamma^*X$, are treated in Sec. III; both pion and antiproton reactions are examined in Sec. IV. Conclusions are summarized in Sec. V.

II. Drell-Yan Model

In the Drell-Yan model [8] for $hN \to \mu\bar{\mu}X$, the $\mu\bar{\mu}$ pair arises from the decay of a massive photon γ^*. This virtual photon is produced in the annihilation of a quark (antiquark) from hadron h with an antiquark (quark) from the target nucleon: $q\bar{q} \to \gamma^*$. The invariant mass M of the γ^* is related to the fractional longitudinal momenta x_i of the annihilating partons by $M^2 = sx_1x_2$. The scaled center-of-mass fractional longitudinal momentum of the pair is $x_F = 2p_L(\mu\bar{\mu})/\sqrt{s} = x_1 - x_2$. Alternatively, one may measure the rapidity y of the lepton pair. In this case $x_1 = \sqrt{\tau}\exp(y)$, and $x_2 = \sqrt{\tau}\exp(-y)$, where $\tau = M^2/s$. The variable s denotes the square of the center of mass energy of the hN process. If the doubly differential cross section $d^2\sigma/dMdx_F$ (or $d^2\sigma/dMdy$) is measured as a function of both M and x_F, one may explore the variation with x_1 and x_2 of the quark and antiquark densities of the colliding hadrons. The Drell-Yan model [8] provides a successful description of data for M $\tilde{>}$ 4 GeV.

For $hA \to \gamma^*X$, I use the notation $\sigma^A(x_1,x_2)$ to denote either $d^2\sigma/dMdx_F$ or $d^2\sigma/dMdy$. For the ratio of the cross sections per nucleon measured on nucleus A and nucleon N, the model specifies that

$$R_{DY}^{hA}(x_1,x_2) \equiv \frac{\sigma^A(x_1,x_2)}{\sigma^N(x_1,x_2)} = \frac{\sum_f e_f^2\{\bar{q}_f^h(x_1,M^2)q_f^A(x_2,M^2) + q_f^h(x_1,M^2)\bar{q}_f^A(x_2,M^2)\}}{\sum_f e_f^2\{\bar{q}_f^h(x_1,M^2)q_f^N(x_2,M^2) + q_f^h(x_1,M^2)\bar{q}_f^N(x_2,M^2)\}} . \quad (1)$$

The sum in (1) is over the different quark flavors f; $\bar{q}^i(x,M^2)$ and $q^i(x,M^2)$ denote the antiquark and quark number density distributions of hadron i. In the case of a nucleus, $\bar{q}^A$ and q^A are densities per nucleon, meaning than an overall factor A has been removed in their definitions.

Higher order radiative contributions [9] in perturbative QCD, whose approximate effect is to multiply numerator and denominator by a constant (K) factor of about 2, should cancel to a good approximation in this ratio. Because of known differences in the x dependences of the up and down flavor distributions, $u^p(x)$ and $d^p(x)$, comparisons should be made of nuclei with similar ratios of the numbers of neutrons and protons. In the following, I use symbol N to denote quantities per "isoscalar" nucleon N. In particular, σ^N is the cross section per nucleon for scattering from a deuteron target, and $u^N(x) = 0.5(u^p(x) + u^n(x))$.

Prior to the observation of the EMC effect, it was generally believed that for large enough Q^2, and for $0.05 \lesssim x \lesssim 0.7$, the quark and antiquark densities of an (isoscalar) nucleus would equal those of free nucleons:

$$q^A(x,Q^2) = q^N(x,Q^2) \quad , \text{ and } \quad \bar{q}^A(x,Q^2) = \bar{q}^N(x,Q^2) \ . \tag{2}$$

For small Q^2 and/or small x, shadowing was expected to invalidate (2). For large enough x, Fermi smearing of the nucleons' momentum distribution also invalidates (2). The EMC effect is the observation that there are ±15% deviations from (2) for all x.

If (2) were correct then we would expect $R^{hA}_{DY}(x_1,x_2) \equiv 1$ for all x_i. The EMC effect leads us to expect deviations whose behavior should provide valuable information on the densities $q^A(x,Q^2)$ and $\bar{q}^A(x,Q^2)$.

If the <u>first</u> terms are dominant in the numerator and denominator of (1), then the ratio $R_{DY}(x_1,x_2)$ is controlled by the A dependence of the <u>quark</u> distribution $q^A(x_2,M^2)$. On the other hand, if the second terms are dominant, the ratio provides information on the antiquark distribution $\bar{q}^A(x_2,M^2)$. Judicious selections on x_1 and x_2 (i.e. on M and x_F) should permit an investigation of $q^A(x,M^2)$ and $\bar{q}^A(x,M^2)$ independently.

III. Proton Induced Reactions

It is instructive to examine the implications of (1) separately for proton and π^- beams (i.e. h = p or π^-).

I begin with $pA \to \gamma^* X$. When x_1 is large enough, e.g. $x_1 > 0.2$, $\bar{q}^p(x_1) \ll q^p(x_1)$ for the u and d flavors, and (1) simplifies to

$$\frac{\sigma^{pA}(x_1,x_2)}{\sigma^{pN}(x_1,x_2)} \to \frac{\sum_f e_f^2 q_f^p(x_1,M^2)\ \bar{q}_f^A(x_2,M^2)}{\sum_f e_f^2 q_f^p(x_1,M^2)\ \bar{q}_f^N(x_2,M^2)} \ . \tag{3}$$

This approximation should be reliable as long as x_2 is not too large. Because $e_u^2 = 4e_d^2$ and $u^p \simeq 2d^p$, the up flavor dominates in (3), and a further simplification yields

$$\frac{\sigma^{pA}(x_1,x_2)}{\sigma^{pN}(x_1,x_2)} \rightarrow \frac{\bar{u}^A(x_2,M^2)}{\bar{u}^N(x_2,M^2)} \qquad (x_1 \gtrsim 0.2) . \tag{4}$$

Results of explicit numerical calculations with both the pion exchange [5] and rescaling [6] models verify that (4) is a good approximation for $x_1 > x_2$, i.e. for $x_F > 0$ (or $y > 0$). Its validity increases with increasing x_F.

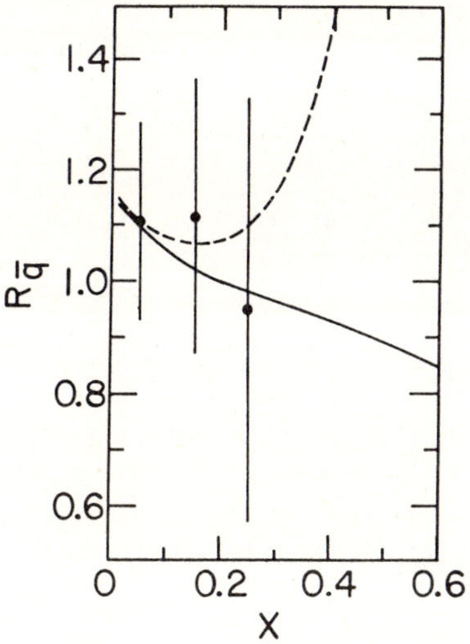

Fig. 1. Data of Abramowicz et al. [3] on the ratio $R_{\bar{q}}(x) = \bar{q}^{Fe}(x)/\bar{q}^{P}(x)$ from neutrino scattering. Here $\bar{q} = (\bar{u} + \bar{d} + 2\bar{s})$. The dashed curve illustrates predictions of the pion exchange model [5], whereas the solid curve shows expectations of the rescaling model [6].

As shown in Fig. 1, current experimental information on the value and x dependence of the ratio $\bar{q}^A(x)/\bar{q}^N(x)$ is of very limited precision. The data in Fig. 1 are extracted from data on deep-inelastic neutrino scattering. They yield an average value $\langle R_{\bar{q}} \rangle = 1.10 \pm 0.10 \pm 0.07$. From these results it is not possible to ascertain whether there is any enhancement of the antiquark density in nuclei at small x. Expectations of models are discussed below.

Equation (4) shows that the reaction $pA \rightarrow \gamma^* X$, with $x_F \geqslant 0$, should be a valuable source of detailed information on the A dependence of the antiquark distribution $\bar{q}^A(x,Q^2)$, notably that of the up flavor. Ideally, $\bar{u}^A(x_2,M^2)$ extracted from massive lepton pair data [10] should be compared directly with predictions. Unfortunately, owing to limited kinematic coverage, model dependent assumptions are introduced in the data analyses concerning both the A dependence of the cross section and the x dependence of the quark structure functions. To eliminate extraneous model dependence, I will therefore compare theoretical predictions with directly measured observables.

The Columbia Fermilab Stony-Brook (CFS) collaboration [10] investigated the A dependence of $pA \rightarrow \gamma^* X$ at 400 GeV/c using platinum and beryllium targets. Their data are concentrated near rapidity $y = 0$, at which $x_1 = x_2 = M/\sqrt{s}$. They parametrize the A dependence by the functional form $\sigma \propto A^{\alpha}$;

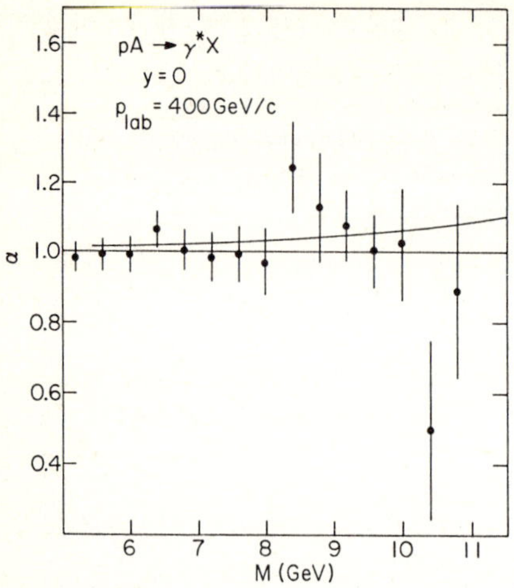

Fig. 2 Nuclear dependence of massive lepton pair production, $pA \to \mu\bar{\mu}X$, as a function of the mass of the lepton pair; α is defined in (5). Data are from Ito et al., [10]. The solid curve is the prediction of the pion exchange model [5]. For the rescaling model, α remains close to unity, varying from $\alpha \simeq 0.99$ at M = 5 GeV to $\alpha \simeq 0.98$ at M = 13 GeV.

$$\alpha = \frac{\ln(\sigma_{Pt}/\sigma_{Be})}{\ln(A_{Pt}/A_{Be})} \quad . \tag{5}$$

In terms of the variable α, absence of A dependence implies that $\alpha \equiv 1$. The CFS results are shown in Fig. 2. The data are consistent with a constant value of α over the mass and transverse momentum ranges covered in the experiment: $\langle\alpha\rangle = 1.007 \pm 0.018 \pm 0.028$ for $5 < M < 11$ GeV, but the error flags are large and do not exclude some variation of α with mass.

The goal of experiments is of course to determine the antiquark content of nuclei, not to test models, all of which are surely naive. However, models consistent with the EMC effect in $\mu N \to \mu' X$ provide a range of possible outcomes for α, and calculations should help to guide further investigations. Shown in Fig. 2 is a curve computed with quark and antiquark densities of a conventional nuclear physics model based on pion exchange [5,7]. The A dependence of the theoretical expression for $R^{pA}_{DY}(x_1 = x_2 = x)$ follows closely the shape of the pion exchange model prediction for $\bar{u}^A(x,M^2)/\bar{u}^N(x,M^2)$. The significant anticipated increase of R_{DY} with mass for $\tau^{1/2} > 0.3$ reflects the broadening of the antiquark distribution characteristic of the pion exchange model [5]. (When α is computed, the magnitude of the predicted deviation from unity is reduced considerably. For example, in the comparison of iron and deuterium, $R_{DY} = 1.1$ reduces to $\alpha = 1.03$; $R_{DY} = 1.4$ reduces to $\alpha = 1.1$; and $R_{DY} = 2.02$ reduces to $\alpha = 1.21$.)

In the pion exchange model [5], the antiquark distribution $\bar{q}^A(x)$ is broadened in a nucleus because exchange pions supply valence antiquarks. In the Q^2-rescaling model [6], QCD evolution leads to a narrowing of $\bar{q}^A(x)$ relative to $\bar{q}^N(x)$. As indicated in Fig. 1, the two models are in agreement at small x, but they provide quite different expectations for the x dependence of antiquark ratio

$R_{\bar{q}}(x)$ measured in deep inelastic neutrino scattering. These differences between models are apparent also in the predicted mass dependences of the massive lepton pair ratio $R_{DY}(x_1 = x_2 = M/\sqrt{s})$. In contrast to the positive value of $\alpha - 1$ and the rise with M shown for the pion exchange model in Fig. 2, a nearly constant value, varying from $\alpha \simeq 0.99$ at $M = 5$ GeV to $\alpha \simeq 0.98$ at $M = 13$ GeV is anticipated in the rescaling model.

Although the expectations of the two models are quite different, neither the neutrino nor the massive lepton pair data shown in Figs. 1 and 2 are adequate to indicate a preference. Owing to the very small cross sections, this situation is unlikely to change in the neutrino case at large Q^2 ($Q^2 \gtrsim 10$ GeV^2). It is noteworthy that the range in x spanned by current lepton pair data is greater than in the neutrino case, e.g., varying from $x = 0.15$ to $x = 0.4$ for $4 < M < 11$ GeV at 400 GeV/c and $y = 0$. This range could be extended by studies at larger values of τ. Lepton pair data appear, therefore, to have more discriminating power, and a precise series of experiments on $pA \to \gamma^* X$ seems advisable.

Although experiments sensitive for $x_2 \gtrsim 0.2$ are required for distinguishing between the pion exchange and Q^2-rescaling approaches, it should be stressed that data in the range $x_2 \lesssim 0.2$ are also crucial. Indeed, good data at small x, where $\bar{q}^N(x)$ is largest, are necessary to establish whether there is any enhancement of the antiquark ocean in nuclei. A proposal under consideration at Fermilab is directed specifically at this important issue [11].

Experimental studies of A dependence in massive lepton pair production have been designed in the past to distinguish between expectations such as $\alpha \simeq 2/3$ (shadowing), $\alpha \simeq 1$ (naive parton model), and $\alpha \simeq 4/3$ (rescattering). The quality of the data in Fig. 2 is sufficient for this purpose. However, the difference between $\alpha \equiv 1$ and the pion exchange model curve in Fig. 2 may be used to estimate the precision needed to make an impact on models of the nucleus. Measurements good to 3% on α are necessary, and the mass and x_F dependences of α are crucial.

IV. Pion and Antiproton Reactions

For $\pi^- A \to \gamma^* X$ and $\bar{p}A \to \gamma^* X$, there are valence antiquarks in the beam to annihilate with the valence quarks in the target. Correspondingly, for most x_F, the first terms are dominant in both the numerator and denominator of (1) (i.e. they provide over 90% of the sum for most values of x_F). Since the up flavor supplies the principal contribution to $\bar{q}q \to \gamma^*$ and to the deep inelastic process $\gamma^* q \to q$, we may expect that

$$R_{DY}^{\pi^- A}(x_1,x_2) \simeq R_{EMC}^{A}(x_2,M^2) \; ; \qquad (6)$$

and

$$R_{DY}^{\bar{p}A}(x_1,x_2) \simeq R_{EMC}^{A}(x_2,M^2) \; . \qquad (7)$$

Here, $R_{EMC}(x,Q^2)$ is the ratio $F_2^A(x,Q^2)/F_2^N(x,Q^2)$, where F_2^A is the deep-inelastic structure per nucleon for nucleus A, and F_2^N is the deuteron structure function per nucleon. Explicit numerical computations with both rescaling [6] and the pion exchange [5] models verify that (6) and (7) are excellent approximations for all $x_1 < 0.9$.

Note the clear differences between (4), on the one hand, and (6) and (7). In $pA \to \gamma^* X$, data may be used to determine the A dependence of the antiquark distribution, whereas in $\bar{p}A \to \gamma^* X$ and $\pi^- A \to \gamma^* X$, the behavior of $R_{DY}(x)$ should follow very closely that of $R_{EMC}(x)$. For M = 4 GeV at $\sqrt{s}$ = 20 GeV, the region $x_F \geq 0$ corresponds to $x_2 \leq 0.2$. This suggests that the difficult interval $x < 0.2$ in which current data on $R_{EMC}(x)$ are both interesting and controversial could be studied with greater resolution in massive lepton pair production than in deep inelastic scattering. A careful study of the A dependence of $\pi^- A \to \gamma^* X$ over a broad range in x_F would provide unique insight into the A dependence of the quark distributions of nuclei. A more extended discussion, including specific calculations and comparisons with data may be found in [7].

V. Conclusions

Careful study of the nuclear dependence of massive lepton pair production is important both for developing models of the nucleus and for detailed tests of perturbative QCD. Among the specific points made in this paper are:

1) Judicious selections of τ and x_F should permit an independent determination of the nuclear dependence of the quark and antiquark distributions, $q^A(x,M^2)$ and $\bar{q}^A(x,M^2)$ as functions of the fractional longitudinal momentum x.
2) In $pA \to \gamma^* X$, for $x_F > 0$, the A dependence of the cross section is determined by that of the antiquark distribution, principally the $\bar{u}$ flavor distribution, $\bar{u}^A(x,M^2)$. As discussed in Sec. III, the range in x over which the antiquark distribution can be explored is greater in lepton pair experiments than in deep-inelastic neutrino experiments. Lepton pair data offer the opportunity for gaining greater insight into nuclear models and more discriminating power.
3) In $\pi^- A \to \gamma^* X$ and $\bar{p}A \to \gamma^* X$, the nuclear dependence of the cross section is determined by that of the quark distribution $q^A(x,M^2)$. It is expected that $R_{DY}^{\pi^- A}(x_1,x_2) \simeq R_{DY}^{\bar{p}A}(x_1,x_2) \simeq R_{EMC}^{A}(x_2,M^2)$.
4) In $\pi^- A \to \gamma^* X$, the region $x_F \geq 0$ corresponds to the small x region in $R_{EMC}(x)$. Massive lepton pair data would shed special light, with perhaps greater resolution in both x and Q^2, on the nuclear physics of an interesting kinematic region in which deep inelastic data are sparse and controversial.
5) Currently popular models such as the pion exchange model [5] and rescaling models [6] are consistent with the limited data available on A dependence in massive lepton pair production. However, their distinct signatures could be distinguished with more precise data from $pA \to \gamma^* X$.
6) Use of (6) shows that nuclear effects are a likely source of error in present attempts to determine the x dependence of the pion structure function $q^\pi(x)$ and the magnitude and kinematic variation of the QCD K factor. The points are discussed in [7].

Acknowledgment

This work was supported by the U.S. Department of Energy, Division of High Energy Physics, under Contract W-31-109-ENG-38.

References

1. J. J. Aubert *et al.*, Phys. Lett. 123B, 275 (1983).
2. A. Bodek *et al.*, Phys. Rev. Lett. 50, 1431 (1983) and 51, 534 (1983); R. G. Arnold *et al.*, Phys. Rev. Lett. 52, 727 (1984); G. Bari *et al.*, Phys. Lett. 163B, 282 (1985).
3. H. Abramowicz *et al.*, Z. Phys. C25, 29 (1984); M. A. Parker *et al.*, Nucl. Phys. B232, 1 (1984); A. M. Cooper *et al.*, Phys. Lett. 141B, 133 (1984).
4. For reviews of experiment and theory, as well as an extensive list of references, see E. L. Berger, Argonne report ANL-HEP-PR-86-69, Proc. Second Conf. on the Intersections between Particle Nuclear Physics, Lake Louise, Canada, May, 1986.
5. E. L. Berger and F. Coester, Phys. Rev. D32, 1071 (1985); E. L. Berger, F. Coester, and R. B. Wiringa, Phys. Rev. D29, 398 (1984). See also C. H. Llewellyn Smith, Phys. Lett. 128B, 107 (1983); M. Ericson and A. W. Thomas, *ibid.* 128B, 112 (1983).
6. E. Close *et al.*, Phys. Lett. 129B, 346 (1983); R. L. Jaffe *et al.*, Phys. Lett. 134B, 449 (1984); F. E. Close *et al.*, Phys. Rev. D31, 1004 (1985).
7. E. L. Berger, Nucl. Phys. B267, 231 (1986), and references therein.
8. S. D. Drell and T. M. Yan, Phys. Rev. Lett. 25, 316 (1970). For reviews, consult E. L. Berger, in *Proc. of the Workshop on Drell-Yan Processes*, Fermilab, 1982, pp. 1-62, and in *Particles and Fields-1982*, ed. by W. E. Caswell and G. A Snow (AIP Conf. Proc. No. 98), pp. 312-342.
9. For a discussion of the QCD "K factor", see W. J. Stirling in *Proc. of the Workshop on Drell-Yan Processes*, *op. cit.*, pp. 131-154.
10. A. S. Ito *et al.*, Phys. Rev. D23, 604 (1981).
11. Fermilab proposal P-772, J. Moss *et al.*

The EMC Effect and the Swelling of Nucleons in Nuclei

M. Ericson

Institut de Physique nucléaire et IN2P3, 43 Bd. du 11 Novembre 1918,
F-69622, Villeurbanne Cedex, France and
CERN, Geneva, Switzerland

The EMC effect has proved to be a stimulus for nuclear physics. Whatever the final word will be, it has triggered a number of investigations of great interest. Two schools of thought have emerged to interpret the EMC effect. The first one links it to some change in scale for a nucleon inside the nucleus, which entails the rescaling phenomenon [1]. This change in scale corresponds to a swelling of the nucleons, interpreted as a partial deconfinement of the valence quarks, a first step towards the total deconfinement of the quark–gluon plasma. This is a very exciting possibility, as it opens the way for a third generation of nuclear physics, after the traditional one and that of the meson degrees of freedom.

The second interpretation sees in the EMC effect a manifestation of phenomena which belong to the by now classical nuclear physics. In this case the influence of quantum chromodynamics (QCD) is only indirect as it is deeply buried in the physics of the nuclear forces or of the meson exchange theory. The most concise example is the description of the EMC effect in terms of the nuclear binding proposed by Garcia Canal et al. [2] and Akoulinitchev et al. [3]. These authors showed how the binding of the nucleons affects the structure function. The potential which binds the nucleons appears on the same footing as their kinetic energy which produces the Fermi motion correction previously considered. This effect can explain the softening of the valence quarks. Such a description asks no questions about the origin of the field which binds the nucleons. This binding is usually thought of as due to the meson exchange NN interaction but there are some attempts to link it more directly to QCD. The binding model does not reproduce the small x enhancement. In this sense, the pionic model is more complete [4]. Here the small x enhancement is attributed to the existence of a pion excess in nuclei and the softening of the valence quark to the loss of momentum of the nucleons to the benefit of the pions. The link between this model and the binding correction is contained [5] in the relation between the pion excess number per nucleon $\langle n_\pi \rangle$ and the expectation value $\langle V_\pi \rangle$ of the one pion exchange potential (OPEP):

$$\langle n_\pi \rangle = \langle V_\pi \rangle / \langle \omega_\pi \rangle,$$

where $\langle \omega_\pi \rangle$ is the average pion energy. The existence of a pion excess follows from the attractive nature of the OPEP potential. According to Ref. [5], pions provide most of the nuclear binding with $\langle V_\pi \rangle \approx -50$ MeV, which leads to a pion excess of $\langle n_\pi \rangle \approx 0.2$ per nucleon.

A recent work by Oka and Amado [6] seems to reconcile these two classes of interpretations. These authors show that the existence of a swelling follows from general grounds for a quantum system and is a consequence of the attractive nature of the binding potential. This step establishes a bridge between binding and swelling and justifies the existence of the swelling. So it seems that everybody should be happy, as all phenomena follow from each other. I am afraid that, in this talk, I will introduce some dissonance in this harmony by addressing the question of the swelling and of its interpretation.

The existence of a swelling has been looked for in low-energy nuclear physics. Are there any indications that a nucleon gets bigger in the nucleus? The answer is positive, there is an indication from inelastic electron scattering, from the Coulomb sum rule. Sum rules are particularly useful to disentangle nucleon properties from the nuclear ones. In the inclusive ee′ scattering, the longitudinal response (charge) has been separated out [7] from the transverse one in the region of the quasi-elastic peak where the photon absorption occurs on one nucleon, a region which is well suited to measuring the nucleonic properties. The area under the longitudinal response should measure the total charge of the nucleus. This is the Coulomb sum rule. The puzzling result of these experiments is that ≈ 40% of the charge is missing (Fig. 1). Noble was bold enough to suggest [8] that this is the signature of a swelling. Indeed, to obtain the charge, the nucleon form factor had been divided out, with the assumption that it is the same as that of a free nucleon. Noble [8] and later Shakin [9] proposed that this is not the case and that for a bound nucleon it drops faster than for a free nucleon, a clear signature of a swelling effect. The increase in the radius needed to recover the nuclear charge is $\Delta\langle r^2\rangle_p \approx 0.4\ \mathrm{fm}^2$, as compared to $\langle r^2\rangle_p = 0.65\ \mathrm{fm}^2$. This is the most natural interpretation of the missing-charge problem.

The questions that I want to address now are: What is the physical interpretation of this swelling? What part of the nucleon structure is concerned? Does it justify the rescaling assumption? These questions lead to the simultaneous discussion of the root mean square (r.m.s.) charge radius and of the electromagnetic (e.m.) polarizability of

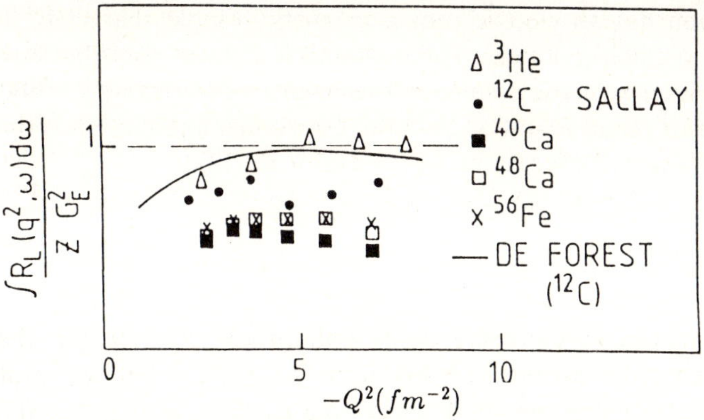

Fig. 1 The Coulomb sum rule for ^{3}He, ^{12}C, ^{40}Ca, ^{48}Ca, and ^{56}Fe. From Ref. [7].

a nucleon inside the nuclear medium. There are certain similarities between these quantities as the understanding of one helps in the understanding of the other. The polarizability is defined from the forward Compton amplitude, whose imaginary part is proportional to the total photoabsorption cross-section. It then becomes interesting to discuss our understanding of nuclear photoabsorption. Consider first the polarizability and the r.m.s. charge radius for a free nucleon and let us then discuss how these quantities are modified in the nucleus [10].

Free nucleon

The e.m. polarizability determines the leading t dependence of the forward and spin-averaged Compton amplitude $f(\omega)$

$$f(\omega) = f(0) + a\omega^2 + \dots , \tag{1}$$

where $a = \alpha + \beta$ is the sum of the electric and magnetic polarizabilities. This is a first similarity to the r.m.s. charge radius which fixes the t dependence of the electric form factor $G_E(t)$

$$G_E(t) = G_E(0) + (1/6)\langle r^2\rangle t + \dots \tag{2}$$

In addition, both quantities a and $\langle r^2\rangle$ arise from the internal structure of the nucleon, in fact from the most peripheral part of this structure which is dominated by the pion cloud. Its influence on the quantity a, for instance, can be gauged from the following dispersive integral

$$a_p = (1/2\pi^2) \int_{m_\pi}^{\infty} d\omega\, \sigma_p(\omega)/\omega^2 , \tag{3}$$

where σ_p is the photoabsorption cross-section on the proton. In the cross-section itself the Δ resonance is very prominent but in the integral the denominator ω^2 strongly favours states near the pion threshold, i.e. πN states in the continuum [11]. Here the photoproduction cross-section has an electric (not a magnetic) dipole character. It occurs via the contact term or pion-pole one. The dominance of these contributions over that of the Δ explains why the electric dominates the magnetic polarizability in the quantity a ($\alpha \approx 10 \times 10^{-4}\ fm^3$ versus $\beta \approx 4 \times 10^{-4}\ fm^3$). A similar analysis based on dispersion relations gives the r.m.s. charge radius of the proton as [12]

$$\langle r^2\rangle_p = (6/\pi) \int_{4m_\pi^2}^{\infty} dt\, \mathrm{Im}\, G_E(t)/t^2 , \tag{4}$$

where the spectral function $\mathrm{Im}\, G_E(t)$ is dominated by two-pion states (Fig. 2). As in the previous case, the denominator t^2 favours the low-mass states near the two-pion threshold, a region where the two pions are nearly uncorrelated, rather than the region around $t = 30m_\pi^2$ where the ϱ resonance occurs.

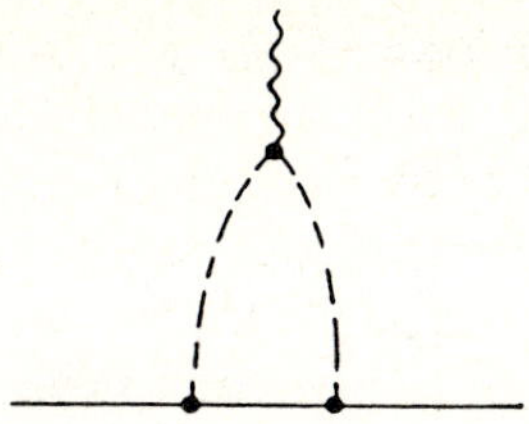

Fig. 2 The pion-cloud contribution to the r.m.s. charge radius of the proton

The importance of the pion cloud in the two structure constants has been confirmed by theoretical studies in the chiral bag model [13–15]. Oset et al. [14] find for the pion contribution to the r.m.s. charge radius: $\langle r^2\rangle_{pion} \gtrsim 0.5\ fm^2$. In the study of Theberge et al. [15] the cloud contribution is not as dominant, i.e. $\langle r^2\rangle_{pion} = 0.2\ fm^2$, but it is by no means negligible.

In fact the analogy between the electric polarizability, which dominates the e.m. one, and the r.m.s. charge radius is more than formal. There is a link between these two quantities, which can be established as follows. The static electric polarizability is

$$\alpha = 2 \sum_n |\langle 0|D|n\rangle|^2/(E_n - E_0) = (1/2\pi^2) \int d\omega\, \sigma_{dip}/\omega^2\,, \tag{5}$$

where D is the electric dipole operator which for a particle bound by a potential is $D = e\vec{\epsilon}\cdot\vec{r}$, and σ_{dip} is the unretarded electric dipole cross-section. On the other hand, the r.m.s. charge radius can be expressed with closure in terms of a sum over intermediate states

$$\langle r^2\rangle = 3/(4\pi^2 e^2) \int d\omega\, \sigma_{dip}/\omega\,, \tag{6}$$

which provides the relation

$$\langle r^2\rangle = (3/2e^2)\, \langle\omega_{dip}\rangle\, \alpha\,, \tag{7}$$

where $\langle\omega_{dip}\rangle$ is a typical E_1 energy.

Applied to the free proton, relation (7) gives an idea of the pion contribution to the proton charge radius. With the value of the intrinsic electric polarizability [16] of the proton $\bar{\alpha} = \alpha - (e^2/3M)\,\langle r^2\rangle_p = 8.6 \times 10^{-4}\ fm^3$ and a typical E_1 energy $\langle\omega_{dip}\rangle \approx$ 200 MeV, relevant for the πN continuum, the relation (7) gives $\langle r^2\rangle_{pion} \approx 0.2\ fm^2$, which is of the right magnitude.

In the nucleus

The pion cloud is the longest range structure of the nucleon, extending in the nucleus in regions where other nucleons are present. Appreciable distortion of this cloud must

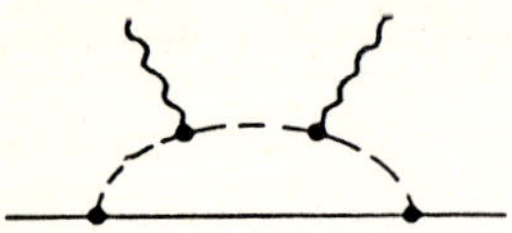

Fig. 3 A contribution to the nucleon polarizability arising from the pion cloud

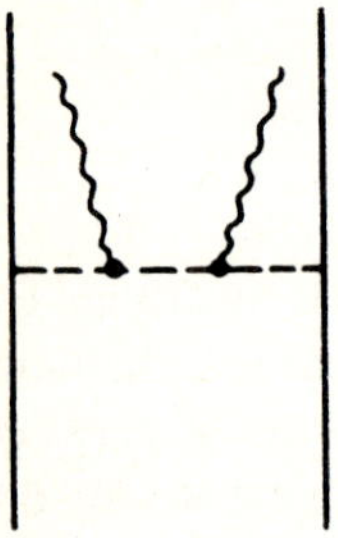

Fig. 4 A meson exchange correction to the nucleonic polarizability

then be expected in the nuclear medium [17]. This should have repercussions both on the polarizability and on the r.m.s. charge radius. As an illustration take a specific contribution to the proton e.m. polarizability, such as the one shown in Fig. 3. Here the pion is absorbed and re-emitted by the same nucleon. In the nucleus it can also be a pion in flight between two different nucleons, as shown in Fig. 4. In this case we obtain a meson-exchange correction to the nucleon e.m. polarizability. As an example, the evaluation of the meson-exchange correction, such as the one shown in Fig. 4, for a free Fermi gas, gives the Pauli correction to the e.m. polarizability. The overall Pauli correction results in a quenching of a_N by $\approx$ 20%, by no means a negligible effect [10]. According to the relation (7) a similar Pauli quenching should apply to the r.m.s. radius. We will now see that this is not the whole story but the message is already clear. A change in the r.m.s. charge radius of the nucleon inside the nucleus is more likely to reflect a distortion of the pion cloud, rather than a swelling of the quark core as is commonly thought. In order to see what further modifications occur, I will first describe the photoabsorption cross-section on nuclei, an ingredient which enters into the expression for the polarizability and for the r.m.s. charge radius.

Nuclear photoabsorption cross-section

This cross-section displays two strong resonances, the giant dipole resonance (GDR) at low energy and the Δ one at $\approx$ 300 MeV (Fig. 5). These two resonances are very different in nature. The GDR is a collective oscillation of the protons against the neutrons. Its position varies from nucleus to nucleus, with the law ω_{GDR} (MeV) $\approx$ 80 $A^{-1/3}$. The Δ instead is an excitation of the individual nucleons. Its position is the same for all nuclei. This difference is conceptually important for the understanding of the question of the e.m. polarization and of the r.m.s. charge radius on nucleons inside the nuclear medium. This appears in the investigation of the nuclear polarizability a_A described below. This quantity can be described by a dispersion relation similar to (3)

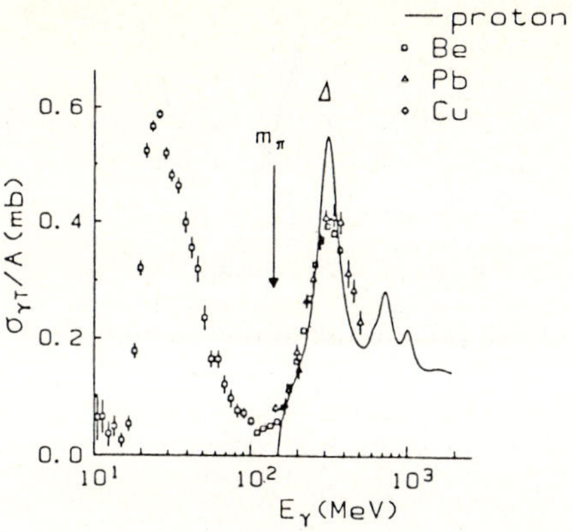

Fig. 5 Experimental values of the nuclear photoabsorption cross-section per nucleon. From Ref. [18].

with the integral starting at $\omega = 0$. We now want to separate out the contribution from the *individual nucleons.* For this we make the following observation. Suppose that in the cross-section σ_A there were only the two resonances, the GDR and the Δ, well separated, then the nucleonic contribution to a_A, $\tilde{a}_N$ could easily be singled out

$$\tilde{a}_N = (1/2\pi^2) \int_{m_\pi}^{\infty} [\sigma_A(\omega)/A]/\omega^2 . \tag{8}$$

The comparison between the free polarizability a_N and that of a bound nucleon $\tilde{a}_N$ would amount to the comparison between the free cross-section and the nuclear one per nucleon above pion threshold, which are close. However, there is more than just these two resonances, which makes this separation more subtle. Above the GDR the photoabsorption cross-section does not vanish but it shows a plateau. This is the so-called quasi-deuteron (QD) region and the name arises from the belief that the absorption is a two-nucleon process which occurs on correlated pn pairs. The quasi-deuteron cross-section has been parametrized as a scaling of the deuteron one with the Levinger factor L′. In addition there is a damping factor $e^{-80/\omega}$, at low energy, which is due to the Pauli blocking that forces the emitted nucleons to be above the Fermi sea:

$$\sigma_{QD}(\omega) = L' (ZN/A) [62.5 (\omega - 2.2)/\omega^{5/2}] e^{-80/\omega} , \tag{9}$$

with σ in mb and ω in MeV.

Microscopically, the quasi-deuteron absorption, up to $\omega \approx 140$ MeV, is a pionic process. The excitation of the 2p2h state occurs essentially via pion exchange, through the contact term (a) or the pionic current one (b) shown in Fig. 6. Some contribution also comes from the correlation term, one example of which is shown in Fig. 7. The important feature is that the quasi-deuteron absorption is an electric dipole mode. This

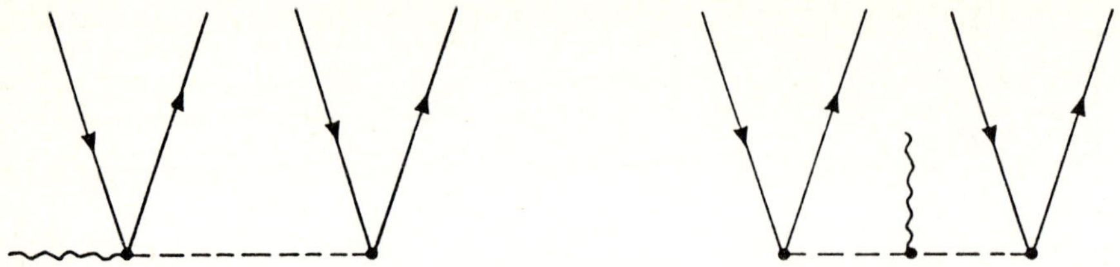

Fig. 6 Excitation of 2p2h states by the contact term (a) and the pionic current (b)

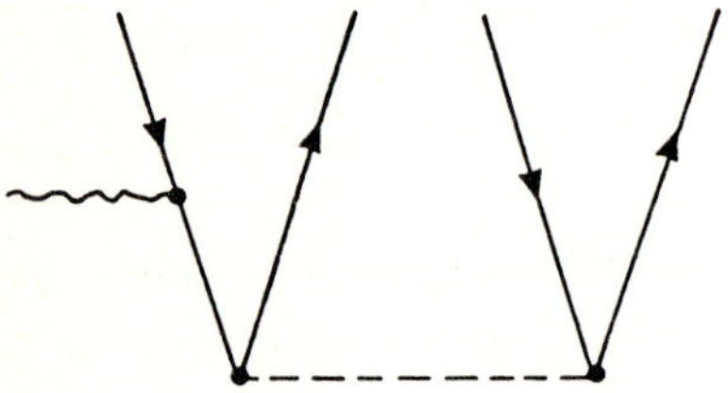

Fig. 7 An example of the excitation of a 2p2h state by the correlation current

does not hold on the nuclear scale but at the level of the pn pairs, which set the relevant scale for this absorption mechanism.

Coming back to the question of the nucleonic polarizability inside the nucleus, how should we consider the contribution to the nuclear polarizability of the quasi-deuteron cross-section? Does it pertain to the nucleus or the individual nucleon? To answer this, we observe that the quasi-deuteron cross-section has a *volume* character, in contradistinction to the GDR one which is a surface effect. It is thus normal to incorporate this part of the cross-section in the expression of the nucleon polarizability and to write $\tilde{a}_N$ as

$$\tilde{a}_N = (1/2\pi^2) \int_0^\infty d\omega \, [\sigma_A'(\omega)/A]/\omega^2, \tag{10}$$

where in σ_A' only the volume part of the cross-section has been retained: $\sigma_A' = \sigma_A - \sigma_{GDR}$. With the expression (10) and the parametrization (9) of σ_{QD} we find that the nucleonic polarizability in the nucleus is greatly enhanced, by a factor of ≈ 3. The origin of this large enhancement is the existence of low-lying ($\omega \lesssim m_\pi$) 2p2h excitations which generate σ_{QD}. Notice that this enhancement concerns mostly the electric polarizability.

We would like now to understand this result in the language of the meson exchange picture, to translate the physical effects in the cross-section into the treatment with virtual pions. In order to illustrate the connection between the two languages, take the specific example of the Pauli blocking effect which was derived in the virtual meson theory. We will restrict ourselves to one single contribution shown in Fig. 8. Can we trace back this effect in the cross-section? To recover the Pauli blocking effect in the cross-section we perform a cut in the graph of Fig. 8 so as to put the intermediate state

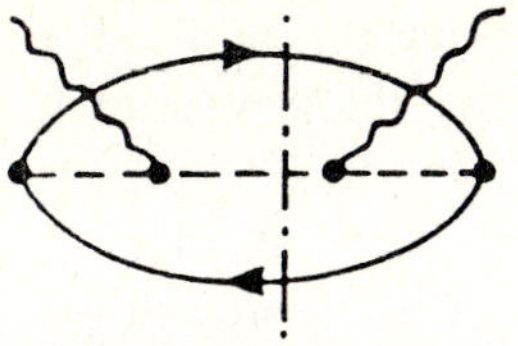

Fig. 8 A contribution to the e.m. polarizability of a nucleon inside a Fermi gas. The dash-dotted line shows the cut which puts the intermediate state on shell.

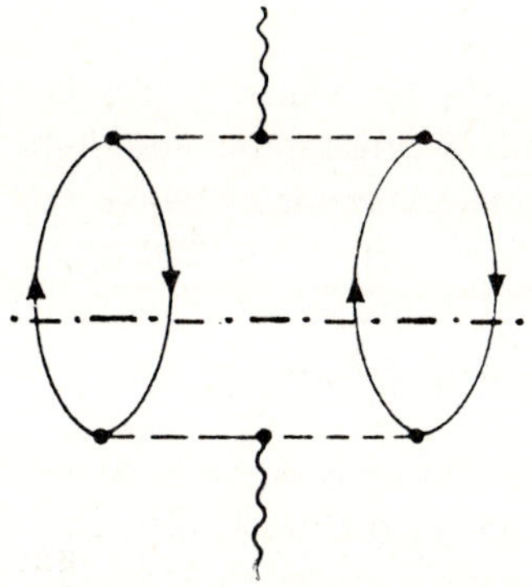

Fig. 9 A contribution to the Compton amplitude. The dash-dotted line shows the cut which leads to the photoabsorption process of Fig. 6b.

on shell. By doing so we obtain a contribution to the photoproduction process in the quasi-elastic region, with 1p1h excitation. Near threshold there is indeed a strong influence of the Pauli blocking effect which should quench the nuclear photoproduction cross-section as compared to the free one [19].

For the quasi-deuteron effect we proceed in the opposite way, going from the cross-section to the meson exchange. The quasi-deuteron graph of Fig. 6b, for instance, is the imaginary part of a contribution to the Compton amplitude shown in Fig. 9 and its physical meaning becomes apparent. It is the same graph as that of Fig. 8, but now the exchanged pion has scattered on a neighbouring nucleon before getting reabsorbed by the first one. The effect of the quasi-deuteron cross-section can therefore be seen as a manifestation of the distortion of the pion cloud by the environment and we have seen that this has a large effect. Similarly we expect also a large influence on the r.m.s. charge radius. To get an idea of the change in the r.m.s. charge radius of the proton inside the nucleus we use the bremsstrahlung weighted sum rule (6). The difference from the corresponding relation for the free proton gives the r.m.s. radius increase. Above the pion threshold the free nucleon cross-section and the nuclear one per nucleon are close. Most of the r.m.s. radius modification then comes from the region below the pion threshold

$$\Delta\langle r^2\rangle = \langle r^2_{eff}\rangle - \langle r^2\rangle_p \approx 3/(\pi^2 e^2) \int_0^{m_\pi} d\omega\, \sigma_{QD}(\omega)/\omega \approx 0.6\ \text{fm}^2$$

[expression (6) has been multiplied by four to account for the fact that the np distance in the quasi deuteron is twice the distance which enters in the electric dipole operator]. This represents a large increase, nearly a doubling of the free value $\langle r^2\rangle_p = 0.65\ fm^2$. Is this value reasonable? Can we understand such a large effect? After all, a large increase is not surprising since the effective r.m.s. charge radius of the proton in the nucleus should feel the internucleon distance. If only 1/10 of the proton charge is transferred to a neighbouring neutron, at a distance of $\approx$ 1.8 fm, the r.m.s. radius increases by $\approx 0.3\ fm^2$.

Is this phenomenon the origin of the missing strength in the ee′ experiments? We cannot be certain because the previous estimate is made in the static and long wavelength limit, which does not strictly apply to the actual experiments where the momenta range from 1 to 2.7 fm^{-1} and the frequencies from zero to $\approx$ 200 MeV. However, it is plausible that an effect of this origin exists and the magnitude should be qualitatively correct. It has nothing to do with a swelling of the quark core.

Strong experimental support in favour of this type of interpretation is provided by the following arguments. The effect that we have suggested concerns the *isovector* radius. The coupling of the photon to two pions is of an isovector nature; the isoscalar coupling proceeds instead through three-pion exchange. An increase in the isovector radius implies that the isovector form factor becomes, at finite momentum, quite different from its free value and hence from the isoscalar one which is not renormalized by this mechanism. Hence the coupling of the neutron to the photon, $G_s(q^2) - G_v(q^2)$, which is small for a free neutron, becomes appreciable in the nuclear medium. The photon *'sees' the neutrons in a nuclear environment.* This phenomenon is apparent in the experimental data, as the longitudinal response of ^{48}Ca is definitely different [7] from that of ^{40}Ca, which would not occur if G_s and G_v were equally renormalized. We have fitted [20] the experimental data, in the region of large momentum transfer, where the influence of collective effects should vanish, with the following inputs. The longitudinal cross-section is proportional to the quantity $Z(G_s + G_v)^2 R_p^0 + N(G_s - G_v)^2 R_n^0$, where R_p^0 and R_n^0 represent the responses of the free proton and neutron gases. The large renormalization of G_v in the nuclear medium provides a natural explanation of the difference between the ^{48}Ca response as compared to that of ^{40}Ca. In order to fit the experimental data, we have taken for the isoscalar form factor the free value $G_s(q^2) = (1 + q^2/\lambda_s^2)^{-2}$ with $\lambda_s^2 \approx 23\ fm^{-2}$, and for the isovector one a similar expression with an adjustable value of λ_v. A fit to the data is obtained with the value $\lambda_v^2 \approx 4\ fm^{-2}$ or equivalently an isovector radius $\langle r^2\rangle_v = 3\ fm^2$. This increase is compatible with our previous estimated increase $\Delta\langle r^2\rangle_v = 2\Delta\langle r^2\rangle_p \approx 1\ fm^2$.

I consider the isotope effect to be an indication in favour of the change in r.m.s. radius associated with the two-pion exchange part of the coupling of the photon. It will be interesting to extend the exploration to the region of lower momenta, where the collective effects are important and the proton and neutron responses mix.

To summarize, I have proposed here a pionic interpretation of the swelling effect. The nucleon acquires an effective r.m.s. charge radius much larger than the free value, from the spreading of the charge to the neighbouring nucleons or to the pion cloud. This type of swelling affects exclusively the isovector swelling. It does not concern the quark core, which remains unchanged. This picture has two merits. Firstly it explains the differences observed in the longitudinal responses of the two isotopes ^{40}Ca and ^{48}Ca. Secondly, a swelling of this nature, which does not touch the quark core, explains why the excitations of this core do not seem to be changed in the nuclear environment. For instance, the Δ resonance is found at the same excitation energy in the nucleus and for a free proton. A large swelling of the core instead should normally lower this energy [21] and one has to invoke compensating effects to justify this fact. Of course our interpretation does not preclude the existence of a genuine (but more modest) modification of the core radius. But it makes unlikely its observation in the ee′ longitudinal response, which is more sensitive to pionic effects.

The question that I want to raise now concerns the rescaling assumption in the EMC effect. Does the pionic swelling, which represents a change in scale, entail the rescaling? I believe it does not, as rescaling implies a change in scale for the valence quarks, which does not take place here.

So in order to have rescaling as the cause of the EMC effect we have to accept the idea that this effect is totally disconnected from the missing strength problem in ee′ scattering. Personally I believe instead that the two effects are two facets of the same physical reality, that of pions in nuclei. The second choice is esthetically more appealing and likely to be closer to reality.

REFERENCES

(1) R. Jaffe, F. Close, R. Roberts and G. Ross: Phys. Lett. **134B,** 449 (1984).
(2) C.A. Garcia Canal, E.M. Santangelo and H. Vucetich: Phys. Rev. Lett. **53,** 1430 (1984).
(3) S.V. Akulinichev, S. Shlomo, S.A. Kulagin and G.M. Vagradov: Phys. Rev. Lett. **55,** 2239 (1985).
(4) C. Llewellyn Smith: Phys. Lett. **128B,** 107 (1983).
M. Ericson and A.W. Thomas: Phys. Lett. **128B,** 112 (1983).
(5) V.R. Pandharipande: Nucl. Phys. **A446,** 189c (1985).
(6) M. Oka and R.D. Amado: University of Pennsylvania preprint UPR-0295T (1986).
(7) P. Barreau et al.: Nucl. Phys. **A402,** 515 (1983).
Z. Meziani et al.: Phys. Rev. Lett. **52,** 2130 (1984) and **54,** 1233 (1985).
(8) J. Noble: Phys. Rev. Lett. **46,** 412 (1981).
(9) C.M. Shakin: Nucl. Phys. **A446,** 323c (1985).
(10) M. Ericson and M. Rosa-Clot: preprint CERN-TH 4420/86 (1986).

(11) M. Rosa-Clot and M. Ericson: Z. Phys. **A320**, 675 (1985).
M. Ericson: Lectures at the School on Nuclear Dynamics, Dronten, 1985, preprint Lycen/8577 (1985).
(12) G. Höhler and E. Pietarinen: Phys. Lett. **53B**, 471 (1975).
(13) P. Weiner and W. Weise: Phys. Lett. **159B**, 59 (1985).
(14) E. Oset, R. Tegen and W. Weise: Nucl. Phys. **A426**, 456 (1984).
(15) S. Theberge, G. Miller and A.W. Thomas: Canad. J. Phys. **60**, 59 (1982).
(16) T. Ericson and J. Hüfner: Nucl. Phys. **B47**, 405 (1972).
(17) J. Delorme, M. Ericson, A. Figureau and C. Thévenet: Ann. Phys. **102**, 273 (1976).
M. Ericson: Progr. Part. Nucl. Phys. **1**, 67 (1978).
(18) J. Ahrends: Nucl. Phys. **A446**, 229c (1985).
(19) W.M. MacDonald, E.T. Dressler and J.S. O'Connel: Phys. Rev. C **10**, 455 (1979).
(20) W.M. Alberico, P. Czerski, M. Ericson and A. Molinari: in preparation.
(21) J. Noble: Abstract A17, p. 82, 11th Europhysics Divisional Conf. Paris, July 1985.

The Colour Conductivity Model and the Shadow Phenomenon in Nuclei

O. Nachtmann

Institut für Theoretische Physik, Universität Heidelberg, Philosophenweg 16, D-6900 Heidelberg, F. R. Germany

The colour conductivity model for deep inelastic lepton-nucleus scattering at high momentum transfer is briefly reviewed. A connection between the shadowing phenomenon and the Bethe-Weizsäcker formula for the nuclear binding energy is made using the energy momentum sum rule.

1. The Basic Hypotheses

This talk is based on work by H. Pirner and myself done already some time ago [1]. I will first recall the basic hypotheses of the colour conductivity model. Then I will make a bridge between low energy nuclear physics and the high energy world where we consider nucleons and nuclei as composed of quarks and gluons. This bridge will have a rather solid footing in the energy momentum sum rule and will lead us to a connection between the Bethe-Weizsäcker formula for the nuclear binding energy and the shadow phenomenon at high energies.

We will be interested in deep inelastic lepton-nucleus scattering, e.g. muon scattering on a nucleus with mass number A:

$$\mu + A \rightarrow \mu + X, \tag{1}$$

where X stands for the final hadronic state and we use the conventional kinematic variables: ν for the energy transfer, Q^2 for the 4-momentum transfer squared, etc.. Let, furthermore, be x and x_A the Bjorken variables defined with the nucleon mass M and the nucleus mass M_A, respectively:

$$x = \frac{Q^2}{2M\nu} \quad , \quad x_A = \frac{Q^2}{2M_A\nu} \tag{2}$$

The kinematic limits are:

$$0 \leqq x \leqq M_A/M \cong A; \quad 0 \leqq x_A \leqq 1$$

The structure functions and parton densities per nucleon for a nucleus A will be denoted by $F_2(x, Q^2, A)$ and $N_j(x, Q^2, A)$, where the index j numbers quarks, antiquarks and gluons ($j = u, d, \bar{u}, \bar{d}, G$). Heavy quarks (s, c, b) will be neglected.

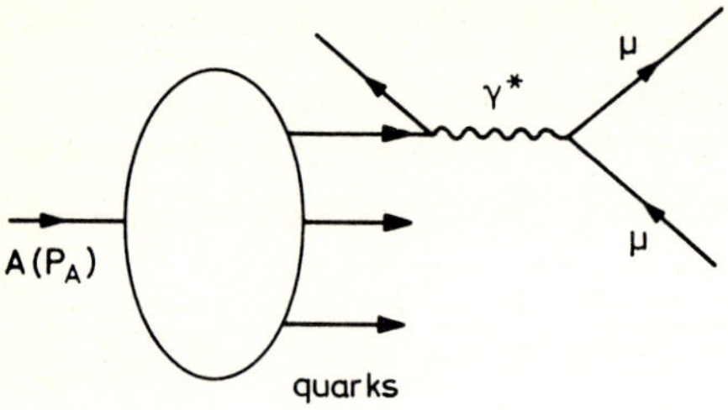

Fig. 1 Deep inelastic muon-nucleus scattering in the Breit-frame

Consider now the reaction (1) at high Q^2 in the Breit-frame (Fig. 1), where the incident and outgoing muons have the same energy and the nucleus travels with high momentum P_A:

$$P_A = \frac{Q}{2x_A} . \quad (3)$$

Our basic assumption is that the quark modes, on which the muon scatters at high Q^2, extend over the whole transverse size of the nucleus. The same is assumed for the gluon modes. Then the nucleonic structure of the nucleus should become unimportant and we should be able to apply simple scaling arguments. Let A and A' be two isoscalar nuclei with radii R_A and $R_{A'}$ (Fig. 2). Scattering a lepton on these nuclei probes them with a resolution Q^{-1}, where Q is the momentum transfer. If there were no other scales in the problem except the radius and the resolution distance, we could conclude immediately that the structure functions of nuclei should at fixed x only depend on the ratio $R_A/Q^{-1} = R_AQ$. This means that going from a small nucleus A' to a bigger one, A, at the same Q^2 should be equivalent to a change from Q^2 to Q'^2, keeping the target nucleus A' fixed, where

$$Q' = Q \frac{R_A}{R_{A'}} . \quad (4)$$

Thus, the structure functions of the two nuclei should be related by

$$F_2(x, Q^2, A) = F_2(x, Q'^2, A'). \quad (5)$$

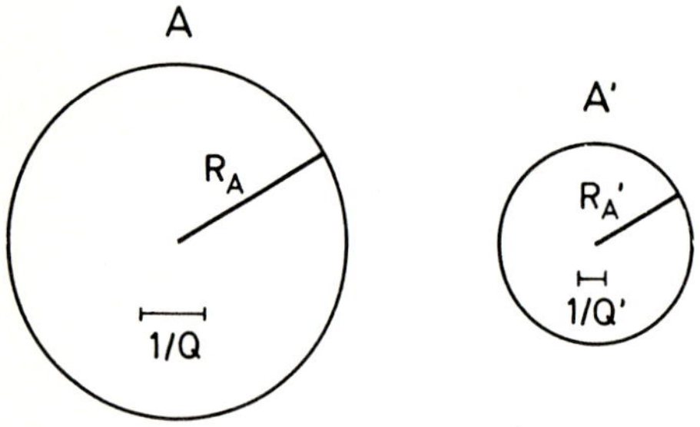

Fig. 2 Transverse view of two nuclei A and A', radii R_A and $R_{A'}$, probed with resolutions 1/Q and 1/Q', respectively

Of course, in quantum chromodynamics (QCD) we have another scale in the problem, the inverse of the Λ-parameter, or, if you want, the confinement radius of the free nucleon. This makes the simple rescaling formula (4) incompatible with standard QCD, but it is easy to remedy this defect [1]. Instead of (4), we find in the framework of QCD the rescaling relation

$$Q' = Q\left(\frac{R_A}{R_{A'}}\right)^{\bar{\alpha}_s/\alpha_s(Q^2)} \tag{6}$$

where $\alpha_s(Q^2)$ is the usual running coupling strength of QCD, and $\bar{\alpha}_s$ is the effective coupling parameter of the quark and gluon modes extending over the whole transverse size of the nucleus.

Here we will not discuss further the comparison of this ansatz to the data on deep inelastic lepton-nucleus scattering. All we will use in the following is that there is a rescaling relation of the type (5) for the structure function respectively quark and gluon densities:

$$N_j(x,\, Q^2,\, A) \quad = \quad N_j(x,\, Q'^2,\, A') \tag{7}$$

with Q' given by (6). We would like to emphasize that the colour conductivity model unambiguously requires (7) to hold at fixed Bjorken variable $x = Q^2/(2M\nu)$ defined with the nucleon mass. (Any fixed mass, independent of A and A' would, of course, be as good.) The essential reason being that (7) derives from our nuclear evolution equation which conserves momentum [1].

2. Shadowing and Colour Conductivity

If our assumptions of chapter 1 were true down to x = 0, we would have quark and gluon modes extending over the whole transverse and longitudinal space of the nucleus. In other words, we would have a nucleus as one big bag filled with quarks and gluons. This cannot be true, and there must be some critical value x_c below which our assumptions of chapter 1 are no longer valid. We will now argue that

$$0 \leqq x \leqq x_c$$

is the shadowing region and estimate x_c from the Bethe Weizsäcker formula, which for isoscalar nuclei reads as follows [2]:

$$M_A = A\,(M - b_V + b_S\, A^{-1/3}). \tag{8}$$

Here b_V = 15.56 MeV is the volume term, b_S = 17.23 MeV determines the surface term, and we have neglected the Coulomb-term as an electromagnetic correction.

The key observation is that energy and momentum are conserved quantities. Thus we can measure the energy of the nucleus in the conventional way or - if we are high energy freaks - by adding up the energy carried by the quarks and gluons seen at high Q^2. This leads to the well-known energy-momentum sum rule of parton physics which is in fact a rigorous consequence of Wilson's operator product expansion in the limit $Q^2 \to \infty$. It reads as follows:

$$\sum_j \int_0^1 dx_A \, x_A N_j^A (x_A, Q^2) = 1 \tag{9}$$

where x_A, defined in (2), is the Bjorken variable appropriate for nucleus A and N_j^A are the appropriate parton densities which are related to our previous ones by

$$N_j^A (x_A, Q^2) \, dx_A = A \, N_j (x, Q^2, A) \, dx. \tag{10}$$

We want to emphasize that these parton densities are (at high enough Q^2 where $1/Q^2$ terms can be neglected) directly measurable quantities to be obtained from structure functions and their scaling violations. From (9) and (10) we find now

$$\int_0^{M_A/M} dx \, x \sum_j N_j(x, Q^2, A) = \frac{M_A}{AM} = 1 - \frac{b_V}{M} + \frac{b_S}{M} A^{-1/3} . \tag{11}$$

For $x > x_c$ we will use our colour conductivity ansatz (7) to relate the parton densities of nucleus A to the ones of deuterium. This ansatz implies essentially no A-dependence apart from logarithms for the parton densities per nucleon in this region. For very small values of x, on the other hand, shadowing should occur. The virtual photon can either be directly absorbed at some point $\vec{y}$ in the nucleus or it can produce coherently a hadronic state upstream in the nucleus which then is absorbed at the same point $\vec{y}$ (Fig. 3). The two processes of Fig. 3a,b interfere destructively and imply that only the quarks of the surface region of the nucleus contribute to deep inelastic scattering at very small values of x (cf. [3] and references cited therein). Then the cross section should be proportional to $A^{2/3}$, the parton densities per nucleon correspondingly to $A^{-1/3}$. A simple ansatz interpolating between the two above regimes is as follows:

$$N_j(x, Q^2, A) = N_j(x, Q'^2, d) \left[\frac{x}{x_c} + \left(\frac{A}{2}\right)^{-1/3} \left(1 - \frac{x}{x_c}\right)\right] \tag{12}$$

for $0 \leq x \leq x_c$ and

$$N_j(x, Q^2, A) = N_j (x, Q'^2, d) \tag{13}$$

for $x \geq x_c$, where we have related the nuclear parton densities to those of the deuteron (d, A = 2) using (6) and (7).

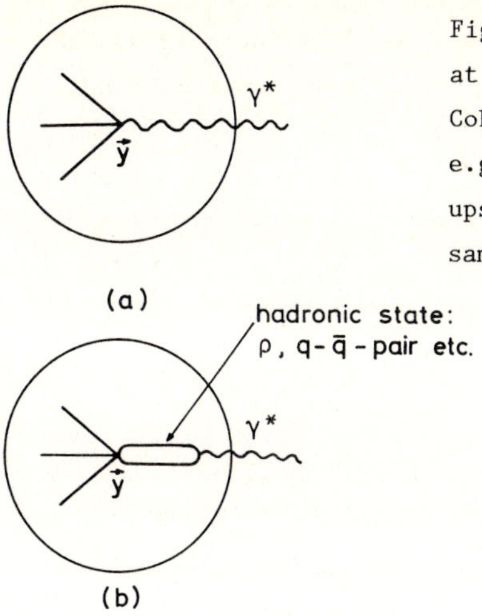

Fig. 3 Absorption of a virtual photon at a point $\vec{y}$ in the nucleus (a). Coherent production of a hadronic system, e.g. a vector meson ρ or a q-$\bar{q}$-pair, upstream which is then absorbed at the same point $\vec{y}$ (b)

It is now a simple matter to insert our ansatz into (11). This leads to

$$\int_0^{M_A/M} dx\, x \sum_j N_j\,(x,\, Q'^2,\, d) - [1 - (\tfrac{A}{2})^{-1/3}] \int_0^{x_c} dx\, x \sum_j N_j(x, Q'^2,\, d)(1 - \tfrac{x}{x_c})$$

$$= \quad 1 - \frac{b_V}{M} + \frac{b_S}{M}\; 2^{-1/3} - [1 - (\tfrac{A}{2})^{-1/3}]\, \frac{b_S}{M}\, 2^{-1/3} \tag{14}$$

Comparing the right and left hand sides, we find the following results:

(1) The first integral on the l.h.s. corresponds to the energy momentum sum rule for the deuteron and equals 1, neglecting the small deuteron binding energy for the moment. Using the near equality of b_V and $b_S \cdot 2^{-1/3}$ (8) we conclude then

$$\int_0^{x_c} dx\, x \sum_j \; N_j(x,\, Q'^2,\, d)(1 - \frac{x}{x_c}) \quad \cong \quad \frac{b_S}{M}\; 2^{-1/3} \;. \tag{15}$$

Thus x_c must be essentially independent of A and Q^2. Shadowing must, for consistency, be a leading twist effect in the colour conductivity model, i.e. it must persist at arbitrarily high Q^2.

(2) The behaviour of the parton densities in the nucleus A is shown schematically in Fig. 4. A strict colour conductivity relation down to x = 0 would correspond to the dash-dotted line. For the energy momentum sum rule (11) of nucleus A the area below the dash-dotted line is taken away - this corresponds to the negative volume term b_V of the binding energy - and instead a piece proportional to $A^{-1/3}$ is added, corresponding to the positive surface energy $b_S \cdot A^{-1/3}$.

(3) We obtain a relation between b_S, b_V and the binding energy of the deuteron ε_B:

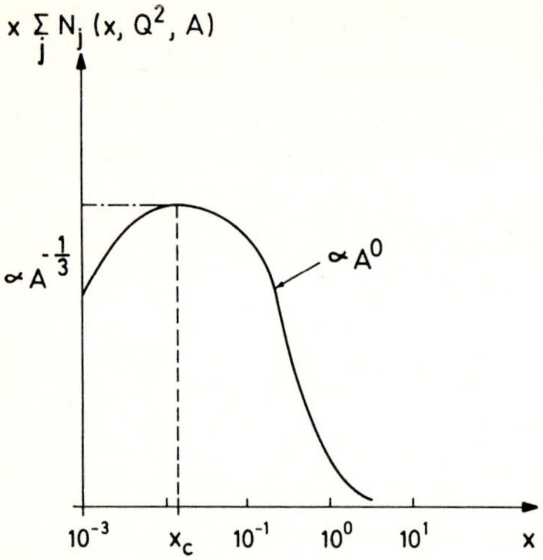

Fig. 4 Schematic behaviour of the parton densities per nucleon of a nucleus A. The dash-dotted line corresponds to applying the colour conductivity relation (7) down to x = 0. The critical value of x below which shadowing appears, is denoted by x_c.

$$b_S = (b_V - \frac{\varepsilon_B}{2})\ 2^{1/3}. \tag{16}$$

Using ε_B = 2.2 MeV and b_V = 15.6 MeV, (16) predicts b_S = 18.2 MeV, which agrees with experiment within 6 %.

(4) From the known parton densities of the deuteron at $x \cong 0$ we can estimate the critical value x_c to be

$$x_c \cong \frac{5}{18} \frac{1}{F_2(0,Q'^2,d)} \frac{b_S}{2^{1/3} M} \cong 0.012\ . \tag{17}$$

This appears to be quite reasonable. A similar estimate for the upper limit of the shadowing region has e.g. been given by Mueller on completely different grounds [3].

We hope to have shown in this contribution that the colour conductivity model makes through consistency arguments definite predictions for the shadowing phenomenon. We have also connected the time-honoured Bethe-Weizsäcker formula to the modern quark-gluon concepts. Hopefully, experimentalists will be able to check our relations. Existing data (e.g. [4]) do not reach high enough Q^2. But this should be the case for the forthcoming EMC-experiment.

References

1 O. Nachtmann and H. J.Pirner: Z. f. Physik C21, 227 (1984); Colour Conductivity at High Resolution: A New Phenomenon of Nuclear Physics, to appear in Ann. Phys. (Leipzig);

O. Nachtmann: "Theoretical Ideas concerning the EMC Effect", in Proceedings of the XIth International Conference on Neutrino Physics and Astrophysics, ed. K. Kleinknecht and E. A. Paschos (World Scientific, Singapore 1984)

2 A. Bohr and B. Mottelson, Nuclear Structure Vol. I, 169 (W. A. Benjamin, New York-Amsterdam 1969)

3 A. H. Mueller, in Quarks, Leptons and Supersymmetry, ed. J. Tran Thanh Van (Edition Frontieres, Gif-sur-Yvette 1982)

4 M. S. Goodman et al., Phys. Rev. Lett. 47, 293 (1981)

Nuclear Effects in Deep-Inelastic Lepton-Nucleus Scattering

S. Shlomo[1], *S.V. Akulinichev*[2], *S.A. Kulagin*[2], *and G.M. Vagradov*[2]

[1]Cyclotron Institute, Texas A & M University, College Station, TX 77843, USA
[2]Institute for Nuclear Research, Academy of Sciences of the USSR, Moscow 117312, USSR

INTRODUCTION

Deep-inelastic lepton-nuclear scattering (DILNS) experiments [1-3] have focused the attention to many important aspects of nuclear and particle physics. The main question raised by these experiments is: Where is the boundary between the interpretation in terms of quarks and that in terms of effective nucleons and mesons degrees of freedom? The deviation between the free nucleon structure function (SF), $F_2^N(x)$, and the nuclear SF, $F_2^A(x)$, [the European Muon Collaboration (EMC) effect] has led to many theoretical models which explain the EMC effect by assuming: The nucleon's quark structure changes inside a nucleus, the appearance of multiquark configurations, the enhancement of the "sea", etc [4].

We consider the DILNS within the conventional nuclear theory taking into account the effects of nuclear binding and two-body short range correlations. We show that the EMC effect is mainly due to nuclear binding. It is also shown that the DILNS data for ^{12}C [3] in the $x > 1$ region are well reproduced when two-body short range correlations are taken into account; x is the Bjorken variable for a free nucleon (see refs. 5-7 for details). We add that these nuclear effects were introduced in some papers [8-11] through parameters fitted to the data. In our work we carry out the calculations on the basis of known nuclear data with no free parameters.

We assume that a nucleus is the bound system of nucleonic and mesonic fields and at asymptotically large $-q^2 \gg m_N^2$ the virtual photon is scattered from the nuclear consistuents incoherently. We also take the bound nucleon's SF to coincide with that of the free nucleon.

THE EMC EFFECT

We consider the deep-inelastic scattering (DIS) from single nucleons occupying the nuclear shell characterized by the separation energies, ϵ_λ (negative) and the single particle wave functions ϕ_λ. Writing the tensor $W_{\mu\nu}$, which describes the hadronic part of the DIS cross section, in terms of single-nucleon electromagnetic currents, we find [5-7] that

$$F_2^A(x) = \int_x^\infty dz\, f^A(z)\, F_2^N(x/z) \tag{1}$$

$$f^A(z) = \sum_\lambda \int \frac{d\vec{p}}{(2\pi)^3}\, \delta\left(z - \frac{p_\lambda q}{m_N q_o}\right) \left| \phi_\lambda(\vec{p}) \right|^2, \tag{2}$$

where $\phi_\lambda(\vec{p})$ is the Fourier transform of $\phi_\lambda(\vec{r})$, m_N is the nucleon mass and p_λ and q are the four vector momenta of the nucleon and the virtual photon, respectively. The sum in (2) is over all occupied single particle states. The physics of the binding energy effect is in taking $p_\lambda = (m + \epsilon_\lambda, \vec{p})$ since the off-mass shell nuclear nucleon has energy ϵ_λ and momentum $\vec{p}$, with scaling variable $x_\lambda = -q^2/(2p_\lambda q)$. Taking the z axis in the direction of $\vec{q}$ we have $p_\lambda q = (m+\epsilon_\lambda)q_o - p_z q_z$ with $q_z/q_o = 1$ in the scaling limit. The well-known Fermi smearing effect is obtained from (1) and (2) using $\epsilon_\lambda = 0$. Figure 1 compares the results of our calculation for the ratio $R(x) = F_2^A(x)/[A\, F_2^N(x)]$ with the data of ref. 2. We have used harmonic oscillator wave functions (HOWF) for ϕ_λ with $\hbar\omega = 45/A^{1/3} - 25/A^{2/3}$. The values of ϵ_λ are taken from a compilation [12] of experimental data for the (e,e'p) reaction. We take the nucleons in the deuteron to be free and use the parameterization of ref. [13] for $F_2^N(x)$. We note that the low x behaviour of R(x) is due to nuclear binding and the rise of R(x) for x > 0.6 is due to Fermi motion.

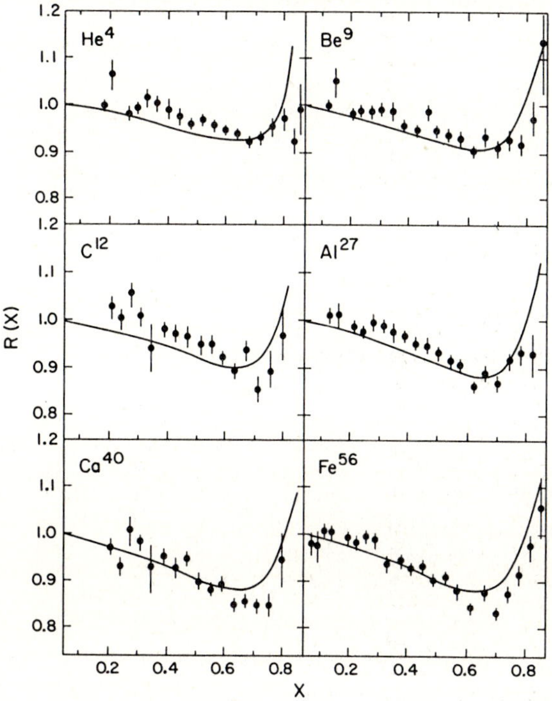

FIG. 1. $R(x) = F_2^A/(AF_2^N)$ as a function of x for various nuclei. The data are taken from Ref. 2. The curves are the results including nuclear binding. See text for details.

THE x > 1 REGION

For sufficiently large x, ϵ_λ can be neglected so that $f^A(z)$ can be written in terms of N(p), the momentum distribution of nucleons in the nucleus. We thus write $f^A(z) = f^A_{mf}(z) + f^A_c(z)$ where the mean field contribution $f^A_{mf}(z)$ was calculated above using HOWF and experimental values of ϵ_λ and

$$f^A_c(z) \simeq \int_{p_z}^{\infty} \frac{dp\; pm_N}{(2\pi)^2} \frac{(2\pi)^3}{4\pi} N_c(p) \qquad (3)$$

is due to short-range correlation. For the high momentum tail, $N_c(p)$, we use the results of Brueckner theory of finite nuclei with realistic two-body interaction [14], i.e.,

$$N_c(p) = N_c(p_c) \exp[-\beta(p-p_c)] \qquad \text{for } p > p_c, \tag{4}$$

with $\beta = 1.5$ fm, $p_c = 2\ \text{fm}^{-1}$ and $N_c(p_c) = 0.021A\ \text{fm}^3$. In (3), $p_z = \max[p_c, |m_N(z-1) - \langle\epsilon_\lambda\rangle|]$. We note that $f^A(z)$ is normalized according to $\int_0^\infty f^A(z)dz = A$. The results of our calculation for ^{12}C is compared with the experimental data [3] in fig. 2. This clearly demonstrates the important effect of short-range correlation.

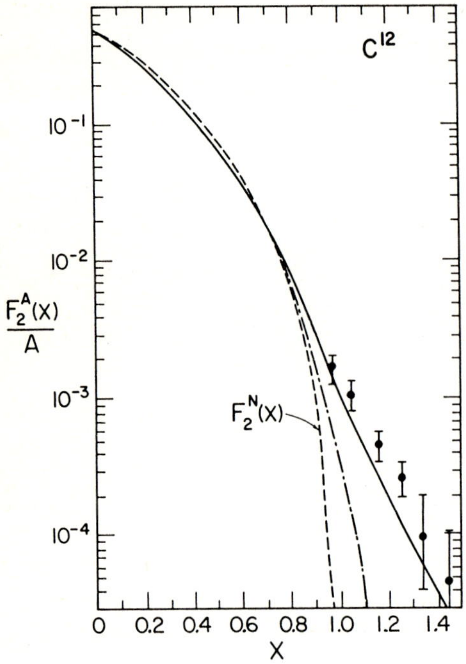

FIG. 2. Plot of structure functions as a function of x. Data are from the Dubna experiment (Ref. 3). The solid line corresponds to our full result, the dash-dotted line to that obtained when the correlation part of the distribution function $f_c^A(z)$ is neglected.

CONCLUSIONS

Before closing, we now mention very briefly other possible effects which should be taken into account in a refinement of the calculations presented above. Using (1) and (2) we have that

$$\frac{1}{A}\int_0^A dx\, F_2^A(x) = \frac{1}{A}\int_0^A dz\, z\, f^A(z) = 1 + \frac{\langle\epsilon_\lambda\rangle}{m_N} \approx 0.97. \tag{5}$$

This violation of the momentum sum rule can be restored by including the contributions of mesons [5] and the simultaneous scattering of the electron from two nucleons [15], which affect R(x) for $x \leq 0.3$, improving the agreement with data. The relatively small effects of the deuteron SF and the $Q^2 = -q^2$ dependence of the neutron and proton SF should be included. Also the problems related to off-mass shell analytic continuation of the scattering amplitude, final state interaction and relativistic corrections to the nuclear wave function should be investigated. However, these effects are not expected to change our main conclusion concerning the significance of nuclear binding, Fermi motion and short-range correlations. Our considerations indicate the surprising stability of the nucleon inside the nucleus and restrict the domain of pure quark effects.

References

1. J. J. Aubert, et al.: Phys. Lett. 105B, 322 (1982).
2. R. G. Arnold, et al.: Phys. Rev. Lett. 52, 727 (1984).
3. I. A. Savin: International Seminar on High Energy Physics, Joint Institute for Nuclear Research, Report No. D1, 2-81-728, 1981.
4. J. P. Vary: Nucl. Phys. A418, 195c (1984).
5. S. V. Akulinichev, S. Shlomo, S. A. Kulagin and G. M. Vagradov: Phys. Rev. Lett. 55, 2239 (1985).
6. S. V. Akulinichev and S. Shlomo: Phys. Rev. C33, 1551 (1986).
7. S. V. Akulinichev, S. A. Kulagin and G. M. Vagradov: Phys. Lett. 158B, 485 (1985).
8. C. H. Llewellyn Smith: Phys. Lett. 128B, 107 (1983).
9. C. A. Garcia Canal, E. M. Santangelo and H. Vucetich, Phys. Rev. Lett. 53, 1430 (1984).
10. E. L. Berger and F. Coester: Phys. Rev. D32, 1071 (1985).
11. H. Araseki and T. Fujita: Nucl. Phys. A439, 681 (1985).
12. S. Frullani and J. Mougey: Adv. Nucl. Phys. 14, 1 (1984).
13. J. G. H. deGroot et al.: Phys. Lett. 82B, 456 (1979) and Z. Phys. C1, 143 (1979).
14. J. Van Orden, W. Truex and M. K. Banerjee: Phys. Rev. C21, 2628 (1980).
15. B. L. Birbrair et al.: Phys. Lett. 166B, 119 (1986).

Electromagnetic Response in Nuclei in Terms of the Quark Structure

C.M. Shakin

Department of Physics, Brooklyn College of the City University of New York, Brooklyn, NY 11210, USA

1. Introduction

In the short space available for this presentation, I would like to emphasize that three topics of current interest in nuclear physics: the EMC effect [1], the quenching of the longitudinal response seen in inclusive (e,e') reactions [2], and the use of the Dirac equation to describe nucleon motion in nuclei [3-5], are all related to a single effect, the modification of the gluon condensate in nuclei [6]. We will argue that, in all three cases mentioned above, we are seeing the effects of a change of a QCD mass (or length) scale away from its vacuum value. The order parameter describing this scale change is an order parameter of the gluon condensate [7]. In the absence of current quark masses, there is only a single mass scale developed dynamically in QCD. It is possible to construct an effective Lagrangian for QCD [7] which only contains a single dimensional order parameter. A change in this parameter will lead to a corresponding change of all dimensional quantities. In particular, we have shown [6,7] how a length scale may be specified for QCD by making use of the gauge and Lorentz invariant parameter,

$$\langle vac|\frac{g^2}{4\pi^2} G^a_{\mu\nu} G^{\mu\nu}_a |vac\rangle = \frac{3}{32\pi^2} g^2 \phi_o^2 \quad . \tag{1}$$

Here g^2 is the QCD coupling constant renormalized at the mass scale, μ^2. (Equation (1) may be taken as a definition of the quantity $g^2\phi_o^2$.) We remark that a value of 0.012 $(\text{GeV})^4$ has been obtained for the left-hand side of (1) in work on QCD sum-rules [8]. (Note that this quantity is a renormalization group invariant in a physical gauge.)

As we have seen in other works [6,7], various dynamical masses are given in terms of the quantity, $g^2\phi_o^2$. We obtain a <u>dynamical</u> gluon mass,

$$m_G^2 = \frac{3}{8} g^2 \phi_o^2 , \tag{2}$$

and a <u>dynamical</u> quark mass,

$$(m_q^{G\ell})^2 = \frac{1}{6} g^2 \phi_o^2 , \tag{3}$$

as well as a number of other mass parameters, all of which are proportional to the same order parameter. The quark also obtains a dynamical mass via the formation of a chiral condensate, however, in our model the chiral condensate order parameters do not define an independent mass scale. Therefore, if we take the current quark mass to be zero for the up and down quarks, there is only a single mass scale in our effective Lagrangian, which we assume describes QCD at large length scales [6,7].

Now the presence of quarks tends to break down the gluon condensate and in nuclear matter we claim that ϕ_o should be replaced by $\phi_{NM} < \phi_o$. Indeed, we want to show that if $\phi_o/\phi_{NM} \cong 1.25$, we can understand the various phenomena mentioned at the beginning of the introduction.

2. The EMC effect

There have been a very large number of theoretical papers dealing with the EMC effect [9]. We prefer the explanation of this effect based upon the rescaling model of Close, Ross, Roberts and Jaffe [10]. However, we do not use the specific dynamical model which these authors have used to calculate the change of scale in nuclei. We can summarize their model as follows. Moments of structure functions are assumed to exhibit "rescaling". With A specifying a nucleus of mass number A, and N denoting the nucleon, it is assumed that moments are related by the following expression [10],

$$M_A(Q^2) = M_N(\xi_{NA}(Q^2)Q^2), \tag{4}$$

where the quantity $\xi_{NA}(Q^2)$ evolves with Q^2 as follows,

$$\xi_{NA}(Q^2) = \xi_{NA}(Q_o^2)^{\alpha_s(Q_o^2)/\alpha_s(Q^2)} \quad . \tag{5}$$

Here $\alpha_s(Q^2)$ is the running coupling constant and Q_o^2 is the momentum scale for which a valence quark model (such as the bag model) is supposed to give a good description of the nucleon structure function, $Q_o^2 \leq 1$ GeV^2). The essential assumption is that $\xi_{NA}(Q_o^2)$ is given by a length scale modification,

$$\xi_{NA}(Q_o^2) = \left(\frac{\lambda_A}{\lambda_N}\right)^2 > 1. \tag{6}$$

Here λ_N is a length scale appropriate to the nucleon in vacuum, and λ_A is the length scale appropriate for the nucleus. In the papers dealing with the rescaling model one finds calculations of the ratio λ_A/λ_N based upon models of the nucleon-nucleon correlation functions [10]. We prefer to make the identification

$$\frac{\lambda_A}{\lambda_N} = \frac{\phi_o}{\langle\phi\rangle_A} \quad , \tag{7}$$

where the brackets denote the average value of $\phi(r)$ in the nucleus. For example, we can write, using a local-density approximation,

$$\phi(r) \cong \phi_o\left[1 - \frac{1}{5}\,\frac{\rho(r)}{\rho_{NM}}\right] , \tag{8}$$

where $\rho(r)$ is the matter density of a nucleus and ρ_{NM} is the density of nuclear matter. Therefore, we have

$$\langle\phi\rangle_A \cong \phi_o\left(1 - \frac{1}{5}\,\frac{\langle\rho\rangle_A}{\rho_{NM}}\right) \tag{9}$$

and

$$\frac{\lambda_A}{\lambda_N} \cong \frac{1}{\left(1 - \frac{1}{5}\frac{\langle\rho\rangle_A}{\rho_{NM}}\right)} \quad . \tag{10}$$

Thus, the A dependence of the ratio λ_A/λ_N is here related to the fact that nuclei of different mass number have different percentages of surface nucleons. (A naive extrapolation of (10) to higher densities would indicate a deconfining phase transition at about five times nuclear matter density.)

We have noted that in the effective Lagrangian we have suggested to model QCD at large length scales there is only a single dimensional quantity, if we neglect the small current masses of the up and down quarks [7]. Dimensional quantities will then scale with the value of this order parameter. For example, the radius of a nucleon in nuclear matter will be given by,

$$\frac{R_{NM}}{R_{vac}} = \frac{\phi_o}{\phi_{NM}} , \tag{11}$$

where R_{vac} is the nucleon radius in vacuum. The average radius of a nucleon in a nucleus is then given by,

$$\frac{\langle R\rangle_A}{R_{vac}} = \frac{\phi_o}{\langle\phi\rangle_A} . \tag{12}$$

Thus, using (7), we can also identify

$$\frac{\lambda_A}{\lambda_N} = \frac{\langle R\rangle_A}{R_{vac}} . \tag{13}$$

This result is consistent with the fact that in the rescaling model one "rescales" the moments (or the structure function) of the nucleon itself.

3. Modification of nucleon electromagnetic form factors in nuclei

We have published a number of papers on this topic and have shown that the momentum transfer dependence and the mass number dependence of the quenching of the longitudinal response in nuclei may be understood in terms of the medium-modified form factors we calculated in earlier work [11]. For example, consider the usual phenomenological expression for the electromagnetic form factor of the proton,

$$G_E^p(q^2) \cong \frac{1}{\left(1 - \frac{q^2}{a_{vac}^2}\right)} . \tag{14}$$

Here $a_{vac}^2 = 0.71\ \mathrm{GeV}^2$ is the value of this quantity in vacuum. In nuclear matter we have

$$\frac{a_{vac}}{a_{NM}} = \frac{\phi_o}{\phi_{NM}} , \tag{15}$$

or alternatively,

$$\frac{[r_p^2]^{\frac{1}{2}}_{NM}}{[r_p^2]^{\frac{1}{2}}_{vac}} = \frac{\phi_o}{\phi_{NM}} , \tag{16}$$

where $[r_p^2]^{\frac{1}{2}}$ is the r.m.s. radius of the proton calculated from the slope of the form factor at $q^2=0$. We have shown in an earlier work [12] that the ratio

$$\frac{\langle [r_p^2]^{\frac{1}{2}} \rangle_A}{[r_p^2]^{\frac{1}{2}}_{vac}} = \frac{\phi_o}{\langle \phi \rangle_A} \quad , \tag{17}$$

which we have calculated using a soliton model of the nucleon [11], reproduces the values of (λ_A/λ_N) which are required to fit the EMC effect. That is, the electromagnetic radius of the nucleon (calculated using a soliton model of the nucleon) scales with the inverse of the dynamical quark mass and that dynamical mass scales as the ratio ϕ/ϕ_o [6,7].

The medium-modified form factors obtained earlier [11] have been used to explain a large body of data dealing with the longitudinal response in nuclei [13-15], the charge distribution of ^{208}Pb [16], and the charge density difference of ^{206}Pb and $^{205}T\ell$ [17]. The situation with respect to the transverse response is more complicated since there appears to be a large amplitude for two-nucleon processes which is important in the region of the quasi-elastic peak [13]. More theoretical and experimental work is needed to clarify the situation in the case of the transverse response.

4. Dirac phenomenology and the relativistic Brueckner-Hartree-Fock theory

In the work of Noble [18] one finds the first attempt to relate the quenching of the longitudinal response to a change of nucleon size. Noble uses the scaling relation,

$$\frac{R_{NM}}{R_{vac}} = \frac{m_N}{\tilde{m}_{eff}} \quad , \tag{18}$$

where

$$\tilde{m}_{eff} = m_N + U_s \ . \tag{19}$$

Here U_s is the scalar potential felt by a nucleon in nuclear matter. Dirac phenomenology [4,5] yields $U_s \cong -400$ MeV so that $R_{NM}/R_{vac} \cong 1.74$, or in ^{56}Fe $\langle R\rangle/R_{vac} \cong 1.4$. This represents an increase of the average nucleon radius in ^{56}Fe of about 40 percent, while to explain the EMC effect (or to explain the quenching of the longitudinal response) in iron, the radius increase needs to be only 15 percent [10,14]. At first sight there might appear to be a problem with the rescaling analysis, however, as we will discuss below, an understanding of the relativistic Brueckner-Hartree-Fock theory [3] allows us to clarify this situation and to see the applicability of the rescaling analysis.

The problem with the simple analysis of (18) and (19) is that U_s contains a number of effects which have nothing to do with the change of mass scale. In particular the various contributions to U_s include <u>exchange</u> (Fock) terms arising from the exchange of omega, rho and pi "mesons" between nucleons. We must remove these terms from U_s before we calculate a value for the modified mass parameter, $\tilde{m}$. In our analysis we found $U_s \cong -350$ MeV, however, only about 60 percent of this scalar potential was due to sigma exchange. (This may be seen from inspections of Figs. (2.8) - (2.12) of [3], for example.) Therefore, $\tilde{m} \cong 938\text{-}210 = 728$ MeV, and

$$\frac{R_{NM}}{R_{vac}} = \frac{m_N}{\tilde{m}} \cong 1.29 \ . \tag{20}$$

There is certainly some theoretical error to be associated with the estimate in (20), but the result is quite close to that obtained from our previous analysis.

More precisely, we can see that in the theory of covariant soliton dynamics [11], the mass and radius of a nontopological soliton are given by [19],

$$m = f(g_\chi, \eta)\, m_q^{dyn} , \tag{21}$$

$$R = \frac{h(g_\chi, \eta)}{m_q^{dyn}} , \tag{22}$$

where m_q^{dyn} is a dynamical quark mass arising from the coupling of the quark to the QCD condensates and f and h are dimensionless functions of a coupling constant, g_χ, and a mass ratio, η. The mass ratio does not change upon rescaling, so that

$$\frac{R_{NM}}{R_{vac}} = \frac{m_q^{dyn}}{\tilde{m}_q^{dyn}} \tag{23}$$

$$= \frac{\phi_o}{\phi_{NM}} \tag{24}$$

as noted earlier.

5. Summary

In summary, we can say that if we use the order parameter of the gluon condensate to set the mass and length scale both in vacuum and in nuclei, we can understand several interesting phenomena from a unified point of view. Either theoretical analysis or phenomenological considerations lead to the conclusion that the gluon condensate order parameter is reduced by about 25 percent in nuclear matter. This effect may be considered as a precursor of a deconfining phase transition [20].

6. References

1. J.J. Aubert et al. (EMC collaboration) Phys. Lett. 123B, 275 (1981); A. Bodek et al., Phys. Rev. Lett. 50, 1431 (1984); 51, 534 (1984).
2. Z.E. Meziani et al., Phys. Rev. Lett. 52, 2130 (1984).
3. M.R. Anastasio, L.S. Celenza, W.S. Pong, and C.M. Shakin, Phys. Rep. 100, 327 (1983).
4. L.S. Celenza and C.M. Shakin, Relativistic Nuclear Physics: Theories of Structure and Scattering (World Scientific, Singapore, 1986)-in press.
5. B.D. Serot and J.D. Walecka, in Advances in Nuclear Physics, Vol. 16, edited by J.W. Negele and E. Vogt (Plenum Press, New York, 1985).
6. L.S. Celenza and C.M. Shakin, Description of the gluon condensate, Brooklyn College Report B.C.I.N.T. 85/021/149 (1985). To be published in Physical Review D.
7. L.S. Celenza and C.M. Shakin, Effective Lagrangian Methods in QCD, Brooklyn College Report B.C.I.N.T. 86/011/150 (1986). To be published in the International Review of Nuclear Physics (World Scientific, Singapore).
8. M.A. Shifman, Ann. Rev. Nucl. Part. Sci. 33, 199 (1983).
9. For a review see R.L. Jaffe, Lectures presented at the 1985 Los Alamos School on Quark Nuclear Physics, June 10-14, 1985; M.I.T. preprint CTP 1261 (July 1985).
10. F.E. Close, R.J. Jaffe, R.G. Roberts and G.G. Ross, Phys. Rev. D31, 1004 (1985).
11. L.S. Celenza, A. Rosenthal and C.M. Shakin, Phys. Rev. C32, 232 (1985).
12. L.S. Celenza, A. Rosenthal and C.M. Shakin, Phys. Rev. Lett. 53, 892 (1984).
13. L.S. Celenza, A. Harindranath and C.M. Shakin, Phys. Rev. C32, 248 (1985).

14. L.S. Celenza, A. Harindranath, C.M. Shakin and A. Rosenthal, Phys. Rev. C32, 650 (1985).
15. L.S. Celenza, A. Harindranath and C.M. Shakin, Phys. Rev. C33, 1012 (1986).
16. L.S. Celenza, A. Harindranath, C.M. Shakin and A. Rosenthal, Phys. Rev. C31, 1944 (1985).
17. L.S. Celenza, A. Harindranath and C.M. Shakin, Phys. Rev. C32, 2173 (1985).
18. J. Noble, Phys. Rev. Lett. 42, 412 (1981).
19. L.S. Celenza, V.K. Mishra, and C.M. Shakin, Mass scales in QCD and nuclear dynamics, Brooklyn College Report B.C.I.N.T. 86/051/155 (1986). Submitted for publication.
20. See for example, M. Satz, Ann. Rev. Nucl. Part. Sci. 35, 245 (1985).

Many Quark Effects in Electron-Nucleus Scattering

P.J. Mulders

National Institute for Nuclear Physics and High Energy Physics, NIKHEF-K, P.O. Box 41882, NL-1009 DB Amsterdam, The Netherlands

1 Introduction

In the hierarchy that exists in the structure of matter the scales for nuclei and nucleons are remarkably close, less than one order of magnitude apart. The scale for nucleons is of the order of $R_N \approx 1$ fm, related to the scale of Quantum Chromodynamics (QCD). The scale for nuclei, say the typical internucleon distance in nuclei is only about twice as large, $R_A \approx 2$ fm. This is the range of the nuclear strong force, the van der Waals force of QCD.

In spite of the proximity of these scales, the role of subnucleonic degrees of freedom in the nucleus at low energies is not very important. The reason is the difference in excitation energies. QCD prefers to build color singlets with masses of the order of $M_N \approx 1$ GeV, leading to excitation energies in nuclei of the order of $1/(M_N R_A{}^2) \approx 10$ MeV. For a nucleon the excitation energies are much higher; for massless quarks they are of the order $1/R_N \approx 200$ MeV. Of course, the understanding of the (effective) nucleon-nucleon interaction does involve quarks, but except for the short range part of the interaction it is well-known that mesonic degrees of freedom, especially the pion, are the relevant ones.

As soon as the nuclear wave function is relevant, however, the effects of quark degrees of freedom are extremely important. It is not clear, for instance, that the factorization of form factors in a nucleon form factor, with or without modifications, and a body form factor is meaningful. The reason is that the wave function of a nucleus is an antisymmetrized many quark wave function. In this context one could also question the applicability of the impulse approximation in electron-nucleus scattering used for the calculation of structure functions in the quasi-elastic scattering (QES) region at intermediate momentum transfer and for the application of smearing corrections [1] in deep-inelastic scattering (DIS).

The observation of scaling in inclusive electron scattering off a target allows the study of the substructure of that target. One must keep in mind that the scaling arises mainly because of kinematics and is not very sensitive to the wave function; the scaling functions, however, depend on the detailed wave function of the target and may be very sensitive to the interplay of nucleon and quark degrees of freedom. The cross section for lepton-nucleus scattering is proportional to the imaginary part of the forward Compton scattering amplitude for virtual photons,

$$\sigma_{eA} \propto 2M\, W_{\mu\nu} = \mathrm{Im}\,[\, i\,\pi^{-1} \int d^4x\, e^{iq.x} \langle A|T\, J_\mu(x)\, J_\nu(0)\, |A\rangle\,]\,. \tag{1}$$

In the case of an electromagnetic scattering process this tensor can be expanded:

$$W_{\mu\nu}(p,q) = \left(\frac{q_\mu q_\nu}{q^2} - g_{\mu\nu}\right)W_1 + \left(p_\mu - \frac{p.q}{q^2}\, q_\mu\right)\left(p_\nu - \frac{p.q}{q^2}\, q_\nu\right)\frac{W_2}{M^2}\,, \tag{2}$$

where W_1 and W_2 are the structure functions which depend on the invariants $p.q \equiv M\nu$ and $q^2 \equiv -Q^2$. For deep-inelastic scattering (momentum transfer squared $Q^2 \geq 2$ GeV^2) one can use the operator product expansion in QCD for the Compton scattering operator to obtain the dominant contribution, which constitutes the familiar parton model. The scattering process can be described as the incoherent sum of quasi-elastic scattering off the individual partons, quarks and gluons, inside the target. The final state interactions, gluon exchange between the struck quark and other quarks in the target, are suppressed by powers of Q^2. For lower momentum transfers the perturbative expansion discussed above breaks down. The color final state interactions become strong, but because these forces saturate for color singlets, leading to baryons (N, Δ, ..) one can intuitively understand the appearance of a scaling region where the process is dominated by the quasi-elastic scattering off colorless hadrons in the target, which is considerably above the scale for nuclear excitations. The "QCD final state interactions" become the hadronic form factors.

In this paper the emphasis is on the quasi-elastic scattering off colorless hadrons in the nucleus. Unlike the case of DIS, there is no rigorous framework available. In QES the momentum squared satisfies $Q^2R_N^2 >> 1$ (R_N is the radius of the nucleon) and one should not expect that an effective field theory of nucleons and mesons is able to describe the process, since short range nucleon-nucleon correlations are involved. We want to discuss a phenomenological approach resembling the quark parton model. The dominant contribution in this case is well-known, namely quasi-free single nucleon knock-out. For this one needs the formalism of West or y-scaling [3] that is discussed in section 2, and which includes the effects of nuclear binding. For the recently measured longitudinal and transverse structure functions on ^3He [4], ^{12}C [5], ^{40}Ca [6,7], ^{48}Ca [7], and ^{56}Fe [8,7], y-scaling seems to work well. Because the internal nucleon structure is still important in y-scaling as a multiplicative factor, namely the form factor, the data can be used to extract 'experimental' single nucleon properties in nuclei. These properties can be obtained with less ambiguities from exclusive single nucleon knock-out processes, but only very few data are available for such processes [9].

It is not surprising that inconsistencies arise in the single nucleon knock-out description of the inclusive scattering process, like the problem with the Coulomb sum rule [10], which yields only about 70 % of the expected result. One of the consequences of our present understanding of hadrons as composite extended objects is a huge probability of finding two overlapping nucleons. In order to get an idea of the order of magnitude, the following examples: In the deuteron the probability of finding two nucleons within 1 fm of each other is almost 5 % . This percentage can be obtained from the scattering data for the NN $^3S_1$ - $^3D_1$ waves using the P-matrix formalism and the 'known' long range (> 1 fm) part of the NN interaction [11]. For ^{12}C it is simple to estimate the overlapping volume per nucleon, which yields about 40 %. This number, obtained from ref.[12], was calculated to obtain an estimate of the increase of the quark confinement scale in nuclei. Correspondingly one can expect a significant contribution of two nucleon knock-out in electron-nucleus scattering because the struck quark finds itself in a 'six quark' environment. If the momentum transfer in the reaction is sufficiently high, say at least twice the fermi momentum, reabsorption of a nucleon in the nucleus is small, and the process of two nucleon knock-out can be added incoherently to the single nucleon knock-out.

2 Quasi-elastic Scattering and y-scaling

Assuming the impulse approximation to be valid, the contribution of a hadron h to the nuclear structure function is given by

$$W_{\mu\nu}^A(P,q) = \sum_h \int d^4p \; S_h(E,p) \; W_{\mu\nu}^h(p,q) = \sum_h \int d^3p \; n_h(p) \; W_{\mu\nu}^h(p,q) \,. \qquad (3)$$

We will assume that the average three-momentum distribution n(p) in the nucleus rest frame (LAB-frame) is spherically symmetric. The kinematics for the quasi-elastic scattering process is shown in figure 1. The recoil momentum of the final-state, of which the invariant mass is denoted as M_{A-1}, is opposite to the momentum of the struck hadron. The energy of the struck hadron thus is determined and can be integrated over. Assuming for $W_{\mu\nu}^h(p,q)$ the (on-shell) result in eq. (2) one obtains for the longitudinal and transverse structure functions, $W_L = -W_1 + (1+\nu^2/Q^2)W_2$ and $W_T = W_1$, the result

$$W_{L/T}(\nu, Q^2) \approx (M_A/|q|) \sum_h F_{L/T}^h(Q^2) \, P_h(y_h), \qquad (4)$$

where F^h are the form factors, for example for a nucleon one has $F_L^N(Q^2) = G_E^2(Q^2)$ and $F_T^N(Q^2) = Q^2 G_M^2(Q^2)/4M^2$. The function P_h is the longitudinal momentum distribution of hadrons in the

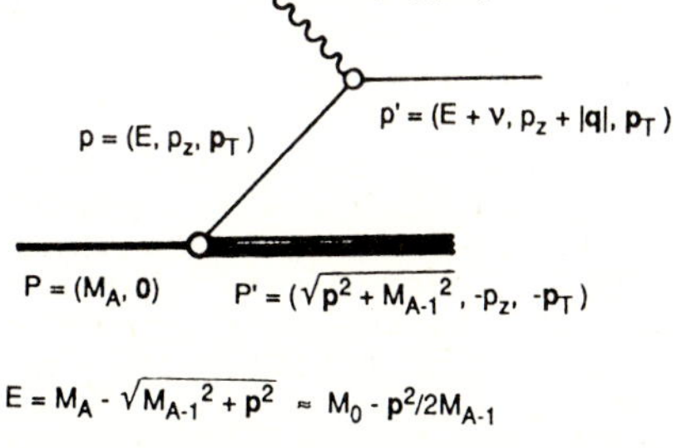

Fig. 1: Kinematics for the quasi-elastic scattering process

nucleus. The function P_h in eq. (4) is evaluated at [2,13]

$$y = \frac{\sqrt{|q|^2 + \alpha(2M_0\nu - Q_{eff}^2)} - |q|}{\alpha} \approx \frac{2M_0\nu - Q_{eff}^2}{2\,|q|} + O(\frac{1}{|q|^3}), \qquad (5)$$

where $O(p^4)$ terms have been neglected, $M_0 = M_A - M_{A-1}$, $\alpha = (M_A + \nu)/M_{A-1}$, and $Q_{eff}^2 = Q^2 + M_h^2 - M_0^2$. The effects of nuclear binding, $M_0 \neq M_h$, are taken into account. Equation (4) is analogous to DIS where the structure functions become functions of the Bjorken scaling variable $x = Q^2/2M\nu$. When single nucleon knock-out is the dominant contribution one can obtain a scaling function by dividing $W(\nu,Q^2)$ by the single nucleon form factor $F^N(Q^2)$; if only one type of hadrons, say nucleons, contribute, this scaling function then can be interpreted as the longitudinal momentum distribution for them. For the longitudinal structure function of ^{12}C [4] the scaling function is shown in Figs. 2a-c for a unmodified nucleon form factor ($\rho = 0$) and for modifications corresponding to an increased nucleon charge radius of 5 and 10 % (ρ = 1.05 and 1.10) respectively. The excellent scaling indicates that the single nucleon contribution dominates with, at least under the assumption that the impulse approximation is valid, no appreciable (< 10 %) increase in the charge radius [14,15]. From the ratio of W_T and W_L one obtains (with appropriate factors) the magnetic moment of the struck hadron (see Fig. 3). The approximately constant result in the region of the quasi-elastic peak indicates that the 'magnetic moment of a nucleon' in the nucleus is increased by about 25 %. The above interpretation of the inclusive data has recently been confirmed by an exclusive scattering experiment $^{12}C(e,e'p)^{11}B$ at NIKHEF [9] . The importance of these experiments is that one *exclusively* measures the single nucleon

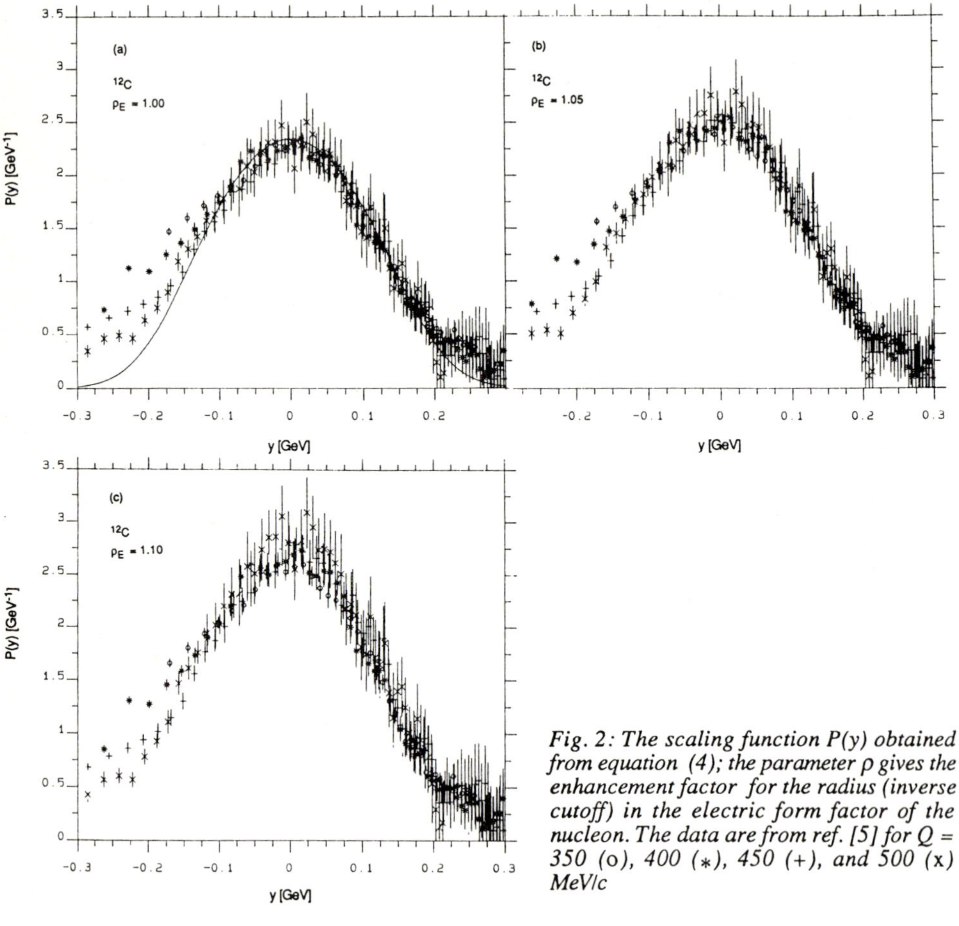

Fig. 2: The scaling function P(y) obtained from equation (4); the parameter ρ gives the enhancement factor for the radius (inverse cutoff) in the electric form factor of the nucleon. The data are from ref. [5] for Q = 350 (o), 400 (∗), 450 (+), and 500 (x) MeV/c

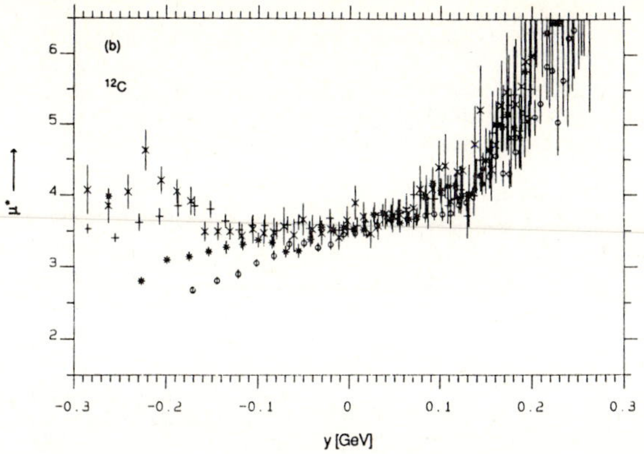

Fig. 3: The 'proton magnetic moment' μ^* in a nucleus determined from the ratio of W_T and W_L for ^{12}C. The data and the symbols used are as in Fig. 2

knock-out. When the proton is observed in the direction of the photon momentum (parallel kinematics), one easily obtains the longitudinal and transverse structure functions. In principle these depend also on the spectroscopic factors, but by fixing the momentum of the struck proton only a common factor is left. The result of the analysis of van der Steenhoven et al. [9] for four data points at y = -90, -70, -40, and + 100 MeV is shown in Fig. 4, and it indicates a similar enhancement for the 'magnetic moment'.

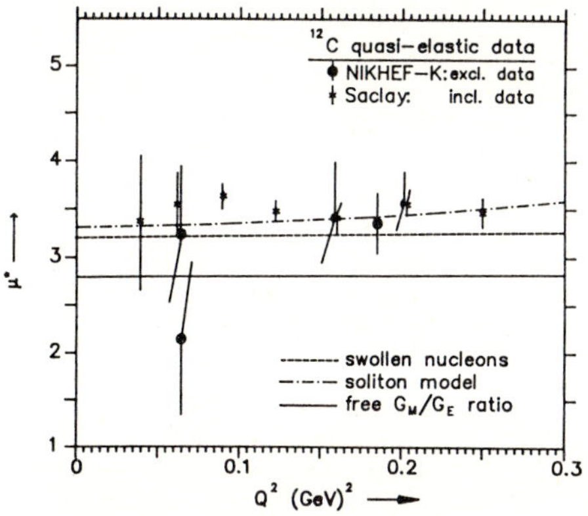

Fig. 4: The 'proton magnetic moment' μ^* in a nucleus determined from the ratio of W_T and W_L for ^{12}C using (e,e'p) data. The figure has been taken from ref. [9]. The exclusive scattering data (•) are compared with inclusive scattering data from ref. [5]

3. Many quark effects

The need for other contributions is evident from the value of the integral over P(y) in Fig. 3a, $\int dy\, P(y) = 0.66$. Although use of a correct antisymmetrized many-quark wave function may also yield a result that deviates from that using the factorization implied by the impulse approximation, one would not expect an effect of 35 %. The influence of multiquark knock-out as mentioned in the introduction, is a promising possibility. The larger radius of six quark states and elementary counting rules lead to a more rapidly vanishing form factor for six quark states compared to three quark states. In this way the suppression in the longitudinal structure function can be understood.

In Fig. 5, I show the result of an - admittedly oversimplified - calculation for ^{12}C using eq. (4) with a 40 % probability of finding a quark in a six quark cluster. The relative masses of clusters with different spin and isospin have been estimated using the color hyperfine interaction. The quark wave function in a single nucleon is modified consistent with (e,e'p) data, to give a 5 % larger nucleon radius and a 30 % larger nucleon magnetic moment. This differs from e.g. modifications in the M.I.T. bag model, where the two numbers are coupled to the increase of the bag radius [16]. The calculation is able to explain features in the data like the suppression in the longitudinal structure function and the cross section in the dip region. Of course accurate data over a larger Q^2 range are needed to confirm the usefulness of eq. (4). Also semi-inclusive processes, where one or two nucleons are detected in coincidence with the scattered electron, will yield valuable information.

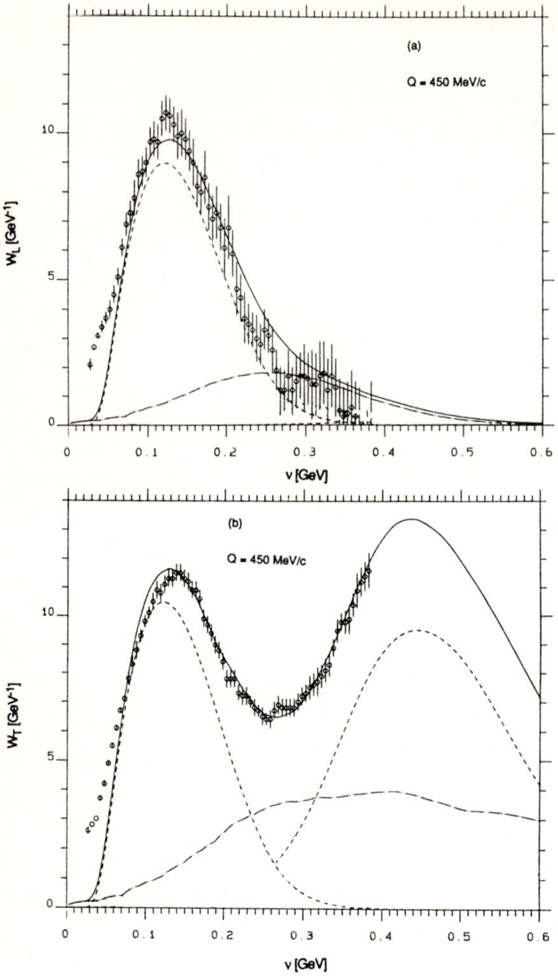

Fig. 5: The longitudinal (a) and transverse (b) structure functions for ^{12}C in a simple calculation including three (dashed) and six (dot-dashed) quark contributions, added incoherently. The data again are from ref. [5]

This work is included in the research program supported by the Foundation for Fundamental Research on Matter (FOM) and the Netherlands Organization for the Advancement of Pure Research (ZWO).

[1] S.V. Akulinichev et al., Phys. Rev. Lett. **55** (1985) 2239; G.V. Dunne and A.W. Thomas, Phys. Rev. **D33** (1986) 2061 and University of Adelaide preprint
[2] G.B. West, Phys. Rep. **18C** (1975) 264
[3] P.J. Mulders, NIKHEF report P-8 (1986), to be publ. in Nucl. Phys. A
[4] C. Marchand et al., Phys. Lett. **153B** (1985) 29
[5] P. Barreau et al., Nucl. Phys. **A402** (1983) 515
[6] M. Deady et al., Phys. Rev. **C28** (1983) 631
[7] Z. Meziani et al., Phys. Rev. Lett. **52** (1984) 1233 and 2130
[8] R. Altemus et al., Phys. Rev. Lett. **44** (1980) 965; A. Hotta et al., Phys. Rev. **C30** (1984) 87
[9] G. van der Steenhoven et al., NIKHEF preprint (1986)
[10] T. de Forest, Nucl. Phys. **A132** (1969) 305
[11] B.L.G. Bakker and P.J. Mulders, NIKHEF report P-16 (1985), to be publ. in Adv. Nucl. Phys.
[12] F.E. Close, R.L. Jaffe, R.G. Roberts and G.G. Ross, Phys. Rev. **D31** (1985) 1004
[13] E. Pace and G. Salme, Phys. Lett. **110B** (1982) 411
[14] The first to suggest a radius increase was J.V. Noble, Phys. Rev. Lett. **46** (1981) 412
[15] The use of y-scaling in such analyses was suggested by I. Sick, Phys. Lett. **157B** (1985) 13
[16] P.J. Mulders, Phys. Rev. Lett. **54** (1985) 2560

Medium Effects on Nucleon Size

I. Sick

Department of Physics, University of Basel, CH-Basel, Switzerland

In order to determine an eventual change of size of the nucleon due to the nuclear environment, we investigate 3 observables that involve the least uncertainties of interpretation. From elastic magnetic and charge electron-nucleus scattering, and from inclusive scattering at large q, we conclude that the nucleon radius change is less than a few percent.

A number of observations and calculations suggest that nucleons could change as they are put into the nuclear medium. Among the observations most cited is the EMC effect, which indicates that the momentum distribution of quarks in bound nucleons differs from the one of free ones[1]. The lack of the strength in the longitudinal sum rule measured in quasielastic electron nucleus scattering also has been taken as evidence that the bound- and free-nucleon form factors differ [2]. The q-dependence of the nucleon form factor measured in (e,e'p) experiments seems to point in the same direction [3]. An increase of the bound-nucleon size of ~20% has been advocated.

A priori, such a change of the nucleon due to the nuclear medium is quite plausible.The nucleon has internal degrees of freedom, and these can play a role in a medium where the average distance to the next nucleon is only 20% larger than the sum of nucleon radii. It therefore would be astonishing to find bound and free nucleons to be identical.

The experimental evidence cited above is far from conclusive, however. A number of different explanations for deep inelastic muon-nucleus scattering data have been advanced; these are discussed in various contributions to this meeting. The longitudinal sum rule at the momentum transfers investigated has a number of problems, that make an interpretation in terms of nucleon form factors doubtful; nucleon FSI cannot be neglected at the momentum transfers of $\lesssim 2K_F$ presently studied, and the removal of strength from the main quasielastic peak due to short-range correlations needs to be accounted for.

The results of (e,e'p) experiments are not yet conclusive given the uncertainties due to FSI, even if these have been minimized by working at constant recoil nucleon energy. In addition, appreciable effects on the (e,e'p) cross section can be expected due to (e,e'n) plus subsequent (n,p) change exchange; the effect of the different q-dependence of p and n cross sections remains to be investigated.

In this paper, we use a number of different observables of electron-nucleus scattering to obtain information on the bound-nucleon form factor. These observables are chosen such as to allow a determination of the bound-nucleon form factor which is as clean as possible, and least subject to complication due to nuclear structure, FSI, etc. The observables we will discuss concern both charge and magnetic elastic scattgering off selected nuclei, and inclusive electron-nucleus scattering at very large momentum transfer.

Elastic magnetic scattering on ^{51}V

The relation [4] between elastic magnetic form factors, nuclear wave functions and nucleonic form factors in general is a fairly involved one. Configuration mixing, which is not yet understood in a quantitative way, plays a major role. This relation does, however, become quite simple for one special case, where we consider scattering from a valence nucleon with the largest j-value of all occupied states, and where we look at the maximum multipolarity $\Lambda = 2j$ only. In this case the relation becomes a simple one

$$F_{M\Lambda}(q) = F_N(q) \cdot \text{const} \cdot \psi^{\Lambda}_{jj} \cdot \int R_j^2(q) \cdot j_{\Lambda-1}(qr) \cdot r^2 dr \qquad (1)$$

Here F_N is the magnetic form factor of the bound nucleon, ψ is the one-body density matrix element of multipolarity Λ, R is the valence nucleon radial wave function.

The simplifications alluded to above stem from the (dominant) one-body nature of the electromagnetic interaction. Multipolarity $\Lambda = 2j$ necessarily involves states with $j' \geq j$. If contributions from states $j' > j$ are small since these states are "empty" (and they are small in calculations that do account for configuration admixtures including those states), then only shell j contributes, and eq (1) is correct. Two body meson exchange currents (MEC), which also would complicate eq (1), have been calculated [5] and have produced small corrections to the observables employed in the following. For the case of elastic magnetic scattering off ^{51}V, a nucleus with an unpaired $f_{7/2}$ proton, one finds $\psi = 0.93 \pm 0.05$. Fig. 1 shows the data [6] employed. We know that ψ must

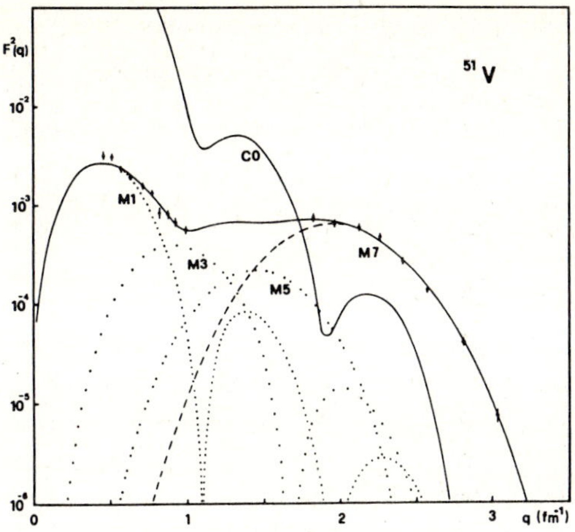

Fig. 1 Data for magnetic elastic form factor of ^{51}V together with WS-fit

be $\leqq 1$. Since nucleons in nuclei are subject to short-range correlations that move strength to very large energies, we might expect ψ to be $\leqq 0.9$, say.

Above experimental value of $\psi = 0.93$ has been determined using for F_N the magnetic form factor of the free nucleon. If the bound nucleon has a size that exceeds the one of the free one by 20%, F_N would be 30% smaller in the region 2.5-3 fm^{-1} where ψ is determined. To explain the data, one then would need a ψ 30% larger than the value of 0.93. This clearly contradicts both the rigorous bound ($\leqq 1$) and realistic expectations ($\leqq 0.9$). The magnetic radius of the bound nucleon therefore cannot be significantly larger than the one of the free nucleon.

<u>Elastic charge scattering from ^{205}Tl and ^{206}Pb</u>

From a measurement of the difference of ^{205}Tl and ^{206}Pb charge density we obtain [7] accurate information on the radial wave function of the added proton, and its absolute spectroscopic factor. In general, it is rather difficult to determine from $\Delta\rho(r)$ these two quantities; configuration mixing and core polarization lead to an involved relationship that cannot be disentangled without extensive theoretical input. This is not a problem, however, for the particular case studied here. The 3s radial wave function has a unique behaviour, a $j_o(q_o r)$-like shape (Fig. 2), the radial dependence of which is strikingly different from other radial wave functions, or core polarization. The Fourier transform of $\Delta\rho(r)$, measured via the cross section difference of Pb and Tl, shows a clear signal of the 3s-contribution, a spike at $q_o \cong 2\text{fm}^{-1}$ (which becomes a $\delta(q_o)$-function if $\Delta\rho \equiv j_o(q_o r)$). Hartree-Fock calculations have shown that effects due to core polarization and configuration mixing are small in $F(\sim 2\text{fm}^{-1})$.

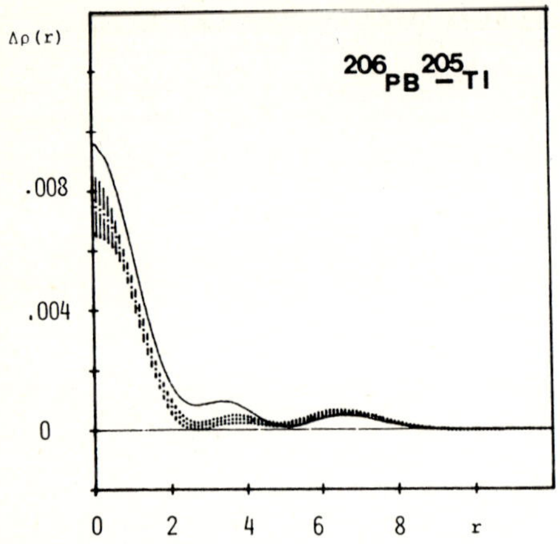

Fig. 2 $^{206}Pb-^{205}Tl$ charge density difference together with prediction for HF calculation.

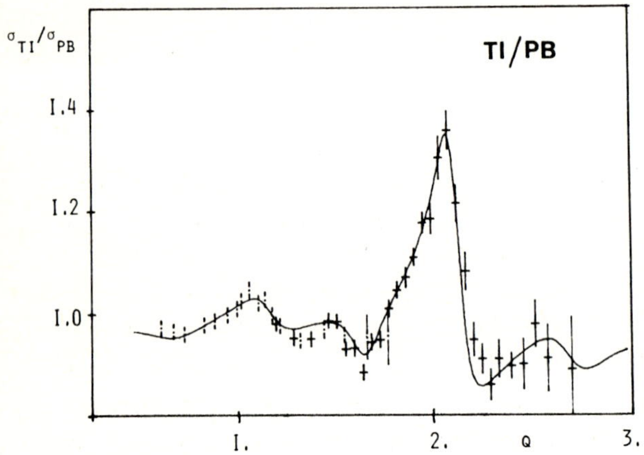

Fig. 3 $^{206}Pb/^{205}Tl$ cross section ratio together with HF-predictions (S_{3s} = 0.7).

Experimentally, one measures precise cross section ratios σ^{Pb}/σ^{Tl}, as shown in Fig. 3. The difference of this ratio to one basically determines near $2fm^{-1}$

$$F^N(q)\cdot S \cdot \int R_{3s}^2(r)\, \sin(qr)\cdot r^2 dr \;/\; \int \rho_{Tl}(r)\, \sin(qr)\, r^2 dr \qquad (2)$$

Here F_N is the bound-nucleon charge form factor, s ist the 3s proton spectroscopic factor, R ist the 3s radial wave function, and ρ_{Tl} the experimental charge density of Tl determined accurately from the Tl(e,e) cross sections we also measured in the same experiment.

Fig. 2 shows that the experimental data very nicely determine the 3s radial wave function, with its 3 maxima and 2 nodes. It is basically the peak in σ^{Pb}/σ^{Tl} near $2fm^{-1}$ which measures the 3s spectroscopic factor. Provided we use the charge form factor of the free proton, we find [7)] S = 0.7 ± 0.07.

The 3s spectroscopic factor of 0.7 respresents an absolute measurement of the difference between the ^{205}Tl and ^{206}Pb occupation numbers. Pb(e,e'p) and Pb(d,^{3}He) experiments [3,9] have been performed for both ^{206}Pb and ^{208}Pb, and for all (experimentally observable) final $\ell=0$ states in Tl. These experiments allow to determine the ratio of the 3s occupation in ^{208}Pb to the 3s spectroscopic factor of ^{206}Pb/^{205}Tl; this ratio is [1] 0.77 ± 0.02. Combining it with the absolute 3s spectroscopic factor from (e,e) yields $n = 0.91\pm0.1$ for the 3s occupation in ^{208}Pb.

Theoretically, we know that $n\leq1$. Given the presence of short range correlations between nucleons, we have good reasons to expect $n\lesssim0.85$. The experimental value of $n\simeq0.91\pm0.1$ agrees with both these values. It would disagree if we were to assume that the bound proton charge radius were 20% larger; this would increase the 3s occupation by 20%, and make it incompatible with both the rigorous upper bound, and a sensible expectation. We therefore conclude that the proton charge radius in the interior of Pb ($\approx$ nuclear matter) is not significantly larger than the free proton radius.

Inclusive scattering at large transfer

The most direct way [10] to obtain information on the bound-nucleon form factor is to scatter incoherently electrons from individual nucleons in the medium. This can be done by measuring the inclusive response function for quasi-elastic electron-nucleus scattering. This response function is related in a simple way to the nuclear momentum distribution and the bound-nucleon form factor provided a number of complications can be eliminated.

- Pauli blocking reduces the cross section; the momentum transfer must be much bigger than $2K_F$ to eliminate it.
- The final state interaction of the recoil nucleon distorts the scattered-electron spectrum. To eliminate its effect, one needs large recoil-nucleon energy (hundreds of MeV), which again can be obtained if the momentum transfer is much larger than 500 MeV/c $\sim 2K_F$.
- We must expect short-range correlations to lead to a decrease of the strength at low K by ~15%, and this strength probably will be missing under the main quasielastic peak. The data should be interpreted such as to not depend on this poorly known reduction.

The first two complications can be taken care of by using experimental data at appropriately large q. The last one can be overcome by analyzing the data in terms of y-scaling [11].

In quasielastic scattering we deal with a process where the electron scatters elastically off an individual bound nucleon, and ejects it from the nucleus. The kinematics of this e-N scattering process define a relation between the observables $\vec{q},\omega$ and the initial nucleon momentum $\vec{K}$ and energy E. This purely kinematical relation is correct provided the nucleon FSI can be neglected (i.e. if ω is large) and provided MEC are small. In the limit of very large q,ω ($q^2 >> K_F^2$) this relation simplifies: the nucleon momentum component $K_\perp^2$ perpendicular to q can be neglected relative to terms of order $K_\parallel \cdot q, q^2$, and the initial nucleon energy E is negligible relative to ω. In this case kinematics defines a relation between $q,\omega,K_\parallel$. The observables q,ω thus are no longer the independent variables they usually are. A relation known from kinemtatics ties them to $K_\parallel$. The cross section shows a scaling behaviour in terms of the varialbe $y = K_\parallel$ introduced in ref. [11).]

If such a scaling behaviour is observed, the inclusive cross section can be expressed in terms of

$$\frac{d\sigma}{d\Omega}(q,\omega)\, d\omega \,/\, \Sigma\, \sigma_{eN}(q) = F(y)\cdot dy \qquad y = y(q,\omega) \tag{3}$$

Here $\Sigma\, \sigma_{eN}$ is the in-medium electron nucleon cross section summed over all nucleons, F(y) is the probability to find nucleons with momentum $y = K_\parallel$ in the nucleus.

The scaling property of F(y) allows to investigate a number of points concerning the reaction mechanism and nucleon wave function.

- FSI and MEC do not lead to a scaling of $d\sigma/d\Omega$. The kinematics is more complicated, the cross section acquires an additional dependence on ω, and vertex form factors different from F_N occur. If the data do scale with a certain accuracy, FSI and MEC are smaller than this accuracy.
- The function F(y) yields the momentum distribution of nucleons in nuclei. This is of particular interest for large momenta, where reactions involving the detection of hadrons in initial or final state have failed due to multi-step processes.
- The function F(y) will not scale if in eq.(3) one uses an in-medium cross section that does not have the right q-dependence. This observation can be exploited to measure the in-medium form factor of nucleons, the object of interest is the present context.

Before discussing the quantitative consequences of eq.(3), let me emphasize again that eq. (3) can be derived only if $q >> K_F$, and if $\omega >> E$. Following the first application [11)] of y-scaling to data up to $q=10\text{fm}^{-1}$, a number of analyses have used cross sections for $q \leq 2\text{fm}^{-1}$. At these transfers, the conditions required to derive scaling are not fulfilled, in which case no significant information can be extracted.

In order to apply the y-scaling concept to a determination of σ_{eN}, we use the inclusive ^{56}Fe(e,e) data we recently measured [12)] at SLAC at energies 2-3.6 GeV and momentum transfers 3-12 fm^{-1}. These cross sections cover many decades, and the quasielastic peak exhibits a pronounced change of position and width as q increases. The function F(y) calculated using the free e-N cross section, is shown in Fig. 4.

At negative values of y (energy loss smaller than for the maximum of the quasielastic peak), the data show a striking scaling behaviour;the cross sections from different q,ω define a unique curve. At y >-50 MeV/c, no scaling is observed, in accordance with expectation that cross sections dominated by Δ- excitation, MEC or inelastic e-N scattering do not scale in y.

In order to judge the quality of the scaling, one can fit the values of F(y) for y <-50 MeV/c using a flexible parametrization. Scaling implies good χ^2, bad scaling bad χ^2.

In Fig. 5 we show the χ^2 of the fit obtained by using different e-N cross sections corresponding to different bound-nucleon size. This can be achieved easily by using for σ_{eN} the dipole form factor, which gives a good description of the e-N data with only one parameter, the nucleon radius. Since most quark models predict a linear relationship between nucleon radius and mass (given the occurence of only one scale parameter), $\Delta m/m = -\Delta r/r$ is used to calculate Fig. 5.

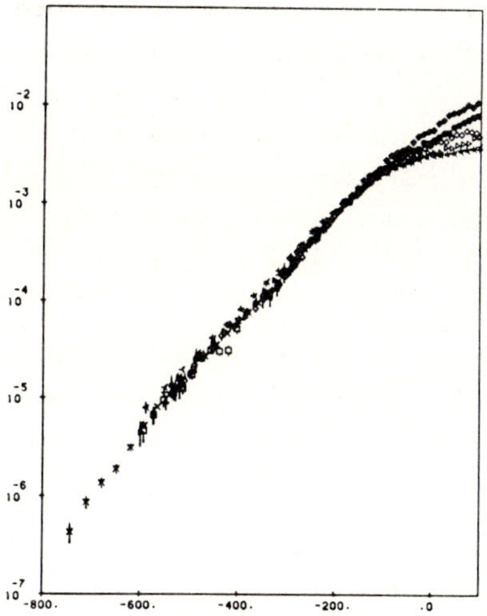

Fig. 4 ^{56}Fe(e,e') data (q=3-12fm^{-1}) plotted in terms of F(y) as a function of the scaling variable y.

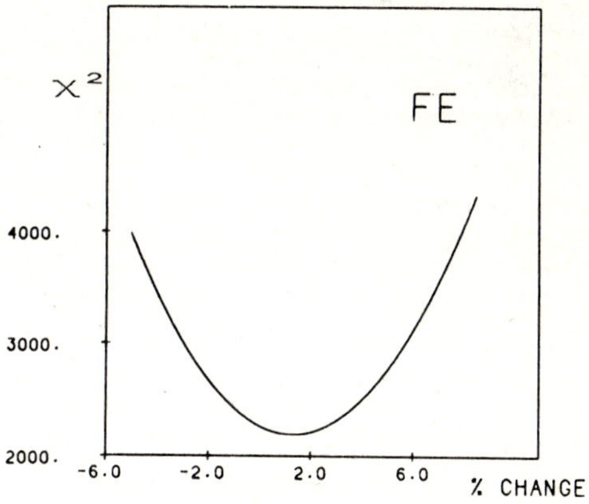

Fig. 5 χ^2 of fit to F(y) as function of the assumed change of radius of the bound nculeon

Fig. 5 shows a minimum of χ^2 , corresponding to best scaling, for a nucleon radius that differs only slightly from the free one. Systematical errors, due to the choice of different form factor parametrizations, add a ±2% uncertainty to the location of the minimum. The fact that the lower-q data must be expected to show systematic deviation from scaling due to $q \neq \infty$ adds another uncertainty, which could displace the minimum towards smaller $\Delta r/r$ (but not larger) by 2-3%.

From Fig. 5 we conclude that the bound- and free nucleon form factors have the same q-dependence. Their radius can differ by less than 3%.

Conclusion

From the 3 observables discussed above we conclude that for both the charge and magnetic radius of the nucleon the nuclear medium leads to an increase of, at most, a few percent.

How is this finding compatible with an expectation that the internal degrees of freedom of the nucleon indeed should show up? In short-range NN collisions, when the quark wave functions of two nucleons do overlap, these internal degrees of freedom certainly play a role, and we must expect that under these conditions the two nucleons are very different from free ones. However, nucleons in short-range interaction with others account for a small fraction of the total wave functions only,10%,say. Long-range interactions dominate, and the experimental evidence discussed above shows that this long-range interaction does not lead to an average size change. A search for these short-range effects of the quark degrees of freedom thus is needed.

References

1) J.J. Aubert et al., Phys. Lett. 123B (83) 123
2) Z.E. Meziani et al., Phys. Rev. Lett. 52 (84) 2130
3) P.K.A. DeWitt-Huberts, Nucl. Phys. A446 (85) 301
4) T.W. Donnelly, I. Sick, Rev. Mod. Phys. 56 (84) 461
5) T. Suzuki et al., Z. Phys. A293 (79) 5, and T. Suzuki, PhD Thesis, unpubl.
6) S.K. Platchkov et al., Phys. Rev. C25 (82) 2318
7) J.M. Cavedon et al., Phys. Rev. Lett. 49 (82) 1978
8) B. Frois et al., Nucl. Phys. A396 (83) 409
9) P. Grabmayr et al., Phys. Lett. 164B (85) 15
10) I. Sick,Phys. Lett. 157B (85) 13
11) I. Sick, D. Day, J.S. McCarthy, Phys. Rev. Lett. 45 (80) 871
12) D. Day et al., to be publ.

Spin-Isospin Response in Nuclei

H. Toki

Department of Physics, Tokyo Metropolitan University, Setagaya, Tokyo 158, Japan

Recently it became available at Los Alamos (LAMPF) to polarize high energy proton beams and measure the polarization of outgoing protons in any direction. Using this facility, spin responses in nuclei were measured at $E_P = 500\,\mathrm{MeV}$ on Pb at the momentum transfer $q = 1.75\,\mathrm{fm}^{-1}$, which corresponds to the momentum of large collectivity in pion channel [1]. It was found that the ratios of the longitudinal spin response ($\boldsymbol{\sigma}\cdot\boldsymbol{q}e^{iqr}$) against the transverse spin response ($\boldsymbol{\sigma}\times\boldsymbol{q}e^{iqr}$) were close to 1; $S_L/S_T{\sim}1$ [2]. This finding was totally unexpected, since the effective interaction is attractive in the longitudinal channel, while it is repulsive in the transverse channel in the $\pi + \varrho$ model [3]. This result led the experimentalists to conclude that the pion field is not enhanced in nuclei [2]. Hence, out of the quark and pion interpretations of the EMC effect, the latter was rejected.

This interpretation influences also the phenomena discussed with the $\pi+\varrho$ model. We notice, however, that the high energy proton sees nucleons only at the nuclear surface due to strong nucleon-nucleon interaction [4]. Therefore we would have to consider the surface effect on the response functions before we would make any comparison with data.

Response functions are driven by particle-hole interactions. The $\pi + \varrho$ model provides the particle-hole interaction in the spin-isospin channel as

$$W_{\mathrm{ph}} = [W^L_{\mathrm{ph}}(q)\boldsymbol{\sigma}_1\cdot\boldsymbol{q}\;\boldsymbol{\sigma}_2\cdot\boldsymbol{q} + W^T_{\mathrm{ph}}(q)\boldsymbol{\sigma}_1\times\boldsymbol{q}\cdot\boldsymbol{\sigma}_2\times\boldsymbol{q}]\boldsymbol{\tau}_1\cdot\boldsymbol{\tau}_2$$

with the longitudinal part

$$W^L_{\mathrm{ph}}(q) = \frac{f_\pi^2}{m_\pi^2}\left[g' - \frac{q^2}{m_\pi^2 + q^2 - \omega^2}\right]$$

and with the transverse part

$$W^T_{ph}(q) = \frac{f_\pi^2}{m_\pi^2}\left[g' - C_\varrho\frac{q^2}{m_\varrho^2 + q^2 - \omega^2}\right].$$

To be specific, we use $g' = 0.7$ and $C_\varrho = 2.18$ [3]. The surface effect is taken into account in three methods; local Fermi gas model [5], semi-infinite slab

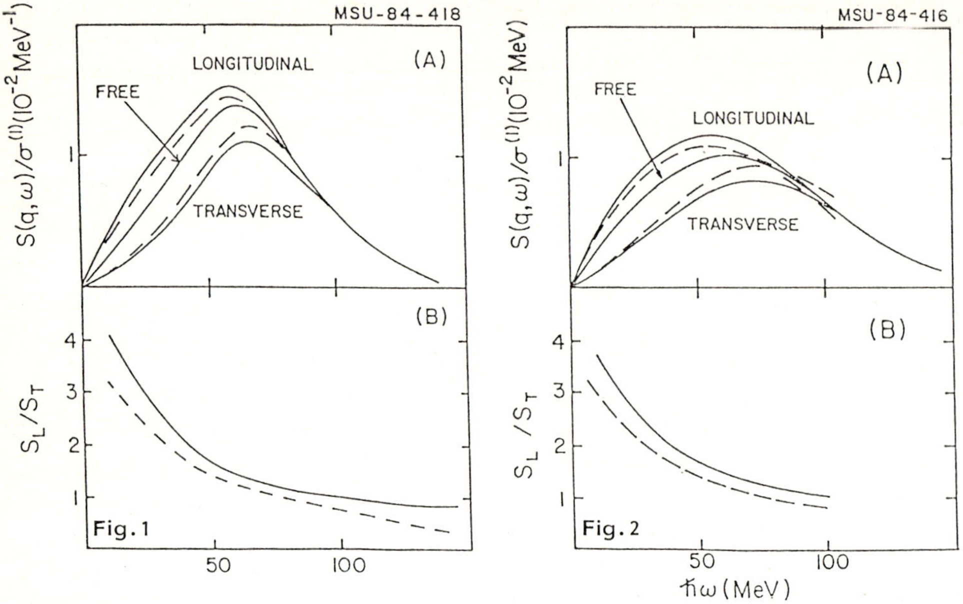

Fig. 1. Spin response functions in the local Fermi gas model as a function of excitation energy. The solid lines are the results with delta and the dashed lines are without delta excitations. Taken from [5]

Fig. 2. Spin response functions in the semi-infinite slab model as a function of excitation energy. Taken from [5]

model [5,6] and the finite nucleus model [7,8]. The distortion effect on the scattering proton is considered under the eikonal approximation.

Numerical results are shown first for the local Fermi gas model and the semi-infinite slab model in Figs. 1 and 2 [6]. The longitudinal spin response is enhanced and the transverse spin response is suppressed from the free response function. The dashed lines denote the results with only the nucleon-hole excitations. The results with the inclusion of delta isobar excitations are shown by the solid lines. The shapes of the response functions are found slightly different between the two models as can be seen by comparing Figs. 1 and 2. However, the ratios are very similar. Figure 3 shows the ratios between the longitudinal and transverse responses as compared with experiment. It is clearly seen that the ratios come down largely from the ones at the normal nuclear matter density $\varrho_0 = 0.17\,\text{fm}^{-3}$. For example, the ratio at the excitation energy $\omega = 30\,\text{MeV}$ comes down to $R = 2.3$ from $R = 8$ due to the surface effect.

An additional effect has to be considered for the case of proton scattering, since the spin responses are caused not only by the isovector excitation but also by the isoscalar excitation. Spin response functions induced by protons become now linear combinations of the isovector and isoscalar spin responses;

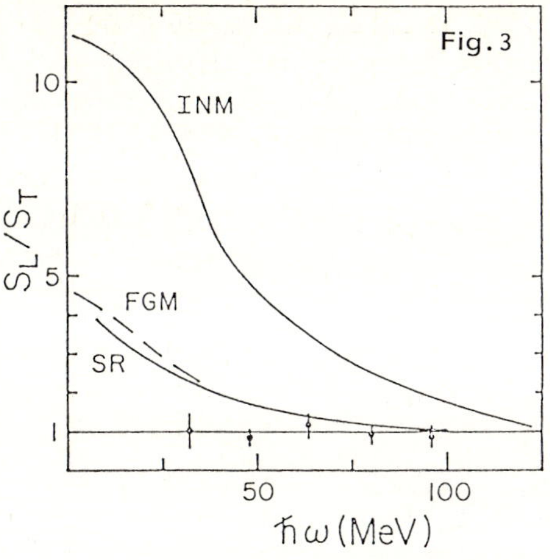

Fig. 3. The ratio of the longitudinal and the transverse spin response functions, calculated in the semi-infinite slab model (SR), the local Fermi gas model (FGM) and infinite nuclear matter model (INM) are compared with experimental data [2]. Taken from [5]

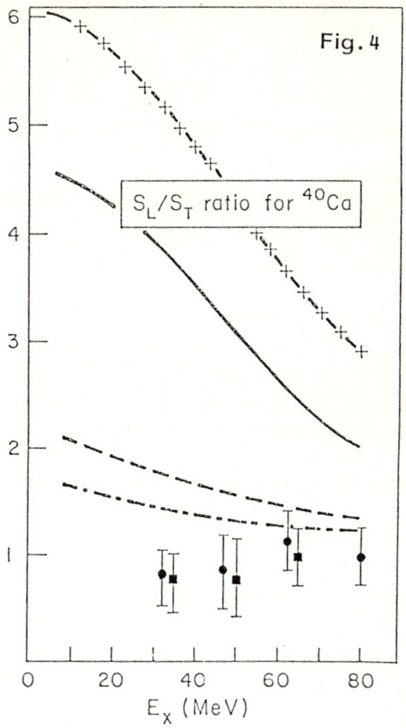

Fig. 4. The ratio of the longitudinal and the transverse spin response functions as a function of excitation energy. The results without the distortion effects are shown for the cases with delta (dash-cross) and without delta (solid). The results with the distortion effects are shown by dash-dot line for the full calculation and by dashed line for the case without the longitudinal-transverse mixing effect. Taken from [7]

$$(S_L/S_T)_{\text{proton}} = \frac{aS_L^{T=1} + bS_L^{T=0}}{a'S_T^{T=1} + b'S_T^{T=0}} \ .$$

The coefficients a, b, a' and b' are fixed by the nucleon-nucleon interaction [9]. Assuming further that the isoscalar spin responses are given by the free ones, we can estimate the isospin mixing effect. The ratio is found $(S_L/S_T)_{\text{proton}} = 1.8$ at $\omega = 30\,\text{MeV}$.

We shall now come to the finite nucleus model. Of interest here is not only the surface effect but also the mixing of the longitudinal and the transverse (L–T) responses arising from the momentum nonconservation, which is bigger for smaller nuclei. The results are shown in Fig. 4 for the case of ^{40}Ca. The dashed line denotes the numerical results with the L–T mixing, while the dash-dotted line corresponds to those without the L–T mixing. The L–T mixing effect is found quite large; the ratio at $\omega = 30\,\text{MeV}$ is brought down to $R = 1.4$ from $R = 1.8$. If we add further the isospin mixing effect as described above on top of the results with the L–T mixing, we find $R_{\text{proton}} = 1.2$. This value is very close to experiment.

In conclusion, we have found that the surface effect reduces largely the ratios between the longitudinal and the transverse spin-isospin responses. The isospin mixing effect to be considered for proton scattering and the longitudinal and the transverse mixing effect further cut down the ratios.

Acknowledgements. I am grateful to Y. Okuhara and B. Castel for fruitful discussions on the longitudinal and transverse mixing effect on the ratio of the spin response functions.

References

1. E. Oset, H. Toki, W. Weise: Phys. Report **83**, 281 (1982)
2. T.A. Carey et al.: Phys. Rev. Lett. **53**, 144 (1984)
3. W.M. Alberico, M. Ericson, A. Molinari: Nucl. Phys. A**379**, 429 (1982)
4. G.F. Bertsch, O. Scholten: Phys. Rev. C**25**, 804 (1982)
5. H. Esbensen, H. Toki, G.F. Bertsch: Phys. Rev. C**31**, 1816 (1982)
6. H. Esbensen, G.F. Bertsch: Ann. Phys. **157**, 255 (1984)
7. Y. Okuhara, B. Castel, I.P. Johnstone, H. Toki: Queen's preprint (1986)
8. W.M. Alberico et al.: To be published in Phys. Rev. C
9. T.A. Carey, J.B. McClelland: Private communication

On the Longitudinal Charge Response in the Quasielastic Peak Region

U. Stroth [1], *R.W. Hasse* [2], *and P. Schuck* [3]

[1] Institut Laue-Langevin, F-38042 Grenoble-Cedex, France
[2] GSI, Postfach 11 05 41, D-6100 Darmstadt 11, F. R. Germany
[3] Institut des Sciences Nucléaires, F-38026 Grenoble-Cedex, France

Abstract . We calculate for ^{12}C, ^{40}Ca, and ^{56}Fe the charge response for momentum transfers $1.0 < q < 2.\ fm^{-1}$ within a RPA theory which is pushed to highest sophistication presently possible. This is achieved using a semiclassical theory and the finite range and density dependent Gogny force which compares well with microscopic G-matrix calculations up to $q \cong 1.5\ fm^{-1}$; antisymmetrisation is taken into account in the meanfield and residual interaction. Remaining differences with experiment should then originate from effects not contained in RPA theory. We find good agreement with experiment for ^{12}C for the momentum transfers considered but for ^{40}Ca and ^{56}Fe the missing charge problem persists. We tentatively conclude that it is the missing coupling to 2p-2h states in the S = 0, T = 1 part of the response which is responsible for this and give detailed arguments for our speculation.

1. Introduction

The longitudinal charge response excited in inelastic electron scattering exhibits a by now longstanding problem : at least for heavier nuclei the data points overshoot considerably the theoretical predictions which have been essentially based on RPA or TDA theory using however additional more or less severe approximations like Fermi gas, effective mass, etc.

In order to account for the discrepancy between theory and experiment Noble, and, later Shakin [1] proposed the idea of a swelling of the nucleon in the nucleus due to in medium corrections of the nucleon form factor.

Before adopting such a spectacular explanation we thought it worthwhile to push RPA theory to highest possible sophistication and to investigate the conclusions one can draw from such a study. We therefore used the finite range Gogny force [2] which compares well with G-matrix calculations up to $q \cong 1.5\ fm^{-1}$ and has proven to be very successfull for the description of groundstate properties and giant resonances [2,3]. Since over the exchange terms momenta around k_F are involved even in these low energy phenomena we think that, together with the agreement with the G-matrix, this force should be quite reliable for transferred momenta up to $q \cong 1.5\ fm^{-1}$ but we tentatively used it for q-values up to $2.\ fm^{-1}$.

Full antisymmetrisation in the mean field and residual interaction was retained, a feature which has been made possible by the use of a recently developed semiclassical theory [4].

2. Formalism

Our formalism is basically a local momentum approximation to the nuclear response which in a model study, has turned out to be very accurate [4] :

$$\pi(q,\omega) = \int d^3 R \; \frac{\pi^0(q,\omega,\, k_F(R))}{1 - v_{eff}(q,\omega,k_F(R))\; \pi^0(q,\omega,k_F(R))} \qquad (1)$$

where $\pi^0(q,\omega,k_F)$ is the free nuclear matter response to a plane wave excitation calculated with the antisymmetrised mean field corresponding to the Gogny force. The Fermi momentum is then replaced by its local value

$$k_F(R) \qquad (\varepsilon_F - V(R))^{1/2} \qquad (2)$$

where we take for V(R) the phenomenological Woods Saxon potential of [5]. The effective interaction v_{eff} is constructed in such a way that (1) agrees in nuclear matter up to first order perturbation theory in the p-h force with the fully antisymmetrised response function. This defines v_{eff} uniquely, includes direct and exchange term on an equal footing and represents in fact the lowest order continued fraction expansion of the response function [6]. This procedure is valid since it turns out that v_{eff} is relatively weak altering the response with respect to its non interacting values by not more than ∿25%. It should be emphasised that our theory outlined above does not contain any free parameter.

3. Results

In Fig. 1 we show our results for ^{12}C for four momentum transfers between 1.0 and 2.0 fm^{-1}. We see that experiment is described quite satisfactorily. However, some overshooting of the theoretical results on the high energy side of the quasielastic peak is observed for all momenta.

In Fig. 2, we display the results for ^{40}Ca and ^{56}Fe for two momenta between 1.5 and 2.0 fm^{-1} (the former is the lowest one measured). With absolutely the same theoretical ingredients a dramatic increase of the discrepancy between theory and experiment with respect to the ^{12}C situation shows up. This is the well known missing charge problem [7]. Let us study the results in more detail : one of the main features in the heavier nuclei is the appearance of a plateau like structure in the response which is absent in the ^{12}C case. This is most evident for ^{56}Fe at q = 1.52 fm^{-1} but it is more or less pronounced also in the other cases. Inspection of our results reveals that this plateau is already present in the T = 0 part of the response whereas in ^{12}C there is no plateau. Let us discuss where this qualitative change comes from. Our effective force is moderately repulsive in the bulk but strongly attractive in the surface (similar to the Landau parame-

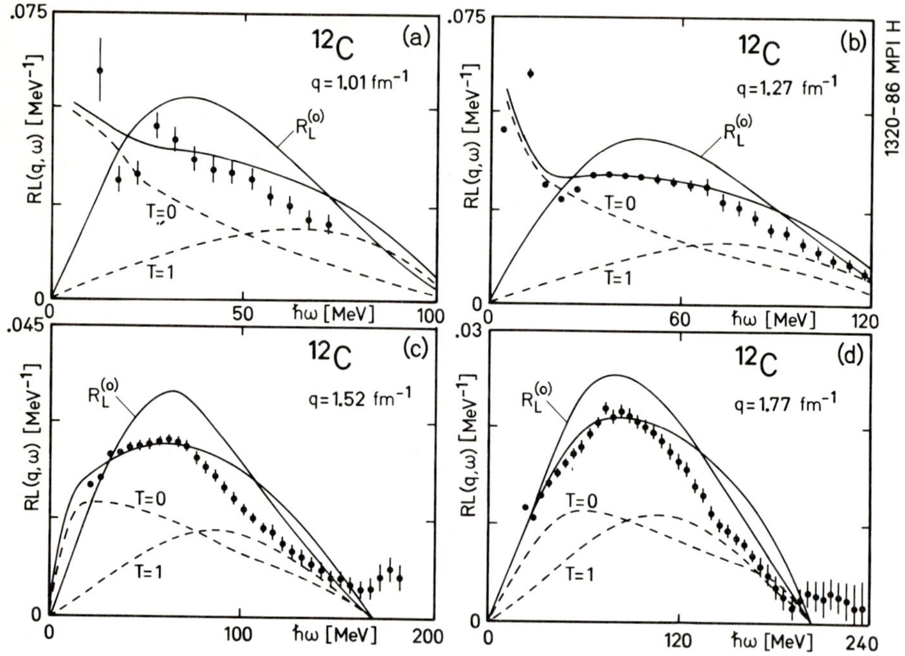

Fig. 1 Response function of ^{12}C for different momentum transfers. Also shown are the individual responses T = 0,1 and the non interacting response $R_L^{(0)}$.

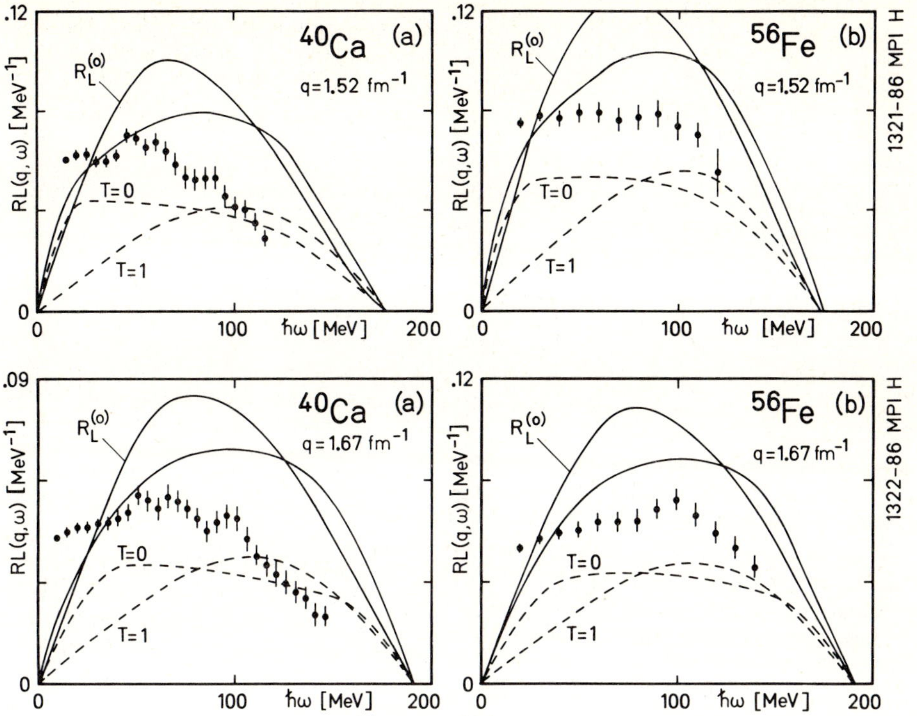

Fig. 2 Same as Fig. 1 but for ^{40}Ca and ^{56}Fe.

ter at q = 0) one effect pushing strength down the other up in energy. These opposite trends built up the plateau in heavier nuclei. In ^{12}C there is no bulk and therefore no plateau ! What happens to the T = 1 part which destroys agreement in shape with experiment ? Recent studies in the giant resonance region [8] revealed that there can be a dramatic difference of the influence of 2p-2h states on the T = 0 and T = 1 part of the response the latter being very much washed out whereas the width of the former is affected only very little. Should this feature prevail at the momenta considered here the isovector part would contribute essentially a flat background to the isoscalar one and it is easily conceivable that quantitative agreement with the data points could be achieved in this way. This also could explain the strong A - dependence of the effect since the density of 2p-2h states very strongly increases with mass number as shown in Fig. 3.

4. Conclusion

In this work we pushed the RPA theory for the longitudinal charge response in the quasielastic peak region to highest sophistication presently possible. The finite range Gogny force taking fully account of antisymmetrisation was employed. Our study suggests that the inclusion of 2p-2h states could resolve the problem of the missing charge in spreading essentially the T = 1 part of the response and not the T = 0 part. We think that this should be studied in detail before accepting spectacular proposals like the swelling of the nucleon in nuclei.

Acknowledgements

We are grateful to W. Alberico, M. Ericson and A. Molinari for discussions.

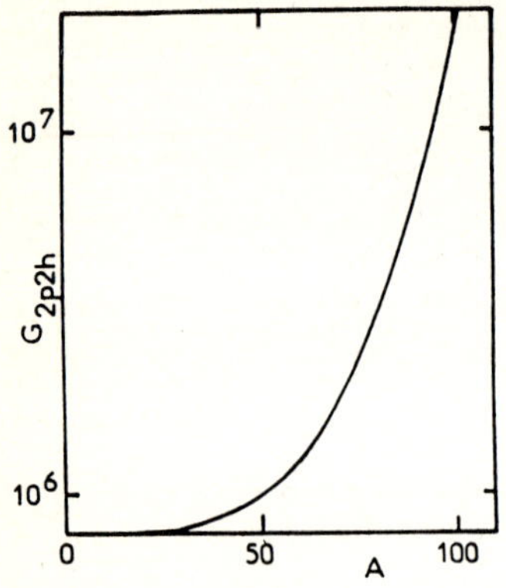

Fig. 3 Density of 2p-2h states at excitations energy E = 50 MeV as a function of mass number for a harmonic oscillator potential [9]

References

[1] J. Noble, Phys. Rev. Lett. 46, 412 (1967)
C.M. Shakin, Nucl. Phys. A446, 323 c (1985)
[2] D. Gogny , J. Déchargé, Phys. Rev. 21, 1568 (1980)
[3] J. Déchargé, L. Sips, Nucl. Phys. A407, 1 (1983)
[4] P. Schuck, Lecture Notes on Random Phase Approximation (Trieste, Feb. 1984)
U. Stroth et al., Phys. Lett. 156B, 291 (1985)
[5] W.D. Myers, Nucl. Phys. A145, 387 (1974)
[6] P. Schuck, J. Low Temp. Phys. 7, 459 (1972)
[7] Z.E. Meziani et al, Phys.Rev. Lett. 52,2130 (1984) and Phys.Rev.Lett.54,1283(1985)
[8] S. Drozdz et al., Nucl. Phys. A451, 11 (1986)
[9] G. Gosh et al., Phys. Rev. Lett. 50, 1250 (1983)

Semi-Classical Calculation of the Nuclear Spin-Isospin Response Functions

G. Chanfray

Institut de Physique Nucléaire (and IN 2P3), Université Claude Bernard Lyon-I, 43, Bd du 11 Novembre 1918, F-69622 Villeurbanne Cedex, France

Abstract : Semi-classical theory has been proven to be a very fruitful tool for calculating the response functions in the quasielastic peak. In particular a Thomas-Fermi RPA calculation (TF-RPA) can give a good description of the response functions measured in electron scattering. This may be no longer the case for surface responses measured with hadronic probes (p,p') for which the (TF-RPA) calculation predicts a sizeable contrast between transverse and longitudinal spin responses in contradiction with the Los Alamos $(\vec{p},\vec{p}')$ experiment. We show that $\hbar^2$ corrections to the (TF-RPA) theory significantly improve the agreement with the Los Alamos data.

The nuclear matter [1] and Thomas-Fermi [2] spin-isospin response functions show marked collective effects due to the presence of a residual p-h interaction constructed from pion and rho exchanges together with a short range repulsive interaction usually characterized by a Landau-Migdal parameter g'. The transverse response R_T is expected to be quenched even at momenta as high as 2 fm^{-1} because the g' interaction still dominates over rho exchange. The transverse structure function measured in (e,e') scattering in Saclay [3] confirms this collective behaviour [2,4]. On the other hand, the longitudinal spin-isospin response R_L should show inverse collective effects [1] as a signature of the pionic field enhancement in nuclei. This mechanism [6,7] has been proposed as an explanation of the EMC effect [5]. Thus, a net contrast between the enhanced longitudinal response and the quenched transverse response should be visible at momenta of order 2 fm^{-1}. For this purpose a $(\vec{p}, \vec{p}')$ experiment has been performed in Los Alamos [8]. The resulting data do not show contrast and seem to contradict the existence of the pionic enhancement. It is the purpose of this paper to demonstrate that there is no contradiction since the protons essentially probe the nuclear surface where specific effects depending on the gradient of the density appear [4]. Those effects cannot be simulated by simply lowering the density. They are computed by performing a semi classical Wigner-Kirkwood expansion of the response up to order $\hbar^2$. Keeping only the $\hbar^0$ term, we would recover the Thomas-Fermi (TF) response. The TF theory is a very accurate tool as far as we are concerned with average volume properties. The TF mean-field p-h response compares extremely well with the exact quantum mechanical result [9] and the TF - RPA calculation gives a good description of both longitudinal [10,11] and transverse [2,4] structure functions measured in (e,e') experiment. This is no longer the case where the nucleus is probed by protons because the surface peaked $\hbar^2$ terms acquire a much more important relative weight.

I.Calculation of the semi-classical spin-isospin responses up to order $\hbar^2$

I.a - The pure p-h (or mean-field) response function

We consider a spherical nucleus with N = Z where the nucleons move in a local single particle potential V(R). The pure particle-hole response is independent on the particular spin-isospin structure of the excitation operator and writes very generally ($\vec{q}$ and ω are the transferred momentum and energy) :

$$R^0(q;\omega) = \int \frac{d^3\vec{R}\, d^3\vec{p}}{(2\pi)^3} \int \frac{dt}{2\pi}\, e^{i\omega t}\, [\theta(\hat{H}-\varepsilon_F)e^{-i\hat{H}t}]_W (\vec{R}, \vec{p}+\vec{q}) \cdot [\theta(\varepsilon_F-\hat{H})e^{i\hat{H}t}]_W (\vec{R},\vec{p}) \tag{1}$$

where $\hat{H}$ is the single particle Hamiltonian ε_F is the Fermi energy and the index W refers to the Wigner transforms (WT) of the operators under consideration [12]. We expand the r-h-s of Equ.1 up to order $\hbar^2$. We recover the Thomas-Fermi response [9] as the $\hbar^0$ part of R^0 $(q;\omega)$. In the case of ^{40}Ca (Wood-Saxon potential, $q \sim 2\ fm^{-1}$) the $\hbar^2$ contribution gives a few percent correction which is positive at low energy and negative at high energy. One should mention that, at very high energy, the response may become negative or divergent although the S_1 sum rule remains satisfied. However, this unpleasant feature, which could be cured by partial resummation methods, does not occur in our region of interest (ω < 100 MeV).

I.b The RPA spin-isospin response functions

The response R_L (R_T) to a longitudinal (transverse) spin-isospin excitation $\vec{\sigma}.\hat{q}\ \tau^\alpha$ $(\vec{\sigma} \times \hat{q}\ \tau^\alpha)$ is calculated in the RPA ring approximation. The residual p-h interaction V_L (V_T) in the longitudinal (transverse) channel consists of pion (rho) exchange plus the short range g' interaction. R_L (R_T) is proportionnal to the imaginary part of the full longitudinal (transverse) polarization propagator which is the solution of a system of coupled integral equations. These equations are solved semi-classically by taking their Wigner transforms and then solving the resulting equations order by order up to $\hbar^2$. The final expression of the response is quite complicated and we give here only the main terms:

$$R_L(q;\omega) = -\frac{1}{\pi} \mathrm{Im} \Big[\int d^3R \Big\{ \alpha^0(q,R;\omega)\, D_L(q,R;\omega) - \frac{2\hbar^2}{3q^2} (\nabla_{\vec{R}}\, \alpha^0(q,R;\omega))^2\, D_L^2(q,R;\omega)\, (V_L(q;\omega)\, D_L(q,R;\omega) - V_T(q;\omega)\, D_T(q,R;\omega)) + \text{"other terms"} \Big\} \Big]$$

$$D_{L,T}(q,R;\omega) = [1 - V_{L,T}(q;\omega)\, \alpha^0(q,R;\omega)]^{-1} \tag{2}$$

Here $\alpha^0(q,R;\omega)$ is formally identical to the nuclear matter expression once one has introduced a local Fermi-momentum $k_F(R) = [2M(\varepsilon_F - V(R))]^{1/2}$. The first term of Equ. 2 is the enhanced RPA response of TF theory. The second term, containing the longitudinal-transverse coupling, is the main $\hbar^2$ contribution. It is surface peaked and proportionnal to the tensor force $V_L - V_T$, to leading order in the residual p-h interaction. Its effect is to reduce the enhancement of R_L. Although this term remains relatively small for the full volume response, it becomes important for the surface response probed by protons. The "other terms" contain, among others, contributions involving gradients of α^0 with respect to $\vec{q}$ and $\vec{R}$. The transverse response R_T is simply obtained by exchange of L and T in Equ. 2. The effect of the $\hbar^2$ surface term is reversed ie the quenching of R_T is reduced. These surface effects are less important than in the longitudinal case.

II.Results and discussion

We apply our formalism to interpret the Los Alamos experiment on the ratio of the longitudinal and transverse spin-responses probed through polarized proton scattering:

$$X = (1/4.6)\,(3.6\,R_L(\sigma\tau) + R(\sigma))\,/\,(R_T(\sigma\tau) + R(\sigma)) \qquad (3)$$

The absorption effect is taken into account through a position-dependent weight factor C(R) multiplying the integrand of Equ. 2 . C(R) is derived in ref. [2] within an eikonal approximation. In this preliminary calculation we have approximated the isoscalar response $R(\sigma)$ by the free one R^0. We have used a Wood-Saxon potential with parameters taken from MYERS [12]. The results are displayed on figure 1 for ^{40}Ca with $g' = 0.7$. The $\hbar^2$ contribution is seen to considerably improve the agreement with data.

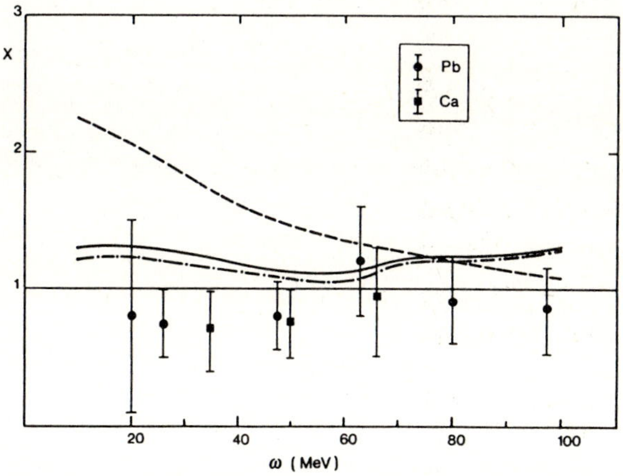

Figure 1 : The ratio X as a function of the transferred energy ω at $q = 350$ MeV/C. The dashed curve refers to the TF-RPA ($\hbar^0$) calculation for ^{40}Ca and the dot-dashed curve is the RPA result including $\hbar^2$ corrections. The full curve is the same as the dot-dashed curve but for a system with $Z = N = 104$. The experimental data are from ref. [8]

Also shown is the calculated ratio for a big system with Z = N = 104 which is supposed to simulate ^{208}Pb. The results are very similar, reflecting the fact that the surface properties do not depend very much on the mass number. Although our results are still above the data at low energy, they are certainly very encouraging. Many improvements may be advocated like renormalization of the isoscalar response, 2p - 2h contributions and refinement of the calculation by use of partial resummation methods.

References

[1] WM. Alberico, M. Ericson, A.Molinari, Nucl. Phys. **A 379** 429 (1982).

[2] U. Stroth et al., Phys. Lett. **156 B**, 291 (1985).

[3] ZE. Meziani et al., Phys. Rev. Lett. **54** 1233 (1985).

[4] WM. Alberico et al., Preprint Lyon, LYCEN/8575, To be published.

[5] JJ. Aubert et al., Phys. Lett. **123 B** 275 (1983).

[6] CH. LLewellyn-Smith, Phys. Lett. **128 B** 107 (1983).

[7] M. Ericson, A.W. Thomas, Phys. Lett. **128 B** 112 (1983).

[8] TA. Carey et al., Phys. Rev. Lett. **53** 144 (1984).

[9] U. Stroth, RW. Hasse, P. Schuck, Jl. de Physique **C6** 343 (1984).

[10] WM. Alberico, P. Czersky, M. Ericson, A. Molinari, Preprint Lyon, LYCEN/8614, To be published in Nucl. Phys. A

[11] U. Stroth, RW. Hasse, P. Schuck, To appear in Phys. Lett. B.

[12] P. Ring, P. Schuck, The Nuclear Many Body Problem, Springer Verlag, Berlin(1980).

[13] WD. Myers, Nucl. Phys. **A 145** 387 (1970).

Nucleon Form Factors from Elastic Scattering of Polarized Leptons (e, μ, τ) from Polarized Nucleons

R. Tegen

Institute of Theoretical Physics and Astrophysics, University of Cape Town, Private Bag, Rondebosch 7700, Republic of South Africa

1. Introduction

Recently nucleon form factors have regained much interest, both theoretical and experimental. Future electron scattering facilities (BATES, CEBAF) will devote considerable efforts to the study of the neutron charge form factor $G_E^n(q^2)$. In all these experiments, however, the lepton beam is an *electron* beam. At the "meson factories" (SIN, LAMPF, TRIUMF) the pion beams produce a longitudinally polarized muon (μ) beam (through spin rotation techniques a transverse polarization of the muon beam can be achieved) which can favourably be used to remove persistent uncertainties in our knowledge of the nucleon form factors: i) the size of the proton (measured by the slope of the charge form factor $G_E^p(q^2)$ near zero) is 0.86 ± 0.01 fm, known only with a rather large uncertainty [1,2]; the corresponding neutron charge form factor $G_E^n(q^2)$ is badly known away from the rather precise thermal neutron on atomic electrons scattering data (at very low 4-momentum transfer squared $(-q^2)$ [3]); ii) the magnetic form factors $G_M^{p,n}(q^2)$ are badly known for very low $(-q^2)$ (due to a kinematical suppression $\sim q^4$ in the cross section, for a *massless* lepton beam) where the charge form factor $G_E^p(q^2)$ dominates (at *high* $(-q^2)$, $G_M^p(q^2)$ dominates by far the cross section).

Since the deviation of the magnetic form factor $G_M(q^2)$ from constancy describes both the charge *and* spin distribution inside the nucleon, the corresponding *magnetic* r.m.s. radius, as compared to the charge r.m.s. radius, is of interest for the understanding of the microscopic spin structure of the nucleon; iii) neither sign of $G_E(q^2)G_E(q^2)$ nor $G_M(q^2)$ can be determined by using the standard technique of analysis via Rosenbluth plots [1,2]; iv) although the dipole fit for $G_E^p(q^2)$ and $G_M^p(q^2)$ is an impressive overall fit for $0 \leq -q^2 \leq 10 (\mathrm{GeV/c})^2$ the deviations from this purely phenomenological fit show up for both small and very high $(-q^2)$ [2,4]. The proton r.m.s. charge radius obtained from the dipole fit is $\langle r_p^2 \rangle^{1/2} = 0.81$ fm whereas the data prefer a somewhat larger radius for the proton; furthermore, the dipole fit does not describe data well around $-q^2 \approx (0.3\text{–}0.5)(\mathrm{GeV/c})^2$. For the modelling of the nucleon in terms of a quark core and the meson cloud distribution it is important to better understand experimentally the charge from factor $G_E^p(q^2)$ in this region. It has been demonstrated that in chiral quark models the region $-q^2 \approx 0.3(\text{GeV}/c)^2$ is the transition region for probing the nucleon as a quasiparticle (a quark core with a surrounding pion cloud) and probing just the quark core [5]. Beyond $(-q^2) \approx 0.4(\text{GeV}/c)^2$ the lepton beam does not probe the pion cloud any more. This is the region then where perturbative QCD starts to become applicable and the power law behaviour q^{-4} for $G_{E,M}(q^2)$ emerges [4] (without, however, explaining the "magic number" 0.71 GeV^2 in the dipole fit).

We have shown in [6] that, using a *muon*[1] instead of an electron beam has two advantages, i) one has a highly polarized μ beam (both longitudinal *and* transverse polarization) which allows us to measure the asymmetry for scattering of a polarized target, ii) one circumvents the kinematical q^4 suppression of the magnetic form factor contribution to the cross section near $q^2 \approx 0$; this is a distinct feature of the *muon* due to its non-negligible rest mass (the lepton-nucleon mass ratio squared is $(3\times10^{-7}, 1.3\times10^{-2}, 3.6)$ for (e, μ, τ)).

2. Nucleon Electromagnetic Form Factors

We consider both a polarized lepton beam *and* a polarized target (proton). To lowest order in the fine structure constant α, knowing the polarization of the incoming lepton beam alone does not produce any asymmetry[2] Let $k = (E, \boldsymbol{k})$ and $k' = (E', \boldsymbol{k}')$ be the 4-momenta of the incoming and outgoing leptons, s and s' are the corresponding lepton polarization 4-vectors, S the proton polarization 4-vector. We have $s = \frac{\lambda}{m}(|\boldsymbol{k}|, E\hat{k})$ or $s = (0, \hat{n})$ for longitudinal $(\lambda = \hat{k} \cdot \hat{s} = \pm 1)$ or transverse polarization $(\hat{k} \cdot \hat{n} = 0)$ respectively (m is the lepton mass), and $S = (0, \hat{S})$ in the rest system of the incoming proton. We report here only our result [6]:

Longitudinal polarization $(\lambda = \pm 1)$:

$$d\sigma/d\Omega_{||} = [d\sigma/d\Omega]^{\text{n.s.}} [R + \lambda \hat{S} \cdot \boldsymbol{A}_{||}] \quad \text{where} \tag{1}$$

$$[d\sigma/d\Omega]^{\text{n.s.}} = \frac{\alpha^2}{4E^2} \frac{1 - (-q^2)/4EE'}{[-q^2/4EE']^2} \frac{t^{-1}}{[1 + 2Et \sin^2 \frac{\theta}{2}/M + E(1-t)/M]}$$

is the "no-structure" cross section with $t = \sqrt{(1 - m^2/E^2)/(1 - m^2/E'^2)}$ and $\Theta = \sphericalangle(\boldsymbol{k}, \boldsymbol{k}')$ the scattering angle, and

[1] Theoretically a tau (τ) beam, if available, would be optimal. However, the mean life $\sim(>10^{27}, 2\times10^{-6}, 3\times10^{-13})$sec of (e, μ, τ) leaves only the $e-$ and $\mu-$ beams as a practical option.

[2] An asymmetry of order 10^{-5} can only arise from the interference between the electromagnetic amplitude (associated with the exchange of one photon) and the parity violating weak amplitude (associated with the exchange of the Z° boson) [7].

$$q^2 = -4EE' \sin^2\frac{\theta}{2}\sqrt{(1-m^2/E^2)(1-m^2/E'^2)} + 2m^2$$
$$- 2EE'(1-\sqrt{(1-m^2/E^2)(1-m^2/E'^2)})$$

and

$$R = (G_E^2 + \eta G_M^2)/(1+\eta)$$
$$+ (2\eta - m^2/M^2)G_M^2(q^2)(-q^2)/4EE'[1-(-q^2)/4EE']^{-1}$$
$$\boldsymbol{A}_{\|} = \eta G_M^2(q^2)(\hat{k}a_1 + \hat{k}'a_2) + G_E(q^2)G_M(q^2)(\hat{k}a_3 + \hat{k}'a_4)$$

with $\eta = -q^2/4M^2$ and $a_i(i = 1,\ldots,4)$ being energy and angle dependent functions, see [6]. Note that $(\hat{S}\cdot\boldsymbol{A}_{\|})$ vanishes identical for $\hat{S}$ perpendicular to the scattering plane, independent of the lepton mass.

For $m = 0$ we have $-q^2/4EE' = \sin^2\Theta/2, t \equiv 1$, and $[d\sigma/d\Omega]^{\text{n.s.}}R$ reduces to the wellknown Rosenbluth formula [2]. The asymmetry due to the presence of $\boldsymbol{A}_{\|}$ in (1) has been discussed in the limit $m = 0$ in [8].

Transverse polarization $(\hat{k}\cdot\hat{n} = 0)$:

$$d\sigma/d\Omega_{\perp} = [d\sigma/d\Omega]^{\text{n.s.}}[R + \hat{S}\cdot\boldsymbol{A}_{\perp}] \quad \text{where} \tag{2}$$
$$\boldsymbol{A}_{\perp} = \eta G_M^2(q^2)\boldsymbol{q}b_1 + G_E(q^2)G_M(q^2)(\hat{n}b_2 + \boldsymbol{q}b_3)$$

with $b_i{\sim}m/M$ energy and angle dependent functions [6] $i = (1,2,3)$.

Since $b_2 \neq 0$ the asymmetry does *not* vanish for $\hat{S}$ perpendicular to the scattering plane.

Let p be the polarization degree of the right-handed lepton beam; i.e. $p \equiv p_R - p_L$, then the asymmetry is given by

$$A_{\|} = p\frac{d\sigma(\lambda=+1,\hat{S}) - d\sigma(\lambda=+1,-\hat{S})}{d\sigma(\lambda=+1,\hat{S}) + d\sigma(\lambda=+1,-\hat{S})}$$

and correspondingly for $A_{\perp}$. Before we discuss the asymmetry let us come back to the unpolarized cross-section, viz. R, Eq. (1). For $m = 0$ the Θ-dependent term (for fixed q^2) $2\eta G_M^2 \tan^2\Theta/2$ serves to disentangle $G_E(q^2)$ and $G_M(q^2)$. Now, this term is very small in the interesting region of small $(-q^2)$, leading consequently to large error bars for $G_M(q^2)$ close to $q^2 = 0$ [2]. According to (1) a muon beam would lead to much less suppression of the G_M^2-term near $q^2 = 0$, due to $(2\eta - m^2/M^2)$ instead of 2η. Thus the magnetic form factor near $q^2 = 0$ could in principle be better determined with a muon beam [6]. The longitudinal asymmetry from a μ beam can presently not compete with the e-asymmetry (only effects $\sim m^2$). The situation changes if one looks at the transverse asymmetry which is *linear* in the lepton mass. It is possible to choose the kinematics such that $\hat{S}\cdot\boldsymbol{A}_\perp$ is proportional to *either* $G_M^2(q^2)$ *or* $G_E G_M(q^2)$, thus determining the *relative* sign of G_E and G_M for *low* $(-q^2)$[3]. For $\hat{S}\cdot\hat{k}=\hat{S}\cdot\hat{k}'=0$ and $\hat{S}\cdot\hat{n}\neq 0$ (i.e. target polarization perpendicular to the scattering plane) we find [6]

$$(1+\eta)\hat{S}\cdot\boldsymbol{A}_\perp = -\frac{mM}{EE'}2\eta(1+\eta)\Big[\frac{\hat{S}\cdot\hat{n}}{1-(-q^2)/4EE'}\Big]G_E(q^2)G_M(q^2).$$

For $-q^2\approx 10^4\,\mathrm{MeV}^2/\mathrm{c}^2$ and at backward angles $\Theta\approx 150^\circ$ the differential cross-section is of the order $\mu b/sr$ for muons and of the order $1/10\,\mu b/sr$ for an electron beam. This is presently measurable at the "meson factories". The transverse asymmetry for (muons, electrons) is $\approx(-0.7,0)$ for $-q^2\approx 10^5\,\mathrm{MeV}^2/\mathrm{c}^2$ and $\Theta\approx 150^\circ$, whereas the longitudinal asymmetry is maximal $\sim(-0.5,-0.95)$ for (μ,e) and decreases with decreasing scattering angle Θ [6].

3. Chiral Quark Models

In chiral quark models one finds for the quark core contributions

$$G_M^p(q^2) = \frac{M}{E_0}G_E^p(q^2) - \frac{2M}{3E_0}\int_0^\infty dr\, f^2(r)\Big[\frac{3j_1(qr)}{qr} - 3j_2(qr)\Big]$$

where $f(g)$ are the lower (upper) components of the quark spinors [5], which, in the non-relativistic limit, leads to the empirical "isoscaling law" [2]

$$G_M^p(q^2) = \mu_p G_E^p(q^2)$$

with μ_p the proton magnetic moment. Since pionic contributions to the *axial* form factor $G_A(q^2)$ from a $f_\pi\partial_\mu\phi$ term vanish [10], pionic and quark core contributions can be separated here and the size of the quark core can be constrained by comparison to available data [11], see Fig. 1. It turns out that the axial radius (0.54 fm) is somewhat smaller than the charge radius, the difference being of relativistic origin [10]. The pseudo-scalar radius (0.63 fm) (related to $G_{\pi NN}(q^2)$) is always *larger* than the axial radius, due to purely relativistic effects [10]. The situation is more complicated with respect to $G_{E,M}^{p,n}$ as the meson cloud contributes sizably. The length scale is set by the pion Compton wavelength 1.4 fm, which is more than double the quark core size. This marked difference between quark and pion length scales is responsible for the *positive* slope of the neutron charge form factor near $-q^2=0$, see *Oset* et al. [5]. The

[3] The asymmetry in e-p elastic scattering has been used at SLAC to determine the sign of $G_E G_M$ to be positive at one kinematical point ($E = 6.47\,\mathrm{GeV}$, $\Theta = 8^\circ$, and $-q^2 = 0.76(\mathrm{GeV/c})^2$ [9].

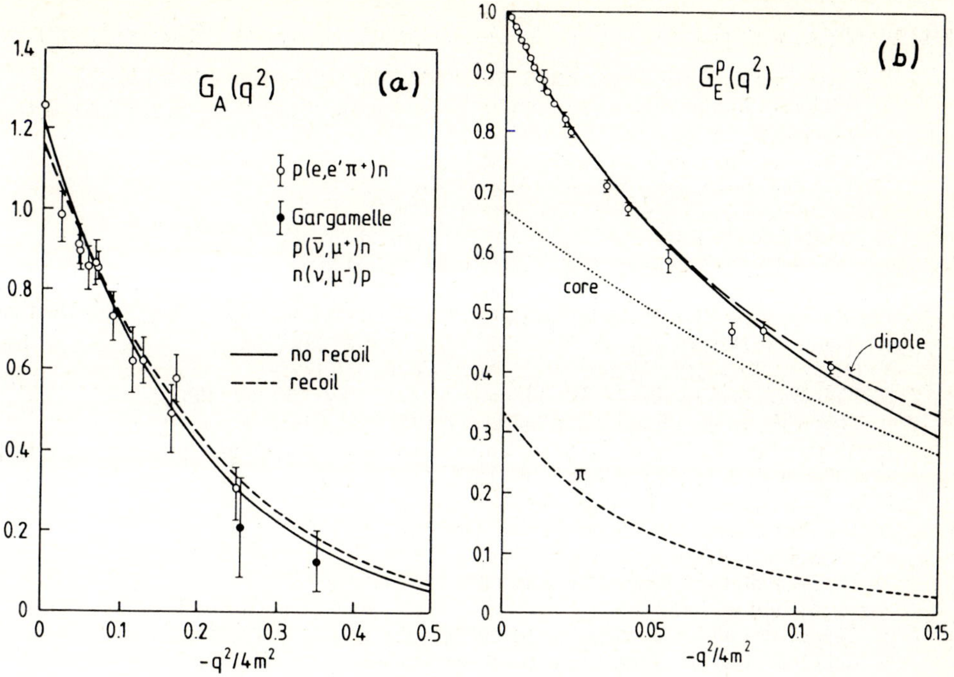

Fig. 1. **a)** The axial form factor $G_A(q^2)$ (see [10]) calculated with a confining potential $M(r) = cr^3$, $c = 930\,\mathrm{MeV\,fm^{-3}}$. **b)** Proton charge form factor $G_E^p(q^2)$ as in Oset, Tegen, and Weise [5]; the quark core parameter is the same as in a). The dashed (π) and dotted (core) curves show the separate contributions of the pion cloud and the quark core. The sum of both (solid curve) is compared to data [2] and to the standard dipole fit $G_E = [1 - q^2/(0.71\,\mathrm{GeV}^2)]^{-2}$

small mass of the pion has the effect that pionic contributions to $G_E^p(q^2)$ die out very fast with increasing $(-q^2)$, so that beyond $-q^2 \approx 0.4\,\mathrm{GeV}^2/c^2$ effectively only the *quark core* contribution to G_E^p survives, see Fig. 1b. That is the perturbative domain of QCD. It is here where data start to deviate from *both* the dipole fit *and* the chiral quark models [2, 5].

4. Conclusion

We have discussed the experimental and theoretical possibility of probing the size etc. of nucleons with a "heavy" lepton beam. The quark core size can be constrained by the (remeasured) axial form factor $G_A(q^2)$ whereas the meson cloud can be constrained by comparison to $G_{E,M}^{p,n}(q^2)$. Better data (possibly from a muon scattering experiment) for $G_{E,M}(q^2)$ will considerably improve our understanding of the nucleon in terms of its spin-carrying (quark, Skyrmion) and spinless constituents (π, σ).

References

1. F. Borkowski et al.: Z. Phys. A**275**, 29 (1975); Nucl Phys. B**93**,461 (1975)
2. G. Höhler: In LandoltBörnstein, New Series, Vol. 9 (Springer, Berlin, Heidelberg 1982) pt. b2, ch 2.5;
   M. Gourdin: Phys. Rep. **11**C, 29 (1974);
   M.N. Rosenbluth: Phys. Rev. **79**, 615 (1950);
3. V.E. Krohn et al.: Phys. Rev. D**8**, 1305 (1973);
   L. Koester et al.: Phys. Rev. Lett. **36**, 1021 (1976)
4. S.J. Brodsky, G.P. Lepage: Phys. Scripta **23**, 945 (1981)
   G.P. Lepage, S.J. Brodsky: Phys. Rev. D**22**, 2157 (1980)
   For a review see: S.J. Brodsky: Springer Tracts in Modern Physics **100**, 81 (1982)
5. R. Tegen, R. Brockmann, W. Weise: Z. Phys. A**307**, 339 (1982);
   E. Oset, R. Tegen, W. Weise: Nucl. Phys. A**426**, 456 (1984);
   S. Théberge, G.A. Miller, A.W. Thomas: Can. J. Phys. **60**, 59 (1982);
   S. Théberge, A.W. Thomas: Nucl. Phys. A**393**, 252 (1983)
6. B.M. Preedom, R. Tegen: Preprint Univ. South Carolina (1986)
7. R.H. Cahn, F.J. Gilman: Phys. Rev. D**17**, 1313 (1978);
   V.W. Hughes: AIP Conf. Proc. **51**, 171 (1978)
8. N. Dombey: Rev. Mod. Phys. **41**, 236 (1969);
   M. Gourdin: Ref. [2]
9. M.J. Alguard et al.: Phys. Rev. Lett. **37**, 1258 (1976);
   See also Hughes: Ref. [7]
10. R. Tegen, W. Weise: Z. Phys. A**314**, 357 (1983)
11. A. del Guerra et al.: Nucl. Phys. B**107**, 65 (1976);
    E. Amaldi, S. Fubini, G. Furlan: Springer Tracts in Modern Physics, Vol. 83 (1979)

Y-Scaling, FSI and the Choice of a Scaling Variable

A.S. Rinat

Service de Physique Nucléaire, Haute Energie and
Laboratoire National Saturne, CEN Saclay, F-91191 Gif-sur-Yvette Cedex, France

Consider cross sections for the inclusive scattering of electrons from nuclei, properly reduced by the Mott cross section. The resulting quantities are responses $R(q\omega)$ ($q\omega$ are momentum and energy transfer) : we address below some questions related to y-scaling of R which naturally emerge from a number of observations :

1) It is impossible to obtain exact expressions for R for truly interacting systems.

2) Approximations $R_i(q\omega)$ to R usually neglect (parts of) the final state interaction (FSI) between the ejected nucleon with a core. For instance, instead of

$$R^{exact}(q\omega) \sim R_{IA}(q\omega) + R_{FSI}(q\omega) \tag{1}$$

one retains R_{IA}, (some form of) the impulse approximation (IA).

3) For several approximations one constructs :

$$\xi_i(q\omega)\ R_i(q\omega) = F_i(y_i) \sim F(y_i) \tag{2a}$$

E.g. for the IA, $\xi_{IA} = q$. Thus [1] :

$$q\ R_{IA}(q\omega) = F(y_{IA}) \tag{3}$$

with

$$F(y) = \alpha \int_y^\infty n(k)k\ dk \tag{4}$$

in terms of the single nucleon momentum distribution n(k). Eq.(3) and all other approximations (2a) clearly exhibit perfect scaling in variables

$$y_i = y_i(q\omega) \tag{5}$$

Responses may also be expressed in terms of y and q. "Inversion" of equations like (5) will assign energy loss variables $\omega_i = \omega_i(yq)$ depending on "i". Thus

$$\xi_i(q\omega_i)\ R_i(q\omega_i) = F(y) \tag{2b}$$

4) For approximations R_i satisfying (2a) or (2b), one has

$$n(k) = -\ (\alpha\ y_j)^{-1}\ dF(y_j)/dy_j \tag{6}$$

5) R^{exact} nor actual data will scale. Instead consider functions ϕ_i [2] :

$$\phi_i(yq) \equiv \xi_i(q\omega_i)\ R^{exact}(q\omega) \tag{7}$$

which will of course not scale perfectly. We mention two of its properties :

a) $y_i - y_j \underset{q \to \omega}{\to} \theta(q^{-1})$ $\phi_i(yq) \underset{\substack{y \text{ fixed} \\ q \to \infty}}{\to} F(y)$ (8)

b) $\phi_i(yq) \underset{y \to \infty}{\sim} F(0)$ (9)

Eq.(8) implies respectively that differences between various y_i vanish asymptotically, and that all ϕ_i tend to one and the same limit, viz the IA. Finally, Eq. (9) expresses that also in the quasi-elastic peak ($y \to 0$) FSI are negligeable.

6) Experimental responses $q\, R_{exp}(q\omega)$ [3], for instance plotted against y_{IA}, cluster in bands [4]. That y_{IA} scaling must be imperfect : the same data $q\, R_{exp}(y_{IA}q)$ for fixed y_{IA} show appreciable dependence on q [1]. Note, that the observation of approximate scaling does not imply eq.(4), i.e. association of the scaling function with n(k).

7) At least for ^{3}He [3], $q\, R_{exp}$, shows (approximate) scaling of comparable quality in more than one variable [5]. It is obviously impossible to interpret for both F(y) as in eq.(4). It is then mandatory to find a criterion, which is able to distinguish between various imperfect scaling representations.

We approached the choice of the criterion in an exactly solvable model [2]. For it R^{exact} and $\phi_i(yq)$ may be computed. All will for $q \to \infty$ tend to the asymptotic limit F(y), but there will be differences how? The sooner (for growing q) $\phi_i(yq) \to F(y)$, the better the approximation.

In the figure we show ϕ_i for y = -0.4, -0.6 GeV (in the contribution to this conference, curves shown for $|y|$ have erroneously been assigned $-|y|$!). The manifestly superior approximation is the one where the knocked-out particle is fast and the full FSI is treated in an eikonal approximation. That approximation leads to a scaling variable

$$y_{eik.} = |\,(-\,q + \xi - m\,\langle V\rangle/\xi)\,| \qquad \xi^2 = 2m\,(\omega - \varepsilon_0) \qquad (10)$$

with ε_0 the binding energy of the ground-state and $\langle V\rangle$ some average of the interaction. The figure also clearly shows that n(k) is best determined from $\phi_{eik}(yq)$.

The above shows how actual data ought to be handled in principle : ϕ_i, eq.(7) with R^{exact} replaced by $R^{exp.}$ ought to be determined for several approximations i. The criterion above will decide which y_j is a preferred one and n(k) will then optimally be determined. One is obviously interested in seing a generalization of $y_{eik.}$ with FSI somehow included, to become paramount. As yet we did not succeed in proving a generalization of (10) for actual nuclei.

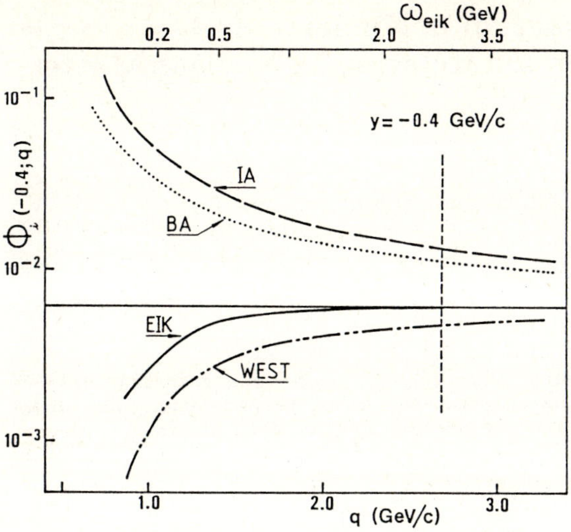

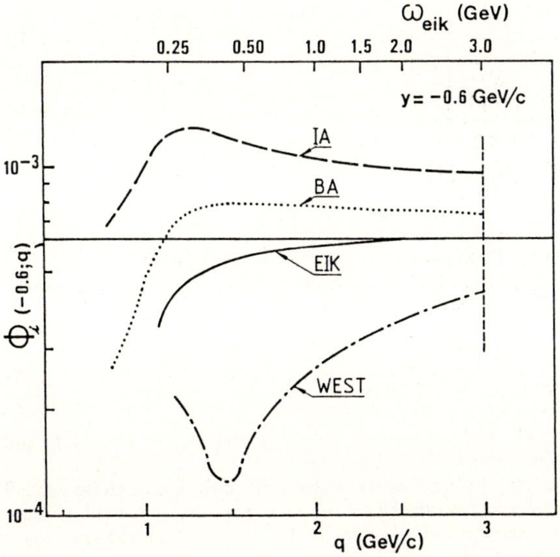

Fig. - The approach to the scaling limit (indicated by an horizontal line) for y =- 0.4, - 0.6 GeV/c.

We conclude by remarking, that imperfect scaling around the quasi elastic peak, and before asymptotic q values are reached is a property of nuclei. It does not seem to be related to any fundamental property of the internucleon force.

References

[1] E.g. P. Bosted et al. : Phys. Rev. Lett. 49 (1982) 1380.

[2] S.A. Gurvitz and A.S. Rinat : in preparation.

[3] E.g. D. Day et al.: Phys. Rev. Lett. 43 (1979) 1143.

[4] E.g. I. Sick et al. : Phys. Rev. Lett. 45 (1980) 981.

[5] S.A. Gurvitz, J. Tjon and S. Wallace, submitted for publication.

The Inclusive $(\gamma,\pi^+\pi^-)$ Reaction in Nuclei as a Test of the Pion Dispersion Relation in Nuclear Matter

E. Oset[1*] *and M.J. Vicente-Vacas*[2]

[1]CERN, Geneva, Switzerland
[2]Physics Department, University of Valladolid, Spain

The poles of the free pion propagator, $(\omega^2-\vec{q}^2-\mu^2)^{-1}$, give the free pion dispersion relation, $\omega = (\vec{q}^2+\mu^2)^{\frac{1}{2}}$, relating the pion energy to its momentum. Inside of nuclear matter, the pion interacts with the nuclear medium and the pion propagator gets modified to

$$D(\omega,q) = \left[\omega^2 - \vec{q}^2 - \mu^2 - \Pi(\omega,q)\right]^{-1} \tag{1}$$

where $\Pi(\omega,q)$ is the pion self-energy, $(\Pi \equiv 2\omega\, V_{opt})$, which depends as well upon the nuclear density. The poles of (1) give the pion dispersion relation in the nuclear medium;

$$\omega^2 - \vec{q}^2 - \mu^2 - \mathrm{Re}\,\Pi(\omega,q) = 0 \quad \rightarrow \quad \omega = \tilde{\omega}(q) \tag{2}$$

in addition, the pion acquires a width due to strong interaction given by $\Gamma = -\mathrm{Im}\,\Pi/\omega$, which takes into account the loss of elastic pion flux due to quasi-elastic scattering or pion absorption.

The knowledge of the pion optical potential provides the medium dispersion relation immediately by means of (2). The experimental determination of the optical potential at low pion energies has, however, severe ambiguities. Indeed, omitting unnecessary details, the optical potential at low energies is parametrized as [1]

$$V_{opt}(r) \propto b_0\rho(r) + B_0\rho^2(r) + \vec{\nabla}\left[c_0\rho(r) + C_0\rho^2(r)\right]\vec{\nabla} \tag{3}$$

which contains an s-wave (b parameters) and a p-wave part (c parameters). On the other hand, it contains a part proportional to ρ and another one proportional to ρ^2. However, as shown in Ref. [2], the pionic atom data and the low-energy pion nucleus scattering data do not provide the parameters of the optical potential but there is a correlation between them, such that for values of the parameters fulfilling a relationship of the type

$$\begin{aligned} b_0 + \alpha B_0 &= \beta \\ c_0 + \gamma C_0 &= \delta \end{aligned} \tag{4}$$

one obtains equally good fits to the data. This indicates that there is an effective density felt by the pions in these physical situations and, as a consequence, one cannot determine the density functional of the potential.

A possibility of learning about the needed extra information is through pion production processes in the nucleus, in particular those having three bodies in the final state, since the phase space is very sensitive to the energies of the particles involved. We have thus looked at the $\pi A \rightarrow \pi\pi A'$ and $\gamma A \rightarrow \pi\pi A'$ inclusive

*) Permanent address: Departamento de Fisica Teorica, Facultad de Ciencias Fisicas, Universidad de Valencia, Burjassot (Valencia), Spain.

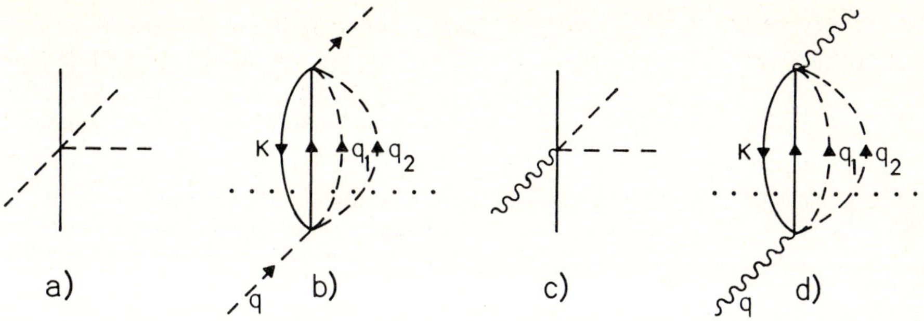

Fig. 1: b) Pion self-energy diagram containing the physical process $\pi N \to \pi\pi N$ of Fig. 1a) in the analytical cut shown by the dotted line. c) and d) same for the photon.

reactions in nuclei. In order to calculate the cross-section for these reactions, we follow the procedure outlined in Fig. 1 and construct a pion (or photon) self-energy diagram containing the $\pi N \to \pi\pi N$ amplitude (or $\gamma N \to \pi\pi N$) at each vertex. The cross-section for the reaction is related to the imaginary part of $\Pi'(q)$ from the analytical cut shown in the figure by means of:

$$d\sigma = -\frac{1}{q} \operatorname{Im} \Pi'(q,\rho) d^3r \tag{5}$$

In order to calculate σ in a finite nucleus, we use the local density approximation, evaluate $\Pi'(q,\rho)$ in infinite matter and integrate (5) over the nuclear volume with ρ being the local density of the nucleus at any integration point. Following the standard Feynman rules and performing some trivial integrations one obtains, assuming free pion propagators for the intermediate lines of Fig. 1:

$$\sigma = \frac{\pi}{q} \int d^3r \int \frac{d^3k}{(2\pi)^3} \int \frac{d^3q_1}{(2\pi)^3} \int \frac{d^3q_2}{(2\pi)^3}\, n(\vec{k})\left[1-n(\vec{q}+\vec{k}-\vec{q}_1-\vec{q}_2)\right]$$
$$\Sigma_{s_i,s_f} |T|^2 \frac{1}{2\omega(\vec{q}_1)} \frac{1}{2\omega(\vec{q}_2)}\, \delta(q^0+\varepsilon(\vec{k})-\omega(\vec{q}_1)-\omega(\vec{q}_2)-\varepsilon(\vec{q}+\vec{k}-\vec{q}_1-\vec{q}_2)) \tag{6}$$
$$R_{\substack{\pi\\ \gamma}}(\vec{b},z)\ \tilde{R}_{\pi 1}(\vec{r},q_1)\ R_{\pi 2}(\vec{r},q_2)$$

where T is the $\pi N \to \pi\pi N$ reaction matrix (or $\gamma N \to \pi\pi N$) and $R_\pi (R_\gamma)$ takes into account the probability that the original pion (photon) reaches the point $\vec{r}$ without absorption or quasielastic scattering (in which step the pion loses some energy hence reducing the phase space for the $\pi N \to \pi\pi N$ reaction), while $\tilde{R}_\pi$ takes account of the possibility that the pion goes out of the nucleus without absorption. In addition, $n(\vec{k})$ is the occupation number for the local Fermi sea at the integration point.

The next step requires the evaluation of the elementary $\pi N \to \pi\pi N$ and $\gamma N \to \pi\pi N$ reaction matrix. A thorough job has been done for the first reaction [3] which requires the evaluation of 32 Feynman diagrams contributing to the process, by means of which a fair description of the experimental data is reached. For the $\gamma N \to \pi\pi N$ reaction some simplified models have been used [4], but on the light of the findings of Ref. [3], a fresh look at this amplitude would be very interesting although it has not yet been done.

If one wishes to take now into account in (6) the effects of the renormalization of the pions in the medium, this is easily accomplished by using renormalized pion propagators as in (1) instead of the free pion propagators in the evaluation of (6). This is trivially implemented and reduces to substitute the residue $(2\omega(\vec{q}))^{-1}$ by $(2q^0 - \partial\Pi/\partial q^0)^{-1}$ [calculated at $q^0 = \tilde{\omega}(\vec{q})$] and $\omega(\vec{q})$ by $\tilde{\omega}(\vec{q})$ inside the δ function. Since the pion self-energy is essentially attractive in all the

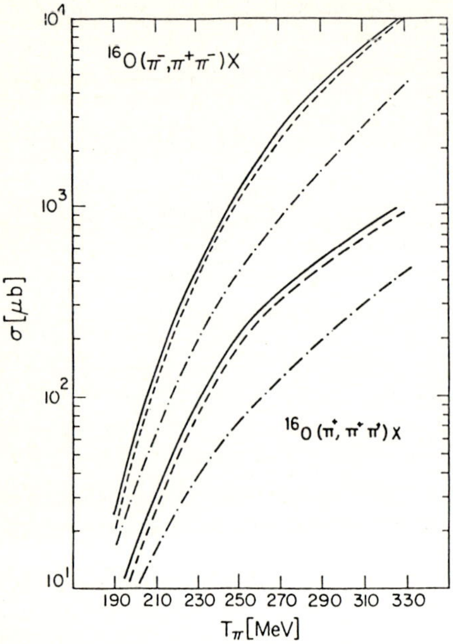

Fig. 2: Predicted cross-section for the $(\pi^-,\pi^+\pi^-)$ and $(\pi^+,\pi^+\pi^+)$ reactions in ^{16}O. The dashed-dotted curves give the results with the free pion dispersion relation. The dashed curves are the results with the pion medium corrections accounting for the new pion dispersion relation. The full lines take into account additional medium renormalization of the amplitudes, Ref. [5].

phase space, $\tilde{\omega}(q) < \omega(q)$ and this increases the range of pion momenta allowed, thus increasing the value of the phase space at the local level. The results of this renormalization can be seen in Fig. 2. The effect of the renormalization is an increase of about a factor 2-2.5 in the integrated cross-section for the reaction.

The method can be applied in a straightforward way to the study of the $(\gamma,\pi^+\pi^-)$ inclusive reaction in nuclei, by using (6) with the proper T elementary amplitude and $R_\gamma \equiv 1$ for photons. As we mentioned, the elementary $\gamma N \rightarrow \pi^+\pi^- N$ amplitude awaits a thorough study like the one in Ref. [3]. However, in spite of that, we can still assess the relevance of the medium effects on the pions for this reaction, if not give absolute values of the reaction cross-section. This is accomplished by dividing two results with and without the pionic medium effects, in which the average $|T|^2$ magnitude cancels. Hence we can get the renormalization factor due to these medium correction. This is shown in Fig. 3.

The cross-section for this reaction gets increased by a factor of around 2.5-3 in the range of energies of the figure. Note that because the initial pions in

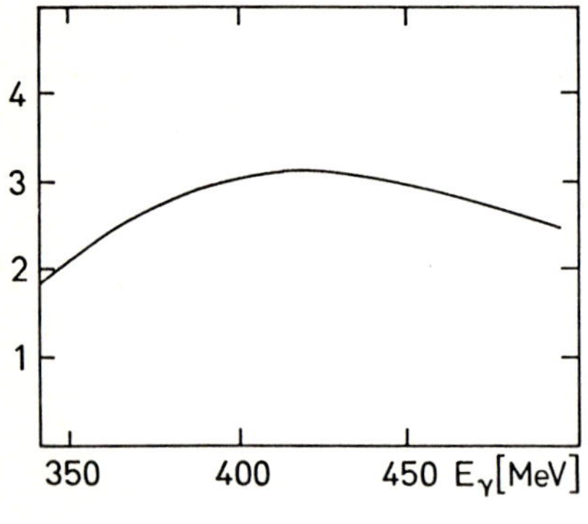

Fig. 3: Renormalization factor in $(\gamma,\pi^+\pi^-)$ inclusive reaction in ^{16}O due to the proper use of the medium pion dispersion relation.

Fig. 2 are around the resonance region, they are strongly absorbed and the reaction takes place in the periphery of the nucleus, thus testing a region of low density. Instead, the $(\gamma,\pi^+\pi^-)$ reaction takes place in the whole nuclear volume since the low energetic final pions are not much absorbed. Hence this last reaction tests a region of high densities.

In conclusion, we can say that because of the sensitivity of these two reactions to the pion dispersion relation in the nuclear medium, together with the fact that the two reactions test a different density regime of the nucleus, they can be very valuable tools to provide information on the density dependence of the optical potential not provided by scattering or pionic atoms experiments.

REFERENCES

[1] M. Ericson and T.E.O. Ericson, Ann. of Physics 36, 383 (1966).

[2] R. Seki and K. Masutani, Phys. Rev. C27, 2799 (1983) ;
R. Seki, K. Masutani and K. Yazaki, Phys. Rev. C27, 2817 (1983).

[3] E. Oset and M.J. Vicente-Vacas, Nucl. Phys. A446, 584 (1985).

[4] J.M. Laget, Physics Reports 69, 1 (1981).

[5] E. Oset and M.J. Vicente-Vacas, Nucl. Phys. A454, 637 (1986).

Part 3

Status and Test of Electroweak Theories and GUT's

3.1 Status of Electroweak Theory

Status of Electroweak Theory for Heavy Quark Decays and CP Violation

L.-L. Chau

Department of Physics, University of California, Davis, CA 95616 and
Department of Physics, Brookhaven National Laboratory, Upton, NY 11973, USA

Abstract: The phenomena of quark mixing in weak interaction are described. The current experimental status of the mixing matrix, charm and beauty particle decays, and CP violations are reported. Future interesting experiments are pointed out, especially in the event that higher generations of quarks exist.

Introduction: Some Historical Remarks

It is a great honor for me to participate in the celebration of the 600-year anniversary of Heidelberg University, and to talk at this Symposium on Weak and Electromagnetic Interactions in Nuclei.

The subject I am going to survey for you all started more than 50 years ago in nuclear reactions. It is 52 years since Fermi proposed the Fermi coupling for β decay, 51 years since Yukawa formulated his theory, and 32 years since Yang and Mills constructed their theory. Nature's answer to these human endeavors has been most elegant and rewarding. The electroweak unified theory of Glashow, Salam, and Weinberg has been confirmed by the recent discovery of the long-awaited intermediate bosons $W^{\pm}$ and Z^{0}. Their masses,[6] and even their production and decay characteristics, were predicted. (The observed lepton-forward-backward asymmetry, i.e., ℓ^{-} from the W^{-} moves in the proton direction, and ℓ^{+} from the W^{+} moves in the antiproton direction, provides a truly remarkable confirmation of the quark-parton description of the reaction.) This was the climax of the strikingly beautiful development of more than 50 years of history of weak interactions, as we have heard from many of the preceeding talks in this conference, and many of you in the audience have made major contributions.

In the light of this milestone of the great success of electroweak unified theory, I am going to review for you some particular areas of research[22-24] in weak interactions and their outlook in our ever ongoing pursuit to understand the constituents of matter and the forces governing their interactions.

For review articles on the subject see Refs. (1) to (4).

I. Status of the Quark Mixing Matrix

As well known, the mixing matrix of the three left-handed generations of quarks are defined as follows,

$$\begin{pmatrix} d' \\ s' \\ b' \end{pmatrix} = \begin{pmatrix} V_{ud} & V_{us} & V_{ub} \\ V_{cd} & V_{cs} & V_{cb} \\ V_{td} & V_{ts} & V_{tb} \end{pmatrix} \begin{pmatrix} d \\ s \\ b \end{pmatrix} \qquad (1.1)$$

Here I would like to emphasize to you a type of parametrization of the Kobayashi-Maskawa matrix, that our accumulated wisdom has shown to be the most convenient in describing the current physical situation (of course physics does not depend upon ways of parametrization, but Nature seems to have chosen a most convenient way to describe her phenomena). The 3×3 K-M matrix can be parametrized by three consecutive rotations, with the phase attached to the further most corner, where the matrix element is the smallest, e.g. V_{ub} :

$$V = \begin{pmatrix} 1 & 0 & 0 \\ 0 & c_y & s_y \\ 0 & -s_y & c_y \end{pmatrix} \begin{pmatrix} c_z & 0 & s_z e^{-i\phi} \\ 0 & 1 & 0 \\ -s_z e^{-i\phi} & 0 & c_z \end{pmatrix} \begin{pmatrix} c_x & s_x & 0 \\ -s_x & c_x & 0 \\ 0 & 0 & 1 \end{pmatrix}$$

$$s_x, s_y, s_z \sim 0 \qquad \begin{array}{c} \\ u \\ c \\ t \end{array} \!\!\!\! \begin{array}{c} \begin{array}{ccc} d \qquad\qquad & s \qquad\qquad & b \end{array} \\ \begin{pmatrix} c_x c_z & s_x c_z & s_z e^{-i\phi} \\ -s_x c_y - c_x s_y s_z e^{i\phi} & c_x c_y - s_x s_y s_z e^{i\phi} & s_y c_z \\ s_x s_y - c_x c_y s_z e^{i\phi} & -c_x s_y - s_x c_y s_z e^{i\phi} & c_y c_z \end{pmatrix} \end{array} \qquad (1.2)$$

In this order of rotation, each angle is related to a direct experimental measurement:

$$|V_{ud}| = c_x[1 + O(10^{-4})] = 0.9737, \quad \text{from nuclear } \beta \text{ decays;}^{4,5} \tag{1.3}$$

$$|V_{us}| = s_x[1 + O(10^{-4})] = 0.225, \quad \text{from strange particle decays;}^{4,5}$$

$$|V_{cb}| = s_y[1 + O(10^{-4})] = 0.059, \quad \text{from } \tau_b = 10^{-12} \text{ sec;}^{6,8}$$

$$|V_{ub}| = s_z \lesssim 0.0082, \quad \text{from } \Gamma(b \to u)/\Gamma(b \to c) < 0.05^{7,8};$$

Recent news on the ancient Cabibbo fit is that the old experimental discrepancy in the lepton-momentum asymmetry in Σ decays, which gave g_1/f_1 a wrong sign, has now disappeared. A recent high-statistics experiment[9] showed $g_1/f_1 = -0.29 \pm 0.07$, which is consistent with the Cabbibo theory. In addition, a somewhat smaller value of V_{us}, $V_{us} = 0.225$ is obtained fitting this experiment, which is is in better agreement with older fits.[4] A recent theoretical study on K_e3 gives even smaller values of V_{us} with much smaller errors.[10] One interesting experimental uncertainty is still the neutron lifetime.[11] For recent detailed analysis see talks in Ref. (2).

Our knowledge of the quark mixing matrix has dramatically improved recently owing to the b-lifetime measurement[4] and the $\Gamma(b \to u)/\Gamma(b \to c) < 0.05$[5] bound from the $b \to \ell X$ measurements, which indicated that X is mainly made out of states with charm. At present the errors on the b lifetime are still rather large. Here we use $\tau_b = 10^{-12}$ sec as an example to obtain Eq. (1.2). Using unitarity and a phase convention[12 14] such that the phase factor of the matrix elements appears with the smallest matrix elements $\lesssim 10^{-3}$, the values of the other matrix element can be obtained through unitarity

$$V = \begin{pmatrix} 0.9737 & 0.225 & V_{ub} \lesssim 0.0082e^{i\phi} \\ -0.23-0.059\,V_{ub} & 0.98 & 0.059 \\ 0.01-V_{ub} & -0.06+0.228\,V_{ub} & 1\pm O(10^{-2}) \end{pmatrix} \tag{1.4}$$

The interesting features of the mixing matrix are that the gap between the second and third generations has been widened: $|V_{cb}|$ is about a quarter of $|V_{ud}|$. Because of such severe suppression, the 2 × 2 GIM matrix is almost unitary, i.e., $V^*_{cs}/V_{cd} \cong -V^*_{us}/V_{ud}$. It is very important to check this relation independently. Significant deviation from it can have serious implications, e.g., breakdown of the model or the existence of the fourth generation. The nice feature of using the parameterization of Eq. (1.3) is that the imaginary parts appear at matrix elements with magnitudes $<10^{-3}$. Because the phase is now with the smallest matrix element, it can be ignored in discussing physics phenomena unrelated to CP violation. Also, it is clear in this parametrization that V_{ub} cannot be zero in order to explain CP violation. Actually one can show that in order to give CP violation none of the matrix elements can be zero.[3] The analysis of CP violation in the K system can give an estimate of the lower bound on V_{ud}, which is not too far below the experimental upper bound, as we shall discuss later.

The CP-violation effects from this 3×3 KM matrix have a quite striking feature. It was shown in Ref. (13) that if there are only three generations of quarks, the CP-violation phenomena in all different decays of three generations of quarks are characterized by one single parameterization-invariant parameter

$$X_{cp} = s_x s_y s_z s_\phi c_x c_y c_z^2 \sim 2\cdot 10^{-4}, \tag{1.6}$$

as determined from ε in K decay.[13] This nice unique phase-convention-independent CP violating parameter feature was also discovered much later in other ways.[5,16] Now this analysis has been carried out for more than three generations of quarks in Ref. (16), where we find that the number of invariant CP violation parameters $X_{cp,i}$ jumps to nine in the four-generation case, saturating the parameter space for generation numbers higher than three. Thus the unique X_{cp} in the three-generation case can serve as a base for searching for signals of the existence of higher generations. I shall come back to this in Section VI.

Further, in this parametrization the mixing-matrix relation to ideas in quark mass matrix can also be described in a more transparent way.[17] See also discussions in Section VII.

II. CP Violation in the Kaon and the Hyperon Decays

Weak interactions mediate $K^0 \rightleftarrows \bar{K}^0$, i.e., an off-diagonal term in the mass-matrix exists, so that the physical states are no more K^0, $\bar{K}^0$ but $|K_{S,L}> = N_{\bar{\varepsilon}}[(1+\bar{\varepsilon})\, K^0> \pm (1-\bar{\varepsilon})|\bar{K}^0>]$, where $N_{\bar{\varepsilon}}$ is a normalization factor. Such mixing gives unequal masses and decay width of the $K_{S,L}$ system, $\delta m \equiv m_S - m_L$, $\delta\Gamma \equiv \Gamma_S - \Gamma_L$, $\tilde{\Gamma} \equiv (\Gamma_S + \Gamma_L)/2$. The mixing parameters δm, $\delta\Gamma$ can be measured in an interference experiment involving regeneration of K_S from a K_L beam. Because of $\Gamma_S \gg \Gamma_L$, the K^0, $\bar{K}^0$ is a maximally mixed system, since no matter whether we have K^0 or $\bar{K}^0$ to begin with, soon K_S decays away and leaves only K_L which is almost an equal mixture of K^0, $\bar{K}^0$. The parameter $\eta \equiv |1-\bar{\varepsilon}|/|1+\bar{\varepsilon}| \neq 1$ is a phase-convention independent indicator of the mass-matrix CP violation, and was directly measured in K_L decay

$$\delta = \frac{\Gamma(K_L \to \pi^- \ell^+ \overset{\rightarrow}{\nu}) - \Gamma(K_L \to \pi^+ \ell^- \overset{\leftarrow}{\nu})}{\ell \quad '' \quad + \quad ''} = \frac{2\mathrm{Re}\bar{\varepsilon}}{1+|\bar{\varepsilon}|^2} = \frac{1-\eta^2}{1+\eta^2} = (3.3 \pm 0.12)\times 10^{-3}. \tag{2.1}$$

The mass-matrix CP violation can be calculated through the diagram[18], Fig. II.1. After calculating the box graph in the middle, which contains X_{CP}, we still need to calculate hadronic matrix element

$$B_K = \langle K^o|[\bar{d}\gamma_\mu(1-\gamma_5)s]^2|\bar{K}^o\rangle(-4/3\, f_K{}^2 M_K)^{-1},$$

which describes how the K^0, $\bar{K}^0$ are made of quarks. This is the main uncertainty in the calculation. There is still no reliable way to calculate this B_K parameter. For a discussion on the evaluation of the values of B_K, see Ref. (19). However, we shall see later, by fitting ε, the range of the value B_K is rather limited in the model of three generations of quarks.

The historically first measured CP violation[20] in K decays are

$$\eta_{+-} = \frac{A(K_L \to \pi^+\pi^-)}{A(K_S \to \pi^+\pi^-)} = \frac{(A_{+-}/\bar{A}_{+-}) - (1-\bar{\varepsilon})/(1+\bar{\varepsilon})}{'' \quad + \quad ''} = \varepsilon + \varepsilon', \tag{2.2a}$$

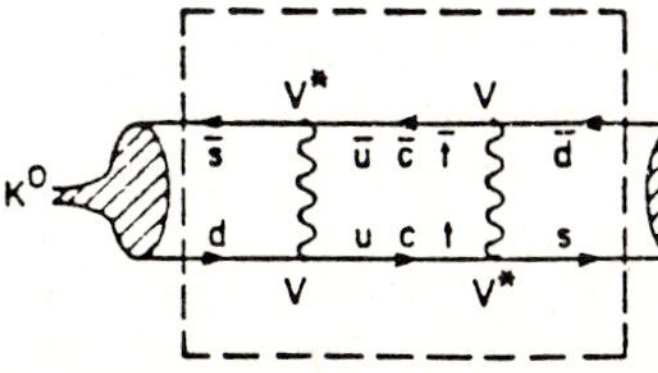

Fig. II.1 Box diagram for $K^0 \rightleftarrows \bar{K}^0$ transition, the B_K parameter describes the transition without the enclosed part; for details, see Ref. (19).

$$\eta_{00} = \frac{A(K_L \to \pi^0\pi^0)}{A(K_S \to \pi^+\pi^-)} = \frac{(A_{00}/\bar{A}_{00}) - (1-\bar{\varepsilon})/(1+\bar{\varepsilon})}{'' \quad + \quad ''} = \varepsilon - 2\varepsilon' . \tag{2.2b}$$

One can easily show that[21,3] $\varepsilon' = 0$ only if $A_{+-}/\bar{A}_{+-}$ as well as $A_{00}/\bar{A}_{00}$ can simultaneously be made to be one, i.e., no decay-amplitude CP violation. In that case and phase convention, $\eta_{+-} = \eta_{00} = \varepsilon = \bar{\varepsilon}$. Thus $\eta_{00}/\eta_{+-}{}^2 - 1 \neq 0$ is a measurement of decay-amplitude CP violation. The current experimental data give

$$|\eta_{00}/\eta_{+-}|^2 - 1 = -6\,\mathrm{Re}(\varepsilon'/\varepsilon) = \begin{cases} -0.0046 \pm 0.0053 \pm 0.0024,^{23} \\ +0.0017 \pm 0.0082,^{22} \end{cases} \tag{2.3}$$

which are still consistent with zero. It was noted by Gilman and Wise[24] that the KM model can give a nonvanishing number of ε'. The W-loop diagram shown in Fig. II.2 is essential in giving such an effect. The one-gluon exchange approximation (the "Penguin") has been used to estimate the effect. Again the main uncertainty comes from the hadronic amplitude of B_K', which describes how the K and 2π are related through quark fields, Fig. II.2. For a discussion of the values of B_K' from various calculation, see Refs. (19 and 25).

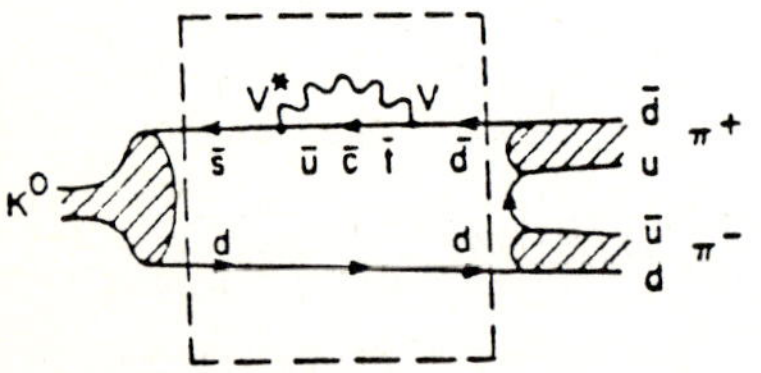

Fig. II.2 W-loop diagram for $K^0 \to \pi^+\pi^-$ Its one-gluon-exchange approximation is the "Penquin" diagram. The B_K' parameter describes the transition without the enclosed part; for details see Ref. (19,25).

Since the KM matrix is designed to give CP violation, the CP-violation measurements ε, ε'/ε must give further constraint on the quark mixing matrix. After the determination of $s_y = 0.059$ from the b lifetime, and $s_z < 0.0082$ from $|V_{ub}/V_{cb}| < 0.14$, we still have s_z, s_ϕ, m_t as independent parameters. After fitting ε, assuming some values for B_K, B_K', we obtain fixed m_t, or fixed ε'/ε contours in the s_z, ϕ plane, Fig. II.3. From such analyses, we learn the following: 1) The mass of the t quark m_t cannot be too small, depending on the values of B_K', use of $m_t \sim$ 40-GeV current data indicates $B_K > 0.33$. 2) Other than all the other calculable factors, $|\varepsilon'/\varepsilon|$ depends upon B_K', and its sign depends upon the relative sign of B_K, B_K'; current data indicate $B_K' < 1$. 3) The lower limit on s_z from CP violation is already only a factor of 2 or 3 of the upper limit given by the experimental $|V_{ub}|$ bound.

We can see that the three-generation quark model is quite narrowed down by the current experiments.

There are now two experiments[26] being carried out measuring $|\eta_{00}/\eta_{+-}|$ of the K_S, K_L system with higher sensitivity. Another new type[27] of experiment is being planned to measure ε'/ε via the tagged $K^0, \bar{K}^0 \to 2\pi$ decays at LEAR, CERN, through the following reaction $p\bar{p} \to \pi^+K^-K^0$, or $\pi^-K^+\bar{K}^0$;[28] we eagerly await the results from these experiments, and most importantly the observation of the top quark and its mass.

Here I would like to point out an interesting historical irony. Because of the $\Delta = 1/2$ - dominance phenomenon from $\Gamma(K^+ \to \pi^+\pi^0)/\Gamma(K^0 \to \pi^+\pi^-) \cong 1/670$, ε'/ε is

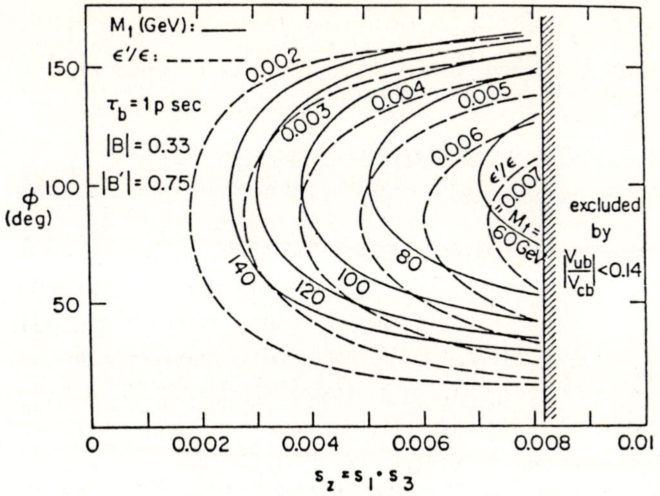

Fig. II.3 Fixed m_t, or ε'/ε contours in the s_z-ϕ plane, using the b lifetime to be 1 psec, and the parameters B_K = 0.33, B_K = 0.75; for discussions on long-range effects and details see Ref. (25).

automatically suppressed by a factor of 20. This same factor, which is now causing difficulty for experimentalists in their attempts to measure ε'/ε, actually made it possible for Professor Dalitz to have some fifteen τ events ($K^+ \to \pi^+\pi^+\pi^-$) in the 1950's. If there is no such suppression of $\Delta I = 3/2$ decays of $K^+ \to \pi^+\pi^0$, the θ, τ puzzle would not have been established before 1956, and the discovery of parity violation would have been delayed for many years. Now this $\Delta I = 3/2$ suppression is making the measurement of ε'/ε difficult. So we began to ask the question whether there are decay channels of the kaon, such that this suppression rule does not apply. Indeed we found that in $K_L \to \gamma\gamma$, the decay amplitude CP-violation effect can be as big as the mass-matrix (superweak) CP-violation effect.[29] It is possible to observe the CP-violation effect in the K_L system only if it is not swamped by $K_S \to \gamma\gamma$. Interestingly, we found that $A(K_S \to \gamma\gamma) \cong 2.4\ A(K_L \to \gamma\gamma)$, i.e., Br $(K_S \to \gamma\gamma) \cong 5 \times 10^{-6}$. The CERN ε'/ε experiment NA31 can soon measure this branching ratio. If it turns out that $K_L \to \gamma\gamma$ is not overshadowed by $K_S \to \gamma\gamma$, as predicted by our calculation, it is possible to measure the CP violation in $K_L \to \gamma\gamma$. To measure such effects in K_L, one <u>must</u> do the K^0, $\bar{K}^0$ tagging experiment as provided by the LEAR experiment.

Other decay-amplitude CP violation effects are the partial decay rate difference in $K^\pm \to (3\pi)^\pm$, and in Λ, $\bar{\Lambda}$ decays.[30] Convenient quantities to measure are

$$\frac{\Gamma(K^+\to\pi^+\pi^+\pi^-)/\Gamma(K^+\to\pi^+\pi^0\pi^0)}{\Gamma(K^-\to\pi^-\pi^-\pi^+)/\Gamma(K^-\to\pi^-\pi^0\pi^0)} = 1 + 2\Delta_{K\to3\pi}, \quad \frac{\Gamma(\Lambda\to\pi^-p)/\Gamma(\Lambda\to\pi^0n)}{\Gamma(\bar{\Lambda}\to\pi^+\bar{p})/\Gamma(\bar{\Lambda}\to\pi^0\bar{n})} = 1 + 2\Delta_\Lambda. \quad (2.4)\&(2.5)$$

Current estimates give $\Delta_{K\to3\pi} \sim 0.74\ \varepsilon'$,[31] and $\Delta_\Lambda \gtrsim 10^{-5}$.[32] The advantages of measuring such ratio of ratios rather then the percentage partial decay difference Δ itself are twofold: first the effects are two times as large, secondly some experimental errors tend to cancel. Another interesting CP-violation effect to look for is the difference of the momentum asymmetry parameter of the pion with respect to the polarization of Σ^+ in $\Sigma^+ \to \pi^+ n$ decay,

$$A(\Sigma^+_+) \cong 20X_{CP} \sim 4 \times 10^{-3}, \tag{2.6}$$

according to an estimate by Chau and Cheng in Ref. (32). Such experiments will be done at BNL, LEAR, or another kaon, hyperon-rich laboratory.

III. Mass-Matrix Mixing and CP Violation in Heavy Quark Decays

A. Mass-matrix mixing and CP violation -- Tagged-Neutral-Particle-Decay Experiments:

We are fortunate with the kaon system to have the K_S, K_L system to study mass-matrix mixing (the neutral particle-antiparticle mixing) and the superweak CP-violation parameters $\eta \equiv |1 - \bar{\varepsilon}|/|1 + \bar{\varepsilon}|$ in that K_L indeed lives rather long because the three-pion mass is very close to the kaon mass. However, heavier quark systems like, D^0, $\bar{D}^0$; B^0, $\bar{B}^0$, both mass eigenstates D^0_S, D^0_L; B^0_S, B^0_L will be short lived. It is very difficult or impossible, to do K_L, K_S-type of interference experiments.[32] So the question is whether we still can measure the mass-matrix parameters δm, $\delta\Gamma$, and CP-violating parameter η separately. The answer is yes, as discussed in Refs. (3, 21). However, one has to do neutral-meson-tagged experiments. The D^0 can be tagged in $e^+e^- \to \psi\ (4030) \to \pi^+ D^- D^0$, $\pi^- D^+ \bar{D}^0$. The three parameters δm, $\delta\Gamma$, η_D can be measured by the following three experiments: one to measure $D^0\bar{D}^0$ mixing parameter,

$$\gamma_D = \frac{"D^0 \to \bar{D}^0"}{"D^0 \to D^0"} = \eta_D^2 \frac{(\delta m/\tilde{\Gamma})^2 + (\frac{1}{2}\delta\Gamma/\tilde{\Gamma})^2}{(\delta m/\tilde{\Gamma})^2 + 2 - (\frac{1}{2}\delta\Gamma/\tilde{\Gamma})^2} , \tag{3.1}$$

which can be measured by a tagged D^0, $\bar{D}^0$ decays, $\gamma = "D^0 \to \ell^- x"/"D^0 \to \ell^+ x"$. Or it can manifest itself in the same sign dipletons productions. That $\delta m \neq 0$, $\delta\Gamma \neq 0$ are indications of mixing. Maximal mixing can happen in two ways: one way is when $\tilde{\Gamma} = (\Gamma_S + \Gamma_L)/2$ is dominated by either Γ_S or Γ_L so that $(1/2)\ \delta\Gamma/\tilde{\Gamma} \tilde{\sim} 1$, which is the case of neutral kaon decays; the other way is $\delta m/\tilde{\Gamma} >> 1$, and $\delta m/\tilde{\Gamma} >> (1/2)\ \delta\Gamma/\tilde{\Gamma}$. We shall see that the latter situation may happen in the heavy quark systems.

Two more experiments are to measure the $\ell^\pm$ partial decay rate differences:

$$\Delta(\ell^+) = \frac{"D^0 \to \ell^+ X" - "\bar{D}^0 \to \ell^+ X"}{" \quad " + " \quad "} = -\frac{2\mathrm{Re}\bar{\varepsilon}_D}{1 + |\bar{\varepsilon}_D|^2} + \frac{\Gamma_S \Gamma_L}{(\delta m)^2 + (\tilde{\Gamma})^2} , \tag{3.2}$$

and

$$\Delta(\ell^-) = \frac{"D^0 \to \ell^- X" - "\bar{D}^0 \to \ell^+ X"}{" \quad " + " \quad "} = -\frac{2\mathrm{Re}\bar{\varepsilon}_D}{1 + |\bar{\varepsilon}_D|} + \frac{\Gamma_S \Gamma_L}{(\delta m)^2 + (\tilde{\Gamma})^2} , \tag{3.3}$$

therefore,

$$\Delta(\ell^-) + \Delta(\ell^+) = -\frac{4\mathrm{Re}\bar{\varepsilon}_D}{1 + |\bar{\varepsilon}_D|^2} \equiv 2\,\frac{1 - \eta_D{}^2}{1 + \eta_D^2} , \tag{3.4}$$

is a pure CP violation effect, and

$$\Delta(\ell^-) - \Delta(\ell^+) = \frac{2\Gamma_S \Gamma_L}{(\delta m)^2 + (\tilde{\Gamma})^2} , \tag{3.5}$$

is a pure mass-matrix mixing effect. From Eqs. (3.1)-(3.5), knowing either Γ_S, or Γ_L, δ_m, $\delta\Gamma$, are measurable. The mass-matrix CP-violation effects $\eta_D \neq 1$ can be measured using Eq. (3.4).

It has been quite firmly established[8] that in the case of three generations of quarks, the only appreciable neutral meson $P^0\bar{P}^0$ (mass-matrix) mixing and CP-violation effects is in the B_S^0, $\bar{B}_S^0$ mixing. Observing deviation from this prediction can provide evidence for the new physics of higher-than-three generations of quarks.

B. Partial-decay-rate differences

Studying the partial-decay-rate differences is a convenient way to study the decay-amplitude CP violations.[2,30,34] As we mentioned previously, all CP-violation effects in the case of three generations of quarks are from a single CP-violation parameter X_{CP}, which with other quark-mixing matrix elements are rather well known. Given the same uncertainties in the hadronic decay amplitudes, we can search for the most likely channels where partial-decay-rate differences can be big. Consistently, we find in the B decays, charged[35] as well as neutral, that the partial[35,36] decay rates can be big (few x 10%), though the branching ratios of these exclusive decays are very small ($10^{-3} \sim 10^{-4}$). Here we have interestingly reversed the situation, as in kaon decays, where the partial-decay-rate differences are very small ($\sim 10^{-3}$), but branching ratios large ($\sim$10%), since the number of events needed is inversely proportional to the square of partial-decay-rate difference, but inversely proportional to the branching ratio. Thus we have a more advantageous situation in the B decay.

Here we list some very encouraging examples,[35,36] e.g.

	Δ tree	Δ W-loop	Br	# of events needed
$B_u^- \to D^- D^{0*}$	-1.6×10^{-2}	1×10^{-3}	3×10^{-3}	1.3×10^{6}
	(-0.86)	(7.3×10^{-4})	(4.1×10^{-3})	(3.3×10^{2})
$B_c^- \to \pi^- D^0$	4×10^{-2}	3.7×10^{-3}	2.3×10^{-4}	2.7×10^{6}
	(-0.86)	(1×10^{-3})	(8.5×10^{-4})	(1.6×10^{3})
$B_d^0 \to K^- \pi^+$	0	10×10^{-2}	1.6×10^{-4}	6.3×10^{5}
$B_S^0 \to D^- F^+$	0	4×10^{-2}	9.6×10^{-4}	5.6×10^{5}
$B_d^0 \to \pi^+ \pi^-$	0	25×10^{-2}	6.1×10^{-4}	2.6×10^{4}
$B_S \to D^+ D^-$	0	-27×10^{-2}	7.0×10^{-4}	2.0×10^{4}

The changed B partial rate differences are mainly from the tree graphs, e.g.,

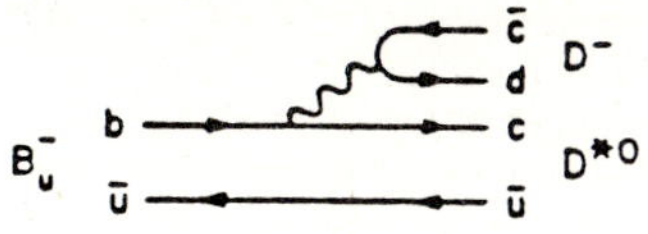

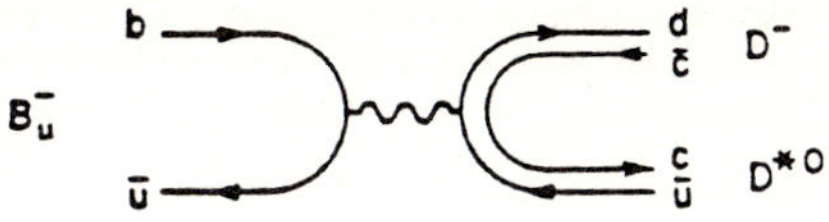

The neutral B^0 partial-rate differences can only come from interference between the tree graph and the loop group, e.g.,

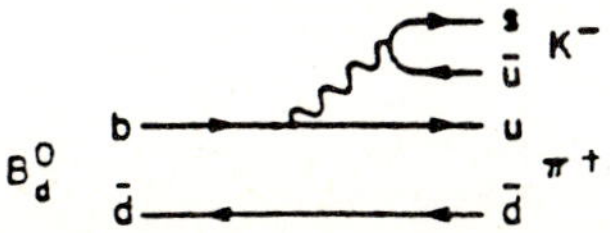

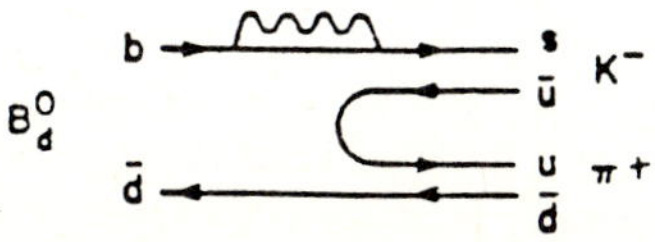

For final state with no definite CP, the partial rate differences are only decay-amplitude CP violation. However, for final states with definite CP, the partial rate differences are a combined result of mass-matrix mixing, and decay-amplitude CP violation. For details, see Refs. (25,36).

Such large partial-decay-rate differences can be searched for at CLEO II, Fermilab. LEP, SLC of current luminosity design gives barely enough B particles to do such experiments.

IV. Comparison of CP-Violating Mechanisms

Of course there is the left-right symmetric theories;[37] they have one more phase parameter than the KM model so that they give an easier fit to the current data.

CP violations can also come from the multiple Higgs fields. The more specific model of Weinberg[38] with three Higgs doublet at one point was thought to be ruled out,[39] i.e., $|\varepsilon'/\varepsilon| \gtrsim 1/20$ too large comparing even to the old experimental bound. Recently, after reexamining the calculations,[40] $|\varepsilon'/\varepsilon|$ becomes small enough to be compatible with present experimental measurement. But the neutron electric dipole moment d_n ironically is found to be much larger[41] than previous calculations. However, its violation of the current experimental bound[42] of $d_n = (0.3 \pm 4.8) \times 10^{-25}$ cm depends on an incalculable parameter. H.-Y. Cheng in Ref. (40) gave a very thorough analysis. For other CP-violation effects, see the talk by G. Segre in Ref. (1).

V. Nonleptonic Decays

Ever since the observation of the suppression of $K^+ \to \pi^+\pi^0$ in 1956,[43] $\Gamma(K^+ \to \pi^+\pi^0)/\Gamma(K^0 \to \pi^+\pi^-) = 1/670$, i.e., the $\Delta I = 1/2$ dominance rule theorists are still struggling to understand such phenomena in nonleptonic decays. The uncertainties in the calculation of the parameters of B_K and $B_{K'}$ reflect the same difficulties. The "penguin" mechanism[44] is as elusive as ever. Many brave attempts in model calculations to estimate the nonleptonic charm decays are challenged again and again as more data become available, and the theory is forced to be amended and modified. For recent impressive efforts in calculating these non-leptonic decays see Ref. (2). Here I would like to point out that this process in furthering understanding of nonleptonic decays can be guided by a model-independent analysis, i.e., using the quark-diagram-formulation, of the available experimental data, in the same spirit as Fermi introduced his coupling where only the then-known model-independent information was used.

One fortunate feature of charm and heavier quark decays is that there are many channels open, unlike in the K decays which have only 2π, 3π decays. As has been known for some time now,[45,2] all meson → 2 meson decays can be generally described in a model-independent way by six quark diagrams,[46] as shown in Fig. V., multiplied by quark-mixing matrix elements, which are now quite well determined as discussed in Section I. There are twenty-plus channels of decays in D^+, D^0, F^+ to two-pseudoscalars, PP, decays, and double that number in pseudoscalar-vector, PV, decays. Thus here the model-independent quark-analysis can be very useful in such a quark-diagram approach. Here I shall demonstrate what we have learned from the present charm-meson-decay data, and what their implications are in this approach.

Recently, many two-body decays of charm particles have been beautifully measured,[49,50] which we list in Tables V.1 and V.2. Here we shall put the quark-diagram formalism to use, analyzing all existing charm two-body decay data and discussing their implications for various theoretical model calculations.

A. Charm meson → pseudoscalar-vector decays

I begin with the PV decays because of the relative simplicity in presenting the discussion, though the data of PV decays are not yet as good as those for some

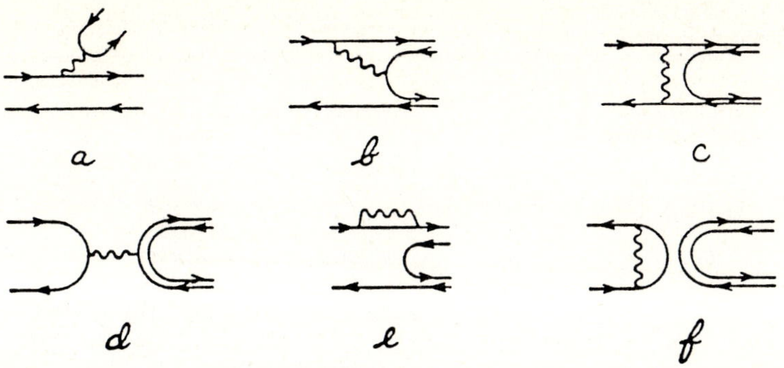

Fig. V The six quark diagrams for a meson decaying to two mesons.

TABLE I. Charm meson decays into a vector boson and a pseudoscalar meson.

	Experimental branching ratio (%)	Amplitudes with SU(3) symmetry[f]	Amplitudes with SU(3) breaking and final-state interactions[g]
		D^+ decays	
$\bar{K}^{*0}\pi^+$	$3.0 \pm 1.9 \pm 1.7$[a]	$(c_1)^2\{a'+b'\}$	$(c_1)^2\{a'+b'\}\exp(i\delta^{\bar{K}^*\pi}_{3/2})$
$\rho^+\bar{K}^0$	$12.2 \pm 2.8 \pm 1.9$[a]	$(c_1)^2\{a+b\}$	$(c_1)^2\{a+b\}\exp(i\delta^{\rho\bar{K}}_{3/2})$
$\phi\pi^+$	$0.93 \pm 0.26 \pm 0.17$[a]	$(s_1c_1)\{b'\}$	$(s_1c_1)\{b'\}\exp(i\delta^{\phi\pi})$
$\bar{K}^{*0}K^+$	$0.53 \pm 0.24 \pm 0.14$[a]	$(s_1c_1)\{a'-\tilde{d}\}$	$(s_1c_1)\{a'-\tilde{d}+\delta e\}\exp(i\delta^{\bar{K}^*K}_1)$
		D^0 decays	
$\phi\bar{K}^0$	1.4 ± 0.5[b]	$(c_1)^2\{\tilde{c}'\}$	$(c_1)^2\{\tilde{c}'\}\exp(i\delta^{\phi\bar{K}})$
$\omega\bar{K}^0$	$3.8 \pm 1.5 \pm 1.0$[a]	$(1/\sqrt{2})(c_1)^2\{b+c\}$	$(1/\sqrt{2})(c_1)^2\{b+c\}\exp(i\delta^{\omega\bar{K}})$
$K^{*-}\pi^+$	$7.8 \pm 1.2 \pm 0.9$[a]	$(c_1)^2\{a'+c'\}$	$(c_1)^2\{(a'+c')-\frac{1}{3}(a'+b')[1-\exp(i\Delta_{\bar{K}^*\pi})]\}\exp(i\delta^{\bar{K}^*\pi}_{1/2})$
	$7.1 \pm 1.6 \pm 1.3$		
$\bar{K}^{*0}\pi^0$	$2.1 \pm 0.9 \pm 0.6$[a]	$(1/\sqrt{2})(c_1)^2\{b'-c'\}$	$(1/\sqrt{2})(c_1)^2\{(b'-c')-\frac{2}{3}(a'+b')[1-\exp(i\Delta_{\bar{K}^*\pi})]\}\exp(i\delta^{\bar{K}^*\pi}_{1/2})$
ρ^+K^-	$13.7 \pm 1.3 \pm 1.5$[a]	$(c_1)^2\{a+c\}$	$(c_1)^2\{(a+c)-\frac{1}{3}(a+b)[1-\exp(i\Delta_{\rho\bar{K}})]\}\exp(i\delta^{\rho\bar{K}}_{1/2})$
$\rho^0\bar{K}^0$	$1.3 \pm 0.4 \pm 0.3$[a]	$(1/\sqrt{2})(c_1)^2\{b-c\}$	$(1/\sqrt{2})(c_1)^2\{(b-c)-\frac{2}{3}(a+b)[1-\exp(i\Delta_{\rho\bar{K}})]\}\exp(i\delta^{\rho\bar{K}}_{1/2})$
		F^+ decays	
$\phi\pi^+$	3.3 ± 1.1[c]; 4.4[d]	$(c_1)^2\{a'\}$	$(c_1)^2\{a'\}\exp(i\delta^{\phi\pi})$
	$13.0 \pm 3.0 \pm 4.0$[e]		

[a]Reference 49.
[b]Reference 50; see also Refs. 49 and 52.
[c]Reference 51.
[d]Reference 52.
[e]Reference 53.
[f]$V_{us}V^*_{cs} \cong -V_{ud}V^*_{cd} \cong s_1c_1$ used.
[g]$\delta e \equiv \tilde{e}-e$; $\delta f \equiv \tilde{f}-f$; $\delta c \equiv \tilde{c}-c$; the amplitudes with tildes have ΔS.

of the PP decays. The simplicity in discussing the PV decays comes from the purity of the quark contents in ϕ and ω. Many PV decays are given by one type of amplitude, as shown in Table I: e.g., $F^+ \to \phi\pi^+$ ($\propto$ a'), $D^+ \to \phi\pi^+$ ($\propto$ b'), $D^0 \to \phi\bar{K}^0$ ($\propto$ $\underline{c}$'), respectively. Thus from the decay rates, we can determine their absolute values:

$$|a'| = (2.50\pm0.42)10^{-6},\quad |b'| = (3.67\pm0.51)10^{-6},\quad |\underline{c}'| = (1.68\text{-}2.10)10^{-6}. \qquad (5.1)$$

The only theoretical assumption used here is that $|e^{i\delta\phi\pi}| = 1 = |e^{i\delta\phi\bar{K}}|$. To obtain the decay widths from the measured branching ratios as given in Table I, the following charm-decay lifetimes are used:[72] $\tau(D^+) = (8.8^{+1.0}_{-0.8})\times10^{-13}$ sec, $\tau(D^0) = (4.3^{+0.4})\times10^{-13}$ sec, $\tau(F^+) = (2.8^{+1.4})\times10^{-13}$ sec. Among the three measurements of $Br(F^+ \to \phi\pi^+)$ we used (3.3 ± 1.1)%. To obtain the amplitudes from the rates, we have made use of the phase-space factor $p_c^3/(8\pi m_V^2)$, where p_c is the

TABLE II. Charm meson decays into two pseudoscalars. The same notations a to f are used for amplitudes, but in general they have no relations to those in the PV decays in Table I.

	Experimental branching ratio (%) (Ref. 1)	Amplitudes with SU(3) symmetry[a]	Amplitudes with SU(3) breaking and final-state interactions[b]
		D^+ decays	
$\bar{K}^0\pi^+$	$3.5 \pm 0.5 \pm 0.4$	$(c_1)^2\{a+b\}$	$(c_1)^2\{a+b\}\exp(i\delta^{\bar{K}\pi}_{3/2})$
$\bar{K}^0K^+$	$1.11 \pm 0.34 \pm 0.21$	$(s_1c_1)\{a-d\}$	$(s_1c_1)\{a-\tilde{d}+\delta e\}\exp(i\delta^{\bar{K}K})$
$\pi^0\pi^+$	$\leqslant 0.53$	$(1/\sqrt{2})(s_1c_1)\{a+b\}$	$(1/\sqrt{2})(s_1c_1)\{a+b\}\exp(i\delta^{\pi\pi}_2)$
		D^0 decays	
$K^-\pi^+$	$4.9 \pm 0.4 \pm 0.4$	$(c_1)^2\{a+c\}$	$(c_1)^2\{(a+c)-(a+b)\frac{1}{3}[1-\exp(i\Delta_{\bar{K}\pi})]\}\exp(i\delta^{\bar{K}\pi}_{1/2})$
$\bar{K}^0\pi^0$	$2.2 \pm 0.4 \pm 0.2$	$(1/\sqrt{2})(c_1)^2\{b-c\}$	$(1/\sqrt{2})(c_1)^2\{(b-c)-(a+\bar{b})\frac{2}{3}[1-\exp(i\Delta_{\bar{K}\pi})]\}\exp(i\delta^{\bar{K}\pi}_{1/2})$
$\bar{K}^0\eta$	$1.8 \pm 0.8 \pm 0.3$	$\cos\theta A(D^0 \to \bar{K}^0\eta_8) + \sin\theta A(D^0 \to \bar{K}^0\eta_0)$	
$K^0\bar{K}^0$	$\leqslant 0.62$	$(s_1c_1)0$	$(s_1c_1)\{(-\delta e - 2\delta f) + (a+\tilde{c}-\delta e)\frac{1}{2}[1-\exp(\Delta_{\bar{K}K})]\}\exp(i\delta^{K\bar{K}}_0)$
K^-K^+	$0.60 \pm 0.10 \pm 0.08$	$(s_1c_1)\{a+c\}$	$(s_1c_1)\{(a+c)+(\delta e+2\delta f)-(a+\tilde{c}-\delta e)\frac{1}{2}[1-\exp(i\Delta_{\bar{K}K})]\}\exp(i\delta^{\bar{K}K}_0)$
$\pi^+\pi^-$	$0.16 \pm 0.09 \pm 0.03$	$-(s_1c_1)\{a+c\}$	$-(s_1c_1)\{(a+c)+(\delta e+2\delta f)-(a+b)\frac{1}{3}[1-\exp(i\Delta_{\pi\pi})]\}\exp(i\delta^{\pi\pi}_0)$
$\pi^0\pi^0$		$\frac{1}{2}\sqrt{2}(s_1c_1)\{b-c\}$	$\frac{1}{2}\sqrt{2}(s_1c_1)\{(b-c)+(\delta e+2\delta f)-(a+b)\frac{2}{3}[1-\exp(i\Delta_{\pi\pi})]\}\exp(i\delta^{\pi\pi}_0)$

[a] $V_{us}V^*_{cs} \simeq -V_{ud}V^*_{cd} \simeq s_1c_1$ used.
[b] $\delta e \equiv \tilde{e} - e$; $\delta f \equiv \tilde{f} - f$; $\delta c \equiv \tilde{c} - c$; the amplitudes with tildes have $s\bar{s}$.

center-of-mass momentum. Note here that neither amplitude b' nor amplitude $\underline{c}'$ is negligible, as preferred by some model calculations.

$D^+ \to \bar{K}^{*0}\pi^+$ [$\propto$(a' + b')] is an exotic channel which implies elastic and small $\delta^{\bar{K}^*\pi}_{3/2}$. Therefore, the rate gives

$$|a' + b'| = (1.16 \pm 0.37) \times 10^{-6}. \tag{5.2}$$

From Eqs. (5.1) and (5.2), it is evident that a' and b' are of opposite signs, and the $|a'|$, and $|b'|$ obtained from $F^+ \to \phi\pi^+$, $D^+ \to \phi\pi^+$ are in excellent agreement with $|a' + b'|$ determined from $D^+ \to \bar{K}^{*0}\pi^+$. Thus we have

$$a' = (2.50 \pm 0.42) \times 10^{-6}, \quad b' = -(3.67 \pm 0.51) \times 10^{-6}. \tag{5.3}$$

Note that a' and b' are severely destructive, consistent with the decay width of D^+ being smaller than D^0. Since $D^0 \to K^{*-}\pi^+$, $\bar{K}^{*0}\pi^0$ are given in terms of (a' + b'), and (a' + c')/(a' + b') [= 1 - (b' - c')/(a' + b')], and the phase shift $\Delta_{\bar{K}^*\pi} = \delta^{\bar{K}^*\pi}_{1/2} - \delta^{\bar{K}^*\pi}_{3/2}$, and assuming $\left|e^{i\delta^{\bar{K}^*\pi}_{1/2}}\right| = 1$, we obtain the following two solutions,

$$(a'+c')/(a'+b') = 2.36 \pm 0.67, \to c' = -(5.25 \pm 0.39) \times 10^{-6}, \quad \Delta_{\bar{K}^*\pi} = (52^{+30}_{-52})°; \tag{5.4a}$$

$$(a'+c')/(a'+b') = -1.70 \pm 0.67, \to c' = -(0.53 \pm 0.39) \times 10^{-6}, \quad \Delta_{\bar{K}^*\pi} = 180° - (52^{+30}_{-52})°. \tag{5.4b}$$

We note that the errors on $\Delta_{\bar{K}^*\pi}$ are so large that the data can be accomodated by real amplitudes without final-state interactions. The amplitude c' is quite different from $\underline{c}'$ in Eq. (5.1). (Underlined amplitudes involve strange quark and antiquark pair production.) To make definite conclusions, we need better measurements.

One nice prediction from this analysis is

$$Br(D^0 \to \phi\pi^0) = (1/2)Br(D^+ \to \phi\pi^+)\Gamma(D^+)/\Gamma(D^0) \cong 0.21\%. \tag{5.5}$$

This will be an important measurement if we are to test this scheme.

We next proceed to determine the unprimed amplitudes from $D \to \rho\bar{K}$ and $D^0 \to \omega\bar{K}$

decays. From Table V.1 it follows that $D^+ \to \rho^+\bar{K}^0$, $D^0 \to \omega\bar{K}^0$, determine

$$|a + b| = (2.18 \pm 0.25) \times 10^{-6}, \quad |b + c| = (2.57 \pm 0.51) \times 10^{-6}. \tag{5.6}$$

From the measurements of $D^0 \to \rho^0\bar{K}^0$, ρ^+K^-, we find two solutions:

$$(a + c)/(a + b) = 1.55 \pm 0.16, \quad \Delta\rho\bar{K} = (24^{+25}_{-24})^\circ, \tag{5.7a}$$

$$(a + c)/(a + b) = -0.89 \pm 0.16, \quad \Delta\rho\bar{K} = 180^\circ -(24^{+25}_{-24})^\circ \tag{5.7b}$$

These give the following three possible solutions for amplitudes a, b, c:

$$a = (4.06 \pm 0.38)10^{-6}, \quad b = -(1.88 \pm 0.28)10^{-6},$$

$$c = -(0.68 \pm 0.28)10^{-6}, \quad \Delta_{\rho\bar{K}} = (24^{+25}_{-24})^\circ; \tag{5.8a}$$

$$a = (1.50 \pm 0.38)10^{-6}, \quad b = (0.68 \pm 0.28)10^{-6},$$

$$c = (1.89 \pm 0.28)10^{-6}, \quad \Delta_{\rho\bar{K}} = (24^{+25}_{-24})^\circ; \tag{5.8b}$$

$$a = (1.41 \pm 0.38)10^{-6}, \quad b = (0.77 \pm 0.28)\times 10^{-6},$$

$$c = -(3.34 \pm 0.28)\times 10^{-6}, \quad \Delta_{\rho\bar{K}} = 180^\circ - (24^{+25}_{-24})^\circ; \tag{5.8c}$$

Again, because of the large errors, the data are compatible with real amplitudes. The measurement of $D^+ \to \bar{K}^{*0}K^+$, Table V.1, gives us

$$|a' - \underline{d} + \delta e| = (2.70 \pm 0.61) \times 10^{-6} \tag{5.9}$$

Future measurements of $\bar{K}^{*0}\eta_8$ $[\propto(b' + c' - 2\underline{c})]$ and $\bar{K}^{*0}\eta_0$ $[\propto(b' + c' + \underline{c})]$, can help to determine amplitudes $\underline{c}$ and $(b' + c')$, and then c', since b' is known. From future measurements of $D^0 \to \bar{K}^{*0}K^0$, $K^{*0}\bar{K}^0$ $[\propto(c - c')]$, and the known information on c' we can determine the amplitude c. Then we can check which solution of equation (5.8) will be picked, and thus determine a and b individually. From $F^+ \to \rho^+\pi^0$ $[\propto(d - d')]$, $F^+ \to \omega\pi^+$ $[\propto(d + d')]$, we can determine d, d'. We then know all the amplitudes a, b, c, d, and a', b', c', d', and their relative signs. The rest of the PV decays are predictable, up to SU(3) breaking and final-state interactions. These results must be conformed to by any theoretical calculations.

Next we discuss the case of charm meson decay into two pseudoscalars, $P_c \to PP$. (Note that the amplitudes a to f here for PP decays have no relation to those for the PV decays. When needed for clarity, we use subscript PP to denote the distinction). Here, the data are of greater accuracy than for the $P_c \to PV$ case, Table V.2. From $D^+ \to \pi^+\bar{K}^0$, $D^0 \to K^-\pi^+$, $\bar{K}^0\pi^0$, we can conclude definitely that real amplitudes $(a, b, c)_{PP}$, (without including effects like final-state interactions) cannot fit the data. From $D^+ \to \pi^+\bar{K}^0$, we obtain

$$|a + b_{PP}| = (1.66 \pm 0.11)10^{-6} \text{ GeV}. \tag{5.10}$$

Then from $D^0 \to K^-\pi^+$, $\bar{K}^0\pi^0$, we obtain the following two solutions,

$$[(a + c)/(a + b)]_{PP} = 1.95 \pm 0.14, \quad \Delta_{\bar{K}\pi} = (79^{+10}_{-14})^\circ; \tag{5.11a}$$

or $$[(a + c)/(a + b)]_{PP} = -1.28 \pm 0.14, \quad \Delta_{\bar{K}\pi} = 180 -(79^{+10}_{-14})^\circ. \tag{5.11b}$$

We want to caution about the interpretation of the phase shift difference $\Delta_{K\pi} \equiv \delta^{K\pi}_{1/2} - \delta^{K\pi}_{3/2}$ obtained here from charm decays. Its relation to the hadronic scattering phase shifts is complicated by the other competing channels, e.g. $\pi\pi\bar{K}$, $\pi\pi\pi K$ (not including $P\bar{K}$, πK^*, which do not communicate with $\bar{K}\pi$ through strong interactions). Only if all the strong-interaction communicating channels are negligible. $\delta^{\bar{K}\pi}$ here is $\bar{K}\pi \to \bar{K}\pi$ scattering phase shift.

Unlike the PV decays, it is much harder here to determine individual amplitudes, since none of the decays is given by a single amplitude. It is interesting to point out that the nonspectator-diagram amplitudes c, $\underline{c}$ and d can be measured in a model-independent way by observing the following decay modes,

$$\frac{\Gamma(D^0\to\bar{K}^0\eta_8)}{\Gamma(D^0\to\bar{K}^0\eta_0)} = \frac{1}{2}\left|\frac{b+c-2\underline{c}}{b+c+\underline{c}}\right|^2_{pp}; \quad \frac{\Gamma(F^+\to\eta_8\pi^+)}{\Gamma(F^+\to\eta_0\pi^+)} = 2\left|\frac{a-d}{a+2d}\right|^2_{pp}. \tag{5.12}$$

From the absolute rates of these decays we can determine $[(b+c), \underline{c}; \text{ and } a, d]_{pp}$. Combining with the solutions Eq. (5.10) and (5.11), we shall determine all amplitudes $(a, b, c \text{ and } d)_{pp}$ and their relative signs.

Next we go to the mixing-matrix singly suppressed measurement of $\Gamma(D^0 \to K^+K^-)/\Gamma(D^0 \to \pi^+\pi^-) \neq 1$ first observedly the Mark II collaboration.[55] From Table V.2b, we see that such differences can be attributed to the SU(3) breaking effect of $(\delta e + 2\delta f)_{pp}$, which contributes with opposite sign to $D^0 \to K^+K^-$, $\pi^+\pi^-$ and/or to the final-state-interaction effect (e.g., $\delta_0^{\pi\pi}$ has a larger absorptive part than $\delta_0^{K\bar{K}}$). To clarify these mechanisms it is of paramount importance to measure $D^0 \to \pi^0\pi^0$ (see Table II.b), since the same unknown $(\delta e + 2\delta f)_{pp}$, $\delta_0^{\pi\pi}$ are present, but the rest of the amplitudes $(b-c)_{pp}$, $(a+b)_{pp}$ are known [from Eqs. (5.10) and (5.11)].

Using the known relation between η, η' and η_8, η_0 (here the mixing angle of (-10°) is used. In the analysis on future measurements of decays involving η', care should be taken to subtract any component in η' that is not η_0 or η_8), the current measurement of $D^0 \to \bar{K}^0\eta$, Table V.2b, gives $|1.23b - 0.49c|_{pp} = (4.57 \pm 1.01) \times 10^{-6}$ GeV, if there is no SU(3) breaking, i.e. $\underline{c} = c$; or it gives $|b + c|_{pp} = (3.71 \pm 0.83) \times 10^{-6}$ GeV, if SU(3) breaking is maximal, i.e., $\underline{c} = 0$. Future measurements of $D^0 \to \bar{K}^0\eta_8$ $[\propto(b + c - 2\underline{c})]$, or $\bar{K}^0\eta'$, will help to determine amplitudes $[(b+c), \text{ and } \underline{c}, \text{ thus } b, c]_{pp}$ individually when combining with the results of Eqs. (5.10) and (5.11). The measurement of $D^+ \to \bar{K}^0K^+$ gives $|a - \underline{d} + \delta e|_{pp} = (4.29 \pm 0.66) \times 10^{-6}$ GeV. Future measurements of $F^+ \to \pi^+\eta_8[\propto(a-d)_{pp}]$, and $\pi^+\eta_0[\propto(a+2d)_{pp}]$, can give $[a, \text{ and } d]_{pp}$; $\bar{\Gamma}(D^+ \to \bar{K}^0K^+)$ is predicted to be equal to $(s_1/c_1)^2\ \bar{\Gamma}(F^+ \to \pi^+\eta_8)$ if there is no SU(3) breaking. The measurements of $D^0 \to K^0\bar{K}^0$, $\eta_0\eta_0$, which are nonzero only from SU(3) breaking, can give a direct modification of SU(3) breaking effects. The long-predicted relation $\bar{\Gamma}(D^+\to\pi^+\pi^0)/\bar{\Gamma}(D^+\to\bar{D}^0\pi^+) = 1/2\ |V_{cd}/V_{cs}|^2$ should be checked by experiments.

As discussed in the previous sections future measurements of $\bar{K}^{*0}\eta_8$, $\bar{K}^{*0}\eta_0$, $D^0\to\bar{K}^{*0}K^0$, $K^{*0}\bar{K}^0$, $F^+\to\rho^+\pi^0$, $\omega\pi^+$; and $D^0\to\bar{K}^0\eta_8$, $\bar{K}^0\eta_0$, $F^+\to\pi^+\eta_8$, $\pi^+\eta_0$, will give definite and model-independent results about individual amplitudes, to which theoretical calculations must conform.

I hope that I have demonstrated to you that the general quark-diagram approach provides a framework in which experimental results can be analyzed in a model-independent way, and new experiments can be pointed out to further test certain specific model calculations.[56] The current experimental results have already provided much information on non-leptonic decay mechanism in term of the quark diagram amplitudes, which can be used for comparison with theoretical calculations. Many interesting predictions have resulted. The story of nonleptonic decay is very complex, but also very interesting. It will take our persistent effort, both theoretically and experimentally, to find the conclusion of the story.

VI. Beyond the Three Generations of Quarks

As we see the experiments measuring the b lifetime, $|V_{ub}/V_{cb}|$, ε, ε'/ε, and the hint of m_t from UA1 experiment at CERN have really pushed the three-generation model to a corner. Future measurements of ε'/ε, $|V_{ub}/V_{cb}|$ with improved error bars, and with m_t pinpointed can really imply very specifically about B_K, $B_K{}'$, and the likelihood of the three generation model. So it's a good time to think ahead if there are four generations of quarks.

If there is a fourth generation of quarks b', t', the number of parameters increase from the three generation 4 to 9, six angles and three phases. We generalize the parameterization of Eq. (1.3) as follows,[36]

$$V_4 = \begin{bmatrix} 1 & 0 & 0 & 0 \\ 0 & 1 & 0 & 0 \\ 0 & 0 & c_u & s_u \\ 0 & 0 & -s_u & c_u \end{bmatrix} \begin{bmatrix} 1 & 0 & 0 & 0 \\ 0 & c_v & 0 & s_v e^{-i\phi_3} \\ 0 & 0 & 1 & 0 \\ 0 & -s_v e^{i\phi_3} & 0 & c_v \end{bmatrix} \begin{bmatrix} c_w & 0 & 0 & s_w e^{-i\phi_2} \\ 0 & 1 & 0 & 0 \\ 0 & 0 & 1 & 0 \\ -s_w e^{i\phi_2} & 0 & 0 & c_w \end{bmatrix} \begin{bmatrix} & & & 0 \\ & & & 0 \\ & V_3 & & 0 \\ 0 & 0 & 0 & 1 \end{bmatrix} \tag{6.1}$$

where V_3 is the 3×3 matrix given in Eq. (1.3). The reason we put the additional phases ϕ_2, ϕ_3 where there are in Eq. (6.1) is to anticipate that the widening of the generation gap will keep its pace. Of course physical consequences are independent of ways of parameterization. What are the different physical phenomena if fourth generation do exist?

(1) If the masses are low enough, we should search for them in e^+e^-, $p\bar{p}$, pp and cosmic rays.

(2) $V^*_{cd}/V_{cs} \cong V^*_{us}/V_{ud}$ does not have to be true. More accurate measurements of V_{cd}/V_{cs} should be studies from semileptonic decays of $D \to \ell^{\pm}X$ containing s or not, and from the long predicted,[61] $\Gamma(D^+ \to \pi^+\pi^0)/\Gamma(D^+ \to \bar{K}^0\pi^+) = 1/2|V_{cd}/V_{cs}|^2$.

(3) As mentioned in Sect. I, we showed a surprising result in Ref. (16), that rather than the unique phase-convention-invariant CP violation parameter X_{CP} in the three-generation case, there are now nine of them $X_{CP,i}$, $i = 1,\ldots 9$, for generation number n = 4, and always saturating the whole parameter space for $n \geq 4$. So the consequences of CP violations are much more varied then the three generation case, e.g. ε'/ε can be zero without making $B_K' = 0$; V_{ub} can be zero without annihilating the CP-violation effect ε. Due to the presence of more $X_{CP,i}$ variables, many more appreciable partial rate differences can happen. Those effects from the tree graphs are actually only dependent on the presence of the higher generation, not on their masses. For example the partial-decay-rate difference Δ_F in mixing-matrix singly suppressed decay $F \to K^0\pi$ can be increased to

10^{-1}, and Δ_T for t quark decays can be very large too. This makes the improvement of SPEAR experiment, and BEPC extremely important. For details, see Ref. (16).

(4) There are new contributions in the mass matrix from the new generation of b', t', so that the mass-matrix mixing (i.e. the neutral meson-antimeson mixing results from the three-generation model can be drastically changed. However such effects are from loop diagrams and very sensitive to the b', t' quark masses. As shown in Ref. (57). If one picks suitable values of the quark-mixing matrix and masses of the new quarks, e.g. D^0, $\bar{D}^0$ mixing can be increased from $\ll 10^{-4}$ to 10^{-2} if $m_{b'} \gtrsim 200$ GeV; $B_d^0, \bar{B}_d^0$ can also be appreciable; the mass-matrix CP violation in D, B mesons can also be large so that $(``\ell^+\ell^+" - ``\ell^-\ell^-")/(``\ell^+\ell^+" + ``\ell^-\ell^-")$, $(KK-\bar{K}\bar{K})/(KK+\bar{K}\bar{K})$ are reasonably large. Such possibilities are extremely interesting to search for experimentally.

VII. Quark Mass Matrix

So far in the electroweak unified gauge theory, the most obscure part is the mass generating mechanism for gauge particles and for quarks. Here I would like to reiterate, the intimate relations between the quark mass matrix M' in the bases of quark eigenstate in weak interaction and the mixing matrix V. The mass term in the interaction Lagrangian can be written in the strong-interaction eign quark state u, d, or equivalently the weak-interaction eigenstate u', d',

$$\text{mass} = \bar{u}_L M(2/3)\, u_R + \bar{d}_L\, M(-1/3)\, d_R + \text{H.C.},$$

$$= \bar{u}_L M(2/3)\, u_R + \bar{d}'_L\, M'(-1/3)\, d'_R + \text{H.C.}, \tag{7.1}$$

where $V_L M'(-1/3)\, V_R^+ = M(-1/3)$, and $V_R^+ V_R = 1 = V_L^+ V_L$.

We can show that M'(-1/3) can always either choose to be Hermitean or symmetric:[58] From

$$M'(-1/3) = V_L^+\, M(-1/3)\, V_R, \tag{7.2}$$

we multiply a unitary matrix W on the right hand side of Eq. (7.2),

$$M'' \equiv M'\, W = V_L^{\;+} M\, V_R W. \tag{7.3}$$

If we require M'' to be Hermitean $M''^+ = M''$ i.e. $W^+V_R^+MV_L = V_L^{\;+}MV_RW$ or $V_LW^+V_R^+M = MV_R\, WV_L^+$. A sufficient solution for W is $V_RWV_L^+ = 1$, i.e. $W = V_R^+V_L$, and $M'' = V_L^+ MV_L$; or if we want M'' to be symmetric $M''^T = M''$, i.e. $W^TV_R^TMV_L^* = V_L^+\, MV_RW$. A sufficient solution for W is $V_RWV_L^T = 1$, i.e. $W = V_R^+V_L^*$, and $M'' = V_L^+M\, V_L^* = V_L^+MV_L^{+T}$. All this says that we can choose a base such that M'(-1/3) is either Hermitean or symmetric. Now the charged weak current is

$$J_\mu = \bar{u}_L\, \Gamma_\mu\, d'_L = \bar{u}_L\, \Gamma_\mu\, V\, d_L. \tag{7.4}$$

This V is the quark mixing matrix. So in the d'_L base the charge - (-1/3) mass-matrix is

$$M'_H\,(-1/3) = V^+\, M(-1/3)\, V, \text{ if in the Hermitial base} \tag{7.5}$$

$$M'_S\,(-1/3) = V^T M(-1/3)\, V, \text{ if in the symmetric base;} \tag{7.6}$$

using the V matrix given in Eq. (1.3), we obtain M'_H (-1/3) and $M'_S(-1/3)$. Picking the hermitian case, the mass matrix becomes:

$$M_H \equiv V\, M^D\, (\text{diag.})\, V^+ =$$

$$= \begin{bmatrix} m_d + m_s s_x^2 & m_s s_x + m_b s_y s_z e^{-i\phi_1} & m_b s_z e^{-i\phi_1} - m_s s_x s_y \\ m_s s_x + m_b s_y s_z e^{i\phi_1} & m_s c_x^2 + m_b s_y^2 & (m_b - m_s) s_y - m_s s_x s_z e^{-i\phi_1} \\ m_b s_z e^{i\phi_1} - m_s s_x s_y & (m_b - m_s) s_y - m_s s_x s_z e^{i\phi_1} & m_b c_y^2 \end{bmatrix} \qquad (7.7a)$$

To get (7.7) we have used the smallness of s_y, s_z and $m_d/m_b \lesssim \lambda^4$, $m_s/m_b \lesssim \lambda^2$. Similarly we can get the symmetric mass matrix VM^D (diag.) V. Equation (7.7) is accurate up to order $m_b\lambda^4$. These are the mass matrices the weak decay phenomenology has told us. They should be conformed by model building. We also known the down-quark mass matrix in its symmetric form:

$$M_S = VM^D V^T. \qquad (7.7b)$$

We can see that the quark mixing matrix and its complex phase, the source of CP violation, is intimately related to the quark mass matrix in the base of weak-interaction quarks. Thus the mass-generating sector, i.e. the Higgs sector, is truly now the reservoir of our ignorance in the gauge theory description of the particle world. It is of great importance to make progress on it.

VIII. Concluding Remarks and Outlook

As we can see there are many interesting experiments that can be done to shed light on our understanding of the physics of heavy-quark decays and CP violation. Here I list them briefly.

a. The Measurements of the b Particle Lifetime and $\Gamma(b\to u)/\Gamma(b\to c)$ have well constrained the quark mixing matrix. It is important to further study the quark mixing matrix elements: V_{ub} via $B\to\tau\nu_\tau$, and the charmless b decays; V_{cs}, V_{cd} via $D\to\ell\, X_{s,d}$. The fitting of the CP-violation effects in K decays, ε and ε'/ε, has cornered the model with three generations of quarks, and has constrained the nonleptonic dynamical parameter B_K, B_k', and the t quark mass. If V_{ub} becomes smaller than needed for CP violating effects, it's an indication of trouble with the KM scheme for CP violation, or of the exciting possibility of the existence of higher than three generations of quarks.

b. There are still many important and interesting CP violating effects to be measured in the kaon and hyperon decays: ε'/ε, ϕ_{00}; η_{+-0}; $Br(K_S^0\to\gamma\gamma)$; R_+/R_-, where $R_\pm = \Gamma(K^\pm\to\pi^\pm\pi^\pm\pi^\mp)/\Gamma(K^\pm\to\pi^\pm\pi^0\pi^0)$; $R_\Lambda/R_{\bar\Lambda}$, where $R_\Lambda \equiv \Gamma(\Lambda\to\pi\, p)/\Gamma(\Lambda\to\pi\, n)$, $R_{\bar\Lambda} = \Gamma(\bar\Lambda\to\pi^+\bar p)/\Gamma(\bar\Lambda\to\pi^0\bar n)$; Pion momentum asymmetry difference in $\Sigma^\pm\to\pi^\pm n$ decays. The tagged $K^0,\bar K^0$ experiments for partial rate difference $\Delta(K^0,\bar K^0\to 2\pi)$, $\Delta(K^0,\bar K^0\to\gamma\gamma)$ are very interesting and will mark the beginning of this new type of CP experiments.

c. Some partial decay rate differences in beauty particle decays can be very large, such that only some tens of restructured beauty decays are needed:

$$\Delta(B_u^\pm\to D^\pm \overset{(-)}{D}{}^{*0}),\ \Delta(\overset{(-)}{B}{}_d^0\to K^\mp\pi^\pm),\ \Delta(\overset{(-)}{B}{}_s^0\to D^+D^-),\ \Delta(\overset{(-)}{B}{}_d^0\,\pi^+\pi^-),$$

$\Delta(B_c^\pm\to\pi^\pm \overset{(-)}{D}{}^0)$, etc. See Table in Section IIIb, and Refs. (35, 36).

In the three-generation case, such partial decay rate differences for charm and top quarks are in general an order of magnitude smaller. This information can serve as a base for looking for hints of the existence of higher generations of quarks.

d. Look for the B_S^0, $\bar{B}_S^0$ mixing effects, e.g. same-sign dileptons, which has been consistently predicted to be substantial in the case of three generations of quarks while mixing in B_d^0, $\bar{B}_d^0$ is small and sensitive to parameters, and mixing in $D^0,\bar{D}^0$ and $T^0,\bar{T}^0$ are extremely small.

e. Experiments tagging neutral mesons can provide new ways to study CP-violation effects that the conventional interference experiments can not do, e.g. measuring decay-amplitude CP violation effects in $K_L \to \gamma\gamma$; by measuring $\Delta(K^0,\bar{K}^0 \to \gamma\gamma)$; separately measuring the neutral meson $P^0,\bar{P}^0$ (mass-matrix) mixing parameters δm, $\delta\Gamma$, and the mass-matrix CP violation $\eta_p \equiv 1-\varepsilon_p / 1-\varepsilon_p$ for short lived heavy quark systems.

f. Experiments should be planned to look for indications of higher then three generations of quark: the most important are the direct observations of the top particles, and the heavier ones; since the phenomenological consequences in the case of three generations of quarks have been quite well determined, any observation of substantial deviations from them are possible indications of the existence of higher generations, e.g. large partial decay rates in charm and top quark decays, in addition to the beauty particle decays which are the only ones calculated to be substantial in the three-generation case; large D^0, $\bar{D}^0$, B_d^0, $\bar{B}_d^0$ mixing in addition to B_s^0, $\bar{B}_s^0$ which is the only one calculated to be substantial in the three-generation case; large mass-matrix CP-violation in B^0, D^0 decays, e.g. $\Delta(\ell^+\ell^+ - \ell^-\ell^-)=0$; "long" lived top quark ($\tau_\tau \gtrsim 10^{-18}$ sec).

g. The well determination of the quark mixing matrix also helps for a systematic study on nonleptonic decays via the many channels available in charm and beauty particle decays. Recent charm decay data analyzed via the quark-diagram scheme already have shed some light on the structure of charm decays. The quark-diagram amplitudes for charm meson $\to$ pseudoscalar-vector-meson decays are quite well determined in magnitudes and relative signs. This leads to many predictions, e.g. $Br(D^0 \to \phi\pi^0) \cong 0.21\%$; future measurements of $D^0 \to \bar{K}^{*0}K^0$, $K^{*0}\bar{K}^0$ can shed light on amplitude c'-c; $F^+ \to \rho^+\pi^0$, $\rho^0\pi^+$ can give information on amplitude d-d'; and $F^+ \to \omega\pi^+$ can give information on d+d'; and many more. For charm meson $\to$ pseudoscalar-pseudoscalar meson decays, measurements of the following few charm decays will be most helpful: $(D^+ \to \pi^+\pi^0)$, so that the long predicted relation $\Gamma(D^+ \to \pi^+\pi^0)/\Gamma(D^+ \to \bar{K}^0\pi^+) = 1/2\ V_{cd}/V_{cs}\ ^2$ can be checked. Measurements on $D^0 \to \bar{K}^0\eta'$ in addition to the recently measured $D^0 \to \bar{K}^0\eta$ can test the importance of W-exchange amplitude c; the measurement of $D^0 \to \pi^0\pi^0$ can help us to understand the details of the mechanism for the ratio $\Gamma(D^0 \to K^+K^-)/\Gamma(D^0 \to \pi^+\pi^-) \neq 1$; the observation of $D^0 \to K^0K^0$, $\eta_0\eta_0$ can give a clear indication and measurements of SU(3) breaking.

h. Measurement of F, $D \to \tau\upsilon_\tau$, $\mu\upsilon_\mu$ can help to determine g_τ/g_μ and the υ_τ mass.[77]

i. Measurements of rare decays can put standard models to stringent tests, and provide windows for a glimpse of possible new physics. The case of $K_L \to \mu^+\mu^-$ ($Br = 9.1 \times 10^{-9}$) was such an example. Now there are four rare decay experiments being carried out at Brookhaven National Laboratory:[60]

Exp. 777, BNL - U. of Washington-Yale Collaboration,

$K^+ \to \pi^+e^-\mu^+$, sensitivity $Br \sim 10^{-11}$,

Exp. 780, BNL-Yale Collaboration,

$K_L \to \mu e$, sensitivity $Br \gtrsim 10^{-10}$,

Exp. 787, BNL-Carnegie-Mellon-Columbia-Princeton-TRIUMF Collaboration,

$K^+ \to \pi^+ \nu\bar{\nu}$, π^+ "?", sensitivity Br $\sim 10^{-10}$,

Exp. 791, UCLA-Los ALamos-U. of Pennsylvania-Princeton-Stanford-Temple Collaboration,

$K_L \to \mu e$, Sensitivity 10^{-12},

$K_L \to \pi^0 e^+ e^-$, Sensitivity 10^{-12},

μ polarization $K_L \to \mu^+ \mu^-$, sensitivity 10 $\sim$ 20%.

Further tightening of the already very impressive bound on the neutron electric dipole moment will help to test various CP-violating models.

We can see that despite the milestone of the observation of the long awaited intermediate bosons, and the many beautiful recent experimental results on charm and beauty particle decays, many fundamental questions in electroweak interactions still remain to be answered: how are the masses of gauge bosons and quarks generated? What is the reason for the V-A nature of the weak current? Where is the origin of CP noninvariance? Why do the quarks and leptons mix and repeat themselves? How many more are there? The apparent complexity of my discussions presented here reflect the immaturity of the field. However, this is the very nature of our pursuit: once a good question is answered, more new meaningful questions can be asked. There will be frontier research as long as there is life itself. I look forward to many new results from accelerators and as well as those non-accelerator experiments discussed at this conference.

Acknowledgement: I would like to thank Professor H.V. Klapdor for inviting me to give this talk, and to thank him and other organizers, especially Dr. J. Metzinger for this splendid conference. I would also like to thank the Rector of Heidelberg University, Dr. J. zu Putlitz, for his warm hospitality.

The advances in the subject discussed here have been made by many theorists and experimentalists, many of them are present here in the audience. Their contributions are cited in the references. My contributions to the subject have been made in recent years in collaboration with H.-Y. Cheng (Indiana University, Bloomington), W.-Y. Keung (University of Illinois, Chicago), and also recently with F. Botella, a Fulbright Fellow from Valencia, Spain.

REFERENCES

1. For recent development on the subject, see talks in Proceedings of "Flavor Mixing in Weak Interactions," Europhysics Conference, March 5-10, 1984, Erice, Italy, Ed. L.-L. Chau, Plenum 1985.
2. Proceedings of the International Symposium on Production and Decay of Heavy Flavours, May 20-23, 1986, Heidelberg.
3. For a general survey, see L.L. Chau, Phys. Rept. 95 (1983) 1, and L.L. Chau, Physics of Heavy Quark Decays and CP Violation, Proceedings of the Kyoto International Symposium, the Jubilee of the Meson Theory, Aug. 15-27, 1985, Kyoto, Japan, Prog. The Phys. Sup. No. 85, 1985.
4. L.L. Chau, "Comments on Heavy Quark Decays and CP Violation," Proceedings of the G.F. Chew Jubilee (Sept. 29, 1984), to be published by the World Scientific Pub. Co.
5. For earlier fits to find V_{ud}, V_{us}, R. Shrock, and L.L. (Chau) Wang, Phys. Rev. Lett. 41 (1978) 1692; for more recent fits, see the next reference. J.F. Donoghue and B.R. Holstein, Phys. Rev. D25 (1982) 2015; A Garcia and P. Kielanowski, Phys. Lett. 110B (1982) 498; and the most recent fits, WA2 experiment at CERN, M. Bourquin et al., "IV. Tests of the Cabibbo Model," CERN preprint (1983). See talk by H.W. Siebert in Ref. 22; J.F. Donoghue, B.R. Holstein, Phys. Lett. 160B (1985) 173; W.J. Marciano and A. Sirlin, Phys. Rev. Lett. 56 (1986) 22, found $V_{ud} = 0.9729 \pm 0.0012$; and A Bohm, M. Kmiecik, Phys. Rev. D31, (1985) 3005, include new Fermilab hyperon data, Ref. 30, and found $V_{us} = 0.225 \pm 0.002$ to be in better agreement with the older

Ke3 fit of Ref. 25.
6. For the b lifetime, E. Fernandez, Phys. Rev. Lett. 51 (1984) 1022; N.S. Lockyer et al., Phys. Rev. Lett. 51 (1983) 1316; for more up to date information, see talks by W.T. Ford, and G.H. Trilling in Ref. 22. See P. Ginsparg, S. Glashow, M. Wise, Phys. Rev. Lett. 50 (1983), 1415.
7. For $\Gamma(b \to u)/\Gamma(b \to c)$, C. Klopfenstein et al., Phys. Lett. 103B, (1983) 444; A. Chen et al., Phys. Rev. Lett. 111B, (1984) 1084; and talks by J. Lee-Franzini, and P. Avery in Ref. 22.
8. L.-L. Chau, W.-Y. Keung, and M.D. Tran, Phys. Rev. D27, (1983) 2145, L.-L. Chau and W.-Y. Keung, Phys. Rev. D29, (1984) 592.
9. Fermilab Polarized Σ^- experiment, S.Y. Hsueh et al., Phys. Rev. Lett. 54, (1985), 2399.
10. H. Leutwyler, M. Roos, Z. Phys. C 25, 91 (1984).
11. See H.W. Siebert's talk in Ref. (22).
12. L. Wolfenstein, Phys. Rev. Lett. 51, (1984) 1945.
13. L.-L. Chau and W.-Y. Keung, Phys. Rev. Lett. 53, (1984) 1802.
14. L. Maiani, Phys. Lett. 62B (1976) 183; R. Mignami Lett. Al Nuovo Cimento, 28 (1980), 529.
15. C. Jarlskog, Phys. Rev. Lett. 55, 1039 (1985), O.W. Greenberg, Phys. Rev., D32, 1841 (1985); D.D. Wu, Phys. Rev., D33, 860 (1986).
16. F.J. Botella and L.-L. Chau, "Anticipating the Higher Generations of Quarks from Rephasing Invariance of the Mixing Matrix," Brookhaven preprint (1985).
17. H. Fritzsch, Phys. Rev. D32 (1985), 3058.
18. M.K. Gaillard and B.W. Lee, Phys. Rev. D10 (1974) 897.
19. See L.-L. Chau, H.-Y. Cheng and W.-Y. Keung, "A KM Model Study for ε, ε/ε, and M_t", BNL preprint BNL-35163, July '84 (unpublished); and Phys. Rev. 32, (1985) 1837; the uncertainties in the B_K parameter and the numerical findings of these papers were presented by L.-L. Chau in a talk immediately following B. Winstein at the APS Meetings at Washington, DC, 23-26, April '84. J. Bijnens, H. Sonoda and M.B. Wise, Phys. Rev. Lett. 53, (1984) 2367. A. Pich and E. de Rafael, Phys. Lett. 158B, (1985) 477.
20. J.H. Christenson, J.W. Cronin, V.L. Fitch and R. Turlay, Phys. Rev. Lett. 13, (1964) 138; and the classical theoretical discussions on CP violation. T.D. Lee, R. Oehme and C.N. Yang, Phys. Rev. 106, (1957) 340; T.T. Wu and C.N. Yang, Phys. Lett. 13, (1964) 380.
21. L.-L. Chau, "Comments on CP Violation" Brookhaven preprint, (1984).
22. J.K. Black et al., Phys. Rev. Lett. 54, (1985) 1628.
23. R.H. Bernstein et al. ibid, 54, (1985) 1631.
24. F.J. Gilman and M.B. Wise, Phys. Lett. 83B, (1979) 83.
25. L.-L. Chau, H.-Y. Cheng, and W.-Y. Keung, "CP Violation in the Kaon Systems", Brookhaven preprint 1985.
26. The current experiments on ε'/ε: Chicago, -Fermilab, -Orsay-Princeton collaboration, Fermilab experiment #731; CERN-Dortmund-Edinburg-Orsay-Pisa-Siegen collaborations, CERN NA31.
27. For discussion partial rate differences in K decays see Sect. 3.2.2 of Ref. (6); L.-L. Chau, talk at Erice Conference on "Electroweak Effects at High Energies," Feb. 1-12 '83; C. Kounnas, A.B. Lahanas and P. Pavlopoulos, Phys. Lett. 127B, 381 (1983).
28. "Physics at LEAR with Low-Energy Cooled Antiprotons," edited by U. Gastaldi and R. Klapisch, Ettore Majorana Int. Sci. Series, ed. A. Zichichi (Plenum); P. Pavlopoulos, Talk in Ref. (22); tagged K^0 experiment: Athen-Basel-ETH-Zurich-Liverpool-Saclay-Sin collaboration, CEAR, LEAR, p. 82.
29. L.-L. Chau, H.-Y. Cheng, Phys. Rev. Lett. 54, (1985) 176B.
30. L.-L. Chau Wang, AIP Conf. Proc. No. 72, Particle and Fields, Subseries No. 23, Virginia Polytechnic Inst. 1980, eds. G.B. Collins, L.N. Chang and J.R. Ficenec; L.-L. Chau, Proceedings of Workshop on Weak Interactions and Neutrinos, Javea, Spain, Sept. 5-11, 1983.
31. C. Avilez, Phys. Rev. D23, (1981) 1124. See also B. Grinstein, S.-J. Rey and M.B. Wise, CALT-68-1286 (1985).
32. L.-L. Chau and H.Y. Cheng, Phys. Lett. 131B, (1983) 202; T. Brown, S.F. Tuan and S. Pakvasa, Phys. Rev. Lett. 51, (1983) 1823. I would like to thank Dr.

K. Kilian for illuminating discussion on the possibility of doing this experiment at LEAR.

33. For a review of CP violation in the K system see K. Kleinknecht, Ann. Rev. Nucl. Sci. 26, 26 (1976). R.G. Sachs, Ann. Phys. 22 (1963), 239. T.D. Lee, and C.S. Wu, Ann. Rev. Nucl. Sci. 15, 381 (1966); "Theory of Weak Interactions in Particle Physics", R.E. Marshak, Riazuddin, C.P. Ryan, Publisher Wiley-Interscience (1969).
34. L.M. Sehgal, and L. Wolfenstein, Phys. Rev. 162 (1976), 1362, O.E. Overseth and S. Pakvasa, Phys. Rev. 189 (1969) 1663. A. Pais and S.B. Treiman, Phys. Rev. D12, 2744 (1975); L.B. Okun, V.I. Zakharov and B.M. Pontecorvo. Lett. Nuovo Cim. 13, 218 (1975). J. Barshay and J. Geris, Phys. Lett. 84B, (1979), 319; M. Bander, D. Silverman and A. Soni, Phys. Rev. Lett. 43, (1979) 242; J. Barnabeu and C. Jarlskog, Z. Phys. C8, (1981) 233. I.I. Bigi and A.I. Sanda, Nucl. Phys. B193, (1981) 81. L. Wofenstein, Nucl. Phys. B200 (1984), 45 and talk by K.-C. Chou in Ref. 22.
35. L.-L. Chau, H.-H. Cheng, Phys. Rev. Lett. 53, (1984) 1037.
36. L.-L. Chau, H.-Y. Cheng, "CP Violation in Decay Rates of Neutral B Mesons", Brookhaven preprint, BNL 36915 (1985), Phys. Lett. 165B (1986).
37. For left-right symmetric theory, see talk by R.N. Mohapatra in Ref. (22). J.C. Pati and A. Salam, Phys. Rev. D10, (1974) 275; R.N. Mohapatra and J.C. Pati, Phys. Rev. D11, (1975) 566, 2558.
38. See e.g. S. Weinberg, Phys. Rev. Lett. 37, (1976) 657, Physica 96A, (1979) 327.
39. A.I. Sanda, Phys. Rev. D23 (1981) 2647; N.G. Deshpande, ibid D23 (1981) 2654.
40. Y. Dupont and T.N. Pham, Phys. Rev. D28, (1983) 2169. J.F. Donoghue and B.R. Holstein, UMHEP-213 (1984). H.-Y. Cheng, "Weinberg CP Violation Model Revisited", Brandeis preprint (1984).
41. A.A. Anselm, V.E. Bunakov, V.P. Gudkov, and N.G. Uraltsev, Leningrad preprint (1984).
42. N.F. Ramsey, Rep. Prog. Phys. 45, (1982) 95; J.M. Pendlebury et al., Phys. Lett. 136B, (1984) 327; I.S. Alterev et al., ibid 102B, (1981) 13; and B. Heckel's talk in Ref. (22).
43. Observation of $K^+ \to \pi^+ \pi^0$ suppression R.W. Birge et al., Nuovo cimento 4 (1956) 834; G. Alexander et al., Nuovo cimento 6 (1957) 478, for theoretical discussions, see M. Gell-Mann and A. Pais, Proc. Int'L Conf. of High Energy Physics, Glasgow (Pergamon Press, London 1955); Proc. Fifth Annual Rochester Conf. on High Energy Physics (NY 1955) p. 136.
44. M.A. Shifman, A.I. Vainshtein, V.I. Zakharov and Nucl. Phys. B120, 316 (1977); Sov. Phys. JETP 45 (1977) 670.
45. L.-L. Chau, talk in the Proceedings of the 1980 Quangzhou Conference (Jan. 5-10, 1980), Science Press, Beijing, Reinhold Comp., China; Van Nostr. and Proc. of VI Int'l. Conf. on Meson Spectroscopy, BNL, April 24-25, 1980; AIP.
46. Here I would like to emphasize that the graphs given here include all strong interaction gluon clouds. They are not Feynman graphs. For example the graph recently given by J. Donoghue in Phys. Rev. (figure on the left below) actually is an exchange diagram, which can be easily seen if we redraw as on the right below.

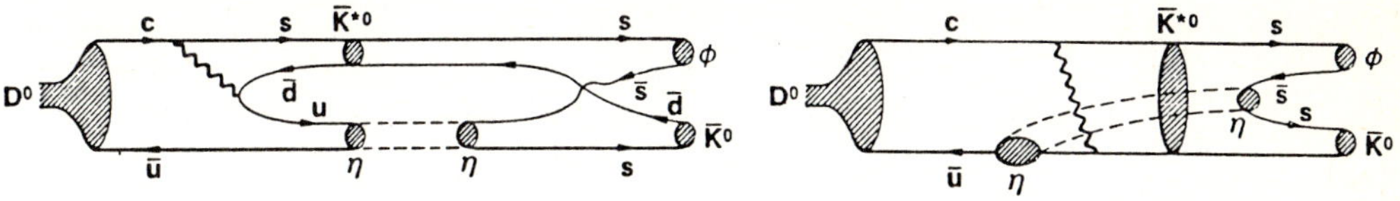

47. For charm → PP, see Table 2.5, Ref. (2), and footnote (2) of Ref. (3); For charm → PV see Table 2.6, Ref. (2) and M. Gorn, Nucl. Phys. B191, (1981). X.Y.-Y. Li, S.-F. Tuan (to be submitted to Zeits für Physik C). For treatment of SUB(3) breaking and final-state-interaction, see next reference.
48. For charm → PP, PV including SU(3) breaking and final-state-interaction effects, see L.-L. Chau and H.-Y. Cheng, "Exclusive Two Body Decays of Charm

Meson", Brookhaven-Brandes preprint 1985; and L.-L. Chau and H.-Y. Cheng, "Quark Diagram Analysis of Charm Meson Decays", Brookhaven-Indiana preprint 1985.
49. MARK III collaboration, R. Baltrusaitis, Phys. Rev. Lett. 55, (1985) 150. R.H. Schindler, "New Results on Charmed D Meson Decay", invited talk at the 1985 SLAC Summer Institute for Particle Physics.
50. CLEO collaboration, A. Chen et al., Phys. Rev. Lett. 51, (1983) 634; see also talk by S. Stone, Int. Sym. or Lepton and Photon Interaction at High Energies, Cornell July 83; and paper contributed to the 1985 Int'l. Lepton and photon conference at Kyoto, Japan, Aug. '85.
51. ARGUS collaboration C. Darden et al., DESY preprint 84-04-3, H. Albrecht et. al., Phys. Lett, 158B (1985) 525.
52. TASSO collaboration, M. Althoff et al., Phys. Lett. 136B, (1984) 130.
53. HRS collaboration, M. Derrick et al., Phys. Rev. Lett. 54, (1985) 2568.
54. For a review on charm lifetime see talk by K. Niu in Ref. (22). I would lik to thank Professor N. Reay for very informative discussions on the subject.
55. R. Schindler et al., Phys. Rev. D24, (1981) 78. Talk by G. Goldhaber at 18th Moriond Conference, March 13-19, 1983.
56. For current comparisons with various model calculations, see discussions in Ref. (66); and for other ways of analyzing the data and comparison with calculations. See the rather exhaustive references given there in, and talk by R. Rückel in Ref. (22), and talk by B. Stech at Moriond Conference, Spring 1985.
57. X.-G. He and S. Pakvasa Phys. Lett. 156B, (1985) 236. A.A. Anselm et al., Phys. Lett. 156B, (1985) 102. M. Gronau and J. Schechter Phys. Rev. D31, (1985) 1668. U. Türke et. al., DO-TH 84/26. I.I. Bigi PITHA 84/19.
58. P.H. Frampton and C. Jarlslog Phys. Lett. 154B (1985) 421, R.N. Mohapatra, Private communication.
59. Chao Kuang-ta, Chau Ling-Lie, Huang Tao, Du Dong-Sheng, Wu Dan-di, Chinese Phys. Lett. 2, (1985) 117.
60. See Talk by L. Littenberg in Ref. (22).
61. L.-L. Chau Wang, and F. Wilczek, Phys. Rev. Lett. 43 (1979), 816.

Experimental Determination of the Kobayashi-Maskawa Matrix Elements

K.R. Schubert

Institut für Hochenergiephysik der Universität Heidelberg,
Schröderstraße 90, D-6900 Heidelberg, F. R. Germany

In the standard elektroweak theory, the charged current is given by

$$\mathbf{J}_\lambda = (\,\overline{\nu}_e \quad \overline{\nu}_\mu \quad \overline{\nu}_\tau\,)\,\gamma_\lambda(1-\gamma_5)\,\mathbf{1}\begin{pmatrix} e \\ \mu \\ \tau \end{pmatrix} + (\,\overline{u} \quad \overline{c} \quad \overline{t}\,)\,\gamma_\lambda(1-\gamma_5)\,\mathbf{V}\begin{pmatrix} d \\ s \\ b \end{pmatrix} + h.c.$$

and describes the couplings of the charged vector bosons $W^\pm$ to the lefthanded leptons and quarks. The unit matrix **1** in the lepton part reflects our present knowledge that emission of a W^- by an electron always leads to an electron neutrino, never to a μ or τ neutrino, and that muons and τ-leptons obey the corresponding rule. In the quark sector the rules are different, emission of a W^- by a d quark may may result in any of the three charge 2/3 quarks u, c, or t, and the same holds for the s and the b quark. The coupling strengths are however strictly connected to each other, the standard theory requires **V** to be a unitary matrix.

The nine coupling strengths $g(q_i W q_j)$ between charge -1/3 quarks q_j and charge +2/3 quarks q_i are products of the universal weak coupling constant g and the unitary matrix elements V_{ij}:

$$g(q_i W q_j) = g \begin{pmatrix} V_{ud} & V_{us} & V_{ub} \\ V_{cd} & V_{cs} & V_{cb} \\ V_{td} & V_{ts} & V_{tb} \end{pmatrix}$$

with $g = g(\nu_e W e)$ and $G_F = g^2/M_W^2$. The matrix **V** for six quarks was introduced by Kobayashi and Maskawa [1] in 1973 generalizing the quark mixing ideas of Cabibbo (with u, d, and s) and Glashow - Iliopoulos - Maiani (with u, d, s, and c), realizing that a world with only four quarks cannot have CP violation within the standard model.

After the discovery of the b quark in 1977 [2] six of the nine Kobayashi - Maskawa (KM) matrix elements are experimentally accessible. This report describes the methods and results of the experiments and demonstrates that unitarity of the KM matrix is well fulfilled. Phase conventions and parametrizations of the matrix elements will be discussed in the last chapter. I will use a phase convention in which V_{ud}, V_{us}, V_{cb}, and V_{tb} are real and positive.

Determination of V_{ud}

The weak coupling strength between u and d quarks is determined by "superallowed" nuclear β decays, and V_{ud} is obtained by comparing this strength with muon decay. Superallowed β decays are $0^+ \to 0^+$ transitions within the same isospin triplet. The elementary process is $u \to d\,W^+$, $W^+ \to e^+\,\nu_e$. One u quark inside the nucleus transforms into a d quark and leaves the nuclear structure essentially unchanged. In the V-A theory, $0^+ \rightarrow 0^+$ transitions are pure vector, and the conserved vector current principle (CVC) guarantees that the weak coupling strength is not renormalized by the strong interaction. The β half life t is given by [3]:

$$f \cdot t \cdot (1 - \delta_c)(1 + \delta_R)(1 + \Delta_\beta) = \frac{2\pi^3 \cdot \ln 2 \cdot \hbar^7}{m_e^5 \cdot c^4 \cdot 2 \cdot G_V^2},$$

where f is the phase space factor including influences of the nuclear charge distribution on the Dirac wave function of the positron and the recoil of the nucleus, t is not the observed half life but excludes electron capture, δ_c is the nuclear structure correction of 0.3% to 0.8% taking into account isospin symmetry breaking effects like protons being slightly less bound in the nucleus than neutrons, δ_R is the outer radiative correction of 1.3% to 2.2% depending on the positron energy and on Z of the nucleus, Δ_β is the inner radiative correction which modifies the elementary process $u \rightarrow dW^+$, $W^+ \rightarrow e^+\nu_e$ through the presence of photons and Z^0 bosons, and G_V is the Fermi constant G_F multiplied by V_{ud}. The mean life τ_μ of muon decay is given by:

$$\frac{\hbar}{\tau_\mu} = \frac{G_F^2(m_\mu c^2)^5}{192\pi^3(\hbar c)^6}[1 - 8\frac{m_e^2}{m_\mu^2}][1 + \frac{\alpha}{2\pi}(\frac{25}{4} - \pi^2)][1 + \Delta_\mu],$$

where the first bracket is the recoil correction, the second bracket is the outer ("QED") radiative correction, and the last bracket is the inner ("elektroweak") radiative correction. The world avarage [4] for τ_μ is:

$$\tau_\mu = (2.197\ 033 \pm 0.000\ 039)\mu s,$$

giving a Fermi coupling constant of

$$G_F\sqrt{1 + \Delta_\mu} = (1.166\ 347 \pm 0.000\ 013) \cdot 10^{-5}\ GeV^{-2}(\hbar c)^3.$$

The ft values of the eight most precisely known superallowed β^+ transitions are shown in fig. 1a. They are not constant because of small nuclear structure effects. Correcting for these effects [3] leads to a rather constant value of $ft(1 - \delta_c)$ as shown in fig. 1b. Because of the energy and Z dependent QED correction δ_R, only fig. 1c should show a constant result. This is unfortunately not the case. An unbiased average of all eight values gives [3]:

$$\mathcal{F}t = ft(1 - \delta_c)(1 + \delta_R) = (3080.1 \pm 2.4)\ s,$$

where the error has been increased to cover the not-understood step. This leads to:

$$G_V\sqrt{1 + \Delta_\beta} = (1.148\ 09 \pm 0.000\ 45) \cdot 10^{-5} GeV^{-2}(\hbar c)^3.$$

The inner radiative corrections Δ_β and Δ_μ have recently been evaluated by Marciano and Sirlin [5] using the standard electroweak model with one Z^0 boson. They find $\Delta_\beta - \Delta_\mu = (2.3 \pm 0.2)\%$ which leads to:

$$V_{ud} = 0.9729 \pm 0.0010,$$

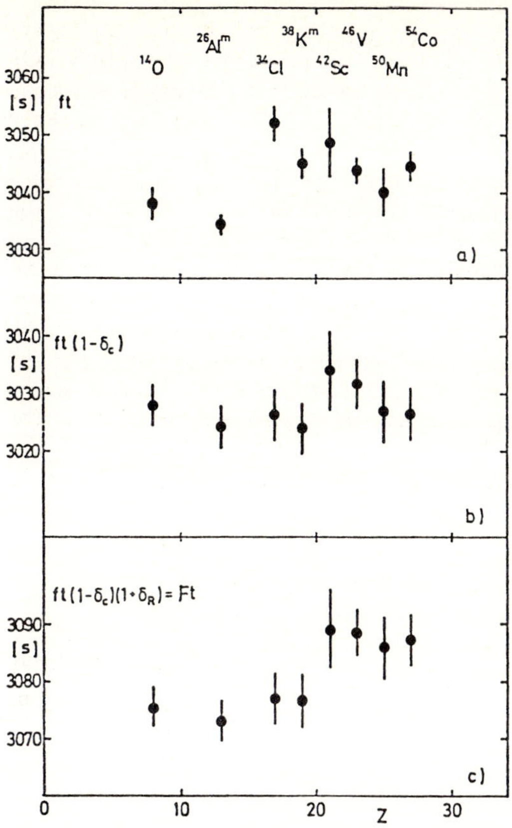

Fig.1: ft Values of eight superallowed β^+ decays as used in refs. 3 and 5 to determine V_{ud}. a) experimental values, corrected for electron capture, b) including nuclear structure corrections, c) including outer radiative corrections.

where the error is dominated by the uncertainty in the inner radiative corrections (± 0.0009). Another precise source for V_{ud}, not influenced by nuclear structur corrections and less influenced by outer radiative corrections, would be pion β - decay. At present, the best determination of its partial rate [6] has an error of 3.8% which is more than 10 times worse than the present correction uncertainties in the nuclear experiments.

Determination of V_{us}

The *us* coupling strength is obtained from hyperon β decays and K_{e3} decays. Hyperon decays are discussed by H.W.Siebert at this conference. The basic process is $s \to uW^-$, $W^- \to e^-\overline{\nu}_e$. It is observed in four hyperon decays $\Sigma^- \to ne\overline{\nu}$, $\Xi^- \to \Lambda e\overline{\nu}$, $\Xi^- \to \Sigma^0 e\overline{\nu}$, and $\Lambda \to pe\overline{\nu}$ by the CERN-WA2 group [7]. Applying radiative corrections and taking into account SU(3) breaking effects, the group finds $V_{us} = 0.229 \pm 0.003$. The same data have been reanalyzed by Leutwyler and Roos [8] with larger uncertainty in the SU(3) breaking, they give

$$V_{us}(hyperon) = 0.231 \pm 0.005 \; .$$

K_{e3} decays are the two K meson decays $K^+ \to \pi^0 e^+ \nu$ and $K^0_L \to \pi^- e^+ \nu$ where the basic process $\overline{s} \to \overline{u}W^+$, $W^+ \to e^+\nu_e$ takes place in a pure vector ($0^- \to 0^-$) transition. As in superallowed nuclear β decays, CVC allows a simple expression for the partial rates:

$$\Gamma_i(K \to \pi e \nu) = \frac{G_F^2 V_{us}^2}{192\pi^3} \cdot C_i^2 \cdot F \cdot |f_+(0)|^2(1+\delta),$$

where C_i^2 is a Clebsch-Gordon isospin coefficient, 1/2 for K^+, 1 for K^0, 1/2 for K_L^0, F is the phase space integral including the q^2 dependence of the $K \to \pi$ transition form factor $f_+(q^2)$, $f_+(0)$ is the value of this formfactor at $q^2 = 0$ which can be interpreted as the overlap integral between the quark wave functions of the K meson and the π meson, and δ is the radiative correction. Good data on these partial rates are old [4], $f_+(0)$ and radiative corrections have recently been evaluated by Leutwyler and Roos [8]. They find $f_+(0, K^0) = 0.961$, $f_+(0, K^+) = 0.982$, and

$$V_{us}(K_{e3}) = 0.2196 \pm 0.0023 \ .$$

The two V_{us} values are slightly incompatible. I average them following standard rules and obtain

$$V_{us} = 0.222 \pm 0.004 \ .$$

Determination of $|V_{cd}|$

The only information on the cd coupling comes from dimuon production in high energy neutrino reactions. The basic processes are $\nu_\mu \to \mu^- W^+$, $dW^+ \to c$ and its charge conjugate. All relevant graphs for muon and dimuon production by neutrinos and antineutrinos on an isoscalar target are shown in fig.2 together with their relative strengths within the obtainable precision. α is the fraction of $\bar{u}$ and $\bar{d}$ in the sea of the target, β the fraction of s and $\bar{s}$ in the sea. With the approximations $V_{ud}^2 \approx |V_{cs}|^2 \approx 1$ and $|V_{cd}|^2 \ll 1$, the total μ^- production by ν_μ is propotional to $(1 + \alpha/3 + \beta)$, the total μ^+ production by $\bar{\nu}_\mu$ proportional to $(1/3 + \alpha + \beta)$. Whenever a c (or $\bar{c}$) quark is produced, it has the probability B to decay into a μ^+ (or μ^-) through the process $c \to s\mu\nu_\mu$. Therefore one observes $\mu^+\mu^-$ production by ν_μ proportional to $B(\beta + |V_{cd}|^2)$ and by $\bar{\nu}_\mu$ proportional to $B(\beta + \alpha|V_{cd}|^2)$. Combining the four observations, one obtains:

$$\frac{\sigma(\nu \to \mu\mu) - \sigma(\bar{\nu} \to \mu\mu)}{\sigma(\nu \to \mu^-) - \sigma(\bar{\nu} \to \mu^+)} = \frac{3}{2} B |V_{cd}|^2 \ ,$$

and the CDHS group [9] at CERN finds $B|V_{cd}|^2 = (0.41 \pm 0.07)\%$. The muon decay branching fraction B are very different for the different charmed hadrons [4], $B(D^0 \to$

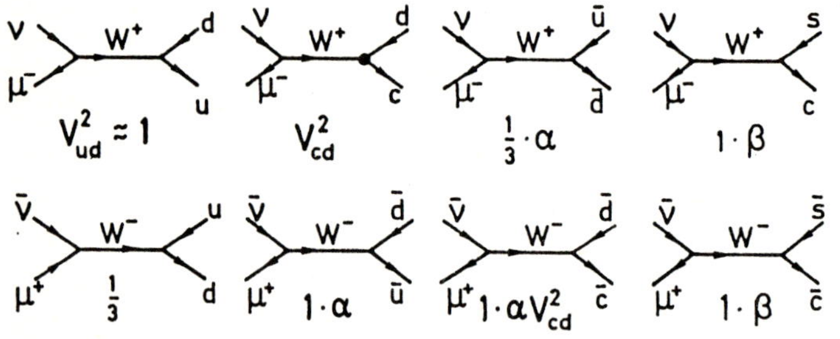

Fig.2: Graphs of muon and dimuon production by neutrinos and antineutrinos on an isoscalar target with their relative strength for the determination of $|V_{cd}|$

$\mu X) = (7.0 \pm 1.1)\%$, $B(D^+ \to \mu X) = (18.2 \pm 1.7)\%$, $B(\Lambda_C \to \mu X) = (4.5 \pm 1.7)\%$. $B(D_S \to \mu X)$ is unknown, but $(6 \pm 2)\%$ is a good estimate. Using these branching fractions and taking [10] the relative abundance of $D^\pm$, D^0, Λ_C and D_S production in high energy neutrino reactions from a recent emulsion experiment [11] to be $(0.32 \pm 0.04) : (0.48 \pm 0.05) : (0.13 \pm 0.03) : (0.07 \pm 0.02)$, the relevant average B is $(10.2 \pm 1.2)\%$. This results in:

$$|V_{cd}| = 0.200 \pm 0.021 \ .$$

Determination of $|V_{cs}|$

The cs coupling is taken from the rate of exclusive semileptonic decays of charmed mesons. The basic process in $D \to Ke\nu$ decays is $c \to sW^+$, $W^+ \to e^+\nu$. The partial rates $\Gamma(D^0_{e3})$ and $\Gamma(D^+_{e3})$ are given by:

$$\Gamma(D^{0,+}_{e3}) = \frac{1}{\tau(D^{0,+})} \cdot B(D^{0,+} \to e^+ \ \nu_e \ K^-, K^0) \ .$$

Values for the branching ratios B have only be obtained by the Mark III group [12], they produce pairs of D mesons by e^+e^- annihilation at the resonance energy ($3770\ MeV$) of the Ψ'' meson. Tagging D^+ decays with fully reconstructed D^- decays (into $K^+\pi^-\pi^-$) or D^0 with reconstructed $\overline{D}^0$ decays (into $K^+\pi^-$ or $K^+\pi^-\pi^+\pi^-$), they determine the fraction in which the tagged D^+ or D^0 decays into an electron and a K meson without extra visible particles. They find $B^0 = (4.1 \pm 0.6 \pm 0.5)\%$ and $B^+ = (10.2 \pm 3.3 \pm 1.0)\%$. The lifetimes $\tau^{+,0}$ of D^+ and D^0 mesons have been determined by a large number of experiments, the most precise one being E691 at Fermilab [13]. This experiment reconstructs production vertex, decay vertex and momentum of up to now 675 D^0 and 480 $D^\pm$ mesons using silicon microstrip detectors. A fit to the distribution of flight times between production and decay yields $\tau(D^0) = (4.4 \pm 0.3) \cdot 10^{-13}\ s$ and $\tau(D^+) = (10.9 \pm 1.0) \cdot 10^{-13}\ s$. These results are nearly as precise as the preent world averages [14] $\tau^0 = (4.34 \pm 0.24) \cdot 10^{-13}\ s$ and $\tau^+ = (10.1 \pm 0.7) \cdot 10^{-13}\ s$. Using the world averages and averaging B^0/τ^0 and B^+/τ^+, one obtains:

$$\Gamma(D \to Ke\nu) = (9.6 \pm 1.1) \cdot 10^{10}/s \ .$$

In complete analogy to K_{e3} decays used for the determination of V_{us}, theory expects:

$$\Gamma(D \to Ke\nu) = 15.4 \cdot 10^{10}/s \cdot |V_{cs}|^2 \cdot |f_+^D(0)|^2 \ ,$$

where, as before, $f_+^D(0)$ is the overlap between the quark wave functions of D and K mesons. Aliev et al [15] find $f_+^D(0) = 0.6 \pm 0.1$, whereas Wirbel et al [16] give $f_+^D(0) = 0.76$. Using $f_+^D(0) = 0.7 \pm 0.1$, I obtain:

$$|V_{cs}| = 0.94 \pm 0.08.$$

Determination of $|V_{ub}|$ and V_{cb}

Couplings of the b quark to c and u are obtained from the rate and the momentum spectrum shape of inclusive semileptonic decays of B mesons. The basic processes are $b \to cW^-$ and uW^-, $W^- \to e^-\nu$ and $\mu^-\nu$. Lifetimes and semileptonic branching fractions of the two lightest B mesons, $B^- = b\overline{u}$ and $\overline{B}^0 = b\overline{d}$ are not known separately, and no exclusive semileptonic decays have been observed so far. The only known values are $\overline{B}(B \rightarrow e\nu X)$, $\overline{B}(B \rightarrow \mu\nu X)$, and $\overline{\tau}(B)$ as averages of $B^{\pm}$ and B^0. The average of electrons and muons allows * to determine $\overline{\Gamma}(B \rightarrow l\nu X) = \overline{B}/\overline{\tau}$ which is expected to be

$$\overline{\Gamma}(B \rightarrow l\nu X) = \Gamma(b \rightarrow l\nu X) = \frac{G_F^2\, m_b^5}{192\, \pi^3} \left(|V_{ub}|^2 + 0.45\, V_{cb}^2 \right)$$

in the spectator model. $\overline{B}(B \rightarrow \mu\nu X)$ and $\overline{B}(B \rightarrow e\nu X)$ are measured by CLEO, CUSB, and ARGUS [17] in e^+e^- annihilation at the resonance energy (10577 MeV) of the $\Upsilon(4S)$ meson. The muon electron average of the three groups is

$$\overline{B}(B \rightarrow l\nu X) = (11.8 \pm 0.7)\%.$$

B meson lifetime measurements have not yet reached the precision in the charm sector. There are only two known candidates for B meson decays with reconstructed production and decay vertices [18] with an estimated mean life of $\overline{\tau} = (3^{+13}_{-2})\ 10^{-13}\ s$. There are, however, eight additional experiments, all with e^+e^- annihilation at CMS energies around 30 GeV using various indirect methods for the estimation of flight lengths between B meson production and decay. Ref.14 gives a recent review of these methods, fig.3 summarizes the results, and the average is

$$\overline{\tau}(B) = (10.8 \pm 2.0)\ 10^{-13}\ s.$$

Using $m_b = (5.0 \pm 0.2)\ GeV$, one obtains the result

$$|V_{ub}|^2 + 0.45\, V_{cb}^2 = (1.01 \pm 0.28)\ 10^{-3}$$

which is represented in fig.4.

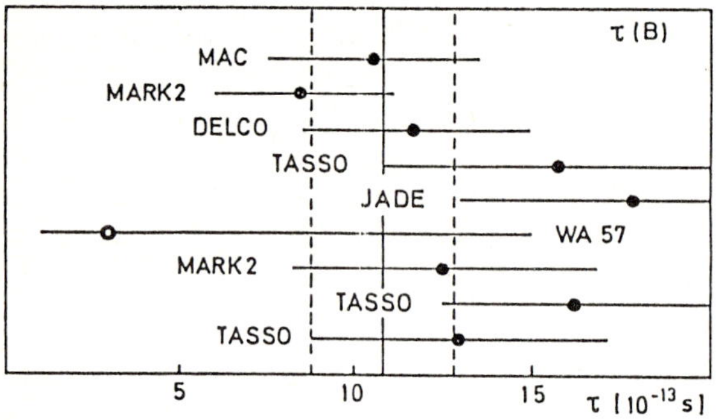

Fig.3: Present experimental results for the lifetime of B mesons. Error bars are combined statistical and systematic errors. The vertical band indicates the average of all experiments.

The decomposition of the observed semileptonic decays into their two components with charm ($\propto V_{cb}^2$) and without charm ($\propto |V_{ub}|^2$) has not yet been successful. No correlations between leptons and charmed mesons are observed so far; the only tool for studying the decomposition has been the shape of the lepton spectrum. To illustrate the method, fig.5 shows the electron momentum spectrum in the inclusive reaction

* If $B^+/\tau^+ = B^0/\tau^0$ as observed in the case of charm semileptonic decays and as expected in the spectator model, then $\overline{\Gamma} = \overline{B/\tau} = \overline{B}/\overline{\tau}$ independent of the two individual lifetime values. Bars denote the average of $B^{\pm}$ and B^0 mesons.

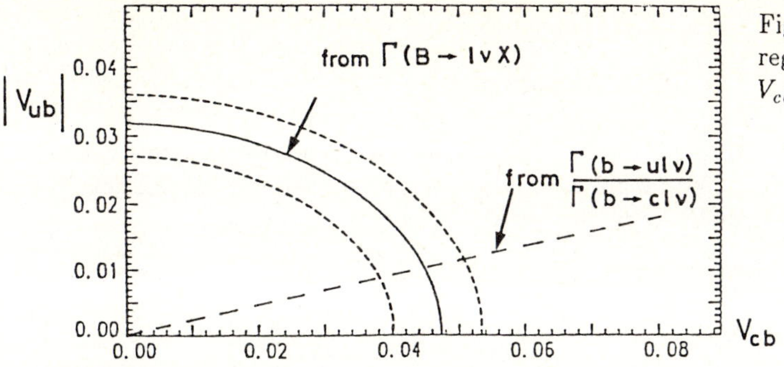

Fig.4: Experimental regions for $|V_{ub}|$ and V_{cb} .

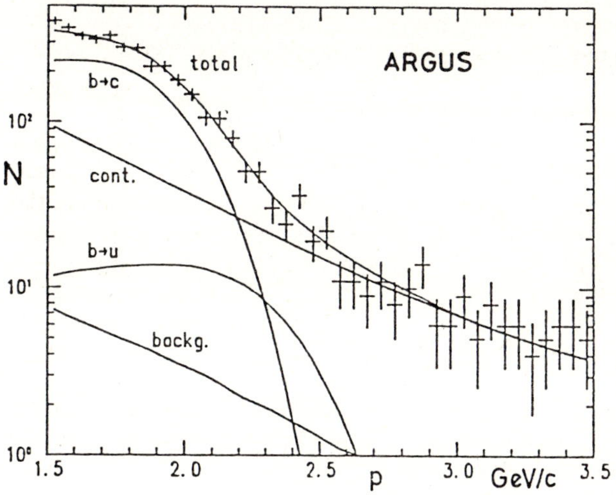

Fig.5: Momentum spectrum of inclusive electrons in the reaction e^+e^- (10577 MeV) $\rightarrow$ $e^{\pm}$ + $hadrons$ as measured by ARGUS [19]. The curves give the components of the spectrum; the ratio $b \rightarrow ue\nu / b \rightarrow ce\nu$ determines $|V_{ub}|/V_{cb}$, cont = from nonresonant decays, backg = misidentified hadrons.

$e^+e^-(10577\ MeV) \rightarrow e^{\pm} + hadrons$ as obtained by ARGUS [19]. Electron candidates originate from semileptonic decays of B mesons produced as decay products of the $\Upsilon(4S)$, from nonresonant decays like $e^+e^- \rightarrow c\bar{c}$, $c \rightarrow DX$, $D \rightarrow e\nu K$, or from hadrons misidentified as electrons. The nonresonant contribution is measured by running the experiment at an off-resonance energy, and hadron misidentification is studied by running on the $\Upsilon(1S)$ and $\Upsilon(2S)$ resonances where many extra hadrons and no extra electrons are produced. The signal electrons originate from $b \rightarrow ce\nu$ and $b \rightarrow ue\nu$ with two different momentum spectra because of the higher mass of the c quark. The signal curves in fig.5 are obtained from a spectator model where the b quark motion in the B meson is taken into account but the b quark decay is treated like that of a free particle, using a typical set of model parameters.

Previous determinations of $\Gamma(b \rightarrow ue\nu)/\Gamma(b \rightarrow ce\nu)$ had used a too narrow range of model parameters and found zero with small upper limits [20]. The strong dependence of the method on model parameters is now generally accepted. Allowing for an as wide parameter range as the data points in fig.5 allow, ARGUS [19] finds $\Gamma(b \rightarrow ue\nu)/\Gamma(b \rightarrow ce\nu) < 0.12$ with 90% confidence, which translates into

$$|V_{ub}|\ /\ V_{cb}\ <\ 0.23.$$

This 90% confidence limit is shown as straight line in fig.4. $|V_{ub}|$ and V_{cb} are essentially uncorrelated, the results are

$$V_{cb} = 0.047 \pm 0.007,$$

$$|V_{ub}| < 0.012.$$

More data and more theoretical understanding on B mesons are needed to obtain a smaller confidence interval for $|V_{ub}|$. With more data, also alternative methods like the search for the exclusive decays $B \to pe\nu$, $B \to p\pi$, and others will become possible.

Summary of Results

The experiments discussed lead to the following matrix element values:

$$|V_{ij}| = \begin{pmatrix} 0.9729 \pm 0.0010 & 0.222 \pm 0.004 & < 0.012 (90\% C.L.) \\ 0.200 \pm 0.021 & 0.94 \pm 0.08 & 0.047 \pm 0.007 \\ ? & ? & ? \end{pmatrix}$$

Check of Unitarity

In the absence of any t quark data, only three out of six unitarity relations may be checked: $\sum |V_{uj}|^2 = 1$, $\sum |V_{cj}|^2 = 1$, and $\sum V_{uj} V_{cj}^* = 0$. The sum of squares in the first row is $0.9958 \pm 0.0039 = 1 - 1.1\sigma$, and in the second row $0.93 \pm 0.16 = 1 - 0.4\sigma$. The row product is $0.014 \pm 0.027 = 0 + 0.5\sigma$ using the phase convention of the next chapter. Unitarity is obeyed with a surprisingly good precision. This allows to draw the following conclusions:

1. The coupling of the $W^{\pm}$ gauge boson to quarks is as perfectly described by the standard theory as its coupling to leptons, with the only distinction that (d, s, b) does not enter the charged current like (e, μ, τ) but $(d', s', b') = \mathbf{V}(d, s, b)$.

2. There is no fourth family of quarks, or its KM-like mixing with the known three families is very weak.

3. Since radiative electroweak corrections are imortant for the evaluation of especially V_{ud}, there is no indication for the existence of neutral gauge bosons in addition to γ and Z^0.

Parametrization and Phase Convention

Since unitarity is well fulfilled within experimental errors, let me present $\mathbf{V}$ in a strictly unitary form. In general, a unitary 3x3 matrix has nine free parameters. Since the quark field phases in the charged current are unobservable, the KM matrix has only four observable parameters, three angles and one phase. The standard theory allows this phase to be different from 0 or π, i.e. $\mathbf{V}$ is allowed to be complex (even with arbitrary quark phases) and therefore to create CP and T violation. The experiments discussed in this report do not put any constraint on the phase. Its value can be determined from CP violation experiments if the KM matrix is the only source of the observed CP violation, but I will not discuss this determination here.

Unfortunately there are several different parametrizations of the KM matrix in the literature. It is very convenient to work with a form of $\mathbf{V}$ which is "nearly real", i.e. where the observable phase is shifted to that matrix element which is the smallest experimentally. This form differs strongly from the original one [1], it was introduced by Stech [21] and

Chau et al [22] and is very close to an earlier form introduced by Maiani [23].

$$\mathbf{V} = \begin{pmatrix} 1 & 0 & 0 \\ 0 & c_\gamma & s_\gamma \\ 0 & -s_\gamma & c_\gamma \end{pmatrix} \begin{pmatrix} c_\beta & 0 & s_\beta e^{-i\delta'} \\ 0 & 1 & 0 \\ -s_\beta e^{i\delta'} & 0 & c_\beta \end{pmatrix} \begin{pmatrix} c_\theta & s_\theta & 0 \\ -s_\theta & c_\theta & 0 \\ 0 & 0 & 1 \end{pmatrix}$$

$$= \begin{pmatrix} c_\theta c_\beta & s_\theta c_\beta & s_\beta e^{-i\delta'} \\ -s_\theta c_\gamma - c_\theta s_\beta e^{i\delta'} s_\gamma & c_\theta c_\gamma - s_\theta s_\beta e^{i\delta'} s_\gamma & c_\beta s_\gamma \\ s_\theta s_\gamma - c_\theta s_\beta e^{i\delta'} c_\gamma & -c_\theta s_\gamma - s_\theta s_\beta e^{i\delta'} c_\gamma & c_\beta c_\gamma \end{pmatrix},$$

where θ, β, γ are the three angles, δ' is the phase, and $s_\theta = sin\ \theta$, $c_\beta = cos\ \beta$ etc. It is possible to choose the quark field phases in such a way that all three angles are in the first quadrant,

$$0 \leq \theta \leq \pi/2, \quad 0 \leq \beta \leq \pi/2, \quad 0 \leq \gamma \leq \pi/2.$$

The phase δ' can take any value between 0 and 2π. In this parametrization, the complex phase factor $e^{i\delta'}$ is always attached to $sin\beta$. In the limit $\beta = 0$ (which is still possible experimentally), **V** becomes real and cannot be the origin of CP violation. In the limit $\beta = \gamma = 0$ (which is experimentally excluded), **V** is real and coincides with the Cabibbo matrix.

Fitting the parametrization above to the experimental results leads to

$$\theta = (13.05 \pm 0.16)^o, \quad \beta < 0.6^o, \quad \gamma = (2.7 \pm 0.4)^o.$$

With these values, the unitarity - constrained matrix becomes

$$\mathbf{V} = \begin{pmatrix} .9742 \pm .0006 & .2258 \pm .0027 & .0000 \pm .0100 \\ -.2256 \pm .0028 & .9731 \pm .0007 & .0471 \pm .0070 \\ .0106 \pm .0100 & -.0459 \pm .0072 & .9989 \pm .0003 \end{pmatrix}$$

$$+ i \begin{pmatrix} 0 & 0 & 0 \pm .0100 \\ 0 \pm .0005 & 0 \pm .0001 & 0 \\ 0 \pm .0100 & 0 \pm .0024 & 0 \end{pmatrix}.$$

Only future experiments may be able to decide if the angle β is different from zero, if the observed CP violation has its origin or part of it in the KM matrix, and what is the value of the phase δ'.

References

1 M.Kobayashi and T.Maskawa, Progr.Theor.Phys. 49(1973)652

2 S.W.Herb et al, Phys.Rev.Lett. 39(1977)252

3 I.S.Towner and J.C.Hardy, Proc. 7th Int. Conf. on Atomic Masses and Fundamental Constants, Darmstadt 1984, ed. by O. Klepper, p.564, and references therein

4 Review of Particle Properties by the Particle Data Group, Phys.Lett. 170B(1986)1, and original references therein

5 W.J.Marciano and A.Sirlin, Phys.Rev.Lett. 56(1986)22

6 W.K.MacFarlane et al, Phys.Rev. D32(1985)547

7 M.Bourquin et al (WA2), Z.Physik C21(1983)27

8 H.Leutwyler and M.Roos, Z.Physik C25(1984)91

9 H.Abramovicz et al (CDHS), Z.Physik C15(1982)19
10 following a procedure given by K.Kleinknecht and B.Renk in "A New Analysis of Weak Mixing Angles between three or four Quark Generations", Proc.Int.Symp. on Production and Decay of Heavy Hadrons, Heidelberg 1986
11 N.Ushida et al (E531), Phys.Rev.Lett. 56(1986) 1767 and 1771
12 D.Hitlin (MARK-III), presented at the Int. Symposium on Production and Decay of Heavy Hadrons, Heidelberg 1986
13 M.Witherell (E691, TPS), presented at the Int. Symposium on Production and Decay of Heavy Hadrons, Heidelberg 1986
14 V.Lüth, presented at the Int. Symposium on Production and Decay of Heavy Hadrons, Heidelberg 1986
15 T.M.Aliev et al, Yad.Fiz. 40(1984)823, Sov.J.Nucl.Phys. 40(1984)527
16 M.Wirbel, B.Stech, and M.Bauer, Z.Physik C29(1985)637
17 A.Chen et al (CLEO), Phys.Rev.Lett. 52(1984)1084
C.Klopfenstein et al (CUSB), Phys.Lett. 130B(1983)444
G.Levman et al (CUSB), Phys.Lett. 141B(1984)271
S.Weseler (ARGUS), presented at the Int. Symposium on Production and Decay of Heavy Hadrons, Heidelberg 1986
18 J.P.Albanese et al (WA75), Phys.Lett. 158B(1985)186
19 Unpublished. I thank my colleagues in the ARGUS collaboration, especially S.Weseler, for their permission to present this result.
20 C.Klopfenstein et al (CUSB), Phys.Lett. 130B(1983)444
A.Chen et al (CLEO), Phys.Rev.Lett. 52(1984)1084
21 B.Stech, Proc. Europhysics Topical Conference on Flavor Mixing in Weak Interactions, Erice 1984, ed. by L.-L. Chau, p.735
22 L.-L. Chau and W.-Y.Keung, Phys.Rev.Lett. 53(1984)1802
23 L.Maiani, Proc.Int.Symp. on Lepton and Photon Interactions at High Energies, Hamburg 1977, p.867

Massive Neutrinos and Gauge Theories

*S.T. Petcov**

Institute of Theoretical Physics, University of Heidelberg,
Philosophenweg 16, D-6900 Heidelberg, F. R. Germany

Abstract: The status of massive neutrinos in the gauge theories of electroweak interactions is reviewed.

1 Introduction

There has been an unceasing interest in the properties of massive neutrinos and the physics they are associated with ever since the idea of the existence of neutrinos was proposed by PAULI [1] in 1930. It increased remarkably after the appearance of the gauge theories of electroweak interactions when it was realized that massive neutrinos and neutrino mixing may arise quite naturally in the gauge theories [2]. The following circumstances seem to have stimulated this interest throughout the years after 1930. First, no profound principle excluding the possibility of massive neutrinos has been discovered. Second, nonzero neutrino masses and neutrino mixing were found to imply an extremely rich spectrum of possible neutrino properties. And third, it was realized that the neutrino mass problem is intimately connected with the problem of the basic symmetries and the structure of the theory of electroweak interactions.

This latter connection is particularly transparent in the modern gauge theories of electroweak interactions. Indeed, the status of massive neutrinos and neutrino mixing in the electroweak gauge theories is closely related to the status of the symmetries associated with the conservation of the additive lepton charges L_e, L_μ, L_τ and of the total lepton charge $L = \sum_{l=e,\mu,\tau} L_l$ in these theories [2]. Now, the symmetries playing a fundamental role in the construction of the gauge theories - the local gauge symmetries - are dynamical. They fix unambiguously the character of the dynamics governing the basic particle interactions (electroweak, strong,..) and ensure renormalizability of the theory once the symmetry group and the particle content of the theory are specified. It has been known for a long time [3] that the conservation of the lepton charges L_l and L could not possibly be associated with unbroken local symmetries.**) It may reflect the existence of exact global symmetries which, however, are not inherent to and have to be imposed as an additional

---

*) Permanent address: Institute of Nuclear Research and Nuclear Energy, Bulgarian Academy of Sciences, 1784 SOFIA, Bulgaria

**) Otherwise the resulting gauge interactions would introduce discrepancy in the Eötvös experiment unless they were characterized with ultrasmall coupling constants.

constraint on the gauge theories. In this sense the global symmetries implying lepton charge conservation cannot be considered as fundamental in the context of the gauge theories. The latter admit violations of these symmetries whenever the requirement of local gauge invariance (and renormalizability) and the relevant multiplet content permit it. As a consequence, lepton number nonconservation, finite neutrino masses and neutrino mixing arise quite naturally in the gauge theories of electroweak interactions [4] and especially in the grand unified theories (GUTs) [5]. In the present talk we shall review some of the mechanisms of neutrino mass generation in the gauge theories. As an introduction to the subject the possible types of neutrino mass terms will be briefly considered and the properties of the massive Dirac and massive Majorana neutrinos will be compared.

2 Possible Types of Neutrino Mass Terms and Massive Neutrinos

The ordinary flavour neutrinos ν_l, $l = e,\mu,\tau$, possess certain well-known specific features which distinguish them from the charged leptons and quarks and predetermine the possible diversity in the properties of the massive neutrinos. Let us recall them once again. i) Neutrinos are electrically neutral: $Q_\nu = 0$. ii) The neutrinos ν_l (antineutrinos $\bar{\nu}_l$) taking part in the standard weak interaction have a definite left-handed (right-handed) helicity (at least to a very good approximation); correspondingly, the neutrino fields $\nu_{lL}(x)$, $l = e, \mu, \tau$, entering into the weak interaction Lagrangian are left-handed (LH): $\frac{1}{2}(1+\gamma_5)\nu_{lL}(x) = \nu_{lL}(x)$, which implies $\frac{1}{2}(1-\gamma_5)\nu_{lL}(x) = 0$. iii) As the existing experimental upper limits on the neutrino masses [6]-[8]

$$m_{\nu_e} < 18 \text{ eV}, \quad \mu_{\nu_\mu} < 250 \text{ keV}, \quad m_{\nu_\tau} < 56 \text{ MeV} \tag{1}$$

indicate (the notations are obvious), if massive, the neutrino ν_l should be much lighter than the charged lepton and quarks belonging to the same family.

The neutrinos with definite mass appearing in the electroweak gauge theories may be Dirac particles (ν_k) [9]; several varieties of massive Dirac neutrinos may exist [10]. Being electrically neutral, the massive neutrinos may also be Majorana particles (χ_k) [11, 12], that is particles which possess no distinctive antiparticles. As a rule, the flavour neutrinos ν_l (weak eigenstate neutrinos) do not coincide in the gauge theories with the mass eigenstate neutrinos and one has

$$\nu_{lL}(x) = \begin{cases} \sum_k U^D_{lk}\, \nu_{kL}(x), \text{ or} & \text{(2a)} \\ \\ \sum_k U^M_{lk}\, \chi_{KL}(x) & \text{(2b)} \end{cases}$$

where $\nu_{kL}(x)$ ($\chi_{kL}(x)$) is the LH component of the field of the Dirac (Majorana) neutrino ν_k (χ_k) with definite mass m_k and $U^{D(M)}$ is the corresponding lepton mixing matrix. The relations (2) imply the existence of neutrino oscillations [12].

The type of massive neutrinos arising in a gauge theory of electroweak interactions is specified by the neutrino mass term L_m^ν, more precisely, by the symmetries L_m^ν and the total Lagrangian of the theory have. By definition, fermion mass term is any piece of the Lagrangian, invariant under the proper Lorentz transformations, formed only by fermion fields and bilinear in them. Note that it is not required to be C, P and/or CP invariant. The neutrino mass term originates usually in gauge theories of the electroweak interactions from Yukawa type couplings of the lepton fields with Higgs scalar fields, some components of which develop non-zero vacuum expectation values [4]. In order not to spoil the renormalizability of the theory these couplings have to be gauge invariant. However, they may not conserve the lepton charges L_l and L.

Let us consider now the possible types of neutrino mass terms. It is convenient to divide them into three categories. For simplicity, we shall discuss first the case of one neutrino flavour.

A) <u>Dirac mass term</u>. Let us assume [13] that the neutrino mass term has the same form as the mass terms of the charged leptons and quarks:

$$L_D^\nu = - m \bar{\nu}_L \nu_R + \text{h.c.} \tag{3}$$

(m is real). Here $\nu_L(x)$ is the LH flavour neutrino field and $\nu_R(x)$ is the field of hypothetical right-handed (RH) neutrino ν_R and LH antineutrino $\tilde{\nu}_L$. The existing data on the weak processes suggest [2] that if ν_R and $\tilde{\nu}_L$ exist, their interactions with the charged leptons and quarks should be much weaker than the interaction of the ordinary flavour neutrinos. Therefore RH neutrinos are often called inert or sterile neutrinos [12]. Note that the number of postulated sterile neutrinos which mix (i.e. form mass terms) with the ordinary flavour neutrinos in a given theory may not coincide with the number of the latter.

Obvioulsy, L_D^ν is invariant with respect to the global transformations $\nu_{L(R)} \to e^{i\alpha}\nu_{L(R)}$, where α is a real parameter. This implies that L_D^ν conserves an additive quantum number, which is the fermion number F, coinciding with the lepton charge L in this case. The neutrino with definite mass ν ($\nu(x) = \nu_L(x) + \nu_R(x)$) is a Dirac particle, distinguised from its antiparticle by the value of the lepton charge L.

B) <u>Majorana mass term</u>. One can construct a neutrino mass term using only the field ν_L [14]:

$$L_M^\nu = - \frac{1}{2} m \bar{\nu}_L C \bar{\nu}_L^T + \text{h.c.} = - \frac{1}{2} m \bar{\nu}_L \nu_R^c + \text{h.c.} \tag{4}$$

where C is the charge conjugation matrix ($C^{-1}\gamma_\mu C = - \gamma_\mu^T$, $C^T = -C$, $C^+ = C^{-1}$). It is not difficult to show that the field

$$\nu_R^c \equiv C \bar{\nu}_L^T \tag{5}$$

i) is a RH field ($\frac{1}{2}(1-\gamma_5)\nu_R^c = \nu_R^c$) and

ii) transforms as ν_L under the proper Lorentz transformations. It is evident from (5) that at the same time ν_R^c transforms as the Dirac conjugate field of ν_L with respect to the global gauge transformations associated with the conservation of additive quantum numbers (such as the fermion number F, lepton charge L, weak hypercharge Y_w, etc.)

Expressed in terms of the combination of ν_L and ν_R^c

$$\chi(x) = \frac{1}{\sqrt{2}} (\nu_L(x) + \nu_R^c(x)) \tag{6}$$

L_M^ν takes the standard form

$$L_M^\nu = - \frac{1}{2} m \bar{\chi}(x)\chi(x)$$

so m is the mass of the field $\chi(x)$. It follows from eqs. (5) and (6) that $\chi(x)$ satisfies the Majorana condition [11]

$$C\bar{\chi}^T(x) = \chi(x) \tag{7}$$

i.e., $\chi(x)$ is a massive Majorana field.

Note that the mass term (4) differs substantially from the Dirac mass term (3) as no charge carried by ν_L is conserved by the former. Even the fermion number is not preserved and, e.g., the transition of a LH neutrino into a RH antineutrino in one space-time point becomes possible due to L_M^ν. The mass term (4) is called Majorana mass term. Obviously, no charge particle can have such a mass term.

C) Dirac and Majorana mass term . In the most general case the neutrino mass Lagrangian may include [15] both L_D^ν, L_M^ν, and a Majorana piece formed by ν_R:

$$-L^\nu_{D+M} = \kappa_1 \bar{\nu}_R^c \nu_L + \kappa_2 \bar{\nu}_R \nu_L + \kappa_3 \bar{\nu}_R \nu_L^c + \kappa_4 \bar{\nu}_R^c \nu_L^c + \text{h.c.} \tag{8}$$

where $\nu_L^c = C\bar{\nu}_R^T$ and κ_i are, in general, complex constants. Obviously, in its most general form L^ν_{D+M} does not conserve any charge assigned to ν_L and/or ν_R.

Since $\bar{\nu}_R^c \nu_L^c = \bar{\nu}_R \nu_L$ we have $\kappa_4 = \kappa_2$. Introducing the column of fields $n = \binom{\nu}{\nu^c}$, one can rewrite L^ν_{D+M} in the form

$$-L^\nu_{D+M} = \bar{n}_R^c M n_L + \text{h.c.} \tag{9}$$

where

$$M = \begin{pmatrix} \kappa_1 & \kappa_2 \\ \kappa_2 & \kappa_3 \end{pmatrix} \tag{10}$$

is a complex symmetric matrix: $M^T = M$. Any complex symmetric matrix can be reduced to a diagonal matrix with real nonnegative elements via the transformation:

$$U^T M U = m, \quad (m)_{ik} = \delta_{ik} m_k, \quad m_k \geqslant 0, \quad i,k = 1,2 \tag{11}$$

where in our case U is a 2x2 unitary matrix. The neutrino fields (χ_i) possessing definite masses (m_i) can easily be found:

$$\chi_i = U_{i\alpha} n_{\alpha L} + U^*_{i\alpha} n^c_{\alpha R}, \; i = 1,2 \tag{12}$$

They satisfy the Majorana condition

$$\chi_i = C \bar{\chi}_i^T, \quad i = 1,2 \tag{13}$$

and hence correspond to massive Majorana neutrinos.*)

Note that in this example we have one flavour neutrino but two neutrinos with definite mass.

Our simplified discussion of the possible types of neutrino mass terms can be easily generalized to the case of arbitrary numbers of flavour neutrinos (n) and sterile neutrinos (m) [2]. Any neutrino mass term can be reduced to one of the three types we have considered.

In general, massive Dirac neutrinos may arise in theories in which at least one additive lepton charge is conserved. This may be the total lepton charge L, or, for example, a generalization of the Zeldovich-Knopinsky-Mahmoud (ZKM) lepton charge**) [17,18], which, e.g., in the case of three lepton flavours can have the form [19] $L' = L_e - L_\mu + L_\tau$. The properties of the massive Dirac neutrinos arising in a given theory depend strongly on the type of the lepton charge conserved in the theory. Massive Majorana neutrinos may arise in theories in which the total lepton charge L is not conserved. Their couplings preserve no lepton charge. Finally, the number of neutrinos with definite mass can exceed the number of LH flavour neutrinos only if sterile neutrinos (neutral fermions) with nonzero masses which mix (i.e. form mass terms) with the flavour neutrinos ν_1 exist [15, 20]. It follows from our considerations that the observation of neutrinos with nonzero mass would imply either that the total lepton charge L is not conserved or that sterile neutrinos exist, or both.

3 Massive Dirace and Massive Majorana Neutrinos: Comparison of the Properties

Let us consider now in more detail the properties of a Majorana neutrino having a field $\chi(x)$,

$$C\bar{\chi}^T(x) = \xi\chi(x), \; \xi = \begin{cases} 1, & \text{if CP invariance does not hold in the leptonic sector} \\ \pm 1, & \text{if CP invariance holds,} \end{cases} \tag{14}$$

---

*) Obviously, under certain conditions L^{ν}_{D+M} may lead to massive Dirac neutrinos (e.g., if $\kappa_{1,3} = 0$).

**) The ZKM lepton charge L' was introduced [16] assuming that there exists one four-component (Dirac) neutrino ν associated with both e and μ and that $L'(e^-) = L'(\nu) = L'(\mu^+) = 1$ (i.e., $L' = L_e - L_\mu$). The charge L' was supposed to be conserved by the weak interaction.

and compare them with the properties of a Dirac neutrino ν (whose field we shall denote by $\nu(x)$). In (14) ξ is an unphysical sign factor which in certain cases turns out to be convenient to introduce by changing the phase of the Majorana field [21].

First, it is not difficult to show using (14) that the particle and the antiparticle creation and annihilation operators in $\chi(x)$ coincide. Therefore, $\chi(x)$ is the field of a particle identical with its antiparticle, i.e., a truly neutral spin-$\frac{1}{2}$ object which cannot carry additive quantum numbers as lepton charge, fermion number, weak hypercharge, etc.. Hence, in contrast to the Dirac neutrino fields, the Majorana fields cannot be subject to continuous global phase transformations $\chi(x) \rightarrow e^{i\alpha}\chi(x)$ (they cannot "absorb" phases). Such transformations are incompatible with the Majorana condition (14).

Second, besides the standard lepton (frmion) number conserving propagator $\overline{\chi_\alpha(x)\ \bar{\chi}_\beta(y)}$ the Majorana neutrino has a non-trivial lepton (fermion) number non-conserving propagator:

$$\overline{\chi_\alpha(x)\chi_\beta(y)} = \xi\ C_{\beta\delta}\ \overline{\chi_\alpha(x)\bar{\chi}_\delta(y)} \neq 0. \tag{15}$$

Obviously, $\overline{\nu_\alpha(x)\nu_\beta(y)} = 0$.

Third, it is easy to show, using (14) that

$$\bar{\chi}(x)\ \sigma_{\mu\nu}\ \chi(x) = 0, \quad \bar{\chi}(x)\ \sigma_{\mu\nu}\gamma_5\ \chi(x) = 0, \quad \bar{\chi}(x)\ \gamma_m\ \chi(x) = 0 \tag{16}$$

The first two equations in (16) imply that the massive Majorana neutrinos cannot possess intrinsic magnetic or electric dipole moments, which is not valid, in general, for the massive Dirac neutrinos.

Finally, the CPT and CP transformation properties of the massive Dirac and the massive Majorana neutrinos are very different. Since Majorana particles have no distinctive antiparticles, the CPT and the CP transformations change only the kinematical characteristics of the Majorana particle states*). In contrast, they transform a state of a Dirac neutrino into a state of its antiparticle. Furthermore, in the case of CP conservation, Majorana neutrinos possess a definite CP parity:

$$|\chi(r,p)> \xrightarrow{CP} \eta_{CP}\ (\chi)|\chi(-r,p_p)>, \ p_p = (-\vec{p},\ ip_o), \eta_{CP}(\chi) = \pm i \tag{17}$$

where $|\chi(r,p>$ denotes a state of χ with definite four-momentum $p = (\vec{p}, ip_0)$ and projection r of the spin on the momentum $\vec{p}$. As was noticed first by Wolfenstein [23], the relative CP parity of Majorana neutrinos is an observable quantity and may play a fundamental role in the processes involving real or virtual neutrinos such as $(\beta\beta)_{0\nu}$ -decay, neutrino radiative decays, etc.. Obviously, massive Dirac neutrinos have no such characteristic.

---

*) The indicated specific CPT and CP transformation properties of the Majorana particle states have interesting implications for, e.g., the production of Majorana fermions in e^+-e^- annihilation and p-$\bar{p}$ collisions [22].

4 Massive Neutrinos in the Gauge Theories

We shall consider next some examples of neutrino mass generation in the gauge theories of electroweak interactions. All three types of neutrino mass terms we have discussed can appear in the gauge theories.

i) $SU(2)_L xU(1)$ theories. As is well known, in the standard (minimal) electroweak theory the fields of the LH flavour neutrinos $\nu_{lL}(x)$ and the LH components of the charged lepton fields $l_L(x)$ form $SU(2)_L$ doublets, while the RH components of the charged lepton fields $l_R(x)$ are assumed to be $SU(2)_L$ singlets:

$$\begin{pmatrix} \nu_{eL} \\ e_L \end{pmatrix} \qquad \begin{pmatrix} \nu_{\mu L} \\ \mu_L \end{pmatrix} \qquad \begin{pmatrix} \nu_{\tau L} \\ \tau_L \end{pmatrix}$$
$$e_R \qquad \mu_R \qquad \tau_R \tag{18}$$

No RH neutrinos are present in the theory. Therefore, neutrinos might acquire a mass term of Majorana type, (4), only. However, it is impossible to generate such a mass term in a gauge invariant manner preserving the renormalizability of the theory as the product $(\bar{\nu}_{lL} C \ \bar{\nu}^T_{l'L})$ changes the weak isospin by one unit and the only Higgs field available in the theory (Φ) is isodoublet ($\Phi = \begin{pmatrix} \Phi^{(+)} \\ \phi^{(o)} \end{pmatrix}$).

Neutrino mass term of Dirac type arises rather naturally in the minimally extended standard theory which includes $SU(2)_L$ singlet RH neutrino fields $\nu_{lR}(x)$ $l = e,\mu,\tau$ as counterparts of $\nu_{lL}(x)$ [24]. In this case the most general gauge invariant fermion-Higgs boson interaction Lagrangian of Yukawa type contains the term:

$$L_{l-\phi} = - \sum_{l,l'} G_{l'l} \ \bar{\nu}_{l'R} \ \phi^{c+} \begin{pmatrix} \nu_{lL} \\ l_L \end{pmatrix} + \text{h.c.} \tag{19}$$

where $\Phi^c = i\,\tau_2\,\Phi^*$ is the charge conjugate of the standard Higgs doublet ϕ, whose neutral component has a nonzero vacuum expectation value $<\Phi^{(o)}>_o = \frac{\lambda}{\sqrt{2}} \neq 0$, $\lambda \simeq 250$ GeV, and $G_{ll'}$ are, in general, complex constants. The coupling (19) gives rise to a Dirac mass term L^ν_D similar to (3)

$$L^\nu_o = - \sum_{l,l'} \bar{\nu}_{lL} \ M^D_{ll'} \ \nu_{l'R} + \text{h.c.},$$

with a mass matrix $M^D_{l'l} = <\phi^{(o)}>_o \ G_{l'l}$.

L^ν_D is the only neutrino mass term possible in the theory if one assumes that the total lepton charge L is conserved (obviously, L^ν_D does not conserve L_e, L_μ and L_τ, but preserves L). In this case neutrinos are treated on equal footing with the other fermions in the theory and there exists a complete analogy between leptons and quarks [13], [24]. The neutrinos with definite mass ν_i are Dirac particles. Their number is equal to the number of the flavour neutrinos ν_l and (2a) is valid, U^D being an unitary matrix with the help of which the neutrino mass matrix is diagonalized. The neutrino mass spectrum as well as the lepton mixing matrix can be arbitrary.

The lepton charges L_e, L_μ, and L_τ are not conserved only by the term (19) in the total Lagrangian of the theory associated with the generation of neutrino masses. Therefore the probabilities of the lepton number nonconserving reactions and decays $\mu^- + N \to e^- + N$, $\mu^+ \to e^+\gamma$, $\mu^+ \to e^+e^-e^+$, etc. predicted in the theory, depend strongly on the neutrino masses and vanish in the limit of massless neutrinos ($G_{ll'} \to 0$). For the values of neutrino masses allowed by the existing experimental bounds (1) in the case of three lepton families, these probabilities can be shown [24] to be unobservably small. As a consequence, besides the nonzero neutrino masses, the only predicted new phenomena that might lead to observable effects in this case are, in essence, the oscillations of neutrinos.

There are two basic mechanisms of generation of neutrino mass term of Majorana type L_M^ν in the $SU(2)_L xU(1)$ theories with minimal fermionic content (no $\nu_{lR}(x)$) and enlarged Higgs sector [4].

If a triplet of neutral, charged and doubly charged Higgs fields

$$H = \begin{pmatrix} -H^+/\sqrt{2} & H^{++} \\ H^o & H^+/\sqrt{2} \end{pmatrix} \qquad (21)$$

whose neutral component has a nonzero vacuum expectation value (i.e., $\langle H^o \rangle_o = \frac{v}{\sqrt{2}} \neq 0$) is introduced, the gauge invariant coupling

$$L_{l-H} = \frac{1}{\sqrt{2}} \sum_{l,l'} h_{ll'} (\bar{\nu}_{l'L} \ \bar{l}'_L) \ H^+ i\tau_2 \begin{pmatrix} C \ \bar{\nu}_{lL}^T \\ C \bar{l}_L^T \end{pmatrix} + \text{h.c.} \qquad (22)$$

where $h_{ll'}$ are complex (symmetric) constants, leads to L_M^ν

$$L_M^\nu = -\frac{1}{2} \sum_{l,l'} \bar{\nu}^c_{l'R} M^L_{l'l} \nu_{lL} + \text{h.c.} \qquad (23)$$

with $M^L_{ll'} = v h^*_{ll'}$. In this case the neutrinos with definite mass are Majorana particles and the mixing (26) takes place, U^M being the unitary matrix which diagonalizes the neutrino mass matrix M^L.

If we assign a lepton charge to H ($L_H = -2$) and assume that the lepton charge L is conserved by the Lagrangian of the theory (note that the coupling (22) conserves it), the global symmetry corresponding to this conservation law will be spontaneously broken [25] (i.e., the vacuum state will not possess this symmetry of the Lagrangian) in case $\langle H^o \rangle_o \neq 0$. The resulting model is due to GELMINI and RONCADELLI [26] and has been widely discussed [27]. Its most remarkable feature is the presence of a physical massless neutral scalar particle (the Goldstone boson of the broken global symmetry) called Majoron, which couples predominantly to neutrinos and extremely weakly to the charged leptons and quarks. This theory has an interesting phenomenology [27]. In particular, in addition to the nonzero neutrino masses and nonconservation of the lepton charges L_l and L, the theory predicts the existence of three neutral, one charged and one doubly charged, physical

Higgs scalar particles which are relatively light on the scale of electroweak symmetry breaking breaking $\lambda \cong 250$ GeV. Together with the Majoron, they should take part in or mediate a number of characteristic processes at low energies. For instance, the $\mu^+ \to e^+\gamma$ and $\mu^+ \to e^+e^-e^+$ decay branching ratios may be close to the existing experimental upper limits and the second decay may be faster than the first by a factor of π/α. The neutrinoless double β-($(\beta\beta)_{o\nu}$-) decay $(A,Z) \to (A,Z+2) + e^+e^-$ is allowed and may have a rate close to the existing experimental limits. The cosmological implications of the model are unconventional as well. According to it, neutrinos in contrast to the photons could not survive during the evolution of the Universe to form cosmic background. They are predicted to disappear annihilating pairwise into pairs of Majorons.

An alternative mechanism for generating L_M^ν within the $SU(2)_L \times U(1)$ theories with a minimal fermionic content and an enlarged Higgs sector, including at least two Higgs doublets ϕ_1 and ϕ_2, was suggested by ZEE [28]. This mechanism relies on the fact that there exist more than one lepton families. The left-handed neutrinos ν_{lL} acquire a radiatively induced Majorana mass as a result of the introduction of an $SU(2)_L$ singlet charged Higgs field H'. It couples to $SU(2)_L$ singlet combinations of two lepton doublets which are antisymmetric in the flavour indices:

$$\sum_{l,l'} f^o_{ll'} (\bar{\nu}_{lL}\ \bar{l}_L)\, H'^+\, i\,\tau_2 \begin{pmatrix} \nu^c_{l'R} \\ l'^c_R \end{pmatrix} + h.c. = 2 \sum_{l,l'} f^o_{ll'}\, \bar{l}'_L\, H'^+ \nu^c_{lR} + h.c. \qquad (24)$$

where $f^o_{ll'} = - f^o_{l'l}$ are, in general, complex constants. According to (24) H' can be assigned two units of the lepton charge L. The lepton number violation effects originate then from trilinear couplings of H' to ϕ_i, which have to be antisymmetric in the indices of the Higgs doublets in order to preserve the gauge symmetry:

$$\sum_{j,k} c_{jk}\, \phi_j^+\, \phi_k^c\, H'^+ + h.c. \qquad (25)$$

and $c_{jk} = - c_{kj}$ are constants (for the case of two Higgs doublets c_{12} is real). The interactions (24) and (25) together with the standard Yukawa couplings of the lepton doublets and l_R with ϕ_i, which give rise to the charged lepton mass matrix, combine at one loop level to produce a finite Majorana mass term of the type given by (23). The corresponding mass matrix $M^{L(Z)}$ is calculable and takes a particularly simple form in the case of three families if only one of the Higgs doublets, say ϕ_1, couples to the leptons and the ϕ_1-lepton couplings are flavour diagonal [29,30]:

$$M^{L(Z)}_{ll'} = g f^o_{ll'} \frac{m_l^2 - m_{l'}^2}{M_W} C', \quad l,\ l' = e,\ \mu,\ \tau \qquad (26)$$

where g is the $SU(2)_L$ gauge coupling constant, m_l and M_W are the masses of the charged lepton l and the $W^\pm$-bosons and C' is a flavour-independent real dimensionless constant, $|C'| \ll 1$. The value of C' is determined [30] by the values of

i) the masses of the two physical charged Higgs particles present in the theory, ii) the vacuum expectation values $<\phi_{1,2}>_o$, and iii) the constant c_{12} in (25). The most interesting feature of the mass matrix (26) is that it has zero diagonal elements.

If, e.g., the H'-lepton coupling constants $f^o_{ll'}$, $l,l' = e, \mu, \tau$ exhibit weak flavour dependence, the character of the resulting neutrino mass spectrum as well as the CP-parities of the Majorana mass eigenstate neutrinos (CP is conserved in the leptonic sector) are determined by the ratios of the squares of the charged lepton masses. In this case two of the Majorana neutrinos with definite mass, say χ_2 and χ_3, are almost degenerate in mass, have opposite CP-parities, and are much heavier than the third one χ_1: $|m_2 - m_3| \sim m_1 \sim m_2\, m^2_\mu/m^2_\tau \ll m_{2,3}$. In particular, values of $m_{2,3}$ as large as $(10 \div 20)$ eV are possible. However, even for $m_{2,3} \sim 20$ eV the $(\beta\beta)_{o\nu}$-decay rate is strongly suppressed [23],[30] as a consequence of the specific form of the neutrino mass matrix (26) ($M^{L(Z)}_{ee} = 0$). At the same time the $\mu^+ \to e^+\gamma$ and $\mu^+ \to e^+e^-e^+$ decay probabilities can be close to the existing experimental upper limits [30].

Finally, the neutrino mass Lagrangian of (Dirac + Majorana) type (8) arises, e.g., in the SO(10) GUTs [5], and that is crucial for generating neutrino masses compatible with the observations. In the minimal SO(10) theory the neutrino of a given flavour ν_l (i.e. family) acquires at the three level a Dirac mass of the order of m_q, where m_q is the mass of the charge 2/3 quark in the same family. So, $\kappa_2 \sim m_q$ and $\kappa_1 = 0$. As follows from (1), neutrino masses of this order are excluded experimentally. A possible solution to this problem is achieved by assuming that the RH neutrino fields ν_{lR} are superheavy, having a Majorana mass M_R, e.g., of the order of the unification scale M_{GUT} of the electroweak and strong interactions $M_R \sim M_{GUT} \sim 10^{14}$ GeV. It turns out to be possible to generate $M_R \sim M_{GUT}$. Neglecting for simplicity the possible interfamily mixing, one obtains the following matrix for the neutrinos of each family (see (8) and (10)):

$$M^\nu_{SO(10)} = \begin{pmatrix} 0 & m_q \\ m_q & M_R \end{pmatrix}, \quad M_R \gg m_q .$$

Since m_q and M_R are real, the neutrino mass term is CP conserving. The eigenvalues $m_{1,2}$ of $M^\nu_{SO(10)}$ and the corresponding eigenvectors $\chi_{1,2}$ can be easily found:

$$m_1 \simeq \frac{m_q^2}{M_R}, \quad m_2 \simeq M_R, \quad \eta_{CP}(\chi_1) = -\eta_{CP}(\chi_2) = i$$

$$\chi_1 \simeq (\nu_L - \nu^c_R) + \frac{m_q}{M_R}(\nu^c_L - \nu_R), \quad \chi_1 = -C\,\bar{\chi}_1^T$$

$$\chi_2 \simeq -\frac{m_q}{M_R}(\nu_L + \nu^c_R) + (\nu^c_L + \nu_R), \quad \chi_2 = C\,\bar{\chi}_2^T. \tag{27}$$

Obviously, the neutrino χ_1 takes part in the weak interactions (as $\chi_{1L} \simeq \nu_L$) and it is light enough to satisfy the existing limits on the neutrino masses. The second neutrino χ_2 is superheavy and it practically decouples at low energies. This is the famous mechanism for generation of small neutrino masses in SO(10) GUTs suggested by GELL-MANN, RAMOND, and SLANSKY, by YANAGIDA and by STECH [31].

The mass M_R is not fixed unambiguously in the theory. It can exceed or be smaller than the unification scale by few orders of magnitude. Using, e.g., for the current masses of the u, c, and t quarks the values 5 MeV, 1.5 GeV, and 40 GeV, respectively, and $M_R = 10^{14}$ GeV, we get the following light neutrino mass spectrum*):

$$m_1 \simeq 10^{-11}\ \text{eV},\quad m_2 \simeq 10^{-6}\ \text{eV},\quad m_3 \simeq 7 \times 10^{-4}\ \text{eV} \tag{28}$$

Variations by few orders of magnitude are possible due to the uncertainty in the value of M_R as well as due to the effects of flavour mixing. However, for the range of values of the neutrino masses predicted by the SO(0) theory, the oscillations of neutrinos may change dramatically the flux of the electron neutrinos from the Sun [32]. Therefore, neutrino masses in the indicated range will be probed in the future solar neutrino experiments [33].

References

1. W. Pauli, Letter (December 1, 1930) to a meeting of physicists in Tübingen (see, e.g., Brown, Physics Today, September 1978).
2. See, e.g., the review articles:
   S. M. Bilenky and B. Pontecorvo, Phys. Rep. 41, 226 (1978);
   J. D. Vergados, Phys. Rep. 133, Vol. 1,2 (1986);
   G. Costa and F. Zwirner, Univ. of Padova preprint DFPP 10/85;
   M. Doi, T. Kotani, and E. Takasugi, Progr. Theor. Phys. Suppl. N83 (1985).
3. T. D. Lee and C. N. Yang, Phys. Rev. 98, 101 (1955).
4. See, e.g.: T. P. Cheng and L.-F. Li, Phys. Rev. D22 2860 (1980).
5. For a review see, e.g.: P. Langacker, Phys. Rep. 72, 185 (1981).
6. M. Fritschi et al., SIN preprint, March 1986.
7. R. Abela et al., Phys. Lett. 146B, 431 (1984).
8. S. Abachi et al., Talk given at the VI Moriond Workshop on Massive Neutrinos in Particle Physics and Astrophysics, Tignes, France, 1986.
9. P. A. M. Dirac, Proc. Roy. Soc. A117, 610 (1928); ibid. A118, 351 (1928); ibid. A126, 360 (1930).
10. See, e.g.: S. T. Petcov, preprint CERN-TH. 4399/86.
11. E. Majorana, Nuovo Cimento 14, 10 (1937).
12. B. Pontecorvo, Zh. Eksper. Teor. Fiz. 33, 549 (1957); ibid. 34, 247 (1958).
13. S. M. Bilenky and B. Pontecorvo, Phys. Lett 61B, 248 (1976).
14. V. Gribov and B. Pontecorvo, Phys. Lett. 28B, 463 (1969).
15. S. M. Bilenky and B. Pontecorvo, Lett. Nuovo Cim. 17, 569 (1976).
16. Ya. B. Zeldovich, DAN SSSR 86, 505 (1952);
    E. J. Konopinsky and H. Mahmoud, Phys. Rev. 92, 1045 (1953).
17. S. M. Bilenky and B. Pontecorvo, Phys. Lett. 102B, 32 (1981).
18. L. Wolfenstein, Nucl. Phys. B186, 147 (1981).

*) The u, c and t quarks are assumed to have the masses quoted above at the scale of 1 GeV, while the neutrino masses are determined by the values of the quark masses evaluated at the unification scale [5]. The latter are approximately by a factor of 4.7 smaller than the former.

19. S. T. Petcov, Phys. Lett. 110B, 245 (1982);
C. N. Leung and S. T. Petcov, Phys. Lett. 125B, 461 (1983).
20. J. Schechter and J. F. W. Valle, Phys. Rev. D22, 2227 (1980).
21. S. M. Bilenky, N. P. Nedelcheva, and S. T. Petcov, Nucl. Phys. B247, 461 (1984);
B. Kayser, Phys. Rev. D30, 1023 (1984).
22. S. T. Petcov, Phys. Lett. 139B, 421 (1984);
S. M. Bilenky, E. Ch. Christova, and N. P. Nedelcheva, Phys. Lett. 161B, 397 (1985);
S. T. Petcov, preprint CERN-TH. 4442/86.
23. L. Wolfenstein, Phys. Lett. 107B, 77 (1981).
24. S. T. Petcov, Sov. J. Nucl. Phys. 25, 340 (1977); errata ibid. 25, 698 (1977).
25. Y. Chikashige, R. N. Mohapatra, and R. D. Peccei, Phys. Lett. 98B, 265 (1981).
26. G. B. Gelmini and M. Roncadelli, Phys. Lett. 99B, 411 (1981).
27. See, e.g.: H. Georgi, S. L. Glashow, and S. Nussinov, Nucl. Phys. B193, 297 (1981).
28. A. Zee, Phys. Lett. 93B, 389 (1980).
29. L. Wolfenstein, Nucl. Phys. B175, 93 (1980).
30. S. T. Petcov, Phys. Lett. 115B, 401 (1982).
31. M. Gell-Mann, P. Ramond, and R. Slansky, in Supergravity, eds. P. van Nieuwenhuizen and D. Freedman, North Holland (1979), p. 315;
T. Yanagida, Progr. Theor. Phys. B135, 66 (178);
B. Stech, in Unification of Fundamental Particle Interactions, eds. J. Ellis, S. Ferrara, and P. van Nieuwenhuizen, Plenum Press, New York (1980) p. 23.
32. S. P. Mikheyev and A. Yu. Smirnov, these Proceedings.
33. W. Hampel, these Proceedings.

Constraints on the Left-Right Symmetric Models of Weak Interactions

R.N. Mohapatra †

Department of Physics and Astronomy, University of Maryland, College Park, MD 20742, USA

We review the present experimental status of left-right symmetric models of weak interactions. The limits on the masses of the new gauge bosons, $W_R^{\pm}$ and Z_R as well as the right-handed neutrinos, N_R that are needed in this model are presented. Additional features of the model such as baryon and lepton number violation, CP-violation are outlined. Finally, latest SO(10) grandunification scenerio of this model is discussed.

§1. Introduction

Left-right symmetric models of weak interactions were developed [1] in 1974 in order to provide a more satisfactory theoretical framework for discussing the origin of parity violation in low energy weak interactions. The new feature of these models is the introduction of V+A (or Right-handed) current interactions into the weak Lagrangian. As such, they provided a theoretically consistent scheme to study the extent of validity of the V-A theory of weak interactions, which forms the basis of the standard electroweak model of Glashow, Weinberg and Salam. Since then a number of attractive features of these models have been realized such as predictions of quark mixing angles [2], understanding a possibly small but non-vanishing neutrino mass and connect it to the suppression of V+A interactions [3], natural suppression [4] of the strong CP-violating parameter θ, connecting the smallness of CP-violating amplitudes in K-decays to suppression of V+A current weak interactions [5] and more recently, a possible new setting for understanding weak isospin violation [6] such as the mass splitting between the up and down quarks. Moreover, these attractive features remain unaffected in the process of embedding this group in grandunified schemes such as SO(10) [7] or E_6 [8] or superstring [9]. It is therefore likely that nature will exhibit the parity symmetry of all fundamental interactions at some distance scale. In this brief review, we discuss the various phenomenological

† Work supported by a grant from National Science Foundation.

constraints on the crucial parameters of the model, which distinguish it from the standard model i.e. masses of the right-handed charged and neutral gauge bosons, heavy right-handed Majorana neutrino, their mixings with their left-handed counterparts i.e. W_L and ν_L respectively. We then discuss the violation of B-L symmetry and its implications for lepton number violating processes such as [10] $\mu^-e^+ \rightarrow \mu^+e^-$, $\mu^-(Z,A) \rightarrow e^+(Z-2,A)$ etc. and $\Delta B=2$ baryon violating processes such as $N-\bar{N}$ oscillation [11]. We discuss the model of CP-violation which relates the smallness of CP-violation to the suppression right-handed weak interactions, which also provides an upper limit on the mass of the W_R-boson. Finally we discuss the embedding of this group in the SO(10) grandunified model.

§2. Motivation and Description of the Model

First I want to say a few words about why I believe left-right symmetry is a fundamental and natural symmetry of the quark-lepton world. It is well-known that weak interactions are symmetric between quarks and leptons. We can use this symmetry to write a unified formula for electric charge in terms of weak isospin and B-L quantum number, in a manner analogous to the Gell-Mann-Nishijima formula for the hadronic world i.e.

$$Q = I_{3W} + \frac{B-L}{2} \tag{1}$$

Since at low energies electroweak processes involve only the left-handed weak isospin, we must rewrite the eqn. (1) as

$$Q = I_{3L} + I_{3R} + \frac{B-L}{2} \tag{2}$$

It is then natural to assume that Nature realizes the full right-handed weak isospin in the same way as it realizes the left-handed one. This then leads to the left-right symmetric model of weak interactions, based on the gauge group $SU(2)_L \times SU(2)_R \times U(1)_{B-L}$. The quarks and leptons are assigned to the gauge group as follows: (denote $Q \equiv (u,d)$ and $\psi \equiv (\nu,e^-)$)

$$\begin{aligned} &Q_L: (2,\ 1,\ 1/3); \quad && Q_R: (1,\ 2,\ 1/3) \\ &\psi_L: (2,\ 1,\ -1); \quad && \psi_R: (1,\ 2,\ -1) \end{aligned} \tag{3}$$

It is clear that gauge interactions are symmetric between left and right-handed fermions; therefore, prior to spontaneous breaking of gauge symmetry, weak interactions like strong, electromagnetic and gravitational interactions conserve parity. This can be seen explicity, by writing the gauge interactions

$$\mathcal{L}_{WK} = \frac{g}{2}\,[\vec{J}_{\mu,L}\cdot\vec{W}^{\mu}_{L} + \vec{J}_{\mu,R}\cdot\vec{W}^{\mu}_{R}] + g'V^{B-L}_{\mu}\cdot B^{\mu} \tag{4}$$

As long as the W_L and W_R have the same mass, the weak interactions conserve parity. The observed parity violation at low energies is then attributed to vacuum being parity asymmetric, which manifests itself in $m_{W_R} \gg m_{W_L}$. The symmetry breaking pattern responsible for the parity violation observed at low energies can be obtained using the following generic pattern for the Higgs multiplets: (we drop the reference to color gauge group from now on) ϕ: (2,2,0), H_L (a, 1, b) + H_R (1, a, b) where a and b will be fixed later. The first stage of the symmetry breaking is implemented by the neutral component of the right-handed multiplet H_R acquiring a v.e.v. i..e. $\langle H_R^o\rangle = V_R$. This reduces the electroweak symmetry to $SU(2)_L \times U(1)_Y$, which is subsequently broken by ϕ acquiring a v.e.v.

$$\langle\phi\rangle = \begin{pmatrix} \kappa & 0 \\ 0 & \kappa' e^{i\alpha} \end{pmatrix} \tag{5}$$

The second stage of the symmetry breaking can induce [3] a v.e.v. for $\langle H_L^o\rangle = v_L \simeq \gamma\, n^2/v_R$ assuming $\kappa' \gg \kappa$; The quarks and leptons acquire a mass at the second stage, as do the W_L and Z-bosons. The form of the low energy charged current interaction in this model is as follows:

$$H_{WK}^{c.c.} = \frac{G_F}{\sqrt{2}} \left[(\cos^2\zeta + \eta \sin^2\zeta) J^+_{\mu_L} J_L^{\mu-} + (\eta \cos^2\zeta + \sin^2\zeta) J^+_{\mu_R} J_R^{\mu-} \right.$$

$$\left. + e^{i\alpha}\cos\zeta \sin\zeta (1-\eta)\, J^+_{\mu_L} J_R^{\mu-} + h.c.\right] \tag{6}$$

where $\eta = \left(\frac{m_{W_L}}{m_{W_R}}\right)^2$ and ζ is the mixing angle between the charged W-bosons and is given in terms of symmetry breaking parameters as $\zeta \simeq (\kappa\kappa'/m_{W_R}^2)$. The currents $J_{\mu L,R}$ are given as follows:

$$J^+_{\mu L} = \bar{P}\gamma_\mu(1+\gamma_5)U_L N + \bar{E}^o\gamma_\mu(1+\gamma_5)V_L E^- \tag{6a}$$

where P, N, E^o and E^- denote the up quark, down quark, neutrino and electron column vector involving all generations and U_L and V_L denote the charged current mixing matrices for the quark and lepton sector. The corresponding right-handed currents are obtained by replacing L by R and γ_5 by $-\gamma_5$. We would now like to confront the Hamiltonian in eqn. (6) with experiment; and since weak processes involving leptons are less complicated by strong interactions, we would like to study purely leptonic and semileptonic processes to gain information about η and ζ. For this purpose, we need to know the nature of the neutrino and the structure of the weak right-handed current (specifically the quark and lepton mixing matrix in the right-handed sector).

§3. Dirac vs. Majorana Neutrino:

The nature of the neutrino depends on the nature of the first stage breaking of left-right symmetry i.e. the nature of the Higgs multiplets H_L and

H_R. Two possible choices of H_L and H_R can be considered: the first one which we will call the canonical choice was introduced in ref. 3 (case i) and it leads to Majorana neutrinos and provides an understanding of the smallness of the neutrino mass being related to the suppression of V+A currents. The second one we mention here leads to light Dirac neutrinos (case ii) at the price of expanding the model to include singlet leptons.

Case (i): Here, H are chosen to be triplets under the weak gauge group and are denoted by $\vec{\Delta}_L(3,1,2) + \vec{\Delta}_R(1,3,2)$. They couple to leptons only as follows:

$$\mathcal{L}_Y' = h\,\{\psi_L^T C^{-1}\tau_2\vec{\tau}\cdot\vec{\Delta}_L\psi_L + \psi_R^T C^{-1}\tau_2\vec{\tau}\cdot\vec{\Delta}_R\psi_R\} + \text{h.c.} \tag{8}$$

Choosing $\langle\Delta_R^o\rangle = V_R$ and assuming that the neutrino Dirac mass coming from the $SU(2)_L$ breaking is m_e, we obtain the following 2×2 mass matrix for $(\nu_L, N_R)'$

$$\begin{array}{ccc} & \nu_L & N_R \\ \nu_L & \multicolumn{2}{c}{\left(\begin{array}{cc} h\gamma\,\kappa^2/V_R & m_e \\ m_e & gV_R \end{array}\right)} \\ N_R & & \end{array} \tag{7}$$

This mass matrix provides a qualitative explanation from small neutrino mass, which comes out to be $m_\nu \simeq (h\gamma\kappa^2 - m_e^2)/V_R$; according to this formula, $m_\nu \to 0$ as $V_R \to \infty$ in the absence of right-handed weak interactions; it is, therefore, aesthetically pleasing that if neutrino has a small mass, it is not just a freak accident of nature but is related to a new intrinsic property of weak interactions being ultimately parity conserving. Turning this question around, we could say that once mass of the neutrino is known, that would provide a rough idea about the mass scale for right-handed interactions. There is, however, one technical difficulty with eqn. (5) i.e. to get acceptable values for neutrino mass (i.e. ev or less), we will have to tune the parameter $\gamma \simeq 10^{-6}$ or so. It turns out that such a small value arises automatically [12], if parity symmetry is broken at a scale higher than the breaking scale of $SU(2)_R$ gauge symmetry [13]. This has the impact that at $\mu \simeq m_{W_R}$, $g_L \neq g_R$ and further the multiplet Δ_L decouples from low energies. Alternatively, it is worth pointing out if γ is set to zero at the tree level, then it receives contributions at the one loop level only from Yukawa couplings, which are much smaller than one ($\sim 10^{-2}$ or so). So if we assume that such one loop graphs are cut off in momentum at $\mu \simeq M_{Planck}$, we would expect $\gamma \simeq \frac{h^4}{16\pi^2}\,\ln(M_P/m_W) \lesssim 10^{-8}$ or so, which may justify the finetuning adopted earlier.

Turning to eqn. (4), we note that the right-handed leptonic current involves the right-handed Majorana neutrino. As a result, whether the right-handed leptonic currents manifest themselves in low energy decay processes depends on the mass, $m_{N_R} \simeq hV_R$, which for natural values of coupling parameters is expected to be in the tens to hundreds of Gev range. We will derive constraints on this mass from neutrinoless double beta decay.

A further point worth discussing in connection with the mass matrix in eqn. (7) is that if the scale V_R is high, then for arbitrary γ, we get $m_{\nu_i} \simeq h\gamma\cdot\kappa^2/v_R$. We find $m_{\nu_i} \approx 10\text{eV}$ requires for $h \simeq 10^{-3}-10^{-4}$, $\gamma \simeq 10^{-1}$, $V_R \simeq 10^8-10^9$ Gev. An important characteristic of this point of view is that all light neutrinos can be nearly degenerate in mass, which could manifest themselves in substantial neutrino oscillations. This approach will be relevant in discussion of solar neutrino oscillations using resonant amplification of neutrino oscillations in the solar interior [14].

Case (ii): We now show that a very mild extension of the fermion sector of the model and an alternative Higgs content can lead to light left-handed neutrinos which are Dirac particles. To do this, we extend the model by including two gauge singlet fermions θ_1 and θ_2 which transform to one another under parity symmetry and replacing the $\Delta_L + \Delta_R$ Higgs bosons by doublets $\chi_L(2,1,+1)$ and $\chi_R(1,2,+1)$. The Yukawa couplings of the leptonic sector can be written as:

$$\mathcal{L}'_Y = f(\bar{\psi}_L\chi_L\theta_1 + \bar{\psi}_R\chi_R\theta_2) + \mu\bar{\theta}_1\theta_2 + \text{h.c.} \tag{9}$$

If we implement the first stage of the symmetry breaking by $\langle\chi_R^o\rangle = V_R$ with $\langle\chi_L^o\rangle = 0$, we find a neutrino mass matrix of the form [15,16]

$$\begin{array}{c} \\ \nu_L \\ \theta_1 \\ \nu_R \\ \theta_2 \end{array} \begin{array}{c} \begin{array}{cccc} \nu_L & \theta_1 & \nu_R & \theta_2 \end{array} \\ \begin{pmatrix} 0 & 0 & m_D & 0 \\ 0 & 0 & 0 & \mu \\ m_D & 0 & 0 & fV_R \\ 0 & \mu & fV_R & 0 \end{pmatrix} \end{array} \tag{10}$$

This leads to two Dirac neutrinos, one heavy with mass $fV_R\cdot\bar{\nu}_R\theta_2$ and one light one with mass $(m_D\mu/fV_R)\bar{\nu}_L\theta_1$. If μ is in the MeV range, this can give rise to a Dirac neutrino with mass in the electron volt range. A slight modification of this mass matrix by the additon of a non-vanishing Majorana mass for θ_2 can provide an understanding of the solar neutrino puzzle [17].

A further point to note here is that the right-handed neutrino is heavy being in the Gev range and will, therefore, prevent the V+A currents from manifesting themselves at low energies even for Dirac neutrinos. In discussing the experimental bounds on the mass of m_{W_R}, we will, however, shed all theoretical prejudices and consider the limits from muon and beta-decays.

§4. Limits on the Masses of the Z_2, W_R and the Right-Handed Neutrino

To search for the signatures of new physics associated with the right-handed symmetry of weak interactions, we must determine to what extent known low energy physics restricts the mass of Z_2, W_R and ν_R. The most model independent limit arises on the mass of the Z_2-boson. The form of the neutral current interaction of Z_1 and Z_2 can be written in these models in the form [18,19,20]

$$\mathcal{L}_{WK}^{N.C.} = \frac{h}{\cos\theta_W} [Z_{1\mu} \{J^{\mu}_{z_L} - \eta_Z J_{\mu} + \frac{1}{(\cos 2\theta_W)}^{1/2} Z_{2\mu} J_{\mu}] \qquad (11)$$

where $\eta_Z = (M_{Z_1}/M_{Z_2})$ and $J = \sin^2\theta_W J_{ZL} + \cos^2\theta_W J_{Z_R}$; $J_{ZL,R} = (I_{3L,R} - Q\sin^2\theta_W)$. We, then, see that the effects of right-handed bosons vanish as $M_{Z_2} \to \infty$ and we obtain the neutral current interaction of the standard model. Typical precision of neutral current experiments is about 10%, which allows one to have a rather light Z_2; a recent analysis [10,20] yields $M_{Z_2} \geq 275$ Gev for these models.

As far as the charged W_R-boson is concerned, we consider both the light Dirac and Majorana cases. And, we will further assume that the left and right handed mixing matrices are equal. This is the case of manifest left-right symmetry [21] which emerges for the simplet choice of Higgs multiplets ϕ, In this case, if neutrino is a Dirac particle with ν_L and ν_R forming this particle or is a Majorana particle such that $m_{\nu_R} \leq 10$ Mev, the most model independent bounds on m_{W_R} and ζ come from the analysis of a recent muon decay experiment [22]. In this experiment the endpoint spectrum of the positrons moving along a direction opposite to the spin of a stopped 100% polarized μ^+ is measured. In pure V-A theory, this spectrum vanishes at the end point, thus any non-vanishing would be an indication of the presence of V+A currents. The quantity measured this way, to be called R is given in terms of the muon decay paramters δ, ξ and ρ as

$$R = 1 - \frac{\delta\xi}{\rho} P_{\mu} \qquad (12)$$

where in terms of the η and ζ defined after eqn. (6) we get

$$R = 4\eta^2 + 2\zeta^2 + 4\eta\zeta \qquad (13)$$

The experimental result $R > .0031$ implies $\eta < .039$ for arbitrary ζ and $\zeta < .056$. The corresponding lower limit on m_{W_R} is 420 Gev. These analysis have recently been extended [23] to the case of non-manifest left-right symmetry [24].

Turning now to the case of the heavy right-handed neutrino (i.e. m_{ν_R} in the Gev range), the only bound comes from non-leptonic decays such as K-decays [25] and K_L-K_S mass difference [26] (ΔM_K), the latter case leading to the most stringent bound. The reason for this bound is that calculation of the one loop $\Delta S=2$ effective Hamiltonian arising from the W_L-W_R exchange leads to an enhanced contribution to ΔM_K as follows

$$\Delta m_K = \Delta m_{K_{LL}} [1-430\eta] \qquad (14)$$

Thus we see that unless $\eta \leq 2.3 \times 10^{-3}$ or $m_{W_R} \geq 1.6$ Tev , eqn. (14) will give the wrong sign and contradict observations. This is the most stringent bound on m_{W_R}.

Coming now to the bound on the mixing parameter ζ, we have already seen that for Dirac neutrinos muon decay experiments provide a bound $\zeta < 5.6\%$. For heavy right-handed neutrinos, the most model independent bound comes from the study [27] of y-distribution in deep inelastic scattering of antineutrinos off nuclei. In the absence of V+A current, the valence quarks lead to $(1-y)^2$ type distribution, whereas the left-right mixing leads to flat y-distribution proportional to ζ. Analysis of existing experiments [28] lead to $\zeta < .1$. Other more model dependent bounds can be obtained by considering deviations from current algebra relations in non-leptonic decays [25], ($\zeta \leq 4 \times 10^{-3}$) as well as from beta decays [29] of nucleus combined with unitarity of the KM-matrix ($\zeta < .005$). Finally, we note that with the simplest Higgs structure (one ϕ and $\Delta_{L,R}$), one has the following theoretical upperbound [30] on the mixing parameter $\zeta \leq \left(\frac{m_{W_L}}{m_{W_R}}\right)^2$. If we use the bound from K_L-K_S mass difference, it implies $\zeta \leq 2 \times 10^{-3}$, which is the most stringent, though model dependent bound on this parameter.

Bound on the mass of Majorana ν_R:

As mentioned earlier, the most natural way to understand small neutrino mass is to assume that the right-handed neutrino is a heavy Majorana particle. While the smallness of m_ν implies that m_{ν_R} is in the tens of Gev range, we would like to discuss the limits on m_{ν_R} from phenomenological constraints. An interesting limit that correlates $(m_{\nu_R})_{min}$ with m_{W_R} arises from present experimental limits [31] on neutrinoless double β-decay [32]. The point is that a heavy right-handed neutrino contributes to $(\beta\beta)_{0\nu}$ decay via the exchange of W_R boson with an amplitude proportional to $G_F^2\eta^2 m_{\nu_R} \langle e^{-m_{\nu_R}r}/r\rangle_{nuc.}$ and this contribution adds incoherently with the usual light neutrino contribution. Using the available nuclear matrix element calculations [33], we obtain the bound shown in fig. 1. an upper limit on $m_{\nu_R} \leq m_{W_R}$ arises from considerations of vacuum stability, which is the right-hand line in fig. 1. The limits on ν_R mass in other mass regions have been discussed earlier [34].

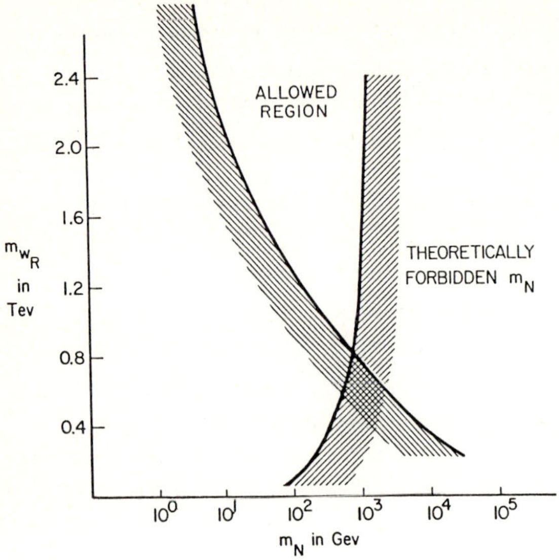

Fig. 1. Correlated bounds on m_{ν_R} and m_{W_R} from neutrinoless double beta-decay and vacuum stability.

§5. Neutrino Mass Spectrum and a Possible Light Neutral Higgs Boson

We saw in the previous section that the neutrino masses in these models scale with the square of the charged lepton masses i.e. $m_{\nu_e}:m_{\nu_\mu}:m_{\nu_\tau} \simeq m_e^2:m_\mu^2:m_\tau^2$. If it turns out that the electron neutrino has a mass in the range of 1-5ev, we would then expect $m_{\nu_\mu} \simeq$ 40-200 Kev and $m_{\nu_\tau} \simeq$ 10-50 Mev. The immediate question to worry about is whether ν_μ decays fast enough to be consistent with cosmological mass density constraints. For decaying neutrinos, the upper limit on the mass depends on its lifetime as $m_{\nu_H} \leq (t_U/\tau_{\nu_H})^{1/2}$ 100 ev, from which it follows that for $m_{\nu_\mu} \simeq$ 100 Kev (which we take as a typical value for the purposes of discussion), $\tau_{\nu_\mu} \lesssim 10^{12}$ sec. the photonic decay mode of the ν_μ is much too long for this purpose [35] ($\tau_{\nu_\mu} \rightarrow \nu_e + \gamma \gtrsim 10^{15}$ sec. for $m_{\nu_\mu} \simeq 200$ Kev as the most optimistic estimate). We, therefore, resort to the $\nu_\mu \rightarrow 3\nu_e$ decay [36] mediated by Δ_L^o-Higgs component. It, however, turns out that the same weak multiplet as Δ_L^o contains a doubly charged Δ_L^{++} which can mediate $\mu \rightarrow 3e$ decay [37], whose branching ratio is highly constrained i.e. $B(\mu\rightarrow 3e) \leq 3 \times 10^{-11}$. Combining these two constraints implies that $(m_{\Delta_L^{++}}/m_{\Delta_L^o})^2 \geq 2 \times 10^4$. Since the masses of these two particles can only be split by $SU(2)_L$-breaking we would expect $m_{\Delta_L^{++}} - m_{\Delta_L^o} \simeq (m_{W/g}) \simeq 250$ Gev or so. Assuming that this splitting can be generously of order of a Tev, we predict that $m_{\Delta^o} \lesssim$ 1-5 Gev [38]. Such a light Higgs boson would not have been visible in any of the exiting experiments since it couples only to left-handed neutrinos. However the Z-boson will decay to $\Delta_L^o\bar{\Delta}_L^o$ with $(B(Z\rightarrow\Delta_L^o\bar{\Delta}_L^o)/B(Z\rightarrow\nu\bar{\nu}))$

= 2 so that an accurate measurement of the Z-width can test this picture. As far as the ν_τ is concerned, it is heavy enough to decay fast to $e^+e^-\nu_e$ without causing any cosmological havoc.

§6. CP-Violation and Neutron-Anti-Neutron Oscillation:

Oscillation: In this section I will briefly touch on the connection between CP and P violation. it is well known that in gauge theories with only V-A weak interactions, one needs three generations to introduce a single nontrivial CP-phase which gives the Kobayashi-Maskawa model of CP-violation [39]. The viability ofthis model hinges on one crucial parameter, θ_{bu}, which measures the semileptonic decay $b \to ue\bar{\nu}_e$. At present, it has an upper limit [40] of about .008; if however, this value is reduced to about .001, KM model will fail to explain CP-violation in $K_L^o \to 2\pi$ decays. Moreover, if the upper limit on other CP-violating parameter ε'/ε is also reduced to the level of 10^{-4}, regardless of θ_{bu}, KM model will be in serious trouble and as the left-right symmetric models will provide an attractive alternative. The interesting point about CP-violation arising from the right-handed currents is that in the limit of $m_{W_R} \to \infty$, CP-violating effects vanish. As a result, if there was no CP-violating effect from the left-handed currents, we would have a dynamical connection between CP- and P-violation i.e. $\varepsilon_{K_L^o \to 2\pi} = (m_{W_L}/m_{W_R})^2 \sin\delta$, where δ is the CP-phase. One can also put an upper limit on $m_{W_R} \leq 35$ Tev from these considerations [41]. A test of right-handed currents as the origin of CP-violation is provided by a connection between ε'/ε with electric-dipole moment of the neutron [42], and a muon polarization in $K_L^o \to \mu\bar{\mu}$ decay at the level of a few percent [43]. The extension of these models to three generations [43a] leads to an estimate of $\varepsilon'/\varepsilon \simeq 10^{-3}$-$10^{-4}$.

Another interesting implication of left-right symmetric models follows from the electric charge formula in eqn. (2) once we switch on parity violation. Since $\Delta Q = 0$ and $\Delta I_{3_L} = 0$ for $\mu \geq m_{W_L}$, we obtain [44] $\Delta(B-L) = 2|\Delta I_{3_R}|$. This, of course, implies Majorana neutrino as mentioned before; and for processes involving no leptons, implies $\Delta B = 2$ transitions such as neutron - anti-neutron oscillations. The strength of the $\Delta B = 2$ transitions, however, depends on the scale of $SU(4)_C$ quark-lepton unification. For $M_x \simeq 10^5$-10^6 Gev as may arise in a class of models with SO(10) grandunification [7], we estimate N-$\bar{N}$ oscillation time of about 10^8-0^9 sec. An important point is that this is accessible to presently ongoing reactor experiments [45]. Another test of the intermediate $SU(4)_C$ unification is the decay $K_L^o \to \mu\bar{e}$, which can proceed with a branching ratio of $\simeq 10^{-10}$.

§7. SO(10) Grandunification

So far we have discussed left-right symmetry group as a partial unification group of electroweak and nuclear forces. In this section, we discuss the prospects for grandunification and its impact on the low energy predictions of the model. The natural grandunification group in our case is SO(10) which has the maximal subgroup $SU(2)_L \times SU(2)_R \times SU(4)_C \times D$ where D is a discrete symmetry which changes from left to right SU(2) groups. In the original analysis of the breaking of SO(10) symmetry down to its subgroups, the role of D-parity was not recognized; as a result, several problems were encountered. To understand these problems, we start by discussing the predictions for proton decay in SO(10) models. Ignoring Higgs contributions and assuming that the left-right symmetric group $SU(2)_L \times SU(2)_R \times U(1)_{B-L} \times D$ appears as an intermediate symmetry group above a scale M_R, it was shown [46] that,

$$(\tau_p)_{SO(10)} = (\tau_p)_{SU(5)} \left(\frac{M_U}{M_R}\right)^2 \tag{15}$$

where M_U is the GUT-scale for the minimal SU(5) model. This, therefore, clearly indicated that to reconcile the present life times on $p \to e^+\pi^o$ decay mode, we must have $M_R < (M_U/10)$. On the other hand, it was noted that if D-parity and $SU(2)_R$ was broken together as is done when one employs a 126-dimensional representation by $SU(2)_R$, domain walls appear [47] whose decay probabilty goes like $\exp - (M_U/M_R)^6$; eqn. (15) would then imply that these domain walls would have no time to disappear, thus giving too much mass to the universe in conflict with observations. This problem could, of course, be solved by the inflationary scenario but then it would require that M_R must be bigger than the reheating temperature of 10^{12}-10^{13} GeV, making right-handed symmetry unobservable at low energies.

It has also been observed [48] that baryon generation in SO(10) model acquires an additional suppression factor of $(M_R/M_U)^4$ over SU(5) model essentially due to the fact that D-parity acts like charge conjugation on fermions. One would have problems in generating an adequate baryon number of the universe unless $M_R \approx M_U$, which would conflict with eqn. (15). Thus, a low right-handed boson mass would appear to be precluded in SO(10) and the conventional symmetry breaking scenario would have some problems.

All the above problems are avoided in a scenario proposed in Ref. 7, where it was shown that parity and the $SU(2)_R$ symmetries need not be broken at one scale by explicit choice of Higgs bosons in both the left-right symmetric as well as SO(10) theories. One can employ a 210-dimensional Higgs boson to break D-parity but leaving $SU(2)_R$-group intact. The scale (to be called M_P) can then be made equal by M_U thereby avoiding both the domain wall and baryosynthesis problems. Furthermore, since this scenario leads to $g_L \neq g_R$ at low energies, it can also accomodate an M_{W_R} in the TeV range without conflict with observed $\sin^2\theta_W$.

In conclusion, we have summarized the salient implication of the idea that weak interactions may become parity conserving at a distance scale $\simeq 10^{-21}$ cm or so, and discussed the constraints on the parameters of the model from observations at low energies. Some crucial tests of the idea are noted and the proper way to embed in SO(10) grandunified group is discussed.

References

1. J. C. Pati and A. Salam, Phys. Rev. D10, 275 (1974);
   R. N. Mohapatra and J. C. Pati, Phys. Rev. D11, 566, 2558 (1975);
   G. Senjanovic and R. N. Mohapatra, Phys. Rev. D12, 1502 (1975);
   For a recent review and more complete reference see R. N. Mohapatra, "Quarks, Leptons and Beyond", ed. by H. Fritzsch et al., Plenum (1985). p.
2. H. Fritzsch, Phys. Lett. 73B, 317 (1978);
   B. Stech, Phys. Lett. 130B, 189 (1983);
   G. Ecker, W. Grimus and W. Konetschny, Nuc. Phys. B177, 489 (1981);
   R. N. Mohapatra, Erice Lecutres, 1984; ed. L. L. Chau (Plenum, N.Y. 1985).
3. R. N. Mohapatra and G. Senjanovic, Phys. Rev. Lett. 44, 912 (1980) and Phys. Rev. D23, 165 (1981).
4. M. A. B. Beg and H. S. Tsao, Phys. Rev. Lett. 41, 278 (1978);
   R. N. Mohapatra and G. Senjanovic, Phys. Lett. 79B, 283 (1978).
5. R. N. Mohapatra and J. C. Pati, Phys. Rev. D11, 566 (1975).
6. D. Chang, R. N. Mohapatra, P. Pal and J. C. Pati, Phys. Rev. Lett. 55, 2756 (1985).
7. D. Chang, R. N. Mohapatra and M. K. Parida, Phys. Rev. Lett. 52, 1072 (1984);
   D. Chang, R. N. Mohapatra, J. Gibson, R. E. Marshak and M. K. Parida, Phys. Rev. D31, 1718 (1985).
8. F. Gursey, P. Sikivie and P. Ramond, Phys. Lett. 60B, 117 (1976);
   Y. Achiman and B. Stech, Phys. Lett. 77B, 389 (1976);
   P. K. Mohapatra, R. N. Mohapatra and P. Pal, Phys. Rev. D33, 2010 (1986)
   J. Rosner, Comments on Nuc. and Part. Phys. 15, 195 (1986).
   T. Rizzo, Ames Laboratory preprints (1986).
9. E. Witten, Nuc. Phys. B258, 75 (1985).
10. Riazuddin, R. E. Marshak and R. N. Mohapatra, Phys. Rev. D24, 1310 (1981).
11. R. N. Mohapatra and R. E. Marshak, Phys.Rev. Lett. 44, 1316 (1980).
12. D. Chang and R. N. Mohapatra, Phys. Rev. D32, 1248 (1985).
13. D. Chang, R. N. Mohapatra and M. K. Parida, ref. 7.
14. S. Mikheyev and Y. Smirnov, Moscow Preprint (1985);
    H. Bethe, Phys. Rev. Lett. 56, 1305 (1986);
    L. Wolfenstein, Phys. Rev. D17, 2369 (1978).
15. R. N. Mohapatra and J. W. F. Valle, Phys. Rev D (1986) (to appear);
16. M. Roncadelli and D. Wyler, Phys. Lett. 133B, 325 (1983).
    P. Roy and O. Shankar, Phys. Rev. Lett. 52, 713 (1984).
    G. Ecker, W. Grimus and M. Gronau, Vienna Preprint (1986).
17. R. N. Mohapatra and J. W. F. Valle, Phys. Lett. B (to appear) (1986).
18. V. Barger, E. Ma and K. Whisnant, Phys. Rev. D26, 2378 (1982).
19. L. Durkin and P. Langacker, Phys. Lett. B (to appear) (1986).
20. For earlier analysis see, J. Rosner, Comments in Nuclear and Particle Physics, (1985).
21. M. A. B. Beg, R. Budny, R. N. Mohapatra and A.Sirlin, Phys. Rev. Lett. 38, 1252 (1977); For further discussion J.Maalampi, K. Mursula and M. Roos, Nuc. Phys. B207, 233 (1982).

22. J. Carr et al., Phys. Rev. Lett. 51, 627 (1983);
D. P. Stoker et al., Phys. Rev. Lett. 54, 1887 (1985).
23. P. Herczeg, Los Alamos Preprint LA-UR-85-2761 (1985).
24. R. N. Mohapatra, New Frontiers in High Energy Physics, ed. A. Perlmutter and L. Scott (Plenum, New York, 1978), p. 337.
25. J. F. Donoghue and B. R. Holstein, Phys. Lett. 113B, 382 (1982).
26. G. Beall, M. Bander and A. Soni, Phys. Rev. Lett. 48, 848 (1982).
27. I. I.Bigi and J. M. Frere, Phys, Lett. 110B, 255 (1982).
28. H. Abramowicz et al, Z. Phys. C12, 225 (1982)
29. L. Wolfenstein, Phys. Rev. D29, 2130 (1984).
30. E. Masso, Phys. Rev. Lett. 52, 1956 (1984).
31. D. O. Caldwell et al., Phys. Rev. D33, 2737 (1986).
T. Ejiri et al., Nuc. Phys. A448, 27 (1986).
E. Fiorini et al., Phys. Lett. 146B, 450 (1984)
F. T. Avignone III et al., Phys. Rev. Lett. 54, 2309 (1985).
32. R. N. Mohapatra, Phys. Rev. D34 (1986).
33. For reviews see, W. Haxton and G. Stephenson, Jr., Progress in Nuclear and Particle Physics 12, 409 (1984);
M. Doi, T. Kotani and E. Takasugi, Prog. Theor. Phys. Suppl. 83, 1 (1985); J. Vergados, Phys. Rep. 133, 1 (1986).
34. M. Gronau and R. Yahalom, Nuc. Phys. B236, 233 (1984);
M. Gronau, C. Leung and J. Rosner, Phys. Rev. D29, 2539 (1984).
35. U. Chattopadhyaya and P. Pal, Univ. of Maryland Preprint (1986).
36. M. Roncadelli and G. Senjanovic, Phys. Lett. 107B, 59 (1983).
37. P. Pal, Nuc. Phys. 227B, 237 (1983).
38. R. N. Mohapatra and P. Pal, Univ. of Maryland Preprint (1986).
39. For a review, see L. L. Chau, this volume.
40. K. Shubert, this volume.
41. R. N. Mohapatra, ref 24.
H. Harari and M. Leurer, Nuc. Phys. B233, 221 (1984).
42. G. Beall and A. Soni, Phys. Rev. Lett. 47, 552 (1981);
G. Ecker, W. Grimus and H. Neufeld, Nuc. Phys. B229, 421 (1983).
43. D. Chang and R. N. Mohapatra, Phys. Rev. D30, 2005 (1984).
43a. G. Branco, J. M. Frere and J. M. Gerand, Nucl. Phys. B221, 317 (1983);
D. Chang, Nuc. PHys. B214, 435 (1983);
R. N. Mohapatra, PHys. Lett. 159B, 374 (1985).
44. R. N. Mohapatra and R. E. Marshak, Phys. Rev. Lett. 44, 1316 (1980).
45. M. Baldoceolin, this volume.
46. Y. Tosa, G. Branco and R. E. Marshak, Phys. Rev. D28, 1731 (1983).
47. T. Kibble, G. Lazaridis and Q. Shafi, Phys. Rev. D26, 435 (1982).
48. V. Kuzmin and M. Shaposnikov, Phys. Lett. 92B, 115 (1980).

Determination of the Electro-Weak Mixing Angle in Neutrino Interactions

U. Dore

Dipartimento di Fisica, Università "La Sapienza", Roma, INFN, Sezione di Roma, P. le Aldo Moro, 2, I-00185 Roma, Italy

All the experimental values of the electro-weak mixing parameter $\mathrm{sen}^2\theta_W$ = S2 obtained so far in a large variety of physical processes are well consistent with a single value of this constant as predicted by the Glashow-Salam-Weinberg theory at the lowest order. Precision measurements of this parameter in different processes allow a test of the theory at the level of first order radiative electroweak corrections.

In this report measurements of S2 in neutrino interactions will be reviewed. The determination of S2 can come from the study of neutrino-nucleon deep inelastic scattering (Sect. 1) and from the study of ν_μ + e and $\bar{\nu}_\mu$ + e elastic scattering (Sect. 2). Section 3 will be devoted to the comparison of the neutrino results on S2 with the ones obtained from the measurements of the W and Z° boson masses at the CERN proton-antiproton Collider.

1. Determination of S2 in ν_μ-N Deep Inelastic Scattering

S2 can be extracted (neglecting small corrections that will be discussed later) by the relation /1/

$$R = \frac{\phi NC}{\phi CC} = \frac{1}{2}\,\mathrm{sen}^2\theta_W + \frac{5}{9}\,\mathrm{sen}^2\theta_W\,(1 + r) \qquad (1.1)$$

where R is the ratio of the neutral (NC) to the charged current (CC) ν_μ-N cross section and r is the ratio of neutrino and antineutrino CC cross sections. Here and everywhere in this report we have assumed $\rho = 1$. ρ = ratio of NC and CC currents strenghts.

Several experiments have been performed and S2 has been obtained from (1.1). The corresponding results are given in table 1.

The crucial point in these experiments is a clean unbiased separation of charged and neutral current events i.e. of events containing a muon and of muonless events. This separation is based either on the topology or on the length of the events. Muons have a much bigger penetration in the detectors than hadronic showers. Systematic error are mainly due to the following causes:

1) a muon from a π or K decay changes a NC into a CC event.

Table 1

Experiment	S2	stat error	syst error	tot exp error
CCFR /2/	0.239	0.008	0.006	(0.010)
CDHS /3/	0.225	(0.004)	(0.003)	0.005
CHARM /4/	0.236	(0.003)	(0.004)	0.005
FMMF /5/	0.246	0.012	0.011	(0.016)

2) a low energy muon is hidden by the hadronic shower.

3) ν_e interactions (from the ν_e contamination of the beam) are interpreted as ν_μ NC interactions.

Statistical and systematic error due to the above corrections are shown in table 1 together with the total experimental error. Numbers within parentheses have been deduced from the published values.

Theoretical corrections need to be applied to the S2 value obtained using relation (1.1). These corrections take into account the fact that formula (1.1) is valid in the presence of u and d quarks and isoscalar targets. So corrections must be made to take into account the presence of s and c quarks in the nucleon. Radiative corrections have also to be applied.

The values of these corrections are similar in all the experiments quoted in table 1 and have been extimated to be affected by an error of the order of $\pm$ 0.005.

Combining all the previous results we obtain

$$S2 = 0.232 \pm 0.003 \pm 0.005$$

where the first error contains statistical and systematic experimental uncertainties and the second one is the theoretical uncertainty.

No large improvement can be obtained in this sector if our knowledge of the hadronic part will not improve.

2. Determination of S2 in Purely Leptonic Neutrino interactions

The ratio of the cross sections for the processes

$$\nu_\mu + e \to \nu_\mu + e \tag{2.1}$$

$$\bar{\nu}_\mu + e \to \bar{\nu}_\mu + e \tag{2.2}$$

is a function of S2

$$R_V = \frac{\sigma(\nu_\mu + e)}{\sigma(\bar{\nu}_\mu + e)} = 3 \times \frac{1 - 4\,\mathrm{sen}^2\theta_W + \frac{16}{3}\,\mathrm{sen}^4\theta_W}{1 - 4\,\mathrm{sen}^2\theta_W + 16\,\mathrm{sen}^4\theta_W} \tag{2.3}$$

In the region where S2 $\simeq$ 0.2 this ratio changes very steeply with S2. Hence

$$\Delta sen^2_W = \frac{1}{8}\frac{\Delta R_V}{R_V} \tag{2.4}$$

So a measure of R_V at the 1% level will give a measurement of S2 at 0.001. Other advantages of this method of measuring S2 are that all systematic effects in experimental acceptances will tend to cancel out in the ratio as well as effects due to radiative corrections. The difficulty of this type of experiments comes from the smallness of the above processes cross section:

$$\phi(\nu_\mu + e) \sim 10^{-4} \;\; \phi(\nu_\mu + N) \sim 10^{-42}\ cm^2 \tag{2.5}$$

This implies that: i) Very large detectors are needed. ii) One must separate the events due to reactions (2.1) and (2.2), that are events with a single electromagnetic shower, from a large majority of events having an hadronic shower. iii) The signal must be extracted from all background processes that give also an electromagnetic shower in the final state. These events, that have a much larger cross section are ν_e quasi elastic processes from the ν_e contamination of the beam and π° coherent production on nuclei.

Concerning this last point the qualifying characteristic of any experiment is its angular resolution, since the angular distribution of the processes (2.1) and (2.2) is much narrower than the one of the background processes. The ultimate limitation to the precision of the method is given by the error on the knowledge of the flux ratios.

Two experiments have measured these reaction: CHARM at CERN /6/ and the E734 experiment at BNL /7/. Their results are summarized in table 2.

Table 2

Experiment	# events	S2	stat error	syst error
CHARM	195 $\pm$ 26	0.215	0.032	0.012
E734	104 $\pm$ 14	0.209	0.029	0.013

The average value that can be obtained from the two previous results is

$$S2 = 0.212 \pm 0.023 \pm 0.08$$

The CHARM result refers to the full sample of events, whereas experiment E34 has collected a total of 290 events. The data collection is just finished and the final error that will be obtained will be of the order of $\pm$ 0.015. New dedicated experiments are under way or now being proposed. The CHARM2 /8/ experiment is ready to take data this year and the final statistics should be enough to achieve an error of ~0.005.

3. Determination of S2 from the Mass of the W and Z° particles

As said in the introduction the most precise measurements outside the neutrino field come from the CERN Collider measurements of the W and Z° masses. At present tests of the standard model at the level of the electro-weak radiative corrections coming from this kind of information are not significant but the situation will improve in the years to come.

Let us review the present experimental information on the masses that comes from the UA1 /9/ and UA2 /10/ experiments.

Z° mass (GeV)	$93.0 \pm 1.4 \pm 3.2$	UA1	$91.5 \pm 0.9 \pm 1.5$	UA2
W mass (GeV)	$83.5^{+1.1}_{-1.0} \pm 2.7$	UA1	$80.3 \pm 0.8 \pm 1.3$	UA2

The first error is statistical, the second is systematical and comes mainly from the energy scale calibration.

Following the definitions of Marciano and Sirlin /11/ we have

$$S2 = 1 - \frac{MW^2}{MZ^{\circ 2}} \tag{3.1}$$

$$MW^2 = \frac{A^2}{(1-Dr)\, S2} \tag{3.2}$$

$$MZ^{\circ 2} = \frac{A^2}{(1-Dr)\, S2\, (1-S2)} \tag{3.3}$$

$$A = \frac{\pi \alpha}{\sqrt{2}\, G_F} = 37.28 \text{ GeV} \tag{3.4}$$

$$Dr = 0.0696 \pm 0.002 \tag{3.5}$$

The factor (1-Dr) takes into account first order radiative corrections. Using (3.1) one obtains

$S2 = 0.194 \pm 0.031$ UA1 $\qquad$ $S2 = 0.230 \pm 0.020 \pm 0.009$ UA2

It must be noted that the systematic errors cancel out in large part in the ratio. Combining (3.2) and (3.3) and assuming the value (3.5) for Dr to be known one can compute S2 with great precision:

$S2 = 0.214^{+0.005}_{-0.006} \pm 0.015$ UA1 $\qquad$ $S2 = 0.232 \pm 0.004 \pm 0.008$ UA2

Viceversa eliminating S2 from (3.2) and (3.3) the UA2 coll. gives for Dr the value

$Dr = 0.08 \pm 0.10 \pm 0.03$ UA2

The errors do not allow any significant check of (3.5).

We can use the neutrino value of S2 and we obtain

$$Dr = 0.07 \pm 0.02 \pm 0.04$$

so the radiative corrections are tested but with a 50% error. The dominant error is the systematic one. To this contribute the systematic errors on the mass scale (main contribution) and the theoretical errors on the neutrino result.

The increase in statistics in the Collider data due to ACOL and better energy calibrations can bring down the error to $\pm$ 0.01 $\pm$ 0.02. When a Z° mass measurement will come out of LEP (the mass of the Z° will be measured with an error of $\pm$ 50 MeV) and values of S2 will be given from purely leptonic channels (Sect. 2) the systematic errors will become negligible. The error on the final results will depend from the statistics of the neutrino experiments. A measurement of S2 at $\pm$ 0.005 will determine Dr with a 20% error.

References

1 C.H.Llewellyn-Smith: Nucl. Phys. B228, 205 (1983).

2 CCFR Coll., F.Merritt: Neutrino-86 Conf. Sendai, Japan, June 1986.

3 CDHS Coll., G.Guyot: Neutrino-86 Conf. Sendai, Japan, June 1986.
H.Abramowicz et al.: submitted to Physics Letters.

4 CHARM Coll., J.Panman: Neutrino-86 Conf. Sendai, Japan, June 1986.
J.Abt et al.: submitted to Physics Letters.

5 FMMF Coll., R.Brock: Neutrino-86 Conf. Sendai, Japan, June 1986.
D.Bogert et al.: Phys. rev. Lett. 55, 1969 (1985).

6 CHARM Coll., F.Bergsma et al.: Phys. Lett. B147, 481 (1984).

7 E734 expt. Y.Suzuki: Neutrino-86 Conf. Sendai, Japan, June 1986.
L.A.Aharens et al.: Phys. rev. Lett. 54, 18 (1985).

8 CHARM2: CERN/SPSC/83-24.

9 UA1 Coll., S.J.Winpenny: XXI Rencontres de Moriond, Les Arcs, France, March 1986.

10 UA2 Coll., A.Clark: Neutrino-86 Conf. Sendai, Japan, June 1986.

11 W.J.Marciano and A.Sirlin: Phys. Rev. D29, 945 (1984).

Neutrinoless Double β-Decay and Lepton Flavor Violation*

G.K. Leontaris and J.D. Vergados

The University of Ioannina, Department of Physics, Ioannina, Greece

We discuss the phenomenological implications of gauge theories on lepton number and lepton flavor non-conservation. In particular we compare neutrinoless double β-decay to muon violating processes.

1. INTRODUCTION

Lepton number (L) and lepton flavor (L_f) appear to be conserved in all experiments performed thus far [1]. Such a conservation is understood in the context of the phenomenologically successful standard model as a consequence of a global symmetry of the Lagrangian. This does not completely settle the issue, however, since only (local) gauge symmetries are accepted as exact. Beyond the standard model one can achieve both L and L_f non-conservation. The most familiar mechanism is that which involves massive intermediate neutrinos. Thus the presence of Dirac mass terms which enter when the right handed neutrino exists, lead to L_f violation. Furthermore the presence of Majorana mass terms , which connect a neutrino with an antineutrino, may lead to $\Delta L = 2$ transitions. In any case the leptonic currents are no longer diagonal but they can take the form [2]

$$\begin{aligned} j_\mu^L &= 2(\bar{e}_L\gamma_\mu U^{(11)}\nu_L + \bar{e}_L\gamma_\mu U^{(12)}N_L + \text{h.c}), \\ j_\mu^R &= 2(\bar{e}_R\gamma_\mu U^{(21)}\nu_R + \bar{e}_R\gamma_\mu U^{(22)}N_R + \text{h.c}) \end{aligned} \tag{1.1}$$

The conjugate currents are easily expressed [2] as:

$$\begin{aligned} (j_\mu^L)^c &= 2(\bar{\nu}_R e^{i\alpha}(U^{(11)})^T e_R^c + \bar{N}_R e^{i\varphi}(U^{(12)})^T\gamma_\mu e_R^c) + \text{h.c.} \\ (j_\mu^R)^c &= 2(\bar{\nu}_L e^{i\alpha}(U^{(21)})^T e_L^c + \bar{N}_L e^{i\varphi}(U^{(22)})^T\gamma_\mu e_L^c) + \text{h.c.} \end{aligned} \tag{1.2}$$

Less familiar mechanisms involve exotic Higgs scalars [3] or even supersymmetric partners of known particles [4].

2. GROSS FEATURES OF L AND L_f VIOLATING PROCESSES IN NUCLEI

The oldest lepton violating process related to the nature of the neutrino (Majorana or Dirac) is neutrinoless ($o\nu$) $\beta\beta$-decay

---

* Presented at the Conference by J.D. Vergados

$$(A,Z) \to (A,Z \pm 2) + e^{\mp} + e^{\mp}, \quad e_b^- + (A,Z) \to (A,Z-2) + e^+ \tag{2.1}$$

which together with the allowed (2ν) $\beta\beta$-decay

$$(A,Z)\to(A,Z\pm2)+e^{\mp}+e^{\mp}+ \begin{cases} 2\tilde{\nu}_e \\ 2\nu_e \end{cases}, \quad e_b^- + (A,Z)\to(A,Z-2)+e^- +\nu_e + \tilde{\nu}_e \tag{2.2}$$

are the only decay modes of some otherwise absolutely stable nuclei [2].

One finds that

$$T_{1/2}(o\nu) = K_{o\nu}/|\eta|^2 |ME|^2_{o\nu}, \quad T_{1/2}(2\nu) = K_{2\nu}/|ME|^2_{2\nu} \tag{2.3}$$

where η is a suitably defined lepton violating parameter and $|ME|^2$ are the corresponding nuclear matrix elements. The quantities $K_{o\nu}$ and $K_{2\nu}$ are functions of (A,Z) which can take the following values [2]

$$1.5 \cdot 10^{13}y < K_{o\nu} < 2.5 \cdot 10^{17}y, \quad 2.5 \cdot 10^{9}y < K_{2\nu} < 3.3 \cdot 10^{26}y \tag{2.4}$$

Thus if η, which contains all information about the gauge models, is not much smaller than 10^{-5}, the present experimental limit [1], detection of $o\nu$ $\beta\beta$-decay is within the capabilities of present experiments.

Another interesting process is the (μ^-,e^+) conversion [2,5]

$$\mu^- + (A,Z) \to e^+ + (A,Z-2)^*, \quad T_{1/2}(\mu^-,e^+) = K'_{o\nu}\ |\eta'|^2 |ME'|^2_{o\nu} \tag{2.5}$$

where

$$K'_{o\nu}(\mu^-,e^+)=2.2 \cdot 10^{10}y\ A^{2/3}/(Z^4_{eff}/Z) \quad \text{or} \quad 5 \cdot 10^{7}y < K'_{o\nu} < 6 \cdot 10^{8} y \tag{2.6}$$

for nuclei ranging from ^{58}Ni to ^{12}C. Even though this process is 10^{10} faster than its sister (e^-,e^+) conversion, it must compete against ordinary muon capture $\mu^- + A \to A^* + \nu_\mu$. Thus one gets a branching ratio

$$R = \Gamma(\mu^-,e^+)/\Gamma(\mu^-,\nu_\mu) = 1.5 \cdot 10^{-21}|\eta'|^2|ME'|^2/|Z(1.62(Z/A)-0.62)| \tag{2.7}$$

where η' is the corresponding lepton violating parameter. Thus for $|\eta'| \approx |\eta| \le 10^{-5}$ this process is unobservable even if transitions to all final nuclear states are considered $|ME'|^2 \approx 0.1Z^2 = 100$. (The present experimental limit is $R < 9 \cdot 10^{-9}$, see A. BADERTSCHER ref. [1]).

Among the various L_f violating processes the most interesting for this audience is the process

$$\mu^- + (A,Z) \to e^- + (A,Z)^* \tag{2.8}$$

For ground state transitions we get [2]

$$R = \frac{\Gamma(\mu^- \to e^-)}{\Gamma(\mu^- \to \nu_\mu)} = |\tilde{\eta}_\alpha|^2 \rho_\alpha \frac{E_e p_e}{m_\mu^2} \frac{h_\alpha(A,Z)}{Z[1.62(Z/A)-0.62]} |F_{ch}(q^2)|^2, \quad \alpha = \nu,N \tag{2.9}$$

In the special case of ^{58}Ni with $F_{ch}(-m_\mu^2) = 0.5$ we obtain

$$R = \begin{cases} 2 \cdot 10^{-21} |\tilde{\eta}^{L}_{\nu}|^2 \\ 2 \cdot 10^{-2} \; |\tilde{\eta}^{L}_{N}|^2 \end{cases} \text{(L-L)}, \; R = \begin{cases} 1.0 \cdot 10^{-8}(|\tilde{\eta}^{+}_{\nu}|^2 + |\tilde{\eta}^{-}_{\nu}|^2) \\ 3.0 \cdot 10^{-2}(|\tilde{\eta}^{+}_{N}|^2 + |\tilde{\eta}^{-}_{N}|^2) \end{cases} \text{(L-R)} \quad (2.10)$$

where the upper row refers to light neutrinos while the lower to heavy ones.

Comparison with the corresponding branching ratios [2] for $\mu \to e\gamma$

$$\frac{R(\mu \to e)}{R(\mu \to e\gamma)} = \begin{cases} 6 \cdot 10^3 \\ 1 \cdot 10^3 \end{cases} \text{(L-L)}, \quad \frac{R(\mu \to e)}{R(\mu \to e\gamma)} = \begin{cases} 0.14 \\ 0.14 \end{cases} \text{(L-R)} \quad (2.11)$$

shows that (μ^-, e^-) is favorable in left handed theories.

3. L AND L_f VIOLATING PARAMETERS

It is now straightforward to compute the amplitude for L and L_f violating processes. Typical diagrams are shown in figs. (1) and (2).

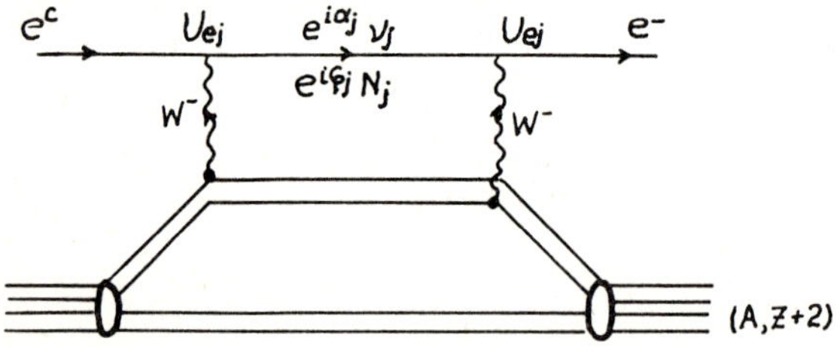

Fig. 1. Neutrinoless $\beta\beta$-decay mediated by Majorana neutrinos. The relevant submatric for the coupling depends on the type of leptonic current (L or R) and the neutrino type (ν_j or N_j). The diagram for (μ^-, e^+) is analogous.

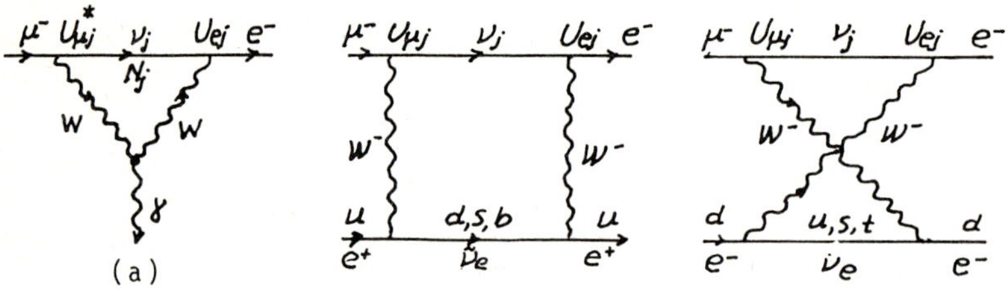

Fig. 2. Typical diagrams which lead to flavor change. Diagram (a) can also lead to (μ^-, e^-) and $\mu \to ee^+e^-$ if the photon is virtual. Diagrams with W replaced by Higgs scalars must also be considered when appropriate.

We are now in a position to obtain expressions for the lepton violating parameters. We distinguish the following cases

i) Both leptonic currents of fig. (1) and (2) are left-handed (j_L-j_L). Then for light neutrinos ($m_j \ll m_W$) we get

$$\eta_\nu = \eta^L_\nu = \frac{\langle m_\nu \rangle}{m_e}, \; \langle m_\nu \rangle = \sum_j (U^{(11)}_{ej})^2 e^{i\alpha_j} m_j \quad (0\nu \;\; \beta\beta\text{-decay}) \qquad (3.1a)$$

$$\eta' = \eta'^{L}_{\nu} = \sum_j U^{*(11)}_{ej} U^{*(11)}_{\mu j} e^{-i\alpha_j} m_j \qquad (\mu^- \to e^+) \qquad (3.1b)$$

$$\tilde{\eta} = \tilde{\eta}^L_\nu = \sum_j U^{(11)}_{ej} U^{*(11)}_{\mu j} (m_j/m_e)^2 \qquad (\mu \to e\gamma,\ \mu^- \to e^-,\ \mu \to \bar{e}ee) \qquad (3.1c)$$

Note the presence of CP eigenvalues in (3.1a) and (3.1b) which distinguish between Majorana and Dirac particles, and the unfavorable mass dependence in case of (3.1c). In the case of heavy neutrinos ($m_j \geq m_W$) one finds

$$\eta = \eta^L_N = \sum_j (U^{(12)}_{ej})^2 e^{i\varphi_j} (m_p/m_j) \qquad (m_p\text{: proton mass}) \qquad (3.2a)$$

$$\eta' = \eta'^L_N = \sum_j U^{*(12)}_{ej} U^{(12)}_{\mu j} e^{-i\varphi_j} m_p/M_j \qquad (3.2b)$$

$$\tilde{\eta} = \tilde{\eta}^L_N = \sum_j U^{(12)}_{ej} U^{(12)}_{\mu j} e^{-i\varphi_j} (m_W/M_j)^2 [a\ln(M_j/m_W)^2 + b] \qquad (3.2c)$$

where a and b are constants of order unity [2] which depend on the details of the graph of fig. (2). Note that $U^{(12)}$ is expected to be small.
ii) Both leptonic currents are right-handed (j_R-j_R). The light neutrino contribution is now negligible. For heavy neutrinos one gets

$$\eta = \eta^R_N = (\epsilon^2 + \kappa^2) \sum_j (U^{(22)}_{ej})^2 e^{i\varphi_j} m_p/M_j \qquad (3.3a)$$

$$\eta' = \eta'^R_N = (\epsilon^2 + \kappa^2) \sum_j U^{*(22)}_{ej} U^{*(22)}_{\mu j} e^{i\varphi_j} m_p/M_j \qquad (3.3b)$$

$$\tilde{\eta} = \tilde{\eta}^R_N = (\epsilon^2 + \kappa^2) \sum_j U^{(22)}_{ej} U^{(22)}_{\mu j} (m_W/M_j)^2 [a\ln\frac{M_j^2}{m_W^2} + b - a\ln\frac{\epsilon^2}{\kappa^2+\epsilon^2}] \qquad (3.3c)$$

where $\kappa = (m_W/m_{WR})^2$ and $\epsilon = \tan\zeta$, $\zeta = w_L - w_R$ mixing angle. The suppression now is due to the mass of the vector boson which mediates the right-handed interaction and/or the small w_L- w_R mixing.
iii) The leptonic current is j_L- j_R type. In this case the helicities are such that one picks the momentum instead of the mass of the intermediate neutrinos. One finds

$$\eta = \eta_{RL} = \sqrt{\kappa^2+\epsilon^2} \sum_j U^{(11)}_{ej} U^{*(21)}_{\mu j} e^{i\alpha_j},\quad \eta' = \eta'_{RL} = \sqrt{\kappa^2+\epsilon^2} \sum_j U^{*(11)}_{ej} U^{*(11)}_{\mu j} e^{-i\alpha_j} \qquad (3.4a)$$

$$\tilde{\eta} = \tilde{\eta}^{\pm}_\nu = (\kappa \text{ or } \epsilon) \sum_j (U^{(11)}_{ej} U^{*(21)}_{\mu j} \pm U^{(21)}_{ej} U^{*(11)}_{\mu j})\ m_j/m_e \qquad (3.4b)$$

$$\tilde{\eta} = \tilde{\eta}^{\pm}_N = (\kappa \text{ or } \epsilon) \sum_j (U^{(12)}_{ej} U^{*(22)}_{\mu j} \pm U^{(22)}_{ej} U^{*(12)}_{\mu j})\ m_w/M_j \qquad (3.4c)$$

We note the favorable explicit dependence on the neutrino mass but we caution that this may be offset by the smallness of $U^{(21)}$ and $U^{(12)}$. (The heavy neutrino contribution in η and η' is negligible).

4. INTERMEDIATE HIGGS SCALARS AND SUPERSYMMETRIC PARTICLES

One can have L and L_f violation, even if the neutrino mass mechanism is suppressed, via intermediate Higgs scalars [2]. Typical L_f violating diagrams are shown in fig. (3). ($o\nu$) $\beta\beta$-decay and (μ^-, e^+) cannot occur directly since the

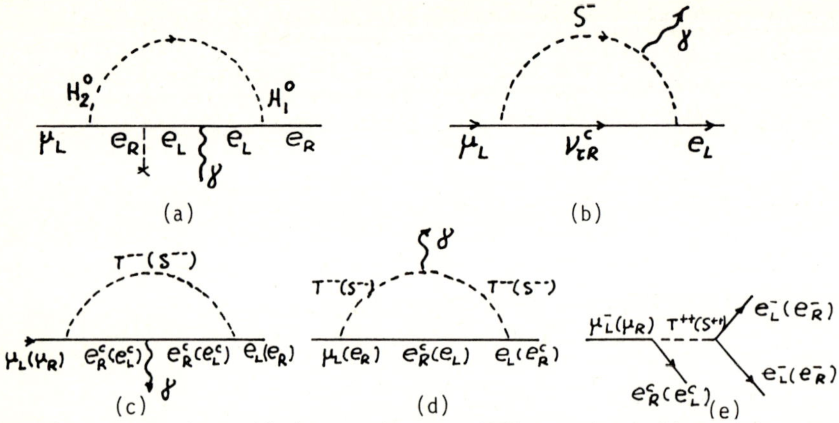

Fig.3. $\mu \to e\gamma$ decay if there exist two different isodoublets (a), a singly charged isosinglet (b) (which always changes flavor) and doubly charged isosinglet (S^{--}) and isotriplet (T^{--}) (c) and (d). In the last case $\mu \to ee^+e^-$ can proceed faster at the tree level (e).

exotic Higgs scalars (S^-,S^{--} and T^{--}) do not directly couple to the quarks. They can proceed, however, via the couplings of such Higgs with the ordinary Higgs isodoublets or with the right-handed vector bosons [2,5]. Thus it is natural to expect that L_f will be favored over L. The contribution of supersymmetric particles is discussed elsewhere [4].

5. CONCLUSIONS

The most conventional L and L_f violating mechanisms involve the neutrino mass. Since the neutrinos produced in weak interactions are linear combinations of mass eigenstates the mass combination measured in various experiments [2] are:

i) Decay experiments producing flavor α measure

$$\sum_j |U_{\alpha j}|^2 m_j \qquad \text{or} \qquad |U_{\alpha j}|^2 \quad \text{and} \quad m_j$$

depending in whether the masses can be resolved.

ii) (oν) $\beta\beta$-decay measures the quantity

$$|\langle m_\nu \rangle| = |\sum_j U_{ej}^2 e^{i\alpha_j} m_j|$$

Thus it is possible that [1] $\langle m_\nu \rangle = 0$ even if the neutrinos are massive Majorana particles.

iii) Neutrino oscillations in principle can disentangle [2] the neutrino mixing from the mass and measure the quantity

$$\delta m^2 \frac{L}{E_\nu} \quad , \qquad \delta m^2 = m_\kappa^2 \quad m_e^2$$

where E_ν = neutrino energy and L = source - detector distance.

The present limits on lepton violating parameters as well as the predictions of some models are given in Table I. From the table we see that the most likely

Table I The Lepton violating parameters in some gauge models which are completely analyzed in ref. [2]. The experimental limits are obtained (a) from (oν) $\beta\beta$-decay, T. KIRSTEN et al. (b) from (μ^-,e^-) BRYMAN et al., (c) from $\mu \rightarrow e\gamma$ KINNISON et al., ref. [1].

Parameter \ model	Witten	SO(10)	M - $S_{(A)}$	M - $S_{(B)}$	Experimental limit	
η_ν^L	$2x10^{-6}$	$4x10^{-9}$	$1.5x10^{-6}$	$4.0x10^{-6}$	$1.0x10^{-6}$	(a)
η_N^L	$5x10^{-14}$	$6.0x10^{-34}$	$2.0x10^{-8}$	-	$6.0x10^{-7}$	(a)
η_N^R	$1.0x10^{-7}$	$2x10^{-11}$	$3.0x10^{-12}$	$4.0x10^{-7}$	$6.0x10^{-7}$	(a)
η_{RL}	$5.0x10^{-7}$	$6x10^{-12}$	-	-	$1.0x10^{-6}$	(a)
$\tilde{\eta}_\nu^L$	$1.0x10^{-7}$	$1.4x10^{-13}$	$1.3x10^{-7}$	$2.1x10^{-5}$	$1.0x10^{5}$	(b)
$\tilde{\eta}_N^L$	-	$2.5x10^{-38}$	$9.4x10^{22}$	-	$3.0x10^{-5}$	(b)
$\tilde{\eta}_N^{(R)}$	-	0	$1.8x10^{-11}$	$4.2x10^{-7}$	$3.0x10^{-5}$	(b)
$\tilde{\eta}_\nu^{(+)}$	$5.0x10^{-12}$	$1.6x10^{-17}$	$2.2x10^{-22}$	$8.9x10^{-6}$	$4.0x10^{-2}$	(c)
$\tilde{\eta}_\nu^{(-)}$	$5.0x10^{-12}$	$2.0x10^{-17}$	0	0	$4.0x10^{-2}$	(c)
$\tilde{\eta}_N^{(+)}$	-	$2.0x10^{-24}$	$5.8x10^{-22}$	-	$3.0x10^{-7}$	(c)
$\tilde{\eta}_N^{(-)}$	-	$1.8x10^{-24}$	0	-	$3.0x10^{-7}$	(c)

process for observing lepton non-conservation is the (oν) $\beta\beta$-decay. The best candidate for observing lepton flavor non-conservation appears to be the neutrino oscillation since nature allows a variety of L/E_ν so that the smallness of δm^2 can be overcome. Admittedly these experiments are very hard and they must be done against pessimistic theoretical predictions. We hope however, that, since such experiments are truly fundamental and the theoretical predictions not completely reliable, they will continue unhindered.

REFERENCES

1. F.T. Avignone, Phys. Lett. 54, 2309 (1985); T. Kirsten, et al., Phys. Rev. Lett. 50, 474 (1983); W. Bertl et al., Nucl. Phys. B260, 1 (1985); D.A. Bryman et al., Phys. Rev. Lett. 55, 465 (1985) and Private Communication; W. Kinnison et al., Phys. Lett. D25, 2846 (1986); A. Badertscher et al., Nucl. Phys. A377, 406 (1982).
2. J.D. Vergados, Phys. Rep. 133, 1 (1986).
3. S.T. Petcov, Phys. Lett. 115B, 401 (1982), G.K. Leontaris, K. Tamvakis and J.D. Vergados, Phys. Lett. 162B, 153 (1986).
4. G.K. Leontaris, K. Tamvakis and J.D. Vergados, Phys. Lett. 171B, 412 (1986).
5. G.K. Leontaris and J.D. Vergados, Nucl. Phys. B224, 137 (1983).

3.2 Electroweak Interactions and Symmetries in Baryons, Nuclei and Atoms

3.2.1 Nuclear Beta Decay and Weak Coupling Constants

The Beta Decay of the Neutron

D. Dubbers

Institut Laue-Langevin, 156 X, F-38042 Grenoble-Cedex, France

Recent developments in neutron beta decay experiments are described, and the accuracy of the data needed for various applications is discussed.

1. Introduction

Neutron beta decay has been investigated since the late 1940's in a number of beautiful experiments; several reviews on these experiments were given on the occasion of the neutron's 50th anniversary in 1982 /1, 2, 3/. In recent years there seems to be a new rush for better neutron decay data; I want to discuss the reason for this renaissance, and list the projects that are underway.

2. The observables in neutron beta decay

One main object of elementary particle physics is to determine the structure and strength of all interaction vertices appearing in nature. Neutron beta decay experiments test the structure and strength of the weak hadronic vertex for the quarks of the first generation in the charged current sector.

In the three body decay: $n \to p\, e^- \bar{\nu}_e$, Q = 1.293 MeV, the neutron, usually in a cold or thermal beam, is essentially at rest, there are two detectable particles in the final state, and all particles involved have spin 1/2; the neutron can very efficiently be polarized, but final state spins are, as usual, difficult to detect.

From neutron decay one can, in principle, obtain a larger number of weak interaction observables. In the allowed approximation, when we assume the validity of the V-A law, and with the ratio of axial vector to vector coupling written as $g_A/g_V = \lambda e^{i\varphi}$, these are /4/:

i. The decay rate

$$\tau^{-1} = \text{const} \times \cos^2\theta_C\,(g_V{}^2 + 3\,g_A{}^2); \qquad (1)$$

when we assume Conservation of the weak Vector Current (CVC), take g_V from superallowed $0^+ \to 0^+$ nuclear transitions, and the Cabbibo angle θ_C from hyperon decay, then we can write /5/ the neutron half life as:

$$t_{\frac{1}{2}} = \tau \ln 2 = \frac{(3596.2 \pm 1.7)}{1 + 3\lambda^2} \text{ sec} ; \qquad (2)$$

ii. The beta-neutrino correlation coefficient

$$a = \frac{1 - \lambda^2}{1 + 3\lambda^2} ; \qquad (3)$$

iii. The beta-decay asymmetry from polarized neutron decay

$$A = -\frac{2\lambda (\lambda + \cos\phi)}{1 + 3\lambda^2} ; \qquad (4)$$

iv. The corresponding neutrino asymmetry coefficient

$$B = \frac{2\lambda (\lambda - \cos\phi)}{1 + 3\lambda^2} ; \qquad (5)$$

v. The triple correlation coefficient

$$D = \frac{2\lambda \sin\phi}{1+3\lambda^2} , \qquad (6)$$

which tests time reversal invariance in neutron decay.

V-A with time reversal invariance requires $\cos\phi = -1$. The sensitivity to λ is highest for a and A, and lowest for B. The coupling constant g_A so far can only be obtained from neutron decay; because g_A is quenched in nuclei in a hardly controllable way /6/. The precise measurement of both the lifetime τ and one correlation coefficient could also give g_V independent of nuclear structure. Finally, precise measurements of all four observables τ, a, A, B would allow tests beyond the V-A model.

These foregoing coefficients τ,a,A,B,D have been measured, see /4/, and the newer data below. The following observables have not yet been measured but are accessible with present day techniques:

vi. The precise shapes of four different correlated proton and electron energy spectra both from unpolarized and polarized neutron decay contain information on additional "induced" coupling constants. Measurement of two such spectra should allow the determination of weak magnetism and second class current form factors, separately.

vii. Measurement of internal bremsstrahlung spectra in neutron decay allows to check one of the radiative correction terms directly.

The measurement of the following observables does not seem to be impossible, but would require a larger effort, which may or may not be justified:

viii. The spin of the outgoing electron.

ix. The decay into bound hydrogen and its excited states, which is sensitive to V + A admixtures /7/.

3. Possible competitors to neutron decay experiments

i. Most of what is known about weak interactions of the quarks of the first generation comes from nuclear beta decay; some of the coupling constants derived from nuclear beta decay need only minor corrections for nuclear structure effects, whereas others are strongly changed in the nuclear medium, and it seems desirable to know all coupling constants from single particle decays.

ii. The only other single particle decay involving merely first generation quarks is pion decay. In the two body decay of the spinless pion the main observable is the decay rate, which depends on a strong interaction vertex. The three body decay of the pion is very interesting, though extremely rare.

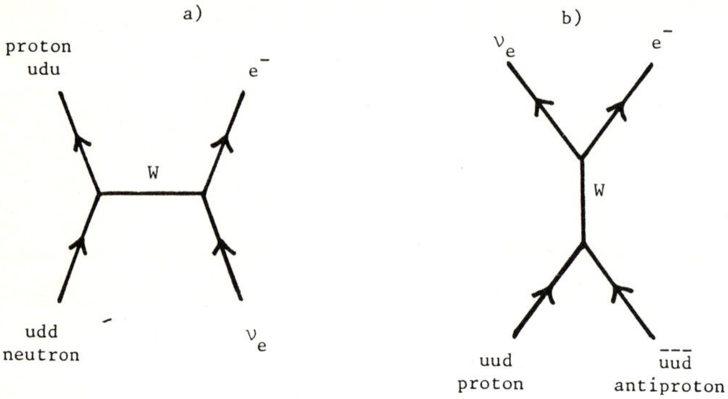

Fig. 1. Feynman diagrams for neutron decay, and W-boson decay after $p\bar{p}$ collision

iii. Interestingly, W-boson decay after production in $p\bar{p}$ collisions has the same Feynman diagram as neutron decay, see Fig. 1, and weak interaction parameters will be derived from this process; however, there is only one particle in the final state, and the spin variable is not as easy to handle as in the case of the neutron. Further, the data will rather be used to test the validity of the quark parton model.

iv. Weak interactions studied in nucleon-nucleon parity violation experiments may one day give valuable additional information on strong interactions, as do non-leptonic particle decays, but weak interaction parameters will enter mainly as an input.

4. Experiments in neutron decay

Thus neutron decay experiments can give unique information on weak interaction vertices. On the other hand, neutron decay experiments are notoriously difficult: only one out of 10^7 neutrons in a beam transversing a typical neutron decay instrument will actually decay there, but every neutron lost in or near the apparatus and every gamma in the beam may contribute to the background. Thus, until recently, high energy experiments on the weak decays of higher generation quarks gave, even for very exotic particles, cleaner energy spectra than their first generation counterpart, i.e. neutron decay experiments.

In recent years, however, at the High Flux Reactor of the Institut Laue-Langevin, new neutron sources relevant for neutron decay experiments were developed. First of all, for "in beam" experiments, high intensities of cold neutrons at very low background are available at the exit of the ILL's neutron guides far away from the nuclear reactor; further, very efficient and easy to use "super-mirror" neutron polarizers have been developed at ILL, nowadays with 99 % polarization. For neutron storage experiments, a new high intensity Ultra Cold Neutron (UCN) source is available in 1986, which increased the available UCN flux by two orders of magnitude; magnetic storage of UCN is being further improved, and new fluid walled UCN-storage boxes give very long storage times. Most of this work is still unpublished.

Substantial progress has also been made in neutron decay instrumentation. Protons from neutron decay can successfully be stored before counting, a procedure by which the effective detector background is substantially reduced /8/. Electrons from neutron decay can nowadays be detected at high count rates (1300 s^{-1}) and low background (70 s^{-1}) with an instrument named

PERKEO[1]. This instrument utilizes a large neutron decay volume within a magnetic field produced by an array of altogether nine superconducting coils which gives an effective detector solid angle of 4π to better than 10^{-3}, and avoids all losses due to electron backscattering, see Fig. 2 and Ref. /9/ and /10/. Beta decay spectra measured with PERKEO are displayed in Fig. 3 and 4.

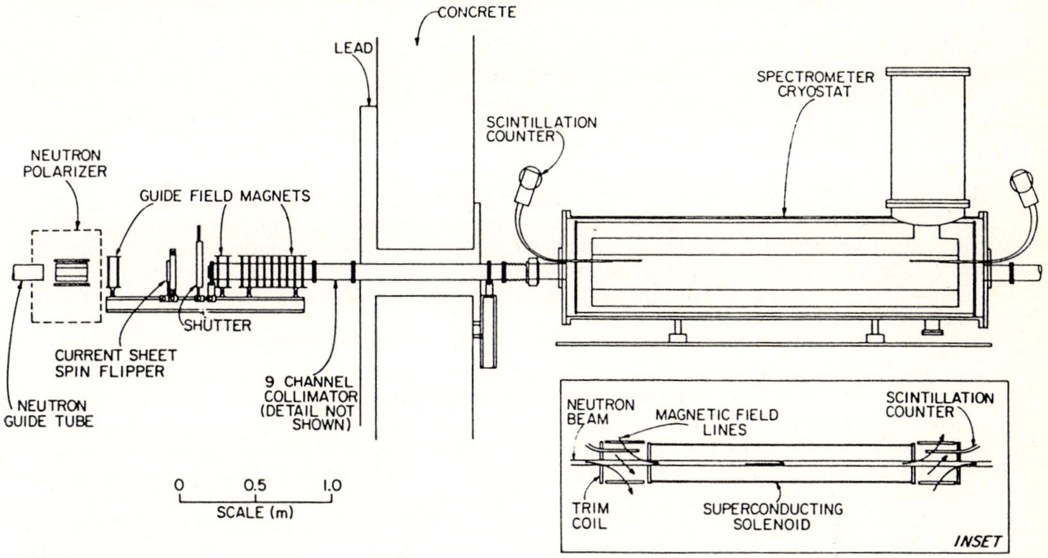

Fig. 2. Arrangement of the experiment. The inset shows details of the inner region of the superconducting solenoid

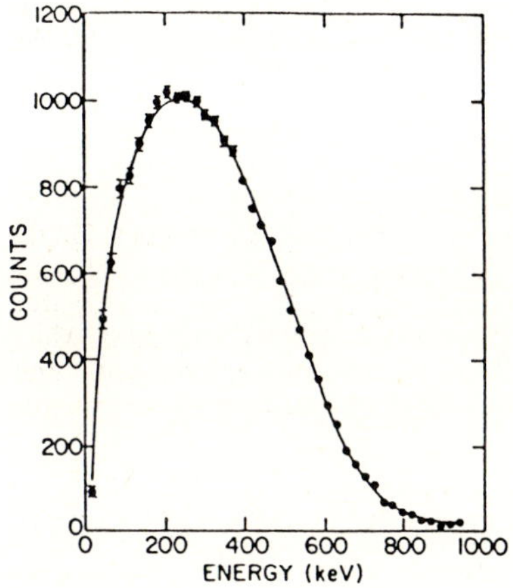

Fig. 3. β-decay energy spectrum from one run

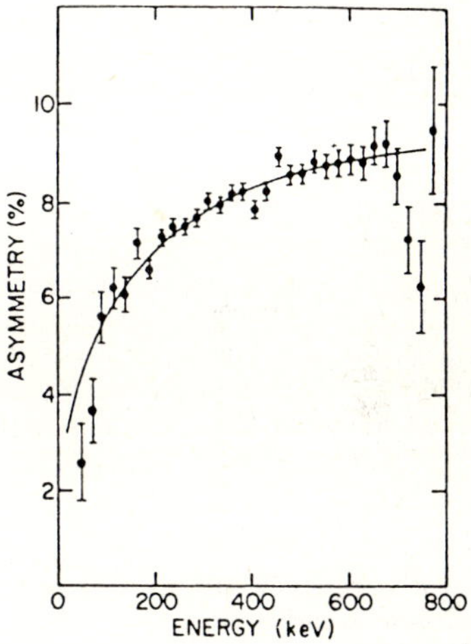

Fig. 4. Experimental β asymmetry as a function of β energy

1) the PERKEO project is a collaboration between the University of Heidelberg (at present: J. Last, M. Arnold, J. Doehner, E. Klemt, and P. Barker from the University of Auckland), the Argonne National Laboratory (S.J. Freedman), and the Institute Laue Langevin (D. Dubbers).

New ways to determine the neutron lifetime "in beam", but without need of precise neutron capture data, have recently been employed both with PERKEO and with a newly developed neutron decay drift chamber detector /11/. This method uses a chopped neutron beam, a black neutron detector, and the elementary equation linking the rate of betas to the number of neutrons in a bunch: $n_\beta = N_n/\tau$, and circumvents many of the difficult neutron beam calibration problems.

Further projects are being prepared, like a 4π detector for e^- and p^+ track reconstruction /12/, proton storage with variable neutron decay volumes /13/, calorimetric neutron beam calibration /14/, and use of secondary electrons for proton detection in PERKEO.

Recent neutron beta decay results can be summarized as follows:

The neutron lifetime still is most difficult to assess experimentally, and results remain controversial. The measured values show a larger spread, see also the contribution by J. Byrne. The upper part of Fig.5 shows an ideogram for the λ values determined from different lifetime measurements "in beam". Measurements with bottled neutrons, on the other hand, have not yet given significant neutron lifetime results. - However, in view of the efforts invested in current neutron lifetime projects, the situation may clear up in the nearer future.

The beta decay asymmetry has recently been measured /9/ with PERKEO to $A = -0.1146 \pm 0.0019$, after application of radiative and recoil corrections. From this the ratio of coupling constants $\lambda = -1.262 \pm 0.005$ is derived with (4), which is more accurate than but in agreement with the values of λ derived from other measurements of A and a (see fig. 5). A good overall agreement is reached when the error bars of the individual lifetime measurements are increased by a factor of about 2.5: from τ, one then has $\lambda = 1.257 \pm 0.012$, from a: $\lambda = -1.256 \pm 0.015$, and from A: $\lambda = -1.262 \pm 0.004$. The overall value is $\lambda = 1.2612 \pm 0.0037$. New lifetime experiments should have errors of about 3% to be competitive with previous lifetime measurements.

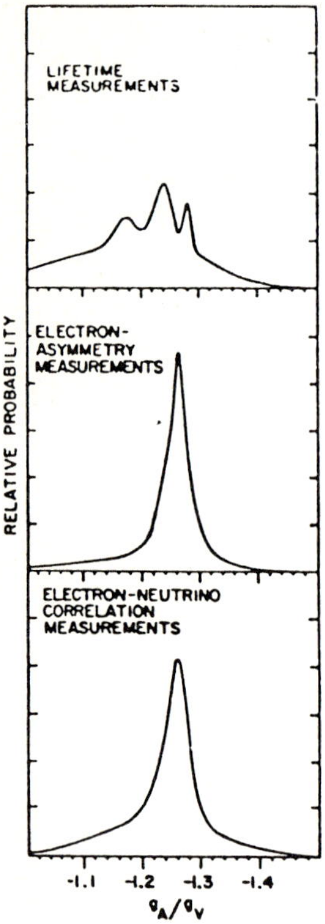

Fig. 5. Ideograms of experiments that determine g_A/g_V. The three ideograms have equal areas

There is no new measurement of B, the current value being /4/: $B = 1.001 \pm 0.026$, giving $\lambda = 1.0 \pm 0.4$, nor of the time reversal violating coefficient D, its current value being $D \leq 2.10^{-3}$ or $\phi=(180.11 \pm 0.17)^{o}$, though there are new proposals for experiments.

When the beta decay asymmetry from Fig. 2b is fitted with the weak magnetism form factor g_M as a free parameter one obtains $g_M = (+ 2.2 \pm 3.4).10^{-3}$ /MeV, while CVC predicts $g_M = + 6 \cdot 10^{-3}$/MeV. This is not yet meant to be a serious measurement of neutron's weak magnetism, but it shows that also this quantity is within reach.

Finally, a new instrument has been assembled /15/ to measure inner bremsstrahlung from neutron decay, with a predicted intensity of 100 photons per frequency octave and per meter neutron beam length.

5. How well do we need to know neutron decay parameters?

As a general rule, we want to know nature's coupling constant as precisely as possible. To be more specific, however, we want to discuss in what fields the present level of accuracy of neutron decay data still poses a problem, and how well we would like to know them.

The lowest requirements come, as often, from quark theory. The simple non-relativistic quark model prediction for $\lambda = |g_A/g_V|$ is $\lambda = 5/3$, and relativistic corrections /16/ give $\lambda = 1.35$. QCD lattice calculations /17/ give $\lambda = 1.40 \pm 0.14$, and QCD sumrules with chiral symmetry breaking /18/ give $\lambda = 1.22$. The Goldberger-Treiman relation, corrected for chiral symmetry breaking /19/ gives $\lambda = 1.25$, and the Adler-Weisburger sum rules /20, 21/ values between $\lambda = 1.16$ and $\lambda = 1.24$. The errors of the theories are difficult to evaluate, but the large spread of the results shows that they are larger than present experimental errors.

Higher requirements on neutron decay data must be posed if we want to obtain not only g_A, but also other weak coupling constants from neutron decay data. If we want to improve on the g_V value derived from nuclear data, then we must know both τ and, say, A with an accuracy of better than 10^{-3}. If we want to check on the existence of a fourth quark generation via the unitarity of the Kaboyashi-Maskawa (K.-M.) matrix /22/, then a similar accuracy is required. Limits on the V-A structure of weak interactions could be improved by improved neutron decay data, as g_S/g_V, g_T/g_V etc. with S = scalar, T = Tensor, etc., enter into (2) to (6) on the same footing as $\lambda = |g_A/g_V|$.

When we take into consideration the very newest (1986) developments in cold neutron sources and neutron polarizers at the ILL, this 10^{-3} accuracy can be achieved for the beta decay asymmetry A with a rebuilt PERKEO instrument. As for the accuracy of future lifetime τ measurements, one cannot yet make reliable predictions.

It is interesting to compare neutron beta decay data with the corresponding data from hyperon decay, within the framework of Cabbibo theory, or its K.-M. extension. Neutron decay data are accurate enough for this only since recently, as older neutron data gave agreement with Cabbibo predictions /23/, whereas our newer neutron results show a significant deviation. More on this in the following contribution by H. W. Siebert.

The "strong" CVC hypothesis states that weak and electromagnetic vector currents form an isovector triplet, which is the starting point of all electroweak unified theories. It thus predicts that the weak magnetism form factor in beta decay is related to the ordinary particle magnetic moments by the Wigner Eckert theorem. This has so far not yet accurately been tested with elementary particles, and our neutron measurements are just a beginning.

The most urgent requirements on the accuracy of neutron decay data, however, do not come from elementary particle theory, but from other fields, like cosmology, astrophysics, solar models, and neutrino detection. As J. Byrne will cover this field in his talk, I will only state that a true 1% accuracy of the neutron lifetime data would, at present, be highly welcome, but requirements on the quality of neutron data will also here probably increase in the course of time.

In summary neutron beta decay is, still today, a promising source for better weak interaction parameters, and there seems at present to be a new desire for improved neutron decay data by a growing number of groups.

References:

/1/ J.M. Robson, Contemp. Phys. 24, 129 (1983)

/2/ A.I. Frank, Sov. Phys. Usp. 25, 280 (1982)

/3/ J. Byrne, Rep. Progr. Phys. 45, 115 (1982)

/4/ E.G. Commis, P.H. Bucksbaum, "Weak interactions of leptons and quarks", Cambridge, 1983

/5/ D.H. Wilkinson, Nucl. Phys. A377, 474 (1982)

/6/ E. Oset, M. Rho, Phys. Rev. Lett. 42, 47 (1979)

/7/ L.L. Nemenov, Sov. J. Nucl. Phys. 31, 115 (1980)

/8/ J. Byrne et al., Phys. Lett. 92B, 274 (1980)

/9/ P. Bopp et al., Phys. Rev. Lett. 56, 919 (1986)

/10/ P. Bopp et al., Nucl. Instr. Meth. A, submitted 1986

/11/ K. Schreckenbach, private communication

/12/ T. Bowles, private communication

/13/ G. Greene et al, private communication

/14/ N. Jarmi et al., IEEE Trans. on Nucl. Sci NS-30, 1508 (1983)

/15/ M.Beau, Diplomarbeit, University of Heidelberg, 1986 (unpublished)

/16/ C. Hayne, N. Isgur, Phys. Rev. D25, 1944 (1982)

/17/ F. Fucido et al., Phys. Lett. 115B, 148 (1982)

/18/ R. Koniuk, R. Tarrach, Z. Phys. C18, 179 (1983)

/19/ C.A. Dominguez, Phys. Rev. D25, 1937 (1982)

/20/ W.I. Weisberger, Phys. Rev. Lett. 14, 1047 (1965)

/21/ S.I. Adler, Phys. Rev. Lett. 14, 1051 (1965)

/22/ W.J. Marciano, A. Sirlin, Phys. Rev. Lett. 56, 22 (1986)

/23/ M. Bourquin et al., Z. Phys. C21, 27 (1983)

The Neutron Lifetime

J. Byrne

School of Mathematical and Physical Sciences, University of Sussex,
Brighton BN1 9QH, Sussex, UK

1 Introduction

1.1 Weak Decay of the Neutron

The free neutron is a β-active nucleus which decays weakly according to the scheme

$$n \rightarrow p + e^- + \tilde{\nu}_e$$

with a lifetime at the level of 10^3 [sec]. The process was predicted by CHADWICK and GOLDHABER [1], and was first detected by SNELL et al who observed positive ions extracted electrically from an intense neutron beam [2,3]. At about the same time ROBSON verified that the positive ions were indeed protons and estimated a half-life of 9-25 [min]. [4]. The first accurate (20%) measurement of the neutron lifetime was also carried out by ROBSON, and in the same experiment he measured the electron spectrum [5]. The current best estimate of the end point kinetic energy of the β-spectrum as derived from the neutron-hydrogen atom mass difference is 781.567±0.017 [keV] [6]. The corresponding maximum kinetic energy of the protons is 750.0 [eV], which is too low to be recorded easily in conventional charged particle detectors.

The neutron is a member of the lowest SU(3) baryon octet and, by comparison, the lifetimes of corresponding weak decays all have values at the level of 10^{-10} [sec]. All the couplings and decay rates of these baryons are related by CABIBBO theory [7] extended to six quarks and three mixing angles by KOBAYASHI and MASKAWA [8], and the anomalously long neutron lifetime is solely a reflection of the near mass degeneracy of neutron and proton.

1.2 Neutron Lifetime Measurements

In the thirty-eight years which have elapsed since neutron β-decay was first detected, ten direct measurements [5,9-17] of the lifetime have been reported in the literature and these are listed in Table 1. Three of these measurements, those carried out by CHRISTENSEN et al [12-13]. BONDARENKO et al [15] and BYRNE et al [16], are quoted to accuracies at the level of 1-2%. In 1982 the Particle Data Group (PDG) adopted the lifetime value τ = 925±11 [sec] [18] based on the first and third of these results but in their 1984 review they also included the second result in their compilation arriving at the value τ = 898±16 sec [19]

Apart from these direct determinations there exist indirect means of arriving at a value of τ, and the most important route is through a combination of the polarization and angular correlation coefficients in neutron decay and the decay rates of $0^+ \rightarrow 0^+$ superallowed Fermi transitions in mirror nuclei [6]. These results allow one to estimate τ = 898±6 [sec] [20]. On the other hand an analysis of the β-decay of tritium $^3H \rightarrow {}^3He + e^- + \tilde{\nu}_e$, which in many important respects closely resembles neutron decay, leads to the value τ = 911±8 [sec] [21].

The apparent agreement between the PDG value τ = 898±16 [sec] and the 'derived' value τ = 898±6 [sec] is deceptive for the following reasons:

Table 1. Measured values of the neutron lifetime. The results τ = 909±69 [sec] [14] and τ = 874±95 [sec] [17] were obtained using bottled ultra-cold neutrons; all the other results were found by beam techniques.

τ[sec]	Reference
1108±216	[5] ROBSON (1951)
1039±130	[9] SPIVAK et al (1956)
1100±165	[10] D'ANGELO (1959)
1013±26	[11] SOSNOVSKI et al (1959)
935±14	[12] CHRISTENSEN et al (1967)
919±14	[13] CHRISTENSEN et al (1972)
909±69	[14] PAUL and TRINKS (1978)
877±8	[15] BONDARENKO et al (1978)
937±18	[16] BYRNE et al (1980)
874±95	[17] KOSTVINTSEV et al (1980)

(i) there is a 7% difference between the extreme values of the three 'precision' measurements on which the PDG value is based and it is evident that there are systematic errors in all three experiments which have not been taken fully into account

(ii) as compared with the old (1982) value, the new (1984) value is weighted downwards by the 0.9% error claimed for the measurement of BONDARENKO et al [15]. This experiment used the same apparatus and technique as in an earlier study by SOSNOVSKI et al [11] who had reported the result τ = 1013±26 [sec]. Given that, unlike the other two comparable measurements [13,16], this study was carried out in conditions of low collection solid angle and high background, and no results were reported showing the effect of systematic variations in the experimental conditions; given also that current neutron standards hardly permit a determination of neutron density to much better than 1%, the claimed 0.9% precision on this experiment is very difficult to justify.

Broadly speaking then one may say that the neutron lifetime has been directly determined to an accuracy of about 3%, a result which may be compared with an accuracy of 0.8% on the lifetime of the Λ°, 0.09% for the pion and 2.5 x 10^{-3}% for the muon. One should also be aware that, in the absence of SU(3)-breaking, the decay data derived from other members of the SU(3) baryon octet would tend to favour a neutron lifetime longer than current evidence would suggest [22]. Hence there is a real need to improve the accuracy of the direct measurement of this most important parameter.

2 Theoretical Significance of the Neutron Lifetime

2.1 Importance for Weak Interaction Theory

The Fermi coupling constant G_F which characterises the weak interaction is known from studies of muon decay to a precision of 4 x 10^{-3}% and, within the framework of the Cabibbo theory, three additional parameters are sufficient at zero momentum transfer to determine the weak decay amplitudes of baryons in the SU(3) octet. These are the Cabibbo angle θ_C and the parameters F and G which define the strengths of the anti-symmetric and symmetric couplings of two SU(3) octets to form a third. Thus, from the point of view of weak interaction physics, the primary purpose in undertaking a measurement of the neutron lifetime to a precision $\leqslant$1% is the determination of the polar vector and axial vector weak form factors whose ratio $\lambda = g_A(0)/g_V(0) = F + G$ in the Cabibbo scheme.

The parameter λ has additional significance in that, according to the Goldberger-Treiman relation, it is given by the formula

$$\lambda = f_\pi g_{\pi NN}/M_N$$

where f_π is the charged pion decay constant, $g_{\pi NN}$ the pion-nucleon coupling constant and M_N the mean nucleon mass. This relation is currently viewed as a consequence of a spontaneously broken exact chiral symmetry of the strong interactions whose associated conserved currents are identified with the strangeness conserving isovector and isoaxial weak currents. Since f_π does not vanish, conservation of the axial current would require that the pion mass be zero which is not the case. Thus the chiral symmetry cannot be exact and the quantity $\Delta_\pi = 1- \lambda M_N/f_\pi g_{\pi NN}$ provides a measure of the degree of chiral symmetry breaking which in principle is calculable [23].

2.2 Importance for Astrophysical Theory

Although theorists can and do argue about the validity of these notions, and about the reality and significance of apparent discrepancies, there is one area of contemporary physics where there is complete unanimity as to the crucial importance of establishing a reliable value for the neutron lifetime. This special area is the realm of astrophysics and cosmology and there are two major problems where the precise value of the neutron lifetime is of direct relevance.

The first problem concerns the rate of helium production in the early universe [24] a question which is central as a test of the hot big bang theory. According to this theory the helium abundance is determined essentially by the ratio of the neutron lifetime to the expansion time of the universe from that epoch at which the neutron/proton ratio goes out of thermal equilibrium with the radiation field, to the point at which deuterium becomes stable in the thermal field and primordial nucleosynthesis commences. A related question concerns the number of neutrino species; adding one more species increases the density and the expansion rate and hence increases the helium abundance for a fixed neutron lifetime.

The second problem concerns the rate at which energy is produced by synthesis of hydrogen into helium in the sun. It is well known that the measured rate of neutrino emission from the sun is less than the theoretical prediction by a factor of about 3; this is the 'solar neutrino problem' [25]. Solar energy generation is governed by the weak interaction process

$$p + p \rightarrow d + e^+ + \nu_e$$

whose rate is proportional to the square of the axial vector coupling constant in neutron β-decay. If, for example, the neutron lifetime were to be reduced, the p-p reaction would proceed faster in the same proportion. But, since the total solar luminosity is fixed, it would be necessary to assume a lower collision rate to compensate for the increased weak interaction cross-section. The net consequence would be a drop in temperature and a drop in emission of those neutrinos to which present detectors are most sensitive. Hence one contribution to solving the solar neutrino problem would be the demonstration that the neutron lifetime had been substantially overestimated. One might add that the uncertainty in the neutron lifetime contributes approximately one third of the normalization error for neutrino counting in reactor searches for the neutrino oscillation process which has long been regarded as a strong candidate for solving the solar neutrino problem [26].

3 Experimental Studies of the Neutron Lifetime

3.1 Techniques of Measurement

If sources of confined neutrons were available off-the-shelf then the neutron lifetime $\tau = \lambda^{-1}$ could be determined directly from the exponential law of decay

$$N(t) = N(0)e^{-\lambda t}$$

Alternatively one could visualize observing the reduction of neutron number with distance along a collimated beam, but, even for very cold neutrons, the distances involved turn out to be prohibitive. It has even been suggested [27] that the necessary distances could be achieved by exploding a nuclear device in outer space although it is unlikely that such an experiment would ever be approved. Two experiments have been reported to observe the loss by β-decay of neutrons confined by magnetic fields [14] or material bottles [17]. Although the statistical accuracy reached to date does not approach that achieved by other methods, mainly due to the low intensity of the ultra-cold neutrons available, the technique is a promising one and new experiments in both variants are currently under way [28,29].

'Beam' methods for measuring the neutron lifetime are based on the differential equation

$$dN(t)/dt = -\lambda N(t)$$

where $dN(t)/dt$ is determined from the measured rate of emission of decay particles from a known volume of neutron beam and $N(t)$ is determined by measuring the neutron density within that known volume. To measure $dN(t)/dt$ one must first decide whether to count electrons or protons, which detectors to use and how to distinguish signal from background. The special importance attached to background discrimination follows from the fact that only 0.1% of neutrons in a beam decay per second and that weak signal can be swamped by (n,γ) events. Until the development of neutron guides one also had to cope with γ-ray contamination of the beam itself. To know the volume one must determine the efficiencies and collection solid angles of all detectors involved.

All these requirements present great problems to the experimenter but, with the development of special detectors operating in 4π-geometry and the evolution of various techniques for beam handling and background suppression, a number of solutions have emerged which suggest that, aside from the special problem of determining neutron density, the accumulated error on the remaining parameters can be reduced to about 0.1%.

In all beam measurements to date the neutron density 'n' has been determined by measuring the neutron capture rate w_C in materials obeying the 1/v law of absorption [30]. In this case for a thin target of area A and surface density ρ the capture rate is

$$w_C = A\rho\sigma_T v_T n$$

where v_T (= 2200 [m/sec]) is the velocity and σ_T the capture cross-section of a thermal neutron. There are three reactions which involve charged particles in the final state, $^{10}B(n,\alpha)^7Li$, $^6Li(n,\alpha)^3H$ and $^3He(n,p)^3H$ which have been used for this purpose, but in each case there is a problem in making an accurate assay of the target mass. An alternative procedure is to measure the activity following (n,γ) capture in mononuclidic targets of which the reactions $^{197}Au(n,\gamma)^{198}Au$, $^{59}Co(n,\gamma)^{60}Co$ and $^{55}Mn(n,\gamma)^{56}Mn$ are the most important. Utilizing some combination of these techniques it is probably possible to achieve an overall accuracy of 0.5% but the neutron density determination sets an absolute limit to the accuracy of the measurement since the neutron lifetime is ultimately determined as a multiple of the mean time between capture collisions. It is therefore necessary to scrutinize in the most minute detail the experimental procedures employed.

3.2 Current Programme

The experimental research in this field carried out over the period 1950-80 has been reviewed elsewhere [31] and the latest generation of experiments has evolved directly from this work. Currently there are six neutron lifetime experiments either in progress or in various stages of preparation. The principles and prospects of each were debated at a recent Workshop on the Investigation of Fundamental Interactions with Cold Neutrons, held at the NATIONAL BUREAU OF

STANDARDS (NBS) in GAITHERSBURG MD., and the details are available in the published Proceedings [32]. Five of these experiments, two 'storage' and three 'beam' experiments will be performed at the INSTITUT LAUE-LANGEVIN (ILL) GRENOBLE. The sixth experiment, also a 'beam experiment', will be carried out at the NBS, and has the particular feature that the neutron flux is determined by a cryogenic calorimeter [33] which seems capable in the long term of delivering a measurement of neutron density accurate to 0.1%. The ultimate objective of all these experiments would be to achieve accuracies better than 1% on the neutron lifetime, thereby providing an independent direct check on the most accurate result obtained to date via the indirect route [20].

References

1 J. Chadwick and M. Goldhaber, Proc. Roy. Soc. A151, 479 (1935)
2 A.H. Snell and L.C. Miller, Phys. Rev. 74, 1217 (1948)
3 A.H. Snell, F. Pleasonton and R.V. McCord, Phys. Rev. 78, 310 (1950)
4 J.M. Robson, Phys. Rev. 78, 311 (1950)
5 J.M. Robson, Phys. Rev. 83, 349 (1951)
6 D.H. Wilkinson, Nucl. Phys. A377, 474 (1982)
7 N. Cabibbo, Phys. Rev. Lett. 10, 531 (1963)
8 M. Kobayashi and T. Maskawa, Prog. Theor. Phys. 49 652 (1973)
9 P.E. Spivak, A.N. Sosnovski, A.Y. Prokofiev and V.S. Sukolov, Proc. Int. Conf. on Peaceful Uses of Atomic Energy, Geneva 1955. Vol. 2, P. 3 (U.N., New York 1956).
10 N. D'Angelo, Phys. Rev. 114, 285 (1959)
11 A.N. Sosnovski, P.E. Spivak, Y.A. Prokofiev, I.E. Kutikov and Y.P. Dobrinin, Nucl. Phys. 10, 395 (1959).
12 C.J. Christensen, A. Nielsen, A. Bahnsen, W.K. Brown and B.M. Rustad, Phys. Lett. B26, 11 (1967)
13 C.J. Christensen, A. Nielsen, A. Bahnsen, W.K. Brown amd B.M. Rustad, Phys. Rev. D5, 1628 (1972).
14 W. Paul and U. Trinks, La Recherche 9, 1008 (1978)
15 L.N. Bondarenko, V.V. Kurguzov, Y.A. Prokofiev, E.Y. Rogov and P.E. Spivak, JETP Lett. 28, 303 (1978)
16 J. Byrne, J. Morse, K.F. Smith, K. Green and G.L. Greene, Phys. Lett. B92, 274 (1980)
17 Y.Y. Kostvintsev, Y.A. Kushnir, V.I. Morozov and G.I. Terekhov, JETP Lett. 31, 326 (1980)
18 Review of Particle Properties, Phys. Lett. B111, 1 (1982)
19 Review of Particle Properties, Rev. Mod. Phys. 56, Part II, S1 (1984)
20 P. Bopp, D.Dubbers, L. Hornig, E. Klemt, J. Last and H. Schütze, Phys. Rev. Lett. 56, 919 (1986)
21 J. Byrne, Nature 310, 212 (1984)
22 M. Bourquin et al, Z. Phys. C21, 27 (1983)
23 C.A. Dominguez, Phys. Rev. D5, 1937 (1982)
24 R.J. Tayler, Nature 282, 559 (1979)
25 J.N. Bahcall, W.F. Huebner, S.H. Lubow, P.D. Parker and R.K. Ulrich, Rev. Mod. Phys. 54, 767 (1982)
26 H.A. Bethe, Phys. Rev. Lett. 56, 1305 (1986)
27 F.J. Dyson, General Dynamics Corporation, General Atomic Division Report GAMD-957 (1959)
28 K.G. Kügler, K. Moritz, W. Paul and U. Trinks, Nucl. Instr. and Meth. 228, 240 (1985)
29 P. Ageron, W. Mampe, J.C. Bates and J.M. Pendlebury, Nucl. Instr. and Meth. (to be published)
30 N.E. Holden, Neutron Capture Cross Sections for BNL325, Fourth Edition BNL-NCS-51388 UC-346 (1981)
31 J. Byrne, Rep. Prog. Phys. 45, 115 (1982)
32 G.L. Greene, Editor: The Investigation of Fundamental Interactions with Cold Neutrons, NBS Special Publication 711, (1986)
33 R.G.H. Robertson and P.E. Koehler, Nucl. Instr. and Meth. (to be published)

Constraints on General $SU(2)_L \times SU(2)_R \times U(1)$ Electroweak Models from Nuclear Beta Decay

P. Herczeg

Los Alamos National Laboratory, Los Alamos, NM 87545, USA

1. Introduction

The minimal standard model of the electroweak interactions is consistent with all available data. Nuclear β-decay experiments contribute to this conclusion through the absence of evidence for deviations from the V-A structure of the underlying charged-current quark-lepton interaction [1]. New contributions to the β-decay interaction are expected at some level in many extensions of the minimal standard model, motivated by the problems and the shortcomings of the latter.

An attractive class of extensions of the minimal standard model, which sheds a new light on the apparent V-A structure of the charged-current weak interactions, is the class of left-right symmetric models based on the gauge group $SU(2)_L \times SU(2)_R \times U(1)$ [2]. A characteristic feature of these models is the presence of right-handed charged currents. Among the sensitive probes of right-handed currents are some observables in nuclear beta decay. Except for the time-reversal odd correlation [3,4] $\langle\vec{J}\rangle\cdot\vec{p}_e\times\vec{p}_\nu$ ($\vec{J}$ ≡ nuclear spin) and some preliminary remarks on $e^\pm$ polarization [4], the implications of the corresponding measurements have been considered [5-9] so far only for models with manifest left-right symmetry [5] and no mixing in the leptonic sector. Here we shall analyze the implications of beta-decay experiments for more general versions of $SU(2)_L \times SU(2)_R \times U(1)$ models, including the most general one which allows for CP-violation, unequal left- and right-handed quark mixing angles, and mixing in the leptonic sector. For each scenario we shall compare the constraints on the pertinent parameters from beta-decay measurements with the constraints provided on them by other data.

2. The Beta-Decay Interaction in $SU(2)_L \times SU(2)_R \times U(1)$ Models

In $SU(2)_L \times SU(2)_R \times U(1)$ models there are two distinct charged gauge boson fields W_L and W_R. Their coupling to the fermions is described by the Lagrangian*

$$L = \frac{g_L}{2\sqrt{2}} W_L (\bar{P}\Gamma_L U_L N + \bar{N}^{(0)}\Gamma_L U^\dagger E) + \frac{g_R}{2\sqrt{2}} W_R (\bar{P}\Gamma_R U_R N + \bar{N}^{(0)}\Gamma_R V^\dagger E) + \text{H.c.} \quad , \tag{2.1}$$

where g_L and g_R are the gauge coupling constants, $\Gamma_L \equiv \gamma_\lambda(1 - \gamma_5)$, $\Gamma_R \equiv \gamma_\lambda(1 + \gamma_5)$ (the Dirac indices have been suppressed), $\bar{P} \equiv (\bar{u},\bar{c}, \ldots)$, $\bar{N} \equiv (\bar{d},\bar{s}, \ldots)$, $\bar{E} \equiv (\bar{e},\bar{\mu}, \ldots)$, and $\bar{N}^{(0)} \equiv (\bar{\nu}_1,\bar{\nu}_2, \ldots)$. U_L, U_R and U, V are the quark and lepton mixing matrices, respectively. The fields W_L and W_R are linear combinations of the mass-eigenstates W_1 and W_2

$$W_L = \cos\zeta W_1 + \sin\zeta W_2$$
$$W_R = e^{i\omega}(-\sin\zeta W_1 + \cos\zeta W_2) \quad , \tag{2.2}$$

where ζ is a mixing angle and ω is a CP-violating phase.

---

*A brief review of the relevant aspects of $SU(2)_L \times SU(2)_R \times U(1)$ is contained in Ref. 10.

The Hamiltonian responsible for nuclear beta decay resulting from (2.1) is given by

$$H^{(\beta)} = a_{LL}[\bar{e}\Gamma_L\nu_e^{(L)}\ \bar{u}\Gamma_L d + \eta_{RR}\bar{e}\Gamma_R\nu_e^{(R)}\ \bar{u}\Gamma_R d + \eta_{LR}\bar{e}\Gamma_L\nu_e^{(L)}\ \bar{u}\Gamma_R d + \eta_{RL}\bar{e}\Gamma_R\nu_e^{(R)}\ \bar{u}\Gamma_L d] + \text{H.c.} \quad , \tag{2.3}$$

where $\nu_e^{(L)} = \Sigma_j U_{ej}\nu_j$, $\nu_e^{(R)} = \Sigma_j V_{ej}\nu_j$. Assuming that m_1^2/m_2^2 can be neglected relative to one and that $\tan^2\zeta$ can be neglected relative to m_1^2/m_2^2, the constants a_{LL}, η_{RR}, η_{LR}, and η_{RL} are given by

$$\begin{aligned} a_{LL} &\simeq g_L^2\cos^2\zeta/8m_1^2 \\ \eta_{RR} &\simeq e^{i\alpha}(g_R^2m_1^2/g_L^2m_2^2)\cos\theta_1^R/\cos\theta_1^L \\ \eta_{LR} &\simeq -e^{i(\alpha+\omega)}(\cos\theta_1^R/\cos\theta_1^L)(g_R\tan\zeta/g_L) \\ \eta_{RL} &\simeq -e^{-i\omega}g_R\tan\zeta/g_L \quad , \end{aligned} \tag{2.4}$$

where m_1, m_2 are the masses of W_1, W_2, and α is a CP-violating phase from U_R ($U_{ud}^L = \cos\theta_1^L$, $U_{ud}^R = e^{i\alpha}\cos\theta_1^R$).

A Hamiltonian of the form (2.3) with arbitrary constants would be determined by seven real parameters (four complex numbers minus an overall phase). In $SU(2)_L \times SU(2)_R \times U(1)$ models the number of independent parameters is six, in view of the relation $\eta_{RR}\eta_{RL}^*/|\eta_{RR}||\eta_{RL}| = \eta_{LR}/|\eta_{LR}|$. One of these, associated with an interference term between left-handed and right-handed leptonic currents, can appear only through contributions proportional to the neutrino masses and will be ignored in the following. As the neutrinos are not detected, the observed β-decay probability is a sum of the probabilities of decays into energetically allowed neutrino mass-eigenstates. We shall assume in the following that the effects of the masses of the neutrinos that can be produced in the decay can be neglected. Taking all the above into account, the following six parameters are available in β-decay: $a_{LL}^{(e)} \equiv a_{LL}\sqrt{u_e}$, $|\eta_{RR}^{(e)}| \equiv |\eta_{RR}|\sqrt{\tilde{v}_e}$, $|\eta_{LR}|$, $|\eta_{RL}^{(e)}| \equiv |\eta_{RL}\sqrt{\tilde{v}_e}|$, η_{LR} and $\eta_{RR}^{(e)}\eta_{RL}^{(e)*}$, where $u_e = \Sigma_i'|U_{ei}|^2$, $v_e = \Sigma_i'|V_{ei}|^2$, $\tilde{v}_e = v_e/u_e$; the summation is over the neutrino states produced in the decay. Only five of the above parameters are independent, due to the mentioned relation.

For a measurement to yield significant constraints on new interactions the expression for the chosen observable must be free of quantities with large theoretical uncertainties or experimental errors. This restricts the choice to allowed decays. With the exception of the coefficient of the T-odd correlation $\langle\vec{J}\rangle\cdot\vec{p}_e\times\vec{p}_\nu/E_eE_\nu$ (D-coefficient) we shall consider from these only pure transitions, since generally the ratio of the Gamow-Teller and Fermi matrix elements is not known with sufficient accuracy (an exception is neutron decay,* where the matrix elements are known exactly, and which provides the value of the axial-vector coupling constant g_A). For the D-coefficient, which vanishes (up to electromagnetic final-state effects) in the minimal standard model, the precise knowledge of the nuclear matrix elements is not essential.

The parameter $a_{LL}^{(e)}$ appears only in the decay rate (in the $O^{14} \to N^{14*}e^+\nu$ superallowed Fermi transition it is involved in the combination of parameters which define $G\cos\theta_c$, where G is the μ-decay coupling constant). The normalized spectrum depends only on the η_{ik}'s. In pure transitions (ignoring recoil-order terms, higher-forbidden contributions and electromagnetic effects) all the observables (except for the rates) are either independent of the η_{ik}'s or proportional to the quantities.

$$x_V = \frac{|1+\eta_{LR}|^2 - |\eta_{RR}^{(e)}+\eta_{RL}^{(e)}|^2}{|1+\eta_{LR}|^2 + |\eta_{RR}^{(e)}+\eta_{RL}^{(e)}|^2} \simeq 1-2|\eta_{RR}^{(e)}+\eta_{RL}^{(e)}|^2 \quad \text{(Fermi transitions)} \tag{2.5}$$

and

*Another case is ^{19}Ne-decay (see B. R. Holstein and S. B. Treiman, Ref. 6), where constraints can be obtained on manifestly symmetric $SU(2)_L \times SU(2)_R \times U(1)$ models by eliminating the unknown matrix elements using other data.

$$x_A = \frac{|1-\eta_{LR}|^2 - |\eta_{RR}^{(e)} - \eta_{RL}^{(e)}|^2}{|1-\eta_{LR}|^2 + |\eta_{RR}^{(e)} - \eta_{RL}^{(e)}|^2} \simeq 1 - 2|\eta_{RR}^{(e)} - \eta_{RL}^{(e)}|^2 \quad \text{(Gamow-Teller transitions)} \tag{2.6}$$

In both cases they are independent of the nuclear matrix elements. It follows that information can be obtained only on the parameters $|\eta_{RR}^{(e)}|$, $|\eta_{RL}^{(e)}|$, and $\mathrm{Re}\,\eta_{RR}^{(e)}\eta_{RL}^{(e)*}$. In addition, the D-coefficient provides information on $\mathrm{Im}(\eta_{LR} + \eta_{RR}^{(e)}\eta_{RL}^{(e)*})$. CP-conserving observables in mixed transitions would be generally sensitive also to $\mathrm{Re}\,\eta_{LR}$ and $|\eta_{LR}|$.

3. Constraints from Beta-Decay Measurements

The average value of x_A from experimental results on Gamow-Teller transitions is [7]

$$(x_A)_{expt} = 1.001 \pm 0.012 \quad . \tag{3.1}$$

A recent accurate measurement of the positron longitudinal polarization in a Fermi transition ($P_e^F = x_V$) yielded [9]

$$(x_V)_{expt} = 0.99 \pm 0.04 \quad . \tag{3.2}$$

An approach followed in recent and in ongoing experiments [11] involves a comparison of positron longitudinal polarizations (P_e^F, P_e^{GT}) in a Fermi and a Gamow-Teller transition for positrons of the same energy.* The present experimental result on P_e^F/P_e^{GT} is [9]

$$(P_e^F/P_e^{GT})_{expt} = 0.986 \pm 0.038 \quad . \tag{3.3}$$

The accuracy for $(P_e^F/P_e^{GT})_{expt}$ is expected to be improved by 1-2 orders of magnitude [11].

The experimental value of the D-coefficient from a recent experiment [13], which has the smallest error, is

$$(D)_{expt} = 0.0004 \pm 0.0008 \quad . \tag{3.4}$$

The results (3.1) and (3.2) imply at 90% confidence level $|\eta_{RR}^{(e)} - \eta_{RL}^{(e)}| < 0.085$ and $|\eta_{RR}^{(e)} + \eta_{RL}^{(e)}| < 0.18$, yielding the bounds

$$|\eta_{RR}^{(e)}| < 0.13 \quad \text{for any } |\eta_{RL}^{(e)}| \text{ and } \cos(\alpha + \omega) \quad , \tag{3.5}$$

$$|\eta_{RL}^{(e)}| < 0.13 \quad \text{for any } |\eta_{RR}^{(e)}| \text{ and } \cos(\alpha + \omega) \quad . \tag{3.6}$$

The result (3.3) implies the limit [note that $(1-P^F/P^{GT})/8 \simeq \mathrm{Re}\,\eta_{RR}^{(e)}\eta_{RL}^{(e)*}$]

$$|\mathrm{Re}\,\eta_{RR}^{(e)}\eta_{RL}^{(e)*}| < 10^{-2} \quad \text{(90\% confidence).} \tag{3.7}$$

A slightly better limit ($|\mathrm{Re}\,\eta_{RR}^{(e)}\eta_{RL}^{(e)*}| < 8 \times 10^{-3}$) follows from (3.1) and (3.2).

The D-coefficient has been discussed previously in Refs. 3 and 4. Barring a cancellation, the result (3.4) sets the constraints**

$$|\mathrm{Im}\,\eta_{LR}| \lesssim 2 \times 10^{-3} \tag{3.8}$$

$$|\mathrm{Im}\,\eta_{RR}^{(e)}\eta_{RL}^{(e)*}| \lesssim 2 \times 10^{-3} \quad . \tag{3.9}$$

4. Constraints on the Beta-Decay Parameters from Other Sources

Among other data the most stringent constraints on the parameters of $SU(2)_L \times SU(2)_R \times U(1)$ models come from muon-decay measurements, and from data which include some nonleptonic transitions. It should be noted that the latter are less reliable, in view of the uncertainties in calculations of nonleptonic

*A brief account of the conclusions regarding P_e^F/P_e^{GT} reported here is given in Ref. 12.

**We note that $|\mathrm{Im}\,\eta_{RR}^{(e)}\eta_{RL}^{(e)*}| < |\mathrm{Im}\,\eta_{LR}|$, provided that $g_R^2 m_1^2/g_L^2 m_2^2 < 1$.

amplitudes. The implications of muon-decay data on the β-decay parameters depend on whether $\tilde{v}_\mu$ is arbitrary or $\tilde{v}_\mu = 1$.* We shall consider three classes of models, distinguished by the values of $\tilde{v}_\mu$ and $\tilde{v}_e$.

(A) Models with $\tilde{v}_e = \tilde{v}_\mu = 1$. Examples are models where U = V (such as $SU(2)_L \times SU(2)_R \times U(1)$ models with Dirac neutrinos and a discrete left-right symmetry). $\tilde{v}_e = \tilde{v}_\mu = 1$ also if all the neutrinos are sufficiently light to be produced in β-decay.

Constraints from μ-decay.** The μ-decay Hamiltonian resulting from (2.1) is given by

$$H^{(\mu)} = c_{LL}[\bar{e}\Gamma_L\nu_e^{(L)}\bar{\nu}_\mu^{(L)}\Gamma_L\mu + \kappa_{RR}\bar{e}\Gamma_R\bar{\nu}_\mu^{(R)}\Gamma_R\mu$$
$$+ \kappa_{LR}\bar{e}\Gamma_L\nu_e^{(L)}\bar{\nu}_\mu^{(R)}\Gamma_R\mu + \kappa_{RL}\bar{e}\Gamma_R\nu_e^{(R)}\bar{\nu}_\mu^{(L)}\Gamma_L\mu] + \text{H.c.} \quad , \tag{4.1}$$

where $\nu_\mu^{(L),(R)}$ are defined as $\nu_e^{(L),(R)}$ except for $e \to \mu$, and $c_{LL} = a_{LL}/\cos\theta_1^L$, $\kappa_{RR} = \eta_{RR}(\cos\theta_1^L/\cos\theta_1^R)e^{-i\alpha}$, $\kappa_{LR} = e^{-i\alpha}\eta_{LR}(\cos\theta_1^L/\cos\theta_1^R)$, and $\kappa_{RL} = \eta_{RL}$.

Since $|\cos\theta_1^R/\cos\theta_1^L| \lesssim 1$, we have $|\eta_{RR}| \lesssim |\kappa_{RR}|$ and $|\eta_{LR}| \lesssim |\eta_{RL}| = |\kappa_{RL}| = |\kappa_{LR}|$. The best limit on $|\kappa_{RR}|$ from leptonic and semileptonic processes comes from the quantity $R = 1 - \delta\xi P_\mu/\rho$, (δ, ξ, and ρ are the usual muon spectrum parameters), related to the end point of the positron spectrum in polarized muon decay. The present experimental limit implies $|\kappa_{RR}| < 0.039$, and therefore

$$|\eta_{RR}^{(e)}| = |\eta_{RR}| \lesssim 0.039 \quad . \tag{4.2}$$

The best limit on $|\kappa_{RL}|$ is provided by the experimental value of the ρ-parameter, implying $|\kappa_{RL}| < 0.033$, so that

$$|\eta_{RL}^{(e)}| = |\eta_{RL}| < 0.033 \quad , \tag{4.3}$$

$$|\eta_{LR}| \lesssim 0.033 \quad . \tag{4.4}$$

The bounds (4.2) and (4.3) are to be compared with the constraints (3.5) and (3.6) obtained from β-decay data. For $|\mathrm{Re}\,\eta_{RR}^{(e)}\eta_{RL}^{(e)*}|$ the bounds (4.2) and (4.3) imply***

$$|\mathrm{Re}\,\eta_{RR}^{(e)}\eta_{RL}^{(e)*}| \lesssim 1.3 \times 10^{-3} \quad , \tag{4.5}$$

to be compared with the bound (3.7) resulting from the direct measurement.

Constraints from information involving nonleptonic transitions. In models where $\theta_i^R = \theta_i^L$ and there is no CP-violation (models with "manifest left-right symmetry" [5]) the K_L-K_S mass difference Δm_K sets a limit [14]

$$|\eta_{RR}^{(e)}| = |\eta_{RR}| \simeq m_1^2/m_2^2 \lesssim 3 \times 10^{-3} \tag{4.6}$$

on $|\eta_{RR}^{(e)}|$ (we have set $g_R = g_L$ as appropriate for such models), and an analysis of nonleptonic K-decays yields [15]

$$|\eta_{LR}| \simeq |\zeta| \lesssim 4 \times 10^{-3} \quad . \tag{4.7}$$

If CP-violation is present in the nonleptonic sector, but still $\theta_i^R = \theta_i^L$ (so-called "pseudomanifest left-right symmetry"), the limit from Δm_K and the CP-violating parameter ε imply again the bound (4.6) [3,16]. The bound (4.7) is also recovered, combining the limit from nonleptonic K-decays [now proportional to

---

*u_μ, v_μ, are defined as u_e, v_e, except for $e \to \mu$; $\tilde{v}_\mu \equiv v_\mu/u_\mu$.
**A study of the implications for general $SU(2)_L \times SU(2)_R \times U(1)$ models of measurements of the positron momentum spectrum end point in polarized μ-decay, which we use here, is given in Ref. 10.
***Inspection shows that the constraint $|\eta_{LR}| < 0.033$ improves the bound on κ_{RR} from R only slightly.

$\cos(\alpha + \omega)]$ with the limit (3.7) [proportional to $\sin(\alpha + \omega)$] (see Ref. 10). From (4.6) and (4.7) one obtains the stringent bound

$$|\mathrm{Re}\,\eta_{RR}^{(e)*}\eta_{RL}^{(e)*}| \lesssim 2 \times 10^{-5} \quad . \tag{4.8}$$

For models where θ_i^R and θ_i^L are unrelated, the $K^o \to \bar{K}^o$ amplitude sets no constraints on η_{RR} or κ_{RR} (see Ref. 10). The constraints from nonleptonic K-decays and the D-coefficient takes the form

$$|\eta_{LR}| \simeq |g_R\zeta\cos\theta_1^R/g_L\cos\theta_1^L| \lesssim 4 \times 10^{-3} \quad . \tag{4.9}$$

Observing that $\eta_{RR}\eta_{RL}^* = \kappa_{RR}\eta_{LR}$, we obtain from the limit on $|\kappa_{RR}|$ from R and from (4.9) the bound

$$|\mathrm{Re}\,\eta_{RR}\eta_{RL}^*| \lesssim 2 \times 10^{-4} \quad . \tag{4.10}$$

$|\mathrm{Im}\eta_{LR}|$ is constrained by the CP-violating parameter ε' in $K_L \to 2\pi$ decays, and also by the electric dipole moment of the neutron (D_n) to be less than $\sim 10^{-4}$ [4]. These constraints are, of course, less reliable than the constraints (3.8) from the direct measurement.

(B) Models with arbitrary $\tilde{v}_e$ and $\tilde{v}_\mu$. In this case muon decay does not provide a constraint on $\eta_{RR}^{(e)}$. The ρ-parameter yields

$$|\eta_{RL}^{(e)}| < 0.047 \quad (90\%\ \text{confidence}), \tag{4.11}$$

which is the best limit on $|\eta_{RL}^{(e)}|$ from leptonic and semileptonic data. Combining (4.11) with the bound $|\eta_{RR}^{(e)} - \eta_{RL}^{(e)}| < 0.085$ from data on Gamow-Teller β-decays (3.1) yields

$$|\eta_{RR}^{(e)}| < 0.12 \tag{4.12}$$

i.e., the same limit as from x_V and x_A (3.5). The limits (4.11) and (4.12) imply

$$|\mathrm{Re}\,\eta_{RR}^{(e)}\eta_{RL}^{(e)*}| < 6 \times 10^{-3} \quad . \tag{4.13}$$

We note that here a bound on $\eta_{RL}^{(e)}$ does not imply a bound on $|\eta_{LR}|$. The best limit on $|\eta_{LR}|$ in this case from leptonic and semileptonic processes is $|\eta_{LR}| < 0.1$ provided by data on inclusive neutrino and antineutrino scattering [17].

Since $\tilde{v}_e \lesssim 1$ (see Ref. 10), for models with manifest or pseudomanifest left-right symmetry the limits (4.6) and (4.7) hold for $|\eta_{RR}^{(e)}|$ and $|\eta_{LR}|$, implying the bound (4.8), as for models of class (A). In models where θ_i^R and θ_i^L are unrelated only the limit (4.9) from nonleptonic K-decays and the D-coefficient holds. The implication for $\mathrm{Re}\,\eta_{RR}^{(e)}\eta_{RL}^{(e)*}$ is

$$|\mathrm{Re}\,\eta_{RR}^{(e)}\eta_{RL}^{(e)*}| \lesssim 4 \times 10^{-3} \quad , \tag{4.14}$$

provided that $g_R^2 m_1^2/g_L^2 m_1^2 < 1$ (since $|\eta_{RR}^{(e)}\eta_{RL}^{(e)*}| = |\kappa_{RR}\sqrt{\tilde{v}_e}\,\eta_{LR}\sqrt{\tilde{v}_e}| \lesssim |\kappa_{RR}||\eta_{LR}|$). The limits on $|\mathrm{Im}\eta_{LR}|$ from ε' and D_n are, of course, the same as in class (A).

(C) Models with $\tilde{v}_\mu = 1$, and arbitrary $\tilde{v}_e$. This scenario would arise if all the neutrinos could be produced in μ-decay but not in β-decay. The conclusions regarding the limits on the beta-decay parameters from other sources are the same as for class (A), except for that, as in class (B) $|\eta_{LR}|$ is not bounded by a limit on $|\eta_{RL}^{(e)}|$.

5. Conclusions

Excluding from consideration mixed transitions and not counting the parameter involved only in the decay rates, β-decay measurements are sensitive to three combinations of parameters of $SU(2)_L \times SU(2)_R \times U(1)$ models: the constants $|\eta_{RR}^{(e)}|$, $|\eta_{RL}^{(e)}|$, and $\mathrm{Re}\,\eta_{RR}^{(e)}\eta_{RL}^{(e)*}$.

In $SU(2)_L \times SU(2)_R \times U(1)$ models where $\tilde{v}_\mu = 1$ the available muon-decay data set more stringent limits on the β-decay parameters than the existing β-decay measurements. In particular, the upper limit from muon decay data on the parameter $|Re\eta_{RR}^{(e)}\eta_{RL}^{(e)*}|$ (which is measured by the ratio P_e^F/P_e^{GT} of beta-ray polarizations) is smaller by an order of magnitude than the limit from the existing direct measurement. The limit on $|Re\eta_{RR}^{(e)}\eta_{RL}^{(e)*}|$ derived from data involving nonleptonic processes is better than the limit from μ-decay data by an order of magnitude.

In $SU(2)_L \times SU(2)_R \times U(1)$ models where $\tilde{v}_\mu$ is arbitrary $|\eta_{RR}^{(e)}|$ is not constrained if beta decay data are not included. The best limit on $|Re\eta_{RR}^{(e)}\eta_{RL}^{(e)*}$ from leptonic and semileptonic processes (obtained by combining the information from the ρ-parameter and Gamow-Teller β-decay data), as well as the limit from data involving nonleptonic processes are only slightly better than the present limit from a direct measurement.

In all models where $\theta_i^R = \theta_i^L$ the constraints on the β-decay parameters derived from nonleptonic processes are much more stringent than those implied by other data.

Searches for a nonzero D-coefficient provide constraints on $Im\eta_{LR}$ and $Im\eta_{RR}^{(e)}\eta_{RL}^{(e)*}$. The best limits on these from leptonic and semileptonic processes come from the direct measurement. The constraints derived from nonleptonic processes are more stringent by an order of magnitude, but less reliable.

I would like to thank Professors J. Deutsch, A. Rich, and M. Skalsey for informative conversations. This work was performed under the auspices of the U.S. Department of Energy.

References

[1] A. I. Boothroyd, J. Markey, and P. Vogel, Phys. Rev. C29, 603 (1984).

[2] J. C. Pati and A. Salam, Phys. Rev. Lett. 31, 661 (1973), Phys. Rev. D10, 275 (1974); R. N. Mohapatra and J. C. Pati, Phys. Rev. D11, 566, 2558 (1975); G. Senjanovic and R. Mohapatra, Phys. Rev. D12, 1502 (1975).

[3] P. Herczeg, Phys. Rev. D28, 200 (1983).

[4] P. Herczeg, in Neutrino Mass and Low-Energy Weak Interactions, Telemark 1984, ed. by V. Barger and D. Cline, World Scientific Publishing Co., 1985, p. 288.

[5] M. A. Bég et al., Phys. Rev. Lett. 38, 1252 (1977).

[6] B. R. Holstein and S. B. Treiman, Phys. Rev. D16, 2369 (1977).

[7] J. van Klinken, F. W. J. Kobs, and H. Behrens, Phys. Lett. 79B, 199 (1978).

[8] M. Skalsey et al., Phys. Rev. Lett. 49, 708 (1982).

[9] J. van Klinken et al., Phys. Rev. Lett. 50, 94 (1983).

[10] P. Herczeg, "On Muon Decay in Left-Right Symmetric Electroweak Models," Los Alamos National Laboratory preprint LA-UR-85-2761, to be published.

[11] A. Rich and M. Skalsey, private communication; J. van Klinken et al., Phys. Rev. Lett. 50, 94 (1983); J. Deutsch, private communication, and in these Proceedings ; see also Ref. 8.

[12] P. Herczeg, in Proceedings of the Second Conference on the Intersections Between Particle and Nuclear Physics, Lake Louise, Canada, May 26-31, 1986. To be published by the AIP.

[13] A. L. Hallin et al., Phys. Rev. Lett. 52, 337 (1984).

[14] G. Beall, M. Bander, and A. Soni, Phys. Rev. Lett. 48, 848 (1982).

[15] J. F. Donoghue and B. R. Holstein, Phys. Lett. 113B, 382 (1982).

[16] H. Harari and M. Leuver, Nucl. Phys. B233, 221 (1984).

[17] H. Abramowicz et al., Z. Phys. C12, 255 (1982).

Search for Anomalous "V+A" Currents in Nuclear Beta Decay

A.S. Carnoy, J. Deutsch, T.A. Girard[+]*, and R. Prieels*

Institute de Physique Nucléaire, Université Catholique de Louvain,
B-1348 Louvain-la-Neuve, Belgium

1) Introduction

The high precision searches for deviations from the standard $SU(2)_L \times U(1)$ model in muon decay are well known [1]. In a companion contribution P. Herczeg discusses the interest and complementarity of similar searches in beta decay to constrain the parameters of most general $SU(2)_L \times SU(2)_R \times U(1)$ models.

Such deviations from the 100% parity-violation predicted by the standard model have been searched for by measuring the absolute Gamow-Teller decay polarization [2] or the ^{19}Ne decay asymmetry parameter [3]. It was recognized also that the comparison of beta-polarizations from Fermi- and Gamow-Teller decays not only eliminates many potential systematic errors inherent to absolute measurements [4] but also allows one to significantly constrain a specific combination of the model-parameters [5].

Such a comparison measurement was previously performed to a precision of 3.6% [6], and efforts are underway to improve this limit significantly [7].

In the following, we shall

- describe the Positron Polarization Comparator installed at our cyclotron and the progress of our experiments,
- comment on the nuclear physics limitation of these searches,
- compare the three approaches underway.

2) The Positron Polarization Comparator of the Louvain University

A schema of the comparator is shown in Fig. 1 and cannot be discussed here in detail. Basically, we compare the polarization of 1.2 MeV positrons emitted in ^{14}O (Fermi) and ^{10}C (Gamow-Teller) decay by observing the time-spectrum of their annihilation with the electrons of fine-grained MgO powder placed in a 12 kG magnetic field of alternating direction.

The distinct feature of this experiment is the production of both activities in the same targets so as to avoid possible systematic errors arising from depolarization differences in two targets (cfr. 4). The target is self-sustaining boron nitride (BN) enriched (to 90 %) in ^{10}B and the two activities

---

[+]Now at Univ. of South Carolina, Columbia, S.C. 29208, U.S.A.

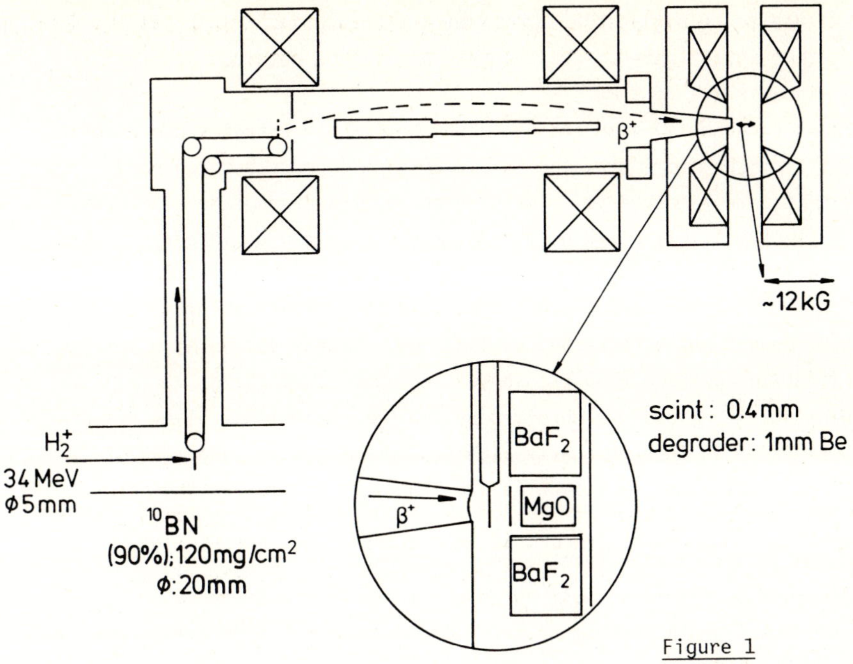

Figure 1

are distinguished using their distinct lifetimes (20 sec and 70 sec). This is illustrated in Fig. 2 which shows the beta decay time spectrum starting 7 sec after the end of a 40 sec irradiation period. Window A contains typically 75 %, and window B 15 %, of of ^{10}C-activity; the complement is practically ^{14}O-decay alone. (A small amount of the mixed-superallowed ^{11}C activity is present and can be taken into account). The slight loss of sensitivity one has to pay comparing the polarizations measured in the A- and B-windows is easily obtained: if we define, with P. Herczeg, $P^F/P^{GT} - 1 = 8r$, where F(GT)) = Fermi (Gamow-Teller), one obtains $P^B/P^A - 1 = 8x0.6xr$.

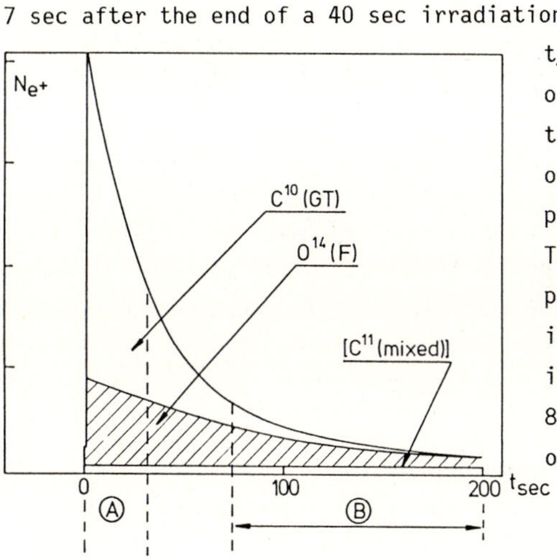

Figure 2

The beta-spectrometer has a transmission of 12 % of 4π; for a 5 mm diameter active spot, the injection efficiency into the polarimeter is 40 % and the momentum-resolution is 16 % FWHM. The end-points of the two positron-emitters are sufficiently similar (cfr. 4); as we focus positrons near the maximum of their intensity distribution, differences in scattering on the spectrometer walls are believed to be negligible at our level of precision. This point will be however further investigated. Typical beta-activities obtained at the beginning of window A are of 2 Mc.

Positron pass through a 0.4 mm scintillator, have their energies degraded in a 1 mm beryllium (Be) foil and stop in a 0.35 g/cm^3 compressed MgO powder of 20 nm grain-size kept in vacuum. The presence of the Be-degrader increases the typical asymmetry to be observed (pseudo triplet-to-triplet intensity ratios compared for two opposite field directions) by about 20 % of its value. The positronium formation ratio observed was typically 25 % (rather similar to comparable SiO_2 powder but 4 times better than the 7 % observed in Al_2O_3); the triplet lifetime observed (140 ns) showed negligible quenching.

The annihilation photons are observed by two BaF_2 detectors (51 mm diameter and 38 mm thick) coupled through quartz light-guides to quartz window XP2020 photomultipliers. The time resolution is 1.5 FWHM: count rates at the origin of the A window are 150 kHz in each detector and 40 kHz in their coincidence.

Positron annihilation time spectra were constructed in each irradiation cycle separately for both detectors as well as for their coincidences in both observation windows A and B. The magnetic field direction was randomly inverted between cycles. Typical time spectra obtained both for MgO and Al are shown in Fig. 3. Since no positronium is formed in Al, the spectrum is used to determine the response function: the slight tailing observed is interpreted as due to positrons stopped in the start scintillator. The coincidence spectra had a cleaner response function than the single ones.

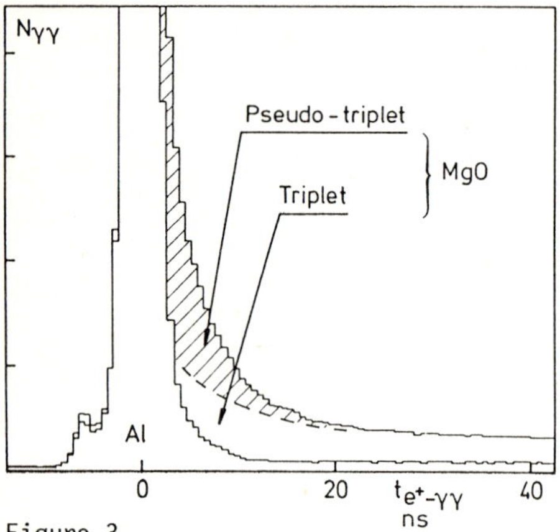

Figure 3

In the preliminary analysis we fitted the MgO-spectra with two exponentials (pseudo triplet and triplet) after allowing for the effect of the response function. The ratio of the two components was 0.3 for the single and 0.4 for the coincidence spectra, in reasonable agreement with expectation.

On the basis of the observed count rates and the expected asymmetry, a 1 % statistical precision should be achievable in a couple of hours.

3) "Nuclear" Effects: Contributions to P_F/P_{GT}

Even in the absence of any deviation from the standard model, P_F/P_{GT} will deviate from unity due to recoil effects [8,9]. This deviation is dominated practically by the weak magnetism contribution to the GT transition and is proportional to the ratio b/c, where b is the weak magnetism form factor and c the main axial amplitude. As $c = K/(ft)^{1/2}$, it is advantageous to choose fast GT decays for the comparision so as to make the recoil contribution negligible.

In the case of the $^{26m}Al(F)$-^{30}P (GT) comparison, envisioned by both the Michigan-Princeton-Toronto (MPT) and Groeningen groups, an evaluation of the b/c-term, using nuclear models, leads to an expected deviation of $P_F/P_{GT} - 1 = (6 \pm 2)10^{-4}\ E_e^{-1}$(MeV) [9]; an alternative evaluation [10], using the gamma width of the relevant gamma transition and CVC, yields an expectation of $(11 \pm 1)10^{-4}$ E_e^{-1}(MeV). If one seeks a determination of the r-parameter better than a couple of 10^{-3}, it may be interesting to choose a faster GT transition than that of ^{30}P.

^{30}P has a beta decay of log ft = 4.83. The log ft of the ^{10}C decay is 3.0: in this case, the contribution of the recoil terms is expected to be a factor of 10 smaller, i.e. completely negligible. We shall see below, however, that this case raises some experimental problems.

(4) Comparision of the three Ungoing Experiments

The Groeningen comparator uses Bhabba-scattering on magnetized iron, detecting in coincidence the scattered positron and the electron knocked out [6]. Taking into account the fraction of polarized electrons in iron, the highest asymmetry for an ideal lay-out (point detectors, perfect energy resolution, etc.) would be about 5 %. We do not know how severely the actual running conditions degrade this effect: the quoted 3.6 % precision [6] being based on 2×10^7 coincidences, the mean asymmetry is estimated to be around 1 %. This result was obtained in 8 days of running, but a simultaneous use of four polarimeters is envisioned [7] to increase the acceptance considerably.

The polarimeters based on time resolved annihilation spectroscopy [4], used by the MPT- and Louvain-groups do not require concidence measurements and can obtain in principle asymmetries as high as 10 %. It should be noted, however, that only about 5 % of the stopped positrons contribute to the pseudo triplet fraction to be observed and that a further factor of 5 to 10 is lost in detecting the annihilation photon. Moreover, the positron rate incident into the polarimeter must be limited to a couple of MHz so as to avoid a prohibitive number of accidental photons. Assuming a conservative mean incoming rate of 0.4 MHz, the relevant photon detection rate would be about 2 KHz; a measurement to 1 % relative precision of a 10 % asymmetry would then require only a couple of hours.

The choice of the decays to be compared, and hence the targets to be activated, is an important difference between the three groups. The main concern of both the MPT- and the Groeningen-groups was to choose decay doublets of as near an end-point energy as possible so as to avoid any eventual systematic errors due to scattering on the walls. The $^{26m}Al(F)$-$^{30}P(GT)$ doublet has end-point energies identical to 0.2 %; the Louvain choice of the $^{14}O(F)$-$^{10}C(GT)$ doublet realizes this condition only to 2.8 %. It should be noted that due to target preparation difficulties the $^{30}P(GT)$ decay is for the time being replaced with ^{25}Al (superallowed mixed) (MPT) [7] and ^{27}Si (superallowed mixed) (Groeningen) [7]. For these cases the end-point equality condition is realized only to 1.4 % and 16 %, respectively. Clearly all three groups will have to investigate the real impact of these differences for their particular geometry.

As said above (cfr. 2), an important particularity of the Louvain comparator is the use of the same target for both comparison decays. The depolarization estimates derive from ref. 11 and - to our knowledge - were never checked experimentally with the required accuracy. According to our estimate based on ref. 11, the difference in depolarization of a positron crossing 20 mg/cm^2 of Si or Mg amounts to 0.2 %(0.1 %) for 1 MeV(2 MeV) positrons. If we wish to push the precision to the 0.1 %-level with different targets, an experimental investigation of the depolarization mechanism may be desirable.

A fringe benefit of the single target is to be able to work with thicker targets and so obtain easily the required statistical accuracy. The draw-back of a reduced sensitivity (cfr. 2: a factor of 0.6) does not seem prohibitive; in this respect one notes that the use of mixed transitions by the MPT- and Groeningen-groups (cfr. above) also reduces the sensitivity by a factor of about 0.75. Possible systematic effects arising from rate-differences in the compared A- and B windows of our experiment have to be given however, serious consideration.

In conclusion, it seems rather fortunate that three different groups progress in parallel in these important precision experiments in which systematic errors are so easily overlooked.

We wish to thank A. Rich, M. Skalsey, J. Van Klinken and V.A. Wichers for many illuminating discussions and for having informed us on the progress of their experiments.

References:

1. J. Carr et al.: Phys. Rev. Lett. 51, 627 (1983) and D.P. Stoker et al.: Phys. Rev. Lett. 54, 1887 (1985).
2. J. Van Klinken: Nucl. Phys. 75, 145 (1966).
3. F.P. Calaprice et al.: Phys. Rev. Lett. 35, 1566 (1975).
4. M. Skalsey et al.: Phys. Rev. Lett. 49 708 (1982).
5. e.g.: M.A. Beg et al.: Phys. Rev. Lett. 38, 1252 1997 and P. Herczeg, communication to this conference.
6. J. Van Klinken et al.: Phys. Rev. Lett. 50, 84 (1983).
7. Private communication from M. Skalsey and A. Rich (Michigan-Princeton-Toronto collaboration), private communication from J. Van Klinken and V.A. Wichers (KVI, Groeningen).

8. B.R. Holstein: Phys. Rev. C16 1258 (1977).
9. T.A. Girard: Phys. Rev. C27, 2418 (1983).
10. H.P.C. Rood: private communication quoted in W.Z. Venema, Groeningen thesis, 1984, unpublished.
11. C. Bouchiat and J.M. Lévy-Leblond: Nuovo Cimento 33, 193 (1964) and M. Skalsey: Univ. Michigan thesis, 1982, unpublished).

Weak Interaction Studies of Oriented Nuclei Far from Stability

L. Vanneste, N. Severijns, D. Vandeplassche, E. van Walle, and J. Wouters

Instituut voor Kern- en Stralingsfysika, Leuven University, B-3030 Leuven, Belgium

Recent progress in different experimental areas has led to a new situation with regard to prospects of weak interaction studies of nuclei in extreme conditions. An immense variety of new nuclei has become available for β-decay investigations through the coupling of heavy ion accelerators to electromagnetic isotope separators. At the same time the progress in (and understanding of) implantation phenomena allows to prepare near ideal β-sources. Also we demonstrated since 1980 that almost all of these nuclei can be polarised to a very high degree by introducing new cryogenic "on line" orientation techniques [1]. Last but not least we developed detectors for particles emitted from a sample at m K temperatures without any window inbetween. The technique of measuring A_1 parameters using this equipment was tested off-line: the results correspond with theoretically predictable A_1 parameters within a few percent, even without any scattering correction.

We applied this combination of techniques first to α-decay in order to study shape variations [2]. In β-decay two fundamental topics were tackled first. The first is the systematic determination of A_1 parameters of mirror transitions in nuclei. Only two cases of nuclei have been measured previously, both using the atomic beam technique (^{19}Ne and ^{35}Ar): the results are conflicting with regard to an eventual vanishing of the Cabibbo angle in nuclear matter (see ref. 3 and the references contained therein). The A_1 of the bare nucleon n has been remeasured recently to high accuracy [4]. Therefore we started measurements of A_1 parameters of mirror nuclei systematically combining cryogenic and dynamic polarisation. As a first case we measured ^{17}F, with a surprisingly large A_1 as result (for discussion see separate contribution to this conference).

It has been argued at several occasions that a C_{Ae} (effective axial vector constant) can be predicted esp. in the case of "good single-particle" systems without major reliance on wave functions. In mirror nuclei it is possible to deduce the value of C_{Ae} from the β decay and the isoscalar part μ_0 of the magnetic moments. Another recent even more direct approach is possible using sum rules [5], and even more convincing since it does not rely on the intrinsically small μ_0. The surprising result is that C_{Ae} = 1.00(2) with a very high correlation coefficient for all mirror decays. How does this reconcile with the two nuclear cases for which A_1 has been measured? In the neutron the A_1 asymmetry is needed to obtain a highly accurate value of C_A/C_V: how is it possible that no drastic effect shows up esp. in the very accurately measured ^{19}Ne? The latter has been used effectively to look for effects in the promille range connected with time-reversal invariance, while a 27% deviation in C_A/C_V goes unnoticed. Although C_A and C_V come always in combination with the respective matrix elements, it is expected - esp. after the large shell-model calculation of Wildenthal for the sd-shell - that one should be able to calculate the ratio $\langle\sigma\rangle/\langle 1\rangle$ to much better accuracy than 25%.
Other factors may be uncertainties in the ft-value (as in the neutron case). In any case the A_1 coefficient is strongly dependent on the C_A/C_V ratio: in the

allowed approximation for the neutron one obtains $A_1 = -\frac{2\lambda(\lambda-1)}{1+3\lambda^2}$.
In the case of ^{19}Ne the A_1-asymmetry is small (of the order 4% vs. 12% for the neutron). In ^{35}Ar it is much larger (22%) but the extracted A_1 has raised controversy. Indeed, when combined with the ft-value, the value for C_V is anomalous (about 4% too large). This has led to the suggestion that the Cabibbo angle may vanish in nuclei for high magnetic fields (~10^{16} gauss). The Cabibbo angle - more exactly a quark mixing angle - turns out to have the normal value in a supposed high field nucleus (^{24}Al) according to a recent measurement [3]. Is it possible that the field is just large enough to induce a phase transition by which the quark mixing angle disappears in one nucleus and not in another?

From the J→J (J=4) analog transition in the β^+ decay of ^{24}Al→^{24}Mg, one can indeed derive the F matrix element if the GT one is known. Relying on the extracted life-time of $t_{1/2} = 2.053 \pm 0.004$ sec., and on the assumption that GT is vanishingly small, a G_V is extracted in good agreement with the value from ^{19}Ne and the superallowed (0^+)-(0^+) transitions.

From this, the authors of ref. 3 claim that the ^{35}Ar result is due to an experimental error. Considering the energy put into checking all parts of the ^{35}Ar puzzle, this is puzzling. The most suspected value is the one of the asymmetry parameter $A_1 = 0.22(3)$. Indeed it was remeasured by the same method with results varying from 0.16(4) to 0.33(6), but in order to obtain agreement a value of 0.43 is required. The problem is that only one method (AB) gives access to A_1 information of mirror decays. That method is restricted to two cases, both - not by coincidence - noble gases. In terms of nuclear structure both are less than ideal. The normal behaving one (^{19}Ne) gives no problems at first glance but has a small A_1: the one with a large A_1 leads to problems.
The recent A_1 value from the neutron also leads to disagreement with Cabibbo theory, in as far that the derived C_A/C_V disagrees with the result from equally recent semileptonic hyperon decays.

Obviously there is an urgent need for additional measurements, preferentially with - a different technique
- on mirror nuclei with "good" nuclear structure, i.e. closed shell ± one nuclei.

The problems concern both anomalies in C_V and C_A, as extracted from neutron decay and "nuclear" mirror transitions.

In addition to the analysis of Buck and Perez [5] we performed an analysis using improved shell model estimates and precise magnetic moments (not the isoscalar part). Surprisingly we obtain again a high correlation fit with

$$0.90 < g_A/g_V < 1.05$$

instead of

$$1.00 < g_A/g_V < 1.60$$

from an analysis of Ramar et al. using the intrinsically small (but probably inaccurate) isoscalar part. This and other "circumstancial evidence" makes it plausible that C_A might be renormalized in multinucleon systems but not in nucleons. We intend to corroborate this assertion by a systematic measurement of all available mirror decays, thus establishing a series of superallowed A_1 of which only two candidates are available now.

Another exciting prospect is the possibility of measuring isospin impurities in heavy nuclei with $N \simeq Z$. In this case the famous Lane-Soper argument, on which the claim for isospin (im)purity is based despite large Coulomb interactions, after the unexpected discovery of isobaric analogue states by Anderson et al., is no longer valid. Hence large Fermi contributions must develop in these nuclei, providing in turn again - through interference - excellent candidates for T-violation tests in nuclear matter. Evidence for at least one case, i.e. ^{75}Se, has been obtained previously, close to the stability line but in a region of possible strong shape changes. This work is repeated and extended to a series of nuclei far from stability.

New possibilities arise also, in connection with the "missing" Gamow-Teller strength, combination with electric dipole manifestation in light actinides (many closely lying opposite parity states), the influence of sudden deformation change in the new found regions of extremely large deformation and many others.

References:

1. Proceedings of the Intern. Symp. on Nuclear Orientation and Nuclei far from Stability (1985), Eds. B.I. Deutch and L. Vanneste, Baltzer AG.
2. J. Wouters, D. Vandeplassche, E. van Walle, N. Severijns and L. Vanneste, Phys. Rev. Lett. 56 (1986) 1901.
3. E.G. Adelberger, P.B. Fernander, C.A. Gossett, J. L. Osborne, V.J. Zeps, Phys. Rev. Lett. 55 (1985) 2129.
4. P. Bopp, D. Dubbers, L. Hornig, E. Klemt, J. Last, H. Schütz, S.J. Freedman, O. Schärpf, Phys. Rev. Lett. 56 (1986) 919.
5. Buck and Perez, Phys. Rev. Lett. 50 (1983) 1975.

Beta-Decay Asymmetry Measurements of the Mirror Nuclei

N. Severijns, D. Vandeplassche, E. van Walle, J. Wouters, and L. Vanneste

Instituut voor Kern- en Stralingsfysika, Celestijnenlaan 200 D, B-3030 Leuven, Belgium

We started a systematic determination of the beta-decay asymmetries of the mirror nuclei. Up to now only three such measurements exist, viz. the neutron, ^{19}Ne and ^{35}Ar [1-3]. For the neutron and for ^{19}Ne, the values for the vector weak coupling constant G_V which can be extracted from the measured asymmetries yield values for the Cabibbo angle that are in good agreement with the ones derived from the Ft-values for superallowed $0^+ \to 0^+$ beta-transitions between T=1 states and from hyperon beta-decays. ^{35}Ar on the other hand is anomalous for many years already, and suggests that the Cabibbo angle would be zero in nuclear beta-decay.

For the $0^+ \to 0^+$ beta-transitions, only the vector current contributes and no magnetic fields are present. For odd nuclei however the axial-vector current contributes also and magnetic fields inside the nucleus could be high. This has led Salam and Strathdee [4] to the suggestion that the Cabibbo angle would vanish in magnetic fields of $\sim 10^{16}$ gauss. They furthermore made the prediction that if CP-symmetry is violated spontaneously, it should be restored above a critical magnetic field which they calculate to be 10^{11} to 10^{14} gauss. Suranyi and Hedinger [5] have shown that the magnetic field inside an odd nucleus could indeed be of the order of 10^{16} gauss.

Apart from the Cabibbo angle problem, beta-decay asymmetry measurements of the mirror nuclei could also give more information on the possible renormalization of the axial-vector current. This renormalization was suggested by Wilkinson [6] already 15 years ago and more recently by Buck and Perez [7]. In order to extract C_A/C_V from a measured asymmetry however, one needs the value of the Gamow-Teller matrix element for the transition concerned. The calculation of this is in general a difficult problem, but, as Jensen and Mayer [8] pointed out, for mirror nuclei there exists a definite relationship between the GT-matrix element and the nuclear magnetic moment. Using this relation, together with the most recent ft-values we found $C_A/C_V = 1.01$ (this will be published), which is to be compared with $C_A/C_V = 1.25$ for the neutron.

To shed more light on these problems we decided to extend the measurements to other mirror nuclei. Instead of the atomic beam technique, which was used for the ^{19}Ne and ^{35}Ar measurements, we use two different nuclear orientation techniques, both in combination with ion implantation at low temperature [9]. One is the conventional static low temperature nuclear orientation, which is based on the thermal equilibrium between the implanted ions and the host lattice. This method is used for those nuclei experiencing a sufficiently large hyperfine field in the host lattice. For the other ones we will use a new dynamic nuclear orientation method which we developped during the past two years and which is based on the capture of spin-polarized electrons (SPEC). With this method, radioactive ions are reflected under a grazing incidence angle of typically 1° from a magnetized Ni 110-surface. During this reflection they get neutralized and as they approach the surface only to a distance of 1 to 2 Å, only spin-polarized 3d-electrons from the Fermi-surface can be captured. The nuclear polarization is then achieved by a partial transfer to the nucleus of this electronic polarization, by hyperfine interaction.

We already measured a first nucleus, viz. ^{17}F (64.5 sec). The nucleus was oriented cryogenically and the anisotropy of the 1.75 MeV (I = 100%) $5/2^+ \rightarrow 5/2^+$ beta-transition (Fig. 1) was detected with thin solid state detectors working at 4K. Without any scattering correction being necessary we find the value for the asymmetry to be in the range -0.64 to -0.68 implying a value for sin Θ_V between 0.63 and zero and for C_A/C_V between 1.55 and 1.07. The accuracy can still be enhanced as we had to work with a rather low counting rate, but because in the case of ^{17}F the asymmetry varies very slowly with $y = C_V M_F / C_A M_{GT}$, it will be difficult to reach an accuracy being sufficiently high to draw definite conclusions concerning sin Θ_V and C_A/C_V. Nevertheless, since the results extracted from this ^{17}F experiment are of the good order of magnitude, this measurement proves that with our technique we can determine beta-decay asymmetries without any corrections being necessary. Therefore we are now concentrating ourselves on the more favorable cases ^{15}O, ^{19}Ne, ^{29}P, ^{35}Ar and ^{39}Ca.

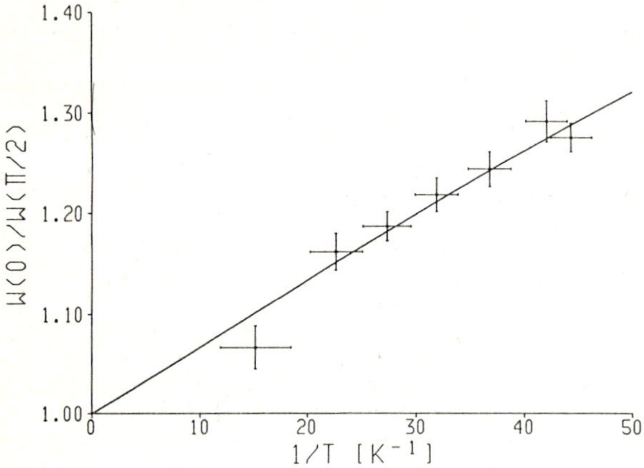

Fig.1. Asymmetry as a function of temperature for the 1.75 MeV beta-decay of ^{17}F.

References.

1. P. Bopp, D. Dubbers, L. Hornig, E. Klemt, J. Last, H. Schütze, S.J. Freedman and O. Schärpf: Phys. Rev. Lett. 56, 919 (1986)
2. F.P. Calaprice, E.D. Commins, H.M. Gibbs, G.L. Wick and D.A. Dobson: Phys. Rev. 184, 1117 (1969)
3. W.G. Mead: Princeton University thesis, 1974.
   F. Calaprice: University of California thesis, UCRL-17551, 1967.
   F. Calaprice, E. Commins and D. Dobson: Phys. Rev. 137B, 1453 (1965)
4. A. Salam and J. Strathdee: Nature 252, 569 (1974)
5. P. Suranyi and R.A. Hedinger: Phys. Lett. 56B, 151 (1975)
6. D.H. Wilkinson: Phys. Rev. C7, 930 (1973); Nucl. Phys. A209, 470 (1973); Nucl. Phys. A225, 365 (1974)
7. B. Buck and S.M. Perez: Phys. Rev. Lett. 50, 1975 (1983)
8. M. Mayer and J. Jensen: Elementary Theory of Nuclear Shell Structure (John Wiley and Sons, Inc., New York 1955)
9. D. Vandeplassche, L. Vanneste, H. Pattyn, J. Geenen, C. Nuytten and E. van Walle: Nucl. Instrum. Methods 186, 211 (1981)

Recent Calculations of Isospin-Mixing Corrections to the Fermi Matrix Element in Superallowed β-Decay and the Determination of the Weak Vector Coupling Constant

W.E. Ormand and B.A. Brown

National Superconducting Cyclotron Laboratory, Michigan State University, East Lansing, MI 48824, USA

In this report, corrections to the Fermi matrix element for superallowed β-transitions due to isospin impurities are presented. An important feature of these decays is that, since they are purely vector, their ft values are given by

$$f_R t = K/G_V^2 \; |M_F|^2 \; , \tag{1}$$

where K is a constant, f the statistical rate function (which includes the nucleus -dependent "outer" radiative correction δ_R of Sirlin [1], i.e. $f_R = f(1+\delta_R)$), t the partial half-life, G_V the effective vector coupling constant for single nucleon β-decay, and M_F the Fermi matrix element for the transition. Once all nucleus-dependent corrections have been applied to (1), it is possible to extract empirical values of G_V. This is important because the effective vector coupling constants for nucleon and muon β-decay are related by

$$G_V^2 = G_\mu^2 \cos^2 \theta_C (1 + \Delta_\beta - \Delta_\mu), \tag{2}$$

where θ_C is the Cabibbo angle, and Δ_β and Δ_μ are the "inner" radiative corrections to nucleon and muon β-decay, respectively. With G_V/G_μ and θ_C determined from experiment, it is then possible to test current theoretical estimates for $\Delta_\beta - \Delta_\mu$.

Although many superallowed transitions have been observed experimentally, at present ft values for only eight transitions have been measured with sufficient accuracy to permit a test of (1) [2]. For this reason, this work concentrates on corrections for the ground state β-decay of ^{14}O, ^{34}Cl, ^{42}Sc, ^{46}V, ^{50}Mn, and ^{54}Co and the decay of the metastable state in ^{26}Al and ^{38}K.

If the initial and final states are perfect analogs, then the Fermi matrix element is model-independent, and values for G_V could be extracted from measured ft values with (1). The most accurately determined ft values, however, are not constant within experimental uncertainty [2], suggesting the breaking of analog symmetry between the initial and final nuclei due to the presence of isospin nonconserving (INC) interactions. The extent to which analog symmetry is broken is embodied in the correction factor δ_C, defined by $|M_F|^2 = |M_{F0}|^2(1-\delta_C)$, where M_{F0} is the Fermi matrix element between states with analog symmetry.

The starting point of our calculations is the usual spherical shell-model wave functions that possess definite isospin. Due to the wide range of nuclei under investigation, four separate configuration spaces were used. These were: the 0p shell, the 1s-0d shell, the 0d3/2 and 0f7/2 orbits, and the 0f-1p shell (for more details, see [3]). Within each basis, it is necessary to consider the effects of isospin mixing between states within the model space as well as with those outside. Isospin mixing within the model space is calculated by adding the INC interaction onto the isospin-conserving Hamiltonian, while isospin mixing outside the active shell is taken into account by allowing the proton radial wave functions to be pushed out relative to the neutron wave functions due to the Coulomb potential. These two effects can be factored [4] into components due to the radial overlaps (RO) and isospin mixing within the configuration space (IM), $\delta_C = \delta_{RO} + \delta_{IM}$.

Previously [5], δ_{RO} was evaluated using radial wave functions obtained with a central Woods-Saxon (WS) plus Coulomb potential. This procedure overestimates the difference between the proton and neutron radial wave functions by neglecting an induced isovector interaction that arises from differences in the proton and neutron densities. To account for this induced interaction we have performed self-consistent Hartree-Fock (HF) calculations with a Skyrme-type interaction (SGII, [6]). δ_{RO} obtained with the Hartree-Fock wave functions are compared with those of Towner and Hardy [5], and are given in Table I. Values obtained with HF wave functions are systematically reduced relative to those given in [5]. This reduction is due to both the Coulomb and nuclear central potentials used in each calculation. In Hartree-Fock calculations, we find that protons are effectively in a potential well, which is both deeper at the origin and has a higher barrier at the nuclear surface relative to the WS procedure. This additional potential tends to draw in the proton radial wave functions, and thus reduce the value of δ_{RO}.

Table I Comparison of the corrections to the Fermi matrix element obtained in the present work with those of Towner and Hardy [4,5]. Values of δ are given in %.

Decaying nucleus	$f_R t$ [2,4]	Present Work				Previous Values [4,5]			
		δ_{IM}	δ_{RO}	δ_C	$G_V G_\mu$	δ_{IM}	δ_{RO}	δ_C	G_V/G_μ
^{14}O	3085.7(22)	0.010	0 134	0.144	0.9842(4)	0.05	0.23	0.33	0.9851(4)
^{26}Al	3083.3(14)	0.012	0.255	0.267	0.9852(3)	0.07	0.27	0.34	0.9855(3)
^{34}Cl	3103.3(28)	0.056	0.432	0.488	0.9831(5)	0.23	0.62	0.85	0.9849(5)
^{38}K	3098.1(26)	0.111	0.453	0.564	0.9843(4)	0.16	0.54	0.70	0.9849(4)
^{42}Sc	3104.2(63)	0.109	0.209	0.318	0.9821(10)	0.13	0.35	0.48	0.9829(10)
^{46}V	3100.9(19)	0.013	0.230	0.243	0.9822(3)	0.04	0.36	0.40	0.9830(3)
^{50}Mn	3099.2(38)	0.004	0.296	0.300	0.9828(6)	0.03	0.40	0.43	0.9834(6)
^{54}Co	3105.9(23)	0.005	0.359	0.364	0.9821(4)	0.04	0.56	0.60	0.9832(4)
^{22}Mg	3016(63)	0.018	0.191	0.209	-	0.06	0.29	0.35	-
^{26}Si	3037(49)	0.035	0.186	0.221	-	0.04	0.38	0.42	-
^{30}S	3077(80)	0.053	0.636	0.689	-	0.34	0.87	1.21	-
^{34}Ar	3000(51)	0.006	0.369	0.375	-	0.13	0.91	1.04	-

The contribution δ_{IM} was evaluated with isospin-mixed wave functions obtained by adding an INC interaction onto the isoscalar Hamiltonian. V_{INC} for each configuration space was determined empirically [3] by requiring that the parameters of a Coulomb plus phenomenological isovector and isotensor potential reproduce experimental b- and c-coefficients of the isotopic mass multiplet equation. The isospin-mixing corrections are compared with those of Towner and Hardy [4], and are also listed in Table I. The values of δ_{IM} calculated here are much smaller than those of Towner and Hardy. This difference is due to both the zeroth-order wave functions and the INC interactions used in each work.

A comparison between values of G_V/G_μ determined with the isospin-mixing corrections presented here and those of Towner and Hardy are shown in Fig. 1 (the error bars reflect the limits of experimental precision). From the figure, it is clear that extracted values of G_V/G_μ are not constant within experimental error. In fact, the results of Towner and Hardy give two inconsistent values: one for $Z < 21$, and another for $Z \geq 21$. The results reported here represent an improvement in this sense, as values of G_V/G_μ are consistent for $Z = 17$, 21, and 25.

As was mentioned above, the uncertainties in G_V/G_μ examined thus far have been purely experimental. There are, however, some theoretical uncertainties as well. In particular, the effects of the model-space truncation on the 0p- and 0f-1p-shell nuclei are unknown. In the case of the 0p shell, no particle-hole excitations into the low-lying 0d5/2 and 1s1/2 orbits were allowed. As for the 0f-1p-shell nuclei, no more than one particle was allowed outside the 0f7/2 orbit. At present, there exists no isoscalar Hamiltonian for the nuclei in question which can adequately account for these effects. The situation concerning the 1s-0d-

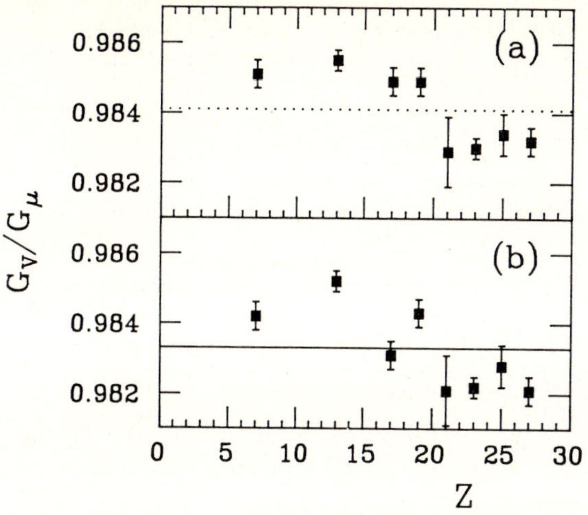

Figure 1 Comparison between values of G_V/G_μ determined in the present work (b) with those of Towner and Hardy(a). The line in each figure represents the average value obtained in each work.

shell, however, is somewhat better. A full-space calculation for nuclei in the middle of the sd-shell is feasible within present limits of computing technology; and, recently, an improved isoscalar Hamiltonian that includes the full sd-shell has been developed [7]. With this in mind, perhaps the best experimental strategy for the future would be to concentrate on understanding the discrepancies that lie within the sd-shell. This course should also include more precise measurements of the ft values for the superallowed β-decays of ^{22}Mg, ^{26}Si, ^{30}S, and ^{34}Ar. Corrections for these transitions are also presented in Table I.

Finally, as an illustration of what can be learned from (2), we conclude with some remarks regarding the "inner" radiative corrections Δ_β and Δ_μ. The difference $\Delta_\beta - \Delta_\mu$ is given by [8]

$$\Delta_\beta - \Delta_\mu = \frac{\alpha}{2\pi}\,[3\ln(M_Z/M_p) + 6\bar{Q}(M_Z/M_A) + 2C + \ldots], \tag{3}$$

where M_Z is the Z-boson mass, M_p, the proton mass, α the fine-structure constant, and $\bar{Q}$ is the average quark charge (for u and d quarks $\bar{Q}$ is 1/6). The quantity C is the least understood part of (3), and present estimates are C = 1 [9] and C=-0.5 [10]. Using values of G_V/G_μ obtained in the present work, we find the quantity C to be 1.64(17) and -0.26(39) for the decay ^{26}Al and ^{34}Cl, respectively.

1. A. Sirlin, Phys. Rev. 164, 1767 (1967)
2. V.T. Koslowsky, et al., AMCO-7, p. 60
3. W.E. Ormand, Ph.D. thesis, Michigan State University (1985); W.E. Ormand and B.A. Brown, Nucl. Phys. A440, 274 (1985)
4. I.S. Towner and J.C. Hardy, Nucl. Phys. A205, 33 (1973)
5. I.S. Towner, J.C. Hardy, and M. Harvey, Nucl. Phys. A284, 269 (1977)
6. Nguyen Van Giai and H. Sagawa, Nucl. Phys. A371, 1 (1981)
7. B.H. Wildenthal: Progress in Particle and Nuclear Physics, edited by D.H. Wilkinson (Pergamon Press, Oxford, 1984) vol. 11, p. 5.
8. A. Sirlin, Rev. Mod. Phys. 50, 573 (1978)
9. A. Sirlin, Nucl. Phys. B71, 29 (1974)
10. E.S. Abers et al., Phys Rev. 167, 1461 (1968)

Axial-Vector Weak Coupling Constant g_A and Quark Confinement in Nucleons

R. Tegen

Institute of Theoretical Physics and Astrophysics, University of Cape Town, Private Bag, Rondebosch 7700, Republic of South Africa

Within the $V - A$ theory for the charged weak current the β decay of the free neutron, $n \to pe\overline{\nu}_e$, is determined by momentum-independent vector and axial-vector weak coupling constants g_V and g_A. The neutron-spin-electron-momentum angular correlation is sensitive to g_A/g_V; a recent experiment at the reactor in Grenoble obtains $g_A/g_V = 1.262 \pm 0.005$ [1]. For $q^2 = 0$ the weak current couples only to those nucleon constituents which build up the helicity of the nucleon [2]; the strength of this coupling is g_A. In confining potentials which follow a simple power law (including the MIT bag for an infinite power) g_A/g_V is independent of the scale parameter set by the (phenomenological) potential. In such models g_A/g_V depends only on the surface thickness and decreases monotonically with decreasing surface thickness, see Fig. 1 [3]. Whereas all confining models (except the *vector* potentials) give the static SU(4) value for $n \to 0$, different classes of potentials converge to different "asymptotic" values. A purely scalar potential stays within the bounds [1.09, 5/3] whereas the admixture of a *vector* potential allows for even *lower* values of g_A/g_V. For an *equal* amount of *scalar* and *vector* potential the limit $n \to \infty$ coincides with the SLAC-bag value 5/9 [4] although the quark density is *not* concentrated in the surface region. The MIT bag core contributes 1.09 to g_A/g_V (without recoil/center-of-mass corrections). Recoil/c.m. corrections to (g_A/g_V) were found by several groups to be small [5–9]. In view of our helicity argument [2], see below, it is obvious that the spurious center-of-mass motion should contribute very little (if at all) to the nucleon helicity.

The problem with all bag models is the *sharp* bag boundary which cuts off the quark and (in some models) the pion fields. If the pion is cut off as well, g_A/g_V becomes much too large, $3/2(g_A/g_V)_{\text{quark}} \approx 5/3$ [10–12]. In a more realistic picture where the pion field "dies out" continuously over a thin surface and hardly exists in the interior the pionic contribution to (g_A/g_V) vanishes [8, 13] and $(g_A/g_V) = (g_A/g_V)_{\text{quark}}$ results, independent of the bag radius/quark core size.

Such a surface transition region is simulated in confinement potential models [14]. An ansatz of the form $M(r) = c_n r^n$ (with $n = 2, 3$ which is favoured by Fig. 1) for a scalar potential has been used previously to obtain a successful description of *both* the axial- and charge form factors of the nucleon [8, 15].

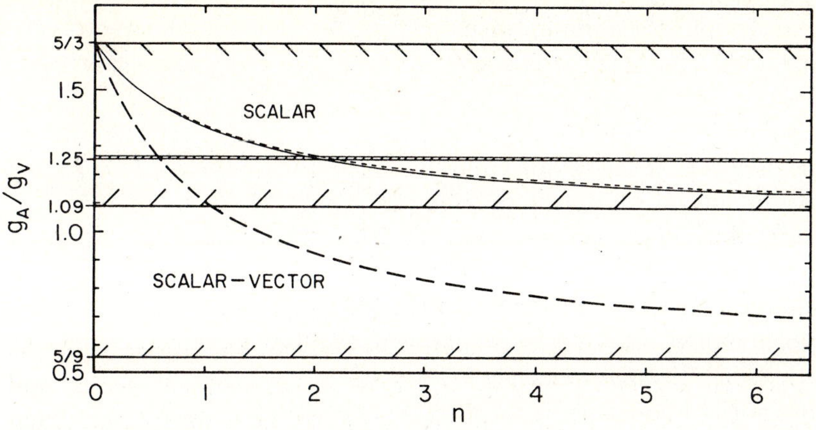

Fig. 1. g_A/g_V as a function of $n = r\,d/dr \log M(r)$, compared with recent data, $g_A/g_V = 1.262(5)$ [1]; solid curve: scalar confinement $M(r) = c_n r^n$ for massless quarks; short-dashed: scalar confinement for a current quark mass $m_q = 10\,\mathrm{MeV}$ [27]; long-dashed: scalar-vector confinement $\frac{1}{2}(1+\gamma_0) = c_n r^n$; a pure vector confinement gives $g_A \equiv 0$, see text

The expectation value of the axial-vector current between free nucleon states $\langle P_f, s_f | A^\mu_\lambda(q^2) | P_i, s_i\rangle$ is a combination of $G_A(q^2)$, $G_p(q^2)$ and $G_T(q^2)$ [8]. We have shown [2] that it is possible to choose a Lorentz frame (the "infinite momentum frame" (IMF) [16]) which disentangles contributions from G_A, G_p and G_T, namely

$$\langle P_f, s_f | A^{\mu=0}_\lambda(q^2) | P_i s_i\rangle \underset{\text{(IMF)}}{\longrightarrow} G_A(q^2=0)\left\langle \boldsymbol{\sigma}_N \cdot \hat{P}_N \frac{\tau^\lambda_N}{2}\right\rangle$$

where $\hat{P}_N = \boldsymbol{P}_i/|\boldsymbol{P}_i| = \boldsymbol{P}_f/|\boldsymbol{P}_f|$ in the IMF.

Equation (1) says that the weak hadronic current couples to the helicity $\langle \boldsymbol{\sigma}_N \cdot \hat{P}_N\rangle$ of the nucleon with a strength given by $g_A \equiv G_A(q^2 = 0)$. This means that (g_A/g_V) receives contributions only from fields contributing to the helicity of the nucleon. This enables us to understand the smallness of c.m./recoil corrections to (g_A/g_V) as well as the strong surface dependence of (g_A/g_V).

In terms of upper (lower) components of the quark spinors ($1s_{1/2}$ orbits) $g(f)$ we obtain [14]

$$g_A/g_V = 5/3\left(1 - (4/3)\int_0^\infty dr\, f^2(r) \Big/ \int_0^\infty dr (g^2 + f^2)\right) \quad . \tag{2}$$

For a pure *vector* potential one simply has[1] [3]

$$\lim_{R\to\infty}\left[\int_0^R dr\, f^2(r) \Big/ \int_0^R dr(g^2+f^2)\right] \equiv 3/4 \tag{3}$$

[1] As a consequence of Klein's paradox [17] the individual terms diverge, only the ratio gives a finite value.

for *arbitrary*, confining vector potentials; thus $g_A/g_V = 0$ for *any* vector potentials as expected. The MIT values are ~1/3 of (3), while the experiment requires ~3/16.

Does (in general) the meson cloud contribute to (g_A/g_V)? The answer clearly depends on whether the meson cloud builds up the nucleon spin (Skyrme models [18–20]) or not (chiral quark models [5,6,10,11,21–24]).

The spin distribution in the Skyrmion [18,19] is not optimal as evidenced by a too small (g_A/g_V). *Rho* et al. have demonstrated [20] that the spin distribution is indeed improved by supplementing the Skyrmion with a spin-carrying quark core (hybrid model). It will be instructive to evaluate in detail the helicity argument for (hybrid) Skyrme models.

In the remainder we concentrate on chiral quark models in which the meson fields do *not* contribute to the nucleon spin and hence do not contribute to g_A, see (1). The simplest chiral quark model includes only the pion field $\underset{\sim}{\phi}$ and gives rise to a term $-f_\pi \delta^\mu \underset{\sim}{\phi}$ in the axial current. As long as the pion field is continuous in all space it does not contribute to g_A. This is the result found in [11]; it is consistent with Eq. (1). In chiral soliton models where $\sigma(r) \neq f_\pi$ inside the nucleon, the *scalar* σ-field does *not* contribute to the nucleon spin and hence (following our helicity argument, (1) [2]) also not to g_A.

The mean field contribution to g_A from a scalar field $\sigma(r)$ depends strongly on its surface behaviour. We find [2]

$$(g_A)_{\text{mesons}} \sim \int_0^\infty dr\, r^2 h(r) d/dr \sigma(r) \tag{4}$$

where $h(r)$ is the radial pion field $\underset{\sim}{\phi}(r) = \langle \sigma_N \cdot \hat{r} \underset{\sim}{\tau}_N \rangle h(r)$ with the asymptotic Yukawa form, and $M(r)$ is identified with $g\sigma(r)$ [2,21,24]. The vanishing of (4) required by the helicity argument, Eq. (1), is satisfied in models where $\sigma(r) \equiv f_\pi$. If the scalar σ is treated as a dynamical field, however, this is not the case any more. The reason is that Lorentz covariance [which demands $(g_A)_{\text{IMF}} = (g_A)_{\text{c.m.}}$] is lost in the mean field treatment of the mesons, so that g_A acquires an *artificial* contribution by transforming axial-vector matrix elements from the IMF to the C.M. system. Recently unacceptably large contributions to g_A from the σ-field have been reported [21]. The fact that $(g_A)_{\text{mesons}}$ does not automatically vanish might be a strong indication that the mean field calculation is inherently flawed. On the other hand the constraint $(g_A)_{\text{mesons}}=0$, Eq. (4), derived from Eq. (1) has interesting consequences, as it requires a *maximum* for $\sigma(r)$ in the surface region, see Fig. 2, implying a deviation from the "chiral circle" $\sigma^2 + \phi^2 = f_0^2$ in the surface region, consistent with other findings [21,24,25]. We find [26] for a potential form as in Fig. 2, constrained by $(g_A)_{\text{mesons}}=0$, Eq. (4), the following:

i) The constraint $(g_A)_{\text{mesons}}=0$ is not easy to fulfill and really *constrains* possible shapes.

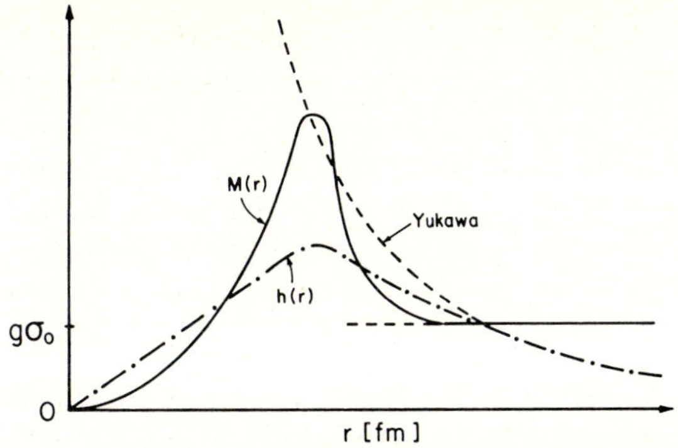

Fig. 2. Solid curve: Dirac-scalar potential $M(r)$; dashed-dotted: radial pion field $h(r)$; dashed: Yukawa's radial pion field for a pointlike nucleon, see text

ii) g_A/g_V depends strongly on the "inside" and weakly on the "outside" of $M(r)$.

iii) The position and the magnitude of the maximum determine the quark eigenenergy E_0 which is ~ inversely proportional to $\langle r^2\rangle^{1/2}_{\text{quark}}$. The r.m.s. radius is independently constrained by comparison to $g_A(q^2)$ [8] leaving E_0 in the "allowed" range (500–600)MeV. E_0 depends on the asymptotic value of $M(r)$ and decreases with increasing $g\sigma_0$. Reasonable values are obtained with $g\sigma_0 \approx (700\text{–}1000)$MeV [2,26].

iv) The pion field is suppressed inside the confinement region (a factor $\gtrsim 2$ smaller than the "cloudy bag" pion field [5]), see Fig. 2, and brings this model closer to bag models which exclude the pions from the bag [10,12,28].

v) Rather drastic changes of the scalar σ field are unavoidable if one wants g_A to be correctly predicted at the mean field level. While these changes affect the σ-field in the surface region, they do not spoil low energy properties of the nucleon.

References

1. P. Bopp et al.: Phys. Rev. Lett. **56**, 919 (1986)
2. R. Tegen, P. Zimak, R.D. Viollier: Preprint University of Cape Town-TP 37/86
3. R. Tegen: Phys. Lett. **172B**, 153 (1986)
4. W.A. Bardeen et al.: Phys. Rev. D**11**, 1094 (1975)
5. A. Thomas: Adv. Nucl. Phys. **13** (1986)
6. C. DeTar: Phys. Rev. D**24**, 752, 762 (1981)
7. C.W. Wong: Phys. Rev. D**24**, 1416 (1981)
   K.F. Liu, C.W. Wong: Phys. Lett. **113**B, 1 (1982)
8. R. Tegen, W. Weise: Z. Phys. A**314**, 357 (1983)
9. P.A.M. Guichon: Phys. Lett. **129**B, 108 (1983)
   M. Betz, R. Goldflam: Phys. Rev. D**28**, 2848 (1983)
10. G.E. Brown, M. Rho: Phys. Lett. **82**B, 177 (1979)
    G.E. Brown, M. Rho, V. Vento: Phys. Lett. **84**B, 383 (1979)
    V. Vento, M. Rho, E.M. Nyman, J.H. Jun, G.E. Brown: Nucl. Phys. A**345**, 413 (1980)
    G.E. Brown, F. Myhrer: Phys. Lett. **128**B, 229 (1983)

11. R.L. Jaffe: Phys. Rev. D**21**, 3215 (1980); Acta Phys. Austr. Suppl. **22** (1980)
12. S.A. Chin: Nucl. Phys. A**382**, 355 (1982)
13. R. Tegen, M. Schedl, W. Weise: Phys. Lett. **125**B, 9 (1983)
14. R. Tegen, R. Brockmann, W. Weise: Z. Phys. A**307**, 339 (1982)
15. E. Oset, R. Tegen, W. Weise: Nucl. Phys. A**426**, 456 (1984)
16. V. de Alfaro, S. Fubini, G. Furlan, C. Rosetti: Currents in Hadron Physics (North-Holland, Amsterdam 1973)
17. O. Klein: Z. Physik **53**, 157 (1929)
18. T.H.R. Skyrme: Proc. Roy. Soc. A**260**, 127 (1961)
N.K. Pak, H.C. Tze: Ann. Phys. **117**, 164 (1979)
J. Gipson, H.C. Tze: Nucl. Phys. B**183**, 524 (1981)
A.P. Balachandran et al.: Phys. Rev. Lett. **49**, 1124 (1982); Phys. Rev. D**27**, 1153 (1983)
E. Witten: Nucl. Phys. B**223**, 422, 433 (1983)
19. G. Adkins, C.R. Nappi, E. Witten: Nucl. Phhys. B**228**, 552 (1983)
G. Adkins, C.R. Nappi: Nucl. Phys. B**233**, 109 (1984); Phys. Lett. **137**B, 251 (1984)
20. M. Rho, A.S. Goldhaber, G.E. Brown: Phys. Rev. Lett. **51**, 747 (1983)
21. M.C. Birse, M.K. Banerjee: Phys. Lett. **136**B, 284 (1984); Phys. Rev. D**31**, 118 (1985)
B. Golli, M. Rosina: Phys. Lett. **165**B, 347 (1985)
22. R. Friedberg, T.D. Lee: Phys. Rev. D**15**, 1694 (1977); D**16**, 1096 (1977); D**18**, 2623 (1978)
23. R. Goldflam, L. Wilets: Phys. Rev. D**25**, 1951 (1982); Comm. Nucl. Part Phys. **12**, 191 (1984)
24. S. Kahana, G. Ripka, V. Soni: Nucl. Phys. A**415**, 351 (1984)
G. Ripka, S. Kahana: Phys. Lett. **155**B, 327 (1985)
25. K. Goeke et al.: Phys. Lett. **164**B, 249 (1985)
26. E.J.O. Gavin et al.: In preparation, University of Cape Town (1986)
27. J. Grasser, H. Leutwyler: Phys. Rep. **87**, 77 (1982); Ann. Phys. **158**, 142 (1984)
28. A. Chodos, C.B. Thorn: Phys. Rev. D**12**, 2733 (1975)

Exchange Currents and Configuration Mixing Effects in the $^{16}O(0^+)$-$^{16}N(0^-)$ Transitions

S. Nozawa[1] *and K. Kubodera*[1,2]

[1]Swiss Institute for Nuclear Research, CH-5234 Villigen, Switzerland
[2]Department of Physics, Sophia University, Tokyo, Japan

The 0^+-0^- transitions in A=16 nuclei have been studied intensively with a view of testing the meson exchange current A_0^{ex} in the time component of the axial current [1-4]. The dominant part of A_0^{ex} comes from a one-soft-pion exchange process whose amplitude can be determined in an almost model-independent way using the soft-pion theorem. (We shall denote by A_0^{soft} the piece corresponding to the one-soft-pion exchange process.) In this sense, this test has a fundamental bearing upon the working of chiral symmetry within the nucleus. Experimental information is available both on the β-decay rate Γ_β for $^{16}N(0^-) \to {}^{16}O(0^+)+e^-+\nu_e$ [6,7] and on the μ-capture rate Γ_μ for $\mu^-+{}^{16}O(0^+)\to\nu_\mu+{}^{16}N(0^-)$ [8].

The ratio $R=\Gamma_\mu/\Gamma_\beta$ is known to be less sensitive to nuclear models. To obtain definitive evidence for A_0^{ex} is a very challenging problem, which involves careful studies of configuration mixing effects as well as the examination of possible non-soft-pion exchange processes (We shall denote by $A_0^{non\text{-}soft}$ the pieces corresponding to the non-soft-pion processes.) Regarding to the latter, we present a reasonably systematic calculation of the possible non-soft-pion processes and show how the dominance of A_0^{soft} should hold in this system. As far as the configuration-mixing effects are concerned even the most elaborate existing calculations [3,4] have one definite point to be improved upon; viz, the second-order perturbation terms should be summed up over all relevant intermediate states rather than within limited configurations. We present these results here.

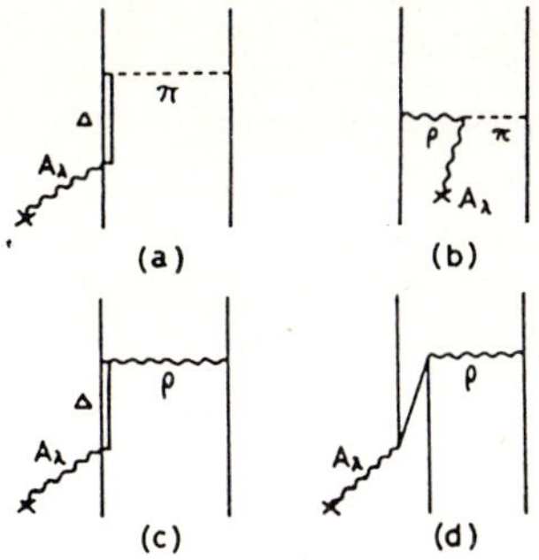

Fig.1 The non-soft-pion processes.

The estimation of $A_0^{non\text{-}soft}$ can be made only by introducing a model hamiltonian. Since, however, we expect at least qualitatively that A_0^{soft} should be dominant, it will be reasonable to use models to estimate $A_0^{non\text{-}soft}$. Assuming the standard interaction hamiltonians, we here consider the processes shown in Fig. 1: (a) Δ-excitation due to π-exchange; (b) ρπ diagram; (c) Δ-excitation due to ρ-exchange; (d) ρ-meson exchange recoil term. Table 1 and Table 2 show the relative contributions of various non-soft-pion processes in comparison with A_0^{soft} to the β-decay and μ-capture, respectively. In this estimation we use harmonic oscillator wave functions and cut off the relative wave functions at r_c=0.6fm in order to take into account the internucleon

Table 1. The relative values of the non-soft-pion exchange current to the soft-pion exchange current for the β-decay.

(%) x	-0.10	0.0	0.10
(1) $M_{\beta}^{\pi N\Delta}$	9.7	9.5	9.4
(2) $M_{\beta}^{\rho N\Delta}$	6.5	6.6	6.6
(3) $M_{\beta}^{\rho\text{-recoil}}$	-10.0	-10.3	-10.7
(4) $M_{\beta}^{\rho\pi}$ (ref. [9])	-5.0	-4.9	-5.2
(5) $M_{\beta}^{\rho\pi}$ (ref. [10])	-16.5	-16.5	-17.4
$M_{\beta}^{\text{non-soft}}$ =(1)+(2)+(3)+(4)	1.2	0.9	0.1
$M_{\beta}^{\text{non-soft}}$ =(1)+(2)+(3)+(5)	-10.3	-10.7	-12.1

Table 2. The relative values of the non-soft-pion exchange current to the soft-pion exchange current for the μ-capture.

(%) x	-0.10	0.0	0.10
(1) $M_{\mu}^{\pi N\Delta}$	10.1	10.0	9.6
(2) $M_{\mu}^{\rho N\Delta}$	6.5	6.6	6.7
(3) $M_{\mu}^{\rho\text{-recoil}}$	-10.3	-10.5	-10.8
(4) $M_{\mu}^{\rho\pi}$ (ref. [9])	-4.9	-4.9	-5.1
(5) $M_{\mu}^{\rho\pi}$ (ref. [10])	-16.8	-17.0	-17.4
$M_{\mu}^{\text{non-soft}}$ =(1)+(2)+(3)+(4)	1.4	1.2	0.4
$M_{\mu}^{\text{non-soft}}$ =(1)+(2)+(3)+(5)	-10.5	-10.9	-11.9

short-range correlations. (In these tables, x represents a mixing parameter between two configurations of the $^{16}N(0^-)$ wave function: viz $|^{16}N(0^-)_{16}\rangle=\sqrt{1-x^2}\cdot|\,0p_{1/2}^{-1}1s_{1/2};\,T=1,\,J=0\rangle + x|0p_{3/2}^{-1}\,0d_{3/2};\,T=1,\,J=0\rangle$. The $^{16}O(0^+)$ wave function is assumed to be a closed shell.) Because of cancellations between the non-soft-pion contributions, the net effect of all non-soft-pion processes considered here is to weaken A_0^{soft} by about 0~12%. Thus, the dominance of A_0^{soft} in the exchange current A_0^{ex} seems to hold well, substantiating the prediction in ref. [5].

In order to get a definitive conclusion on A_0^{soft} we need to investigate the interplay between the exchange current and the configuration-mixing at full length, and also we need to check the results by the existing experimental data. In the previous work [3] we presented the second-order perturbation calculation in terms of a reasonable residual interaction, where we limited ourselves to the estimation of the normalization terms which would be dominant among the second-order quantities. Here this is improved by taking into account second-order terms depicted in Fig.2. The

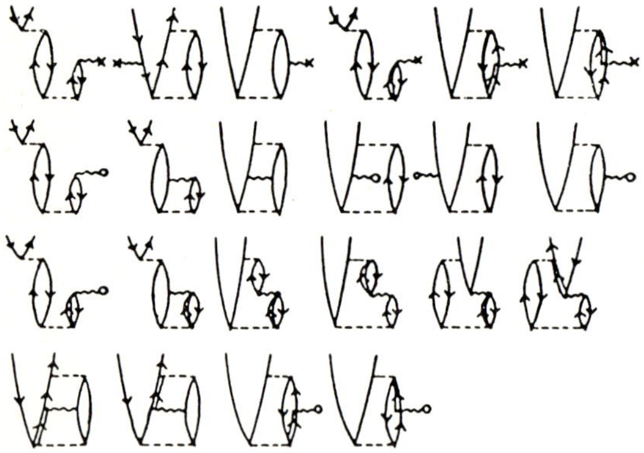

Fig.2 Diagrams of second-order terms other than normalization terms. The wavy lines stand for A_0^{IA} or A_0^{soft}, and the broken lines stand for the residual interaction; the double lines represent Δ-particles and the lines without arrow-heads correspond to both hole-lines and particle-lines.

intermediate state is summed up to $20\hbar\omega$ as the excitation energy, which gives a reasonable convergence of the matrix elements. (In order to illustrate the convergence we have plotted in Fig.3 the matrix elements corresponding to the diagrams of Fig.2 (the A_0^{soft} case) as a function of the excitation energy E_{excit}.) In the case of the normalization terms we found that the intermediate state added up constructively [4]. On the other hand the second-order terms in Fig.2, similarly to the case of the first-order terms, have partial cancellation among themselves because of the mixture of signs which occurs in the matrix elements. Table 3 shows the μ-capture rate Γ_μ, β-decay rate Γ_β and the ratio $R=\Gamma_\mu/\Gamma_\beta$ in each order of the perturbation expansion of the effective interaction. The cases with and without A_0^{soft} are shown. (Here the mixing parameter x is fixed at x=0 because x is known to be small $|x|<0.08$ [4] and the rates Γ_μ and Γ_β are less sensitive to this value [5]).

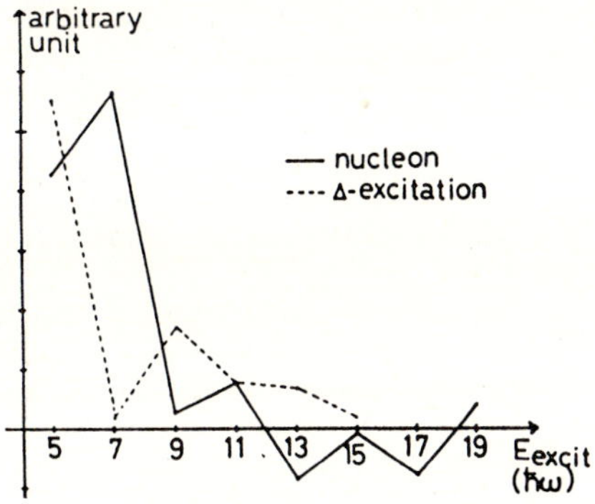

Fig.3 Convergency properties

The experimental values are denoted on the bottom line of the table. The results support the previous conclusion [4,5] that the soft-pion plays an important role in explaining the experimental value of $R=(3.23\pm0.34)\cdot10^3$ $((2.6\pm0.47)\cdot10^3)$. The full calculation gives $R=4.12\cdot10^3$, whereas the calculation without A_0^{soft} gave $R=10.8\cdot10^3$. Even the absolute rates for Γ_μ and Γ_β are within 30% of the existing experiments.

Table 3. The results of the perturbation theory and the experimental values.

	IA			IA + soft-pion		
	$\Gamma_\mu(\times10^3 s^{-1})$	$\Gamma_\beta(s^{-1})$	$R(\times10^3)$	$\Gamma_\mu(\times10^3 s^{-1})$	$\Gamma_\beta(s^{-1})$	$R(\times10^3)$
zeroth order	1.89	0.316	5.98	2.80	0.956	2.93
+first order	1.88	0.167	11.3	2.82	0.710	3.97
+second order	1.46	0.135	10.8	2.12	0.515	4.12
exp.	$\Gamma_\mu=(1.57\pm0.10)\times10^3 s^{-1}$ [8], $\Gamma_\beta=0.486\pm0.02\ s^{-1}$ [7] or $0.60\pm0.07\ s^{-1}$ [6] $R=(3.23\pm0.34)\times10^3$ [8,7] or $(2.6\pm0.47)\times10^3$ [8,6]					

References:

[1] P. Guichon, H. Giffon and C. Samour, Phys. Lett. 74B, 15 (1978)

[2] S. Nozawa, Y. Kohyama and K. Kubodera, Phys. Lett. 140B, 11 (1984)

[3] S. Nozawa, K. Kubodera and H. Ohtsubo, Nucl. Phys. A453, 645 (1986)

[4] I.S. Towner and F.C. Khanna, Nucl. Phys. A372, 331 (1981)
[5] K. Kubodera, J. Delorme and M. Rho, Phys. Rev. Lett. 40, 755 (1978)
[6] T. Minamisono et al., Phys. Lett. 130B, 1 (1983)
[7] G. Garvey, Proc. Int. Symp. on nuclear spectroscopy and nuclear interactions, Osaka, March 1984, p193.
[8] P. Guichon et al., Phys. Rev. C19, 987 (1979)
[9] W.K. Cheng, B. Lorazo and B. Goulard, Phys. Rev. C21, 374 (1980)
[10] H. Jager, M. Kirchbach and E. Truhlik, Nucl. Phys. A404, 456 (1983)

Measurements of the Longitudinal Electron Polarization in Nuclear Beta-Decay

R. Gauder, O. Boslau, A. Hilscher, K.-W. Hoffmann, J. Kayser, E. Speller, and U. Zierer

Institut für Strahlenphysik, Allmandring 3,
D-7000 Stuttgart 80, F. R. Germany

One important consequence of the maximum violation of parity in nuclear ß-decay and the pure V-A-interaction is the resulting longitudinal polarization P of the emitted electrons with P = -v/c. Deviations from this value can be caused by:

a.) the mechanism of the weak interaction itself (non-maximum violation of parity, Fierz-interference etc.),
b.) nuclear structure - and renormalization effects (admixtures of the strong interaction).

According to b.) measurements of the longitudinal polarization of ß-decay electrons can yield information about the magnitude of these effects (described by form-factors) and in conjunction with a proper nuclear model about the ß-matrix elements. In contrast to measurements for a.) which have to be absolute our measurements concerning b.) could be done relative to the well-known transition of ^{60}Co. Keeping in mind the relativeness of our results one can, however, deduce values for Fierz-interference or contributions from right-handed currents for different types of ß-transitions.

Our measurements are based on the well-known Mott-scattering technique: longitudinally polarized electrons emitted by a radioactive sample are velocity-selected by a magnetic short lens spectrometer and focussed by a magnetic quadrupole doublet to guide them through a Wien-filter (crossed electric and magnetic fields) which transforms the polarization from longitudinal to transverse. Afterwards the electrons are scattered by a thin gold-foil and are detected by a classical 4-detector scheme [1,2] . One left-right detector pair is placed at 45° with respect to the beam where the analyzing power of the scattering foil is zero thus enabling a measurement of the instrumental asymmetry whereas the other pair is placed at 110° (maximum analyzing power) to measure the Mott-asymmetry which is proportional to the initial degree of polarization. To avoid a calibration of the Mott-polarimeter and to reduce systematical errors all measurements were carried out relative to a ^{60}Co-source for which the polarization is known with a very low statistical error (≈1%) [1,3] . Then the ratio of the measured counting rate asymmetries equals the ratio of the longitudinal polarizations.

The radioactive samples were carefully prepared in order to reduce the intrinsic depolarization in the source. For this purpose various preparation techniques were applied, such as molecular plating, electrospraying and ordinary electrolysis. The samples always showed good homogeneity. To apply MÜHLSCHLEGEL's formula [4] for depolarization corrections the sample thickness was restricted stringently thus resulting in a very low effective activity (see sample activities A in next chapter).

In contradiction to Ryu [5] we didn't see any deviation from -v/c for ^{45}Ca. Also the discrepancies of some measurements [1,6] concerning ^{147}Pm could be removed. ^{204}Tl also showed no deviation from -v/c as predicted by theory [7]. The ß-decay of ^{147}Pm was evaluated more quantitatively because of the very accurate results. For Fierz-interference we found an upper limit of 0.01±0.012. Therefore S- and T-admixtures to V and A are less than 1%. Though ^{147}Pm has six form factors we could show that $^{V}F_{101}$ and $^{V}F_{110}$ are dominant and of the order of 0.12. The para-

Table 1: Electron polarization P

Energy [keV]	^{45}Ca	^{147}Pm	^{204}Tl
80	1.055±0.084	0.996±0.025	
100	1.017±0.048	1.006±0.026	
130	0.993±0.039	1.012±0.020	
150			0.96±0.09
160	0.987±0.046	1.018±0.02	
200			1.006±0.062
$\bar{P}$	1.013±0.023	1.008±0.011	
measuring time [h]	3160	3790	1860

meter g_o and g_1 according to BEHRENS [7] which describe the deviations from -v/c were both in the range of 0.01. According to [8] we computed the contributions from right-handed currents and found a lower limit for the mass of the right-handed boson W_{V+A}: $m(W_{V+A}) \geqq 3.7\ m\ (W_{V-A})$.

Literature references:

1: J. van Klinken: Nucl. Phys. 75, 145 (1966)
2: J. Kessler: Polarized Electrons (Springer, Berlin, Heidelberg, New York 1985)
3: A.I. Boothroyd et al.: Phys. Rev. C 29 603 (1984)
4: B. Mühlschlegel: Z. Phys. 155, 69 (1959)
5: N. Ryu: Memoirs of the Faculty of Engineering Hiroshima University, Vol. 5, No. 3 (1975)
6: H. Hasai: J. Phys. Soc. Jpn 47, 9 (1979)
7: H. Behrens, W. Bühring: Electron Radial Wave Functions and Nuclear Beta-Decay (Int. Series of Monographs on Physics 67, Clarendon Press, Oxford 1982)
8: J. van Klinken, F.W. Koks, H. Behrens: Phys. Letters 79B, 199 (1978)

A Universal Source of Polarized Cold and Ultracold Neutrons at the LNPI WWR-M Reactor

A.P. Serebrov

Leningrad Nuclear Physics Institute, Gatchina,
Leningrad district, 188350, USSR

A universal source of polarized cold and ultracold neutrons is mounted at the LNPI WWR-M reactor. The liquid hydrogen moderator (volume about 1 litre) is placed at the center of the reactor core. The thermal neutron flux at the source is (1.5 - 2) • 10^{14} n/cm^{2}s. The heat release in the source material and in the liquid hydrogen is ~2 kW. The heat power is removed by the circulation of liquid hydrogen in the loop with a heat-exchanger which is connected with the helium refrigerator. The neutron guide system consists of the neutron guide for polarized cold neutrons and the neutron guide for ultracold neutrons (Fig. 1).

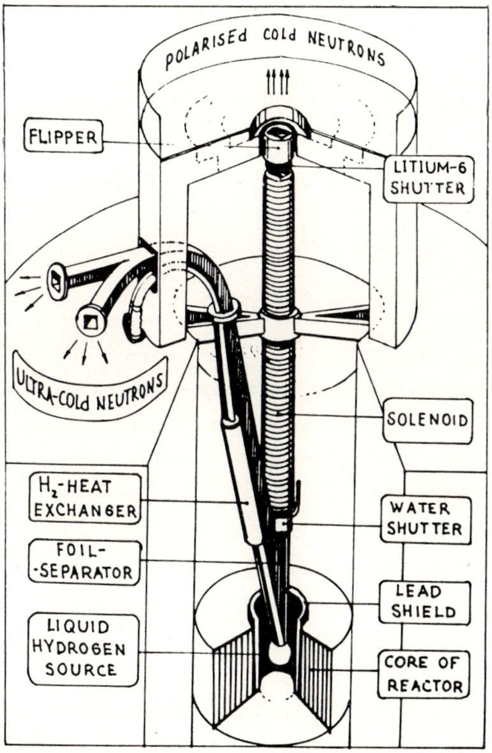

Fig. 1. The scheme of the cold source in the WWR-M reactor

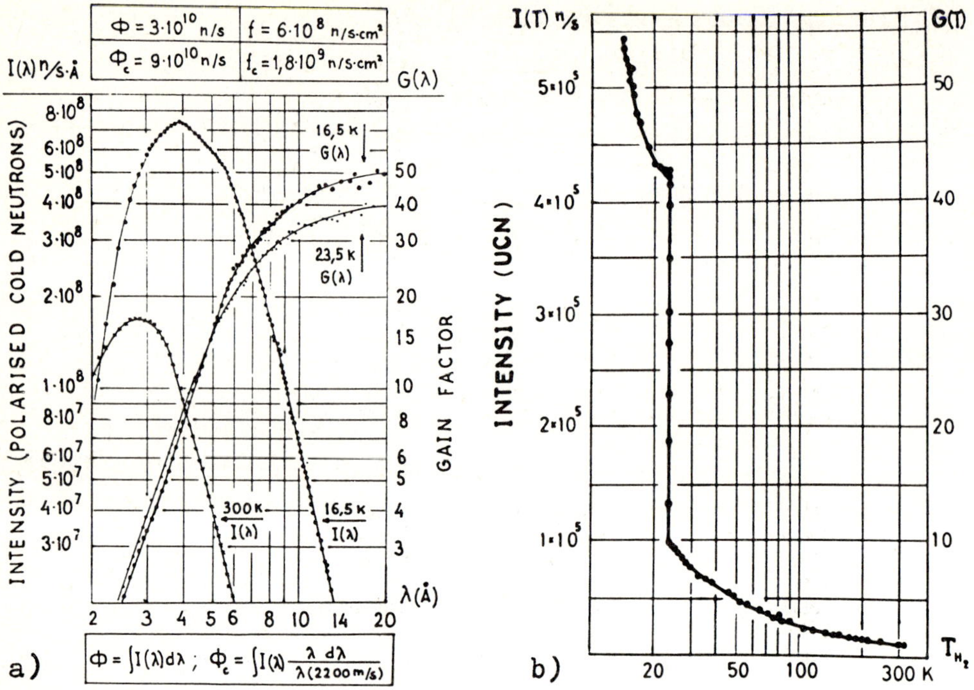

Fig. 2.

The time-of-flight spectra of polarized cold neutrons which were measured before and after condensation of hydrogen in the source are shown in Fig. 2a. Their ratio gives the gain factor of the cold neutron yield due to the liquid hydrogen moderator. The gain factor has the spectrum dependence and amounts to 40 to 50 at the great wavelengths (10-20 Å). The temperature dependence for the yield of ultracold neutrons is shown in Fig. 2b. The total (for both beams) intensity of ultracold neutrons with velocity components V_x, V_y, $V_z < 7.8$ m/s is equal to $2 \times 2.5 \cdot 10^5$ n/s with a flux density $6 \cdot 10^3$ n/cm^{2}s. The intensity of polarized cold neutrons $3 \cdot 10^{10}$ n/s with a flux density of $6 \cdot 10^8$ n/cm^{2}s is a record. The result is achieved by placing the source at the maximum thermal neutrons flux of the reactor (it became possible due to the liquid hydrogen circulation system) and by the multislit neutron guide which simultaneously polarize the neutron beam. The polarization of the cold neutron beam is about 90%.

The research programme for the new source intends to continue the experiment on the search for the electric dipole moment of the neutron [1,2,3] and to arrange new precision experiments on neutron β-decay. One of these experiments is the measurement of the correlative coefficients for neutron β-decay, another is the measurement of neutron lifetime by means of storage of ultracold neutrons. The aim of the experiments is to determine independently the ratio of axial and vector constants for the neutron β-decay at a level of a few units 10^{-3} and the determination of the neutron lifetime with a precision less than 0.5%.

References

1. I.S. Altarev et al.: Nucl. Phys. A341, 269 (1980)
2. I.S. Altarev et al.: Phys. Lett. 80A, 413 (1980)
3. V.M. Lobashev, A.P. Serebrov: J. de Physique C3, Suppl. 45, C3-11 (1984)

3.2.2 Hyperons and Hypernuclei

Semileptonic Hyperon Decays

H.W. Siebert

Physikalisches Institut, Universität Heidelberg, Philosophenweg 12,
D-6900 Heidelberg 1, F. R. Germany

1. Introduction

More than twenty years ago, in 1963, N. Cabibbo published his paper on Unitary Symmetry and Leptonic Decays [1]. At that time, only very little information on semileptonic hyperon decays (SLYD) was available, but two decades of experimental effort have brought a wealth of data. The Cabibbo model in its elegant simplicity accomodated all new data without serious problems. Only recently experiments have reached a level where we begin to see SU(3) symmetry breaking.

Until about ten years ago, the bulk of information on SLYD came from bubble chamber experiments in K^-beams. Recently, experiments in high-energy hyperon beams have increased the experimental basis considerably. For the first time, five different SLYD have been measured with greatly increased precision in a single experiment at CERN [2], and a comprehensive test of the Cabibbo model could be performed without the inconsistencies inevitably arising in comparisons of very different experiments and analysis procedures. This problem had plagued many earlier "Cabibbo fits". Other high-statistics experiments on the decay $\Lambda \rightarrow pe\bar{\nu}$ at BNL [3] and on $\Sigma^- \rightarrow ne\bar{\nu}$ at FNAL [4] will also be discussed together with the experimental result on neutron decay [5] obtained at ILL and described by D. Dubbers in these Proceedings.

In section 2 the theoretical framework and the relevant experimental quantities are outlined, the experimental results are described in section 3 and discussed in section 4.

2. Theoretical framework

Assuming the usual V,A structure, the matrix element M for the semileptonic decay $B \rightarrow B' + l + \bar{\nu}$ is given by the product of the matrix elements of the baryon and lepton weak currents:

$$M = \frac{G}{\sqrt{2}} \langle B' | J_h^\mu | B \rangle \, \bar{u}_l(p_l) \gamma_\mu (1+\gamma_5) u_\nu(p_\nu) \, ,$$

where G is the universal weak coupling constant. The baryon matrix element is expressed in terms of six form factors which take into account the effects of the strong interaction:

$$\langle B' | J_h^{\mu} | B \rangle = C\bar{u}_{B'}(p') \left\{ f_1(q^2)\gamma^{\mu} + i \frac{f_2(q^2)}{M} \sigma^{\mu\nu} q_{\nu} + \frac{f_3(q^2)}{M} q^{\mu} + \left[g_1(q^2)\gamma^{\mu} + i \frac{g_2(q^2)}{M} \sigma^{\mu\nu} q_{\nu} + \frac{g_3(q^2)}{M} q^{\mu} \right] \gamma_5 \right\} u_B(p),$$

with $C = \cos\theta_c$ for $\Delta S = 0$ and $C = \sin\theta_c$ for $\Delta S = 1$ transitions. θ_c is the Cabibbo angle, p,M and p',M' are the four-momenta and masses of the initial and final baryon and $q = p - p'$. In the Kobayashi-Maskawa six-quark mixing scheme [6], $\cos\theta_c$ corresponds to $U_{ud} = \cos\theta_1$ and $\sin\theta_c$ to $U_{us} = \sin\theta_1 \cdot \cos\theta_3$.

For a meaningful experimental analysis, the number of parameters must be reduced drastically. Kinematics allow us to drop the terms with f_3 or g_3, because they are multiplied with a factor $(m_l/M)^2$, and are therefore completely negligible in electronic decays, which are discussed here. T-invariance implies that f and g are real. In the Cabibbo model, SU(3) symmetry implies that matrix element terms with the same form factor, but from different decays within the baryon octet, are determined by two reduced matrix elements F and D. The CVC hypothesis relates the vector form factors f_1 and f_2 to the electric charges and the anomalous magnetic moments of protons and neutrons. CVC also predicts $f_3 = 0$, but this term is negligible anyhow. In the limit of exact SU(3), the absence of "second class currents" implies $g_2 = 0$ and $f_3 = 0$.

Thus, in the Cabibbo model, all semileptonic decays in the baryon octet should be described by three parameters only: the parameters F and D determining the form factors g_1 in the different decays, and the Cabibbo angle θ_c. It should be noted that these conclusions are not changed in the framework of the standard model of electroweak interactions.

Experiments provide three types of information: decay rates, Dalitz plot distributions and asymmetries. In good approximation, the decay rates are $\Gamma = \text{const} \cdot C^2 \cdot (1 + 3g_1^2/f_1^2)$, so they provide information on the Cabibbo angle and on the absolute value of g_1/f_1.

The Dalitz plot, the two-dimensional distribution of the baryon recoil energy T' and the electron energy T_e, depends strongly on $|g_1/f_1|$, especially in the T' projection. The formfactors depend on the four-momentum transfer q^2, and because of the relation between q^2 and T', the assumptions on the q^2-dependence enter into the determination of $|g_1/f_1|$. The electron-neutrino asymmetry $a_{e\nu}$ is correlated with the Dalitz plot and provides no independent information. The electron energy spectrum depends weakly on g_1/f_1 and can be used to determine its sign.

Both the magnitude and the sign of g_1/f_1 can be determined from the lepton asymmetries with respect to the polarization of either the mother or the daughter baryon. Experimentally, these two cases are very different. Polarized hyperons can be produced at large p_T in proton-nucleus collisions, and this has been exploited in the FNAL experiment on $\Sigma^- \to ne\bar{\nu}$ decay. In the CERN experiment, Ξ^- decays provided a sample of Λ with known longitudinal polarization $P = \alpha_{\Xi^-}$, and this sample was used to study $\Lambda \to pe\bar{\nu}$ decay. On the other hand, if the daughter baryon is a Λ, then its polarization is accessible through the decay proton asymmetry. This has been used in the study of $\Sigma^- \to \Lambda e\bar{\nu}$ and $\Xi^- \to \Lambda e\bar{\nu}$ decays.

All quoted distributions depend also on the form factors f_2 and g_2, but to a much lesser extent.

3. Experimental results

SLYD have branching ratios ranging from below 10^{-4} to 10^{-3}. This poses two problems: Firstly, the total number of hyperons available in experiments must be very high to obtain reasonable statistics, and secondly, these decays must be identified amongst a much larger background of nonleptonic hyperon decays, which produce a π^- instead of the $e\bar{\nu}$ pair ($\Sigma^- \to \Lambda e\bar{\nu}$ is an exception). The first problem was overcome by the large hyperon fluxes available in the latest generation of hyperon beams (for a review, see for example [7]). The second problem was solved at lower energies by the use of gas-filled threshold Cherenkov counters, and at higher energies by the

Table 1: Recent experimental results on semileptonic hyperon decays. Numbers are from the CERN experiment unless noted otherwise. BR = branching ratio.

Decay	$10^4 \cdot$ BR	f_1/g_1
$\Sigma^- \to \Lambda e\bar{\nu}$	0.561 ± 0.031	+ 0.03 ± 0.08
		g_1/f_1
$\Sigma^- \to ne\bar{\nu}$	9.6 ± 0.5	- 0.34 ± 0.05 - 0.29 ± 0.07 FNAL
$\Xi^- \to \Lambda e\bar{\nu}$	5.64 ± 0.31	+ 0.25 ± 0.05
$\Xi^- \to \Sigma^0 e\bar{\nu}$	0.87 ± 0.17	
$\Lambda \to pe\bar{\nu}$	8.57 ± 0.36 8.47 ± 0.17	+ 0.70 ± 0.03 + 0.715 ± 0.026 BNL
$\Omega^- \to \Xi^0 e\bar{\nu}$	56 ± 28	
$n \to pe\bar{\nu}$		+ 1.262 ± 0.005 ILL

combined use of transition radiation counters and of electromagnetic calorimeters based on leadglass counters.

The results of the three latest experiments are listed in table 1. At the CERN-SPS charged hyperon beam, five SLYD were measured in a single experiment running from 1977 to 1979 [2]. These results provide the backbone of the existing data. We also include data on $\Lambda \to pe\bar{\nu}$ from a hyperon beam experiment at BNL [3] and on $\Sigma^- \to ne\bar{\nu}$ from a recent hyperon beam experiment at FNAL [4]. The latter experiment used polarized Σ^- and provided an unambiguous determination of the sign of g_1/f_1 in this decay, a subject of much earlier controversy. For earlier results, the reader is referred to the reviews in [8] and [9], here we will just state that there is no discrepancy between t he earlier and the new experiments.

The measurement of the muonic decay modes $B \to B'\mu\bar{\nu}$ was, unfortunately, not possible in these experiments because of the large background from $\pi \to \mu\bar{\nu}$ decays.

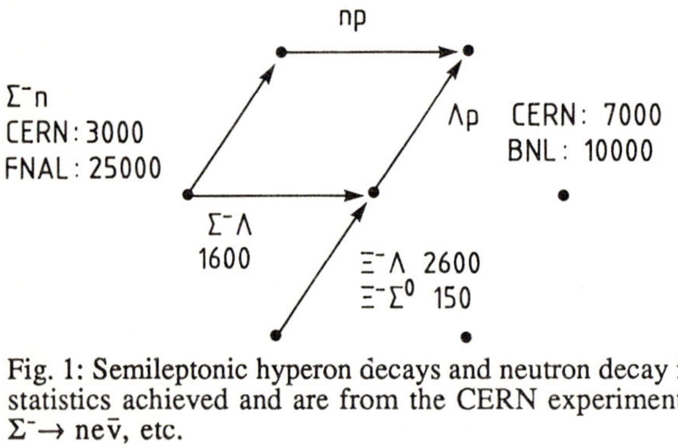

Fig. 1: Semileptonic hyperon decays and neutron decay in the baryon octet. Numbers indicate the statistics achieved and are from the CERN experiment unless noted otherwise. Σ^-n stands for $\Sigma^- \to ne\bar{\nu}$, etc.

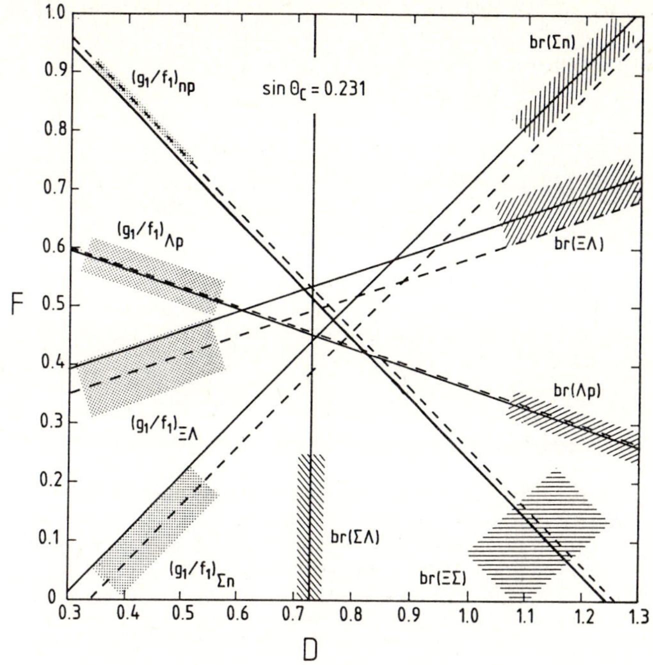

Fig. 2: F and D values from branching ratios (solid lines) and from g_1/f_1 ratios (dashed lines). Shaded bands indicate the experimental errors.

In fig. 1, we present schematically the SLYD in the ground-state baryon octet, with the statistics achieved in hyperon beam experiments. The decay $\Omega^- \to \Xi^0 e\bar{\nu}$ has also been measured in the CERN experiment [10], as a decuplet-octet transition, however, it is not described in the Cabibbo model.

In figure 2 we show the F, D values derived from the CERN experiment. Each branching ratio and each g_1/f_1 ratio corresponds to one line in this plot, with the experimental uncertainties indicated by shading. The decay $\Sigma^- \to \Lambda e\bar{\nu}$ contributes only through its branching ratio, as CVC predicts $f_1 = 0$ for this decay, in good agreement with experiment (see table 1). In order to derive F, D values from the measured branching ratios, we have used $\sin \theta_c = 0.231$, as obtained in overall fits.The good agreement with the Cabibbo model is reflected in the fact that all bands meet in a zone at, roughly, F = 0.45, D = 0.73. We have also plotted a line for the g_1/f_1 ratio in neutron decay. The plot does not change significantly if the new BNL and FNAL results are included.

An overall fit of the CERN, BNL and FNAL hyperon data to the Cabibbo model was made, taking into account radiative corrections and assuming a q^2-dependence of the formfactors derived from electroproduction and ν scattering. For f_2, the values given by CVC were used, and g_2 was assumed to be zero for all decays. For a discussion of these assumptions, see [2,8]. The result is given as FIT 1 in table 2. Here F + D and $\alpha = D/(F+D)$ are quoted, because their fit errors are almost uncorrelated. These numbers demonstrate excellent agreement of SLYD with the Cabibbo model.

Problems arise, however, if one wants to include neutron decay into this fit. The latest value for g_1/f_1 in this decay comes from the experiment at ILL, $g_1/f_1 = 1.262 \pm 0.005$ [5] in good agreement with earlier correlation measurements. As the Cabibbo model predicts $g_1/f_1 = F + D$ for neutron decay, we have a striking disagreement with hyperon decays. Indeed, including the neutron data into the Cabibbo fit leads to the values given by FIT 2 in table 2, which has a bad χ^2. The disagreement comes mainly from the $\Sigma^- \to \Lambda e\bar{\nu}$ and $\Xi^- \to \Lambda e\bar{\nu}$ branching ratios, where the measurements are 2.5 standard deviations below and above the fit values, respectively.

Table 2: Results of the Cabibbo fits to the CERN, BNL and FNAL data. The results of these experiments have been averaged for the fit, where appropriate. $\alpha = D/(F+D)$. FIT 3 is a two-angle fit with different values of θ_c for vector and axialvector amplitudes.

	FIT 1	FIT 2	FIT 3
$\sin\theta_c$	0.231 ± 0.003	0.224 ± 0.002	0.231 ± 0.003 V 0.215 ± 0.004 A
F + D	1.183 ± 0.021	1.258 ± 0.005	1.256 ± 0.005
α	0.619 ± 0.008	0.617 ± 0.008	0.607 ± 0.009
χ^2/NDF	6.0/5	19.1/6	11.0/5

4. Discussion

The discrepancy between hyperon and neutron decays may indicate SU(3) symmetry breaking, caused by the different wavefunctions of s quarks and u,d quarks. Donoghue and Holstein have calculated such effects on the basis of the MIT bag model, and conclude that in $\Delta S = 1$ decays the vector form factors should decrease by 3 % and the axialvector formfactors increase by 11% [11]. We can imitate this effect by fitting two independent effective Cabibbo angles for vector and axialvector amplitudes. The results are given as FIT 3 in table 2. The χ^2 is somewhat better, but not quite satisfactory, and we note that the data prefer shifts opposite to those estimated by Donoghue and Holstein. Eeg and Lie-Svendsen have repeated these calculations [12], including recoil corrections, and find that g_1 should be reduced by about 10 % in $\Delta S = 1$ decays, in better agreement with the data. Detailed calculations for each decay have yet to be performed.

So far, $g_2 = 0$ has been assumed in the analysis. If one uses instead the values estimated in [11], the overall fit hardly changes, as the experiments are not yet sensitive enough to establish the presence of g_2 terms.

In principle, the Cabibbo angle derived from K decays should be equal to the value derived from hyperon decays. In both cases, the bare weak angle may be changed by strong interaction effects leading to SU(3) symmetry breaking. The Ademollo-Gatto theorem predicts that these effects should be small in vector transitions. Leutwyler and Roos have determined the bare weak Cabibbo angle from K_{e3} decays, which are pure vector transitions, and find $\sin\theta_c = V_{us} = 0.2196 \pm 0.0023$. This value is closer to the axialvector effective Cabibbo angle measured in hyperon decays.

In view of these problems, it is not easy to decide which is the value of the "bare weak" Cabibbo angle, and therefore of the Kobayashi-Maskawa matrix element $|V_{us}| = \sin\theta_c$. Because of the Ademollo-Gatto theorem, we are inclined to believe that the answer is closer to the value derived from K_{e3} decay.

The matrix element V_{ud} can be determined from a comparison of superallowed Fermi β-decays and muon decay. In a recent analysis, Marciano and Sirlin [14] find $|V_{ud}| = 0.9729 \pm 0.0008 \pm .0009$, where the second error reflects the theoretical uncertainties. The matrix element V_{ub} is quite small, about 0.01. If we put $|V_{us}| = 0.220$ or 0.230, we obtain $|V_{ud}|^2 + |V_{us}|^2 + |V_{ub}|^2 = 0.9950$ or 0.9995, resp. The persisting theoretical uncertainties in the determination of V_{ud} and V_{us} do not allow a statement on whether this sum deviates from 1, the value required by three-generation unitarity. This topic is also discussed by K. Schubert in these Proceedings.

To summarize, the Cabibbo model still holds at the level of precision achieved in measurements of semileptonic hyperon decays at hyperon beams. Observed discrepancies between results from hyperon, kaon and neutron decays may be the first indication of SU(3) symmetry breaking.

References

1. N. Cabibbo, Phys. Rev. Lett. 10 (1963) 531

2. M. Bourquin et al., Z. Physik C21 (1983)27 and earlier references quoted therein

3. J. Wise et al., Phys. Lett. 91B, 165 (1980); Phys. Lett. 98B, 123 (1981).
   D. Jensen et al., Proc.Int. Europhysics Conf. on High-Energy Physics, Brighton 1983, p. 255.

4. S.Y. Hsueh et al., Phys. Rev. Lett. 54 (1985) 2399

5. D.Dubbers et al., Phys. Rev. Lett. 56 (1986) 919

6. M. Kobayashi and T. Maskawa, Progr.Theor. Physics 49 (2973) 652.

7. M.Bourquin and J.-P. Repellin, Phys. Rep. 114 (1984) 99

8. J.-M. Gaillard and G. Sauvage, Ann.Rev.Nucl.Sci. 34 (1984) 351

9. H.W. Siebert, Proc. European Topical Conf. on Flavor Mixing in Weak Interactions, Erice, March 1984, p. 37

10. M. Bourquin et al., Nucl. Phys. B 241 (1984) 1

11. J.F. Donoghue and B.R. Holstein, Phys. Rev. D 25 (1982) 206

12. J.O. Eeg and Ø. Lie-Svendsen, Z. Physik C27 (1985) 119

13. H. Leutwyler and M. Roos, Z. Physik C25 (1984) 91

14. W.J. Marciano and A. Sirlin, Phys. Rev. Lett. 56 (1986) 22

Electroweak Properties of the Baryons in QCD*

J. Pasupathy

Centre for Theoretical Studies, Indian Institute of Science, Bangalore 560 012, India

The method of QCD sum rules has been applied to the study of the properties of the Nucleon and other Baryons by a number of authors [1,2]. Ioffe and Smilga [3] and independently Balitsky and Yung [4] have computed the magnetic moments by evaluating the nucleon current correlation in the presence of an external magnetic field. This was extended to study the axial vector renormalization constants in ref. [5,6]. The analysis of ref. [3,4] was extended in ref. 7 by including additional terms in the operator product expansion (OPE) and adopting a different procedure for analyzing the sum rules. In particular, it was pointed out that expressed in appropriate units [cf. Table 1] the magnetic moments have a simple pattern. We have extended the calculation to Λ hyperon [8]. Our results are summarised in ref. [6,7,10]. The following is a brief outline of the calculations.

The nucleon current defined by Ioffe [1] is

$$\eta(x) = u^a(x) c\gamma^\mu u^b(x)\gamma_\mu\gamma^5 d^c(x)\varepsilon^{abc} \tag{1}$$

u,d, are the up and down quark fields, a,b,c are color indices. Introduce the correlation function

$$\pi(p) = i\int d^4x\, e^{ip\cdot x} \langle 0|T(\eta(x)\bar{\eta}(0))|0\rangle \tag{2}$$

The invariant coefficients of the decomposition

$$\pi(p) = \pi_1(p^2)p + \pi_2(p^2)1 \tag{3}$$

are assumed to satisfy the dispersion relations

$$\pi_{1,2}(p^2) = \frac{1}{\pi}\int \frac{\mathrm{Im}\pi_{1,2}(s)}{s-p^2-i\varepsilon}\,ds \tag{4}$$

Sum rules arise when eq.(2) is calculated using OPE on the one hand and while eq. (4) is evaluated using Physical intermediate states. The nucleon contribution to Im (s) is

$$\mathrm{Im}\pi\,(s)\Big|_{\mathrm{Nucleon}} = \lambda_N^2\delta(p^2 - m_N^2)\,(\hat{p} + m_N) \tag{5}$$

where

$$\langle 0|\bar{\eta}(0)|p\rangle = \lambda_N u(p) \text{ and } \bar{u}u=2m_N \tag{6}$$

After a Borel transformation Ioffe and Belyaev [1] derived the following two sum rules

$$\frac{M^6}{8L^{4/9}} + \frac{bM^2}{32L^{4/9}} + \frac{1}{6}a^2L^{4/9} - \frac{a^2m_o^2}{24M^2} = \beta_N^2 e^{-m_N^2/M^2} + \text{excited state contributions} \tag{7}$$

---

*Work done in collaboration with C.B. Chiu, J.P. Singh and S.L. Wilson

$$\frac{aM^4}{4L^{4/9}} - \frac{ab}{72} + \frac{17}{81}\frac{\alpha_s}{\pi}\frac{a^3}{M^2} = m_N \beta_N^2 e^{-m_N^2/M^2} + \text{excited state contributions} \quad (8)$$

Here

$$a=-(2\pi)^2 <0|\bar{q}q|0> \quad b=<0|g_c^2 G_n^{\alpha\beta} G^{n\alpha\beta}|0> \quad am_o^2=+(2\pi)^2 <0|\bar{q}\sigma\cdot Gq|0>$$

$$\alpha_s=g_c^2/4\pi \quad L=\ell n(M^2/\Lambda_{QCD}^2)/\ell n(\mu^2/\Lambda_{QCD}^2) \quad \mu=500 \text{ MeV} \quad \Lambda_{QCD}=100 \text{ MeV}$$

$$\beta_N^2=(2\pi)^4\lambda_N^2/4 \quad a=0.45 \text{ (GeV)}^3 \quad b=0.5 \text{ (GeV)}^4 \quad m_o^2=0.8 \text{ GeV}^2 \quad (9)$$

Using these sum rules Ioffe [1] computed the nucleon mass ≈ 1 GeV. The constant $\lambda_N^2 = 0.64\ 10^{-3}$ (GeV)^6 has an important bearing in Nucleon and Particle Physics. Marshak and Pasupathy pointed out [9] that λ_N^2 is a measure of the size of the proton. In particular it will be impossible to reconcile bag models which have small radii like 0.5 fm with the QCD value of λ_N^2.

The static properties of the Nucleon can be computed by studying the current propagator eq.(2) in the presence of an external field [3-8]. With a constant external axial vector field Z_μ (Isovector I_3=0) we can write

$$\pi(p) = i\int d^4x\, e^{ip\cdot x} <0|T\{\eta(x)\bar{\eta}(0)\}|0>|Z_\mu$$

$$= f_1(p^2)p\cdot Z\ \hat{Z}\gamma_5 + f_2(p^2)\hat{Z}\gamma_5 + F_3(p^2)p\cdot Z\gamma_5 + f_4\sigma^{\alpha\beta}Z_\alpha\chi_\beta\gamma_5 \quad (10)$$

The external field Z_μ couples directly to the valence quarks as well as affects their propagation by polarizing the QCD vacuum ie. we now have external field induced vacuum correlations like

$$<0|\bar{q}\gamma_\mu\gamma_5 q|0> = g_q\chi Z_\mu <0|\bar{q}q|0> \ ; \ <0|\bar{q}G_{\mu\nu}\gamma_\nu q> = Z_\mu K <0|\bar{q}q|0>.$$

Using [5] PCAC $\chi<0|\bar{q}q|0> = -f_\pi^2$ while $K<0|\bar{q}q|0> = -\frac{1}{3}\xi f_\pi^2 0.2$ GeV^2. In ref. [5] the value ξ=1 is adopted while in [6] ξ is retained as a free parameter. The sum rules for the odd structures f_1 and f_2 [5,6] read

$$\frac{M^6}{8L^{4/9}} + \frac{KaM^2}{6L^{68/81}} + \frac{bM^2}{32L^{4/9}} + \frac{5}{18}a^2L^{4/9} = \beta_N^2(G_A+Am^2)e^{-m_N^2/m^2} + \text{excited state terms} \quad (11)$$

$$\frac{M^6}{8L^{4/9}} - \frac{M^4\chi a}{2L^{4/9}} - \frac{3}{2}\frac{M^2ka}{L^{68/81}} + \frac{bM^2}{32L^{4/9}} + \frac{a^2}{18}L^{4/9} = \beta_N^2[G_A(\frac{-2m_N^2}{M^2}+1)+B]e^{-m_N^2/m^2}$$

$$+ \text{ excited state terms} \quad (12)$$

These sum rules have the remarkable property that the asymptotic terms in eq.(7) (11) and (12) are identical. This of course arises from the fact that the external field coupling is also a gauge coupling and preserves chirality. The second important point to note [6] is the sharp change in the nucleon pole coefficient in eq.(11) and (12) and the coefficients A and B to take into account non-diagonal terms involving transition between the nucleon and excited states. It was shown in ref. [6] that this structure has a natural explanation. Comparison of Eq. (7) and eq.(11) suggests that G_A for excited states tends to 1 while its associated A tends to zero. This latter is possible when there is parity doubling for highly excited states ie. chiral symmetry is realized in its Wigner-Weyl mode as compared to the Nambu-Goldstone mode at low energies. Results are summarised in [6]; discussion of the Ratio method of analysis of sum rules can be found in [10].

For the magnetic moment of the proton, the sum rule corresponding to its odd chiral structure reads [3]

Table 1 Baryon magnetic moments [8]. The last column uses $\chi=-3\text{GeV}^{-2}$ $K=0.75$ $\xi=-1.5$ Λ(low momentum cut off)=500 MeV.

	in nuclear magneton	Definition of $(1+\delta_B)$	$1+\delta_B$	Sum rule value in Nuclear magneton
p	2.793	$4e_u \frac{eK}{2m_N c}(1+\delta_p)$	1.047	2.92
n	-1.913	$4e_d \frac{eK}{2m_n c}(1+\delta_N)$	1.435	-1.72
Σ^+	2.38	$4e_u \frac{eK}{2m_\Sigma c}(1+\delta_{\Sigma^+})$	1.13	2.72
Σ^-	-1.12	$4e_d \frac{eK}{2m_\Sigma c}(1+\delta_{\Sigma^-})$	1.06	-1.26
Ξ^-	-0.69	$4e_s \frac{eK}{2m_\Xi c}(1+\delta_{\Xi^-})$	0.73	-0.74
Ξ^o	-1.25	$4e_s \frac{eK}{2m_\Xi c}(1+\delta_{\Xi^o})$	1.32	-1.12
Λ	-0.61	$\frac{2}{3}(e_u+e_d+4e_s)(1+\delta_n)$	1.19	-0.72

$$\frac{e_u M^6}{8L^{4/9}} + \frac{bM^2}{L^{4/9}}\{\frac{(e_u+\frac{1}{4}e_d)}{48} - \frac{1}{288}(e_u+e_d)[\ln\frac{M^2}{\Lambda^2} -1-\gamma_E] - \frac{1}{432}e_u[\ln\frac{M^2}{\Lambda^2} + \frac{1}{6} -\gamma_E - \frac{M^2\cdot}{2\Lambda^2}]\}$$
$$+ \frac{a^2}{72}L^{4/9}[-(2e_u+3e_d)+e_u(2k-\xi)] - \frac{e_u\chi a^2}{12L^{4/27}}[M^2-m_o^2/8L^{4/9}]$$
$$= \frac{\beta_N^2}{4}(\mu_p+A^1M^2)e^{-m_N^2/M^2} + \text{excited state terms} \qquad (13)$$

The second term computed in ref. [8] is numerically small. Dividing eq.(15) by e_u it is seen that the leading coefficients in the l.h.s. is identical to eq.(7) as in the G_A sum rule and moreover in the right hand side we have 3/8 μ_p which is close to unity. This sum rule can then be analyzed in exactly the same way as the G_A sum rule and the results are displayed in Table 1.

1. B.L. Ioffe, Nucl. Phys. B188, 317 (1981); B191, 591 (E) (1981); V.M. Belyaev and B.L. Ioffe, Zh. Eksp. Teor. Fiz. 83, 876 (1982).
2. Y. Chung, H.G. Dosch, M. Kremmer, and D. Schall, Phys. Lett. 102B 175 (1981); Nucl. Phys. B197, 55 (1982).
3. B.L. Ioffe and A.V. Smilga, Nucl. Phys. B232, 109 (1984); Phys. Lett. 133B, 436 (1983).
4. I.I. Balitsky and A.V. Yung, Phys. Lett. 129B 328 (1983).
5. V.M. Belyaev and Y.I. Kogan, Pis'ma Zh. Eksp. Theor. Fiz. 37, 611 (1983); Phys. Lett. 136B 274 (1984); V.M. Belyaev, B.L. Ioffe and Ya.I. Kogan, Phys. Lett. 151B, 290 (1985).
6. C.B. Chiu, J. Pasupathy and S.L. Wilson, Phys. Rev. D32, 1786 (1985).
7. C.B. Chiu, J. Pasupathy and S.L. Wilson, Phys. Rev. D33, 1961 (1986); and preprints.
8. C.B. Chiu, J. Pasupathy, J.P. Singh and S.L. Wilson preprint June 1986.
9. J. Pasupathy and R. Marshak, Phys. Rev. D30, 265 (1984).
10. J. Pasupathy and S. Wilson, to be published.

The Weak Decay of Hypernuclei

G.B. Franklin

Department of Physics, Carnegie Mellon University,
Pittsburgh, PA 15213, USA

Hypernuclei whose ground states are stable against strong decay are used to study two-baryon weak interactions. A review of the existing experimental data, including recent results from the AGS on ${}^{12}_{\Lambda}C$ and ${}^{11}_{\Lambda}B$, shows that the lifetimes and branching ratios can be used to test the effective weak Hamiltonians used in the rate caculations.

1. Introduction

Binding a Λ hyperon in a nucleus provides an opportunity to study the strangeness-changing weak interaction. A particle stable hypernucleus can decay through the mesonic channels, $\Lambda \to n+\pi^0$ and $\Lambda \to p+\pi^-$, which are analagous to the decay modes of a free Λ. However, the bound Λ may also interact with nucleons through the weak nonmesonic channels $\Lambda+p \to n+p$ (proton stimulated) and $\Lambda+n \to n+n$ (neutron stimulated). In all but the lightest of hypernuclei, it is expected that the nonmesonic channels dominate since the mesonic $\Lambda \to N+\pi$ modes are supressed by Pauli-blocking of the final state nucleon due to its low momentum.

These nonmesonic channels, not available to the free Λ, may be viewed as a conventional meson exchange with one weak vertex or as a W exchange between quarks as shown in Fig.1. In fact, the hybrid-quark-hadron model (HQH)[1] uses both; a π exchange is used at large distances and quark wavefunctions are used for $r_0 < \sim 1$ fm. Since a significant fraction of the transition rate appears to be due to short range interaction the study of nonmesonic decay rate Γ_{nm} provides an opportunity to explore the transition from conventional nuclear physics to quark dynamics.
The effective weak quark interaction arises from W exchange diagrams with strong interaction corrections from QCD perturbation theory and has a Hamiltonian of the form:

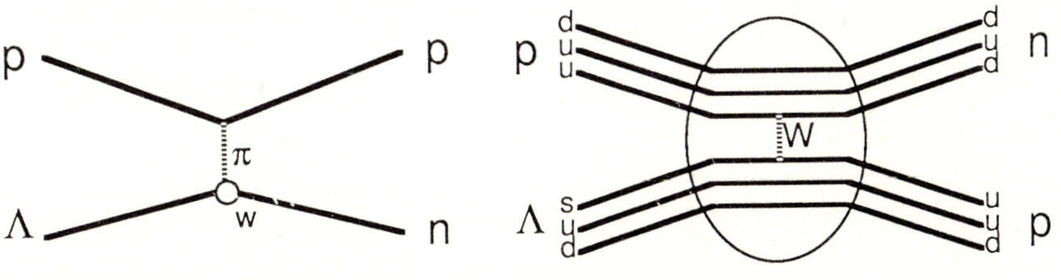

Fig. 1. Meson exchange and quark models of weak decay

$$H^{weak}_{(\Delta S=1)} = \frac{G}{\sqrt{2}} \sin\theta_c \cos\theta_c \sum_{i=1}^{6} C_i Q_i \qquad (1)$$

The local operators Q_i connect strange and nonstrange quark lines. The coefficints C_i have been calculated by Gilman and Wise [2] and it is found that only the first two terms, C_1Q_1 and C_2Q_2, contribute significantly. However, this effective weak Hamiltonian gives rise to both $\Delta I=1/2$ and $\Delta I=3/2$ transitions, in contradiction of the empirical $\Delta I=1/2$ rule observed in strangeness changing decay. The ratio of $\Delta I=1/2$ to $\Delta I=3/2$ transition rates can be increased somewhat by considering strong corrections to C_1 and C_2 but is still smaller than experimental data [3]. Heddle and Kisslinger [1], who use this Hamiltonian in their HQH hypernuclear lifetime calculation, find that by varying a cutoff parameter used in the calculation of C_1 and C_2 the $\Delta I=1/2$ rule can be obtained while simultaneously providing a closer agreement with the measured nonmesonic weak decay rate of $^{12}_{\Lambda}C$.

2. Existing Data

Early data on the decay of the light ($A\leq5$) hypernuclei come from bubble chamber and emulsion studies. There is also one $^{16}_{\Lambda}O$ measurement obtained by measuring the decay distances of hypernuclei created with an ^{16}O ion beam. In ongoing experiments at the AGS, the decays of $^{12}_{\Lambda}C$, $^{11}_{\Lambda}B$, and $^{5}_{\Lambda}He$ are being studied. The hypernuclei are first created and tagged with the reaction $K^-+Z\rightarrow\pi^-+_{\Lambda}Z$. The decay products are then detected in a pion/proton range spectrometer located beneath the target and a time-of-flight neutron detector located overhead as shown in Fig.2. The lifetime is measured using high precision timing scintillators to measure the delay between the formation and decay of the hypernucleus. In parallel, the time resolution of the apparatus is constantly measured by detecting prompt protons in the range spectrometer from the reaction $\pi^-+Z\rightarrow p+X$ and the comparison of the time distributions of these prompt protons with the protons from the tagged hypernuclear decay gives a direct measurement of the lifetime. The preliminary results from the recent $^{5}_{\Lambda}He$ measurement are shown in Fig.3.

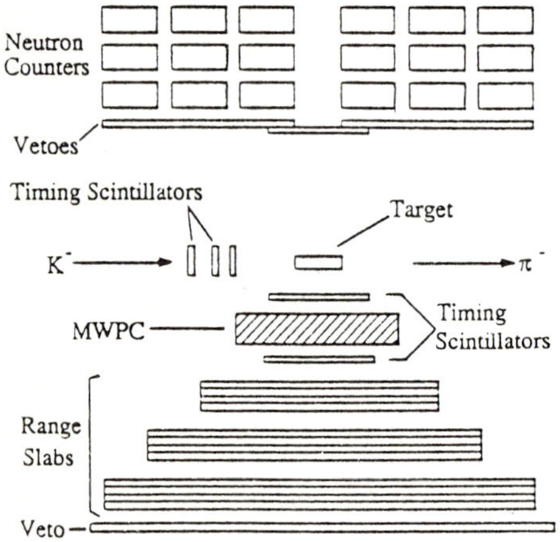

Fig. 2. AGS experimental setup

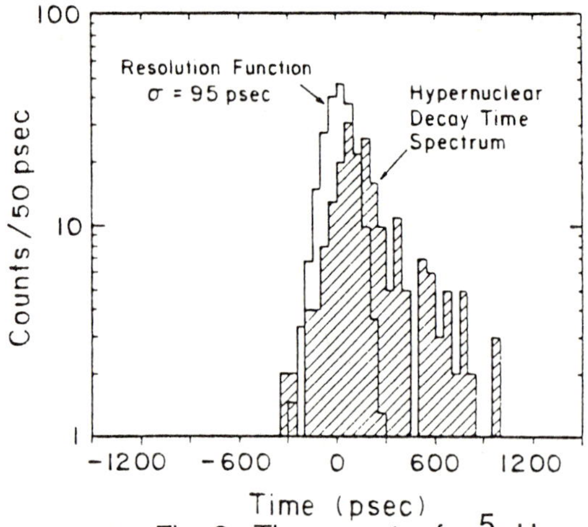

Fig. 3. Time spectra for $^{5}_{\Lambda}He$

Figure 4 shows the hypernuclear lifetime data reviewed in references [4] and [5], along with the recent ${}^{12}_{\Lambda}C$ and ${}^{11}_{\Lambda}B$ data from the AGS [6]. Preliminary results on ${}^{5}_{\Lambda}He$, which give a total decay rate somewhat larger than the decay rate of a free Λ, have not been included. Note that no data for A>16 exists and the data for A=4 hypernuclei are of limited value due to their large errors. To rectify this situation, the AGS program will be extended to include measurements of the lifetime and partial decay rates of ${}^{4}_{\Lambda}He$. The total lack of data on heavy hypernuclei has been a considerable problem since most of the detailed caculations have used nuclear matter approximations. However, a new experiment at LEAR, which will be reviewed by Polikanov in this session, has observed the decay of heavy hypernuclei and will be able to provide this important data when their analysis is completed [7].

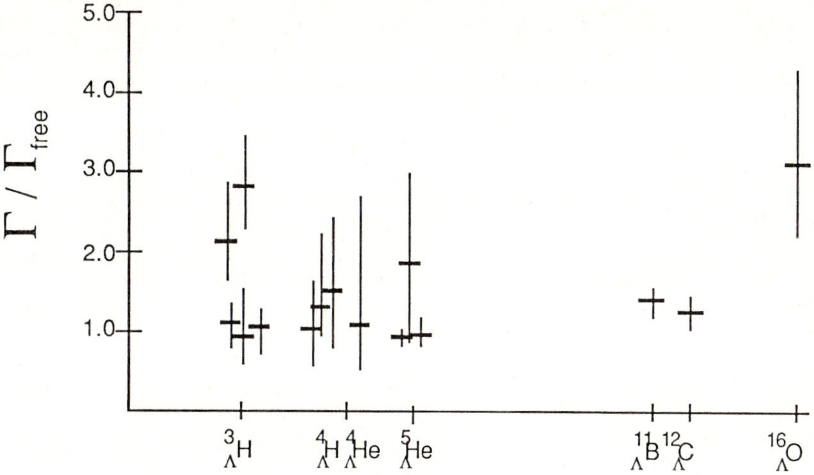

Fig. 4. Total transition rates

Measurements of the neutron stimulated fraction $n=\Gamma_n/\Gamma_{nm}$ and the ratio of the nonmesonic decays to $\pi-$ decays, $Q^-=\Gamma_{nm}/\Gamma_{\pi^-}$, have also been obtained for numerous nuclei. Tables of the data available prior to the recent AGS and LEAR experiments can be found in a review by Dover and Walker [4]. Most of the A≤5 data has large error bars and lacks sufficient information to extract the individual partial rates and it should be noted that many of the quoted numbers in the literature for Q^- and n are not direct experimental measurements. It is advisable to refer to the original publications before using these data for a detailed comparison to theory. For example, the neutron stimulated fraction numbers are often obtained by fitting the theoretical proton energy distributions generated by intranuclear cascade calculations (INC) for neutron and proton stimulated decays to the measured proton spectrum. The neutron stimulated fractions for large A hypernuclei obtained in this manner are as large as 0.9, a result which is in serious disagreement with existing nuclear matter calculations. It is not clear whether this represents a serious discrepency or is merely an artifact of the analysis procedures.

3. Nonmesonic Decay Rates

A number of detailed calculations of Γ_{nm} have been performed in an attempt to understand the weak ΛN interaction. The calculation of Adams [8] is based on one-pion exchange, as is the calculation of Oset and Salcedo [9] although the latter include modifications to the pion propogator due to nuclear medium effects. McKellar and Gibson [10], use both π and ρ exchange while Dubach [5] includes π,ρ,η,ω,K, and K^* exchanges. Oset et al. have used a local density approximation to calculate the transition rate for ^{12}C while Heddle and Kisslinger [1] have expanded the ΛN clusters via spectroscopic factors to handle the nuclear strucutre.

Since Dubach will discuss these calculations in more detail later in the proceedings of this conference, I wish to show how the nuclear structure aspects of the problem, often an unwelcomed detail of a theory attempting to understand basic two-body interactions, can in this case be used to select out components of the effective weak Hamiltonian. A carefully selected set of measurements will provide a rigorous testing grounds for the models.

Consider the neutron stimulated decay rate Γ_n. The Hamiltonian for a one pion exchange has the form:

$$V(r) = [V_0(r)\sigma.\sigma + V_1(r)\sigma.r + V_2(r)S_{12}]\ \tau_1.\tau_2 \qquad (2)$$

The use of this Hamiltonian in a nuclear matter calculation gives a transition rate which is dominated by the $^3S_1 \rightarrow {}^3D_1$ parity conserving transition. Since the $^3D_1$ state must have isospin I=0, a two neutron final state is excluded. This gives a neutron stimulated fraction $n=\Gamma_n/\Gamma_{nm} \ll 0.5$ in contradiction of existing data which has $n \sim 0.5$ and larger. Dubach obtains an increase in the neutron stimulated channel when he includes the exchange of heavier mesons. The HQH model, which provides an alternate method of treating the short range effects, does not get a significant neutron stimulated rate when applied to nuclear matter. However, their model gives $n \sim 0.5$ when applied to ${}^{12}_{\Lambda}C$ since in this case the strength comes from the parity nonconserving term in the Hamiltonian. This term acts on the ΛN P-stat components of the ${}^{12}_{\Lambda}C$ wavefunction to induce P→S and P→D transitions.

These alternative explanations will be tested when analysis on the ${}^{5}_{\Lambda}He$ AGS data is complete. In this case there are no initial P-state components so the neutron decay rate should be reduced in the HQH calculation. The effect shoul be quite different if the $^3S_1 \rightarrow {}^3P_1$ transition is responsible for a significant fraction of the neutron stimulated strength. The future ${}^{4}_{\Lambda}He$ measurements will allow extraction of the partial rates for this A=4 system. Since the Λn are coupled as a $^1S_0$ pair in the ground state while the Λp couples to both $^1S_0$ and $^3S_1$, the ratio of Γ_n/Γ_p should be particularly interesting.

These calcualtions indicate that this series of light hypernuclei, ${}^{4}_{\Lambda}He$, ${}^{5}_{\Lambda}He$, ${}^{12}_{\Lambda}C$, systematically select out components of the effective weak Hamiltonian. When completed, the data from these experiments will be combined with the heavy hypernuclear lifetime data of Polikanov [7] to provide a rigorous test of our understanding of the weak nonleptonic interaction.

1 D.P. Heddle and L.S. Kisslinger, Phys. Rev. C33, 608 (1986)
2 F.J. Gilman and M.B. Wise, Phys. Rev.D20, 2392 (1979)
3 B. Desplanques, J. Donoghue and B.R. Holstein, Ann. Phys. 124 449 (1980)
4 C.B. Dover and G.E. Walker, Phys. Rep. 89, 1 (1982)
5 J.F. Dubach, Nucl. Phys. A450, 71c (1982)
6 R.R. Grace, P.D. Barnes, R.A. Eisenstein, G.B. Franklin, C. Maher, R. Reider, J. Seydoux, J. Syzmanski, W Wharton, S. Bart, R.E. Chrien, P. Pile, Y. Xu, R. Hackenburg, E.V. Hungerford B. Bassaleck, M. Barlett, E.C. Milner and R.L. Stearns, Phys. Letts. 55, 1055 (1985)
7 S. Polikanov; this volume
8 J.B. Adams, Phys. Rev. 150, 1611 (1967)
9 E. Oset and L.L. Salcedo, Nucl. Phys. A443, 704 (1985)
E. Oset and L.L. Salcedo, Nucl. Phys. A450, 371 (1986)
10 B.H.J. McKellar and B.F. Gibson, Phys. Rev. C30 322 (1984)

Theoretical Aspects of the Weak Decay of Hypernuclei

J. Dubach[1]

Department of Physics and Astronomy, University of Massachusetts, Amherst, MA 01003, USA

The present status of theoretical descriptions of mesonic ($\Lambda \to N\pi$) and non-mesonic ($\Lambda N \to NN$) decay modes of hypernuclei is reviewed. Calculations for the non-mesonic mode are discussed in some detail within the context of a model which describes the strangeness-changing, parity-violating, $\Lambda N \to NN$ "transition potential" in terms of π, ρ, ω, η, K, K^*, and "σ" exchanges. Results are presented for the total decay rate, the ratio of proton- to neutron-stimulated decay rates, and the as yet unmeasured ratio of parity-violating to parity-conserving decay rates. Calculations for nuclear matter, which are in reasonable agreement with experiment, suggest that these rates (particularly the two ratios) can provide important tests of the form of the transition potential. Considerations of finite hypernuclei are also discussed. Finally, other theoretical approaches and the present experimental situation are briefly summarized.

1 Introduction

A study of the weak decay modes of hypernuclei offers a unique insight into weak interactions. When a Λ hyperon is embedded in a nucleus, the Λ's free-space (mesonic) decay mode ($\Lambda \to p\pi^-$ (64.2%) and $\Lambda \to n\pi^0$ (35.8%)) will be suppressed due to "Pauli-blocking" of the outgoing nucleon. A Λ at rest inside a hypernucleus decaying via the mesonic mode will release only 5.7(5.4) MeV of energy to the neutron(proton) with a corresponding nucleon momentum of 104(100) MeV/c. This energy release to the nucleon is smaller than typical binding energies of the Λ in the ground state of a hypernucleus. Hence the nucleon resulting from the mesonic decay mode must remain bound. However, nuclei have Fermi momenta, k_F, of approximately 270 MeV/c and nucleon states in the vicinity of 100 MeV/c will already be mostly occupied. Thus, the mesonic mode can be expected to be strongly suppressed inside all but perhaps the lightest hypernuclei and other decay processes may become significant or even dominate the hypernuclear decay.

This was recognized as early as 1953 by CHESTON and PRIMAKOFF[1] who suggested that the two-hadron, non-mesonic decay mode $\Lambda N \to NN$ would be the most significant decay mode. This mode might be viewed as the same as the free-space decay mode $\Lambda \to N\pi$ except that the pion is now virtual and is absorbed on a second nucleon in the (hyper)nucleus. As we shall see below, however, this process also receives contributions from the exchange of more massive mesons, mesons whose production is below threshold for free-space Λ decay. It is the two-hadron nature of this decay mode and the involvement of these other mesons that makes hypernuclear decays such an intriguing topic for the study of weak interactions.

The experimental signal of the non-mesonic mode is fairly obvious. Since the full Λ-N mass difference is available, "fast" nucleons each with energies of roughly 80 MeV will be produced. Unfortunately, the hypernuclear lifetimes are so short that direct measurement of the rate for the non-mesonic decay mode has, until recently, proven extremely difficult. Nonetheless, the dominance of the non-mesonic over the mesonic decay mode could be verified in a series of experiments (mostly emulsion and bubble-chamber experiments)[2]. These measurements of the ratio of non-mesonic to mesonic rates are summarized in Fig. 1.

---

[1]Work supported in part by the U. S. Department of Energy.

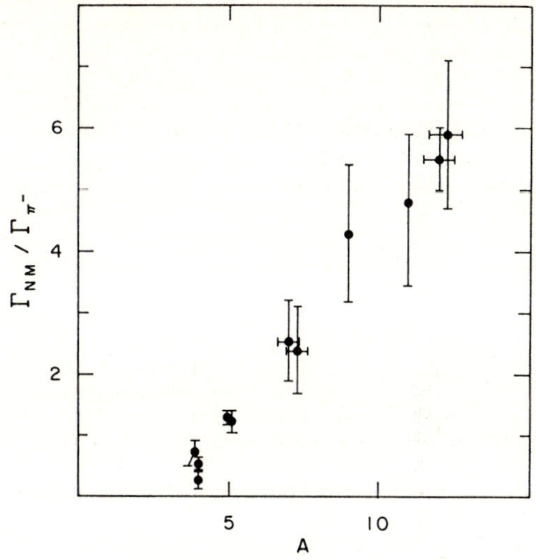

Fig. 1 Selected experimental data for Γ_{NM}/Γ_{π} as a function of hypernuclear mass number.

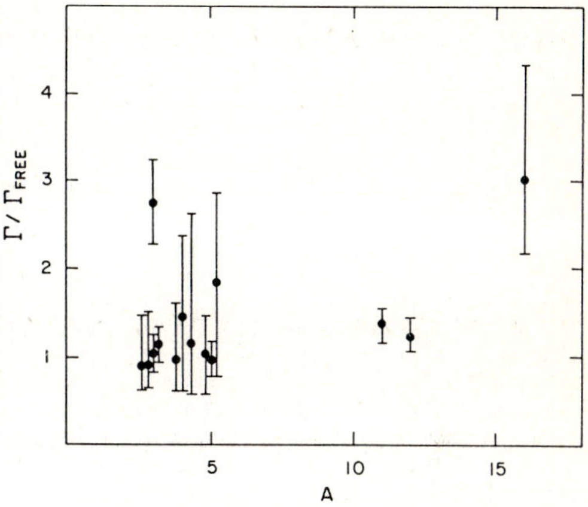

Fig. 2. Experimental values for the hypernuclear decay rates.

The mesonic decay rates may be estimated from a simple model of the Pauli-blocked suppression of the free-space decay. Within the context of a single-particle shell model for the hypernucleus and assuming free-space kinematics, i.e., neglecting differences between the lambda and nucleon single-particle energies and recognizing the fact that the pion recoils against the nucleus as a whole instead of a single nucleon (these assumptions are equivalent to a closure approximation with an appropriately chosen average binding energy for the decay nucleon), one obtains the simple expression:

$$\frac{\Gamma[\pi^-(0)]}{\Gamma^{free}[\pi^-(0)]} = 1 - \frac{1}{2} \sum_{nlj} N_{nlj} |\langle nlj | j_\ell(k_\pi r) | 1s_{1/2} \rangle|^2 \tag{1}$$

where N_{nlj} is the number of protons(neutrons) in the nlj shell and k_π is the pion momentum, i.e., $k_\pi \simeq 100$ MeV/c. Γ_{free} is the rate for the free-space decay mode, corresponding to the experimental value for the total Λ lifetime of 260 psec (most of the results quoted hereafter will be in units of this rate). BANDO and TAKAKI[3,4] have recently performed much more sophisticated calculations of the mesonic decay mode including the use of density-dependent Hartree-Fock wave functions, more realistic kinematics, pion distortion and absorption, and correlation effects. Their basic conclusions are similar to those obtained from the simple expression above, viz., the suppression expected for the mesonic decay mode is verified with suppressions by factors of 6-10 for hypernuclei with $A \simeq 12$ and by as much as 1000 for $A \simeq 100$.

From these calculations and the measurements shown in Fig. 1, we may infer that the rate for the non-mesonic mode in hypernuclei for $A<20$ is roughly equal to the free-space rate. The direct measurements[5] of the hypernuclear lifetimes are shown in Fig. 2. Included are the recent data of GRACE et al.[6,7] for A=11 and 12. These authors measured the lifetime directly by a delayed coincidence between the fast π from the (K,π) reaction which produces the hypernucleus and the fast nucleon from the non-mesonic decay. Their measurements are the first accurate measurements of lifetimes of hypernuclei with $A>5$ and have provided much of the impetus for the renewed theoretical interest in this process. As can be seen, the expectation that $\Gamma \simeq \Gamma_{free}$ is well borne out by these measurements.

2 Theory

Much of the underlying physics for the non-mesonic process has been understood for many years due to the work of PRIMAKOFF and collaborators[1] and DALITZ and co-workers[8]. More recent calculations have been made by a number of authors [9,10,11,12,13,14,15] and I will restrict my attention to a discussion of these results. The most common approach has been to treat the decay of a Λ at rest inside nuclear matter. In that case, the non-mesonic decay rate may be written:

$$\Gamma_{NM} = \frac{1}{8\pi}\,\mu_{\Lambda N}^{3}\int_{0}^{\mu_{\Lambda N}k_F} q^2 dq\; t_0 m_N \sum_{\alpha,\beta} |\langle\beta|V|\alpha\rangle|^2 \;\equiv\; \sum_{\alpha,\beta}\Gamma(\beta \leftarrow \alpha) \qquad (2)$$

where q is the relative Λ-N momentum, $\mu_{\Lambda N} = m^3/(m_\Lambda + m_N)^3$, and $t_0^2 = m_N(m_\Lambda - m_N) + 0.5q^2(1 + m_N/m_\Lambda)$ is the relative momentum of the final NN pair. Since q is fairly small, initial state configurations, α, are restricted to s-waves, $^1S_0$ and $^3S_1$. Partial waves considered for the relative NN final state, β, include $^1S_0$, $^3S_1$, $^3P_0$, $^3P_1$, $^1P_1$, and $^3D_1$. Values chosen for k_F vary somewhat from author to author; we[10] have chosen to use k_F = 268 MeV/c. As we shall see in the next section, corrections due to the physical structure of the finite hypernucleus can be important for a detailed understanding of the experimental results. It remains instructive, however, to consider in some detail the nuclear matter results before considering the finite nucleus calculations.

The most important physics input in such a treatment is the "transition potential" V. I report here on one model[10] which includes π, ρ, ω, η, K, and K^* exchanges in forming the potentials. We consider explicitly the one-meson exchange processes depicted by diagrams (a) and (b) of Fig. 3. The full strangeness-changing weak interaction which is responsible for the transformation of a hyperon into a nucleon is indicated by a circled cross. The remaining vertices represent strong interactions. Relying on earlier successes of such models for the weak parity-mixing[16] nucleon-nucleon interaction, we consider only the pseudoscalar (π,η,K) and vector (ρ,ω,K^*) exchanges.

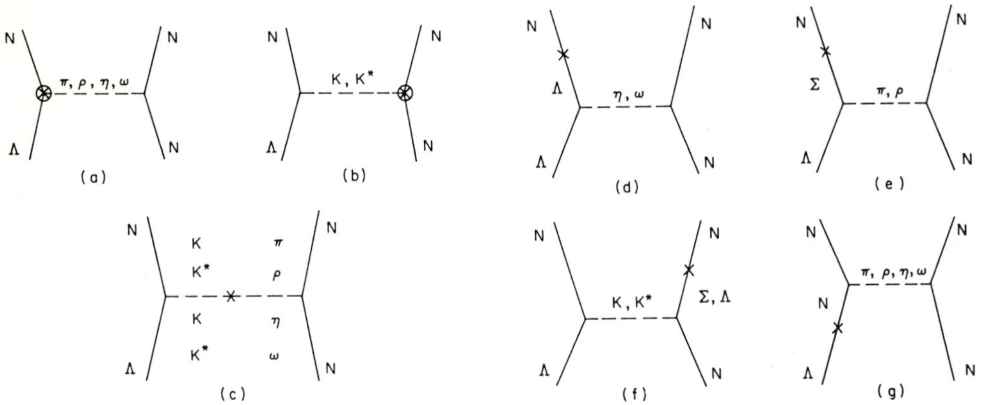

Fig. 3. Meson exchange diagrams used to evaluate the transition potential in the present model.

The full parity-conserving weak vertices are calculated assuming a pole model as depicted in diagrams (c)-(g) of Fig. 3 where the simple cross indicates a weak meson $\rightarrow$ meson or baryon $\rightarrow$ baryon transition amplitude. Using the $SU(6)_W$ symmetry and enforcing the empirical $\Delta I = 1/2$ rule one obtains $\langle M'|H_W|M\rangle \sim \langle\pi|H_W|K\rangle$, i.e., that all the necessary meson $\rightarrow$ meson amplitudes can be obtained from the K $\rightarrow$ π amplitude. Using PCAC this amplitude can be related to the physical K $\rightarrow$ $\pi\pi$ decay rate, thus determining all of the necessary $\langle M'|H_W|M\rangle$ amplitudes. Similarly, using PCAC, the baryon $\rightarrow$ baryon amplitudes needed can be related to lambda decay and sigma decay: $\langle N|H_W|\Lambda\rangle \sim \langle\pi^0 n|H_W|\Lambda\rangle$ and $\langle N|H_W|\Sigma\rangle \sim \langle\pi^0 p|H_W|\Sigma^+\rangle$. The strong couplings are obtained using SU(3) symmetry and PCAC and the Goldberger-Treiman relation as necessary.

The above pieces can then be combined to construct the full Λ-N-meson vertices. For the one measurable amplitude, $A^{\Lambda N\pi}$, we construct:

$$A^{\Lambda N\pi} = \bar{N}\, \underset{\sim}{\tau}\cdot\underset{\sim}{\pi} \begin{bmatrix} 0 \\ 1 \end{bmatrix} \{ g^{\Lambda\Sigma\pi} \frac{1}{m_N - m_\Sigma} A_{N\Sigma} + $$

$$+ g^{NN\pi} \frac{1}{m_\Lambda - m_N} A_{N\Lambda} + g^{\Lambda NK} \frac{1}{m_K^2 - m_\pi^2} A_{K\pi} \frac{2m_\pi^2}{m_K^2 + m_\pi^2} \} \tag{3}$$

This is evaluated as described above to give 1.35×10^{-6} for the $\Lambda n \pi^0$ vertex which may be compared to an experimental value of 1.61×10^{-6}. This then gives us some confidence that our model is capable of producing correct weak Λ-N-meson couplings at the level of perhaps 25% accuracy.

In a similar fashion, all parity-conserving baryon-baryon-meson couplings can be related by the $SU(6)_W$ symmetry to the $A^{\pi^- p\Lambda}$ and $A^{\pi^0 p\Sigma^+}$ couplings where again we enforce the $\Delta I = 1/2$ rule. The full details of our procedure for constructing both the parity-conserving and parity-violating potentials along with the values of all couplings used may be found in [10].

Other authors have taken somewhat simpler models for the transition potential including only π exchange[9,13,14,15] or π and ρ exchanges[12]. KISSLINGER and collaborators[13,14] have also considered calculation of this decay mode directly in terms of W exchange between quarks for Λ-N separations of less than 1 fm. As we shall see below, the mesons more massive than the π seem to play an essential role in understanding the experimental results. The extent to which the same physics is described by the shorter-ranged exchanges in our model and the direct treatment of quark degrees of freedom in [13,14] remains to be seen.

Finally, it must be realized that nucleon-nucleon correlations will play a very important role in the evaluation of (2) and we consider a number of different possibilities for both initial and final state relative wave functions. For the initial state these include i) a simple (plane-wave) spherical Bessel function, ii) an approximate form[9,13,14] for a solution to the Bethe-Goldstone equation with a hard-core radius of 0.4 fm, and iii) a simple phenomenological correlation function of the form $f(r) = 1 - e^{-\alpha r^2}$ applied to the spherical Bessel function of form i)[12]. For final state wave functions we have considered both the simple spherical Bessel function and the much more complicated solution to the Schrodinger equation when a Reid soft core nucleon-nucleon interaction is introduced [12]. KISSLINGER et al.[13,14] consider the Bethe-Goldstone form for their initial lambda-nucleon system and introduce an eikonal approximation to obtain the final nucleon-nucleon wave function in the presence of a nuclear optical potential. The correlation function is somewhat less critical for these authors since their premise is that the short- distance behavior of the lambda-nucleon and nucleon-nucleon systems is more properly described in terms of six-quark systems.

3. Results

Results obtained from our model are summarized in Table 1, where partial-wave contributions and the total rate are given for a number of variations of the model. The first column gives the results obtained from considering only pion exchange and taking spherical Bessel functions for both initial and final state relative wave functions. The $S \leftarrow S$ potential formally involves a $\delta(r)$ piece which dominates the $S \leftarrow S$ contributions; however, since any realistically correlated wave function can be expected to vanish at the origin, the δ-function contribution has been omitted (care must be taken when comparing different calculations because of different treatments of this term). The second column shows the rates for only pion exchange but now with fully correlated wave function (using the phenomenological form for the initial state and the Reid correlation for the final state). Since the tensor piece of the Reid interaction mixes $^3S_1$ and $^3D_1$, the notation $^3S_1 \leftarrow {}^3S_1$ and $^3D_1 \leftarrow {}^3S_1$ should be understood to mean the two mixed $^3S_1/^3D_1$ states in both final states. As expected, once the correlations are turned on the $S \leftarrow S$

Table 1. Decay rates for a number of variations of the present model.

	π no corr	π corr	$\pi + \rho$	$\pi, \rho, \eta, \omega, K, K^*$	$+$"σ"
$^1S_0 \leftarrow {^1S_0}$	.01	--	.001	.001	.004
$^3P_0 \leftarrow {^1S_0}$	.156	.037	.052	.018	.018
$^3P_1 \leftarrow {^3S_1}$	.312	.117	.113	.456	.456
$^1P_1 \leftarrow {^3S_1}$	.468	.128	.100	.110	.110
$^3S_1 \leftarrow {^3S_1}$	.01	.789	.589	.202	.203
$^3D_1 \leftarrow {^3S_1}$	2.93	.751	.693	.444	.444
Total	3.89	1.82	1.55	1.23	1.23

transition is negligible regardless of the δ-function term. The third column shows the fully correlated calculation including both pion and rho exchange. The fourth column shows the full results of the present model including in the fully correlated calculation the contribution of π, ρ, η, ω, K, and K^* mesons. Finally, since 'correlated' two-pion exchanges are known to be so important in nucleon-nucleon strong interaction potentials, we have included as well the effects of the exchange of a scalar "σ meson" of mass 760 MeV. The strong σNN coupling is obtained phenomenologically[10], while the weak σ coupling is described in a two-pion exchange model where one baryon-baryon-pion coupling is taken to be the weak coupling as described above. The results including such a model σ-exchange are given in the last column of Table 1. As can be seen, the σ plays only a very modest role by slightly modifying the $S \leftarrow S$ transitions.

These results are to be compared with the experimental value for A=12 of 1.14 ± 0.20[7]. Clearly the correlations play a critical role by reducing the theoretical predictions. (In fact, the original motivation of the work of ADAMS[9] was to exploit the sensitivity of this process to the correlations to learn something about the correct form of the correlations.) The other authors [12-15] observe similar results when the correlations are included in their models.

While the total rates change little when the shorter-ranged η, ω, K, and K^* exchanges are added, one observes that the various partial rates (particularly the $^3S_1/^3D_1 \leftarrow {^3S_1}$ and $^3P_1 \leftarrow {^3S_1}$) change drastically. This suggests that additional measurements for which the partial rates enter in different combinations should be considered. The (measurable) ratios

$$\frac{\Gamma_{pv}}{\Gamma_{pc}} = \frac{\Gamma(^3P_0 \leftarrow {^1S_0}) + \Gamma(^3P_1 \leftarrow {^3S_1}) + \Gamma(^1P_1 \leftarrow {^3S_1})}{\Gamma(^1S_0 \leftarrow {^1S_0}) + \Gamma(^3S_1 \leftarrow {^3S_1}) + \Gamma(^3D_1 \leftarrow {^3S_1})}$$

$$\frac{\Gamma_{\Lambda p \to np}}{\Gamma_{\Lambda n \to nn}} = \frac{3\Gamma_{T=0} + \Gamma_{T=1}}{2\Gamma_{T=1}} \tag{4}$$

with

$$\Gamma_{T=0} \equiv \Gamma(^1P_1 \leftarrow {^3S_1}) + \Gamma(^3S_1 \leftarrow {^3S_1}) + \Gamma(^3D_1 \leftarrow {^3S_1})$$

$$\Gamma_{T=1} \equiv \Gamma(^1S_0 \leftarrow {^1S_0}) + \Gamma(^3P_0 \leftarrow {^1S_0}) + \Gamma(^3P_1 \leftarrow {^3S_1})$$

may be particularly useful. These are the ratio of the parity-violating to parity-conserving non-mesonic decay rates and the ratio of the "proton-stimulated" to "neutron-stimulated" non-mesonic rates. The latter has been measured in the A=12 experiments[6] to be .8 (with an error bar encompassing .4 to 1.9)[7].

Table 2 shows the results for these ratios for a number of variations of the present work (these are, of course, easily obtained from the values given in Table 1). A simple model suggests that the pion-only calculations should give 9.0 for the proton/neutron ratio. Indeed with or without correlations the pion exchange models are far removed from the experimental values. Including the ρ exchange

Table 2. The ratios of (4) for the present model.

	Γ_{pv}/Γ_{pc}	$\Gamma_{\Lambda p\to np}/\Gamma_{\Lambda n\to nn}$
π (no corr)	0.14	11.2
π (corr)	0.18	16.6
$\pi+\rho$	0.21	13.1
$\pi,\rho,\eta,\omega,K,K^*$	0.90	2.9
+"σ"	0.90	2.9

does not change these ratios much. However, adding the remaining mesons brings the parity-violating/parity-conserving ratio to nearly 1.0 and it reduces the proton/neutron ratio to about 3.0. This reflects the decrease in the (parity-conserving, T=0) $^3D_1 \leftarrow {}^3S_1$ channels and the increase in the (parity-violating, T=1) $^3P_1 \leftarrow {}^3S_1$ channel as discussed above. None of the other authors whose work has been discussed above have quoted values for these ratios.

The value for the measured proton/neutron ratio certainly suggests that the simple pion-exchange calculations have overlooked a significant portion of the physics. Inclusion of the heavier meson exchanges seems to provide some of the reduction necessary to bring the predicted proton/neutron ratio into agreement with experiment. While these predictions arise from complicated interferences between a number of terms, it is apparently the contribution from kaon exchange that has the largest effect on the predicted ratio. Since the kaon exchange is the longest-ranged exchange after the pion exchange and since the weak vertex for the K exchange is roughly three times that of the pion (in the T=1 channel) this may not be too surprising.

Since these ratios are clearly sensitive to the partial-wave decomposition of the (hyper)nuclear structure, one should question at this point the validity of the nuclear matter treatment of this structure. Indeed, in a simple shell model for $^{12}_{\Lambda}C$, the $1s_{1/2}$ Λ "sees" four nucleons in $1s_{1/2}$ orbits and seven in $1p_{3/2}$ orbits so there is ample opportunity for the initial ΛN pair to be in a relative P state as well as in the relative S state to which the nuclear matter calculations were limited. In order to examine such effects we have undertaken to repeat our calculations using a shell model to describe the (hyper)nuclear structure. To date we have only considered an extreme single-particle model (with no configuration mixing) and only phenomenological forms for the correlation functions. But our preliminary results, as shown in Table 3, are most intriguing.

As can be seen, the experimental numbers for both the total decay rate and the proton/neutron ratio in $^{12}_{\Lambda}C$ can be reproduced within the shell model if all of the exchanges are included. The shell model description of the (hyper)nuclear structure provides a mechanism for decreasing the proton/neutron ratio from that obtained in nuclear matter. It is still the case, however, that the pion exchange only transition potential seriously over-predicts this ratio and that inclusion of the shorter-ranged exchanges improves the situation considerably. Experimental results for $^5_{\Lambda}He$ are expected within a few months[17]. For this case we again see a drastic change in the proton/neutron ratio (resembling what was seen in nuclear matter) when the shorter-ranged exchanges are introduced. Measurement of this ratio and the total decay rate for $^5_{\Lambda}He$ should provide strong constraints on the various ingredients used in these calculations.

Table 3. Results using a shell model description of the hypernucleus

	$\Gamma_{NM}/\Gamma_{free}$	Γ_{PV}/Γ_{PC}	$\Gamma_{\Lambda p\to np}/\Gamma_{\Lambda n\to nn}$
$^5_{\Lambda}He$			
π (no corr)	1.6	0.1	15.
π (corr)	0.9	0.1	19.
$\pi,\rho,\eta,\omega,K,K^*$	0.5	0.8	2.1
$^{12}_{\Lambda}C$			
π (no corr)	3.4	0.1	4.6
π (corr)	2.0	0.1	5.0
$\pi,\rho,\eta,\omega,K,K^*$	1.2	1.1	1.2

4 Conclusions

The experimentally observed total non-mesonic decay rates of hypernuclei can be obtained in a variety of theoretical models. Given the sensitivity of these total rates to nucleon-nucleon correlations and other details of the models, this is probably not surprising. However, the measured ratio of the rates for proton- to neutron-stimulated decay is less sensitive to the correlations and much more difficult to obtain with the various models. It appears that no model based solely on a one-pion-exchange transition potential can obtain this ratio. Instead, both heavier-meson exchange and detailed consideration of the (hyper)nuclear structure must apparently be included. These ratios can thus be used to distinguish between models and perhaps to define the correct interpretation of a two-hadron weak interaction process.

The experimental results for ${}^{5}_{\Lambda}He$ are anxiously awaited. As seen above these results should confirm or deny the present interpretation. Measurements of the parity-violating to parity-conserving ratios would provide additional "orthogonal" evidence for or against this interpretation. On the theoretical side, the shell model calculations must be explored more thoroughly, the interplay between the various partial waves and the various meson exchanges must be understood in detail as should the role of the correlations in the finite hypernucleus models. Finally, it would be extremely useful if the theorists could cogently select one or two additional hypernuclear systems for future experimental study that would provide additional enlightenment about the weak interactions responsible for hypernuclear decay.

Acknowledgements

The results discussed above have been obtained in collaboration with my colleagues at the University of Massachusetts: L. delaTorre, J. F. Donoghue, B. R. Holstein, and M. Kimura.

1 W. Cheston and H. Primakoff, Phys. Rev. 92 (1953) 1537.
2 A. Montwill, P. Moriarty, D.H. Davis, T. Pniewski, T. Sobczak, O. Adamovic, U. Krecker, G. Coremans-Bertrand, and J. Sacton, Nuc. Phys. A234 (1974), 413; and references therein.
3 H. Bando and H. Takaki, Prog. Th. Phys. 72 (1984) 106.
4 H. Bando and H. Takaki, Phys. Lett. 150B (1985) 409.
5 K.J. Nield, T. Bowen, G.D. Cable, D.A. DeLise, E.W. Jenkins, R.M. Kalbach, R.C. Noggle, and A.E. Pifer, Phys. Rev. C13 (1976) 1263; G. Keyes, J. Sacton, J.H. Wickens, and M.M. Block, Nucl. Phys. B67 (1973) 269; G. Bohm, J. Klabuhn, U. Krecker, F. Wysotzki, G. Bertrand-Coremans, J. Sacton, J. Wickens, D.H. Davis, J.E. Allen, and K. Garbowska-Pniewska, Nucl. Phys. B23 (1970) 93; R.E. Phillips and J. Schneps, Phys. Rev. 180 (1969) 1307; G. Keyes, M. Derrick, T. Fields, L.G. Hyman, J.G. Fetkovich, J. McKenzie, B. Riley, and I.T. Wang, Phys. Rev. D1 (1970) 66; R.J. Prem and P. H. Steinberg, Phys. Rev. 136 (1964) B1803; and further references in Ref. 8.
6 R. Grace, P.D. Barnes, R.A. Eisenstein, G.B. Franklin, C. Maher, R. Rieder, J. Seydoux, J. Szymanski, W. Wharton, S. Bart, R.E. Chrien, P. Pile, Y. Xu, R. Hackenburg, E. Hungerford, B. Bassalleck, M. Barlett, E.C. Milner, and R.L. Stearns, Phys. Rev. Lett. 55 (1985) 1055.
7 P.D. Barnes, Nucl. Phys. A450 (1986) 43c.
8 See, e.g., R.H. Dalitz, Phys. Rev. 112 (1958) 605; R.H. Dalitz and G. Rajasekharan, Phys. Lett. 1 (1962) 58; M.M. Block and R.H. Dalitz, Phys. Rev. Lett. 11 (1963) 96.
9 J.B. Adams, Phys. Rev. 156 (1967) 1611.
10 L. de la Torre, J.F. Donoghue, J. Dubach, and B.R. Holstein, to be submitted to Ann. Phys.
11 J. Dubach, Nucl. Phys. A450 (1986) 71c.
12 B.H.J. McKellar and B.F. Gibson, Phys. Rev. C30 (1984) 322.
13 C.Y. Cheung, D.P. Heddle, and L.S. Kisslinger, Phys. Rev. C27 (1983) 335.
14 D.P. Heddle and L.S. Kisslinger, to be published.
15 E. Oset, L.L. Salcedo, and L. N. Usmani, Nucl. Phys. A450 (1986) 67c.
16 See, e.g., B Desplanques, Contr. to 8th Int. Workshop on Weak Interactions and Neutrinos, Javea, Spain (1982).
17 J. Szymanski, private communication.

Heavy Hypernuclei

J.P. Bocquet[1], *M. Epherre-Rey-Campagnolle*[2], *G. Ericsson*[3], *T. Johansson*[3], *J. Konijn*[4], *T. Krogulski*[5], *M. Maurel*[1], *E. Monnand*[1], *J. Mougey*[6], *H. Nifenecker*[1], *P. Perrin*[1], *S. Polikanov*[7], *C. Ristori*[1], *and G. Tibell*[3]

[1]Centre d'Etude Nucléaires de Grenoble, B.P. 85X, F-38041 Grenoble Cedex, France
[2]Laboratoire Rene Bernas du CSNSM, B.P. 1, F-91406 Orsay, France
[3]Gustab Werner Inst., Univ. of Uppsala, P.O. Box 531, S-752 21 Uppsala, Sweden
[4]Nat. Inst. f. Nucl. Phys. and H.E. Phys. (NIKHEF), Kruislaan 409, P.O. Box 41882, NL-1009 DB Amsterdam, The Netherlands
[5]Univ. of Warsaw, Ul. Hoza 69, PL-00-681 Warszawa, Poland
[6]Centre d'Etudes Nucléaires de Saclay, F-91191 Gif-sur-Yvette Cedex, France
[7]Ges. f. Schwerionenforschung (GSI), Postfach 11 05 41, D-6100 Darmstadt 11, F. R. Germany

Information on the lifetimes of lambda hypernuclei has mainly been obtained in experiments with kaons. The most accurate measurement has recently been carried out for ${}^{12}_{\Lambda}C$ and ${}^{11}_{\Lambda}B$ hypernuclei [1]. It was found that the lifetime for those hypernuclei is somewhat shorter than for the free lambda hyperon. It was also shown that for ${}^{12}_{\Lambda}C$ the mesonic decay $\Lambda \rightarrow N+\pi$ is strongly suppressed in comparison with the nonmesonic reaction $\Lambda+N \rightarrow N+N$. Nothing is known about the lifetimes of hypernuclei with masses substantially exceeding that of ${}^{12}C$, but one can expect that for those nuclei nonmesonic decay predominates even more. Residual nuclei produced in the nonmesonic decay are excited, and in the case of hypernuclei with masses close to 200 they may fission.

In the present work for the first time an antiproton beam was used for production of hypernuclei, and for the first time the decay of hypernuclei in the region of uranium has been observed. What we registered was delayed fission resulting from the decay of hypernuclei produced in the annihilation of antiprotons in ${}^{238}U$. We assume that the primary antiproton annihilation produces kaons which in turn interact with residual nuclei to form these hypernuclei. However, the high yield of hypernuclei also indicates a possibility that the antiproton annihilates on two nucleons (CUGNON and VANDERMEULEN [2]).

The present experiment has been carried out in the antiproton beam from the Low Energy Antiproton Ring (LEAR) at CERN. The experimental method is based on the recoil-distance technique which was developed in the study of short-lived fission isomers [3]. The apparatus used in our experiment is shown schematically in Fig. 1a, and the principle of the recoil-distance technique is explained in Fig. 1b.

Antiprotons from LEAR, after passing through the beryllium window 1 of the storage ring, the kapton window 2 of our experimental setup and slowing down in degrader 3 and scintillator 4, were stopped in a 0.1 mg/cm^2 thick ${}^{238}U$ target 5 of the size 2.5×5.0 mm^2. The degrader 3 was of variable thickness, and the scintillator 4 which

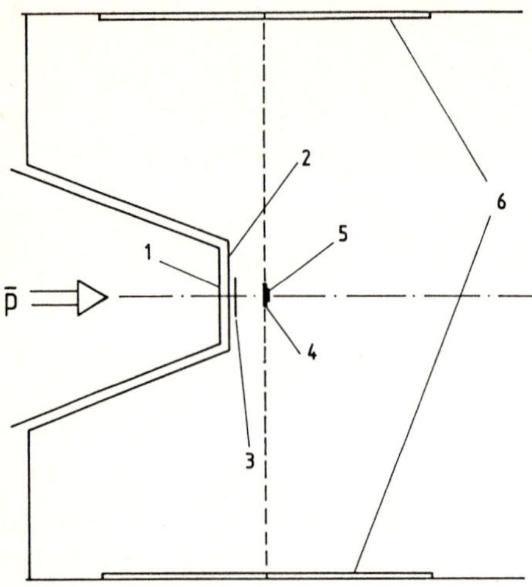

Fig. 1a.
Schematics of the apparatus:
1-Be window, 2-kapton window, 3-degrader
4-scintillator, 5-uranium target,
and 6-fission detectors

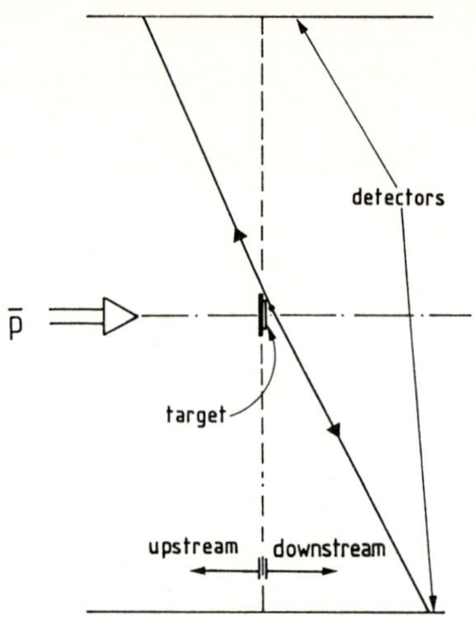

Fig. 1b.
Principle of the
recoil-distance technique

was 100 μ thick and 3 (or 10) mm wide served simultaneously as a backing for the target and an antiproton detector. The thickness of degrader 3 was chosen to provide maximum stopping rate of antiprotons in the target. Fission fragments were registered by the two parallel-plate avalanche counters 6. For each fission-event the velocities of fragments, positions and pulse heights (ionization) in the counters were measured.

Since prompt fission induced by the annihilation occurs in a short time ($\simeq 10^{-18}$ sec) the fissioning nucleus does not leave the target. Therefore, prompt fissions are registered only in the downstream hemisphere of the detector. On the contrary, hypernuclei having longer lifetimes decay at some distance from the target (Fig. 1b), and fission fragments (shown in Fig. 1b schematically by the arrows) can also be registered upstream. Accordingly, a correlation exists between the position distribution of fission fragments in the counters and the distances which hypernuclei travel before they decay. These distances obviously depend on the lifetime of the hypernuclei and on their velocities.

In total, 102 events which were identified as delayed fission have been observed in experiment with a 3-mm wide scintillator. Fig. 2 shows the distribution in angle between the fragments from delayed fission. This distribution is dependent on the momentum distribution of fissioning nuclei and on the moments of fission fragments. In the region of uranium the latter are close to 4000 MeV/c, and are weakly dependent on the fission mode. The dashed histogram in Fig. 2 is the result of a calculation by a Monte Carlo method on the assumption that the momentum distribution for hypernuclei is described by the relation $p^2/(1+\exp((p-p_o)/\Delta p))$, where $p_o = 600$ MeV/c, and

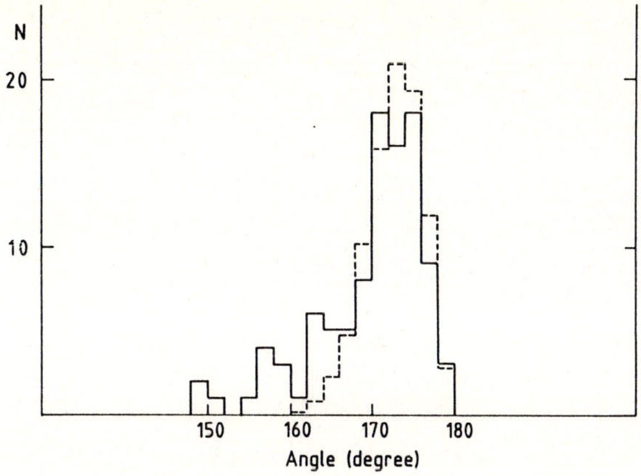

Fig. 2. Distribution in angles between the fragments from delayed fission. The solid line-experiment, the dashed line-calculation

Δp = 150 MeV/c. We can see that the calculation reasonably well reproduces the experimental results except for a few events at small opening angles. Since the evaluation of the lifetime for delayed fission is based on a comparison between the experimental and calculated distribution we have excluded from further analysis 15 events with an angle less than 160°.

Fig. 3 shows the position distribution in the counters for fission fragments registered in the upstream hemisphere and also for complementary fragments recorded downstream. In Fig. 3 we also have shown the result of a Monte Carlo calculation for

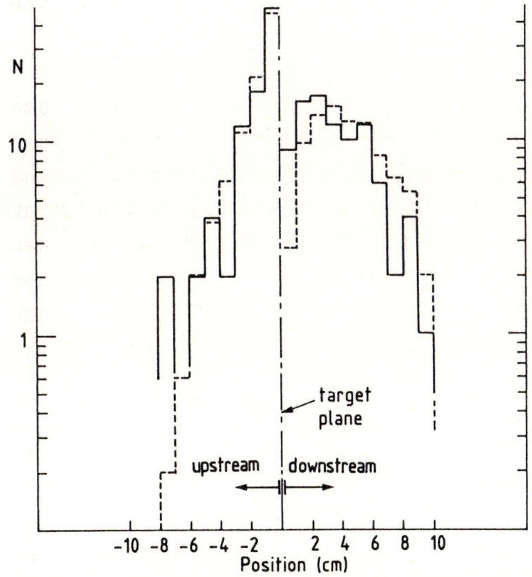

Fig. 3 Position distribution of delayed-fission events: the solid line-experiment, the dashed line-calculation for the lifetime 1.0×10^{-10} sec.

the lifetime 1.0×10^{-10} sec. In the experiment with a 10-mm wide scintillator only 5 events were recorded within a 1-cm bin near the target plane, and this number is close to the expected one for the lifetime 1.0×10^{-10} sec.

Therefore, our results show that some products of the annihilation of antiprotons in ^{238}U undergo delayed fission with the lifetime of about 10^{-10} sec. The yield of these products is about 10^{-3} per stopped antiproton, and the measurement of fission-fragment velocities indicates a symmetric mode of fission. The only plausible explanation of our results is that we observe delayed fission resulting from the decay of hypernuclei produced at the annihilation of antiprotons in ^{238}U nuclei. Presumably these hypernuclei are isotopes of thorium and of neighboring elements. We consider our result as preliminary, and postpone detailed comparison with theoretical predictions until better statistics will have been collected and a refined analysis carried out. Nevertheless we can conclude that the lifetime of hypernuclei in the region of uranium is shorter than the lifetime for $^{12}_{\Lambda}C$ by a factor of about 2.

REFERENCES

1. R. Grace et al., Phys. Rev. Lett. *55* 1055 (1985).
2. J. Cugnon, and J. Vandermeulen, Preprint P.T.M.-84/16, Universite de Liege, 1984.
3. V. Metag et al., Nucl. Instr. and Meth. *114* 445 (1974).

Electromagnetic Transitions in Hypernuclei

R.E. Chrien

Physics Department, Brookhaven National Laboratory, Upton, NY 11973, USA

The interaction of electromagnetic radiation with the nucleus has provided classical nuclear physics with an indispensible tool for illuminating the structure of nuclei. Similarly, for the study of hypernuclei, the observation of γ-ray transitions is essential in the interpretation of hypernuclear level schemes.

Hypernuclei are those nuclear systems which contain one or more hyperons in addition to the usual nucleons. They are usually produced in strangeness exchanging reactions from kaon beams. The detailed comparison of hypernuclear level schemes with those of nuclei provides a valuable insight into the nature of the strong force.

The importance of this information can be placed in perspective by noticing that the spin-dependent hyperon-nucleus potentials--and here, and in what follows one concentrates exclusively on the Λ-hyperon--are weaker than the corresponding nuclear ones. The spin-spin and spin-orbit splittings of hypernuclear level multiplets will be smaller and harder to resolve than in the nuclear case.

High resolution spectroscopy is therefore essential in unravelling hypernuclear decay schemes, and it must be carried out under extremely adverse conditions. The difficulties include 1) relatively poor magnetic resolution for in-flight kaon measurements--at best 0.2%; 2) relatively low beam intensities compared to conventional nuclear physics, about 8 orders of magnitude lower; 3) relatively poor beam quality--typically 90% pion contamination in a kaon beam; 4) extremely unfavorable background conditions from meson decays and background reactions; and, finally, 5) a world-wide scarcity of suitable facilities.

In spite of these obstacles, electromagnetic transitions from hypernuclei have been observed, and have contributed, in some cases, decisively to our knowledge of the effective two-body hyperon-nucleon forces in the nucleus. The object of the review is to survey these contributions and make some evaluation of possible future research.

1. Predictions for Electromagnetic Transitions in Light Hypernuclei

All of the available γ-ray data presented are limited to the p-shell hypernuclei and the A=4 systems. An extensive analysis of p-shell hypernuclear transitions have been presented by MILLENER et al. [1] and DALITZ and GAL [2]. These are generally restricted to particle-bound states of the form $p_N s_\Lambda$. We will consider transitions between the core states in which the Λ is a spectator and the core configuration changes, and between members of the same multiplet, where a Λ spin flip occurs. Schematically this is shown in Fig. 1.

The ΛN effective interaction is characterized by three spin-dependent matrix elements: Δ (spin-spin), S_Λ (Λ spin-orbit), and T (tensor). In addition the energy separation of states built on different core states involves S_N, the induced nucleon spin-orbit interaction.

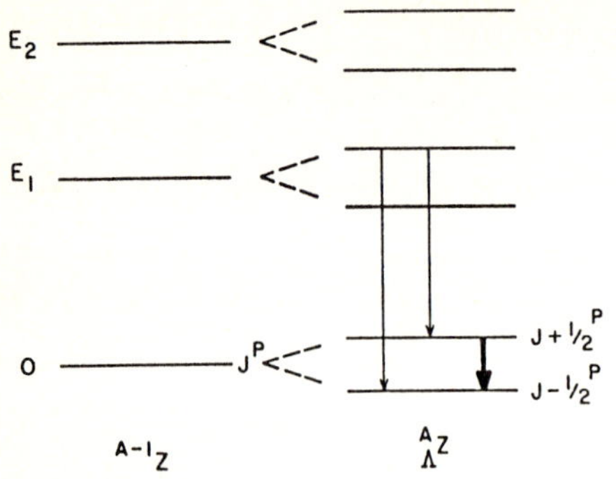

Fig. 1 Doublet states of Λ-hypernuclei in which an S-shell Λ is coupled to core states of the ^{A-1}Z target

The spin-doublet splittings are determined by Δ, S_Λ and T, which can be expressed in terms of radial integrals over the ΛN potentials. These are taken as constants across the p-shell. An approximate representation of the doublet splittings in terms of the parameters δ, δ', for the $p_{3/2}$ and $p_{1/2}$ shells respectively, is as follows:

$$\delta = 2/3\ \Delta + 4/3\ S_\Lambda - 8/5\ T \quad ; \quad \delta' = -\ 1/3\ \Delta + 4/3\ S_\Lambda + 8\ T \qquad (1)$$

As is apparent from the above, the effect of the tensor interaction is relatively large at the end of the p-shell. A systematic measurement across the shell together with the assumption of constant coefficients would determine the matrix elements uniquely. This is a major goal of Λ hypernuclear spectroscopy.

2. Experiments

The data base for γ rays in hypernuclei is extremely limited. All of the published observations to date are listed in Table I. These do not include results described in conferences, proceedings, or work in progress.

The experiments of Table I fall into three classes. Those of references 3-6 referring to stopping kaons in a thick target; reference 8 refers to an in-flight experiment with no energy selection on the incoming kaon or outgoing pion. In ref. 7, the energy differences between incoming and outgoing particles are measured in a spectrometer with a precision of about 1%. Thus only in ref. 7 is it possible to associate the observed γ ray with the excitation of the hypernucleus, and to associate γ-ray prediction with the region of hypernuclear particle bound states, or the production of specific hypernuclear fragments. Thus the assignment of γ rays to specific transitions is unambiguous only in that experiment.

The first observations were reported by BAMBERGER et al. [3] and subsequently refined by the CERN-LYON-WARSAW collaboration [4-6]. Gamma rays were observed by stopping kaons in thick targets of lithium. The observed radiation were ascribed to the hyperfragments $_\Lambda^4H$ and $_\Lambda^4He$. It is important to note that for these light systems the weak decay mode is primarily mesonic, and because of the large binding of the α particle, phase space favors the following modes:

$$\begin{aligned} &{}_\Lambda^4H^* \to {}_\Lambda^4H+\gamma \ ; \ {}_\Lambda^4H \to {}^4He+\pi^- \ [E_{\pi^-} = 53\ MeV] \ ; \quad E_\gamma = 1.04\ MeV \\ &{}_\Lambda^4He^* \to {}_\Lambda^4He+\gamma \ ; \ {}_\Lambda^4He \to {}^4He+\pi^\circ \ [E_{\pi^\circ} = 57\ MeV] \ ; \quad E_\gamma = 1.15\ MeV \end{aligned} \qquad (2)$$

In these experiments the γ rays were not resolvable with the sodium iodide detectors used. The energy assignments of 1.04 MeV for $_\Lambda^4H$ and 1.15 MeV for $_\Lambda^4He$ were made on the basis of examining the spectra in coincidence with the π^-

Table I: Table of Reported Hypernuclear γ Rays

Target	E_γ	Identification	Ref.
^{6}Li, ^{7}Li	1.09	$1^+\rightarrow0^+$; $^4_\Lambda$H; $^4_\Lambda$He	3
^{7}Li	1.09±0.03	$1^+\rightarrow0^+$; $^4_\Lambda$H	4
^{6}Li, ^{7}Li	1.04±0.04	$1^+\rightarrow0^+$; $^4_\Lambda$H	5
	1.15±0.04	$1^+\rightarrow0^+$; $^4_\Lambda$He	5
^{7}Li	1.108±0.01	$1^+\rightarrow0^+$; $^4_\Lambda$H $^4_\Lambda$He	7
^{7}Li	2.034 ± 0.023	$5/2^+\rightarrow1/2^+$; $^7_\Lambda$Li	7
^{9}Be	3.079±0.04	$5/2^+,3/2^+\rightarrow1/2^+$; $^9_\Lambda$Be	7
^{7}Li	0.789±0.04	? ($^7_\Lambda$Li)	8
^{6}Li, ^{7}Li	1.42	?	3
^{9}Be	0.31±0.02	?	3
^{9}Be	1.22±0.02	? ($^8_\Lambda$Li)	6

and π°, respectively. These 1 MeV γ rays are attributed to the $1^+\rightarrow0^+$ spin-flip transitions in these systems. These transitions can be used to evaluate the spin-spin interaction matrix Δ appropriate to the p shell if three-body forces are neglected and a nuclear size correction is made. The result is $\Delta \approx 0.5$ MeV.

These stopped kaon experiments are useful only for the lightest systems because of their inability to specify the initial state of the hypernucleus unambiguously. The next step in the development of hypernuclear γ spectroscopy was to place γ-ray detectors at the target position of the hypernuclear spectrometer, Moby Dick, at the AGS, and record radiation in coincidence with the kaon and pion in the strangeness exchange reaction (K^-,π^-). The simultaneous measurement of the K,π momentum difference permits a specification of the energy spectrum of the hypernuclear states. In practice the energy resolution for a thick target, while poor, is in the order of 6-7 MeV. This is adequate to separate the region of particle bound states (where γ transitions are expected to occur) from the quasi-free (unbound or continuum) region. Further, the continuum can be divided into different regions corresponding to the thresholds for production of various hyperfragments.

The technique is illustrated in the observation of a transition in $_\Lambda{}^9$Be of 3.079 MeV, as shown in Fig. 2. Two γ rays of approximately equal intensities are expected to depopulate the first excited $5/2^+$, $3/2^+$ doublet built on the 2.94 2^+ core excited state of ^{8}Be. The spacing of this doublet is almost exclusively determined by the matrix element of the spin-orbit term, S_Λ. The failure to resolve this doublet places a tight constraint on S_Λ; namely $|S_\Lambda| < 0.04$ MeV. This constraint is the most stringent one on the Λ spin-orbit interaction, which is known to be small from previous (K^-,π^-) studies in, for example, ^{13}C.

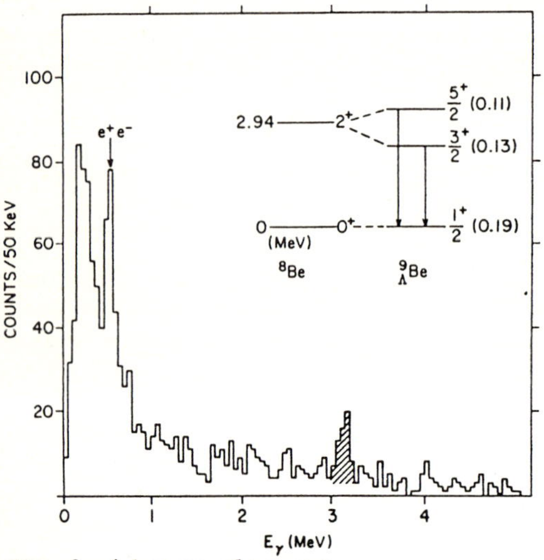

Fig. 2 Λ hypernuclear γ-ray transitions in ${}_\Lambda^9$Be.

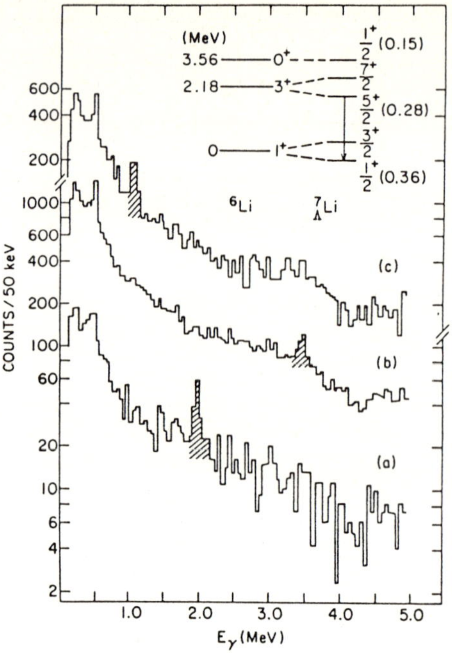

Fig. 3 The γ-ray transitions observed for ${}_\Lambda^7$Li

The same experiment also reported results for ${}_\Lambda^7$Li as shown in Fig. 3. The observation of the γ ray at 2.034 MeV is attributed to the transition for the $5/2^+$ member of the doublet built on the 2.18 MeV 2^+ ^{6}Li core to the $1/2^+$ gs of ${}_\Lambda^7$Li. The reduction of the energy from the core value of 2.18 MeV is attributed by Millener et al. to the induced spin-orbit matrix element S_N. This matrix element can be affected by 3-body ΛNN contributions. Finally, one very important result of the experiment is the confirmation of the 1 MeV γ rays reported by the stopped kaon experiments. These appear at an excitation above break up of ${}_\Lambda^7$Li into the hyperfragments ${}_\Lambda^4$He and ${}_\Lambda^4$H.

As a final remark on previous experiments, I list in Table I, below the horizontal line, those experiments reporting γ rays which have not been verified, and which do not fit into the expected level schemes of refs. 1 and 2.

3. Present and Future Plans

It is natural to consider the extension of hypernuclear spectroscopy to high resolution Ge diode detectors. There is actually a limited energy range of usefulness for such detectors as illustrated in Fig. 4. The lower limit of energy (≈50 keV) is determined by the competition with weak decay, and for γ's of higher than 300 keV, Doppler broadening becomes significant and detector efficiency falls. Fortunately this range is well-suited to the region estimated for the spin-flip M-1 γ rays for p shell hypernuclei.

Valuable experience with high resolution γ spectroscopy was gained in AGS experiment 781, where an array of 6 Ge coaxial detectors, in mini (0.4ℓ)-cryostats was assembled to view targets of ^{10}B and ^{16}O (as water) [9]. With present facilities, an estimated sensitivity is obtained, for 100 keV of γ-ray energy, of about 10 μb/sr. This is comparable to predicted transition rates. Thus the experiments are marginal at best. Peaks which are plausible candidates for spin-flip M-1's were observed at 160 keV for ^{10}B and near 80 keV for ^{16}O in the above experiment.

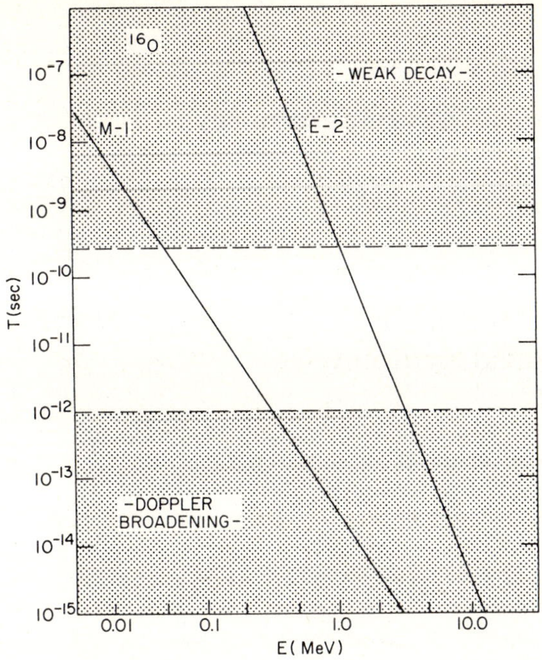

Fig. 4 The window of utility for high resolution spectroscopy

Detailed analyses, which are still underway, have however failed to establish statistically significant confidence levels for those peaks.

For future facilities such as the proposed new "kaon factories" it will be possible to continue a productive γ-ray spectroscopy program for hypernuclei. The most helpful advances would be the provision of highly separated beams, preferably with K to π ratios near one to one, sufficient intensities ($> 10^6$) to allow well shielded detectors to be placed outside the beam halo, and 100% duty factor. These will provide the order-of-magnitude increase in rate that is desired.

Acknowledgements

I appreciate the many useful discussions and suggestions from Henryk Piekarz, Morgan May, Martin Deutsch, and John Derderian. Research has been performed under contract DE-AC02-76CH00016 with the U.S. Department of Energy.

1. R. H. Dalitz and A. Gal: Ann. Phys. (N.Y.) 116, 167 (1978)
2. D. J. Millener, A. Gal, C. B. Dover, R. H. Dalitz: Phys. Rev. C31, 499 (1985)
3. A. Bamberger, M. Faessler, U. Lynn, H. Piekarz, J. Piekarz, J. Pniewski, B. Povh, H. Ritter, and V. Soergel: Nucl. Phys. B60, 1 (1973)
4. M. Bedjidian, A. Filipowski, J. Grossiord, A. Guichard, M. Gusakow, S. Majewski, H. Piekarz, J. Piekarz, J. R. Pizzi: Phys. Lett. 62B, 469 (1976)
5. M. Bedjidian, E. Descroix, J. Grossiord, A. Guichard, M. Gusakow, M. Jacquin, M. Kudla, H. Piekarz, J. Piekarz, J. Pizzi, and J. Pniewski: Phys. Lett. 83B, 252 (1979)
6. M. Bedjidian et al.: Phys. Lett. 94B, 480 (1980)
7. M. May, S. Bart, S. Chen, R. Chrien, D. Maurizio, P. Pile, Y. Xu, R. Hackenburg, E. Hungerford, H. Piekarz, Y. Xue, M. Deutsch, J. Piekarz, P. Barnes, G. Franklin, R. Grace, C. Maher, R. Rieder, J. Szymanski, W. Wharton, R. L. Stearns, B. Bassaleck, B. Budick: Phys. Rev. Lett. 51, 2085 (1983)
8. J. C. Herrera, J. J. Kolata, H. W. Kraner, C. L. Wang, R. Allen, D. Gockley, M. Hasan, A. Kanofsky, and S. Lazo: Phys. Rev. Lett. 40, 158 (1978)
9. M. May: Nucl. Phys. A450, 179 (1985); R. E. Chrien: Proc. Intern. Symp. on In-Beam Nuclear Spectroscopy, Debrecen (Akademiai Kiado, Budapest, 1984)

3.2.3 Parity and CP Violation, Charge Symmetry

Nuclear Probes of Fundamental Symmetries

E.G. Adelberger

Nuclear Physics Laboratory, GL-10, University of Washington, Seattle, WA 98195, USA

1. Introduction

Nuclei can be used in many ingenious studies of fundamental symmetries and my topic is so broad that it cannot be covered in a 30 minute talk. Hence I will restrict myself to giving an experimentalists quick overview of nuclear probes of three symmetries: isospin (I), parity (P) and time-reversal (T). I omit many currently active areas of research. Some of these (lepton nonconservation, neutrino masses and right-hand charged weak currents) are discussed elsewhere in this conference; others (such as axion searches) are not.

2. The I-Violating NN Interaction

We can distinguish three levels of I-violation in hadronic systems.

- the electromagnetic interaction itself (direct EM effects)
- the interplay between electromagnetism and strong interactions (indirect EM effects)
- the strong interaction (u, d quark mass difference)

In complex nuclei isospin violation is easily detectable - being a ~1 to ~10% effect. The rich and interesting phenomena are dominated by the simplest direct EM effect - the Coulomb interaction between nucleons. We can view I-violation in complex nuclei as another way (more subtle than electron scattering) of using EM to study nuclear structure. I shall not consider it further in this talk.

To observe what happens when Coulomb effects are absent (or calculable with some degree of reliability) we must turn to the NN system. HENLEY and MILLER [1] clarified possible I-violating NN interactions as follows. They express a general NN interaction as

$$V_{ij}=V_{ij}^{I}+V_{ij}^{II}+V_{ij}^{III}+V_{ij}^{IV}$$

where $V_{ij}^{I}=V^0+V^1\tau_i\cdot\tau_j$ is an isospin scaler ($\Delta I=0$)

$V_{ij}^{II}=V^2(\tau_{3i}\tau_{3j}-\tau_i\cdot\tau_j/3)$ is an isospin tensor ($\Delta I=2$)

$V_{ij}^{III} = V^3\left(\tau_{3i}+\tau_{3j}\right)$ is a symmetric isospinvector ($\Delta I=1$)

$V_{ij}^{IV}=V^4(\tau_{3i}-\tau_{3j})+V^5(\tau_i\times\tau_j)_3$ is an antisymmetric isospin vector ($\Delta I=1$)

and V^0, V^1, V^2, and V^3 are symmetric functions of spins and momentum while V^4 and V^5 are antisymmetric. Two different NN isospin violating observables have been studied.

1) <u>Differences in the $^1S_0$ scattering lengths in the nn, np and pp systems.</u>

The results have been available for some time

a_{nn} = −16.4±1.2 fm

a_{np} = −23.715±0.015 fm

a_{pp} = −7.823±0.011 fm

= −17.2±3.0 fm (Coulomb effects subtracted)

The scattering length differences are sensitive to the V^{II} and V^{III} interactions. The evidence for $V^{II}\neq0$ is quite clear, while that for $V_{III}\neq0$ is marginal. The existence of the V_{II} interaction has been explained as an indirect effect of pion Coulomb energies on the nuclear force. All three NN systems (pp, np and nn) receive contributions for π^0 exchange. The np system also experiences $\pi^{\pm}$ exchange. MILLER and ERICSON [2] have shown that the ~5 MeV mass difference between charged and neutral pions accounts for most of the observed scattering length difference.

2) <u>S=0 ⟷ S=1 transitions in np scattering</u>

The Pauli principle tells us that S+I+L of the NN system must be odd. Since the oddness or eveness of L is determined by the parity (which we assume to be exact) we see that $|\Delta S|=|\Delta I|$ so that S=0 ⟷ S=1 transitions necessarily violate I by one unit.

Angular momentum conservation forbids ΔS=1 transitions for L=0 states; to see them we need L≥1. This implies that only experiments at bombarding energies ≥200 MeV will have much sensitivity to the I-violating interaction. In the NN system I-violation occurs only for the I=0 and I=1 states of the np system. Since V^{III} vanishes in the np system and V^{I} and V^{II} cannot couple an I=0 to an I=1 NN system $\Delta S=1$ transitions in np scattering are a direct probe of the V^{IV} interaction.

To detect $\Delta S=1$ transitions one clearly needs polarized neutron beams and polarized proton targets. If $V^{IV}=0$ then $A_n(\theta)$, the analyzing power for polarized neutron scattering by unpolarized protons, must equal $A_p(\theta)$, the analyzing power for unpolarized neutron scattering by polarized protons. It is very difficult to make an accurate measurement of $A_n(\theta)-A_p(\theta)$ by comparing absolute measurements of A_n and A_p. The sensible strategy is to compare the angles at which $A_n(\theta)$ and $A_p(\theta)$ cross through zero, since this does not require accurate knowledge of the beam or target polarizations. Groups at TRIUMF and IUCF have undertaken V^{IV} experiments; I will discuss results from TRIUMF [3] at $E_n^{lab} = 477$ MeV. The TRIUMF apparatus is shown in Fig. 1. The experiment ran in two modes, in which a polarized (unpolarized) proton beam on a D_2 target produced polarized (unpolarized) neutrons incident on a depolarized (polarized) target of the frozen-spin type. Scattered neutrons and recoil protons from the frozen-spin target were detected in coincidence in two left-right symmetric detectors. The only differences between the two interleaved modes were the polarization of the beam and target. The difference in zero-crossing angles for A_n and A_p was found to be $\theta_{0n}(A_n)-\theta_{0n}(A_p) = +0.13°\pm0.06°\pm0.03°$ which may be converted into $\Delta A \equiv A_n(\theta_{0n})-A_p(\theta_{0n}) = 0.0037\pm0.0017\pm0.0008$.

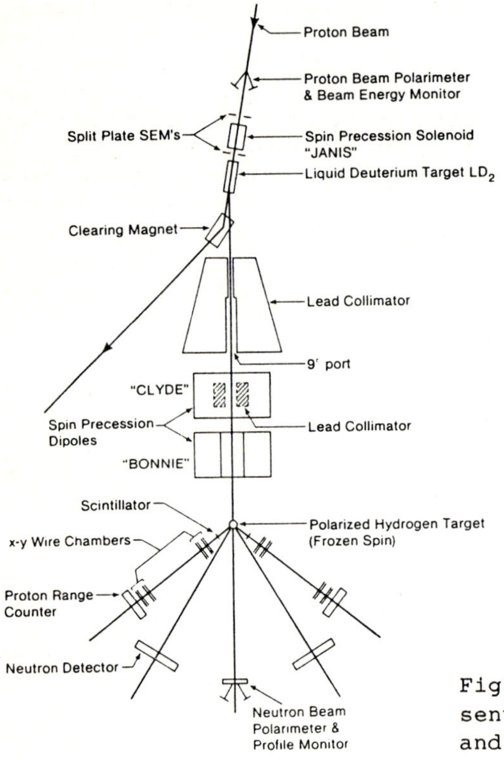

Fig. 1 Schematic representation of beam line and apparatus.

How does this result compare to theoretical expectations? MILLER, THOMAS and WILLIAMS [4] have recently reexamined the theory. They considered the V^{IV} effects arising from one γ exchange, one π exchange (OPE), ρ and mixed ρ-ω exchanges and the π exchange plus explicit quark effects. (Note that charge-dependent meson-nucleon coupling constants and π-η mixing do not lead to V^{IV} interaction. Miller, Thomas and Williams conclude that the dominant effect in the TRIUMF experiment is the influence of the n-p mass difference on the OPE potential; and their result for ΔA agrees within error with the TRIUMF measurement. So it seems as if I symmetry violation in the np system, both for the scattering length differences and for ΔA arises primarily from mass splitting effects on the OPE process - for scattering lengths it is the pion mass splitting, while for ΔA it is the nucleon mass splitting.

3. The P-Violating NN Interaction

Haxton and I recently completed an extensive review [5] of this topic. Consequently, I shall focus on new results and refer the reader to our review for more details.

There are two motives for undertaking difficult measurements of P-violating hadronic interactions. First, in principle, one may learn something new about the fundamental weak interaction. The standard theory is not complete and may require extensions, possibly extra Z^o weak bosons. If these "extra" Z's have very small couplings to leptons they would have probably escaped detection in conventional particle physics experiments. On the other hand, if the extra Z's coupled with normal or greater strength to the quarks then they would have detectable consequences in the P-violating NN interaction since flavor-conserving P-violating hadronic interactions provide a unique window on Z^o exchange between the quarks. We can only study weak interactions among hadrons when the strong interaction is suppressed by a symmetry, normally flavor violation as in the decay $\Lambda \rightarrow N\pi$. However, the neutral current contribution to flavor-changing processes is highly suppressed by the GIM mechanism. To probe the neutral current hadronic interaction we must study flavor-conserving processes which violate parity - of which the P-violating NN interaction is the only existing practical example.

The second motive is more applied. Even if we had a complete theory of electroweak interactions of quarks we still could not compute the weak interactions of hadrons because we do not have exact knowledge of the structure of the hadrons. This problem of computing weak matrix elements between hadronic states is both important and interesting. Important, for example, because it limits our ability to interpret the lovely ϵ'/ϵ data in kaon decay, and interesting because of the appearance of dynamical symmetries (e.g., the $\Delta I=1/2$ rule) which are still not well understood. By studying a new class (flavor-conserving) of hadronic weak interactions we may get new insight into the problem of weak matrix elements.

Under I rotations, the P-violating NN interaction is expected to transform as a mixture of $\Delta I=0$, 1, and 2 with the $\Delta I=1$ contribution being especially sensitive to Z^o exchange. The standard theoretical analysis of the P-violating NN interaction employs a one-meson exchange model in which one of the meson (M)-nucleon-nucleon vertices is a weak process. Bag-model estimates of the weak MNN vertices by DESPLANQUES, DONOGHUE and HOLSTEIN [6] (DDH) predict that the $\Delta I=0$, 1, and 2 components of the P-violating NN interaction should have roughly equal strengths.

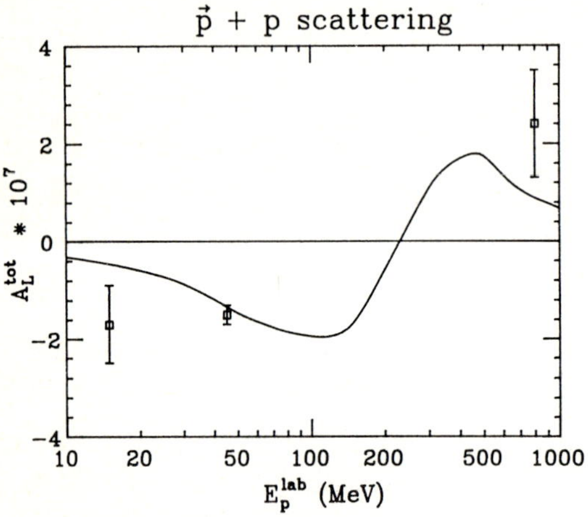

Fig. 2. Measured values of A_L in $\vec{p}$+p scattering. The curve is a P-violating meson exchange theory calculation described in Ref. 5.

How do these predictions compare to the experimental results? Unfortunately, the data is sparse because it is difficult to perform sufficiently precise experiments. In principle, one could completely determine the low-energy P-violating NN interaction from six different measurements in the NN system, involving p+p, n+p and n+n initial states. However, the predicted P-violating effects range between $\sim 10^{-7}$ to $\sim 10^{-8}$ and definite effects have been seen only in the p+p system. An impressively precise new SIN result for A_L(pp) (the longitudinal analyzing power) in p+p at $E_p^{lab} \approx 45$ MeV has recently been reported [7]. The new SIN result is shown in Fig. 2 along with previous results from the Los Alamos tandem and LAMPF. Also shown is a prediction for A_L (see Ref. 5 for details) based on the DDH estimates of the MNN vertices. The agreement is encouragingly good. The other new datum in the NN system is an ILL measurement [8] of A_γ (np) (the asymmetry when polarized cold neutrons are captured by protons). This quantity is particularly interesting because it is sensitive primarily to F_π, the $\Delta I=1$ amplitude for weak $\pi^\pm$ exchange. (Note that pp scattering is completely insensitive to F_π in lowest order. The experimental value [8] $A_\gamma=(-4.7\pm4.7)\times10^{-8}$ is consistent with the prediction (see Ref. 1) $A_\gamma=-5\times10^{-8}$. BLAYNE HECKEL [9] has proposed [5] measuring A_L(np) for cold neutrons on a parahydrogen target. It seems that a measurement of A_L(np), which is mainly sensitive to the ρ and ω exchange amplitudes, can be made with enough precision to test the theory.

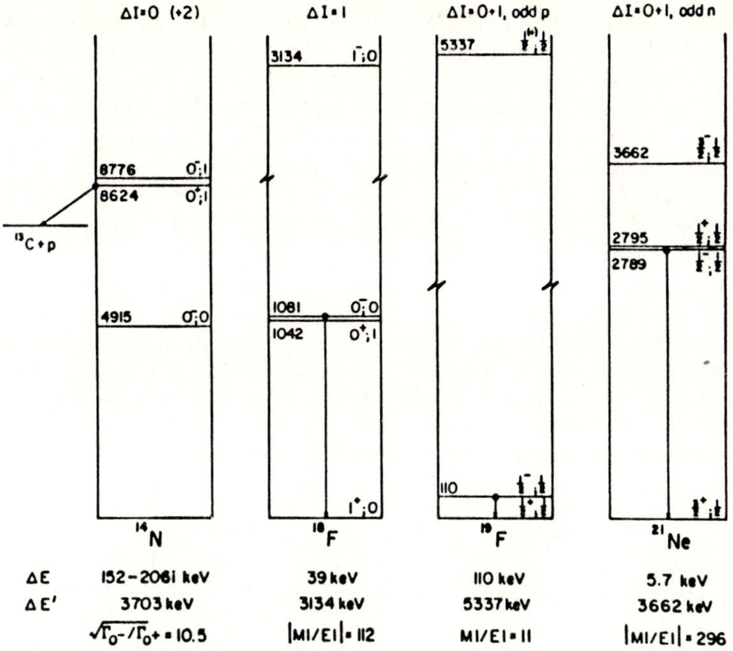

Fig. 3. Parity doublets in light nuclei. The transitions displaying the amplified P-violating effect are indicated. The

On the other hand, it is currently impossible to do an adequately precise $A_L(nn)$ measurement and quite difficult to make a large improvement in the errors on $A_\gamma(np)$. However, one can study nn and np scattering indirectly, by investigating parity impurities in nuclear states. The most interesting cases of nuclear parity mixing are the "parity doublets" - nuclei with closely spaced pairs of levels of the same spin and opposite parity. Some important examples of parity doublets are shown in Fig. 3. In each of these quantities ΔE and $\Delta E'$ are the smallest and next smallest energy denominators governing the parity mixing. The quantities in the bottom row are "amplification factors."

examples one member of the doublet has a much longer decay lifetime than the other. As a result pseudoscalar observables associated with the decay of the longer lived states will be considerably enhanced - because of the small energy dominator and the fact that a little admixture of a rapidly decaying level has a relatively big effect.

The doublets also act as "isospin filters", the various doublets isolate specific isospin components of the P-violating NN interaction. For example the ^{18}F doublet involves the mixing of I=1 and I=0 levels; it probes weak $\pi^{\pm}$ exchange which has $\Delta I=1$. Therefore $P_\gamma(^{18}F)$ and $A_\gamma(np)$ both are sensitive to the same underlying physics. They provide a dramatic illustration of the gain of the nuclear P-violating amplifier. The DDH "best value" predictions for the observables are $A_\gamma(np)=-5\times10^{-8}$ and $|P_\gamma(^{18}F)|=1.5\times10^{-3}$ - which corresponds to a gain of 3×10^4. In this case the gain is also nearly "noise-free". The ^{18}F nuclear matrix element needed to compute $P_\gamma(^{18}F)$ from F_π is determined by the measured $^{18}Ne\rightarrow{}^{18}F(0^-)$ β decay rate (see Ref. 5).

Recently groups at Queens University [10] and Firenze [11] have published results for $P_\gamma(^{18}F)$ with considerably smaller errors than the previous world average. To obtain this precision the experimenters had to develop high-rate data acquisition systems and produce targets that would survive under prolonged intense bombardments. When the new ^{18}F results are added to previous data the new world average is $P_\gamma=(1.2\pm3.8)\times10^{-4}$, which corresponds to $|F_\pi|<3\times10^{-7}$.

The $A_\gamma(^{19}F)$ and $A_L(\vec{p}+{}^4He)$ data (both ^{19}F and $p+{}^4He$ are odd proton systems) yield similar and consistent constraints on a linear combination of $\Delta I=1$ and $\Delta I=0$ amplitudes. The $P_\gamma(^{21}Ne)$ result (an odd neutron system) corresponds to a different linear constraint on the $\Delta I=1$ and $\Delta I=0$ amplitudes. The $\pm1\sigma$ constraints, shown in Fig. 4, do not overlap. However, the ^{21}Ne constraint is shown as dashed because it was not possible to "calibrate" the shell model matrix elements against corresponding first-forbidden β-decay rates. I therefore discount the ^{21}Ne constraint and provisionally conclude that F_π is considerably smaller and F_0 somewhat larger than the DDH best values.

How can we test these conclusions? It seems difficult to make significant improvements in either the $A_\gamma(np)$ or $P_\gamma(^{18}F)$

TWO PARAMETER ANALYSIS OF PARITY DOUBLETS

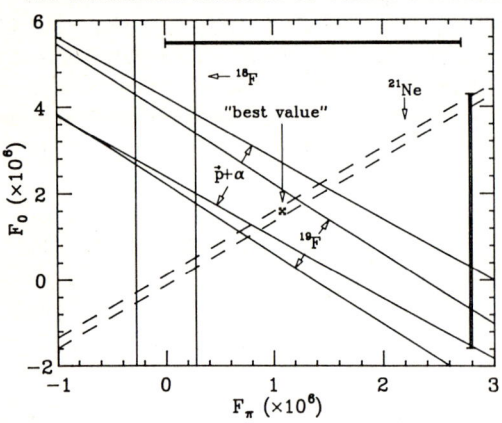

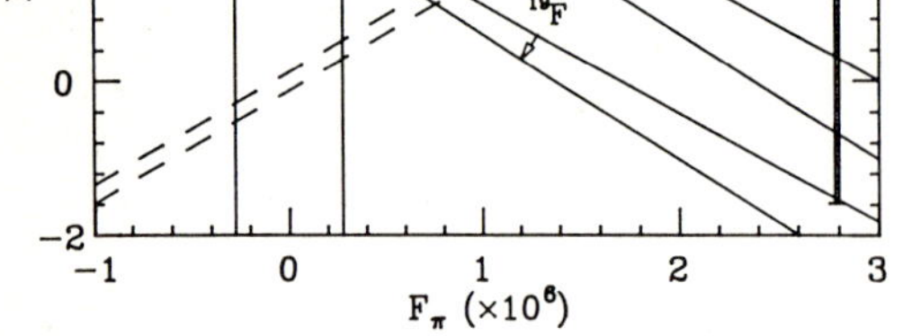

Fig. 4. Analysis of P-violation in the light systems $p+{}^4He$, ^{18}F, ^{19}F and ^{21}Ne in terms of F_0 and F_π as defined in Ref. 5. The linear $\pm1\sigma$ constraints imposed by the experimental results are shown. The DDH "best value" is shown as a cross and the DDH "reasonable ranges" are shown as double bars on the top and right.

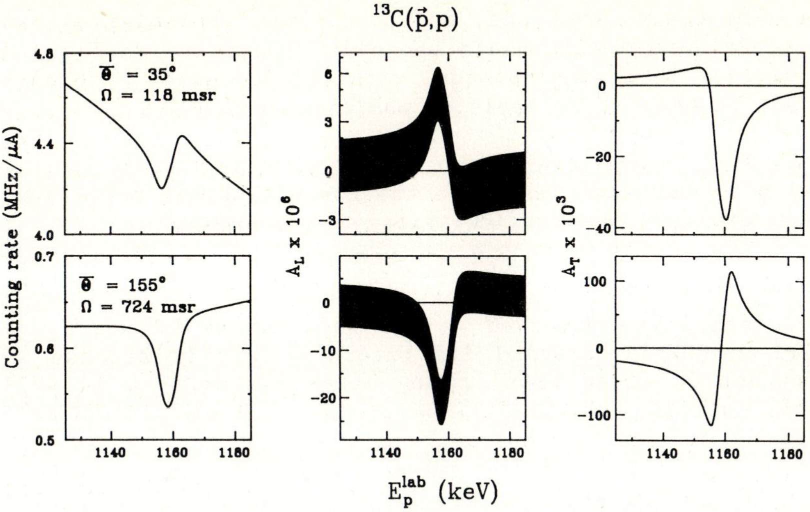

Fig. 5. Predicted counting rates, P-violating A_L, and P-conserving A_T for the 0^+ resonance in $^{13}C+p$. Predictions are based on the DDH "best values." The solid region in the A_L plot represents a $\pm 1\sigma$ band where σ is the statistical error after running for 1 μA-day of integrated beam.

experiments. A possible test would be to measure a pure $\Delta I=0$ transition that imposes a horizontal constraint in Fig. 4. The mixing [12] of I=1 0^+ and 0^- levels in ^{14}N (see Fig. 3) provides just such a constraint. One could then check for internal consistency among the ^{18}F, ^{19}F, $p+\alpha$ and ^{14}N results. We are currently trying to study the P-mixing in ^{14}N by a novel technique - measuring $A_L(\vec{p}+^{13}C)$ at the elastic-scattering resonance corresponding to the narrow 0^+ state [13]. The P-violating effect predicted using the DDH best values and Haxton's matrix element [12] is shown in Fig. 5. We hope to have results within a year.

What conclusions can be drawn at this point?

1) There are no strikingly large anomalies - cases where the measured effects are much larger than can be accounted for by the standard model. Previous anomalies, such as the earlier $P_\gamma(np)$ and $A_\gamma(nd)$ results have been superceded by more precise data which is in general agreement with expectations. Hence there is nothing that lends strong support to the existence of "extra" Z^0's.

2) There is a suppression of F_π (the $\Delta I=1$ amplitude for weak $\pi^\pm$ exchange) compared to bag model calculations. DONOGHUE and HOLSTEIN [14] have considered the implications of small values for F_π and argued that it teaches us about quark masses and SU(3) breaking effects. It would be good to update their analysis.

3) The one-boson exchange model of the P-violating NN interaction at low and intermediate energies appears to work quite well (see Fig. 1). This is consistent with the successes of boson-exchange models in many other areas of nuclear physics [15].

4) It would be very interesting to extend the $A_L(pp)$ results to considerably higher energies where the boson exchange model must break down and explicit quark calculations are required.

4. Time-Reversal Invariance

Although T-violation was observed in the kaon system 22 years ago [16] we still do not understand its origin - largely because T-violation has not yet been seen in any other system. It is well known that the existing limits [17] on the neutron electric dipole moment provide a sensitive constraint on theories of T-violation. What is the status of T-violation tests in nuclei? In a nutshell none of the existing nuclear searches for T-violation are sensitive enough to see the expected effects which in any case will be no larger than 10^{-3} of the weak interaction. For example, detailed balance in compound nuclear reactions [18] sets a limit at roughly 5×10^{-4} of the strong interaction, correlations in γ-decay [19] set a limit at $\sim 10^{-3}$ of the electromagnetic (or strong) interactions, and correlations in β-decay [20] restrict T-violation to be $\sim 10^{-3}$ of the weak interaction.

I will briefly discuss an intriguing new possibility for sensitive searches for T-violation in neutron scattering. One of the remarkable surprises in neutron physics was the discovery that when low energy ($\lesssim$eV) neutrons scatter from heavy nuclei one occasionally sees very large P-violating effects. The record so far occurs on a p-wave resonance in ^{139}La+n where a 7% P-violating effect [21] has been observed! The effects are so large because low-energy penetrabilities are much larger for s-waves than for p-waves and because of fortuitous combination of PNC matrix elements, energy splittings and reduced widths. Unfortunately one cannot use these strikingly large effects to learn more about the P-violating NN interaction because the interesting issues are quantitative and we do not (and may never) know the wavefunctions of the resonances well enough to make a quantitative analysis of the results. On the other hand, it has been pointed out by Blayne Heckel and others that one might be able to exploit the large P-violating effects to do an interesting and sensitive qualitative search for T-violation. The idea is to compare a T-odd, P-odd observable on a given resonance to a T-even, P-odd observable on the same resonance. The ratio of the two observables gives a measure of the T-odd fraction of the weak matrix element. The ratio could be a very sensitive probe of T-violation for cases in which the P-violating effect is large enough (say $\geq 10^{-2}$). One first searches for a case where there is a large P-violating effect with polarized neutrons on an unpolarized target, for example by looking for P-violating A_L's or P-violating neutron spin rotation. This amounts to studying a $\vec{\sigma}\cdot\vec{p}$ term in the n-target interaction. Then one polarizes the target spin $\vec{J}$ and searches for a $\vec{\sigma}\cdot\vec{p}$ x $\vec{J}$ term in the interaction. This can be done in several ways. (see STODOLSKY [22] for a discussion of this point) One possibility is to polarize $\vec{J}$ along a direction perpendicular to the neutron momentum, $\vec{p}$. Now compare the transmission without spin-flip of neutrons polarized along $\hat{n}$ and $-\hat{n}$ where $\vec{n}=\vec{J}$ x $\vec{p}$. A difference in transmissions is unambiguous evidence for T-violation, and cannot be an artifact of final-state interactions as can occur in the γ-and-β-decay correlations [19,20] mentioned above. This type of measurement (which Blayne Heckel and I propose to do at ILL) is practical with cold neutrons which can be polarized and analyzed with high efficiency. On the other hand it requires a bit of luck to find a case where a polarizable nucleus has a suitable p-wave resonance close enough to threshold to give a large P-violating effect (and hence high sensitivity to T-violation).

Another possibility is to use "white" epithermal neutrons from a spallation source and let time-of-flight select neutrons that are right on the peak of a p-wave resonance. In this case one expects to find a number of favorable cases. However polarization and analysis are no longer very efficient so one cannot afford to analyze as well as polarize the beam. As a result one must bombard a polarized target with polarized neutrons and measure the transmission summed over both spin directions of the transmitted neutrons. Now one must contend with "false signals" arising from a combination of $\vec{\sigma}\cdot\vec{p}$ and $\vec{\sigma}\cdot\vec{J}$ terms in the n-target interaction [see ref 22]. An ingenious scheme for dealing with such problems is discussed by C.D. BOWMAN [23] elsewhere in this conference.

It is interesting that searches for T-violation in neutron scattering could, in the next few years, conceivably reach a level of sensitivity comparable to or, even better than, that obtained from the neutron electric dipole moment.

References

1. E.M. Henley and G.A. Miller in Mesons in Nuclei eds., M. Rho and D.H. Wilkinson, (North-Holland, Amsterdam, 1979) p. 405.
2. T.E.O. Ericson and G.A. Miller, Phys. Lett. B 132, 32 (1983).
3. R. Abegg et al., Phys. Rev. Lett. 56, 2571 (1986).
4. G.A. Miller, A.W. Thomas, and A.G. Williams, Phys. Rev. Lett. 56, 2567 (1986).
5. E.G. Adelberger and W.C. Haxton, Ann. Rev. Nucl. Part. Sci. 35, 501 (1985).
6. B. Desplanques, J.F. Donoghue, and B.R. Holstein, Ann. Phys. 124, 449 (1980).
7. M. Simonius, reported at Lake Louise Conference on Intersections between Nuclear and Particle Physics, 1986, Lake Louise, Canada.
8. R. Wilson et al., The Investigation of Fundamental Interactions with Cold Neutrons, ed., G.L. Greene, NBS Special Publ. 711 p. 85.

9. B. Heckel, The Investigation of Fundamental Interactions with Cold Neutrons, ed., G.L. Greene, NBS Special Publ. 711 p. 90.
10. H.G. Evans et al., Phys. Rev. Lett. 55, 791 (1985).
11. M. Bini et al., Phys. Rev. Lett. 55, 795 (1985).
12. E.G. Adelberger, P. Hoodbhoy, and B.A. Brown, Phys. Rev. C 30, 456 (1984); Phys. Rev. C 33, 1840 (1986).
13. E.G. Adelberger et al., Annual Report, Univ. Washington, Nuclear Physics Lab, 1985, 1986 (unpublished).
14. J.F. Donoghue and B.R. Holstein, Phys. Rev. Lett. 46, 1603 (1981).
15. G.E. Brown and M. Rho, Comm. Nucl. Part. Phys. 15, 245 (1986).
16. J.H. Christenson et al., Phys. Rev. Lett. 13, 138 (1964).
17. see talks by N.E. Ramsay and V.M. Lobashov at this conference.
18. E. Blanke et al., Phys. Rev. Lett. 51, 355 (1983) and D. Boose, H.L. Harney and H.A. Weidenmüller, Phys. Rev. Lett. 56 , 2012 (1986).
19. see J.L. Gimlett et al., Phys. Rev. C25 1567 (1982) and references therein.
20. A.L. Hallin et al., Phys. Rev. Lett. 52 337 (1984).
21. V.P. Alfimenkov et al., Phs'ma Zh. Eksp. Teor. Fiz. 35, 42 (1982) [JETP Lett. 35, 51 (1982)].
22. L. Stodolsky, MPI München preprint MPI-PRE/PTH 84/85.
23. C.D. Bowman, talk presented at this conference.

Parity Violation in Atoms

C.-A. Piketty

Laboratoire de Physique Théorique de l'Ecole Normale Supérieure,
24 rue Lhomond, F-75231 Paris Cedex 05, France

I INTRODUCTION

1 Limits of the Talk

The purpose of this talk is to give a concise review of experiments searching for parity violation (PV) in atoms [1] (time-reversal invariance being assumed). Stress will be laid on guiding principles rather than on technical details. Due to lack of time, we are going to concentrate on those proposals which, to our knowledge, have materialized in actual experiments now completed. In practice we will leave aside the microwave experiments on hydrogen, deuterium and He^+ which are in progress [2] and only discuss experiments on heavy atoms (HA). Furthermore we will concentrate on the recent experiments [3,4] on cesium for the following reasons: i) at present they constitute the only case of uncontested agreement between measurements performed by different groups using different methods; ii) they are now approaching the precision (<10%) of the best high-energy tests of the electroweak theory; iii) experimental results can be reliably interpreted in an atom as simple as cesium. In fact, quite different approaches obtain consistent results within the theoretical uncertainty which is now lowered down to 5% [5-8].

2 Motivations to search for PV in HA

Let us briefly recall some motivations to search for PV in heavy atoms, despite already abundant experimental support in high energy physics [9] for the standard electroweak model [10]. Different physics are involved. First in high energy (HE), nucleons are broken and quarks act incoherently while in atomic physics the nucleus remains intact so that quarks act coherently. Both types of experiments appear to be complementary: the weak vector charges they measured are almost "orthogonal". Let us give the explicit form of the weak vector charge Q_W involved in atomic physics, where u and d quarks contribute coherently:

$$Q_W = -2((2Z + N)\ C_u^{(1)} + (Z + 2N)\ C_d^{(1)}) \qquad (1)$$

Since in HA $N \simeq 1.5Z$, we obtain the same linear combination of $C_u^{(1)}$ and $C_d^{(1)}$ for all HA:

$$Q_W \simeq -7Z(C_u^{(1)} + 1.14\ C_d^{(1)}) \qquad (2)$$

Secondly, different momentum ranges are involved, larger than 1 GeV/c in HE, between 1 to 10 MeV/c in HA.

In fact, it turns out that the bounds on Q_W give the only constraints [11] on some alternatives to the standard electroweak model involving an extra U(1) gauge group (see for instance [12]).

II HOW IS IT POSSIBLE TO MEASURE PV IN ATOMIC PHYSICS? MAIN STEPS

0 The starting Point

PV occurs through the interference of a pure electromagnetic amplitude A_{em} and a weak one A_w:

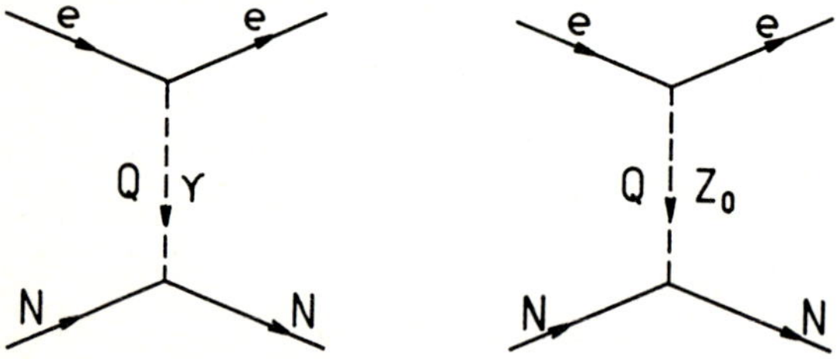

Fig. 1: Electromagnetic and weak amplitudes

In the standard model of EW interactions, the asymmetry $\mathcal{A} = 2\,A_w/A_{em}$ is simply given by:

$$\mathcal{A} = 2\ Q^2/(Q^2 + M_Z^2\ c^2) \qquad (3)$$

where Q is the momentum transfer and M_Z the Z_0 mass: $M_Zc^2 = 93$ GeV [9]. In atoms $Q \simeq m\ c\ \alpha$ (where m is the electron mass and $\alpha = 1/137$). Thus one gets $\mathcal{A} \simeq 2\alpha^2 m^2/M_Z^2 \simeq 3 \times 10^{-15}$. This looks desperately small! Fortunately in 1973 M.A. BOUCHIAT (MAB) and C. BOUCHIAT (CB) found enhancement factors [13].

1 The Z^3 Enhancement Factor of Heavy Atoms

The first one comes when working with HA, MAB and CB exhibited the so-called Z^3 enhancement factor which is in fact Z^2Q_W : Z^2 comes from the point-like structure of the interaction while the weak vector charge Q_W comes from the coherent contribution of the nucleons, if PV occurs at the leptonic vertex. Let us remark that if PV occurs at the hadronic vertex, then the nucleons act coherently through their nuclear spin I which is of order one, so there is a factor Z lost. A further suppression is expected, in the standard model, due to the leptonic weak vector coupling constant which is almost zero. Therefore in HA we will mostly be dealing with the vector nucleonic - axial electronic term of the weak interactions. Careful computations give another enhancement K_r which is a relativistic one. Typically K_r equals 2.8 for Cs, 9 for Bi. We get the resulting enhancement factor:

$$Q_W Z^2 K_r \simeq \begin{cases} 6 \cdot 10^5 & \text{for Cs} \\ 6 \cdot 10^6 & \text{for Tl} \end{cases} \tag{4}$$

2 Forbidden Transitions. First Level of Interdiction

A Allowed Magnetic Transitions

The second step is to decrease the electromagnetic amplitude, that is to work with forbidden transitions. In fact two levels of interdiction are involved.

The best physical quantity, and maybe the only one, to search for PV is to look for an off-diagonal matrix element of the electric dipole between two states of "same" parity. If weak interactions are off, then $\vec{E}_1 = \langle a|e\vec{r}|b\rangle$ vanishes. If weak interactions are on, then each state $|a\rangle$ becomes admixed with opposite parity states: $|\tilde{a}\rangle$, so that $\vec{E}_1^{PV} = \langle\tilde{a}|e\vec{r}|\tilde{b}\rangle$ is non-vanishing. $\vec{E}_1^{PV}$ is the so-called PV electric dipole amplitude. The parity conserving (PC) magnetic dipole transition M_1 is non zero:

$$M_1 = \langle a|\mu/c|b\rangle \simeq \langle\tilde{a}|\mu/c|\tilde{b}\rangle \simeq \frac{\alpha}{2}\, e\, a_0 \tag{5}$$

where a_0 is the Bohr radius; $\alpha/2$ is the first level of interdiction involved in allowed magnetic dipole transitions. From the time reversal invariance, E_1^{PV} is pure imaginary. Final estimate is the following:

$$E_1^{PV} = i \operatorname{Im} E_1^{PV} = -i\, 2.5 \times 10^{-13}\, (Z\alpha)^2\, Q_W\, K_r\, e\, a_0 \tag{6}$$

PV effects are proportional to $X_{PV} = \operatorname{Im}E_1^{PV}/M_1$, in allowed magnetic dipole transitions, typically:

$$X_{PV} \simeq 10^{-7} \quad \text{for Tl, Pb and Bi} \tag{7}$$

This corresponds to the first generation of experiments. They began around 1974 in Novosibirsk, Oxford, Seattle and in 1979 in Moscow.

B Optical Rotation in Atomic Vapour

The PV signal is the following: a linearly polarized resonant light passes through an atomic vapour. One measures the angle of rotation ϕ_{PV} of its polarization. In other words, the measured pseudo-scalar is $X_{PV} = \vec{\Omega} \cdot \hat{k}$ where $\vec{\Omega}$ is the rotation axial vector and $\hat{k}$ the unit vector of the laser beam. ϕ_{PV} is of the order of 10^{-7}, so there is no problem with statistics. ϕ_{PV} has a sharp dispersive behaviour with the light frequency around the resonant one. This behaviour is crucial to discriminate ϕ_{PV} from the background.

Background is the first severe problem of these experiments. For the first line studied in Bi (648 nm) the observed pattern was dominated by the molecular dimmer absorption [14-16]. It is less severe in other experiments on Bi (876 nm) [17] and Pb [18].

The second, even more severe problem, is with systematic effects. They may originate in purely electromagnetic effects, which are 10^7 larger than the weak one which is of interest, due to imperfect reversal in the handedness of the experimental configuration. In fact, the problem of systematics became clear in 1982 after the publication of the results of the patient and delicate work on Cs of the Paris group. It is clearly illustrated, by the first null results obtained by the three experimental groups, now denied and rejected, as well as, in the evolution of quoted uncertainties in the recent and best available numbers of each experimental group which are summarized in Table I, (in particular see the comments). Note also that systematics are the likely interpretation for the remaining discrepancy between experimental results. As for the theoretical estimates, we only kept those where many-body (MB) corrections (such as shielding) have been included either phenomenologically or in the framework of a Hartree-

Table I: Recent results [10^8 $\mathrm{Im}E_1^{PV}/M_1$] for PV Optical Rotation in Atoms

EXPERIMENT		THEORY[1]
	BISMUTH$^{209}_{83}$	Q_W = -114.3
	λ = 648nm	
-20.2 ± 2.7* Novosibirsk (1979)[14]	-12.7	parametric potential with shielding[19]
	-16.3	semi-empirical [20]
- 9.3 ± 1.5** Oxford (1984)[15]	-17	semi-empirical [21]
- 7.8 ± 1.8*** Moscow (1984)[16]	-10.1	relativistic HF method with shielding[22]
	λ = 876nm	
-10.4 ± 1.7** Seattle (1984)[17]	-10.9	parametric potential with shielding[19]
	-12.7	semi-empirical [20]
	- 7.5	relativistic HF method with shielding[22]
	LEAD$^{207}_{82}$	Q_W = -113.4
- 9.9 ± 2.5** Seattle (1984)[18]	-11.2	semi-empirical [23]
	-14.3	relativistic HF method with no shielding[24]

*Quoted error is statistical only

**Quoted error is predominantly systematic

***Quoted error includes possible systematic (0.8), calibration uncertainty (0.8) and statistical one (0.2).

1) Results of the various authors have been renormalized to Q_W with radiative corrections included in the zero momentum limit [25], $Q_W = -2(ZC_{2p}^{(1)} + NC_n^{(1)})$ with $C_p^{(1)} = -0.4882 + 1.9518 \sin^2\theta_W$, $C_n^{(1)} = -0.4883$, using $\sin^2\theta_W = 1 - M_W^2/M_Z^2 = 0.223$ [26]; $\sin^2\theta_W(M_W)$ used in [25] is $\sin^2\theta_W/1.006$.

Fock treatment. The dispersions among the theoretical predictions might be taken as indications of the uncertainties coming from atomic physics. (They are a consequence of the complicated structure of Pb (and Bi) : 4 (and 5) electrons in the outmost shell).

3 Forbidden Transitions. Second Level of Interdiction: Forbidden Magnetic Transitions

Forbidden magnetic transitions occur between single particle states which differ by their radial numbers only. In the non-relativistic limit, M_1 is strictly zero due to the orthogonality of the wave functions. Experiments are performed on the forbidden magnetic transitions: either the 6S-7S of Cs, or the $6P_{1/2}$-$7P_{1/2}$ of Tl. For both: $M_1 \simeq 10^{-5} \frac{\alpha}{2} ea_0$ (8). So the level of interdiction is $\frac{\alpha}{2} \times 10^{-5} \simeq 10^{-7}$

$$X_{PV} \text{ is } \simeq \begin{cases} 10^{-4} & \text{for Cs} \\ 10^{-3} & \text{for Tl} \end{cases} \tag{9}$$

Therefore systematics are accordingly reduced. This corresponds to the second generation of experiments for which systematics are a fundamental concern in the design and execution of the experiments. Particularly they are all performed in a static electric field $\vec{E}$.

4 External Static Electric Field $\vec{E}$

The electric dipole $\vec{d}$ gets extra pieces which are proportional to $\vec{E}$:

$$\vec{d} = -\alpha\vec{E} - i\beta\vec{\sigma}\wedge\vec{E} + M_1\vec{\sigma}\wedge\vec{k} - i \text{ Im } E_1^{PV} \vec{\sigma} \tag{10}$$

α and β are the scalar and vector polarizabilities; in Cs : $\alpha \simeq -10\beta$, in Tl : $\alpha \simeq \beta$. In all experiments the magnitude of $\vec{E}$ is such that the inequalities $|\alpha E|, |\beta E| \gg |M_1| \gg |E_1^{PV}|$ hold. So E_1^{PV} is detected through the interference with a stark induced electric dipole amplitude E_1^{st} (rather than M_1). Let us point out some of the related advantages: i) the favoured direction of the electric field labels the PV signal with a particular symmetry which provides an excellent discrimination against collisional or molecular effects, isotropic on the average: all background is eliminated; ii) the magnitude of $\vec{E}$ turns out to be a very convenient parameter to optimize the signal/noise ratio; iii) E_1^{st} E_1^{PV} interference effects are even under reversal of the direction of the beam $\vec{k}$, while M_1 E_1^{PV} effects are odd. Consequently the former can be amplified by multiple forward-backward passes of the beam, which improve the statistics and reduce systematics; iv) reversal or modulation of $\vec{E}$ is used to distinguish the PV interference term.

Three types of experiments, originally proposed by MAB and CB [29] have now been completed, we shall analyze them separately.

III THE THREE TYPES OF EXPERIMENTS IN FORBIDDEN MAGNETIC TRANSITIONS

1 In Zero External Magnetic Field: PV Electronic Polarization of the Excited State

The PV signal is an electronic polarization of the final state $\vec{P}_{PV}$ which behaves like a true vector (instead of an axial one):

$$\vec{P}_{PV} \propto \xi \, \vec{k} \wedge \vec{E} \tag{11}$$

where ξ and $\vec{k}$ are the helicity and the momentum of the resonant light. The PC polarizations are $\vec{P}_1 \propto \vec{k} \wedge \vec{E}$ which is parallel to $\vec{P}_{PV}$, and $\vec{P}_2 \propto \xi \vec{k}$ which is along the beam. The basic experimental set-up is given on Fig. 2.

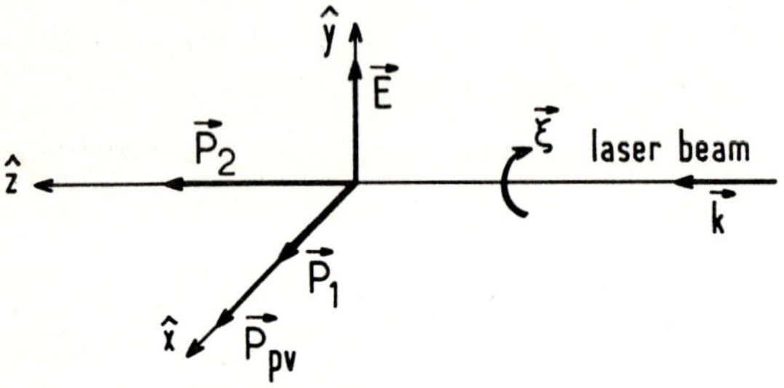

Fig. 2: Principle of measurement of PV electronic polarization in an external electric field $\vec{E}$: $\vec{P}_{PV}$; $\vec{P}_1$ and $\vec{P}_2$ are the PC electronic polarizations; $\xi = \pm 1$.

Experiments on Cs in Paris [3] and Tl in Berkeley [27] began around 1974. We will discuss some details of the first experiment performed in Paris on the $\Delta F = 0$ $6S_{1/2} - 7S_{1/2}$ transition of Cs, since one could find in it all the essential points leading to mastering the systematics; these have ever since been extensively exploited in all other experiments on atoms (including light ones). The experimental conditions are such that $|\alpha E| \gg |\beta E| \gg |M_1|$. The origin and magnitude of the PV polarization $\vec{P}_{PV}$, and of the PC ones, $\vec{P}_1$ and $\vec{P}_2$, as well as their signatures under the reversal of ξ, ξ & $\vec{k}$, $\vec{E}$ are summarized on Table II.

Table II: Properties of the final electronic polarizations in an external electric field for the Paris $\Delta F = 0$ experiment on Cs

Origin	Orders of magnitude	Signatures under reversal of		
		ξ	ξ & $\vec{k}$	$\vec{E}$
$\vec{P}_{PV}$: $\dfrac{\mathrm{Im}\, E_1^{PV}(\alpha E)}{(\alpha E)^2}$	2 to 3 x 10^{-6}	-	+	-
$\vec{P}_1$: $\dfrac{M_1(\alpha E)}{(\alpha E)^2}$	3 to 5 x 10^{-2}	+	-	-
$\vec{P}_1$: $\xi \dfrac{(\beta E)\ (\alpha E)}{(\alpha E)^2} = \xi \dfrac{\beta}{\alpha}$	0.1	-	+	+

$\vec{P}_2$, which is along the beam, is independent of $\vec{E}$ and large, so $\vec{P}_2$ is used as a calibration of the polarization. Therefore the measured quantity is $P_{PV}/P_2 \simeq \mathrm{Im}E_1^{PV}/(\beta E)$, which gives $\mathrm{Im}E_1^{PV}/\beta$.

The PC polarization $\vec{P}_1$ which is parallel to $\vec{P}_{PV}$, is 10^4 larger than $\vec{P}_{PV}$, but under the reversal of the laser beam ξ & $\vec{k} \to -\xi$ & $-\vec{k}$, $\vec{P}_1 \to -\vec{P}_1$ while $\vec{P}_{PV} \to +\vec{P}_{PV}$. The laser beam follows a zigzag pass between two spherical mirrors, $\vec{P}_1$ is then reduced by more than two orders of magnitude with respect to $\vec{P}_{PV}$, systematics are of course accordingly reduced. This is the beginning of a long and delicate work, the final result of which is to reduce each potential systematic effect below a few per cent of the PV effect so as to avoid the need of corrections. As it would easily be verified (see Table II), in the ideal situation when both geometry and apparatus are perfect, the PV signal differs from each present PC ones by at least two features (e.g. orthogonal directions, or opposite behaviour under reversal of a certain parameter). Consequently, in the real situation, systematics appear only as the product of at least two small imperfections $\delta_1\ \delta_2$, which makes them second order small. To achieve a reliable estimation of these imperfections, the atoms themselves are used as probe during the experimental run. To this purpose, they measured, continuously or periodically, atomic quantities, other than the PV signal, sensitive to these imperfections. In many cases they deliberately increased one imperfection (say δ_1) up to a large known value, making the systematic $\delta_1\ \delta_2$ measurable, so that δ_2 could be determined. In addition, numerous atomic controls were continuously available, owing to the clear labeling and modulation of the polarization which gives continuous access to the response of the atoms to all possible states of the incident polarization. In the second experiment performed on the $\Delta F = 1$ transition of Cs in Paris [3], the physical quantities involved in the PV measurement were different as well as those involved in most systematics. Consequently the two experiments can provide a genuine cross-check. Their final results (after net systematic corrections < 2%) are:

$$[\mathrm{Im}\ E_1^{PV}/\beta]_{\Delta F=0} = -1.33 \pm 0.22 \pm 0.11 \text{ mv/cm} \tag{12}$$

$$[\mathrm{Im}\ E_1^{PV}/\beta]_{\Delta F=1} = -1.75 \pm 0.26 \pm 0.06 \text{ mv/cm} \tag{13}$$

where the quoted uncertainties come from statistics (the largest ones) and systematics, respectively. Within experimental uncertainties the two results are in fair agreement, so they can be combined if the theoretical expected values are the same for the two different hyperfine transitions; theoretical estimates indicate that this is true at the level of 10^{-3} [28]. Following the philosophy adopted in related experiments[2)] one gets:

---

2) Each result is affected with a net uncertainty (quadratic sum of the statistical and systematic uncertainties) and weighted by the square reciprocal of that uncertainty.

$$\mathrm{Im}\ E_1^{PV}/\beta = -1.52 \pm 0.18\ \mathrm{mv/cm} \quad (14)$$

The detailed discussion is postponed until we give the second type of experiments on Cs in a non zero magnetic field. A similar experiment has been performed by the Berkeley group on the ΔF=1 $6P_{1/2}$ - $7P_{1/2}$ forbidden transition of Tl (Exp.I) [27] βE and M_1 both contribute, the measured quantity was E_1^{PV}; after a net systematic correction of 50%, the result was:

$$\mathrm{Im}E_1^{PV}/M_1 = 1.40 \pm 0.35_{\mathrm{stat.}} \ {}^{+0.15}_{-0.10}{}_{\mathrm{syst.}} \quad (15)$$

2 In Crossed Electric $\vec{E}$ and Magnetic $\vec{H}$ Static Fields

The interference is observed directly in the transition rate by applying a magnetic field to break the degeneracy of the Zeeman levels. Two experiments have been performed using different E_1^{PV} E_1^{st} interference effects, the first one at Berkeley on the $6P_{1/2}$ - $7P_{1/2}$ Tl transition [30], the second one at Boulder on the 6S - 7S Cs transition [4]. We will first discuss the cesium experiment.

B1 Circular Dichroism in Crossed $\vec{E}$ and $\vec{H}$ Fields

The circular dichroism is associated with the presence of the pseudoscalar:

$$X_{PV} = \xi\ \vec{k} \cdot \vec{E} \wedge \vec{h} \quad (16)$$

in the resonant absorption cross-section. The basic experimental set-up, which maximizes X_{PV}, is described in Fig. 3a

The experiment was performed at Boulder on pure Δm = +1 and Δm = -1 components of the ($6S_{F=4}$ - $7S_{F'=3}$) and ($6S_{F=3}$ - $7S_{F'=4}$) hyperfine lines of Cs. The unique feature of the experiment is the use of crossed laser and atomic beams (the latter being along $\vec{H}$). This yields narrow transition line widths which allow the use of a small magnetic field. Other desirable features of an atomic beam exper-

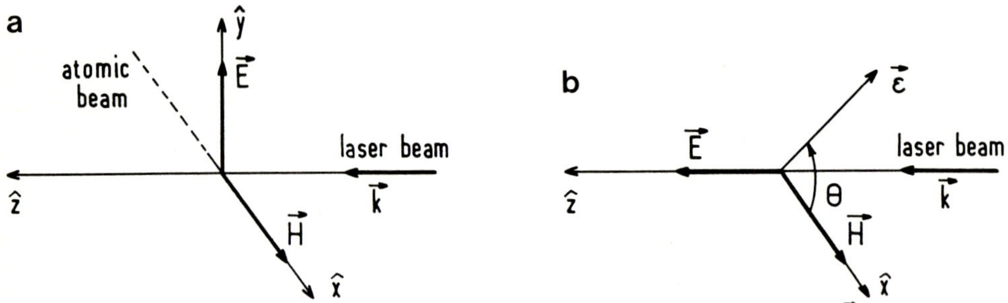

Fig. 3: a) b) Basic experimental set up, in crossed static electric ($\vec{E}$) and magnetic ($\vec{H}$) fields, to search a) for circular dichroism, b) for handed absorption of plane polarized light ($\vec{\epsilon}$).

iment include the reduction of collision, radiation trapping, and molecular background. This is important in view of the physical quantity monitored in this experiment. The PV signal is amplified by a multipass of the beam. It is isolated by observing the modulation in the transition rate with reversal of $\vec{E}$, $\vec{H}$ and ξ. An additional Δm reversal is achieved by a change of the laser frequency from the pure $\Delta m = +1$ transition to the $\Delta m = -1$ of a given multiplet. The measured quantity is $\mathrm{Im}E_1^{PV}/\beta$. Identification and measurement of systematic effects are similar to that used in the Paris experiment [3]. Some systematics turn out to be quite large, up to 50% of the PV signal but are expected to be well controlled and measured. Their average over all the data-runs amounts to a net systematic correction of 14(1)% in contrast to the ones of the two Paris experiments which are less than 2%. Results of both experiments are in fair agreement:

$$\mathrm{Im}E_1^{PV}/\beta = -1.52 \pm 0.18 \text{ mv/cm} \qquad \text{Paris exp. I + II} \qquad (14)$$

$$\mathrm{Im}E_1^{PV}/\beta = -1.63 \pm 0.13 \text{ mv/cm} \qquad \text{Boulder exp.} \qquad (17)$$

In both experiments the statistical error dominates the systematic one. Both groups have determined β semi-empiricallly [31,32], within errors (<3%) their

Table III: Forbidden $6S_{1/2}$ - $7S_{1/2}$ Transition in $\mathrm{Cesium}_{55}^{133}$. Parity-Violating Amplitude in Units of $[10^{-11}\ |e|\ a_0]$: Experiment and Theory.
The quantity $\mathrm{Im}E_1^{PV}/(Q_W/N)$ is purely atomic. The weak charge Q_W is computed in the framework of the standard electroweak model with radiative corrections included (see footnote 1 of Table I).

EXPERIMENT		THEORY	
Experiment $\beta = 26.8\ a_0^3$[31]	with standard model $Q_W = -70.35$	Atomic theory	with standard model $Q_W = -70.35$
$\mathrm{Im}\ E_1^{PV}$	$\mathrm{Im}\ E_1^{PV}/(Q_W/N)$	$\mathrm{Im}\ E_1^{PV}/(Q_W/N)$	$\mathrm{Im}\ E_1^{PV}$
-0.79 ± 0.10 [3]	0.88 ± 0.11	0.97 ± 0.10* [5]	-0.87
-0.86 ± 0.07 [4]	0.95 ± 0.08	0.88 ± 0.03**	-0.79
combined***:		0.888 ± a few %**** [7]	-0.797
-0.84 ± 0.06	0.93 ± 0.07		
		0.93 ± 0.02 ± 0.03 [8] exp. theor.	-0.84
		combined: 0.91 ± 0.04	

* Relativistic corrections included and shielding effect in a semi-empirical way; the dominant neglected non local MB corrections are of the order of -5%.
** Perturbative MB computation with a relativistic HF treatment of the ion - 1st order corrections included as well as dominant higher ones involving correlations.
*** See footnote 1.
**** See [6]. It contains a complete summation of single particle excitations but no correlation effects.

values agree with each other and with the theoretical estimates [5,6]; $\beta = 26.8 \pm 0.8\ a_0^3$ will be used [31]. Results for $\mathrm{Im}E_1^{PV}$ are summarized in Table III.

Among all available predictions of E_1^{PV} in Cs, we only retain the recently published evaluations including many-body (MB) corrections, either semi-empirically [5,8] or from first principles [6,7]. We would like to make a brief comment on a new semi-empirical method [8] which uses the information on the atomic wave-functions near the nucleus contained in the empirical hyperfine splittings (hfs). It deals with the PV matrix elements rescaled by the geometrical mean of S and P states hfs constants which is very slightly affected by many-body corrections (<5%). Results are summarized in Table III.

Before going to the theoretical implications of the Cs results, we would like to present the second experiment performed on Tl in a non zero magnetic field.

B2 Chiral Absorption of Linear Polarized Resonant Light in Crossed Electric and Magnetic Fields

The measured pseudoscalar X_{PV} is:

$$X_{PV} = (\vec{\epsilon} \cdot \vec{H})\ (\vec{\epsilon} \cdot \vec{H} \wedge \vec{E}) \tag{18}$$

where $\vec{\epsilon}$ is the linear polarization of the incident laser beam. One can get rid of the PC scalar X_{PC}:

$$X_{PC} = (\vec{\epsilon} \cdot \vec{H})\ (\vec{\epsilon} \cdot \vec{H} \wedge \vec{E}) \tag{19}$$

by choosing the laser beam momentum $\vec{k}$ along $\vec{E}$. So we get the basic experimental set-up illustrated in Fig. 3b. The signatures of X_{PV} are a characteristic laser frequency dependence and well defined symmetries under the reversal of Θ, $\vec{E}$, $\vec{H}$ and $\vec{k}$. The multipass of the beam could have been used to reduce X_{PC}/X_{PV} but it has not been in the second experiment performed in Berkeley [33]. Absorptions corresponding to well separated $\Delta m = 0$ and $\Delta m = 1$ transitions between Zeeman levels of the $6P_{1/2} - 7P_{1/2}$ have been studied. The measured quantity is $\mathrm{Im}E_1^{PV}/\beta$:

$$[\mathrm{Im}E_1^{PV}/\beta]_{\mathrm{exp\ II}} = -1.73 \pm 0.26 \pm 0.07\ \mathrm{mv/cm} \tag{20}$$

Using β/M_1, $\mathrm{Im}E_1^{PV}/\beta$ can be derived from the first experiment (see (15)) one gets:

$$[\mathrm{Im}E_1^{PV}/\beta]_{\mathrm{exp\ I}} = -1.80 \pm 0.45 \pm {}^{0.20}_{0.15}\ \mathrm{mv/cm} \tag{21}$$

Both results agree within errors. β has been measured recently [34]. The discrepancy of 60% with theoretical estimates [36,37] clearly shows the importance of MB corrections in Tl. New theoretical computations are now in progress [38].

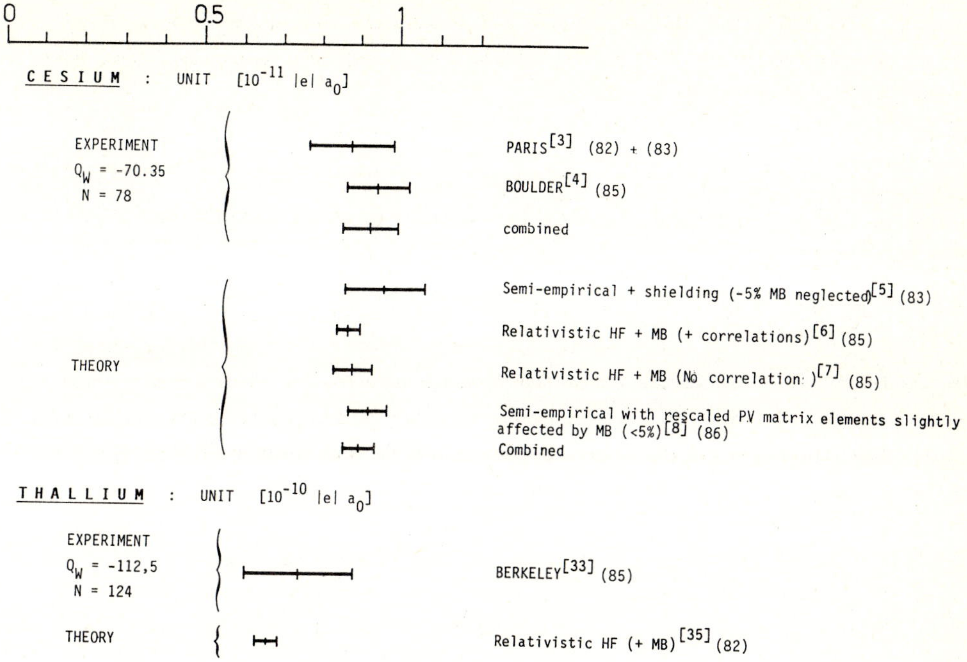

Fig. 4: Summary for Im $E_1^{PV}/(Q_W/N)$: Experiment and Theory, for Cs (unit $10^{-11}|e|a_0$), for Tl (unit $10^{-10}|e|a_0$)

Results for $\mathrm{Im}E_1^{PV}$ and for the pure atomic quantity $\mathrm{Im}E_1^{PV}/(Q_W/N)$ are summarized in Table IV. We only keep the theoretical estimate with many-body corrections included. Experimental and best theoretical results for $\mathrm{Im}E_1^{PV}(Q_W/N)$ for Cs are summarized in Fig. 4 and compared with Tl results.

IV Theoretical implications of the Cs results

As we have already pointed out at the beginning, these results on PV in atoms constitute the only case of uncontested agreement between measurements performed by

Table IV Forbidden $6P_{1/2}$ - $7P_{1/2}$ transition in Tl^{205}_{81}

Parity violating amplitude in units of $[10^{-10}\ |e|\ a_0]$, experiments and theory; see Table III for detailed explanations.

EXPERIMENT		THEORY	
Experiment $\beta = 198\ a_0^3$[34]	with standard model $Q_W = -112.5$	Atomic theory	
Im E_1^{PV}	Im $E_1^{PV}/(Q_W/N)$	Im $E_1^{PV}/(Q_W/N)$	
-0.69 ± 0.13 [33]	0.76 ± 0.14	0.68 ± 0.03[35]	Relativistic HF method MB corrections included

different groups using different methods and with theoretical estimates using different approaches. They both can be combined, and one deduces an empirical determination of the weak charge of the Cs

$$Q_W(Cs) = -72 \pm 6 \tag{22}$$

Assuming the standard model, (radiative corrections are included, see footnote 1), one deduces:

$$\sin^2\theta_W(Cs) = 0.23 \pm 0.03 \tag{23}$$

in agreement with the values extracted from HE experiments of different types [26] (see Fig. 5). The interest of this agreement lies in the considerable difference in momentum transfer between HE experiments (> 1GeV/c) and atomic physics (typically 2.4 MeV/c in Cs). This illustrates one aspect of the complementarity of the two types of experiments.

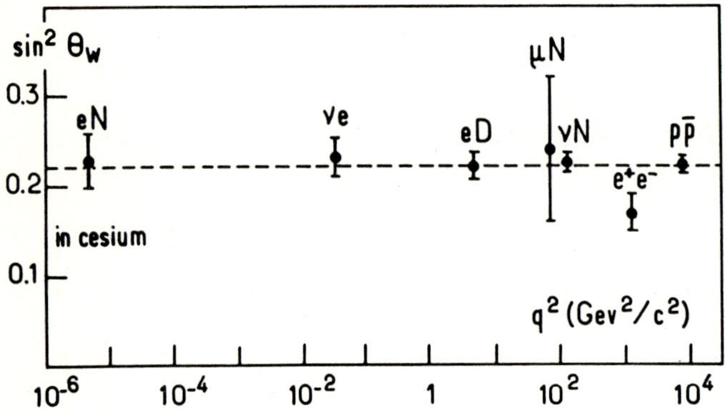

Fig. 5: Values of $\sin^2\theta_W$ from Cs results and from totally different types of experiments at high energy in the scattering of various partners. The dashed line represents the average of all data: $\sin^2\theta_W$ 0.223 ± 0.004, (including radiative corrections).

Another aspect of this complementarity is illustrated by the model-independent interpretation of $Q_W(Cs)$ coming from coherent contribution of quarks in HA (see equation 1). The linear combination of $C_u^{(1)}$ and $C_d^{(1)}$ is almost orthogonal to the one measured in the SLAC-YALE experiment in which quarks contribute incoherently. This is illustrated on Fig. 6. The results of both experiments allow both $C_u^{(1)}$ and $C_d^{(1)}$ to be determined while neither experiment can achieve this alone. One gets:

$$C_u^{(1)} = -0.25 \pm 0.08 \qquad ; C_d^{(1)} = 0.40 \pm 0.08 \tag{24}$$

(where uncertainties are correlated).

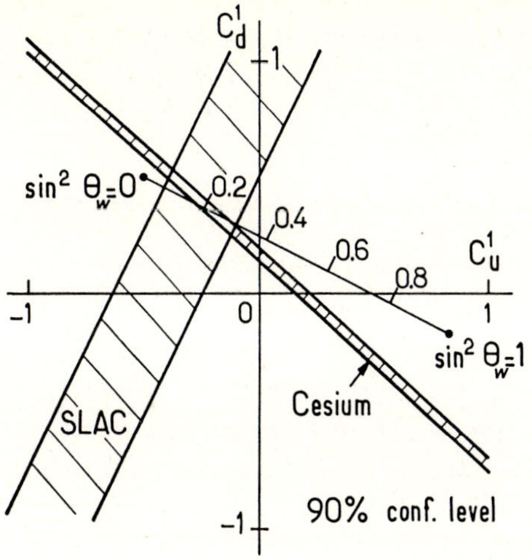

Fig. 6: Regions of $C_u^{(1)}$ and $C_d^{(1)}$ allowed by Cs analyses (see (24)): $C_u^{(1)} + 1.122\ C_d^{(1)} = 0.191 \pm 0.016$ and the SLAC results: $C_u^{(1)} - 0.5\ C_d^{(1)} = -0.45 \pm 0.12$ when error bars are multiplied by 1.64 to obtain 90% confidence level. The graduated segment is the prediction of the standard model.

The agreement of the above model-independent interpretation with the standard model prediction is note worthy. The corresponding value of $\sin^2\theta_W \gtrsim 0.2$ is precisely the value obtained from various HE experiments.

Finally, up to now, we have neglected the contribution (expected to be 1/Z smaller at least) where PV occurs at the hadronic vertex. The parameter involved in the sum is the weak axial moment:

$$A_W(Z,N)\ \frac{\vec{I}}{I} = 2\Big(\sum_p C_p^{(2)}\vec{\sigma}_p + \sum_n C_n^{(2)}\ \vec{\sigma}_n\Big) \tag{25}$$

where $\hbar\vec{\sigma}_{p,n}/2$ represent the spins of protons and neutrons. The PV electric dipole amplitude is found to be of the form:

$$\langle 7S,\ m_s'|\vec{d}^{PV}|6S,\ m_s\rangle = -i\ \mathrm{Im}\ E_1^{PV}\ \langle m_s'|\vec{\sigma} + \eta\ \vec{I}/I + i\ \eta'\ \vec{\sigma}\wedge\vec{I}/I|m_s\rangle \tag{26}$$

where η and η' are real coefficients. The ratio η'/η is a purely atomic quantity which has been computed for the 6S - 7S Cs transition: $\eta'/\eta \simeq 2$ [29]. The parameter η can be expressed in terms of a nuclear matrix element of the Gamow-Teller type:

$$\vec{m}_A = \Big\langle\sum_p C_p^{(2)}\ \vec{\sigma}_p + \sum_n C_n^{(2)}\ \vec{\sigma}_n\Big\rangle$$

$$\eta\ \frac{\vec{I}}{I} = \frac{2(2\sqrt{1 - Z^2\alpha^2} + 1)}{3(-Q_W)}\ \vec{m}_A \tag{27}$$

If one assumes that the even number of neutrons couple with a zero total angular momentum (as it is suggested by shell model of nuclei), $\vec{m}_A$ can be computed in terms of the magnetic moments of the Cs: μ_{Cs} and of the proton μ_p:

$$\vec{m}_A = \frac{\mu_{Cs} - I}{\mu_p - 1/2} C_p^{(2)} \frac{\vec{I}}{I} \tag{28}$$

Within the standard model, including radiative corrections (see footnote 1), $C_p^{(2)}$ is quite sensitive to $\sin^2\Theta_W$ value), one gets:

$$C_p^{(2)} = - 0.065 \tag{29}$$

Using either the experimental value of μ_{Cs} : $\mu_{Cs} = 2.579$ (or the shell-model estimation $\mu_{Cs} = 1.72$) with $\mu_p = 2.79$, one gets $\eta = +0.7 \times 10^{-3}$ (or $\eta = +1.4 \times 10^{-3}$).

But (26) represents the most general description of $\vec{d}^{PV}$, provided time reversal invariance holds. For this reason, it is worthwhile to consider also a purely phenomenological discussion of the results starting from (26) without any assumption concerning the value of η. The two Paris experiments on the F=4 → F'=4 and F=3 → F'=4 transitions in zero magnetic field are not very sensitive to η. The recent Boulder measurements which involve some components of the F=3 → F'=4 and F=4 → F'=3 transitions are more sensitive to the nuclear spin-dependent interaction through η'. Their present results support $\eta \ll 1$. Within the standard model, assuming the value of the shell-model for μ_{Cs}, they derive $C_p^{(2)} = -2 \pm 2$; if the experimental value is used instead, the bounds become: $C_p^{(2)} = -4 \pm 4$. We may note that comparison of the two $\Delta F = 0$ hf components would provide as sensitive a test with the additional advantage of being independent of η' and thereby independent of the atomic calculations.

V <u>Prospects</u>

In Cs experiments, statistics are much larger than systematics. Besides the obvious improvement of statistics, significantly higher precision is expected to be achieved with future refinements of the experiments. The Boulder group will use a spin-polarized atomic beam. As for the Paris group, the most promising gain expected is in detection efficiency. The electronic polarization is detected through the analysis of the polarization of the light emitted in the $7S_{1/2}$ -$6P_{1/2}$ transition. In a new project the Paris group intends to monitor the stimulated emission in the 7S - 6P transition induced by a probe beam, rather than the spontaneous fluorescence light. Because all photons are emitted in the direction and at the frequency of the probe beam, high efficiency is expected; furthermore new atomic quantities will become measurable [40]. A different experimental approach is also investigated by the Zurich group [41].

Using the new semi-empirical approach in which the physical quantity studied turns out to be very slightly affected by MB corrections, we expect to lower the uncertainty down to the 1% level, provided improved empirical oscillator strengths become available. We would like to mention also a very extensive computation program concerning weak interactions in HA which has been undertaken by

W.R. JOHNSON, in which various single particle potentials are used as starting point. Only the first part has been published in which MB have not been included [37].

If precision of the Cs analyses reaches the per cent level, new domains of physics will become accessible: test of radiative corrections in the low momentum regime, information on the weak neutral interaction involving nucleon spins (measurement of $C_p^{(2)}$). Of course, such measurements could by no means replace experiments in hydrogen.

Acknowledgements

We thank M.-A. Bouchiat and C. Bouchiat for fruitful and enjoyable discussions.

References

1 Review articles on PV in Atomic Physics:
a. M.A. Bouchiat: New Trends in Atomic Physics, eds. G. Grynberg and R. Stora (1982) pp. 887;
b. L. Pottier: Proceedings of the International School of Physics of Exotic Atoms (Erice 1984), Plenum Press.
c. M.A. Bouchiat and L. Pottier: Atomic Physics 9, eds. R.S. Van Dyck Jr. and E.N. Fortson (World Scientific, Singapore 1984).
d. E.N. Fortson and L.L. Lewis: Physics Reports 113, 289 (1984).
Theory : C. Bouchiat: Atomic Physics 7, eds. D. Kleppner and F.M. Pipkins (Plenum, New York 1981) pp. 83.

2 L.P. Levy and W.L. Williams: Phys. Rev. Lett. 48, 607 (1982), and L. P. Levy: Ph.D. Thesis, Univ. of Michigan (1982);
E.G. Adelberger et al.: Nucl. Instr. and Meth. 179, 181 (1981); See also Ref. 1 c and d;
E.A. Hinds and V.W. Hughes: Phys. Lett. 67B, 487 (1977); E.A. Hinds, Phys. Rev. Lett. 44, 374 (1980);
R.W. Dunford: Phys. Lett. 99B, 58 (1981).

3 M.A. Bouchiat, J. Guéna, L. Hunter and L. Pottier: Phys. Lett. 117B, 358 (1982); 121B, 456 (1983) and 134B, 463 (1984); J. Physique 46, 1897 (1985), J. Physique (to be published, July and October 1986);
J. Guéna, Thesis for the Doctorat d'Etat, Paris 1985, unpublished.

4 S.L. Gilbert, M.C. Noecker, R.N. Watts and C.E. Wieman: Phys. Rev. Lett. 55, 2680 (1985); see also Boulder preprint (1986) to appear in Phys. Rev. A.

5 C. Bouchiat, C.A. Piketty, D. Pignon: Nucl. Phys. B221, 68 (1983).

6 V.A. Dzuba, V.V. Flambaum, P.G. Silvestrov and O.P. Sushkov: J. Phys. B18, 597 (1985).

7 A.M. Martensson-Pendrill: J. Physique 46, 1949 (1985).

8 C. Bouchiat and C.A. Piketty (1986), to be published in Euro-Phys. Lett.

9 C.Y. Prescott et al.: Phys. Lett. 77B, 347 (1978) and 84B, 524 (1979);
For a discussion of the information extracted from this experiment, see also J.E. Kim et al.: Rev. Mod. Phys. 53, 211 (1981);
UA1 Coll.: Phys. Lett. 122B, 103 (1983) and 126B, 398 (1983);
UA2 Coll.: Phys. Lett. 122B, 476 (1983) and 129B, 130 (1983).
For a recent review article, see [26].

10 S. Weinberg: Phys. Rev. Lett. 19, 1264 (1967); A. Salam: Proc. 8th Nobel Symposium (Almquist and Wilsell, Stockholm 1968) pp. 169; S.L. Glashow: Nucl. Phys. 22, 579 (1961); S.L. Glashow et al.: Phys. Rev. D2, 1285 (1970).

11 R.W. Robinett and J.Rosner: Phys. Rev. D25, 3036 (1982); C. Bouchiat and C.A. Piketty: Phys. Lett. 128B, 73 (1983).

12 P. Fayet: Phys. Lett. 84B, 416 (1979) and 96B, 83 (1980).

13 M.A. Bouchiat and C. Bouchiat: Phys. Lett. 48B, 111 (1974); J. Physique 35, 899 (1974) and 36, 493 (1975).

14 L.M. Barkov and M.S. Zolotorev: Phys. Lett. 85B, 308 (1979) and Comments At. Mol. Phys. 8, 79 (1979).

15 J. Taylor: D. Phil. Thesis, Oxford 1984;
P.G.H. Sandars and D. Stacey: private communication.

16 C.N. Birich et al.: Zh. Ehsp. Tcor. Fiz. 87, 776 (1984).

17 J.H. Hollister et al.: Phys. Rev. Lett. 46, 643 (1981).

18 T.P. Emmons, J.M. Reeves and E.N. Fortson: Phys. Rev. Lett. 51, 2089 (1983).

19 P.G.H. Sandars: Phys. Scr. 284, 21 (1980).

20 L. Novikov, O. Sushkov, I. Khriplovich: Sov. Phys. JETP, 44, 872 (1976).

21 Results quoted in [14].

22 A.M. Martensson, E. Henley, L. Wilets: Phys. Rev. A24, 308 (1981).

23 S.L. Carter and H.P. Kelly: Phys. Rev. Lett. 42, 966 (1979).

24 P.G.H. Sandars, quoted in [18].

25 W.J. Marciano and A. Sirlin: Phys. Rev. D27, 552 (1983) and D29, 75 (1984); B.W. Lynn: in "1983 Trieste Conf. on Radiative Corrections in SU(2) x U(1)", Eds. B.W. Lynn and J.F. Wheater (World Scientific, Singapore 1984) p. 311.

26 P. Langacker: Proceedings of 1985 Int. Sympos. on lept. photon int. at HE, Kyoto August 1985, eds. M. Konuma and K. Takahashi (Nissha Printing Co. Ltd., Kyoto).

27 P. Bucksbaum, E. Commins and L. Hunter: Phys. Rev. Lett. 46, 640 (1981) and Phys. Rev. D24, 1134 (1981).

28 M.A. Bouchiat, M. Poirier and C. Bouchiat: J. Physique 40, 1127 (1979) and 43, 729 (1982); M. Poirier: Thesis (1979) unpublished.

29 C. Bouchiat and C.A. Piketty: in preparation.

30 P.S. Drell and E.O. Commins: Phys. Rev. Lett. 53, 968 (1984).

31 M.A. Bouchiat, J. Guéna, L. Hunter and L. Pottier: Opt. Comm. 45, 35 (1983).

32 Results quoted in [4].

33 P.S. Drell and E.O. Commins: Phys. Rev. A32, 2196 (1985).

34 C.E. Tanner and E.D. Commins: Phys. Rev. Lett. 56, 332 (1986).

35 B.P. Das et al.: Phys. Rev. Lett. 49, 32 (1982).

36 D. Neuffer and E.D. Commins: Phys. Rev. A16, 844 (1977).

37 W.R. Johnson, D.S. Guo, M. Idrees and J. Sapirstein: Phys. A32, 2093 (1985).

38 W.R. Johnson and J. Sapirstein, A.M. Martensson-Pendrill, B.P. Das: private communication, quoted in [34].

39 M.G. Mayer and J.H.D. Jensen: Elementary Theory of Nuclear Shell Structure, eds. John Wiley and Sons, New York 1957, pp. 79.

40 M.A. Bouchiat, Ph. Jacquier, M. Lintz and L. Pottier: Optics Communications 56, 100 (1985).

41 P.P. Hermann, J. Hoffnafle, N. Schlumpf, V.L. Telegdi and A. Weiss: to be published in J. Physique B.

Charge Symmetry and Charge Independence*

K.K. Seth

Northwestern University, Evanston, IL 60201, USA

1 INTRODUCTION

Symmetry principles occupy a hallowed place in physics, indeed in all human culture. At the level at which we believe in a friendly and benevolent Nature, we believe in symmetries. Symmetries appeal to our sense of logic, order, and beauty. Nuclear physicists are very fond of pointing out that isospin (T) invariance was the first 'internal' symmetry (as distinguished from space-time symmetries) to be postulated[1], initially for nucleons, and soon after for mesons as well. Isospin invariance leads to the principle of *charge independence* (CI) which may be stated as:

Hadronic forces are invariant under rotations in isospin space, or

$$[H_{hadr}, \vec{T}] = 0. \tag{1}$$

Obviously, we can formulate a weaker principle, i.e., *charge symmetry* (CS), which is contained in charge independence, but states:

Hadronic forces are invariant under rotations by 180° in isospin space, or,

$$[H_{hadr}, e^{i\pi T_2}] = 0. \tag{2}$$

This is equivalent to the statement that a system with hadronic forces only is invariant if mesons and baryons in it $(\pi^-, \pi^+, \cdots, n, p, \cdots)$ are all replaced by their charge symmetric counterparts $(\pi^+, \pi^-, \cdots, p, n, \cdots)$ in the same space-spin states. It may be noted[2] that since apparent charge symmetry effects (e.g., equality of π^+ and π^- masses) may arise as a consequence of another unrelated invariance principle (TCP, in this case), it is clearly not enough to just test CS breaking (CSB). Charge independence breaking (CIB) must be also studied.

Recently there has been a renewed interest in isospin invariance. The reason for this is the development of QCD (quantum chromodynamics) as the fundamental candidate- theory of strong interactions. Before the advent of QCD it was generally believed that all observed departures from CS and CI were directly or indirectly electromagnetic in origin, even though many of them (e.g., n-p mass difference) were never successfully explained as such. It was almost an article of faith that hadronic interactions are charge independent. QCD tells us that the up- and the down- quarks have intrinsically different masses (at the level of current algebra masses, $m_d/m_u = 1.79$)[3]. Thus for the first time we have isospin breaking at a purely hadronic level (i.e., if quarks themselves do not have any structure). We are told that because the constituent (or dressed) quarks acquire about 300 MeV additional mass, the effective masses m_d and m_u become remarkably equal, with $m_d/m_u \approx 1.01$. According to QCD, it is to this 'accident' or divine providence that we owe isospin invariance as we know it in hadronic interactions. Thus a new dimension has been added to our interest in studying CSB and CIB. Now we must remove electromagnetic effects from our observables to examine if what is left is consistent with the known differences in up and down quark masses. Indeed we may aspire to 'determine' quark masses from our observables.

* *This work was supported in part by the U. S. Department of Energy.*

This talk is primarily intended to provide a review of the latest experiments relating to CSB and CIB. We believe that because of nuclear structure uncertainties it is not possible to draw conclusions about isospin invariance at the level of two-body interactions from experiments involving nuclei heavier than ^{4}He. For this reason I will not discuss masses of nuclear isospin multiplets (A$\geq$4), nuclear Coulomb energies, isospin mixing in nuclear levels, or isospin forbidden β-decays. For detailed discussions of these experiments reference is made to the earlier reviews of the subject[2,4-8]. Similarly, for details of the meson-theoretic origin of charge dependent nuclear forces we refer to the excellent reviews of HENLEY and his colleagues[2,4,5].

Let me first make two general observations:

a) A perusal of the literature shows that most experiments done to date only address the question of charge symmetry breaking. Few experiments examine the wider symmetry implied by charge independence. It is clear that this situation needs to be improved.

b) Tracking the energy and momentum transfer dependence of an observable is a time-honored way of disentangling interfering contributions to it. However, in the past most experiments relating to CI and CS were done at low energies; little or no data existed at higher energies ($\geq$ 100 MeV). This situation needs to be improved, and is indeed improving as more and more experiments are being done at intermediate energy facilities such as TRIUMF, IUCF, LAMPF, Saturne and others.

c)The scale of background effects, in presence of which CSB and CIB effects must be detected, is set by the electromagnetic interaction which breaks both CS and CI, and is essentially omnipresent. This scale is that of the fine-structure constant, α, i.e., $\approx$ 1%. All existing evidence points out that hadronic CSB and CIB effects are of the same order, i.e., $\leq \approx$ 1%. Therefore one has to design experiments in which direct electromagnetic effects can be kept below the level of a few percent and the errors in the measurements can be kept at sub- 1% level.

2 NUCLEON–NUCLEON INTERACTION

Before we review the experimental data it is useful to restate the isospin classification of the nucleon-nucleon interaction which was first introduced by HENLEY[2]. The nucleon(i)-nucleon(j) interaction can be classified as follows:

Class I : Isoscalar, CI (and CS) conserving:

$$V_1(ij) = A[1 + \vec{\tau}(i) \cdot \vec{\tau}(j)]. \tag{3}$$

This is the dominant part of the nuclear interaction.

Class II : Isotensor, CS conserving, CI breaking:

$$V_2(ij) = B[\tau_3(i)\tau_3(j) - \frac{1}{3}\vec{\tau}(i) \cdot \vec{\tau}(j)]. \tag{4}$$

This interaction makes the np interaction different from the nn and pp interactions.

Class III : Isovector, CI and CS breaking, symmetric under particle exchange:

$$V_3(ij) = C[\tau_3(i) + \tau_3(j)]. \tag{5}$$

This interaction is zero for the np system.

Class IV : Isovector, CI and CS breaking, antisymmetric under particle exchange:

$$V_4(ij) = D[\tau_3(i) - \tau_3(j)][\vec{\sigma}(i) - \vec{\sigma}(j)] \cdot \vec{L}_{ij} + E[\vec{\tau}(i) \times \vec{\tau}(j)]_3[\vec{\sigma}(i) \times \vec{\sigma}(j)] \cdot \vec{L}_{ij}. \tag{6}$$

This interaction is zero for the nn and pp systems. Its space-spin part has the form of a spin-orbit interaction and its manifestations should be most pronounced in polarization observables.

A similar isospin classification of pion-nucleon interaction is rendered quite difficult because of the presence of numerous pion-nucleon resonances. One makes the implicit assumption that π-nucleon coupling constants are isospin invariant.

3 EXPERIMENTAL EVIDENCE FOR CHARGE SYMMETRY

3.1 *Nucleon–Nucleon Scattering*

Nucleon-nucleon scattering experiments at low energies have been the forerunners of all CI and CS studies. This is because for s-wave scattering the phase shift $\delta_0[\sigma = (4\pi/k^2)\sin^2\delta_0]$ is simply given by the effective range formula

$$k\cot\delta_0 = (-1/a) + (1/2)rk^2. \tag{7}$$

The scattering length a is a very sensitive measure of the interaction because the NN system is very close to binding in its T=1 state. A 1% change in potential produces about 18% change in a.

The value of the pp scattering length obtained directly from the data, which has remained unchanged since 1966, is $a_{pp} = -7.828 \pm 0.008$ fm. To this a large electromagnetic correction (≈ -9.4 fm) has to be applied. Time was when this correction was assigned only an error of 0.2 fm[2]. However, ever since SAUER and WALLISER[9] showed that this correction depends sensitively on the unknown short-range part of the nuclear force, a great amount of uncertainty has been introduced into the problem. Even if one tempers the highly pessimistic conclusions of Sauer and Walliser with criteria of reasonableness[10] and keeping relativistic effects small[11] one is forced to assign[5] at least an error of ± 2 fm to the electromagnetically corrected a_{pp}. Thus our best result is $a_{pp} = -17.2 \pm 2.0$ fm.

The story of a_{nn} is quite the opposite. Here one has no problems of electromagnetic corrections, but definite problems with experimental measurements. So far all measurements depend on an analysis of the final state interaction between two neutrons produced (along with other particles) in some reaction. Clearly the best such reaction is that in which no strongly interacting particles other than the two neutrons are produced. The $\pi^- d \to nn\gamma$ reaction is therefore everybody's favourite. The latest result, $a_{nn} = -18.6 \pm 0.5$ fm, is due to GIBIOUD et al.[12] in which the shape of the γ-ray spectrum near its maximum energy was analyzed. This result barely overlaps with $a_{nn} = -16.7 \pm 1.3$ fm obtained in an earlier, but more complete experiment, in which the γ's as well as the neutrons were detected[13]. There also remain some questions about the formalisms used for analysis of the two data sets. Thus at least this auther remains rather skeptical about the ± 0.5 fm error quoted by GIBIOUD et al.[12]

What is needed is a true neutron-neutron scattering experiment. This long standing dream appears to be closer to realization today than ever before. A Los Alamos- Oak Ridge collaboration[14] is attempting to directly measure the nn scattering length by means of two colliding neutron beams produced in two synchronized nuclear explosions. A feasibity test was actually performed on April 8, 1986. In this experiment neutrons from the simultaneous (within 1 - 2 nanoseconds) underground explosion of two fission-fusion devices are collimated by $\approx$565 tons of Cu-Fe-Pb shielding into two channels at 3.8° angle. Even though the kinetic energy of each neutron beam peaks at $\approx$14 MeV, their relative energy is $E_{cm} \approx$ 38–20 eV. About 10^{15} interactions take place in a collision volume of $\approx$12.4 cm^3. The scattered neutrons are essentially confined on the surface of a 'Jacobian cone'. They are detected in scintillators located on the surface of the cone. Background is similarly measured just beyond the surface of the cone. Integral charge from the phototubes, rather than counts are measured. In principle, a simultaneous n-p scattering experiment can also be done with CH_2 and C targets intercepting the outgoing neutron 'beams'. The entire arrangement is schematically shown in Fig. 1. It is hoped that nn scattering cross sections can be measured with errors of the order of $\pm 10\%$. This should lead to the determination of a_{nn} with errors of ± 0.5 fm.

As mentioned earlier, the feasibility experiment was indeed done on April 8. However, no results are so far available. Due to a radiation leak into the area of experimental equipment, the data can not be retrieved yet—it is literally 'too hot to handle'!

From the existing results we find that

$$|a_{nn}| - |a_{pp}| = [18.6 \pm 0.5(?)] - [17.2 \pm 2.0(?)] \text{ fm} = 1.4 \pm (\geq)2.0 \text{ fm}. \tag{8}$$

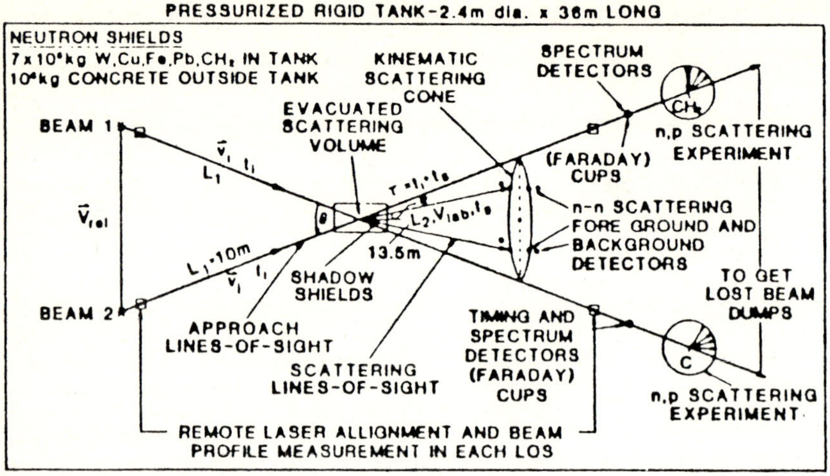

Figure 1: Geometry of experiment to measure a_{nn}[14].

This implies that V_{nn} is about 0.5% $\pm$ 0.7% stronger than V_{pp}. It is often claimed that this result demonstrates that charge symmetry holds in NN interactions. We feel that the uncertainties in the experimental results are still too large to justify this conclusion. We note that several authors[15] have calculated contributions due to electromagnetic ρ-ω and π-η mixing which lead to Class III and Class IV forces. They find that meson mixing contributes $\approx$1 fm of the above difference of scattering lengths. Recently THOMAS, BICKERSTAFF and GERSTEN[16] have claimed that as a result of the up- and down- quark mass difference π^0 coupling constant to the neutron is about 0.4% bigger than to the proton. This contributes another 0.3 fm to $|a_{nn}| - |a_{pp}|$. Thus the magnitude of the experimental difference is just about correct. If only the errors were smaller!

3.2 *Masses of ^3H and ^3He*

The difference in binding energies of ^{3}H and ^{3}He is 763.75 $\pm$ .01 keV. The best calculation of the electromagnetic differences (Coulomb plus all the subtle ones) explains only 683$\pm$29 keV of this difference, leaving a discrepancy of 81$\pm$29 keV or $\approx$ 1.5% of the binding energy as unexplained[17]. This is often referred to as one of the best measures of charge symmetry breaking. It is claimed that meson-mixing ($\pi\eta$ or $\rho\omega$) and photon+meson exchanges can not explain this discrepancy[18]. It has also been shown that Δ contributions to the mass difference are too small (only $\approx$ 12 keV)[19]. Since np mass difference and mass differences between the different charge states of the Δ's are already taken into account in the last calculation[19], one can only try more exotic explainations. These include three-body forces[20] and six-quark clusters[21]. At present these suggestions must be considered highly speculative, and the problem must be considered still open.

3.3 *$\pi^\pm d$ Total Cross Sections*

PEDRONI et al.[22] have made precision measurements of $\pi^\pm$ total cross sections on deuterium and analyzed them to obtain mass difference between two charge states of the Δ. Their result

$$\Delta m(T = 3/2) \equiv m(\Delta^0) - m(\Delta^{++}) = 2.7 \pm 0.3 MeV \tag{9}$$

is one of the best in the entire field of charge symmetry experiments. It can be directly used to determine if Coulomb effects and quark mass differences are entirely sufficient to explain the apparent charge symmetry violation. Quite schematically,

$$\Delta m(T = 3/2) \equiv m(ddu)_{3/2} - m(uuu)_{3/2} \cong 2(m_n - m_p) - (1/9)\langle e^2/r\rangle + \delta M(T) \tag{10}$$

where δ M(T) is the additional postulated T-dependent mass difference. From the known value of $m_n - m_p = 1.29$ MeV and the value of Δm above we immediately get

$$\delta M(T) = 0.17 \pm 0.30 MeV. \quad (11)$$

In other words, there is no evidence of any T-dependent contribution to the mass besides that implied in the quark construct of the Δ's.

3.4 $\pi^{\pm} d$ *and* $\pi^{\pm}\,{}^4He$ *Elastic Scattering*

Asymmetries A $\equiv [d\sigma(\pi^-) - d\sigma(\pi^+)]/[d\sigma(\pi^-) + d\sigma(\pi^+)]$ have been measured for pion elastic scattering from deuterium at 65 MeV[23], 142 MeV[24] and 256 MeV[24], and from ^{4}He at 24, 51, and 75 MeV. The asymmetries are large (as much as 50%) at low energies and small angles, but they seem to be quite satisfactorily explained in terms of external Coulomb corrections[25,26].

3.5 $\pi^{\pm}$ *Elastic Scattering From* ^3H *and* ^3He

Considerable stir was caused by a recent report of the measurement of ratios:

$$r_1 \equiv d\sigma(\pi^+ + {}^3H)/d\sigma(\pi^- + {}^3He), \qquad r_2 \equiv d\sigma(\pi^- + {}^3H)/d\sigma(\pi^+ + {}^3He) \quad (12)$$

and the super-ratio

$$R \equiv r_1 r_2 = [d\sigma(\pi^+ + {}^3H) \cdot d\sigma(\pi^- + {}^3H)]/[d\sigma(\pi^+ + {}^3He) \cdot d\sigma(\pi^- + {}^3He)]. \quad (13)$$

It is claimed that if CS is valid the super-ratio R = 1 at all angles and energies. Experimentally R is attractive because it is unaffected by uncertainties in several experimental parameters like beam flux, spectrometer acceptance, etc. Theoretically several corrections for Coulomb effects are also expected to cancel in R in the first order. Thus it was indeed very surprising when NEFKENS et al.[27] reported considerable angular variation in R, with values as large as R = 1.31 $\pm$ 0.09 ($\theta_{cm} = 65°$, T(π) = 180 MeV). This report generated a flurry of excitement in theoretical circles. KIM[28] proposed that the observation could be explained as manifestation of multiquark compound resonances. BARSHAY and SEHGAL[29] proposed a model of a specific three-nucleon correlation which distorts in going from ^{3}H to ^{3}He.

All the frantic activity of theorists has come to naught because the experimentalists have repeated their measurements and discovered that their earlier results were apparently wrong. In the new experiments[30] the super-ratio R was measured at several angles between $\theta_l = 40°$ and 110° for T(π) = 140, 180 and 220 MeV. No value larger than 1.13 was measured, and the average of eleven data points was found to be R = 1.08$\pm$0.04. At this level it appears that external Coulomb effects will be quite adequate to explain the results.

3.6 *The Reaction* $dd \rightarrow \alpha\pi^0$

This reaction is isospin forbidden. However, as HENLEY[2] has pointed out because only self conjugate particles and nuclei are involved, this reaction only tests charge symmetry. The experimental upper limits have been improved by more than two orders of magnitude since the original experiment[31]. The latest result is that due to a collaboration at Saturne[32] at Saclay. They find:

$$\begin{aligned} &T_d = 800\ MeV,\ \theta_{cm} = 100°,\ \ d\sigma/d\Omega < 0.8 \text{ pb/sr}, \\ &T_d = 1350\ MeV,\ \theta_{cm} = 77°,\ \ d\sigma/d\Omega < 5.0 \text{ pb/sr}. \end{aligned} \quad (14)$$

These results can be compared to the theoretical predictions of CHEUNG[33] (T_d = 700 MeV, θ = 0°, dσ/d$\Omega \approx$ 0.03 pb/sr) and of COON and PREEDOM[34] (T_d = 1.95 GeV, θ = 6°, dσ/dΩ $\approx$ 0.12 pb/sr), or to the experimental results for the isospin allowed reaction dd $\rightarrow \alpha\gamma$. For this reaction the Saturne collaboration reports dσ/dΩ = 4.7 $\pm$ 1.0 pb/sr at 800 MeV and dσ/dΩ = $3.5^{+2.8}_{-1.4}$ pb/sr at 1350 MeV. Both comparisons show that the experimental upper limits in the dd

$\rightarrow \alpha\pi^0$ experiments have to be improved by yet another factor of 10 or so before this reaction can provide useful information about charge symmetry. This is obviously a quite difficult undertaking.

3.7 *The Reaction* $np \rightarrow d\pi^0$

The differential cross sections for the T=1 → T=1 reaction pp → $d\pi^0$ are symmetric about θ_{cm} = 90°. If isospin is conserved the reaction np → $d\pi^0$ can only proceed through the T=1 initial state and its cross sections should be exactly half of that for the pp → $d\pi^+$ reaction everywhere. In particular, charge symmetry demands that it should also be symmetric about 90°. At T(n) = 795 MeV HOLLAS et al.[35] have measured the front-back asymmetry, $A_{fb} \equiv (F - B)/(F + B)$, where $F \equiv \int_0^{\pi/2} \sigma \, d\Omega$ and $B \equiv \int_{\pi/2}^{\pi} \sigma \, d\Omega$. They find that $|A_{fb}| \leq 0.15\%$. Theoretical predictions for the asymmetry range from +0.16% at 577 MeV[36] and −0.11% at 800 MeV[37]. The primary contribution is from π^0-η mixing. It is once again clear that the precision of the experiment has to be improved by at least a factor five before meaningful conclusions about CSB breaking mechanisms can be drawn from this reaction. Since this experiment involves no external Coulomb corrections it appears to be a worthwhile undertaking to make the required improvements in the experiment.

3.8 *Polarization in np Elastic Scattering*

In the purely nucleonic sector, where theoretical interpretation is the least ambiguous, the only test of charge symmetry so far is that provided by the difference in s-wave nn and pp scattering lengths. It only tests Class III CSB. Earlier we pointed out that Class IV CSB interaction can be best explored in measurements of polarization observables at intermediate energies. Indeed, as early as 1956, WOLFENSTEIN[38] noted that in np elastic scattering the polarization of the two outgoing particles should be equal, i.e., $\Delta P = P_n(\theta) - P_p(\pi - \theta) = 0$. This follows from the generalized Pauli principle or the overall antisymmetry of the space-spin-isospin wave function. The same relation holds for the (more convenient to measure) analyzing powers, if alternately polarized beam or polarized target are used:

$$\Delta A = A(\vec{n}p, \theta) - A(n\vec{p}, \theta) = 0. \tag{15}$$

In fact, it can be shown that ΔA is directly proportional to the isospin forbidden triplet to singlet transition amplitude. Several theoretical calculations of ΔA have been made[39-41] for np scattering at intermediate energies and at least two ambitious experiments, which differ only in details, have been undertaken. The IUCF experiment[42] is not yet operational, but the first results from the very elaborate TRIUMF experiment, whose schematic is shown in Fig. 2, have been just published[43].

In the TRIUMF experiment advantage is taken of the fact that if the measurements are made in the vicinity of the angle at which A(np) crosses zero ($\theta_{cm} \approx 71°$ at intermediate energies) then the experimental results are relatively unaffected by errors in the precise knowledge of beam and target polarizations, and one can equivalently measure $\Delta\theta$, the difference in the zero crossing angles.

ABEGG et al.[43] have made their first measurement at T(p) = 477 MeV. Extreme care taken in controlling systematic errors and the authors report

$$\begin{aligned} &\Delta\theta = -0.13° \pm 0.07° \pm (0.03°) \\ or, \quad &\Delta A = 0.0034 \pm 0.0017 \pm (0.0008) \end{aligned} \tag{16}$$

in the vicinity of the zero-crossing angle. This compares quite well with the post-experiment theoretical result due to MILLER, THOMAS, and WILLIAMS[44]

$$\Delta A = 0.0054 \pm 0.0040. \tag{17}$$

In this calculation[44] charge symmetry breaking terms due to one photon exchange, n-p mass difference affecting π, ρ, and 2π exchanges, ρ-ω mixing, and effects of up-down quark masses were included. We note that an earlier prediction for ΔA by GE and SVENNE[41] made a negative prediction ($\Delta A = -0.0035$). It is claimed that the earlier calculations of Refs. 39 and 41 had a sign error in the calculation of the effect of n-p mass difference on the one pion exchange contribution.

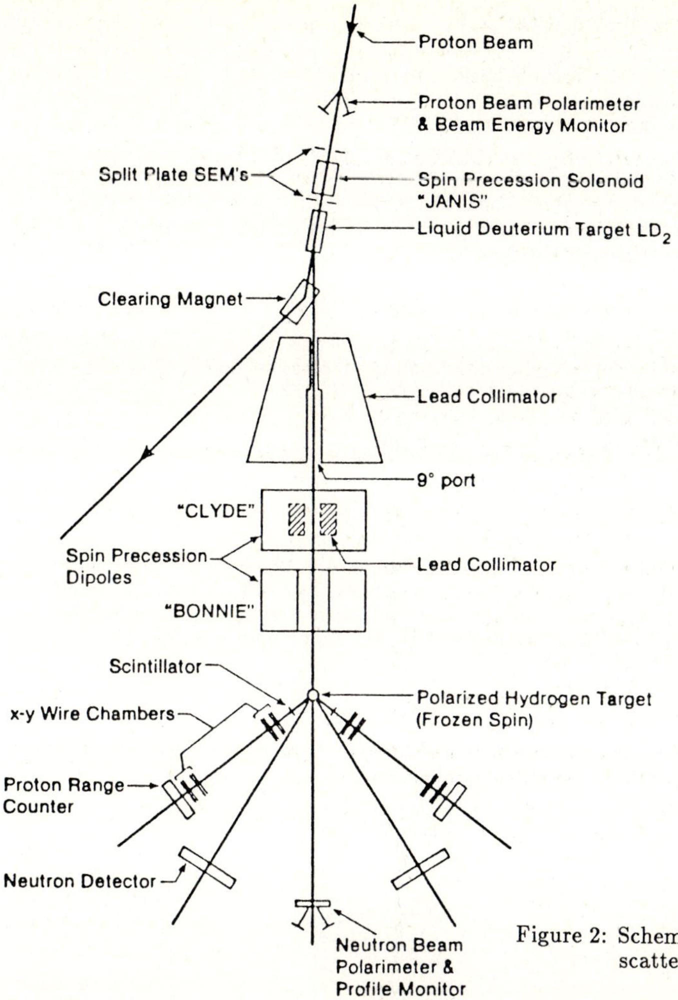

Figure 2: Schematic of the TRIUMP *np* scattering experiment[43].

The collaboration of this first experiment at TRIUMF[43] plans to make measurements with improved errors at T(n) = 350 MeV in the near future.

4 EXPERIMENTAL EVIDENCE FOR CHARGE INDEPENDENCE

4.1 *Masses of Baryon and Meson Isospin Multiplets*

The masses of baryon and meson isospin multiplets are found to be nearly degenerate. The differences range from 0.14% (for n-p) to 0.81% ($K^{\pm} - K^0$), with an average of about 0.5%. The notable exception is the pion with a $\pi^{\pm} - \pi^0$ mass difference of 3.35%. These mass differences are reasonably well explained in bag model[45] and potential model[46] calculations whose basic ingredients are Coulomb energies and 'known' differences in up-, down- and strange- quark masses.

4.2 *Nucleon–Nucleon Scattering*

Neutron proton s-wave scattering length has been measured to be $a_{np} = -23.72 \pm 0.02$ fm[47]. This measurement does not suffer from the uncertainties of Coulomb corrections or of final state interactions. This value of a_{np} and the average of a_{pp} and a_{nn} from section 3.1 lead to

$$\Delta a = a_{np} - \langle a_{nn}, a_{pp} \rangle = -5.4 \pm 0.5(?) \text{ fm}. \tag{18}$$

This large difference, amounting to 26±3% is the basis of the often quoted statement that CI is quite obviously broken. We must of course remember that this CI violation amounts to the statement that the space averaged np potential is about (2.1±0.5)% more attractive than the pp and nn potentials[2]. It has been known for a long time that the $\pi^{\pm}, \pi^0$ mass difference alone accounts for about half of this differene via the one pion exchange potential. When a more consistent model of NN interactions is made, including one and two pion exchanges, $\pi\rho, \pi\sigma, \pi\omega$ exchanges, and nucleons as well as deltas are considered, CHEUNG and MACHLEIDT[48] are able to account for $\Delta a = 4.4$ fm, or essentially for the entire experimental difference.

4.3 *Pion-Nucleon Total Cross Sections*

An upper limit of CIB is provided by the difference of π^+p and π^-p cross sections. As is well known, in term of cross sections for pure isospin states

$$\sigma(\pi^- p) = (1/3)\sigma_{3/2} + (2/3)\sigma_{1/2}, \qquad \text{and} \quad \sigma(\pi^+ p) = \sigma_{3/2}. \tag{19}$$

It follows that

$$\Delta\sigma \equiv 3\sigma(\pi^- p) - \sigma(\pi^+ p) = 2\sigma_{1/2}. \tag{20}$$

The cross section $\sigma_{1/2}$ is of course not equal to zero, but at the peak of the (3,3) resonance its contribution is small. In any case the quantity, $\Delta\sigma/\sigma_{3/2}$ at the peak of the (3,3) resonance establishes the upper limit of CI violation in pion-nucleon interaction. In Fig. 3, constructed from the precision data of PEDRONI et al.[22], we can see that this upper limit is $\approx$ 4%.

4.4 *Pion Production Reactions*

As HENLEY has pointed out[5], among the few tests of charge independence which appear possible, perhaps the potentially most accurate one is the comparision of the two reactions:

$$\begin{aligned} p + d &\rightarrow {}^3H + \pi^+ \\ p + d &\rightarrow {}^3He + \pi^0. \end{aligned} \tag{21}$$

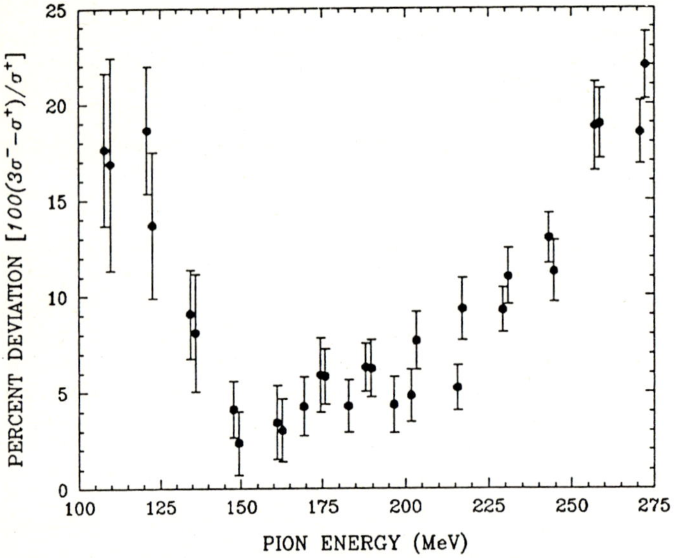

Figure 3: Percent deviation between $3\sigma(\pi^- p)$ and $\sigma(\pi^+ p)$.

Indeed a comparision of the differential cross sections for these two reactions for testing CI was suggested as early as 1952[49]. Charge independence predicts that

$$R \equiv d\sigma(\theta, {}^3H)/d\sigma(\theta, {}^3He) = 2. \tag{22}$$

Many experiments in which this ratio is measured have been reported over the last thirty years[50]. By far the best of the old experiments is that done by HARTING et al.[51]. We have just completed a high precision measurement of R at LAMPF in which a further improvement of almost a factor ten in statistical and systematic errors has been achieved.

The primary emphasis in the Northwestern experiment[50,52] was however not on R, but on the measurement of the analyzing power difference

$$\Delta A_{y0} \equiv A_{y0}(\theta, {}^3H) - A_{y0}(\theta, {}^3He), \tag{23}$$

which is predicted to be zero by charge independence. It may be expected that just as in the case of the polarized neutron-proton scattering experiment of section 3.8, measurement of ΔA_{y0} will be specially sensitive to Class IV charge independence breaking forces. For T($\vec{p}$) = 730 MeV, $\theta_{cm}(\pi)$ = 130° we obtain

$$\begin{aligned} R &= 2.193 \pm 0.007 \pm (0.024) \\ -\Delta A_{y0} &= 0.0065 \pm 0.0040 \pm (0.0018), \end{aligned} \tag{24}$$

where the first errors are statistical and second (in parantheses) are systematic.

The theoretical analysis of these results is obviously more complicated than for the np scattering experiment. No entirely satisfactory theory of pion production is available so far. However, if we combine KOHLER's[53] estimates of external Coulomb corrections (dominated by differences in ^{3}H and ^{3}He wave function) with LAGET and LECOLLEY's[54] estimates of effects of mass differences of exchanged pions and intermediate state Δ's, we obtain the 'predictions'

$$\begin{aligned} R &= 2.014 \\ -\Delta A_{y0} &= 0.003. \end{aligned} \tag{25}$$

Admittedly these 'predictions' are rather unreliable. However, since ΔA_{y0} is expected to have only a weak dependence on wave function differences, its prediction may not be as bad as that of R.

We hope that the precision results obtained in this experiment will spur the theorists to make new calculations. We look forward to experts in three-body problems and experts in pion-production reactions joining hands to solve the challenging problem posed by these new results.

References

1. W. Heisenberg: *Z. Phys.* 77 (1932) 1.
2. E. M. Henley in *Isospin in Nuclear Physics*, ed. D. Wilkinson, (North Holland, Amsterdam 1969) Ch. 2.
3. S. Weinberg: *Trans. N.Y. Acad. Sci.* 38 (1977) 185; see however, D. Kaplan and A. V. Manohar: *Phys. Rev. Lett.* 56 (1986) 2004.
4. E. M. Henley, G. A. Miller in *Mesons in Nuclei*, eds. M. Rho and D. Wilkinson, (North Holland, Amsterdam, 1979) vol. I, p. 405.
5. E. M. Henley in *Proc. Nucl. Theory Summer Workshop*, Sante Barbara, (1981) ed. G. F. Bertsch, (World Scientific, Singapore, 1981) Ch. 1.
6. E. G. Adelberger in *Symmetries in Nucl. Structure*, eds. K. Abrahams, K. Allart and A. E. L. Dieperink, (Plenum Press, N.Y. , 1982), p. 55.
7. W. T. H. van Oers, *Comments Nucl. Part. Phys.* 10 (1982) 251.
8. S. Schlomo: *Rep. Prog. Phys.* 41 (1978) 66.

9. P. U. Sauer and H. Walliser: *J. Phys.* G3 (1977) 1513.
10. J. M. Allen and H. Fiedelday: *Nucl. Phys.* A260 (1976) 213.
11. M. Rahman and G. A. Miller: *Phys. Rev.* C27 (1983) 917.
12. B. Gibioud et. al.: *Phys. Rev. Lett.* 42 (1979) 1508; *Phys. Lett.* 103B (1981) 9. At the symposium we were informed that this group at SIN has done a new experiment in which the γ's as well as neutrons were detected. The results of the new experiment are stated to be consistent with those of the older one.
13. R. M. Salter et al.: *Nucl. Phys.* A254 (1975) 241.
14. D. W. Glasgow et al: *Proc. Int. Conf. on Nucl. Data for basic and applied Science*, Santa Fe (1985) , to be published by in *Radiation Effects*; also D. W. Glasgow, *priv. comm.*
15. P. C. McNamee et al.: *Nucl. Phys.* A249 (1975) 483; *ibid.*, A287 (1977) 381; J. L. Friar and B. F. Gibson: *Phys. Rev.* C17 (1978) 1752.
16. A. W. Thomas, P. Bickerstaff and A. Gersten: *Phys. Rev.* D24 (1981) 2539.
17. R. A. Brandenburg, S. A. Coon and P. U. Sauer: *Nucl. Phys.* A294 (1978) 305.
18. S. A. Coon and M. D. Scadron: *Phys. Rev.* C26 (1982) 562; S. A. Coon in *Proc. Workshop on Charge Symmetry, TRIUMF Report* TRI-81-3 (1981).
19. H. Baier, W. Bentz, Ch. Hajduk and P. U. Sauer: *Nucl. Phys.* A386 (1982) 460.
20. J. L. Friar and B. F. Gibson: *Ann. Rev. Nucl. Sci.* 34 (1984) 403.
21. J. M. Greben and A. W. Thomas: *Phys. Rev.* C30 (1984) 1021.; V. Koch and G. A. Miller: *Phys. Rev.* C31 (1985) 602.
22. E. Pedroni et al.: *Nucl. Phys.* A300 (1978) 321.
23. B. Balestri et al.: *Nucl. Phys.* A392 (1983) 217.
24. T. G. Masterson et al.: *Phys. Rev. Lett.* 47 (1981) 220; *Phys. Rev.* C26 (1982) 2091; *Phys. Rev.* C30 (1984) 2010.
25. J. Frölich, and B. Saghai: *Nucl. Phys.* A435 (1985) 738.
26. M. Th. Kankhasayev et al.: *Phys. Lett.* (submitted), Dubna preprint, E4-85-612 (1985).
27. B. M. K. Nefkens et al.: *Phys. Rev. Lett.* 52 (1984) 735.
28. Y. E. Kim: *Phys. Rev. Lett.* 53 (1984) 1508.
29. S. Barshay: *Phys. Rev.* C31 (1985) 2133.
30. C. Pillai et al.: *Bull. Amer. Phys. Soc.* 31 (1986) 800, abst. DH2.
31. Yu. K. Akimov et al.: *Sov. Phys. JETP* 14 (1962) 512.
32. The Saturne Collaboration *(unpublished)*, as reported by A. Stetz, *Bull. Amer. Phys. Soc.* 31 (1986) 848, abst. Hb4.
33. C. Y. Cheung: *Phys. Lett.* 119B (1982) 47.
34. S. A. Coon,B. M. Preedom: *Phys. Rev.* C33 (1986) 605.
35. C. L. Hollas, et al.: *Phys. Rev.* C24 (1981) 1561.
36. C. Y. Cheung, E. M. Henley and G. A. Miller: *Phys. Rev. Lett.* 43 (1979) 1215; *Nucl. Phys.* A348 (1980) 365.
37. C. Y. Cheung: *Private Comm.* quoted in Ref. 38.
38. L. Wolfenstein: *Ann. Rev. Nucl. Sci.* 6 (1956) 43.
39. C. Y. Cheung, E. M. Henley and G. A. Miller: *Nucl. Phys.* A348 (1980) 365; *ibid.* A305 (1978) 342.
40. A. Gersten: *Phys. Rev.* C24 (1981) 2174; *ibid.* C18 (1978) 2252.
41. L. Ge and J. P. Svenne: *Phys. Rev.* C33 (1986) 417.
42. S. E. Vigdor et al.: *Proc. Vth Internat. Symp. on Polarization Phenomena in Nucl. Phys.* (Santa Fe, 1980), *AIP Conf. Proc.* 69 (1981) 1455; also L. D. Knutson et al.: *IUCF Scientific and Technical Report* (1985) 12.
43. R. Abegg, et al.: *Phys. Rev. Lett.* 56 (1986) 2571; *Proc. 6th Internat. Symp. on*

Polarization Phenomena in Nucl. Phys., Osaka, (1985), *J. Phys. Soc.* Japan, 55 (1986) 369.

44. G. A. Miller, A. W. Thomas, A. G. Williams: *Phys. Rev. Lett.* 56 (1986) 2567.
45. R. P. Bickerstaff and A. W. Thomas: *Phys. Rev.* D25 (1982) 1869.
46. S. Godfrey and N. Isgur: *University of Toronto Preprint* TRI-PP-85-110 (Dec. 1985).
47. H. P. Noyes: *Ann. Rev. Nucl. Sc.* 22 (1972) 465.
48. C. Y. Cheung and R. Machleidt: *contributed paper PANIC X*, Heidelberg (1984) C13.
49. A. M. L. Messiah: *Phys. Rev.* 86 (1952) 432; M. Ruderman: *Phys. Rev.* 87 (1952) 383.
50. For references to these experiments and a critical discussion of their limitations, see M. Artuso: *Ph.D. dissertation, Northwestern University* (1986), *unpublished.*
51. D. Harting et al.: *Phys. Rev. Lett.* 3 (1959) 52; *Phys. Rev.* 119 (1960) 1716.
52. M. Artuso, K. K. Seth, B. Parker, R. Soundranayagam and D. Barlow, *to be published.*
53. H. S. Kohler: *Phys. Rev.* 118 (1969) 1345.
54. J. M. Laget and J. F. LeColley: *contributed paper, PANIC X*, Heidelberg (1984) E21; J. M. Laget: *priv. comm.*

Electric Dipole Moment of ^{3}He

Y. Avishai[1] *and M. Fabre de la Ripelle*[2]

[1]Department of Physics, Ben Gurion University of the Negev, Beer Sheva, Israel
[2]Division de Physique Theorique, Inst. de Physique Nucléaire, F-91406 Orsay Cedex, France

In the present work we estimate the contribution of CP violating nucleon-nucleon interaction to the electric dipole moment of ^{3}He. We use two models of CP violating interactions in combination with a Reid soft core strong nucleon-nucleon interaction. In the Kobayashi-Maskawa model of CP violation the order of magnatude is 10^{-30} e-cm while the presence of the θ term in the QCD lagrangian contributes an order of magnitude $10^{-16}\bar{\theta}$ e-cm.

The increasing interest in CP violation[1] together with the ever decreasing limit on the electric dipole moment (EDM) of the neutron[2] ($p(n) < 6 \times 10^{-25}$ e-cm) revived the interest in measuring the EDM of finite nuclei. Like the situation encountered in parity violation, it is expected that CP violation effects will be larger for heavy nuclei than for light ones. However, in order to relate the observed violation to the fundamental CP violating force it is much better to study light nuclei. In fact, there is a recent suggestion[3] to detect a macroscopic EDM in the polarized Al phase of superfluid ^{3}He. Therefore, theoretical attempts to evaluate the EDM of nuclei seem timely, and some rough estimates have recently been reported.[4,5,6] Such EDM can arise from several reasons (EDM of nucleons, CP odd electron nucleon interaction etc.), but here we consider the possible occurence of P and T non-conserving (PTNC) nucleon-nucleon (NN) interaction, and compute its contribution to the EDM of ^{3}He.

At low energy it is reasonable to model the PTNC NN interaction in terms of meson exchange, just as is successfully done in modeling the strong and the weak (but T conserving) NN interaction.[7] The PTNC nucleon-nucleon-meson coupling constants at the vertices depend on the mechanism responsible for the CP violation in the pertinent subnuclear Hamiltonian. In the Kobayashi-Maskawa (KM) model[8] it arises in the fermion sector. Another possible mechanism of CP violation is the occurrence of the θ term in the QCD Lagrangian which violates CP.[9,10]

The KM mechanism of CP violation received much attention in the last decade. As has been shown,[4-6] it leads to a PTNC-NN isovector interaction arising from K meson exchange whose nonrelativistic form is

$$V_{KM}(\mathrm{PTNC}) = -i(A/M)(\vec{\tau}_1-\vec{\tau}_2)_z(\vec{\sigma}_1+\vec{\sigma}_2)\cdot[\vec{p}; v_{m_k}(r)] \quad . \tag{1}$$

In (1), M is the nucleon mass, $\vec{\tau}_i$, $\vec{\sigma}_i$ are respectively isospin and spin operators of nucleon i $(i=1,2)$ while $\vec{p} = -i\hbar\vec{\nabla}$ and $v_{m_k}(r) = \exp(-m_k r)/r$ are enclosed in a commutator. The small number A is proportional to the coupling constants at the two nucleon-nucleon-meson vertices (one of them is CP odd while the other is CP even). In terms of the Fermi constant G and a CP violating parameter $\eta \approx 0.67 \times 10^{-8}$, the numerical value of A is $A = Gm_k^2\,\eta/16\pi\sqrt{2} = 2.7 \times 10^{-16}$.

As for the CP violating θ term in the QCD Lagrangian it leads in the non-relativistic limit to PTNC NN isoscalar interaction of the one pion exchange form[11]

$$V_\theta(\mathrm{PTNC}) = i(g\bar{g}/8\pi M)\vec{\tau}_1\cdot\vec{\tau}_2\ (\vec{\sigma}_1-\vec{\sigma}_2)\cdot[\vec{p},v_{m_\pi}(r)] \tag{2}$$

where $g \approx 13.5$ is the strong $NN\pi$ coupling constant and $\bar{g}$ is the CP odd $NN\pi$ constant. Using SU(3) results and quark mass ratios,[12] $\bar{g}$ is related to $\bar{\theta}$ as $\bar{g} = 0.027\bar{\theta}$.

The ground state wave function of ^{3}He is the solution of the Schrödinger equations with both strong and PTNC NN interactions. Thus

$$[H_0 + V(\text{strong}) + V(\text{PTNC})]\psi = E\psi\ . \tag{3}$$

The wave function is then divided into strong component ψ^S and weak CP odd component ψ^W. To first order in V(PTNC) the equations satisfied by these components are

$$[H_0 + V(\text{strong})]\psi^S = -B\psi^S \tag{4}$$

$$[H_0 + V(\text{strong}) + B]\psi^W = -V(\text{PTNC})\psi^S \tag{5}$$

For $V(\text{strong}) = \sum_{i<j} v_{ij}$ we take v_{ij} = Reid soft core NN force. First, (4) is solved yielding the strong component of the ground state wave function and the ^{3}He binding energy B. The solution is then used as a source term in (5) which is an inhomogeneous equation for the weak component ψ^W. Finally, the matrix element of the dipole operator $\vec{p} = (e/2)\sum_{i=1}^{3}(1-\vec{\tau}_{iz})\vec{r}_i$ is taken in the ground state wave function. To first order in V(PTNC), $p(^3\text{He}) = 2 < \psi^W|p_z|\psi^S >$.

In the KM model of CP violation (the interaction (1)) we have obtained $p_{KM}(^3\text{He}) = 0.168\times10^{-29}$ e-cm. This value is greater almost by two orders of magnitude from the corresponding EDM of the neutron in the KM model.[13] The fact that in this model the EDM of finite nuclei is much larger than that of the neutron has already been noticed elsewhere.[4] It is mainly a nuclear physics effect, which is estimated[4,14] to be ≈ 60.

The contribution of the θ term (the interaction (2)) to the EDM of ^{3}He is found to be $p_\theta(^3\text{He}) = 3.68\times10^{-11}\bar{\theta}$ e-cm which should be compared with the EDM of the neutron $p_\theta(n) = 3.60\times10^{-16}\bar{\theta}$ e-cm. In this case, no enhancement is obtained. The reason for that is not very clear and may be due to some subtle cancellation of various terms in the CP odd part of the ^{3}He wave function.

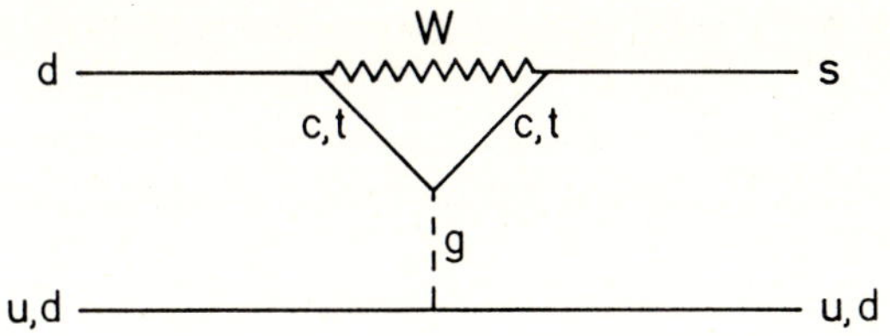

Fig. 1. Penguin term in the Kobayashi-Maskawa mechanism of CP violation (the dashed line refers to the gluon)

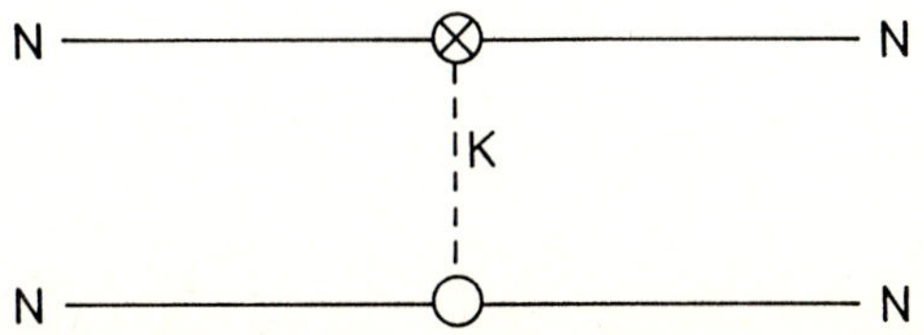

Fig. 2. K meson exchange contributing to CP odd NN interaction. The upper vertex is CP odd (arising from the interaction described in Fig. 1). The lower vertex is weak but CP even

References

1. J.W. Cronin: Rev. Mod. Phys. 53, 374 (1981).
2. I.S. Altarev et al.: Phys. Lett. 102B, 13 (1981); N.F. Ramsey: Phys. Rep. 43C, 409 (1977).
3. I.B. Khriplovich: JETP Lett. 35, 485 (1982). (Pizma Zh. Exp. Teor. Fiz. 35, 392 (1982)).
4. O.P. Sushkov, V.V. Flambaum and I.B. Khriplovich: Soviet Phys. JETP 60, 873 (1984). (Zhur. Exp. and Thor. Fyz. 87, 1521 (1984)).
5. Y. Avishai: Phys. Rev. D32, 314 (1985).
6. Y. Avishai and M.F. de la Rippele: Phys. Rev. Lett. 56, 2121 (1986).
7. B. Des planques, J.F. Donoghue and B. Holstein: Ann. Phys. 124, 449 (1980).
8. M. Kobayashi and T. Maskawa: Prog. Theor. Phys. 49, 652 (1973).
9. A.A. Belavin et al.: Phys. Lett. 59B, 85 (1975). G. t'Hooft: Phys. Rev. Lett. 37, 8 (1976).
10. V. Baluni: Phys. Rev. D19, 2227 (1979).
11. W.C. Haxton and E.M. Henley: Phys. Rev. Lett. 51, 1937 (1983).
12. R.J. Crewther et al.: Phys. Lett. 88B, 123 (1979); Phys. Lett. 91B, 487 (1980).
13. I.B. Khriplovich and A.R. Zhitinitsky: Phys. Lett. 109B, 490 (1982).
14. O.P. Sushkov, V.V. Flambaum and I.B. Khriplovich: Nucl. Phys. A449, 750 (1986).

A Proposal for a High Sensitivity Search for T-Violation in Slow Neutron Resonances

C.D. Bowman, J.D. Bowman, and V.W. Yuan

Los Alamos National Laboratory, Los Alamos, NM 87545, USA

An experiment is proposed using eV neutrons from the Los Alamos Proton Storage Ring (PSR) to perform a search for direct evidence of violation of time-reversal invariance. Over twenty years ago CP-violation was discovered in the decay of the neutral kaon[1] and subsequently verified in a number of experiments. The resulting CPT symmetry implication that time reversal invariance violation exists (referred to hereafter as T-violation) has never been directly observed. Our lack of knowledge of the level of T-violation is a major barrier to further improvements in the standard model for weak interactions.

The purpose of this experiment is to search for T-violation by studying the resonance transmission of polarized neutrons through polarized targets exhibiting p-wave resonances in the 1-100 eV range[2]. The T-violation interaction studied is $\sigma \cdot k \times I$ where σ is the spin of the neutron, k, the momentum vector of the neutron, and I the direction of sample polarization. At the same time the relatively easily observed P-violating interaction $\sigma \cdot k$ can be studied. The result is a search for T-violation relative to P-violation in the same slow neutron resonance. Amplification effects in the slow neutron resonances along with the extraordinary pulsed neutron intensity of the PSR allow sensitivity for a T-violation component of the weak force which may approach 3×10^{-5} (3σ).

A comparison of the sensitivity of the experiment with the neutron EDM experiments is highly model dependent, but it appears that this experiment may readily match or exceed the sensitivity of the present limit on the EDM. In addition, a number of favorable resonances can be studied thereby eliminating any accidental insensitivity that might obtain in a single resonance or in the neutron system itself.

Transmission measurements on specially selected slow neutron resonances offer extraordinary sensitivity for both P-odd and T-odd studies because of the amplification of the weak interaction relative to the strong interaction for low energy p-wave resonances. The amplification arises from two sources[3]. First, the close spacing of a few eV for s- and p-wave resonances in heavy nuclei promotes mixing of s-wave amplitude from an s-wave resonance through the weak force into a neighboring p-wave resonance. This effect, which depends on the level separation, is typically a factor 10^4 more favorable for slow neutron resonances than for the most favorable cases for study of parity violation in the decay of low-lying levels of lighter nuclei. The second effect is the angular momentum barrier which strongly suppresses the p-wave amplitude relative to the s-wave amplitude in slow neutron resonance permitting the small s-wave admixture in a p-wave resonance to be observed. The largest example of parity violation found[4] is in the 0.75 eV resonance of ^{139}La where an effect of 14% was observed and we believe even larger effects in the 50% range can be expected.

The PSR is now the world's most intense pulsed laboratory neutron source and therefore the site where this experiment could be performed with the best sensitivity. The facility accepts a full 850 microsecond-long LAMPF macropulse, compresses it to 0.27 microseconds and dumps it onto a tungsten target. Each burst produces about 10^{15} neutrons[5] and the repetition rate presently is 12 Hz

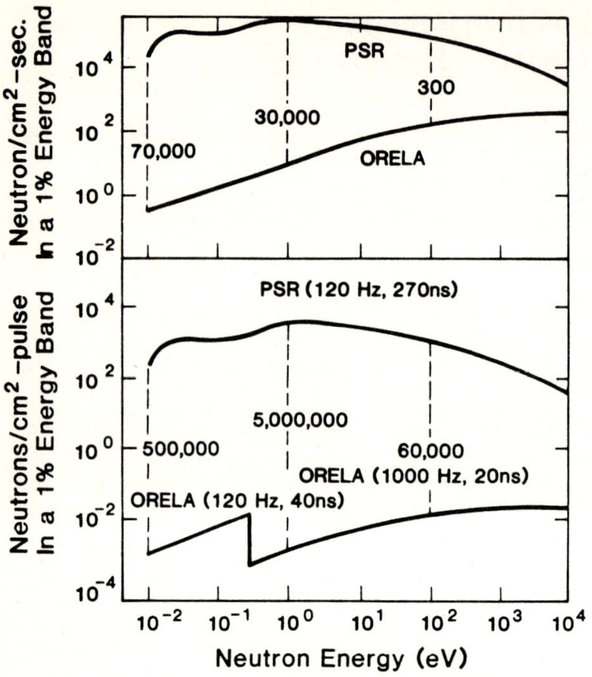

Fig. 1 Comparison of PSR and ORELA Neutron Intensities. a) the average neutron intensity per second in a 1% energy band is shown with flight path and repetition rate adjusted for greatest intensity and no overlap. The figures give the ratio of intensities. b) The neutron intensity per pulse is given to emphasize the advantages for experiments requiring low duty cycle. Practical repetition rates and burst widths are included. The curve demonstrates the effectiveness of the PSR for the epithermal range; ORELA is competitive in the higher kilovolt and lower MeV range.

but might go higher. These neutrons are moderated and energy resolved along paths which range in length from 5 to 200 meters. The intensity is compared in Fig. 1 with ORELA[6] at the Oak Ridge National Laboratory which is the most intense neutron source extensively used for neutron physics in the eV range. Figure 1a shows the average neutron flux using a flight path appropriate to maintain a 1% energy resolution at the detector. Figure 1b emphasizes the neutron intensity per burst which is an important advantage for many experiments. The factor of about 10^4 in flux of Fig. 1a in the 1-30 eV range can be translated into a factor of 100 in sensitivity advantage for the PSR.

While the experiment is somewhat complex, the individual components require only the replication of routine techniques. Our proposed schedule calls for the installation of modest equipment in time to allow confirmation of the P-violation reported for 139 La resonance by 12/15/86 and to search for other favorable resonances. T-violation studies would begin in earnest in September 1987.

[1] J. H. Christenson, J. W. Cronin, V. L. Fitch and R. Turlay, Phys. Rev. Lett. **13**, 138 (1984).

[2] V. E. Bunakov and V. P. Gudkov, Nucl. Phys. **A401**, 93 (1983).

[3] For example O. P. Sushkov and V. V. Flambaum, Sov. Phys. Usp. 25, 1 (1982).

[4] V. P. Alfimenkov, et al., Conf. Proc. Nuclear Data for Science and Technology, Antwerp, p. 773, D. Reidel Publishing Co. (1982).

[5] G. J. Russell, C. D. Bowman, E. R. Whitaker, H. Robinson and M. M. Meier, High Power (200μA) Target Moderator-Reflector Shield Eighth Meeting of the International Collaboration on Advanced Neutron Sources ICANS VIII, Oxford, England (1985).

[6] D. K. Olsen, J. A. Martin and D. J. Horen, "Possible ORELA Replacement Option", Report ORNL/TM8669 (1984).

P-Violating Effects in the Integral Gamma-Ray Spectrum of nγ-Reactions on Nuclei

V.A. Nazarenko

Leningrad Nuclear Physics Institute, Gatchina,
Leningrad District, 188350, USSR

Two years ago, at a Workshop on Reactor based fundamental physics in Grenoble, our group reported on the first results of an experimental search for P-violating effects in the integral γ-ray spectrum of the nγ-reaction on nuclei /1/. We measured the asymmetry of γ-emission in polarized thermal neutron capture and the circular γ-polarization in radiative capture of unpolarized reactor neutrons. It turned out that in some cases no noticeable attenuation of the effects in the total γ-ray spectrum is evident contrary to what could be expected basing on the fact that the effects from partial γ-transitions occurring from the capture state to states of different spin and parity should have opposite signs and thus cancel largely one another. An asymmetry at a level of $2 \cdot 10^{-5}$ was found to exist for the Cl, Br and La nuclei, as well as a circular polarization of $(2 - 16) \cdot 10^{-5}$ for natural tin, bromine, chlorine and lanthanum.

Meaningful theoretical analysis of the observed effects required obtaining additional experimental data. Therefore the circular polarization measurements where interpretation appears to be simpler because of the absence of the so-called spin factor were continued.

The general scheme of the experiments was not changed. Targets of the nuclei to be studied were placed into a water cavity in the core of the LNPI WWR-M reactor. The cavity was surrounded by a 90 mm lead shield to protect against the core γ-rays polarized because of the contributions of bremsstrahlung from electrons emitted in the β-decay of uranium fission fragments in the fuel elements. The γ-rays from the radiative neutron capture on the target nuclei were led out via a collimating channel through the reactor tank water to a transmission-type polarimeter described in /2/ and detected with CsI(Tl) scintillation counters operating in integral mode. The experimental effect δ_{exp} was determined, as usual, from the relative variation of the γ-ray intensity as the polarimeter magnetization was reversed once a second.

Since the nγ-reaction spectrum under study has a composite character, the effect in question is actually weighted over all the γ-transitions involved, the weight being the product of the energy by the partial intensity of a given γ-transition. A noticeable contribution to the effect comes only from γ-rays above 1.0 - 1.5 MeV, since the low-energy part of the spectrum is absorbed by the polarimeter absorber which in our case is made up of 7 cm of iron. The experimental effect is related to the circular polarization Pγ by the expression $P_\gamma = \delta_{exp}\, \epsilon$, where ϵ is the polarization efficiency. Generally speaking, ϵ is the energy dependent, however, in range of interest to us (1.5 - 6.0 MeV) ϵ varies only by about 10%.

An important point in the selection of nucleus for study is the presence of β-activity in irradiated target material, since the bremsstrahlung of the β-decay electrons is circularly polarized and this may imitate or mask the effect under study. The possible contribution of this process was determined by calculation. Apart from this, the effect was measured in all cases immediately after reactor shutdown, when there are no radiation capture γ-rays and the effect is totally determined by bremsstrahlung.

The measurements were carried out on natural abundance samples of Na, K, Sc, Fe, Co, J, Cs, and Ho having mono - or nearly monoisotopic composition. The control experiment, described in /2/, did not reveal any noticeable contribution to the effect from the nγ-reaction on the shield and target structural materials (lead and zirconium) and residual water in the cavity. The depression of the thermal neutron flux in the cavity with the targets in place was found by calculation. The results of the measurements together with the earlier published data /1,3/ (marked with an asterisk) are presented in Table 1 (column 3).

Table 1

Nucleus	E_p, eV	$P_\gamma^{exp} \cdot 10^5$	$P_\gamma^{th} \cdot 10^5$	v_{exp}, meV	v'_{exp}, meV	v_{st}, meV
Na		-0,27±0,50				
Cl*	398	6,4 ±0,50		78 ± 22	250 ± 80	25
K		0,26±0,20				
Sc		-0,08±0,09				
Fe	1147	2,5 ± 0,4		55 ± 15	32 ± 8	30
Co	1380	0,0 ±0,12		30		12
Br*	0,88	3,1 ± 0,2	2,0	4,6 ± 0,4	3,0 ± 0,5	3,2
Cd*	7	≤ 0,3	0,26	≤ 0,8	≤ 0,2	1,3
Sn*	1,33	1,9 ± 0,5	0,41	3,7	2,5 ± 0,5	2
J	7,6	0,18±0,20		≤ 2	0,5 ± 0,4	0,8
Cs	9,5	-0,24±0,08		0,9 ± 0,3		1,7
La*	0,75	-16 ± 2,5	21,7	0,96	1,3 ± 0,1	4
Ho	10,3	0,15 ±0,12		≤ 2,4		0,73

The theoretical analysis /4,5/ of the first experimental results published in /3/ is given in column 4. It was based on the concept of P-violating effect enhancement near p-wave compound-resonances involving a statistical approach which assumes complete statistical independence of the partial γ-transition matrix elements. Within this approach $P_\gamma = 2v/E_p \cdot A_\gamma$, where $v = \langle s|V_W|p\rangle$ is the matrix element of weak interaction mixing the s- and p-resonances, E_p is the p-resonance position. A_γ is a factor governing the spectrum shape and radiative widths of the levels. If the partial γ-ray amplitudes are unknown, then the factor A_γ can be estimated only in terms of its dispersion /4,5/. The value of the "weak" matrix element required to calculate P_γ^{th} is derived from other independent experiments.

A satisfactory description of the experimental data /3/ permitted BUNAKOV et al. /4,6/ to derive from integral type measurements /3,7/ the values of the matrix elements v_{exp} for Co, Br, Cd, J, Cs, La, and Ho listed in column 5 of Table 1. The values v_{exp} for the Cl, Fe and Sn nuclei to which the statistical approach is inapplicable were obtained basing on the available information on the intensities of partial γ-ray transitions from the s- and p-resonances under study and on the quantum characteristics of the low-lying excited states.

For comparison, Table 1 presents values of v'_{exp} derived from independent experiments on individual transitions or on the transmission of neutrons with different helicity. On the whole, these values agree fairly well with those of v_{exp} obtained from integral measurements, which supports the validity of using the statistical approach in describing the radiative decay characteristics.

The last column of Table 1 contains the values of v_{st} calculated from standard deviations /8/ (in the compound-compound mixing mechanism, the resonance-averaged value is $\bar{v} = 0$). On the whole, the theory is seen to agree satisfactorily with experiment which suggests the dynamic mechanism of parity mixing in compound nucleus stage used in calculating v_{st} to provide a dominant contribution.

Thus integral-type experiments provide meaningful information on the weak interaction matrix elements for the nuclei to which the statistical description of radiative capture is applicable. Note that no additional data on partial transition amplitudes are required and no knowledge of radiative level widths is assumed. In addition, integral γ-ray spectrum experiments are simpler to carry out than measurements on individual γ-ray transitions.

The present report draws primarily upon the data of refs /6,7/.

References

1. V.M. Lobashev and V.A. Nazarenko: J. de Physique. 45, 103 (1984).
2. V.A. Knyazkov et al.: Nucl. Phys. A417, 209 (1984).
3. V.A. Vesna et al.: Pisma Zh. Eksp. Teor. Fiz. 36, 169 (1982).
4. V.E. Bunakov et al.: Yad. Fiz. 40, 188 (1984).
5. V.V. Flambaum and O.P. Sushkov: Nucl. Phys. A435, 352 (1985).
6. V.E. Bunakov et al.: LNPI Preprint-1101, Leningrad (1985).
7. A.I. Egorov et al.: LNPI Preprint-1067, Leningrad (1985).
8. S.G. Kadmenskii et al.: Yad. Fiz. 37, 581 (1983).

Some Macroscopic Effects of P- and T-Violation in Atoms

A.N. Moskalev

Leningrad Nuclear Physics Institute, Gatchina,
Leningrad 188350, USSR

The methods of atomic physics and optics may be successfully used to investigate the P- and T-violating electron-nucleus interactions and to search the electric dipole moments of electron or nucleon. As an illustration we shall consider the influence of such interactions on the electromagnetic properties of the substances.

The parity violating interactions produce the mixing of the atomic states of different parities and T-violation changes the phases of the admixture coefficients. As a result, the material equations which relate the electric and magnetic "displacement vectors" to the electric and magnetic fields are modified. For the monochromatic fields and low density substances these equations may be written as

$$\begin{aligned} \vec{D} &= \varepsilon\vec{E} + \gamma \mathrm{rot}\vec{E} + \delta\vec{H} + i\vec{E} \times \vec{g}_E + \vec{H} \times \vec{G}_p + \mathrm{rot}\vec{E} \times \vec{G}_\tau , \\ \vec{B} &= \mu\vec{H} + \gamma \mathrm{rot}\vec{H} + \delta\vec{E} + i\vec{H} \times \vec{g}_H - \vec{E} \times \vec{G}_p - \mathrm{rot}\vec{H} \times \vec{G}_\tau . \end{aligned} \tag{1}$$

The first three terms of these expressions correspond to the isotropic substances, the other terms correspond to the substances with oriented angular momenta of atoms (the so-called gyrotropic substances). The vectors $\vec{g}_E$ and $\vec{g}_H$ are the electric and magnetic gyration vectors. The terms which contain γ and $\vec{G}_p$ describe the parity violating effects. The terms with δ and $\vec{G}_T$ characterize the effects of simultaneous P- and T-violation. One may readily express all parameters ε, μ, γ, δ etc. in terms of matrix elements of atomic electric and magnetic dipole transition moments. Some applications of (1) will be discussed below.

1. Electromagnetic Wave Propagation

Let us consider the monochromatic electromagnetic wave in an isotropic homogeneous medium. For the right (+) and left (-) circularly polarized components of this wave (1) take the form

$$\begin{aligned} &\vec{D}_\pm = \varepsilon_\pm \vec{E}_\pm , \quad \vec{B}_\pm = \mu_\pm \vec{H}_\pm , \\ &\varepsilon_\pm = \varepsilon \pm (\gamma\frac{\omega}{c} - i\delta), \quad \mu_\pm = \mu \pm (\gamma\frac{\omega}{c} + i\delta), \quad n_\pm = \sqrt{\varepsilon\mu} \pm \gamma\frac{\omega}{c} . \end{aligned} \tag{2}$$

It follows from (2) that P-violating γ-term leads to a difference in refractive indices n_+ and n_-. Some consequences of this optical activity, such as the rotation of the plane of polarization, the circular dichroism, the splitting of circularly polarized wave bv reflection into two waves etc. have been discussed by many authors (see e.c. Ref. [1]). Note that the refractive indices $n_\pm$ do not depend on δ. Hence, T-violating interactions lead only to a phase shift δ between $\vec{D}_\pm$ and $\vec{E}_\pm$, $\vec{B}_\pm$ and $\vec{H}_\pm$. In principle, this shift may be separated from the shifts arising from the dissipative effects and relating to the imaginary parts of ε, μ, and γ.

Similarly one may consider the wave propagating in a gyrotropic medium. In this case the effects of P- and T-violation are rather complicated and will not be discussed here.

2. Static Fields

For static electric and magnetic fields (1) reduces to

$$\vec{D} = \varepsilon\vec{E} + \delta\vec{H} + \vec{H} \times \vec{G}_p, \quad \vec{B} = \mu\vec{H} + \delta\vec{E} - \vec{E} \times \vec{G}_p. \tag{3}$$

The second terms at the right-hand sides of (3) describe T-odd correlation between the magnetic and electric dipole moments of atoms. The experimental searches of this correlation with the help of quantum superconducting interferometer have been reported in Ref. [2]. The experiment has been based on detecting the change of magnetic flux in ferrite induced by the application of electric field. It is seen from (3) that the same method may be used to study the effects of P-violation (the last terms of (3)) when the ferrite is magnetized, i.e. $\vec{G}_p \neq 0$.

References

1. E.L. Altshuler, A.N. Moskalev, V.A. Ryzhov, R.M. Ryndin, P.G. Silvestrov, V.N. Fomichev and I.B. Khriplovich: J. Techn. Phys., 54, 1956 (1984) (in Russian).

2. B.V. Vasiliev and E.V. Kolycheva: JETP, 74, 466 (1978) (in Russian).

Parity Nonconserving NN Interaction in $SU(3)_C \times SU(2)_L \times U(1)$ Theory

V.M. Dubovik[1] *and S.V. Zenkin*[2]

[1]Joint Institute of Nuclear Research, Laboratory of Theoretical Physics, Dubna 141980, USSR
[2]Institute for Nuclear Research, Ac. Sc., Moscow, USSR

In the framework of the standard electroweak and colour interaction theory we calculate the PNC constants in MNN vertices: $h_M \sim \langle MN|H^{PNC}|N\rangle$. In the effective four-quark Hamiltonian H^{PNC} the colour interactions are taken into account through the summation of leading logs within the renormalization group approach /1a/. All the contributions in the matrix elements h_M are calculated without any arbitrary (fitting) parameters. In the cases when matrix elements are not reduced to experimentally known ones of the hadronic currents, so-called non-factorizing contributions, the latter are estimated in the MIT-bag model. The constants in the vertices VNN ($V = \rho,\omega$) with changing isospin $\Delta T = 0,1,2$ are equal to (in units 10^{-7} /1b/: $h^0_\omega = -8.3$, $h^2_\rho = -6.7$, $h^0_\omega = -3.9$, $h^1_\rho = 0.39$, $h^1_\omega = -2.2$ and correspond qualitatively to the earlier suggested "best values" (b.v.) /2/. The constant $h_\pi = 1.3 \approx 1/3h_\pi^{b.v.}$, that does not contradict the recent experimental restrictions (see e.g. /3/). The most accurate experiments provide the corridors of possible values h^1 and h^0 shown in Fig. 1. A larger picture of the possibilities for describing the low-energy PNC data by our h_M is demonstrated in Table 1 /1c/. Emphasize that the agreement of the theoretical and experimental results follows from manifestation of all the components of the standard theory: charge and neutral currents and the colour interactions.

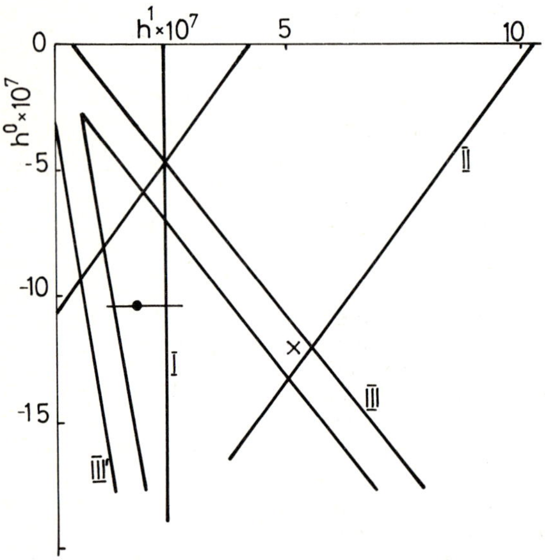

Fig. 1 Restrictions on the combinations $h^1 = h_\pi - 0.12(h^1_\rho + h^1_\omega)$ and $h^0 = h^0_\rho + 0.56\ h^0_\omega$ following from the data on $P_\gamma(^{18}F)$ /3/, $A_\gamma(^{19}F)$ and $P_\gamma(^{21}Ne)$ (90% c.l.). The black circle corresponds to our values of h_M; the segment, to the interval $0.6 \cdot 10^{-7} \le h_\pi \le 3.0 \cdot 10^{-7}$; the cross, to $h_M^{b.v.}$. $P_\gamma^{exp}(^{18}F)$ gives the upper bound I on h^1; $A_\gamma^{exp}(^{19}F)$ provides the corridor II, whereas $P_\gamma^{exp}(^{21}Ne)$ the corridors III or III' depending on the used nuclear model /1c/.

Table 1:

The contributions of the charge and neutral currents are given separately. Value at an effect with colour interactions switched off is given in brackets. For the experimental papers see refs. [1c,3].

process	Q	Q Q Q	Q	Q^{exp}
$\vec{p}p \to pp$	$A_L \cdot 10^7$			
E_p = 15 MeV		-1.2-0.1=-1.3	(-0.43)	-1.2±0.6
45 MeV		-2.1-0.2=-2.3	(-0.76)	2.31±0.89
$np \to d\vec{\gamma}$	$P_\gamma \cdot 10^7$	0.49-0.14=0.35	(0.28)	1.8±1.8
$\vec{n}p \to d\gamma$	$A_L \cdot 10^7$	0.00-0.15=-0.15	(-0.054)	0.6±2.1
$\vec{\gamma}d \to np$	$A_L \cdot 10^7$			
E = 0-0.1 MeV		0.145-0.120=0.25	(0.14)	-
1 MeV		-0.025		-
10 MeV		-0.068		-
30 MeV		-0.17		
$\vec{n}d$	$\varphi \cdot 10^7 (\frac{zad}{m})$	-0.98+3.7=2.7	(1.6)	-
$\vec{n}d \to T\gamma$	$P_\gamma \cdot 10^7$	0.7-3.7=3.0	(0.44)	-
$\vec{p}d \to pd$	$A_L \cdot 10^7$			
E_p = 15 MeV		-2.4-4.1=-6.5	(-1.2)	-3.5±8.5
$\vec{p}\alpha \to p\alpha$	$A_L \cdot 10^7$			
E_p = 46 MeV		-2.0-1.5=-3.5	(-0.39)	-3.3±0.9
$\vec{n}\, {}^4He$	$\varphi \cdot 10^7 (\frac{zad}{m})$	2.6-0.82=1.8	(-1.3)	
${}^{16}O\ (2^-) \to {}^{12}C\alpha$	$\sqrt{\Gamma_\alpha} \cdot 10^5 (MeV)^{1/2}$	0.79-0.31=1.1	(-0.11)	1.0±0.1
${}^{41}K \left(\frac{7}{2}^- \to \frac{3}{2}^+ \right)$	$P_\gamma \cdot 10^5$	0.61+0.69=1.3	(0.13)	2.0±0.4
${}^{175}Ta \left(\frac{5}{2}^+ \to \frac{7}{2}^+ \right)$	$P_\gamma \cdot 10^5$	1.7+2.3=4.0	(0.28)	5.5±0.5
${}^{181}Lu \left(\frac{9}{2}^- \to \frac{7}{2}^+ \right)$	$P_\gamma \cdot 10^6$	-1.5-2.0=-3.5	(-0.22)	-5.2±0.5

1. Dubovik V.M., Zenkin S.V. Comm. JINR, a) E2-83-611, b) E2-83-615, c) E2-83-922, Dubna, 1983.
2. Desplanques B., Donoghue J., Holstein B., Ann. Phys. 1980, 124, 449.
3. Bini M. et al. Phys. Rev. Lett., 55, 794, (1985); Evans C. et al. Phys. Rev. Lett. 55, 791, (1985).

Measurement of the Parity Violation in Quasi-Elastic Electroweak Electron-Scattering from ^{9}Be

W. Achenbach[3], *J. Ahrens*[1], *H.G. Andresen*[2], *A. Bornheimer*[2], *D. Conrath*[3], *K.-J. Dietz*[3], *W. Gasteyer*[3], *H.-J. Gessinger*[3], *W. Hartmann*[3], *W. Heil*[2], *J. Jethwa*[4], *H.-J. Kluge*[3], *H. Kessler*[3], *T. Kettner*[2], *L. Koch*[3], *F. Neugebauer*[3], *R. Neuhausen*[2], *E.W. Otten*[3], *E. Reichert*[3], *F.P. Schäfer*[4], and *B. Wagner*[2]

[1]Max-Planck-Institut für Chemie, Universität Mainz, D-6500 Mainz, F. R. Germany
[2]Institut für Kernphysik, Universität Mainz, D-6500 Mainz, F. R. Germany
[3]Institut für Physik, Universität Mainz, D-6500 Mainz, F. R. Germany
[4]Max-Planck-Institut für Biophysikalische Chemie, Universität Göttingen, D-3400 Göttingen, F. R. Germany

In the energy range of about 300 MeV, available at the Mainz- Linac, quasi-elastic scattering dominates the total cross section at backward scattering angles. This process can therefore be detected efficiently by a gas Cerenkov-counter with large solid angle. This is a prerequisit for experiments on parity violation due tc the very small asymmetry effect being of the order of $1 \cdot 10^{-5}$ at these energies [1]. The counting system built, consists of 12 elliptical mirrors, imaging the Cerenkov photons seen in target direction onto photomultipliers. The mirrors cover the full azimuth for polar angles $115° \leq \vartheta \leq 145°$, thus covering 20 % of 4π (Fig. 1). The detector has been proved to yield a statistical asymmetry error limit of $1 \cdot 10^{-5}$ within half an hour running time at an average current of 20 μA. Under stable Linac conditions the error histogram is Gaussian and its width is compatible with the number of scattered electrons (Fig. 2).

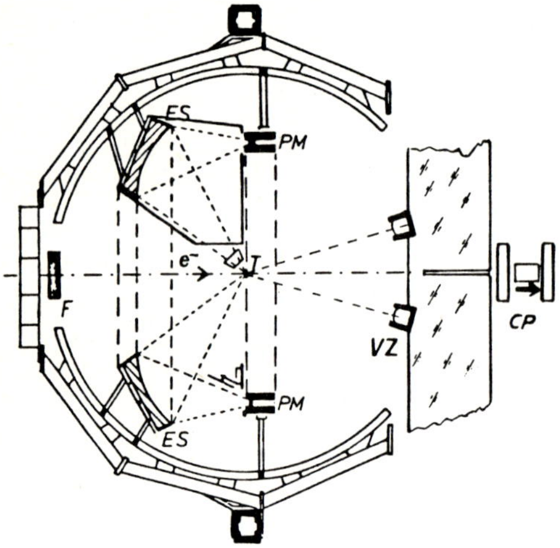

Fig. 1: Side-view of the gas Cerenkov-detector system. ES - elliptical mirror, PM - photomultiplier, T - target, F - ferrit, VZ - forward-angle counter, CP - Compton-polarimeter

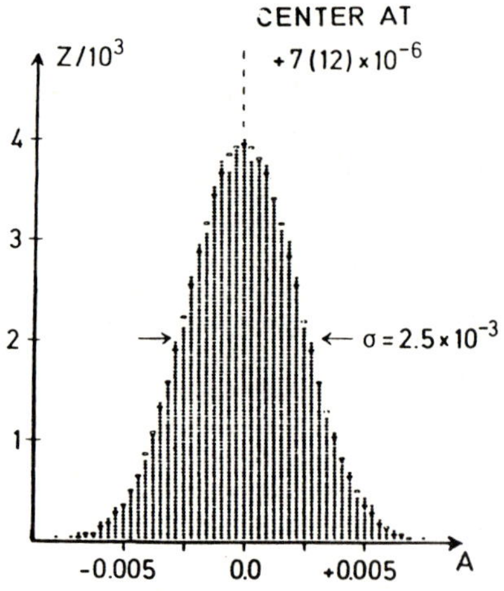

Fig. 2: Distribution of the measured single asymmetries $A = \frac{N^+ - N^-}{N^+ + N^-}$ with unpolarized, ordinary e^- source. Polarization (+,-) defined at will

The polarized electron source is a GaAs photocathode illuminated by a circular-polarized pulsed laser beam from a flash lamp driven dye laser. The construction is analogue to the source used in the SLAC experiment [2]; however, the requirements regarding pulse stability and lifetime are much harder, since the asymmetry effect to be measured is a factor of 10 smaller in our experiment. At present the source reaches the necessary current and pulse width. Its stability and lifetime during operation could be increased up to 200 hours. To avoid apparative asymmetries due to systematic drifts we have reversed the electron helicity statistically between two pulses by a Pockels-cell. With the positions (+,-) of a λ/2-plate in front of the Pockels-cell one changes the sign of the helicity of the circular-polarized light. Therefore keeping all parameters unchanged an observed helicity asymmetry must change its sign by switching the position of the λ/2-plate. Fig. 3 shows the sequence of asymmetry runs (λ/2+) for a total running time of 50 hours. The distribution of the measured asymmetries is shown in Fig. 4 for both λ/2-positions. They are centered at $A = (3.2 \pm 1.6) \cdot 10^{-6}$ for λ/2+ and $A = (-4.2 \pm 1.6) \cdot 10^{-6}$ for λ/2-, respectively. The averaged asymmetry is

$$|\overline{A_{exp}}| = (3.9 \pm 1.1) \cdot 10^{-6}.$$

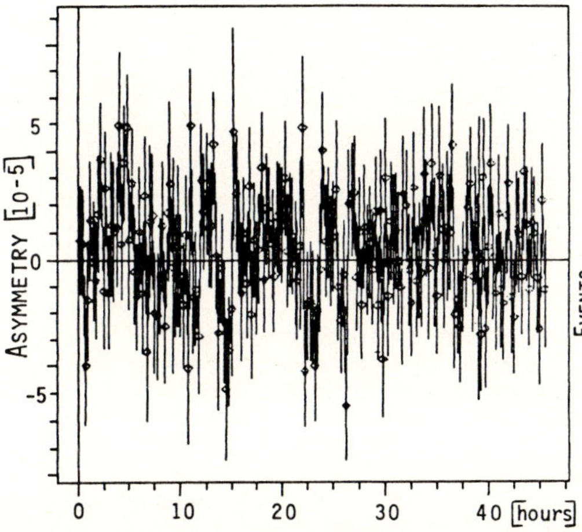

Fig. 3: Sequence of asymmetry runs (15 min each) at an average current of 7 μA, λ/2-position (+)

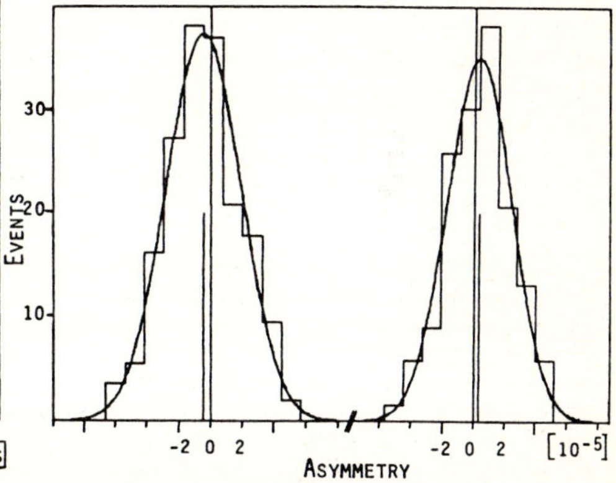

Fig. 4: Distribution of the 15 min asymmetry runs for both λ/2-positions

In order to identify and correct apparative asymmetries correlated to the switching of the polarization by the Pockels-cell, there are control systems, like microwave cavities etc., recording energy and intensity variations, beam position changing and inhomogenities of the target thickness. A correlation between the measured asymmetries of the Cerenkov-counters and the energy variation is shown in Fig. 5. In a separate experiment the polarization of the electrons has been measured at full energy by Møller-scattering. This measurement served also for calibrating a Compton-polarimeter positioned behind the detector system. The Compton-polarimeter provided an on-line control of the longitudinal polarization of the electron beam at target position during the measurement (Fig. 6).

The interest in the parity violating effect in quasi-elastic scattering lies in the fact that both pv-amplitudes of the weak neutral current contribute, the vector part of the nucleon as well as its axial vector part, given in the form [3]:

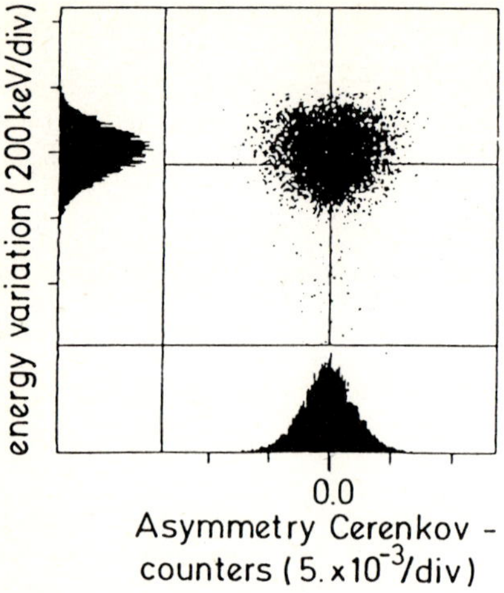

Fig. 5: Correlation plot between the asymmetry of the Cerenkov-counters and the energy variation measured with a microwave cavity in the beam line

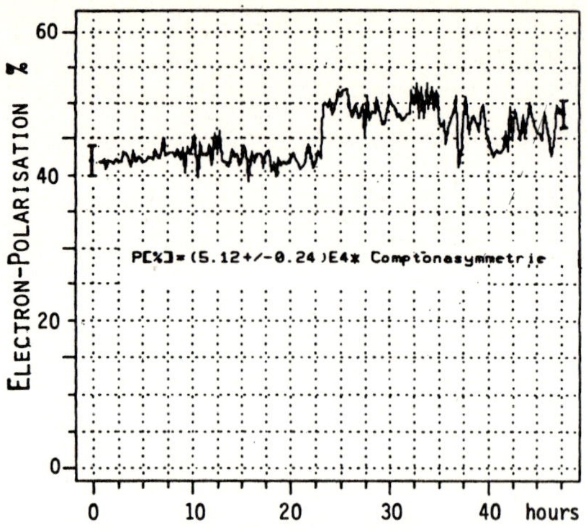

Fig. 6: Electron polarization at target position during experiment

$$L^{eH}_{p.v.} = - \frac{G_F}{2} [\bar{e} \gamma_\mu \gamma_5 e(\alpha V^3_\mu + \gamma V^0_\mu) + \bar{e} \gamma_\mu e(\beta A^3_\mu + \delta A^0_\mu)]$$

The Slac-asymmetry-experiment together with atomic parity violating experiments succeeded in determining $\tilde{\alpha},\tilde{\gamma}$ in a model-independent way [4]. We included their results in calculating the $\tilde{\beta},\tilde{\delta}$ coupling constants. This is seen from the diagram (Fig. 7) defining in a plane of coupling constants the allowed area which is the cross section of the different experimental results (neutrino experiments, SLAC, ...). Measurements at different energies in the medium energy region will allow us to determine $\tilde{\alpha}$, $\tilde{\beta}$, $\tilde{\gamma}$, $\tilde{\delta}$ in a model-independent way. Regarding future experiments at the Mainz microtron Δexcitation by polarized electrons would be another interesting medium energy experiment in this context; it would isolate the isovector part of the neutral current [5].

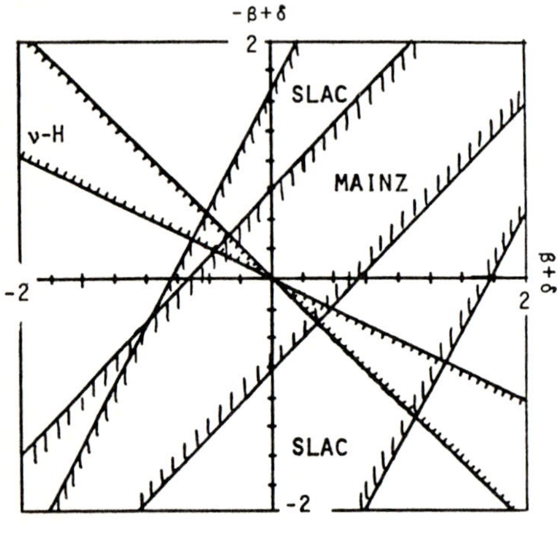

Fig. 7: Array of the axial vector coupling constants $(\tilde{\beta},\tilde{\delta})$ of the hadronic current defined by present experiments (90 % C.L.)

References:

[1] E. Hoffmann and E. Reya: Phys. Rev. 18D, 3230 (1978)

[2] E. Reichert: AIP Conference Proceedings, High Energy Spin Physics-1982 Brookhaven

[3] P.Q. Hung and J.J. Sakurai: Phys. Lett. 63B, 295 (1976)

[4] P.Q. Hung and J.J. Sakurai: Annual Rev. of Nucl. and Part. Sience 31 (1981)

[5] D.R.T. Jones and S.T. Petcov: Phys. Lett. 91B, 137 (1980)
L.M. Nath, K. Schilcher and M. Kretzschmar: Phys. Rev. D25, 2300 (1982).

Strong Interaction Effects in Parity Violation in p-p Elastic Scattering

G. Roy[1], *J. Birchall*[2], *and W.T.H. van Oers*[2]

[1]Physics Dept. University of Alberta, Edmonton, Alberta, Canada T6G 2J1
[2]Cyclotron Lab., University of Manitoba, Winnipeg, Manitoba, Canada R3T 2N2

Parity violation (PV) in the hadronic interaction has been the focus of experimental and theoretical studies for many years. While the strong and electromagnetic interactions conserve parity, the weak interaction does not, and thus parity violation measurements serve as a probe of the weak part of the hadronic interaction. One example of an observable which is only sensitive to the weak interaction between hadrons is the longitudinal polarization asymmetry A_z in proton-proton scattering[1)]. At low energies it has become standard practice to relate PV effects to an interaction between nucleons based on one and two vector-meson exchanges between a parity conserving (strong interaction) vertex and a parity violating (weak interaction) vertex. The strong interaction is presently well understood at these energies and is described by meson-exchange potentials, parametrized by phase-shift analysis. The PV interaction is parametrized by a set of meson-nucleon coupling constants. An extensive review paper by Desplanques et al.[2)] synthesized various approaches based on the quark model and SU(6) to calculating these coupling constants. The purpose of the present work is to concentrate on the strong interaction effects in PV in p-p scattering. Calculations have been performed using Thorndike's[3)] and Woodruff's[4)] approach: PV phase shift transitions are calculated involving known phases and a PV mixing angle is used as a parameter to set the magnitude of A_z. The PV phase shift mixings considered here are ${}^1S_0-{}^3P_0$, ${}^3P_2-{}^1D_2$, ${}^1D_2-{}^3F_2$ etc. The calculations deal only with the strong interaction effects (i.e. the phases). They do not determine by themselves the magnitude of A_z, but do show the angular dependence which is set by the strong interaction. In order to set the magnitude of A_z, we normalise to the experimen-tal results at 15 MeV[6)], 45 MeV[7)] and 800 MeV[8)].

Our calculations are shown in the figures. In both cases, A_z has been multiplied by 10^7. Figure 1 shows our results for energies of 15 MeV and 45 MeV. The solid points are our results, while the open points are taken from Brown, Henley and Krejs[5)]. They obtained weak interaction information from a potential model using the Born approximation. Their results have been multiplied by one-half to take into account their definition of A_z. They obtained the wrong sign for A_z; the meson-nucleon coupling constants they used were at the limits of the acceptable values found by DDH[2)]. Figure 2 shows our calculations at 800 MeV, for three different cases of mixing: the curve labelled J = 0 signifies ${}^1S_0-{}^3P_0$ mixing, while J = 0+2 means J = 0 plus $^3P_2-^1D_2$ plus $^1D_2-^3F_2$ mixing, and J = 0+2+4 means J = 0+2 plus $^3F_4-^1G_4$ and $^1G_4-^3H_4$ mixing. The sharp drop at forward angles in our results is due to Coulomb effects; these were excluded by Brown, Henley and Krejs[5]. As expected, at low energies, the angular dependence is mostly isotropic; however at 800 MeV, considerable variation occurs with angle. Note that all the experimental measurements actually are of the angle-integrated A_z, and that at 800 MeV, the three possible mixings shown all yield an angle-integrated result that is consistent with the LAMPF experimental value, once the angular distribution is folded in. Thus the positive sign for A_z at 800 MeV is due to the strong interaction vertex. It is unfortunate that angle-dependent information was not obtained in the 800 MeV measurement, in order to help unfold the different J-contributions to A_z.

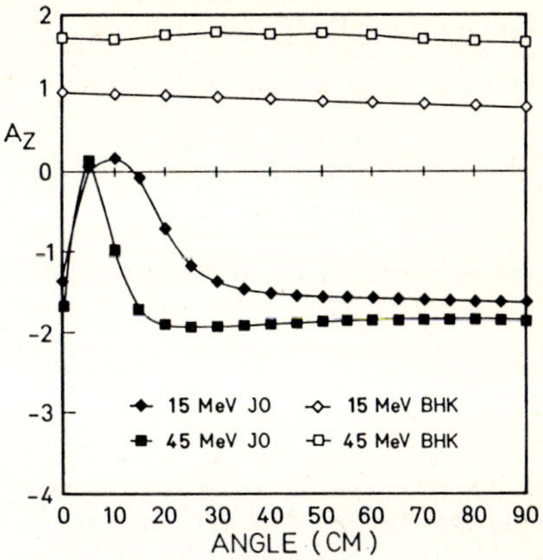

Fig. 1: A_z at 15 and 45 MeV

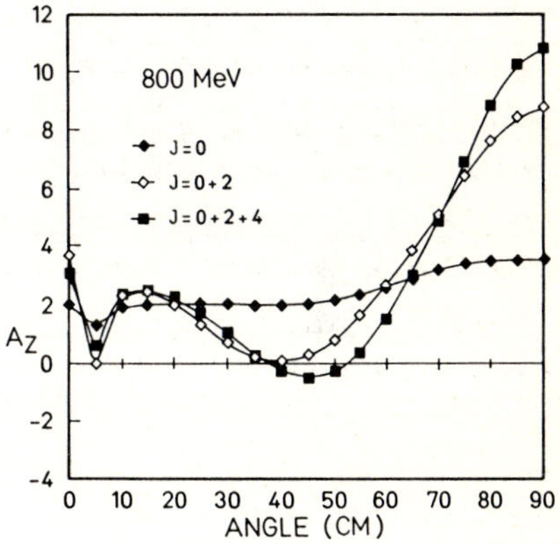

Fig. 2: A_z at 800 MeV

References

1. M. Simonius, Phys. Lett. 41B, 415 (1972).
2. B. Desplanques, J.F. Donoghue and B. Holstein, Ann. Phys. (N.Y.) 124, 449 (1980).
3. E.H. Thorndike, Phys. Rev. 138B, 586 (1965).
4. A.E. Woodruff, Ann. Phys. (N.Y.) 7, 65 (1959).
5. V. Brown, E. Henley, and F. Krejs, Phys. Rev. C9, 935 (1974).
6. J.M. Potter et al., Phys. Rev. Lett. 33, 1307 (1974).
7. R. Balzer et al., Phys. Rev. C30, 1409 (1984).
8. V. Yuan, Proc. of the 18th LAMPF Users Group Mtg., 1984 LA-10370-C.

Progress Report on an Experiment to Measure $\Delta I=0$ Parity Mixing in ^{14}N

H.E. Swanson[1], V.J. Zeps[1], E.G. Adelberger[1], C.A. Gossett[1], J. Sromicki[2], W. Haeberli[2], and P. Quin[2]

[1]University of Washington, Seattle, WA 98195, USA
[2]University of Wisconsin-Madison, Madison, WI 53706, USA

We are attempting to detect parity mixing of the $J^{\pi}=0^+;1 - 0^-;1$ doublet in ^{14}N at $E_x \approx 9$ MeV by forming the narrow ($\Gamma^{cm} = 3.8$ keV $E_p = 1.16$ MeV) 0^+ level as a resonance in $^{13}C(p,p)$ and measuring the longitudinal analyzing power A_z[1]. This mixing is primarily sensitive to the $\Delta I=0$ component of the PNC NN force. Its interpretation has fewer nuclear structure uncertainties than the α decay of $^{16}O(2^-)$ [2]. The calculated $^{13}C(p,p)$ scattering cross sections and transverse analyzing powers A_t [3] agree favorably with our data. This gives us confidence in our model of the reaction. The PNC signal is $A_z(b) - A_z(f)$ where b and f refer to counters placed at $\Theta_b = 155^o$ and $\Theta_f = 35^o$. Using recently measured values for E_x and Γ of the doublet [4], along with Haxton's PNC matrix element and the standard DDH "best value" PNC interaction we predict a signal of -2.2×10^{-5}.

Our apparatus is shown in figure 1. A 25 $\mu g/cm^2$ ^{13}C target is viewed by 8 scintillation counters. The beam is held along the symmetry axis of the apparatus by high-bandwith feedback loops. The mean spin direction on target is held longitudinal by feedback systems which null the transverse asymmetries in the scintillation counters. The apparatus was constructed and tested with unpolarized beam in Seattle and then taken to Madison to run with their polarized proton beam.

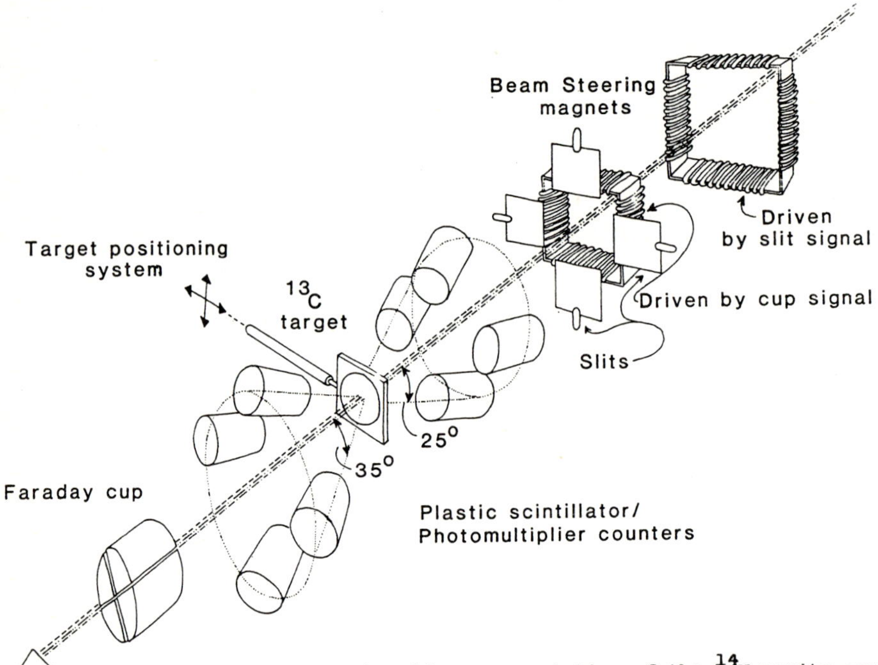

Fig. 1. Schematic representation of the ^{14}N parity experiment.

We are currently investigating systematic contributions to the measurement of A_z. These fall in two categories, spin independent beam modulations and effects from transverse polarization gradients. Tests with 1.16 MeV unpolarized beam on ^{13}C targets and 1.7 MeV polarized beam on an Au ($A_t \approx 0$) target show that spin independent modulations give a negligible contribution to our result. Tests with polarized beam on a ^{12}C target ($A_t \approx 0.8$) show significant false asymmetries arising from residual transverse components of polarization. To investigate these spurious spin dependent effects, we use the overlaping $3/2^-$, $5/2^+$ resonances in $^{12}C(p,p)$ at $E_x \approx 1.7$ MeV which cannot be parity mixed but which have large transverse analyzing powers. When we scan a strip target (0.3 mm wide 150 $\mu g/cm^2$ C strip evaporated on a 2 $\mu g/cm^2$ C backing) through the beam at our target position, we observe polarization gradients $\partial P_y/\partial x$ and $\partial P_x/\partial y$ which correspond to a transverse polarization of 1 - 2% at the edge of the beam. The measured beam parameters, sensitivities of our apparatus and corresponding false asymmetries are shown in Table 1.

Table 1. Systematic contributions to A_z from coherent beam modulations.

Quantity	Measured Value	Sensitivity*	Resulting False Asymmetry
	Spin Independent Effects		
Intensity ($\Delta I/2I$)	$\lesssim 5\text{x}10^{-4}$	$-2.1\text{x}10^{-3}$	$\lesssim 1.0\text{x}10^{-6}$
Position	$(5\pm2)\text{x}10^{-4}$ mm	$-1\text{x}10^{-4}/\text{mm}^2 \times \epsilon$	$\lesssim 1.2\text{x}10^{-8}$
Angle	-	$-1.7\text{x}10^{-7}/\text{mrad}^2 \times \alpha$	-
Width ($\Delta\Gamma/\Gamma$)	$<2\text{x}10^{-3}$	$7.5\text{x}10^{-4}$	$<1.5\text{x}10^{-6}$
Energy	0.3 ± 0.6 eV	$2\text{x}10^{-6}/\text{eV}$	$(0.6\pm1.2)\text{x}10^{-6}$
	Spin Dependent Effects		
$(\langle\vec{\epsilon}\rangle \times \langle\vec{p}\rangle)\cdot\hat{z}$	$<2.5\text{x}10^{-4}$ mm	$-5.5\text{x}10^{-4}/\text{mm}$	$<1.4\text{x}10^{-7}$
$\langle\vec{\alpha}\rangle \cdot \langle\vec{p}\rangle$	$<6\text{x}10^{-4}$ mrad	$2.4\text{x}10^{-4}/\text{mrad}$	$<1.4\text{x}10^{-7}$
$\langle\epsilon_x p_y\rangle - \langle\epsilon_y p_x\rangle$	≈ 0.02 mm	$-6.7\text{x}10^{-5}/\text{mm}$	$\approx 1.3\text{x}10^{-6}$
$\langle\alpha_x p_x\rangle + \langle\alpha_y p_y\rangle$	-	$2.0\text{x}10^{-5}/\text{mrad}$	-

$\hat{z}$ = nominal chamber axis, $\vec{\epsilon}$ = displacement of beam on target from z-axis, $\vec{\alpha}$ = angle of beam on target from z-axis, $\vec{p}$ = polarization vector.

* Sensitivities are for E_p(lab)=1158 keV, 25 $\mu gm/cm^2$ ^{13}C target.

Recent studies suggest that the fringing fields of the spin precessor are responsible for the observed transverse polarization gradients. Since known systematic errors are small compared to the predicted A_z, we are planning to take data on the $^{13}C(p,p)$ 0^+ resonance to see if the $\Delta I=0$ PNC NN interaction has the expected strength.

References

1. Adelberger, E.G., Hoodbhoy, P., Brown, B.A., Phys. Rev. C30 456 (1984)
2. Neubeck, K, Schober, H., Waffler, H., Phys. Rev. C 10:340(1974)
3. Adelberger, E.G., Hoodbhoy, P., Brown, B.A., Phys. Rev. C33 1840 (1986)
4. Fernandez, P.B., Gossett, C.A., Osborne, J.L., Zeps, V.J., Adelberger, E.G., Bull. Am. Phys. Soc. 30 1161 (1985)

3.3 Lepton Number Violation and Neutrino Mass

3.3.1 Double Beta Decay

Double Beta Decay and Nuclear Structure

K. Grotz and H.V. Klapdor

Max-Planck-Institut für Kernphysik, D-6900 Heidelberg, F. R. Germany

Abstract: The influence of different properties of nuclear structure on nuclear matrix elements for two-neutrino and neutrinoless double beta decay is studied. We further present nuclear structure calculations for all potential 35 double beta emitters with $A \geq 70$. New and more stringent limits on the neutrino mass are deduced from existing double beta decay experiments. It is found that two-neutrino decay is strongly influenced by pairing as well as by various long-range forces. Neutrinoless decay on the other hand is found to be determined mainly by pairing, allowing more reliable predictions for this decay mode. As a consequence of the different effects of the long-range forces on the matrix elements the often-used "scaling procedure" leads to overestimates of the neutrino mass. It is discussed to what extent interference between different neutrinos affects the obtained mass limits.

1. Introduction

The exploration of the nature of the neutrino at the lower end of the mass hierarchy of elementary particles is a great challenge. Nuclear physics can contribute to the solution of the question of neutrino masses and mixing by the investigation of neutrinoless double beta decay ($0\nu\ \beta\beta$). The latter is a sensitive probe for a Majorana mass of the neutrino [1], which might be a unique property of the neutrino among the elementary fermions.

Neutrinos with Majorana masses would correspond to the breaking of B-L symmetry and should manifest themselves in the appearance of neutrinoless double beta decay. We shall briefly repeat the motivation from grand unified theories (GUTs) for searching for Majorana neutrinos by double beta decay experiments.

Starting from the standard SU(5) model, the prediction for the neutrino mass is zero, because on one hand in minimal SU(5) the right-handed neutrino ν_R (not to be confused with the normal antineutrino conventionally denoted by $\bar{\nu}; \bar{\nu} = (\nu_L)^C$) does not exist at all (fig. 1); hence no Dirac mass is possible. On the other hand

SU(5)   SO(10)

$$\bar{5} = \begin{bmatrix} d_g^c \\ d_r^c \\ d_b^c \\ e^- \\ -\nu \end{bmatrix}_L \quad 10 = \begin{bmatrix} 0 & -u_b^c & u_r^c & u_g & d_g \\ & 0 & -u_g^c & u_r & d_r \\ & & 0 & u_b & d_b \\ & \text{anti-} & & 0 & e^+ \\ & \text{symmetric} & & & 0 \end{bmatrix}$$

$$\begin{array}{|ccccc|cccc|} \hline \nu_L & d_g^c & d_r^c & d_b^c & u_b & u_r & u_g & e^+ \\ \hline e^- & u_g^c & u_r^c & u_b^c & d_b & d_r & d_g & \nu_R^c \\ \hline \end{array}$$

$$16_{SO(10)} = 10_{SU(5)} + \bar{5}_{SU(5)} + 1_{SU(5)}$$
$$\searrow \nu_R$$

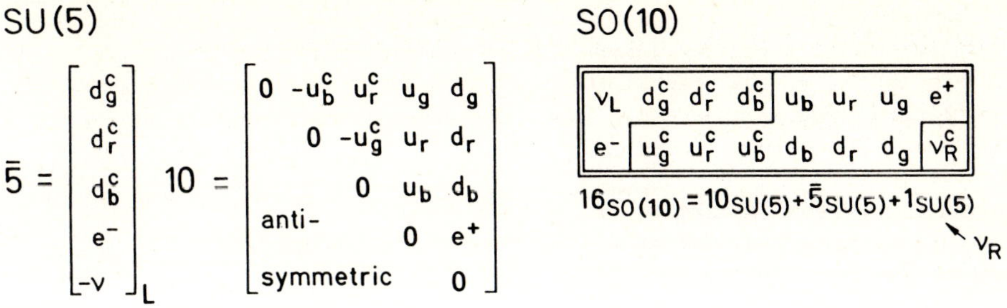

Fig. 1. Multiplets of the fermions of the first generation in the SU(5) and in the SO(10) model. The right-handed neutrino (left-handed antineutrino), not existing in SU(5), is a natural consequence in SO(10).

SU(5) invariant Majorana couplings are not possible with the Higgs content of the minimal model. However, it is also well known that this result should not be taken too seriously, because there are many shortcomings of this model, for example it does not naturally include CP violation.

Extending the standard model leads to nonvanishing neutrino masses. In SO(10), for example, the elementary fermions of one family are placed in a 16-dimensional spinor representation, which consists of the 5- and 10-dimensional SU(5) representations and one additional neutral fermion, which has to be interpreted as the right-handed neutrino ν_R.

Although there is a possibility of making the Dirac mass term for the neutrino vanish by a very delicate choice of model parameters, the most natural prediction is a neutrino Dirac mass in the MeV range. An attractive model for the generation of fermion masses has recently been considered by STECH [2].

It is the exceptional position of the neutrinos that could solve this apparent conflict with reality. An interplay between the large Dirac mass terms, which can hardly be avoided, and an additional Majorana mass term could lead to the phenomenologically desired small neutrino mass eigenstate. As considered by GELL-MANN, RAMOND and SLANSKY [3], in the presence of a right-handed Majorana term $\overline{m_R^M (\nu_R)^C} \nu_R$ much larger than m^D, related to the breaking of a global B-L symmetry, the neutrino mass matrix simplified to the one-flavor case (for n flavors m^D and m_R^M have to be replaced by n x n matrices) would have the form

$$(\overline{\nu_L (\nu_R)^C}) \begin{bmatrix} 0 & m^D \\ m^D & m_R^M \end{bmatrix} \begin{bmatrix} (\nu_L)^C \\ \nu_R \end{bmatrix} \quad \begin{array}{l} m_R^M \gg m^D \\ m^D = 0 \text{ (MeV)} \end{array} \quad (1)$$

Diagonalization yields one very light Majorana neutrino with

$$m_1 \simeq \frac{(m^D)^2}{m_R^M} \quad (2)$$

consisting mostly of ν_L and one very heavy with $m_2 \simeq m_R^M$. For $m_R^M \geq 10^3$ GeV $m_1 \lesssim 0$ (eV) would result. The weak point in this simple model is the fact that the left-handed Majorana mass term must be zero or at least very small. However, the model illustrates the general mechanism of generating small Majorana neutrino masses by the mixing of the unavoidable (except in SU(5)) large Dirac mass terms with ever larger Majorana mass terms. As a consequence the occurrence of the neutrinoless double beta decay is expected.

2. Neutrinoless Double Beta Decay

Let us first briefly repeat the main properties of the two decay modes of double beta decay. The so-called 2ν decay,

$$ {}^{A}_{Z}X \rightarrow {}^{A}_{Z+2}Z + 2e^{-} + 2\bar{\nu} \tag{3} $$

can be understood simply as a reduplication of single beta decay (Fig. 2a).

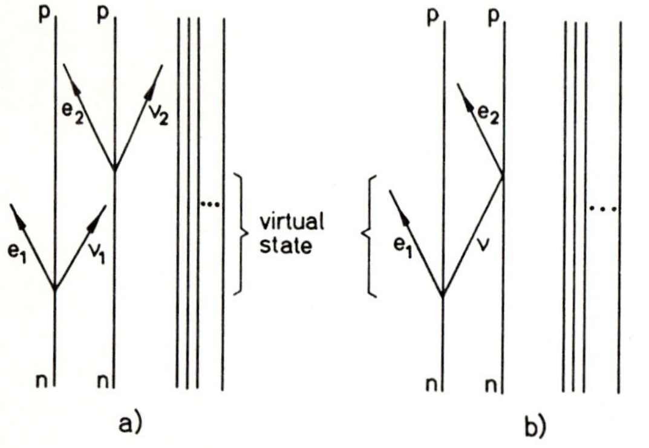

Fig. 2. Schematic representations of the 2ν and 0ν ββ decay modes, both as dominating two-nucleon processes.

Of course, this is an interesting process for itself, because it is a second-order effect of the weak interaction with extremely large half-lives (10^{20} years typically), and only this decay mode has been observed, though indirectly, by geochemical methods (see Table 1). On the other hand 2ν double beta decay is generally thought of as being a test for the nuclear structure calculations needed to analyze the neutrinoless decay mode

$$ {}^{A}_{Z}X \rightarrow {}^{A}_{Z+2}X + 2e^{-} \tag{4} $$

in respect to B-L violating parameters (however, as we will show, this is not so much the case). In the neutrinoless decay mode the neutrino serves only as an exchange particle between the two vertices.

It is easy to see that this process is impossible with a Dirac neutrino. In the first decay step, which can be thought of as a single beta decay, an antineutrino is emitted together with an electron; however the second vertex corresponds to neutrino capture. This process is therefore only possible if the neutrino is its own antiparticle, which means it must be a Majorana particle. If this condition is fulfilled, the still existing mismatch of the handedness of the neutrino and the handedness of the second decay vertex can be removed only if

a) the weak interaction is not purely left-handed

or

b) the neutrino has a finite (Majorana) mass and, therefore, no definite handedness.

0ν $\beta\beta$ decay is therefore a sensitive probe for the nature of the neutrino as well as for the nature of the weak interaction.

Writing down the effective (low energetic) form of a generalized weak interaction [4]:

$$H_W = \frac{G_B}{\sqrt{2}} [j_{L\mu} J^{\mu +}_L + \kappa j_{L\mu} J^{\mu +}_R + \eta j_{R\mu} J^{\mu +}_L + \lambda j_{R\mu} J^{\mu +}_R] \qquad (5)$$

with the left- (right-) handed weak leptonic and hadronic currents $j_{L(R)}$ and $J_{L(R)}$, respectively, we may parametrize the decay rate $\omega^{0\nu}$ of neutrinoless double beta decay of a given nucleus:

$$\omega^{0\nu} = A\langle m_\nu\rangle^2 + B\langle\lambda\rangle^2 + C\langle\eta\rangle^2 + D\langle m_\nu\rangle\langle\lambda\rangle + E\langle m_\nu\rangle\langle\eta\rangle + F\langle\lambda\rangle\langle\eta\rangle \qquad (6)$$

The brackets around the neutrino mass and the interaction parameters indicate that these parameters are effective parameters. For the neutrino mass we have (see discussion in section 4)

$$\langle m_\nu\rangle = \sum_i U^2_{e_i} m_i$$

with the mixing coefficients U_{e_i}. The coefficients A, B, ... F depend on the decay energy, the atomic number and the nuclear structure of the considered nucleus.

We will not discuss the terms involving $\langle\lambda\rangle$ and $\langle\eta\rangle$ further. This is done in the contribution by Tomoda to this conference [5]. We restrict ourselves to the case $\kappa = \lambda = \eta = 0$, where we can write:

$$\omega^{0\nu} = A\langle m_\nu\rangle^2 = f^{0\nu} |1 - X_F|^2 |R_0 M^{0\nu}|^2 \langle m_\nu\rangle^2 \qquad (7)$$

Here we have split the coefficient A into a factor $f^{0\nu}$, containing the weak coupling constant and the leptonic phase space corrected for the Coulomb interaction (using the relativistic Fermi function), and the nuclear structure matrix elements $M^{0\nu}$ and X_F. The form of $M^{0\nu}$ is given by:

$$M^{0\nu} = \langle f|\Sigma\sigma\tau^-\sigma\tau^- H(r)|i\rangle \qquad (8)$$

X_F is the relative fraction of Fermi-type contributions

$$X_F = \frac{1}{c_A^2} \frac{\langle f|\Sigma\tau^-\tau^- H(r)|i\rangle}{M^{0\nu}} \qquad (9)$$

with $c_A = 1.25$.

H(r) is a correlation function depending on the distance of the two decay vertices, arising from the propagation of the virtual neutrino. H(r) is usually called the neutrino "potential" and is approximately given by

$$H(r) = \int \frac{e^{i\vec{q}\vec{r}}}{\langle E\rangle + q_o} d^3q \qquad (10)$$

with q_μ being the four momentum of the intermediate neutrino and $\langle E\rangle$ the mean nuclear excitation energy. Using in (10) the mean value $\langle E\rangle$ instead of the excitation energy E_m of the individual nuclear levels leads to the so-called closure approximation. For the 0ν decay this approximation is well justified because important contributions to H(r) arise only for $q_o \gg \langle E\rangle$. In the case of 2ν decay, however, the intermediate 1^+ spectrum must be included explicitly [6].

The major problem in applying eq. (7) to deduce limits for the neutrino mass arises from the fact that there is no way of deducing the nuclear matrix elements $M^{0\nu}$ experimentally. They can only be calculated and there exists also no obvious way of testing these calculations. What can be tested by experiment are predictions for the conventional allowed 2ν $\beta\beta$ decay. However such calculations up to now predicted much too large rates for this decay mode as compared to the geochemical data for ^{82}Se and ^{128,130}Te. Table 1 shows the predictions of some nuclear structure calculations for 2ν $\beta\beta$ half-lives in comparison with experimental results. For the sometimes used calculations by HAXTON et al. [13] this discrepancy with experiment ranges up to a factor of 150 for ^{130}Te [10]. Since the calculated matrix elements for 2ν decay were too large, up to now it was widely believed (see, for example, [4,10,13]) that similarly the matrix elements for the 0ν decay $M^{0\nu}$ resulting from the same nuclear structure calculations are too large, too, and cannot be used without modification in analyzing experimental data.

Table 1. Results from various nuclear structure calculations in comparison with experimental data for 2ν $\beta\beta$ decay.

	$T^{2\nu}_{1/2\ exp}$	$T^{2\nu}_{1/2\ calc}$
^{48}Ca	$>3.6 \times 10^{19}$ [7]	0.61×10^{19} [12] 9×10^{19} a) [11]
^{76}Ge	$>8 \times 10^{19}$ [8]	3.7×10^{20} b) [13] 1.18×10^{21} [14] 2.2×10^{20} [6]
^{82}Se	$>1.0 \times 10^{20}$ [9] $(1.45 \pm 0.15) \times 10^{20}$ [10]	0.17×10^{20} b) [13] 0.33×10^{20} [14] 0.15×10^{20} [6]
^{128}Te	$>8 \times 10^{24}$ [10]	0.08×10^{24} b) [13] 0.11×10^{24} [14] 0.57×10^{24} [6]
^{130}Te	$(2.55 \pm 0.20) \times 10^{21}$ [10]	0.023×10^{21} c) [15] 0.017×10^{21} b) [13] 0.023×10^{21} [14] 0.12×10^{21} [6]

a) Result extrapolated from restricted model space.
b) Closure treatment.
c) Corrected for phase space.

A usual procedure was therefore to scale $M^{0\nu}$ down according to the observed discrepancy on the 2ν sector.

$$M^{0\nu}_{scaled} = M^{0\nu}_{calc} \left| \frac{\omega_{2\nu_{exp}}}{\omega_{2\nu_{calc}}} \right|^{1/2}$$

Implicitly this scaling assumes that the 2ν and 0ν matrix elements are reduced proportional to each other by some unknown reduction mechanism. As will be discussed below, we find that such a scaling procedure underestimates systematically the 0ν matrix element and leads to too large neutrino mass limits.

3. Nuclear Structure Studies of 2ν and 0ν $\beta\beta$ Decay

There have been a lot of attempts to calculate nuclear matrix elements for double beta decay, especially for the 2ν decay mode. The straightforward method is to use conventional shell model techniques [11-13,16,17]. This method, however, is limited to light double beta emitters. For heavy nuclei the model space and configurations have to be seriously truncated and the collective effects important in double beta decay can only partly be included. It is difficult or even impossible to treat explicitly the intermediate 1^+ spectrum, which could be compared with results from (p,n) experiments, within this approach. This has only been done for ^{48}Ca [11,12]. In the other calculations [13,16,17] the detailed structure of the intermediate states was neglected and the closure approximation was used. In this approximation one obtains the analogue of eq. (7) for the 2ν decay:

$$\omega_{2\nu} = f^{2\nu} \frac{|M^{2\nu}|^2}{\langle E\rangle^2} \qquad \text{with } M^{2\nu} = \langle F|\sum_{ij} \vec{\sigma}(i)\tau^-(i)\vec{\sigma}(j)\tau^-(j)|I\rangle \tag{11}$$

This greatly simplifies the calculations, but is not justified for 2ν decay [6,11,12].

The most important collective effect in double beta decay comes from pairing correlations. They lead to the fact that many shells contribute all coherently to double beta decay. This effect is included in calculations using BCS wave functions [17,18], or in an HFB approach [5]. However, taking into account only pairing forces yields much too large matrix elements for the 2ν decay [6,19]. The next step is to include besides pairing also neutron-proton forces, the most important of these for double beta decay being the spin-isospin part, which leads to the concentration of the GT strength for single β^- decay in the Gamow-Teller giant resonance and also to a reduction of the total β^- strength and more important of the β^+ strength. The spin-isospin (GT) force together with the pairing forces can be included in an RPA approach based on BCS wave functions [15]. It is important to note that in this approach the closure approximation is avoided.

We have performed such RPA calculations using a deformed Nilsson potential for all double beta emitters with $A \geq 70$ [20]. A similar approach was later used in ref. [14] for several nuclei. However, in our calculations the BCS wave functions were separately optimized for parent and daughter nucleus (taking into account the change in proton and neutron numbers). The results are given in Table 2. Unfortunately this approach is not suitable to calculate explicitly the 0ν matrix elements. One step beyond this RPA treatment is to include besides the spin-isospin part also other neutron-proton forces. We have done this for the strongest part of the neutron-proton interaction, the quadrupole-quadrupole force [6,19]. This force especially in the isotopes $^{128,130}Te$ and $^{128,130}Xe$ leads to low-energetic 2^+ phonons, which are admixed to the 0^+ ground states and give rise to destructive contributions to the 2ν double beta matrix elements. We have performed these calculations for several nuclei using particle number projected BCS wave functions (PBCS) in large model spaces (12 subshells). The strength of the pairing interaction was deduced from existing pickup and stripping data. The neutron-proton Hamiltonian

$$H_{np} = \chi\Sigma\vec{\sigma}(i)\tau^-(i)\vec{\sigma}(j)\tau^-(j) - \kappa QQ \tag{12}$$

was then diagonalized in this model space. Ground state correlations involving four quasiparticles arising from the spin-isospin part were included by exact diag-

Table 2. Results of the RPA calculations for double beta decay. Given are the calculated half-lives for the 2ν and 0ν decay modes and calculated closure matrix elements $M^{2\nu}$. The 2ν half-lives were calculated taking into account the full intermediate 1^+ spectrum. The 0ν half-lives were obtained as described in the text. The 0ν half-lives are neutrino mass dependent and the given figures correspond to $\langle m_\nu \rangle = 1$ eV. The $T^{2\nu *}_{1/2}$ are the half-lives obtained with the PBCS treatment. The 0ν half-lives obtained using the matrix elements given in the third column of Table 4 are also marked with an asterisk. T_0 denotes the $\beta\beta$ Q-value, δ the adopted deformation of parent and daughter nucleus.

	T_0 (MeV)	QRPA δ	$M^{2\nu}$	$T^{2\nu}_{1/2}$ (years)	$T^{2\nu *}_{1/2}$ (years)	$T^{0\nu}_{1/2} \times \langle m_\nu \rangle^2$ (years x eV^2)
^{70}Zn	1.00	0	6.79	1.9×10^{22}		7.6×10^{23}
^{76}Ge	2.04	0.2	5.54	1.1×10^{20}	2.2×10^{20}	2.6×10^{23}*
^{80}Se	0.136	0.2	5.33	9.0×10^{28}		6.7×10^{25}
^{82}Se	3.01	0.2	4.47	4.5×10^{18}	1.5×10^{19}	9.5×10^{22}*
^{86}Kr	1.25	0	1.63	3.5×10^{22}		5.0×10^{24}
^{94}Kr	1.15	-0.1	4.63	4.1×10^{21}		6.2×10^{23}
^{96}Zr	3.35	-0.12	4.83	5.2×10^{17}		1.6×10^{22}
^{98}Mo	0.111	-0.19	4.33	1.6×10^{29}		7.3×10^{25}
^{100}Mo	3.03	-0.24	3.81	1.8×10^{18}		3.3×10^{22}
^{104}Ru	1.30	-0.26	3.81	1.8×10^{21}		5.0×10^{23}
^{110}Pd	2.01	-0.23	3.56	5.0×10^{19}		1.3×10^{23}
^{114}Cd	0.54	0.14	1.80	2.7×10^{24}		1.7×10^{25}
^{116}Cd	2.81	0	1.63	8.3×10^{18}		1.7×10^{23}
^{122}Sn	0.36	0	1.75	1.4×10^{26}		3.6×10^{25}
^{124}Sn	2.28	0	1.65	9.3×10^{19}		3.1×10^{23}
^{128}Te	0.87	0.15	2.36	1.2×10^{23}	5.7×10^{23}	9.8×10^{23}*
^{130}Te	2.53	0.10	2.55	1.9×10^{19}	1.2×10^{20}	4.6×10^{22}*
^{134}Xe	0.84	0	3.18	5.1×10^{22}	2.5×10^{23}	8.7×10^{23}*
^{136}Xe	2.48	0	1.22	6.0×10^{19}	3.3×10^{19}	3.0×10^{23}*
^{142}Ce	1.41	0	1.21	2.8×10^{21}	4.1×10^{20}	4.7×10^{23}*
^{146}Nd	0.061	0	3.38	2.9×10^{30}		5.6×10^{25}
^{148}Nd	1.93	0.18	2.63	2.5×10^{19}		1.1×10^{23}
^{150}Nd	3.37	0.24	2.17	4.8×10^{17}		2.4×10^{22}
^{154}Sm	1.25	0.28	3.10	9.5×10^{20}		2.4×10^{23}
^{160}Gd	1.73	0.29	3.46	4.4×10^{19}		6.4×10^{22}
^{170}Er	0.66	0.27	2.80	6.6×10^{22}		9.1×10^{23}
^{176}Yb	1.08	0.26	2.69	1.1×10^{21}		2.5×10^{23}
^{186}W	0.49	0.20	2.46	3.2×10^{23}		1.2×10^{24}
^{192}Os	0.41	-0.15	2.32	1.7×10^{24}		1.6×10^{24}
^{198}Pt	1.04	-0.10	0.78	1.2×10^{22}		1.6×10^{24}
^{204}Hg	0.41	0	0.47	4.6×10^{25}		2.6×10^{25}
^{232}Th	0.85	0.23	3.57	1.6×10^{20}		3.8×10^{22}
^{238}U	1.15	0.24	2.87	2.2×10^{19}		2.4×10^{22}

onalization. The spin-isospin force was also applied to the intermediate 1^+ spectrum and the force strength χ was adjusted to reproduce the known position of the GTGR.

The 2ν decay rates resulting from this more refined (compared to the RPA) treatment are also given in table 2. It is seen that inclusion of the κQQ term in some cases leads to a further strong reduction of the decay rates (especially for

Table 3. Closure matrix elements $M^{2\nu}$ calculated in the PBCS treatment in various approaches of increasing complexity from left to right. The first column gives the large matrix elements in the simple pairing model. These are reduced step by step including the spin-isospin force first in the nucleonic and then also in the Δ nucleon hole space and further by including the 2^+ phonon correlations in lowest order. The matrix elements in the last column include also higher order g.s. correlations from the quadrupole-quadrupole forces (see [19]).

	only pairing	pairing + $H_{\sigma\tau}$ (no Δ-h mixing)	pairing + $H^{\Delta}_{\sigma\tau}$ (with Δ-h mixing)	pairing + $H^{\Delta}_{\sigma\tau}$ + H_{QQ}, without higher order phonon contributions	pairing + $H^{\Delta}_{\sigma\tau}$ + H_{QQ} including higher order phonon contributions
^{128}Te	6.34	2.00	1.77	0.61	0.52
^{130}Te	5.86	1.75	1.55	0.56	0.48
^{82}Se	5.52	2.83	2.33	1.84	1.38
^{76}Ge	7.05	3.76	3.26	2.64	1.93

$^{128,130}Te$). The discrepancy with the experiments on the 2ν sector is considerably reduced compared to the calculations of [13]. In fig. 3 the step-by-step reduction of the matrix elements by including the various correlations is demonstrated . The important point is now that the influence of these long-range forces on the 0ν matrix elements is surprisingly weak [21]. Table 4 shows together the results for the 2ν as well as for the 0ν matrix elements, calculated using the same wave functions. Shown are also the matrix elements with simple BCS wave functions. It is seen that in contrast to the 2ν matrix elements the 0ν matrix elements are determined mainly by pairing. 0ν and 2ν matrix elements are, therefore, not proportionally reduced as assumed by the scaling procedure, but $\underline{M^{0\nu} \text{ remains large}}$.

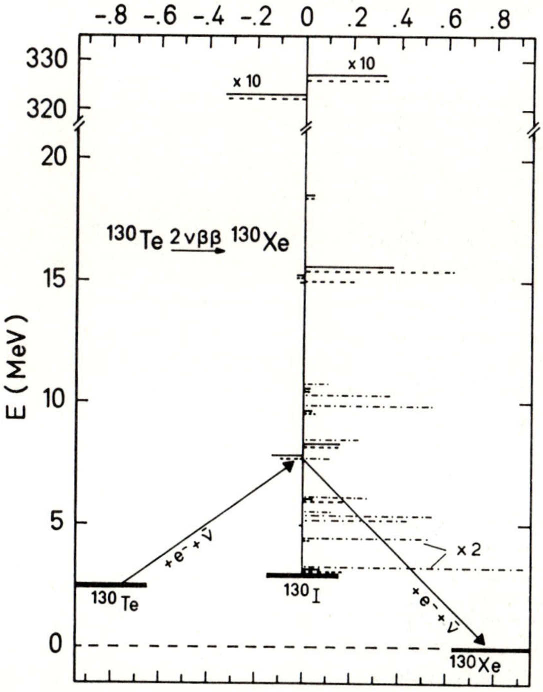

Fig. 3. 2ν ββ amplitudes for the transition $^{130}Te \xrightarrow{\beta\beta} {}^{130}Xe$ calculated with pairing alone (dashed-dotted lines) and pairing plus Gamow-Teller forces including Δ-h mixing (dashed lines). The solid lines show the final result including also quadrupole-quadrupole forces.

Table 4. Nuclear closure matrix elements for 0ν and 2ν (compare Table 2) double beta decay calculated a) using only particle number projected BCS wave functions (pairing), and b) with additional spin-isospin and quadrupole-quadrupole forces. Given is also the ratio $\xi = |R_o M^{0\nu}/M^{2\nu}|$ for the case b).

	$R_o\|M^{0\nu}\|$		$\|M^{2\nu}\|$		
	Pairing	Pairing $+\sigma\tau\sigma\tau$ +QQ	Pairing	Pairing $+\sigma\tau\sigma\tau$ +QQ	ξ
^{76}Ge	12.5	10.4	7.1	1.93	5.4
^{82}Se	9.7	8.2	5.5	1.38	5.9
^{128}Te	12.3	10.0	6.3	0.52	19.3
^{130}Te	11.9	9.4	5.9	0.48	19.7
^{134}Xe	14.7	11.2	8.0	0.77	14.6
^{136}Xe	6.0	3.9	3.6	0.08	50.8
^{142}Ce	7.7	6.2	3.3	1.30	4.8

Instead of scaling we propagate, therefore, to use the calculated 0ν matrix elements without modification. The reason for the different behaviour of the two types of matrix elements lies in the neutrino potential H, which favors the decay of neutrons close to each other in the case of the 0ν mode. Long-range correlations are, therefore, less important. One can express this result also by the ratio of closure matrix elements $\xi = R_o M^{0\nu}/M^{2\nu}$. As can be seen from table 4 we find for ξ a strong dependence on the considered nucleus and generally large values. Using these results for $M^{0\nu}$ we obtain the 0ν half-lives given in table 2. For the nuclei which we have treated only in RPA and for which we did not calculate $M^{0\nu}$ explicitly, a mean value for ξ was used, deduced from the calculated $M^{0\nu}$ values (see [20]).

4. Limits for the Neutrino Mass

According to the findings discussed above, we have analyzed existing experimental data using the matrix elements from various calculations to obtain limits for the neutrino mass. The results are shown in table 5. It is seen that using the unmod-

Table 5. Limits for the neutrino mass $\langle m_\nu \rangle$ deduced from experimental results for the 0ν $\beta\beta$ decay and calculated matrix elements $M^{0\nu}$.

	Experimental $T^{0\nu}_{1/2}$ [years]	Ref	Limit for m_ν [eV] Unscaled	Scaled	Matrix elements used Ref.
^{48}Ca	$>2 \times 10^{21}$	[7]	<40		[13]
			<50		[12]
^{76}Ge	$>3.9 \times 10^{23}$	[8]	< 0.8	< 2.5	[19]
			< 1.6	< 3.3	[13]
			< 1.0	< 2.2	[5]
^{82}Se	$>1.1 \times 10^{22}$	[9]	< 2.9	<9.0	[19]
			<7.5	<17	[13]
^{128}Te	$>8 \times 10^{24}$	[10]	< 0.35	< 1.6	[19]
			< 0.7	< 8.8	[13]
^{130}Te	$>2.2 \times 10^{21}$	[10]	< 4.5	<21	[19]
			< 8.7	<108	[13]

ified matrix elements leads to improved mass limits. The effect is most clearly seen by comparing the mass limit for ^{128}Te based on the data of [10] with the analyses using the scaled Haxton matrix elements. Depending on whether we use directly the calculated $M^{0\nu}$ or scale according to the remaining discrepancy on the 2ν sector, we obtain a limit of 0.35 eV or 1.6 eV for $\langle m_\nu \rangle$. In the case that the still-remaining discrepancy for the 2ν rates could be explained by some refined treatment of long-range nuclear structure correlations, we would expect the lower limit of 0.35 eV to be the more realistic one.

5. Interference Effects in 0ν $\beta\beta$ Decay

The true (electron) neutrino mass might still be substantially larger than the conservative bound of 1.6 eV obtained above. This could be the case, if different neutrino types give rise to interference effects in the neutrino exchange amplitude. One can distinguish two classes of such interferences [22-25], which however can mix with each other.

a) Interference Between Neutrinos of Different Flavor

If neutrino mass eigenstates are not identical with the interaction states one has a nondiagonal mixing matrix. For two flavors (ν_e, ν_μ), it reads

$$\begin{pmatrix} \cos\Theta & \sin\Theta\, e^{i\beta} \\ -\sin\Theta\, e^{-i\beta} & \cos\Theta \end{pmatrix} \tag{13}$$

While the observables in neutrino oscillation experiments depend only on the mixing angle Θ, double beta decay is sensitive also to the phase β, which is connected to CP violation. The CP-conserving cases are $\beta = 0$ and $\beta = \pi/2$. In the second case one could have two Majorana neutrinos with opposite CP eigenvalues leading to destructive interference in neutrinoless double beta decay:

$$\langle m_\nu \rangle = |m_1 \cos^2\Theta - m_2 \sin^2\Theta| \tag{14}$$

The effective mass in $\beta\beta$ decay $\langle m_\nu \rangle$ could be considerably smaller than the true neutrino masses m_1 and m_2. In the extreme case $\tan^2\Theta = m_1/m_2$ one obtains $\langle m_\nu \rangle = 0$.

Restrictions on Θ and $\Delta m^2 = |m_2^2 - m_1^2|$ exist from neutrino oscillation experiments [26,27] which, taken into account, restrict the possible solutions with $m_2 \geq m_1 \gg \langle m_\nu \rangle = 1.6$ eV to

$$\begin{aligned} &\text{a)} \quad m_1 \simeq m_2 \ (\Delta m^2 \lesssim 1\ \mathrm{eV}^2), \quad \text{and } |\Theta| \simeq 45^\circ \\ &\text{or} \\ &\text{b)} \quad m_2 \gtrsim 10^3\, m_1 \ (\Delta m^2 \gtrsim 10^6\ \mathrm{eV}^2), \quad \text{and } |\Theta| \lesssim 3^\circ \end{aligned} \tag{15}$$

b) Interference Between Light and Heavy Neutrinos of the Same Flavor

The mechanism for generating small neutrino masses discussed above requires the existence of very heavy neutrinos. In the case of one light ν_1 and one heavy neutrino ν_2 (per generation), the neutrino state coupling to the left-handed electronic current is given by

$$\nu_{eL} = \nu_{1L} \cos\Theta + \nu_{2L} \sin\Theta \tag{16}$$

If the two mass eigenstates ν_{1L} and ν_{2L} have opposite CP eigenvalues η_1 and η_2, they give rise to a destructive interference in neutrinoless double beta decay, the effective mass in this process being given by:

$$m_{eff}(A) = |\eta_1 m_1 \cos^2 \Theta + \eta_2 m_2 (\sin^2 \Theta) F(m_2,A)| \tag{17}$$

In the case $\eta_1\eta_2 = -1$, the situation is quite similar to the effect discussed above. However, eq. (17) involves a nuclear structure-dependent correction factor $F(m_2,A)$, the reason for which is a modified neutrino exchange potential H for the heavy neutrino. Approximately one has [22]

$$H(m_2,r_{ij}) \simeq \frac{e^{-m_2 r_{ij}}}{r_{ij}} \quad \text{for } 10 \text{ MeV} \lesssim m_2 \lesssim 1 \text{ GeV} \tag{18}$$

The matrix element $M^{0\nu}_{heavy}$ for 0ν decay mediated by heavy neutrinos can be written

$$M^{0\nu}_{heavy} = F(m_2,A) M^{0\nu}_{light} \tag{19}$$

The correction $F(m_2,A)$, arising from the exponential behaviour of eq. (18), in the crudest approximation simply depends on the nuclear radius and is therefore a monotonic function of A.

From this A dependence one gets a restriction for possible values of the light neutrino mass m_1. While there may be perfect cancellation in one nucleus, to be specific let's say in ^{128}Te (corresponding to the condition $m_1 \cos^2 \Theta = m_2(\sin^2 \Theta)F(m_2,128)$), this is no longer the case in the other nuclei.

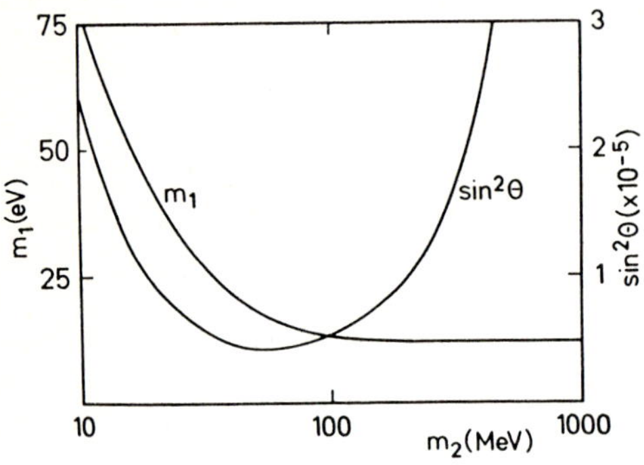

Fig. 4. The upper limit of the Majorana mass m_1 of the light neutrino consistent with the double beta decay results for ^{76}Ge and ^{128}Te (see text) in the case of interference effects with one heavy neutrino (of same flavour) with mass m_2 and coupled to the electron with $\sin^2 \Theta$. For $m_2 \gtrsim 100$ MeV a limit of $m_1 \lesssim 13$ eV is obtained.

Combining the experimental results $m_{eff}(76) < 3.1$ eV and $m_{eff}(128) < 1.6$ eV the limit on m_1 shown in fig. 4 is obtained as function of the heavy neutrino mass m_2. Also shown is the value for $\sin^2 \Theta$ needed for complete cancellation in ^{128}Te. F(m,A) was calculated from a simple uniform nucleon-nucleon correlation function with hard core [25].

For $m_2 \gtrsim 100$ MeV the mass of the light neutrino is restricted to $m_1 \lesssim 13$ eV. For $m_2 \lesssim 100$ MeV still larger values for m_1 could be consistent with the observed upper limits for neutrinoless double beta decay.

In the simplest case naturally arising in the SO(10) model, given by the mass matrix (1), m_1, m_2, and $\sin^2 \Theta$ are related to each other and the heavy neutrino does not lead to any measurable interference effect, because one has $m_2 \gtrsim 100$ GeV and $\sin^2 \Theta \lesssim 10^{-10}$.

In an extended model involving three neutrinos of the same flavor, in addition to ν_L and ν_R introducing an SO(10) singlett N_L [28], the following mass matrix could be naturally expected:

$$(\overline{\nu_L}, (\overline{\nu_R})^C, \overline{N_L}) \begin{pmatrix} 0 & m^D & \epsilon \\ m^D & 0 & M^D \\ \epsilon & M^D & \eta \end{pmatrix} \begin{pmatrix} (\nu_L)^C \\ \nu_R \\ (N_L)^C \end{pmatrix} \qquad (20)$$

with $m^D \simeq 0$ (MeV)

$M^D >> m^D$

$\epsilon, \eta << m^D$

Again interference effects between the light and the two heavy neutrinos resulting from (20) should influence the neutrino mass observed in 0ν $\beta\beta$ decay:

$$m_{eff}(A) = |m_1 |U_1^2| + m_2 |U_2^2| F(m_2, A) - m_3 |U_3^2| F(m_3, A)| \qquad (21)$$

The U_1 are the mixing amplitudes calculable from the mass matrix (20). In this model strong cancellation effects are possible for $M^D \simeq 0$ (100 MeV). With reasonable approximations one arrives again at the same constraints on m_1 as shown in fig. 4 for the case of two neutrinos of the same flavour.

5. Conclusion

Nuclear structure is found to be more in favour of neutrinoless double beta decay than of the conventional 2ν decay. While the latter depends sensitively on nuclear structure details and is strongly reduced by long-range correlations, the former is determined mainly by pairing correlations. Therefore the resulting large matrix elements for the neutrinoless decay can be calculated with much more confidence than the tiny matrix elements for the 2ν decay. Consequently the previously used "scaling" yields incorrect (too small) matrix elements for the neutrinoless mode. Applying the matrix elements from our nuclear structure calculations improved limits for the neutrino mass are obtained. The most stringent limit from "direct" experiments we find at present from $\beta\beta$-decay of ^{76}Ge yielding an effective mass of 0.8 eV. Besides the detailed calculations including both decay modes for the up-to-now most important $\beta\beta$ nuclei also RPA predictions for all potential $\beta\beta$ emitters with $A \geq 70$ are presented. Considering interference between different types of neutrinos, the decay rate for neutrinoless decay could be retarded allowing for a larger mass of the lightest neutrino. In the case of mixing with a heavy neutrino with a mass around 100 MeV a conservative limit for the light neutrino mass could be as large as 13 eV.

References

1. H. Primakoff, S.P. Rosen: Ann. Rev. Nucl. Part. Sci. 31, 145 (1981)
2. B. Stech: Phys. Lett. 130B, 189 (1983)
3. M. Gell-Mann, P. Ramond, S. Slansky: In Supergravity, eds. P. van Nieuwenhuizen and D.Z. Freedman (North-Holland, Amsterdam 1979)
4. M. Doi, T. Kotani, H, Nishiura, K. Okuda, E. Takasugi: Prog. Theor. Phys. 66, 1739 (1981)
5. T. Tomoda: Contr. to this volume.
6. H.V. Klapdor, K. Grotz: Phys. Lett. 142B, 323 (1984)
7. R.K. Bardin, P.J. Gollon, J.D. Ullmann, C.S. Wu: Nucl. Phys. A158, 337 (1970)

8. D.O. Caldwell et al. Phys. Rev. Lett. 54, 281 (1985) and contrib. to this volume.
9. S.R. Elliott, A.A. Hahn and M.K. Moe, contrib. to this volume.
10. T. Kirsten, H. Richter, E. Jessberger: Phys. Rev. Lett. 50, 474 (1983); T. Kirsten: In Proc. Workshop on Science Underground (Los Alamos 1982)
11. A. Brown; preprint 1985.
12. T. Tsuboi, K. Muto, H. Horie: Phys. Lett. 143B, 293 (1984)
13. W.C. Haxton, G.J. Stephenson, Jr., D. Strottmann: Phys. Rev. Lett 47, 153 (1981); W.C. Haxton: Phys. Rev. D 25, 2360 (1982); Comm. Nucl. Part. Phys. 11, 41 (1983); Prog. Part. Nucl. Phys. 12, 409 (1984)
14. P. Vogel and P. Fisher: Phys. Rev. C 32, 1362 (1985)
15. A.H. Huffmann: Phys. Rev. C 2, 741 (1970)
16. V.A. Khodel: Phys. Lett. 32B, 583 (1970)
17. L. Zamick and N. Auerbach: Phys. Rev. C 26, 2185 (1982)
18. W.C. Haxton, G.J. Stephenson, Jr.: Phys. Rev. C 28, 458 (1983)
19. K. Grotz, H.V. Klapdor: Nucl. Phys. A, in press.
20. K. Grotz, H.V. Klapdor: Phys. Lett. 157B, 242 (1985)
21. K. Grotz, H.V. Klapdor: Phys. Lett. 153B, 1 (1985)
22. C.W. Kim, H. Nishiura: Phys. Rev. D 30, 1123 (1984)
23. M. Doi, M. Kermoka, T. Kotani, H. Nishiura, E. Takasugi: Prog. Theor. Phys. 70, 1331 (1983);
24. C. Wolfenstein: Phys. Lett. 107B, 77 (1981)
25. A. Halprin, S.T. Petcov, S.P. Rosen: Phys. Lett. 125B, 335 (1983)
26. V. Zacek et al., Phys. Lett. 164B (1985) 193 and contrib. to this volume.
27. J.F. Cavaignac et al. Phys. Lett. 148B (1984) 387 and D.H. Koang, contrib. to this volume.
28. H. Georgi, D.V. Nanopoulos, Phys. Lett. 82B, 392 (1979)

Neutrinoless Double Beta Decay and a Limit on the Right-Handed Leptonic Current

T. Tomoda

Institut für Theoretische Physik, Universität Tübingen,
D-7400 Tübingen, F. R. Germany

1 Introduction

With the recent development of gauge theories, a growing interest has been focused on the nature of the neutrino. Since an observation of neutrinoless double beta ($0\nu\beta\beta$) decay would be practically the only possible evidence for the neutrino to be a Majorana particle, much effort has been devoted to the problem of nuclear $\beta\beta$ decay [1-4].

In most of the theoretical investigations of the $0\nu\beta\beta$ decay the non-relativistic limit of the nuclear current was used. It is justified in a type of $0\nu\beta\beta$ decay caused by a non-vanishing mass of the neutrino since this is an "allowed" decay. In another type of $0\nu\beta\beta$ decay caused by a specific admixture of a right-handed current, however, the contributions of "second forbidden" processes become important because those of "allowed" processes cancel each other to a great extent. One kind of these "second forbidden" $0\nu\beta\beta$ transition consists of those processes in which one of the two electrons is emitted in a P-wave. There is, however, another kind of "second forbidden" $0\nu\beta\beta$ transitions in which both of the electrons are emitted in S-waves but a relativistic correction to the nuclear current plays a role [3,5]. It turns out that the process of this type even dominates over all others [6].

The relativistic correction becomes essential because the neutrino exchanged between nucleons is virtual and the momentum transfer to it can be very large ($\sim$ twice the Fermi momentum $\sim$ 540 MeV/c). This is in sharp contrast to the single β-decay, where both the electron and the neutrino are real particles and the momentum transfer from a nucleon to these leptons is restricted by the Q-value (a few MeV) of the decay.

In the present talk I will report on the calculation of the $0\nu\beta\beta$ decay rates of ^{76}Ge [6] and on the results of comparison with an experimental lower bound for the decay half-life. It will be shown that because of the large enhancement of the nuclear matrix element due to the above-mentioned correction one can deduce a very stringent limit on the right-handed leptonic current.

2 Formalism

We start from the following effective weak interaction Hamiltonian density [2,3]

$$H_W(x) = (G\cos\theta_C/\sqrt{2})(j_{L\mu}J_L^{\mu\dagger} + \kappa j_{L\mu}J_R^{\mu\dagger} + \eta j_{R\mu}J_L^{\mu\dagger} + \lambda j_{R\mu}J_R^{\mu\dagger}) + \text{h.c.}, \tag{1}$$

where $G = 1.16637\times10^{-5}$ GeV^{-2}, $\cos\theta_C = 0.9737$ with the left and right-handed leptonic currents

$$j_L^{\mu}(x) = \bar{e}(x)\gamma^{\mu}(1-\gamma_5)\nu_{eL}(x),$$
$$j_R^{\mu}(x) = \bar{e}(x)\gamma^{\mu}(1+\gamma_5)\nu'_{eR}(x), \tag{2}$$

and the current neutrinos

$$\nu_{eL}(x) = \sum_{i=1}^{2n} U_{ei}N_{iL}(x), \tag{3}$$

$$\nu'_{eR}(x) = \sum_{i=1}^{2n} V_{ei}N_{iR}(x).$$

N_i is the eigenstate of the Majorana neutrino mass matrix with the eigenvalue m_i, n the number of generations. The nuclear current is assumed to be

$$(J_L^{o\dagger}(\vec{x}),\vec{J}_L^{\dagger}(\vec{x})) = \sum_{n=1}^{A} \tau_n^{+}\delta(\vec{x}-\vec{r}_n)(g_V - g_A C_n, -g_A\vec{\sigma}_n + g_V\vec{D}_n), \quad (4)$$

where $\langle p|\tau^{+}|n\rangle = 1$, $g_V = 1$, $g_A = 1.254$, and the right-handed current $J_R^{\dagger}(x)$ is assumed to be obtained by replacing g_A with $-g_A$. C_n and $\vec{D}_n$ are the relativistic correction terms [7] (see also ref. [3])

$$C_n = (\vec{p}_n + \vec{p}_n')\cdot\vec{\sigma}_n/2M,$$
$$\vec{D}_n = [\vec{p}_n + \vec{p}_n' - i\mu_\beta\vec{\sigma}_n \times (\vec{p}_n - \vec{p}_n')]/2M, \quad (5)$$

with

$$\mu_\beta = \frac{1}{2}(g_p - g_n) = 4.7, \quad (6)$$

where $\vec{p}_n$ and $\vec{p}_n'$ are the initial and final nucleon momenta, M the nucleon mass, and g_p, g_n the spin g-factors of the proton and neutron. The effect of the term proportional to κ in (1) will be neglected in the following because κ contributes to the $0\nu\beta\beta$ decay amplitude only in the combination $1\pm\kappa$ and we expect $|\kappa| \ll 1$.

We calculate the decay amplitude in second-order perturbation theory and employ the closure approximation [8] in taking a summation over intermediate excited nuclear states. This approximation means that the energy of the intermediate nuclear state, E_N, in the energy denominator is replaced by some average value $\langle E_N\rangle$, i.e.

$$(\omega + E_j + E_N - E_I)^{-1} \rightarrow (\omega + E_j + \langle E_N\rangle - E_I)^{-1}, \quad (7)$$

where ω, E_j, E_I are the energies of the neutrino, the j-th electron (j = 1,2), and the initial nuclear state, respectively. This approximation is expected [8] to be good for the $0\nu\beta\beta$ decay because the neutrino exchanged between two nucleons is virtual and its typical energy is [2] $\omega \gtrsim k \sim 1/\bar{r}_{NN} \sim 200 m_e$ (where k is the neutrino momentum and $\bar{r}_{NN}$ the mean internucleon distance), which is much larger than the typical nuclear excitation energy $\langle E_N\rangle - E_I \sim 20 m_e$.

The wave functions of the electrons emitted in the $0\nu\beta\beta$ decay are expanded in terms of the solutions of the Dirac equation in a spherical basis. The leading contributions for a $0^+\rightarrow 0^+$ decay come from the S-(g_{-1}, f_1) and P-wave (g_1, f_{-1}) radial wave functions with j=1/2. These will be included in the present work. These radial wave functions are then expanded in powers of r, and the leading terms (a constant for g_{-1}, f_1; a linear term for f_{-1}, g_1) will be retained [9] (using by this the long-wave-length approximation).

The half-life for the $0\nu\beta\beta$ decay $\tau_{1/2}^{0\nu}$ is expressed as

$$(\tau_{1/2}^{0\nu})^{-1} = C_{mm}(\langle m_\nu\rangle/m_e)^2 + C_{\lambda\lambda}\langle\lambda\rangle^2 + C_{\eta\eta}\langle\eta\rangle^2 + 2C_{m\lambda}(\langle m_\nu\rangle/m_e)\langle\lambda\rangle$$
$$+2C_{m\eta}(\langle m_\nu\rangle/m_e)\langle\eta\rangle + 2C_{\lambda\eta}\langle\lambda\rangle\langle\eta\rangle, \quad (8)$$

where

$$\langle m_\nu\rangle = \sum_i{}' m_i U_{ei}^2,$$
$$\langle\lambda\rangle = \lambda\sum_i{}' U_{ei}V_{ei}, \quad (9)$$
$$\langle\eta\rangle = \eta\sum_i{}' U_{ei}V_{ei}.$$

The summation in (9) should be taken over light neutrinos ($m_i \ll 200\ m_e$). We neglect possible contributions from heavy neutrinos for which the coefficients C´s in (8) would become dependent on the neutrino mass m_i. We assume CP invariance so that the three parameters $\langle m_\nu\rangle$, $\langle\lambda\rangle$ and $\langle\eta\rangle$ which characterize lepton number violation are all real. The coefficients C_{mm} etc. are the sums of the products of electron phase space integrals and nuclear matrix elements. There appear nine nuclear matrix elements [6] $M_{GT}^{(0\nu)}$, χ_F, $\tilde{\chi}_{GT,F}$, $\chi_{GT,F,T,P,R}$. Of these

$$\chi_R' = (g_V/g_A)M_R'^{(0\nu)}/M_{GT}^{(0\nu)} \quad (10)$$

with

$$M_R^{(0\nu)} = (\mu_\beta/3m_eM)\langle F|\frac{1}{2}\sum_{nm}[4\pi\delta(\vec{r}_{nm}) - \frac{2}{\pi}\bar{A}r_{nm}^{-2}]\vec{\sigma}_n\cdot\vec{\sigma}_m\tau_n^+\tau_m^+|I\rangle + \text{small correction} \quad (11)$$

originates from the relativistic correction to the nuclear current (the term $\vec{D}_n$ in (5)). $\bar{A}=\langle E_N\rangle-E_I+m_ec^2+Q_{\beta\beta}/2$ is the sum of the average electron and nuclear excitation energies.

We obtained the zero-range operator in (11) using the relation $\vec{\nabla}\cdot(\vec{r}/r^3)=4\pi\delta(\vec{r})$. If the short-range NN correlations are taken into account, the matrix element of this part vanishes completely. This follows, however, from our assumption of a point nucleon in (4). If we also take into account the finite extension of the nucleon, we should obtain an operator which has a finite range and the matrix element of which is not affected so drastically by the short-range correlations. Thus we replace the vector and axial vector coupling constants in momentum space with the dipole form factors [2,4]

$$g_V \rightarrow g_V(1+(k^2/\Lambda^2))^{-2},$$
$$g_A \rightarrow g_A(1+(k^2/\Lambda^2))^{-2}, \quad (12)$$

where Λ=850 MeV. The δ-function in (11) is then modified into

$$(2\pi)^{-3}\int d\vec{k}e^{i\vec{k}\cdot\vec{r}}(1+(k^2/\Lambda^2))^{-4} = (\Lambda^3/64\pi)e^{-\Lambda r}[1 + \Lambda r + (\Lambda r)^2/3]. \quad (13)$$

Similarly, $1/r$ in (11) is replaced by

$$(4\pi)^{-1}\int d\vec{k}e^{i\vec{k}\cdot\vec{r}}k^{-1}(1+(k^2/\Lambda^2))^{-4}. \quad (14)$$

For all the other transition operators we neglect the effect due to the finite nucleon size since either they are relatively long-ranged or their matrix elements do not contribute very much to the total decay rate even without the short-range correlations.

The nuclear wave functions of the initial and the final states are assumed to be of the Hartree-Fock-Bogoliubov type and obtained by minimizing the expectation value of the nuclear Hamiltonian after particle-number and angular-momentum projection [10]. A model space consisting of active $1p_{3/2}$, $0f_{5/2}$, $1p_{1/2}$ and $0g_{9/2}$ orbitals is assumed and the modified surface delta interaction [11] is used as an effective NN interaction.

The nuclear wave functions obtained in this way lack the short-range repulsive correlations. Their effect is especially important for the evaluation of the nuclear matrix element χ_R since the operator is still relatively short-ranged even after the modification (13) due to the finite nucleon size. We multiply [2,4] the two-nucleon wave functions by $f(|\vec{r}_n-\vec{r}_m|)$, where [13]

$$f(r) = 1 - e^{-ar^2}(1-br^2), \quad (15)$$

with a=1.1 fm^{-2} and b=0.68 fm^{-2}.

3 Results

The first column of Table 1 gives the calculated nuclear matrix elements. Aside from the difference in nuclear wave functions the present calculation differs from that of HAXTON and STEPHENSON [2] in the following points:
In the present work
1) the matrix element $\tilde{\chi}_R$ arising from the relativistic correction is calculated;
2) the approximation $\chi_F=\tilde{\chi}_F$, $\chi_{GT}=1$ is not used;
3) the sign of χ_P' is opposite to that of ref.[2].

Table 2 gives the calculated coefficients C_{mm} etc. Using these values in (8) and comparing with the experimental lower limit $\tau_{1/2}^{0\nu}$ > 3.9×10^{23}y by CALDWELL et al. [14], one obtains the upper limits on $\langle m_\nu\rangle$, $\langle\lambda\rangle$ and $\langle\eta\rangle$ shown in Table 3. While our limits on $\langle m_\nu\rangle$ and $\langle\lambda\rangle$ are similar to those of ref.[2], the present limit on the right-handed current coupling strength $\langle\eta\rangle$ is by two orders of magnitude more stringent than that of ref.[2]. The coefficient $C_{\eta\eta}$ is dominated by the contribution from χ_R', where the electron phase space integral associated with this matrix element involves only S-waves. The enhancement of the P-wave components relative to the S-wave components of the electron wave functions due to the Coulomb field (called a "P-wave effect" by DOI et al.[3]) does not affect the $0\nu\beta\beta$ decay rates very much because the contribution from the matrix element χ_P' is much smaller

Table 1 The nuclear matrix elements for the $0\nu\beta\beta$ decay of ^{76}Ge. The values are taken from refs.[6] and [2]

	present	Haxton
$M_{GT}^{(0\nu)}[fm^{-1}]$	-0.563	-0.411
χ_F	-0.219	-0.200
$\tilde{\chi}_{GT}$	0.857	1.000
$\tilde{\chi}_F$	-0.182	-0.200
χ'_{GT}	1.143	1.141
χ'_F	-0.255	-0.231
χ'_T	-0.026	-0.013
χ'_P	-0.218	0.269
χ'_R	31.9	-----

Table 2 The coefficients C's. The values are taken from refs.[6] and [2]

	present	Haxton
$C_{mm}[10^{-12}y^{-1}]$	0.315	0.156
$C_{m\lambda}[10^{-12}y^{-1}]$	-0.0560	-0.0350
$C_{m\eta}[10^{-12}y^{-1}]$	12.9	-0.0296
$C_{\lambda\lambda}[10^{-12}y^{-1}]$	0.326	0.224
$C_{\eta\eta}[10^{-12}y^{-1}]$	2210	0.101
$C_{\lambda\eta}[10^{-12}y^{-1}]$	-0.129	-0.104

Table 3 The upper limits on the neutrino mass $\langle m_\nu\rangle$ and the right-handed current coupling strengths $\langle\lambda\rangle$ and $\langle\eta\rangle$ deduced from the experimental lower bound $\tau_{1/2}^{0\nu}>3.9\times10^{23}y$

	present	Haxton
$\|\langle m_\nu\rangle\|$[eV]	<1.7	<2.5
$\|\langle\lambda\rangle\|$	$<2.9\times10^{-6}$	$<5.4\times10^{-6}$
$\|\langle\eta\rangle\|$	$<3.9\times10^{-8}$	$<8.1\times10^{-6}$

than that of χ'_R and plays no important role. (This would not be the case if $\chi'_R \lesssim 10\chi'_P$.)

4 Summary

The $0\nu\beta\beta$ decay rate for the transition $^{76}Ge(0_1^+) \rightarrow {}^{76}Se(0_1^+)$ has been calculated taking into account the relativistic correction to the nuclear current including weak magnetism. The upper limit on the parameter $\langle\eta\rangle$ describing an admixture of the right-handed leptonic current is determined dominantly by the "second-forbidden" matrix element χ'_R. This is because i) there occurs a systematic cancellation in "allowed" and other "second-forbidden" processes, and ii) the neutrino exchanged between nucleons is virtual and the momentum transfer to it is limited only by the amount which a nucleon in a nucleus can provide i.e. twice the Fermi momentum. Comparing the present calculation with the experimental lower bound $\tau_{1/2}^{0\nu}>3.9\times10^{23}y$ we obtain $|\langle\eta\rangle|<3.9\times10^{-8}$ which is by one or two orders of magnitude more stringent than those obtained by DOI et al.[3] or HAXTON et al.[2], respectively.

I acknowledge the collaboration with A. Faessler, K.W. Schmid and F. Grümmer. This work was supported by the Bundesministerium für Forschung und Technologie.

References

1 H. Primakoff and S.P. Rosen: Ann. Rev. Nucl. Part. Sci. 31,145(1981)
2 W.C. Haxton and G.J. Stephenson Jr.: Prog. Part. Nucl. Phys. 12,409(1984)
3 M. Doi, T. Kotani and E. Takasugi: Prog. Theor. Phys. Suppl. 83,1(1985)
4 J.D. Vergados: Phys. Rep. 133,1(1986)
5 M. Doi, T. Kotani, H. Nishiura, K. Okuda and E. Takasugi: Prog. Theor. Phys. 66,1739(1981)
6 T. Tomoda, A. Faessler, K.W. Schmid and F. Grümmer: Nucl. Phys. A452,591(1986)
7 T. Tomoda, A. Faessler, K.W. Schmid and F. Grümmer: Phys. Lett. 157B,4(1985)
8 H. Primakoff and S.P. Rosen: Rep. Prog. Phys. 22,121(1959)
9 M. Doi, T. Kotani, H. Nishiura and E. Takasugi: Prog. Theor. Phys. 69,602 (1983)
10 K.W. Schmid, F. Grümmer and A. Faessler: Nucl. Phys. A431,205(1984)
11 P.W.M. Glaudemans: private communication cited in ref.[12]
12 M. Didong, H. Müther, K. Goeke and A. Faessler: Phys. Rev. C14,1189(1976)
13 G.A. Miller and J.E. Spencer: Ann. Phys. 100,562(1976)
14 D.O. Caldwell: talk presented at Int. Symp. on Nuclear Beta Decays and Neutrino, Osaka, June 11-13, 1986

Nuclear Matrix Elements of $^{48}Ca(0_1^+) \rightarrow {}^{48}Ti(0_1^+)$ Double Beta Decay

K. Muto

Department of Physics, Tokyo Institute of Technology, Oh-okayama, Meguro, Tokyo 152, Japan

Uncertainties involved in the nuclear matrix elements of neutrinoless double beta decay are estimated. Uncertainties originating from the average energy of intermediate states, the NN short-range correlations and the nucleon finite size effects are less than 50%, about a factor of 2 in the decay rate, which would possibly be smaller than that arising from the nuclear wave functions.

The neutrinoless double beta decay, which is a lepton-number nonconserving process, is expected to give an answer to the questions of whether neutrinos have a finite mass and whether the right-handed weak charged current exists [1,2]. However, in order to deduce the neutrino mass and the magnitude of the right-handed current from observed decay rates, the relevant nuclear matrix elements have to be evaluated theoretically.

Among a variety of nuclear matrix elements of the neutrinoless mode, two elements are of particular interest. One is for the neutrino mass and the other for the right-handed current. They are given by

$$M_m = \langle {}_f | \sum_{ij} (t_-)_i (t_-)_j \, (g_A^2 \vec{\sigma}_i \cdot \vec{\sigma}_j - g_V^2) \, h_+ (r_{ij}) | {}_i \rangle \tag{1}$$

$$M_\eta = \langle {}_f | \sum_{ij} \frac{1}{3} f_W \, (t_-)_i \, (t_-)_j \, g_V g_A \, \vec{\sigma}_i \cdot \vec{\sigma}_j \, h_R \, (r_{ij}) | {}_i \rangle \tag{2}$$

where $g_V = 1$, $g_A = 1.251$, $f_W = 4.708$, and t_- and σ are isospin lowering and Pauli spin operators, respectively. The neutrino potentials h_+ and h_R are functions of distance between two nucleons which exchange a neutrino. The h_+ behaves as $\sim 1/r$, while the h_R contains a potential singular at the origin which gives a dominant contribution to the matrix element relevant to the right-handed current [3].

We usually utilize the closure approximation, since a sum of intermediate states extends over all J^π nuclear states. It is to replace the intermediate state energies with an average value and then complete the sum by closure. With the approximation we can avoid a complicated calculation, but at the same time we abandon information on the intermediate states. The parametrization [1] $\mu_0 m_e = \langle E_a \rangle - (E_i + E_f)/2$ is taken, where m_e is the electron mass and E_a, E_i and E_f are energies of intermediate, initial and final nuclear states, respectively.

The NN short-range correlations are not taken into account in the nuclear wave functions of initial and final states. Though they depend in nature on states and NN interactions, it is assumed that they are represented by a correlation function. The functional form [4]

$$f(r) = 1 - e^{-ar^2} (1 - br^2) \tag{3}$$

is assumed, where b/a = 0.616, and the correlation length is defined as

$$\ell_c = \int_0^\infty [1 - (f(r))^2]dr \quad . \tag{4}$$

Commonly used values are more or less $\ell_c = 0.7$ fm [2,3].

The finite nucleon size effects are taken into account by introducing the momentum-transfer dependent dipole form factors [2,3].

$$g_{V,A} \to g_{V,A}(1 + q^2/\Lambda^2)^{-2} \quad . \tag{5}$$

If CVC is correct, $\Lambda = 0.84$ GeV for the vector coupling, and experiments do not contradict this value. On the other hand, the axial vector mass is not well established experimentally, the values scattering in the range $\Lambda = 0.6 - 1.1$ GeV [5].

Uncertainties of the nuclear matrix elements coming from uncertainties of the three parameters are estimated for the double beta decay of ^{48}Ca. The nuclear wave functions are the same as used in the calculation of two-neutrino mode decay [6]. One parameter is varied in a reasonable range, while the others are fixed to the typical values, $\mu_0 = 15$, $\ell_c = 0.7$ fm and $\Lambda = 0.85$ GeV. Results are summarized in Table 1. The nuclear matrix elements, in particular the singular potential term, are sensitive to the short-range correlations.

However, more uncertainties would come from nuclear wave functions. These matrix elements, as for the two-neutrino mode, are dominated by contributions from spin-singlet components of the two-nucleon wave functions, and depend largely on the pairing force of NN interaction and also occupations of spin-orbit partners, $j = \ell \pm 1/2$.

Table 1. Uncertainties in the nuclear matrix elements for the $^{48}Ca \to {}^{48}Ti$ double beta decay.

Parameter	Range	M_m	M_η
μ_0	10–50	20%	10%
ℓ_c	0.6–0.8 [fm]	30%	50%
Λ	0.6–1.1 [GeV]	10%	20%

References

[1] M. Doi, T. Kotani and E. Takasugi, Progr. Theor. Phys. Suppl. 83, 1 (1985).

[2] W.C. Haxton and G.J. Stephenson, Prog. Part. Nucl. Phys. 12, 409 (1984).

[3] T. Tomoda, A. Faessler, K.W. Schmid and F. Grümer, Nucl. Phys. A452, 591 (1986).

[4] G.A. Miller and J.E. Spencer, Ann. Phys. (N.Y.) 100, 562 (1976).

[5] N.J. Baker, A.M. Cnops, P.L. Connolly, S.A. Kahn, H.G. Kirk, M.J. Murtagh, R.B. Palmer, N.P. Samios and M. Tanaka, Phys. Rev. C23, 2499 (1981).

[6] T. Tsuboi, K. Muto and H. Horie, Phys. Lett. 143B, 293 (1984).

Double Beta Decay: Experiments and New Techniques

E. Bellotti

Dipartimento di Fisica dell' Universita and INFN, via Celoria 16, I-20133 Milano, Italy

1 Introduction

Double beta decay [1] (d.b.d.) consists in the contemporary emission of two electrons from a nucleus (A,Z) which transforms in the (A,Z+2) isobar, according to one of the following reactions:

$$(A,Z) \rightarrow (A,Z+2) + 2e^- + 2\bar{\nu} \qquad (1)$$
$$(A,Z) \rightarrow (A,Z+2) + 2e^- \qquad (2)$$
$$(A,Z) \rightarrow (A,Z+2) + 2e^- + \chi^\circ \qquad (3)$$

(where χ° is a majoron)

Reactions (2) and (3) imply lepton number violation, while reaction (1) is a double weak process allowed by any known selection rule. Similar processes, like double positron emission, positron emission and electron capture, double electron capture, have been studied theoretically and searched for experimentally (one could also consider d.b.d. of elementary particle, like $K \rightarrow \pi l l$, l is a lepton), however these modes will be not considered in this short review.

Nuclear d.b.d. involves a triplet of isobars (Fig.1), where the two steps decay (A,Z) → (A,Z+1) → (A,Z+2) is forbidden by energy conservation or strongly hindered by large change in angular momentum and small transition energy. Transitions to excited states of the final nucleus can also occur.

The half-life for the various decay modes have been computed by many authors [1].

a) 2ν mode (reaction (1))

According to the diagramm of Fig. 2a, the inverse half-life for the g.s. - g.s. transition can be written as:

$$\left[T_{1/2}^{2\nu}(0^+ \rightarrow 0^+)\right]^{-1} = G_{GT}^{2\nu} \left[M_{GT}^{2\nu}/\mu\right]^2$$

where $G_{GT}^{2\nu}$ is a kinematical factor which also includes the Fermi coupling constant and the Coulomb factor, $M_{GT}^{2\nu}$ is the Gamow-Teller nuclear matrix element and μ has the meaning of a mean excitation energy of the intermediate nucleus. As it is well known, there is disagreement between theoretical and experimental decay rates, the latter ones being slower than predicted.

b) Neutrinoless Mode (reaction (2))

This decay mode is usually described according to diagrams of Fig. 2a) and b) where a down quark emits an electron and a Majorana neutrino, which is absorbed by a second down quark. The amplitude for these processes is different from zero if neutrinos don't carry any lepton number and have a mass (Fig.2a) or right handed currents exist (Fig.2b). Quite a general hamiltonian is:

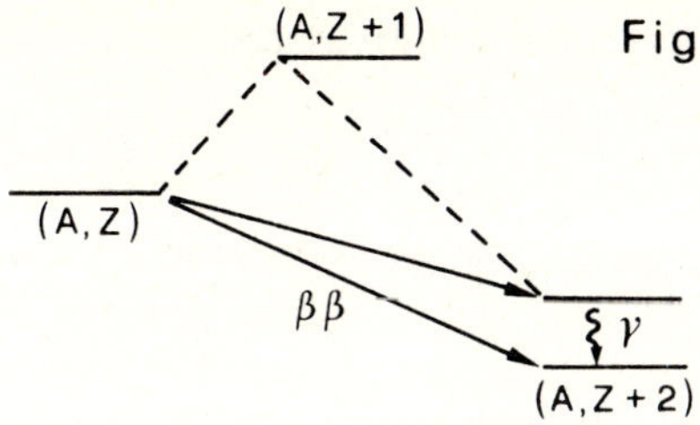

Fig. 1

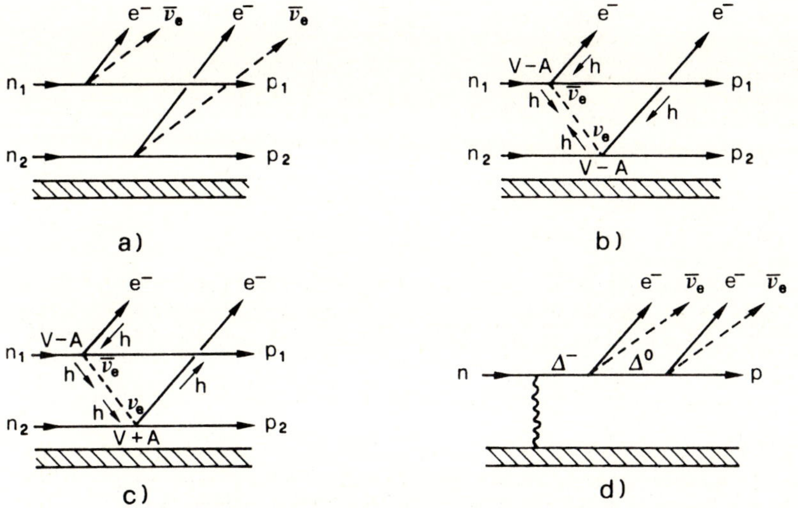

Fig. 2

$$H = G/\sqrt{2}\{j_L J_L^+ + \chi j_L J_R^+ + \eta j_R J_L^+ + \lambda j_R J_R^+\} + h.c.$$

where j (J) are the leptonic (hadronic) currents and L (R) indicates their handness; χ, η and λ can be easily interpreted in the frame of left-right symmetric models [1].

Neglecting contributions from the so called "Δ mechanism" (Fig. 2d is an example of this mechanism), the inverse half-life (with some assumption on CP conservation) can be written as

$$\left[T_{1/2}^{o\nu}(0^+\to0^+)\right]^{-1} = \left[M_{GT}^{o\nu}\right] \cdot \left\{c_1 \frac{\langle m_\nu\rangle^2}{m_e^2} + c_2 \frac{\langle\lambda\rangle\,\langle m_\nu\rangle}{m_e} + c_3 \frac{\langle\eta\rangle\,\langle m_\nu\rangle}{m_e} + c_4 \langle\lambda\rangle^2 + c_5\langle\eta\rangle^2 + c_6\langle\lambda\rangle\,\langle\eta\rangle\right\}$$

where $\langle m_\nu\rangle$, $\langle\eta\rangle$ and $\langle\lambda\rangle$ are "effective" values (for an exact definition see [1]) and c_i's depend on phase space as well on nuclear matrix elements and have been computed by many authors, with quite a different result [2-4].

c) Majoron Mode (reaction(3))

The inverse half-life has been written as

$$\left[T_{1/2}^{\chi^o}(0^+\to0^+)\right]^{-1} = \langle g_{\chi^o}\rangle \left[M_{GT}^{o\nu}\right]^2 c_{\chi^o}$$

where $\langle g_{\chi^o}\rangle$ is the "effective" coupling of the majoron to leptons and c_{χ^o} is a kinematical factor.

Table 1 reports for a few nuclei of interest some of the relevant quantities contained in the above expressions

Table 1 (From DOI et al.[1]), χ_F is the ratio of the Fermi n.m.e. to the Gamow Teller one; E xx means 10^{xx}

Element	T(KeV)	$T_{1/2}^{2\nu}\left[\frac{M_{GT}^{2\nu}}{\mu}\right]^2$	$T_{1/2}^{0\nu}\left[m_\nu(eV)M_{GT}^{0\nu}(1-\chi_F)\right]^2$
^{76}Ge	2040	7.7 E 18	4.1 E 25
^{82}Se	2995	2.3 E 17	9.3 E 24
^{100}Mo	3034	1.1 E 17	5.7 E 24
^{128}Te	868	1.2 E 21	1.4 E 26
^{130}Te	2533	2.1 E 17	5.9 E 24
^{136}Xe	2479	2.1 E 17	5.5 E 24
^{150}Nd	3367	8.4 E 15	1.3 E 24

2 Experimental Results

Recent experimental results will be summarized in this section and some of the apparatuses which should give results in a short time, will be mentioned.

Germanium - 76

Many experiments are devoted to search for neutrinoless d.b.d. of ^{76}Ge on ground state as well as on excited states. There are many reasons which make Germanium very attractive to experimentalists:

- natural Ge contains 7.8 % of ^{76}Ge
- transition energy is reasonably large
- Ge detectors have an excellent energy resolution and are intrinsecally quite free from contamination of other materials

Table 2 contains a list of running experiments (the Milano experiment has been concluded a few months ago) and of results obtained on neutrinoless decay mode [5].

As it will be discussed later, the best limits on lepton number violating parameters are obtained by these experiments.

Selenium - 82

The new experiment by the Irvine group [6] is now running. Data obtained in ~5000 hours are no more in contradiction with the geochemical result on the 2ν mode. A sizeable part of the observed signal has to be due to 2 decay, if the geochemically determined half-life is correct. However a more complete understanding of the background is needed before giving a quantitave estimate of the decay rate. More statistics is also necessary to clarify the origin of the signal (4 events) at 2.75 MeV

Table 2 Experiments on ^{76}Ge

	Volume(cm^3)	Shielding	Bkg*		$T^{0\nu}_{1/2}$ (10^{23} yr)	
			$(0^+\to2^+)$	$(0^+\to0^+)$	$(0^+\to2^+)$	$(0^+\to0^+)$
Milano (M.nt Blanc)	138 155	P P	8.0 3.0	4. 0.8	0.2	2.3
Caltech-SIN Neuchatel (Gotthard)	90	A		0.6		0.6
Osaka (Kamioka)	164	A	0.5	1.0	0.6	0.74
Pacific N.L. South Carolina	135	P	1.9	0.4	0.8	1.4
Zaragoza Bordeaux Strasbourg (Frejus)	4x(~100)	A	0.5	6.6	0.2	0.2
Guelph-Aptec (salt mine)	3x(~190)	A	2	0.5	-	1.2
UCSB-LBL (200 m of rock)	8x(~165)	A		0.3	1.5	4.0

*) counts/(KeV x year x 10^{23} at. ^{76}Ge)
Limits at 68% C.L.
P=passive, A=active shielding

Molybdenum - 100

The Osaka group [7] started a measurement with a new detector made by Si detectors interleaved with thin foils of enriched ^{100}Mo. The result up to now obtained shows the presence of a non negligible background in the energy region of interest for the 2ν mode. However more statistcs, comparison with dummy samples and a more refined data analysis will probably allows this group to measure the 2ν rate and to set significant limit on the neutrinoless modes.

Xenon - 136

Results on ^{136}Xe have been recently presented by a Soviet group [8].

A proportional chamber of 3.14 l operated at 25 atm.s was filled with enriched ^{136}Xe. In a preliminary run, a careful investigation of the background due to materials close to the detectors (mainly the preamplifiers) has been carried out; in the same time interesting limits have been obtained, namely:

$$T^{2\nu}_{1/2}(0^+\to0^+) > 1.8\ 10^{19}\ \text{yr}$$

$$T^{0\nu}_{1/2}(0^+\to0^+) > 1.2\ 10^{21}\ \text{yr}$$

$$T^{0\nu}_{1/2}(0^+\to2^+) > 4.9\ 10^{20}\ \text{yr}$$

these limits will be improved in the future, thank to the use of cleaner materials and better statistics.

The Milano group [9] has constructed a multicell proportional counter of about 100 l of volume, which can be operated at 10 atm.s or more. This detector is ready for installation in the Gran Sasso Laboratory; preliminary results are expected in few months.

The Caltech group [11] is completing the construction of a large TPC and is developing solid Xenon detectors; also in this case, results are expected in a short time.

Neodimium - 150

Neodimium, as results from Table 1, is a very interesting candidate for d.b.d.searches; a Soviet group intends to carry on a new experiment on this nuclide at Baksan.

3 Discussion

From the above results the following conclusions can be tentatively drawn:

- d.b.d. has not yet been observed in direct experiments
- the Irvine group and the Osaka group do not seem to be too far from reaching sensitivity sufficient to observe the 2ν mode, respectively for Se and Mo; this point seems to me great importance, in order to give more inputs and cheks of theorethical calculations
- from the UCSB results on Ge it is possible to extract interesting limits on $\langle m_\nu \rangle$ and on the right-handed currents contributions; they are reported in Table 3, according to the presently available theoretical estimates on half-lives.

Table 3 - Limits from ^{76}Ge

$\langle m_\nu \rangle$		$\langle \eta \rangle$		$\langle \lambda \rangle$	
<2.0 eV	[2]	$<5.10^{-6}$	[2]	$<3.4.10^{-6}$	[2]
<1.5 eV	[4]	$<3.4.10^{-7}$	[1]	$<3.9.10^{-6}$	[1]
<0.8 eV	[3]	$<3.6.10^{-8}$	[4]	$<2.9.10^{-6}$	[4]

In view of the fact that the Lyubimov result on m_ν from ^{3}H is contradicted by other recent experiments, d.b.d. appears to be the best way to reach the lowest, although model dependent, limits on m_ν .

4 Particle Detection by Calorimetric Devices

The use of calorimetric detectors in d.b.d. searches has been proposed some time ago [11]. The principle on which these detectors are based is simple. The specific heat at the absolute temperature T of a covalent crystal of mass m (g),mass number A, Debye temperature Θ_D is:

$$C_V = 1944\ (m/A)\ (T/\Theta_D)^3\ J\ K^{-1}$$

It can be small enough to allow a measurable increase in temperature when an energy of a few KeV or MeV is deposited in the crystal. In the assumption (not yet proved experimentally) that all the energy is converted in heat, the energy resolution is:

$$\Delta E = \xi\ (k_B\ T^2\ C_V)^{1/2}$$

k_B being the Boltzmann constant and ξ is in general of the order of a few units. Encouraging results have been obtained by CORON et al. [12], MCCAMMON et al. [13], NIINIKOSKI et al. [14]. Owing to the fact that many elements can be used

(Ge, Mo, Nd ...) which are also sources of d.b.d. ,the energy resolution could be better than in most of other types of detectors and sample mass can be quite large, this technique could bring, in a few years from now, at extremely interesting results.

References

1 M.Doi et al.: Progr.Theor.Phys.Supp. 83,1 (1985).
This review contains a complete list of theoretical as well experimental papers.

2 W.C Haxton and G.J.Stephenson jr.: Prog. Part. Nucl. Phys. 12, 409 (1984).

3 K.Grotz and H.V. Klapdor: Phys.Lett. 153B,1 (1985) and also this conference

4 T.Tomoda et al.: Nucl.Phys. A452, 591 (1986).

5 Most of the results reported in this Table will be reported in Proc. of the Int. Symposium on Nuclear Beta Decays and Neutrinos - June 11-13, 1986 Osaka to be published and in the Proc. of The ν'86 Int. Conference June 2-8, 1986 Sendai , to be published.

6 M.K.Moe: Proc. of the Int. Symposium on Nuclear Beta Decays and Neutrinos -June 11-13, 1986 Osaka - to be published.

7 N.Kamikubota et al.: Proc. of the Int. Symposium on Nuclear Beta Decays and Neutrinos - June 11-13, 1986 Osaka - to be published.
H.Ejiri -this Conference.

8 A.S. Barabash et al.:Proc. of The ν'86 Int. Conference - June 2-8,1986 Sendai.

9 E.Bellotti et al.: Proc. of the Int. Symposium on Nuclear Beta Decays and Neutrinos - June 11-13, 1986 Osaka - to be published.

10 F.Boehm et al.: Proc. of the Int. Symposium on Nuclear Beta Decays and Neutrinos - June 11-13, 1986 Osaka - to be published.

11 E. Fiorini and T. Niinihoski: Nucl. Instrum. and Meth. 244, 83 (1984).

12 N. Coron et al.: Nature 314, 75 (1985).

13 D.C. McCammon et al.: I.E.E.E. NS 33, 236 (1986).

14 T. Niinihoski et al., Europhys. Lett. 3, 449 (1986).

Ultralow Background Searches for $\beta\beta$-Decay, Cold Dark Matter and Solar Axions

F.T. Avignone III[1], S.P. Ahlen[2], R.L. Brodzinski[3], S. Dimopolous[4], A.K. Drukier[5], G. Gelmini[6*], B.W. Lynn[7], H.S. Miley[1], J.H. Reeves[3], D.N. Spergel[6], and G.D. Starkman[4]

[1]Department of Physics, University of South Carolina, Columbia, SC 29208, USA
[2]Department of Physics, Boston University, Boston, MA 02215, USA
[3]Pacific Northwest Laboratory, Richland, WA 99352, USA
[4]Department of Physics, Stanford University, Stanford, CA 94305, USA
[5]Harvard-Smithsonian Center for Astrophysics, Cambridge, MA 02138, USA
[6]Department of Physics, Harvard University, Cambridge, MA 02138, USA
[7]Stanford Linear Accelerator Center, Stanford, CA 94305, USA

An ultralow background Ge detector in the Homestake gold mine is applied to searches for 0ν $\beta\beta$-decay, cold dark matter candidates and invisible axions from the sun. A large body of low background Ge data imply $\langle m_\nu \rangle \lesssim 2.3$ eV. Particles with spin independent Z° exchange interactions and masses between 20 GeV and 5 TeV are excluded as dominant in the galactic halo. Finally, axions with $F/2x_e' \lesssim 0.5 \times 10^7$ GeV are also ruled out.

1. Introduction

Recently achieved low background levels in the PNL/USC 135cm³ prototype Ge detector in the Homestake gold mine are beginning to permit sensitive searches for weakly interacting massive particles (WIMPs) in the halo of our galaxy and Dine-Fischler-Srednicki (DFS) axions from the sun, as well as 0ν $\beta\beta$-decay of ^{76}Ge, for which it was originally intended. The status of these experiments and their interpretation are reported here.

While the program of background reduction continues, the present low levels represent a new technology applicable to these searches. The major improvements, which impact the WIMP and axion searches, have occured in the low energy end of the spectrum (~ 5 keV) and result from several generations of cryostat construction and shielding configurations. The details of our initial background identification and elimination are published elsewhere [1]. A new prototype has just been installed in the mine which does not have indium in direct contact with the Ge crystal, nor any electrical solder connections near the detector. Naturally occurring ^{115}In has a half life of 6 X 10^{14} yr and β-decays with an endpoint energy of 490 keV which significantly affects the low energy portion of the spectrum. The Pb in solder normally contains ^{210}Pb which β-decays to ^{210}Bi which, in turn, β-decays to ^{210}Po contributing background to the spectrum up to about 1.2 MeV. The ^{210}Po α-decays with an energy of 5.3 MeV. Polonium appears to come to the surface of melted solder, and a broad peak at this energy, as well as an energy degraded continuum, were evident in our recently replaced prototype, with which the present data were obtained. The new prototype represents a significant reduction in background in some regions of the spectrum; however, a complete assessment must await further counting. Spectra collected under a variety of cryostat and shielding configurations are shown in Fig. 1.

*On leave from the University of Rome II, Rome, Italy 00173.

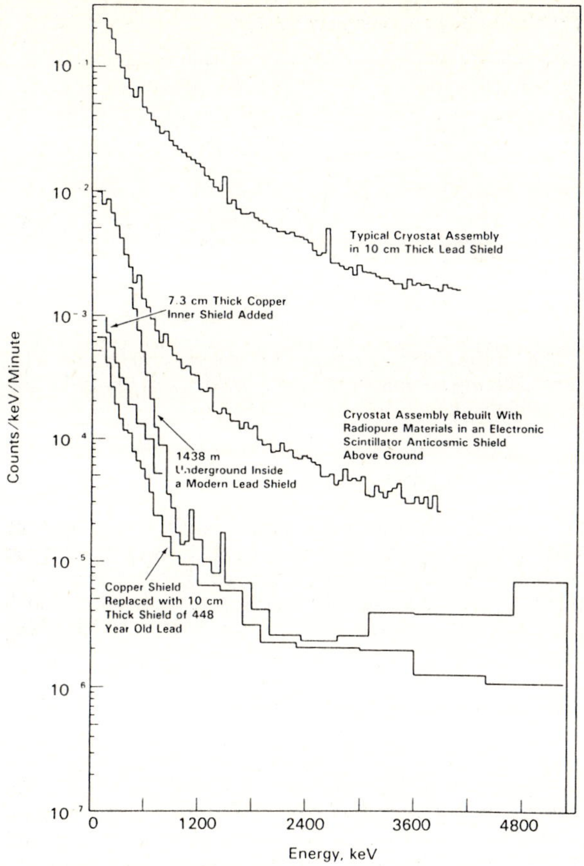

Figure 1. Spectra of 135cm³ Ge prototype detector under various conditions

2. Neutrinoless ββ-Decay

This process can be generated either by Majorana ν-mass or by right-handed ν-currents and has been extensively reviewed [2]. Since the early experiments of PROFESSOR FIORINI and his co-workers, ^{76}Ge experiments have become popular in a major international effort to study the fundamental properties of neutrinos. The purpose of this paper is to present the results of recent background reduction efforts discussed above and to interpret the sum of data from five experiments of similar background levels. The nuclear structure calculations used to obtain limits on the Majorana ν-mass are those of HAXTON and STEPHENSON [2]. A summary of the experimental data is given in Table I from the articles given in ref. [3].

The combined data, added by energy bins, are equivalent to that of a 0.4 yr count with the total Ge (∿ 1770 cm³) used in the experiments of ref. [3]. The most optimistic interpretation of the corresponding likelihood function yields $T_{1/2}^{0\nu} \gtrsim 5 \times 10^{23}$ yr. If the slight depression in the spectrum at 2041 keV is neglected, $T_{1/2}^{0\nu} \gtrsim 3.3 \times 10^{23}$ yr which corresponds to $\langle m_\nu \rangle \lesssim 2.7$ eV. To achieve a sensitivity of $T_{1/2}^{0\nu} \gtrsim 10^{24}$ yr, and $\langle m_\nu \rangle \lesssim 1.6$ eV, the current total effort would have to continue for 3.7 yr. If continued for 20 yr, a sensitivity of $T_{1/2}^{0\nu} \gtrsim 2.3 \times 10^{24}$ yr and $\langle m_\nu \rangle \lesssim 1.0$ eV can be achieved. Therefore, experiments with this level of background, though impressive, are reaching the point of

TABLE I. Summary of five recent ^{76}Ge ββ-decay experiments [3]

EXPERIMENT	(BG/Nt)X10^{23}	Nt(10^{23}yr)	$T_{1/2}^{0\nu}$ limit (10^{23}yr)
CALTECH	0.50	2.08	0.55
GUELPH	0.50	6.56	2.10
MILANO	0.68	2.64	0.74
PNL/USC	0.40	4.07	1.40
UCSB/LBL	0.35	9.19	2.50
TOTAL	0.45	24.45	4.90

diminishing returns. To probe the important regime of $\langle m_\nu \rangle \lesssim 0.1$ eV will require a background reduction of between one and two orders of magnitude. Our recent studies imply that this is feasible.

3. Search For Dark Matter Candidates of the Galactic Halo

In this section, we present the results of our laboratory search for weakly interacting massive particles (WIMPs) which interact coherently with Ge nuclei in the detector via spin independent interactions mediated by Z^0 exchange [4]. The differential cross section with respect to recoil energy, of a WIMP of mass m_X and velocity v, is given by:

$$\frac{d\sigma}{dT} \cong \frac{G_F^2 m_N c^2}{8\pi v^2} [Z(1-4 \sin^2 \theta_W) - N]^2 \, [1 + (1 - T/E)^2 - \frac{m_N T + m_X^2}{E^2}]$$
$$\times \exp(-m_N TR^2/3), \tag{1}$$

where m_N is the mass of the nucleus, Z and N are the proton and neutron numbers respectively, E is the total WIMP energy, R is the nuclear radius and the exponential factor is an approximate nuclear form factor.

The hypothesized particles of the halo are assumed to have a velocity distribution function f(v) with an r.m.s. of 250 km/s and a maximum of 550 km/s, while the halo, like the galactic spheroid, rotates with a velocity of 80 km/s. The count rate was calculated as a function of WIMP mass, halo density, total Dirac neutrino-mass to baryonic-mass ratio, and the coupling constant.

The low energy portion of the pulse height spectrum was used in the search for WIMPs. The data from 10 weeks of counting at high gain were analyzed. The results exclude a halo dominated by particles with $\sigma_{s.i} = \sigma_{weak}$ and with 20 GeV $\leq m_X \leq 5$ TeV. The data also exclude the existence of stable Dirac neutrinos with masses $\geq$ 20 GeV, except in a narrow range near the Z^0 resonance at $m_\nu = m_{Z^0}/2$. To arrive at this result, the cosmological ratio of these neutrinos to baryons was calculated using an analytical solution to the Boltzmann equation [5] in order to find ρ(ν; cosm.) and the largest ρ(baryon; cosm.) consistent with big-bang nucleosynthesis [6].

Our main results are shown in Fig. 2, where the ranges of mass and cross-section for particles excluded as main components of the halo, lie above the curves. The ratio g/g_W is defined as $(\sigma/\sigma_W)^{1/2}$ where σ_W is the cross section for standard Dirac (s.i.) or Majorana (s.d.) neutrinos. A far more detailed discussion, with proper reference to earlier work, can be found in reference [7].

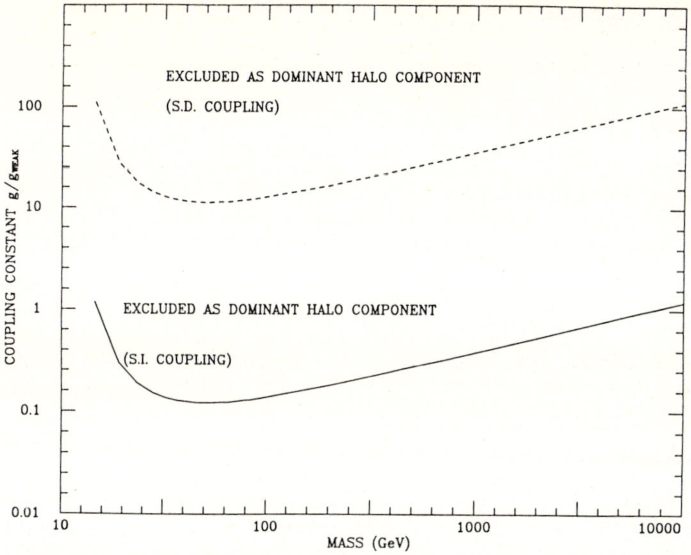

Figure 2. Excluded regions of coupling constant/mass plane lie above the curves

4. Laboratory Limits on Solar Axions

Atomic enhancements can be exploited to detect axion ionization of atoms by the axioelectric effect, analogous to the photoelectric effect. The photoelectric effect is enhanced by factors of $\sim 10^3$ when the photon energy is near the electron binding energy [8]. This should also apply to solar axions because T(sun) $\sim$ 1 keV which is approximately equal to atomic binding energies. In the dipole approximation

$$\sigma_a = (\alpha_a/\alpha)\,(\omega/2m_e)^2\,\sigma_{pe}, \qquad (2)$$

where $\alpha_a = (2X_e'\,m_e/F)^2/4\pi$. The coupling constant is defined by the Lagrangian which describes the direct interaction of DFS axions directly to electrons,

$$L=(2X_e'm_e/F)\,\bar{a}ei\gamma_5 e\,. \qquad (3)$$

The solar flux was calculated for T(sun) = 1 keV using only the bremsstrahlung emission process [9]. In Fig. 3, the number of events/kg/day in Ge are plotted against incoming axion energy for $F/2X_e' = 0.5 \times 10^7$ GeV (solid line) and $F/2X_e' = 10^7$ GeV (dashed line). The major contribution to the event rate would come from a narrow band between 1 and 6 keV of recoil energy. Also plotted in Fig. 3 are a few experimental points (crosses) for $\omega \cong 4$ keV. From this, the bound $F/2X_e' \lesssim 0.5 \times 10^7$ GeV was deduced. The axion mass can be related to F by

$$m_a \cong 7.2\text{ eV}\,(10^7\text{ GeV}/F), \qquad (4)$$

or $m_a < 15$ eV from the data [10].

If $F/2X_e' \cong 0.5 \times 10^7$, the axion luminosity would be about four times the photon luminosity and the sun would burn too fast to be consistent with many observations. To be really meaningful, our bound must be improved by a factor of 2 to 3. The prognosis is excellent that this goal can be met in the not too distant future. For a more complete discussion, with proper reference to earlier work, see ref. [11].

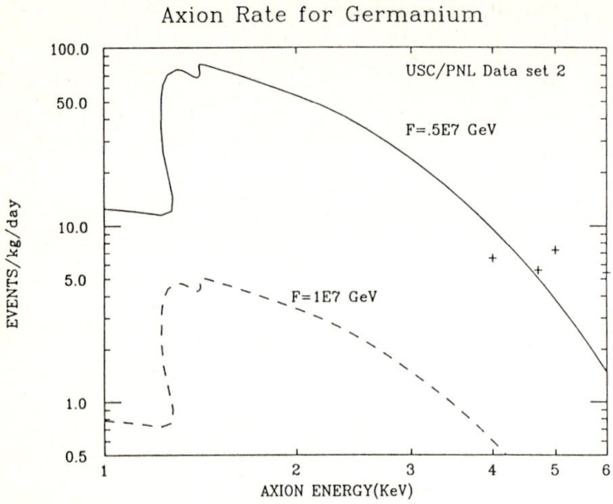

Figure 3. Solar axion events per kg per day for Ge, $F/2X'_e$ = 0.5 X 10^7 GeV (solid), 1.0 X 10^7 GeV (dashed); the crosses are data points

References

[1] R.L. Brodzinski, D.P. Brown, J.C. Evans, Jr., W.K. Hensley, J.H. Reeves, N.A. Wogman, F.T. Avignone III and H.S. Miley, Nucl. Instr. and Meth. A239, 207 (1985).

[2] H. Primakoff and S.P. Rosen, Ann. Rev. Nucl. Part. Sci 31, 145 (1981); W.C. Haxton and G.J. Stephenson, Jr., Progress in Particle and Nuclear Physics 12, 409 (1984); M.G. Shchepkin, Sov. Phys. Usp. 27, 555 (1984); Masuru Doi, Tsuneyuki Kotani and Eijchi Takasugi, Progress in Theoretical Phys. Suppl. 83, 1 (1985); J. Vergados, Phys. Repts. 133, 1 (1986).

[3] F.T. Avignone III, R.L. Brodzkinski, D.P. Brown, J.C. Evans, Jr., W.K. Hensley, J.H. Reeves and N.A. Wogman, Phys. Rev. Lett. 54, 2309 (1985); E. Bellotti, O. Cresmonesi, E. Fiorini, C. Liguori, A. Pullia, P. Sverzellati and L. Zanotti, Phys. Lett. 146B, 450 (1984); P. Fisher, Proc. Moriond Conf. (1986) (in Press); J.J. Simpson (private communication); D.O. Caldwell, R.M. Eisberg, D.M. Grumm, D.L. Hale, M.S. Witherell, F.S. Goulding, D.A. Landis, N.W. Madden, D.F. Malone, R.H. Pehl and A.R. Smith, Phys. Rev. D33, 2737 (1986).

[4] M.W. Goodman and E. Witten, Phys. Rev. D31, 3059 (1985).

[5] J. Bernstein, L.S. Brown, and G. Feinberg, Phys. Rev. D32, 3261 (1985).

[6] J. Yang, M.S. Turner, G. Steigman, D.N. Schramm, and K.A. Olive, Astrophys. J. 281, 492 (1984).

[7] S.P. Ahlen, F.T. Avignone III, R.L. Brodzinski, A.K. Drukier, G. Gelmini, and D.N. Spergel, Harvard Smithsonian Center for Astrophysics Preprint 2292, (1986).

[8] S. Dimopoulos, B.W. Lynn and G.D. Starkman, SLAC preprint 3850, October 1985.

[9] G.G. Raffelt, Phys. Rev. D33, 97 (1986).

[10] D.B. Kaplan, Nucl. Phys. B260, 215 (1985). M. Srednicki, Nucl. Phys. B45, 689 (1985).

[11] F.T. Avignone III, R.L. Brodzinski, S. Dimopoulos, A.K. Drukier, G. Gelmini, B.W. Lynn, D.N. Spergel and G.D. Starkman, SLAC-PUB-3872 (1986).

Limits on Lepton Number Non-Conservation Studied by Double Beta Decays of ^{76}Ge and ^{100}Mo*

*H. Ejiri, N. Kamikubota, Y. Nagai**, T. Nakamura, K. Okada, T. Shibata, T. Shima, N. Takahashi, and T. Watanabe*

Dept. Phys. & Lab. Nucl. Studies, Osaka University, Toyonaka, Osaka 560, Japan

The neutrinoless double beta decay ($0\nu\beta\beta$), which violates the lepton number (L) conservation, provide a very sensitive probe for the conservation rule to the order of 10^{-30} with respect to the L conserving single beta decay. The $0\nu\beta\beta$ gives an evidence for the Majorana neutrino, and a finite mass $<m_\nu>$ and/or a finite right handed weak current. General aspects of the $0\nu\beta\beta$ have been already discussed by previous speakers: Details of the $0\nu\beta\beta$ theories and experiments may be found in recent reviews [1,2,3] and references therein.

This report gives our recent results on the $0\nu\beta\beta$ from ^{76}Ge to both the 0^+ ground and 2^+ first excited states in ^{76}Se and the present status of the $0\nu\beta\beta$ from ^{100}Mo. Some of them have preliminarily been reported [3,4,5].

The weak interaction Hamiltonian is written as [1]

$$H = \frac{G}{\sqrt{2}} (j_L J_L + \eta j_R J_L + \lambda j_R J_R + \kappa j_L J_R) \tag{1}$$

where j_i and J_i are lepton and hadron currents, respectively, with i = L and R standing for the left-handed and right handed ones. Then the $0\nu\beta\beta$ transition rate $t^{0\nu} \equiv \ln 2/T^{0\nu}_{1/2}$ is

$$t^{0\nu} = GM^2_{\beta\beta}[|<m_\nu>|^2/m_e^2 + C_\lambda<\lambda>^2 + C_\eta<\eta>^2] \tag{2}$$

where G is the phase factor, $M_{\beta\beta}$ is the nuclear matrix element, $<m_\nu> = \Sigma m_j U^2_{ej}$, $<\lambda> = \lambda\Sigma U_{ej}V_{ej}$, and $<\eta> = \eta\Sigma U_{ej}V_{ej}$. The $0\nu\beta\beta$ shows up as a sharp peak among huge continuum background in the electron sum energy $E_\beta + E_\beta$ spectrum. Then the fluctuation of the background counts (N_B) gives the lower limits on the peak yield (N_t) which can be identified, namely $N_t > \sqrt{N_B}$. Then

$$0.76\, t^{0\nu} N_o t\, k/\sqrt{N_{BG}\cdot\Delta E\cdot t} > 1 \tag{3}$$

where N_o is the number of source nuclei, t is the measurement time in units of year, ΔE is the energy resolution, k is the detection efficiency, N_{BG} is the

---

* Presented by H. Ejiri

** Present address, Dept. Phys. Tokyo Inst. Technology, Meguro, Tokyo 152 background count rate per keV per year. The coefficient 0.76 is the probability that the peak falls within the energy window ΔE. Thus $\langle m_\nu \rangle$, $\langle \lambda \rangle$, and $\langle \eta \rangle$ are written in terms of the nuclear sensitivity S_N and the detection sensitivity S_D as

$$\{|\langle m_\nu \rangle|/m_e,\ C_\lambda \langle \lambda \rangle,\ C_\eta \langle \eta \rangle\} = 1/S\,t^{1/4} \tag{4a}$$

$$|\langle m_\nu \rangle|^2/m_e^2, \ldots\ldots = t^{0\nu}/S_N,\quad t^{0\nu} > 1/(S_D\sqrt{t}) \tag{4b}$$

$$S = \sqrt{S_N \cdot S_D}\ ,\quad S_N = GM^2_{\beta\beta},\quad S_D = 0.76\ N_o k/\sqrt{N_{BG}\cdot \Delta E} \tag{4c}$$

S is the overall sensitivity. Since G is nearly proportional to the $\beta\beta$ Q value $(Q_{\beta\beta})^5$, a nucleus with large $Q_{\beta\beta}$ and $M_{\beta\beta}$ has a large S_N. A detector with large N_o and k and small N_{BG} and ΔE has a large S_D.

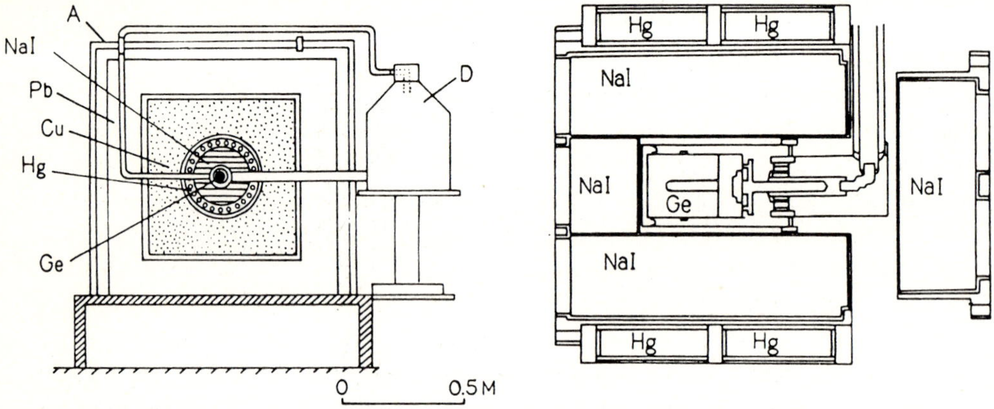

Fig. 1. Left: Side view of ELEGANTS. Right: Top view. Ge: 171 cc active volume intrinsic Ge. NaI: 10"ϕ·12" NaI segmented into 5 and 8"ϕ·3" NaI. Cu: 15 cm thick OFHC. Pb: 15 cm thick lead. D: Liq. N_2 dewar.

A detector system ELEGANTS (ELEctron GAmma-ray NeuTrino Spectrometer) has been developed as shown in Fig. 1. For ^{76}Ge $\beta\beta$ the 171 cc Ge serves as both the $\beta\beta$ detector and the ^{76}Ge source [6]. N_o (^{76}Ge) is $5.7\cdot 10^{23}$. It is surrounded by a 4π geometry 6-NaI crystals. Recording all energy and time signal from the Ge and 6 NaI in a list mode, being followed by the online-offline analysis, is very useful to identify the $0\nu\beta\beta$ $0^+ \to 0^+$ (no γ ray) and $0\nu\beta\beta$ $0^+ \to 2^+$ ($2^+ \to 0^+$ γ ray) events among huge electron events followed by many γ-rays (including Compton ones). [3,5,7]. Thus such β-γ spectroscopic device of the ELEGANTS gives high detector sensitivities S_D because of the low N_{BG} for both the $0^+ \to 0^+$ and $0^+ \to 2^+$ $\beta\beta$. Advantages of using the Ge detector are high resolution (small ΔE) and high efficiencies (k = 1 for $0^+ \to 0^+$ and k = 0.34 for the $0^+ \to 2^+$).

In case of ^{100}Mo a mosaic type source detector system consisting of 11 layers of Si(Li) detectors with 10 ^{100}Mo foils with 50 mg/cm^2, sandwiched between the Si

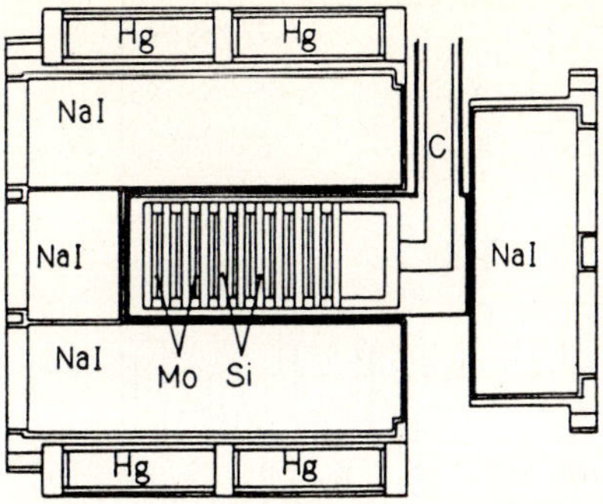

Fig. 2. Si(Li) detectors and ^{100}Mo foils set in place of the Ge in Fig. 1.

detectors with 1500 mm^2 × 4 mm, is used in place of the Ge detector, as shown in Fig. 2. The total ^{100}Mo nuclei is $N_o = 0.5\cdot10^{23}$. Major advantages of using ^{100}Mo are firstly the large S_N because of the large $Q_{\beta\beta}$ ($Q_{\beta\beta}(0^+)$ = 3.03 MeV) in comparison with the $Q_{\beta\beta}(0^+)$ = 2.0407 MeV for ^{76}Ge, and secondary the large S_D (small N_{BG}) because of requirement that two β-rays emitted into opposit directions are detected in coincidence by the adjacent two Si detectors. The energy resolution of the Si detector cooled down to Liq. N_2 temperature is as good as 5 keV and the overall one of around $\Delta E \approx 100$ keV is entirely due to the source thickness. The efficiency k, however, is small. k is evaluated for the $0\nu\beta\beta$ process due to $<m_\nu>$ by a Monte Carlo calculation.

The electron energy spectra were measured at the underground laboratory located in the 1000 m deep (2700 m we) Kamioka mine. Run 1 for ^{76}Ge was carried out for 1600 hr. with the mercury shield around the NaI. Run 2 was done for 7021 hrs. without the mercury shield to avoid the 2044 keV line from ^{194}Hg contained in the shield. The energy spectra around the regions of $Q_{\beta\beta}(0^+)$ and $Q_{\beta\beta}(2^+)$ are shown in Fig. 3. No appreciable $0\nu\beta\beta$ peaks at $Q_{\beta\beta}(0^+)$ and $Q_{\beta\beta}(2^+)$ are found.

The sensitivities and half-life limits are evaluated as follows. For the ^{76}Ge $0^+ \to 0^+$ transition N_{BG} = 5/keV·y, ΔE = 3 keV, k = 1 and $S_D(0^+) = 1.1\cdot10^{23}y^{1/2}$, while for the ^{76}Ge $0^+ \to 2^+$ transition N_{BG} = 2/keV·y, ΔE = 2.5 keV, k = 0.35, and $S_D(2^+) = 0.6\cdot10^{23}y^{1/2}$. The value for $S_D(0^+)$ are in the same order of magnitude as other sensitive detectors [8,9,10], and the $S_D(2^+)$ is the best. The half-life limits are $T^{0\nu}_{1/2}(0^+) \gtrsim 7.4\cdot10^{22}$ y and $T^{0\nu}_{1/2}(2^+) \gtrsim 6.0\cdot10^{22}$y, both on 68 % CL. The $0\nu\beta\beta$ process accompanied by the Majoron and the $2\nu\beta\beta$ process are examined at $E_\beta + E_{\beta'} = 0.9\ Q_{\beta\beta}(0^+)$ and $0.5\ Q_{\beta\beta}(0^+)$, respectively. The lower limits on the half lives are $T^{0\nu M}_{1/2} \gtrsim 2.5\cdot10^{20}$y and $T^{2\nu}_{1/2} \gtrsim 0.2\cdot10^{20}$y.

The nuclear sensitivity for ^{76}Ge is rather small, S_N for $0^+ \to 0^+$ is $7.0\cdot10^{-13}$ by using the $M_{\beta\beta}$ in ref. 11. Then the overall sensitivity is $S = \sqrt{S_N\cdot S_D} = 3\cdot10^5$. The present limit on $T^{0\nu}_{1/2}(0^+)$ gives $<m_\nu> \lesssim 1.9$ eV on the 68 % CL. This is in accord with the value $m_e/S \approx 1.8$ eV. The upper limit on the Majoron coupling is $<g_B> \lesssim 6.4\cdot10^{-4}$ by using the $M_{\beta\beta}$ [11]. The nuclear parameters in ref. 12 lead to

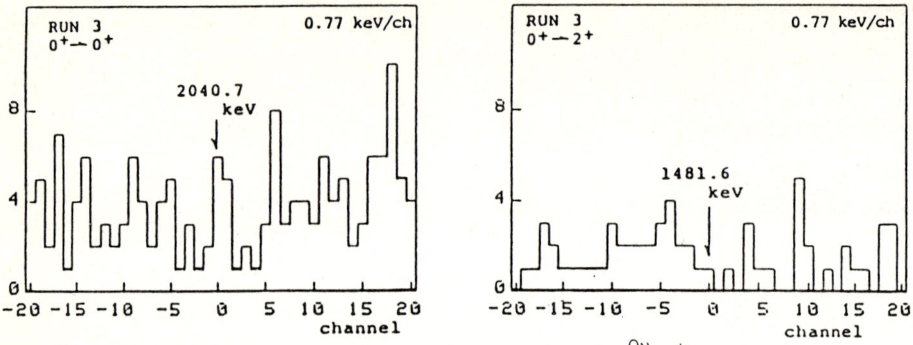

Fig. 3. Enlarged spectra for the Ge detector at the $Q^{0\nu}_{\beta\beta}(0^+)$ and $Q^{0\nu}_{\beta\beta}(2^+)$ regions. Left: for events followed by no γ signals from the 4π NaI crystals. Right: for events followed by one 559 keV $2^+ \to 0^+$ γ signal from one of the NaI segments.

$\langle m_\nu \rangle \lesssim 4.8$ eV, $\langle \eta \rangle \lesssim 1.1\cdot10^{-5}$, $\langle \lambda \rangle \lesssim 0.8\cdot10^{-5}$, and $\langle g_B \rangle \lesssim 1.6\cdot10^{-3}$, while those in ref. 13 are $\langle m_\nu \rangle \lesssim 3.6$ eV, $\langle \eta \rangle \lesssim 8.7\cdot10^{-8}$ and $\langle \lambda \rangle \lesssim 5.7\cdot10^{-6}$. The recent limit of $T^{0\nu}_{1/2}(0^+) \gtrsim 2.5\cdot10^{23}$y [10] leads smaller limits on these quantities by a factor 2.

The sum energy spectrum for ^{100}Mo was obtained by adding signals from two adjacent Si detectors. The result of the preliminary run for 1391 hr is shown, together with the Monte Carlo calculation for the $0\nu\beta\beta$ (0^+), in Fig. 4. Here the result obtained by only the 6 sets of the Si·Mo units out of ten is shown because of α activity contaminants in other sets. The background rate is indeed very small, but no appreciable $0\nu\beta\beta$ peak appears. The sensitivity for the 6 operating Si-detectors is obtained as $\tilde{S}_D = 5\cdot10^{20}y^{1/2}$ from $\tilde{N}_{BG} = 0.13$/keV·y, $\Delta E = 200$ keV, $N_o = 0.28\cdot10^{23}$ and $k = 0.11$. It is estimated to be $S_D = 1.7\cdot10^{21}$ for the present 11 Si-detector system with $N_o = 0.5\cdot10^{23}$ and $k = 0.22$. The half-life limit for $0\nu\beta\beta$ process is $\tilde{T}^{0\nu}_{1/2}(0^+) \gtrsim 1.7 \sim 1.0\cdot10^{20}$y, depending on the way of analysis. The half life limit measurable for the one year run with the 11 Si detector is

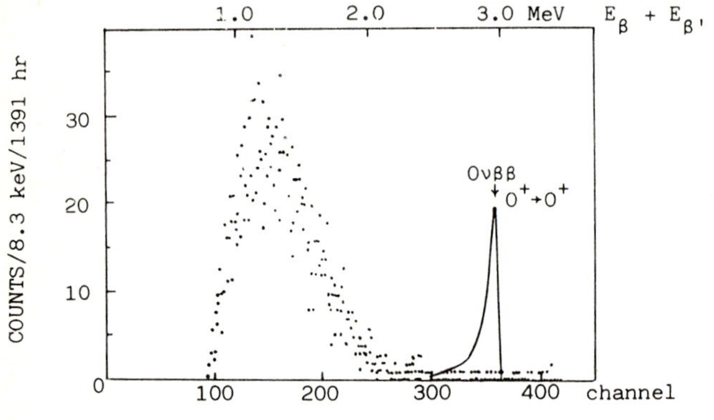

Fig. 4. Energy spectrum of $E_\beta + E_{\beta'}$ for 6 pairs of the adjacent two Si detectors with ^{100}Mo. The solid line is the Monte Carlo calculation.

$T^{0\nu}_{1/2} \approx S_D \sim 1.7\cdot10^{21}$y. The lower limit for the 0νββ with the Majoron is $\tilde{T}^{0\nu M}_{1/2} \gtrsim 7\cdot10^{18}$y on the 68 % CL.

The ^{100}Mo has a large nuclear sensitivity of $S_N = 55\cdot10^{-13}$ by using the $M_{\beta\beta}$ in ref. 10, and the overall sensitivity is $S = \sqrt{S_N \cdot S_D} = 1.0\cdot10^5$. The present half-life limit gives $<\tilde{m}_\nu> \lesssim 14 \sim 18$ eV on the 68 % CL, and the $<m_\nu>$ measurable with the 11 Si detector is $<m_\nu>/m_e \sim S^{-1} \sim 10\cdot10^{-6}$ ($<m_\nu>$ measurable: ~5 eV). The upper limit on the Majoron coupling is $<\tilde{g}_B> \lesssim 10^{-3}$. Because of low background rate one year run with the 11 Si-detector array may elucidate the $<g_B>$ in the order of $3\cdot10^{-4}$ and $T^{2\nu}_{1/2}$ in the order of $10^{17} - 10^{18}$y.

A detector system "ELEGANT V" with drift chambers surrounded by NaI scintillators is now under progress. The two β-ray tracks and the vertex are measured by the drift chamber in order to select the true ββ event. It is used to search for 0νββ, 0νββ with Majoron and 2νββ processes of ^{100}Mo and ^{150}Nd with fairly large S_N. Evaluated sensitivities are $S_N = 75 \sim 55\ 10^{-13}$, $S_D = 5\cdot10^{23}$ and $S \approx 20\cdot10^5$. Thus this can access to $<m_\nu>$ in the order of 0.3 eV.

References

1. M. Doi, T. Kotani and E. Takasugi: Prog. Theor. Phys. Supp. No.83 (1985)
2. S.P. Rosen: Proc. Int. Conf. Neutrino '86 (Univ. Hawaii 1982).
3. H. Ejiri: Proc. Int. Workshop Grand Unification/ICOBAN, Toyama, (World Scientific, Singapore 1986)
4. H. Ejiri et al. : Proc. Int. Symp. Nucl. Spectroscopy & Nucl. Interactions (Osaka, World Scientific, Singapore 1984) p.284
5. H. Ejiri et al. : Nucl. Phys. A448 (1986) 271
   N. Kamikubota et al. : Nucl. Instr. Methods A245 379 (1986)
6. E. Fiorini et al. : Nuovo Cim. 13A 747 (1973)
7. F. Leccia et al. : Nuovo Cim. 78A 50 (1983)
8. E. Bellotti et al. : Phys. Lett. 121B 72 (1983)
9. F.T. Avignone et al. : Phys. Rev. Lett. 54 2309 (1985)
10. D.O. Caldwell et al. : Phys. Rev. Lett. 54 281 (1985), Preprint 1986.
11. K. Grotz and H.V. Klapdor: Phys. Lett. 157B 242 (1985)
12. W.C. Haxton et al. : Phys. Rev. D25 2360 (1982)
13. T. Tomoda et al. : Phys. Lett. 157B 4 (1985)

New Limits on Neutrino Masses and Right-Handed Currents from Double Beta Decay

D.O. Caldwell[1], *R.M. Eisberg*[1], *D.M. Grumm*[1], *D.L. Hale*[1], *M.S. Witherell*[1], *F.S. Goulding*[2], *D.A. Landis*[2], *N.W. Madden*[2], *D.F. Malone*[2], *R.H. Pehl*[2], *and A.R. Smith*[2]

[1]Physics Department, University of California, Santa Barbara, CA 93106, USA
[2]Lawrence Berkeley Laboratory, Berkeley, CA 94720, USA

Data obtained for different time periods from 4, 6, and 8 Ge detectors in an active NaI shield yield a maximum likelihood 68% C.L. lower limit of 3.9×10^{23}yr for the half-life of the $0^+\to0^+$ transition, and 1.5×10^{23}yr for the $0^+\to2^+$ transition, for zero-neutrino double beta decay of ^{76}Ge. They also set 90% C.L. lower limit of 8×10^{19}yr for the two-neutrino decay, and 6×10^{20}yr for the mode in which a light boson (Majoron) is emitted with no neutrinos. These results are used to set limits on light and very heavy neutrino masses, on Majoron-neutrino coupling, and on right-handed currents.

Because double beta decay provides a sensitive probe of lepton conservation, Majorana neutrino mass (m_ν), and right-handed currents (RHC), it is important to push limits for the zero-neutrino decay ($\beta\beta_{0\nu}$) as far as possible. We report here new limits for both the $0^+\to0^+$ (ground state) and $0^+\to2^+$ (first excited state) transitions for this decay in ^{76}Ge. The relative rates of these two transitions can resolve the ambiguity between m_ν and RHC as the source of the decays.[1] A new limit is also set on the less conventional decay which could occur by the emission of a Majoron. The lepton-number conserving two-neutrino decay ($\beta\beta_{2\nu}$) serves as a check on calculations of the $\beta\beta_{0\nu}$ rate; hence the improved limit we quote for this process is also useful.

Double beta decay can occur because pairing energy prevents the decay of an even-even nucleus into the adjacent odd-odd nucleus, whereas decay to the next even-even nucleus is possible, such as in the case of ^{76}Ge$\to$^{76}Se studied here. The second-order weak decay, $^{76}\text{Ge}\to{}^{76}\text{Se}+2e^-+2\bar{\nu}_e$ certainly occurs, but a decay involving the emission of no neutrinos would be favored by a phase space factor $\sim10^8$ were it not that such a decay requires lepton number nonconservation and either m_ν or RHC to provide an abnormal helicity admixture. The $0^+\to2^+$ transition can occur only via RHC.

Descriptions have been published [2,3] of earlier versions of our apparatus. The first version of our experiment, [3a] the data from which will not be used here, was done above ground with two Ge detectors inside the NaI anticoincidence detector background suppression shield. Then the apparatus was moved to the power station of the Oroville Dam in Northern California, about 200m below ground. The average background rate was approximately seven times smaller in the present experiment, per kg of Ge,

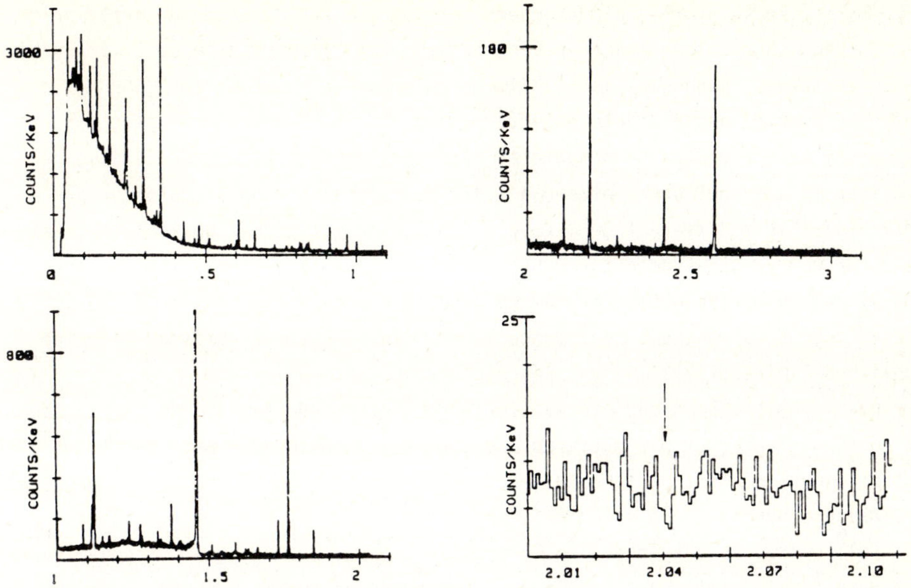

Fig. 1a-d. Energy spectra accumulated in four Ge detectors for 3550 hours of running. The total spectrum is broken into three 1-MeV intervals, and the region near 2.041 MeV (indicated by the arrow) is shown in detail. Small corrections for gain shifts between runs are the cause of fractional events in each bin.

than in the earlier one. At first, four Ge detectors were used for 3550 hours, and those results were published. [3b] Then a total of six detectors were installed and used for about 4000 additional hours. Now eight detectors are in operation.

Figure 1 shows spectra obtained with the 4 and 6 detector systems. The peaks seen have all been identified and are more prominent than in some other double beta decay systems. These are largely due to the NaI, but the NaI also drastically suppresses the Compton scattering associated with these peaks, and hence they do not affect our results. Since no peak is seen at 2.041 MeV, the energy that would characterize the sum of the two electron energies in the $0^+ \to 0^+$ $\beta\beta_{0\nu}$ decay, we can set a lower limit on the half-life for this process. The limit is obtained from the measured average background of 0.27 counts per keV per year per 10^{23} ^{76}Ge atoms in the interval from 2000 keV to 2100 keV, surrounding the energy where a peak would be. (Note: The background has been gradually decreasing since the equipment was first set up underground; it dropped by a factor of 2 in the first 6 months.) For 7.1×10^6 hour·cm^3 (about 43,000 detector hours) of counting with the various detectors giving an averaged 3.4 keV resolution in this energy region, the half-life limit is 3.9×10^{23} yr, using the square root of the number of background counts expected in a region equal to one FWHM resolution with a 68% confidence level. This is to be compared with the two best published [4,5] results of 1.2×10^{23}yr (0.8×10^{23} using the $\sqrt{N}$ method). In giving our results we have not used the data from our above-ground experiment,[3a] which gave a limit of 0.5×10^{23}yr.

The primary purpose of the NaI detectors surrounding the Ge detector array is to act as a Compton shield. An additional purpose is to search for the $0^+ \to 2^+$ transition in $\beta\beta_{0\nu}$ decay. This is done by looking for a coincidence between the 0.559 MeV de-excitation γ, seen by one of these detectors, and the 1.482 MeV summed electron energy deposited in one of the Ge detectors. Another possibility, which is also searched for, is a coincidence between a 1.482 MeV signal in one Ge detector and a 0.559 MeV signal in a different Ge detector. Allowing for either possibility, data collected for the 4- and 6-detector running (6.1×10^6 hr·cm^3) were analyzed, using a measured 90 keV FWHM resolution for the NaI, to give a half-life limit of 1.5×10^{23}yr from the $\sqrt{N}$ method, at the 68% confidence level. This is to be compared with the best published limits of 2.2×10^{22}yr, [4] and 4.4×10^{22}yr. [6]

After peaks due to known radioactivities were subtracted from the Ge energy spectrum, a continuum spectrum was obtained and then searched for the $\beta\beta_{2\nu}$ decay. A significant component of this background spectrum is due to γ-rays from the ^{40}K peak at 1.46 MeV which convert so close to the insensitive region of the detector that less than the full energy is detected. We have collected spectra using a potassium source near the detector to study this component. The shape of the background spectrum was described by adding the potassium component to a smooth function decreasing with energy. A fit to the sum of this background and the calculated $\beta\beta_{2\nu}$ spectrum gives a total of 500±800 for the number of $\beta\beta_{2\nu}$ events. Because of the broad shape of the $\beta\beta_{2\nu}$ spectrum, however, the statistical error on the $\beta\beta_{2\nu}$ contribution is small compared to the systematic error due to uncertainty in the shape of the background. Methods used to estimate this systematic error include studying the stability of the fit to different background shapes and using the calculated value of chi square to see how much the statistical errors underestimate the total errors. After scaling up the errors to include these effects, the 90% confidence level upper limit on the size of the $\beta\beta_{2\nu}$ contribution is 5500 events. The spectrum used for this analysis represents 2456 hours of data in the four detectors, and the corresponding lower limit on the $\beta\beta_{2\nu}$ half-life is 8×10^{19}yr. Recent calculations [7,8] predict a half-life of 2×10^{20}yr.

The same procedure was used to search for neutrinoless double beta decay induced by emission of a scalar boson ($\beta\beta_{0\nu.B}$), which can be a familon or a Majoron. This spectrum, which is shown in Ref. 9, would peak at about 1.55 MeV, while the $\beta\beta_{2\nu}$ spectrum peaks at about 0.65 MeV, and hence the two processes are quite distinguishable. In the $\beta\beta_{0\nu,B}$ case the limit on the half-life is 6×10^{20}yr. From the evaluation of DOI, KOTANI, and TAKASUGI, [9] our result gives a limit on the coupling of the boson to the electron neutrino of $<1.1\times10^{-3}$ using the matrix elements of Ref. 10 and $<4.1\times10^{-4}$ using those of Ref. 11. A somewhat lower limit [9] from ^{130}Te depends on additional assumptions, but the present direct result is more stringent than those from ^{48}Ca and ^{150}Nd; see Refs. 12 and 13.

Interpretation of these several results is difficult at this time because of uncertainties relating half-life to neutrino mass, right-handed currents, or Majoron-neutrino coupling. In fact, the measurement of the $\beta\beta_{2\nu}$ rate is mainly motivated by the need to have a result which can in principle be calculated. Unfortunately, present calculations do not agree by at least an order of magnitude with geochemical results [14] for $\beta\beta_{2\nu}$ half-lives, which are now supported by a laboratory measurement [15] for ^{82}Se.

Even if the matrix elements for the $\beta\beta_{2\nu}$ rate are wrong, those for $\beta\beta_{0\nu}$ might still be correct. [7] The higher energy, shorter range virtual neutrino process is quite different from the regular second-order weak interaction, and for the latter low-lying 1^+ intermediate states play a bigger role. Thus at this stage one can simply take the $\beta\beta_{0\nu}$ matrix elements at face value and get neutrino mass and right-handed current limits, keeping in mind that these could well be low by at least a factor of three.

Table 1

Neutrino Mass and RHC Parameter Limits: $T_{1/2} > 3.9\times10^{23}$yr

	Heidelberg[11]	Osaka[9,16]	Los Alamos[10]	Tübingen[8]
m_ν	0.8 eV	-----	2.0 eV	1.5 eV
$\langle\eta_{RL}\rangle$	---	3.3×10^{-7}	5.0×10^{-6}	3.6×10^{-8}
$\langle\eta_{RR}\rangle$	---	3.9×10^{-6}	3.4×10^{-6}	2.9×10^{-6}

The first row of Table 1 shows mean neutrino mass limits, m_ν, for a half-life at our measured limit of 3.7×10^{23}yr, assuming that there are no right-handed currents, and employing the calculations of GROTZ and KLAPDOR [11], of HAXTON, STEPHENSON, and STROTTMAN [10], and of TOMODA, FAESSLER, SCHMIDT, and GRÜMMER. [8] Assuming $m_\nu = 0$, the data yields the limits shown in the second and third rows of Table 1 for one of the right-handed current parameters, $\eta_{RL}\langle 1\rangle_{LR}$ or $\eta_{RR}\langle 1\rangle_{LR}$, assuming the other to be zero. These limits employ calculations of DOI, KOTANI, and TAKASUGI [9,16], as well as of the authors just mentioned. In the notation used, the first subscript refers to the handedness of the leptonic current and the second to that of the hadronic current, while the subscript LR on the second factor indicates interference between left-handed and right-handed neutrino fields. Clearly, combinations of m_ν, $\langle\eta_{RL}\rangle$, and $\langle\eta_{RR}\rangle$ can co-exist. The matrix element uncertainty applies to all of these limits.

While the numbers in the third row are in rough agreement, those in the second are not. The value using Ref. 8 is one or two orders of magnitude smaller than the others. This most recent calculation includes a relativistic correction to the nuclear current, including weak magnetism, combined with the inclusion of short-range nucleon-nucleon correlations and finite nucleon size effects.

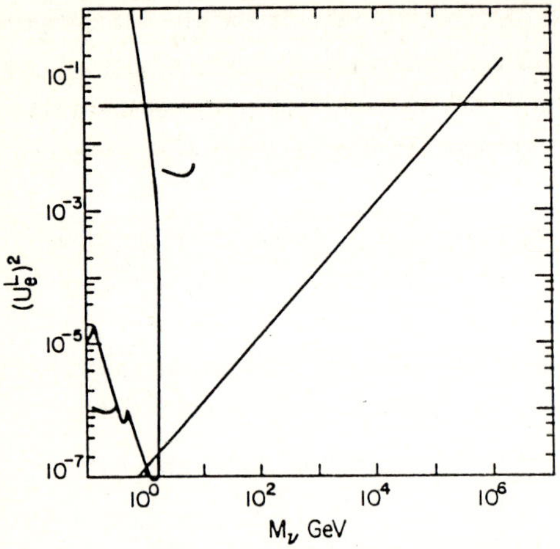

Fig. 2. Limit on the mass of a left-handed Majorana neutrino as a function of its probability for mixing with an electron neutrino.

The $0^+ \to 0^+$ half-life limit may be used also to set a lower limit on the mass of a heavy, left-handed neutrino which couples to the electron neutrino, since this could also provide the abnormal helicity admixture needed to induce neutrinoless double beta decay. What is actually determined for this massive Majorana neutrino is a product of the mixing coefficient, U_e^L, and its mass, M_ν. The limits for these quantities are given as a straight line in Fig. 2, based on a calculation of HAXTON and STEPHENSON. [10] GILMAN and RHIE [17] have recently compiled limits from other experiments, and the relevant portion of those limits is shown in the figure. Values below and to the right of the lines are allowed. The limit from this experiment is seen to be much more stringent than those from others, except near 1 GeV. However, the other limits apply to Dirac neutrinos also. If the heavy neutrino were from a fourth generation of leptons, it could be either a Dirac or a Majorana particle, but extra neutrinos, such as are usually required in left-right symmetric models and particularly in the low-energy limit (E6) of string theories, are most frequently Majorana particles. Thus extending these mass limits can supply important restrictions.

The remarkable sensitivity of this second-order weak process allows one to set limits (with the usual caveats regarding nuclear matrix elements and the possible cancellation caused by the mixing of neutrinos with opposite CP phases) on Majorana neutrino masses at very small values ($\langle m_\nu \rangle < 1\text{-}2$ eV) or at very large values ($M_\nu \gtrsim 10^2$ GeV if its mixing probability with the electron neutrino is $>10^{-5}$, for example). The results given here also allow the setting of stringent limits on the Majoron's coupling to the electron neutrino of $\lesssim 10^{-3}$ and on the mixing of right-handed currents, which may be as small as 10^{-7}.

Work supported in part by the U.S. Department of Energy.

REFERENCES

1. H. Primakoff and S.P. Rosen, Ann. Rev. Nucl. Sci. 31, 145 (1981).
2. F.S. Goulding, C.P. Cork, D.A. Landis, P.N. Luke, N.W. Madden, D.F. Malone, R.H. Pehl, A.R. Smith, D.O. Caldwell, R.M. Eisberg, D.M. Grumm, D.L. Hale, and M.S. Witherell, IEEE Trans. Nuc. Sci. NS-32, No. 1, 463 (1985); IEEE Trans. Nuc. Sci. NS-31, No. 1, 285 (1984).

3(a) D.O. Caldwell, R.M. Eisberg, D.M. Grumm, D.L. Hale, M.S. Witherell, F.S. Goulding, D.A. Landis, N.W. Madden, D.F. Malone, R.H. Pehl, and A.R. Smith, Phys. Rev. Lett. 54, 281 (1985); (b) Phys. Rev. D33, 2737 (1986).

4. E. Belloti, O. Cremonesi, E. Fiorini, C. Liguori, A. Pullia, P. Sverzellati, L. Zanotti, Physics Lett. 146B, 450 (1984).
5. F.T. Avignone, R.L. Brodzinski, D.P. Brown, J.C. Evans, W.K. Hensley, H.S. Miley, J.H. Reeves and N.A. Wogman, Phys. Rev. Lett. 54, 2309 (1985).
6. H. Ejiri, N. Takahashi, T. Shibata, Y. Nagai, K. Okada, N. Kamikubota, T. Watanabe, T. Irie, Y. Itoh, and T. Nakamura, Nuc. Phys. A448, 271 (1986).
7. H.V. Klapdor and K. Grotz, Physics Lett. 142B, 323 (1984).
8. T. Tomoda, A. Faessler, K.W. Schmidt, and F.Grümmer, Phys. Lett. 157B, 4 (1985), Nucl. Phys. A452, 591 (1986).
9. M. Doi, T. Kotani, and E. Takasugi, Osaka Univ. preprint OS-GE 85-03 (1985).
10. W.C. Haxton, G.J. Stephenson, and D. Strottman, Phys. Rev. Lett. 47, 153 (1981); W.C. Haxton and G.J. Stephenson, Prog. Part. Nucl. Phys. 12, 409 (1984).
11. K. Grotz, and H.V. Klapdor, Physics Lett. 153B, 1 (1985).
12. J.D. Vergados, Physics Lett. 109B, 96 (1982); erratum Physics Lett. 113B, 513 (1982).
13. A.A. Klimenko, A.A. Pomansky, and A.A. Smolnikov, Proc. "Neutrino-84" Conference, Dortmund, W. Germany, 161 (1984).
14. T. Kirsten, H. Richter, and E. Jessberger, Phys. Rev. Lett. 50, 474 (1983); B. Srinivasan, B.C. Alexander, and O.K. Manuel, Econ. Gelo. 67, 592 (1972).
15. M.K. Moe, Univ. Calif. at Irvine Neutrino Report No. 133, 323 (1984).
16. M. Doi, T. Kotani, and E. Takasugi, Osaka Univ. preprint OS-GE 85-07 (1985).
17. F.J. Gilman and S.H. Rhie, Phys. Rev. D32, 324 (1985).
18. J. Bagger and S. Dimopoulos, Nucl. Phys. B244 (1984); J. Bagger, S. Dimopoulos, E. Masso, and M.H. Reno, Phys. Rev. Lett. 54, 2199 (1985).
19. R.N. Mohapatra and G. Senjanovic, Phys. Rev. D23, 165 (1981).

An Experimental Search for Double Beta Decay in ^{82}Se

S.R. Elliott, A.A. Hahn, and M.K. Moe

Physics Department, University of California, Irvine, CA 92717, USA

A time projection chamber with a selenium double beta decay source as the central electrode has yielded a lower limit of 1.0×10^{20} years at the 68% confidence level for the two-neutrino half life of ^{82}Se. For the neutrinoless mode we find a corresponding limit of 1.1×10^{22} years, also at the 68% confidence level.

This experiment is searching for three frequently considered modes of double beta decay. The 2 neutrino mode,

$$(A,Z) \rightarrow (A,Z+2) + 2e^- + 2\bar{\nu} \ ,$$

the neutrinoless mode,

$$(A,Z) \rightarrow (A,Z+2) + 2e^-$$

and the Majoron mode,

$$(A,Z) \rightarrow (A,Z+2) + 2e^- + B.$$

The 2 neutrino mode ($\beta\beta(2\nu)$) is expected in the standard model as a second-order weak process whereas the neutrinoless mode ($\beta\beta(0\nu)$) violates Lepton number conservation. The Majoron mode ($\beta\beta(0\nu,B)$) is possible only if the Majoron (B) exists.[1]

The detector is a gas time projection chamber (TPC) which consists of two 10cm drift regions on either side of a planar central electrode. It is octagonal in shape with an 80cm diameter and filled with 92.5% helium, 7.5% propane mixture. The detector is contained in a lead house which itself is within a 4π gas proportional counter cosmic ray veto system. This entire setup is immersed in a 713 gauss magnetic field and is located at sea-level. Fourteen grams of selenium, 97% enriched in isotope 82, is deposited on the central electrode to form a source which 7.1mg/cm^2 thick. A detailed description of the apparatus has been reported[2,3] elsewhere.

Each electron from double beta decay forms a helix in the magnetic field of the TPC. The TPC displays these two individual helices and allows the identification and removal of many background processes.

The trigger rate of 2.5 per second is reduced by on-line event stripping by about a factor of 4. This sample is further reduced by off-line analysis to a group of about 200 events per day which must be scanned by a physicist. The measured $\beta\beta(2\nu)$ detection efficiency is $.154 \pm .013$ which includes loss of 47% of the spectrum[4] due to a sum energy threshold at 1.1 MeV. Backscattering and trajectories too parallel or too perpendicular to the magnetic field for measurement account for most of the remaining losses. Dead time in the system is about 10%.

The energy calibration was determined by the internal conversion lines following the beta decay of ^{208}Tl and ^{207}Bi. The thallium was introduced into the TPC by a ^{220}Rn injection. The radon daughters have short half lives and so are not a worry after a few days. The bismuth was sealed between two mylar sheets and suspended in the TPC.

The energy resolution is a function of the angle the electron makes with the magnetic field. In the best cases the two-electron sum energy resolution is about 4.5% full width at half maximum (FWHM), and averages about 13% FWHM for double beta decay candidates within the range of acceptance.

Eighty-eight double beta decay candidates were recorded in 5426 hours of live time. The energy spectra shown in Fig. 1a,b include all observed events with two electrons over 1.1 MeV sum energy, originating from a common point on the selenium, and having no further activity at the vertex during the following millisecond. Below 1.1 MeV the sum energy spectrum is swamped by the beta decay, internal conversion sequence in ^{214}Pb. This contaminant is deposited on the outer surface of the source by decaying ^{222}Rn in the TPC gas. A sum threshold at 1.1 MeV excludes the ^{214}Pb contribution from spectra (a) and (b) in the figure.

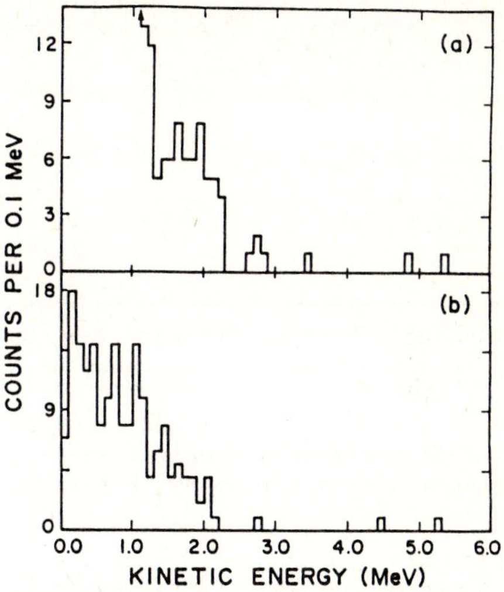

Fig. 1 a) The sum energy histogram of the 88 ββ candidates above a 1.1 MeV sum threshold. Identified backgrounds over threshold contribute mostly below 1.5 MeV.

b) The 2x88=176 electrons taken singly.

Above 1.1 MeV however, the background is less severe. The makeup of the identified background in this region consists of beta decays followed by internal conversion and Compton scattering of gamma rays incident on the source (see Table I). Each of these background mechanisms can be associated with some other detectable process and therefore its intensity can be estimated.

For example, the isotopes ^{208}Tl and ^{214}Bi both may have an internal conversion follow their respective beta decay. The number of thallium 2e$^-$ events can be determined by the rate of the easily detected ^{212}Bi - ^{212}Po beta - alpha sequence in a competing branch of the thorium series. The background from ^{214}Bi can be removed by the observance of the 164 μsec ^{214}Po alpha particle within the millisecond following the primary event. Both thallium and bismuth are daughters of radon isotopes. Tests have implicated the TPC wires as the source of at least part of this radon gas.

Gamma rays incident on the Se source can produce background by Compton scattering. The second e$^-$ comes from a Möller scatter of the Compton electron, or from a second scattering or photoelectric absorption of the photon. This contribution can be deduced from the rate of single electrons (not members of pairs) observed to leave the source. An upgrade of the TPC, intended to reduce these backgrounds, will begin this summer.

To report a conservative limit for the ^{82}Se half life, the background rate is assumed to be the minimum of Table I. This leaves 62 ± 9 counts as possible

TABLE 1. Identified backgrounds between 1.1 and 3.0 MeV

Source of Background		How Measured	Rate (5426 HR) Min.[a]	Max.[b]
β Decay with Internal Conversion				
Thorium	^{228}Ac	β-α sequence ^{212}Bi-^{212}Po	0	21
Series	^{212}Bi	"	0	0
	^{208}Tl	"	7	14
Uranium	^{234m}Pa	β-α sequence ^{214}Bi-^{214}Po	0	7
Series	^{214}Bi	"	0	7
Gamma Induced $2e^-$ events		Single Electrons	19	54
		TOTAL	26	103

[a]Assumes U and Th daughters come from Rn in TPC gas, and that the single electrons are mostly β particles without gamma rays.

[b]Assumes U and Th daughters are imbedded in the Se, and that the single electrons are mostly Compton recoil electrons. "Min" and "max" are the one-sigma limits accompanying these assumptions.

ββ(2ν) events, from which the ^{82}Se half life is limited to > 1.0×10^{20}yr at 68% confidence. This result puts to rest the suggestion of a much shorter half life by an earlier cloud chamber experiment.[5] It also disagrees with the relatively short half life predicted for ^{82}Se by shell model calculations.[6] The proximity to the geochemical half life of $(1.26 \pm .04) \times 10^{20}$yr reported by Kirsten[7] provides some hope that a clean double beta decay spectrum is within reach.

Very little of the predicted sum energy spectrum for ββ(2ν) survives beyond 2.1 MeV. However, the spectrum predicted for the Majoron mode is at its strongest in this region.[8] If we attribute the 12 events between 2.1 and 3.0 MeV to ββ(0ν,B) we find the half life for this mode to be > 4.3×10^{20} yr at 68% confidence.

The sum energy for a ββ(0ν) ground-state transition in ^{82}Se is 3.0 MeV. The probablity that a neutrinoless event would be detected in a 300 keV window centered on 3 MeV is estimated to be 0.21 ± 0.02. the absence of counts in this window during 7106 hours[9] corresponds to a ββ(0ν) half life of > 1.1×10^{22}yr at 68% confidence.

We are indebted for advice and encouragement to Professor F. Reines. This work is supported by the U. S. Department of Energy through Contract No. DE AT03-76SF00010.

References

1. Y. Chikashige, R. N. Mohapatra, and R. D. Pecci, Phys. Lett. 98B, 265 (1981).

2. M. K. Moe, A. A. Hahn, and H. E. Brown, _The Time Projection Chamber_, Edited by J. A. Macdonald (AIP Conference Proceedings No. 108, 1984) p. 37.
3. M. K. Moe, A. A. Hahn, and S. R. Elliott, _Proceedings of the Conference on Neutrino Mass and Low Energy Weak Interactions_, Telemark, Wisconsin, 1984, edited by Vernon Barger and David Cline (World Scientific, Singapore, 1985).
4. H. Primakoff and S P. Rosen, Repts. Progr. Phys. _22_, 121 (1959).
5. M. K. Moe and D. D. Lowenthal, Phys. Rev. C _22_, 2186 (1980).
6. W. C. Haxton and G. J. Stephenson, Jr., Prog. Part. Nucl. Phys. _12_, 409 (1984).
7. T. Kirsten, _International Symposium on Nuclear Beta Decays and Neutrino_, Osaka, June 1986.
8. M. Doi, T. Kotani, and E. Takasugi, Osaka Univ., Prog. Theor. Phys. Supp., _83_, (1985)
9. Includes 1680 hours with the unimproved shield.

Neutrinoless Double Beta Decay of ^{76}Ge. Preliminary Results of an Experiment in the Frejus Tunnel

A. Morales[1], *J. Morales*[1], *R. Nuñez-Lagos*[1], *J. Puimedón*[1], *J.A. Villar*[1], *D. Dassie*[2], *Ph. Hubert*[2], *F. Leccia*[2], *P. Mennrath*[2], *M.M. Villard*[2], *J. Chevallier*[3], *and B. Haas*[3]

[1]Departamento de Fisica Nuclear, Universidad de Zaragoza
[2]Centre d'Etudes Nucléaires de Bordeaux-Gradignan, Université de Bordeaux I, F-Bordeaux, France
[3]Centre de Recherches Nucléaires de Strasbourg, F-Strasbourg, France

I. Introduction

As is well known the detection of the neutrinoless double beta decay of nuclei would indicate the violation of lepton number and would provide significant information about the character, possible mass and chirality couplings of the neutrino. Several direct experiments aiming to detect such a rare event have been recently carried out or are being done. We present here the preliminary results of an attempt to search for the neutrinoless double beta decay of ^{76}Ge, currently being done in the Frejus tunnel. The experiment (which uses as source and detector an assembly of 4 Germanium crystal [1]) investigates the transition from the ground state 0^+ of the ^{76}Ge to the ground state 0^+ of ^{76}Se, as well as to the excited state 2^+ (559.1 KeV), of ^{76}Se, looking for a gaussian shaped peak at $E_{2\beta}$=2040.7 KeV in the anticoincidence spectrum, and at E=1481.6 KeV in the coincidence spectrum $2\beta/\gamma$ (E_γ=559.1 KeV), respectively, by surrounding the 4-Ge detector with a 4π assembly of NaI scintillators. Transition to the 0_2^+ (1122.3 KeV) excited state is also investigated.

The neutrinoless double beta decay is a low energy, very rare process which needs a proper matching between the helicities of the neutrinos emited and absorbed, respectively, by two nucleons within a nucleus. In the neutrinoless $0^+\to0^+$ transitions that matching is provided by a neutrino mass term and/or by a right handed admixture in the neutrino coupling, whereas in the $0^+\to2^+$ transitions, only the right handed admixture is involved. The $0^+\to2^+$ transition although is even more rare than the transition to the ground state, has the bonus of the deexcitation photon signal which help to clean the background (provided that the NaI crown be also extremely clean), and gives unambigous information about the presence of the right handed couplings.

II. Experimental set-up

The experimental set-up (Fig.1) consists in a set of 4 hyperpure Germanium detectors of 456 cm^3 total volume (417 cm^3 total active volume), manufactured by PGT Europe with low background specifications. The common cryogenic container is a cylindrical box of dimensions 160 Ø x 120 cm and 2 mm thick, where the four Germanium detectors are placed symmetrically with their z-axis parallel to the vertical axis of the cyllindrical box. This common container as well as the end caps of the detectors and other mechanical parts holding the assembly are made in Cu(OFHC). The averaged energy resolutions in 2478 hours are 2.7 KeV and 3.1 KeV in the $0^+\to2^+$ and $0^+\to0^+$ regions, respectively. The 4-Ge detector is surrounded by a 4π system of polycristalline NaI hexagonal scintillators (dimensions of the crystals 135 Ø x 204 cm). Fourteen of these detectors (Bicron), manufactured with low background specifications are wrapped in teflon, have quartz windows and are packaged in low activity stainless-steel (0.5 mm thick). The other five NaI (Harshaw), of the same dimensions, are packaged in aluminium and are located below the 4-Ge box. The averaged energy resolutions of the scintillators is about 11% at 560 KeV. The detector system is contained in a box of 65 x 65 x 105 cm , in Cu(OFHC), 2.5 cm thick, shielded with 10 cm of lead, and is placed for operation at the Modane Underground

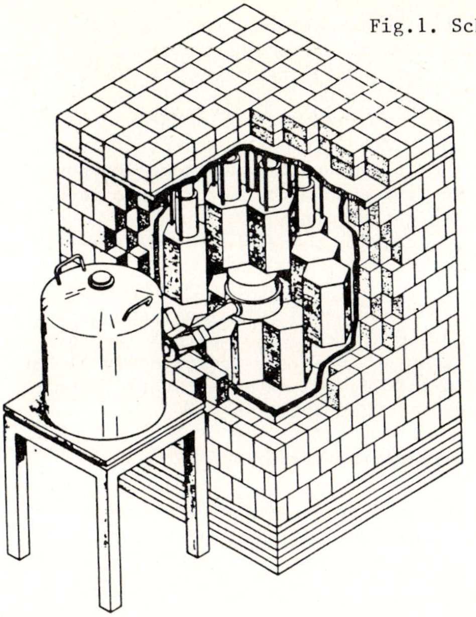

Fig.1. Schematic view of the experimental set-up

Laboratory, in the Frejus tunnel (4700 mwe deep). Associated to the 23 detectors one has conventional read out electronics. The multiparametric adquisition system is controlled by a PDP 11/23 microcomputer through a CAMAC crate. The trigger is the logical OR of the Germanium detectors. A logical discriminator rejects high multiplicity events (in the 4-Ge and/or the 19 NaI detectors) due to external background. Each event candidate is stored in a standard magnetic tape (with 2 Kw buffer), recording the time, the configuration mask, energies and associated TACs. Coincidence events ($0^+\rightarrow$ excited states transitions) as well as anticoincidence events ($0^+\rightarrow$ ground state) are thus simultaneously recorded.

The experimental procedure is as follows: For each fast signal coming out of any one of the 4-Ge detectors, we look at the number N of the NaI fired in a 100 nsec window (N=0 anticoincidence, N=1, 2, 3, 4, 5 - fold coincidences), and record the event candidate as indicated before. Sets of data recorded every $\sim$250 hours ($\sim$1500 blocks of 2 Kw) are then combined by taking properly into account slight gain shifts, in order to add only events of the same energy. At each tape replacement the NaI energy calibration and energy resolution are checked by taking data with a ^{22}Na source. The 4-Ge energy resolution and the energy calibration were checked every two or three days by using the background lines of ^{234}Pa (1001 KeV), ^{60}Co (1173, 1332 KeV), ^{40}K (1460 KeV), ^{214}Bi (1764 KeV) and ^{208}Tl (2614 KeV), as well as by measuring with the ^{22}Na source. The overall energy resolution is that of the total summed data. Broadening after 2478 hours gives, Γ_{FWHM}(1482 KeV)=2.7 KeV and Γ_{FWHM}(2041 KeV)=3.1 KeV as quoted before.

As is well known, the reduction of the background is the main problem in the search of such a rare double beta transitions. The anticoincidence 4π-Sodium iodine detectors are essential to reduce both the internal background of the 4-Ge detector and that of the charged cosmic particles. It is clear that efficient detection of gamma rays in coincidence with the beta signal from the Ge detector is of paramount importance to identify internal background of the Ge detector. That implies that the materials surrounded the Ge crystal have to be kept at a minimum and to be made of low-Z elements. That is also essential for having a reasonable probability for the 560 KeV deexcitation photon to escape from the Ge crystals (and from the cryogenic enceinte), without interaction, to have a chance to be absorbed by the 4π NaI scintillators. To reduce external background (outside the Ge crystal) low radioactivity materials have been used, and passive lead and

Cu(OFHC) shields employed as described before. On the other hand, the operation in a laboratory deep underground (4700 mwe) assures the dropping of the cosmic background, giving so a essential reduction of the induced neutrons. In our spectra we have found the natural decay chains of ^{235}U, ^{238}U, ^{232}Th as well as ^{40}K, ^{132}Ba, ^{22}Na, ^{60}Co, ^{137}Cs. On the other hand, we also find a sizable background due to α particles of 5.2 MeV from the decay of the ^{210}Pb [2]. The sources of the background obtained so far are being traced in order to reduce it, and some improvements are being considered. In particular, the five old NaI detectors have been replaced by new ones, made with low activity specifications.

III. Sensitivity and experimental parameters

The double beta decay experiment look for a gaussian shaped peak (corresponding to the summed two electrons energy signal) at about $Q_{2\beta}$ (or $Q_{2\beta}-E_\gamma$), with a few KeV width at half maximum, superimposed on a background. So far no positive signal has been seen, and only a lower limit for the half-life $T_{1/2}$, can be extracted from the flat background spectrum in the region of interest. The limit is given by

$$T_{1/2} \geqslant \frac{\ln 2}{A} \; \frac{V(cm^3).N_A.\rho(gr/cm^3).f}{P_A} \; \frac{t(h)}{365x24} \; \varepsilon \quad \text{years}$$

where A is the upper limit of the double beta "peak" area (at a given confidence level) and ε is the overall efficiency. For the ^{76}Ge case one has

$$T_{1/2} \geqslant \frac{2.71 \times 10^{17} \times V_f \times t}{A} \times \kappa \times P \quad \text{years}$$

where V_f is the fiducial volume, $P \equiv P_{00}=1$ for $0^+\to0^+$ transitions and $P\equiv P_{02}=P^\gamma_{esc} x P^\gamma_{abs}$ for the $0^+\to2^+$ coincidence experiments. P^γ_{esc} is the probability for the 559.1 KeV photon to escape without interaction from the Ge-detector assembly and P^γ_{abs} is the probability of being totally absorbed in the 4π-NaI scintillators. The factor κ takes into account the fraction of the peak area of double beta true events falling inside the energy bin chosen, (gaussian distribution assumed), i.e., κ is the probability for the double beta event to be in the bin chosen. In the $0^+\to0^+$ transition $\kappa=\kappa_{Ge}$ refers to the bin chosen around $E(0^+\to0^+)=2040.7$ KeV. In the $0^+\to2^+$ transitions $\kappa=\kappa_{Ge} \times \kappa_{NaI}$, refers both to the bin around $E(0^+\to2^+)=1481.6$ KeV, and to the energy window in NaI around $E_\gamma=559.1$ KeV, chosen for the assumed double beta/gamma coincidence.

The sensitivity is defined as the simple gaussian rule (square root of the flat background), in a bin centered in the region of interest and with a width equal to the FWHM

$$S_{1/2} = 0.76 \times 1.37 \times 10^{20} \sqrt{\frac{V_f(cm^3).t(h)}{B(c/KeV.y.10^{23}).\Gamma(KeV)}} \times (\kappa_{NaI} \times P_{02}) \quad \text{years}$$

where B is the background and $\kappa_{NaI} \times P_{02}$ is to be used in $0^+\to2^+$ transitions only.

To improve the sensitivity, one needs (besides a large counting time) to increase simultaneously $\sqrt{V_f}$ and P_{02} (which in $0^+\to2^+$ transitions amounts to use a set of Ge detectors) and to improve both B and Γ. The materials surrounding the 4-Ge detector, mechanical parts, caps, etc, as well as the geometry, must be chosen in order to have a not too big damping factor P_{02}. In fact, in the current coincidence experiments P_{02} is always less than $\sim$0.30 - 0.35 [3] [4].

We have computed the fiducial volume of our 4-Ge assembly by Monte Carlo simulation by using our own code [5] for transport of low energy electrons as well as the well known EGS4 code [6]. We have distinguished between $0^+\to0^+$ and $0^+\to2^+$ by using the single electron kinetic energy spectrum and angular correlation as given by Doi et al [7], obtaining respectively $V_f(0^+\to0^+)=389$ cm³ and $V_f(0^+\to2^+)=403$ cm³.

The P_{02} probability defined above, has been also calculated with Monte Carlo simulation EGS4. By allowing respectively, ±31 KeV (2.35σ) and ±40 KeV (3σ) for the

560 KeV gamma ray, we obtain the escape probability $P^{\gamma}_{esc}(0^+\rightarrow 2^+)=0.32$ (0.33). Notice that the probability of escape from our 4-Ge bare crystal (assuming no surrounding material at all) would have been 0.44 (0.45).

The total absorption probability in the 4π-system of NaI scintillators (packaged in 0.5 mm of stainless-steel) arranged in the geometry depicted in Fig.1, computed also by Monte Carlo EGS4, is $P_{tot.abs.}(0^+\rightarrow 2^+)=0.70$ (0.71) (one fold coincidence). In the analysis of our preliminary data, we shall use as overall probability $P_{02}=0.23$ for a 3σ NaI window. (σ=26.4 KeV).

Finally, as far as the upper limit of the peak area, A, is concerned, we shall use the probability density function method [8] as well as the commonly used maximum likelihood function method.

IV. Preliminary results

1) $0^+\rightarrow 0^+$ transition

In Fig.2 we show the enlarged region of the anticoincidence spectrum, centered around E=2040.7 KeV (channel 2721). Live time of measurement t=2475 hours. The background , averaged over 33.75 KeV is B=6.4 c/KeV.y.10^{23} at. of ^{76}Ge. The sensitivity is $S_{1/2}=2.3\times 10^{22}$ years. We obtain the following lower limit of the half-life - $T_{1/2}$ $(0^+\rightarrow 0^+)$:

i) Probability density function method
   Energy bin ΔE=5.25 KeV (7 ch)=$\pm 1.99\sigma$, $\kappa_{Ge}=0.953$
   The background averaged over ±14.25 KeV above and below the bin is B=24.2 c/KeV
   Result: $T_{1/2}(0^+\rightarrow 0^+) \geq 2.4\times 10^{22}$ years (68% C.L.)

ii) Maximum likelihood method.
   Result: $T_{1/2}$ $(0^+\rightarrow 0^+) \geq 2.4\times 10^{22}$ years (68% C.L.)

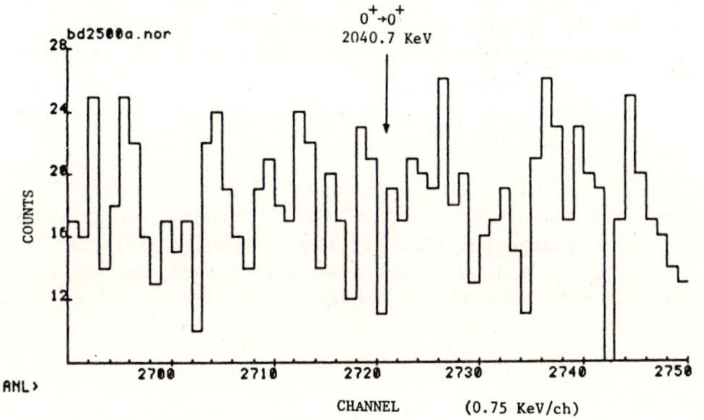

Fig.2. Enlarged region of the $0^+\rightarrow 0^+$ (anticoincidence) spectrum

2) $0^+\rightarrow 2^+$ transition

In Fig.3 we show the enlarged region of the $2\beta/\gamma$ coincidence spectrum, centered around $E=Q_{2\beta}-E_{\gamma}$=1481.6 KeV (channel 1976), gated by a 3σ window in NaI (559.1±39.6 KeV). The background averaged over 33.75 KeV is B=0.46 c/KeV.y.10^{23} at. of ^{76}Ge. The sensitivity is $S_{1/2}=1.9\times 10^{22}$ years. We obtain the following lower limit of - the half-life $T_{1/2}(0^+\rightarrow 2^+)$:

i) Probability density function method
   Energy bin (Ge) ΔE=3.75 KeV (5 ch)= $\pm 1.63\sigma$, $\kappa_{Ge}=0.897$
   (NaI) $\Delta E=3\sigma$, $\kappa_{NaI}=0.866$
   The background averaged over ±15 KeV above and below the bin is B=1.8 c/KeV
   Result: $T_{1/2}(0^+\rightarrow 2^+) \geq 2.1\times 10^{22}$ years (68% C.L.)

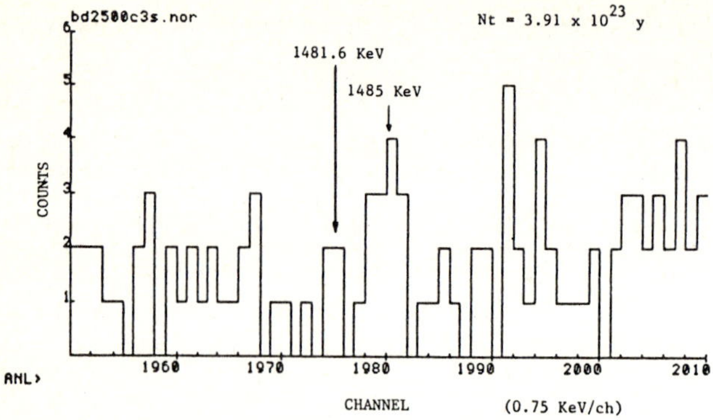

Fig.3. Enlarged region of the $0^+\to2^+$ (coincidence) spectrum

ii) Maximum likelihood method. Result: $T_{1/2} > 2 \times 10^{22}$ years.

A two dimensional analysis (various NaI windows an Ge energy bins) does not - change, essentially, the results.

3) $0^+\to0_2^+$ (1123.3 KeV)

Half-life limit: Result: $T_{1/2}(0^+\to0_2^+) \geqslant 1 \times 10^{22}$ years

Finally, we would like to point out the appearance of an accumulation of counts, as it is shown in the coincidence $0^+\to2^+$ spectrum (Nt=3.91 x 10^{23} y) of Fig.3. This "peak" is located at about 1485 KeV, i.e., 3.4±0.5 KeV above the currently acepted double beta Q-value for $0^+\to2^+$ transition (1481.6 KeV). A simple statistical analysis shows that this accumulation of counts (obtained in about 2500 hours) would co rresponds to a "peak" of 8.0±3.8 counts. We do not have yet enough statistics to - draw any definite conclusion, neither about its established presence nor about - its origin, but it is interesting to notice, that Avignone et al, have reported [9] also an accumulation of counts (5.7±2.8 counts) in a $0^+\to0^+$ ^{76}Ge spectrum (Nt=1.8 x 10^{23} y), just at an energy of 2044.7 KeV, i.e., shifted also 4 KeV above the $Q_{2\beta}$ value. This peak has been, however, somehow weaken when the statistics has increased (Nt=4 x 10^{23} y) [10].

A new run has started, with improved conditions, in order to rise the lower limits of the half-lifes obtained so far, and to investigate the evolution of the - shifted accumulation of counts reported above.

The authors thank the IN2P3 (France) and CAICYT (Spain) for financial support as well as R. Barloutaud, S. Jullian, D. Lalanne and Ph. Roy for providing all the facillities in the Laboratoire Souterrain de Modane. They also appreciate the collaboration of B. Gerona and A. Larrea as well the able technical assistance of A. Guiral.

References

1. E. Fiorini et al: Nuovo Cimento, 13A (1973) 747
2. Ph. Hubert et al: "α-rays induced background in ultra low level counting with Ge spectrometers", submitted to Nucl. Instr. & Meth., Jun 1986
3. D.O. Caldwell et al: Phys. Rev. Lett. 54 (1985) 281
4. H. Ejiri et al: Nucl. Phys. A448 (1986) 271
5. J.A. Villar: "Métodos de simulación en experiencias de muy bajas actividades". Thesis Univ. of Zaragoza. Publ. FAN/UZ, Jun 1985
6. W.R. Nelson, H. Hirayama, D.W.O. Rogers: "The EGS4 Code System". SLAC Report 265 (Dec 1985)
7. M. Doi, T. Kotani, E. Takasugi: Progr. of Theor. Phys., Japan, Suppl. 83, (1985)
8. O. Helene: Nucl. Instr. & Meth. 212 (1983) 319
9. F.T. Avignone et al: Phys. Rev. Lett. 54 (1985) 2039
10. F.T. Avignone: private communication

Searching for $\beta\beta$ Decay of ^{150}Nd. Next Step

A.A. Klimenko, S.B. Osetrov, A.A. Pomansky, A.A. Smolnikov, and S.I. Vasilyev

Institute for Nuclear Research, Academy of Sciences, Moscow 117312, USSR

Results of preliminary measurements performed with a modified experimental set-up to search for $\beta\beta$ decay of some isotopes are described. The sensitivity of the experiment to different modes of ^{150}Nd and ^{100}Mo $\beta\beta$ decay is estimated for a measuring time of one year.

Recently the first step of measurements to search for double beta decay of ^{150}Nd has been completed [1]. Data obtained for 4000 h measuring time yield some nontrivial limits on the ^{150}Nd half-lives for neutrinoless $\beta\beta$ decay ($T_{1/2}>2.3\times10^{21}$ y for the $\langle m_\nu \rangle$-mechanism) and for the mode in which a Majoron is emitted with no neutrinos ($T_{1/2}>1.3\times10^{20}$ y). Hereafter all experimental limits are given at 90% c.l. . Taking into consideration some recent theoretical estimations of the rate of two-neutrino $\beta\beta$ decay for ^{150}Nd [2,3] we can suppose that our experimental limit $T_{1/2}(2\nu, 0^+ \to 0^+)>2.4\times10^{19}$ y is close to the real half-life. In such a case one can hope to detect at least this $\beta\beta$-mode for ^{150}Nd using the method of multidimensional analysis applied in our experiment. Moreover, our previous experience showed there are some ways to increase the detection sensitivity through possible improvements of certain parameters of the experimental set-up.

In continuing our $\beta\beta$ decay studies we have begun a modified version of our experiment. In principle, both the method of event acquisition and the main design of the detector system have been unchanged on the whole. But the electronic equipment as well as the detector have been replaced by new ones. The new detector as the old one [4] consists of 4 plastic scintillators, only the design of the light-pipes is changed. However, the optical and scintillating properties of the detector meet the heightened requirements to improve the scintillation response. A new type of low-noise phototubes is used as well. So, the energy resolution of the detector has been improved from 20% to 17% (FWHM) at 1 MeV. The number of components from metal is strictly limited in the detector assembly and this metal is titanium only. The careful control of the radioactivity contamination of the construction materials for the detector assembly and the same care in the process of counter production allowed us to achieve a background reduction of about a factor of 10 compared to the old detector in the energy region of 2ν-mode (0.5–2.0 MeV) and of a factor of 5 in the region of the 0ν $\beta\beta$ decay (2.7–3.7 MeV) of ^{150}Nd and ^{100}Mo. In connection with the recent interest in the ^{100}Mo nuclear matrix elements [3] we plan to measure this isotope in turn with Nd. It is an essential feature of the assembly that one can replace an exposed sample by another one. The complete set of the samples includes the isotopically enriched sources, the sources for energy calibration and the samples applied for background recognition. An exposed sample is sandwiched between two inner scintillation counters to measure energy-energy correlations. The IN 96B multichannel analyser, based on a quick and powerful minicomputer, is used in the modified version of equipment. This system handles 6 independent ADC inputs; two of these are used for event acquisition from two outer scintillators and the four others are used for inner ones. The latter enables us to compare the position of an event along the length of each scintillator providing an additional selection of the two-electron events. This in turn gives an additional background reduction at least of a factor of 3.

As a result of the above mentioned improvements of the resolution and background the sensitivity of the experiment to search for $\beta\beta$ decay is increased considerably. Preliminary measurements have shown that we can achieve the following limits on $\beta\beta$ half-lives for ^{150}Nd and ^{100}Mo, respectively, during a measurement of about one year:

2ν-mode – 1×10^{20} y and 5×10^{19} y,
$0\nu\chi^{\circ}$-mode – 5×10^{20} y and 3×10^{20} y,
0ν-mode – 1×10^{22} y and 6×10^{21} y (all at 90% c.l.).

References

1. A.A. Klimenko, A.A. Smolnikov, A.A. Pomansky: "Low background scintillation installation for double beta decay experiments", in "Low Radioactivities '85" Int. Conf. Proc., Bratislava, 1985
2. H.V. Klapdor: "Nuclear Beta Strength and the Neutrino Mass", Invited Talk given at Int. Symp. on Nuclear Beta decays and Neutrino, Osaka, June 11–13, 1986;
K. Grotz, H.V. Klapdor, Phys. Lett. **157**B (1985) 242
3. P. Vogel: "Nuclear Structure and Double Beta Decay" in Proc. of same Symposium
4. M.P. Baskov et al.: Proc. Int. Conf. "Neutriono '82", v. 1 (1982), p. 202

New Possibilities in a Double Beta Decay Experiment Using Enriched ^{76}Ge Inside of an Active Si(Li) Shielding

L.A. Popeko[1], *A.V. Derbin*[1], *I.A. Kondurov*[1], *V.V. Martynov*[1], *H.V. Klapdor*[2], *and J.Metzinger*[2]

[1]Leningrad Nuclear Physics Institute, Gatchina, Leningrad district, 188350, USSR
[2]Max-Planck-Institut für Kernphysik, D-6900 Heidelberg, F. R. Germany

At present some direct neutrinoless double beta-decay experiments are in progress at a sensitivity level of 10^{23} y. The presently best limit for ^{76}Ge of $T_{1/2}^{0\nu} \geq 3.9 \cdot 10^{23}$ y [1] corresponds to a Majorana neutrino mass $m_\nu \leq 1$ eV [2].

Common feature of present experiments is the application of germanium detectors inside low background passive or active shielding. A decisive quantity for such experiments is the background counting rate of a germanium detector per energy resolution interval at the energy of 2.041 MeV. Measurements in underground laboratories at levels deeper than 4000 m water equivalent (w.e.) eliminate practically the cosmic rays. The residual background of $\sim 10^{-4}$ (keV • h • 100 cm^3)$^{-1}$ is due to the natural radioactivity of the low background shielding.

As possibilities to improve the experimental sensitivity have been discussed:

- increase of the active volume of the germanium detector used;
- use of isotopically enriched ^{76}Ge as detector material;
- use of a passive shielding of semiconductor purity.

The possibilities of active scintillation shielding [3] are practically exhausted.

A new way to improve the sensitivity is to construct an active silicon shielding which consists of a great number of Si(Li) detectors.

Recently a Si(Li) multidetector for ($\bar{\nu}$,e) scattering experiments was proposed [4]. The detector contains 400 kg of Si(Li) modules with an active volume of a 100 cm^3 each. The modules are packed closely without any support materials and are surrounded by an 80 mm thick mercury shielding in titanium containers. These containers are placed inside a vacuum cryostat at liquid nitrogen temperature. The outside preamplifiers are connected with the Si(Li) modules by HP copper conductors. The main feature of the detector is application of semiconductor purity components (the contaminations of ^{238}U and ^{232}Th are less than 10^{-20} Ci/g).

A background test has been made of a 25 kg detector model consisting of 305 modules. Figure 1 (curves 5 and 6) shows the experimental data at a level of 30 m w.e. The background index is 10^{-4} (keV • h • 100 cm^3)$^{-1}$ [4]. This result is not changed when testing at the level of 1100 m w.e., i.e., the background counting rate is caused by the materials nearest to the detector.

The result obtained was used for the calculation of the background index of a 200 cm^3 HP Ge detector placed in the centre of the 400 kg Si(Li) multidetector

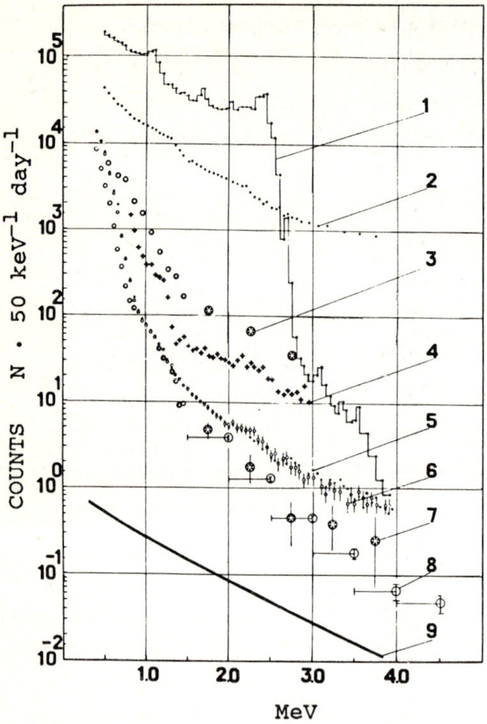

Fig. 1. The results of the background test of the 25 kg Si(Li) active shielding:
1 - the calibration spectrum of ^{24}Na for the whole detector;
2 - the background spectrum of the whole detector at ground level;
3 - the background of the central module;
4 - the background spectrum at a level of 30 m w.e., the preamplifier components are inside the first layer (45 mm) of the passive shielding;
5,6 - the background with the preamplifier components outside the 200 mm passive shielding and after increasing the active shielding;
7 - the same as curve 6 but for the central module;
8 - the results of the Reines's $(\bar{\nu},e)$ scattering experiment;
9 - the calculated recoil electron spectrum of the $(\bar{\nu},e)$ scattering experiment for a neutrino flux of $6 \cdot 10^{12}$ cm^{-2} s^{-1}.

which consists of 2000 modules. The index calculated is 10^{-5} (keV · h · 100 cm^3)$^{-1}$. In case of using 7.6 kg enriched ^{76}Ge as HP detector material inside this active shielding the background counting rate will be 30 counts/y, which corresponds to a sensitivity level of $T_{1/2} \geq 10^{25}$ y for the neutrinoless double beta-decay process.

References

[1] D.O. Caldwell et al., this conference
[2] H.V. Klapdor, Proceed. VI Moriond Workshop on Massive Neutrinos in Astrophysics and Particle Physics, Tignes, France, 25.1-1.2.1986, Editions Frontiéres, Gif-sur-Yvette, p. 597, K. Grotz, H.V. Klapdor, Phys. Lett. 153B (1985) 1 and Nucl. Phys. A, in press (1986)
[3] F.T. Avignone et al., Nucl. Instr. Meth. A239, 207 (1985)
[4] A.V. Derbin, L.A. Popeko, A.V. Cherny and G.A. Shishkina, Pisma v GETF 43, 164 (1986) (in Russian)

3.3.2 Solar Neutrinos

Solar Neutrinos: Theory

J.N. Bahcall

Institute for Advanced Study, Princeton, NJ 08540, USA

1. Overview

The solar neutrino problem can be stated simply. The observed capture rate is [1,2]:

$$\text{Observed} = (2 \pm 0.3)\text{SNU}. \tag{1}$$

The predicted capture rate for the ^{37}Cl experiment is [3,4]:

$$\text{Predicted} = (7.5 \pm 2.5)\text{SNU}. \tag{2}$$

[The convenient unit for discussing solar neutrinos is the product of a flux times a cross section and is: 1 SNU = 10^{-36} events per target particle per second. I give here preliminary values of event rates predicted by the standard solar model that have been obtained by an improved set of calculations with updated parameters. The final values will be published in reference 4.] The observational error quoted above is 1-σ(by tradition); the error I quote on the predicted rate corresponds to an *effective* 3-σuncertainty, defined in [3]. The fact that the theoretical and the experimental numbers differ significantly *is* the solar neutrino problem. There is no generally accepted solution to the problem although neutrino oscillations in matter provides a plausible and attractive hypothesis [5,6].

For physics, the Sun provides a beam of collimated low energy (MeV) neutrinos that traverse a large distance before they are detected. Hence, solar neutrino experiments provide a means for exploring, for large mixing angles, neutrino mass matrix elements as small as 10^{-12} eV^2 and, for small mixing angles, more modest neutrino masses ($\lesssim 10^{-4}$eV^2). For astronomy, solar neutrino experiments constitute rigorous tests of our understanding of how stars generate energy and evolve. The long-standing discrepancy between calculation and observation in the ^{37}Cl experiment could be caused by new weak interaction physics or by our lack of understanding of the simplest stage of stellar evolution.

I will summarize the theory of stellar evolution as it refers to the Sun and describe the theoretical expectations - and their uncertainties - for solar neutrino experiments. The central question for solar neutrino research is easily stated. *Is the solar neutrino problem caused by a lack of understanding of the interior of the Sun or by the discovery of new phenomena in the propagation of neutrinos?* Is the origin of the problem in the *production* or the *propagation* of neutrinos? This question can only be answered experimentally and I will therefore discuss crucial experiments involving deuterium, electron scattering, ^{40}Ar, and ^{71}Ga .

2. The Standard Solar Model

An observation of solar neutrinos is a *critical* test of the theory of stellar evolution. We know much more about the Sun than about any other star. We know its mass, its luminosity, its surface chemical composition, and its age much more accurately than we can ever hope to determine these crucial parameters for any other stellar object. Moreover, the Sun is undergoing the simplest stage of stellar evolution (otherwise we wouldn't be here now). The Sun is sitting quietly on the main sequence, perking along by burning hydrogen without (according to the standard theory) any violent or rapid evolution. Thus we ought to be able to calculate what the Sun is doing more accurately than we can predict what more distant, less well-behaved stars are doing.

The theoretical basis for the calculations of solar neutrino fluxes is described in detail in reference [3].

The Sun is assumed to be spherical and to have evolved quasi-statically for a period of 5×10^9 years. The evolution is caused primarily by the loss of photons from the surface of the star, which is compensated for by the burning of (four) protons into α-particles. Energy transport in the interior is mainly by radiation (which means that the radiative opacity is important) and the pressure is largely thermal.

The physical conditions are different from those we experience in everyday life but they are not so different as to suggest that the relevant physics will contain important surprises. The central temperature is a little more than a keV and the central density is somewhat more than 100 gm cm^{-3}. So far as we know, the phyiscs of the gaseous (largely non-degenerate) solar interior is relatively simple. The primordial chemical composition is assumed equal to the present-day surface composition. The surface of the Sun is too cool for nuclear reactions to have altered significantly the primordial composition.

In order to calculate the neutrino fluxes, we make numerical models of the Sun, beginning with an initial guess for the primordial ratio of hydrogen to heavy elements. The surface measurements give only the ratios of the abundances of elements heavier than helium. The model parameters are evolved quasi-statically, taking account of the composition changes and the energy released by nuclear reactions. The accuracy with which the interior calculations must be carried out is much higher than for most other applications of stelllar evolution theory because the measurements relate directly to processes occurring in the solar interior and because the calculated fluxes are sensitive to the precise conditions.

The theory of stellar evolution has had a number of remarkable successes. The most basic achievement is the prediction of a relation between the mass and the photon luminosity of stars that is in agreement with observation over almost two-orders of magnitude in mass (six orders of magnitude in luminosity). In addition, the theory successfully accounts for the positions of all known stars in the luminosity-temperature or luminosity-color plane. Since most of the observed plane (known technically as an H-R diagram) is empty, the representation of the positions of the known stars by conventional models is a major triumph. Perhaps the greatest acheivement of the theory is that it is used constantly by astronomers in applications to many different kinds of traditional astronomical problems without encountering obvious inconsistencies.

3. Nuclear Energy Generation

The Sun shines by converting protons into alpha-particles. The main nuclear burning reactions are well known [3,7].

Over the past two decades, many difficult experiments and detailed calculations have been performed to determine the rates of individual reactions in the p-p chain and the

CNO cycle. This work has determined the nuclear reaction rates to an accuracy that shows that the solar neutrino problem cannot be accounted for by recognized uncertainties in the nuclear fusion parameters. This collective achievement is the work of many people in institutions distributed around the world.

There have been many improved experiments over the past several years [7] which have refined nuclear reaction rates. I record here only those changes from the best-estimates (and uncertainties) of the nuclear parameters discussed in [3].

The rate of the p-p reaction is now known more accurately because of the improved determination of the axial vector coupling constant from assymetry measurements in neutrino decay [8]. The best estimate now is $S_{p-p} = 4.07(1 \pm 0.017) \times 10^{-25}$ MeV-b. The rate of the pep reaction is changed only because it is proportional to S_{p-p}. There are many new accurate data for the ^{3}He - ^{4}He reaction [7], which results in $S_{3-4} = (0.54 \pm 0.02)$ keV-band $S'_{3-4} = -3.1 \times 10^{-4} b$. A careful review of the ^{7}Be - p reaction yields $S_{1-7} = (0.0243 \pm 0.0018)$ keV-b, 16 % smaller than the value of 0.029 keV-b adopted in [3]. The experimental errors quoted above are 1-σ values and should be multiplied by 3 in calculating uncertainties in neutrino fluxes according to the perscription in [3]. An improved ^{8}B neutrino spectrum has been derived [9] and forbidden corrections included in the capture cross sections, but the net change is small (1-3 %) for all the solar neutrino experiments discussed here.

The challenge for the next two decades is to reduce the uncertainties in each of the reaction rates to a level where the errors in determining the fusion parameters do not influence the interpretation of the next-generation solar neutrino experiments, which roughly translates into a 3-σ uncertainty in the predicted neutrino fluxes of less than 10 % . Refinements in our knowledge of low energy cross sections that will permit this accuracy in the predictions will be difficult and expensive in terms of time, thought, and accelerator facilities, but are required in order to keep the uncertainties resulting from nuclear fusion parameters from interfering with the ultimate precision of the solar neutrino experiments.

The most important reaction for solar neutrino astronomy, the ^{7}Be (p, γ)^{8}B reaction, is, with the existing hard-won data, the most susceptible to systematic uncertainties [8]. The reason that this crucial reaction has not been investigated even more thoroughly in the laboratory is that it is difficult to study because the target is radioactive (with an inconveniently short lifetime of 52 days) and the reaction rate is very small at the enrgies of interest. More experimental work is required because the predictions for a number of solar neutrino experiments (e. g., ^{37}Cl, ^{2}H, ^{40}Ar , electron scattering, and ^{98}Mo) are proprotional to, or approximately proportional to, the value of the low energy cross section factor for this reaction. We should aim at reducing the 1-σ uncertainty of about 3 % so that it will not confuse the interpretation of the solar neutrino experiments.

The 3-σ uncertainty in the p-p rate still causes an appreciable uncertainty, $\simeq$ 14%, in the predicted ^{8}B neutrino flux. The main contributor is the meson exchange correction which has a significant effective 3-σ uncertainty of 4 % (based upon the spread among the results of different authors [3], although individual authors have estimated uncertainties that are a factor of two smaller. The nuclear matrix element is rather well determined (effective 3-σ uncertainty of only 2.5 %). The recent determinations of the ratio of axial vector to vector weak coupling constants [10] correspond to a total 3-σ uncertainty of only 1.5 %.

4. Crucial New Experiments

More experiments are required, with different techniques and sensitivities. The number of suggested theoretical explanations for the solar neutrino problem is very large, including: resonant neutrino oscillations (with mixing angles and mass matrix elements between at least three flavors of neutrinos), non-resonant neutrino oscillations, a variety of non-standard solar models (i. e., ones in which non-standard physics is introduced), and a multitude of entertaining suggestions like a central massive black hole or a large neutrino dipole moment. Several new carefully chosen experiments must be performed to in order to select uniquely the correct explnation.

The ^{71}Ga Experiment

A gallium experiment is the consensus next step. The reasons are both theoretical and experimental: 1). ^{71}Ga is sensitive to neutrinos from the basic proton-proton reaction; and 2). the detection scheme is simple and well established. For the standard solar model, we obtain [4] as a preliminary value:

$$\Sigma(\phi_i \, \sigma_i) = (134 \pm 17) \text{ SNU}, \tag{3}$$

The largest uncertainties are from the heavy element abundance (11 SNU) and the neutrino absorption cross sections (9 SNU). For non-standard solar models that are consistent with the ^{37}Cl experiment, transitions to excited states of ^{71}Ge are unimportant, eliminating for this class of models the uncertainty that is largest for the predictions of the standard solar model.

The ^{71}Ga experiment will be discussed in detail at this conference by W. Hampel.

Some Other Experiments

The most promising or advanced detectors include: ^{2}H, ^{40}Ar, ^{81}Br, ^{98}Mo, and electron scattering. The ^{98}Mo experiment is the furtherest along and may yield results within one year [11,12]. It will provide unique information about the ^{8}B neutrino flux over the past several million years.

The ^{81}Br detector is unique in being primarily sensitive to ^{7}Be neutrinos. However, the neutrino absorption cross sections are not accurately determined [13] (p,n measurements on this target are essential) and the counting is difficult but feasible [14].

The other experiments are senstive only to ^{8}B neutrinos. With ^{2}H and ^{40}Ar detectors, one can test for both the ν_e flux and the total solar neutrino flux, using electron scattering (sensitive to neutrinos of all flavors) to measure the total flux. The ^{2}H experiment will be discussed at this conference by E. D. Earle. I will therefore say a few words about the ^{40}Ar experiment.

The Liquid Argon Experiment

A liquid argon time projection chamber can be used [15] to make separate measurements of electron neutrinos (by absorption) and neutrinos of different flavors, to determine the incident neutrino spectrum, and to verify that neutrinos come from the direction of the Sun. *Resonant neutrino oscillation has a dramatic effect on the shape of the predicted spectrum of recoil electrons.*

The cross sections for neutrino capture to the analogue state of ^{40}K can be calculated accurately and are: $\sigma(^8B) = 7.8 \times 10^{-43} cm^2$ and $\sigma(^3He - p) = 3.1 \times 10^{-42} cm^2$. The standard solar model implies a capture rate of 4.2 SNU or about 1×10^3 events per year in a kiloton detector.

The most important aspect of this detector is its ability to measure the shape of the recoil spectrum of electrons produced by neutrino capture, thus determining directly the electron neutrino spectrum after it leaves the Sun. If the MSW effect is operating in the Sun (see talks by A. Smirnov and S. P. Rosen at this conference), then the ^{40}Ar detector may be able to determine mixing angles and mass-differences by observing the recoil electron spectrum.

5. Summary

The first quarter of a century of solar neutrino astronomy has produced a well defined "solar neutrino" problem. I believe that experiments to be performed in the next 25 years will reveal the solution to this problem and point us toward either a more complete theory of stellar energy generation or of neutrino propagation. If we are fantastically lucky, solar neutrino experiments could even do both.

This work was supported in part by the National Science Foundation.

[1] R. Davis, in *Proceedings Informal Conference on Status and Future of Solar Neutrino Research* edited by G. Friedlander (BNL Report 40879, 1978), **Vol. 1**, p. 1.

[2] J.N. Bahcall, B. T. Cleveland, R. Davis, Jr. ,and J. K. Roweley, Ap. J. Letters **292**, L79-L82 (1985).

[3] J.N. Bahcall, W.R. Huebner, S.H. Lubow, P.D. Parker, and R.K. Ulrich, Rev. Mod. Phys., **54**, 767 (1982).

[4] J. N. Bahcall and R. K. Ulrich (in preparation) (1986).

[5] S.P. Mikheyev and A. Yu. Smirnov, 10th International Workshop, Savonlinna, Finland, June 16-25, 1985; L. Wolfenstein, Phys. Rev. **D20**, 2634 (1979).

[6] S. P. Rosen and J. M. Gelb, Phys. Rev. D (submitted) (1986); H. A. Bethe, Phys. Rev. Letters **56**, 1305 (1986).

[7] P. D. Parker, in *Physics of the Sun*, **Vol. 1**, p. 15 (D. Reidel, New York, 1986).

[8] S. J. Freedman, Argonne preprint PHY=4682 (1986).

[9] J. N. Bahcall and B. Holstein, Phys. Rev. C, **33**, 2121 (1986).

[10] F. C. Barker and R. H. Spear, Ap. J. (to be published) (1986).

[11] G. A. Cowan and W. C. Haxton, Science **216**,51 (1982).

[12] G. A. Cowan, private communication (1986).

[13] J. N. Bahcall, Phys. Rev. C, **24**, 2216 (1981).

[14] G. S. Hurst et al., Phys. Rev. Letters **53**, 1116(1984).

[15] J. N. Bahcall, M. Baldo-Ceollin, D. Cline, and C. Rubbia, Phys. Letters (submitted) (1986).

Neutrino Oscillations in Matter

S.P. Mikheyev and A.Yu. Smirnov

Institute for Nuclear Research, Academy of Sciences of the USSR, 60-October Anniversary prospect 7a, Moscow 117312, USSR

1 Introduction

Neutrino Oscillations ... in vacuum/1/ are determined by mass difference and mixing angle.

... in Matter the effect of coherent forward ν-scattering must be taken into account/2/. It results in modification of evolution equations for ν-states and, consequently, in changing of the length and the depth of oscillations/2,3/.

The influence of matter on weakly mixed neutrinos has a resonance character (resonance in density or energy)/4/. Matter may amplify ν-oscillations resonantly. Moreover in matter with varying density qualitatively new effects appear/4/, in particular, at adiabatic condition almost complete trasformation of initial type neutrino into another one is possible in a wide energy region/4/. Neutrino transformations in matter give new insight into solar neutrino puzzle/4/.

Recently many aspects of neutrino osgillations in matter have been elaborated in details.

1)Basis and background of phenomena. Matter effect in terms of effective neutrino masses (energy levels) have been described/5,6/ and new insight into neutrino system resonance was done/5/. Many analogies of the considered effects have been found/2, 12-15,22/.

2)Neutrino oscillations in different regimes. Oscillations at adiabatic condition are considered thoroughly/4,7,8,10,12,15-16,22/. Some results for nonadiabatic regime have been obtained/4,8,10,12,17-19,22/.

3)Divergence of neutrino wave packets. Properties of this divergence and its consequencies for ν-oscillations are studied/8,12,14,15,22/.

4)Applications. ν-oscillations in the Sun/4,5,9-12,16-19/, in the Earth/8,12,13/, in cores and envelopes of collapsing stars/4,7,8/, in Early Universe/8,20/ have been considered.

5)The implication to neutrino mass spectrum and particle physics/5,20,21/.

In this paper we describe united formalism of ν-oscillations for different regimes, which is immediate generalization of vacuum oscillations theory. Adequate graphical representation of this formalism is given. We summarize main properties of ν-oscillations for different density distributions.

2 Remarks on WOLFENSTEIN's Equations

Evolution equations for two mixed neutrinos with definite flavours $\nu_f = (\nu_\alpha, \nu_\beta)$ can be written as follows:

$$i \frac{d}{dt} \vec{\nu}_f = \hat{M} \vec{\nu}_f \tag{1}$$

The evolution matrix M contains two parts

$$\hat{M} = \hat{M}^V + \hat{M}^I, \quad \hat{M}^V = \begin{vmatrix} M_\alpha^V & \overline{M} \\ \overline{M} & M_\beta^V \end{vmatrix}, \quad \hat{M}^I = \begin{vmatrix} \Sigma_\alpha & 0 \\ 0 & \Sigma_\beta \end{vmatrix} \tag{2}$$

First - M^V is "vacuum part" with $M_\alpha^V + M_\beta^V = (m_1^2 + m_2^2)/2k$, $M_\alpha^V - M_\beta^V = (\Delta m^2/2k) \cdot \cos 2\theta$ and $2\overline{M} = -(\Delta m^2/2k) \cdot \sin 2\theta$. Here m_i are the masses of admixtures, k is momentum of neutrinos, θ is vacuum mixing angle. Second - M^I is the "matter part", describing the interactions of neutrinos:

$$\Delta\Sigma=\Sigma_\alpha-\Sigma_\beta=\sum_i \frac{\Delta f^i(0)\cdot n_i}{k} - \sqrt{2}\cdot G_F\cdot \frac{\rho^{\alpha\beta}_{eff}}{m_N} \quad (3)$$

where $\rho_{eff} = \sum_i \Delta f_i(0)\, n_i\, m_N/\sqrt{2}\, G_F$ (4)
is the effective density of matter. $\Delta f_i(0) = f^i_\alpha(0) - f^i_\beta(0)$; $f^i_\alpha(0)$ is the amplitude of forward ν_α-scattering on i-component of matter, n_i is the concentration of this component, m_N is nucleon mass, G_F is Fermi constant.

The interest phenomena take place when $\Delta\Sigma$ compensites at some density ρ_R the difference between $M^V_f(f=\alpha,\beta)$:

$$M = \Delta M^V + \Delta\Sigma = 0 \quad (5)$$

Vacuum and matter may be considered on one step here. In general both vacuum and matter are nonsymmetrical mediums respect to ν_α and ν_β. Moreover, vacuum may be represented as matter with a constant density $\rho^V = -\rho_R(E)$ (see sect.3). Then (4) corresponds to $\rho^m = \rho^V$. At transition from flavour oscillations to mass oscillations ($\nu_1 - \nu_2$), vacuum and matter change their roles. For flavour oscillations vacuum mixes neutrino, matter is diagonal respect to ν_e, ν_μ. For mass oscillations inversely: vacuum is diagonal $\hat{M}^V = \mathrm{diag}(m_1,m_2)$, matter gives mixing/2/.

Instead of vacuum "second matter" may be considered. The corresponding interaction should be nondiagonal in flavours/2/. The effects of varying density may be reproduced here if the effective densities (4) of these two matters depend on concentrations differently and if the relative concentration (composition of matter) varies with a distance (time). Note that in contrast with (vacuum plus matter) - case here effects do not depend on neutrinos energy.

Futher on we will consider "usual" situation - vacuum and matter - and $\nu_e - \nu_\mu$ oscillations for definiteness.

3 Neutrino Oscillations, General Formalism

§1 Neutrino mixing in matter

In matter neutrino mixing can be represented as follows:

$$\vec{\nu}_\alpha = \hat{S}_m \vec{\nu}_m; \quad \hat{S}_m = \begin{pmatrix} \cos\theta_m & \sin\theta_m \\ -\sin\theta_m & \cos\theta_m \end{pmatrix} \quad (6)$$

Here $\vec{\nu}_m = (\nu_{1m}, \nu_{2m})$ are neutrino eigenstates in matter/4,8,10/, that is the eigenstates of evolution matrix $\hat{M}$ (2). θ_m is the mixing angle, determined from diagonalization condition

$$\hat{S}^{-1}\hat{M}\hat{S} = \hat{M}^{diag}; \quad \hat{M}^{diag} = \mathrm{diag}(M^d_1, M^d_2) \quad (7)$$

$$\sin 2\theta_m = 2\bar{M}/(M^2 + 4\bar{M}^2)^{1/2} \quad (8)$$

M is introduced in (5). The eigenvalues of $\hat{M}$ are

$$M^d_{1,2} = \frac{1}{2}\left(M_e + M_\mu \pm (M^2 + 4\bar{M}^2)^{1/2}\right) \quad (9)$$

According to (6) mixing angle θ_m determines the flavours (ν_e, ν_μ-content) of neutrino eigenstates. In particular,

$$\langle \nu_e | \nu_{1m} \rangle = \cos\theta_m, \quad \langle \nu_e | \nu_{2m} \rangle = \sin 2\theta_m \quad (10)$$

ν_{im}, θ_m and M^d_i generalize ν_i, θ and m_i correspondingly for vacuum case. Relations are straitforward; if $\rho \to 0$ $\nu_{im} \to \nu_i$, $\theta_m \to \theta$, $M^d_i \to m_i^2/2k$.

In matter θ_m depends on density or energy. From (8) one has:

$$\sin^2 2\theta_m = \sin^2 2\theta \cdot R(\rho/\rho_R, \theta) \quad (11)$$

where R is the resonance factor

$$R = 1/\cos^2 2\theta \left((1-\rho/\rho_R)^2 + \tan^2 2\theta\right) \quad (12)$$

with $\rho_R = m_N \Delta m^2 \cos 2\theta / 2\sqrt{2}\, G_F E$. Two properties are important:
1) The dependence of $\sin^2 2\theta_m$ on ρ (see (11)) or E has a resonance character/4/ (fig.1). $\sin^2 2\theta_{m\,max} = 1$ at resonant density $\rho = \rho_R$, the half width of resonance:

$$\Delta\rho_R = \rho_R \tan 2\theta \quad (13)$$

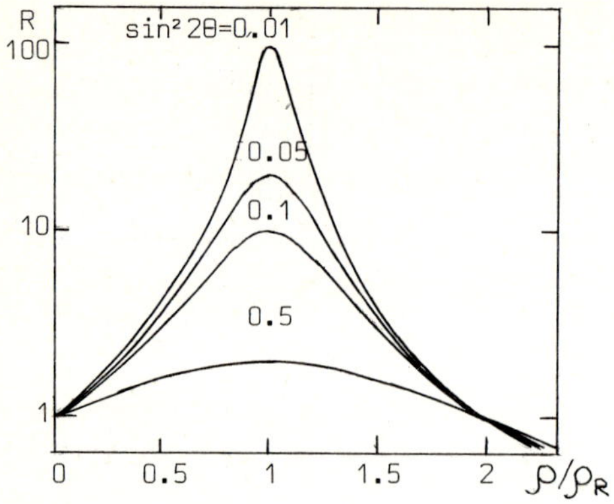

Fig.1 The resonant factor as a function of density

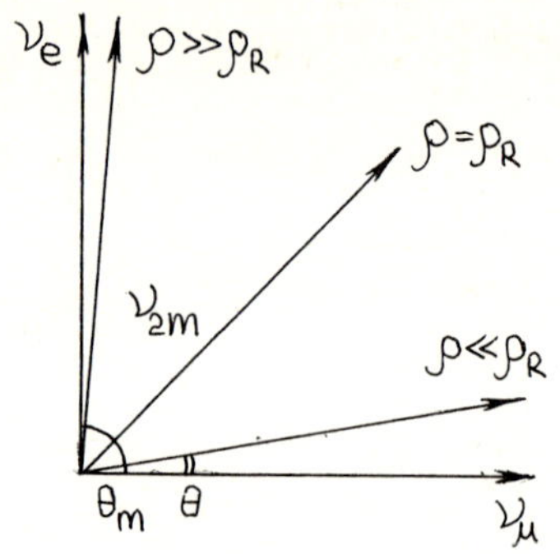

Fig.2 The dependence of flavour of neutrino eigenstate ν_{2m} on density

2) Flavours of ν-eigenstates are changed with density (see (10)). When ρ diminishes from $\rho \gg \rho_R$ to $\rho \ll \rho_R$, the mixing angle θ_m decreases from $\pi/2$ to θ. Correspondingly, flavours of ν_{im} at small θ change almost completely. For example, if $\nu_{1m}(\rho \gg \rho_R) \approx \nu_\mu$, then $\nu_{1m}(\rho \ll \rho_R) \approx \nu_e$.

The same result follows from eigenvalues consideration/5,6/. Eq's (9) give: if $M_1^d(\rho \gg \rho_R) \approx M_\mu$, then $M_1^d(\rho \ll \rho_R) \approx M_e$.

§2 Oscillations

Generalizing vacuum consideration futher, we may represent arbitrary neutrino state as follows

$$|\nu(t)\rangle = \cos\theta_a |\nu_{1m}\rangle + \sin\theta_a e^{i\varphi} |\nu_{2m}\rangle \tag{14}$$

Here θ_a is the angle which fixes the admixtures of ν-eigenstates in $\nu(t)$. $\varphi = \varphi_1 - \varphi_2$ is the phase difference between ν_{im}; general factor $\exp(-i\varphi_1)$, which is unessential for oscillation picture, have been omitted. Monotonic changing of φ results in flavour oscillations in $\nu(t)$. Using (10) one has the amplitude of probability to find ν_e in moment t:

$$\langle \nu_e | \nu(t) \rangle = \cos\theta_a \cos\theta_m + \sin\theta_a \sin\theta_m e^{i\varphi} \tag{15}$$

For the probability it follows from (15):

$$P(t) = \bar{P}(t) + \frac{1}{2} A_p(t) \cdot \cos\varphi \tag{16}$$

where the average probability is

$$\bar{P}(t) = \cos^2\theta_a \cos^2\theta_m + \sin^2\theta_a \sin^2\theta_m \tag{17}$$

the depth of oscillations equals to

$$A_p(t) = \sin 2\theta_a \sin 2\theta_m \tag{18}$$

and quasiperiod is

$$T_m = L_m = 2\pi / \dot{\varphi} \tag{19}$$

In contrast with vacuum case $\bar{P}$, A_p and T_m are changed in time now. In (16-19) θ_m is known function of density (see (11)) and the task is to determine $\theta_a(t)$ and $\varphi(t)$.

§3 Graphical representation of oscillations

Describe the graphical picture which reflects the relations (14-19) immediately/8/. It gives not only clear representation of ν-state evolution, but enables to reproduce results of analitic consideration.

Introduce basis $\{\nu_m\} = \{\nu_{1m}, \nu_{2m}^R, \nu_{2m}^I\}$ where axes ν_{2m}^R and ν_{2m}^I correspond to real and imaginary parts of ν_{2m} wave function. Then ν-state (14) is described by unit vector $\{\cos\theta_a, \sin\theta_a \cos\varphi, \sin\theta_a \sin\varphi\}$. Vectors $\vec{\nu}_e = \{\cos\theta_m, \sin\theta_m, i \sin\theta_m\}$ and $\vec{\nu}_\mu = \{-\sin\theta_m, \cos\theta_m, i \cos\theta_m\}$ correspond to ν_e and ν_μ-states (see(6)). Projection of $\nu(t)$

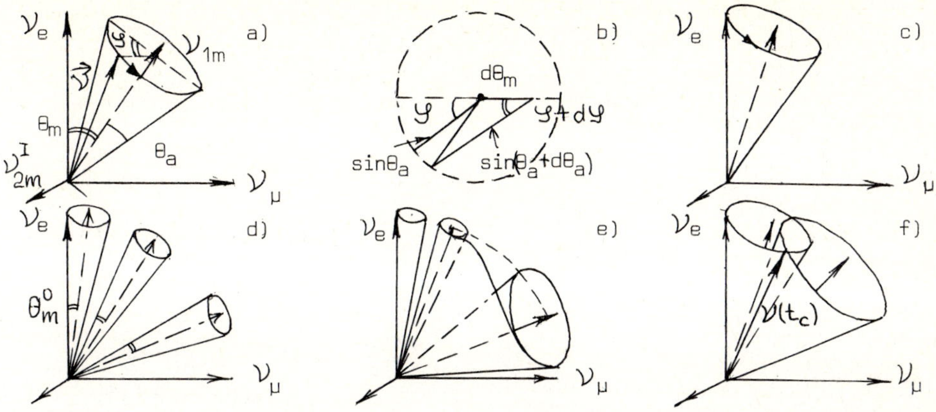

Fig.3 Graphical representation of ν-oscillations a) representation of ν-state; b) changing of phase and cone angle; c) vacuum and constant density; d) adiabatic regime; e) adiabatic violation in resonant layer; f) jump of density.

on given axis ν_x equals to the amplitude of probability to find ν_x in $\nu(t)$. ν-changing in (14) corresponds to rotation of $\vec{\nu}(t)$ around $\vec{\nu}_{1m}$. $\vec{\nu}(t)$ moves on the cone surface with angle θ_a (θ_a is the angle between $\vec{\nu}_{1m}$ and $\vec{\nu}(t)$) and with quasiperiod $L_m = 2\pi/\dot{\varphi}$ (fig3a).

In matter with varying density the evolution consists of two rotations: 1) rotation of $\vec{\nu}_{1m}$ with respect to $\vec{\nu}_e$, $\vec{\nu}_\mu$ vectors. The angle between ν_{1m} and ν_e is θ_m, ν_{1m}-rotation is determined uniquely by density changing. 2) Rotation of $\vec{\nu}(t)$ around $\vec{\nu}_{1m}$. The corresponding angular velocity is $\dot{\varphi}$. Moreover cone angle θ_a is changed in time also.

§4 Evolution of ν-eigenstates in matter. Equations for φ and θ_a

Coefficients in front of $|\nu_{im}\rangle$ in (14)

$$\nu_{1m}(t) = \cos\theta_a, \quad \nu_{2m}(t) = \sin\theta_a \exp(i\varphi) \tag{20}$$

are obviously the wave functions of ν_{1m} and ν_{2m} normalized so that $|\nu_{1m}|^2 + |\nu_{2m}|^2 = 1$. Evolution equations for ν_{im} follow from (6) and (1)

$$i\dot{\vec{\nu}}_m = \hat{M}_m \vec{\nu}_m; \quad \hat{M}_m = (\hat{M}^{diag} - i\frac{d\hat{S}}{dt}\hat{S}^{-1}) = \begin{vmatrix} M_1^d & -i\dot{\theta}_m \\ i\dot{\theta}_m & M_2^d \end{vmatrix} \tag{21}$$

where

$$\frac{d\theta_m}{dt} = \frac{\sin^2 2\theta \cdot R}{2\Delta\rho_R} \cdot \frac{d\rho}{dt} \tag{22}$$

(R - resonant factor (see(12)). Inserting (20) in (21) we have

$$\dot{\theta}_a = \dot{\theta}_m \cos\varphi, \quad \dot{\varphi} = \dot{\varphi}^d - 2\dot{\theta}_m \sin\varphi \cot 2\theta_a \tag{23a,b}$$

Here $\dot{\varphi}^d = M_1^d - M_2^d$. If the initial state coincides with one of eigenstates of weak interactions, for example, $\nu(0) = \nu_e$, then it follows from (14):

$$\theta_a^\circ = \theta_m^\circ, \quad \varphi_o = 0 \tag{24}$$

Remark some properties of (23). Infinitesimal changing of cone angle θ_a is due to ν_{1m}-rotation only, $\nu(t)$-rotation ($\dot{\varphi}$) does not participate. $\dot{\varphi}$ has two contributions. First term in (23b) describes the changing of phase due to $\vec{\nu}$ rotation around $\vec{\nu}_{1m}$, second term corresponds to $\vec{\nu}_{1m}$-rotation. Relations (23) can be obtained graphically (fig.3b).

Finally, introducing in (23a) the integration over φ one has

$$\theta_a = \theta_a^\circ + \int_0^\varphi (\dot{\theta}_m/\dot{\varphi}) \cos\varphi' d\varphi' = \theta_a^\circ + \int_0^\varphi (\dot{\theta}_m/\dot{\varphi}^d)(1 - (\dot{\theta}_m/\dot{\varphi}^d) 2\sin\varphi' \cot 2\theta_a)^{-1} \cos\varphi' d\varphi' \tag{25}$$

which will be usefull for futher analizes.

4 Neutrino Oscillations in Different Regimes

The properties of ν-oscillations depend on character of density changing. One can single out several regimes: 1)constant density, 2)adiabatic regime (slow changing of density),3)violation of adiabacity in resonant layer (fast changing of density), 4)strong violation of adiabacity with jump of density as a limit case.

§1 Constant density

Using results of sect.2 we have

$$\dot{\theta}_m = 0,\ \dot{\theta}_a = 0 \text{ or } \theta_a = \theta^\circ_a,\ \dot{\varphi} = \dot{\varphi}^d \tag{26}$$

(see(23)). 1) ν_{im} diagonalize system (21); there is no $\nu_{1m} \leftrightarrow \nu_{2m}$ transitions, ν_{im} evolute independingly; admixtures of ν_{im} in given $\nu(t)$ do not change ($\dot{\theta}_a$=0). 2)Flavours of ν_{im} are fixed (θ_m=const). Consequently, the picture of ν-oscillations in matter with ρ=const is similar to that in vacuum. Values of oscillation parameters are changed only. If $\nu(0)=\nu_e$ (see(24)) then $\theta_a=\theta^\circ_m=\theta_m$. Substituting this θ_a into (17), (19) we reproduce WOLFENSTEIN's results/2/:

$$\bar{P} = 1 - (1/2) \sin^2 2\theta_m,\ A_p = \sin^2 2\theta_m = \sin^2 2\theta \cdot R,\ L_m = 1/(M^d_1 - M^d_2) = L_v \cdot R^{1/2} \tag{27}$$

Evolution of $\nu(t)$ state corresponds to rotation of vector $\vec{\nu}(t)$ around $\vec{\nu}_{1m}$ with a constant frequency. Position of $\vec{\nu}_{1m}$ does not change.

The manifestation of resonance is the following. In matter with $\rho=\rho_R$ the depth of oscillation is maximal A_p=1. If neutrinos with continuous energy spectrum are generated then in interval $E_R - E_R \div E_R + E_R$ oscillations will be amplified resonantly. Here E_R is the resonant energy, determined from condition $\rho_R(E_R)=\rho_R$, $\Delta E_R = E_R \cdot \tan 2\theta$.

§2 Adiabatic regime

Let the density changes so slowly that/4,7,10,15,16/

$$\dot{\theta}_m \ll \dot{\varphi}^d = M^d = 2\pi/L_m \tag{28}$$

(remember that $\dot{\theta}_m \propto d\rho/dt$). According to (25), at adiabatic condition (28):

$$\theta_a = \theta^\circ + O(\dot{\theta}_m/M^d),\ \dot{\varphi} = M^d + O(\dot{\theta}_m/M^d) \tag{29}$$

In first (adiabatic) approximation admixtures of ν_{im} in ν-state do not change and equal to those in initial moment: $\theta_a=\theta^\circ_m$= const. In other words the transitions $\nu_{1m} \leftrightarrow \nu_{2m}$ can be neglected, ν_{im} evolute independingly (see(21)), and thus turns out to be quasieigenstates of Hamiltonian. This is quite similar to ρ=const- case. In contrast with the case of ρ=const, flavours of ν-eigenstates change now depending on density variation $\theta_m=\theta_m(\rho(t))$. It is this changing results in a strong transformations of ν-states.

In geometrical terms the adiabatic condition means that $\vec{\nu}$ should rotate around $\vec{\nu}_{1m}$ much quickly than $\vec{\nu}_{1m}$ to do itself (fig.3d). The evolution of ν-state is the rotation of ν-cone axis (that is $\vec{\nu}_{1m}$) without changing of cone angle. This angle is determined by initial mixing.

Both $\dot{\theta}_m$ and $\dot{\varphi}$ depend on $\rho(t)$ and therefore inequality (28) should be satisfied for any moment t. But in resonance layer ($\rho=\rho_R$) it becomes the most crucial/4,10,15/: on one hand $\dot{\theta}_m$ contains resonant factor R (22), on the other hand L_m is maximal in in resonance (27). Moreover it is in resonance layer the most strong transformation of ν-state takes place and consequently adiabacity violation beyond the resonance region does not change results significantly. From (22,27,28) for $\rho=\rho_R$ one has

$$2\Delta r_R \gg L_m/2\pi \tag{28a}$$

where $\Delta r_R = (d\rho/dr)^{-1} \Delta\rho_R$ is the spatial width of resonance layer/4/

Substituting $\theta_a=\theta^\circ_m$ in (17,18) one has

$$\bar{P} = \cos^2\theta^\circ_m \cos^2\theta_m + \sin^2\theta^\circ_m \sin^2\theta_m,\ A_p = \sin^2 2\theta^\circ_m \sin^2 2\theta_m,\ \varphi = \int M^d\, dt \tag{30}$$

Properties of adiabatic solution (30) are widely discussed. Main of them are: 1)Universality/7,8/. $\bar{P}$ and A_p are the functions of densities in initial and in a given moment and do not depend on density distribution. In terms of dimensionless variable $n=(\rho-\rho_R)/\Delta\rho_R$, which is the distance (in density scale) from resonance in units $\Delta\rho_R$, we have $\sin 2\theta_m=(n^2+1)^{-1/2}$ (see(11)) and from (30):

$$\bar{P}(n,n_o)=0.5(1+n\cdot n_o \cdot ((n^2+1)\cdot(n_o^2+1))^{-1/2}),\ A_p=((n^2+1)\cdot(n_o^2+1))^{-1/2} \tag{31}$$

2)Monotonic changing of $\overline{P}$ with density. If neutrinos are produced at $\rho>\rho_R$ ($n_o>0$) then $\overline{P}$ diminishes with ρ monotonically. $\overline{P}$ follows ρ. This gives strong transformation of initial type neutrinos into another ones when they cross the layer with sufficiantly large difference $\rho_o-\rho_f$ adiabatically. If at the exit $\rho_f=0$, then:

$$\overline{P} = 0.5(1-n_o\cos 2\theta/\sqrt{(n_o^2+1)}) \text{ and } A_p = \sin 2\theta/(n_o^2+1)^{-1/2}$$

The smaller angle θ (in contrast with vacuum case) and the greater ρ_o, the stronger suppression of initial neutrinos flux can be reached.

3)Oscillationless limit/8/. At $n_o \to \infty$ (see(31)) the depth of oscillations goes to zero and $\overline{P}(n)$ converges to asymptotic dependence $P_{as}=0.5\ (1+ n\cdot(n^2+1)^{-1/2})$. At $n_o \to \infty$ cone angle $\theta_a \to 0$ and $\nu(t)$ coincides with ν_{1m}.

4)Resonance manifestation. In resonance the depth of oscillations is maximal: $A_p(n=0) > A_p(n\neq 0)$ and average probability $\overline{P}$ is equal 1/2 independently on n_o. It is in resonance layer the most strong transformations of ν-beam take place: $P^o= P(-1,n_o) - P(1,n_o) = n_o\cdot(2\ (n_o^2+1))^{-1/2}$ increases from 1/2 to $1/\sqrt{2}$ when n_o rises from $n_o=1$ to $n_o \to \infty$.

§3 Adiabacity violation in resonant layer

If the adiabatic condition is violated then the transitions $\nu_{1m} \leftrightarrow \nu_{2m}$ become essential. The admixtures of ν_{im} in a given ν-state do not conserve. $\dot{\theta}_a\neq 0$ - cone angle is changed. $\dot{\varphi} \neq M^d$. $\overline{P}$ and A_p depend on density distribution and phase of oscillation now.

If $\dot{\theta}_m/M^d<1$ corrections to adiabatic results can be found according to (25):

$$\theta_a = \theta_a^o + \int_0^{\varphi} (\dot{\theta}_m/M^d) \cos\varphi'\, d\varphi'.$$

As it have been pointed out in §2 adiabatic condition is the most crucial in resonance. So if the density distribution becomes more and more steep first of all adiabacity is broken in resonance layer. In this case before and after it neutrinos evolute adiabatically (fig.3e). Cone angle after adiabacity restoration θ_a' depend not only on density distribution but on phase of oscillations at the beginning of violation (fig.4). The conditions are possible at which θ_a diminishes, that is nonadiabatic crossing of resonance may results in increasing of transition probability respect to adiabatic one. But the cone angle, averaged on φ_{in} rises. Because of $\theta_m^R = 45°$ in the limit case of very fast density changing $\overline{\theta}_a \to 45°$ and $P \to 1/2$.

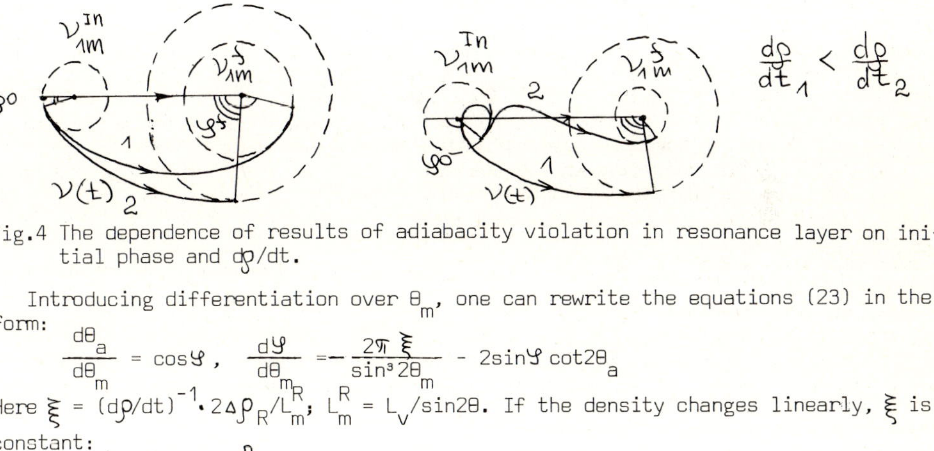

Fig.4 The dependence of results of adiabacity violation in resonance layer on initial phase and $d\rho/dt$.

Introducing differentiation over θ_m, one can rewrite the equations (23) in the form:

$$\frac{d\theta_a}{d\theta_m} = \cos\varphi\,, \quad \frac{d\varphi}{d\theta_m} = -\frac{2\pi\,\xi}{\sin^3 2\theta_m} - 2\sin\varphi\cot 2\theta_a \tag{32}$$

Here $\xi = (d\rho/dt)^{-1}\cdot 2\Delta\rho_R/L_m^R$; $L_m^R = L_v/\sin 2\theta$. If the density changes linearly, ξ is constant:

$$\xi = 2\Delta r_R/L_m^R \tag{33}$$

It turns out to be the parameter which determines the edge of adiabacity for resonance layer (see(28a)). For linear density distribution the solution can be obtained in special functions/18/. If initial $\rho_o \gg \rho_R$ and final $\rho_f \ll \rho_R$, then for small θ the integration limits for (32) are $\theta_m^o = \pi/2$ and $\theta_m^f = 0$. In this case: 1) the solution of system (32) depends on parameter ξ only, $\theta_a^{fm} = \theta_a(2\Delta r_R/L_m^R) = \theta_a(\sin 2\theta\cdot\Delta m^2/E)$, correspondingly, $\overline{P}_f = \overline{P}_f(\sin 2\theta\cdot\Delta m^2/E)$ (firstly this result have been obtained in /10/ in another terms); 2) Solution can be written in a simple form/18/

$$\sin^2\theta_a = \exp(-\pi^2 2\Delta r_R/2L_m^R) \qquad (34)$$

Results 1), 2) are in fact good approximation to arbitrary density distribution for small $\sin^2 2\theta$. Indeed for small $\sin^2 2\theta$ the width of resonance layer is small and linearity is quite acceptable. Moreover it is in resonance layer the most strong transitions take place.

§4 Very fast changing of density, jump of density

If θ_m varies very rapidly, the first term in (23b) can be neglected. Mainly phase changes due to ν_{1m}-rotation in ν_e,ν_μ space. In this case system (23) has analitical solutions. It corresponds in first approximation to jump of density in some moment t_c. Solution coincides with that obtained graphically/8/. Vector $\vec{\nu}$ is at rest when ν_{1m} changes its position. After jump the rotation angle θ_a and phase φ are determined by $\vec{\nu}(t_c)$ and new position of $\vec{\nu}_{1m}$ (fig.3f).

5 Applications to the Sun

We have performed new calculations of matter effect to ν-oscillations in the Sun, using standard solar model/23/. Some results are shown on fig.5.

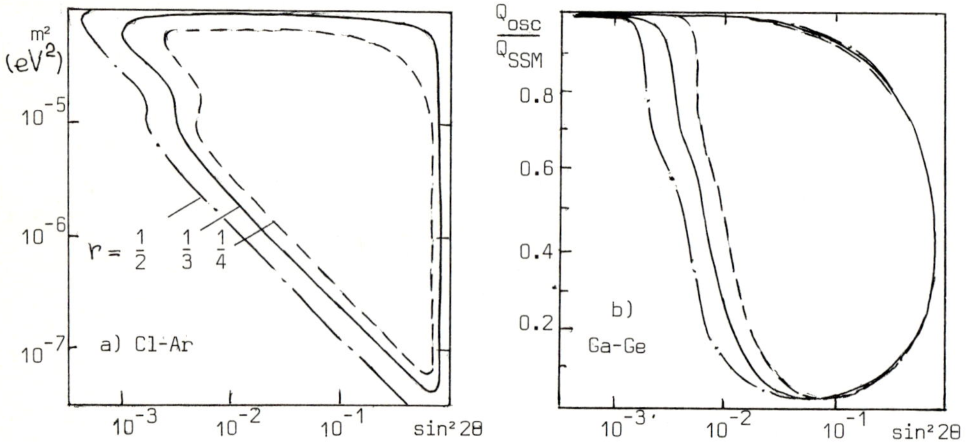

Fig.5 a)Parameters of neutrino oscillations for which rate of ν-cature in Cl-Ar experiment is r of standard one; b)Predictions for Ga-Ge experiment as a function of $\sin^2 2\theta$ (standard value is 110 SNU's).

References

1.B.M.Pontecorvo, ZhETF 23, 549 (1958); S.M.Bilenky and B.M.Pontecorvo, Phys.Rep. C41, 225 (1978).

2.L.Wolfenstein, Phys.Rev. D17, 2369 (1978); ibid D20, 2634 (1979).

3.V.Barger et al, Phys.Rev. D22, 2718 (1980).

4.S.P.Mikheyev and A.Yu.Smirnov, Talk given at 10th Int. Conf. on Weak Interactions, Savonlinna, Finland (1985); Yadernaja Fizika 42, 1441,(1985); Nuov.Cim. 9C, 17 (1986)

5. H.Bethe, Phys.Rev.Lett. 56, 1305 (1986).

6. N.Cabibbo, Summary talk at 10th Int. Conf. on Weak Interactions (see in ref.4)

7.S.P.Mikheyev and A.Yu.Smirnov, ZhETF 91, 7, (1986)

8.S.P.Mikheyev and A.Yu.Smirnov, Talk given at VIth Moriond Workshop on Massive Neutrinos in Particle Physics and Astrophysics, Tignes, France (1986) (to be published)

9.S.P.Rosen and J.M.Gelb, Talk given at VIth Moriond Workshop (see ref.8)

10.A.Messiah, Talk given at VIth Moriond Workshop (see ref.8); preprint PhT/86-46 Sacl.

11.W.Hampel, Max-Planck-Inst. Preprint, Heidelberg (1986)

12.M.Spiro, Talk given at VIth Moriond Workshop (see ref.8); J.Bouchez et al, Preprint DPhPE/86-10, Saclay (1986); J.Bouchez Talk given at "Neutrino'86", Sendai,Japan(1986

13.V.A.Chechin et al, Preprint 45 Lebedev Institute (1986).

14. S.P.Mikheyev and A.Yu.Smirnov Talk given at Int. Seminar Quarks'86, Tbilisi, USSR.
15.S.P.Mikheyev and A.Yu.Smirnov Talk given at 7th-WOGU/ICOBAN'86, Toyama, Japan.
16.V.Barger et al, Preprint MAD/PH/280 (1986).
17.W.C.Haxton, Preprint Univ. of Washington, Seatle, Washington (1986)
18.S.J.Parke, Preprint Fermilab Pub-86/67-T (1986)
19.E.W.Kolb et al, Preprint Fermilab Pub-86/69-A (1986)
20.P.Langacker et al, Preprint CERN-TH.4421/86 (1986)
21.L.Wolfenstein, Talk given at Int. Conf. "Neutrino'86", Sendai, Japan (1986).
22.S.P.Mikheyev and A.Yu.Smirnov Talk given at Int. Conf. "Neutrino'86", Sendai, Japan.
23.J.N.Bahcall et al, Rev.Mod.Phys. 54, 767 (1982).

The Signal from the Gallium Solar Neutrino Detector: Implications for Neutrino Oscillations and Solar Models

W. Hampel

Max-Planck-Institut für Kernphysik, D-6900 Heidelberg, F. R. Germany

In order to interpret the experimental result of the Gallium Solar Neutrino Experiment, to be performed by the European GALLEX Collaboration [1-3], it has to be compared to predicted capture rates. These predicted rates depend on the spectrum and intensity of solar neutrinos incident at the detector and on the cross section for neutrino capture in ^{71}Ga. After a short summary of the experimental procedure and the present status of the GALLEX experiment, we discuss (a) our present knowledge on the ^{71}Ga neutrino capture cross section, and (b) the impact of solar models and neutrino oscillations on the expected signal from the gallium detector.

Experimental procedure and present status

The gallium detector is designed to detect the pp neutrinos generated in the primary nuclear fusion reaction in the sun. It is based on the neutrino capture reaction $^{71}\mathrm{Ga}(\nu_e,e^-)^{71}\mathrm{Ge}$. The detector will consist of 30t gallium as $GaCl_3$-HCl solution, to be contained in a single large tank ($\sim$ 80 m^3). In this solution the neutrino produced ^{71}Ge ($T_{1/2}$=11.43d) as well as the Ge carrier added before the extraction will be present as $GeCl_4$, a rather volatile compound which simply can be swept out of the solution by gas purge. The extracted $GeCl_4$ is then converted to GeH_4 which together with xenon is introduced into a small proportional counter in order to observe the ^{71}Ge decay. The whole experimental procedure was tested in a pilot experiment [2-4] and has been demonstrated to be feasible.

The GALLEX experiment is funded since December 1985. It will be carried out in the European Underground Physics Laboratory at the Gran Sasso in Italy. The scientific committee for this underground facility has approved installation of the experiment in the south wing of hall A. The time schedule of the experiment is mainly determined by the gallium delivery. According to the contract with the gallium supplier, the whole amount of gallium will be available in October of 1989. Measurements of the solar neutrino flux will then last for 4 years, eventually interrupted for the calibration of the detector with an artificial ^{51}Cr neutrino source. The experimental conditions achieved so far allow to measure a solar neutrino signal of 90 SNU to within 10% [3].

The ^{71}Ga neutrino capture cross section

In order to calculate the cross section for neutrino capture in ^{71}Ga one needs the Gamow-Teller strength B(GT) connected with all states in ^{71}Ge (up to the neutron separation energy at 7.4 MeV) which can be populated by neutrino capture in allowed transitions. In any case, the largest contribution to the total rate for the gallium detector comes from the transition to the ^{71}Ge ground state. We have redetermined the B(GT) value for the ground state transition by remeasuring the ^{71}Ge halflife [5] and the ^{71}Ge Q_{EC} value [6]. The resulting B(GT) (and thus the neutrino capture cross section) is 6% larger than the old one, the uncertainty is reduced from 4% to 1%.

Information on the B(GT) strength for excited states in ^{71}Ge must come either from shell model calculations [7-9] or from forward scattering (p,n) experiments [10], since the corresponding ^{71}Ge-^{71}Ga β decays cannot be observed in the laboratory. The B(GT) values used for our cross section calculation are listed in Table I. Except for the ground state transition (see above) they are based on the (p,n) data of Krofcheck et al. [10]. For states below 1.5 MeV the adopted values are slightly larger than those inferred from the (p,n) data in order to account for the difference in the ground state B(GT) between the (p,n) measurement and the direct determination from β decay. We have used the usual formula for allowed neutrino capture and have applied corrections to it for overlap and exchange effects as evaluated by Bahcall [11]. The Fermi function $F(Z,p_e)$ was taken from the tables of Behrens and Jänecke [12].

Table I : Gamow-Teller strength B(GT) for the ground state and excited states in ^{71}Ge (see text).

^{71}Ge states	B(GT)
Ground state	0.0886 ± 0.0010
175 keV	0.005 ± 0.005
500 keV	0.011 ± 0.006
708 keV	0.023 ± 0.012
0.9-1.5 MeV	0.087 ± 0.026
1.5-7.4 MeV	4.1 ± 0.7

The resulting cross section is plotted in Fig. 1. Folding this cross section with the neutrino fluxes from the new Standard Solar Model (SSM) of Bahcall and Ulrich [13] and a so-called "Consistent" model [14] (a model which predicts 2 SNU for the Cl solar neutrino detector, in agreement with the measured value of 2.0 ± 0.3 SNU [15]) yields the capture rates listed in Table II. The contribution from the ground state transition in the total rates given in Table II is 86% for the SSM and 95% for the consistent model.

Solar models and neutrino oscillations: Impact on the Ga and Cl detector rates

A large number of so-called non-standard solar models have been suggested during the last 20 years in order to solve the solar neutrino problem posed by the Cl experiment. In most cases these models were invented in order to reduce the strongly temperature-dependent ^{8}B and ^{7}Be neutrino fluxes to a level consistent with the Cl result. In Fig. 2 we have plotted the expected rate for the Cl detector versus the corresponding Ga detector rate for the SSM and for a variety of these non-standard solar models (see [2]). It follows that models in agreement with the measured 2σ range for the Cl experiment (1.4 to 2.6 SNU [15]) predict capture rates between 85 and 100 SNU for the Ga detector.

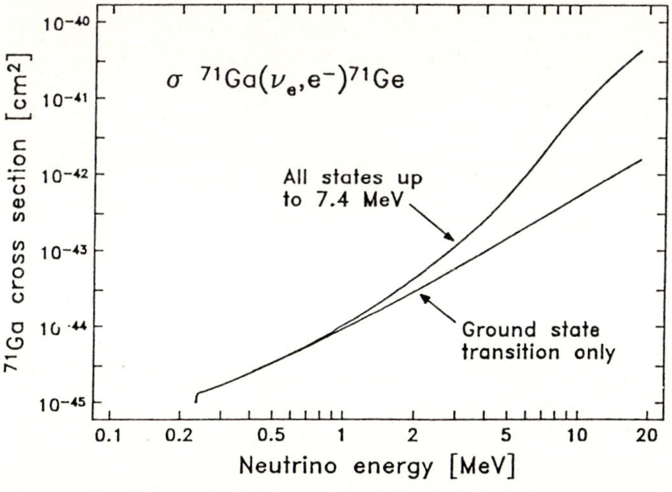

Fig. 1. The cross section for neutrino capture in ^{71}Ga.

Table II : ^{71}Ge production rates including excited state contributions for the Standard Solar Model and a "Consistent" Model.

Neutrino source and energy [MeV]	Standard Solar Model†		Consistent Model‡	
	Flux on earth [10^{10}cm^{-2}sec^{-1}]	Production rate [SNU]	Flux on earth [10^{10}cm^{-2}sec^{-1}]	Production rate [SNU]
pp (0–0.42)	6.0	69.3 ± 0.6	6.19	71.5 ± 0.6
pep (1.44)	0.015	3.2 ± 0.2	0.015	3.2 ± 0.2
^{7}Be (0.38, 0.86)	0.475	34.5 ± 1.5	0.21	15.2 ± 0.7
^{8}B (0–14.06)	0.00054	16.2 ± 2.6	0.00011	3.3 ± 0.5
^{13}N (0–1.20)	0.06	3.7 ± 0.1	0.0048	0.3 ± 0.01
^{15}O (0–1.73)	0.05	5.9 ± 0.3	0.0064	0.8 ± 0.04
Total		132.8 ± 5.3		94.3 ± 2.1

†Bahcall (1986); Bahcall and Ulrich (1986), to be published.
‡Model D3 of Demarque et al. (1973).
All errors in this table include only the uncertainty in the cross sections.

Besides non-standard solar models, the only other explanation to the solar neutrino puzzle seriously discussed is a reduced ν_e flux incident at a terrestrial detector due to neutrino oscillations. However, for vacuum ν oscillations large mixing between 2 neutrino flavors with a squared mass difference Δm^2 near 10^{-10} eV^2 or almost maximal mixing between 3 neutrino flavors is required in order to account for the discrepancy between theory and experiment in the case of the Cl detector [16]. This situation has changed dramatically since the discovery of Mikheyev and Smirnov [17] following the fundamental work of Wolfenstein [18] on the modification of vacuum ν oscillations through matter effects. With the Mikheyev-Smirnov-Wolfenstein (MSW) effect a large fraction of the ν_e generated in the solar core can be converted to ν_μ or ν_τ during their passage through the sun, even if the vacuum mixing angle is small (see also [19]).

Two groups within the GALLEX Collaboration (Saclay, Heidelberg) have performed extensive numerical calculations on the MSW effect (assuming 2ν mixing) for the Ga and Cl solar neutrino detectors [20]. We have updated these calculations with the new SSM neutrino fluxes of Bahcall and Ulrich [13]. The effect introduced by neutrino passage through the earth at night has been included. For large mixing angles and for $E_\nu/\Delta m^2$ values near $3\cdot10^6$ MeV/eV^2

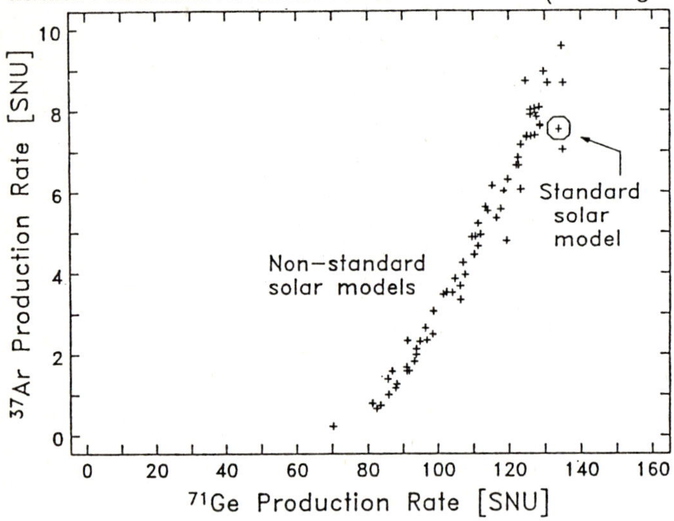

Fig. 2. Cl detector signal versus Ga detector signal: Expectations from the Standard Solar Model and from non-standard solar models.

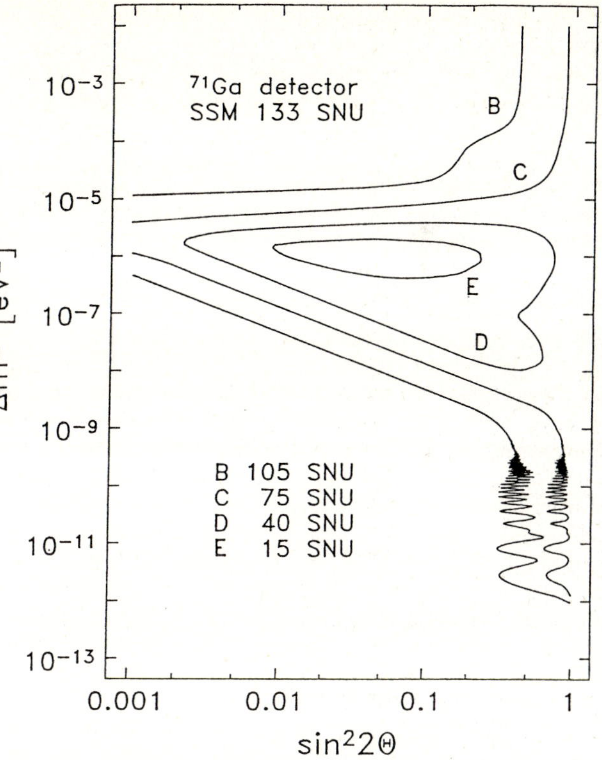

↑**Fig. 3.** Δm^2-$\sin^2 2\theta$ plot for the Ga detector. The curves indicate the allowed values for Δm^2 and $\sin^2 2\theta$ if the rate measured with this detector is 105, 75, 40 or 15 SNU.

the earth partly reverses the MSW effect caused by the solar matter, thus reducing the overall effect (see also [20]). With the radiochemical detectors under consideration one cannot observe day/night effects. Also, because of the statistical uncertainties, one is not able to resolve small annual variations of the production rate. We therefore have averaged the earth effect over day and night as well as over the course of a year, assuming a detector located at a latitude λ=42.5°N (Gran Sasso) which is similar to that of the Cl detector at the Homestake mine (λ=44°N).

In Figs. 3 and 4 we present plots of Δm^2 versus $\sin^2 2\theta$ for both the Ga and the Cl detector. The contours correspond to different possible reductions of the Ga and Cl detector signals as compared to the SSM expectation. The earth effect is visible for large mixing angles in both Fig. 3 (at $\Delta m^2 \simeq -10^{-7}$ eV^2) and Fig. 4 (at $\Delta m^2 \simeq -3 \cdot 10^{-6}$ eV^2), respectively.

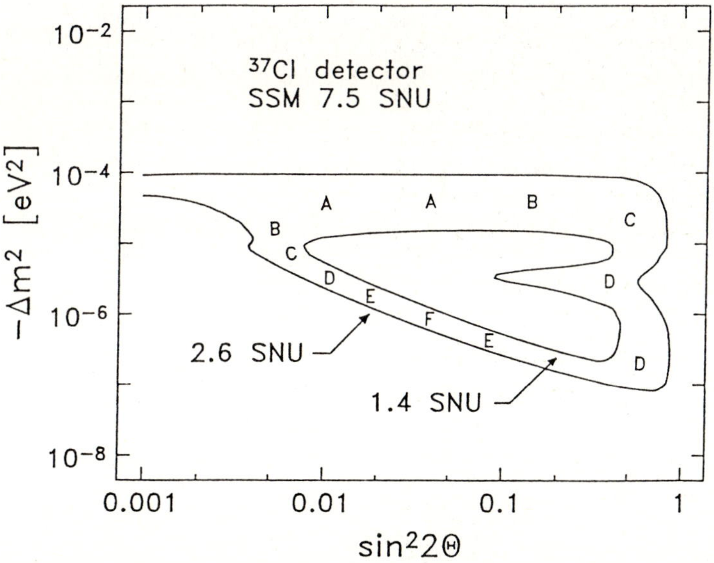

If the Cl detector result is interpreted in terms of the MSW effect, then the Ga detector can measure any signal between 10 and 115 SNU, depending on the actual values of Δm^2 and $\sin^2 2\theta$ (see Fig. 4). The Ga detec-

Fig. 4. Δm^2-$\sin^2 2\theta$ plot for the Cl experiment. The curves represent the 2σ range of the experimental result (1.4-2.6 SNU). Letters denote the corresponding Ga detector signal (in SNU): A (115), B (105), C (75), D (40), E (15), F (10).

tor result thus puts important constraints on these neutrino oscillation parameters. There are, however, two regions in the allowed Cl band of Fig. 4 (between the points labelled "B" and "C") where the Ga detector would measure a rate between 75 and 105 SNU, the range of values to be expected from non-standard solar models (see above). This ambiguity is the prize which has to be paid for the enhanced sensitivity of the Ga detector to neutrino oscillations due to the MSW effect.

The ambiguity cannot be resolved with other radiochemical solar neutrino detectors, since the integral fluxes of the different solar neutrino sources are affected in about the same manner for both explanations. To distinguish between these two possibilities a direct counting detector with some ability in energy spectroscopy is required in order to observe whether or not the ^{8}B neutrino spectrum incident at the earth is disturbed by matter oscillation effects.

References

1 **GALLEX Collaboration**: W. Hampel, G. Heusser, J. Kiko, T. Kirsten (spokesman), A. Lenzing, E. Pernicka, B. Povh, S. Richter, K. Schneider, M. Schneller, H. Völk, R. Wink (**MPIK Heidelberg**); R. von Ammon, K. Ebert, E. Henrich (**KfK Karlsruhe**); R. L. Mössbauer (**TU München**); M. Cribier, G. Dupont, B. Pichard, J. Rich, M. Spiro, D. Vignaud (**CEN Saclay**); G. Berthomieu, E. Schatzman (**Univ. of Nice**); E. Bellotti, O. Cremonesi, E. Fiorini, C. Liguori, S. Ragazzi, L. Zanotti (**Univ. of Milano**); R. Bernabei, S. d'Angelo, L. Paoluzi, R. Santonico (**Univ. of Rome**); I. Dostrovsky (**WIS Rehovot**).

2 W. Hampel, AIP Conf. Proc. **126**, Editors M.L. Cherry, W.A. Fowler, and K. Lande, 100 (1985).

3 T. Kirsten, Proc. VIth Moriond Workshop on "Massive Neutrinos in Particle Physics and Astrophysics", Tignes, France (1986).

4 **Pilot experiment collaboration**: Brookhaven Nat. Lab., MPIK Heidelberg, IAS Princeton, WIS Rehovot.

5 W. Hampel and L.P. Remsberg, Phys. Rev. **C31**, 666 (1985).

6 W. Hampel and R. Schlotz, Proc. 7th Int. Conf. on Atomic Masses and Fundamental Constants (AMCO-7), Darmstadt-Seeheim, 89 (1984).

7 G.J. Mathews, S.D. Bloom, G.M. Fuller, and J.N. Bahcall, Phys. Rev. **C32**, 796 (1985).

8 K. Grotz, H.V. Klapdor, and J. Metzinger, Phys. Rev. **C33**, 33 (1986).

9 T. Oda (1986), contribution to this symposium.

10 D. Krofcheck, E. Sugarbaker, J. Rapaport, D. Wang, J.N. Bahcall, R.C. Byrd, C.C. Foster, C.D. Goodman, I.J. Van Heerden, C. Gaarde, J.S. Larsen, D.J. Horen, and T.N. Taddeucci, Phys. Rev. Lett. **55**, 1051 (1985); E. Sugarbaker (1986), contribution to this symposium.

11 J.N. Bahcall, Rev. Mod. Phys. **50**, 881 (1978).

12 H. Behrens and J. Jänecke, Numerical Tables for Beta-Decay and Electron Capture, Springer-Verlag, Berlin (1969).

13 J.N. Bahcall and R. Ulrich (1986), to be published; J.N. Bahcall, priv. comm. (1986).

14 P. Demarque, J.G. Mengel, and A.V. Sweigart, Ap. J. **183**, 997 (1973).

15 J.K. Rowley, B.T. Cleveland, and R. Davis Jr., AIP Conf. Proc. **126**, Editors M.L. Cherry, W.A. Fowler, and K. Lande, 1 (1985).

16 W. Hampel, Proc. Neutrino 84, Nordkirchen, Editors K. Kleinknecht and E.A. Paschos, World Scientific Publishing, Singapore, 530 (1984).

17 S.P. Mikheyev and A.Yu. Smirnov, Yad. Fiz. **42**, 1441 (1985); Sov. J. Nucl. Phys. **42**, 913 (1986); A.Yu. Smirnov (1986), contribution to this symposium.

18 L. Wolfenstein, Phys. Rev. **D17**, 2369 (1978).

19 H.A. Bethe, Phys. Rev. Lett. **56**, 1305 (1986).

20 J. Bouchez, M. Cribier, J. Rich, M. Spiro, D. Vignaud, and W. Hampel, to be published in J. Phys. C. (1986).

The Effective Interaction Dependence of the ^{8}B Neutrino Capture Rate of the Ga Solar Neutrino Detector

T. Oda and K. Muto

Department of Physics, Tokyo Institute of Technology, Oh-Okayama, Meguro-ku, Tokyo 152, Japan

The full-scale solar neutrino experiment using ^{71}Ga as a detector has been proposed and discussed [1 - 5] in order to settle the discrepancy between the observation in the ^{37}Cl experiment [6] and the prediction of the standard theory of stellar evolution [7]. The Ga detector, first suggested by KUZMIN [8], has a sufficiently low threshold energy of 0.236 MeV and is considered to be sensitive to low-energy neutrinos from the basic reaction $p + p \rightarrow {}^2H + e^+ + \nu_e$ of the pp chain. The neutrino flux from this basic reaction contributes more than 90% of the solar neutrino flux on the earth and can be calculated practically independent of the solar model [5]. The ^{37}Cl detector, on the other hand, is insensitive to this low-energy pp neutrino due to its high threshold energy of 0.814 MeV, and sensitive to high-energy neutrinos from ^{8}B decays. The flux of ^{8}B neutrinos is highly temperature-dependent, and thus dependent on the solar model.

Nuclear structure studies of the Ga solar neutrino detector have been intensively performed recently from both sides of experiment [9,10] and theoretical calculation [11,12,13]. The neutrino capture rate for the Ga detector is determined by the Gamow-Teller (GT) strength (B(GT)) distribution in ^{71}Ge up to a neutron emission threshold of 7.4 MeV.

The ground state ($3/2^-$ in ^{71}Ga) to ground state ($1/2^-$ in ^{71}Ge) GT strength can be obtained reliably from the ft value for ^{71}Ge (β^+) ^{71}Ga to be 0.09. One can see the role of Gamow-Teller strength distribution in ^{71}Ge for each solar neutrino source from the excitation energy dependence of $\phi \langle\sigma_{av}(E_{exc})\rangle$, where ϕ is the neutrino flux and $\langle\sigma_{av}(E_{exc})\rangle$ is the weighted average phase space factor for the neutrino absorption cross section for the fixed final state. From this excitation energy dependence of $\phi \langle\sigma_{av}(E_{exc})\rangle$, one can see the following properties. First, the pp neutrino capture rate is determined only by the Gamow-Teller strength to the ground state. Therefore, the pp neutrino capture rate is reliably estimated by using the GT strength from the β decay ft value of ^{71}Ge. Secondly, the ^{7}Be neutrino capture rate is determined by the Gamow-Teller strength to the lowest three levels in ^{71}Ge. ORIHARA et al. [9] performed the (p,n) reactions with 35 MeV protons and deduced the GT matrix elements to the

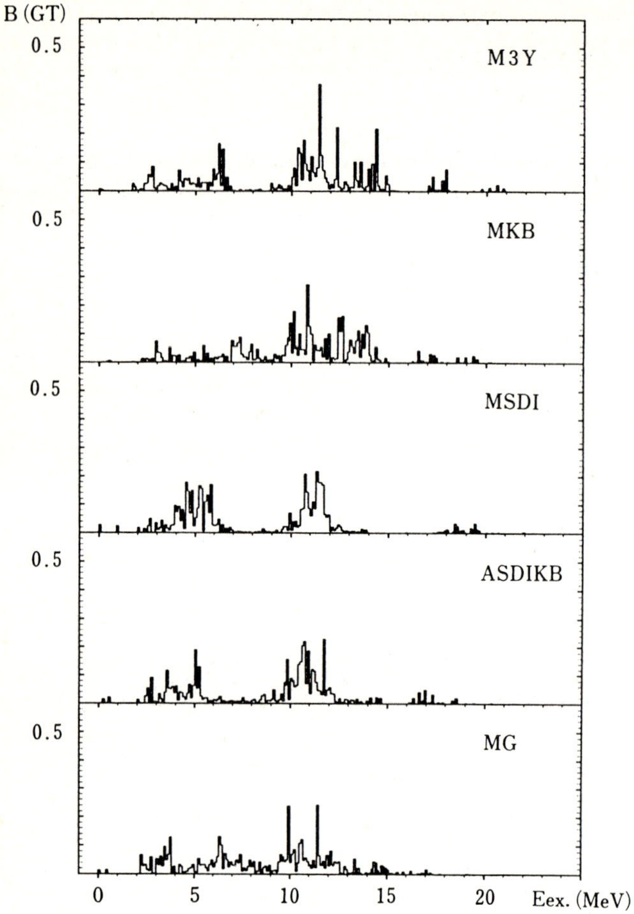

Fig. 1 Calculated GT strength distributions in ^{71}Ge.

first three states in ^{71}Ge, assuming the proportionality of the (p,n) forward-angle cross sections to B(GT). Their deduced B(GT) to the first $5/2^-$ state is comparable to that to the ground state and gives the ^{7}Be neutrino capture rate comparable contribution as B(GT) to the ground state. BALTZ et al. [12] discussed the (p,n) reaction mechanism on the proportionality above mentioned for a low energy projectile. Thirdly, the ^{8}B neutrino capture rate is very small if only those Gamow-Teller strength to the lowest three levels are taken into account. However, if there is large Gamow-Teller strength in the 2 ~ 5 MeV region, the ^{8}B neutrino capture rate could be increased to the comparable order as the ^{7}Be neutrino capture rate. KLAPDOR et al. [11] first pointed out by BCS calculation that there is a bump of GT strength at 2 ~ 6 MeV in ^{71}Ge and it increases considerably the ^{8}B neutrino capture rate. KROFCHECK et al. [10] studied the ^{71}Ga (p,n) ^{71}Ge reaction at proton bombarding energies of 120 and 200 MeV, and deduced the GT strength distribution in ^{71}Ge up to ~14 MeV. Their adopted upper limit of B(GT) to the first $5/2^-$ state is 0.009, which is one order smaller than that to the ground ($1/2^-$). MATHEWS et al. [13] performed shell model studies within truncated model spaces and with KALLIO-KOLLTVEIT interaction [14], and obtained similar conclusion as that of GROTZ et al. [11] and KROFCHECK et al. [10] that the GT strength distribution for the excited states increases the ^{8}B neutrino cap-ture rate by a factor of 10 compared with the estimate of BAHCALL [2] which takes into account only the lowest three levels.

We have performed detailed shell model calculations of the Gamow-Teller strength distribution in ^{71}Ge within $(f_{5/2},p_{1/2},p_{3/2})^{15} + f^{-1}_{7/2}$ $(f_{5/2},p_{1/2},p_{3/2})^{16}$ configuration space and with five sets of interaction in order to study the effective interaction dependence of this ^{8}B neutrino capture rate by the Ga detector. The effective interactions used are M3Y [15], KUO-BROWN (MKB) [16], modified surface delta interaction (MSDI) and adjusted surface delta interaction (ASDIMKB) of KOOPS-GLAUDEMANS [17], and MOOY-GLAUDEMANS (MG) [18].

The single particle energies and the monopole parts of diagonal matrix elements of the effective interactions are modified to reproduce the lowest three levels in ^{71}Ga and the Gamow-Teller giant resonance at 11 ~ 12 MeV in ^{71}Ge. The calculated GT strength distributions in ^{71}Ge are shown in fig. 1. These are the calculated results in the truncated space, in which the summed value of seniority in each orbit for each configuration is limited in isospin formalism as equal to or smaller than 3. However, the total B(GT) calculated in this truncated space exhausts about 90% of that in the full space of $(f_{5/2},p_{1/2},p_{3/2})^{15} + f^{-1}_{7/2}$ $(f_{5/2},p_{1/2},p_{3/2})^{16}$.

We have made the following modification on this strength distributions. For the ground state, we have normalized the calculated strength to be 0.09 which is reliably obtained from the β decay ft value of ^{71}Ge. We have multiplied this normalization factor to the calculated strength for the first $5/2^-$ and $3/2^-$ states. For the other excited states, we have multiplied the quenching factor 0.5.

The solar neutrino capture rates thus obtained are shown in table 1. The percentage of the pp neutrino capture rate in the total capture rate is almost constant (47 ~ 51%). Our calculated B(GT) values to the first $5/2^-$ state are smaller than 0.004, and the ^{7}Be neutrino capture rate does not have the strong effective interaction dependence as shown in table 1. On the other hand, the ^{8}B neutrino capture rate depends on the GT strength distribution up to 7.4 MeV. Our calculated capture rate has a strong effective interaction dependence, and varies from 13 SNU to 26 SNU. According to this calculations, the solar model dependent "background" from ^{8}B neutrinos in a Ga solar neutrino experiment ranges - in the standard solar model - between 19.5 and 39% of the pp-rate - , i.e. a factor of 10 larger than the Bahcall expectation. It is to be noted that Gamow-Teller strength distribution should be carefully calculated with a reliable effective interaction in a reliable model space for the prediction of solar neutrino capture rate. It might be the same situation in the case of ^{205}Tl experiment.

Table 1 Calculated solar neutrino capture rates (in SNU) in five sets of interactions

Neutrino source	M3Y	MKB	MSDI	ASDIMKB	MG
pp	65.5	65.5	65.5	65.5	65.5
pep	3.1	4.2	3.4	4.0	3.6
^{7}Be	32.6	39.4	34.4	38.4	35.5
^{8}B	18.8	12.8	25.6	21.0	21.7
^{13}N	3.1	3.6	3.2	3.5	3.3
^{15}O	4.7	5.9	5.0	5.7	5.2
Total ($\Sigma\sigma\phi$)	127.7	131.4	137.1	138.2	134.7

References:

1. J.N. Bahcall et al.: Phys. Rev. Lett. 40, 1351 (1978)
2. J.N. Bahcall: Rev. Mod. Phys. 50, 881 (1978)
3. J.N. Bahcall et al.: Rev. Mod. Phys. 54, 767 (1982)
4. W. Hampel: in Proc. Int. Conf. Neutrino Physics and Astrophysics, Erice, Italy 1980, ed. E. Fiorini (Plenum, New York, 1982), p. 61.
5. H.V. Klapdor: Prog. Part. Nucl. Phys. 10, 131 (1983)
6. R. Davis, Jr.: in Proc. Conf. Status and Future Solar Neutrino Research, ed. G. Friedlander (BNL 50879), Vol. 1, p.1
7. J.N. Bahcall et al.: Phys. Rev. Lett. 45, 945 (1980)
8. V. A. Kuzmin: Zh. Eksp. Teor. Fiz. 49, 1532 (1965) [Sov. Phys. JETP 22, 1051 (1966)]
9. H. Orihara et al.: Phys. Rev. Lett. 51, 1328 (1983)
10. D. Krofcheck et al.: Phys. Rev. Lett. 55, 1051 (1985)
11. K. Grotz et al.: in Capture Gamma-Ray Spectroscopy and Related Topics - 1984, ed. S. Raman, AIP Conf. Proc. No. 125, p.793, Phys. Rev. C33, 1263 (1986)
12. A.J. Baltz et al.: Phys. Rev. Lett. 53, 2078 (1984)
13. G.J. Mathews et al.: Phys. Rev. C32, 796 (1985)
14. A. Kallio and K. Kolltveit: Nucl. Phys. 53, 87 (1964)
15. W.G. Love: in The (p,n) Reaction and the Nucleon-Nucleon Force, ed. C.D. Goodman et al., p. 23
16. T.T.S. Kuo and G.E. Brown: Nucl. Phys. A114, 241 (1968)
17. J.E. Koops and P.W.M. Glaudemans: Z. Physik A280, 181 (1977)
18. R.B.M. Mooy and P.W.M. Glaudemans: Z. Physik A312, 59 (1983)

Microscopic Calculation of Neutrino Capture Rates in ^{69,71}Ga and the Detection of Solar and Galactic Neutrinos

H.V. Klapdor, K. Grotz, and J. Metzinger

Max-Planck-Institut für Kernphysik, D-6900 Heidelberg, F. R. Germany

1. Introduction

A measurement of the practically solar model independent main part of the solar neutrino flux, namely those coming from the reaction $p + p \to d + \nu_e + e^+$ (pp neutrinos) using a gallium detector is expected to help to answer the question of the origin of the observed ^{8}B neutrino deficit [1,2]. One of the explanations of this so-called solar neutrino problem could be the occurrence of neutrino oscillations, complicated by effects of neutrino oscillations in the solar matter [3]. A less exotic explanation is given e.g. by [4].

Although in principle extremely small ν-mass differences lead to sizable effects, even in the favourable case of maximum 2ν-mixing with $\Delta m^2 \gtrsim 2 \times 10^{-8} eV^2$ corresponding to a mixing length L smaller than R_0, the solar radius, the expected reduction in the ν_e-flux would only be a factor 2. For smaller than maximum mixing and also for $L \gtrsim R_0$ the reduction could be considerably smaller. Therefore, a comparison of the experimental capture rate with theoretical expectations requires accurate knowledge of the neutrino capture cross section $\sigma_\nu(E_\nu)$. We shall show that earlier phenomenological estimates [5] for this quantity in ^{69}Ga and ^{71}Ga are insufficient. A reliable prediction of $\sigma_\nu(E_\nu)$ requires a careful treatment of the beta strength distribution in ^{69,71}Ge.

2. Calculation of the neutrino capture cross section $\sigma_\nu(E_\nu)$

The capture cross section $\sigma_\nu(E_\nu)$ is dominated by the isospin invariance determined Fermi transition to the isobaric analog state (IAS) and the Gamow Teller (GT) transitions to the daughter nucleus:

$$\sigma_\nu(E_\nu) = \frac{g_A^2}{\pi c^3 h^2} \int_0^{E_\nu - Q} p_e \, E_e \, F(Z,E_e) \, S_\beta(E')dE' \qquad (1)$$

with

$$S_\beta(E) = \frac{1}{dE} \sum_{E'=E}^{E+dE} \left[B_{GT}(E') + B_F(E') \left(\frac{g_V}{g_A} \right)^2 \right]$$

and

$$E_e = E_\nu - Q - E' + m_e c^2$$

In the above formula g_V and g_A are the vector and axial vector coupling constants respectively, p_e and E_e are the electron momentum and energy, respec-

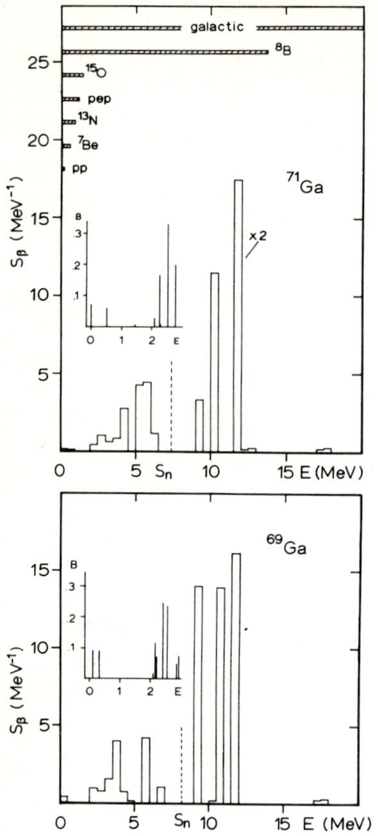

tively and F(Z,E) is the usual Fermi function. The beta strength function $S_\beta(E)$ here is defined as reduced transition strength B_i (in natural units) per energy unit.

Refining earlier studies [6] we have calculated $\sigma_\nu(E_\nu)$ for the first time (see [7]) based on a microscopic nuclear model. For details of this calculation see [8]. The calculated distributions of GT strength for ^{71}Ga and ^{69}Ga are shown in Fig. 1. Our calculation reproduces quite well (see Table 1) the reduced transition probability B for the transition $^{71}Ga_{g.s.} \rightarrow {}^{71}Ge_{g.s.}$ which is known to be 0.09 from the EC decay of ^{71}Ge. The calculation shows also that the large GT strength for the first excited $5/2^-$ state claimed by [9] basing on their (p,n) experiment at 35 MeV proton energy, could hardly be correct (see Table 1). The investigation by [10] cleared up this

Fig. 1: Microscopically calculated Gamow Teller (GT) part of the beta strength distribution $S_\beta(E)$ $(= \frac{1}{\Delta E} \Sigma_{\Delta E} B_{GT}(E)$; $\Delta E = 0.5$ MeV in this figure) for ^{71}Ga and ^{69}Ga. The inserts show the calculations for the low-energy part on an expanded scale. The energy range important for the detection of solar neutrinos from different sources is indicated.

Table 1. Calculated Gamow Teller strength for β^- decay of ^{71}Ga to low-lying states in ^{71}Ge compared with values from (p,n) data and the g.s. β-decay value.

Final state in ^{71}Ge (MeV)	this work [7,8]	Baltz et al. [10]	Mathews et al. [11]	g.s. β-decay	(p,n) Orihara et al. [9]
g.s.	0.071	0.238	0.052	0.09	0.083
0.175 ($5/2^-$)	0.0001	0.011	0.012	---	0.080
0.500 ($3/2^-$)	0.061	0.058	7.5×10^{-4}	---	0.019

problem in detail. The latter authors calculated the $^{71}Ga(p,n)^{71}Ge$ cross section for 35 MeV bombarding energy at zero degree and also the GT strength in an (fp) model space and found that there is only little correspondence between these two quantities at such a low bombarding energy, since the (p,n) cross section for the $3/2^- \rightarrow 5/2^-$ transition at this energy is totally dominated by higher multipoles.

The neutrino capture cross sections as function of E_ν calculated by eq. (1) on the basis of the strength distributions of Fig. 1 are shown in Fig. 2. For typical energies of the ^8B neutrinos (intensity maximum at $E_\nu \simeq 7$ MeV) the microscopically calculated capture cross section is about an order of magnitude larger than the ground state contribution.

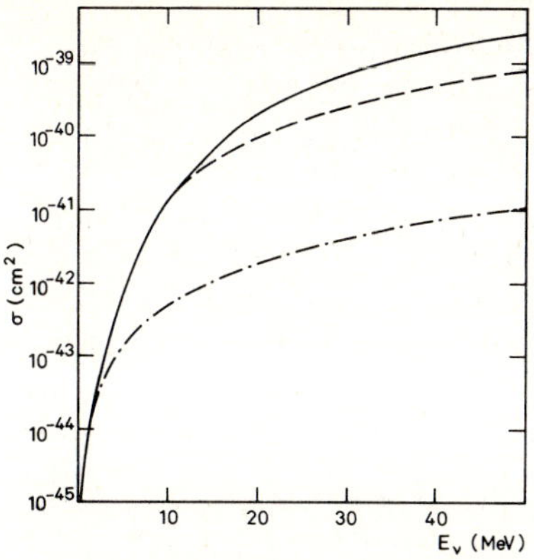

Fig. 2a: Neutrino capture cross section as deduced from the calculated S_β of fig. 1. The solid line includes all strength (also the Fermi strength in the isobaric analogue state), the dashed line only the strength lying below S_n. The dashed-dotted line is the contribution from the ^{71}Ge ground state alone, on which previous estimates were based.

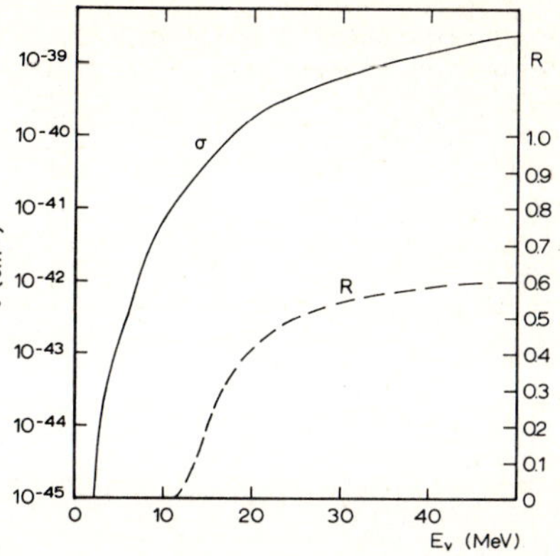

Fig. 2b: $\sigma_\nu(E_\nu)$ for ^{69}Ga including all strength (also the Fermi strength). Shown is also R, the fraction of the total capture cross section resulting from population of states above the neutron separation energy S_n (process (3)).

The enhanced cross section for neutrino energies $\gtrsim$ 4 MeV is the result of a sizable fraction of the total GT strength between 2 and 6 MeV excitation energy in ^{71}Ga (see Fig. 1a). It is not an effect of the IAS nor of the GT giant resonance, which start to dominate $\sigma_\nu(E_\nu)$ for $E_\nu \gtrsim 15$ MeV but do not lead to detectable events, because neutron emission from these states leads to the stable ^{70}Ge.

3. Consequences for a solar neutrino experiment

We find, based on the standard solar model flux prediction and the calculated $\sigma_\nu(E_\nu)$ for ^{71}Ga, that ^8B neutrinos produce a background of the order of about 15 SNU (to be compared to the pp neutrino signal of 71 SNU). This value is still dependent on the degree of quenching of GT strength in the 'nuclear' region of excitation (Table 2).

With a quenching factor of $= 1 - B_{GT}^{quench}/B_{GT} = 0.35 \pm 0.15$ for the strength below the GTGR (for discussion of this point see [12]) this would lead to the capture rates given in the last column of Table 2. We compare this result in Table 2 also with the more recent calculation by the Livermore group [11]. It is remarkable, that this different theoretical approach leads practically to the same result when making the same assumption on the quenching factor in both cases.

A recent study of the effective interaction dependence of the ^8B neutrino capture rate performed by Oda and Muto [13] leads - with a quenching factor of 0.5 - to values between 12.8 and 21.7 SNU.

The Gamow-Teller strength distribution in ^{71}Ge has been measured by the $^{71}Ga(p,n)^{71}Ge$ reaction at $E_p = 120$ MeV [14,15]. The authors deduce a capture rate of 13 SNU from this experiment for the standard solar model. This result

Table 2. Calculated ratios $\sigma_{tot}/\sigma_{g.s.}$ for the capture rates of ^8B solar neutrinos as function of quenching of low-lying GT strength and the predicted capture rates for solar ^8B neutrinos (by states in ^{71}Ge below S_n) in SNU (1 SNU = 1 neutrino capture per second in 10^{36} target atoms; capture in the g.s. state corresponds to 1.2 SNU). The quenching factor q is defined as $q = 1 - B_{GT}^{quench}/B_{GT}$. The last line gives the result of the (p,n) experiment of Krofcheck et al. [14,15].

	$\sigma_{tot}/\sigma_{g.s.}$	q	$\sigma_{^8B-\nu}$ (SNU)	$\sigma_{^8B-\nu}$ (SNU) q=0.35±0.15
this work [7,8]	22.0	0	26.4	
	14.3	0.35	17.2	17.2±4
	11.0	0.5	13.2	
Mathews et al. [11]	19.5	0	23.4	
	12.7	0.35	15.2	15.2±4
	10.4	0.5	12.8	
Bahcall [5]	1.86		2.2	2.2
Krofcheck et al. [14,15]	10.8±30%		13±4	13±4

agrees well with the prediction (see also [7]) of the present calculation. It should be recognized in this context, that according to the ^{37}Cl experiment [16] the ^8B neutrino flux might be smaller than the standard solar model prediction on which our capture rates are based. In a model yielding 1.9 SNU in ^{37}Cl [16] the ^8B as well as the ^7Be ν flux would be strongly reduced. The solar model dependent contribution to the Gallium detector signal (mainly ^8B + ^7Be neutrinos) could therefore range from 22-68% of the pp signal (for a more detailed discussion see [8].

4 Consequences for the detection of collapsing star neutrinos

In hydrodynamical models of the gravitational collapse of a massive star neutrino-interactions are an essential ingredient. The trapping of neutrinos in the core and also the neutrino-pair production rate depends not only on hydrodynamical model parameters but also on the Weinberg angle Θ_W, the number of generations of quarks and leptons and on whether neutrinos possess Majorana masses or not [17,18]. The observation of the neutrino-burst emitted by a collapsing star is therefore of fundamental interest.

When estimating the feasibility of a Gallium detector for detection of galactic neutrinos, three different mechanisms have to be considered (see also [19]): (1) Capture by ^{71}Ga leading to states in ^{71}Ge below the neutron separation energy S_n. These events can be distinguished only statistically from solar neutrino events. (2) Capture by ^{69}Ga to states below S_n in ^{69}Ge. (3) Capture by ^{69}Ga to ^{69}Ge states above S_n, decaying to the ^{68}Ge ground state.

In Table 3 we give the calculated number of nuclei produced in these three channels assuming two different neutrino spectra as calculated by [20] and [21]. The figures hold for a 30t detector. Probably the most important point is the

Table 3. Expected number of nuclei produced in a 30 t gallium detector by a 1.8×10^{53} erg ν-burst at 1 kpc distance.

process	capture product	$T_{1/2}$	number of nuclei produced by: solar neutrinos	collapsing star neutrinos ν-spectrum from Wilson [20]	Roberts et al. [21]
(1)	^{71}Ge	11.2 d	18	120	255
(2)	^{69}Ge	39 h	0.3	147	370
(3)	^{68}Ge	288 d		67	275

high expectation for process (1). Also given in Table 3 is the normal solar contribution. For the latter the saturation value is given, which is the number of nuclei being present in statistical equilibrium between production and decay.

Our expectation of 120 ^{71}Ge nuclei (255 for the ν-spectrum of [21] is much larger than the solar saturation value and of the same order as the production rate by processes (2) and (3). Since the IAS in ^{69}Ge lies below S_n, process (2) has the largest cross section, but has the disadvantage of the short half life of ^{69}Ge, therefore needing a trigger signal. The disadvantage of process (3) is the large half life of ^{68}Ge of 288 d, which, despite the large production rate, would lead only to a very small decay rate of 0.7 per day for the assumed hypothetical neutrino burst.

5. Conclusion

The nuclear beta strength is found to have impact on questions ranging from the decay heat of nuclear reactors to the determination of the neutrino mass and to astrophysical and cosmological problems [22]. The present paper adds one more example in this respect.

The neutrino capture cross section for $^{69,71}Ga$ has been calculated as function of the neutrino energy. We find that a non-negligible solar model dependent 'background' of ^{8}B neutrinos of about 15 SNU has to be expected in a gallium solar neutrino experiment besides the pp signal.

The calculations yield a larger sensitivity of the Gallium detector than assumed previously for galactic neutrinos.

References

[1] V.A. Kuzmin, Sov. Phys. JETP 22 (1966) 1051

[2] W. Hampel, this volume.

[3] A. Smirnov, this volume.

[4] G. Marx, Proceed. Internat. Symp. on Nuclear Beta Decays and Neutrino, Osaka, June 11-13, 1986.

[5] J.N. Bahcall et al., Rev. Mod. Phys. 50 (1978) 881 and 54 (1982) 767

[6] H.V. Klapdor, Progr. Part. Nucl. Phys. 10 (1983) 131

[7] K. Grotz, H.V. Klapdor, J. Metzinger, Proceed. Internat. Sympos. Capture Gamma Ray Spectroscopy and Related Topics, Knoxville, USA, Sept. 1984, publ. in AIP Conf. Proceed. No. 125 (1985) p. 793, New York

[8] K. Grotz, H.V. Klapdor, J. Metzinger, Astron. Astrophys. 154 (1986) L1 and Phys. Rev. C 33 (1986) 1263
[9] H. Orihara, C.D. Zafiratos, S. Nishihara, K. Furukawa, M. Kabasawa, K. Maeda, K. Miura, H. Ohnuma, Phys. Rev. Lett. 51 (1983) 1328
[10] A.J. Baltz, H. Weneser, B.A. Brown, J. Rapaport, Phys. Rev. Lett. 53 (1984) 2078
[11] G.J. Mathews, S.D. Bloom, G.M. Fuller, J.N. Bahcall, 1985, Phys. Rev. C 32 (1985) 796
[12] K. Grotz, H.V. Klapdor, J. Metzinger, Phys. Lett. 132B (1983) 22
[13] T. Oda and K. Muto, this volume.
[14] D. Krofcheck, E. Sugarbaker, J. Rapaport, D. Wang, J.N. Bahcall, R.C. Byrd, C.C. Foster, C.D. Goodman, I.J. Van Heerden, C. Gaarde, J.S. Larsen, D.J. Horen, T.N. Taddeucci, Phys. Rev. Lett. 55 (1985) 1051
[15] E. Sugarbaker, this volume.
[16] R. Davis, B.T. Cleveland, J.K. Towley, Conf. on the Interactions between Particle and Nuclear Physics, Steamboat Springs, May 23-30, 1984
[17] E.W. Kolb, D.L. Tubbs, D.A. Dicus, Astrophys. J. 255 (1982) L57 and E.W. Kolb, Proc. XI Int. Conf. Neutrino Phys. and Astrophys., Nordkirchen, 1984, p. 243
[18] E.W. Kolb, this volume.
[19] W. Hampel. Proc. Int. Conf. Neutrino Phys. and Astrophys., Erice, Italy 1980, p. 61.
[20] J.R. Wilson, Astrophys. J. 163 (1971) 209
[21] A. Roberts, H. Blood, J. Learned, F. Reines, Proc. Int. Neutrino Conf., Aachen 1976, p. 688
[22] H.V. Klapdor, Invited lecture presented at Internat. School of Nucl. Phys.: The Early Universe and its Evolution, Erice, Sicily, 2-14 April 1986, in press in Progr. Part. Nucl. Phys.

Gamow-Teller Strength Functions via (p,n) and the Ga Solar Neutrino Detector

E. Sugarbaker

Department of Physics, The Ohio State University, Columbus, OH 43210, USA

The process of neutrino absorption or capture via inverse beta decay requires a transition in which a neutron in the ground state of the detecting nucleus is transformed into a proton and an electron is emitted. If the transition is an allowed $\Delta L = 0$, spin-independent one, the total transition strength which is not Pauli-blocked will be concentrated in the isobaric analog state (the IAS) of the final nucleus [1]. This total Fermi (F) strength function is characterized by the sharp IAS resonance located significantly higher in energy than the initial state due to the coulomb energy difference. Expressing the Fermi strength B(F) in units such that B(F) = 1 for a free neutron, we have B(F) = N - Z = 9 assigned to the IAS in ^{71}Ge. However, only neutrinos above a threshold energy of 9.18 MeV could be captured into this IAS, which in turn is above the 7.4 MeV particle-emission threshold in ^{71}Ge. It is therefore apparent that the Fermi strength function is not relevant to detectable neutrino capture by the proposed ^{71}Ga detector.

In addition to the above Fermi transition, an allowed $\Delta L = 0$, spin-dependent Gamow-Teller (GT) transition may also participate in neutrino absorption. Due to the additional spin interaction, the GT strength function is not characterized by a sharp resonance degenerate with the IAS. Rather it is extremely fragmented, and even has small but non-zero overlap matrix elements for states at very low excitation. In particular, electron capture results [2] for ^{71}Ge decay yield a ground state B(GT) value of 0.091 (a small fraction of the total expected B(GT) = 3(N - Z)). However, with a threshold energy of only 0.233 MeV, neutrino capture into the ground state of ^{71}Ge will still be significant due to the large flux of low-energy solar neutrinos from the p + p reaction. Excited states at 0.175, 0.500 MeV and many others are also likely to have significant GT matrix elements with the ground state of ^{71}Ga and therefore contribute to the total neutrino capture cross section. Large-scale nuclear structure calculations (see columns 2 - 4 of Table 1) do not agree well on the magnitudes to expect in this low-excitation region of the GT strength function. Due to the limits imposed by the energetics of beta-decay processess, direct experimental information was not available on the B(GT) of the excited states until our recent (p,n) measurements.

GOODMAN et al. [7] argued that the 0 deg (p,n) differential cross section $\sigma(0^{\circ})$ at intermediate energies can be represented as

Table 1 Gamow-Teller strength for ^{71}Ga(β^-)^{71}Ge calculated and measured

E_{ex}	Calculated			β-decay	(p,n)	
[MeV]	Ref. [3]	Ref. [4]	Ref. [5]	[2]	Ref. [6]	Present
0.0	0.238	0.052	0.071	0.091	0.083	0.085(0.015)
0.17	0.011	0.012	0.0001	-	0.080	<0.009
0.50	0.058	0.000075	0.061	-	0.019	0.010(0.005)
0.7-7.4	-	~5.96[a]	~7.25[a]	-	-	4.21 (0.62)

[a] Excluding estimates of "quenching"

$$\sigma(0^{\circ}) = K_{\alpha}\, N_{\alpha}\, |J_{\alpha}|^{2}\, B(\alpha) \tag{1}$$

on the assumption that these cross sections are dominated by $\Delta L = 0$ transitions by action of a primarily direct one-step DWIA reaction mechanism. The α designates either a $\Delta S = 0$, $\Delta J = 0$ Fermi transition or a $\Delta S = 1$, $\Delta J = 1$ Gamow-Teller transition. The K's are kinematic factors, the N's are distortion factors, and the J's are the Fourier transform of the relevant free N-N t-matrix at momentum transfer $q \simeq 0$. The $B(\alpha)$ represent the *same* nuclear structure matrix elements as derived from the inherently $\Delta L = 0$, allowed beta decay processes *if* the reaction mechanism successfully suppresses other transition strength involving higher $\Delta L > 0$ multipoles. This is a reasonable approximation for studies at $E_p > 100$ MeV.

Use of (1) in the extraction of B(GT) is hampered by uncertainties in measuring absolute cross sections, in extracting empirical J's at a given bombarding energy and in calculating the N's, suggesting the need for an internal calibration. TADDEUCCI et al. [8] showed that (1) applied to Fermi and Gamow-Teller transitions within the same 0° spectrum yields a double ratio of

$$\frac{\sigma_{GT}(0^{\circ})\; /\; B(GT)\; K_{GT}}{\sigma_{F}(0^{\circ})\; /\; B(F)\; K_{F}} = \frac{N_{\sigma\tau}\; |J_{\sigma\tau}|^{2}}{N_{\tau}\; |J_{\tau}|^{2}} \simeq (E_p/E_0)^{2} \tag{2}$$

where E_0 is found by fitting the results at E_p between 80 and 200 MeV for states with B(F) and B(GT) known from β decay. The E_0's obtained are given in Fig. 1.

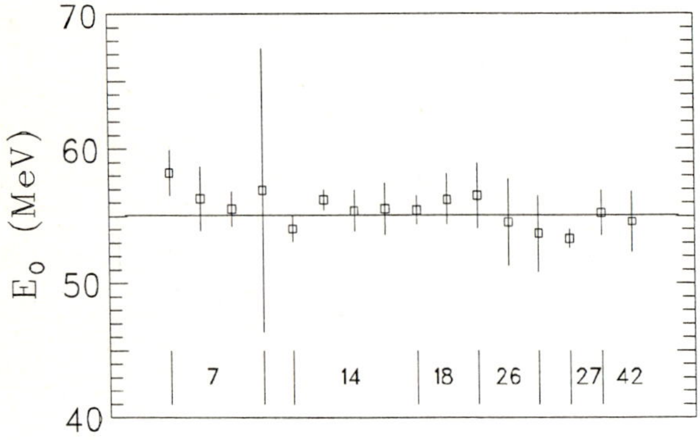

Fig. 1 Empirical determination of E_0 parameter in (2) using (p,n) data for transitions having known beta decay matrix elements (from Ref. 8)

We measured the ^{71}Ga(p,n)^{71}Ge reaction at E_p of 120 and 200 MeV using a neutron flight path of 130 meter at the Indiana University Cyclotron Facility (IUCF).[9] The lower bombarding energy was selected to permit resolution of the first excited state from the ground state while still maintaining as much as possible of the desired $\Delta L = 0$, Gamow-Teller selectivity associated with the higher energies. The higher resolution (~215 keV FWHM) data at 120 MeV proved more useful, but required subtraction of estimated non-$\Delta L = 0$ contributions at 0° to yield agreement with the more $\Delta L = 0$ selective 200 MeV data. Such analysis required a few-point angular distribution and is outlined in KROFCHECK et al. [10]. It resulted in the B(GT) strength function shown in Fig. 2b, derived from the 0° spectrum shown in Fig. 2a. The specific values are listed in the last column of Table 1. A large background of low-energy neutrons from the previous beam burst (about 120 ns earlier) has been subtracted from the spectrum of Fig. 2a, and is the primary basis for the large uncertainties in the B(GT) values we have quoted.

There has been much interest in the potential contribution of the 175 keV first excited state in ^{71}Ge to the neutrino capture efficiency. The low energy (p,n) study of ORIHARA et al. [6] reported nearly identical 0° cross sections for both the ground state and first excited state and suggested a B(GT) of about 0.080 assuming that (1) is still valid at $E_p = 35$ MeV. These results would imply that

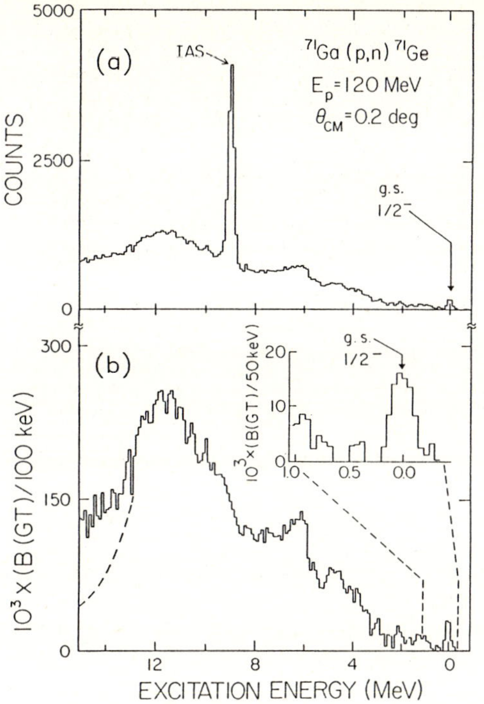

Fig. 2 (a) The neutron time-of-flight spectrum after subtraction of background of low-energy neutrons and conversion to energy bins
(b) The GT strength function in ^{71}Ge (from KROFCHECK et al. [10])

an additional 32 SNU must be added to that of the ground state solar neutrino capture rate of 107 SNU, due primarily to enhanced capture of neutrinos from the ^{7}Be source reaction into the first excited state. BALTZ et al. [3] recently showed that at such energies the $\Delta L > 0$, GT forbidden contributions to the 0° cross sections can be significant. They find this to be particularly important for this state, calculating that only 18% of the 0° cross section at 35 MeV is attributable to $\Delta J^\pi = 1^+$, of which only a fraction is allowed GT strength. The situation at 120 MeV is greatly improved, with about 70% of the 0° cross section arising from $\Delta J^\pi = 1^+$ components. Our initial data which was reported in KROFCHECK et al. [10] established only an upper limit on the *small* 0° cross section to the first excited state. However, significant $\Delta L > 0$ contributions appear in the angular distribution for this state. After the subtraction process described above, the B(GT) spectrum (see insert in Fig. 2b) is consistent with negligible strength at this excitation energy.

Our initial data is adequate to show that the contribution from the first excited state to the total neutrino capture process is not more than 3 SNU. However, the higher-lying excited states do contribute significantly, primarily via capture of the higher energy neutrinos from the ^{8}B source reaction. This increases the predicted capture rate by an additional 12.5 ± 3.2 SNU over that due to capture into only the ground and first excited states. Table 2 presents the correction factors Q as defined by BAHCALL [11] based on our initial B(GT) values and also shows a comparison of the predicted capture rates per neutrino source using various GT strength functions and assuming the neutrino flux given by the Standard Solar Model of BAHCALL [11]. It is obvious that even at the higher E_p, an attempt must be made to identify with as small uncertainty as is feasible the allowed GT components of the 0° spectrum. To that end we recently obtained additional data at 0° and 2.5° using the newly installed stripper loop at IUCF, which provides intense beam bursts separated by about 1 μsec. This removes much of the uncertainty associated with the earlier data by significantly reducing the amount of background. Preliminary analysis of these new data suggests that it is now possible to identify

Table 2 Calculated correction factors $Q = \sigma_{total}/\sigma_{gs}$ and capture rates in ^{71}Ga

Neutrino source	E_{ν}^{max} [MeV]	Q Present	SNU gs only[a]	SNU Ref. [4]	SNU Ref. [5]	SNU Ref. [6]	SNU Present
p + p	0.420	1.00	70.2	70.2	71.3	67.5	70.2
pep	1.442	1.27	2.5	3.0	2.5	5.4	3.2
^7Be	0.862	1.065	27.0	31.2	31.2	58.7	28.7
^7Be	0.384	1.00					
^8B	14.02	11.1	1.2	11.6	17.2	4.4	12.6
^{13}N	1.198	1.08	2.8	3.3	2.9	5.4	3.1
^{15}O	1.737	1.23	3.9	4.6	4.0	8.2	4.8
Total			107.5	124.	129.	149.	122.

[a] Ref. [11]

the first excited state in these forward-angle cross-section spectra. These new data should also yield B(GT) values with uncertainties under 10% at the 1-σ level for all but the most weakly populated states.

In conclusion, the(p,n) reaction studies at intermediate energies provide unique knowledge of the excited GT strength which could participate significantly in capture of higher-energy solar neutrinos. Such information may prove critical in light of the recent predictions (MIKHEYEV and SMIRNOV [12], BETHE [13]) that electron neutrinos above a certain energy might be converted into muon neutrinos while passing through the sun's interior.

The material discussed represents the work of our (p,n) group at IUCF, which includes R. Byrd, T. Carey, C. Gaarde, C. Goodman, D. Horen, D. Krofcheck, J. Rapaport, J. Larsen, and T. N. Taddeucci. I am especially indebted to J. Bahcall for his help in the application of our (p,n) results to this interesting problem.

References

1. J.D. Anderson, C. Wong and J.W. McClure: Phys. Rev. 126, 2170 (1962)
2. W. Hampel and L.P. Remsberg: Phys. Rev. C31, 666 (1985)
3. A.J. Baltz et al.: Phys. Rev. Lett. 53, 2078 (1984)
4. G.J. Mathews et al.: Phys. Rev. C32, 796 (1985)
5. H.V. Klapdor, K. Grotz and J. Metzinger: in Proceedings of the First Symposium on Underground Physics, Saint Vincent, Italy, April 25-28, 1985 (in press)
6. H. Orihara et al.: Phys. Rev. Lett. 51, 1328 (1983)
7. C.D. Goodman et al.: Phys. Rev. Lett. 44, 1755 (1980)
8. T.N. Taddeucci et al.: Phys. Rev. C25, 1094 (1982) and to be published
9. C.D. Goodman et al.: IEEE Trans. Nucl. Sci. 26, 2248 (1979)
10. D. Krofcheck et al.: Phys. Rev. Lett. 55, 1051(1985)
11. J.N. Bahcall: Rev. Mod. Phys. 50, 881 (1978) and private communication
12. S.P. Mikheyev and A. Yu. Smirnov: in Proceedings of the Tenth International Workshop on Weak Interactions, Savonlimma, Finland, June 16-25, 1985 (unpublished)
13. H.A. Bethe: Phys. Rev. Lett. 56, 1305 (1986)

The Sudbury D_2O Neutrino Detector

E.D. Earle[1], *G.T. Ewan*[2], *H.W. Lee*[2], *H.-B. Mak*[2], *B.C. Robertson*[2], *R.C. Allen*[3], *H.H. Chen*[3], *P.J. Doe*[3], *D. Sinclair*[4], *W.F. Davidson*[5], *C. Hargrove*[5], *R.S. Storey*[5], *G. Aardsma*[6], *P. Jagam*[6], *J.J. Simpson*[6], *E.D. Hallman*[7], *A.B. McDonald*[8], *A.L. Carter*[9], and *D. Kessler*[9]

[1]AECL, Chalk River, Ontario, K0J 1J0, Canada
[2]Queen's University, Kingston, Ontario, K7L 3N6, Canada
[3]University of California, Irvine, CA 92717, USA
[4]University of Oxford, Oxford, OX1 3NP, UK
[5]National Research Council of Canada, Ottawa, Ontario, K1A 0R6, Canada
[6]University of Guelph, Guelph, Ontario, N1G 2W1, Canada
[7]Laurentian University, Sudbury, Ontario, P3E 2C6, Canada
[8]Princeton University, Princeton, NJ 08544, USA
[9]Carleton University, Ottawa, Ontario, K1S 5B6, Canada

The construction of a solar neutrino observatory in a nickel mine near Sudbury, Canada, has been proposed[1]. This proposal is both attractive and timely because of the temporary surplus of over 1000 Mg of D_2O stockpiled for the Canadian-designed nuclear power reactors, of the large rock overburden at the INCO mine and of the scientific interest in the solar neutrino flux and neutrino oscillations. The primary objective of the experiment is to measure the incident electron neutrino (ν_e) spectrum above 5 MeV in real time with energy and directional resolution. It is also expected that the total neutrino (ν_x) flux above 2.2 MeV can be measured thereby providing important information concerning neutrino oscillations. The observatory will also be a sensitive instrument for measuring neutrinos produced by cosmic rays, solar flares and collapsing stars.

The ν_e flux above 2 MeV, dominated by the decay of ^8B in the sun, is predicted[2] to be 4×10^6 $cm^{-2}s^{-1}$ by the standard solar model. This is more than a factor of two larger than the experimental flux reported[3] by ROWLEY et al. This shortfall in the ν_e flux is referred to as the solar neutrino problem (SNP). Lower theoretical estimates of the ν_e flux involve extreme changes in the solar model. Alternatively, neutrino vacuum oscillations could cause a factor of three reduction in the flux between sun and earth. Recently it has been proposed[4] that the occurrence of neutrino oscillations can be enhanced in the dense centre of the sun, thereby reducing the ν_e flux and changing its energy spectrum.

Two Ga-Ge radiochemical experiments[5,6] have been funded and will complement the Cℓ-Ar experiment[3]. The Ga experiments, with a detection threshold of 0.23 MeV, will be sensitive to the ν_e flux from the $p + p \rightarrow d + e^+ + \nu_e$ reaction, $\nu_e(pp)$, the dominant and first nuclear reaction occurring in the sun. Until recently it was argued that a $\nu_e(pp)$ experimental flux consistent with solar model predictions would suggest no neutrino oscillations. The SNP would then be due to errors in that part of the model used to predict the $\nu_e(^8B)$ rate. On the other hand a low $\nu_e(pp)$ experimental flux would suggest vacuum oscillations. Neutrino matter oscillations[4,7] now allow the possibility that there is no $\nu_e(pp)$ flux reduction but a significant $\nu_e(^8B)$ flux reduction. A measurement of the $\nu_e(^8B)$ energy spectrum would then be required to resolve the SNP.

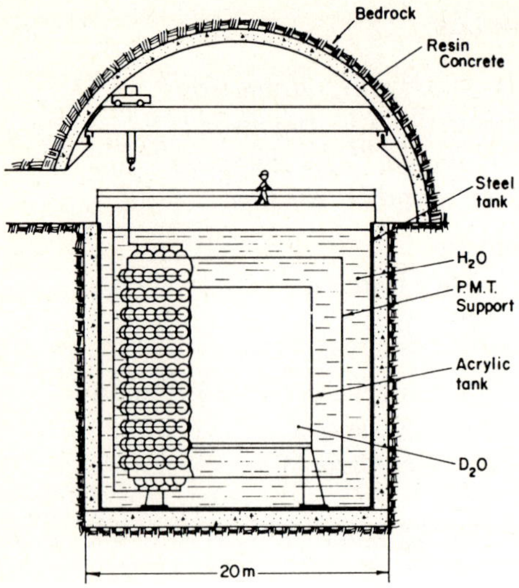

Fig. 1 Sketch of the SNO detector

The proposed D_2O detector, sketched in Fig. 1, includes 1000 Mg of 99.8% D_2O enclosed in an acrylic tank. This 10.5 m × 10.5 m cylinder is surrounded by 3.5 m of H_2O and 1 m of low-background concrete. Twenty-four hundred 50 cm diameter photo-multipliers mounted uniformly in the H_2O, 2.5 m from the acrylic tank, view the D_2O and provide 40% coverage. The detector, located 2000 m underground, will measure the Cerenkov light from the electrons produced in the reactions 1) $\nu_e + e \rightarrow \nu_e' + e'$, 2) $\nu_e + d \rightarrow p + p + e$ and 3) $\nu_x + d \rightarrow p + n + \nu_x'$, $n + d \rightarrow t + \gamma$, $\gamma + e \rightarrow \gamma' + e'$. Reaction 3 measures the total neutrino flux and tests the standard solar model independently of neutrino oscillations. Reaction 2 measures the ν_e spectrum, the shape of which may be affected by neutrino oscillations. The ratio of the rate of reaction 2 to the rate of reaction 3 provides a test for neutrino oscillations independently of solar models. The calculated rates[1] for each reaction are listed in Table 1 assuming a $\nu_e(^8B)$ flux of 2×10^6 $cm^{-2}s^{-1}$, a $\nu_x(^8B)$ flux of 5×10^6 $cm^{-2}s^{-1}$ and a conservative detection threshold of 7 MeV. Doping the D_2O with Gd will improve the efficiency and sensitivity for reaction 3 by a factor of three. To determine the yield for the various reactions sequential runs involving H_2O, D_2O and D_2O doped with ^{10}B are required. The flux sensitivities are estimates of the minimum ^8B neutrino flux that could be observed with confidence (3σ) in a period of six months running with D_2O after running a similar time with H_2O. All backgrounds associated with the (ν_e,e) and (ν_e,d) reactions are included. The background from radioactive decay of trace amounts of Th in the acrylic and D_2O are not included in the flux sensitivity calculation for the (ν_x,d) reaction.

Table 1 Reaction rates per day and sensitivity to the $\nu(^8B)$ flux

Reaction	Total events (day^{-1})	Total events × detector efficiency (day^{-1})	Flux sensitivity ($cm^{-2}s^{-1}$)
1. $\nu_e e$	3.5	0.7	3×10^5
2. $\nu_e d$	12	8	7×10^4
3. $\nu_x d$	14	3 (undoped) 10 (doped)	4×10^5 (undoped)

The proposal[1] is supported by extensive Monte Carlo calculations[8] on the performance of D_2O Cerenkov detectors, by background measurements in the mine and of detector materials, by light attenuation measurements[9] in D_2O, by timing measurements of 50 cm photomultipliers, by a determination of the availability of suitable D_2O, by negotiations with the mine management concerning the site of the laboratory and by discussions with an acrylic tank manufacturer.

The rock overburden provides an adequate muon shield. The concrete and H_2O stop γ-rays and neutrons from the rock and phototube materials so that the background above 7 MeV for reactions 1 and 2 is less than 2 per day and is from γ-rays external to the D_2O not rejected after reconstruction. In addition to the two background events per day from external γ-rays, reaction 3 has an internal background due to the presence of ^{232}Th and its daughters which decay via the 2.614 MeV first excited state of ^{208}Pb. These 2.614 MeV γ-rays cause deuteron photodisintegration which produces neutrons indistinguishable from those from deuteron neutrino disintegration. Th in the acrylic at 10^{-11} g/g and in the D_2O at 10^{-14} g/g will each produce one background event per day, events not included in the 4×10^5 value listed in Table 1. It is expected that acrylic without dye and ultraviolet absorber will have less Th than the 4×10^{-11} g/g measured[10]. The Th content of a 10 L D_2O sample was determined[11] to be $(4\pm1) \times 10^{-14}$ g/g by removing the Th from the D_2O by adsorption on silica-immobilized 8-hydroxyquinoline and then measuring the concentrate by inductively coupled plasma mass spectrometry. It is expected that on-line filtration systems will reduce the Th content of the water to 10^{-14} g/g.

These projected and measurable concentrations of Th will not solve the background problems if Th is in disequilibrium with its daughters, in particular ^{228}Ra. Accordingly the ability of the detector to detect (ν_x,d) events has not yet been demonstrated. Procedures to determine the concentration of Th daughters are being investigated for both the acrylic and D_2O.

The good event rate and detection efficiency of the D_2O Cerenkov detector as compared to existing detectors indicates that a measurement of the $\nu_e(^8B)$ energy spectrum will be practical. Doping the D_2O with ^{10}B will remove any residual background from neutrons due to deuteron disintegration. ROSEN and GELB[7] have indicated the importance of this spectrum as a means of choosing between oscillations and the solar model as the cause of the SNP and also as a means of distinguishing between different sets of oscillation parameters. The predicted spectrum in our D_2O detector from a $\nu_e(^8B)$ flux of 2×10^6 cm^{-2} s^{-1} is shown in Fig. 2 as a function of detected photoelectrons, assuming 40% surface

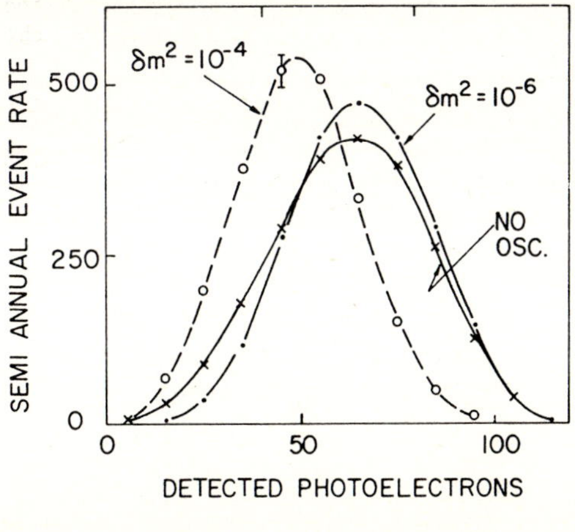

Fig. 2 Predicted $\nu_e(^8B)$ spectra

coverage. Eighty photoelectrons will be detected from a 10 MeV electron. The Monte Carlo calculated resolution function has been folded with the expected $\nu_e(^8B)$ energy spectrum for no neutrino oscillations, for matter oscillations in the body of the sun ($\delta m^2 = 10^{-6}$ eV^2, $\sin^2 2\theta = 0.04$) and for matter oscillations in the sun's core ($\delta m^2 = 10^{-4}$ eV^2, $\sin^2 2\theta = 0.04$). With a 5 MeV detection threshold then a measured $\nu_e(^8B)$ spectrum similar to the $\delta m^2 = 10^{-4}$ curve will suggest that matter oscillations in the sun's core is the reason for the SNP. If the spectrum is similar to the other curves then it will be more difficult, but not impossible, to distinguish between matter oscillations in the body of the sun and non-standard solar models. In such a case, the entire yield in the Cℓ experiment is due to the $\nu_e(^8B)$ thereby enhancing the Cerenkov detector yield and, in addition, the Ga radiochemical experiments[5,6] are likely to measure a low $\nu_e(pp)$ yield if oscillations are the reason for the SNP.

1. G.T. Ewan et al., SNO-85-3, July 1985 (NRC, Ottawa, Canada).
   D. Sinclair et al., Il Nuovo Cimento C, Vol. 9, Issue #2, 308 (1986)
2. J.N. Bahcall, AIP Conf. Proc. 126, 60 (1985).
3. J.K. Rowley, B.T. Cleveland, R. Davis, Jr., AIP Conf. Proc. 126, 1 (1985).
4. S.P. Mikheyev and A. Yu Smirnov, Sov. J. Nucl. Phys. 42, 913 (1985).
5. W. Hampel, AIP Conf. Proc. 126, 162 (1985).
6. I.R. Barabanov et al., AIP Conf. Proc. 126, 175 (1985).
7. S.P. Rosen and J.M. Gelb, submitted to Phys. Rev. D.
8. R.C. Allen, H.H. Chen, P.J. Doe, K. Roemheld, UCI-Neutrino #162, SNO-85-5, Nov. 1985, (Irvine, California).
9. L.P. Boivin, W.F. Davidson, R.S. Storey, D. Sinclair, E.D. Earle, Applied Optics 25, 877 (1986).
10. G. Aardsma, P. Jagam and J.J. Simpson, private communication.
11. J.W. McLaren, R.E. Sturgeon, S.N. Willie and W.F. Davidson, Physics in Canada 42, 38 (1986).

Low Energy Neutrino Detection with the Mont Blanc LSD Experiment

M. Aglietta[1], *G. Badino*[1], *G.F. Bologna*[1], *C. Castagnoli*[1], *A. Castellina*[1], *W. Fulgione*[1], *P. Galeotti*[1], *O. Saavedra*[1], *G.C. Trinchero*[1], *P. Vallania*[1], *S. Vernetto*[1], *V.L. Dadykin*[2], *F.F. Khalchukov*[2], *V.B. Korchagin*[2], *P.V. Korchagin*[2], *E.V. Korolkova*[2], *A.S. Malgin*[2], *V.G. Ryassny*[2], *O.G. Ryazhskaya*[2], *V.P. Talochkin*[2], *V.F. Yakushev*[2], and *G.T. Zatsepin*[2]

[1]Istituto di Cosmogeofisica del CNR, Torino, Italy
Istituto di Fisica Generale dell'Università, Torino, Italy
[2]Institute for Nuclear Research, Academy of Sciences, Moscow, USSR

A 90 tons Liquid Scintillation Detector (LSD) is fully running since October 1984 in the Mont Blanc Laboratory, at a depth of 5200 hg/cm^2 under ground. The detection of cosmic neutrinos of different origin is the main goal of this experiment. The detector is very well shielded against the local radioactivity and is extremely sensitive to detect low energy particles with a threshold of 5 MeV.

We discuss here the status of the experiment and present data on: 1) An experimental limit on the rate of neutrino bursts from galactic stellar collapses, 2) The possibility to detect solar neutrinos either from the ^{8}B decay or correlated with large solar flares, and 3) An experimental limit on the flux of $\overline{\nu}_e$ in the energy range 5 to 60 MeV.

1. Introduction

Since the main purpose of the LSD experiment is to search for neutrino bursts from galactic stellar collapses, the detector was designed to be sensitive to very low energy, and with a very good signature to electron antineutrino interactions. We payed attention to reduce any source of background, both from the local radioactivity of the rock and from the material of the detector itself. Background is the most stringent problem in detecting solar neutrinos.

Finally, from the data so far analyzed we give an experimental limit on the flux of $\overline{\nu}_e$ produced in the atmosphere by muon decay.

2. The LSD Experiment

The LSD experiment and its main characteristics are described in detail elsewhere (Ref. [1, 2] and references therein). Briefly, LSD was designed in order to be a very low background experiment, in which both products of the $\overline{\nu}_e + p \rightarrow n + e^+$ interaction can be detected: the e^+ at the energy threshold 7 MeV, and the neutron, in a delayed coincidence within 500 μs, though the gammas from (n, p) capture at the energy threshold 0.8 MeV.

The active mass (90 tons of liquid scintillator) is contained in 72 stainless steel counters (1.5 m^3 each), shielded with iron slabs (total mass 200 tons), surrounding each counter and the whole detector to reduce the local radioactivity background. The nature of the background is still under an extensive and systematic study, and some preliminary results have already been published [3].

Each counter is watched by 3 FEU-49B photomultipliers (15 cm diameter), whose pulses are fed into a 3-fold coincidence. The whole detector is triggered whenever a pulse with $E > 7\,\mathrm{MeV}$ is detected in any of the 72 counters. This trigger opens a gate, $500\,\mu s$ wide, within which pulses with $E \geq 0.8\,\mathrm{MeV}$ in any counter are recorded. This method allows to measure γ's from the (n,p) reaction with an efficiency of 70% [4].

Present Status of the Experiment

LSD is running since October 1984. The experiment was interrupted at the end of 1985 to improve the electronics and to install 2 cm of Fe at the roof of the installation, so reducing the trigger rate by a factor of 2. Since April 1986 the energy thresholds for 16 of the internal counters are reduced to 5 MeV.

The total trigger rate is $1.3\,\mathrm{min}^{-1}$, 3.5 muons/hour being detected. The trigger rate of internal counters (events/counter) is in average 7 times less than the average value. The counting rates at the threshold of 0.8 MeV are shown in Fig. 1.

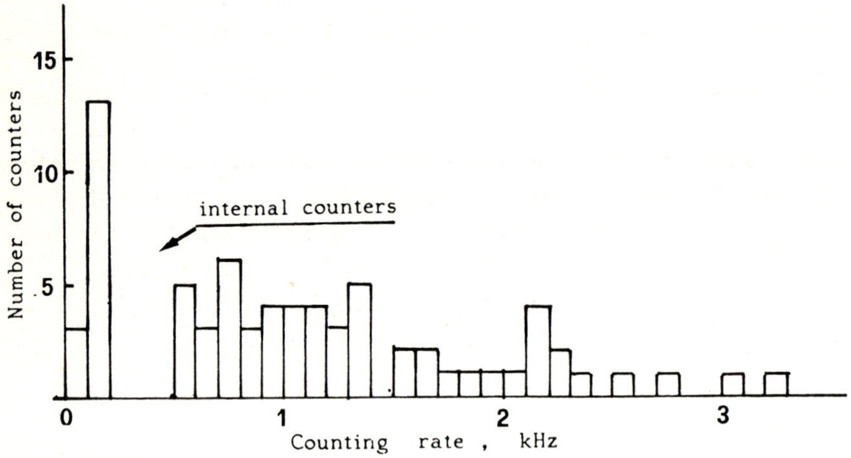

Fig. 1. Background conditions for the energy range ≥ 0.8 MeV

Regarding the delayed coincidence pulses in the $500\,\mu s$ gate, we detect in average 7% of time-correlated events. 14% of them appeared in the ground layer counters, while in the middle layer internal counters no such event was detected during 75 days. Taking the detection efficiency for prompt γ's from uranium decay of 10.5% and that for decay neutrons of 70% [4], we can derive the effective uranium contamination (averaged over the scintillator, PM materials, steel, rock etc.) of $\leq 1.8 \times 10^{-10}$ g/g at 90% c.l.

3. Search for $\bar{\nu}$-Bursts from Collapsing Stars

The analysis of the data is based on counting the number of pulses during a time interval δt in the energy range 7 to 60 MeV; events with more than 1 counter fired during 200 ns will be rejected. The rate of selected pulses is $6.2\times10^{-3}\,s^{-1}$. We look for a burst of such pulses during δt =1 to 200 s. For every burst we will count, then, the number of triggers which are accompanied by 0.8 MeV pulses in 500 μs gate and in the same counter. The probability to imitate this configuration of pulses by a background fluctuation can easily be calculated. We reject all configurations which are imitated with the frequency of $\geq$1/20 days. No high-multiplicity burst has been recorded, and from the present running time of LSD of $\sim$250 days the limit on the frequency of galactic stellar collapses is $fc \leq 3.7\,\text{year}^{-1}$ at 90% c.l.

The $\bar{\nu}$ burst candidates were compared with the Artyomovsk Scintillator Detector (ASD) data [5]. No coincident candidates have been obtained for the overlapping runs in 1985–1986.

4. Atmospheric Low-energy $\bar{\nu}_e$

The analysis of 75 days of LSD running time with the improved shielding does not show any signal in the 12 to 60 MeV energy range in the inner counters, whose total mass is 15 tons. Therefore the atmospheric $\bar{\nu}_e$ flux from $\mu - e$ decay is

$$j \leq 5\times10^4\,\text{cm}^{-2}\text{s}^{-1}\ , \quad \text{at} \quad 90\%\,\text{c.l.}$$

This limit does not contradict to the estimates of *Gaisser* and *Stanev* [6], which lead to $\leq$1 event in 15 tons of scintillator during 75 days.

5. Possibility to Detect Solar Neutrinos in LSD

^8B solar neutrinos can be detected by neutrino-electron scattering in the scintillator. The response of LSD to the solar neutrino flux was obtained by Monte Carlo simulation, taking the energy distribution of the ν_e computed by *Bahcall* et al. [7].

During 75 days 2 events have been recorded with $E \geq 12$ MeV in the 6 inner counters (scintillator mass of 7.2 tons) with the lowest background counting rates. Since we expect 0.4 such events, the very low background in the internal counters seems to allow the detection of solar neutrinos with LSD after further shielding to reduce the total counting rate by a factor of 5 to 10.

6. Possible Correlation Between Solar Activity and Solar Neutrinos

Bazilevskaya et al. [8] have suggested a correlation between the ^{37}Ar production rate and powerful solar flares. Two large solar proton flares were observed on the 17th and 20th July, 1985, and we, therefore, analyzed LSD data taken in the period from 16–20 July, 1985. The counters were selected to have a total counting rate of $\leq 5\times 10^{-4}\,s^{-1}$ in 3 energy intervals: ≥ 7, ≥ 10 and ≥ 13 MeV, and corresponding groups of 19, 32 and 64 counters were obtained. At July 15th the LSD started a new run after a 4-day interruption. During 16–17 July the counting rates of selected groups reasonably exceeded the average value [Neutrino 86]. The counter groups with different thresholds seem to behave in a practically independent way due to the steep decrease of the background with the energy. Assuming a statistical independence among the groups, the probability of the background fluctuation will be 3×10^{-4}, which corresponds to in average 0.004 such events in the observational interval. We think that this single event can hardly be interpreted in definite terms, but anyway shows the capability of LSD to search for solar neutrino bursts associated with powerful solar flares.

7. Conclusions

No candidate for a $\bar{\nu}_e$-burst from collapsing stars has been observed during 250 days of LSD lifetime. The data collected were used to obtain an upper limit on the flux of atmospheric neutrinos, and to examine the possibility to detect solar neutrinos and the correlation between their flux and the solar activity.

References

1. G. Badino et al.: Nuovo Cimento C**7**, 573 (1984)
2. G. Badino et al.: Neutrino '84, Proc. 11th Int. Conf. on Neutrino Phys. and Astrophys., Dortmund 1984, p. 556
3. V.L. Dadykin et al.: Nuovo Cimento, in press
4. M. Aglietta et al.: Proc. 19th Int. Cosmic Ray Conf. 1985, Vol. 8, 108
5. V.L. Dadykin et al.: Paper presented at Neutrino '86, to be published
6. T.K. Gaisser, T. Stanev: Proc. 19th Int. Cos. Ray. Conf., 1985, Vol. 8, 156
7. J.N. Bahcall et al.: Astroph. J. **184**, 1 (1974)
8. G.A. Bazilevskaya et al.: Sov. JETP Lett. **35**, 341 (1982)

The Solar Neutrino Problem as a Probe for Nuclear Astrophysics

H.J. Haubold[1] *and A.M. Mathai*[2]

[1]Akademie der Wissenschaften der DDR, Zentralinstitut für Astrophysik, DDR-1502 Potsdam-Babelsberg, GDR

[2]McGill University, Department of Mathematics and Statistics, Montreal, P.Q., Canada H3A 2K6

In a recently published paper KLAPDOR [1] has shown the central importance of methods for obtaining the shape of the beta strength function in nuclear physics and astrophysics. In addition to the beta strength function determined by the weak interaction there is the reaction rate leading the synthesis of elements in thermonuclear processes determined mainly by strong and electromagnetic interactions of nuclei. In the light of the solar neutrino problem we started the development of a new mathematical approach to the analytic representation of nuclear reaction rates, cp. HAUBOLD and JOHN [2]. The intention to come up with closed form representations of nuclear reaction rates is to take into account first physical principles of a reaction in a thermonuclear plasma as consistent as possible and to avoid mathematical approximations, cp. HAUBOLD and MATHAI [3]. The present status of our programme of the mathematical approach to the representation of nuclear reaction rates is as follows.

For reactions mainly involving n, p, and α with heavier nuclei in a nonrelativistic, nondegenerate plasma the reaction rate is defined by

$$r_{ij} = (1 - \frac{1}{2}\delta_{ij}) n_i n_j \langle\sigma v\rangle, \qquad (1)$$

where $\langle\sigma v\rangle$ denotes the reaction probability of particles i and j. The reaction probability can be calculated as an average over the *MAXWELL-BOLTZMANN distribution* for nonresonant and resonant cross sections, respectively. The strong energy dependence of the nuclear cross section is caused by the *GAMOW penetration factor* which is based on the solution of the *SCHROEDINGER equation* for the *COULOMB wave functions*. Representing the intrinsically nuclear parts of the probability for the occurrence of a nonresonant nuclear reaction by *SALPETER's astrophysical cross section factor* S(E) expressed in terms of a *MACLAURIN series expansion* up to the second power of the energy we have

$$\langle\sigma v\rangle = \left(\frac{8}{\pi\mu}\right)^{1/2} \sum_{\nu=0}^{2} \frac{1}{(kT)^{-\nu+1/2}} \frac{S^{(\nu)}(0)}{\nu!} \int_0^d dy\, e^{-ay}\, e^{-by^{\delta}}\, y^{\nu}\, e^{-z/(y+t)^{1/2}}, \qquad (2)$$

where $y = E/kT$, $z = 2\pi(\mu/2kT)^{1/2} Z_i Z_j e^2/\hbar$, and $t = Z_i Z_j e^2 \kappa$ denotes the electron screening parameter including the *DEBYE-HUECKEL length* κ. In the standard case of the reaction rate ($d = \infty$, $b = 0$, $t = 0$) the closed form representation can be found as

$$N_1(z;a,\nu) = a^{-(\nu+1)} \frac{1}{\pi^{1/2}} G^{3,0}_{0,3}\left(\left[\frac{z}{2}\right]^2 a \,\middle|\, 0, 1/2, 1+\nu\right), \qquad (3)$$

where $G^{m,n}_{p,q}(.)$ denotes *MEIJER's G-function*, cp. MATHAI and SAXENA [4]. If due to collective plasma effects a depletion of the *MAXWELL-BOLTZMANN distribution* must be considered in the reaction rate we have ($d = \infty$, $b = 1$, $t = 0$)

$$N_2(z;\delta,a,\nu) = \sum_{k=0}^{\infty} \frac{(-1)^k}{k!} a^{-(\nu+1+k\delta)} \frac{1}{\pi^{1/2}} G^{3,0}_{0,3}\left(\left[\frac{z}{2}\right]^2 a \,\middle|\, 0, 1/2, 1+\nu+k\delta\right). \qquad (4)$$

Considering dissipative collision processes in a thermonuclear plasma a cut off of the high energy tail of the *MAXWELL-BOLTZMANN distribution* could occur, thus we obtain (b = 0, t = 0)

$$N_3(z;d,a,\nu) = d^{(\nu+1)} \sum_{k=0}^{\infty} \frac{(-ad)^k}{k!} \frac{1}{\pi^{1/2}} G^{3,0}_{1,3}\left([\tfrac{z}{2}]^2 \tfrac{1}{d}\Big|^{2+\nu+k}_{1+\nu+k,0,1/2}\right). \quad (5)$$

In a dense plasma each nucleus attracts neighboring electrons and repels neighboring nuclei, thus forming a screening cloud of electrons in some sense characterized by the *DEBYE-HUECKEL length* κ which is a measure of the size of the ion cloud. We have ($d = \infty$, $b = 0$)

$$N_4(z;t;a,\nu) = t^{\nu+1} e^{at} \sum_{r=0}^{\nu} \binom{\nu}{r}(-1)^r\{N_1(zt;at,\nu-r) - N_3(zt;1,at,\nu-r)\}, \quad (6)$$

where $N_1(.)$ and $N_3(.)$ are given in (3) and (5), respectively.

For a nuclear reaction proceeding via a single resonance the reaction probability taking into account the *BREIT-WIGNER single level formula* is given by

$$\langle\sigma v\rangle = ([2\pi]^5/\mu)^{1/2} \frac{Z_iZ_je^2}{(kT)^{3/2}} \frac{R_0\omega\Gamma_{k1}D}{1+(\frac{1}{2}\Gamma_1)^2} \int_0^{\infty} dx\, \frac{e^{-ax}\, e^{-z/(ax)^{1/2}}}{(b-x)^2 + g^2}, \quad (7)$$

where $x = E(1+[\frac{1}{2}\Gamma_1]^2)$, $a = \{kT(1+[\frac{1}{2}\Gamma_1]^2)\}^{-1}$, $b = E_r - \frac{1}{4}\Gamma_0\Gamma_1$, and $g = \frac{1}{2}(\Gamma_0 + E_r\Gamma_1)$. The closed form representation of the resonance integral in (7) can be written

$$R_1(z,a,b,g) = \frac{1}{g^2 a} \sum_{k=0}^{\infty} \frac{(-1)^k}{(g^2)^k} \sum_{k_1=0}^{2k} \binom{2k}{k_1} \frac{(-1)^k}{a^{k_1}}\, b^{2k-k_1} \times \frac{1}{\pi^{1/2}} G^{3,0}_{0,3}\left([\tfrac{z}{2}]^2 \tfrac{1}{a}\Big|_{0,1/2,1+k_1}\right). \quad (8)$$

At least the results given in (1) to (8) should be the mathematical foundation of *FOWLER's reaction rate systematics.*

References

1 H.V. Klapdor: Fortschr. Phys. 33, 1(1985)

2 H.J. Haubold and R.W. John: "A New Approach to the Analytic Evaluation of Thermonuclear Reaction Rates", in Fundamental Problems in the Theory of Stellar Evolution, eds. D. Sugimoto, D.Q. Lamb, and D.N. Schramm (D. Reidel Publishing Company, Dordrecht 1981), p. 317

3 H.J. Haubold and A.M. Mathai: Fortschr. Phys. 33, 161(1985)

4 A.M. Mathai and R.K. Saxena: Generalized Hypergeometric Functions with Applications in Statistics and Physical Sciences, Lecture Notes in Mathematics, Vol. 348 (Springer-Verlag, Berlin - Heidelberg -- New York 1973)

Exchange Currents in the Neutrino-Deuteron Reaction and the Solar Neutrino Problem

S. Nozawa[1], *Y. Kohyama*[2], *and K. Kubodera*[2]

[1]Swiss Institute for Nucelar Research, CH-5234 Villigen, Switzerland
[2]Department of Physics, Sophia University, Tokyo, Japan

In spite of intensive studies on the solar neutrino flux over fifteen years, there still exists a significant discrepancy between the observed capture rate and the expected value from the standard solar model. Recently, there is augmenting interest in the direct observation of the solar neutrinos with large water Cerenkov counters, such as developed at KAMIOKANDE[1-4]. The interest is enhanced even more by the project[5] to construct a KAMIOKANDE-type detector which uses D_2O instead of H_2O. These combined experimental data are expected to provide us a more fruitful information on the properties of the solar neutrinos. Motivated by this situation, we have made a detailed evaluation of the cross section for the following two reactions[2]

$$\nu_e + d \rightarrow e^- + p + p \tag{1}$$

$$\bar{\nu}_e + d \rightarrow e^+ + n + n \tag{2}$$

The existing estimates[6] based on the impulse approximation (IA) are improved by including the exchange currents (EC). The ρ-exchange as well as π-exchange effects are considered. We shall show that the uncertainties due to nuclear physics are small enough to use these reactions in order to investigate the properties of solar neutrinos. The model dependence of the EC effects is examined by considering the two typical treatments: model A due to Guichon and Samour[7], and model B due to Bargholtz[8]. Tables 1 and 2 show the average cross sections calculated with and without EC. Here, the average cross sections are defined as

$$\sigma_i = \int_{|Q|}^{E_\nu^{max}} dE_\nu \phi(E_\nu) \int^{E_e^{max}} dE_e (d\sigma_i/dE_e) \Big/ \int^{E_\nu^{max}} dE_\nu \phi(E_\nu) \tag{3}$$

where i represents the reactions (1) and (2), and $\phi(E_\nu)$ is the differential solar neutrino flux coming from the ^8B decay. (For the reaction (2) the same flux function has been used.) One can see from the tables that the EC enhances the cross sections by about 2∼6%.

Table 1. Cross section for the reaction $\nu_e + d \rightarrow e^- + p + p$.

σ ($\times 10^{-42} cm^2$)	IA	IA+π	IA+π+ρ+FF
model A	1.10	1.22	1.16
model B	1.10	1.11	1.10

Table 2. Cross section for the reaction $\bar{\nu}_e + d \rightarrow e^+ + n + n$.

σ ($\times 10^{-42} cm^2$)	IA	IA+π	IA+π+ρ+FF
model A	0.420	0.466	0.444
model B	0.420	0.421	0.421

The relevance of the present results to the neutrino oscillation problem has been discussed by Koshiba et al.[1], whose argument may be sketched as follows. Let us suppose that one wants to check the possible neutrino oscillation of the following two types: (a) $\nu_e \rightarrow (1-x_a)\nu_e + x_a(\nu_\mu + \nu_\tau)$ (b) $\nu_e \rightarrow (1-x_b)\nu_e + x_b\bar{\nu}_e$, where x_a and x_b are the mixing parameters. Then, one measures the yields of the $e^\pm$ in the KAMIOKANDE-type counters, one with ordinary water and the other with heavy water, as the functions of E^{min}, the minimum detected energy of $e^\pm$. If one defines R^α by $R^\alpha = (Y^\alpha(D_2O) - Y^\alpha(H_2O))/Y^\alpha(H_2O)$, where α=a or b, and the yields $Y^\alpha(H_2O)$ and $Y^\alpha(D_2O)$ are assumed to be normalized to the same number of water molecules, R^α turns out to be sensitive to x_a and x_b. Fig. 1 shows R^α for some representative values of x_a and x_b; the shaded bands indicate the variations due to the different treatments of EC[2]. It is noteworthy that (i) the cases (a) and (b) give rather markedly different R's as functions of E_e^{min}, (ii) the results are very sensitive to the small admixture of x_b, and (iii) the ambiguities in the nuclear transition amplitudes are small enough to encourage a serious pursuit of experiments to measure R^α.

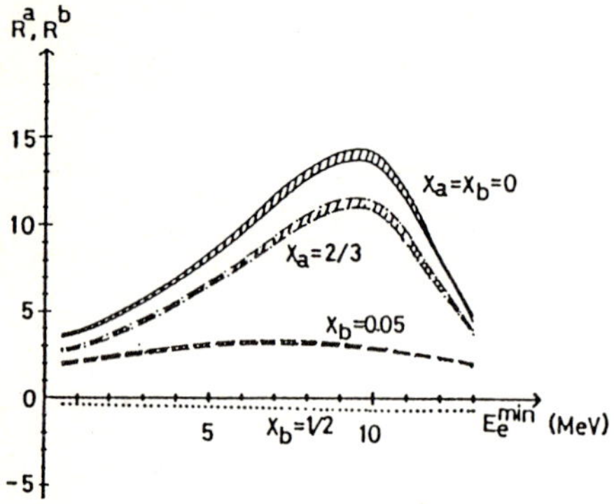

Fig.1 The values of R^α as the function of E_e^{min}.

We wish to express our sincere thanks to Prof. M. Koshiba and M. Nakahata for helpful discussions.

References:

[1] M. Koshiba, M. Nakahata, K. Kubodera and S. Nozawa, UT-ICEPP-85-04, 1985, University of Tokyo.

[2] S. Nozawa, Y. Kohyama, T. Kaneta and K. Kubodera, INS-Rep.-568, Feb. 1986, University of Tokyo.
[3] T. Kajita et al., J. Phys. Soc. Jpn. 54 (1982) 767.
[4] H. Cheng, Phys. Rev. Lett. 55 (1985) 1534.
[5] G. Ewan et al., Sudbury Neutrino Observatory, SNO-85-3, July 1985, Queen's University.
[6] S. Ellis and J. Bahcall, Nucl. Phys. A114 (1968) 636.
[7] P. Guichon and C. Samour, Nucl. Phys. A382 (1982) 461.
[8] C. Bargholtz, Phys. Lett. B81 (1979) 286.

3.3.3 Reactor Neutrino Oscillation Experiments

Neutrino Oscillation Experiments at Nuclear Power Reactors

V. Zacek

CERN, CH-1211 Geneva 23, Switzerland

1. Introduction

The question whether the neutrino possesses a finite rest mass remains one of the most important and challenging issues in todays physics. Unfortunately, there is little guidance from theory. Whereas in the minimal version of Grand Unified Theories – the SU(5) model – there is no room for neutrino mass generating processes, extended models beyond SU(5) easily allow neutrino masses in the range 10^{-6} eV to 10 eV. A finite neutrino mass together with violation of lepton number conservation offers the possibility of neutrino oscillations [1]. The underlying assumption for the existence of ν – oscillations is, that the neutrino ν_ℓ being created and detected by weak interaction processes are coherent superpositions of neutrino states ν_i with definite mass m_i, in analogy to the KM – mixing of hadronic charged currents:

$$\nu_\ell = \Sigma\, U_{\ell,i} \nu_i \quad \ell = e,\mu,\tau ... \quad i = 1,2,3... \tag{1}$$

where $U_{\ell i}$ are the mixing amplitudes. In the simplest picture, where only two neutrino species are involved, neutrino oscillations are characterized by two parameters: the mixing angle Θ, denoting the degree of admixture of the neutrino mass eigenstates to the weak eigenstates and the mass parameter $\Delta m^2 = |m_1{}^2 - m_2{}^2|$ defined by the difference of the squared corresponding mass eigenvalues. The probability for a weak neutrino type ν_ℓ ($\ell = e,\mu,\tau$) with energy E_ν to be still found in the state ν_ℓ at a distance L from the source can then be expressed as:

$$P(E_\nu, L, \Delta m^2, \Theta) = 1 - \sin^2 2\Theta \, \sin^2[1.27 \Delta m^2(eV^2)\, L(m)/E_\nu(MeV)] \tag{2}$$

The figure of merit according to which different experiments can be classified is the ratio $L(m)/E_\nu(MeV)$, since maximum sensitivity to the mass parameter is obtained for $\Delta m^2 \approx E_\nu(MeV)/L(m)$.

With typical distances in the order of L≈40m and with neutrino energies between 2MeV and 8MeV, reactor experiments especially investigate the domain of small massparameters of about $0.01 eV^2$ to $1 eV^2$.

In a nuclear reactor most of the produced fission products end up in an unstable neutron rich configuration and further undergo β^- –decays. The great number of these decays, each accompanied by an electron antineutrino thus make a nuclear reactor an intense neutrino source providing a flux of 1.9×10^{14} $\bar{\nu}_e$ sec^{-1} $GWth^{-1}$. Oscillation experiments at reactors search for an intensity reduction of the primary $\bar{\nu}_e$ –flux. The sensitivity to the mixing angle is limited by statistics, the knowledge of the primary $\bar{\nu}_e$ –flux and uncertainties in the neutrino detection process (detector efficiency, detection cross section).

2. The Experiment at Gösgen

In an oscillation experiment by a CALTECH–SIN–TU Munich collaboration at Gösgen (Switzerland), flux and energy spectrum of the $\bar{\nu}_e$ were monitored at three distances from the core of a 2.8GWth power reactor (37.9m, 45.9m, 64.7m) [2] and about 10^4 $\bar{\nu}_e$ were collected in each position. The detector system was based on the reaction $\bar{\nu}_e + p \rightarrow e^+ + n$ and an alternating array of scintillation counters, filled with a proton rich liquid scintillator, and ^3He–filled wire chambers, provided the neutrino target and served for positron and neutron detection. As signature for a good neutrino event a time and position correlation of the detected neutron and positron was required. From the measured positron energy spectrum the energy spectrum of the incident $\bar{\nu}_e$ can directly be inferred.

For a commercial power reactor, like the one at Gösgen, the reactor working period is about 10 months followed by a scheduled shut–down of about one month to allow the replacement of one third of the fuel elements. At the beginning of a cycle the number of fissions divides up in 69% ^{235}U, 21% ^{239}Pu, 7% ^{238}U and 3% ^{241}Pu. During operation the fissionable isotopes ^{239}Pu and ^{241}Pu are bred continuously from ^{238}U. The neutrino spectra following the fission of the plutonium and the uranium isotopes differ among each other ([3], [4], [5]), and therefore the time dependent fission contributions have to be registered over the entire measuring period. Since the measurements in the three positions covered slightly different time spans and burnup conditions of the reactor fuel cycle, differences between the corresponding antineutrino spectra result. They did not exceed 5% and were taken into account in the data analysis.

By comparing the positron spectra at the three distances, the data were analyzed without relying upon the precise knowledge of the neutrino source spectrum, detection efficiency and detection cross section. The result is compatible with the absence of neutrino oscillations. Starting from their three data sets the Gösgen group also calculated "backwards" to deduce the reactor source spectrum, underlying their measurements. This spectrum was in good agreement with the reactor neutrino spectrum predicted in an independent way, where essentially β–spectroscopic data from the fissioning isotopes were used [3]. Incorporating this independently predicted reactor source spectrum into the analysis allowed to exclude a large area in the oscillation parameter plane. The resulting limits (90% c.l.) on the oscillation parameters are $\Delta m^2 < 0.019 eV^2$ for maximum mixing and $\sin^2 2\Theta < 0.18$ in the limit of large mass parameters ($\Delta m^2 > 5 eV^2$).

3. The Experiment at Bugey

Essentially the same detector concept was used by an Annecy – Grenoble collaboration at the 2.8GWth power reactor in Bugey (France) [6]. The energy spectra of the $\bar{\nu}_e$ were measured at two distances (13.6m, 18.3m) from the core, with 4×10^4 and 2.3×10^4 events recorded in the close and far position respectively. By taking the ratio of the observed integrated positron yields between 1.5MeV and 6.5MeV a value $R = 1.102 \pm 0.01$(stat.)± 0.028(syst.) was obtained, after correcting for solid angle and different fuel compositions in both positions. This observation was interpreted as indication for neutrino oscillations above the 3σ level. A best fit to the data was obtained for $\Delta m^2 = 0.2 eV^2$ and $\sin^2 2\Theta = 0.25$.

By comparing the results of the two reator experiments performed under similar conditions. it turns out that the two experiments are in serious conflict if one asks for a common reactor neutrino source spectrum that simultaneously is compatible with all the individual measurements at Gösgen and Bugey: Inferring the reference ($L = 0$m) spectrum from the Bugey 13.6m measurement under the assumption of neutrino oscillations, described by the parameters corresponding to the Bugey best fit solution, the result differs at 4MeV by 20% from neutrino yield predictions. A more detailed discussion is found in ref.[2].

4. The Experiment at Rovno

Two other experiments use as neutrino detector a Gd loaded liquid scintillator, which simultaneously provides the proton target and also serves as positron and neutron detection medium. Selection criterium for a valid neutrino event is a delayed coincidence between a positron and a gamma ray following neutron capture in Gd.

A group from Kurchatov Institute (Moscow) published results obtained from a sample of 1.5×10^4 $\bar{\nu}_e$ detected with a 240 ℓ detector in 18m distance from the core of the 1.38 GWth reactor in Rovno (USSR) [7]. Limits on the oscillation parameters were derived by comparing the measured neutrino rate with prediction. A ratio of measured over predicted yield of $R = 0.95 \pm 0.11$ was obtained. Though the outcome of the analysis is not very restrictive in the $\Delta m^2 - \sin^2 2\Theta$ plane the experiment does not show evidence for neutrino oscillations (fig.1). In the meantime data taking continued in a second 25m location and a ratio of the integrated rates found in the far position to those in the close position of $R = 0.986 \pm 0.039$(stat.)± 0.029(syst.) is given [8].

5. The Experiment at Savannah – River

Preliminary results were reported from a UCLA – Irvin group which measured the neutrino flux with a detector also based on a Gd loaded scintillator at 18.5m

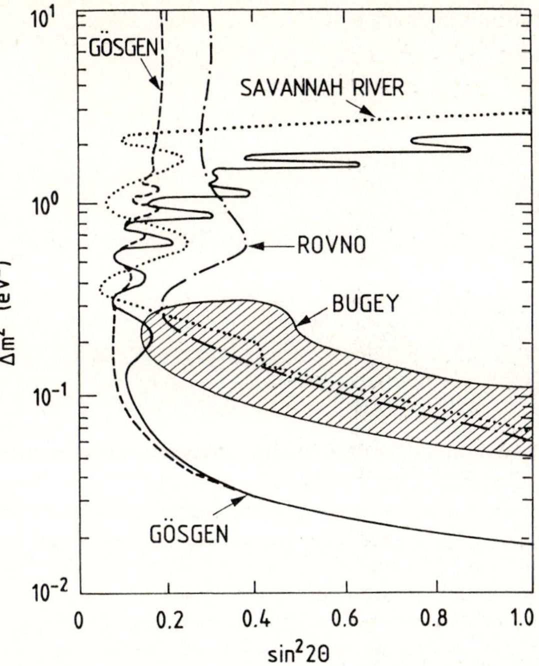

Figure 1: Results from $\bar{\nu}_e \rightarrow \bar{\nu}_x$ disappearance studies at reactors. Solid curve: 90% c.l. derived from the three Gösgen experiments, without relying on external data. Dashed line: 90% c.l. obtained from a comparison of the expected reactor $\bar{\nu}_e$ spectrum and the Gösgen data. Dot – dashed line: Rovno single position (18m) analysis. Dotted line: 90% c.l. from the two position analysis of the Savannah River experiment. In each case the areas to the right of the curves are excluded. Shaded area: 90% c.l. of oscillation parameters allowed by the Bugey data.

and 23.8m from the core of the Savannah River 2.2GWth reactor [9]. Compared to the rest of the reactor experiments, performed at commercial power reactors here the core composition is particularily simple, since it consists by more than 85% of ^{235}U. Around 3×10^4 $\bar{\nu}_e$ were collected at both detector sites and a near to far position integral ratio of $\bar{R} = 1.04 \pm 0.02$(statistical error only) was found. Although a final evaluation of the systematic error contribution does not exist for the moment the deviation from 1 is considered as not significant. Limits on the oscillation parameters (90% c.l.) obtained in a comparison of the two spectra are shown in fig.1.

6. Conclusions

In summarizing, no convincing evidence for neutrino oscillations has been found in the present round of reactor experiments set up between 13m and 66m distance to the respective reactor cores (table 1). The only claim of an oscillation effect (Bugey) is strongly contradicted by the Gösgen experiment and partially invalidated by the experiments at Rovno and Savannah River. As an upper limit for the mixing angle a value of $\sin^2 2\Theta < 0.18$ (90% c.l.) can be stated for $\Delta m^2 > 5eV^2$ and the lowest bound on the mass parameter is $\Delta m^2 < 0.019eV^2$ (90% c.l.) assuming full mixing.

Table 1: List of reactor neutrino oscillation experiments

experiment	reactor power [GWth]	distances [m]	oscillation effect
Gösgen (CH)	2.8	37.9, 45.9, 64.7	NO
Bugey (F)	2.8	13.6, 18.3	YES
Savannah River (USA)	2.3	18.5, 23.8	NO
Rovno (USSR)	1.38	18., 25.	NO

Since the existing experiments are already pushed close to the practical limits a further improvement on the oscillation parameter limits cannot be expected in the near future. However a new generation of larger and more efficient neutrino counters should be capable of lowering the present mass parameter limits by still one order of magnitude.

References:

1. S.M.Bilenky et al.: Phys.Rep. C41, 225 (1978)
2. V.Zacek et al.: Phys.Lett. 164B, 193 (1985)
3. F.v.Feilitzsch et al.: Phys.Lett. 118B, 162 (1982)
4. P.Vogel et al.: Phys.Rev. C24, 1543 81981)
5. H.V.Klapdor et al.: Phys.Rev.Lett. 48, 127 (1982)
6. J.F.Cavaignac et al.: Phys.Lett. 148B, 387 (1984)
7. A.Afonin et al.: JETP Lett. 42, 285 (1985)
8. A.Pomansky: Proc. of the ν'86 Conference, Sendai, to be published
9. H.Sobel: Proc of the VIth Moriond Workshop on "Massive Neutrinos in Particle and Astrophysics", edition Frontieres, (Feb.1986)

The Bugey Neutrino Oscillation Experiment [1] Status and a New Neutrino Detector

D.-H. Koang

Institut des Sciences Nucléaires, Grenoble, France

Introduction

In a previous experiment at Bugey, positron energy spectra from the $\overline{\nu}_e p \rightarrow ne^+$ reaction have been measured at two distances, 13.6 and 18.3 m from the reactor core. A significant difference in the counting rates of $\overline{\nu}_e$ events has been observed (Fig. 1) and has been explained in a simple two neutrino oscillation scheme with the solution $\sin^2 2\theta = 0.25$ and $\Delta m^2 = 0.2\,\text{eV}^2$ [1]. This solution is in fair agreement with recent results from Savannah river plant [2] where high statistics data have been collected at distances of 18.2 and 23.8 m. However it is not supported by the Gösgen experiment [3] in particular when limits on the oscillation parameters were deduced on the basis of an assumed known reactor neutrino spectrum. In the Gösgen experiment, data were collected at three different distances 37.9, 45.9 and 64.7 m but with a relatively poorer statistics. Results from the experiment at the Rovno reactor have also been communicated [4] but did not lead to any further clarification.

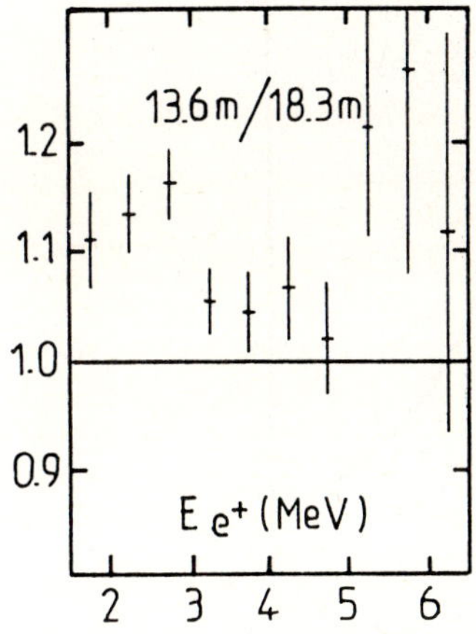

Fig. 1. Ratio between event yields at 13.6 m and 18.3 m as a function of positron energy

[1] The Bugey experiment is undertaken by a collaboration of LAPP (Annecy), ISN (Grenoble), CPPM (Marseille), CdF (Paris) et DPHPE (Saclay

Status of the Experiment

Background studies and complementary measurements have been undertaken at Bugey. Improved shielding conditions and installation of external cosmic counters had lead to a reduction of the background rate by a factor 4. The main cause of the background suppression comes from the removal of a layer of internal lead [1] in which neutral components of cosmic rays did induce correlated neutron-gamma events. A new measurement of the energy spectrum at 18.3 m where we observed disappearence of $\overline{\nu}_e$ events has given results compatible with previous one.

Optimal Conditions for a New Experiment

The main constraint for the results of the previous Bugey experiment comes from the integrated counting difference between the two positions. Large uncertainties on the absolute response of the detector and on the initial neutrino spectra preclude any conclusion to be drawn from single position spectra (Fig. 2).

In the new experiment now undertaken by a larger collaboration we intend to measure precisely the *relative shape* of the energy spectra at two positions so as to be able to unambiguously prove or eventually disprove the observed oscillation effect from the shape distorsion.

Monte-Carlo simulations have been carried out to define the most favorable conditions. For a $\Delta m^2 = 0.2\,\mathrm{eV}^2$, the optimal distance is found to be around 35 m if the first position is keep at 13.6 m (Fig. 3). These studies also show that even with a reduced background, an increase of the counting rate by an order

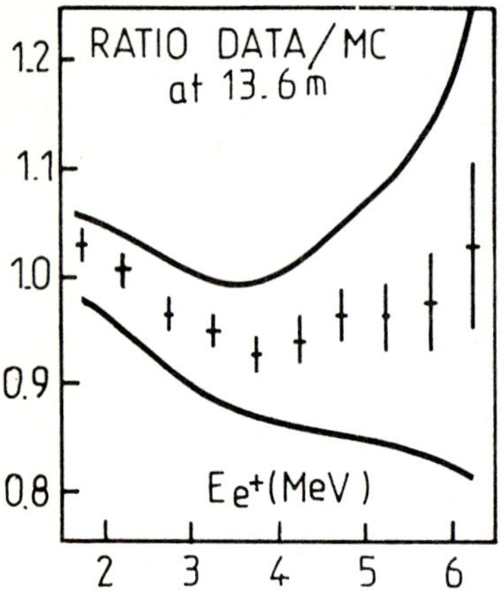

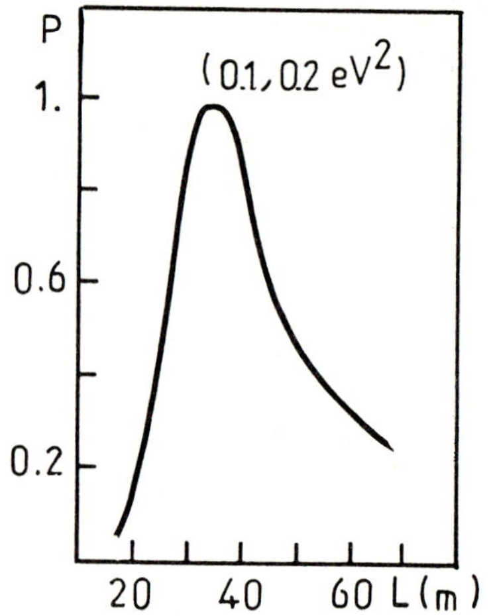

Fig. 2. Solid lines show changes in the ratio with different predicted ^{241}Pu and ^{238}U spectra. A systematic shift in energy calibration of 100 keV induces changes of about the same amount

Fig. 3. Probability of excluding no-oscillation by shape analysis as a function of distance of the second position

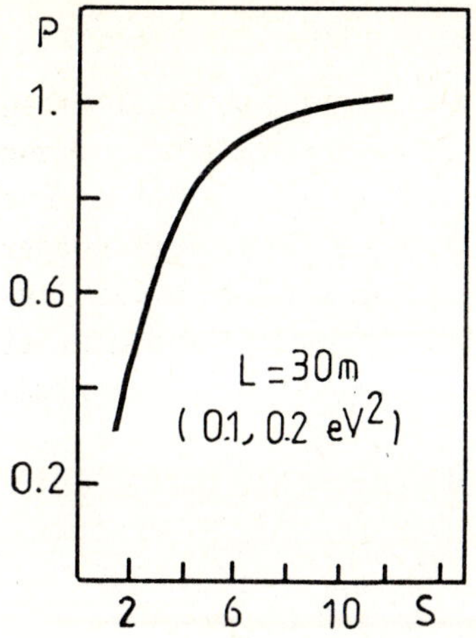

Fig. 4. Probability of excluding no-oscillation by shape analysis as a function of the increase in detector sensitivity

of magnitude is necessary if one wants to exclude the no oscillation case at 95% confidence level (Fig. 4), for one year of data taking.

A New $\overline{\nu}_e$ Detector of High Efficiency

In the detector that we have developed for the ILL and Bugey experiments, the detection functions of the positron and the neutron are independent. The ^{3}He wire counters have small sensitivity to γ rays and provide very clean neutron signature. However the mean efficiency is relatively low (25%). This is due to neutron absorption by hydrogen nuclei of the liquid scintillator and also due to neutron escape at the borders of the detector.

The new detector is homogeneous and is based on liquid scintillator loaded with ^{6}Li. The thermal neutron is captured through the reaction $^6\mathrm{Li} + n \to \alpha + t + 4.8\,\mathrm{MeV}$ with a cross section $\sigma = 940$ barns. A concentration of ^{6}Li of few tenth of percent is enough to reduce to a negligeable portion the capture by hydrogen nuclei. This detector has the following advantages:

- the signature of a thermal neutron is given by the couple of $\alpha - t$ particles with defined energy
- the small ranges of the $\alpha - t$ particles ($\sim$100 μm) in the liquid scintillator allows more precise location of the interacting point
- escapes at the periphery of the detector are reduced
- the mean duration for neutron capture is shorter (30 μs instead for 120 μs) than in the old detector
- the expected efficiency is about 75%.

A prototype of an elementary cell of 6 ℓ is now working. The identification of $\alpha - t$ particles is performed using a pulse shape discrimination method.

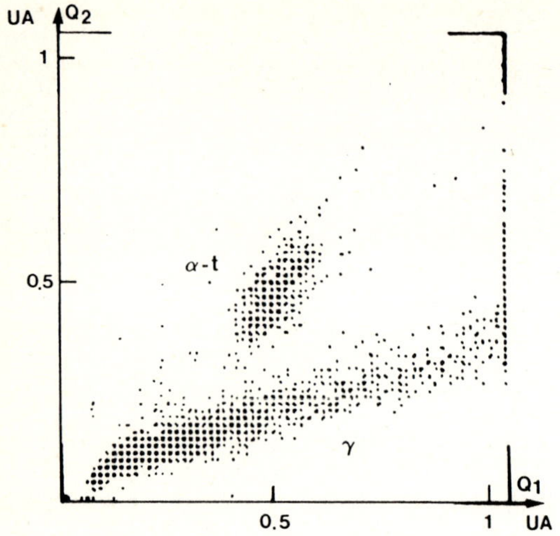

Fig. 5. Identification of thermal neutron capture events in a ^{6}Li loaded liquid scintillator. *Q1* is proportional to the energy, *Q2* is a discrimination parameter

In Fig. 5, one can distinguish clearly the neutron capture events with defined energy well separated from γ-events.

A module of 100 cells is expected to have a counting rate five times higher than the old detector at the same distance. One module will be installed at 13.6 m and two such modules at 35 m with the possibility of mutual permutation.

The construction of the first module is scheduled for the end of this year.

References

1. J.F. Cavaignac et al.: Phys. Lett. **148**B, 387 (1984)
2. V. Zacek et al.: Phys. Lett. **164**B, 193 (1985)
3. N. Baumann et al.: Neutrino 86 Sendai (Japan)
4. A.I. Afonin et al.: Neutrino 86 Sendai (Japan)

Reactor Core Antineutrino Spectra

K. Schreckenbach[1], *A.A. Hahn*[2], *W. Gelletly*[3], *F. von Feilitzsch*[4], *G. Colvin*[1], *and B. Krusche*[1]

[1]Institute Laue-Langevin, F-38042 Grenoble, France
[2]University of California, Irvine, CA 92717, USA
[3]University of Manchester, Manchester M13 9PL, Great Britain
[4]Physik Department, Technische Universität München, D-8046 Garching, F. R. Germany

The accurate knowledge of the antineutrino spectrum emitted from a nuclear reactor is of major importance for a variety of experiments in low energy neutrino physics. Fundamental questions in charged and neutral weak current reactions can be investigated under the favourable condition of a high source strength (see discussions in [1,2]). In fact a nuclear reactor emits 2 x 10^{17} $\bar{\nu}_e$ per megawatt of thermal power, with energies up to $\sim$ 10 MeV, steming from the beta-decaying fission products in the core.

Recent neutrino oscillation experiments with reactor $\bar{\nu}_e$ have achieved a high statistical accuracy [3,4,5]. These measurements were carried out at different distances from the reactor core searching for possible $\bar{\nu}_e$ flux variations. The knowledge of the $\bar{\nu}_e$ source spectrum has a strong impact on the interpretation of these $\bar{\nu}_e$ detector data. Short neutrino oscillations would show up in a $\bar{\nu}_e$ flux deficit independent of the distance from the core. For a longer oscillation length the inclusion of the $\bar{\nu}_e$ source spectrum improves significantly the sensitivity and the confidence level for oscillation parameters [3,4]. Part of the controversy between the neutrino group at the Bugey reactor, claiming evidence for neutrino oscillations [3], and the Gösgen group, excluding such oscillation parameters [4] is still based on the confidence in the neutrino source spectrum used in the evaluation. In the following we would like to discuss how this spectrum was deduced and present new data on the $\bar{\nu}_e$ spectrum from ^{241}Pu fission.

Having the origin of reactor $\bar{\nu}_e$ in mind the following procedure allows the experimental determination of the source spectrum:

- measurement of the cumulated beta spectrum of fission products from neutron induced fission for each reactor fuel isotope

- conversion of the beta spectra into the correlated $\bar{\nu}_e$ spectra.

With the knowledge of the number of fissions per time unit for each fissile isotope in the core, the total composite $\bar{\nu}_e$ source spectrum can be determined.

In power reactors the dominant contributions to the $\bar{\nu}_e$ spectrum stem from the thermal neutron induced fission of ^{235}U and ^{239}Pu. Contributions from fission of ^{238}U and ^{241}Pu are below 10% of the total fission rate, but ^{241}Pu is a strongly varying component as a consequence of breeding.

We have measured the beta spectra of the isotopes ^{235}U [6,8], ^{239}Pu [7] and ^{241}Pu, which are fissile by thermal neutrons. These experiments were performed at the magnetic beta-spectrometer BILL of the ILL High Flux Reactor in Grenoble. The in-pile target arrangement is shown in Fig. 1. The target represents a small reactor which is transparent to electrons but will contain the fission products. The amount of target material is limited by self-heating in the high neutron flux of 3 x 10^{14} ncm^{-2} s^{-1} (at the full reactor power of 57 MW). Typically 1 mg of fissile material in an area of 2 x 6 cm^2 is used. The target is exposed to a constant neutron flux and the emerging electrons are analysed in the spectrometer. The elec-

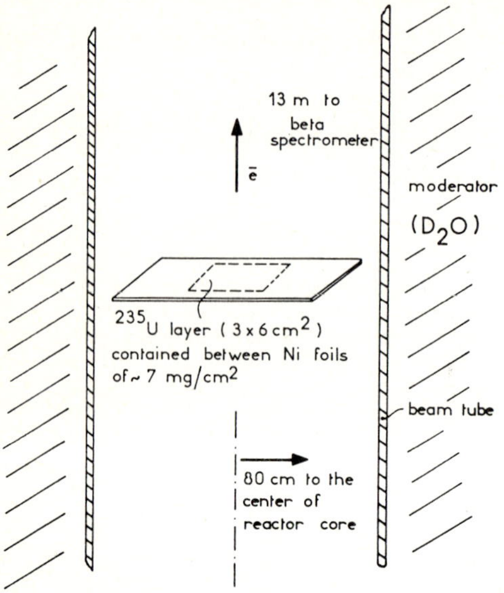

Fig.1 - Schematic view of the target site at the BILL spectrometer

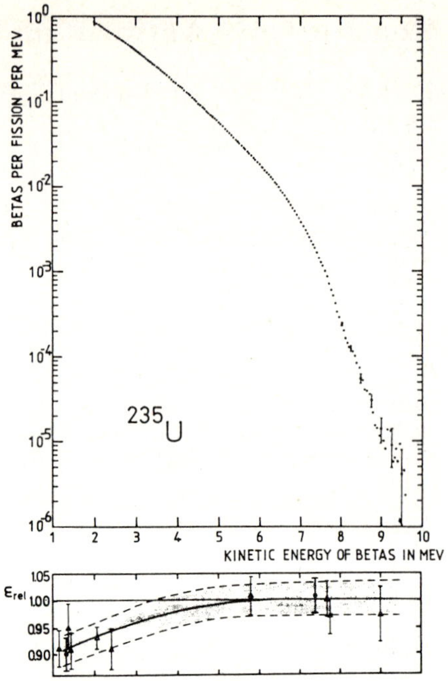

Fig.2 - Experimental beta spectrum of ^{235}U fission products. Below the spectrometer response function is shown [8]

trons are finally detected by multi-wire proportional counters. In our more recent studies [8] a 32-wire detector with a rear mounted plastic scintillator was used.

The background from the beam tube is determined from a target without a fissile material, but of the same geometry and mass. For the most important case of ^{235}U a measurement was performed at 4 MW reactor power with 18 mg target material. Thus the signal to backgrund ratio could be improved by nearly an order of magnitude to 6,2.5, 0.5 and 0.12 at 3,6,8 and 9 MeV, respectively [8]. The result was consistent with our earlier study [6]* and give us confidence as to the correctness of the background determination.

The absolute intensity calibration per fission of the beta spectrum was based on the proportionality of the reaction rate in the target to $\sigma_n \cdot \phi_n \cdot m$, where σ_n denote the relevant thermal neutron capture cross section, ϕ_n the neutron flux and m the mass of the target. A comparative measurement with target isotopes of known (n,e^-) cross sections provided a calibration over a wide range of beta energies. Figure 2 shows the result on ^{235}U with such a measured intensity calibration.

First results from our recent measurement on ^{241}Pu are given in Table 1. The measurement was performed at full reactor power with a target mass of 1.4 mg ^{241}Pu. The intensity calibration was performed similarly to that shown in Fig. 2. The exposure time to neutrons was more than 36 hours for the given values. The absolute rates are preliminarly normalised at 3 MeV to the theoretical values of ref. [1] and [9], as the target masses in our measurements have still to be

*For comparison, the values in ref. [6] have to be increased by 3% since the non 1/v neutron capture cross section of ^{235}U was not taken into account.

Table 1: Experimental beta spectrum from ^{241}Pu fission products and deduced $\bar{\nu}_e$ spectrum. The exposure time to a constant thermal neutron flux was 36 hours.

E_β (MeV)	N_β a)	ΔN_β b) (%)	$N_{\bar{\nu}}$ c)
2.0	9.3-1	1.5	1.30
2.5	6.3		9.0-1
3.0	4.15		6.4
3.5	2.59		4.4
4.0	1.52		2.75
4.5	8.6-2		1.65
5.0	4.85	2.5	9.2-2
5.5	2.56		5.3
6.0	1.27	4.0	2.85
6.5	6.1-3		1.50-3
7.0	2.41	7	7.0
7.5	8.5-4	10	2.55-4
8.0	2.32	15	8.9
8.5	5.9	25	2.2
9.0	9.3-6	70	5.0-5

a) Normalised to 0.415 betas per fission and MeV at E_β = 3 MeV.

b) Includes only statistical error and the variations between the different runs. The uncertainty in the response function of the spectrometer (∿ 3% uncertainty in the ratio 2 MeV to 8 MeV) is not included.

c) Deduced from N_β. Conversion error ∿ 4% below 7.5 MeV. Total uncertainty in the shape is $\sim(\Delta N_\beta^2 + 25)^{1/2}$ [%].

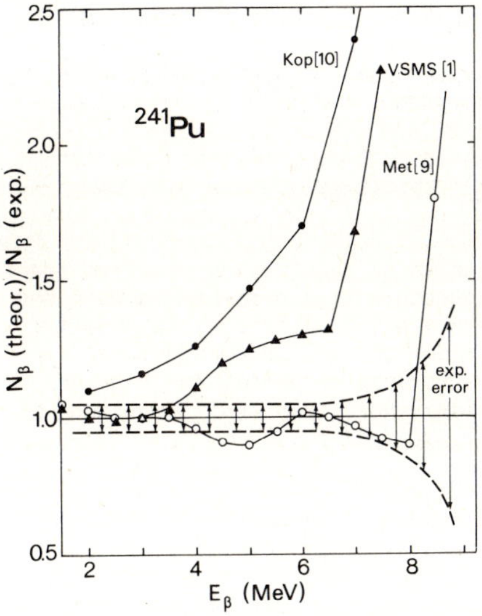

Fig.3 - Ratio between different calculated values and the present experiment for the beta spectrum from ^{241}Pu fission products. The experimental data are normalized to 0.415 betas per fission per MeV at E_β = 3 MeV

determined by mass spectroscopy. In Fig. 3 our results are compared with various theoretical predictions. The best agreement is observed with the theoretical approach of METZINGER [9], at least below 8 MeV. As in the cases ^{239}Pu and ^{235}U all theories overestimate the rates above that energy. In particular our ^{235}U data [8] are statistically very accurate even at these high energies.

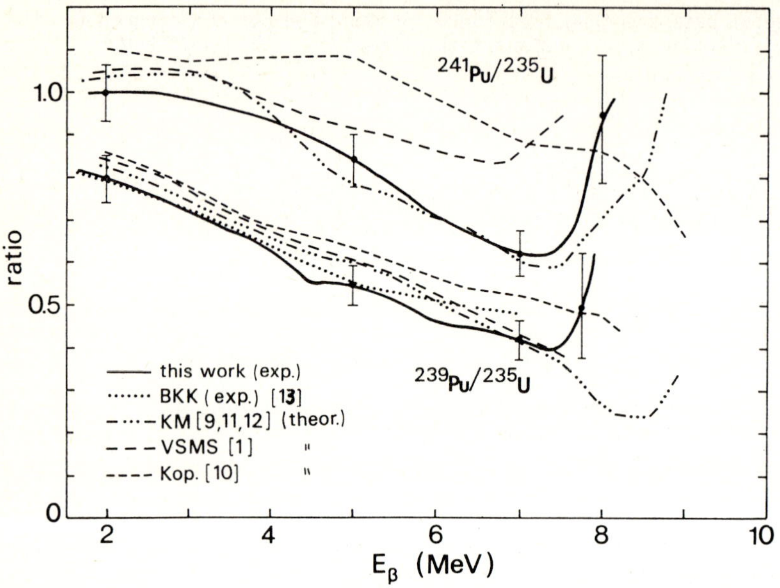

Fig.4 - Ratios of beta spectra from fission products. The experimental ratios of [13] are normalized at 3 MeV to our result.

In Fig. 4 the ratio of the beta spectra $^{239}Pu/^{235}U$ and $^{241}Pu/^{235}U$ are given. BOROVOI et al. [13] performed a comparative measurement of the shape of the beta spectrum of ^{239}Pu and ^{235}U. The ratio $^{239}Pu/^{235}U$ agrees well with our data (see Fig. 4), but the individual shapes are significantly harder (30% at 7 MeV compared to 3 MeV) than our results. A similar trend can be observed for other measurements which used a ΔE-E plastic scintillator telescope [14]. Only the much earlier measurement of CARTER et al. [15] on ^{235}U agrees well with our data. In that study a proportional counter was used as ΔE detector in front of a plastic scintillator to suppress the γ-ray sensitivity of the system.

The conversion of the measured beta spectrum into the correlated $\bar{\nu}_e$ spectrum was discussed in our earlier publications [6,7,8] in some detail. We have compared two methods.

In the first method an end-point distribution $\{a_i, E_0^{(i)}\}$ is extracted from the experimental beta spectrum, where a_i denote the strength of a beta-branch with end point energy $E_0^{(i)}$. Only allowed beta branches were assumed. An energy dependent mean proton number $\bar{Z}(E_0^{(i)})$ was used in the Fermi function. A Z-independent radiative correction term [16] and a correction due to weak magnetism and high order Coulomb terms [17] were also included. The individual beta branches i were then transformed into the corresponding $\bar{\nu}_e$ replacing E_β by $E_0^{(i)}-E_\nu$.

In a second method theoretical spectra which agree reasonably well with our beta spectrum are used for the conversion. The differences between calculated and measured beta spectra can be converted into the $\bar{\nu}_e$ spectra by the relation for relativistic cases $\Delta N_\nu(E_\nu) \approx \Delta N_\beta (E_{kin}+m_e c^2)$. This method is therefore equivalent to modifying the theory until a perfect agreement with the measured beta spectrum is achieved.

The resulting $\bar{\nu}_e$ spectra from the two methods agreed within a few percent [8]. The $\bar{\nu}_e$ spectrum from ^{241}Pu fission products was deduced in this way and is given in the last column of Table 1. The total uncertainty in shape is estimated as 5%, apart from the statistical error at high energies.

To conclude we have experimentally determined the beta and neutrino spectra of ^{235}U, ^{239}Pu and ^{241}Pu fission products. Apart from the ^{238}U component (fast neutron fission), the $\bar{\nu}_e$ spectrum of a reactor can now be determined with an absolute precision of typically 5%. The uncertainty in the conversion methods remains one of the major limitation.

The authors acknowledge the valuable help of the ILL reactor staff and the technical assistance of G. Blanc. Stimulating discussions with V. Zacek, R.L. Mössbauer, J.G. Cavaignac, P. Vogel, H.V. Klapdor and J. Metzinger are much appreciated.

References

1. P. Vogel, G.K. Schenter, F.M. Mann and R.E. Schenter: Phys. Rev. C24, 1543 (1981)
2. F.T. Avignone III and Z.D. Greenwood: Phys. Rev. C22, 594 (1980)
3. J.F. Cavaignac et al.: Phys. Lett. 148B, 387 (1984)
4. V. Zacek et al.: Phys. Lett. 164B, 193 (1985)
5. H.W. Sobel et al.: 6th Moriond Workshop on Massive Neutrinos (Tignes 1986, France)
6. K. Schreckenbach et al.: Phys. Lett. 99B, 251 (1981)
7. F. von Feilitzsch et al.: Phys. Lett. 118B, 162 (1982)
   K. Schreckenbach et al.: J. de Phys. 45, C3-135 (1984)
   K. Schreckenbach et al.: 4th Moriond Workshop on Massive Neutrinos (La Plagne 1984)
   ed. J. Tran Thanh Van (Editions Frontieres, Dreux) p. 125
8. K. Schreckenbach et al.: Phys. Lett. 160B, 325 (1985)
9. J. Metzinger: Thesis 1984, University of Heidelberg, W.Germany
10. V.I. Kopeikin: Sov. J. Nucl. Phys. 32, 780 (1980)
11. H.V. Klapdor and J. Metzinger: Phys. Lett. 112B, 22 (1982)
12. H.V. Klapdor and J. Metzinger: Phys. Rev. Lett. 48, 127 (1982)
13. A.A. Borovoi, Yu. Klimov and V.I. Kopeikin: Sov. J. Nucl. Phys. 37, 801 (1983)
14. N. Tsoulfanidis et al.: Nucl. Sci. Eng. 43, 42 (1971)
    J.W. Kutscher and W.E. Wyman: Nucl. Sci. Eng. 26, 435 (1966)
    U. Keyser: Z. Phys. A322, 529 (1985)
15. R.E. Carter et al.: Phys. Rev. 113, 280 (1959)
16. A. Sirlin: Phys. Rev. 164, 1767 (1969)
17. P. Vogel: Phys. Rev. D29, 1918 (1984)

Absolute Measurement of the Sum Beta-Spectra of all Fission Products from $^{235}U(n_{th},f)$ and $^{239}Pu(n_{th},f)$*

U. Keyser and F. Münnich

Institut für Metallphysik und Nukleare Festkörperphysik,
Technische Universität Braunschweig, Mendelssohnstr. 3,
D-3300 Braunschweig, F. R. Germany

The aim of this experiment was to determine the integral beta-spectra of $^{235}U(n_{th},f)$ and $^{239}Pu(n_{th},f)$ with an accuracy as high as possible, because these spectra are of importance for experiments performed at power reactors and related to weak interactions processes [1].

The sum beta-spectra of ^{235}U and ^{239}Pu fission products have been measured with a plastic scintillator telescope at the external guide tube H22 for thermal neutrons ($\Phi_n = 3{,}2\cdot10^8\ cm^{-2}s^{-1}$) at the high flux reactor of the ILL in Genoble. The highly enriched targets of various mass depletion from 150 $\mu g\cdot cm^{-2}$ up to 800 $\mu g\cdot cm^{-2}$ were placed in a fission chamber at a distance of approximately 110 m from the reactor core to reduce γ-background.

The accumulation of the data was started after a constant β-counting rate had been reached in order to obtain the near-equilibrium beta-spectra for $^{235}U(n_{th},f)$ and $^{239}Pu(n_{th},f)$. The background measurements have been performed under comparable conditions without any change of geometry. The absolute determination of the beta counting rate is easily obtained since by using a surface barrier detector simultaneously, only the solid angles of the fission chamber and of the scintillator telescope have to be known. But in order to obtain the absolute shapes of the beta-spectra, the response matrix of the detector must be known too. This matrix was derived from the measurement and parametrization of the response function determined from electron conversion lines up to 9 MeV at the double focussing spectrometer BILL of the ILL [2]. The precision and accuracy of the matrix inversion necessary for the deconvolution of the sum beta-spectra from ^{235}U and ^{239}Pu has been proved for the spectrum of the conversion lines from ^{207}Bi measured with the plastic scintillator telescope. In Figure 1a,b the deconvoluted sum beta-spectra of ^{235}U and ^{239}Pu derived in the present work are compared to those of other recent measurements and calculations [3-11]. For better distinction of the graphs no error bars are given in the Figures. For the present study, the total errors of the number N_β of electrons per MeV and fission have been determined to be in the range $2{,}5\% < \varepsilon < 5{,}4\%$ für ^{235}U and $3{,}1\% < \varepsilon < 47\%$ für ^{239}Pu respectively for electron energies between 1 and 8,5 MeV.

In Figure 2, the ratio V of the number N_β for ^{235}U and ^{239}Pu is shown as a function of beta-energy for 3 different experimental and 2 calculated spectra. As can be seen, V is always greater than 1 and raises with electron energy. Up to 7 MeV, all graphs derived either from experiments or from calculations [3-11] agree surprisingly well. Above this energy, however, large discrepancies are observed. For the experimental studies, these are mainly caused by the low counting rates. For the calculated sum beta-spectra the reason is due to the lack of knowledge in the decay data for the far unstable nuclei [9,10,12]. A more detailed discussion and the whole set of numbers will be given in a forthcoming publication.

---

* Supported in part by the Bundesministerium für Forschung und Technologie, Germany.

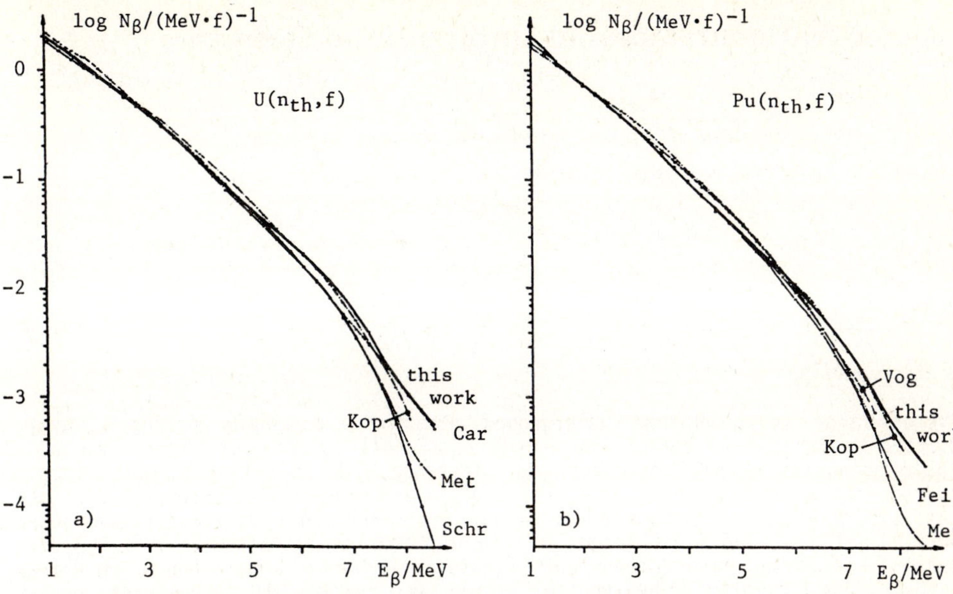

Fig.1: Comparison of measured and calculated sum beta-spectra for a) $^{235}U(n_{th},f)$ and b) $^{239}Pu(n_{th},f)$ with the result of this work.

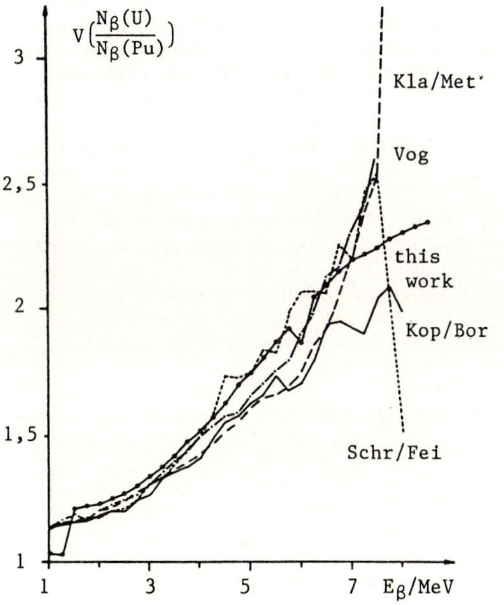

Fig.2: Ratio V of the number of electrons per MeV and fission for $^{235}U(n_{th},f)$ and $^{239}Pu(n_{th},f)$

References

[1] U.Keyser: Habilitationsschrift, TU Braunschweig (1985)

[2] W.Mampe, K.Schreckenbach, P.Jeuch, B.P.K.Maier, F.Braumandl, J.Larysz, T.v.Egidy: Nucl.Inst.Meth,154,127 (1978)

[3] R.E.Carter, F.Reines, I.I.Wagner, M.E.Wyman: Phys.Rev.113,280(1959)

[4] V.I.Kopeikin: Sov.J.Nucl.Phys.32, 1507(1980) and ORNL-tr-4842,June 1982

[5] K.Schreckenbach, H.R.Faust, F.v. Feilitzsch, A.A.Hahn, K.Hawerkamp J.L.Vuilleumier: Phys.Lett.99B, 251(1981)

[6] K.Schreckenbach, G.Colvin, W. Gelletly, F.v.Feilitzsch: Phys. Lett.1608,325(1985)

[7] P.Vogel, G.K.Schenter, F.M.Mann, R.E.Schenter: Phys.Rev.C24,1543 (1981)

[8] F.v.Feilitzsch, A.A.Hahn, K. Schreckenbach: Phys.Lett.118B, 162(1982)

[9] H.V.Klapdor: MPI H-1982 V35 and MPI H-1984 V15

[10] A.A.Borovoi, Yu.V.Klimov, V.J. Kopeikin: Yad.Fiz.37,1345(1983)

[11] J.Metzinger: Dissertation Univ. Heidelberg (1984)

[12] K.-L.Kratz: JKMZ 83-5 Univ.Mainz und Nucl.Phys.A417,447(1984)

Reactor Antineutrinos and Underground Detectors

P.O. Lagage

Service d'Astrophysique, Centre d'Etudes Nucléaires de Saclay,
F-91191 Gif-sur-Yvette Cedex, France

1. Introduction

Several large ($\gtrsim 1$ kiloton) underground detectors, currently proposed, are expected to be able to detect low energy ($\lesssim 10$ MeV) $\bar{\nu}_e$ and ν_e. One kiloton of liquid scintillator will be used, for instance, in the LVD (Large Volume Detector) experiment, planned at Gran Sasso [1] and which follows smaller (≈ 100 tons) experiments of the same kind, already in operation [2,3,4 and references therein]. The use of 6.5 kilotons of Liquid Argon is proposed for the ICARUS (Imaging Cosmic And Rare Underground Signals) experiment, also planned at Gran Sasso [5]. A 1 kiloton of heavy water is required for the SNO (Sudbury Neutrino Observatory) experiment [6].

These detectors have various objectives: search of proton decays, of monopoles, detection of neutrinos, study of cosmic rays. . . As concerns neutrinos, two major goals have been drawn: the detection of ^{8}B solar ν_e (flux $\lesssim 2\ 10^6\ \nu_e\ cm^{-2}\ s^{-1}$) and of neutrinos from a supernova (SN) explosion in our galaxy. The interest in the knowledge of the solar ν_e spectrum has been recently accentuated when Mikheyev and Smirnov [7] realized that the solar ν_e puzzle could be solved by invoking ν_e oscillations inside the Sun, with as a consequence a change in the ν_e energy spectrum. The detection of SN neutrinos is also very interesting as they are the principal witnesses of fundamental processes occurring during a SN explosion. But, the value generally adopted for the SN explosion rate in our galaxy is very low: $\sim$one every 30 years [8 and references therein], with a large uncertainty.

In the same time, the fluxes expected from other continuous sources of low energy $\bar{\nu}$ or ν have been calculated ; the sources considered are all past SN explosions which have generated an universal diffuse SN ν and $\bar{\nu}$ background [9,10,11], the Earth which emits $\bar{\nu}_e$ via the β-decay of isotopes inside the earth lithosphere [10] and all the nuclear reactors of the planet, via the β-decay of fission products [12]. The following sections deal mainly with the reactor $\bar{\nu}_e$ background.

2. Antineutrino Background from Far Away Nuclear Reactors

In 1985, the 374 nuclear power stations in operation on the planet have delivered a total power of 250 GWe [13]. The $\bar{\nu}_e$ background generated by these reactors depends, of course, on the point on Earth considered. The mean $\bar{\nu}_e$ flux induced during 1985 in various underground laboratories is indicated in Table 1. It varies from $2.6\ 10^4\ \bar{\nu}_e\ cm^{-2}\ s^{-1}$ at Kolar Gold Field to $2.1\ 10^6\ \bar{\nu}_e\ cm^{-2}\ s^{-1}$ at the Mont-Blanc tunnel. The minimum flux that can be found on Earth is $\approx 5\ 10^3\ \bar{\nu}_e\ cm^{-2}\ s^{-1}$.

If we now take into account the nuclear power stations under construction, the fluxes increase by less than a factor 2 in all the laboratories considered except at Fairport Harbor. There, the flux increases by a factor ≈ 40 because a nuclear power station will be soon in operation at about 10 km from the laboratory. Then, a flux of $\approx 4\ 10^7\ \bar{\nu}_e\ cm^{-2}\ s^{-1}$ is expected.

Table 1: Mean antineutrino fluxes induced in various underground laboratories by far away nuclear power stations in operation during 1985. The mean distance covered by these $\bar{\nu}_e$ from their source to the laboratory is also indicated. The symbols ●(o), ■(□), ▲(△) indicate the main purpose of the experiments placed in the laboratories: respectively study of proton decays, supernova explosions, solar neutrinos; full symbol: in operation, open symbol: proposed or planned. The data on the nuclear power stations have been extracted from the french CEA/DPg ELECNUC data base informations

Location of laboratories	Mean $\bar{\nu}_e$ fluxes $[cm^{-2}s^{-1}]$ during 1985	Mean distance [km] covered by these $\bar{\nu}_e$
Mont-Blanc (France) ●,■	$2.1\ 10^6$	300
Fréjus (France) ●	$1.9\ 10^6$	320
Kamioka (Japan) ●,o	$1.2\ 10^6$	290
Fairport-Harbor (Ohio, USA) ●	$8.1\ 10^5$	480
Sudbury (Canada)△	$5.3\ 10^5$	600
Gran Sasso (Italy) o,□,△	$3.3\ 10^5$	1020
Artemosk (USSR)■	$2.7\ 10^5$	1000
Soudan (Minesota, USA) ●,o	$1.9\ 10^5$	1190
Baksan (USSR) ■,△	$9.6\ 10^4$	1970
Lead (Dakota, USA) ■,△	$7.6\ 10^4$	2180
Kolar Gold Field (India) ●	$2.6\ 10^4$	3370

I turn now to smaller reactors: those which equip the fleet with nuclear propulsion. These moving sources are as numerous as the nuclear power stations but, as they are less powerful (from ≈ 45 MWth to ≈ 300 MWth), the total power available is only ≈ 1/20 of the power generated by the nuclear power stations. Then, it can be guessed that the $\bar{\nu}_e$ background generated by these sources is less important that the one due to nuclear power stations, at least in the underground laboratories considered.

3. Counts Expected in Large Underground Detectors

The total number of counts induced per year by reactor $\bar{\nu}_e$ charged current interactions, in one kiloton of liquid scintillator, is indicated in Fig. 1a) as a function of the energy threshold of detection; with a positron energy threshold of 3 MeV, about 1 event per year would be expected in a detector located at Kolar Gold Field, ≈ 20 events at Gran Sasso and ≈ 1600 events at Fairport Harbor. Figure 1b) contains the reactor $\bar{\nu}_e$ electron elastic scattering interactions in 6.5 kilotons of liquid Argon; with an energy threshold of 3 MeV, less than 1 event per year would be counted by a detector located at Kolar Gold Field, 3 at Gran Sasso and 250 at Fairport Harbor. Although the cross-section of electron elastic scattering interactions is much lower than that of charged current interactions, these interactions have the interesting possibility of keeping some statistical memory of the arrival direction of the neutrinos.

A comparison of the number of counts expected from reactor $\bar{\nu}_e$ with the number of counts from other neutrino sources (see fig.1) shows that reactor $\bar{\nu}_e$ have to be taken into account at low energies (1-7 MeV).

4. Reactor $\bar{\nu}_e$ oscillations and Large Underground Detectors

A positive use of such a background would be the study of neutrino oscillations as it is done with small detectors near one nuclear power station. The mean oscillation lengths tested could be large: several hundred of kilometers (see Table 1); but the number of counts expected is low compared to the numbers

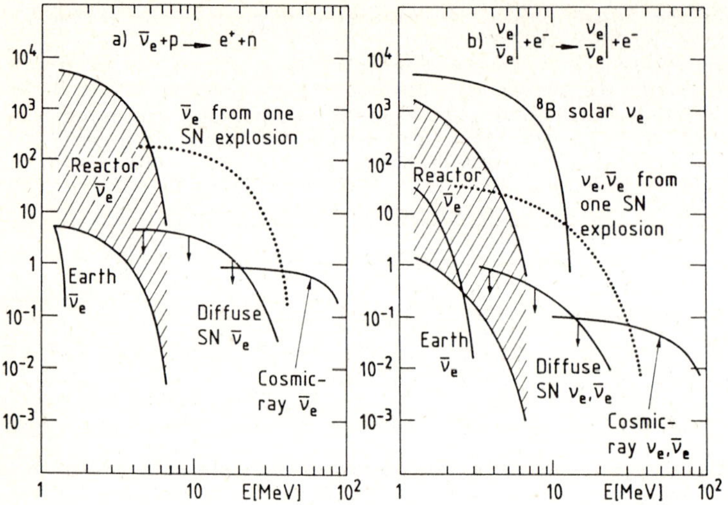

Figure 1a): Number of e^+ with energy (taken equal to E_ν-1.8 MeV) greater than E, expected per year from charged current interactions of various $\bar{\nu}_e$ in one kton of liquid scintillator (9.4 10^{31} free protons), as a function of E. The e^+ from reactor $\bar{\nu}_e$ lie inside the shaded region, according to the location of the laboratory considered; bottom line: mean $\bar{\nu}_e$ flux at Kolar Gold Field in 1985; upper line: expected flux in 1988 at Fairport Harbor (assuming a reactor efficiency of 70%). The SN explosion has been assumed to occur at 10 kpc from us and to have expelled 2 10^{53} ergs, distributed in the 3 types of $(\nu, \bar{\nu})$. The upper limit for the diffuse SN $\bar{\nu}_e$ background is from [11 and the references therein]; this limit corresponds to a flux of 250 $\bar{\nu}_e$ cm^{-2} s^{-1}; (larger limits can appear, but the corresponding $\bar{\nu}_e$ are at lower energies). The earth $\bar{\nu}_e$ are from [10]. For the calculations of the cosmic-ray $\bar{\nu}_e$ interactions, only the cosmic-ray $\bar{\nu}_e$ with energy < 100 MeV have been taken into account from [14]

1b): Number of e^- with energy greater than E, expected per year from electron elastic scattering interactions of various $\bar{\nu}_e$ or ν_e in 6.5 ktons of liquid Argon (1.8 10^{33} e^-), as a function of E. In addition to the neutrino sources of Fig.a), the ^{8}B solar ν_e have been considered with a standard energy spectrum and a flux of 2 10^6 ν_e cm^{-2} s^{-1}, compatible with the Davis experiment [15]

obtained with small detectors located near a nuclear power station. Furthermore, these $\bar{\nu}_e$ originate from several nuclear power stations, which are not in operation at the same time; so that it will be difficult to put in evidence neutrino oscillations, if any. These inconveniences could be, at least partly, circumvented at Fairport Harbor in the near future, when the nuclear power station under construction will be in operation. It is also amusing to notice that a 1 kiloton detector located at 700 meters from a small "moving reactor" of 100 MWth would be the place of the same number of $\bar{\nu}_e$ interactions, as a 0.3 ton detector located at 65 meters from a nuclear power station [16].

But the detection of low energy neutrinos is a difficult task because the (non-neutrino) background increases a lot below 10 MeV. Careful analyses of this background in large underground detectors are essential.

Acknowledgement: It is a pleasure to thank J.C. Le Ralle for kindly providing information on nuclear power stations. I am also grateful to the organisers of the symposium for the kind invitation.

References

1 C. Alberini et al.: Proc. of the UP85 Conf. on underground Physics, St-Vincent (Italy), April 1985, in press in Il Nuovo Cimento
2 F.F. Khalchukov, V.G. Ryassny, O.G. Ryashskaya, G.T. Zatsepin: 19th Int. Cosmic Ray Conf., La Jolla (USA), 8, 140 (1985)
3 G. Badino, G. Bologna, C. Castagnoli, W. Fulgione, P. Galeotti, O. Saavedra, V.L. Dadykin, V.B. Korchaguin, P.V. Korchaguin, A.S. Malguin, O.G. Ryazhskaya, A.L. Tziabuk, V.P. Talochkin, G.T. Zatsepin, V.F. Yakushev: Il Nuovo Cimento, 74, 6, 573 (1984)
4 M.L. Cherry, S. Corbato, T. Daily, E.J. Fenyves, D. Kieda, K. Lande, C.K. Lee, 19th Int. Cosmic Ray Conf., La Jolla (USA), 8, 246 (1985)
5 ICARUS collaboration: preprint, INFN/AE-85/7
6 Simpson and the SNO collaboration: Proc. of the Moriond Workshop on Massive Neutrinos in Particle Physics and Astrophysics, Tignes (France), Jan. 1986, J. Tran Thanh Van Editor, in press
7 S.P. Mikheyev, A.Yu. Smirnov: Proc. of the 10th Workshop on Weak Interactions and Neutrinos, Savonlinna (Finland), June 1985, in press in Il Nuovo Cimento
8 V. Trimble: Rev. of Modern Physics 54, 1183 (1982)
9 G.S. Bisnovatyi-Kogan, Z.F. Seidov: Ann. N.Y. Acad. Sci. 422, 319 (1984)
10 L.M. Krauss, S.L. Glashow, D.N. Schramm: Nature 310, 191 (1984)
11 S.E. Woosley, R. Wilson, R. Mayle: Astrophys. J. 302, 19 (1986)
12 P.O. Lagage: Nature 316, 42 (1985)
13 Nuclear Power Reactors in the World (Ref. data Ser. N°2, IAEA, Vienna, Austria 1986)
14 T.K. Gaisser, T. Stanev: Proc. Neutrino 84 Conf., Dortmund, 370 (World Scientific Publishing, 1984)
15 R. Davis, B.T. Cleveland, J.K. Rowley: AIP Conf. Proc. 96 (1983)
16 P.O. Lagage: Proc. of the Moriond Workshop on Massive Neutrinos in Particle Physics and Astrophysics, Tignes (France), Jan. 1986, J. Tran Thanh Van Editor, in press

Present Knowledge of the Lepton Mixing Matrix from Neutrino Oscillation Experiments

K. Kleinknecht

Institut für Physik der Universität Mainz, D-6500 Mainz, F. R. Germany

An analysis of experimental results in the framework of simultaneous oscillations of three neutrino flavours is used, in a combined fit, to constrain the elements of the 3 x 3 unitary lepton mixing matrix and to obtain limits on the three lepton mixing angles $\alpha_{e\mu}$, $\alpha_{e\tau}$ and $\alpha_{\mu\tau}$.

If the lepton number is not conserved, neutrinos of different flavour may mix. In analogy to quark mixing, this lepton mixing can be described by a 3 x 3 unitary matrix. In fact, the weak eigenstates ν_e, ν_μ and ν_τ are related to the mass eigenstates ν_1, ν_2 and ν_3 by a unitary matrix U_{ik} [1]:

$$|\nu_\ell\rangle = \Sigma_k U_{\ell k} |\nu_k\rangle \qquad (\ell = e,\mu,\tau;\ k = 1,2,3) \tag{1}$$

and this matrix $U_{\ell k}$ can be parametrized in analogy to the quark mixing matrix. We use the parametrization of Maiani [2], given in table 1.

Table 1

Maiani parametrization of lepton mixing matrix

$$U = \begin{pmatrix} C_\beta C_\theta & C_\beta S_\theta & S_\beta \\ -S_\gamma C_\theta S_\beta e^{i\delta'} - S_\theta C_\gamma & C_\gamma C_\theta - S_\gamma S_\beta S_\theta e^{i\delta'} & S_\gamma C_\beta e^{i\delta'} \\ -S_\beta C_\gamma C_\theta + S_\gamma S_\theta e^{i\delta'} & -C_\gamma S_\beta S_\theta - S_\gamma C_\theta e^{-i\delta'} & C_\gamma C_\beta \end{pmatrix}$$

Up to now, the scarcity of data and the complexity of the problem have prevented a general analysis of experimental data without restrictive assumptions, i.e. with three free angles and neutrino masses. Instead, experimentalists searching for neutrino oscillations have usually made the ad hoc assumption that only one angle and one neutrino mass difference is different from zero, and have given their results in terms of these two quantities.

We present here a general analysis of available experimental data with five free parameters, i.e. three angles and two neutrino mass differences [3].

The time evolution of an initially pure ν_ℓ state can be expressed in terms of mass differences and the neutrino energy E. If this energy is large compared to the neutrino mass, the probability to find a $\nu_{\ell'}$ after a distance L(m) from the production point of an initially pure ν_ℓ state of energy E(MeV) is [4]:

$$P_{\ell\ell'} = \sum_{k,k'=1} U_{\ell k} U_{\ell k'} U_{\ell' k} U_{\ell' k'} \cos (2.54 \, \Delta m^2_{k'k} \, L/E) \tag{2}$$

where $\Delta m^2_{k'k} = |m^2(\nu_{k'}) - m^2(\nu_k)|$ is measured in (eV^2).

As a simplification, the elements $U_{\ell k}$ are assumed to be real. This leaves us with five free parameters to be determined, because for three neutrino masses only two Δm^2 values are independent.

For the Maiani parametrization, in the limiting case of only one angle being finite, this angle can be directly related to oscillations between two flavours of neutrinos. The Maiani angles θ, β and γ are thus uniquely related to the oscillations channels $\nu_e \leftrightarrow \nu_\mu$, $\nu_e \leftrightarrow \nu_\tau$ and $\nu_\mu \leftrightarrow \nu_\tau$ respectively. We call these angles $\theta = \alpha_{e\mu}$, $\beta = \alpha_{e\tau}$ and $\gamma = \alpha_{\mu\tau}$. We note that this decoupling of angles does not occur in the Kobayashi-Maskawa notation.

The experimental data on neutrino oscillation can be divided into two classes: 1. Disappearance experiments, where the flux of neutrinos of one flavour is measured at two distances from their production point. 2. Appearance experiments, where neutrinos of flavour ℓ are searched in a beam of neutrinos of flavour $k \neq \ell$.

For the simultaneous fit of oscillation experiments of different kinds, we have used all data from reactor and accelerator experiments available at this time. They are listed in table 2 [4]. For each experiment, the original data were used, such that for a given neutrino energy each experiment gives a contribution to the global

Table 2

Oscillation experiments included in the fit [4]. The last column is the χ^2 contribution of the experiment for the hypothesis of no oscillation.

experiment	measured quantity	Δm^2 range $[eV^2]$	$\chi^2_{\text{no osc.}}$/NDF
(1) CCFRR	$p_{\mu\mu}$ ratio	30. - 1000.	15.5/14
(2) CDHS	$p_{\mu\mu}$ ratio	0.24 - 90.	15.3/14
(3) CHARM	$p_{\mu\mu}$ ratio	0.60 - 20.	1.8/3
	$p_{e\mu}$ difference	1. - 10.	0/0
(4) GOESGEN	p_{ee} absolute	> 0.01	15.5/15
(5) GOESGEN	p_{ee} ratio	0.03 - 3.	7.4/15
(6) BUGEY	p_{ee} ratio	0.02 - 5.	20.9/8
(7) ν_τ Exp.'s	$p_{\mu\tau}/p_{\mu\mu}$	> 0.2	6.3/5
(8) BNL	$p_{\mu e}$ absolute	> 0.43	2.4/6

χ^2 in the fit. For each experiment we checked that, making the restrictive assumptions of the authors, the procedure used here gave back the quoted result of the authors in the two-parameter model with only one θ and one Δm^2.

A total chisquare function was constructed from the measured quantities Q^m (ratios of oscillation probabilities at different L/E or absolute probabilites from a comparison to calculated initial fluxes) and the corresponding computed values Q^c. The data points of each experiment i are allowed to vary together by a scale factor N_i within the quoted normalization uncertainty $\sigma(N_i)$. The expression

$$\chi^2 = \sum_{i=1} \left[\sum_j \left(\frac{N_i Q^m_{ij} - Q^c_{ij}}{\sigma(Q^m_{ij})} \right)^2 + \left(\frac{N_i - 1}{\sigma(N_i)} \right)^2 \right] \tag{3}$$

is minimized by fitting the mixing angles and mass differences, where the index i labels the experiments, and the index j gives the bin in L/E for a specific experiment.

The results of the best fit are listed in table 3. Limits on the oscillation parameters can be derived from (3) in three parameter planes ($\sin^2 2\alpha_{e\mu}$, Δm^2_{12}), ($\sin^2 2\alpha_{e\tau}$, Δm^2_{13}), and ($\sin^2 2\alpha_{\mu\tau}$, Δm^2_{23}). The curves in Fig.1 give the largest allowed $\sin^2 2\alpha$ values (for 90 % C.L.) as a function of the corresponding Δm^2. Each such point has been obtained under the condition that in the other two planes all combinations ($0. < \sin^2 2\alpha < 1.$, $0.01 < \Delta m^2 < 1000\ eV^2$) are possible.

There are two best fit solutions with a $\chi^2_{min} = 78$ for 84 D.F. One of them is (solution a) reached for the parameters $\Delta m^2_{21} = 34\ eV^2$ and $\Delta m^2_{31} = 0.2\ eV^2$. For this solution, the angles are constrained to be $\sin^2 2\alpha_{e\mu} < 4 \times 10^{-3}$, $\sin^2 2\alpha_{e\tau} < 0.13$

Table 3a

90 % C.L limits on mixing matrix elements for best fit values $\Delta m^2_{12} = 34\ eV^2$ and $\Delta m^2_{13} = 0.2\ eV^2$

$$U = \begin{pmatrix} 1.00 - 0.98 & 0. - 0.03 & 0. - 0.18 \\ -0.04 - 0. & 1.00 - 0.99 & 0. - 0.07 \\ -0.18 - 0. & -0.07 - 0. & 1.00 - 0.98 \end{pmatrix}$$

$\sin^2 2\alpha_{e\mu} < 0.004$; $\sin^2 2\alpha_{e\tau} < 0.13$; $\sin^2 2\alpha_{\mu\tau} < 0.02$

Table 3b

90 % C.L. limits on mixing matrix elements for fit values $\Delta m^2_{12} = 0.2\ eV^2$ and $\Delta m^2_{13} = 38\ eV^2$

$$U = \begin{pmatrix} 1.00 - 0.97 & 0. - 0.19 & 0. - 0.15 \\ -0.21 - 0. & 1.00 - 0.98 & 0. - 0.07 \\ -0.14 - 0. & -0.10 - 0. & 1.00 - 0.98 \end{pmatrix}$$

$\sin^2 2\alpha_{e\mu} < 0.15$; $\sin^2 2\alpha_{e\tau} < 0.09$; $\sin^2 \alpha_{\mu\tau} < 0.02$

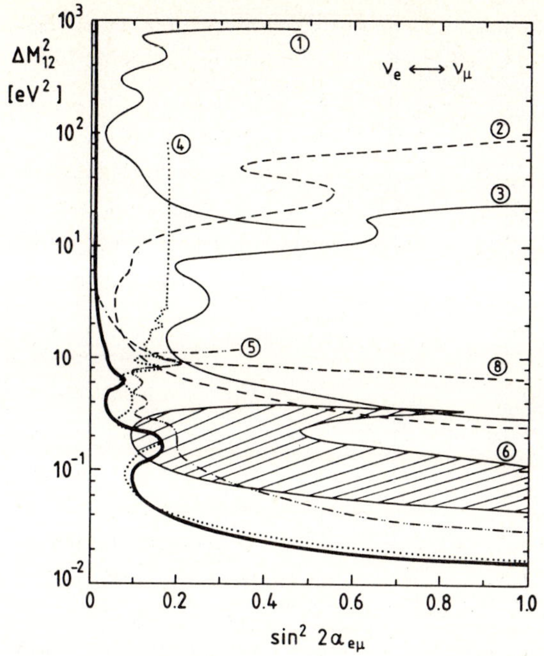

Fig.1a: Limits on mixing parameter $\sin^2 2\alpha_{e\mu}$ vs. neutrino mass difference ΔM^2_{12}. The thin lines are 90 % C.L. upper limits on $\sin^2 2\alpha_{e\mu}$ from individual experiments 1) 2) 3) 4) 5) and 7) in ref. [4] assuming $\alpha_{e\tau} = \alpha_{\mu\tau} = 0$, the shaded area is the allowed range from the Bugey experiment (6) in ref. [4]. The broad line is the 90 % C.L. upper limit on $\sin^2 2\alpha_{e\mu}$ from the three flavour oscillation analysis, allowing the other two mixing angles to vary over the whole range $0 \leq \sin^2 2\alpha_{e\tau} \leq 1$ and $0 \leq \sin^2 2\alpha_{\mu\tau} \leq 1$.

and $\sin^2 2\alpha_{\mu\tau} < 0.02$, and the mixing matrix is given in Table 3a, with 90 % C.L. limits. The values for Δm^2 are different from zero only at the 1.5 standard deviation level, i.e. the finite values are not significant.

A second local minimum of χ^2 is found at $\Delta m^2_{21} = 0.2$ eV^2 and $\Delta m^2_{31} = 39$ eV^2, i.e. interchanging the two values from solution a. Here also $\chi^2_{min} = 78$, the 90 % C.L. limits on the angles are $\sin^2 2\alpha_{e\mu} < 0.15$, $\sin^2 2\alpha_{e\tau} < 0.09$ and $\sin^2 2\alpha_{\mu\tau} < 0.02$, and the range of matrix elements is given in Table 3b.

Since there is no way of discriminating between the two solutions, global limits on angles for mass differences in the range $\Delta m^2_{12} > 0.06$ eV^2, $\Delta m^2_{13} > 0.04$ eV^2 and $\Delta m^2_{23} > 2$ eV^2 are $\sin^2 2\alpha_{e\mu} < 0.15$, $\sin^2 2\alpha_{e\tau} < 0.13$ and $\sin^2 2\alpha_{\mu\tau} < 0.02$ while for any specific mass difference $\Delta m_{k'k}$, the 90 % C.L. limit on the corresponding $\sin^2 2\alpha_{\ell\ell'}$ is given in Figs. 1a through 1c. For the hypothesis of no oscillation, i.e all mixing angles fixed to zero, a χ^2 of 85 for 87 D.F. is obtained, i.e. the data are consistent with this hypothesis.

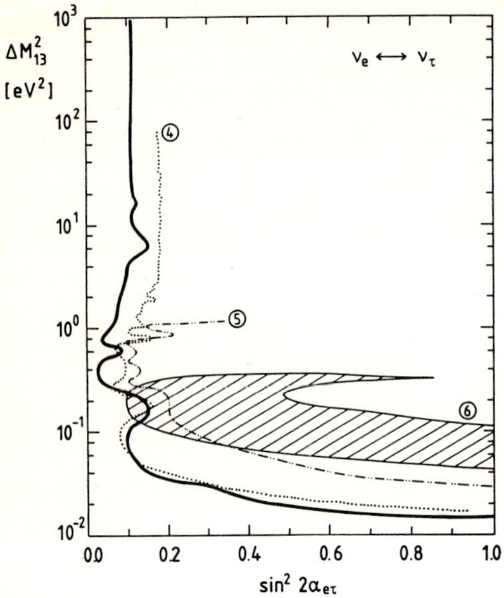

Fig.1b: Limits on mixing parameter $\sin^2 2\alpha_{e\tau}$ vs. ΔM^2_{13}. Thin lines are 90 % C.L upper limits from individual experiments assuming $\alpha_{e\mu} = \alpha_{\mu\tau} = 0$. Broad line from three-flavour analysis.

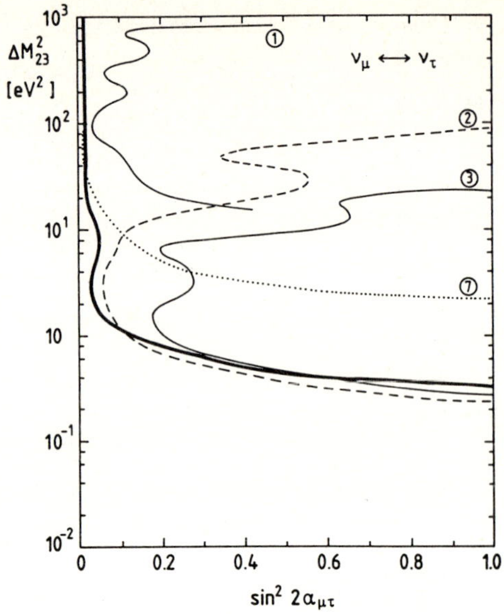

Fig.1c: Limits on mixing parameter $\sin^2 2\alpha_{\mu\tau}$ vs. ΔM^2_{23}.

In conclusion, we have shown that an analysis of existing data on oscillations of three neutrino flavours yields limits on the mixing matrix and on the three mixing angles each corresponding to one channel of oscillations. While this method does not in general allow stringent constraints on the mixing parameters to be extracted form <u>one</u> experiment, the combination of the different experimental data gives limits on angles and the mixing matrix elements which are about as restrictive as earlier two-neutrino analyses on separate oscillation channels. The ensemble of data is consistent with no oscillation occuring.

If neutrino masses are finite [5], then mixing angles of the order of $\sqrt{(m_e/m_\mu)}$ or $\sqrt{(m_\mu/m_\tau)}$ are expected in some models invoking family symmetries [6]. Experiments have nearly reached the level of sensitivity needed for testing such models. As in the case of quark mixing, the problem of family mixing is still waiting for a solution.

<u>References</u>

[1] S.M.Bilenky and B.Pontecorvo, Phys. Rep. 41C (1978) 225
A.deRujula et al., Nucl. Phys. B 168 (1980) 54
S.P.Rosen, Preprint Los Alamos LA-UR-84-1789 (1984)

[2] L.Maiani, Proc. Intern. Symp. on Lepton and Photon Interactions at High Energies (Hamburg 1977), o.877

[3] H.Blümer and K.Kleinknecht, Phys. Lett. 161B (1985) 407 and references quoted therein.

[4] Experimental results are taken from the following references:

CCFRR	(1)	C.Haber et al., Phys. Rev. Lett 52 (1984) 1384
CDHS	(2)	F.Dydak et al., Phys. Lett. 134B (1984) 281
CHARM	(3)	F.Bergsma et al., Phys. Lett. 142B (1984) 103
Goesgen	(4)	J.L.Vuilleumier et al., Phys. Lett. 114B (1982) 298
Goesgen	(5)	K.Gabathuler et al., Phys. Lett. 138B (1984) 449
Bugey	(6)	J.F.Cavaignac et al., Phys. Lett. 148B (1984) 387
ν_τ search	(7)	N.J.Baker et al., Phys. Rev. Lett. 47 (1981) 1576
		N.Armenise et al., Phys. Lett. 100B (1981) 182
		O.Errique et al., Phys. Lett. 102B (1981) 73
		N.Ushida et al., Phys. Rev. Lett. 47 (1981) 1694
		G.N.Taylor et al., Phys. Rev. D28 (1983) 2705
		H.C.Ballagh et al., Phys. Rev. D30 (1984) 22/1
BNL	(8)	L.A.Ahrens et al., Phys. Rev. D31 (1985) 2732

[5] V.A.Lubimov et al., Phys. Lett. 94B (1980) 226; Proc. EPS Conf. on High Energy Phys., Brighton (1984) p.386
K.Bergkvist, Phys. Lett. 154B (1985) 224
J.J.Simpson, Phys. Rev. Lett. 54 (1985) 1891
T.Altzitzoglou et al., Phys. Rev. Lett 55 (1985) 799

[6] Z.G.Berezhiani, Phys. Lett. 150 B (1984) 177
H.Fritzsch and P.Minkowski, Phys. Rep. 73 (1981) 67
H.Fritzsch, Nucl. Phys. B155 (1979) 189
B.Stech, Phys. Lett. 130B (1983) 89
M.Shin, Phys. Lett. 145B (1984) 285

Mixing Among Three States: A Practical Approximation and Its Application to Neutrino Oscillations

T. Sauerland

Institut für Experimentalphysik I, Ruhr-Universität Bochum, D-4630 Bochum, F. R. Germany

The mixing matrix, leading from the pure states to the physical states, has been parametrized in the form of three real rotations:

$$U = R(\alpha,0,0) \cdot R(0,\beta,0) \cdot R(0,0,\gamma).$$

Applying this representation to the isospin mixing among triplets in ^{16}O and ^{14}N I found [1] that neglecting the mixing across the middle state and thus setting $\beta = 0$ is a very good approximation.

Also from the measured flavor mixing in quarks [2] through the weak interaction I extracted an angle β compatible with zero:

$$\alpha = 2.56^{\circ} \pm 0.26^{\circ}, \qquad \beta = 0^{\circ} \text{ to } 0.3^{\circ},$$

$$\gamma = 13.36^{\circ} \pm 0.15^{\circ} \qquad \text{(Cabibbo angle)}.$$

Here the additional phase factor $e^{i\delta}$ in the unitary mixing matrix [3], essential to CP violation, has been set to one.

The suggested mixing among the three neutrino mass eigenstates can be described in the same manner. Formulas connecting the eigenmasses m_1, m_2, m_3 with the members of the mass matrix $\langle\nu_i|m|\nu_j\rangle = m_{ij}$ $(i,j = e,\mu,\tau)$ and also formulas for the expected conversion probabilities in neutrino oscillations in terms of α, β, γ are too long to be reproduced here.

The assumption $\beta = 0$ means an important simplification. For this approximation I am able to give parameter-free expressions for the eigenmasses:

$$\left.\begin{matrix} m_1 \\ m_2 \end{matrix}\right\} = \frac{1}{2}\left\{ m_{ee} + m_r \mp \left[(m_r - m_{ee})^2 + 4 \cdot (m_{e\mu}^{\ 2} + m_{\tau e}^{\ 2}) \right]^{1/2} \right\} ,$$

$$m_3 = (m_{\tau e}^{\ 2} \cdot m_{\mu\mu} + m_{e\mu}^{\ 2} \cdot m_{\tau\tau} - 2 \cdot m_{e\mu} m_{\mu\tau} m_{\tau e})/(m_{e\mu}^{\ 2} + m_{\tau e}^{\ 2})$$

with

$$m_r = (m_{e\mu}{}^2 \cdot m_{\mu\mu} + m_{\tau e}{}^2 \cdot m_{\tau\tau} + 2 \cdot m_{e\mu} m_{\mu\tau} m_{\tau e})/(m_{e\mu}{}^2 + m_{\tau e}{}^2) \quad .$$

The frequently discussed case of maximum neutrino mixing [4], characterized by a probability matrix with 1/3 in the three diagonal positions after averaging over the various momenta of the neutrinos or over the various distances from their origins, cannot be represented by $\beta = 0$ and also not by $\beta \neq 0$, $\alpha = \gamma$. The angles corresponding to this special situation ought to be determined numerically. But I proved the finding that no solution with real angles exists and so this paradigm lacks any physical reality. The upper left element 1/3 already determines $\sin^2\beta = 1/3$ and $\sin^2\gamma = 1/2$ and fixes the total matrix

$$\overline{P} = \begin{pmatrix} 1/3 & 1/3 & 1/3 \\ 1/3 & 1/2 & 1/6 \\ 1/3 & 1/6 & 1/2 \end{pmatrix} \quad .$$

For $\alpha = \gamma = 45^0$ the choices $\beta = 45^0$ or 0^0 give

$$\overline{P} = \frac{1}{32} \begin{pmatrix} 12 & 10 & 10 \\ 10 & 19 & 3 \\ 10 & 3 & 19 \end{pmatrix} \quad \text{or} \quad \begin{pmatrix} 1/2 & 1/4 & 1/4 \\ 1/4 & 3/8 & 3/8 \\ 1/4 & 3/8 & 3/8 \end{pmatrix}$$

respectively.

1. T. Sauerland: Verhandl. DPG (VI) 20, 562 (1985) and
   Jahresbericht DTL 11, 22 (1984)
2. K. Kleinknecht: Comments Nucl. Part. Phys. 13, 219 (1984)
3. M. Kobayashi and T. Maskawa: Progr. Theor. Phys. 49, 652 (1973)
4. H. Primakoff, S. P. Rosen: Ann. Rev. Nucl. Part. Sci. 31, 145 - 192 (1981) .

3.3.4 Tritium Decay and Electron Capture

An Upper Limit for the Electron Antineutrino Mass

W. Kündig, M. Fritschi, E. Holzschuh, J.W. Petersen[a]*, R.E. Pixley, and H. Stüssi*

Physics Institute, University of Zürich, CH-8001 Zürich, Switzerland

The endpoint region of the tritium β-spectrum has been measured with 27 eV resolution, using a magnetic spectrometer. The tritium activity was implanted into a thin layer of carbon. The neutrino mass determined is consistent with zero with an upper limit of 18 eV, which includes instrumental and statistical uncertainties as well as uncertainties due to the energy loss in the source and the final electronic states.

Since 1980 a group at ITEP (Institute for Theoretical and Experimental Physics, Moscow) has been publishing a series of papers [1] claiming evidence for a nonzero mass of the electron antineutrino as determined from the tritium β-spectrum. The experiment has been improved continuously, the latest result being $20 < m < 35$ eV with a central value of 35 eV. However, this result has been subject to some critisisme [2,3]. Up to now no other experiment has reached a sufficient sensitivity to be a true, independent test. Here we report first results of our measurement of the endpoint region of the tritium β-spectrum.

The instrument employed in the present investigation consists of a toroidal field, magnetic spectrometer of the Tretyakov type [4] with 2662 mm source-detector distance, modified with a radial, electrostatic retarding field around the source. The main advantage of this spectrometer is its high absolute resolution together with a high luminosity [5]. The reason for this is that due to the decceleration of the electrons only a very modest relative resolution in the magnetic spectrometer is required. This allows the use of a large source (157 cm^2) and guarantees that the spectrometer can be built with sufficient accuracy. Since the electric and magnetic fields of the spectrometer are known analytically, the various spectrometer parameters, especially the resolution function can be reliably calculated. Detailed conversion electron measurements [5] have confirmed the calculated shape of the spectrometer resolution function (SRF). With a magnetic field setting for 2.2 keV electrons the basic spectrometer parameters are: 0.6 % transmission, 1.1 eV/mm dispersion, and 27 eV FWHM resolution. The resolution of the spectrometer is proportional to the magnetic field setting, the counting rate inversely proportional to the square of the field setting.

[a] Present address: CERN, CH-1211 Geneva

Three sources were prepared by implantation of molecular tritium ions with about 300 eV into carbon. The implanted sources may be assumed to be uniform, this in contrast to evaporated sources, which may form islands. The extreme stability of the sources and various indications in the literature strongly suggest that C-T bonds are formed. The depth profiles were obtained by bombarding the sources with 400 keV C and measuring the energy distribution of the recoil tritons under 45° [6].

The well known [7] energy loss in carbon is mainly due to plasmon excitation and is easily calculable from the measured profiles. Independently the spectra of various conversion electron sources covered with evaporated carbon have been measured. This test confirmed the validity of the procedure described. The sum of the spectrometer resolution function and the energy loss function folded with the SRF gives the total resolution function (TRF). Backscattering due to multiple elastic scattering and plasmon excitations in the source substrate was investigated by Monte Carlo simulation. It was found that this effect can be adequately represented by a constant distribution below the noloss line.

The data reported here were taken in four runs with three sources, totalling 27 days of effective measuring time. Figures 1 and 2 show the measured spectrum plotted in a linear scale and in the form of a Kurie-plot. The line through the datapoints is a least squares fit to the assumed spectrum

$$N(E) = A\,F(E)\,p\,E_t\,(1 + \alpha\varepsilon_0) \sum_i W_i \varepsilon_i^2 \left[1 - m^2/\varepsilon_i^2\right]^{1/2} + BG \qquad (1)$$

where $\varepsilon_i = E_{oi} - E$ is the neutrino energy, F(E) the Fermi function, and p, E, and E_t are the electron momentum, kinetic and total energy, respectively. The sum runs over all final states (FS) with branching fractions W_i and endpoints E_{oi}. The E_{oi} are defined by $E_{oi} = E_{oo} - E_{ex,i}$, where E_{oo} correspondes to the electronic groundstate and $E_{ex,i}$ are the excitation energies measured from the groundstate. Backscattering is taken into account by the parameter   and enters in this approximate form by a convolution of the backscattering distribution with the β-spectrum. The fit function is obtained by convoluting Eq. (1) with TRF, and by including a simple correction factor for the varying spectrometer acceptance with changing retarding voltage. The free parameters are α, E_{oo}, the normalization A, background BG, and the neutrino mass squared m^2. The later was allowed to take on nonphysical, negative values, in which case the square root in Eq. (1) was replaced by its first order expansion $|1 - m^2/(2\varepsilon_i|\varepsilon_i|)|$. This gave symmetric, parabola-like curves for   versus m^2. Our final result does not depend on this assumption.

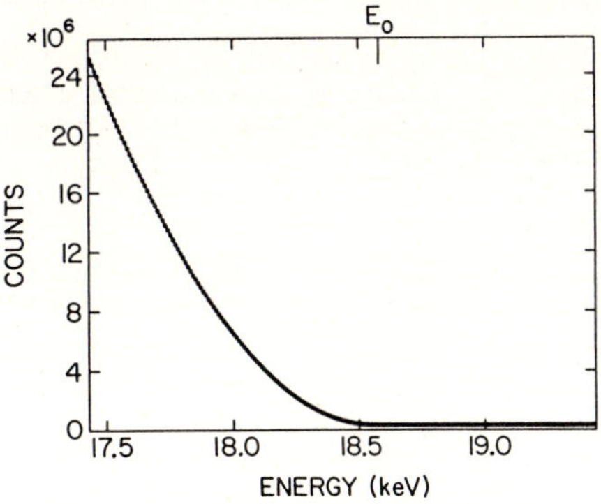

Fig. 1. β-spectrum of tritium. The vertical scale is correct for the region around E_o. Outside this region the scale is a factor 10 less.

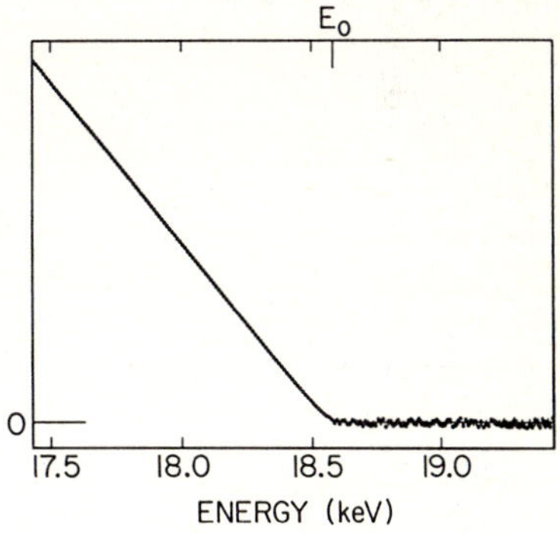

Fig. 2. Kurie-plot of the tritium β-spectrum shown in Fig. 1. The fitted curve corresponds to m = 0.

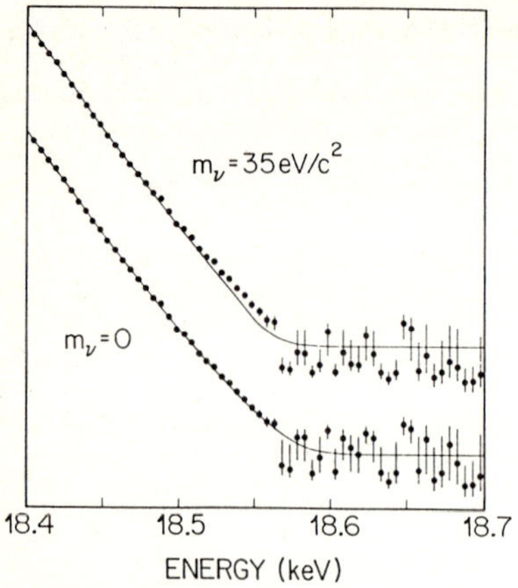

Fig. 3. Kurie-plot of the data from run 4 with fits corresponding to m = 0 and m = 35 eV. The final states of CH_3T were assumed. Notice the systematic deviations of the 35 eV fit below E_o and in the strongly mass correlated background.

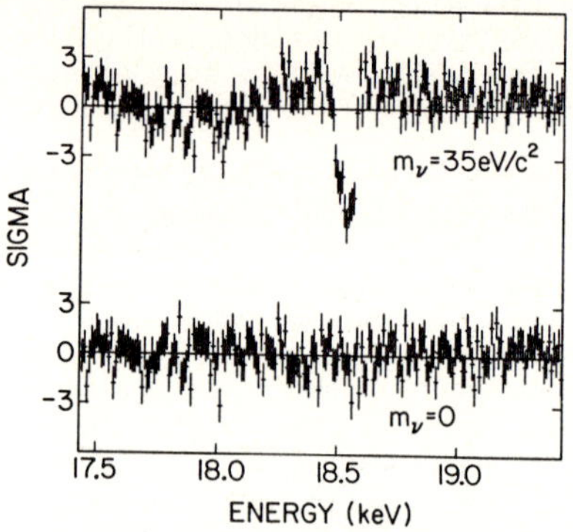

Fig. 4. Plot of the difference between the fitted function and the data divided by the standard deviation for two fixed values of m. Notice the deviations not only in the critical endpoint region but also in the background region and well below E_o.

The data were fitted to various final state distributions. Kaplan et al. [8] calculated the FS for a variety of molecules with C-T bonds and found very little variation in terms of their averaged properties. In this paper CH_3T is used as an adequate representation for the FS of our sources.

Figure 3 shows the Kurie-plot of the data from run 4 near the endpoint with fits corresponding to m = 0 and m = 35 eV. Of course all the other free parameters (A, α, E_o, BG) being fitted. The difference between the measured and fitted datapoints is shown in Fig. 4. The two figures clearly indicate that a mass in the order of 35 eV can not fit our data. The resultes for the four runs with three different sources are given in Table I. The statistical result is: $m^2 = -11 \pm 63$ eV^2. The 95 % upper limit is:

$$m^2 < 95 \text{ eV}^2$$

Table I. Fit results for four runs with three different sources and the combined set of data, using CH_3T final states. Errors indicated are one standard deviation.

Data set	1	2	3	4	all
source	T1	T2	T2	T3	-
no loss fraction	0.573	0.589	0.589	0.662	-
χ^2	328.8	345.6	349.9	343.7	1370.2
degr. of freedom	337	356	356	317	1369
E_{oo} - 18500 (eV)	82.6±.3	84.5±.2	84.4±.2	77.6±.2	- *)
α (keV^{-1})	0.018±.004	0.020±.002	0.022±.002	0.021±.002	-
m^2 (eV^2)	140±130	-9±140	-22±140	-85±93	-11±63
95 % UL (eV^2)	356	218	202	70	95

*) The systematic deviations in E_{oo} are due to calibration errors and inaccuracies in the source and detector position. They have no influence on the neutrino mass

Table II. Fitted values of m^2 and the corresponding shifts m^2 when the input parameters of the fitted function are changed. Shown are only changes which increase m^2. The complete data set with 1369 degrees of freedom was fitted.

Input parameter	relativ change	χ^2	m^2 (eV^2)	Δm^2 (eV^2)
Best set	-	1370.2	- 11	-
Spectrometer resolution	+ 10 %	1370.2	52	63
mean source thickness	+ 15 %	1373.8	120	131
Ground-state fraction	- 2.5 %	1370.5	30	41
Mean excitation energy	+ 6 %	1373.2	111	122
Width of excited states	+ 30 %	1386.3	52	63

In Table II conservative estimates of the systematic errors are given. Adding the statistical and systematical errors linearly one arrives at the result, $m^2 < 310\ eV^2$ or:

$$m < 18\ eV.$$

In conclusion we find no indication of a nonzero mass for the electron antineutrino, which is in strong contradiction to the result of Ref. [1]. Although the various systematic errors are still under investigation, we see at present no possible source of error in our experiment large enough to account for this discrepancy. Additional details are given in Ref. [9].

References:

1 V.A. Lubimov et al., Phys. Lett. 94B, 266 (1980); Sov. Phys. JETP 54,616 (1981); Phys. Lett. 159B, 217 (1985).

2 J.J. Simpson, Phys. Rev. 30D,1110 (1984).

3 K.E. Bergkvist, Phys. Lett. 154B, 224 (1985); Phys. Lett. 159B, 408 (1985).

4 E.F. Tretyakov, Izv. Akad. Nauk SSSR Ser. Fiz. 39, 583 (1975).

5 W. Kündig, J.W. Petersen, R.E.Pixley, H. Stüssi, and M. Warden, Proc. of the Fourth Moriond Workshop, ed. by J. Tran Thanh Van (Edition Frontières, Paris 84)

6 G.G. Ross, B. Terreault, G. Gobeil, G. Abel, C. Boucher, and G. Veilleux, J. Nucl. Mat. 128, 730 (1984).

7 R.E. Burge and D.L Misell, Phil. Mag. 18, 251 (1968).

8 I.G. Kaplan, G.V. Smelov, and V.N. Smutnyi, DAN USSR 279,1110 (1984).

9 M. Fritschi, E. Holzschuh, W. Kündig, J.W. Petersen, R.E. Pixley, and H. Stüssi, Phys. Lett. 173B, 485 (1986); Proc. of the Sixth Moriond Workshop, ed. by J. Tran Thanh Van (1986, in press).

A Limit on the $\bar{\nu}_e$ Mass in Free Molecular Tritium Beta Decay

T.J. Bowles[1], *J.F. Wilkerson*[1], *J.C. Browne*[1], *M.P. Maley*[1], *R.G.H. Robertson*[1], *D.A. Knapp*[2], *and J.A. Helffrich*[3]

[1]Physics Division, Los Alamos National Laboratory, Los Alamos, NM 87545, USA
[2]Princeton University, Princeton, NJ 08544, USA
[3]University of California at San Diego, La Jolla, CA 92093, USA

The question of a nonzero neutrino mass has received considerable attention since the claims of Lyubimov et al [1] in 1980 were published which showed evidence for an electron antineutrino mass between 14 and 46 eV, with a best fit value of 35 eV. However, there are still considerable concerns about possible systematic problems in their experiment. Many of these concerns revolve around the use of a tritiated valine source, in which the energy given up in final state excitations of the molecule following the beta decay of one of the tritium atoms is comparable to the size of the neutrino mass observed. The effect of these final state effects is difficult to calculate in a molecule as complex as valine. In addition, ionization energy loss and backscattering of the betas in traversing the solid source are appreciable and must be very accurately accounted for. These concerns have led us to carry out an experiment using free molecular tritium as the source material. The final state effects have been accurately calculated for the tritium molecule [2-4] and the uncertainties in these calculations cannot generate a spurious neutrino mass greater than 1 eV. In addition, the energy loss in the source is small because the source consists of tritium only and there is no backscattering.

The apparatus has been discussed in detail elsewhere [5] and will only be briefly described here. Molecular tritium is passed through a palladium leak and enters a 3.8-m long, 3.8-cm inner diameter aluminum tube at the center and is pumped away and recirculated at the ends. The tube is held at approximately 130 K to increase the source strength and is uniformly biased to typically -8 kV. The source tube is inside a superconducting solenoid so that betas from the decay of tritium spiral along the field lines without scattering from the tube walls. At one end, the betas are reflected by a magnetic pinch and at the other end are accelerated to ground potential. Also located at the magnetic pinch is a hot filament that emits thermal electrons that neutralize the trapped positive atoms in the source. This keeps the change in source potential due to space charge buildup to less than a volt. The betas are guided through the pumping restriction where the tritium is differentially pumped away and the betas are then focused by nonadiabatic transport through a rapidly falling magnetic field to form an image on a 1 cm diameter collimator at the entrance to the spectrometer. The collimator projects an image down the center of the source tube so that decays originating on or close to the walls of the source tube are not viewed by the spectrometer. A small Si detector is located at a position in front of the collimator where it intercepts a small fraction of the betas from decays in the source tube. This beta monitor serves to normalize the source strength from point to point. The spectrometer is a 5-m long, 2-m diameter, 72-coil toroidal beta spectrometer similar in design to the Tretyakov instrument, but with a number of improvements. Betas from a 1.7 cm^2 area in the source tube are transmitted with 25% efficiency through the spectrometer entrance collimator and form a cone of 30° half angle into the spectrometer. Betas between 19.5^o and 29.5^o are transmitted through the spectrometer to a position sensitive gas proportional counter at the focal plane of the spectrometer. The focal plane detector is 2 cm in diameter with a 2 mm wide entrance slit. The energy resolution for 26 keV betas is 20% and the position resolution is 4 mm FWHM (position information is used to reject backgrounds outside of the slit acceptance). The earth's magnetic field is cancelled to a level of ±10 mG in the spectrometer volume by a set of cosine coils wound around the spectrometer and the zero field setting is determined by fluxgate magnetometers mounted in the spectrometer. The event rate in the last 100 eV was typically 0.10 counts/sec.

The beta spectrum is scanned by changing the voltage applied to the source tube so that betas of constant energy are analyzed by the spectrometer. By accelerating the betas by several keV, not only is the emittance of the source improved, but the betas of interest from the source are raised in energy well above backgrounds from betas originating from decay elsewhere in the pumping restriction or spectrometer. The beta monitor is biased at the same voltage as the source tube, which results in constant energy betas being detected by the beta monitor.

In order to determine the overall source and spectrometer resolution, we introduce ^{83m}Kr into the source tube in the same manner as tritium is injected. The krypton produces a 17.835(20)-keV conversion line and the shakeup and shakeoff effects are known [6-9] so that their contribution can be accurately removed from the resolution function. The contribution from scattering of the conversion electrons from nitrogen molecules in the source gas (which builds up due to the recirculation of the krypton) has been calculated using existing experimental data [10-11]. These measurements yield a spectrometer resolution function which has a skewed Gaussian shape with a FWHM of 52 eV for the first data set and 38 eV for the last two data sets. The change in resolution between the data sets was due mainly to improved cancellation of residual magnetic fields from the source magnets in region of the spectrometer. The total resolution function for the complete source and spectrometer consists of the skewed Gaussian optical resolution function determined from the krypton measurements which is folded in with the energy loss spectrum of betas scattering from tritium molecules in the source. This energy loss spectrum was determined using the measured tritium density in the source tube together with Monte Carlo calculations of electron scattering on molecular hydrogen [12-13] including tracking the betas along the magnetic field lines in the source region. Approximately 10% of the betas are trapped in local magnetic minima in the source region and must scatter several times to escape from the source region, while approximately 5% of the untrapped betas scatter before leaving the source region.

Measurements of backgrounds from the source and tritium contamination of the spectrometer have been made and we do not observe any backgrounds originating from the source walls, extraction region, or from tritium contamination of the spectrometer at a level less than 1 count/500 sec. The background rate in the focal plane detector has remained constant at 1 count/270 sec and is primarily due to cosmic ray muons traversing the detector.

Three data sets were taken, each of 3-4 days duration, with operating conditions (given in Table I) varied somewhat between runs to check systematic effects. The first two runs were taken with the spectrometer set to analyze 26.0-keV betas and the beta spectrum was scanned from 16.44 to 18.94 keV in 10 eV steps. Two randomly selected data points were taken for 600 seconds each, followed by a 200 second data run at 16.44 keV in order to check for time dependent systematic errors. The third data set was taken in a similar manner, except that the spectrometer was set to analyze 26.5 keV betas in order to check for any systematic effects in varying the extraction voltage (and therefore the extraction efficiency). Extra data points were taken in 5-eV steps near the endpoint in the last run. Several other data sets taken were not used because resolution measurements were not available, or the runs were incomplete.

To analyze the data, a predicted beta spectrum is generated which includes the molecular final states, Coulomb corrections, screening corrections, nuclear recoil effects, weak magnetism, and acceleration gap effects (the last three are negligible). The total system resolution and energy loss in the source are folded in with the calculated spectrum. A five-parameter fit (varying the

TABLE I. Summary of Parameters for each run and results from fitting procedure for each run. Uncertainties in Δm_ν^2 are 1σ.

	RUN 3	RUN 4-A	RUN 4-B	COMBINED
E_{SPECT} (keV)	26.0	26.0	26.5	
ΔE(FWHM,eV)	52.1 ± 1.7	32.0 ± 1.5	32.4 ± 1.3	
Skewness	.133	.153	.173	
Total Events	5,081,270	944,353	567,581	6,593,204
Counts in 100 eV	170	93	273	536
Background in 100 eV	36	28	53	117
Quadratic Term ($10^{-8}/eV^2$)	-1.28	-1.80	-0.64	
Number of Data Points	254	250	220	
E_o(eV)	18584.8	18585.7	18584.4	18585.0
$\Delta m_\nu 2$ (statistical eV^2)	1126	1720	688	
$\Delta m_\nu 2$ (resolution, eV^2)	70	364	52	638
$\Delta m_\nu 2$ (e_2 loss, eV^2)	50	28	25	
$m_\nu 2$ (eV^2)	-1190	1880	-63	-186

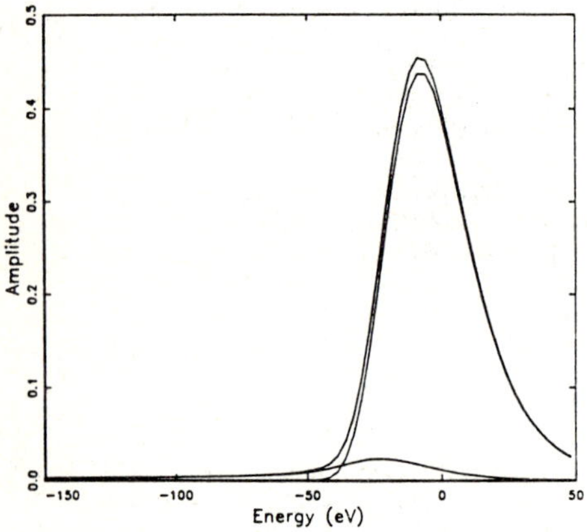

Fig. 1. Resolution function for Run 4-B showing optical resolution, energy loss component and sum of both.

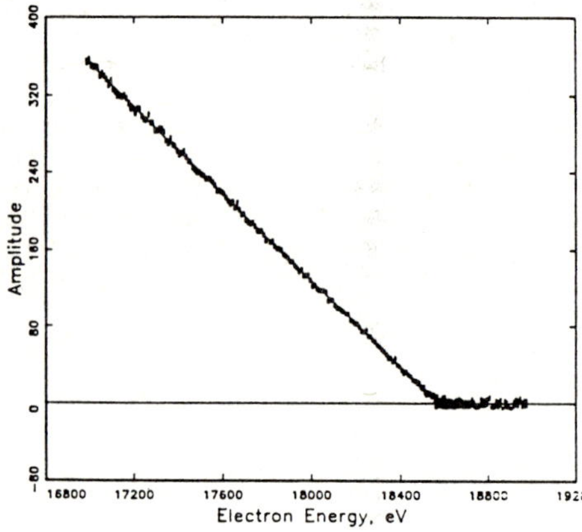

Fig. 2. Kurie Plot for Run 4-B.

amplitude (determined by total number of events), endpoint energy, neutrino mass, background level, and a quadratic extraction efficiency term) in a maximum likelihood procedure with Poisson statistics is then performed. Extensive Monte Carlo calculations were carried out in order to study systematic effects and correlations between variables, and to verify the unbiased character of the fit estimator.

In Table I we summarize run parameters and fit results for the four data sets. The consistency between the measured endpoint energies is good, notwithstanding the large change in spectrometer resolution between runs 3 and 4A, and the change to 26.5 keV operation in run 4B. The overall uncertainty in the endpoint energy is dominated by the 20-eV uncertainty in the energy of the ^{83m}Kr calibration line, however. The quadratic correction term varies from run to run owing both to changes in focus coil excitation and (in run 4B) to normalization of the source intensity by interpolation between calibration points rather than by the Si detector, which had become excessively contaminated. A linear term was tried in place of the quadratic one and gave similar results but with larger variations as the fitting interval was successively truncated. Such variations were within statistics with the (fixed) quadratic term when the fitting interval was varied over the range 2500 to 300 eV. There was no statistical evidence that both linear and quadratic terms were required.

Statistical errors in m_ν^2 were extracted from the Ξ^2 plots (which were closely parabolic in m_ν^2). A conservatively estimated systematic error arising from imperfect knowledge of the resolution function in each run was then added linearly to the statistical error. The resolution-function uncertainties have both systematic and statistical components, but are in any case believed to be largely uncorrelated from run to run. Finally, a systematic uncertainty from the measurement of the density of the source gas and the Monte Carlo simulation of multiple scattering was added linearly to the weighted average of all runs. These were the only systematic uncertainties considered to be non-negligible.

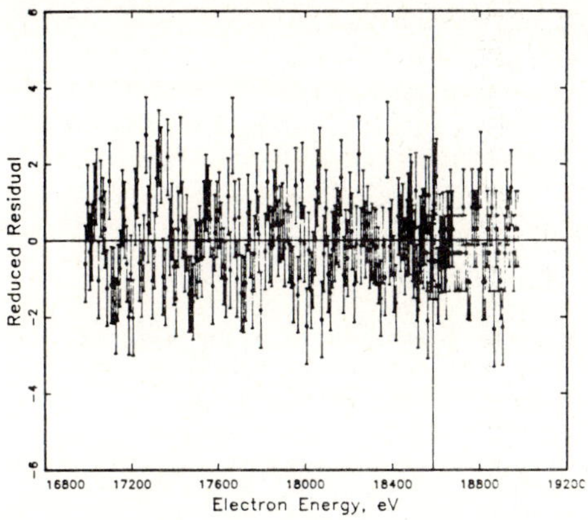

Fig. 3. Residual plot for Run 4-B with m_ν=0.

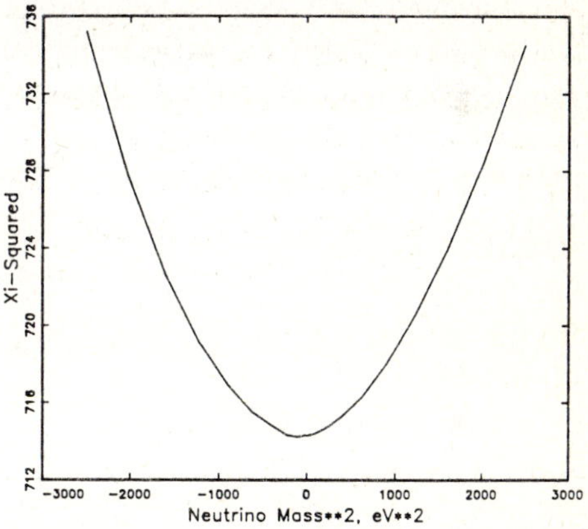

Fig. 4. Ξ^2 plot for all three runs.

The uncertainty in the final result is predominantly statistical. An upper limit the mass of the electron antineutrino is found to be 29.3 eV at the 95% confidence level (C.L.) or 25.4 eV at the 90% C.L. It does not support the central value reported by Lyubimov (1) of 30(2) eV, but neither does it exclude the lower part of the range 17 to 40 eV. The present result is, for all practical purposes, model independent. Improvements to the apparatus transmission and resolution now in progress are expected to result in a sensitivity to neutrino mass in the vicinity of 10 eV.

1. S. Boris et al., Phys. Lett. 159B, 217 (1985)
2. R.C. Martin and J.S. Cohen, Phys. Lett. 110A, 95 (1985)
3. W. Kolos et al., Phys. Rev. A31, 551 (1985)
4. O. Fackler et al., Phys. Rev. Lett. 55, 1388 (1985)
5. J.F. Wilkerson, Proc. XIX Recontre de Moriond (1985) to be published
6. T.A. Carlson and C.W. Nestor, Phys. Rev. A8, 2887 (1973)
7. D.P. Spears et al., Jour. Chem. Phys. 60, 103 (1974)
8. J.S. Levinger, Phys. Rev. 90, 11 (1953)
9. X. Bambynek et al., Rev. Mod. Phys. 44, 716 (1972)
10. E.N. Lassettre, Can. Jour. Chem. Phys. 47, 1733 (1969)
11. T.C. Wong et al., Phys. Rev. A12, 1846 (1975)
12. R.C. Ulsh et al., Jour. Chem. Phys. 60, 103 (1974)
13. J.W. Liu, Jour. Chem. Phys. 59, 1988 (1973)

Measurement of the Mass of the Electron Neutrino Using Electron Capture in ^{163}Ho

S. Yasumi[1,2], M. Ando[2], H. Maezawa[2], H. Kitamura[2], T. Ohta[2], F. Ochiai[2], A. Mikuni[2], M. Maruyama[3], M. Fujioka[4], K. Ishii[4], T. Shinozuka[4], K. Sera[4], T. Omori[4], G. Izawa[4], M. Yagi[4], K. Masumoto[4], K. Shima[5], T. Mukoyama[6], Y. Inagaki[6], I. Sugai[7], A. Masuda[8], and O. Kawakami[8]

[1]Teikyo University, Hachioji, Tokyo 192-03, Japan
[2]National Laboratory for High Energy Physics, KEK, Ibaraki-ken 305, Japan
[3]Osaka University, Toyonaka, Osaka 560, Japan
[4]Tohoku University, Sendai 980, Japan
[5]University of Tsukuba, Niihari, Ibaraki 305, Japan
[6]Kyoto University, Kyoto 606, Japan
[7]Institute for Nuclear Study, University of Tokyo, Tanashi, Tokyo 188, Japan
[8]University of Tokyo, Tokyo 113, Japan

1. Introduction

The mass of the electron neutrino has been recently investigated using electron capturing isotopes including 163Ho [1,2,3], 193pt [4] and so on. There are two ways to measure $m_{\nu e}$; one proposed by A. De Rújula [5] of CERN which is essentially based on three body phase space in radiative electron capture process, and another that utilizes the $m_{\nu e}$-dependence of the electron capture rate, which was discussed by Bennett et al. [6]. Considerations of the intensity of internal bremsstrahlung at electron capture in a 163Ho source, indicate that the first approach is very difficult [7,8]. Therefore, for the time being, we decided to pursue the second approach [9].

2. Principle of the Method

If $Sp^{163}Ho$ stands for a photon spectrum from 163Ho, where the number of photons per atom per second is plotted as a function of the energy of the photons, we then have

$$Sp^{163}Ho(x) = \lambda M1 \cdot SM1(x) + \lambda M2 \cdot SM2(x) \quad (1)$$

where

SMi(x)(i=1,2) : M X-ray spectrum of dysprosium in the case where there is one vacancy in the Mi subshell only,

x : channel number.

Equation (1) tells us that when we reconstruct $Sp^{163}Ho$ using the SM1 and SM2 spectra, the coefficients of SM1 and SM2 correspond to λM1 and λM2, respectively.

The experimental procedures for obtaining the SM1 and SM2 spectra for dysprosium are as follows. If SEα denotes an M X-ray fluorescence spectrum from Dy ions excited by monochromatic photons having an energy Eα, SEα is represented by the following equation,

$$SE\alpha(x) = N\alpha m \sum_i \sigma_i \alpha \cdot SMi(x) \quad (i = 1 \sim 5,\ \alpha = a \sim e) \quad (2)$$

where

SMi (i = 1 ∿ 5) : the same as in eq. (1),

$\sigma_i\alpha$: photoinonization cross section of Mi subshell for a photon of energy Eα,

Nα : total number of incident photons of energy Eα per second,

m : number of dysprosium atoms in a target per cm^2.

For Eα, we take five different energies: Ea, Eb, Ec, Ed and Ee where Ea > EM1 > Eb > EM2 > Ec > EM3 > Ed > EM4 > Ee > EM5, and EMi (i = 1 ∿5) stands for the binding energy of the Mi subshell of the dysprosium atom. Then equation (2) becomes five equations. If $\sigma_i\alpha$, Nα and m are known, one can obtain SMi (i = 5 ∿ 1) by using these five equations. Spectra SM1 and SM2 thus obtaind, can be used to reconstruct the $Sp^{163}Ho$ spectrum for obtaining λM1 and λM2. On the other hand, λt (total decay constant) can be determined by isotope-dilution mass spectrometry. Using λM1, λM2 and λt, the three quantities, $m_{\nu e}$, the Q-value, and the nuclear matrix element relevant to the decay of 163 Ho, can be determined.

3. Experiments and Results

Undulator radiation, with a double reflection crystal monochromator, from the 2.5 Gev Electron Storage Ring of the Photon Factory in KEK was used as a light source for the fluorescence measurement on dysprosium. We measured SEα spectra using monochromatic photon beams with five different energies; Ea = 2.250 KeV, Eb = 1.944 KeV, Ec= 1.758 Kev, Ed = 1.504 KeV and Ee = 1.314 KeV.

Despite much effort, we have not yet succeeded in obtaining reliable data on σ_i^α and N^α in eq. (2). We therefore analysed the data using theoretical SMi spectra calculated by us, which were modified so as to fit a 163Ho photon spectrum and the SEα dysprosium spectra. The reconstruction using these SM1 and SM2 thus obtained, was performed as shown in Fig. 1. Results of the reconstruction are:

$$\lambda M1 = 0.9740 \pm 0.0041 \times 10^{-12}\ sec^{-1},$$
$$\lambda M2 = 0.0817 \pm 0.0035 \times 10^{-12}\ sec^{-1}.$$

On the other hand, the half life of 163Ho was determined by measuring the production rate of 163Dy due to electron capture in 163Ho with isotope-dilution mass spectroscopy. Our result was

$$T1/2 = 4569 \pm 40\ yr\ (\lambda t = 4.807 \pm 0.042 \times 10^{-12} sec^{-1})$$

which is in excellent agreement with Baisden et al.'s [10].

λM1 and λM2 divided by λt are plotted in Fig. 2. From this figure we conclude

$m_{\nu e}$ < 550 eV (68% CL).

Furthermore, we found 4.98±0.01 in terms of log ft-value for the matrix element relevant to the decay 163Ho → 163Dy. This value lies in between CERN-Aarhus-group's value [1] and Princeton-Livermore group's one [3,10].

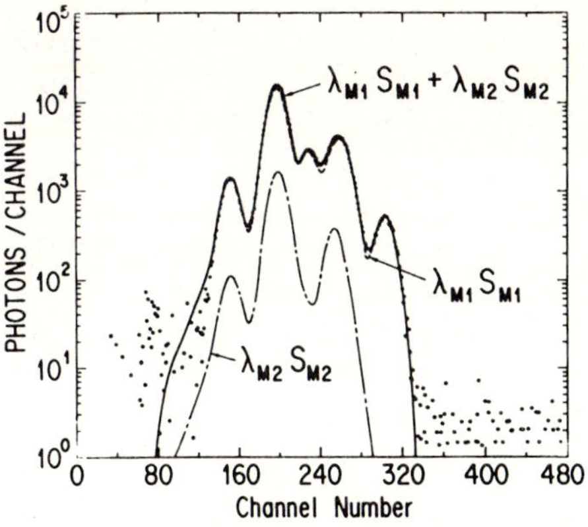

Fig. 1 Reconstruction of 163Ho spectrum

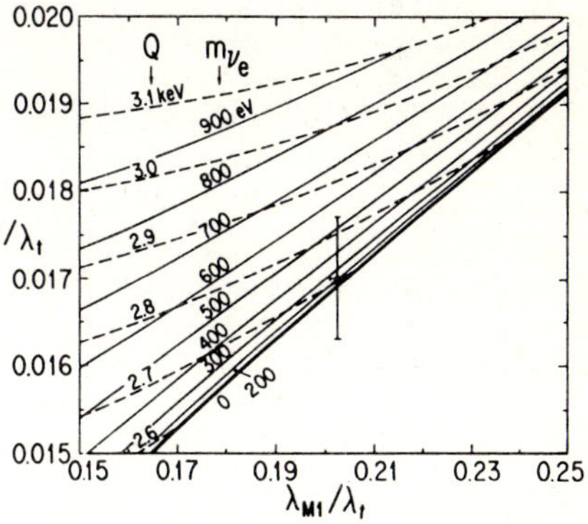

Fig. 2 Summary of the results

References:

1 J.U. Andersen et al.: Phys. Lett. 113B, 72 (1982)

2 S. Yasumi et al.: Phys. Lett. 122B, 461 (1983)

3 F.X. Hartmann and R.A. Naumann: Phys. Rev. C31, 1594 (1985)
4 B. Jonson et al.: Nucl. Phys. A396, 479C (1983)
5 A. De Rujula: Nucl. Phys. B188, 414 (1981)
6 C.L. Bennett et al.: Phys. Lett. 107B, 19 (1981)
7 S.Yasumi: Proc. the International Europhysics Conference on High Energy Physics, Brighton, UK, July 1983. p. 391.
8 K. Riisager et al.: Physica Scripta 31, 321 (1985)
9 S. Yasumi et al.: Proc. Neutrino '84, Nordkirchen near Dortmund, June 11-16, 1984. p. 202.
10 P.A. Baisden et al.: Phys. Rev. C28, 337 (1983)

3.4 Muon Physics

3.4.1 Muon Decay and Lepton-Flavor Conservation

Study of Rare and Forbidden μ- and π-Decays

R. Engfer

Physik-Institut der Universität, Schönberggasse 9,
CH-8001 Zürich, Switzerland

1. Introduction

The standard Glashow-Salam-Weinberg model [1-2] combines weak and electromagnetic interactions on the basis of a spontaneously broken $SU(2)_L xU(1)$ symmetry. The experimental verification [3-5] of the existence of its most noticeable prediction, the W and Z bosons led to profound trust in this model as the correct description of electroweak processes at "low energies" up to tens of GeV. The phenomenon of maximal parity violation, i.e. the V-A structure of weak interactions and the masslessness of neutrinos is built into the theory by assigning lefthanded fermions to SU(2) doublets and righthanded ones to singlets and requiring lepton number conservation.

However, there is a widespread mainly aesthetical aversion against the acceptance of the standard model as the ultimate theory concerning the many arbitrary parameters (coupling constants, mixing angles, masses, and potential parameters), the left-right asymmetry, the family replication, the lack of unification with other interactions like QCD and gravity, and the existence of elementary Higgs scalars. Candidates for a more complete theory are grand unification models [6], supersymmetric [7,8] and superstring theories [9], technicolor [10-12], horizontal symmetries [13-16] as well as composite models [17]. In most of these models neutrinos can acquire masses [18-20] which have either direct effects on the normal μ decay spectrum [21-24] or can influence it by the presence of Lorentz structures different from V-A [25-27]. In addition, these models do not require exact conservation of μ-lepton number and thus predict decays like $\mu \to e\gamma$ and $\mu \to 3e$. Already Lee and Yang [28] have discussed that baryon- and lepton-number are not necessarily conserved as "gauge charges" on the basis of an exact unbroken gauge symmetry. Specific models predict various ratios between the branching ratios of different μ-lepton number violating processes, however, they do not predict any lower limit.

The success of the Glashow-Salam-Weinberg model [1,2], and the hope that physics beyond this standard model could also be described by gauge theories revived the interest in this "new physics" for the last 9 years. Experimentators were stimulated to increase as much as possible their sensitivities to detect small deviations from V-A interaction in normal μ decay or to detect a finite branching ratio for "forbidden" μ decays such as $\mu \to e\gamma$. Apart from μ decay, also other low energy experiments get very important to test physics beyond the standard model: the search for finite neutrino masses by endpoint spectroscopy [29,30], neutrinoless double beta decay [31], neutrino oscillations [32]. neutrino decays [33], secondary peaks in pseudoscalar meson decays [34], exotic pion-, tau-, and kaon-decays [35] as well as to improve on (g-2) for the muon [36]. These experiments have to be regarded as complementary to direct searches for new particles in high energy experiments.

Spectrometers to study e.g. the forbidden decay $\mu \to 3e$ are also suitable to study rare but allowed decays of μ's and π's with several charged particles in the decay channel. Therefore, such reactions have been studied partly simultaneously or after minor modifications of the existing spectrometers. Examples discussed are the $\mu \to 3e2\nu$ decay (Chap. 2), which is observed as a background in a $\mu \to 3e$ search and measures the μ decay coupling constants of weak interaction and the $\pi \to 3e\nu$ decay (Chap. 4) which is sensitive to the weak form factors of the pion. In addition, these data put experimental limits for short-lived neutral particles postulated to cure the CP-problem in QCD (Chap. 5).

2. Normal and Rare μ Decays

At present the maximal parity violation, the V-A structure of the weak interaction and the masslessness of neutrinos are built into the theory by hand by assigning lefthanded fermions to SU(2) doublets and righthanded ones to singlets and requiring lepton number conservation. Decay parameters determined from muon decay experiments set the best constraints on non-standard Lorentz structure of purely leptonic charged weak interactions. These data are free of uncertainties from hadronic contributions as observed in semileptonic or hadronic weak interactions. As shown in [37,38] they agree with the predictions of the standard model. V, A and T couplings different from pure V-A are zero within 4-11%, whereas S and P couplings could be as large as 45% and a mixed right-lefthanded scalar coupling ($h_{12} \equiv g^S_{RL}$ in the helicity projection form) is unconstrained. It should be mentioned that new data of the rare decay $\mu^+ \to e^+\nu\bar{\nu}e^+e^-$ yield constraints on the weak coupling constants of about a factor 4 less accuracy [39,40]. The experimental branching ratio $(3.4\pm0.4)\times10^{-5}$ agrees with the V-A value of $(3.5916\pm0.0022)\times10^{-5}$ and also the invariant mass distribution of the observed events agrees with the calculation (Fig. 1).

It's remarkable that from specific μ decay parameters alone constraints on parameters of non-standard models can be given. The mass of a righthanded boson in left-right symmetric theories with massless neutrinos is limited to $m_{W_R} > 432$ GeV/c^2 for arbitrary mixing by the decay parameter $P_\mu\xi\delta/\rho$ [41]. Masses of supersymmetric gauge bosons are constrained to $m_{\tilde{W}} > 350$ GeV/c^2 [42] and charged superleptons to $m_{\tilde{l}} > 150$ GeV/c^2 [43] by the decay parameters ρ, δ and ξ requiring masses of less than

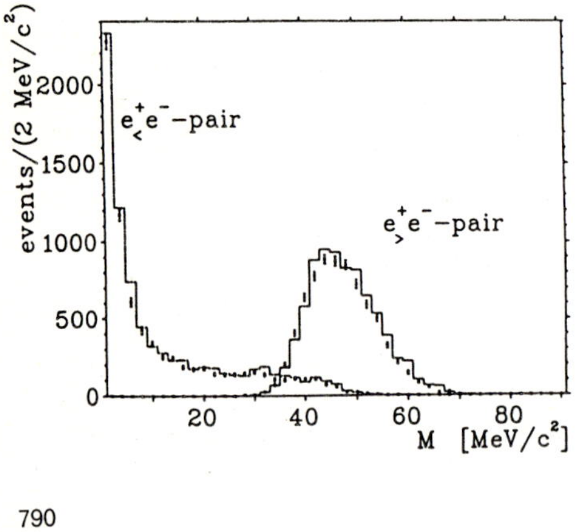

Fig. 1: Distribution of the invariant mass M of the e^+e^- pairs in the rare decay $\mu^+ \to e^+\nu\bar{\nu}e^+e^-$ as measured with the SINDRUM spectrometer [39,40]. The two pairs e^+e^- with the direct positron and the Dalitz pair e^+e^- are distinguished with high confidence by their higher or lower invariant mass $(Mc^2)^2 = \Sigma E_i{}^2 - |\Sigma \vec{p}_i c|^2$. The Monte Carlo simulation (histogram) is calculated for a pure V-A interaction and takes into account all kinematic constraints and efficiencies of the spectrometer. Small accidental background and background from Bhabha scattering has been subtracted.

20 MeV/c^2 for superneutrinos and photinos. Other limits to non-standard models can be found in [41].

3. Search for Forbidden μ Decays

The present status of experiments and new proposals to search for processes violating the conservation of muon number is summarized in Table 1. If no candidate event has been found in an experiment, the upper limit for the branching ratio is B (90% C.L.) = 2.3 / ($\dot{N}$ T ω ε) given by the stop rate $\dot{N}$ in a measuring time T a detector solid angle ω and a total specific efficiency ε. Background contributions are prompt, which are independent of the stop rate and duty cycle and which is suppressed by kinematic detector resolution, and accidental background which is reduced by time resolution and high duty cycle of the beam. As an example: The decay $\mu^+ \rightarrow e^+e^+e^-$ is kinematically characterized by $\sum_{i=1}^{3} E_i = m_\mu c^2$ and $\sum_{i=1}^{3} \vec{p}_i = 0$ and the three particles are prompt and have a common vertex within the target. The background consists of the rare decay $\mu^+ \rightarrow e^+\nu\bar{\nu}e^+e^-$ and of two- and threefold accidental coincidences. The experimental problem is the reduction of a stop-rate of $10^7\,s^{-1}$ to a tolerable rate of about 1 event s^{-1} to be written on tape by a fast on-line trigger system [e.g. 61].

Table 1. Experiments and proposals for processes violating the conservation of muon number. The upper limits for the branching ratios corresponds to 90% C.L. K-decays are added for completeness.

Process	Present branching ratio	Ref.	Expected sensitivity of ongoing experim. or new proposals	Ref.
$K_L^o \rightarrow \mu e$	$< 10^{-8}$ a)	[44]	10^{-10}	[52]
			$5 \cdot 10^{-13}$	[53]
			10^{-11}	[54]
$K^+ \rightarrow \pi^+\mu e$	$< 4.8 \cdot 10^{-9}$	[45]	10^{-11}	[55]
$\mu \rightarrow e\gamma$	$< 4.9 \cdot 10^{-11}$	[46]	10^{-13}	[56,57]
$\mu \rightarrow e\gamma\gamma$	$< 7.2 \cdot 10^{-11}$	[47]	10^{-12}	[56]
$\mu \rightarrow 3e$	$< 3.1 \cdot 10^{-11}$	[47]		
	$< 2.4 \cdot 10^{-12}$	[48]	$8 \cdot 10^{-13}$	
$\mu^-N \rightarrow e^-N$	$< 1.6 \cdot 10^{-11}$	[49]		
	$< 4 \cdot 10^{-12}$	[50]	10^{-13}	[58]
$\mu^+e^- \rightarrow \mu^-e^+$	< 0.04	[51]	10^{-5}	[59,60]

a) multiplied by a factor of five (see Ref. 35)

The offline analysis rejects accidental background - mainly twofold coincidences between a normal positron and an e^+e^- pair from pair production and Bhabha scattering in the target - by requiring timing (640 ps FWHM)and a vertex on the target surface (2 mm FWHM). The result of the final analysis is shown in Fig. 2. No $\mu \rightarrow 3e$ event is found which yields an upper limit for the branching ratio of $B(\mu \rightarrow 3e) < 2.4 \times 10^{-12}$ (90% C.L.) using the total of 7.3×10^{12} stopped μ^+ and an overall efficiency of (13.8 ± 0.5)% [48]. This ratio will be improved by a factor of about 3 this year.

Details on the LAMPF experiments to search for $\mu \rightarrow e\gamma$ and μ-$e\gamma\gamma$ decays are presented at this conference by R. Mischke, therefore, as a second example, the search for the process $\mu^-N \rightarrow e^-N$ with a hexagonal time projection chamber (TPC) at TRIUMF [49] as shown in Fig. 3 will be discussed. This neutrinoless reaction is characterized by the emission of a single electron with an energy $E_{e^-} \cong m_\mu c^2$. The main background are μ^-

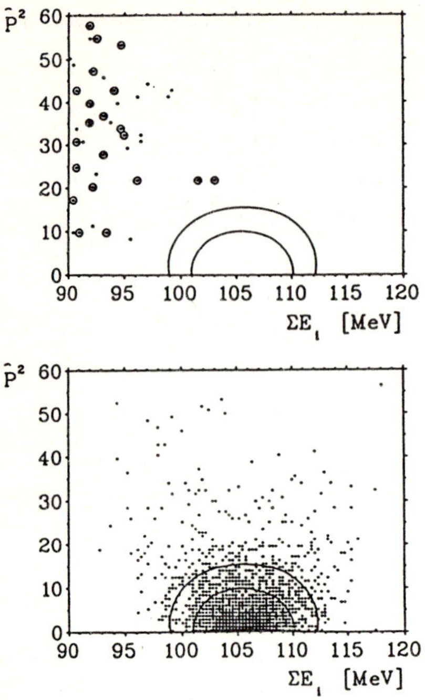

Fig. 2. Total normalized momentum squared versus total energy for $e^+e^+e^-$ events from the SINDRUM $\mu \to 3e$ search [48], for measured 42 prompt events (top), 2312 Monte Carlo simulated $\mu \to 3e$ events (bottom]. Encircled events survive an additional small angle cut. No candidate for the decay $\mu \to 3e$ is found within the 90% and 68% contours indicated.

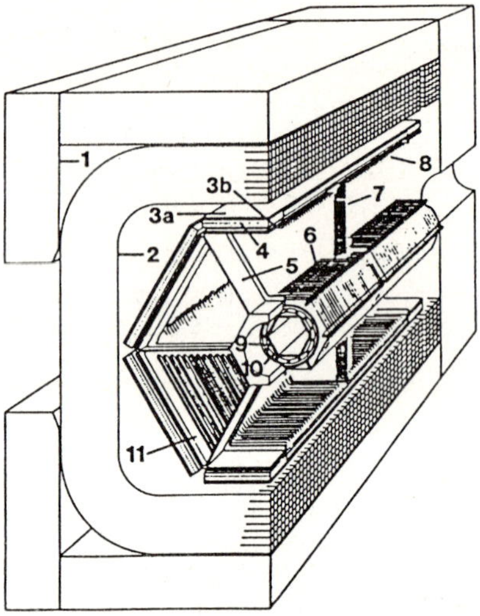

Fig. 3. Perspective view of the time projection chamber used for the search of anomalous muon conversion at TRIUMF [49]. 1: Magnet iron, 2: Coil, 3: Outer trigger scintillators, 4: Outer trigger proportional counters, 5: Support, 6: Inner field wires, 7: Central high voltage plane, 8: Outer field wires, 9: Inner trigger scintillators, 10: Inner trigger proportional chamber, 11: End cap proportional wire modules.

decays in orbit with an electron energy ranging up to $m_\mu c^2$, radiative capture of muons and beam contaminating pions. The result given in Ref. [49] of $B_{\mu-e} = \Gamma(\mu^-Ti \to e^-Ti) / \Gamma(\mu^-Ti\ \text{capture}) < 1.6 \times 10^{-11}$ (90% C.L.) has already been improved to 4×10^{-12} [50]. A sensitivity of $< 10^{-13}$ is planned for a proposed experiment at SIN [58] with the SINDRUM II spectrometer (Fig. 4) for which the final data taking is expected in 1990.

Various limits on parameters of specific non-standard models are deduced from the forbidden μ decays. Superleptons must be degenerate in mass in the percent region [62] to meet the stringent bounds from $\mu \to e\gamma$ etc. Alternatively, if these transitions are mediated by hypothetical horizontal gauge bosons [13,14] or Extended Technicolor gauge bosons [12] their masses must be larger then 50-100 TeV/c^2, whereas, if leptons are composed out of preons the compositeness scale must be larger than ~ 500 TeV [63] corresponding to a size of the muon of $< 4 \cdot 10^{-20}$ cm. The limits quoted for exotic contributions are subject to large uncertainties because of unknown coupling constants and mixing angles. On the other hand, the same kind of uncertainties also apply to limits deduced from high energy experiments as long as the corresponding particles are not produced as real particles. As a further example limits on the masses of particles postulated by specific models derived from the upper limit of the $\mu \to 3e$ decays are given in Table 2. A more detailed review on the motivation and on experiments of normal and forbidden μ decays is given in [38].

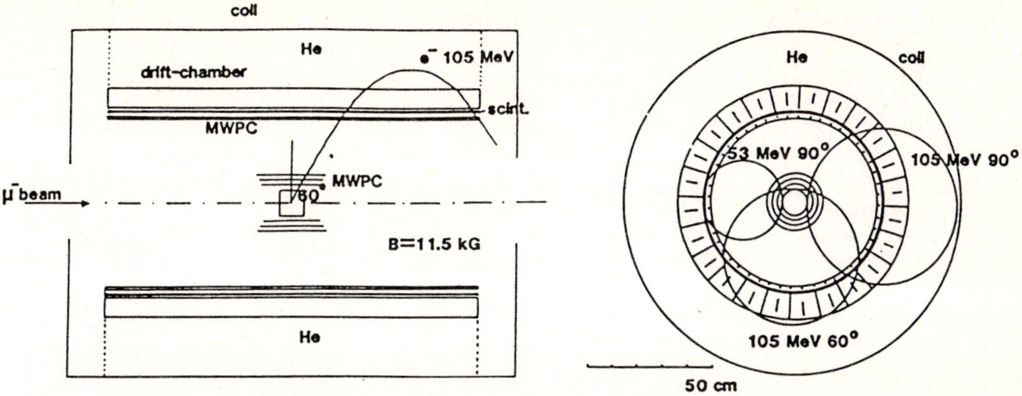

Fig. 4. Apparatus proposed by the SINDRUM II collaboration [58] to search for the anomalous muon conversion $\mu^- N \to e^- N$ with a sensitivity of ~ 10^{-13}.

Table 2: Lower limits on masses of particles postulated by specific non-standard models quoted in the references. The values are derived from the upper limit $B(\mu \to 3e) < 2.4 \times 10^{-12}$ [48] under the assumptions: any mixing is maximal, the coupling constant of any new interaction is equal to the Fermi coupling constant. All masses are given in TeV/c^2.

Theory	Assumptions	Limits	Ref.
Extension of the Standard-Model			
Heavy Leptons	$M_W = 0.08$	$\Delta M<0.016$	[64]
Righthanded Boson		$M_R>0.6$	[20]
Neutral scalar Higgs	$m_\tau = 1.78 \times 10^{-3}$	$M_H>2$	[15]
Heavy gauge boson	$B(K_L\text{-}\mu e) < 2 \times 10^{-9}$	$M>50$	[14]
ETC gauge boson		$M>75$	[12]
Other Interactions			
Milliweak interaction	y = CP-viol. param.	$y<2.2\times10^{-5}$	[65]
Composite Models			
Chiral symmetric		$M>500$	[66]
μ = excited e	via IC($\gamma \to e^+e^-$)	$M>150$	[67]

4. The rare radiative decays $\pi^+ \to e^+\nu_e\gamma$ and $e^+\nu_e e^+e^-$

The two radiative decays $\pi^+ \to e^+\nu\gamma$ and $\pi^+ \to e^+\nu e^+e^-$ are sensitive to both the vector and the axial-vector weak hadronic currents |68|. The normal pion decay $\pi^+ \to \mu^+\nu$ has contributions from the axial current alone. The contributions to the amplitude of the radiative decays can be seperated in an axial-vector current part of inner bremsstrahlung (IB) and a structure dependent part (SD_A), the vector current only leads to a structure dependent term (SD_V). The IB term is parametrized by the pion decay constant f_π = 128 MeV [69]. The terms SD_V and SD_A are parametrized by the vector form factor F_V and by the axial-vector form factors F_A and R or, alternatively, by F_V, $\gamma = F_A/F_V$ and $\xi = R/F_V$ [70,71]. F_V is predicted from the CVC hypothesis, which connects F_V with the π^0 lifetime, $|F_V|$ = 0.0255 [69,72].

The sign of F_V has not been determined. The PCAC hypothesis relates R with the electromagnetic radius of the pion, $R = 1/3\ m_\pi\ f_\pi\ \langle r_\pi^2 \rangle$ [68,70,71]. With $\langle r_\pi^2 \rangle = 0.45\ fm^2$ [73], this gives ξ=2.7. Recent relativistic quark models give $\gamma = \xi/2 = 1$ [74-76]. Replacing the calculated pion radius by the measured one, these models obtain $\gamma = 1.4$ [74,75]. Small values for γ are given by a modified bag model ("non zero but very small") [77] and a non-perturbative QCD calculation, $\gamma = 0.0 \pm 0.3$ [78].

Experimental information on γ has been obtained for the decay $\pi^+ \to e^+\nu\gamma$. Unfortunately, the experiments measured mainly a SD term proportionally to $(1+\gamma)^2$, giving two values for γ, $\gamma = 0.50 \pm 0.12$ or $\gamma = -2.42 \pm 0.12$ [79] and $\gamma = 0.52 \pm 0.06$ or $\gamma = -2.48 \pm 0.06$ [80]. In the first experiment the large negative value gave a better fit and in the second experiment the positive value was favored by a likelihood ratio of 8.5, thus an ambiguity of the sign existed. The recent LAMPF data [81] yield $\gamma = 0.25 \pm 0.12$ with a likelihood ratio of 2000 for the positive value.

From the $\pi^+ \to e^+\nu e^+e^-$ decay an unambiguous determination of γ is possible. Also the second axial-vector form factor R, which contributes to the decay amplitude for massive virtual photons only, can be determined. The IB contribution can be suppressed effectively by requiring the opening angle of any e^+e^- pair to be larger than 10°. In an experiment done at SIN with the SINDRUM spectrometer the $\pi^+ \to e^+\nu e^+e^-$ decay was observed for the first time [82]. A total of $(3.8 \pm 0.7) \times 10^{12}\ \pi^+$ were stopped in the target. The online and offline data reduction was done in a similar way as in the previous muon decay experiments [40,48]. The distribution of $\pi^+ \to e^+\nu e^+e^-$ events remaining after a final offline analysis [82] are shown in Fig. 5. The $\mu^+ \to e^+\nu\bar{\nu}e^+e^-$ events are clearly separated. From the data a branching ratio of $(3.4 \pm 0.5) \times 10^{-9}$ has been determined, they are in good agreement with the distribution of Monte Carlo generated $\pi^+ \to e^+\nu e^+e^-$ events (Fig. 5). A maximum likelihood analysis of the 79 events in the five dimensional phase space yielded $F_V = 0.029 \pm 0.019$, $F_A = 0.018 \pm 0.015$, $R = 0.063 \pm 0.026$, and $\gamma = F_A/F_V = 0.7 \pm 0.5$ in agreement with the small positive value of γ and $\xi = R/F_V = 2.3 \pm 0.6$ which is in agreement with the PCAC prediction.

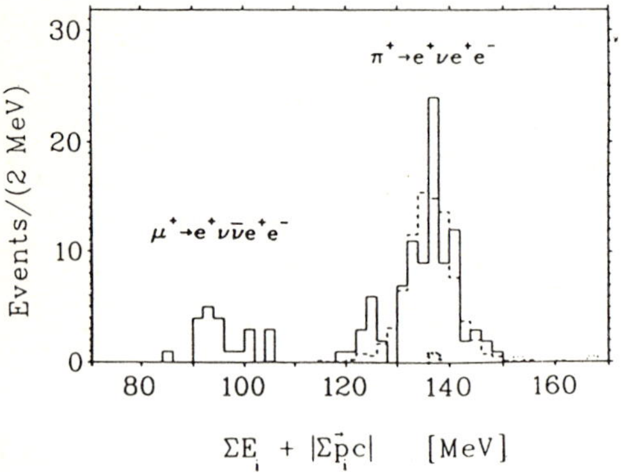

Fig. 5. $\Sigma E_i + \Sigma|\vec{p}_i c|$ distribution of 79 $\pi^+ \to e^+\nu e^+e^-$ events [82]. A cut was applied on the e^+e^- opening angle between 18° and 32° and $E_{e^+} < 56$ MeV for the other positron which reduces background from the $\pi^+ \to e^+\nu\gamma$ or $\mu^+ \to e^+\nu\bar{\nu}$ decays followed by external conversion or Bhabha scattering, respectively. The dashed line is the distribution of Monte Carlo generated $\pi^+ \to e^+\nu e^+e^-$ events.

5. Limits for short lived neutral particles in μ^+ or π^+ decay

Specific non-standard theories have been proposed to solve the strong CP problem; they introduce new neutral particles ϕ e.g. the axion [83,84]. All axion searches have been negative so far [85,86], however, a new interest in axions arose after the observation of positron and electron peaks at 300 keV in heavy ion collisions at GSI [87-90]. Proposing a variant axion an isovector transition $\pi^+ \to e^+\nu\phi$ could occur in the model [90] with a branching ratio of 2×10^{-6} [91]. If the ϕ is decaying fast enough into an e^+e^- pair this decay should have been observed in the search for the $\pi^+ \to e^+\nu e^+e^-$ decay [82].

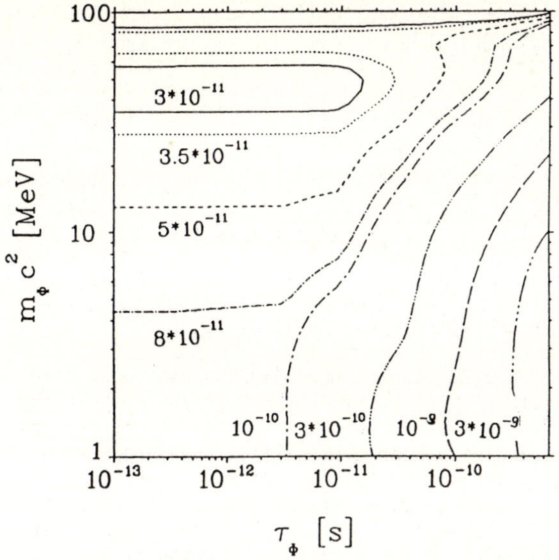

Fig. 6. Contour plot of the upper limit for the branching ratio $\Gamma(\pi^+ \to e^+\nu\phi) / \Gamma(\pi^+ \to \text{all})$ (90% C.L.) as a function of the assumed mass and lifetime [92].

The new $\pi^+ \to e^+\nu e^+e^-$ data have been analysed with respect to the $\phi \to e^+e^-$ decay [92]. The experimental upper limit for the existence of the $\phi \to e^+e^-$ decay in the π^+ decay depends on the ϕ-mass and its lifetime, it is given in Fig. 6. A similar analysis has been done with the data of the search for the $\mu \to 3e$ decay yielding upper limits for the decay $\mu^+ \to e^+\phi$ with the successive decay $\phi \to e^+e^-$. The result is given in [92]. Summarizing no evidence has been seen for a neutral particle emitted in μ^+ or π^+ decay of a mass between 1.02 and 100 MeV/c^2 and decaying with a lifetime of less than 10^{-9}s into e^+e^-. Combining this result with a bound for a $\Delta T = 0$ transition in ^{10}B exclude the variant axion [91,93].

References

1. S.L. Glashow, Nucl. Phys. 22(1961)579
2. S. Weinberg, Phys. Rev. Lett. 19(1967)1264,
   A. Salam, Proc. 8th Nobel Symp. Stockholm, Almquist and Wiksells (1968) p.367
3. G. Arnison et al. Phys. Lett. 122B(1983)103
4. M. Banner et al. Phys. Lett. 122B(1983)476
5. G. Arnison et al. Phys. 126B(1983)398
6. P. Langacker, Phys. Rep. 72C(1981)185
7. H. Haber, G. Kane, Phys. Rep. 117(1985)75
8. H.P. Nilles, Phys. Rep. C110(1984)1
9. J.H. Schwarz, report CALT-68-1290 (1985) "Introduction to Superstrings" Lectures at Trieste Work Shop 1985.
10. R.K. Kaul, Rev. Mod. Phys. 55(1983)449
11. E. Farhi, L. Susskind, Phys. Rep. C74(1981)278
12. S. Dimopoulos, J. Ellis, Nucl. Phys. B182(1981)505
13. R.N. Cahn, H. Harari, Nucl. Phys. B176(1980)135
14. G.L. Kane, R. Thun, Phys. Lett. 94B(1980)513
15. O. Shanker, Nucl. Phys. B185(1981)382 and B206(1982)253
16. D.R.T. Jones et al., Nucl. Phys. B198(1982)45
17. L. Lyons, Progr. Part. Nucl. Phys. V10(1983)227
18. M. Gell-Mann et al. In Supergravity, eds. Nieuwenhuizen, P.V., and Freedman, D. Amsterdam. North Holland, 1979. p. 317
19. T. Yanagida, In Proc. Workshop on Unified Theory and Baryon Number in the Universe, eds. O. Sawada, A. Sugamoto, Tsukuba: KEK. 1979

20. R.N. Mohapatra, G. Senjanovic, Phys. Rev. D23(1981)165
21. P. Kalyniak, J.N. Ng, Phys. Rev. D24(1981)1874
22. J. Missimer et al. Nucl. Phys. B188(1981)29
23. M. Doi et al. Progr. Theor. Phys. 71(1984)1440
24. M.S. Dixit et al. Phys. Rev. D27(1983)2216
25. J. Maalampi et al. Nucl. Phys. B207(1982)233
26. K. Mursula, F. Scheck, Nucl. Phys. B253(1985)189
27. R.E. Shrock, Phys. Rev. D24(1981)1275
28. T.D. Lee, C.N. Yang, Phys. Rev. 98(1955)1501
29. F. Boehm, P. Vogel, Ann. Rev. Nucl. Part. Sci 34(1984)125
30. M. Fritschi et al. Phys. Letters 173B(1986)485
31. W.C. Haxton, G.J. Stephenson, Progr. Part. Nucl. Phys. 12(1984)409
32. H. Blümer, Proc. Int. Conf. on High Energy Physics, Bari, Italy, July 18-24, 1985. Laterza Bari. p. 429
33. F. Vanucci, Proc. 18. Renc. de Moriond, La Plagne, March 13-19 1983. p. 63
34. J. Deutsch, see Reference 27. p. 438
35. H.K. Walter Nucl. Phys. A434(1985)409c
H.K. Walter "The Future of Medium- and High-Energy Physics in Switzerland", Les Rasses, Switzerland, May 17-18, 1985, p. 87
36. V.W. Hughes, Kinoshita, T. Comm. Nucl. Part. Phys. 14(1985)341
37. W. Fetscher et al. Phys. Lett. 173B(1986)102
38. R. Engfer, H.K. Walter, Ann. Rev. of Nucl. and Part. Sc. Vol.36(1986)
39. N. Kraus, A. Kersch, Nucl. Phys. to be published and contribution to this conference
40. N. Kraus thesis University Zürich 1985
41. J. Carr et al. Phys. Rev. Lett. 51(1983)627 51(1983)1222. See also D.P. Stoker theses LBL-20324 1984. A.E. Jodidio thesis LBL. 1986
42. W. Buchmüller, F. Scheck Phys. Lett. 145B(1984)421
43. J. Barber, R.E. Shrock Phys. Lett. 139(1984)427
44. A.R. Clark et al. Phys. Rev. Lett 26(1971)1667
45. A. Diamant-Berger et al. Phys. Lett. 62B(1976)485
46. R.D. Bolton et al. Draft paper LAMPF, Jan. 1986
47. R. Bolton et al. Phys. Rev. Lett. 53(1984)1415 and Ref. 32 p. 447
48. W. Bertl et al. Nucl. Phys. B260(1985)1
49. D.A. Bryman et al. Phys. Rev. Lett 55(1985)465
50. D. Bryman, Workshop on Fundamental Muon Physics, LAMPF, Jan.20-22, 1986 and private communication.
51. G.M. Marshall et al. Phys. Rev. D25(1982)1174
52. R.C. Larsen et al. AGS exp. E780, Brookhaven (1983)
53. S.G. Wojcicki AGS exp. E791, Brookhaven (1984)
54. T. Inagaki et al. KEK Internal 85-1, Ibaraki (1985)
55. M. Zeller AGS exp. E77, Brookhaven (1982)
56. M.D. Cooper LAMPF exp. 969, LAMPF-Los Alamos (1985)
57. H.K. Walter et al. SIN letter of intent R-85-15.0, SIN-Villingen (1985)
58. A. Badertscher et al. SIN letter of intent R-85-07.0, SIN-Villingen (1985)
59. V.W. Hughes et al. LAMPF exp. 985, LAMPF-Los Alamos (1985)
60. K.P. Arnold et al. SIN exp. R-85-08.1., SIN-Villingen (1985)
61. W.H. Bertl et al., Nucl. Instr. Meth. 217(1983)367
62. J. Ellis and D.V. Nanopoulos Phys. Lett. 110(1982)44
63. H. Terazawa et al. Phys. Lett. 112B(1982)387
64. S.M. Bilenky, B.Pontecorvo, Phys. Rep. C41(1978)225
65. S. Barshay, Phys. Lett. 58B(1975)86
66. M. Peskin, Proc. Int. Symp. "Lepton-Photon Interact. at High Energies", Bonn 1981
67. R. Barbieri et al. Phys. Lett. 96B(1980)63 and TH-2850-CERN(1980)
68. F. Scheck and A. Wullschleger Nucl. Phys. B67(1973)504
69. D.A. Bryman et al. Phys. Rep. 88(1982)152
70. O.Yu. Bardin et al. Sov. Nucl. Phys. 14(1982)239
71. A. Kersch, Diplomarbeit, Univ. Mainz, 1984, unpublished;
A. Kersch and F. Scheck Nucl. Phys. B263(1986)475

72. H.W. Atherton et al. Phys. Lett. 158B(1985)81
73. S.R. Amendolia et al. Phys. Lett. 146B(1984)116
74. N. Paver and M.D. Scadron, Nuovo Cim. 78A(1983)159
75. L. Ametller et al. Phys. Rev. D29(1984)916
76. C.Y. Lee, Phys. Rev. D32 (1985) 658
77. Q. Ho-Kim and H.C. Lee, Phys. Rev. D29(1984)1017
78. N.F. Nasrallah et al. Phys. Lett. 113B(1982)61
79. A. Stetz et al. Nucl. Phys. B138(1978)285
80. A. Bay et al. Phys. Lett 174B(1986)455
81. L. Piilonen etal. LAMPF preprint 1986, Lake Louise Conf. May 1986
82. S. Egli et al. Phys. Lett. 175B(1986)97
83. S. Weinberg, Phys. Rev. Lett. 40(1978)223,
F. Wilczek, Phys. Rev. Lett 40(1978)279
84. R.D. Peccei and H.R. Quinn, Phys. Rev. Lett. 38(1977)1440; Phys. Rev. D16(1977)1791
85. S. Yamada, Proc. of the 1983 Int.Symp. on Lepton and Photon Interactions at High Energies, Cornell University, Aug. 4-9, 1983.
Y. Asano et al. Phys. Lett. 107B(1981)159
86. For a review of axion bounds see A. Zehnder, in Fundamental Interactions in Low-Energy Systems, Plenum Press, New York, 1985 and references therein.
87. J. Schweppe et al. Phys. Rev. Lett. 51(1983)2261;
M. Clemente et al. Phys. Rev. Lett. 137B(1984)41;
T. Cowan et al. Phys. Rev. Lett. 54(1985)761.
88. T. Cowan et al. Phys. Rev. Lett. 56(1986)444
89. A. Shafer et al. J. Phys. G:Nucl. Phys. 11(1985)L69
A. B. Balantekin et al. Phys. Rev. Lett. 55(1985)461
J. Reinhardt et al. Phys. Rev. C33(1986)194
L. Kraus and F. Wilczek, NSF-ITP-86-18
90. R. D. Peccei et al. DESY 86-013(1986)
91. W.A. Bardeen et al. DESY 86-054 (1986)
92. R. Eichler et al. Phys. Lett 175B(1986)101
93. L.M. Krauss et al. report YTP 86-13 /CALT-68-1356, May 1986

Search for Muon - to - Electron Conversion in Titanium

P. Depommier[1], *R. Poutissou*[1], *S. Ahmad*[2], *G. Azuelos*[2], *D.A. Bryman*[2], *R.A. Burnham*[2], *E.T.H. Clifford*[2], *M. Hasinoff*[2], *J.A. Macdonald*[2], *T. Numao*[2], *J.-M. Poutissou*[2], *J. Summhammer*[2], *M.S. Dixit*[3], *C.K. Hargrove*[3], *H. Mes*[3], *M. Blecher*[4], and *K. Gotow*[4]

[1]Université de Montréal, Montréal, Québec, Canada, H3C 3J7
[2]TRIUMF, 4004 Wesbrook Mall, Vancouver, B.C., Canada, V6T 2A3
[3]National Research Council, Ottawa, Canada, K1A 0R6
[4]Virginia Polytechnic Institute and State University, Blacksburg, VA 24061, USA

1 Introduction: The question of muon – to – electron conversion is as old as the muon itself. When it was realised that the muon does not decay into electron and photon the concept of separately conserved muonic and electronic lepton numbers was introduced as an ad hoc hypothesis, without profound justification. But it is only recently that this question has become of great importance, with the advent of the gauge theories of fundamental interactions. The "standard model" based on the electroweak group SU(2) x U(1) with the minimal particle content (no right – handed neutrino, only one Higgs doublet) does not allow any muon number (or any partial lepton number) violation. But the "standard model" is not believed to be the final theory of nature. It has to be modified, generalised, and hopefully embedded in a more complete unification scheme. To do this there are many possibilities which lead to muon number violation in a very natural and inescapable way. Therefore the study of muon number (and other lepton number) violation is one of the most important issues in particle physics. Processes which violate muon number have already been searched for extensively at the meson factories. They have not been observed but very stringent upper limits for their existence have been obtained, which put strong constraints on theoretical models, in particular on mixing parameters and masses of hypothetical particles on the TeV scale. All muon number violating processes are important and must be studied. They will be discussed at length by other speakers at this conference. This paper describes an experiment which has been carried out at TRIUMF, the Canadian meson facility, to search for muon – to – electron conversion in a nucleus with a Time – Projection – Chamber (TPC): μ^- nucleus $\rightarrow$ e^- nucleus. This process has some advantages. There is a coherent contribution of the quarks for the ground – state to ground – state transition, leading to an enhancement of the reaction rate, whereas inelastic excitations are strongly inhibited by Pauli blocking. Experimentally one has to search for a single electron of well – defined energy (the muon mass less the binding energy of the muonic atom) in an energy region which is almost background – free.

Preliminary results have already been published with upper limits down to 1.6 x 10^{-11} for the branching ratio R of muon – to – electron conversion to total muon capture[1].

2 The Experimental Apparatus: The TRIUMF Time – Projection – Chamber which is used as an electron spectrometer has been described elsewhere[2]. It consists of two large (hexagonal) drift volumes separated by a central plane which is connected to a negative high voltage. A magnetic field B, parallel to the electric field E, provides a measurement of the particle momentum. The target is located at the center of the TPC. A charged particle coming from the target produces ionisation electrons in the gas, which drift along the electric fields (towards the end – caps of the TPC) and spiral around the magnetic field. When these electrons reach the end – caps they give rise to an avalanche on the anode wires which induce positive charges on the cathode pads which are placed behind the anode wires. In this way one is able to determine the circular projection of the trajectory (an helix) on a plane (called xy). The third coordinate z is obtained by measuring the drift time, which is defined as the time between the triggering of internal and external scintillation counters (placed inside and outside the drift volume) and the arrival of the ionisation electrons at the anode wire. Table 1 gives the various characteristics of the TRIUMF TPC.

Table 1: TPC Characteristics

TPC length (along beam direction)	69 cm
TPC width	100 cm
Gas mixture	Argon(80%) – Methane(20%)
Gas pressure	Atmospheric
Electric field	25 KV/m.
Magnetic field	0.9 Tesla for $\mu^- \rightarrow e^-$ runs
	0.6 Tesla for $\pi^+ \rightarrow e^+\nu$ calibration runs
Drift velocity for electrons	7 cm/μs.
Number of sectors	12 (6 upstream, 6 downstream)
Number of anode wires	144 (12 per sector)
Number of pads	636 per sector
Number of anode ADC's	144
Number of anode TDC's	12
Number of cathode ADC's	636 (multiplexed)
Number of internal scintillators	6
Number of external scintillators	12 (2 layers)
Number of internal wire chamber	1 segmented in 6
Number of external wire chambers	6 (each covers 2 sectors)

The spatial resolution of the TPC has been discussed in a previous paper[3]. The energy resolution is better for electrons than for positrons due to the E x B effect near the anode wires. The energy resolution for positrons has been measured many times during the experiment by using the monochromatic positrons from the $\pi^+ \rightarrow e^+\nu$ decay (70 MeV) at a magnetic field value of 0.6 Tesla. The observed energy resolution agrees with a Monte – Carlo calculation which also predicts the energy resolution for conversion electrons to be σ = 2.3 MeV at an energy of 101 MeV (the expected energy of the conversion electron after energy losses in various materials) at a magnetic field value of 0.9 Tesla. The TPC acceptance has been monitored during the experiment with the $\pi^+ \rightarrow e^+\nu$ decay, a process with a well-known branching ratio. This acceptance is 11% and it agrees with the result of a Monte-Carlo calculation.

The target is made of natural titanium. The choice of the target material is somewhat arbitrary. Medium-mass nuclei are a good compromise[4]. Titanium is easy to handle. The target is made of shredded titanium metal to minimise the target thickness seen by the electrons. Along the beam direction the target thickness is 2 g/cm^2. The beam consists of 73 MeV/c negative muons. A RF separator is used to reduce the pion contamination to about 10^{-4} and the electron contamination to 10^{-2}. The muons were stopped in the target at a rate of about 10^6/s. The total number of stops in the whole experiment was 1.1 x 10^{13}. The trigger was made to accept particles firing an internal and an external scintillation counters in the same sector or in two adjacent sectors appropriate for a negative particle. Events with momenta larger than 70 MeV/c were accepted in a time gate from 5 to 1000 ns. after a stop signal in the beam telescope. The trigger rate was a few events per second. Protection against cosmic rays was provided by drift chambers and scintillation counters placed on top and around the TPC magnet and by segmentation of

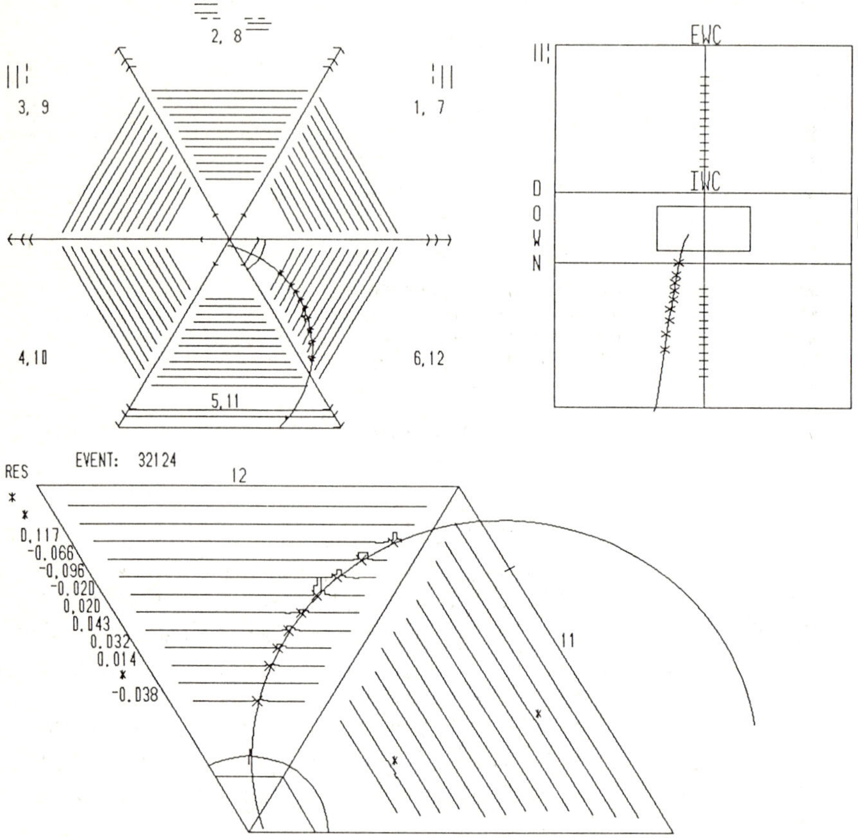

Figure 1: A typical event in the TPC. Electron with p = 94.2 MeV/c

trigger counters and TPC. Figure 1 shows a typical event (electron from muon decay in orbit). The elements which have fired are shown: the internal scintillator and wire chamber in sector 12, the external scintillators and wire chamber in sector 11.

There are several sources of background:
– muon decay in orbit which produces electrons with energies up to the muon mass. Using theoretical expressions[5] a Monte–Carlo calculation has been used to predict the contribution of this process under our experimental conditions (see Figure 2, dashed curve).
– radiative muon capture followed by the creation of a very asymmetric electron–positron pair. This is negligible at the present level of precision.
– residual pions in the beam producing gamma rays, and pair creation. Software cuts are applied to eliminate prompt events (in coincidence with a beam particle).
– cosmic rays. Between beam periods we have accumulated considerable data on cosmic rays. Some events are expected and observed above 106 MeV/c.

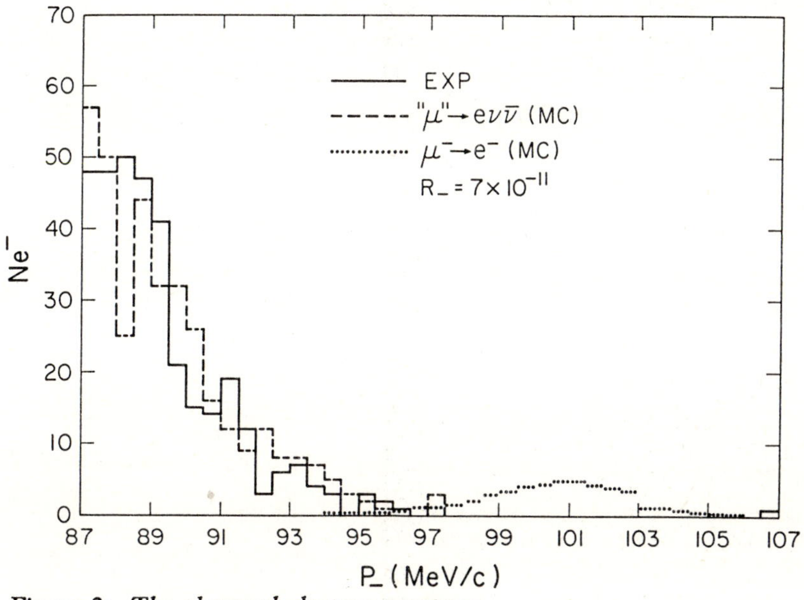

Figure 2: The observed electron spectrum

3 Results: Various cuts have been applied to the data (minimum number of points per track, dE/dx, chi squares for the xy and z fits, etc...) and the corresponding losses of efficiency have been determined. No events were seen in the energy window 96.5–106 MeV (which should contain 85% of the conversion events). For a total number of stops (live–times) of 1.1×10^{13} and a branching ratio of 0.83 for muon capture the total number of muon captures was 9.1×10^{12}. With an overall acceptance (preliminary number) of 6% we obtain from Poisson statistics a preliminary upper limit for the branching ratio of muon–to–electron conversion to total muon capture:

$$R < 4 \times 10^{-12} \text{ (90\% confidence limit)}$$

Figure 2 shows the observed electron spectrum. The dotted curve shows the result of a Monte–Carlo calculation for muon–to–electron conversion assuming a branching ratio $R = 7 \times 10^{-11}$.

References

1 D.A. Bryman et al, Nucl. Phys. A434, 469c (1985)
D.A. Bryman et al., Phys. Rev. Lett. 55, 465 (1985)

2 C.K. Hargrove et al., Physica Scripta 23, 668 (1981)
H. Mes et al., Nucl. Instr. Meth. 225, 547 (1984)
D.A. Bryman et al., Nucl. Instr, Meth. A234, 42 (1985)

3 C.K. Hargrove et al., Nucl. Instr. Meth. 219, 461 (1984)

4 S. Weinberg and G. Feinberg, Phys. Rev. Lett. 3, 111, 244(E) (1959)
O. Shanker, Phys. Rev. D20, 1608 (1979)

5 F. Herzog and K. Alder, Helv. Phys. Acta 53, 53 (1980)

New Results for Rare Muon Decays

R.E. Mischke[1], *R.D. Bolton*[1], *J.D. Bowman*[1], *M.D. Cooper*[1], *J.S. Frank*[1], *A.L. Hallin*[1a], *P.A. Heusi*[1b], *C.M. Hoffman*[1], *G.E. Hogan*[1], *F.G. Mariam*[1], *H.S. Matis*[1c], *D.E. Nagle*[1], *L.E. Piilonen*[1], *V.D. Sandberg*[1], *G.H. Sanders*[1], *U. Sennhauser*[1d], *R. Werbeck*[1], *R.A. Williams*[1], *S.L. Wilson*[2e], *R. Hofstadter*[2], *E.B. Hughes*[2], *M.W. Ritter*[2f], *D. Grosnick*[3], *S.C. Wright*[3], *V.L. Highland*[4], and *J. McDonough*[4]

[1]Los Alamos National Laboratory, Los Alamos, NM 87545, USA
[2]Stanford University, Stanford, CA 94305, USA
[3]University of Chicago, Chicago, IL 60637, USA
[4]Temple University, Philadelphia, PA 19122, USA

1. Abstract

Branching-ratio limits obtained with the Crystal Box detector are presented for the rare muon decays $\mu \to eee$, $\mu \to e\gamma$, and $\mu \to e\gamma\gamma$. These decays, which violate the conservation of separate lepton-family numbers, are expected to occur in many extensions to the standard model. We found no candidates for the decay $\mu \to eee$, yielding an upper limit for the branching ratio of $B_{\mu 3e} < 3.1 \times 10^{-11}$ (90% C.L.). A maximum-likelihood analysis of the $\mu \to e\gamma$ candidates yields an upper limit of $B_{\mu e\gamma} < 4.9 \times 10^{-11}$ and an analogous analysis of $\mu \to e\gamma\gamma$ candidates gives an upper limit of $B_{\mu e\gamma\gamma} < 7.2 \times 10^{-11}$. These results strengthen the constraints on models that allow transitions between lepton families.

2. Introduction

The muon has been an enigma since its discovery. Originally mistaken for the pion, it behaves as if it were a heavy electron. The only known decay mode of the muon is $\mu \to e\nu_\mu\bar{\nu}_e$ (and its radiative correction). Many experiments have searched for lepton-family-nonconserving decays such as $\mu \to e\gamma$, $\mu \to 3e$, and $\mu \to e\gamma\gamma$. Figure 1 shows how the limits for these decays have been steadily

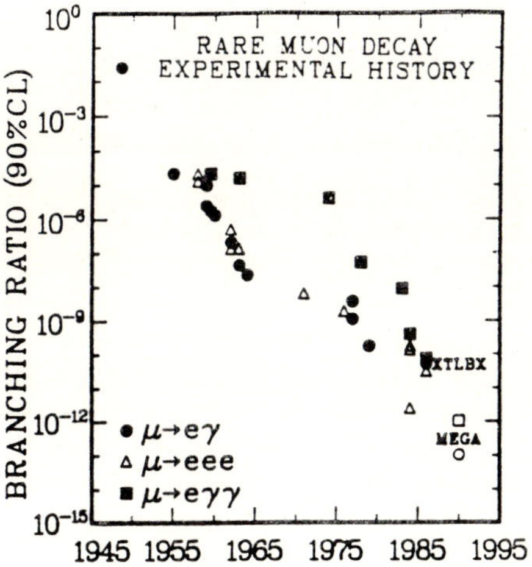

Fig. 1 Branching ratio limits for rare muon decays versus the year of the measurement

---

[a] Present address: Princeton University, Princeton, NJ 08544
[b] Present address: ELEKTROWATT Ing. AG., Zurich, Switzerland
[c] Present address: Lawrence Berkeley Laboratory, Berkeley, CA 94720
[d] Present address: SIN, CH-5234 Villigen, Switzerland
[e] Present address: Los Alamos National Laboratory, Los Alamos, NM 87545
[f] present address: Lockheed Missiles and Space Company, Palo Alto, CA 94304

improved. The absence of the decay $\mu \to e\gamma$ led to the discovery of separate muon and electron neutrinos and to the hypothesis of separate conserved quantum numbers for electrons and muons. Recently it has become widely accepted that because lepton number does not relate to a space-time symmetry as do energy and momentum conservation, nor is it associated with a massless gauge boson, as is electric charge, there is no reason to think it is a conserved quantity. The standard model[1] has been very successful but it does not address the question of lepton number. Because the standard model is incomplete, many extensions have been considered and several of these can lead to lepton-family nonconservation.[2] The existing experimental upper limits[3-5] impose model-dependent constraints on the theoretical parameters, like mixing angles or gauge-boson masses, that describe such processes. In general, there are too many free parameters in these theories to predict absolute decay rates. However, they frequently predict the ratios of rates for these decays. It is important for experiments to consider all channels because it is impossible to predict in which decay mode lepton-number nonconservation will first be seen.

3. Detector and Data Acquisition

The Crystal Box detector,[6] shown in Fig. 2, is located at the Stopped Muon Channel of LAMPF. A large (6.7-cm effective radius), thin (52 mg/cm^2) polystyrene target stops 26 MeV/c positive muons. Surrounding the target is a 728-cell, eight-plane, large-stereo-angle drift chamber,[7] which determines the three dimensional trajectories of charged particles. The drift chamber is surrounded by a 36-section plastic scintillator hodoscope, which provides discrimination between charged and neutral particles and timing resolution for charged particles of 290 ps (FWHM). Energy information is provided by a large-solid-angle 396-element NaI(Tℓ) array. Ninety crystals form each of four quadrants and nine crystals are arranged in each corner between two quadrants. The detector has an energy resolution of ~8% FWHM at 50 MeV for both positrons and photons, and a timing resolution of 1.2 ns for photons. The position resolution of the origin of a charged particle on the target is 2 mm. The photon conversion point is determined to 4.1 cm from energy sharing between NaI crystals.

The absolute gain of each NaI crystal was set to 10% using a Pu-α-Be source (4.43 MeV γ) and calibrated using photons from the reactions $\pi^- p \to n\pi^0$ ($\pi^0 \to \gamma\gamma$) ($55 \leq E_\gamma \leq 83$ MeV) and $\pi^- p \to n\gamma$ ($E_\gamma = 129.4$ MeV). The pion data were taken with a liquid-hydrogen target replacing the drift chamber. The gain stability of each NaI channel was monitored every two hours using a Xe flashtube and fiber optics cables connected to each photomultiplier and by the end-point of the positron energy distribution from normal μ decay.

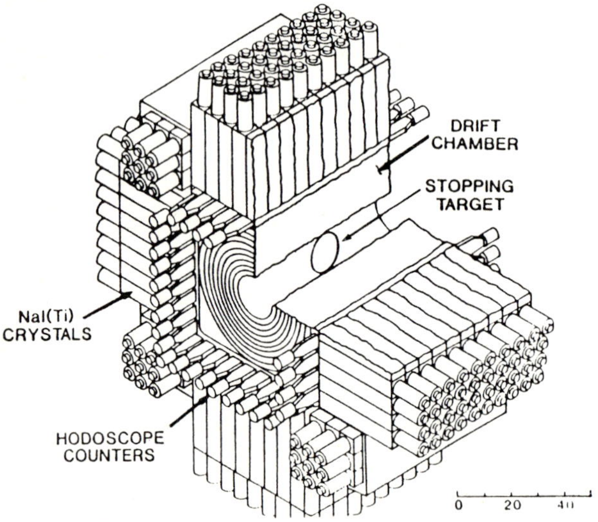

Fig. 2 The Crystal Box detector

The sensitivity limit of a rare μ decay experiment is determined by the number of μ decays examined and by how well the backgrounds are identified or suppressed. The sources of background are random coincidences between positrons from normal μ decay and bremsstrahlung photons, and the prompt processes $\mu \to ee e\nu\bar{\nu}$, $\mu \to e\gamma\nu\bar{\nu}$, and $\mu \to e\gamma\gamma\nu\bar{\nu}$. Random coincidences dominate the background for the Crystal Box in all three decay modes. However, for the $e\gamma$ mode the prompt background contributes about 10%. To identify a rare decay event the energy, time, and position resolutions of the detector must be adequate to show that the particles are in time, that the total energy is equal to that of the muon, and that the vector sum of the momenta is zero. In addition, for 3e events, all tracks must have a common origin on the target.

The data acquisition system collected candidates for all three decay modes simultaneously. The trigger defined a "positron quadrant" as a signal in a hodoscope scintillator with more than 5 MeV of energy in a crystal in one of the three rows of crystals directly behind the scintillator within 15 ns of the hodoscope signal. A "photon quadrant" was defined by requiring energy in the NaI with no scintillator firing within 20 ns in front of it, or in the nearest scintillator in the adjacent two quadrants. The 3e trigger required that there be signals in three non-adjacent plastic scintillators, that the scintillators be in a geometric pattern kinematically consistent with a 3e decay, and that three scintillators fire within 5 ns of each other. The $e\gamma$ trigger required a coincidence within ±5 ns of a positron in one quadrant and the opposite photon quadrant, and that each have an NaI energy greater than 30 MeV. The $e^+\gamma\gamma$ trigger required a time coincidence within ±12 ns of a positron quadrant and two photon quadrants, with at least 70 MeV deposited in the NaI calorimeter. These trigger requirements generated a trigger rate of about 20 Hz with 7.7 MHz instantaneous of muons stopping in the target at a 7% duty factor (500 kHz average).

The trigger generated a signal to start all the TDC's, a gate for the ADC's, and a start signal for the readout of the event. For each event all the scintillator ADC and TDC data were recorded. Distributed processors performed a sparse data scan for the drift-chamber TDC information and the NaI pulse-height and timing information. In addition, a second ADC with a different gate was used on the NaI crystals to detect pileup. A computer acquired and filtered the events by making on-line cuts before taping the data.

Approximately 10^6-10^7 events were recorded for each trigger from 1.2×10^{12} muons that were stopped. About 30% of the data were taken without benefit of the additional ADC system to reject pileup in the NaI. All the data were processed by a multistage filtering process. The data remaining after the first two passes consisted of 10^3-10^4 events in each of the data streams. These were carefully investigated to look for a prompt signal and any candidates for lepton-family-nonconserving decays.

4. $\mu \to e\gamma$ Analysis

The data sample containing possible $\mu \to e\gamma$ events consisted of 17 073 events satisfying $|\Delta t_{e\gamma}| < 5$ ns, $\theta_{e\gamma} \geq 160^\circ$, $E_e \geq 44$ MeV, and $E_\gamma \geq 40$ MeV. An ideal $\mu \to e\gamma$ event would have $|\Delta t_{e\gamma}| = 0$, $\theta_{e\gamma} = 180^\circ$, and $E_e = E_\gamma = 52.8$ MeV. Fig. 3a shows $\Delta t_{e\gamma}$, the photon-positron relative timing, for a subset of these events. The broad distribution is due to random coincidences, while the prompt peak is due to $\mu^+ \to e^+\nu\bar{\nu}\gamma$ and possible $\mu^+ \to e^+\gamma$ events.

The number of each component in the data sample was determined by maximizing the likelihood

$$L(n_{e\gamma}, n_{IB}) = \prod_{i=1}^{N} \left[\frac{n_{e\gamma}}{N} P(\vec{x}_i) + \frac{n_{IB}}{N} Q(\vec{x}_i) + \frac{n_R}{N} R(\vec{x}_i) \right]$$

with respect to the parameters $n_{e\gamma}$, n_{IB}, and $n_R = N - n_{e\gamma} - n_{IB}$ that estimated the number of $\mu^+ \to e^+\gamma$, $\mu^+ \to e^+\nu\bar{\nu}\gamma$, and random events in the total sample of N events. The vector $\vec{x}$ had components $\theta_{e\gamma}$, $\Delta t_{e\gamma}$, E_e, and E_γ. P, Q, and R were the probability distributions for $\mu^+ \to e^+\gamma$, $\mu^+ \to e^+\nu\bar{\nu}\gamma$, and random events, respec-

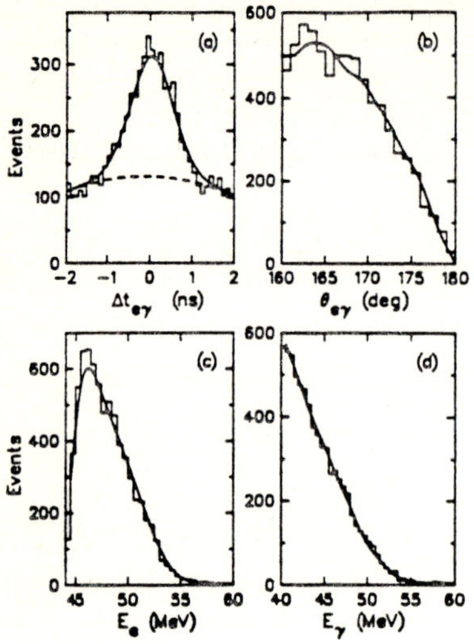

Fig. 3 Spectra for each of the quantities used in the $\mu \to e\gamma$ data analysis

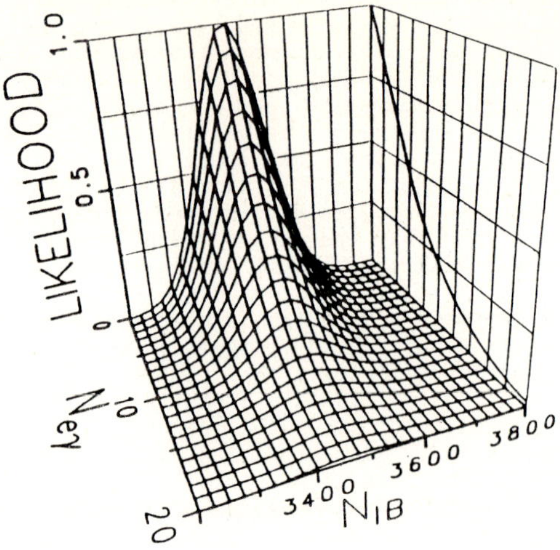

Fig. 4 The likelihood function versus the number of $\mu \to e\gamma$ and $\mu \to e\gamma\nu\bar{\nu}$ in the data

tively. The distributions for P and Q and the acceptance of the apparatus were determined with a Monte Carlo simulation, based on the shower code EGS3,[8] that accurately reproduced the response of the detector to photons and positrons.

Fig. 4 shows the normalized likelihood function. It peaks at $n_{e\gamma} = 0$ and $n_{IB} = 3470 \pm 80 \pm 300$ events. The latter agrees well with the $3960 \pm 90 \pm 200$ $\mu^+ \to e^+\nu\bar{\nu}\gamma$ events expected in the data. The likelihood function distribution implies $n_{e\gamma} < 11$ events (90% C.L.). Using the number of muons stopped, the apparatus acceptance for $\mu^+ \to e^+\gamma$, 0.305, and the detection efficiency, 0.613, we obtain $B_{e\gamma} < 4.9 \times 10^{-11}$. Fig. 3 shows the agreement between the data (histogrammed) and the best mix of $\mu^+ \to e^+\nu\bar{\nu}\gamma$ and randoms (smooth curve) as determined by the likelihood analysis.

5. $\mu \to e\gamma\gamma$ Analysis

The number of events surviving the first analysis from the $e\gamma\gamma$ trigger was 41 656; an additional 968 candidates were found in the $e\gamma$-triggered events, where the positron and one photon occupied the same quadrant. Fig. 5 shows the relative timing distribution for some of these events, the majority being backgrounds from triple random coincidences or two-particle prompt events in random coincidence with a third particle (e.g., $\mu^+ \to e^+\nu\bar{\nu} + \gamma$). The cuts applied during the data analysis removed most of the double- and triple-random coincidences while retaining most of the $\mu^+ \to e^+\gamma\gamma$ events, making minimal assumptions for the $\mu^+ \to e^+\gamma\gamma$ matrix element.[9] Events with one particle showering and appearing as two hits in the trigger in coincidence with another particle were removed by energy cuts.

The number of $\mu^+ \to e^+\gamma\gamma$ events in the remaining sample of nine events was estimated by maximizing the likelihood

$$L(n_{e\gamma\gamma}) = \prod_{i=1}^{N} \left[\frac{n_{e\gamma\gamma}}{N} P(\vec{x}_i) + \frac{n_R}{N} R(\vec{x}_i) \right]$$

with respect to the parameters $n_{e\gamma\gamma}$ and $n_R = N - n_{e\gamma\gamma}$ that estimated the number of $\mu^+ \to e^+\gamma\gamma$ and background events in the sample of N events. The components of $\vec{x}$ were E_{tot}, $\tau = 2t_e - t_{\gamma 1} - t_{\gamma 2}$, $p_\parallel = |\vec{p}_a + \vec{p}_b + \vec{p}_c \times \hat{p}_{ab}|$ and $\cos\alpha = \hat{p}_c \cdot \hat{p}_{ab}$, where $\vec{p}_a$ and $\vec{p}_b$ were the momenta most nearly perpendicular to each other, $\hat{p}_{ab}$ was the unit vector normal to the p_a-p_b plane, and $\vec{p}_c$ was the third particle's momentum. P and R were the probability distributions for $\mu^+ \to e^+\gamma\gamma$ and back-

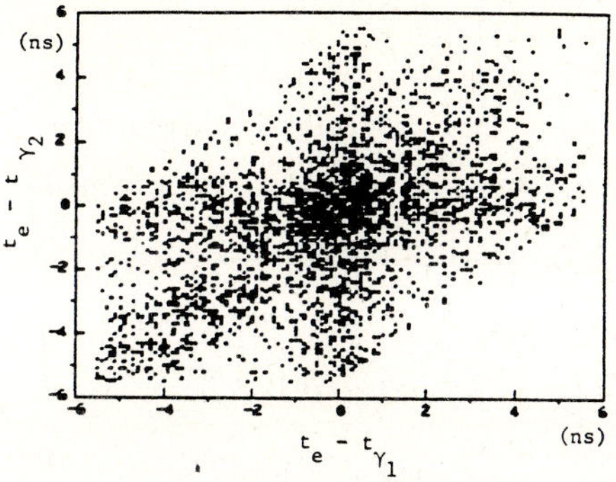

Fig. 5 Timing scatter plot for $\mu \to e\gamma\gamma$ data

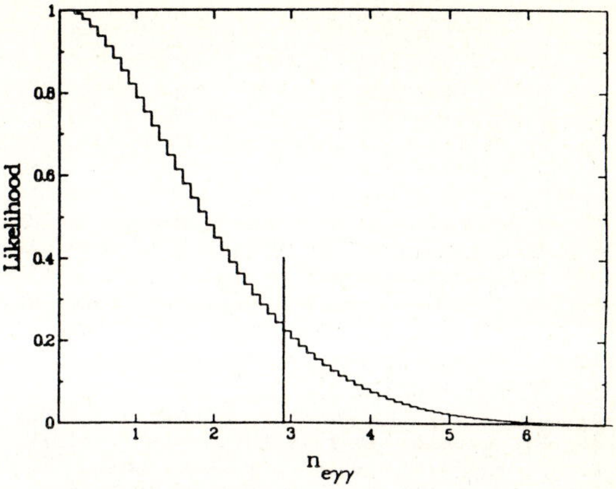

Fig. 6 Maximum likelihood for $\mu \to e\gamma\gamma$ candidates

ground, respectively. The Monte Carlo program gave the distributions for P and the distributions for R were taken from data with $\tau \neq 0$.

The likelihood function distribution in Fig. 6 implies $n_{e\gamma\gamma} < 2.9$ (90% C.L.). Using the number of muons stopped, the apparatus acceptance for $\mu^+ \to e^+\gamma\gamma$, 0.064, and the detector efficiency, 0.524, we obtain $B_{e\gamma\gamma} < 7.2 \times 10^{-11}$ (90% C.L.).

6. Summary

The results of the experiment for the three decay modes are as follows:

$$\frac{\Gamma(\mu\to e\gamma)}{\Gamma(\mu\to e\nu\bar{\nu})} < 4.9 \times 10^{-11} \qquad \frac{\Gamma(\mu\to eee)}{\Gamma(\mu\to e\nu\bar{\nu})} < 3.1 \times 10^{-11} \qquad \frac{\Gamma(\mu\to e\gamma\gamma)}{\Gamma(\mu\to e\nu\bar{\nu})} < 7.2 \times 10^{-11}$$

We conclude that there is no evidence for nonconservation of lepton family number. As examples of theoretical constraints imposed by our result, we show how these new values of $B_{\mu e\gamma}$ and $B_{\mu e\gamma\gamma}$ limit the parameters in a few models. Using the formula of Tomozawa[10] for the mass of the constituents of muons and electrons, where the muon is taken to be a 2S excited state of the electron, $B_{\mu e\gamma}$ can be combined with $B_{\mu e\gamma\gamma}$ to yield a lower limit on the mass of the constituents of 7.15×10^{7} GeV. In a composite model[11] based on the inclusion of heavy vector weak isodoublets, $\mu \to e\gamma$ constrains the mass scale of the heavy leptons to be $\Lambda > 2.4 \times 10^{2}$ TeV. In supersymmetric theories,[12] where the symmetry is broken by gravity,[13] the mass of the supersymmetric partner of the muon must be greater than 36 GeV. A model[14] based on O(18) that contains the standard model and a discrete family symmetry with four generations includes massive neutrinos with mass less than 40 GeV. Taking the estimate that the mixing angle with e and μ is $|\theta^{*}_{\mu}\theta_e| = 10^{-3}$, the limit on $\mu \to e\gamma$ constrains the neutrino mass to be at the upper end of its range: $M_N > 40$ GeV. An effective Lagrangian analysis of possible deviations from the standard model shows that unless there are unnatural suppressions, $\mu \to e\gamma$ provides a stringent limit on the scale of new interactions of $\Lambda = 10^{4}$ TeV.[15] In all cases, the mass limits vary as $[B_{\mu e\gamma}]^{-1/4}$.

We acknowledge the extraordinary assistance from the many people at each of our institutions and from the operations staff at LAMPF. This work was supported in part by the U.S. Department of Energy and the National Science Foundation.

7. References

1. S. L. Glashow, Nucl. Phys. 22, 579 (1961); A. Salam, in Elementary Particle Theory: Relativistic Groups and Analyticity (Nobel Symposium No. 8), edited by N. Svartholm (Almqvist and Wiksell, Stockholm, 1968), p. 367; S. Weinberg, Phys. Rev. Lett. 19, 1264 (1967).

2. C.M. Hoffman, in Fundamental Interactions in Low-Energy Systems, ed. P. Dalpiaz et al., (Plenum Press, New York, 1985), p. 138; R. Engfer and H. K. Walter, Ann. Rev. Nucl. and Part. Sci. Vol. 36 (1986).
3. W. W. Kinnison et al., Phys. Rev. D 25, 2846 (1982)
4. W. Bertl et al., Nucl. Phys. B260, 1 (1985).
5. G. Azuelos et al., Phys. Rev. Lett. 51, 164 (1983).
6. R.D. Bolton et al., Phys. Rev. Lett. 56, 2461 (1986) and references therein.
7. R.D. Bolton et al., Nucl. Instr. and Methods 241, 52 (1985).
8. R.L. Ford and W.R. Nelson, Stanford Linear Accelerator Center No. SLAC-210, 1978 (unpublished).
9. J. Dreitlein and H. Primakoff, Phys. Rev. 126, 375 (1962).
10. Y. Tomozawa, Phys. Rev. D25, 1448 (1982).
11. P. Esposito, Università "La Sapienza" - Roma Preprint N. 479, Nov. 1985 (unpublished).
12. J. Ellis and D. V. Nanopoulos, Phys. Lett. 110B, 41 (1983).
13. E. Cremmer et al., Phys. Lett. 122B, 41 (1983).
14. J. Bagger et al., Phys. Rev. Lett. 54, 2199 (1985).
15. W. Buchmüller and D. Wyler, Nucl. Phys. B268, 621 (1986).

Measurement and Analysis of the Rare Muon Decay $\mu^+ \to e^+\nu_e\overline{\nu}_\mu e^+e^-$

U. Bellgardt, W. Bertl, S. Egli, R. Eichler, R. Engfer, Ch. Grab, E.A. Hermes, A. Kersch, N. Kraus, N. Lordong, J. Martino, C. Niebuhr, H.S. Pruys, A. van der Schaaf, and H.K. Walter

IMP der ETH Zürich, Physik-Institut der Universität Zürich, CH-8001 Zürich, Switzerland
Swiss Institute for Nuclear Research, CH-5234 Villigen, Switzerland
Laboratoire National Saturne, CEN Saclay, F-91191 Gif-sur-Yvette Cedex, France
RWTH-Aachen, D-5100 Aachen, F. R. Germany
Institut für Physik der Universität Mainz, D-6500 Mainz, F. R. Germany

The SINDRUM-Spectrometer has measured the kinematical distribution of 7443±148 events of the rare muon decay $\mu \to 3e2\nu$. The muons were (80 ± 2) % polarized and only the high energy region was investigated. This measurement was a by-product in the search for the lepton number violating decay $\mu \to 3e$. The phase space for the $\mu \to 3e2\nu$ decay was reduced by a factor of $\sim 5.8 \times 10^{-5}$ by the following cuts on the charged particle momenta:

- each momentum transverse to beam direction > 16 MeV/c
- $36° < \Theta < 144°$, where Θ is the azimuthal angle
- total transverse momentum $\leq$ 45 MeV/c.

Reference [1] gives detailed description of the detector and of the data reduction.

The measured branching ratio to the ordinary decay $B = (3.4 \pm 0.4) \times 10^{-5}$ is in agreement with the calculated V-A value of $(3.5916 \pm 0.0022) \times 10^{-5}$ in [2]. Figure 1 shows the distributions of the kinematical variable $K = \Sigma E_i + |\Sigma \vec{p}_i c|$ of prompt events (solid line), normalized accidental events (dotted line) and of Monte Carlo simulated $\mu \to 3e2\nu$ events (with error bars) assuming V-A, which is normalized to the measured spectrum.

To perform an analysis of the data, the most general differential decay rate corresponding to a local, lepton-number conserving, derivative-free

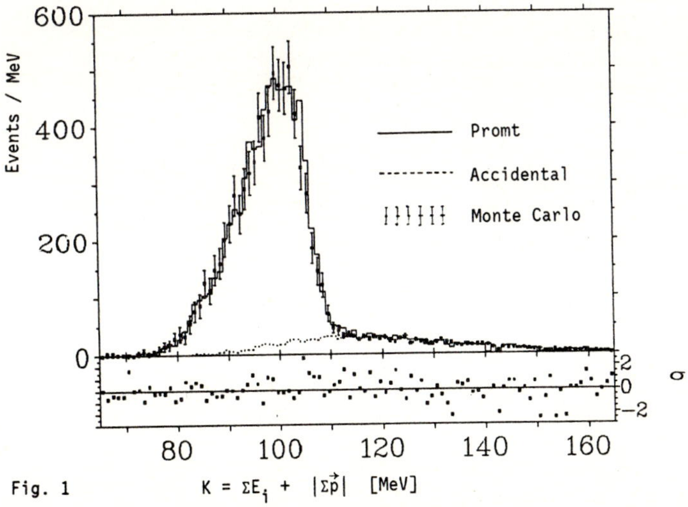

Fig. 1

four-fermion interaction has been calculated in the charge changing form. The matrix element is the one of the ordinary muon decay supplied with additional electromagnetic interaction of the charged leptons and antisymmetrized final state. See [3] for description of the weak coupling constants h_{ij}, g_{ij}, f_{ii}, ij = 1,2. In the standard model g_{22} = 1, all other equal to zero.

In the work of Pratt [4] it is shown that all possible kinematical distributions depend on 10 real combinations of the 10 complex coupling constants of the above matrix element. 4 combinations factorize with time reversal violating parts and have as a factor either m_e, the electron mass, or σ_e^T, the transverse polarization of the electron. In our case only the high energy part and no electron polarization was measured. Hence our data are only sensitive to the other 6 combinations a,b,c,a',b',c' of [3], but not to time reversal violation. The h_{12} (h_{21}) distribution alone and the g_{22} (g_{11}) distribution alone are indistinguishable in this experiment. Hence our data are sensitive to 6 independent combinations of the coupling constants h_{11}, h_{22}, g_{11}, g_{12}, g_{21}, g_{22}, f_{11}, f_{22}. In particular, our data contain information in the kinematical distributions about h_{22}, g_{12}, f_{22}, which in ordinary muon decay is contained in the longitudinal polarization of the electron and measured with the ξ' parameter. Figure 2 shows the calculated distributions of the cosine of the angle between muon spin and total transverse momentum of a g_{11}, g_{12}, g_{21}, g_{22}, interaction respectively, with-out experimental cuts.

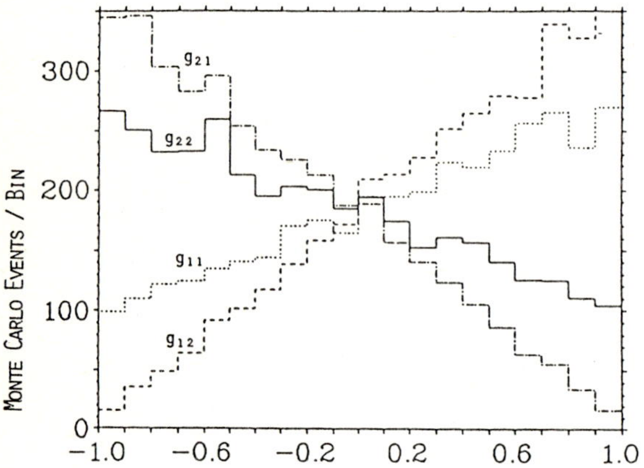

FIG. 2 COSINE OF ANGLE BETWEEN $\Sigma\vec{p}_i$ AND μ-SPIN

The preliminary analysis of the data includes the following restrictions:

- two of the coupling constants respectively have been set to zero not to exceed the number of independent parameters
- the muon polarization has not been determined in an independent experiment
- the Monte Carlo integration of the total rate, needed in the maximum likelihood analysis, has a still too large error.

Table 1 gives the preliminary results. The limits for the couplings are compared with the values [5], where a complete model independent analysis of all muon decay experiments is given. In a final model independent analysis (following [5] and limiting positive semi-definite combinations of couplings) we do not expect much different results.

We conclude:

- measured rate and spectra can be explained with the standard model
- in the $\mu \to 3e2\nu$ decay the value of several coupling constants can be constrained in a single experiment
- the couplings related to the ξ' parameter are as well limited as the other ones

table 1

coupling constant	h_{11}	h_{22}	g_{11}	g_{12}	g_{21}	f_{11}	f_{22}
limit from $\mu \to 3e2\nu$ decay (68% CL)	<0.50	<1.30	<0.25	<0.47	<0.59	<0.25	<0.53
limits from [5]	<0.116	<0.371	<0.074	<0.192	<0.104	<0.116	<0.325

- since the limits of the couplings scale with the quartic root of measured decays, it will hardly be possible to surpass the ordinary muon decay experiments.

Literature References

1 W. Bertl et al.: Nucl.Phys. B260, 1 (1985)
2 P.M. Fishbane, K.-J. Gaemers: Phys.Rev. D33, 159 (1986)
3 Particle Data Group: "Review of Particle Properties", (1986)
4 R.H. Pratt: Phys.Rev. 111, 649 (1958)
5 W. Fetscher, H.J. Gerber, K.F. Johnson: Phys.Lett. B173, 102 (1986)

Complete Determination of the Charged Leptonic Weak Interaction in Muon Decay

W. Fetscher, H.-J. Gerber, and K.F. Johnson

Institut für Mittelenergiephysik, ETHZ, CH-5234 Villigen, Switzerland

The decay of the muon is well suited to investigate the structure of the charged leptonic weak interaction. Surprisingly enough the interaction had, up to now, not been determined conclusively without including semileptonic data or theoretical assumptions restricting the most general set of decay parameters.

We have based a general analysis on data from normal and inverse muon decay. Although only eleven independent experiments have been performed, it will be shown that these experiments suffice to determine the interaction to be "V-A", and to give upper limits to all other possible couplings. Our analysis is described in more detail in ref. [1].

The decay of the muon $\mu \rightarrow e\bar{\nu}_e\nu_\mu$ can be described by the most general, local, derivative-free, lepton-number conserving four-fermion interaction hamiltonian [2]. It contains ten complex coupling constants and therefore, neglecting on arbitrary common phase, 19 independent parameters. The hamiltonian thus is general enough to allow deviations from universality and violation of T-invariance. We have used the "helicity projection form" [3,4,5] (abbreviated HPF) to analyze the experimental results. This form uses fields of definite handedness.

Our matrix element is given by

$$M \sim \sum_{\substack{S,V,T \\ \varepsilon,\mu=L,R \\ (n,m)}} g^{\gamma}_{\varepsilon\mu} \langle \bar{e}_\varepsilon | \Gamma^{\gamma} | (\nu_e)_n \rangle \langle (\nu_\mu)_m | \Gamma_\gamma | \mu_\mu \rangle$$

γ indicates the type of interaction: Γ^S, Γ^V, Γ^T (scalar, vector, tensor). Parity violation is indicated by the coupling to handed particles. The indices ε and μ label the chiral projections (left-handed, right-handed) of the spinors of the experimentally observed particles, $\varepsilon \mathrel{\hat{=}}$ electron, $\mu \mathrel{\hat{=}}$ muon. n indicates the helicity of the electron antineutrino, m that of the muon neutrino. n and m are uniquely determined for given γ, ε and μ.

The decay rate determines the strength of the interaction. Since we are interested in the relative amount of each coupling, we have normalized the $g^{\gamma}_{\varepsilon\mu}$ by setting

$$\begin{aligned} A \equiv\ & 4(|g^S_{RR}|^2 + |g^S_{LR}|^2 + |g^S_{RL}|^2 + |g^S_{LL}|^2) \\ & + 16(|g^V_{RR}|^2 + |g^V_{LR}|^2 + |g^V_{RL}|^2 + |g^V_{LL}|^2) \\ & + 48(|g^T_{LR}|^2 + |g^T_{RL}|^2 = 16 \end{aligned} \quad (1)$$

"V-A" corresponds to $g^V_{LL} = 1$ and all other $g^\gamma_{\varepsilon\mu} = 0$. By adding the quantities in eq. (1) with the same combination of handedness, but different coupling, we get the four quantities $Q_{\varepsilon\mu}$:

$$Q_{RR} = |g^S_{RR}|^2/4 + |g^V_{RR}|^2$$

$$Q_{LR} = |g^S_{LR}|^2/4 + |g^V_{LR}|^2 + 3\,|g^T_{LR}|^2$$

$$Q_{RL} = |g^S_{RL}|^2/4 + |g^V_{RL}|^2 + 3\,|g^T_{RL}|^2$$

$$Q_{LL} = |g^S_{LL}|^2/4 + |g^V_{LL}|^2$$

We note: $0 \leqq Q_{\varepsilon\mu} \leqq 1$ and $\Sigma Q_{\varepsilon\mu} = 1$. A $Q_{\varepsilon\mu}$ is thus the probability for a transition from a muon of handedness μ to an electron of handedness ε.

It now turns out that numerical values for the $Q_{\varepsilon\mu}$ can be derived from existing measurements on μ decay. The method is described in [1,6], where also the references for the experimental data are given. We find upper limits for Q_{RR}, Q_{LR} and Q_{RL}, which yield upper limits for the absolute values of eight coupling constants $g^\gamma_{\varepsilon\mu}$ [see Table 1]. Since Q_{LL} is bounded by a lower limit, it is not possible to deduce an upper limit for $|g^S_{LL}|$. In fact, with the data from normal muon decay we can not tell if

$$g^S_{LL} = 0, \quad g^V_{LL} = 1 \quad \text{(V-A)} \qquad \text{or}$$

$$g^S_{LL} = 2, \quad g^V_{LL} = 0 \quad !$$

The first combination corresponds to left-handed, the second to right-handed neutrini, with the charged leptons being left-handed in either case.

This ambiguity is resolved by the data from inverse muon decay:

$$\nu_\mu + e^- \rightarrow \mu^- + \nu_e$$

Table 1
Complete set of coupling constants $g^\gamma_{\varepsilon\mu}$ of the muon decay interaction determined from experiments without model assumptions. The "standard model" sets $g^V_{LL} = 1$, all others to zero. An additional "V+A interaction" is represented by g^V_{RR}. The $Q_{\varepsilon\mu}$ are the relative rates. 68 (90%) C.L. is given.

Handednesses of electron (ε) and muon (μ) $\varepsilon\mu$	Relative abundance $10^3 \times Q_{\varepsilon\mu}$	Type of interaction γ	Handednesses of ν_μ(m) and ν_e(n) mn	Coupling constant $10^3\lvert g^\gamma_{\varepsilon\mu}\rvert$	Main input
RR	<1.4(2.0)	S	LR	<74(91)	$\xi\delta/\rho$
		V	RL	<37(45)	
LR	<2.7(3.9)	S	LL	<116(137)	
		V	RR	<52(62)	$\xi\delta/\rho$, ρ, δ
		T	LL	<34(40)	
RL	<36(45)	S	RR	<378(448)	
		V	LL	< 97(114)	$P_L(\xi')$
		T	RR	< 95(112)	
LL	>960(949)	S	RL	<743(961)	S, h_{ν_μ}
		V	LR	>928(877)	Q_{RR}, Q_{LR}, Q_{RL}

The total rate S, normalized to the rate predicted by V-A, has been measured as

$$S = 0.98 \pm 0.12 \quad [7]$$

The influence of eight of the $g^{\gamma}_{\varepsilon\mu}$ on S is found to be negligible. Thus

$$S \approx |g^V_{LL}|^2 \cdot (1-h)/2 + (3/32)|g^S_{LL}|^2 \cdot (1+h)/2,$$

where h is the helicity of the ν_μ from pion decay. The deviation of $|h|$ from 1 is known very precisely: $1 - |h| < 4.1 \times 10^{-3}$ [8]; the sign of h is taken for ν_μ and $\bar{\nu}_\mu$ from [9,10]. With $h(\nu_\mu) \approx -1$ we finally obtain

$$|g^V_{LL}| \approx S \quad \text{and} \quad |g^S_{LL}|^2 \leq 4\,(1-S).$$

The limits obtained for the $g^{\gamma}_{\varepsilon\mu}$ are shown in Fig. 1.

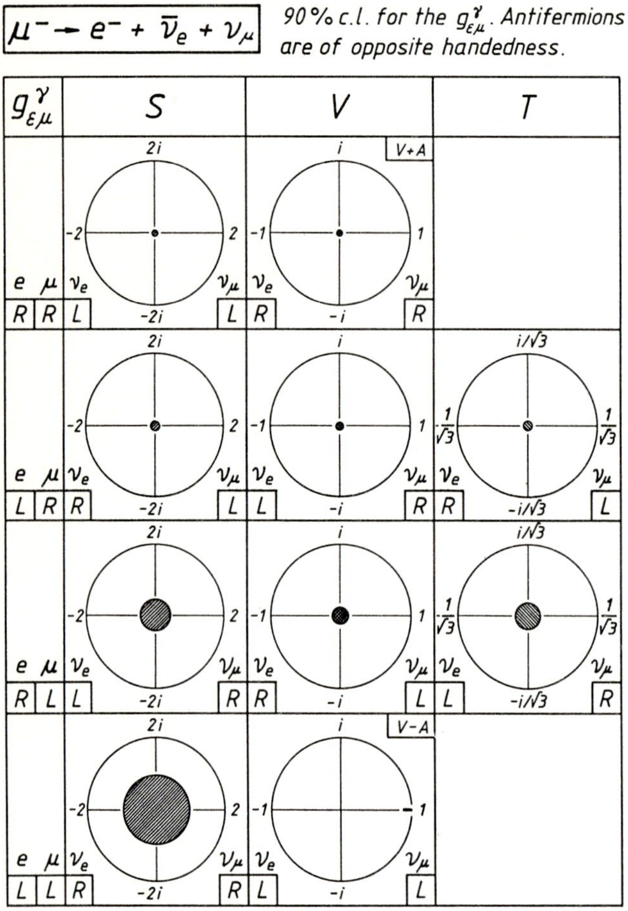

Fig. 1 90% c.l. limits for the coupling constants $g^{\gamma}_{\varepsilon\mu}$. Each coupling is uniquely determined by the handednesses ε and μ of the electron and the muon, respectively, and the type of interaction γ = S, V or T.

In summary, we have completely determined the weak interaction in muon decay from existing experiments. Present experimental errors, however, still allow substantial contributions from interactions other than V-A.

References

[1] W. Fetscher, H.-J. Gerber and K.F. Johnson, Phys. Lett. 173B, 102 (1986)
[2] L. Michel. Proc. Phys. Soc. A63, 514 (1950)
[3] F. Scheck, Leptons, Hadrons and Nuclei (North-Holland, Amsterdam, 1983)
[4] K. Mursula and F. Scheck, Nucl. Phys. B253, 189 (1985)
[5] Review of Particle Properties, Phys. Lett. 170B, 1 (1986)
[6] H. Burkard, F. Corriveau, J. Egger, W. Fetscher, H.-J. Gerber, K.F. Johnson, H. Kaspar, H.J. Mahler, M. Salzmann and F. Scheck, Phys. Lett. 160B, 343 (1985)
[7] F. Bergsma et al., Phys. Lett. 122B, 465 (1983)
[8] W. Fetscher, Phys. Lett. 140B, 117 (1984)
[9] L.Ph. Roesch et al., Helv. Phys. Acta 55, 74 (1982)
[10] A.I. Alikhanov et al., JETP 11, 1380 (1960);
G. Backenstoss et al., Phys. Rev. Lett. 6, 415 (1961);
M. Bardon et al., Phys. Rev. Lett. 7, 23 (1961);
A. Possoz et al., Phys. Lett. 70B, 265 (1977);
R. Abela et al., Nucl. Phys. A39, 413 (1983)

3.4.2 Muon Capture and μ-Atoms

Muon Capture in Nuclei and Determination of Weak Coupling Constants

H. Ohtsubo

Department of Physics, Osaka University, Toyonaka, Osaka 560, Japan

1. Introduction

Beta decay and muon capture in nuclei have provided us information on the semi-leptonic weak interaction of a nucleon[1]. At present, it is known that the semileptonic processes are described in a good accuracy, by a product of current and current as

$$H_I = (V_\lambda + A_\lambda)L_\lambda/\sqrt{2} + h.c., \quad (1)$$

where L_λ is the lepton current, and V_λ and A_λ are, the vector and axial-vector nuclear currents, respectively. Then, our main interest is to determine details of the currents V_λ and A_λ, i.e., various coupling constants involved in weak nucleonic current in order to test the working hypotheses so far proposed. The matrix elements of the vector and axial-vector currents of a free nucleon are expressed as

$$(n|V_\lambda|p) = i\bar{u}(n)[g_V\gamma_\lambda + g_W\sigma_{\lambda\mu}q_\mu - ig_Sq_\lambda]u(p), \quad (2)$$

and

$$(n|A_\lambda|p) = i\bar{u}(n)\gamma_5[g_A\gamma_\lambda + ig_Pq_\lambda - g_T\sigma_{\lambda\mu}q_\mu]u(p) \quad (3)$$

with $q = n - p$.

The coupling constants g_V, g_W, g_S, g_A, g_P and g_T are called the vector, weak magnetism, induced scalar, the axial-vector, induced pseudoscalar and induced tensor coupling constants. They are generally functions of the square of momentum transfer q^2. The vector and axial-vector coupling constants at $q^2=0$ are determined by the ft-values of the 0^+– 0^+ beta decays[2] and the neutron beta decay[3] with a high precision.

$$g_V = (1.4122 \pm 0.0043)\times 10^{-45}\text{erg}\cdot\text{cm}^3, \quad (4)$$

and

$$g_A/g_V = -(1.2533 \pm 0.00071). \quad (5)$$

The coupling constant g_V is expressed in terms of the universal weak coupling constant and the Cabbibo angle. The absolute value of the axial-vector coupling constant g_A is consistent with the famous Adler-Weisberger sum rule[4] with the current algebra and the PCAC hypothesis[5]. The other four coupling constants are predicted by the CVC[6] and PCAC hypotheses, and also the definite G-parity of these currents.

From the CVC hypothesis, we have

$$g_W/g_V = -(\mu_p - \mu_n)/2M = -3.706/2M, \quad (6)$$

and

$$g_S = 0, \quad (7)$$

M being the mass of a nucleon. Their validity is confirmed by the detailed study of beta-ray spectra from ^{12}B and ^{12}N, which will be briefly discussed.

The scalar term in the vector current and the induced tensor term in the axial-vector current are transformed differently from others under the G-parity transformation[7], and are called the second-class current, while the other four terms, the first-class current. Equation (7) means that the vector current is of the first-class. Whether the axial-vector current is of the first-class or not, is essentially important on constructing a unified theory of elementary particles. In fact, during the last decade, it has been studied experimentally and theoretically[8-13]. As a result we have found that the induced tensor coupling constant g_T is also vanishingly small. The latest situation of this problem will be discussed in section 2.

A remaining coupling constant to be determined is the induced pseudoscalar coupling constant in the axial-vector current, which is directly related with a test of the PCAC theory. It predicts

$$g_P/g_A = 2M/(q^2 + m_\pi^2), \quad (8)$$

if we assume pion pole dominance. It, however, cannot be determined in the beta decay, because the induced pseudoscalar interaction is proportional to the momentum transfer, which is vanishingly small in the beta decay. Therefore, determination of the induced pseudoscalar coupling constant should be performed through the muon capture reaction.

The most favored process for this purpose is the muon capture in the hydrogen, because it is a fundamental process of weak interaction and is free from uncertainties of nuclear structure. This reaction, however, involves problems concerning atomic and molecular physics. Therefore, the previous experimental values scatter relatively in a wide range[14]. Recently, there appeared a new measurement of muon capture rate by detecting electrons from muon decay[15]. We hope that this type of experiment will exclude ambiguities of molecular physics and determine induced pseudoscalar coupling constant with a high precision.

Muon capture in nuclei has been studied theoretically and experimentally to determine the induced pseudoscalar coupling constant[16]. There are two types of muon capture, i.e., radiative muon capture and nonradiative muon capture. Due to very small branching ratio of the former process, it has been difficult to draw a definite conclusion on the coupling constant g_P. Recently, the SIN group has attacked this difficult problem successfully. Since Dr. Truöl will present its details, we shall concern ourselves with the nonradiative muon capture, more specifically, polarization phenomena in muon capture on ^{12}C in connection with the beta decay.

2. Beta-ray energy spectra and angular distributions in ^{12}B and ^{12}N

The beta decay of ^{12}B and ^{12}N into the ground state of ^{12}C is the Gamow-Teller transition. To show explicitly roles of individual terms involved in the weak nuclear current, we shall adopt a simplified form of the beta-decay rate as[8]

$$\begin{aligned} W(\theta,E)dEd\cos\theta = {} & (2\pi^3)^{-1} pE(E_0-E)^2 F_0(\pm Z,E)|g_A \textstyle\int\boldsymbol{\sigma}|^2[1 + R(E,E_0)]dEd\cos\theta \\ & \times[1 \pm (8/3)aE] \\ & \times\{1 - P(p/E)[1 + \alpha_\pm E]P_1(\cos\theta) + A\alpha_\pm E P_2(\cos\theta)\} \end{aligned} \quad (9)$$

with

$$\alpha_\pm = (2/3)[\pm(a-b_T) - y/2M] \quad (10)$$

$$a = -\textstyle\int\mathbf{r}\times\mathbf{V}/\int\mathbf{A} \quad \rightarrow \quad -(g_V/g_A)(1 + \mu_p - \mu_n)/2M \quad (11)$$

$$b_T = \textstyle\int\mathbf{A}_{(2)}/\int\mathbf{A} \quad \rightarrow \quad g_T/g_A \quad (12)$$

and

$$y = -2M\textstyle\int A_0\mathbf{r}/\int\mathbf{A} \quad \rightarrow \quad -2M\int i\gamma_5\mathbf{r}/\int\boldsymbol{\sigma}. \quad (13)$$

Here, the upper(lower) sign refers to the electron(positron)decay. p, E and E_0 are the momentum, energy and its maximum value of the electron, respectively. $\mathbf{A}_{(2)}$ stands for the second class current. The first line of the righthand side shows the phase space part with the Fermi function $F_0(\pm Z,E)$ and the radiative correction $R(E,E_0)$. Expressions (11-13) indicated by arrows are only for eye guide.

The second line is called the spectra shape factor. It was measured for the first time by Lee, Mo and Wu[17], and, later, independently by the Heidelberg group[18]. It was investigated in detail theoretically by Koshigiri et al.[19] and by Behrens et al.[20]. They showed validity of the CVC prediction (11).

Information on the second-class current is involved in the coefficient $\alpha_\pm$, which can be determined from the beta-ray angular distribution from the beta decay of polarized nuclei($P \neq 0$) or of aligned nuclei($A \neq 0$)[8]. The latter experiment is the best for determination of $\alpha_\pm$ owing to the low background.

Physical importance of these coefficients can be seen from the following combinations:

$$\alpha_- - \alpha_+ = (4/3)(a - b_T), \tag{14}$$

and

$$\alpha_- + \alpha_+ = - (2/3M)y. \tag{15}$$

Equation (14) suggests that we can obtain information on the second-class current, since the term a is determined by the CVC theory, while (15) determines the time component of the axial-vector current, on which the exchange current suggested by Kubodera, Delorme and Rho[21] affects significantly. And it serves a test of soft pion theorem in nuclei. At the same time it strongly depends on details of nuclear structure pointed out by Koshigiri et al.[22] and, later, by Gichon et al.[23], since the time component of the axial vector current behaves differently from other components in the weak current under the time reversal operation. Measurement of the angular distribution of beta-ray from the aligned nuclei has been done by the ETH group[10,11] and the Osaka group[12,13,26]. From the data in Refs.[11], and [26], we obtain

$$g_T/g_A = -(0.13 \pm 0.48)/2M$$
$$= -(0.08 \pm 0.20)/2M$$

and

$$y = 3.79 \pm 0.48$$
$$= 3.83 \pm 0.11,$$

respectively, by using the ft-values and M1 transition rate from ^{12}C(15.1MeV). From this result, the second class current is vanishingly small. As for the time

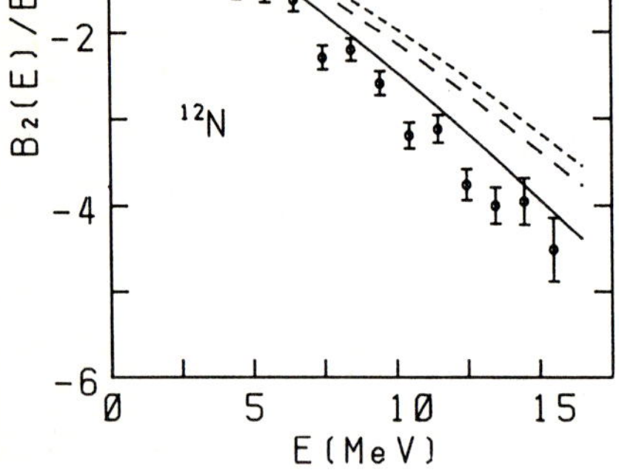

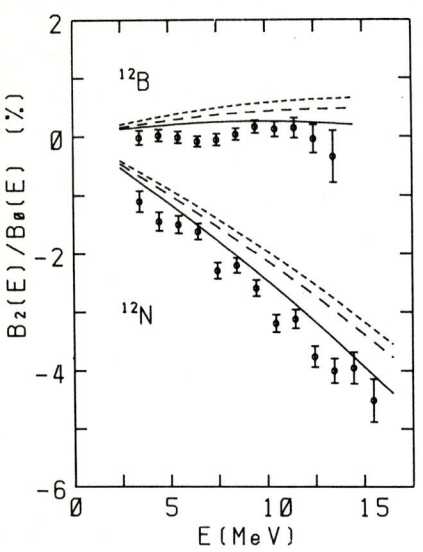

Fig.1 Alignment coefficient for ^{12}B and ^{12}B

Short and long dashed curves are the impulse calculations with and without the core polarization. Solid curve includes both the core polarization and exchange current.

component of axial-vector current, the impulse approximation predicts y = 3.17, if we adopt the p-shell model of Hauge and Maripuu[24] for the nuclear wave functions. The exchange current of Kubodera, Delorme and Rho enhances it about 1.09, while the core polarization effect in the first order perturbation reduces it about -0.56, if we assume the Sussex interaction[25]. Final result is y = 3.70, which is in agreement with the experimental data. In Fig.1, our result is compared with the latest experimental data[26].

3. Muon Capture and Nuclear polarization

Absolute value of the capture rate is governed by the absolute values of the relevant nuclear matrix elements, and is nuclear model-dependent. Therefore, it is difficult to determine the coupling constant g_P precisely. One exception is to study the muon capture in the few-body system owing to the rapid progress of the few body-problem. Another way for our purpose is to study physical quantities less sensitive to the nuclear structure and sensitive to the induced pseudoscalar coupling constant. Candidates are i) ratio of the muon capture rate to the beta decay rate and ii) polarizations of residual nucleus after muon capture.

The most typical example in case i) is the $0^+ - 0^-$ transitions in the A=16 system, which are completely governed by the axial-vector current and depend strongly upon the induced pseudoscalar coupling constant. At early days this ratio was thought to be independent of nuclear model[27]. However, it is found that this is not the case. The exchange current strongly enhances the time component of the axial-vector current[21,28], and the nuclear core polarization changes the ratio of matrix elements of the spatial and time components of the axial-vector current[29,30]. Recently, the extensive study of this reaction has been done[31] and its details will be presented later by Dr. Nozawa.

The celebrated case ii) is polarization of ^{12}B in muon capture on ^{12}C. The matrix elements determined from the beta decay of ^{12}B and ^{12}N, are reproduced fairly well by the shell model. Nevertheless, the absolute value of the muon capture rate depends more strongly upon the details of nuclear model than a ratio of the induced pseudoscalar and Gamow-Teller coupling constants g_P/g_A. On the other hand, polarizations of the recoil nucleus such as average and longitudinal polarizations are much less sensitive to the nuclear model, because polarizations are expressed by combinations of ratios of the various types of matrix elements to the Gamow-Teller matrix element in the same transition[32,33], and they are insensitive to the adopted nuclear model.

The average and longitudinal polarizations, P_{AV} and P_L, are defined and expressed by one parameter X as follows:

$$P_{AV} = \int d\Omega_{\boldsymbol{\nu}} \mathrm{Tr}\langle(\mathbf{P}_\mu\cdot\mathbf{J})\rho/P_\mu\rangle / \int d\Omega_{\boldsymbol{\nu}} \mathrm{Tr}\langle\rho\rangle = 2(1 + 2X)/3(2 + X^2) \tag{16}$$

and

$$P_L = -\int d\Omega_{\boldsymbol{\nu}} \mathrm{Tr}\langle(\boldsymbol{\nu}\cdot\mathbf{J})\rho/\nu\rangle / \int d\Omega_{\boldsymbol{\nu}} \mathrm{Tr}\langle\rho\rangle = -2/(2 + X^2), \tag{17}$$

where ρ and $\mathbf{J}$ are density matrix and angular momentum operator of ^{12}B, respectively, and $\mathbf{P}_\mu$ is the muon polarization vector. $\boldsymbol{\nu}$ is the neutrino momentum. By factoring out the muon wave function, the parameter X can be expressed in terms of matrix elements as

$$X = [1 + (\langle\rho_A\rangle + \langle ps\rangle + \langle\mathbf{A}'\rangle)/\langle\mathbf{A}\rangle]/(1 + \langle\mathbf{V}\rangle/\langle\mathbf{A}\rangle) \tag{18}$$

with

$$\langle\mathbf{A}\rangle = \langle\sqrt{2/3}\,\mathbf{Y}_{101}\cdot\mathbf{A}j_0 - \sqrt{1/3}\,\mathbf{Y}_{121}\cdot\mathbf{A}j_2\rangle, \tag{19}$$

$$\langle\mathbf{A}'\rangle = \sqrt{3}\langle\mathbf{Y}_{121}\cdot\mathbf{A}j_2\rangle, \tag{20}$$

$$\langle\mathbf{V}\rangle = \langle i\mathbf{Y}_{111}\cdot\mathbf{V}j_1\rangle, \tag{21}$$

$$\langle\rho_A\rangle = \langle i\sqrt{2}\rho_A Y_1 j_1\rangle, \tag{22}$$

and

$$\langle ps\rangle = \langle -i(m_\mu/2M)g_P(\boldsymbol{\sigma}\cdot\boldsymbol{\nabla})Y_1 j_1\rangle. \tag{23}$$

In contrast to the beta decay, we encounter the matrix element $\langle\mathbf{Y}_{121}\cdot\boldsymbol{\sigma}\rangle$, a correction to the main term, sensitive to the nuclear structure. At the same time the time component of the axial-vector current $\langle \rho_A \rangle$ is sensitive to both the nuclear structure and the exchange current.

In our previous study[34], we showed that the first order core polarization reduces the matrix element $\langle \rho_A \rangle$, while the exchange current enhances. Net effect enhances the matrix element about 28 %, and reduces the average polarization about 4.7%. Recently, we have investigated this problem by taking account of the second order core polarization[35], starting from the 1p-shell model with the Sussex interaction. In the first order perturbation, we included spin-orbit splitting of single particle states consistent with the residual interaction, and the intermediate states with upto $6\hbar\omega$. In the second order calculation, we evaluated the cross term between the first order core polarization and the exchange current, and effects of the second order core polarization by keeping the intermediate states with $2\hbar\omega$ carefully. Contributions of higher energy configurations were estimated by using the jj-coupling model. They improve the absolute value of capture rate, while they scarcely affect polarizations. Finally, we found that reduction of the average polarization is about 3.4% instead of 4.7% in the first order calculation. From our results on the beta decay and muon capture, we can determine g_P/g_A with good accuracy. Possible uncertainty in our calculation may be the parameter y, which brings at most 3% error in g_P/g_A.

There is, however, ambiguity to obtain the polarization relevant to determination of g_P/g_A[32]. Experimentally, we observe the recoil nuclei of the ground state of ^{12}B, which are produced directly by the muon capture on ^{12}C, or which is subsequently produced by the deexcitation of the excited states of ^{12}B produced by the muon capture. We must therefore subtract the latter effect from the experimental data. A dominant correction comes from the 1^- state, and therefore we need the branching ratio of the partial muon capture rates of the 1^- state and the ground state. Unfortunately, we have different values of the branching ratio obtained by different groups[36,37]. If we adopt the experimental ratio of the average and longitudinal polarizations[38] and the latest value of the branching ratio[36], we obtain

$$R(1^+) = P_{AV}(1^+)/P_L(1^+) = -0.499 \pm 0.044,$$

which is shown by the shaded area in Fig.2. Our result is plotted as a function of g_P/g_A. From this figure, we obtain $g_P/g_A = (9.5 \pm 1.7)/m_\mu$, which is consistent with the PCAC prediction $g_P/g_A = 7$ within the experimental error. The same result is obtained from the experimental data on the average polarization[39,40].

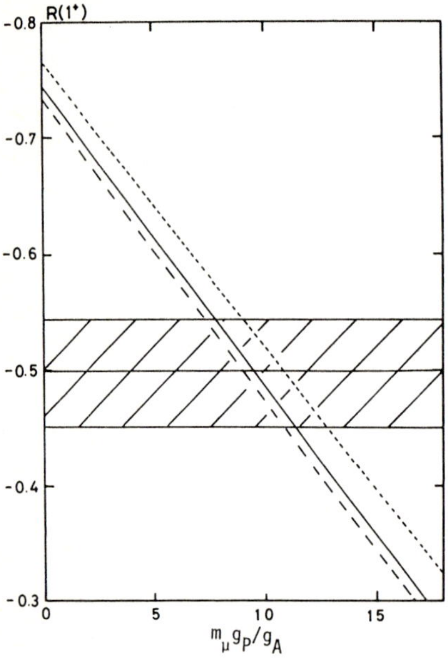

Fig.2 Ratio $R(1^+)$ of ^{12}B in muon capture on ^{12}C

Short dashed curve is the impulse approximation with the p-shell wave function. Long dashed and solid curves include corrections of the first and second order, respectively.

4. Remark

We discussed the beta-ray angular distribution and nuclear polarization in muon capture in order to determine the coupling constant ratio g_P/g_A. Here it should be emphasized that g_P/g_A cannot be determined by only the muon capture, unless nuclear matrix elements of vector and axial-vector currents are confirmed to be correct. Therefore, the beta-decay plays an essentially important role in testing these matrix elements. This standpoint is also found in Ref.[41] where the same problem is studied by the particle treatment of nuclei.

Finally, we should mention the nuclear polarizations after muon capture on spin-nonzero nuclei[42]. We can expect that muon capture in the different muonic hyperfine states will give us much information on the nuclear matrix element. Among them, we found that the average alignment of residual nuclei in the muon capture on ^{13}C and ^{14}N in the muonic hyperfine states is rather insensitive to the detail of the nuclear model, and is suitable for determination of induced pseudoscalar coupling constant.

1 M. Morita: Hyperfine Int. 21, 143 (1985)
2 D. H. Wilkinson and D. E. Alburger: Phys. Rev. C13, 2517 (1976)
3 Particle Data: Rev. Mod. Phys. 52, S1 (1980)
4 S. L. Adler: Phys. Rev. Letters 14, 14 (1965)
W. I. Weisberger: Phys. Rev. Letters 14, 1047 (1965)
5 M. Gell-Mann and M. Lévy: Nuovo Cim. 16, 703 (1960)
6 R. P. Feynman and M. Gell-Mann: Phys. Rev. 109, 193 (1958)
7 S. Weinberg: Phys. Rev. 112, 375 (1958)
8 M. Morita et al.: Prog. Theor. Phys. Suppl. 60, 1 (1975)
9 K. Kubodera et al.: Phys. Letters 58B, 402 (1975)
10 P. Lebrun et al.: Phys. Rev. Letters 40, 301 (1978)
11 H. Brändle et al.: Phys. Rev. Letters 40, 306 (1978); 41, 299 (1978)
12 K. Sugimoto et al.: J. Phys. Soc. Japan, Suppl. 44, 801 (1978)
13 Y. Masuda et al.: Phys. Rev. Letters 43, 1083 (1979)
14 N. C. Mukhopadhyay: Phys. Reports C30, 1 (1977)
15 G. Bardin et al.: Phys. Letters 104B, 320 (1981)
16 W. Hughes and C. S. Wu: Muon Physics, vol.2, (Academic, New York 1975)
17 C. S. Wu et al.: Phys. Rev. Letters 10, 253 (1963); 39, 72 (1977)
18 W. Kaina et al.: Phys. Letters 70B, 411 (1977)
19 K. Koshigiri at al.: Nucl. Phys. A319, 301 (1979)
20 H. Behrens et al.: Inst. für Kernphysik, Teck Hochschule Darmstadt Report IKDA 78/5, (1978)
21 K. Kubodera, J. Delorme and M. Rho: Phys. Rev. Letters 40, 755 (1978)
22 K. Koshigiri, H. Ohtsubo and M. Morita: Prog. Theor. Phys. 66, 358 (1981); J. Phys. Soc. Japan, 55, 1014 (1986)
23 P. A. M. Guichon and C. Samour: Nucl. Phys. A382, 461 (1982)
24 P. S. Hauge and S. Maripuu: Phys. Rev. C8, 1609 (1973)
25 J. P. Elliott et al.: Nucl. Phys. A121, 241 (1968)
26 T. Minamisono et al.: J. Phys. Soc. Japan, Suppl. 55, 382 (1986)
27 A. Maksymowicz: Nuovo Cim. 48A, 320 (1967)
28 P. A. M. Guichon, M. Giffon and C. Samour: Phys. Letters 74B, 15 (1978)
29 K. Koshigiri, H. Ohtsubo and M. Morita: Prog. Theor. Phys. 62, 706 (1979)
30 I. S. Towner and F. C. Khanna: Nucl. Phys. A372, 331 (1981)
31 S. Nozawa et al.: Prog. Theor. Phys. 74, 926 (1985)
S. Nozawa, K. Kubodera and H. Ohtsubo: Nucl. Phys. A453, 645 (1986)
32 M. Kobayashi et al.: Nucl. Phys. A312, 377 (1978)
33 H. Ami et al.: Prog. Theor. Phys. 65, 632 (1981)
34 M. Fukui et al.: Prog. Theor. Phys. 70, 827 (1983)
35 M. Fukui et al.: J. Phys. Soc. Japan, Suppl. 55, 222 (1986)
36 L. Ph. Roesch et al.: Phys. Letters 107B, 31 (1983)
37 M. Giffon et al.: Phys. Rev. C24, 241 (1981)
38 L. Ph. Roesch et al.: Phys. Rev. Letters 46, 1507 (1981)
39 A. Possoz et al.: Phys. Letters 70B, 265 (1977)
40 Y. Kuno et al.: Phys. Letters 148B, 270 (1984)
41 S. Nozawa et al.:Prog. Theor. Phys. 70, 892 (1983)
42 K. Koshigiri et al.: Prog. Theor. Phys. 74, 736 (1985)

Radiative Muon Capture and the Induced Pseudoscalar Coupling in Nuclei*

M. Döbeli[1], M. Doser[1], L. van Elmbt[1], M. Schaad[1], P. Truöl[1], A. Bay[2], J.P. Perroud[2], and J. Imazato[3]

[1]Physik-Institut der Universität Zürich, Schönberggasse 9, CH-8001 Zürich, Switzerland
[2]Institut de Physique Nucléaire, Université de Lausanne, CH-1015 Lausanne, Switzerland
[3]Meson Science Laboratory, University of Tokyo, Bunkyo-Ku, Tokyo 113, Japan

The experimental attempts to exploit the sensitivity of the radiative muon capture (RMC) reaction[1]

$$\mu^- N(A,Z) \rightarrow N^*(A,Z-1)\nu_\mu \gamma$$

to the induced pseudoscalar coupling g_P until now are limited to the two nuclear targets ^{16}O[2] and ^{40}Ca [2,3,4]. The observables are the photon energy spectrum above a cutoff energy of 57 MeV, below which the muon decay electron induced bremsstrahlung prohibits the measurement, and the muon spin photon angular correlation. The latter quantity has only been measured with poor statistical accuracy for ^{40}Ca[3,5,6]. The data are scarce because the low rate (the radiative branch above 57 MeV contributes only $2 * 10^{-5}$ to the total muon absorption rate in ^{40}Ca), the modest resolution attainable for medium energy photons and the high background environment of fast neutrons severely trouble the measurement. But even this limited amount of data has lead to surprising results, when compared to theoretical calculations [7,8,9], which proceed in the impulse approximation with a Hamiltonian derived from the elementary $\mu^- p \rightarrow n\nu_\mu\gamma$ reaction and a nuclear response function, adapted to describe the results of other medium energy weak or electromagnetic processes. The interpretation of the integrated photon yield and the photon spectrum in terms of the pseudoscalar coupling yields $g_P = (4.5 \pm 1.5) g_A$ for ^{40}Ca[4] and g_p exceeding $12 g_A$ for ^{16}O [1,2,9]. This values are to be compared with $g_P = (7.0 \pm 1.5) g_A$ extracted from the orthomolecular capture rate for muonic hydrogen [10] and $g_P = (10.0 \pm 2.5) g_A$ [11] (and $g_P \sim 12 g_A$ [12]) deduced from the β-decay and the muon capture rate of the $^{16}N(0^-, 0.12\ MeV)$ to $^{16}O(g.s)$ transition. Prompted by these findings we started a series of experiments with two principal goals. The measurement of photon spectra for various other targets ranging from ^{12}C to ^{209}Bi should shed light on the question, whether the induced pseudoscalar coupling is indeed enhanced in lighter and suppressed in heavier nuclear systems. Secondly the photon-muon spin angular correlation asymmetry parameter Γ_γ seems to be less dependent on the model used to describe the initial and the

*presented by P. Truöl

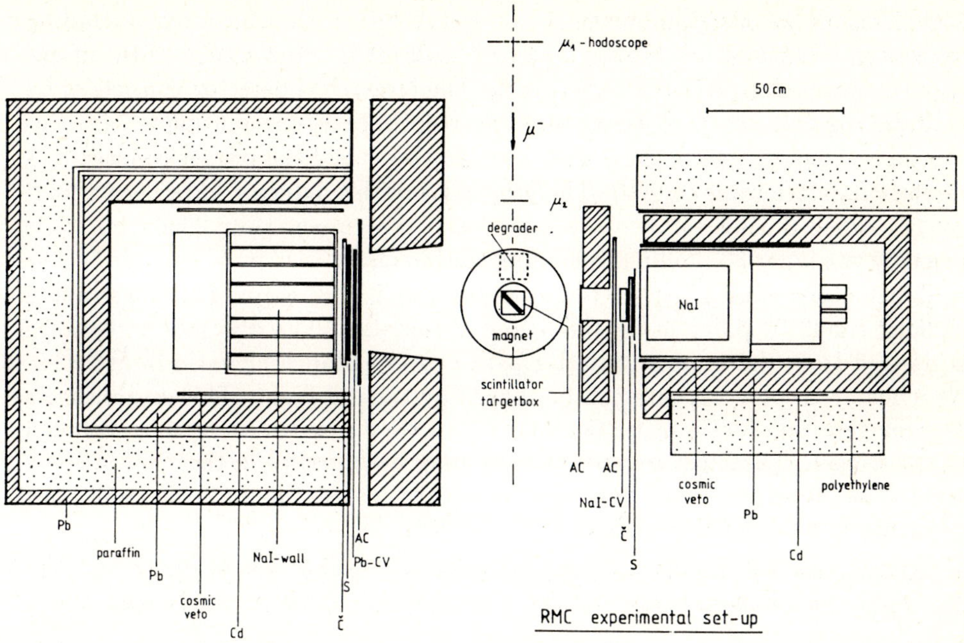

Figure 1: Experimental Set-Up

nonobserved final nuclear states then the photon yield. Thus a more precise determination of this quantity aims at eliminating the uncertainties arising from the assumed nuclear response in ^{40}Ca.

The experimental setup at the SIN superconducting muon channel is shown in Figure 1. Muons with 125 MeV/c momentum were brought to rest in the following target materials : ^{12}C, H_2, ^{16}O, ^{27}Al, ^{40}Ca, ^{nat}Fe, ^{165}Ho and ^{209}Bi. The targets were placed in a hermetically closed scintillator box, when muon stopping rates were determined at low intensities. For the correlation experiment the ^{40}Ca target is placed in a transverse magnetic field produced by two Helmholtz coils. The initial muon spin direction is antiparallel to the beam and with a field of 0.374 Tesla all arriving muons precess in phase, since the precession frequency equals the RF-frequency of the SIN-accelerator (50.7 MHz). Thus the time difference between the RF-signal and a signal from either of the two photon detectors provides a measure of the muon spin direction at the moment of absorption. This stroboscopic method [6] allows to use the full flux of $30 * 10^6\ sec^{-1}$ muons. During the measurement of the photon spectra both the magnet and the scintillator box are removed to reduce target non-associated background. The two NaI-spectrometers placed perpendicular to the beam are equipped with electron-positron pair converters followed by a scintillator and a lucite Čerenkov counter to tag the pair. This renders both arms sufficiently insensitive to neutrons. The smaller NaI-detector has cylindrical shape (27 cm diameter, 33 cm length) and an active disk-shaped NaI-converter of 2 cm thickness. This allows a measurement of the energy loss in the converter. Including conversion efficiency its total acceptance is $\Delta\Omega/4\pi = 1.2 * 10^{-3}$ with an energy resolution $\Delta E/E(FWHM) = 11\%$. The larger NaI detector consists of 64 modules ($6x6x40\ cm^3$). Without the converter its energy resolution is 7.0%, including the energy loss in the 0.4 cm lead converter and the trigger counters the resolution deteriorates to 32%. The properties of this detector have been extensively studied during our $\pi^+ \rightarrow e^+\nu_e\gamma$ decay experiment [13]. The solid angle given by the inner 36 modules including conversion is $\Delta\Omega/4\pi = 3.8 * 10^{-3}$. The calibration procedures rely on the $\pi^- p \rightarrow n\gamma$ and $\pi^- p \rightarrow n\pi^o$ reactions (muon channel tuned to pions using CH_2- and C-targets). The absolute acceptance is derived from the known radiative pion capture branching ratio in ^{12}C [14]. Neutrons from pion capture can be tagged by time of flight. This establishes a neutron rejection factor of better then 10^4, if the Čerenkov counter is included in the trigger. The most important remaining background in the photon spectrum stems from the pion contamination in the beam. Figure 2 shows the raw spectrum from radiative muon capture in ^{40}Ca.

Apart from the bremsstrahlung contribution visible below $m_\mu/2$ and the $(\mu^-, \nu_\mu\gamma)$ contribution below m_μ the pion induced signal is clearly seen above

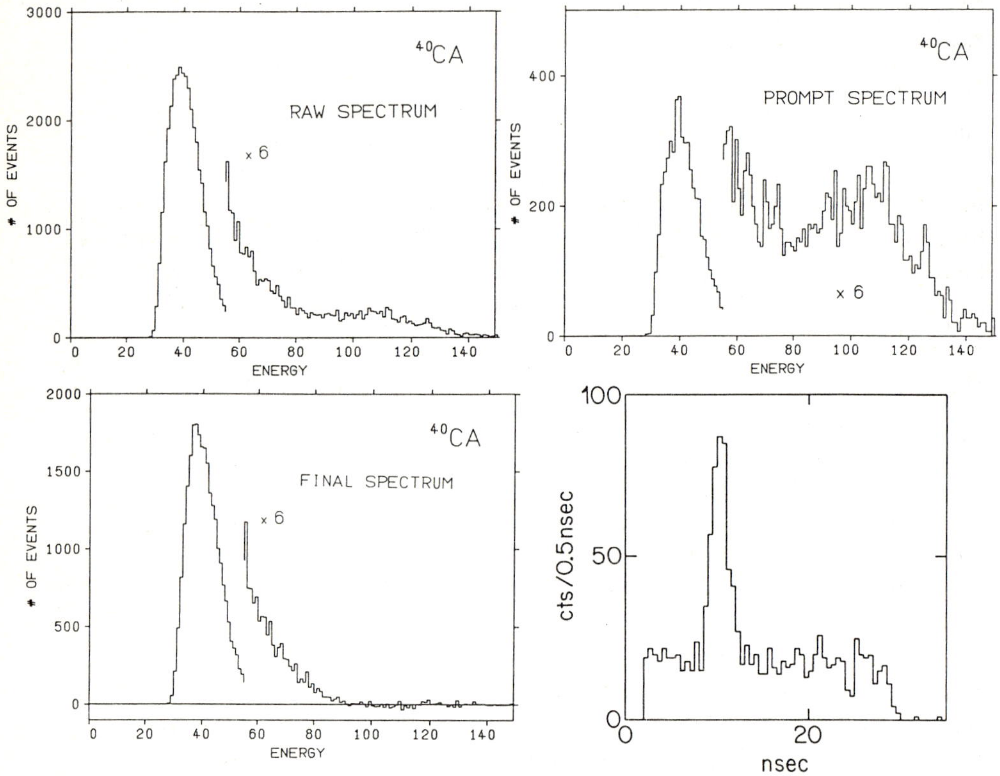

Figure 2: Photon spectra from radiative muon capture in ^{40}Ca. Bottom right: Event time spectrum, the peak contains the prompt events.

Table 1: Radiative Muon Capture Branching Ratios

Target	Yield[1] [*10^-5]	Yield[2] [*10^-5]	g_P/g_A	Theoretical Model
^{12}C		10.6 ± 3.0		
^{16}O	8.5 ± 1.7	9.3 ± 1.9	> 16[3]	shell modell [8]
		6.2 ± 0.8	> 16 [3]	giant resonance [9]
^{27}Al		2.5 ± 0.7	8.7 ± 3.4[4]	giant resonance [9]
^{40}Ca	2.3 ± 0.4	2.2 ± 0.4		
		2.11 ± 0.16 [3]		shell modell [8]
	1.86 ± 0.16 [4]	2.07 ± 0.20 [4]	4.0 ± 1.5	giant resonance [7]
^{56}Fe		1.6 ± 0.3	2.2 ± 2.0[5]	Fermi gas [15]
^{156}Ho		0.89 ± 0.18	0.9 ± 1.9 [5]	Fermi gas [15]
^{209}Bi		0.76 ± 0.15	2.0 ± 2.0 [5]	Fermi gas [15]

[1]Yield extracted using theoretical curve

[2]Yield extracted using parametrisation in energy or model-independent unfolding technique

[3]calculations for $g_P > 16 g_A$ not available

[4]interpolating ^{16}O and ^{40}Ca

[5]interpolating from other nuclei

m_μ. With a cut in the TDC-spectrum measuring the time difference between a beam particle and a photon (insert of Figure 2) the prompt pion spectrum can be removed. The number of good events lost by this cut can be determined from the number of μ-decay events in the prompt spectrum below $m_\mu/2$. Applying a similar procedure to the target out data and subtracting this contribution one obtains the final spectrum shown in Figure 2. Fitting this spectrum with a theoretical spectrum folded with the line shape and the relative acceptance and then integrating the spectrum above the cut-off energy of 57 MeV yields the branching ratioes given in Table I.

To arrive at the entries in Table I one must account properly for the muons stopping in the target, the fraction of muons decaying, the dead-time losses, accidental vetos of photons caused by decay electrons, photon absorption in the target and the absolute acceptance. Not in all cases has the full data sample been used, which is why the results must still be considered preliminary. It is comforting however, that for the previously measured targets the results are reproduced. Judging from the published analysis of previous ^{40}Ca-data [3,4] the differential yield depends little on the particular theoretical model, since the photon spectrum is monotonically decreasing with energy and largely determined by phasespace factors. Both microscopic calculations [7,8] as well as simple polynomial ansatz used in reference 3 lead to the same branching ratio. For those nuclei where there are no theoretical calculations existing only the polynomial ansatz has therefore been used.

The treatment of asymmetry data is similar. With prompt and target non-associated background removed, the photon-RF time spectra are fitted with the function

$$dN/dt = N_0 e^{-t/\tau}(1 + A\cos(\omega t + \delta))$$

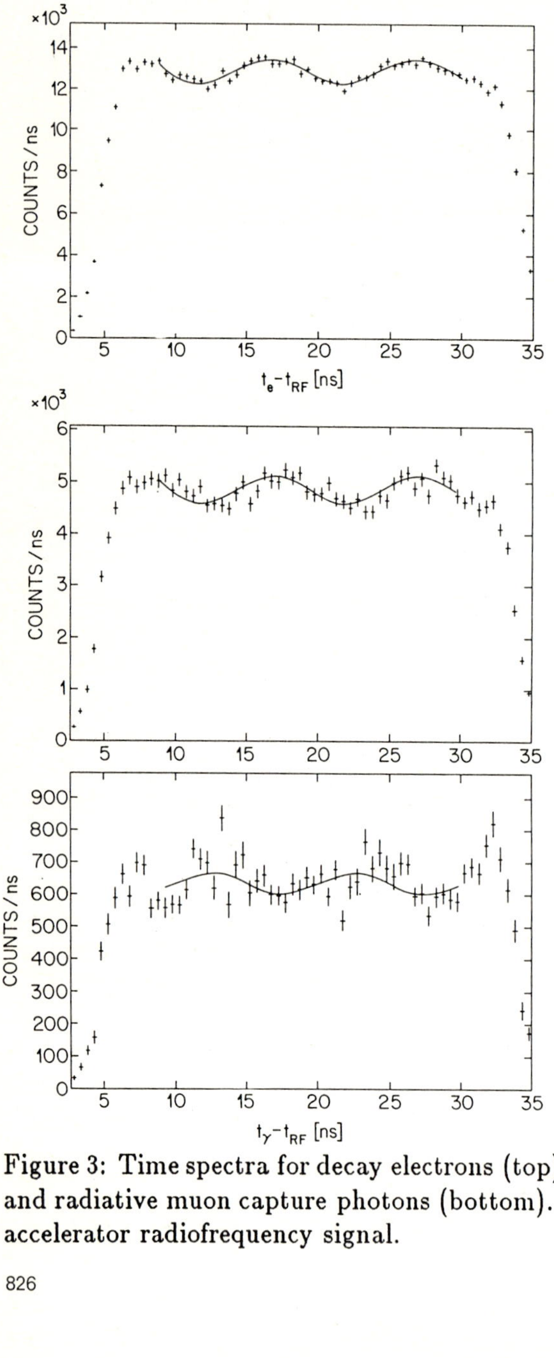

Figure 3: Time spectra for decay electrons (top), bremsstrahlung events (center) and radiative muon capture photons (bottom). t_{RF} is the time marker from the accelerator radiofrequency signal.

Table 2: Muon-Spin Angular Correlation Data

Reaction	Experimental Asymmetry[1] %	Number of events
$\mu \to e^- \overline{\nu}_e \nu_\mu$	-5.49 ± 0.29	750000
$\mu \to e^- \overline{\nu}_e \nu_\mu + \gamma$	-6.51 ± 0.75	280000
$^{40}Ca(\mu^-, \nu_\mu \gamma)$	6.9 ± 1.5	40000
	Γ_γ[2]	Reference
$^{40}Ca(\mu^-, \nu_\mu \gamma)$	0.80 ± 0.17	this experiment
	0.92 ± 0.26	this experiment[3]
	0.90 ± 0.50	Ref. [3]
	0.82 ± 0.76	Ref. [6]
	0.85 ± 0.25	Ref. [5][4]
	0.84 ± 0.12	average

[1] large NaI-data

[2] Third entry divided by first entry times expected electron asymmetry $\overline{\alpha(E)} = -0.64$

[3] small NaI-data

[4] Value corrected for the experimental depolarisation factor 0.137 ± 0.010 (instead of 0.166 used in reference [5])

folded with a Gaussian of 6 ns width to account for the time spread of the muon beam burst. The free parameters are N_o, δ and A.

The asymmetry parameter A represents the energy average correlation parameter $\Gamma_\gamma(E\gamma > 57\ MeV)$ multiplied with a μ^- depolarisation factor and the beam polarisation. The latter two factors are determined from time-differential decay electron spectra, where the correlation parameter is known. Further calibration is afforded by the bremsstrahlungs events which also exhibit the decay electron asymmetry. Time spectra for electrons and photons are shown in Figure 3, the results for the asymmetries and Γ_γ are given in Table II. Some $\mu^+ \to e^+$ decay data were taken also to measure the beam polarisation and verify the stroboscopic field.

Within their limited precision all measurements of Γ_γ agree with each other. In Figure 4 we compare the average of all measurements with theoretical results.

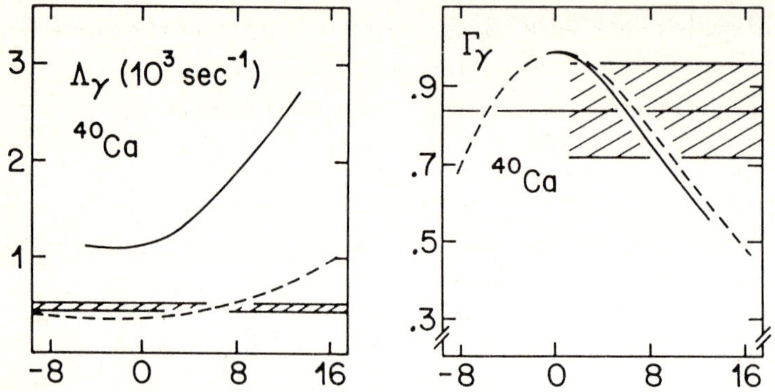

Figure 4: Yield and asymmetry in ^{40}Ca versus g_P/g_A. Shadowed regions indicate the experimental result. Dashed curve: Ref. [17]; solid curve: Ref. [7]

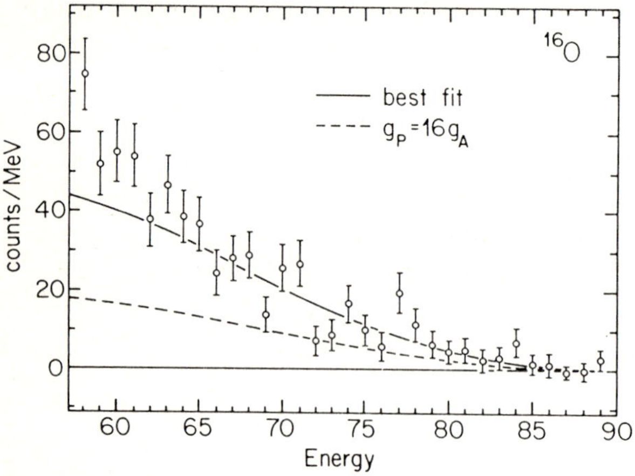

Figure 5: Photon-spectrum for ^{16}O. Dashed curve: Ref. [8](properly normalized). Solid line: Fit to polynomial in energy.

Though the value of g_P extracted from this Figure, $g_P/g_A = 6.8 \pm 3.4$ is not yet sufficiently accurate, one may expect, that a further improvement of the measurement may provide the key to understand the puzzling yield data. Two different models for the nuclear response, the simple shell model [16] and the giant resonance model which lead to quite different results for the yield produce nearly identical asymmetries. The inclusion of the constraints provided by the continuity equation on the electromagnetic vertices (Siegert's theorem), which bring the shell model results in line with the giant resonance model [8] for the yield and lead to $g_P/g_A = 4.0 \pm 1.5$ [4] essentially leave the asymmetry unaffected. A recently discovered new technique [20] to repolarize the muon through its hyperfine interaction with a polarized nucleus in ^{209}Bi may provide the means to measure Γ_γ in a heavy nucleus, too. Without such results the possibility, that the low yields found in heavy elements as compared to the prediction of the Fermi gas model can be ascribed to a deficiency of the nuclear model, cannot be completely discarded. Yet that same technique on average represents the measured total muon capture rate within a few percent.

The discrepancy between experiment and theory (using the standard Goldberger-Treiman or nucleonic value for g_P) is particularly striking in ^{16}O. The wave functions, which enter into the model calculation [8], have been successfully tested in predicting the radiative pion capture and inelastic electron scattering data. Only a few dipole and magnetic quadrupole transitions contribute. Yet the experimental rate exceeds the predictions even for g_P/g_A equal to twice the nucleonic value by more than a factor of two. Even accepting an enhancement of g_P as evident also from the exclusive μ-capture data to the 0^- state [11] the situation is quite unsatisfactory. Again the solution might be found in a - yet unattempted - measurement of Γ_γ in this nucleus.

Figure 6 shows the radiative muon capture yield versus Z. The experimental results are marked with □ (bottom: Z-values indicated in first line; •△• : theoretical results for $g_P/g_A = 16, 8, 0$, respectively (Z-values indicated on lower line); $Z > 42$ [15], $Z = 20$ [7], $Z = 8$ [8,9], $Z = 2$ [18], $Z = 1$ [19]; top: g_P extracted from yield measurements). In conclusion it is observed, that the smooth dependance of the radiative muon capture yield upon nuclear charge (Figure 6), approximately proportional to Z^{-1}, is in contradiction with theoretical calculations and would lead to unexpected variation of the induced pseudoscalar coupling with nuclear mass.

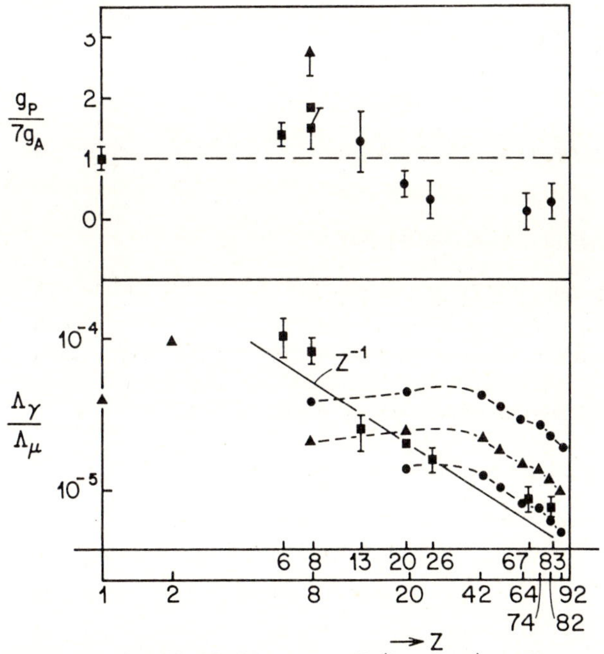

Figure 6: RMC Yield versus Z (bottom) and g_P extracted from yield measurements (top).

References

[1] For a recent review see M. Gmitro, P. Truöl, to appear in Adv. Nucl. Phys.

[2] A. Frischknecht et al.: Czech. J. Phys. B32, 270 (1982)

[3] R.D. Hart et al.: Phys. Rew. Letters 39, 399 (1977)

[4] A Frischknecht et al.: Phys. Rev. C32, 1506 (1985)

[5] M.D. Hasinoff et al.: Proc. Xth Int. Conf. on Particles and Nuclei, Heidelberg, Abstracts Vol. II, p. H4, 1984

[6] A. Frischknecht et al.: Hel. Phys. Acta 53, 647 (1980)

[7] P. Christillin: Nucl. Phys. A362 391 (1980)

[8] M. Gmitro and A.A. Ovchinnikova, T.V. Tetereva: Nucl. Phys. A453 685 (1986)

[9] P. Christillin and M. Gmitro: Phys. Letters 150B 50 (1985)

[10] G. Bardin et al.: Phys Letters 104B 320 (1981)

[11] C.A. Gagliardi et al.: Phys. Rev. C28 2423 (1983)

[12] A.R. Heath and G.T. Garvey: Phys. Rev. C31 2190 (1985)

[13] A. Bay et al.: Phys. Letters 174 455 (1986)

[14] J.P. Perroud et al.: Nucl. Phys. A453 542 (1986)

[15] P. Christillin and M. Rosa-Clot and S. Servadio: Nucl. Phys. A345 331 (1980)

[16] M. Döbeli et al.: Czech. J. Phys. B36 386 (1986)

[17] M. Gmitro et al.: Czech. J. Phys. B31 499 (1981)

[18] L. Klieb and H.P.C. Rood: Phys. Rev. C29 223 (1984)

[19] M. Gmitro and A.A. Ovchinnikova: Nucl. Phys. A356 323 (1981)

[20] R. Kadono et al.: Proc. 6th Int. Symp. Polarisation Phenomena in Nuclear Physics, Osaka, Japan, (1985)

Possibilities to Measure Electroweak Effects in Muonic Atoms

L.M. Simons

Schweizerisches Institut für Nuklearforschung, CH-5234 Villigen, Switzerland

At first glance, muonic atoms offer several attractive features for parity violation studies compared to the proposals in atomic physics with electronic atoms. First, the overlap of the muon wave function with the nucleus is $(m_\mu/m_e)^2$ times more than in electronic atoms with correspondingly larger weak interaction matrix elements. Also, the structure of these atoms is as simple as the hydrogen atom. This facilitates the calculation of the relevant matrix elements, and hence makes the interpretation of an eventual experiment feasible to a high degree of accuracy. Next, the preparation of a pseudoscalar observable is possible either via the natural polarization of a muonic atom produced with a polarized muon beam or via the detection of the decay electron from the ground state.

Because of the interplay of the finite size effect and the vacuum polarization; there exist several near degeneracies of the 2s and 2p states. The mixing of these parity-odd states produces an interference of electric dipole with the highly forbidden magnetic dipole transitions. The resulting asymmetry of the direction of emission of the M1 transition radiation with respect to a direction of polarization are several percent[1,2]. It is to be noticed, however, that the observation of a M1 transition is exceedingly difficult because of the low M1 branching ratio of the only weakly populated 2s level. In solid targets and gas targets of pressures high enough to stop a sufficient number of muons, the dominant decay mode of the 2s level is the conversion of the 2s-2p transition with the emission of an electron from the atomic L shell for muonic atoms with $Z \gtrsim 3$. A search for a metastable 2s level therefore was not sucessfull[3].

Because of new techniques available to form more muonic atoms in gases at pressures at and below 1 atm compared to conventional set-ups, muonic atoms can be formed without contact with neighbouring atoms during the muonic cascade time. Since the cascade begins at main quantum numbers $n > 14$, and because each muonic Auger transition initiates a vacancy cascade in the electron shell, there will be almost complete ionization even for atoms with $Z \sim 18$[4]. In measurements using certain intensity ratios of circular transitions to indicate the number of remaining electrons, this picture was verified. In muonic Ne a single electron is still bound at the time the muon reaches the $n = 5$ level in less than 3 % of the muonic atoms. In muonic argon the probability for one electron remaining is less than 50 %. Cascade calculations show that these relations also approximately hold for the 2s level.

In order to keep the muonic atom free of electrons during the lifetime of the 2s-level measured electron capture cross sections have been used to calculate the maximum allowed pressure for Ne to be 1 atm[5]. Here a velocity of the muonic atom of 10^5 cm/s' has been assumed. For Argon the corresponding pressure is several atm.

An additional requirement for measuring electroweak effects is to distinguish between the 2s - 1s and 2p - 1s transitions by their energy without going to high Z where the effect to be measured and the efficiency of solid state detectors drop rapidly. Hence the conclusion is, that gaseous compounds of Al or Si should be examined next to determine whether enough electrons remain stripped at feasible pressures.

It is worth-while to mention that a second window is open for a possible observation of a parity violation effect at $Z = 5$. Here it is assumed, that more than 10^6 muonic boron atoms can be formed in diborane at pressures below 10^{-4} atm. Using an absorption edge in Zn at 9.66 keV the two 2s hyperfine components can be filtered out so that in coincidence with the 2s - 1s transition the different coupling constants of the weak neutral current effects could be determined.

References

1. J. Bernabeu, T.E.O. Ericson and C. Jarlskog, Phys. Lett. 50B (1974) 467; G. Feinberg and M.Y. Chen, Phys. Rev. 10 (1974) 190; A.N. Moskalev, JETP Lett. 19 (1974) 216.

2. J. Missimer and L. Simons, Phys. Rep. 118C (1985) 179.

3. A.L. Carter, et al. , Phys. Lett. 124B (1983) 465.

4. R. Bacher, et al., Phys. Rev. Lett. 54 (1985) 2087.

5. C.R. Vane, M.H. Prior and R. Marrus, Phys. Rev. Lett. 46 (1981) 107.

Heavy Muonic Atoms and Muon Capture

P. David

Institut für Strahlen- und Kernphysik, Universität Bonn, Nußallee 14–16, D-5300 Bonn 1, F. R. Germany

The study of muonic actinide atoms and the subsequent fission of these deformed heavy nuclei has revealed several aspects about the electromagnetic interaction of the muon with the nucleus at low energies and about the weak interaction. Information on nuclear ground state shapes, nuclear structure and on capture mechanisms has been obtained.

For the muon-nucleus interaction, Wheelers prediction |1| has been proven that radiationless (r.l.) transitions between the lower muonic levels can lead to prompt fission by direct electromagnetic excitation of the nucleus and that the muon in the atomic ground state may be captured into the nucleus by weak interaction, thus exciting the nucleus with the possible consequence of delayed fission. In fact this process may be considered as partial muon capture on one of the bound protons in the weak process $\mu^- + p \rightarrow n + \nu_\mu$. The spectroscopy of the fission channel results in measuring the total muon capture rate in averaging over the spectrum of nuclear excitation energies. The prompt fission process takes place with the muon in its 1s orbit strongly overlapping with the nucleus. The binding energy of the muon depends appreciably on the nuclear deformation and thus strongly influences the nuclear potential energy. As a consequence of this, changes of static fission properties, like the height of the fission barrier, the characteristics of isomeric states and the rate of their population may be expected. It also seemed worthwhile investigating the spectator role of the muon while probing fission dynamics.

1. Muonic X-rays and nuclear structure effects

For ^{237}Np a γ - γ -measurement was performed at SIN with a 10g NpO_2 target, with a 30% intrinsic Ge-detector with BGO shield and with two large volume scintillation detectors |3|.

The 2p-1s, 3d-2p and 4f-3d complexes in the muonic ^{237}Np X-ray spectrum show several low lying excited states belonging to the nuclear rotational band of the Nilsson ground state configuration 5/2 + [642]↑ to be mixed appreciably with the muonic 3d, 2p and 1s states |2,3|. The nuclear spectroscopic quadrupole moment was determined by fitting the observed hyperfine splitting of the $5g_{9/2}$-$4f_{7/2}$ and $5g_{7/2}$-$4f_{5/2}$ muonic X-rays, giving an average value of Q ($^{237}Np_{g.s.}$) = (3.63 ± 0.02)b. This method has the advantage of not being sensitive to the finite nuclear size and to the dynamical mixing of nuclear states. The K, L and M X-ray data therefore were only used to determine the nuclear charge distribution parameters. The Fermi-parameters obtained from the experimental X-ray transition energies are c=7.06 fm and t=2.3 fm |2,3|.

2. The probability of non radiative decay of the 3d level in muonic ^{237}Np

The role of quadrupole 3d-1s and of dipole 2p-1s r.l. transitions in nuclear excitation has been investigated for ^{238}U, where the r.l. 3d-1s transition probability has been measured to be (14 ± 5)% |4|. A similar measurement was performed for ^{237}Np. Normalizing the ratio of the gated to non gated ^{237}Np muonic L X-ray complexes to unity and basing on the 4f-3d and 5g-4f intensity ratios, the 2p level is bypassed on the average by (12 ± 4)%. With the relative radiative width of the 3d level in muonic ^{237}Np measured to be (3 ± 1)% |6| the probability for a r.l. 3d-1s transition in muonic ^{237}Np is (9 ± 4)%. These results for ^{238}U and ^{237}Np show the phenomenological model predictions of Teller and Weiss |5| to be quite good.

3. Mean life times of negative muons bound to actinide nuclei and capture rates

Muon-fission fragment time spectra T(μ^-,f) have been measured for the target nuclei ^{235}U, ^{238}U, ^{237}Np, ^{242}Pu and ^{244}Pu with a resolution of 1.2 ns (FWHM) |7|. The mean life times of the muons in the 1s orbit and the ratios of prompt to delayed fission yields (p/d-ratio) have been determined from them. The figure presents mean life times of muons

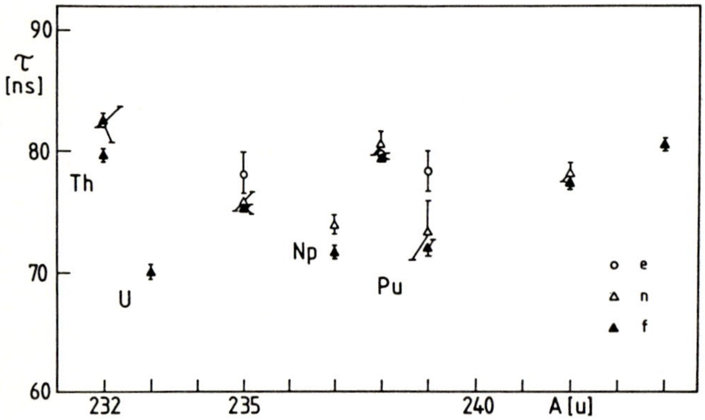

Fig.1

bound to actinide nuclei as determined from the fission decay channel in comparison with values obtained from other decay channels. Correcting the experimental data for systematic errors due to muon capture by fission fragments and by surrounding materials, the corrected values from the various decay channels do not show a significant time difference which would indicate the presence of an isomeric state. Also the evaluation of the time spectra of ^{237}Np, ^{242}Pu and ^{244}Pu by directly including a second time component, did not give a statistically significant result.

The measured muon capture rates evidence an isotope effect, also predicted by Primakoff |8|. The results obtained within this model |9| underestimate, the results obtained within the original giant resonance model of Kozlowski and Ziglinski |10,11| overestimate the measured capture rates. Evidently, more systematic and precise data on separated isotopes are required for further conclusions on microscopic absorption mechanisms |12|. Such measurements virtually free of systematic errors are prepared to be performed at SIN.

4. The ratios of prompt and delayed fission yields and average excitation energy in delayed fission

The ratios of prompt to delayed fission yields have been determined from the time spectra for the nuclei ^{235}U, ^{238}U, ^{237}Np, ^{242}Pu and ^{244}Pu. From these ratios the fission probabilities per stopped muon and the average excitation energy of delayed fission processes have been calculated. Some of the experimental values of the fission probabilities per stopped muon show differences up to a factor five, as compared to an average value, possibly due to systematical errors in the measurements. Therefore calculations were performed using nuclear excitation energy distributions from Hadermann and Junker |14| and from Singer |15| and using excitation functions of the respective fission probability calculated up to 40 MeV of excitation energy. With few exceptions the calculated fission probabilities per stopped muon agree nicely with the experimental values.

The same excitation energy distributions of refs. |14,15| have been used to calculate also the average excitation energy in delayed fission. For ^{237}Np a value of 19 MeV has been derived from the mass yield distribution of the delayed fission processes. The calculated and the experimental value for ^{237}Np agree nicely. The value of the average excitation energy in delayed fission is of interest, since it allows, when knowing the energy dependence of the fission probability, to discriminate distribution functions of the excitation energy calculated in microscopic models.

5. The barrier augmentation

In prompt fission the energy to raise the muonic 1s level during the deformation is taken from the nuclear excitation energy. The nuclear barrier height increased by this could most directly be measured in identifying a fission isomer of a muonic atom by its life time. No clear indication of such a state is presently available, however. Another way is offered in measuring the fission probability $P_f(E_x)$ at the energies of the sharp radiationless transitions in prompt muonic fission. The following table gives the probabilities for radiationless transitions per muon stop in ^{238}U and ^{237}Np

	3d-1s	2p-1s
^{238}U	14 ± 5 %	18.6 ± 3.2 %
^{237}Np	9 ± 4 %	13.2 ± 2.2 %

For ^{238}U the fission probability per muon stop is 15 %, giving a prompt fission probability of 1.3 % and from this P_f(3d-1s) = 6.9 ± 2.8 %. In comparison to the value of about 22 % from the (γ,f) reaction |16|, this is a clear indication of a barrier augmentation.

In a constant temperature model and considering a double humped barrier an augmentation of 0.75 MeV for the outer barrier of ^{238}U has been calculated from the above P_f(3d-1s)-value |6|. For ^{237}Np an augmentation of the outer barrier by 0.75 MeV results in a fission probability of about 30% for the r.l. 3d-1s transition. Such the fission probability per muon stop for this transition is 3.6 ± 1.7 %. The calculations performed in the same model for ^{237}Np give only about one third of all prompt fission events being induced by a r.l. 3d-1s transition. The calculated value does not change much for different barrier augmentations assumed. The result is in strong contrast to (74 ± 15)% that have been measured for ^{238}U |4|. The reason may be that in ^{237}Np both the $2p_{3/2}$-$1s_{1/2}$ and $2p_{1/2}$-$1s_{1/2}$ hyperfine complexes have energies above the barrier height, whereas in ^{238}U the $2p_{1/2}$-$1s_{1/2}$-complex is below the barrier. The measurement $^{237}Np(\mu^-,\gamma f)$ is in preparation.

6. Fission fragment spectroscopy

Mass yield and total kinetic energy release (TKE) distributions of fragments from prompt and delayed muon induced fission, separately, have been measured for the isotopes ^{235}U, ^{238}U, ^{237}Np and ^{242}Pu. Similarly the reactions ^{238}U, $^{237}Np(\pi^-,f)$ were studied.

The mean compound nucleus excitation energy in delayed muon and in pion induced fission processes is determined from the symmetric mass yields to be (19 ± 1) MeV and about 80 MeV, respectively. In comparing the TKE at asymmetric mass splits $m_H \gtrsim 135$ u, in prompt and delayed muon induced fission, i.e. at different nuclear excitation energies, these TKE do not differ appreciably. This is probably due to permanent deformations, which do not change much by increasing the excitation energy. The TKE released in prompt muon induced fission of ^{235}U and ^{237}Np is smaller by 1 to 2 MeV, if compared to the results from the $(\alpha,\alpha'f)$ reaction on the same targets. This can be ascribed to the partial screening of the nuclear charge by the muon.

Both processes, prompt and delayed muon induced fission, fit very well to the general picture that mass and kinetic energy distributions of fission fragments are dominantly influenced by the excitation energy deposited in the fissioning nucleus. This supports the spectator role of the muon and the weak coupling of static and dynamical properties of the fissioning muonic system. The muon can thus be treated as an ideal probe for studying fission dynamics, e.g. by measuring the muon attachment probability to the fission fragments, to investigate the transition of the nucleus from saddle to scission point.

I am thankful to my collegues
J.F.M. d'Achard van Enschut, W. Duinker, C. Gugler, H. Hänscheid, J. Hartfiel, H. Janszen, T. Johansson, J. Konijn, T. Krogulski, C.T.A.M. de Laat, T. Mayer-Kuckuk, W. Müller, R. von Mutius, C. Petitjean, S. Polikanov, H.W. Reist, R. Reuter, F. Risse, Ch. Rösel, L.A. Schaller, L. Schellenberg, A.K. Sinha, W. Schrieder, A. Taal, J.P. Theobald, G. Tibell, N. Trautmann
for their collaboration and stimulating and refreshing conversation. It is a pleasure to thank Prof. J.P. Blaser and his staff for the excellent working conditions at SIN and the ever given support to our collaboration.

This work was supported by the Bundesministerium für Forschung und Technologie der Bundesrepublik Deutschland.

References

|1| J. Wheeler, Rev. Mod. Phys. 21 (1949) 133

|2| C.T.A.M. de Laat, A. Taal, W. Duinker, J. Konijn, J.F.M. d'Achard van Enschut, P. David, J. Hartfiel, H. Janszen, T. Mayer-Kuckuk, R. von Mutius, C. Gugler, L.A. Schaller, L. Schellenberg, T. Krogulski, C. Petitjean, H.W. Reist, W. Müller, to be published in Phys. Lett.

|3| P. David, J. Hartfiel, H. Janszen, T. Mayer-Kuckuk, R. von Mutius, C.T.A.M. de Laat, A. Taal, W. Duinker, J. Konijn, J.F.M. d'Achard von Enschut, C. Gugler, L.A. Schaller, L. Schellenberg, T. Krogulski, C. Petitjean, H.W. Reist, W. Müller, to be published in Phys. Lett.

|4| T. Johansson, J.Konijn, T. Krogulski, S. Polikanov, H.W. Reist, G. Tibell, Phys. Lett. 116 B (1982) 402 and Phys. Lett. 97B (1980) 29

|5| E. Teller, M.S. Weiss, Trans. N.Y. Acad. Sci. 40 (1980) 222

|6| R. von Mutius, Thesis 1985, Institut f. Strahlen- und Kernphysik, University of Bonn

|7| P. David, J. Hartfiel, H. Janszen, T. Mayer-Kuckuk, R. von Mutius, C. Petitjean, H.W. Reist,S.M. Polikanov, W. Duinker, J. Konijn, C.T.A.M. de Laat, A. Taal, T. Krogulski, T. Johansson, G. Tibell, J.F.M. d'Achard van Enschut, J.P. Theobald, N. Trautmann, L.A. Schaller, L. Schellenberger, C. Gugler, to be published in Z. Physik

|8| H. Primakoff, Rev. Mod. Phys. 31 (159) 802
B. Goulard, H. Primakoff, Phys. Rev. C10 (1974) 2034, C11 (1975)

|9| W.W. Wilcke, M.N. Johnson, W.N.Schröder, J. R. Huizenga, D.G. Perry, Phys. Rev. C18 (1978) 1452

|10| T. Kozlowski, A. Ziglinski, Phys. Lett. 50B (1974) 222, Nucl. Phys. A 305 (1978) 368

|11| W.W. Wilcke, M.W. Johnson, W.U. Schröder, D. Hilscher, J.R. Birkelund, J.R. Huizenga, J.C. Browne, D.G. Perry, Phys. Rev. C21 (1980) 2019

|12| N. Auerbach, A. Klein, Nucl. Phys. A 422 (1984) 480

|13| P. David, J. Hartfiel, H. Janszen, T. Mayer-Kuckuk, R. von Mutius, C. Petitjean, H.W. Reist, S.M. Polikanov, J. Konijn, T. Johansson, G. Tibell, J.F.M. d'Achard van Enschut, J.P. Theobald, N. Trautmann, L. Schellenberg, L.A. Schaller, C. Gugler, to be published in Z. Physik

|14| J. Hadermann, K. Junkër, Nucl. Phys. A 256 (1976) 521

|15| P. Singer, Springer Tracts in Modern Physics 71 (1974) 39

|16| J.T. Caldwell, E.J. Dowdy, B.L. Berman, R.A. Alvarez, P. Meyer, Phys. Rev. C21 (1980) 1215
A. Veyssière, H. Beil, R. Bergère, P. Carlos, A. Lepretre, K. Kernbath, Nucl. Phys. A 199 (1973) 45

A Measurement of the Muon Capture Rate in Liquid Deuterium by the Lifetime Technique

J. Martino

Service de Physique Nucléaire, Haute Energie, CEN Saclay, F-91191 Gif-sur-Yvette Cedex, France

Since neutron β-decay and muon capture on the proton define the nucleon weak form factors, muon capture on the deuteron allows to study the two-nucleon system with the weak interaction probe, investigating problems like the deuteron wave function, meson exchange currents and final state interactions.

Muon capture on the deuteron*, according to the reaction $\mu^- + d \rightarrow n + n + \nu_\mu$, occurs after the formation of a μd atom and leads to the same difficulties as on the proton : i) the capture rate is very small and strongly depends on the hyperfine state of the μd atom ; ii) the deuterium must be extremely pure because of the high cross section for muons to be transferred to impurities ; iii) in the two previous experiments the capture rate was measured by counting the capture neutrons : major limitations then come from neutron identification and efficiency knowledge ; iv) in high deuterium density $d\mu d$ molecules catalyse fusion reactions which give an intense neutron background and poison the μd muon source with μ^3He atoms.

For the two previous experiments, at a low deuterium density to avoid $d\mu d$ formation, the interpretation of the capture rates relied on an assumption, that of all μd were in the Doublet hyperfine state. This puzzle of muonic chemistry in deuterium has been partially cleared recently, leading to the following consequences : i) the two previous experiments can only give a lower and an upper limit for the Doublet capture rate : $325\ s^{-1} < \lambda_C^D < 557\ s^{-1}$; ii) the initial μd population can be defined unambiguously only in ultra-pure dense deuterium.

A new measurement was therefore performed at Saclay in ultra-pure liquid deuterium and the neutron detection difficulties have been avoided by using the lifetime technique : the capture rate is obtained by an accurate comparison of the μ^+ and μ^- lifetimes : $\lambda_C = 1/\tau_{\mu^+} - 1/\tau_{\mu^-}$. This method takes advantage on the pulsed (3000 Hz) beam of the Saclay Linac : the muons stop in the target during the 3 μs beam burst and their lifetime is measured after the burst by the decay electron time distribution. The capture rate can thus be measured without the systematic uncertainties due to neutron counting. After subtraction of the cosmic and room backgrounds, correction for a small dead time distortion proportional to the acquisition rate, and extrapolation to zero hydrogen concentration, the mean value of the decay electron time distribution is : $\tau_{\mu^-} = 2194.53 \pm 0.11$ ns. Taking for τ_{μ^+} the world average value $\tau_{\mu^+} = 2197.03 \pm 0.04$ ns, the average capture rate is : $\lambda_c = 518 \pm 24\ s^{-1}$.

To get the Doublet capture rate λ_c^D, the distortion due to μ^3He captures has to be corrected out. This correction, calculated from the muonic chemistry parameters, is equal to : $\Delta\lambda_c = -60 \pm 16\ s^{-1}$. The major ingredient in this calculation is the product of the muon sticking parameter to ^3He by the dμd Doublet formation rate. It happens that the dμd fusion neutron time distribution directly gives this product. Such a measurement was therefore also done at Saclay by replacing the electron telescopes by neutron counters. After a very careful elimination of the huge γ-ray contamination by pulse shape discrimination, subtraction of cosmic and photo-neutron backgrounds, the fusion neutron time distribution yields $\Delta\lambda_c = -59 \pm 23\ s^{-1}$. The validity of the muonic chemistry calculation is thus confirmed and, after including a final correction of 12 s^{-1} due to the μ-atomic bond, the following value is obtained for the capture rate in the μd Doublet spin state :

$$\lambda_c^D = 470 \pm 29\ s^{-1}\ .$$

This result is higher than the most recent theoretical prediction, $\lambda_c^D(th) \sim 413\ s^{-1}$, by two standard deviations. A careful investigation has been made of the possible systematic errors. Auxiliary measurements showed in particular that : i) a residual impurity effect is less than 5 s^{-1} ; ii) even in non optimal stopping conditions, muon transfer from μd atoms to the Cu nuclei of the target wall is not significant ; iii) a residual muon polarization of 1 %, possible according a recent measurement, gives a distortion smaller than 2 s^{-1}.

This difference of two standard deviations might be a statistical fluctuation or even due to some unknown systematic error. The answer to this question requires further efforts to increase the experimental precision. On the other hand, it might also be a theoretical problem. But it could hardly be explained by only adjusting the weak coupling constant. Further investigations on the nuclear side, taking for instance into account short range effects in the deuteron, could perhaps lead to a better agreement and would be of great interest.

---

*The details of the present experiment and the related references, in particular to previous measurements, to theoretical calculations and to the recent progress in muonic chemistry, can be found in the paper by G. Bardin et al., Nucl. Phys. A453 (1986) 591.

Pion Exchange Current Effects in $\nu_\mu + d \to \mu^- + p + p$

S.K. Singh and H. Arenhövel*

Institut für Kernphysik, Johannes Gutenberg-Universität,
D-6500 Mainz, F. R. Germany

Abstract: The effect of pion exchange currents in the process $\nu_\mu + d \to \mu^- + p + p$ is estimated using realistic wave functions for the deuteron and closure over the final dinucleon states. Its implications on the determination of axial vector form factor $F_A(q^2)$ are discussed.

The neutrino reaction $\nu_\mu + n \to \mu^- + p$ provides a direct method to determine the weak form factor $F_A(q^2)$ of the nucleon which is parametrized in the dipole form, i.e.,

$$F_A(q^2) = F_A(0)/(1 + q^2/M^2)^2 \tag{1}$$

and the value of M_A is determined from the experiments. In general, the neutrino reactions are done with nuclear targets, where nuclear effects become important. In order to minimize the uncertainties in the determination of $F_A(q^2)$ associated with the nuclear effects, it is essential to do experiments with the deuteron target, where reliable estimates of these effects can be made |1|. However, no serious attempts have been made to include the various corrections due to the relativistic and meson exchange current (MEC) effects in the analysis of neutrino deuteron scattering experiments which have been recently used to determine the dipole mass M_A |2-4|. Keeping in mind the fact that these experiments provide most accurate value of M_A, it is important to estimate the size of these effects.

We have calculated the effect of the D-state and MEC on the differential cross section $d\sigma/dq^2$ for the reaction $\nu_\mu + d \to \mu^- + p + p$ using realistic wave functions for the deuteron and closure over the final dinucleon states. The effect of MEC is estimated using one pion exchange as shown in fig. 1.

The MEC operators are evaluated in the static limit using quark model values for the various couplings involving Δ excitations, i.e., $g_{\Delta N\pi}$, $g_{\Delta N\gamma}$ etc. and PCAC with vector dominance to evaluate $g_{\rho\pi A}$ |5|.

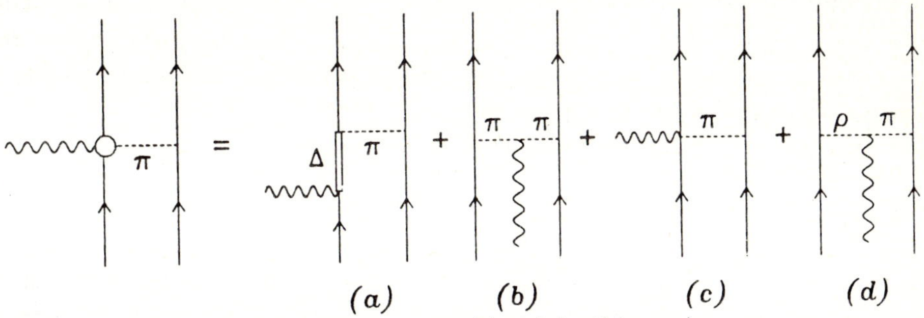

Fig. 1.: Exchange current diagrams considered in this work.

*) Supported by the Deutsche Forschungsgemeinschaft (SFB 201)

The numerical calculations are made for the ratio $R(q^2)$ defined as

$$R(q^2) = d\sigma/dq^2(\nu_\mu + d \to \mu^- + p + p)/d\sigma/dq^2(\nu_\mu + n \to \mu^- + p) \qquad (2)$$

as a function of q^2 using various wave functions for the deuteron as obtained from Paris, Reid soft core and Reid hard core potentials. We find following results.

1. The effect of the D-state is to decrease $R(q^2)$ while MEC increase $R(q^2)$. The net effect is to increase the value of $R(q^2)$ by 5-15% in various regions of q^2 (fig. 2).

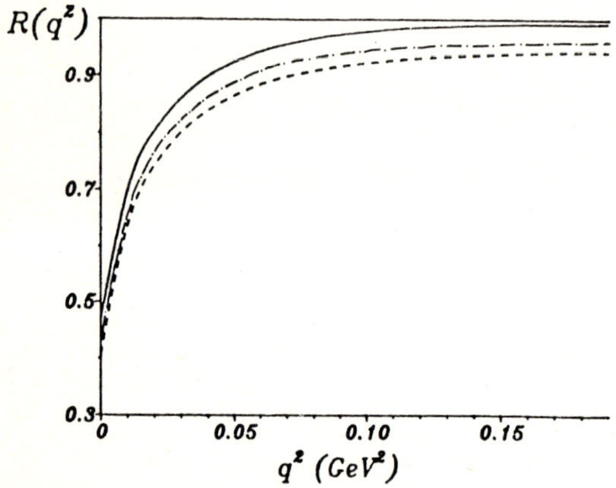

Fig. 2.: $R(q^2)$ vs. q^2 at E = 1 GeV for deuteron wave function with Paris potential: S waves - . - ., S + D waves - - -, S + D + MEC ———.

2. $R(q^2)$ does not depend significantly upon the choice of wave function, once realistic wave functions are used.

3. $R(q^2)$ is independent of neutrino energy E within 1-2% even when the effect of MEC are taken into account.

4. $R(q^2)$ at $q^2 = 0$ is independent of the deuteron structure in the absence of MEC. When the effect of MEC is taken into account, the deuteron structure gives 10-12% increase depending upon the deuteron wave function.

With respect to the determination of M_A it has been found by TARRACH and PASCUAL |6| that inclusion of the D-state leads to a smaller value. However, with the net increase through additional MEC contribution one finds a higher value of M_A.

The world average of M_A as determined from the neutrino deuteron experiments is 1.03 ± 0.04 GeV |4|. However, the value of M_A as determined from the high energy neutrino experiments ($5 \le E_\nu < 200$ GeV) is not consistent with this value if the events at low q^2 are also included which have been left out in the analysis of this experiment |4|. This discrepancy will be removed if the new values of $R(q^2)$ are used in the analysis of this experiment. Moreover, an overall increase of $R(q^2)$ will also affect the flux determinations performed in the Argonne experiment |2|.

Present calculations of MEC effects in $\nu_\mu + d \to \mu^- + p + p$ can be improved by taking into account the effect of momentum dependence in the isobar propagator and doing a partial wave expansion of the final dinucleon states instead of using closure approximations. The MEC effects are underestimated in the closure approximation where quasifree kinematics is used, while neglect of momentum dependence in the isobar propagator results in an overestimation of this effect. It will be interesting to see the interplay of these two effects in an improved calculation.

References:

1. S.K. Singh, Nucl. Phys. B36, 419 (1972); Phys. Rev. C10, 988 (1974)
   J. Bernabeu, P. Pascual, Nuovo Cim. 10A, 61 (1972)
   J. Bernabeu, Phys. Lett. 39B, 313 (1972)
2. W.A. Mann et al., Phys. Rev. Lett. 31, 844 (1973)
   S.J. Barish et al., Phys. Rev. D16, 3103 (1977)
   K.L. Miller et al., Phys. Rev. D26, 537 (1982)
3. N.J. Baker et al., Phys. Rev. D23, 2499 (1981)
4. T. Kitagaki et al., Phys. Rev. D28, 436 (1983)
5. W. Müller and M. Gari, Phys. Lett. 102B, 389 (1981)
6. R. Tarrach and P. Pascual, Nuovo Cim. 18A, 760 (1973)

Search for the Lambshift in Muonic Helium at Low Helium Pressures

H.P. von Arb[1], *C. Brandes*[1], *F. Dittus*[1], *P. Egelhof*[1,2], *H. Hofer*[1], *F. Kottmann*[1], *C. Lüchinger*[1], *R. Schaeren*[1], *F. Studer*[1], *D. Taqqu*[1,3], and *J. Unternährer*[1]

[1]Inst. für Hochenergiephysik der ETH-Zürich, CH-8093 Zürich, Switzerland
[2]Inst. für Physik der Universität Basel, CH-4056 Basel, Switzerland and Inst. für Physik der Universität Mainz, D-6500 Mainz, F. R. Germany
[3]Schweizerisches Inst. f. Nuklearforschung, CH-5234 Villigen, Switzerland

The measurement of the Lambshift, i.e. the 2s-2p energy difference in light muonic atoms leads to precise tests of the vacuum polarization and (by looking for different isotopes) to improvements in the determination of nuclear charge radii[1)]. A good experimental precision can be attained by using laser resonance techniques.

The Lambshift in muonic ^{4}He has been measured a decade ago in a pioneering experiment at CERN[2)], using a gaseous ^{4}He target at 40 atm helium pressure. The experimentally observed long lifetimes of the metastable 2s-state (τ_{2s} > 1 μs up to pressures of 40 atm)[3)], which were necessary to perform the laser resonance experiment, are up to now not understood satisfactorily from the theoretical point of view. Recent investigations[4)] of our collaboration on the 2s-lifetime in the pressure range 30 torr < p < 6 atm show a quadratic pressure dependence of the 2s-decay rate resulting in a 2s-lifetime < 150 ns above 400 torr in accordance with theoretical predictions. These data as well as recent investigations at pressures up to 40 atm[5)] are in strong contradiction to the data of Ref. 3.

In order to shed more light on this unclear situation the aim of the present experiment is to measure the Lambshift in muonic ^{4}He at a pressure of 30 torr where the 2s-lifetime has been detected to be larger than 1 μs[4)]. The principle of the measurement is to form metastable (μHe) 2s-ions using the muon-bottle technique[6)] which has been especially developed for the formation of muonic atoms at low pressures, to induce (after a delay of approximately 1 μs) a 2s-2p transition by means of a short laser pulse and to detect an 8 keV x-ray from the 2p-1s transition in coincidence with the laser pulse. The laser is triggered by the scintillation light which is produced during the slowing down procedure of the muon. The muonic x-rays are detected in 2 large area gas-scintillation-proportional-chambers. The laser system (Fig. 1) consists of a Q-switched Nd:Yag oscillator, a Nd:Yag preamplifier and 3 parallel working Nd:Yag amplifiers, which are followed by frequency doubling in KD*P crystals. The green light pumps a dye laser system consisting of an oscillator and several amplifiers in series. The observed output energy is about 160 mJ/pulse at λ = 812 nm with a repetition rate of 7 Hz. In order to illuminate the total effective volume of the muon bottle an optical cavity consisting of 4 highly reflective mirrors and a system of laser beam diagnostics is used. With the present set-up a transition probability of about 7% can be reached.

A first short run using the full laser power has been performed just recently at a laser-wavelength of λ = 811.7 nm (c.f. Ref. 2) taking events with laser on and off respectively. On the basis of a preliminary data analysis no evidence for the resonance signal has

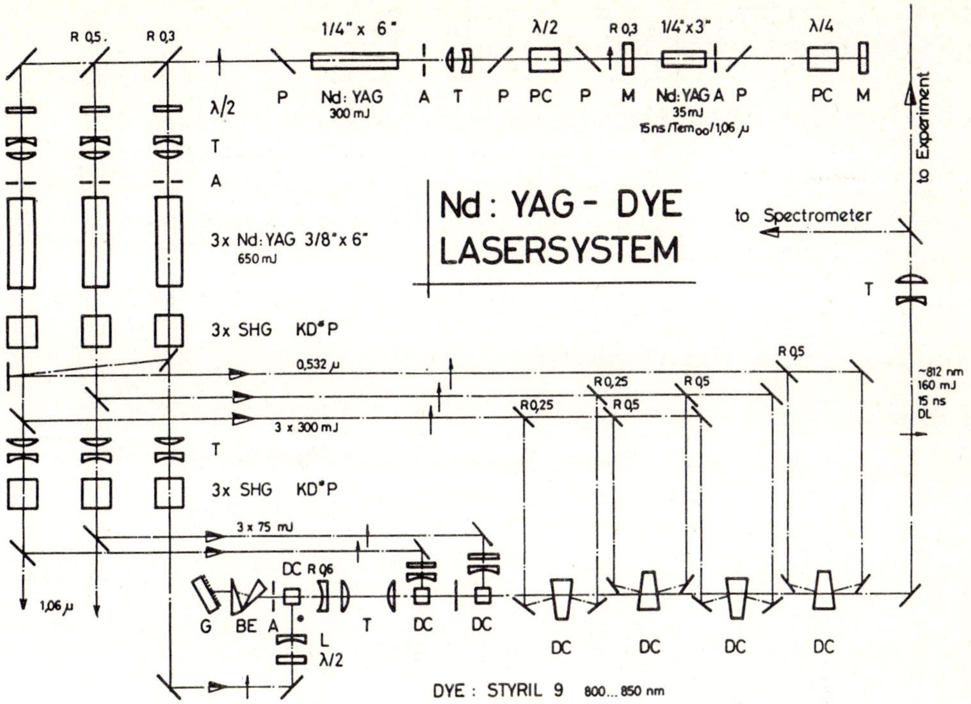

Fig. 1 Lasersytem for the Lambshift-Experiment in μ^4He

been found within the limited statistics. After improvement of the apparatus we plan to search for the resonance signal in the wavelength range corresponding to predictions using the measured nuclear rms-radius from electron scattering.

References

1 E. Zavattini: in Exotic Atoms, ed. by G. Fiorentini and G. Torelli, Frascati (1977) p. 43
2 G. Carboni et al.: Nucl.Phys. A278, 381 (1977)
3 A. Placci et al.: Nuo. Cim. 1A, 445 (1971)
A. Bertin et al.: Nuo.Cim 26B, 433 (1975)
4 H.P. von Arb et al.: Phys.Lett. 136B, 232 (1984) and references therein
5 M. Eckhause et al.: Phys. Rev. A33, 1743 (1986)
6 H. Anderhub et al.: Phys. Lett. 101B, 151 (1981)

3.5 GUT's, SUSY's, Superstrings

3.5.1 Further Basic Experiments for GUT's

Nucleon Decay Experiments

H. Meyer

FB-Physik der Universität*, D-5600 Wuppertal 1, F. R. Germany

1 OVERVIEW of the EXPERIMENTS

At the time of this conference five experiments (KOLAR II[1], NUSEX, FREJUS, IMB III [2] and KAMIOKANDE II [3] designed to discover nucleon decay take data, located all around the world but only in the northern hemisphere, as it is shown in figure 1. One more experiment is being constructed (SOUDAN II[4]) and two projects are under serious discussions (SUPERKAMIOKANDE[5], ICARUS[6]), however nucleon decay has not been dicovered yet. The effort goes to both larger and more sophisticated experiments, table 1 containes the relevant information. As can been seen three experiments have recently either been reconstructed (KOLAR II) or significantly been improved (KAMIOKANDE II, IMB III). The total useful luminosity of the five experiments adds up to ~6,5 kty[1], that is ~4×16^{33} nucleons (proton + neutrons) have been looked at for one year. From this year on KOLAR, IMB and KAMIOKANDE are much improved and add luminosity at better sensitivity for some of the potential nucleon decay modes and in general with better background rejection. Improvements concern the light collection of the new IMB III setup, which has all 2048 5" PMT[2] replaced by new 8" PMT with a wavelength shifter plate on each of them giving a factor 4 in light collection efficiency. KAMIOKANDE II has now time and pulse height measurement on each of the 1024 20"PMT furthermore an active veto region(H_2O+PMT looking outwards)surrounding the whole detector. KOLAR II is larger than KOLAR I (260 to vs 140 to) has shorter Fe sampling length (6mm vs 10 mm) and has been moved to a new location somewhat higher up in the same mine. The Fréjusdetector will continue unmodified, while the Nusexdetector may well be turned off at the end of 1986. SOUDAN II is expected to come into operation with a fraction of the total mass sometimes in 1987. Exptrapolating now assuming continuous running up to the end of 1988 the integrated luminosity for each experiment is shown again in table 1 providing additional 14 kty in total. For some of the more easy decay modes (e.g. $e^+\pi^o$) the luminosity integrated up to now is as useful as the new one and a maximum of ~20kty (~$1.3 \cdot 10^{34}$ nucleons for 1 year) could be available by 1989. It should be mentioned however that continuous running has not always been achieved in the past at all the experiments which should not be too surprising in view of the very remote places of the experiments and - sometimes - not to easy working conditions. Next generation experiments have to be very large if the level of nucleon decay candidates remains to be small, at

1 kty = 1 kiloton year   2 PMT = Photomultipliertube

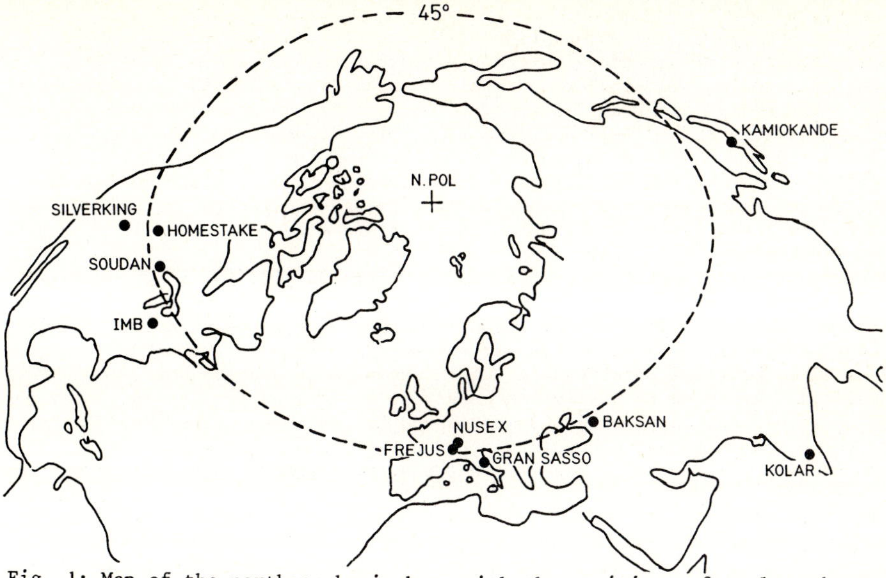

Fig. 1: Map of the northern hemisphere with the positions of nucleon decay detectors

Table 1

EXPERIMENT	STATE	START	FID.MASS M_f (to)	M_f/M_{tot}	L(6/86) (kty)	L(12/88) (kty)
KOLAR I KOLAR II	INDIA	Oct. 80 Dec. 85	65 120	0.46	.325	.600
NUSEX	ITALY	July 82	120	0.86	.384	
IMB I IMB III	USA	Aug. 82 June 86	3300	0.47	3.770	8.000
KAMIOKANDE I KAMIOKANDE II	JAPAN	July 83 Jan. 86	880 800	0.29	1.470	2.500
FREJUS	FRANCE	March84	750	0.83	.700	2.500
SOUDAN II	USA	1987	900	0.83		1.000
SUPER KAMIOKANDE	JAPAN	project	22000	0.46		
ICARUS	ITALY	project	4000	0.66		

least as large as SUPERKAMIOKANDE (22 kt fiducal mass) to make a real impact on the field with further improvements on detection quality for individual events of course very wellcome. I will come back to this point at the end of this talk.

2 NEUTRINO BACKGROUND

Only events due to neutrino interactions in the detector are of relevance as a source of background for nucleon decay. Other sources like e.g. neutral hadrons (n, K_L^o) from hadronic interactions of high energy muons in the material surrounding the detector are considered (sometimes known) be to negligible. The neutrino flux originates from the decay of pions, kaons and muons produced in interactions of primary cosmic rays in the earth atmosphere. ν_μ and $\bar{\nu}_\mu$ are more frequent than ν_e and $\bar{\nu}_e$ by ~a factor of 1,5 in contrast to standard accelerator neutrino beams that have an ν_e component at the % level. The energy spectrum falls off rather quickly and the absolute flux is known with moderate accuracy (~20-30%)[7]. It is also dependent on the geomagnetic location on earth and on the solar spot cycle. The flux is not isotropic, the horizontal direction being favoured by ~a factor 2 as compared to the upward or downward direction. The nucleon decay experiments will certainly provide rather accurate measurements on the atmospheric neutrino flux as more and more luminosity is integrated by the experiments.

The observed (and expected) rate of neutrino events is ~130 for 1 kty, because of the steeply falling neutrino spectrum this depends sensitively on the trigger threshold and somewhat on the detector position on earth, it is lowest in the Kolardetector due to the highest geomagnetic cut-off there.

Based on data (mostly from work with bubble chambers in accelerator ν-beams) Monte Carlo Simulations are used to predict the neutrino background for nucleon decay detectors. Only one calibration run has been attempted by the Nusex group [8]. Similar experiments for H_2O-detectors are considered to be a very major effort, since the detector has to be of similar size as the original one. Furthermore accelerator ν-beams are not free of n, $K_L 0$ background, which is at a very low level in the nucleon decay experiments. In addition the ν_e-component of the atmospheric neutrino flux needs to be specifically simulated. Additional difficulties arise from simulating the influence of the nucleus (O, Fe) on the final appearance of neutrino events in the detector [9]. Despite all those reservations the agreement of the simulation with real data is generally considered to be very satisfactory, which, at this stage of the development partly may just reflect the dominance of simple final states in the total sample, viz. quasi elastic scattering (ν_μ+N $\rightarrow$ μ+N; or single π-production). The M.C. studies continue to be of very high importance as more luminosity is being collected for example the Fréjuscollaboration is reanalysing the Aachen-Padova-neutrino experiment[10] to provide a large data basis for comparison with the simulation in a detector of very similar structure.

3 SEARCH for NUCLEON DECAY CANDIDATES

Candidate events for nucleon decay are being searched for by applying the decay kinematics of a particle at rest. This implies:

$$\sum_i P_i \simeq P_f \qquad 1$$

$$\sum_i E_i = E_N \qquad 2$$

$$E_N^2 - (\sum_i P_i)^2 = M_N^2 \qquad 3$$

P_f is the Fermimomentum of nucleons in the respective nucleus (O, Fe) and M_N the nucleon mass. The basic experimental information is Čerenkov rings for H_2O-detectors and sample points on the particle tracks for Fe-detector. A Ĉ-ring uniquely determines the direction of a track with further details from the time development of the Ĉ-rings. The number of photoelectrons (pulseheight) determines the Ĉ-energy loss of the particle. To convert this information into momentum and energy of the particle, it's nature has to be assumed. For the KAMIOKANDE experiment and for simple event patterns, e.g. 1 or 2 ring events, can a distinction between a showering and a non showering particle be made. Further information comes from time delayed signals they do indicate the decay of muons. Particles with $\beta \lesssim 0.72$ are not detected, e.g. protons in neutrino events in general, or pions and muons below ~200 MeV/ĉ momentum. This causes a serious loss of particles in events more complicated than just the two body decay modes of nucleons.

In fine grain tracking detectors - like the Fréjus experiment with 3mm Fe sampling-tracks are very well defined and showering and non-showering tracks can be distinguishe with high reliability. The detection threshold for particle tracks is well below 100 MeV/ĉ in general, e.g. a μ-decay electron (35 MeV) produces 6 hits on average. The track direction can not always be determined, except for showering tracks and in favourable cases using the increase of multiple scattering towards the end of a track. Muon decay is seen with moderate (~30%) efficiency detecting time delayed hits (NUSEX, FREJUS). Containement of the events is defined by requiring the particles to stop well inside the detector surface. For H_2O detectors this requires ~2m cut and a considerable loss in detector volume (~50%), in Fe-detectors one can go closer to the detector face (~(10-50)cm) with correspondingly reduces losses (10-20%) of useful detector mass.

IMB I

Since particle identification on an event by event basis is not available, the data is compared to M.C. predictions on the assumption that the detected light originates from a showering particle [11]. For each event then the total visible C'renkov-energy E_C is calculated. The momentum balance is approximated by a quantity called anisotropy A_c it measures the energy weighted vectorsum of the directions of PMT's fired w.r.t. the vertex of the event. Single prong events have~ A_c~0.75, momentum balanced events have $0<A_c \lesssim 0.4$. Various nucleon decay modes cover regions in E_c vs. A_c depending on the assumed final state. If all energy appears in showering particles like $p \to e^+ \pi^o$, $p \to e^+ + \omega^o \to e^+ + \pi^o + \gamma$, the visible energy corresponds to the nucleon mass M_N, while for modes with nonshowering particles like $n \to \mu^+ \pi^-$ the visible energy is in general much lower. In the real data sample there are candidates events in the nucleon decay regions defined by M.C. calculation for almost all decay modes searched for.

Neutrino events simulated on the basis of accelerator data, account for essentially all of the candidate events and therefore only lower limits on nucleon lifetime have been derived. Figure 2 shows the data (a) the neutrino simulation (b) and the simulation of a particular nucleon decay modes $p \to \mu^+ \pi^o$ (c)
In general IMB I has reached the background level for the majority of the interesting decay modes of nucleons [12].

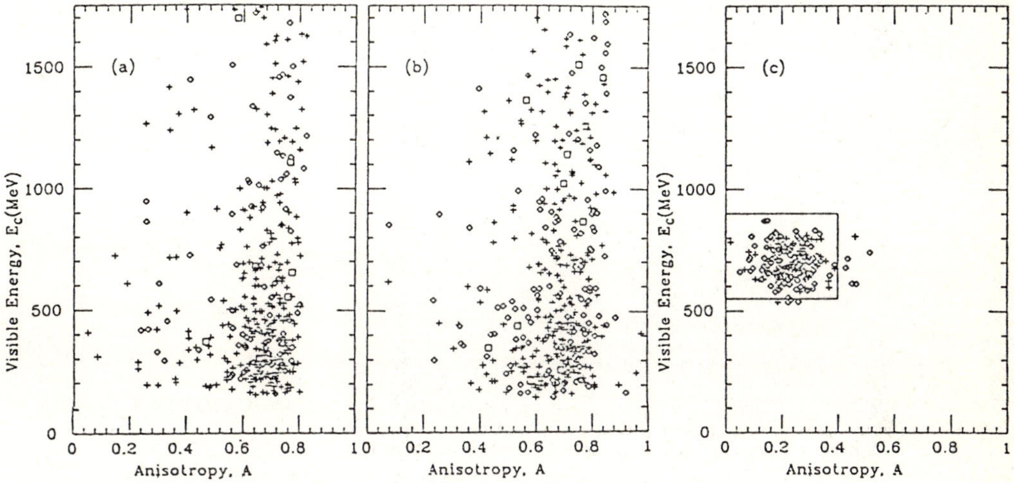

Fig. 2: Data and simulation for ν-events and nucleon decay for the IMB detector

KAMIOKANDE I

KAMIOKANDE has much higher light collection efficiency for each event due to the very large diameter (20") of the phototubes. This gives good energy measurement and in particular allows distinction between showering and nonshowering particles in favourable cases (e.g. 1 ring event). Also individual particle tracks are identified by observing their Cerenkovring pattern. Therefore more specific kinematic tests can be applied. The neutrino background is again simulated on the basis of accelerator data by M.C. [13]. The comparison is shown in figure 3a (2 rings) and figure 3b(≥ 3 rings [14].

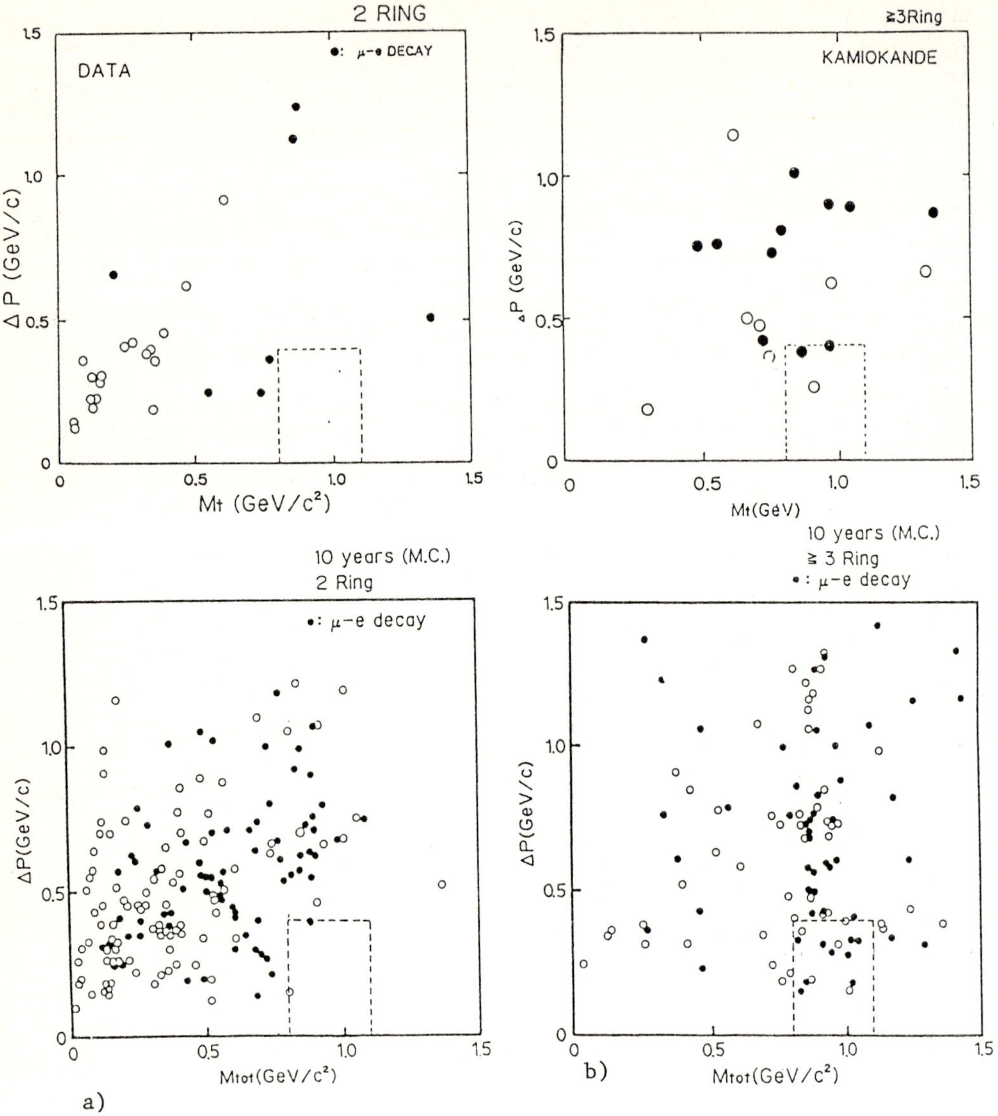

Fig. 3: Data and simulation for ν-events for the KAMIOKANDE detector

In the two ring sample no candidate appears in the nucleon decay region, but also the M.C. calculation (with 7times more statistics) does not predict a background event. Clearly an improvement over the IMB data.

In the 3 ring data sample some events appear in the nucleon decay region, but also the M.C. simulation predicts ν-events as nucleon decay candidates. The background level therefore is reached also in KAMIOKANDE for more complicated events and only lower limits for nucleon life time can be quoted.

KOLAR I

The KOLAR I experiment claimes several nucleon decay candidates [15]. A general kinematical analysis of all contained events has however not been presented [16], rather individual events, considered compatible with nucleon decay have been discussed in considerable detail [1]. If true nucleon decays, they should have by now been confirmed by the larger detectors, which is clearly not the case.

NUSEX

The NUSEX group has described several nucleon decay candidates [17], however the neutrino background expectation is at a similar level. Therefore only lower limits for nucleon lifetime have been quoted [18]. This is in (mild)disagreement with the KOLAR I experiment since the respective luminosities are similar.

FREJUS

The FREJUS experiment is both larger (6times NUSEX) and has finer granularity, e.g. a 1GeV showering particle has ~18 hits (KOLAR I) and 35 hits (NUSEX) as compared to 145 hits (FREJUS). Showering from nonshowering particles can be distinguished with high reliability, event vertices can be determind with mm precision. For the two, three and more prong contained events momentum balance (P_{mis}) and visible energy E_{vis} are determined as shown in figure 4 for E_{vis}<1,7 GeV. Only one event can be considered as a nucleon decay candidate. It has four prongs and two vertices as shown in figure 5.

Under the assumption that vertex B is the primary vertex (3 prongs) a nucleon decay hypothesis is unlikely and vertex A is a secondary scattering e.g. $\pi p \rightarrow \pi p$. If vertex A is the primary vertex 4 particles emerge and vertex B is a secondary π-nucleus

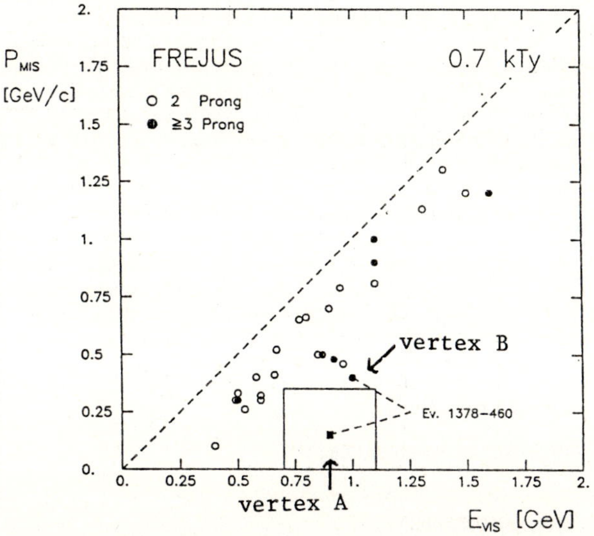

Fig. 4: The visible energy and the momentum balance for the Fréjus detector

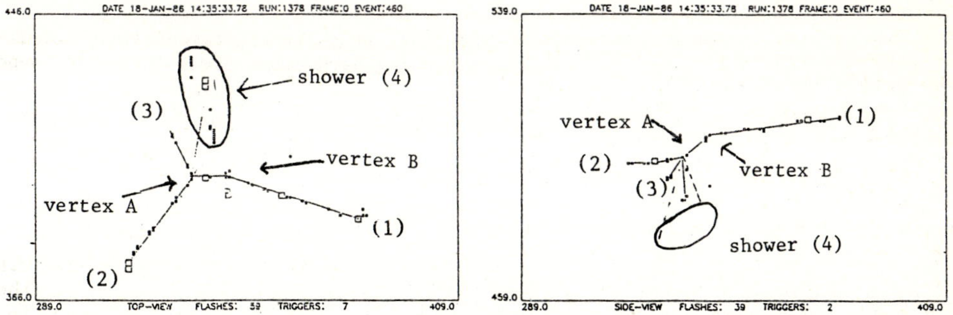

Fig. 5: The nucleon decay candidate from Fréjus

elastic scattering. The event then is rather well balanced (<200 MeV/$_c$) and has a total energy-from range measurements- close to the nucleon mass. In general a nucleon decay with four particles in the final state is strongly influenced by the nucleus (Fe) like in a final state $e\pi\pi\pi$, the probability to observe all three pions without loss through π interactions in the nucleus is clearly very small, however the event nominally is balanced. A balanced nucleon decay event with all energy visible is rather easily obtained in final states with 3 leptons like $e\mu\mu\pi$. More events of this kind need to be observed to consider a case for nucleon decay, one event is just not enough it could very well be a (very unlikely thought) ν-background event. On the other hand this type of event is very difficult to observe in H_2O Cerenkov detectors, since in almost all configurations one or two particles would be below Cerenkov-threshold and the event appears strongly unbalanced.

4 LIFETIME LIMITS

A lifetime limit for a particular decay mode is calculated according to

$$\tau_N \cdot \frac{1}{B} = \frac{X \cdot 10^{29} \cdot L \cdot 10^3}{N} \cdot \varepsilon \qquad 4$$

X is 2,9 if N is the proton and 3.2 if N is the neutron for Fe as the detector material, L is the luminosity in kiloton year, N the number of obșerved events (2.3 if no candidate is seen, 3.8 if one candidate is considered, at 90% confidence level), B the nucleon decay branching ratio for the decay mode considered and the efficiency ε can vary considerably from experiment to experiment. It is certainly high for a decay mode like $p \to e^+\pi^o$ and rather low for $p \to \nu\nu\nu\pi^+$. Also the background level for a given decay mode can be rather high therefore reducing the lifetime limit accordingly. A detailed comparison of limits given by IMB and KAMIOKANDE (background subtracted on the basis of M.C. simulation of ν-events) is shown in figure 6 [19]. They are based on 1.47 kty (KAMIOKANDE) and 3.77 kty (IMB) and they are very similar. The channels $p \to e^+\pi^o$, $e^+\eta^o$ are hardly background limited and the limits reflect the luminosities. The difference for the channels $e^+\pi^-$ and $e^+\rho^-$ reflect very different event efficiencies. The mode $n \to \nu\pi^o$ has very high background for IMB and rather low background for KAMIOKANDE. The limits from NUSEX are in general lower by the ratio of luminosities (factor 4). From the FREJUS experiment no detailed information is available yet, since the M.C. for ν-background as well as nucleon decay detection efficiency has not been completed. The limits will however be $\sim 3 \cdot 10^{31}$ years at least.

For two much discussed decay modes, $p \to e^+\pi^-$ the expected dominant mode in SU5-GUT and $p \to \nu K^+$ in SUSY-GUT the combined limits are

$$\tau/_B\ (p \to e^+\ \pi^o) > 4 \cdot 10^{32}\ y \quad \text{(IMB,FREJUS,KAMIOKANDE)} \qquad 5$$

and

$$\tau/_B\ (p \to \nu k^+) > 6 \cdot 10^{31}\ y \quad \text{(IMB,KAMIOKANDE)} \qquad 6$$

They will (probably) be improved by about a factor 4 by the combined impact of the third generation detectors (KAMIOKANDE II, IMB III, FREJUS, SOUDAN II, KOLAR II) by 1.1.1989 (see table 1). Limits are of considerable interest e.g. SU5 may already be ruled out with the high upper limit for $p \to e^+\pi^o$ however it is much more interesting to ask for the chances to observe nucleon decay within the next 3 years with the new detectors. A look at table 2 provides a guideline, it shows the luminosities, ν-event rates and the candidate level and the corresponding background level of the running experiments. One observes:

1. The candidate level is only $\sim$(2-5)% of the ν-event rate, that is <5 events/kty
2. The rate for good candidates is even smaller only one each from NUSEX, FREJUS and KAMIOKANDE, this means $\sim$1 event/kty.

This allows then by 1989 for $\lesssim$10 good candidates from all experiments taken together in my mind hardly enough (leaving aside some lucky circumstances) to convincingly prove nucleon decay.

5 FUTURE EXPERIMENTS

4th generation nucleon decay detectors need to be very large >50.000 t otherwise the rate of interesting events is to small. Since one is then approching the neutrino

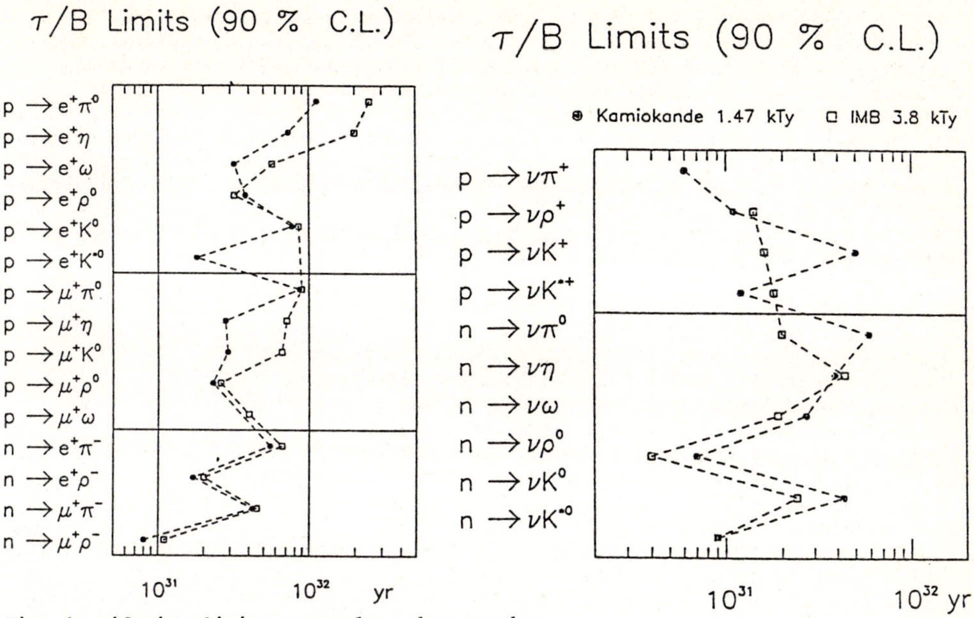

Fig. 6: Lifetime limits on nucleon decay modes

Table 2

EXPERIMENT	$\int$Ldt (kty)	# of contained events*	CANDIDATES	BACKGROUND
KOLAR I	.325		5	.6
NUSEX	.380	37	3 (1)	1.6 (0.15)
FREJUS	.610	51	1	prob. small
KAMIOKANDE I	1.470	173	5 (1)	3.2 (0.03)
IMB I	3.770	401	20	17

* Not corrected for trigger threshold !

background limit (at least for some channels) the detector needs to be also more sophisticated then the third generation detectors. The proposal to build a SUPERKAMIOKANDE certainly is a very interesting attempt in this direction, especially if it proves to be possible to perform other fundamental experiments simultanously (like ^{8}B ν's froms the sun). I would prefer however a tracking calorimeter, since nucleon decay could be complicated and simply not accessible for H_2O-Cerenkov detectors.

Acknowledgement: I would like to thank my colleges from the Fréjus experiment for numerous discussions on the subject of this talk.

* Supported by BMFT, Bonn

REFERENCES

1 M.R. Krishnaswamy et.al.: "K.G.F. Proton Decay Experiment" Contribution to the ν-86 Conference, Sendai, Japan

2 J.M. Lo Secco: Contribution to the ν-86 Conference, Sendai, Japan
3 KAMIOKANDE II Collaboration (Tokyo, Niigata, KEK, Pennsylvania, Caltech): Contribution to the 7th workshop on Grand Unification, ICOBAN'86, Toyama,Japan
4 D. Ayres et.al.: "The Soudan II Nucleon Experiment" Argonne Report ANL-HEP-PR-84-30 (1984)
5 Y. Totsuka: "SUPERKAMIOKANDE", Contribution to the 7th Workshop on Grand Unification, ICOBAN'86, Toyama, Japan
6 ICARUS-Proposal, CERN, HAVARD, MILANO, PADOVA, ROMA, TOKYO, WISCONSIN Collaboration: INFN/AE-85-7, September '85
7 T.K. Gaisser et.al.: Phys. Rev. Lett 51, 223 (1983)
8 G. Battistoni et.al.: Nucl. Inst. Meth. 219, 300 (1984)
9 C.H.Q. Ingram: Nucl. Phys. A374, 319 C (1982)
10 H. Faissner et.all.: Phys. Rev. Lett. 41, 213 (1978)
11 G. Blewitt: Thesis, CALT-68-1327
12 G. Blewitt et.al.: Phys. Rev. Lett. 55, 2114 (1985) and Ref. 11
13 M. Nakahata et.al.: KAMIOKANDE Collaboration: Preprint UT-ICEPP-86-05
14 T. Suda, KAMIOKANDE Collaboration: Contribution to the 7th Workshop on Grand Unification, ICOBAN'86, Toyama, Japan and Ref. 13
15 M.R. Krishnaswamy et.al.: Phys. Lett. 106B, 339 (1981); 115B, 349 (1982) and Ref.1
16 See however, M.R. Krishnaswamy et.al.: ICOMAN'83, Frascati January 1983 and Pramana 19,525 (1982)
17 G. Battistoni et. al: Phys. Lett. 133B, 454 (1983)
18 G. Battistoni et.al.: Contribution to the XXII Int. Conf. on High Energy Physics, Vol. I, 246, Leipzig, July 1984
19 T. Haines, IMB Collaboration: Contribution to the 7th Workshop on Grand Unification, ICOBAN'86, Toyama, Japan and Ref. 13, 14

Neutron-Antineutron Oscillation Experiments

M. Baldo-Ceolin

Dipartimento di Fisica "G. Galilei" and
Istituto Nazionale di Fisica Nucleare, Sezione di Padova, Via Marzolo, 8,
I-35131 Padova, Italy

1. Physics Motivation and Experimental Status

In the hypothesis that baryon number (B) conservation law is only approximate every proton as well as every neutron bound in nuclei, must eventually decay into lighter particles[1].

Then, due to the fact that the particles carrying non vanishing B are fermions, conservation of the angular momentum requires that any possible decay mode satisfays the selection rule: $\Delta(B+L)$ be an even number.

Processes with $\Delta B=2$ give also rise to matter-antimatter oscillations which would occur as a first-order process.

In the following we shall be mainly concerned with $n-\bar{n}$ oscillations, which seem to be the most interesting process among these obeying the selection rule $\Delta B=2$.

In the quark-lepton picture the simplest term in the effective Lagrangian that can induce $\Delta B=2$ processes is of the form

$$L_{eff} \simeq G\ (qqq \cdot qqq) + h.c.$$

where $G \simeq e^4/M_x$ and M_x, the mass of the bosons associated with such processes, come in with a power of five, because the six quarks operator has dimension nine. The above effective interaction can be generated through the coupling of the quark pairs with appropriate bosons: it is to be noted that in order to get detectable effects, these bosons should be relatively light, $M_x < 10^7$ GeV.

The $n-\bar{n}$ oscillations are characterized by *the oscillation time* τ_{osc}, which contains all the relevant parameters

$$(\tau_{osc})^{-1} = \delta m = \langle \bar{n} | H' | n \rangle \qquad (1)$$

The theoretical models leave a large incertitude in the value to be expected for τ_{osc}. The so-called "left-right symmetric" models, however, suggest for the neutron oscillation time a value τ_{osc} $(10^8 \div 10^9)$sec[2].

Experimentally: $\tau_{osc} > 10^6$sec[3] and correspondingly $\delta m < 6 \cdot 10^{-28}$ MeV, thus by dimensional arguments

$$M_x > 5 \cdot 10^5 \text{ GeV}$$

Deep underground experiments, measuring nuclear stability lifetimes[4], give for $\Delta B=2$ processes $T_{osc} > 10^{31}$yr and this can be interpreted in terms of $n-\bar{n}$ oscillations to give a lower limit to τ_{osc} of $(5 \div 10) \cdot 10^7$sec.

It is to be stressed that if neutron-oscillation processes are observed, this besides being a manifestation of the baryonic number non conservation will open a new physics in the mass range $M_x \simeq (10^4 \div 10^6)$GeV and provide new experimental input regarding ways to extend the Standard Model.

2. Phenomenology and experimental methods

As a consequence of a $\Delta B=2$ interaction there will be a $n-\bar{n}$ mixing. Thus in vacuum an initially pure neutron state (B=+1) will in time acquire an antineutron component (B=-1) with probability[5] $P(\bar{n},t)$

$$P(\bar{n},t) = \sin^2(\delta m t) = (t/\tau)^2$$

When external perturbations introduce an energy difference between the $n-\bar{n}$ states, the $n-\bar{n}$ transition probability becomes:

$$P(\bar{n},t) = [\delta m^2/(\delta m^2+\Delta E^2)]\sin^2[(\delta m^2+\Delta E^2)^{\frac{1}{2}}\, t] \tag{1}$$

where $2\Delta E$ is the $n-\bar{n}$ state energy difference. These external interactions may be magnetic, acting through the equal but opposite n and $\bar{n}$ magnetic moments - $\Delta E \simeq 10^{-18}$ MeV in the hearth magnetic field - or nuclear through the differing n and $\bar{n}$ strong interaction properties - $\Delta E \simeq 10^2 \div 10^3$ MeV -.

From eq.(1) it appears that for $\Delta E \gg \delta m$, both the oscillation amplitude and period decrease and tend to zero respectively as $(\delta m/\Delta E)^2$ and $1/\Delta E$, so that $n-\bar{n}$ transitions are suppressed and neutrons appear stable.

Two situations, however, are experimentally interesting: the two limiting cases:

a) $\Delta E\cdot t \ll 1$ and b) $\Delta E\cdot t \gg 1$

The first case corresponds to neutron beams propagating in vacuum. Under the condition $\Delta E\cdot t \ll 1$, eq.(1) reduces to

$$P(\bar{n},t) = (t/\tau_{osc})^2 \tag{2}$$

which coincides with the free neutrons - no external field - case. The "quasi free condition", $\Delta E\cdot t \ll 1$, which allows the optimization of experimental conditions, is of basic importance in designing experiments aiming at detecting free neutron oscillations.

The second case corresponds to neutrons bound in nuclei. Here $P(\bar{n},t)$ oscillates very rapidly, and its average value becomes extremely small $\langle P(\bar{n})\rangle_{av} \simeq (\delta m/\Delta E)^2 < 10^{-60}$. Thus the nuclei would become unstable against $n-\bar{n}$ oscillations with $\Gamma_{osc} \simeq (\delta m/\Delta E)^2\, \Gamma_{n\bar{n}}$, where $\Gamma_{n\bar{n}}$ is the $\bar{n}$-nucleus annihilation rate. So if the nuclear lifetime T_{osc} has been measured it might be related to the interesting parameter τ_{osc} through the relation

$$T_{osc} = \tau_{osc}^2\, T_r$$

where T_r takes into account the nuclear contributions and corresponds according to the nuclear model, to values $T_r \simeq (1\div 2)10^{23}$ sec^{-1}[6].

Thus the sensitivity of an experiment, defined as the maximum value of the oscillation time the measurement can detect, depends on the total number of available neutrons, the effective time along which neutron oscillation can develop, and the frequency of background events. Therefore, in order to attain high sensitivities, first of all very intense neutron sources are required, such as nuclear matter or artificial sources such as nuclear reactors or accelerators.

Depending on the choice between the two possibilities, whether the observed neutrons are bound in nuclear matter or artificially produced, two types of measurements can be made: a static or a dynamical one. In both cases the signal is a $\bar{n}$-nucleon annihilation, i.e. an energy release of $\sim$2 GeV, distributed over several pions, 5 in average, and a total momentum $\bar{p} = 0$.

3. Static measurements

Static measurements take advantage of the fact that a very large number of neutrons are present in nuclear matter. They require very massive set-ups acting as source and detector at the same time in order that a large amount of matter can be observed over a long period to detect events due to annihilation processes. A high spatial and energy resolution is needed.

The main and unavoidable source of background in this type of experiments, which must be carried out deep underground in order to avoid cosmic ray interactions, are the neutrino interactions (∿120 events per Kton per year), since when the annihilation process takes place in the core of a heavy or medium heavy nucleus its characteristic signature results practically destroyed due to the high absorption probability of the produced pions in the nucleus.

The evaluation of τ_{osc} from measurements of the static type is not straightforward; there are substantial uncertainties in relating free and bound $n-\bar{n}$ mixing depending on nuclear model assumption[7], furthermore it has to be stressed that the strength of $n\rightarrow\bar{n}$ transitions needs not to be the same for "free neutrons" or neutrons bound in nuclear matter, (consider i.e. the free and bound neutron β decay probability) and consequently T_{osc} does not constrain the value of the free neutron oscillation time[8].

The main characteristics of the static experiments and their results are reported in Table I[9].

As it may be seen from Table I, this generation of experiments has already reached, due to the presence of background events, its maximum sensitivity on τ_{osc} but newer experiments are in preparation

TABLE I

Experiment	Detector Type and Fiducial Mass (tons)	Selection Criteria and Evaluated Efficiency	Candidates and Backgrounds Events	T_{osc} in $>10^{31}$ years measured at 90% C.L.	τ_{osc} (10^7 sec) evaluated according to DGR	or ABM
KGF	Tracking Detector M=65	Visible Energy =n.e.	N_C=2 N_B n.e.	0.34	2.8	1.6
NUSEX	Tracking Detector M=120	$1 \leq E_{vis} \leq 2$GeV ≥1 e.m. Shower =0.6	N_C=4 N_B n.e.	1.0	4.7	2.5
FREJUS	Tracking Detector M=750	Not yet analyzed				
IMB	Cerenkov H_2O M=3300	Energy and Isotropy =0.6	N_C=15 N_B=8.6	3.0	10.0	5.5
KAMIOKANDE	Cerenkov H_2O M=880	Energy p<600 MeV/c =0.33	N_C=0 N_B=1.17	4.3	12.0	6.8

or in project such as Soudan II, Superkamiokande and Icarus where better resolution will allow to reach higher sensitivities.

4. Experiments of the dynamical type

Experiments of the dynamical type benefit by the "quasi free neutron condition" $\Delta E \cdot t \ll 1$, so providing a more straightforward way for measuring τ_{osc}. They require a very intense source of moderated and possibly cooled neutrons (from nuclear reactor or accelerators) and need a region adequately free from all kind of perturbing interactions - for t=1 sec the residual gas pressure has to be $p<10^{-6}$ torr and the residual magnetic field $B<10^{-4}$ gauss -.

The constraint which defines the quality of a neutron-oscillation experiment may be deduced from eq.(2) as

$$\tau_{osc} \propto (N \cdot \varepsilon)^{\frac{1}{2}}\, t = (I \cdot T \cdot \varepsilon)^{\frac{1}{2}}\, L/v \qquad (3)$$

where: I, the neutron current in $n \cdot sec^{-1}$, depends on the power of the neutron source; ε, the fraction of annihilation events which can be unambiguously identified, depends on the properties ("quality") of the detector; $t = L/v$, is the "quasi free propagation" time in sec; v, the neutron velocity in m/sec; T, the data recording time; Nt^2, for a given source, depends upon the neutron energy and the annihilation target area.

Therefore in order to reach high sensitivities are needed: a very intense neutron beam travelling a long distance at low velocity; a target at the end of the flight path, where the antineutron component annihilate; a detector large and massive enough as to accept and stop the annihilation products with high spatial and energy resolution, and an efficient Cosmic Ray veto and shield.

It is to be stressed that a very important feature of this experiments consists in the fact that the possible $\bar{n}$ component can be suppressed by the application of an external magnetic field - $(P(\bar{n}) \propto B^{-2})$, which allows to verify the validity of any observed signal.

Background for this type of experiments will consist primarily of events from cosmic rays into the target that simulate an $\bar{n}$-nucleus event, their rate being proportional to the target mass and less than 1 event per kg per year[3] depending on the detector quality. Very thin annihilation target can be used - $\sigma_n \propto 1/v$ - of the order of $\sim 0,25$ kg per 1 m^2 target area.

Moreover, source and target associated reactions give rise to a large radiation flux on the detector reducing the detection efficiency. Increasing the path length L for a fixed neutron source and a fixed annihilation target, reduces the neutron beam intensity at the target and the related background in proportion, while the sensitivity remains constant since the loss of solid angle is compensated by the increase in t^2. Furthermore, the radiation effect can be conveniently reduced transporting the neutrons from the source to the experimental area by means of neutron beam guides slightly bent so as to switch off the radiation coming along with the neutron beam.

In Table II are reported the main characteristics of the dynamical experiments[10], in Fig. 1 the lay-out of the Grenoble II experiment is shown as an illustration of an experimental apparatus.

TABLE II

Experiment	Neutron Source Type and Power	Neutron $\langle E \rangle$ in eV and Intensity $I_0 = \phi_0 S_0$ ($n\ sec^{-1}$) at the Source	Neutron I ($n\ sec^{-1}$) Target Area (m^2) and Detector Effifiency	Drift Length in m and Drift Time in sec	Status of Experiment	τ_{osc} in 10^7 sec at 90% C.L.
GRENOBLE I	Nuclear Reactor - 57 MW	Cold $\langle E \rangle = 10^{-4}$ - $1.5 \cdot 10^9$	$1.5 \cdot 10^9$ - A=0.06 $\varepsilon = 0.3$	$\langle L \rangle = 4.5$ $\langle t \rangle = 0.027$	complet-ed	0.1
GRENOBLE II	Nuclear Reactor - 57 MW	Cold $\langle E \rangle = 10^{-3}$ - $3.3 \cdot 10^{11}$	$4.4 \cdot 10^{11}$ A=0.65 $\varepsilon = 0.8$	$\langle L \rangle = 50$ $\langle t \rangle = 0.1$	in prepa-ration	20.0
PAVIA	Nuclear Reactor - 0.25 MW	Thermal $2.5 \cdot 10^{-2}$ - $1.1 \cdot 10^{14}$	$2.5 \cdot 10^{11}$ - A=0.8 $\varepsilon = 0.3$	L=16 $\langle t \rangle = 0.007$	Taking Data	0.5
MMF MOSCOW	Meson Factory - 500 μA	Cold $2.5 \cdot 10^{-3}$ - $1.2 \cdot 10^{15}$	$6 \cdot 10^{12}$ - A=25.0 ε n.e.	L=70 $\langle t \rangle = 0.1$	Project	90.0

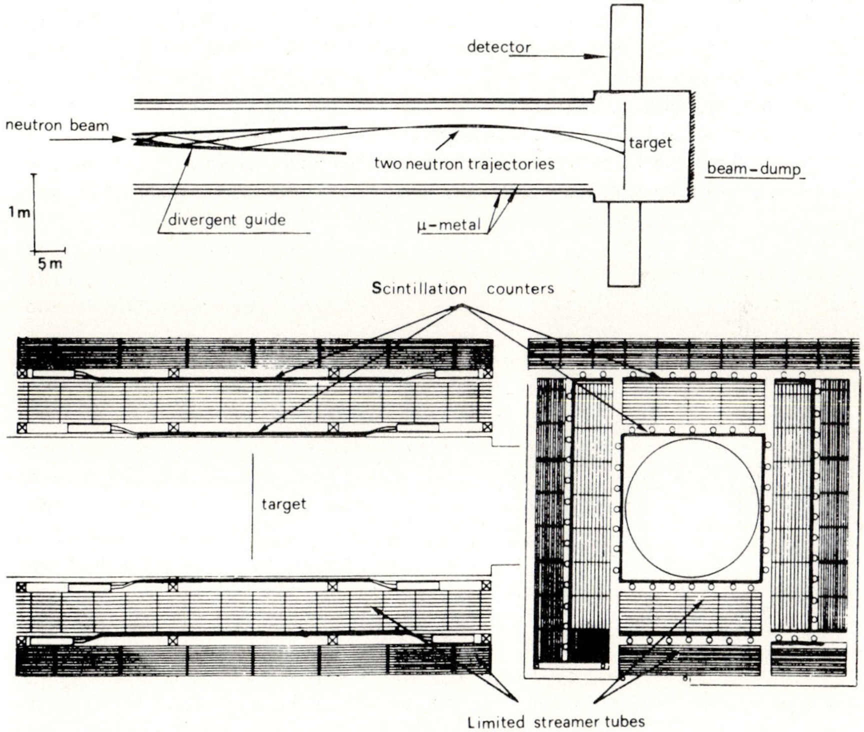

Longitudinal and transversal view of the detector

Fig. 1 - The lay-out of the Grenoble II experiment.

References

(1) See the P.Langacker: Contribution to this Conference.
(2) See R.N. Mohapatra: this Conference.
(3) G. Fidecaro et al.: Phys.Lett. 156B, 122 (1985).
(4) For a review on Proton Decay Experiments see H. Meyer: this Conference.
(5) M. Baldo-Ceolin: Proceed. of the "Conference on Astrophysics and Elementary Particles: Common Problems", Roma (1980), p. 251.
R.E. Marshak and R.N. Mohapatra: Phys.Lett. 94B, 183 (1980).
(6) W.M. Alberico, A. Bottino and A. Molinari: Nucl.Phys. A429, 445 (1984).
C.B. Dover, A. Gal and M. Richard: Phys.Rev. D27, 1090 (1983).
(7) See for example Ref.(6).
(8) P.K. Kabir: Phys.Rev.Lett. 51, 231 (1983).
J. Basecq and L. Wolfenstein: Nucl.Phys. B224, 21 (1983).
(9) KGF : M.R. Krishnaswami et al.: Phys.Lett.106B, 339 (1981)
NUSEX : G. Battistoni et al.: Phys.Lett. 133B, 454 (1983) and E. Bellotti Private Communication.
FREJUS : H. Meyer: "Neutrino 86 Sendai".
IMB : T.W. Jones et al.: Phys.Rev.Lett. 52, 720 (1984) and J.M. Losecco: Private Communication.
KAMIOKANDE: M. Koshiba: 1986 Aspen Winter Physics Conference.
(10) GRENOBLE I: CERN-ILL-PADOVA-RHEL-SUSSEX Collaboration, G. Fidecaro et al.: Phys.Lett. 156B, 122 (1985).
GRENOBLE II: HEIDELBERG-ILL-PADOVA-PAVIA Collaboration. See for example M. Baldo-Ceolin: "Workshop on Reactor Based Fundamental Physics", Grenoble 1983, Journal de Physique, Colloque C3, Suppl.n.3, Tome 45, C3 - 173-183 (1984).
PAVIA : PAVIA-ROMA Collaboration. See S. Ratti: Proceed. of the "ICOBAN" Meeting, Bombay 1982, p.197.
MMF : MOSCOW MESON FACTORY: see V.A. Kuzmin, Proceed. of the "ICOBAN" Meeting, Bombay 1982, p.197.

Search for a Neutron Electric Dipole Moment

N.F. Ramsey

Harvard University, Cambridge, MA 02138, USA

1. Early Experiments

In 1950 at a time when parity conservation was almost universally believed, PURCELL and RAMSEY [1] pointed out that such an assumption must be based on experiment and that there was little experimental evidence for the assumption at that time. We further noted that a search for a neutron electric dipole moment would provide such a test since a dipole moment is forbidden under parity symmetry and the test would be particularly sensitive since the neutron has no electric charge and is consequently not accelerated out of the observation region when a strong external electric field is applied.

SMITH, PURCELL and RAMSEY [2] constructed a neutron beam magnetic resonance apparatus with a strong electric field applied in the region between the two separated oscillatory fields. If the resonance frequency should change when the relative orientations between the electric and magnetic fields were changed from parallel to antiparallel, the change would have to be attributed to a neutron electric dipole moment provided spurious effects were eliminated, such as those arising from changes in magnetic field due to leakage currents associated with the electrostatic field. When we started the experiment, the lowest limit tht could be set by any existing observations was 3×10^{-18} e.cm and in the experiment we lowered the limit to 5×10^{-20} e.cm, but found no electric dipole moment.

LEE and YANG [3] in 1956 proposed parity non-conservation in the weak interaction as an explanation of an anomally in K meson decay. The proposed parity non-conservation in the weak interaction was confirmed in 1957 by WU, et al. [4] in the angular distribution of the electrons in the decay of polarized ^{60}Co. Although the discovery of the failure of parity symmetry eliminated that symmetry principle as a fundamental argument against the possibility of a particle electric dipole moment, LANDAU [5] and others [6] argued that with the failure of parity (P) symmetry there should still be conservation of CP and of T (where C is the charge exchange operator and T is the time reversal operator) and that T symmetry would eliminate the possibility of a particle electric dipole moment. However, at that time I wrote a letter [7] pointing out that the T symmetry assumption must be based on experiment and that a search for a neutron electric dipole moment was a particularly sensitive test of T conservation; JACKSON, et al. [8] also emphasized the need for an experimental basis for the assumptions of T and CP symmetry. In 1964 CHRISTENSON, CRONIN, FITCH and TURLAY [9] showed that there was a failure of CP conservation in the decay of the K_L^0 and hence a failure of T is CPT were conserved. The subsequent theories that accounted for the observed CP violation in the K_L^0 usually predicted a non-zero value for the neutron electric dipole moment, with the initial predictions being usually larger than 10^{-23} cm. As a result there were a series of experiments [6], mostly by the neutron beam magnetic resonance method, which successively lowered the experimental limit on the neutron electric dipole moment. The most sensitive of the neutron beam experiments was that of DRESS, MILLER, PENDLEBURY, PERRIN and RAMSEY [11,12]. Neutrons from the Grenoble reactor at 180 m/s were polarized and analyzed by reflection from magnetized iron mirrors between which there was a 180 cm long region with a 17 G magnetic field and a 100 kV/cm electric field. Neutron reorientation transitions were induced by the method of separated oscillatory fields [1]. If the neutron had an electric dipole moment, the neutron resonance frequency would be different when the electric and magnetic fields were parallel and antiparallel; such a frequency shift was sought. None was found and the result was equivalent to an experimental neutron electric dipole moment of $(4 \pm 15) \times 10^{-25}$ cm.

2. Recent Experiments

The most recent experiments on the neutron electric dipole moment involve neutrons of less than 7 m/s velocity trapped by neutron total reflection at the surface of the trap, which typically consists of beryllium and beryllium oxide. Figure 1 shows the apparatus used by the Leningrad group of ALTAREV, et al. [6,12] and Fig. 2 shows that of the Harvard-Sussex-Rutherford-ILL collaboration of PENDLEBURY, et al. [6,13] working at the ILL in Grenoble.

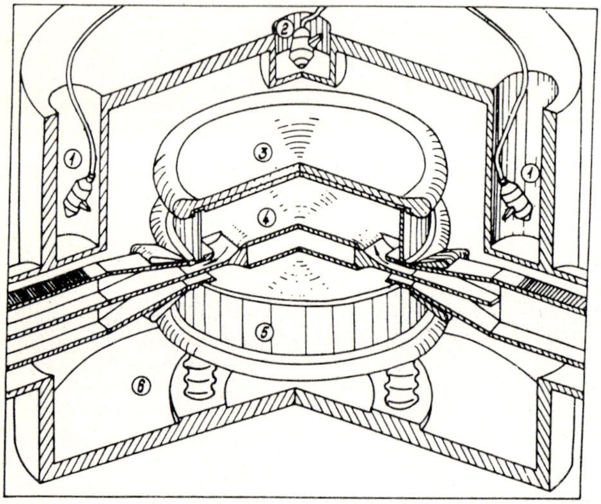

Fig. 1 -- General view of the Leningrad double-bottle neutron magnetic resonance spectrometer: (1) indicates a magnetometer for control of the magnetic field (2) a magnetometer for field stabilization, (3) the top electrode, (4) the central electrode, (5) the chamber wall made of beryllium-oxide coated fused quartz, and (6) the bottom electrode. The initial and final oscillatory field coils are shown on the neutron pipes at the left and right hand sides of the figure. The spectrometer is surrounded by three layers of magnetic shielding.

In the Leningrad experiment polarized neutrons enter through a pipe on the right and stay in one of the two storage volumes for approximately 5 seconds before exiting through one of the two pipes on the left. They are excited by separated oscillatory fields with an adiabatic excitation [12] in the entrance and exit pipes. The electric potential is applied at the central electrode of the double bottle so that the electric fields in the two bottles are in the opposite direction. This helps to cancel out the effects of drifting magnetic fields but does not fully compensate for local magnetic effects due to sparks and leakage currents from the applied electric fields. The electric dipole moment was detected by a change in the neutron magnetic resonance frequency when the electric fields were changed from parallel to anti-parallel in the magnetic field.

In the experiment of the Harvard-Sussex-Rutherford-ILL collaboration at the ILL, we provided a guide to bring the ultra cold neutrons from the left of Fig. 2. The neutrons are stored in a cylinder approximately 25 cm in diameter and 10 cm high

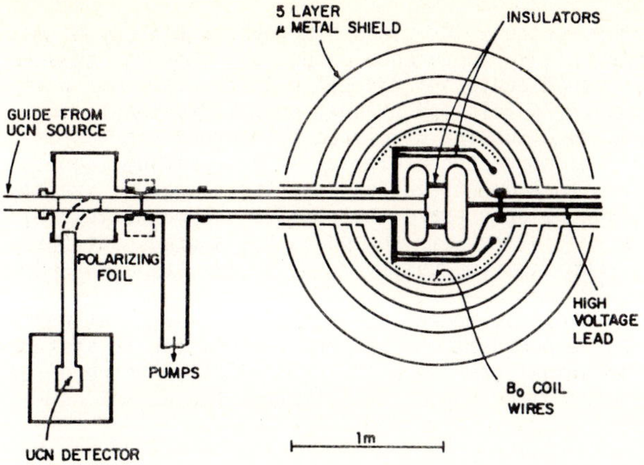

Fig. 2 -- Schematic diagram of apparatus used for measuring electric dipole moment with bottled neutrons at ILL

with the end plates being of metallic beryllium and the sides of the cylinder being of beryllia. After the neutron bottle is filled, a shutter is closed, storing the neutrons for approximately 80 s. The oscillatory field is applied as initial and final coherent pulses. The resonance is observed in a fashion similar to the neutron beam experiment: the neutrons are polarized on passage through the indicated polarizing foil, analyzed during their return passage through the foil, and counted at the indicated ultra-cold neutron detector. The observations are at the steepest point of the resonance curve. The change in beam intensity correlated with the application of an electric field is then examined to set a limit to the neutron electric dipole moment. The magnetic field was monitored with three rubidium magnetometers. So far the only neutron beam available to the experiment has been one from a moderator at room temperature or above. As a result, only a very small fraction of the beam is at velocities less than 7 m/s so the stored neutron density is very low (about 0.05 neutron/cm^3). As will be discussed later, this beam intensity has been markedly increased for the current experiments.

The use of stored ultra-cold neutrons offers two particularly important advantages. The resonance curve with stored neutrons is 7,000 times narrower than in the beam experiment. Furthermore, in the beam experiment a large fraction of the running time must be devoted to eliminating the E x v/c effect, which effectively vanishes in the bottle experiment since the average value of v is very small when the neutrons enter and exit through the same port with a storage time of 80 s.

The published results for the electric dipole moment of the neutron as measured by the Leningrad [12] group is $(2.3 \pm 2.3) \times 10^{-25}$ e.cm with an informal later value of $(-2 \pm 1) \times 10^{-25}$ e.cm given at a subsequent meeting [27].

The latest published results of the ILL group [13] are $(0.3 \pm 4.8) \times 10^{-25}$ e.cm. with a late informal value of $(-1.8 \pm 2.9) \times 10^{-25}$ e.cm. given at a subsequent meeting [14]

Current Improvements

In the next two years there should be a significant lowering of the above experimental errors, since both the Leningrad and ILL groups have made major improvements in their experiments and are working on further improvements.

The Leningrad group has revised its apparatus to increase the neutron density, to eliminate the velocity effect of a net flow of neutrons through the apparatus, and to diminish the danger of systematic errors.

Our group at the ILL now has approximately 100 times the neutron density as was available before the recent improvements in the Grenoble reactor. Most of this improvement comes from the use of a liquid deuterium moderator which could not previously be used with neutrons below 7 m/s velocity. Additional improvements in neutron density were obtained by using a turbine developed by Steyerl and his associates which slows one component of the neutron velocity. The rubidium magnetometers have been improved and other improvements to decrease the danger of systematic errors are being made. Plans are being prepared to monitor the magnetic field with a low pressure magnetometer vapor in the same region as the stored neutrons [20]. Possible vapors are cesium, rubidium, mercury, ^{3}He or ^{129}Xe.

The limits on the neutron electric dipole moment are already so low, that any further decrease comes slowly with the necessity to pay great attention to systematic errors, particularly those which might provide a small correlation between the strength of the magnetic field and the direction of the electric field.

Theoretical Predictions

The theoretical predictions [6] for the value of the neutron electric dipole moment range from 10^{-19} e.cm. to 10^{-33} e.cm. Most of the early predictions which attributed the electric dipole moment to T non-conservation in electromagnetic theory or in a milliweak force predicted values larger than 10^{-23} e.cm. and have been abandoned, due in part to the disagreement between the predictions and the above experimental values. Currently the three most popular theories for predicting a neutron electric dipole moment are (a) theories [6] that attribute CP violations to the exchange of Higgs bosons and predict values of 10^{-24} e.cm., (b) theories [17,18] that attribute CP violation to the exchange of quarks and predict values usually smaller than 10^{-29} e.cm. and (c) cosmological theories [19] that account for the baryon-anti-baryon asymmetry and relate the neutron electric dipole moment to the entropy of the universe generated since bariosynthesis.

Parity Non-Conserving Spin Rotation of the Neutron

Although my talk is primarily on experiments on the neutron electric dipole moment, in view of this being a conference on weak interactions, I should like to conclude my talk by giving the results of our ILL group on parity non-conserving spin rotations when neutrons are transmitted through different. These rotations are presumably due to the weak force.

The possibility of detecting parity non-conserving spin rotations of the neutron due to the weak interaction when neutrons pass through various samples of matter was independently suggested by MICHEL [23] and STODOLSKY [24], but for many years the predicted rotations were considered to be too small to be observable. However, when I first heard of their proposals, I realized that with the sensitivity of the neutron magnetic resonance techniques which we used in measuring the electric dipole moment we had plenty of sensitivity. The principle difficulty would be in distinguishing the parity non-conserving spin rotation from the spin rotation due to extraneous magnetic fields.

The arrangement of our experiment for measuring the spin rotation has been described by HECKEL et al. [25]. Our results for the parity non-conserving rotations, ϕ_{pnc}, of the neutron in passing through a number of substances are:

$^{124}_{50}$SN: $\phi_{pnc} = -(0.48 \pm 1.49) \times 10^{-6}$ rad/cm

$_{50}$SN: $= -(3.19 \pm 0.40)$

$^{117}_{50}$SN: $= -(37.0 \pm 2.5)$

$_{82}$Pb: $= +(2.24 \pm 0.33)$

$_{57}$La: $= -(219 \pm 29)$

References

1 E.M. Purcell and N.F. Ramsey, Phys. Rev. 78, 807 (1950); N.F. Ramsey, Physics Today 33, 25 (1980).
2 J.H. Smith, E.M. Purcell and N.F. Ramsey, Phys. Rev. 108, 120 (1957).
3 T.D. Lee and C.N. Yang, Phys. Rev. 104, 254 (1956).
4 C.S. Wu, E. Ambler, R.W. Hayward, D.D. Hoppes, and R.R. Hudson, Phys. Rev. 105, 1413 (1957).
5 L. Landau, Nucl. Phys. 3, 127 (1957).
6 See references [10] and [11] for a complete list of references.
7 N.F. Ramsey, Phys. Rev. 109, 225 (1958).
8 J.D. Jackson, S.B. Treiman and H.W. Wyld, Jr., Phys. Rev. 106, 517 (1957).
9 J.H. Christernson, J.W. Cronin, V.L. Fitch, and R. Turlay, Phys. Rev. Lett. 13, 138 (1964).
10 W.B. Dress, P.D. Miller, J.M. Pendlebury, P. Perrin, and N.F. Ramsey, Phys. Rev. D 15, 9 (1977).
11 N.F. Ramsey, Ann. Rev. Nucl. Part. Science 32, 211 (1982).
12 I.S. Altarev, et al., Phys. Lett. 102B, 13 (1981).
13 J.M. Pendlebury, et al., Phys. Lett 136B, 327 (1984).
14 A. Steyerl, and W. Mampe, private communications.
15 R. Golub and J.M. Pendlebury, Rep. Prog. Phys. 42, 439 (1979).
16 S. Weinberg, Phys. Rev. Lett. 37, 367 (1976).
17 M. Kobayashi and K. Maskawa, Prog. Theor. Phys. 49, 652 (1973).
18 D.V. Nanopoulos, A. Yildiz and P. Cox, Annals of Physics 127, 126 (1980).
19 J. Ellis, et al., Nature 293, 41 (1981).
20 N.F. Ramsey, Acta Physica Hungarica 55, 117 (1984).
21 R. Dombeck, et al., Rep. Tr-79-85, pp. 79-153, Univ. of Maryland Physics Dpt.
22 J. Morse, R. Golub, and others, private communication.
23 F.C. Michel, Phys. Rev. 133, B329 (1964).
24 L. Stodolsky, Phys. Lett. 50B, 352 (1974).
25 B. Heckel, et al., Phys. Rev. C 29, 2389 (1984).
26 A. Steyerl, et al., Condensed Matter 50, 281 (1983) and ICSU/AB:07.
27 A.P. Serebrov, et al., Journal de Physique 45, C3-286 (1984), J. Morse, et al., Journal de Physique 45, C3-286 (1984).

An Experimental Search for the Neutron Electric Dipole Moment

V.M. Lobashev

Leningrad Nuclear Physics Institute of the USSR Academy of Sciences, Gatchina, Leningrad district, 188350, USSR and Institute for Nuclear Research, Academy of Sciences, Moscow 117312, USSR

Abstract: An experiment searching for the neutron electric dipole moment (EDM) as a measure of CP-violation has been carried out by the method of containment of ultracold neutrons (UCN). The measurements performed yielded for the neutron EDM a value $d_n=-(1.4\pm0.6)\cdot10^{-25}$ e·cm which is equivalent to establishing an upper limit of $d_n\leq2.6\cdot10^{-25}$ e·cm at a 95% confidence level.

The mystery of the CP-violation nature has been remaining one of the most actual and intriguing problems of elementary particle physics for more than 20 years. At present besides K_L-meson decay the neutron electric dipole moment seems to be the physical phenomenon which is most sensitive to CP-violation searching. The best restrictions on the value of the neutron EDM has been obtained in ref./1-3/ by means of a magnetic resonance spectrometer with applied electric field using ultracold neutrons.

This communication reports on the results of a new experimental search for the neutron EDM carried out at the WWR-M reactor in the LNPI of the Academy of Sciences of the USSR by the group including I.S. Altarev, Yu.V. Borisov, N.V. Borovikova, A.B. Brandin, A.I. Egorov, S.N. Ivanov, E.A. Kolomensky, M.S. Lasakov, A.N. Pirozhkov, A.P. Serebrov, Yu.V. Sobolev, R.R. Taldaev, E.V. Shulgina.

The distinction of previous experiments /1-3/ was the usage of a socalled flow-through type spectrometer with an average UCN containment time of about 5 sec. The use of two UNC containment chambers with oppositely directed electric field and common magnetic and high frequency fields (differential method for measuring the EDM) permitted one to compensate fluctuations of the magnetic field and to control systematic effects involved by the reversal of electric field.

The UCN source represented a helium cooled liquid hydrogen converter which was placed in a beryllium reflector of the WWR-M reactor and a vertical mirror neutron guide. Information about characteristics of the source has been given in ref. /4/.

The measurements performed yielded for the neutron EDM a value $d_n=(-2\pm1)\cdot10^{-25}$ e·cm, which is equivalent to establishing an upper limit of $d_n \leq 4.10^{-25}$ e·cm at the 95% confidence level. A maximum sensitivity which is an error of EDM per day of measurements amounted to $\pm8\cdot10^{-25}$ e·cm.

Further increase of the sensitivity in a flow-through mode requests a long-term containment of UCN in a locked chamber. In recent experiments conducted in the Institute Laue-Langevin (ILL, Grenoble) in the storage mode with one chamber a containment time of about 80 sec was achieved, and $d_n=(0.3\pm4.8)\cdot10^{-25}$ e·cm /5/ was measured.

We used a differential method involving two chambers for neutron containment with oppositely directed electric field (see Fig.1) and with common (static) magnetic and radio-frequency fields. The central high voltage electrode separating the chambers was put between two grounded electrodes (lids), the upper and lower lids being connected by arcs which formed some sort of grounded screen around the central electrode. All connections were made in such a way as to decrease the influence of magnetic field components induced by recharge and leakage currents on the resonance conditions.

The system for producing a reversible electric field is described in ref. /6/. The average electric field strength was ±12-15 kV/cm at leakage currents smaller than 30 nA.

The walls of the chambers were formed of quartz rings coated with BeO and Be_3N_2. Shutters at the input and output neutron guides ensured opening and shutting in 0.3 sec. Like in previous measurements /1-3/ we used a system for simultaneous detection of oppositely polarized UCN.

The necessity of stabilizing the spectrometer magnetic field up to $\sim 10^{-12}$T for 5-10 min which is necessary for a containment time of 50 sec has proved to be a very difficult task. A three-layer permalloy screen which was used in ref. /1-3/ provided a stability approximately by an order of magnitude worse than mentioned above. To stabilize the resonance position to the desired level we placed quantum self-generating Caesium magnetometers near to upper and lower chambers as it is shown in Fig. 1. Since the ratio of resonant frequency of Cs magnetometers to that of neutron resonance is equal to 119.9 which is close to 120, a relatively simple system for stabilization was developed. It was based on scaling of a frequency of one magnetometer (one-channel mode) or averaged frequency of two magnetometers (the two channel mode) by the factor of 120 and applying a signal from a scaler to a radio-frequency coil enveloping both UCN-storage chambers /7/. Owing to this system the resonant conditions were maintained with stability corresponding to $\sim 2\cdot 10^{-12}$T for 5-10 min. The effective stabilization factor was about 7 in the one-channel mode and 15 in the two-channel mode.

The residual fluctuations of the resonant conditions were suppressed due to the differential method of measurements, so the difference between the root mean-square (rms) spread and the statistical one amounted only to 5-10%.

In primary experiments as a UCN source a liquid hydrogen converter in a beryllium reflector /4/ was used. In main series of measurements a new universal source of cold and ultracold neutrons, put in service by the end of 1985, was utilized.

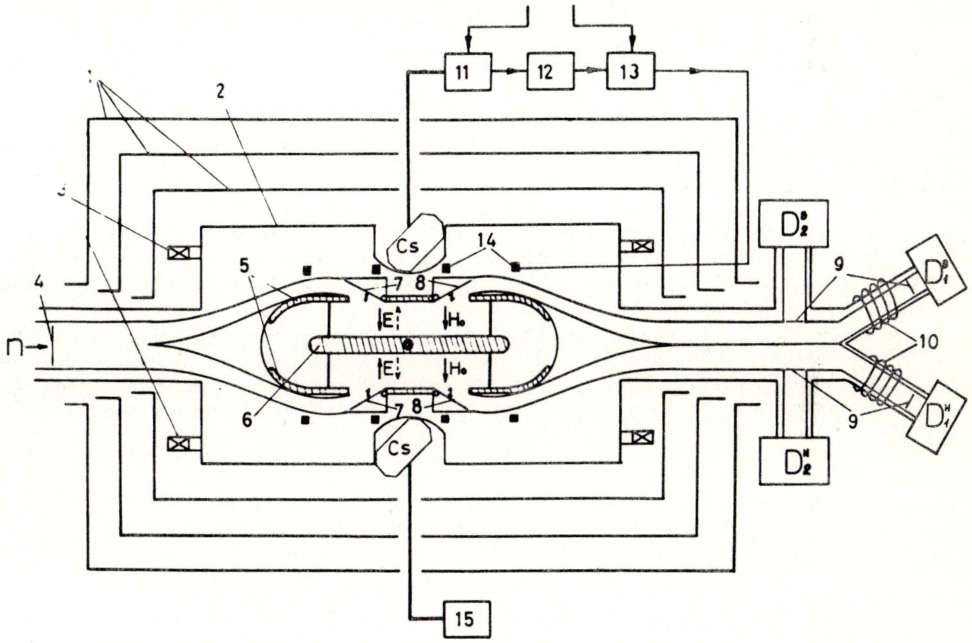

Fig.1 Layout of the magnetic resonance spectrometer of ultracold neutrons for measuring the neutron EDM. 1-permalloy screens, 2-vacuum chamber, 3-Helmholtz coils for producing static magnetic field, 4-UCN polarizer, 5-grounded electrodes, 6-high voltage electrode, 7-input shutters, 8-output shutters, 9-analyzers, 10-radio-frequency flipper, 11-Cs-magnetometer controlled by computer, 12-frequency scaler, 13-r.f. pulse unit controlled by computer, 14-coil for producing oscillating field, 15-Cs-magnetometer. D-UCN detectors (B and H correspond to the upper and lower chambers, respectively, 1 and 2 - to opposite polarizations of UCN).

It contained 1ℓ of liquid hydrogen at 18-19K and was placed in the center of the WWR-M reactor active zone, in the lead shielding.

The increase in intensity of UCN by 3-4 times and development of the storage type spectrometer with a UCN containment time of ~50 sec permitted us to obtain a sensitivity for measuring the EDM of $\sim(2.5\text{-}2.7)\cdot 10^{-25}$ e·cm per day.

The measurement cycle included: filling of chambers by UCN- -30 sec, containment - 50-55 sec, yield and counting of UCN - 30 sec. Radio-frequency signals were applied during 1.2 sec at the beginning and at the end of a containment time.

The electric field polarity was reversed in an interval of 40 sec between measurement cycles. All the rest measurement procedures were similar to that described in /2/. For each detector the EDM was calculated as:

$$d_i = - \frac{h(N_i^+ - N_i^-)}{4E\ \partial N/\partial \nu} ,$$

N_i^+, N_i^- being the number of detected neutrons at different polarities, E - electric field strength, $\partial N/\partial \nu$ - dependence of a neutron count on a shift of oscillating field frequency, which determines the resonant curve slope at the point of resonance being measured. The results of measurements for four detectors indicated as in fig. 1 were presented as follows:

$$d_n = \frac{1}{4}\left[(d_1^B + d_2^B) + (d_1^H + d_2^H) \right]$$

$$R = \frac{1}{4}\left[(d_1^B + d_2^H) - (d_1^B + d_2^H) \right]$$

$$P = \frac{1}{4}\left[(d_1^B - d_2^B) - (d_1^H - d_2^H) \right]$$

$$C = \frac{1}{4}\left[(d_1^B - d_2^B) + (d_1^H - d_2^H) \right]$$

where d_n determines the neutron EDM; R characterizes a synchronous shift of neutron resonance in both chambers (one-chamber effect); P may serve as a criterion of the pick-up from a high voltage to detector circuits and C is a criterion of the compensation of these systematic and random factors.

Table 1 summarizes results of the measurements of the EDM including those of /3/ with a flow-through type system.

One may notice that results for EDM obtained by a flow-through method and by the method of containment are in good agreement. R seems to be rather large in some runs of measurements, however, the precision of synchronous effects compensation for two chambers gives a factor of about 0.05 which resulted in a possible contribution to d_n on the average less than $0.3\cdot 10^{-25}$ e·cm for each run.

Table 2 shows that the removal of those measurements that have large values of one chamber effect (R) made no substantial influence upon the value of d_n. As estimated, other possible systematic effect can not make a sizable contribution to d_n. Thus, for example, the effect of non-parallelism of electric and magnetic fields accounts for $d_n \lesssim 10^{-26}$ e·cm and the leakage current effect (at I<30 nA) doesn't exceed $d_n < 10^{-26}$ e·cm. So, the final result is the following:

$$d_n = -(1.4 \pm 0.6) \cdot 10^{-25} \text{ e}\cdot\text{cm}$$

Though the value of EDM obtained appeared to be at variance from zero this fact cannot be considered yet as a conclusive evidence for existence of nonzero EDM. Therefore, it seems to be reasonable to explain the result as a new upper limit

$$d_n < 2.6 \cdot 10^{-25} \text{ e}\cdot\text{cm}$$

(at a 95% confidence level).

TABLE 1

Results of the measurements of the neutron EDM (in units of 10^{-25} e·cm)

Measurement conditions	E D M	One-chamber effect	Pick-up effect	Compensation effect
	d_n(rms/stat)	R(rms)	P(rms)	C)rms
Flow-through spectrometer	2.1(2.4/2.4)	16.6(4.8)	5.3(3.5)	4.8(2.4)
	-1.0(2.8/2.7)	1.5(5.8)	-5.0(4.1)	-2.6(2.7)
	-3.4(1.3/1.3)	1.1(2.8)	0.4(2.2)	-3.6(1.3)
A v e r a g e	-2.0(1.0)	4.5(2.2)	0.6(1.7)	-1.8(1.0)
Storage type spectrometer				
UCN-source in a beryllium reflector, one-channel stabilization	0.97(2.60/2.55)	-3.7(4.5)	1.3(2.9)	0.8(2.6)
UCN-source:				
H^0 ↓ two-channel stabilization	10.4(4.9/4.2)	-12.5(8.2)	5.0(5.7)	2.7(4.8)
H^0 ↓ two-channel stabilization	-1.9(1.1/1.1)	8.9(1.7)	-0.7(1.6)	0.9(1.1)
H^0 ↓ one-channel stabilization	-0.6(2.3/2.2)	9.4(4.4)	-0.8(3.1)	-1.2(2.1)
H^0 ↑ one-channel stabilization	-3.9(2.2/2.1)	5.2(4.4)	-0.5(2.5)	1.0(2.1)
H^0 ↑ two-channel stabilization	-0.5(1.4/1.3)	-1.9(2.3)	5.4(1.7)	0.3(1.3)
A v e r a g e	-1.1(0.7)	4.0(1.2)	1.7(1.0)	0.4(0.7)
General average	-1.4(0.6)	4.1(0.9)	1.4(0.9)	0.3(0.6)

TABLE 2 The results of data analysis with cut-off on value of R (one-chamber effect) in the storage type spectrometer

Cut off threshold	E D M	One-chamber effect	Pick-up effect	Compensation effect
	d_n(rms/stat)	R(rms)	P(rms)	C)rms
R/stat < 10	-1.3(0.7/0.7)	3.5(1.3)	1.8(1.0)	0.4(0.7)
R/stat < 7	-1.1(0.7/0.7)	4.0(1.2)	1.7(1.0)	0.4(0.7)
R/stat < 5	-1.1(0.7/0.7)	4.1(1.1)	1.7(1.0)	0.3(0.7)

Comparison with predictions of the most popular theoretical models for CP-violation makes it possible to abandon with no doubt only the Weinberg one /8,9/. For some other models definite restrictions should be put on the values of parameters involved.

The authors are grateful to V.L. Ryabov for handling of results, V.L. Varentzov for help in measurements, V.G. Muratov, L.A. Grigorjeva and V.N. Slyusar for assistance in equipment adjustment.

References:

1. Altarev I.S. et al., Nucl. Phys. 1980, A341, 269.
2. Altarev I.S. et al., Phys. Lett. 1981, 102B, 13.
3. Lobashev V.M. and Serebrov A.P., J. Physique Collog., 1984, 45, C3-11.
4. Altarev I.S. et al., Phys. Lett. 1980, 80A, 413.
5. Pendlebury T.J. et. al., Phys. Lett. 1984, 136B, 327.
6. Borisov Yu.V., Ezhov V.F., Ivanov S.N., Lobashev V.M., Serebrov A., Preprint LNPI 1985, N 1148.
7. Altarev I.S., Borisov Yu.V., Brandin A.B., Ivanov S.N., Lobashev V.M., Serebrov A.P., Taldaev R.R., Preprint LNPI 1985, N1117.
8. Khriplovich I.B. and Zhitnitsky A.R. Yad. Fiz. 1981, 34., 167.
9. Anselm A.A., Bunakov V.E., Gudkov V.P., Uraltsev N.G., Phys. Lett. 1985, B152, 116-120.

Search for Short-Lived Axions Emitted from Neutron Capture on Protons

S.J. Freedman[1], *M. Arnold*[2], *J. Doehner*[2], *J. Last*[2], *and D. Dubbers*[3]

[1]Argonne National Laboratory, Argonne, IL 60439, USA
[2]Physik. Institut der Universität Heidelberg, D-6900 Heidelberg, F.R. Germany
[3]Institut Laue-Langevin, 156X, F-38042 Grenoble Cedex, France

We have searched for a light neutral particle like the axion, which decays rapidly to positron-electron pairs, produced in the reaction $n + p \rightarrow D + a$. Instead of axions we found the e^+e^- signal expected from direct internal pair conversion. A preliminary analysis gives $(\Gamma_p/\Gamma_\gamma)_{expt}/(\Gamma_p/\Gamma_\gamma)_{cal} = 1.06 \pm 0.12$, for the measured to expected pair decay branching ratio. These results exclude existing axion models over the mass range $2\ m_e < m_a < 2.22$ MeV/c^2. The axion mass would fall in this range if the axion was the explanation for the narrow positron lines observed in recent heavy-ion experiments at the GSI.

1. Introduction

The beautiful explanation of strong P- and CP-conservation based on the Peccei-Quinn symmetry [1] is so appealing that its most dramatic physical consequence, the axion, continues to be the object of experimental search, despite numerous failures. Recently, exciting indications of narrow positron peaks observed in heavy-ion collisions at the GSI [2] revived the hope that the axion may indeed exist, but with properties that allow it to evade detection in previous direct searches. The "standard axion model" [3] describes the axion as a pseudoscalar with mass $m_a \approx 0.025\ N\ (1/X + X)$ MeV/c^2, where N is the number of quark generations and X is a free parameter equal to the ratio of two Higgs field vacuum expectation values. Depending on its mass, the axion decays into two gammas or into positron-electron pairs. The coupling strength to quarks and leptons is a function of X, and thus the axion's production and decay properties are predicted in a myriad of physical processes. The possibility of discovering the axion has led to a number of fruitless searches. The most compelling evidence against the standard axion comes from searches for $\psi \rightarrow \gamma + a$ and $\Upsilon \rightarrow \gamma + a$ [4]. These experiments are complimentary, the rate for $\psi \rightarrow \gamma + a$ goes like X^2 and $\Upsilon \rightarrow \gamma + a$ like $1/X^2$, and the limits from the two experiments essentially rule out the axion for all X. The interpretation of these experiments is complicated slightly if the lifetime is very short and the axion decays near the interaction point in these high energy e^+e^- annihilation searches [5]. Failure to observe axions coming from nuclear transitions provides additional negative evidence [6]. The nuclear experiments are each sensitive only to limited regions of X, and to relatively long lifetimes, say $\tau > 10^{-10}$ sec. In addition, most nuclear searches concentrate on axions with masses less than 2 m_e.

If the axion explains the narrow peaks in the GSI experiments it would have a mass around 1.7 MeV. Experiments are compatible with a lifetime in the range 10^{-19}-10^{-10} sec, but theoretical arguments favor $\tau \approx (1\text{-}5) \times 10^{-13}$ sec. The standard axion at $m_a \approx 1.7$ MeV/c^2 would have $\tau \approx 5 \times 10^{-12}$ sec, and this is ruled out by limits on $\Upsilon \rightarrow \gamma + a$. Motivated by the GSI results and negative axion searches Peccei, Wu and Yanagida [7], and Krauss and Wilczek [8] have considered variations of the standard axion model. Briefly these models have two main characteristics: (1) The axion couples essentially only to light quarks, mitigating the negative evidence from vector meson decay searches; and (2) The axion has a very short lifetime for decay into positron-electron pairs, explaining why it was not seen in nuclear decay searches, and making it a candidate for the GSI particle. With these models as a guide one is lead to search for axion emission from nuclear transitions, but in experiments sensitive to axions that decay to e^+e^- with a short lifetime ($\tau < 10^{-10}$ sec). All the models predict that axion decay will compete favorably with particular M1 nuclear γ-ray transitions [3]. Thus we have searched for axions emitted in the decay of the 2.22 MeV state of the deuteron. This simple isovector M1 transition is an ideal place to search for axion-like particles of the type suggested by the GSI result. The various axion models predict the branching ratio, Γ_a/Γ_γ, for axion decay relative to γ-decay to be at the 10^{-3} level.

2. The Experiment

The experiment is straightforward. We attempt to detect e^+e^- pairs from the sequence of reactions $n + p \rightarrow D^*(E_x = 2.2245 \text{ MeV}) \rightarrow D + a$ and $a \rightarrow e^+ + e^-$. Figure 1 shows the experimental arrangement. A cold neutron beam from the Institute Laue-Langevin reactor intersects a polyethelene target (≈ 0.21 mm thick) at the center of 2 m long superconducting solenoidal spectrometer (PERKEO). The spectrometer has previously been used to measure the β-asymmetry of the neutron [9] and more recently to measure the neutron lifetime. In the present experiment e^+e^- pairs are detected in coincidence with plastic scintillator detectors (5 mm thick) at either end of the spectrometer. The 15 kgauss solenoidal field causes low energy electrons and positrons to move in helical paths with diameters less than 1 cm. Trim coils at the ends of the solenoid pervert the field causing the particle trajectories to intersect the scintillators which are outside of the neutron beam. The system is calibrated with internal conversion sources and the energy resolution corresponds to about 120 photoelectrons/MeV.

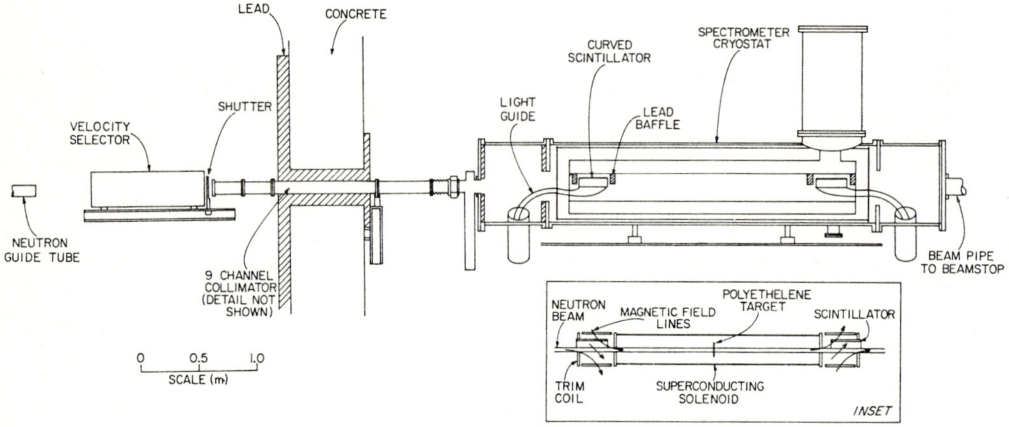

Figure 1. Layout of the axion search experiment. The insert shows the details of the inner volume of PERKEO

The expected signal for axion decay is a peak in the two detector sum energy spectrum at ≈ 1.2 MeV ($E_x - 2m_ec^2$). The main background in the experiment is direct internal pair conversion. Because of the low transition energy in the deuteron the internal pair conversion probability is small, $\Gamma_p/\Gamma_\gamma = 3.33 \times 10^{-4}$. Moreover, the rate is reliably predicted, and since $Z = 1$ the Born approximation is adequate [10]. There are three additional sources of coincidence background: (1) Cosmic rays liberating δ-rays from one scintillator that moves through the spectrometer causing a coincidence; (2) Electrons from neutron decay that backscatter off a detector; and (3) Background γ-rays that liberate electrons from one detector by Compton scattering. These additional backgrounds can be partially identified by detector timing information but this feature is not used in the present preliminary analysis. Figure 2 is the sum energy spectrum from an 18 hour run of the experiment. The neutron beam capture flux is $\phi_n = 3.9 \times 10^7$ neutrons/sec, determined by Cobalt activation analysis and observation of the neutron β-decay rate. The peak in Fig. 2 is at the right energy for either axion decay or internal pair conversion after accounting for elec-

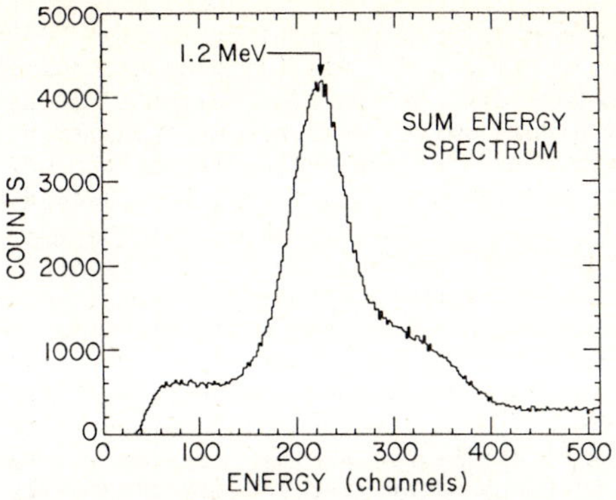

Figure 2. Sum energy spectrum for an 18 hour run. The distortion to the high energy side of the peak at 1.2 MeV is from positron annihilation in the detector

tron (and positron) energy loss in the target foil. The distortion to the peak on the high energy side is also expected. Since one of the detected particles is a positron that annihilates in a detector, one or both annihilation photons can Compton scatter, increasing the observed energy of the event.

At low energy (below ≈ 1/2 MeV) the background in Fig. 2 is primarily from neutron β-decay backscattering. The background from cosmic rays and γ-rays (mostly from Al capture γ-rays coming from target scattered neutrons) is rather smooth up to the largest detectable energy. To account for internal pair decay, we performed a Monte Carlo calculation to determine the experimental acceptance corrected for backscattering in the target foil. The effective backscattering coefficient was measured with a ^{207}Bi conversion source. The error from background subtraction, backscattering corrections, and uncertainties about the exact target composition dominates the error the present experiment error. A more elaborate analysis, now underway, and a chemical analysis of the target promises to reduce the error significantly.

3. Results

The e^+e^- detection rate we expect for pair decay is given by the expression, $N_p = \phi_n \cdot \sigma_o \cdot (\Gamma_p/\Gamma_\gamma) \cdot N_H \cdot \varepsilon_p$, where σ_o is the thermal capture cross section for $H(n,\gamma)D$, N_H is the number of target hydrogen atoms per cm^2, and ε_p (≈ 0.37) is the calculated detection efficiency which accounts for the angular correlation of the pairs and target backscattering. We obtain $N_p = 2.42 \pm 0.24$; the measured rate is $N_p(expt) = 2.58 \pm 0.16$. These results are consistent within errors and we construct a 2σ upper limit for the axion emission branching ratio, Γ_a/Γ_γ. Figure 3 shows the experimental limit compared to various theoretical expectations. The expected rate for axion decay is given by the expression,

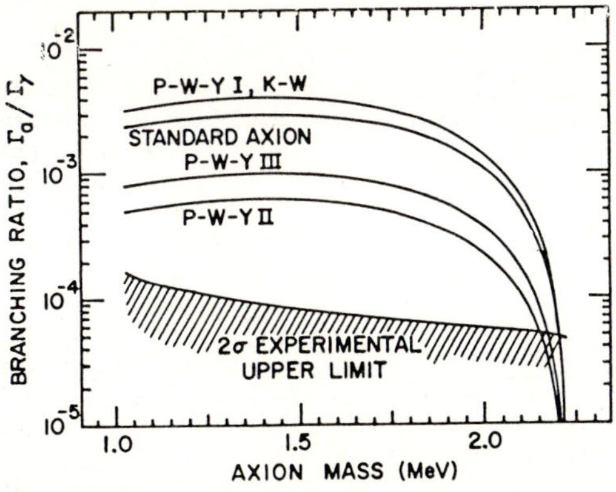

Figure 3. The preliminary experimental upper limit on Γ_a/Γ_γ as a function of axion mass. The theoretical curves are obtained from the various axion models: K-W Ref. 8; P-W-Y I-III Ref. 7; and the standard axion with N=3 from Ref. 3. There is some loss of experimental sensitivity just above 2 m_e because the axion lifetime would be longer than 10^{-10} sec, but this mass region is not significant

$$\Gamma_a/\Gamma_\gamma \approx 1/2\ \tilde{\alpha}/\alpha\ (k_a/k_\gamma)^3(\rho^{(1)}/4.71)^2\ [3]\ ,$$

where k_i are the momenta, $\tilde{\alpha}$ is a coupling constant, the analog of α, and $\rho^{(1)}$ is a measure of the axions isovector coupling to nucleons which is a function of X that depends on the details of the particular model. The predictions corresponding to the models described in Refs. 3, 7 and 8 are plotted in Fig. 2 as a function of axion mass. The mass dependence of the experimental limit reflects the kinematic dependence of the acceptance. All the models are ruled out for the range of masses of interest for the GSI experiments. In general the present experiment is sensitive to hypothetical particle emission in the range $2m_e < m_a < E_x/c^2$. We would also detect particles with lifetimes up to $\approx 10^{-10}$ sec. Hypothetical particles with longer lifetimes would begin to leave the volume of the spectrometer before decaying.

This research supported by the U.S. Department of Energy, Nuclear Physics Division, under contract W-31-109-ENG-38 and by the Bundesministerium für Forschung und Technologie.

References

[1] R. D. Peccei and H. R. Quinn, Phys. Rev. Lett. 38, 1140 (1970); Phys. Rev. D16, 1991 (1977); S. Weinberg, Phys. Rev. Lett. 40, 223 (1978); F. Wilczek, Phys. Rev. Lett. 40, 249 (1978).

[2] J. Schweppe et al., Phys. Rev. Lett. 51, 2261 (1983); M. Clemente et al., Phys. Lett. 137B, 41 (1984); T. Cowan et al., Phys. Rev. Lett. 54, 761 (1986); see also a. B. Balantekin et al., Phys. Rev. Lett. 55, 461 (1985) and A. Schäfer et al., J. Phys. G11, 169 (1978).

[3] T. W. Donnelly et al., Phys. Rev. D18, 1607 (1978); W. A. Bardeen and S. H. H. Tye, Phys. Lett. 74B, 229 (1978).

[4] C. Edwards et al., Phys. Rev. Lett. 48, 903 (1982); M. Sivertz et al., Phys. Rev. D26, 717 (1982); M. S. Alam et al., Phys. Rev. D27, 1665 (1983); Niczyoruk et al., Z. Phys. C17, 197 (1983).

[5] G. Mageras et al., Phys. Rev. Lett. 56, 2672 (1986); T. Bowcock et al., Phys. Rev. Lett. 56, 2676 (1986).

[6] For a review see A. Zehnder, in Fundamental Intreractions in Low-Energy Systems, Eds. P. Dalpiaz, G. Fiorentini and G. Torelli (Plenum Press, New York 1985), p. 337.

[7] R. D. Peccei, T. T. Wu and T. Yanagida, DESY preprint 86-013 (1986).

[8] L. M. Krauss and F. Wilczek, Yale Physics Dept. preprint YTP 86-03 (1986).

[9] P. Bopp et al., Phys. Rev. Lett. 56, 919 (1986).

[10] M. E. Rose, Phys. Rev. 76, 678 (1949).

Search for Short-Lived Axions in a Nuclear Isoscalar Transition

F.W.N. de Boer [a], *K. Abrahams* [b], *A. Balanda* [c], *H. Bokemeyer* [d], *R. van Dantzig* [e], *J.F.W. Jansen, B. Kotlinski, M.J.A. de Voigt, and J. van Klinken*

Kernfysisch Versneller Institut, University of Groningen, NL-9747 AA Groningen, The Netherlands

Axions, if existing, can cause a strong signal of positron-electron pairs in isoscalar M1 transitions in competition with gamma-ray emission. We have searched for such a signal in the 3.59 MeV transition in ^{10}B with a fourfold Mini-Orange spectrometer. No axion events were found within one percent of the prediction for the standard axion- to γ-ray branching ratio.

A light pseudoscalar Higgs particle, the axion, has been introduced by Weinberg[1)] and Wilczek[2)], as a consequence of the global U(1) symmetry proposed by Peccei and Quinn[3)] to explain the absence of large P and CP violations in strong interactions. In a treatise of possible axion search experiments, Donnelly et al.[4)] discussed the probability for nuclear deexcitation by axion emission in competition with magnetic γ-radiation. Subsequent axion searches provided such stringent upper bounds on the pseudoscalar particle mass and lifetime that the non-existence of axions was anticipated [5)]. However, the interest in an observable axion was revived recently[6)] after the puzzling observation in superheavy collisional systems at GSI of a sharp positron peak at 330 keV, which was later[7)] found to be in coincidence with a similar electron peak. Could the observed (e^+e^-) pairs arise from a "visible "axion with a mass of about 1.68 MeV and a lifetime $\leq 10^{-13}$ s (refs 8,9,10)?

The axion mass and lifetime can be related to one free parameter X, the ratio of vacuum expectation values for two Higgs field components f.sin λ and f.cos λ with $f = (G_F 2^{1/2})^{-1/2} =$ 250 GeV (G_F being the Fermi coupling constant) :

$$m_a = 75\,(X + 1/X) \qquad \text{(keV)} \tag{1}$$

and for $m_a > 2m_e$ when the axion will predominantly decay in a positron-electron pair:

$$\tau(a \to e^+e^-) = 8\pi f^2 X^2 m_e^{-2}\,(m_a^2 - 4m_e^2)^{-1/2} \qquad \text{(sec)} \tag{2}$$

Searches for axions decaying into two photons with radioactive sources, reactors and beam dumps constrain[9)] the mass to > 1 MeV and the lifetime to < 10^{-6}s, whereas a search[11)] for decay into positron-electron pairs limits the lifetime to < 2.10^{-10}s. Searches with pseudo scalar- and vector mesons (K^+, J/ψ and Y) were sensitive to lifetimes of 10^{-11}s due to time dilation of the order of 10^4. The anomalous magnetic moment of the electron constraints the lifetime to > 5.10^{-13}s [9,10)]. In summary, a rather narrow window of lifetimes between $10^{-13}$and 10^{-10}s cannot be excluded on the basis of existing experimental evidence . If the observed positron peak is due to an axion with mass 1.68 MeV, its X value according to eq. (1) is either X = 0.045 or X = 0.0422 leading to a lifetime of about 5 x 10^{-12}s (allowed) or 2.10^{-6}s (excluded). Either of these two X-values would lead to contradictions for branching ratios of meson decays in the standard

---

a. Present address: NIKHEF-K, P.O. Box 4395, 1009 AJ Amsterdam, The Netherlands
b. Energieonderzoek Centrum Nederland (ECN), Petten, The Netherlands
c. Physics Laboratory of the Free University , Amsterdam, The Netherlands and Jagiellonian University , Krakow, Poland
d. Gesellschaft für Schwerionenforschung(GSI), D-6100 Darmstadt, Federal Republic of Germany
e. NIKHEF-K, Amsterdam, The Netherlands

axion model. Quite recently, however, Peccei et al. [12] have succeeded to reconcile the theory with results from previous unsuccessful searches with an axion of 1.68 Mev in a "viable" axion model and stressed together with a.o. Brodsky et al. [13] the need of new experiments.

In this letter we report on a search for axions with lifetimes in the unexplored region ($\tau < 10^{-9}$ s). We looked for the $e^+ e^-$ decay of such axions in competition with a magnetic γ-ray transition. Specifically for isoscalar transitions, the branching ratio for (e^+e^-) pair creation from a pseudoscalar decay particle and M1 gamma- decay can be as large as 40%. This is three orders of magnitude larger than the internal pair creation (IPC). As isoscalar transition the 3.59 MeV M1 + E2 transition in ^{10}B was chosen.

The 2^+, 0 state at 3.59 MeV in ^{10}B decays to the 3^+, 0 groundstate with a 19% branching ratio. The transition has a mixing ratio[14] δ(M1/E2) of 1.5 ± 0.6 while the parallel $3^+,0 \rightarrow 1^+,0$ transition of 2.87 MeV seems [14] to be predominantly E2 with ($\delta^{-1} < 0.2$). With the transition energy fixed at 3.59 MeV, the energy and angular correlations of (e^+e^-) pairs is fixed kinematically. In our experimental arrangement the M1 and E2 IPC is detected with total efficiency ε of typically O(10^{-4}), see below. The IPC angular correlation between the e^+ and e^- is maximum at relative angle $\theta = 0^o$, while in case of the axions the maximum will occur around $\theta = 45^o$. The angle between the axes of neighbouring Mini-Oranges is 60° (fig. 1), which causes that an axion decay would be seen with somewhat higher efficiency than IPC; a numerical estimate gave $\varepsilon_A \approx 1.5\varepsilon$, assuming that the axions are emitted isotropically. The nuclear axion-decay rate(in case of existing axions) is predicted by Donnelly et al. [4] to be as large as $I_a = 0.39\, I_\gamma$(M1), whereas the theoretical IPC coefficients α_π have been obtained from ref. 15. The number of observed pairs can now be written as:

$$N(e^+e^-)/\varepsilon = \{0.39 \times 1.5 + \alpha_\pi(M1)\}\, I_\gamma(M1) + \alpha_\pi(E2) I_\gamma(E2) \qquad (3)$$

where ε is the efficiency. If axions indeed exist then the first term (= 0.585) will dominate since the others are of O(10^{-3}).

In the experiment the 3.59 MeV state in ^{10}B was excited in inelastic scattering of 7 MeV protons by a ^{10}B target of 1 mg/cm^2. Using a fourfold Mini-Orange system[16] with two detector systems for e^+ and two for e^- detection, we registered prompt (e^+e^-) signals and their energy sums. The four spectrometers were set at a backward angle of 45° with respect to the incident proton beam. This is shown by a view in the horizontal plane in fig. 1(left). The axes of opposite Mini-Oranges (with the same polarity) made an angle of 90° with each other, those of neighbouring ones (with alternating polarity) 60° with each other. For clarity fig. 1 (right) shows the four equal Mini-Orange systems, each composed of four magnets (e^+ and e^- systems being only different by inversion of their magnetic fields). In reality the present experiment was limited by the available

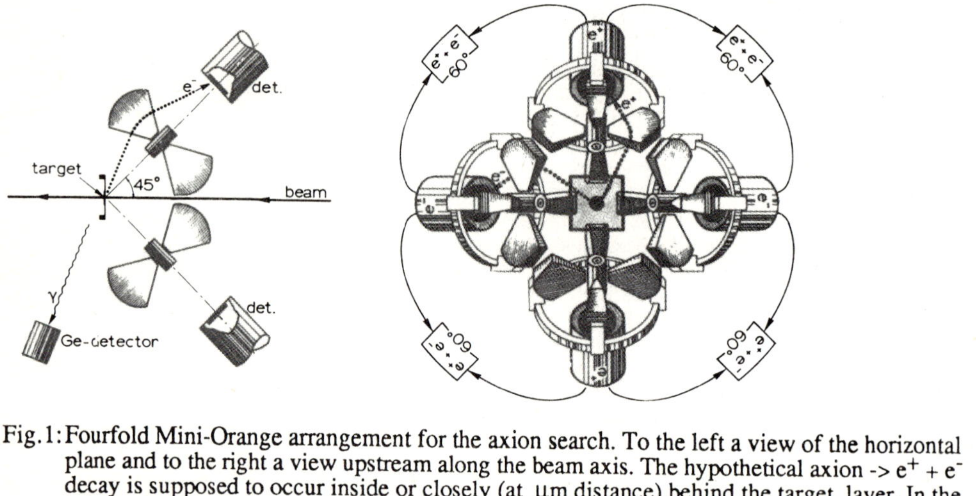

Fig.1: Fourfold Mini-Orange arrangement for the axion search. To the left a view of the horizontal plane and to the right a view upstream along the beam axis. The hypothetical axion -> $e^+ + e^-$ decay is supposed to occur inside or closely (at μm distance) behind the target layer. In the left part of the figure an e^+ is indicated, while the corresponding e^- particle will be detected by a neighbouring e^- detector.

magnets so that two four-gap and two five-gap systems had to be used. Since the latter magnets were relatively weaker, the overall features of the four magnet systems were similar, with maximum transmissions around 1.3 MeV.

The Mini-Orange spectrometers were equipped with 300 mm^2 Si(Li) detectors: two with a thickness of 3 mm and two with a thickness of 5 mm. With the KVI data acquisition system (VAX 750) the twofold coincidences were registered on magnetic tape for off-line analysis.

Figure 2 (top) displays the essential result: the sum of the (e^+e^-) pair spectra from the four neighbouring detector combinations . This spectrum is compared with a simultaneously recorded spectrum of gamma rays (bottom) using a Ge detector. The internal pair peaks of the two isoscalar transitions at 2.87 MeV and 3.59 MeV depopulating the 3.59 MeV level in ^{10}B are only weakly visible.

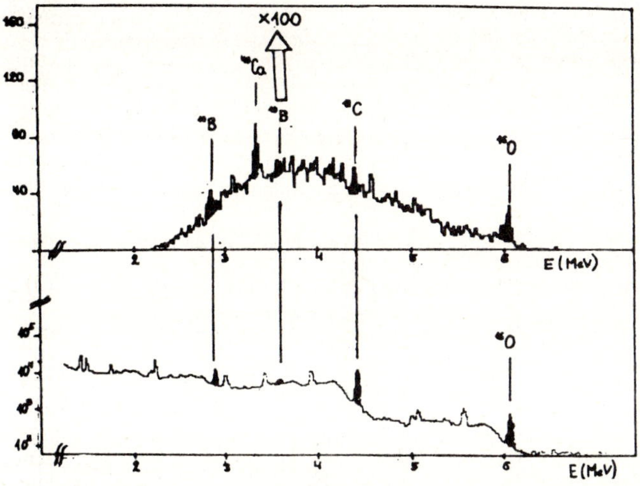

Fig.2: Sum of $e^+ + e^-$ pair spectra (top) compared with simultaneously recorded γ-ray spectrum (bottom) using a Ge detector, following the ^{10}B (p,p') reaction at 7 MeV. In case of an existing axion with mass of 1.68 MeV and life time 5×10^{-12}s, the prediction by Donnelly et al [4)] corresponds to a peak at 3.59 MeV with height of over 1000 counts.

Interestingly, two rather prominent peaks at 3.35 and 6.05 meV show up in the sum spectrum. They had no counterpart in the γ-rays spectrum and belong to totally converted E0 transitions in ^{40}Ca and ^{16}O respectively, which are to our knowledge the only two existing $E0(0^+_2 - 0^+_1)$ transitions between 3 and 7 MeV in nature. The spectra of scattered protons, recorded separately, indeed showed a significant content of ^{16}O (and ^{12}C) impurity in the target. A tracer impurity causes the appearance of the ^{40}Ca, E0 transition in the pair spectra. The 6.05 MeV transition in ^{16}O occurs just above energy threshold and also in the extreme upper transmission range of the Mini-Oranges. Furthermore it is observed despite the fact that the two 3 mm Si(Li) detectors had insufficient stopping power for most of the e^+ and e^- pulses. Evidently, the arrangement is hyper sensitive for detection of such totally converted E0 transitions from first-excited 0^+ states.

The background of the pair spectrum is complex, containing a continuum distribution following $^{10}B(p,\alpha)$, distributions from pairs created inside the vacuum walls, e^+ and e^- particles from e.g. the E0(^{16}O) transitions which pass through the detectors. Yet, the main part of the spectrum belongs to IPC type coincidences and not to e.g. Compton scattered γ-rays since the yield of $e^+ + e^+$ and $e^- + e^-$ combinations is less by an order of magnitude than the $e^+ + e^-$ pairs. In the γ-ray spectrum all transitions of sufficiently short half-life are Doppler broadened and/or shifted. The ^{10}B 2.87 and 3.59 MeV γ-rays show up strongly . The spectrum also shows a number of ^{27}Al γ-rays produced in the walls of the Mini-Orange vacuum chamber, together with the 4.43 MeV E2 transition in ^{12}C and the 6.13 MeV E3 transition in ^{16}O (target impurities).

A comparison of the intensity of inelastically scattered protons from the 4.43 MeV level in ^{12}C with those from the 3.59 MeV level in ^{10}B showed us that only a fraction of the corresponding

γ-ray intensity in ^{12}C is produced in the target and apparently most of it in the walls of the chamber. From the ratio the target γ-ray intensity was determined using experimental data for differential and total cross sections for ^{10}B (p,p') (ref. 16) and ^{12}C (p,p') (ref. 17) inelastic scattering at 7 MeV under 135°. It was used together with the intensity of the internal pair creation and the theoretical [14] IPC of 1.3×10^{-3} for an E2 transition of 4.43 MeV to internally determine the efficiency of the M.O. system at 3.41 MeV. At the low energy side the system was internally calibrated at 1.85 MeV from the 2.87 MeV transition in ^{10}B ($\geq$ 98% E2) and externally at 1733 keV using the 2755 keV E2 transition in ^{24}Na with theoretical [14] IPC's of 7.4×10^{-4} respectively 7.0×10^{-4}. All three efficiency calibrations are equal to a value of $\varepsilon = 2.3 \times 10^{-4}$ within their error bars of 35% .

The isoscalar 3.59 MeV transition has in the (e^+e^-) pair spectrum roughly the same intensity as the parallel 2.87 MeV transition used for internal calibration. The branching ratio of the 3.59 pair peak was determined to be $(2.1 \pm 1.1) \times 10^{-3}$ which is consistent with internal pair creation, α_π being 9.9×10^{-4} for a 28% M1 and 72% E2 transition. The full strength of axion decay with the theoretical value of $I_a = 0.39\ I_\gamma$ (M1) for an axion of 1.68 MeV would have caused a peak height of over 1000 counts or two orders of magnitude (a factor 136) larger than the observed value.

Generally, our result shows that any axions with masses between 1 and 3 MeV and lifetimes between 2×10^{-11} and 2×10^{-12}s (see formulas 1 and 2), are at least hindered with respect to the prediction by Donnelly et al. [4] with factors between 180 respectively 30. Consequently our search yields no evidence for the existence of a viable and visible standard axion.

After evaluation of our experimental data, we became aware of IPC measurements on ^{10}B in the early decades of nuclear physics, when axions were not yet advocated, but when ($e^+ e^-$) angular correlations were used as a probe for distinguishing multipolarities of high-energy γ-rays. We were both surprised and impressed by the angular correlations reported for the 3.59 MeV transition in the clean reaction $^9Be(d,n)^{10}B$ by Gorodetsky et al.[18].They used scintillation counters with inherently two orders of magnitude less energy resolution than our Si(Li) detectors (partly spoiled by Doppler broadening), and could hardly separate a strong 3.37 MeV transition in ^{10}B from the 3.59 MeV isoscalar transition in ^{10}B. Yet, their angular distribution shows no enhancement near 45° where the standard axion would have given a one to two orders of magnitude larger signal than measured, as discussed above. This report in old literature can be interpreted as testifying against the existence of axions, but (to our knowledge) it has been overlooked in recent surveys on existing evidence with respect to axions.

This investigation was supported by the Netherlands Organization for Pure Scientific Research (ZWO) through the Foundation for Fundamental Research on Matter (FOM).

1. S. Weinberg, Phys. Rev. Lett. 40 (1978) 223
2. F. Wilczek, Phys. Rev. Lett. 40 (1978) 279
3. R.D. Peccei and H.R. Quinn, Phys. Rev. Lett. 38 (1977) 1440
4. T.W. Donnelly et al., Phys. Rev. D18 (1978) 1607
5. A. Zehnder, SIN report PR 84-08A
6. T. Cowan et al., Phys. Rev. Lett. 54 (1985) 1761
7. T. Cowan et al., Phys. Rev. Lett. 56 (1986) 444
8. A. Schäfer et al., J. Phys. G11 (1985) 169
9. A.B. Balantekin et al., Phys. Rev. Lett. 55 (1985) 461
10. N. Mukhhopadhyay and A. Zehnder, Phys. Rev. Lett. 56 (1986) 206
11. F.P. Calaprice et al., Phys. Rev. D20 (1979) 2708
12. R.D. Peccei, Tai Tsun Wu and T. Yanagida, Phys. Lett. B172 (1986) 435
13. E.K.Warburton et al., Phys. Rev. 17 (1968) 1178
14. P. Schlüter and G. Soff, Atom. Data and Nucl. Data Tables 24 (1979) 509
15. J. van Klinken, S.J. Feenstra and G. Dumont, Nucl. Instr. and Meth. 151 (1978) 433
16. B.A.Watson et al., Phys. Rev. 187 (1969) 1351
17. J.B. Swint et al., Nucl. Phys. 86 (1966) 119
    A.C.L. Barnard et al., Nucl. Phys. 86 (1966) 130
18. S. Godoretsky, P. Chevallier, R. Armbruster and G.Sutter, Nucl. Phys. 12 (1959) 349

3.5.2 Theory of GUT's, SUSY's, Superstrings

The Present Status of Proton Decay and Baryon Number Nonconservation

P. Langacker

Department of Physics, University of Pennsylvania,
Philadelphia, PA 19104, USA

The present status of ordinary and supersymmetric grand unified theories is reviewed. The theoretical motivations, successes, and difficulties, and the expectations for proton decay and other baryon number nonconserving processes are described. Although the minimal SU_5 model is ruled out by the nonobservation of proton decay, many plausible modifications and extensions lead to longer lifetimes and/or different decay modes, often in an experimentally accessible range.

The Standard Model [1,2]

The standard $SU_3 \times SU_2 \times U_1$ model (QCD plus the Glashow-Weinberg-Salam [3] electroweak theory) has been spectacularly successful. It is a mathematically consistent renormalizable field theory that either predicts or is compatible with all known facts in particle physics except possibly for the anomalous positron events recently observed at Darmstadt. In particular, the standard model successfully predicted the existence and detailed form of the neutral current interaction, the W and Z boson masses, and the existence of the charm quark (in order to avoid strangeness changing neutral currents). The standard model supplemented with classical general relativity is almost certainly an approximately correct theory of nature, with a range of validity including ordinary terrestrial and astrophysical conditions and accelerator energies up to several tens of GeV. Despite these successes, the standard model cannot be considered the ultimate description of Nature - it is simply too complicated and arbitrary and leaves too many fundamental questions unanswered.

These difficulties can be summarized under five headings.

a) The gauge problem: the standard model gauge group is a complicated direct product of three groups with three distinct coupling constants. Furthermore, because of the U_1 factor the average electric charges for the particles in an SU_2 multiplet are essentially arbitrary except for two constraints from the cancellation of anomalies. Hence there is no fundamental explanation for the observed quantization of fermion and boson charges in multiplets of e/3 (or therefore for the equality of the magnitudes of the electron and proton charges.)

b) The fermion problem: the fermions are assigned to a complicated reducible representation of the $SU_3 \times SU_2 \times U_1$ group. No explanation is given for the existence or number of fermion families. Furthermore, neither the fermion masses, which are observed to vary over a range of five orders of magnitude, nor the fermion mixing angles are predicted by the theory: they must be taken from experiment.

c) The Higgs problem: $SU_2 \times U_1$ symmetry is broken by the introduction of one or more fundamental Higgs multiplets ϕ, which should have mass^2 parameters μ_ϕ^2 that are not too much larger in magnitude than M_W^2. However, μ_ϕ^2 receives quadratically divergent corrections from gauge, Higgs, and fermion loop diagrams of order

$$\delta\mu_\phi^2 = O(g^2, \lambda, h^2)\Lambda^2, \tag{1}$$

where g, λ, and h represent gauge, quartic Higgs, and Yukawa couplings, respectively, and Λ is the next highest mass scale in the theory above the weak scale (at which the otherwise divergent integrals are presumably cut off.) In the standard model the only higher mass scale is the Planck mass $m_p = G_N^{-1/2} \cong 10^{19}$ GeV. Hence an incredibly accurate cancellation (fine-tuning) between the bare value of μ_ϕ^2 and the correction is needed.

d) The strong CP problem: CP and T violation associated with nonperturbative instanton effects in the strong (QCD) sector of the theory will lead to an unacceptably large value for the neutron electric dipole moment unless a parameter θ_{QCD} is fine-tuned to a value $< 10^{-9}$.

e) The graviton problem: finally, I come to the graviton problem, which has several aspects. First, gravity is not unified with the other interactions in the standard model in a fundamental way. Secondly, even though general relativity can be incorporated into the model by hand, we have no idea how to achieve a mathematically consistent theory of quantum gravity: attempts to quantize gravity within the standard model framework lead to horrible divergences and a non-renormalizable theory.

Finally there is yet another fine-tuning problem associated with the cosmological constant. The vacuum energy density associated with the spontaneous symmetry breaking of $SU_2 \times U_1$ generates an effective renormalization of the cosmological constant $\delta\Lambda = 8\pi G_N\langle V\rangle$, which is about 50 orders of magnitude larger than the observed value. One must fine-tune the bare cosmological constant against the correction to this incredible degree of precision.

Another characterization of many of these problems is that the standard model with massless neutrinos, three fermion families, and the minimal Higgs structure (one doublet) has nineteen free parameters. These are three gauge couplings, two CP-violating θ parameters (θ_{QCD} and an analogous parameter in the weak sector of the theory), nine fermion masses, three Kobayashi-Maskawa-Cabibbo (KMC) mixing angles, one CP-violating KMC phase, and the W and Higgs scalar masses, minus one overall mass scale. If one includes classical gravity one must add the Planck mass and the (observationally tiny) cosmological constant to the list.

As successful as the standard model may be, it is almost certainly not the ultimate description of Nature!

Grand Unified Theories (GUTs) [4]

Grand unified theories are models in which $SU_3 \times SU_2 \times U_1$ is embedded in a simple group G with a single gauge coupling constant g_G. One hopes that the additional symmetries in G will constrain some of the arbitrary features of the standard model subgroup.

The simplest and most popular GUT is the Georgi-Glashow SU_5 model. [5] The fermion representations are still rather complicated in SU_5: each family is assigned to a reducible $5^* + 10$ dimensional representation:

$$
\begin{array}{ccc}
5^* & & 10 \\
{\scriptstyle \uparrow W^{\pm} \downarrow}\begin{pmatrix} \nu_e & \\ & \bar{d} \\ e^- & \end{pmatrix}_L & &
\begin{pmatrix} & u & \\ e^+ & & \bar{u} \\ & d & \end{pmatrix}_L \\
\leftarrow X,Y \rightarrow & & \leftarrow X,Y \rightarrow
\end{array}
\qquad (2)
$$

where the color indices have been suppressed. In addition to the twelve generators and corresponding gauge bosons ($W^{\pm}, Z, \gamma$, 8 gluons) of the standard model, there are twelve new bosons associated with transformations between adjacent columns in (2). These are the color antitriplet bosons X and Y, which form a nearly degenerate SU_2 doublet with charges 4/3 and 1/3, respectively, and their antiparticles. They can mediate proton and bound neutron decay, as shown in Fig. 1, and therefore should have masses $\gtrsim 10^{14}$ GeV.

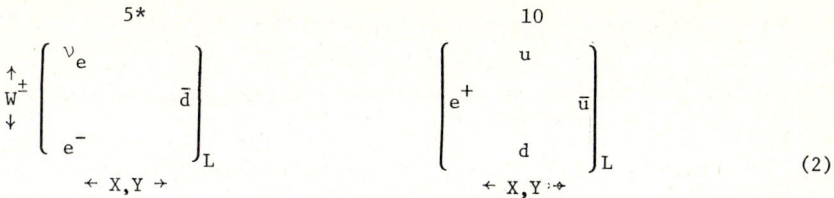

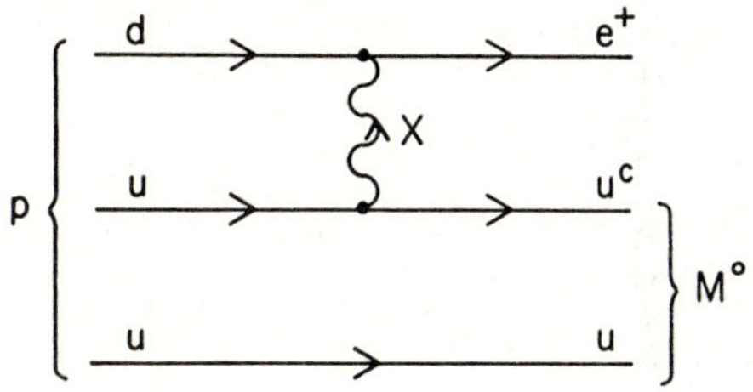

Fig. 1. A typical proton decay diagram.

In SO_{10} the $5^* + 10$ is combined with a Majorana neutrino $\bar{N}_L$ to form a 16-plet. In E_6, each family is placed in a 27-plet, which decomposes into a $16 + 10 + 1$ of SO_{10}. The singlet is a Majorana neutrino S^o and the 10 consists of $(D, \bar{D}, E^{\pm}, E^o, \bar{E}^o)_L$, where D, E^- and E^o have the $SU_3 \times U_1$ quantum numbers of the d, e^-, and ν, respectively. All of the new fermions are presumably heavy (e.g. in the 100 GeV range).

GUTs are the most successful in dealing with the gauge problem, because there is only one underlying interaction. Charge quantization follows from the fact that the electric charge operator Q is a (traceless) generator of the gauge group. Hence, the sum of the charges of all particles in a multiplet (which generally includes both quarks and leptons) must vanish. In the simpler GUTs this implies the desired relation between the quark and lepton charges. However, the correct relation does not follow uniquely in all theories because there may be more than one way to embed Q in the theory (for some amusing recent speculations see Okun et al.[6]) or because there may be additional fermions in the multiplet.

Coupling Constants

Grand unified theories have only one underlying gauge coupling constant. Therefore, the properly normalized running coupling constants g_s, g, and $\sqrt{5/3}\, g'$ of SU_3, SU_2 and U_1, respectively, should all come together at the unification scale M_X, above which spontaneous symmetry breaking can be ignored, as shown in Fig. 2. If there are only two mass scales (M_W and M_X) in the theory then one can use the observed ratio α/α_s at low energy to predict M_X^2 and the weak angle $\sin^2\theta_W \sim g'^2/(g^2 + g'^2)$ + h.o.t. In addition to the assumption of only two scales, one needs to know: (a) the quantum numbers of all of the light (mass << M_X) particles in the theory (to compute the renormalization group coefficients), and (b) the quantum numbers of a full G multiplet (in order to properly normalize the generators - i.e. to calculate the $\sqrt{5/3}$ factor multiplying g').

Given these assumptions the biggest uncertainty is in the value of the strong scale parameter $\Lambda_{\overline{MS}}$ related to α_s by

$$\alpha_s(Q^2) = \frac{6\pi}{33-2n} [\ell n(Q/\Lambda^{(n)}_{\overline{MS}})]^{-1} + \text{h.o.t.} \tag{3}$$

where n is the number of quark flavors that are light compared to Q. There are a number of ways to determine $\Lambda_{\overline{MS}}$ (e.g. deep inelastic scattering, $e^+e^- \to$ jets, Υ widths) but each has serious uncertainties [7]. For example, Lepage [8,9] has given the average

$$\Lambda^{(4)}_{\overline{MS}} = 100 \begin{smallmatrix}+100 \\ -50\end{smallmatrix} \text{ MeV} , \tag{4}$$

while Duke and Roberts [7] have found the somewhat larger range 200 - 250 MeV.

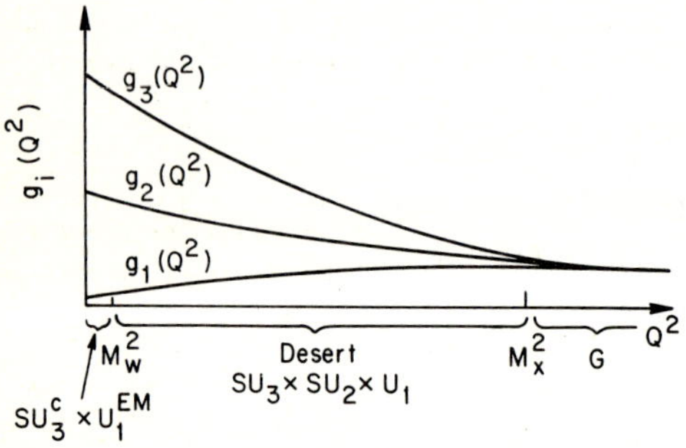

Figure 2. The momentum dependent coupling constants $g_3 \equiv g_s$, $g_2 \equiv g$, and $g_1 \equiv \sqrt{5/3}\, g'$.

(Other recent determinations include 186 ± 60 MeV from deep inelastic scattering [10], 193 ± 43 MeV from the photon structure function F_2^γ [11] and 300 ± 60 MeV from heavy quarkonium states [12]). For definiteness I will use the value for $\Lambda^{(4)}_{\overline{MS}}$ given in (4), but larger values are not entirely excluded.

For three fermion families, a single Higgs doublet, and m_t = 50 GeV one predicts [9]

$$M_X \simeq 1.3 \times 10^{14} \text{ GeV} \left[\frac{\Lambda^{(4)}_{\overline{MS}}}{100 \text{ MeV}} \right] (1.5)^{\pm 1} \tag{5}$$

and

$$\sin^2\theta_W = .218 \pm 0.003 + 0.006\, \ell n \left[\frac{100 \text{ MeV}}{\Lambda^{(4)}_{\overline{MS}}} \right] \tag{6}$$

where the logarithmic term in (6) is $\mp$ 0.004 for $\Lambda_{\overline{MS}}$ in the range (4). The prediction in (6) is in excellent agreement with the experimental value[1] 0.229 ± 0.003. This is a major success for a wide class of models. Unfortunately, most of these models also predict an unacceptably short proton lifetime of order 10^{29} years.

In many extended or modified models with a longer proton lifetime $\sin^2\theta_W$ is a free parameter, so that the success of (6) would be an accident. I will take the pragmatic point of view that maintaining the approximate validity of (6) is a constraint on modified models.

[1] This is a preliminary value of a global fit to existing neutral current data and the W and Z masses [13]. The quoted error is experimental only. Theoretical uncertainties have not yet been calculated in detail, but are likely to be around 0.005.

In many of the simplest GUTs (such as minimal SU_5) m_τ and m_b are generated by the same Yukawa coupling and are equal when evaluated at M_X. One then successfully predicts [14] m_b (5 GeV) $\simeq$ 5 GeV for three families, while m_b is too large for four or more families.

Unfortunately, these models also predict $m_s \simeq 500$ MeV, which is larger than most phenomenological determinations [4] (100 - 200 MeV). Even worse, one has $m_s/m_d = m_\mu/m_e \simeq 200$, while current algebra determinations yield $m_s/m_d \simeq 20$. One must hope that new physics (e.g. associated with the Planck scale) can perturb m_s and m_d sufficiently without upsetting the m_b prediction by too much.

Proton Decay

The most dramatic prediction of grand unification is proton and bound neutron decay. Unfortunately, even within minimal SU_5 the expected proton lifetime has large uncertainties both from $\Lambda_{\overline{MS}}$ (which determines M_X) and also from the poorly known hadronic matrix elements of the effective four-fermion operators. There have been many theoretical calculations [15], based on relativistic and non-relativistic quark models, the MIT bag, PCAC, vector meson dominance, chiral Lagrangians, QCD sum rules, etc. Even for the same value of M_X these calculations vary in their predictions for the proton lifetime by as much as two orders of magnitude.[2] A reasonable range for the partial lifetime (τ/B) into $e^+\pi^o$ is

$$\begin{aligned}\tau(p\to e^+\pi^o) &\simeq 6.6 \times 10^{28\pm0.7}\ [M_X/1.3 \times 10^{14}\ \text{GeV}]^4\text{yr} \\ &\simeq 6.6 \times 10^{28\pm1.4}\ [\Lambda^{(4)}_{\overline{MS}}/\ 100\ \text{MeV}]^4\ \text{yr} \quad ,\end{aligned} \tag{7}$$

where the theoretical uncertainty $10^{\pm0.7} \sim (5)^{\pm1}$ is a guess based loosely on the spread of theoretical estimates. Equation (7) implies that $\tau(p\to e^+\pi^o)$ is less than 3×10^{31} yr. for $\Lambda_{\overline{MS}}$ in (4), while larger reasonable estimates for $\Lambda_{\overline{MS}}$ could allow $\tau(p\to e^+\pi^o)$ as large as $\simeq 10^{32}$ yr. The present experimental limit [17] $\tau(p\to e^+\pi^o) > 3.3 \times 10^{32}$ yr. therefore essentially rules out minimal SU_5. However, it is hard to make an absolute statement to that effect because the uncertainties in $\Lambda_{\overline{MS}}$ and the matrix elements are large and difficult to quantify.

Other gauge groups G, such as SO_{10}, E_6, etc., give identical predictions as SU_5 for $\sin^2\theta_W$ provided they break directly to $SU_3 \times SU_2 \times U_1$ at M_X and have no additional light particles. They typically predict slightly faster decays into $e^+\pi^o$ and enhanced branching ratios into neutrino modes because of the additional superheavy gauge bosons.

Although the simplest SU_5 model is apparently ruled out, the basic ideas of grand unification are still alive and well. Many relatively simple modifications or extensions of minimal SU_5, some of which are natural and some ad hoc, lead to a longer lifetime and/or different decay modes. Unfortunately, they also tend to have less predictive power for $\sin^2\theta_W$.

The models with longer proton lifetimes fall into three principal classes. The first possibility is to add structure in the desert between M_W and M_X (e.g. in the form of new gauge, Higgs or fermion mass scales), so that the renormalization group equations which determine M_X and $\sin^2\theta_W$ are modified. Some examples of this mechanism are: (a) it is possible to have more than two symmetry breaking (i.e. gauge boson mass) scales [18]. For example, SO_{10} might first break to $SU_3 \times SU_{2L} \times SU_{2R} \times U_1$, at a mass $M_X >> 10^{14}$ GeV, and then to $SU_3 \times SU_2 \times U_1$ at a scale M_R. Then one has [19] $M_X \sim M_5(M_5/M_R)^{\frac{1}{2}} >> M_5$, where $M_5 \simeq 1.3 \times 10^{14}$ GeV is the SU_5 scale. $\sin^2\theta_W$ depends on the mass ratios in such schemes, so one has no real prediction. If one requires $\Delta\sin^2\theta_W < 0.01$ then $M_R > 10^{12}$ GeV and $M_X < 2 \times 10^{15}$ GeV, so that τ_p can be increased by about 4 orders of magnitude. (b) Another reasonable [20]

[2] Nuclear effects may modify the bound proton lifetime by an additional factor of 0.5-2. See ref. [16].

possibility is that there are heavy Higgs multiplets (e.g. in the 75 or 45 of SU_5) that are not quite degenerate with M_X. (c) Approximately degenerate new G multiplets of fermions or scalars do not affect M_X to lowest order. On the other hand, split fermion [21] or Higgs [22] multiplets which have some light (e.g. $O(M_W)$) and some heavy (e.g. $O(M_X)$) particles can affect τ_p and $\sin^2\theta_W$ significantly in either direction. Such multiplet splitting models are usually quite ad hoc, and most splittings lead to unacceptable values of M_X and $\sin^2\theta_W$. Also, it is hard to understand how the mass splittings come about, although one splitting (of a 5 dimensional Higgs representation in minimal SU_5) is needed in any case to generate the $SU_2 \times U_1$ scale without having very fast proton decays mediated by the partners of the ordinary Higgs doublet. (d) Theories with a low energy supersymmetry are considerably less arbitrary, and will be discussed below.

A very different possibility is that the proton decay rate is suppressed by mixing angle effects [4]. It is possible, though not very natural, that in models with complicated Higgs structures the light quarks and heavy leptons, for example, are associated together in multiplets. In this case the dominant amplitudes could be into energetically forbidden channels such as $p \to \tau^+\pi^o$. In most cases the lifetime is expected to increase by no more than $\sin^{-2}\theta_c \sim 20$, however. One exception to this limitation is models in which the proton is made absolutely stable by imposing an ad hoc new quantum number on the theory. Such models generally involve a doubling of the fermions and can lead to the dramatic signal of baryon number violation at accelerators.

Finally, effective non-renormalizable operators [23] such as $\mathrm{Tr}(F_{\mu\nu}F^{\mu\nu}\phi)/M$, where ϕ is the adjoint Higgs representation, could be generated if a GUT is embedded in a larger theory, such as supergravity or a Kaluza-Klein theory, with mass scale M. Such operators modify the gauge kinetic energy terms and have the ultimate effect of increasing M_X.

It is clear that grand unified theories allow a wide range of reasonable possibilities for the proton lifetime. Experiments on τ_p and $\sin^2\theta_W$ should be pushed as far as is reasonably possible.

Supersymmetry (SUSY) [24]

Grand Unified Theories are very successful with the gauge problem, but they shed little light on the fermion, Higgs/hierarchy[3], strong CP, or graviton problems. One attractive extension is supersymmetry, in which there is a symmetry between bosons and fermions.

In addition to their theoretical elegance, supersymmetric models partially solve the hierarchy problem by cancelling the worst divergences in the renormalization of the Higgs mass scale. Furthermore, local supersymmetric models (supergravity) involve a nontrivial unification of the other interactions with gravity. Supergravity does not by itself render quantum gravity finite or renormalizable, but recent developments with superstring theories [25] in ten space-time dimensions are promising.

In supersymmetric GUTs all of the light and superheavy particles acquire supersymmetric partners. The calculable modification of the $\sin^2\theta_W$ and m_b predictions as well as the requirement of perturbative unification are very strong constraints.

In the minimal SUSY-GUT model with three fermion families, two Higgs doublets, and superpartners for all of the ordinary particles, one has

$$M_X^{SUSY} \simeq 4.8 \times 10^{15}\ \mathrm{GeV}\left[\frac{\Lambda_{\overline{MS}}^{(4)}}{100\ \mathrm{MeV}}\right] , \qquad (8)$$

[3] GUTs even aggravate the Higgs problem by introducing difficulties at the tree level.

about 30 times larger than (5), so that the proton lifetime due to X and Y exchange is increased by about 10^6 to the probably unobservable level of $\sim 10^{35}$ yr. $\sin^2\theta_W$ is increased to about 0.24, which is still acceptable. The m_b predictions are essentially unchanged. (One can modify these models to obtain predictions similar to minimal SU_5 by the ad hoc addition of new chiral supermultiplets that are split into light and superheavy components.)

Some SUSY-GUTs (e.g. in which supersymmetry and the grand unification symmetry are broken in the same sector [26] require new chiral supermultiplets with the same $SU_3 \times SU_2 \times U_1$ quantum numbers as the gluons and electroweak bosons (these are in addition to the gauginos). In the simpler models these lead [27] to unacceptable $\sin^2\theta_W$ and/or m_b. More complicated versions with extra supermultiplets may give [26] acceptable $\sin^2\theta_W$, however, with M_X^{SUSY} close to the super Planck scale $m_p/\sqrt{8\pi} \sim 2.4 \times 10^{18}$ MeV.

Finally, there are supergravity theories with non-minimal kinetic energy terms, leading to non-universal gauge couplings at the unification scale. A class of such models yields [28] acceptable $\sin^2\theta_W$ and m_b for M_X^{SUSY} in the range $(2.4 \times 10^{16} - 3.5 \times 10^{17})$ GeV.

Proton Decay in SUSY-GUTs

In SUSY-GUTs the proton decay rate from X and Y exchange is probably unobservably slow because of the increase of the unification scale (eqn. 8). However, there are new mechanisms for proton decay, as shown in Fig. 3.

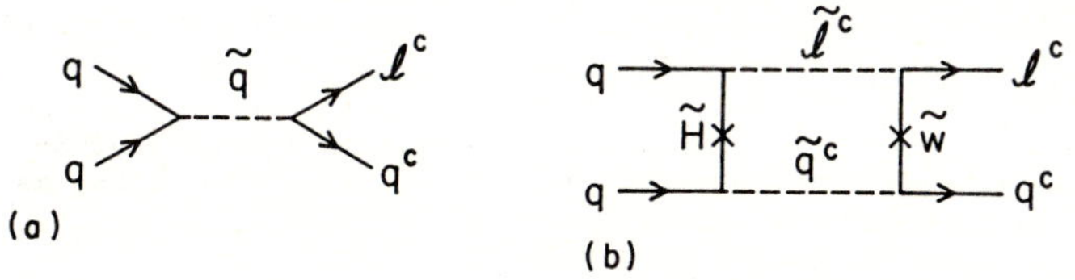

Figure 3. New proton decay diagrams that can occur in supersymmetric GUTs. (a) a dimension 4 diagram that must be forbidden to avoid a disastrously short lifetime. (b) a dimension 5 diagram.

The diagram of Figure 3a involves the exchange of a light scalar quark. It is not suppressed by any power of M_X^{SUSY} and, if present, would lead to a disastrously short proton lifetime. Fortunately, it can be forbidden by imposing discrete symmetries on the Lagrangian [24].

The colored Higgsino ($\tilde{H}$) exchange in the left half of the box diagram in Figure 3b leads to the dimension 5 operator

$$L \sim h^2 qq\tilde{q}\tilde{l}/M \tag{9}$$

where h is a Yukawa coupling and $M \sim M_X^{SUSY}$ is the Higgsino mass. After exchanging a light wino or gluino[4] these operators lead to proton decay. The favored mode is usually $\bar{\nu}K$, but models can be constructed [30] in which $\bar{\nu}\pi$ dominates because of cancellations.

The predicted lifetime is of order M^2 (rather than the usual M_X^{4}). This appears disastrous, but in fact there are a number of suppressions to the decay rate ($M_X^{SUSY} > M_X$, $1/16\pi^2$ from the loop, two small Yukawas, smaller higher order enhance-

[4] Gluino exchange is important if there are off-flavor-diagonal gluino vertices and a splitting between the u and d scalar quarks, both of which are expected in most models [29,30].

ment factors). The expected lifetime is typically in the range 10^{26} - 10^{31} yr (with large uncertainties from the superpartner masses) for models with hidden sector breaking. In fact, Enqvist et al. [31] have recently claimed that a large class of minimal supergravity models are ruled out by proton decay. However, some models with non-minimal kinetic energy terms [28] or in which supersymmetry and GUTs are broken in the same sector [26] typically have larger unification masses and are still viable, as are the models in which $\bar{\nu}\pi$ dominates because of cancellations in the $\bar{\nu}K$ amplitude [30]. It is possible to find models in which the dimension 5 operators are absent, but for most supersymmetric models that are not already ruled out, proton decay should be observable.

Superstrings [25]

Supergravity theories are very attractive, but they contribute little insight into the fermion problem and do not yield sensible theories of quantum gravity. An extremely exciting recent development is that superstring theories in 10 space-time dimensions may generate completely finite theories of quantum gravity (and the other interactions) with no arbitrariness.

The predictions for $\sin^2\theta_W$ in superstring theories are similar to those in other supersymmetric GUTs, except that one must take into consideration the exotic new particles predicted by such theories. The correct value $\sin^2\theta_W$ is therefore a constraint on model building and on assumptions as to whether there are intermediate thresholds in the desert, at which the extra Z's or exotic fermions in such models could possibly acquire masses.

Unfortunately proton decay is probably unobservable in superstring theories. This is because the basic unification scale is typically of order 10^{18} GeV so that gauge boson mediated decays are unobservably slow. Models typically have proton decays via Higgs that are either much too fast, in which case those models are excluded experimentally, or much too slow to observe. (The necessary Yukawa interactions can in principle be calculated in terms of topological considerations, but we are unable to do that at present.) However, it is always possible that in some models proton decay may occur at an observable rate [32].

Other Models of Proton Decay

Proton decay may also be mediated by the exchange of colored Higgs particles. Because of the smaller couplings of Higgs particles to fermions this is only likely to be important for Higgs boson masses around 10^{12} GeV (there is no particular reason for the masses to actually be in this range). Also, the existence of appropriate Higgs masses does not by itself lead to any suppression of the ordinary X and Y gauge boson mediated processes. The proton lifetime from Higgs exchange is completely arbitrary. In the simplest models the μ^+K^o mode should be very important, with a branching ratio from 35-80%. Other important modes are likely to be $\mu^+\omega$ and $\bar{\nu}K^+$.

Too rapid proton decay is a very serious constraint on models with composite quarks and leptons. It is hard to generalize about the rates and branching ratios in the viable models.

There are many variations on the Pati-Salam models. In the integer-charge-quark versions there are quark decay diagrams that can lead to $p \to 3\nu\pi^+$, $3\nu+3\pi$, $\nu\nu e^-\pi^+\pi^+$, etc. The fractional-charge-quark versions usually have proton decay by some very model dependent Higgs exchange diagrams. The lifetime is almost arbitrary in such models.

Neutron Oscillations

Another possibility for baryon number nonconservation is $\Delta B = 2$ operators that can cause neutrons to oscillate into antineutrons. The current reactor limit [33] on the free neutron oscillation time $\tau_{osc} > 10^6$ sec. may ultimately be improved to 10^8 sec. A more stringent but less direct limit may be obtained from the nonobservation of $\Delta B = 2$ processes in nuclei. Dover, Gal, and Richard [34] have recently done a thorough optical model study utilizing existing analyses of $\bar{p}$-atom level shifts and widths as well as new low energy $\bar{p}$-^{12}C scattering data from LEAR. They find that the limit [35] $\tau_{nuc} > 2.4 \times 10^{31}$ yr from IMB implies $\tau_{osc} > (1 \pm 0.3) \times 10^8$ sec., where the ± 0.3 is their estimate of the theoretical uncertainties.

Most models leading to $\Delta B = 2$ interactions involve heavy Higgs particle exchange. In many cases the Higgs particles are introduced in a rather ad hoc manner. Furthermore, τ_{osc} is proportional to M^5, where M is the typical mass of the particles mediating the interaction. Only for M in the rather narrow range $10^5 - 10^6$ GeV are neutron oscillations experimentally relevant. This range has not emerged in a natural way in any theory, and in particular a number of authors [36] have shown that it does not occur for any viable SO_{10} symmetry breaking pattern if one makes the reasonable requirement that M be associated with a gauge boson mass scale. The theoretical estimates are therefore not very optimistic, but nevertheless it seems well worth pushing the experiments as far as possible.

<u>The Cosmological Baryon Asymmetry</u> [37]

One of the most attractive features of grand unification is that it can account for the observed baryon asymmetry

$$\frac{n_B}{n_\gamma} \simeq 10^{-10} \quad , \; n_{\bar{B}} \ll n_B \quad , \tag{10}$$

where n_B, $n_{\bar{B}}$ and $n_\gamma \simeq 400/cm^3$ are the present average number densities of baryons, antibaryons, and microwave photons, respectively. If baryon number were exactly conserved, as in the perturbative standard model, then most likely the asymmetry $(n_B - n_{\bar{B}})/n_\gamma \sim 10^{-10}$ would have to be imposed as an initial condition on the big bang. This is certainly possible but not very attractive. Alternatively, one can imagine that there is no net asymmetry but that there is a large scale separation of baryons and antibaryons in the Universe. This view runs into severe difficulties, however.

In GUTs, the baryon asymmetry can be generated dynamically in the first 10^{-35}s after the big bang when the temperature was comparable to M_X. In addition to B violation, the necessary ingredients are C and CP violation, to distinguish baryons from antibaryons, and nonequilibrium. The latter is necessary because otherwise $n_B = n_{\bar{B}}$ since B and $\bar{B}$ are degenerate by CPT. (Kuzmin [38] has recently made the interesting suggestion that one could utilize CPT breaking rather than nonequilibrium).

In most models the actual mechanism is the decay of super-heavy Higgs particles H and their antiparticles $\bar{H}$. CP violation allows the relative rates for $H \to qq$ and $\bar{q}\bar{l}$ to differ from those for $\bar{H} \to \bar{q}\bar{q}$ and ql. In the minimal SU_5 model with three fermion families the asymmetry requires a three loop diagram and is too small ($\sim 10^{-20}$). However, an adequate asymmetry can be generated for ≥ 4 families or if additional superheavy Higgs multiplets are added to the theory. [37,39].

<u>Other Sources of Baryon Number Violation</u>

All grand unified theories predict the existence of superheavy (mass $\simeq M_X/\alpha_G \simeq 10^{16}$ GeV) magnetic monopoles. [40] In addition to their electromagnetic interactions, GUT monopoles may catalyze baryon number violating processes such as $Mp \to Me^+\pi^o$ with strong interaction strength [41] (much larger than the naive geometric estimate $\sigma \sim M_X^{-2} \sim 10^{-56} cm^2$). The search for relic monopoles in principle gives a very useful window not only on grand unification but also on the early Universe.

There may well be baryon number violation associated with quantum gravity, but the lifetime is likely to be around 10^{50} years.

Finally, Kuzmin et al. [42] have suggested the possibility of baryon number violation at the time of the $SU_2 \times U_1$ phase transition. B violation associated with weak instantons is normally negligibly small. However, it could be unsuppressed at $T \sim M_W/\alpha_W \sim 10$ TeV (or in the decays of very heavy fermions). This could conceivably generate the baryon asymmetry if the transition is first order. However, the mechanism raises a disturbing problem: It could also destroy any existing (e.g. GUT) asymmetry without adequately creating a new one. Clearly, the issue needs further consideration.[5]

Summary

The standard $SU_3 \times SU_2 \times U_1$ model is almost certainly approximately correct up to around 100 GeV. However, the standard model with classical gravity has 21 free parameters and is plagued by many unexplained features. Therefore, new underlying physics is necessary.

Most extensions of the standard model predict baryon-number violating processes, such as proton decay, $n \to \bar{n}$ oscillations, or monopole catalysis of baryon-number violation at some level, and many can naturally explain the baryon asymmetry of the Universe. In particular, the non-observation of proton decay rules out the simplest grand unified theories, but many variations and extensions are alive and well, as shown in Table 1.

Table 1. Predictions for proton decay in various models

Model	Typical lifetime (years)	Important Modes
Simple GUTs (2 scale)	$10^{27} - 10^{31}$	$e^+\pi^0$, $\bar{\nu}\pi^+$, $e^+\omega$, μ^+K^0
Large M_X GUTs	10^{31}-10^{35} (may be longer)	same
Mixing angle suppressions	10^{28}-10^{32} (may be longer)	e^+M, μ^+M, $\bar{\nu}M$, $M = \pi, \omega, K$, etc.
Higgs exchange	arbitrary	μ^+K^0, $\mu^+\omega$, $\bar{\nu}K^+$
Supersymmetry	10^{26}-10^{31} (10^{35} in variants)	$\bar{\nu}K$, $\bar{\nu}\pi$
Superstrings	10^{35} (less in variants)	-
Pati-Salam	arbitrary	$3\nu\pi$'s, $\nu\nu e^-\pi^+\pi^+$
Compositeness	arbitrary	-

Proton decay and other baryon number violating processes provide a unique window on physics beyond the standard model. They should be pushed as far as possible.

Acknowledgement

It is a pleasure to thank Prof. Klapdor and the organizers of this meeting for their invitation and travel support.

References

1. For reviews of QCD, see W. Marciano and H. Pagels: Phys. Rep. 36, 137 (1978) and A. H. Mueller: Proc. of the 1985 Int. Symp. on Lepton and Photon Interactions at High Energies, ed. M. Konuma and K. Takahashi (Nisha, Kyoto, 1986), p. 162.
2. For reviews of the standard electroweak model, see P. Langacker: 1985 Int. Symp. on Lepton and Photon Interactions, p. 186; W. Marciano: First Aspen Winter Physics Conference, ed. M. Block, (N.Y. Acad. Sci., N.Y., 1986) p. 367.

[5] Other implications are considered in [43].

3. S. Weinberg: Phys. Rev. Lett. 19, 1264 (1967); A. Salam: in Elementary Particle Theory, ed. N. Svartholm (Almquist and Wiksells, Stockholm, 1969) p. 367; S. L. Glashow: Nucl. Phys. 22, 579 (1961); S. L. Glashow, J. Iliopoulos, and L. Maiani: Phys. Rev. D2, 1285 (1970).
4. For reviews of GUTs, see P. Langacker: ref. [2]; Comm. on Nucl. and Part. Phys. 15, 41 (1985); and Phys. Rep. 72, 185 (1981); G. C. Ross: Grand Unified Theories (Benjamin, Menlo Park, 1984).
5. H. Georgi and S. L. Glashow: Phys. Rev. Lett. 32, 438 (1974).
6. L. B. Okun, M. B. Voloshin, and V. I. Zakharov: Phys. Lett. 138B, 115 (1984).
7. For a recent review, see D. W. Duke and R. G. Roberts: Phys. Rev. 120, 275 (1985).
8. G. P. Lepage: 1983 Int. Lepton-Photon Symposium, eds. D. G. Cassel and D. L. Kreinick (Cornell, 1983) p. 565.
9. W. J. Marciano: Fourth Workshop on Grand Unification, ed. H. A. Weldon et al., Birkhäuser, Boston, 1983) p. 13 and ref. [2].
10. F. Sciulli: 1985 Int. Symp. on Lepton and Photon Interactions, p. 8.
11. H. Kolanoski: 1985 Int. Symp. on Lepton and Photon Interactions, p. 90.
12. Y. J. Ng, J. Pantaleone, and S. H. Tye: Phys. Rev. D33, 777 (1986).
13. U. Amaldi, A. Böhm, S. Durkin, P. Langacker, A. Mann, W. Marciano, A. Sirlin, and M. H. Williams: to be published.
14. D. V. Nanopoulos and D.. Ross: Phys. Lett. 108B, 351 (1982); J. Oliensis and M. Fischler: Phys. Rev. D28, 194 (1983).
15. For reviews see ref. [4] and M. Goldhaber and W. J. Marciano: Comm. on Nucl. Part. Phys. 16, 23 (1986); W. Lucha: Fortschr. Phys. 33, 547 (1985); K. Enqvist and D. V. Nanopoulos: CERN-TH 4066/84. A recent calculation is P. Falkensteiner et al.: Z. Phys. C27, 477 (1985) and Wien UW Th Ph-1984-6.
16. A. Axelrod: Phys. Rev. D33, 741 (1986); C. Dover et al.: Phys. Rev. D24, 2886 (1981), and references therein.
17. For a recent review, see Y. Toksuka: 1985 Int. Symp. of Lepton and Photon Interactions, p. 120.
18. For an interesting recent model, see S. Dimopoulos and L. J. Hall: Nucl. Phys. B255, 633 (1985).
19. See for example, Y. Tosa et al.: Phys. Rev. D28, 1731 (1983) and references therein.
20. T. Hübsch et al.: Phys. Rev. D31, 2958 (1985) and references therein.
21. A. Yu. Smirnov: JETP Lett. 31, 737 (1980); P. Frampton and S. L. Glashow: Phys. Lett. 131B, 340 (1983), 135B, 515(E), (1984); S. Nandi, Phys. Lett. 142B, 375 (1984).
22. W. J. Marciano: ref. [9]; K. Hagiwara et al.: Phys. Lett. 141B, 372 (1984); K. S. Babu and E. Ma: Phys. Lett. 144B, 381 (1984); J. C. Wu: Phys. Rev. D32, 1253 (1985).
23. C. T. Hill: Phys. Lett. 135B, 47 (1984); Q. Shafi and C. Wetterich: Phys. Rev. Lett. 52, 875 (1984); J. Ellis et al.: Phys. Lett. 156B, 189 (1985); T. Hübsch et al.: ref. [20]; F.-X. Dong et al.: Beijing BIHEP-TH-85-8; C. Panagiotakopoulos and Q Shafi: Phys. Rev. Lett. 52, 2336 (1984).
24. For reviews of supersymmetry, see H. E. Haber and G. L. Kane: Phys. Rep. 117, 75 (1985); H. P. Nilles: Phys. Rep. 110C, 1 (1984); S. Dawson, E. Eichten, and C. Quigg: Phys. Rev. D31, 1581 (1985); J. Ellis: 1985 Int. Symp. on Lepton and Photon Interactions, p. 850.
25. For reviews of superstrings, see M. B. Green: 1985 Int. Symp. on Lepton and Photon interactions, p. 372; G. Segre: NATO Advanced Study Inst. on Particle Physics, Cargese, France, 1985; V.S. Kaplunovsky and C. R. Nappi: Princeton preprint.
26. B. Ovrut and S. Raby: Phys. Lett. 138B, 72 (1984).
27. J. E. Bjorkman and D.R. Jones: Nucl. Phys. B259, 533 (1985).
28. J. Ellis et al.: Phys. Lett. 156B, 189 (1985).
29. J. Milutinovic et al.: Phys. Lett. 140B, 324 (1984); S. Chadha et al.: Phys. Lett. 149B, 477 (1984); J. McDonald and C. E. Vayonakis: Phys. Lett. 163B, 148 (1985), 144B, 199 (1984).
30. R. Arnowitt et al.: Phys. Lett. 156B, 215 (1985) and Phys. Rev. D32, 2348 (1985); T.-C. Yuan: Northeastern NUB #2682.
31. K. Engvist et al.: Phys. Lett. 156B, 209 (1985).

32. See, for example, G. Lazarides et al.: Phys. Rev. Lett. 56, 557 (1986) and Thessaloniki preprint UT-ST PT-3/86; R. N. Mohapatra and J. W. F. Valle: Maryland preprint 86-127.
33. G. Fidecaro et al.: Phys. Lett. 156B, 122 (1985).
34. C. B. Dover, A. Gal, and J. M. Richard: Phys. Rev. C31, 1423 (1985).
35. T. W. Jones et al.: Phys. Rev. Lett. 52, 720 (1984).
36. For example, see P. Majumdar et al.: Phys. Lett. 137B, 181 (1984) and references therein. For E_6 models, see P. K. Mohapatra et al.: Phys. Rev. D33, 2010 (1986).
37. For a review, see E.W. Kolb and M. S. Turner: Ann Rev. Nucl. Part. Sci. 33, 645 (1983).
38. V. A. Kuzmin: Moscow Qu. Gravity 1984:270.
39. For a recent study, see G. Branco and A. I. Sanda: Phys. Lett. 135B, 383 (1984).
40. For a review, see M. Turner: First Aspen Winter Physics Conference, p. 639 and ref. [4].
41. V. A. Rubakov: JETP Lett. 33, 644 (1981); Nucl. Phys. B203, 311 (1982); C. G. Callan: Phys. Rev. D26, 2058 (1982); 25, 2141 (1982), Nucl. Phys. B212, 391 (1983). For a recent discussion, see W. Bernreuther and N. S. Craigie: Phys. Rev. Lett. 55, 2555 (1985).
42. V. A. Kuzmin et al.: Phys. Lett. 155B, 36 (1985); V.. Rubakov: Nucl. Phys. B256, 509 (1985); M. Fukugita and V. A. Rubakov: Phys. Rev. Lett. 56 988 (1986).
43. M. Evans: Rockefeller preprint RU/86/144.

Quasi Standard Model Physics

R.D. Peccei

Deutsches Elektronen-Synchrotron DESY, D-2000 Hamburg, F.R. Germany

Possible small extensions of the standard model are considered, which are motivated by the strong CP problem and by the baryon asymmetry of the Universe. Phenomenological arguments are given which suggest that imposing a PQ symmetry to solve the strong CP problem is only tenable if the scale of the PQ breakdown is much above M_W. Furthermore, an attempt is made to connect the scale of the PQ breakdown to that of the breakdown of lepton number. It is argued that in these theories the same intermediate scale may be responsible for the baryon number of the Universe, provided the Kuzmin Rubakov Shaposhnikov (B + L) erasing mechanism is operative.

1. Big and Little Excursions from the Standard Model

The standard SU(3) x SU(2) x U(1) model of the strong and electroweak interactions works extremely well phenomenologically. Particle theorists, however, are unhappy because they do not understand the deep reasons behind some of the structural aspects of the standard model. Putting it succinctly /1/, theorists would like to know:

i) why these are the forces we see in nature?
ii) why the matter we see are quarks and leptons?
iii) what fixes the dynamics which generates masses for all the elementary excitations?

Elaborate theoretical constructs exist which try to address these deep structural queries. Composite models, technicolor, supersymmetry, GUTs, supergravity and superstrings are some of the key concepts employed to try to provide an answer to the above deep questions. However, it is not my intention here to speak of any of these beautiful theoretical ideas (whose common link, alas, is that of having as yet no evidence for their validity!) Rather than looking at these rather large extrapolations beyond the standard model, I want to concentrate on two points which require only modest excursions beyond the standard model and for which one can adduce some experimental/theoretical evidence in their favor.

The first of these little excursions concerns the strong CP problem. Due to the structure of the QCD vacuum /2/ and the presence of an ABJ anomaly for chiral rotations /3/, one has an effective CP violating term in the standard model Lagrangian:

$$\mathcal{L}_{CP\ viol.} = \frac{\alpha_s}{4\pi} (\Theta + \text{Argdet } M)\, F_a^{\mu\nu} \tilde{F}_{a\mu\nu} = \frac{\alpha_s}{4\pi} \bar{\Theta}\, F_a^{\mu\nu} \tilde{F}_{a\mu\nu} \tag{1}$$

Here Θ is the QCD vacuum angle and M is the quark mass matrix. However, the parameter $\bar{\Theta}$ is very strongly bounded by the absence of a neutron electric dipole moment /4/

$$\bar{\Theta} \lesssim 10^{-8} - 10^{-9} \tag{2}$$

It is totally ununderstandable theoretically why $\bar{\Theta} \simeq 0$, unless some new physics forces a cancellation between the SU(3) and SU(2) x U(1) pieces in $\bar{\Theta}$. Augmenting the standard model by an extra global $U(1)_{PQ}$ chiral symmetry locks automatically the phase of the quark mass matrix to the vacuum angle giving $\bar{\Theta} = 0$ /5/. However, such a global symmetry also implies the existence of a pseudo Goldstone boson, the axion /6/.

The question of axions, as a price to pay to solve the strong CP problem, will be discussed in the next two sections. In Sec. II I will review the most recent bounds on visible axions, particularly those for the recently proposed variant axions /7/. The conclusion which will emerge is that no window is left for axions to exist, if the scale of the $U(1)_{PQ}$ breakdown is connected to that of the weak scale. In Sec. III, the astrophysical and cosmological bounds on invisible axions will be discussed, narrowing the range for an allowed scale of the $U(1)_{PQ}$ breakdown to a window from 10^8 to 10^{12} GeV. Physical arguments will be presented in this section which will connect the scale of the $U(1)_{PQ}$ breakdown to that of lepton number. If this connection really obtains then light neutrinos, such as those needed to solve the solar neutrino puzzle, may indeed point to the same dynamical scale as that needed to solve the strong CP puzzle! /8/

The second little excursion from the standard model which I would like to discuss concerns the Universe baryon asymmetry. The observed ratio /9/

$$(n_B - n_{\bar{B}})/n_\gamma \simeq 10^{-10} \tag{3}$$

could be a peculiar initial condition. However, if the Universe started in a symmetric way, to obtain (3) it is necessary that there should be baryon number violating interactions at some level. Baryon number, classically, is a global symmetry of the standard model and so it is natural to presume that for the Universe's asymmetry to obtain it is necessary to go beyond the standard model. However, at the quantum level, baryon number is violated in the standard model /2/ because the baryon number current has an ABJ anomaly /3/. Thus one must check first that these effects are irrelevant, before invoking physics beyond the standard model to explain the ratio in (3).

The physics of baryon number violation in the standard model is quite analogous to that of the strong CP problem. For baryon number violation what is relevant also is the presence of a non trivial vacuum structure, connected with the appearance of an electroweak vacuum angle Θ_{EW}. However, under normal circumstances, baryon violating processes are suppressed by a factor of $\exp(-4\pi\sin^2\Theta_W/\alpha)$ and are totally negligible. However, as pointed out by Kuzmin, Rubakov and Shaposhnikov /10/, and as discussed in Sec. IV, in the early Universe the non trivial electroweak vacuum structure may lead to significant baryon number violation. Indeed if the KRS mechanism is operative it is quite possible that any previously produced baryon number (more precisely B + L) could get erased at temperatures of order of M_W. Thus, there exist the exciting possibility that a lepton asymmetry in the Universe, generated at temperatures of the order of the $U(1)_{PQ}$ symmetry breaking, may ultimately be responsible for the observed ratio (3). In this way, the strong CP problem, light neutrinos and the baryon asymmetry of the Universe are mutually interconnected phenomena.

2. The Last Hurrah for Visible Axions

Technically to solve the strong CP problem, by imposing an additional global $U(1)_{PQ}$ symmetry /5/, requires at least two doublets of Higgs in the theory. When these doublets ϕ_i get vacuum expectation value not only does SU(2) x U(1) break down to $U(1)_{em}$, but also $U(1)_{PQ}$ is broken down. Fortunately, the resulting Goldstone boson - the axion /6/ - is not totally massless because the $U(1)_{PQ}$ symmetry is anomalous. However, the axion is still very light with a mass of order $m_\pi f_\pi/V$, with V being the scale of the electroweak symmetry breaking ($V \sim \langle\phi_i\rangle$) and hence ought to be visible.

Two classes of axion models have been invented: standard and variant. In the standard axion model /5/ /6/ all quark flavors are treated in a symmetric fashion under $U(1)_{PQ}$, while in variant models /7/ /11/ the quarks are treated asymmetrically under $U(1)_{PQ}$ (For instance, only the first generation of quarks effectively have PQ charges). In terms of the couplings to the Higgs field ϕ_i, the distinction between standard and variant models is that in the standard axion case all charge 2/3 (all charge -1/3) right handed quark fields couple to $\phi_1(\phi_2)$, while in the variant case some charge 2/3 right handed quark fields couple to $\tilde{\phi}_2$ rather than ϕ_1.

Standard axions have been ruled out experimentally already a number of years ago by a combination of experiments, including quarkonia radiative decays ($\psi \to \gamma a$; $\Upsilon \to \gamma a$), beam dump experiments and a variety of nuclear deexcitation experiments /12/. Variant axion models were (re)invented this year to try to explain the positron peaks and e^+e^- correlated signals seen at GSI /13/. Since these phenomena require kinematically that the axion mass be near 1.7 - 1.8 MeV it follows that one of the Higgs expectation values is much greater than the other (see below). Because of this, unless one has asymmetric couplings to quarks of different families, either the rate for $\psi \rightarrow \gamma a$ or that for $\Upsilon \rightarrow \gamma a$ is very enhanced. Variant axion models, very neatly avoid this problem /7/. Furthermore, these models also have a very short lifetime for $a \rightarrow e^+e^-$ and can avoid in this way the old beam dump and nuclear deexcitation bounds. Hence the GSI signals raised the exciting possibility that perhaps some kind of visible variant axion might really exist.

Unfortunately variant axion models are now also ruled out experimentally. First of all, the appearance of a second correlated peak in the EPOS experiment /13/, plus the failure of finding any convincing production mechanism /14/ have considerably weakened the axion interpretation for the GSI phenomena. Most importantly, however, new experiments contradict the expectations of the most general variant axion models and thus, independently of the GSI observations, eliminate this remaining option for an axion model, where the scale of breaking of $U(1)_{PQ}$ is the same as that of the electroweak theory.

Variant axion models have essentially two free parameters: the ratio $x = \langle\phi_2\rangle/\langle\phi_1\rangle$ of Higgs vacuum expectation values and the number, N_{PQ}, of quark doublets which have a PQ symmetry. The axion mass fixes, however, one combination of these parameters. One has /11/

$$m_a \simeq \frac{m_\pi f_\pi}{V} \frac{(m_u m_d)^{1/2}}{(m_u + m_d)} N_{PQ}(x + \frac{1}{x}) \simeq 25\ N_{PQ}(x + \frac{1}{x})\ \mathrm{keV} \qquad (4)$$

If $m_a \simeq 1.7$ MeV, $N_{PQ}(x + \frac{1}{x})$ must be large. Further, to avoid the quarkonia bounds, x not x^{-1} must be large. Hence the GSI identification implies $N_{PQ}x \simeq 70$. Since one can measure experimentally three characteristics of variant axions, as a function of the remaining free parameter: the axion coupling to electrons, the isovector content of the axion and the isoscalar content of the axion, these models are testable.

Electron beam dump experiments, recently performed at KEK /15/ and Orsay /16/, rule out any coupling of variant axions to e^+e^- which are consistent with the g-2 bound on this coupling. In particular, values of $x \lesssim 70$ are clearly excluded. Experiments measuring axion deexcitation in hadronic transitions also rule out variant axions /1/. Isovector transitions like $np \rightarrow da$ /17/, $\pi^+ \rightarrow a\, e^+ \nu_e$ /18/ or the decay of the 2^+1 9.17 MeV state in ^{14}N to its 1^+0 ground state /19/ measure the mixing parameter

$$\lambda_3 \simeq \frac{x}{2} \left[1 - N_{PQ} \frac{(m_d - m_u)}{(m_d + m_u)} \right] \simeq \frac{x}{8} (4 - N_{PQ}) \qquad (5)$$

The most stringent bound comes from the π^+ decay experiment where one predicts /20/ /21/

$$B(\pi^+ \rightarrow ae^+\nu_e) \simeq 3 \times 10^{-9} (\lambda_3)^2 \qquad (6)$$

to be compared to the SIN bound /18/

$$B(\pi^+ \rightarrow ae^+\nu_e) \lesssim (1\text{-}2) \times 10^{-10} \tag{7}$$

yielding $|\lambda_3| \lesssim 0.25$. Such a value is only compatible with (5) and $xN_{PQ} \simeq 70$ if $N_{PQ} = 4$. However, such a value is in conflict with the recent result of an isoscalar, axion induced, nuclear deexcitation experiment in ^{10}B. The axion to photon rate for the 3.59 MeV $2^+0 \rightarrow 3^+0$ transition is predicted to be /1/ /20/

$$\frac{\Gamma_a}{\Gamma_\gamma} = 7.9 \times 10^{-4} (\lambda_s)^2 \tag{8}$$

while experimentally /22/ one observes

$$\frac{\Gamma_a}{\Gamma_\gamma} \lesssim 7.2 \times 10^{-3} \tag{9}$$

implying a bound for the isoscalar mixing: $|\lambda_s| \lesssim 3$. However, in variant axion models one has a constraint

$$\lambda_3 - \lambda_s \simeq \frac{3}{8} \times N_{PQ} \simeq 26 \tag{10}$$

which is clearly violated by the above bounds.

These results have dashed all hopes to prove experimentally that the solution of the strong CP problem is due to having an extra $U(1)_{PQ}$ in the theory, which then begets a visible axion. That is not to say, however, that this may not be still the solution to the strong CP problem. If the scale of the break down of $U(1)_{PQ}$, V_{PQ}, is much greater than the weak interaction scale $V \sim 250$ GeV, then the concomitant axion is superlight ($m_a \sim \frac{1}{V_{PQ}}$) and very weakly coupled ($\sim\frac{1}{V_{PQ}}$) and hence essentially invisible.

3. Invisible Axions and Elusive Neutrinos - is there a Connection?

Invisible axion models /23/ make use of an additional SU(2) x U(1) singlet Higgs field σ which carries $U(1)_{PQ}$ charge and which has a very large vacuum expectation value

$$\langle\sigma\rangle = V_{PQ} >> V \tag{11}$$

Because of the extra $U(1)_{PQ}$ symmetry, the strong CP problem is solved ($\bar{\Theta} = 0$). However, now there are no experimental problems (or tests!) for the resulting axion because it is very light, very weakly coupled and extremely long lived. The real question then becomes how can one tell that this is right? Remarkably, invisible axions have some astrophysical and cosmological constraints and, under certain circumstances, can even be searched for experimentally.

Astrophysics gives a lower bound on the symmetry breaking scale V_{PQ}. This bound follows because invisible axions can efficiently cool stars by a Compton-like process $\gamma + e \to e + a$, with the resulting axions escaping from the star because they are so weakly coupled. The energy loss due to axions is proportional to the cross section for the above process and thus is inversely proportional to $V_{PQ}{}^2$. Only if V_{PQ} is sufficiently big, axions would not have affected the known life cycle of stars. Detailed investigations /24/ give a bound for V_{PQ}

$$V_{PQ} \gtrsim 10^8 \text{ GeV} \tag{12}$$

Cosmology gives actually an upper bound for V_{PQ}. This comes about because in the early Universe, at temperatures below those of the scale of the $U(1)_{PQ}$ breakdown but much above the scale of QCD, the locking mechanism which fixed the phase of the quark mass matrix is not operative. In this temperature regime this phase is arbitrary and not fixed to be $-\theta$. This unrestricted Yukawa phase corresponds to a coherent axion field of magnitude of order V_{PQ} (i.e. a phase of $O(1)$):

$$a(x) \sim V_{PQ} \tag{13}$$

As the temperature decreases locking takes place and the phase oscillates about its final value $-\theta$. These phase oscillations (axion oscillations) contribute to the Universe energy density and one finds /25/ that if V_{PQ} is too big, the energy density today in the oscillating axion field exceeds the critical density needed to close the Universe. This gives an upper bound for V_{PQ} of order /25/:

$$V_{PQ} \lesssim 10^{12} \text{ GeV} \tag{14}$$

We see that the strong CP problem can be solved by having an extra $U(1)_{PQ}$ symmetry in the theory only if the scale of the breakdown lies between:

$$10^8 \text{ GeV} \lesssim V_{Pq} \lesssim 10^{12} \text{ GeV} \tag{15}$$

In particular, if V_{PQ} is near the upper end of the above range then the Universe's energy density is dominated by axions! The axion mass, if $V_{PQ} \sim 10^{12}$ GeV, is $m_a \sim 10^{-5}$ eV which corresponds to a gigahertz frequency. Such axions, because they would permeate our galaxy, may be amenable to direct detection /26/. Since an axion has a coupling to two photons, axions in the Milky Way halo would give a tiny, but perhaps measurable, Q shift in an appropriate electromagnetic cavity.

Irrespective of whether invisible axions will ever be seen experimentally, there remains the theoretical question of why V_{PQ} should lie in the range (15). With Langacker and Yanagida /27/, I have recently put forward the suggestion that this scale could be the same as the scale at which lepton number breaks. If a Majorana mass for right handed neutrinos indeed has a scale V_{PQ}, then the numerology of neutrino masses is quite nice. Using the standard see-saw mechanism /28/, one expects

$$m_{\nu_1} \sim \frac{m_1^2}{m_{\nu_R}} \sim \frac{m_1^2}{V_{PQ}} \tag{16}$$

which for $V_{PQ} \sim 10^{10}$ GeV gives $m_{\nu_\tau} \sim 0.1$ eV, $m_{\nu_\mu} \sim 10^{-3}$ eV, $m_{\nu_e} \sim 10^{-7}$ eV. One sees therefore that the mass difference $\delta(m_{\nu_\mu}^2 - m_{\nu_e}^2)$ is of the right order of magnitude to appeal to the Mikheyev Smirnov Wolfenstein /29/ $\nu_e \to \nu_\mu$ conversion mechanism in the sun to understand the solar neutrino puzzle.

Can one argue for this connection between V_{PQ} and the scale of breakdown of lepton number theoretically? The answer is yes, if one demands that the breakdown of the $U(1)_{PQ}$ augmented standard model leads to axions as the <u>only</u> approximate Goldstone bosons /27/. With ϕ_1, ϕ_2 and σ one can write down 5 possible Yukawa couplings (per family)

$$\bar{Q}_L\phi_1 u_R;\quad \bar{Q}_L\phi_2 d_R;\quad \bar{L}_L\phi_1\nu_R;\quad \bar{L}_L\phi_2 e_R;\quad \nu_R^T\, C\, \sigma\, \nu_R \tag{17}$$

These preserve 9 - 5 = 4 U(1) symmetries, corresponding to weak hypercharge, baryon number, lepton number and $U(1)_{PQ}$. When σ and ϕ_i acquire vacuum expectation value, B remains unbroken but the other U(1) quantum numbers break down. Hypercharge is gauged, so its Goldstone boson is eaten, but there remain two other Goldstone excitations. One is the axion and the other, corresponding to the breakdown of L, is a Majoron /30/. To avoid this extra Goldstone boson one must break one of the above four U(1)'s explicitly in the Higgs potential via, for example, a term like

$$V = K\, \phi_1\phi_2\sigma \tag{18}$$

In this case the Majoron acquires a mass proportional to $K\langle\sigma\rangle$ and one is left with an invisible axion model where the scale of the $U(1)_{PQ}$ breaking $\langle\sigma\rangle$ is precisely the same as that which gives the right handed neutrino a large Majorana mass. Of course, one must ultimately argue for the reason for the appearance of (18) and the necessary fine tuning needed to keep $\langle\sigma\rangle >> \langle\phi_i\rangle$, which requires K to be very small.

4. <u>The KRS catastrophe and the great L ↔ B switch</u>

I would like to discuss finally another, quite speculative, reason for having an intermediate scale of lepton number violation of the order of V_{PQ}, connected to baryon number violation in the standard model. Although baryon number is a classical global symmetry of the standard model, the baryon number current has an ABJ anomaly:

$$\partial^\mu B_\mu = \frac{2n_f\alpha}{\pi \sin^2\Theta_W}\, W_a^{\mu\nu}\, \tilde{W}_{a\mu\nu} \tag{19}$$

Here $W_a^{\mu\nu}$ is the SU(2) field strength and n_f is the number of families. Eq. (19) implies that there exist baryon number violating Green's functions in the theory /2/. The real vacuum state for the electroweak theory $|\Theta_{EW}\rangle$ is a superposition of distinct gauge variant vacua $|n\rangle$

$$|\Theta_{EW}\rangle = \sum_n e^{in\Theta_{EW}} |n\rangle \tag{20}$$

As a result of this the vacuum amplitude from which all Green's functions are generated, contains a superposition of terms involving transitions which change $|n\rangle$

$$_+\langle\Theta_{EW}|\,\Theta_{EW}\rangle_- = \sum_\nu e^{i\nu\Theta_{EW}} \left\{ \sum_n {}_+\langle n+\nu\,|n\rangle_- \right\} \tag{21}$$

One can show that terms of non zero ν involve a change in B /2/. However, $\nu \neq 0$ amplitudes are ridiculously small, of order $\exp(-2\pi \sin^2\Theta_W \nu/\alpha)$, where the exponential factor is essentially the probability for tunnelling between different $|n\rangle$ vacua. Although there exists baryon number violation in the standard model, it appears to be of no practical importance.

Recently, however, Kuzmin, Rubakov and Shaposhnikov (KRS) /10/ pointed out that the electroweak vacuum structure may in fact be important in the early Universe. Their idea is very simple. If $T \neq 0$ one can imagine going over from an $|n\rangle$ vacuum to an $|n+1\rangle$ vacuum by thermal fluctuations instead of tunnelling, thereby bypassing perhaps the exponential suppression factor discussed above. KRS specifically made use of a saddle point solution for the free energy, found by Klinkhamer and Manton /31/, which allows one to connect an $|n\rangle$ to an $|n+1\rangle$ vacuum. The rate for this transition, near the temperature T_c where the electroweak phase transition takes place, computed by KRS is not exponentially damped and can in fact exceed the Universe expansion rate. As a result, it may be possible for this phenomena to erase all previously accumulated (B + L) asymmetry in the Universe at a very low temperature (since (B - L) has no anomaly in the standard model only (B + L) really gets affected). Furthermore, when these B + L violating processes finally go out of equilibrium, it is too late to reestablish any significant baryon asymmetry /10/.

The KRS result is somewhat of a catastrophe, if true. If it is effective, it means that the baryon number asymmetry after KRS is just

$$B = \frac{1}{2}\,(B - L)_{\text{early Universe}} \tag{22}$$

Thus for any GUTs where (B - L) is conserved, like SU(5), one is left with no final asymmetry! It is of course possible that the KRS estimate for the baryon violating transition rate is a gross overestimate. This could be because the Klinkhamer Manton path is only a very particular path /32/ or, more likely, because near T_c the whole vacuum structure itself begins to lose meaning altogether. In this case there would be no low temperature erasure of any baryon number established at high temperature. More interesting, however, is to suppose that the KRS mechanism does work and that the final baryon asymmetry is a reflection of an earlier established (B - L) asymmetry. An obvious candidate for this latter asymmetry would be a lepton number asymmetry established at some intermediate scale /33/. Clearly having the possibility of breaking lepton number at a scale V_{PQ} would be very interesting in this connection /27/. Out of equilibrium decays of ν_R at $T \sim V_{PQ}$ would create an initial lepton number asymmetry which, by the KRS mechanism, would then at $T \sim T_c \sim V$ be turned into a baryon number asymmetry /27/ /33/. So a simple extension of the standard model which incorporates a $U(1)_{PQ}$ symmetry plus a singlet Higgs field could interrelate invisible axions with light neutrinos and, at the same time, be the trigger for the baryon asymmetry of the Universe!

5. Conclusions

I hope to have given you an impression that small excursions from the standard model are well worth exploring. Although certainly not compelling, it is possible to envisage models where apparently disconnected phenomena, like small neutrino masses, invisible axions and the Universe's baryon asymmetry, in fact have a common origin. These theoretical musing suggests, if nothing else, that besides exploring the high energy frontier experimentally it is very worthwhile to look for subtle effects in low energy experiments.

Acknowledgements

I would like to thank W. Bardeen, P. Langacker and T. Yanagida for some of their insights.

References

1 For a somewhat more extended discussion, see for example R.D. Peccei in the Proceedings of the 12th International Conference on Neutrino Physics and Astrophysics, Sendai, Japan, June 1986

2 G. 't Hooft: Phys. Rev. Lett. 37 8 (1976); Phys. Rev. D14 3432 (1976)

3 S.L. Adler: Phys. Rev. 177 2426 (1969); J.S. Bell and R. Jackiw: Nuovo Cimento 60A 49 (1969); W.A. Bardeen: Phys. Rev. 184 1848 (1969)

4 V. Baluni: Phys. Rev. D19 2227 (1979); R. Crewther, P. di Vecchia, G. Veneziano and E. Witten: Phys. Lett. 89B 123 (1979)

5 R.D. Peccei and H.R. Quinn: Phys. Rev. Lett. 38 1440 (1977); Phys. Rev. D16 1791 (1977)

6 S. Weinberg: Phys. Rev. Lett. 40 223 (1978); F. Wilczek: Phys. Rev. Lett. 40 279 (1978)

7 R.D. Peccei, T.T. Wu and T. Yanagida: Phys. Lett 172B 435 (1986); L.M. Krauss and F. Wilczek: Phys. Lett 173B 189 (1986)

8 For somewhat different arguments for this connection see M.S. Turner and B.J. Carr: Fermilab-Pub-86/66-A (May 1986)

9 J. Yang, M.S. Turner, G. Steigman, D.N. Schramm and K. Olive: Ap. J. 281 493 (1982)

10 V.A. Kuzmin, V.A. Rubakov and M.E. Shaposhnikov: Phys. Lett. 155B 36 (1985)

11 W.A. Bardeen and S.-H-H. Tye, Phys. Lett. 74B 229 (1978)

12 For a review see A. Zehnder, Proceedings of the 1982 Gif-sur-Yvette Summer School

13 J. Greenberg, these proceedings

14 A. Schäfer et al.: J. Phys. G, Nucl. Phys. 11 L69 (1985); A. Chodos and L.C.R. Wijewardhana: Phys. Rev. Lett. 56 302 (1986); J. Reinhardt et al.: Phys. Rev. 33C 194 (1986)

15 A. Konaka et al.: KEK preprint 86-9 May 1986

16 M. Davier, private communication

17 S.J. Freedman, these proceedings

18 R. Eichler et al.: SIN preprint PR 86-07 May 1986

19 M.J. Savage et al.: Phys. Rev. Lett. 57 178 (1986)

20 W.A. Bardeen, R.D. Peccei and T. Yanagida: DESY 86-054, Nucl. Phys. B to be published

21 L.M. Krauss and M.B. Wise: Yale preprint YTP 86-13 May 1986

22 F.W.N. de Boer et al.: Groningen preprint, Phys. Lett. B to be published

23 J.E. Kim: Phys. Rev. Lett. 43 103 (1979); M.A. Shifman, A.I. Vainshtein and V.I. Zakharov, Nucl. Phys. B166 493 (1980); M. Dine, W. Fischler and M. Srednicki: Phys. Lett. 104B 99 (1981)

24 D.A. Dicus et al.: Phys. Rev. D18 1829; D22 839 (1980); M. Fukugita, S. Watamura and M. Yoshimura: Phys. Rev. D26 1840 (1982); N. Iwamoto: Phys. Rev. Lett. 53 1198 (1984)

25 J. Preskill, M.B. Wise and F. Wilczek: Phys. Lett. 120B 127 (1983); L.F. Abbott and P. Sikivie: Phys. Lett. 120B 133 (1983); M. Dine and W. Fischler, Phys. Lett. 120B 137 (1983)

26 P. Sikivie: Phys. Rev. Lett. 51 1415 (1983); L.M. Krauss, J. Moody, F. Wilczek and D. Morris: Phys. Rev. Lett. 55 1797 (1985)

27 P. Langacker, R.D. Peccei and T. Yanagida, in preparation

28 T. Yanagida in Proceedings of the Workshop on Unified Theory and Baryon Number of the Universe KEK Japan 1979; M. Gell-Mann, P. Ramond and R. Slansky in Supergravity (North Holland 1979)

29 L. Wolfenstein: Phys. Rev. D16 2369 (1978); S.P. Mikheyev and A. Smirnov, Nuovo Cimento 9C (1986) 17

30 Y. Chikashige, R.N. Mohapatra and R.D. Peccei; Phys. Lett. 98B 265 (1981)

31 R.F. Klinkhamer and N.S. Manton: Phys. Rev. D30 2212 (1984)

32 E.W. Kolb and M.S. Turner, private communication

33 M. Fukugita and T. Yanagida: Phys. Lett. 174B 45 (1986)

Hierarchy and Mass Spectrum from Minimal Supergravity

N. Dragon

Institut für Theoretische Physik, Universität Hannover, Appelstraße 2,
D-3000 Hannover, F. R. Germany

The successful explanation of the strong, weak and electromagnetic forces as gauge interactions of the group SU(3) × SU(2) × U(1) has encouraged attempts to unify all these interactions including, if possible, gravity. This unification, however, can occur only at energy scales which are enormously above the weak scale M_W - the scale set by the mass of the W- and Z-bosons $\sim 10^2$GeV. Typical models which unify the gauge interactions deal with mass scales of about 10^{15}GeV. If gravity is included it introduces the Planck-scale into the game

$$M_{Pl} = 2.4 \cdot 10^{18}\ \mathrm{GeV} = (8\pi G_N)^{-1} \quad (c=\hbar=1,\ G_N = \text{Newton's constant})\ .$$

Therefore each unification model raises the question why such disparate mass scales occur. In terms of the standard SU(3) × SU(2) × U(1) model the unification scale is hierarchically large while in terms of scales set by the gauge unification or by gravity the weak scale is unnaturally small.

This hierarchy problem or naturalness problem becomes even more puzzling if one considers quantum corrections to hierarchically different scales. These scales are determined by vacuum expectation values (vev)of scalar (Higgs) fields which in turn depend on the mass parameters of the scalar fields. These masses get quantum corrections by diagrams as  which scale like Λ^2, if a cutoff Λ is used to make the loop integration finite. To cancel these contributions one has to introduce counterterms which again have to be finely tuned to yield hierarchically different mass scales.

A reasonable solution to this problem seems to be the suggestion that the large scale is the natural scale and that the weak scale vanishes due to some symmetry in lowest order. The breakdown of this symmetry enables the scalar fields to develop masses and thereby sets the weak scale M_W. Moreover the symmetry has to tame quantum corrections to the masses in such a way that the quadratically divergent contributions  are cut off at about $\Lambda^2 = M_W^2$.

Under mild restrictions (e.g. finite number of particles) all symmetries compatible with a local relativistic quantum field theory have been classified |1|. Apart from one candidate no symmetry can accomplish the above requirements. The only candidate left is supersymmetry: and in fact, it does the job. Supersymmetry is generated by infinitesimal transformations D_α which change bosonic fields into fermionic fields and vice versa.

$$D_\alpha(\text{Boson}) = \text{Fermion} \qquad\qquad D_\alpha(\text{Fermion}) = \text{Boson} \tag{1}$$

If one combines this simple statement with the fundamental fact (of relativistic local quantum field theory) that fermions carry half integer spin and anticommute while bosons have integer spin and are quantized by commutation relations, then the main peculiarities of supersymmetry result.

D_α has to carry half integer spin i.e. α is a spin index and D_α behaves like a fermionic object, it changes the algebraic properties of the fields. Its Lie-algebra therefore involves anticommutators. The decisive relation is given by |2|

$$\{D_\alpha, \bar{D}_{\dot\alpha}\} = -2i\sigma^m_{\alpha\dot\alpha}\frac{\partial}{\partial x^m} \tag{2}$$

($\bar{D}_{\dot{\alpha}}$ is the complex conjugate of D_α, $\sigma^m_{\alpha\dot{\alpha}}$, m = 0,1,2,3 are the matrix elements of the 4 hermitian 2 × 2 matrices α = 1,2, $\dot{\alpha}$ = 1,2). The relation states that supersymmetry transformations close into space-time translations $\partial/\partial x^m$, or stated slightly different that supersymmetry transformations behave like the square root of translations. Therefore one can understand why gauged supersymmetry will contain gravity, the gauge interaction of gauged translations. Gauged supersymmetry is therefore called supergravity.

The algebraic relations of D_α are unique (up to doubling, tripling etc. of the D_α yielding extended supersymmetry) as shown by Haag, Sohnius and Łopuszanski |1|. The relations which we did not list (confer e.g. |3,4|) imply that supertransformations commute with the mass operator and charge operator. So if supersymmetry is exact the masses and charges of susy partners, i.e. particles related by supersymmetry, are equal. Spin and statistics of susy partners, however, are different.

Experimentally no supermultiplet has been observed in the accessible energy range. This is neither surprising nor discouraging. Supergravity is governed by the scale of gravitational interactions

$$M_{Pl} = 2,4 \cdot 10^{18} \text{ GeV}$$

and one would not expect supergravity to be relevant below this scale. If the second scale, present in supersymmetric models, the scale of supersymmetry breaking M_S would be much below the weak scale M_W one would have to understand anew how M_W orginates.

After these reasuring considerations we proceed and introduce the systematic notation for particles contained in supermultiplets. The quarks q and leptons l of the standard model enter a supersymmetric model with spin 0 partners sq and sl with identical gauge interactions. There are as many scalar fields as fermionic ones, i.e. each lefthanded quark or lepton field q_l, l_l is partner of a complex scalar field sq_l, sl_l, the same aplies to righthanded fields q_R, l_R occurs with sq_R and sl_R. The Higgs field H of the standard model is accompanied by a chiral fermion $\tilde{H}$. Generically superpartners of standard bosons are denoted by the final syllable -ino, so the Higgs particle is partner to the Higgsino $\tilde{H}$, the photon γ comes together with $\tilde{\gamma}$ the photino etc. These fermionic partners have spin $\frac{1}{2}$. Only the gravitino, the partner of the graviton, carries spin $\frac{3}{2}$. The multiplets are classified according to the highest spin as 1) chiral multiplet ϕ containing a complex scalar and a chiral spin $\frac{1}{2}$ fermion 2) gauge multiplet $\tilde{\gamma}$ containing gaugino and Yang-Mills spin 1 field and 3) gravity multiplet consisting of gravitino and graviton. If the Higgs mechanism makes a gauge multiplet massive, a chiral multiplet H, $\tilde{H}$ combines with a gauge multiplet Z, $\tilde{Z}$ to yield a massive vector boson Z, a massive Dirac fermion consisting out of $\tilde{H}$ and $\tilde{Z}$ and a real scalar partner sZ (the remnant of the complex Higgs) which is needed to make number of degrees of freedom in the bosonic sector match the number in the fermionic sector of the massive gauge multiplet. The physical properties of the members of the multiplets are determined by the Lagrangian which is resticted by supersymmetry and the requirement that Bose-fields enter with two derivatives, Fermi-fields with one derivative at most. This fixes the Lagrangian to be of the form

$$L_{Susy} = D^2[\bar{D}^2 K(\phi,\bar{\phi}) + \tfrac{1}{4} f(\phi)\, \tilde{\gamma}\, \tilde{\gamma} + g(\phi)] \quad (3)$$

D and $\bar{D}$ generate infinitesimal supertransformations (1,2) $\tilde{\gamma}$ denotes collectively the gaugino fields and ϕ the complex scalar fields of the chiral multiplets. $K(\phi,\bar{\phi})$ is a real function which determines the kinetic energies of the chiral multiplets. $f(\phi)$ is an analytic function which multiplies the kinetic energies of the gauge multiplets. If one requires L_{Susy} to be renormalizable and kinetic energies to be positive then K and f are completely fixed and g is a 3rd order polynomial

$$K(\phi,\bar{\phi}) = -\frac{1}{3}|\phi|^2 + \text{const}$$
$$f(\phi) = \frac{1}{e^2} = \frac{1}{(\text{gauge-coupling})^2} \quad (4)$$

The additive constant in K can be dropped as it does not enter any physical quantity. $g(\phi)$ is called the superpotential because the potential $V(\phi,\bar{\phi})$ is determined by $g(\phi)$ if one spells out the shorthand notation (3) of the Lagrangian

$$V(\phi,\bar{\phi}) = \left|\frac{\partial g(\phi)}{\partial \phi}\right|^2 \tag{5}$$

The superpotential also determines masses and Yukawa-couplings of $\tilde{\phi}$, the ferminonic partner of ϕ

$$L_{Yukawa} = -\frac{1}{2}\tilde{\phi}\,\tilde{\phi}\,\frac{\partial^2 g}{\partial\phi\partial\phi} + \text{h.c.} \tag{6}$$

Couplings and masses of fermions and bosons are therefore related. Masses of scalars are tied to masses of fermions and can therefore be forced to vanish if a chiral symmetry keeps fermion masses zero. So supersymmetry contains a mechanism to make scalar fields massless. In fact, it accomplished more: it tames the quantum corrections and cancels the quadratically divergent contributions from scalar loops by fermionic loops

(7)

This cancellation works because the ϕ^4 coupling of the first diagram is the square of the Yukawa-coupling of the second diagram, this follows from (5,6), the propagator of the Bose-field is the square of the Fermi-propagator and ultimately the Fermi-loop comes with a relative minus sign because Fermions anticommute. More generally one can show that no quantum corrections require counterterms for the superpotential - wich is the non-renormalization property of supersymmetric models. So supersymmetry tames quantum corrections.

Supersymmetry has more mechanisms in stock to provide massless scalars at tree level . O'Raifeartaigh |5| has shown that spontaneous breakdown of supersymmetry is always accompanied not only by a massless Goldstone-fermion field (as required by Goldstone's theorem) but also its superpartner call it y, remains massless. In fact its vev cannot be fixed at tree level because the potential is constant if the remaining fields are fixed to their vev.

$$V_{tree}(y, \phi = \langle\phi\rangle) = v(y) \equiv \text{const} \tag{8}$$

Because $\langle y\rangle$ cannot be fixed at tree level, y is called a sliding field. This property (8) is not enforced by a symmetry : $\langle y\rangle$ can be fixed by quantum corrections. They can therefore generate a scale which is different from the natural scale M_S of the model set by the breakdown of supersymmetry |6|. There is even a further possibility in supersymmetric models to make the Higgs doublets masslesss which are to break the weak gauge symmetry: if the Higgs doublets enter a gauge unification model, take (SU(5) for example, they are part of a larger multiplet. The breakdown of SU(5) would normally provide masses for all components of the multiplet

$$M_{Doublet} = c_2 M_{GUT} \qquad\qquad M_{Triplet} = c_3 M_{GUT} \tag{9}$$

with different (i.e. SU(5) breaking) coefficients for the doublet part and the remaining components (triplet) of the SU(5)-Higgs multiplet. M_{GUT} is some SU(5)-breaking vev. One can introduce a gauge singlet field S into the model and change the mass relations to

$$M_{Doublet} = c_2 M_{GUT} - \langle S\rangle \qquad\qquad M_{Triplet} = c_3 M_{GUT} - \langle S\rangle \tag{9a}$$

This does not look more promissing than the former relation: if $\langle S\rangle$ is fixed by the parameters of the Lagrangian a cancellation $M_{Doublet} = 0$ needs finetuning of the parameters. The situation changes completely in supersymmetric models where symmetries can enforce $\langle S\rangle$ to be undetermined at tree level. If S is a sliding singlet then quantum corrections can, and in fact will, fix $\langle S\rangle$ so as to cancel the tree level Higgs mass | 6,7|. Let us now briefly describe the situation if one gauges supersymmetry, i.e. if one introduces gravity and the gravitino. Apart from technical details the supergravity Lagrangian is still given by (3). One has to take D and $\bar{D} = (D)^*$ to be the covariant versions of supersymmetry transformations |4| moreover a volume element has to be included to make the integration $\int d^4x$ covariant. Supergravity models are specified by the choice of a real function $K(\phi,\bar{\phi})$ determining the kinetic energies of the chiral multiplets and the gravity multiplet, an analytic function $f(\phi)$ entering the kinetic energies of the gauge multiplets and an analytic superpotential $g(\phi)$. Renormalizability can no longer be imposed to restrict K, f and g but requiring dilatational invariance does the same job. If in addition all resulting kinetic energies are to be positive then K, f and g are fixed to be

$$\begin{aligned} K &= |\phi^\circ|^2 - \frac{1}{3}\sum_i |\phi^i|^2 \\ f &= \frac{1}{e^2} = \frac{1}{(\text{gauge-coupling})^2} \\ g &= c_{ijh}\,\phi^i\phi^j\phi^h \qquad \text{(cubic)} \end{aligned} \tag{10}$$

Only if the vev of K is positive, the gravitational kinetic energy (the Einstein action) is positive. This fixes the sign of $|\phi^\circ|^2$ in K and requires $|\langle\phi^\circ\rangle|^2 > \frac{1}{3}\Sigma|\phi^2|^2$. The Lagrangian in fact has a larger symmetry than gauged supersymmetry: it is also invariant under gauged dilatations which can be used to transform ϕ° from whatever value it has to

$$\phi^\circ = M_{Pl} = 2.4 \cdot 10^{18}\ \text{GeV} \tag{11}$$

So ϕ° drops out of the model and the functions K, f, and g are now

$$K = M_{Pl}^2 - \frac{1}{3}|\phi|^2 \qquad f = \frac{1}{e^2} \qquad g \quad 3^{rd} \text{ order polynimial} \tag{12}$$

This is exactly the same input as for globally supersymmetric renormalizable models (4), the constant term in K now aquires a physical meaning: it is the inverse of Newton's gravitational constant.

Dilatationally invariant supergravity Lagrangians (with spontaneous breakdown of this dilatational invariance (11)) therefore coincide with the Lagrangian obtained by minimal substitution from renormalizable globally supersymmetric models. This is why we call such models minimal. The kinetic energies of the scalar fields still reflect the SU(n,1) symmetry of K (1o), which is spontanenously broken (11). They form a non-linear σ-model $(SU(n,1)|(SU(n)\times U(1))$. This symmetry of the kinetic energies however, is an accidental byproduct of dilatational invariance and is broken by the superpotential.

Because of the comparatively complicated kinetic energies resulting from the minimal choice (12) one investigated mainly the case that canonical kinetic energies emerge. This requires as input

$$K = M_{Pl}^2 \cdot \exp\left(-\frac{|\phi|^2}{3M_{Pl}^2}\right) \tag{13}$$

which we call canonical. Introducing supergravity with this canonical choice spoils all the mechanisms described so far to generate massless Higgs fields and to explain the smallness of the weak scale M_W.

On the other hand minimal supergravity models allow to understand why hierarchically different mass scales emerge: One finds |7,4|: 1) the cosmological constant, generated by $\langle V(\phi,\bar{\phi})\rangle$ is never positive 2) if the cosmological constant is arranged to vanish (as required by astromomical observations $|\langle V(\phi,\bar{\phi})\rangle| \leq (10^{-30}M_{Pl})^4$) and if supersymmetry is spontaneously broken, then as in (8) a sliding scalar field y emerges, i.e.

$$V_{tree}(y,\phi = \langle\phi\rangle) = v(y) \equiv 0 \tag{8'}$$

3) all scales including the scale of supersymmetry breaking, the gravitino mass $M_{3/2}$, depend on $\langle y\rangle$ and slide. 4) all couplings of y to the remaining fields ϕ are surpressed by inverse powers of M_{Pl}, so y is hidden.

The dependence of the masses M_ϕ on <y> is devers depending on the coupling of the y in the superpotential.

Introducing

$$\varepsilon(\langle y\rangle) = \left|\frac{M_{Pl} + \sqrt{3}\langle y\rangle}{\sqrt{M_{Pl}^2 - 3|\langle y\rangle|^2}}\right| \tag{14}$$

the gravitino mass is found to scale like

$$M_{3/2}(\langle y\rangle) = \mu \cdot \varepsilon^3(\langle y\rangle) \tag{15}$$

while scalar masses can scale as

$$M_\phi = (\frac{N}{2\varepsilon}\ , \ \varepsilon \cdot M'\ , \ \varepsilon^3 \cdot M\ ) \tag{16}$$

μ, N, M', M are fixed parameters of the superpotential and can all be of the natural scale

$$\mu,\ N,\ M',\ M \quad \sim 0(\ M_{Pl}\ ) \tag{17}$$

If the sliding vev <y> is fixed by quantum corrections so as to make ε small, hierarchically different mass scales emerge.

At tree level the masses in the light chiral multiplets are split, so supersymmetry is broken in this sector

$$M_{Scalar}^2 = M_{Fermion}^2 \pm 2\, M_{3/2} \cdot M_{Fermion} \qquad = \varepsilon^6(M^2 \pm 2\mu M) \tag{18}$$

(The ± spin corresponds to the two real fiels contained in the complex scalar). Because of this breakdown of supersymmetry quantum corrections to the masses of gauginos and scalar partners of quark and leptons no longer vanish. They are finite however and therefore calculable in terms of gauge coupling and M . Working through a typical model |8| one finds a Higgs potential

$$V(\phi) = \lambda\phi^4 + p\, M_{3/2}^2\phi^2 + q\, M_{3/2}^2\, \mathrm{Re}\phi^2 \ln \frac{M^2}{M_{Pl}^2} \tag{19}$$

The ϕ^4 coupling λ is an input parameter, a sliding singlet has canceled whatever tree mass the Higgs had (9a). $pM_{3/2}^2$ is the finite quantum correction to the mass showing that $M_{3/2}$ works as a cutoff Λ. Actually the coefficient p comes from a 2-loop diagram containing gauge couplings only and is therefore no free parameter. The sliding singlet which cancelled the tree mass obligingly contributes the third term via a 3-loop diagram, containing unspecified Yukawa-couplings. Because of the higher loop order q is naturally smaller than p. The potential still contains the sliding vev of y via the gravitino mass (15) and therefore has to be extremal with respect to φ and $M_{3/2}$. This fixes $M_{3/2}$

$$\frac{\partial V}{\partial M_{3/2}} = 0 \quad \Longleftrightarrow \quad M_{3/2}^2 = M_{Pl}^2\ e^{-(\frac{p}{q}+1)} \tag{20}$$

and a hierarchy emerges naturally because q is smaller than p. Using the experimentally determined mass

$$M_W = g_2 \langle\phi\rangle = 82 \text{ GeV} \tag{21}$$

to fix λ and (20) to eliminate q, the potential (and therefore the complete low energy Lagrangian) contains only $M_{3/2}$ as parameter which is not yet experimentally fixed.

In particular the masses of all superpartners of the known quarks and leptons can be calculated in terms of the gravitino mass |8|. In the range of

$$3\ \text{TeV} < M_{3/2} < 11\ \text{TeV} \tag{22}$$

the model is consistent with the absence of experimental evidence for supersymmetry.

The next generation of accelerators, the Stanford Linear Collider,however, will decide whether the explanation of the hierachy problem is tenable. It can pairproduce charged particles with mass $M \leq 50$ GeV. Depending on the magnitude of supersymmetry breaking, however, either the scalar partners of right handed electrons, muons, or taus or the lighter fermionic partner of the W have mass below 40 GeV. They cannot all escape detection at the SLC.

Supersymmetry had soon been realized to be the only remedy against quantum corrections which spoil hierarchical mass differences of scalar fields. Supergravity can even generate hierarchically different mass scales. Within the near future experiments will tell whether supersymmetry is in fact the solution to the hierarchy problem.

References

1 R. Haag, J. Łopuszanski and M. Sohnius, Nucl. Phys. B 88, 257 (1975)
2 J. Wess and B. Zumino, Nucl. Phys. B 70, 39-50 (1974), Phys. Lett. 49 B 52-54 (1974)
3 J. Wess and J. Bagger, Supersymmetry and Supergravity, Princeton University Press (1983)
4 N. Dragon, U. Ellwanger and M.G. Schmidt, Supersymmetry and Supergravity, Heidelberg preprint HD-THEP-86-3 to appear in "Progress in Particle and Nuclear Physics"
5 L. O'Raifeartaigh,Nucl. Phys. B 96, 331-352 (1975)
6 E. Witten, Phys. Lett. 105 B, 267 (1981)
7 N. Dragon, M.G. Schmidt and U. Ellwanger, Phys. Lett. 145 B, 192-196 (1984)
8 N. Dragon, U. Ellwanger and M.G. Schmidt, Phys. Lett. 154 B, 373-380 (1985)

Superstrings

J.-P. Derendinger

Laboratoire de Physique, Théorique de l'Ecole Normale Supérieure*,
24 rue Lhomond, F-75321 Paris Cedex 05, France

1. INTRODUCTION

The Standard Model [1] gives a satisfactory description of strong and electroweak interactions, at least in the range of energies accessible with present experiments. It is a Yang-Mills theory and is then fully consistent at the quantum level. At larger energies, there may be a further step of unification, for instance in a simple gauge group (SU(5), SO(10),...), introducing new interactions. This new level of synthesis would not in principle require a fundamental change in the theoretical framework, as long as it occurs at energies smaller than the gravity scale (the Planck scale). One would have again a gauge theory with a new, unified gauge group and an enlarged set of matter fields.

When one reaches the energy at which gravitational interactions have a strength comparable to the other interactions, a new theoretical framework has to be adopted for the description of particle interactions. A hypothetical quantum theory with gravity cannot contain only particle with spin 0, 1/2 and/or 1, in addition to the spin 2 graviton. Such a theory is known to be non renormalizable. Supergravity theories [2] have a more satisfactory ultraviolet behaviour. The presence of supersymmetric partners allows the cancellation of some divergences. For supersymmetric Yang-Mills theories, this mechanism is the origin of important non-renormalization or even finiteness theorems. For supergravity, explicit cancellation of some gravitational divergences has been observed, but there are however no reasons to believe that supergravities are either renormalizable or finite. Moreover, realistic supergravity models, with acceptable gauge groups and particle content, are divergent. A common belief is now that supergravity could be the effective field theory resulting from a more fundamental unified theory of all interactions, which would also make sense at the quantum level, and at energies larger than the Planck scale, $M_P \simeq 10^{19}$ GeV.

The only known candidate for this fundamental theory is a small class of superstring theories [3], [4]. Each of them contains an infinite set of particles, with arbitrarily high spins, but with specific masses and couplings. They also possess

---

* Laboratoire Propre du Centre National de la Recherche Scientifique, associé à l'Ecole Normale Supérieure et à l'Université de Paris-Sud.

unique quantum properties [5], [6]. For specific choices of the gauge group, these superstrings are finite at one-loop level [5]. There are arguments, but no proof, suggesting that they are indeed finite to all orders. String theories exist only for fixed space-time dimensions. For superstrings, this critical dimension is ten. In ten dimensions however, a quantum theory is in general anomalous in the Yang-Mills and in the gravity sector. It was however found that the mechanism which cancels the divergences also removes the anomaly [6]. These two important results strongly suggest that these superstrings may be consistent quantum theories with gravity.

The list of anomaly-free, one-loop finite superstrings is rather short. Three models are known: Type I superstrings [3] with gauge group SO(32), and heterotic strings [7] with gauge group SO(32) or $E_8 x E_8$. At present, it is not clear whether this list is complete.

These notes are divided in two parts. An elementary review of string and superstring theories is first given. Since many aspects of all string theories are already present in the simplest models, I will mainly concentrate on bosonic strings. The second part is devoted to the connection of superstrings with the gauge theory describing the interactions of quarks and leptons, the Standard Model or a more unified gauge theory. I will shortly discuss some implications of this string unification program at lower energies.

2. STRINGS AND SUPERSTRINGS

As indicated by the name, strings are one-dimensional objects of finite length, living in a D-dimensional space-time. We will certainly have to distinguish between the two possible configurations, open and closed strings. These two cases possess different vibration modes, and it is in fact the spectrum of these vibrations which gives the particle content of the string theory.

To describe the string, we give its space-time coordinates in terms of two parameters, σ and τ. At any given time τ, the string is a curve (in space-time) parametrized by σ. We arbitrarily choose $0 \leq \sigma \leq \pi$. Thus, for each time τ and for each position σ along the string, we give a full Lorentz vector of coordinates,

$$X^{\mu}(\sigma, \tau) \quad , \qquad \mu = 0, 1, \dots, D-1. \tag{1}$$

These coordinates, which describe the evolution (with time τ) of the string, define a surface in space-time, the world-sheet, analogous to the path of a pointlike particle in space-time. For closed strings, the world-sheet will look like a tube (Fig. 1). The description of the string with a full vector of coordinates at each point (σ,τ) of the world-sheet is useful for a Lorentz covariant treatment of string dynamics. We have however too many degrees of freedom. It is for instance clear that the two time-like quantities τ and $X^0(\sigma,\tau)$ are redundant. This difficulty is eluded by the requirement of reparametrization invariance of the world-sheet, which means that the dynamics of the string does not depend on the choice of parametrization of

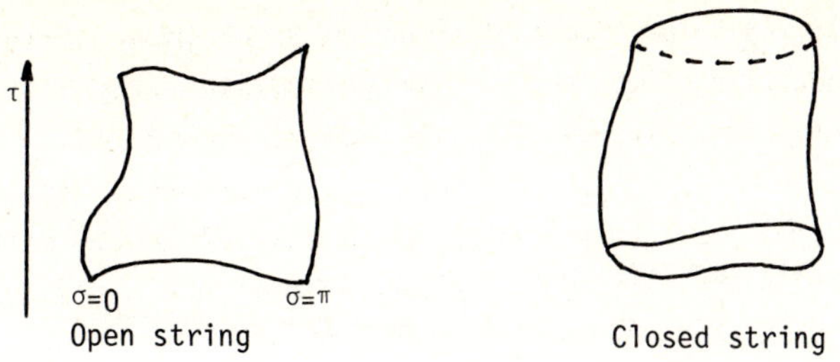

Fig. 1 : The world-sheet

the world-sheet, in terms of σ and τ. This invariance principle allows us, for instance, to choose a parametrization for which $\tau \propto X^0$. A covariant formalism should then be invariant under the transformations

$$\sigma , \tau \quad \rightarrow \quad \sigma' (\sigma , \tau) \;, \; \tau' (\sigma , \tau) \; . \tag{2}$$

The requirement of invariance under general coordinate transformations in the two-dimensional space (σ,τ) suggests that string theories should share some resemblance with gravity theories.

The appropriate equation of motion, describing vibrating strings, is the wave equation

$$- \left(\frac{\partial^2}{\partial\tau^2} - \frac{\partial^2}{\partial\sigma^2} \right) X^\mu = 0 \; , \tag{3}$$

which can be rewritten in the more compact form

$$\eta_{\alpha\beta} \, \partial^\alpha \partial^\beta \, X^\mu = 0 \; . \tag{4}$$

The notations

$$\partial_1 = \partial_\tau = \frac{\partial}{\partial\tau} \quad , \quad \partial_2 = \partial_\sigma = \frac{\partial}{\partial\sigma} \tag{5}$$

are used and the metric $\eta_{\alpha\beta}$ is

$$\eta_{\alpha\beta} = \begin{pmatrix} -1 & 0 \\ 0 & 1 \end{pmatrix} \; . \tag{6}$$

The equation of motion is supplemented by boundary conditions, specifying whether the string is open or closed. For open strings one requires

$$\partial_\sigma X^\mu (\sigma=0 , \tau) = \partial_\sigma X^\mu (\sigma=\pi , \tau) = 0 \; , \tag{7a}$$

and for closed strings, the condition is

$$X^\mu (\sigma=0 , \tau) = X^\mu (\sigma=\pi , \tau) \; , \tag{7b}$$

for all values of τ. The wave equation (3) will allow vibrations of the string with a spectrum of modes characteristic of the boundary conditions (7). Each of these vibration modes will carry a fixed amount of energy. In the quantum string theory, these modes will correspond to an infinite series of particle states with a mass related to the energy level of the modes.

To construct further the theory and in particular to quantize it, we need an action. It is very simple to obtain the equation of motion (3) from a variational principle in the two-dimensional space (σ,τ). The corresponding action is

$$S_1 = -\frac{1}{4\pi\alpha'} \int_0^\pi d\sigma \int d\tau \left(\eta_{\alpha\beta}\, \partial^\alpha X^\mu\, \partial^\beta X^\nu\, \eta_{\mu\nu} \right) . \tag{8}$$

The quantity α' sets the energy scale and will further on be identified with the Regge slope, or the inverse string tension. The action S_1 does not however fulfill explicitely the requirement of reparametrization invariance. Consider now the action

$$S = -\frac{1}{4\pi\alpha'} \int_0^\pi d\sigma \int d\tau \sqrt{-g}\, g^{\alpha\beta}\, \partial_\alpha X^\mu \partial_\beta X^\nu\, \eta_{\mu\nu}. \tag{9}$$

We have introduced a two-dimensional metric $g_{\alpha\beta}(\sigma,\tau)$, with $g_{\alpha\beta} = g_{\beta\alpha}$, and its inverse $g^{\alpha\beta}$. These degrees of freedom are new dynamical variables. In eq. (9), g is the determinant of $g_{\alpha\beta}$, which is chosen with one positive and one negative eigenvalue. The action S possesses a large invariance, under:

i : global Poincaré transformations in D dimensions,

ii : local reparametrizations of the world-sheet, Eq. (2),

iii: Weyl rescalings $g_{\alpha\beta} \to \exp \lambda(\sigma,\tau) \cdot g_{\alpha\beta}$, with X^μ inert.

This large invariance can be used to obtain a variety of actions related to S by specific gauge choices. For instance we can go to the conformal gauge in which

$$g_{\alpha\beta} = \exp \phi(\sigma,\tau) \cdot \eta_{\alpha\beta} , \tag{10}$$

with the help of transformations ii. In fact this choice does not exhaust the invariance under reparametrization of the world-sheet: there is a residual conformal invariance. Substituting (10) into the action S leads directly to the action S_1. Since S is invariant under the Weyl rescalings iii, ϕ does not appear in S_1. The decoupling of ϕ is only true at the classical level [8]. In the quantized version the decoupling will persist only in the critical dimension $D = 26$.

The equations of motion of the dynamical variables $g_{\alpha\beta}$ can be used to obtain a more geometrical form of the action. One easily obtains that S is proportional (classically) to the area of the world-sheet. The analogy with point particles is now fully clear: the action for point particles is the length of the path, and trajectories are paths of minimal length.

The first form S_1 of the action corresponds to the gauge choice (10), substituted in S. S_1 is however subject to additional constraints due to the equation of motion for $g_{\alpha\beta}$. Using these constraints, one can go to a light-cone gauge in which

$$X^+(\sigma , \tau) = \frac{1}{\sqrt{2}} (X^0 + X^{D-1}) = x^+ + p^+ \tau , \tag{11}$$

i.e. time τ is identified with X^+. The action depends only on physical degrees of freedom and reads

$$S_{LC} = \frac{1}{4\pi\alpha'} \int_0^\pi d\sigma \int d\tau\, \eta^{\alpha\beta}\, \partial_\alpha X^i\, \partial_\beta X^i \quad , \quad i = 1, \dots , D-2 , \tag{12}$$

in terms of transverse string coordinates only. Lorentz transformations have become non linear in that gauge.

The various actions we have described could in principle be the starting point of the quantum theory. There are two alternative approaches. First, one can adopt a Lorentz covariant quantization procedure, with an action containing the full vector of coordinates and linear Lorentz transformations. The explicit Lorentz covariance makes the quantization simpler and more elegant. The problem is the unphysical degrees of freedom contained in X^μ . The theory contains ghost states with negative probalities. These unphysical states have to be removed by imposing constraints on the physical spectrum. This procedure can be achieved only in the critical dimension D = 26.
The second possibility is to start with an action containing only physical degrees of freedom. The light-cone action (12) is the starting point. The difficult issue is Lorentz invariance. The SO(D-1,1) algebra spanned by the non-linear Lorentz transformations is particularly difficult to check, and this problem becomes worse when string interactions are introduced. Here, the critical dimension D = 26 arises from the fact that Lorentz invariance of the quantized theory holds only in that dimension. The light-cone gauge on the other hand exhibits directly the complete physical spectrum of the strings, which we will now describe. Let us first consider open strings. The equation of motion for the transverse coordinates,

$$- (\partial_\tau^2 - \partial_\sigma^2) X^i = 0 , \tag{13}$$

supplemented by boundary conditions (7a) leads to the following mode expansion:

$$X^i (\sigma,\tau) = x^i + 2\alpha' p^i \tau + i \sum_{n \neq 0} \frac{1}{n} \alpha_n^i \cos n\sigma \; e^{-in\tau} . \tag{14}$$

To quantize, impose canonical relations for the transverse coordinates and their conjugate momenta. These commutation relations lead to the following algebra for the operators α_n^i:

$$[\alpha_n^i , \alpha_m^j] = m \, \delta^{ij} \, \delta_{m+n,0} . \tag{15}$$

This algebra corresponds to an infinite set of harmonic oscillators, with raising operators $\alpha_{-n}^i/\sqrt{n}$, n = 1, 2, ... Furthermore, one can easily calculate the mass-shell condition

$$\alpha' M^2 = N - 1 = \sum_{n=1}^{\infty} \alpha_{-n}^i \alpha_n^i - 1 . \tag{16}$$

Thus, starting with a ground state |0>, annihilated by all lowering operators, which forms the N=0 level, we build the full set of string states acting with the raising operators. The ground state, N = 0, is a scalar tachyon with $M^2 = - \alpha'^{-1}$. The first level consists of a massless vector state:

$$N = 1 : \quad \alpha_{-1}^i \; |0\rangle . \tag{17}$$

The second level, with $\alpha' M^2 = 1$, contains the states

$$N = 2 : \quad \alpha_{-2}^i \; |0\rangle \; , \; \alpha_{-1}^i \, \alpha_{-1}^j \; |0\rangle \quad : \tag{18}$$

These states are a vector and a symmetric tensor of SO(24). They can be collected into a symmetric traceless rank two tensor of SO(25). All massive states appearing at higher levels will form representations of SO(25), which is the angular momentum in 26 dimensions. All higher levels can be analyzed in the same way. Their states lie on a linear Regge trajectory, with slope α'.

An analogous discussion can easily be given for closed strings. The basic difference comes from the boundary conditions (7b), which leads to the following mode expansion for the transverse string coordinates:

$$X^i(\sigma,\tau) = x^i + 2\alpha' p^i \tau + \frac{i}{2}\sum_{n\neq 0}\frac{1}{n}\left(\alpha_n^i e^{-2in(\sigma-\tau)} + \beta_n^i e^{-2in(\sigma+\tau)}\right) . \tag{19}$$

We now have two independent sets of oscillators, constructed with the α_n^i's and the β_n^i's, corresponding to harmonic waves moving along the closed string in the direction of increasing σ or decreasing σ. The algebra (15) holds for both sets and

$$[\alpha_n^i, \beta_m^j] = 0 . \tag{20}$$

The particle content of the closed bosonic string is dictated by the mass formula

$$\frac{1}{4}\alpha' M^2 = N_\alpha - 1 = N_\beta - 1 , \quad \text{where} \tag{21}$$

$$N_\alpha = \sum_{n=1}^{\infty} \alpha_{-n}^i \alpha_n^i \quad ; \quad N_\beta = \sum_{n=1}^{\infty} \beta_{-n}^i \beta_n^i . \tag{22}$$

We have again a scalar tachyon, $N_\alpha = N_\beta = 0$. The massless sector consists of

$$\alpha_{-1}^i \ \beta_{-1}^j \ |0\rangle . \tag{23}$$

The part which is symmetric in i and j is a symmetric SO(24) tensor, containing a scalar and a graviton.

Let us summarize the main aspects of free bosonic strings. They live only in 26 dimensions and suffer a fatal disease: the ground state is a tachyon. Bosonic strings cannot give consistent quantum theories due to this basic problem. Open bosonic strings have a massless vector state which will be convenient to generalize Yang-Mills theories, while closed strings contain the massless symmetric tensor to be identified with the graviton.

The problem of the tachyon is elegantly solved in the case of superstring theories. In addition to the bosonic (transverse) coordinates X^i, superstrings also contain fermionic coordinates

$$S_\alpha^a , \quad \alpha = 1, 2 . \tag{24}$$

Since the critical dimension is now ten instead of twenty-six, a = 1, ... , 8 is an SO(8) spinor index. Accordingly, α is an SO(1,1) spinor index, in the (σ,τ) space. The corresponding equation of motion is a two-dimensional Dirac equation

$$\begin{aligned}(\partial_\tau + \partial_\sigma)\, S_1^a &= 0 ,\\ (\partial_\tau - \partial_\sigma)\, S_2^a &= 0 ,\end{aligned} \tag{25}$$

and appropriate boundary conditions have to be imposed for open or closed superstrings. The light-cone bosonic action (12) can be easily extended to describe the dynamics of the (free) fermionic coordinates. One only needs to add a light-cone Dirac action in two dimensions. An important new ingredient of superstrings is that they possess ten-dimensional supersymmetry. All string states can then be collected at each level in supersymmetry multiplets.

From the equation of motion (25) and the boundary conditions, one gets the mode expansion of S^a_α. Quantization now proceeds through imposition of anti-commutation relations. The effect of this procedure is to remove the tachyon state and the fundamental level is now a multiplet of massless states which is, for open strings, a Yang-Mills supermultiplet (with gauge fields A_μ and gaugino spinors χ) and, for closed superstrings, a supergravity multiplet (containing the graviton, the gravitino and some additional bosons and fermions). Closed superstrings also offer the possibility of extended supersymmetry. In fact, each set of oscillators (α^i_n and β^i_n in (19)) carries an independent supersymmetry. We then have three different classes of closed superstrings models. One can restrict the theory in such a way that only states built symmetrically in α^i_n and β^i_n are retained: this subsector is a consistent model, with only one supersymmetry (Type I closed superstrings). Then, theories with N = 2 supersymmetry are of two species, depending whether the two supersymmetries have opposite (Type IIa) or same (Type IIb) handedness. Type I and Type IIb superstrings are chiral models, in which all spinors have a definite (left or right) handedness.

Before discussing which superstring models could be physically relevant, let us sketch how do strings interact. Interactions can occur when two strings meet at some point. There are two different basic interactions. First, if two string endpoints meet. This results in either a trilinear open string vertex (i.e. two open strings joining to form an open string or one open string splitting into two pieces), or a transition vertex between closed and open strings (with an open string closing onto itself, or a closed string opening at some point). The second interaction occurs when two strings meet at a point other than endpoints. The corresponding vertices can involve open strings only, closed strings only, or both kinds of strings.
The first interaction generates a trilinear open string vertex, which in particular contains a three vector coupling in the massless sector. Its strength will then correspond to the Yang-Mills coupling constant. The second interaction generates a trilinear closed string vertex, and a trilinear graviton interaction in the massless sector; its strength will then be the gravitational constant. The full set of possible interactions is then:

open + open	$\leftrightarrow$	open	:	coupling g.
open	$\leftrightarrow$	closed	:	coupling g.
closed + closed	$\leftrightarrow$	closed	:	coupling κ.
closed + open	$\leftrightarrow$	open	:	coupling κ.
open + open	$\leftrightarrow$	open + open	:	coupling κ.

Clearly, an interacting theory with open strings has to contain all five interactions, while a theory with closed strings only possesses the third interaction only. In superstrings, the fermionic degrees of freedom interact in a manner fixed by supersymmetry.

To close this part, let us give a brief list of string theories, emphasizing their possible relevance as a unified theory of all interactions. Bosonic strings were considered long ago as possible theories of hadrons. Open bosonic strings have the scattering amplitudes of the Veneziano model. The necessity of introducing fermions suggested the spinning string model. This model was in fact the first theory possessing a two-dimensional supersymmetry. These old models contain tachyons and cannot be consistent quantum field theories. The spinning string is the prototype of superstrings, which are of three classes: Type I (open and closed strings, N = 1 supersymmetry) with possible gauge groups SO(n) or Sp(2n), Type IIa and Type IIb (closed strings only) with N = 2 supersymmetry, but no gauge group. Type I supestrings possess all the ingredients necessary to a unified theory of all interactions. Only one model however is possibly consistent at the quantum level. It was shown [5,6] that if the gauge group is SO(32), Type I superstrings are one-loop finite and free of anomalies. This result also holds for a gauge group $E_8 x E_8$, which is not admissible with Type I superstrings. A new class of "hybrid" superstring theories was constructed, with SO(32) and $E_8 x E_8$ gauge groups only: these models are called heterotic strings [7]. In fact, $E_8 x E_8$ heterotic strings offer the most promising features as a realistic unified theory of all interactions. It is not yet clear whether the list of string theories is really complete and if more promising models are still to be constructed.

3. FROM SUPERSTRINGS TO LOW-ENERGY PHYSICS

Even if we assume that anomaly-free superstrings are really consistent quantum theories with gravity, the existence of strong and electroweak interactions as well as the presence of quarks of leptons has to be demonstrated before the model can be considered as a possible unified theory. The problem of relating a fundamental string theory with the Standard Model is very complicated. The low-energy limit, $E \ll M_P$, has first to be taken. Since superstrings live in ten dimensions, six dimensions have to be compactified. This process should be derived from string dynamics, the choice of compactification background (and the number of compact dimensions) arising spontaneously from equations of motion. Compactification will give masses to most of the light string states. The necessity of keeping massless states and gauge symmetries compatible with the Standard Model is a very strong constraint on compactification.

The first question is the choice of the fundamental string theory. $E_8 x E_8$ is a more attractive gauge group than SO(32). The gauge group of the Standard Model is obviously a subgroup of both E_8 and SO(32), but it is a formidable task to break spontaneously the huge group SO(32) (dimension = 496, rank = 16) into G_{SM} = SU(3)x

SU(2)xU(1).Moreover the quantum numbers of the light states will appear to be closer to those of quarks and leptons when E_8 is the unified group. The existence of a second gauge group E_8' will have interesting implications. It could be the origin of a possible breaking of low-energy supersymmetry in the effective gauge theory [9,10]. We will then assume that $E_8 x E_8$ heterotic string is the fundamental theory.

The massless sector of Type I superstrings contains the following fields: a graviton $g_{\mu\nu} = g_{\nu\mu}$, an antisymmetric tensor $B_{\mu\nu}$, a real scalar ϕ, a gravitino spinor ψ_μ and a spinor λ form the supergravity multiplet, gauge fields A_μ and gaugino spinors χ form the Yang-Mills multiplet. Heterotic strings have the same massless sector, they differ only at higher, massive levels. These massless states generate upon compactification the fields to be identified with quarks, leptons, gauge fields and scalar bosons. The Lagrangian describing their interactions is a N = 1 chiral supergravity in ten dimensions. It is only partially known [11] and is the starting point to study the problem of compactification [12].

Compactification of six dimensions involves to find a space K_6 with a geometry compatible with string invariances (and in particular with conformal symmetry). The most attractive compactifications lead to unbroken supersymmetry in four dimensions: supersymmetry is probably desirable if one wants to explain the smallness of the weak interaction scale with respect to the Planck scale [13]. Low-energy predictions are also more reliable when these supersymmetric compactifications are considered. Two classes of candidate spaces K_6 have been considered: Calabi and Yau spaces [12] and orbifolds [14]. They lead basically to the same predictions at lower energies. The compactification process induces a first step of gauge symmetry breaking (through vacuum expectation values of gauge field strengths, $\langle F^\alpha_{\mu\nu} \rangle$). $E_8 x E_8'$ is broken into $E_6 x E_8'$. Massless states remaining after compactification are obviously classified in N = 1 supersymmetry multiplets in four dimensions. They will include the gauge multiplets of $E_6 x E_8'$ (with gauge vector bosons and spin 1/2 gauginos), and a set of chiral multiplets (with left-handed spin 1/2 states and complex scalars) transforming under E_6 according to

$$N\ 27 \quad + \quad \delta\ (\ 27 + \overline{27}\ ) \tag{26}$$

where N and δ are fixed by the geometry of the compact space K_6. Each 27 contains one generation of quarks and leptons, completed by twelve states possessing vector-like strong and electroweak interactions. The number N should then be three or four (larger N's would give problems with asymptotic freedom). In most cases of interest, δ = 1 : the 27 + $\overline{27}$ states are useful and necessary to complete the symmetry breaking.

The chiral multiplets (26) do not transform under E_8'. This means that there is no light state which feels simultaneously strong and electroweak interactions and E_8' interactions. E_8' is thus hidden and could only be detected using gravitational interactions. The existence of such a hidden world is certainly not in contradiction with any experimental data. Cosmology may give some constraints.

Compactification predicts that the unified gauge group is E_6 at the Planck scale (with, in addition, a hidden E_8'). It is however necessary to break E_6 further during compactification: if not, the scalar fields contained in chiral multiplets (26) are unable to complete the symmetry breaking into SU(3)xSU(2)xU(1). If the internal space K_6 is not simply connected (Calabi-Yau spaces with small values of N and orbifolds are not simply connected), then vacuum expectation values of gauge potentials $\langle A_\mu^\alpha \rangle$ become physically relevant and break further the symmetry. The dynamics of the corresponding Wilson loop operators [12] is not understood yet. One can however give the list of the resulting gauge groups and discuss whether they can give rise at low energies to a physics compatible with the Standard Model. This is essentially done by requiring that the scalar fields available among the chiral multiplets (26) are then able to break this resulting effective group G' into SU(3)xSU(2)xU(1) [15]. There are several possible models, giving acceptable values of the low-energy gauge coupling constants, provided some assumptions on the light spectrum are accepted. Two general results emerge:

- G' contains at least a new U(1) gauge symmetry with fixed quantum numbers on quarks and leptons; the scale of its breaking is not theoretically determined; this scale will however correspond to the mass of the twelve new states completing each quark and lepton generation, each 27 of E_6.
- G' cannot be a simple (or semi-simple) group like SU(5), SO(10)...

Many phenomenological implications of superstrings have been tentatively discussed. They strongly depend on the explicit form of the effective supersymmetric Lagrangian. Attempts to obtain this Lagrangian, by a "string inspired" dimensional reduction of ten-dimensional supergravity [16], or with the help of classical symmetries of superstrings [17], have led to the conclusion that the effective theory has the characteristics of a "no scale" model [18]. Two kinds of contributions should modify the Lagrangian: higher derivative terms (see for instance [19]) and loop corrections which will in general break classical symmetries [20-22]. The full modifications and their implications are yet unknown.

Another point of interest is supersymmetry breaking, which will arise from two sources: vacuum expectation values of the $B_{\mu\nu}$ tensor, and condensation of E_8' gauginos [9,10]. E_8' is a confining force which will generate gaugino condensation at a scale Λ, which should be 10^{10-11} GeV to induce the right scale of SU(2)xU(1) breaking. This condition implies that E_8' has to be broken in a quite contrived way by Wilson loop expectation values [20].

The supergravity Lagrangian derived by Witten [16], supplemented with supersymmetry breaking induced by gaugino condensates, has been used in several articles to study phenomenological implications of superstrings. The results of these preliminary investigations will not be reported here (see for instance [23]).

4. CONCLUSIONS

The study of superstring theories will probably bring decisive progresses in our understanding of quantum gravity. Their exceptional properties suggest that they are in fact consistent quantum theories with gravitation. A complete proof of this statement would by itself be a remarkable achievement.

Ultimately , a superstring theory could be a viable fundamental unified theory of all interactions. Heterotic strings contain most of the necessary ingredients. At present, there are still many theoretical mysteries, which have to be solved before phenomenology of superstrings can be safely analyzed. Superstring unification can work: all investigations have shown that no basic obstruction exists in that direction. But we are still clearly unable to discuss whether string dynamics will choose to realize the scheme, or prefer to follow a physically unrealistic road to low energies.

REFERENCES

1) S.L. Glashow, Nucl. Phys. 22 (1961) 579.
S. Weinberg, Phys. Rev. Lett. 19 (1967) 1264.
A. Salam, in "Elementary Particle Theory", ed. N. Svartholm (Almqvist and Wiksell, Stockholm, 1968).

2) S. Ferrara, D.Z. Freedman and P. van Nieuwenhuizen, Phys. Rev. D13 (1976) 3214.
S. Deser and B. Zumino, Phys. Lett. 62B (1976) 335.
For reviews, see: P. van Nieuwenhuizen, Phys. Rep. 68 (1981) 189.
S.J. Gates, M.T. Grisaru, M. Rocek and W. Siegel, "Superspace", Benjamin, Reading, 1983.

3) M.B. Green and J.H. Schwarz, Nucl. Phys. B181 (1981) 502; B198 (1982) 252; B198 (1982) 441.

4) For reviews, see: J.H. Schwarz, Phys. Rep. 89 (1982) 223;
M.B. Green, Surveys in High Energy Physics 3 (1983) 127;
L. Brink, CERN preprint TH.4006 (1984).
See also the collection of reprints in: "Superstrings", ed. by J.H. Schwarz, World Scientific, Singapore, 1985.

5) M.B. Green and J.H. Schwarz, Phys. Lett. 151B (1985) 21.

6) M.B. Green and J.H. Schwarz, Phys. Lett. 149B (1984) 117; Nucl. Phys. B255 (1985) 93.

7) D.J. Gross, J.A. Harvey, E. Martinec and R. Rohm, Phys. Rev. Lett. 54 (1985) 502; Nucl. Phys. B256 (1985) 253; Nucl. Phys. B267 (1986) 75.

8) A.N. Polyakov, Phys. Lett. 103B (1891) 207.

9) J.-P. Derendinger, L.E. Ibanez and H.P. Nilles, Phys. Lett. 155B (1985) 65.

10) M. Dine, R. Rohm, N. Seiberg and E. Witten, Phys. Lett. 156B (1985) 55.

11) E. Bergshoeff, M. de Roo, B. de Wit and P. van Nieuwenhuizen, Nucl. Phys. B195 (1892) 97.
G.F. Chapline and N.S. Manton, Phys. Lett. 120B (1983) 105.

12) P. Candelas, G.T. Horowitz, A. Strominger and E. Witten, Nucl. Phys. B258 (1985) 46.

13) N. Dragon, these Proceedings.
For a review, see also: H.P. Nilles, Phys. Rep. 110 (1984) 1.

14) L. Dixon, J.A. Harvey, C. Vafa and E. Witten, Nucl. Phys. B261 (1985) 651.

15) S. Cecotti, J.-P. Derendinger, S. Ferrara, L. Girardello and M. Roncadelli, Phys. Lett. 156B (1985) 318.
M. Dine, V. Kaplunovsky, M. Mangano, C. Nappi and N. Seiberg, Nucl. Phys. B259 (1985) 519.
J.D. Breit, B.A. Ovrut and G. Segré, Phys. Lett. 158B (1985) 33.

16) E. Witten, Phys. Lett. 155B (1985) 151.

17) C.P. Burgess, A. Font and F. Quevedo, Univ. of Texas preprint UTTG-31-85/CTP-046 (1985).

18) E. Cremmer, S. Ferrara, C. Kounnas and D.V. Nanopoulos, Phys. Lett. 133B (1983) 61.
J. Ellis, C. Kounnas and D.V. Nanopoulos, Nucl. Phys. B241 (1984) 406; B247 (1984) 773; Phys. Lett. 143B (1984) 410.

19) S. Cecotti, S. Ferrara, L. Girardello and A. Pasquinucci, UCLA preprint 85/TEP/24 (1985).

20) J.-P. Derendinger, L.E. Ibanez and H.P. Nilles, Nucl. Phys. B267 (1986) 365.

21) Y.J. Ahn, J.D. Breit and G. Segré, Univ. of Pennsylvania preprint (1985).
M. Quiros, CERN preprint TH.4363/86 (1986).

22) L.E. Ibanez and H.P. Nilles, CERN preprint TH.4355/86 (1986).

23) J. Ellis, CERN preprint TH.4439/86 (1986).

Baryon and Lepton Number Violation in Superstring Motivated Models

Q. Shafi *†

Bartol Research Foundation, University of Delaware, Newark, DE 19716, USA

We discuss how realistic low energy physics can arise from the $E_8 \times E_8$ superstring. A key role is played by the discrete symmetries which typically arise after compactification. Cosmological problems are avoided because the discrete symmetry we employ is shown to effectively be embedded in a Peccei-Quinn symmetry. The models we present seem to satisfy the known phenomenological constraints. In particular, they possess a harmless axion. Implications for baryon and lepton number violation are discussed.

1. Superstring Model Building

The discovery of anomaly free superstring theories [1] and the subsequent development of the $E_8 \times E_8$ heterotic theory [2] have fueled expectations that we may have at hand a theory which provides a consistent unification of gravity with the other three forces. Compactification of the $E_8 \times E_8$ theory on Calabi-Yau (C-Y) spaces can lead to chiral fermion families, a must for any realistic theory, and even delivers an unbroken $N = 1$ supersymmetry [3]. However, attempts to obtain a realistic low energy phenomenology and a consistent cosmology must face up to some formidable problems. For instance, there are potential dangers from sources such as rapid proton decay, $\sin^2\theta_W$, flavor changing neutral currents, neutrino masses, domain walls, visible axions, etc. Moreover, the model independent axion in superstring theories [4] does not satisfy the standard astrophysical and cosmological constraints.

In this talk I wish to present a unified approach to these and some other related questions. The aim is to come up with models which, in addition to satisfying all of the above constraints, make some predictions.

Before embarking on the actual model building and the ensuing consequences, we need to clarify a number of points. First, let us state some of our main assumptions:

i) The compactification of the ten-dimensional $E_8 \times E_8$ superstring theory on a suitable C-Y space gives rise to three generations of chiral massless

---

* Based on work done in collaboration with G. Lazarides and C. Panagiotakopoulos.

† Supported in part by the Department of Energy under Contract Grant Number DE-AC02-78ER05007.

fermions (contained in the $\underline{27}$'s of E_6). In addition, there will be a certain number of chiral matter multiplets arising from incomplete $\underline{27}$'s and $\overline{\underline{27}}$'s of E_6 [5],[6].

ii) The C-Y space is non-simply connected thus allowing for non-trivial Wilson loops [3]. The effective gauge group below the compactification scale M_C is a subgroup G of E_6 and, depending on the details of the symmetry breaking, possesses rank five or six [5]. Clearly, G must contain $SU(3)_C \times SU(2)_L \times U(1)_Y$ (3-2-1) as a subgroup.

iii) There exists some mechanism of spontaneous supersymmetry breaking, presumably triggered by the hidden E_8 sector [7], which gives rise to the weak scale M_W. All scalars are assumed to get $(mass)^2$ terms of order $M_W{}^2$ through gravitational interactions.

Next, in order to obtain a phenomenologically acceptable theory, it seems necessary, even in the case of a three family model, that many of the extra fields in the $\underline{27}$ acquire an intermediate mass M_I, which is much larger than M_W. Some reasons for such an intermediate scale are:

a) The requirement of perturbative unification.

b) The possibility of creating some mixing between the Higgs doublets $\bar{H}$ and H which give masses to the up quarks and the down quarks (and the charged leptons). The mixing is important in order to obtain tree level electroweak breaking without the need for large Yukawa couplings that certainly are absent for H.

c) The possibility of having acceptable proton lifetime without imposing exact baryon number conservation, which helps in the creation of baryon asymmetry in the universe.

The most plausible intermediate scale in superstring models appears to be $-\sqrt{M_W M_C} \equiv M_I \sim 10^9 - 10^{10}$ GeV for $M_C \sim 10^{17}$ GeV. Its appearance requires the existence of at least one light 3-2-1 singlet from a $\underline{27}$, accompanied by a corresponding singlet from a $\overline{\underline{27}}$, plus the existence of D- and F- flat directions in the potential of this singlet [8],[9].

Having convinced ourselves (!) that superstring models ought to possess an intermediate scale M_I at which a part of the gauge group G is broken, we will argue that at least one independent global symmetry also be broken at M_I in order that the models be not phenomenologically and cosmologically unacceptable. The reasons for this include:

1) The absence of large flavor changing neutral currents which are generally present in models with several light Higgs doublets. In superstring models there are at least six such doublets. On the basis of their quantum numbers with respect to the local symmetries there is no reason why some of them are heavy and the others light, or why some of them couple to quarks and leptons and the others do not. If however, there is a suitable global symmetry which differentiates between them the problem can be overcome.

2) The absence of large masses for the known neutrinos. Unless an extra global symmetry is present neutrino masses generally turn out to be much too large [6].

3) Acceptable proton lifetime. The presence of a global symmetry eliminates some of the undesirable couplings which otherwise would lead to a much too rapid proton decay [5],[6],[8],[9].

An obvious candidate for the global symmetry in a superstring model is the discrete symmetry that typically arises after compactification of a superstring theory on a C-Y space [5]. Such a symmetry may help in avoiding large neutrino masses and a rapid proton decay. It should be remembered, however, that discrete symmetries create severe domain wall problems if they are spontaneously broken.

The domain wall problem would be neatly resolved if the discrete symmetry could effectively be embedded in some continuous (global) symmetry. Before proceeding to a discussion of what we consider an attractive candidate for the continuous symmetry, let us remark that the continuous symmetry could not reasonably be expected to be a symmetry of the complete four dimensional theory, including all the non-renormalizable interactions. It can, at best, be expected to be the symmetry of all those field operators that could affect the relevant physics between M_C and the QCD scale.

In order to motivate the nature of the continuous global symmetry, let us recall the model independent axion in superstring models [4]. The decay constant for such an axion is close to M_C, much larger than the upper bound of 10^{12} GeV allowed by standard cosmological arguments [10]. The presence of an additional global U(1) Peccei-Quinn (PQ) symmetry [11] (in the sense mentioned above) broken at an intermediate scale resolves the axion problem in a neat way. The true axion is now an appropriate linear combination of the two fields, the model independent axion and the PQ axion, and couples to $F\tilde{F}$ with a decay constant characterized by $M_I \sim 10^{10}$ Gev [12]. It is only natural to identify the desired continuous global symmetry with the $U(1)_{PQ}$ symmetry.

To summarize, our task now is to construct an acceptable axion model based on the field content suggested by the $E_8 \times E_8$ superstring. We also should identify the relevant discrete symmetry which is effectively embedded in $U(1)_{PQ}$. The problem of constructing C-Y spaces which lead to such a discrete symmetry and, in addition, have all the other desirable features is, of course, a formidable one and is not addressed in this paper.

In the following we first present a model based on the subgroup $G \equiv SU(3)_C \times SU(2)_L \times U(1)_L \times U(1)_R$ [5] which appears to have all the desirable features. The standard hypercharge generator is proportional to the direct sum of the generators of the U(1)'s.

In order for this model to develop an intermediate scale the C-Y space must have other (1,1) harmonic forms besides the Kahler form [8]. We also need two 3-2-1 singlets with different global charges in order to break $U(1)_{PQ}$ at M_I.

These singlets come from $\underline{27}$'s and are accompanied by their mirrors from $\overline{27}$'s. We denote them S_1 and S_2. Their mirrors are denoted $\tilde{S}_1$, $\tilde{S}_2$. The fields in $\underline{27}$'s with the same local quantum numbers as S_1 and S_2, but which do not have light mirrors are denoted by N.

It turns out that we need additional fields which come from incomplete $\underline{27}$'s (and $\overline{27}$'s). For reasons that will become clear later we introduce four doublets, H'', H''', $\bar{H}''$, $\bar{H}'''$ and their mirrors $\tilde{H}''$, $\tilde{H}'''$, $\tilde{\bar{H}}''$, $\tilde{\bar{H}}'''$ as well as two color triplets g' and their mirrors $\tilde{g}'$. In Table I are given the quantum numbers of the various fields with respect to G and a global Z_9 symmetry which we impose on the model. Also listed is the multiplicity N_m of each field. F denotes the number of families (three).

Table I: Field content, multiplicity and quantum numbers. For the Z_9 the (global) charges are clearly defined only mod9.

	$SU(3)_C$	$SU(2)_L$	$U(1)_L$	$U(1)_R$	global	N_m
Q	3	2	1	0	0	F
L	1	2	-1	-2	8	F
H	1	2	-1	-2	-12	1
$\bar{H}$	1	2	-1	4	14	1
U_c	$\bar{3}$	1	0	-4	-14	F
D_c	$\bar{3}$	1	0	2	12	f
E_c	1	1	2	4	4	F
g	3	1	-2	0	-10	F
g_c	$\bar{3}$	1	0	2	7	F
H'	1	2	-1	-2	-2	F-1
$\bar{H}'$	1	2	-1	4	-1	F-1
N	1	1	2	-2	-2	2F
S_1	1	1	2	-2	3	1
S_2	1	1	2	-2	-7	1
$\tilde{S}_1$	1	1	-2	2	-3	1
$\tilde{S}_2$	1	1	-2	2	7	1
H''	1	2	-1	-2	-2	1
$\tilde{H}''$	1	2	1	2	2	1
$\bar{H}''$	1	2	-1	4	-1	1
$\tilde{\bar{H}}''$	1	2	1	-4	1	1
H'''	1	2	-1	-2	-2	1
$\tilde{H}'''$	1	2	1	2	2	1
$\bar{H}'''$	1	2	-1	4	-1	1
$\tilde{\bar{H}}'''$	1	2	1	-4	-9	1
g'	3	1	-2	0	-10	2
$\tilde{g}'$	$\bar{3}$	1	2	0	20	2

The superpotential of the model contains (identifying fields with the same quantum numbers) the following dimension three (d = 3) terms: $\bar{H}QU_c$, HLE_c, QD_cH, $H\bar{H}N$, gg_cS_1, $g_cg_cU_c$, gD_cN, $H'H'E_c$, $\bar{H}'LS_2$, $H'\bar{H}'S_1$, $\tilde{H}''\tilde{\bar{H}}''\tilde{S}_1$, $\tilde{H}'''\tilde{\bar{H}}'''\tilde{S}_2$. The d = 4 terms are: $\tilde{H}''\tilde{H}''H'H'$, $\tilde{H}''\tilde{H}''LH$, $\tilde{H}''\tilde{\bar{H}}''gg_c$, $\tilde{H}''\tilde{\bar{H}}''$, $H'\bar{H}'$, $\tilde{H}''\tilde{\bar{H}}'''L\bar{H}'$, $\tilde{\bar{H}}''\tilde{\bar{H}}''\bar{H}'\bar{H}'$, $\tilde{H}''\tilde{g}'gH$, $\tilde{\bar{H}}'''\tilde{g}'g\bar{H}'$, $\tilde{H}''\tilde{S}_1H'S_1$, $\tilde{H}''\tilde{S}_1LS_2$, $\tilde{H}''\tilde{S}_2HS_1$, $\tilde{H}''\tilde{S}_2H'S_2$, $\tilde{\bar{H}}''\tilde{S}_1H'E_c$, $\tilde{H}''\tilde{S}_1\bar{H}'S_1$, $\tilde{\bar{H}}'''\tilde{S}_2H'E_c$,

$\tilde{\bar{H}}'''\tilde{S}_2\bar{H}'S_1$, $\tilde{\bar{H}}''\tilde{S}_2HE_c$, $\tilde{\bar{H}}''\tilde{S}_2\bar{H}'S_2$, $\tilde{\bar{H}}'''\tilde{S}_1QD_c$, $\tilde{\bar{H}}'''\tilde{S}_1LE_c$, $\tilde{g}'\tilde{S}_1gS_2$, $\tilde{S}_1\tilde{S}_1S_1S_1$, $\tilde{S}_1\tilde{S}_2S_1S_2$, $\tilde{S}_2\tilde{S}_2S_2S_2$, $\tilde{S}_1\tilde{S}_2NN$. Although we have imposed only a discrete Z_9 global symmetry, the important thing is to know the actual symmetries of the theory. It turns out that, if we restrict ourselves to terms in the superpotential with $d \leq 4$, the theory possesses a larger global symmetry than Z_9, namely a $U(1)_{PQ}$ to which the Z_9 is embedded. The maximal symmetry of the model is $G \times U(1)_{PQ}$. The $U(1)_{PQ}$ charges of the various fields are the same as the Z_9 ones given in Table I. The Z_9 symmetry is also sufficient to guarantee the absence of $d > 4$ terms (e.g. $S_1S_1S_1\tilde{S}_2\hat{S}_2\tilde{S}_2$) which violate the $U(1)_{PQ}$ symmetry and whose absence is important for the PQ mechanism.

Through the couplings gg_cS_1, $H'\bar{H}'S_1$, $\bar{H}'LS_2$, $\tilde{H}''\tilde{\bar{H}}''\tilde{S}_1$, $\tilde{H}'''\tilde{\bar{H}}'''\tilde{S}_2$, the 3-2-1 singlets S_1, S_2, $\tilde{S}_1'$, $\tilde{S}_2$ can develop a negative $(\text{mass})^2$ and acquire vev's $\langle S_1\rangle = \langle\tilde{S}_1\rangle$, $\langle S_2\rangle = \langle\tilde{S}_2\rangle$, all of the order of M_I. These vev's break $U(1)_{L-R} \times U(1)_{PQ}$ down to a Z_{40} generated by $(e^{-3\pi i/40}, e^{2\pi i/10})$. The couplings $NN\tilde{S}_1\tilde{S}_2$ and $H\bar{H}N$ create mixing between H and $\bar{H}$ $(M_I{}^2M_C{}^{-1}H\bar{H}N^*)$ necessary for tree level electroweak breaking. For suitable values of the parameters of the theory, H, $\bar{H}$ and N acquire vev's of order M_W [9]. These vev's break the Z_{40} down to a Z_{20} generated by $(e^{-3\pi i/20}, e^{2\pi i/5})$. No other field is supposed to develop a vev.

The model has baryon and lepton number non-conservation due to the presence of the couplings $g_cf_cU_c$ and $H'H'E_c$. For reasonable values of the couplings $(\sim 10^{-1}\text{-}10^{-2})$ the proton lifetime is acceptable and possibly experimentally accessible. The dominant modes are of the type $\ell^+\ell^-M^+\nu(\bar{\nu})$ (M denotes a meson and the rest are leptons) (see Fig. 1).

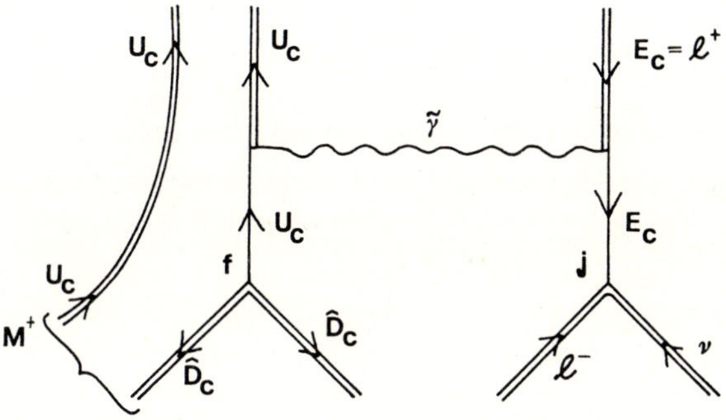

Fig. 1: One of the dominant diagrams for proton decay

The $U(1)_{PQ}$ also ensures the absence of unacceptably large neutrino masses. Their actual values depend on the unknown parameters of the underlying theory. Electron volt neutrino masses are, in principle, possible.

The (one loop) renormalization group equations are consistent with $M_C \simeq 10^{17}$ GeV, $M_I \simeq 10^{10}$ GeV, $\alpha_s(M_W) \simeq 0.11$ and $\sin^2\theta_W(M_W) \simeq 0.22$. For $F = 3$, $\alpha_G \simeq 0.14$ and the calculation therefore is reliable.

2. Models with n-$\bar{n}$ Oscillations

In this section we wish to further pursue this phenomenolofical (experimental) approach to superstring model building. Following the strategy outlined above, we construct a new class of phenomenologically and cosmologically acceptable models which have some interesting experimental consequences. These include a stable proton, massless or extremely light (<< eV) neutrinos, and experimentally accessible n-$\bar{n}$ oscillations. The models also lead to a baryon asymmetry in accord with observations. A new feature here, absent in previous models, is the appearance of an (accidental) unbroken global U(1) symmetry which corresponds to lepton number for the "known" particles.

We base our model once again on the rank five subgroup $G \equiv SU(3)_c \times SU(2)_L \times U(1)_L \times U(1)_R$ of E_6. We also impose a (discrete subgroup of a) global $U(1)_{PQ}$ symmetry. The $U(1)_L \times U(1)_R \times U(1)_{PQ}$ symmetry spontaneously breaks down to $U(1)_Y$ at an intermediate scale $M_I(\sim 10^9$ GeV). The model consists of three 27's of E_6, two $SU(3)_c \times SU(2)_L \times U(1)_Y$ singlet fields S_1, S_2 (plus their mirrors $\tilde{S}_1,\tilde{S}_2$)which acquire vev's of order M_I, a set of extra doublets H", $\bar{H}$" (plus mirrors $\tilde{H}$, $\tilde{\bar{H}}$), and an extra color triplet g' (plus mirror $\tilde{g}$) with mass on the order of the electroweak scale.

The superpotential of the model contains the following degree three and four terms (For quantum numbers of the fields see Table II. Note that we identify H" with H'; $\bar{H}$" with $\bar{H}$' and g' with g): $\bar{H}QU_c$, HQD_c, HLE_c, $\bar{H}HN$, $g_cD_cU_c$, gg_cS_1, $\bar{H}'LS_2$, $H'\bar{H}'S_1$, gD_cN, $\tilde{H}\tilde{\bar{H}}\tilde{S}_2$, $\tilde{H}\tilde{\bar{H}}H'H'$, $\tilde{H}\tilde{\bar{H}}L\bar{H}'$, $\tilde{\bar{H}}\tilde{g}g\bar{H}$, $\tilde{H}\tilde{S}_1H'S_1$, $\tilde{H}\tilde{S}_1LS_2$, $\tilde{H}\tilde{S}_2H'S_2$, $\tilde{\bar{H}}\tilde{S}_2\bar{H}'S_1$, $\tilde{g}\tilde{S}_1gS_2$, $\tilde{S}_1\tilde{S}_1S_1S_1$, $\tilde{S}_2\tilde{S}_2S_2S_2$, $\tilde{S}_1\tilde{S}_2S_1S_2$, $\tilde{S}_1\tilde{S}_2NN$.

The above superpotential possesses two global U(1) symmetries (in addition to the local symmetry G) which can be identified with the anomalous $U(1)_{PQ}$ and the non-anomalous $U(1)_\ell$ where the subscript ℓ denotes generalized lepton number. The quantum numbers pertaining to these two symmetries are listed in Table II. Note the important point that $U(1)_\ell$ is an automatic symmetry although we only imposed $G \times U(1)_{PQ}$ (more precisely $G \times Z_n$, where Z_n is an appropriately large subgroup of $U(1)_{PQ}$) on the superpotential to begin with.

The axion domain wall problem can be taken care of in essentially the same manner as previously discussed and will not be repeated here.

The one loop renormalization group equations for the low energy couplings are consistent with the compactification scale $M_c \simeq 10^{17}$ GeV, $M_I \simeq 10^9$ GeV, $\sin^2\theta_w \simeq 0.22$, $\alpha_s(M_W) \simeq 0.11$ and $\alpha_G \simeq 0.09$.

The out of equilibrium condition requires that the couplings f and j (f stands for the gD_cN coupling and j for the $U_cD_cg_c$ coupling) be smaller than or equal to about 1/5. Taking into account the dilution factor of about 10^6 the final baryon asymmetry turns out to be roughly of order 10^{-10} which is highly satisfactory.

Table II: Field content and quantum numbers

	$SU(3)_C$	$SU(2)_L$	$U(1)_L$	$U(1)_R$	$U(1)_{PQ}$	U(1)
Q	3	2	1	0	0	0
L	1	2	-1	-2	2	1
$\bar{H}$	1	2	-1	-2	-2	0
$\bar{H}$	1	2	-1	4	-1	0
U_c	$\bar{3}$	1	0	-4	1	0
D_c	$\bar{3}$	1	0	2	2	0
E_c	1	1	2	4	0	-1
g	3	1	-2	0	-5	0
g_c	$\bar{3}$	1	0	2	-3	0
H'	1	2	-1	-2	-8	1
$\bar{H}'$	1	2	-1	4	0	-1
N	1	1	2	-2	3	0
S_1	1	1	2	-2	8	0
S_2	1	1	2	-2	-2	0
$\tilde{S}_1$	1	1	-2	2	-8	0
$\tilde{S}_2$	1	1	-2	2	2	0
H"	1	2	-1	-2	-8	1
$\tilde{H}$	1	2	1	2	8	-1
$\bar{H}''$	1	2	-1	4	0	-1
$\tilde{\tilde{H}}$	1	2	1	-4	-10	1
g'	3	1	-2	0	-5	0
$\tilde{g}$	3	1	2	0	15	0

If we assign baryon number (1/3(-1/3) to a color triplet (antitriplet), the sole baryon number violating vertex involving light quarks (mass eigenstates) is $q\, d_c d_c u_c$, where the effective coupling constant $q \sim fj \langle N\rangle/M_g$. Here $\langle N\rangle \sim$ electroweak (SUSY) breaking scale, $M_g(\sim M_I)$ denotes the mass of the heavy colored fields g and g_c, and f and j (both $\lesssim 1/5$) are the coefficients of gD_cN and $g_cU_cD_c$ couplings.

An important constraint on q arises from considerations related to the baryon asymmetry in the Universe. We must require that the baryon asymmetry produced at $T \gtrsim M_s$ is not washed out due to the presence of the baryon number violating coupling $q\, d_c d_c u_c$. The rate of $2 \leftrightarrow 2$ baryon number violating scatterings at temperature $T \sim M_s$ is of order q^2T. For these scatterings to be out of thermal equilibrium, their rate must be smaller than the expansion rate of the Universe. In a radiation dominated Universe this gives $q^2T \lesssim 30T^2/M_P$, where $M_P \simeq 1.2 \times 10^{19}$ GeV is the Planck mass. For $T \sim M_s \sim$ few hundred GeV, $q^2 \lesssim 10^{-15}$. This constraint on q is readily satisfied in our class of models because of the suppression factor $\langle N\rangle/M_I \sim M_s/M_I$ which appears in the definition of q.

Due to the absence of lepton number violating couplings, the proton is effectively stable in these models provided the gauginos and higgsinos are heavier than a GeV. The proton may decay through the exchange of superheavy ($\sim 10^{17}$ GeV) particles in which case its lifetime would be $\sim 10^{38-40}$ yrs., well beyond the scope of any forseeable experiment.

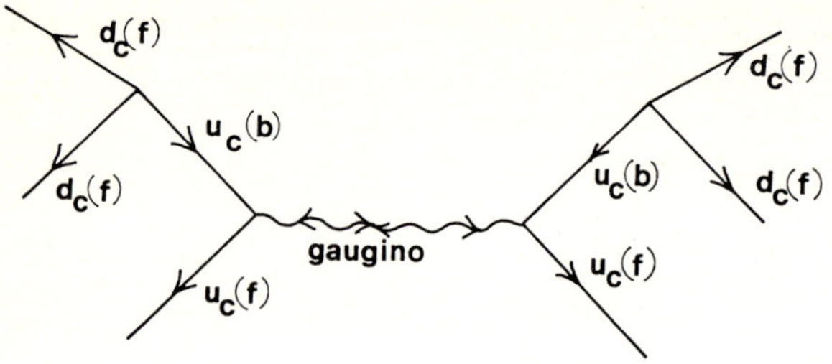

Fig. 2: Diagram for n-$\bar{n}$ oscillations

A prediction of the model which may be most amenable to experimental searches concerns n-$\bar{n}$ (neutron-antineutron)oscillations. A relevant diagram is shown in Fig. 2. using standard estimates, the oscillation time is expected to be given by

$$\tau^{-1}_{n-\bar{n}} \simeq q^2 \alpha^2 M_s^{-5} |\psi(0)|^4 \ .$$

Here ψ denotes the nuclear wave function. For $|\psi(0)|^4 \simeq 10^{-3}$ GeV^6, $q^2 \sim 10^{-16}$ and $M_s \sim 250$ GeV,

$$\tau_{n-\bar{n}} \sim 10^8 \text{ sec}$$

This may be within the reach of ongoing experiments searching for n-$\bar{n}$ oscillations. We should emphasize, however, that the value of $|\psi(0)|^4$ is not accurately known and could even be one or two orders of magnitude smaller.

The unbroken $U(1)_{\mathcal{L}}$ symmetry implies the absence of any neutrino masses. Higher order (degree >4) terms in the superpotential need not conserve lepton number. The neutrinos may acquire a small (<<eV) mass as a consequence.

References:

1 M.B. Green and J.H. Schwarz, Phys. Lett. 149B, 117 (1984).

2 D.J. Gross, J. Harvey, E. Martinec and R. Rohm, Phys. Rev. Lett. 54, 502 (1985).

3 P. Candelas, G. Horowitz, A. Strominger and E. Witten, Nucl. Phys. B258, 46 (1985).

4 E. Witten, Phys. Lett. 149B, 351 (1984); Phys. Lett. 153B, 243 (1985).

5 E. Witten, Nucl. Phys. B258, 75 (1985).

6 J.D. Breit, B.A. Ovrut and G.C. Segre, Phys. Lett. 158B, 33 (1985).

7 M. Dine, R. Rohm, N. Seiberg and E. Witten, Phys. Lett. 156B, 55 (1985);
J.P. Derendinger, L.E. Ibanez and H.P. Nilles, Phys. Lett. 155B, 65 (1985).

8 M. Dine, V. Kaplunovsky, M. Mangano, C. Nappi and N. Seiberg, Nucl. Phys. B259, 549 (1985).

9 M. Mangano, "Low energy aspects of superstring theories", Princeton preprint (1985).

10 J. Preskill, M. Wise and F. Wilczek, Phys. Lett. 120B, 127 (1983);
L. Abbott and P. Sikivie, ibid. 133;
M. Dine and W. Fischler, ibid. 137.

11 R. Peccei and H. Quinn, Phys. Rev. Lett. 38, 1440 (1977);
S. Weinberg, Phys. Rev. Lett. 40, 223 (1978);
F. Wilczek, Phys. Rev. Lett. 40, 279 (1978).

12 K. Choi and J. Kim, Phys. Lett. 154B, 393 (1985).

Expectations for Neutrino Mass and Baryon Number Violation in Superstring Models

J.W.F. Valle

Dept. de Fisica Teorica, Univ. Autonoma de Barcelona, Bellaterra, Barcelona, Spain

The neutrino mass problem of superstring models is discussed and possible solutions are identified. I focus on models with intermediate scales where the gauge group below the Planck scale is restricted. Models in which light E_6 singlets survive after compactification are shown to give rise to new mechanisms of understanding small neutrino masses while avoiding fast proton decay. I also briefly discuss a recently proposed solution to the solar neutrino problem involving the oscillation of the electron neutrino into a new sterile lepton predicted in the superstring theory.

1. Introduction

In the standard electroweak gauge theory neutrinos are massless and all lepton numbers are exactly conserved global symmetries. This follows in the model with minimal Higgs content from the absence of right-handed neutrinos. Nonzero neutrino mass can of course be accomodated in various ways [1, 2]. The question is, if neutrino masses turn out to be nonzero, why are they so small on the scale of the charged fermion masses which is set by the Higgs doublet vacuum expectation value (vev) $\langle\varphi\rangle$. An attractive explanation is given by the see-saw mechanism which is described by the mass matrix [3]

$$\begin{array}{c c} & \begin{array}{cc} \nu & \nu^c \end{array} \\ \begin{array}{c} \nu \\ \nu^c \end{array} & \begin{pmatrix} 0 & D \\ D & M_R \end{pmatrix} \end{array} \tag{1}$$

where the Dirac mass component is D and thus proportional to $\langle\varphi\rangle$ while M_R is an SU(2)×U(1) singlet and thus expected to be much bigger.

The details of the physics associated with M_R depend on the nature of B-L breaking: explicit versus spontaneous and, in the second case, whether B-L is a global [4] or a gauge symmetry [5] as in left-right symmetric theories such as SO10. In both cases there is new dynamics for neutrinos; either a Majoron or gauge bosons dominantly coupled to right-handed neutrinos. In any event, regardless of the precise nature of the underlying physics, as long as $M_R \gg D$ (1) has a light eigenstate of mass

$$m_\nu \simeq D^2/M_R \tag{2}$$

which is mostly along the SU(2) doublet two-component neutrino ν with a small admixture of the singlet ν^c. I will now consider what happens in superstring models.

A very appealing approach towards unification of all forces along with all matter is found in the superstring context [6]: physics at the Planck scale is that of an extended object – the string – which can vibrate in any of an infinite number of modes. The excited modes are at the Planck scale and are believed to provide a consistent treatment of quantum gravity. The massless modes or ground state describe the physics normally accounted for by conventional field theory: this is the low energy limit of the superstring. It consists of a $N = 1$ supergravity grand unified theory (GUT) in 10 dimensions with $E_8 \times E_8'$ or SO(32) as gauge groups.

Compactification on a suitable six dimensional internal space K_6 is possible while preserving $N = 1$ supersymmetry in four dimensions so as to keep stable the gauge hierarchy. For the case of $E_8 \times E_8'$ a spectrum of *chiral* fermion (denoted ψ) is produced after compactification, the number of generations being related to the Euler number of the internal manifold and the E_8 symmetry being reduced to E_6 close to the Planck scale. The E_8' is believed to play the role of a hidden sector responsible for supersymmetry breaking. In addition a number $b_{1,1}$ (Betti-Hodge number of K_6) of pairs of *mirror* multiplets H, $J \sim 27$ and $\overline{H}$, $\overline{J} \sim \overline{27}$ (in some models these may play the role of Higgs fields) arise from the decomposition of the 248 of E_8

$$248 \rightarrow [78,1] + [27,3] + [\overline{27},\overline{3}] + [1,8] \tag{3}$$

where the first number in brackets gives the transformation under the E_6 while the second gives the SU(3) transformation (which is the "holonomy" of the manifold).

In this approach much of low energy physics has its origin in the process of compactification and the topology of K_6; the GUT gauge group, E_6 and its ("flux") breaking pattern, as well as the field representation content in four dimensions and even the Yukawa couplings are in principle determined.

It is useful to decompose the 27 of E_6 with respect to the [SO10,SU5] subgroups

$$27 \rightarrow [16,10] + [16,\overline{5}] + [16,1] + [10,5] + [10,\overline{5}] + [1,1] \ . \tag{4}$$

In one family of E_6 there are five neutral leptons which will be relevant to discuss the issue of neutrino mass (see below)

$$\nu,\ \nu^c,\ N,\ N^c,\ n_0 \ . \tag{5}$$

The components of the chiral multiplets ψ remain light as a consequence of the index theorem while the components of the mirror multiplets may or may not acquire mass at compactification, depending on the details of the flux breaking. In particular, in some models the Weinberg-Salam doublets must come from the chiral multiplets since the corresponding doublets in the mirror multiplets do not survive the flux breaking.

2. Phenomenological Problems in Superstring Models

Many attempts have been made to extract phenomenology out of the superstring [7] but among others, the following difficulties have been recognized: too fast proton decay mediated by the exotic quarks contained in the [10,5] and [10,$\bar{5}$] piece of the 27 (Fig. 1) and unacceptably large neutrino masses.

Unlike the situation in the standard model, in the massless sector of the string there is no Higgs representation with the correct quantum numbers to provide the large Majorana mass M_R of the right-handed neutrinos (1).

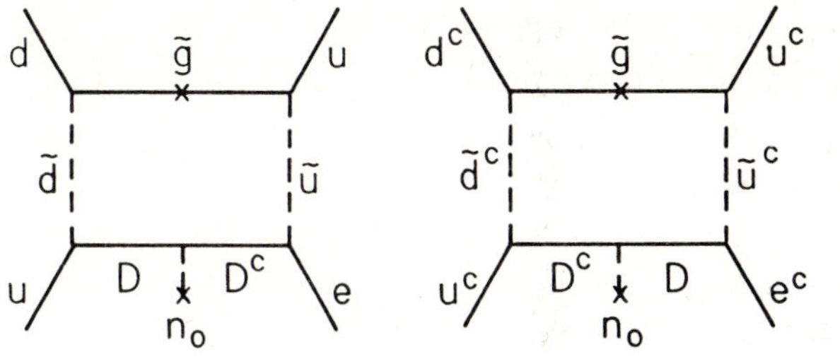

Fig. 1. Diagrams for proton decay

3. Possible Solutions

One possible way out of these problems would be simply to postulate that the couplings that cause rapid proton decay and large neutrino mass are absent as a result of some suitable discrete symmetry of the internal manifold K_6. Such simple "solution", although we can not discard, has several possible criticisms. First the required discrete symmetry may for all we know not be compatible with details of compactification. In addition, in these models where there is no mass scale between the Planck mass and the weak interaction scale, there is in general at least one new Z' gauge boson with mass not much above the TeV scale [8]. In this case there are too many light neutrinos contributing to the energy density of the universe during nucleosynthesis and the prediction for the primordial ^{4}He abundance will be upset unless $m_{z'}>500$ GeV [9] which however still leaves an allowed region.

One can also have a solution to the neutrino mass problem in models with no intermediate scale if R parity is broken at the TeV scale. However, this removes only one neutrino and a discrete symmetry is still needed to solve the neutrino mass problem of the other two families [10].

4. Models with Two Intermediate Scales

We prefer a more general solution to the problems of proton stability and neutrino mass of superstring models which is possible in models which have intermediate scales. First note that the existence of such scales is not necessarily incompatible with a stable gauge hierarchy because large vevs can arise along flat directions of the potential which do not break supersymmetry at that scale, but only in the TeV region. Second, there do exist Higgs fields which are singlets under the standard model and which can remain light after E_6 breaking and thus acquire large scale vevs. More interestingly the requirement that such singlets survive the E_6 breaking actually determines the gauge structure which follows that breaking, depending on the assumed discrete symmetry for the internal space [11]. We assume the existence of two intermediate scales, possible in models with $b_{1,1} = 2$ for which case we have given the allowed gauge symmetry groups below the Planck scale when Z_n is the manifold discrete symmetry (for the case of only one such scale again additional discrete symmetry is still needed [12]).

As can be seen from Table 1 some discrete symmetries such as Z_2, Z_4 and Z_5 are unacceptable because they lead to low energy baryon number violation through light gauge bosons.

Table 1

Discrete symmetry	Gauge group below Planck mass
Z_2	$SU(6)\times SU(2)_L$
Z_3	$SU(3)_C\times SU(3)_L\times SU(3)_R$
Z_4	$SU(5)\times SU(2)_L\times U(1)$
Z_5	$SU(5)\times SU(2)_N\times U(1)$
Z_6	$SU(3)_C\times U(1)_4\times SU(2)_L\times SU(3)_R$
$Z_n, n>6$	$SU(3)_C\times SU(2)_L\times SU(2)_N\times U(1)\times U(1)$

A general comment about models with two intermediate scales is that the Weinberg Salam Higgs doublet must come from the chiral multiplets. The only exception is the Z_3 model with gauge group SU(3)×SU(3)×SU(3) where the weak doublet can come from the mirror multiplets [11]. In all cases one needs to fine tune a Yukawa coupling to be very small so as to guarantee that a light doublet survives intermediate breaking.

In models with two intermediate scales it is possible to revive the see-saw mechanism [13] if the superpotential contains non renormalizable couplings of generic form

$$hM_P^{-1}\psi^2\overline{J}^2 \qquad (6)$$

and, say, the J field acquires a large vev V_{BL} along the [16,1] component. In this case the right-handed neutrino will acquire a Majorana mass component

$$M_R = hV_{BL}^2/M_P \tag{7}$$

which typically gives a neutrino mass of order

$$m_\nu \simeq h^{-1}(D/V_{BL})^2 M_P \ . \tag{8}$$

If the vev along the component [1,1] is high enough proton decay will be suppressed but because of the large value of V_{BL} we expect observable $n\overline{n}$ oscillations through the gluino exchange mechanism (shown in Fig. 2) as was first noticed in [11].

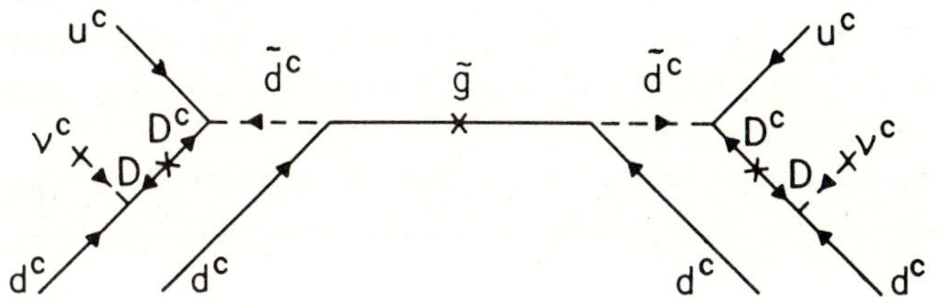

Fig. 2. Tree diagram for $n - \overline{n}$ oscillation

We now turn to the possibility suggested by Witten [14] that some E_6 singlet fermions may survive massless the compactification from 10 to 4 dimensions. Models of this type were given in [11] which I will now summarize.

By choosing a high enough value for the [1,1] vev V_6 of one of the Higgs fields (H) the neutral fermions in the 10 of SO10,N and N^c, will acquire superheavy mass by virtue of the $27_\psi 27_\psi 27_H$ coupling and will decouple from the others along with the new quarks that would otherwise cause fast proton decay. Our model is so constructed that there are no observable $n\overline{n}$ oscillations, unlike the case described above. In order to take care of the neutrino mass problem for the other three fermions we resort to the use of the light E_6 singlets: if they have a nonzero coupling to the matter multiplets, we may combine the right handed neutrino with one such singlet to make up a heavy Dirac particle of mass $0(V_{BL})$. First note that in our model the SO10 singlet fermion n_0 may remain massless or light without any conflict with experiment and cosmology because it only couples to superheavy gauge bosons. (It could also acquire a superheavy mass by coupling with a second singlet, if present.)

In this case we identify two different scenarios described by the following mass matrices (for simplicity we consider only the case of one generation but the models apply to any number of generations, provided there are as many E_6 singlets).

5. Model 1

$$\begin{array}{c} \\ \nu \\ \nu^c \\ S \end{array} \begin{array}{c} \begin{array}{ccc} \nu & \nu^c & S \end{array} \\ \begin{pmatrix} 0 & D & 0 \\ D & 0 & M_{BL} \\ 0 & M_{BL} & 0 \end{pmatrix} \end{array} \tag{9}$$

Here the state which is mostly SU(2) doublet neutrino remains massless while the right-handed neutrino ν^c combines with the E_6 singlet fermion S to form a heavy Dirac fermion of mass M_{BL}. This mass arises from the [16,1] vev of one of the mirror multiplets, J, which we choose to couple to the E_6 singlet. This mass matrix has an exactly conserved symmetry, and the neutrino is massless. In this respect this model is the superstring inspired analog of the standard model. In the three generation case this model has interesting phenomenological signatures for the case where M_{BL} is small [15].

Symmetry breaking effects may arise as a result of supersymmetry breaking and/or radiative corrections which could, for example, generate a Majorana mass for S. Now the massless eigenstate neutrino acquires a small Majorana mass

$$m_\nu \simeq (D/M_{BL})^2 M_S \tag{10}$$

so that lepton number is now violated and processes such as neutrinoless double beta decay become allowed.

6. Model 2

In this case the two-component SO10 singlet fermion n_0 acquires a small mass by combining with the two-component SU(2) doublet neutrino ν to form a Dirac particle [1] of mass

$$m_\nu \simeq \mu D/M_{BL} \tag{11}$$

which for $D \simeq 10\,\text{MeV}$, $\mu \simeq 10\,\text{TeV}$, $M_{BL} \simeq 10^{10}\,\text{GeV}$ is of order $10\,\text{eV}$. Here μ is an effective vev of J along the [1,1] direction. The full mass matrix is 4×4

$$\begin{array}{c} \\ \nu \\ n_0 \\ \nu^c \\ S \end{array} \begin{array}{c} \begin{array}{cccc} \nu & n_0 & \nu^c & S \end{array} \\ \begin{pmatrix} 0 & 0 & D & 0 \\ 0 & 0 & 0 & \mu \\ D & 0 & 0 & M_{BL} \\ 0 & \mu & M_{BL} & 0 \end{pmatrix} \end{array} \tag{12}$$

It is easy to see that the mass eigenstates are two Dirac neutrinos, a light one of mass given in (11) and a superheavy one of mass M_{BL} formed by combining the right-handed neutrino with the E_6 singlet fermion S.

The light mass eigenstates are

$$i(\nu - n_0)/\sqrt{2} \tag{13a}$$

and the orthogonal combination,

$$(\nu + n_0)/\sqrt{2} \ . \tag{13b}$$

The symmetry which enforces the two-component neutrinos in our model to combine into Dirac neutrinos [1] may be broken in various ways, for example by a non-zero Majorana mass for the field S as discussed above. In this case the model predicts that neutrinos are Quasi-Dirac particles [16] with a mass splitting given by

$$\delta m = (D/V_{BL})^2 M_S \tag{14}$$

which is much smaller than the actual mass given (11). In this model it is natural for neutrinos to be light and Quasi-Dirac: the smallness of the mass (11) and that of the mass splitting (14) are governed by different powers of the large symmetry breaking scale V_{BL}.

In the past many attempts were made to construct models of Dirac and Quasi-Dirac neutrinos in an attempt to reconcile the null results from neutrinoless double beta decay with the hint of a finite neutrino mass from the ITEP group [16]. This was achieved in an ad hoc manner either by increasing the gauge group of the weak interactions or by introducing new fermions [17].

7. Solar Neutrino Oscillations in Superstring Models

Recently there has been a revived interest in the long standing puzzle posed by Davis experiment on solar neutrinos and its possible implications for particle physics. Using an earlier suggestion due to Wolfenstein [18] Mikheyev and Smirnov [19] have shown that in the presence of matter the charged current weak interactions enhance the *ee* component of the neutrino mass matrix due to resonant processes in the interior of the sun so that a substantial depletion of the solar ν_e flux measured on earth is possible as a consequence of flavor oscillations, even for small mixings. Barring the possibility that neutrino mass differences are much smaller than neutrino masses for which in general there is no reason, this would indicate masses far below levels of detectability even in the most refined tritium decay experiments.

I now discuss how in model 2 we have a natural solution to the solar neutrino puzzle in terms of large oscillations between the electron neutrino and the sterile lepton n_0 discussed in (12). In weak interaction processes one produces the weak eigenstate ν which, in the course of time, acquires a component n_0, proportional to

$$P(n_0, L) = \sin^2(m_{\nu_e} \delta m L/E) \tag{15}$$

where E is the energy with which the neutrino is emitted (of a few MeV for the ^{37}Cl experiment) and L is the earth-sun distance. Since the field n_0 is inert with respect to weak interactions, there will be a depletion of the active SU(2) doublet component from the solar neutrino flux. This depletion can become significant for an electron neutrino mass of 10 eV or so (11) and a very tiny mass splitting of 10^{-12} eV (14).

Another amusing consequence of the present model is the existence of neutral current oscillations due to a large off diagonal coupling of the Z boson with the light eigenstates (13). The persistence probability for an electron neutrino against neutral current oscillation is also given in (16).

Finally another way to break the lepton number in this model is discussed in [20] and also leads to the possibility of substantial depletion in the solar neutrino flux.

8. Conclusion

I have summarized various possible scenarios for understanding small neutrino masses in the context of superstring models, while avoiding catastrophic proton decay. The gauge structure which follows the breaking of E_6 is restricted according to Table 1. Neutrinos could be: (i) massless, (ii) light Majorana particles, (iii) light Dirac particles or (iv) light Quasi-Dirac particles.

Finally I discussed how in the last case a natural solution can be given to the solar neutrino puzzle in terms of oscillations between "active" and "sterile" neutrinos which offers an alternative to the possibility of resonant matter oscillations.

Acknowledgements. This talk is largely based on work done in collaboration with R. Mohapatra and supported by Spanish Ministry of Science and National Science Foundation.

References

1. J. Schechter, J.W.F. Valle: Phys. Rev. D**22**, 2227 (1980)
2. T.P. Cheng, L.F. Li: Phys. Rev. D**22**, 2860 (1980);
   S. Bilenky, J. Hosek, S.T. Petcov: Phys. Lett. **94**B, 495 (1980)
3. M. Gell-Mann, P. Ramond, R. Slansky: In *Supergravity*, ed. by P. van Nieuwenhuizen and D. Freedman (North Holland, 1979);
   T. Yanagida: Proc. of Workshop on Unified Theory and Baryon Number of the Universe, ed. by O. Sawada et al. (KEK, Japan 1979)
4. Y. Chikashige, R. Mohapatra, R. Peccei: Phys. Lett. **98**B, 265 (1980)
5. R. Mohapatra, G. Senjanovic: Phys. Rev. Lett. **44**, 912 (1980)

6. D. Gross et al.: Phys. Rev. Lett. **54,** 502 (1985); Nucl. Phys. B**256,** 251 (1985); P. Candelas et al.: Nucl. Phys. B**258,** 46 (1985); E. Witten: Nucl. Phys. B**258,** 75 (1985)
7. E. Witten: Proc. of the Symposium on Anomalies, Geometry and Topology, ed. by W.A. Bardeen and A. White (World Scientific, 1985); M. Dine et al.: Nucl. Phys. B**259,** 549 (1985); J. Breit, B. Ovrut, G. Segre: Phys. Lett. **158**B, 33 (1985); J. Derendinger, L. Ibanez, H. Nilles: Nucl. Phys. B**267,** 365 (1986); F. del Aguila et al.: CERN preprint 1985; S. Cecotti et al.: Phys. Lett. **156**B, 318 (1985); P. Binetruy et al.: LBL 20317 (1985)
8. J. Ellis et al.: Phys. Lett. **167**B, 457 (1986); A. Masiero et al.: Rockefeller preprint (1986)
9. G. Steigman, K. Olive, D. Schramm, M. Turner: Phys. Lett. B (in press)
10. R. Mohapatra: Phys. Rev. Lett. **56,** 561 (1986)
11. R.N. Mohapatra, J.W.F. Valle: Maryland Physics Publ. 86–127, Phys. Rev. D (in press)
12. C. Nappi, V. Klapunovsky: Comm. Nucl. Part Phys. (in press); R. Mohapatra, J.W.F. Valle: In preparation
13. S. Nandi, U. Sarkar: Phys. Rev. Lett. **56,** 566 (1986)
14. E. Witten: Princeton preprint (1985)
15. J. Bernabeu, A. Santamaria, A. Mendez, J.W.F. Valle: In preparation
16. J.W.F. Valle: Phys. Rev. D**27,** 1672 (1983); S.T. Petcov: Phys. Lett. **110**B, 245 (1982); M. Doi et al.: Prog. Theor. Phys. **70,** 1331 (1983)
17. J.W.F. Valle, M. Singer: Phys. Rev. D**28,** 540 (1983); D. Wyler, L. Wolfenstein: Nucl. Phys. **218,** 205 (1983); P. Roy, O. Shanker: Phys. Rev. Lett. **52,** 713 (1984)
18. L. Wolfenstein: Phys. Rev. D**17,** 2369 (1978); D**20,** 2634 (1979)
19. S. Mikheyev, A.Yu. Smirnov: INR preprint, Moscow, 1985; H. Bethe: Phys. Rev. Lett. **56,** 1305 (1986); P. Langacker, S.T. Petcov, G. Steigman, S. Toshev: CERN TH4421 (1986); S.P. Rosen, J. Gelb: Los Alamos LA UR **86,** 804 (1986)
20. R. Mohapatra, J.W.F. Valle: Maryland Physics Publ. 86–164 (1986), Phys. Lett. **177**B, 47 (1986)

Part 4

Weak Interaction in Astrophysics and Cosmology

Supernovae and High Density Nuclear Matter

S. Kahana[1]

Physics Department, Brookhaven National Laboratory, Upton, NY 11973, USA

The role of the nuclear equation of state (EOS) in producing prompt supernova explosions is examined. Results of calculations of Baron, Cooperstein, and Kahana incorporating general relativity and a new high density EOS are presented, and the relevance of these calculations to laboratory experiments with heavy ions considered.

I. INTRODUCTION

It has been known for a considerable time that the properties of nuclear matter play a significant part in stellar evolution. If the density of normal, saturated, nuclear matter, $\rho_0 \approx 2.6 \times 10^{14}$ g/cm^3, is taken as a standard, then for most of a star's life one needs only consider nuclear material at quite low density. During the gravitational collapse of the cores of many massive stars, however, densities several times greater than ρ_0 are achieved. It is such highly compressed matter and its properties that particularly concern me here. The progenitors of type II supernovae, i.e. those supernovae whose spectra contain hydrogen lines, are believed to possess initial total mass in excess of nine or ten solar masses. I will discuss hydrodynamic simulations of collapse, carried out by Edward Baron, Jerry Cooperstein and myself [1], which are the first to produce a prompt shock-explosive mechanism for type IIs, beginning with the $12M_{\odot}$ and $15M_{\odot}$ models of Weaver and Woosley [2].

Over the years, two major classes of mechanism have been proposed. One approach seeks an ejection of the stellar mantle and envelope by an explosive shock created in a core bounce after collapse. A second approach would have the exterior regions blown off by the energetic neutrinos produced somewhat after collapse. Both of these scenarios are present in the classic papers of BURBIDGE, BURBIDGE, FOWLER, and HOYLE [3], and in the intervening years early work by COLGATE and collaborators [4] was carried out along both lines. Most recently, several authors (MAZUREK, COOPERSTEIN, KAHANA [5]; WILSON [6], ARNETT [7]; HILLEBRANDT [8]) have examined the collapse and subsequent shock formation with apparently negative results. Artificial initial models (COOPERSTEIN [9]; COOPERSTEIN, BETHE, and BROWN [10]; KAHANA, BARON, and COOPERSTEIN [11]) for the degenerate iron core can lead to explosions, but attempts to begin with the realistic evolutionary models of Weaver and Woosley and collaborators (WWZ [12]; WWF [13]; WW [2]) have uniformly ended with stalled, accreting shocks.

---

[1]The submitted manuscript has been authored under contract DE-AC02-76CH00016 with the U.S. Department of Energy.

Initial quiet evolution in the more massive stars, taking some 10-100 million years, ends with a core consisting only of elements near iron in atomic number, supported by degenerate electron pressure. Collapse is first triggered by photodisintegration of iron, then by the pressure drop following rapid β-capture, and finally halted by the stiffening of nuclear matter at high density. Bounce occurs when densities at the outer edge of the homologously collapsing part of the core reach nuclear saturation values. A shock is inevitably formed, at or near the radius where infall and sound velocities match, i.e. at the sonic point (see Fig. 4). In question then is the outward propagation of this shock through the remaining dense material in the core, while additional material continues to rain in.

The eventual fate of the shock is determined by the initial shock energy and by the losses suffered along its path. The initial energy in the shock is clearly borrowed from the gravitational well, and the losses result from dissociation of nuclei and from neutrino escape. Interestingly, although the collapsing core mass is almost all non-relativistic nuclei or nuclear matter, it is supported by the pressure of a highly relativistic, degenerate, gas. The adiabatic index C_p/C_v for such a system is $\Gamma = 4/3$, and the non-relativistic virial theorem tells us that the total core energy starts close to zero. The entropy is initially low and remains low throughout collapse (BETHE and collaborators [14]); thus the process is essentially adiabatic, and the total energy stays close to zero throughout. At maximum compression, this zero core energy is divided between approximately 100×10^{51} ergs (100 foes[1]) of negative gravitational energy and 100×10^{51} ergs of internal energy. Since only one foe emerging in the shock is sufficient to unbind the mantle and envelope (bound by a few tenths of a foe) as well as produce the required visual display, one must be prepared to carry out an accurate calculation on a delicately balanced system.

The early work of MAZUREK, COOPERSTEIN, and KAHANA [5] began with the 10-25 $M_\odot$ models of WEAVER, WOOSLEY, and ZIMMERMAN [12] and found stalled shocks. It was argued that excessive neutrino escape and dissociation were responsible for the enfeebled shock. However, WWZ [12] models ascribed a mass of 1.51 $M_\odot$ to the initial iron core. The shock could simply not traverse so much dense material in the highly compact core. In further work by WEAVER, WOOSLEY, and FULLER (WWF) [13] the omission of important β-capture channels was corrected, the initial electron fraction Y_e^i lowered, and hence the effective core mass (recall M(Chandresakhar) ~5.6 Y_e^2) was reduced to 1.36 $M_\odot$. We shall see that, though the WWF core is more diffuse than that of WWZ, we were still unable to successfully simulate type II supernovae. Our development (BARON, COOPERSTEIN, and KAHANA [1]) depends critically on the equation of state used for nuclear matter at high density and on the introduction of relativistic gravitation into the collapse phase. By theoretically altering the equation of state, corresponding to a softening of high density matter, we are able to increase the energy initially pumped into the shock. In a purely Newtonian calculation this change in not sufficient. Paradoxically, it is the introduction of general relativity, despite the apparent harm caused by the strengthening of gravitation, in combination with the softer matter that leads to a successful theoretical mechanism.

II. THE EQUATION OF STATE

Role in Supernovae Studies

Driven by the exigencies of the supernova problem, we have in earlier work employed as simple an equation of state as permitted [1,11,15]. Such an approach is justified not only from calculational necessity, i.e. reduction in computing time, but also from informational reality. Considerable theoretical knowledge and experimental constraint can be brought to bear on the nuclear equation of state at densities below the saturation value $\rho_0(x)$, x=Z/A, for normal nuclear matter. At higher density, however, little is known from laboratory experiment. A second point in favour of simplicity is the restricted sensitivity of the collapse calculations to various components in the equation of state. Baldly put, the collapsing stellar core near maximum compression senses some average compressibility at saturation and some average adiabatic index over a range of densities from ρ_0 to perhaps 3-4 ρ_0, and little else.

[1]G. E. Brown is responsible for the unit 1 foe = 10^{51} ergs.

Another relevant issue to keep in mind, missed by many earlier workers, is the actual environment in the stellar core at collapse. In particular, the charge to mass ratio, $Z/A = Y_e$ drops to the value $x = .32$ because of β-capture. For densities greater than $\rho_0(Z/A)$, equal to 2.4×10^{14} gm/cm^3 for $Z/A \approx 1/3$, we then suggested the simple form

$$P_N(\rho) = \frac{K_0(Z/A)\ \rho_0(Z/A)}{9\gamma}\left[u^\gamma - 1\right], \quad u = \rho/\rho_0(Z/A) \tag{1}$$

for the cold nuclear pressure. Here K_0 is the nuclear modulus of incompressibility at saturation, i.e. $u = 1$, and at the charge to mass ratio Z/A, while γ is the limiting adiabatic index at high density. The corresponding nuclear energy per baryon is

$$E/A = -16.0 + 29.3\left(1 - \frac{2Z}{A}\right)^2 + E_\rho + \frac{\pi^2 T^2}{4\varepsilon_{F,0}} \cdot u^{-2/3}\ \text{MeV} \tag{2}$$

with the cold compressional energy

$$E_\rho = \int_{\rho_0}^{\rho} d\rho\, \frac{P_N}{\rho^2} = \frac{K_0}{9\gamma(\gamma-1)}\left[u^{\gamma-1} + \frac{\gamma-1}{u} - \gamma\right]. \tag{3}$$

The thermal energy in (1) is that of a non-interacting Fermi gas, incorporating correlation effects only through the effective mass in the Fermi energy $\varepsilon_{F,0} = p_F^2/2m^*(p_F)$. Thus, our equation of state for dense nuclear matter can be viewed as containing three unknown parameters, $K_0(Z/A)$, γ, and the effective mass. Figure 1 displays P_N from (1) as a function of density and parametrically of γ.

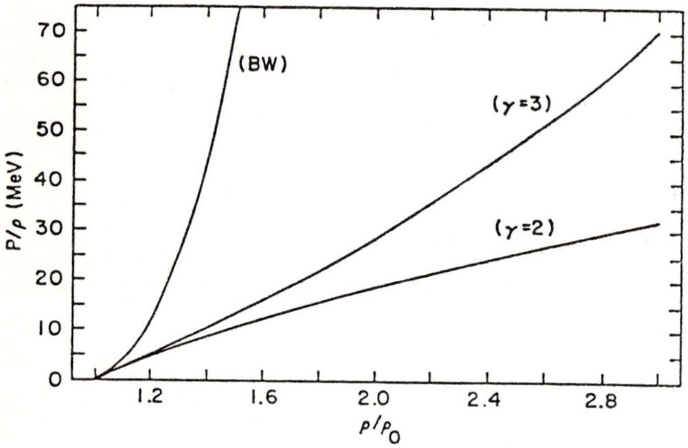

Fig. 1. Pressure versus densities are given for the Baron, Cooperstein, Kahana high density equation of state for two values of the parameter γ (the high density adiabatic index) in (1). For comparison, the same curve is shown for a very stiff equation of state (Brick Wall) in an artificially constructed model.

At the high densities reached in the collapsing cores of supernovae, and certainly at the perhaps even higher densities of the final compact remnant (presumably a neutron star), two forms of matter may compete for dominance: the hadronic matter I have considered and quark matter. Expressing the baryon equation of state in terms of the thermodynamic parameters, volume, chemical potential (μ) and temperature, and then recalling the phenomenological equation of state (at T=0)

$$P_{quark} = a' \cdot \mu^4 - B\ , \text{ with B the bag pressure}\ , \tag{4}$$

allows one to get some handle on the transition point. The main point I make here is to note the ambiguities inherent in proceeding with hydrodynamical simulations based on too elaborate an equation of state and in assuming only hadrons are present, even though the transition density is not achieved during collapse. Equation (1) should then be thought of as playing a phenomenological, symbolic role.

Asymmetric Matter

Recall also, as I have indicated above, that in the collapse environment β-capture has reduced the electron fraction, i.e. Y_e to near 1/3. The incompressibility K_0 is a strong function of the asymmetry of nuclear matter, a point first made in BARON, COOPERSTEIN, and KAHANA [15]. Using knowledge of the symmetry energy, BCK arrived at the simplified expressions

$$K_0(Z/A) = K_0(1/2)\ [\ 1-a(Z/A-1/2)^2\ ]$$
$$\rho_0(Z/A) = \rho_0(1/2)\ [\ 1-b(Z/A-1/2)^2\ ] \tag{5}$$

which for a=2, b=3/4 yield a good representation of the calculations of KOLEHMAINEN and collaborators [16]. Since

$$\frac{K(\rho)}{9} = \left[\frac{\partial P}{\partial \rho}\right]_{\rho_0} = \rho_0^2 \left[\frac{\partial^2 E}{\partial \rho^2}\right]_{\rho_0}, \quad \left[\frac{\partial E}{\partial \rho}\right]_{\rho_0} = 0 \quad , \tag{6}$$

it is clear that somewhere between Z/A = 1/2 and completely asymmetric neutron matter, Z/A = 0, $K_0(Z/A)$ must vanish; the saturation minimum becomes a point of inflection and then disappears. Most of the softening we require, to achieve theoretical explosions, from the BLAIZOT [17] value for $K_0(1/2)$ that I later discuss, results from the extrapolation exhibited in (5). The development in this section runs contrary to the oft-stated opinion that asymmetric matter is "stiffer" than symmetric matter. This statement is presumably based on the true but misleading changes in effective forces expected as one goes towards neutron-rich matter. The only meaningful density at which to cite a compression modulus is the saturation value $\rho_0(Z/A)$. Clearly from (5) this incompressibility decreases as $Z/A \to 0$, i.e. asymmetric matter is softer at saturation than is symmetric matter; this effect is heightened by the $[\rho_0]^2$ factor in (6) coupled to the behaviour of $\rho_0(Z/A)$ exhibited in (5).

Nuclear Compressibility Information from Low-lying Excitations

Since the changes in the properties of nuclear matter we require are somewhat controversial, I would now like to consider briefly just what information we have about $\Gamma(\rho)$ and $K_0(Z/A$ (and $m^*(p_F)$) at the densities achieved in the central core at bounce. We can of course get K_0, γ, and m* directly from nuclear matter calculations, but it is also interesting to bring whatever empirical evidence exists directly to bear. In terms of the Landau parameters $F_\ell(\ell=0,1,...)$ characterising the effective particle-hole interaction in nuclear matter, one can write

$$K_0 = 6\varepsilon_F(1 + F_0)$$
$$m^*(p_F) = m[1 + F_1/3]\ . \tag{7}$$

In the simplified case of Skyrme forces BLAIZOT [17], whose development I follow here, derives the result

$$K_0 = \frac{a}{\rho_0} - \frac{9\varepsilon(\rho_0)}{\rho_0} + d\left[\ \frac{-9\varepsilon(\rho_0)}{\rho_0} + \frac{3a}{\rho_0}\ \right] \tag{8}$$

where $\varepsilon(\rho_0)$ is the energy density functional at saturation and $a = \frac{3\hbar^2 k_{F_0}^2}{10m}\rho_0$,

while the force is given by

$$v(\underline{r}\cdot\underline{r}') = 1/V(t_0 + t_3\ \rho^d)\ \delta(\underline{r}-\underline{r}') \tag{9}$$

For the choice $K_F = 1.35\ fm^{-1}$ and $\varepsilon(\rho_0)/\rho_0 = -16$ MeV, Blaizot [17] obtains $K_0 = 167 + 212d$ MeV. K_0 then varies between 238 MeV and 96 MeV for d varying between 1/3 and -1/3, corresponding effectively to an increase in the range of the force. To relate these quantities in infinite matter directly to finite nuclear information, Blaizot [17] performed a random phase calculation (RPA) for the energy of the breathing mode in heavy nuclei. His results are summarized in Fig. 2.

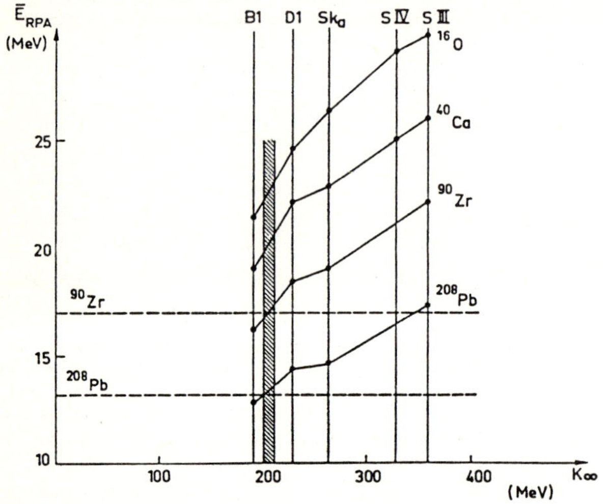

Fig. 2. The RPA predictions of Blaizot for the breathing mode in ^{208}Pb and ^{90}Zr compared to the experimental values (dashed lines). The vertical lines indicate the calculations for different forces used in the RPA. B1 and D1, which give results closest to experiment, are both finite range forces, while Skyrme forces Sk_a, SIV, and SIII all give larger compressibilities and disagree with experiment. The infinity in K_∞ (and $\rho_0(\infty)$ above) refers to the infinite matter limit.

A major problem in this standard treatment of finite nuclear compressibility is apparent in the relation for $K_0(1/2)$ in terms of F_0 and m^*. The effective mass is evaluated at the Fermi surface, whereas the single particle (hole) states involved in the finite nucleus RPA are appreciably removed from P_F. The well publicized difficulty in reproducing known single particle levels above and below the closed shell nucleus ^{208}Pb, with the quasi-zero range forces like Skyrme and its variants, is a strong warning in this regard. The two particle correlations in the RPA shift the monopole resonance some 14 MeV downwards, and are an added complication. I personally feel we are a long way from pinning down the incompressibility of nuclear matter at saturation; nevertheless, I feel Blaizot's result, $K_0(1/2) = 210 \pm 30$ MeV, is an honest and useful estimate.

Relativistic Heavy Ion Collisions

The analyses of relativistic heavy ion experiments performed at Lawrence Berkeley Laboratory by STOCK, HARRIS, and collaborators [18,19] and GUSTAFSSON, GUTBROD and collaborators [20] are also viewed as a commentary on the nuclear EOS. Two differing phenomena: (1) pion production in heavy ion collisions, and (2) the transverse momentum distributions of collision fragments, are used to deduce information on the state of matter. The extraction of an infinite matter compressibility from such analyses is complicated in this instance by the non-equilibrium nature of the actual collision and perhaps also by the fairly high temperatures achieved.

In the case of pion production (HARRIS and collaborators [18,19]; KITAZOE and collaborators [21] the reasoning is simple: energy generated in the nuclear medium by collision is divided into compressional and thermal components, with only the latter available for pions. One imagines a division of nuclear matter into two regions, one collision-free or unshocked, and one shocked, in which all pions are generated. Presupposing chemical equilibrium for the two baryon species N and Δ, one obtains for the pion multiplicity

$$\langle m_\pi \rangle = \frac{\rho_\pi + \rho_\Delta}{\rho_N + \rho_\Delta} = \frac{\rho_\pi + \rho_\Delta}{\rho} \tag{10}$$

The relativistic Rankine-Hugoniot equations applied for a one-dimensional geometry are then used to derive the conditions in the shocked region 2 from the known state of normal nuclear matter $\rho_1 = \rho_0(1/2)$ in the unshocked region 1.

The form of the compressional energy used in the LBL analyses [18,19] is the so-called parabolic dependence

$$P_\rho = \frac{K_0 \rho_0}{9} u^2 (u - 1) \ , \ \Gamma(\rho) = 3 + \frac{1}{u-1} \ . \tag{11}$$

Eq. (11) can be compared to the BCK equation of state, i.e., to eq. (1). From (1) it appears the parabolic form, for equal K_0, generally is stiffer than BCK with $\gamma \lesssim 3$.

Baron, Brown, Cooperstein, and Prakash [22] have attempted to pursue a fundamental derivation of the high density EOS by introducing necessary relativistic effects. Figure 3 from these authors presents comparative results for $\langle m_H \rangle$ from BCK and from the parabolic form. Clearly with K_0 = 400 MeV, an even stiffer EOS than afforded by γ = 4 in BCK, for which the parabolic and BCK forms are roughly equivalent, is required by the data. The stiffer the EOS employed, the less energy energy available for pions. A cold compressibility of K_0 = 600-800 MeV seems necessary to fit the pion data [18,22].

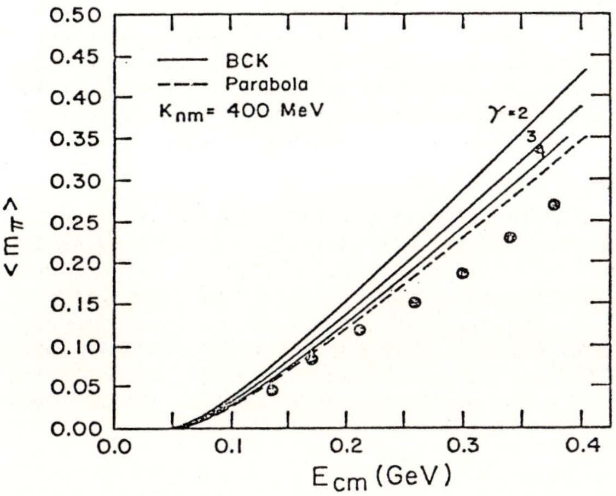

Fig. 3. Heavy-ion pion production. Comparisons of BCK for γ=2,3,4 and of the "parabolic"EOS with the data for the number of pions/baryons, $\langle m_\pi \rangle$, produced in heavy ion collisions, from Baron and collaborators [22]. The data is best fitted by an even stiffer EOS with $K_0(1/2) = K_{nm} \approx$ 600-800 MeV.

BARON and collaborators [22] argue that the nuclear EOS is indeed stiffer at higher temperature, e.g. at the T ≈ 70 MeV reached in the experiments. Their argument exploits the weakening in the attractive tensor force occasioned by higher average excitation energies in colliding matter. Similar conclusions, but not so extreme with respect to the degree of stiffness, follow from an analysis of sideways flow in heavy ion collisions (GUSTAFFSON and collaborators [20]; MOLITORIS and STOCKER [23]).

This disagreement between compressibilities, already present between the two different analyses of nuclear properties considered, is again exhibited in the hydrodynamic evolution of stellar collapse I now turn to. Ironically, it is only in the distant stellar laboratory that truly uniform nuclear matter obtains in conditions of chemical and thermal equilibrium.

III. HYDRODYNAMIC SIMULATION: NEWTONIAN

The hydrodynamic scheme I discuss here is based on a Newtonian code developed by Cooperstein [9] and used extensively in earlier work with my collaborators [11,15]. The force equation being studied for a spherically symmetric distribution of matter with velocity field U(r,t) is

$$\rho\frac{dU}{dt}(r,t) = - G\,\frac{\rho(r)\ m(r)}{r^2} - \frac{dP}{dr}\quad , \text{ with } P = P_N + P_{lept}\ . \tag{12}$$

The mass m(r) within the sphere radius r provides a Lagrangian coordinate for the problem. Supplementing (12) one must have some treatment of the only non-equilibrium aspect of this problem, neutrino transport. Wilson [6] has included a more complete diffusion approach in his calculations, and these provide a normalizing standard. In BCK [1,11,15] two approximate treatments are employed: (1) trapping at some density plus free-streaming, or 2) leakage (EPSTEIN and PETHICK [24]). The leakage scheme permits neutrinos to escape from all zones in the star, and was conservatively designed to overestimate the escape. Energy is explicitly conserved in all these Lagrangian schemes.

The Newtonian simulations performed following the early work of MAZUREK, COOPERSTEIN, and KAHANA [5] were intended for answering one question: What is the sensitivity of shock stalling to the nuclear EOS above (and just below) saturation density? We recognized early on that little constraint is provided here by laboratory experiment, and also we felt that the shock energy would be clearly enhanced by a softening of the high density matter. Increased gravitational compression of the "spring" in the stiffening matter would leave the final hydrostatic core more bound, and could result in the increased transfer of energy to the shock. The realistic initial models used were those of WEAVER, WOOSLEY, and FULLER [13] with central entropies of 0.67 per baryon (in units of k_B) and with a core mass of approximately 1.36 $M_\odot$, much to be preferred to the 1.51 $M_\odot$ core of WWZ [12].

Other, artificially generated, models were examined to study the effects of a softer EOS and also to study shock energy loss during outward propagation. Table I, reproduced from BCK [15], illustrates the role of the EOS, while Fig. 4 from the

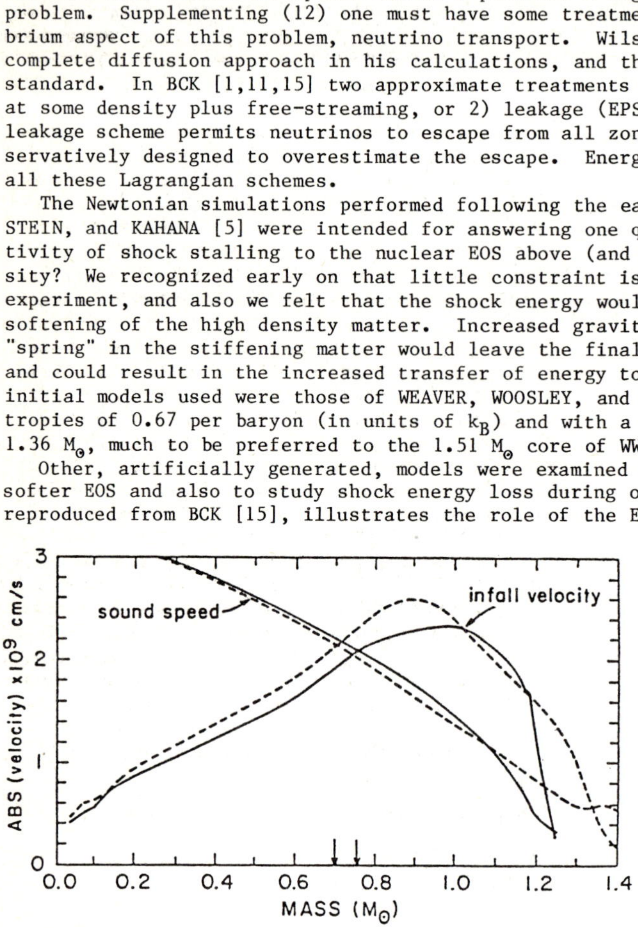

Fig. 4. The sonic point is displayed for two artificially constructed Newtonian models, FAIL and EXT. The dotted lines refer to the model FAIL, which produced a stalled shock, and the solid lines refer to model EXT(TREME), which ended in a successful explosive shock. Note the small mass difference between the formation of the sonic point in the two models (indicated by the arrows on the abscissa). This small mass difference is quite important to the shock propagation.

TABLE 1: The Effect of Equation of State on Shock Parameters for Two Models

Initial Model	$K_0(1/3)$	γ	R_{max} (km)	ρ^c_{max} $(g(cm^3)^{14})$	E_S (foes)
(BCK) I (a) 0.6	220	2	300	4.67	0.30
(b) "	150	2	835	5.54	0.64
(c) "	120	2	(1200)	6.08	0.96
(WWF) II (a) 0.67	220	2	160	4.80	1.47
(b) 0.67	120	2	320	5.60	1.80

Model I is artificially constructed with core mass $M_{core} = 1.32\ M_\odot$; Model II is that of WWF [13]. R_{max} is the largest radius reached by the shock before stalling. In I(c) the shock is marginally successful in ejecting the mantle. ρ_{max}, the maximum central density reached during collapse, is to be compared with $\rho_0[(Z/A) = .32] = 2.3 \times 10^{14}$ gm/cm^3. The energy E_S, a measure of shock energy discussed in BCK [15], is evaluated when the shock is at the Lagrangian coordinate $M = 1.21\ M_\odot$.

same source shows the relationship to shock propagation of the mass coordinate for the sonic point just after the last good homology. The pressure waves reaching the edge of the homologous core at this time must generate a shock at the sonic point. The shock is more likely to stall if it has to traverse more mass during its passage outwards.

We concluded from our studies of the EOS in a Newtonian setting that indeed shock energies are enhanced. We mapped out the losses per unit mass traversed by the shock, but were forced to accept the unlikeliness of a prompt explosion in the realistic evolutionary models. One interesting point, however, was the rapid increase in central density associated with the softening of higher density nuclear matter.

IV. HYDRODYNAMIC SIMULATION: GENERAL RELATIVISTIC GRAVITATION

The Schwarzschild radius of some 2 km for a homologous core of 0.7 $M_\odot$, which collapses to 15-20 km in radius, coupled to the cancellation of core internal and gravitational energies, alluded to above, suggests that the effects of general relativity cannot be ignored. The softer EOS further motivates the introduction of relativity, since non-linear gravitational effects are larger for the increased central densities. At first sight the strengthening of gravity inherent in general relativity will lead to deeper digging into the gravitational potential. However, this strengthening may also produce a smaller homologous core, and hence force the shock to traverse more matter on its way out. Indeed, early work by VAN RIPER [25] and TAKAHARA and SATO [26] suggested general relativity might be harmful to shock health, but VAN RIPER noted the possible sensitivity to the EOS. It is the unique combination of the softer EOS and relativistic gravitation present in our calculations that changes the picture.

The Newtonian force equation (12) is replaced by (see MISNER and SHARP [27])

$$\Delta_t\ U = e^{-\phi}\frac{\partial U}{\partial t} = -\frac{4\pi GRP}{c^2} - \frac{G\tilde{m}}{R^2} - \frac{4\pi R^2}{\omega}\tilde{\Gamma}\frac{\partial P}{\partial m} \tag{13}$$

with a gravitational mass

$$\tilde{m}(m,t) = \int_0^{R(m)} 4\pi R^2\rho(1 + E/c^2)dR\ , \tag{14}$$

and where m is a comoving Lagrangian coordinate in terms of which the metric is

$$d\sigma^2 = g_{\mu\nu}\ dx^\mu dx^\nu = -\ e^{2\phi}\ c^2dt^2 + e^{2\lambda}\ dm^2 + R^2d\Omega^2\ . \tag{15}$$

The factor $\tilde{\Gamma}$ in (13) is related to the radial metric coefficient by $e^{2\lambda} = \left(\frac{\partial R}{\partial m}\right)^2 \tilde{\Gamma}^{-2}$, while the specific enthalpy is $\omega = 1 + E/c^2 + \frac{P}{\rho c^2}$.
The metric must be matched to an exterior Schwarzschild solution at some low-density surface $m = m_g(t)$, thus introducing the total mass-energy within this surface. The stress energy tensor is taken to be that of a perfect fluid. Eq. (13) must be solved using the EOS for the pressure together with the comoving energy equation

$$\frac{dE}{dt} = -\frac{d(1/\rho)}{dt} - e^{\phi}\dot{S} \qquad (16)$$

Here $\dot{S}$ is the rate of energy loss in escaping neutrinos and e^{ϕ} a (small) neutrino redshift.

V. RESULTS

Our general relativistic calculations have been performed using the initial models of WEAVER and WOOSLEY [2], with dramatic results. I have already referred to these latest models in the introduction, but note here they resemble WWF [13] in structure for $M \lesssim 15\ M_{\odot}$, having iron cores with masses near $1.36\ M_{\odot}$. However, between $M = 15\ M_{\odot}$ and $20\ M_{\odot}$ a sudden change in core mass appears in the WW [2] analysis, a jump to core masses $\gtrsim 2.0\ M_{\odot}$. This discontinuity results from faster rates used for $^{12}C(\alpha,\gamma)^{16}O$ (ROLFS [28]) and depends on the formation or non-formation of a carbon-burning shell. If the carbon-shell forms, its large entropy creates a barrier to further burning ending in iron. This is the case for the lighter initial stellar masses. Table 2 is a summary of numerical experiments carried out for several initial models differing in their parametrization of the EOS. Figure 5 contains the radius vs time profiles for the $12\ M_{0}$, model 41 of Table 2. Reasonable choices for $K_0(1/2)$, very close if not within the range obtained by BLAIZOT [17], and for γ values near those expected from purely theoretical considerations, result in viable prompt explosions with energies adequate to account for any visual display. I wish to reemphasize that most of the softening from the near 200 MeV value of $K_0(1/2)$ comes from the well understood extrapolation to asymmetric matter. Although one continues to characterize the nuclear EOS with $K_0(1/2)$ it is the value near $Z/A = 1/3$ that is most important for supernova calculations and not the history between $Z/A = 1/2$ and $1/3$.

"Successful" explosions in these calculations are most succinctly characterized by the maximum central density achieved during collapse (see Table 2). This may well be misleading, since phase changes to quark-gluon or other exotic forms of

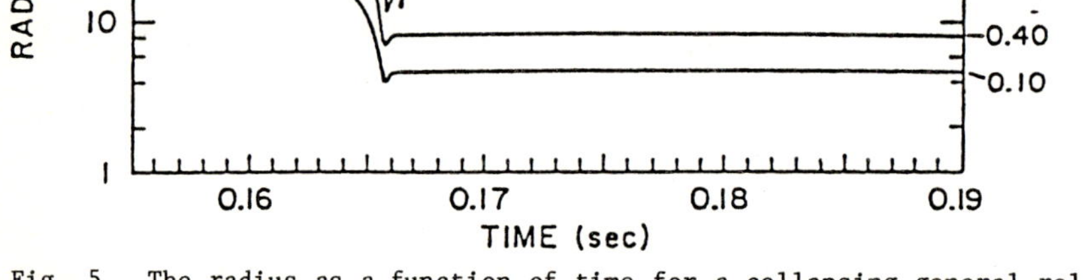

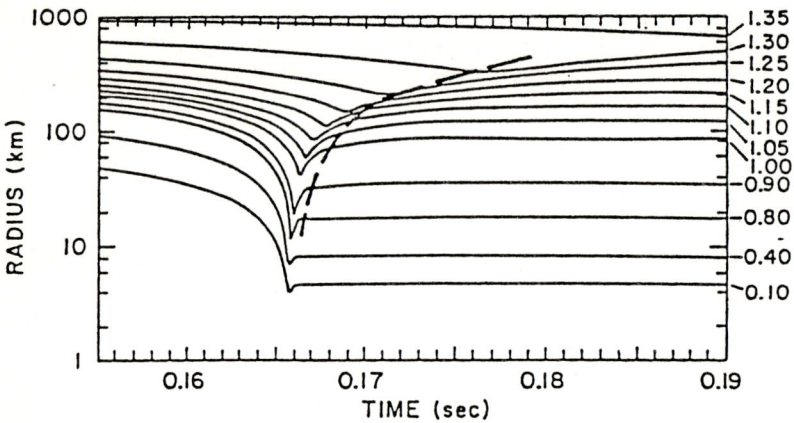

Fig. 5. The radius as a function of time for a collapsing general relativistic model. ($12\ M_{\odot}$ with $\gamma = 3$, $K_0(1/3) = 140$ MeV). The numbers on the right hand side indicate the total enclosed mass in units of $M_{\odot}$. The dotted line denotes the approximate position of the shock.

Table 2 Description of Calculations

Model	Mass ($M_\odot$)	K_0^{sym} (MeV)	$K_0(0.33)$ (MeV)	γ	GR	$\frac{\rho_c^{max}}{\rho_0(0.33)}$	E_{exp} (foes)	E_{lost} (foes)
32	12	220	220	2	no	1.7	-	2.6
33	12	220	170	2	no	2.0	-	2.1
38	12	180	140	2	no	2.3	0.1	3.2
40	12	180	140	2	yes	12.0	3.2	2.2
41	12	180	140	3	yes	3.1	0.8	3.3
29	15	220	220	2	no	1.7	-	2.1
42	15	180	140	3	yes	3.1	-	2.5
43	15	180	140	2.5	yes	4.1	1.7	3.4
44	15	140	120	3	yes	3.3	-	3.2
45	15	90	90	3	yes	4.0	0.8	3.2

The column labeled GR refers to whether general relativity or Newtonian gravity was assumed. $K_0(0.33)$ refers to the value of the nuclear incompressibility at saturation density when there are twice as many neutrons as protons (Z/A =0.33). For model 33 the form of the incompressibility was taken to be that of (3) with K_0^{sym} = 220 MeV, while for models 38 and 40-43 K_0^{sym} was chosen to be 180 MeV. Model 44 had $K_0(x) = 140\left(1-1.3(1-2x)^2\right)$ and the other models (29, 32, 45) had K_0 = constant at the value given in this table. γ refers to the high-density adiabatic index discussed in the text. ρ_c^{max} is the maximum central density achieved just prior to bounce. E_{exp} is an estimate of the explosion energy, neglecting the binding energy of the mantle and envelope. No entry in this column means the shock failed to reach the edge of the iron core. E_{lost} is the total neutrino loss when the calculation was stopped (roughly 50 milliseconds after bounce).

matter are excluded in our work, e.g. at sufficiently high density one expects hadron matter to coexist with two-flavour quark matter. Nevertheless, a high central density signals that the unfavourable aspects of general relativity, viz. the smaller homologous core and the perhaps stronger gravitational force the outward-moving shock must fight against, have been offset by an increase in the energy transferred to the shock from the gravitational well. An unexpected, but important, further advantage of stronger gravitation is occasioned by the faster collapse of the central core. The shock forms faster and will thus meet the infalling outer core zones at times when the latter are at lower density and falling slower. The passage outward is thus eased considerably relative to that of the Newtonian shock. This effect is illustrated in the density profiles in Fig. 6.

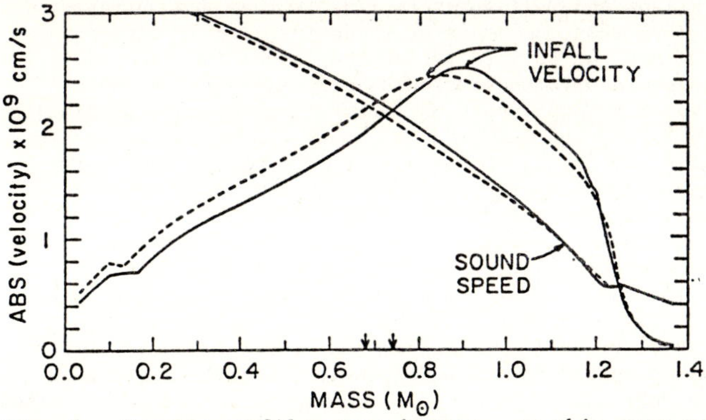

Fig. 6. Density profiles near bounce. In this case general relativity (dashed line) clearly produces a more favourable configuration for shock propagation.

To summarize the results: In Table 2 for the reasonable choices $\gamma = 2.5$ and $K_0(.33) = 140$, corresponding to $K_0(1/2) = 180$ MeV, one finds explosion energies near 2.0 foes in the general relativistic calculations. Lower shock energies are indeed acceptable, and one could probably produce prompt explosions for the 12 $M_\odot$ and 15 $M_\odot$ WW [2] models with $K_0(0.50) \approx 200$ MeV, i.e. very close to the BLAIZOT [17] central value. These shock energies are not to be taken too seriously at this point, partly because of the uncertainties in our parametrizations and partly because the calculations should be followed further in time. Nevertheless, sufficient energy has been generated to eject the non-core regions of the star and to account for the visual display of a supernovae. Equally certain is that the remnant of such an explosion will be a neutron star, although its actual observable mass remains in some doubt.

VI. NEUTRON STAR MASSES

A possible constraint on the K_0, γ used in the nuclear EOS may result from a study of the hydrostatics of neutron stars. Inordinately soft matter would be unable to sustain the 1.4 $M_\odot$ mass observed for known neutron stars. For example, dropping the constant term in (1) yields as an appropriate non-saturating EOS for neutron matter:

$$P = \frac{K_0 \rho_0}{9\gamma} u^\gamma , \qquad (16)$$

From calculations based on (16) one concludes $\gamma \gtrsim 2.5$ is required, a not very severe constraint in light of Table 2. However, one could easily imagine a density variation in the adiabatic index which kept γ low in the regions important in collapse while having γ rise sufficiently for $\rho \gtrsim 3$ to accommodate known neutron star masses. Such a choice would also go a long way towards explaining the Berkeley heavy ion experiments mentioned earlier. More work on the EOS is required for both the collapse and neutron star phases, and certainly more fundamental theoretical work incorporating amongst other things, the nuances of phase changes.

VII. CONCLUSIONS

In conclusion, we note, no Newtonian calculation with WW initial models yields sufficient energy to disrupt the star. The most favourable case for $K_0(.33) = 140$ MeV and $\gamma = 2$ (model 38) yields a shock which propagates off the mathematical grid, but whose .1 foe energy cannot overcome the few tenths of a foe binding of the mantle and envelope. There is a threshold of softness in nuclear matter beyond which the effects of general relativity become helpful to shocks. Shock energies are enhanced over Newtonian values by an order of magnitude reasonably close to this threshold in $K_0(.33)$ and γ.

The WW initial models [2] for stars more massive than, perhaps, 18 $M_\odot$, possessing iron cores greater than 2.0 $M_\odot$ in mass, will not explode; there is almost no chance for a hydrodynamic shock to be viable in such a massive core. Perhaps the delayed neutrino-heating scenario of Wilson and collaborators [29] is then appropriate. However, the use by nature of more than one means of producing type II supernova appears inelegant. Arnett [30] has put forward yet another mechanism for reviving stalled shocks: convection generated by the gradients in Y_e and entropy behind the shock. Baron, Bethe, Brown, and Cooperstein [31] have examined this possibility for the model 42 of BCK (Table 2) and conclude that although convection is indeed possible sufficiently far behind the shock, the details of the matter and energy distribution render such convection unfavourable to further progress of the shock.

Undoubtedly we have not yet heard the end of this tale, although I believe an important watershed has been reached. Initial models still retain the capacity for terror, for example a drop in core masses below 1.25 $M_\odot$ would dissolve all difficulties, while an appreciable increase in core mass would prevent any prompt shock from succeeding. Finally, the details of nucleosynthesis must be foloowed hydrodynamically for a successful shock-produced explosion, and thus the initial stellar mass range required to partake in supernova for an adequate discription of the known elemental abundances, would be defined.

REFERENCES

1. E. Baron, J. Cooperstein, and S.H. Kahana: Phys. Rev. Lett. 55, 126 (1985).
2. S. E. Woosley and T.A. Weaver: Bull. Am. Astr. Soc. 16 971 (1984).
3. E.M. Burbidge, G.R. Burbidge, W.A. Fowler, and F. Hoyle: Rev. Mod. Phys. 29 547 (1957).
4. S.A. Colgate and H.J. Johnson: Phys. Rev. Lett. 5 235 (1960). S.A. Colgate and R.H. White: Ap. J. 142 626 (1966).
5. T.J. Mazurek, J. Cooperstein, and S.H. Kahana: DUMAND '80, ed. V.J. Stenger (DUMAND Center, Honolulu, 1981). T.J. Mazurek, J. Cooperstein, and S.H. Kahana: Supernovae: a Survey of Current Research, ed. M. Rees and R.J. Stoneham (Reidel, Dordrecht, 1982).
6. J.R. Wilson: Ann. N.Y. Acad. Sci. 336 358 (1980).
7. W.D. Arnett: Supernovae: a Survey of Current Research, ed. M. Rees and R.J. Stoneham (Reidel, Dordrecht, 1982).
8. W. Hillebrandt: Supernovae: a Survey of Current Research, ed. M. Rees and R.J. Stoneham (Reidel, Dordrecht, 1982).
9. J. Cooperstein: PhD thesis, State University of New York at Stony Brook (unpublished).
10. J. Cooperstain, H.A. Bethe, and G. E. Brown: Nucl. Phys. A429 527 (1984).
11. S.H. Kahana, E. Baron, and J. Cooperstein: in Problems of Collapse and Numerical Relativity, ed. D. Bancel and M. Signore (Reidel, Dordrecht, 1984) p. 163.
12. T.A. Weaver, B. Zimmerman, and S.E. Woosley: Ap. J. 225 1021 (1978).
13. T.A. Weaver, S.E. Woosley, and G.M. Fuller: Bull. Am. Astr. Soc. 14 No.4 957 (1982) and in Numerical Astrophysics, ed. J. Centrella, J. Leblanc, and R. Bowers (Jones and Bartlett, Boston, 1982).
14. H.A. Bethe, G.E. Brown, J. Applegate, and J. Lattimer: Nucl. Phys. A324 487 (1979).
15. E. Baron, J. Cooperstein, and S.H. Kahana: Nucl. Phys. A440 744 (1985).
16. J-P. Blaizot: Phys. Rep. 64 171 (1980).
17. K. Kolehmainen, M. Prakash, J. Lattimer, and J. Treiner: Nucl. Phys. A (to be published).
18. J.W. Harris, R. Bock, R. Brockmann, A. Sandoval, R. Stock, H. Stroebele, G. Odyniec, H.G. Pugh, L.S. Schroeder, R.E. Renfordt, D. Schall, D. Bangert, W. Rauch, and K. Wolf: Phys. Lett. 153 377 (1982).
19. R. Stock, R. Bock, R. Brockman, J.W. Harris, A. Sandoval, H. Stroebele, K.W. Wolf, H.G. Pugh, L.S. Schroeder, M. Maier, R.E. Renfordt, A. Daca, and M.E. Ortiz: Phys. Rev. Lett. 49 1236 (1982).
20. H.A. Gustafsson, H.H. Gutbrod, B. Kolb, H. Löhner, B. Ludewigt, A.M. Poskanzer, T. Renner, H. Riedesel, H.G. Ritter, A. Warwick, F. Weik, and H. Wieman: Phys. Rev. Lett. 52 1590 (1984).
21. Y. Kitazoe, M. Gyulassy, P. Danielewicz, H. Toki, Y. Yamamura, and M. Sano: Phys. Lett. 138B 341 (1984).
22. E. Baron, G.E. Brown, J. Cooperstein, and M. Prakash: SUNY at Stony Brook, preprint.
23. J.J. Molitoris and H. Stöcker: Phys. Rev. C32 346 (1985).
24. R. Epstein and C.J. Pethick: Ap. J. 243 1003 (1981).
25. K. Van Riper: Ap. J. 232 558 (1979).
26. M. Takahara, and K. Sato: Prog. Theor. Phys. 72 978 (1984).
27. C. Misner and D. Sharp: Phys. Rev. 136 571 (1964).
28. C. Rolfs: See for example the proceedings of the April 1986 Erice symposium on Nuclear Physics.
29. J.R. Wilson: in Numerical Astrophysics, ed. J. Centrella, J. Leblanc, and R. Bowers (Jones and Bartlett, Boston, 1985).
30. W.D. Arnett: to be published, (1985).
31. E. Baron, H.A. Bethe, G.E. Brown, and J. Cooperstein: Private communication (1986).

Electron Capture in Stellar Collapse

J. Wambach

Department of Physics, University of Illinois at Urbana-Champaign, Urbana, IL 61801, USA

Massive stars in the range $M > 9\ M_{\odot}$ are thought to end in Type II supernova explosions. Once the elements up to iron have been reached in the thermonuclear burning, further fusion reactions are halted. The iron core, supported by degenerate electron pressure, develops an instability induced by a combination of nuclear photodisintegration and electron capture and implodes. The evolution of the capture rates on free protons and nuclei during the collapse phase is discussed. These change the two most important quantities in the collapse dynamics, namely the number of leptons per nucleon and the entropy. It is believed that high lepton fraction and low entropy are essential for shock wave formation and propagation, which eventually expels the outer parts of the star. Dynamical collapse calculations are discussed which indicate that the electron fraction per nucleon and the entropy at neutrino trapping densities are not changed significantly from their initial values.

1. Introduction

Massive stars at the end of their thermonuclear burning may die explosively as Type I or Type II supernovae. While in both cases the energy release is quite comparable ($\sim 10^{51}$ ergs), the light curves, the hydrogen content in their spectra and the progenitor masses are quite different. It is currently believed that Type I supernovae are the outcome of white dwarfs, presumably composed of carbon and oxygen, accreting mass from a binary compagnion, while Type II supernovae originate from more massive stars ($M > 9\ M_{\odot}$) which undergo gravitational collapse. The explosion mechanisms for Type I's and Type II's are also quite different. Currently favoured models[1)] indicate that white dwarfs explode via thermonuclear "combustion waves" (deflagration) while Type II's result from prompt or "delayed" hydrodynamical shock waves formed at the sonic point.

At several stages of the evolution of massive stars nuclear electron capture plays a decisive role. To name some:

- detailed network calculations[2)] of explosive nucleosynthesis in the core of mass accreting white dwarfs and the resulting element abundances especially in the Fe-Ni region are sensitive to capture rates.
- presupernova conditions, in particular core sizes, of more massive stars dependent on the degree of neutronization. Recent rate estimates[3)] for example lead to a downward revision of iron core masses compared to previous estimates, a welcome precondition for successful explosions.
- the gravitational instability of Type II progenitors is partly triggered by e^--capture on the iron group nuclei in the core once the Chandrasekhar limit has been reached. The subsequent collapse dynamics up to neutrino trapping densities depends sensitively on rates for neutron rich nuclei in the $60<A<100$ mass region.

- hydrodynamical shock propagation in the post-bound phase and "delayed" explosions[4] depend on the neutrino luminosity behind the shock, which is governed by capture in the homologous core.

An adequate review covering all aspects of electron capture in massive stars goes certainly beyond the scope of this presentation. To unravel the relevant nuclear physics, which is common to most of the phenomena listed above, we will restrict ourselves to the neutronization of the stellar core starting from the initial collapse phase up to or a little above neutrino trapping densities. This covers a density range $0.1 \leqslant \rho_{11} \leqslant 10$ ($\rho_{11} = 10^{11}$ g/cm^3).

2. Presupernova Conditions

Supernova progenitors evolve through various thermonuclear burning stages of the star which depend on the stellar mass. For $M \geqslant 12\ M_\odot$ all six stages ignite hydrogen, helium, carbon, neon, oxygen and silicon burning. The structural evolution can be quite complex, in particular for light stars, characterized by off center ignition and sensitivity to electron capture etc.[1]. Figure 1 displays the composition of a 15 $M_\odot$ presupernova star at the time when the edge of its iron core begins collapsing[5].

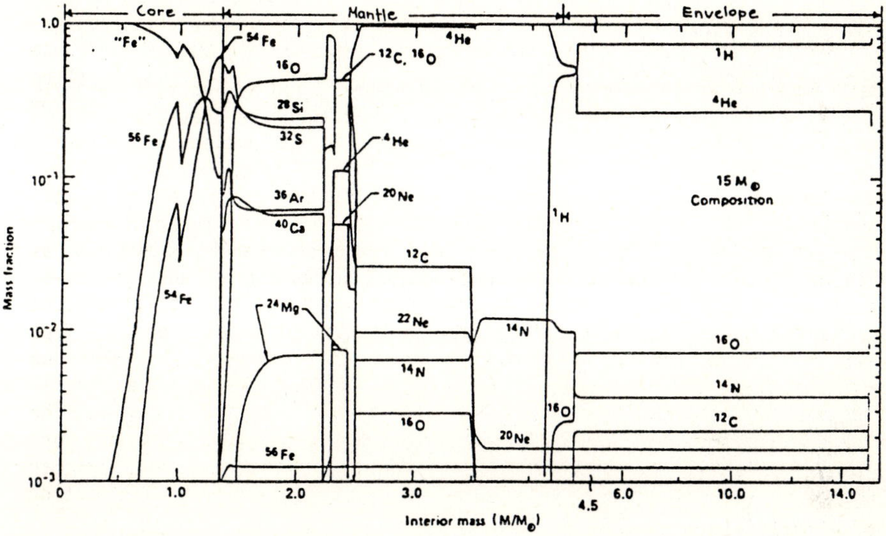

Fig. 1 Element composition of a 15 $M_\odot$ presupernova star at the time when the edge of iron core starts collapsing. The central temperature is 7.62×10^9 K (.66 MeV) and the central density .1 ρ_{11} (reproduced from ref. 5).

In the collapse phase the core evolution is essentially decoupled from the envelope and the mantle[6]. Before collapse the core of radius R and mass M_c is in hydrostatic equilibrium

$$\frac{1}{\rho}\frac{\partial P}{\partial R} + \frac{GM_c}{R^2} = 0 \tag{2.1}$$

$$M(r) = 4\pi \int_0^R r^2\, dr\, \rho(r) \tag{2.2}$$

with the pressure P being dominated by relativistic, essentially degenerate electrons with an adiabatic index of 4/3, i.e. $P \propto \rho^{4/3}$. The number of electrons/nucleon Y_e is typically .42 for Type II cores. The core temperature at this point is about 0.8-1.0 MeV. Once the Chandrasekhar limit

$$M_c \geqslant M_{Ch} \sim 5.8\ Y_e^2 \sim 1.02\ M_\odot$$

is reached through Si-burning ($\rho_c \sim .1\ \rho_{11}$), collapse is triggered by

- electron capture from the iron group which reduces Y_e and thus the degeneracy pressure.
- photodisintegration of iron group elements into α's and neutrons which also leads to a pressure decrement since the breakup costs energy. This process is dominant for $M \geqslant 20\ M_\odot$).

The presupernova structure is best understood in terms of the entropy per nucleon of S/k, particularly in the center. Massive stars are born with a nearly constant entropy profile with $S/k \sim 25$ throughout the star. Near the collapse stage, however, $S/k \sim 1$ in the core, while that in the envelope has increased, principally by radiation transport to ~ 40. It is the low value of the central entropy which makes the core of the star sensitive to the Chandrasekhar limit.

3. Electron Capture Rates

The two most important quantities during collapse are Y_e and S/k as first pointed out by Bethe et al.[6] (BBAL). Low entropy ensures that the core does not halt and bounce at $\rho_{11} \sim 3\times10^2$ because of the thermal pressure of hot nucleons from photodisintegrated matter, but will continue to densities beyond nuclear matter, $\rho_{11} \sim 3\times10^3$. Both Y_e and S/k change in the collapse phase due to electron capture. A change in entropy is introduced by the fact that weak interactions are not equilibrated on the collapse time scale of a few milliseconds. β-equilibrium is only achieved after the collapse becomes adiabatic because of neutrino trapping ($\rho_{11} \sim 5$).

Two capture processes are relevant in the collapse scenario: capture on free protons

$$p+e^- \to n+\nu_e \tag{3.1}$$

or nuclei

$$(N,Z)+e^- \to (N+1,Z-1)+\nu_e\ . \tag{3.2}$$

While free proton capture always leads to a reduction in entropy, as long as neutrinos escape, ground state capture on nuclei goes to excited states which increases the entropy. To quantify the importance of both effects a detailed study of the rates is necessary.

For given proton and nuclear concentrations X_p and X_A and capture rates λ_p and λ_A the time evolution of Y_e is determined by

$$\dot{Y}_e = \dot{Y}_e^P + \dot{Y}_e^A = (X_p\lambda_p + X_A \langle\lambda_A\rangle)Y_e\ . \tag{3.3}$$

At finite temperature there is a statistical distribution $\phi(A)$ of nuclei to be determined from thermodynamics (sect. 4). Thus the nuclear rates for given mass number A have to be averaged according to

$$\langle\lambda_A\rangle = \int \phi(A)\lambda_A dA\ . \tag{3.4}$$

It is a standard exercise in β-decay physics to evaluate λ_p and λ_A[7]. In the relativistic electron limit ($P_e = \varepsilon_e$) and for a given temperature T we have

$$\lambda_p = -\frac{G_W^2}{2\pi^3}(g_V^2+3g_A^2)\int_{M_{pn}}^{\infty} d\varepsilon_e \varepsilon_e^2(\varepsilon_e - M_{pn})^2 n(\varepsilon_e,\varepsilon_\nu,T) \tag{3.5}$$

and

$$\lambda_A = -\frac{G_W^2}{2\pi^3}\int_Q^{\infty} d\varepsilon_e \varepsilon_e^2(\varepsilon_e - Q)^2 F(\varepsilon_e,Z) S_A(Q,\varepsilon_e,T) n(\varepsilon_e,\varepsilon_\nu,T) \tag{3.6}$$

with the Pauli blocking factor

$$n(\varepsilon_e,\varepsilon_\nu,T) = f\left(\frac{\varepsilon_e-\mu_e}{kT}\right)\left[1 - f\left(\frac{\varepsilon_\nu-\mu_\nu}{kT}\right)\right] \tag{3.7a}$$

$$f(x) = (1-e^x)^{-1} . \tag{3.7b}$$

From energy conservation ε_e and ε_ν are not independent but related to the Q-value of the capture reaction

$$Q = \varepsilon_e - \varepsilon_\nu = \varepsilon_n - \varepsilon_p = \varepsilon^* + \hat{\mu} + M_{pn} \tag{3.8}$$

Here we define $\hat{\mu} = \mu_n - \mu_p$ as the difference in the neutron-proton chemical potentials and $M_{pn} = M_n - M_p = 1.29$ MeV as the neutron-proton mass difference. ε^* denotes the excitation energy in the daughter nucleus (N+1,Z-1). $F(\varepsilon_e,Z)$ is the Coulomb penetration factor.

For a given capturing nucleus (N,Z) the detailed nuclear structure information is contained in the "shape factor" S_A which describes the transition strength per proton. At finite T it is determined by an ensemble average of transition matrix elements of the weak Hamiltonian H_W between nuclear states $|i\rangle$ and $|f\rangle$

$$S_A(\varepsilon_e,Z,T) = \frac{1}{Z}\sum_{i,f} W_i(N,Z;T)\,|\langle (N,Z)i|H_W(\varepsilon_e)|(N+1,Z-1)f\rangle|^2\,\delta(E-Q) . \tag{3.9}$$

The weight factor W_i gives the probability for finding the initial state $|i\rangle$ in the grand canonical ensemble

$$W_i(N,Z;T) = \frac{e^{-\frac{1}{kT}(E_i-\mu_N N-\mu_Z Z)}}{\sum_j e^{-\frac{1}{kT}(E_j-\mu_N N-\mu_Z Z)}} . \tag{3.10}$$

Specific nuclear models for S_A will be discussed in sect. 5.

The change in Y_e (3.3) specifies the change in entropy. According to BBAL[6])

$$\frac{dS}{dY_e} = -(\mu_e - \hat{\mu} - M_{pn} - \langle\varepsilon_\nu\rangle)/T \tag{3.11}$$

The average neutrino energy released in the capture process is defined as[8])

$$\varepsilon_\nu = [(\dot{Y}_e)_p \langle\varepsilon_\nu\rangle_p + (\dot{Y}_e)_A \langle\varepsilon_\nu\rangle_A]/\dot{Y}_e \tag{3.12}$$

The proton part $\langle\varepsilon_\nu\rangle_p$ and the nuclear part $\langle\varepsilon_\nu\rangle_A$ is easily calculated from the corresponding rate expressions by including in the integrands an extra weight factor

$$\varepsilon_\nu = \varepsilon_e - Q . \tag{3.13}$$

4. Stellar Core Composition

To obtain the time dependence of Y_e we have to determine the concentrations X_p and X_A and the nuclear distribution function $\phi(A)$. At the relevant densities the core composition is specified by photons which form a Stephan-Boltzmann gas, relativistic degenerate electron-positron gases, a neutrino gas, free nucleons, α-particles and heavier nuclei. At collapse time scales strong interactions are equilibrated. Thus, once, T, ρ and Y_e are specified the concentrations of nuclear species are determined from minimization of the Helmholtz free energy. It is convenient to deduce the heavy nucleus part from the liquid drop model[11]. Neglecting compressional and curvature effects the binding energy per nucleon is

$$B_A(Y_e,\rho) = W_b + W_s\, A^{-1/3} + W_c\, A^{2/3} \, . \qquad (4.1)$$

Since the nuclei are immersed in a degenerate electron gas and occupy only a fraction of space W_s and W_c are functions of Y_e and ρ. The statical distribution function

$$\phi(A) = \frac{1}{\sigma_A\sqrt{\pi}}\, e^{-\frac{(A-A_o)^2}{\sigma_A^2}} \qquad (4.2)$$

is specified by B_A. The most abundant nucleus with mass number A_o is obtained from the equilibrium condition

$$\left.\frac{\partial B_A}{\partial A}\right|_{Y_{e,\rho}} = 0 \qquad (4.3)$$

which leads to the "virial-like" expression

$$A_o = \frac{W_s}{2W_c} \, . \qquad (4.4)$$

The dispersion σ_A is calculated from the second order derivative

$$\sigma_A^{-2} = A_o \left.\frac{\partial^2 B_A}{\partial A^2}\right|_{A_o, Y_e, \rho} /kT \qquad (4.5a)$$

which yields

$$\sigma_A = A_o \sqrt{\frac{3kT}{W_s A_o^{2/3}}} \qquad (4.5b)$$

Finally the proton-neutron chemical potential difference $\hat{\mu}$ is determined by the Y_e-dependence of B_A

$$\hat{\mu} = -\frac{1}{A_o} \left.\frac{\partial B_A}{\partial Y_e}\right|_{A_o, Y_e, \rho} \, . \qquad (4.6)$$

5. Nuclear Models

To evaluate the shape factors, several methods have been applied. Statistical treatment based on an average nuclear partition functions have been proposed by Fuller et al.[3]. BBAL used a shell model description subsequently refined by Zaringhalam[9], K. Kar and A. Roy[10] and J. Cooperstein and J. Wambach[8]. We shall review this approach in more detail. BBAL considered ^{56}Fe as the most probable nucleus. The shell occupation is indicated in Fig. 2.

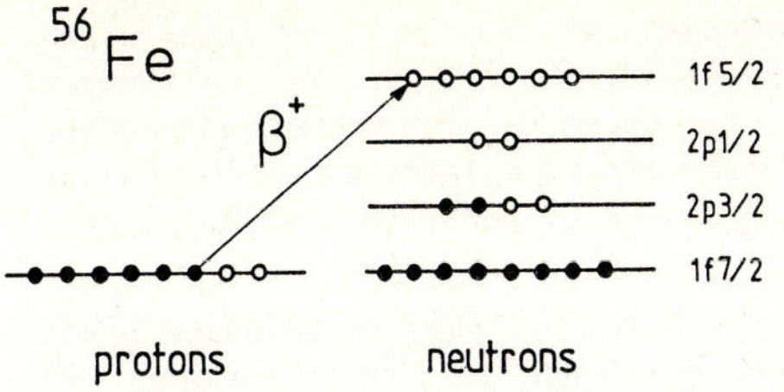

Fig. 2 Shell-model description of e⁻-capture in ^{56}Fe. In the capture, protons go to the $1f_{5/2}$-level of the daughter nucleus, which then γ-decays to the 2p-orbitals.

In the original work[6], temperate effects have been neglected and H_W was taken in the long wavelength limit. Thus only Fermi(F)- and Gamow-Teller(GT)-transitions survive. Under these assumptions S_A becomes

$$S_A(Q,\varepsilon_e,0) = \frac{1}{Z}\sum_{i,f} \{g_V^2|\langle j_i\|1\|j_f\rangle|^2 + g_A^2|\langle j_i\|\vec{\sigma}\|j_f\rangle|^2\} n_{ph}(i,f)\delta(E-Q) \quad (5.1a)$$

with

$$n_{pn}(i,f) = \frac{n_p^i}{2j_i+1}\left(1 - \frac{n_n^f}{2j_f+1}\right) \quad (5.1b)$$

where n_p^i and n_n^f are the number of protons and neutrons in the initial and final j-shells respectively. For ^{56}Fe the resulting rate is easily calculated

$$\frac{\lambda_A}{\lambda_P} = \frac{2g_A^2}{g_V^2+3g_A^2}\,\frac{6}{26} = .13 \ . \quad (5.2)$$

Because of Pauli blocking the nuclear rate λ_A is considerably reduced over the free proton rate λ_p. Using the iron rates through the entire collapse phase BBAL found a considerable drop in Y_e (Y_e^f = .32 at ρ_{11} = 5). Subsequently Fuller et al.[6] pointed out an inconsistency. As Y_e drops nuclei become more neutron rich and allowed capture is quickly Pauli blocked since the $1f_{7/2}$ neutron shell fills up. On the other hand, as the matter density goes up, μ_e increases and electrons become more energetic. Thus "parity forbidden" transitions become important. In the nonrelativistic limit these are represented by the operators[7]

$$\Delta J^\pi = 0^- \qquad D = g_A\left[\frac{\vec{\sigma}\cdot\vec{p}}{M} + i\,\frac{\alpha Z}{2R}\,\vec{\sigma}\cdot\vec{r}\right]\tau_+ \quad (5.3a)$$

$$\Delta J^\pi = 1^- \qquad D = [g_V \frac{\vec{p}}{M} + \frac{\alpha Z}{2R}\,[g_A\,\vec{\sigma}\times\vec{r} + i g_V \vec{r}]\tau_+ \quad (5.3b)$$

$$\Delta J^\pi = 2^- \qquad D = i g_A/\sqrt{3}\,[\vec{\sigma}\times\vec{r}]^2_m\,\sqrt{\varepsilon_e^2+(\varepsilon_e-Q)^2}\,\tau_+ \quad (5.3c)$$

$$R = 1.12\,A^{1/3} \quad (5.4)$$

In the "unique first forbidden" case ($\Delta J^\pi = 2^-$) D depends on the electron energy and the Q-value.

Zaringhalam[9] included first forbidden transitions in the capture calculation and found that the rates are uniformly faster by a factor of ~ 20 than the BBAL estimates which results in a higher Y_e at trapping.

In Zaringhalam's work, however, two effects have been left out which were analysed in ref. 8:

- Nuclei immersed in the presupernova environment are not at zero temperature and hence nuclear excited states are thermally populated. In the shell model picture of BBAL the ensemble average (3.9) reduces to temperature dependent shell occupation numbers. Obviously, finite temperature leads to unblocking of GT-strength which is typically of the order of 1-2 %, consistent with Fuller's estimates from averaged partition functions[6].

- e^--capture proceeds mostly via low Q-values which are favoured by phase space (3.6). At low Q both g_V and g_A are renormalized due to ph-correlations. In terms of the isospin and spin-isospin susceptibilities κ_τ and $\kappa_{\sigma\tau}$ one obtains effective coupling constants as given by

$$g_V^{eff}/g_V = (1 + \kappa_\tau)^{-1} \tag{5.6a}$$

$$g_A^{eff}/g_V = (1 + \kappa_{\sigma\tau})^{-1} \ . \tag{5.6b}$$

Since the ph-interactions V_τ and $V_{\sigma\tau}$ are both repulsive κ_τ and $\kappa_{\sigma\tau}$ are > 0, i.e. g_V and g_A are quenched. This quenching has been measured by Eijiri et al.[12] as a function of neutron excess and more recently for unique first forbidden β-decay by A. Richter et al.[13]. The results are summarized in Fig. 3.

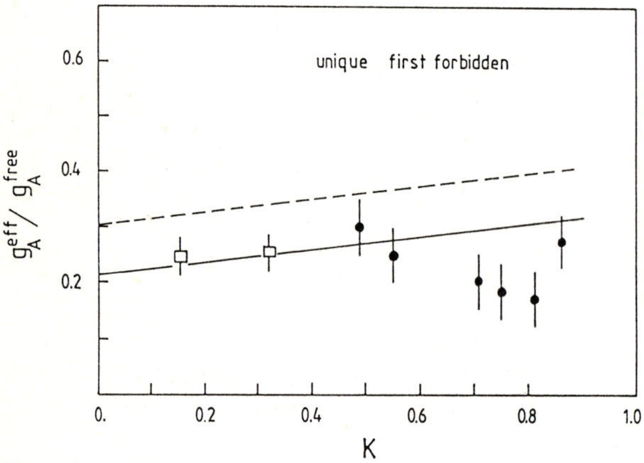

Fig. 3 Effective axial vector coupling constants for unique first forbidden β-decay. The data are taken from refs. 12 and 13. K denotes the neutron excess in units of the number of nucleons in a single $\hbar\omega$-shell.

Typically

$$g_i^{eff}/g_i \sim .3 \qquad (i=V,A) \tag{5.7}$$

Theoretically the measured susceptibilities are well reproduced by standard RPA adjusting V_τ and $V_{\sigma\tau}$ to the giant dipole and GT-resonances respectively[8].

The influence of quenching on the rates in Type I supernovae and the resulting Fe-Ni-abundances have been analysed recently[2]. The refined rates are helpful to reduce the overproduction of ^{54}Fe, ^{58}Ni and ^{62}Ni but do not solve the problem at present.

6. Electron Capture and Core Collapse

To have a complete picture of the collapse, the neutronization of the inner core during this phase has to be calculated. Following ref. 8, the implosion can be simulated by a simplified "one-zone" treatment[6]. In this model the state of the core matter is described by integrating the equation for the internal energy per unit mass

$$\frac{dE}{dt} = \frac{P}{\rho} \log \dot{\rho} + \langle \varepsilon_\nu \rangle \dot{Y}_e \tag{6.1}$$

where P/ρ is the pressure per nucleon. P/ρ is obtained from the equation of state of Bethe et al.[14] suitably extended to cover lower densities, including free protons, drip neutrons and α-particles together with a liquid drop description of nuclei and relativistic electron-positron and photon gases.

The lower part of Fig. 4 displays the evolution of the mean nucleus mass number A_0 together with Z and N.

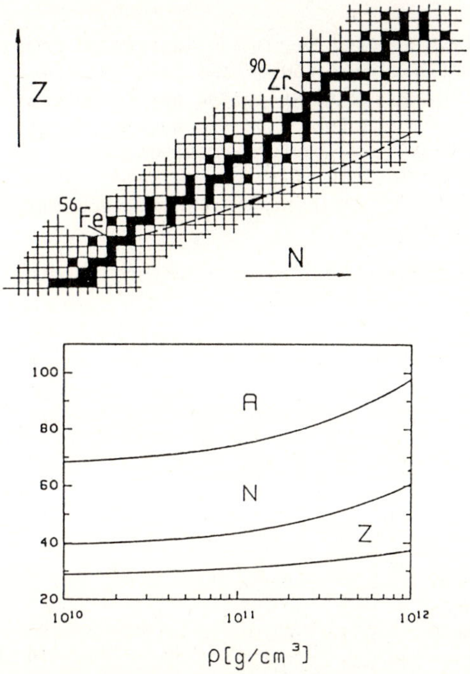

Fig. 4 Element trajectory (upper part) and evolution of the mean nucleus as a function of the matter density in the inner core (lower part).

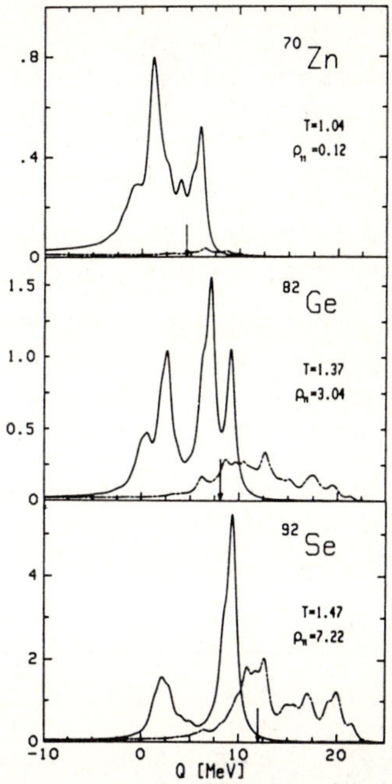

Fig. 5 Q-value dependence of the capture rates $\langle \lambda_a \rangle$ (3.4) at three stages of the collapse. The full lines give thermally unblocked GT-transitions while the dashed-dotted lines denote contributions from first forbidden capture. The arrows denote the T=0 capture thresholds.

Starting from the iron region the element trajectory quickly moves away from the stability line (upper part of Fig. 4). The Q-value distribution of rates at various points on this trajectory is given in Fig. 5.

At low A they are dominated by thermally unblocked GT-transitions (full lines) which mostly occur below the T=0 threshold (indicated by the arrows). With increasing A forbidden capture takes over (dashed-dotted lines), mostly above threshold.

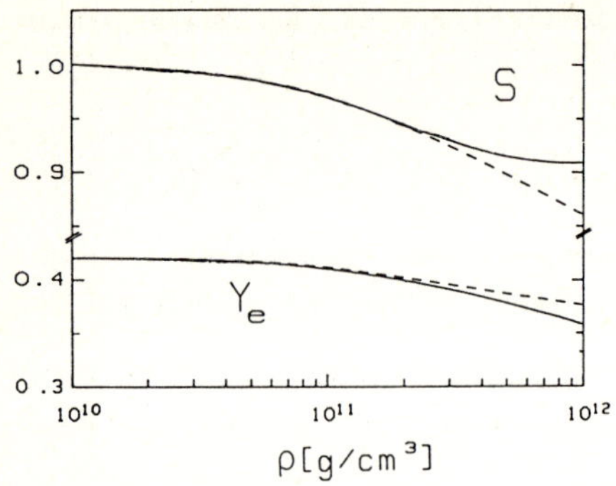

Fig. 6 Evolution of the entropy per nucleon and Y_e as a function of core density. The dashed lines include capture from free protons only, while the full lines allow for capture on nuclei as well.

The most relevant quantities Y_e and S are displayed in Fig. 6. Capture on free protons reduces the entropy since this process always cooles the core. However, also capture from nuclei does not lead to entropy increase in the early collapse stages. This is a finite temperature effect. For capture from the nuclear ground state the entropy has to increase. In the beginning, capture proceeds, however, via unblocked GT's mostly below the T=0 threshold (Fig. 5), i.e. we have cooling since thermally excited states are deexcited. At later stages, where forbidden capture takes over, the entroy begins to increase.

Table 1 summarizes the evolution of the relevant collapse parameters in more detail.

The overall entropy decrease below trapping has important consequences for successful explosions. Any entropy gain in the inner core during the collapse is harmful. An increase of 0.2 in the central core for instance can reduce the shock energy by ~ 10^{51} ergs! Detailed neutrino transport will modify the collapse dy-

Table 1 One-zone-collapse results for the relevant collapse parameters

ρ_{11}	Y_e	T	S	$\hat{\mu}$	μ_e	$X_p \times 10^4$	X_H
0.100	0.420	0.885	1.000	3.09	8.09	2.69	0.964
0.126	0.420	0.911	1.000	3.11	8.68	3.17	0.962
0.159	0.419	0.941	0.998	3.14	9.39	3.75	0.959
0.200	0.419	0.972	0.996	3.18	10.17	4.34	0.956
0.251	0.419	1.00	0.994	3.25	11.00	4.91	0.954
0.316	0.418	1.04	0.992	3.34	11.90	5.41	0.952
0.398	0.417	1.07	0.989	3.46	12.87	5.82	0.951
0.501	0.416	1.10	0.985	3.63	13.91	6.04	0.951
0.631	0.414	1.14	0.981	3.85	15.02	6.04	0.951
0.794	0.412	1.18	0.976	4.13	16.22	5.78	0.952
1.00	0.410	1.21	0.969	4.48	17.50	5.29	0.953
1.26	0.406	1.25	0.962	4.89	18.87	4.64	0.954
1.59	0.403	1.28	0.954	5.39	20.34	3.82	0.956
2.00	0.399	1.31	0.946	5.97	21.92	2.99	0.957
2.51	0.394	1.34	0.936	6.63	23.60	2.23	0.958
3.16	0.389	1.37	0.927	7.35	25.40	1.59	0.958
3.98	0.384	1.39	0.920	8.17	27.33	1.07	0.956
5.01	0.378	1.42	0.914	9.07	29.38	0.694	0.954
6.31	0.372	1.44	0.910	10.06	31.57	0.431	0.951
7.94	0.365	1.46	0.908	11.04	33.92	0.274	0.947
10.0	0.358	1.49	0.909	12.18	36.40	0.162	0.941

namics above trapping density. This has been neglected in ref. 8. However, there are two opposing effects with a tendency to cancel. On the one hand, as trapping sets in, the capture phase space is severely reduced since μ_ν builds up (3.7). Thus the rates decrease. On the other hand, there will be diffusive heating which has a tendency to increase the entropy. Therefore no significant entropy changes above trapping are expected.

Acknowledgement

This work was supported by grants NSF-PHY-84-15064 and NATO RG.85/0093. I would like to thank J. Cooperstein for a fruitful collaboration and F. Thielemann for useful discussions especially on Type I supernovae.

References

1. S.E. Woosley and T.A. Weaver: Livermore preprint (1986)
2. F.K. Thielemann et al.: Astron. Astrophys. 158, 17 (1986)
3. G.M. Fuller et al.: Ap. J. Suppl. Ser. 48, 279 (1982); Ap. J. 252, 715 (1982)
4. J.R. Wilson et al.: Twelfth Texas Symp. on Rel. Ap. Proc. (Ann. NY Acad. Sci. in press)
5. S.E. Woosley and T.A. Weaver: Nucleosynthesis and its Implications on Nuclear and Particle Physics, Fifth Moriand Astrophysics Conf. Proc., ed. J. Audouze and T. van Thuan (D. Reidel, Dordrecht, in press)
6. H.A. Bethe et al.: Nucl. Phys. A324, 487 (1979
7. E.J. Konopinsky: The Theory of Beta Radioactivity (Oxford Univ. Press 1966); J.N. Bahcall: Ap. J. 139, 318 (1964)
8. J. Cooperstein and J. Wambach: Nucl. Phys. A420, 591 (1984)
9. A. Zaringhalam: Nucl. Phys. A404, 599 (1983)
10. K. Kar and A. Ray: Phys. Lett. 94A, 322 (1983)
11. G. Baym et al.: Nucl. Phys. A175, 225 (1971)
12. H. Ejiri: Phys. Rev. C26, 2628 (1982)
13. A. Richter et al.: Prog. Part. and Nucl. Phys. 13, 1 (1984)
14. H.A. Bethe et al.: Nucl. Phys. A403, 625 (1982); J. Cooperstein: PhD Thesis 1982, Stony Brook

Neutron Star Formation and the Weak Interaction

A. Burrows

Department of Physics and Astronomy, University of Arizona,
Tucson, AZ 85721, USA

The collapse of the core of a massive star has no photon signature because this core is completely obscured by its opaque envelope. The supernova that might attend collapse will announce itself to an external optical observer long after the implosion/explosion of the core itself, as it takes about a day for the internally generated shock wave to traverse the giant envelope.

The only known direct diagnostic of the central event is its neutrino emission. The imprint of the entire internal evolution is stamped on the spectrum, mix of flavors, luminosities, and features of the accompanying neutrino burst. Detection and scrutiny of this neutrino signal will test theories concerning stellar collapse, Type II supernovae, and the formation of neutron stars in ways impossible by other means. Despite the fact that an incredible 3×10^{53} ergs may be emitted in neutrinos after the initiation of collapse, the very weakness of the neutrino/matter interaction that allows them to penetrate the stellar envelope and escape makes their detection at the Earth very difficult. Though neutrino astronomy is not yet a mature discipline, the physical theories of collapse have progressed to a sufficient degree that specific and detailed predictions can be made about the neutrino emissions that with future detector technology might be tested. The time seems propitious to summarize and review what is known and suspected about the neutrino signature of collapse, the potential for its detection, and how it can be used to test our ideas about the death of massive stars and the birth of neutron stars.

1. Introduction

Statistical, observational, and theoretical arguments strongly suggest that OB star death, pulsar formation, supernovae, and α- nuclei nucleosynthesis are correlated [1]. The catastrophic implosion and subsequent bounce at nuclear densities of either white dwarfs or the white dwarf-like heavy element (Fe or O-Ne-Mg) cores in the centers of massive stars ($60 M_\odot \geq M_\star \gtrsim 8\ M_\odot$) when they exceed the Chadrasekhar mass is implicated in both supernovae and the birth of neutron stars. Indeed, it is thought that neutron stars can form in the context of only such a gravitational collapse. Figure 1 summarizes the parent population. Neutron star formation is one of the few contexts in which the weak interaction plays a crucial role in macroscopic events. Whether it is by accretion-induced collapse or after the thermonuclear exhaustion of the cores of massive stars, the transition from electron-rich ($Y_e \sim 0.4$-0.5) "white dwarf" to neutron-rich ($Y_e \sim 0.05$) "neutron star" involves the same dynamics and physical processes.

An electron degenerate core of mass $\sim 1.4 M_\odot$ and radius ~ 2000 km, made unstable by photodisintegration or electron capture, collapses five orders of magnitude in central density and two orders of magnitude in radius within ~ 0.5 seconds. When

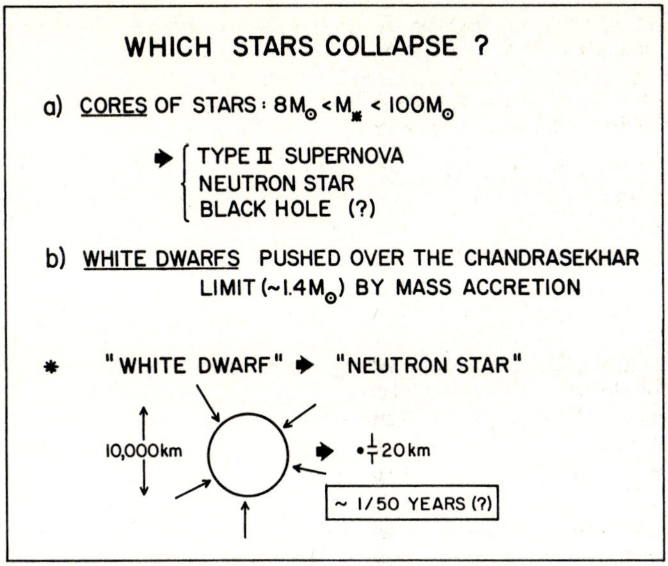
WHICH STARS COLLAPSE ?
a) CORES OF STARS : 8M⊙ < M* < 100M⊙
TYPE II SUPERNOVA
NEUTRON STAR
BLACK HOLE (?)
b) WHITE DWARFS PUSHED OVER THE CHANDRASEKHAR LIMIT (~1.4M⊙) BY MASS ACCRETION
* "WHITE DWARF" ➧ "NEUTRON STAR"
10,000km
20km
~ 1/50 YEARS (?)

Fig. 1: Parent population

nuclear density is reached and the matter stiffens, the inner core rebounds and drives a shock wave into the infalling outer core. Within a few milliseconds of bounce, the still lepton-rich core settles into hydrostatic equilibrium [2]. The shock, now in the outer envelope of the core, for most massive star core collapses, is a Type II supernova in its infancy. The proto-neutron star that obtains just subsequent to bounce does not resemble the cold, compact pulsar with which most are familiar. It is only marginally bound ($O(10^{51}$ ergs)), has a hot, extended envelope, and its quite electron-rich. Only after the tens of seconds of quasihydrostatic thermal, structrual, and compositional settling is the residue a neutron star. It is during this"long-term" cooling and neutronization phase, not the early dynamical phase, that the few x 10^{53} ergs of binding energy that must be lost during neutron star formation is radiated as neutrinos of all species (ν_e, $\bar{\nu}_e$, ν_μ, $\bar{\nu}_\mu$, ν_τ , $\bar{\nu}_\tau$).

2. The Dynamical Phase and the Weak Interaction

The duration of the birth process and the persistence of high electron fractions is a consequence of the fact that the residue is quite opaque ($\tau \sim 10^{3-5}$) to the neutrinos whose loss regulates its evolution. Neutronization is slaved to the slow pace of lepton transport (by ν_e's). The lepton fraction ($Y_e + Y_{\nu e}$) is "frozen" at values (0.35-0.4) significantly above neutron star values early during the collapse when densities have reached only one thousandth of nuclear density. Figure 2 depicts the competing timescales that determine this early "trapping" [3] during collapse. When the loss timescale exceeds the dynamical timescale (τ(free-fall)), collapse phase neutronization ceases, despite the fact that the electron capture timescale has become very short. A sea of degenerate ν_e's is generated and electron capture's inverse process, $\nu_e + n \rightarrow e^- + p$, puts the matter into a "chemical" equilibrium. Note that electron-type lepton number is a conserved quantum number. Figure 3 summarizes the major weak and neutrino processes important in stellar collapse and neutron star formation.

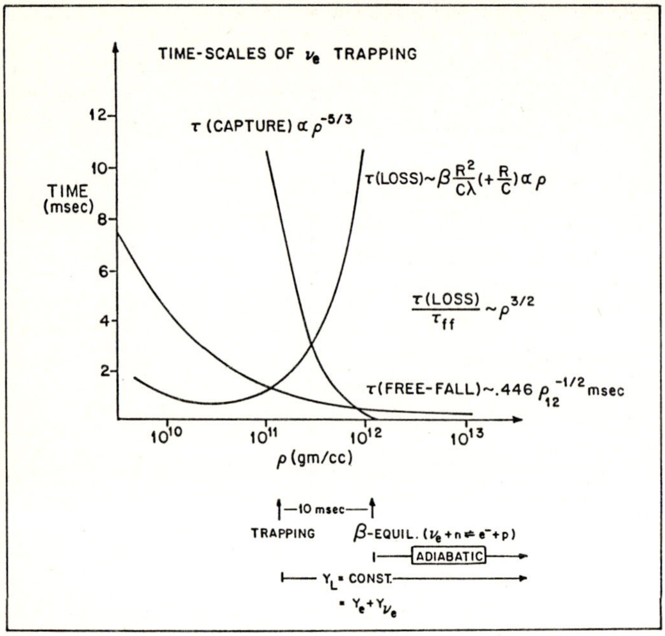

Fig. 2: The time-scales of ν_e trapping. τ_{ff} is the dynamical time ($\rho/[d\rho/dt]$), τ (loss) is a measure of the ν_e escape time, and τ (capture) is a measure of the electron capture time. After $\sim 10^{12}$ gm/cm^3, β-equilibrium and adiabaticity obtain.

* ν SCATTERING AND ABSORPTION: OPACITY SOURCES

$\nu + e^- \to \nu + e^-$
$\nu_e + n \to e^- + p$ ←— ABSORPTION
$\nu + (n,p) \to \nu + (n,p)$
$\nu + A \to \nu + A$ } NEUTRAL CURRENTS
  ↳ FREEDMAN SCATTERING (COHERENT) $\propto A^2$

* $\lambda_{\nu_e} \sim 5\text{-}10\,\text{km}\,\frac{1}{\rho_{12}}(\frac{10\,\text{MeV}}{\epsilon_\nu})^2$, STIFF DEPENDENCE ON ϵ_ν

* λ_{ν_e} (at bounce) $\sim$ 10-100 cm R $\sim 10^{6\text{-}7}$ cm, $\underline{\tau \sim 10^5!!}$

* ELECTRON CAPTURE: $e^- + p \to n + \nu_e$ (NEUTRONIZATION)

* PAIR PRODUCTION: $\gamma_{plasmon} \to \nu \bar{\nu}$ (ALL SPECIES)
      $e^+ + e^- \to \nu \bar{\nu}$ ("   "   )

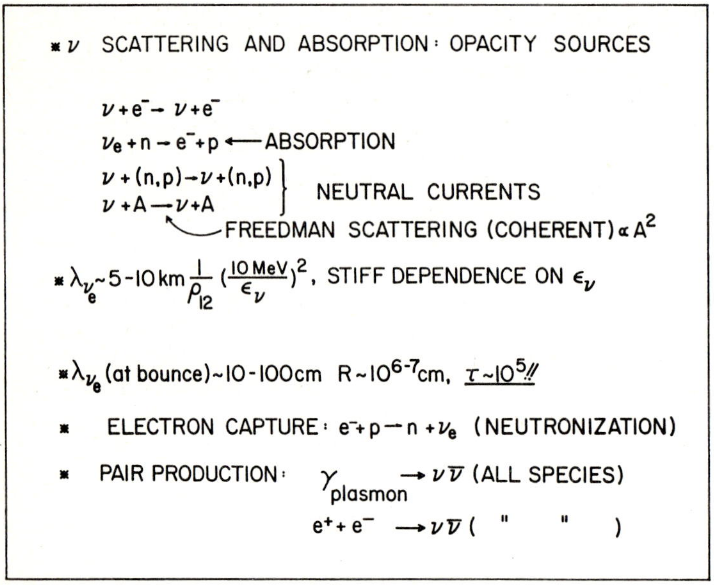

Fig. 3: Synopsis of major weak and neutrino processes in collapse and neutron star birth.

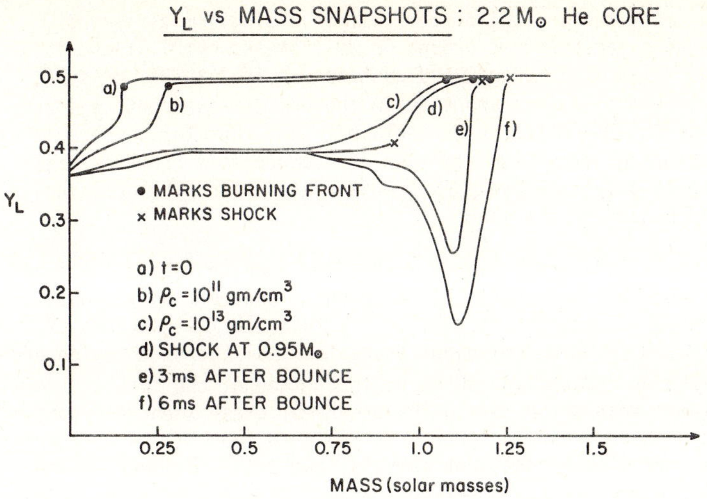

Fig 4: Dynamical phase neutronization of a O-Ne-Mg core. ρ_c is the central density.

Figure 4 depicts an evolution of the lepton fraction profile during collapse and shock formation. This sequence was calculated by BURROWS and LATTIMER [2] for the implosion of the O-Ne-Mg core of Nomoto's 2.2$M_\odot$ helium star, but in its gross features, is representative of all early core composition histories. Note that Y_L does not plummet in the center to 0.05 during the dynamical phase. Most of the residue's neutronization and energy loss occurs after the phase represented in Figure 4 during the long-term, quasi-static phase.

3. The Neutrino Signature of Stellar Collapse and Neutron Star Birth

What follows is a short, quantative description of the neutrino signature that we believe accompanies stellar collapse and neutron star formation. We are handicapped, however, by the lack of direct data. A stellar collapse occurs only about once every forty years in our galaxy and we have yet to be graced with a detection that would confirm or confound us. Ergo, caveat lector.

During collapse, electron neutrinos from electron capture on protons, in and out of nuclei, dominate the emission. As the collapse accelerates, the average energy of the neutrinos, which indirectly represents the electron fermi energy, increases from ∿1 MeV to ∿10 MeV. The average energy of emitted electron-type neutrinos does not much exceed 10 MeV because the matter becomes opaque to the higher energy neutrinos generated when higher densities are reached. Through the ν_e fermi energy is ∿ 200 Mev when the core bounces around nuclear density, these high energy neutrinos do not stream out, but are degraded and downscattered in energy as they diffuse or are convected out to the "neutrinosphere." There they finally escape with, not 100 MeV, but ∿10 MeV. This spectral softening has profound consequences for detectability, since neutrino cross-sections are stiffly increasing functions of energy.

Most of the collapse capture emission occurs in the last ∿10 milliseconds before bounce and involves only $\sim 10^{51}$ ergs. The shock wave is formed 20 kms from the center and moves out many ten's of kilometers within a single millisecond. When the shock hits the neutrinosphere at around 80 km ("breaks-out"), there is a burst of ν_e's that lasts $\sim$ 1-3 milliseconds and contains up to 3×10^{51} ergs [4]. The shock continues to move out in mass and stay in the semi-transparent region, whether the supernova starts immediately or is delayed [5]. Therefore, there is a continuing, though, on average, declining emission of capture ν_e's from the still collapsing electron-rich envelope. These $\underline{\nu_e}$'s are generated after the accreted matter is compressed and heated by the shock.

The break-out burst has a luminosity near 0.25-0.5 $M_\odot$/s, which rivals the total optical output of the observable universe! It represents, however, only about 1% of the total energy emission of neutron star birth. Within $\sim$10 milliseconds of break-out, the ν_μ, ν_τ, $\bar{\nu}_\mu$, $\bar{\nu}_\tau$, and $\bar{\nu}_e$ emissions start to rival the ν_e emission and the long-term, neutronization and cooling phase begins. Though, the luminosities during this phase are lower than those during the break-out phase when ν_e's dominated, most of the neutron star's binding energy is radiated over its 1-10 seconds. Curiously, most of this energy comes out, not as ν_e's, but as "ν_μ's," making the detection of neutron star birth by charge-current detectors more difficult. Figures 5 and 6 show schematically the time dependence of the luminosities and average neutrino energies, respectively. The long-term "hydrostatic," as opposed to hydrodynamic, phase involves fascinating physics and processes, too involved to delve into here. The interested reader is referred to BURROWS and LATTIMER [6] for further discussion. However, Figures 7 and 8 from BL, depicting the early history of the temperature and lepton fraction profiles, respectively, of a representative residue are reproduced below. They are intended to give the reader some sense of the rich range of phenomena encountered during the early life of a neutron star. Note that in Figure 8 are the <u>neutronization</u> curves. The salient neutrino signature facts are summarized in Figure 9.

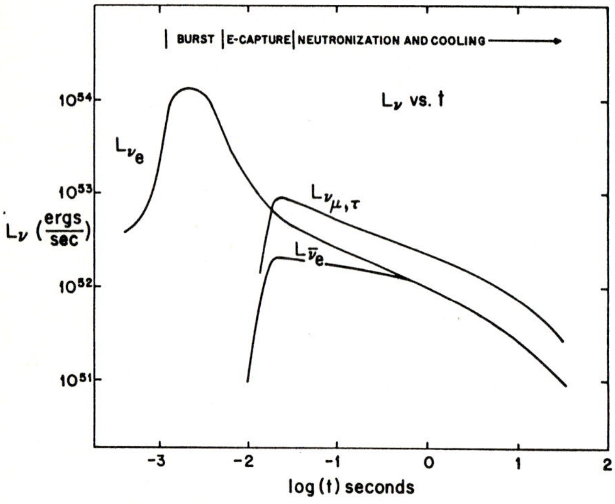

Fig. 5 A schematic of the luminosities of the various neutrino species versus time. Note the logarithmic time axis. The initial spike in the electron neutrino luminosity occurs at shock "break-out."

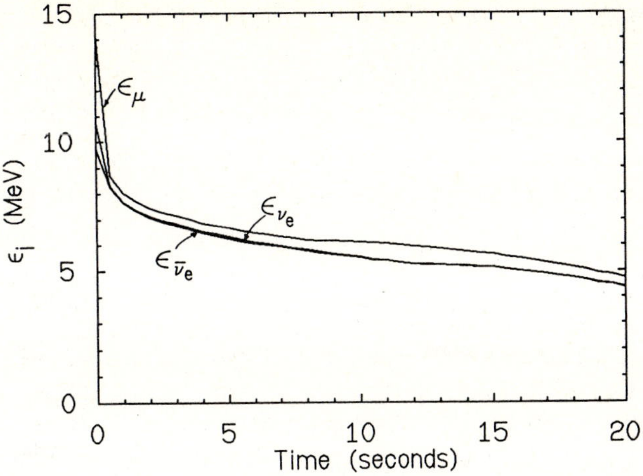

Fig. 6 ε_i versus coordinate time. ε_i is the average emitted neutrino energy, where $i = (\nu_\mu , \nu_e , \bar{\nu}_e)$ and ν_μ represents mu and tau neutrinos and antineutrinos. ε_i is in MeV's and Time is in seconds.

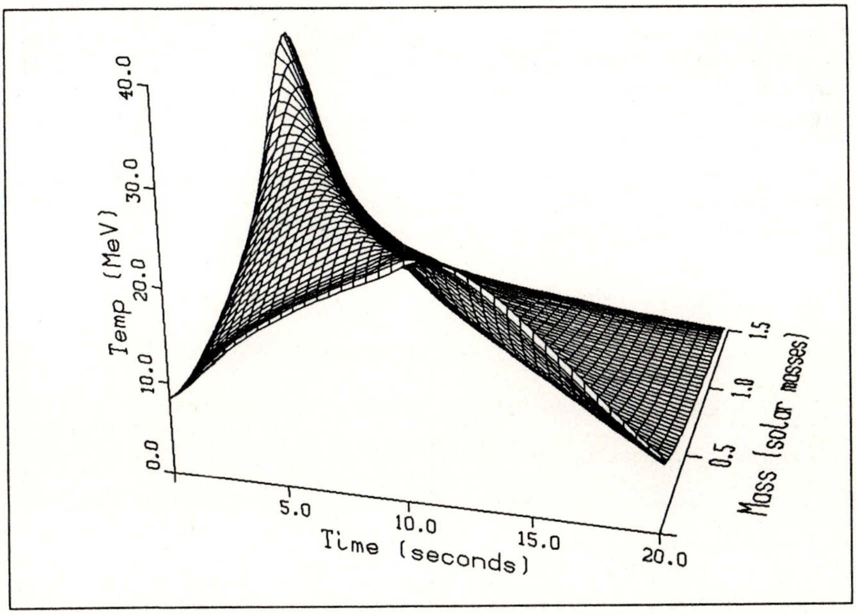

Fig. 7 Temperature in MeV versus enclosed baryon mass in solar masses versus time in seconds. The early stages are to the left and the center of the star is in the foreground. The total baryon mass of the residue is $1.4M_\odot$. This surface is from BL and represents the post-collapse thermal evolution of a neutron star.

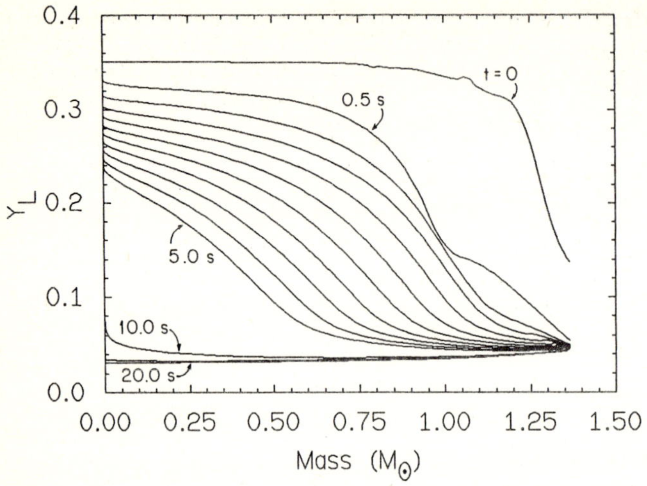

Fig. 8 Y_L versus enclosed baryon mass profiles at various times for the baseline simulation of Figure 7. Snapshots are taken every 0.5 seconds for 5 seconds and then every 5.0 seconds until 20.0 seconds is reached. The last line is the neutron star. The t=0 line is a representative post-collapse lepton profile. These are the neutronization curves.

NEUTRINO COMPONENTS

ν_e

	DURATION	ENERGY
COLLAPSE	10-10^2 MILLISECONDS	$\sim 10^{51}$ ergs
SHOCK BREAK-OUT	1-3 MILLISECONDS	1-3×10^{51} ergs
NEUTRONIZATION & COOLING (PAIRS)	1-10 SECONDS	$\sim 6 \times 10^{52}$ ergs

"ν_μ" $(\nu_\mu, \overline{\nu}_\mu, \nu_\tau, \overline{\nu}_\tau)$

COOLING (PAIRS)	1-10 SECONDS	$\sim 2 \times 10^{53}$ ergs

- TOTAL # ν's RADIATED $\sim 10^{58}$, MOSTLY "ν_μ's" (PAIRS)
- TOTAL ENERGY RADIATED $\sim 3 \times 10^{53}$ ERGS, $\langle \epsilon_\nu \rangle \sim$ 10 MeV

* NEUTRINO SIGNATURE IS THE DIAGNOSTIC OF THE DYNAMICS OF STELLAR COLLAPSE, TYPE II SUPERNOVAE, NEUTRON STAR BIRTH.

Fig. 9 A crude synopsis of the neutrino signature of stellar collapse and neutron star birth.

4. Detection

A neutronization event 1 kpc from the earth hurls ∿3 tonnes of neutrinos through the earth. Despite this abundance, the notoriously low interaction cross-sections of neutrinos with matter require that the mass of all neutrino detectors of "collapse" be at least in the kiloton range. Otherwise, the number of events/detector will be trivally small. These "neutrino telescopes" are frequently made of water or scintillator and are always collapse detectors as an afterthought. The primary motivations for their construction are cosmic ray, proton decay, or solar neutrino studies. This is as it should be, collapse events being so infrequent. However, unambiguous detection, especially coincident detections at many sites around the world, would be one of the most important scientific events of the decade. Many "underground" experimenters know this and have included paragraphs in their proposals that concern collapse, although a few still require tuition concerning the current ideas about the neutrino signal and mix of neutrino types.

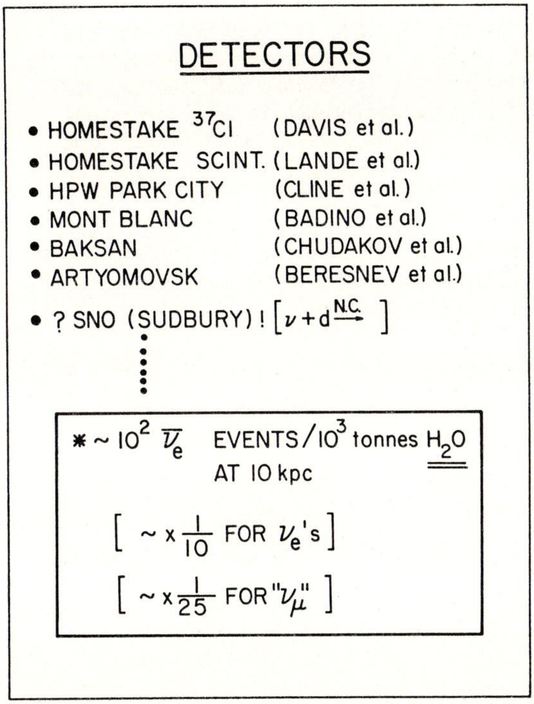

Fig. 10 Detectors and a few numbers

A summary of extant and planned experiments that are relevant to neutron star birth detection is given in Figure 10. Included is a scaling relation the layman can use to estimate the integrated number of events in H_2O expected from the canonical collapse. It is essential in this enterprise to have good time resolution and a low energy threshold ($\sim$5 MeV). Further, a neutral current capability (for the "ν_μ's") would be useful. The proposed Sudbury heavy water detector (SNO) is particularly relevant in this regard. Lacking data, this section ends. I sincerely hope my next review of this subject can include a comparison of theory with experiment.

5. Sundries

The reader can see that the physics of neutron star birth is quite exotic. A curious fact of the early life of the residue is that it is convective. Indeed, a neutron star follows what is known as a Hayashi track, similar in broad outline to the physical path of a protostar evolving towards the main sequence. The idea of convecting nuclear matter may take a while to adjust to and is just now getting theoretical attention. Convection may influence not only the birth of neutron stars, but also the structure of pulsar magnetic fields and the mechanism of Type II supernovae.

A black hole would form in the context of stellar collapse if the core accreted mass after bounce in excess of the maximum mass possible for a stable neutron star ($\gtrsim 2.0\ M_\odot$ (baryon)). However, the neutrino signature of black hole formation is expected to be only slightly different from that of neutron star formation [7]. Collapse, bounce, and the initial settling are the same for both proto-objects. However, in black hole formation, once a critical mass has been accreted, the fat core implodes dynamically (in ∼0.1 milliseconds). It "blinks out" and the neutrino emissions effectively cease. Whether this difference in behavior is detectable is problematic.

References

1. G. A. Tammann: in "Supernovae: A Survey of Current Research," ed. by M. J. Rees and R. J. Stoneham (Dordrecht: Reidel), p. 371.

2. A. Burrows and J. M. Lattimer: Astrophys. J. (Letters) 299, L19 (1985), and references therein.

3. T. J. Mazurek: Nature 252, 287 (1974).

4. A. Burrows and T. J. Mazurek: Astrophys. J. 259, 330 (1982).

5. J. R. Wilson: in "Numerical Astrophysics", ed. by J. Centrella, J. LeBlanc, and R. L. Bowers Jones and Bartlett: Boston), p. 422 (1983).

6. A. Burrows and J. M. Lattimer: Astrophys. J. 307, in press (1986) (BL).

7. A. Burrows: Astrophys. J. 300, 488 (1986).

Baryon and Lepton Number Violation in Astrophysics

E.W. Kolb

Fermi National Accelerator Laboratory, Batavia, IL 60510, USA

The cosmological and astrophysical significance of baryon and lepton number violating process is the subject of this talk. The possibility of baryon-number violating processes in the electroweak transition in the early universe is reviewed. The implications of lepton-number violation via Nambu-Goldstone bosons are discussed in detail.

1. Introduction

In spite of heroic experimental efforts in the search for neutrinoless double beta decay or for proton decay, there is no conclusive evidence for non-conservation of baryon or lepton number. The standard low-energy model $SU_3 \times SU_2 \times U_1$ has conservation of baryon (B) and lepton (L) number in the perturbative sector, but violation of B and L (with conservation of B-L) in the non-perturbative sector. The fact that B and L are violated by non-perturbative effects suggests that there is nothing sacred about B and L conservation and that they might be violated in the perturbative sector as well. Indeed, grand unified models predict perturbative B and L violation, and many extensions of the standard model to include masses for neutrinos predict L violation.

In this talk I will examine some astrophysical consequences of B and L violation. In the second section I will discuss the cosmological significance of baryon number violation in grand unified theories (GUTs) and in the electroweak transition. In the third section I will review the cosmological and astrophysical consequences of lepton number non-conservation, both non-conservation of the total lepton number, and non-conservation of the individual lepton numbers, L_e, L_μ, and L_τ. The cosmological and astrophysical effects are strongly dependent on the particle physics, and in this talk I will only consider a particular class of particle physics models, those with lepton-number violation through the interactions of neutrinos with Nambu-Goldstone bosons. The Nambu-Goldstone bosons I will consider arise either from the breaking of the total lepton number (Nambu-Goldstone boson = Majoron) [1,2,3] or from the breaking of a global family symmetry (Nambu-Goldstone boson = familon [4,5]).

2. Baryon Number Violation and the Baryon Asymmetry

One of the most striking observational facts about the large-scale structure of the Universe is the apparent asymmetry between the number of baryons and the number of antibaryons. [6] In the absence of any convincing evidence for large amounts of antimatter in the Universe, it is usually assumed that the number density of antibaryons in the Universe is zero, and the baryon density is given by

$$n_B = n_b - n_{\bar{b}} = 1.15 \times 10^{-5}\ \Omega_B h^2 \mathrm{cm}^{-3}\ , \tag{2.1}$$

where Ω_B is the fraction of the critical density ($\rho_c = 3H_o^2/8\pi G = 1.88 \times 10^{-29}\ h^2$ gcm^{-3}) due to baryons. A variety of observations, including the requirement of consistency of primordial nucleosynthesis, age of the Universe, observation of galactic mass density, etc., requires $10^{-2} < \Omega_B h^2 < 1$. So long as baryon number is conserved, n_B simply scales with the cosmic scale factor R, as $n_B \propto R^{-3}$. Note that it is the difference in n_b and $n_{\bar{b}}$ that is defined as the baryon density.

It is convenient to express the baryon asymmetry as the ratio of n_B to the entropy density, $B = n_B/s$. The entropy density, $s = (\rho+p)/T$ contributed by relativistic particles is $s = (2\pi^2/45)g_* T^4$, where g_* counts the effective massless ($m < T$) degrees of freedom $g_* = \Sigma g_{BOSONS} + (7/8)\, \Sigma g_{FERMIONS}$. Since the expansion of the Universe should be adiabatic during much of its history, $s \propto R^{-3}$, and $B = n_B/s$ is constant so long as baryon number and entropy is conserved.

One of the most interesting applications of modern developments in particle theory to the early Universe is the connection between Grand Unified Theories and the development of a non-zero baryon number from an initially symmetric state. [7] It is by now well known that the drift and decay of the baryon-number violating X-bosons of GUTs fulfill the three requirements (first stated by Sakharov in 1967) for generation of the baryon asymmetry. In the simplest GUT based upon SU_5, a baryon number and lepton number is generated, but B-L is zero. In more complicated models, a non-zero value of B-L might also be generated. This scenario for baryon number generation occurs when the temperature of the Universe is about $10^{14} - 10^{15}$ GeV.

This simple and beautiful picture for the generation of the baryon asymmetry might be altered due to anomalous electroweak baryon-number non-conservation. The basic point was made by KUZMIN, RUBAKOV, and SHAPOSHNIKOV. [8] They noticed that topological electroweak field configurations with indefinite baryon number exist. These configurations have the form (A^a_μ is the SU_2 gauge field and ϕ the SU_2 Higgs field)

$$A^a_o = 0,\; A^a_i = \varepsilon_{iak} \frac{x^k}{r^2}\, g(r/r_o)\;,\; \phi = \frac{vi}{\sqrt{2}} \frac{\vec{\tau}\cdot\vec{x}}{r} \begin{pmatrix} 0 \\ 1 \end{pmatrix} f(r/r_o) \tag{2.2}$$

where $\vec{\tau}$ are the Pauli matricies, $r_o^{-1} = g_W v$ with v the vacuum expectation value of ϕ, and g_W is the SU_2 gauge coupling constant. For technical reasons it is convenient to ignore the U_1 part of $SU_2 \times U_1$. The functions f and g in (2.2) have simple asymptotic forms for small r/r_o ($f(r/r_o) = r/r_o + \ldots$, $g(r/r_o) = (r/r_o)^2 + \ldots$) and large r/r_o ($f(r/r_o) \to 1$, $g(r/r_o) \to 1$). The free energy of the classical field configuration of (2.2) is finite, and given by ($\alpha_W = g_W^2/4\pi$)

$$F = 2M_W/\alpha_W\, B(\lambda/\alpha_W) \tag{2.3}$$

where λ is the coefficient of ϕ^4 in the Higgs potential. Although λ/g_W is unknown, $B(\lambda/g_W)$ only varies from 1.5 to 2.7 as λ/g_W varies from 0 to ∞.

These field configurations correspond to local maxima of the free energy. They separate different Θ-vacua with different topological (and baryon) number, and hence represent field configurations that must exist in the transition between states of different baryon number in different Θ-vacua. It is reasonable to assume that the free energy of the field configuration (2.2) determines the action in the calculation of the finite-temperature tunneling rate. KUZMIN, etal. [8] find the rate for the baryon number non-conservation to be

$$\dot{B} = -BTC \exp\left(-\frac{2M_W}{T\alpha_W} B\right)\;, \tag{2.4}$$

where C is a constant (which they assume to be unity) and the pre-factor T is introduced on dimensional grounds. If (2.4) is the rate for baryon number violation, then it exceeds the expansion rate of the Universe, $\dot{R}/R \simeq T^2/m_{pl}$ for temperatures larger than $T^* = 2M_W\alpha_W^{-1} B\, \ln(m_{pl}/T^*)$, or about 200 GeV. This period would be long enough to eradicate any baryon number that has a zero projection on

B-L. The only way to generate a baryon asymmetry is to have a non-zero B-L before the electroweak transition, or to generate the baryon asymmetry after the electroweak transition.

The potential importance of these effects in the electroweak transition justify a detailed analysis of the problem. One aspect of the problem deserving careful scrutiny is the effects of fermions on the fluctuations of the bosonic fields. To consider the effects of fermions, consider the field configurations of (2.2) as a particle denoted as X, with mass M = F, with baryon number violation in its decay. In this approximation [7],

$$\dot{n}_B + 3Hn_B = -n_B n_X^{eq} \Gamma_D \qquad (2.5)$$

where n_X^{eq} is the equilibrium abundance of X ($n_X^{eq} = (FT)^{3/2} e^{-F/T}$ for F>>T), and Γ_D is the decay width of the field configuration.

If the Universe is a perfect conductor, the field configuration (2.2) will be frozen-in and have an infinite lifetime (Γ_D=0). If it takes an infinite time to decay, then by detailed balance it takes an infinite time to establish the field configuration and there will be no baryon number violation at the electroweak transition. A detailed calculation of Γ_D is in progress [9]. If Γ_D is large enough, i.e. $\Gamma_D \geq T$, then the fermions will have no effect, and the scenario of KUZMIN, etal. [8] should be obtained.

3. Lepton Number Violation

If there exists a Nambu-Goldstone boson with couplings to neutrinos that are non-diagonal in family space, then they can mediate flavor-changing neutrino decay $\nu_H \rightarrow \nu_L \phi$, where ν_H is a massive neutrino, ν_L is a light (assumed massless) neutrino and ϕ is the Nambu-Goldstone boson. The decay width for the process is $\Gamma = h_{HL}^2 M/16\pi$, where h_{HL} is the non-diagonal coupling of ϕ to $\nu_H \nu_L$, and M is the mass of ν_H. One expects h_{HL} = M/V, where V is the scale of symmetry breaking that gives rise to the Nambu-Goldstone boson. The lifetime of ν_H will be

$$\tau = 16\pi V^2/M^3 = 10^{-1}(V/10^{10}\text{GeV})^2 \ (1 \text{ MeV}/M)^3 \text{ years} . \qquad (3.1)$$

The scale of symmetry breaking is very model dependent. In the original Majoron model of CHICKASHIGE, MOHAPATRA, and PECCEI [1], the Nambu-Goldstone was the Majoron, and it arose from the breaking of global lepton number. In the original model the Majoron coupling is diagonal [3,10], and h_{HL} = 0. It is possible to construct more complicated models to give $h_{HL} \neq 0$ by introducing additional Higgs representations [11,12]. Such extensions are only viable if $V \geq 10^{10}$ GeV. Majoron models have also been constructed with V ~ 1 keV for the SSB scale [2]. In these models ν_H annihilation will be dominant, and ν_H decay will not be important. In attempts to understand the observed family structure, a global family symmetry is often assumed. The breaking of this symmetry leads to a Nambu-Goldstone boson known as the familon. Experimental bounds on $\mu \rightarrow ef$ and $K \rightarrow \pi f$ (f is the familon) imply $V \geq 10^{10}$ GeV. If the family group also contains the Peccei-Quinn symmetry, then astrophysical arguments suggest $V \leq 10^{12}$ GeV. [13]

The neutrino lifetime as a function of mass for several values of V is shown in Fig. 1. Curves marked 10^9, 10^{10}, 10^{11} correspond to the lifetime given by (3.1) with V = 10^9, 10^{10}, 10^{11}.

One interesting feature of the violation of lepton family number in $\nu_H \rightarrow \nu_L \phi$ is the possibility of avoiding cosmological bounds on neutrino masses.

Neutrinos decouple in the early Universe at a temperature of a few MeV. [14] If we consider neutrinos of mass less than a few MeV, then they were relativistic at decoupling, and their number density was $n_\nu = g_\nu (3\zeta(3)/4\pi^2)T_D^3$, where g_ν is the

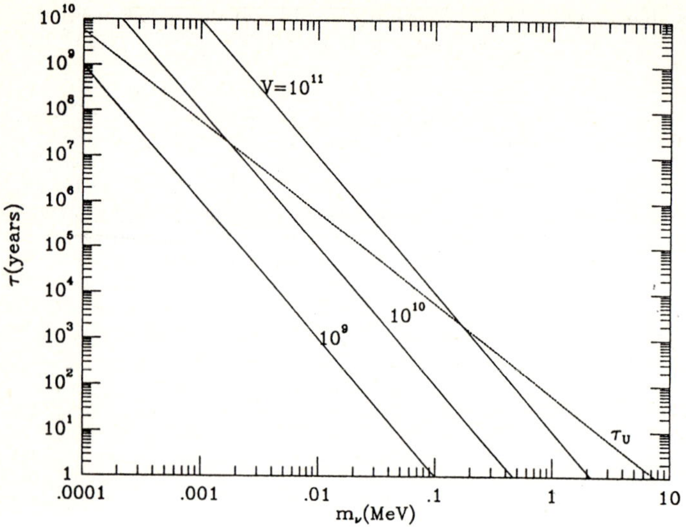

Figure 1

number of neutrino spin states (which we shall take as $g_\nu = 2$) and T_D is the decoupling temperature. The photon number density at decoupling is given by $n_\gamma = (2\zeta(3)/\pi^2)T_D^3$, which results in $n_\nu/n_\gamma = 3/4$ at neutrino decoupling. Subsequent to neutrino decoupling, e^+e^- annihilation heats the photons, increasing their temperature relative to the photon temperature by a factor of $(T_\nu/T_\gamma)^3 = 4/11$. Therefore, after e^+e^- annihilation, $n_\nu/n_\gamma = (3/4)(4/11)$. Today, the number density of photons in the Universe is accurately known from the microwave background radiation. If we take the temperature of the microwave background radiation to be $T_\gamma = 2.7$K, then today $n_\nu = 399$ cm^{-3}, and $n_\nu = (3/4)(4/11)n_\gamma = 109$ cm^{-3}.

If neutrinos have a mass greater than $T_\nu = 1.9$K, then the present mass density due to the relic massive neutrinos would be $\rho_\nu = m_\nu n_\nu = (m_\nu/1\text{eV})\ 1.09 \times 10^2$ eV cm^{-3}. It is convenient to express this density in terms of $\Omega_\nu = \rho_\nu/\rho_c$. In terms of the neutrino mass, Ω_ν is given by $\Omega_\nu h^2 = 1.05 \times 10^{-2}\ (m_\nu/1\text{eV})$. A conservative upper limit on $\Omega_\nu h^2$ would be $\Omega_\nu h^2 \le 1$, which requires $m_\nu \le 100$ eV. Up to numerical factors this is the Cowsik-McClelland bound on neutrino masses [15].

The Cowsik-McClelland bound can be avoided by making the neutrino mass greater than the decoupling temperature of a few MeV, thereby suppressing the number density at decoupling (hence the number density today) by a factor of e^{-m_ν/T_D} relative to the photon number density. For neutrino masses above about 5 GeV the decrease in the number density results in $\Omega_\nu h^2 \le 1$. [16] However, in this paper I will assume that the neutrinos have a mass much less than 5 MeV, in fact, I will assume the masses are less than 1 MeV.

A second way to avoid the cosmological neutrino mass limit is to have the neutrinos unstable, with a lifetime less than the age of the Universe [17]. If a massive neutrino decays to relativistic particles, the energy density of the relativistic decay products will decrease in the expansion as R^{-4} (where R is the Robertson-Walker scale factor) compared to a R^{-3} decrease in the energy density of a massive particle. This "extra" decrease in the energy density of the decay products results in the massless decay products having an energy density today of

$$\Omega h^2 = 1.05 \times 10^{-2} \left(\frac{m_\nu}{1\text{eV}}\right) \frac{R_D}{R_o}\ , \tag{3.2}$$

where R_D/R_o is the ratio of the scale factor at decay to the scale factor today.

It is customary to express this ratio of scale factors in terms of a cosmological red-shift z_D, $R_o/R_D = 1+z_D$, where z_D is the "red-shift" of the neutrino decay ($z_D = 0$ corresponds to today).

If we require $\Omega h^2 \leq 1$, then there is a relation between z_D and m_ν that must be satisfied for the decay products to have an energy density less than ρ_c: $(1+z_D) \geq 10^{-2}\ (m_\nu/1\text{eV})$.

If we assume that the microwave background is primordial in origin and not the result of some other mechanism (such as photons produced in neutrino decay), then $T \propto R^{-1} \propto 1+z$, and the temperature of the Universe at decay is given by $T_D = 2.7\text{K}(1+z_D)$, which must satisfy the inequality $T_D \geq 2.7 \times 10^{-2}\text{K}\ (m_\nu/1\text{eV})$.

This limit on the temperature of the Universe at neutrino decay can be translated into a limit on the neutrino lifetime. The massive neutrinos decay when the age of the universe ($t_U = (2/3)\ H^{-1}$ if the Universe is matter dominated when the neutrinos decay) is equal to the particle lifetime, τ. The expansion rate at the time of neutrino decay is given by ($m_{pl} = G^{-1/2}$)

$$H^2 = \frac{8\pi G}{3}\rho = \frac{8}{3m_{pl}^3}\, m_\nu n_\nu\ , \tag{3.3}$$

if the massive neutrino dominates the Universe at the time of its decay. Since at the temperature of neutrino decay $n_\nu = (3/4)\ 2\zeta(3)\ T_D^3\ (4/11)$, equating $t_u(T_D) = \tau$ results in the relation between τ and T_D:

$$\tau = \frac{m_{pl}}{m^{1/2}T_D^{3/2}}\ . \tag{3.4}$$

Therefore the limit on T_D results in the limit [17]

$$\tau \leq 5.9 \times 10^3 \left(\frac{100\text{keV}}{m_\nu}\right)^2\ \text{yr}. \tag{3.5}$$

This upper limit as a function of m_ν is shown as the dashed line in Fig. 1, and compared to the lifetime expected for neutrino decay into a light neutrino plus a Nambu-Goldstone particle. In order for the massless decay products to give $\Omega h^2 \leq 1$, the lifetime must be below the dashed line. From the figure we see that if $V \leq 10^9$ GeV the lifetime will always be short enough and there would be no limit on m_ν. If $V = 10^{10}$ GeV, then the lifetime would be short enough only if $m_\nu \geq 1.7$ keV, while if $V = 10^{11}$ GeV the lifetime will be short enough only if $m_\nu \geq 170$ keV. If $V \geq 10^{12}$ GeV, the lifetime will be too long, and the neutrino decay products will have $\Omega h^2 > 1$. Again, I emphasize that I am only considering neutrinos with masses less than a few MeV.

Other decay mechanisms for ridding the Universe of massive neutrinos have been considered. Most other mechanisms have a large fraction of the decay products in photons. It is very difficult to hide the photons from detection today. The advantage of the particle physics models considered here is that the decay products are "invisible" (i.e. the final state contains no detectable photons or charged particles). For neutrinos of mass less than a few MeV, the decays into Nambu-Goldstone particles with $V \leq 1010$ GeV are the only models to give a short enough lifetime.

Saturation of the limit in (3.5) is interesting, because it would result in a radiation-dominated Universe, i.e. the Universe would be dominated by ν_L and ϕ decay products of ν_H. Thus it is possible to have a flat Universe as suggested by inflation and a clustered component of $\Omega = .2 - .3$ suggested by observation. This possibility has generated a great deal of recent interest. [18]

Although in the Gelmini-Roncadelli model the Majoron coupling to neutrinos is diagonal and the neutrino is stable at the tree-level against decay into a neutrino plus Majoron (M), there is another way to rid the Universe of massive neutrinos and allow neutrinos of mass greater than 100 eV to exist but not to contribute to the present mass density. GEORGI, GLASHOW and NUSSINOV [3] pointed out that neutrino annihilation through Majoron exchange will reduce the neutrino density below the level necessary to give $\Omega_\nu h^2 < 1$. [3,19]

In the Gelmini-Roncadelli (GR) model, neutrinos recieve a Majorana mass by the introduction of a SU(2) Higgs triplet $\vec{\chi}$, which couples to the usual lepton doublets. The introduction of a Yukawa term in the Lagrangian of the form

$$L_\chi = -h_3 \bar{\psi}_L^c \vec{\tau} \cdot \vec{\chi} \Psi_L , \tag{3.6}$$

where Ψ_L is one of the usual lepton doublets (e.g. $\Psi_L^T = (\nu_e, e)_L$ for the first generation), $\vec{\chi}$ is the Higgs triplet, and $\vec{\tau}$ are Pauli matrices, results in a Majorana mass for the neutrino $m_\nu = h_3 V_T$, where V_T is the VEV of the neutral component of the triplet (constrained to be << 1 MeV).

In the GR model the Higgs potential is constructed to conserve a global lepton number. A lepton number of L = -2 can be assigned to χ, and when χ gets a VEV the global lepton number is spontaneously broken. The resulting Nambu-Goldstone boson is the Majoron. The Majoron coupling to neutrinos is of the form

$$L_{M\nu} = ih_3 \bar{\nu}\gamma_5 \nu M = i(m_\nu / V_T) \bar{\nu}\gamma_5 \nu M \tag{3.7}$$

where ν is the neutrino Majorana field $\nu = \nu + \nu^c$. The Majoron coupling to charged particles is suppressed by a factor of V_T/V_D, where V_D is the VEV of the usual Higg doublet field, $V_D \sim 250$ GeV. Since this ratio is expected to be small, the Majorons effectively decouple from charged particles and only interact with neutrinos.

Neutrinos can scatter by the exchange of the massless Majoron, and the neutrino-neutrino scattering cross section at energies greater than m_ν will be $\sigma_M \sim h^4/s$ where $\sqrt{s}$ is the center-of-mass energy and h is a Yukawa coupling. This is larger (for energies less than m_W) than the usual neutrino-neutrino scattering cross section mediated by the exchange of a massive W or Z, $\sigma_{W,Z} \sim G_F^2 s \sim g^4 s/m_W^4$ (g is a gauge coupling), by the ratio $\sigma_M/\sigma_{W,Z} \sim h^4 m_W^4/g^4 s^2$. In the GR model neutrino-neutrino scattering is much larger than weak.

Through Majoron interactions, the neutrinos will interact amongst themselves long after they have decoupled from the rest of the particles in the Universe at the usual neutrino decoupling temperature of $T \approx 1$ MeV. Neutrinos can annihilate into Majorons, $\nu\nu \to MM$, with a cross section of $\sigma_{\nu\to M} = h^4 |\vec{v}|/48\pi s$, where $|\vec{v}|$ is the relative velocity and h is the coupling related to m_ν and V_T. The p-wave annihilation is responsible for the funny velocity dependence. The number density of neutrinos change due to annihilation and expansion according to

$$\dot{n}_\nu = -[n_\nu^2 - (n_\nu^{eq})^2] \sigma_A |\vec{v}| - 3 \dot{R}/R \, n_\nu \tag{3.8}$$

where σ_A is the total neutrino annihilation cross section (for the lightest mass neutrino $\sigma_A = \sigma_{\nu\to M}$), and $\dot{R}/R$ is the expansion rate. If one solves (3.8) for V_T less than the red-giant limit, neutrino annihilation into Majorons decreases the relic neutrino density enough to give $\Omega_\nu h2$ today less than one.

In conclusion, in the GR model, as in the original Majoron model and the familon model, it is possible to evade the cosmological bound on the mass of neutrinos. The Nambu-Goldstone boson in the GR model also has other interesting astrophysical consequences.

The Majoron in the GR model couples to neutrinos and is responsible for lepton-number violating reactions such as the $|\Delta L| = 2$ reaction $\nu_L \nu_L \leftrightarrow MM$ or the $|\Delta L| = 4$ reaction $\nu_L \nu_L \leftrightarrow M \leftrightarrow \nu_R \nu_R$ (in this model the anti-neutrino is the right handed neutrino). Furthermore, the $\Delta L \neq 0$ ν-ν scattering cross sections are much larger than the usual weak interaction result. This suggests that one way to test the GR model is to do ν-ν scattering. Neutrino beams have been used since the 1950's, but neutrino targets are not yet available, at least terrestrially. The best place today in the Universe to study ν-ν scattering is during the latter stages of the collapse of massive stars. If the GR model is correct it would have profound implications on models of gravitational collapse and Type II supernovae. [20,21,22]

The standard theory of Type II supernovae is described in these proceedings by several speakers. The crucial point with regard to Majorons is that during the gravitational collapse of massive iron cores, electron neutrinos are copiously produced in electron capture and are "trapped" in the core, i.e. the mean free path between neutrino scatterings is much smaller than the size of the core and the time it takes for neutrinos to diffuse out of the core is longer than the time for core collapse.

At the point of neutrino trapping the mass density is about 10^{12} g cm^{-3}, the temperature is about 1.5 MeV, and the neutrino number density is about 10^{34} cm^{-3}! The existence of a large neutrino density means that in the standard collapse picture, β-equilibrium is established in reactions of the type $e^- + p \leftrightarrow n + \nu_e$ where the protons and neutrons are either free or bound in nuclei. The β-equilibrium condition is $\mu_e = \mu_\nu + \hat{\mu} + \Delta m$ where μ_e (μ_ν) is the electron (electron neutrino) chemical potential, Δm is the neutron-proton mass difference ($\Delta m = 1.293$ MeV), and $\hat{\mu} = \mu_n - \mu_p$ is the difference of the neutron and proton chemical potentials. The existence of a large density of trapped neutrinos is necessary to allow ν-capture on neutrons to establish β-equilibrium.

The rapid establishment of β-equilibrium results in little entropy production, and the standard collapse scenario is a low-entropy collapse. In a low entropy collapse the nucleons remain bound in nuclei and the pressure is supplied by the relativistic electrons, which is insufficient to stop the collapse before reaching a few times nuclear matter density.

In the GR model however the scenario for collapse is much different. The Majorons mediate $\Delta L \neq 0$ reactions amongst the neutrinos. The rate for lepton-number violating reactions is greater than the weak rates trying to establish β-equilibrium. The lepton-number violating reactions will drive the neutrino chemical potential to zero so the β-equilibrium condition becomes $\mu_e = \hat{\mu} + \Delta m$. Therefore the value of μ_e in β-equilibrium is much less in the Majoron collapse than in the standard collapse. The decrease in μ_e (i.e., in number of electrons) leads to a large entropy increase ($Tds = \mu_e dN$). The large entropy increase causes the nuclei to melt, increasing the number of non-relativistic particles and eventually leading to non-relativistic nucleons dominating the pressure. Thus, Majoron interactions will change the equation of state in the collapse.

In the standard collapse, after neutrino trapping the total lepton number of the core is conserved. The equilibrium obtained in the standard collapse is subject to the constraint of conservation of lepton number. The existence of a large neutrino chemical potential blocks electron capture, and the equilibrium conditions obtained in collapse are not too different than the conditions at neutrino trapping, resulting in a small increase in entropy. If lepton number is not conserved in the collapse, and the maximum entropy equilibrium state will be obtained with the lepton number zero. The de-leptonization of the core produces a large amount of entropy, changing the equation of state during collapse. Until detailed numerical calculations are done it is impossible to know whether the de-leptonization of the core helps or prevents the supernova explosion.

Finally, it should be emphasized that Majorons were only the agents for mediating lepton-number violation. The change in the collapse scenario only depended upon lepton number violation occurring rapidly during collapse. Any particle physics model with "fast" lepton number violating reactions amongst neutrinos would lead to the same effect.

This work was supported in part by NASA and the Department of Energy. I would like to thank F. Accetta, D. A. Dicus, R. Pisarski, V. L. Teplitz, D. L. Tubbs and M. S. Turner for collaboration in the work reported here.

References

1 Y. Chikashige, R.N. Mohapatra, and R.D. Peccei: Phys. Lett. 98B, 265 (1981).

2 G.B. Gemini and M. Roncadelli: Phys. Lett. 99B, 411 (1981).

3 H.H. Georgi, S.L. Glashow, and S. Nussinov: Nucl. Phys. B193, 297 (1981).

4 F. Wilczek: Phys. Rev. Lett. 49, 1549 (1982).

5 D.B. Reiss: Phys. Lett. 115B, 217 (1982).

6 G. Steigman: "Observational Tests of Antimatter Cosmologies", in Ann. Rev. Ast. Astro. 14, 339 (1976).

7 E.W. Kolb and M.S. Turner: "Grand Unified Theories and the Origin of the Baryon Asymmetry", in Ann. Rev. Nucl. Part. Sci. 33, 645 (1983).

8 V.A. Kuzmin, V.A. Rubakov and M.E. Shaposhnikov: Phys. Lett. 155B, 36 (1985).

9 F. Accetta, E.W. Kolb, R. Pisarski, and M.S. Turner, work in progress.

10 J. Schechter and J.W.F. Valle: Phys. Lett. D25, 774 (1982).

11 J.W.F. Valle: Phys. Lett. 131B, 87 (1983).

12 G.B. Gelmini and J.W.F. Valle: Phys. Lett. 142B, 181 (1984).

13 D.A. Dicus and V.L. Telpitz: Phys. Rev. D28; 1778 (1983).

14 S. Weinberg: Gravitation and Cosmology (Wiley, New York, 1972).

15 R. Cowsik and McClelland: Phys. Rev. Lett. 29, 669 (1972).

16 B.W. Lee and S. Weinberg: Phys. Rev. Lett. 39, 165 (1977).

17 D.A. Dicus, E.W. Kolb, V.L. Teplitz: Phys. Rev. Lett. 39, 168 (1977).

18 D.A. Dicus, E.W. Kolb, V.L. Teplitz: Ap. J. 221, 327 (1978); D.A. Dicus, E.W. Kolb, V.L. Teplitz, and R.V. Wagoner; Phys. Rev. D18, 1829 (1978); J. Gunn, B. Lee, I. Lerch, D. Schramm, and G. Steigman: Ap. J. 223, 1015 (1978); T. Goldman and J. Stephenson: Phys. Rev. D16, 2256 (1977); A. DeRujula and S. Glashow, Phys. Rev. Lett. 45, 942 (1980); E.W. Kolb, in Proceedings of 1980 Neutrino Mass Conference, Ed. V. Barger and D. Cline (1980): M.S. Turner, G. Steigman, and L.L. Krauss: Phys. Rev. Lett. 52, 2090 (1984); G. Gelmini, D.N. Schramm and J.W.F. Valle: Phys. Lett 146B, 311 (1984); D.A. Dicus and V.L. Teplitz: preprint DOE/ER 140200-599.

19 E.W. Kolb and M.S. Turner: Phys. Lett. 159B, 102 (1985).

20 E.W. Kolb, D.A. Dicus, an D.L. Tubbs: Ap. J. Lett. L57, 25 (1982).

21 G.B. Gelmini, S. Nussinov, and M. Roncadelli, Nucl. Phys. B209, 157 (1982).

22 D.A. Dicus, E.W. Kolb, D.L. Tubbs: Nucl. Phys. B223, 532 (1983).

The Stellar Beta-Decay Rate of ^{79}Se

N. Klay and F. Käppeler

Kernforschungszentrum Karlsruhe GmbH, Institut für Kernphysik,
P.O.B. 3640, D-7500 Karlsruhe, F. R. Germany

At the very high temperatures in the interior of massive stars the beta half-life of ^{79}Se will be reduced compared to the terrestrial value. This is of relevance for the slow neutron capture process (s-process) which occurs presumably during the Helium burning phase at the late stages of stellar evolution. The enhancement of the beta decay rate can be understood as the result of thermal excitation of nuclear levels with strongly different beta decay properties. For ^{79}Se the most important level in this respect is the isomeric state at 96 keV, because it has the lowest excitation energy and can undergo allowed beta decay whereas the ground state decay is a unique first forbidden transition.

Theoretical calculations of the ^{79}Se half-life as a function of temperature have been performed by several authors [1-3]. It turned out that the largest uncertainty in those calculations results from the estimated log ft-value for the 96 keV isomer ^{79m}Se. We report in the following on an experimental determination of this quantity via direct observation of the weak beta decay branch of ^{79m}Se.

The total half live of this isomer is 3.91 minutes. Hence, a sufficient amount of ^{79m}Se can be produced via (n,γ) activation of ^{78}Se, in order to measure the beta decay immediately after irradiation. The reactor of the ILL Grenoble provided the necessary combination of a high neutron flux and a fast rabbit system. In order to minimize backgrounds, isotopically pure ^{78}Se was implanted by a mass separator into an extremely clean graphite sample backing of only 0.1 mm thickness and 7 mm diameter.

The observation of the beta spectrum with a Si(Li) detector required suppression of the competing strong conversion lines from the 96 keV E3 transition to the ground state of ^{79}Se. We used a mini-orange filter built with permanent magnets [4] whose low energy cutt-off property was improved by shielding of field inhomogeneities near the magnet surfaces (Fig. 1). For the present study this filter was optimized for transmission in the energy range from 120 keV up to the endpoint energy of 256 keV, while the conversion electrons at 83 keV and 94 keV are suppressed by a factor of $\sim$ 60.

The measurement of the beta spectrum could be started typically 4 minutes after the end of the irradiation in the reactor. Subsequent spectra were recorded for 16 m in intervals of 1 m in order to verify our background correction via reproduction of the 3.9m half-life of ^{79m}Se. With this technique, background from longer-lived beta emitters (mainly ^{41}Ar and ^{38}Cl) could be subtracted accurately. Backgrounds from shorter-lived contaminations could be excluded from simultaneous gamma spectroscopy of the activated samples with a Ge(Li) detector. After $\sim$ 30 minutes the conversion lines were still strong enough to be measured by removing the magnetic filter. Figure 2 shows the experimental sum spectra for one activation.

From the observed count rates in the beta spectrum and of the conversion electrons the branching ratio between electromagnetic transitions and beta decays was calculated. We obtained $\lambda_\gamma/\lambda_\beta$ = 1970 $\pm$ 350 as a result of five activations. This branching ratio leads to <u>log ft = 4.80 $\pm$ 0.08</u> for the beta decay of ^{79m}Se.

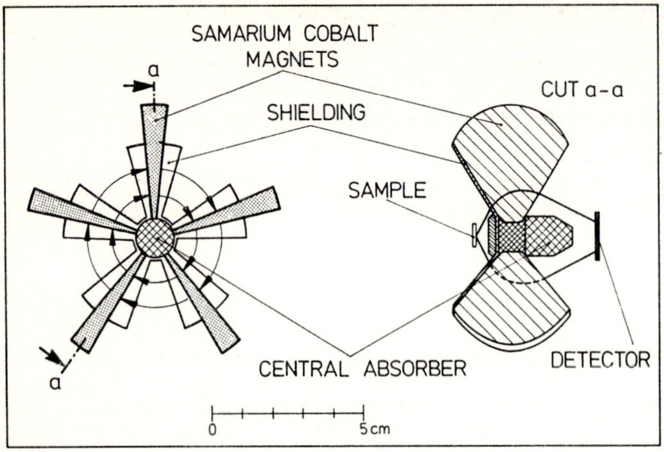

Fig. 1 Mini-Orange spectrometer setup for suppression of conversion electrons below 96 keV. Idealized trajectories for electron energies of ∿ 140 keV are indicated in the right part of the figure.

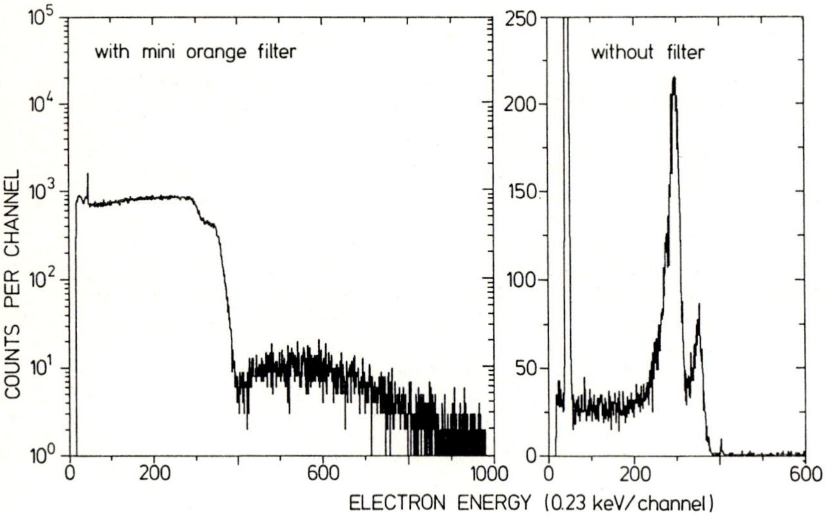

Fig. 2 Experimental spectra: Electrons from the beta decay of ^{79m}Se (left, counts beyond channel 400), and conversion electrons from the ground state transition (right).

With this experimental result we calculated the temperature dependence of the ^{79}Se half-life according to the formalism of YOKOI and TAKAHASHI [3], who also took bound state beta decay into account. The calculated function $t_{1/2}(T)$ (Fig. 3) is particularly important for stellar nucleosynthesis in the slow neutron capture process. From s-process analyses in the mass range A = 80-90 one can deduce the effective stellar half-life of ^{79}Se to range between 2 y and 13 y. For this estimate, Fig. 3 yields immediately an allowed range of temperatures between 200 and 320 million degrees. This rather narrow range of s-process temperatures depends completely on the small uncertainty of the present experimental result. For a detailed discussion of the unterlying s-process model the reader is referred to Refs. [5] and [6].

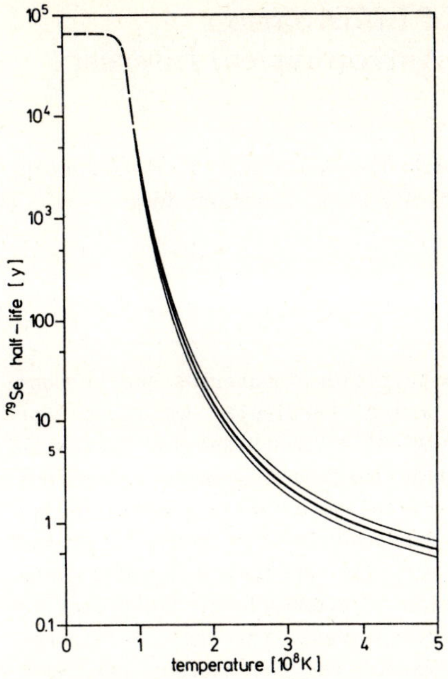

Fig. 3 The stellar beta half-life of ^{79}Se as a function of temperature. The error band reflects the uncertainty due to the measured log ft value.

Acknowledgements: We thank G. Rupp for his excellent technical assistance, as well as A. Hanser and B. Feurer for the mass separation of ^{78}Se.

References:

1. K. Cosner and J.W. Truran: Astrophys. Space Sci. 78, 85 (1981)
2. H.J. Conrad: Ph.D. Thesis, University of Heidelberg (1976)
3. K. Yokoi and K. Takahashi: Report KfK-3849, Kernforschungszentrum Karlsruhe (1985)
4. J. van Klinken, S.J. Feenstra, G. Dumont: Nucl. Instr. Meth. 151, 433 (1978)
5. H. Beer, G. Walter, R.L. Macklin, P.J. Patchett: Phys. Rev. C30, 464 (1984)
6. G. Walter, H. Beer, F. Käppeler, G. Reffo, F. Fabbri: Astr. Astrophys. (in print).

Coulomb Dissociation as a Source of Information on Radiative Capture Processes of Astrophysical Interest

G. Baur[1], C.A. Bertulani[1], and H. Rebel[2]

[1]Inst. für Kernphysik, KFA Jülich, Postfach 1913, D-5170 Jülich, F. R. Germany
[2]Kernforschungszentrum Karlsruhe, Inst. für Kernphysik, Postfach 3640, D-7500 Karlsruhe, F. R. Germany

The cross sections for radiative capture of α-particles, deuterons and protons by light nuclei at very low relative energies are of particular importance for the understanding of the nucleosynthesis of chemical elements and for determining the relative elemental abundances in stellar burning processes at various astrophysical sites. As examples we quote the reactions $\alpha+d\rightarrow{}^6Li+\gamma$, $\alpha+{}^3He\rightarrow{}^7Be+\gamma$, or $\alpha+{}^{12}C\rightarrow{}^{16}O+\gamma$. As an alternative to the direct experimental study of these processes we consider [1,2] the inverse process, the photodisintegration, by means of the virtual photons provided by a nuclear Coulomb field. The radiative capture process $b+c\rightarrow a+\gamma$ is related to the inverse process, the photodisintegration $\gamma+a\rightarrow b+c$ by the detailed balance theorem. Except for the extreme case very close to the threshold the phase space favours the photodisintegration cross section as compared to the radiative capture. The disintegration by means of the virtual photons in a Coulomb collision

$$Z+a \rightarrow Z+b+c \tag{1}$$

makes use of the high virtual photon number and also of possible kinematical advantages. The fragments b and c have high energies in the lab system and can be conveniently detected. From the kinematics the relative energy E_{bc} can be accurately deduced (see Ref. 2). Care will have to be taken for "post-acceleration" effects, possibly more serious for fragments with different charge to mass ratios.

The double differential cross section for the excitation of the projectile into the continuum with excitation energy E_γ and angle ϑ of the c.m. of the broken up pair is given by

$$\frac{d^2\sigma}{d\Omega E_\gamma} = \frac{1}{E_\gamma}\sum_\lambda \frac{dn_{E\lambda}}{d\Omega}\,\sigma^{photo}_{E\lambda} \tag{2}$$

where $\sigma^{photo}_{E\lambda}$ denotes the photodisintegration cross section of multipolarity $E\lambda$ (magnetic excitations are neglected). The virtual photon numbers for the multipolarity $E\lambda$ are denoted by $dn_{E\lambda}/d\Omega$ and are calculated from the kinematics of the process. We are especially interested in the case where the scattering angle is small, $\vartheta \ll 1$, i.e. $\varepsilon = \frac{1}{\sin\vartheta/2} \gg 1$. In this case we have

$$\frac{dn_{E1}}{d\Omega} = \frac{Z^2}{4\pi^2}\,\alpha\,\varepsilon^2\left(\frac{c}{v}\right)^2 x^2\left[K_0^2(x)+K_1^2(x)\right]^2 = \frac{Z^2}{4\pi^2}\,\alpha\,\varepsilon^2\left(\frac{c}{v}\right)^2 \varphi_1(x) \tag{3}$$

where $x = \frac{\omega b}{v}$ is the adiabaticity parameter. The impact parameter is given by b, v is the velocity of the projectile and $E_\gamma = \hbar\omega$. Corrections due to Coulomb repulsion depend on $\xi = \frac{\omega a}{v}$, where a is half the distance of closest approach in a head on collision, they are small and easily evaluated. For the important E2 case we obtain

$$\frac{dn_{E2}}{d\Omega} = \frac{Z^2\alpha}{\pi^2}\frac{1}{\xi^2}\left(\frac{c}{v}\right)^4 \cdot \varphi_2(x) \qquad (4)$$

where

$$\varphi_2(x) = x^2[K_1^2 + x^2(K_1^2+K_0^2) + xK_0K_1] \qquad (5)$$

The functions φ_1 and φ_2 are given in Fig. 1.

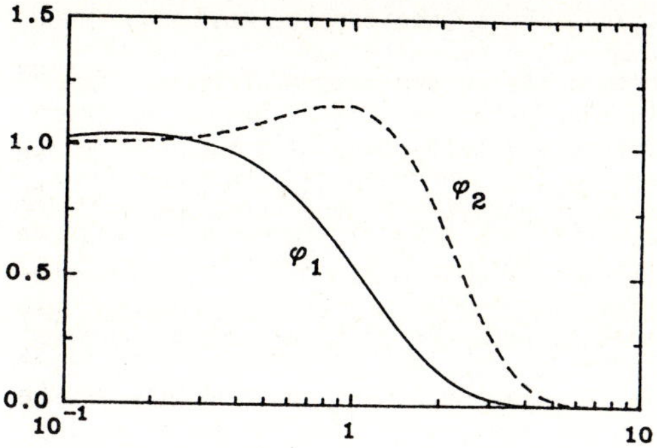

Fig. 1: The shape of the virtual photon spectrum as a function of the adiabaticity parameter x for the multipolarities E1 and E2.

It can be seen that the E2 virtual photon numbers are in many interesting cases much larger than the corresponding E1 ones. From various experimental conditions with different relative E1 and E2 virtual photon numbers the quantity $\sigma_{E\lambda}^{photo}$ can be individually determined. In coincidence studies of reaction (1) interference effects between different multipoles will show up in general, which can in principle help to disentangle the various multipole contributions. A selective population of magnetic substates of the system b+c is expected, it can be directly calculated from the theory of Coulomb excitation.

From the virtual photon numbers (3), (4) and (5) and the estimated cross sections $\sigma_{E\lambda}^{photo}$ (from the astrophysical S-factor measured at higher relative energies) it is concluded that dedicated experiments are feasible in the astrophysically relevant region. It is especially noticeable that even at modest energies (~ 60 MeV/A) the severe adiabaticity condition in the $^{16}O \rightarrow \alpha ^{12}C$ dissociation (Q = -7.162 MeV) can be overcome.

1 H. Rebel: Proc. Workshop on Nuclear Reaction Cross Sections of Astrophysical Interest, unpublished report, Kernforschungszentrum Karlsruhe, February 1985

2 G. Baur, C.A. Bertulani and H. Rebel: Preprint KFK, December 1985, Nucl. Phys. A, in press

Competition of Neutron Capture and Beta Decay at the ^{85}Kr and ^{151}Sm Branchings, a Means to Estimate the s-Process Pulse Conditions

H. Beer

Kernforschungszentrum Karlsruhe, Institut für Kernphysik III,
P.O.B. 3640, D-7500 Karlsruhe 1, F. R. Germany

1. Introduction

s-process abundances of heavy elements are successfully reproduced using an exponential distribution of neutron exposures [1,2]. This behavior has been explained as the result of a pulsed irradiation of s-process seed material in the He-shell of red giant stars [3]. A possibility to ascertain the pulsed nature of the s-process is offered by the analysis of branchings in the synthesis path. A branching is the result of a competition between neutron capture and beta decay of a radioactive isotope situated on the path. The pulsed nature of the s-process can show up due to the interpulse decay of the radioactive branch point nucleus and the necessary reformation of its abundance during the next pulse.

2. The Sm151 and Kr85 branchings

In this work it was found that the branchings at Kr85 and Sm151 are especially sensitive to the pulse conditions. Sm151 can provide a lower bound for the pulse duration whereas Kr85 can give an upper limit. For the analysis, the pulsed s-process model of WARD and NEWMANN[4] assuming a constant irradiation during the pulse was used. The classical s-process appears as a limiting case for very large pulse durations ($\Delta t \to \infty$). In Fig.1 the analysis of the Sm151 branching is shown. Comparison of the solar Gd152 abundance with the calculation exhibits that a pulse duration $\Delta t \geq 3$yr is required. A similar calculation on the Kr85 branching yields an upper limit $\Delta t \leq 50$yr. Fig.2 shows a typical s-process calculation from Fe56 to Bi209 using a pulse duration of 10yr where all empirical data are described satisfacto-

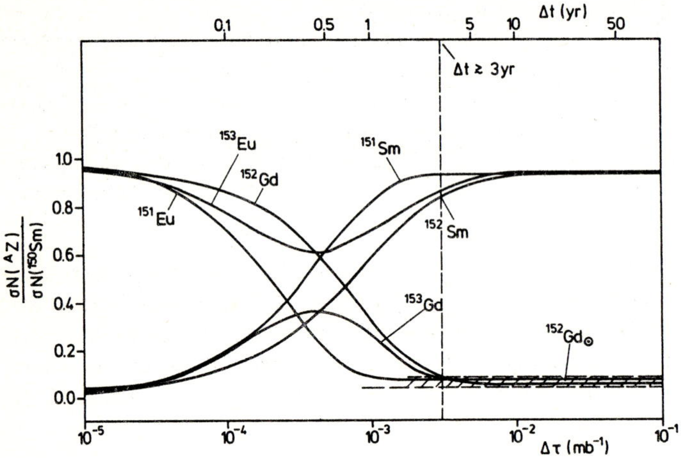

Fig.1 The analysis of the isotopes located in the Sm151 branching is shown. The σN (capture cross section times s-process abundance) values normalized to Sm150 are plotted as a function of the exposure $\Delta\tau$ (neutron density: 1.3 10^8 /cm^3 and temperature: 2.7 10^8 K). For Gd152 an empirical value can be given (hatched area). The intercept of the theoretical curve with the empirical value yields $\Delta t \geq 3$yr. Note that a sufficiently small Δt practically cancels the branch to Sm152.

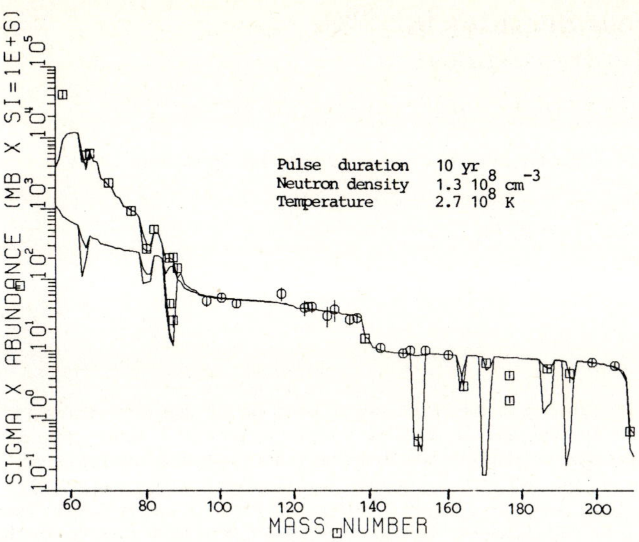

Fig.2 The product σN as a function of mass number. The symbols correspond to empirical data. Significant branchings were identified due to low empirical values on one of the branches. Note that the Kr86 and Rb87 data located in the Kr85 branching are not corrected for r-process contributions. Therefore, the theoretical calculation must remain below the empirical value's to avoid overproduction. Below A=90 and above A=205 s-process components with single flux exposures are superimposed [5]. The course of the pulsed s-process in these regions is indicated, too.

rily. The figure also demonstrates that for nuclei below the Kr85 branching an additional weak s-process becomes more and more influential. This weak s-process was previously modeled as an additional exponential exposure component [1,2] signifying that it has to be treated also as a pulsed s-process. But calculations using an unpulsed single flux neutron exposure result in an improved agreement with the empirical data (Fig.2) [5]. This type of s-process is in agreement with current concepts of stellar nucleosynthesis in this mass region [6,7].

1. F. Käppeler, H. Beer, K. Wisshak, D.D. Clayton, R.A. Ward: Ap. J. 257 , 821 (1982)
2. H. Beer: 5th Moriond Astrophysics Meeting, Nucleosynthesis and its Implications on Nuclear and Particle Physics, Les Arcs, Savoie, March 17-23, 1985, eds. J. Audouze, N. Mathieu (D. Reidel Publishing Company 1986) p. 263
3. K. Cosner, I. Iben jr., J.W. Truran: Ap. J. 238 , L91 (1980)
4. R.A. Ward, M.J. Newman: Ap. J. 219 , 195 (1978)
5. H. Beer: 2nd IAP Rencontre on Nuclear Astrophysics, Paris July 7-11 , 1986
6. S. Lamb, W.M. Howard, J.W. Truran, I. Iben jr.: Ap. J. 217 , 213 (1977)
7. W.D. Arnett, F.-K. Thielemann: Ap. J. 295 , 589 (1985)

Neutron Capture Cross Sections for ^{86}Sr and ^{87}Sr at Stellar Temperatures

R.W. Bauer, G.J. Mathews, J.A. Becker, R.E. Howe, and R.A. Ward

Lawrence Livermore National Laboratory, Livermore, CA 94550, USA

1. Introduction

Recent work on s-process nucleosynthesis has focused attention on the investigation of capture cross sections for nuclei in the mass region near the N=50 closed neutron shell [1,2,3]. Of special astrophysical interest are (i) the analysis of the s-process branching through ^{85}Kr as a monitor of stellar neutron density and temperature and (ii) the investigation of the possible chronometric pair ^{87}Rb-^{87}Sr as an independent measure of the age of the galaxy. For both problems the capture cross sections of the two pure s-process nuclei ^{86}Sr and ^{87}Sr have to be known to an accuracy of 5% or better. The current investigation of the neutron capture cross sections for ^{86}Sr and ^{87}Sr was undertaken to extend recent measurements by WALTER and BEER [2] to energies below 3.5 keV, where strong resonances are known to exist, and to explore the discrepancy in the results of the Maxwellian averaged capture cross section of ^{87}Sr at kT = 30 keV as reported by previous investigators [2,4,5].

2. Experiment and Analysis

The neutron capture cross sections for ^{86,87}Sr have been measured from 100 eV to 1 MeV at the Livermore Electron Linear Accelerator. Neutrons with a continuous energy distribution were produced in a tantalum target bombarded by 100-MeV electrons. The capture events and their flight times were recorded by detecting the prompt gamma-ray cascade with two C_6D_6 scintillators located 11 m from the neutron source. A ^{6}Li-glass scintillator was used to monitor the neutron flux. The background was determined experimentally utilizing the "black resonance" technique. Details of the experimental setup have been presented in previous reports [6,7]. We applied a weighting function to our data such that the resultant efficiency of the capture gamma-ray detectors is independent of the gamma-ray spectrum. Corrections have also been applied for neutron multiple scattering and self-shielding, and for gamma-ray attenuation. The strontium cross sections have been normalized to a standard gold cross section revised to agree with the latest measurements by MACKLIN et al [8]. Figure 1 gives an example of our cross section results for ^{86}Sr.

3. Results

The Maxwellian averaged neutron capture cross sections have been calculated for stellar temperatures ranging from kT = 10 to 100 keV. Our cross sections at 20, 30, 40 and 50 keV are found to be in excellent agreement with those reported by WALTER and BEER [2]. At kT = 30 keV we obtain 70 $\pm$ 4 mb for ^{86}Sr, and 97 $\pm$ 5 mb for ^{87}Sr. Combining our results with those reported previously [2,4,5], we recommend Maxwellian averaged capture cross sections at kT = 30 keV of 70 $\pm$ 3 mb for ^{86}Sr, and 93 $\pm$ 4 mb for ^{87}Sr. These latter values have been used to analyze the branching in the s-process flow at the unstable nucleus ^{85}Kr. This branching can be used as a possible measure of the neutron density during the s-process by comparing the $\sigma \bullet N$ values for ^{86}Sr and ^{87}Sr with the corresponding value for ^{88}Sr. There exists also the additional possibility to use the ^{87}Rb-^{87}Sr isobaric doublet

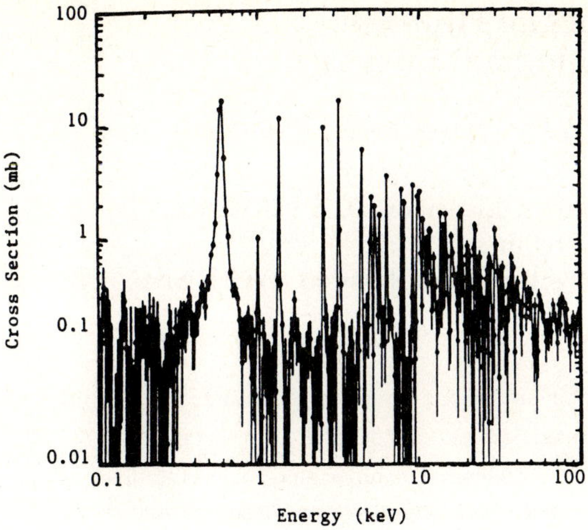

Fig. 1

Measured capture cross sections of ^{86}Sr, displaying strong resonances at 0.588, 1.370, 2.592, 3.247 and 4.496 keV, and an approximate 1/v decrease above 20 keV

as a chronometric pair based on the long half-life of ^{87}Rb. Utilizing analyses of the capture flow based on an exponential distribution of neutron exposures including the temperature dependence of all beta decays and neutron captures, we find a good fit to the branching through ^{85}Kr can be obtained for all temperatures. The optimum conditions correspond to a mean neutron exposure of $\tau_o = 0.40(\pm 0.06)$ $(kT/30)^{1/2}$ mb^{-1} (where kT is in keV), and an average neutron density of roughly $n_n = 4.7(\pm 0.7)x10^7$ (kT/30) cm^{-3}. It appears that this branch requires a slightly larger exposure and a lower-density neutron source than the heavier s-process nuclei. This might be attributed to production in low-mass AGB stars [9].

The data are still too uncertain to be used for a reliable evaluation of the ^{87}Rb–^{87}Sr chronometric pair. However, we can infer from these data an upper limit (95% confidence) to the age of the universe of $\leq$ 14x10^9 years (for a constant rate of nucleosynthesis) which is consistent with other chronometers.

This work was performed under the auspices of the U. S. Department of Energy by the Lawrence Livermore National Laboratory under Contract No. W-7405-ENG-48.

References:

1. F. Käppeler, H. Beer, K. Wisshak, D. D. Clayton, R. L. Macklin and R. A. Ward, Ap. J. 257, 821 (1982)
2. G. Walter and H. Beer, Astron. Astrophys. 142, 268 (1985)
3. G. Walter, B. Leugers, K. Käppeler, G. Reffo, and F. Fabbri, Nucl. Sci. Eng. (to be published)
4. G. C. Hicks, B. J. Allen, A. R. de L. Musgrove, and R. L. Macklin, Austral. J. Phys. 35, 267 (1982)
5. R. L. Macklin and J. H. Gibbons, Ap. J. 149, 577 (1967)
6. B. L. Berman and J. C. Browne, Phys. Rev. C 7, 252 (1973)
7. J. C. Browne and B. L. Berman, Phys. Rev. C 23, 1434 (1981); 26, 969 (1982)
8. R. L. Macklin, private communication (1982). See also Z. Y. Bao and F. Käppeler, KFK Report (February 1986)
9. G. J. Mathews, R. A. Ward, K. Takahashi, and W. M. Howard, Proc. Fifth Moriond Workshop on Astrophysics, Les Arcs, ed. J. Audouze and N. Mathieu, publ. Reidel, Amsterdam, p. 277 (1986)

Laboratory Determination of the Half-Life of ^{187}Re, a Nuclide of Cosmological Interest*

M. Lindner[1], *D.A. Leich*[1], *R.J. Borg*[1], *G.P. Russ*[1], *J.M. Bazan*[1], *D.S. Simons*[2], *and A.R. Date*[3]

[1]Lawrence Livermore National Laboratory, Livermore, CA 94550, USA
[2]National Bureau of Standards, Gaithersburg, MD 20899, USA
[3]British Geological Survey, 64-78 Gray's Inn Road, London, WCIX 8NG, UK

In recent years there has been great interest among cosmologists and geochemists in an accurate laboratory-measured value for the half-life of ^{187}Re. The reason is that the ^{187}Re-^{187}Os isobaric couple has potential use as a terrestrial and cosmic chronometer. We have recently completed experiments which have yielded a value of $(4.35 \pm 0.13)10^{10}$ y. for this half-life. We used micro and radiochemical techniques combined with mass spectrometry to observe the growth of ^{187}Os into a large source of osmium-free rhenium. When our value is compared with the ^{187}Re lifetime recently determined from meteorite analysis, we can set limits on the possible variation of the fine-structure "constant", α, over the past 4.5 billion years.

*This work was performed under the auspices of the U.S. Department of Energy by Lawrence Livermore National Laboratory under contract no. W-7405-Eng-48. Further details may be obtained from Nature, Vol. 320, No. 6059, pp. 246-248, 20 March 1986.

Interpretation of the Solar $^{48}Ca/^{46}Ca$ Abundance Ratio and the Correlated Ca-Ti-Cr Isotopic Anomalies in Inclusions of the Allende Meteorite

W. Hillebrandt[1], *K.-L. Kratz*[2], *F.-K. Thielemann*[1], *and W. Ziegert*[1,2]

[1]Max-Planck-Institut für Physik und Astrophysik, Institut für Astrophysik, D-8046 Garching b. München, F. R. Germany

[2]Institut für Kernchemie, Universität Mainz, D-6500 Mainz, F. R. Germany

In the past, astrophysical models encountered severe difficulties in explaining the solar $^{46,48}Ca$ abundances or the correlated Ca-Ti-Cr isotopic anomalies observed in inclusions of the Allende meteorite [1-3]. Among the various attempts, SANDLER et al. [4] suggested the production of neutron-rich stable Ca-Ti-Cr isotopes in a high neutron density environment of $\sim 10^{-7}$ mol/cm^3 with a neutron-exposure time of 10^3 s. Assuming the initial abundances to be solar and applying Hauser-Feshbach neutron-capture cross sections, the above authors have calculated a $^{48}Ca/^{46}Ca$ abundance ratio which is only a factor of 2.6 smaller than the observed solar value of 56. However, the predicted isotopic anomalies for ^{46}Ca and ^{49}Ti were too large by factors of 13 and 5, respectively, compared to those in the EK-1-4-1 inclusion of the Allende meteorite [1,2]. Recently KÄPPELER et al. [5] have estimated the steady-state abundance ratio of $^{46,48}Ca$ in a high neutron-density environment, but could not reproduce the solar-system value. In a different approach, HARTMANN et al. [6] have calculated the composition of matter that has gone through a phase of nuclear statistical equilibrium in the deep interior of a supernova and is ejected with large neutron excess. With this model it is possible to fit the solar $^{46}Ca/^{46}Ca$ ratio and also the Ti-Cr abundances in the inclusion G1 [3], but the Ca-Ti correlation [1,2] is not well reproduced. Additional difficulties arise from the fact that also other neutron-rich isotopes are overproduced by large factors.

Motivated by those difficulties and by the nuclear structure related large variations of β-decay properties of neutron-rich isotopes near the double magic nucleus ^{48}Ca, we have investigated possible implications of β-decay half-lives ($T_{1/2}$) and β-delayed neutron (βdn) emission probabilities (P_n) on the Ca-Ti-Cr abundances by a systematic study of neutron-capture processes at different neutron densities. Improving our first simple n-process approach [7], we have performed complete network calculations which include stable and neutron-rich radioactive isotopes from S to Cr, (n,γ)- and (γ,n)-reactions, β-decay and βdn-emission.

Since we intended to perform a parameter study in order to find the right conditions for both the solar system $^{48}Ca/^{46}Ca$ ratio and the isotopic anomalies mentioned earlier we have not used a specific astrophysical model but rather have computed the abundances as functions of time for various constant temperatures and neutron densities. Solar system initial abundances were always chosen. We found that the results did not vary much with temperature if T was chosen between a few times 10^8 K and 10^9 K, but were strongly dependent on neutron densities and exposure times.

Figure 1 shows a typical example obtained for $T = 8\cdot 10^8$ K. Such a temperature is expected, e.g., in explosive He-burning. It can be seen that the solar $^{48}Ca/^{46}Ca$ ratio is only reproduced if the exposure time is less than 50 ms and the neutron exposure is at least $5\cdot 10^{-5}$ mol cm^{-3} s, translating into a neutron density larger than 10^{-3} mol cm^{-3}. It should be noted that at present no astrophysical scenario predicts such high neutron densities in matter ejected from a star, but these neutron densities are in the same range as required for the r-process. The strong time dependence seen in fig. 1 is mainly caused by the

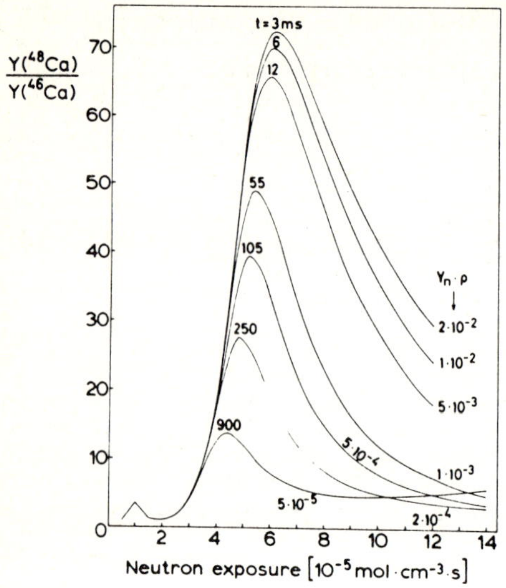

Figure 1:
Abundance ratio $^{48}Ca/^{46}Ca$ as a function of neutron exposure for various combinations of neutron densities $(Y_n \cdot \rho)$ and exposure times in milliseconds. Given is the exposure time at which the maximum abundance ratio was obtained. β-decay half-lives from ref. [8] have been used.

short β-decay half-life of ^{44}S for which we have used the value (310 ms) predicted by KLAPDOR et al. [8]. If neutron-exposure times become comparable to $T_{1/2}(^{44}S)$ nuclei are fed into the Cl isotopic chain and thus the abundances of ^{46}Cl and ^{47}Cl increase which both contribute to the final yield of ^{46}Ca [7]. If, on the other hand, the $T_{1/2}(^{44}S)$ would be considerably longer less neutrons would be needed in our model in order to explain the solar $^{48}Ca/^{46}Ca$ ratio. Since ^{44}S has a closed neutron shell a longer $T_{1/2}$ is certainly possible and not out of experimental range.

A maximum of the abundance ratio shown in fig. 1 exists for neutron exposures of about $(5\pm1)\cdot10^{-5}$ mol cm^{-3} s, nearly independent of neutron density and exposure time. For fixed exposure time the maximum value is mainly determined by the P_n-values of ^{46}Cl and ^{49}K. The experimentally known large P_n-value of ^{49}K (≃ 86%) yields a large ^{48}Ca abundance, whereas tho (so far unknown) P_n-value of ^{47}Cl strongly influences the ^{46}Ca production. Even if $P_n(^{47}Cl)$ would turn out to be large (≈ 100%) the solar $^{48}Ca/^{46}Ca$ ratio (≃ 56) can be reproduced for certain neutron exposures, but with the restrictions on neutron densities and exposure times shown in fig. 1. $P_n(^{47}Cl)$ values around 30% would be more favourable. Again this important physical quantity should be measured in order to restrict the astrophysical parameter space.

In order to derive meteoritic Ca, Ti and Cr abundances, the composition obtained from our network calculations was mixed into a reservoir with solar system abundances. Typically a fraction of 10^{-5} of processed material gave the best fits. At a neutron-exposure of $\sim7\cdot10^{-5}$ mol cm^{-3} s, best overall agreement with the observed Ca-Ti isotopic anomalies in EK-1-4-1 was obtained (see fig. 2). Furthermore, our calculated abundance ratio of $[^{48}Ca]/[^{50}Ti] = 3.55$, where square brackets denote abundances relative to their solar system values, is in excellent agreement with the measured value of 3.6±0.2 [2]. For comparison, the models of SANDLER et al. [4] and HARTMANN et al. [6] predict ratios of 2.6 and 0.8, respectively. In addition, we predict isotopic anomalies in EK-1-4-1 (see the table below) which could, in principle, be measured.

Finally, we have also tried to explain the excess found in ^{54}Cr and ^{50}Ti, in the Allende inclusion G1 by BIRCK and ALLEGRE [3]. Our best fit as well as the observed anomalies are shown in fig. 3. This fit was obtained for a rather moderate neutron exposure of $1.2\cdot10^{-7}$ mol cm^{-3} s and an admixture of processed matter of again roughly 10^{-5}. It can be seen that our model is able to reproduce the observed anomalies within the experimental errors as well.

Predictions for EK-1-4-1

$[^{48}Ca]/[^{54}Cr]$	$[^{54}Cr]/[^{52}Cr]$	$[^{53}Cr]/[^{52}Cr]$
56^{+15}_{-25}	$4.5^{+3.0}_{-1.0}$	$2.9^{+1.0}_{-0.5}$

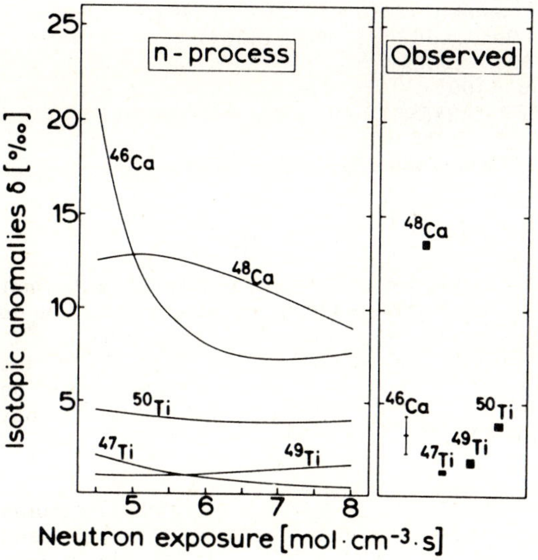

Figure 2:
Isotopic anomalies in calcium and titanium. The figure on the right shows the observed anomalies in the Allende inclusion EK-1-4-1 [2]. The figure on the left shows the computed anomalies obtained from our n-process model as a function of neutron exposure.

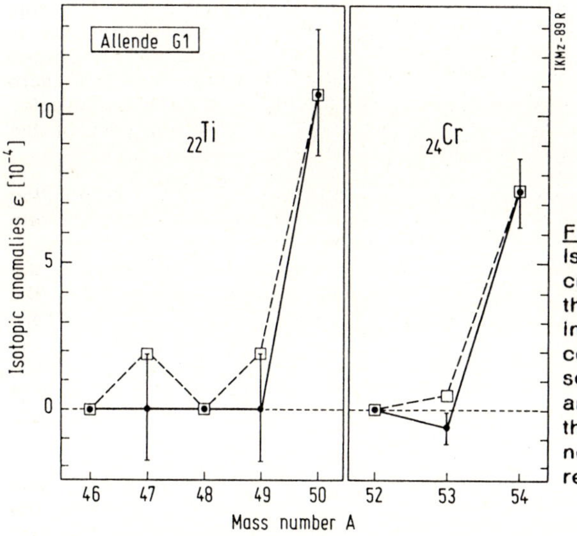

Figure 3:
Isotopic anomalies in titanium and cromium. Observed anomalies in the Allende inclusion G1 [3] are indicated by dots, the anomalies computed from our model by squares. The computed abundances are normalized in such a way that the observed anomalies in the most neutron-rich isotope are exactly reproduced.

So, in conclusion, it seems that several isotopic anomalies show abundance patterns which are characteristic for moderate neutron exposures expected, e.g., in explosive He-burning in supernovae. In order to obtain the solar $^{48}Ca/^{46}Ca$ ratio, on the other hand, a significantly higher neutron density is required. Future experimental determinations of relevant β-decay properties of neutron-rich isotopes will help to make more definite predictions about the astrophysical sites.

References:

1. T. Lee et al.: Ap. J. 220, L21 (1978), and 228, L93 (1979)
2. F.R. Niederer et al.: Geochim. Cosmochim. Acta 45, 1017 (1981)
3. J.-L. Birk, C.J. Allegre: Geophys. Res. Lett. 11, 943 (1984)
4. D.G. Sandler et al.: Ap. J. 259, 908 (1982)
5. F. Käppeler et al.: Ap. J. 291, 319 (1985)
6. D. Hartmann et al.: Ap. J. 297, 837 (1985)
7. W. Ziegert et al.: Phys. Rev. Lett. 55, 1935 (1985)
8. H.V. Klapdor et al.: At. Data Nucl. Data Tables 31, 81 (1984).

The Inflationary Universe : Progress and Problems

R.H. Brandenberger

Department of Applied Mathematics and Theoretical Physics,
University of Cambridge, Cambridge CB3 9EW, England

I briefly review the main features of inflationary universe models. Inflation provides a mechanism which produces energy density fluctuations on cosmological scales. In the original models it was not possible to obtain the correct magnitude of these fluctuations without fine tuning the particle physics models. I discuss a modified model, "chaotic inflation", in which the fine tuning problem is less severe. I also mention other approaches to early universe cosmology.

1. Introduction

Until recently, most cosmological models were based on coupling classical general relativity as the theory of space-time with an ideal gas as a matter source. In the most popular of these models, the "big bang" theory, the universe starts out at an initial time t=0 with infinite temperature and energy density. Obviously, at very high temperatures, i.e. very early times, the classical description of matter as an ideal gas will be very misleading.

In the new cosmological theories the ideal gas as matter source for classical gravity is replaced by a description of matter which will be valid until much higher temperatures, namely in terms of quantum fields. In inflationary universe models matter is taken to be given by a grand unified theory. This may lead to deviations from the standard big bang model at very early times.

The standard big bang model leads to various puzzles for cosmologists, the first of which is the horizon problem. The universe looks homogeneous and isotropic on scales much greater than the causal hor-

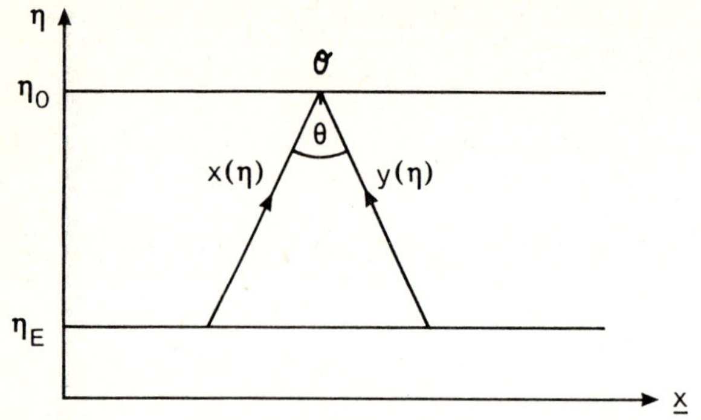

Figure 1 A conformal space-time diagram showing two light rays $X(\eta)$ and $Y(\eta)$ seen by the observer σ and subtending an angle Θ. The last scattering surface is $\eta = \eta_E$. η_0 is the present conformal time.

izon at the time the features we observe were established. The prime example is the homogeneity and isotropy of the microwave background radiation (Figure 1). If we look back at the last scattering surface along two light rays subtending an angle of more than 1°, we are looking at points which were causally disconnected at last scattering. Why should the microwave temperature be the same?

The second puzzle is the flatness problem. We observe no spatial curvature of the universe. Hence the energy density today must be equal to the critical density (the density which is required for a spatially flat solution of Einstein's equations) to within one order of magnitude. In an expanding universe flat space is unstable in the sense that any deviation from the critical energy density grows in time as T^{-2} (T is the temperature). In order to explain the present day flatness of the universe, the energy density ρ would have had to be equal to the critical energy density ρ_0 so one in 10^{55} at a temperature of 10^{17} Gev, a temperature at which we imagine setting up initial conditions. This is a completely unnatural fine tuning of initial conditions.

Finally, the standard big bang model does not explain why inhomogeneities such as galaxies and clusters of galaxies are correlated on scales which initially were outside the causal horizon.

The key to resolving these cosmological puzzles is to change the expansion rate of the universe at very early times. In the standard model the space-time metric is given by

$$ds^2 = -dt^2 + a(t)^2 d\underline{x}^2 \quad (1)$$

a(t) is the scale factor and determines the physical distance between two points at rest. The time dependence of a(t) is determined by the equation of state of matter. For relativistic matter the pressure p equals $1/3\rho$ (ρ is the energy density), and $a(t) \sim t^{1/2}$. For non-relativistic matter p=0 and $a(t) \sim t^{2/3}$. In both cases the Hubble radius

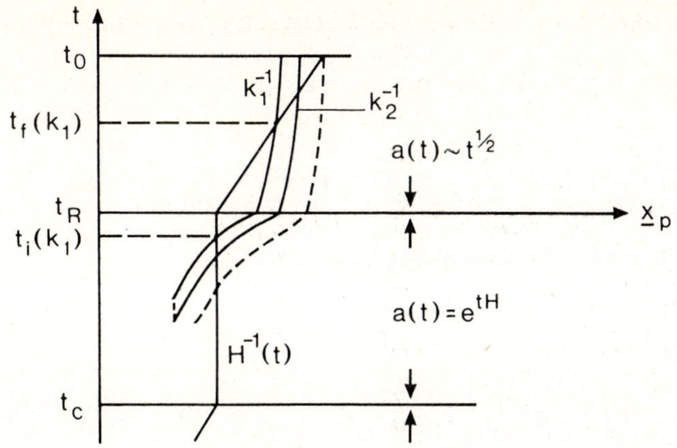

Figure 2 Evolution of the comoving scale corresponding to the present day ($t = t_0$) horizon (dotted line) in the inflationary universe. The solid line $H^{-1}(t)$ is the Hubble radius which is proportional to t in the radiation dominated phases $t < t_c$ (t_c the time when inflation starts) and $t > t_R$ (t_R is the reheating time). Also shown are the scales corresponding to fluctuations of wave numbers k_1 and k_2.

$$H^{-1}(t) = \left[\frac{\dot{a}(t)}{a(t)}\right]^{-1} \tag{2}$$

increases faster than the distance between two points at rest. $H^{-1}(t)$ is the maximal distance microphysical forces can act at time t and can thus be called the causal horizon.

What changes if there is a period in the early universe in which the scale factor increases exponentially (hence the physical size of the universe expands exponentially, i.e. inflates)? Now the Hubble radius is constant while physical scales increase. Distances outside the causal horizon at later times were initially inside the Hubble radius, and thus a solution of the horizon problem is at hand. Provided the period of exponential expansion lasted sufficiently long, our entire observed universe started out inside the causal horizon (Figure 2). In addition, correlations of energy density fluctuations can be created inside the Hubble radius in the exponential phase (the de Sitter phase), expand to be far outside the causal horizon at the beginning of the radiation-dominated Friedmann-Robertson-Walker (FRW) period, and reenter the Hubble radius later as the fluctuations we observe on the scales of galaxies and clusters of galaxies. Finally, the flatness problem is resolved. Our observed universe stems from an exponentially small patch P of the universe at the beginning of the de Sitter phase. Any reasonable curvature initially on scales of the Hubble radius is exponentially small compared to the scale of P. The inflationary universe predicts $\Omega = |\rho-\rho_0|\rho_0^{-1} \simeq 1$.

The crucial observation by Guth [1] was that an equation of state $p = -\rho$ for matter which gives rise to an exponentially increasing scale factor emerges naturally when considering a class of grand unified field theories in the early universe. In section 2 I will describe the basic mechanism of inflation (see Ref 2 for more detailed expositions). I then discuss the causal mechanism by which energy density fluctuations required to form galaxies and clusters of galaxies are generated. The mechanism is very successful in explaining the qualitative features of the spectrum of density perturbations, but in the initial models [3] fails to produce the correct amplitude [4, 5]. In section 4 I outline a possible solution, chaotic inflation [6]. In the final section I will compare the inflationary universe with other approaches to early universe cosmology.

2. The New Inflationary Universe

To illustrate the main ideas I will consider a simple toy model field theory, a double well $\lambda\phi^4$ model with potential (Figure 3)

$$V(\phi) = \frac{\lambda}{4}(\phi^2 - \sigma^2)^2 \tag{3}$$

σ is the scale of symmetry breaking, and is of the order 10^{15} Gev for grand unified theories. It is very natural to consider such a potential, since in any grand unified theory in which the symmetry is spontaneously broken there will be a Higgs field with a similar potential.

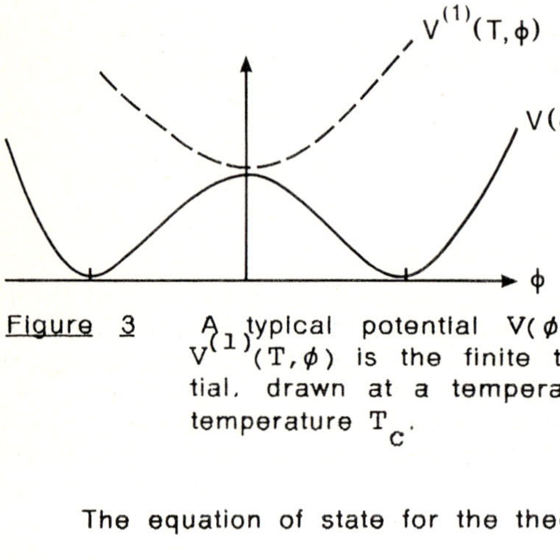

Figure 3 A typical potential $V(\phi)$ for the new inflationary universe $V^{(1)}(T,\phi)$ is the finite temperature one-loop effective potential, drawn at a temperature T much larger than the critical temperature T_c.

The equation of state for the theory with Lagrangian

$$L(\phi) = \tfrac{1}{2}\partial_\mu\phi\partial^\mu\phi - V(\phi) \tag{4}$$

is

$$\rho = \tfrac{1}{2}\dot{\phi}^2 + \tfrac{1}{2}a^{-2}(t)(\underline{\nabla}\phi)^2 + V(\phi) \tag{5}$$

$$P = \tfrac{1}{2}\dot{\phi}^2 - \tfrac{1}{6}a^{-2}(t)(\underline{\nabla}\phi)^2 - V(\phi)$$

We consider a modified big bang model in which there is ordinary matter plus the scalar field ϕ. The energy-momentum tensor is

$$T_{\mu\nu} = T_{\mu\nu}(\phi) + T_{\mu\nu}(\text{rad}). \tag{6}$$

$T_{\mu\nu}$(rad) is the energy-momentum tensor of ordinary matter and gives an equation of state $p = \frac{1}{3}\rho$. $T_{\mu\nu}(\phi)$ is the contribution to $T_{\mu\nu}$ from the scalar field.

The standard lore of new inflation is as follows. After the big bang $\phi(\underline{x},t) = 0$. At high temperatures $T_{\mu\nu}$ is dominated by radiation, the effective equation of state is $p = 1/3\rho$ and the scale factor grows as $a(t) \sim t^{1/2}$. Once the radiation gas has cooled to below the critical temperature T_c determined by

$$\lambda\sigma^4 = \frac{\pi^2}{30} N T_c^4 \qquad (7)$$

(where N is the number of spin degrees of freedom at temperature T_c), $T_{\mu\nu}$(rad) becomes unimportant and the equation of state is given by the scalar field. If $\phi(\underline{x},t) = 0$ then $p = -\rho$, and by the Einstein equations

$$\left(\frac{\dot{a}}{a}\right)^2 = \frac{8\pi G}{3}\rho \qquad (8)$$

$$\dot{\rho} = -3H(t)(\rho + p)$$

we see that a(t) expands exponentially

$$a(t) = e^{Ht}\ , \quad H = \left(\frac{8\pi G}{3}\rho\right)^{1/2} . \qquad (9)$$

The equation of motion for a spatially homogeneous scalar field configuration in an expanding universe is

$$\ddot{\phi} + 3H(t)\dot{\phi} = -\partial V/\partial\phi \qquad (10)$$

$\phi(t) = 0$ is an unstable fixed point; any deviations from it will grow in time. Thus, starting near $\phi(\underline{x}) = 0$, the scalar field will slowly roll down the potential slope (Figure 3). As long as the rolling is slow, the equation of state is still approximately $p = -\rho$, and inflation continues. When $\phi(t)$ approaches $\phi = \pm\sigma$, the scalar field will begin to roll fast, it will oscillate about $\phi = \pm\sigma$ and its energy is converted into thermal energy. In this period, which generally is much shorter than a Hubble expansion time H^{-1}, the universe reheats to close to the critical temperature T_c. The phases of the new inflationary universe are sketched in Figure 4.

Why does the scalar field start out with $\phi(\underline{x},t(T)) \simeq 0$ for $T > T_c$? The answer depends on whether the scalar field ϕ is weakly or strongly coupled to other fields. Consider first the case in which ϕ is strongly coupled to a second field ψ, which for simplicity we take to be a second scalar field. The interaction Lagrangian is $\frac{1}{2}\tilde{\lambda}\phi^2\psi^2$ and will give rise to an effective mass term in the equation of motion for ϕ

$$\ddot{\phi} + 3H\dot{\phi} = -\frac{\partial V}{\partial\phi} - \tilde{\lambda}\psi^2(T)\phi \qquad (11)$$

At high temperatures the mass term will drive $\phi(\underline{x})$ towards $\phi = 0$. An alternate way to see this effect would be to compute the one loop finite

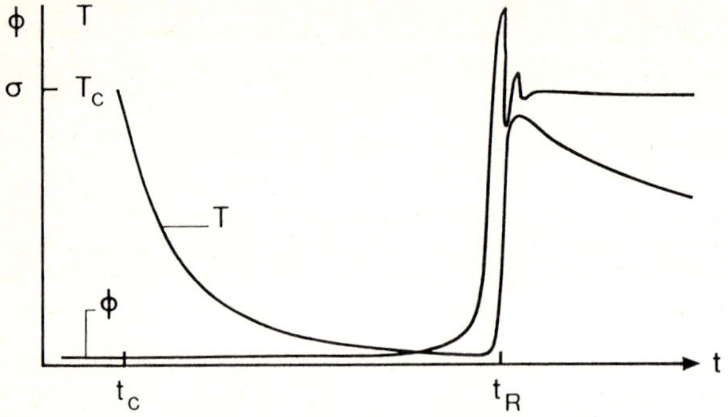

Figure 4 Sketch of the phases in the new inflationary universe. During most of the inflationary (de Sitter) phase ϕ remains close to the origin and T decreases exponentially. At reheating ϕ increases to its ground state value σ and the universe reheats to almost T_c.

temperature effective potential, which will give a similar mass term [7] (see Figure 3).

In the case of a scalar field ϕ only very weakly coupled to other fields a different approach is necessary. In fact, it has been argued [8] that in this case inflation does not occur. At high temperatures there will be large spatial fluctuations in the scalar field $\phi(\underline{x})$. As long as $T > T_c$, $\phi(\underline{x})$ can oscillate in time from large positive to large negative values (large compared to σ). It was conjectured that when T falls below T_c, the scalar field will have lost enough energy so that it "gets stuck" in one or the other potential well, and that thus for $T \leqslant T_c$ spatial domains with $\phi(\underline{x}) = \underline{+}\sigma$ will form. Looking back at the equation of state (5), we see that this would imply an equation of state with $p > -1/3\rho$, and that there would be no inflation.

The above argument, however, misses an essential point, namely the effects of the Hubble damping term $3H\dot{\phi}$. In the presence of inhomogeneities, the equation of motion for the scalar field is

$$\ddot{\phi} + 3H(t)\dot{\phi} - a^{-2}(t)\nabla^2\phi = -\partial V/\partial\phi \qquad (12)$$

The spatial gradient term leads to oscillations of $\phi(\underline{x},t)$. The Hubble term will damp the oscillation, thus producing a configuration $\phi(\underline{x},t) \simeq 0$. In contrast, the nonlinear force $\partial V/\partial\phi$ will tend to drive $\phi(\underline{x})$ towards $\underline{+}\sigma$ and thus produce a domain structure. For a scalar field theory with weak self coupling, the Hubble damping force will be much stronger than the nonlinear force, and inflation will occur [9].

To make the above arguments more precise, we shall for a moment replace the nonlinear force by a small mass term :

$$\frac{\partial V}{\partial\phi} \rightarrow \frac{1}{6}R\phi \qquad (13)$$

where R is the Ricci scalar. With this substitution, Eq (12) is the equation of motion of a conformally coupled free scalar field. Any solution

$\phi(\underline{x},t)$ can be obtained by conformal transformation from a solution $\tilde{\phi}(\underline{x},t)$ of the flat space-time Klein-Gordon equation for a free massless scalar field.

$$\phi(\underline{x},t) = a^{-1}(t)\tilde{\phi}(\underline{x},\tau) , \tag{14}$$

where τ is conformal time given by $dt^2 = a^2(t)d\tau^2$. In flat space-time the solution $\tilde{\phi}$ will be an oscillating wave with constant amplitude. Hence in an expanding universe ϕ will be a wave oscillating in conformal time with amplitude damped as $a^{-1}(t)$.

At this stage the result is no surprise, since we have thrown away the domain forming force. For weak coupling we can analyse the effects of $\partial V/\partial\phi - \frac{1}{6}R\phi$ using a perturbative Green function method [9]. We Fourier expand $\phi(\underline{x},t)$ and study the equation of motion for each Fourier mode $q_k(t)$. Demanding that the effect of the nonlinear force be small for a period of time long enough for sufficient inflation to occur gives an estimate of the maximal value of the coupling constant λ. We can guess this value by comparing the damping force and the nonlinear force at an initial temperature $T = \sigma$ and demanding that the damping force dominates. This gives

$$\lambda < \frac{8\pi^3}{90} N\left(\frac{\sigma}{m_{pl}}\right)^2 \tag{15}$$

where m_{pl} is the Planck mass. This result has been confirmed in numerical work [10].

In order to solve the cosmological puzzles mentioned in the introduction a period of inflation longer than 60 Hubble expansion times is required. This brings us to the first serious problem with the new inflationary universe. Potential (3) gives insufficient inflation. The initial value of $\phi(t)$ will be of the order H, given by the r.m.s. value of quantum fluctuations of a free scalar field in an exponentially expanding universe [11]. The period of inflation will then only be of the order one Hubble expansion time.

A way out of this problem is to abandon the simple toy model of Eq (3) and to use potentials which are very flat near the origin. Such potentials arise in many supergravity theories [12]. The question, however, remains : have we not merely replaced a cosmological fine tuning problem by a particle physics fine tuning?

3. Origin of Energy Density Fluctuations

Probably the most astonishing success of the inflationary universe is that it provides a mechanism which in a causal way generates a spectrum of primordial energy density fluctuations necessary to produce galaxies and clusters of galaxies. The first crucial observation is that in inflationary universe models all scales of interest today originate inside the causal horizon in the de Sitter period [13] (see Figure 2).

We can also derive the shape of the perturbation spectrum on very general grounds. Consider the two scales indicated in Fig 2 and assume that there is some mechanism which produces perturbations at all times on a fixed physical scale. Then the evolution of the perturbations k_1 and k_2 between when they were formed and when they left the Hubble radius at time $t_i(k)$ are related by time translation. Hence the magnitude of the perturbation is independent of k when evaluated at time $t_i(k)$:

$$\frac{\delta M}{M}(k,t_i(k)) = \text{const} \tag{16}$$

$\delta M/M(k,t)$ is the r.m.s. mass excess in a ball of radius k^{-1} at time t. While the perturbation is outside the Hubble radius microphysical processes can not influence it. Hence its physical size will not change until it reenters the horizon at time $t_f(k)$. Therefore,

$$\frac{\partial M}{M}(k,t_f(k)) = \text{const} \tag{17}$$

This shape of spectrum is called scale-invariant or Harrison-Zel'dovich spectrum. It was postulated a long time ago [14] as a reasonable spectrum for galaxy and cluster formation.

Now the second crucial observation [4,5]. Quantum fluctuations in the scalar field $\phi(\underline{x})$ can produce classical energy density perturbations. This problem has recently been studied semiclassically [15], i.e. without quantizing gravity, and in linearized quantum gravity [16,17]. The main idea in the semiclassical analysis is very simple. In quantum mechanics, we all know that the expectation value of the operator Q^2 in a state $|\psi\rangle$ of a simple harmonic oscillator can be interpreted as the square of its classical amplitude q in state $|\psi\rangle$

$$q^2 = \langle\psi|Q^2|\psi\rangle \tag{18}$$

Translating this procedure to field theory we can define a classical field

$$\phi(\underline{x},t) = \phi_0(t) + \delta\phi(\underline{x},t) \tag{19}$$

(ϕ_0 is homogeneous, $\delta\phi(\underline{x})$ has vanishing spatial average) in terms of the expectation values of the operators $\Phi^2(\underline{x})$ and $\tilde{\Phi}^2(\underline{k})$ ($\tilde{\Phi}$ is the Fourier transform of Φ) [15]

$$\phi_0(t) = \langle\psi|\Phi^2(\underline{x})|\psi\rangle_{ren} \tag{20}$$

$$\delta\tilde{\phi}(\underline{k}) = \langle\psi|\tilde{\Phi}^2(\underline{k})|\psi\rangle ,$$

where the subscript ren indicates a renormalized expectation value. The expectation value of $\Phi^2(\underline{x})$ for a massless free scalar field is both IR and UV divergent. The ultraviolet renormalization is done in analogy to the flat space-time procedure, the infrared divergence is cured by inserting a physical cutoff. Since fluctuations with wavelength greater than the Hubble radius act like a change in the background on scales smaller than H^{-1}, we consider them as part of the background in computing $\langle\psi|\Phi^2(x)|\psi\rangle$.

If we choose the state $|\psi\rangle$ to be the state empty of particles at the beginning of the de Sitter phase, then its wave functional is the product of the ground state wave functions for each Fourier mode of ϕ. The expectation values can easily be evaluated. For a free, massless, minimally coupled scalar field we obtain

$$\phi_0(t) = (2\pi)^{-1} H^{\frac{3}{2}} t^{1/2} \tag{21}$$

$$\delta\tilde{\phi}(\underline{k},t) = V^{1/2}(2\pi)^{\frac{3}{2}} a^{-\frac{3}{2}}(t) H^{-1/2} ,$$

where V is the cutoff volume. It is now easy to evaluate the r.m.s. mass perturbations at initial Hubble radius crossing $t_i(k)$. For spectra with reasonable k dependence we can express $(\delta M/M)(k)$ in terms of the Fourier transform of the energy density perturbation

$$\left(\frac{\delta M}{M}\right)^2(k) \simeq V^{-1} k^3 \left(\frac{\delta\tilde{\rho}}{\rho}\right)^2 (k) \tag{22}$$

Since $T_{\mu\nu}$ for a free scalar field is well known, we can immediately compute the mass perturbations

$$\left(\frac{\delta M}{M}\right)(k,t_i(k)) = O(1)\frac{H^4}{\rho} \sim \lambda^{-1}\left(\frac{H}{\sigma}\right)^4 \sim \lambda\left[\frac{\sigma}{m_{pl}}\right]^4 \tag{23}$$

For typical grand unified theories $\sigma \sim 10^{15}$ Gev and hence the mass perturbation at initial Hubble radius crossing is of the order 10^{-16}. The third crucial point was the observation that the amplitude of these fluctuations is amplified by a large factor as a consequence of the change in the equation of state from de Sitter-like $(p \simeq -\rho)$ to radiation dominated $(p = 1/3\rho)$.

In linear perturbation theory fluctuations with different wavelengths do not couple and hence can be analysed independently. While the wavelength of a given perturbation is inside the Hubble radius, microphysical processes determine the amplitude. Once the wavelength is greater than the Hubble radius, microphysical processes can no longer act coherently, and the evolution of the perturbation is determined by gravity alone.

The analysis of the evolution of perturbations outside the Hubble radius is not straightforward. The main problem is to separate the physical degrees of freedom from pure gauge modes. For a detailed explanation, we refer the reader to [5] and [18]. The main point is the following : if we restrict our attention to scalar modes, we can form a gauge invariant variable Φ_H by combining the components of the tensor of the metric perturbation in a clever way [19]. When the wavelength equals the Hubble radius (at time t_H), Φ_H is essentially equal to the energy density perturbation (evaluated in comoving coordinates, which we indicate by a subscript c)

$$\Phi_H(t_H) = \alpha\left(\frac{\delta\rho}{\rho}\right)_c(t_H) \tag{24}$$

Here α is a constant of the order 1. The next step is to consider the linearised Einstein equations and to transform them into an equation of motion for Φ_H. The upshot of a lot of algebra is a fairly simple equation

$$\ddot{\Phi}_H + (4+3c_s^2)H\dot{\Phi}_H + 3(c_s^2-w)H^2\Phi_H = I(t) \tag{25}$$

where $w = p/\rho$ and $c_a^2 = \dot{p}/\dot{\rho}$ determine the equation of state and its change in time, and $I(t)$ is a combination of matter source terms which is negligible for scales outside the Hubble radius.

If the equation of state does not change in time, then Φ_H is constant, as can easily be verified from (25). In inflationary universe models, the equation of state changes during reheating. Given that $\dot{\Phi}_H$ vanishes before and after reheating, Eq (25) can be recast into the form

$$\frac{\Phi_H}{1+w} = \text{const} \tag{26}$$

This and (24) allows us to determine the amplitude of the energy density fluctuations when they enter the Hubble radius at time $t_f(k)$ in the radiation dominated period in terms of the amplitude at initial Hubble radius crossing $t_i(k)$ and the net change in the equation of state

$$\left(\frac{\delta M}{M}\right)(k,t_f(k)) = \frac{1+w(t_f)}{1+w(t_i)}\left(\frac{\delta M}{M}\right)(k,t_i(k)) = \frac{4\rho}{3\dot{\phi}_0^2(t_i(k))}\left(\frac{\delta M}{M}\right)(k,t_i(k)) \tag{27}$$

using in the final step

$$1+w(t) = \frac{\dot{\phi}_0^2}{\rho} \tag{28}$$

for the equation of state of a homogeneous scalar field.

If we evaluate (27) for a scalar field $\phi(\underline{x},t)$ slowly rolling down the slope of the potential in Figure 3 we find

$$\left(\frac{\delta M}{M}\right)(k,t_f(k)) = O(1)\frac{H\delta\phi}{\dot{\phi}_0} \tag{29}$$

For a potential which near the origin can be approximated by [3]

$$V(\phi) = -\lambda\phi^4 \tag{30}$$

we find

$$\left(\frac{\delta M}{M}\right)(k,t_f(k)) \sim \lambda^{1/2} \tag{31}$$

This brings us to the second main problem I would like to mention. The amplification of the perturbations during inflation is too efficient. Unless λ is fine tuned to a value smaller than 10^{-8}, the primordial perturbations produced by inflation are too large. They would have generated observable anisotropies in the microwave background radiation. Again it looks like cosmological fine tuning has been replaced by particle physics fine tuning. There exist, however, many models which give sufficiently small energy density perturbations. Many of these are supergravity models [12], many others employ a new Higgs singlet to produce inflation [20].

4. Recent Progress : Chaotic Inflation

Linde [6] has put forward a new approach to inflation which promises to alleviate the two fine tuning problems mentioned at the end of sections 2 and 3. The basic idea is simple. Consider a scalar field ϕ with a general renormalizable potential

$$V(\phi) = \frac{\lambda}{4}\phi^4 + \beta\phi^3 + \gamma\phi^2 + \delta\phi + \epsilon \tag{32}$$

If we assume that ϕ is weakly coupled to all other fields, then at a high initial temperature T_O when initial conditions are set up (for simplicity we shall take $T_O = m_{pl}$), all values of ϕ with $V(\phi) < \frac{\pi^2}{30} m_{pl}^4$ should arise.

Consider now a region of space in which the scalar field takes on a value close to its maximal value ϕ_{max}. If βm_{pl}, γm_{pl}^2, δm_{pl}^3 and ϵm_{pl}^4 are small, then

$$\phi_{max} \sim \lambda^{-1/4} m_{pl} \tag{33}$$

If $\phi(\underline{x})$ is fairly homogeneous so that the spatial gradient term in the equation of motion (12) for ϕ is negligible compared to the nonlinear force term $\partial V/\partial\phi$, and if $\dot{\phi}$ is small initially, then the equation of motion can be approximated by

$$3H\dot{\phi} = -\frac{\partial V}{\partial \phi} = -\lambda\phi^3 \tag{34}$$

with the Hubble "constant"

$$H(t) = \left[\frac{2\pi\lambda}{3}\right]^{1/2} \frac{\phi^2}{m_{pl}} \tag{35}$$

The solution of (34) is [21]

$$\phi(t) = \exp\left\{-\left[\frac{\lambda}{6\pi}\right]^{1/2} m_{pl} t\right\} \phi(0) \tag{36}$$

where we have taken the initial time to be t=0. As a consistency check, we note that neglecting $\ddot{\phi}$ is self consistent until $\phi(t) = (6\pi)^{-1/2} m_{pl}$.

It can easily be verified [21] that the spectrum of energy density fluctuations is again given by Eq (31). The equation of state is inflationary as long as (36) is valid. We conclude that provided $\lambda < 10^{-8}$ chaotic inflation works in the sense that we have a sufficient period of inflation, and a spectrum of energy density fluctuations consistent with the bounds on microwave background anisotropies.

Chaotic inflation works for fairly general potentials, which is an improvement over the new inflationary universe. However, we had to assume a certain degree of homogeneity in the scalar field configuration. This may lead to a severe problem for the model.

Chaotic inflation has recently received considerable attention. It has been shown (using "naive" [22] and "sophisticated" [23] measures) that almost all initial configurations for the scalar field lead to inflation. Again, spatial inhomogeneities have not been considered in these analyses.

5. Conclusions

The inflationary universe has led to a major change and improvement in cosmological models. The description of matter as an ideal gas can never be the complete answer, and whatever may happen to particular particle physics models, quantum field theoretical processes in the early universe will have to be considered.

I would like to stress two main achievements of inflationary models. Firstly, quantum field theory effects in the early universe can lead to a period in which the universe expands exponentially. In this case the entire observable universe starts out inside the causal horizon in the de Sitter period. This solves the horizon and flatness puzzles of the standard big bang model.

Secondly, inflation provides a mechanism which from first principles in a causal way gives rise to primordial energy density perturbations required for the formation of galaxies and clusters of galaxies.

Inflation is clearly not the answer to all questions. It does not address the question why the cosmological constant today is so small. From a mathematical point of view inflationary models are not consistent in that they treat matter quantum mechanically while retaining the classical description of space-time. This inconsistency leads to the necessity of an ad-hoc prescription when computing the spectrum of energy density perturbations : at some point it is necessary to make the transition between quantum mechanical expectation values and classical quantities. I tried to convince you of the plausibility of this prescription, but it is still only a prescription. Another weak point of the technical analysis of perturbations is that it was necessary to prescribe an initial state for the quantum field. Finally, a certain amount of fine tuning in the particle physics sector is required in order to make inflation work at all.

Some of the technical problems mentioned above are remedied in a new approach, quantum cosmology (see [24] for recent reviews). In this approach, both gravity and the scalar matter field are treated quantum mechanically. In one approach [16], gravity is analysed by linearizing about a Friedmann-Robertson-Walker universe. Using a single prescription, the Hartle-Hawking proposal [25] for the wave function of the universe, and applying the analysis to a $\lambda\phi^4$ model, one gets both an inflationary background and a scale invariant spectrum of perturbations. This approach is mathematically consistent in that it treats matter and gravity on a similar footing, it requires less fine tuning in terms of the particle physics model (in this respect it is similar to chaotic inflation), and it needs only one initial state prescription, not a separate one for the matter sector and the gravity sector.

A completely different approach to early universe cosmology and the question of the origin of energy density fluctuations is the theory of cosmic strings (see [26] for a recent review). Like the inflationary universe, the cosmic string model is based on considering quantum field theories in the early universe. In models with a strongly coupled scalar field (thus no inflation) and with a non-simply connected vacuum manifold, topological defects arise by very general causality arguments [27]. These defects form seed masses about which galaxies and clusters of galaxies accrete (see [28] for some recent progress). The flatness and horizon puzzles are not addressed in this theory, but cosmic strings predict the correct correlation function of clusters [29]. Maybe the next cosmological models will combine ideas from inflation and from the cosmic string theory.

Acknowledgement

Thanks to Professors Doug Eardley and Bill Press for continued support and encouragement.

References

1. A. Guth, Phys. Rev. D23, 347 (1981).
2. A. Linde, Rep. Prog. Phys. 47, 925 (1984);
   R. Brandenberger, Rev. Mod. Phys. 57, 1 (1985).
3. A. Linde, Phys. Lett. 108B, 389 (1982);
   A. Albrecht and P. Steinhardt, Phys. Rev. Lett. 48, 1220 (1982);
   S. Hawking and I. Moss, Phys. Lett. 110B, 35 (1982).
4. A. Guth and S. Y. Pi, Phys. Rev. Lett. 49, 1110 (1982);
   S. Hawking, Phys. Lett. 115B, 295 (1982);
   A. Starobinsky, Phys. Lett. 117B, 175 (1982).
5. J. Bardeen, P. Steinhardt and M. Turner, Phys. Rev. D28, 679 (1983).
6. A. Linde, Phys. Lett. 129B, 177 (1983).
7. L. Dolan and R. Jackiw, Phys. Rev. D9, 3320 (1974);
   S. Weinberg, Phys. Rev. D9, 3357 (1974);
   C. Bernard, Phys. Rev. D9, 3313 (1974).
8. G. Mazenko, R. Wald and W. Unruh, Phys. Rev. D31, 273 (1985).
9. A. Albrecht and R. Brandenberger, Phys. Rev. D31, 1225 (1985).
10. A. Albrecht, R. Brandenberger and R Matzner, Phys. Rev. D32, 1280 (1985).
11. A. Vilenkin and L. Ford, Phys. Rev. D26, 1231 (1982);
    A. Linde, Phys. Lett. 116B, 335 (1982);
    S. Hawking and I. Moss, Nucl. Phys. B224, 180 (1983).
12. A. Linde, Phys. Lett. 132B, 137 (1983);
    A. Linde, JETP Lett. 37, 724 (1983);

D. Nanopoulos, K. Olive, M Srednicki and K. Tamvakis, Phys. Lett. 123B, 41 (1983) and ibid 124B, 171 (1983); B. Ovrut and P. Steinhardt, Phys. Lett. 133B, 161 (1983); R. Holman, P. Ramond and C. Ross, Phys. Lett. 137B, 343 (1984).

13. W. Press, Physica Scripta 21, 702 (1980); K. Sato, Mon. Not. R. Astron. Soc. 195, 467 (1981); V Lukash, JETP 52, 807 (1980); G. Chibisov and V. Mukhanov, JETP Lett. 33, 532 (1981).
14. E. Harrison, Phys. Rev. D1, 2726 (1970); Ya. Zel'dovich, Mon. Not. R. Astron. Soc. 160, 1p (1972).
15. R. Brandenberger, Nucl. Phys. B245, 328 (1984).
16. J. Halliwell and S. Hawking, Phys. Rev. D31, 1777 (1985).
17. W. Fischler, B. Ratra and L. Susskind, Nucl. Phys. B262, 159 (1985).
18. R. Brandenberger and R. Kahn, Phys. Rev. D29, 2172 (1984).
19. J. Bardeen, Phys. Rev. D22, 1882 (1980); R. Brandenberger, R. Kahn and W. Press, Phys. Rev. D28, 1809 (1983).
20. Q. Shafi and A. Vilenkin, Phys. Rev. Lett. 52, 691 (1984); S. Y. Pi, Phys. Rev. Lett. 52, 1725 (1984).
21. R. Kahn and R. Brandenberger, Phys. Lett. 141B, 317 (1984).
22. V. Belinsky, L. Grishchuk, I. Khalatnikov and Ya. Zel'dovich, Phys. Lett. 155B, 232 (1985); T. Piran and R. Williams, Phys. Lett. 163B, 331 (1985).
23. G. Gibbons, S. Hawking and J. Stewart, DAMTP preprint (1986).
24. S. Hawking, 'Quantum Cosmology', in Relativity, Groups and Topology II, Proceedings of the Les Houches Summer School 1983, edited by B. DeWitt and R. Stora (North Holland, 1984); J. Hartle, 'Quantum Cosmology', proceedings of the 1985 Theoretical Advanced Study Institute on Elementary Particle Physics, to be published.
25. J. Hartle and S. Hawking, Phys. Rev. D28, 2960 (1983).
26. A. Vilenkin, Phys. Rep. 121, 263 (1985).
27. T. Kibble, J. Phys. A9, 1387 (1976).
28. N. Turok and R. Brandenberger, Phys. Rev. D33, 2175 (1986).
29. N. Turok, Phys. Rev. Lett. 55, 1801 (1985).

The Age of the Universe in Inflationary Cosmology

H.-J. Blome

Institut f. Astrophysik der Universität Bonn*, Auf dem Hügel 71, D-5300 Bonn 1, F. R. Germany

The age of the oldest objects in the universe has obviously to be smaller than the age of the space-time arena itself.

Lower bounds for the age of the universe are provided by the age of the galaxy and the age of globular clusters. Progress in a better understanding of element synthesis by r-process has led to a new age of the galaxy: $20(+2,-5)*10^9$ [yr]whereas the globular cluster ages are in the range of $14*10^9$[yr]to $25*10^9$[yr].

There exist age-limit problems for the inflationary universe because this scenario predicts a density parameter $\Omega \sim 1$ which implies an age of $t_0 = 2/3 * H_0^{-1} \leq 13*10^9$ [yr] if $H_0 \geq 50$ [km s^{-1} Mpc^{-1}] assuming that the vacuum energy ρ_V associated with Higgs- and matter-fields does not gravitate.

Thus if the cosmological constant ($\Lambda \sim \rho_V$) vanishes, the age limits suggest that the universe must have low density ($\Omega \ll 1$) contradicting the inflationary flatness prediction, or the Hubble constant must be ≤ 50, opposed to present trends, or that we had a power-law inflation instead of an exponential inflation.

We analyze cosmological models with non-vanishing vacuum energy and arrive at Lemaître models with an age $\geq 17*10^9$[yr]and euclidean space structure which also may contain dark matter.

Introduction

The basic questions which cosmology tries to answer concern: the geometrical structure of space, whether time had a beginning, the origin and composition of matter and the formation and distribution of galaxies.

In homogeneous and isotropic Friedmann-Lemaître cosmology space structure, expansion and age of the universe are connected with the density of the universe. The geometrical properties of space depend on density and motion of the matter present ($\rho = \rho_B + \rho_{DM}$, ρ_B baryonic matter, ρ_{DM} dark matter, ρ_V vacuum matter) and is characterized by its gaussian curvature

$$C_G = \frac{k}{R^2} = \frac{8\pi G}{3c^2} (\rho + \rho_V - \rho_c) = \frac{8\pi G}{3c^2} (\rho + \rho_V) - \frac{1}{c^2} \left(\frac{\dot{R}}{R}\right)^2 \quad (1)$$

*Present address: Institut f. Theoretische Physik (SFB 301), Universität Köln

(Where the Hubble constant $H_0 = \dot{R}(t_0)/R(t_0)$ characterizes the present expansion rate and R(t) is the cosmic scale factor and/or curvature radius). If $\rho_0 + \rho_v < \rho_c = 3H_0^2/8\pi G$ the curvature is negative; $\rho_0 + \rho_v > \rho_c$ means positive curvature and $\rho_0 + \rho_v = \rho_c$ implies spatially flat universe.

The age of the Friedmann universes ($\rho_v = o$) can be calculated exactly in terms of H_0 and Ω_0 to give

$$t_0 = H_0^{-1} f(\Omega) \tag{2}$$

where f is an algebraic function with $f'(\Omega) < o$, $f(o) = 1$ and $f(1) = 2/3$. Luminous, baryonic matter is mainly concentrated in galaxies but there is observational evidence that large quantities of non-luminous material in and around galaxies are probably non-baryonic, see PEEBLES [1]. On the contrary BAHCALL and CASERTANO [2] give arguments for baryonic dark matter.

Inflationary models for the early universe ($t < 10^{-34}$[s]) e.g. TURNER [3], predict a spatially flat (k=o), zero curvature universe with $\Omega = 1$ in standard cosmology with $\rho_v = 0$. This implies an age of the universe of $t_0 = 2/3\ H_0^{-1}$. If the present expansion rate $H_0 = \dot{R}_0/R_0$ is greater than $H_0 = 50$ [km s^{-1} Mpc^{-1}], as observations indicate, we arrive at an age of the universe $t_0 \leq 13*10^9$ [yr]. This value is barely compatible with evolutionary ages of globular clusters of the order $(16 \pm 3)*10^9$ [yr]([14]) and an age of the galaxy $t_G = 20.8\ (+2,-5)*10^9$ [yr] [8], or $t_G = (11.6 - 24.6)*10^9$ [yr] as preferred by THIELEMANN [27].

Three different ways are possible to remove this contradiction:

1) $H_0 < 50$[km s^{-1} Mpc^{-1}], e.g. SHANKS [4],
2) power law inflation instead of exponential inflation, LUCCIN et.al. [5],
3) the energy of the final asymmetric vacuum, which may contain contributions from all stages of spontaneous symmetry breakings (SSB) is not zero: BLOME, PRIESTER, [6], [7]; KLAPDOR, GROTZ [8].

A constant term in the scalar potential $U(\phi)$ of the self coupled Higgs field ϕ left over after SSB is equivalent to a cosmological term in the Friedmann equation. Friedmann-Lemaître models with a non-zero cosmological constant would allow a value of $\Omega(t_0) \sim 0.2$ while still maintaining zero spatial curvature and increase the cosmic age to avoid the afore mentioned contradiction in face of $H_0 \geq 50$.

I Cosmological model

Einstein equations relate the time variation of R(t) to the matter density in the universe:

$$\left(\frac{\dot{R}}{R}\right) = \frac{8\pi G}{3}\,(\rho(t) + \rho_v) - \frac{kc^2}{R^2} \tag{3a}$$

$$\frac{\ddot{R}}{R} = -\frac{4\pi G}{3}\,(\rho(t) - 2\rho_v) \tag{3b}$$

Here $\rho(t)$ is the density of incoherent matter ($p = 0$) and $\rho_V = \varepsilon_V/c^2 = \Lambda c^2/8\pi G$ denotes the vacuum energy.

Modern developments in field theory suggest that there may be vacuum contributions to ρ and p which may obey an equation of state like: $P_V = -\rho_V\ c^2$ (ZELDOVICH [9] , GLINER [10]). They could come either from the broiling sea of quantum fluctuations connected with quantized matter fields or from self-interaction terms of the Higgsfield $U(\phi)$ which imply an additional contribution

$$T_{\mu\nu} - U(\phi)g_{\mu\nu} \tag{4}$$

to the energy momentum tensor of matter.

The observed expansion rate of the universe limits the total vacuum (energy) density of the present universe to be

$$\rho_V < 2.3 * 10^{-29} \quad [g \cdot cm^{-3}]$$

or $\Lambda \leq 5 * 10^{-56}$ [cm^{-2}](BLOME, PRIESTER [7]).

In standard cosmology (e.g. WEINBERG [10], or KUNDT [12]) one makes the a priori assumption that the quantum vacuum does not gravitate: $\rho_V = 0$.In this case the observable quantities of the Friedmann-models are at time $t=t_o$:

present expansion rate $\quad H_o = \dot{R}_o/R_o \quad$ (5)

deceleration parameter $\quad q_o = \dfrac{-\ddot{R}_o R}{\dot{R}_o^2} \quad$ (6)

density parameter $\quad \Omega_o = 2q_o = \rho_o/\rho_{cr} \quad$ (7)

From this we derive the curvature of 3-space

$$C_G = \frac{kc^2}{R_o^2} = H_o^{\ 2}\ (3/2\ \Omega_o - q_o - 1) = (\rho_o - \rho_c)\ \frac{8\pi G}{3} \tag{8}$$

A (singular) origin of this model must have occured at a past time which is less than a Hubble time if $\varepsilon + 3p \geq 0$ and $\dot{R}(t_o) > 0$, see Hawking and Ellis (1969)[26].

Inflation, i.e. a phase of exponential growth in the cosmic scale factor, arises in the so-called Inflationary Scenario in which standard Friedmann models are combined with Grand Unified Theories. In these models the dynamics of the very early universe may be controlled by the self-energy of the Higgs field associated with the separation of strong and electroweak forces during symmetry breaking, i.e.:

GU-symmetry $\rightarrow SU(3)_{color} \otimes [SU(2) \otimes U(1)]_{electroweak}$. This occurs when the cosmic temperature drops to $T_{GUT} \sim 10^{14}$ [GeV] at about 10^{-35} [s] after "Big Bang".

The scale factor blows up exponentially if $\rho_V(\phi)$ = const. dominates in (3a):

$$R(t) \sim \exp\left[\sqrt{\frac{8\pi G}{3}\rho_V}\, t\right] \qquad (9)$$

For the presently observable part of the universe to be in causal contact at some earlier time the universe must have inflated by a factor $F_{min} = H_{inf}/H_o$ $z_{inf} \sim 10^{25} - 10^{30}$ (where H is the Hubble constant and redshift z_{inf} corresponds either to Z_{GUT} or Z_{Planck}). Besides removing this horizon problem [3] the inflationary scenario provides monopole dilution and explains the flatness of the universe ($\Omega = 1$) in spite of the possibility that at $t < 10^{-43}$ [s] the curvature of space may have been arbitrarily large. Unfortunately, observations suggests that $\Omega \ll 1$. For inflation to yield $\Omega \sim 0.2$ the inflation factor would have just to be F_{min}, and the pre-inflation curvature radius just comparable to Hubble radius then, making the present radius of curvature just of order the Hubble radius. This demands a fine-tuning which one would prefer to avoid.

II Observational results and the age conflict

Density parameter

Observational determinations of Ω_o indicate that the luminous parts of spiral galaxies give only $\Omega_{olum} = (2\text{-}6)*10^{-3}$. On larger scales, those of binaries and small groups of galaxies, one finds $\Omega_o = 0.05$ to 0.15. Even on largest scales where determinations of Ω have been made one finds that Ω_o is probably not larger than 0.4. Estimates based on modelling the infall of the local group into the local supercluster, or on applying a cosmic virial theorem to pairs of galaxies or to cluster-galaxy pairs suggest $\Omega_o \sim 0.1 - 0.3$ [25]. Calculations of the primordial synthesis of light elements constrain the baryonic density: $0.01 \leq \Omega h^2 \leq 0.04$ [13], where $\Omega H^2 = H^2 \rho/\rho_c = (8\pi G/3)\rho$ and H = 100 h. We would like to stress that the corresponding values of $\rho_B = (0.5^{+0.7}_{-0.3})\ 10^{-30}$ [g $\cdot$ cm^{-3}] are really independent of H_o! This means that most of the mass density is composed of non-luminous (non-baryonic) dark matter [1]. To fulfil the inflationary prediction of $\Omega_o = 1$ one needs non-baryonic dark matter and/or $\rho_V = 0$, but cold matter is constrained to $\Omega_{oDM} \leq 0.3$ [D. SCHRAMM, this volume].

Hubble constant

The Hubble parameter $H = \dot{R}/R$ characterizes the rate of expansion of the universe at any epoch. The data for the present Hubble parameter H_o span the range from 50 ± 7 (SANDAGE, TAMMANN [14]) up to 100 ± 10 (de VAUCOULEURS, PETERS [15]). In the following table we summarize some recent results. For a complete review and critical discussion we refer to the book by M. ROWAN-ROBINSON [16].

Hubble parameter H_o [km s^{-1} Mpc^{-1}]	Authors	Expansion time $t_o = 2/3\ H_o^{-1}\ 10^9$[yr] (flat cosmos)
50 $\pm$ 7	(SANDAGE, TAMMANN [14])	13
67	(ROWAN-ROBINSON [16])	9.8
75	(BORGEST, REFSDAL [19])	8.5
74.3 $\pm$ 11	(VISVANATHAN [17])	8.8
100 $\pm$ 10	(de VAUCOLEURS [15])	6.5

Age of astronomical objects

Mainly two methods have been used to estimate the age of the universe from astronomical objects.

The first involves globular clusters. These clusters are presumed to be at least as old as galaxies because they are distributed nearly spherically around the galaxies. One generally assumes that galaxies formed quite early in the history $t_{form} \sim 10^9$[yr]. If so, then the age of globular clusters would be a good indicator for the age of the universe [G. TAMMANN, this volume]. The ages of the oldest globular clusters are in the range of $(14 - 25) * 10^9$[yr]. A second method of estimating the age of the galaxy is through nucleocosmochronology. In 1981 E.M.B. SYMBALISTY and D. SCHRAMM [20] concluded that the age of our galaxy lies in the range of $(8.7 - 15.8)*10^9$ years. Recent progress, applying the new results for the actinide chronometer production ratios, including ß-delayed fission and neutron emission in an r-process calculation, leads to an age of the galaxy of $t_G = (20.8^{+2}_{-5})*10^9$ [yr], when assuming an exponential model for galactic nucleosynthesis (THIELEMANN, METZINGER, KLAPDOR [21]).

Recently THIELEMANN [27] showed that the age of the Galaxy determined by nuclear cosmochronology was only constrained to lie in the range $(11.6 - 24.6)*10^9$[yr].

With $H_o = 50$ [km s^{-1} Mpc^{-1}] the lifetime of a flat standard Friedmann ($\Omega_o = 1$) - universe is only $13 * 10^9$ [yr] and with $H_o = 100$ [km s^{-1} Mpc^{-1}] it is only $6.5 * 10^9$ [yr]. Thus even with $H_o = 50$ the age of the universe is too short to accomodate old globular clusters and an age of the galaxy of $15.8 * 10^9$ [yr] [8]. Thus we have:

$$(t_G \geq 17 * 10^9 \text{ yr}) \wedge (\text{inflation: } \Omega = 1,\ k = 0) \wedge (H_o \geq 50) \rightarrow t_o < t_G.$$

As a consequence the $\Omega_o = 1$ Friedmann models are ruled out if H_o is greater than 50 [km s^{-1} Mpc^{-1}].

If the universe is dominated by relativistic particles then the age of an inflationary universe is even less than $0.5 * H_0^{-1}$. Only $H_0 < 21$ [km s^{-1} Mpc^{-1}] could solve the contradiction in this case.

Only in a low-density universe, say $\Omega_0 = 0.12$, is it possible to reconcile the age of the universe with ages of the galaxy $t_G \sim 17 * 10^9$ [yr] with ($H_0 = 50$).

In any case, the globular cluster and cosmochronology ages are marginal consistent with an $H_0 = 50$ [km s^{-1} Mpc^{-1}], $\Omega_0 = 1$ standard cosmology, and are certainly incompatible with an $H_0 = 100$[km s^{-1} Mpc^{-1}]Friedmann universe.

Figure 1a shows the relation between density parameter Ω, Hubble parameter $h = H_0/100$ and the age of the universe.

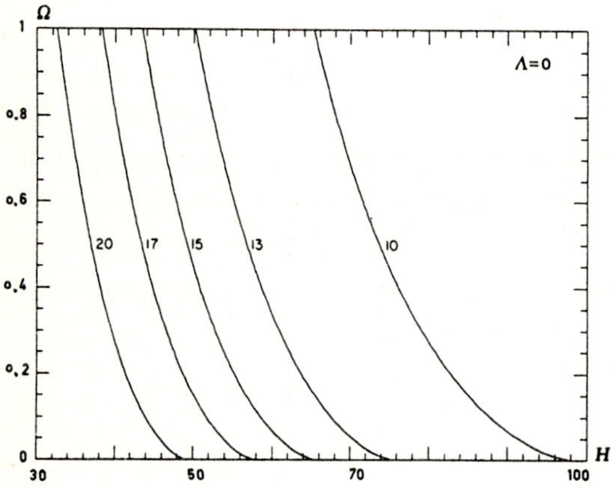

Fig. 1a:

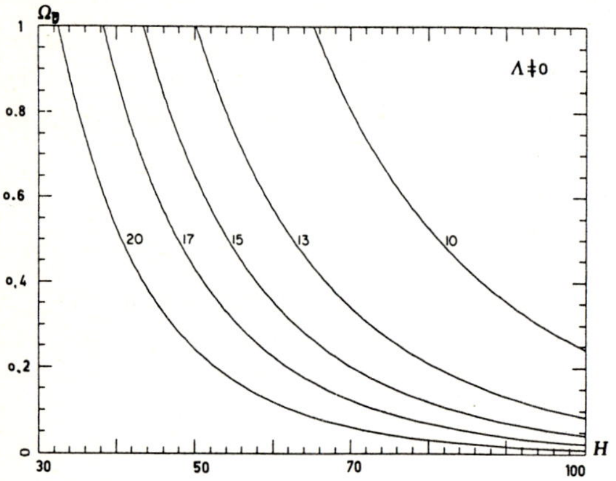

Fig. 1b:

Figure 1:

Relation between Hubble-parameter H_0 and the density parameter Ω_0 or Ω_{B0} (only baryonic contribution) for different ages of the world (in the units of 10^9 years): a) $\Lambda = 0$, b) $\Lambda \neq 0$ and euclidean metric.

Solution of the age conflict

One way the inflationary prediction could be reconciled with observation ($\Omega_0 \ll 1$) is by the existence of an unclustered sea of non-baryonic weakly interacting particles, neutrinos or cold dark matter (see D. SCHRAMM, this volume). But this doesn't resolve the age problem, on the contrary, it shortens the cosmic age!

Another escape route which has been suggested (BLOME, PRIESTER [7]; PEEBLES [20]; KLAPDOR, GROTZ [8]) as a way of rescuing inflation in the face of observation that $\Omega \leq 0.4$ and to avoid the age conflict is the possibility that the cosmological constant associated with the final asymmetric vacuum state after SSB is non-zero:

$$\Lambda = \frac{8\pi G}{c^2} \rho_V(\phi) \leq 5 * 10^{-56} \; [\mathrm{cm}^{-2}]$$

Moreover there may be contributions from all possible quantum fields to vacuum energy. We know for example from quantum electrodynamics that the lowest state of the field configuration (vacuum) consists of particle-antiparticle pairs in a constant state of creation and destruction which has observable effects: Delbrück-scattering, Lamb shift, Casimir effect. Thus it can't be excluded that zero-point energy contributes to ρ_V and acts as a source for gravity. If we allow for $\rho_V \neq 0$ then we arrive at Friedmann-Lemaître models, which lead to the following equations, if inflation is supposed to hold:

$$C_G = \frac{8\pi G}{3c^2} (\rho_0 + \rho_V - \rho_c) = 0 \leftrightarrow \rho_V = \rho_c - \rho_0 \tag{10}$$

$$\Omega(t_0) = 1 - \frac{\rho_V}{\rho_c} \tag{11}$$

$$q(t_0) = \frac{1}{2}\Omega_0 - \frac{\rho_V}{\rho_c} \tag{12}$$

$$t = H_0^{-1} f(\rho_0, \rho_V, \rho_c) \tag{13}$$

During inflation the universe is driven to euclidean structure with a flatness condition which constrains the sum of ordinary energy density and vacuum energy. If $\rho_V = 0$, equation (11) of course implies $\Omega_0 = 1$, but $\rho_V \neq 0$ can compensate for $\Omega_0 < 1$.

Figure 1b displays the connection between Ω_0, t_0 and H_0 if $\rho_V > 0$ for a universe with euclidean structure.

Discussion

Models with $\rho_V \neq 0$ allow a value of $\Omega_0 = (\rho_B(t_0) + \rho_{DM}(t_0)) / \rho_c < 1$ while still maintaining zero spatial curvature. Figure 1b shows the corresponding values of cosmological density parameter and Hubble-parameter H_o for various values of cosmic age

$$t_0 = 1.41 * 10^9 \; \rho_V^{-1/2} * \mathrm{ArCosh}\left(1 + \frac{2\rho_V}{\rho_0}\right) \qquad [\mathrm{yr}] \qquad (14)$$

with ρ_V in units of 10^{-30}[g · cm^{-3}].

It can be seen that a value $\rho_o = 0.4 * 10^{-30}$[g · cm^{-3}] can be combined with a vacuum density $\rho_V = 2.8 * 10^{-30}$ [g/cm^3] to yield an age of $17 * 10^9$ [yr] for $H_o = 50$ [km s^{-1} Mpc^{-1}]. The time dependence of the scale factor in the epoch of non-relativistic matter domination with k = 0 is given by

$$\frac{R(t)}{R(t_0)} = \sqrt[3]{\frac{\rho_0}{2\rho_V}\left(\mathrm{Cosh}\,\frac{t}{\tau} - 1\right)} \qquad (15)$$

where ρ_V denotes vacuum density and $\rho_o = \rho_B + \rho_{DM}$ and $\tau = \sqrt{(24\pi G\rho_V)}^{-1}$

Figures 2 show the scale factor versus cosmic time t for models with $\rho_V \neq 0$, k = ± 1,0 and for present matter density $\rho_o = 0.5 * 10^{-30}$[g/cm^3] with $H_o = 75$ and $H_o = 50$. The age of our galaxy from Th/U is marked in a lower right corner.

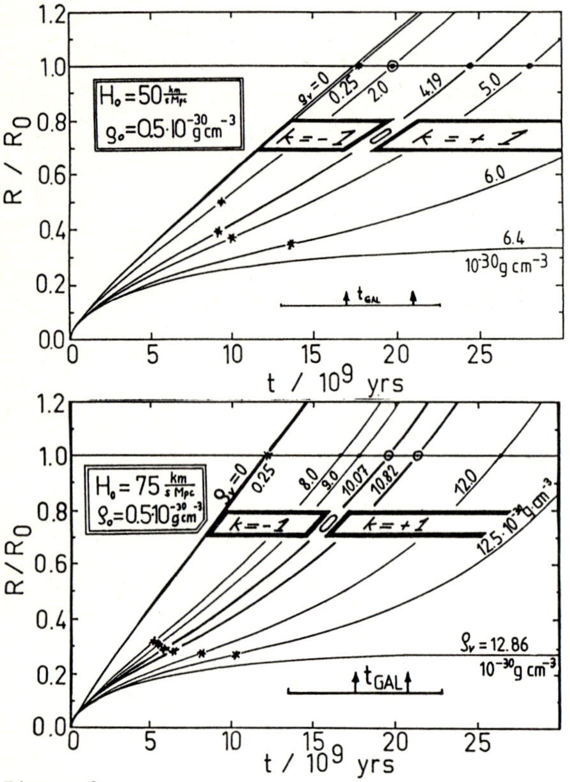

Figure 2:

Cosmic scale factor R/R_o versus time of Friedmann-Lemaître models for two different Hubble-parameters: $H_o = 50$ and $H_o = 75$.

Pure baryonic models are not consistent with present upper limits on the fine-scale anisotropy of background radiation even if recourse is made to a non-vanishing vacuum density [23]. But consistency is found in a model with baryonic- and cold dark-matter ($\rho_{CDM} \geq \rho_B$) and $\rho_V \neq 0$. In this case inflation implies a density parameter:

$$\Omega_0 = \frac{\rho_B(t_0) + \rho_{CDM}(t_0) + \rho_V}{\rho_c} = 1 \qquad (16)$$

In Figures 3 (a: H_0=75[km s^{-1} Mpc^{-1}];b: H_0=50[km s^{-1} Mpc^{-1}])the relation between t_0 and ρ_0 is displayed with ρ_V as parameter.

The allowed region (dotted) follows from $\rho_0 \geq 0.5 * 10^{-30}$ [g cm^{-3}] $t_0 \geq 17 * 10^9$ [yr] and $q_0 > -1.3$ [23].

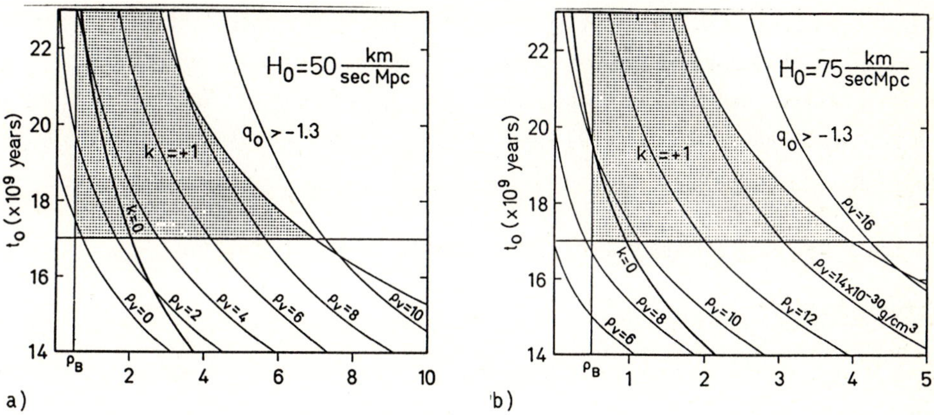

Figure 3(from [8]):
Relation between t_0 and ρ_0 (= ρ_B + ρ_{DM}) and vacuum density $\rho_V = \varepsilon_V/c^2$ (both densities in units of 10^{-30} g/cm^3). In Fig. 3a, H_0 = 50 and in Fig. 3b, H_0 = 75 [km $s^{-1}Mpc^{-1}$]. The allowed region is dotted (see text).

Friedmann models (ρ_V = 0) with euclidean structure are excluded for $H_0 > 50$ [km s^{-1} Mpc^{-1}]. Even for $t_0 > 14 * 10^9$ [yr] one still needs $\rho_V \geq 10^{-30}$ [g cm^{-3}]. In the case H_0 = 75 [km s^{-1} Mpc^{-1}] only models with $\rho_V > 0$ are allowed. A solution of Friedmann equations with ρ_V = 0 is compatible with the lower limit of the cosmic age of $14 * 10^9$ [yr] only if $H_0 < 70$. If we require $\Omega = 1$, we need a contribution of dark, non-baryonic matter: $\rho_{DM}(t_0) \sim 4 * 10^{-30}$ [g cm^{-3}] and would obtain a cosmic age of $13 * 10^9$ years.

Because the inflationary scenario does not make a definite prediction about the sign of the gaussian curvature it may be that we live in a nearly flat hyperbolic or spherical universe. For this reason we calculated Friedmann-Lemaître models with k = ± 1, too. One example is a model with $\Omega = (\rho_0 + \rho_V)/\rho_c = 1.07$. It represents a closed, ever expanding model with

$\rho_V = 10.81 * 10^{-30}$ [g cm^{-3}] and an age $t_0 = 21.4 * 10^9$ years. This example was selected since it yields $q_0 = -1.0$.

Conclusion

Inflation predicts a cosmos with nearly euclidean space-structure. Big Bang nucleosynthesis requires $\Omega(t_0) \approx 0.02$, if $H_0 = 50$ and $\rho_B = 0.5 * 10^{-30}$[g cm^{-3}], but a possible amount of cold dark matter is restricted to $\Omega_{DM} \leq 0.3$. Observations yield a Hubble constant $H_0 \gtrsim 50$, implying an age $t_0 \leq 13 * 10^9$[yr] in a standard ($\rho_V = 0$) model. This age is less than the age of the Galaxy given before.

The solution to this obvious conflict is to take into account the vacuum energy associated with Higgs "matter" and/or zero-point energy of matter fields. The resulting Friedmann-Lemaître cosmos with a (nearly) euclidean metric solves the age conflict and admits dark matter. For $H_0 \gtrsim 50$ km s^{-1} Mpc^{-1}, $k = 0$ and $t_0 > 17 * 10^9$[yr] one gets (Fig. 3): $\rho_{Bo} \leq \rho_0 \leq 4\rho_B$ allowing an amount of dark matter $\rho_{DM} \leq 3\rho_B$.

I thank Wolfgang Priester and Thomas Schmutzler for discussion.

References

[1] P.J. Peebles: Nature 321, 27 (1986)
[2] J. Bahcall, S. Casertano: Astrophys. J. 293, L7 (1985)
[3] M. Turner: Proceed. of. Cargése Summer School on Fundam. interactions and Cosmology 1984 (ed.: J. Audouze, J. Tran Thank Van)
[4] T. Shanks: Vistas in Astron 28, 595 (1985)
[5] F. Luccin, S. Matarrese, N. Vittorio: Astron.Astrophys. 162, 13 (1986)
[6] H.J. Blome, W. Priester: Astrophys.Space Sc. 117, 327 (1985)
[7] H.J. Blome, W. Priester: Naturwissenschaften 71, 528 (1984)
[8] H.V. Klapdor, K. Grotz: Astrophys. J. 301, L39 (1986)
[9] Y. Zeldovich: Sov. Phys. Usp. 24, 216 (1981)
[10] E.B. Gliner: Sov. physics Doklady 15, 559 (1970)
[11] S. Weinberg: Gravitation and Cosmology, New York 1972
[12] W. Kundt: Springer Tracts Mod. Phys. 58, 1 (1971)
[13] J. Yang, M.S. Turner, G. Steigman, D.N. Schramm, K.A. Olive: Astrophys. J. 281, 493 (1984)
[14] A. Sandage, G.A. Tammann: Proceedings of I.ESO-CERN Symposium, 127 (1984)
[15] G. de Vaucoleurs, W.L. Peters: Astrophys. J. 287, 1 (1984)
[16] M. Rowan-Robinson: The cosmological distance ladder, New York 1985
[17] N. Visvanathan: Astrophys. J. 275, 430 (1983)

[18] U. Borgeest, S. Refsdal: Astron. Astrophys. 141, 318 (1984)
[19] E.M.B. Symbalisty, D. Schramm: Rep.Prog.Phys. 44, 293 (1981)
[20] P.J.E. Peebles: Astrophys. J. 284, 439 (1984)
[21] F.K. Thielemann, J. Metzinger, H.V. Klapdor: Z. Phys. A309, 301 (1983)
[22] M. Turner, G. Steigman, L.M. Krauss: Phys.Rev.Lett. 52, 2090 (1984)
[23] N. Vittorio, J. Silk: Astrophys. J. 297, L1 (1985)
[24] G.A. Tammann: Landolt-Börnstein VI/2c, Berlin (1982)
[25] M. Davis, P.J.E. Peebles: Astrophys. J. 267, 465 (1983)
[26] S. Hawking, G.F.R. Ellis: Astrophys. J. 152, 25 (1968)
[27] F. Thielemann: NATO Workshop; Kona, Hawaii (1986).

Constraints on the Age of the Universe from Globular Clusters and the Cosmic Expansion Rate

G.A. Tammann

Astronomical Institute of the University of Basel, Venusstraße 7, CH-4102 Binningen, Switzerland and European Southern Observatory, D-8046 Garching, F. R. Germany

I. Introduction

It is appropriate here to note the 60th birthday Allan Sandage has celebrated last month and who - in 1968 - first pointed out in a paper entitled "The Time Scale of Creation" [1] the amazing agreement of three totally independent time scales, i.e. the age of the oldest stars, the age of the radioactive elements, and the expansion age of the Universe. During the last almost 20 years these ages have been revised numerically, but the agreement persists.

The following pages deal with the oldest stars and the expansion age only, while reference is made to nucleosynthetic dating to which our host at this Symposium, Hans-Volker Klapdor, has contributed so significantly.

Clearly, the radioactive clock started at the epoch when the ejecta of the first dying, short-lived stars in our Galaxy begun to contaminate the interstellar medium. On the other hand, the oldest stars so far known in our Galaxy all contain <u>some</u> "metals"[1)], which must have been built up in a hypothetical first generation of stars (sometimes called "Population III"). Since the metal enrichment proceeded very fast during the earliest phases of our Galaxy, as seen below, the age of the known oldest stars refers also to the time when the first Galactic stars had died. Radioactive elements and oldest stars hence reflect nearly the same epoch, which is only slightly antedated by the formation of the very first stars and of our Galaxy. Circumstantial evidence suggests that our Galaxy was formed during a period of preferred galaxy formation, and then the age of the Universe is obtained by adding the estimated, relatively short interval between the Big Bang and the epoch of preferred galaxy formation to the age of the Galaxy. The global age thus obtained is to be compared with the expansion age of the Universe as derived from the present value of the Hubble constant H_o and of the deceleration parameter q_o.

II. The Age of the Globular Clusters

The search for the oldest objects in our Galaxy is guided by three selection criteria:

- Since massive stars evolve rapidly, the oldest stars have necessarily low mass ($\lesssim$ 1 solar mass).
- The earliest Galactic stars were formed before the Galaxy collapsed into a smoothly rotating disc; the oldest stars therefore do not

1) In astrophysics the term "metals" refers to all elements heavier than ^{4}He.

partake of the Galactic rotation and exhibit large motions out of the Galactic disk.

- Since all metals have been built up by stellar nucleosynthesis ("astration"), low metal content is an indicator of high age.

These criteria all lead to the globular clusters in the Galactic halo (i.e. outside the disk) as candidates of high age. The fact that these candidates are clusters is fortunate, because their multitude of member stars - which are equidistant and presumably coeval and chemically homogeneous - render them particularly suited to observe the effects of stellar evolution and hence to determine their ages.

The theory of stellar evolution determines the locus in the luminosity(L)-effective temperature(T_{eff})-plane of stars which convert H into He in their centres. The locus is a narrow strip - the so-called main sequence - along which individual stars are positioned according to their mass, the highest masses at the highest luminosities and highest temperatures. As stars exhaust their central H supply - and this happens first for the overluminous, most massive stars - they move away from the main sequence to become red giants. For a star cluster the degree of main-sequence depletion, as quantified by the position of the so-called turn-off point, is therefore a measure of age. Stellar evolution theory of low-mass stars, which are relevant for globular clusters, is sufficiently advanced mainly through the work of Ciardullo and Demarque [2], VandenBerg [3], VandenBerg and Bell [4] and others, to date a stellar ensemble from its turn-off point (TO) position within the L-T_{eff} plane to within better than a factor of 1.3, provided the initial chemical composition Y and Z are reasonably known[2)].

Through observations globular cluster stars can be placed into an apparent magnitude(m_V)-colour(B-V)-plane (cf.Fig.1)[3)]. If the cluster distance is known, the apparent magnitudes m_V can be converted into absolute magnitudes M_V. The latter still refer to a special spectral band (V) and must be transferred into absolute bolometric magnitudes M_{bol} by means of empirical bolometric corrections, in order to obtain finally luminosities L. Moreover, model atmospheres are needed to substitute the observed colours into T_{eff}. While these transformations alone are a difficult task, the observational problems in general can hardly be overemphasized. The photometry in highly crowded star fields must be carried down to magnitude ~20 in order to reach the main-sequence. A.Sandage has contributed here ground breaking work since the early 1950's. The advent of new techniques, like image processing of photographic plates and charge-coupled devices (CCD), will eventually improve on the available data.

The TO position in the M_V-(B-V) diagramme (cf.Fig.1) can in principle be used in three different ways to derive ages. (1) The TO luminosity L_{TO}, derived from M_V(TO), can be compared with stellar models. (2) The effective temperature of the TO, T_{eff}(TO), can be fitted to stellar models. But this method is not only sensitive to the transformation of $(B-V)_{TO}$ into T_{eff}(TO), but also on the correction for interstellar reddening, to the assumed metallicity, and to the mixing-length parameter α in the stellar interior, and will therefore not be considered further. (3) Theoretical evolutionary tracks off the main sequence in the L-T_{eff} plane can be used to combine points of equal

---

2) Y is the fractional mass of He, and Z is the fractional mass of "metals".

3) The visual and blue apparent magnitudes, m_V and m_B, are defined (mainly for technical reasons) to lie at particular wave lengths and to have specified passbands. $(B-V) = m_B - m_V$.

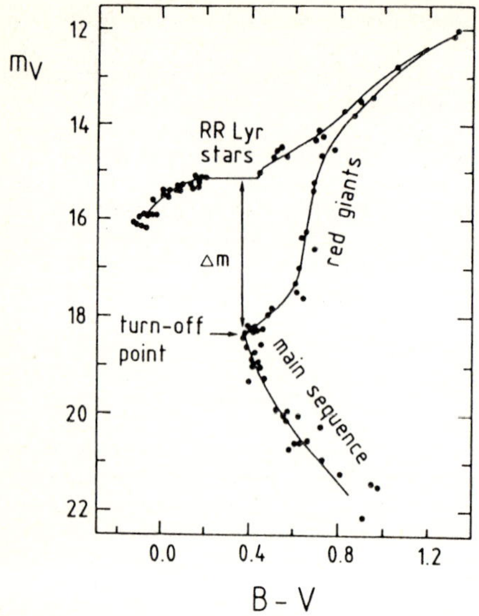

Fig.1. The magnitude-colour diagramme of the globular cluster M92 [5]. The significance of the turn-off point and of Δm is explained in the text. The individual (coeval) stars above the turn-off point do not trace their evolutionary paths, but the isochrones of the cluster.

age, but of different mass to yield isochrones in function of Y, Z, and in addition of α. To the extent that it is possible to translate these isochrones into the M_V-(B-V) diagramme they can be directly fitted to the observed magnitude-colour diagramme of a globular cluster.

Methods (1) and (3) depend both - but in different ways - on the adopted distances, the interstellar reddening, the chemical composition Y and Z, the bolometric correction and the mixing length. Weighing these various soures of error leads to a slight advantage of method (1) [7].

For method (1) the most crucial parameter is the distance. A 5% error in distance corresponds to an age error of 10%. Distances of globular clusters can be obtained either by fitting the cluster main sequence to a fiducial main sequence or through an adopted absolute magnitude of the RR Lyrae variables. The former route requires high-accuracy photometry at very faint levels; moreover, the position of the fiducial main sequence depends on the appropriate chemical composition. RR Lyr stars have the advantage to be much brighter (cf.Fig.1), but their mean absolute magnitude is not well known (not better than within 0.15 mag. corresponding to 7% in distance) and the magnitude requires a poorly understood empirical correction for metallicity. - Several authors have published TO ages for individual clusters during the last years. Mean ages of several clusters have been determined by Carney [8], Sandage [9], and Iben and Renzini [10]. The perfect agreement of their results, i.e. 17, 17 and 16 gigayears (Gyr), respectively, is somewhat fortuitous, because a different combination of the various input parameters used by these authors could alter the result by ~30%.

The isochrone fitting of method (2) yields in principle simultaneously - through the translation and deformation of the isochrones - distance, reddening, age, Y, Z, and α of a cluster. Of course, some of these parameters can be taken from external evidence. Sandage [11] obtained by using the RR Lyr star distances and only the very best observed stars near the TO point an excellent fit with the isochrones of VandenBerg for M15 and M92 (cf.Fig.2), which corresponds to an age of 18±2 Gyr of these two clusters. For about 10 well

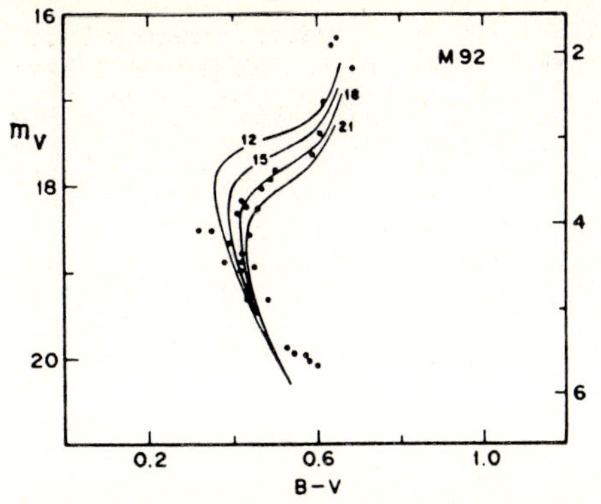

Fig.2. The magnitude-colour diagramme of M92 in the region of the turn-off point. Only the very best observed stars are shown. The isochrones of VandenBerg for Y=0.2, Z=0.0001, α=1.5, and an interstellar reddening of E_{B-V}=0.04 are labelled in Gyr. From Sandage [9].

observed clusters VandenBerg [3] has adopted distances and reddening values from the literature and determined the remaining parameters; his ages range from 15 to 18 Gyr.

The conclusion of these results is that the age of the globular clusters, obtained in two different ways, lie between 15 and 18 Gyr. Following [12] a mean age of 16±3 Gyr is adopted here.

As stated before the age of the globular clusters is clearly a lower limit of the age of the Galaxy, but the significance of their age for the Galactic history and cosmology is greatly enhanced by the following observations.

1. The age of the globular clusters is amazingly uniform [3, 9]. From the constancy of the time-dependent, but distance-independent magnitude difference Δm between the RR Lyr stars on the horizontal giant branch and the TO (cf.Fig.1) - after correction to constant metallicity - Sandage has concluded that the ages of the well observed clusters agree to within 10% [9]. These clusters were therefore formed during a special epoch of the Galaxy, presumably during its early collaps [13].

2. No field stars older than the globular clusters are known. Stars with good parallaxes, which are candidates for high age due to their large space velocities, find their lower bound in the M_V-(B-V) diagramme by the giant branches of the globular cluster 47 Tuc if metal-poor, or even by the younger open cluster NGC 188 if somewhat metal-richer [14].

3. The considerable range in metallicity of the globular clusters does not require a prolonged Galactic prehistory of nucleosynthesis, because the enrichment in metals was very fast during the collaps phase of the Galaxy [14] (cf.Fig.3).

4. The near agreement between the onset of nucleosynthesis and the age of the globular clusters, as shown in Fig.3, calls for a comparison between the mean cluster age of 16±3 Gyr and of nucleo-chronometers. The radioactive pairs $^{232}Th/^{238}U$ and $^{187}Re/^{187}Os$ yield model-independent (i.e. independent of the history of the Galactic production rate) lower and upper limits to the time since the beginning of nucleosynthesis of 9 and 28 Gyr, respectively [15]. If a specific model of the Galactic production rate is adopted, then the β-decay of the actinides gives 17-24 Gyr [16] or 14-22 Gyr [17], and the Re/Os clock 9-19 Gyr [18, 15]. The firm conclusion from this is that the ages of the globular clusters and of the radioactive elements agree to within a factor of $\lesssim$ 2, and they may well be identical.

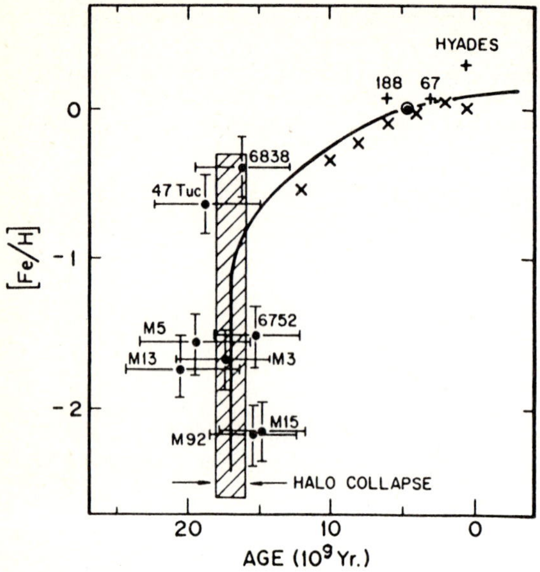

Fig.3. The metal content of various globular and open clusters plotted against their respective ages. The parameter [Fe/H] is the logarithm of the Fe-to-H ratio minus the logarithm of the solar Fe-to-H ratio. From Sandage [14].

5. Also globular clusters outside the Galaxy share the age of their Galactic counterparts. Particularly the Large Magellanic Cloud (LMC), a galaxy of moderate size whose gas richness and stellar population make it appear to be "young", contain - besides many relatively young clusters - very old globulars. For instance the age of the LMC cluster NGC2257 is given as 14±2 Gyr [19], and two other LMC clusters are found to have the same age as the Galactic globular M92 [20]. On the other hand the oldest stars in the "old looking", gas-free dwarf elliptical galaxies Ursa Minor and Draco correspond to M92 and to 16±2 Gyr, respectively [21,22]. RR Lyr stars are known in several other nearby galaxies; their presence alone sets a lower age limit of 10 Gyr [23].

6. The fact that sufficiently close galaxies have a common age gains additional weight from the evidence for a cosmic epoch in which the galaxy formation rate had a maximum. This evidence comes from the constancy of the colours of comparable galaxies. Comparable galaxies are for instance first-ranked galaxies in clusters of galaxies. The first-ranked galaxies within a fixed redshift interval have surprisingly similar colours (Fig.4). While galaxies evolve they become redder; the exact reddening rate is not known, but a possible range can be specified. The corresponding colour distributions are calculated for three different values of the colour decay under the assumption that first-ranked cluster galaxies formed continuously. The skewed distributions in Fig.4 contrast markedly with the observed colours. The conclusion is that first-ranked galaxies formed more or less simultaneously, presumably also 16±3 Gyr ago when the nearby galaxies began to shine [24].

7. The Universe is older than its oldest galaxies. The gestation time of galaxies since the Big Bang could be estimated if the process of galaxy formation were understood. Current theories have difficulties to form galaxies fast. On the other hand the most luminous galaxies, i.e. quasars, are observed backward in space and time to redshifts of $z \lesssim 4$. The time between $z=\infty$ (Big Bang) and $z=4$ depends on the Hubble constant H_o and the poorly known deceleration parameter q_o (see below); the best compromise value is $\Delta t=2.3$ Gyr [25]. This value is an upper limit if galaxies formed at $z > 4$. For convenience we adopt here

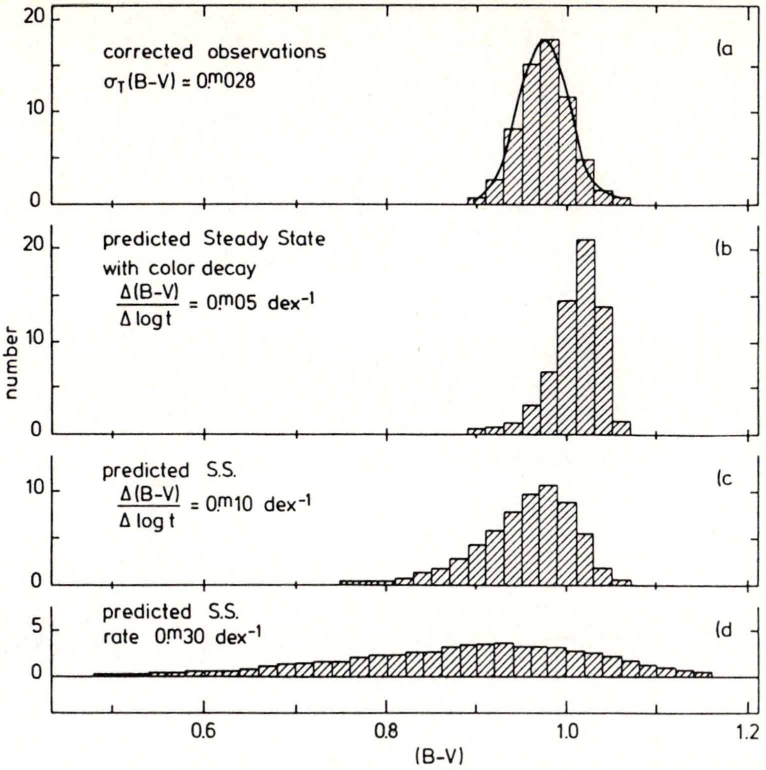

Fig.4. The observed colour distribution of first-ranked cluster galaxies (upper panel). Note the narrow range in (B-V)-colours. The lower panels show the expected colour distribution, if galaxies were formed with constant rate (steady-state situation). Three different cases of colour evolution are assumed. Considering that the assumed colour evolution is improbably small in panel b), first-ranked cluster galaxies must have had a preferred formation epoch in the past. From Sandage [24].

$\Delta t=2$ Gyr and obtain for the age of the Universe $T_o=(16\pm3)+(2\pm1)=18\pm3$ Gyr.

A completely independent dating method will eventually become available from the dynamics of the Local Group of galaxies. The way has been shown by Lynden-Bell [26], who derived on the basis of admittedly unsatisfactory data a dynamical age of 16 Gyr.

III. The Expansion Time Scale

The tangentially linear expansion of the Universe is measured by the present value of the (time-dependent) Hubble constant $H_o \equiv \dot{R}_o/R_o$, where R_o is the radius of curvature. H_o has the dimension of [1/time]; it is normally expressed, however, in km per sec per Megaparsec. The quantity $1/H_o$ is a time, and it gives in fact the expansion time T_o since the Big Bang, provided the expansion proceeds undecelerated (Fig.5). Any realistic value of T_o must take into account, however, the decelerating effect of any matter in the Universe. In all non-

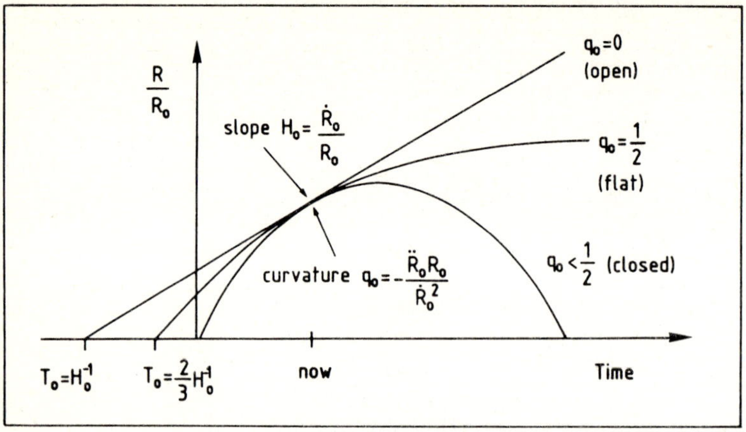

Fig.5. The time dependence of the radius of curvature R (in arbitrary units of its present value R_o) in a Friedmann-type Universe for three different values of the deceleration parameter q_o. The Friedmann time T_o since the Big Bang decreases with increasing deceleration q_o.

empty Friedmann models (which are characterized by a cosmological constant of $\Lambda=0$) T_o is <u>smaller</u> than $1/H_o$. The function $T_o=f(H_o, q_o)$ - where the present value of the deceleration parameter q_o is defined by $\ddot{R}_oR_o/\dot{R}_o^2$ - has first been given and tabulated by Sandage [27].

All attempts so far to measure directly the deceleration q_o have failed [cf.28]. The way out is to measure the present mean mass density ρ_o in volumes large enough to be representative for a homogeneous and isotropic Universe. One disadvantage of the remedial measure is that any undetectable matter/energy remains unaccounted for. The density ρ'_o is generally expressed by the density parameter $\Omega_o = \rho_o/\rho_{o\ crit}$, where $\rho_{o\ crit}$ corresponds to a flat (Euclidean) Universe with $q_o=0.5$. While the ensuing relation $\Omega_o=2\ q_o$ (for $\Lambda=0$!) is trivial, the distinction between q_o and Ω_o is helpful to remember that the deceleration itself has not been measured.

Three independent arguments (the decelerating effect of galaxy clumpings on the expansion in their neighbourhood; the gravitational binding of pairs and clusters of galaxies; the primordial nucleosynthesis of the light elements) give very consistently a value of $\Omega_o=0.1$ [28 and references therein]. Only 10% of this matter is luminous; the remaining 90% are required to account for local deviations from an ideal Hubble expansion field and to bind clusters of galaxies. Some of the dark matter is also "seen" in the flat rotation curves of spiral galaxies. Hence, the dark matter has been detected only because of its association with luminous matter. From the primordial-nucleosynthesis argument it seems that all matter yielding $\Omega_o=0.1$ is baryonic. It can be hidden in a wide range of objects (e.g. pebbles, Jupiters, white dwarfs, black holes). With $\Omega_o=0.1$ the Universe is open and expanding forever.

The determination of the Hubble constant follows from $H_o \equiv \dot{R}_o/R_o = v/r$, where v is the recession velocity and r the distance of a galaxy. The recession velocity v is easily measured as the redshift z of a galaxian spectrum ($v \approx cz$). Good distances r can be obtained from Cepheid variables. The period of these pulsationally unstable stars is a function of mass and hence of luminosity; the calibration of the semi-theoretical relation is provided by Galactic Cepheids with known distances. Unfortunately Cepheids reach at present only to distances (~3 Mpc), where the recession velocities are still influenced by non-Hubble motions induced by density fluctuations. Secondary and tertiary distance indicators must therefore be called upon to determine distances out to recession velocities of 3000-5000 km s^{-1}. At these velocities peculiar motions amount to $\lesssim$ 15%, because the largest peculiar motion so far reliably measured – except for the high virial velocities in clusters of galaxies - is the absolute (three-dimensional) motion of ~ 600 km s^{-1} of the Galaxy with respect to the microwave background.

Brightest blue and red stars in a dozen of spiral galaxies have proven to be good secondary distance indicators [29]. In two of these spirals supernovae of type I have been observed, and their absolute magnitude at maximum light, M_B(max), can thus be calibrated [30]; this calibration finds additional support from supernova models and from historical data of supernovae in our Galaxy [31 and references therein]. Because supernovae of type I are very good standard candles [cf.31], i.e. M_B(max) shows little intrinsic scatter, the calibration can be applied to distant supernovae with v < 10 000 km s^{-1}. The result is H_o=43(+10, -7) [km $s^{-1}$$Mpc^{-1}$] [31].

An independent route is to rely on the empirical relation between maximum rotation velocity of spiral galaxies, their mass and hence their luminosity. The maximum rotation velocity can easily be observed from the width of the 21cm hydrogen line. The line width-luminosity relation can be calibrated using spirals with known Cepheid distances. The relation has considerable intrinsic scatter and is therefore vulnerable to selection bias (see below). If this statistical problem is circumvented one arrives at a Virgo cluster distance of 21.6 Mpc and a Hubble value of 50±9 [25, 32, 33].

Relations like the 21cm line width-luminosity relation are beset by statistical difficulties. Their finite intrinsic scatter leads to a systematic error in the sense that the distance of distant galaxies is underestimated and H_o overestimated. This so-called Malmquist bias is caused by existing galaxy catalogues which are necessarily limited by apparent magnitude. Distant galaxies enter the catalogues only because they are overluminous; their luminosity surpasses easily the mean relation, calibrated locally, by 3σ [34]. It is mainly the neglect of this bias that Hubble values as high as 90-100 are still found in the literature [35, 36].

A completely independent route to H_o is to compare the luminosity distribution of Galactic globular clusters with that of the globulars surrounding M87, an elliptical giant galaxy in the Virgo cluster. The result is 19.3 Mpc [37], and since the Virgo cluster is well tied into the general expansion field [33] it follows H_o=56±10.

A value of H_o=50 is also in good agreement with all circumstantial evidence. To give only one example: If H_o were 100, our Galaxy and our neighbour, the Andromeda galaxy, would both have the largest diameters among several thousand surrounding galaxies; this paradox is avoided only if H_o < 70 [38].

The conclusion is that the convenient value of H_o=50±7 is in excellent agreement with all presently available data. The reciprocal is H_o^{-1}=19.6(+3.1, -2.4) Gyr. With the reduction factor of 0.898, corresponding to Ω_o=0.1 [27], one obtains T_o=17.6 Gyr and a 3σ range of 11.1 < T_o < 25.9 Gyr. (With H_o=100 and Ω_o=0.1 T_o would be reduced to 8.8 Gyr!). The agreement of the time scale of the expansion with that required by the age of globular clusters and of the radioactive elements is most remarkable indeed.

If much matter remained still undetected in intergalactic space, say Ω_o=0.6 and hence T_o=14.3 Gyr, the time scale agreement would be marginally destroyed. The situation could be remedied by abandoning a Friedmann model and by introducing the repulsive force of a positive cosmological constant Λ. If $\Lambda c^2/3H_o^2$=1.3 (a dimensionless number), then T_o would still be as high as 19.5 Gyr with H_o=50 [39, p.180]. This model has the pleasant feature to be finite and closed (k=1), but still to expand forever [28]. Yet the density parameter would be time-dependent (as in open Friedmann models) and it would remain unexplained why Ω was arbitrarily close to 1 at the very early epochs of the Universe.

The last-mentioned question has led to the classical prejudice of $\Omega_o \equiv 1$, which is the only time-independent value of Ω. This high value has gained momentum through the scenario of an Inflationary Universe and it would help - if due to cold matter - the theory of galaxy formation. From the previous discussion it is clear that in this case ~90% of all matter were yet undetected, dark, non-baryonic, and not clustered like galaxies. With Ω_o=1 and H_o=50 the Friedmann time is 13.1 Gyr, which is disfavoured by the Galactic evidence at almost the 2σ level. A cosmological constant of $\Lambda c^2/3H_o^2$=1.5 would bring the expansion age back to T_o=17.3 Gyr. However, models with Ω_o=1 and $\Lambda > 0$ have a coasting phase (i.e. $R/R_o \sim$ constant) at redshift $z \sim 0.8$ [28]. Complete counts of objects versus redshift do not yet preclude a corresponding maximum at z=0.8, but they may eventually provide a powerful test.

IV. Conclusions

Globular clusters and radioactive elements require a total age of the Universe of 18±3 Gyr. The expansion age of a Friedmann Universe with H_o=50 km $s^{-1}Mpc^{-1}$ and Ω_o=0.1 - i.e. with the presently observed values - is 17.6 Gyr with a 1σ uncertainty of ~2.5 Gyr. The agreement between the independent time scales is - to say the least - suggestive.

If Ω_o=1 is postulated, the expansion time becomes precariously short, unless $H_o \leq 43$ (i.e. $T_o \geq 15$ Gyr). Such a low value of the Hubble constant is not excluded, but it is certainly not favoured in the current literature. For $H_o \geq 50$ it seems necessary to take rescue in a Lemaitre Universe with a cosmological constant $\Lambda > 0$. This additional free parameter allows to adjust the time scales at will.

Acknowledgement: The author thanks the Swiss National Science Foundation for financial support. The manuscript was edited by Mrs.M. Saladin.

References

1. A.Sandage: in Galaxies and the Universe, ed.L.Woltjer, Columbia University Press, p.75 (1968)
2. R.B.Ciardullo and P.Demarque: Trans.Yale Univ.Obs., vol.33 (1977)
3. D.A.VandenBerg: Astrophys.J.Suppl.51, 29 (1983)
4. D.A.VandenBerg and R.A.Bell: Astrophys.J.Suppl.58, 561 (1985)
5. A.Sandage: Astrophys.J.162, 851 (1970)
6. A.Sandage: Astron.J.88, 1159 (1983)
7. A.Renzini: in Galaxy Distances and Deviations from Universal Expansion, ed.B.F.Madore, Dordrecht: Reidel, in press
8. B.Carney: Astrophys.J.Suppl.42, 481 (1980)
9. A.Sandage: Astrophys.J.252, 553 (1982)

10. I.Iben Jr. and A.Renzini: Physics Reports 105, 329 (1984)
11. A.Sandage: Astron.J.88, 1159 (1983)
12. A.Sandage and G.A.Tammann: in Inner Space - Outer Space, eds. E.W. Kolb et al., Chicago: University of Chicago Press, p.41 (1986)
13. O.J.Eggen, D.Lynden-Bell and A.Sandage: Astrophys.J.136, 748 (1962)
14. A.Sandage: Astrophys.J.252, 574 (1982)
15. B.S.Meyer and D.N.Schramm: preprint Fermilab-Pub-86/71-A (1986)
16. H.V.Klapdor: Prog.Part.Nucl.Phys.10, 131 (1983); H.V.Klapdor, J. Metzinger and T.Oda: Atomic Data Nucl.Data Tables 31, 81 (1984)
17. F.-K.Thielemann: in Stellar Nucleosynthesis, eds. C.Chiosi and A. Renzini, Dordrecht: D.Reidel, p.389 (1984)
18. K.Hainebach and D.N.Schramm: Astrophys.J.212, 347 (1977)
19. L.L.Stryker: Astrophys.J.266, 82 (1983)
20. J.Andersen, A.Blecha, and M.F.Walker: Astron.Astrophys.150, L12 (1985)
21. E.W.Olszewski and M.Aaronson: Astron.J.90, 2221 (1985)
22. B.W.Carney and P.Seitzer: Astron.J.92, 23 (1986)
23. L.L.Stryker, G.S.Da Costa, and J.R.Mould: Astrophys.J.298, 544 (1985)
24. A.Sandage: Astrophys.J.183, 711 (1973)
25. A.Sandage and G.A.Tammann: in Astrophysical Cosmology, eds. H.A. Bruck, G.V.Coyne, and M.S.Longair, Rome: Pontifical Acad.Sci. p.23 (1982)
26. D.Lynden-Bell: in Astrophysical Cosmology, eds. H.A.Bruck, G.V. Coyne, and M.S.Longair, Rome: Pontifical Acad.Sci. p.85 (1982)
27. A.Sandage: Astrophys.J.133, 355 (1961)
28. A.Sandage and G.A.Tammann: in Large-Scale Structure of the Universe, Cosmology and Fundamental Physics, eds. G.Setti and L.Van Hove, Garching/Genf: ESO-CERN, p.127 (1984)
29. A.Sandage: in press
30. A.Sandage and G.A.Tammann: Astrophys.J.256, 339 (1982)
31. R.Cadonau, A.Sandage, and G.A.Tammann: in Supernovae as Distance Indicators, ed. N.Bartel, Berlin: Springer-Verlag, p.151 (1985)
32. A.Sandage and G.A.Tammann: Nature 307, 326 (1984)
33. A.Sandage and G.A.Tammann: Astrophys.J.294, 81 (1985)
34. G.A.Tammann: in Highlights of Astronomy 6, 301 (1983); Galaxy Distances and Deviations from Universal Expansion, ed. B.F.Madore, Dordrecht: Reidel, in press.
35. G.de Vaucouleurs: Mon.Not.R.astr.Soc. 202, 367 (1983)
36. M.Aaronson and J.Mould: Astrophys.J. 303, 1 (1986)
37. S.van den Bergh, C.Pritchet, and C.Grillmair: preprint (1986)
38. P.C.van der Kruit: Astron.Astrophys., in press (1986)
39. S.Refsdal, R.Stabell, and F.G.de Lange: Mem.Roy.Astron.Soc. 71, 143 (1967)

Evidence for a Nonvanishing Energy Density of the Vacuum or Cosmological Constant

H.V. Klapdor and K. Grotz

Max-Planck-Institut für Kernphysik, D-6900 Heidelberg, F. R. Germany

The better understanding of the synthesis of heavy elements by the r-process - based on an improved description of nuclear beta decay far from stability -, and the investigation of globular clusters led to a larger age of the universe of $\geq 15 \times 10^9$ a. It will be shown that with the assumption of inflationary expansion at the beginning of the evolution of the universe such a number leads to a nonvanishing cosmological constant Λ in the Friedmann equation for a Hubble constant $H_0 \gtrsim 45$ $Mpc^{-1}sec^{-1}$. Consequences of a nonvanishing Λ (corresponding to a nonvanishing energy density of the vacuum) are more stringent limits on the amount of dark matter in the universe and on the neutrino mass.

1. Introduction

An "experimental" value of the cosmological constant Λ or the energy density of the vacuum is of large value for getting deeper insight into the structure of gauge theories, and also for the understanding of the dark matter problem of the universe.

The expansion of a homogeneous and isotropic universe with a Robertson-Walker metric can be described simply by the radius of curvature R(t) of the three-dimensional space manifold at time t after the big bang:

$$\left[\frac{\dot{R}(t)}{R(t)}\right]^2 = \frac{8\pi G}{3}\,\rho(t) - \frac{kc^2}{\left[R(t)^2\right]} + \frac{1}{3}\,\Lambda c^2$$

and (1)

$$\frac{\ddot{R}(t)}{R(t)} = -\frac{4\pi G}{3}\left[\rho(t) + \frac{3p(t)}{c^2}\right] + \frac{1}{3}\,\Lambda c^2$$

Here, $\rho(t)$ is the mean density of matter and p(t) is the radiation pressure in the universe. k is the metric parameter describing spherical, Euclidean and hyperbolic spaces for k = +1, 0 or -1, respectively. Λ is the cosmological constant, which cannot be fixed by general relativity. In most cosmological models Λ is assumed to be zero (for simplicity or for esthetic reasons).

In the context of quantum field theory Λ can be interpreted [1] as the energy density of the vacuum ϵ_v

$$\Lambda = \frac{8\pi G}{c^4}\,\epsilon_v \tag{2}$$

It is convenient to introduce an equivalent vacuum matter density $\rho_v = \epsilon_v/c^2$ In order for inflation (see, e.g., [2,3]) to develop, one must have $\rho_v \gtrsim \rho(t)$

at some very early time t ($t \approx 10^{-35}$ sec). Such large vacuum energies may result from the Higgs fields introduced in gauge theories. On the other hand, contributions to the vacuum energy in our present world must be many orders of magnitude smaller in order not to completely dominate the universe. Simple gauge theory estimates and cosmological limits deviate by at least 50 orders of magnitude (see, e.g. [4,5]). To solve this problem one has probably to include gravitation into the quantum description [6-8]. In this context it is an extremely interesting question whether the cosmological constant is exactly zero or only very small (compared to gauge theory expectation) but finite.

2. The age of the universe and the cosmological constant Λ

The progress made in recent years in the theoretical predictability of β decay properties of nuclei far from stability [9-13] has led to a solution of the problem of the astrophysical site of the r-process being - together with the s-process - responsible for the production of the heavy elements in the universe. Recent work [14,15] in which the authors succeeded to produce a prompt shock-explosive mechanism for type II supernovae using realistic evolutionary supernova models [16] with 12 and 15 solar masses, gave support to this scenario. Understanding of the r-process is a prerequisite for putting the method of cosmochronology on a reliable basis. The age of the universe obtained by this method is $t_o = (21.8^{+2}_{-5}) \times 10^9$ yr [9,10,12,17,18], consistent with and supported by most recent determinations of the age of globular clusters which lead to an age of the universe of 18±3 billion years [19]; see also [20,21]. Earlier work on the actinide chronometers [22,23] suffers severely from neglecting the effect of β-delayed fission [10,18] on the production ratios of the cosmochronometers by the r-process. This strong effect has been confirmed recently also by the work of [24] - even though these authors in their estimates of β-delayed fission rates do not yet include the effect of tunneling through the barrier.

We shall investigate in this paper the consequences for some cosmological questions which would arise from an age of the universe $t_0 \gtrsim 15 \times 10^9$ yr. It will be shown in this section (see also [25]) that with the assumption of inflationary expansion the larger age of the universe leads to a nonvanishing cosmological constant Λ in the Friedmann equation for a Hubble constant $H_0 \gtrsim 45$ km $Mpc^{-1}sec^{-1}$. (Observations cover the range $H_0 \approx 40$ to 100 km $Mpc^{-1}sec^{-1}$). This point is discussed also by Blome and Priester [4,26]. The possibility of a positive nonvanishing cosmological constant was also pointed out by [27,28]. These authors suggested that with $\Lambda > 0$ a flat universe, as favored by inflation, could be obtained despite the fact that the observed matter density is much smaller than the critical value. Our findings are in accordance with these considerations.

In order to obtain a unique solution for R(t) three boundary conditions are needed, if $\rho(t)$ is assumed to be pressure-free matter density and p(t) is neglected (good approximation for $t >> 10^6$ years). Three such boundary conditions, for which experimental information exists, are (1) the Hubble constant $H_0 = (\dot{R}/R)$ today, (2) the matter density ρ_o today and (3) the age of the universe t_o. Although these quantities are fixed experimentally only within relatively wide limits ($H_o \approx 40$-100 km $Mpc^{-1}s^{-1}$ [29-31], $t_o > 15 \times 10^9$ yr [18,19] and $\rho_0 \geq \rho_B \approx (0.5^{+0.7}_{-0.3}) \times 10^{-30}$ g cm^{-3} [4,32,33] already the large value of

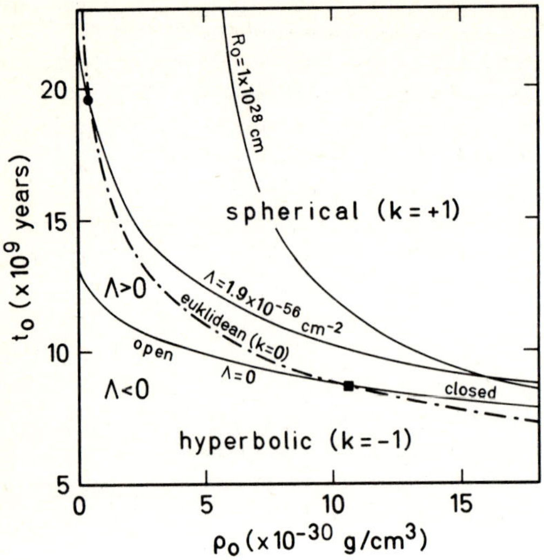

Fig. 1: Relations between the age of the universe t_0 and the present matter density ρ_0 for various cosmological model types (H_0 is fixed at 75 km $Mpc^{-1}s^{-1}$).

t_o for almost the whole range of admittable values of H_o and curvatures of the universe implies a positive Λ. Fig. 1 shows the relation between t_o and ρ_o for the average value H_o = 75 km $Mpc^{-1}sec^{-1}$ in different models. With Euclidean metric (favored by inflation) and Λ = 0, the age t_o would only be 8.7 x 10^9 yr (for H_o = 75 km $Mpc^{-1}sec^{-1}$). But if one allows for Λ > 0 (Friedmann-Lemaître models) the Euclidean models give t_o just in the right range (see also fig. 2).

The interesting region is shown in detail in fig. 3a again with the assumption H_o = 75 km $Mpc^{-1}sec^{-1}$ and in fig. 3b for H_o = 50 km $Mpc^{-1}sec^{-1}$. The allowed regions are shaded. The constraint excluding the upper right corner comes from the experimental limit on the deceleration parameter

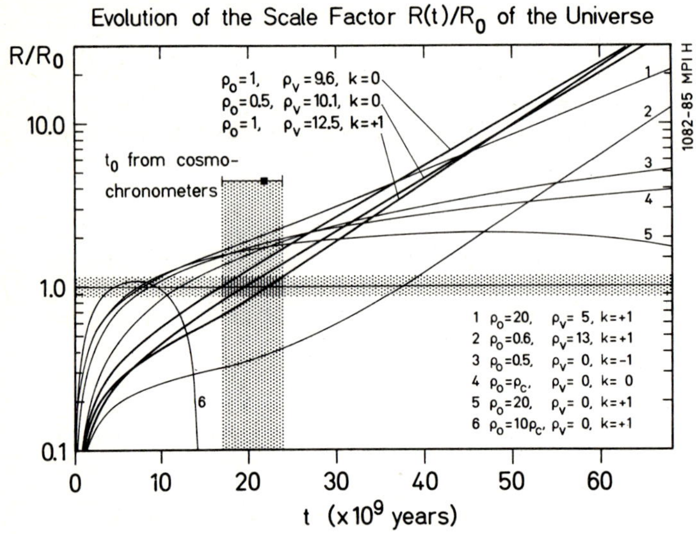

Fig. 2: Some examples for the evolution of the scale parameter $R(t)/R_0$ for various cosmological models with H_0 = 75 km $Mpc^{-1}s^{-1}$.

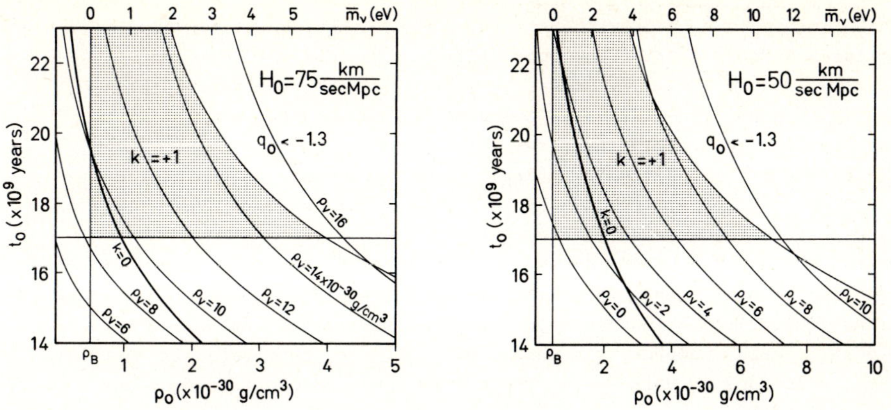

Fig. 3: The relation between t_0 and ρ_0 as in Fig. 1; however, the parameter range of interest shown in more detail. In Fig. 3a, H_0 = 75 km $Mpc^{-1}s^{-1}$ and in Fig. 3b H_0 = 50 km $Mpc^{-1}s^{-1}$ is assumed.

$q_0 = - [\ddot{R}(t)R(t)/\dot{R}(t)^2]_{t=t_0}$ $(-1.27 < q_o < 2$ [34,35].

In the case of Euclidean metric (inflation) and for $H_0 \geq 50$ km $Mpc^{-1}s^{-1}$ non-vanishing values of Λ and ρ_v are found down to ages $t_0 \gtrsim 13 \times 10^9$ yr (for $H_0 \geq 43$ km $Mpc^{-1}s^{-1}$ down to 15.2×10^9 yr).

So in the case of Euclidean metric the larger age of the universe implies a very small but finite positive cosmological constant for a Hubble constant H_0 > 43 km $Mpc^{-1}sec^{-1}$. It may be noted that $\rho_v > 0$ may lead to a universe expanding forever even in the case of a spherical metric (see examples in fig. 2). Spherical metric (k = +1) no more necessarily implies that the universe will contract again in the future. (This would mean $R < R_{max}$ for all times; compare curves 1 and 5 of fig. 2).

3. Constraints on the present matter density of the universe and on the neutrino mass in models with finite Λ.

In models with $\Lambda > 0$ one obtains sharper limits for the total matter density ρ_0 and therefore for the amount of dark matter ρ_D. First, it is clear that a large ρ_0 tends to decrease t_0 for any metric (see Fig. 1). On the other hand, the critical matter density ρ_c, needed for Euclidean metric in the $\Lambda = 0$ case, would be more than 10 times larger than the observed baryonic density $\rho_B \approx 0.5 \times 10^{-30}$ g cm^{-3}. So, in models with Euclidean metric and $\Lambda = 0$ one is faced with the problem of a large amount of missing dark matter ρ_D. There is, of course, no lack of possible candidates for dark matter (see, e.g. [42]). However, they could solve the problem only at the cost of a much too small age t_0. As can be seen from Figures 1 and 3, the situation changes drastically in models with Λ > 0. The positive vacuum energy density to a large amount "replaces" the dark matter needed to obtain Euclidean metric with $\Lambda = 0$.

Concluding, a consistent model can be obtained with the following features: a) positive nonvanishing Λ, b) only a small amount of dark matter, c) Euclidean (or nearly Euclidean) metric d) a) and b) leading to a large t_o consistent with observation. In the Euclidean case we obtain for $t_0 \geq 17 \times 10^9$ yr

and $H_0 \geq 50$ km $Mpc^{-1}s^{-1}$ the limit (see fig. 3): $\rho_B \lesssim \rho_0 \lesssim 4\rho_B \rightarrow \rho_D \lesssim 3\rho_B$. Allowing for spherical metric, the upper limit is: $\rho_0 \lesssim 7 \times 10^{-30}$ g cm^{-3}, $\rho_D \lesssim 6.5 \times 10^{-30}$ g cm^{-3} = $13\rho_B$. For $t_o \geq 14 \times 10^9$ yr the result would be $\rho_D \lesssim 15 \times 10^{-30}$ g cm^{-3}. Attributing the possible dark matter completely to neutrinos the mean neutrino mass (three flavors) $\bar{m}_\nu = 1/3\sum_{i=1}^{3} m_\nu^i$ is $\bar{m}_\nu \lesssim 3$ eV for Euclidean metric (condition $\rho_D \lesssim 3\rho_B$). If no assumption on the metric is made, for $t_0 \geq 17 \times 10^9$ yr and $H_0 \gtrsim 50$ km $Mpc^{-1}s^{-1}$ the less stringent limit $\rho_D \lesssim 6.5 \times 10^{-30}$ g cm^{-3} corresponds to $\bar{m}_\nu \lesssim 11$ eV.
These values which would be the average values of neutrinos with three different flavors (e^-, μ, τ neutrinos) lie well in the order of magnitude, which follows from recent evaluations of double beta decay experiments [12,36-39], which yield $m_\nu < 0.8$ eV. Would, however, the masses of ν_e, ν_μ, ν_τ be related as those of the e^-, μ-, τ-leptons, a value of $m_{\nu_e} \lesssim 10^{-3}$ eV would be obtained from the age of the universe.

4. Conclusion

Recent progress in nuclear physics (beta decay of nuclei far from stability) is found to have impact on questions such as the age of the universe and by this on questions of cosmology and gauge theories. The deduced larger age of the universe ($t_0 \gtrsim 15 \times 10^9$ yr) leads for a range of $H_0 \approx 45$ - 100 km Mpc^{-1} s^{-1}, to a nonvanishing value for the cosmological constant Λ. The deduced value seems consistent with recent conclusions from the observed distribution of quasars [40,41].

With an age of $t_0 \gtrsim 17 \times 10^9$ yr and Euclidean metric a limit for the density ρ_D of dark matter in the universe of $\rho_D \lesssim 3\rho_B$ is obtained. Dark matter candidates fall according to [42] mainly into two categories: hot (neutrino-like) and cold (axion or massive photino-like). Attributing the dark matter density of the above model completely to neutrinos would correspond to a mean neutrino mass of $\bar{m}_\nu < 3$ eV.

It should be mentioned that a nonvanishing Λ would lead to a simpler solution of the dark matter problem in [42], since models with Euclidean metric and $\Lambda > 0$ (in which $\Omega = \rho_0/\rho_{crit} < 1$) are much less faced with this problem.

On the other hand an $\Omega < 1$ is supported by observational dynamical arguments as well as by big bang nucleosynthesis arguments [42], while arguments for an $\Omega = 1$ are at present mainly of philosophical nature.

Finally it should be mentioned that a positive Λ would be of great influence on the structure of SUSY-theories [43].

References

1 Zeldovich, Ya.B., Usp. Fiz. Nauk 95 (1968) 209.
2 Linde, A.D., Phys. Lett. 108B (1982) 389.
3 R. Brandenberger, this volume.
4 Blome, H.J. and Priester, W., Naturwissenschaften, 71 (1984) 528.
5 Bludman S.A., and Ruderman, M.A., Phys. Rev. Lett. 38 (1977) 255.
6 Baum, E., Phys. Lett. 133B (1983) 185.
7 Hawking, S.W., Phys. Lett. 143B (1984) 403.
8 Henneaux, M. and Teitelboim, C., Phys. Lett. 143B (1984) 415.
9 Klapdor, H.V., Progr. Part. Nucl. Phys. 10 (1983) 131.
10 Klapdor, H.V., Fortschr. der Physik 33 (1985) 1.
11 Klapdor, H.V., Metzinger, J. and Oda, T., At. Data Nucl. Data Tables 31 (1984) 81.

12 Klapdor, H.V., Invited lecture presented at Internat. School of Nuclear Physics: The Early Universe and its Evolution. Erice, Sicily, 2-14 April 1986. In press in Progr. Part. Nucl. Phys.
13 Klapdor, H.V., Grotz, K., Metzinger, J., this volume.
14 Baron, E., Cooperstein, J., Kahana, S., Phys. Rev. Lett. 55 (1985) 126.
15 Kahana, S.N., this volume.
16 Woosley, S.E., Weaver, T.A., Bull. Am. Astr. Soc. 16 (1984) 971.
17 Klapdor, H.V., Sterne und Weltraum 3 (1985) 132.
18 Thielemann, F.K., Metzinger, J. and Klapdor, H.V., Z. Phys. A 309 (1983) 301; Astron. Astrophys. 123 (1983) 162.
19 Tammann, G.A., this volume.
20 Sandage, A., Astrophys. J. 252 (1982) 553.
21 Nissen, P.E., ESO Messenger 28 (1982) 4.
22 Fowler, W.A., Proceed. of the Welch Foundation Conferences on Chemical Research, XXI, Cosmochemistry, ed. W.D. Milligan, Houston (1977), p. 61.
23 Fowler, W.A. and Meisl, C.C., preprint OAP-660, Caltech, March 1985.
24 Meyer B.S. and Schramm D.N., Progr. Part. Nucl. Phys., in press.
25 Klapdor, H.V., Grotz, K., Astrophys. J. 301 (1986) L39.
26 Blome, H.J., this volume.
27 Peebles, P.J.E. Ap. J. 284 (1984) 439.
28 Turner, M.S., Steigman, G., and Krauss, L.M. Phys. Rev. Lett. 52 (1984) 2090.
29 Sandage, A. and Tammann, G.A., Astrophys. J. 256 (1982) 339.
30 Sandage, A. and Tammann, G.A., Nature 307 (1984) 326
31 Van den Berg, S., Nature 299 (1982) 297.
32 Blome, H.J. and Priester, W., Naturwissenschaften 71 (1984) 456.
33 Blome, H.J. and Priester, W., Naturwissenschaften, 71 (1984) 515.
34 Ehlers, J., Mitt. Astron. Ges. 38 (1976) 41.
35 Tammann, G.A., Cosmology, Landolt-Börnstein VI/2c, Springer (1982).
36 Grotz, K. and Klapdor, H.V., Phys. Lett. 153B (1985) 1.
37 Grotz, K., and Klapdor, H.V. Phys. Letters 157B (1985) 242.
38 Grotz, K. and Klapdor, H.V., Proceed. 1st Symposium on Underground Physics, Nuovo Cimento in press.
39 Grotz, K. and Klapdor, H.V., Nucl. Phys. A, in press.
40 Fliche, H.H. and Souriau, J.M., Astron. Astrophys. 78 (1979) 87.
41 Fliche, H.H., Souriau, J.M. and Triay, R., Astron. Astrophys. 108 (1982) 256.
42 Schramm, D.N., this volume.
43 Dragon, N., Ellwanger, U., Schmidt, M.G., in press in Progr. Part. Nucl. Phys.

Weak Interaction and the Large Scale Structure of the Universe

D.N. Schramm

The University of Chicago and Fermilab, Chicago, USA

The combined problems of large scale structure, the need for non-baryonic dark matter if $\Omega = 1$, and the need to make galaxies early in the history of the universe seem to be placing severe constraints on cosmological models. In addition, it is shown that the bulk of the baryonic matter is also dark and must be accounted for as well. The nucleosynthesis arguments are now strongly supported by high energy collider experiments as well as astronomical abundance data. The arguments for dark matter are reviewed and it is shown that observational dynamical arguments and nucleosynthesis are all still consistent at $\Omega \sim 0.1$. However, the inflation paradigm requires $\Omega = 1$, thus, the need for non-baryonic dark matter. A non-zero cosmological constant is argued to be an inappropriate solution. Dark matter candidates fall into two categories, hot (neutrino-like) and cold (axion or massive photino-like). New observations of large scale structure in the universe (voids, foam, and large scale velocity fields) seem to be most easily understood if the dominant matter of the universe is in the form of low mass ($9eV \lesssim m_\nu \lesssim 35eV$) neutrinos. Cold dark matter, even with biasing, seems unable to duplicate the combination of these observations (of particular significance here are the large velocity fields, if real). However, galaxy formation is difficult with hot matter. The potentially fatal problems of galaxy formation with neutrinos may be remedied by combining them with either cosmic strings or explosive galaxy formation. The former naturally gives the scale-free correlation function for galaxies, clusters, and superclusters. The latter requires fine tuning and percolation to get the large scales and the scale-free correlation function. However, combining hot matter and strings reduces the ability of the hot matter to give some of the large scale features and still yield $\Omega = 1$. Questions to be examined are raised.

1. Introduction

The major confrontation of early universe studies with the "real" universe now focuses on the problems of galaxy formation, dark matter, and the generation of large scale structure. The observable aspects of these problems came into being shortly after recombination; however, the condition of the universe as it approaches recombination are determined by events taking place much earlier, when nuclear and particle physics effects dominated. Since the recombination epoch is the limiting epoch for direct observations, it is only natural that this epoch serve as the interface between early universe cosmologists and astronomers.

The problems are to produce initial conditions and types of matter which will yield the observable universe, the large scale structure. In particular, the observable universe now appears to have large scale structure on scales of $\sim 40Mpc$ that looks like foam or at least intersecting sheets and filaments with large voids[1,2,3]. In addition, there appear to be large, coherent motions of 40 Mpc clumps with velocities of $\sim 600km/sec$[4]. To this very large scale structure must be added the apparent fact that clusters of galaxies cluster with each other more strongly than galaxies cluster[5], or to use the analysis of Szalay and Schramm[6], the clusters and galaxies appear to cluster in a scale-free manner as if laid out in some fractal pattern.

2. The Dynamical Arguments

To these large scale observations must be added the dynamical measurements of mass and the so-called dark matter problem. In particular, the dynamics of the visible parts of galaxies imply an Ω of ≤ 0.01 (where $\Omega \equiv \frac{\rho}{\rho_{crit}}$ is the critical density of the universe). However, when galaxies interact with other galaxies in binary pairs or in small groups, they interact with $\sim$ 10 times as much mass, implying an $\Omega \sim 0.1$. When galaxies interact with one another in large clusters they interact with possibly even more mass, implying $\Omega \sim 0.1$ to 0.3. (*No well studied system gives anything near* $\Omega = 1$.)

3. Big Bang Nucleosynthesis

To the dynamical arguments we can add the arguments from Big Bang nucleosynthesis (Yang et al.) which show that observed abundances are consistent only if $\Omega_b \sim 0.1$ (where $\Omega_b \equiv \frac{\rho_b}{\rho_{crit}}$ and ρ_b is the density of baryons).

Thus as Gott et al.[7] pointed out over ten years ago, direct astronomical evidence points towards $\Omega \sim 0.1$ with the dark halos being baryonic and no need for exotic stuff. In particular, it should be noted that the lower bound on Ω_b is $\Omega_b \geq 0.03$[8]. Since this is > 0.01, it implies that the bulk of the baryons are dark. (Note that because of this point, dark halos for dwarf spheroidal galaxies are no problem since they can be baryonic.) Also, it is important to remember that nucleosynthesis contrains $\Omega_b < 0.15$. (This is lower than the 0.19 from Yang et al.[9] due to better current upper limits on the microwave background temperature.) Thus, if $\Omega \sim 1$, the bulk of the universe would be non-baryonic *and* could not cluster with the light emitting galaxies and clusters.

The nucleosynthesis arguments are gaining even greater credence now that their predition[9,10,] that the total number of neutrino types (generations) is small (three or at most four) is being verified by collider experiments[11] with current experimental limits at < 5. From particle physics theory alone any number of generations might be possible. The preliminary verification of the cosmological prediction is the first time that cosmology has made a prediction which has been verified by a high energy accelerator experiment.

4. Baryonic Halos?

Can halos of galaxies and dwarf spheriodals really be baryonic? While the coincidence of $\Omega_b \sim 0.1$ and $\Omega_{dynamic} \sim 0.1$ is suggestive, it is certainly not compulsory. Different forms of dark matter can mix with baryons in different ways depending on the mechanism of galaxy formation.

With cold dark matter the halos must be a mixture of $\sim 90\%$ cold matter and 10% baryons whereas in hot matter models the halo mixture depends on the galaxy formation scenario.

If the halos do contain significant baryonic materials, what form can it be? Hegyi and Olive and Schramm have argued that most baryonic things do not work. However, they leave two very important loopholes:

1. Black holes left from an early generation of massive stars with the bulk of the stellar material falling into the hole and not producing excess heavy elements. Such black holes are contrained by Big Bang nucleosynthesis baryon limits since they were baryons then (so they count as baryonic material).
2. Low mass objects too dim to be seen in telescope searches. Jupiter-like clumps or even 0.01 $M_\odot$ stars would work. In order for the abundance of such objects to be sufficient, the abundance spectrum for these objects would probably be above the low mass extrapolation of the Salpeter initial mass function. However, that function is strictly empirical and there could certainly be a low mass excess if the initial stellar generation with pure H and He, but more objects low than currently occurs with heavy elements present. (Option 1., of course, requires exactly the opposite behavior for the early stellar mass function.)

5. The Flatness Arguments

If everything agrees so well with $\Omega \sim 0.1$, why do people continue to think $\Omega = 1$? The only astrophysical evidence for large Ω is clearly weak at the present time. It consists of the following:

1. With Gaussian adiabatic initial density fluctuations of the type described by Zel'dovich and expected from simple inflation models, it is impossible to make galaxies rapidly enough when constrained by limits on microwave background anisotropies unless $\Omega > 0.2$[12,13].
2. The velocity field of IRAS galaxies on scales of $\sim$ 200Mpc implies a virial mass on these large scales of $\Omega \sim 1$[14].
3. The density of galaxy counts versus redshift is optimally consistent with $\Omega = 1$ geometry[15].

The first of these is clearly removable if galaxies form by something other than Gaussian adiabatic fluctuations with a Zel'dovich spectrum. In particular, string models which are also derivable from grand unified gauge models do not yield such a stringent requirement on Ω, nor do, for that matter, models where galaxy formation is stimulated by early explosions[16].

The second argument has the problem that a reliable way to determine distances to IRAS galaxies has not been established and a complete redshift survey of IRAS galaxies remains to be done. In addition, IRAS counts may have a significant north–south bias due to induced instrumental variations in sensitivity of the satellite in the northern and southern hemispheres.

The third argument, while potentially the strongest, still requires a more detailed analysis of galactic evolution effects and normalization of distant galaxy counts to nearby where different techniques are used.

Thus, while suggestive, these arguments do not yet establish $\Omega = 1$. However, there is a Copernican-like argument which is sufficiently powerful that most theoretical physicist believe $\Omega = 1$. The argument was best articulated by Dicke and Peebles and later provided Guth with a strong motivation for inflation which gave a physical mechanism for yielding the desired Ω. The argument, simply stated, is that Ω is a time changing quantity going to $\Omega < 1$ and to ∞ if $\Omega > 1$, and only remaining constant if $\Omega = 1$. The timescale of change is the expansion rate of the universe. Thus, the only long-lived values are 0, 1, and ∞. Since we are here, Ω is neither 0 nor ∞. The only other long-lived value is 1. To have any finite value below unity today would require that we live at a very special time, the early epoch in cosmic time when Ω was not 1 or 0. Such a value would require the extraordinary fine tuning at the Planck time of $\sim$ 60 decimal places, or at least 17 decimal places at the time of Big Bang nucleosynthesis. Thus, unless we live at a special time and some unknown mechanism tunes Ω to exactly the right amount to fantastic accuracy, Ω is probably unity.

Since any early deSitter phase for the universe produces a flat universe ($\Omega = 1$ if the cosmological constant $\Lambda = 0$) and since inflation means an early deSitter phase, and since most scalar fields yield inflation, it is reasonable to believe $\Omega = 1$. While many have recently focused on the problems many models of inflation have been producing, the right sized initial fluctuations[17] any inflation model which solves the horizon problem, getting a nearly constant background temperature, will also solve the flatness problem.

6. The Cosmological Constant

Some astrophysicists (who shall remain nameless) have focused on the formal mathematical loophole that flatness can also be obtained with a non-zero Λ and $\Omega < 1$. However, such a solution is missing the philosophical motivation (like killing for pacifism). If today we have $\Omega \sim 0.1$ and non-zero Λ yields flatness, that is an epoch- dependent solution since the contribution of Ω and Λ vary differently with epoch. Such a solution would imply that we live at the only epoch where Λ and Ω contributions to curvature are comparable, again requiring amazing fine tuning (tuning Λ to ≥ 120 decimal places). Unfortunately we don't as yet have a nice physically motivated mechanism like inflation to set $\Lambda = 0$, but if we buy the philosophy, I believe we should also assume Λ is negligible. Of course both arguments are philosophical (or theological) rather than based on physical observation, but the Copernical principal of us not being special has held up well for several hundred years.

7. Dark Matter and Galaxy and Structure Formation

As mentioned before, if Ω is 1, then we need non-baryonic dark matter. Such matter has been classified as either hot (neutrino-like with high velocities just prior to the epoch of matter-radiation equality) or cold (low velocities prior to matter-radiation equality).

Initially, hot, low mass, neutrinos were quite popular as candidates for solving the cosmological dark matter problem, since they were the least exotic of the non-baryonic options, and they naturally clustered only on large scales where the dark matter was needed, rather than on the small scales where the contribution of dark matter was known to be minimal[18]. They received a major boost with the preliminary reports of measured mass[19] for ν_e (although probably only the most massive ν is cosmologically important, and that might well be ν_τ (or a nucleosynthesis-allowed 4th generation) which could still have a $\sim 10eV$ mass, even if $m_{\nu_e} \ll 1eV$). Also, they gained strength when it was shown[3] that the neutrino Jean's mass was

$$M_J \sim \frac{3 \times 10^{18} M_\odot}{m_\nu^2(eV)} \text{ or } \lambda_J \sim \frac{1300Mpc}{m_\nu(eV)}$$

which for $m_\nu \sim 30eV$ yielded $M \sim 3 \times 10^{15} M_\odot$, and $\lambda \sim 40Mpc$, the mass and scale of large clusters.

Unfortunately, massive neutrinos fell into disrepute as dark matter when it was emphasized[20] that in the standard adiabatic model of galaxy formation with a random phase, Zel'dovich fluctuation spectrum of the type expected by inflation, and with $\delta T/T$ constrained by microwave observations, galaxies did not form until redshift $z \lesssim 1$. This occurred because the initially formed pancakes with mass M_J took a while to fragment down to galaxy size. This contradicted the observations which showed that quasars existed back to $z \sim 3.5$. In addition, if baryons stay in gas form in the potential wells of the large ν pancakes, they light up in the x-rays beyond what is observed[21].

While some[22] have appealed to statistical tails, etc., to escape these conclusions, most cosmologists began abandoning neutrinos and adopting cold dark matter[23], which could enable rapid galaxy formation[24,25].

Cold matter also had its problems[26]. In the standard model, it would all cluster on small scales, and thus be measured by the dynamics of clusters, such as the Virgo infall. Since such measurements implied that $\Omega \sim 0.2 \pm 0.1$ on cluster scales, this meant that $\Omega_{cold} \lesssim 0.3$, and not unity. Remember that $\Omega \sim 0.1$, so observationally, non-baryonic dark matter is not required unless one wants an Ω of unity, so cold matter wasn't naturally solving one problem for which it was postulated. This constraint on cold matter could be escaped if it were *also* assumed that galaxy formation was biased[25,27] and did not occur everywhere. Thus, there could be many clumps of cold matter and baryons that did not shine for some ad hoc reason. Biasing ran into problems when it could not explain the observation[5] of a very large cluster–cluster correlation function, ξ_{cc}, relative to the galaxy–galaxy correlation function[26,27], ξ_{gg}. With biasing $\xi_{cc} \propto \xi_{gg}$ but in all models $\xi_{gg} < 0$ for a few 10's of Mpc, whereas ξ_{cc} was observed to be positive out to scales $\gtrsim 50Mpc$. Hardcore cold matter lovers had to argue that the ξ_{cc} data might be wrong, although no one has been able to disprove it.

A way out of the ξ_{cc} problem was proposed by Szalay and Schramm. There we noted that the correlation functions appear to be scale free, thus implying that large-scale structure is dominated by something other than random noise and gravity, say either percolated explosions or strings. In fact, the scale-free structure is characterized by a fractal of dimension $D \sim 1.2$, not too different from the $D \sim 1$ that naive string theory might yield. String calculations[28] of galaxy formation indeed found support for such a fractal process with the appropriate dimension being valid from galaxy to supercluster scales.

Thus, there were already strong hints that something was wrong with the previous, in vogue, picture of biasing and cold matter with random noise initial fluctuations. To this we now add the new observations of many large voids[1,2] of diameter $50h_{1/2}Mpc$ ($h_{1/2} \equiv H_0/50km/sec/Mpc$), with most galaxies distributed on the walls of the voids, and the observation[4] that our local 40 Mpc region of space is moving with a coherent velocity field of $\sim 600km/sec$ toward Hydra-Centaurus. While at least one large void (in Böotes) had been observed before[3], using a pencil beam approach, until the Harvard redshift[1] survey work, it was not known how ubiquitous voids were. In fact, the Harvard data shows that almost all galaxies are distributed along the "walls" of voids; galaxies and clusters are not randomly distributed, but fit onto a well-ordered pattern.

While the Harvard work only goes out to $\sim 100Mpc$, there is substantial evidence that this sort of pattern persists to redshifts $z \sim 1$ from the Koo and Kron survey[2]. A simple explanation for the peaks and valleys in the distribution of galaxies and quasars with redshift is that one is looking through filaments or shells with voids in between, once again demonstrating that galaxies and clusters are not laid out randomly on the sky, but follow a pattern.

While statistical fluctuations with cold matter might yield a few large voids as well as many small voids[21,25], it is difficult to get all of space filled with large voids and have galaxies appear only at the boundaries unless some special form of "biasing" is used. However, the real killing blow for the cold matter plus biasing scheme comes from the velocity field work. Even if the biasing could be selected so as to give ubiquitous large voids, the velocities of a $40Mpc$ region of galaxies would be relatively small and random, rather than large and coherent[29]. In fact, the more extreme the biasing used to get large voids, the *lower* the large scale velocities. Thus, it appears that the large-scale structure is telling us that we need something that gives us $\sim 40Mpc$ coherent patterns, and cold matter doesn't appear the way to go. (Unless, of course, the large scale velocity field work is in error. In other words, cold matter with gaussian Zel'dovich fluctuation requires *both* ξ_{cc} and the velocity to be completely wrong.

Since neutrinos naturally gave us patterns on this scale, maybe they should be reexamined. In addition, since the voids look rather spherical, and since explosions tend to produce spherical holes after a few expansion times even if the initial explosion is asymmetric, perhaps an explosive mechanism should be considered also.

Since the Ostriker–Cowie[16] explosion mechanism by itself cannot yield such large voids, the only way it could work is via a high density network of explosions which percolated[25,30]. However, to get $\Omega = 1$ with an exploding scenario would still require non-baryonic matter that did not cluster with the light emitting stuff. In principal, this could be either neutrinos or cold matter but at least with neutrinos an $\sim 40Mpc$ scale might still be naturally imposed.

8. Neutrinos plus Strings or Explosions

Of course, in order for neutrinos to work as the dominant matter, some mechanism to rapidly form galaxies must be imposed both to enable galaxies to exist at $z \sim 5$, and to condense out the gas before it falls into the forming deep potential wells, and emits x-rays. Two ways that might achieve this rapid formation are either via the aforementioned explosion scheme within the collapsing ν-pancakes, or via cosmic strings[31] which would act as nucleation sites for galaxy formation. Since strings are not free-streamed away by the relativistic neutrinos[32], the galaxy scale fluctuations remain within the ν-pancakes. Notice that since neutrinos are not used by themselves simple arguments based on relating their primordial fluctuation spectrum to observed galaxy velocity and distribution features are not necessarily valid and must be reexamined in the more complete scenario.

It should be noted that even with strings as seeds so that cold matter can cluster in a scale-free way fitting ξ_{cc}, the large scale velocity fields for cold matter are small, and it is difficult to get $\Omega = 1$ while observing $\Omega_{cluster} \sim 0.2$. However, we have the additional problem that the strings might mess up the nice large scale neutrino features and background of ν's will still slow galaxy growth around the strings over how cold matter would form on the strings.

It is interesting that two surviving galaxy formation options, strings and explosions, involve the same two options that the scale-free cluster–cluster correlation function arguments point towards. Let us look at each of these scenarios in a little more detail and see if there might be ways of resolving whether either of them might actually be correct. Also, let us see what each requires for the physics of the early Universe.

Both of these scenarios seem to need hot matter if we want to solve the velocity field, $\Omega = 1$, and large scale problems.If $\Omega = 1$, as is necessary to avoid our living at a special epoch, and as agrees with the recent large-scale galaxy count arguments of Loh and Spillar[15] (but disagrees with the direct dynamical arguments on scales of clusters and smaller, and with the baryonic measurements from nucleosynthesis), then $m_\nu \lesssim 35eV$. Since with $\Omega = 1$ the age of the Universe $t_0 = \frac{2}{3H_0}$, and since globular clusters and nucleochronology require $t_0 \gtrsim 11 \times 10^9 yr$ (with a best fit of $t_0 \sim 15 \times 10^9 yr$) we must say that $H_0^{-1} \gtrsim 17 \times 10^9 yr$. Thus, $H_0 \lesssim 60 km/sec/Mpc$, or $h_{1/2} \lesssim 1.2$. From the number of neutrinos and photons in the Universe, we know that the most massive neutrino is bounded by (see ref. 18 and references therein)

$$m_\nu \lesssim (25eV)\Omega h_{1/2}^2 \lesssim 35eV.$$

It is curious that the requirement that we want the neutrinos to give us the large-scale structure, $\lambda_J \sim 40Mpc$, or $M_J \sim 10^{16}M_{\odot}$, also gives us $m_\nu \sim 30eV$, a mass about what is necessary to get $\Omega \sim 1$. Also, we have a lower bound from the nucleosynthesis argument[26] that the number of neutrino species with $m_\nu \lesssim 10MeV$ is three or at most four. Since the sum of all neutrino masses cannot exceed the $35eV$ limit mentioned above, and since the lowest mass for the most massive one occurs when they are all equal, then if $N_\nu \leq 4$,

$$m_\nu \gtrsim 9eV.$$

The first scale to be able to condense and thus have their density grow will be the horizon scale when the neutrinos become non-relativistic, which is M_J. However, in the string option, loops of string will exist down to scales of galaxy size (scales smaller than galaxy size gravitationally radiate away[31]). So as the neutrinos become non-relativistic they can be trapped on smaller scales. The baryons will not be able to begin clustering until after recombination. However, the slow-moving baryons will rapidly fall on to the pre-existing loops of string plus neutrinos. Thus, galaxies will be able to form shortly after recombination, and well before $z \sim 1$.

9. Problems with Strings?

Unfortunately, just after matter domination the bulk of the neutrinos will still have relatively high velocities so their Jean's mass, while dropping, will not be low enough for most neutrinos to cluster on the galaxy size loops. Even after recombination the characteristics Jean's mass for the bulk of the neutrinos will still be much larger than galaxy size, so there will be a relatively smooth background of neutrinos which will slow the rate of growth of baryons falling onto the loops of string. Thus, strings plus neutrinos do not grow galaxies as rapidly as strings plus cold matter; however, strings definitely help the neutrino picture along. The quantitative question of whether the neutrino-string picture can form rapidly enough remains to be worked out in detail, since quick and dirty calculations indicate that the results are marginal[33]. With neutrinos, the dimensionless string tension 6μ needs to be higher than for strings with cold matter where $6\mu \sim 10^{-6}$. Unfortunately, it cannot be arbitrarily raised since high values ($\geq 10^{-5}$) cause problems in microwave anisotropy and in radiating too much energy at the time of nucleosynthesis, thus running into the equivalent of the neutrino country bound[34].

Also, it is not clear how the combination of ν's and strings deals with the very large scale structure. While strings by themselves give the scale- free correlation function out through the scales of Abell clusters[28], if neutrino pancaking is too strong, it could mess this up. On the other hand, string perturbations existing on scales smaller than $\sim$ 40Mpc may prevent pancaking from ocurring at all. Horizon length strings at matter-radiation equality will produce large scale adiabatic fluc-turations that could induce pancake formation in the neutrinos, going non-linear at redshift $z \sim 1$. However, the strength of the fluctuations relative to the normal string fluctuations needs to be checked to see which, if any, dominates.

If they really do not go non-linear until $z \sim 1$, they might not mess up the more rapidly forming galaxy and cluster scale fluctuations, so the smaller scale correlation functions might be retained while the neutrino pancake collapse might induce the very large scale velocity field and pancakes, filaments, and voids. Obviously the whole combined picture needs to be examined in much greater detail to see if it really can retain the best features of both models, rather than the two components destroying each others better features.

Because the string picture looks like the current front runner, people have begun looking at it in far greater detail, to see if it really can yield the observable universe. In particular, Peebles has privately circulated a "screed", stating possible problems. At a workshop held at the Aspen Center for Physics, these problems were examined and possible ways out were found. Let us now summarize the Peebles problems and possible solutions.

Problems not previously mentioned:

1. Strings produce loops following a power spectrum $\sim M^{-5/2}$, whereas galaxies are observed from their light to have a much flatter spectrum, up to $\sim 10^{12} M_{\odot}$ and then exponential fall off. Thus, at first glance, it appears that strings give too many small *and* large galaxies if their spectrum is normalized to fit the L^* galaxies at $\sim 10^{12} M_{\odot}$.
2. Strings are small relative to their separation distances. Thus, collapse onto static strings appears unlikely to give large quadropole moments, and thus tidal interactions will not produce the angular momentum observed in galaxies.
3. With strings as seeds, both cold and hot dark matter will cluster on small scales so that Ω measured for clusters should be a good estimate of Ω_{total} which would yield ~ 0.2, not 1. Biased suppression of galaxy formation with strings as seeds is evn more ad hoc than normal cold-matter biasing, so is not a convenient escape.

The possible solutions to these problems are:

1.1. Excess amounts of small strings forming galaxies can be supressed in a variety of ways.

a. For larger 6μ, such as in the neutrino models, gravitational radiation eliminates the excess low mass loops.

b. Vilenkin[35] has shown that global strings rather than gauge strings radiate Goldstone bosons in addition to gravitational radiation. Thus few mass global strings would also not be a problem.

c. Strings do not radiate symmetrically. The differential radiation for small strings results in a rocket effect[36] which supresses their ability to acrete.

d. More fragmentation of the small loops which form early could lower their abundance as the smaller are radiated away.

1.2. The excess amounts of large loops may be a more complex problem and more work needs to be done here. Possible solutions include:

a. Finite velocity may affect accretion.

b. Fragmentation of large loops will reduce their numbers.

c. Big loops may yield CD galaxies at centers of clusters with velocity curves rising as $r^{1/4}$ rather than normal flat rotation curves.

2. Angular momentum may be formed by tidal interactions because accretion is not spherical but sausage-like, due to the finite velocity of loops. Distances moved are comparable to separations so quadropole moments will be approximately large.
3. The solution to the Ω problem requires that somehow clusters don't sample a standard segment of the universe. One way to accomplish this would be if galaxies correlated more with clusters than randomly. Such could occur if large, cluster-producing strings fragment to produce smaller galaxy-producing strings, and the resultant small strings didn't get too far from the clusters. Clearly, this does occur to some degree; however, can it quantitatively yield a factor of three of more enhancement in Ω between its cluster measured value and the true value remains to be shown. The dynamical range of string simulations has not yet enabled such quantitative tests between small and large loops. Note that if galaxy strings are strongly correlated with clusters, then many regions in space will be without loops of strings, and so will not form galaxies even though they have baryons and either hot or cold dark matter.

Another possible problem is that, while the string scenario may naturally yield $D \sim 1$, it does not so naturally give $D = 1.2$. Fine tuning[39] of string parameters may enable such variation on the scale of the galaxy–galaxy correlation function, or some modification of the criteria for the formation of light-emitting regions around the strings may be necessary.

In this regard it should be remembered that because of possible systematic errors, not everyone agrees that 1.2 is significantly different from 1.0, even for the galaxy–galaxy correlation function, which is the best determined[38]. The uncertainties in the exponent of the cluster–cluster correlation functions are *far* larger, thus problems in trying to explain variations from $D = 1$ fractals are not serious at the present time. With strings there is the additional problem of tuning the primordial phase transition so as to inflate first, and then produce strings[39]. While not impossible, this is constraining.

10. Explosive Galaxy Formation

The second way to get neutrinos to work involves explosive galaxy formation. Here we need initial seeds to lead to condensations which produce massive baryonic objects which explode. As mentioned before, such a model does not naturally give us $40Mpc$ structure. If we use neutrinos then the seeds must be in a form which does not get free-streamed away by the relativistic neutrinos. Strings don't work well here because the string scales that might lead to rapidly evolving baryonic objects are radiated away gravitationally. Thus, the seeds must come in some other isothermal-like form. Perhaps the best option would be condensates from the quark–hadron transition, either planetary mass black holes[40] or Witten nuggets[41]. Both have formation problems[42] and the latter have survival problems[43] also. If such objects could form and survive, they do lead naturally[44] to very massive

($\sim 1000 M_{\odot}$) baryonic objects which would explode on rapid timescales. Another option is cold dark matter clumps, in which case small strings work as seeds, but the large scale problems are aggravated.

The scale affected by explosions of single galaxy size[45] is at most a few Mpc; however, it has been shown[30] that at sufficiently high densities and high trigger rates, the explosions can percolate at least out to scales of a few 10's of Mpc. The fractal dimension of such percolated ensembles is quite sensitive to parameter assumptions and usually varies with scale, thus showing that it is not a true scale-free fractal. If it is made to fit the small scale (few Mpc) with $D \sim 1$ it is usually larger ($D \sim 2$) on scales of $\gtrsim 10 Mpc$. Since, as mentioned above, the exponent of the cluster–cluster correlation function is not, at present, well determined, such models cannot be ruled out. With such explosions percolating within ν-pancakes, we might naturally have their pattern superimposed on the $\sim 40 Mpc$ neutrino scale. In addition, although percolated explosions will initially be highly non-spherical, their shape will evolve towards sphericity with the smaller axes catching up in length to the largest one. In order for large-scale percolation to occur, several generations[21] of explosions must occur; however, cooling arguments and time to initial explosions, plus the need for condensed objects by $z \sim 4$ and the need to hide from present observers, the radiation produced by the explosions, severely restrict the possibility of such percolation and thus quite a bit of fine tuning is required to escape the constraints.

11. Conclusion

Thus, while we cannot explicitly rule out this latter case, unless some new physics can be developed to show how the fine-tuned parameters are natural for other reasons, we must lean towards the string option as the present frontrunner. Strings, of course, would have other observational consequences[32] like gravitational double lensing of distant objects and shifts in the 3° background across such a line of lenses, and a background of gravitational radiation from the evaporation of small-scale strings which might affect the millisecond pulsar. Thus, observations should eventually be able to confirm or deny this frontrunner. Table 1 gives a summary of current proposed models and their ability to solve the problems. Note that the location of dark baryons may eventually be detectable and a discriminator of models. No model is yet a clear winner. Some require more calculations to see if they can be made to work. Others require some key bit of observational data to be proven wrong.

In summary, we have come full circle and once again massive neutrinos are looking good. However, with them comes the need for galaxy and structure formation triggered by something other than random phase adiabatic fluctuations. The non-random phase fractal initial conditions such as produced by strings[46] or fractal generating explosions[16,30] seem to be the way to go. It is comforting that the exotica of cosmic strings do seem to be a natural consequence[47] of the current, in vogue, superstring Theories of Everything (T.O.E.).

Table I: Models and Problems

	Hot & Adiabatic	Cold & Adiabatic	Cold & Strings	Hot & Strings	Cold & Explosions	Hot & Explosions
$\Omega = 1$ with $\Omega_{cluster} \sim 0.2$	o.k.	requires ad hoc biasing	requires large cluster–galaxy correlation	requires large cluster–galaxy correlation	requires special biasing	o.k.
Large Cluster–Cluster correlation function	difficult	no	o.k.	o.k. if not destroyed by pancaking	requires fine tuning	requires fine tuning
Filaments, sheets, and voids structure at $\sim$ 40Mpc	o.k.	difficult	not easy	maybe	difficult	o.k.
Large scale high velocities	o.k.	no, worse with biasing	no	maybe	difficult	o.k.
Galaxy formation by $z \gtrsim 4$	no	o.k.	o.k.	marginal	o.k.	depends on seeds
Galaxy mass spectrum	pancake fragmentation	depends on biasing scheme	maybe	probably o.k.	maybe	maybe
Galaxy angular momentum	pancake fragmentation	o.k.	probably o.k.	probably o.k.	probably o.k.	probably o.k.
Contents of voids	mostly hot stuff	$\sim$ 90% cold $\sim$ 10% baryons	$\sim$ 90% cold $\sim$ 10% baryons	$\gtrsim$ 90% hot $\lesssim$10% baryons	$\sim$ 90% cold $\sim$ 10% baryons	mostly hot stuff
Halos of galaxies (including dwarfs)	mostly baryonic	$\sim$ 90% cold $\sim$ 10% baryonic	$\sim$ 90% cold $\sim$ 10% baryonic	> 10% baryonic < 90% hot	$\sim$ 90% cold $\sim$ 10% baryonic	mostly baryonic

Acknowledgements to co-workers J. Charlton, K. Olive, A. Melott, G. Steigman, A. Szalay, and M. Turner are gratefully given. I also acknowledge many useful discussions with N. Turok and P.J.E. Peebles. This work was supported in part by NSF AST 85-15447, and by DOE DE-FG02-85ER40234 at the University of Chicago, and was prepared at the Aspen Center for Physics.

1. deLapparant, V., Geller, M. and Huchra, J. 1986, Center for Astrophysics preprint
2. Koo, D. and Kron, R. 1986, in preparation.
3. Kirschner, R., Oemler, G., Schecter, P., and Shectman, S. 1982, *Ap.J.* **248**, L57.
4. Faber, S., Aaronson, M., Lynden-Bell, D. 1986, *Proc. of Hawaii Symposium on Large-Scale Structure.*
5. Bahcall, N. and Soniera, R. 1983, *Ap.J.* **270**, 20; Klypin and Khlopov 1983, *Soviet Astron. Lett.* **9**, 41.
6. Szalay, A. and Schramm, D. 1985, *Nature* **314**, 718.
7. Gott, J.R., Gunn, J., Schramm, D.N., and Tinsley, B.M. 1974, *Ap.J.* **194**, 543.
8. Freese, K. and Schramm, D. 1984, *Nucl. Physics* **B233**, 167.
9. Yang, J., Turner, M., Steigman, G., Schramm, D., and Olive, K. 1984 *Ap.J.* **281**, 493.

10. Steigman, G., Schramm, D.N., and Gunn, J.E. 1977, *Phys.Lett.* **B66**, 502.
11. Cline, D. 1986, The 6th Proton–Anti-proton Conference, Aachen, West Germany, review talk.
12. Vittorio, N. and Silk, J. 1984, *Ap.J.* **L39**.
13. Bond, J., Efstathiou, G., and Silk, J. 1980, *Phys.Rev.Lett.* **45**, 1980.
14. Rowan-Robinson, M. 1986, in The Proc. ESO/CERN Symposium on Cosmology.
15. Loh, E. and Spillar, E. 1986, Princeton University preprint
16. Ostriker, J. and Cowie, L. 1980, *Ap.J.* **243**, L127.
17. Olive, K. and Schramm, D.N. 1986 *Comments on Nuclear and Particle Physics*, in press.
18. Schramm, D. and Steigman, G. 1981, *Ap.J.* **243**, 1.
19. Lubimov, A. 1986, in this volume.
20. Frenk, C., White, S., and Davis, M. 1983, *Ap.J.* **271**, **417**.
21. Davis, M. 1986 *Proc. 1984 Inner Space/Outer Space*, University of Chicago Press.
22. Melott, A. 1986 *Proc. 1984 Inner Space/Outer Space*, University of Chicago Press.
23. Blumenthal, G., Faber, S., Primack, J., and Rees, M. 1984, *Nature* **311**, 517.
24. Melott, A., Einasto, J., Saar, E., Suisalu, I., Klypin, A., and Shandarin, S. 1983, *Phys.Rev.Lett* **51**, 935.
25. Efstathiou, G., Frenk, C., White, S., and Davis, M. 1985 *Ap.J.Suppl.* **57**, 241.
26. Schramm, D. 1985, *Proc. 1984 Rome Conf. on Microwave Background.*
27. Bardeen, J., Bond, J., Kaiser, N., and Szalay, A. 1985, submitted to *Ap.J.*
28. Turok, N. 1985, U.C. Santa Barbara preprint
29. Melott, A. 1986, Univ. of Chicago preprint.
30. Charlton, J. and Schramm, D. 1986, submitted to *Ap.J.*.
31. Vilenkin, A. 1985, *Physics Reports* **121**, 1.
32. Vittorio, N. and Schramm, D. 1985, *Comments on Nuclear and Particle Physics* **15**, 1.
33. Turok, D.N. and Schramm, D.N. 1986, in preparation
34. Bennet, D. 1986, SLAC preprint
35. Vilenkin, A. 1986, Tufts Univcsity preprint
36. Rashiputi 1986, preprint
37. Pagels, H. 1986, Rockefeller University preprint.
38. Peebles, P.J.E. 1981, *The Large Scale Structure of the Universe*, Princeton University Press.
39. Olive, K. and Seckel, D. 1986, FNAL preprint.
40. Crawford, M. and Schramm, D. 1982, *Nature* **298**, 538.
41. Witten, E. 1984, *Phys.Rev.* **D30**, 272.
42. Applegate, J. and Hogan, C. 1985, *Phys.Rev.* **D31**, 3037.
43. Alcock, C. and Farhi, J. 1985, MIT preprint.
44. Freese, K., Price, R., and Schramm, D. 1983, *Ap.J.* **275**, 405.
45. Vishniac, E., Ostriker, J., and Bertschinger, E. 1985, Princeton University preprint.
46. Turok, N. and Schramm, D. 1984, *Nature* **312**, 598.
47. Witten, E. 1985, *Physics Letters* **B153**, 243.

Part 5

Summary

Summary

A. Faessler

Universität Tübingen, Institut für Theoretische Physik,
D-7400 Tübingen, F. R. Germany

1. Introduction

Recently a friend of mine gave me an envelope which he had inscribed as shown in Fig. 1. I soon convinced myself that this was again an example for a simple and elegant theory which does not need to be correct.

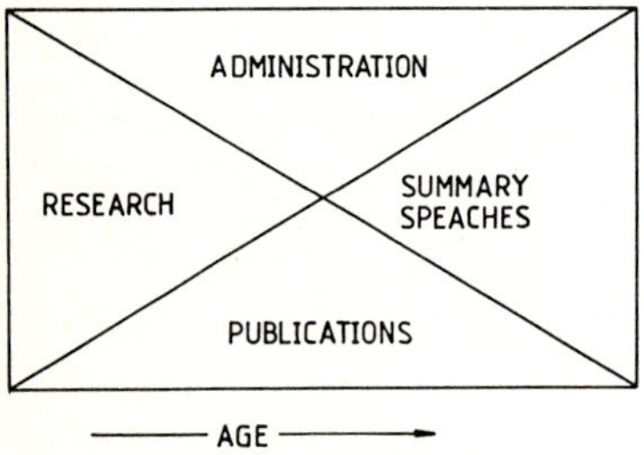

Fig. 1: Example of a simple and elegant theory which does not need to be correct

This conference had a wide range of interesting talks. The conference topics really covered three conferences: One on nuclear physics, one on particle physics and one on astrophysics. Altogether not including the talks of politicians and this summary talk we had 171 invited and submitted contributions. Since it is not possible to summarize in 50 minutes all these topics I want to follow the advice of Adelberger which he gave in his talk citing Weisskopf:

"It is good to cover a whole topic but much better to uncover a part of it".

I want to modify it in the following way:

"It is good to cover a whole conference, but much better to uncover a part of it".

The parts I want to uncover are the following:

(i) A new collective nuclear state 1^+,
(ii) A superdeformed nucleus at extremely high spins,
(iii) Momentum distribution of quarks in nuclei,
(iv) The solar neutrino problem and neutrino properties,
(v) Even mass "super"-heavy elements,
(vi) The positron peaks at GSI.

I will not summarize talks of this morning session since I assume they are still in all the details in your memory.

2. A new Collective State

People who do not work in the field of nuclear structure often believe that collective nuclear states have been detected in the fourties and been studied and understood in the fifties. But the situation is totally different: Most of the different types of collective excitations of the nucleus have been found in the seventies.

- E2 giant resonances of isoscalar and isovector character
- Low and high-lying octupole resonances
- Monopole E0 resonances
- Giant magnetic dipole resonances
- Gamov-Teller resonances

In the talks of Bohle, Nojarov and Gelberg a new type of low-lying collective isovector 1^+ excitation has been described. In Darmstadt, the team of Achim Richter (see BOHLE et al. in these conference proceedings) made inelastic electron scattering for example at ^{156}Gd. In many deformed nuclei of the rare earth region they found around 3 MeV a 1^+ state which one can describe as angle vibrations of the deformed proton against the deformed neutron distribution as indicated in Fig. 2.

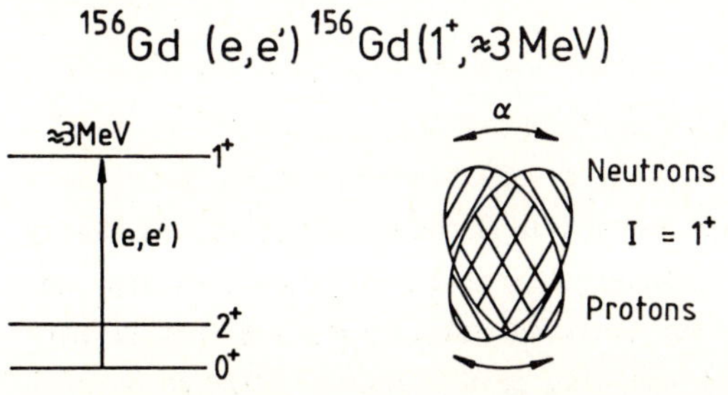

Fig. 2: Excitation of the low-lying 1^+ isovector state around 3 MeV by inelastic electron scattering at the example of ^{156}Gd. This collective isovector state with angular momentum 1^+ is explained as a scissor vibration of the deformed neutron against proton distribution.

Nojarov showed that the restoring force of these scissor vibrations can be calculated from the symmetry energy of the Bethe-Weizsäcker mass formula using a collective model of Amand Faessler from 1966 in which protons and neutrons can be excited to quadrupole vibrations and rotations independent of each other.

The simple picture of the scissor vibrations (see Fig. 2) assumes that these 1^+ states which are excited by rotating the proton distribution against the neutron distribution are pure orbital excitations.

$$R\,\psi = e^{i(I_{xp}-I_{xn})\alpha}\,\psi \approx [1 + i(I_{xp}-I_{xn})\alpha]\psi \qquad (1)$$

From (1) it is obvious that the so excited collective states have the angular momentum 1^+. The central question is now: Are the experimentally found low-lying 1^+ states indeed of this simple character? One essential test could be to find out experimentally that this state is indeed a pure orbital excitation and excitations due to spin flip are of minor importance.

$$\psi = \psi(\text{orbital}) + \psi(\text{spin}) \tag{2}$$

Inelastic electron scattering (e,e') excites both the orbital part and the spin part of the wave function (2). Wesselborg et al. (see contribution of Bohle to this conference) argued that inelastic proton scattering is mainly exciting the spin part of the wave function. Thus if there is no spin part for these low-lying excited isovector 1^+ states they should not be seen in inelastic proton scattering.

$$^{156}\text{Gd}(p,p')^{156}\text{Gd}\ (1^+;\ \gtrsim 3\ \text{MeV})$$

$$V_{NN}(\vec{r}_{0i}) = V_0(r_{0i}) + V_{\sigma\tau}(r_{0i})(\vec{\tau}_0\cdot\vec{\tau}_i)(\vec{\sigma}_0\cdot\vec{\sigma}_i)$$

$$+ V_{\ell s}(r_{0i})(\vec{\tau}_0\cdot\vec{\tau}_i)\vec{\ell}\cdot(\vec{\sigma}_0+\vec{\sigma}_i) + \ldots \tag{3}$$

$$\text{with: } |V_0| >> |V_{\sigma\tau}| >> |V_{\ell\cdot s}|.$$

The experiment was performed at Jülich with 26 MeV protons. The argument runs like this: The state is of isovector nature and therefore can only be excited by terms in the proton-proton interaction which contain the scalar product $\vec{\tau}_0\cdot\vec{\tau}_i$ denoting the isospin Pauli matrices acting on the projectile 0 and on a nucleon in the target i. If the state is indeed of pure orbital nature it can only be excited by the two-body spin orbit term proportional to $\vec{\ell}\cdot\vec{\sigma}_0$. But this term is extremely small so that the state could only be seen if it has also spin character and can be excited by the second term in (3). Since one indeed does not see the low-lying 1^+ states in inelastic proton scattering, one concludes that the state is of pure orbital character. But this conclusion might be a little premature. In inelastic electron scattering one can excite nuclear states only in a one-step-process due to the weak character of the electromagnetic interaction. In inelastic proton scattering a strong central part V_0, which is extremely large at 26 MeV, excites in combination also with the other terms in the nucleon-nucleon interaction (3) many states in multi-step processes. Near 3 MeV we have only few states which can be excited in a one-step-process but one has many states in a nucleus at this energy which can be excited by multi-step processes. Thus the background in proton scattering is much larger than for inelastic electron scattering as indicated in Fig. 3. Thus the fact that one can not see in inelastic proton scattering the 1^+ state around 3 MeV might be solely connected with the larger background due to multi-step processes in inelastic proton scattering.

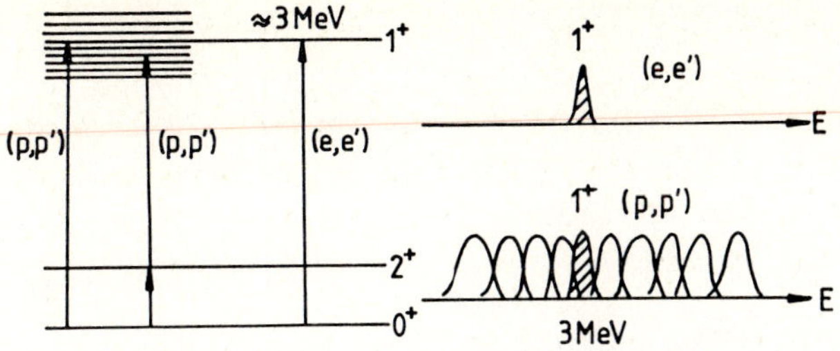

Fig. 3: Inelastic electron scattering excites nuclear states in a one-step process. Inelastic proton scattering excites due to its strong nature also states in multi-step processes. Near 3 MeV are many states which can not be excited in a one-step process starting from the ground state but which can be excited by multi-step processes in inelastic proton scattering essentially also involving the strong central part at 26 MeV bombarding energy. The fact that one does not see the 1^+ state at around 3 MeV might perhaps be connected solely with the larger background in inelastic proton scattering than in inelastic electron scattering.

One possibility to improve the test would be to do the inelastic proton scattering at energies of about 300 MeV. This energy is in the so-called "window" in which the central part $V_0(r_{0i})$ is weak and one-step processes are dominant. An even better possibility would be to use polarized protons at TRIUMF and to measure only those inelastically scattered protons which show a spin flip. Such a set-up was described by Otto Häusser at this conference.

$$^{156}Gd(p\uparrow\ 300\ MeV,\ p\downarrow)^{156}Gd(1^+,\ \approx 3\ MeV) \quad . \tag{4}$$

This condition would exclude one-step contributions from the central part V_0 and thus reduce the background. If in such a reaction performed at TRIUMF the low-lying isovector 1^+ state is not found one can indeed conclude that the spin contribution to the total wave function (2) is negligible. This would strongly confirm the model shown in Fig. 2 for the explanation of this low-lying collective isovector 1^+ state.

3. Superdeformed Nucleus at Extremely High Spins

Several participants at this conference indicated to me that the contribution of Sharpey-Schafer from yesterday afternoon was the high point of this conference. He was looking with the TESSA III collaboration at Daresbury by the reaction

$$^{108}Pd(^{48}Ca\ 205\ MeV,\ 4n)^{152}Dy \tag{5}$$

to high spin states in ^{152}Dy. He measured the de-exciting gamma rays of ^{152}Dy with TESSA III with 12 GeLi's in BGO supressors. The spectrum which he saw is indicated in Fig. 4.

The nature of the excitations of ^{152}Dy up to angular momentum I=36 is well-known. The nucleus is oblate and forms these angular momentum states by independent single-

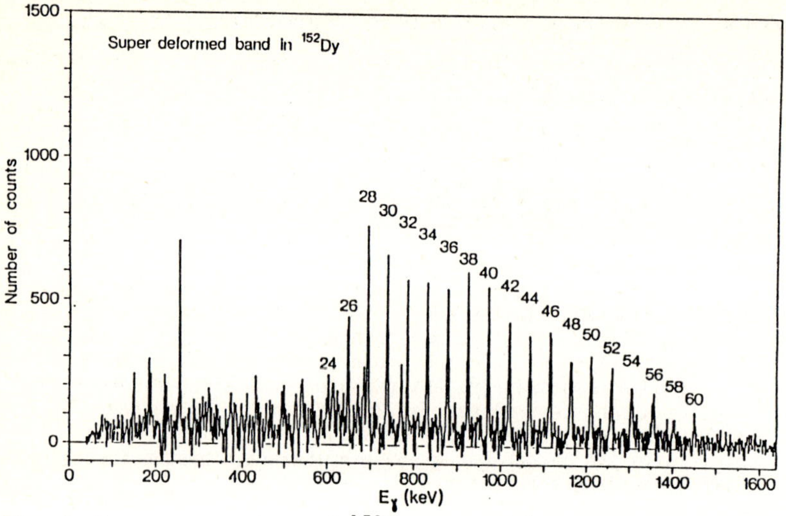

Fig. 4: Gamma ray spectrum of ^{152}Dy measured by the Liverpool group with TESSA III at Daresbury in the reaction (5). The figure shows a regular rotational pattern with 16 transitions between 693 and 1449 keV γ-ray energies.

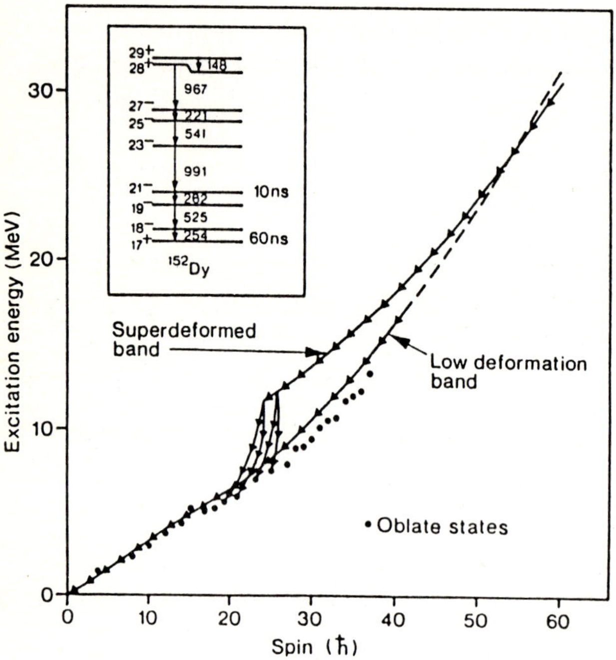

Fig. 5: Excitation spectrum of ^{152}Dy plotted as the excitation energy as a function of the angular momentum. Up to angular momentum 36 the yrast line is a pure statistical one formed by independent single-particle excitation with the angular momenta aligned along this oblate symmetry axis. The 16 collective gamma rays seen in Fig. 4 are indicated here as the superdeformed band with a deformation β=0.6 indicating a ratio of the axis 2:1. The inset at the upper left indicates the statistical yrast states in which the superband is feeding. The major side-feeding goes to the 18^-, 19^-, 21^-, 23^-, 25^- states by 24%, 27%, 10%, 23%, 18%, respectively.

particle excitations which have their angular momenta aligned along the symmetry axis of the oblate deformed nucleus. Since there is no collective rotation possible around a symmetry axis this nucleus forms only statistic yrast states due to the single-particle excitations. This type of excitation yields many isomeric states. These states up to angular momentum 36 are indicated in Fig. 5.

The measurement of the Liverpool group at Daresbury indicates that with higher angular momenta the oblate shape with statistical single-particle excitation gets unstable and deforms in a strongly prolate nucleus with a "superdeformation" of an axes ratio 2:1 with β=0.6. Such a superdeformation at angular momenta 50 to 60 has been predicted in cranked Hartree-Fock-Bogoliubov calculations in the seventies. But it is the first time that this superdeformation is found experimentally.

4. Momentum Distribution of Quarks in Nuclei

The difference in the momentum distribution of quarks in a free nucleon and in the nucleon embedded in a nucleus has first been found by the European Myon Collaboration (EMC-effect). At this conference this topic was treated in talks by Rith, Close, M. Ericson, Nachtmann, Schlomo, Shakin, Mulders, Sick, Milsztajn and others.

Figure 6 indicates the difference of electron- or myon-scattering on a free nucleon and on a nucleus.

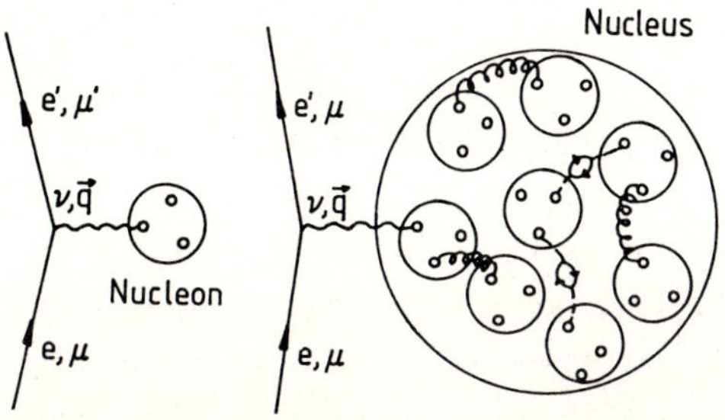

Fig. 6: High momentum transfer scattering of electrons or myons on a free nucleon (left) or on a nucleus (right). The four momentum transfer $q=(\nu,\vec{q})$ is space-like and thus one defines for the four momentum transfer squared the quantity $Q^2 = -q^2$ which is larger than 0.

For small angles only the structure function F_2 is important.

$$\frac{d^2\sigma}{dq^2 d\nu} \propto \frac{\cos^2\vartheta/2}{\sin^4\vartheta/2}\,\frac{1}{\nu}\,F_2\left(\frac{Q^2}{2P_N\cdot q}\right)$$

$$x = \frac{Q^2}{2P_N\cdot q} = \frac{Q^2}{2[E_N\nu-\vec{P}_N\cdot\vec{q}]} \rightarrow \frac{Q^2}{2M_N\nu} = x \qquad (6)$$

$$\text{with: } Q^2 = -q^2 = \vec{q}^{\,2} - \nu^2$$

$$F_2(x) \approx \sum_i c_i^2\, q_i(x)x$$

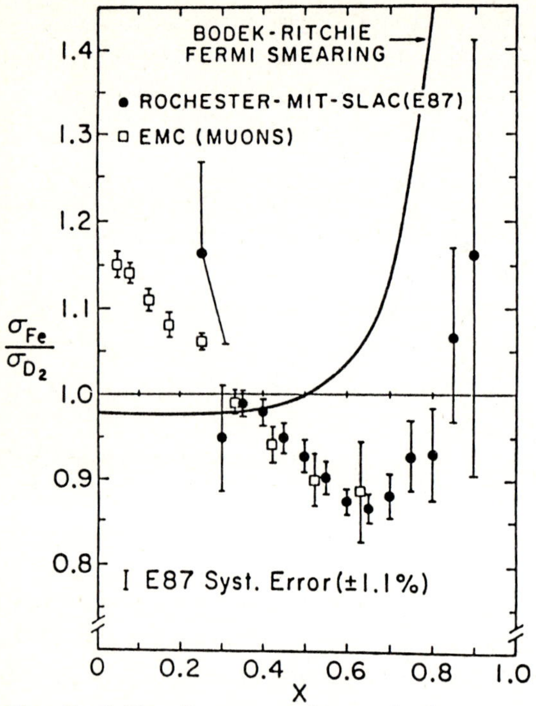

Fig. 7: Ratio of cross sections of electron or myon scattering on iron over the cross section on deuterium as a function of the scaling variable x (EMC-effect).

Here $q_i(x)$ is the momentum distribution of the quark i depending on the scaling variable x which is on the light cone the fraction of the total momentum of a nucleon which resides on quark i. Figure 7 shows the ratio of the cross section on iron and on deuterium as a function of the scaling variable x. Rith gave a nice survey on the data but most of the other talks were devoted to the explanation of the EMC-effect. In the chapters 4a to 4c I want critically summarize the different possible explanations.

4a. Binding of Nucleons in Nuclei

Shlomo proposed in his talk an explanation of the EMC-effect which is put forward by him and several russian groups. He claims that one should not take the scaling variable $x=Q^2/(2M_N\nu)$ in its abbreviated form. The full expression shown on the left hand side in (6) includes the total energy and the momentum $\vec{P}_N$ of the nucleon before it absorbes the exchanged photon. In first order the threemomentum averages to zero but the binding of the nucleon remains.

$$x^A = \frac{Q^2}{2(M_N-|\varepsilon|)\cdot\nu} = \frac{M_N}{M_N-|\varepsilon|}\, x^N \qquad (7)$$

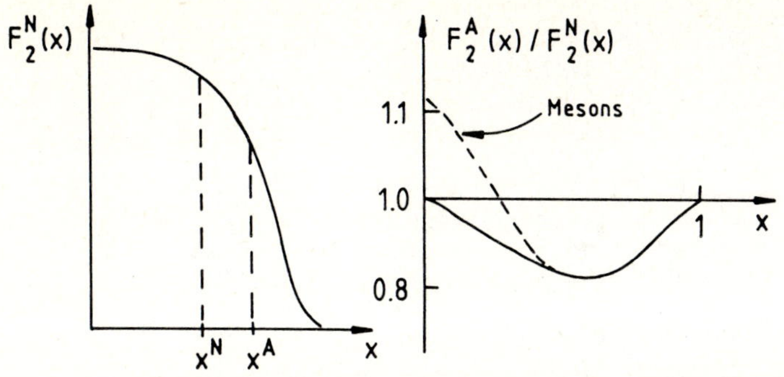

Fig. 8: Qualitative sketch of the nuclear structure function of a free nucleon $F_2^N(x)$ as a function of the scaling variable x. x^N and x^A indicate according to (7) the change of the scaling variable due to the binding effect in a nucleus A. To the right the ratio of the nuclear structure functions F_2 in a nucleus A and in a free nucleon N is sketched. Due to momentum conservation at small x a contribution from the additional mesons which are responsible for the binding effect has to be added.

Due to the binding of the nucleon $|\varepsilon|$ the scaling variable x^A in a nucleus A is increased compared to its value in a free nucleon x^N. Fig. 8 indicates qualitatively the dependence of the free structure function $F_2(x)$ as a function of the scaling variable x. One sees that the scaling indicated in (7) can explain a large part of the EMC-effect. In Fig. 8 one also sees that momentum is missing. This is shifted to the mesons which are responsible for the binding of the nucleons. They give a contribution (dashed curve in Fig. 8) at small x.
This explanation of the EMC-effect is called by Frank Close x-scaling.

An explanation which is connected with this is given by L. Smith, M. Ericson and A. Thomas. They say that due to the interaction (binding) between the nucleons part of the momentum is shifted to additional mesons which move momentum from high x to low x. This explanation has a large overlap with the one given by Schlomo.

4b. Q^2-Scaling

Close and Nachtmann suggested at this conference that the EMC-effect can be explained by the Q^2-scaling suggested by QCD. The theoretical basis is the fact that at higher and higher momentum transfer one sees more and more details. Thus one sees more and more the quark-antiquark pairs and therefore smaller momenta. The hard experimental facts behind Q^2-scaling are qualitatively sketched in Fig. 9.
Figure 9 shows that the structure function in a nucleus can be obtained from the one in a nucleon by using Q^2-scaling.

$$F^A(x, Q^2) = F_2^N(x, \xi Q^2) \quad \text{with: } \xi \approx 2 \qquad (8)$$

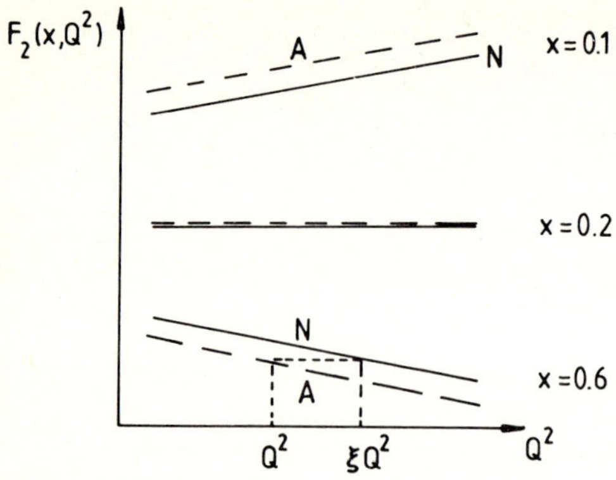

Fig. 9: Sketch of the structure function $F_2(x,Q^2)$ as a function of the fourmomentum transfer $Q^2=-q^2$ for different scaling variables x=0.1, 0.2 and 0.6. The structure function is given for a free nucleon N (solid line) and a nucleon in a nucleus A (dashed lines). One sees that the dashed lines indicating the structure function in a nucleus A can be brought to agreement with the one in a free nucleon by shifting them to the right.

Frank Close argued in his talk that the fact that x-scaling and Q^2-scaling can both explain the data is not an accident of two simple and elegant theories (see introduction) but indicates that both explanations are intrinsically correct. By equating x-scaling with Q^2-scaling he derives that the pion consists essentially out of glue (with an isospin T=1) and that the momentum distribution of the pions in a nucleus must have a definite form. In nuclear physics we know that the binding effects of nucleons in nuclei are not mainly due to pions but due to the so-called σ-mesons of scalar and isoscalar character. Thus the conclusions he draws should not be applied to pions but to the σ-mesons which might be resonance states out of two pions. As a whole the argument of Frank Close sounds to me like the following: The Cheops pyramid is 150 meters high and the sun has a distance from the earth of 150 million kilometers. Thus the old Egyptians did know the distance between the earth and the sun. Some people believe in such arguments. They are hard to disprove.

4c. Nucleons get larger in Nuclei

Immediately after the EMC-effect has been detected it was obvious that it could be explained if the radius of the nucleons is enlarged in nuclei. Such an enlargement of the nuclear radius in nuclei has been discussed at this conference by Shakin and by M. Ericson. Shakin is invoking a general QCD-scale change going from the normal vacuum into nuclei. Magda Ericson favoured meson exchange between neighbouring nucleons for the fact that the nucleon radius is effectively enlarged (see Fig. 10). New compared to former discussions explaining the EMC-effect by enlarging the nucleon radius is the explanation of the longitudinal and transversal response function which can be explained by the same increase of the radius by about 25%.

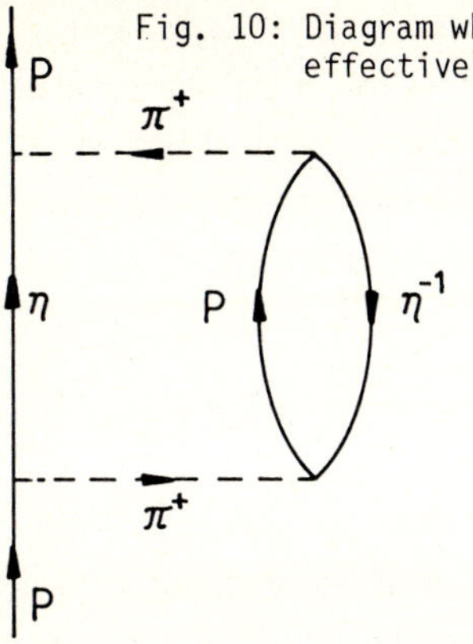

Fig. 10: Diagram which admixes to a proton a two-particle two-hole state which is effectively enlarging the proton radius according to M. Ericson.

For larger scattering angles the differential cross-section (6) has to be written by two structure functions.

$$\frac{d^2\sigma}{dq^2 d\nu} \propto \frac{\cos^2 \vartheta/2}{\sin^4 \vartheta/2} \left[\frac{1}{\nu} F_2(Q^2,\nu) + \frac{2}{M_N} F_1(Q^2,\nu) tg^2 \vartheta/2\right]$$

$$R_L(Q^2,\nu) \propto F_2 - 2x\, F_1 \tag{9}$$

$$R_T(Q^2,\nu) \propto F_2(Q^2,\nu)$$

The longitudinal R_L and the transversal response function R_T can be easily calculated using the free nucleon formfactor multiplying it with the nucleon number and smearing it out according to the Fermi motion. Figure 11 shows that this yields a result which is by about a factor two too large. If one increases the radius of the free

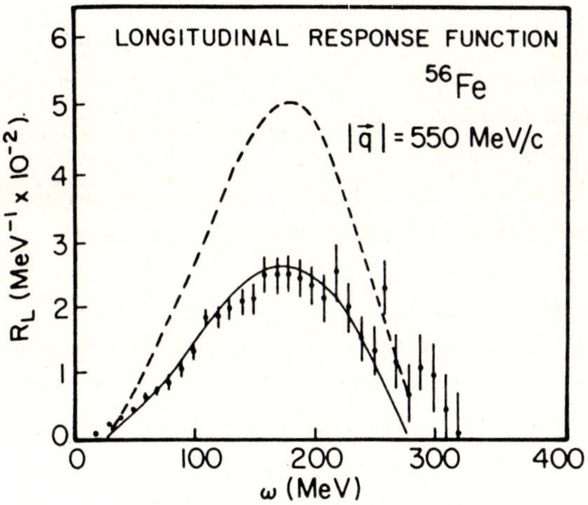

Fig. 11: Longitudinal response function (9) for the threemomentum transfer $|\vec{q}|$= 550 MeV/c measured by electron scattering in ^{56}Fe as a function of the energy transfer $\nu=\omega$ in MeV. The dashed curve is calculated from the free nucleon formfactor. The solid line is obtained by scaling the free nucleon form factor in such a way that the nucleon radius is enlarged by 25%. One obtains with such an increased nucleon radius good agreement with the data.

nucleon by about 25% and uses the corresponding formfactor one obtains agreement (Fig. 11) with the data. Opposite to the longitudinal response function one obtains a good agreement of the transversal response function with the calculated value. But one can argue that if one subtracts the tail from the Δ-resonance one obtains also a too large value compared to the data. Then one again obtains agreement by enlarging the radius of the free nucleon by 25%.

4d. Measurement of the Size of a Nucleon in a Nucleus

If the size of a nucleon inside a nucleus is indeed larger than for a free nucleon one should be able to see this in quasi-elastic electron scattering on a nucleus. Ingo Sick showed us in his talk that in the quasi-elastic scattering région the electron-nucleus cross section divided by the electron-nucleon cross section should be a function of one variable alone.

$$\frac{(d^2\sigma/dq^2d\nu)^A}{(d^2\sigma/dq^2d\nu)^N} = F(Q^2,\nu) = F(y) \tag{10}$$

The scaling variable y is defined as the negative projection of the momentum of the struck nucleon before the interaction onto the momentum transfer from the electron. Thus y is measured in MeV/c and is negative.

Figure 12 shows the ratio (10) solely as a function of the scaling variable y. One clearly sees that all the data which have quite different cross-sections scale excellently as a function of y.

Fig. 12 shows that the differential electron scattering cross sections can only be explained if one assumes that the nucleon has inside the nucleus the same radius as a free nucleon. A 15% increase of the nucleon radius yields a much worse scaling as shown in Fig. 13. The χ^2 fit of the radius change of the nucleon to the scattering data yields a minimum at no change (Fig. 14). One clearly sees from Fig. 14 that an increase of more than 2 to 4% is not tolerable.

But what can then be responsible for the reduction of the longitudinal (and perhaps also the transversal) response function? Several effects have not yet been carefully taken into account which would yield such a reduction:

- -Pauli blocking for the outgoing nucleons due to the rest of the nucleus
- -final state interaction of the outgoing nucleon with the other nucleons in the nucleus
- -short range Brueckner correlations which yield a 15% reduction according to theory

The above three facts seem to be able to explain the needed reduction. Ingo Sick indicated that a new preprint of Fantoni and Pandheripande using the hypernetted chain approach to describe the above effects can reproduce the data.

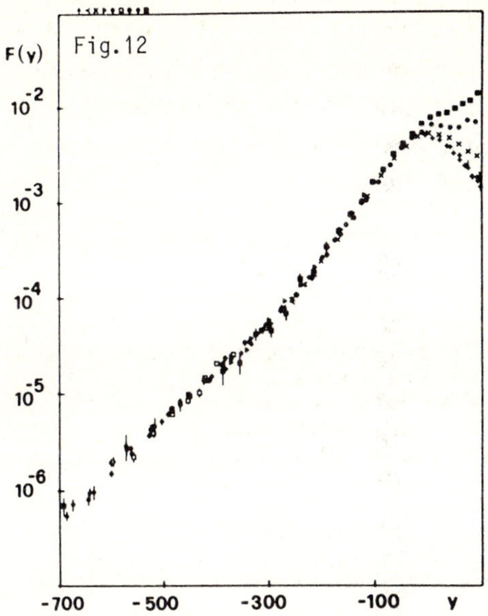

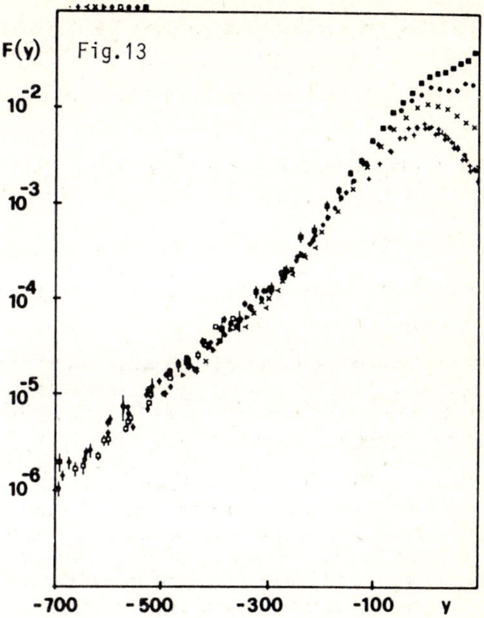

Fig. 12: Ratio of the electron scattering cross-section on ^{3}He divided by the corresponding cross-section on a free nucleon plotted as a function of the scaling variable y which is defined as the momentum of the struck nucleon before the interaction projected onto the negative momentum transfer direction in units of MeV/c. y is negative since it is defined as the negative projection on the momentum transfer.

Fig. 13: The same ratio of the differential electron cross-section on a nucleus Λ and on a free nucleon N as a function of the scaling variable y defined in the figure caption of Fig. 12. Opposite to Fig. 12 the radius of the nucleon inside the nucleus is increased by 15%. One clearly sees that the data are not scaling as well as in Fig. 12.

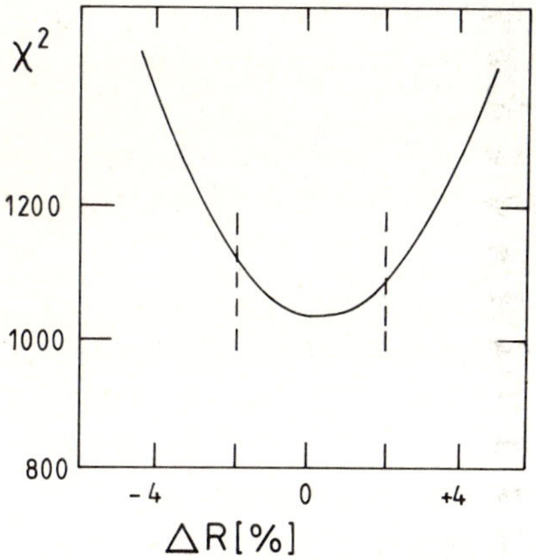

Fig. 14: Chi squared for the fit to the ratio of the differential electron scattering cross section on a nucleus A and on a free nucleon N (10) as a function of the assumed change of the radius of a nucleon inside the nucleus compared to a free nucleon. One sees that it is hard to accomodate a change of the nucleon radius inside the nucleus by more than 2 to 4%.

4e. Explanation of the EMC effect?

If the nucleons inside the nucleus have the same radius as the free nucleons and if the quenching of the longitudinal (and also transversal) response function can be explained without an increase of the nucleon radius, what is then the explanation of the EMC effect?

Keeping in mind that we have at the moment no clear proof about the definite cause of the EMC effect, let me make an educated guess what according to my conviction is causing this effect:

- I am convinced that the main cause for the EMC effect comes from the binding of the nucleons inside the nucleus as discussed in section 4a. This explanation which is often called x-scaling is simple and straight-forward. But is seems not to be able to explain some details, especially the mass dependence of the EMC effect.
- I expect that details of the EMC effect are influenced by the collisions of the nucleons inside the nucleus. During that collision the spatial part of the nucleon wave functions are polarized. One obtains what is often called a six quark bag. But why do then the nucleons look as if they have the same size inside the nucleus? If an electron hits two nucleons which are in the process of colliding (six quark bag) one is not in the scaling region in Fig. 12 but in the deep inelastic scattering region on the right-hand side where the different cross sections do not scale. A dynamical treatment of the collisions of the two nucleons using the resonating group approach has been given by the Tübingen group. Results including this effect are shown in Fig. 15. The solid line includes only the interaction with the next neighbours while the dashed line includes the interaction of all the nucleons in Fe (see Fig. 15). At the moment we prepare in Tübingen a calculation which includes at the same footing the binding effect and the dynamical deformation of the nucleons during the collision inside a nucleus.

4f. The Drell-Yan process

The final decision about the different possible explanations of the EMC effect has to be done by careful experimental measurements. Since all the competing theories are constructed in a way that they all explain the data one needs to do a different type of experiment. A possible way out is the Drell-Yan process which has been discussed in the talk of Berger.

$$\frac{d^2\sigma}{dx_1 dx_2} \propto \sum_i \left[\bar{q}_i^N(x_1) F_2^A(x_2)/x_2 + e_i^2 \, q_i^N(x_1) \bar{q}_i^A(x_2) \right] \qquad (11)$$

with: i = flavor

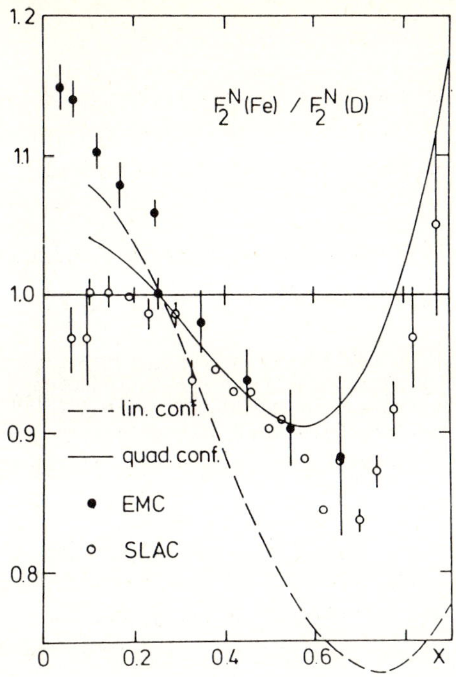

Fig. 15: Structure function of a nucleon inside iron over the structure function of a nucleon in deuterium as a function of the scaling variable x calculated from the spatial polarisation during the collision of two nucleons inside Fe and D. The solid line is for quadratic and the dashed line for linear confinement.

The Drell-Yan process can be performed with any hadron beam on a nucleus. This yields interesting information about the momentum distribution of the quarks in different hadrons. But to get information on the EMC effect we consider here only the nucleon-nucleus collision. The cross section (11) depending on the two scaling variables x_1 from a quark or an anti-quark in the nucleon and on x_2 of a quark or anti-quark in the nucleus consists of two parts: In the first part the anti-quark comes from the sea of the projectile and the momentum distribution of the quarks in the nucleus is described by the structure function F_2^A. This is the same information as one obtains from the EMC effect. The second term contains the momentum distribution of a quark in the projectile N and the momentum distribution of an anti-quark from the sea of the nucleus $\bar{q}_i^A(x_2)$. This last term gives new information. One can now go into different kinematical regimes. If one chooses the scaling variable x_1 around zero, the second term can be neglected since the valence quark distribution in the nucleon $q_i^N(x_1)$ goes to zero as x_1 approaches this value. So one sees that for small x_1 values one obtains the structure function $F_2^A(x_2)$ which is the same information as the one obtained in the EMC effect.

But if one goes to large values of $x_1 \sim 0.7$ the first term can be neglected compared to the second one since the anti-quark distribution $\bar{q}_i^N(x_1)$ of the sea of the

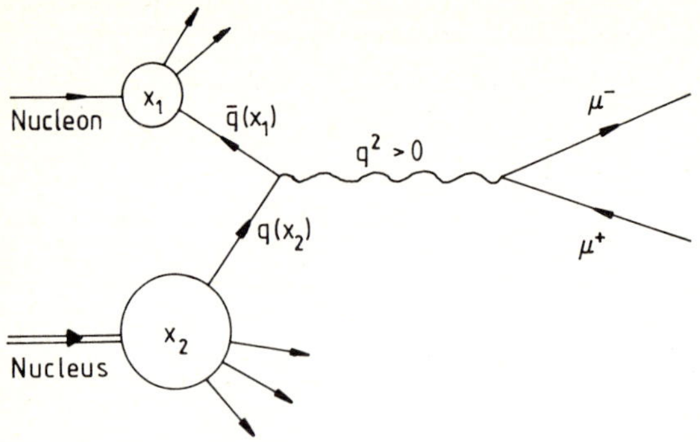

Fig. 16: Drell-Yan process in the collision of a nucleon with a nucleus producing a lepton pair. Either a quark from the nucleus is annihilated with the anti-quark from the sea of the nucleon or a quark from the nucleon is annihilated with an anti-quark from the sea of the nucleus in a timelike photon which produces the lepton pair.

projectile gets zero for larger values x_1. Thus one obtains as a new information the anti-quark distribution of the sea of the target $\bar{q}_i^A(x_2)$.

G. Miller et al. from Seattle have studied theoretically the Drell-Yan process for the proton nucleus collision on Fe and D using different models which give roughly the same results for the EMC effect. Figure 16 shows that the different models yield for the ratio of the Drell-Yan reaction cross sections on iron over the one on deuterium as a function of the scaling variable x_2 quite different results for large x_1=0.7. But in the EMC regime ($x_1 \gtrsim 0.1$) they produce roughly the same dependence. Thus one can expect that the Drell-Yan process is able to distinguish between the different models which can all explain the EMC effect.

In addition one should keep in mind that leptons are blind to flavor while the Drell-Yan process sees flavor. In the Drell-Yan process one can also get in second order an interaction with the gluons inside the projectile and inside the nucleus. This naturally yields additional complications but also the chance to get additional information.

5. Solar Neutrino Problem and Neutrino Properties

This topic was handled in a large number of talks among others by Bahcall, Smirnow, Valle, Hampel, Oda, Sugarbaker, Earle, Zacek, Koang, Kleinknecht, Schreckenbach, Lagage, Kündig, Belotti, Avignone, Ejiri, Grotz, and Tomoda.

5a. The Solar Neutrino Problem

We are all convinced that our sun runs on nuclear power mainly on the p-p cycle which describes the fusion of the protons into α particles. But what is the experimental proof of that? The reaction happens deep in the sun and we are not able to do their measurements. The only experimental information which comes from this region in the sun are neutrinos emitted in the nuclear reactions. So the only possibility to get information about these reactions is to measure these neutrinos on the earth. The p-p cycle is shown in equation 12.

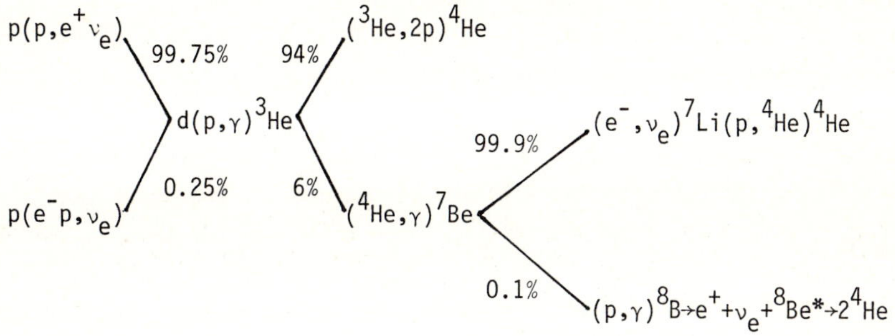

$$
\begin{array}{c}
p(p,e^+\nu_e) \xrightarrow{99.75\%} \\
p(e^-p,\nu_e) \xrightarrow{0.25\%}
\end{array}
d(p,\gamma)^3\text{He}
\begin{array}{c}
\xrightarrow{94\%} (^3\text{He},2p)^4\text{He} \\
\xrightarrow{6\%} (^4\text{He},\gamma)^7\text{Be}
\begin{array}{c}
\xrightarrow{99.9\%} (e^-,\nu_e)^7\text{Li}(p,^4\text{He})^4\text{He} \\
\xrightarrow{0.1\%} (p,\gamma)^8\text{B}\to e^+ +\nu_e + ^8\text{Be}^* \to 2\,^4\text{He}
\end{array}
\end{array}
\quad (12)
$$

The main source of the neutrinos are the reactions:

$$
\begin{aligned}
&p(p,e^+\nu_e)d \\
&^7\text{Be}(e^-,\nu_e)^7\text{Li} \\
&^8\text{B} \to {}^8\text{Be}^* + e^+ + \nu_e
\end{aligned}
\qquad (13)
$$

The neutrinos from the first reaction give a continuous spectrum up to about an energy of 410 keV, while the second reaction yields a discrete neutrino line at around 870 keV. The ^{8}B neutrinos are again a continuum stretching from about 0.7 to 15 MeV.

The ^{37}Cl solar neutrino detector of Davis has a lower threshold of 814 keV and sees therefore essentially only the ^{8}B neutrinos if one takes into account the energy dependence of the cross section for the inverse β-decay. Thus one measures the p-p cycle (12) only in a minor branch which is not very important for the whole energy production. Therefore the result that the ^{37}Cl solar neutrino detector measures only about one third of the expected neutrino flux is not a very valuable test to check or to falsify the present understanding of the physics of the sun.

We would have a more decisive test if we could measure the neutrinos from the first reaction of equation (13). 99.75% of the p-p cycle runs over this reaction. The inverse β-decay of ^{71}Ga

$$^{71}\text{Ga}(\nu_e,e^-)^{71}\text{Ge} \; ; \; Q = -233 \text{ keV} \qquad (14)$$

has a $|Q|$ value of only 233 keV. Thus a ^{71}Ga solar neutrino detector with a lower threshold of 233 keV would measure at least the upper energy and of the neutrinos coming from the first reaction of (13). Such a ^{71}Ga solar neutrino detector is prepared at the moment by a german, italian and french collaboration. In addition also a russian group is building up such a detector.

The measurement of the solar neutrino flux by such a detector will be a decisive step forward in testing the solar energy production and the transportation of the neutrinos from inside the sun to the earth. But let us keep in mind that this will yield us only one number. Naturally, together with the number from the ^{37}Cl detector it will give us new insight. But independently if the number we get from the ^{71}Ga experiment agrees or disagrees with theory we will still have no decisive answer about the energy production in the sun and the transportation of the neutrinos from inside the sun to the earth. Such a final answer could only be given by measuring the full neutrino spectrum coming from the sun. Thus one needs a detector with a very small threshold and the possibility to measure the energy of each neutrino producing the inverse β-decay. This would perhaps be possible with the ^{115}In detector using the reaction

$$^{115}In(\nu_e,e^-)^{115}Sn \; ; \quad Q = -120 \text{ keV} \tag{15}$$

In principle one is able to measure the energy of the electron produced in the inverse β-decay and by that determine the energy of the neutrino for each event by adding to the electron energy 120 keV. But one is not able to distinguish these events from the tremendous amount of background events expected.
The measurements of one or two gamma rays from the inverse β-decay of ^{115}In (see Fig. 17) in coincidence with the electrons makes it perhaps possible to pick out the electrons of the inverse β-decay induced by solar neutrinos from the background and measure their energy. In this way one should obtain the solar neutrino spectrum above 120 keV. Such a spectrum would be a stringent test for any model describing the solar energy production and the transportation of the neutrinos from inside the sun to the earth.

But let us assume that the ^{37}Cl solar neutrino experiment is correct. We have no reason to believe that something is wrong with the experiment. Why do we then observe too few ^{8}B neutrinos? Several possibilities for reducing the number of ^{8}B

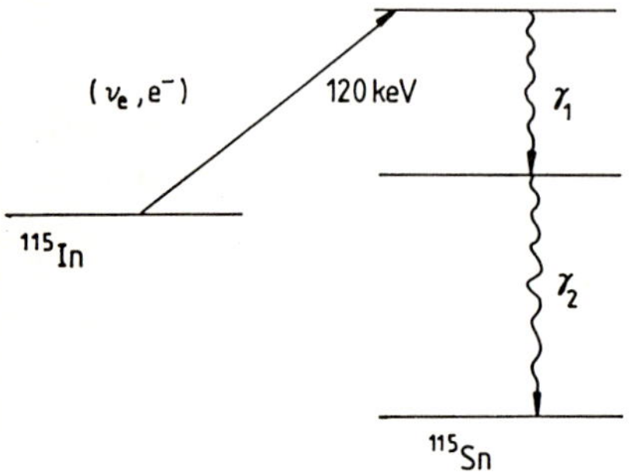

Fig 17: Level scheme of a ^{115}In solar neutrino detector with a threshold for the inverse β-decay of 120 keV and two gamma rays which can be measured in coincidence with the electron. This might allow to pick out the electrons from the background and measure in this way the neutrino spectrum as the electron spectrum plus 120 keV.

neutrinos have been discussed:

- solar models and the temperature in the sun
- nuclear physics
- neutrino oscillations

The astrophysicists insist that the solar models and the temperature they calculate for the sun are correct. There seems to be no modification possible without changing also drastically other results which are in good agreement with observations.

There were some difficulties in the cross sections of several reactions but this has been cleared up and at the moment there seems to be no possibility to change appreciably the ^8B neutrino production if one wants not to get in disagreement with well measured cross sections.

Thus one remains with the question to the particle physicists, what about neutrino oscillations. This is a question which has been discussed in the talks of Smirnow, Zacek, Koang and Kleinknecht. Especially Smirnow discussed a possibility of the neutrino oscillations which seems very promising, to explain the missing electron neutrino flux of ^8B neutrinos.

We assume with different grand unified theories, especially those built on SO10, that as in the quark sector there exists also a mixing between the different families in the lepton sector for the weak interaction. More in detail: The mass matrix of the neutrinos is not diagonal in the weak interaction eigenstates.

$$\begin{pmatrix} \boxed{\begin{matrix} M_{ee} & M_{e\mu} \\ M_{\mu e} & M_{\mu\mu} \end{matrix}} & \begin{matrix} M_{e\tau} \\ M_{\mu\tau} \end{matrix} \\ \begin{matrix} M_{\tau e} & M_{\tau\mu} \end{matrix} & M_{\tau\tau} \end{pmatrix} \tag{16}$$

If we consider for a moment only the mixing between the electron and the myon weak eigenstates as indicated by the box in (16) one obtains for the probability of finding at a given distance L and electron neutrino ν_e if at L=0 the reaction has produced a ν_e:

$$P(\nu_e) = 1 - \frac{1}{2}\sin^2(2\vartheta)[1 - \cos(2.5\Delta^2 L/E_\nu)] \tag{17}$$

with: $\Delta^2 = (m^2_{\nu\mu} - m^2_{\nu e})[eV^2]$

L = distance [m]

E_ν = neutrino energy [MeV]

In the Gösgen experiment, as reported here by Zacek, one obtains an upper limit for the mixing angle and in the Bugey experiment one obtains a value (although in contradiction to the Gösgen experiment):

$$\begin{aligned} &\sin^2 2\vartheta \lesssim 0.18 \quad \text{Gösgen} \\ &\sin^2 2\vartheta \approx 0.25 \quad \text{Bugey} \end{aligned} \tag{18}$$

From (18) and (17) one sees that one never can obtain a reduction to 1/3 of the probability of finding an electron neutrino ν_e.

Smirnow according to a suggestion of Wolfenstein assumes that the interaction of the electron neutrinos and the myon neutrinos with the electrons in the sun are responsible for the reduction of the electron neutrino flux. The electron neutrinos interact with the electron gas in the sun by the charged and the neutral weak current, while the myon neutrinos can only interact via the neutral current. Thus this interaction of the electron gas yields different diagonal matrix elements to the mass matrix of the neutrinos.

$$\begin{pmatrix} M_{ee} + M^m_{ee} & M_{e\mu} \\ M_{\mu e} & M_{\mu\mu} + M^m_{\mu\mu} \end{pmatrix} \tag{19}$$

The diagonalization of this 2×2 matrix yields for the mixing angle of the mass eigenstates of the neutrinos of the first two generations the following value:

$$\sin^2(2\vartheta) = \frac{2M_{e\mu}}{[(M_{ee}+M^m_{ee}-M_{\mu\mu}-M^m_{\mu\mu})^2 + 4M^2_{e\mu}]^{1/2}} \tag{20}$$

The matter contribution M^m_{ee} and $M^m_{\mu\mu}$ to the diagonal elements of the mass matrix due to the interaction with the electron gas in the sun can make the diagonal elements of the mass matrix equal. In this case the mixing angle is $\theta=45^0$. One has a maximal mixing between the weak interaction eigenstates of electron and myon neutrinos. The matter contributions to the diagonal elements of the mass matrix (19) depend on the density of the electron gas in the sun. If the situation is such that the resonance defined by equal diagonal elements in (19) lies somewhere inside the sun one obtains neutrino oscillations qualitatively shown in Fig. 18.
This could at least qualitatively explain the loss of ^8B electron neutrino flux. If this explanation is correct we should have a strong flux of solar myon neutrinos here on earth. Thus as a test for this hypothesis one should measure the solar myon neutrino flux. Thus one should put a high priority for building a solar myon neutrino detector.

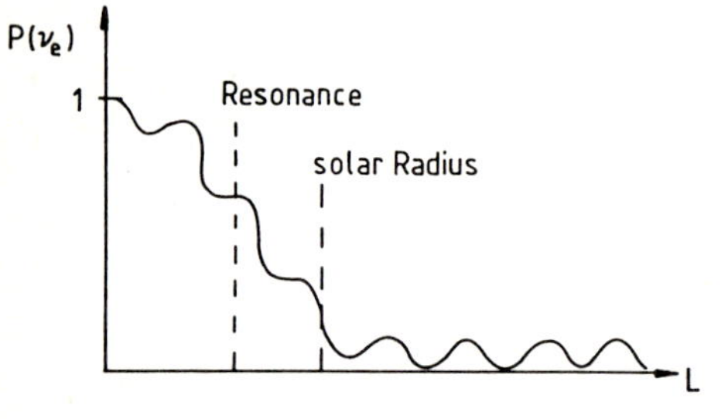

Fig. 18: Probability for finding an electron neutrino as a function from the center of the sun if the matter contribution to the neutrino mass matrix is so that the diagonal elements of (19) are equal somewhere inside or near the surface of the sun.

But this consideration is only a hypothesis. To make it to a theoretical prediction one needs to know the neutrino masses of the electron and the myon neutrino and the mixing angle θ. These quantities one can only find by measurements here on earth in the laboratory. Interest in such measurements are increasing. Some of them we will discuss in the next chapter.

5b. Neutrino Properties

Kündig reported in his talk about the measurement of the electron neutrino mass in the triton beta decay.

$$^3H \rightarrow {}^3He + e^- + \bar{\nu}_e + 18.6 \text{ keV} \tag{21}$$

A finite mass of the electron neutrino could be seen at the upper end of the electron spectrum. This upper end of an electron spectrum around 18.6 keV is shown in Fig. 19. The solid line is a theoretical curve adjusted to the data using a zero neutrino mass or a mass of 35 eV as adjusted by Lubimov et al. to their measurements. By looking to the statistical error one can even see without being a specialist in the field that the Kündig measurement has appreciably smaller statistical errors than competitive measurements from Moscow and Munich. The final analysis of the Kündig data yields including systematic errors:

$$m_{e\nu}\, c^2 \leq 18 \text{ eV} \tag{22}$$

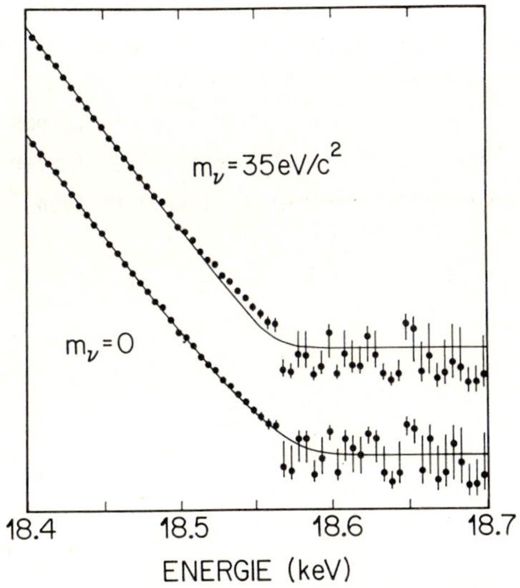

Fig. 19: Upper end of the electron spectrum of the Kündig measurement. The solid lines are fits to the data for the neutrino mass m_ν=0 and the value of Lubimov m=35 eV. One clearly sees that with the data of Kündig a finite value of the neutrino mass with m_ν=35 eV is not in agreement.

If the neutrino is a Majorana particle then the double neutrinoless β-decay can serve for determining its mass. Most of the grand unified theories which are built on SO10 predict that the neutrinos have a mass and that they show also a right-handed weak interaction. This is all interconnected with the fact that neutrinos are Majorana particles, that means loosely speaking that they are identical with their antiparticles.

If the neutrino is a Majorana particle and therefore has a finite mass and a weak right-handed interaction the double neutrinoless β-decay of for example ^{76}Ge into ^{76}Se can happen in essentially two ways indicated in Fig. 20.

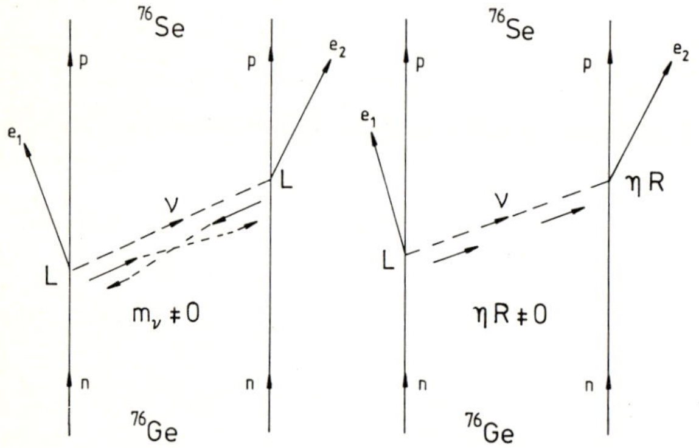

Fig. 20: The double neutrinoless β-decay for example from ^{76}Ge to ^{76}Se can occur by two different mechanisms. On the left-hand side we assume that the neutrino mass is different from zero and thus the helicity of the neutrino emitted by the left-handed interaction L is not a good quantum number and can therefore be absorbed by a left-handed interaction L by a second neutron which is transformed into a proton. The right-hand diagram is only possible if the weak interaction has also a right-handed current R with the strength η. In this way the positive helicity neutrino created at a left vertex by a left-handed interaction L can be absorbed on the right vertex with a positive helicity due to the right-handed interaction R with the strength η.

Tomoda, Faessler, Grümmer and Schmidt did show that in the second diagram of Fig. 20 one has to include the small relativistic amplitudes of the nucleons. In the usual β-decay the relativistic corrections in the transition probabilities go with the neutrino momentum over the nucleon mass squared. This is typically (23) 10^{-6}. Thus one calls β-decays going through this mechanism as double forbidden.

$$\begin{aligned} \beta &: \left[\frac{p\nu}{M_N c} \approx \frac{1}{938}\right]^2 \approx 10^{-6} \\ 2\beta &: \left[(g_{sp} - g_{sn})\,\frac{p\nu}{M_N c} \approx 10\,\frac{2k_F}{M_N c}\right]^2 \approx 25 \end{aligned} \qquad (23)$$

In the double neutrinoless β-decay the intermediate neutrino is virtual. Thus its momentum is not limited by the Q value but by the maximum momentum transfer of a nucleon which stays in a nucleus. This maximum momentum transfer is twice the

Fermi momentum. If one includes weak magnetism one finds that the double neutrinoless β-decay probability is proportional to the second expression in (23) which is about 25 compared to 1 which one obtains if one uses the allowed Fermi or Gamov-Teller transitions. In reality the factor turns out to be even larger. Already the matrix element is by a factor 200 larger than the ones obtained only with the large amplitudes of the nuclear wave functions. If one uses a lower limit for the lifetime or an upper limit for the double neutrinoless β-decay transition probability one can get in this way upper limits for the neutrino mass and the right-handedness of the weak interaction η compared to the left-handedness.

$$w = |M_m m_\nu + M_\eta \eta|^2 \leq w_{upper} \tag{24}$$

If one takes the lower limit for the half life of the double neutrinoless β-decay of ^{76}Ge presented by Caldwell et al. at the Osaka meeting in June 1986 of

$$\tau^{o\nu}_{1/2}(^{76}Ge \rightarrow {}^{76}Se) > 3.9 \times 10^{23} \text{ years} \tag{25}$$

the Tübingen group obtains the following upper limits for the neutrino mass $<m_\nu>$ and the right-handedness parameter $<\eta>$ averaged by the mixing coefficients over the three families:

$$|<m_\nu>| < 1.7 \text{ eV}; \; |<\eta>| < 3.9 \times 10^{-8} . \tag{26}$$

6. Even Mass "Super"-Heavy Elements

Talks about "super"-heavy elements have been given at this conference by Herrmann and by Hofmann. We are aquainted to hear good news from the group of Armbruster at GSI about the confirmation or the detection of new heavy elements

$$\begin{aligned} 1981/82 &: \; Z = 107, 109 \\ 1985 &: \; Z = 108 \text{ (odd A)} \\ 1986 &: \; Z = 108 \text{ (even A)} \end{aligned} \tag{27}$$

The new result reported here in the talk of Hofmann is the detection of even mass Z=108 elements found in the reaction:

$$^{207}Pb(^{58}Fe \text{ 5.1 MeV/A, n})^{264}108 \tag{28}$$

The experimental set-up with which on May 8th, 1986, the first even mass Z=108 element has been detected is the same as the one used in earlier experiments (27). The velocity filter SHIP is used to remove all nuclei which do not correspond to the element Z=108 with the mass A=264. These nuclei are then implanted into a position sensitive detector in which the α decays and the fission is measured. Fig. 21 shows such a decay of an even mass element Z=108, A=264.

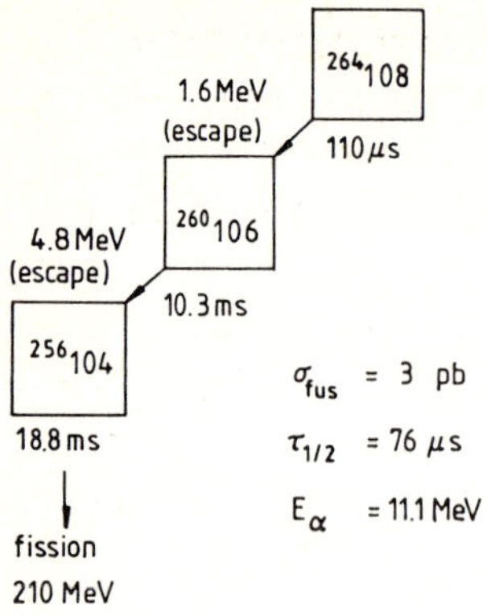

Fig. 21: Decay pattern of one event of the decay of element Z=108, A=264. It decays by two α-particles which escaped out of the detector and therefore the α-decay energy could not be measured in this event. Element Z=104 with mass A=256 decays after 18.8 ms by fission with 210 MeV. From all the measured events one derives a fusion cross section of 3 pb and a half-life for Z=108, A=264 of 76 μs. The first α-decay energy is 11.1 MeV.

The first measurement of an even mass isotope of Z=108 is important since it shows that the relative stability of these elements is not due to some odd-even effect but is a shell effect valid for even and odd mass isotopes. The stabilization of these nuclei to a relative long half-life of 76 μs is due to shell effects. Thus if one defines super heavy nuclei as such nuclei which are solely stabilized by shell effects one would need to include also Z=108 into the super heavies.

7. The Positron Peaks at GSI

Greenberg reported in his talk about the measurements of the EPOS collaboration at GSI of the positron spectrum in the collision between two heavy nuclei. The new result measured already last year is that one sees discrete positron lines in coincidence with discrete electron lines with a total kinetic energy of 760 keV with the total width of not more than 70 keV. The collision between the uranium projectile and a thorium target is shown in Fig. 22. In the electron and positron spectrum one finds also discrete lines at 375 and 380 keV, respectively.

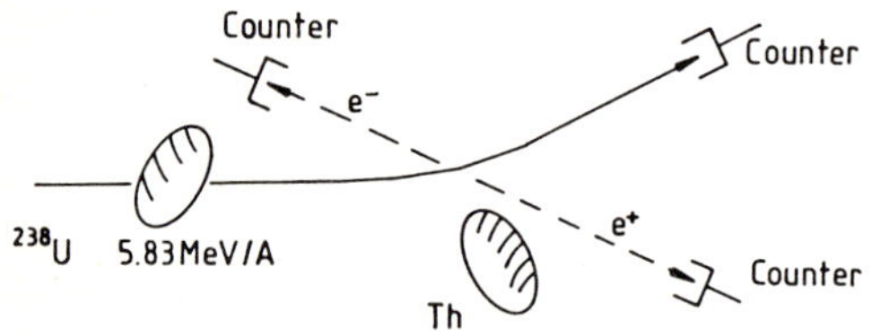

Fig. 22: Bombardment of a thorium target with ^{238}U nuclei of 5.83 MeV/A by the EPOS collaboration. In this experiment the positrons have been measured in coincidence with electrons. The sum of the kinetic energies of the positrons and the electrons in coincidence are 760 keV. The separate electron-positron energies correspond to discrete lines at 375 and 380 keV, respectively.

What are the possible explanations for these coincidences between positrons and electrons? To explain the small width of the sum energy and the discrete lines in each spectrum one must assume that it is something like the decay of a particle almost at rest in the center of mass system. The spontaneous positron production by the diving of the K-shell into the lower Dirac sea is out since this would not yield electrons and positrons in coincidence. In addition, one has seen the positron production also in subcritical systems ($Z<172$). Furthermore, one finds a wrong dependence on the united charge of the projectile and the target. The energy of the positrons is almost constant independent of the united charge while one would expect from the hypothesis of the spontaneous positron production an extremely strong dependence with the power of the order of 20 on the united charge.

Recently Peccei proposed that the electrons and positrons come from the decay of an axion. This can be excluded as shown in the talks of Engfer, Friedmann and de Boer. Engfer showed that the extremely small branching ratios

$$\begin{array}{l} \pi^+ \to e^+ \nu_e a \\ \qquad\quad \llcorner\!\!\longrightarrow e^+e^- \\ \mu^+ \to e^+ a \\ \qquad\quad \llcorner\!\!\longrightarrow e^+e^- \end{array} \tag{29}$$

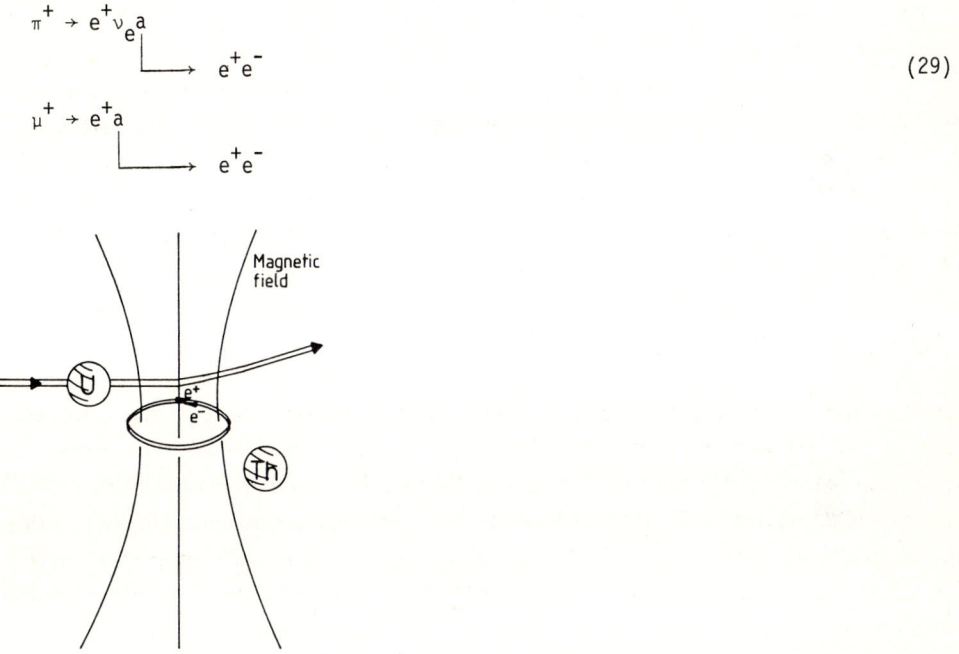

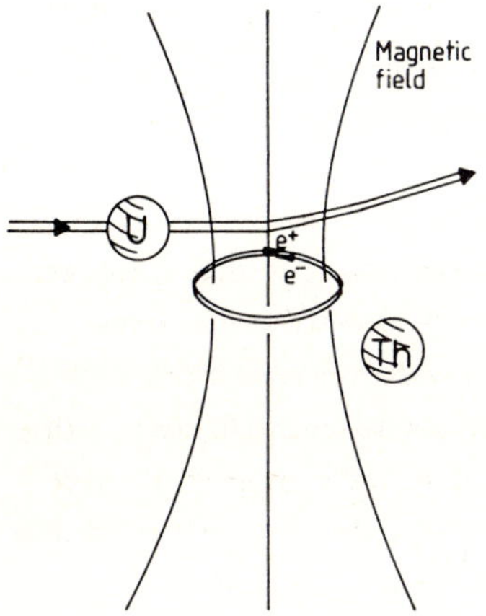

Fig. 23: In the collision of two heavy nuclei like U on Th one obtains by the currents of the two heavy nuclei a strong magnetic field up to 10^{18} Gauss. In this magnetic field an electron positron pair can form a Landau state. Such a ring current has no charge since it consists out of an electron positron pair, but it represents a strong current since the electron and the positron move in opposite directions along the ring. One can obtain not only one but perhaps several such electron-positron pair Landau states. These ring currents may create their own magnetic field which stabilizes them even more. In most cases the electron-positron pairs annihilate adiabatically and the energy goes into the motion of the two heavy ions. But in some cases an electron-positron pair may survive as a Landau state in rest in the center of mass frame and decay after the two heavy ions already moved away. Such a decay could explain the measured data.

of pions and myons exclude the existence of an axion as requested by Peccei. This morning Peccei himself gave in his invited talk a first-class funeral for the axions.

One probably should try harder to search for non-exotic explanations and not invent always new elementary particles which are then proved to be non-existent by other experimental data. Let me paint a classical scenario according to discussions I had with Rafelski and Walter Greiner.
Figure 23 shows such a classical scenario only as a possible example without making the statement that that could be a correct explanation. The idea behind this discussion given in the figure caption of Fig. 23 is to stimulate ideas which try to explain the GSI positrons without invoquing new elementary particles.

8. Conclusions

Professor Brix told us at the dinner in the castle about the history of physics in Heidelberg. Let me conclude here with a story which links the physics in Heidelberg and the one in Tübingen.

When Lenard retired in 1932 the offer for his succession went first to Geiger in Tübingen. Professor Geiger took his oldest assistent, now Professor Haxel here in Heidelberg, and went by train to Heidelberg. He did ring at Philosophenweg 12, the Physikalisches Institut, and the "Hausmeister" opened the door. Lenard was standing up the staircase and when he realized that Geiger wanted to come in he cried: "Don't let him in, close the door." Geiger then said to Haxel: "This solves the problem", and they walked to the railroad station and went back to Tübingen.

Although it is always the privilege of the last chairman to thank the organizers, I think I do not take this privilege away if I thank Professor Klapdor and also Dr. Metzinger and their whole crew from the Max-Planck-Institute for Nuclear Physics and from the University for the excellent organization. Professor Klapdor organized a very wide and interesting scientific program. Together with his team he had also the technical organization of this conference so strongly under control that one had the feeling that nothing could go wrong. I am sure you agree with me when I thank Professor Klapdor and all the people who contributed to the organization for their excellent work.

Part 6

Appendix

Large Scale Computing in Theoretical Physics: Example QCD

K. Schilling

University of Wuppertal, Gaußstraße, D-5600 Wuppertal, F. R. Germany

Evening lecture given on July 1, 1986

The limitations of the classical mathematical analysis of Newton and Leibniz appear to be more and more overcome by the power of modern computers. Large scale computing techniques - which resemble closely the methods used in simulations within statistical mechanics - allow to treat nonlinear systems with many degrees of freedom such as field theories in nonperturbative situations, where analytical methods do fail. The computation of the hadron spectrum within the framework of lattice QCD sets a demanding goal for the application of supercomputers in basic science. It requires both big computer capacities and clever algorithms to fight all the numerical evils that one encounters in the Euclidean world. The talk will attempt to describe both the computer aspects and the present state of the art of spectrum calculations within lattice QCD.

1. INTRODUCTION

This year we celebrate the 40th anniversary of the electronic computer. The first computers carried the frightful names of ENIAC and MANIAC, and they were by today's standards monsters full of old fashioned electronic valves. Their architecture though would still be considered as modern; for it was structured according to the ideas of John von Neumann, just as most of today's computers. As you remember, a von-Neumann-machine consists of a central processing unit, which carries out coded instructions which are stored together with the data in a storage unit.

In the past forty years the computers have been boosted by several orders of magnitude in their performance, mainly due to great technological progress in microelectronics. This has of course changed the notion of supercomputer over the years, since a supercomputer by definition is the most powerful machine at a given time. So today, you can buy for a physicists salary a desk computer, that would have been the pride of the director of your university computer center twenty years ago!

Just to warm up on the topic, let me mention a few technical data which enlighten the speed of miniaturisation just over the last five years:

> Take the supercomputers of CDC and ETA: The Cyber 205, supercomputer of CDC from 1980 contains chips in emitter coupled logic (ECL), with a packing density of 250 gates/chip, each chip producing a heat loss of 5 Watts. Roughly speaking, one needs a million gates to make a supercomputer, which amounts to a total heat production of 20 kW! The chips are interconnected by nearly 800.000 wire connections.

Now the supercomputer to come out this fall, the ETA10:

> It has 20.000 gates/chip in complementary metal oxide technology (CMOS). With a heat loss of 2 Watts/chip, one needs only cool 100 Watts in total. With the miniaturisation of 2400 gates/cm^2 , one can integrate a complete CPU onto one platine. The clocktime of the CPU can be decreased down to 3 nsec.

2. Where the Analysis Ends: Computer and Basic Science

The ever increasing impact of the dramatic progress in microelectronics on human labour has been noticed long ago. It is a common concern for the trade unions all over the world. But who is aware of the implications of the rapid progress in computing technologies onto the natural sciences themselves? A turnover in science history maybe? - I would like to convince you in the rest of my talk, that the present generation of supercomputers - and the ones to come in the near future - will allow us to tackle new problems which were so far not amenable to us. In fact, the rate of qualitative progress that the computers thus offer to the basic sciences is so substantial that we might be tempted to talk about a new era of sciences opening up.

It might be useful to put the short history of electronic computing into perspective to the timescale of modern science. You all know, that the invention of the analysis has very strongly influenced the development of physics. It took about two hundred years from Leibniz and Newton to develop the analysis to its full blossom in the last century. The analysis is a very mighty tool both for the formulation and the solution of physical problems. Just think of the beautiful solution of the two body problem in celestial and quantum mechanics - 17th century and 20th century physics! Yet, at the same time, we notice the limits of analytical methods in solving problems. It ends as early as with the three body problem!

So you arrive soon at perturbation theory. And perturbation theory even of the three-body-problem becomes notoriously complicated in the higher orders! My reverence to such industrious people as the astronomer Delauney, who worked for 20 years on the perturbation of the moon orbit up to the seventh order!

As soon as you leave the regime of linear phenomena, you are totally lost with your analytical tools. This is to say, that the larger part of physics is out of the reach of classical analytical methods. Well-known examples are the anharmonic oscillator or the Navier-Stokes equations in hydrodynamics.

The clever technicians have invented long ago the analog experiment to circumvent the problem. Based on their intuition about simplified hydrodynamical situations, the engineers e.g. simulate aerodynamical flow in a wind tunnel in order to optimize airplane wings. This way mankind was able to build airplanes without fully understanding such complex phenomena as turbulences.

3. Computer Simulation Illustrated With the ISING Model

In basic science we have no wind tunnel, yet we are also faced with highly non linear system with very many degrees of freedom. These systems are defined in the form of models, which might be easy to formulate, but they are in most instances extremely difficult or impossible to solve. Take as an example the ISING model, which you might remember from your textbook in statistical mechanics. ISING formulated it back in 1924 as an extremely simplified mathematical model in order to understand the collective behaviour of a system of elementary magnets on a cubic lattice (1). In the defining Hamiltonian, you start only with nearest neighbour interactions:

$$H = -\frac{1}{2} \sum_{\text{next neighbours}} s_i \, s_k \qquad (1)$$

where s denotes the spin position at the lattice site i which might be either +1 or -1.

The physicists worked for twenty years on the analytical solution of the thermodynamics of this toy model which evidently is of great theoretical interest because it can serve as an illustration about what happens in a phase transition. Onsager became famous for his 1944 analytical solution just for the two-dimensional case. He was able to demonstrate, that the system undergoes a first order phase transition at the critical temperature T = 2.27 . But nobody up to now knows how to solve the model analytically in three dimensions.

The model is however easy to treat with the help of a computer: In order to study the thermodynamics of such a system, you normally start from the partition function

$$Z(\beta) = \sum_{\text{all conf.}} e^{-\beta H} \quad ; \qquad \beta = 1/kT \quad . \tag{2}$$

If you work on a 3dimensional cubical lattice with $V = N^3$ sites, there is a myriad of 2^{ν} different configurations contributing to the sum, most of which though are very unimportant, because they are exponentially suppressed at reasonable temperatures T. So you have no chance to do a straightforward calculation on a computer. What you need is a method to pick out the important contributions, in other word you should have an importance sampling technique. This was achieved by Metropolis et al. (2), who set up a Markov process with transition probability

$$w(\text{state}\{\alpha\} \rightarrow \text{state}\{\beta\}) = \max(1, \exp(-\beta(-H_\alpha - H_\beta))) \quad . \tag{3}$$

You let your computer go over all lattice sites, changing (updating)the spins site by site according to this transition probability. In this way you perform a Markov process, sweeping again and again over the lattice.

It is very instructive to watch this process develop with computer graphics, as exemplified in Fig. 1, obtained with an ATARI PC.

After sufficiently many sweeps, you produce configurations which are distributed according to the well-known Boltzmann probability

$$P\{\alpha\} = c\, e^{-\beta H_{\{\alpha\}}} \tag{4}$$

So you find, that you can compute the average of any observable <A> as an average over n such thermalized configurations

$$\langle A \rangle = \frac{1}{n} \sum_{\nu=1}^{n} A_\nu \tag{5}$$

Note that one needs not compute the partition function (or free energy, entropy) at all in order to extract physics this way.

<u>We summarize at that stage</u>: it is indeed possible nowadays to solve the 3dimensional ISING model numerically just by using the enormous computer power available for Monte-Carlo simulation. The techniques used in these simulations resemble in many respects the methods of experimentalists. In fact, some people like to refer to these computer experiments as '<u>the third method</u>' in science, supplementing the two classical ones: experiment and theory. The analog to a good detector to the computer scientist is the efficient algorithm...on a fast computer. So physicists working in the field have been very competitive in creating fast codes: the world record in the ISING business e.g. appears to be held by Gyan Bhanot at the Supercomputer

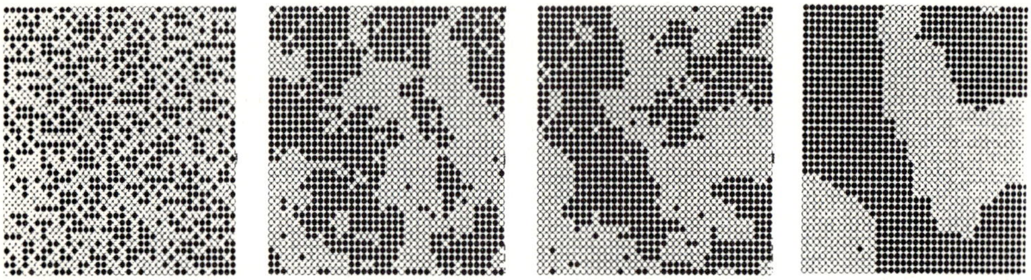

Fig.1.Pictures of a 2d simulation of the ISING model with free boundary conditions. The lattice is cooled stepwise. From left to right: T = infinity, after 20 sweeps at T = 2.3, after another 20 sweeps at T = 1.5, after another 20 sweeps at T = .8. The configurations are clearly still correlated! Black= spin up, white = spin down.

Computations Research Institute in Tallahassee, whose code on a 2-pipe Cyber 205 reached a speed of 100 Megaupdates/second (3). There is of course lots of interesting physics to be learned from these studies, like the critical exponents, finite size scaling etc.etc. But this is not my topic. I have to refer you to the literature (4). I merely mentioned the ISING model as a pedagogical introduction to my main point: Lattice field theory.

4. Lattice Field Theory

While physicists working in solid state physics look at continuum formulations just as an approximation to their lattice problems, Ken Wilson suggested back in 1974 to go the other way around in Quantum Field Theory (QFT) and to consider field theories on a discrete space-time lattice (5). As you know, the evaluation of QFT implies the computation of n-point functions or correlation functions which contain all the physics in the respective theory. They can formally be written in terms of functional integrals, which in most cases again (!) cannot be solved analytically. The traditional way to tackle these integrals is to use perturbation theory. Perturbative field theory is notoriously plagued by infinities which arise due to nonconvergence of loop integrals in the infrared and/or ultraviolet regimes. The ultraviolet problems can be blamed to the fact that in the naive field theory with point particles you pretend to know the correct physics down to all submicroscopic length scales. This is of course not justified. The ultraviolet divergences can be avoided by introducing cut-offs. A common regularisation procedure to get rid of them works by introduction of a high-momentum cut-off P_{cut} in loop integrals. This amounts to injecting a minimal length $a = 1/P_{cut}$. Or to put it differently: you discretize the continuum field theory onto a 4dimensional space-time lattice and thus turn it into a well defined lattice field theory. At the same time, you convert the partition function, which originally had the form of a formal functional integral

$$Z = \int \delta\,\{\text{Fields}\}\; e^{i\int dt \int d^3x\, \mathcal{L}\{\text{Fields}\}} \qquad (6)$$

into a normal integral. Even on a finite lattice this integral is still difficult enough: its dimension equals the number of degrees of freedom of the system, i.e. it is of the order of the lattice volume V.

We are of course finally interested in the limit $a \to 0$ of the theory. This limit can be realized by refining the lattice in sequential steps $\{j\}: a \to \frac{a}{2} \to \frac{a}{4} \ldots$

There is a technique due to Callan and Symanzik (6) known as renormalization group theory that tells you how to do it. You just tune the coupling constants of the theory in each step , such that physical predictions become independent of the actual lattice spacing.

Such procedure should intuitively be the correct one, as long as the physical quantity to be computed, call it mass M, implies a length which is big compared to the actual lattice spacing

$$\xi = \frac{1}{M} \gg a \qquad (7)$$

To phrase it differently: the lattice should be fine enough to possess many lattice points within the physical correlation length. In this manner, the lattice method promises to be a useful technique to compute the previously unaccessible long-range-behaviour of the theory.

We shall see in a minute, that field theory is able to predict the relationship between coupling constant and cut-off in the case of Quantumchromodynamics, the theory describing the strong interactions of quarks, that bind together into hadrons.

Let me just point out at this stage the close formal analogy between the partition function of statistical mechanics and the lattice field theory expression for the lattice approximation of the functional integral equ(6) after passing to Euclidean space-time, $t \to i*t$:

$$Z = \int \prod_{\text{sites } i} d\{\text{Fields}\} e^{-\text{Action}} \qquad (8)$$

The latter formula demonstrates very clearly the practical procedure to evaluate a field theory: you put it onto a sufficiently large lattice, go to the Euclidean world and simulate it on a computer by a Monte-Carlo process as demonstrated some minutes ago for the ISING model.

5. Commercial Number Crunchers

Before I shall elaborate on lattice QCD, I would like to introduce you to some major design features of present supercomputers which are of great importance for the practical realization of large scale computer experiments. All supercomputers by now are built according to the so-called vector principle, which was pioneered on the market ten years ago by Seymour Cray with his CRAY 1 machine. The vector principle helps to speed up the execution of programs that allow for a substantial degree of concurrency. The gain factor is in the order of 10 to 20.

The vector principle results out of a design optimisation, that aims to fully exploit the production capacity of the processors. First you analyze what sort of number crunching is done most of the time. You quickly find, that matrix multiplication is the bulk of the work in most standard applications of large scale computation. Think about partial differential equations: the solution proceeds numerically on a discrete grid of space-time points. The Greensfunction G on this grid is obtained by solving a huge system of linear equations. The matrix involved is a sparse matrix, from the discretised differential operators. The numerical methods typically proceed by iteration. E.g. like this: You want to compute a Greensfunction G to a latticized differential operator D:

$$DG = \delta = \text{localized source}$$

trivially extended
with invertible matrix B:

$$BG + (D-B)G = \delta$$

or converted into iterative scheme,
step (i):

$$BG(i+1) + (D-B)G(i) = \delta$$

which gives you the recursion for the Greens function approximation G(i+1) !

The linear algebra operations involved in this scheme are highly stereotyped. This holds all the more, once you are dealing with matrices of rank 1.000.000 or more as in continuum problems. This structure is exploited in the design of a vector processor. Such a processor is segmented into a number of stations. Once a vector instruction is issued from the instruction processor these stations are able to work concurrently on the various microinstructions involved at different stages of completion of the macrotask. The whole process reminds very much of an assembly line in car industry. Many uncompleted products are simultaneously under work. Once the production line is filled, you have a finished product - here a completed floating point operation - from each assembly line after every cycle of your clock.

The different products of commercial supercomputers vendors, CRAY, CYBER/ETA and the Japanese producers FUJITSU, HITACHI and NEC offer various realizations of this principle, on various stages of hardware technology:

- all of them house different such assembly lines - they are called pipelines, within one floating point unit. If there are more than two such pipelines in a vector unit, they are specialized (to floating point multiplication, addition,...)

- some of them bundle several such parallel vector units into one machine: this has been done by CRAY, that puts up to 4 vector units together, or by ETA, that offers you parallelism from two to eight vector units. These parallel units, of course, can share memory.

A vector machine reaches maximum performance, once all the elements of the vector unit are continuously busy on a steady stream of computations. This stream will of course be interrupted whenever an old production line is closed down or a new one opened up. The startup time is an overhead which becomes negligible with increasing vector length. In order to increase the fexibility of the vector machines, designers have taken great care to cut down the overhead. For the CYBER 205 it still amounts to 50 machine cycles so that the efficiency of the vectorization outpasses the 50 % line only at a vector length of 200. I understand, that this number has been reduced to about 40 with the ETA10. Both CRAY and the Japanese have improved performance on short vector operations from the very beginning. The VP200 of FUJITSU for instance is able to overlay two consecutive vector processes in their closing and opening phases ('pipeline linkage'). It is clear that the full speed number crunching demands a continuous stream of data entering and leaving the 'factory' <u>at the CPU speed</u>. This is of course not a trivial matter, because the CPU is very fast compared to the memory chips that you can afford to use for your random access mass storage. This problem has been solved by structuring the storage both vertically and horizontally:

* Most supercomputers have a fast buffer storage (register) of at least several hundred words which are accessible within one machine cycle. As far as I can see, the most sophisticated vector register setup has been devised by FUJITSU: the VP200 is equipped with a 32 kByte register set that can be dynamically

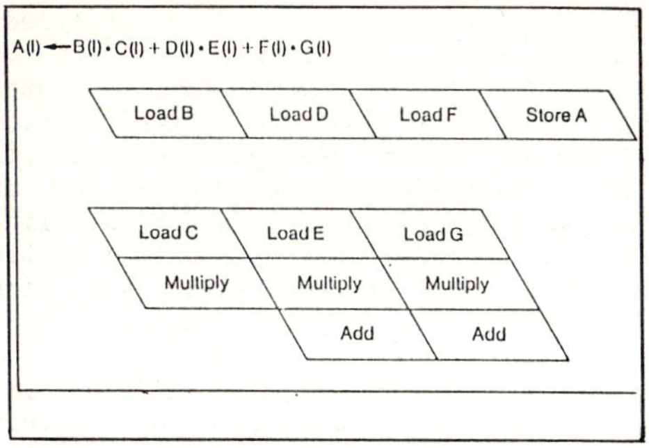

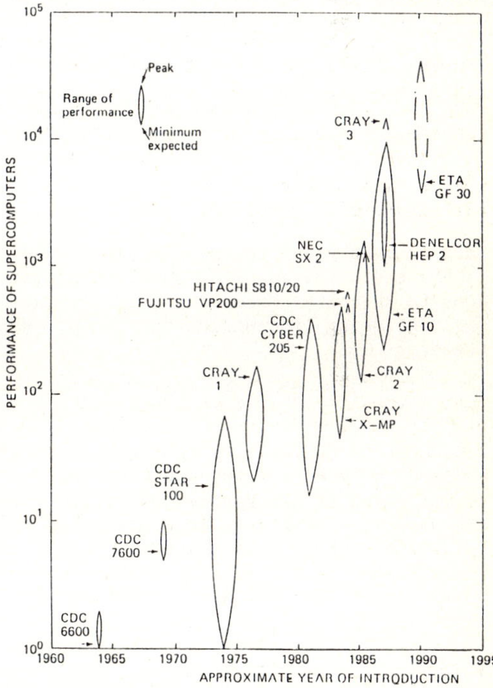

Fig.2 (left). Pipeline parallelism with 'pipeline linkage' as realized in the VP200 of Fujitsu. The loss due to startup times is reduced!
Fig.3 (right) The rise in performance of supercomputers in Megaflops = Mflop/sec vs. the year of introduction, from ref.(8). Indicated are the ranges up to theoretical peak performance.

restructured into different registers of varying length according to the temporary needs. The partitioned register is the hardware basis for the pipeline linkage feature mentioned above.

* The main memory typically has a clock time four times as long as the one of the CPU. Therefore it is common to subdivide it down into 16 to 64 memory banks that are independent, but phaselocked with each other; consecutive elements of a vector being distributed over different banks, it is possible to introduce pipelining into LOAD/STORE operations as well. As a result vector components in natural order can be transferred at a rate of eight 8Byte words per CPU cycle both in the CYBER 205 and in the VP200. When the code requires the components in some other than the natural order, one has to reshuffle the vector, which decreases the speed somewhat (about 15% for the 4dimensional lattice connections).

We summarize: vector computers love to crunch vector components in their natural order. You necessary lose speed, once you cannot avoid in your program

- indirect adressing like G(J) = B(L(J))
- branching like IF...
- recursive structures like G(J) = FUNCTION (G(J-1))

With these caveats, I would like to quote a plot (see Fig. 3) that displays the increase in performance of supercomputers over the years (8). I apologize to the Japanese, that their machines have so minute entries: the plot has evidently been prepared by an American!

After this short excursion on vector machines, let me come back to physics...

6. Quantumchromodynamics Established?

QCD has been proposed in the early seventies as the candidate theory for strong interaction physics . It has the form of a local non-abelian gauge theory with gluons carrying the interaction between the fermionic quarks with three colours. The theory could be nicely worked out for short-distance phenomena, i.e. for situations involving large momentum transfers.

This is mainly due to the fact, that the theory has an ultraviolet fixed point. That is to say, the coupling constant g goes to zero, as the momentum-space cutoff 1/a goes to infinity. More precisely, it has been found (7), that in SU(3) gauge theory, the following connection does hold, which is frequently alluded to as asymptotic freedom:

$$a(g) = \Lambda_L^{-1} \exp\{-\frac{24\pi^2}{33} g^{-2} + \frac{51}{121} \ln \frac{48\pi^2}{33} g^{-2}\} \quad . \qquad (9)$$

Contrary to the history of quantum mechanics of Heisenberg and Schrödinger, which enjoyed an immediate triumph with the correct computation of the Balmer formula, the verification of QCD has seen only slow progress so far. The main testing grounds for the validity of QCD in fact have been in experiments at high energies and momentum transfers, where perturbation theory is expected to work. Detailed experimental checks along this line are greatly hampered by the fact, that perturbation theory is done for quarks which are only indirectly observable, after their hadronization, which is again a nonperturbative low energy phenomenon. The computation of hadron mass spectra, however, is a low energy physics problem and therefore clearly out of reach of perturbation theory; any progress to compute the analogue of the Balmer formula in hadronic physics necessarily requires new theoretical techniques beyond the tools of perturbation theories.

This is precisely the point where lattice methods come into the game: Computer simulation are hoped to supercede analysis. Sic!

7. A Taste of Lattice QCD

Research activities on lattice gauge theory have almost exploded after the discovery in 1979 by Mike Creutz (9), that the static quark-antiquark-potential in pure gauge theory without fermionic matter, as obtained from simulations on 8^4 lattices appeared to rise linearly for large distances (confinement property) with a slope (string tension), that shows asymptotic scaling according to eq.(9). It was then quickly recognized that vectorization could also be applied to Monte Carlo lattice simulations with local interactions.

Therefore, in the past seven years, numerical lattice methods have been applied to tackle a large variety of problems in gauge theories like

* quark potentials
* QCD thermodynamics
* hadron masses
* instanton effects
* chiral condensates
* Higgs phenomenon
* .....

I think it is fair to say that all the previous work in the field must be considered as exploratory in the sense, that systematic errors of the computations are not yet fully under control (12). But the results are so encouraging and motivate QCD investigators to proceed more and more into large scale computing, and even into machine building...

In this talk, I will not even make the attempt to qualitatively review all these topics listed above. Since I have been asked by the organizers to expose the large-scale-computing aspects and implications of the field, I rather aim to give you an idea about its prospects in the present computer scenario. My guideline will be the research activitiy of our lattice group in Wuppertal, so I shall be selective and exemplify matters on hadron mass computations.

How to write the QCD Lagrangian (11)

$$\mathcal{L} = \sum_{\text{flavor } f} \bar{\psi}_f (i\not{D} - m_f)\psi - \frac{1}{4} F^A_{\mu\nu} F^A_{\mu\nu}$$

$$F^A_{\mu\nu} = \partial [\mu A^A_\nu] - g f_{ABC} A^A_\mu A^B_\nu \qquad (10)$$

$$D_\mu = \text{covariant derivative} = \partial\mu + ig \lambda^A/2 A^A_\mu$$

on a lattice?
(Notation: A,B,C=color indices, f=SU(3) structure constants,λ^A=Gell-Mann matrices), g=bare coupling 'constant')

Let's start with the action of matterless QCD! The basic ingredient is the interaction of the local gauge field. This field is living on the links between the lattice points and has values in the gauge group SU(3). We denote it by $U(i,\mu)$, where i stands for the site and μ for the direction of the link starting from i. The interaction is then given by a closed loop construct

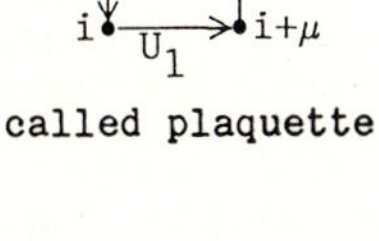

$$\Box_i := U_1 U_2 U_3 U_4 \in SU(3) \qquad (11)$$

called plaquette. It can be shown, that the plaquette action

$$S_g := \beta \sum_i (1 - \frac{1}{3} \text{Tr} \;\Box_i) \qquad (12)$$

has the correct formal continuum limit of the pure gauge sector of interacting gluons. It is gauge invariant under the local gauge transformation

$$U(x,\mu) \to V^{+}(x)U(x,\mu)V(x+\mu) \quad . \tag{13}$$

The combination $6/g^2$ is denoted by ß, stressing the close relationship to the statistical mechanics situation. In order to inject quarks as fermions into the lattice version of the theory, one has to discretize the Dirac equation. For the sake of local gauge invariance the Dirac operator in continuum theory implies covariant derivatives instead of the normal derivatives, as we remember from our electrodynamics class. On the lattice, the derivative operator is replaced by a difference

$$\partial\mu \ \psi(x) \to \frac{1}{a} \{\psi_{x+\mu} - \psi_x\} \ . \tag{14}$$

So you will have contributions of the form $\bar{\psi}_x\psi_{x\pm\mu}$ to the fermionic part of the Lagrangian. These spinorial bilinears can be renderd gauge invariant by inserting the intermediate link operator U(i,i+) in between. In this manner, Wilson 1974 finally arrived at his version of the fermionic action (5)

$$S = S_g + S_f$$

$$S_f = -\sum \psi_x^A \ \{(1-\gamma_\mu)U^{AB}(x,\mu)\}\psi^B_{x+\mu} + \bar{\psi}^A_{x+\mu} \ \{(1+\gamma_\mu)U^{+AB}(x,\mu)\}\psi_x^B - \frac{8+2ma}{4} \ \bar{\psi}_x^A \ \psi_x^A \tag{15}$$

We introduce a "fermionic" matrix Δ^{-1}, which implies a nearest neighbour interaction and reads as follows:

$$\Delta^{-1}_{AB}(x,y) := \kappa_f\{(1-\gamma_\mu)U^{AB}(x,\mu) \ \delta_{y,x+\mu} + (1+\gamma_\mu)U^{+AB}(x,\mu) \ \delta_{x,x+\mu} - \delta_{AB}\delta_{xy}\} \tag{16}$$

The capitals A,B stand for SU(3) or colour degrees of freedom, while the γ_μ denote the usual Dirac matrices, κ_f the hopping parameter.

Hadron propagators will be composed out of quark propagators, which can be written in the form of integrals over the gauge fields and quark fields:

$$Z\langle \psi_A(y) \ \bar{\psi}_B(0)\rangle = \int\{dU\} \ e^{-S_g} \int\{d\psi d\bar{\psi}\}\psi_A(y)\bar{\psi}_B(0) \ e^{-S_f} \tag{17}$$

The integration over the quark fields can be done explicitely with the result

$$Z\langle\psi_A(y) \ \bar{\psi}_B(0)\rangle = \int\{dU\} \ e^{-S_g} \ \det \Delta^{-1} \ \Delta_{AB}(y,0) \tag{18}$$

This means, that one has to average the inverse fermion matrix over appropriately many gauge configurations {U} in order to compute the quark propagator. These gauge configurations should be weighted both with the Boltzmann factor and the fermion determinant.

6. How Large is Large?

So let's assess the numerical problem, once you have a 'thermalized' background configuration: you just have to compute a column of the inverse fermion matrix Δ , which is a sparse matrix carrying colour, spin and spacetime indices. To figure out the numbers for a 16**3*28 lattice: you have to deal with a complex matrix with L = 1.4 Megarows and -columns to be computed out of a "background SU(3) field" with 5.5 million d.o.f. or 22 Mbytes information; the quark propagators necessary to compose hadrons carry 11 Mbyte information, with storage requirements given in real*4 accuracy. So this is not a problem for your local MicroVAX any more! To be honest, it even outpasses most of the supercomputer installations available this summer. Since a word of memory costs about 1 US $, most present configurations around in Germany have just 16 or 32 MByte main memory.

It is important at this stage to acquire some feeling on the necessary size of the lattice. The physical requirements are obvious. In order to reliably compute, say, the proton mass, we require that

* the lattice is fine enough as to have many sites within the proton
* the spatial lattice is 2 or 3 fm in extent
* the typical correlation lengths measured in lattice units are large
* the time-extent of the lattice should be sufficiently large to guarantee observation of asymptotic time-behaviour

There is agreement that the regime of asymptotic scaling does not set in before ß = 6.0. At this ß-value the lattice spacing has been found to be about .1 fm (12). Therefore, the above quoted lattice size corresponds to a world with a volume 5 fm^3, which is not overly large. Yet we are certainly faced with a large-scale computing problem! As an experimentalist in the Minkowski world, you have the pleasure of observing hadron masses as beautiful peaks in some mass distributions. As a theorist doing computer experiments in the Euclidean world, you have to do it the hard way: you typically observe hadron masses - as ground states in the corresponding channels - by analyzing asymptotic slopes of exponentially decreasing hadron propagators (correlation-functions):

$$G(t,0) = e^{-M_{Hadron}\cdot t} \tag{19}$$

That means, that you must fight with the noise of small signals in your computer data! The propagators in our case are deduced from 'hadron channel operators' that are bilinear and trilinear constructs of quark operators. For instance, the pion channel couples to the combination $\bar{\psi}\gamma_5\psi$ of quark operators.

In this theory the mass of the u and d quarks are free parameters that should be tuned to reproduce the empirical rho mass/pion mass ratio. The fermion matrix has zero modes for zero quark masses. Therefore, it is much easier to compute on heavy quark masses, i.e. close to the limit of the static quark model. Most of the computations in the past have therefore been done in this region. For with the very light u and d quarks, you would anticipate convergence problems with the iterative inversion methods for the fermion matrix, due to its nearby zero mode. - For illustration, I would like to mention, that one needs, on a 16**3*28 lattice at ß = 6.0, with 'light' quarks of mass about 100 MeV (corresponding to a pion of mass 600 MeV), about 450 iterations of a conventional conjugate gradient algorithm in order to reach an accuracy of 10^{-5} in the quark propagator. This amounts to 6*10**13 floating point operations. Therefore, in the past most of the the direct numerical computations worked with quark masses in the hundred Mev region and extrapolated down to light quark masses.

6. Reduction of 'Unnecessary' Degrees of Freedom

Much of the computational work goes into the computation of small-distance aspects, which are just lattice artifacts and no continuum physics. It is therefore very natural to concentrate on those degress of freedom which are dominating the infrared content of the lattice physics. In this spirit our Wuppertal group has developed an approximative blockdiagonalization algorithm, that allows to reduce the quark field degrees of freedom by factors 16 (13). This saving is achieved by blocking the lattice into hypercubes, which carry 16 effective mass modes; since only the zero mode among them is important for the long range behaviour of the theory, we are able to cut down the problem to the numerical computation within the zero mode sector. We applied this blocking procedure twice, thus reducing degrees of freedom by the factor 256. This way, we gain a factor 12 in computer time per iteration step of the conjugate gradient. Moreover, since we squeeze the spectrum of the matrix, we gain the benefit of an additional factor 2.2 in convergence of the conjugate gradient.

So altogether we save a factor 256 on storage and a factor 25 on CPU time. This puts us into the position to proceed to hadron propagator computations on the unprecedently large lattice 24**3*48, using just the 16 Mbyte 2 pipe Cyber 205 in Karlsruhe. The project so far needed some five hundred CPU hours. From analyzing the hadron propagators over lattice distances up to 24 lattice units, we were able to extract the hadron masses (14).

7. The Status of Quenched Hadron Mass Calculations

Figure 3 represents the state-of-the-art 1986 on large lattice mass computations. It shows the computed mass ratios m(rho)/m(pi) versus m(nucleon)/m(rho). You see both the static limit and the physical point and the results of a variety of calculations performed by the groups around David Wallace in Edinburgh (15)on the array processor DAPP from ICL and Claudio Rebbi in Brookhaven (16)(on Cyber 205). These groups worked with the socalled staggered fermions (18) whose Dirac components are distributed over neighbouring lattice sites. On a lattice of given size, the staggered fermionic degrees of freedem are reduced by a factor 4, but you buy a very odd interference between flavor and Lorentz symmetries on the lattice. Our measurements are performed with the Wilson fermion action on a larger lattice and down to much lower quark masses. Other measurements with wilson fermions at higher quark masses and on a 16**3*48 lattice have been done by a group in Tsukuba University in Japan on a HITACHI machine(17).

All of these calculations neglect the impact of the fermion determinant onto the equilibrium distribution of background fields - the so-called quenched approximation. The reason is simple: you need at least 100 more in computer time to compute the determinant during the Monte-Carlo updating (19). Therefore, a direct detailed study of the effects of dynamical fermions so far has only been attempted for the (unrealistic) SU(2) case (20).

Physically, the effect of the determinant incorporates the effects of quark-antiquark loops onto the vacuum of the theory, i.e. dynamical fermions. Therefore, the quenched approximation should be o.k. for heavy quarks. Decreasing the quark mass, dynamical quark effects should enter the game; in fact they become necessary to stabilize the quenched approximation. Indeed, we do observe at our lowest quark mass a very striking deviation of the quenched propagators from the normal Gaussian distribution (on three exceptional out of 28 configurations). This can be attributed to fluctuations within the eigenvalue spectrum of the fermionmatrix towards very low eigenvalues. Therefore, the exceptional configurations are expected to be strongly suppressed by the determinant. On the remaining, nonexceptional configurations ('improved quenched approximation'), our theoretical mass ratios for light quarks are in fair agreement with the experimental ones (see Fig.4). So we feel confident, that the lattice computations will finally reach their goal, once we have strong enough machines and clever enough algorithms to treat dynamical fermions.

Our hadron mass project is an illustrative example for the increasing scope of large scale computing projects:

1. We could not have done from the beginning without the help of Philippe deForcrand at CRAY RESEARCH, Inc. in Chippewa Falls, who supplied us with the background fields, that he computed with a CRAY XMP4 with solid state disk. He developed an extremely fast code, that managed to reach an update time of < 6 μsec on a 4 processor machine, attaining a computational speed of 490 Mflops (at 840 Mflops peakrate).

2. We needed substantial computer time on the Cyber 205 machines in Bochum and Karlsruhe in order to develop and apply our method. I think I should mention that thanks to the highly professional und unbureaucratic management of the Deutsche Forschungsgemeinschaft and of the Karlsruhe computer center, we were able to extend our hadron mass calculations to unprecedently large lattices, small quark masses and large statistics. The good old 2-pipe Cyber 205 performed beautifully,

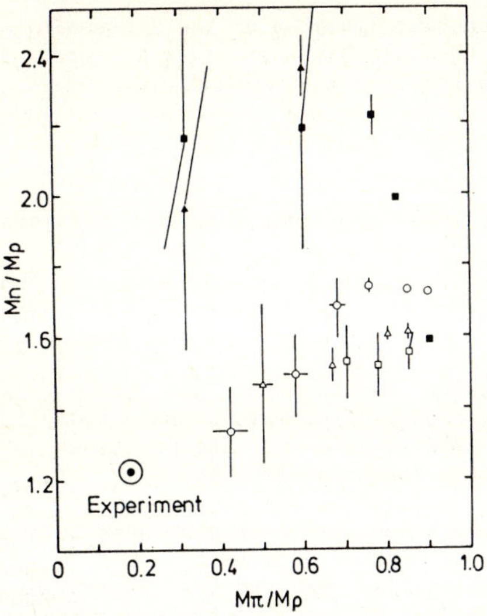

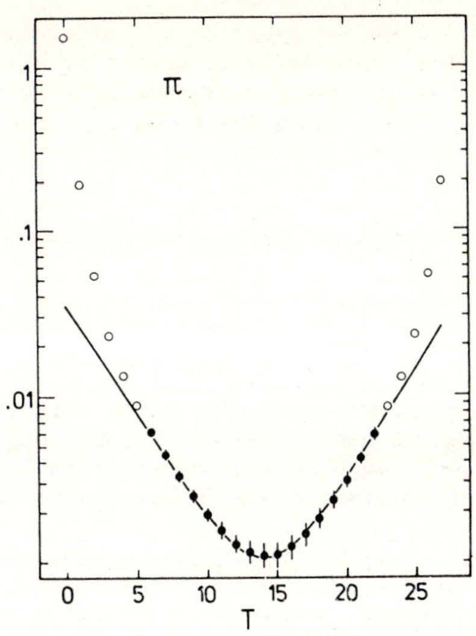

Fig.4(left) Hadron Massratios from lattice computations in quenched approximation: m(nucleon)/m(rho) vs. m(rho)/m(pi). These mass ratios are a function of the quark mass. The plot contains results on 16**3*32 from ref.(16) -symbol △- and 16**4 from ref(15) with Kogut-Susskind fermions with ß = 6.0 -symbol ▲ ■. The Tsukuba group (ref.(17)) -symbol □ - works with a modified Wilson action on a 16**3*48 lattice. Symbol ○ refers to our preliminary data (see ref(14)) obtained with the blocking method and Wilson fermions on a 24**3*48 lattice at ß = 6.3 on a 16 MByte Cyber 205. Configurations with very small values of the fermion determinant were excluded.
Fig.5.(right) The pion propagator as obtained directly (without blocking) from a 16**3*28 lattices by using a VP200. (see ref(21)).

delivering us 65 Mflops or 65% peak rate on the conjugate gradient algorithm (64 bit accuracy).

3. On the other hand, we could not trust our results, if we were not able to get a handle finally on the systematic errors of our blocking approximation: in order to test our method, we are in the process of performing direct hadron mass computations on a FUJITSU VP200 machine on a 16**3*28 lattice. Collaborators in this project are O.Haan and E.Schnepf from the Siemens Company in Munich, which moreover generously supplies us with the necessary computer time. Figure 5 shows you some preliminary data of ours on the pion propagator, which you can very nicely over trace over 14 lattice spacings (20).

You might wonder why we bothered to rewrite our code for another machine. Well, the reason for choosing a Fujitsu machine was twofold:

1. FUJITSU VP200 is presently the machine with the largest available memory (64 Mbyte) in Germany, and
2. We were just curious to study the performance of a non-US product against the Cyber 205.

In case you are interested to learn about our experience: compared to the Cyber 205 the Fujitsu has a beautiful vector compiler, that does all the work for you. So: the user just develops his FORTRAN 77 code on his homely VAX, sends it to the friendly front end Fujitsu computer (with his unfriendly operating system!) and quickly becomes productive on the vector machine without much change to his code. To be specific: our program is mainly a conjugate gradient algorithm, which performed 300 Mflops on the VP200 with 520 Mflops peakrate (to be compared with the 65 Mflops - double precision - reached with a similar program on the 2 pipe CYBER 205 in

Karlsruhe with 100 Mflops peakrate). But sadly enough: inspite of the impressingly high Megafloprate, the VP200 is kept busy with the huge lattice: it still needs 6 hours to compute one full quarkpropagator (with the 12 spin-colour d.o.f. of the source) at an intermediate quark mass.

8. Quo Vamus?

I think I have given you a taste that lattice QCD is quite a challenge and provides motivation, both

*to the theorists to develop better algorithms and organize themselves more efficiently and

*to the computer people to build more powerful machines for the money. It appears that we need a factor 10 more computer power from what we have today in order to pin down the quenched approximation and to look into the questions of dynamical fermions...and a factor 100 to settle dynamical fermions.

In Germany, the Deutsche Forschungsgemeinschaft has just been convinced to set up a major research focus (Forschungsschwerpunkt) to support university research on lattice gauge theory, and federal research centers like DESY and KFA Jülich are launching, together with the universities the project of a computational research center for theoretical physics, which will provide computer time to larger research projects just like CERN offers beam time at its accelerators. The center is supposed to join both computer scientists and theoretical physicists. This development in Germany follows the recent setup of a impressing number of supercomputer centers in the US.

So things are moving. With the commercial supercomputers, we are just entering the Gigaflops era with CRAY2 (1.9 Gflops), VP400 (1.14 Gflops) and ETA10. In the immediate future, the machine with the highest nominal peak rate will be the ETA10, which is a parallel computer with 2 to 8 processors performing 1.2 Gflops each. As I mentioned at the beginning, the machine is based on advanced CMOS technology . For this machine main memory storage will probably not be the bottleneck for quite a while, since there will be 32 Mbytes main memory for each processor, and an additional 2 Gigabytes common memory. Clocktime and memory bandwidth will be improved by a factor 4 with respect to the CYBER 205. As a user I can only hope that we will have soon such a machine available, with many processors. As I am told by the ETA people, it is the cheapest machine ever around, if you scale on the peak rate. But I have not yet seen the pricelist, so I do'nt know what bargain you can make by ordering the maximal configuration. I'm afraid that an eight processor ETA machine will not so quickly be installed. Moreover, it remains to be seen, how frequently the director of your computer center will allow you to use all CPU's simulataneously. So it remains to be seen, who in real life can ever run such a machine at more than, say, 5 Gigaflops.

While the big computer vendors move rather cautiously into parallel computing, ICL has built some years ago a distributed array processor (4096 processing elements), that our colleagues in Edingburgh have used very successfully to do lattice physics on. Quite a few 'lattice physicists' feel, that one should not rely on the most sophisticated hardware, but rather build massively parallel machines (mostly single instruction, multiple data) using 'cheap' components out of more or less off-the shelf.

* K. Wilson is experimenting on a configuration out of many FPS 164 and 264 processors hooked to an IBM 3084

* Our italian colleagues (around G.Parisi and N.Cabibbo) are about to build APE (21), an array processor with a cycle time of 120 nsec. Since they want to reach a one-Gigaflops machine, they chose a design with eight vector pipes for addition/multiplication per processor and bundling 16 such vector units together with a one-dimensional communication channel.

* A Caltech group (with G. Fox) has pioneered a concurrent computation program working with 32,64 and 128 nodes connected like hypercubes in a 5,6 or 7 dimensional world (22). They have produced nice physics results with their first experimental setup.

* A group at Southampton (with T. Hey) is building a parallel machine based on the transputer, a programmable hardware switch for the processor communication.

* D. Weingarten is building with a small group at the IBM Watson Research Center a massive parallel system with 576 processors of 20 Mflops each (23). In 'old fashioned' technology with 200 kW heating power.

I am sure, that all this effort will pay off and that we can report about substantial progress, say in five years time from now. Which is very soon, as seen on the historical time-scale of analysis!

Acknowledgements. I thank my young collaborators, Drs. A.König, R.Sommer, E.Laermann, and E.Schnepf, who are the real experts on the art of vector processing in our group and who let me participate on much about their wisdom about the Cyber 205 and VP200 machines. Thanks also to Dr.H. Gietl (Siemens) for his interest in and support for lattice QCD. I am also grateful to Dr. Soboll (CDC) for his help during the preparation of this talk.

References

1. E.Ising: Z.Phys.31 (1925)253
2. N.Metropolis, A.W.Rosenbaum, M.N.Rosenbluth, A.H.Teller, E.Teller: J.Chem.Phys. 21 (1953)1087
3. G. Bhanot, D.Duke, R.Salvador: 'A Fast Algorithm for the CDC Cyber 205 to Simulate 3-d ISING Model, Tallahasse preprint FSU-SCRI 86-04
4. see e.g. K.Binder:in Phase Transitions and Critical Phenomena, C.Domb and M.S.Green, eds., Vol.5B, Academic Press, New York 1976), and K.Binder(ed.) Monte Carlo Methods, Springer, Berlin-Heidelberg-New York 1979
5. K.G.Wilson: Phys.Rev. D10(1974) 2445
6. see e.g. the textbook of C.Itzykson and J.-B. Zuber: Quantum Field Theory McGraw Hill 1980
7. for the history of asymptotic freedom, see G.t'Hooft: Proceedings of the 'Colloquium in Memoriam Kurt Symanzik', held in Hamburg, Febr.1984, North Holland, Amsterdam
8. S.Fernbach: Future Generations Computer Systems 1(1984) 23
9. M.Creutz: Phys.Rev. D21(1980) 2308
10. for the most recent status,see e.g. the Proceedings of the workshop "Lattice Gauge Theory - A Challenge in Large - Scale Computing", held in in Wuppertal, Nov.1985, R.Bunk,K.H. Mütter, K.Schilling (eds.), Plenum Press, Lodon-New York 1986
    for an introduction, see the lectures of G.Schierholz given at the 27th Summer School of the Scottish Universities in Physics, St.Andrews, August 1984
11. for an introduction see e.g. the textbook of I.J.R Aitchison: An Informal Introduction to Gauge Theories, Cambridge University Press, 1982
12. see e.g. Proceedings of the Workshop 'Advances in Lattice Gauge Theory', held in Tallahassee, April 1985, World Scientific, Singapore 1985
13. K.H.Mütter, K.Schilling: Nucl.Phys. B230 (FS10) (1984) 275
    A.König, K.H.Mütter, K.Schilling: Phys.Lett. 147B(1984) 145
    A.König, K.H.Mütter, K.Schilling: J.Smit, Phys.Lett. 157B (1985) 421

14. Ph. deForcrand, A.König, K.H.Mütter, K.Schilling, R.Sommer, in the Proceedings of the Wuppertal workshop , ibid.,
R.Sommer, contribution to the 2nd Dutch-German Symposion held in Bad Honnef, April 1986
and Wuppertal preprint WU B 86/12, to be published
15. R.D.Kenway, in the Proceedings of the Wuppertal workshop, ibid.
16. D.Barkai, K.Moriarty, C.Rebbi, Phys.Lett. 156B(1985) 385
17. S.Itoh, Y.Iwasaki, T.Yoshie, Tsukuba preprint UTHEP-155, May 1986
18. J.Kogut, L.Susskind, Phys.Rev D11(1975) 395
19. There is a long and still open discussion in the literature about the most efficient algorithm to treat dynamical fermions. We refer the reader to the contributions of J.B.Kogut, G.G.Batrouni, F.Fucito, and D.Weingarten in the Proceedings of the Tallahasse Workshop ,ibid.
20. E.Laermann, F.Langhammer, I.Schmitt, and P.M.Zerwas:' , Masses and chiral Symmetry Breaking: SU(2) Cloour Gauge Theory with Dynamical Fermions', CERN preprint TH 4394/86
21. O.Haan, E.Laermann, K.Schilling, and E.Schnepf: Wuppertal preprint in preparation
22. E.Marinari:'The APE Computer and Lattice Gauge Theories', in the Proceedings of the Wuppertal workshop, ibid.
23. G.C.Fox, S.W.Otto:'Caltech Concurrent Computation Program: A Status Report', Caltech preprint HM 157 B , CALT-68-1317
24. J.Beetem, M.Denneau, D.Weingarten : in the IEEE Proceedings of the 12th International Symposium on Computer Architecture held in Boston, June 1985

Remarks on the History of the University of Heidelberg

G. zu Putlitz

University of Heidelberg, Grabengasse 1, D-6900 Heidelberg, F. R. Germany

Speech given by the Rector, Magnifizenz Prof. G. zu Putlitz at the Concert and Reception in the Alte Aula of the University, on July 2, 1986

Dear Colleagues, let me welcome you at this room, the Alte Aula of the University. This room represents to a very large extent the history of this university and I will take a few moments to explain something to you about this history. The University was founded in 1386, a time when there was big rivalry in Europe between two religious factions, the Pope of Rome, Urban VI, and the Anti-Pope in Avignon, Clement VII. As you know, this was the time of the great schism. In this time the University of Heidelberg was founded. The University of Heidelberg was one of the first foundations not growing out of religious schools but was rather a foundation by a king or a duke, in this case the Prince Elector of the Palatine area, Ruprecht I - you can see him here on the right side in this medallion picture. Many scholars and professors alike had to leave Paris, the Sorbonne, because the Sorbonne adhered to the Pope in Avignon, Clement VII, came to Heidelberg and became part of the faculty or the student body of this university.

On October 23, 1385, Pope Urban VI signed the bull entitling Ruprecht I to found the University. On June 24, 1386, the bull was handed over to Ruprecht I. Then on October 1, Ruprecht I signed his own documents, the constitution of the University and on October 18 a holy mass was celebrated in the Church of the Holy Spirit to start the studies in Heidelberg. The next day, October 19, was the day of the first lectures; among those that I mentioned already in my welcome address yesterday the physics of Aristotle given by Heilmann von Wunnenberg. You see that all this development makes it very logical to use these historical dates to celebrate a 600th anniversary. So, essentially we started on October 23 last year and we will be finishing this year on October 19.

As I said, Heidelberg was one of the first universities founded under the new regime. The sequence is essentially 1348 Prague, 1364 Cracow, 1365 Vienna and 1386 Heidelberg. I'm only mentioning those which have survived. In this sequence of university foundations, Prague became also a place where there was a big struggle between the different nations attending the schools and, as a consequence of this, many professors and students from Prague came to Heidelberg in the succeeding years after the foundation here. In this period there was no language barrier in Europe. Latin was the only language used in science and also there was a free movement between the different countries, something which is unbelievable according to our present-day standards. You can see the big influence from the different regions by noting that among the first 60 Rectors of Heidelberg University, 28 were Magisters from Prague. So at that time faculty exchange was no issue at all in Central Europe.

The history of the University, to sketch it in a few brief terms, is of course very much determined by ups and downs, by wars and other things. For example, in the first century of the University it had to leave many times because of plague. Plague came to the city, and the University had to move to little cities outside Heidelberg. As a consequence of this the University celebrates this year with many little villages a reunion in commemoration of the fact that we once got a refuge there. These little villages are for example Eberbach, Eppingen, Oppenheim, Weinheim, Frankfurt, even Neustadt an der Weinstraße. At all these places we have been once displaced for a period of half a year or a year.

The first century was not really so important academically, but the next one, the 16th century, was quite important for the University and for its academic life because on April 26, 1518, Martin Luther came here to Heidelberg to give his first important disputation in Germany and so Reformation received a fresh start here. Actually there was a big influence of Reformation in Heidelberg in the 16th century and in the second half of the 16th century also Calvinism became here quite important. There was a time in Heidelberg around 1560, when Johann Kasimir was ruling this area as a Prince Elector of the Palatine area, where Calvinism was more important in Heidelberg than at any other place. So Heidelberg was called the third Genf, the third Geneva, because the strongholds of Calvinism were Geneva, Amsterdam, and Heidelberg. So the academic life at that time was flourishing, the discussion about the right religion was going on and was fought with big arguments. For the poor people it was sometimes quite cumbersome because if you pick a certain century in that time, you can recognize that the people living here had to change their religions seven times in a hundred-year period in order to be in accordance all the time with the Prince Elector. This was quite a big change, all the time back and forth; at that time to change your religion had much more influence on your daily life than you can imagine now.

The 17th century then brought two big wars which were quite important for the University. At this site where you are presently sitting, in this Alte Aula, this old lecture hall, the University was located from the very beginning. There was the first building erected here after they tore down some of the living quarters which used to be here. In the beginning of the University time there was a Dionysianum for a hundred years, the so-called Armenburse, essentially a living quarter for students and professors. Then this Armenburse was pulled down in 1560. In 1591, here the Kollegium Casimirianum was erected and this burned down in 1693. A few years later between 1712 and 1716 the present building was erected here as a Baroque building, as the main building of the University. Everything you see in this building is essentially in style Baroque except for this particular room which was redecorated a hundred years ago on the occasion of the 500th anniversary in the style of historism, the style of that time.

Between 1618 and 1648 there was a 30-years' war in Germany and in particular this area was stricken several times. In 1622, Tilly, a general from Bavaria, conquered the city and he took away the very precious library, the most precious library in Central Europe at that time, the Biblioteca Palatina with 3500 handwritings and 5500 printed books, altogether 9000 books. This library was handed over to Maximilian I of Bavaria, who gave it - in exchange for money to carry on the war - to the Pope of Rome, Boniface XV. The Pope made this library part of the Biblioteca Vaticana. If you look back upon the history, this library may not have survived at all in this city because soon after the 30 years' war was finished there was an inheritance war, the so-called Orleans succession war between France and the Palatine area, and in 1693 the whole city was burned down including this building. Also the castle was destroyed; it got its present shape so to speak, and all the buildings in the city were destroyed, including the Church of the Holy Spirit; so the library may have been destroyed at that moment if it hadn't already been a part of the Biblioteca Vaticana in Rome. I mention this also because this very precious library with very world-famous books and handwritings came back for a period of three months in some volumes. We will exhibit 500 volumes of the most precious pieces between July 9 and October 31st in the city on the original site of the library, the balcony of the Church of the Holy Spirit. And if you have a chance to pass Heidelberg in the next three months you should not miss the opportunity to see this library. The books are really fantastic. I have seen a few of those already in the last few days. They have arrived here on a cargo plane of the German Federal Armed Forces.

Well, that was the 17th century. The 18th century was under the influence of Anti-Reformation. The University was without any importance. The influence of the Catholic Church was very strong, academic life was very much suppressed so to speak, and there was again another war which changed the situation. The Napoleonic Wars caused the old Palatine area to dissolve. The part left of the Rhein, on the other side, became part of Bavaria and this part became the Grand Dukedom of Baden. The Grandduke of Baden in 1803, Karl Friedrich von Baden, you see him here on the left side, gave the University a new constitution which made it essentially a state university. As a consequence of this, the University was financed on a regular basis and started to grow and to flourish. In the middle of the last century, many famous scientists of all fields were actually part of this University. Kirchhoff and Bunsen developed spectral analysis here. Leopold Gmelin taught chemistry, Helmholtz was in physiology. These are people known to us as scientists, but also in all the other fields, law and history and philosophy and so on, we had very famous professors here. So at the end of the last century the University had academically a very high standard and this time was its second really important period. Something to mention also: there were small discussion groups all the time, like for example the group around Max Weber, a sociologist, or the group around Henry Thode, a history of arts professor, or the group around Stefan George and Gundolf, a philology group. These groups had a profound influence on German intellectual life but also brought about a very liberal attitude of the University. This University at the turn of the last century was not really Wilhelminic and that is quite important to note because usually at that time the universities were all very much nationalistic, but Heidelberg was not. I think this has been part of the problem which came later when the National Socialists took over, that the very liberal body of professors and students suffered severely from the laws extended by the National Socialists to the University. Well, at the end of the last century the first woman in Germany got a PhD here, actually in philosophy, in the philosophical faculty. The thesis was on partial differential equations. Also in 1885 Shapiro, for example, a mathematics professor, wrote down the first ideas for the Hebrew University in Jerusalem here. The First World War of course was a big incision again particularly because so many professors and students were lost. The time after the First World War was determined by famous names. I mentioned Gundolf and George already. Others to note were for example Gustav Radbruch, he became Minister of Justice in the Weimar Republic, and Karl Jaspers who was a famous philosopher in Heidelberg. But all this ended when the National Socialists came to power. The University lost one-third of its teaching body, of the professors, and a large number of other members. The University came essentially very much under the influence of National Socialists because all the empty positions were filled with people loyal to the National Socialists and also unqualified to quite a large extent. So at the end of the last war the University had to be re-founded so to speak and this was done by our Professor of Surgery, Karl Heinrich Bauer. On January 7, 1946, the University could open again with large help and impetus by the American university officers, and it started to grow again. When I started to study in 1953, the University was pretty much in order again and you could study here regularly because big losses already had somehow been covered up.

At present the University has 28,000 students, it has 10,000 employees including the medical department and the clinics, it has 600 professors, 248 of whom are full professors, it has a budget of 650 million Marks, it is in other words a big enterprise. Half of the students are still in humanities which shows a strong tradition in these fields. We have two campuses, one here in the city for humanities and one in the Neuenheimer Feld, which is about 5 kilometers from here, for the sciences. In the course of our anniversary we have 100 international meetings in Heidelberg which hopefully get many people here to be in contact with the University, at least for a short time. And last not least we have founded an international science forum, a place where some of you have the privilege to live in the Hauptstraße 242 which we hope in the future will be a place of intense scientific discussion in continuing a tradition which was very influential here in Heidelberg at the turn of the century. I should mention also that this coming weekend, from Friday morning on, we celebrate a big festival for all the former students and the present students and any other guests in the city and I hope you take the advantage during this meeting here and sneak away from Mr. Klapdor and participate in our celebrations. Thank you very much.

Six Hundred Years of Physics at Heidelberg

P. Brix

Max-Planck-Institut für Kernphysik, Postfach 10 39 80,
D-6900 Heidelberg, F. R. Germany

Before-Dinner Speech at the Banquet of the International Symposium on Weak and Electromagnetic Interactions in Nuclei in the Königssaal of Heidelberg Castle, on July 3, 1986.

Magnifizenz, ladies and gentlemen, dear colleagues:

Six hundred years of physics at Heidelberg: I shall hurry through them with a speed of two seconds per year, but that will be an average speed.

Physics started here from the very beginning. Already the second rector lectured mainly physics. His name was Heilmannus Wunnenberg de Worms, elected 1387. We have no picture of him. The first portrait available of a physicist at Heidelberg is that of Theophilus Mader de Frauenfeld, a Swiss, teaching 200 years later (Fig. 1). This physicist was elected in 1593 as the 369th Rector Magnificus of Hei-

Fig. 1 Theophilus Mader de Frauenfeld

Fig. 2 The big comet over Heidelberg (1618)

delberg university. You may calculate that in those times a rector could only stand less than a year in office. Our present anniversary rector, the physicist Gisbert Frhr. zu Putlitz, has to serve four years.

Studying physics in the 16th century amounted to reading and discussing the books of Aristotle; early physics experiments - mainly on magnetism - started outside of Heidelberg in Nuremberg and in England.

Figure 2 shows Heidelberg in 1618. The big comet announced to the city the beginning of the 30 years-war and decades of great suffering. But the comet also points to the site of the Max-Planck-Institute for Nuclear Physics which was to be founded by Wolfgang Gentner 340 years later. And who could have imagined at that time that this Max-Planck-Institute would take an active part in a mission "Giotto" to Halley's comet when the university was to celebrate its 600th year of existence?

Another document of the 17th century even forecasts our present conference on weak and electromagnetic interactions. Honestly, only the weak interactions have a specific relation to the castle of Heidelberg where we have our banquet tonight. First I have to explain a little bit to the non-physicists what we experimental nuclear physicists are doing. We are mainly shooting projectiles at targets. If we hit, we call that an "event", and we are happy and wait for the next event. If nothing happens, however, physicists are not disappointed. We just say: well, the cross section is very low. But if we try again and again and nothing happens at all, we conclude that we have discovered a weak interaction. However, very very rarely even in a weak interaction an event may happen. Then there is great joy; a paper is written and all the authors eagerly wait for the next conference on weak interactions.

It will probably never happen again that the highlight - the banquet - of such a conference is taking place at the site where a totally improbable event was first recorded in history. If you have time to walk to the "thick tower" of Heidelberg castle (still intact on Fig. 2, blown up 1689), you will find a monument. There is carved in stone (loosely translated):

Anno 1681, on the 22nd of January
Kurfürst Karl against any reasonable chance
hit a (flying?) bullet with a bullet.

("Hat wieder alles Hoffen mit Kugel Kugel troffen"). This really shows how cleverly Prof. Klapdor has planned everything for this conference.

Only a few years after 1681, the university was destroyed, the city burned, the people had to flee. When the university of Heidelberg was exactly half as old as it is now (1686), Isaac Newton presented his manuscript on the theory of gravity to the Royal Society of England. This may be taken as the end of the physics of Aristotle. A new age for natural science began, but it did not dawn at Heidelberg. Here we had to wait till 1752 before the first professor of experimental physics was appointed. His name was Christian Mayer. Figure 3 shows his picture; he was then 33 years old. Mayer was a man of good humour throughout his life. His salary was one third of that of a professor of law at that time. He also got every seven years the money to buy himself a gown for his lectures, and he received 15 bottles of wine per year.

Fig. 3 Christian Mayer (1719-1783)

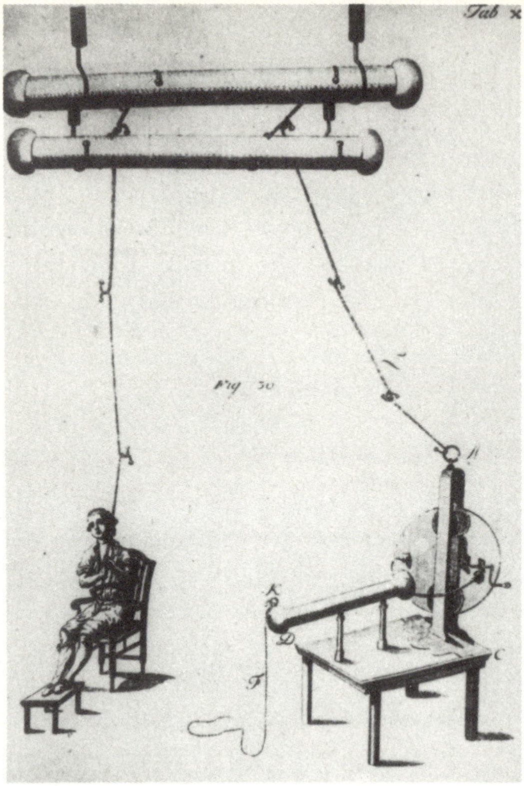

Fig. 4 From a publication of the Mannheim Academy of Sciences 1780

Mayer's auditorium first was on the upper floor and then on the ground floor of the "Alte Aula" where we had this wonderful concert last night, and the reception by the rector. Physics in the 18th century was dominated by studies and discoveries of the effects of electricity. I am happy to show in Fig. 4 a Heidelberg student studying the electromagnetic interaction 200 years ago.

In 1854 a world-famous physicist came to Heidelberg: Gustav Robert Kirchhoff (Fig. 5). Together with the chemist Robert Wilhelm Bunsen he developed here in 1859 "the chemical analysis by spectral observation" as their first paper was called. He applied spectral analysis to the sun and the stars. Kirchhoff was a very careful man. Only after he had calculated a probability of 10^{-12} that 60 lines observed in the sun agreed by chance with 60 lines of the iron spectrum in the laboratory, he dared to state that he had found iron on the sun. Modern physicists may be well advised to think of this occasionally. A memorial plaque at the "Haus zum Riesen" in the "Hauptstrasse" identifies the building where Kirchhoff and Bunsen made their discovery. It housed the first physics laboratory in Germany where students could do experimental work.

Figs. 5 and 6 Gustav Robert Kirchhoff (1824-1887) and Hermann von Helmholtz (1821-1894)

Opposite to that house stands the Friedrichsbau (Fig. 7). From 1863 to 1913 it housed the physics institute of Heidelberg university. It also was the home of other natural sciences. Here Kirchhoff had his official living quarters.

The man who is considered the greatest German physicist of the last century worked in Heidelberg from 1858 to 1871: Hermann von Helmholtz (Fig. 6). However,

Fig. 7 "Friedrichsbau", housing the Institute of Physics from 1863 to 1913

he belonged to the faculty of medicine. Incidentally, Helmholtz got the highest salary that had ever before been paid to a Heidelberg professor. When he went to Berlin in 1871, they built the biggest German institute of physics for him. At the 500th anniversary of our university, Helmholtz was asked to speak in honor of Heidelberg at the ceremonial banquet of August 4, 1886 (taking place in the museum where now the "New University" building stands). Helmholtz asked: Is it by chance that from these green hills mankind looked for the first time into the unthinkable depth of the universe with an insight into the chemical nature of the stars, an endeavour which immediately before must have appeared as ridiculously impossible? ("Ist es ein Zufall, daß von diesen grünen Hügeln aus der geistige Blick des Menschen zum ersten Male mit der Einsicht, wie die chemische Natur der Weltkörper zu ergründen ist, in die unermesslichen Himmelsräume drang? Ein Unterfangen, welches unmittelbar vorher noch als die abenteuerlichste Unmöglichkeit erscheinen musste! Ich glaube das Gegenteil. Etwas vom Schauen des Dichters muss der Forscher in sich tragen.") Helmholtz remarked that the scientist ought to have something of a poet's vision, and he closed with a toast to

"Alt- Heidelberg, die feine
Die Stadt an Ehren reich,
sie wachse, sie blühe, sie lebe!"

"Old-Heidelberg, dear city,
With honors crowned, and rare,
O'er Rhine and Neckar rising,
None can with thee compare."

We owe this English version of the song by Viktor von Scheffel (who died 100 years ago, 1886) to a great American friend of Heidelberg, Jacob Gould Schurman. He was a student here for one year only, 1878/79. But 50 years later, as Ambassador to Berlin of his country, the United States of America, he came back to Heidelberg and presented the university with over half a million dollars which he had collected to provide a new "Hall of Instruction". The white "New University" building is still our central lecture hall. When Schurman got an honorary doctor's degree from the university of Heidelberg, he surprised the audience by expressing his thanks with his translation of Viktor von Scheffel's song in praise of Heidelberg, the "city of merry fellows, with wisdom lad'n and wine".

When Heidelberg university was 550 years old (1936), Schurman declined to take part in the celebration. The "living spirit", that was written over the entrance of "his" building had left the university, it was later replaced by the so called "German spirit". Even the donor's plaque and Schurman's bronze bust were removed. These were dark years for Germany, and they were especially dark and shameful for physics in Heidelberg, due to Philipp Lenard. Lenard, the 1905 Physics Nobel Laureate (for his research on cathode rays), became director of the Heidelberg physics institute in 1907. A new, big institute building (Fig. 8) was built for him,

Fig. 8 Physikalisches Institut der Universität Heidelberg, 1914

and according to his plans wonderfully situated on the other side of the Neckar at the Philosopher's path. It was completed in 1913 and is still the central home of physics at Heidelberg. Our regular Friday colloquia take place in the auditorium, the windows of which may be seen to the right in the picture. In the 1920's Lenard became an ardent National Socialist. He hated Jews and what he called "Jewish science", and even as an emeritus exerted a terribly bad influence beyond physics and far beyond Heidelberg.

The miracle that the physics institute of Heidelberg university, named "Philipp Lenard Institut" in 1935, became an internationally respected center of physics soon after the war is first of all owed to one man: Walther Bothe (Fig. 9). He had been appointed successor of Lenard in 1932. But when the Nazis came to power, life became unbearable for him at the institute. He resigned from the university, but by good fortune the Kaiser-Wilhelm-Society (now: Max-Planck-Society) could in 1934 offer him the position of Director of Physics at the Max-Planck-Institute for Medical Research in Heidelberg. There he was able to continue his research, mainly in the field of nuclear physics, together with Gentner, Maier-Leibnitz and others.

Wolfgang Gentner (Fig. 10) built a Van de Graaff accelerator which was used to generate high energy gamma rays from proton capture on lithium 7. With these gamma rays, Bothe and Gentner discovered in 1937 the nuclear photo effect for complex nuclei (a Heidelberg landmark of electromagnetic interactions in nuclei!). Only the photo-disintegration of the deuteron (Chadwick and Goldhaber 1934) and of beryllium (Szilard and Chalmers, Gentner 1934) had been found before.

Figs. 9 and 10 Walther Bothe (1891-1957) and Wolfgang Gentner (1906-1980)

Figure 11 presents the life spans of some Heidelberg physicists. When Kirchhoff went to Berlin in 1875, Georg Quincke became his successor (till 1907). He was the teacher of Lenard.

After the last war had ended, Bothe resumed his duties as professor of physics at Heidelberg university. He succeeded in getting Jensen (Fig. 12) to Heidelberg in 1949, and together they persuaded Haxel to join them. I have not included Otto Haxel and Christoph Schmelzer in Fig. 11 because they are fortunately still with us (and all the other living famous Heidelberg physicists have also been omitted!).

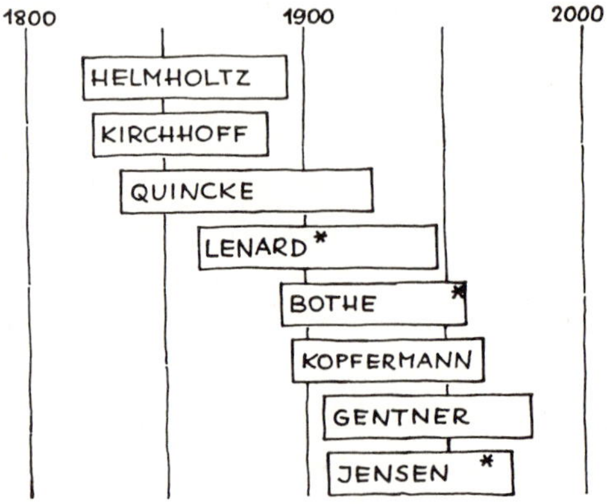

Fig. 11 Life spans of Heidelberg physicists (star: year of Nobel prize)

Figs. 12 and 13 Hans Daniel Jensen (1907-1973) and Hans Kopfermann (1895-1963)

Hans Kopfermann (Fig. 13) came to Heidelberg in 1953 from Göttingen. He was a pioneer in hyperfine spectroscopy, author of the well known book "Nuclear Moments" ("Kernmomente"), a great "chamber musician" of physics, as Weisskopf called him in his obituary. Time runs out, so I could only briefly mention those physicists to whom we owe the rebirth of physics at Heidelberg university after 1945.

But I wish to pay special tribute tonight to Wolfgang Gentner. He could have celebrated his 80th birthday this month (23.7.1986). When Gentner died on September 4, 1980, the Journal Physics Today wrote that "the international community of physicists has lost one of its few great exemplars. He should be remembered not only because of his contributions to physics but also because of his moral character, which enabled him to live through the most tragic period of his country without ever compromizing with the evil forces of those days". In this Königssaal in the castle of Heidelberg a memorial ceremony and colloquium in his honor took place on 1st of April 1981. Weisskopf was one of the speakers; it should be mentioned at this conference that he stressed Gentner's decisive contribution to the creation and development of CERN. Figure 14 shows Gentner when he became a member of the Board of Governors of the Weizmann Institute of Science at Rehovot (now partner city of Heidelberg). We are very grateful to Gentner because he did so much to establish human and scientific relations with Israel.

Ladies and Gentlemen, physics in Heidelberg is now entering the 7th century of its existence. At this moment one may feel like the monk who just lives "when the century goes" and the page is turned in the big book of history. His thoughts have been expressed in 1899 by the German poet Rainer Maria Rilke:

Fig. 14 Wolfgang Gentner (right) with President Meyer Weisgal (left) of the Weizmann Institute of Science and Israel Foreign Minister Abba Eban, 1975

"The lustre of the new-turned page one senses,
Where everything may yet unfold,
The silent forces measure their expanses;
Each other dimly they behold".

With these lines Hans Daniel Jensen (Fig. 12) closed his Nobel Lecture in Stockholm on the 12th of December 1963 after receiving the prize for his part in formulating the nuclear shell model of Haxel, Jensen and Suess, and Maria Goeppert-Mayer. Jensen read the lines in German:

"Man fühlt den Glanz von einer neuen Seite,
auf der noch Alles werden kann.
Die stillen Kräfte prüfen ihre Breite
und sehn einander dunkel an."

"The lustre of the new-turned page one senses" - as one of the older Heidelberg physicists I want to say how much we all appreciate your being with us in this historic moment when the new page is opened in the Heidelberg book of physics.

Thank you so much for listening patiently. Prosit W.E.I.N!

References may be found in P. Brix and G. zu Putlitz: "Rückblicke auf die Heidelberger Physik im Jubiläumsjahr der Universität", Physikalische Blätter 42, 65 (1986). - In preparing this talk I have greatly profited from lectures that Prof. Joachim Heintze gave on the history of physics in Heidelberg, and I thank him very much for Figs. 1,3 and 4.

List of Participants

A. Abbas, Darmstadt, Germany
K. Abrahams, Petten, The Netherlands
E.G. Adelberger, Seattle, USA
S. Akari, Heidelberg, Germany
W. Andrejtscheff, Sofia, Bulgaria
E. Arai, Tokyo, Japan
P. Armbruster, Darmstadt, Germany
A. Arriaga, Lisboa, Portugal
F.T. Avignone, Columbia, USA
Y. Avishai, Beer Sheva, Israel
J.N. Bahcall, Princeton, USA
M. Baldo-Ceolin, Padova, Italy
P.H. Barker, Heidelberg, Germany
R.W. Bauer, Livermore, USA
G. Baur, Jülich, Germany
F. Beck, Darmstadt, Germany
H. Beer, Karlsruhe, Germany
H. Behrens, Karlsruhe, Germany
E. Bellotti, Milano, Italy
W. Bentz, Tokyo, Japan
E.L. Berger, Argonne, USA
K.E. Bergkvist, Stockholm, Sweden
W. Bernreuther, Heidelberg, Germany
F.E. Bertrand, Oak Ridge, USA
P.G. Bizzeti, Firenze, Italy
H.-J. Blome, Bonn, Germany
P. Bock, Heidelberg, Germany
F.W.N. de Boer, Amsterdam, Netherlands
D. Bohle, Darmstadt, Germany
T. Bowles, Los Alamos, USA
C.D. Bowman, Los Alamos, USA
R. Brandenberger, Cambridge, UK
P. von Brentano, Köln, Germany
P. Brix, Heidelberg, Germany
W. Bühring, Heidelberg, Germany
A. Burrows, Stony Brook, USA
A.A. Bykov, Gatchina, USSR
J. Byrne, Brighton, United Kingdom
A.S. Carnoy, Louvain-la-Neuve, Belgium
G. Chanfray, Villeurbanne, France
L.-L. Chau, Davis, USA
R.E. Chrien, Brookhaven, USA
F. Close, Didcot, United Kingdom
F. Coester, Argonne, USA
H.E. Conzett, Marseille, France
L. Crepinsek, Maribor, Yugoslavia
P.K.A. David, Bonn, Germany
P. Depommier, Montreal, Canada
J.-P. Derendinger, Paris, France
B. Desplanques, Orsay, France
J. Deutsch, Louvain-la-Neuve, Belgium
M. Döbeli, Zürich, Switzerland
P. Doll, Karlsruhe, Germany
U. Dore, Roma, Italy
H.G. Dosch, Heidelberg, Germany
N. Dragon, Hannover, Germany
J.F. Dubach, Amherst, USA
D. Dubbers, Heidelberg, Germany
J.-P. Dufour, Bordeaux, France
E.D. Earle, Chalk River, Canada
P. Egelhof, Mainz, Germany
S. Egli, Villigen, Switzerland
H. Ejiri, Osaka, Japan
S.R. Elliott, Irvine, USA

G.T. Emery, Bloomington, USA
H. Emling, Darmstadt, Germany
R. Engfer, Zürich, Switzerland
M. Ericson, Villeurbanne, France
A. Faessler, Tübingen, Germany
F. von Feilitzsch, München, Germany
W. Fetscher, Villigen, Switzerland
E. Fiorini, Milano, Italy
G. Franklin, Pittsburgh, USA
S. Freedman, Argonne, USA
C. Gaarde, Copenhagen, Denmark
S. Gabrakov, Sofia, Bulgaria
R. Gauder, Stuttgart, Germany
A. Gelberg, Köln, Germany
H. Gemmeke, Karlsruhe, Germany
H.J. Gessinger, Mainz, Germany
C. Geweniger, Heidelberg, Germany
C. Giusti, Pavia, Italy
P.W.M. Glaudemans, Utrecht,Netherlands
M. Gmitro, Rez, Czechoslovakia
K. Goeke, Jülich, Germany
F. Gönnenwein, Tübingen, Germany
M. Goldhaber, Brookhaven, USA
D. Greaves, Copenhagen, Denmark
J.S. Greenberg, New Haven, USA
E. Grosse, Darmstadt, Germany
K. Grotz, Heidelberg, Germany
C. Günther, Bonn, Germany
O.F. Häusser, Vancouver, Canada
W. Heil, Mainz, Germany
H. Heiselberg, Aarhus, Denmark
W. Henning, Darmstadt, Germany
P. Herczeg, Los Alamos, USA
A. Hermanni, Mainz, Germany
G. Herrmann, Mainz, Germany
B. Hersman, Durham, USA
E. Heusser, Heidelberg, Germany
G. Heusser, Heidelberg, Germany
K. Heyde, Gent, Belgium
R.W. Hoff, Livermore, USA
R.D.U. Hoffmann, Heidelberg, Germany
S. Hofmann, Darmstadt, Germany
J. Honkanen, Jyväskylä, Finland
J. Hüfner, Heidelberg, Germany
H. Jänsch, Heidelberg, Germany
B. Jonson, Göteborg, Sweden
F. Käppeler, Karlsruhe, Germany
S.N. Kahana, Brookhaven, USA
J. Kayser, Stuttgart, Germany
U. Keyser, Braunschweig, Germany
T.L. Khoo, Argonne, USA
J. Kiko, Heidelberg, Germany
T. Kirchner, Heidelberg, Germany
T. Kirsten, Heidelberg, Germany
V. Kitipova, Sofia, Bulgaria
H.V. Klapdor, Heidelberg, Germany
N. Klay, Karlsruhe, Germany
P. Kleinheinz, Jülich, Germany
K. Kleinknecht, Mainz, Germany
E. Klemt, Heidelberg, Germany
O. Klepper, Darmstadt, Germany
K.T. Knöpfle, Heidelberg, Germany
D.-H. Koang, Grenoble, France
E.W. Kolb, Batavia, USA
F. Kottmann, Zürich, Switzerland
K.-L. Kratz, Mainz, Germany
N. Kraus, Villigen, Switzerland
S. Krewald, Jülich, Germany
A.T. Kruppa, Debrecen, Hungary
W. Kündig, Zürich, Switzerland
P.O. Lagage, Gif-sur-Yvette, France
P. Langacker, Philadelphia, USA
J.F. Last, Heidelberg, Germany
G.A. Leander, Oak Ridge, USA
F. Lichtenberg, Heidelberg, Germany
K.P. Lieb, Göttingen, Germany
W. Lieberz, Köln, Germany
M. Lindner, Livermore, USA
R. Liotta, Stockholm, Sweden
V.M. Lobashev, Moscow, USSR
T.K. Lönnroth, Jyväskylä, Finland
W. Luck, Heidelberg, Germany

R. Madey, Kent, USA
L.A. Malov, Moscow, USSR
W. Mampe, Grenoble, France
P. Manakos, Darmstadt, Germany
N. Mansour, Heidelberg, Germany
B. Martin, Heidelberg, Germany
S. Mattsson, Göteborg, Sweden
T. Matulewicz, Darmstadt, Germany
T. Mayer-Kuckuk, Bonn, Germany
P. Mennrath, Bordeaux, France
J. Metzinger, Heidelberg, Germany
H. Meyer, Wuppertal, Germany
R.A. Meyer, Livermore, USA
A. Milsztajn, Gif-sur-Yvette, France
R.E. Mischke, Los Alamos, USA
R.N. Mohapatra, College Park, USA
A. Morales, Zaragoza, Spain
J. Morales, Zaragoza, Spain
M. Morita, Osaka, Japan
H.-P. Morsch, Jülich, Germany
A.N. Moskalev, Gatchina, USSR
R.L. Mößbauer, München, Germany
A. Moussavi, Heidelberg, Germany
A.C. Mueller, Caen, France
P.J. Mulders, Amsterdam, Netherlands
O. Nachtmann, Heidelberg, Germany
J. Napolitano, Argonne, USA
V.A. Nazarenko, Gatchina, USSR
L. Nilsson, Uppsala, Sweden
R. Nojarov, Tübingen, Germany
S. Nozawa, Villigen, Switzerland
R. Nunez-Lagos, Zaragoza, Spain
E. Nyman, Helsinki, Finland
T. Oda, Tokyo, Japan
W. von Oertzen, Berlin, Germany
H. Ohtsubo, Osaka, Japan
E. Ormand, East Lansing, USA
E. Oset, Valladolid, Spain
A. Osipowicz, Mainz, Germany
F. Osterfeld, Jülich, Germany
E.W. Otten, Mainz, Germany
F.D. Pacati, Pavia, Italy
T. Pacher, Heidelberg, Germany
J. Pasupathy, Bangalore, India
R.D. Peccei, Hamburg, Germany
F. Pellegrini, Padova, Italy
D. Pelte, Heidelberg, Germany
S.T. Petcov, Sofia, Bulgaria
M. Pfützner, Warszawa, Poland
A. Piepke, Heidelberg, Germany
C.-A. Piketty, Paris, France
S. Polikanov, Darmstadt, Germany
M. Potokar, Ljubljana, Yugoslavia
B. Povh, Heidelberg, Germany
G. zu Putlitz, Heidelberg, Germany
N. Pyatov, Dubna, USSR
S. Raman, Oak Ridge, USA
N.F. Ramsay, Cambridge, USA
E. Rathke, Darmstadt, Germany
A. Richter, Darmstadt, Germany
S. Richter, Heidelberg, Germany
A. Rinat, Gif-sur-Yvette, France
A. Ringwald, Heidelberg, Germany
K. Rith, Heidelberg, Germany
G. Roy, Edmonton, Canada
M.L. Rustgi, Buffalo, USA
L.L. Salcedo, Valladolid, Spain
H. Sanchez, Heidelberg, Germany
P.U. Sauer, Hannover, Germany
T. Sauerland, Bochum, Germany
M. Schaad, Zürich, Switzerland
G. Schatz, Karlsruhe, Germany
L. Schellenberg, Fribourg, Switzerland
K. Schiffer, Köln, Germany
K.D. Schilling, Wuppertal, Germany
P. Schlosser, Heidelberg, Germany
U. Schmidt-Rohr, Heidelberg, Germany
K. Schneider, Heidelberg, Germany
M. Schneller, Heidelberg, Germany
D.N. Schramm, Chicago, USA
K. Schreckenbach, Grenoble, France
R. Schuch, Heidelberg, Germany
P. Schuck, Grenoble, France
K.R. Schubert, Heidelberg, Germany

D. Schwalm, Heidelberg, Germany

A.P. Serebrov, Gatchina, USSR

K.K. Seth, Evanston, USA

N. Severijns, Leuven, Belgium

Q. Shafi, Newark, USA

C.M. Shakin, Brooklyn, USA

J.F. Sharpey-Schafer, Liverpool, UK

S. Shlomo, College Station, USA

I. Sick, Basel, Switzerland

H.-W. Siebert, Heidelberg, Germany

S. Sieler-Hornke, Heidelberg, Germany

L.M. Simons, Villigen, Switzerland

S. Singh, Mainz, Germany

S. Skoda, Köln, Germany

A.Yu. Smirnov, Moscow, USSR

A.A. Smolnikov, Moscow USSR

K.A. Snover, Seattle, USA

D. Sober, Washington, USA

A. Sobiczewski, Warszawa, Poland

P. Sona, Firenze, Italy

A.M. Sona-Bizzeti, Firenze, Italy

F. Soramel, Padova, Italy

H.-J. Specht, Heidelberg, Germany

B. Stech, Heidelberg, Germany

E. Steffens, Heidelberg, Germany

T. Stephan, Heidelberg, Germany

R. Stokstad, Heidelberg, Germany

U. Stroth, Grenoble, France

E. Sugarbaker, Columbus, USA

E. Swanson, Seattle, USA

Z. Szeflinski, Warszawa, Poland

I. Talmi, Rehovot, Israel

G.A. Tammann, Binningen, Switzerland

P. Taskinen, Jyväskylä, Finland

R. Tegen, Rondebusch, South Africa

H. Toki, Tokyo, Japan

T. Tomoda, Tübingen, Germany

J. Trümper, München, Germany

P. Truöl, Zürich, Switzerland

J.W.F. Valle, Barcelona, Spain

L. Vanneste, Leuven, Belgium

J.D. Vergados, Ioannina, Greece

M.J. Vicente-Vacas, Valladolid, Spain

M.J.A. de Voigt, Groningen,Netherlands

S.J. Wagner, Heidelberg, Germany

G. Walter, Strasbourg, France

J. Wambach, Urbana, USA

T. Weber, Gießen, Germany

W. Weise, Regensburg, Germany

H.R. Weller, Groningen, Netherlands

C. Wessleorg, Köln, Germany

B.H. Wildenthal, Philadelphia, USA

W. Wilke, Gießen, Germany

R. Wink, Heidelberg, Germany

R. Wirowski, Köln, Germany

P. Wurm, Heidelberg, Germany

I.F. Wright, Manchester, UK

S. Yasumi, Ibaraki-ken, Japan

A. Yokoyama, Tokyo, Japan

V. Zacek, Geneve, Switzerland

S. Zenner, Heidelberg, Germany

H.F.K. Zingl, Graz, Austria

I. Zychor, Darmstadt, Germany

J. Zylicz, Darmstadt, Germany

Index of Contributors